D1156064

DISCARDED

HANDBOOK OF ENGINEERING
POLYMERIC MATERIALS

HANDBOOK OF ENGINEERING POLYMERIC MATERIALS

edited by

Nicholas P. Cheremisinoff, Ph.D.

MARCEL DEKKER, INC.

NEW YORK · BASEL · HONG KONG

Library of Congress Cataloging-in-Publication Data

Handbook of engineering polymeric materials / edited by Nicholas P.
  Cheremisinoff.
       p.   cm.
    Includes index.
    ISBN 0-8247-9799-X (hc: alk. paper)
    1. Polymers. 2. Polymeric composites. 3. Elastomers.
  I. Cheremisinoff, Nicholas P.
  TA455.P58H36   1997
  668.9—dc21
                                                97-20896
                                                  CIP

The publisher offers discounts on this book when ordered in bulk quantities. For more information, write to Special Sales/Professional Marketing at the address below.

This book is printed on acid-free paper.

Copyright © 1997 by Marcel Dekker, Inc. All Rights Reserved.

Neither this book nor any part may be reproduced or transmitted in any form or by any means, electronic or mechanical, including photocopying, microfilming, and recording, or by any information storage and retrieval system, without permission in writing from the publisher.

Marcel Dekker, Inc.
270 Madison Avenue, New York, New York 10016

Current printing (last digit):
10 9 8 7 6 5 4 3 2 1

PRINTED IN THE UNITED STATES OF AMERICA

Preface

Polymers and, in particular, elastomers have evolved into mature products over the last thirty years, with well-established applications, most of which center around the consumer. Despite these materials being mature, and market applications well established, this is by no means a stagnant industry or subject. Driven by fierce competition, product quality improvements, and new applications in such market segments as health care, automotive parts, construction and building materials, mechanical goods, and an enormous range of composites with specialty or niche-type applications, suppliers and manufacturers of elastomers, plastics blends, and polymeric alloying materials continue to strive for new products and variations of feedstock polymers. This volume provides a compendium of some of the latest technological advancements in product applications, new elastomers, blends, alloys, and functionalized materials. Discussions are also included on some of the more conventional polymers to provide a balance to the discussions of alternative materials for end-use applications.

As already noted, although the elastomers and plastics industries may be mature, they are still very active. Products and raw materials are constantly evolving; new ones are introduced while others are dropped. Thermoplastic rubbers began to emerge in the late 1980s and have now made substantial market penetrations into automotive applications, mechanical goods, construction materials, and other areas. Perhaps more than plastics, it is the rubbers or elastomers that continue to show the greatest areas of product innovation because of their versatility in properties, but also their ability for compatibility with certain plastic materials. In general, the demand for synthetic rubber is strong throughout the world, particularly in North America. We should not forget, however, that natural rubber continues to account for 25 to 30% of total elastomer demand, and will likely remain strong in the automotive market sector as the move to steel radial truck tires from bias-ply tires continues. This volume concentrates on the plastic-elastomeric types of blends and alloys and examines both new and conventional products. Emphasis in discussions is given to product characterization and performance attributes, as well as the structural properties of these materials. In this regard the volume is probably most useful to the product development specialist and the applications engineer. The book represents the efforts of a large number of experts from both the industrial and academic communities. Their efforts in preparing contributions to this volume are to be noted and I express heartfelt gratitude for their time and effort. A special thanks is extended to the publisher for the fine production of this volume.

Nicholas P. Cheremisinoff, Ph.D.

iii

OCT   1998

Contents

Contributors

Elsayed Mohamed Abdel-Bary   Department of Chemistry, Mansoura University, Mansoura, Egypt

Sahar Al-Malaika   Chemical Engineering and Applied Chemistry, Aston University, Birmingham, England

Salah M. Aliwi   Department of Chemistry, College of Science, Mustansiriya University, Baghdad, Iraq

R. Asaletha*   Rubber Research Institute of India, Kerala, India

Bhola Nath Avasthi   Department of Chemistry, Indian Institute of Technology, Kharagpur, India

Susanta Banerjee   Department of Synthetic Chemistry (Polymer Group), Defence Research & Development Establishment, Gwalior, India

Antonio Bello   Instituto de Ciencia y Tecnología de Polímeros (CSIC), Madrid, Spain

Rosario Benavente   Instituto de Ciencia y Tecnología de Polímeros (CSIC), Madrid, Spain

Ishwar Singh Bhardwaj   R. S. Petrochemicals Ltd., Baroda, India

Susmita Bhattacharjee   FB-Physikalische Chemie—Polymere, Philipps Universität, Marburg, Germany

Anil K. Bhowmick   Rubber Technology Center, Indian Institute of Technology, Kharagpur, India

A. K. Bledzki   Institute of Materials Technology, Plastic and Recycling Technology, University of Kassel, Kassel, Germany

Robert Bond   Chicago Rawhide (Americas), Elgin, Illinois

Wei-Xiao Cao   Institute of Polymer Science, Peking University, Beijing, China

Jiong Chen   Department of Technical Development, Shanghai Medical Equipment Research Institute, Shanghai, China

Babur Z. Chowdhry   School of Chemical and Life Sciences, University of Greenwich, London, England

Huseyin Çiçek   Chemical Engineering Department, Hacettepe University, Ankara, Turkey

Claude Daneault   Pulp and Paper Research Center, University of Quebec, Trois-Rivieres, Quebec, Canada

Chapal K. Das   Materials Science Centre, Indian Institute of Technology, Kharagpur, India

Santanu Datta   Rubber Technology Centre, Indian Institute of Technology, Kharagpur, India

S. K. De   Rubber Technology Centre, Indian Institute of Technology, Kharagpur, India

---

Current affiliation: Mahatma Gandhi University, Kerala, India

Pradip Kumar Dutta   Department of Applied Chemistry, Shri G. S. Institute of Technology & Science, Indore, India

Eman Mohamed El-Nesr   National Center for Radiation Research and Technology, Cairo, Egypt

Mustafa Ersöz   Department of Chemistry, Selçuk University, Konya, Turkey

Xin-De Feng   Institute of Polymer Science, Peking University, Beijing, China

J. Gassan   Institute of Materials Technology, Plastic and Recycling Technology, University of Kassel, Kassel, Germany

Jayamol George   School of Chemical Sciences, Mahatma Gandhi University, Kerala, India

Snooppy George   School of Chemical Sciences, Mahatma Gandhi University, Kerala, India

Josephine George   School of Chemical Sciences, Mahatma Gandhi University, Kerala, India

Virendra Kumar Gupta   Research Centre, Indian Petrochemicals Corporation Ltd., Gujarat, India

Masanori Hara   Department of Chemical and Biochemical Engineering, Rutgers University, Piscataway, New Jersey

Sadao Hayashi   Faculty of Textile Science and Technology, Shinshu University, Ueda, Japan

Baki Hazer   Department of Chemistry, Zonguldak Karaelmas University, Zonguldak, Turkey

Markku T. Heino*   Department of Chemical Engineering, Helsinki University of Technology, Espoo, Finland

Eggehard Holler   Institute for Biophysics and Physical Biochemistry, Department of Biology and Preclinics, University of Regensburg, Regensburg, Germany

Cheng-Di Huang   Department of Technical Development, Shanghai Medical Equipment Research Institute, Shanghai, China

Rui Huang   Department of Plastics Engineering, Sichuan Union University, Sichuan, China

Koji Ishizu   Department of Polymer Science, Tokyo Institute of Technology, Tokyo, Japan

Andrzej Jeziorny   Institute of Fibre Physics and Textile Finishing, Technical University of Lodź, Lodź, Poland

B. S. Kaith   Department of Applied Sciences and Humanities, Regional Engineering College, Hamirpur, India

Inderjeet Kaur   Department of Chemistry, Himachal Pradesh University, Shimla, India

Ramesh Keshavaraj†   Department of Chemical Engineering, Texas Tech University, Lubbock, Texas

M. B. Khan   CHEMTEC & Prime Glass, Jhelum, Pakistan

Jayasree Konar   Department of Chemistry, Indian Institute of Technology, Kharagpur, India

Refiga Kurbanova   Department of Chemistry, Selçuk University, Konya, Turkey

Valery F. Kurenkov   Kazan State Technological University, Kazan, Russia

Stephen A. Leharne   School of Earth and Environmental Sciences, University of Greenwich, London, England

Guangxian Li   Department of Plastics Engineering, Sichuan Union University, Sichuan, China

Yu. P. Losev   Department of Chemistry, Belarussian State University, Minsk, Belarus

Sukumar Maiti   Materials Science Centre, Indian Institute of Technology, Kharagpur, India

Current affiliations:
* Nokia Cables Ltd., Espoo, Finland
† Milliken & Company, LaGrange, Georgia

Debesh Maldas* Pulp and Paper Research Center, University of Quebec, Trois-Rivieres, Quebec, Canada

Ajit B. Mathur Research Centre, Indian Petrochemicals Corporation Ltd., Baroda, India

Metwally Shafik Metwally Department of Chemistry, Al-Azhar University, Cairo, Egypt

Ramazan Mirzaoglu Department of Chemistry, Selçuk University, Konya, Turkey

Bhupendra Nath Misra Department of Chemistry, Himachal Pradesh University, Shimla, India

Abd-Alla M. A. Nada Cellulose & Paper Department, National Research Centre, Cairo, Egypt

Susumu Nagai Plastics Technical Society, Osaka, Japan

Mohammad Kazim Naqvi PVC Group, Research and Technology Support, Saudi Basic Industries Corporation (SABIC), Riyadh, Saudi Arabia

Raghu S. Narayan Department of Chemical Engineering, Texas Tech University, Lubbock, Texas

Naoto Oku Department of Radiobiochemistry, School of Pharmaceutical Sciences, University of Shizuoka, Shizuoka, Japan

Zacharia Oommen C.M.S. College, Mahatma Gandhi University, Kerala, India

Raphael M. Ottenbrite Department of Chemistry, Virginia Commonwealth University, Richmond, Virginia

Jayant S. Parmar Department of Chemistry, Sardar Patel University, Gujarat, India

Iain F. Paterson School of Earth and Environmental Sciences, University of Greenwich, London, England

José M. Pereña Instituto de Ciencia y Tecnología de Polímeros (CSIC), Madrid, Spain

Ernesto Pérez Instituto de Ciencia y Tecnología de Polímeros (CSIC), Madrid, Spain

Chennakkattu Krishna Sadasivan Pillai Polymer Division, Regional Research Laboratory (CSIR), Thiruvananthapuram, India

Kun-Yuan Qiu Institute of Polymer Science, Peking University, Beijing, China

Ivo Reetz Organic Chemistry Department, Istanbul Technical University, Istanbul, Turkey

Mohini M. Sain Pulp and Paper Research Center, University of Quebec, Trois-Rivieres, Quebec, Canada

Reiko Saito Department of Polymer Chemistry, Tokyo Institute of Technology, Tokyo, Japan

J. A. Sauer Department of Chemical and Biochemical Engineering, Rutgers University, Piscataway, New Jersey

Yongsok Seo Polymer Processing Laboratory, Korea Institute of Science and Technology, Seoul, Korea

Raj P. Singh Polymer Chemistry Division, National Chemical Laboratory, Pune, India

A. S. Singha Department of Applied Sciences and Humanities, Regional Engineering College, Hamirpur, India

T. Siyam Nuclear Chemistry Department, Hot Laboratory Centre, Atomic Energy Authority, Cairo, Egypt

A. K. Srivastava Chemistry Department, Harcourt Butler Technological Institute, Kanpur, India

Yasuo Suda Department of Chemistry, Osaka University, Osaka, Japan

Sunil Department of Applied Sciences, Regional Engineering College, Hamirpur, India

Sabu Thomas School of Chemical Sciences, Mahatma Gandhi University, Kerala, India

Richard William Tock Department of Chemical Engineering, Texas Tech University, Lubbock, Texas

---

Current affiliation: Michigan State University, East Lansing, Michigan

Ali Tuncel   Chemical Engineering Department, Hacettepe University, Ankara, Turkey

Akira Ueda   Osaka Municipal Technical Research Institute, Osaka, Japan

Grzegorz Urbańczyk   Institute of Fibre Physics and Textile Finishing, Technical University of Lodź, Lodź, Poland

Tommi P. Vainio*   Department of Chemical Engineering, Helsinki University of Technology, Espoo, Finland

E. F. Vainstein   Department of Kinetics and Thermodynamics of Cooperative Processes, N. M. Emanuel Institute of Biochemical Physics, Russian Academy of Sciences, Moscow, Russia

Yusuf Yagci   Organic Chemistry Department, Istanbul Technical University, Istanbul, Turkey

Xiao-Su Yi   Institute of Polymers and Processing, Zhejiang University, Hangzhou, China

Mustafa Yilmaz   Department of Chemistry, Selçuk University, Konya, Turkey

Mohamed Adel Yousef   Department of Chemistry, Helwan University, Cairo, Egypt

G. E. Zaikov   Department of Chemical and Biological Kinetics, N. M. Emanuel Institute of Biochemical Physics, Russian Academy of Sciences, Moscow, Russia

Xiongwei Zhang   Department of Plastics Engineering, Sichuan Union University, Sichuan, China

---

* *Current affiliation*: Valmet Corporation, Jyväskylä, Finland

1

Artificial Neural Networks as a Semi-Empirical Modeling Tool for Physical Property Predictions in Polymer Science

Ramesh Keshavaraj,* Richard William Tock, and Raghu S. Narayan
Texas Tech University, Lubbock, Texas

I. INTRODUCTION

Recently, a new approach called artificial neural networks (ANNs) is assisting engineers and scientists in their assessment of "fuzzy information." Polymer scientists often face a situation where the rules governing the particular system are unknown or difficult to use. It also frequently becomes an arduous task to develop functional forms/empirical equations to describe a phenomena. Most of these complexities can be overcome with an ANN approach because of its ability to build an internal model based solely on the exposure in a training environment. Fault tolerance of ANNs has been found to be very advantageous in physical property predictions of polymers. This chapter presents a few such cases where the authors have successfully implemented an ANN-based approach for purpose of empirical modeling. These are not exhaustive by any means.

Typical applications of neural networks have been largely made in the areas of pattern recognition, signal processing, speech recognition, written character recognition, time series analysis, complex dynamic systems, and process control [1–14]. It is important to note that all of these applications involve parallel information transformation between various processing elements connected through different architectures. The specific application of ANNs to physical property prediction, especially with respect to polymers, has been rather limited.

A artificial neural network is to a great extent, a parallel-processing dynamic system, with the to-

pology of a directed graph, which can carry out information processing, by means of its outputs, on either on-off or continuous inputs—Hecht-Nielsen

II. ANN AS A MODELING TOOL

The computational paradigm in an ANN is based on an idealized model of a biological unit called a neuron. The unique characteristics of this ANN model are the inputs of signal from stimulus in a training environment. It is important to note that each neuron works independently of the other neurons. The specific characteristics of ANN models that attract industrial application are:

1. Learning from examples
2. Continued learning while in operation
3. Ability to distinguish noise in experimental data
4. Ability to generalize
5. Self-organization
6. High speed in conjunction with complex parallel processing
7. Error tolerance

These capabilities of ANNs make them a unique tool for a large number of industrial applications. In this chapter, the authors demonstrate, with case studies, the advantages of using this approach to physical property predictions in polymer science.

III. FUNDAMENTALS OF ANNs

To understand why and how artificial neural networks work as they do, it is helpful to study some of the funda-

---

**Current affiliation*: Milliken & Company, LaGrange, Georgia.

mental characteristics of biological neural networks. A biological neuron is a primary processing unit. In a biological neuron, dendrites carry information impulses toward the neuron and can be considered as an input. Similarly, an axon carries information impulses away from the neuron and can be considered as an output. Hence, artificial neural networks are constructed of mathematical processing elements, also known as neurons, which are connected via information channels called interconnections/synapse, as in a biological neuron. Synapses in a network serve as pathways of varying strength for transmission of information. The neurons operate collectively and simultaneously and are configured in regular architecture. They "learn" by extracting preexisting information from the data that describe the relationship between the inputs and the outputs. Hence, in the learning process, the network actually acquires knowledge or information from the input environment. As a result of the interrelationships, the network assimilates trends and relationships that can be recalled later. Neural networks can identify and learn correlative patterns between sets of input data and corresponding target values. Once trained, neural networks can be used to forecast the outputs expected for new levels of input variables. Each neuron can have multiple inputs, but only one output. Each output may, however, branch out as an input to other neurons. Neural networks that are capable of handling complex and nonlinear problems can process information rapidly and help reduce the engineering effort required in developing highly computation-intensive models. Neural networks also come in a variety of types, and each has its distinct architectural differences and uses [15,16].

The structure of a neural network forms the basis for information storage and governs the learning process. The type of neural network used in this work is known as a feed-forward network; the information flows only in the forward direction, i.e., from input to output in the testing mode. A general structure of a feed-forward network is shown in Fig. 1. Connections are made be-

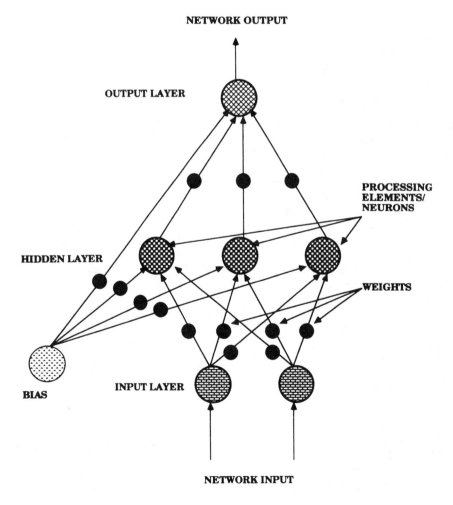

Figure 1   A general structure of a feed-forward neural network.

tween neurons of adjacent layers: a neuron is connected so that it receives signals from each neuron in the immediate preceding layer and transmits signals to each neuron in the immediate succeeding layer. Neural networks that are organized in layers typically consist of at least three layers: an input layer, one or more hidden layers, and an output layer. The input and output layers serve as interfaces that perform the appropriate scaling relationship between the actual and the network data. Hidden layers are so termed because their neurons are hidden from the actual data; the connections provide the means for information flow. Each connection has an associated synaptic weight factor, w_i, expressed by a numerical value that can be adjusted. The weight is an indication of the connection strength between any two neurons.

The neurons in both the hidden and output layers perform summing and nonlinear mapping functions. The functions carried out by each neuron are illustrated in Fig. 2. Each neuron occupies a particular position in a feed-forward network and accepts inputs only from the neurons in the preceding layer and sends its outputs to other neurons in the succeeding layer. The inputs from other nodes are first weighted and then summed. This summing of the weighted inputs is carried out by a processor within the neuron. The sum that is obtained is called the activation of the neuron. Each activated neu-

ron performs three primary functions: (1) receives signals from other neurons; (2) it sums their signals; and (3) it transforms the sum. For example, if the output from the ith neuron with pattern p is designated as $x_{i,p}$, then the input to the jth neuron from the ith neuron is $x_{i,p} w_{i,j}$. Summing the weighted inputs to the jth neuron can be represented as:

$$u_{i,p} = \sum_i x_{i,p} w_{i,j} - w_{B,j} \theta_j \qquad (1)$$

where θ_j is a bias term and $w_{B,j}$ is the weight of the connection from the bias neuron to the jth neuron. The bias neuron is provided to supply an invariant output to each neuron in the hidden layer. If the weighted sum of the scaled inputs into the hidden layers exceeds the weight of the bias hidden layer neuron, the neuron becomes activated or excited. This activation can either be positive, zero, or negative, because the synaptic weightings and the inputs can be either positive or negative. A bias of $+1$ was used in this study. Hence, any weighted input that makes a positive contribution to activation represents a triggering or tendency to turn the neuron on. An input making a negative contribution represents an inhibition, which tends to turn the neuron off. After summing its inputs to determine its activation, the summed total is then modified by a mapping function, also known as a transfer/threshold function. A commonly used transfer function is the "sigmoid," which is expressed as:

$$S_{i,p} = \frac{1}{[1 + \exp(-u_{i,p})]} \qquad (2)$$

A sigmoid (s-shaped) is a continuous function that has a derivative at all points and is a monotonically increasing function. Here $S_{i,p}$ is the transformed output asymptotic to $0 \le S_{i,p} \le 1$ and $u_{i,p}$ is the summed total of the inputs $(-\infty \le u_{i,p} \le +\infty)$ for pattern p. Hence, when the neural network is presented with a set of input data, each neuron sums up all the inputs modified by the corresponding connection weights and applies the transfer function to the summed total. This process is repeated until the network outputs are obtained.

IV. NEURAL NETWORK ARCHITECTURE

The number of neurons to be used in the input/output layer are based on the number of input/output variables to be considered in the model. However, no algorithms are available for selecting a network structure or the number of hidden nodes. Zurada [16] has discussed several heuristic based techniques for this purpose. One hidden layer is more than sufficient for most problems. The number of neurons in the hidden layer neuron was selected by a trial-and-error procedure by monitoring the sum-of-squared error progression of the validation data set used during training. Details about this proce-

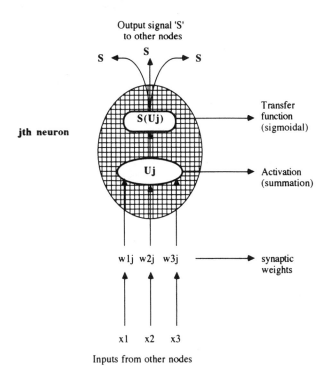

Figure 2 Schematic of typical functions carried out by a neuron.

dure are discussed in the following paragraphs. A feed-forward type of neural network architecture has been used in all the case studies considered in this chapter. The training algorithm used here is based on the familiar Fletcher algorithm for nonlinear optimization used in many thermodynamic property estimations.

V. TYPES OF ANNs

Neural networks can be broadly classified based on their network architecture as feed-forward and feed-back networks, as shown in Fig. 3. In brief, if a neuron's output is never dependent on the output of the subsequent neurons, the network is said to be feed forward. Input signals go only one way, and the outputs are dependent on only the signals coming in from other neurons. Thus, there are no loops in the system. When dealing with the various types of ANNs, two primary aspects, namely, the architecture and the types of computations to be per-

FEED-BACK NEURAL NETWORKS

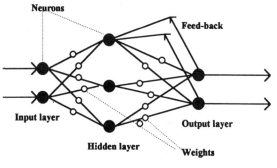

Shown without a Bias

FEED-FORWARD NEURAL NETWORK

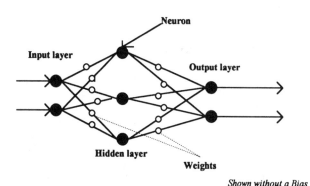

Shown without a Bias

Figure 3 Feed-back and feed-forward artificial neural networks.

formed, have to be understood before their implementation.

Neural networks can also be classified by their neuron transfer function, which typically are either linear or nonlinear models. The earliest models used linear transfer functions wherein the output values were continuous. Linear functions are not very useful for many applications because most problems are too complex to be manipulated by simple multiplication. In a nonlinear model, the output of the neuron is a nonlinear function of the sum of the inputs. The output of a nonlinear neuron can have a very complicated relationship with the activation value.

A. Feed-Back Networks

In a feed-back network, the output signals of neurons directly feed back into neurons in the same or preceding layers, as shown in Fig. 3. Feedback refers to the way in which neurons are connected. Feed-back networks send signals back as inputs to other neurons. This is not same as back-propagation, which describes a training method. With back-propagation, the error at the final output layer is used for correction. Feed-back networks do not use back-propagation for training. Back-propagation networks are not feed-back models.

Feed-back models can be constructed and trained. In a constructed model, the weight matrix is created by adding the output product of every input pattern vector with itself or with an associated input. After construction, a partial or inaccurate input pattern can be presented to the network and, after a time, the network converges to one of the original input patterns. Hopfield and BAM are two well-known constructed feed-back models.

B. Feed-Forward Networks

The second main category of neural networks is the feed-forward type. In this type of network, the signals go in only one direction; there are no loops in the system as shown in Fig. 3. The earliest neural network models were linear feed forward. In 1972, two simultaneous articles independently proposed the same model for an associative memory, the linear associator. J. A. Anderson [17], neurophysiologist, and Teuvo Kohonen [18], an electrical engineer, were unaware of each other's work. Today, the most commonly used neural networks are nonlinear feed-forward models.

Current feed-forward network architectures work better than the current feed-back architectures for a number of reasons. First, the capacity of feed-back networks is unimpressive. Secondly, in the running mode, feed-forward models are faster, since they need to make one pass through the system to find a solution. In contrast, feed-back networks must cycle repetitively until the neuron outputs stop changing. This cycling can typically require anywhere from 3 to 1000 cycles.

Computations in neural networks can also be broadly classified into two other categories, namely, direct throughput and relaxation. Direct throughput is specific to feed-forward neural networks. In this case, the selected network training data are trained so that a certain set of input patterns give rise to desired outputs. Relaxation, to the contrary, is an iterative convergence to a fixed point, which is achieved by feed-back loops constantly feeding the network output into the neurons in the preceding layers. These two modified networks can be used advantageously in optimization problems. Also, relaxation networks, when correctly trained, can also be used as content addressable memory.

VI. LEARNING IN NEURAL NETWORKS

Once the network architecture is selected and the characteristics of the neurons and the initial weights are specified, the network has to be taught to associate new patterns and new functional dependencies. Learning corresponds to adjustments of the weights in order to obtain satisfactory input–output mapping. Since neural networks do not use "a priori" information about the process to be modeled, learning must come from training in the exposed environment. Hence, the selection of training data in any problem becomes very important for neural networks. The weights are learned through experience, using an update rule to change the synaptic weight, w_{ij}.

Learning can be broadly divided into three categories: supervised, unsupervised, and reinforced learning.

1. Supervised learning—A training data set containing a set of inputs and target outputs are provided to the selected network architecture. The synaptic weights are adjusted so as to minimize the error between the desired and the network predicted outputs for each input pattern. A trained network can then be validated with a second independent data set over the same range as the training data set.
2. Unsupervised learning—In this type the network is able to discover statistical regularities in its input space and automatically develops different modes of behavior to represent different types of inputs.
3. Reinforced—This type of network receives a global reward/penalty signal during training. Usually weights are changed so as to develop an input/output behavior that maximizes the probability of receiving a reward, i.e., minimize that of receiving a penalty.

Several different learning rules have been proposed by various researchers [15,19,20], but the aim of every learning process is to adjust the weights in order to minimize the error between the network predicted output and the desired output. The output from each neuron i is $S_{i,p}$, as shown in Eq. (2).

A. Nonlinear Optimization Training Routine

A faster training process is to search for the weights with the help of a optimization routine that minimizes the same objective function. The learning rule used in this work is common to a standard nonlinear optimization or least squares technique. The entire set of weights are adjusted at once instead of adjusting them sequentially from the output to the input layers. The weight adjustment is done at the end of each exposure of the entire training set to the network, and the sum of squares of all errors for all patterns is used as the objective function for the optimization problem. A nonlinear optimization routine based on the Levenberg–Marquardt method [21] is used for solving the nonlinear least-squares problem. The optimization problem can be defined if the model to be fitted to the data is written as follows:

$$F(y) = f(\alpha_1, \alpha_2, \ldots, \alpha_m;$$
$$\beta_1, \beta_2, \ldots, \beta_k) = f(\alpha, \beta) \qquad (3)$$

where $\alpha_1, \alpha_2, \ldots, \alpha_m$ are independent variables, $\beta_1, \beta_2, \ldots, \beta_k$ are the population values of k parameter and $F(y)$ is the expected value of the independent variable y. Then the data points can be denoted by:

$$(Y_i, X_{1i}, X_{2i}, \ldots, X_{mi}) \qquad i = 1, 2, \ldots, n \qquad (4)$$

The problem is to compute those estimates of the parameter that will minimize the following objective function:

$$\phi = \sum_{i=1}^{n} [Y_i - \hat{Y}_i]^2 \qquad (5)$$

where $\hat{Y}_i$ is the value of y predicted by the model at the ith data point. The parameters to be determined is the strength of the connections, i.e., the weights, w_i. If the $(m \times n)$ Jacobian is defined by $J_{ij} = \partial r_i / \partial x_j$ (r_i is the residual $Y_i - \hat{Y}_i$) then each iteration can be written as follows:

$$Y^k = Y^k + \delta^k \qquad (6)$$

where δ is the solution of the set of linear equations of the following form:

$$(A + \lambda I)\delta = -v \qquad (7)$$

where $A = J^T J$ and $v = J^T r$ are evaluated at Y^k, and where λ is an adjustable parameter that is used to control the iteration. More details of this Levenberg–Marquardt method can be found elsewhere [21]. A flow diagram of this original Marquardt algorithm is given in Fig. 4.

This algorithm shares with the gradient methods the ability to converge from an initial guess that may be

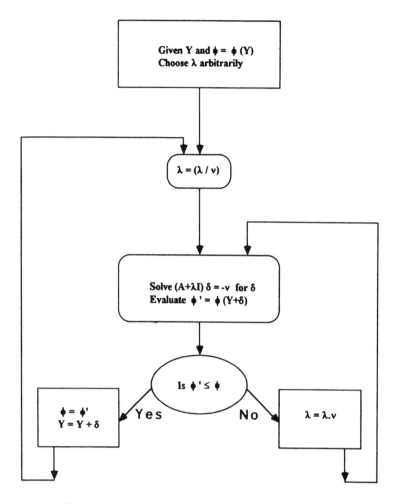

Figure 4 Original Levenberg–Marquardt algorithm.

outside the range of convergence with other methods. It shares with the Taylor series method the ability to close in on the converged values rapidly after the vicinity of the converged values have been reached.

1. Modifications of Levenberg–Marquardt Method: Fletcher's Modification

A number of modifications to eliminate some less favorable aspects of the Levenberg–Marquardt method were considered by Fletcher. For instance, the arbitrary initial choice of the adjustable parameter λ, if poor, can cause an excessive number of evaluations of ϕ, the sum of squared error, before a realistic value is obtained. This is especially noticeable if ν, i.e., $J^T R(x)$, is chosen to be small, i.e., $\nu = 2$. Another disadvantage of the method is that the reduction of λ to λ/ν at the start of each iteration may also cause excessive evaluations, especially when ν is chosen to be large, i.e., $\nu = 10$. The effect of this is that the average number of evaluations

of ϕ per iteration may be about 2, which is unnecessarily inefficient. A further disadvantage of the method is that the tests $\phi' \leq \phi$ or even $\phi' < \phi$ precludes a proof of convergence being made. Finally, when solving problems in which $R = 0$ at the solution, it is possible to achieve a quadratic rate of convergence with the Gauss–Newton method, but only a superfine rate with Marquardt scheme. All of these drawbacks of the original Marquardt method were overcome using Fletcher's modification [21]. Details about this implementation are given in Fig. 5. Where R is the ratio of the actual reduction/predicted reduction. The motivation for Fletcher's modification was that if the ratio R is near 1, then the adjustable parameter λ has to be reduced, and if the ratio is near to or less than zero, than λ has to be increased. Fletcher suggested an arbitrary choice of constants ρ and σ such that $0 < \rho < \sigma < 1$, and a reduction of λ if $R < \rho$, and an increase of λ if $R > \sigma$. Various experiments were performed by Fletcher to study the effect of these

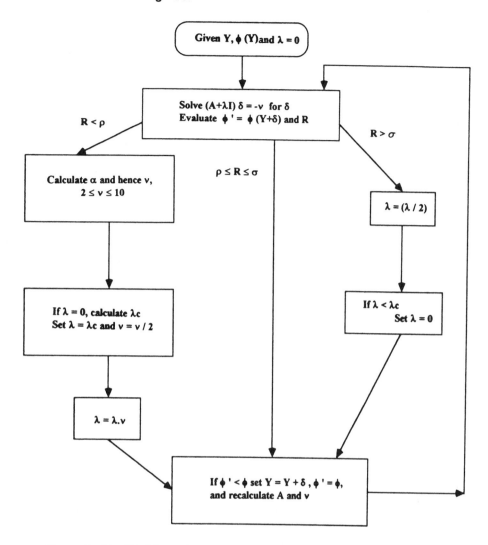

Figure 5 Modified Levenberg–Marquardt algorithm/Fletcher algorithm.

two constants. The rate of convergence was largely insensitive to different choices of ρ and σ, which led to the ultimate choice of the values $\rho = 0.25$ and $\sigma = 0.75$. More details of this modification is given elsewhere [21–26]. The optimization procedure updated weights at every connection and yielded rapid and robust training. The weights were initialized to values in the range ± 0.1 by random assignment.

B. Standard Error–Back-Propagation Training Routine

In a standard back-propagation scheme, updating the weights is done iteratively. The weights for each connection are initially randomized when the neural network undergoes training. Then the error between the target output and the network predicted output are back-propa-

gated through the network. The back-propagation of error is used as a means to update the connection weights [20]. Repeated iterations of this operation results in a final convergence to a set of connection weights.

The general principle behind most commonly used back-propagation learning methods is the "delta rule," by which an objective function involving squares of the output errors from the network is minimized. The delta rule requires that the sigmoidal function used at each neuron be continuously differentiable. This methods identifies an error associated with each neuron for each iteration involving a cause–effect pattern. Therefore, the error for each neuron in the output layer can be represented as:

$$\delta_{i,p} = (T_{i,p} - S_{i,p})f'(U_{i,p}) \tag{8}$$

where $T_{i,p}$ is the desired target output for neuron i and pattern p, and f', is the derivative of the sigmoidal function used for neuron i. The change in the weight of the connection between neuron i and neuron j is given by:

$$\underset{(n+1)}{\Delta} w_{i,j} = \beta(\delta_{i,p}S_{i,p}) + \alpha \underset{n}{\Delta} w_{i,j} \qquad (9)$$

where β is the learning rate, α is the momentum factor, and n indexes the iteration.

The error signal from the neurons in the output layer can be easily identified. This is not so for neurons in the hidden layers. Back-propagation overcomes this difficulty by propagating the error signal backward through the network. Hence, for the hidden layers, the error signal is obtained by:

$$\delta_{i,p} = f'(u_{i,p}) \sum_j (\delta_{j,p}w_{i,j}) \qquad (10)$$

where j represents the neurons to which neuron i in the hidden layer sends the output. Hence, the weights were updated as shown in Eq. (9).

C. Multiple Adaline (Madaline III) Training Algorithm

Back-propagation and Madaline III are useful for training multilayered networks. A gradient-descent technique for minimizing a sum of squared errors is employed in both of these training routines. Gradient-descent techniques function by finding the lowest error point on any error contour in the weight space. This is accomplished by descending down the slope or gradient of the contour plane.

While back-propagation uses an analytical equation to govern the entire training process. Madaline III, a network of multiple adalines, uses a more empirical approach [27]. This approach reduces the system error at each instant in time. Thus, when an input is presented to the network, the resulting output is compared to a target using a sum of the squared error. This approach uses a perturbation method on each neuron's summation of inputs to determine the appropriate connection weight. A function called "perturb" is used in this algorithm. Perturb controls the size of perturbation added to the pre-neuron sum of each neuron. The effect of this added value is propagated through the network. If the network error is reduced as a result of this change, then the perturbation is accepted, and the connection weights are changed. If the error is increased by the perturbation, then a change in the opposite sense is made.

The gradient can be used to optimize the weight vector according to the method of steepest-descent:

$$w_{(k+1)} = w_k - \mu \left(\frac{\Delta(\epsilon_k)^2}{\Delta s}\right) x_k \qquad (11)$$

Here Δs is the perturb, μ is a parameter which controls stability and the rate of convergence, and $\Delta(\epsilon_k)^2/\Delta s$ is the sample derivative.

To some extent, the sample derivative $\Delta(\epsilon_k)^2/\Delta s$ is Madaline III is analogous to the analytical derivative $(\delta E_k/\delta s_k)^2$ used in back-propagation. Hence, these two training rules follow a similar instantaneous gradient, and thus perform nearly identical weight updates. However, the back-propagation algorithm requires fewer operations to calculate gradients than does Madaline III, since it is able to take advantage of a prior knowledge of the sigmoidal nonlinearities and their derivative functions. Conversely, the Madaline III algorithm uses no prior knowledge about the characteristics of the sigmoid functions. Instead it acquires instantaneous gradients from perturbation measurements.

Finally, any training is incomplete without proper validation of the trained model. Therefore, the trained network should be tested with data that it has not seen during the training. This procedure was followed in this study by first training the network on one data set, and then testing it on a second different data set.

VII. DATA PROCESSING/DATA PREPARATION

It is tempting to view ANNS as simplified versions of biological nervous systems. Yet even the most complex neurocomputers, with several million neurons, are unable to mimic the behavior of a fly, which has approximately one million nerve cells. This is because the nerve system of the fly has far more interconnections than are possible with current-day neurocomputers, and their neurons are highly specialized to perform necessary tasks. The human brain, with about 10 billion nerve cells, is still several orders of magnitude more complex.

Even so, artificial neural networks exhibit many "brainlike" characteristics. For example, during training, neural networks may construct an internal mapping/ model of an external system. Thus, they are assumed to "make sense" of the problems that they are presented. As with any construction of a robust internal model, the external system presented to the network must contain meaningful information. In general the following anthropomorphic perspectives can be maintained while preparing the data:

Determine which data are to be used to train and test the selected network. For effective training, the selected data should be relevant, meaningful, and complete.

Divide the available data into training and test data sets (1/3). Test sets are used to validate the trained network and insure accurate generalization.

Scale/map the data to an array of numbers. Transform the real world data into numeric input and target output patterns.

Normalize both the input and the target output data to fit the transfer function range. This implies that the data have to be scaled to fit between the minimum and maximum values of the selected transfer function.

A. Scaling of Experimental/Real-World Data

In many cases, numeric data with a continuous range of values do not need to be encoded prior to being sent to a neural network. In the authors' experience encoding was found to be necessary only in some special cases. In such cases, the analog data had to be scaled to fit within the transfer function range. A maximum-minimum scaling process for standardizing numeric data was used. To do this, the following formula was applied to each data point:

$$X_S = \left[\left(\frac{X_d - X_{MIN}}{d_{MAX} - d_{MIN}} \right) \right. \tag{12}$$
$$\left. \times (TF_{MAX} - TF_{MIN}) \right] + TF_{MIN}$$

where X_S is the Max/Min scaled data point; X_d is the experimental unscaled input data point; d_{MAX} is the minimum value of the raw input data; d_{MIN} is the maximum value of the raw input data; TF_{MAX} is the transfer function maximum and TF_{MIN} is the transfer function minimum.

Similarly, the network predicted data must be unscaled for error estimation with the experimental output data. The unscaling was performed using a simple linear transformation to each data point.

$$Y_d = \left[\left(\frac{Y_S - TF_{MIN}}{TF_{MAX} - TF_{MIN}} \right) \right. \tag{13}$$
$$\left. \times (b_{MAX} - b_{MIN}) \right] + b_{MIN}$$

where Y_d is the actual experimental outcome; Y_S is the neural network predicted output; b_{MAX} and b_{MIN} are the maximum and minimum value of the raw output data.

VIII. OPTIMAL TRAINING

During the training phase, one of the important objective is to achieve optimal training. When training is optimal, the number of training examples required to assimilate the cause–effect relationship between the variables modeled is minimized. The goodness of the internal model built by the neural network depends on this efficiency. Several methods to achieve optimal training have been developed by earlier investigations [15,16,28]. A novel approach, suggested by Weigand et al. [29], uses

a separate validation data set during training. A validation data set is made up of arbitrary number of data points from the original training data set (for example, 10%). Hence, at the end of each exposure, after updating the weights, the validation data are presented to the network and the network's prediction error for the validation is calculated. Training is stopped when this error starts to increase. However, it should be remembered that validation data are not part of the training data once the training begins.

IX. HOW MUCH DATA, NEURONS, AND LAYERS ARE NEEDED WITH ANNs?

In neural network design, the above parameters have no precise number/answers because it is dependent on the particular application. However, the question is worth addressing. In general, the more patterns and the fewer hidden neurons to be used, the better the network. It should be realized that there is a subtle relationship between the number of patterns and the number of hidden layer neurons. Having too few patterns or too many hidden neurons can cause the network to memorize. When memorization occurs, the network would perform well during training, but tests poorly with a new data set.

An easy to apply rule of thumb that the authors have found useful in their experience to determine the number of patterns required for training was:

Training patterns
= 2 × no. of (inputs + hiddens + outputs) to
  10 × no. of (inputs + hiddens + outputs)   (14)

where the inputs, hiddens, and outputs indicated the number of neurons in those layers of a selected architecture. This equation suggests that the number of training data required will be between 2 to 10 times the number of neurons in the selected network.

It should be emphasized that the selected network need not learn every training pattern perfectly. A good network will generalize well for new testing cases. Many users make the mistake of trying to achieve perfect training by adding more and more hidden neurons without realizing that this only leads to a higher possibility of deterioration during testing. There are no hard and fast rules to overcome this problem, but either of the following two guidelines can be used.

Number of hidden neurons
$$= \frac{(\text{inputs} + \text{outputs})}{2} \text{ or}$$

Minimum number of hidden neurons   (15)
$$= \frac{[\text{number of patterns}]}{10} - \text{inputs} - \text{outputs}$$

Maximum number of hidden neurons

$$= \frac{[\text{number of patterns}]}{2} - \text{inputs} - \text{outputs}$$

During the selection of the number of hidden layer neurons, the desired tolerance should also be considered. In general, a tight tolerance requires that the selected network be trained with fewer hidden neurons. As mentioned earlier, cross-validation during training can be used to monitor the error progression, which subsequently serves as a guideline in the selection of the hidden layer neurons,

A network that is too large may require a large number of training patterns in order to avoid memorization and training time, while one that is too small may not train to an acceptable tolerance. Cybenko [30] has shown that one hidden layer with homogenous sigmoidal output functions is sufficient to form an arbitrary close approximation to any decisions boundaries for the outputs. They are also shown to be sufficient for any continuous nonlinear mappings. In practice, one hidden layer was found to be sufficient to solve most problems for the cases considered in this chapter. If discontinuities in the approximated functions are encountered, then more than one hidden layer is necessary.

X. APPLICATION OF ANNs IN CHEMICAL SCIENCE

Applications of neural networks are becoming more diverse in chemistry [31–40]. Some typical applications include predicting chemical reactivity, acid strength in oxides, protein structure determination, quantitative structure property relationship (QSPR), fluid property relationships, classification of molecular spectra, group contribution, spectroscopy analysis, etc. The results reported in these areas are very encouraging and are demonstrative of the wide spectrum of applications and interest in this area.

XI. CASE STUDIES

In all, five case studies are presented in this chapter in an attempt to illustrate the performance of applicability of neural networks in polymer science. The diversity of problems considered in this chapter highlight ANNS as a general tool for problem solving. Although there are numerous other methods that may be just as appropriate for problem solving, ANNS are rapidly becoming the method of choice in certain areas. This is particularly so in those areas where the phenomenological understanding is very limited [23].

This section begins with a very simple case; that of modeling the specific volumes and viscosites of siloxanes. The siloxanes, which are considered here, have been well documented in the literature and the structural

relationships have been well addressed. In addition to this rather simple case, a relatively new application of neural networks namely in QSPR is described in Case V. The overall results obtained from these examples have been very encouraging and demonstrate that ANNS should be considered as a potential tool in polymer science.

A. Case I: Fluid Property Prediction of Siloxanes

An artificial neural network based approach for modeling physical properties of nine different siloxanes as a function of temperature and molecular configuration will be presented. Specifically, the specific volumes and the viscosities of nine siloxanes were investigated. The predictions of the proposed model agreed well with the experimental data [41].

1. Materials

The compounds considered in this work are shown in Table 1. These particular siloxane compounds have been described elsewhere [42]. A system of abbreviation already in use for these compounds is given in the same table. Their abbreviated formulas are expressed on a standard functional basis in Table 1 along with the name of the compound. An explanation of the units in the abbreviated formulas is given below:

$$M, \text{ the monofunctional unit: } (CH_3)_3Si\text{—}O \qquad (16)$$

$$D, \text{ the difunctional unit: } O\text{—}\underset{\underset{CH_3}{|}}{\overset{\overset{CH_3}{|}}{Si}}\text{—}O \qquad (17)$$

2. Specific Volumes of Siloxanes

Specific volumes of polymers generally show a regular decrease with increase in temperature, for both chain and ring structures. Experimental data from earlier investigations [42] indicate that this decrease in specific volume is greater for chain compounds than for ring compounds. For each increase of one difunctional group in the molecule, the specific volume of the chain compound is usually greater than that of a ring compound containing the same number of silicon atoms. It is hypothesized that the ring structure is more compact, and contains stronger interatomic attractive forces holding the molecule than does the chain compounds. With increased molecular size, this type of differentiation becomes much more complicated.

Table 1 List of Siloxanes and Their Abbreviations

No.	Abbreviation	Name of siloxane
1.	MM	Hexamethyldisiloxane
2.	MDM	Octamethyltrisiloxane
3.	MD_2M	Decamethyltetrasiloxane
4.	MD_3M	Dodecamethylpentasiloxane
5.	MD_4M	Tetradecamethylhexasiloxane
6.	D_4	Octamethylcyclotetrasiloxane
7.	D_5	Decamethylcycloentasiloxane
8.	D_6	Dodecamethylcyclohexasiloxane
9.	D_7	Tetradecamethylcycloheptasiloxane

The specific volumes of all the nine siloxanes were predicted as a function of temperature and the number of monofunctional units, M, and difunctional units, D. A simple 3-4-1 neural network architecture with just one hidden layer was used. The three input nodes were for the number of M groups, the number of D groups, and the temperature. The hidden layer had four neurons. The predicted variable was the specific volumes of the silox-anes. A schematic of the typical training process is shown in Fig. 6.

The experimental specific volume data were available in the temperature range of 273K to 353K, with 20K increments. The nine types of siloxanes were arbitrarily divided into two groups, one each for training and testing. The compounds 1, 2, 4, 6, and 8 were utilized in the training phase. The trained network was then

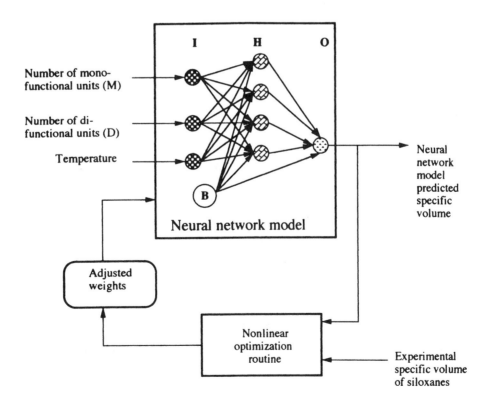

Figure 6 Schematic of a typical neural network training process. I-input layer; H-hidden layer; O-output layer; B-bias neuron.

tested for the specific volume prediction of compounds 3, 5, 7, and 9. The training and testing results are shown in Fig. 7. The root mean squared (rms) error for the model was 0.003.

3. Viscosities of the Siloxanes

Viscosities of the siloxanes were predicted over a temperature range of 298–348 K. The semi-log plot of viscosity as a function of temperature was linear for the ring compounds. However, for the chain compounds, the viscosity increased rapidly with an increase in the chain length of the molecule. A simple 2-4-1 neural network architecture was used for the viscosity predictions. The molecular configuration was not considered here because of the direct positive effect of addition of both M and D groups on viscosity. The two input variables, therefore, were the siloxane type and the temperature level. Only one hidden layer with four nodes was used. The predicted variable was the viscosity of the siloxane.

For all the siloxanes the network was trained at two temperature levels; 25°C and 75°C. The trained network was then tested for its viscosity predictions at 50°C. The network training and testing results are shown in Fig. 8. The rms error for this prediction was 0.002.

4. Relationship Between Viscosity and Specific Molar Volumes

Batschinski's relationship [43] for specific volume and viscosity is probably the best known for siloxanes. This relationship can be written as follows:

$$v = \left(\frac{C}{\eta}\right) + \omega \qquad (18)$$

where v is the specific volume of the siloxane, η is the viscosity of the siloxane and, C and ω are constants that can be obtained from experimental data. Values of these constants for the siloxanes under investigation were reported by Hurd [42]. It should be remembered that C and ω are a function only of the compound and independent of temperature. The experimental viscosity data for the siloxanes under investigation were available at 298 K, 323 K, and 348 K. The proposed neural network model was put to a test by comparing its predictions with Batschinski's equation in this temperature range (298 K, 323 K, and 348 K). Figure 9 shows the comparison between these two predictions. The agreement was very good between the experimental based Batschinski's relationship and the ANN model at least within the limited range of training data.

B. Case II: Prediction of Densities of High Molecular Weight Esters Used as Plasticizers*

To any one concerned with the production of plasticized resins, compatibility has long meant the ability of two or more materials to mix with each other to form a homogeneous composition of useful desired plastic properties [44,45]. A plasticizer, therefore, can be defined as a material that will soften and make inherently rigid and even brittle polymers flexible. Plasticizers are also sometimes

* Reprinted with permission from *Adv. Polym. Tech.*, Vol. 14, No. 3, 215–225 (1995). © 1995 by John Wiley & Sons, Inc.

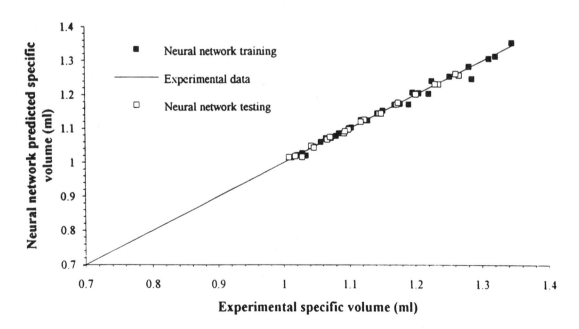

Figure 7 ANN training and testing results for specific volumes of siloxanes.

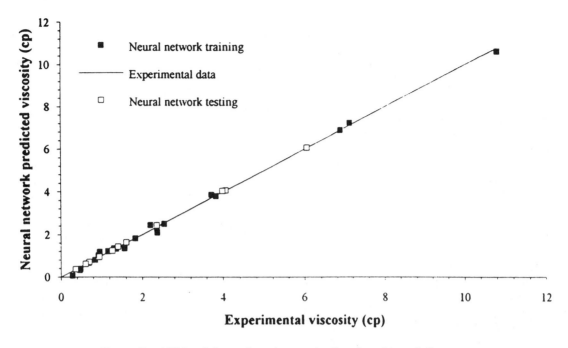

Figure 8   ANN training and testing results for viscosities of siloxanes.

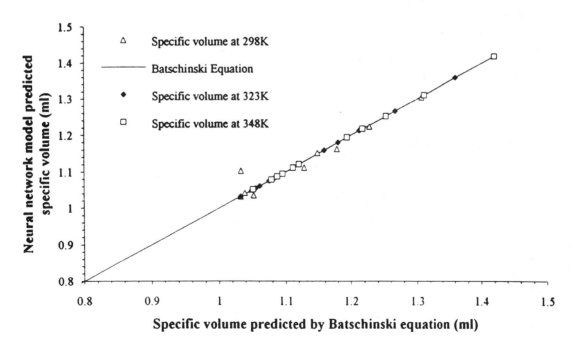

Figure 9   Comparison of viscosity predictions between ANN model and Batschinski's equation.

added to lower the melt viscosity. They can reduce the temperature of the second-order transition for easier processing by virtue of both glass transition temperature (T_g) depression and through internal lubrication. Hence, plasticizers are commonly liquids that are compatible with their parent polymers, but not to the point of complete miscibility.

Primary plasticizers that are used as the sole plasticizing ingredients are typically high molecular weight organic esters. Such esters have found extensive use in the plastic industry. They demonstrate a high degree of permanence and flexibility on a unit weight basis. Also of importance are their low level of toxicity, consistently light color, resistance to degradation, and relatively low cost. Plasticizers generally show a reduction in effectiveness with increasing molecular weight and temperature. As solvents they give solutions that undergo gelation on heating. Lower molecular weight esters, however, can become volatile and/or decompose at elevated processing temperatures.

Since the physical properties of the polymer matrix can be significantly modified by the type of the plasticizer used, these high molecular weight esters have been extensively investigated during the last decade. Liquid density, for example, has been found to be a useful bulk property of these liquid plasticizers, since density helps define the internal molecular flexibility. Most of the time, however, experimental difficulties prevent the accurate determination of density. Moreover, existing empirical models to predict density remain tedious because of many ill-defined parameters. Therefore, a simple neural network based model approach is offered wherein the effect of the number of end groups and temperature on density were incorporated. The proposed model is also capable of differentiating noise or error in the experimental data.

The effects of structure and temperature on the densities of a homologous series of esters were investigated. In all, five different series of esters were considered.

They were: dialkyl sebacate ester (DSEs), 1,10-decanediol diester (DDEs), triglycerides (TGEs), 2-(hydroxymethyl)-2-methyl-1,3-propanediol trialky esters (TTEs), and pentaerythritol tetra alkyl esters (PTEs). Of the five, the first two can be characterized as linear and flexible esters, while the third and fourth are "Y" branched compact molecules. The last series, namely PTEs are "+" shaped and highly rigid. The various esters considered in this study were structurally differentiated by earlier investigators [46–48] who used the number of methylene groups in the molecule. Density measurements should help reveal changes in the compatibility and efficiency of the ester, and relate directly to the plasticizer's permanence. The results of this study offer an insight into the density changes that are produced by structure for these industrially important plasticizers.

1. Materials

The isothermal experimental density data for all the five type of esters were obtained from various literature sources [46–48]. The reported density measurements for DDEs, TGEs, and PTEs were for different temperatures (310–413 K). The chemical–structures of all five esters are shown in Tables 2–6, with the different numbers of methylene groups in the molecules being specified by (X). Numerical results from these references are not presented here.

The isothermal densities within each of the five groups of esters reveal a significant dependence on the molecular structure. At a constant temperature and pressure, the liquid density of the ester decreases with increasing molecular weight and the successive addition of the methylene groups, (X), even though the concentration of the end-groups remained constant.

It should be remembered that the volume requirement for the rotation of a molecule depends on both its shape and the possible directional nature of its interaction with its neighbors. As the chain length in the ester is increased by the successive addition of methylene

Table 2 Dialkyl Sebacate Esters (DSEs)

$$CH_3(CH_2)_X—O—\overset{\overset{\displaystyle O}{|}}{C}—(CH_2)_8—\overset{\overset{\displaystyle O}{|}}{C}—O—(CH_2)_XCH_3$$

Compound	X	Empirical formula	Molecular weight
Di-*n*-butyl sebacate	3	$C_{18}H_{34}O_4$	314
Di-*n*-hexyl sebacate	5	$C_{22}H_{42}O_4$	370
Di-*n*-octyl sebacate	7	$C_{26}H_{50}O_4$	426
Di-*n*-decyl sebacate	9	$C_{30}H_{58}O_4$	482
Di-*n*-dodecyl sebacate	11	$C_{34}H_{66}O_4$	539
Di-*n*-tetradecyl sebacate	13	$C_{38}H_{74}O_4$	595
Di-*n*-hexadecyl sebacate	15	$C_{42}H_{82}O_4$	651
Di-*n*-octadecyl sebacate	17	$C_{46}H_{90}O_4$	707

Table 3 1,10-Decanediol Diesters (DDEs)

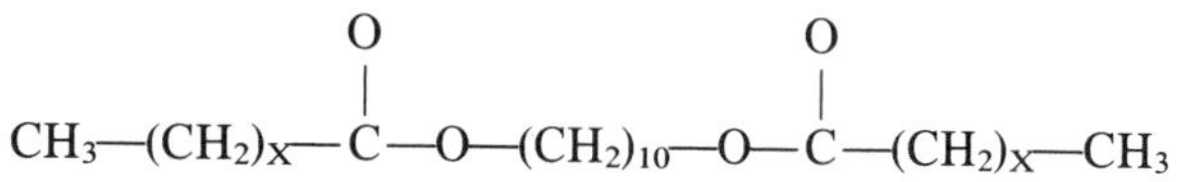

$$CH_3-(CH_2)_X-\overset{\overset{\displaystyle O}{\|}}{C}-O-(CH_2)_{10}-O-\overset{\overset{\displaystyle O}{\|}}{C}-(CH_2)_X-CH_3$$

Compound	X	Empirical formula	Molecular weight
1,10-decanediol dibutyrate	2	$C_{18}H_{34}O_4$	314
1,10-decanediol dioctanaote	6	$C_{26}H_{50}O_4$	426
1,10-decanediol didecanoate	8	$C_{30}H_{58}O_4$	482
1,10-decanediol dipalmitate	14	$C_{42}H_{82}O_4$	650

groups, it tends to loosen the packing by disrupting the large sterically hindered carboxyl group in the ester series. This disruption produces an increase in the specific volume of the ester, hence, the packing efficiencies for small molecules are higher than they are for the larger molecules. The densities of individual groups of esters can also be differentiated by the difference in the number of carboxyl groups. The observed decrease in the density of the esters, with the successive addition of the methylene groups (X), was probably due to the dilution of the —COO— groups. This is because the density of the carboxyl group is greater than the methylene group. The densities of the esters were in the following order:

PTEs > TTEs/TGEs > DSEs > DDEs

This observed order can be explained on the basis of the relative degree of the packing and flexibility of the atoms in the molecules of these esters.

2. Neural Network Performance

A very simple 2-4-1 neural network architecture with two input nodes, one hidden layer with four nodes, and one output node was used in each case. The two input variables were the number of methylene groups and the temperature. Although neural networks have the ability to learn all the differences, differentials, and other calculated inputs directly from the raw data, the training time for the network can be reduced considerably if these values are provided as inputs. The predicted variable was the density of the ester. The neural network model was trained for discrete numbers of methylene groups over the entire temperature range of 300–500 K. The

Table 4 Triglyceride Esters (TGEs)

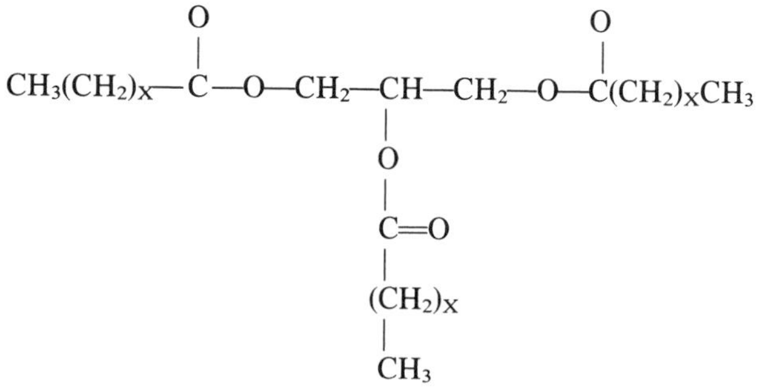

$$CH_3(CH_2)_X-\overset{\overset{\displaystyle O}{\|}}{C}-O-CH_2-\underset{\underset{\displaystyle CH_3}{\underset{\displaystyle |}{\underset{\displaystyle (CH_2)_X}{\underset{\displaystyle |}{\underset{\displaystyle C=O}{\underset{\displaystyle |}{\underset{\displaystyle O}{\underset{\displaystyle |}{CH}}}}}}}}-CH_2-O-\overset{\overset{\displaystyle O}{\|}}{C}(CH_2)_XCH_3$$

Compound	X	Empirical formula	Molecular weight
Tributyrin	2	$C_{15}H_{26}O_6$	302
Tricaproin	4	$C_{21}H_{38}O_6$	386
Trioctanoin	6	$C_{27}H_{50}O_6$	470
Tridecanoin	8	$C_{33}H_{62}O_6$	554
Trilaurin	10	$C_{39}H_{74}O_6$	639
Trimyristin	12	$C_{45}H_{86}O_6$	723
Tripalmitin	14	$C_{51}H_{98}O_6$	807
Tristearin	16	$C_{57}H_{110}O_6$	890

Table 5 2-(Hydroxymethyl)-2-Methyl-1,3-Propanediol Triesters (TTEs)

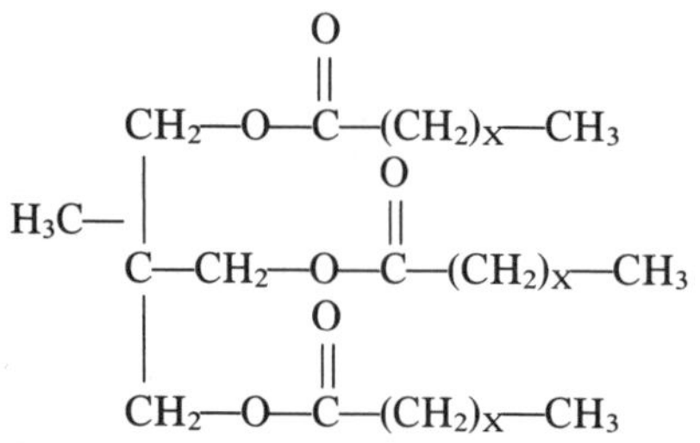

$$
\begin{array}{c}
\text{O} \\
\| \\
\text{CH}_2\text{—O—C—(CH}_2)_\text{X}\text{—CH}_3 \\
\text{O} \\
\| \\
\text{H}_3\text{C—} \ \ \text{C—CH}_2\text{—O—C—(CH}_2)_\text{X}\text{—CH}_3 \\
\text{O} \\
\| \\
\text{CH}_2\text{—O—C—(CH}_2)_\text{X}\text{—CH}_3
\end{array}
$$

Compound	X	Empirical formula	Molecular weight
2-(Hydroxymethyl)-2-methyl-1,3-propanediol tripropionate	1	$C_{14}H_{24}O_6$	288
2-(Hydroxymethyl)-2-methyl-1,3-propanediol tributyrate	2	$C_{17}H_{30}O_6$	330
2-(Hydroxymethyl)-2-methyl-1,3-propanediol trivalerate	3	$C_{20}H_{36}O_6$	372
2-(Hydroxymethyl)-2-methyl-1,3-propanediol triheptanoate	5	$C_{26}H_{48}O_6$	456
2-(Hydroxymethyl)-2-methyl-1,3-propanediol tripoctanoate	6	$C_{29}H_{54}O_6$	498
2-(Hydroxymethyl)-2-methyl-1,3-propanediol trinonanoate	7	$C_{32}H_{60}O_6$	540
2-(Hydroxymethyl)-2-methyl-1,3-propanediol tridecanoate	8	$C_{35}H_{66}O_6$	582
2-(Hydroxymethyl)-2-methyl-1,3-propanediol tripalmitate	14	$C_{53}H_{102}O_6$	834

Table 6 Pentaerythritol Tetra Alkyl Esters (PTEs)

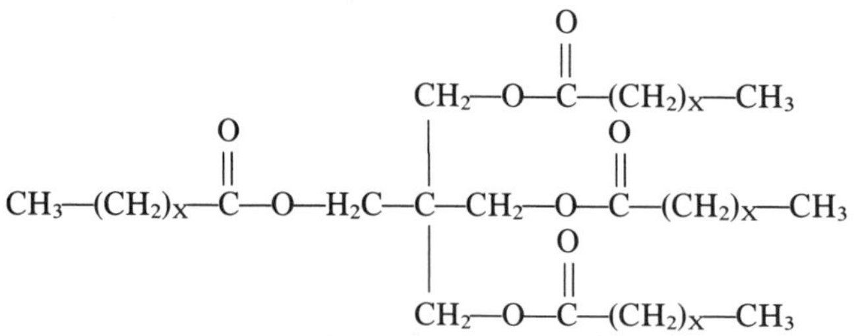

$$
\begin{array}{c}
\text{O} \\
\| \\
\text{CH}_2\text{—O—C—(CH}_2)_\text{X}\text{—CH}_3 \\
\text{O} \ \ \ \ \ \ \ \ \ \ \ \ \ \ \ \text{O} \\
\| \ \ \ \ \ \ \ \ \ \ \ \ \ \ \ \ \| \\
\text{CH}_3\text{—(CH}_2)_\text{X}\text{—C—O—H}_2\text{C—C—CH}_2\text{—O—C—(CH}_2)_\text{X}\text{—CH}_3 \\
\text{O} \\
\| \\
\text{CH}_2\text{—O—C—(CH}_2)_\text{X}\text{—CH}_3
\end{array}
$$

Compound	X	Empirical formula	Molecular weight
Pentaerythritol tetraacetate	0	$C_{13}H_{20}O_8$	304
Pentaerythritol tetrapropionate	1	$C_{17}H_{28}O_8$	370
Pentaerythritol tetrabutyrate	2	$C_{21}H_{36}O_8$	416
Pentaerythritol tetravalerate	3	$C_{25}H_{44}O_8$	472
Pentaerythritol tetraheptanoate	5	$C_{33}H_{60}O_8$	584
Pentaerythritol tetraoctanoate	6	$C_{37}H_{68}O_8$	640
Pentaerythritol tetranonanoate	7	$C_{41}H_{76}O_8$	696
Pentaerythritol tetradecanoate	8	$C_{45}H_{84}O_8$	752

Table 7 ANN Predictions

Compound	"X" at training	"X" at testing	Sum-of-squared-error
DSEs	3, 7, 11, 15 and 17	5, 9 and 13	<0.001
DDEs	2, 6 and 14	8	<0.001
TGEs	2, 6, 10, 14 and 16	4, 8 and 12	0.001
TTEs	1, 3, 6, 8 and 16	2, 5 and 7	<0.001
PTEs	1, 2, 4, 6 and 8	3, 5 and 7	<0.001

trained network was then tested for density predictions using compounds with a different number of methylene groups [49]. The latter were not included in the training phase.

Results for the ANN model are given in Table 7, along with the type of compounds used in the training and testing phases. Each ester group was trained and tested separately. The sum-of-squared error of the proposed network for the density predictions during the testing phase was very low (Table 7). The density predictions were with a high degree of precision considering a $\pm 2\%$ uncertainty in the reported experimental data. This uncertainty can be compared to the noise in a typical process output.

The ANN was able to assimilate the cause–effect relationship of the density of the ester, its structure and temperature. The training and testing results are shown in Fig. 10–14 for individual ester series. The network with the proposed training routine converged in less than 100 iterations for all the esters.

As a comparison, the results from a back-propagation training routine with a gradient search for PTE is shown in Fig. 15. Also, the speed and accuracy of the proposed nonlinear optimization routine are shown as a sum-of-squared error progression plot with the back-propagation training routine in Fig. 16. A sum-of-squared error of 10^{-4} was achieved with just 67 iterations with the proposed nonlinear optimization training routine, whereas with a back-propagating routine only a sum-of-squared error of 10^{-2} was achieved even after 180 iterations.

C. Case III: Prediction of Solvent Activity in Polymer Systems*

In many process design applications like polymerization and plasticization, specific knowledge of the thermodynamics of polymer systems can be very useful. For example, non-ideal solution behavior strongly governs the diffusion phenomena observed for polymer melts and concentrated solutions. Hence, accurate modeling of

---

* Reprinted with permission from *Ind. Eng. Chem. Res.*, Vol. 34, No. 11, 3974–3980 (1995). © 1995 American Chemical Society.

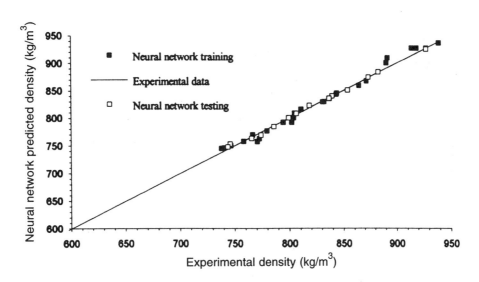

Figure 10 ANN training and testing results for densities of DSEs.

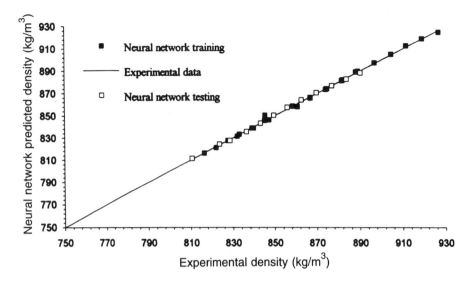

Figure 11 ANN training and testing results for densities of DDEs.

thermodynamic parameters, such as the solvent activity, is a necessary requisite for the proper design of many polymer processes.

It is an arduous task to develop thermodynamic models or empirical equations that accurately predict solvent activities in polymer solutions. Even so, since Flory developed the well-known equation of state for polymer solutions, much work has been conducted in this area [50–52]. Consequently, extensive experimental data have been published in the literature by various researchers on different binary polymer–solvent sys-

tems. When such data are available, modeling solvent activity in polymer solutions can be simplified by the use of the ANN technique.

The well-known Flory treatment [50–52] of the enthropic contribution to the Gibbs energy of mixing of polymers with solvents is still the simplest and most reliable theory developed. It is quite apparent, however, that the Flory–Huggins theory was established on the basis of the experimental behavior of only a few mixtures investigated over a very narrow range of temperature. Strict applications of the Flory–Huggins approach

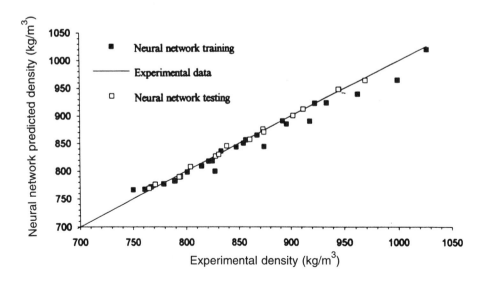

Figure 12 ANN training and testing results for densities of TGEs.

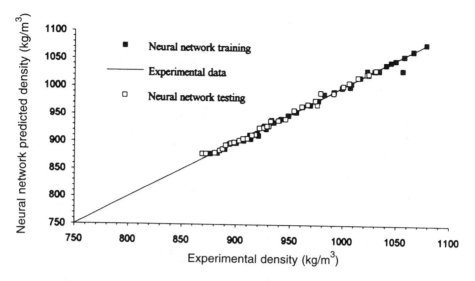

Figure 13 ANN training and testing results for densities of TTEs.

requires "χ" to be constant; yet the temperature dependence of χ can be substantial even in athermal systems. Although this parameter was initially introduced to account for the energetic interactions between the polymer and the solvent, it was later shown that it is convenient to regard it as a free-energy term with both an enthalpic and an entropic contribution.

$$\chi = \chi_h + \chi_s \tag{19}$$

While this approach can quantitatively describe a number of phenomena occurring in many polymer systems, it has some deficiencies.

For more than two decades researchers have attempted to overcome the inadequacies of Flory's treatment in order to establish a model that will provide accurate predictions. Most of these research efforts can be grouped into two categories, i.e., attempts at corrections to the enthalpic or noncombinatorial part, and modifications to the entropic or combinatorial part of the Flory–Huggins theory. The more complex relationships derived by Huggins, Guggenheim, Stavermans, and others [53] required so many additional and poorly determined parameters that these approaches lack practical applications. A review of the more serious deficiencies

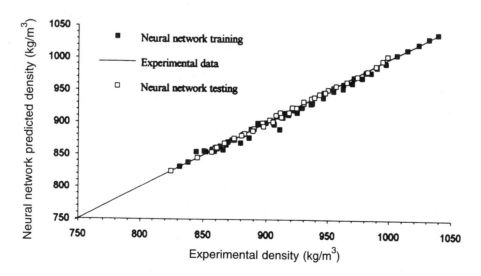

Figure 14 ANN training and testing results for densities of PTEs.

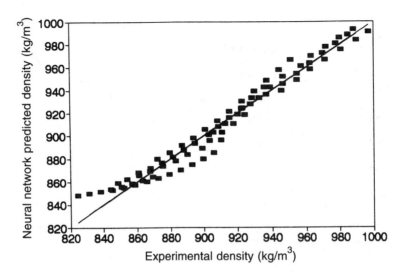

Figure 15 Back-propagation training results for PTEs. ————-Experimental data; ■-Network prediction.

in the existing models was given in a recent article by Kontogeorgis et al. [54]. Despite their differences, the equations which have been developed remain complex and sometimes not fully consistent. Moreover, the predictions of these modified Flory–Huggins equations are especially inadequate for polymer solutions in which strong intermolecular polar forces must be considered.

The purpose of this case study was to develop a simple neural network based model with the ability to predict the solvent activity in different polymer systems. The solvent activities were predicted by an ANN as a function of the binary type and the polymer volume frac-

tion "ϕ." Three different polymer solvent binaries were considered: namely polar binaries, nonpolar binaries, and polar/nonpolar binaries. The proposed neural network based model resulted in good agreement with the experimental data available for the tested polymer–solvent systems. Without this approach, the estimation of a proper correction for the Flory–Huggins equation to give a solution model that will yield a reasonable prediction for all the polymer–solvent systems would be time consuming. This is because considerable efforts are needed to simultaneously optimize the different model parameters estimated from the experimental data.

1. Predicting Polymer Activities in Polymer/ Solvent Binaries

A list of the systems investigated in this work is presented in Tables 8–10. These systems represent 4 nonpolar binaries, 8 nonpolar/polar binaries, and 9 polar binaries. These binary systems were recognized by Heil and Prausnitz [55] as those which had been well studied for a wide range of concentrations. With well-documented behavior they represent a severe test for any proposed model. The experimental data used in this work have been obtained from the work of Alessandro [53]. The experimental data were arbitrarily divided into two data sets; one for use in training the proposed neural network model and the remainder for validating the trained network.

2. Predictions of ANN with Non-Linear Optimization Training Routine

To evaluate the reliability of the proposed neural network model, all the binaries were trained in each of the

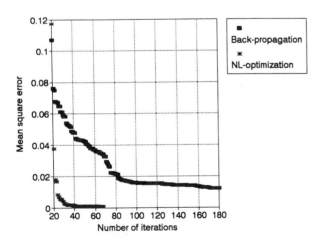

Figure 16 Root-mean-squared error progression plot for Fletcher nonlinear optimization and back-propagation algorithms during training.

Table 8 Experimental Solvent Activity in Nonpolar Binaries

System	ϕ_1	a_1
Cyclohexane/ Polystyrene (PM = 25.900)	0.5150	0.9980
	0.3630	0.9670
	0.3270	0.9696
	0.2340	0.8947
	0.1990	0.8480
	0.1960	0.8442
	0.1290	0.7651
Cyclohexane/ Polystyrene (PM = 440.000)	0.5170	0.6784
	0.5150	0.7479
	0.4320	0.7910
	0.3390	0.8466
	0.3290	0.9128
	0.2420	0.9183
	0.1980	0.9601
	0.1720	0.9823
	0.1410	0.9827
Benzene/ Polystyrene	0.9000	0.995
	0.8005	0.989
	0.8000	0.983
	0.7500	0.975
	0.7000	0.963
	0.6500	0.949
	0.6000	0.930
Toluene/ Polystyrene	0.9000	0.995
	0.8500	0.989
	0.8000	0.983
	0.7500	0.965
	0.7000	0.963
	0.6500	0.949
	0.6000	0.930

Source: Ref. 53.

Table 9 Experimental Solvent Activity in Nonpolar/ Polar Binaries

System	ϕ_1	a_1
Rubber/Acetone	0.154	0.955
	0.132	0.917
	0.084	0.772
	0.053	0.600
	0.045	0.533
Rubber/Ethylacetate	0.3750	0.988
	0.3030	0.932
	0.2600	0.604
	0.2060	0.832
	0.1290	0.684
	0.1200	0.647
	0.0510	0.373
Rubber/ Methylethylketone	0.4210	0.990
	0.3740	0.973
	0.3240	0.955
	0.2570	0.603
	0.2000	0.838
	0.1340	0.709
	0.0900	0.577
Polypropylene/ Diethylketone	0.0070	0.090
	0.0100	0.167
	0.0110	0.209
	0.0220	0.289
	0.0260	0.344
	0.0330	0.374
	0.0350	0.395
	0.0430	0.412
	0.0700	0.565
	0.1210	0.734
	0.1960	0.887
Polyproylene/ Diisopropylketone	0.0150	0.134
	0.0160	0.149
	0.0350	0.236
	0.0660	0.421
	0.0970	0.521
	0.1620	0.699
	0.2300	0.812
	0.3010	0.876
	0.3540	0.923
Polystyrene/Acetone	0.1099	0.608
	0.2174	0.800
	0.3226	0.872
	0.4255	0.900
Polystyrene/ Propylacetate	0.2174	0.672
	0.3226	0.792
	0.4255	0.864
	0.5263	0.896
	0.6250	0.700
Polystrene/Chloroform	0.2174	0.360

(continued)

three widely reported categories (as shown in Tables 8–10) together as one training. A 2-4-1 feed forward neural network architecture was used in the training for all of the three categories of polymer/solvent combinations [56]. As mentioned earlier, the 2-4-1 architecture implies 2 input nodes: namely the binary type, and the volume fraction of the polymer ϕ_1 in the binary. There were 4 hidden nodes and 1 output node, namely the polymer activity, a_1, in the binary. The neutral network model training and testing results are shown in Figs. 17–19 for all three categories of polymer/solvent binaries. As shown, the predictions by the neural network model gave extremely good agreement with the experimental data. The average sum of the squared error for all the three binaries was 8.0×10^{-4}. From the tables and these

Table 9 (*Continued*)

System	ϕ_1	a_1
	0.3226	0.512
	0.4255	0.632
	0.5263	0.744
	0.6250	0.816
	0.7216	0.864
	0.8163	0.896

Source: Ref. 53.

Table 10 Experimental Solvent Activities in Polar Binaries

System	ϕ_1	a_1
Polypropylene glycol/Methanol (PM = 1955)	0.8874	0.997
	0.8315	0.996
	0.6701	0.985
	0.4899	0.958
Polypropylene glycol/Methanol (PM = 3350)	0.8961	0.999
	0.8087	0.996
	0.7233	0.993
	0.5746	0.979
Polyethylene oxide/Chloroform	0.7840	0.912
	0.7140	0.833
	0.6180	0.687
	0.5200	0.526
	0.4950	0.486
	0.4240	0.433
	0.3270	0.431
	0.2010	0.437
	0.1060	0.418
	0.0660	0.407
Cellulose acetate/Acetone	0.9409	0.997
	0.9093	0.995
	0.8761	0.993
	0.8414	0.991
	0.8049	0.989
	0.7666	0.987
	0.7262	0.985
	0.6837	0.981
	0.6388	0.977
Cellulose acetate/ Methylacetate	0.9308	0.998
	0.8945	0.998
	0.8565	0.996
	0.5992	0.994
	0.7352	0.992
	0.6916	0.990
	0.6463	0.987
	0.5992	0.984

Table 10 (*Continued*)

System	ϕ_1	a_1
Cellulose acetate/Dioxane	0.8846	0.996
	0.8440	0.993
	0.8023	0.990
	0.7594	0.986
	0.7152	0.983
	0.6698	0.979
	0.6231	0.974
	0.5749	0.968
Cellulose acetate/Pyridine	0.8895	0.996
	0.8504	0.990
	0.8100	0.982
	0.7683	0.972
	0.7252	0.959
	0.6807	0.943
	0.6346	0.925
Cellulose nitrate/Acetone	0.9495	0.999
	0.9220	0.998
	0.8930	0.996
	0.8623	0.993
	0.8296	0.988
	0.7949	0.977
	0.7579	0.946
	0.7184	0.884
Cellulose nitrate/Methylacetate	0.9406	0.999
	0.9088	0.997
	0.8755	0.993
	0.8406	0.987
	0.8040	0.977
	0.7655	0.955
	0.7251	0.916
	0.6824	0.842

Source: Ref. 53.

figures, it can be concluded that the pure binaries exhibited higher activity than did the polar/nonpolar binaries.

3. Predictions of ANNs with Back-Propagation and Madaline III Training Routines

A 2-6-1 neural network architecture was needed for both of these training routines. Two additional hidden nodes were required with the gradient search based procedures. Since these training routines are extremely slow compared to the nonlinear optimization routine, the error limit on the sum of squared error was fixed at 1×10^{-3} for the termination conditions. The neural network model with back-propagation routine required 29,000 iterations while the Madaline III approach required 57,000 iterations. Compared to these two training routines, the nonlinear optimization routine required less than 100 iterations. The training and testing results for these two algorithms are shown in Figs. 20 and 21 for the polar

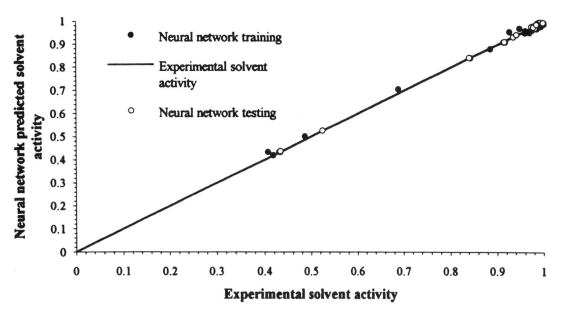

Figure 17 ANN training and testing results for solvent activities of polar/polar binary system.

binaries only. A mean square error progression during the training phase is shown in Fig. 22 for the Fletcher nonlinear optimization routine and a traditional conjugate gradient search routine. Convergence of the Fletcher routine required 35 iterations, compared to several thousand iterations required for gradient search based back-propagation and Madaline III routines. Such comparisons of the proposed training routine were reported for predicting densities of high molecular weight esters in our earlier publication [56].

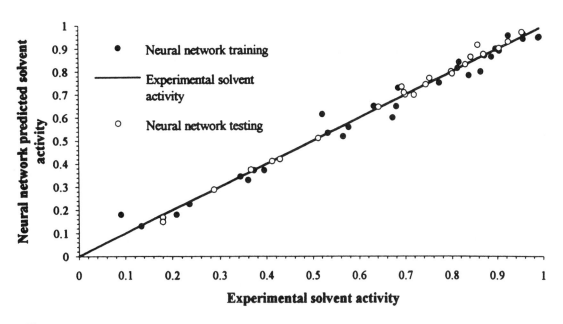

Figure 18 ANN training and testing results for solvent activities of polar/nonpolar binary system.

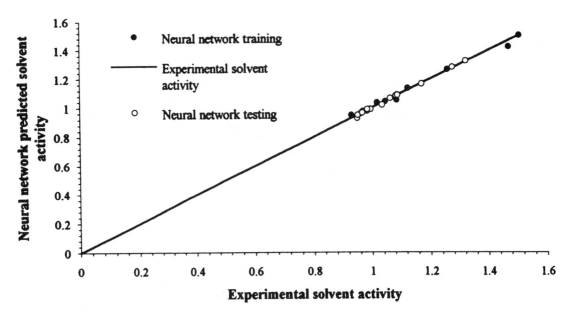

Figure 19   ANN training and testing results for solvent activities of nonpolar/nonpolar binary system.

D. Case IV: Integrating the Concepts of Graphical Theory and ANNs for Polymer Property Predictions in QSPR

As polymer scientists continue to investigate the atomic structure and physical properties of polymers, the need for more accurate models to describe various experi-

mental measurements has become even more apparent. In this section, a new integrated approach is presented that makes use of ANN principals to establish relationships between the atomic structure of a polymer and its physical properties (QSPR). Parameters dependent on atomic structure and particular physical property obtained from graphical theory were utilized in this ap-

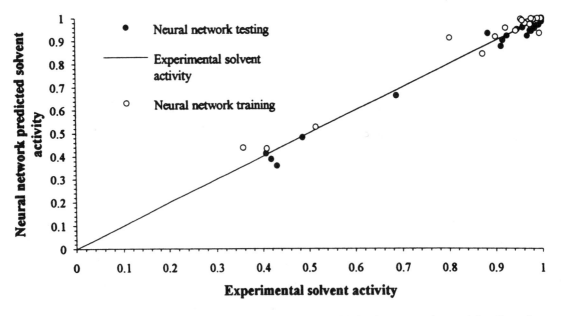

Figure 20   Feed-forward neural network training and testing results with back-propagation training for solvent activity predictions in polar binaries (with learning parameter $\eta = 0.1$).

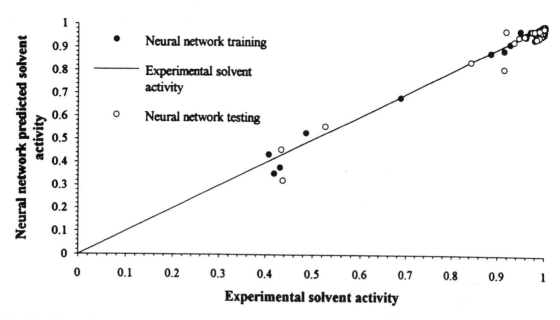

Figure 21 Feed-forward neural network training and testing results with Madaline III training for solvent activity predictions in polar binaries (with learning parameter $\eta = 0.1$ and perturb $\Delta s = 0.1$).

proach. These selected physical properties of various polymers were predicted based solely on their structural attributes. A total of six different properties for each of 45 different polymers were predicted by this integrated approach.

1. Background of QSPR

Physical property prediction in polymer science has evolved from the original basic group contribution meth-ods to the more recently developed graph theory, which is based on topology. With graph theory, connectivity indices for many physical properties can be calculated based on any arbitrary polymer structure. Connectivity indices, therefore, represent numerical parameters that can be calculated from the chemical structures of the constituent atoms and bonds of the mer(s) in the poly-mer. Hence, connectivity indices have been shown to be most useful for polymers with well-defined chemical structure.

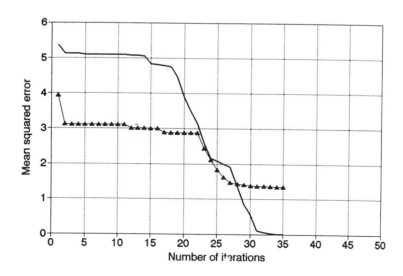

Figure 22 Root-mean-squared error progression plot for Fletcher nonlinear optimization and back-propagation algorithms during training. ————-Fletcher routine; —▲—-Gradient search.

This topological approach with connectivity indices has been extended by Bicerano [58] to a point where many of the physical properties of the polymer can be estimated from empirical predictive equations. Bicerano correlated connectivity indices with group contribution values in order to develop a model equation for a specific property [58]. The usefulness of Bicerano's equations is that they can be extended to predict the same property values for new interested polymers.

A need still exists, however, for ways to simplifying or combine the many correlations that have been created by graph theory. Bicerano generalized some of the correlations by using adjustable parameters to eliminate correction factors that were cumbersome to estimate and difficult to satisfy for the many different polymers that must be considered. Also, higher order indices have been introduced to eliminate the use of many correction factors. However, these higher order indices do not always guarantee substantial improvements. In this chapter, the concepts of graph theory were integrated with a semi-empirical modeling approach based on the neural networks concept. This approach appears to effectively eliminate the above listed shortcomings of graph theory.

2. Graph Theory

Graph theory has been extensively used by earlier investigators to study the various physical properties of polymers [57–64]. From this modest beginning, Bicerano used the topological method of connectivity to develop a set of empirical equations for predicting many polymer properties. Bicerano's process of estimating connectivity indices for a polymer structure based on the principles of graph theory is illustrated in the following paragraphs for a simple molecule of vinyl fluoride.

$$
\begin{array}{ccc}
\text{H} \quad \text{F} & & \text{F} \\
| \quad\quad | & & | \\
\text{C}\!=\!\!=\!\text{C} & \longrightarrow & \text{C}\!-\!\!-\!\text{C} \\
| \quad\quad | & & \\
\text{H} \quad \text{H} & &
\end{array} \qquad (20)
$$

Valence bonds in vinyl fluoride   *Hydrogen suppressed graph*

Starting with the valence bond (Lewis) structure of the vinyl fluoride molecule, the hydrogen atoms are first removed from the structure to construct a hydrogen suppressed graph of the mer unit. Each remaining atom becomes a vertex in the graph and each bond is called an edge. The first atomic index (δ) specifies a connectivity index, which is equal to the number of nonhydrogen atoms to which that particular nonhydrogen atom is bonded.

Information on the electronic configuration of each nonhydrogen atom is incorporated in the second order atomic index (δ^v) of the atom.

$$
\delta^v = \left[\frac{Z^v - N_H}{Z - Z^v - 1} \right] \qquad (21)
$$

In Eq. 21, Z is the atomic number of the atom, Z^v is the number of valence electrons in the atom, and N_H is the number of hydrogen atom bonded to it.

The connectivity indices for vinyl fluoride, therefore, is written as follows:

$$
\begin{array}{ccccc}
1 & & & 7 & \\
| & & & | & \\
1\!-\!\!-\!2 & & & 2\!-\!\!-\!3 & \\
\delta & & & \delta^v &
\end{array} \qquad (22)
$$

Bond indices β and β^v can be assigned to each edge of the hydrogen suppressed graph at the two vertices, i and j.

$$
\beta_{ij} = \delta_i \times \delta_j \qquad (23)
$$
$$
\beta_{ij}{}^v = \delta_i{}^v \times \delta_j{}^v
$$

Hence, the bond indices can be written as follows:

$$
\begin{array}{ccccc}
& 1 & & 7 & \\
2 & | & & 21 & | \\
1\!-\!\!-\!2 & & 2\!-\!\!-\!3 & \\
2 & & & 6 & \\
\beta & & & \beta^v &
\end{array} \qquad (24)
$$

Finally, the atomic and bond indices can be combined to give indices for the whole unit. The zeroth-order connectivity indices O_χ and $O_\chi{}^v$ for the entire molecule can be calculated as a summation over the vertices of the hydrogen suppressed graph, that is:

$$
O_\chi = \sum_{\text{vertices}} \left(\frac{1}{\sqrt{\delta}} \right) \qquad (25)
$$

$$
O_\chi{}^v = \sum_{\text{vertices}} \left(\frac{1}{\sqrt{\delta^v}} \right)
$$

Similarly, the first order connectivity indices 1_χ and $1_\chi{}^v$ for the entire molecule can be written as:

$$
1_\chi = \sum_{\text{edges}} \left(\frac{1}{\sqrt{\beta}} \right) \qquad (26)
$$

$$
1_\chi{}^v = \sum_{\text{edges}} \left(\frac{1}{\sqrt{\beta^v}} \right)
$$

The zeroth and first order indices for vinyl fluoride can be shown to be $O_\chi = 2.7071$, $O_\chi{}^v = 1.6625$, $1_\chi = 1.4142$, and $1_\chi{}^v = 0.06264$, respectively. (Readers are referred to the source book for more details on this procedure.)

In his work Bicerano used only the zeroth and first order indices to develop his correlation model. However, various types of structural and correction parameters were used to correct for underestimation and/or overestimation of the contribution of several structural units. These corrections are cumbersome to use because of the many adjustable parameters that were utilized. Therefore, in this study a different approach was investigated. First, an internal loop for the correlation of six different physical properties represented for a group of 45 different polymers was established using the principles of neural networks. Finally, the prediction by this model was tested for a new set of polymers that the ANN had not seen. This integration of graph theory and ANN procedure will be explained in the following paragraphs.

3. Integrating Graphical Theory with Neural Networks for Establishing QSPR

In this approach, connectivity indices were used as the principle descriptor of the topology of the repeat unit of a polymer. The connectivity indices of various polymers were first correlated directly with the experimental data for six different physical properties. The six properties were: Van der Waals volume (V_W), molar volume (V), heat capacity (C_p), solubility parameter (δ), glass transition temperature (T_g), and cohesive energies (E_{coh}) for the 45 different polymers. Available data were used to establish the dependence of these properties on the topological indices. All the experimental data for these properties were trained simultaneously in the proposed neural network model in order to develop an overall cause–effect relationship for all six properties.

As shown in Fig. 23, five parameters obtained from the hydrogen suppressed graph were used as the input for the ANN. These parameters were, N, the number of vertices in the hydrogen suppressed graph, the zeroth order connectivity indices, O_χ and O_χ^v, and the first order connectivity indices 1_χ and 1_χ^v. Hence, the proposed ANN model had five input nodes/neurons in the input layer. Since, six different properties were to be predicted, six output nodes were used in the output layer. Only one hidden layer was found to be sufficient. The number of hidden layer neurons to be used was determined by a trial-and-error procedure that monitored the rms error progression for all the six output variables. Eight hidden nodes was found to be effective, and, hence, the model had a 5-8-6 architecture. As in the earlier cases, Fletcher's based nonlinear optimization procedure was used. A schematic of the integrated training procedure between graph theory and ANN used is shown in Fig. 24.

The model training and testing results for the six physical properties considered earlier are shown in Figs. 25 and 26, respectively. Because of differences in magnitudes, the results of five of the properties, excluding cohesive energy, are shown separately in Fig. 25. The rms error progression during the training phase is shown in Fig. 27. The speed of the nonlinear optimization algorithm is obvious from this plot. The rms errors obtained during the testing of the different polymers are presented in Table 11 along with their respective input variable ranges. In the authors' experience, the proposed algorithm has been found to be very effective for many physi-

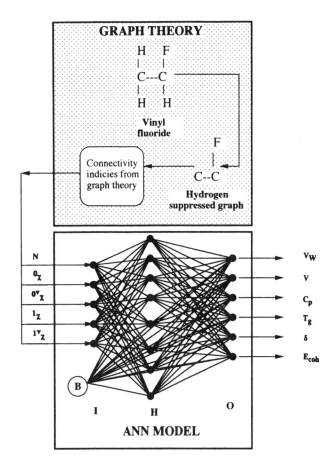

Figure 23 Schematic of the proposed integration of graph theory and neural networks. I-input layer; H-hidden layer; O-output layer; B-bias neuron.

cal property estimations [65]. However, it should be remembered that this algorithm is efficient only as long as the neural network architecture is simple and requires only a single hidden layer for information assimilation. As the ANN architecture becomes more complex, it may be necessary to resort to more traditional back-propagation algorithms that achieve convergence, but at a much slower rate [66]. For all six physical properties considered here, the overall training error compared to the published experimental data was found be within $\pm 0.15\%$, as shown in Fig. 28.

E. Case V: Predicting Crosslink Density Changes in Silicone Elastomers Due to Aging

Most elastomeric sealants used in structural glazing applications are organic polymers or elastomers. Because organic materials degrade, some changes in the properties of the elastomer can be expected to occur with pas-

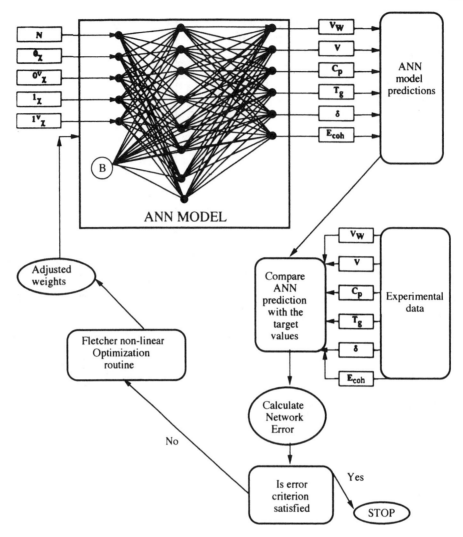

Figure 24   Schematic of the proposed training procedure in the integration.

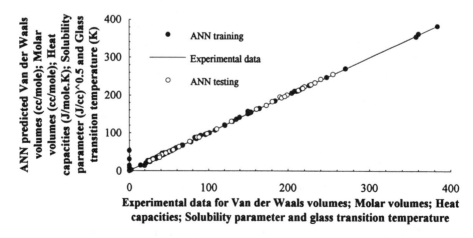

Figure 25   ANN model (5-8-6) training and testing results for van der Waals volume, molar volume, heat capacity, solubility parameter, and glass transition temperature of 45 different polymers.

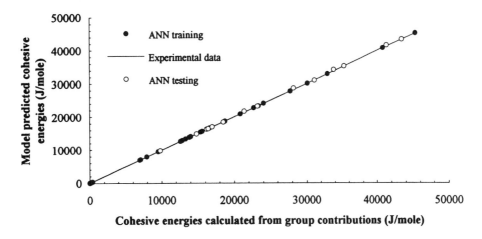

Figure 26 ANN model (5-8-6) training and testing results for cohesive energies of 45 different polymers.

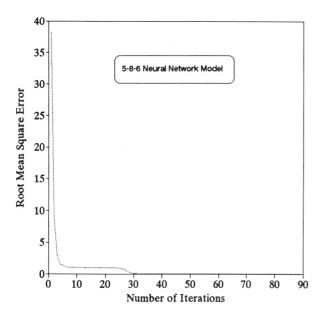

Figure 27 Root-mean-squared error progression plot during training phase.

sage of time, i.e., as the material ages. Of the polymeric sealants used for structural glazing applications, the silicone based elastomers are considered to be highly reliable, and their use is widely prevalent. These structural sealants function as a weather seal and, when used with glazing, also act as a binding agent for the glass to the building substrate. As a weather seal the sealants are exposed to a wide variety of service conditions that include moisture (acid and alkaline), ultraviolet rays, oxidative effects from ozone, buffeting forces of the wind, and the constant live load experienced from gravity and subtle building movements. Many questions arise, therefore, as to the damage caused by these weathering agents and their potential deleterious effects on the life time of a sealant during service. The authors have carried out a wide variety of artificial weathering studies on three different commercially available sealants as part of a project sponsored by the National Science Foundation (NSF) at Texas Tech University. Details of the results of this project can be found in the literature [67–73]. Also, the synergistic effects of several different weather-

Table 11 Comparison of the Proposed Integrated Approach with Traditional Graph Theory for Selective Polymers

Property	N	O_χ	O_χ^v	1_χ	1_χ^v	Percent variation in this work	Percent variation in graph theory
V_w	1–5	0.7–4.07	1.22–5.13	0.5–2.5	0.57–3.99	<0.05%	0.85%
V	1–5	0.7–4.07	1.22–5.13	0.5–2.5	0.57–3.99	<0.05%	0.22%
C_p	1–5	0.7–4.07	1.22–5.13	0.5–2.5	0.57–3.99	<0.05%	1.2%
δ	1–5	0.7–4.07	1.22–5.13	0.5–2.5	0.57–3.99	<0.05%	0.24%
T_g	1–5	0.7–4.07	1.22–5.13	0.5–2.5	0.57–3.99	<0.05%	0.52%
E_{coh}	1–5	0.7–4.07	1.22–5.13	0.5–2.5	0.57–3.99	<0.15%	6.9%

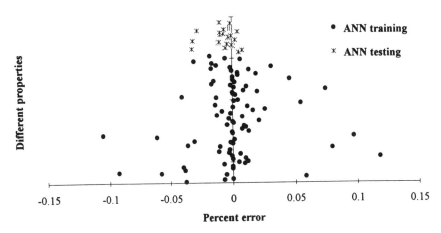

Figure 28 Overall training and testing error for all six properties predicted by the integrated model.

ing agents have been addressed, perhaps for the first time, through these research efforts.

Three different commercial formulations of silicone sealants from Dow Corning was used in the NSF sponsored studies. They were DC-790, DC-995, and DC-983, in the order of increasing modulus. Dumbbell test coupons (samples) were prepared as per the ASTM standards. Some test coupons were maintained at ambient conditions as control and the rest were subjected to simulated weathering. The weathered coupons were removed from the test layout at regular intervals of time and were tested for any changes in crosslink density due to exposure.

If an incompletely cured silicone is exposed to nonneutral pH moisture or other weathering agents, it was observed that the sealant continued to crosslink. This increase in crosslink density was reflected and quantified from the increase of the material's modulus. In general, from the initial mixing of the components in the sealants (polymer, fillers, stabilizers, etc.) to well after structural application, silicone sealants tend to undergo an increase in crosslink density. At some point, however, the available sites for crosslinking will be depleted either by the formation of a crosslink or by deactivation. The continued exposure to weathering agents, therefore, begins to cause degradation of the sealants physical properties.

In our study, the effect of moisture over the non-neutral pH range of 3–11, direct sunlight, ozone at a concentration level of 6000 ppm, and the effects of loading stresses, were investigated for the three commercial sealants. A characteristic variation of crosslink density for the typical silicone sealants is shown in Fig. 29. This figure depicts the results for the coupons exposed to moisture and sunlight. Initially upon exposure, the crosslink density of the sealants exhibit an increase due to the availability of residual uncured crosslink sites

that were still active. Degradation effects, even if they occur during these early stages, go undetected until the vast majority of these active sites have either reacted or been deactivated and the rate of chain scission begins to exceed the crosslink rate. Therefore, after attaining a maximum, a decline in the crosslink density is observed. A decrease in crosslink density is indicative of softening of the sealant due to the scission process.

This transitory behavior was observed to arise from all the weathering agents considered in this study except ozone. Instead, test coupons exposed to ozone exhibited an initial decline in the crosslink density of the silicone with the formation of surface cracks, which were difficult to distinguish with the naked eye. With continued exposure to ozone, however, the material would begin to crosslink. We proposed that ozone's greatest affinity

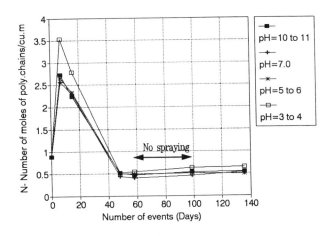

Figure 29 Characteristic variation of crosslink density of silicone elastomers subjected to aging.

for the silicone structure is for any unsaturated carbon-to-carbon double bonds present on the exposed surface. Hence, the weathering process from ozone is such that the bonds at some depth immediately beneath the surface are not attacked until all the double bonds on the surface have been modified. This is true except in the scenario where the intermolecular spaces of the silicone elastomers are filled with adsorbed water. Otherwise, the oxidized surface after reacting with ozone can serve to protect the interior layers. Ozone degradation, therefore, results in discoloration, softening, and a change in the structure of the silicone elastomer. A silicone–ozone complex was observed at the point-of-crack initiation where the maximum discoloration was noticed. The alignment of ozone induced surface cracks was typically perpendicular to the direction of stress loading in the those samples when an initial strain of 10% was induced and maintained on the sample. During the experiments, the rate of uniaxial stress decay was also estimated through relaxation time data. Relaxation times were found be significantly different in those samples where a preload or strain was induced. The coupons were monitored for a period of 138 weathering events. A chemical kinetic model was developed by the authors to model (crosslink density changes) transient crosslinking behavior due to synergistic weathering effects [67,70,72]. This was a semi-empirical model based approach with data reduction and optimization of various variables. In this chapter, however, a simple model based on an ANN is presented for predicting the same crosslink density changes in these sealants due to weathering. The prior experimental data for various weathering parameters were also incorporated in the model and were also used simultaneously in training.

A 3-4-1 architecture was used in this case. The three input nodes were: (1) type of weathering (moisture at four different pH ranges, sunlight, ozone, and loading effects); (2) sealant type (DC-790, DC-995, and DC-983); and (3) number of weathering events. The predicted variable being the crosslink density of the respective sealants. The experimental weathering data for a period of 138 weathering events were used for training the proposed neural network. The experimental data were divided into two data sets, one for training and the second for testing the trained network. The ANN training and testing results are shown in Fig. 30. Predictions were within a 0.1% tolerance. Extrapolation of this trained network was also tested with additional data and was found to be in the same tolerance limit over 218 weathering events. Even though these finding are encouraging, it should be noted that the use of an ANN in the range outside the trained environment (range of variables) is not recommended.

XII. CONCLUSIONS

Literature in the area of neural networks has been expanding at an enormous rate with the development of new and efficient algorithms. Neural networks have been shown to have enormous processing capability and the authors have implemented many hybrid approaches based on this technique. The authors have implemented an ANN based approach in several areas of polymer science, and the overall results obtained have been very encouraging. Case studies and the algorithms presented in this chapter were very simple to implement. With the current expansion rate of new approaches in neural networks, the readers may find other paradigms that may provide new opportunities in their area of interest.

The proposed neural network model with the nonlinear optimization routine is similar to many nonlinear

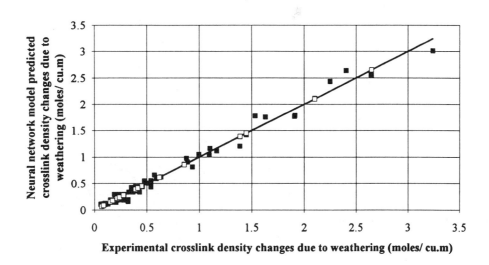

Figure 30 ANN model training and testing results for crosslink density changes due to aging.

least-squares techniques and is computationally simple. While it does not require the choice of additional parameters such as η and α in back-propagation, it is very efficient for predicting various physical properties of the polymers as illustrated in the different case studies. The proposed algorithm was an efficient method for problems considered in this chapter.

The ANN as a predictive tool is most effective only within the trained range of input training variables. Those predictions that fall outside the trained range must be considered to be of questionable validity. Even so, whenever experimental data are available for validation, neural networks can be put to effective use. Since an extensive experimental body of data on polymers has been published in the literature, the application of neural networks as a predictive tool for physical, thermodynamic, and other fluid properties is, therefore, promising. It is a novel technique that will continue to be used, and it deserves additional investigation and development.

REFERENCES

1. M. A. Cohen and S. Grossberg, *IEEE Trans. Syst. Man. Cybern.*, SMC-13: 815–826 (1983).
2. B. Delgutte, *J. Acoust. Soc. Am.*, 75: 879–886 (1984).
3. R. O. Duda and P. E. Hart, *Pattern Classification and Scene Analysis*, John Wiley and Sons, New York (1973).
4. B. Gold, *MIT Lincoln Laboratory Technical Report*, TR-747, AD-A169742 (June 1986).
5. S. Grossberg, *The Adaptive Brain I: Cognition, Learning, Reinforcement, and Rhythm*, and *The Adaptive Brain II: Vision, Speech, Language, and Motor Control*, Elsevier–North-Holland, Amsterdam (1986).
6. T. Kohonen, K. Masisara, and T. Saramaki, "Phonotopic Maps—Insightful Representation of Phonological Features for Speech Representation," Proceedings IEEE 7th International Conference, Montreal, Canada (1984).
7. R. F. Lyon and E. P. Loeb, "Isolated Digit Recognition Experiments with a Cochlear Model," in Proceedings International Conference on Acoustics Speech and Signal Processing—ICASSP-87, Dallas, Texas (April 1987).
8. J. Makhoul, S. Roucos, and H. Gish, *IEEE Proc.*, 73: 1551–1588 (Nov. 1985).
9. T. Martin, Acoustic Recognition of Limited Vocabulary in Continuous Speech, Dept. Electrical Engineering, Univ. Pennsylvania, Ph.D. Thesis (1970).
10. D. E. Rumelhart, G. E. Hilton, and R. J. Williams, *Parallel Processing: Explorations in Microstructure of Cognition. Vol. 1: Foundations*. MIT Press, Cambridge, Massachusetts (1986).
11. D. E. Rumelhart and J. L. McClelland, *Parallel Processing: Explorations in Microstructure of Cognition*, MIT Press, Cambridge, Massachusetts (1986).
12. K. Watanabe, I. Matsuura, M. Abe, M. Kubota, and D. M. Himmelblau, *AICHE. J*, 35: 1803–1812 (1989).
13. V. Venkatasubramanian, R. Vaidyanathan, and Y. Yamamoto, *Comput. Chem. Eng.*, 14: 699–712 (1990).
14. N. Bhat and T. J. McAvoy, *Comput. Chem. Eng.*, 14: 573–583 (1990).
15. S. Haykin, *Neural Networks—A Comprehensive Foundation*, Macmillan College Publishing Company, NY (1994).
16. J. M. Zaruda, *Introduction to Artificial Neural Systems*, West Publishing Co., NY (1992).
17. J. A. Anderson, *Math. Biosci.*, 14: 197–220 (1972).
18. T. Kohonen, *Self-Organization and Associative Memory*, 2nd Ed., Springer-Verlag, New York (1987).
19. R. Battiti, *Neural Comput.*, 4: 121–166 (1992).
20. D. E. Rumelhart and J. L. McClelland, *Psychological and Biological Models*, Vol. 2 M.I.T. Press, Cambridge, MA (1986).
21. R. Fletcher, *AERE-R 6799*, Theoretical Physics Division, Harwell, Berkshire, U.K., 1971.
22. D. W. Marquardt, *J. Soc. Indust. Appl. Math*, 11(2): 431–441 (1969).
23. R. Keshavaraj, R. W. Tock, and G. S. Nusholtz, *Issues in Automotive Safety Technology*, SAE-SP-1072: 55–66 (1995).
24. R. Keshavaraj, R. W. Tock, and G. S. Nusholtz, *J. Appl. Polym. Sci.*, 57(9): 1127–1144 (1995).
25. R. Keshavaraj, R. W. Tock, and G. S. Nusholtz, *Polymer* [in press].
26. R. Keshavaraj, R. W. Tock, and D. Haycook, "Feedforward Neural Network Modeling of Biaxial Deformation of Airbag Fabrics" ANTEC '95 Proceedings, SPE Technical Papers, Modeling of Polymer Properties and Processes, Boston (May 1995).
27. B. Andes, B. Widrow, M. Lehr, and E. Wan, *Proc. Intl. Joint. Conf. on Neural Networks*, 1, 1: 533–536 (1990).
28. Y. Le Cun, J. S. Denker, and S. A. Solla, *Advances in Neural Information Processing Systems*. Morgan Kaufmann, San Mateo, California (1990).
29. A. S. Weigand, B. A. Huberman, and D. E. Rumelhart, *Intl. J. Neur. Syst*, 1(3): 193 (1990).
30. G. Cybenko, *Tech. Rep. Dept. Comp. Sci.*, Tufts University, Medford, MA.
31. J. Gasteiger, J. Zupan, and Angew, *Chem. Int. Ed. Engl.*, 32: 503–527 (1993).
32. J. Zupan and J. Gasteiger, *Neural Networks for Chemists: An Introduction*. New York: VCH (1993).
33. J. Zupan and J. Gasteiger, *Anal. Chim. Acta*, 248: 1–30 (1991).
34. E. W. Robb and M. E. Munk, *Mikro. Chim. Acta*, 1: 131–155 (1990).
35. J. A. Burns and G. M. Whitesides, *Chem. Rev.*, 93: 2583–2601 (1994).
36. D. W. Elrod and G. M. Maggiora, *Proc. Montreux, Int. Chem. Inf. Conf.* (H. Collier, ed.) Infonortics, Clane, UK (1987).
37. Q. C. Van Est, PJ. Schoenmakers, JRM. Smits and WPM. Nijssen, *Vib. Spectrosc.* 4: 263–272 (1993).
38. T. A. Andrea and H. Kalayeh, *J. Med. Chem.*, 34: 2824–2836 (1991).
39. M. J. Lee and J. T. Chen, *Ind. Eng. Chem. Res.*, 32: 995–997 (1993).
40. S. Kito, T. Hattori, and Y. Murakami, *Ind. Eng. Chem. Res.*, 31: 979–981 (1992).
41. R. Keshavaraj, R. W. Tock, R. S. Narayan, and R. A. Bartsch, "Fluid Property Prediction of Siloxanes with the Aid of Artificial Neural Nets," *Polymer-Plastics Technology and Engineering*, 35(6):971–982 (1996).
42. C. Hurd, *J. of Amer. Chem. Soc.*, 68: 364 (1946).
43. Z. Batschinski, *Physik. Chem.*, 84: 643 (1913).
44. J. K. Sears and N. W. Touchette, *Plasticizers, Encyclo-*

pedia of Chemical Technology, 3rd ed., (R. Kirk, D. Orthmer, eds.) John Wiley and Sons: New York, Vol. 18, p. 111 (1982).

45. J. R. Darby, J. K. Sears, *Plasticizers, Encyclopedia of Polymer Science and Technology*, (H. F. Mark, N. G. Gaylord, eds.), John Wiley and Sons, New York, Vol. 10, p. 228 (1969).

46. K. Kishore, H. K. Shobha, G. J. Mattamal, *J. Phys. Chem.*, *94*: 1642 (1990).

47. K. Kishore, G. J. Mattamal, *J. Poly. Sci., Poly. Lett. Ed., 24*: 53 (1986).

48. J. C. Phillips, G. J. Mattamal, *J. Chem. Eng. Data, 23*(1): (1978).

49. R. Keshavaraj, R. W. Tock, R. S. Narayan, and R. A. Bartsch, *Advances in Polymer Technology*, *14*(3): 215–226 (1995).

50. P. J. Flory, *Principles of Polymer Chemistry*, Cornell University Press, London (1953).

51. P. J. Flory, *J. Chem. Phys.*, *10*: 51 (1942).

52. P. J. Flory, J. L. Ellenson, and B. E. Eichinger, *Macromolecules*, *1*: 279–286 (1968).

53. V. Alessandro, *Fluid Phase Equilibria, 34*: 21–35 (1987).

54. G. M. Kontogrorgis, A. Fredenslund, and D. P. Tassios, *Ind. Eng. Chem. Res.*, *32*: 362–372 (1993).

55. J. F. Heil and J. M. Prausnitz, *AIChE J.*, *2*: 678–685 (1966).

56. R. Keshavaraj, R. W. Tock, R. S. Narayan, and R. A. Bartsch, *Ind. Eng. Chem. Res.*, *34*(11): 3974–3980 (1995).

57. R. Keshavaraj, R. W. Tock, and G. S. Nusholtz, *SAE International Congress and Exposition Conference*, Paper # 950343, Detroit (February 1995).

58. J. Bicerano, *Prediction of Polymer Properties*, Marcel Dekker, Inc., NY, (1993).

59. W. C. Forsman, *J. Chem. Phys.*, *65*: 4111–4115 (1976).

60. J. E. Martin, *Macromolecules*, *17*: 1263–1275 (1984).

61. H. Galina, *Macromolecules*, *19*: 1222–1226 (1986).

62. M. R. Surgi, A. J. Polak, and R. C. Sundahl, *J. Polym. Sci., Polym. Chem. Ed.*, *27*: 2761–2776 (1989).

63. J. Bicerano and D. Adler, *Pure and Applied Chemistry*, *59*: 101–144 (1987).

64. D. A. Loy and R. A. Assink, *J. Am. Chem. Soc.*, *114*: 3977–3978 (1992).

65. R. Keshavaraj, R. W. Tock, and R. S. Narayan, [submitted] *Industrial and Engineering Chemistry Research*.

66. R. Keshavaraj and R. W. Tock, Compendium of Results—Chrysler Challenge Fund Project, Chrysler Technology Center (December 1994).

67. R. Keshavaraj and R. W. Tock, *Polymer-Plastics Technology and Engineering*, *32*(6): 579–593 (1993).

68. R. Keshavaraj and R. W. Tock, *Advances in Polymer Technology*, *13*(2): 149–156 (1994).

69. R. Keshavaraj and R. W. Tock, *Journal of Material Science Letters* (UK), *12*: 1627–1629 (1993).

70. R. Keshavaraj and R. W. Tock, *Polymer-Plastics Technology and Engineering*, *33*(5): 537–550 (1994).

71. R. Keshavaraj, R. W. Tock, R. S. Narayan, and C. V. G. Vallabhan, *Journal of Material Science Letters*, (UK), *14*: 964–967 (1995).

72. R. Keshavaraj and R. W. Tock, *Polymer-Plastics Technology and Engineering*, *33*(4): 397–417 (1994).

73. R. Keshavaraj, R. W. Tock, and C. V. G. Vallabhan, *Journal of Construction and Building Materials*, *8*(4): 227 (1994).

2

New Generation High Performance Polymers by Displacement Polymerization

Pradip Kumar Dutta
Shri G. S. Institute of Technology & Science, Indore, India

Susanta Banerjee
Defence Research & Development Establishment, Gwalior, India

Sukumar Maiti
Indian Institute of Technology, Kharagpur, India

I. INTRODUCTION

Much of our modern progress has been based on the use of materials such as metals, ceramics, and polymers in diversified fields. Among those, polymers are the newest member of the materials family. Within a short span of time, the polymer materials are now scientifically mature, and research is moving into increasingly sophisticated frontiers of material applications, such as nonlinear optics (NLO), computer memory cell, piezoelectricity, photoresist, molecular composites, etc. High performance polymers, e.g., polyether etherketone (PEEK), polyimides, Kevlar, etc., exhibit high resistance to heat, solvents, chemicals, etc., as well as superior load-bearing characteristics. Most of these polymers have rigid structures (e.g., semi-ladder or ladder type) that have a high glass transition temperature (T_g) close to their decomposition temperature and poor solubility characteristics. Therefore, they are difficult to process commercially.

In general most of the commercial polymers are not comparable to metals and ceramics in terms of load-bearing property, mechanical strength, and thermal stability. To overcome these difficulties the aromatic ether or sulfide (thioether) linkages in the polymer backbone are inserted. The presence of ether or sulfide linkages in the main chain of the molecules acts as a ball bearing with a significantly lower energy of internal rotation. This imparts chain flexibility, lower glass transition, and melting temperatures (T_m), and enhances solubility while maintaining the desirable high temperature characteristics. This effect is more pronounced when ether or sulfide linkages are present at the "appropriate position" in the polymer molecule [1]. The incorporation of ether or sulfide linkages in high temperature polymers (resulting in polyethers and polysulfides, respectively) has been conveniently made possible by activated displacement polymerization reactions [2–8].

The principle of displacement reaction is shown below:

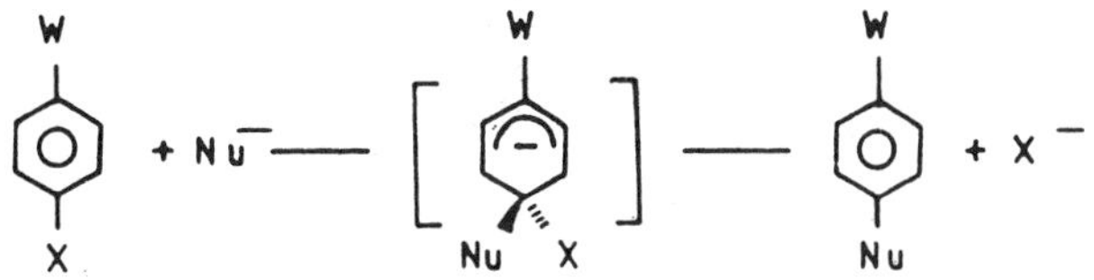

W = Electron withdrawing group, X = Leaving group, Nu = Nucleophile

35

II. AROMATIC SYSTEMS IN DISPLACEMENT POLYMERIZATION

Besides ketone and sulfone groups, it is of interest to investigate other activating groups, such as amide, perfluoroalkyl, etc., that will produce new poly(aryl ethers) [9,10]. The effectiveness of new activating groups can be estimated by Hammett σ values (Table 1) as well as the chemical shift of aromatic protons ortho to the new system in the ^{1}H-NMR spectra. Table 1 shows Hammett σ values of the new activating groups with respect to a keto group.

A comparison of Hammett σ values of the new activating groups, such as amide and —CF$_2$—, with that of conventional activating keto groups shows them to be electronically similar.

Although the sulfone activated biphenyl and the ketone activated naphthalene moiety for the displacement polymerization have been reported by Attwood et al. [11], these were rediscovered by Cummings et al. [12] and Hergenrother et al. [13], respectively, for the synthesis of poly(aryl ethers). Recently, Singh and Hay [14] reported polymers containing O-dibenzoyl benzene (1,2,3) moiety by reaction between bis(O-fluorobenzoyl) benzene or substituted benzene with bisphenates of alkali metal salt in DMAC as follows:

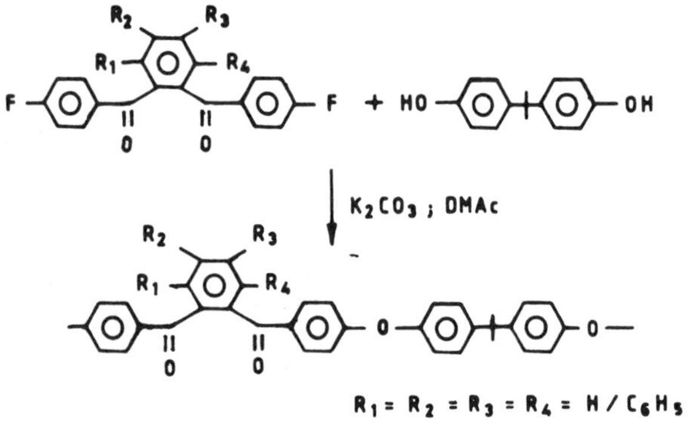

Table 1 Hammett σ Values for Different Activating Groups

Activating group	Hammett σ value	Deshielding of aromatic proton value (δ)	Ref.
—C—$\parallel$O	0.50	7.9	10, 11
—C—N—$\parallel$ $\vert$ O H	0.36	8.1	9
—CF$_2$—	0.54	7.6	10

^{12}F poly(aryl ether ketones) derived from hexafluoro acetone have also been reported [15]. Recently, the synthesis and characterization of fluorinated polyethers prepared from decafluorobiphenyl and bisphenols by displacement reaction have been reported by Irvin et al. [16] and Mercer et al. [17], respectively.

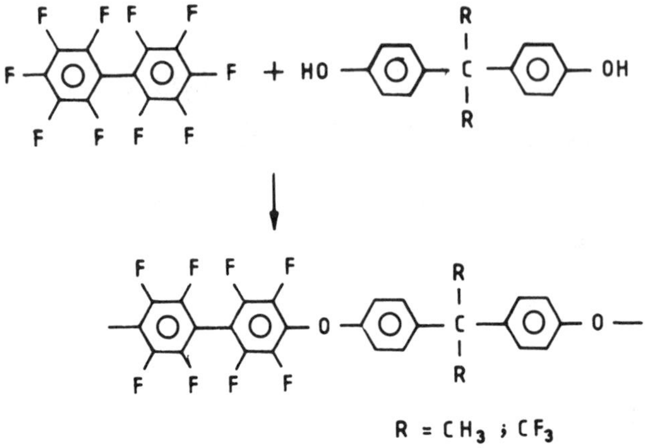

R = CH$_3$; CF$_3$

Although the synthesis of fluorinated polyarylethers by the reaction between decafluorobiphenyl with bisphenols had previously been described by others [18,19], those polymers were not fully characterized and no particular utility was ascribed to them.

The new fluorescent poly(aryl ethers) derived from nonfluorescent monomers have gained significant attention from polymer scientists [20]. These polymers are prepared by the polymerization of phenolphthalein and its derivatives with activated aromatic difluorides.

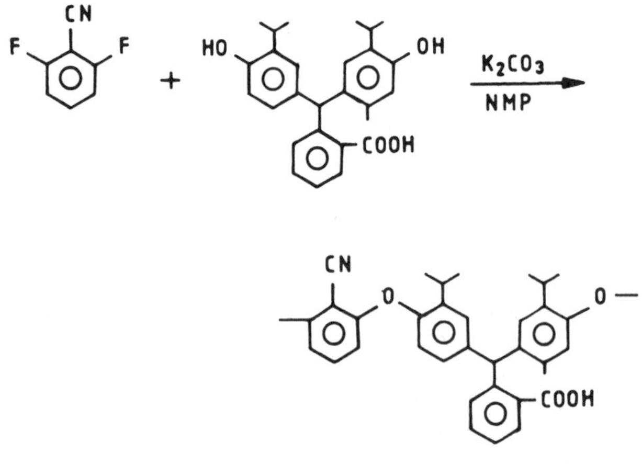

Recently, Dutta and Maiti [21] reported nitro displacement polymerization of the bisphenol dianion with the sulfone activated dinitro aromatic compounds. In addition, there have been recent reports of the development of functionalized PEEK [22] and polyether sulfone ketone (PESK) [23] that are comparable to commercially available high performance polymers.

It is interesting to note that all the new aromatic systems, as described, undergo displacement polymerizations in DMAC solvent by the K_2CO_3 method, except perfluoroalkylene [10] and amide activated polymerization [9], which were performed in NMP solvent. The displacement polymerization in DMAC solvent was carried out at 155–164°C. ^{12}F poly(aryl ether ketones) require less reaction time (3–6 h) than other aromatic systems for synthesis of polyethers [15]. Synthesis of the fluorinated polyether as reported by Irvin et al. [16] was carried out at room temperature for 16 h ($M_w = 75,000$), whereas the same polymer by Mercer et al. [17] was synthesized at 120°C for 17 h ($M_w = 78,970$).

III. SYNTHESIS OF PHENOL/THIOPHENOL BASED POLYMER PRECURSORS

Diphenol/thiophenol is one of the most important polymer precursors for synthesis of poly(aryl ethers) or poly(aryl sulfides) in displacement polymerizations. Commonly used bisphenols are 4,4'-isopropylidene diphenol or bisphenol-A (BPA) due to their low price and easy availability. Other commercial bisphenols have also been reported [7,24,25]. Recently, synthesis of poly(aryl ethers) by the reaction of new bisphenol monomers with activated aromatic dihalides has been reported. The structures of the polymer precursors are described in Table 2. Poly(aryl ether phenylquinoxalines) have been synthesized by Connell et al. [26], by the reaction of bisphenols containing a preformed quinoxaline ring with conventional activated dihalides. An alternative approach, utilized by Hedrick et al. [27] for synthesis of the same polymer, was the reaction of arylfluorides activated by an adjacent heterocyclic ring with conventional bisphenols. In both cases, high molecular weight polymers have been achieved by the K_2CO_3 method. Similarly, a series of bisphenols containing preformed heterocyclic linkages (Table 2) including imidazoles [28], benzoxazoles [29], triazoles, and oxadiazoles [30] are amenable to the nucleophilic aromatic substitution polymerization.

Recently, the pyrazole group containing bisphenols have been synthesized from activated aromatic dihalides and 3,5-bis (4-hydroxy phenyl)-4-phenyl pyrazole or 3,5-bis(4-hydroxy phenyl)-1,4-diphenyl pyrazole. A novel synthesis of imido aryl containing bisphenols has been reported [32]. N-substituted 1,4-bis(4-hydroxy phenyl)-2,3-naphthalimides were prepared from phenolphthalein and copolymerized with aromatic sulfone or ketone difluorides to obtain the poly(imidoaryl ether) sulfones/ketones.

However, heterocycles containing thiophenols have not been reported. It has been observed that the thiophenolate ion undergoes nucleophilic attack by the halo/nitro compounds more easily than the phenolate ion in displacement reactions [37–39]. The experimental result shows that the reactivity of 3-nitro-N-phenyl-phthalimide with 4-methyl-thiophenolate (reaction 1) is 100 times faster than that of 4-methyl phenolate [40] (reaction 2):

$$(1)$$

$$(2)$$

The higher nucleophilic property of sulfur in aromatic nucleophilic displacement reactions is due to the higher polarizability of the sulfur atom and the higher product stability through the transition state, which involves a build-up of a negative charge adjacent to sulfur to be stabilized by the d-orbital resonance. The enhanced reactivity of sulfur nucleophile has also been well documented by various workers using both activated [41,42] and nonactivated systems [4,43–48]. For example, synthesis of the aromatic polysulfide [49] from bis(4-chloro-3-nitrophenyl) sulfone and 4,4'-oxydiphenyldithiol oc-curs under mild conditions, i.e., at a temperature of 15°C for 24 h, whereas the aromatic polyether [50] from bisphenol nucleophile and the same nitro-substituted dichloro monomer may be carried out at 140°C for 24 h. However, the work concerning the nucleophilic property of sulfur in sodium sulfide (Na_2S) to activate halo/nitro compounds in the activated displacement polymerization is rather limited [48,51,52]. Recently, Yoneyama et al. [53] reported the synthesis of polysulfides by the reaction between the thianthrene containing a dichloro monomer with sodium sulfide as follows:

Table 2 New Dihydroxy Aromatic Systems for Activated Displacement Polymerization

Polymer precursor	Year of reporting	Ref.
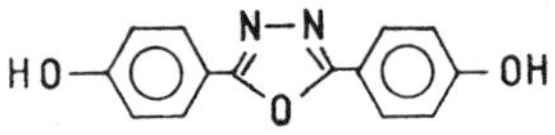	1988	29
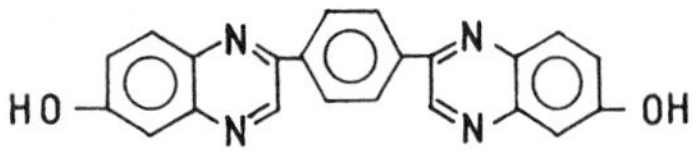	1988	26
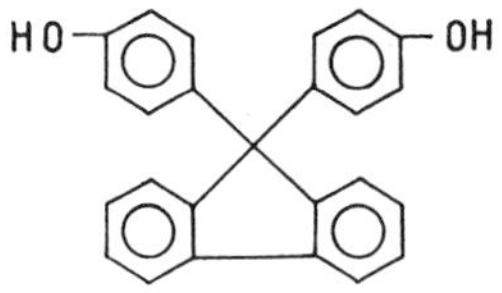	1988	13
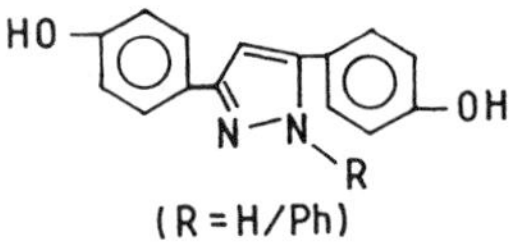 (R = H/Ph)	1991	31
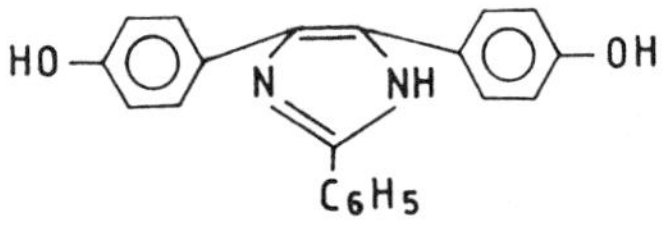	1991	28
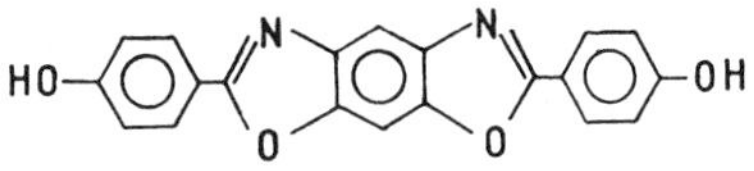	1991	33
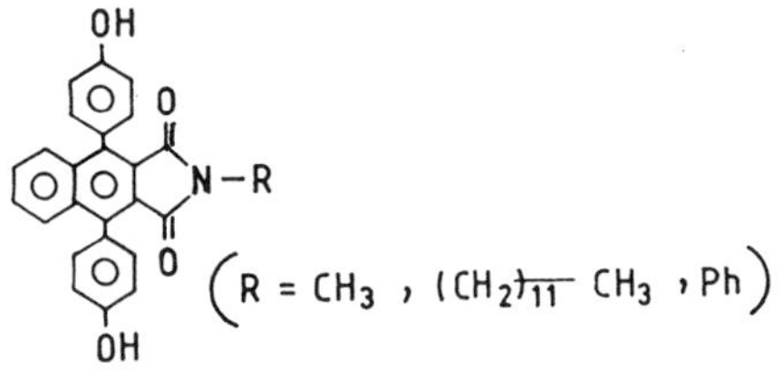 (R = CH$_3$, (CH$_2$)$_{11}$ CH$_3$, Ph)	1991	32
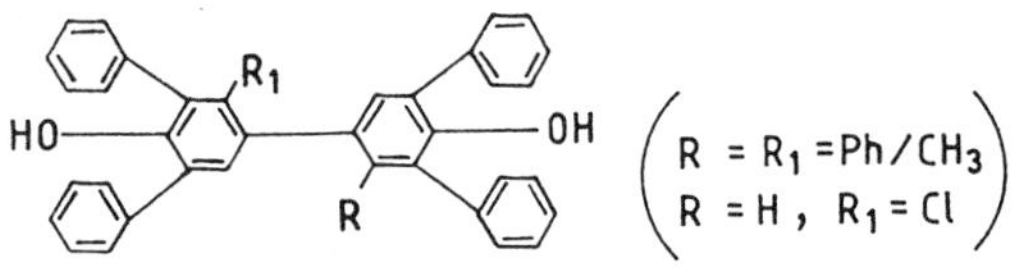 (R = R$_1$ = Ph/CH$_3$; R = H, R$_1$ = Cl)	1991	34
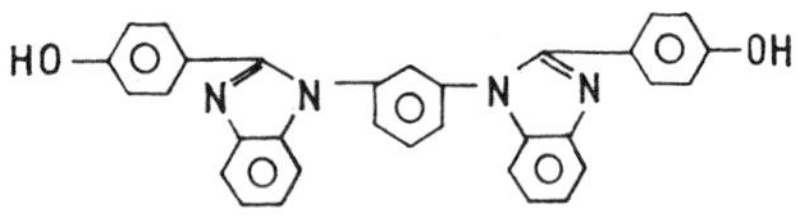	1993	35
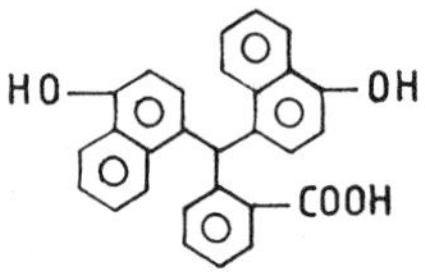	1994	20
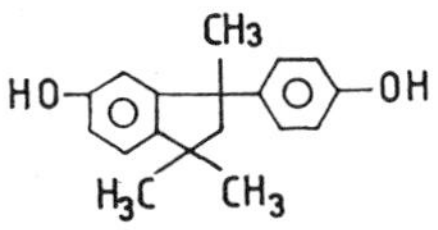	1994	36

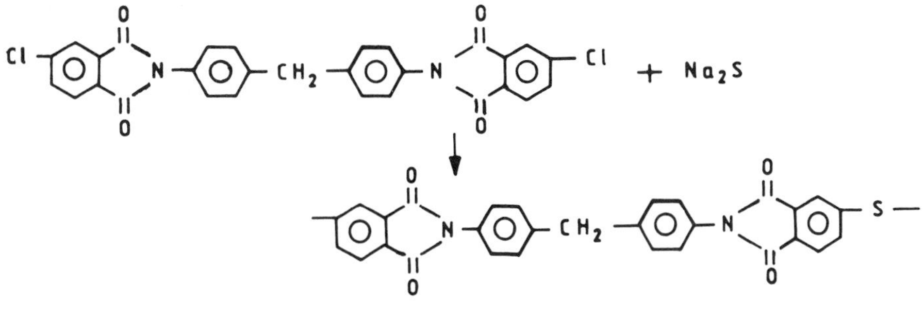

The authors also reported a few polysulfides [54,55] caused the reaction between sodium sulfide and dihalocompounds activated by —SO₂— and amide groups, respectively.

IV. HETEROCYCLIC SYSTEM IN DISPLACEMENT POLYMERIZATIONS

Among organic materials, poly(aryl ethers) and poly(aryl sulfides) have been known, as a class of engineering thermoplastics. The electron withdrawing sulfone and ketone groups usually activate the dihalo or dinitro compounds to facilitate the nucleophilic displacement through the transition state called Meisenheimer-like complex, and, thus, poly(aryl ether or sulfide) sulfones

and poly(aryl ether or sulfide) ketones are formed. The other electron withdrawing groups, i.e., esters, nitrile, azo, azoxy, triazine, pyridazine, phenyl phosphine oxide and fluoroalkyl groups, have also been recognized [7,56]; but these are mainly used for preparation of poly(aryl ethers) only.

The use of other heterocyclic rings in displacement polymerization has been recently reported. Table 3 shows the new dihalo heterocyclic monomers used for synthesis of poly(aryl ethers).

The above mentioned new heterocyclic systems have not been used in the preparation of poly(aryl sulfides). The authors reported [57] a polyheteroarylene sulfide by the reaction between 2,6-dichloropyridine and sodium sulfide.

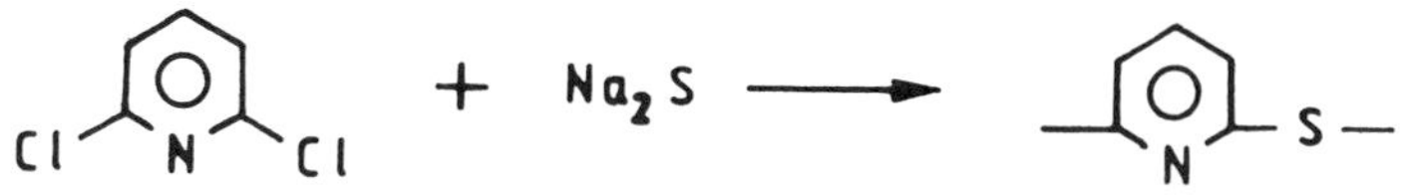

Recently, a poly(heteroarylene sulfide) formed by the reaction between bis(thiophenols) and bis(6-chlorophenyl quinoxalines) has been reported by Hedrick et al. [58].

The common characteristics of the above mentioned heterocycles are electron withdrawing and a site of unsaturation that can stabilize the negative charge developed by the displacement reaction through resonance. For example, the thiazole activated halo displacement is similar to that of a conventional activating group as shown in Scheme 1. The activation is derived from the electron affinity and the stabilization of the negative

charge developed in the transition state through the formation of a Meisenheimer complex, which lowers the activation energy for the displacement reaction [58].

Furthermore, the deshielding effect of the aromatic protons ortho to the new heterocycle, e.g., benzothiazole ring (δ 8.2) is greater than that of a ketone group (δ 7.9) indicating higher electron affinity (Fig. 1). This has resulted in the possibility of facile displacement at the 4-position of the 2-phenyl ring of halogenated benzothiazole compound (Scheme 1). The preparation of heterocycle activated poly(aryl ethers) [58] e.g., poly(benzothiazole ether) is demonstrated as follows:

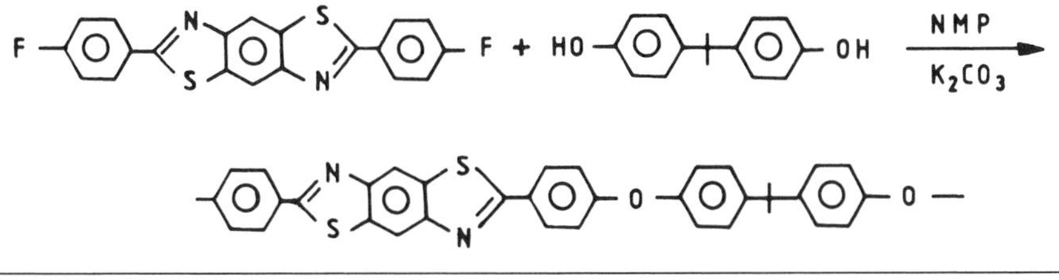

Similarly, the other new heterocycle containing poly(aryl ethers) have been synthesized by many researchers

[13,27,30,59–64]. These syntheses can be considered as heterocyclic analogue of the poly(ether imide) synthesis

Table 3 New Activated Dihaloheterocyclic Systems for Displacement Polymerization

Monomer (X—Ar—X)	Year of reporting	Ref.
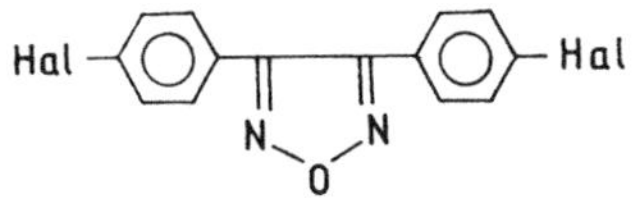	1986	7, 59
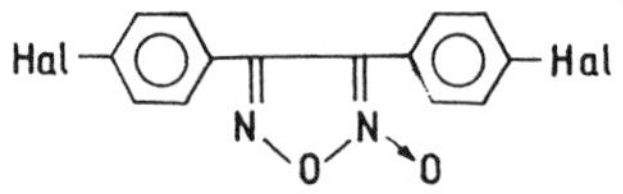	1986	7, 59
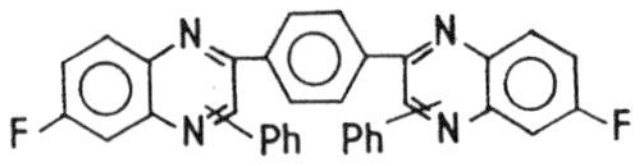	1990	27
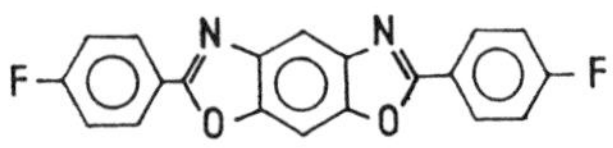	1990	60
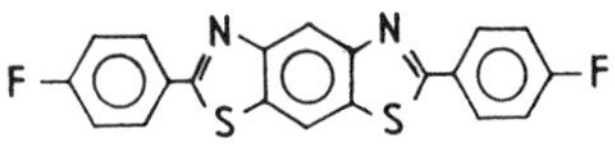	1991	58
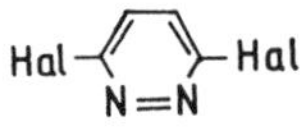	1967	2
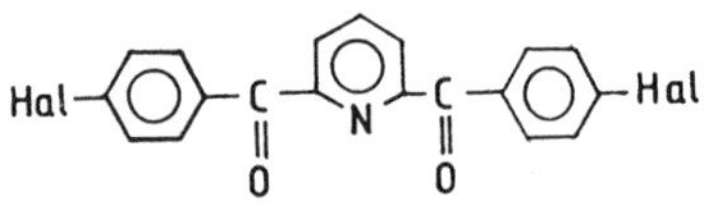	1988	13
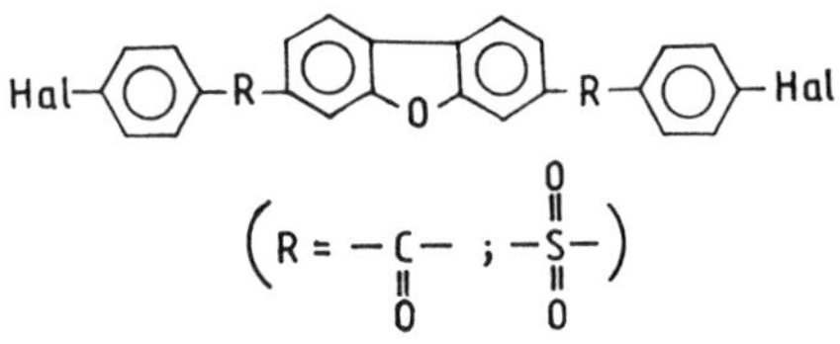	1991	61
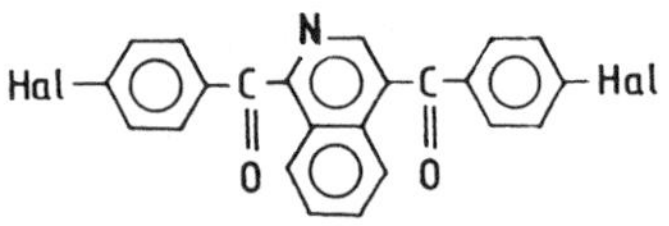	1991	62
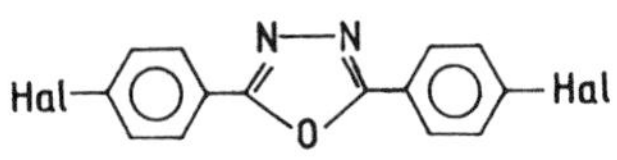	1992	7, 59
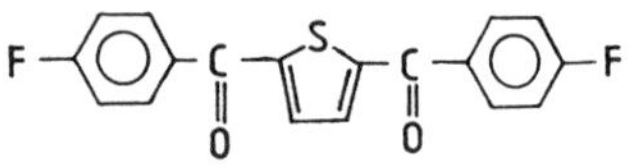	1992	63
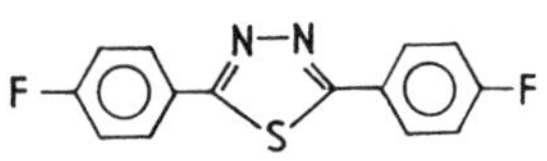	1994	36
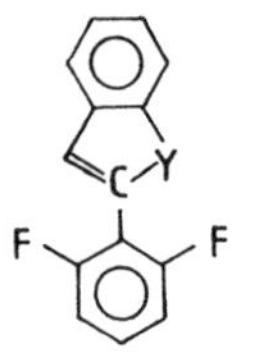	1995	65

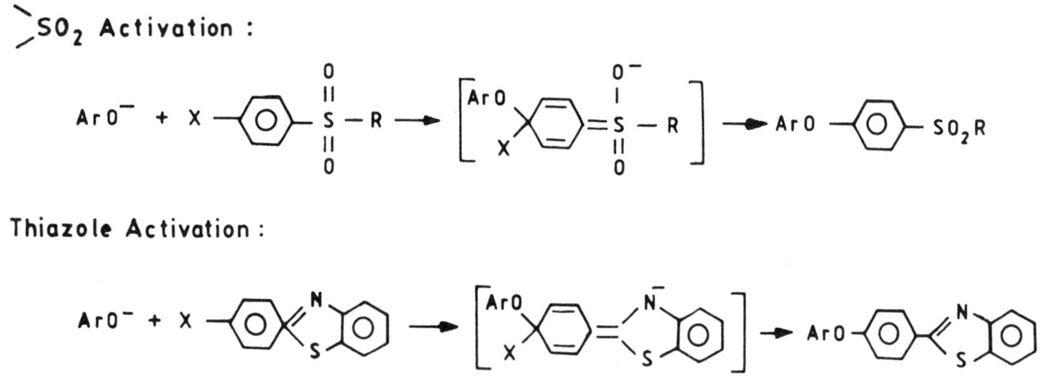

SO₂ Activation :

$>SO_2$ Activation :

Thiazole Activation :

Scheme 1   Activation of sulfone and thiazole groups.

like Ultem [58]. The polyetherimide synthesis requires milder conditions to avoid side reactions associated with the nitrile ion generated during the polymerization reaction. Johnson and coworkers [2] in 1967, first demonstrated the use of heterocycle oxadiazole activated displacement polymerization reaction. Recently, Hedrick et al. [65] used the oxadiazole activated halo and nitro displacement for the synthesis of poly(aryl ether oxadiazoles). Poly(aryl ethers) containing pendent benzoxa-

zole and benzothiazole units have been reported [65]. The use of Reissert compounds for the synthesis of poly-(heteroarylene ether ketones) containing 1,4-isoquinol-inediyl units [62] is a novel approach because the number of such monomer types and the methods for their introduction into polymers appear to be rather limited. Recently, a five membered ring containing thiophene based poly(aryl ethers) by displacement reaction have been reported [36,63,66,67]

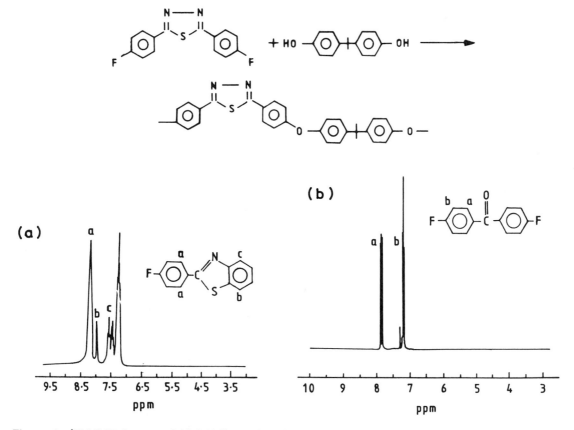

Figure 1   ¹H-NMR Spectra of (a) 2-(4-fluorophenyl)-benzoxthiazole, and (b) 4,4′-difluorobenzophenone.

Synthesis of all these heterocycle activated poly-ethers is carried out in polar aprotic solvents, such as NMPs, by the K_2CO_3 method. The effective displace-ment reactions are reported at varied temperatures (140–190°C) and durations (3–24 h).

V. PHASE TRANSFER CATALYSTS FOR DISPLACEMENT POLYMERIZATIONS

Interests in the phase transfer catalysis (PTC) have grown steadily for the past several years [68–70]. The use of PTC has recently received industrial importance in cases where the alternative use of polar aprotic sol-vents would be prohibitively expensive [71–74]. Thus, the potential application of the phase transfer catalyzed aromatic nucleophilic displacement reactions between phenoxide or thiophenoxide and activated systems has recently received considerable attention for polymer synthesis [75–80]. Nucleophilic aromatic displacement reactions (model reactions) using different phase trans-fer catalysts are summarized in Table 4.

Recently, the above mentioned model reaction has been extended to polycondensation reactions for synthe-sis of polyethers and polysulfides [7,81]. In recent re-ports crown ether catalysts have mostly been used in the reaction of a bifunctional nucleophile with a bifunctional electrophile, as well as in the monomer species carrying both types of functional groups [7]. Table 5 describes the syntheses of aromatic polyethers by the nucleophilic displacement polymerization using PTC.

The thermotropic aromatic main chain liquid crys-talline polymers are also prepared by the phase transfer catalyzed aromatic nucleophilic polymerization [87]. Polyetherification of bis(4-chloro-3-nitrophenyl) sulfone with mesogenic aromatic diols is shown below:

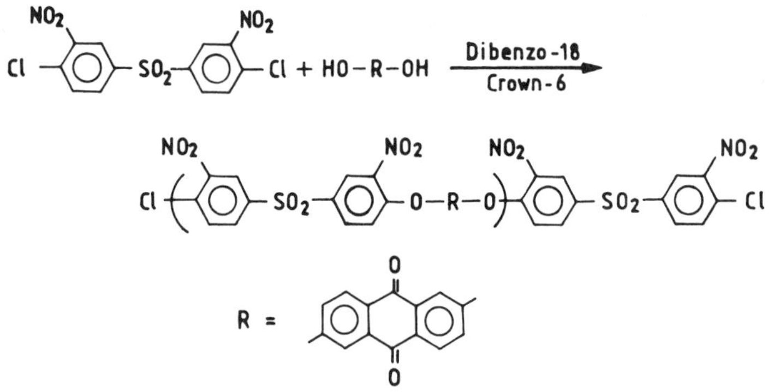

Recently, the scope of these preparations has also been extended to the synthesis of polythioethers and copo-lythioethers.

VI. DISPLACEMENT POLYMERIZATIONS USING SILYLATED MONOMERS

The use of silyl groups at the end of an initiator molecule in chain polymerization was first reported in 1983 [88,89]. This is a novel method of polymer synthesis known as group transfer polymerization (GTP). It in-volves the repeated addition of monomer molecules to a growing polymer chain end that carries a reactive silyl ketene acetal group. The name GTP is derived from the fact that during the addition, the silyl group is transferred to the incoming monomer, regenerating a new ketene acetal functionality ready for reaction with incoming monomer molecules. This process usually needs an or-gano-silicon initiator, a bifunctional catalyst, and α,β-unsaturated esters, ketones, nitriles, and carboxamides monomers.

The reaction of GTP must be strictly conducted under anhydrous conditions. The choice of the solvent depends on the use of catalysts, and the temperature range may vary from $-100°C$ to 150°C, but 0–50°C is preferred. The rate of this polymerization is very fast and can be controlled by the addition rate of the mono-mer. The heat of polymerization is, of course, the same as that generated by other methods, and is removed by refluxing the solvent used. However, the mechanism and the exact nature of the intermediates of the group transfer polymerization have not so far been established.

GTP is a safe operation. A runaway polymerization can be quickly quenched with a protonic solvent. Since the group transfer polymerization goes to completion, no unwanted toxic monomer remains; the silicone group on the living end after hydroxylation is removed as inac-tive siloxane. The "living" polymer in GTP is costlier than traditional polymerization techniques because of the stringent reaction conditions and requirements for pure and dry monomers and solvents. It can be used in fabrication of silicon chips, coating of optical fibers, etc.

The role of silyl groups in condensation polymeriza-tion is different from that in GTP. The use of silylated monomers in condensation polymerization was studied first by Klebe [90–92] in 1964. N-trimethylsilyl-substi-

Table 4 Phase Transfer Catalyzed Aromatic Nucleophilic Displacement Reaction

Substrate	Nucleophile	Catalyst	Solvent	Time (h)	Temp (°C)	Yield (%)	Product
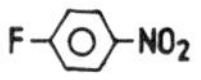 F—⟨O⟩—NO$_2$	PhONa	Dialkylammonium pyridinium salts	Toluene	1/4	110	96	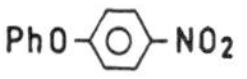 PhO—⟨O⟩—NO$_2$
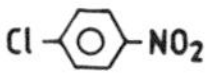 Cl—⟨O⟩—NO$_2$	PhONa	Dialkylammonium pyridinium salts	Chlorobenzene	1/2	125	95	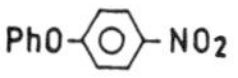 PhO—⟨O⟩—NO$_2$
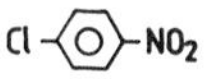 Cl—⟨O⟩—NO$_2$	PhONa	Bu$_4$NBr	Chlorobenzene	8	125	12	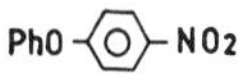 PhO—⟨O⟩—NO$_2$
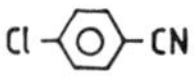 Cl—⟨O⟩—CN	PhSNa	4-dimethyl aminopyridine	0-dichloro-benzene	4	150	86	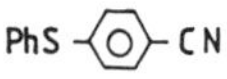 PhS—⟨O⟩—CN
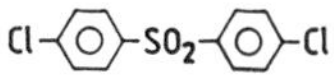 Cl—⟨O⟩—SO$_2$—⟨O⟩—Cl	PhONa	4(4-methyl-piperidinyl) pyridine	0-dichloro-benzene	2	180	87	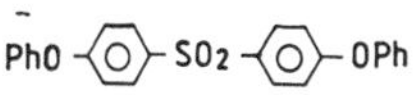 PhO—⟨O⟩—SO$_2$—⟨O⟩—OPh
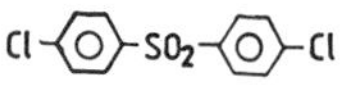 Cl—⟨O⟩—SO$_2$—⟨O⟩—Cl	PhSNa	4-dimethyl aminopyridine	0-dichloro-benzene	2	150	90	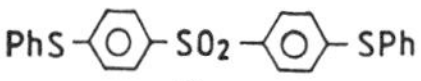 PhS—⟨O⟩—SO$_2$—⟨O⟩—SPh
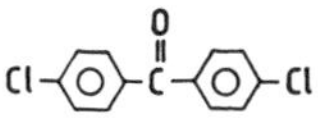 Cl—⟨O⟩—C(=O)—⟨O⟩—Cl	PhSNa	4-dibutyl aminopyridine	Toluene	2	110	90	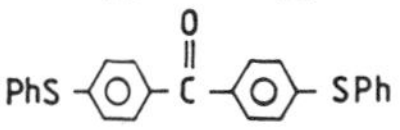 PhS—⟨O⟩—C(=O)—⟨O⟩—SPh
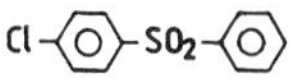 Cl—⟨O⟩—SO$_2$—⟨O⟩	Na$_2$S	PEGM 5000	0-dichloro-benzene	72	180	60	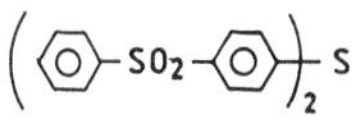 (⟨O⟩—SO$_2$—⟨O⟩)$_2$—S
Cl—⟨O⟩—NO$_2$	(BPA)Na$_2$	1,10-bis(4-dihexyl-amino pyridinium decane dibromide	Chlorobenzene	2	Reflux	99	O$_2$N—⟨O⟩—O—⟨O⟩—⊢—⟨O⟩—O—⟨O⟩—NO$_2$
	Ph(ONa)$_2$			—	—	97	O$_2$N—⟨O⟩—O—⟨O⟩—O—⟨O⟩—NO$_2$
	(PhONa)$_2$S			—	—	98	O$_2$N—⟨O⟩—O—⟨O⟩—S—⟨O⟩—O—⟨O⟩—NO$_2$

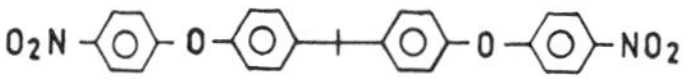

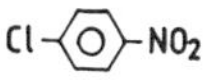

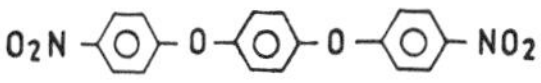

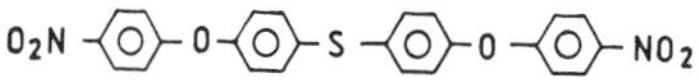

tuted diamines (silylated) and nucleophilic monomers such as, α,α'-dichloroxylene were used for synthesis of condensation polymers, such as polyamines and polyureas, in the presence of catalytic amounts of ammonium chloride. The high polymers were formed with elimination of the halosilane. The reaction of α,α'-dichloro-p-xylene with the bisilyl derivative of piperazine is shown below:

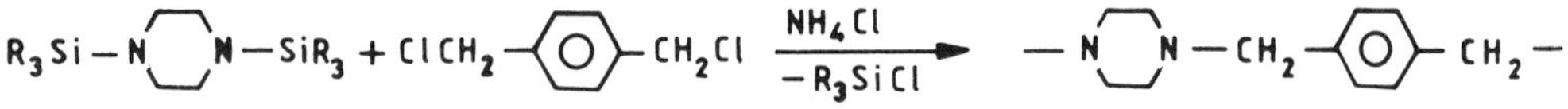

Later, Kricheldorf and coworkers [93,94] extensively demonstrated the use of O-silylated bifunctional monomers, such as diphenols, for synthesis of a wide variety of polycondensation polymers. The silylated oxygen of difunctional phenols may be condensed with activated fluoroaromatics in the molten state following displacement polymerization reaction, to yield poly(ether sulfones), poly(ether ketones), poly(benzonitrile ether), and other polyethers [95–102].

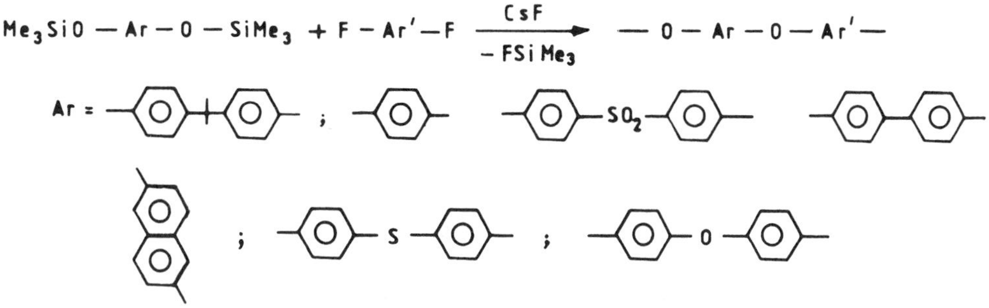

Table 5 Polymers by Phase Transfer Catalysis

Polymer repeat Unit	Monomer(s) Nucleophile	Halocompound	Catalyst	Solvent	Temp (°C)	Time (h)	Yield (%)	n_{inh} (Solv)	Ref.
[triazine structure, R] R = —C₆H₅	BPA	2-phenyl-4,6-dichloro-S-triazine	Cation B	PhNo₂	5	0.5	72	2.58 (ODCB)	82
—OCH₃	BPA	2-methoxy-4,6-dichloro-S-triazine	Cetyldimethyl ammonium choloride	CHCl₃	5	5	—	0.80 (CHCl₃)	80
—OCH₃	BPA + BPA – F	—do—	—do—	CHCH₃	10	5	—	0.90 (CHCl₃)	83
[fluoro-aryl ether structure] X = —C(CH₃)₂—	BPA	Hexafluoro benzene (HFB)	18-Crown-6	DMAC	Reflux	17	85	0.27 (CHCl₃)	75
—S—	BPS	HFB	18-Crown-6	DMAC	80	24	4	insol	75
[sulfone ether structure]	BPA	DCDPS	18-Crown-6	DMAC	80	24	90	0.40	84
[dinitro sulfone structure]	BPA	DNDCPS	18-Crown-6	CH₂Cl₂	20	24	99	0.84 (DMF)	75
[methylene ether structure]	BPA	DCX	Benzyltriethyl-ammonium chloride	CH₂Cl₂	60	7	98	0.45 (NMP)	85
[nitro methylene structure] X = —C(CH₃)₂—	BPA	NXDB	Tetrabutyl-ammonium chloride	CH₂Cl₂	60	10	95	0.44 (DMF)	86
—C(CF₃)₂—	BPA-F	NXDB	—do—	CH₂Cl₂	60	10	85	0.77 (DMF)	86
—C(CH₃)₂—	BPA-F	NXDC	—do—	CH₂Cl₂	60	10	83	0.32 (DMF)	86

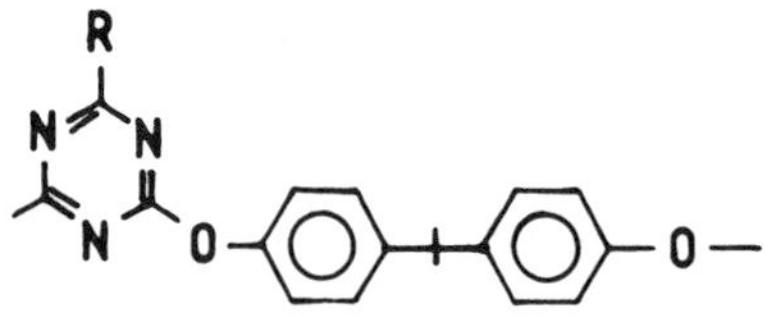

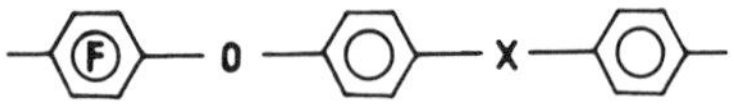

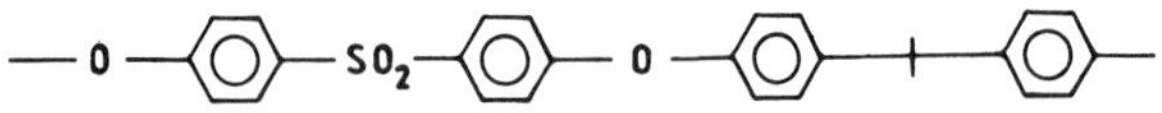

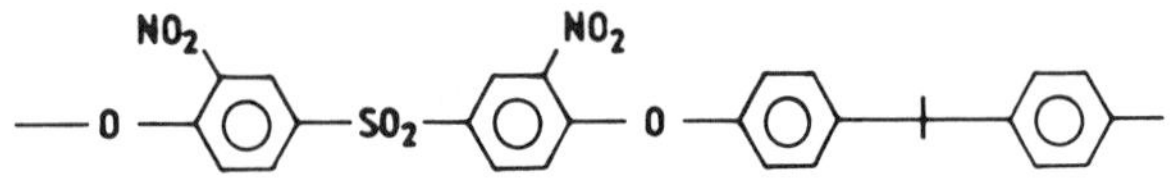

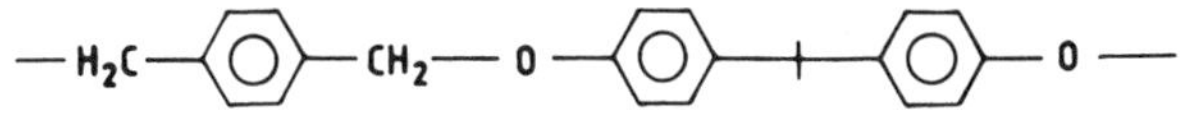

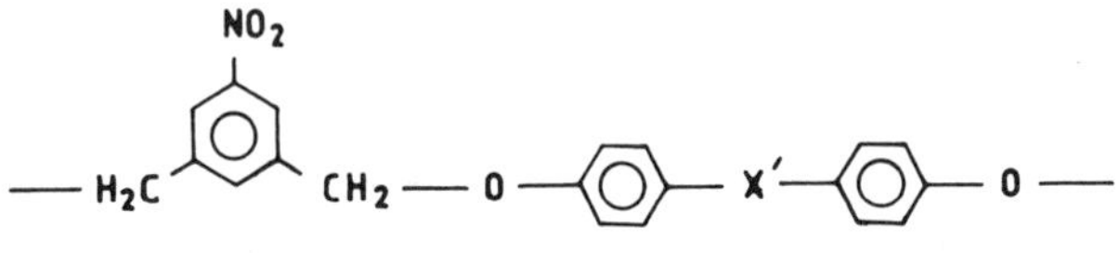

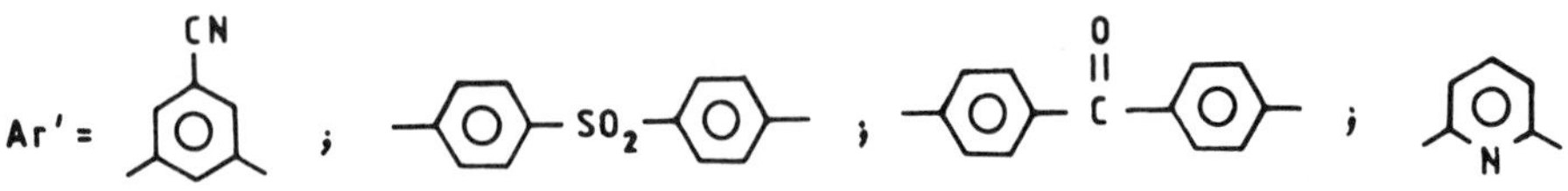

Ar′ = *[structures: cyano-substituted aryl; diphenyl sulfone; benzophenone; pyridine]*

Condensation with activated chloroaromatics with O-silylated bisphenols in NMP in presence of K_2CO_3 has also been reported by Kricheldorf and Jahnke [103]. The procedure is particularly useful for synthesis of copoly-ethers containing N-heterocycles such as pyridine, pyrazine, or pyridazine [93].

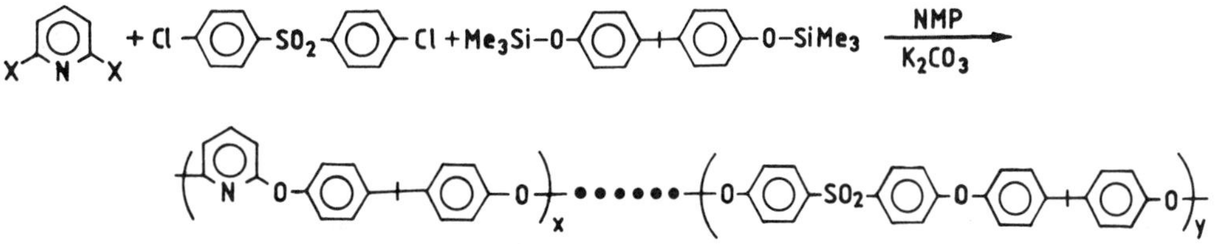

The polymers obtained by this copolymerization [103] show weight average molecular weights upto 2×10^5. Such functionalized copolyethers are of interest for preparation of membranes with variable hydrophilicity and permeability [104].

The utilization of N-silylated diamines as polyamide and polyimide forming monomers has recently been exhaustively reported by Kakimoto et al. [105] and Oishi et al. [106]. But their work is beyond the scope of this present review. It is, at the same time, interesting to note that the first report of the polycondensation of S-silylated aromatic dithiols with activated aromatic dihalides leading to aromatic polysulfides by the use of displacement polymerization was reported by Hara et al. [107]. The solution polycondensation of three S-silylated aromatic dithiols with bis(4-chloro-3-nitrophenyl) sulfone without using any catalyst (such as CsF) gives aromatic polysulfides:

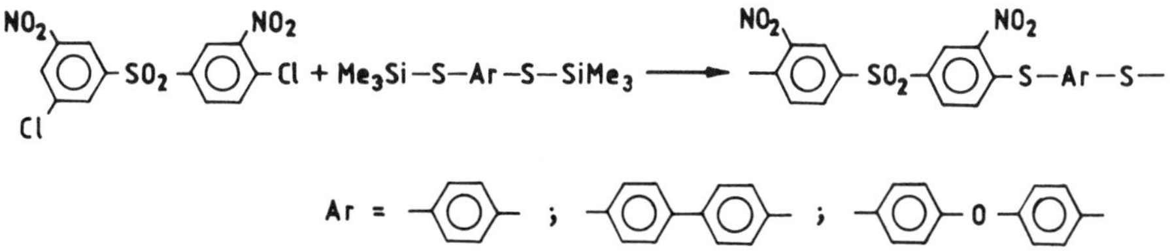

Recently, Kricheldorf and Jahnke [108] reported pyridine and pyridazine containing poly(ether-sulfide)s by polycondensation of silylated 4-mercaptophenol with aromatic dihalide as follows:

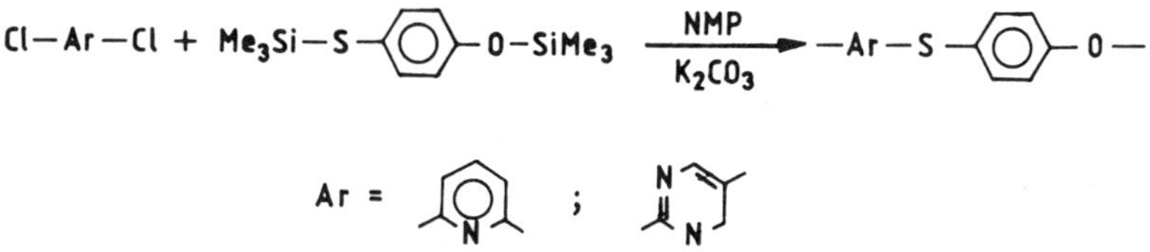

It is observed that the synthesis of various polymers by the silylated method is superior to the conventional diamine, diphenol, or dithiol route particularly with respect to higher inherent viscosity or molecular weight of the resulting polymers.

The strategy of the silylated method has several advantages over the condensation of nonsilylated monomers as shown below:

1. Silylation and distillation of oxidation sensitive or impure technical monomers is a highly effective way of purification.
2. Halosilanes are highly volatile as well as less aggressive, and thus, separation of the resulting polymers from inorganic salts is not required.
3. The O-silylated bisphenol monomer is stable at high temperature, and can act as a good nucleophile beyond certain temperature.
4. The silyl method does not liberate water, like certain conventional methods and, thus, avoids

hydrolysis of functional groups, such as ester, amide, or fluoroaromatic groups.

VII. HIGH PERFORMANCE PHOSPHORUS CONTAINING POLYMERS

Organic polymers provide one of the most versatile groups of materials and have widespread uses. Due to some inherent deficiencies, mainly poor heat and flame resistance, these materials suffer from limitations in certain areas of application. The resistance of polymers to high temperatures and flame may be increased by the incorporation of both aromatic rings and certain chemical elements in the polymer chain. It has been found that phosphorus, present either as a constituent in the polymer chain or incorporated as an additive in the form of a phosphorus compound to the polymer system, can make polymers flame retardant [109].

Polyphosphonates are well-known flame-retardant materials [110] and are generally prepared by melt [111,112], interfacial [113–115] and solution polycondensation methods [116]. A typical example of synthesis is the polycondensation of bifunctional organophosphorus compounds, such as dichlorophenylphosphine oxide, with bisphenols [117,118].

$$RPOCl_2$$
$$+ HO—Ar——OH \longrightarrow (—O—Ar—O—\overset{\overset{\displaystyle O}{\|}}{\underset{\underset{\displaystyle R}{|}}{P}}—)$$

High molecular weight polyphosphonates were successfully synthesized by the interfacial [119] or phase transfer catalyzed methods [120]. Two polymers were prepared by reacting dichlorophenylphosphine oxide (DCPO) with 4,4′-thiodiphenol (TDP) and with 4,4′-sulphonyldiphenol (SDP). The polymers from DCPO/SDP and DCPO/TDP have glass transition temperatures of 146°C and 83°C, respectively. The polymers begin to lose weight at about 395°C in a nitrogen atmosphere. These polymers have good flame retardancy, as indicated by high limiting oxygen indices (LOI) of 50–60. Imai and coworkers also prepared high molecular weight and high T_g polyphosphonates from DCPO and bisphenols having rigid ring structures, such as 4,4′-biphenol ($T_g \sim 120°C$), 3(4-hydroxyphenyl)-1,1,3-trimethyl-5-indanol ($T_g \sim 124°C$), and 9,9-bis(4-hydroxyphenyl)fluorene ($T_g \sim 188°C$) by a phase transfer catalyzed polycondensation method [121]. These polymers were stable up to 300°C in air. Polyphosphonates can be used as matrix resins for heat- and flame-resistant composites [122].

Polyphosphates are also an important class of organophosphorus polymers. In addition to their flame-retardant characteristics, they possess attractive plasticizing properties and can be used as polymeric additives to other polymers [123–128]. In general, polyphosphates can be prepared by interfacial [119,129], melt [130], or solution polycondensation [131,132a,b]. Kricheldorf and Koziel [133] prepared polyphosphates from silylated bisphenols.

Banerjee et al. reported a number of soluble polyimido [134], polyazomethine [135], and polyazoxy phosphonates [136] by the two phase polycondensation method with or without any phase transfer catalyst. Resulting polymers exhibit high thermal stability and fire retardancy.

A number of phosphorus containing polyimides were synthesized by Varma and coworkers and their properties were studied [137–141]. Polyimides have also been synthesized by the reaction of benzophenonetetracarboxylic dianhydride (BTDA) and bis(3-aminophenyl)methylphosphine oxide [142]. Copolyimides were also prepared using bis(3-aminophenyl)methylphosphine oxide,4,2′,4′-triaminobenzanilide and tris(3-aminophenyl) phosphine oxide with pyromellitic dianhydride. The homopolymers and copolymers exhibit good thermal properties. Improved mechanical properties were observed in copolymers.

Cross-linkable phosphorus containing polyimides were also synthesized and characterized [143]. The synthesis was based on the monoamic acid derived from the reaction of tris(3-aminophenyl) phosphine oxide (TAPO) with an equimolar amount of maleic, nadic, or itaconic anhydrides; the product being subsequently reacted with an equimolar amount of pyromellitic dianhydride. Cured resins were obtained by heating the above polymer at higher temperatures. The monomer TAPO may also be used for the synthesis of cross-linkable polyamides and polyureas [143] by reacting the monoamic acid of TAPO with diacid chlorides and diisocyanates, respectively. All these polymers are self-extinguishing and stable up to 300°C and can be used as matrix resins for composites.

Phosphorus containing poly(maleimide-amines) were synthesized from N,N'-bisdichloromaleimido-3,3′-diphenyl alkylphosphine oxides and aromatic diamines or piperazine [144]. The polymers prepared from piperazine are soluble in DMF, DMAC, DMSO, etc., but have poor thermal stability and flame retardancy.

Phosphorus containing poly(anhydride-imides) were synthesized from N,N'-bis(4-carboxyphthalimido)-3,3′-diphenylalkylphosphine oxide, 3,3′-$[N,N'$-bis(4-carboxyphthalimido)] benzophenone, and their mixtures in steps via diacetyl derivatives of the bisimide carboxylic acid [145]. The resulting polymers have low reduced viscosities (0.06–0.14 dl/g) and exhibit improved heat and flame resistance compared to nonphosphorus containing poly(anhydride-imides). Phosphorus containing poly(esterimides) were prepared from N,N'-bis (4 - carboxyphthalimido) - 3,3′ - diphenylphosphine oxide (BCIAP) and various aromatic diacetoxy compounds by acidolysis [146,147]. Poly(amide-imides) containing phosphorus were synthesized from BCIAP and various aromatic diacetamido derivatives by acidolysis [148]. The resulting polymers are fairly soluble in DMAC, DMF, and concentrated H_2SO_4; the reduced viscosities of these polymers are 0.19–0.32 dl/g at 30°C in DMAC. They have good thermal stability and self-extinguishing properties.

In order to achieve better thermal stability, phosphorus containing polymers with phosphorus in a ring system were synthesized. Polyamides containing phenoxaphosphine rings were prepared from 2,8-bis-(chloroformyl)-10-phenoxaphosphine 10-oxide (BCPO) and aromatic diamines in DMAC by low temperature solution polycondensation in the presence of triethylamine as acid acceptor [149]. The resulting polyamide has a reduced viscosity of 0.14–0.40 dl/g in DMAC. Some of the polyamides are soluble in DMF, DMAC, and DMSO. Phenoxaphosphine ring containing polyesters were prepared by the reaction of (BCPO) with bisphenols by interfacial polycondensation [150]. Some of the resulting polyesters were soluble in chloroform, *m*-cresol, and nitrobenzene, and showed reduced viscosities of 0.85–1.92 dl/g in DMAC. Soluble copolyesters have also been synthesized by the reaction of isophthaloyl chloride with bisphenols [151].

Phenoxaphosphine ring-containing poly (1,3,4-oxadiazoles) were synthesized by cyclodehydration of polyhydrazides obtained from (BCPO) and aliphatic and aromatic dihydrazines [152]. All these polymers are soluble in formic acid, *m*-cresol and concentrated H_2SO_4. The polyhydrazides yield transparent and flexible films when cast from DMSO solution under reduced pressure at 80–100°C. The polyhydrazides exhibit reduced viscosities of 0.24–0.40 dl/g in DMAC. Phenoxaphosphine ring-containing oxadiazole polymers showed little degradation below 400°C.

Novel polyimides containing phenoxaphosphine rings were prepared by the reaction of 10-phenylphenoxaphosphine 2,3,7,8-tetracarboxylic dianhydride 10-oxide with diamines via polyamic acid in two steps [150]. The resulting polyimides have reduced viscosities of 0.12–0.84 dl/g in concentrated H_2SO_4 at 30°C. Phenoxaphosphine ring-containing poly(amide-imides) were prepared by cyclodehydration of the poly(amide-acid) derived from 8-chloroformyl 10-phenylphenoxaphosphine 2,3-dicarboxylic anhydride-10-oxide and diamines by low temperature solution polycondensation [153]. The resulting poly(amide-imides) had reduced viscosities of 0.10–0.59 dl/g in DMAC at 30°C. Phenoxaphosphine ring systems containing polyesterimides were prepared by the interfacial polycondensation of bischloroformyl phenoxaphosphine derivative with bisphenols [154] and aminobenzoic acid. Phenoxaphosphine ring-containing polybenzimidazoles and polybenzoxazoles were synthesized by the reaction of 2,8-dicarboxyl 10-phenylphenoxaphosphine 10-oxide with aromatic tetramines and dihydroxydiamines, respectively, at 190–200°C in polyphosphoric acid [155].

It has been observed that all the phenoxaphosphine ring-containing polymers have excellent thermal stability and show better heat resistance than open-chain phosphorus containing polymers. The phenoxaphosphine polymers containing aromatic rings in the backbone show little degradation below 400°C in air.

Phenothiophosphine ring-containing polyamides and polyesters were also prepared by the polycondensation of 2,8-bischloroformyl-10-phenylphenothiophosphine 5,5′,10-trioxide with aromatic diamines such as 4,4′-diaminodiphenyl ether and 4,4′-diaminodiphenylmethane, and bisphenols such as 4,4′-dihydroxybiphenyl and 4,4′-dihydroxydiphenylmethane, respectively [159]. These polymers are soluble in polar aprotic solvents and also exhibit good heat and fire resistance. Phosphorus containing high performance polymers are shown in Table 6.

VIII. HIGH PERFORMANCE POLYIMINES

The imine bond —CH=N— is formed during polycondensation of aromatic/aliphatic diamines with aromatic/aliphatic dialdehydes:

$$H_2N—Ar—NH_2 + HOC - Ar'—COH$$
$$→ = N—Ar—N=HC—Ar'—CH=$$

The literature of polyimines is extensive [164–173]. A number of researchers have tried to synthesize high molecular weight polymers but failed due to poor solubility in organic solvents. Polyimines are of great interest because of their high thermal stability [174–176], ability to form metal chelates [174–177], and their semiconducting properties [178–181]. Due to insolubility and infusibility, which impeded characterization of the molecular structure, the application of these polymers is very limited and of little commercial importance.

The first polyimine was reported by Adams and coworkers [182] from terephthalaldehyde and benzidine and dianisidine. Between 1950 and 1959 Marval and coworkers [174–176] reported a number of polyimines. Suematsu and coworkers [170] reported the first successful synthesis of high molecular weight fully aromatic polyimines by solution polycondensation method using *m*-cresol as reaction medium.

Morgan and coworkers [183] synthesized a variety of fusible aromatic polyimines and copolyimines by solution and melt methods. The polyimines exhibit inherent viscosities of 0.46–4.7 dl/g in concentrated H_2SO_4 at 30°C. However, it is well known that polyimines undergo extensive degradation in the presence of strong protonic acid due to hydrolysis of the —CH=N— linkages [179,184]. Therefore, the observed viscosities of the polyimines in concentrated H_2SO_4 will be viewed with scepticism. These polymers provide a useful range of melting or softening points (200–350°C). This range was attained by ring substitution, copolymerization, and/or introduction of limited chain flexibility. Many of the polyimines yielded liquid crystalline melts, which can be readily spun into oriented, high-tenacity, high-modulus fiber. These fibers can be further strengthened by heat treatment. The polyimines synthesized from methyl-1,4-phenylenediamine and terephthalaldehyde was spun into fiber with tenacity/initial modulus of 7.3/916 g/denier as spun, and the heat treated fiber had tenacity/initial modulus of 38/1012 g/denier.

Wojtkonski [185] has also reported on three series of melt spinnable thermotropic aromatic-aliphatic polyimines. The polyimines were prepared by reaction of 1,2-bis(4-formylphenoxy) ethane, terephthalaldehyde, or 4,4′-biphenyldicarboxaldehyde, respectively, with l,n-bis(4-amino-3-methylphenoxy) alkanes where n = 1–10, 12, 14, and 16 in dry DMAC containing 5% dry lithium chloride. The polymers decomposed at 400°C, and as the length of the flexible aliphatic segments increased, melting points decreased. Polymers with an odd

Table 6 Phosphorus Containing High Performance Polymers

Polymer structure	Year of reporting	Ref.
A. Trivalent phosphorus		
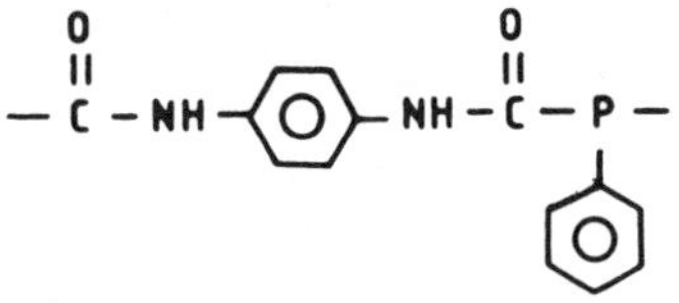	1963	157
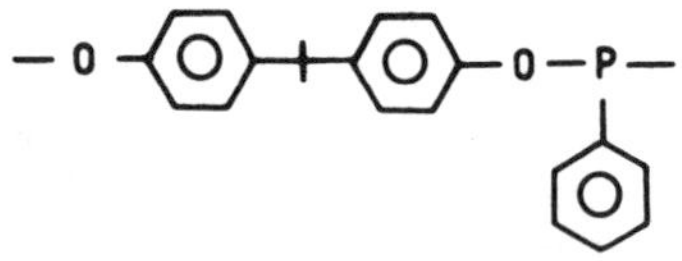	1964	159
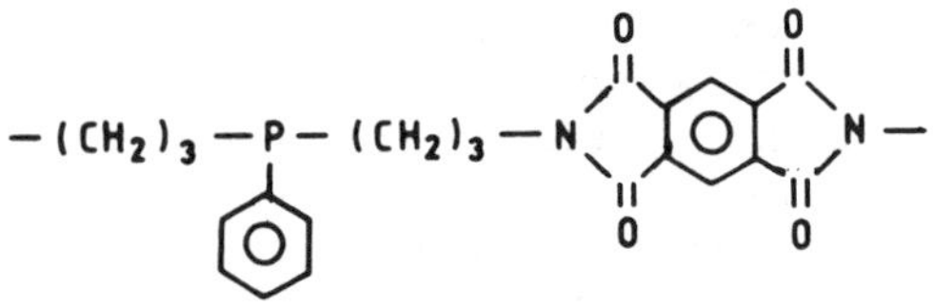	1968	124
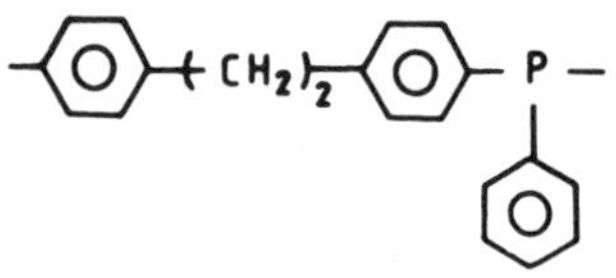	1985	157
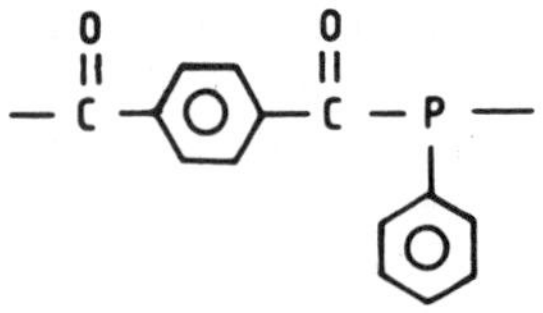	1992	158
B. Pentavalent phosphorus		
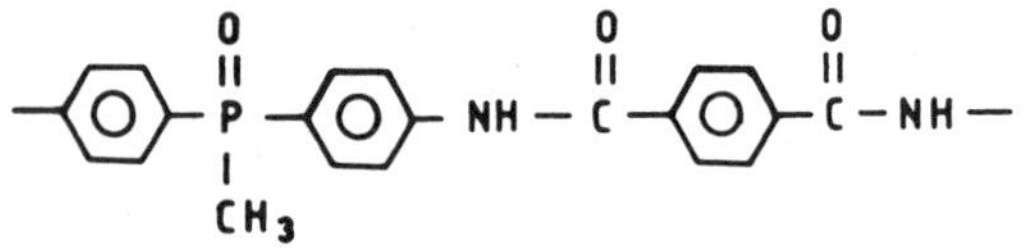	1958	160
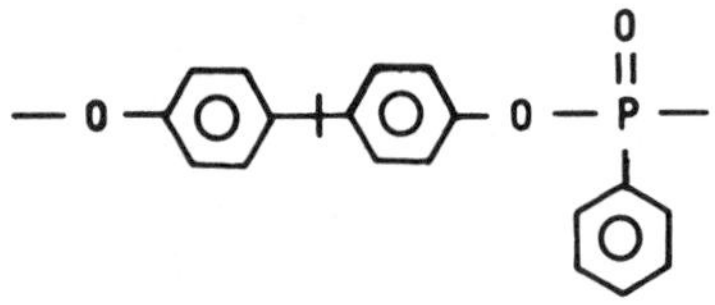	1962	113, 119–120
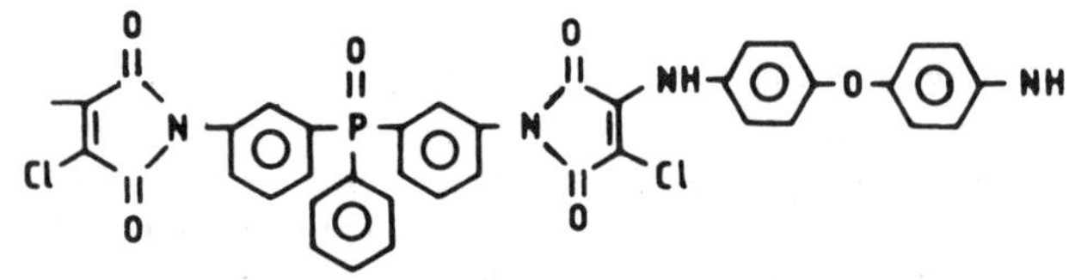	1980	144
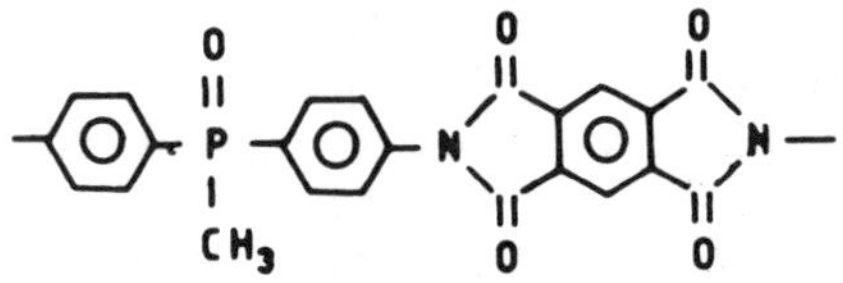	1983	142

Table 6 (*continued*)

Polymer structure	Year of reporting	Ref.
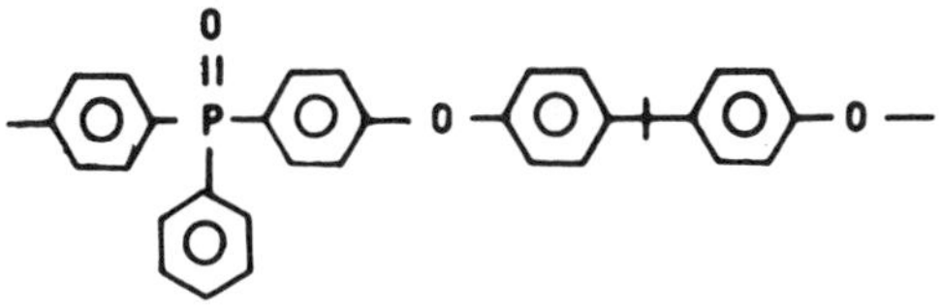	1990	161
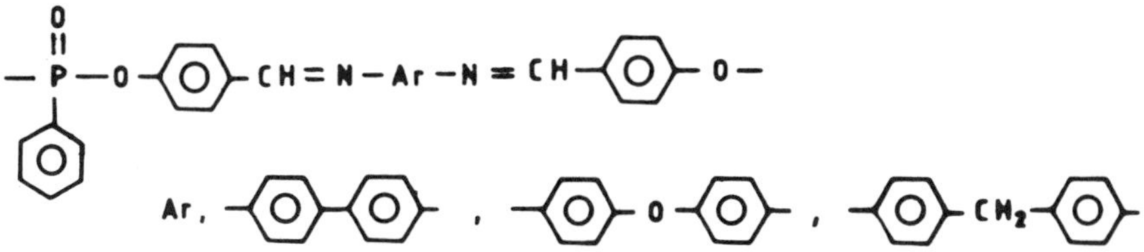	1992	135
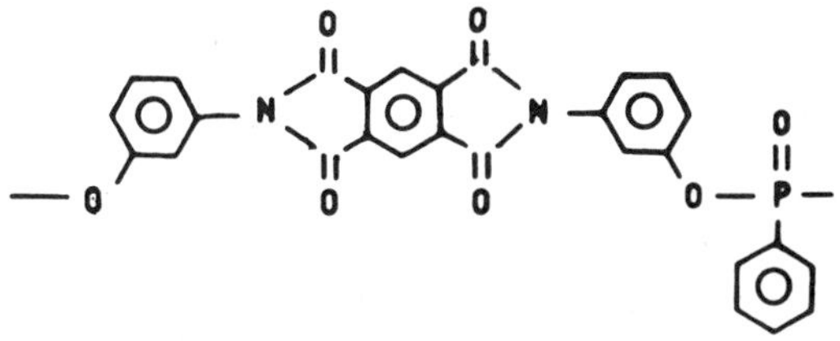	1994	134
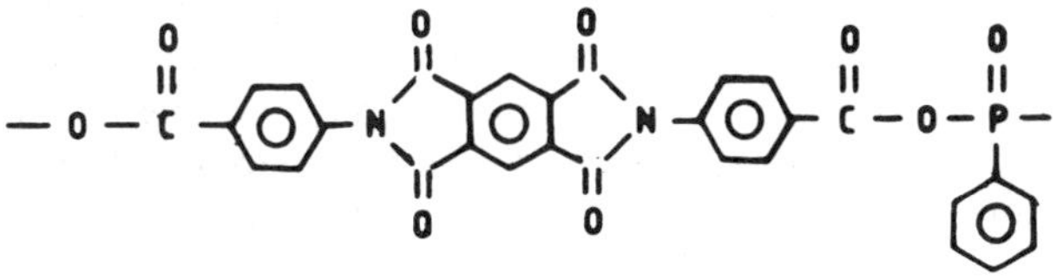	1994	162
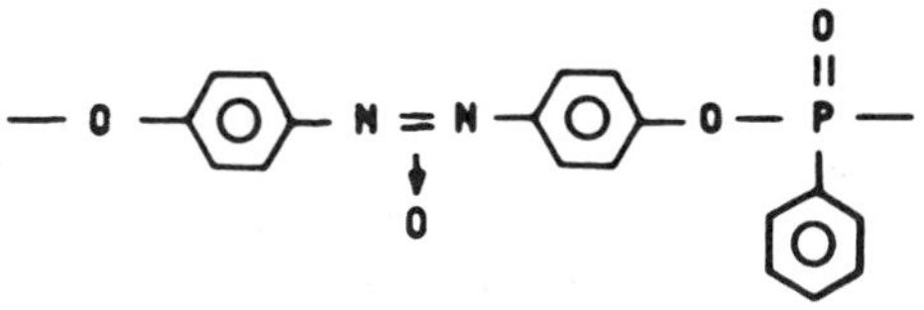	1994	136

C. Phosphorus in the ring

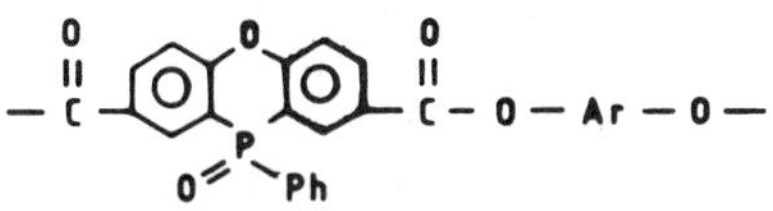	1980	150
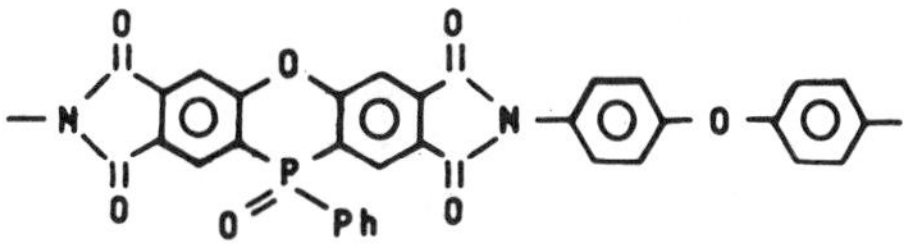	1980	150
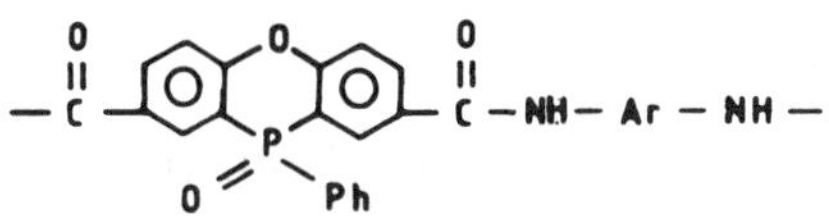	1980	149

(*continued*)

Table 6 (*continued*)

Polymer structure	Year of reporting	Ref.
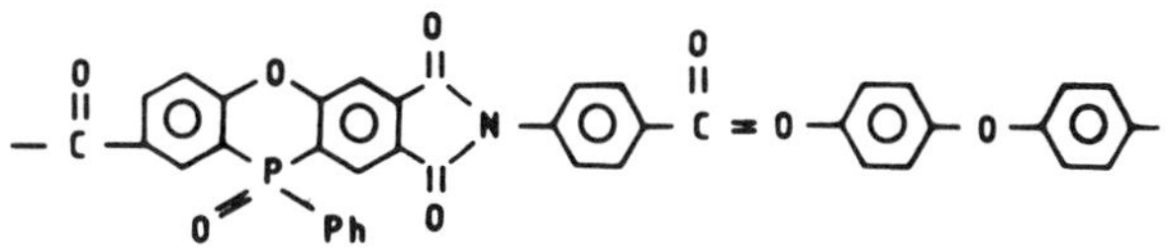 [chemical structure]	1981	154
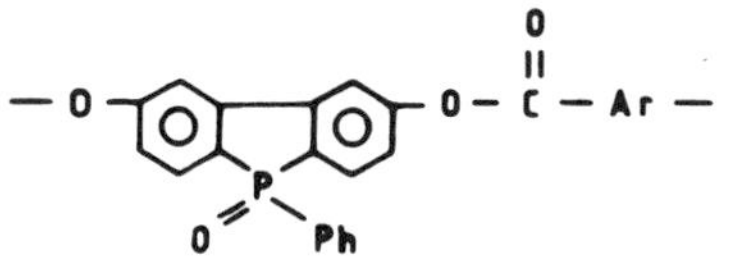 [chemical structure]	1981	163
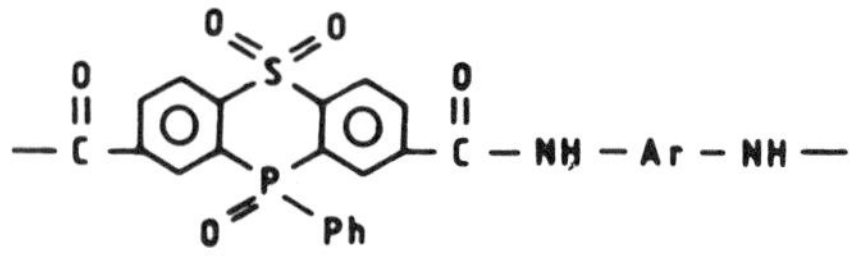 [chemical structure]	1984	156

number of methylene units generally had lower melting and lower modulus (spun fibers) than polymers with an even number of methylene units, indicating poorer molecular packing in the solid and weaker chain alignment in the melt for the odd series.

Yang and Jenekhe [186,187] reported a successful solubilization of aromatic polyimines in organic solvents via their soluble coordination complexes, which facilitated their solution characterization by NMR and processing films and coatings by spin coating and other techniques. This has created opportunities for various studies of the aromatic polyimines.

Recently, polyimines include the synthesis of long alkoxy (C_8–C_{18}) side chain derivatives [188,189], which are presumably soluble to some extent in organic solvents and derivatives containing fluorene cardo unit [190]. Trifluoromethyl groups [191] in the polymer backbone provide solubility in organic solvents. Studies of the electrical conductivity of doped conjugated aromatic polyimines and alkoxy derivatives have been reported [188], and the values are in the range of 10^{-6} to 10^{-2} S/cm.

The presence of ether linkages in the polymer molecule imparts chain flexibility, lowers glass transition temperature, and enhances solubility while maintaining the desired high temperature characteristics [192]. Recently, polyether imines were prepared by the reaction of different diamines with 4,4'-[1,4-phenylene bis(oxy)] bisbenzaldehyde [184]. The polymers synthesized by the solution method were yellow to white in color and had inherent viscosities up to 0.59 dl/g in concentrated H_2SO_4. Some of these polyimines can be considered as

high temperature resistant semiconductors. A number of ketoimines [193] were also synthesized to get soluble polymers by destroying the chain symmetry through pendent methyl groups. Recently, some soluble polyester imines containing pendent ethoxy linkages in the polymer backbone were reported [194]. Some high performance polyimines are summarized in Table 7.

IX. PROPERTIES OF NEW POLYMERS

Polyethers and polysulfides obtained by displacement reaction, in general, are rigid as well as flexible, heat resistant, tough, and melt processable polymers with good electrical, chemical, and durability properties, as well as fire and hydrolysis resistance [195,196].

The advantage of the activated displacement polymerization is the facile incorporation of different and unconventional structural units in the polymer backbone. Most of the heteroarylene activated polyethers prepared by this route are soluble in many organic solvents. The solubility behavior of new polyethers is shown in Table 8. In contrast to many polyphenylenequinoxalines, poly(aryl ether phenylquinoxalines) prepared by the quionoxaline activated displacement reaction are soluble in NMP. Solubility in NMP is important since it is frequently used for polymer processing in the microelectronics industry [27].

A. Mechanical Behavior

The mechanical properties of commercially available polyethers and newly prepared polyethers are shown in

Table 7 High Performance Polyimines

Polymer structure	Year of reporting	Ref.
=N—⟨O⟩—⟨O⟩—N=CH—⟨O⟩—CH= (R, R) R = H, OCH₃, OH	1923	182, 187
=N—⟨O⟩—N=CH—⟨O⟩—O—C(=O)—⟨O⟩—C(=O)—O—⟨O⟩—CH=	1983	170
=N—⟨O⟩—N=CH—⟨O⟩—O—CH₂—⟨O⟩—CH₂—O—⟨O⟩—CH=	1983	170
=N—⟨O⟩—S—⟨O⟩—N=CH—⟨O⟩—S—⟨O⟩—CH=	1983	170
=CH—⟨O⟩—CH=N—⟨O⟩(Me)—N=	1983	170
=N—⟨O⟩(CH₃)—N=CH—⟨O⟩(Cl)—CH=	1987	183
=N—⟨O⟩(CH₃)—N=CH—⟨O⟩⟨O⟩—CH=	1987	183
=N—⟨O⟩(CH₃)—O(CH₂)ₓ—O—⟨O⟩(CH₃)—N=CH—⟨O⟩—CH= X = 1–10, 12, 14, 16	1987	185
=N—⟨O⟩—O—⟨O⟩—N=CH—⟨O⟩(Cl)—CH=	1987	183
=N—⟨O⟩—⟨O⟩—⟨O⟩—N=CH—⟨O⟩(R, R)—CH=N— R = H, OCH₃, OH	1995	187
=N—Ar—N=CH—⟨O⟩—O—⟨O⟩—O—⟨O⟩—CH=	1995	184
=N—Ar—N=C(CH₃)—⟨O⟩—O—⟨O⟩—O—⟨O⟩—CH=	1996	193
=N—Ar—N=CH—⟨O⟩(OCH₂CH₃)—O—C(=O)—⟨O⟩—C(=O)—O—⟨O⟩(OCH₂CH₃)—CH=	1996	194

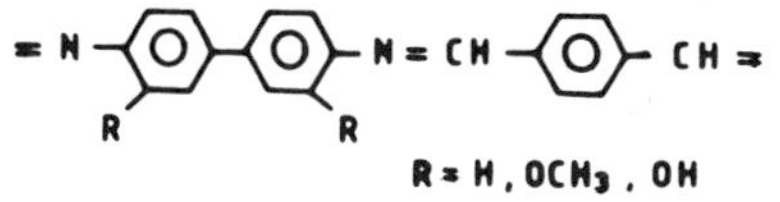

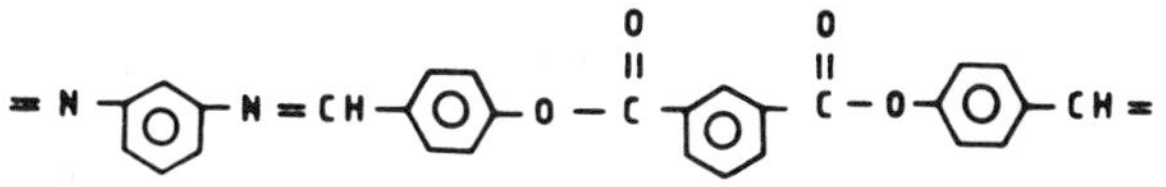

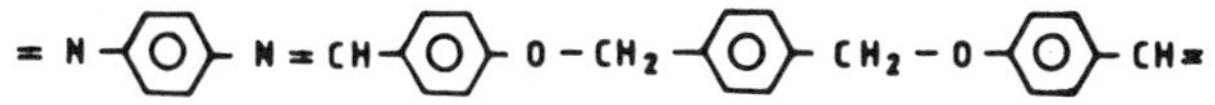

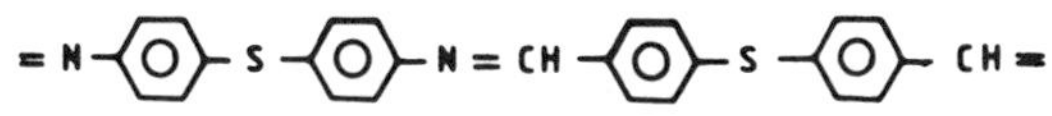

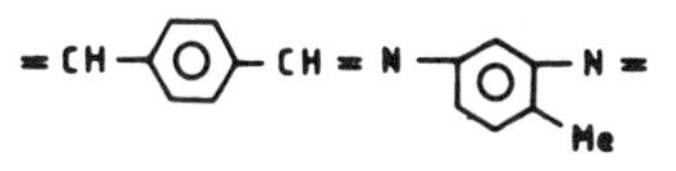

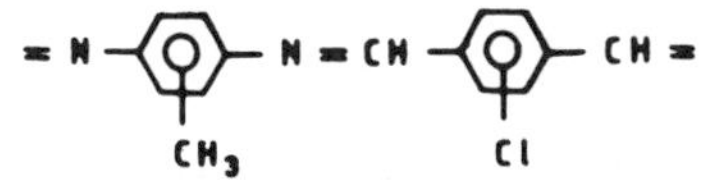

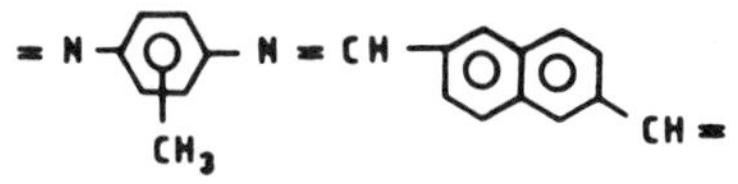

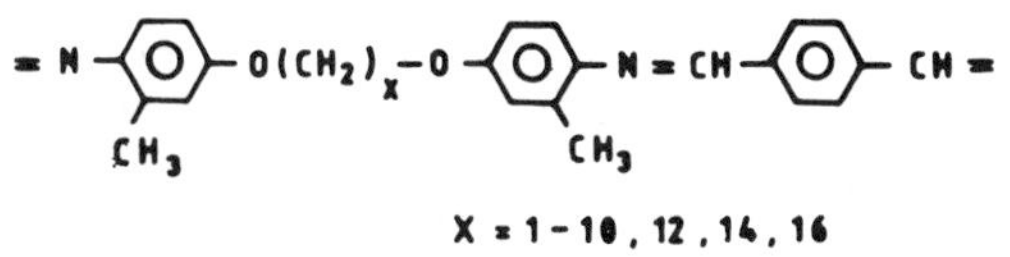

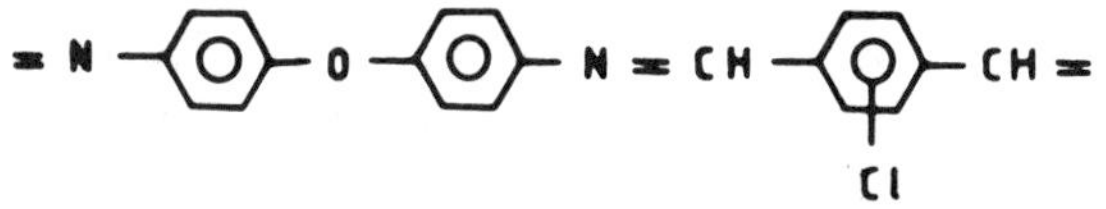

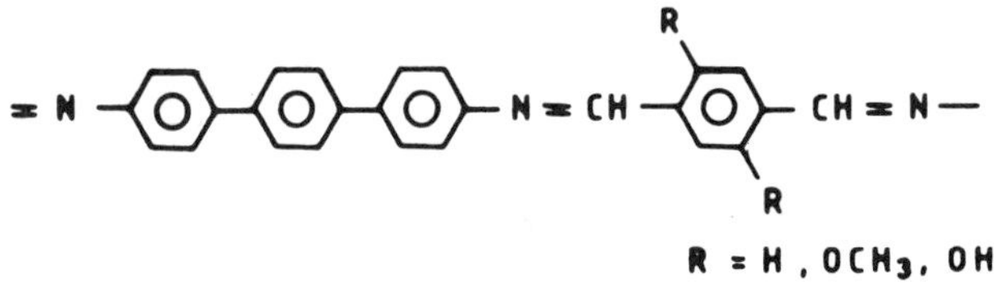

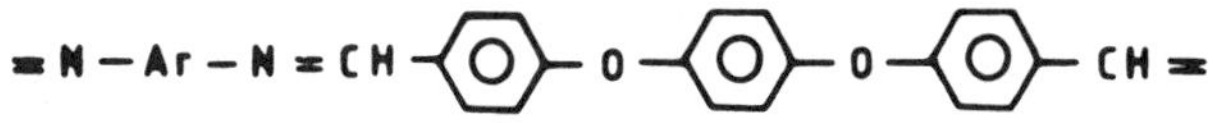

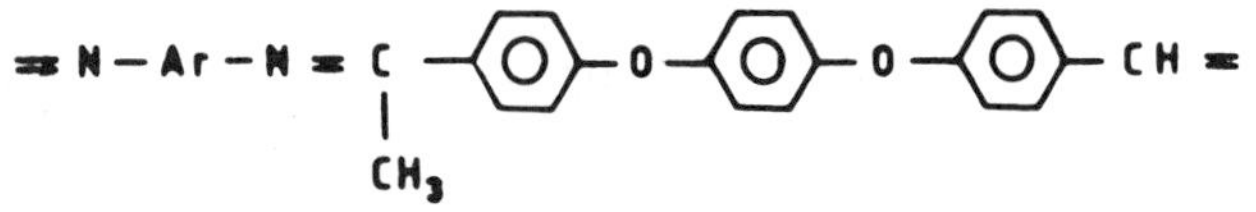

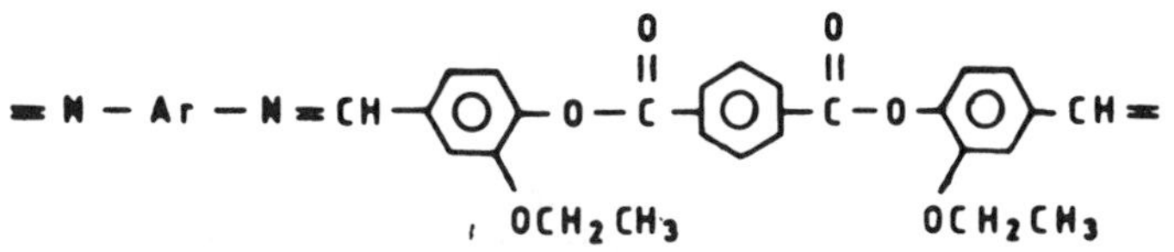

Ar = —⟨O⟩—⟨O⟩— , —⟨O⟩—O—⟨O⟩— , —⟨O⟩—CH₂—⟨O⟩— , (CH₂)₂ , (CH₂)₆

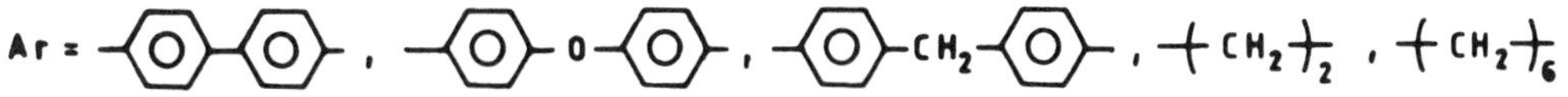

Table 8 Solubility Behavior of New Polyethers

Polymer	NMP	DMF	fMAC	THF	CHCl$_3$	Ref.
Poly(ether bissulfones)	s	s	s	ns	sw	12
Poly(ether imides)	s	s	s	ns	ns	101
12F-Poly(ether ketones)	s	s	s	s	s	15
Poly(ether perfluoroalkylenes)	s	s	s	s	s	10
Poly(ether quinoxalines)	s	ms	ms	ms	ns	26, 27
Poly(ether ketones)	s	s	s	s	s	2, 56
Poly(ether benzoxazoles)	s	ms	ms	ms	ns	29, 60
6F 8F Poly(ether ketones)	s	s	s	s	s	17, 30
Poly(ether oxadiazoles)	ms	ms	ms	ms	ns	33, 63
Poly(ether thiophenes)	s	s	s	s	s	64
Poly(ether dibenzofurans	s	s	s	s	s	61
Poly(ether benzothiazoles)	ms	ms	ms	ms	ns	58
Poly(ether 0-dibenzoyl benzenes)	s	s	s	s	s	14
Poly(ether-1,4-isoquinolinediyl)	s	s	s	s	s	62

s = soluble; sw = swellable; ms = marginal soluble; ns = not soluble.

Table 9. These polymers show excellent mechanical properties, which are comparable to those of commercially available engineering plastics. Ultem [199] possesses the highest tensile modulus, whereas fluoropolyether [10] has lesser modulus. Polyether ketone [200] and polyether ketone isoquinolinediyl [62] have greater breaking elongation, which shows better flexibility than rigid polyether phenylquinoxaline [27]. The yield stress of polyether phenylquinoxaline, Ultem, and perfluoroalkylene are almost the same. Most of the polyethers offer tough as well as flexible transparent films on solution casting or compression molding. Most of the new polyethers mentioned in Table 9 are considered to be promising advanced plastic materials.

B. Thermal Stability

All commercial polyethers, polysulfides, and polyetherimides possess almost comparable thermal properties, but differ in melt processability. The high heat deflection temperature of Kapton limits its processability and use [197]. The ease of melt processability of all displacement polyethers or polythioethers is due to the presence of ether or thioether linkages (C—O—C or

Table 9 Mechanical Properties of Polyethers

Polymer	Tensile strength (MPa)	Elongation at break (%)	Tensile modulus (MPa)	Yield stress (MPa)	Ref.
Polyether sulfone*	—	—	2482	70.3	198
Polyether benzoxazole	74	25	1900	68	60
Polyether phenylquinoxaline	108	13	2800	100	27
Ultem	—	—	3000	103	199
Polyether perfluroalkylene	—	35	1500	100	10
Polyether ketone	100	>100	2700	91	200
Polyether ketone-1,4-iso quinolinedinelene	55	—	1500	—	62
	88	136	2627	—	13
	94	4.6	2606	—	13
12F-Polyether Ketone	74	96	—	—	15
6F 8F-Polyether Ketone	35[a]	12[a]	1528[a]	—	16
	57[b]	85[b]	1689[b]	—	17

* Udel polyether sulfone.

[a] Polymer prepared by Cassidy et al.

[b] Polymer prepared by Mercer et al.

Table 10 T_g and Thermal Stability of Polyarylethers Synthesized by Displacement Reaction

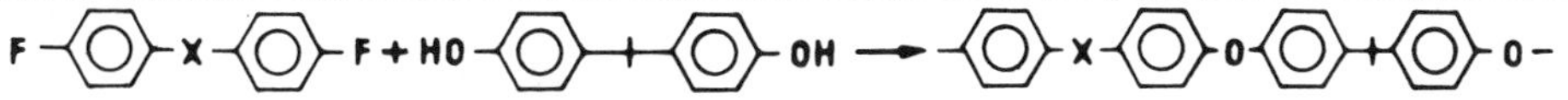

Polyether	T_g (°C)	TGA (10% weight loss in air)	Ref.
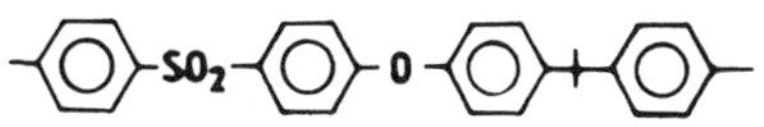 [structure: –⬡–SO₂–⬡–O–⬡–+–⬡–]	190	513	2
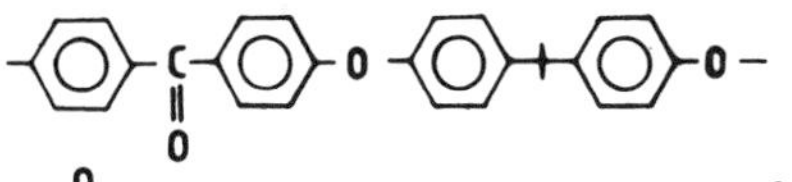 [structure: –⬡–C(=O)–⬡–O–⬡–+–⬡–O–]	155	—	2
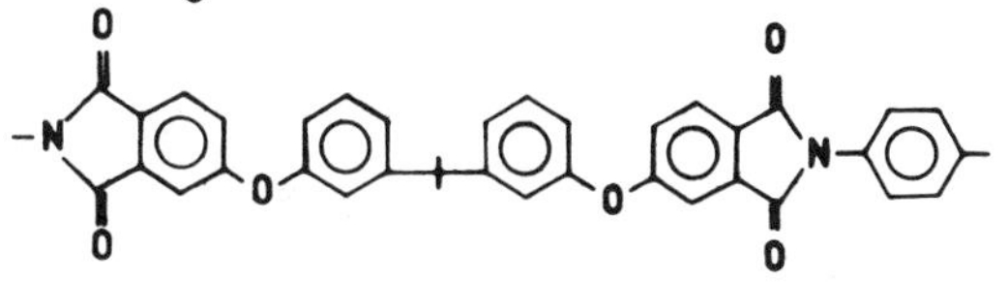 [structure: imide –N...O–⬡–+–⬡–O...N–⬡–]	220	545	56
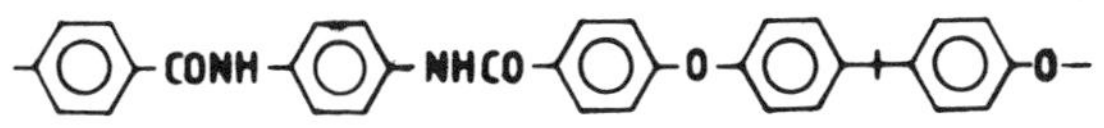 [structure: –⬡–CONH–⬡–NHCO–⬡–O–⬡–+–⬡–O–]	230	420*	9
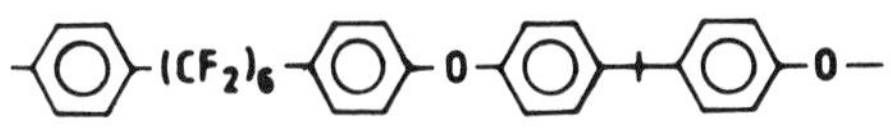 [structure: –⬡–(CF₂)₆–⬡–O–⬡–+–⬡–O–]	93	High thermal stability	10
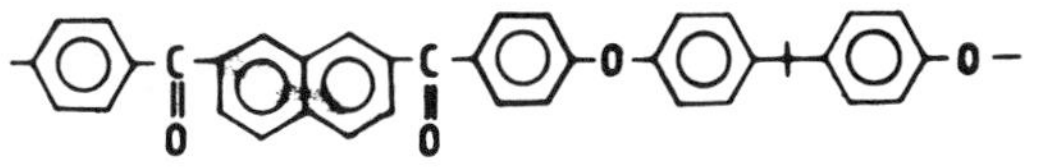 [structure: –⬡–C(=O)–naphthalene–C(=O)–⬡–O–⬡–+–⬡–O–]	185	505*	13
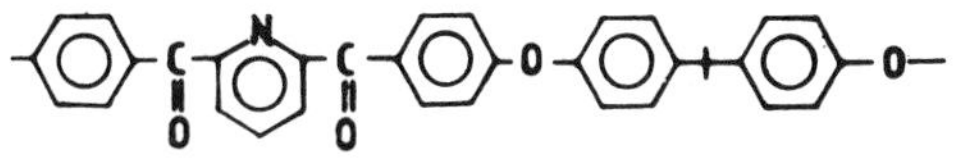 [structure: –⬡–C(=O)–pyridine–C(=O)–⬡–O–⬡–+–⬡–O–]	132	—	13
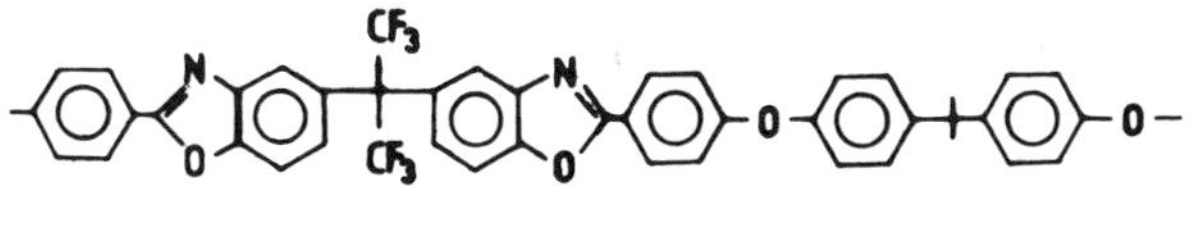 [structure: oxazole C(CF₃)₂ bridge –O–⬡–+–⬡–O–]	241	>450*	60
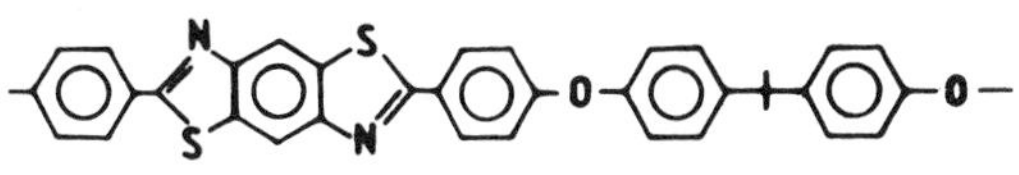 [structure: benzobisthiazole –⬡–O–⬡–+–⬡–O–]	248	~450*	58
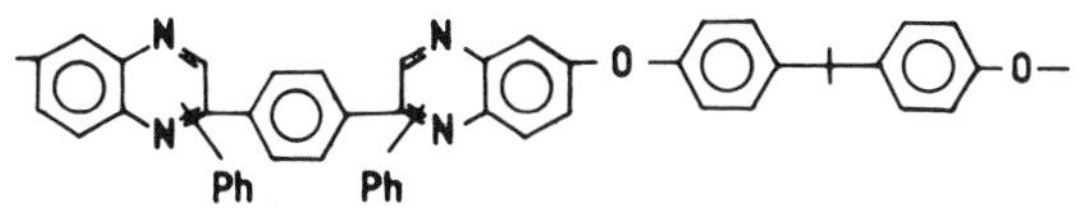 [structure: quinoxaline Ph Ph –O–⬡–+–⬡–O–]	255	—	27
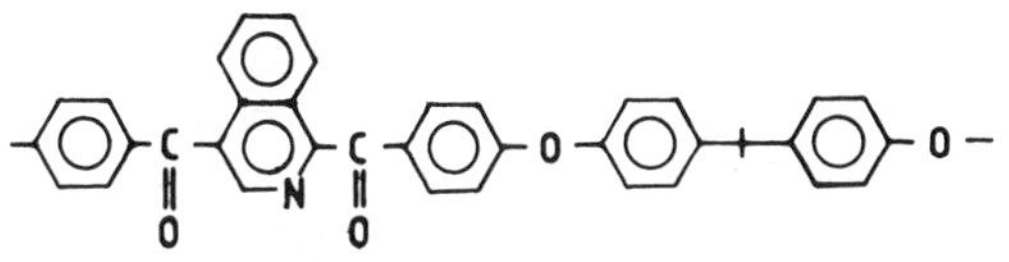 [structure: –⬡–C(=O)–N-ring–C(=O)–⬡–O–⬡–+–⬡–O–]	181	480	62
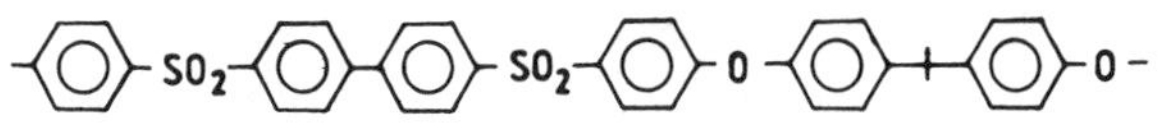 [structure: –⬡–SO₂–⬡–⬡–SO₂–⬡–O–⬡–+–⬡–O–]	241	510	12

(Continued)

Table 10 (*Continued*)

Polyether	T_g (°C)	TGA (10% weight loss in air)	Ref.
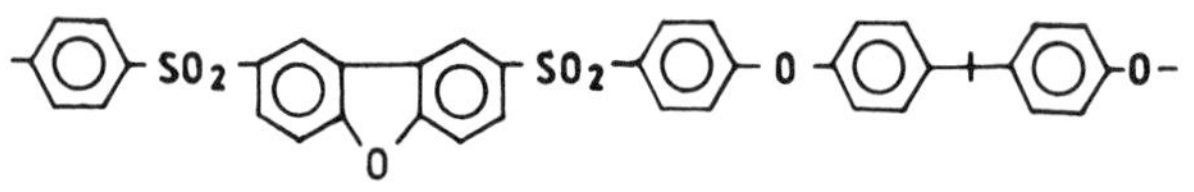	223	475	61
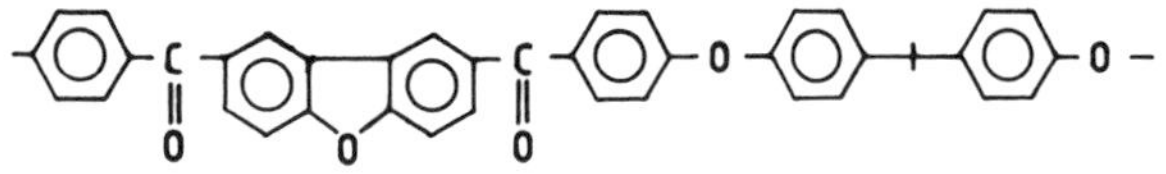	205	505	61
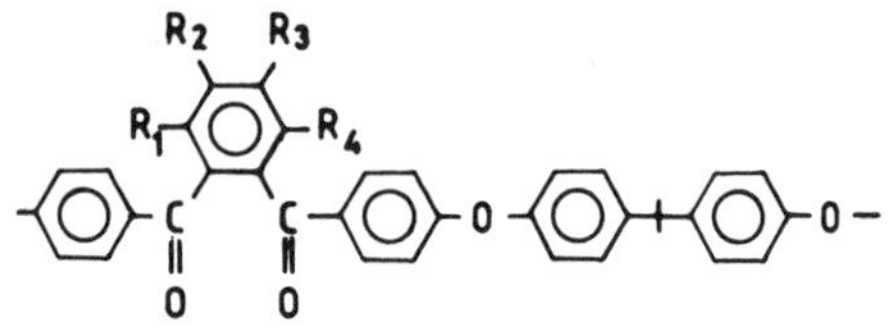	268	520*	14
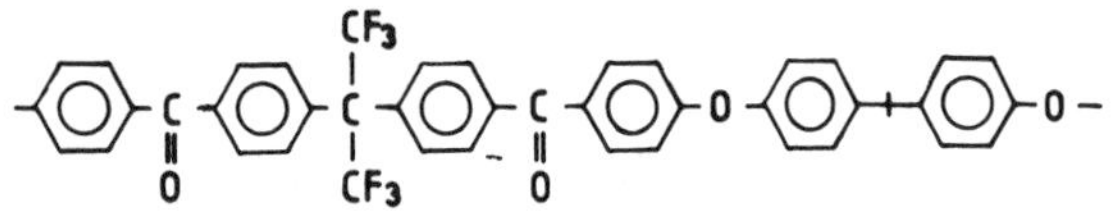	172	485	15
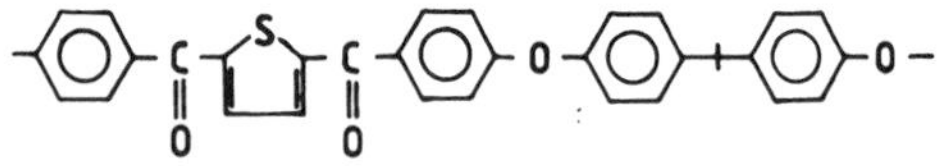	158	>480	64
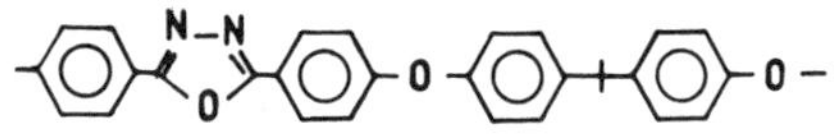	201	>450	63
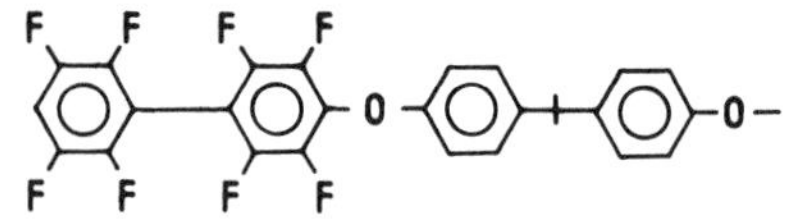	—	460	17, 30

* Polymer decomposition temperature.

C—S—C) in the main chain. The new polyethers prepared either by new heteroarylene activated or by aromatic activated systems have good melt processability. The thermal stability and glass transition temperature of bisphenol-A based new polymers are shown in Table 10.

Table 11 describes the thermal properties of polyether sulfone based on DCDPS and heteroarylenediol. The T_gs range from 230 to 315°C and the decomposition temperature is higher than 450°C. Their thermal stability depends on the bisphenol and activated difluoride used in the polymer synthesis (Tables 10 and 11).

C. Crystallinity

Polyetherimides show no crystallinity as evidenced from calorimetry measurements. The heteroarylene like phenylquinoxaline [27], oxadiazole [30], and benzoxazole [56] activated polyethers show T_gs from DSC thermograms, with no evidence of crystallization, indicating amorphous or glassy morphology. Furthermore, wide angle x-ray scattering measurements show no evidence of crystalline or liquid crystalline type morphologies, consistent with an amorphous structure. ^{12}F polyether

Table 11 T_g and Thermal Stability of Polyether Sulfone Synthesized by Displacement Reaction

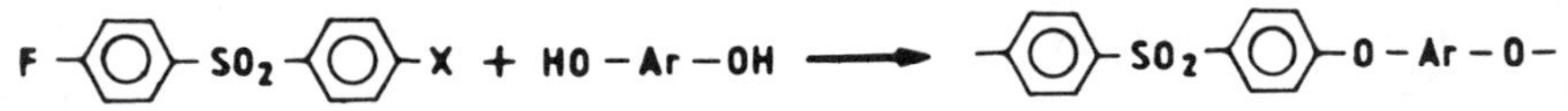

F —◯— SO₂ —◯— X + HO — Ar — OH ⟶ —◯— SO₂ —◯— O — Ar — O —

Polyethersulfone	T_g (°C)	TGA (10% weight loss in air)	Ref.
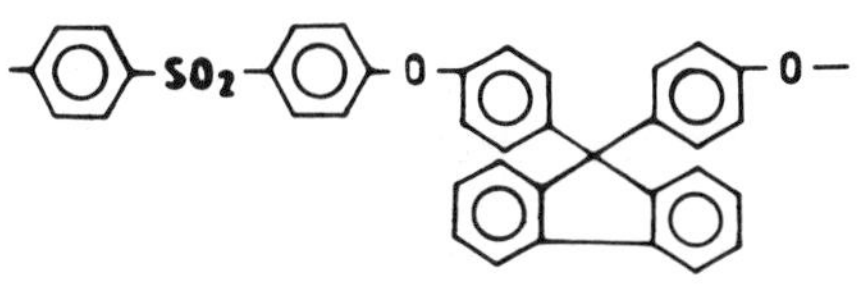	280	504	13
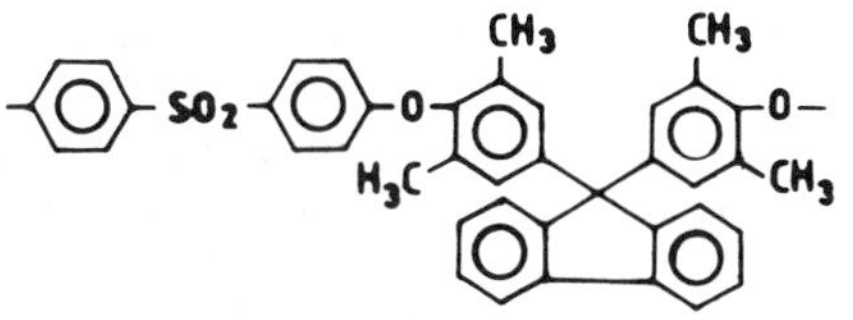	294	—	13
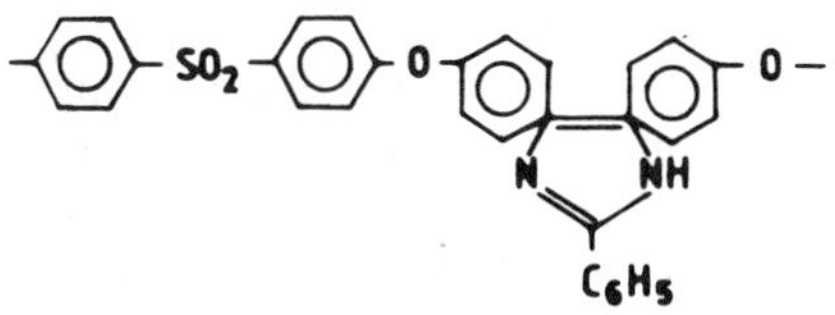	277	~400[a]	28
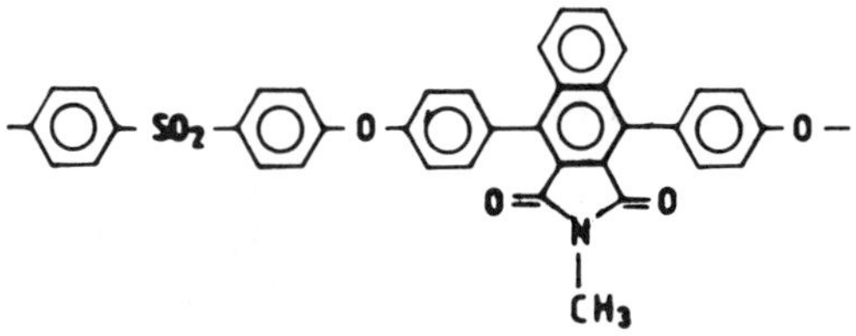	314	503	32
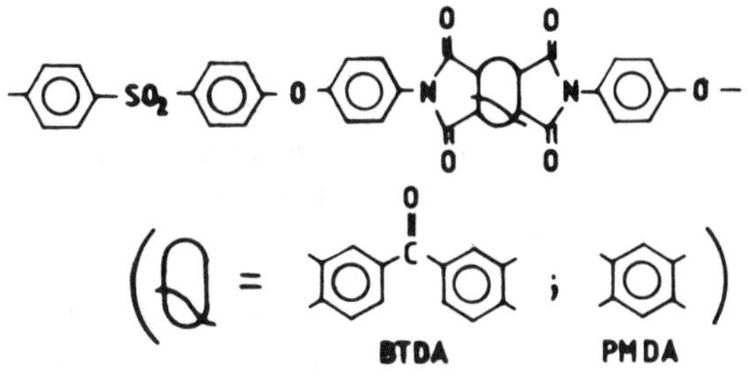	>200	450[b]	201
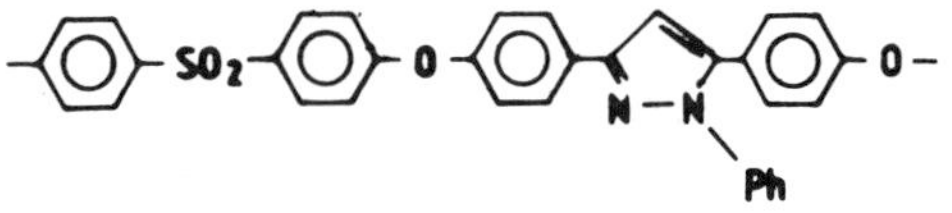	235	474	31

[a] 5% weight loss.

[b] No weight loss in air at 450°C.

ketones are semicrystalline as evidenced by x-ray diffraction study [15]; however, there is no observable crystalline melting point by DSC. The wide angle x-ray diffraction study of polyaryl ethers [13] shows their crystalline nature. Some of these polymers were thought to be liquid crystalline but their chemical structure, melt flow behavior, and appearance under cross-polarized light did not support this assumption. Crystalline morphology of poly sulphide amide is found to be similar to those of PEEK and PPS [55]. The chain symmetry of these polymers leads to close packing of the chains in the polymer crystallites and, hence, shows a fair degree of crystallinity [97]. Since the x-ray patterns of the polymer show the orthorhombic elementary cell similar to that of PEEK, this suggests that the minisegmental poly(sulphide amide)s form a layered supermolecular structure such as PEEK. Such layered structures are known for several thermotropic polyesters with alternating sequence of aliphatic spacers and aromatic mesogens. Thus, poly sulfide amide with an alternating sequence of flexible sulfide spacers and aromatic rigid blocks hydrogen bonded with chains is almost similar to thermotropic polyesters.

Poly(aryl ether O-dibenzoylbenzenes) are amorphous [14]. Poly(aryl ether-1,4-isoquinolinediyls) show a lack of crystallinity, and this can be attributed to asymmetry arising from the presence of nitrogen atoms [62]. The careful annealing experiment of Marand and Prasad [201] have also failed to produce any crystalline system. The unoriented solvent cast films of $^6F\,^8F$ polyether showed a density of 1.57 g/cm^3, and from x-ray diffraction the polymer is found to be approximately 15% crystalline [16]. Recently, the geometric constraints of the aromatic units on the thermal and processing properties of amorphous and semicrystalline polyether materials have been reported by DeSimone and coworkers [63]. The differences between thiophene based polyethers versus 1,3-phenylene and 1,4-phenylene polyethers are studied to obtain the intermediate polymer geometries for larger thermal processing windows without losing the desirable liquid crystalline properties of the latter. In addition, it has been observed that the bilateral asymmetry of the heterocycle influences melting points, rates of crystallization, glass transition temperatures, solubility, miscibility with other polymers, adhesion, etc. For example, new poly(aryl ether isoquinolinediyls) that form

miscible blends with Ultem [62] and yield higher T_gs than comparable PEEK/Ultem blends. The basicity associated with isoquinoline (PK_b = 8.6) is believed to play a role in the miscibility. Poly(aryl ether) bis sulfones exhibit glassy and amorphous morphology [12]. This result is consistent with polyether sulfones because the rigidity of the sulfone group destroys the crystallinity and, therefore, it (sulfone group) becomes unfit to enter into the lattice structure.

D. Environmental Stress Cracking

Many engineering thermoplastics (e.g., polysulfone, polycarbonate, etc.) have limited utility in applications that require exposure to chemical environments. Environmental stress cracking [13] occurs when a stressed polymer is exposed to solvents. Poly(aryl ether phenylquinoxalines) [27] and poly(aryl ether benzoxazoles) [60] show poor resistance to environmental stress cracking in the presence of acetone, chloroform, etc. This is expected because these structures are amorphous, and there is no crystallinity or liquid crystalline type structure to give solvent resistance. Thus, these materials may have limited utility in processes or applications that require multiple solvent coatings or exposures, whereas acetylene terminated polyaryl ethers [13] exhibit excellent processability, high adhesive properties, and good resistance to hydraulic fluid.

E. Radiation Resistance

The interaction of "traditional" (γ-rays, energetic electrons and protons) or "nontraditional" (heavy ions of KeV or MeV energy) ionizing radiation with polymers is, therefore, a topic of interest for both fundamental and technological reasons. The electron irradiation stability of polymers is an important property. It has been shown that PES is particularly stable to electron irradiation compared with other polymer systems [202]. This was proved by its retention of mechanical properties, such as flexural strength and modulus, even after irradiation doses of 200 Mrads. However, electron irradiation produces significant changes in the chemical structure. The experimental observation suggests that the three linkages (I–III), sulfone, ether, and isopropylidine between the phenylene rings are affected by irradiation.

The radiation chemistry has been mainly discussed in terms of degradation reactions (as above) involving the loss of gaseous products and the irreversible change of the stoichiometry [203]. However, more recent results showed that polymers irradiated with radiation depositing a high density of energy (typically because of heavy ions) can be characterized as highly reactive systems in which many simple chemical reactions can occur in competition with each other, producing the modification of the functional groups, modification of the polymer backbone, cross-linking, chain scission, etc. [204]. X-ray photoelectron spectroscopy (XPS) or electron spectroscopy for chemical analysis (ESCA) is generally used to characterize the compositional state of surfaces irradiated in situ. An ESCA study of irradiated PES by various ionizing beams suggests the sulfone groups are dramatically affected by irradiation with low energy projectiles [205,206]. The surface study of PPS powder by heterogeneous reactions suggests that a random terpolymer type of a surface layer is generated consisting of sulfide, sulfoxide, and sulfone groups [207,208].

F. Permeability

Many fluorine containing polymers are being evaluated for possible use as permeable selective membranes. The fluorine containing polymers have been shown to increase permeation rates without decreasing in the selectivity of the membrane. The ^{12}F PEK membrane showed moderate permeation rates with good selectivity ratios for H_2/CH_4 and CO_2/CH_4 gas combinations [15]. The fluorine containing polymers generally show greater flame resistance and lower dielectric constant and water absorption than their nonfluorinated analogues. ^{6}F ^{8}F-polyaryl ether (^{6}F ^{8}F-PAE) as reported by Cassidy et al. [16] exhibits water absorption of 0.3% (by immersion at ambient temperature) with maximum water uptake occurring in 2.5 h whereas ^{6}F ^{8}F-PAE and ^{8}F PAE as obtained by Mercer et al. [17] have water absorption 0.1 and 0.15%, respectively, by immersion of solution-cast films in water for 16 h at 90°C.

The effect of the substituents on chain mobility and chain packing has been related to the gas transport properties [209]. Role of symmetry of methyl group placement on bisphenol rings in PES shows the permeability coefficients in the following order:

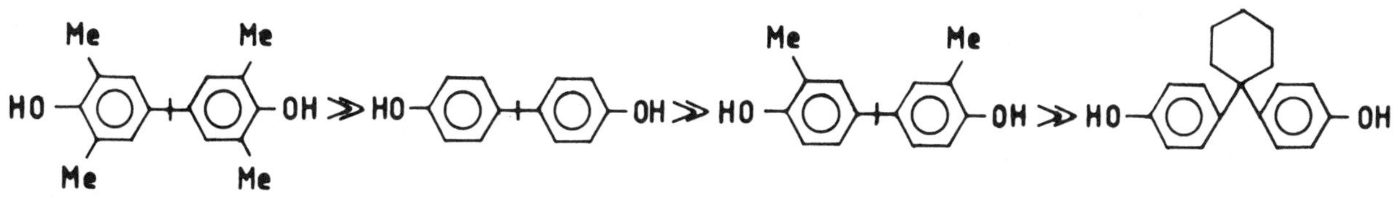

The gas sorption and transport properties also depend on the bisphenol connector groups [210]. The permeability coefficients for all gases rank in the order:

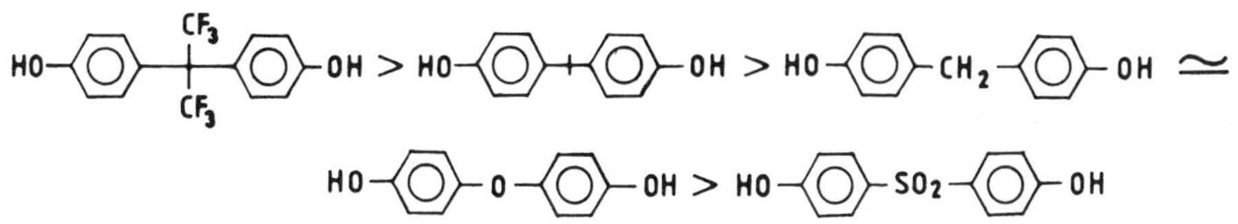

REFERENCES

1. R. S. Irwin, *Polym. Prepr.*, *25*: 213 (1984); S. Maiti, *Special Topics in Polymer Research*. Dibrugarh Univ., Dibrugarh, 1984.
2. R. N. Johnson, A. G. Farnham, R. A. Clendinning, W. F. Hale, and C. N. Merriam, *J. Polym. Sci.*, *A-1*(5): 2375 (1967).
3. A. S. Hay, *Advan. Polym. Sci.*, *4*: 496 (1967).
4. J. T. Edmonds, Jr. and H. W. Hill, Jr., inventors; Phillips Petroleum Co., USA U.S. Patent 3,354,129 (1967).
5. R. Gabler and J. Studinka, German patent 1,909,441 (1969).
6. J. B. Rose, *Chimia*, *28*: 561 (1974).
7. S. Maiti and B. K. Mandal, *Prog. Polym. Sci.*, *12*: 111 (1986).
8. J. E. Harris and R. N. Johnson, *Encycl. Polym. Sci. Engg.*, *2nd ed.*, Vol. 13, (H. F. Mark, N. M. Bikales, C. G. Overberger, G. Menges, eds.), John Wiley and Sons, New York, p. 196 (1988).
9. J. L. Hedrick, *Macromolecules*, *24*: 812 (1991).
10. J. L. Hedrick and J. W. Labadie, *Macromolecules*, *23*: 5371 (1990).
11. T. E. Attwood, D. A. Barr, T. King, A. B. Newton, and J. B. Rose, *Polymer*, *18*: 359 (1977).
12. D. R. Cummings, R. S. Mani, P. B. Balanda, B. A. Howell, and D. K. Mohanty, *J. Macromol. Sci. Chem.*, *A28*: 793 (1991).
13. P. M. Hergenrother, B. J. Jensen, and S. J. Havens, *Polymer*, *29*: 358 (1988).
14. R. Singh and A. S. Hay, *Macromolecules*, *24*: 2637 (1991).
15. G. L. Tullos and P. E. Cassidy, *Macromolecules*, *24*: 6059 (1991).
16. J. A. Irvin, C. J. Neff, K. M. Kane, P. E. Cassidy, *J. Polym. Sci. Part A, Polym. Chem.*, *30*: 1675 (1992).
17. F. Mercer, T. Goodman, J. Wojtowicz and D. Duff, *J. Polym. Sci. Part A, Polym. Chem.*, *30*: 1767 (1992).
18. R. Kellman, R. F. Williams, G. Dimotsis, D. J. Gerbi, and J. C. Williams, *ACS Symp. Ser.*, *326*: 126 (1987).
19. D. J. Gerbi, G. Dimotsis, J. L. Morgan, R. F. Williams, and R. Kellman, *J. Polym. Sci. Polym. Lett.*, *23*: 551 (1985).
20. S. Matsuo, N. Yakoh, S. Chino, M. Mitani, and S. Tagami, *J. Polym. Sci. Part A, Polym. Chem.*, *32*: 1071 (1994).

21. P. K. Dutta and S. Maiti, *Ind. J. Chem. Tech.*, 2: 63 (1994).
22. T. Koch and H. Ritter, *Macromolecules*, 28: 4806 (1995).
23. Y. K. Han, S. D. Chi, Y. H. Kim, B. K. Park and J. J. Jin, *Macromolecules*, 28: 916 (1995).
24. A. D. Rusanov and T. Takekoshi, *Usp. Khim.*, 60: 1449 (1991).
25. B. K. Mandal, Indian Institute of Technology, Kharagpur (1986).
26. J. W. Connell and P. M. Hergenrother, *Polym. Prepr.*, 29: 172 (1988).
27. J. L. Hedrick and J. W. Labadie, *Macromolecules*, 23: 1561 (1990).
28. J. N. Connell and P. M. Hergenrother, *J. Polym. Sci. Part A, Polym. Chem.*, 29: 1667 (1991).
29. P. M. Hergenrother, J. W. Connell, and P. Wolf, Proceedings of Symposium on Recent Advances in Polyimides and Other High Performance Polymers, ACS, San Diego, CA, USA, (1990).
30. J. L. Hedrick and R. Tweig, *Macromolecules*, 25: 2021 (1992).
31. R. G. Bass, K. R. Srinivasan, and J. G. Smith, *Polym. Prepr.*, 32: 160 (1991).
32. M. Strukelj and A. S. Hay, *Macromolecules*, 24: 6870 (1991).
33. J. G. Smith, Jr., J. W. Connell, and P. M. Hergenrother, *Polym. Prepr.*, 32: 646 (1991).
34. W. G. Kim and A. S. Hay, *Polym. Prepr.*, 31: 389 (1991).
35. J. G. Smith, Jr., J. W. Connell, and P. M. Hergenrother, *J. Polym. Sci. Part A, Polym. Chem.*, 31: 3099 (1993).
36. Y. Saegusa, T. Iwasaki, and S. Nakamura, *J. Polym. Sci. Part A, Polym. Chem.*, 32: 249 (1994).
37. J. F. Bunnett and G. T. Davis, *J. Am. Chem. Soc.*, 80: 4337 (1958).
38. M. E. Peach, *The Chemistry of the Thiol Group* (S. Patai, ed.), Wiley and Sons, New York, p. 735 (1974).
39. P. Cogoli, F. Maiolo, L. Testaferri, M. Tingoli, and M. Tiecco, *J. Org. Chem.*, 44: 2642 (1979).
40. F. J. William and P. E. Donahue, *J. Org. Chem.*, 42: 3414 (1977).
41. P. Battistoni, P. Bruni, and G. Fara, *Gazz. Chim. Ital.*, 110: 301 (1980).
42. S. D. Pastor, L. D. Spinvack and D. W. Haughes, *Sulfur Lett.*, 2: 71 (1984).
43. B. Hortling, M. Solder, and J. J. Lindberg, *Angew. Makromol. Chem.*, 107: 163 (1982).
44. I. Haddad, S. Hurley, and C. S. Marvel, *J. Polym. Sci. Polym. Chem.*, 11: 2793 (1973).
45. G. Daccard and P. Sillion, *Polym. Bull.*, 4: 459 (1981).
46. R. W. Lenz, C. E. Handlovits, and H. A. Smith, *J. Polym. Sci.*, 58: 351 (1962).
47. A. B. Port and R. H. Still, *J. Appl. Polym. Sci.*, 24: 1145 (1979).
48. J. R. West, *Adv. Chem. Ser.*, 140: 74 (1975).
49. Y. Imai, M. Ueda, M. Komatsu, and H. Urushibata, *Makromol. Chem. Rapid Commun.*, 1: 681 (1980).
50. Y. Imai and M. Ueda, *J. Polym. Sci. Polym. Chem.*, 24: 2373 (1986).
51. T. L. Evans, F. J. Williams, P. E. Donahue, and M. M. Grade, *Polym. Prepr.*, 25: 268 (1984).
52. R. W. Campbell, (inventors, Phillips Petroleum Co., USA) U.S. patent 4,127,713 (1978).
53. M. Yoneyama, G. Cei, and J. L. Mathias, *Polym. Prepr.*, 32: 195 (1991).
54. P. K. Dutta and S. Maiti, *Angew. Makromol. Chem.*, 211: 79 (1993).
55. S. Maiti and P. K. Dutta, *Ind. J. Chem. Tech.*, 1: 81 (1994).
56. P. K. Dutta, *Ph.D. Thesis*, Indian Institute of Technology, Kharagpur (1993).
57. P. K. Dutta and S. Maiti, *Makrmol. Chem. Rapid Commun.*, 13: 505 (1992).
58. J. L. Hedrick, *Macromolecules*, 24: 6361 (1991).
59. B. K. Mandal and S. Maiti, *Eur. Polym. J.*, 22: 447 (1986).
60. J. G. Hilborn, J. W. Labadie, and J. L. Hedrick, *Macromolecules*, 23: 2854 (1990).
61. L. Cormier, G. Lucotte, and B. Delfort, *Polym. Bull.*, 26: 395 (1991).
62. H. W. Gibson and B. Guilani, *Polym. Commun.*, 32: 324 (1991).
63. J. M. DeSimone, S. Stompel, E. L. Samulski, Y. Q. Wang, and A. B. Brennan, *Macromolecules*, 25: 2546 (1992).
64. S. Matsuo, *J. Polym. Sci. Part A, Polym. Chem.*, 32: 2093 (1994).
65. J. L. Hedrick, H. Jonsson, and K. R. Carter, *Macromolecules*, 28: 4342 (1995).
66. J. L. Hedrick, *Polym. Bull.*, 25: 543 (1991).
67. K. R. Carter, H. Jonsson, R. Twieg, R. D. Miller, and J. L. Hedrick, *Polym. Prepr.*, 33: 388 (1992).
68. W. P. Weber, G. W. Gokel, *Phase Transfer Catalyst in Organic Synthesis*, Springer-Verlag, W. Berlin (1977).
69. C. M. Starks, C. L. Liotta, *Phase Transfer Catalysis*, Academic Press, New York (1978).
70. E. V. Dehmlov, S. S. Dehmlov, *Phase Transfer Catalysis*, Verlag Chimie, W. Berlin (1980).
71. G. H. Alt, J. P. Cheep, U.S. patent 4,371,717 (1978).
72. F. J. Williams, U.S. patent 4,273,712 (1981).
73. D. S. Johnson, U.S. patent 4,163,833, (1979).
74. W. Rieder, U.S. patent 4,430,493 (1984).
75. D. J. Gerbi, R. F. Williams, R. Kellman, and J. L. Morgan, *Polym. Prepr.*, 22: 385 (1981).
76. Y. Imai, M. Ueda, and M. Li, *J. Polym. Sci. Polym. Lett.*, 17: 85 (1979).
77a. V. Percec and B. C. Auman, *Makromol. Chem.*, 185: 617 (1984).
77b. V. Percec and H. Nava, *Makromol. Chem. Rapid Commun.*, 5: 319 (1984).
78. R. Kellman, D. J. Gerbi, R. F. Williams, and J. L. Morgan, *Polym. Prepr.*, 21: 164 (1980).
79. W. P. Reeves, T. C. Bothwell, J. A. Rudis, and J. W. McClusky, *Synth. Commun.*, 12: 1071 (1982).
80. P. P. Shah, *Eur. Polym. J.*, 20: 519 (1984).
81. *Phase Transfer Catalysis*, ACS Symposium Series, 326, C. M. Starks (ed.), Washington, D.C. (1987).
82. Y. Nakamura, K. Mori, K. Tamura, and Y. Saito, *J. Polym. Sci. Part A-1*, 7: 3089 (1969).
83. V. Vangani, S. S. Kansara, and N. K. Patel, *Macromol. Reports, A28*: 193 (1991).
84. R. Kellman, D. J. Gerbi, R. F. Williams, and J. L. Morgan, *Polym. Prepr.*, 21: 2 (1980).
85. N. Yamazaki and Y. Imai, *Polym. J.*, 15: 603 (1983).
86. T. Iizawa, H. Kudou, and T. Nishikubo, *J. Polym. Sci. Part A, Polym. Chem.*, 29: 1875 (1991).
87. T. D. Shaffer and V. Percec, *Makromol. Chem.*, 187: 1431 (1986).
88. O. W. Webster, *Encycl. Polym. Sci. Engg.* 2nd ed., (H.

F. Mark, N. M. Bikales, C. G. Menges, eds.), Vol. 7, John Wiley and Sons, New York, p. 580 (1987).

89. A. D. Jenkins, J. M. Morrison, J. Mykytiuk, L. Trowbridge, I. Tsartolia, and D. R. M. Walton, *Makromol. Chem. Rapid Commun.*, 12: 653 (1991).
90. J. F. Klebe, *J. Polym. Sci.*, A2: 2673 (1964).
91. J. F. Klebe, *J. Polym. Sci.*, B2: 1079 (1964).
92. J. F. Klebe, *Adv. Org. Chem.*, 8: 97 (1972).
93. H. R. Kricheldorf, *Polymer Science (Contemporary Theme)*, Vol. 1, (S. Sivarama, ed.), Tata McGraw Hill, New Delhi, p. 49 (1991).
94. H. R. Kricheldorf and G. Schwarz, *Polym. Bull.*, 1: 383 (1979).
95. H. R. Kricheldorf and G. Bier, *J. Polym. Sci. Polym. Chem.*, 21: 2283 (1983).
96. H. R. Kricheldorf and G. Bier, *Polymer*, 25: 1151 (1984).
97. H. R. Kricheldorf and U. Delius, *Macromolecules*, 22: 517 (1989).
98. H. R. Kricheldorf, G. Schwarz, and J. Erxleben, *Makromol. Chem.*, 189: 2255 (1988).
99. H. R. Kricheldorf, J. Meier, and G. Schwarz, *Makromol. Chem. Rapid Commun.*, 8: 529 (1987).
100. H. R. Kricheldorf, U. Delius, and K. U. Tonnes *New Polym. Mater.*, 1: 127 (1988).
101. H. R. Kricheldorf and U. Delius, *Makromol. Chem.*, 190: 1277 (1989).
102. H. R. Kricheldorf, B. Schmidt, and U. Delius, *Eur. Polym. J.*, 26: 791 (1990).
103. H. R. Kricheldorf and P. Jahnke, *Makromol. Chem.*, 191: 2027 (1990).
104. K. C. O'Brien, W. J. Koros, T. A. Barbari, and E. S. Sanders, *J. Membr. Sci.*, 29: 229 (1986).
105. M. Kakimoto, Y. Oishi, and Y. Imai, *Makromol. Chem. Rapid Commun.*, 6: 229 (1985).
106. Y. Oishi, M. Kakimoto, and Y. Imai, *Polym. Prepr. Jpn.*, 36: 315 (1987).
107. A. Hara, Y. Oishi, M. Kakimoto, and Y. Imai, *Macromolecules*, 29: 1933 (1991).
108. H. R. Kricheldorf and P. Jahnke, *Polym. Bull.*, 28: 411 (1992).
109. S. Maiti, S. Banerjee, and S. K. Palit, *Prog. Polym. Sci.*, 18: 227 (1993).
110. H. E. Stepniczka, *J. Fire Retard. Chem.*, 2: 30 (1975).
111. A. D. F. Toy, U. S. patent 2,435,252 (1984).
112. H. W. Coover and M. A. McCall, U.S. patent 3,719,727 (1973).
113. F. Millich and C. E. Carraher Jr., *J. Polym. Sci. Part A-1*, 7: 2669 (1969).
114. F. Millich and C. E. Corraher Jr., *J. Polym. Sci. Part A-1*, 8: 163 (1970).
115. F. Millich and C. E. Carraher Jr., *Macromolecules*, 3: 253 (1970).
116. Y. Massai, Y. Kato, and N. Fukui, U. S. patent 3,719,727 (1973).
117. A. J. Papa and W. R. Proops, *J. Appl. Polym. Sci.*, 16: 2361 (1972).
118. S. R. Sander and W. Karo, *Polymer Syntheses*, Organic Chemistry Monographs Vol. 29(1), Academic Press, New York, p. 367 (1974).
119. F. Millich and L. L. Lambing, *J. Polym. Sci. Polym. Chem.*, 18: 2155 (1980).
120. K. S. Kim, *J. Appl. Polym. Sci.* 28: 1119 (1983).
121. Y. Imai, H. Kamata, and M. K. Kakimoto, *J. Polym. Sci. Polym. Chem.*, 22: 1259 (1984).
122. Y. Massai, Y. Kato, and N. Fukui, U.S. patent 3,919,185 (1975).
123. M. Sander, and E. J. Steninger, *J. Macromol. Sci. Rev. Macromol. Chem.*, 2: 1 (1968).
124. M. Sander, *Encycl. Polym. Sci. Tech.*, (H. F. Mark, N. G. Gaylord, and N. M. Bikales, eds.), Vol. 10, John Wiley and Sons, New York, p. 123 (1968).
125. W. E. Cars, U.S. patent, 2,616,873 (1952).
126. H. Zenftman, *Chemistry Preprints*, ACS Div. Paint, Plastics, Printing Ink, 18: 361 (1958).
127. L. A. Datskevich, V. D. Maiboroda, and I. P. Losev, *Geterotsepuye Vysokomolek, Soedin*, 6: 243 (1964).
128. H. W. Coover, U.S. patent, 2,95,266 (1960).
129. K. Kishore and P. Kannan, *J. Polym. Sci. Polym. Chem.*, 28: 3481 (1990).
130. R. Yamaguchi, M. Takada, K. Kudo, and Y. Echigo, Japan patent 7,577,618 (1979); *Chem. Abstr.* 846,357 (1976).
131. K. S. Annakutty and K. Kishore, *Polymer*, 29: 756 (1988).
132. (a) K. Kishore, K. S. Annakutty, and J. M. Mallick, *Polymer*, 29: 756 (1988). (b) K. Kishore, K. S. Annakutty, and J. M. Mallick, *Polymer*, 29: 762 (1983).
133. H. R. Kricheldorf and H. Koziel, *J. Macromol. Sci. Chem.*, A3: 1337 (1986).
134. S. Banerjee, S. K. Palit, and S. Maiti, *J. Polym. Sci. Polym. Chem.*, 32: 219 (1994).
135. S. Banerjee, S. K. Palit, and S. Maiti, *J. Polym. Mater.*, 9: 219 (1992).
136. S. Banerjee, S. K. Palit, and S. Maiti, *Coll. Polym. Sci.*, 272: 1203 (1994).
137. I. K. Varma, G. M. Fohlen, and J. A. Parkar, U.S. patent 4,276,344 (1981).
138. I. K. Varma, G. M. Fohlen, and J. A. Parkar, IUPAC 28th *Int. Symp. Macromol.* Abstract of Communication 314, Amherst, MA (1982).
139. I. K. Varma, G. M. Fohlen, and J. A. Parkar, IUPAC 27th *Int. Symp. Macromol.* Abstract of Communication, A-1, 94 (1981).
140. I. K. Varma, G. M. Fohlen, and J. A. Parkar, *J. Macromol. Sci. Chem.*, A19: 39 (1983).
141. I. K. Varma, G. M. Fohlem, Ming-Ta Hsu, and J. A. Parkar, *Contemp. Topics Polym. Sci.*, 4: (1981).
142. I. K. Varma and B. S. Rao, *J. Appl. Polym. Sci.*, 28: 2805 (1983).
143. A. P. Mellaris and J. A. Mikroyannids, *Eur. Polym. J.*, 25: 275 (1989).
144. H. Kondo, M. Sato, and M. Yokoyama, *Eur. Polym. J.*, 16: 537 (1980).
145. M. Sato and M. Yokoyama, *Eur. Polym. J.*, 15: 541 (1979).
146. M. Sato, T. Iijime, T. Uchida, and M. Yokoyama, *Kobunshi Ronbunshu*, 35: 501 (1978).
147. M. Sato, M. Sato, and M. Yokayoma, *Kobunshi Ronbunshu*, 35: 713 (1978).
148. M. Sato and M. Yokoyama, *Eur. Polym. J.*, 15: 75 (1979).
149. M. Sato and M. Yokoyama, *Eur. Polym. J.*, 13: 79 (1980).
150. M. Sato, Y. Tada, and M. Yokoyama, *Eur. Polym. J.*, 16: 671 (1980).
151. M. Sato, H. Kondo, and M. Yokoyama, *J. Appl. Polym. Sci.*, 29: 299 (1984).

152. M. Sato and M. Yokoyama, *J. Polym. Sci.*, *Polym. Chem.*, *18*: 2751 (1980).
153. M. Sato, Y. Tada, and M. Yokoyama, *J. Polym. Sci. Polym. Chem.*, *19*: 1037 (1981).
154. M. Sato, H. Kondo, and M. Yokoyama, *J. Polym. Sci. Polym. Chem.*, *20*: 335 (1982).
155. M. Sato and M. Yokoyama, *J. Polym. Sci. Polym. Chem.*, *19*: 591 (1981).
156. H. Kondo, M. Sato, and M. Yokoyama, *J. Polym. Sci. Polym. Chem.*, *22*: 1055 (1984).
157. D. E. C. Corbridge, *Phosphorus: An Outline of its Chemistry, Biochemistry and Technology*, *Studies in Inorganic Chemistry*, Vol. 6, Elsevier, Amsterdam (1985).
158. S. Banerjee, Md. S. Rahaman, S. K. Palit, and S. Maiti, *Angew. Makromol. Chem.*, *199*: 1 (1992).
159. E. Steinnger and M. Sander, *Kunststoffe*, *54*: 507 (1964).
160. V. V. Korshak, G. S. Kolesnikov, and B. A. Zhubanov, *Izvest. Akad. Nank., U.S.S.R., Otdel. Khim. Nank.*, *613*: (1958).
161. C. S. Smith, D. K. Mohanty, and J. E. McGrath, 35th *Int. SAMPE Symp.*, p. 108 (1990).
162. S. Banerjee, S. K. Palit, and S. Maiti, *Indian J. Chem. 33B*: 43 (1994).
163. H. Kondo, M. Sato, and M. Yokoyama, *Eur. Polym. J.*, *17*: 583 (1981).
164. G. F. D'Alelio, *Encycl. Polym. Sci. Tech.*, (H. F. Mark, N. G. Gaylord and N. M. Bikales, eds.), Vol. 10, John Wiley and Sons, New York, p. 659 (1969).
165. A. D. Delman, A. A. Stein, and B. B. Simms, *J. Macromol. Sci. Chem.*, *A1*: 147 (1967).
166. R. J. Cotter and M. Matzner, *Org. Chem. (N.Y.)*, *13B*(1): 1 (1972).
167. H. A. Goodwin and J. C. Bailer, Jr, *J. Am. Chem. Soc.*, *83*: 2467 (1961).
168. S. Banerjee, S. K. Palit, and S. Maiti, *J. Polym. Mater.*, *9*: 219 (1992).
169. P. W. Morgan, T. C. Pletcher, and S. L. Kwolek, *Polym. Prepr.*, *24*(2): 470 (1983).
170. K. Suematsu, K. Nakamura, and J. Takada, *Coll. Polym. Sci.*, *261*: 470 (1983).
171. K. Suematsu, K. Nakamura, and J. Takada, *Polym. J.*, *15*: 71 (1983).
172. C. J. Yang and S. A. Jenekhe, *Chem. Mater.*, *3*: 878 (1991).
173. Y. Saegusa, K. Sekiba, and S. Nakamura, *Kenkya Hokoku-Asahi Garasu Zaidan*, *56*: 139 (1990).
174. C. S. Marvel and W. H. Hill, *J. Am. Chem. Soc.*, *72*: 4891 (1950).
175. C. S. Marval and N. Tarkoy, *J. Am. Chem. Soc.*, *80*: 832 (1958).
176. C. S. Marval and P. V. Bonsignore, *J. Am. Chem. Soc.*, *81*: 2668 (1959).
177. W. E. Rudzinski, S. R. Guthrie, and P. E. Cassidy, *J. Polym. Sci. Part A*, *Polym. Chem.*, *26*: 1677 (1988).
178. Y. Saegusa, K. Sekiba, and S. Nakamura, *J. Polym. Sci. Part A*, *Polym. Chem.*, *28*: 3647 (1990).
179. M. S. Patel and S. R. Patel, *J. Polym. Sci. Part A*, *Polym. Chem.*, *20*: 1985 (1982).
180. K. B. Al-Jumah, K. B. Wagener, and T. E. HogenEsch, *Polym. Prepr.*, *30*(2): 173 (1989).
181. Y. Saegusa, T. Takashima, and S. Nakamura, *J. Polym. Sci. Part A*, *Polym. Chem.*, *30*: 1375 (1992).
182. R. Adams, R. E. Bullock, and W. C. Wilson, *J. Am. Chem. Soc.*, *45*: 521 (1923).
183. P. W. Morgan, S. L. Kwolek, and T. C. Pletcher, *Macromolecules*, *20*: 729 (1987).
184. S. Banerjee, P. K. Gutch, and C. Saxena, *J. Polym. Sci. Polym. Chem.*, *33*: 1719 (1995).
185. P. W. Wojtkonski, *Macromolecules*, *20*: 740 (1987).
186. C. J. Yang and S. A. Jenekhe, *Chem. Mater.*, *3*: 878 (1991).
187. C. J. Yang and S. A. Jenekhe, *Macromolecules*, *28*: 1180 (1995).
188. B. A. Reinhardt and M. R. Unroe, *Polym. Prepr.*, *31*: 620 (1990).
189. K. S. Lee, J. C. Won, and J. C. Jung, *Makromol Chem.*, *190*: 1547 (1989).
190. S. S. Mohite and P. P. Wadgaonkar, *Polym. Prepr.*, *31*: 482 (1990).
191. R. G. Bryant, *Polym. Prepr.*, *33*: 182 (1992).
192. P. M. Hergenrother, N. T. Wakelyn, and S. J. Havens, *J. Polym. Sci. Polym. Chem.*, *25*: 1093 (1987).
193. S. Banerjee, C. Saxena, P. K. Gutch and D. C. Gupta, *Eur. Polym. J.*, *32*: 661 (1996).
194. S. Banerjee and C. Saxena, *J. Polym. Sci. Polym. Chem.*, 1996 (in press).
195. H. F. Hale, A. G. Farnham, R. N. Johnson, and R. A. Clendinning, *J. Polym. Sci.*, *A-1*(5): 2399 (1967).
196. C. D. Smith, H. J. Grubbs, H. F. Webster, J. P. Wightman, and J. E. McGrath, *Polym. Mater. Sci. Engg.*, *65*: 109 (1991).
197. *Plastics for Electronics*, Cordura Publications, California, (1979).
198. *Polysulfone Design Engineering Data Handbook*, Union Carbide Corporation, Lit., F-47178, 4th Rev., (March, 1979).
199. R. O. Johnson and H. S. Burlhis, *J. Polym. Sci. Polym. Symp.*, *70*: 129 (1983).
200. D. R. Kelsey, L. M. Robeson, R. A. Clendinning, and C. S. Blackwell, *Macromolecules*, *20*: 1204 (1987).
201. H. Marand and A. Prasad, Unpublished results.
202. A. Davis, M. H. Gleaves, J. H. Golden, and M. B. Huglin, *Makromol. Chem.*, *129*: 63 (1969).
203. T. Venkatesan, L. Calcagno, B. S. Elman, and G. Foti, *Ion Implantation in Insulators*, (G. W. Arnold, P. Mazzoldi, eds.), Elsevier, Amsterdam p. 301 (1987).
204. G. Marletta, *Nucl. Instrum. Methods B.*, *46*: 295 (1990).
205. G. Marletta, S. Pignataro, A. Toth, I. Bertoti, T. Szekely, and B. Keszler, *Macromolecules*, *24*: 99 (1991).
206. P. K. Dutta, *Macromol. Reports*, *A31*: 571 (1994).
207. A. Kaul and K. Udipi, *Macromolecules*, *22*: 1201 (1989).
208. S. Maiti and P. K. Dutta, *J. Polym. Mater.*, *10*: 31 (1993).
209. J. S. McHattie, W. J. Koros, and D. R. Paul, *Polymer*, *32*: 840 (1991).
210. J. S. McHattie, W. J. Koros, and D. R. Paul, *Polymer*, *32*: 2618 (1991).

3

Acrylamide Polymers

Valery F. Kurenkov

Kazan State Technological University, Kazan, Russia

I. INTRODUCTION

Acrylamide polymers are widely used and are technically important water-soluble polymers called "polyacrylamides." This family name includes polyacrylamide-nonionic homopolymer, its anionic and cationic derivatives, and also copolymers of acrylamide with ionogenic and nonionogenic monomers. Acrylamide polymers were first developed in 1893, however, they have only been commercially available since the early 1950s, because the raw materials were not readily available. The initial application of acrylamide polymers were as flocculants for the treatment of acid-leached uranium ores and as dry strength agents for paper. Since then numerous other applications have been developed. The polymers are now used for separation and clarification of liquid–solid phases and for thickening, binding, lubrication, and film formation. For the past 30 years the manufacturing of acrylamide polymers has been rapidly increasing due to: (1) the improvement of raw material production, (2) important advances in the controlled polymerization and copolymerization of acrylamide in concentrated aqueous solutions, and dispersions and (3) improvements in the methods used in chemical modification of polymers. The major producers and exporters of acrylamide polymers are large firms in the United States, Japan, and Europe. Exports from Russia, China, and South Africa are of lesser significance. The worldwide manufacturing of acrylamide polymers is being developed. However, the growth rate of the manufacture of polymers does not meet the expected growth rate of 8–10% per year of the polymer market. By the end of the century, the manufacturing of acrylamide polymers is expected to exceed 400 kilotons per year.

II. ACRYLAMIDE

A. Properties

Acrylamide is the most important and the simplest of the acrylic and methacrylic amides. Acrylamide is a colorless crystalline solid. The basic physical properties and solubilities of acrylamide are given in Table 1. Acrylamide is a severe neurotoxin and is a cumulative toxicological hazard.

Acrylamide has two functional groups, a reactive double bond and an amide group. It undergoes many reactions typical of the two functionalities [1,2]. The

Table 1 Physical Properties and Solubilities of Acrylamide

Formula	$CH_2{=}CH$
	$\overset{\mid}{H_2N{-}C{=}O}$
Molecular weight	71.08
Melting point	$84.5 \pm 0.3°C$
Boiling point	87°C (0.27 kPa)
	103°C (0.67 kPa)
	116°C (1.06 kPa)
	125°C (2.10 kPa)
Density	1.122 g/cm^3 (30°C)
Heat of polymerization	82.8 kJ/mol
Solubility (g/100 g at 30°C)	Water, 215.5
	Methanol, 155.0
	Ethanol, 86.2
	Acetone, 63.1
	Benzene, 0.346

amide group undergoes the reactions characteristic of an aliphatic amides: hydrolysis reaction under acidic or alkaline condition, addition of various amines and ammonia, alcohols, and ketones. Under acidic conditions acrylamide with formaldehyde forms N,N'-methylene-bisacrylamide. Under alkaline conditions acrylamide reacts with formaldehyde to form N-methylolacrylamide, which further reacts with alcohol to give N-alkoxymethylacrylamide. Diels-Alder reactions take place at the double bond of acrylamide. The double bond of acrylamide undergoes polymerization and copolymerization reactions.

B. Manufacture

All current industrial production of acrylamide is believed to be by saponification of acrylonitrile [1,2]. Until recently, the saponification of acrylonitrile was done exclusively with sulfuric acid monohydrate.

$$CH_2{=}CHCN \xrightarrow[80\text{--}100°C]{H_2SO_4,\ H_2O}$$
$$CH_2{=}CHCONH_2 \times H_2SO_4 \qquad (1)$$

The reaction is extremely exothermic. Inhibitors such as copper and copper salt are used to suppress side reactions. Acrylamide and sulfuric acid may be separated by adding lime or ammonia to a solution of acrylamide sulfate.

$$CH_2{=}CHCONH_2{\cdot}H_2SO_4 \begin{cases} \xrightarrow{Ca(OH)_2} CH_2{=}CHCONH_2 + CaSO_4 + 2H_2O \\ \\ \xrightarrow{2(NH_4)_2SO_4} CH_2{=}CHCONH_2 + (NH_4)_2SO_4 + 2H_2O \end{cases} \qquad (2)$$

The expensive and difficult step in the process is the recovery of acrylamide from the reaction mixture.

Currently, acrylamide is produced by the hydration of acrylonitrile in the presence of copper-based catalysts.

$$CH_2{=}CHCN \xrightarrow[80\text{--}100°C]{H_2O,\ Cu} CH_2{=}CHCONH_2 \qquad (3)$$

This process eliminates the by-product sulfite and can be run continuously.

Most important in the process is the preparation of acrylamide by the hydration of acrylonitrile in the presence of immunized microorganisms.

$$CH_2{=}CHCN \xrightarrow[<15°C,\ pH\ 7\text{--}8]{H_2O,\ Biocatalyst} CH_2{=}CHCONH_2 \qquad (4)$$

C. Application

The main application of acrylamide is the preparation of water-soluble polymers and copolymers. Smaller quantities of acrylamide are used for the preparation of N-substituted derivatives of acrylamide (N,N'-methylenebisacrylamide, N-methylolacrylamide, etc.). Acrylamide is partially used as an additive to improve the mechanical properties of cement and gypsum plasters as well as to achieve better control over the hardening process of these plasters.

III. POLYACRYLAMIDE

A. Properties

1. Physical Properties

Properties of dry polymer

Polyacrylamide is an odorless, hard glassy white polymer, that exhibits a very low toxicity. Dry polymer is available in a powder or granule form with different particle sizes, depending on the type of polymerization, drying, and grinding processes used. Macromolecules of polyacrylamide are linear with a predominantly head-to-tail structure. Basic physical properties and solubilities of polyacrylamide are given in Table 2. Polyacrylamide is stable to action of oils, fats, and waxs. Dry polymer has a good thermal stability, but with increasing temperatures polyacrylamide decomposes. At temperatures >100°C polyacrylamide may be cross-linked by imide formation. Polyacrylamide starts to decompose at 220°C. In the temperature range 220–335°C ammonia gas is released due to an imidic reaction.

At 335°C a second decomposition range begins, with the decomposition rate reaching its maximum at about 370°C. Attributed to the second range are the breakdown of imide groups to nitrile, completion of the decomposition of amide groups, and the breakdown of the polymer backbone. At about 500°C the weight of the sample becomes constant, and a charlike material remains [3].

Properties of Aqueous Solutions of Polymers

The analysis of the main properties of aqueous solutions of polyacrylamide and copolymers of acrylamide has been reviewed [4,5]. The main characteristics of aqueous solutions of polyacrylamide is viscosity. The viscosity of aqueous solutions increases with concentration and molecular weight of polyacrylamide and decreases with increasing temperature. The relationship between the intrinsic viscosity ($[\eta]$) in cm^3/g and the molecular weight for polyacrylamide follows the Mark–Houwink equations:

Table 2 Physical Properties and Solubilities of Polyacrylamide

Formula	$-[-CH_2-CH-]_n-$ $\quad\quad\quad\quad\quad\mid$ $\quad\quad\quad\quad H_2N-C=O$
Molecular weight	10^4-10^7
Density	1.302 g/cm^3 (30°C)
Glass transition temperature	153°C
Soluble	In water, morpholine, formamide, and ethylene glycol
Partially soluble	In acetic and lactic acids, glycerin, dimethylsulfoxide, and dimethylformamide
Insoluble	In acetone, dioxane, alcohols, hexane, and heptane.

$$[\eta] = 6.8\cdot10^{-2}\overline{M}_n^{0.66} \text{ (in water at 25°C) [6]} \quad (5)$$

$$[\eta] = 3.73\cdot10^{-2}\overline{M}_w^{0.66} \quad (6)$$
$$\times \text{ (in 1 M NaNO}_2 \text{ at 30°C) [7]}$$

$$[\eta] = 6.31\cdot10^{-3}\overline{M}_z^{0.80} \text{ (in water at 25°C) [8]} \quad (7)$$

where $\overline{M}_n$, $\overline{M}_w$, and $\overline{M}_z$ equal the number average, weight average and z average (from sedimentation) molecular weight, respectively.

For copolymers of acrylamide with sodium acrylate, the preexponential factor K and exponent α of $[\eta] = KM^\alpha$ depend on copolymer composition (Table 3).

The coil demention $(\bar{r}^2)^{1/2}$ for polyacrylamide may be obtained from the relation, which is applicable for non-Gaussian coils:

$$[\eta]\cdot M = \phi(\bar{r}^2)^{3/2} \quad (8)$$

where ϕ is the Flory parameter. For polyacrylamide the Huggins equation is used:

$$\frac{\eta_{SP}}{c} = [\eta] + K_H[\eta]^2 c \quad (9)$$

where K_H is the Huggins coefficient. In water for copolymers of acrylamide with ionogenic monomers in comparison with polyacrylamide much higher significances of η_{SP}/c and $(\bar{r}^2)^{1/2}$ are characteristic owing to a polyelectrolyte effect. In this case the Fuoss relation is used:

$$\frac{\eta_{SP}}{c} = \frac{A}{1 + B \times \sqrt{C}} + D \quad (10)$$

where A, B, and D are constants. The parameter D characterizes the interaction of the macromolecular coils with the ingredients of the system; parameters A and B characterize the viscous properties of the polyelectrolyte for the two extreme situations $\mu \to 0$ and $\mu \to \infty$, where μ is the ionic strength. For copolymers of acrylamide with sodium acrylate, the plot $[\eta]/M$ versus $M^{1/2}$ is linear. When an aqueous solution of a polyelectrolyte is diluted, the $\frac{\eta_{SP}}{c}$ of solution increases rapidly over a wide range of concentration. When dilution of the polyelectrolyte occurs in the presence of added salt whose concentration is kept constant, the dependence of $\frac{\eta_{SP}}{c}$ from c agrees with the Huggins equation.

At low shear rates, aqueous solutions of polyacrylamide are pseudoplastic. With increasing shear rates and temperature the viscosity of the solutions decrease. At high shear rates during violent mixing and pumping operations the molecular weight of polyacrylamide decreases by destruction of macromolecules.

When the solution of polyacrylamide is kept at room temperature, intrinsic viscosity decreases because of the conformational change of a single macromolecule with the participation of hydrogen bonds and degradation of the polymer [4]. Degradation can be accelerated by free radicals, ionizing radiation, light, heat, shear, and high speed stirring of aqueous solution of polyacrylamide. The degradation of polyacrylamide in 0.1–0.4% aqueous solution in the presence of $K_2S_2O_8$ at 40–70°C can be used for regulation of molecular weight products [10]. Ozone degradation of polyacrylamide in water at pH 10 was much larger than that at pH 2. A linear relationship was observed under basic conditions between the number of breaks and ozone consumed.

The degradation of polymers may be prevented by the addition of sodium nitrile, thiourea, etc.

2. Chemical Properties

The amide group of polyacrylamide offers a reactive site to change the ionic character or to cross-link the polymer. A polyacrylamide solution undergoes general reactions of the aliphatic amide group [1,2,11]. The impor-

Table 3 Dependence of Constants K and α of $[\eta] = KM^\alpha$ on Mole Fraction of Sodium Acrylate (X_{SA}) for Copolymer Acrylamide with Sodium Acrylate in 0.5 M NaCl at 25°C [9]

X_{SA}, mol%	1.7	9.0	16.8	23.7	27.9	35.1	38.2	41.6	43.7	45.8
$K \times 10^2$	0.69	0.65	0.62	0.62	0.63	0.65	0.67	0.70	0.71	0.72
α	0.78	0.80	0.81	0.82	0.82	0.83	0.83	0.83	0.83	0.82

tant reaction of polyacrylamide is hydrolysis. Hydrolysis can be carried out in an acidic or a basic medium.

$$—CH_2—CH— \xrightarrow{\text{Acid or base}} —CH_2—CH— + NH_3$$
$$\quad\ \big|\qquad\qquad\qquad\qquad\ \big|$$
$$\quad\ CONH_2 \qquad\qquad\qquad\ COOH$$
$$\tag{11}$$

Acidic hydrolysis of the amide group at pH 4.5 is a very slow reaction. Strong acidic conditions leads to a progressive insolubilization of the reaction product because of formation of cyclic imide structures:

$$
\begin{array}{ccc}
O{=}C \quad\quad C{=}O & \quad & O{=}C \quad\quad\quad C{=}O \\
\backslash \quad\ / & & \quad\quad\quad \\
N & & NH \quad HN \\
| & & \\
H & & O{=}C \quad\quad\quad C{=}O
\end{array}
\tag{12}
$$

Cross-linking of the polymer by secondary amine groups is not irreversible since these groups are easily destroyed when the polymer is treated with slightly alkaline solutions:

$$—CONHCO— + NaOH \rightarrow$$
$$—COONa + —CONH_2 \tag{13}$$

Hydrolysis of polyacrylamide under basic conditions is guide rapid and can be used to introduce acrylate groups into macromolecules.

$$—(—CH_2—CH—)_n— + m\ NaOH$$
$$\qquad\qquad\ \big|$$
$$\qquad\qquad\ CONH_2$$
$$\rightarrow—(—CH_2—CH—)_{n-m}$$
$$\qquad\qquad\quad \big|$$
$$\qquad\qquad\quad CONH_2$$
$$—(—CH_2—CH—)_m + m\ NH_3$$
$$\qquad\qquad\ \big|$$
$$\qquad\qquad\ COO^-Na^+ \tag{14}$$

The saponification products have a statistical distribution of acrylate side groups and practically no block structure will be formed. The degree of hydrolysis depends on temperature, reaction time, and the addition of salts (NaCl, KCl). The reactions are influenced by the effects of neighboring groups and the conformation of polyacrylamide. The maximum degree of hydrolysis becomes limited, 70% for polyacrylamide. This may be caused by the decreasing reactivity of the amide groups, depending on the structure of the neighboring groups.

$$\sim CH_2—CH—CH_2—CH—CH_2—CH\sim$$
$$\qquad \big|\qquad\quad \big|\qquad\quad \big|$$
$$\quad\ O{=}C\qquad O{=}C\qquad O{=}C$$
$$\qquad \big|\qquad\quad \big|\qquad\quad \big|$$
$$\quad\ O^-\ ...\ \ H—N—H\ ...\quad O^- \tag{15}$$

At hydrolysis of polyacrylamide in 10 M NaCl at 100°C, the degree of hydrolysis becomes ~95%, but in this case a degradation of the macromolecules is observed. Various modes of hydrolysis of polyacrylamide under basic conditions have been reviewed [2,12].

A Hofmann degradation of polyacrylamide by use of a very small excess of sodium hypochlorite and a large excess of sodium hydroxide at 0°C to 15°C for about 15 h polyvinylamine (95 mol% amine units) is obtainable:

$$\sim CH_2—CH\sim + NaOCl + 2NaOH$$
$$\qquad\quad \big|$$
$$\qquad\quad CONH_2$$
$$\rightarrow \sim CH_2—CH\sim + Na_2CO_3 + NaCl + H_2O$$
$$\qquad\qquad \big|$$
$$\qquad\qquad NH_2 \tag{16}$$

The Mannich reaction of polyacrylamide with formaldehyde and an amine may be used for the obtaining product that contains N-methylol groups (or ethers or ethers thereof).

$$\sim CH_2—CH\sim + CH_2O + HNR_2$$
$$\qquad\quad \big|$$
$$\qquad\quad CONH_2$$
$$\xrightarrow{\text{Base}} \sim CH_2—CH\sim + H_2O$$
$$\qquad\qquad\quad \big|$$
$$\qquad\qquad\quad CONHCH_2NR_2 \tag{17}$$

The reaction rate increases when heated to temperatures up to 40°C. The amino derivatives can then be quaternized if desired. The N-methylol derivatives of polyacrylamide can be made cationic by heating with amines, or they can be made anionic by heating with aqueous bisulfite solution under basic conditions.

Polyacrylamide can be undergone methylolation in the presence of formaldehyde under basic conditions (pH 8–10).

$$\sim CH_2—CH\sim + CH_2O \xrightarrow{\text{Base}} \sim CH_2—CH\sim$$
$$\qquad\quad \big|\qquad\qquad\qquad\qquad\qquad \big|$$
$$\qquad\quad CONH_2 \qquad\qquad\qquad\quad CONHCH_2OH$$
$$\tag{18}$$

The reaction of polyacrylamide with formaldehyde under acid conditions leads to the formation of a methylene bridge ($—CONHCH_2—O—CH_2NHCO—$).

Polyacrylamide may be converted into a sulfomethyl derivative by condensation with formaldehyde and sodium bisulfite under basic conditions.

$$\sim CH_2—CH\sim + CH_2O + NaHSO_3$$
$$\qquad\quad \big|$$
$$\qquad\quad CONH_2$$
$$\xrightarrow{\text{Base}} \sim CH_2—CH\sim + H_2O$$
$$\qquad\qquad\quad \big|$$
$$\qquad\qquad\quad CONHCH_2SO_3Na \tag{19}$$

Under these reaction conditions, polymers will contain significant carboxylate groups, owing to the rapid rate of hydrolysis.

Cross-links of polyacrylamide are formed when polyacrylamide reacts with N,N'-methylenebisacrylamide.

$$2 \sim CH_2 - CH \sim + (CH_2 = CHCONH)_2 CH_2$$

$$\begin{array}{c} | \\ CONH_2 \end{array}$$

$$\rightarrow \sim CH_2 - CH \sim$$

$$\begin{array}{c} | \\ CONHCH_2CH_2CONH \\ \searrow \\ CH_2 \quad (20) \\ \nearrow \\ CONHCH_2CH_2CONH \\ | \\ \sim CH_2 - CH \sim \end{array}$$

B. Manufacture

Acrylamide readily undergoes polymerization by conventional free radical methods, ionizing radiation, ultrasonic waves, and ultraviolet radiation. The base-catalyzed hydrogen transfer polymerization of acrylamide yields poly-β-alanine (Nylon 3) a water insoluble polymer that is soluble in certain hot organics. All current industrial production is believed to be by free radical polymerization.

$$\begin{array}{c} CH_2 = CH \\ | \\ CONH_2 \end{array} \left\langle \begin{array}{c} \xrightarrow{R^\bullet} \quad -(-CH_2 - CH -)_n - \\ \quad\quad\quad | \\ \quad\quad\quad CONH_2 \\ \xrightarrow{A^-} \quad -(-NH - CH_2 - CH_2 - CO -)_n - \end{array} \right.$$

$$(21)$$

where $R^\bullet$ and A^- are radical and anion, respectively.

The commonly used free radical initiators are persulfates, peroxides, perborates, azo compounds, and redox systems. By radical polymerization, polymers with different molecular weights may be produced. However, the polymers with high molecular weight ($\overline{M} = 10^6 - 10^7$) have the greatest practical interest. This requires a high purity of the monomers, very low concentrations of initiators, and the absence of oxygen and of other interfering substances such as metal ions traces. The presence of metal ions traces can interfere with the initiation systems, causing premature polymerization and a reduced molecular weight of the polymer.

Effects of the reaction medium on the radical polymerization of acrylamide have been examined in several reviews [13–15]. It has been shown that a change in the concentration of the monomer, the nature of the initia-

tor, the addition of complexing agents and surfactants, temperature, the nature of the solvent, and other factors influence the rate of polymerization and the molecular weight of polymers.

The principal production methods for acrylamide polymers are polymerization in aqueous solutions, mixed solvent solutions, and various dispersed phases.

1. Aqueous Solution Polymerization

Polymerization of acrylamide is usually performed in aqueous solutions. The principal factors that determine popularity of this polymerization technique are a high rate of polymer formation and the possibility to obtain a polymer with a large molecular weight. The reason for a specific effect produced by water upon acrylamide polymerization lies in protonation of the macroradical, leading to localization of an unpaired electron, which leads to an increase in the reactivity of the macroradical:

$$\begin{array}{c} \quad\quad\quad \overset{\bullet}{C}H \\ \sim CH_2 - \; | \\ \quad\quad\quad C - OH \quad\quad (22) \\ \quad\quad\quad \| \\ \quad\quad +NH_2 \end{array}$$

reflected in high values of the chain growth rate constant k_p. Mutual repulsion of like charges is responsible for reduction of the constant of the rate of bimolecular chain termination k_t. In a nonprotonated radical existing in the case of polymerization in nonaqueous solvents, conjugation of the unpaired electron with α electron of the C=O group brings about stabilization of the radical and reduction of its activity. The high reactivity of acrylamide in aqueous solutions may be associated with supression of autoassociation of this monomer's molecules as they form hydrogen bonds with water molecules. In view of these factors, acrylamide has a rather high value of the ratio of the constants $k_p/k_t^{1/2} = 3.2$–4.4 at 30–60°C, which, along with small values of the constant of chain transfer to monomer and water, stimulates polyacrylamide formation in aqueous solutions at rates and molecular weights unachievable in the case of polymerization in organic solvents.

Other reasons for a wide propagation of polymerization in water include: (1) reduction of energy consumed to separate the initial monomer in crystal form (acrylamide is produced and used in the aqueous solution form), which, in addition, is associated with the probability of its spontaneous polymerization, and (2) recovery of the organic solvents, which results in less environmental pollution and the elimination of the stage of solution of polymer reagents used, as a rule, in the form of the aqueous solutions.

The production of acrylamide polymers by means of

homogeneous polymerization of acrylamide in aqueous solutions is the simplest and the most environmentally clean method. Polymers synthesized by this method are obtained as gels with different concentrations of the principal compound. If necessary, dry polymers can be separated by settling in organic solvents, desiccation of pre-pulverized gel, or azeotropic mixture distillation of water with an organic solvent. In most cases, however, the stage of polymer separation is skipped, and gel of polyacrylamide is used as a commercial product. Various methods of polymerization of acrylamide in aqueous solution were reviewed [15,16].

Polymerization is carried out in 8–10% aqueous solutions of acrylamide in the presence of the redox system in an atmosphere of nitrogen and with pH 8.0–9.5. Polymerization is started at 20–25°C, but in the course of the reaction the temperature rises to 35–40°C. The polymeric product with $\overline{M} = 10^6 - 10^7$ in the gel is obtained.

Polymerization in aqueous solution of acrylamide can also be fulfilled in thin layers (up to 20 mm) applied on a steel plate or a traveling steel band. Polymerization is initiated by persulfates, redox system, UV or γ radiation. Polymerization proceeds in isothermal conditions as the heat of polymerization is dissipated in the environment and, additionally, absorbed by the steel carrier. Nonadhesion of the polymer to the carrier is ensured by the addition of glycerol to isopropyl alcohol or by precoating the steel band with a film based on fluor-containing polymers. This makes polymerization possible at a high concentration of the monomer (20–45%) and in a wider process temperature range. This film of polyacrylamide is removed from the band, crushed, dried, and packed.

In the case of photoinitiated polymerization, an oxygen-free aqueous solution of acrylamide with a concentration of about 50% mixed with a photosensibilizer and other required additives is passed through a column-type apparatus with exterior water-cooling. A thin layer of the solution is exposed to a mercury lamp, acquires the consistency of a plastic film, which then can be passed through a second exposure zone, and is crushed and dried. Acrylamide polymers produced by this method are easily soluble and have a low residual monomer content.

To maintain a high polymerization rate at high conversions, reduce the residual amount of the monomer, and eliminate the adverse process of polyacrylamide structurization, polymerization is carried out in the adiabatic mode. An increase in temperature in the reaction mixture due to the heat evolved in the process of polymerization is conductive to a reduction of the system viscosity even though the polymer concentration in it rises. In this case, the increase in flexibility and mobility of macromolecules shifts the start of the oncoming gel effect into the range of deep transformation or eliminates it completely.

The production process consists of: the stages of preparation of the monomer and additive solutions; elimination of the dissolved oxygen from the solutions; polymerization; compounding (i.e., stabilization and granulation); drying, crushing, and packing of the finished product.

Adiabatic polymerization is carried out in 20–35% aqueous solutions of acrylamide at an initial temperature of 5–20°C in the presence of a redox system in a stainless steel reactor with a stirrer. To reduce the adhesion of the polymer to the reactor walls and to facilitate polymer unloading, the interior surface of the reactor is polished or coated with synthetic materials. In the polymerization process, the temperature of the reacting mass may reach or even exceed 100°C, and the polymer is kept at this temperature for a long period of time (up to 20 h). Therefore, to prevent polymer hydrolysis, polymerization is carried out at pH 4–6 in the presence of ammonium salts and amines. The polymer gel is crushed in extruders and dried. In the process of crushing and drying of the polymer, antisticking agents (glycine, β-alanine, polyethylene glycols, fatty acids, etc.) and destruction stabilizers (2-mercaptobenzimidazol, phloroglucinol, urea, etc.) are added. Easily storable and noncacking polymer gels may be obtained without drying by crushing the polymer particles with a material poorly wetted by water, such as an active silicon acid, starch, dextrine, etc.

The drying agent can be of various designs. When a two-stage converger drier is used, polymer granules with a humidity of 50–90% are continuously fed into the first stage, laid in a layer 10–200 mm thick, and passed through the drier equipped with special agitator. After this, the polymer with a humidity of 5–40% arrives at the second stage where it is dried to the preselected residual humidity. In the first stage of drying, the air temperature can reach 120–130°C. Polyacrylamide granules can also be dried in a drum-type drier if the drum has shelves for agitation of the granules. Drying is performed by hot air at 100–120°C, the preliminary drying proceeding at 120°C, and the final stage at 100°C. Another suggested method is mixer drying. In this method the polymer is dispersed in cyclohexane in the presence of an emulsifying agent (polyoxyethylene phenyl ester) and other additives, and then azeotropic distillation is undertaken at 85–90°C. Finally, polymer granules are filtered and the organic solvent is dried up.

The polymerization of acrylamide in aqueous solutions in the presence of alkaline agents leads to the obtainment of partially hydrolyzed polyacrylamide. The polymerization process under the action of free radicals R· (formed on the initiator decomposition) in the presence of OH⁻ ion formed on the dissociation of an alkali addition (NaOH, KOH, LiOH), and catalyzing the hydrolysis can be described by a simplified scheme (with Me = Na, K, Li):

$$\text{(23)}$$

Process 2–4 is main. With an increasing concentration of alkali the rate of polymerization increases and the molecular weight decreases. The nature of hydrolyzing agent and salt additions influence the acrylamide polymerization process. Various methods of the acrylamide polymerization in the presence of alkaline agents have been reviewed [12,16,17].

2. Mixed Solvent Solution Polymerization

Polymerization of acrylamide is carried out in water–organic media that serve as solvents for monomers and as precipitants for polymers. For this reason, at the beginning of polymerization the reaction mixture is homogeneous, while during the process the precipitation of polymers takes place and the reaction proceeds under heterophase conditions. Among the organic solvents used as cosolvents are lower aliphatic alcohols (methyl, ethyl, isopropyl, *tert*-butyl), dioxane, acetone, and acetonitrile. The commonly used initiators are persulfates and perborates, benzoyl peroxide, AIBN, and UV and γ radiation. The molecular weights of polymers produced by precipitation polymerization are much lower than those obtained under homogeneous conditions (in aqueous solutions), but the molecular weight distribution is more narrow in the case of precipitation polymerization. The values of the molecular weight of polymers may be varied by choosing organic solvents and the ratio of organic solvent to water. The low viscosity of the precipitated polymer allows for better heat transfer. The polymer separates when formed and is, therefore, easily isolated and purified. However, the powdered polymers obtained by precipitation are readily caked, which hinders their storage and subsequent use.

3. Dispersion Polymerization

Suspension polymerization

Suspension (co)polymerization is carried out in aqueous solutions of monomers dispersed in the form of 0.1–5 mm diameter droplets by stirring in nonmixed water–organic liquids in the presence of initiators. The organic liquids that are not dissolving monomers and (co)polymers are represented by solvents that either form azeotropic water mixtures (toluene, heptane, cy-

clohexane, etc.) or do not form the azeotrope (hexane, paraffins, kerosene, and other unsaturated hydrocarbons), and by mixtures of organic solvents. The initiator can be a water-soluble compound (persulfate, peroxide, azo-compound, redox system, etc.). The suspension stabilizers are represented by surface-active substance possessing a nonionogeneous and a low hydrophilic–lipophilic balance (HLB) value (polyoxyethylated fatty alcohols and alkylphenols, sorbitan ethers of fatty acids, etc.). In practice, mixtures of various stabilizers are often applied that, together with their main function, reduce the scale formation on the stirrer and reactor walls, improve stability and control the molecular weight of the polymer, and stabilize the polymer on drying, etc. Depending on the stabilizer concentration and nature and the stirring conditions, the polymers are obtained in the form of powder or granules. The polymerization occurs in droplets of an aqueous monomer solution acting as microreactors, and its kinetics resemble solution polymerization, although still affected by the stabilizers. The rate polymerization and the molecular weight of polymers obtained in suspensions depend on the nature and concentration of the stabilizer, initiator, and monomer content, aqueous-to-organic phase ratio, and salt additions. The polymerization rate and the molecular weight in suspensions are lower than those observed in aqueous solutions. In the literature [17–19], the features of suspension polymerization of acrylamide are analyzed.

The technological process of obtaining polyacrylamide by suspension polymerization consists of the different stages of raw material preparation (preparation of monomer, initiator, stabilizer, and the other agent solutions; and the removal of oxygen from all these solutions), polymerization, water removal by azeotropic distillation, centrifugation, polymer drying, sieving, weighing, and packing of the final product.

The suspension polymerization of 65% acrylamide aqueous solution dispersed in *n*-hexane (aqueous phase: *n*-hexane = 1:5) in the presence of a stabilizer (sorbitan monostearate, 1.4% with respect to *n*-hexane) and an initiator (2,2'-azo-bis-*N*,*N*'-dimethyleneisobutylamide chloride) carried out at 65°C for 3 h, with subsequent holding at 110°C, yields a powdered product with the granule size of 0.5 mm, while the addition of Na_2SO_4

(50% with respect to monomer) allows the obtainment of a finely dispersed powder with a particle size of $2 \cdot 10^{-3}$ mm. The endothermal phase transition in Na_2SO_4 provides a partial removal of the polymerization reaction heat. This property is also offered by some other compounds: acetates of alkalimetals, borates, carbonates, phosphates, and silicates, which can be used as additions during the suspension polymerization.

To obtain a cation-active acrylamide copolymer with a particle size of 0.1–1 mm, suspension polymerization is carried out in the medium of a hydrophobic organic solvent having a density of 0.95–1.25 g/cm^3 in the presence of a stabilizer mixture containing anhydrosorbit monooleate with HLB 4.3 and cellulose derivatives (0.01–1% of the organic phase). A reactor is loaded with the organic solvent and then the stabilizer is dissolved. The mixture is thoroughly agitated, the oxygen is removed by purging dry nitrogen, and the reactor is heated to 45–70°C. After this, an aqueous solution of acrylamide mixed with an aminoalkylmetacrylate salt is introduced (the total monomer concentration in the aqueous solution is about 80%), an initiator solution is added, and the reaction is carried out during < 1.5 h. After cooling the reaction medium, the polymeric granules are separated by filtering and dried; the product yield reachers ~99%.

Emulsion polymerization

At the (co)polymerization in inverse emulsions, 30–70% aqueous solutions of monomers are dispersed (to provide the particle size of 1 to 10 μm) in a continuous hydrophobic organic phase (aliphatic and aromatic hydrocarbons, etc.) in the presence of the water-in-oil type emulsifier. The emulsifiers are represented by surface active substances with a low HLB (sorbitan ethers of fatty acids, polyoxyethylated fatty alcohols, and alkylphenols, etc.) and by mixtures of surface active substances with low and high HLB. The process is initiated by an oil-soluble (benzoyl peroxide, AIBN, etc.) or water-soluble initiators (persulfates, redox systems, etc.). The pecularities of emulsion polymerization of acrylamide have been analyzed in the literature [17,20–23]. The polymerization process that occurred in a low-viscosity media (the viscous polymer mass is the internal phase) in small particles provides efficient heat removal and has little effect on the chain termination reaction. For these reasons, the rate of polymerization in inverse emulsions is, under comparable conditions, higher, but the molecular weight of the polymer is lower than that obtained in aqueous solutions. The use of an oil-soluble initiator usually results in a greater molecular weight as compared with the water-soluble initiators. The character of the effects of the emulsifier on the process depends on the nature of the dispersion medium. By increasing the polymerization rate, the molecular weight decreases. The temperature variations due to a change of the adsorption saturation of emulsion layers can be accompanied by a decrease in the emulsion stability and phase inversion. The reaction results in the formation of a colloidal emulsion of hydrophilic polymeric particles dispersed in a continuous organic phase. This latex is characterized by a board distribution of the particle size and retains stability during several hours or days. It can be used as the final product or be subjected to azeotropic distillation, solvent removal, and drying.

The emulsion (co)polymerization is usually carried out as a periodic process. The production of emulsion acrylamide (co)polymers involves the following stages: initial component preparation; polymerization; emulsion concentrating and/or replacing the organic solvent; coagulate filtering, weighing, and packing of the final polymer product.

The reactor is loaded with a solution of emulsifier in an organic solvent and the aqueous monomer solution (20–60%) is dispersed in the organic phase by stirring. The obtained emulsion is deoxygenated by purging dry nitrogen or by multiple evacuation and thermostated at 30–60°C. Then, an initiator solution is introduced in the reaction mixture and the process is carried out at 30–60°C for 3–6 h, after which the reaction mixture is aged for 1–5 h.

To accelerate the polymerization process, some water-soluble salts of heavy metals (Fe, Co, Ni, Pb) are added to the reaction system (0.01–1% with respect to the monomer mass). These additions facilitate the reaction heat removal and allow the reaction to be carried out at lower temperatures. To reduce the coagulate formation and deposits of polymers on the reactor walls, the additions of water-soluble salts (borates, phosphates, and silicates of alkali metals) are introduced into the reaction mixture. The residual monomer content in the emulsion can be decreased by hydrogenizing the double bond in the presence of catalysts (Raney Ni, and salts of Ru, Co, Fe, Pd, Pt, Ir, Ro, and Co on alumina). The same purpose can be achieved by adding amidase to the emulsion.

The emulsion polymerization of acrylamide yields a high-molecular polymer (with the molecular weight reaching $2.5 \cdot 10^7$), which can be easily dispersed in water to obtain water-in-oil type latex (containing 30–60% polymer). On prolonged storage, the emulsion exhibits lamination, but subsequent stirring allows easy redispersal of the product.

The latex can be concentrated by different means. One method consists of heating the latex under vacuum conditions to remove excess water and organic solvent by evaporation. In the case of an organic solvent forming azeotrope with water, the final concentration may be higher and the temperature of the treatment can be reduced. Using the concentrating treatment, the polymer content in the latex can be increased to 70% and the water content can be reduced to 2%.

Another method of concentrating the latex is based on the relative instability of latex over time. The concentration is achieved by removal of the organic solvent from latex upon its lamination in storage.

The polymeric latex obtained in a hydrophobic organic solvent is poorly dispersed in water because of the presence of an emulsifier with a low HLB value. For this reason, a wetting agent is added to water or emulsion prior to the dissolution. The wetting agent (a surface active substance with a high HLB value) facilitates the inversion of latex phases to produce a direct type emulsion. Usually, it belongs to oxyethylated alkylphenols, fatty alcohols, or fatty acids.

The solubility of latex in water can be improved by replacing the solvent used in the system. Initially, the water is removed and than a hydrophobic organic solvent is replaced by a hydrophilic solvent, which has a boiling point above 100°C. This last solvent can be ethylene glycol, diethyl ether of diethylene glycol, monoethyl ether of ethylene glycol, or polyethylene glycols. This treatment results in a pastelike composition that can be easily mixed with water and used as a final product.

The polymers obtained by emulsion polymerization are seldom prepared in a powdered form because the material is highly hydroscopic and does not allow easy preparation of aqueous solutions. If it is necessary to obtain a finely dispersed powder, a precipitating agent is added to the latex on continuous stirring (methanol, isopropanol, etc.), and the precipitate formed is then filtered and dried.

4. Copolymerization

Radical copolymerization is used in the manufacturing of random copolymers of acrylamide with vinyl monomers. Anionic copolymers are obtained by copolymerization of acrylamide with acrylic, methacrylic, maleic, fumaric, styrenesulfonic, 2-acrylamide-2-methylpropanesulfonic acids and its salts, etc., as well as by hydrolysis and sulfomethylation of polyacrylamide Cationic copolymers are obtained by copolymerization of acrylamide with N-dialkylaminoalkyl acrylates and methacrylates, 1,2-dimethyl-5-vinylpyridinum sulfate, etc. or by postreactions of polyacrylamide (the Mannich reaction and Hofmann degradation). Nonionic copolymers are obtained by copolymerization of acrylamide with acrylates, methacrylates, styrene derivatives, acrylonitrile, etc. Copolymerization methods are the same as the polymerization of acrylamide.

The different physical properties, the reactivity of comonomers, and the reaction medium affect copolymerization. The majority of the real processes of copolymerization of acrylamide are complicated. Therefore, copolymerization may not be characterized by the classic equations. The following are the main complicating factors in the copolymerization of acrylamide.

1. Acrylamide can form inter- and intramolecular hydrogen bonds. The reactivities of acrylamide are af-

fected by the composition of the initial monomer mixture, the conversion degree, and the change of the overall concentration of the initial monomer mixture. These effects are significant at the copolymerization of acrylamide in organic and water–organic solvents. For such systems, the reactivity ratios r_1 and r_2 should be considered to be the apparent characteristics of the copolymerization.

2. The copolymerization of acrylamide with an ionogenic monomers leads to a nonisoionic condition of the process. Under these conditions, the value of the ionic strength for the reaction medium differs with changing concentrations of ionogenic monomers in the initial reaction mixture or with the addition of neutral salts, as well as a result of the consumption of ionogenic monomers during copolymerization. This leads to a change of r_1 and r_2. The nonisoionic condition of copolymerization causes the difference between experimental and theoretical integral curves of composition distribution. In the presence of salts, the copolymerization takes place under isoionic conditions. It stabilizes the reactivity of ionogenic monomers and establishes satisfactory correlation between experimental and theoretical integral curves of composition distribution. The peculiarities of the radical copolymerization ionogenic monomers in various media have been analyzed in reviews [24,25].

3. Heterogeneous copolymerization of acrylamide causes redistribution comonomers between phases I and II. This leads to a change of copolymer composition in phases I and II. As a result, the values of r_1 and r_2 change. This accounts for anomalous widening of the experimental composition distribution curves as compared with theoretical curves.

These complicating factors influence not only the middle composition and composition distribution curves of copolymers, but also the kinetic parameters of copolymerization and the molecular weight of copolymers. An understanding of these complicating factors makes it possible to regulate the prosesses of copolymerization and to obtain copolymers with different characteristics and, therefore, with various properties.

Random copolymers are more technologically important than the graft and block copolymers, which are less convenient to prepare. Acrylamide can be grafted into numerous substrates for the purpose of increasing hydrophilicity, altering crystallinity, reducing susceptibility to degradation, or providing a reactive site. Acrylamide is grafted using free radical sources, UV light, or x-rays. Substances include polyolefins, polyvinyl chloride, cellulose, polyvinyl alcohol, polyamides, urethanes, etc. Acrylonitrile, acrylic acid, ethylene oxide, vinylchloride, or styrene can be grafted into polyacrylamide substrates. Block copolymers of acrylamide can be obtained by mechanical or UV degradation of polymers in the presence of acrylamide, or by incorporating a reactive end group in the base polymer.

C. Application

Polymers and copolymers of acrylamide (obtained by copolymerization or postreaction of polyacrylamide) with different values of the molecular weight, composition, distribution of molecular weight and compositions, linear and cross-linked have different functions and are used in many fields. The main functions and applications of acrylamide polymers are shown in Table 4.

The effectiveness of the applications of the acrylamide polymers depends on their characteristics. The main use of nonionic polymers is in mineral processing (as flocculation and dewatering agents). Anionic polymers are used in water treatment, mining, paper refining, and enhancement of oil recovery. Cationic polymers are used in the manufacture of paper, by concentrating and dewatering sludges. High molecular weight polymers $[\overline{M} = (2-18) \times 10^6]$ (the effectiveness of which increases with an increase in molecular weight) are used

for flocculation, thickening, binding, and film formation. Low molecular weight polymers $[\overline{M} = (0.005-0.4) \times 10^6]$ are used as dispersants; suspending, coating, and sizing agents; stabilizers of drilling fluids; additives for dilution of oil; and scale deposition inhibitors. In most cases, water-soluble polymers are used, but cross-linked polymers are used in coating, films, adhesives, and water-holding gels. An analysis of the applications of acrylamide polymers has been reviewed [26,27].

Among the basic fields of applications, the major use of acrylamide polymers is liquid–solid separation in water treatment and waste treatment. Smaller quantities are used in the manufacturing of paper and in the processing of minerals in mining. Relatively nonlarge quantities are use as additives for enhanced oil recovery.

Most of water-soluble acrylamide polymers find practical applications as highly efficient flocculents for clarification and treatment of potable water and municipal and industrial effluents; and in the mining, papermak-

Table 4 Applications of Acrylamide Polymers

Fields of application	Functions
Water treatment	Clarification of potable water, industrial effluents, municipal waste water; thickening and dewatering of sludge; filtration of primary sludge, digested sludge; food processing
Pulp and paper	Filler retention, binder pigments, dry and wet strengthening, sizing, coating, clarification of waste water, concentration and dewatering of sludge
Mining and ore processing	Setting slimes, flocculation of flotation waste, clarification of waste water from gravel wash, coal washeries, concentration and dewatering of sludges
Petroleum	Enhancing oil recovery, regulation of filterability and rheological properties of drilling muds, thickening of water, soil structure formation, oil flotation
Agriculture	Soil stabilization; microencapsulation of mineral fertilizers, fungicides, and herbicides
Medicine	Microencapsulation of water soluble pharmacological preparates, nonthrombogenic granulated gels, contact lenses in ophthalmology, high-quality tampons, diapers, etc.
Construction	Soil stabilization; water absorber; water retention aids in cements, grouts, and tiles; improvement of mechanical properties of cement and gypsum plasters and water based paints
Other fields	Friction reduction agents, dispersants, adhesives, fiber dressers

ing, food, metallurgy, coal, and oil industries. Action of high molecular weight flocculents proceeds principally by two mechanisms. In one mechanisms the macromolecules are absorbed on suspension particles that bind them together into large flocs. This is the bridge flocculation mechanism. In the second mechanisms, charged macroions are absorbed on charged dispersed particles. This neutralizes them and lowers the sedimentational stability of the system. This is called the neutralization flocculation mechanism. For high molecular weight polymers, the bridge flocculation mechanism prevails as a rule. In most cases, polymers are used with inorganic coagulants, such as aluminum sulfates. When coagulants are added to the water being treated, they are charged positively and neutralize the negative charge of the colloidal particles. When anionic polymers are added after inorganic coagulants, large flocs by bridge formatting are produced and sedimentation is greatly accelerated. The influence of various factors on the process of acrylamide polymers flocculation has been reviewed [28]. The nature of the second monomer and the composition of the copolymer greatly affect flocculation properties of acrylamide copolymers.

The flocculation activity of polymers increases with increasing molecular weight. Linear macromolecules are better flocculents than branched ones with the same total molecular weight. In the future the largest growth of polymer applications is expected to be in waste and water treatment, because of ecological problems.

In the pulp and paper industry, anionic and cationic acrylamide polymers are used as chemical additives or processing aids. The positive effect is achieved due to a fuller retention of the filler (basically kaoline) in the paper pulp, so that the structure of the paper sheet surface layer improves. Copolymers of acrylamide with vinylamine not only attach better qualities to the surface layer of paper, they also add to the tensile properties of paper in the wet state. Paper reinforcement with anionic polymers is due to the formation of complexes between the polymer additive and ions of Cr^{3+} and Cu^{2+} incorporated in the paper pulp. The direct effect of acrylamide polymers on strength increases and improved surface properties of paper sheets is accompanied by a fuller extraction of metallic ions (iron and cobalt, in addition to those mentioned above), which improves effluent water quality.

Acrylamide polymers are used as multipurpose additives in the oil-producing industry. Introduction of polymers into drilling fluids–drilling muds improves the rheological properties of the fluids in question, positively affects the size of suspended particles, and adds to filterability of well preparation to operation. Another important function is soil structure formation, which imparts additional strength to the well walls. A positive effect is also observed in secondary oil production, where acrylamide polymers additives improve the mobility of aqueous brines injections, which contribute to a more complete oil extraction from porous media. Acrylamide polymers are not adsorbed on the rock in any significant amount, so that the effects arising from pore blocking are less pronounced. From the process viewpoint, water-in-oil emulsions are preferable to aqueous solutions, as this form of polymeric component is easier to inject into a wall, the viscosity of emulsions being notably lower than of aqueous solutions of acrylamide polymers. If the effect of polymer destruction in the course of injection is to be reduced, polyacrylamide should be replaced with methylol derivatives of polyacrylamide. The most efficient way to apply acrylamide polymers in oil production is to use them in oil pools operated by a water drive. The introduction of polymeric additives significantly reduces water content in raw oil, creating a protective screen against a water-bearing stratum.

REFERENCES

1. D. C. MacWilliams, in *Functional Monomers, Vol. 1*, (R. H. Yacum and E. B. Nyquist, eds.) Marcel Dekker, Inc., New York, p. 1 (1973).
2. V. F. Kurenkov, ed., *Polyacrylamide*, Khimiya, Moscow (1992).
3. W. M. Leung, D. E. Axelson, and J. D. Van Dyke, *J. Polym. Sci., Polym. Chem. Ed.*, 25: 1825 (1987).
4. W. M. Kulicke, R. Kniewske, and J. Klein, *Progr. Polym. Sci.*, 8: 373 (1982).
5. V. A. Myagchenkov and V. F. Kurenkov, *Acta Polymerica*, 42: 475 (1991).
6. E. Collinson, F. D. Dainton, and McNaughton, G. *Trans. Faraday Soc.*, 53: 489 (1957).
7. American Cyanamid Co., *Acrylamide*, Wayne, New Jersey (1957).
8. W. Scholtan, *Makromol. Chem.*, 14: 169 (1954).
9. J. Klein and K.-D. Conrad, *Makromol. Chem.*, 179: 1635 (1978).
10. V. F. Kurenkov and E. D. Tazieva, *Z. prikladnoi. khim.*, 67: 1162 (1994).
11. J. D. Morris and R. J. Penzenstadler, *Kirk-Other Encyclopedia of Chemical Technology*, 3rd ed., Vol. 1, John Wiley & Sons, New York, p. 312 (1978).
12. V. F. Kurenkov and T. A. Baiburdov, *Izv. VUZ, Khim. khim. Technol.*, 32: 3 (1989).
13. V. F. Gromov and P. M. Khomikovskii, *Uspekhi khimii*, 48: 1943 (1979).
14. V. F. Kurenkov and V. A. Myagchenkov, *Eur. Polym. J.*, 16: 1229 (1980).
15. V. F. Kurenkov and L. I. Abramova, *Polym.-Plast. Technol. Eng.*, 31: 659 (1992).
16. V. F. Gromov and E. N. Teleshov, *Plastmassy*, 10: 9 (1984).
17. V. F. Kurenkov and V. A. Myagchenkov, *Polym.-Plast. Techn. Eng.*, 30: 367 (1991).
18. M. V. Dimonie, C. M. Boghina, N. N. Marinescu, M. M. Marinescu, C. I. Cincu, and C. G. Oprescu, *Eur. Polym. J. 18*: 639 (1982).
19. D. Hunkeler, A. E. Hamielec, and W. Baade, *Polymer*, 30: 127 (1989).
20. J. W. Vanderhoff, F. V. Distefano, M. S. ElAasser, R. O'Leary, O. M. Shaffer, and D. L. Visioly, *J. Dispers Sci. Technol.*, 5: 323 (1984).

21. W. Baade, K. H. Reichert, *Eur. Polym. J.*, *20*: 5 (1984).
22. C. Graillat, C. Pichot, A. Guyot, and M. S. El Aasser., *J. Polym. Sci.*, *Polym. Chem. Ed.*, *24*: 427 (1986).
23. Ks. O. Sochilina and V. F. Gromov, *Vysokomolek. Soedin.*, *Ser. B*, *37*: 1228 (1995).
24. V. F. Kurenkov and V. A. Myagchenkov, *Acta Polymerica*, *37*: 517 (1986).
25. V. F. Kurenkov and V. A. Myagchenkov, *Polymer Materials Encyclopedia*, CRC Press, Inc., Boca Raton, FL Vol. 10 (1996).
26. A. F. Nikolaev and G. I. Ohrimenko, *Vodorastvorimye Polymery*, Khimiya, Leningrad (1979).
27. V. A. Myagchenkov and V. F. Kurenkov, *Polym.-Plast. Technol. Eng.*, *30*: 109 (1991).
28. M. A. Nagel, V. F. Kurenkov, and V. A. Myagchenkov *Izv. VUZ. Khim. khim. Technol.*, *31*: 1 (1988).

4

Transparent Polyolefins

Yu. P. Losev
Belarussian State University, Minsk, Belarus

An intrinsic defect of polyolefins is the presence of turbidity, which limits their use as an optical material. Only poly-4-methylpentene represents a transparent material. However, polyolefins in melts are transparent and the turbidity is due to the formation of coarse spherulite structures during the process of crystallization. At the same time, rapid cooling of polyolefin melts, i.e., "quenching," results in the preservation of the structure particular to the melt. This fact is confirmed by results obtained in polarization microscopy studies of the structure characteristics of normal and quenched polyolefins (Fig. 1). Quenched polyolefins exhibit fine-grain spherulite structure, while in normal polyolefins a coarse spherulite structure is revealed. In normal polyolefins with coarse spherulite structure, where the sizes are comparable to the wavelength of the transmitted light flux, the scattering process takes place and, consequently, the light permeability is rather low. The sizes of spherulites in quenched polyolefins are much smaller than the relative wavelength of transmitted light flux and the scattering is insignificant. An investigation into the light permeability of polyolefins upon quenching, over the wavelength range of 300–700 nm, has shown that after quenching the structure of the polyolefins has a considerable effect on light transmission. Of the polyethylene samples examined, the highest transmission has been observed for high-pressure polyethylene. Light permeability in the ultraviolet region may be as high as 70%, rising monotonically with the wavelength and reaching 95% relative to glass at the wavelength of 700 nm. Light permeability of low- and medium-pressure polyethylene is practically at the same level. An increase in the wavelength leads to a linear rise in the light permeability, which reaches 70% at 700 nm. Lower light permeability of medium- and low-pressure polyethylene upon quenching is conditioned by slight branching, which re-

sults in the absence of steric barriers for the formation of spherulites in contrast to high-pressure polyethylene. Nonquenched polyethylene samples possess negligible light permeability. The highest light permeability is exhibited by high-pressure polyethylene, the light transmission of which amounts to 15% at 300 nm and increases to as high as 55% at 700 nm. In instances of low- and medium-pressure polyethylene, the light permeability of nonquenched samples increased from 10% at 300 nm to 35% at 700 nm.

Light permeability of quenched polypropylene in the visible spectral region is practically the same as that of quenched high-pressure polyethylene. In the ultraviolet region the light permeability amounts to 35% at 300 nm, then it drastically increases in conjunction with the wavelength and may become as high as 92% at 700 nm. The transmission curve of nonquenched polypropylene is much lower; as the wavelength increases the light transmission also increases from 10% at 300 nm to 50% at 700 nm. Higher light permeability of quenched polypropylene, as compared to medium- and low-pressure polyethylene, is determined by the presence of methyl substituents hindering the formation of coarse spherulite structures (Fig. 2).

Using the quenching procedure, optically transparent films and plates up to 2 mm thick can be produced from high-pressure polyethylene. As thickness is increased the light permeability of quenched polyethylene decreases, but this decrease is insignificant. Light permeability of a nonquenched high-pressure polyethylene 350 μc thick sample (Fig. 3) comes into the visible spectral region only up to 5%, while that of the 350 μc thick quenched sample is 80%. A decrease in the light permeability of polypropylene in relation to the sample thickness has been also observed, and this decrease was also modest. Increasing the sample thickness from 100 μc to

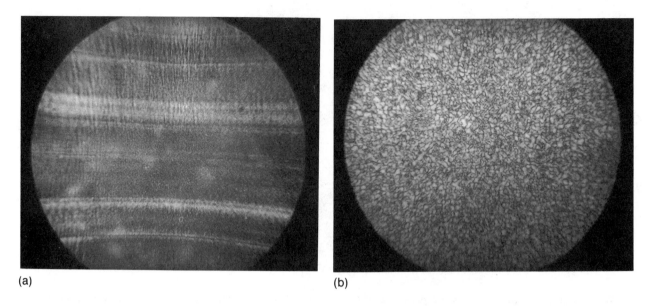

(a)   (b)

Figure 1 Photomicrographs of polyethylene films 1 mm thick: (a) quenched, and (b) nonquenched.

500 μc, in the case of quenched polypropylene, decreases the light permeability in the visible spectral region by 10%, while for normal polypropylene such a decrease was one-fourth as large. A sample of polypropylene 500 μc in thickness displays 10% light

permeability, and quenched samples show 80% light permeability (Fig. 4).

In an effort to obtain optically transparent cured polyethylene, the vulcanization procedure followed by quenching has been developed. Increasing concentration of the curing agent (i.e., dicumyl peroxide) leads to only a small increase in the light permeability (Fig. 5). This increase derives from the fact that an increase in the vulcanization depth with an increasing concentration of the curing agent prevents the formation of coarse

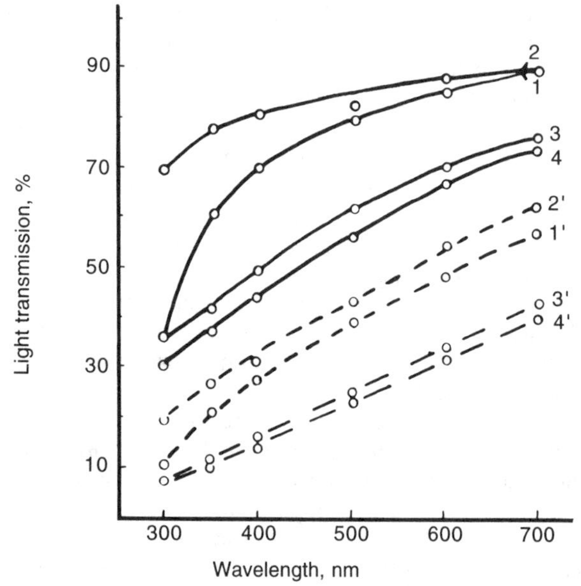

Figure 2 Light permeability of polyolefins after quenching (1–4) and of nonquenched samples (1'–4'): 1,1'-polypropylene (PP); 2,2'-high-pressure polyethylene (HPPE); 3,3'-low-pressure polyethylene (LPPE); 4,4'-medium-pressure polyethylene (MPPE). Film thickness-150 μc; moulding time-10 minutes, moulding pressure: HPPE-160°C; LPPE, MPPE, PP-190–200°C.

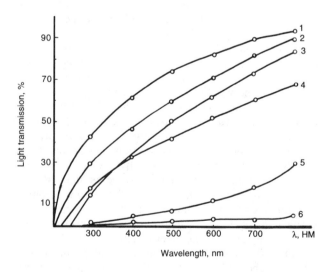

Figure 3 Light permeability of the high-pressure polyethylene sample as a function of thickness: 1–3-quenched polyethylene; 4–6-normal polyethylene; 1 and 4-50 μc thick; 2 and 5-200 μc thick; 3 and 6-350 μc thick.

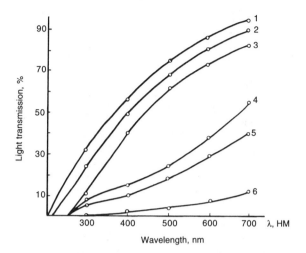

Figure 4 Light permeability of polypropylene as a function of the sample thickness: 1–3-quenched polypropylene; 4–6-normal polypropylene; 1 and 4-100 μc thick; 2 and 5-150 μc thick; 3 and 6-500 μc thick.

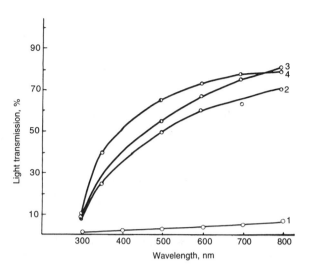

Figure 6 Light transmission of quenched cured polyethylene as a function of the sample thickness (vulcanization temperature 160°C, time 15 min.). 1-sample of normal polyethylene 2 mm thick; 2, 3, and 4-quenched cured samples of different thickness; 2-1 mm, 3-2 mm, 4-4 mm.

spherulite structures. As the sample thickness increases, the light permeability of quenched cured polyethylene decreases. Whereas normal polyethylene 2 mm thick is practically opaque, the light permeability of quenched cured polyethylene is as high as 70%, and for a sample 4 mm thick the light permeability is 60% in the visible spectral region (Fig. 6).

The presence of fine-grain spherulite, optically permeable structure contributes to the improvement of serviceability (Table 1).

Transparent polyethylene is characterized by an enhanced freeze- and heat-aging resistance, and also by a

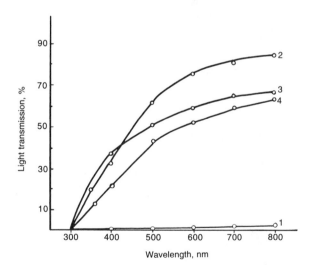

Figure 5 Light transmission of polyethylene as a function of the curing agent concentration (plate thickness 1 mm, moulding temperature 160°C, time 15 min.). 1-noncured polyethylene; 2, 3, and 4-quenched cured polyethylene with different concentrations of the curing agent; 2-0.5%, 3-1%, 4-2%.

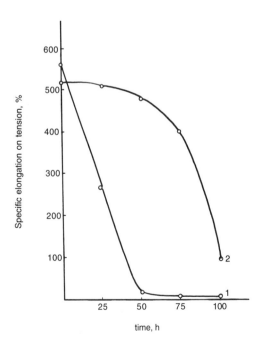

Figure 7 Light resistance of low-density polyethylene upon ultraviolet irradiation. 1-normal; 2-quenched.

Table 1 Serviceability of Transparent and Normal High-Pressure Polyethylene

Quality	Time (hours)	Tensile strength* (kg/cm²)		Specific elongation (%)	
		1	2	3	4
Freeze resistance at liquid nitrogen temperature		356	350	530	480
	36	360	400	530	505
Refrigeration at −54°C	50	343	390	485	495
Heat aging at 95°C	100	312	370	380	400
Weather resistance	150 days	84	97	30	260

* Columns 1 and 3 refer to normal polyethylene; columns 2 and 4 refer to the transparent one.

considerable degree of weather resistance. Fine-grain spherulite transparent structure favors improved light resistance (Fig. 7).

On exposure to ultraviolet irradiation over a period of 25 hours, specific elongation of normal polyethylene decreases by half, and complete decomposition is observed after a 50-hour exposure. At the same time, specific elongation of quenched polyethylene, when exposed to irradiation not greater than 50 hours, is little affected, but when exposed over a period of 100 hours, its specific elongation amounts to 100%.

Useability of a transparent polyethylene film in hotbeds and greenhouses is obvious, since this film, as opposed to glass, displays high light permeability in the ultraviolet spectral region.

Transparent polyethylene can be also applied to the protection of window glass against aggressive media, e.g., the effect of hydrogen fluoride on the plants producing superphosphate fertilizers. The use of transparent polyethylene film for window glass makes it possible to cut down on the heat losses due to the lower thermal conductance of polyethylene as compared to glass.

5

Polyolefin Stabilizers with Intramolecular Synergism

Yu. P. Losev

Belarussian State University, Minsk, Belarus

The synergistic compositions of stabilizers are highly efficient, but they have a number of drawbacks in regard to the selection of stabilizers indifferent to one another, which limits the number used and availability of optimal relations. Drawbacks differences in physico-chemical characteristics—volatility, compatibility with polyolefins—that upsets the optimal relations and, as a result, reduces their efficiency; occasional chemical interactions in the course of processing and service. Stabilizers with intramolecular synergism capable of inhibiting the thermal oxidative breakdown of polyolefins do not have these drawbacks. Such stabilizers simplify considerably the introduction method; they are of high economic efficiency as they render the production of several different stabilizers unnecessary. Moreover, such stabilizers are known to exert a constant stabilizing action over time. In studying stabilizers with intramolecular synergism we have investigated polyamine disulfides produced by the condensation reaction of sulfur monochloride with a number of aromatic amines and diamines. They are high-molecular or oligomeric compounds, and therefore are compatible with polyolefins; they are low-volatile and heat-proof in air at 220–250°C; they have paramagnetic centers with a concentration 10^{16}–10^{20} spin/g.

A study of the thermal oxidative breakdown of polyethylene under static conditions has revealed that polyamine disulfides are stabilizers, yet to varying degrees (Fig. 1).

Aliphatic polyamine disulfides—imino-, carbamide-, thiocarbamide-, ethylenediamine- —are low-efficiency stabilizers except for polythiosemicarbazide disulfide and polysulfide of hydrorubeanic acid. Aromatic polyamine disulfides are more efficient. Their series are more efficient than the synergistic composition of stabilizers—2,2′-bis-methylene (-tert-butyl-4-methylphenol) and dilauryl thiodipropionate in the ratio of I:I. A study of efficient polyamine disulfides and the effect of con-

centration on the induction period has shown that 1% mass is the optimal concentration (Fig. 2). Polythiosemicarbazide disulfide is more efficient. Higher concentrations of polyamine disulfides considerably increase the

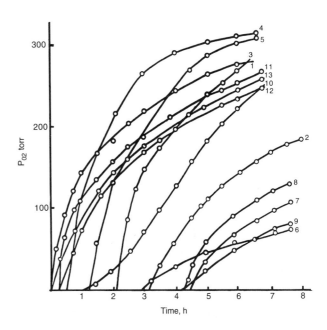

Figure 1 Thermal oxidative breakdown of polyethylene (temperature 200°C; P_{O_2} = 350 Tor; stabilizer concentration 0.5 mass percent). 1-without stabilizer; 2-CaO-6; 3-polydiiminodiphenylmethane disulfide; 4-polydiiminodiphenylsulfon disulfide; 5-polyparaoxydiphenylamine disulfide; 6-polydimethylaniline disulfide; 7-polyaniline disulfide; 8-polydiiminodiphenyloxide disulfide; 9-polythiosemicarbazide disulfide; 10-polyamine disulfide; 11-polycarbamide disulfide; 12-polythiocarbamide disulfide; 13-polyethylenediamine disulfide.

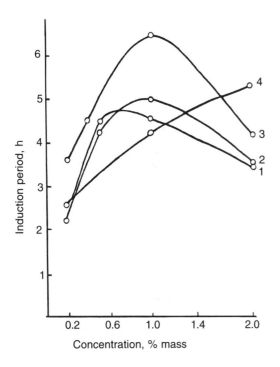

Table 1 Effect of Stabilizer Concentration on Film Polyethylene Aging (temperature 120°C, thickness 0.2 mm)

Stabilizer 1	Concentration (mass %) 2	Breakdown time (hour) 3
Polydiiminodiphenyloxide	0.2	900
	1.0	1200
	2.0	1350
Polyiminoaniline disulfide	0.2	550
	1.0	800
	2.0	870
Polythiosemicarbazide disulfide	0.2	700
	1.0	675
	2.0	750
2,2-thiobis(6-tert-butyl-4-methylphenol) (CaO-6)	0.2	550
	1.0	1100
	2.0	1250

Figure 2 The effect of stabilizer concentration on oxidation induction period. 1-polyamineaniline disulfide; 2-polydiaminodiphenyloxide disulfide; 3-polythiosemicarbazide disulfide; 4-CaO-6.

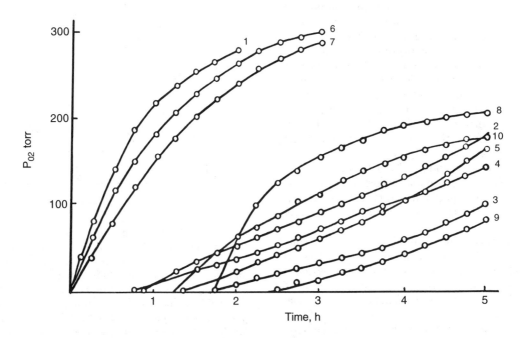

Figure 3 Thermal oxidative breakdown of polypropylene (temperature 200°C; P_{O_2} = 350 Tor; stabilizer concentration 0.5 mass percent). 1-without stabilizer; 2-Santanox®; 3-polythiosemicarbazide disulfide; 4-polyamineaniline disulfide; 5-polydiiminodiphenyloxide disulfide; 6-polyaniline disulfide; 7-polydiiminodiphenylmethane disulfide; 8-hydrorubeanicpoly disulfide; 9-thiocarbamide polysulfide; 10-polyiminoazobenzene disulfide.

time until the complete breakdown of film due to thermal aging (Table 1).

Polythiosemicarbazide disulfide is the most efficient aliphatic polyamine disulfide for inhibiting the thermal oxidative breakdown of polypropylene, while polyiminoaniline disulfide and polydiiminodiphenyloxide disulfide (Fig. 3) are the most efficient aromatic polyamine disulfides. In contrast to polyethylene, the thermal oxidative breakdown period increases as the concentration increases (Fig. 4). Depending on the concentration, the flow–melt index at 230°C increases at a lower rate than in the case of commercial stabilizer Santanox® (Table 2).

Just as in the case of polyethylene, the efficiency of polyamine disulfides can be presented in the series: polythiosemicarbazide disulfide, polydiiminodiphenyloxide disulfide, polyiminoaniline disulfide.

The stabilizing activity of polyamine disulfides is specified by the imine group and the disulfide bridge, which participate simultaneously in the breaking of the chain by suppressing the radicals and in the free radical destruction of peroxides and hydroxides.

The formation of a sulfoxide group is confirmed by infrared spectroscopy; absorption bonds typical of a sulfonic group can be observed in $1120 \div 1160$ and $1310 \div 1350$ cm^{-1} spectral regions. Inhibition of the thermal oxidative breakdown of polyolefins by pre-oxidized polyamine disulfides decreases the induction period of thermal oxidation from 265 minutes to 65 minutes for polydiiminodiphenyloxide disulfide, and from 270 minutes to 250 minutes for polythiosemicarbazide disulfide. A break of alkyl and alkoxy radicals may also be present due to "adhesion" to the stabilizer paramagnetic centers.

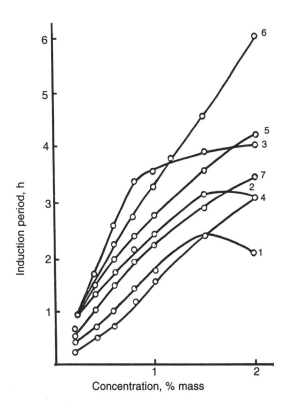

Figure 4 The effect of stabilizer concentration on the induction period of polypropylene thermal oxidative breakdown (temperature 200°C; P_{O_2} = 300 Tor). 1-polyaminoaniline disulfide; 2-polydiiminodiphenyloxide disulfide; 3-polythiosemicarbazide disulfide; 4-Santanox®; 5-hydrorubeanic polydisulfide; 6-thiocarbamidepoly disulfide; 7-polydisulfide.

Table 2 Effect of Polyamine Disulfide Concentration on the Polypropylene Flow–Melt Index During Thermal Oxidative Breakdown

Stabilizer	Concentration (mass %)	Sampling time (min)			
		5	20	30	50
1	2	3	4	5	6
Without stabilizer Santonox®	—	1.40	2.48	2.65	3.20
	0.2	1.00	1.20	1.30	1.40
	0.8	1.04	1.45	1.60	1.80
	2.0	1.15	1.70	2.00	2.40
Polyaminoaniline disulfide	0.2	0.88	0.90	0.90	1.00
	0.8	0.86	0.98	0.99	1.10
	2.0	1.07	1.11	1.10	1.45
Polydiiminodiphenyloxide disulfide	0.2	0.89	0.96	0.94	0.91
	0.8	0.80	0.80	0.80	0.85
	2.0	1.00	1.05	1.09	1.45

6

High-Temperature Stabilization of Polyolefins

Yu. P. Losev
Belarussian State University, Minsk, Belarus

I. INTRODUCTION

Polyolefins represent one of the main types of synthetic polymeric materials. World production of polyolefins in 1980 amounted to 23 million tons, and since then the tendency to further growth has been prevailing [1]. The grave defect of polyolefins is low thermal and heat resistance, which is detrimental to the processing efficiency and limits their useability (Table 1).

Polyethylene displays good heat resistance in the absence of oxygen in vacuum or in an inert gas atmosphere, up to the temperature of 290°C. Higher temperature brings about the molecular-chain scission followed by a drop in the molecular-weight average. At temperatures in excess of 360°C the formation of volatile decomposition products can be observed. The main components are as follows: ethane, propane, n-butane, n-pentane, propylene, butenes and pentenes [7].

At temperatures 400–420°C the decomposition rate is so great that in 1 hour the molecular weight decreases from 27,100 to 3100, and within 5 hours the polymer is transformed into a mixture of liquid products [8].

Thermal stability of polyethylene is greatly affected by the value of molecular weight and molecular-weight distribution. The samples of polyethylene with high molecular weight and wide molecular-weight distribution are less stable. The activation energy on thermal destruction of polyethylene is within the interval 60–70 kcal/mol and pure thermal breaking of such a chain with the formation of two radicals is highly improbable [9]. However, thermal destruction of polyethylene takes place even at 290°C, while hexadecane is stable up to 400°C. This stability is derived from the fact that the molecular structure of polyethylene is more complex than may be expected from the schematic representation. A macromolecule of polyethylene possesses "weak links" that are, first of all, susceptible to thermal destruction. These weak links may be represented by side chains, peroxy, carbonyl, and hydroxyl groups (forming in the process of radical polymerization of ethylene in the presence of peroxide initiators and oxygen), atactic portions of macromolecular chains and branches. To illustrate, the molecular-chain scission of linear polyethylene is random in nature, whereas the behavior of branched samples deviates from this pattern, as evidenced by the curves of destruction rate versus the depth approximating the curves characteristic of depolymerized polymer and by the mechanisms of volatile products formation [10]. Thermal stabilities of branched polyethylene and polypropylene are lower than those of linear polyethylene. Ethylene-propylene copolymers display an intermediate thermal stability between polyethylene and polypropylene and, their stability being dependent on the composition, could influence the stability of polyethylene or polypropylene.

According to current knowledge, the complex process of thermal decomposition of polyethylene is a combination of a number of elementary reactions proceeding through the free radical mechanism [11–13].

1. Initiation-formation of macroradicals due to the polymeric-chain scission under the action of heating.

   $$...—CH_2—CH_2—CH_2—CH_2—...$$
   $$\rightarrow ...—CH_2—CH_2\cdot + \cdot CH_2—CH_2—...$$

2. Chain reaction followed by the detachment of monomer from the end macroradicals.

   $$...—CH_2—CH_2—CH_2—CH_2\cdot$$
   $$\rightarrow ...—CH_2—CH_2\cdot + CH_2{=}CH_2$$

Table 1 Temperature Conditions in Polyolefin Processing

Polyolefin 1	Process 2	Temperature (°C) 3	Reference 4
Low-density polyethylene	Press moulding	140–160	2
	Sheet moulding	120–135	3
	Pipe extrusion	140–150	
	Film extrusion	160–170	
	Centrifugal casting	250	
	Blow moulding	160–170	
	Injection moulding	150–250	
	Welding	250	
	Paper coating by extrusion	up to 260	
High-density polyethylene	Press moulding	150–180	
	Granulation by screw machine	250–260	
	Injection moulding	230–260	
	Deposition on metal by welding	300–380	
	Injection moulding	200–270	
	Sheet moulding	200–270	5
Polypropylene	Extrusion of films and profiled products	190–240	6
	Insulating coating of wires and cables	200–250	
	Filming of paper, steel, aluminum	230–260	
	Blow moulding of hollow products	175–185	

3. Monomolecular chain transfer caused by the detachment of hydrogen macroradicals from other macromolecule accompanied by the breaking of C—C bond.

—CH$_2$—CH$_2$· + —CH$_2$—CH$_2$—CH$_2$—CH$_2$—
→ —CH$_2$—CH$_3$ + . . . —CH=CH$_2$
+ ·CH$_2$—CH$_2$—. . .

4. Intramolecular chain transfer due to the detachment of a hydrogen atom in the macroradical proper and movement of its valence to the end of the chain followed by the breaking of C—C bond.

—CH$_2$—CH$_2$· + CH$_2$=CH—CH$_3$

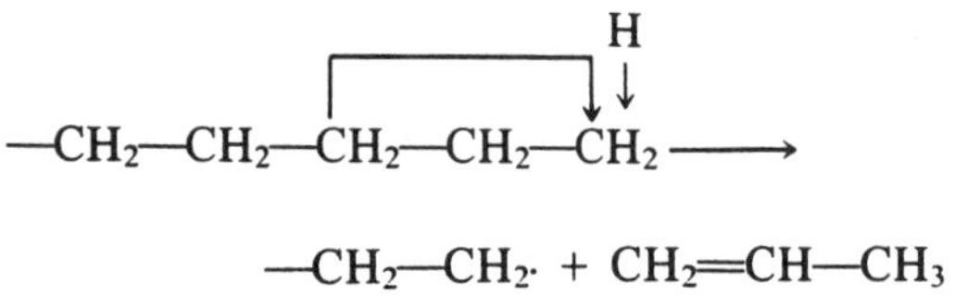

5. Destruction of macroradicals—scission of kinetic chains. A disproportionation reaction is most common at this stage:

. . .—CH$_2$—CH$_2$—CH$_2$—CH$_2$·
+ ·CH$_2$—CH$_2$—CH$_2$—. . .
→ . . .—CH$_2$—CH$_2$—CH=CH$_2$
+ CH$_3$—CH$_2$—CH—. . .

Upon thermal destruction of polyethylene the chain transfer reactions are predominant, but depolymerization proceeds to a much lesser extent. As a result, the products of destruction represent the polymeric chain fragments of different length, and monomeric ethylene is formed to the extent of 1–3% by mass of polyethylene. C—C bonds in polypropylene are less strong than in polyethylene because of the fact that each second carbon atom in the main chain is the tertiary one.

The hydrogen atom attached to the tertiary carbon atom is more reactive and more easily detached than the hydrogen attached to the secondary carbon atom, and methyl substituent being small in size is no hindrance to the hydrogen atom migration. In consequence, the breaking of carbon–carbon bonds in the chain of polypropylene proceeds predominantly with migration of the hydrogen atom according to the following scheme:

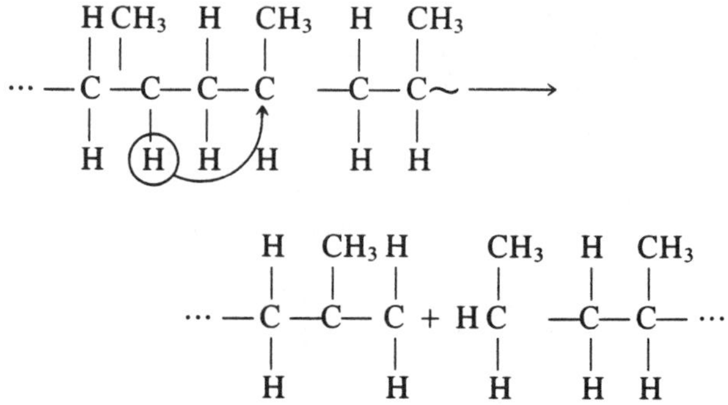

The other pattern of breaking the carbon–carbon bonds which results in the formation of free radicals is observed to much lesser degree and is responsible for an insignificant propylene content upon thermal destruction.

Thus, suppression of the radical-chain thermal destruction reaction of olefins necessitates an addition of substances having the ability to react with active macroradicals and to yield inactive or low-reactivity products.

II. STABILIZATION OF POLYOLEFINS BY HYDROXIDES OF ALKALI METALS

At present, high-temperature stabilization of polyolefins is still misunderstood; besides, this problem presents serious difficulties. Stabilization of thermal oxidation and photoinduced destruction with the use of stabilizers in this case is inefficient, since at high temperatures these stabilizers are easily evaporated out of the polyolefin melt and decomposed with the formation of radicals capable of initiating additional kinetic chains of destruction.

It has been only found that some antioxidants and light stabilizers show the ability for partial inhibition of thermal destruction of the polyolefins.

Among the antioxidants one should point out the alkyl phenols. It has been demonstrated that an increase in the length of an alkyl substituent leads to increased compatibility with polyolefin and to decreased volatility of alkyl phenol [14]. The greatest efficiency is exhibited by bisalkyl phenols, with the substituents in the ortho position being associated with a cage effect [15]. Of the nitrogen-containing stabilizers some may be aliphatic, aromatic amines, and tricarboxylic acids $HOOCCH_2N(CH_2COOR)_2$ [16]. The substances containing nitrogen and sulphur (e.g.,

where R^I and R^{II} represent alkyl) are more efficient [17].

Among the sulphur-containing substances, esters of $3,3^1$-thiopropionic acids $S(CH_2CH_2COOR)_2$, where $R=C_4-C_{20}$ [18], are most efficient.

Organic phosphites $POR(OR^I)OR^{II}$, where $R=C_1-C_{30}$ represents aliphatic, cycloaliphatic, or aromatic radical, are also able of inhibiting the thermal destruction of polyolefin [19]. Of light stabilizers, benzophenone derivatives have the ability for inhibiting thermal destruction of polyolefins, too.

where R—H, OH and R^I—H, C_1-C_{18} means alkyl; R^{II}, R^{III} represent aryl, acyl, and alkyl [20].

Phthalic anhydride also shows the ability to inhibit thermal destruction of polyolefins [21]. Among the organometallic compounds may be quoted organotin compounds $R_2Sr(OR^I)_2$, where R_2 means alkyl, aryl, or cycloalkyl; OR^I means alkoxyl, acyl, or $R_2Sn(CH_2COOR_1)_2$, where $R_1=C_1-C_{18}$ means alkyl, allyl, or benzyl; R_2 represents chloro-, mono-, or triorganotin mercaptans [22,23].

Among the polymeric stabilizers may be listed polycondensed polymers based on alkyl phenols, aldehydes, and ketones of the aliphatic series, where $n = 1 - 8$ and R,R^I means alkyl [24], Na, K, Ca phenolates of polycondensed polymers [25], and also products of epichloride with one or more aliphatic amines C_3-C_{30} [26].

Such inorganic compounds as sulphates of heavy metals and calcium [27], sulphites of calcium and potassium, salts of nickel [28] have been reported as thermostabilizers of polyolefins.

It should be noted that the aforementioned few compounds behave as stabilizers of thermal destruction of polyolefins only at temperatures from 200–250°C.

Thermal stabilization of polyolefins has been first demonstrated for low-molecular models–normal structure alkanes [29]. It has been shown that metallic sodium and potassium hydroxide with absorbent birch carbon (ABC) as a carrier are efficient retardants of thermal destruction of n-heptane during a contact time of 12–15 s up to the temperature of 800°C [130]. Olefins and nitrous protoxide, previously reported as inhibitors of the hydrocarbon thermal destruction, are ineffective in this conditions.

Potassium hydroxide with the use of the carrier is effective as an inhibitor of hexadecane destruction up to the temperature of 600°C.

Cracking of n-alkanes is effectively inhibited by metallic sodium and potassium hydroxide with ABC carrier even under pressure and over a long period of time (Table 2).

Metallic sodium practically completely retards crecking of heptane and hexadecane over a period of 120 min at 500°C, and in presence of potassium hydroxide with ABC carrier a fraction of decomposed hydrocarbon amounts to 2.1% in case of heptane and to 3.1% for hexadecane.

High inhibitive efficiency relative to thermal destruction of n-alkanes is displayed by hydrides and amides of alkali metals [33–35].

In case of polyethylene potassium hydroxide with an ABC carrier is effective as a thermostabilizer up to the temperature of 440°C. Regarding the thermostabiliz-

Table 2 Inhibition of *n*-Heptane and Hexadecane Cracking Under Pressure in Autoclave

Inhibitor	Gassing (% by mass)	Amount of compaction product (% by mass)	Hydrocarbon decomposed (% by mass)
Inhibitor-free heptane	45.2	11.5	56.7
Heptane with sodium	2.1	—	2.1
Heptane with potassium on ABC carrier	2.5	0.1	2.6
Hexadecane	49.3	15.6	64.1
Hexadecane with sodium	2.3	—	2.3
Hexadecane with KOH on ABC carrier	2.8	0.3	3.1

Temperature = 500°C; Time = 120 min.

ing efficiency relative to polyethylene, hydroxides of alkali metals form the following series: KOH, NaOH, and LiOH. Thermal destruction of polyolefins is inhibited by potassium hydroxide to a variable degree (Table 3).

The inhibitive effect of potassium hydroxide with an ABC carrier is observed for all polyolefins, but to a variable degree. Thermal destruction of polypropylene is inhibited only slightly, being associated with the mechanism of thermal destruction that is primarily conditioned by hydrogen transfer. A little portion of polypropylene undergoes thermal destruction following the radical pattern (not inhibited), while destruction of polyethylene proceeds according to the radical mechanism with breaking of the carbon–carbon chain under the laws of chance, and the inhibitive effect is significant.

Hydroxides of alkali metals are effective as inhibitors of thermal destruction of polyolefins even without the carrier, yet at lower temperatures (Table 4).

Thermal destruction of low-pressure polyethylene with molecular weight of 34,800 and of high-pressure polyethylene is completely retarded by potassium hydroxide. The molecular weight of high-molecular polyethylene decreases by a factor of 1.8, and without an

inhibitor, by a factor of 7.5 with less efficient inhibition of thermal destruction.

At 300°C and in the presence of KOH an increase in the molecular weight is observed, i.e., the reaction of macropolymerization is realized [38,39]. Potassium hydroxide is effectively inhibiting thermal destruction of polyethylene at temperatures from 350–375°C. The per cent change in molecular weight is half or one-third as high as that without the use of an inhibitor. At 400°C the efficiency of inhibition is insignificant. Potassium hydroxide with an ABC carrier is effective up to the temperature of 440°C due to the increased contact surface of the inhibitor with macroradicals.

Fiber glass provides effective inhibition of polyethylene thermal destruction up to 400°C. The inhibitive efficiency increases with increased content of sodium oxide from 0.7–16% (Table 5).

A similar situation is observed when studying the effect of temperature on inhibition of thermal destruction of polyethylene by fiber glass of varying composition (Table 6). The molecular weight of polyethylene is practically unchanged when exposed over a period of 6 hours at 350°C with 30% of fiber glass containing 16%

Table 3 Inhibition of Polyolefin Thermal Destruction by KOH with ABC Carrier in Nitrogen Atmosphere

Polyolefin	Molecular weight prior to destruction	Molecular weight after destruction	
		Free of inhibitor	KOH + ABC carrier
Low-pressure polyethylene	34,800	400	18,700
	158,000	1050	9500
High-pressure polyethylene	37,500	2100	15,000
Medium-pressure polyethylene	63,000	5750	35,300
Polypropylene	366,400	900	2300

Temperature = 420°C; time = 120 min; polyolefin/inhibitor = 1:1.

Table 4 Inhibition of Polyolefin Thermal Destruction by Potassium Hydroxide

	Molecular weight		
		After destruction	
Polyolefin	Before destruction	Without inhibitor	With inhibitor
Low-pressure polyethylene	15,800	21,000	86,200
	34,000	12,750	37,000
High-pressure polyethylene	37,500	13,600	37,600
Polypropylene	366,400	161,800	253,400

Temperature = 350°C; time = 120 min; POH concentration = 5%.

of sodium oxide. At the same time, without an inhibitor the molecular mass decreases by a factor of 7. When used as a filler, alkali fiber glass opens the way for the production of fiber glass plastics based on normal and cured polyethylene of enhanced thermal stability [40].

Hydroxides of alkali metals and alkali metals provide for inhibition of polyolefin thermal destruction following the radical-chain pattern. These substances fail to inhibit thermal depolymerization of polystyrene that proceeds according to the molecular mechanism and have little inhibitive effect on thermal destruction of polypropylene following the hydrogen-transfer pattern.

Under the influence of thermal motion and on endothermic electron transition from the OH^- ion to the Me^+ ion in alkali metal hydroxides the formation of the Me. and OH. radicals takes place. As a result, free va-

Table 5 Inhibition of Polyethylene Thermal Destruction by Filler—Fiberglass of Varying Alkalinity

Na_2O content (% by mass) 1	Filler content (% by mass) 2	Time (h) 3	Molecular weight of polyethylene after destruction 4
0.7	10	2	27,500
		4	24,000
		6	22,000
	30	2	34,500
		4	29,000
		6	28,000
11.5	10	2	28,000
		4	25,000
		6	23,500
	30	2	33,000
		4	32,000
		6	30,500
16	10	2	29,500
		4	26,000
		6	24,500
	30	2	34,000
		4	33,500
		6	32,000
Without filler		2	12,000
		4	10,000
		6	5000

Vacuum temperature = 350°C; Polyethylene molecular weight = 35,000.

Table 6　Temperature Effect on Thermal Destruction of Polyethylene Stabilized by Fiberglass of Varying Alkalinity

Na$_2$O content (% by mass) 1	Filler content (% by mass) 2	Temperature (°C) 3	Molecular weight of polyethylene after destruction 4
0.7	10	300	33,000
		350	32,000
		380	15,000
		400	8000
	30	300	33,000
		350	33,000
		380	23,000
		400	14,000
11.5	10	300	33,500
		350	33,000
		380	25,000
		400	9000
	30	300	34,000
		350	33,500
		380	28,000
		400	15,000
16.0	10	300	34,000
		350	33,500
		380	26,000
		400	12,000
	30	300	34,000
		350	33,500
		380	28,000
		400	16,000
Without filler		300	23,000
		350	12,000
		380	8000
		400	3000

Time = 120 min; Vacuum, molecular weight of polyethylene = 35,000.

lences, the concentration of which is determined by the equilibrium:

$$Me^+ + OH^- \rightleftarrows OH\cdot + Me\cdot$$

appear at the surface of an inhibitor.

Different efficiency of alkali metal hydroxides seems to be associated with differing ionization potentials.

According to the ionization potential and electron-transfer work, alkali metals form the following series: Li > Na > K, and their hydroxides are arranged in the sequence KOH > NaOH > LiOH as to their inhibitive efficiency relative to thermal destruction of polyolefins. And the efficiency of alkali metals can be represented by the sequence Na > K > Li. This seems to be due to the fact that the electron-transfer work of sodium is somewhat higher than that of potassium, whereas the atomic dimensions of sodium are much less than those of potassium, and sodium radical is more effective accepting the macroradicals formed upon thermal destruction of polyethylene as compared to the potassium radical.

The inhibitive efficiency of alkali metal hydroxides increases with increased branching of polyethylene. This is confirmed by more pronounced effect of these hydroxides diminishing the yield of propane and propylene than in case of ethane and ethylene. The decreased yield of propane and propylene is also conditioned by more efficient inhibition of the macroradical isomerization stage by alkali metal hydroxides. Upon thermal destruction of polyethylene with the use of inhibitors the number of vinylidene (882 cm^{-1}) and vinylene (965 cm^{-1}) bonds is drastically growing, while the number of vinyl bonds (909 and 993 cm^{-1}) decreases sharply. The overall nonsaturation of destructed polyethylene with the use of an inhibitor is increased by a factor of 1.5 as compared to the initial one, but branching is greatly reduced.

At the first stage of polyethylene thermal destruction the metallizing of polyethylene macroradical by the metal radical takes place.

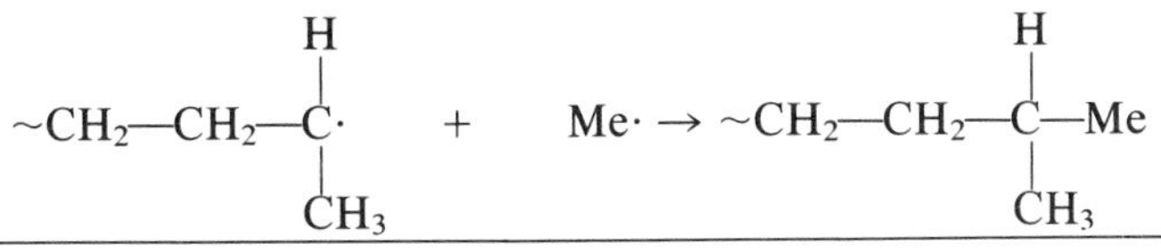

$$\sim CH_2-CH_2-\underset{\underset{CH_3}{|}}{\overset{\overset{H}{|}}{C}}\cdot \quad + \quad Me\cdot \rightarrow \sim CH_2-CH_2-\underset{\underset{CH_3}{|}}{\overset{\overset{H}{|}}{C}}-Me$$

One would think that thermal destruction of polyethylene should be inhibited by hydroxides of alkali metals according to the following scheme, as with phenols:

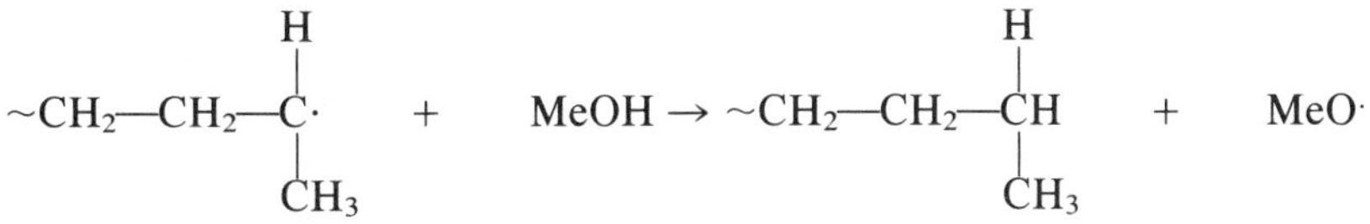

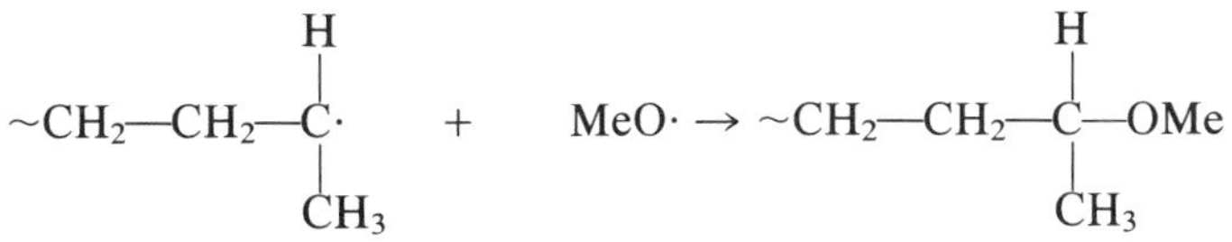

$$\sim CH_2-CH_2-\underset{\underset{CH_3}{|}}{\overset{\overset{H}{|}}{C}}\cdot \quad + \quad MeOH \rightarrow \sim CH_2-CH_2-\underset{\underset{CH_3}{|}}{\overset{\overset{H}{|}}{C}H} \quad + \quad MeO\cdot$$

$$\sim CH_2-CH_2-\underset{\underset{CH_3}{|}}{\overset{\overset{H}{|}}{C}}\cdot \quad + \quad MeO\cdot \rightarrow \sim CH_2-CH_2-\underset{\underset{CH_3}{|}}{\overset{\overset{H}{|}}{C}}-OMe$$

However, this mechanism is at variance with the following factors: (1) high bonding strength of OH-Me (90 kcal/mol) that is much greater than that of OH in phenol molecules thus eliminating the possibility of breaking at temperatures from 350 to 400°C; (2) the absence of the absorption bonds 700–900 cm^{-1} characteristic of the bonding C—O—Me in infrared spectra of polyethylene after destruction; and (3) the absence of EPR-signal in the products after destruction.

Metallizing is supported by the fact that thermal destruction of polyethylene is inhibited by alkali metals.

At the second stage the decomposition of metal alkyl takes place. Metal alkyl is liable to decompose into metal alkyl and olefin causing the increased saturation of polyethylene macromolecules:

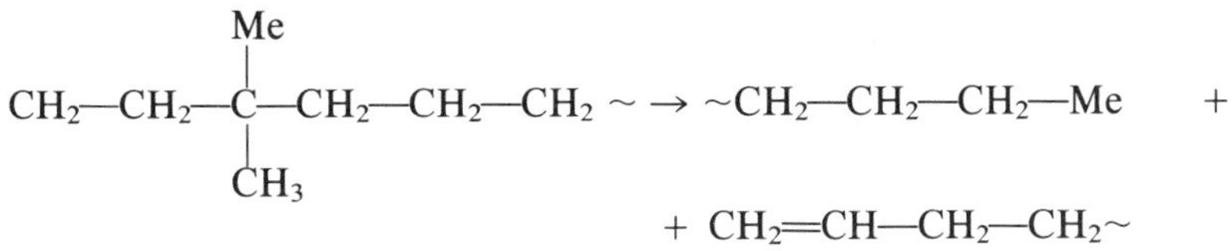

$$CH_2-CH_2-\underset{\underset{CH_3}{|}}{\overset{\overset{Me}{|}}{C}}-CH_2-CH_2-CH_2 \sim \rightarrow \sim CH_2-CH_2-CH_2-Me \quad +$$

$$+ \quad CH_2=CH-CH_2-CH_2\sim$$

Metal alkyl is liable to decompose with breaking of the carbon–carbon bond that is in β-position relative to the C—Me bond:

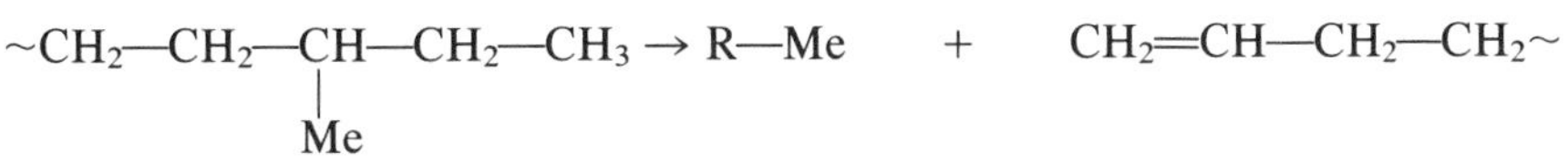

$$\sim CH_2-CH_2-\underset{\underset{Me}{|}}{CH}-CH_2-CH_3 \rightarrow R-Me \quad + \quad CH_2=CH-CH_2-CH_2\sim$$

where

$$R = C_1-C_3.$$

This reaction leads to drastic decrease in the number of methyl groups in polyethylene and to an increase in the number of double bonds.

The resultant metal alkyl is rapidly decomposed at temperatures from 200–250°C with the formation of metal, alkanes, and alkenes C_2-C_3.

Metal alkyl is liable to react with nonsaturated ends of polyethylene molecules:

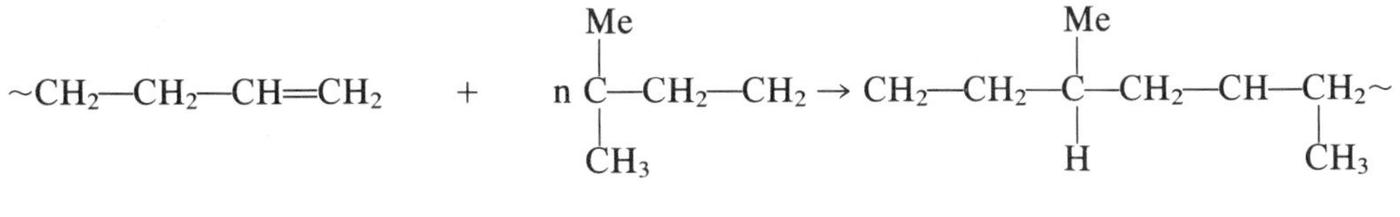

$$\sim CH_2-CH_2-CH=CH_2 \quad + \quad n\,\underset{\underset{CH_3}{|}}{\overset{\overset{Me}{|}}{C}}-CH_2-CH_2 \rightarrow CH_2-CH_2-\underset{\underset{H}{|}}{\overset{\overset{Me}{|}}{C}}-CH_2-CH-CH_2\sim$$

As this takes place, scission of polymeric chains is hindered and the molecular dimensions are even growing up to the temperature of 300°C.

Metal alkyl reacts with nonsaturated ends of macromolecules following the pattern of substitution reaction [and causing a decrease in the number of methyl groups]:

$$\sim CH_2-CH_2-CH=\underset{\underset{CH_3}{|}}{C} \quad + \quad Me-\underset{\underset{CH_3}{|}}{C}-CH_2-CH_2\sim \rightarrow CH_2-C=CH \quad + \quad CH_3-CH_2-CH_2\sim$$
$$\qquad\qquad\qquad\qquad\qquad\qquad\qquad\qquad\qquad\qquad\qquad\qquad H \quad CH_2Me$$

III. INHIBITION OF THERMAL DESTRUCTION BY BORIC ACID ESTERS

Boric acid esters have aroused considerable interest because they are stable up to the temperature of 450°C, display low volatility and good compatibility without migration from olefin.

Boric acid esters provide for thermal stabilization of low-pressure polyethylene to a variable degree (Table 7). The difference in efficiency derives from the nature of polyester. Boric acid esters of aliphatic diols and triols are less efficient than the aromatic ones. Among polyesters of aromatic diols and triols, polyesters of boric acid and pyrocatechol exhibit the highest efficiency. Boric acid polyesters provide inhibition of polyethylene thermal destruction following the radical-chain mechanism, are unsuitable for inhibition of polystyrene depolymerization following the molecular pattern and have little effect as inhibitors of polypropylene thermal destruction following the hydrogen-transfer mechanism.

The inhibitive efficiency of boric acid polyesters differs greatly. The highest efficiency is exhibited by polyesters of boric acid, aromatic diols and triols. This derives from the fact that in this case the radicals are accepted not only by boron, but also by the aromatic nucleus. Among the aromatic polyesters, most efficient is ester of boric acid and pyrocatechin due to the Frank–Rabinovich cage effect. The efficiency of inhibition in case of polyethylene is cymbatically growing with increase in branching, being associated with growing importance of the disproportionation and macroradical-decomposition reactions as compared to the chain transfer reactions, since branching presents hindrance to the detachment and transfer of hydrogen to the macroradical. In presence of polypyrocatechin borate the branching of polyethylene after destruction is greatly extended, whereas nonsaturation is growing only slightly. Interaction between polypyrocatechin borate and macroradicals of polyethylene gives inhibiting radicals that have been detected by the EPR method. After heating under the same conditions, but without an inhibitor the samples of polyethylene revealed no EPR-signal.

In the process of inhibition polypyrocatechin borate interacts with polyethylene macroradicals to form the B—O—C bonds. This is confirmed by the fact that the absorption spectrum of polyethylene inhibited with polypyrocatechin borate revealed the bands in the region of 1350 cm^{-1} characteristic for the B—O—C bond. There is no such a band in the spectrum of pure polypyrocatechin borate after heating under the same conditions. Chemical analysis of boron in polyethylene provides support for the IR-spectroscopy data concerning the presence of chemically bonded boron in polyethylene after destruction.

Inhibition of polyethylene thermal destruction by polypyrocatechin borate could be represented as follows. The initial molecular-chain scission of branched

Table 7 Inhibition of Polyethylene Thermal Destruction by Boric Acid Polyesters

Polyester of	Molecular weight before destruction	Molecular weight after destruction	
		Without inhibition	With inhibition
Ethylene glycol	158,000	1050	9200
Glycerol	158,000	1050	15,100
Pyrogallol	158,000	1050	20,500
Hydroquinone	158,000	1050	16,800
Resorcinol	158,000	1050	14,200
Pyrocatechol	34,800	2650	24,000
Pyrocatechol	31,500	2100	15,000
Polypropylene	366,000	6300	22,200

Temperature = 420°C; Time = 2 hours; Stabilizer concentration = 5%.

polyethylene leads to the formation of radicals R_1 and R_2:

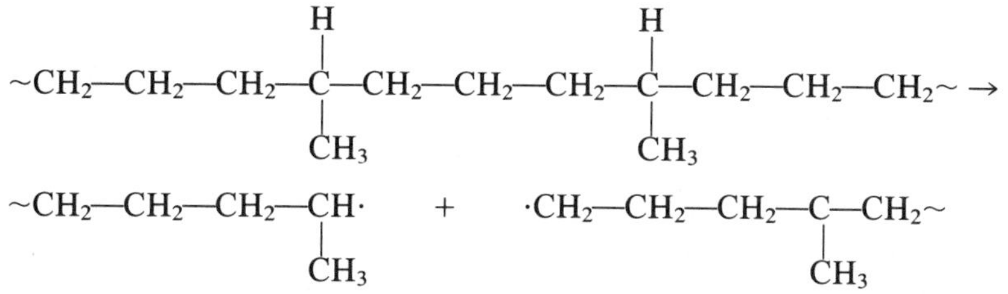

$$\sim CH_2-CH_2-CH_2-\underset{\underset{CH_3}{|}}{\overset{\overset{H}{|}}{C}}-CH_2-CH_2-CH_2-\underset{\underset{CH_3}{|}}{\overset{\overset{H}{|}}{C}}-CH_2-CH_2-CH_2\sim \;\rightarrow$$

$$\sim CH_2-CH_2-CH_2-\underset{\underset{CH_3}{|}}{CH\cdot} \;+\; \cdot CH_2-CH_2-CH_2-\underset{\underset{CH_3}{|}}{C}-CH_2\sim$$

Isomerization, disproportionation, and decomposition reactions of the radical R_1 yield nonsaturated end groups $CH_2{=}CH-CH_2$ and result in the formation of ethane and ethylene. Isomerization and decomposition

of the radical R_2 proceed with the formation of propylene and new end radical that reacts with the macromolecule of polyethylene according to the following scheme:

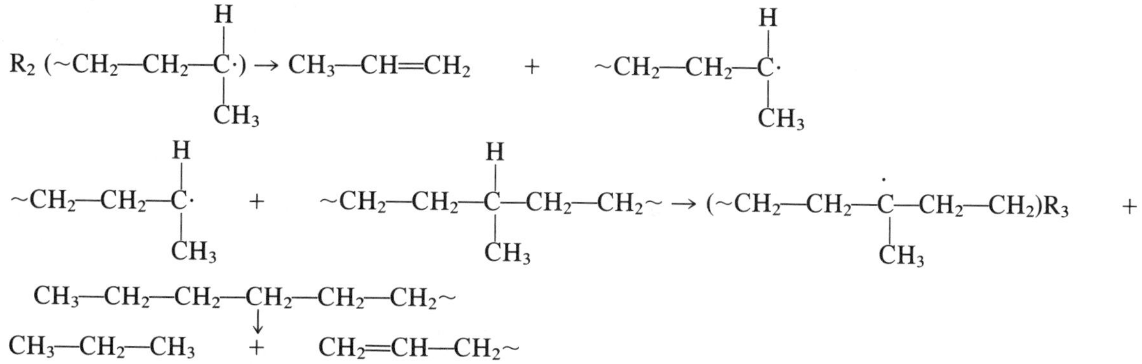

$$R_2\,(\sim CH_2-CH_2-\underset{\underset{CH_3}{|}}{\overset{\overset{H}{|}}{C}\cdot}) \rightarrow CH_3-CH{=}CH_2 \;+\; \sim CH_2-CH_2-\underset{\underset{CH_3}{|}}{\overset{\overset{H}{|}}{C}\cdot}$$

$$\sim CH_2-CH_2-\underset{\underset{CH_3}{|}}{\overset{\overset{H}{|}}{C}\cdot} \;+\; \sim CH_2-CH_2-\underset{\underset{CH_3}{|}}{\overset{\overset{H}{|}}{C}}-CH_2-CH_2\sim \rightarrow (\sim CH_2-CH_2-\underset{\underset{CH_3}{|}}{\overset{\overset{\cdot}{}}{C}}-CH_2-CH_2)R_3 \;+\;$$

$$CH_3-CH_2-CH_2-CH_2-CH_2-CH_2\sim$$
$$\downarrow$$
$$CH_3-CH_2-CH_3 \;+\; CH_2{=}CH-CH_2\sim$$

As a result, the central radical R_3 is formed, and the fragment with the end methyl group breaks down into propane and a new fragment with the end vinyl group.

Polypyrocatechin borate has a higher inhibitive efficiency relative to the reactions of the radical R_2 than to that of the radical R_1, since the latter exerts a more severe decrease in the yield of propane and propylene than in the yield of ethane and ethylene, and favors an insignificant growth of polyethylene nonsaturation in the pro-

cess of destruction. However, this offers no explanation for the diminished number of methyl groups in polyethylene as compared to polyethylene without the use of inhibition.

It is believed that the recombination product is liable (due to the increased mobility of the hydrogen atom following the carbon atom in β-position relative to the C—O—C bond) to isomerization and decomposition along the ester bond:

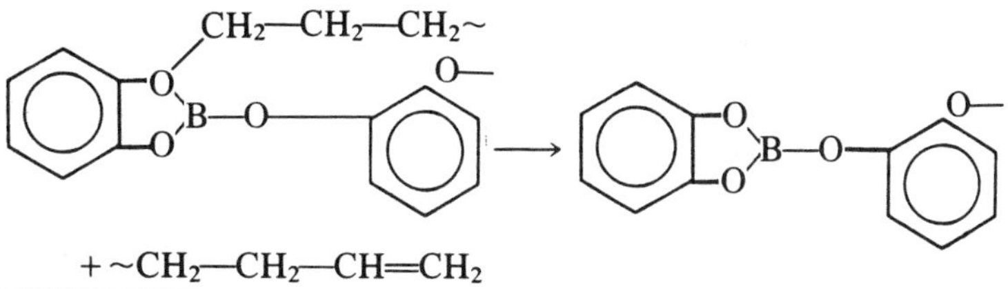

$$+ \sim CH_2-CH_2-CH{=}CH_2$$

As this takes place, the inhibitor molecule is regenerated, and side methyl group is substituted by end vinyl [41].

IV. STABILIZATION OF CURED POLYETHYLENE

Polyethylene cured by the chemical and radiation–chemistry methods undergoes thermal destruction upon heating as in normal polyethylene. Thermostabiliz-

ers should possess good thermal stability up to 300–350°C being no bar to chemical vulcanization, i.e., being inert to the curing agent. Besides, these thermostabilizers should have good compatibility, low volatility without sweating out from the polyethylene mass. Currently used thermostabilizers of polyethylene (alkyl phenols, aromatic amines and diamines, sulphur- and phosphorus-containing stabilizers) exhibit thermal stability up to the temperatures from 200–250°C, decompose at higher temperatures and contribute to the initiation of thermal destruction. They react with a curing agent (most commonly peroxide) causing reciprocal consumption of peroxide and thermostabilizer. To take an example, the use of β-dinaphtyl n-phenylene diamine (diafen NN) as a stabilizer in compositions with dicumyl peroxide involves a decrease in gel fraction by 15–40% depending on the concentration of dicumyl peroxide [42].

At present, very few compounds are used as thermostabilizers for cured polyethylene. Among them may be listed I,3-dihydro-2,2,4-trimethylquinoline; β-dinaphtyl, n-phenylenediamine, zinc mercaptobenzimidazole [43–45].

Polyamine disulphides do not inhibit peroxide vulcanization of polyethylene, are stable in air up to 300–350°C, exhibit good compatibility and show no sweating out from the polyethylene mass. Table 8 gives the comparison between the efficiency of polyamine disulphides as thermostabilizers of cured polyethylene.

A series of polyamine disulphides (polyaniline disulphide, polyamine disulphide, and polyparaphenylenediamine disulphide) represent effective thermostabilizers of cured polyethylene, and provide a decrease in gel fraction 2.5–3 times as large as that in case of inhibited thermal destruction. Stabilizers of normal polyethylene (Neozone "D", Santonox "R") are inefficient as stabilizers of cured polyethylene, these substances decompose and even initiate thermal destruction of cured polyethylene.

Polyamine disulphides are effective thermostabilizers of cured polyethylene up to 400°C. In presence of polyamine disulphides a decrease in gel fraction is one half as large as that of nonstabilized cured polyethylene over the temperature range from 350–380°C [46].

The highest efficiency is exhibited by polyaniline disulphide in presence of which a decrease in gel fraction amounts to 7% at 350°C, 19% at 380°C, and 60% at 400°C, whereas without an inhibitor the decrease in gel fraction becomes 24% at 350°C, 41% at 380°C, and 92% at 400°C. [47].

An investigation into the effect of the concentration of polyaniline disulphide on inhibition of thermal destruction in case of cured polyethylene has demonstrated that polyaniline disulphide is efficient even at the concentration of 0.25%. An increase in the concentration over the range 0.25–1.0% results in the increased efficiency, while further increase in the concentration leads to a slight drop in inhibition.

Polyamine disulphides as inhibitors of thermal destruction of cured polyethylene are effective over a long period of time.

Upon 10-h exposure, a decrease in gel fraction amounts to 40% without the use of inhibition and to 8% in presence of polyaniline disulphide. High efficiency of polyaniline disulphide has been confirmed by the investigation of strength properties for cured polyethylene. Tensile strength of cured polyethylene in vacuum at 300°C decreases insignificantly, and after a 10-h exposure it amounts to 192 kg/cm². Without inhibition tensile strength decreases depending on the exposure time, and over a period of 10 hours it comes to only 58 kg/cm². With the use of polyaniline disulphide a specific elongation at rapture decreases slowly as a function of time, and after 10 hour exposure this parameter amounts to 90% of the initial value. Without an inhibitor a specific elongation decreases sharply with increase in time, amounting to only 40% of the initial value after 10-h exposure. Polyaniline disulphide is an efficient inhibitor of heat aging in case of cured polyethylene (Table 9).

A drop in gel fraction of nonstabilized cured polyethylene amounts to 50% after 25-h exposure, 75% after 50 hours and after a 75-h exposure the complete fall is observed. At the same time, a decrease in gel fraction in presence of polyaniline disulphide is observed only after a 50-h exposure and comes to only 2%, whereas

Table 8 Inhibition of Thermal Destruction of Low-Density Cured Polyethylene in Vacuum (10^{-3} torr)

	Gel fraction (%)		Decrease in gel fraction
	Before destruction	After destruction	
Polyaniline disulphide	84	68	19
Polyparaphenyldiamine disulphide	82	60	27
Polyamine disulphide	64	50	22
Polycarbamide disulphide	70	44	39
Polythiocarbamide disulphide	59	47	24
Polyethylenediamine disulphide	70	52	26
Without stabilizer	87	43	41
Neozone "D"	78	49	62
Santonox "R"	75	39	49

Temperature = 380°C; Inhibitor concentration = 1%; Time = 120 min.

Table 9 Heat Aging of Cured Polyethylene in Air

Time (h)	Decrease in gel fraction (%)	
	Without inhibitor	With inhibitor
25	53	0
50	76	2
75	100	5
100	—	10

Temperature = 150°C; Polyaniline disulphide concentration = 0.8% by mass.

after 100 hours a decrease in gel fraction amounts to 10% [48].

High thermostabilizing efficiency of polyamine disulphides relative to chemically cross-linked polyethylene is conditioned by the ability to accept macroradicals at the disulphide bridge and imine group. Besides, the presence of paramagnetic centers causes the adherence of macroradicals providing for an extra stabilizing effect [49].

REFERENCES

1. A. G. Sirota, *Modification of Structure and Properties of Polyolefins,* Chemistry Publishers, Leningrad, p. 3 (1994).
2. A. K. Wardenburg, *Plastics in Electrotechnical Industry,* Gosenergoizdat Publishers (1957).
3. V. S. Shifrina and N. N. Samosatsky, *High-Pressure Polyethylene,* Goschimizdat Publishers (1958).
4. S. S. Mindlin and N. N. Samosatsky, *Production of Polyethylene Products by the Extrusion Method,* Goschimizdat Publishers (1959).
5. *Low-Pressure Polyethylene* (N. M. Egorov, ed.), Goschimizdat Publishers (1958).
6. F. Klema, *Mitt. Chem. Forsch., 12*: 159 (1958).
7. H. Stepanek, *Plastverarbeiter, 10,* 137 (1959).
8. Madorsky, Straus, Thompson, and Williamson, *J. Polymer. Sci.,* 4: 639 (1949).
9. Hopff *Kunststoffe 42*: 2; 423–426 (1952).
10. H. U. Ellinek, *J. Polymer Sci.,* 4: 850 (1948).
11. R. Simka, L. A. Wall, and R. G. Blatz, *J. Polymer Sci.,* 5: 615 (1959).
12. L. A. Wall and S. Straus, *J. Polymer Sci.,* 44: 113 (1960).
13. N. N. Semenov, *On Some Problems of Chemical Kinetics and Reactivity,* USSR Ac. Sci. Publishers (1959).
14. I. Feugt, *Stabilization of Synthetic Polymers Against Light and Heat Effect,* Chemistry Publishers, Leningrad, p. 26–29 (1972).
15. I. Feugt, *Stabilization of Synthetic Polymers Against Light and Heat Effect,* Chemistry Publishers, Leningrad, p. 184–185 (1972).
16. A. G. Farbenfabriken Bayer, *Neth. Appl., 6*: 5, 15,965 (1966).
17. French patent 1350966, 23, XII (1969).
18. *Badische Anilin Soda-Fabrik A. G. To Adolf-Hrubeschanol,* Hans Moeller Belg., 617, 190, 5, XI (1962).
19. E. Glazence and F. Holstrup, Eastmann Kodak Co., Brit., 972989, 21, X (1964).
20. *Eastman Kodak Co.,* (C. E. Fholstrup, ed.), France I, 366, 533, 10 VII (1964).
21. R. Renneth, Hills and R. G. Walton, Phillips Petroleum Co., U.S. 3, 227, 676 (1969).
22. *Azien de Colori National Affini of CNASPA* (G. Vigailo, ed.), Ital., 630562, 18 (1961).
23. *Badische Anilin Soda-Fabrik A. G.* (Hans Burger and Beinhurd Razinalla, eds.), German I, 159, 645, 19 XII (1965).
24. *Badische Anilin Soda-Fabrik A.G. Neth. Appl. 6,* 413, 754, 31 V (1969).
25. *Montecatini Societa Generall per Industria Mineraria Chemica Ital.* 642, 042, 5 VII (1962).
26. *Hitto Chemical Industry Co. Ltd.* (F. Scki, J. Kanakarni, and F. Katsamura, eds.), Japan 645, 5166, 19 IV (1966).
27. British patent 833853 (1960).
28. *Shell Internationale Research* (T. H. Bouthle and C. C. Gosselik, eds.), German, I, 203, 583, 9 XII (1965).
29. A. V. Topchiev, Ya. M. Paushkin, A. V. Nepryakhina, P. G. Anan'ev, and N. N. Dmitrievsky, *Reports of USSR Ac. Sci., 133*: 134 (1960).
30. A. V. Topchiev, Ya. M. Paushkin, A. V. Nepryakhina, P. G. Anan'ev, and N. N. Dmitrievsky, *Izvestiya USSR Ac. Sci., Chemical Series, 10*: 1838 (1960).
31. Ya. M. Paushkin, Yu. P. Losev, and P. G. Anan'ev, *J. Neftekhimiya, IX*: 60–62 (1969).
32. Ya. M. Paushkin, Yu. P. Losev, and P. G. Anan'ev, *Izvestiya USSR Ac. Sci., Chemical Series, 6*: 1276–1278 (1969).
33. Yu. P. Losev, Ya. M., and Paushkin, V. M. *Khoruzhy Reports USSR Ac. Sci., XVIII,* 1014–1015 (1974).
34. Yu. P. Losev, Ya. M. Paushkin, V. M. Khoruzhy, and D. I. Metelitsa, *Reports USSR Ac. Sci., 218*: 390–392 (1974).
35. Yu. P. Losev, Ya. M. Paushkin, V. M. Khoruzhy, and G. V. Dedovich, *Reports USSR Ac. Sci., 218*: 1365–1367 (1974).
36. Yu. P. Losev, and Ya. M. Paushkin, *Reports Belarussian Ac. Sci. XII,* 522–525 (1968).
37. Ya. M. Paushkin, and Yu. P. Losev, *J. Polymer Sci., Part C,* 501–511 (1968).
38. Ya. M. Paushkin, and Yu. P. Losev, Proceedings of Intern. Symp. on Macromolecular Chemistry, Brussels, 1967, 121–124, (1968).
39. Ya. M. Paushkin, and Yu. P. Losev, *New Petrochemical Products and Processes,* Central Research Institute for Petrochemical Technology, Moscow, pp. 50–54 (1971).
40. Ya. M. Paushkin, Yu. P. Losev, and M. E. Elyamberg, *J. High-Molecular Compounds, IX,* 362–365 (1967).
41. Ya. M. Paushkin, Yu. P. Losev, E. I. Karakozova, and V. N. Isakovich, *High-Molecular Compounds, A, XV,* 2496–2500 (1973).
42. E. I. Evdokimov, I. V. Konoval, Yu. I. Firsov, E. A. Vasilenko, A. S. Glebko, V. N. Chkalova, T. L. Zinevich, and A. N. Kopchenkov, *Plastmassy* (Plastics), 29–31 (1972).
43. U.S. patent no. 4028332.
44. Author's certificate no. 572472 (USSR).
45. Author's certificate no. 606866 (USSR).
46. Ya. M. Paushkin, Yu. P. Losev, and D. M. Bril', *New Petrochemical Products and Processes,* Central Research Institute for Petrochemical Technology, Moscow, pp 38–45 (1971).

47. Yu. P. Losev, Ya. M. Paushkin, and V. N. Isakovich, *J. High-Molecular Compounds, A,* 2502–2505 (1974).
48. Yu. P. Losev, Ya. M. Paushkin, V. N. Maksimenko, E. P. Natal'ina, and V. N. Isakovich, *Reports Belarussian Ac. Sci., v. XVI,* 916–918 (1972).
49. Yu. P. Losev, V. I. Shonorov, S. P. Baranov, T. S., Martynenko, E. S. Savostenko, V. N. Isakovich, and V. I. Fursikov, USSR Author's Certificate, no. 637411, *Inventor's Bulletin,* no. 46 (1978).
50. Yu. P. Losev, Ya. M. Paushkin, and V. N. Isakovich, Abstracts at Intern. Symp. on Destruction and Stabilization of Polymers, 257–258 (1974).

7

Poly(malic Acid) from Natural Sources

Eggehard Holler
University of Regensburg, Regensburg, Germany

I. INTRODUCTION

β-Poly(L-malate) is a "young" polymer. Because of some of its chemistry it is occasionally counted among the poly(hydroxyalkanoates). However, it owns a wealth of properties and structural diversity not shared by any of these polymers.

Poly(L-malic acid) denotes a family of polyesters derived from L-malic acid as the building unit. By chemical synthesis, three kinds of poly(L-malic acid) have been obtained, depending on the molecular position of the ester bond: the α-type(I) [1], the β-type(II) [2], and the α,β-mixed-type(III) [3].

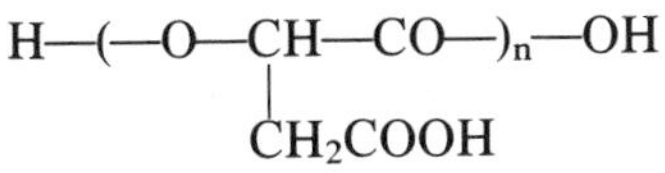

$$H—(—O—CH—CO—)_n—OH \qquad (I)$$
$$\qquad\;\; CH_2COOH$$

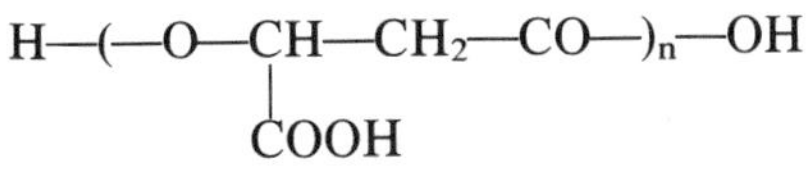

$$H—(—O—CH—CH_2—CO—)_n—OH \qquad (II)$$
$$\qquad\;\; COOH$$

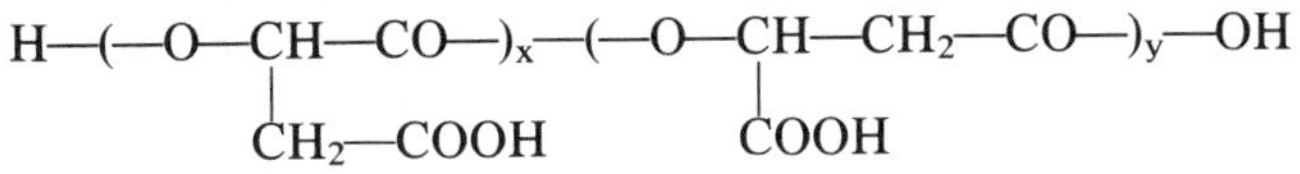

$$H—(—O—CH—CO—)_x—(—O—CH—CH_2—CO—)_y—OH \qquad (III)$$
$$\qquad\;\; CH_2—COOH \qquad\quad COOH$$

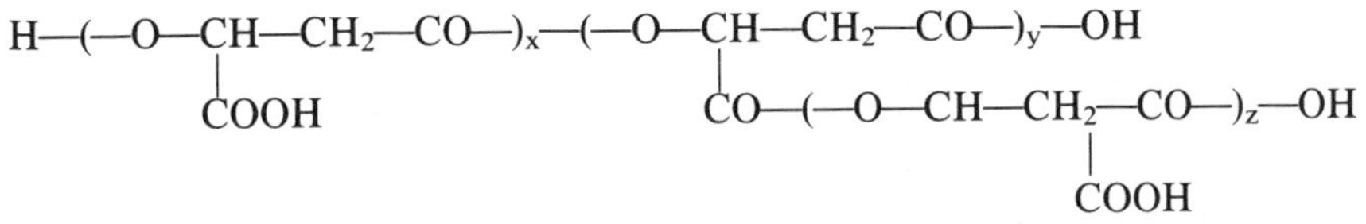

$$H—(—O—CH—CH_2—CO—)_x—(—O—CH—CH_2—CO—)_y—OH \qquad (IV)$$
$$\qquad\;\; COOH \qquad\qquad\quad CO—(—O—CH—CH_2—CO—)_z—OH$$
$$\qquad\qquad\qquad\qquad\qquad\qquad\qquad\quad COOH$$

In recent years poly(malic acid) has been discovered in several fungal strains, notably in *Physarum polycephalum* [4] and in *Aureobasidium* sp. [5,6], and have been analysed as being of the β-type(II) either linear (*P. polycephalum*) or branched (type IV) (*Aureobasidium* sp.).

Poly(malic acid) is of pharmaceutical interest because its chemical derivatives may harbor both tissue-specific homing molecules and therapeutic effectors to be used for tissue (tumor) targeting in chemotherapy [2]. Because of its efficient production by fermentation, its biodegradability and nontoxicity, it is also considered as raw material in the industrial production of detergents, glues, and plastic materials.

The following chapters will be devoted to the production of β-poly(L-malic acid) or its salt by fermentation, its isolation, and physico-chemical characterization. The biosynthesis, degradation, and presumed physiological role will be also considered.

II. PURIFICATION OF β-POLY(L-MALATE) FROM FUNGI

A. Fungal Producers of β-Poly(L-Malate)

The first organism reported to produce poly(L-malic acid) was *Penicillium cyclopium* [7]. An amount of 2.6-

g freeze-dried poly(L-malic acid) from 8.6 liters of crude fungus extract was purified by repeated anion exchange chromatography and 66% acetone (pH 4) precipitation followed by size exclusion chromatography on Sephadex G50 and concersion of the salt to the polymer acid with Amberlite IR-120. (Rohm & Hans). Based on the finding of L-malic acid after hydrolysis (hydrochloric acid) and other criteria, the polymer was assumed to be poly(L-malic acid).

In the year 1989, plasmodia of *P. polycephalum* were shown to synthesize β-poly(L-malate), the identity of which was confirmed by a variety of analytical methods also involving NMR techniques [4,8]. The polymer is isolated from the liquid growth medium of 2-d old cultures (grown as "microplasmodia" during shaking). The polymer can be collected in several batches adsorbed to DEAE-cellulose. The combined batches were subjected to purification by repeated chromatography on DEAE-cellulose, alcohol precipitation, and size exclusion chromatography on Sephadex G25 fine and lyophylization. Average yields were 100 mg of polymer in 100 ml of culture medium. β-Poly(L-malate) can also be prepared from the plasmodia containing at maximum 80 mg of the polymer in 100 g of cells, but this purification was tedious involving acid precipitation (perchloric acid pH 2) of protein in the first step. Highly purified β-poly(L-malic acid) devoid of contaminating salts was prepared by passage of the purified polymer salt over Amberlite IR-120 (H$^+$ form), lyophylization, solubilization in acetone, removal of debris, and evaporation of the solvent. The final polymer was obtained in colorless translucent sheets. A detailed description of the purification is given in Table 1.

During a large-scale screening (Rathberger, Molitoris and Holler, unpublished results), 232 different fungus strains were tested for the production of β-poly(L-malate) among them 53 species of marine *Ascomycetes*, 6 species of yeast-like marine *Ascomycetes*, 9 species of terrestrial *Ascomycetes*, 13 species of marine *Basidomycetes*, 7 species of yeast-like marine *Basidiomycetes*, 44 species of terrestrial *Basidiomycetes*, 54 species of marine *Deuteromycetes*, 15 species of yeast-like marine *Deuteromycetes*, 19 species of terrestrial *Deuteromycetes*, *Dictyostelium discoideum*, *Dictyostelium mucoroides*, *Physaum polycephalum*, and 7 species of terrestrial *Phycomycetes* (all cultures Glc-NS-Y medium, see Table 2 footnotes). The best producers were *Aureobasidium* species of the terrestrial *Deuteromycetes* (4–9 g/liter).

The strongest known producer of β-poly(L-malic acid) has been identified as *Aureobasidium* sp. providing 61 g of polymer from 1 liter of culture medium [5,6]. β-Poly(L-malate, Ca^{2+}-salt) of the culture broth was first separated from accompanying bulk pullulan by methanol precipitation. The water-redisolved precipitate was converted to the polymer acid by passage over Amberlite IR-120B (H$^+$-form). Thus, the best to-day producers of

β-poly(L-malate) are the *Aureobasidiae* followed by *P. polycephalum* (0.5–1.0 g/liter), several *Cladosporium* species of marine *Deuteromycetes* (0.02–0.35 g/liter), and *Corollospora* species of marine *Ascomycetes* (0–0.02 g/liter). A list of the producers is given in Table 2.

B. Methods of Purification of β-Poly(L-Malate) from Fungal Producers

The protocol for the purification of β-poly(L-malic acid) in Table 1 has been worked out for *P. polycephalum*. It allows an efficient removal of contaminating proteins, nucleic acids, and low-molecular mass impurities from the culture medium or from the cellular extracts of this organism. In order to avoid low-molecular mass polymer fractions, the culture medium has to be harvested early during growth, and the purification has to be limited to only the high-ionic strength eluates after chromatography on DEAE-cellulose and to the high-molecular mass fractions after size exclusion chromatography. The mass-average molecular mass value of 53 kDa in Table 2 is the maximum range obtained but preparations of as low as 12-kDa masses with a high value for the polydispersity (in the range of 3.5–4.0) can be obtained. Extremely low mass values were found for the culture medium of macroplasmodia being of the order of 3–5 kDa. A decrease in molecular mass reflects spontaneous hydrolysis, introducing mostly intrachain cuts, and the hydrolytic action of polymalatase, a specific enzyme degrading the polymer from one of its ends to L-malate (see the following). Spontaneous hydrolysis is promoted by the slightly acid pH 4.25–4.53 of the culture medium during the first 3 days of plasmodial growth.

The purification of β-poly(L-malic acid) from *Aureobasiae* has been reported involving methanol precipitation of the polymer in the form of the Ca^{2+} salt [5]. This is possible because a high concentration of CaCO$_3$ is present in the growth medium. Unfortunately, the polymer acid is not soluble in aceton thus missing an additional purification step. In our hands, purification of β-poly(L-malate) from several *Aureobasidiae* strains was unsatisfactory because of low yields and resisting impurities.

III. BIOSYNTHESIS OF β-POLY(L-MALATE)

A. Correlation of Growth and Synthesis of β-Poly(L-Malate) in *P. polycephalum*

The production of the polymer depends on several factors such as the composition of the growth medium, the time of harvest, and the particular stage of the life-cycle of organism under consideration. For *P. polycephalum* only plasmodia are the producers of β-poly(L-malate); neither amoebae nor spherules (specialized cell forms that can survive unfavorable environmental conditions)

Table 1 Purification of β-Poly(L-Malic Acid) and Its Potassium Salt from the Culture Medium of Plasmodia of *Physarum polycephalum*

1. Grow microplasmodia in 25 × indented[1] 2 liter Erlenmeyer flasks, each inoculated with 10-g plasmodia (24 h old) of *Physarum polycephalum* strain M₃CVII in 500 ml growth medium.[2]
2. Growth at 21°C (27°C for strain M₃CVIII) for 3 d in the dark.
3. Harvest conditioned medium (11 liter) by sieving.[3]
4. Stirr into 700-g DEAE-cellulose[4] during 2 h.
5. Wash DEAE-cellulose with 10 liter of buffer A[5]/0.3 M KCl on a Buechner funnel.[6]
6. Elute the polymer with 5 liter of buffer A/0.7 M KCl.
7. Polymer-containing fraction[7]: adjust with 10 mM buffer A to 0.35 M KCl, readsorb to 300-g DEAE-cellulose.
8. Wash loaded DEAE-cellulose on Buechner funnel with 3 liters of buffer A/0.2 M KCl.
9. Pour column ∅ (8 cm). Elute with 1.5 liter of a 0.2–1 M KCl gradient/buffer A.
10. Dilute active fraction with 2.5 vol of buffer A. Stir into 500-g DEAE-Sephacel (Pharmacia) (2 h).
11. Wash loaded DEAE-Sephacel on Buechner funnel with 3 liter of buffer A/0.2 M KCl.
12. Pour column (∅ 8 cm). Elute with 2.5 liter of a 0.2 M KCl–1.5 M KCl gradient/buffer A.
13. Adjust the polymer containing fraction to 70% (v/v) ethanol and keep at minus 20°C overnight.
14. One hour of centrifugation at 16,000 × g to harvest the precipitate.
15. Remove traces of ethanol and dissolve precipitate in a minimum (10 ml) of distilled water. Molecular sieving on a Sephadex G25 fine column (1300 ml, ∅ 5 cm). Assay fraction for β-poly(L-malate)[8] and chloride ions.[9] Use only salt-free fractions.
16. Obtain pure β-poly(L-malate) potassium salt by lyophylization. Store in the freezer.
17. Obtain free β-poly(L-malic acid) after passage over Amberlite IR 120 (H⁺-form) (20 ml bed volume/1 g of polymer salt). Lyophylize, dissolve powder in acetone, remove insoluble material by centrifugation, and evaporate acetone from the supernatant.

[1] Indentation supports aeration during growth on a culture shaker.
[2] The culture medium Glc-NS-Hem (9) contained in 1 liter (adjusted with NaOH to pH 4.5 ± 0.05): 10 g Bacto-Trypton, 1.5-g yeast extract, 11-g D-glucose monohydrate, 3.45-g citric acid, 2-g KH₂PO₄, 0.6-g CaCl₂·2H₂O, 0.085-g FeSO₄·7H₂O, 0.6-g MgSO₄·4H₂O, 0.084-g ZnSO₄·7H₂O, and 0.005-g hemin. The solution of hemin (0.5 g/liter in 1% (w/v) NaOH is autoclaved separately. For strain propagation prepare shaking cultures at 27°C in the dark using indented Erlenmeyer flasks (100 ml medium, 500 ml flask, rotary shaker). Transfer every 2 days 2% of the culture to fresh medium. Macroplasmodia are grown on agar plates (see below): inoculation with 5-ml microplasmodia (grown for 24 h and pelleted for 1 min at 2400 × g, resuspended in 1 vol distilled water). Spherules are activated as follows: Place filter paper with spherules on agar (two-fold diluted culture medium, 2% agar agar, 20 min at 120°C autoclaved. After cooling to 50°C pour into petri dishes, dry them over night in the dark, incubate 4–5 days at 27°C in the dark, allowing a macroplasmodium to grow. Scrape the cell from the agar and transfer it to 100-ml culture medium (500 ml indented Erlenmeyer flask).
[3] The cells may be extracted for cellular β-poly(L-malate).
[4] Whatman DE 52, equilibrated in buffer A (footnote no. 5).
[5] Buffer A contains 10 mM potassium phosphate pH 7.
[6] Steps 4–15 are carried out in the cold room at 4°C.
[7] Assay for β-poly(L-malate) see Fischer et al. 1989 [4].
[8] The assay for ester bonds is more convenient at this stage [4,10].
[9] Nephelometric assay for chloride ions with AgNO₃ under weakly HNO₃-acidic conditions.

nor spores contain the polymer. The polymer synthetic activity of the plasmodia increases together with the cellular mass as a function of the D-glucose concentration in the culture medium (Fig. 1a). At a concentration of 11 g D-glucose/liter routinely used in the growth medium, plasmodia grow for a period of 4–6 days under a concomitant increase in the polymer concentration in the culture medium (Fig. 1b). At the same time the polymer content within the plasmodia stays approximately constant. Polymer production ceases during the arrest of growth. From thereon the polymer content (and the polymer average molecular mass) decreases due to the hydrolytic decomposition to L-malate (Fig. 1b). Since the decomposition begins from the onset of growth, it is advantageous

to harvest plasmodia at an early stage of the culture if a polymer of a high-average molecular mass is desired. The decline in polymer content of the culture medium during prolonged times is also observed with other organisms such as *Corollospora fusca* M-214 and probably refers there as well to the activity of specific hydrolases (see polymalatase from *P. polycephalum*).

B. Biochemistry of β-Poly(L-Malate) Synthesis

The biochemistry of β-poly(L-malate) synthesis has been investigated for *P. polycephalum* but is far from

Table 2 Producers of β-Poly(L-Malate), Growth Conditions, Amount of Produced Polymer, and Selected Properties of Polymer[1]

Organism	Growth medium,[2] days of growth, (temperature,°C)	Polymer content[3] (mg/liter medium)	Polymer content[4] (mg/g cells)	Solubility of free acid in acetone	Molality of KCl for elution from DEAE-cellulose, (overall content)		$\overline{M}_n$ (kDa)[5]	$\overline{M}_w$ (kDa)	Polydispersity
Penicillium[6] *cyclopium*	wheat bran	5.7×10^3						5.0	
Physarum polycephalum[7] M₃CVIII[8]	Glc-NS-Hem,[9] 3 (27)								
Culture medium		1×10^3	—	yes	<0.3	(100%)	25[10]	50	2.0
Cytoplasm[11]			0.35 (0.063)	yes	0.5	(9%)	4.9	7.4	1.5
			macro(micro)- plasmodium		0.6	(33%)	11	14	1.2
					0.7	(54%)	17	29	1.7
					1.0	(4%)	27	100	3.7
Nuclear extract			0.35 (0.19)	yes	0.5	(2%)	4.8	6.5	1.3
			macro(micro)- plasmodium		0.6	(20%)	14	20	1.4
					0.7	(70%)	31	52	1.7
					1.0	(8%)	29	92	3.2
Aureobasidium sp. A-91[12]	Glc-NS,[13] 7 (25)	61×10^3		insoluble				9	
Aureobasidium pullulans[14]									
M-156	Glc-NS-Y,[15] 20	9.2×10^3	0.05–0.15	slightly	0.2–0.5	(75%)	4.0	4.5	1.1
T-207	(21)	8.0×10^3	0.3	slightly	0.5	(70%)	5.0	5.5	1.1
Corollospora fusca[14] M-214	Glc-NS-Y, 8 (21)	22		slightly	0.6	(60%)	6.1	6.8	1.1
Cladosporium cladosporioides[14]									
M-203	Glc-NS-Y, 7 (21)	24							
M-204		92							
TS-03		72							
Cladosporium herbarum[14]									
M-202	Glc-NS-Y, 7 (21)	56							
M-205		350							

[1] Unless mentioned otherwise, the properties listed are for the polymer isolated from the growth medium.

[2] Organisms are grown in the dark.

[3] After saponification of the polymer, L-malalate is assayed according to [4,10].

[4] It may be necessary to purify the polymer on DEAE-cellulose before performing the polymer assay.

[5] Measurement by gel permeation chromatography in 0.2 M phosphate buffer pH 7.0 with polystyrenesulfate or polyethylene glycols ([5] in the case of *aureobasidium* sp. A-91) as molecular weight standards. Data processing as described in Ref. [11].

[6] Ref. [1].

[7] Ref. [4].

[8] Similar results were obtained with *Physarum polycephalum* strain M₃CVII.

[9] See Table 1 footnotes.

[10] The molecular masses may be smaller depending on the time of harvest and on the selection of chromatographic fractions during purification.

[11] After cell fractionation according to Ref. [12].

[12] Ref. [5].

[13] 8% Glucose, 0.3% ammonium succinate, 0.2% succinic acid, 0.04% K₂CO₃, 0.01% KH₂PO₄, 0.01% MgSO₄ × 7H₂O, 5 ppm ZnSO₄ × 7H₂O, 0.05% corn steep liquor, and 2% CaCO₃ in deionized water [5].

[14] From the collection Molitoris, Regensburg, Germany.

[15] 12% Glucose, 0.1% NaNO₃, 0.01% KH₂PO₄, 0.05% KCl, 0.02% MgSO₄ × 7H₂O, 3% CaCO₃, 0.01% yeast extract (Rathberger, Molitoris, Holler unpublished results).

being understood [13, 14, and unpublished results]. As indicated, of the various forms of cells in the life cycle (amoeba, plasmodium, spherule, spore) only the plasmodium produces the polymer. Newly synthesized polymer appears first in cell nuclei, which, however, maintain a constant level of the polymer over time (homeostasis). Homeostasis is accomplished by a high rate of polymer synthesis and a release of surplus polymer via the cytosol into the growth medium. The molecular mass distribution of β-poly(L-malate) during growth

(a)

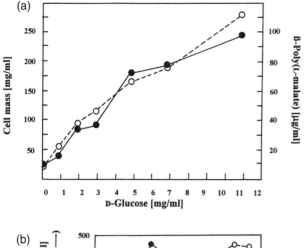

(b)

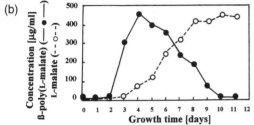

Figure 1 β-Poly(L-malate) released into the culture medium during plasmodial growth of *P. polycephalum*. (a) Effect of the nutrient (D-glucose) on the growth of plasmodia (—●—), and the production of β-poly(L-malate) (−−○−−). Growth conditions are otherwise as indicated in Table 1 footnote 2. (b) Content of β-poly(L-malate) (—●—) and L-malate (−−○−−) in the culture medium during growth of strain M_3CVIII under conditions indicated in Table 1 footnote 2. Inoculation on day 0. Growth termination on day 4.

is such that the nuclei and the cytosol of a plasmodium contain approximately equal amounts of β-polyl L-malate.

β-Poly(L-malate) is produced from D-glucose involving the citric acid cycle for the production of precursors [13]. Plasmodia have a relatively high content in L-malate (0.3–3 mg/g cells). The content of D-glucose in the culture medium can be varied in the range 0–11 mg/ml, and the amount of polymer released into the growth medium varies proportionally (Fig. 1a). In contrast, the content in the cells stays constant due to homeostasis. By using D-[^{14}C]glucose in the culture medium or by injecting L-[^{14}C]malate, the production of radioactively labeled β-poly(L-malate) can be preparatively accomplished.

In the beginning of the investigation it was believed that enzymatic synthesis of β-poly(L-malate) follows a route, which is similar to the synthesis of poly(β-hydroxybutyrate) (for a review see [15]). However, enzymati-

cally synthesized β-L-[^{14}C]malyl-CoA [16] was not used as a substrate and radioactivity was not incorporated into nascent β-poly(L-malate) (Willibald and Holler, unpublished results). Also the desulfoanalogue of coenzyme-A had no effect on the cellular synthesis when injected into macroplasmodia. Homology PCR screening with consensus sequences of bacterial poly(β-hydroxybutyric acid) (PHB) synthases and cDNA from *P. polycephalum* did not show homologous DNA fragments (Windisch, unpublished results). We think that unlike the bacterial synthesis of poly(β-hydroxybutyric acid), the synthesis of β-poly(L-malic acid) does not involve β-L-malyl-CoA. It may rather use adenylate formation for malic acid activation, which is accompanied by the release of inorganic pyrophosphate from ATP as the second substrate. In vivo injection experiments indicate an inhibition of the polymer production by the ATP analogue α,β-methyleneadenosine 5'-triphosphate (AMP-CPP) but not with β,γ-methyleneadenosine 5'-triphosphate (AMP-PCP) (unpublished results) in agreement with this assumption. The polymer synthetic activity seems to be regulated by a signal pathway. The GTP analogue β,γ-methyleneguanosine 5'-triphosphate (GMP-PCP) inhibits the polymer synthesizing activity suggesting that a G-Protein is involved. Inhibition was also observed after injection of tyrosine kinase inhibitors, suggesting that one of the cascade proteins in the signal pathway should not be phosphorylated (we have reasons to assume that this is not the β-poly(L-malate) synthase itself). During plasmodium disruption that might activate the signal pathway, an extremely rapid inactivation of β-poly(L-malate) synthase is observed.

IV. PHYSICO-CHEMICAL PROPERTIES

A. Molecular Mass

Structural information about β-poly(L-malate) is available for the polymers from *P. polycephalum* [4, 8, and unpublished results] and *Aureobasidium* sp. [5,6]. These polymers differ substantially by their molecular mass (Table 2). The *Aureobasidium*-β-poly(L-malate) is of molecular mass below 10 kDa, whereas that of *P. polycephalum* in the absence of degradation is of the order of at least 50 kDa and may be as high as several hundred kDa. These two classes of β-poly(L-malate) can be also distinguished by their elution position from the anion exchanger DEAE-cellulose and by their solubility as free acids in acetone (Table 2). Furthermore, the *Aureobasidium* polymer does not inhibit DNA polymerases (see the following). The producers of the two classes of β-poly(L-malate) are also characteristically different. The high molecular mass polymer is made by *P. polycephalum*, a fragile slime mold with extremely large multinucleate plasmodial cells, whereas the low molecular weight polymer is made by cell walls embedded mononucleate cells which display hyphen/yeast dimorphismus.

B. Spectral Properties

NMR-spectra have been recorded for the free polymer acid, the Na^+, and Ca^{2+} salts (Table 3). As the spectra for α- and β-poly(L-malic acids) or the copolymers are very similar and are, therefore, difficult to distinguish [6,8], a comparison of the corresponding high-resolution ^{1}H and ^{13}C NMR spectra has been carried out confirming that the polymer from *P. polycephalum* was indeed of the poly(β-hydroxy acid) structure [8]. The absence of band fine splitting revealed also that the natural polymer was of extreme enantiomeric purity exceeding that of synthetic β-poly(L-malate). For β-poly(L-malate) from *Aureobasidium* sp. A-91, the observed peculiar variation in methine- and methylene peaks during hydrolytic cleavage into L-malic acid was interpreted as evidence for chain branching according to the structure (IV) (see Introduction) [5,17]. Branching is also thought to be a reason for the absence of a melting point (decomposition above 185°C) and the insolubility in acetone or other organic solvents [5]. In contrast, β-poly(L-malic acid) (mass-average molecular mass of 50 kDa) has a melting point of 192°C, shows decomposition above 200°C

(Gassner and Holler, unpublished results) and is readily soluble in acetone.

Infrared spectra and the degree of specific rotation show typical features of the malic acid polyester (Table 3). Ultraviolet absorbance spectra of β-poly(L-malate) from both *P. polycephalum* and *Aureobasidium* sp. A-91 are similar and are reminiscent of malate itself [4,5]. For a solution of 1.0 mg/ml polymer, absorbance increases from 0.40 units at 230 nm to 10 units at 190 nm wavelength. After saponification and pH-neutralization, the absorbance increased from 8.7 units at 230 nm to 100 units at 190 nm.

According to molecular calculations, β-poly(L-malic acid) is highly flexible since long-range ordering forces are absent (Hendl, Urbani and Cesaro, unpublished results). The situation is similar as found for poly(β-hydroxybutyric acid), which can be considered as the reduced form of β-poly(L-malic acid) with methyl groups in the place of the pending carboxylic groups. Due to charge repulsion in the ionized polymer, an extended conformation is preferred in β-poly(L-malate). Nevertheless, the polyanion probably does not display higher-ordered structures in water.

Table 3 Spectral Properties of β-Poly(malic Acid) and Its Salts

β-Poly(L-malic acid)	^{1}H NMR (ppm)	^{13}C NMR (ppm)	Infrared (KBr) spectra, ν_{max} (cm^{-1})	Specific rotation $[\alpha]_D$
Na salt* (D$_2$O)	2.9 (methylene, doublet), 5.0 (methine, triplett)	36.7 (methylene), 72.3 (methine), 172.3 (ester carbonyl), 176.1 (carboxylate)	3000; 2940 (C—H), 1750–1720 (C=O, ester, carboxylic acid), 1650–1600 (carboxylate)	−17° (c 6.0 H$_2$O) at 20°C
Na salt† (D$_2$O)		36.34 (methylene), 71.95 (methine),‡ 171.98 (ester carbonyl), 175.78 (carboxylate)‡		
Ca^{2+} salt§ (D$_2$O)	3.0 (methylene, doublet), 5.2 (methine, triplet)	38.7 (methylene), 74.2 (methine), 174.2 (ester carbonyl), 178.2 (carboxylate)	3400; 2940; 1740; 1600; 1410; 1280; 1180; 1100; 1050	16°–17° (c 5.0, H$_2$O) at 25°C
Free acid† (CD$_3$COCD$_3$)		36.33 (methylene),‡ 69.28 (methine),¶ 169.07 (ester carbonyl),‡ 170.03 (carboxylic acid)		
Free acid§ (D$_2$O)	3.0 (methylene, doublet), 5.4 (methine, triplet)	38.4 (methylene), 71.9 (methine), 173.1 (ester carbonyl), 174.6 (carboxylic acid)	3400; 2940; 1740; 1410; 1180; 1050	−8° (c 6.0, H$_2$O) at 25°C

* Ref. [4].
† Ref. [8].
‡ Meso diads.
§ Ref. [5].
¶ Isotactic triads.

V. CHEMICAL PROPERTIES OF β-POLY(L-MALATE)

A. Assay of β-Poly(L-Malate)

A specific assay for β-poly(L-malate) is alkaline hydrolysis of the polymer followed by the enzymatic dehydrogenation of L-malate and measurement of the NAD/H$^+$ specific A$_{340}$ absorbance [4]. Hydrolysis is carried out at concentrations of 1–20 μg of polymer in 100-μl samples overnight (12 h) in the presence of 0.45 M NaOH (10 μl of 5 M NaOH) at 37°C. Samples that contain protein (cell extracts) were deproteinized after acidification to pH 2 with 7 M perchloric acid, and removal of the precipitate by centrifugation. The supernatant is then subjected to alkaline hydrolysis. This acid precipitation has the drawback that polymer is lost by adhering to the protein precipitate. A better method is the chromatographic removal of protein on a 1 ml DEAE-cellulose column (sufficient for an extract from 1 g of cells) by washing with 5 ml of 0.2–0.4 M KCl in 10 mM potassium phosphate buffer pH 7.0. The polymer is eluted with 3 ml of buffer containing 1 M KCl. For a low-molecular mass polymer such as of 4–8 kDa, the salt concentration in the washing buffer should not be higher than 0.15 M. After hydrolysis, 100-μl portions are assayed spectrophotometrically by following the A$_{340}$ due to formation of NADH/H$^+$ in the presence of 830 μl of glycin/hydrazine buffer pH 9.0 (0.5 M glycine, 0.4 M hydrazine stock solution), 65 μl of NAD$^+$ (40 mM), and 5 μl of malate dehydrogenase (30 units/μl in 10 mM Tris/HCl pH 7.5) [18]. The L-malate background in the sample is assayed for a sample, which has not been subjected to alkaline hydrolysis. The absorbance at 340-nm wavelength is standardized with known amounts of L-malate. The sensitivity of the assay is approximately 0.5 μg of polymer. A 10-fold higher sensitivity can be obtained with the method of Peleg et al. [19] that couples the malate dehydrogenase reaction with the reduction of 3-(4,5-dimethylthiazol-2-yl)-2,5-diphenyltetrazolium-bromide) (MTT) yielding the corresponding formazan dye. This method has been reported to be applicable as a plate-assay for the screening of L-malic acid secreting microorganisms and may be used for the assay of β-poly(L-malate) after a proper technical adaptation.

Alkaline (and also acidic) ester hydrolysis of β-poly(L-malate) is accompanied by side reactions leading to the formation of fumarate, maleate and/or racemization, especially at elevated temperatures. The above assays thus underestimate the polymer contents due to the formation of small amounts of 2–4% fumarate (unpublished results). This fraction of fumarate increases for the hydrolysis of more concentrated polymer solutions.

A less time consuming alternative to the saponification/malate dehydrogenase assay is the albeit less specific measurement of the carboxylic ester groups according to Kakác and Vejdelek [10]. The sample (160 μl) is mixed with a solution of 10% (w/v) aqueous hydroxylammonium chloride (160 μl) and a solution of 10% (w/v) NaOH (160 μl). After 5–10 min at room temperature, a solution of 4 M HCl (160 μl) is added followed by a solution of 5% (w/v) FeCl$_3$ (160 μl). The absorbance A$_{540}$ is read after 10 min. One unit of A$_{540}$ corresponds to 0.29 mg ± 0.02 mg of β-poly(L-malate, potassium salt).

B. Acid Dissociation

β-Poly(L-malic acid) ionizes readily in water giving rise to a highly soluble polyanion. Thus, a 2% solution of the free acid of the polymer from *Aureobasidium* sp. A-91 showed a pH 2.0 [5]. The ionic constants have been determined to be pK$_a$ = 3.6 for the polymer from *Aureobasidum* sp. A-91 [5] and pK$_a$ (25°C) = 3.45 for β-poly(L-malic acid) of M$_w$ 24 kDa from *P. polycephalum* (Valussi and Cesaro, unpublished results) Thus, the polymer is highly charged under physiological conditions (pH 7.0).

C. Noncovalent Binding of Small Cations

Polyvalent cations such as the protonated forms of spermidine [*N*-(3-amino-propyl)-1,4-butanediamine], spermine [*N*,*N*'-bis(3-aminopropyl)-1,4-butanediamine], poly(ethylenimine), poly(L-lysine), histones, etc., bind noncovalently to β-poly(L-malate) with high affinities. For spermidine and spermine the dissociation constants (pH 7.5, 28°C, 50 mM ionic strength) are K$_{Diss}$ ≈ 0.6 mM and ≈ 0.06 mM, respectively [4]. The dye tetranitro blue tetrazolium binds to the polymer (K$_{Diss}$ = 0.7 mM, pH 7.0, 20°C, 50 mM ionic strength) with a concomitant increase in absorbance at 275 nm and 350 nm wavelength, and can thus be used as an indicator for studying competitive interactions of other ligands with the polymer [13]. By this method, the binding of Ca^{2+} and Mg^{2+} was found to follow dissociation constants (20°C, pH 7.0, 50 mM ionic strength) of 0.11 M and 0.16 M, respectively. These salt complexes are rapidly reversible by the addition of high salt in the molar concentration range. It is of pharmacological interest that cis-diamminepolymalatoplatinum(II) is obtained from a mixture of β-poly(L-malate) and cis-diamminediaquaplatinum(II). The platinum complex inhibits the growth of MCF-7 cell cultures similarly as observed with cisplatin (Natarajan and Holler, unpublished results).

D. Chemical Derivatization of β-Poly(L-Malic Acid)

Molecules of interest that contain free amino groups can be coupled in aqueous solution to β-poly(L-malate) as amides using carbodiimides such as the water-soluble 1-ethyl-3(3-dimethylaminopropyl)carbodiimide hydrochloride (EDC) [2,12,20,21]. By this method, the molecules are attached randomly. A selective amide bond formation at the carboxylate terminus can be achieved without a coupling reagent by a direct attack (aminolysis) of the polymer ester bond, thereby shortening the polymer chain. By using a soluble acyl chloride, the terminal hydroxyl group can be esterified in aqueous solution. The yields of such reactions in water are low and the molecular mass of the polymer is reduced by spontaneous hydrolysis. Several kinds of reactions may be carried out in organic solvents such as N-methylpyrrolidone, dioxane, THF, CH_2Cl_2 or acetone, in which the acid is (partially) soluble [2]. In principle, reporter, effector and targeting molecules can be singly or collectively attached to β-poly(L-malate) in order to serve pharmaceutical or other purposes.

VI. BIOCHEMICAL PROPERTIES

β-Poly(L-malate) from *P. polycephalum* forms specific complexes with DNA polymerases α, δ, and ϵ from the same organism [4,12,13,22,23]. When bound to the polymer these enzymes are completely inhibited. Activity can be recovered by the addition of spermine hydrochloride. Complex formation (inhibition) was also observed with DNA polymerase α from β-poly(L-malate)-producing *Aureobasiae* (unpublished results) but not with DNA polymerases from higher eukaryotes. In contrast, β-poly(L-malate) from *Aureobasidiae* was unable to form such complexes. Other nuclear proteins like histones and HMG-like proteins bind also to the polymer from *P. polycephalum* and have the potential to reactivate DNA polymerases from inhibition with the polymer [4,22]. These proteins have been shown to bind in cell-free systems with dissociation constants in the sub-μmolar concentration range. In agreement with these findings, multicomponent complexes of high stability have been found to be constituents of the nuclei in the plasmodium of *P. polycephalum* [12]. The spontaneous dissociation of DNA polymerase from nuclear complexes is very slow and occurs in minutes to hours. The addition of biogenic amines (like putrescine, spermidine, or spermine), and histones, like high concentrations of salt, to nuclear extracts accelerate the dissociation of these complexes and causes an activation of the DNA polymerases [4]. Because of their small sizes, the amines have an access to the ion pairs of the protein/polymer complexes and provoke the rapid dissociation.

Model calculations have indicated that β-poly(L-malate) displays a certain degree of isosterism with the phosphodeoxyribose backbone of DNA (and probably with the backbone of RNA) regarding the distance between the negative charges [22]. It is, therefore, possible that β-poly(L-malate) mimicks DNA in many of its activities.

VII. EVIDENCE FOR PHYSIOLOGICAL FUNCTION

Evidence for a physiological function of β-poly(L-malate) is available for plasmodia of *P. polycephalum*. The plasmodium is a giant amorphous cell containing billions of diploid nuclei that develop through the mitotic cycle with high synchrony. Since the polymer is only found in the plasmodium and not in other cell types of the organism, β-poly(L-malate) can be assumed to be essential. The growing plasmodium contains the polymer in its nuclei at high, homeostatically controlled concentrations (100–150 mg/ml). The polymer forms large complexes of molcular mass in the range of $1–2 \times 10^6$ Da involving nuclear proteins such as histones, DNA polymerases, and HMG-proteins [12]. The present view of the physiological role of β-poly(L-malate) is a function as a mobile matrix and storage device for certain nuclear proteins among them histones and DNA polymerases. In this role, β-poly(L-malate) provides an equal supply of material to all nuclei that is necessary to maintain the observed synchrony between nuclei. In its role as a complexing agent the polymer can participate in nucleus organization. Due to its polyvalency it probably also functions as a molecular chaperon of nucleosome assembly [22]. By its property as a molecular buffer that binds competitively both histones and DNA polymerases, the synthesis of histones becomes coupled with the availability of free active DNA polymerases at the onset of DNA replication.

The secretion of the polymer into the growth medium is not only a means to control nuclear β-poly(L-malate) concentration but also is a prerequisite for sporulation. L-Malate, which is a co-inductor of sporulation (Renzel and Hildebrandt, personal communication), is obtained by spontaneous and polymalatase-catalyzed hydrolysis of the secreted polymer.

VIII. DEGRADATION

A. Spontaneous Ester Hydrolysis

β-Poly(L-malate) decomposes spontaneously to L-malate by ester hydrolysis [2,4,5]. Hydrolytic degradation of the polymer sodium salt at pH 7.0 and 37°C results in a random cleavage of the polymer, the molecular mass decreasing by 50% after a period of 10 h [2]. The rate of hydrolysis is accelerated in acidic and alkaline solutions. This was first noted by changes in the activity of the polymer to inhibit DNA polymerase α of *P. polycephalum* [4]. The explanation of this phenomenon was that the degradation was slowest between pH 5–9 (Fig. 2) as would be expected if it were acid/base-catalyzed. In choosing a buffer, one should be aware of specific buffer catalysis. We found that the polymer was more stable in phosphate buffer than in Tris/HCl-buffer.

Whereas the cleavage of β-poly(L-malate) at neutral pH is at random [2], alkaline hydrolysis reveals characteristic patterns of the cleavage products, which is due to nonrandom chain scission (Fig. 3). The phenomenon is explained by an autocatalytic ester hydrolysis. Assuming that one (or both) of the polymer ends bends

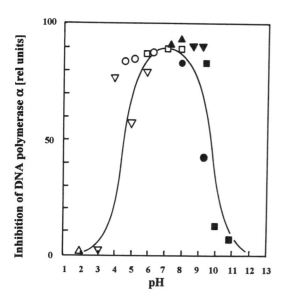

Figure 2 Stability of β-poly(L-malate) measured by its activity to inhibit purified DNA polymerase α of *P. polycephalum*. The relative degree of inhibition is shown (100 rel. units refer to complete inhibition). The DNA polymerase assay was carried out in the presence of 5 μg/ml β-poly(L-malate) as described [4]. The polymer was preincubated for 7 days at 4°C in the following buffer solutions (50 mM): KCl/HCl (—△—). Citrate (—▽—). 2-(*N*-Morpholino)-ethanesulfonic acid, sodium salt (—○—). Sodium phosphate (—□—). *N*-(2-Hydroxyethyl)piperazine-*N'*-(2-ethanesulfonic acid), sodium salt (—▲—). *N,N*-bis (2-Hydroxyethyl)-glycine, sodium salt (—▼—). Tris/HCl (—●—). 3-(Cyclo-hexylamino)-1-propanesulfonic acid, sodium salt (—■—).

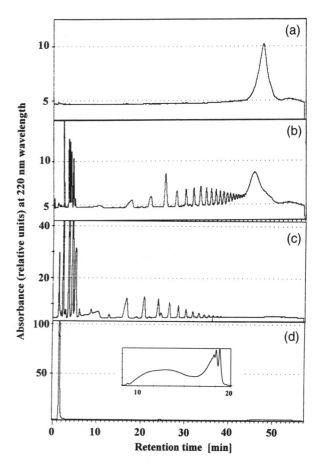

Figure 3 Reversed-phase chromatography of products after alkaline hydrolysis of β-poly(L-malate). Discrete polymer products are formed, which differ in length by several units of L-malate. The absorbance at 220-nm wavelength was measured. (a) β-Poly(L-malate) before hydrolysis. (b) After 10-min incubation in 20 mM NaOH at 37°C. (c) After 15 h in 20 mM NaOH at 37°C. (d) After 1 h in 500 mM NaOH at 100°C. High pressure chromatography (HPLC) on Waters reversed-phase C_{18}-μ-Bondapak. The methanol gradient (in water-trifluoro acetic acid, pH 3.0) was programmed as follows: 0–40 min 0.3–23%, 40–47 min 23–40%, 47–49 min 40%, 49–54 min 40–0%. (d) Inset: size exclusion chromatography after 3-min alkaline hydrolysis at pH 10.2. BioSil SEC 250 column of 300 mm × 7.8 mm size, 0.2 M potassium phosphate buffer pH 7.0.

toward the polymer chain while functioning as the catalyst, the observed cleavage patterns are thought to be generated. An autocatalytic hydrolysis was also postulated for degradation under acidic conditions [17].

B. Enzymatic Hydrolysis by Polymalatase

The stability of the polymer in a conditioned culture medium was strongly decreased by the action of an esterase. The results shown in Fig. 1 indicate that during prolonged times the content in β-poly(L-malate) decreases giving rise to the formation of L-malate. In controls with fresh medium this degradation was not detected over a period of several days, indicating that the phenomenon was enzyme catalyzed. An esterase was isolated after ammonium sulfate precipitation of proteins in the conditioned culture medium, followed by repeated hydrophobic interaction chromatography on butyl-Toyoperl, and size exclusion chromatography on Pharmacia Superdex 200 (exclusion limit for proteins 1.3 × 10⁶ Da) [24]. The hydrolase has a molecular mass of 68,000 and is specific for β-poly(L-malate). Because of its specificity it

is termed "polymalatase." It removes L-malic acid from one of the two termini of the polymer. Due to its activity optimum (pH 3–5) it is active only in the relatively acidic culture medium. The properties of the enzyme are summarized in Table 4. It has been found that the esterase inhibitors of serin proteases do not inactivate polymalatase, and that the enzyme contains an essential sulfhydryl group. It is, thus, distinguished from the "depolymerases" of poly(β-hydroxyalkanoates), that are related to lipases. Judged by the above and several other crite-

Table 4 Properties of β-Poly(L-Malate) Hydrolase from *P. polycephalum*[1]

Molecular mass[2]	68,000, monomeric
Specific activity (munits/mg)[3]	18000 (pH 3.5)[4]; 9000 (pH 4.0)[5]; 230 (pH 6.0)[6]
K_m (mM)	11 (pH 3.5); 3.5 (pH 4.0); 0.11 (pH 6.0)
Maximum activity,[7] pH	3.5
Temperature	27°C
Inhibitors, residual activity[8] (%)	
KCl, 100 mM	50
CaCl$_2$, 100 mM	15
ZnCl$_2$, 100 mM	80
FeCl$_2$; CoCl$_2$; CuCl; CuSO$_4$ (5–100 μM)	100
EDTA, 3 (20) mM	80 (40)
2-Mercaptoethanol 10 mM	80
Glycerol, 15% (50%) (v/v)	50 (10)
p-mercuribenzoate, 0.025 (0.1) μg/ml	50 (0)
N-Bromosuccinimide, 8 (12) μM	50 (0)
Spermine hydrochloride, 0.7 (1.0) mM	50 (0)
0.3 mM	170
Inhibitors of serin proteases[9]	100
D-gluconolactone, 22 (200) mM	50 (0)
Poly(L-aspartate)[10], 10 mM	100
Poly(L-glutamate), 10 mM	100
Poly(vinyl sulfate), 10 mM	100
Poly(acrylate), 10 mM	100

[1] Ref. [24]

[2] SDS-polyacrylamid gelelectrophoresis.

[3] Samples containing appropriate buffers were incubated for varying times with varied concentrations of β-poly(L-malate). In 30 min intervals, 100 μl aliquots were removed and assayed for L-malate content according to Ref. [18].

[4] sodium citrate (20 mM).

[5] potassium acetate (30 mM).

[6] 2-(*N*-morpholino)ethanesulfonic acid (20 mM).

[7] The assay as above was used containing 1.5 μg/ml β-poly(L-malic acid) (9.6 mM in terms of malyl residues).

[8] The same assay as in footnote 7 was used. In cases of no inhibition, assays were repeated in the presence of 0.09–1.8 mM polymer (in terms of malyl residues).

[9] For composition of the inhibitor cocktail see Ref. [24].

[10] For molecular masses Ref. [24].

ria, the catalytic mechanism of polymalatase appears to be closer to that of glycosidases than to that of serin-esterases.

Substantial amounts of polymalatase have been isolated from plasmodial extracts. This may refer to stored enzyme before secretion, because β-poly(L-malate) is not degraded in plasmodia [24]. Several other fungi were found to secrete β-poly(L-malate) degrading activities to L-malic acid (Ratberger, Molitoris and Holler, unpublished results). These enzymes have not yet been purified and characterized.

Polymalatase may be useful for the tayloring of β-poly(malic acid) and its derivatives, and for analytical purposes. If the hydrolase is arrested at points of polymer branches or covalently/physically attached ligands, the hydrolase can be used in studies analogous to those known for DNA and exonucleases.

IX. SYNTHETIC β-POLY(L-MALATE)

Synthetic β-poly(L-malic acid) can be obtained by polymerization of malolactonic acid benzylester followed by hydrogenolysis of the resulting linear β-poly(L-malic acid benzylester) [25], and from L-aspartic acid [26], or L-malic acid [27]. The molecular mass of the linear products is in the range for the natural polymer. The biochemical reactivity of synthetic β-poly(L-malate) has been indistinguishable from the natural polymer [4,22]. Studies with the synthetic polymer have indicated very low levels of toxicity and high immunological tolerance [2]. The natural polymer, which shows a higher degree of optical purity than the synthetic material, may be superior to the synthetic material with regard to properties such as crystallizability or biocompatibility. The extremely low toxicity demonstrated for the synthetic polymer is compatible with L-malic acid being a natural metabolite.

X. FUTURE

β-Poly(L-malic acid) may be visualized to a certain extent as a member of the large poly(hydroxyalkanoate) family, which is wide-spread in many bacterial strains, and is available in relatively large quantities by fermentation (sustainable feedstocks). According to the instability of the ester bond and the susceptibility of the building units to metabolic degradation, the polymers are environmentally safe raw materials, the more since many of these polyesters can be actively degraded by bacteria and fungi. Production costs can be relatively low, and some of the polyesters are, thus, eligible as raw material for manufacturing plastics. As for plastics, β-poly(L-malic acid) is not competitive, because its production costs are much too high.

PHB and related polymers are water insoluble, limiting an application to nonaquaeous systems. β-Poly(L-malate) is extremely water-soluble and is, thus, of complementary nature. Aside qualifying as raw material for the manufacture of water-soluble plastics or tissue, the polyanionic nature allows several other applications, some of which probably justify the relatively high production costs. Such applications are the use as precipitating agent, glue or as a surface film or molecular shuttle for carrying covalently bound drugs, cellular targets and tags. This kind of application should be of value in the fields of pharmacology, medicine, and agriculture. Poly(L-malic acid) may also be applicable in the form of blends with poly(hydroxyalkanoates) or other polymers. As more organisms are discovered as producers, a diversity in structure and composition may become available that renders this material even more interesting.

REFERENCES

1. T. Ouchi, and A. Fujino, *Makromol. Chem. 190:* 1523 (1989).
2. C. Braud, and M. Vert, *Trends Polymer Sci 3*: 57 (1993).
3. T. Fujino, and T. Ouchi, *Polym. Prepr., Jpn. 35:* 2330 (1985).
4. H. Fischer, S. Erdmann, and E. Holler, *Biochemistry 28*: 5219 (1989).
5. N. Nagata, T. Nakahara, and T. Tabuchi, *Biosci. Biotech. Biochem. 57*: 638 (1993).
6. N. Nagata, T. Nakahara, T. Tabuchi, R. Morita, J. R. Brewer, and S. Fujishige, *Polym. J. 25*: 585 (1993).
7. K. Shimada, K. Matsushima, J. Fukumoto, and T. Yamamoto, *Biochem. Biophys. Res. Commun. 35*: 619 (1969).
8. S. Cammas, Ph. Guerin, J. P. Girault, E. Holler, Y. Gache, and M. Vert, *Macromolecules 26*: 4681 (1993).
9. J. H. Daniel, and H. H. Baldwin, *Methods in Cell Physiology* (D. A. Prescott, ed.) Academic Press, New York, pp. 9–13 (1964).
10. B. Kakàk, and Z. J. Vejdèlek, *Handbuch der photometrischen Analyse organischer Verbindungen 1*: 333 (1974).
11. J. F. Johnson, *Encyclopedia of Polymer Science and Engineering 3*: 501 (1985).
12. B. Angerer, and E. Holler, *Biochemistry, 34*: 14741 (1995).
13. E. Holler, B. Angerer, G. Achhammer, S. Miller, and C. Windisch, *FEMS Microbiol. Rev. 103*: 109 (1992).
14. C. Windisch, S. Miller, H. Reisner, B. Angerer, G. Achhammer, and E. Holler, *Cell Biol. Internat. Reports 16*: 1211 (1992).
15. A. J. Anderson, and E. Dawes, *Microbiol. Rev. 54*: 450 (1990).
16. B. Willibald, H. Boves, and E. Holler, *Anal. Biochem. 227*: 363 (1995).
17. S. Fujishige, R. Morita, and J. R. Brewer, *Makromol. Chem., Rapid Commun. 14*: 163 (1993).
18. I. Gutmann, and A. W. Wahlefeld, *Method. Enzymat. Anal. 2*: 1632 (1974).
19. Y. Peleg, J. S. Rokem, and I. Goldberg, *FEMS Microbiol. Letters 67*: 233 (1989).
20. G. E. Means, and R. E. Feeney, *Chemical Modification of Proteins,* Holden-Day Inc., San Francisco (1971).
21. H. Yamada, T. Imoto, K. Fujita, K. Okazaki, and M. Motomura, *Biochemistry 20*: 4836 (1981).
22. E. Holler, G. Achhammer, B. Angerer, B. Gantz, C. Hambach, H. Reisner, B. Seidel, C. Weber, C. Windisch, C. Braud, Ph. Guerin, and M. Vert, *Eur. J. Biochem. 206*: 1 (1992).
23. G. Achhammer, A. Winkler, B. Angerer, and E. Holler, *Current Genetics 28*: 534 (1995).
24. C. Korherr, M. Roth, and E. Holler, *Can. J. Microbiol. 41 (Suppl. 1)*: 192 (1995).
25. M. Vert, and R. W. Lenz, *Polym. Prepr. (Am. Chem. Soc., Div. Polym. Chem.) 20*: 608, 611 (1979)
26. M. Vert, and R. W. Lenz, *Polym. Prepr. (Am. Chem. Soc., Div. Polym. Chem.) 20*: 609 (1979)
27. Ph. Guerin, M. Vert, C. Braud, and R. W. Lenz, *Polym. Bull. 14*: 187 (1985).

8

Stabilization of Polyolefins

Sahar Al-Malaika

Aston University, Birmingham, England

I. INTRODUCTION

Oxidation of organic materials is a major cause of irreversible deterioration for a large number of substances. It is responsible not only for the loss of physical properties of plastics, rubbers, fibers, and other polymeric substrates, but also for rancidity of foodstuffs, deterioration of hydrocarbon lubricating oils, as well as biological ageing and is also implicated in some diseases. Inhibition of this oxidation process is, therefore, very important and the art of stabilization of organic materials against the effect of molecular oxygen has evolved over the past 90 years from an entirely empirical basis to a mature science-based technology, which is in place throughout the polymer industry. Almost all synthetic polymers require stabilization against the adverse effect of their processing, fabrication, storage, and the end use environment. The early intuitive trial-and-error approaches in the development of stabilizers are the roots to the current state-of-the-art stabilization technology of polymers.

II. EARLY DEVELOPMENTS

"Aging" or "perishing" of natural rubber was noted long before Hoffman [1] in 1861 drew attention to the role of oxygen in the deterioration of organic materials. Practical solutions were sought to overcome the problem of loss of properties with aging. Animal foodstuffs were preserved by smoking; this process almost certainly results in the absorption of phenolic materials by the fatty components of the food. Similarly, additives were selected to prolong the "life" span of natural rubber. By the early part of this century, considerable progress was made toward finding practical solutions to combat the problem of oxidation and property deterioration, and a number of chemical compounds including phenols, quinones, and amines were patented and offered as remedies [2–4]. Maureau and Dufraisse [5] in the 1920s believed that the primary role of "antioxygens" (i.e., antioxidants) was to deactivate "peroxides" giving inert products. As such this concept failed to provide a suitable mechanism of antioxidant action; their findings, however, provided a phenomenological basis upon which the modern theory of antioxidant action was founded. This was followed later by the concept of chain reactions involving free radicals and molecular oxygen.

Bateman, Gee, Barnard, and others at the British Rubber Producers Research Association [6,7] developed a free radical chain reaction mechanism to explain the autoxidation of rubber which was later extended to other polymers and hydrocarbon compounds of technological importance [8,9]. Scheme 1 gives the main steps of the free radical chain reaction process involved in polymer oxidation and highlights the important role of hydroperoxides in the autoinitiation reaction, reaction 1b and 1c. For most polymers, reaction 1e is rate determining and hence at normal oxygen pressures, the concentration of peroxyl radical (ROO$^{\cdot}$) is maximum and termination is favoured by reactions of ROO$^{\cdot}$: reactions 1f and 1g.

III. ANTIOXIDANT MECHANISMS AND CLASSIFICATIONS

The development of the autoxidation theory, in which the propagating radicals, alkyl, and alkylperoxyl (R$^{\cdot}$ & ROO$^{\cdot}$), and the hydroperoxide (ROOH) are the key intermediates, has therefore led to a comprehensive theory of antioxidant action; Scheme 2 shows the two major

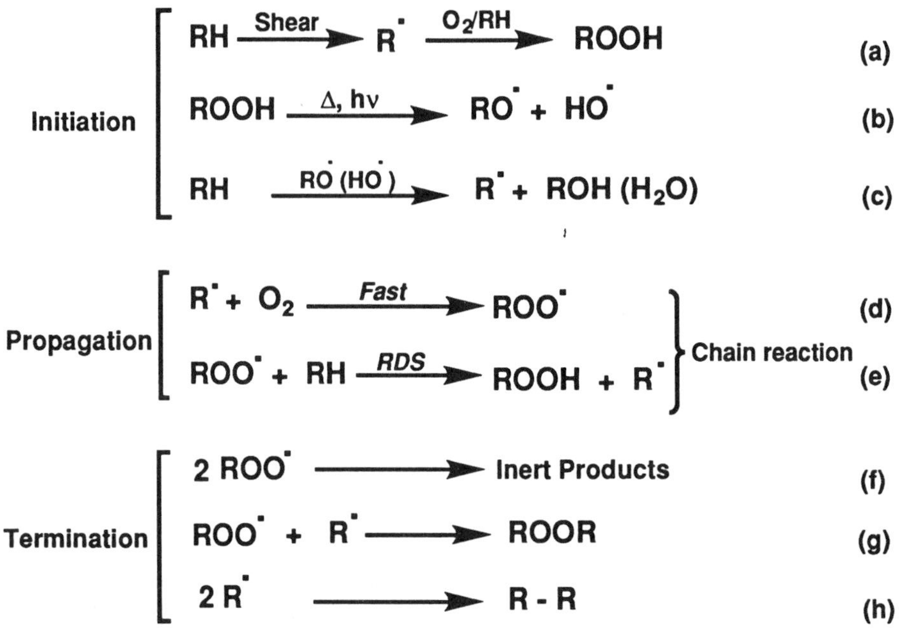

$$\text{Initiation} \begin{cases} RH \xrightarrow{\text{Shear}} R^\bullet \xrightarrow{O_2/RH} ROOH & \text{(a)} \\[2mm] ROOH \xrightarrow{\Delta,\, h\nu} RO^\bullet + HO^\bullet & \text{(b)} \\[2mm] RH \xrightarrow{RO^\bullet\,(HO^\bullet)} R^\bullet + ROH\,(H_2O) & \text{(c)} \end{cases}$$

$$\text{Propagation} \begin{cases} R^\bullet + O_2 \xrightarrow{\text{Fast}} ROO^\bullet & \text{(d)} \\[2mm] ROO^\bullet + RH \xrightarrow{RDS} ROOH + R^\bullet & \text{(e)} \end{cases} \Big\} \text{Chain reaction}$$

$$\text{Termination} \begin{cases} 2\,ROO^\bullet \longrightarrow \text{Inert Products} & \text{(f)} \\[2mm] ROO^\bullet + R^\bullet \longrightarrow ROOR & \text{(g)} \\[2mm] 2\,R^\bullet \longrightarrow R\text{-}R & \text{(h)} \end{cases}$$

Scheme 1 Free radical chain process involved in polymer oxidation.

antioxidant mechanisms: the chain breaking and preventive inhibition processes.

The basis to the chain breaking donor (CB—D) mechanism, which was the first antioxidant mechanism to be investigated, was laid down by the late 1940s [10–12]. Many reducing agents, e.g., hindered phenols and aromatic amines, which reduce the ROO$^\bullet$ to hydroperoxide in a CB—D step have already been empirically selected and used for rubbers and by this time also for the newer plastics industry (e.g., Table 1a, AO 1–8 and 9–12). The major mechanistic landmarks of the antioxi-dant action of hindered phenols are: (1) the importance of the stability of the derived phenoxyl radical (see Scheme 3, reactions 3b) in order to prevent chain transfer reactions (see reactions 3c); and (2) the dependence of the activity of hydrogen donor antioxidants on oxygen pressure (see reaction 3d).

Oxidizing agents, e.g., quinones, which were shown to be able to retard oxidation [13] can function as antioxidants (via a chain breaking acceptor process, CB—A) if they can compete with oxygen for the alkyl radicals (Scheme 4). In the case of polymers, reaction 4a can

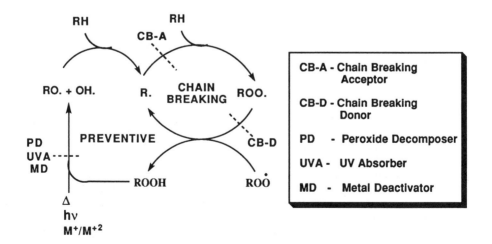

Scheme 2 Autoxidation cycles and antioxidant mechanisms.

Table 1a   Some Commercial Thermal Antioxidants

Antioxidant (AO)	Commercial or common name	Antioxidant (AO)	Commercial or common name
Chain-Breaking Antioxidants		*Peroxide Decomposers*	

Chain-Breaking Antioxidants

(AH)

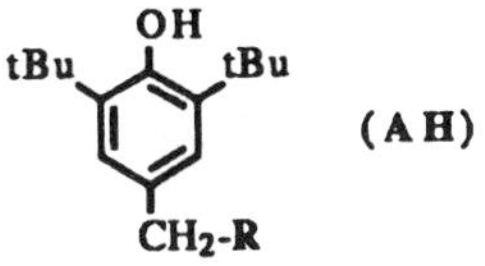

R = -H — 1 BHT
— 2 Topanol O

R = -CH₂CO₂C₁₈H₁₇ — 3 Irganox 1076

R = -(-CH₂CO₂CH₂)₄C — 4 Irganox 1010

R = (trimethylphenyl) — 5 Ethanox 330 / Irganox 1330

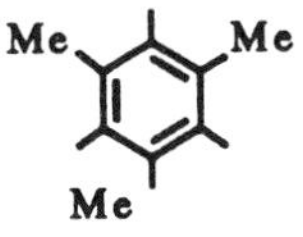

R = (triazine) — 6 Irganox 3114 / Goodrite 3114

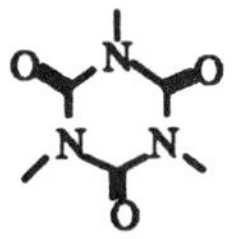

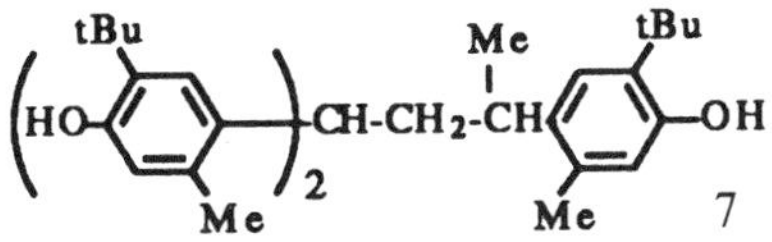

7 Topanol C

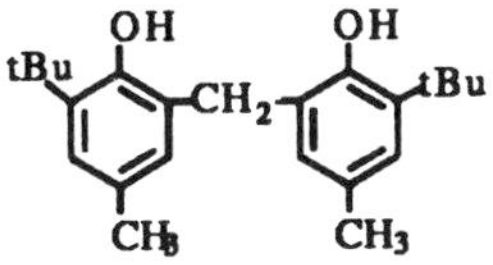

8 Cyanox 2246

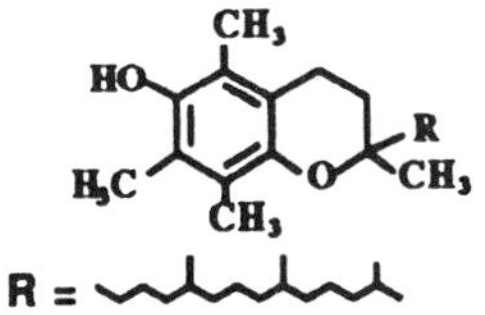

9 α-Tocopherol

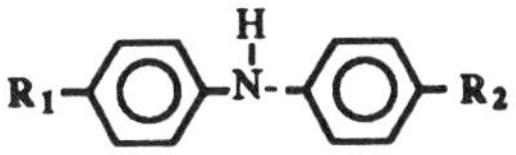

10 R₁ = R₂ = tOct Nonox OD

11 R₁ = H, R₂ = HN⟨⟩ Nonox DPPD

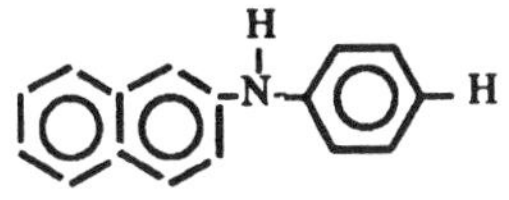

12 Vulkanox PBN

Peroxide Decomposers

$(C_{12}H_{25}O-)_3$-P — 13 Phosclere P312 / Ultranox TLP

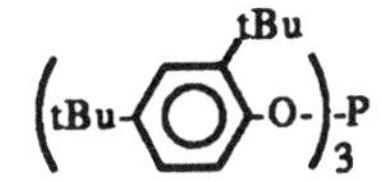

14 Irgafos 168

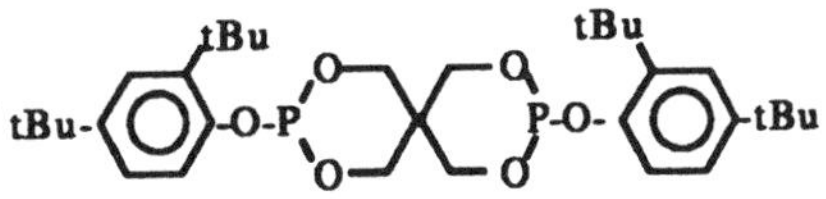

15 Ultranox 626

$[RO\text{-}\overset{O}{C}\text{-}CH_2CH_2\text{-}]_2$-S   R = C₁₈H₃₇ — 16 Irganox PS802
R = C₁₂H₂₅ — 17 Irganox PS800

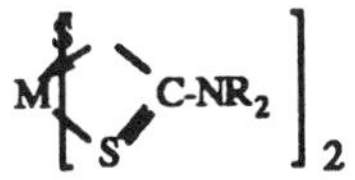

M = Zn, R = C₄H₉ — 18 Robec Z bud (dithiocarbamate)

Metal Deactivators

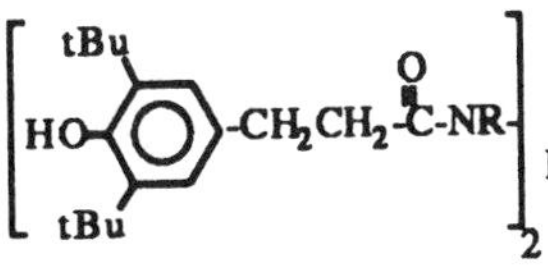

R=H — 19 Irganox MD-1024
R= H-(CH₂)₃ — 20 Irganox 1098

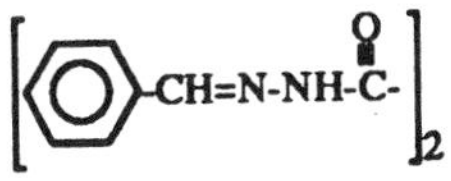

21 Eastman OABH

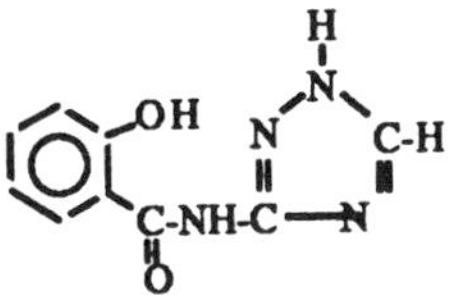

22 Mark 1475

Table 1b   Some Commercial Photoantioxidants (Light Stabilizers)

Antioxidant (AO)	Commercial or common name	AO code
UV-Absorbers		
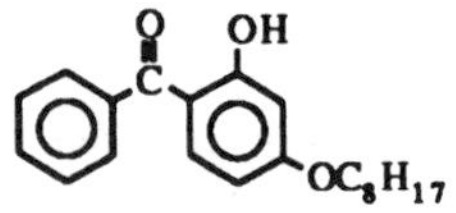	Cyasorb UV531 Chimasorb 81	23
R₁=R₂= tBu, R₃=Cl	Tinuvin 327	24
R₁=tBu, R₂=CH₃, R₃=Cl	Tinuvin 326	25
R₁=H, R₂= CH₃, R₃=H	Tinuvin P	26
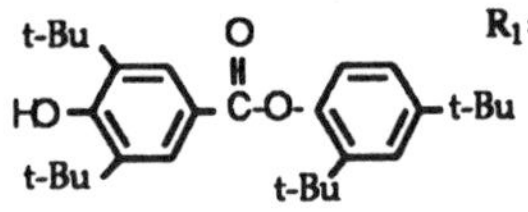	Tinuvin 120	27
Nickel Complexes		
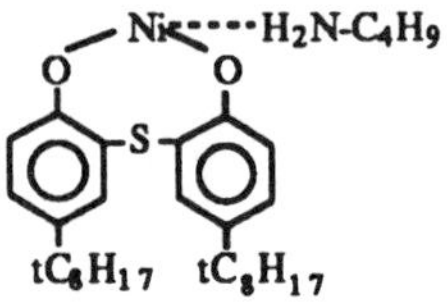	Cyasorb UV1084 Chimasorb N-705	28
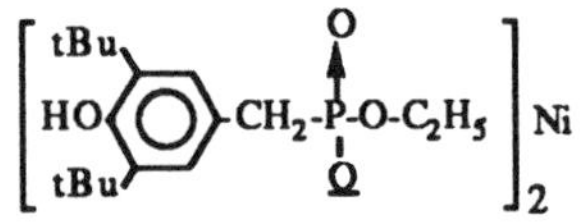	Irgastab 2002	29
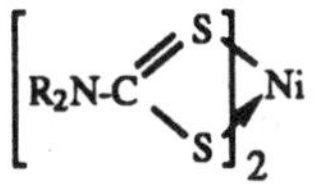	Nickel dialkyldithiocarbamate (NiDRC)	30
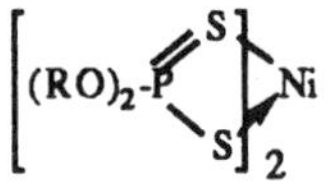	Nickel dialkyldithiophosphate (NiDRP)	31
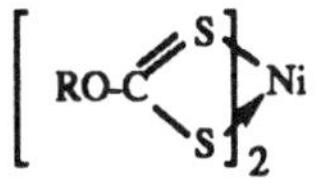	Nickel dialkylxanthate (NiRX)	32
Hindered Amines		
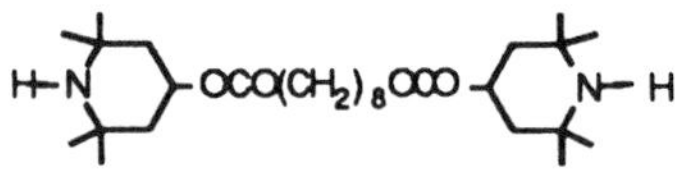	Tinuvin 770	33
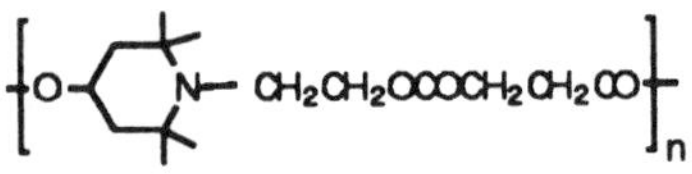	Tinuvin 622	34
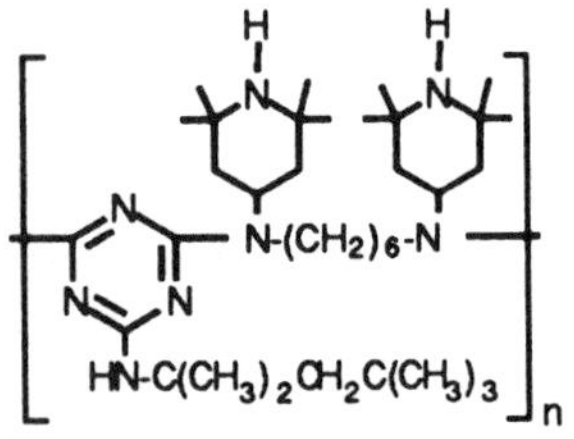	Chimasorb 944	35

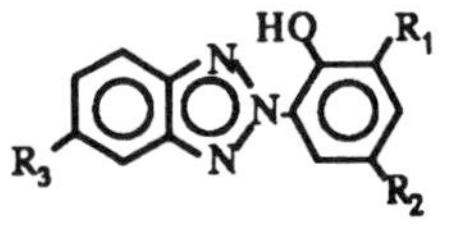

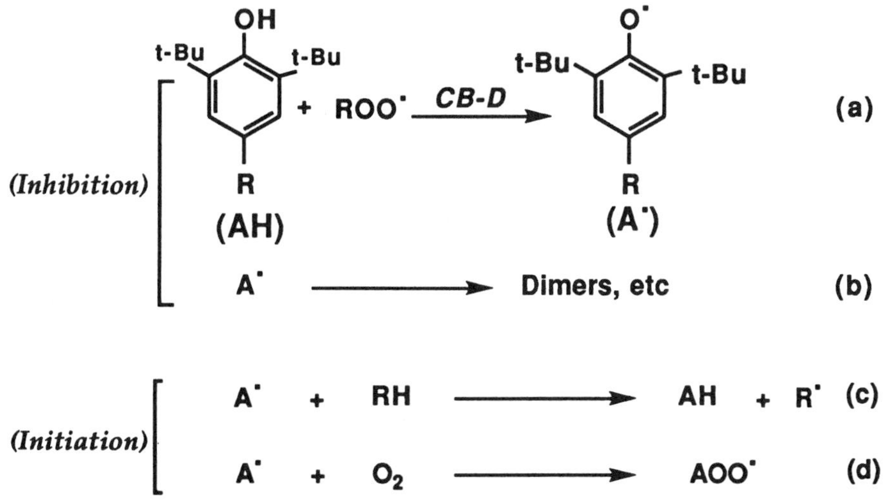

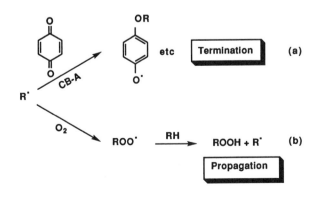

Scheme 3 Antioxidant mechanism of hindered phenols.

compete with reaction 4b during melt processing or fatiguing due to the high concentration of macro alkyl radicals under these conditions, hence low oxygen concentration at the reaction site. Many stable oxidizing radicals and certain metal ions were later shown [14,15] to act by this CB—A mechanism.

The early work of Kennerly and Patterson [16] on catalytic decomposition of hydroperoxides by sulphur-containing compounds formed the basis of the preventive (P) mechanism that complements the chain breaking (CB) process. Preventive antioxidants (sometimes referred to as secondary antioxidants), however, interrupt the second oxidative cycle by preventing or inhibiting the generation of free radicals [17]. The most important preventive mechanism is the nonradical hydroperoxide decomposition, PD. Phosphite esters and sulphur-containing compounds, e.g., AO 13–18, Table 1a are the most important classes of peroxide decomposers.

The simple trialkyl phosphites (e.g., Table 1a, AO13) decompose hydroperoxides stoichiometrically

(PD—S) to yield phosphates and alcohols, see Scheme 5 reaction a. Sterically hindered aryl phosphites (e.g., AO 14) have an additional chain breaking activity, i.e. they react with peroxyl and alkoxyl radicals during their function as antioxidants (reactions 5b and 5c) [18].

Sulphur compounds, e.g., thiopropionate esters and metal dithiolates (Table 1a, AO 16 and 17), decompose hydroperoxides catalytically, i.e., one antioxidant molecule destroys several hydroperoxides through the intermediacy of sulphur acids [19,20]. Scheme 6 shows a simplified scheme for the antioxidant mechanism of simple alkyl sulphides.

Metal deactivators (MD) act, primarily, by retarding metal-catalyzed oxidation of polymers; they are, therefore, important under conditions where polymers are in contact with metals, e.g., wires and power cables. Metal deactivators are normally polyfunctional metal chelating compounds (e.g., Table 1a, AO 19–22) that can chelate with metals and decrease their catalytic activity [21].

UV absorbers (UVA) act by absorbing UV light hence retarding the photolysis of hydroperoxides. Their activity is also associated with hydrogen bonding between the 2-hydroxy group and the carbonyl chromophore [22]. Typical examples are based on 2-hydroxy-benzophenones and 2-hydroxybenztriazoles (e.g., Table 1b, AO 23 and AOs 24–26).

IV. EFFECTS OF PROCESSING AND ENVIRONMENTAL FACTORS ON OXIDATIVE DETERIORATION OF POLYMERS

Thermoxidative degradation of polymers can occur at all stages of their lifecycle (polymerization, storage, fabrication, weathering) but its effect is most pronounced

Scheme 4 Antioxidant mechanism of quinones.

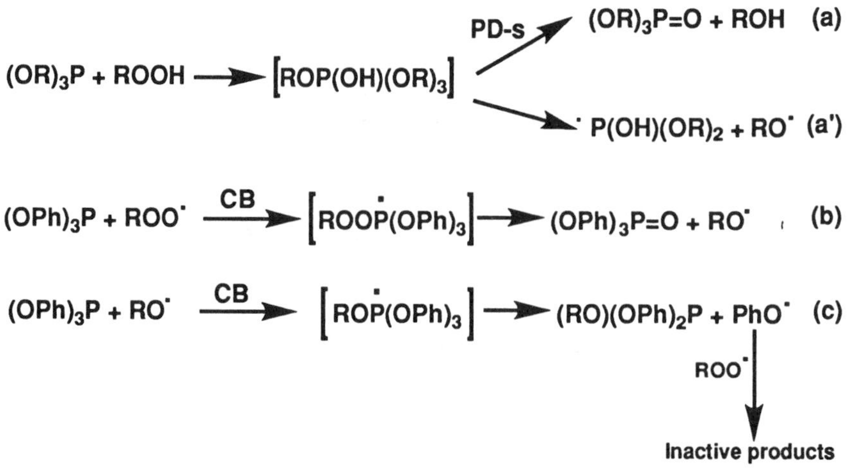

$$(OR)_3P + ROOH \longrightarrow \left[ROP(OH)(OR)_3\right] \underset{\nearrow}{\overset{PD-s}{\longrightarrow}} (OR)_3P=O + ROH \quad (a)$$

$$\overset{\cdot}{P}(OH)(OR)_2 + RO^\cdot \quad (a')$$

$$(OPh)_3P + ROO^\cdot \overset{CB}{\longrightarrow} \left[ROO\overset{\cdot}{P}(OPh)_3\right] \longrightarrow (OPh)_3P=O + RO^\cdot \quad (b)$$

$$(OPh)_3P + RO^\cdot \overset{CB}{\longrightarrow} \left[RO\overset{\cdot}{P}(OPh)_3\right] \longrightarrow (RO)(OPh)_2P + PhO^\cdot \quad (c)$$

$$ROO^\cdot$$

Inactive products

Scheme 5 Antioxidant mechanism of phosphites.

during conversion processes of the polymer to finished products. Polymer fabrication and conversions are normally achieved through the use of high shear mixing machinery (e.g., in extrusion, injection moulding, internal mixing, milling, calendering). Oxidative degradation of polymer articles during outdoor weathering is often exacerbated by the combined effects of the environment, e.g., sunlight, rain, ozone, temperature, humidity, atmospheric pollutants, and micro-organisms.

The prior thermal-oxidative history of polymers determines, to a large extent, their photoxidative behavior in service. Hydroperoxides formed during processing (and to a lesser extent during manufacturing and storage) are the primary initiators during the early stages of photoxidation, while the derived carbonyl-containing products (Scheme 7) exert deleterious effects during later stages of photoxidation [23–25]. The initiating species, e.g., hydroperoxides, and their decomposition products are responsible for the changes in molecular structure and overall molar mass of the polymer that are manifested in practice by the loss of mechanical properties (e.g., impact, flectural, tensile strengths, elongation) and by changes in the physical properties of the polymer surface (e.g., loss of gloss, reduced transparency, cracking, chalking, yellowing).

The extent of oxidative degradation of the macromolecular chain during melt processing and in-service depends ultimately on the nature and structure of the base polymer. Polyolefins exhibit widely different oxidative stabilities due to both chemical and physical effects. Morphological differences [26], for example, are implicated in the greater susceptibility of polypropylene (PP) and high density polyethylene (HDPE) towards photodegradation than for low density polyethylene (LDPE). PP undergoes mainly oxidative chain scission during processing (Scheme 8). Vicinal hydroperoxides are formed during oxidation of the macromolecular (polymer) chain leading to very high efficiency of thermal and photo-initiation. The breakdown of macromolecular hydroperoxides during melt processing, and subsequently in service under the influence of uv light, yields macroalkoxyl radicals (PPO.); these radicals which are

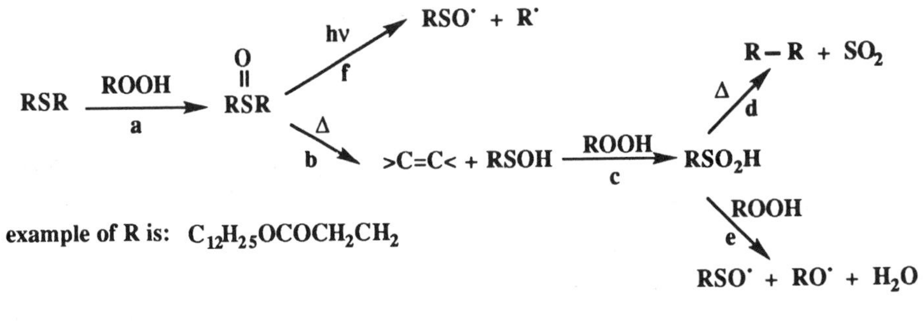

$$RSR \overset{ROOH}{\underset{a}{\longrightarrow}} \overset{O}{\underset{RSR}{\overset{\parallel}{}}} \overset{hv}{\underset{f}{\nearrow}} RSO^\cdot + R^\cdot$$

$$\overset{\Delta}{\underset{b}{\searrow}} > C=C< + RSOH \overset{ROOH}{\underset{c}{\longrightarrow}} RSO_2H \overset{\Delta}{\underset{d}{\nearrow}} R-R + SO_2$$

$$\overset{ROOH}{\underset{e}{\searrow}} RSO^\cdot + RO^\cdot + H_2O$$

example of R is: $C_{12}H_{25}OCOCH_2CH_2$

Scheme 6 Antioxidant mechanism of simple alkyl sulphides.

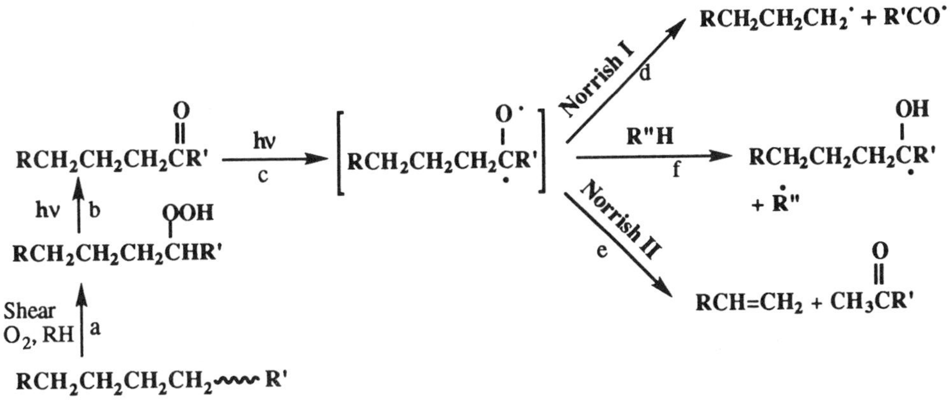

Scheme 7   Polymer hydroperoxidation during processing and further photolysis of derived carbonyl compounds.

key intermediates in the autoxidation cycle are capable of generating new reactive macroalkyl radicals (reaction 5d) via a β-scission process. In practice, the deleterious effects of processing on PP cause a dramatic reduction of its molecular weight as evidenced by a sharp drop in its melt viscosity and high photoxidative instability. In LDPE, however, chain scission is less important whereas cross-linking reactions (associated with increases in both molecular weight and melt viscosity) predominate: these reactions occur by different radical

(alkyl, alkoxyl, alkylperoxyl) combination processess, see reactions 8a and 8c [27].

V.   THERMAL STABILIZATION OF POLYOLEFINS

Polyolefins are exposed to the effects of high temperatures initially during processing and fabrication and subsequently during in-service. Thermal stabilization of po-

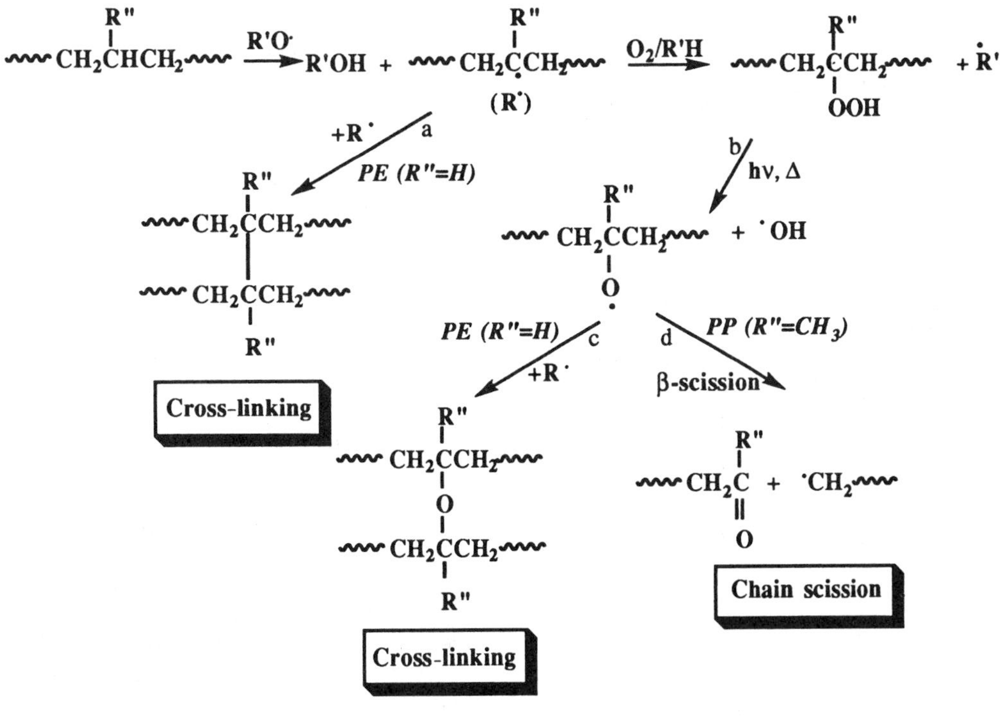

Scheme 8   Oxidative degradation of PE and PP.

lyolefins against mechano-oxidative degradation during high temperature processing is essential in order to stabilize the polymer melt and to minimize the formation of adverse molecular impurities and defects, which may contribute to early mechanical failure of finished product during service. The choice of antioxidants for melt stabilization vary depending on the oxidizability of the base polymer, the extrusion temperature, and the performance target of the end-use application.

The effectiveness of melt processing antioxidants is normally measured by their ability to minimize changes in the melt flow index (MFI) of the polymer that occur in their absence. Chain breaking antioxidants are generally used to stabilize the melt in most hydrocarbon polymers. Hindered phenols (CB—D, e.g., AO-1 to 4, Table 1a) are very effective processing antioxidants for polyolefins, see Table 2. Aromatic amines, on the other hand, have limited use because they give rise to highly colored conjugated quinonoid structures during their antioxidant function. Although hindered phenols do not suffer as much from the problem of discoloring polymers during melt processing, yellowing can occur as a result of intensely colored oxidation products, e.g., stilbene quinone (SQ) from BHT (Scheme 9) [28].

Transformation products of stabilizers formed during melt processing may exert either or both anti- and/or pro-oxidant effects. For example, in the case of BHT, peroxydienones, PxD (reactions 9b, b″) lead to pro-oxidant effects, due to the presence of the labile peroxide bonds, whereas quinonoid oxidation products, BQ, SQ, and G· (reaction 9 b′, c, d) are antioxidants and are more effective than BHT as melt stabilizers for PP [29]. The quinones are effective CB—A antioxidants and those which are stable in their oxidized and reduced forms (e.g., galvinoxyl, G·, and its reduced form, hydrogalvinoxyl, HG) may deactivate both alkyl (CB—A mecha-

nism) and alkylperoxyl (CB—D mechanism) radicals in a redox reaction (reaction 9e and f).

Alkyl sulphides, e.g., Table 1a, AO-16 and 17, which are catalytic peroxide decomposers (PD-C) acting through further oxidation to sulphur acids, are also used as antioxidants during processing of polyolefins (Table 2). However, a major disadvantage of these simple sulphides is that their conversion to sulphur acids (see reaction Scheme 6) involves a parallel series of pro-oxidant reactions involving the formation of propagating radicals (e.g., reaction 6e). Therefore, such antioxidants, e.g., AO-17, are always used in combination with effective chain breaking donor antioxidants, Table 2. Similarly, in the case of phosphites, e.g., AO-14, free radical forming reactions (see Scheme 5, reactions 5a′, 5b) are believed to occur simultaneously with the stoichiometric peroxide decomposition reaction (reaction 5a), hence the normal practice of using phosphites in combination with CB—D antioxidants (e.g., Irganox 1010) for stabilizing polyolefin melts (Table 2).

In the case of long-term thermoxidative stability of polyolefins, stabilizers with high molar masses and lower volatility (e.g., antioxidants 4 and 5 in Table 1) are potentially more effective than those with lower molar masses (and higher volatility) containing the same antioxidant function, e.g., BHT (antioxidant 1 in Table 1). Furthermore, peroxide decomposers, e.g., sulphur-containing compounds, enhance the performance of high molar mass phenols under high temperature in service conditions. For example, dialkyl sulphides such as antioxidants 16 and 17 (see Table 1) are often used as peroxide decomposer synergists in the thermoxidative stabilization of polyolefins.

VI. PHOTOSTABILIZATION OF POLYOLEFINS

The presence of light absorbing impurities, trace level of metals and adventitious species arising from commercial production processes (polymer manufacture, processing and fabrication) renders many commercial polymers (e.g., PE and PP) vulnerable to the deleterious effects of their service environment. The outdoor performance of polymers, however, can be greatly enhanced by a suitable choice of stabilizers. In practical applications, the end-use performance is governed by both physical parameters and chemical factors. Other factors which can affect the ultimate photostability of polymers include sample thickness, polymer crystallinity, coatings, and the presence of other additives, e.g., pigments and fillers. The largest market for light stabilizers is in polyolefins; for example photostabilization of commercial PP is essential for outdoor and indoor end-use applications because of its sensitivity to ultraviolet (UV) light.

An effective photoantioxidant must satisfy not only the basic chemical and physical requirements mentioned

Table 2 Melt Stabilizing Efficiency of Antioxidants in PP (processed in an internal mixer at 190°C). Melt flow index (MFI) Measured at 230°C and 2.16 Kg

Antioxidant	% w/w	MFI, g/10 min
Control (no antioxidant)	0	11.7
Irganox 1010 (AO-4)	0.05	5.9
Irganox 1010 (AO-4)	0.10	4.6
Irganox 1010 (AO-4)	0.20	3.9
Irganox 1076 (AO-3)	0.20	3.7
BHT (AO-1)	0.20	3.5
Irgafos 168 (AO-14)	0.05	7.7
Irgafos 168 (AO-14)	0.10	7.2
Irganox PS 800 (AO-17)	0.10	8.4
α-tocopherol (AO-9)	0.05	3.8
α-tocopherol (AO-9)	0.10	3.6
Irganox 1010+Irgafos 168	0.05+0.05	4.3
Irganox 1010+Irganox PS 800	0.05+0.10	4.9

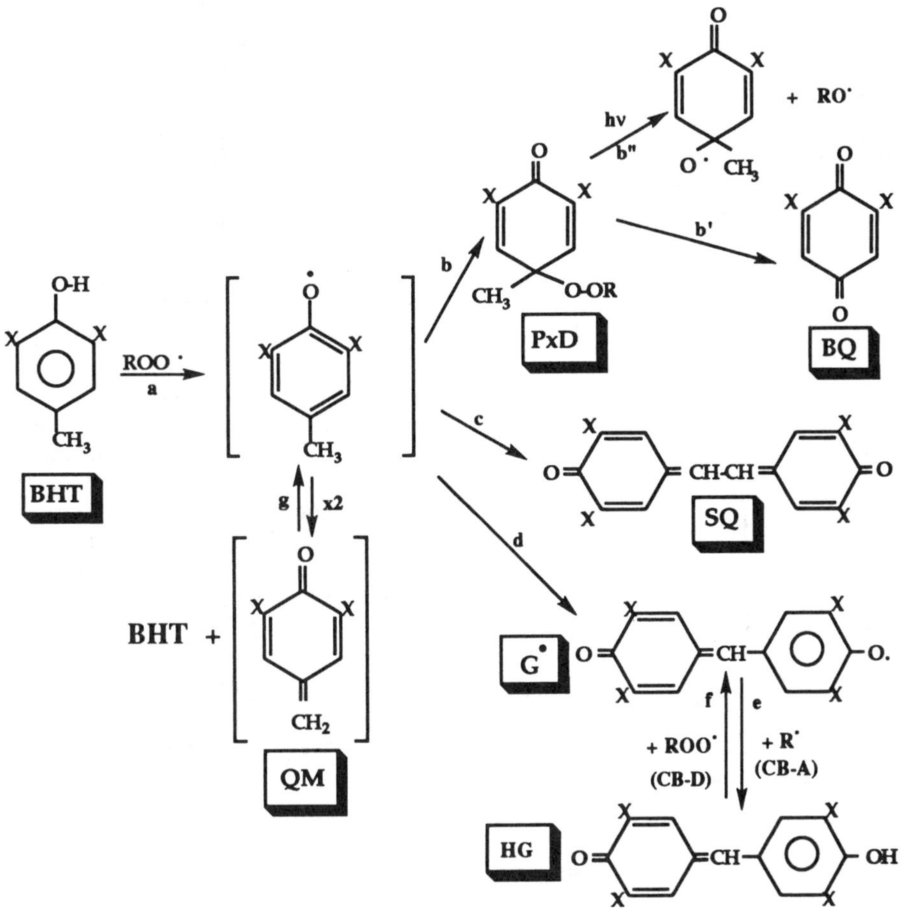

Scheme 9 Oxidative transformation products formed during the antioxidant action of BHT.

above, but must also be stable to UV light and to withstand continuous periods of exposure to UV light without being destroyed or effectively transformed into sensitizing products. Chain breaking donor antioxidants, e.g., hindered phenols, are relatively ineffective under photoxidative conditions as they are generally unstable to UV light and some of their oxidative transformation products are photosensitizing. For example, in the case of the commercial hindered phenol BHT (AO-1, Table 1), which is a good processing antioxidant, the peroxydienones (PxD, reaction 9b) undergo photoisomerization [28] through a radical intermediate and can act therefore as powerful photoinitiators (see reaction 9b″). Similarly, many sulphur containing antioxidants which are very effective peroxide decomposers (PD) and thermal antioxidants are not effective photoantioxidants [30]. For example, the intermediate sulphoxide that is formed during the antioxidant action of thiodipropionate esters, e.g., AO-17 Table 1a, photodissociates readily to free radicals (reaction 6f) with deleterious effects on their photoantioxidant activity. However, both the hind-

ered phenols and these sulphide antioxidants can synergise with UV stabilizers and become much more effective photoantioxidants.

Metal containing sulphur complexes including many metal thiolates, such as dithiocarbamates, e.g., AO-30, dithiophosphates, AO-31, xanthates, AO-32, differ from sulphide antioxidants in that they are generally more stable to UV light. All metal dithiolates are effective processing stabilizers for polyolefins but their activity as thermal and photo antioxidants depends on the nature of the metal ion. Transition metal complexes (e.g., M=Ni, Co, Cu) are more photostable than group II metal complexes, e.g., Zn, hence better overall photostabilizing effectiveness (Table 3). Like most other metal dithiolates, the primary oxidation product formed during the antioxidant function of NiDRP (AO-31) is the corresponding disulphide that undergoes further oxidation to give different sulphur acids (the real catalysts for peroxide decomposition), see Scheme 10 [19].

The UV stabilizing action of nickel and iron complexes (e.g., NiDRC and FeDRC) is strongly concentra-

Table 3 UV-Embrittlement Times (EMT) of PP and LDPE Films Containing Different Concentrations of Antioxidants (processed in an internal mixer at 190°C and 150°C, respectively, and exposed to UV light in an accelerated sunlamp–blacklamp UV aging cabinet)

Antioxidant	Concentration mol/100g x 10⁴	EMT, h PP	EMT, h LDPE
Controls (no antioxidant)	0	90	1200
ZnDEC	2.5	175	1400
NiDEC	0.25	140	—
NiDEC	2.5	740	—
NiDEC	3	790	1800
NiDEC	10	840	—
FeDMC	0.25	85	
FeDMC	2.5	150	
FeDMC	5	336	
CuOctX	3	250	
FeOctX	3	330	
NiOctX	3	700	
CoOctX	3	1600	
Irganox 1076	3	—	1800
UV 531	3	245	1650
Tinuvin 770	3	950	2400
ZnDEC + UV 531	3	—	>4000, Synergism
Tin 770 + UV 531	4	1750	—
NiDEC + Irganox 1076	3	—	1580, Antagonism
NiDEC + Tinuvin 770	3	—	1850, Antagonism

tion dependent; nickel complexes are much more stable to UV light than iron complexes (Table 3). Further, the behavior of the iron complexes (e.g., FeDRC) changes with concentrations; at low concentration (below 0.05%, in polyolefins) the iron complexes photoxidize rapidly eliminating the thiocarbomoyl ligand and freeing the metal ion, which is a very powerful sensitizer of photoxidation through its reaction with light and hydroperoxides (Scheme 11 reactions e, f). Conversely, at high concentration, FeDRC shows a photo-induction period during which the iron complex behaves as a photoantioxidant due to the oxidation of the sulphur ligand to low molar mass sulphur acids (by a mechanism similar to that of the dithiophosphate discussed above), (see Scheme 10 and Scheme 11 reaction a, b and Table 3) [31].

This dual role of FeDRC (an antioxidant during processing, storage and very early stages of exposure to UV light, and an effective photosensitizer on further exposure to light) is suitable for the precise time controlled stabilization required for agricultural applications. Such systems, referred to as the "Scott-Gilead" systems are now being used for agricultural applications in several countries [32].

The UV absorber (UVA) class (e.g., 2-hydroxy-benzophenones and benzotriazoles, see AO 23–27 in Table 1b) are stable to UV light and have high extinction coefficients in the region 330–360 nm. They operate pri-

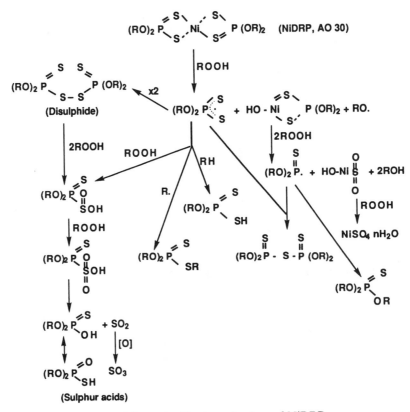

Scheme 10 The antioxidant mechanism of NiDRP.

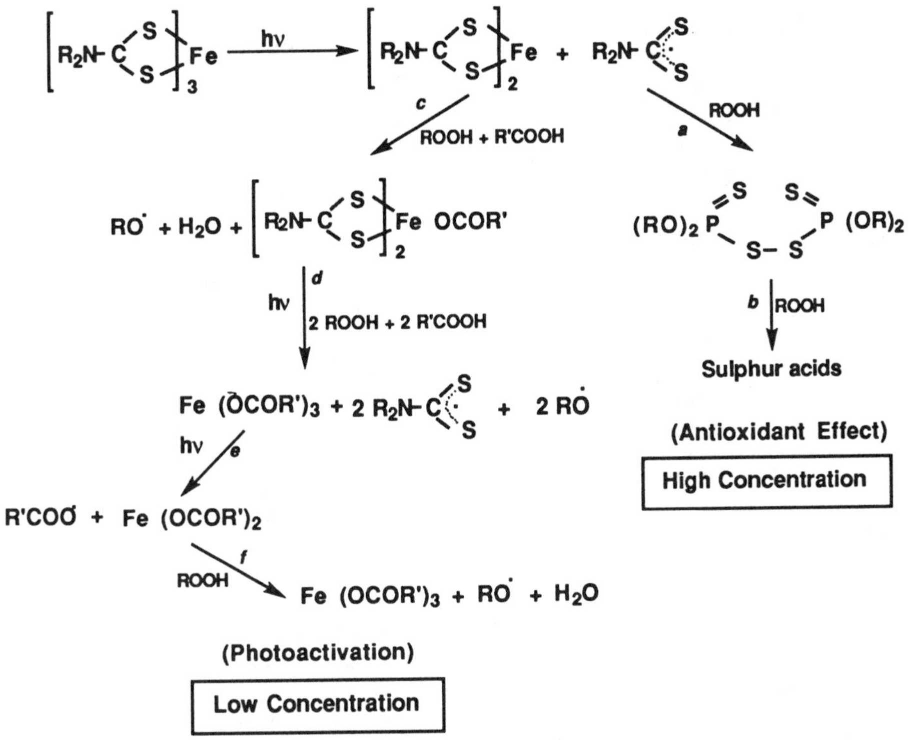

Scheme 11 Mechanism of antioxidant action of FeDRC.

marily by absorbing UV light and dissipating it harmlessly as thermal energy, e.g., via an excited state keto-enol tautomerism. It is important to point out that most known UV stabilizers do not act by a single mechanism; e.g., UV531 (AO 23, Table 1b) function not only as a UV screen but also as a sacrificial antioxidant removing chain initiating radicals (e.g., alkoxyl radicals) via a weak CB—D mechanism. The limited effectiveness of UV531 as a UV stabilizer Table 3) is not due to its photolability but because of its instability toward hydroperoxides and carbonyl compounds under photoxidative conditions. Thus UV531 is a much more effective photostabilizer for mildly processed saturated polyolefins than for severely oxidized polyolefins [33]. UV absorbers such as UV531 synergize with peroxide decomposers (Table 3).

The early recognition of the role of stable nitroxyl free radicals, e.g., 2,2,6,6-tetramethyl-4-oxopiperidine, and their hindered amine precursors, in polymer stabilization soon led to the development of the hindered amine light stabilizer (HALS) class of photoantioxidants. The first HALS, Tinuvin 770, AO-33, (commercialized in 1974) proved to offer much higher UV-stability to polymers than any conventional UV-stabilizer available at the time such as UV-absorbers, nickel compounds and benzoates, Table 3).

Unlike secondary aromatic amines, aliphatic analogues (HALS) are not effective CB—D antioxidants due to the high N—H bond strength, nor are they good UV absorbers or excited chromophore quenchers. Mechanisms involving quenching of singlet oxygen, complexation with important transition metal ions did not appear important. The real antioxidant action of HALS was found to be due to the ability of their generated stable nitroxyl radical to scavenge the alkyl radicals in competition with oxygen. This alone, however, could not account for the very high effectiveness of HALS; nitroxyl radical regeneration mechanism was, therefore, proposed and researches throughout the 1970s and 1980s produced evidence for its regeneration from both the corresponding hydroxylamine and alkylhydroxylamine [34–37], (Scheme 12) [37].

VII. FACTORS CONTROLLING ANTIOXIDANT EFFECTIVENESS AND USE OF REACTIVE AND BIOLOGICAL ANTIOXIDANTS

The effectiveness of antioxidants depends not only on their intrinsic activity but also on their physical retention in the polymer. Migration of antioxidants into the sur-

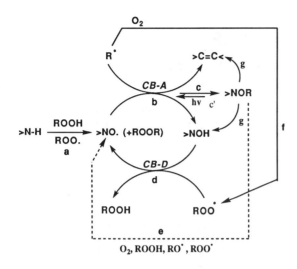

Scheme 12 Antioxidant mechanism of HALS.

rounding environment, e.g., leaching out into food constituents by the extracting action of oils and fats, leads not only to premature failure of the polymer article but also to problems associated with health hazards and toxicological effects. Solutions have been sought for enhancing the permanence of antioxidant in polymers especially when used in applications that require direct contact with aggressive solvent extractive environ-

ments. These include the use of large molar mass antioxidants, copolymerizing antioxidant functions in polymers (during polymer synthesis) and the in-situ grafting of antioxidant functions onto polymer backbones during the fabrication and manufacture of polymer articles using reactive processing methods. Each of these approaches has advantages and disadvantages. The basic tenet here is that if the antioxidant structure can become fully molecularly dispersed in the polymer matrix, and achieve 100% grafting efficiency while retaining its antioxidant function, then the problems of physical loss, migration, and safety of antioxidants will not be an issue.

Grafting of reactive antioxidants on the polymer backbone during processing is an important approach to polymer stabilization. Under reactive processing procedures, very high grafting levels of antioxidants can be achieved, hence virtually eliminating the problems of physical loss and migration of antioxidants. Grafting of antioxidants that contain polymer reactive functions (e.g., Scheme 13, AO-I, II, III) on saturated polymers can be facilitated by the incorporation of small concentration of a free radical initiator during processing. However, the efficiency of the grafting reaction (reaction 13b) was found to be low (about 20–30%) because of competitive homopolymerization of the antioxidant, reaction 13c [38]. Higher levels of grafting can be achieved when the polymer melt reaction is carried out in the presence of non-homopolymerizable antioxidants (e.g., bis-(2,2,6,6-tetramethylpiperidine-4-yl) maleate [39], or by

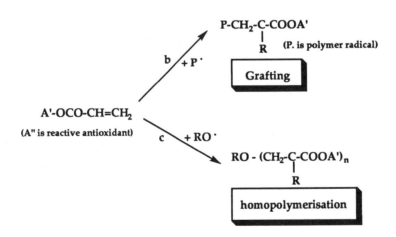

Examples of the reactive antioxidants A" are:

Scheme 13 Grafting reactions of some reactive antioxidants.

processing reactive antioxidants with a co-agent (a multifunctional monomer without an antioxidant function) [40]. In these case, the efficiency of the melt-grafting reaction can be increased to over 90% together with high efficiency of stabilization, albeit after achieving a delicate balance between the chemical, physical, and rheological characteristics of the reactants.

The advantages in adopting a reactive processing approach to polymer stabilization include high retention of antioxidants in the polymer matrix leading to efficient stabilization, reduced risk of migration of antioxidants into the human environment, cost effectiveness and attractiveness of the option of producing highly modified concentrates for use as conventional additives in the same or other polymers, and, site-specific antioxidant grafts can be targeted for premium performance of speciality niche products.

Another approach to safer stabilization is to use a biological antioxidant such as vitamin E (α-tocopherol is the active form of vitamin E, AO-9, Table 1a). It is essentially a hindered phenol which acts as an effective chain breaking donor antioxidant, donating a hydrogen to ROO· to yield a very stable tocopheroxyl radical. α-Tocopherol is a very effective melt stabilizer in polyolefins that offers high protection to the polymer at very low concentration [41], (Table 2).

VIII. SYNERGISM AND ANTAGONISM

A cooperative interaction between two or more antioxidants (or antioxidant function) that leads to an overall antioxidant effect greater than the sum of the individual effects of each antioxidant is referred to as synergism. Synergism can be achieved in different ways. It may arise from the combined action of two chemically similar antioxidants, e.g., two hindered phenols (homosynergism), or when two different antioxidant functions are present in the same molecule (autosynergism); the latter is exemplified by many commercial antioxidants (e.g., Irgastab 2002, AO 29 Table 1b), which have CB and UVA activity.

Synergism can also arise from cooperative effects between mechanistically different classes of antioxidants, e.g., the chain breaking antioxidants and peroxide decomposers (heterosynergism) [42]. For example, the synergism between hindered phenols (CB—D) and phosphites or sulphides (PD) is particularly important in thermal oxidation (Table 2). Similarly, effective synergism is achieved between metal dithiolates (PD) and UV-absorbers (e.g., UV 531), as well as between HALS and UV-absorbers, (Table 3).

Anti-synergistic effects, on the other hand, arise when antioxidants show antagonistic effects and give rise to a reduced net effect when compared to the sum of their individual effects [42]. Antagonism during photoxidation of PP occurs when phenolic antioxidants, e.g.,

Irg 1076, AO-3 (CB), are used in combination with metal dithiolates, e.g., NiDEC, AO-30 (PD), due to the sensitized photoxidation of dithiolates by the oxidation products of phenols, particularly stilbenequinones (SQ, see reaction 9C) (Table 3). Hindered piperidines exhibit a complex behavior when present in combination with other antioxidants and stabilizers; they have to be oxidized initially to the corresponding nitroxyl radical before becoming effective. Consequently, both CB-D and PD antioxidants, which remove alkyl peroxyl radicals and hydroperoxides, respectively, antagonise the UV stabilizing action of this class of compounds (e.g., Table 3, NiDEC + Tin 770). However, since the hindered piperidines themselves are neither melt- nor heat-stabilizers for polymers, they have to be used with conventional antioxidants and stabilizers.

REFERENCES

1. A. W. Hoffman, *J. Chem, Soc.*, *13*: 87 (1861).
2. US Patent 99, 935 (1870).
3. US Patent, 680, 387 (1901).
4. German Patent 221, 310 (1908).
5. C. Moureu, and C. Dufraisse, *Bull. Soc. Chim.*, *31*(4), 1152 (1922).
6. J. L. Bolland and G. Gee, *Trans. Faraday Soc.*, *42*: 236, 244 (1946); J. L. Bolland, *Quart. Rev.*, *3*: 1 (1949).
7. L. Bateman, and G. Gee, *Proc. Roy. Soc.* {a} 195, 376 (1948–9); L. Bateman, and A. L. Morris, *Trans. Faraday Soc.*, *49*: 1026 (1953).
8. A. V. Tobolsky, D. J. Metz, and R. B. Mesrobian, *J. Amer. Chem. Soc.*, *72*: 1942 (1950).
9. J. R. Shelton, *Rubb. Chem. Tech.*, *30*: 1251 (1957).
10. H. L. J. Backstrom, *J. Am. Chem. Soc.*, *49*: 1460 (1927) & *51*: 90 (1929).
11. C. D. Lowry, G. Egloff, J. G. Morrell, and C. G. Dryer, *Ind. Eng. Chem*, *25*: 804 (1933).
12. J. L. Bolland and P. Ten Have, *Trans Faraday Soc.*, *43*: 201 (1947); J. L. Bolland and P. Ten Have, *Discuss Farad Soc.*, *2*: 252 (1947).
13. W. F. Watson, *Trans. IRI 29*: 32 (1953).
14. G. Scott, *Brit. Polym. J.*, *3*: 24 (1971).
15. G. Scott, *Developments in Polymer Stabilization-7,* Elsevier App Sci, London, p. 65 (1984).
16. G. W. Kennerly and W. L. Patterson, *Ind. Eng. Chem.*, *48*: 1917 (1956).
17. S. Al-Malaika, *Atmospheric Oxidation and Antioxidants, vol. 1,* (G. Scott, ed.), Elsevier Science Publishers, Amsterdam, Chap. 5 (1993).
18. K. Schwetlick, *Mechanisms of Polymer Degradation and Stabilization* (G. Scott, ed.), Elsevier Science Publishers, New York, Chap. 2, (1990).
19. S. Al-Malaika, *Mechanisms of Polymer Degradation and Stabilization* (G. Scott, ed.), Elsevier Science Publishers, New York, Chap. 3 (1990).
20. S. Al-Malaika, K. B. Chakraborty and G. Scott, *Developments in Polymer Stabilization-6,* (G. Scott, ed.), App. Sci. Pub., London, Chap. 3 (1983).
21. H. Muller, *Plastics Additives Handbook,* (R. Gachter, and H. Muller, eds.), Hanser, Munich, Chap. 2 (1987).
22. F. Gugumus, *Plastics Additives Handbook,* (R. Gachter, and H. Muller, eds.), Hanser, Munich, p. 128 (1987).

23. K. B. Chakraborty and G. Scott, *Europ. Polym. J., 15*: 731 (1977).
24. G. Scott, *Developments in Polymer Degradation-1,* (N. Grassie, ed.), App. Sci. Pub., London, Chap. 7 (1978).
25. D. J. Carlsson, A. Graton and D. M. Wiles, *Developments in Polymer Stabilisation-1,* (G. Scott, ed.), App. Sci. Pub., London, Chap. 7 (1979).
26. W. L. Hawkins, *Polymer Stabilization,* (W. L. Hawkins, ed.), Wiley Interscience, London and New York, p. 7, (1972).
27. S. Al-Malaika, and G. Scott, *Degradation and Stabilisation of Polyolefins,* (N. S. Allen, ed.), App. Sci. Publ., London, Chap. 6 (1983).
28. J. Pospisil, *Developments in Polymer Stabilisation, vol. 1,* (G. Scott, ed.), App. Sci. Pub., London, Chap. 1 (1979).
29. T. J. Henman, *Developments in Polymer Stabilisation vol. 1,* (G. Scott, ed.), App. Sci. Pub., London, Chap. 2 (1979).
30. J. R. Shelton, and K. E. Davis, *J. Sulfur Chem., 8*: 217 (1973).
31. S. Al-Malaika, A. Marogi, and G. Scott, *J. App. Polym. Sci., 31*: 685 (1986); S. Al-Malaika, A. Marogi, and G. Scott, *J. App. Polym. Sci., 33*: 1455 (1987), S. Al-Malaika, A. Marogi, and G. Scott, *J. App. Polym. Sci., 34*: 2673 (1987).
32. G. Gilead, *Degradable Polymers,* (G. Scott, and D. Gilead, eds.) Chapman and Hall, New York, Chap. 10 (1995).
33. S. Al-Malaika, and G. Scott, *Europ. Polym. J., 19*: 241 (1983).
34. V. YaShlyapintokh, and V. B. Ivanov, *Developments in Polymer Stabilisation-5,* (G. Scott, ed.), Elsevier App. Sci., London Chap. 3 (1982).
35. D. W. Grattan, D. J. Carlsson, and D. M. Wiles, *Polym. Deg. Stab., 1*: 69 (1979).
36. G. Scott, *Developments in Polymer Stabilisation-7,* (G. Scott, ed.), Elsevier App. Sci., London Chap. 2 (1985).
37. S. Al-Malaika, E. O. Omikorede, and G. Scott, *J. Appl. Polym. Sci., 33*: 703 (1987).
38. S. Al-Malaika, G. Scott and B. Wirjosentono, *Polym. Deg. & Stab, 40*: 233 (1993).
39. S. Al-Malaika, A. Q. Ibrahim, and S. Al-Malaika, *Polym. Deg. Stab., 22*: 233 (1988).
40. S. Al-Malaika, and G. Scott, Patent Application Number, PCT/W090/01506 (1990).
41. S. Al-Malaika and S. Issenhuth, *Advances in Chemistry Series,* ACS, Washington, (R. L. Clough, K. T. Gillen, and N. C. Billingham, eds.) in press.
42. G. Scott, *Atmospheric Oxidation and Antioxidants, vol. 2,* Elsevier Science Publishers, Amsterdam, Chap. 9 (1993).

9

Gamma Radiation Induced Preparation of Polyelectrolytes and Its Use for Treatment of Waste Water

T. Siyam
Hot Laboratory Centre, Atomic Energy Authority, Cairo, Egypt

I. INTRODUCTION

Interest in the chemistry of water-soluble polymers (polyelectrolytes) has been continually increasing during the past 45 years. The tremendous scope of utility for water-soluble polymers has led to a vigorous search for new materials and the rapid development of polyelectrolytes into a dynamic field of industrial research. Growth in this field has been especially rapid since 1960; and today, many companies are engaged in synthesis and applications research on polyelectrolytes that are primarily used in four main marketing areas: water treatment, paper, textiles, and oil recovery [1]. Polyacrylamide gel was also used as soil conditioner [2–4].

Polyelectrolytes are classified into three main groups: nonionic, anionic, and cationic depending upon the nature of the residual charge on the polymer in aqueous solution as shown in Table 1.

The most verstatile and useful type of polyelectrolytes are the cationic, which are comprised of three classes: ammonium (primary [I⁰], secondary [II⁰], tertiary [III⁰] amines, and quaternaries), sulfonium, and phosphonium quaternaries, as shown in Table 2.

The extensive industrial and commercial utilization of water-soluble polymers (polyelectrolytes) in water treatment has been developed based on the charge along the polymer chains and the resultant water solubility. The use of water-soluble polymers in water treatment has been investigated by several authors [5–26] in the recovery of metals; radioactive isotopes, heavy metals, and harmful inorganic residues. This allows recycling water in the industrial processes and so greatly saves the consumption of water in industry. This is evidently of great importance to meet the requirements of the population, which is continually increasing as is its water consumption. In addition, there is the need to save water for agricultural use to parry the fear of a food crisis.

Water-soluble polymers such as nonionic, anionic, cationic, and amphoteric are described as shown in Fig. 1.

Polymeric resins such as poly(acrylamide-acrylic acid) [24,25] [cationic resin, pAM-AA], poly(acrylic acid-diallylethylamine-HCl) [20] [amphoteric resin, pAA-DAEA-HCl], and poly(acrylamide-acrylic acid-diallylamine-HCl [26] [amphoteric resin. pAM-AA-DAA-HCl] and poly(acrylamide-acrylic acid-diallylethylamine-HCl) [26] [amphoteric resin, pAM-AA-DAEA-HCl] were also used in water treatment.

II. GAMMA RADIATION-INDUCED PREPARATION OF WATER SOLUBLE POLYMERS

A. Preparation of Nonionic Polymer (polyacrylamide)

Polyacrylamide (pAM) and copolymers of acrylamide are used on a large scale in waste water treatment and other industrial applications. All of these reasons show that the production and use of polyacrylamide (pAM) and copolymers of acrylamide are a material objective.

Acrylamide is polymerized by the conventional free radical initiators, e.g., peroxides [27,28], redox pairs [29–33], and azo compounds [34]. Electro-chemical initi-

Table 1 Polyelectrolytes

Nonionic	Anionic	Cationic
Polyethers	Carboxylic	Amonium
Polyamides	Sulfonic	Amines
Poly(N-vinyl-	Phosphonic	Quaternaries
heterocyclies)		Sulfonium
		Phosphonium

ation [35], ultrasonic waves [36], photo-chemical sensitizers [37–40], ultraviolet radiation [41], radio-isotopes radiation [42–44], x-rays [45], gamma radiation [45–48], and accelerated electrons [49] also produce pAM.

III. KINETICS OF RADIATION-INDUCED POLYMERIZATION OF ACRYLAMIDE IN AQUEOUS SOLUTION

Acrylamide polymerization by radiation proceeds via free radical addition mechanism [37,38,40,45,50]. This involves three major processes, namely, initiation, propagation, and termination. Apart from the many subprocesses involved in each step at the stationary state the rates of formation and destruction of radicals are equal. The overall rate of polymerization (R_p) is so expressed by Chapiro [51] as:

$$R_p = k_p k_t^{-1/2} R_i^{1/2} [M] \tag{1}$$

Table 2 Cationic Polyelectrolytes

I Amonium			
	$\underset{\underset{I^0}{\overset{\displaystyle H}{\mid}}}{\overset{\displaystyle H}{\underset{\displaystyle \mid}{\overset{\displaystyle +}{N}}}}$ H	$\underset{\underset{II^0}{\overset{\displaystyle H}{\mid}}}{\overset{\displaystyle H}{\underset{\displaystyle \mid}{\overset{\displaystyle +}{N}}}}$ R	$\underset{\underset{III^0}{\overset{\displaystyle H}{\mid}}}{\overset{\displaystyle H}{\underset{\displaystyle \mid}{\overset{\displaystyle +}{N}}}}$ R
I^0, II^0, III^0 amines (Protonated)			
Quaternary	$-\overset{\overset{\displaystyle R}{\mid +}}{\underset{\underset{\displaystyle R}{\mid}}{N}}-R$		
II. Sulfonium	$-\overset{}{\underset{\underset{\displaystyle R}{\mid}}{S^+}}-R$		
III. Phosphonium	$-\overset{\overset{\displaystyle R}{\mid +}}{\underset{\underset{\displaystyle R}{\mid}}{P}}-R$		

where k_p, k_t, and R_i are the propagation, termination, and initiation rate constant, respectively, and $[M]$ is the initial monomer concentration.

In homogeneous media acrylamide is terminated by bimolecular termination [51–53]. In this case the degree of polymerization ($\overline{DP}_n$), as defined by Chapiro [51] is:

$$\overline{DP}_n = 2\,k_p k_t^{-1/2} R_i^{1/2}\,[M] \tag{2}$$

The advantage of using radiation-induced polymerization is that the polymer is homogeneous and free from any impurities. Moreover, the molecular weight of the formed polymer is controlled by varying doses and doserates.

IV. PREPARATION OF ANIONIC AND CATIONIC POLYACRYLAMIDE

Anionic polyacrylamide was prepared by gamma radiation-initiated copolymerization of acrylamid with sodium acrylate in aqueous solution at optimum conditions for the copolymerization [17]. The copolymerization process produces water-soluble poly(acrylamide-sodium acrylate [pAM-AANa] of high molecular weight [17,54].

Radical polymerization of diallylamine derivatives produce water-soluble polymers of low molecular weight [22,55–57]. In order to increase the molecular weight, acrylamide has been copolymerized with these diallylamine derivatives to produce cationic polymers with variable charge density depending on the content of the structural units of pyrrolidinium rings and acrylamide in polymeric chains [22,55,58–61].

Poly(acrylamide-diallylethylamine-HCl) (cationic polyacrylamide pAM-HCl) was prepared by gamma radiation-initiated copolymerization of acrylamide with diallylethylamine-HCl in aqueous solution at the optimum composition for copolymerization of acrylamide with diallylamine derivatives [61].

A. Kinetics of Copolymerization Process

1. Influence of Dose-Rate

The dose-rate was varied by lining the irradiation chamber of the gamma cell irradiation by lead foil of uniform thickness [17]. The dose was kept constant at 0.15 and 0.35 KGy for copolymerization of AM-AANa and AM-DAEA-HCl system, respectively. The results are shown in Figs. 2 and 3, which show that the rate of polymerization, R_p increases while the degree of polymerization ($\overline{DP}_n$) and the intrinsic viscosity [η] decrease with the dose-rate. The exponents of the dose-rate for AM-AANa system [17,54] were determined to be:

$$R_p \propto I^{1.3} \tag{3}$$

$$Dp_n \propto I^{-0.5} \tag{4}$$

(a) - Nonionic polymers:

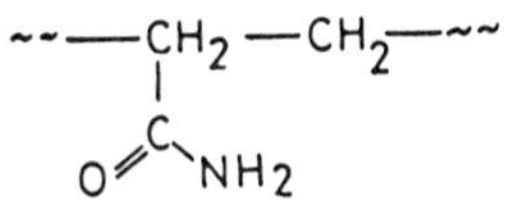

$$\sim\!\!-CH_2-CH_2-\!\!\sim$$

Polyacrylamide (pAM)

"Structure 1"

(b) - Anionic polymers:

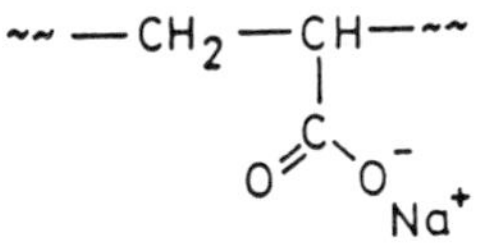

Polysodium acrylate (pAAMa)

"Structure 2"

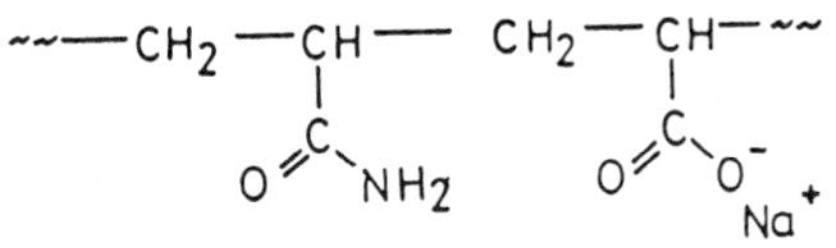

polyacrylamide-sodium acrylate
(anionic polyacrylamide pAM-AANa)

"Structure 3"

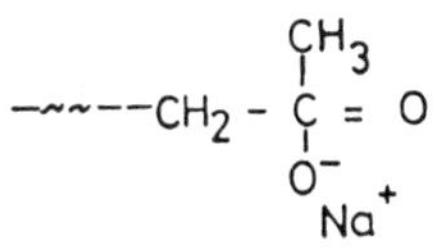

Polysodium methacrylate
(pAAMNa)

"Structure 4"

(c) - Cationic polymers:

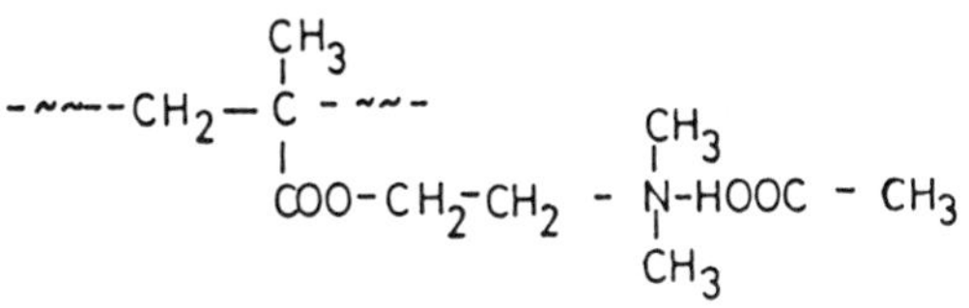

Poly β-dimethylaminoethyl methacrylate hydroacytate
(pDMAEM)

"Structure 5"

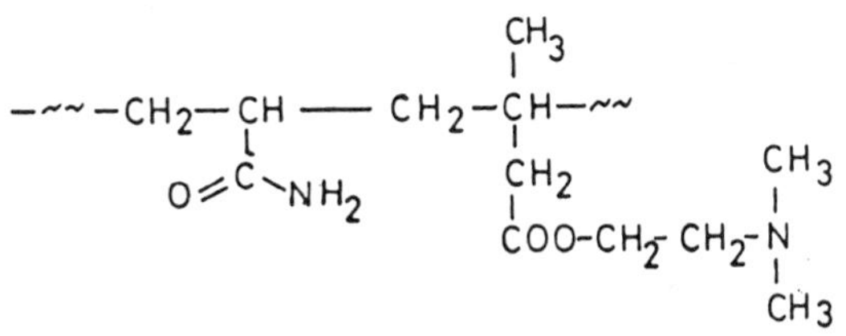

Poly(acrylamide-N,N-dimethylaminoethyl methacrylate
(pAM-DMAEM MC)

"Structure 6"

Figure 1 Structures of water-soluble polymers.

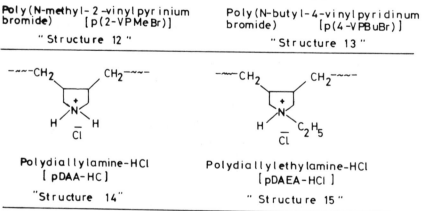

Poly(diallyldimethylammonium choloride-SO₂ [pDADMAm-Cl-SO₂]

"Structure 7"

Poly(diallyldiethylammonium choloride-SO₂ [pDADEAmCl-SO₂]

"Structure 8"

Poly(diallylmethylamine-HCl-SO₂) [pDAMA HCl-SO₂]

"Structure 9"

Poly(diallylamine-HCl-SO₂) [pDAA-HCl-SO₂]

"Structure 10"

Polyethyleneimine [pEI]

"Structure 11"

Poly(N-methyl-2-vinylpyrinium bromide)   [p(2-VPMeBr)]

"Structure 12"

Poly(N-butyl-4-vinylpyridinum bromide)   [p(4-VPBuBr)]

"Structure 13"

Polydiallylamine-HCl [pDAA-HC]

"Structure 14"

Polydiallylethylamine-HCl [pDAEA-HCl]

"Structure 15"

Figure 1   (*Continued*)

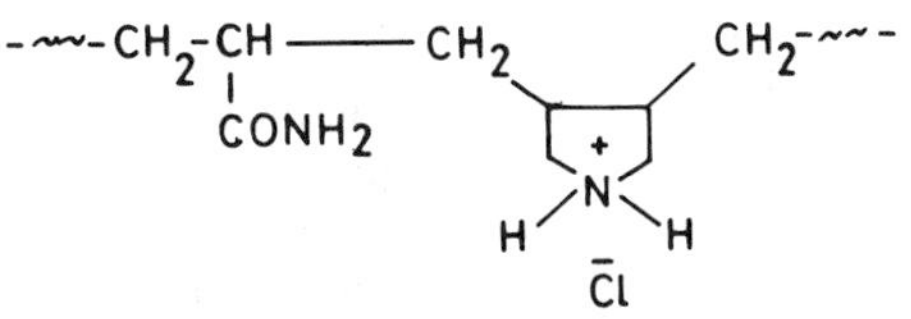

$$-\sim-CH_2-CH———CH_2———CH_2-\sim-$$

CONH$_2$

Poly(acrylamide-diallylamine-HCl)
[Polyamido-amines, pAM-DAA-HCl]

"Structure 16"

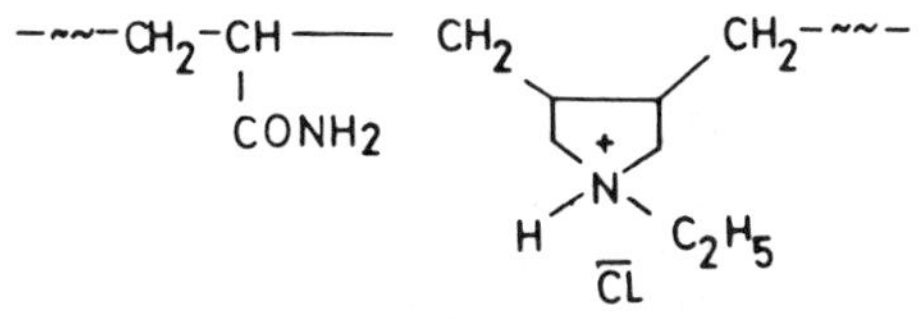

$$-\sim-CH_2-CH———CH_2———CH_2-\sim-$$

CONH$_2$

Poly(acrylamide-diallylethylamine-HCl)
[Polyamido-amines, pAM-DAEA-HCl]
"Structure 17"

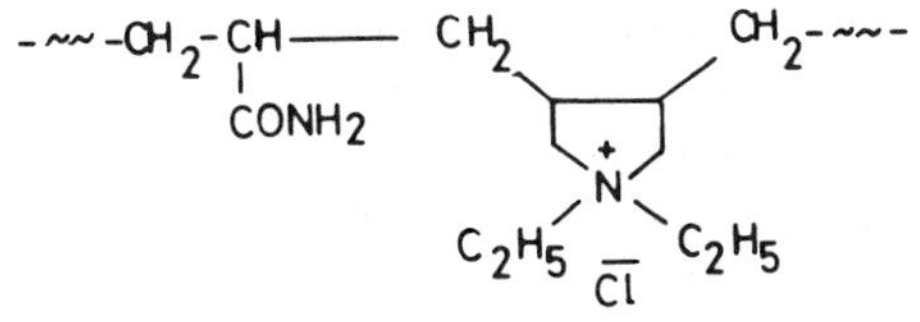

$$-\sim-CH_2-CH———CH_2———CH_2-\sim-$$

CONH$_2$

Poly(acrylamide-diallyldiethylammonium-chloride
[Polyamido-amines, pAM-DADEAm-Cl]
"Structure 18"

(d) **Amphoteric polymers:**

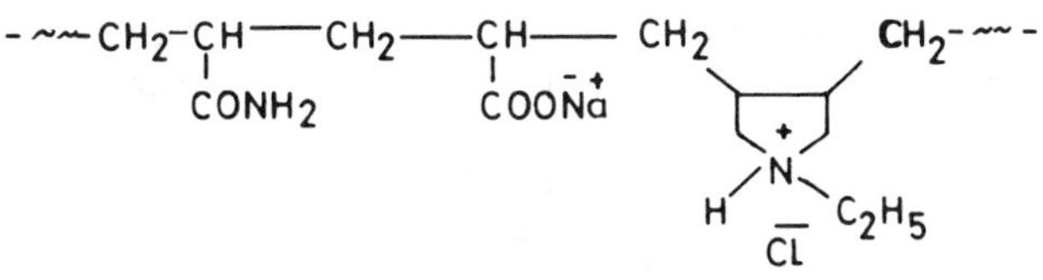

$$-\sim-CH_2-CH———CH_2———CH———CH_2———CH_2-\sim-$$

CONH$_2$ COONa

Poly(acrylamide-sodium acrylate-diallylethylamine-HCl)
[pAM-AANa- DAEA-HCl]
"Structure 19"

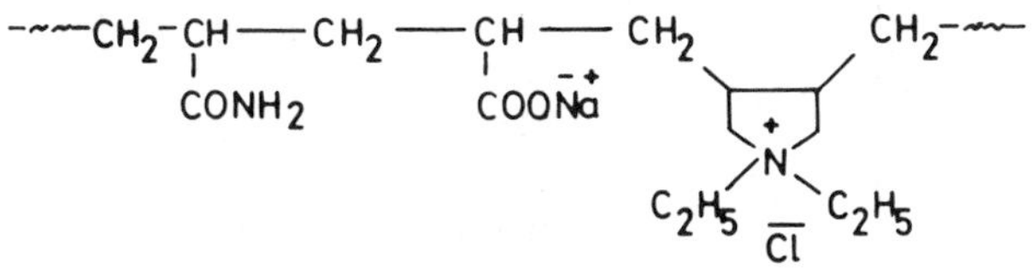

$$-\sim-CH_2-CH———CH_2———CH———CH_2———CH_2-\sim-$$

CONH$_2$ COONa

Poly(acrylamide-sodium acrylate-diallyldiethylammonium-
chloride)
[pAM-AANa-DADEAm-Cl]

"Structure 20"

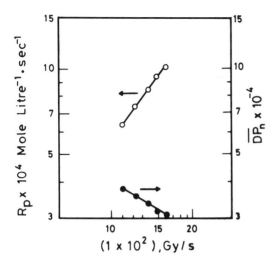

Figure 2 Variation of R_p and $\overline{DP}_n$ with the dose-rate.

However, the exponent of the dose-rate for AM-DAEA-HCl system was also determined to be [22]:

$$R_p \propto I^{0.9} \tag{5}$$

The reported values for the exponent of the dose-rate for the polymerization rate in gamma radiation-induced copolymerization of acrylamide with methyl chloride salt of N,N-dimethylaminoethyl methacrylate (DMAEM-MC) in aqueous solution was found to be 0.8 [16]. However, the dose-rate exponent of the polymerization rate at a lower dose-rate was found to be slightly higher than 0.5 for gamma radiation-induced polymerization of acrylamide in aqueous solution [45,62].

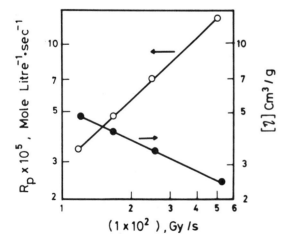

Figure 3 Effect of dose-rate on the copolymerization of acrylamide with DAEA-HCl. $\bigcirc = R_p$; $\bullet = [\eta]$.

It will be known that for the radical polymerization the increase on the rate of initiation would increase the polymerization rate Eq. (1) and decrease the degree of polymerization Eq. (2). In the present systems, the monomer concentration was relatively high so that initiating radicals are formed to some extent from the monomer and solvent, i.e., R_i in Eq. (1) may be represented as follows [51]:

$$R_i = I \, f(\phi_M[M + \phi_S[S]) \tag{6}$$

where ϕ_M and ϕ_S are the rate of production of free radicals in the monomer and in the solvent expressed in mole per liter per unit of radiation dose and I is the dose-rate. Thus, the increase in the dose-rate would increase the rate of initiation, which leads to an increase in the rate of polymerization, R_p, and a decrease in the degree of polymerization, $\overline{DP}_n$.

The observed increase in the expont of the dose-rate of the polymerization rate in our systems can be attributed to the peculiarity of acrylamide, which scavenges almost all the radiolytic products of water [63]. The reduction of termination reactions of two propagating radicals, because the reaction becomes highly viscous as the polymerization proceeds (trapping of alive radicals) [16]. Dainton and Tordoff [40] found that the $k_p/k_t^{0.5}$ for acrylamide exceeds that of any other polymerization system. Electron and charge transfer processes may also play an effective role in multiple initiation of polymeric radicals. These all together leave the system highly sensitive toward radiation. So, the abnormally high value of the dose-rate exponent might be expected [54].

Equation (4) shows that the dose-rate exponent of the degree of polymerization agrees with the theory Eq. (2). However, the degree polymerization and the intrinsic viscosity decrease with increasing dose rate is probably due to increased termination reactions caused by the increasing radical population at high dose-rate [22].

2. Influence of Monomer Concentration

Keeping the composition of copolymerization media constant the total comonomer concentration of which is varied. The absorbed dose was kept constant at 0.14 KGy for the AM-AANa and at 0.35 KGy for the AM-DAEA-HCl systems. The results are shown in Figs. 4 and 5, which show the rate of polymerization, R_p, the degree of polymerization, and the intrinsic viscosity increase with increasing monomer concentration. At comonomer concentration >2.1 M/L, $\overline{DP}_n$ decreases with increasing comonomer concentration. From the logarithmic plots, exponents of the comonomer concentration for the AM-AANa system were determined to be [17,54].

$$R_p \propto [M]^{1.33} \tag{7}$$

$$\overline{DP}_n \propto [M]^{1.34} \qquad [M] < 2.1 \text{ M/L} \tag{8}$$

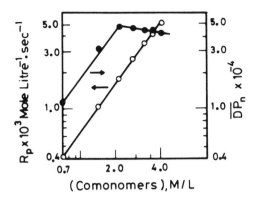

Figure 4 Variation of R_p and $\overline{DP}_n$ with the comonomers concentration.

$$\overline{DP}_n \propto [M]^{-0.15} \qquad [M] < 2.1 \text{ M/L} \qquad (9)$$

In addition, the exponent of the comonomer concentration for AM-DAEA-HCl system was also determined to be [22]:

$$R_p \propto [M]^{1.13} \qquad (10)$$

The reported values for the exponent of the monomer concentration for the rate of polymerization were found to be 1.26[61], 1.3[16] for gamma radiation-induced copolymerization of acrylamide with N,N-diethyldiallylammonium chloride and methyl chloride salt of N,N-dimethylaminoethyl methacrylate (DMAEM-MC). Ishigue and Hamielec [34] have shown that the

monomer exponent varies from 1.0 to 1.5 in acrylamide polymerization by chemical initiator.

In the present systems water is subjected to radiolysis to form radiation products such as: e_{aq}^-, H˙, OH$^-$, H_3O^+, H_2 and H_2O_2 [64]. The three primary radical species react with monomers to give radicals derived from monomers [63]. All these generated radicals are available to contribute to the chain initiation. This increases the exponent of the monomer concentration in these systems.

Figures 4 and 5 show that the degree of polymerization and the intrinsic viscosity increase with increasing the comonomer concentration. Thus, the increase in the comonomer concentration would increase $\overline{DP}_n$ and, consequently [η] Eq. (2). However, the exponent of the monomer concentration for the degree of polymerization decreases at a high comonomer concentration. The abrupt change in the exponent at comonomer concentration >2.1 M/L (Eq. 9) may be ascribed to polydispersity. This was found by others [34] to be at 2.1 and 2.2 M/L, although it starts theoretically at 2.0 M/L.

3. Influence of Radiation Dose

The influence of the radiation dose on the copolymerization of the AM-AANa system was investigated at 1.4 M/L [17,54]. The results are shown in Figs. 6 and 7. The data for the copolymerization of the AM-DAEA-HCl system at a comonomer concentration of 1.344 M/L [22] are also shown in Fig. 8.

Figures 6 and 8 show, generally that the conversion first increases fast then slowly with the dose. This is in agreement with the findings of Azzam [48] and Siyam [61]. The significant increase in the conversion percentage is attributed to the gel-effect [51,65]. In the gel-state, as the conversion percentage increases the viscosity of the medium is highly increased and the growing poly-

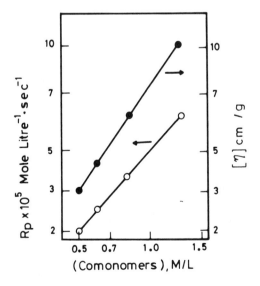

Figure 5 Effect of comonomer concentration on copolymerization of acrylamide with EAEA-HCl. ○ = R_p, ● = [η].

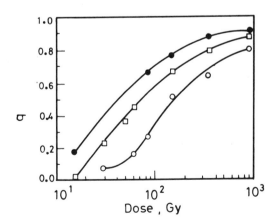

Figure 6 Influence of the absorbed dose on the conversion (q). ○ = 0.7035; □ = 1.407; ● = 2.1105 M/L.

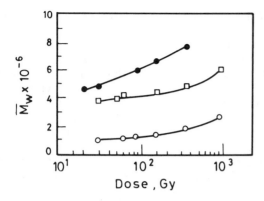

Figure 7 Influence of the adsorbed dose on the weight-average molecular weight. $\bigcirc$ = 0.735; $\square$ = 1.047; $\bullet$ = 2.1105 M/L.

Table 3 Influence of Radiation Dose on the Copolymer Composition and the Swelling Degree

Dos KGy	% Amine	% Acrylamide	Swelling degree
0.06	21.26	78.74	Soluble
0.10	23.52	76.48	Soluble
0.22	26.18	73.82	Soluble
0.60	28.68	71.32	Soluble
1.12	31.25	68.75	Soluble
6.00	41.24	58.76	542
14.00	42.56	57.44	750
29.00	44.41	56.59	360
42.00	45.40	54.60	311

meric chains are trapped in the viscous medium. Thus, the termination by mutual interaction of two growing chains becomes very unlikely while propagation continues. The interactions of high molecular weight growing chains become diffusion-controlled reactions [66,67]. Consequently, the molecular weight and the intrinsic viscosity of the formed copolymer would increase, as shown in Figs. 7 and 8 for AM-AANa and for AM-DAEA-HCl systems, respectively.

The influence of radiation dose on the polymer composition and the swelling degree of (pAM-DAEA-HCl) are shown in Table 3. The results show that the percent of acrylamide in the copolymer is higher than that of the amine. This can be attributed to smaller reactivity ratios of monomers of diallylammonium salts relative to acryl-

amide monomer [59,60]. At low doses water-soluble polymers are produced, the intrinsic viscosity of the resulting copolymers are shown in Fig. 8. At high dose, >6 kGy, the formed polymer is converted into polymeric gel. The gel formation is attributed to polymer branching [51], plurimolecular aggregates [68], and supermolecular architecture [52]. Table 3 also shows that the swelling degree increases then decreases when increasing the exposure dose. The decrease in the value of swelling degree is probably due to the increase in the extent of crosslinking between the polymeric chains as a result of radiation.

It was found that the polymerization of monomer of diallyammonium salts proceed by intra- and intermolecular cyclization to form polypyrrolidine polymers [64–77], which contain pyrrolidinium rings alternating along the polymeric chains [61]. Consequently, the copolymers include the amine as pyrrolidinium rings with acrylamide units alternating along the polymeric chains depending on the composition of the copolymer, as shown in "Structure 21."

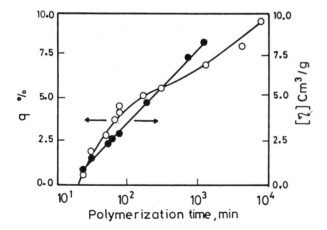

Figure 8 Effect of radiation dose on copolymerization of acrylamide with DAEA-HCl. $\bigcirc$ = q(%); $\bullet$ = [η].

"Structure 21"

4. Influence of Temperature

The influence of temperature on the copolymerization was investigated at constant absorbed dose of 0.12 and 0.16 KGy for copolymerization of AM-AANa [17,54] and AM-DAEA-HCl [22], respectively. The results are shown in Figs. 9 and 10, which show that the R_p values increase while the intrinsic viscosity and the degree of polymerization decrease with increasing the polymerization temperature. However, the increase in the temperature of the polymerization medium increases the swell-

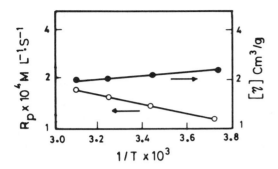

Figure 9 Effect of temperature on copolymerization of acrylamide with DAEA-HCl. $\bigcirc = R_p$; $\bullet = [\eta]$.

ing of the formed copolymer in the comonomer feed. This increases the concentration of the comonomer around the growing chains as the temperature rises. So, the propagation reactions and the polymerization rate should increase [78,79].

The increase in the temperature reduces the viscosity of the polymerization medium which increases the termination reactions. This is attributed to an increase in chain transfer reactions higher than that of propagation reactions [16,51]. Consequently, the weight-average molecular weight of the formed polymer decreases.

From the logarithmic plot of the Arrhenius equation shown in Figs. 8 and 9, the overall activation energy, E_{R_p}, was calculated to be 0.65 and 0.56 Kcal/mol for AM-AANa and AM-DAEA-HCl systems, respectively. However, the corresponding reported values for gamma radiation induced copolymerization of acrylamide with DMAEM-MC in aqueous solution was found to be 2.0 Kcal/mol [16].

In general, the overall activation energy for the polymerization rate is given by:

$$E_{R_p} = \left[\frac{E_i}{2} + \left('E_P - \frac{E_t}{2} \right) \right] \qquad (11)$$

where, E_i, E_p, and E_t are the activation energy of the initiation, propagation, and termination, respectively. In radiation-induced polymerization E_i is considered to be approximately 0. The E_p value for the radical polymerization is known to be 6–8 Kcal/mol [16,51]. The determined values of E_{R_p} indicate that E_t is about 10.7–14.7 and 10.88–14.88 Kcal/mol for AM-AANa and AM-DAEA-HCl systems, respectively. These large values of E_t are presumably due to the diffusion barrier of propagation chains and to the gel-effect.

5. Influence of After Effect

The influence of the after effect was studied at an absorbed dose of 0.12 KGy and a temperature of 0°C for copolymerization of the AM-AANa system [17,54]. The results are shown in Fig. 11.

Figure 11 shows that the conversion and weight-average molecular weight increases after ceasing irradiation. The copolymer yield seemingly come to an equilibrium within 6 h, the molecular weight continues increasing for nearly 50 h. These may be argued to be structural changes [80], possibly via conformational equilibria [81]. Aggregation [82] and cluster formation [52] may not be neglected. In fact, there is ample evidence on segmental motion of these and related copolymers [83,84].

V. USE OF WATER-SOLUBLE POLYMERS IN WATER TREATMENT

Based on the application of the established theory of colloid stability of water treatment particles [8,85–88], the colloidal particles in untreated water are attached to one another by van der waals forces and, therefore, always tend to aggregate unless kept apart by electrostatic repulsion forces arising from the presence of electrical charges on the particles. The aggregation process

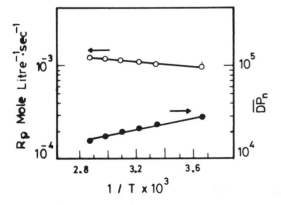

Figure 10 Variation of R_p and $\overline{DP}_n$ with temperature.

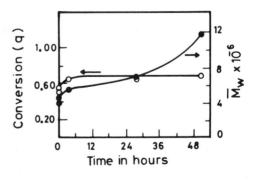

Figure 11 Influence of after-effect on acrylamide-acrylate copolymerization.

is aided by the addition of simple electrolytes, which reduce the electrostatic repulsion. This is commonly called "coagulation." Aggregation can also be brought on by various polymeric substances, which bind particles together with the chain molecules forming interparticle bridges. Following la Mer's suggestion [89,90], the term flocculation, refers specificaly to the latter process.

An understanding of the action of polyelectrolytes in flocculation has evolved from a number of investigations of the interaction of various polymeric flocculants [6–20,91–100] with suspended particles. The bridging theory of flocculation was first postulated by Ruehrewein and Ward [5]. This theory postulates that the polymer flocculation can be analyzed in terms of a primary adsorption process followed by a second flocculation process [6,8,10,11,14,15]. The polymer molecules attach themselves to the surface of the suspended particles at on or more adsorption sites, and that part of the chain extends out into the bulk of the solution. When these extended chain segments make contact with vacant adsorption sites on other particles, bridges are formed. The particles are thus bound into a small floc, which can grow to a size limited by the degree of agitation and the amount of polymer initially adsorbed on the particle surfaces. If too many adsorption sites are occupied, bridging will be hindered and wholly inhibited if all are occupied. If too few sites are occupied, bridging may be too weak to withstand the shearing forces imposed by even mild agitation. Consequently, in the first process of the adsorption the extended chain segments bind particles or ions together with polymer chains. These chain segments interact with each other to form a floc or a sludge in a subsequent flocculation process.

The adsorption of polymer molecule to the adsorbent surface may occur by chemical or physical interaction, depending on the characteristics of the polymer chain and the adsorbent surface. For nonionic polymers, the polymer is adsorbed via hydrogen bonding [8,11,15]. The adsorption of anionic polymers by the suspended particles is considered to be either hydrogen bonding, anion exchange, or chemical reaction with the charged particles [5–7,10,11,19]. However, cationic polymer is adsorbed on the surface of the particles via cation-exchange mechanism [7,11,19].

It was found that [5–7] the rate of flocculation of particles produced by the bridging action of polymer is the slower process and, consequently, the rate-determining step. The primary adsorption of polymer is fairly rapid, but the slow attainment of the adsorption equilibrium under agitation arises at least in part from the breakdown of flocs offering new surfaces for adsorption. Thus, the bridging step is slow because a polymer adsorbed on one particle must find another particle having a free surface available to complete the bridge.

The interparticle bridging mechanism was affected by the charge [7], adsorption sites [5] of particles extending in the flocculation system, the physical characteris-

tics of polymers [7,8,10,11,13,101], the degree of polymer adsorption [7], the valency and charge of ions in solution [12,18], the thickness the nature of adsorbed water layer [14,20], and the degree of agitation [5,11].

In previous works [18–20,23,102] water-soluble polymers such as: polyacrylamide (pAM), polysodium acrylate (pAA Na), poly(acrylamide-sodium acrylate) (pAM-AA Na), poly(acrylamide-diallyethylamine-hydrochloride) (pAM-DAEA-HCl), and poly(acrylamide-sodium acrylate-diallyethylamine-hydrochloride) (pAM-AANa-DAEA-HCl) were used in the recovery of cations and some radioactive isotopes from aqueous solutions. It was found that the floc is formed between the added polymer and ions of the solution in the flocculation process with the formation of a crosslinked structure. The formed cross-linked structure is characterized by [103–105]:

1. The transition bond energy is strong energy.
2. The metal ion is readily removed by treating with an acid.
3. The coordination number and stereostructure of the metal complex are specific for a given metal ion species.
4. The crosslinking reaction is detectable without difficulty by spectroscopic and magnetic measurements.

The efficiency of flocculation of these polyelectrolytes was investigated at different pH and polymer concentrations, valency of cations, and weight-average molecular weight of the polymer. The results are shown in Figs. 12–16.

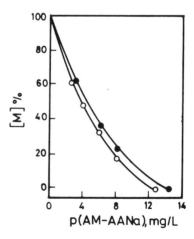

Figure 12   Effect of polymer concentration on the residual metal concentration [M] for p(AM-AANa). ● = Mg^{2+}, ○ = Cu^{2+}.

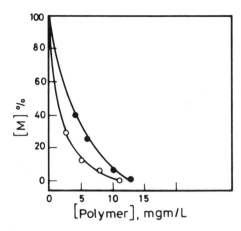

Figure 13 Effect of polymer concentration on the residual metal concentration [M] for p(AM-DAEA-HCl). ● = Mg^{2+}, ○ = Cu^{2+}.

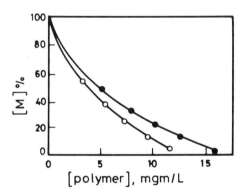

Figure 14 Effect of polymer concentration on the residual metal concentration [M] for p(AM-AANa-DAEA-HCl). ● = Mg^{2+}, ○ = Cu^{2+}.

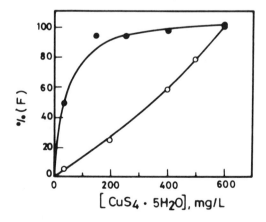

Figure 15 Influence of valency of cations on the polymer efficiency. ● = Eu^{2+}, ○ = Co^{2+}.

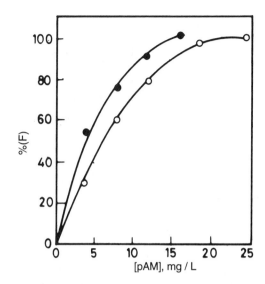

Figure 16 Influence of polymer concentration and weight-average molecular weight ($\overline{M}_w$) on the polymer efficiency. ● = $\overline{M}_w$ = 5.5 × 10^6; ○ = $\overline{M}_w$ = 3.0 × 10^6.

A. Effect of pH

The influence of pH on the residual metal concentration [M] was studied at a constant polymer concentration of 10 mg/l and copper sulphate concentration of 10 gm/l, results are shown in Table 4. It is clear that [M] decreases with increasing pH value. Results are given in Table 4, which shows that the concentration of Cu^{2+} decreases with an increase in the pH value for each polymer used. This is attributed to the effect of the pH value on the active groups, which are distributed along the polymer chains. At a low pH value the amide groups

Table 4 Effect of pH on the Percent of Residual Cu^{2+} Concentration

Anionic polyacrylamide (pAM-AANa)		Cationic polyacrylamide (pAM-DAEA-HCl)		Amphoteric polyacrylamide (pAM-AANa-DAEA-HCl)	
pH	%[Cu]$^{2+}$	pH	%[Cu]$^{2+}$	pH	%[Cu]$^{2+}$
5.25	95.36	5.22	95.42	5.29	80.24
5.37	82.18	5.30	90.51	5.25	60.15
5.40	30.25	5.40	75.21	5.31	55.02
5.50	23.22	5.65	50.50	5.42	36.48
5.56	46.44	5.70	30.40	5.52	38.52
5.70	11.82	5.90	10.21	5.81	30.31
5.80	7.31	5.95	00	6.10	5.25
6.15	00	6.50	00	6.14	00

are protonated [13,105], which can undergo imidization [13], while both the carboxylate groups [7,11] and quaternary ammonium groups [61] are unionized. The polymer chains are often coiled in the solution. Consequently, no interaction can occur between the coiled polymeric chains and the ions that might be present in the solution.

At higher pH values the amide groups are unprotonated [106], while the carboxylate groups [7,11] and quaternary ammonium groups [61] are ionized. Alternatively, the interaction can occur between these cations of the solution and the active groups of the polymer chains for floc formation.

B. Effect of Polymer Concentration

The influence of polymer concentration on the percent of residual metal concentration [M] was studied at a polymer concentration of 10 mg/l, metal sulphate concentration of 10 gm/l, and pH ~ 6, results are shown in Figs. 12–14. It is clear that [M] decreases with increasing polymer dosage. As the polymer concentration increases, the number of extended segments onto the solution increases. These extended segments, which affect the probability of the segment collision and consequently the probability of forming the floc, increase. This is in agreement with the findings of other investigators [7,10,11,13,17]. The change in the polymer dosage required to bring 0% [M] varies from one type of polymer to another. This is due to the difference in the weight-average molecular weight of the polymer used. Figs. 12–14 show that the polymer concentration for complete precipitation of Mg^{2+} is higher than that of Cu^{2+}. This is due to the difference in the hydration of cations. Increases in the hydration decrease the polymer efficiency. This is in agreement with the findings of Dolimore et al. [14]. They found that the extent of flocculation decreases with increasing thickness of the adsorbed water layer around the cations of the clay particles upon treatment with polymeric flocculants.

C. Influence of Valancy of Cations

Different samples of aqueous solution containing radionuclides of Co^{2+} and Eu^{3+} were prepared at different copper sulphate concentrations and constant polymer concentrations (pAM) of 15 mg/l. The addition of salt to the system was done to reduce both the repulsion forces between the radionuclides and the interaction between the polymeric chains [7]. The polymer efficiency for the prepared samples was determined, results are shown in Fig. 15. It is clear that the polymer efficiency for Eu^{3+} is higher than for Co^{2+}. This can be explained by the difference in the tightly bound structured water associated with different cationic species [14,107]. On this basis, we expect that Co^{2+} is more hydrated than Eu^{3+}. This is due to the difference in the ionic size. The hydra-

tion energy decreases with increasing ionic size. This is in agreement with the finding of Dolimore et al. [14]. They found that the extent of flocculation decreases with increasing thickness of the adsorbed water layer around the cations. They also found that the flocculation of Li by polyacrylamide from a kaolinite clay is less than that of K and Na. This can be attributed to the strong hydration of Li than of K and Na.

D. Effect of Polymer Concentration and Weight-Average Molecular Weight

The influence of polymer concentration on the polymer efficiency (F) was studied at constant copper sulphate concentration of 600 mg/l at two different $\overline{M}_w$ of 3.9×10^6 and 5.5×10^6, results are shown in Fig. 16. It is clear that the polymer efficiency (F) increases with increasing the polymer dosage as discussed above.

It is found that at high polymer dosage >100 mg/l, the floc is not formed. This is due to the high viscosity of the solution and the complete interaction of polymer chains with each other.

It was observed that the polymer efficiency (F) increases with increasing molecular weight of the polymer ($\overline{M}_w$). The polymer efficiency (F) reaches 100% at polymer dosage of 24 and 15 mgm/l at $\overline{M}_w$ of 3.9×10^6 and 5.5×10^6, respectively. Consequently, the higher is the value of $\overline{M}_w$ the lower the polymer dosage. This is attributed to increasing the efficiency of the polymer with increasing weight-average molecular weight of the polymer. On increasing $\overline{M}_w$, the extended polymer segments are longer making collision and thus bridging is more probable. This increases the probability of forming the floc. This is in agreement with the findings of other investigators [7,10,11,13,15,17].

VI. MECHANISM OF INTERACTION OF WATER-SOLUBLE POLYMERS WITH IONS IN AQUEOUS SOLUTION

The infrared spectra of the formed floc from copper sulphate with water-soluble polymers such as pAM, pAANa, p(AM-AA-HCl), and p(AM-DAEA-HCl) are summarized in Tables 5 and 6. In the spectrum of pAM the shift in the absorption bands, characteristic of both >NH stretching and >NH bending to a lower wavelength, is due to the hydrogen bonding formation between the polymer chains in plurimolecular aggregates of polyacrylamide as shown in "Structure 22" [79,108,109]. The same shift in the absorption bands, characteristic for —OH group to a lower wavelength, is due to the presence of water molecule, in addition, the appearance of the absorption bands characteristic of the sulphate ion indicate the presence of the sulphate group in the floc. The absorption bands can lead to the conclusion that: (1) polyacrylamide forms complexes with

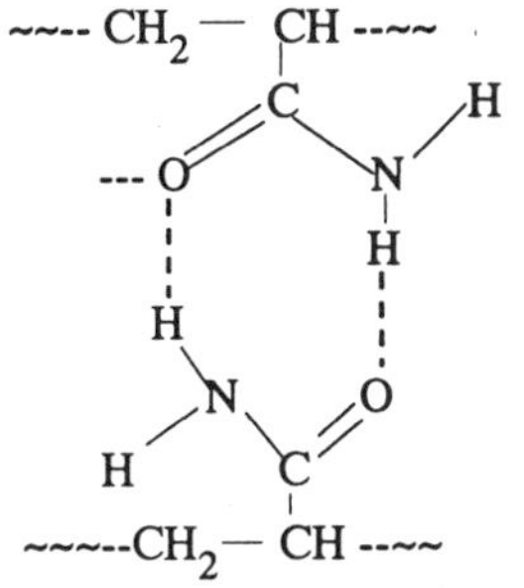

"Structure 22"

Plurimolecular aggregates
of pAM

Cu(II) and these complexes may by an octahederal structure due to formation of coordination bonds between the amide group and water of hydration with Cu(II); (2) the polymer chains form hydrogen bonds with each other; (3) the sulphate ions may be present in the floc. Copper sulphate form crosslinked structures with pAM as show in the possible Structure 23. This is in

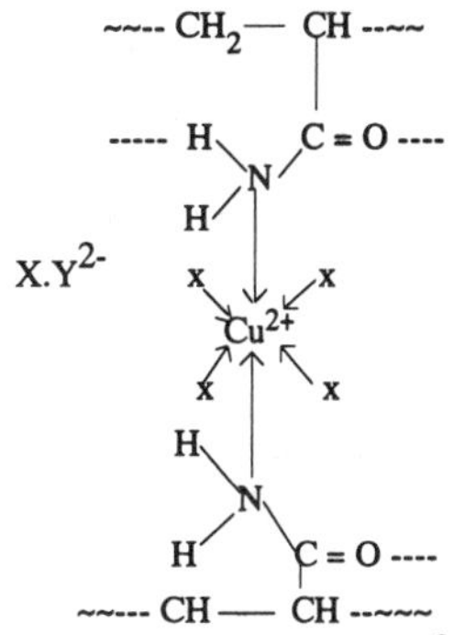

$X.Y^{2-}$

$x = H_2O$ and $Y^{2-} = SO_4^{2-}$

"Structure 23"
Cross-linked structure of pAM-CuSO$_4$·5H$_2$O complexes

Table 5 IR Spectral Data of Polymers with Copper Sulphate

Absorption bands of original groups (cm^{-1})		Absorption bands of	
		p(AM)	p(AANa)
Amide:			
Free $>$NH stret.	3500, 3400		
(Amide II)			
Bonded NH stret.	3350, 3180	3390, 3270	
(Amide II)			
$>$NH bend.	1620, 1590		
$>$C=O stret.	1650	1651	
(Amide I)			
Aliphatic:			
—CH$_2$ stret.	2926–2853	2965	2923
—CH$_2$ bend.	1485–1445	1485	
—CH$_3$ bend.	1470–1430	—	—
Carboxylate:	1610–1550	—	1565
—COO$^-$	1400–1300	—	1419
Ester $>$C=O stret.	1750–1735	—	—
Hydroxyl:			
Free—OH stret.	3650–3590		
Bonded—OH	3400–3200	3583	3394
C—O—stret. and	1420–1300	1125–1091	1118
—OH bend.	−1200 and 920	941	
Amine:			
C—N stret.	1410		
Sulphate:			
Ionic	1140–1070, 480–400	1126–1091	—
		500–489, 420	
Covalent	1450–1350	—	1419
	1230–1150	—	1118
	650–550	—	609

agreement with the suggestion of several authors [110,111].

In the I.R. spectrum of pAANa, which is shown in Table 5, the absorption bands characteristic of the carboxylate group (—COO⁻), the covalent sulphate group (—O—SO₂—O—), and the hydroxyl group (—OH) are due to the —COONa interaction with copper sulphate according to the following mechanisms:

1. Cation exchange between Na⁺ of COONa and Cu(II).
2. Anion exchange between —COO⁻ and SO₄²⁻. This is in agreement with the suggestion of Vreudge et al. [11], Plack et al. [7], Whayman et al. [10], and Siyam et al. [19,20].
3. Water may form coordinate bonds with ions or

with polymer chains. Copper sulphate forms a crosslinked structure with the carboxylate group as shown in the possible "Structure 24."

In the spectrum of p(AM-DAA-HCl) and p(AM-DAEA-HCl), the shift in the absorption bands of >NH stretching of the amide group to a lower wavelength is due to hydrogen bond formation between the amide groups as mentioned previously. The appearance of the absorption bands characteristic for —OH group at a lower wavelength may be also due to the presence of water molecules coordinated with ions or with polymer chains. In addition, the appearance of absorption of ionic sulphate indicates the presence of the sulphate groups in the floc. The spectra do not reveal absorption bands characteristic for >NH stretching (3000–2700

Table 6 IR Spectral Data of Polymers with Copper Sulphate

Absorption bands of original groups (cm⁻¹)		Polymers (cm⁻¹)	
		p(AM-DAA-HCl)	p(AM-DAEA-HCl)
Amide:			
Free >NH stret. (Amide II)	3500, 3400		
Bonded NH stret. (Amide II)	3350, 3180	3359, 3243	3359, 3289
>NH bend.	1620, 1590		
>C=O stret.	1650		
(Amide I)		1654	1654
Aliphatic:			
—CH₂ stret.	2926–2853	2923	2958
—CH₂ bend.	1485–1445	1519–1450	1437
—CH₃ bend.	1470–1430	—	1437
Carboxylate:			
—COO⁻	1610–1550	—	—
	1400–1300	—	—
Ester >C=O stret.	1750–1735	—	
Hydroxyl:			
Free—OH stret.	3650–3590		
Bonded—OH	3400–3200		
C—O—stret and	1420–1300	3359	3386
—OH bend.	−1200 and 920	1110	1122
Amine:			
C—N stret.	1410,	1427	1437
Sulphate:			
Ionic	1140–1070	1110	1122
	480–400	497	424
Covalent	1450–1350	—	—
	1230–1150	—	—
	650–550	—	—

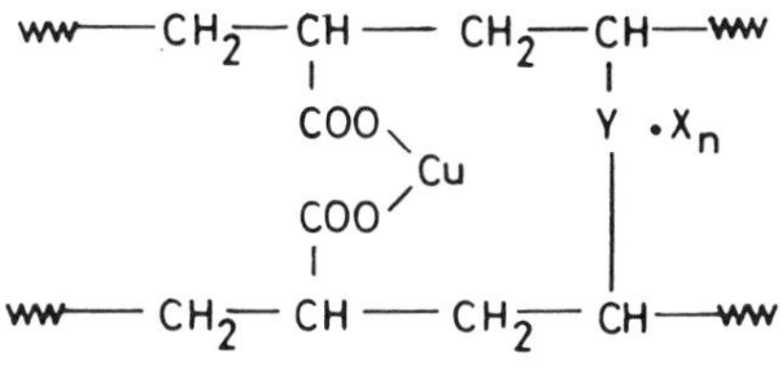

$X = H_2O$, $Y^{--} = SO^{--}$
and n is the number of X

"Structure 24"

Cross-linked structure of
$pAANa-CuSO_4 \cdot 5H_2O$

cm^{-1}) and (2700–2500 cm^{-1}) of amine salts. This can lead to the conclusion that:

1. Amide groups interact with copper sulphate by the complex formation mechanism as mentioned previously.
2. The ammonium groups interact according to the following mechanisms:
   a. Cation exchange between H^+ of ammonium group and Cu(II).
   b. Anion exchange between Cl^- and SO_4^{2-}.

This is in agreement with the suggestion of Vreudge et al. [11], Plack et al [7], and Siyam et al. [19,20]. Copper sulphate forms crosslinked structures with ammonium groups of p(AM-DAA-HCl) and p(AM-DAEA-HCl) are shown in the possible "Structure 25" and "Structure 26", respectively.

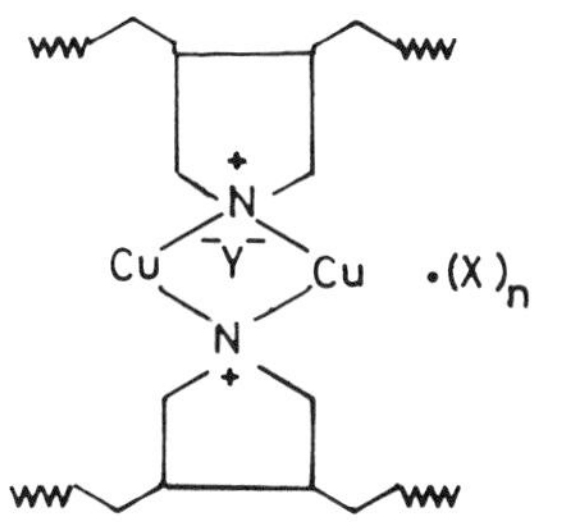

$X = H_2O$, n is the number
of X, $Y = SO_4^{--}$

"Structure 25"

Cross-linked structure of
$pDAA-HCl-CuSO_4 \cdot 5H_2O$

The spectroscopic studies show that:
1. Water-soluble pAM "neutral polymer" interacts with ions of the solution through the complex formation between amide groups and hydrated ions.
2. Water-soluble pAANa "anionic polymer" interacts with ions through cation and anion exchange mechanisms.
3. Water-soluble pAM-DAA-HCl and pAM-DAEA-HCl "cationic polymers" interact with ions by the following mechanisms:
   a. Complex formation between amide groups and ions of Cu^{2+} in solutions.
   b. Ammonium groups of amine salts interact with ions through cation and anion exchange mechanisms.

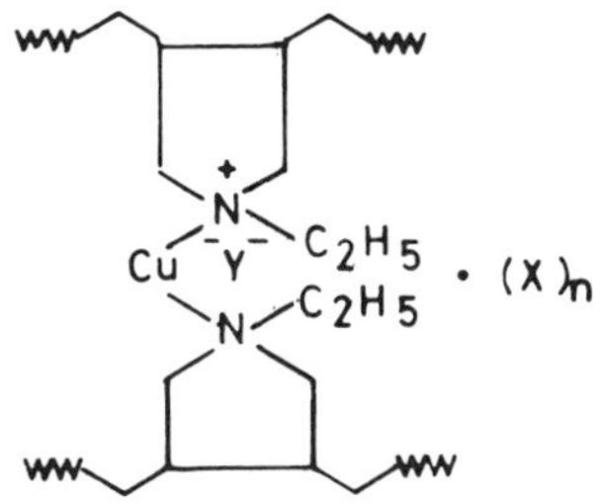

$X = H_2O$, n is the number

of X, $Y = SO_4$

"Structure 26"
Cross-linked structure of
$pDAEA \ HCl \ CuSO_4 \cdot 5H_2O$

ACKNOWLEDGMENT

I wish to express my sincere gratitude to Prof. Dr. H. F. Aly, Chairman of Atomic Energy Authority, Prof. Dr. I. M. El-Naggar, Prof. of Physical Chemistry, Hot Lab. Centre, Atomic Energy Authority and Prof. Dr. R. Azzam (who passed away on November 18, 1993), Prof. of Applied Chemistry, Atomic Energy Authority, Prof. Dr. Ahmady A. Yassin Prof. of Polymer Chemistry, Faculty of Science, Cairo University, for their assistance in carrying out the experimental part of this work in Nucl. Chem. Dept., Hot Lab. Centre, Atomic Energy Authority, Cairo, Egypt.

REFERENCES

1. M. F. Hoover, *J. Macromol. Sci.-Chem., A4*: 1327 (1970).
2. R. Azzam and T. Siyam, *Annals of Agric. Sc., Moshtohor, 13*: 215 (1980).
3. G. Burillo and T. Ogawa, *Radiat. Phys. Chem., 18*: 1143 (1981).
4. R. Azzam, *Sandy Soil Plantation in Semi Arid Zones by Polyacrylamide Gel Preparation by Ionizing Radiation,* "IAEA-Contract No. 2596/RB, Progress Rpt" July 1980–Oct. 1981, Vienna.
5. R. A. Ruehrwein and D. W. Ward, *J. Soil Sci., 13*: 485 (1952).
6. T. W. Healy and V. K. laMer, *J. Phys. Chem. 66*: 1835 (1962).
7. A. P. Plack, F. B. Brrkner, and J. J. Morgan, *J. Ain. Water Works Ass. (JAWWA), 57*: 1547 (1965).

8. M. Pressman, *J. Am. Water Works Ass. (JAWWA), 59*: 169 (1967).

9. T. Uesla and S. Harado, *J. Appl. Polym. Sci., 12*: 2383 (1963).

10. E. Whayman and O. L. Cress, *Sugar J., 20* Nov. (1975).

11. M. J. A. Vreudge and G. W. Polling, *CIM Pull., 68*: 54, (1975).

12. J. E. Unbehend, *J. Technical Assoc. of the Bull. and Paper Industry TAPPT, 59*: 74 (1976).

13. T. Wada, H. Sekiya, and S. Machi, *J. Appl. Polym. Sci., 20*: 3233 (1976).

14. D. Dolimore and T. Horridge, *Power Technology, 17*: 207 (1977).

15. D. W. Royers and J. A. Poling *CIM Bull., 71* (1978).

16. T. Okada, I. Ishigaki, T. Suwa, and S. Machi, *J. Appl. Polym. Sci., 24*: 1713 (1979).

17. T. Siyam, Studies on Gamma Radiation Induced Copolymerization of Acrylamide Sodium Acrylate as Flocculant, M. Sc. Thesis, Fac. Sci., Cairo Univ. (1982).

18. T. Siyam, M. I. El-Dessouky, and H. F. Aly, 2nd Arad. Intern. Conf. on Advanced Materials Science & Engineering (Polymeric Material), 39, 6–9 Sept. Cairo, Egypt (1993).

19. T. Siyam, R. Ayoub, and N. Souka, *Egypt. J. Chem. 37*: 457 (1994).

20. T. Siyam and E. Hanna, *J. Macromol. Sci., Pure Appl. Chem. A 31(3&4)*: 349 (1994).

21. T. Siyam, *J. Macromol. Sci. Pure Appl. Chem., A31 (3&4)*: 383 (1994).

22. T. Siyam, *J. Macromol. Sci. Appl. Chem. A31 (3&4)*: 371 (1994).

23. T. Siyam, *J. Macromol. Sci. Pure Appl. Chem. A32 (5&6)*: 801 (1995).

24. T. Siyam, M. M. Abdel-Hamid and I. M. El-Naggar, *J. Macromol. Sci. Pure Appl. Chem. A32 (5&6)*: 871 (1995).

25. T. Siyam and R. Ayoub, 3rd Arab Intern. Conf. on Advanced Materials Science & Engineering (Polymeric Materials) 4–7 Sept., Cairo, Egypt (1995).

26. T. Siyam, 3rd Arab Intern. Conf. on Advanced Materials Science, (Polymeric Materials 4–7 Sept., Cairo, Egypt (1995).

27. J. P. Riggs, and R. Rodriguez, *J. Polym. Sci. A-1, 5*: 3151 & 3167 (1967).

28. T. Takahashi, Y. Hori and I. Sato, *J. Polym. Sci., A-1, 6*: 2091 (1968).

29. S. P. Rout, A. M. N. Rout, B. G. Single, and Santappa, *Makromol. Chem., 178(7)*: 1971 (1977).

30. R. K. Samal, P. L. Nayale, and T. R. Mohanto, *Macromol., 10(2)*: 489 (1977).

31. S. N. Bhadani and Y. K. Prasad, *Makromol. Chem., 178(6)*: 1841 (1977).

32. T. S. Shukla and D. C. Mirsva, *J. Polym. Sci., A-2, 11*: 751 (1973).

33. N. C. Eevi and V. Mahadevan, *J. Polym. Sci. A-1, 10*: 903 (1972).

34. T. Ishige and A. E. Hamielec, *J. Appl. Polym. Sci., 17*: 1479 (1973).

35. Z. Ogumij, I. Tari, Z. Rakahara, and S. Yoshizawa, *Bull. Chem. Soc. Japan, 47(8)*: 1843 (1974).

36. S. Tazuke, K. Tsukamoto, K. Hayashi, and S. Ohamura, *Kohumshi Kagaku, 24*: 264, 302 (1967).

37. R. Bhaduri and S. Aditya, *Makromol. Chem. 178*: 1385 (1977).

38. G. Oster and G. Prati, *J. Am. Chem. Soc, 79(1)*: 595 (1957).

39. K. Imamuva, M. Asai, S. Tazuka, and S. Okamura, *Makromol. Chem., 174*: 91 (1973).

40. F. S. Dainton and M. Tordoff, *Trans. Farad Soc., 53*: 499 (1957).

41. B. M. Baysal, H. N. Evten and U. S. Ramelsow, *J. Polym. Sci., A-1,9(3)*: 1581 (1971).

42. N. Ootuska and T. Yamamoto, *Oye Butsuri, 42(12)*: 1185 (1973).

43. N. Ootuska and T. Yamamoto, *Oye Butsuri, 43(6)*: 588 (1974).

44. M. Ootuska and T. Yamamoto, *Hiroshima, Univ. Ser. A: Phys. Chem., 40(2)*: 327 (1976).

45. E. Collinson, F. S. Dainton, and G. S. McNughton, *Trans. Farad. Soc., 53*: 476 (1957).

46. A. Chapiro and L. Perec. *Eur. Polym. J., 7*: 1335 (1971).

47. V. V. Guston and A. G. Kazakevich, *Khim. Vys. Emerg., 11(5)*: 26376 (1977) (Russ).

48. R. Azzam and K. Singer, *Polym. Bull., 2*: 147 (1980).

49. G. B. Korneeva, B. B. Mamin, A. P. Sheinker, and A. D. Abkin, *Radiat. Khim, 2*: 27 (1973).

50. T. Wada, H. Sekiga, and S. Macki, *J. Polym. Sci., A-1, 12*: 1858 (1975).

51. A. Chapiro, *Radiation Chemistry of Polymeric System*, Interscience Publishers, New York (1962).

52. V. A. Kabanov, *J. Polym. Sci. Symp., 50*: 71 (1975).

53. A. J. Swallow, *An Introduction To Radiation Chemistry*, Longman, London (1973).

54. R. Azzam, A. Yasin, and T. Siyam, 6th IUPAC, Bratislava Conf. on Modified Polymers, Czechoslovakia, July 2–5, II, 100–106 (1984).

55. J. E. Boothe, H. Flock, and J. Fredhoover, *Macromol. Sci. Chem. A4*: 1419 (1970).

56. G. B. Butler, IUPAC, Inter. Symposium on Polymeric Amines and Ammonium Salts, 125, Ghent, Belgium, 24–26 Sept. (1979).

57. Y. Negi, S. Harada, and O. Hishizuka, *J. Polym. Sci., A3*: 2063 (1967).

58. S. Loan, G. Mocanu, and Maxim, *Eur. Polym. J., 15*: 667 (1979).

59. T. Siyam, *Egypt. J. Chem., 37(1)*: 69 (1994).

60. Ch. Wandrey and W. Jaeger, *Acta Polymeric, 34*: 100 (1985).

61. T. Siyam, "Studies on Gamma Radiation-Induced Preparation of Cationic and Amphoteric Copolymers." Ph.D. Thesis, Fac. Sci. Cairo Univ., Cairo, Egypt (1986).

62. Schalz, Renner, Henglen, and Kern, *Makvomol. Chem., 12*: 20 (1954).

63. K. W. Chambers, E. Collinson, and F. S. Dainton, *Trans. Farad. Soc. 66(1)*: 142 (1970).

64. V. Madhavan, N. N. Lichtin, and E. Hayon, *J. Am. Chem. Soc., 97*: 2889 (1975).

65. G. Odian, *Principles of Polymerization*, McGraw Hill Inc., New York, 3, p. 165 (1970).

66. A. N. Ghauri, and S. R. Palit, *Trans. Farad. Soc., 64*: 1603 (1968).

67. A. M. North and G. A. Reed; *J. Polym. Sci., A-1*: 1311 (1963).

68. A. Chapiro, *Eur. Polym. J., 7*: 1355 (1971) and 417 (1973).

69. S. R. Johns, R. I. Willing, S. Middleton, and A. K. Ong., *J. Macromol. Sci. Chem., A-1(5)*: 875 (1976).

70. D. G. Hawthorne, S. R. Johns, D. H. Solomom, and R. I. Willing, *Aust. J. Chem., 29*: 1955 (1976).

71. D. H. Solomon, and D. G. Hawthorne, *J. Macromol. Sci. Rev., Macromol. Chem. C-15*: 143 (1976).

72. J. H. Hodgkin and D. H. Solomom, *J. Macromol. Sci. Chem., A-10*: 893 (1976).
73. J. H. Hodgkin and R. J. Allen, *J. Macromol. Sci., A-11*: 937 (1977).
74. J. H. Hodgkin, and S. Demearac, *Adv. Chem. Ser., 187*: 211 (1980).
75. J. H. Hodgkin, R. L. Willing, and R. Eibl, *J. Polym. Sci. A-1*: 19, 1239 (1981).
76. J. H. Hodgkin, S. R. Johns, and R. L. Willing, *Polym. Bull., 7*: 353 (1982).
77. R. M. Ottenbrite, and W. S. Ryan Jr., *Ind. Eng. Chem. Prod. Res. Dev., 19*: 528 (1980).
78. S. N. Bhadani, and Y. K. Prasad, *J. Polym. Sci. Polym. Letters Ed., 17*: 493 (1979).
79. A. Chapiro, *Am. Chem. Soc. Symposium Series, 175*: 233 (1981).
80. F. A. Makblis, *Radiation Physics and Chemistry of Polymers,* John Wiley & Sons Jerusalem (1975).
81. P. L. Luisi and F. Ciardelli, *Configuration and Conformation in High Polymers in, Reactivity, Mechanism and Structure in Polymer Chemistry,* (A. D. Jenkins and A. A. Ledwith, eds.), John Wiley & Sons, London (1979).
82. A. Chapiro, *Eur. Polym. J., 9:* 417 (1973).
83. R. Azzam 5th "Tihany" Symp. on Radiation Chemistry Proc., 771 (1982).
84. R. Azzam, et al. Internat. Symp. on Isotope and Radiation Techniques in Soil Physics and Irradiation Studies, Aix-en-Provence, Proc., p. 321, (1983).
85. E. W. Verwer, and J. T. G. Dverbeek, Elsevier Publishing Co, N.Y. (1948).
86. H. R. Kryut, *Colloid Science,* Elsevier Publishing Co, N.Y., 1, (1952), 2, (1949).
87. A. P. Black, *Water and Sewage Works* (1961).
88. T. M. Riddich, *J. Water Works Ass., 53*: 1007 (1961).
89. V. K. LaMer and T. W. Healey, *J. Phys. Chem., 67*: 2417 (1963).
90. V. K. LaMer, *J. Colloid Sci., 19*: 291 (1964).
91. A. S. Michalls, *Ind. Eng. Chem., 46*: 1485 (1954).
92. A. S. Michaels, and O. Morelos, *Ind. Eng. Chem., 47*: 1801 (1955).
93. H. Van Olphen, *An Introduction to Clay Colloid Chemistry,* John Wiley & Sons, Inc., N.Y. (1963).
94. V. K. LaMer, and T. H. Healy, *Rev. and Pure Appl. Chem., 13*: 112 (1963).
95. W. E. Walles, *J. Colloid & Inerfac. Sci., 27*: 797 (1968).
96. T. Ueda, and S. Harada, *J. Appl. Polym. Sci., 12*: 2383 (1968).
97. G. Fanta, R. C. Burv, C. R. Eussell, and C. E. Rist, *J. Appl. Polym. Sci., 14*: 2601 (1960).
98. P. H. King, J. W. Oliver, Randall, and J. A. Ceskey, 4, *Recent Advances in Industrial Pollution Control,* Technonic, Pub. Co. (1971).
99. B. Vicent, *Adv. Coll. Int. Sci., 4*: 193 (1974).
100. A. Van Lierde, *Int. J. Minerol Processing, 81* (1974).
101. N. S. Miyata, I. Sakata, and R. Senyu, *Bull. Chem. Soc. Japan, 48(11),* 3367 (1975).
102. T. Siyam, and E. Hanna, 4th Nat. Phys. Conf., 23, 28–30 Nov., Cairo, Egypt (1992).
103. H. Nishide and E. Tsuchida, *Makromol. Chem., 177*: 2295 (1976).
104. G. Wulff, A. Sarkan, and K. Zabiochi, *Tetrahedron Lett, 4329* (1973).
105. T. Takagushi, and I. M. Rlotz, *Bipolymers, 11*: 483 (1972).
106. V. F. Kuenkov, and V. A. Myagchenkov, *Eur. Polym. J., 16*: 1229 (1980).
107. B. D. Xay, and P. F. Low, *Heats of Compression of Clay Water Mixtures, Clayminer., 23*: 266 (1975).
108. A. Chapiro and Dulieu, *Eur. Polym. J., 13*: 563 (1977).
109. A. Chapiro, *Pure Appl. Chem., 643* (1981).
110. E. J. Goethals, "Polymeric Amines and Ammonium Salts," International Symposium on Polymeric Amines and Ammonium Salts, Ghent, Belgium, 24–26 Sep. (1979), pp. 255–299, Pergamon Press (1979).
111. I. Karatas, and G. Irez, *J. Macromol. Sci. Pure Appl. Chem. A30 (3&4)*: 241 (1993).

10

Polyvinylchloride (PVC)/Thermoplastic Polyurethane (TPU) Polymeric Blends

Cheng-Di Huang and Jiong Chen
Shanghai Medical Equipment Research Institute, Shanghai, China

I. INTRODUCTION

In the field of plastics, the annual production of polyvinylchloride (PVC) is second only to polyethylene. PVC has long been used in various areas, ranging from agriculture and industry to medical equipment and daily life, due to its well-developed production techniques, easy processing, and low price. However, PVC has its own disadvantages, mainly its low stability toward heat and ultraviolet (UV) light. Also, pure PVC is a very hard material that cannot be easily processed and practically used. Common PVC plastics contain various amounts of plasticizers and other additives, including modifiers, stabilizers, and lubricants.

Hard PVC contains little or no plasticizer, thus it is hard and brittle, lacking toughness at low temperatures. Flexible PVC usually contains significant amounts of plasticizer, thus it is soft and maintains good mechanical properties at low temperatures, although these properties usually will not be sustained when the temperature is lower than $-20°C$. Also, the strength of the material under normal conditions is significantly lower than that of PVC itself because of the presence of a large quantity of low-molecular weight liquid plasticizer.

Due to the large differences in size between PVC polymers and the relatively small plasticizer molecules, the attraction (mainly the van der Waals force) between these two is not strong. Therefore, the migration of the plasticizer molecules from inside the material to the outer surface is unavoidable. This phenomenon is particularly severe when PVC products are used with heat, are subjected to light, or come in contact with blood or other organic liquids. These cause the deterioration of the PVC materials by causing hardening, shrinking, and other damage. Also, the plasticizer leeching off the materials will cause contamination to the surrounding medium. This contamination is of a particular concern in the medical field and food industry where flexible PVC materials are wisely used. The most common plasticizer for PVC is dioctylphthalate (DOP). Its possible toxicity, although not confirmed, has already been the cause for concern among various industries [1–3].

Thermoplastic polyurethane (TPU) is a type of synthetic polymer that has properties between the characteristics of plastics and rubber. It belongs to the thermoplastic elastomer group. The typical procedure of vulcanization in rubber processing generally is not needed for TPU: instead, the processing procedure for normal plastics is used. With a similar hardness to other elastomers, TPU has better elasticity, resistance to oil, and resistance to impact at low temperatures. TPU is a rapidly developing polymeric material.

However, TPU has several disadvantages: (1) its high price (currently TPU is at least four times the cost of PVC), (2) its narrow temperature range for processing, and (3) its tendency to be hydrolyzed especially in the case of polyester-based TPU.

Polymer blending is one of the fastest growing areas in polymer science and technology. The idea of combining PVC and TPU to form blends is based on the assumption that these two polymers have a good compatibility and that the different properties of these two will reinforce each other and give better results. In addition, the use of TPU as a modifier to PVC, replacing the bulk of liquid plasticizer, will eliminate the concern about the migration of plasticizers. These kinds of polymeric blends with good properties are also called polymer alloys, which are derived from metallurgy [4].

There are three methods of making polymer blends: mechanical blending, solution mixing, and chemical synthesis. This chapter will focus only on the mechanical blending of polymers.

II. THE COMPOSITION OF PVC/TPU BLENDS

A. Compatibility

1. The Theory of Compatibility

In a thermodynamic sense, the compatibility of polymers is similar to the dissolving solute in a solvent. The thermodynamic standard of solubility is the free energy of mixing ΔG_M. If $\Delta G_M < 0$, then two components are soluble to each other. According to the definition:

$$\Delta G_M = \Delta H_M - T \Delta S_M,$$

since the entropy of mixing ΔS_M is always positive, the enthalpy of mixing ΔH_M must be smaller than a certain value in order to have a negative ΔG_M.

According to Hildebrand and Scott [5], the enthalpy of mixing per unit volume Δh_M is related to the volume fraction Φ and the solubility parameter δ of the two components:

$$\Delta h_M = \Phi_1 \Phi_2 (\delta_1 - \delta_2)^2$$

Thus, two materials with the same solubility parameters will be soluble in one another due to the fact that their mixed enthalpy is zero. This is consistent with the fact that chemically and structurally similar materials tend to be soluble in one another.

The above theories are a necessary condition to judge the compatibility between two polymers, but not a sufficient condition.

2. The Measurement of the Compatibility Between PVC and TPU

The properties of the blending system are closely related to its morphology. Therefore, the measuring of the compatibility of PVC and TPU is important in order to understand the properties of the blending materials.

Differential scanning calorimetry (DSC) is fast, sensitive, simple, and only needs a small amount of a sample, therefore it is widely used to analyze the system. For example, a polyester-based TPU, 892024TPU, made in our lab, was blended with a commercial PVC resin in different ratios. The glass transition temperature (T_g) values of these systems were determined by DSC and the results are shown in Table 1.

These data showed that within the range of $0.5 \leq$ TPU/PVC ≤ 1.075, the T_g value of the blends is very close to the theoretical value calculated from the Fox equation. This suggests that 892024TPU is compatible with PVC within this range.

However, it must be pointed out that there are some limitations when DSC is used to analyze the polymer blends:

1. TPU has a high degree of crystallinity and its content in the blending system is high. Under this situation, the analysis could be distorted by the high crystallinity of the TPU.
2. There is a very small amount of one of the components.
3. When the difference between the T_g of the elasticized PVC and that of the TPU is less than 20°C [6].

In these special situations, other methods, such as electronic microscopy, IR spectrum, and fluorescence spectrum are to be used.

B. TPU Used in Blending with PVC

TPU is usually made from hydroxyl-terminated polyether or polyester diols, diisocyanates, and bifunctional chain extenders. Since the composition, the synthetic method, molecular weight, and its distribution are all changeable, there are numerous types of TPUs available, and their prices and properties vary significantly.

Not all TPU polymers are suitable for blending with PVC. Certain types of TPUs when mixed with PVC produce a hard, deep-yellow colored material that whitens when bent [7]. Only a few types of TPUs are suitable for blending with PVC.

Table 1 The Relationship Between the PVC/TPU Ratio and the T_g of the Blends

TPU/PVC (ratio by weight)	Pure TPU	1.075	0.882	0.692	0.500	Pure PVC*
$T_{g(exp.)}$(K)	223	257	264	273	281	338
$T_{g(the.)}$(K)	—	267	272	279	288	—
ΔT_g(K) = $T_{g(exp.)}$ — $T_{g(the.)}$	—	−10	−8	−6	−7	—
$\Delta T_g/T_{g(exp.)}$ (%)	—	−3.9	−3.0	−2.2	−2.5	—

PVC:TPU:Plasticizer:Additives = 100:(50–107.5):5:3 (by weight).
* Shanghai Tianyuan Chemical Co., SG-3 suspension resin with plasticizer and additives.

There are three major measurements used in judging TPU as a polymeric plasticizer for PVC: the glass transition temperature (T_g), the compatibility with PVC, and the degree of crystallinity.

The main function of a plasticizer is to increase the mobility of the polymer chain, thus decreasing its T_g. The T_g of the PVC/TPU system is mainly determined by the T_g of TPU. It could be calculated by the Fox equation:

$$\frac{1}{T_g} \approx \frac{W_{PVC}}{T_{gPVC}} + \frac{W_{TPU}}{T_{gTPU}}$$

here, W_{PVC} and W_{TPU} are the weight fractions of PVC and TPU [8]. It is clear that only when T_{gTPU} is sufficiently low, the T_g of the blends could be lower than room temperature. Usually the T_g of TPU used in blending with PVC should be lower than $-30°C$ if it is used as a plasticizer.

The compatibility is also a very important fact. Good compatibility means that the two kinds of polymers in the blends are mixed at the molecular level to form an apparently single phase. This kind of blending system shows only one T_g. If the two polymers have only partial compatibility, the entire system will maintain two different T_gs and TPU only serves as a modifier not a plasticizer.

Another fact that affects the properties of the blends is the degree of crystallinity of TPU. TPUs with a high degree of crystallinity cannot serve as plasticizer for PVC.

1. The Selection of TPU

The δ values of both PVC and TPU are available. The selection of TPU should also be based on the application itself. For example, in the case of modifying PVC, if TPU is used as the substituent of liquid plasticizer, then it must have a good compatibility with PVC in order to make a homogeneouslike system. If TPU is just used to improve the toughness of PVC, then the amount of TPU in the blends is usually less than 15 parts (PVC as 100 parts), therefore, partial compatibility is enough.

Both polyester- and polyether-based TPU could be used to blend with PVC, although the former constitutes the majority of the commercial products. All of the blends should meet the following requirements: (1) they must have good or relatively good compatibility with PVC, (2) their processing temperature should be close to or lower than that of PVC, (3) they have to meet the specific requirements of the products, for example, TPUs used for medical purposes should be colorless (if possible), transparent, nontoxic, and able to be sterilized, and (4) they should not be expensive.

2. The Synthesis of TPU

In some cases, satisfactory TPU cannot be found from commercial sources and synthesis is necessary. Since the polyester-based TPU is more frequently used for blending with PVC, it is selected as the representative in the following discussion. The hydroxyl-terminated polyesters used for synthesizing TPU are usually from commercial sources. They are composed of aliphatic diacids and diols with four or more carbons, their average molecular weight is between 500–3000. The water content should be lower than 0.03% and the hydroxyl value must be tested before use. The amount of catalyst that is remaining in the polyester should be as low as possible. A catalyst is necessary in polyester synthesis but is a nuisance in the blends. Its existence could cause the hydrolysis of the TPU products.

Aromatic diisocyanates such as toluene 2,4-diisocyanate (TDI) and methylene di-p-phenylene isocyanate (MDI) are usually used. Aliphatic diisocyanate such as hexanediisocyanate (HDI), although it has the advantage that the TPU synthesized from it is softer and not prone to turning yellow, is seldom used due to its high cost.

A chain extender could be selected from the following compounds: ethylene glycol, 1,2-propanediol, 1,4-butanediol, 1,3-butanediol, neopentanediol, hexanediol, methyl-dihydroxylethyl amine (MDEA) [9], etc. These should be as dry as possible.

Increasing the molecular weight of polyester (or polyether) or changing its chemical composition could lower the T_g of the TPU and decrease the crystallinity of the polymer. For example, a TPU composed of poly(δ-lactone), MDI, and 1,4-butanediol was found to have the lowest degree of crystallinity and, therefore, the best compatibility with PVC when the hard segment in it is 36% by weight [10].

Also, the ratio of isocyanate and hydroxyl functional groups in the system has a certain optimal value's range. A smaller value than the optimal value's range will cause banding during the processing and the blending product would have a low strength. A higher value than the optimal value's range will also cause certain problems during the processing. In both cases, the final material will have a rough surface. These kinds of defects are most prominent in the case of forming blown film.

When both the ratio of the reactants and the method of synthesis are fixed at their optimal conditions, the control of reaction time has the decisive role in the completion of the reaction and the property of the product.

TPU could be synthesized by either a one-step or a two-step method. If the twin (or multi) screw extruder is used, the continued production is possible.

Heat treatment of TPU after its synthesis could increase its molecular weight, thus increase the strength of the blending products. It also decreases the hardness of the products. For example, the polymer blends from PVC and 881014TPU, a TPU made in our lab, showed a 10% increase in its tensile strength and 5% increase in its elasticity when the TPU was treated at 105°C for 7–9 hours compared with no treatment.

C. PVC Resins Used in Blending with TPU

PVC has been widely used in every aspect of life. Its production has been refined and its quality is relatively stable. Unlike TPU, PVCs used in blending with TPU are almost exclusively from commercial sources. Most of them are made by suspension polymerization.

The molecular weight is the major concern when selecting a PVC for blending with TPU. But the exact effect on the properties of the blends also depends on the relative amount of PVC and TPU in the blends itself. Experimental data from our lab suggested that when PVC is the major component in the blending system (PVC/TPU > 1), within a certain range of molecular weight, the PVC is compatible with the particular type of TPU, but the elasticity decreased with the increase of the PVC molecular weight. The strength of the material did not show any significant difference.

When TPU was the major component (PVC/TPU < 1), the experimental data [11] showed that within a certain range of molecular weight, the strength of the material increased with the increase of PVC molecular weight and the elasticity did not show any significant difference. However, the melt viscosity increased, and this caused difficulties in the processing.

Thus, the impact of molecular weight of a PVC on the properties of the blends is complicated. It is closely related to the ratio of the two polymer components in the blending system. It is also related to the properties of TPU. Therefore, it should be considered in combination with other facts.

D. Plasticizers

Small amounts (usually <10%) of plasticizer could be used in the blending system to improve the processing properties of the material by lowering the melting and glass-transition temperatures. The addition of liquid plasticizer also makes the material soft but at the same time, the strength and toughness of the material decreases.

The effects of added DOP on the properties of PVC/TPU blending material are shown in Table 2.

When selecting a plasticizer, the overall properties are the major concern. The processing properties of the blending material and cost should also be considered.

More than one plasticizer could be used. One is the main plasticizer and the other (or others) is the secondary plasticizer. The compatibility between the plasticizers and the polymers in the blending system, especially the major component, is necessary to obtain the best plasticizing effect. Combining the main plasticizer and secondary plasticizer(s) usually gives better results.

The most common types of plasticizer are: the phthalic acid esters of C_6, C_8, C_{10} alcohols; citric acid esters; and epoxy aliphatic esters.

E. Additives

1. Stabilizers

The most commonly used stabilizers are: barium, cadmium, zinc, calcium and cobalt salts of stearic acid; phosphorous acid esters; epoxy compounds and phenol derivatives. Using stabilizers can improve the heat and UV light resistance of the polymer blends, but these are only two aspects. The processing temperature, time, and the blending equipment also have effects on the stability of the products. The same raw materials and compositions with different blending methods resulted in products with different heat stabilities. Therefore, a thorough search for the optimal processing conditions must be done in conjunction with a search for the best composition to get the best results.

2. Lubricants

Lubricants used in processing can be divided into inner and outer lubricants. The former is slightly soluble in the melted polymer, thus it lowers the melt viscosity of the polymer; the latter forms a thin film between the surfaces of the melted polymer and the hot metal surface of the processing machine, thus it does not allow the polymer to stick to the surface of the machine.

The amount of a lubricant used in processing usually does not exceed 0.5–2 parts. Overuse can cause a decrease in the transparency of the material and the lubri-

Table 2 The Effect of Added DOP on the Mechanical Properties of PVC/TPU Blends

DOP (% by weight)	Tensile strength (MPa)	Ultimate elongation (%)	Permanent deformation (%)	Hardness (°ShA)	T_g (K)
0	17.8	497	37.1	86.0	273.0
4.2	16.1	461	46.6	78.7	264.5
6.6	12.1	454	45.2	72.8	262.5
8.8	11.6	439	47.6	69.6	257.0

PVC:TPU:Plasticizer:Additives = 100:78.8:changeable value:fixed value.

cant will eventually migrate to the surface of the blending material (lubricant-bloom).

The most widely used lubricants are stearic acid, its esters, and amides. They usually have a long aliphatic chain and a polar functional group. The polar group has certain compatibility with PVC and TPU. The long aliphatic chain serves as the lubricating agent. Increasing the length of the aliphatic chain will help the lubricating effect but its compatibility decreases at the same time.

III. THE PROCESSING PROPERTIES OF THE PVC/TPU SYSTEM

A. The Viscous Flow Temperature (T_f) of the PVC/TPU System

Zhong et al. [12] have studied the processing properties of the PVC/TPU system. The T_f of the PVC/TPU system changes with different PVC/TPU ratios. Pure PVC has a T_f of 161°C, and with a ratio of 1:1, the system has a T_f of 202°C. Before the ratio of TPU/PVC reaches 0.5, the T_f of the system decreases in conjunction with the increase of the ratio of TPU/PVC; but after that point, the T_f of the system increases when the ratio increases. This phenomenon can be explained by the following: TPU polymer molecule consists of both soft and hard chain segments, therefore, it has a certain flexibility. When the amount of TPU is small, it exists as a dispersed phase in the blending system. This phase resembles the plasticizer. The result, therefore, is a decrease of the T_f of the blending system. When the TPU amount increases, it becomes a continuous phase, and the blends become more and more similar to TPU. Furthermore, as the amount of TPU increases, the polar groups in the TPU chain have more attraction to the chlorine atoms on the PVC chain. This also causes an increase of the T_f of the system.

B. The Plasticizing Properties of the TPU/PVC System

Zhong et al. [12] also studied the plasticizing properties of the PVC/TPU blending system by measuring the torque (TQ) during plasticating.

Whatever processing temperature was used, the system always had an initial period of time during which the TQ reached a maximum. As time went on, the TQ decreased until plasticizing occurred, at which time the TQ had reached its minimum. After that, the TQ showed a gentle increase. Both physical and chemical changes occurred during processing. During the initial period of time, the polymer molecules do not move alot, this is shown by the high TQ; after the system reaches equilibrium, melting and plasticizing occurs. At this time, the polymer molecules move violently as viscous flow occurs. This causes both viscosity and TQ to drop. After reaching equilibrium, another increase of the TQ suggests that possible chemical reactions have occurred during the processing, such as grafting and crosslinking. This actually improves the properties of the material.

The fact that these kinds of chemical reactions occur during the processing were further demonstrated by the testing of the insoluble material in the blends. The data showed that after dimethylformamide (DMF) treatment and after processing, the solubles significantly increased. It also showed that the insoluble content is also related to the TPU content in the blending system. It peaks at the TPU content of 15%–20%, after which the amount drops dramatically. Accordingly, the mechanical properties of the material also showed the same trend.

IV. THE PRODUCTION OF PVC/TPU BLENDS

An illustrative flow chart of the entire procedure of the blending processing is shown in Scheme 1.

A. Raw Materials

1. PVC Resins, Plasticizers, and Additives

The quality control of PVC, plasticizers, and additives must be done according to certain standards. In the case of making the blending material for medical uses, medical grade PVC, which contains <3 ppm vinylchloride monomers, should be used.

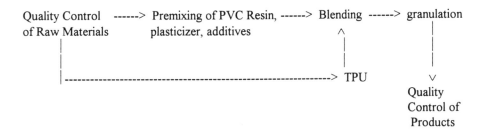

Scheme 1

Table 3 The Mechanical Properties of the 566 TPU/PVC Polymeric Blends

TPU/PVC (ratio by weight)	Tensile strength (MPa)	Ultimate elongation (%)	Permanent deformation (%)	Hardness (°ShA)
0.500	22.6	304	25	98
0.821	26.3	423	15	88
1.08	24.3	466	13	79
1.34	19.8	553	13	71
2.00	17.1	720	26	64

According to Chinese National Standards GB 1040-79, GB 528, GB2411-80.
PVC:TPU:Plasticizer:Additives = m:n:5:4 (by weight, m + n + 5 + 4 = 200).

2. TPU

Since TPU is much more expensive than PVC, and its quality is usually not as stable as that of PVC, especially when TPU is synthesized instead of a commercial product, the quality inspection of TPU must be done under strict control. One thing that needs to be paid attention to is the water content of TPU. TPUs with higher than allowed amounts of water must be preheated to remove the bulk of the water.

It is always necessary to check the compatibility between the selected TPU and PVC before large scale production. A simple and practical method is to blend a small sample in the lab with a twin-roller calender or a small extruder.

A compatible blend should have good processing properties along with a smooth surface and cross-section. If needed, further tests on mechanical properties should be carried out on testing samples made from 2-mm films produced by compression molding.

B. The Blending

PVC resin, plasticizers, and additives are weighed according to the recipe and mixed in a high-speed mixer. This mixture is then processed at 140–180°C for a certain period of time to form the preplasticized PVC. The TPU is then added and further blending is performed until completion, depending on each case.

If TPU is the major component, then the TPU should be processed first, then added to the preplasticized PVC to form the final blend.

The processing (blending) can be done on either calenders or extruders. Usually, one-step processing is feasible on calenders, but for two-step processing extruders are necessary. The first step is making semi-plasticized PVC granules by the extruder. Then this is mixed with TPU in the proper ratio and added to the extruder once again for final processing. By using a twin-screw extruder, one-step processing is also possible.

Since quite a bit of difference exists between raw materials, the recipe, and the equipment, the processing procedure and conditions vary a lot. Also, the processing procedures of commercial products are usually not available to the public. Thus, much work needs to be done to find the best procedure and condition for each individual system. In general, a good procedure is a combination of optimal processing time, temperature, and rotating speed of the screw (in the case of extruder use) or the roll nip (in the case of calender use).

V. THE PROPERTIES OF THE PVC/TPU BLENDS

A. General Remarks

In general, the properties of the PVC/TPU blends reflect the ratio of PVC and TPU in the mixture.

Table 4 The Low-Temperature Impact Test

Temperature (°C)	Commercial soft PVC, plasticized by DOP, medical grade	566TPU/PVC with 39.3% (by weight) TPU	566TPU/PVC with 54.7% (by weight) TPU
−25	no pass	—	—
−30	no pass	—	—
−40	—	pass	—
−60	—	pass	pass

According to Chinese National Standards GB5470-85.

Table 5 Mechanical Properties of the Modifier/PVC Blends

Blends	P83/PVC	E741/PVC	566TPU/PVC	PVC control
Hardness (°ShA)	86.2	82.2	74.2	93.4
Tensile strength (MPa)	13.5	12.0	22.1	29.2
Ultimate elongation (%)	362	381	460	23
Permanent deformation (%)	53	47	30	23

PVC:Modifier:DOP:Additives = 100:75:20:4 (by weight).

PVC modified by TPU has improved abrasive resistance, improved impact resistance (especially at low temperatures), improved oil resistance, improved heat and UV light resistance, and improved adhesive capability with other materials.

TPU modified by PVC has improved processability, improved moisture resistance, improved flame resistance, improved heat resistance, lower costs, and lower abrasion coefficiency.

B. Commercial PVC/TPU Polymeric Blends

Several commercial products of PVC/TPU blends are available. The BF Goodrich Chemical Group has a PVC/TPU blend based on their Estane series TPUs. For example, their Estane 54620, a polyester-based TPU with a °ShA 85 hardness, shows excellent compatibility with flexible PVC. The blends are produced by mixing PVC, TPU, plasticizer, stabilizer, and lubricant in a twin-screw extruder. These polymeric blends show intermediate mechanical properties between PVC and TPU.

Dianippon Ink & Chemical Company (DIC) manufactures the Pandex series of TPUs that are used to make polymeric blends with PVC. These polyblends show comparable mechanical properties to others. Germany's Beyer Chemical Company also has similar products. The related information about these commercial products can be obtained from the manufacturers.

C. Others

A few noncommercial PVC/TPU systems have been reported. Makarof et al. [13] reported a polymer blend of PVC and polyester-based TPU. The TPU series were made from a polyesterglycol (composed of butanediol and hexanediacid), MDI, and 1,4-butanediol as chain extender. The polymeric blends were tested on various mechanical properties and generally showed good properties.

A series of polyester-based TPU (566TPU series) were synthesized in our lab and used to blend with PVC to manufacture a modified PVC material for medical uses [14]. Morphological studies showed that 566TPU has very good compatibility with PVC. Detailed mechanical and electronic property tests were also conducted. Some of the data are provided in Tables 3 and 4.

These data show that the 566TPU/PVC polymeric blend has good mechanical properties, especially at low temperatures. Other tests showed very good oil resistance of this material. Also, the migration rate of plasticizer is only one-fourth of that of commercial medical grade flexible PVC material.

D. The Comparison Between TPU and Other Polymeric Modifiers of PVC

A series of tests [15] were conducted to compare three types of polymeric modifiers for PVC: Du Pont's Elvaloy 741, a copolymer of ethylene, vinylacetate, and carbon monoxide; Goodyear's Chemigum P83, a copolymer of butadiene and acrylonitrile; and 566TPU from our lab, a polyester-based TPU. Some of the results are provided in Tables 5 and 6.

These results showed that TPU, represented here by 566TPU, has the best overall results in modifying PVC among those tested. It has good compatibility with PVC, and the resulting polymeric blends have good mechanical properties suitable for various processing methods.

Table 6 DSC Results of Modifiers and Their Blends with PVC

	P83	P83/PVC	E741	E741/PVC	566TPU	566TPU/PVC	PVC control
T_g (°C)	−23.7	$T_{g1} = -17.9$ $T_{g2} = 52.3$	−24.4	11.3	−32.8	1.7	50.1
T_m (°C)	—	—	57.7	47.7	—	—	—

PVC:Modifier:Plasticizer:Additives = 100:75:20:4 (by weight).

VI. APPLICATIONS AND DEVELOPMENTS

Although relatively new, PVC/TPU polymeric blends have already found substantial applications in various fields. This is evident by the numerous patents applied for in this area.

Since it possesses good properties of both PVC plastics and polyurethane elastomers, it has been used in those areas where PVC and polyurethane have traditionally played dominant roles. For example, it is a very promising replacement for flexible PVC used for medical purposes and in the food industry [16,17], because it essentially eliminates the concern regarding plasticizer contamination. It has been used in combination with the copolymer of butadiene and acrylonitrile (NBR) to make the abrasion-resistant aprons and rolls used on textile machines [18]. A PVC/TPU/ABS blend serves as a substitute for leather [19]. This could have a tremendous impact on the shoe industry. It has also been found to have an application as a building coating [20,21]. This trend will certainly grow and more applications will be found. This in turn should bring new developments in the material itself.

REFERENCES

1. G. Pastuska, *Polymer Processing and Properties* (G. Astarita, L. Nicolais, eds.) Plenum Press, New York, p. 295 (1984).
2. C. D. Huang and J. Chen, *PVC (China), 1*: 44 (1992).
3. P. W. Albro, *Environ. Health Perspect. 65*: 293 (1986).
4. L. A. Utracki, D. J. Walsh, and R. A. Weiss, *Multiphase Polymers: Blends and Ionomers* (L. A. Utracki and R. A. Weiss, eds.), ACS, Washington, DC, p. 1 (1989).
5. J. H. Hildebrand and R. L. Scott, *Solubility of Non Electrolytes,* Reinhold, New York (1950).
6. M. Jiang, *Physical Chemistry of Polymer Alloys,* Sicuan Educational Press, China (1988).
7. C. D. Huang and H. L. Zheng, *Plastics Industry, 1*: 39 (1991).
8. L. E. Nielsen, *Mechanical Properties of Polymer and Composites,* Marcel Dekker, Inc., New York (1974).
9. X. Q. Jiang, B. Ha, and C. Z. Yang, *Synthetic Rubber Industry (China), 16(4)*: 220 (1993).
10. *The Soviet Union Chemistry Abstract,* 7C135 (1981).
11. G. Q. Mei and Y. C. Yan, *Liming Chemical Engineering (China), 3*: 14 (1995).
12. Z. D. Zhong, B. F. Wang, and Y. Jin, *Plastics Processing (China), 4*: 19 (1992).
13. A. C. Makarof, et al. *Plastics (the Soviet Union), 1*: 26 (1989).
14. C. D. Huang, H. L. Zheng, J. Chen, J. Zhao, and B. Ding, unpublished work.
15. C. D. Huang and J. Chen, *Polym.-Plast. Technol. Eng. 33(5)*: 615 (1994).
16. Sako Eiji (The Green Cross Corp.) British Patent, No. 1550260 (1979).
17. Sako Eiji (The Green Cross Corp.) Austrian Patent, No. 357013 (1980).
18. C. D. Huang and Z. H. Wu, unpublished work.
19. Luo Lukovsky. Czech Patent, No. CS 246466 (1987).
20. Y. F. Chen and Z. D. Cao, *Polyurethane Industry (China), 1*: 22 (1992).
21. Y. F. Chen and W. H. Gu, "PU/PVC Building Coating," presented at the 7th Chinese National Polyurethane Meeting, Tai Yuan City, 1994.

11

Mechanical Properties of Ionomers and Ionomer Blends

Masanori Hara and J. A. Sauer
Rutgers University, Piscataway, New Jersey

I. NATURE OF IONOMERS

Ionomers are polymers into which ionic groups have been introduced, usually at a concentration of up to 10–15 mol%, onto some of the monomeric units of the hydrocarbon macromolecular chains, or, in the case of telechelic polymers, at one or both of the chains ends [1–3]. Ionomers can be synthesized by copolymerization of one monomer with another that contains acrylic, methacrylic, sulfonic or other type of acid group, or they may be prepared directly from the homopolymer by chemical modifications. The acid groups, whether carboxylic or sulfonic, are subsequently neutralized, partially or fully, by the addition of appropriate neutralizing agents, such as metal hydroxides.

The ionic groups of the ionomer interact with one another and thereby alter the cohesion and the physical and mechanical properties. Although the precise microstructure of ionomers is still an open question, it is generally accepted that "multiplets," consisting of small groupings of ion pairs, and "clusters," consisting of an ion-rich second phase containing both multiplets and portions of hydrocarbon chains, are present [4]. Strong evidence for the presence of the ionic cluster "phase" comes from studies of ionomers by means of small-angle x-ray scattering (SAXS) and small-angle neutron scattering (SANS). Both of these techniques show the presence of a peak in the scattered intensity at low scattering angles, corresponding to a Bragg spacing of 2–5 nm. This so-called ionic peak is attributed to the presence in the ionomer of a dispersed second phase consisting of ionic clusters.

As the ion content of an ionomer increases, the proportion of the cluster phase to the multiplet-containing matrix phase rises. Then, at some critical ion content, or ion content range, the ionic cluster "phase" becomes dominant over the multiplet-containing matrix phase. One way of determining the critical ion content for a given ionomer is to measure the dynamic mechanical properties, i.e., the storage modulus (E') and the mechanical loss ($\tan \delta$), as a function of temperature. Measurements of this type have been made on a number of ionomers. The mechanical loss, as a function of temperature, usually shows two peaks, one corresponding to the glass transition temperature of the matrix phase, T_{g1}, and the other, at a higher temperature, to the glass transition of the more tightly bound cluster "phase," T_{g2}.

This type of plot, for the Na-salt of a sulfonated polystyrene (SPS) ionomer is shown in Fig. 1 [5]. As the ion content rises, the maximum intensity of the matrix peak falls while that of the cluster peak increases. For this ionomer, the critical ion content at which the cluster "phase" becomes dominant is about 6–7 mol%. A comparable value of the critical ion content has been found in another ionomer, viz. poly(styrene-co-sodium methacrylate). However, for a poly(methyl methacrylate) (PMMA) ionomer, formed from neutralization of a copolymer of methyl methacrylate and methacrylic acid, the critical ion content appears to be about 12 mol% [6]. One reason for the higher critical ion content for the PMMA-based ionomer is that the PMMA has a higher dielectric constant than polystyrene (PS); hence, at any given ion content, the tendency for multiplets to form clusters is reduced. Thus, a higher ion content is required before the cluster "phase" becomes dominant over the matrix phase.

The mechanical properties of ionomers, such as their modulus or stiffness, tensile strength and energy-

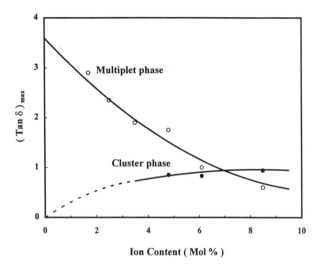

Figure 1 (tan δ)$_{max}$ versus ion content for the multiplet phase and the cluster phase of Na-SPS ionomers.

to-fracture, induced deformation modes, and lifetime-to-fracture under alternating loading, depend on many factors. These include type of ionomer, ion content, processing conditions and thermal treatment, percent conversion (neutralization), nature of the counterion, and extent of blending, if any [7]. In this chapter, our primary interest is in the mechanical performance of solid ionomers under ambient temperature conditions and in the influence of various variables on mechanical properties. It is pertinent, however, to first note the very extensive changes in properties that occur at elevated temperatures in the rubbery region above the glass transition

temperature of the parent homopolymer. An illustration of the profound changes that can be produced is evident from Fig. 2, which shows how the elastic storage modulus of PMMA-based ionomers depends on temperature [6]. Note the dramatic increase of several decades or more in the rubbery plateau modulus as the ion content, and the extent of the ionic cluster "phase," rises. There is also a shift of the multiplet-containing matrix T_g to higher temperatures. Both of these results are indications that the ionic groups are acting rather like covalent crosslinks, although ionic bonds in ionomers are generally weaker than covalent bonds. Ionomers, however, have an advantage over covalently crosslinked polymers; i.e., they can be processed by conventional molding techniques provided the ion content is not too high.

II. INFLUENCE OF ION CONTENT

As the ion content of an ionomer is increased, the proportion of the ionic cluster "phase" that is present rises. This leads to an increase in "entanglement strand density." The significance of strand density to mechanical behavior has been pointed out in the literature [8]. One consequence is that a low-strand density polymer, like polystyrene, deforms in tension only by crazing and is susceptible to early fracture by craze fibril breakdown and subsequent crack propagation through the already crazed material. In contrast, a high-strand density polymer, like polycarbonate, readily undergoes shear deformation and the presence of shear leads to enhanced ductility and toughness.

For a brittle polymer, such as polystyrene, it is known that the entanglement density can be increased by radiation crosslinking. This produces a change in deformation mode from crazing only to combined crazing and shear, and the presence of shear deformation hinders the growth of crazes. Also, as a result of the crosslinking, and the increase in entanglement density, a higher stress is now required to initiate and propagate crazes. Somewhat similar effects on deformation modes may be anticipated to occur in ionomers as a result of ionic interaction and ionic crosslinking.

As an indication of the changes in deformation modes that can be produced in ionomers by increase of ion content, consider poly(styrene-co-sodium methacrylate). In ionomers of low ion content, the only observed deformation mode in strained thin films cast from tetra hydrofuran (THF), a nonpolar solvent, is localized crazing. But for ion contents near to or above the critical value of about 6 mol%, both crazing and shear deformation bands have been observed. This is demonstrated in the transmission electron microscope (TEM) scan of Fig. 3 for an ionomer of 8.2 mol% ion content. Somewhat similar deformation patterns have also been observed in a Na-SPS ionomer having an ion content of 7.5 mol%. Clearly, in both of these ionomers, the presence of a

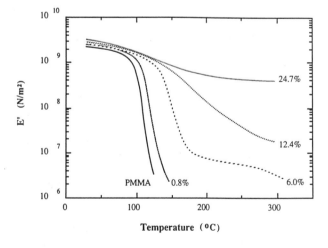

Figure 2 Modulus versus temperature for Na salts of PMMA ionomers of various ion contents.

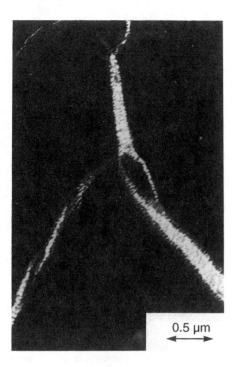

Figure 3 TEM micrograph of a deformed thin film of an 8.2 mol% poly(styrene-co-sodium methacrylate) ionomer cast from THF.

dominant cluster "phase" and a significantly higher strand density has led to the development of shear deformation. The changes observed in strained films are a promising indication that beneficial changes in mechanical properties may also be realized in bulk specimens of ionomers.

Relatively few investigations of the mechanical behavior of solid or glassy ionomers have been reported in the literature [7]. In one study of the tensile properties of Na-SPS ionomers, a modest increase of tensile strength was observed at low ion contents; and a marked increase, to values about 60% higher than for the PS homopolymer, was found as the ion content approached and exceeded the critical range of about 6 mol% [9]. The energy-to-fracture, or toughness, followed a similar trend with the maximum value being about 100% above that of the homopolymer.

In general, as the ion content is raised, the modulus or stiffness of the ionomer is increased, as shown by the data in Fig. 2. While the increase is much greater in the elevated temperature range, where the polymer is acting more like a crosslinked rubber, there is still a significant increase in the glassy modulus below T_g. For example, for the PMMA-based ionomer of Fig. 2, the modulus at 30°C is almost 20% above that of the homopolymer for an ionomer having an ion content of 12.4 mol%. For the

ionomer with ion content of 24.7 mol%, the increase in stiffness is about 45%.

Another property of glassy ionomers, which has been investigated as a function of ion content, is its resistance to fracture under applied alternating stress. In tests made on Na-SPS ionomers, the fatigue life was observed to increase with ion content and to reach a maximum value at about 6 mol%. Thus, such mechanical properties as modulus, tensile strength, energy-to-fracture in simple tension, and resistance to fracture under alternating stress, are all enhanced by introduction of sufficient ionic groups into polystyrene so that microstructure becomes dominated by the ionic cluster "phase." The clusters provide more effective crosslinking than simple multiplets and the resulting increase in ionic entanglement strand density, as the ion content is raised to the critical value, is then sufficient to induce some of the same beneficial changes in properties that are known to occur as a result of radiation crosslinking of PS.

In nonrigid ionomers, such as elastomers in which the T_g is situated below ambient temperature, even greater changes can be produced in tensile properties by increase of ion content. As one example, it has been found that in K-salts of a block copolymer, based on butyl acrylate and sulfonated polystyrene, both the tensile strength and the toughness show a dramatic increase as the ion content is raised to about 6 mol% [10]. Also, in Zn-salts of a butyl acrylate/acrylic acid polymer, the tensile strength as a function of the acrylic acid content was observed to rise from a low value of about 3 MPa for the acid copolymer to a maximum value of about 15 MPa for the ionomer having acrylic acid content of 5 wt% [11]. Other examples of the influence of ion content on mechanical properties of ionomers are cited in a recent review article [7].

III. EFFECTS OF PROCESSING CONDITIONS

The mechanical properties of ionomers can be appreciably altered by the manner in which the ionomer is prepared and treated prior to testing. Some of the factors that are influential are: the degree of conversion (neutralization) from the acid form to the salt form, the nature of the thermal treatment or aging, the type of counterion that is introduced, the solvent that is used for preparation of thin films, and the presence and nature of any plasticizers or additives that may be present. In the scope of this chapter, it is not possible to provide a complete description of the influence of each of these variables on the wide variety of ionomers that are now commercially available or produced in the laboratory. Instead, one or more examples of the changes in properties that may be induced by each of the processing variables is presented and discussed.

A. Degree of Conversion (Neutralization)

In most ionomers, it is customary to fully convert to the metal salt form; but, in some instances, particularly for ionomers based on a partially crystalline homopolymer, a partial degree of conversion may provide the best mechanical properties. For example, as shown in Fig. 4, a significant increase in modulus occurs with increasing percent conversion for both Na and Ca salts of a poly(-ethylene-co-methacrylic acid) ionomer; and in both cases, at a partial conversion of 30–50%, a maximum value, some 5–6 times higher than that of the acid copolymer, is obtained and this is followed by a subsequent decrease in the property [12]. The tensile strength of these ionomers also increases significantly with increasing conversion but values tend to level off at about 60% conversion.

The best combination of properties of polyethylene-based ionomers, such as stiffness, strength, transparency, and toughness, are realized at partial degrees of conversion of about 40–50% [13]. The initial increase in properties is a result of the presence of ionic interactions, which strengthen and stiffen the polymer. There is, however, some loss of crystallinity as a result of the presence of the ionic groups. When the loss of crystallin-ity becomes great enough to begin to counter the beneficial effects of the ionic crosslinking, a maximum in performance is reached; and this is followed by a decline in properties at higher degrees of neutralization due to the loss of the strengthening effect of the crystalline phase.

In some ionomers, an excess of degree of neutralization may be helpful to mechanical properties. For example, in a Zn-salt of a poly(butyl acrylate-co-acrylic acid) ionomer (5.4 mol%), tensile strength was observed to increase steadily with degree of neutralization and to reach a maximum value, some 2.5 times that of the butyl acrylate homopolymer, when there was about 60% excess of neutralizing agent [11]. As another example, at 100% excess of neutralizing agent in a Na-salt of an SPS ionomer, the tensile strength was increased by about 20%, and the average lifetime-to-fracture under an alternating stress of 16.5 MPa, was found to be over 4 times that of a fully neutralized sample [14]. Additional examples are cited in a recent review [7]. Possible reasons for the increase in performance when an excess of neutralizing agent is present are: strengthening of the ionic aggregates by the excess ions, neutralization of any unneutralized acid groups that were initially present, and the presence of the excess neutralizing agent as small, reinforcing second phase particles.

B. Thermal Treatment

The properties of ionomers are strongly influenced by the manner in which they are produced and by the thermal treatment that is provided. For a given ionomer to reach its equilibrium microstructure, there must be sufficient time at an elevated temperature for ionic groups to interact and for ionic aggregates to form. Various examples to illustrate the possible effects of thermal treatment and physical aging are presented. For example, in a compression molded sample of a Na-SPS ionomer of 2.5 mol% ion content, the storage modulus was found to drop sharply above the matrix glass transition temperature and no rubbery plateau modulus was achieved. However, when this same ionomer was given an additional heat treatment (24 h at 200°C), an extended rubbery plateau region was developed and a small mechanical loss peak, indicative of the presence of a cluster "phase," was observed [15].

Thermal treatment and the nature of the casting solvent can also affect the deformation modes achieved in strained films of ionomers. For example, in films cast from polar dimethylformamide (DMF), the solvent interacts with ion-rich clusters and essentially destroys them, as is evident form absence of a second, higher temperature loss peak in such samples. As a result, even in a cast DMF sample of Na-SPS ionomer of high ion content (8.5 mol%), the only deformation mode observed in tensile straining is crazing. However, when these films are given an additional heat treatment (41 h at 210°C), shear

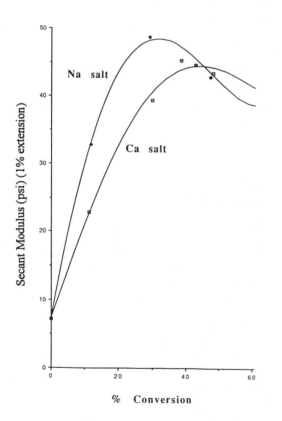

Figure 4 Secant modulus versus percent conversion for Na and Ca salts of an ethylene/methacrylic acid ionomer.

bands are seen in addition to crazes [7]. This is an indication that the ionic clusters have been restored. Additional confirmation comes from dynamic mechanical measurements, which show the presence of a second elevated temperature loss peak.

Studies of PMMA-based ionomers also demonstrate the influence of thermal treatment on deformation modes (16). For Na salts of PMMA-based ionomers of 6 and 12 mol% that were cast from DMF, only crazes were observed on straining. However, after an additional heat treatment (48 h at 160°C), which also removes any DMF solvent that is present, shear deformation zones are induced. Hence, the ionic cluster "phase," which was destroyed by the polar solvent, has been restored by the heat treatment.

For partially crystalline ionomers, such as those based on copolymers of ethylene and methacrylic acid, even time or aging at room temperature can have an effect on mechanical properties. For example, upon aging at 23°C, the modulus of the acid form of the copolymer increased 28%, while in the ionomer form, the increase ranged up to 130%, with the specific gain in modulus depending on the degree of conversion and on the counterion that was present [17].

C. Nature of Counterion

The type of counterion present in an ionomer may, or may not, have a significant effect of properties. For polyethylene-based ionomers, where the presence of crystallinity has an appreciable effect on properties, the type of counterion present does not appear to have a significant effect on either modulus or tensile strength, as Fig. 4 indicates. However, in amorphous ionomers, the effects of changing the counterion from a monovalent one, as in Na or K, to a divalent one, such as Ca, may be appreciable.

As one example, in thin films of Na or K salts of PS-based ionomers cast from a nonpolar solvent, THF, shear deformation is only present when the ion content is near to or above the critical ion content of about 6 mol%; and the TEM scan of Fig. 3, for a sample of 8.2 mol% demonstrates this; but, for a THF-cast sample of a divalent Ca-salt of an SPS ionomer, having only an ion content of 4.1 mol%, both shear deformation zones and crazes are developed upon tensile straining in contrast to only crazing for the monovalent K-salt. This is evident from the TEM scans of Fig. 5. For the Ca-salt, one sees both an unfibrillated shear deformation zone, and, within this zone, a typical fibrillated craze. The Ca-salt also develops a much more extended rubbery plateau region than Na or K salts in storage modulus versus temperature curves; and this is another indication that a stronger and more stable ionic network is present when divalent ions replace monovalent ones. Still another indication that the presence of divalent counterions can enhance mechanical properties comes from

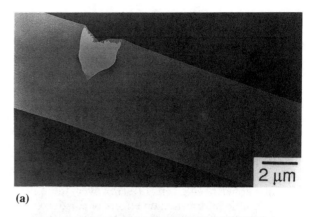

(a)

(b)

Figure 5 TEM micrographs of deformed thin films of an SPS ionomer having an ion content of 4.1 mol% and cast from THF K salt (a), and Ca salt (b).

study of the lifetime-to-fracture under alternating stress conditions. For SPS ionomer samples, with an ion content of 4.1 mol%, divalent Ca-salts exhibited an average lifetime-to-fracture that was 2.8 times greater than that achieved by monovalent K or Cs salts of the same ion content [18].

The combined effects of a divalent Ca counterion and thermal treatment can be seen from studies of PMMA-based ionomers [16]. In thin films of Ca-salts of this ionomer cast from methylene chloride, and having an ion content of only 0.8 mol%, the only observed deformation was a series of long, localized crazes, similar to those seen in the PMMA homopolymer. When the ionomer samples were subject to an additional heat treatment (8 h at 100°C), the induced crazes were shorter in length and shear deformation zones were present. This behavior implies that the heat treatment enhanced the formation of ionic aggregates and increased the entanglement strand density. The deformation pattern attained is rather similar to that of Na salts having an ion content of about 6 mol%; hence, substitution of divalent Ca for monovalent Na permits comparable deformation modes, including some shear, to be obtained at much lower ion contents.

D. Influence of Plasticizers

In the preparation and processing of ionomers, plasticizers may be added to reduce viscosity at elevated temperatures and to permit easier processing. These plasticizers have an effect, as well, on the mechanical properties, both in the rubbery state and in the glassy state; these effects depend on the composition of the ionomer, the polar or nonpolar nature of the plasticizer and on the concentration. Many studies have been carried out on plasticized ionomers and on the influence of plasticizer on viscoelastic and relaxation behavior and a review of this subject has been given [19]. However, there is still relatively little information on effects of plasticizer type and concentration on specific mechanical properties of ionomers in the glassy state or solid state.

Two recent studies in this laboratory have examined the effects of a polar and a nonpolar plasticizer on the properties of a PS-based ionomer and on a PMMA-based ionomer each of which was given an ion content in the critical range where the matrix phase and the cluster "phase" had comparable magnitudes. Results of this study on the Na-SPS ionomer (5.5 mol%) are in accordance with the general concept that a nonpolar plasticizer, such as dioctyl phthalate (DOP), selectively plasticizes the multiplet-containing matrix phase, thereby reducing T_{g1}, but has only a modest effect on the ion-rich cluster "phase." The polar plasticizer, glycerol, acted in an opposite fashion. It selectively plasticized the cluster "phase," and essentially destroyed it, while having only a small effect on the matrix phase.

The effects of the same two plasticizers on a Na-PMMA based ionomer (12.4 mol%) are somewhat different. For a concentration of 22.4 wt%, the nonpolar DOP reduced the matrix T_{g1} about 60°C but the cluster T_{g2} was reduced by only 30°C. Thus, the nonpolar DOP acts in a rather similar fashion for the PMMA ionomer as for the PS-based one. This is not the case though for the polar plasticizer, glycerol. It, at comparable concentration (19.4 mol%) to that of the nonpolar DOP, acts as a "dual" plasticizer for the PMMA ionomer, decreasing both the T_{g1} of the multiplet-containing matrix phase and the T_{g2} of the ionic cluster "phase" by essentially the same amount, viz. ~60°C. Evidently, the higher dielectric constant of PMMA vs that of PS, together with the polar nature of the ester side group in PMMA, causes the polar glycerol to interact more strongly with the backbone chains of PMMA than of PS.

The presence of plasticizers not only shifts the glass transition temperatures to lower values but it can cause significant changes in the glassy modulus. As one example, it has been observed in the Na-PMMA ionomer (12.4 mol%), that the modulus in the solid state (measured at 30°C) decreased at the rate of 2.5% per wt% of DOP but at a higher rate of 7.7% per wt% of glycerol. Thus, for essentially comparable amounts of plasticizer, the polar glycerol has a much greater effect on the properties of

PMMA ionomers in the glassy state than does the nonpolar DOP.

The effects of glycerol and DOP, at 10 wt% concentration, on tensile properties of a glassy Na-SPS ionomer have been examined. Addition of DOP decreased the modulus, the tensile strength, and the toughness. Glycerol at the same concentration, increased the elongation to fracture and the toughness by about 30%, while slightly reducing the stiffness and the tensile strength. The influence of plasticizer concentration on the tensile properties of the same ionomer have also been explored, over the range from 0–30 wt%. With increasing concentration, the modulus and tensile strength steadily decreased but the strain-to-fracture and the toughness rose and reached maximum values at 15 wt%. At this concentration, the energy-to-fracture rose over 70%. These examples illustrate the wide range in properties that can be realized in ionomers, not only by control of ion content and heat treatment, but by selection and control of an appropriate plasticizer.

IV. IONOMER BLENDS

A. Ionomer/Homopolymer

The mechanical properties of blends of ionomers with homopolymers have not been extensively studied but a discussion has been given of the available information [7]. The purpose of blending is to enhance one or more properties of the homopolymer. It has been discovered that, in several blends of an ionomer with a suitable polymer, synergistic effects can be realized, i.e., the measured property of the blend becomes significantly higher than values expected on the basis of the simple rule of mixtures. As one example, the influence of blend composition on the tensile strength of a blend of a sulfonated butyl rubber (SBR) ionomer and polypropylene (PP) is shown in Fig. 6 [20]. At all blend compositions the tensile strength lies above the rule of mixtures value and it reaches a maximum value of 38 MPa at a composition containing about 30 wt% of the SBR ionomer component. The enhanced performance of the blends may result from the presence of two separate interpenetrating networks, one arising from the presence of crystallites and the other from the ionic crosslinking of the ionomer component. Another factor may be good adhesion between two phases resulting from some degree of affinity as a result of common methyl groups in both components. In support of this, in similar blends of SBR with high-density polyethylene (HDPE), which lacks methyl side groups, no synergistic effects are found.

Another example of favorable synergistic effects in ionomer/homopolymer blends is evident from a study of the tensile properties of blends of an SPS ionomer with PS. Over most of the composition range these two polymers are incompatible. For small additions of the SPS ionomer to PS, TEM studies of cast thin films show that

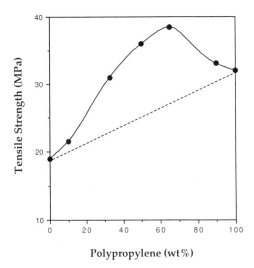

Figure 6 Tensile strength versus percent polypropylene in SBR ionomer/PP blends.

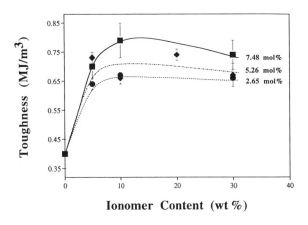

Figure 7 Toughness versus ionomer content (wt%) for Na-SPS ionomer/PS blends.

the ionomer is present as well-bonded, spherical-shaped, small, dispersed particles whose size increases with ion content. Upon straining such films, crazes develop in the PS matrix and the SPS particles in the crazed regions undergo plastic deformation and become appreciably elongated [21]. Also, a higher stress is required to produce craze fibrillation within the ionomer particles as a result of the presence of ionic crosslinking.

A reinforcing effect of dispersed ionomer particles in similar blends of Na-SPS ionomers and PS is also manifest from measurements of the energy to fracture (toughness) of bulk, compression molded specimens of the blends. This is evident from the toughness vs ionomer content data shown in Fig. 7 for specimens having various ion contents in the ionomer component [22]. For all three ion contents, there is a marked rise in toughness with increasing ionomer content with maximum values, which rise with increasing ion content, being reached at about a 10% addition of ionomer. The measured tensile strengths follow a similar trend. An example of this for a blend in which the ionomer component has an ion content of 5.26 mol% is given in Fig. 8. For both tensile strength and toughness, the maximum values achieved are well above values anticipated on the basis of the rule of mixtures (dotted lines in Figs. 6 and 8).

As another example, a synergistic enhancement of tensile strength has been reached in blends of a sulfonated polyacrylonitrile terpolymer (SPAN) with a polyurethane (PU) cationomer [23]. Maximum enhancement was achieved at a blend composition of 30/70 (PU/SPAN). At this blend composition, the tensile strength was raised from an initial value of 78.5 MPa to 196 MPa and the strain-to-fracture was at its highest. The en-

hancement of mechanical properties is attributed to the strong ionic interactions between ionic groups with opposite sign, which are attached to the respective backbone chains.

In a partially crystalline homopolymer, nylon 6, property enhancement has been achieved by blending with a poly(ethylene-co-acrylic acid) or its salt form ionomer [24]. Both additives proved to be effective impact modifiers for nylon 6. For the blends of the acid copolymer with nylon 6, maximum impact performance was obtained by addition of about 10 wt% of the modifier and the impact strength was further enhanced by increasing the acrylic acid content from 3.5 to 6%. However, blends prepared using the salt form ionomer (Surlyn®: 9950-Zn salt) instead of the acid, led to the highest impact strength, with the least reduction in tensile

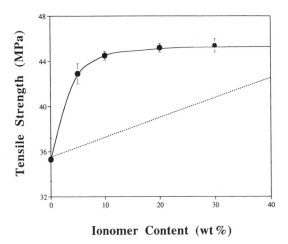

Figure 8 Tensile strength versus ionomer content (wt%) for a blend of an Na-SPS ionomer (5.26 mol%) and PS.

strength, at a blend composition containing 25 wt% of the ionomer.

B. Ionomer/Small Additives

Blending of certain additives to an ionomer can result in both a plasticizing action, thereby enhancing elevated temperature processing of the blend, and also in a reinforcing action, thereby enhancing the strength and stiffness properties of the composite at ambient temperatures. This dual action occurs, for example, when zinc stearate ($ZnSt_2$), which is a crystalline material with a low melting temperature, is added to the Zn salt of a sulfonated ethylene-propylene-diene terpolymer (SEPDM) rubbery type ionomer. Upon melt mixing the two components, the zinc stearate penetrates and attacks the ionic aggregates of the SEPDM ionomer. This lowers the melt viscosity and permits easier processing. In this respect, the zinc stearate additive acts rather as a conventional plasticizer.

However, when the stress–strain properties of the blends are examined at ambient temperature, the zinc stearate, which is present in the form of dispersed crystallites, acts as a strong reinforcing agent. This is evident from Fig. 9 where the stress–strain response of a 50/50 blend is compared to that of the EPDM homopolymer, a sulfur vulcanized EPDM (X-EPDM) and the Zn-SEPDM ionomer [25]. While the ionomer itself shows significant increases in modulus and strength as compared to the homopolymer or to the vulcanized EPDM, it is evident that the greatest enhancement of strength, stiffness, and toughness (area under the stress–strain curve) occurs for the $ZnSt_2$/Zn-SEPDM blend. The beneficial effects on mechanical properties are a result of the reinforcing action of the $ZnSt_2$ crystalline phase and the network of strong ionic interactions that are present.

V. SUMMARY

The mechanical properties of ionomers are generally superior to those of the homopolymer or copolymer from which the ionomer has been synthesized. This is particularly so when the ion content is near to or above the critical value at which the ionic cluster phase becomes dominant over the multiplet-containing matrix phase. The greater strength and stability of such ionomers is a result of efficient ionic-type crosslinking and an enhanced entanglement strand density.

Aside from ion content, a wide range of properties is available in ionomers by control of various processing variables, such as degree of conversion (neutralization), type of counterion, plasticizer content and thermal treatment. Various examples illustrating possible effects of these variables on mechanical relaxation behavior and on such mechanical properties as stiffness, strength, and time- or energy-to-fracture have been given.

Blending of ionomers with other homopolymers is also one means of enhancing mechanical performance. Frequently, in ionomer/polymer blends, synergistic effects are realized and properties may be significantly increased over anticipated values based on the rule of mixtures. This area of study has not been extensively explored and the probability clearly exists that new materials and new blends, having even a greater degree of property enhancement, will become available in the near future.

ACKNOWLEDGMENTS

Our research studies on ionomers and on ionomer blends has received financial support from the U.S. Army Research Office and ACS-PRF, whose assistance is gratefully appreciated. Our appreciation is also expressed to former graduate students, Drs. P. Jar, M. A. Bellinger, and X. Ma, who made significant contributions to some of the research results reported herein.

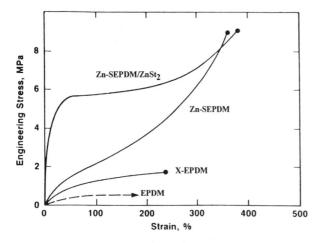

Figure 9 Stress–strain curves for EPDM, vulcanized EPDM, Zn-SEPDM ionomer, and 50/50 blend of Zn-SEPDM and $ZnSt_2$.

REFERENCES

1. L. Holliday (ed.), *Ionic Polymers,* John Wiley & Sons, New York (1975).
2. A. Eisenberg and M. King, *Ion Containing Polymers,* Academic Press, New York (1977).
3. L. A. Utracki and R. A. Weiss (eds.) *Multiphase Polymers-Blends and Ionomers,* ACS Symp. Ser., *395* (1989).
4. A. Eisenberg, B. Hird, and R. B. Moore, *Macromol., 23*: 4098 (1990).
5. M. Hara, P. Jar, and J. A. Sauer, *Polymer, 32*: 1622 (1991).

6. X. Ma, J. A. Sauer, and M. Hara, *Macromol., 28*: 3953 (1995).
7. M. Hara and J. A. Sauer, *Rev. Macromol. Chem. Phys., C34(3)*: 325 (1994).
8. E. J. Kramer and L. L. Berger, *Adv. Polym. Sci., 91/92*: 1 (1990).
9. M. Bellinger, J. A. Sauer, and M. Hara, *Macromol., 27*: 1407 (1994).
10. R. D. Allen, I. Volger, and J. E. McGrath, *ACS Symp. Ser., 302*: 79 (1986).
11. H. Xie and Y. Feng, *Polymer, 29*: 1216 (1988).
12. S. Bonotto and E. F. Banner, *Macromol., 1*: 510 (1968).
13. R. W. Rees, *Polyelectrolytes* (K. C. Frisch, D. Klemper and A. V. Paris, eds.), Technomic Publ. Co., Westport, Connecticut, p. 177 (1976).
14. M. Hara, P. Jar, and J. A. Sauer, *Macromol., 23*: 4964 (1990).
15. M. Hara, P. Jar, and J. A. Sauer, *Polymer, 32*: 1380 (1991).
16. X. Ma, J. A. Sauer and M. Hara, *Macromol., 28*: 5526 (1995).
17. E. Hirisawa, Y. Yamamoto, K. Tadano, and S. Yano, *Macromol., 22*: 2776 (1989).
18. M. Hara, P. Jar, and J. A. Sauer, *Macromol., 23*: 4465 (1990).
19. C. G. Bazuin, *ACS Symp. Ser., 395*: 476 (1989).
20. H. Xie, J. Xu, and S. Zhao, *Polymer, 32*: 95 (1989).
21. M. Hara, M. Bellinger, and J. A. Sauer, *Polymer Intern, 26*: 137 (1991).
22. M. Bellinger, J. A. Sauer, and M. Hara, *Macromol., 27*: 6147 (1994).
23. Y. Oh, Y. Lee and B. Kim, J., *Macromol. Sci.-Phys., B33*: 243 (1994).
24. R. D. Deanin, S. A. Orroth and R. I. Bhagat, *Polym. Plast. Technol. Eng., 29*: 289 (1990).
25. I. Duvdevani, R. D. Lundberg, C. Wood-Cordova, and G. L. Wilkes, *ACS Symp. Ser., 302:* 184 (1986).

12

Metallocene-Based Polyolefins: Product Characteristics

Virendra Kumar Gupta

Indian Petrochemicals Corporation Ltd., Gujarat, India

I. INTRODUCTION

An enormous interest in the synthesis of polyolefins has been created in recent years due to the discovery of a new class of catalysts called metallocenes [1–3]. These catalysts produce new polymers, such as syndiotactic polypropylene, and cycloolefin copolymers, or can be used in the field of already existing polyolefins, such as linear low-density polyethylene (LLDPE), high-density polyethylene, isotactic polypropylene, and ethylene-propylene rubber. The most remarkable feature of these emerging catalyst systems is the fact that all metallocene sites produce polymer chains with virtually the same architecture (Fig. 1) as compared with the different structures of polymer chains obtained with traditional multiple site heterogeneous catalysts. This characteristic opens up many possibilities for producing tailored polyolefin materials with controlled molecular weights, consistent comonomer contents, desired molecular weight and chain branching distribution, and control tacticity pattern, etc. New molecular architecture of homogeneous polymer molecules coupled with the compatability of metallocene systems with existing polyolefin production processes has led to faster commercialization of metallocene-based polyolefins (Table 1) [4]. To date, most efforts have been devoted to catalyst development and their evaluation for polyolefin synthesis. Presently, attention has also been focused on other aspects of the new generation of polyolefin materials, such as processability and additive package for various end use applications [5]. Most early applications have been in speciality polymer markets where value-added and higher priced new polyolefins can compete. The metallocene-based polyolefins are expected to compete in a broader thermoplastic market as more understanding of the different commercially significant aspects are realized.

II. METALLOCENE CATALYST SYSTEMS

Metallocene consists of a transition metal atom sandwiched between ring structures to form a sterically hindered site [6]. The main interest in the use of metallocenes as catalysts for polyolefins was started in the late 1970s when Kaminsky's group discovered that a small addition of water in a metallocene–trimethylaluminum system gives high activity for ethylene polymerization [7]. Since then various types of metallocene–alumoxane systems have been investigated for polyolefin synthesis (Table 2) [1]. These systems show high activities, good copolymerization behavior, and excellent control of stereoregularity. However, homogenous systems produce polymers with poor morphology leading to their nonutilization in gas or slurry phase processes. Supported metallocenes are being developed to overcome these drawbacks. Inorganic supports, such as silica or magnesium dichloride, have been used for the incorporation of metallocenes. Another class of catalysts, termed cationic metallocenes, have also been used [1,8]. Such catalysts provide a hope for development of alumoxane-free commercial catalysts.

III. POLYETHYLENE

Novel polyethylene materials have been synthesized by copolymerization of ethylene with 1-butene, 1-hexene, and/or 1-octene using metallocene catalyst systems [9].

(a) Metallocene-Single-Site

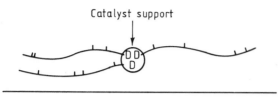

(b) Ziegler- Natta-Multi-Site

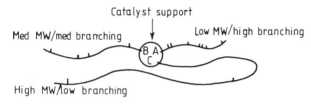

Figure 1 Metallocene-single-site versus Ziegler–Natta multi-site catalysts. (From Ref. 34. Reprinted with permission from *Chemical Engineering,* McGraw-Hill, Inc., New York, 1993.)

Table 1 Global Capacity for Metallocene-Based Polyolefins

Company	Location	Capacity (millions of pound per year)
Polyethylene		
Dow Plastics	U.S.	250
Dow Plastics	Spain	125
Exxon Chemical	U.S.	253
Mitsubishi	Japan	220
Nippon Petrochemicals	Japan	110
Ube Industries	Japan	44
Total		1,002
Polypropylene		
BASF	Germany	26
Chisso	Japan	44
Exxon Chemical	U.S.	220
Hoechst	Germany	220
Mitsui Toatsu	Japan	165
Total		675
Polycyclicolefins		
Dow Plastics	U.S.	Pilot
Hoechst	Germany	Pilot
Mitsui Petrochemicals	Japan	7

Source: Ref. 4. (Reprinted with permission from Chem. & Eng. News, Sept 11, 1995, p15; American Chemical Society.)

Metallocene and monocyclopentadienyl-amide derivatives of titanium complexes (Fig. 2) [2] randomly incorporate ethylene and alpha olefins to give polymers with narrow molecular weight distribution, higher comonomer contents, and good compositional homogeneity [3]. A wide variation in copolymerization efficiency of metallocenes has been observed. Stereorigid metallocenes readily incorporate alpha olefins as compared with nonstereorigid metallocene systems. Dimethyl silyl bridge amido cyclopentadienyl titanium complexes are found to be less restricted in terms of the nature of the comonomer and incorporate a much higher amount of even higher alpha olefins. These homogenous polyethylene polymers are structurally, compositionally, thermally, morphologically, and optically distinct materials from traditional heterogenous catalyst-based products. The solution, slurry, and gas phase production processes have been utilized for polymer synthesis [9,10]. The so-

lution phase process has mainly been used commercially for the production of different grades of polymers (Table 3). The properties of metallocene-based LLDPE by solution and gas phase processes are shown in Tables 4 and 5 [9].

LLDPE with narrow molecular weight distribution exhibits a lower, a sharper melting point [3], better hot tack and heat seal properties as well as higher clarity and better impact resistance (Fig. 3), tensile strength (Fig. 4) [11], and lower levels of alkane-soluble components. The most distinguishing characteristic of metallocene-based LLDPEs is that they are not restricted by the current immutable property relationships that are

Table 2 Metallocene Catalyst Systems

Catalyst	Cocatalyst	Products
Nonstereorigid metallocene	Alumoxanes	PE, atactic PP
Stereorigid metallocene	Alumoxanes	PE, PP, cyclic polyolefins
Supported metallocene	Trialkyl aluminum/Alumoxanes	PE, PP
Cationic metallocene	—	PE, PP
Monocyclopentadienyl metallocenes	Alumoxanes	PE, PP

PE = polyethylene; PP = polypropylene.

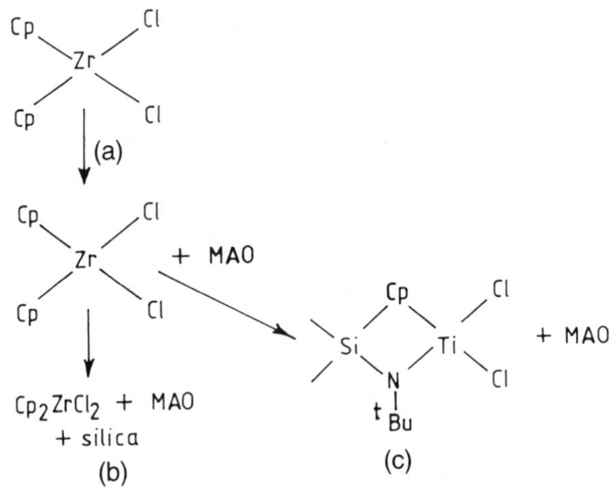

Figure 2 Different metallocene types for polyethylene. (From Ref. 2.)

Table 4 Comparison of Properties of LLDPEs Produced by Metallocene Catalyst and Conventional Ziegler–Natta Catalyst

Property	Exxpol EX-101	Standard polyethylene
Melt Index	50	55
Density	0.923	0.926
Spiral flow (cm)	69.9	78.1
Tensile strength (MPa)		
Yield	12.1	16.1
Break	7.9	7.4
Elongation (%)		
Yield	25	21
Break	275	77
Tensile impact (J/m)	715	502
Total energy		
Impact (J)	12.1	4.2
Flex modulus (MPa)	363	459
Shrinkage (%)	2.0	2.2

Source: Reprinted from Ref. 9, with permission from Elsevier Science Ltd., Kidlington, U.K.

commonly applied for Ziegler–Natta polymeric products [2,12]. It includes the control of melt tension of polymer independent of melt flow rate (Fig. 5) and nonexistence of a relationship between melt flow characteristic and molecular weight distribution. It is possible to synthesize one density grade of metallocene polyethylene with different melt flow rates (Fig. 6) [13]. These developments indicate that metallocene catalysts have provided the opportunity to tailor polymer performance to specific applications—both in terms of processability and end use characteristics.

Narrow molecular weight distribution, which is characteristic of metallocene-based polyethylene (Fig. 7), causes processing difficulty in certain applications due to increased melt pressure, reduced melt strength, and melt fracture [14,15]. This problem can be overcome by blending the metallocene polymer with other prod-

ucts, such as LDPE, processing aids, etc. Recently, Dow Chemical has introduced a new family of ethylene-based branched homogenous polymers using insite technology [2,3,16]. Using this technology, the monocyclopentadienyl metallocene catalyst and process parameters hold the short chain branching level at a precise target while increasing the long chain branching level. This results in the creation of a narrow molecular weight and the homophasic product characteristics of a metallocene catalyst, but with lower melt viscosity and higher melt strength than observed in the homogenous linear

Table 3 Metallocene-Based Polyethylene Processes

Process (company)	MI range (g/10 min)	Density (g/cc)	Comonomer (%)
Solution (Dow)	0.5–30	0.880–0.950	Octene (0–20%)
Solution (Exxon)	1–100	0.865–0.940	Propylene, Butene, Hexene
Gas phase (Mobile)	1	0.918	Hexene
Gas phase (BP Chem.)	2.3	0.916	Hexene

Source: Reprinted from Ref. 9, with permission from Elsevier Science Ltd., Kidlington, U.K.

Table 5 Properties of Gas Phase LLDPE (Hexene-1 Comonomer) Using Supported Metallocenes

Property	BP chemicals	Mobile	Mobil's standard polymer*
Density (g/cc)	0.916	0.91	0.918
Melt index	2.3	1	1
MFR (g/10 min)	—	18	28
Dart drop (g)	900	>800	180
Tear strength (g/mm)			
Machine direction	14.1	14.6	13.8
Transverse direction	18.6	19.3	29.5
Extractables (%)	0.8	0.6	3.5
Haze (%)	—	5.7	18

* Ziegler–Natta synthesized hexene polymer.
Source: Reprinted from Ref. 9, with permission from Elsevier Science Ltd., Kidlington, U.K.

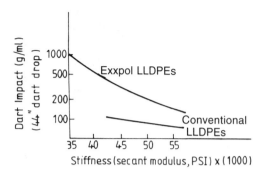

Figure 3 Impact strength of metallocene LLDPE (Ethylene-hexene resin, 1 melt index). (From Ref. 11.)

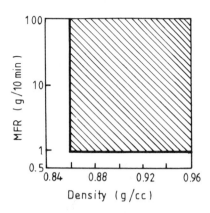

Figure 6 Product profile of LLDPE with respect to density and MFR. (From Ref. 13.)

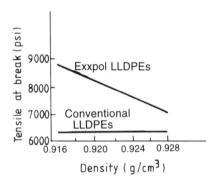

Figure 4 Tensile properties of metallocene LLDPE (Ethylene-hexene resin, transverse direction). (From Ref. 11.)

ethylene–alpha olefin resins. This new class of ethylene-based polymers couples the advantages of metallocene catalyzed compositional purity with the processing behavior of highly branching low-density polyethylene materials. Polymer design parameters (density, molecular weight, short chain branching, long chain branching, etc.) of polymeric materials can be controlled independently with Dow's technology. This breaks the existing set of rules regarding process, structure, and property relationship.

Metallocene-based polyethylene products are expanding the polyethylene market by taking the polymer into new and existing applications [11,17–22]. Metallocene–LLDPE (Tables 6 and 7) has been targeted for film and packaging applications while high-density and medium-density polyethylene grades are used for injection moulding [23] (Table 8; applications: thin-walled containers and stadium cups) and rotational moulding (Table 9; applications: agriculture and industrial containers, trash containers, tanks, etc.), respectively. Commercial applications of LLDPE are notably in the blown and

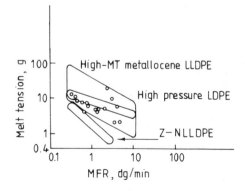

Figure 5 High melt tension LLDPEs from supported metallocenes. (From Ref. 2.)

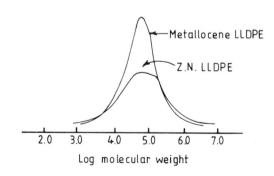

Figure 7 GPC curve of LLDPE obtained with metallocene catalyst system. (From Ref. 13.)

Table 6 Metallocene-Based Linear Low Density Polyethylene

Property	Metallocene system
Density (g/cc)	0.915–0.925
MI (dg/min)	0.5–3.0
Molecular weight distribution	2.1–2.2
Melting point (°C)	114–115
Processability	Good
Dart impact (g)	140–800
MD tear (g)	150–400
TD tear (g)	350–500
% Haze	3.8–6.0
% Gloss	100–135

Source: Reprinted from Ref. 23, with permission of Phillips Petroleum Company, U.S.A.

Table 7 Metallocene-Based Very Low Density Polyethylene

Property	Metallocene catalyst
Density (g/cc)	0.908–0.913
MI (dg/min)	1.8–2.4
Molecular weight distribution	2.2
Melting point (°C)	100–107
Dart impact (g)	600–1400
MD tear (g)	200–240
TD tear (g)	380–410
% Haze	4–11

Source: Reprinted from Ref. 23, with permission of Phillips Petroleum Company, U.S.A.

Table 8 Metallocene-Based High-Density Polyethylene

Property	Metallocene catalyst
Density (g/cc)	0.953
MI (dg/min)	46
Molecular weight distribution	2.6
Melting point (°C)	130
Spiral flow index (190°C, 1500 psi, dg/min)	4.50
Flexural modulus (psi. 10)	19
Tensile strength @ Yield (psi)	4000
Drop impact (ft–lbs)	2.7

Source: Reprinted from Ref. 23, with permission of Phillips Petroleum Company, U.S.A.

Table 9 Metallocene-Based Medium-Density Polyethylene

Property	Metallocene Catalyst
Density (g/cc)	0.936
MI (dg/min)	3.3
Molecular weight distribution	2.1
Melting point (°C)	126
ESCR, Condition A, F-50 hrs	>1000
Odor	Low
Part impact	Good

Source: Reprinted from Ref. 23, with permission of Phillips Petroleum Company, U.S.A.

cast film use, such as stretch film, as well as can liners and heavy duty sacks. Film applications for metallocene polyethylene are intended to maximize physical, optical, and heat-sealing attributes. Metallocene–LLDPEs have exceptional toughness, low levels of extractables, excellent optical properties, and outstanding heat seal and hot tack characteristics. Film manufacturers have incorporated this unique set of properties into a variety of film-making processes. It has provided end users with many advantages such as: (1) increased packaging speeds due to lower seal initiation temperature, higher hot tack, and reduced blocking; (2) reduced package failures due to greater toughness and superior resistance to abuse; (3) improved package aesthetics due to lower haze and higher gloss; and (4) improved packaged product quality due to reduced package–product interactions, lower odor, and extractables, etc. Most often the value of metallocene-based polyethylene is not derived from one characteristic but from a superior group of properties.

IV. STEREOREGULAR POLYPROPYLENE

Propylene polymerizes by metallocene catalysts to give a variety of polymeric structures (Fig. 8) [13,24]. Stereostructure and other characteristics of polymers are determined by symmetry of metallocene, cocatalyst nature, and polymerization parameters. In addition to atactic and isotactic polypropylenes, new stereoregular polymers such as syndiotactic polypropylenes and hemi-isotactic polypropylenes have been obtained. Stereoblock isotactic polypropylene containing isotactic blocks of alternating orientation and stereoblock polypropylene consisting of isotactic blocks alternating with atactic blocks could also be synthesized [25]. Among these polypropylene structures, isotactic polypropylene and syndiotactic polypropylene have presently gained much commercial significance. Atactic polypropylene is also gaining importance as an additive and polymeric material for blends.

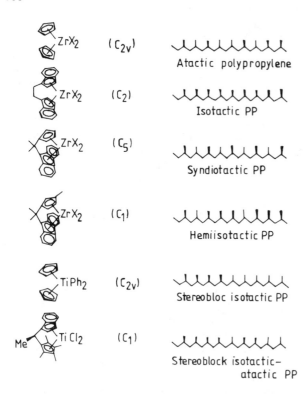

Figure 8 Stereostructures of polypropylene and its relationship with metallocene structures. (From Ref. 3, with permission from Elsevier Trend Journals, Cambridge, U.K.)

Polypropylene can be synthesized by metallocene catalysts with tailored characteristics [26,27] such as: (1) different degree of tacticity; (2) higher comonomer content with uniform distribution; (3) controlled molecular weight, width, and shape of molecular weight distribution; and (4) lower isotactic or atatic product fractions. Presently, only a part of these options are used in synthesizing new generation polypropylene resins. These polymeric materials show great potential for their

use in a wide range of product applications. Potential areas of metallocene polypropylene application in the existing polymer market are indicated in Table 10 [28].

A. Isotactic Polypropylene

Many modifications in metallocene structures have been incorporated, as shown in Fig. 9, to synthesize isotactic polypropylene with a range of properties including molecular weight, isotacticity, mechanical properties, etc.

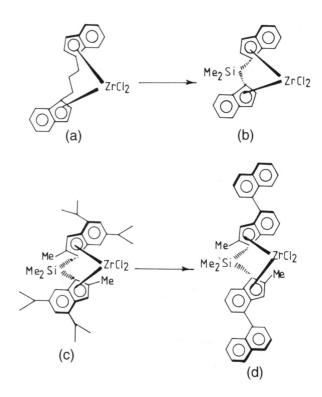

Figure 9 Metallocene structures for polypropylene synthesis. (From Ref. 2.)

Table 10 Potential Areas of Application for Polypropylene

Molecular features of polypropylene	Expected process/product effects	Potential application
Narrow molecular weight distribution	Difficult extrusion except where NMWD resins are used.	Fiber/nonwovens
	Low melt strength, good melt drawability	Fiber/nonwovens
Narrow tacticity distribution	Low extractables	Food packaging
Narrow composition distribution	Absence of high comonomer fraction ease processesing and comonomer effective in reducing m.p.; sharp m.p.	Heat seal layers in composites films
New copolymers with higher alpha olefins	Expect superior ultimate properties	Tough films, flexible moulding

Source: Reprinted from Ref. 28, with permission of Exxon Chemical Company, U.S.A.

[1–3]. Earlier discovered metallocenes, such as ethylene bridged indenyl zirconocenes, showed high activity for isotactic polypropylene (i-PP) synthesis as compared to heterogenous catalyst systems. However, the properties comparison reveal that zirconocenes-based i-PP has less stereoregular, lower melting points, narrow molecular weight distribution, and low-molecular weights (Table 11) [3]. Replacement of ethylene bridge with a silylene bridge improved activity of these systems to at least ten times that of modern Ziegler–Natta catalysts. However, molecular weights, crystallinities, and the melting point of isotactic polypropylene are found to be well below those of conventional isotactic polypropylenes. Further modifications in metallocenes have resulted in the formation of isotactic polypropylene at a commercially useful molecular weight and isotacticity [29]. The performance data of some metallocene structures (Fig. 10) suitable for high-molecular weight isotactic polypropylene are shown in Table 12.

Some of the important features of metallocene-based isotactic polypropylene are as follows:

1. Chemical nature and high activity of metallocene results in the production of isotactic polypropylene with a very low level of chlorine (<3 ppm) and without any or a very small titanium content. Metallocene polypropylene also shows higher resistance to free radical attack due to higher chemical purity. These characteristics help in reducing the additive requirement as compared with those used in conventional isotactic polypropylene. Furthermore, such a polymer does not induce corrosion of the processing machines and will be useful in many applications where a low generation of free radical initiators are essential for product longevity (outdoor applications and medical applications).

2. Metallocene isotactic polypropylenes (MET.PP) are accessible with different melting points under the commercial range of melt flow rate (Fig. 11). The variation of melting points in these polymers is linked with the presence of different lengths of isotactic sequences. Mechanical properties of polypropylene

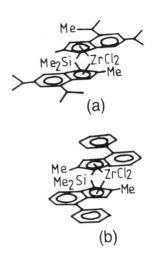

Figure 10 Metallocene structures for making high-molecular weight isotactic polypropylene. (From Ref. 29, with permission from Sterling Publications Limited.)

Table 12 Performance Data of Metallocenes in Polymerization of Liquid Propylene

Zirconocene (ligand)	Activity (kg PP/g catalyst/h)	Weight average molecular weight (kg/mol)	Melting point (°C)
A	170	500	146
[2-Me,4,6-(iPr)$_2$Ind]			
B	1000	1100	160
[2-Me,4-PhInd]			

Source: Reprinted from Ref. 29, with permission from Sterling Publications Limited.

Table 11 Properties of Isotactic Polypropylene

Property	Ethylene-bridged zirconocenes	Ziegler–Natta
%mmmm	82.7	95.1
%rrrr	0	0.4
Mw	38000	350000
Mw/Mn	3	7.5
M.P. (°C)	133	163

* Ethylene-bridged bis-indenyl zirconocene dichloride-methylalumoxane system.
Source: From Ref. 3, with permission of Elsevier Trends Journals, Cambridge, U.K.

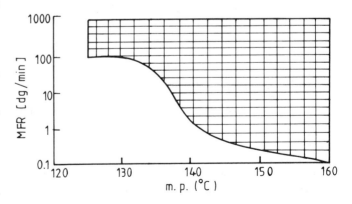

Figure 11 Homopolymer polypropylene product range with metallocenes. (From Ref. 29, with permission from Sterling Publications Limited.)

Table 13 Comparison of Metallocene-Based Isotactic Polypropylene

Property	MET.PP1	MET.PP2	MET.PP3	Conventional PP
Melting point	139	151	160	162
MWD	2.2	2.3	2.5	5.8
Modulus (N/mm.mm)	1060	1440	1620	1190
Hardness (N/mm.mm)	59	78	86	76
Impact resistance Izod (mJ/mm.mm)	128	86	100	103
Light transmission (%)	56	44	35	34

* All samples MFR = 2 dg/min; MET.PP = metallocene based polypropylenes.
Source: Reprinted from Ref. 29, with permission from Sterling Publications Limited.

show a variation with a change of the melting point (Table 13). MET.PP 1 and MET.PP2, having a lower isotacticity, show enhanced transparency and more hardness and modulus as compared with transparent random copolymers. High-melting MET.PP 3 has more hardness and 35% higher modulus than the conventional isotactic polypropylene. Higher mechanical strength is found to be true for the whole MFR range (Fig. 12). By introducing a modeled MWD, further increase in the modulus level is possible (Met.PP 4 in Fig. 12). An isotactic polypropylene of melting point 145°C has the same mechanical properties as a conventional isotactic polypropylene with a melting point of 160°C. Despite the low melting point of metallocene polypropylene, the heat deflection temperature is similar to conventional isotactic polypropylene resulting in the same service temperature. Use of metallocene-based isotactic polypropylene in a melt–blown fiber process gives a very high throughput and enhanced fiber strength. The new range of polypropylene also offers the option of heat-sealable homopolymers having better optical and mechanical properties.

3. Metallocene catalysts produce random copolymers [29–31] with different property profiles (Table 14). These data show that random copolymers have higher stiffness and higher transparency at certain melting point levels. A very low content of extractables in low-melting

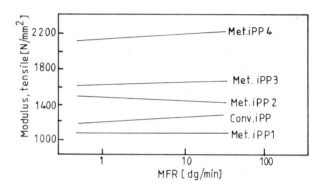

Figure 12 Modulus versus MFR in metallocene versus conventional homo-iPP. (Met iPP 1–3-narrow MWD; Met iPP 4-modeled MWD). (From Ref. 29, with permission from Sterling Publications Limited.)

point products is observed as compared with conventional copolymers (Table 15). This results in nonstickiness of low-melting random copolymers, allowing their synthesis in conventional bulk or gas phase processes. These polymers can be used in fiber and nonwoven applications.

4. Metallocene catalysts produce high-comonomer content products, such as polypropylene block

Table 14 Comparison of Conventional Random Copolymner with Metallocene Copolymer and Homopolymer

Property	Conventional copolymer	Metallocene copolymer	Metallocene homopolymer
Melting point	141	140	142
Modulus (N/mm.mm)	620	940	1120
Hardness (N/mm.mm)	41	59	65
Impact resistance notched (mJ/mm.mm)	23.1	11.3	7.3
Light transmission (%)	57	65	48
Extractables (%)	7.9	1.1	0.7

Source: Reprinted from Ref. 29, with permission from Sterling Publications Limited.

Table 15 Low Melting Metallocene Versus Conventional Propylene-Ethylene Copolymer

Property	Conventional copolymer	Metallocene copolymer
Melting point	134	135
Extractables (%, hexane, 50°C)	5.6	0.5
Extractables (%, hexane, 69°C)	14.2	1.7

Source: Reprinted from Ref. 30, with permission from Sterling Publications Limited.

Table 16 Properties of Syndiotactic Polypropylene

Polymer	Crystallinity (%)	% RRRR	Melting temperature
SPP1	21	76.5	120/130
SPP2	22	78.0	121/130
SPP3	29	91.1	146/151

Source: Reprinted from Ref. 35, with permission from Fina Oil and Chemical Company, U.S.A.

copolymer and reactor blends, with uniform composition distribution. Thus, the overall crystallinity pattern becomes better as compared with that in a conventional copolymer with the same comonomer content. This results in better hardness–impact resistance ratio and high transparency of the product.

B. Syndiotactic Polypropylene

Syndiotactic polypropylene [1–3,24,32–34] has pendant methyl chains arranged on alternating sides of the polymer backbone in symmetrical patterns. Such a structural feature provides greater impact, higher strength, and more flexibility than conventional isotactic polypropylene, as well as lower haze, heat deflection temperature, and residual monomer content. Unlike metallocene technology for isotactic polypropylene, only one basic catalyst structure (with minor variations) has so far been found for syndiotactic polypropylene. The most commonly cited structure is a dimethyl methylene-bridged cyclopentadienyl fluroneyl zirconium dichloride [35,36]

(Fig. 13). Property evaluation of polymers indicates that tacticity variation results in a different crystallinity and melting point profile (Table 16). The increase in resins melting points as a function of increasing percentage of racemic pentads indicates that more stereospecific catalysts can produce resins with higher melting points. Polydispersities of metallocene-based resins are lower than isotactic polypropylene produced with Ziegler–Natta catalyst (Table 17). Narrower molecular weight distribution causes lowering in die swell behavior and influences rheological properties. These polymers also show a slower crystallization rate than isotactic polypropylene.

Processing properties of syndiotactic polypropylene have been improved by blending with isotactic polypropylene [37]. It has allowed a syndiotactic polypropylene-based polymer to process well with conventional processing machines in conventional operating conditions. The transparent articles with moderate rigidity can be injection-molded without the sticking problem of polymer to the mold cavity. Film and sheet are also cast with conventional machine settings and even fabrication of blown film of unusual transparency is achieved by quenching with water or air.

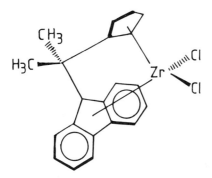

Figure 13 Metallocene structure for syndiotactic polypropylene. (From Ref. 35 with permission from Fina Oil and Chemical Company, U.S.A.)

Table 17 Properties of Syndiotactic Polypropylene

Property	SPP1	SPP2	I-PP (Z.N.)
MFR	5.34	8.90	1.2
Mn	59300	51700	52000
Mw	157000	134000	417700
Mz	579000	482000	1356000
MWD	2.6	2.6	8.0
MWD′	3.7	3.6	3.24
% Xylene	2.08	3.07	2.6
Soluble die swell	1.21	1.13	2.39
Density (g/cc)	0.827	0.870	0.907

Source: Reprinted from Ref. 35, with permission from Fina Oil and Chemical Company, U.S.A.

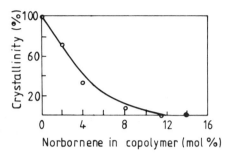

Figure 14 Crystallinity of ethylene/norbornene copolymers. (From Ref. 25. Reprinted with permission from Elsevier Science.)

V. CYCLO-OLEFIN POLYMERS

New polymeric materials are synthesized by polymerizing cyclic olefin (cyclobutene, cyclopentene, cycloheptene, norbornene) or copolymerizing cyclic olefin with linear olefin using metallocene catalysts [1,24,25]. These materials are not yet accessible sofar with the classical Ziegler–Natta catalyst. Homopolymers consist of rigid polymer chains with a high melting point (>400°C). It causes difficulty in the processing of the material. On the other hand, copolymers are amorphous in nature. The degree of amorphousness can be controlled by the amount of cyclic monomer incorporated into the product. It has been observed, in the case of ethylene-norbornene copolymers, that 12–14 wt% incorporation of norbornene gives amorphous polymer (Fig. 14) [25]. Other properties of copolymers are low melting point, high transparency, better optical characteristics, good heat stability, and high chemical resistance. Mechanical strength of the polymers is retained up to their plasticization temperature. Property variation can also be achieved by alloying cycloolefin copolymers with engineering plastics. Such a properties profile of cyclic olefin copolymers opens up many applications in areas such as optoelectronic data transmission and data storage on a new generation of optical disks [38,39]. These copolymers can effectively compete with existing acrylontrile-butadiene-styrene polymer, polycarbonates, and acrylics products [9]. This will push polyolefins into areas of specialty applications.

VI. FUNCTIONALIZED POLYOLEFINS

A long-standing goal in polyolefins is the synthesis of polymers bearing polar functional groups such as acrylate, esters, or vinyl ethers, etc [24,40]. These copolymers might endow polyolefins with useful properties such as adhesiveness, dyeability, paintability, and printibility. Advances have recently been made in polymerizing polar monomers with cationic metallocene catalysts

[24,37]. These catalysts are considerably more tolerent of functional groups than other types of metallocene systems. Functionalized polyolefins are synthesized by using monomers such as 4-tert-butyl dimethylsiloxy-1-pentene, 5-N,N diisopropylamino-1-pentene, or 4-trimethylsiloxy-1,6-heptadiene. Propylene oligomers with an olefinic end group, synthesized by using metallocene, have been converted to various other functional groups. Thiol-terminated oligopropylene has been used as a chain transfer reagent in methylmethacrylate polymerization to form polypropylene-block-methylmethacrylate polymer. A new class of polymers containing pendant polypropylene chains are also derived from copolymerization of methacrylate-terminated oligopropylene macromonomers with acrylic esters, acrylonitrile, or styrene.

Polypropylene block and graft copolymers are efficient blend compatibilizers. These materials allow the formation of alloys, for example, isotactic polypropylene with styrene-acrylonitrile polymer or polyamides, by enhancing the dispersion of incompatible polymers and improving their interfacial adhesion. Polyolefinic materials of such types afford property synergisms such as improved stiffness combined with greater toughness.

VII. CONCLUSIONS

Novel polyolefins with superior properties are a result of the remarkable abilities of metallocene catalyst systems to control the insertion of monomer into a growing polymer chain. This feature provides the opportunity to make a polyolefin product with tailored characteristics, such as better film clarity, good radiation resistance, higher flexural modulus, improved impact strength and toughness, improved melt rheology, etc. A wide properties profile and processing advantages of metallocene-based polyolefins are expected to significantly increase the growth of a new generation of polyolefin materials in the polymer market.

REFERENCES

1. V. K. Gupta, S. Satish, an I. S. Bhardwaj, *J. Macromol. Sci. Revs. Macromol. Chem. Phys., C34:* 439 (1994).
2. K. B. Sinclair and R. B. Wilson, *Chem. & Ind.,* 857 (1994).
3. A. D. Horton, *Trends in Polymer Science, 2:* 158 (1994).
4. A. M. Thayer, *Chem & Eng. News, Sept 11:* 15 (1995).
5. R. Simmons, W. P. Chatham, R. Criswell, M. Whitlock, and J. Ward, "Additive requirments for mPOs," MetCon 95 Proceedings, USA, May 1995.
6. R. L. Halterman, *Chem. Rev., 92:* 965 (1992).
7. H. Sinn and W. Kaminsky, *Adv. Organomet. Chem., 18:* 99 (1980).
8. R. F. Jordan, *Adv. Organomet. Chem., 32:* 325 (1991).
9. S. S. Reddy and S. Sivaram, *Prog. Polym. Sci., 20:* 309 (1995).
10. U. Moll and M. Lux, "Manufacture of ethylene/alpha olefin copolymers with metallocene catalysts in slurry loop

and high pressure processes,'' MetCon 95 Proceedings, USA, May 1995.

11. J. Baker, *European Chemical News,* Aug 7–13: 39 (1995).
12. M. Ohgizawa, M. Takahashi and N. Kashiwa, "Metallocene catalyzed polyethylene with unique rheological property", MetCon 95 Proceedings, USA, May 1995.
13. A. Akimoto and A. Yano, "Production of ethylene copolymers with metallocene catalysts at high pressure and its properties," MetCon 94 Proceedings, USA, May 1994.
14. B. C. Childress, "Properties of homogeneous and heterogeneous polyolefins: Metallocene catalyzed versus Ziegler-Natta catalyzed resins," MetCon 94 Proceedings, USA, May 1994.
15. R. Halle, "Structure, properties and blown film processing of a new family of linear ethylene polymers," SME Blown Film Technology Seminar, Oct 1993, USA.
16. G. Lancaster, J. Damen, C. Orozco, and J. Moody, "Global product and application development utilizing Insite technology," MetCon 94 Proceedings, USA, May 1994.
17. P. C. Wu, "Metallocene ethylene based resins in cast film extrusion and application," MetCon 94 Proceedings, USA, May 1994.
18. D. J. Michiels, "Advance performance terpolymers for blown film applications," MetCon 94 Proceedings, USA, May 1994.
19. *Modern Plastic International, Aug*: 55 (1994).
20. A. A. Montagna and J. C. Floyd, *Hydrocarbon Processing, March*: 57 (1994).
21. *Modern Plastic International, June*: 18 and 38 (1995).
22. S. Shang, "Expectation of medical packing industry from metallocene based polyolefins," MetCon 95 Proceedings, USA, May 1995.
23. M. B. Welch, S. J. Palackal, R. L. Geerts, and D. R. Fahey, "Polyethylene produced in Phillips slurry loop reactors with metallocene catalysts," MetCon 95 Proceedings, USA, May 1995.
24. H. H. Brintzinger, D. Fischer, R. Mulhaupt, B. Rieger, and R. M. Waymouth, *Angew. Chem. Int. Ed. Engls., 34*: 1143 (1995).
25. W. Kaminsky, *Catalysis Today, 20*: 257 (1994).
26. *Modern Plastics International, March*: 48 (1995).
27. *European Chemical News, 20–26: March*: 31 (1995).
28. J. J. McAlpin and G. A. Stahl, "Applications potential of Exxpol metallocene based polypropylene," MetCon 94 Proceedings, USA, May 1994.
29. W. Spalek, B. Bachman, and A. Winter, *Hydrocarbon Technology International, Spring*: 117 (1995).
30. W. Spaleck, A. Winter, B. Bachmann, V. Dolle, F. Kuber and J. Rohrmann, "New isotactic polypropylenes by metallocene catalysts," MetCon 93 Proceedings, USA, May 1993.
31. F. Langhauser, J. Kerth, M. Kersting, P. Kolle, D. Lilge, and P. Muller, *Angew. Makromol. Chem. 223*: 155 (1994).
32. *Modern Plastic International, Oct*: 34 (1991).
33. *Modern Plastic International, April*: 67 (1993).
34. J. Chowdhury and S. Moore, *Chem. Eng., April*: 34 (1993).
35. E. S. Shamshoum, L. Sun, B. Reddy, and D. Turner, Properties and applications of low density syndiotactic polypropylene," MetCon 94 Proceedings, USA, May 1994.
36. *European Chemical News, May 24*: 43 (1993).
37. T. Shiomura, M. Kohno, N. Inoue, Y. Yokote, M. Akiyama, T. Asanuma, R. Sugimoto, S. Kimura, and M. Abe, *Studies in Surface Science and Catalysis Design for Tailor-Made Polyolefins,* Kodansha, 327 (1994).
38. *European Chemical News, April 11*: 27 (1994).
39. *Chemicalweek, Aug 31/Sept 7*: 14 (1994).
40. T. C. Chung, "New utilities of metallocene catalysts and borane reagents in the functionalization and block/graft reactions of polyolefins," MetCon 95 Proceedings, USA, May 1995.

13

Preparations and Properties of Porous Poly(vinyl alcohol)–Poly(vinyl acetate) Composites

Sadao Hayashi
Shinshu University, Ueda, Japan

I. INTRODUCTION

Poly(vinyl acetate) (PVAc) latexes prepared in the presence of poly(vinyl alcohol) (PVA) as a protective colloid are widely used in adhesives, paints, textile finishes, coatings, and so on. The most important properties required for those uses are that the PVAc latexes have a high viscosity and a characteristic viscosity behavior. The physical and application properties of the PVAc latexes will be generally affected by the amount, degree of hydrolysis, or molecular weight of the PVA or the degree of blocking of vinyl acetate (VAc) units in the PVA chain. However, it is known that the properties are easily controlled by polymerization conditions except the temperature of polymerization and the stirring speed; e.g., the kind, amount, and addition time of the initiator, and the addition method of VAc under the same charge of ingredients for polymerization [1].

The PVAc latexes generally contain graft copolymers between PVA and PVAc formed during the polymerization. The graft copolymers are either water-soluble, benzene-soluble, or in water and benzene, and, because of the graft copolymers, the dried-down PVAc latex films are not easily dissolved in solvents. It has already been shown from the method of solvent extraction that the grafting during polymerization occurs to a greater extent with partially hydrolyzed PVAc (P-PVA) (the degree of hydrolysis of about 88 mol%) rather than with PVA (the degree of hydrolysis over 98 mol%), and persulfate is more effective than hydrogen peroxide (HPO) as a grafting initiator [2,3]. In addition, the presence of graft copolymers in the PVAc latexes that were formed using P-PVA as a stabilizer and ammonium persulfate (APS) as an initiator have been established by the techniques of turbidimetric titration and adsorption and partition chromatography on paper [4]. Furthermore, the number of grafting reactions in the emulsion polymerization of VAc in the presence of PVA increases with the rise in the initiator concentration [5]. Therefore, judging from the standpoint of these investigations, it seems to be difficult to expect a high degree of extraction of PVAc from the film of PVAc latexes prepared in the presence of PVA. However, it has been reported that in the PVAc latexes containing PVA as a protective colloid, no graftings were found [6]. In most cases, a thorough mixture of two polymers, of which one is soluble and the other is insoluble in the extraction solvent, is incompletely separated after an extraction time of 24 h [7]. Consequently, even if the residual polymers that were not extracted with acetone and boiling water were obtained, it is not possible to conclude whether the polymers were graft copolymers.

We have found that in the system of presulfate initiator, the PVAc latexes are not dissolved transparently in the methanol–water mixture [8], and in the system of HPO initiator, the extraction of the polymer from the PVAc latex films with acetone greatly depends on the polymerization condition [9]. These results suggest that if a polymerization method can be found in which the grafting polymerization of VAc onto PVA is controlled to the minimum, a large portion of PVAc in the latex film will have a chance of extraction with solvents. In this Chapter, the preparations of the unique porous films from the PVAc latexes containing PVA as a protective colloid by an extraction of the PVAc particles with acetone and the characteristic properties of the porous films are summarized.

II. PREPARATIONS OF PVAC LATEXES IN THE PRESENCE OF PVA

A. Polymerization Methods

There are three important processes for preparing PVAc latexes in the presence of PVA as a protective colloid: batch, semi-continuous, and delayed addition of monomer [10]. In this Chapter, the effects of the addition of VAc and initiators on the properties of PVAc latexes are discussed using the three methods under the same charge of ingredients for polymerization as shown in Fig. 1 [1,11].

Polymerizations were performed in a 1-L five-necked separable flask equipped with a thermometer, a reflux cooler, a VAc dropping funnel, an initiator dropping funnel, and a stirrer. In the standard recipe, the flask was first charged with 250 g of aqueous solution containing 25 g of PVA (the degree of polymerization was approximately 1700 and the degree of hydrolysis was 98.8%) and 5 g of water (in the case of the HPO-tartaric acid [TA] system, 5 g of aqueous solution containing TA was the activator). It was then immersed in a water bath thermostated at 85°C. After all VAc has been added, the temperature of the water bath was heated to more than 95°C for 30 min. The stirring speed was maintained at 240 rpm. When other recipes and methods were adopted, written information was maintained on all conditions.

Polymerization methods [I], [II], and [III] (Fig. 1) indicate, respectively, the dropwise addition of VAc and initiator; the dropwise addition of VAc and the stepwise addition of initiator; the batch method, in which all ingredients of water, VAc, PVA, and initiator were put into the reaction vessel before starting polymerization. In method [I], when the temperature of the PVA solution in the flask attained 70°C, dropwise additions of 20 g of an aqueous solution containing initiator and 250 g of VAc were started. In method [II], the process was similar to method [I], except the initiator was added stepwise. When the temperature of the contents in the flask was raised to 70°C, 24 g of an aqueous solution containing half the prescribed amount of initiator was first added,

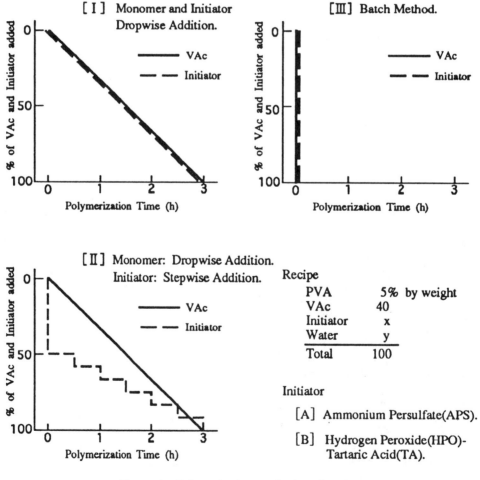

Figure 1 Polymerization methods and recipe.

and then 1 g of an aqueous solution containing one-sixth of residual initiator was added at intervals of 30 min. In method [III], when the temperature of the PVA solution in the flask reached 70°C, 200 g of VAc and 30 g of an aqueous solution containing the initiator were added. Thereafter, the temperature of the water bath was adjusted to 70–75°C to prevent bumping during polymerization. If the temperature of the contents in the flask went above 70°C, the temperature of the water bath was instantly raised to 85°C. After VAc reflux was over, the reaction mixture was permitted to stand in a water bath more than 95°C for 30 min. The obtained latexes were cooled below 35°C, and conversions were determined gravimetrically. The PVAc latex viscosity was measured at 30°C by a Type BH Brookfield viscometer after keeping the latex in a water bath at 30°C for 1 h. The dispersion quantity was determined from the ratio of the extinction at 400 nm (D_{400}) to that at 600 nm (D_{600}) at 30°C, where the D_{400} was fixed within 0.40–0.45.

B. Influence of Initiators

In methods [I] and [III] using APS, the PVAc latexes coagulated during polymerization or cooling. In the HPO-TA system, stable PVAc latexes were formed using high conversion in every polymerization method. The results of the polymerization and properties of the PVAc latexes obtained are summarized in Table 1 [11].

The PVAc particle size, which was evaluated from the dispersion quantity, was in the order of [I] > [II] > [III] in the method using HPO-TA. The PVAc particle size in method [II] using APS was larger than that in any method using HPO-TA. The transmittance of the PVAc latex films, which were formed above the minimum film-forming temperature, varied from a high value in methods [II] and [III] using HPO-TA to a low value in method [II] using APS and method [I] using HPO-TA. The difference in these values depends on the size of PVAc particles. In the dropwise addition of VAc, the viscosity of the PVAc latex is greatly influenced by the polymerization condition, especially by the amount of initiator added before the start of polymerization. In general, the viscosity of a latex is dependent upon the particle size. In the presence of PVA, the viscosity of PVAc latexes will also be greatly affected by the interaction of PVA and the PVAc particles, the content of the graft copolymer between PVA and PVAc, and the molecular weight of PVA in addition to the size of PVAc particles. This explains why the viscosity of the PVAc latex contained PVA is independent of the particle size. The PVAc latex with a high viscosity was very convenient for casting the latex and for forming the latex film. This will have a very important impact on the greater uses of the PVAc latexes in the fields of adhesives, paints, and so on.

In order to prepare the PVAc latexes with Newtonian character or dilatancy behavior, the polymerization

Table 1 Preparations and Properties of PVAc Latexes

	Polymerization method[a]							
	[A] Ammonium persulfate		[B] Hydrogen peroxide-tartaric acid					
	[II]	[II]	[I]	[I]	[II]	[II]	[III]	[III]
Initiator (wt%)	0.20	0.40	0.06	0.12	0.06	0.12	0.06	0.12
Activator, TA (wt%)	—	—	0.10	0.10	0.10	0.10	0.10	0.10
Conversion (%)	99.6	99.3	99.4	99.2	99.4	99.9	98.4	98.3
Viscosity of latex (CP at 10 rpm)	7,560	134,000	13,400	30,400	15,800	61,500	5,100	3,700
Dispersion quantity (D_{400}/D_{600})	1.08	1.19	1.26	1.22	1.39	1.31	1.53	1.54
Extraction from latex film with acetone (%)[b]	28.4 (32.0)	35.5 (40.0)	21.5 (23.6)	25.7 (29.9)	62.4 (70.2)	71.7 (80.7)	83.0 (93.5)	85.6 (96.5)
Relative viscosity of polymer extracted with acetone[c]	1.15	1.13	1.24	1.17	1.11	1.09	1.59	1.30
Swelling of latex film in acetone[d]	Very large	Very large	Very large	Very large	Large	Large	Small	Small

[a] See Figure 1.
[b] Values in brackets indicate percentage of polymers extracted with acetone to total PVAc.
[c] Acetone was used as solvent. Determination of viscosity was carried out at 30°C.
[d] Very large, large, and small imply increases of 70–90%, 20–30%, and 2–3% in area, respectively.

in method [I] using a very small amount of initiator is adopted. These latexes are very good as adhesives for high-speed application to paper. Although the PVAc latexes in the presence of PVA show Newtonian character, independent of the PVA amount, the PVAc latexes in the presence of P-PVA show only thixotropic viscosity, in spite of the P-PVA amount. This seems to be the reason that P-PVA has a long sequence of VAc units in the polymer molecules, so that it plays an important role in a water solution as a surfactant. The sequence distribution of VAc units in P-PVA chains, which is the important factor governing the adsorption of the PVA onto the PVAc particles, is qualitatively or quantitatively evaluated using infrared spectroscopy [12], the color reaction with iodine–iodide [13], and differential thermal analysis [14]. In the ^{13}C-NMR spectra of P-PVA, the block characters and the mean run lengths of the VAc units in the polymer chain are calculated from the three (OH, OH), (OH, OAc), and (OAc, OAc) [15–18].

In the polymerization in method [II], the PVAc latexes have thixotropic characteristics, which are good for general adhesives and paints. The viscosity of PVAc latexes is not affected by the total amount of initiator, but is governed mainly by the initiator amount added at the initial stage of the polymerization. With an increasing initiator amount, the latex viscosity increases, reaches a maximum, and then decreases. This tendency changes more greatly with the addition of the initiator at the initial stage than during the course of the polymerization. This fact appears to offer very powerful insights concerning the polymerization of VAc in the presence of PVA. For example, under the same charge of ingredients for polymerization, the properties of PVAc latexes can be controlled by the amount and the addition time of the initiator.

The polymerization in method [III] will probably be unsuitable for industrial production due to the heat of polymerization, but it can be used to produce a freeze–thaw stable adhesive with rapid drying and good adhesion to paper, which cannot be obtain by other polymerization methods. However, the water resistance of the latex film is not improved.

As mentioned previously, the properties of the PVAc latexes prepared in the presence of PVA have something to do with the reaction between PVA and initiators during the polymerization.

III. TRANSFORMATION OF LATEX FILMS TO POROUS MATERIALS

A. Transformation to Porous Film

The PVAc latex was cast 1.8 mm in thickness on a poly(ethylene) plate and dried at room temperature. The latex film was dried further in a vacuum desiccator for 24 h, and extracted in Soxhlet extractor with acetone for 20

h. After extraction, the obtained white film was immersed in n-hexane to exchange acetone and quickly dried with blowing air, which had been passed through calcium chloride to prevent a destruction of PVA skeletons by sucking up moisture from air that adheres on the film accompanying the heat of vaporization of acetone [11,19]. The porous white film was further dried in vacua for 10 h.

By acetone extraction, the latex films were transformed from semi-transparent to white. The swelling degree of the latex film in acetone showed the opposite relation, compared to the extraction percentage of PVAc, as shown in Table 1. In the latex films of the HPO-TA system, the percentage of extracted polymers to total PVAc was about 25% in method [I], 70–80% in method [II], and more than 90% in method [III]. This tendency is closely associated with the solubility of the PVAc latex in the mixed solvent of methanol and water (5.0:1.0 in volume). Comparing the latex films prepared in method [II], the PVAc amount extracted with acetone was greater in the HPO-TA system than in the APS system. In the polymerization systems using the same initiator, even if the same amount of initiator is used, the amount of the extracted polymer is greatly dependent on the polymerization method. In the polymerization systems using different initiators, even if the same polymerization method is adopted, the amount of the extracted polymer relates to the nature of initiators.

In the emulsion polymerization of VAc, microgels are formed in particles as a consequence of a chain transfer reaction onto PVAc [20], and especially, they increase as the pH value of latex decreases [5]. Actually, the pH value of the PVAc latex was approximately 2.4–2.6 in the persulfate system and 3.4–3.6 in the HPO-TA system. If persulfate is used as an initiator, because of sulfuric acid produced by the side reaction of persulfate, the latex will be acidic, as compared with latex from the HPO-TA system. As a result, it is interpreted that the polymers containing microgels in the persulfate system did not dissolve in the mixed solvent of methanol and water. Emulsion polymerization proceeds mainly in particles. In methods [I] and [II] in which VAc is added dropwise for 3 h, since refluxing of VAc during polymerization is slight, the composition in the particles is predicted to be PVAc-rich. This situation is similar to a high conversion in mass polymerization. Moreover, if a small amount of the HPO-TA redox initiator, as in method [I], is added dropwise for a length of polymerization, the radical concentration in the particles will result in a very low state from the early stage of polymerization. Accordingly, the formation of polymers with high-branched structures in the emulsion polymerization of VAc is a natural result. This fact may also have bearing on the lower extraction percentage with acetone from the PVAc latex films containing PVA.

Judging from the extraction percentage of PVAc and from the swelling degree of latex films in acetone, it is concluded that method [III] using HPO-TA is the most suitable method for making the porous polymer.

B. Structure of Porous Film

From the results in Table 1, the PVAc latex prepared in method [III] using HPO-TA was adopted for making porous polymer film. The extraction with acetone proceeded smoothly holding constant the original form of the film. The latex film, which was initially semi-transparent, changed to a white film just as paper would after extraction. This change indicates the occurrence of many holes in extracted film. This supports the hypothesis that the holes, which were formed by taking away polymer particles from the latex film, pass through the film. PVA used in this study showed a blue color with an iodine–iodine solution, while the white film developed a red-violet color on reaction with the iodine–iodide solution, which proves the existence of PVAc, i.e., the PVAc graft copolymers to PVA [13]. The residual PVAc could be eliminated by hydrolyzing in methanol solution containing sodium methylate.

As shown in Fig. 2a, the size of PVAc particles was found to have a range of approximately 0.1–1.0 μm from electron micrograph of the crack surface of the latex film. The electron micrograph of the crack surface after extraction is shown in Fig. 2b. It is clear that the white film is very porous and the holes are made in every direction. Figures 2c and 2d are the upper surface views of the latex film before and after extraction. Large PVAc particles were distinctly seen on the upper surface before extraction. The result is different from the observation on the lower surface of the PVAc latex film with a low PVA concentration [21]. This confirms that a size gradient of PVAc particles from top to bottom in the film did not take place because of the high viscosity of the latex. It is obvious from Fig. 2b that the substantial holes passing through the porous film are not the size of PVA cells formed by an extraction of PVAc particles, but are the small size bored through the wall between the PVA cells. Although the PVAc particles were not observed on the bottom surface of the latex film, after acetone extraction, many holes were observed. This suggests that PVAc particles, swelled with acetone, penetrated through a thin bottom PVA layer, and the acetone solution of the PVAc broke through the PVA layer and extended to the outside of the film. In the film after extraction, the size of holes on the bottom surface, on the average, were smaller than those on the upper surface.

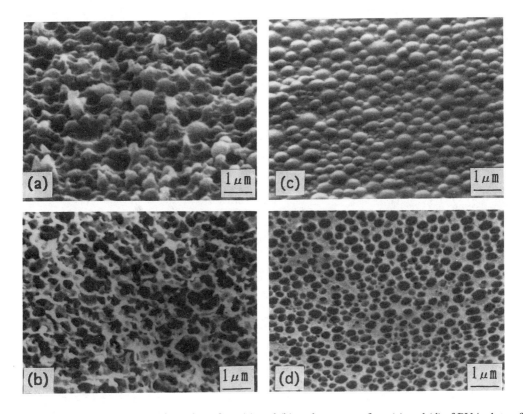

Figure 2   Electron microphotographs of crack surface (a) and (b) and upper surface (c) and (d) of PVAc latex film. Acetone extraction: (a) and (c) before, (b) and (d) after.

From the electron micrographs, assuming that PVAc particles in the latex are the same size, the formation model of the porous film from the latex film can be illustrated as in Fig. 3 [19]. When the latex forms a dried film over minimum film-forming temperature, it is concluded that PVA coexisted in the latex and is not excluded to the outside of the film during filming, but is kept in spaces produced by the close-packed structure of PVAc particles.

Since the porous polymer consists of skeletons of PVA cells with grafting PVAc and has an enormous interface, it can be expected to be utilized for various purposes in the future.

IV. PROPERTIES OF POROUS PVA-PVAc COMPOSITES

A. Permeability to Organic Solvents

The PVAc latex containing PVA as a protective colloid prepared in method [III] using the HPO (0.12%)-TA (0.10%) system as an initiator in Table 1 was cast to about 1.8 mm in thickness on a poly(ethylene) plate and dried at room temperature. The dried latex films were 0.7–0.9 mm in thickness and were semi-transparent. The porous film after acetone extraction changed to a white color without a change in the film size.

The permeability of the porous film to n-hexane, cyclohexane, and benzene was investigated at room temperature under various pressures [11,22]. As shown in Fig. 4, n-hexane and cyclohexane were able to permeate through the porous film at rate of 5.50×10^{-2} and 1.93×10^{-2} ml/cm^2·s at 1.5 kg/cm^2, respectively; but the rate for benzene was only 1.33×10^{-2} ml/cm^2·s

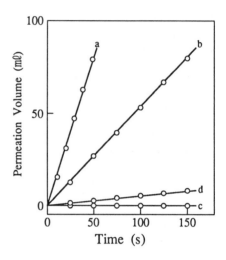

Figure 4 Permeability of porous PVA-PVAc composite film to organic solvents. (a) n-Hexane at 0.5 kg/cm^2; (b) cyclohexane at 0.5 kg/cm^2; (c) benzene at 0.5 kg/cm^2; (d) benzene at 60 kg/cm^2.

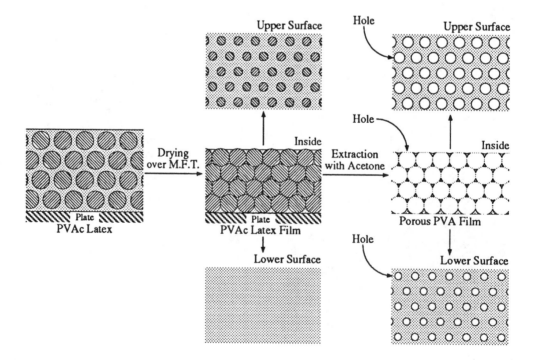

Figure 3 Formation model of porous PVA film from PVAc latex.

at 60 kg/cm². The difference in permeability to *n*-hexane and cyclohexane is explained by the viscosities. A strange phenomenon for benzene is due to the grafting PVAc or to the insoluble PVAc-like microgels inside the PVA cells of the porous film. As the PVAc swells in benzene and shrinks in *n*-hexane or cyclohexane, it seems to act in organic solvents as valves of small pores passed through and between PVA cells. If the porous film is dipped in benzene, the film size is barely affected by the solvent because the swelling of the PVAc occurs only in PVA cells. The porous PVA film that the grafting PVAc was hydrolyzed in a methanol solution of sodium methylate, indicated nearly the same flux as that observed for nonsolvents for PVAc. Thus, it is concluded that the grafting PVAc localized on the surface of PVA cells in the porous film plays a critical role and behaves as a valve that closes with benzene or a good solvent for PVAc and opens with *n*-hexane or a poor solvent.

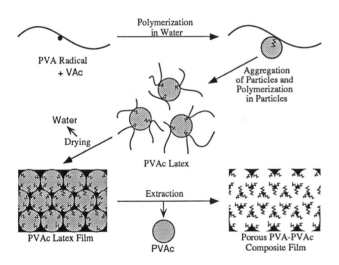

Figure 5 Model preparation of porous PVA-PVAc composite film. ∼ : PVA molecule grafted to PVAc; ■: PVA; ∖∖ : PVAc molecule grafted to PVA; ▨ : PVAc.

This prediction is drawn according to the following model. Figure 5 illustrates that in the latex state the grafting PVA protects hydrophobic PVAc particles in water by concentrating on the surfaces of PVAc particles, but in the porous film after acetone extraction, the insoluble grafting PVAc conversely exists as an important component on the inner surface of spherical cells of PVA.

Therefore, the porous film obtained by using the polymerization of VAc in the presence of PVA as a protective colloid is certainly the porous PVA–PVAc composite film.

B. Chromism in Organic Solvents

1. Influence of PVAc Amounts

The PVAc latexes initially were prepared in method [III] using the HPO-TA system as an initiator. The polymerization conditions and the results of the porous PVA–PVAc composites are summarized in Table 2 [23]. The extraction ratio of PVAc from the latex films increased with an increase in the HPO amount in the polymerization. The residual PVAc in the porous films was approximately 20–45%, relative to the total weight of the film before acetone extraction. This residual PVAc is considered to be polymers grafted onto PVA or highly branched polymers existed in the spherical cells of PVA formed by acetone extraction, which act as valves in the pores passing between the PVA cells for solvents and nonsolvents [19,22]. This interesting structure produces a satisfactory result by allowing penetration of organic solvents into the inner parts of the microporous films.

If paraffin permeates a white opaque paper, one can read letters through the paper because the paper becomes transparent. This phenomenon is based on the simple principle that micropores in the paper are filled with paraffin, which has a refractive index that is close to that of cellulose. If the porous PVA–PVAc composite film is soaked in organic solvents having the same refractive indices as that of PVA, the porous film is expected to become transparent again, according to the same principle as the phenomenon between paraffin and cellulose. On the basis of this consideration, subsequent experi-

Table 2 Results of Polymerization and Acetone Extraction

Expt no.	HPO amount (%)	Conversion (%)	Average particle size (μm)	Residual PVAc in porous film (%)[a]	Extraction ratio vs. total PVAc (%)
1	0.035	99.5	0.64	44.9	90.1
2	0.070	99.9	0.61	35.1	93.2
3	0.140	99.7	0.48	26.0	95.6
4	0.210	99.6	0.44	22.8	96.3
5	0.280	99.3	0.43	17.9	97.2

[a] PVAc ratio vs. total weight (100 × PVAc/(PVAc + PVA)).

ments on the porous films were carried out in various organic solvents [23].

One can read letters through the porous PVA–PVAc film in benzene, but one cannot do so in cyclohexane nor in the case of the blank. This is supported by the fact that the refractive indices of benzene are close to that of PVA, but the refractive index of cyclohexane is far from that of PVA. When the porous film was dipped in a mixed solvent of benzene and cyclohexane (8.0:2.0 in weight), it became semi-transparent. To make this point clearer, the refractive index and the dispersive power of polymers and organic solvents were measured. The results are shown in Table 3, which shows that the refractive index of PVA is near that of benzene and that the dispersion power of aliphatic compounds is lower than that of aromatic compounds.

When the porous films were dipped in organic solvents, they are colored at high transmittance over 80%, the wavelength of transmitted light of the porous film in chlorobenzene shifted from the short to the long side with increase in the residual amount of PVAc, as shown in Table 4. The scattered color of the porous film, which was complementary to that of the transmitted light, did not change with the observing direction and shifted from yellow to green through orange, red, purple, and blue as the residual amount of PVAc increased. This interesting phenomenon was more remarkable in aromatic than in aliphatic solvents, and was not observed in the hydrolyzed porous PVA film because sufficient penetration of the aromatic solvents into the film did not take place. To check the different coloration of the porous film in aliphatic and aromatic solvents, the transmittance spectra of the porous film in solvents were measured. Figure 6 displays the effects of the mixed solvents of chloroform and bromoform on the transmission curve of the porous

Table 4 Influences of PVAc Content on Color of Porous Film (Expt. Nos. 1–5) in Chlorobenzene

Expt. no.	Residual PVAc in porous film (%)[a]	Maximum wavelength of transmittance (nm)	Color of scattered light
1	44.9	478.4	Yellow
2	35.1	509.4	Purple
3	26.0	578.8	Bluish purple
4	22.8	612.4	Blue
5	17.9	624.4	Blue

[a] PVAc ratio vs. total weight (100 × PVAc/(PVAc + PVA)).

film (Expt. No. 3 in Table 3). In those spectra, the maximum wavelength of the transmittance was scarcely observed. This was common to all samples in Table 2. Figure 7 is the transmittance spectra of the porous film (Expt. No. 2 in Table 3) in mixed solvents of benzene and bromobenzene where the maximum in the spectra is clearly observed. This phenomenon was found in all porous films listed in Table 2. As a result, the transmittance spectra made some predictions about the coloration phenomena of the porous film in organic solvents.

The mechanism of the color development of the porous film and the effect of the PVAc amount in the porous film on the color development are discussed again in the next section.

2. Solvatochromism

The color of the porous PVA–PVAc composite films in organic solvents did not change with the observing

Table 3 Refractive Index of Materials Used for Experiments

Material	Refractive index (n_D^{20})	Dispersive power (DP)[a]
PVA	1.5214	0.0173
PVAc	1.4665[b]	0.0131[c]
Bromoform	1.5690	0.0255
Chloroform	1.4438	0.0176
Cyclohexane	1.4256	0.0172
Benzene	1.5014	0.0299
Chlorobenzene	1.5239	0.0310
Bromobenzene	1.5590	0.0344
Carbon disulfide	1.6272	0.0542

[a] DP = $(n_F - n_C)/(n_D - 1)$, where n_F, n_D, and n_C are refractive indices at λ = 486.1, 589.3, and 656.3 nm.
[b] From Ref. [24].
[c] Calculated value, cf. Ref. [23].

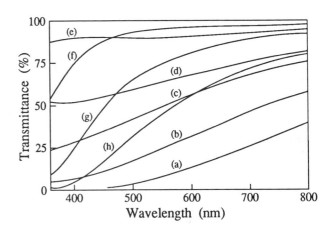

Figure 6 Transmittance spectra of porous film (expt. no. 3) in mixed solvents of chloroform and bromoform. Mixing ratio of chloroform and bromoform (w/w): (a) 8.0:2.0; (b) 6.0:4.0; (c) 5.0:5.0; (d) 4.0:6.0; (e) 3.0:7.0; (f) 2.0:8.0; (g) 1.0:9.0; (h) 0:10.0.

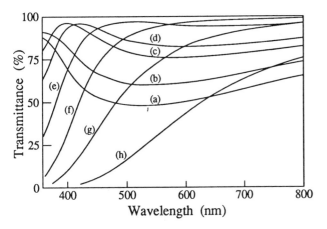

Figure 7 Transmittance spectra of porous film (expt. no. 2) in mixed solvents of benzene and bromobenzene. Mixing ratio of benzene and bromobenzene (w/w): (a) 10.0:0; (b) 9.0:1.0; (c) 8.0:2.0; (d) 7.0:3.0; (e) 6.0:4.0; (f) 4.0:6.0; (g) 2.0:8.0; (h) 0:10.0.

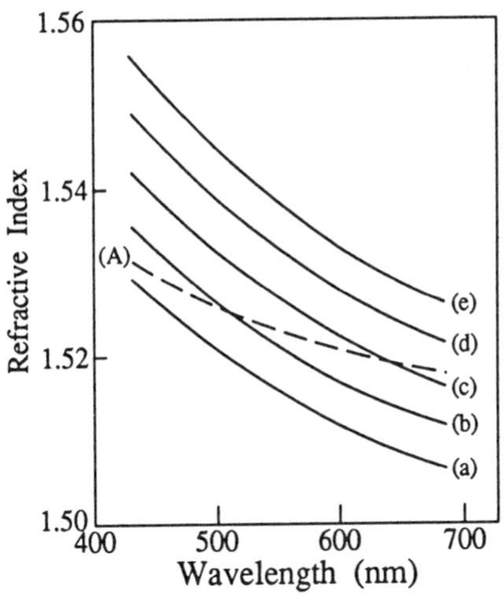

Figure 8 Dispersion curves of PVA and mixed solvents of benzene and bromobenzene. Mixing ratio of benzene and bromobenzene (w/w): Solid lines: (a) 7.0:3.0; (b) 6.0:4.0; (c) 5.0:5.0; (d) 4.0:6.0; (e) 3.0:7.0. Broken line: (A) PVA.

direction except for transmitted light and was complementary to the color of the transmitted light. This suggests that the coloration results from light scattering based on the difference between the refractive indices of the polymer and the solvents. Because it is reasonable to assume that PVAc grafted on PVA forms a viscous solution by swelling with solvents penetrated in the spherical cells of PVA, this light scattering may be concentrated on the refractive indices of PVA and PVAc solutions. It is thought that the refractive indices of PVA agree with that of the PVAc solution at the maximum wavelength of the transmission curve.

Figure 8 shows the light dispersion curves for PVA (broken line) and mixed solvents of benzene and bromobenzene is (solid lines). Because the negative inclination of the dispersion curve for PVA is slower then those of

the solvents, the PVA curve necessarily crosses some curves of the solvents. If the crosspoint exists in the visible region, PVA transmits the color light of the wavelength at the crosspoint and refracts color light except for the transmitted light. The refraction of the light is repeated in all directions passing through numerous polyhedra of PVA in the porous film and the scattered light is observed as the same color regardless of the observing direction. The coloration results of the porous film dipped in the mixed solvents of benzene and bromobenzene are shown in Table 5 [23]. The color shifted from yellow to green through orange, red, purple, and blue with the increase in the refractive index of the mixed solvents. The color development of the porous film in aliphatic solvents was not clearly revealed. This is because the dispersion power of the aliphatic solvents is

Table 5 Influence of Refractive Index on Color of Porous Film (Expt. No. 3) in Mixed Solvents of Benzene and Bromobenzene

Mixed solvent (in weight)		Refractive index (n_D^{20})	Maximum wavelength of transmittance (λ_{max}, nm)	Color of scattered light
Benzene	Bromobenzene			
7.0	3.0	1.5341	416.0	Yellow
6.5	3.5	1.5312	439.2	Orange
6.0	4.0	1.5286	499.8	Reddish purple
5.5	4.5	1.5255	533.3	Purple
5.0	5.0	1.5228	590.1	Bluish purple
4.5	5.5	1.5202	638.7	Blue

lower than that of the aromatic solvents. Consequently, to develop more beautiful colors in the porous PVA–PVAc composite films in organic solvents, it is concluded that the use of halogenated aromatic solvents is preferred.

One of the characteristics of the porous film is that there is no effect on the film size in the solvents, despite the existence of PVAc, because of the enormous space taken up by the PVA cells versus the PVAc amount. If the porous film is dipped in a solvent, the PVAc concentration in the PVA cells may be appreciated by the residual PVAc amount. Because the refractive index of the PVAc solution in contact with PVA cells becomes lower as the amount of PVAc with a low-refractive index increases, the wavelength of the transmitted light for the porous film shifts to the short side, and the color of the scattered light shifts to the yellow side. This consideration successfully explains the experimental results in Table 4.

This principle is applied not only to the PVA–PVAc composites but to other polymer composites. The composite structure does not always need to be porous but may be powders and gels designed for the wettability by solvents and the extension of the surface area in soluble polymers. From this point-of-view, the present work sheds a new light on the research on composite materials related to graft polymers and copolymers.

3. Thermochromism

The effects of temperature on the color development of the porous film in chlorobenzene were shown in Table 6 [23]. The coloration was reversible thermochromism. The refractive index of the materials generally decreases as the temperature increases, and the temperature dependence of the liquid is greater than that of the solid. For example, the temperature dependence ($\Delta n_D/°C$) of PVA and chlorobenzene was found to be 3.0×10^{-3} and 4.5×10^{-4} at 589.3 nm. Consequently, it is interpreted that the wavelength of the crosspoint between the dispersion curves of PVA and chlorobenzene shifts from the long side to the short side with increasing tem-

Table 6 Influence of Temperature on Color of Porous Film (Expt. No. 3) in Chlorobenzene

Temperature (°C)	Maximum wavelength of transmittance (λ_{max}, nm)	Color of scattered light
10	736.4	Greenish blue
20	578.8	Bluish purple
30	495.2	Reddish purple
40	435.2	Yellow
50	397.2	Pale yellow

perature, and from the short side to the long with decreasing temperature. This is why the coloration phenomena of the porous films in solvents showed reversible thermochromism.

4. Piezochromism

There have been only a few investigations of piezochromism because piezochromic compounds in general have particular structures and exhibit performance only under certain strict conditions such as grinding or high pressure [24,27]. There have been no more than 10 reports on piezochromism of polymers since 1967 [28–37]. Therefore, if piezochromism is observed in popular compounds and can be revealed in a simple way, it is very interesting. It was recently found that the porous PVA–PVAc composite does not show a color change in solvents, but displays an interesting color change in a mixture of a solvent and a nonsolvent for PVAc with increasing pressure. In this section, the piezochromism of the porous PVA–PVAc composite film in various mixed solvents for PVAc is discussed.

In order to compare the color of a porous film before and after increasing the pressure, two oblong glass plates were used to press the sample. The porous film colored in solvents was pressed from both sides of the two oblong glass plates using paper clips, and the color photographs were taken in the same mixed solvent to prevent evaporation of the solvents at 20°C. Both sides of the porous film show the color before the pressure was increased and the middle narrow part of the porous film shows the color after the pressure was increased. This method provides the advantage of photographically observing the color change in the porous film before and after increasing the pressure.

When chlorobenzene as a single solvent or a mixture of benzene and bromobenzene as a mixed solvent for PVAc was used, the color of the porous film did not change with pressure; it was purple in both solvents. However, when the mixture of bromobenzene as solvent and cyclohexane as nonsolvent for PVAc was used, the color of the porous film changed remarkably from yellow to blue with pressure. Taking into account the fact that the coloration of the porous film in organic solvents is based on the intersection between the dispersion curve of PVA and that of PVAc solution swollen with solvents in PVA cells in the visible region, the color change from yellow to blue suggests that the intersection between both the dispersion curves shifted to the long wavelength side; that is, the composition of bromobenzene in the mixed solvent in the PVA cells was increased by pressure. It seems reasonable that the PVAc in the PVA cells selectively absorbed the solvent from the mixed solvent, and the nonsolvent remaining in the PVA cells was forced out of the porous film by pressure. In order to inspect the results mentioned previously, whether or not PVAc selectively absorbs the solvent from the mixture of solvent and nonsolvent for PVAc was examined by a gas chromatograph. The composition of solvent and nonsolvent was in the range in which PVAc remains insoluble, although it swells in the mixed solvent. The ratio of the solvent to nonsolvent in the mixed solvent after dipping PVAc obviously decreased in comparison with that before dipping PVAc. This can be considered proof that PVAc in the microcells of PVA selectively absorbs the solvent from the mixture of solvent and nonsolvent.

Furthermore, to elucidate the piezochromism more clearly, some experiments were carried out with various combinations of solvents and nonsolvents for PVAc. One was the A-group in which the refractive indices of the solvent were higher than that of the nonsolvent, and another was the B-group in which the refractive index of the solvent was lower than that of the nonsolvent for PVAc as listed in Table 7 [39]. In the A-group, the three combinations of chlorobenzene or bromobenzene as solvent and cyclohexane as nonsolvent for PVAc were used. In A-(1), because chlorobenzene is a solvent for PVAc, cyclohexane, which is a nonsolvent for PVAc remains in the PVA cells in the porous film, is forced out by pressure. Consequently, the color of the porous film after increasing the pressure changes to purple in chlorobenzene. The color in A-(2) and in A-(3) using bromobenzene changed from yellow to blue and from purple to blue with pressure. The color difference in A-(2) and A-(3) before the pressure was increased relates to the refractive indices based on the weight fraction bromobenzene in the mixed solvents. In the A-group, it seems reasonable that the color of the porous films shifted with pressure from the color in the mixed solvent having a low-refractive index to that in the solvent hav-ing a high-refractive index because cyclohexane as the nonsolvent, having a low-refractive index, was forced out of the porous film. In the B-group, the three combinations of benzene or chlorobenzene as solvents and carbon disulfide as nonsolvent for PVAc were used. In B-(2), the color of the porous film changed from blue to purple with pressure. This indicates that the weight fraction of chlorobenzene in the spherical cells of PVA increased from 0.9 to nearly 1.0 because of the exclusion of carbon disulfide from the PVA cells with pressure. In the case where benzene was used as the solvent, the color of the porous film changed with pressure from purple to yellow in B-(1) and from blue to orange in B-(3). In the B-group, the color of the porous films changed from the color in the mixed solvent having a high-refractive index because carbon disulfide as a nonsolvent, having a higher refractive index, was forced out of the porous film.

So far as we know, this piezochromism is the simplest and has the widest range of color change in comparison with all reported results.

V. CONCLUSIONS

There has been a conventional sense that the PVAc latexes prepared in the presence of PVA as a protective colloid contain the graft copolymer of PVA and PVAc, so that PVAc particles in the dried latex film are not extracted at a high ratio with solvents. In this Chapter, it has been defined without an influence by the usual sense that the porous PVA–PVAc composite can be prepared from the PVAc latex film with acetone extraction. The porous film consists of the spherical cells of PVA

Table 7 Relationships Between Solvent Composition and Color Change of Porous Film Before and After Pressure

Expt. no.	Combination of solvents	Mixing ratio (w/w)	Refractive index (n_D^{20})	Color change with pressure Before	Color change with pressure After	Piezo-chromism
Single solvent						
	chlorobenzene	—	1.5239	Purple	Purple	×
Mixture of two solvents						
	benzene:bromobenzene	5.0:5.0	1.5228	Purple	Purple	×
Mixture of solvent and non-solvent						
A-group						
(1)	chlorobenzene:cyclohexane	9.0:1.0	1.5141	Yellow	Purple	○
(2)	bromobenzene:cyclohexane	8.0:2.0	1.5160	Yellow	Blue	○
(3)	bromobenzene:cyclohexane	8:4:1.6	1.5238	Purple	Blue	○
B-group						
(1)	benzene:carbon disulfide	7.0:3.0	1.5252	Purple	Yellow	○
(2)	chlorobenzene:carbon disulfide	9.0:1.0	1.5289	Blue	Purple	○
(3)	benzene:carbon disulfide	6.5:3.5	1.5297	Blue	Orange	○

Refractive index; A-group: solvent > nonsolvent, B-group: solvent < nonsolvent.
Mark for piezochromism; ○: positive, ×: negative.

from which PVAc particles were extracted, and grafting PVAc exists on the inside surface of the cells. From this characteristic structure, the porous film shows an interesting permeability to organic solvents, which benzene passes through, but cyclohexane does not. The porous film, furthermore, shows chromism in aromatic organic solvents with the refractive index near that of PVA. Especially in the mixed solvent of solvent and nonsolvent for PVAc, the porous film reveals piezochromism. This piezochromism is the simplest, and the color change with pressure covers all the ranges in the visible portion by the course from short wavelength to long wavelength or conversely from long wavelength to short wavelength.

REFERENCES

1. S. Hayashi, *Polym.-Plast. Technol. Eng., 27:* 61 (1988).
2. S. Okamura, and T. Motoyama, *Kobunshi Kagaku, 15:* 165 (1958).
3. S. Okamura, T. Motoyama, and T. Yamashita, *Kobunahi Kagaku, 15:* 171 (1958).
4. F. D. Hartley, *J. Polym. Sci., 34:* 397 (1959).
5. I. Gavat, V. Dimonie, D. Donescu, C. Hagiopol, M. Munteanu, K. Gpsa, and T. Deleanu, *J. Polym. Sci., Polym. Symp., 64:* 125 (1978).
6. A. H. Traaen, *J. Appl. Polym. Sci., 7:* 58 (1963).
7. H. Dexheimer, and O. Fuchs, *Makromol. Chem., 96:* 172 (1966).
8. S. Hayashi, and K. Nakamura, *Kogyo Kagaku Zasshi, 71:* 127 (1968).
9. S. Hayashi, T. Yanagisawa, and N. Hojo, *Nippon Kagaku Kaishi,* 402 (1973).
10. H. Lamont, *Adhes. Age, 16:* 24 (1973).
11. S. Hayashi, T. Hirai, and N. Hojo, *J. Appl. Polym. Sci., 27:* 1607 (1982).
12. E. Nagai, and N. Sagane, *Kobunshi Kagaku, 12:* 195 (1955).
13. S. Hayashi, C. Nakano, and T. Motoyama, *Kobunshi Kagaku, 20:* 303 (1963).
14. R. K. Tubbs, *J. Polym. Sci.,* Part A-1, *4:* 623 (1966).
15. T. Moritani, and Y. Fujiwara, *Macromolecules, 10:* 532 (1977).
16. G. van der Velden, and J. Beulen, *Macromolecules, 15:* 1072 (1982).
17. S. Toppet, *Polymer, 24:* 507 (1983).
18. D. C. Bugada, and A. Rudin, *Polymer, 25:* 1759 (1984).
19. S. Hayashi, M. Takagi, and N. Hojo, *J. Colloid and Interface Sci., 77:* 6 (1980).
20. M. Schmit, D. Nerger, and W. Burchard, *Polymer, 20:* 582 (1979).
21. W. A. Cote, Jr., A. C. Day, G. W. Wilkes, and R. H. Marchessanlt, *J. Colloid and Interface Sci., 27:* 32 (1968).
22. S. Hayashi, T. Hirai, F. Hayashi, and N. Hojo, *J. Appl. Polym. Sci., 28:* 304 (1983).
23. S. Hayashi, J. Xu, K. Fuse, K. Asada, and T. Hirai, *J. Colloid and Interface Sci., 163:* 315 (1994).
24. J. Brandrup, and E. M. Immergut (eds.), *Polymer Handbook, 3rd ed.,* Wiley-Interscience, New York (1989).
25. C. Reichardt, *Chem. Soc. Rev., 21:* 373 (1992).
26. K. L. Bray, and H. G. Drickamer, *J. Phys. Chem., 94:* 2154 (1990).
27. K. L. Bray, and H. G. Drickamer, *J. Phys. Chem., 93:* 7601 (1989).
28. K. L. Bray, H. G. Drickamer, E. A. Schmitt, and D. N. Hendrickson, *J. Am. Chem. Soc., 111:* 2848 (1989).
29. Y. Ogo, Y. Nishino, and T. Sato, *High Temp.-High Pressures, 14:* 319 (1982).
30. H. Tamura, N. Mino, and K. Ogawa, *J. Photopolym. Sci. Technol., 2:* 158 (1989).
31. K. Song, H. Kuzmany, G. M. Wallraff, R. D. Miller, and J. F. Rabolt, *Macromolecules, 23:* 3870 (1990).
32. J. F. Rabolt, K. Song, H. Kuzmany, R. Sooriyakumaran, G. Fickes, and R. D. Miller, *Polym. Prepr. (Am. Chem. Soc. Div. Polym. Chem.), 31:* 262 (1990).
33. R. A. Nallicheri, and M. F. Rubner, *Macromolecules, 24:* 517 (1991).
34. K. Song, R. D. Miller, G. M. Wallraff, and J. F. Rabolt, *Macromolecules, 24:* 4084 (1991).
35. B. F. Variano, C. J. Sandroff, G. L. Baker, *Macromolecules, 24:* 4376 (1991).
36. F. C. Schilling, A. J. Lovinger, D. D. Davis, F. A. Bovey, and J. M. Zeigler, *J. Inorg. Organomet. Polym., 2:* 47 (1992).
37. K. Song, R. D. Miller, G. M. Wallraff, and J. F. Rabolt, *Macromolecules, 25:* 3629 (1992).
38. K. Song, R. D. Miller, and J. F. Rabolt, *Macromolecules, 26:* 3232 (1993).
39. S. Hayashi, K. Asada, S. Horiike, H. Furuhata, and T. Hirai, *J. Colloid and Interface Sci., 176:* 370 (1995).

14

Hydrophobization of Polyanionic Polymers to Achieve Higher Biological Activity

Naoto Oku
University of Shizuoka, Shizuoka, Japan

Raphael M. Ottenbrite
Virginia Commonwealth University, Richmond, Virginia

Yasuo Suda
Osaka University, Osaka, Japan

I. INTRODUCTION

Since the pioneer work by Merigan in 1967 [1], many kinds of synthetic or natural polyanionic polymers have been examined for their biological activities [2–8], such as cytotoxicity, antiviral activity, antitumor activity, and immunomodulating activity. Although the biological results were interesting, the extent of activity for clinical application was still low.

We hypothesize that the low activity of polyanionic polymers is related to the uneasiness of the first stage of interaction between target cell membranes and polyanionic polymers because the electro-repulsion between both negatively charged substances hinders the interaction. Therefore, the most important factor for producing the high activity of synthetic or natural polyanionic polymers may be to increase the affinity of polymers for the cell membrane. Once the polymer attaches to the cell surface, the subsequent uptake of the polymer into the cell may readily occur and high biological activity is expected. This hypothesis is supported by the following two facts.

Several mechanisms between drug-encapsulated liposomes and cells was presented [9,10]. Sunamoto and coworkers [11] applied the liposomal technique for improving the biological activity of polyanionic polymers.

They encapsulated poly(MA-CDA) into mannan-coated liposomes and evaluated superoxide production from mouse macrophages. The activity was three- to five-fold high compared with uncapsulated poly(MA-CDA) itself [5,11], suggesting that an increased incorporation of the polymer by the receptor-mediated endocytosis mediated the higher biological activity.

The second fact is the existence of biological response modifiers (BRMs) from bacterial cell surface. Figure 1 shows a structure of typical BRM, lipopolysaccharide (LPS) from gram-negative bacteria, which is a macromolecular and negatively charged compound [12]. LPS contains a very hydrophobic region in the so-called lipid A part. The hydrophobic region may have a high affinity for the cell membrane at the earliest stage of interaction with cells. This was a good reference point for our strategy of increasing the biological activity of synthetic polyanionic polymers.

We modified polyanionic polymers by use of a grafting reaction of hydrophobic groups onto the polymers. After an extensive evaluation for the affinity of the hydrophobically modified (hydrophobized) polymers to cell membrane, the immuno-stimulating activity of polymers was investigated by in vitro or ex vivo experiments. Consequently, the increased biological activity was found in the hydrophobized polymer, indicating that

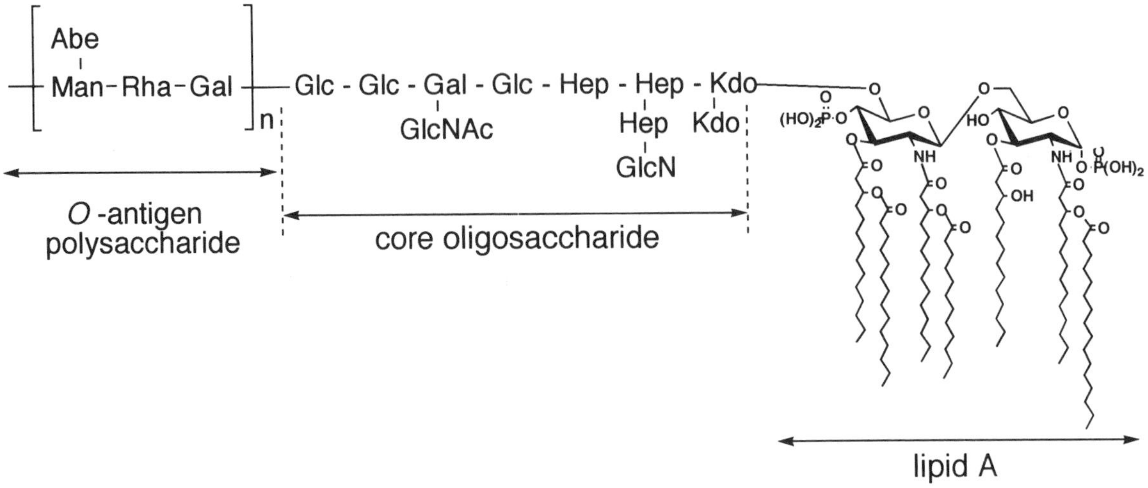

Figure 1 Structure of lipopolysaccharide (LPS) from gram-negative *Salmonella* species.

the higher activity was due to the increased membrane affinity of polymers. In addition, the immuno-stimulating biological activity was affected by the molecular weight of polymers.

II. HYDROPHOBIZATION OF POLY(MALEIC ACID-*ALT*-3,4-DIHYDROXYPHENYLPROP-1-ENE) [POLY(MA-DP)]

A. Preparation of Hydrophobized Poly(MA-DP)

In order to obtain the moderate property of membrane affinity and water solubility many kinds of hydrophobic groups were first tested using poly(maleic acid-*alt*-3,4-dihydroxyphenylprop-1-ene)[poly(MA-DP)] [13]. The preparation of polymer and the hydrophobization were done according to Scheme 1. Hydrophobic groups were grafted onto the maleic anhydride residue in the polymer by changing the contents (Table 1). From their solubility in water and the partition coefficient examined by *n*-octanol/buffer system, the hexyl group and phenyl group seemed to have better properties for the use of hydrophobization.

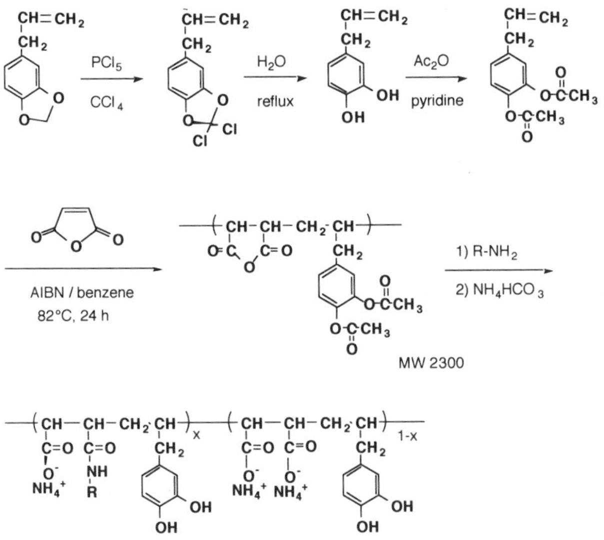

B. Evaluation of Cell Membrane Affinity In Vivo: Interaction with Rat Intestinal Cells

Two hydrophobized polymers, MA-DP-A20 (phenyl group grafted in 20 mol%) and MA-DP-H68 (hexyl group grafted in 68 mol%), were selected for the evaluation of the interaction between polymers and cell membranes

Scheme 1

because they have better water solubility and a partition coefficient close to 1.0. The in situ loop method was first applied to estimate the extent of interaction between the modified polymers and cell membranes (Fig. 2) (14). Compared with unmodified polymer (MA-DP), the recovery of phenyl group or hexyl group grafted modified polymers (MA-DP-A20 or MA-DP-H68) from rat small intestine was 10–20% less. This lower recovery may be derived from the adhesion of the polymer to the lipid bilayer of the intestinal epithelial cells or to the partial absorption of the polymer into the cells. That was the

Table 1 Abbreviation and Partition Coefficient of Unmodified or Modified Poly(Maleic acid-*alt*-3,4-dihydroxyphenylprop-1-ene)

Abbreviation	Grafting group	% Grafted	P.C.
MA-DP	none	0	0.002
MA-DP-P23	$CH_3(CH_2)_2NH-$	23	0.026
MA-DP-P100	$CH_3(CH_2)_2NH-$	100	0.058
MA-DP-B35	$CH_3(CH_2)_2NH-$	35	0.050
MA-DP-B66	$CH_3(CH_2)_3NH-$	66	0.302
MA-DP-B100	$CH_3(CH_2)_3NH-$	100	2.204
MA-DP-H30	$CH_3(CH_2)_5NH-$	30	0.088
MA-DP-H68	$CH_3(CH_2)_5NH-$	68	0.723
MA-DP-H100	$CH_3(CH_2)_5NH-$	100	n.d.
MA-DP-A8	C_6H_5NH-	8	0.012
MA-DP-A13	C_6H_5NH-	13	0.243
MA-DP-A20	C_6H_5NH-	20	0.868
MA-DP-O29	$CH_3(CH_2)_7NH-$	29	0.071
MA-DP-O59	$CH_3(CH_2)_7NH-$	59	0.309
MA-DP-O100	$CH_3(CH_2)_7NH-$	100	n.d.
MA-DP-D19	$CH_3(CH_2)_9NH-$	19	0.144
MA-DP-D74	$CH_3(CH_2)_9NH-$	74	n.d.

P.C. = partition coefficient by *n*-octanol/phosphate buffer (pH 6.5).

n.d. = not determined because of their insolubility into water.

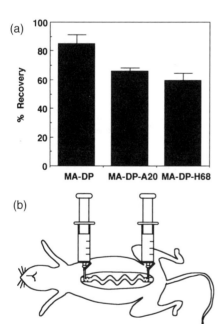

Figure 2 The percent recovery of the modified polymer from rat small intestine after 1 h (a), and the schematic representation of in situ loop method using rat small intestine (b).

first evidence that hydrophobized polymer showed higher affinity for the cell membranes than the unmodified polymer.

C. Evaluation of Membrane Affinity In Vitro: Interaction with Liposomal Lipid Bilayer

The in situ method using rat living intestine was simple and qualitative. However, it was difficult to evaluate the weak interaction between polymers and cell membranes quantitatively. Therefore, the lipid bilayer of liposome was used as a model of cell membranes for the quantitative evaluation for the affinity of the hydrophobized polymers (15).

The experimental principle is illustrated in Fig. 3. The interaction of the polymer with the liposomal membranes causes the perturbation of the bilayer. This perturbation follows the leakage of calcein from the liposome. Calcein in high concentration in the liposome is self-quenched, but has strong fluorescence intensity by the leak from the liposome. Therefore, the extent of the membrane interaction can be estimated quantitatively from the fluorescence spectroscopy.

Figure 4 shows the results of the interaction between the negatively charged liposomes and the prepared polyanionic polymers. The release of calcein was dependent on the concentration of polymers. The unmodified polymer (MA-DP) had very little effect on the permeability of liposomal membrane. This fact suggests that the unmodified polyanionic polymer is so hydrophilic that the interaction with cell membranes is very weak. On the contrary, the two modified polymers, phenyl group grafted MA-DP-A20 and hexyl group grafted MA-DP-H68, caused membrane perturbation even at low concentrations. The difference was the 2 to 4 order in concentration compared with the unmodified polymer. These results indicate the stronger interaction between polymers and cell membranes in the case of hydrophobized polymers.

Based on the data in Figs. 2 and 4, it appears that the affinity of the polyanionic polymer for cell membrane can be increased by the hydrophobization of polymers.

III. HYDROPHOBIZATION OF POLY(MALEIC ACID-*ALT*-7,12-DIOXASPIRO-[5,6]-DODEC-9-ENE) [POLY(MA-CDA)]

A. Preparation of Hydrophobized Poly(MA-CDA)

Based on the results of the hydrophobization of poly(MA-DP), we applied the hydrophobically grafting technique to poly(maleic acid-*alt*-7,12-dioxaspiro-[5,6]-

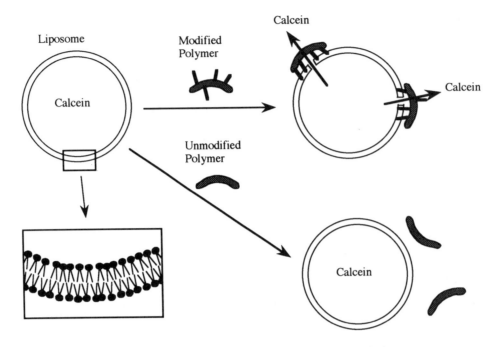

Figure 3 The principle for the evaluation of the interaction between modified or unmodified polyanionic polymer and liposomal membrane.

dodec-9-ene) [poly(MA-CDA)] (16), which had been previously reported to have a significant antitumor activity at a very high dose (100 mg/kg) (17).

Poly(MA-CDA) was synthesized by a free radical copolymerization of maleic anhydride and 7,12-dioxaspiro-[5,6]-dodec-9-ene, followed by hydrolysis in aqueous ammonium bicarbonate. As shown in Scheme 2, higher (MW 20,000) and lower (MW 3,400) molecular weight polymers were prepared by changing the concentration of initiator (AIBN) to evaluate the molecular-weight dependency on the biological activity of polymers. The structure of the polymer was characterized

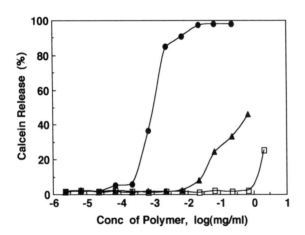

Figure 4 Effect of modified and unmodified poly(MA-DP)s on the stability of the negatively charged liposome. □ = unmodified MA-DP; ▲ = MA-DP-A20; ● = MA-DP-H68.

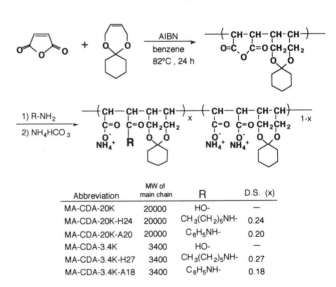

Abbreviation	MW of main chain	R	D.S. (x)
MA-CDA-20K	20000	HO-	—
MA-CDA-20K-H24	20000	$CH_3(CH_2)_5NH-$	0.24
MA-CDA-20K-A20	20000	C_6H_5NH-	0.20
MA-CDA-3.4K	3400	HO-	—
MA-CDA-3.4K-H27	3400	$CH_3(CH_2)_5NH-$	0.27
MA-CDA-3.4K-A18	3400	C_6H_5NH-	0.18

Scheme 2

by ^{1}H NMR, IR, and elemental analysis. The molecular weight (peak molecular weight) was determined by GPC-HPLC. The hydrophobization was performed according to the similar amidation to that in the previous study on MA-DP where a hexyl or phenyl group was selected for the hydrophobic group.

B. Evaluation of Membrane Affinity In Vitro

The affinity of hydrophobized polymers to cell membranes was examined using negatively charged liposomes. Figure 5 shows the results of high-molecular weight poly(MA-CDA) (MW 20,000). In the case of unmodified polymer (abbreviated as MA-CDA-20K), no calcein was released even over the concentration of 10 mg/ml, indicating that the interaction of unmodified poly(MA-CDA) with liposomal membrane is very weak. This may be the reason that a very high dose of this polymer was necessary for the antitumor activity in vivo [17]. However, hydrophobized polymers, phenyl group modified (MA-CDA-20K-A20) and hexyl group modified (MA-CDA-20K-H24), interacted with negatively charged liposomes even at low concentrations. The results of the low-molecular weight polymer (MW 3,400) is shown in Figure 6. In addition, the unmodified polyanionic polymer (MA-CDA-3.4K) had little interaction, but modified polymers (MA-CDA-3.4K-A18, MA-CDA-3.4K-H27) interacted with liposomal membrane even at low concentrations.

There was no distinct difference between polymers having different molecular weights. From the data in Figures 2, 4, 5, and 6, it seems to be established that the affinity of polyanionic polymers for cell membrane is adjustable by a simple grafting of hydrophobic groups.

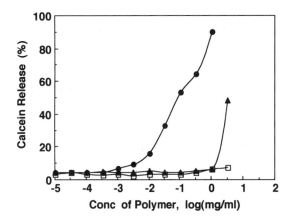

Figure 6 Effect of low-molecular weight (3,400) poly(MA-CDA)s on the stability of negatively charged liposome. □ = unmodified MA-CDA-3.4K; ▲ = MA-CDA-3.4K-A18; ● = MA-CDA-3.4K-H27.

C. Evaluation of Biological Activity In Vitro: Superoxide Release from Cultured Cells

The biological activity of the polymers was evaluated in vitro by the ability to stimulate the release of superoxide from DMSO-differentiated HL-60 cells [18–20]. The released superoxide was monitored by the cytochrome C method [21,22]. As shown in Figure 7, when the differen-

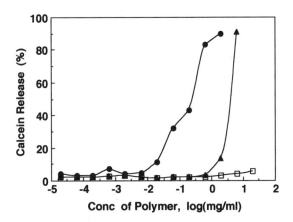

Figure 5 Effect of high-molecular weight (20,000) poly(MA-CDA)s on the stability of negatively charged liposome. □-unmodified MA-CDA-20K; ▲ = MA-CDA-20K-A20; ● = MA-CDA-20K-H24.

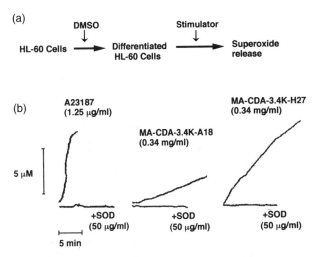

Figure 7 Differentiation of HL-60 cells by the incubation with dimethylsulfoxide [DMSO] (a), and the assay of superoxide release in the DMSO-differentiated HL-60 cells by cytochrome C method (b).

tiated cells were incubated with A23187 (1.25 μg/ml as a final concentration), the optical density at 550 nm increased. Since this increase was inhibited by the addition of superoxide dismutase (50 μg/ml, final concentration), it was obvious that the increase in the optical density was related to the superoxide released from DMSO-differentiated HL-60 cells.

The effect of poly(MA-CDA)s having high-molecular weight on the release of superoxide in DMSO-differentiated HL-60 cells is shown in Fig. 8. Each value was the average from at least three experiments and is shown as a relative activity to the positive control, A23187. The stimulating activity of unmodified polymer (MA-CDA-20K) was observed only at the high concentration (2mg/ml) examined. In contrast, the hydrophobically modified polymer (MA-CDA-20K-A20 and MA-CDA-20K-H24) stimulated the superoxide release even at low concentrations. The distinct difference was observed at a concentration of greater than 0.027 mg/ml.

Figure 9 shows the results of low-molecular weight polymers. Compared to Figure 8, the low-molecular weight modified polymers were found to have a much higher stimulating effect on the superoxide release. At the highest concentration examined, the hexyl group grafted polymer (MA-CDA-3.4K-H27) stimulated superoxide release approximately 40 times greater than the blank level, and the phenyl group grafted polymer (MA-CDA-3.4K-A18) stimulated 15 times. At the lower concentrations, the superoxide release stimulation was demonstrated only in the case of modified polyanionic polymers.

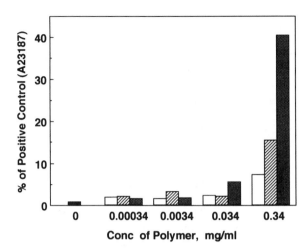

Figure 9 Effect of low-molecular weight (3,400) poly(MA-CDA)s on the superoxide release in the DMSO-differentiated HL-60 cells. The data are the mean of three experimental results. ■ = control; ☐ = unmodified MA-CDA-3.4K; ▨ = MA-CDA-3.4K-A18; ■ = MA-CDA-3.4K-H27.

Based on these observations in Figs. 8 and 9, it is suggested that the hydrophobic group modification (hydrophobization) is an effective method for improving the immuno-stimulating activity of polyanionic polymers.

D. Cytotoxic Action on Cultured Cells

Qualitatively, it was found that these modified and unmodified polymers exhibited almost no cytotoxicity against DMSO-differentiated HL-60 cells during the experimental period of superoxide release (20 min to 1 h). The long-term influence of the polymers on cells was evaluated by the use of J774.1 cultured cells derived from macrophage [23]. Figure 10 shows the effect of high-molecular weight polymers on the growth of the cells for 2 days. All three kinds of polymers showed no effect at concentrations less than 0.1 mg/ml. In the case of hexyl group grafted polymer (MA-CDA-20K-H24), cytostatic effect was observed at concentrations greater than 0.3 mg/ml, and the other two polymers showed cytostaticity at levels greater than 1 mg/ml. The cytotoxic and cytostatic effects of the low-molecular weight polymers are shown in Fig. 11. Compared with the results shown in Fig. 10, the cytotoxic action of all low-molecular weight polymers seemed to be greater. The hexyl group grafted polymer (MA-CDA-3.4K-H27) showed cytotoxicity at the concentration of 0.3 mg/ml. The strongest cytotoxic action of hexyl group grafted polymers having both high- and low-molecular weight, MA-CDA-20K-H24 and MA-CDA-3.4K-H27, among the

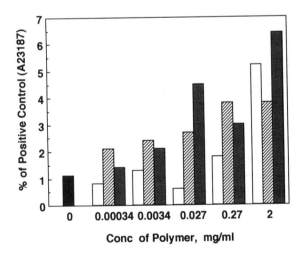

Figure 8 Effect of high-molecular weight (20,000) poly(MA-CDA)s on the superoxide release in the DMSO-differentiated HL-60 cells. The data are the mean of three experimental results. ■ = control; ☐ = unmodified MA-CDA-20K; ▨ = MA-CDA-20K-A20; ■ = MA-CDA-20K-H24.

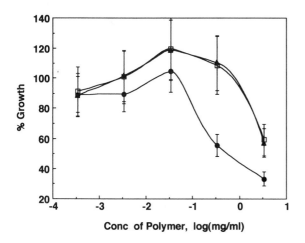

Figure 10 Effect of high-molecular weight (20,000) poly(MA-CDA)s on the growth of J774 cells for 2 days. □ = MA-CDA-20K; ▲ = MA-CDA-20K-A20; ● = MA-CDA-20K-H24.

comparative polymers, might be due to too strong an interaction with the membranes of J774.1 cells.

Since it might be possible that the perturbation of membrane directly stimulated the NADPH-oxidase located on the cell membrane, which is the enzyme for the production of superoxide [24], the possibility was examined by the assay using detergent (Triton X-100) instead of polymers. At 0.001% of Triton X-100, no stimulation of superoxide release from DMSO-differentiated HL-60 cells was observed. At 0.01% of Triton X-100, a

very strong cytotoxicity was found for the differentiated HL-60 cells, where most of the cells were lysed immediately after the addition of the detergent (data not shown), suggesting that the membrane perturbation is involved in the cytotoxicity, but does not mediate the superoxide production.

IV. THE EFFECT OF THE HYDROPHOBIZATION AND THE MOLECULAR WEIGHT OF POLY(MA-CDA) ON CYTOKINE-INDUCING ACTIVITY

A. Preparation and Characterization of Polymers

For additional evaluation of the effect of hydrophobization and the molecular weight of the polymers on the biological immuno-stimulating activity, we investigated the ex vivo cytokine (interleukin-6 [IL-6], and tumor necrosis factor [TNF]-inducing activity from human peripheral whole blood cells of hydrophobized polymers by use of fractionated poly(MA-CDA) with narrow polydispersity. Since this assay uses the intact human cells, it shows more accurate results than in vitro assay using cultured cell line [25].

Poly(MA-CDA) was synthesized as described previously by a free radical copolymerization followed by hydrolysis in aqueous solution. By the fractional precipitation of the copolymerization product (MW = 14,200, $\overline{MW}/\overline{Mn}$ = 3.1) different average-molecular weight poly(MA-CDA)s with narrow polydispersity were obtained as shown in Table 2.

From the previous data regarding especially cytotoxic assay using J774.1 cells, the hexyl group seemed not to be the better of the hydrophobic groups. We, therefore, selected the phenyl group for the hydrophobization with a content of approximately 10 mol%. According to the similar procedure described above, the hydrophobization of poly(MA-CDA) was done by grafting aniline onto the polymer backbone through amidation between the anhydride group of poly(maleic anhydride-alt-7,12-dioxaspiro-[5,6]-dodec-9-ene) and amino functionality of aniline. The degree of grafting was determined by ^{1}H NMR. These results are also listed in Table 2.

B. Evaluation of Biological Activity Ex Vivo: Cytokine Induction

The ex vivo IL-6 and TNF-inducing activities of fractionated and modified or unmodified poly(MA-CDA) were performed according to the method reported [26] and shown in Figs. 12 and 13, respectively. A similar tendency was shown in IL-6 and TNF induction from peripheral whole blood cells by those of poly(MA-CDA).

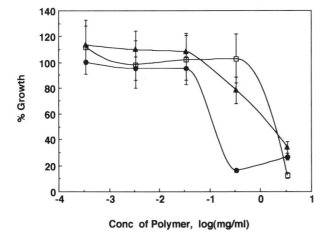

Figure 11 Effect of low-molecular weight (3,400) poly(MA-CDA)s on the growth of J774 cells for 2 days. □ = MA-CDA-3.4K; ▲ = MA-CDA-3.4K-A18; ● = MA-CDA-3.4K-H27.

Table 2 Molecular Weights, Hydrophobization and *Limulus* Activity of Fractionated poly(MA-CDA)s

Fraction	Molecular weight			Unmodified polymer	Modified polymer		
	$\overline{M}w(10^3)$	Mp(10^3)	$\overline{M}w/\overline{M}n$	Abbreviation	Content (mol%)	Abbreviation	*Limulus** activity
#1	26.6	21.4	1.83	MA-CDA-21K	9.4	MA-CDA-21K-A	<1
#2	22.8	19.2	1.67	MA-CDA-19K	9.2	MA-CDA-19K-A	3
#3	15.3	15.1	1.74	MA-CDA-15K	9.5	MA-CDA-15K-A	4
#4	8.88	8.74	1.69	MA-CDA-9K	9.9	MA-CDA-9K-A	<1
#5	6.06	5.35	1.98	MA-CDA-5K	10.5	MA-CDA-5K-A	63
#6	2.78	2.34	1.77	MA-CDA-2K	10.7	MA-CDA-2K-A	16

* Equivalent to a reference standard LPS derived from *Escherichia coli* 0111:B4 (Sigma Chem. Co.) in ng/mg.

The hydrophobized polymers demonstrated higher activity than unmodified polymers. The molecular weight dependency on the activity was also found. In particular, MA-CDA-5K-A (MW 5,000) hydrophobized with 10 mol% aniline induced both cytokines remarkably higher than unmodified MA-CDA and modified MA-CDA with different molecular weights.

The *Limulus* activity of hydrophobized polymers was measured by means of an Endospecy test™, which is known to have the highest sensitivity for the detection of LPS. The MA-CDA-5K-A, which showed the highest

ex vivo cytokine-inducing activity, was almost negligible in *Limulus* activity, but slightly higher than those of other modified MA-CDAs (Table 2). The 3-hydroxymyristic acid, which is a major fatty acid in LPS, was not detected in the fatty acid analysis of the polymer using GC-MS (Shimadzu QP-5000, Kyoto, Japan; the method was as reported in reference 26). Further, the polymer was treated with LPS (Endo-toxin)-adsorbed gel (Kuttuclean, Maruha, Japan) to confirm that no contamination of even trace amounts of LPS were in the polymer. Figure 14 shows the IL-6 activity of MA-CDA-5K-A with

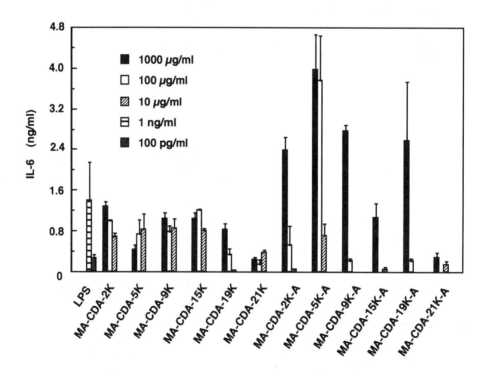

Figure 12 IL-6 inducing activity of poly(MA-CDA)s from human peripheral whole blood cell culture. The doses of poly(MA-CDA)s were 1 mg/ml, 100 μg/ml, and 10 μg/ml. The doses of the LPS were 1 ng/ml and 100 pg/ml.

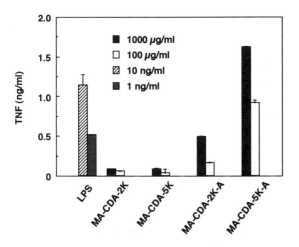

Figure 13 TNF inducing activity of poly(MA-CDA)s from human peripheral whole blood cell culture. The doses of poly(MA-CDA)s were 1 mg/ml and 100 μg/ml. The doses of the LPS were 10 and 1 ng/ml.

and without treatment of the gel. The activity was the same for each, which strongly suggests that there was no LPS contamination in the polymer.

From these facts, we concluded that the cytokines from peripheral whole blood cells were induced by the action of the synthetic polycarboxylic polymer itself. Hydrophobization may contribute to the higher affinity of the polymer to the target cells in human peripheral whole blood. The effect of molecular weight is currently unclear. However, it may relate to the action mechanism inside the cell after the polymer is incorporated into the cell. It is obvious from our data shown in Figs. 5 and 6 that the hydrophobized polymers having similar contents of modified groups possess similar affinity for the liposomal membranes, even though the molecular weights were different. The polyanionic polymer having an appropriate molecular weight (5,000) acts as a more effective immuno-stimulator to the organ or enzyme in the cell contributing to the cytokine induction than the polymer having different molecular weights.

V. CONCLUSION

The affinity of polyanionic polymers for cell membranes could be increased and controlled by the simple hydrophobization through grafting of the hydrophobic group. Thus, the biological activity of these polymers could be enhanced as demonstrated by in vitro tests using cultured cell lines and ex vivo test using peripheral whole blood cells. Especially from the last ex vivo test, it was clear that the hydrophobized polymer, having both appropriate affinity for cell membrane and molecular weight, possessed higher biological immuno-stimulating activity. Additional experiments to clarify the action mechanism of poly(MA-CDA) in the cells and to test the anticancer activity in vivo using animals are planned. Our current data will result in a new field of polymer drugs.

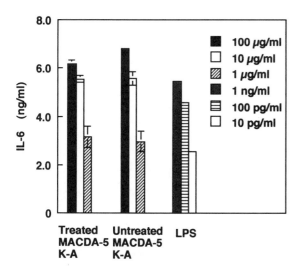

Figure 14 IL-6 inducing activity of Kuttu-clean treated and untreated hydrophobized poly(MA-CDA)s from human peripheral whole blood cell culture. The doses of poly(MA-CDA)s were 100, 10, and 1 μg/ml. The doses of the LPS were 1 ng/ml, 100 pg/ml, and 10 pg/ml.

REFERENCES

1. T. C. Merigan, *Nature 214*: 416 (1967).
2. W. Regelson, A. Munson, and W. Wooles, "Interferon and Interferon Inducers," *International Symposium Standards, London, 1969,* Symposium Series Immunobiological Standards, S. Kager, Basel, pp. 14, 227 (1978).
3. L. G. Donaruma, R. M. Ottenbrite, and O. Vogel, (eds.), *Anionic Polymeric Drugs,* John Wiley & Sons, NY (1980).
4. R. M. Ottenbrite and G. B. Butler, (eds.) *Anticancer and Interferon Agents,* Marcel Dekker, Inc., NY (1984).
5. M. Akashi, H. Iwasaki, N. Miyauchi, T. Sato, J. Sunamoto, and K. Takemoto, *J. Bioact. Compat. Polymers 4*: 124 (1989).
6. P. S. Bey, R. M. Ottenbrite, and R. R. Mills, *J. Bioact. Compat. Polymers 2*: 312 (1987).
7. M. J. Han, K. B. Choi, J. P. Chae, and B. S. Hahn, *J. Bioact. Compat. Polymers 5*: 80 (1990).
8. T. Espevik, M. Otterlei, G. Skjak-Braek, L. Ryan, S. D. Wright, and A. Sundan, *Eur. J. Immunol. 23*: 255 (1993).
9. P. Machy and L. Leserman, *Liposomes in Cell Biology and Pharmacology,* John Libbey Eurotext, London and Paris (1987).
10. J. N. Weinstein and L. D. Leserman, *Pharmacol. Therapeut. 24*: 207 (1984).
11. T. Sato, K. Kojima, T. Ihda, J. Sunamoto, and R. M. Ottenbrite, *J. Bioact. Compat. Polymers 1*: 448 (1986).

12. E. Th. Rietschel, L. Brade, B. Lindner and U. Zähringer. *Bacterial Endotoxic Lipopolysaccharides. Vol. I. Molecular Biochemistry and Cellular Biology* (D. C. Morrison, and J. L. Ryan, eds.) CRC Press, Boca Raton, Florida, p. 3–41 (1992).

13. Y. Suda, H. Yamamoto, M. Sumi, N. Oku, F. Ito, S. Yamashita, T. Nadai, and R. M. Ottenbrite, *J. Bioact. Compat. Polymers, 7*: 15 (1992).

14. J. T. Doluisio, N. F. Billups, L. W. Dittert, E. T. Sugita, and J. V. Swintosky, *J. Pharma. Sci., 58*: 1196 (1969).

15. N. Oku, N. Yamaguchi, N. Yamaguchi, S. Shibamoto, F. Ito, and M. Nango, *J. Biochem. 100*: 935 (1986).

16. Y. Suda, S. Kusumoto, N. Oku, H. Yamamoto, M. Sumi, F. Ito, and R. M. Ottenbrite, *J. Bioact. Compat. Polymers, 7*: 275 (1992).

17. A. M. Kaplan and R. M. Ottenbrite, *Ann. NY Acad. Sci. 169*: (1986).

18. S. J. Collins, R. C. Gallo, and R. E. Gallagher, *Nature 270*: 347 (1977).

19. J. Nath, A. Powledge, and D. G. Wright, *J. Biol. Chem. 264*: 848 (1989).

20. S. J. Collins, R. W. Ruscetti, R. E. Gallagher, and R. C. Gallo, *Proc. Natl. Acad. Sci. USA. 75*: 2458 (1978).

21. B. M. Babior, R. S. Kipnes, and J. J. Curnutte, *J. Clin. Invest. 52*: 945 (1973).

22. B. Meier, H. H. Radeke, S. Selle, M. Younes, H. Sies, K. Resch, and G. G. Habermehl, *Biochem. J. 263*: 539 (1989).

23. P. Raiph, J. Prichard, and M. Cohn, *J. Immuno. 114*: 898 (1975).

24. H. Sumitomo, K. Takeshige, and S. Mizukami, *ENSHOU 5*: 89 (1985).

25. Y. Suda, M. Hashimoto, J. Yasuoka, S. Kusumoto, N. Oku, and R. M. Ottenbrite, *J. Bioact. Compat. Polym. 11*: 100 (1996).

26. Y. Suda, H. Tochio, K. Kawano, H. Takada, T. Yoshida, S. Kotani, and S. Kusumoto, *FEMS Immunology and Medical Microbiology 12*: 97 (1995).

15

Uniform Latex Particles

Ali Tuncel and Huseyin Çiçek
Hacettepe University, Ankara, Turkey

I. INTRODUCTION

Various novel applications in biotechnology, biomedical engineering, information industry, and microelectronics involve the use of polymeric microspheres with controlled size and surface properties [1–3]. Traditionally, the polymer microspheres larger than 100 μm with a certain size distribution have been produced by the suspension polymerization process, where the monomer droplets are broken into micron-size in the existence of a stabilizer and are subsequently polymerized within a continuous medium by using an oil-soluble initiator. Suspension polymerization is usually preferred for the production of polymeric particles in the size range of 50–1000 μm. But, there is a wide size distribution in the product due to the inherent size distribution of the mechanical homogenization and due to the coalescence problem. The size distribution is measured with the standard deviation or the coefficient of variation (CV) and the suspension polymerization provides polymeric microspheres with CVs varying from 15–30%.

Typically, polymer latices containing uniform polymeric particles in submicron-size range are produced by emulsion polymerization. The first studies relating to emulsion polymerization were started at the beginning of the 1900s. A German patent was issued on emulsion polymerization in 1909 [4]. In 1927, a procedure was patented by I. G. Farben Company on the emulsion polymerization of butadiene leading to a synthetic latex with the use of soaps as emulsifiers and hydrogen peroxide as the initiator [5]. But, the product of emulsion polymerization, latex, was first observed in 1947 by using electron microscopy. The researchers at the Dow Chemical Company discovered that the latex product comprised the excellent spherical particles of identical diameters in submicron size [6]. Today, the emulsion process is usually used for the production of uniform polymeric microspheres in the size range of 0.05–1.5 μm.

A new process, from Norway, has filled the size gap between emulsion and suspension polymerization techniques [7,8]. This novel polymerization method, the so-called swollen emulsion polymerization has been developed by Ugelstad for producing uniform polymeric particles in the size range of 2–100 μm. This process comprises successive swelling steps and repolymerizations for increasing the particle size of seed polymer particles by keeping the monodispersity of the seed latex.

In the early 1970s, a novel method, dispersion polymerization was proposed as a single-step method for the production of large uniform latex particles in the micron-size range [9]. In this process, the reaction mixture starts out as a homogeneous solution containing dissolved monomer in an inert medium and resulting polymer precipitates as solid spherical particles, stabilized by a steric barrier of dissolved polymer. The uniform latex particles have also been produced on shuttle flights in space starting in the early 1980s. Emulsion polymerization runs in space provided large uniform latex particles up to 15 μm in size.

In this chapter, the polymerization methods used for the production of uniform latex particles in the size range of 0.1–100 μm are described. Emulsion, swollen emulsion, and dispersion polymerization techniques and their modified forms for producing plain, functionalized, or porous uniform latex particles are reviewed. The general mechanisms and the kinetics of the polymerization methods, the developed synthesis procedures, the effect of process variables, and the product properties are discussed.

II. EMULSION POLYMERIZATION

Most of the currently available uniform latex particles are produced by emulsion polymerization, which is best suited to submicron (<1 μm) particles. The emulsion process can be extended to the average size value of 1.5 μm. The CVs of the latex particles produced with emulsion polymerization are typically around 1%. In the emulsion polymerization process, the polymerization reaction can be conducted within a wide temperature range (0–80°C), depending on the selected initiation system. The heat transfer and agitation in the polymerization medium are easier relative to the other polymerization methods (i.e., solution or suspension polymerizations). Emulsion polymerization is a very suitable process for continuous operation. The liquid form of the reactants and product allows transfer by means of pumps and pipelines. The use of expensive solvent and solvent recovery problems are eliminated since water is the main component of the emulsion medium. The polymeric product (i.e., latex) having high-molecular weights can be produced with high polymerization rates. The viscosity of latex is independent of the molecular weight of the polymer. It also possible to change the form of the product by converting the liquid latex dispersion into a solid product by means of filtering and drying processes.

A. General Mechanism

The mechanism of emulsion polymerization was explained by the definition of an ideal emulsion polymerization system [10]. The main constituents of an ideal emulsion polymerization system are the water-insoluble monomer, the free radical initiator, the emulsifier, and water. When certain chemical compounds, called surface active agents or emulsifiers including either polar or apolar groups, are dissolved within water in concentrations above a critical level aggregates form. These aggregates, called micelles, contain 100 or so emulsifier molecules. In these roughly spherical structures, the hydrophobic portion of each molecule is directed toward the center of the micelle, while the hydrophilic or polar portion is placed at the outer side, which is adjacent to water. The micelles are in colloidal dimensions and about 10^{16}–10^{18} micelles/ml of the aqueous phase at the emulsifier concentrations usually used in the emulsion polymerization. When a hydrophobic monomer-like styrene is mixed with an aqueous medium containing an emulsifier, a small amount of monomer (around 1%) enters into the micelle structure and the dimension of the micelles swollen by the monomer become about two-fold of their original dimension. However, an insignificant amount of hydrophobic monomer is solubilized within the water phase. Therefore, prior to the initiation step, most of the monomer is found in the form of droplets in an emulsion medium. The typical diameter of

monomer droplets is around 1 μm. In a conclusion, an ideal emulsion polymerization system prior to the initiation step includes: (1) an external water phase, (2) monomer droplets dispersed within the aqueous phase, (3) monomer swollen emulsifier micelles.

Water-soluble free radical initiators (i.e., potassium persulfate, $K_2S_2O_8$) are used in the emulsion polymerization process. Upon heating, the persulfate ion decomposes into two sulfate ion free radicals according to the following reaction:

$$S_2O_8 \rightarrow 2SO_4^{-}*$$

In the ideal system, the primary radicals are generated in the water phase. In the Smith–Ewart theory, it is assumed that the free radicals generated within the aqueous phase immediately enter monomer swollen micelles and initiate the polymerization of monomer within these structures [10]. But, the adsorption of these usually ionic and highly water-soluble radicals by the monomer swollen micelles having a relatively apolar structure is reasonably difficult. This point was emphasized by different researchers in recent years [9,11,12]. According to the proposed mechanism by these researchers, the primary radicals interact first with the solubilized monomer molecules within the water phase to form relatively apolar and low-molecular weight oligomer units. Therefore, the oligomeric radicals produced within the water phase can be adsorbed by the monomer swollen micelles due to their apolar characters relative to the primary ones. The existence and the molecular weights of these oligomers formed from primary ionic radicals were analyzed by using gel permeation chromatography (GPC) techniques [13–15]. These analyses showed that the degree of initial polymerization within the water phase changed from 1 to 60 depending upon the conditions of emulsion medium.

The progression of an ideal emulsion polymerization is considered in three different intervals after forming primary radicals and low-molecular weight oligomers within the water phase. In the first stage (Interval I), the polymerization progresses within the micelle structure. The oligomeric radicals react with the individual monomer molecules within the micelles to form short polymer chains with an ion radical on one end. This leads to the formation of a new phase (i.e., polymer latex particles swollen with the monomer) in the polymerization medium.

In the next stage (Interval II), the polymerization inside the micelle structure proceeds rapidly and the latex particles begin to grow. The emulsifier is adsorbed onto the growing surface of the latex particles. Thus, the forming particles are stabilized by emulsifier molecules on their surfaces. The presence of emulsifier molecules on the surface of forming particles prevents the flocculation of latex particles. The emulsifier concentration in the water phase decreases continually by the progressing adsorption process. This causes a change in the balance between the solubilized emulsifier molecules and those within the micelles. Therefore, the unacti-vated micelles begin to disintegrate by the progressing adsorption to restore this balance. When the emulsifier concentration decreases below the critical micelle concentration value, the unactivated micelles disappear completely. This occurs after 10–20% of monomer conversion. In Interval II, the polymerization rate increases with the increasing monomer conversion due to the increase taking place in the formation of oligomeric radicals. After this interval, no new particle formation occurs since the initiation takes place only within the swollen micelle structure. Therefore, the number of latex particles present in the medium is fixed, unable to disappear from the micelle structure.

In the next stage (Interval III), the polymerization progresses within the monomer swollen latex particles in the presence of monomer diffusion through the aqueous phase from the monomer droplets to the forming latex particles. The monomer concentration within the forming particles is kept constant by means of the monomer diffusion. Therefore, the polymerization rate is constant during this step and does not change with the increasing monomer conversion. At about 60% conversion, all of the monomers in the water phase (i.e., in the form of droplets) are transferred into the forming particles, then the monomer droplets in the aqueous medium disappear. After this point, the monomer within the forming particles is consumed by the polymerization reaction and the monomer concentration within the forming particles decreases. This results in a significant decrease in the polymerization rate. This step is termed as the polymerization of the monomer within the monomer swollen latex particles. The last stage is the termination of the polymerization reaction in which all the monomer is reacted within the latex particles. At this stage, the ion-monomer free radical reacts with the another free radical to terminate the reaction. Therefore, the polymerization is terminated when each polymer chain has an ionic end on both ends. The variation of polymerization rate with the monomer conversion at the different steps of ideal emulsion polymerization process is shown in Fig. 1.

The stable aqueous dispersion of uniform polymeric microspheres in submicron size obtained as the product of emulsion polymerization is called "latex." The term latex is used because the process evolved from synthetic rubber production and the emulsion has a milky appearance. A typical latex particle is composed of a large number of polymer chains. Each chain has a molecular weight in the range of about 10^5–10^7. Based on the rearrangement of polymer chains within the particle, the latex particles can be amorphous, crystalline, rubbery, or glassy. The colloidal behavior of the latex is determined by its surface properties, which are determined by the polymerization conditions. A typical latex particle includes: (1) charged groups coming from the initiator, and (2) physically adsorbed or chemically grafted stabilizer molecules on its surface. The surface charge of the

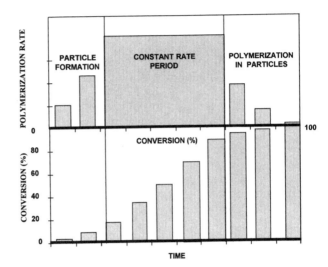

Figure 1 The typical tendencies for the variation of monomer conversion by the polymerization time and for the variation of polymerization rate by the monomer conversion in the ideal emulsion polymerization process.

latex particles usually arises from the negatively charged acidic sulfate groups formed by the decomposition of the initiator. The charge density of the latex particles can be determined after removal of the physically adsorbed emulsifier molecules and water-soluble ions by ion exchange treatment. The potentiometric or conductometric titration methods can be used for the determination of charge density (CD_w) expressed in the equivalents of negatively charged groups per unit weight of particles by the following expression [16]:

$$CD_w = (C_B \times V_B)/W_p \qquad (1)$$

In this expression, W_p is the weight of particles titrated (g), C_B and V_B are the concentration of the base (i.e., titrant) and the volume of the base at equivalence point. The surface charge density can be calculated for the particles having a known diameter by means of the following expression:

$$CD_s = 1.004 \times CD_w \times \rho_b \times d \qquad (2)$$

Where CD_s is the surface charge density (number of charged groups/A^2), ρ_b is the polymer bulk density (g/cm^3), and d is the particle diameter (μm). In the aqueous medium, the surface charge of the latex particle is balanced with the counterions having opposite charge within the solution. The electro-neutrality obtained by this balance is called as electrical double layer in which an equilibrium is established between electro-static forces and diffusion forces [17]. The latex particle surface has an electro-static surface potential, which can be positive or negative depending upon the charged groups on the particle surface. A typical scanning electron micrograph of the submicron size, uniform polystyrene microspheres produced by the emulsion polymerization of styrene is given in Fig. 2.

In this characterization, the following properties of uniform latex particles should be considered: (1) total solid content of the latex, (2) specific gravity of the latex, (3) pH of the latex, (4) viscosity of the latex, (5) average size and size distribution of the latex particles, (6) charge density of the latex particles, (7) average molecular weight and molecular weight distribution of the latex particles, (8) bulk and surface chemistry of the latex particles, (9) crystallinity and glass transition temperature of the latex particles, (10) crosslinking density of the latex particles, and (11) porosity of the latex particles.

B. Kinetic Theory of Emulsion Polymerization

The kinetic mechanism of emulsion polymerization was developed by Smith and Ewart [10]. The quantitative treatment of this mechanism was made by using Harkin's Micellar Theory [18,19]. By means of quantitative treatment, the researchers obtained an expression in which the particle number was expressed as a function of emulsifier concentration, initiation, and polymerization rates. This expression was derived for the systems including the monomers with low water solubility and partly solubilized within the micelles formed by emulsifiers having low critical micelle concentration (CMC) values [10].

An emulsion polymerization reaction follows three conventional steps, namely, initiation, propagation, and termination. These steps can be described by the conventional reactions that are valid for any free radical polymerization. Smith and Ewart [10] proposed that a forming latex particle in an ideal emulsion polymeriza-

tion system contained either one or no free growing radical. It was assumed that a mutual combination occurred instantaneously when a free radical diffused into a forming particle including a growing free radical since the polymerization volume was very small. Therefore, the average number of radicals per forming particle is equal to one-half. The number of particles (N) in an emulsion polymerization system is given by Eq. (3). In this expression, ρ is the radical production rate, μ is the rate of volume increase of a particle (i.e., the ordinary derivative of particle volume with respect to time and which is assumed to be constant), a_E is the area occupied by one emulsifier molecule, $[E]$ is the concentration of emulsifier, and F is a constant between 0.37 and 0.53 [10].

$$N = F(\rho/\mu)^{2/5} \times (a_E [E])^{3/5} \qquad (3)$$

The rate of an ideal emulsion polymerization is given by Eqn (4). In this expression $[I]$ is the initiator concentration, $[E]$ is the emulsifier concentration, and $[M]$ is the concentration of monomer within the forming latex particles. This value is constant for a long reaction period until all the monomer droplets disappear within the water phase.

$$R_P = K[I]^{2/5}[E]^{3/5}[M] \qquad (4)$$

As a conclusion, the basic kinetic features of the emulsion polymerization system may be summarized as follows:

1. The rate of termination reaction is slower than that observed in the homogenous bulk or solution polymerization since the limited number of free radicals exists in the polymerization loci having a reasonably small volume (i.e., monomer swollen forming latex particle). Higher degree of polymerizations can be achieved in an emulsion system relative to the homogenous polymerization due to the existence of this limitation.

2. The monomer concentration within the forming latex particles does not change for a long period due to the diffusion of monomer from the droplets to the polymerization loci. Therefore, the rate of the propagation reaction does not change and a constant polymerization rate period is observed in a typical emulsion polymerization system.

3. The overall polymerization rate depends on the number of forming particles within the emulsion medium.

The Smith–Ewart theory has been modified by several researchers [13,20–24]. These researchers argued against the Smith–Ewart theory that (1) the particle formation also occurs in the absence of micellar structure, (2) the predictions on particle number with the Smith–Ewart theory are higher relative to actual case,

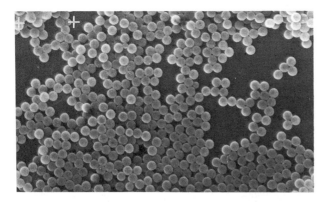

Figure 2 A typical scanning electron micrograph of the submicron size (0.2 μm), uniform polystyrene microspheres produced by the emulsion polymerization of styrene. Magnification: 20,000×.

(3) more polar monomers do not follow the theory, (4) the maximum polymerization rate at the end of nucleation period in the theory is not usually observed in experimental systems [25].

In most of the studies on the kinetics of emulsion polymerization, the role of monomer droplets in the particle formation step is usually neglected. This approach usually arises from the smaller surface area of the monomer droplets relative to that of monomer swollen micelles and the uniform character of the final particles. After development of the Smith–Ewart model, the detailed investigations on particle formation step indicated that the particle formation can occur by different mechanisms rather than that proposed in the Smith–Ewart model. Based on the results of these investigations, a growing radical within continuous phase can be adsorbed by the existing forming particles, or can enter into the monomer droplets, or can precipitate within the continuous phase to form a new stable particle. Fitch and Tsai [13] proposed the ''homogeneous nucleation mechanism'' involving the formation of primary particles from the growing radicals within the continuous medium. According to the proposed model, the hydrophobicity of growing radicals increases with the increasing chain length. Therefore, the growing radicals precipitate in the continuous medium when they reach a certain molecular weight to form primary particles. The interaction of growing radicals with the micellar structure is not taken into account in the homogenous nucleation mechanism. Therefore, in a real system, the formation of primary particles by the interaction between the micellar structure and the growing radicals accompanies this mechanism. However, Ugelstad and coworkers [26,27] clearly demonstrated that even the monomer droplets can compete with the monomer swollen micelles or with the forming particles for capturing of radicals from the aqueous phase when they can be made sufficiently small in size.

Various kinetic models on particle formation were proposed by different researchers. These may be classified as follows: (1) radical absorption mechanisms by Gardon [28–34] and Fisch and Tsai [13], (2) micellar nucleation: newer models by Nomura et al. [35,36] and by Hansen and Ugelstad [37], (3) homogeneous nucleation by Fistch and coworkers [13,38,39].

C. Polymerization Procedure and Equipment

1. Typical Procedure

Following is a typical procedure for the emulsion polymerization of hydrophobic monomers in batch fashion. The first step in this process is the preparation of the continuous phase solution. As mentioned before, water is usually selected as the base material of the continuous phase for the emulsion polymerization of hydrophobic monomers. The continuous phase is prepared by dissolving a certain amount of emulsifier in a concentration that is slightly higher than the CMC value. After charging the emulsifier solution into the reactor, the proper amount of monomer is added and mixed with the continuous phase. The monomer-to-water volumetric ratio is kept at about one-half or one-third in conventional emulsion polymerization. This ratio can be extended to 1:1 for the preparation of high solid latexes. The stirring and heating were started for the emulsification of the monomer within the water phase. Low stirring rates in the range of 50–250 rpm are usually preferred. When the reactor is attained to the polymerization temperature, the stirring of emulsion medium is continued for a certain period at the polymerization temperature for complete emulsification of the monomer. Thus, the monomer is distributed within the emulsion medium in the form of fine droplets. A purge operation may be applied at this step by passing nitrogen through the emulsion medium for a certain period. This purge operation may also be applied during the heating period before initiation of the polymerization. In some cases, nitrogen is passed through the polymerization medium continuously for the whole polymerization period. After completion of the emulsification step, the polymerization is started by the injection of the initiator solution, including the dissolved initiator within a small volume of water, into the polymerization medium.

The polymerization reaction is conducted at the desired temperature with a slow stirring regime for a certain period. A typical recipe for the emulsion polymerization of styrene is exemplified in Table 1 [40]. As seen here, potassium persulfate and sodium dodecyl sulfate were used as the initiator and the stabilizer, respectively. This recipe provides uniform polystyrene particles 0.22 μm in size.

2. Polymerization Equipment

The emulsion polymerization process is usually carried out within batch stirred reactors. The polymerization method is also suitable for the continuous operation of the reactor. The schematical representation of a typical laboratory-scale stirred reactor system for emulsion polymerization is given in Fig. 3 [41]. Sealed reactors are usually preferred since the process is usually conducted under nitrogen atmosphere. The nitrogen is supplied from a pressurized cylinder and sent to the reactor with a certain flow rate prior to and during the polymerization. In some cases, the reactor is purged with nitrogen for only a certain period before the initiation of polymerization to remove the dissolved oxygen through the continuous medium. Then, the reactor is sealed and the nitrogen flow is stopped after adding of the initiator. A heating jacket is located around the reactor. Temperature control of the polymerization medium is achieved by circulating a hot fluid (usually water) through the jacket around the reactor. The heating fluid is withdrawn

Table 1   A Typical Recipe for the Emulsion
Polymerization of Styrene [40]

Ingredients	Amount
Styrene (ml):	75
Distilled water (ml):	225
Potassium peroxydisulfate (g):	0.30 g in 10 ml of distilled water
Sodium dodecyl sulfate (g):	0.80
Nitrogen purge	
Time:	After the addition of initiator solution, for 10 minutes
Flow rate:	100 ml/min
Polymerization conditions	
Heating time to the polymerization temperature (min):	40
Stirring at the polymerization temperature for complete emulsification (min):	20
Polymerization temperature (°C):	70
Polymerization time (h):	3
Agitation Rate (rpm):	200

Reactor properties

Glass, jacketed and cylindrical, sealed reactor (56 mm in diameter and 200 mm in height) equipped with a blade type stirrer in 30-mm length and 10-mm width.

from a thermostatic bath in which the temperature of the heating fluid was controlled according to the desired reactor temperature. The total pressure of the polymerization medium is a measure of monomer conversion and can be monitored by a sensitive pressure gauge. The reactor should include proper inlets for feeding of the continuous medium, monomer, and initiator solutions and proper outlets. Zero valves are usually preferred to prevent plugging during the discharge of the latex. The use of a blade type stirrer is usually enough to agitate the polymerization medium with relatively slow agitation rates. The size of the final particles is not affected by the agitation rate in the emulsion polymerization. Relatively slow agitation rates (100–250 rpm), which are sufficient to obtain a homogenous temperature distribution within the reactor and to prevent the coalescence of the individual monomer droplets, are selected. The surface of the reactor coming into contact with the polymerization medium should be reasonably smooth. The formation of coagulum can occur on the reactor walls when reactors with relatively rough inner surfaces are used. The laboratory-scale reactors are usually made of glass. In industrial scale reactors, using polished stainless steel as an inner surface or placing a glass liner within a metal reactor can solve this problem. The sealed glass bottles placed in a shaking water bath can also be used as the reactors in laboratory-scale emulsion polymerizations.

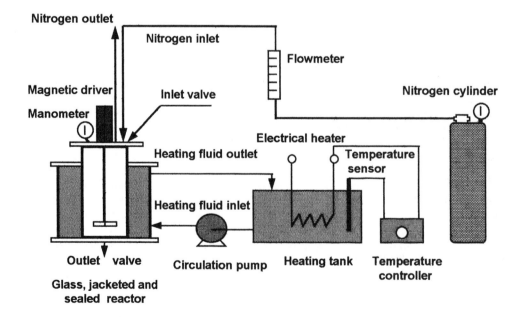

Figure 3   The schematical representation of a typical laboratory scale stirred reactor system for emulsion polymerization.

D. Effect of Process Variables

1. Initiator

Water-soluble initiators that can generate active free radicals are used in emulsion polymerization. The generation of active free radical can occur by two different mechanisms: (1) thermal decomposition, and (2) chemical interaction.

Thermal Initiators

Water-soluble peroxydisulfates, hydroperoxides, and hydrogen peroxide are the most widely used initiators in the emulsion polymerization. The overall polymerization rate increases with the increasing initiator concentration since the free radical production rate increases. The effect of initiator on the polymerization rate is exemplified in Fig. 4 for the emulsion polymerization of styrene. The average molecular weight of the latex particles decreases and the average particle size increases with the increasing initiator concentration. As a result of the interaction between the monomer molecules and the sulfate free radicals formed by the thermal decomposition of the initiator, the sulfate groups are covalently bound to the polymer chains and are located on the surface of latex particles. The presence of sulfate groups on the surface provides a negative charge to the particles. The schematical view of the latex particles obtained as the product in the emulsion polymerization is given in Fig. 5.

Kolthoff et al. [42] showed that the thermal decomposition of peroxydisulfate ion yielded two free sulfate radicals according to the following reaction:

$$S_2O_8{}^{2-} \rightarrow 2SO_4^{\bar{\cdot}}$$

In some cases, due to the highly polar character of the sulfate radicals, peroxydisulfate initiators can provide slow polymerization rates with some apolar monomers since the polar sulfate radicals cannot easily penetrate into the swollen micelle structures containing apolar monomers. The use of mercaptans together with the peroxydisulfate type initiators is another method to obtain higher polymerization rates [43]. The mercaptyl radicals are more apolar relative to the free sulfate radicals and can easily interact with the apolar monomers to provide higher polymerization rates.

Redox Initiators

Since the peroxydic compounds are strongly oxidizing agents, these compounds can generate free radicals in the existence of a reducing agent due to the oxidation–reduction reaction occurring between these two compounds. The initiator systems containing an oxidizing agent and a reducing agent are termed redox initiators. The redox system was reviewed by Warson [44]. These systems are usually used at low temperatures (i.e., between 4–25°C) in the emulsion polymerization processes. Sodium metabisulphite, Fe^{2+} complexes, diamines (i.e., ethylene diamine, tetramethylethylene di-

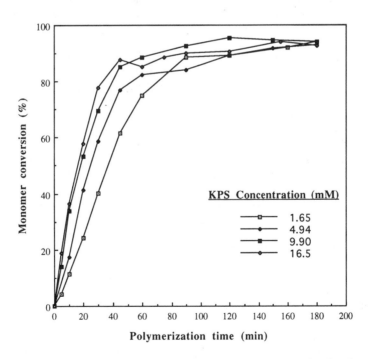

Figure 4 The effect of initiator concentration on the variation of monomer conversion by the polymerization time in the emulsion polymerization of styrene. Styrene–water = 1/3; SDS = 15.4 mM; reaction volume = 300 ml; stirring rate = 250 rpm; temperature = 70°C.

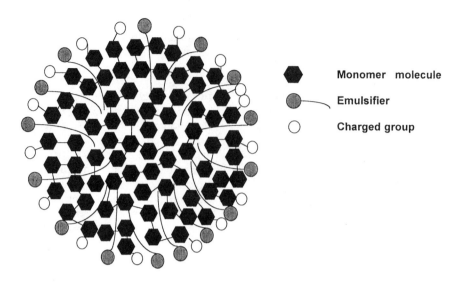

Figure 5 The schematical view of the latex particles obtained as the product of emulsion polymerization.

amine) are the most common reducing agents that can be used with the peroxydic compounds to generate free radicals. For example, a typical redox system including a free radical forming agent, ammonium persulfate, and a reducing agent, sodium metabisulphide, was used in the emulsion polymerization of vinyl acetate carried out in the aqueous medium including copper sulfate and buffered with sodium bicarbonate. The polymerization reaction was initiated at 25°C and continued in nonisothermal conditions [44].

The following properties should be considered in the selection of the proper redox initiator system: (1) the generated radicals should interact chemically with the monomer molecules for producing oligomeric radicals at the selected polymerization temperature, (2) a chemical interaction between the constituents of the selected redox system and the emulsifier should not occur in the emulsion polymerization conditions, (3) the generated radicals should diffuse into the swollen micelle structure and interact with the monomer molecules with a proper rate. Therefore, the polarity of the generated radicals should be suitable to the polarity of the monomer molecules. In the existence of a large polarity difference, the generated radicals cannot easily diffuse into the monomer swollen micelle structure and the polymerization progresses very slowly. However, when the polarity of the generated radicals and monomer is very close, more radicals diffuse into the micelle structure, and this leads to a decrease in the degree of polymerization. Thus, the latices with high-molecular weights cannot be achieved. Organic hydroperoxides are usually preferred as the oxidizing agents of redox systems in the emulsion polymerizations of relatively apolar monomers like butadiene and styrene. Relatively polar monomers (i.e., acryl-

amide, acrylonitrile, 2-hydroxyethylmethacrylate, or methylmethacrylate) can be initiated with the inorganic peroxydic compounds as the oxidizing agents, and (4) the residues forming in the decomposition reaction should not interact with the polymer particles after production. When the color of the product is important, the initiator systems, which can release colored metallic contaminants, should not be used.

2. Emulsifier

The function of emulsifier in the emulsion polymerization process may be summarized as follows [45]: (1) the insolubilized part of the monomer is dispersed and stabilized within the water phase in the form of fine droplets, (2) a part of monomer is taken into the micel structure by solubilization, (3) the forming latex particles are protected from the coagulation by the adsorption of monomer onto the surface of the particles, (4) the emulsifier makes it easier the solubilize the oligomeric chains within the micelles, (5) the emulsifier catalyzes the initiation reaction, and (6) it may act as a transfer agent or retarder leading to chemical binding of emulsifier molecules to the polymer.

One of the most important characteristics of the emulsifier is its CMC, which is defined as the critical concentration value below which no micelle formation occurs. The critical micelle concentration of an emulsifier is determined by the structure and the number of hydrophilic and hydrophobic groups included in the emulsifier molecule. The hydrophile-lipophile balance (HLB) number is a good criterion for the selection of proper emulsifier. The HLB scale was developed by W. C. Griffin [46,47]. Based on his approach, the HLB number of an emulsifier can be calculated by dividing the percentage of ethylene oxide in the emulsifier by 5 [48]. Davies [49] developed some structural parameters for the determination of the HLB number for a given emulsifier. Greth and Wilson [50] applied the HLB scale on the selection of emulsifiers for the emulsion polymerizations of styrene and vinyl acetate. The best stability and the highest polymerization rate for polystyrene latexes were obtained with the emulsifiers having HLB numbers between 13–16.

Based on the Smith–Ewart theory, the number of latex particles formed and the rate of polymerization in Interval II is proportional with the 0.6 power of the emulsifier concentration. This relation was also observed experimentally for the emulsion polymerization of styrene by Bartholome et al. [51]. Dunn and Al-Shahib [52] demonstrated that when the concentrations of the different emulsifiers were selected so that the micellar concentrations were equal, the same number of particles having the same size could be obtained by the same polymerization rates in Interval II in the existence of different emulsifiers [52]. The number of micelles formed initially in the polymerization medium increases with the increasing emulsifier concentration. This leads to an increase in the total amount of monomer solubilized by micelles. However, the number of emulsifier molecules in one micelle is constant for a certain type of emulsifier and does not change with the emulsifier concentration. The monomer is distributed into more micelles and thus, the amount of monomer used by each forming particle decreases. This leads to a decrease in the average size of the final product. The polymerization progress in the existence of more particles formed depends on the number of micelles existing initially. Therefore, the overall polymerization rate increases with the increasing emulsifier concentration since the polymerization rate is directly proportional to the number of forming particles as described in the basic kinetic relationship of the ideal emulsion polymerization. A typical graph showing the variation of monomer conversion by the polymerization time at different emulsifier concentrations is given for the emulsion polymerization of styrene in Fig. 6.

The Smith–Ewart theory was developed by referring to the ionic emulsifiers. The studies performed by Maron et al. [53] and Piirma and Chen [54] indicated that a_E for anionic emulsifiers decreased with the increasing alkyl chain length of the emulsifier and increased with the increasing temperature. The dependency of a_E on the polarity of the polymer surface was shown by Piirma and Chen [54] and Paxton [55], although the variation of a_E with the type of polymer was ignored in the early studies [56]. Therefore, the value of a_E should be determined or selected carefully in the light of these findings for the estimation of particle number and polymerization rate. It should be noted that the effects of these findings are not taken into account in the Smith–Ewart model. Ionic emulsifiers are the most widely used ones in the

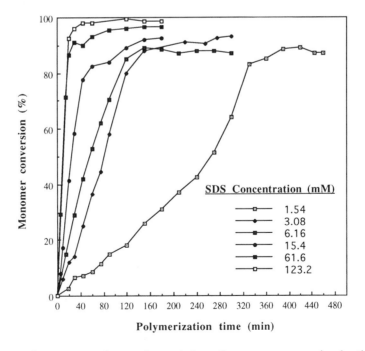

Figure 6 The effect of emulsifier concentration on the variation of monomer conversion by the polymerization time in the emulsion polymerization of styrene. Styrene–water = 1/3; KPS = 1.65 mM; reaction volume = 300 ml; stirring rate = 250 rpm; temperature = 70°C.

emulsion polymerization studies. The types and stability ranges of these emulsifiers are given in Table 2.

Medvedev et al. [57] extensively studied the use of nonionic emulsifiers in emulsion polymerization. The emulsion polymerizations in the presence of nonionic emulsifiers exhibited some differences relative to those carried out with the ionic ones. Medvedev et al. [57] proposed that the size of latex particles remained constant during the reaction period, but their number increased continually with the increasing monomer conversion. The use of nonionic emulsifiers in emulsion polymerization usually results in larger sizes relative to those obtained by the ionic emulsifiers. It is possible to reach a final size value of 250 nm by the use of nonionic emulsifiers in the emulsion polymerization of styrene [58].

3. Monomer

The free radical initiators are more suitable for the monomers having electron-withdrawing substituents directed to the ethylene nucleus. The monomers having electron-supplying groups can be polymerized better with the ionic initiators. The water solubility of the monomer is another important consideration. Highly water-soluble (relatively polar) monomers are not suitable for the emulsion polymerization process since most of the monomer polymerizes within the continuous medium. The detailed emulsion polymerization procedures for various monomers, including styrene [59–64], butadiene [61,63,64], vinyl acetate [62,64], vinyl chloride [62,64,65], alkyl acrylates [61–63,65], alkyl methacrylates [62,64], chloroprene [63], and isoprene [61,63] are available in the literature.

Table 2 Ionic Emulsifiers and Their Stability Ranges

Ionic emulsifier type	Stable latex	Unstable latex
Anionic emulsifiers		
Alkyl sulfates	From fairly acidic to alkaline values	—
Alkylarylsulphonates	From fairly acidic to alkaline values	—
Alkali salts of the carboxylic acids	Alkaline pH range	Acidic pH range
Cationic emulsifiers		
Quaternary ammonium salts, Salts of alkyl amines, Alkyl chain substituted cyclic amines	Acidic pH range	Alkaline pH range

The effect of monomer concentration on the emulsion polymerization process is strongly related to the type of monomer and emulsifier. The number of forming particles is mainly controlled by the emulsifier concentration according to the kinetic model of ideal emulsion polymerization. But, in practice, the monomer concentration is an effective parameter on the number of forming particles due to the following reasons. The number of monomer droplets in the aqueous phase increases with the increasing monomer concentration. The increased surface area of the droplets requires more emulsifier for the stabilization of them within the aqueous phase. This leads to a decrease in the amount of emulsifier that is used for the formation of micelles. The decrease occurring in the number of micelles decreases the number of forming particles. Therefore, the increase in the monomer concentration usually causes the formation of larger final beads since more monomer is used for fewer forming particles.

The increase in the monomer-to-water ratio causes an increase in the monomer concentration within the forming particles during the constant rate period. This is an effect involving an increase in the overall polymerization rate. But, as explained previously, the number of forming particles decreases with the increasing monomer concentration and this effect involves a decrease in the polymerization rate. The dominant one from these two opposite cases determines the apparent effect of monomer concentration on the polymerization rate. The effect of the monomer-to-water ratio on the polymerization rate for the emulsion polymerization of styrene is exemplified in Fig. 7. As seen here, the polymerization rate at the earlier times increased with the decreasing monomer content of the emulsion medium since the number of micelles increased with the decreasing monomer concentration. At the extended reaction times, the polymerization rate was controlled by either the number of forming particles or the styrene concentration within the forming particles. Therefore, the polymerization rates with lower initial styrene concentrations were relatively slow for the step in which the polymerization occurred within the forming particles since the monomer concentration within the particles was lower due to the existence of more particles in the polymerization medium. In addition to apolar (hydrophobic) monomers, relatively polar (hydrophilic) monomers can be polymerized within water by applying the emulsion polymerization procedure. Rembaum et al. [66,67] produced uniform latex microspheres in the size range of 0.03–0.34 μm by the emulsion copolymerization of hydrophilic monomers. In their method, the uniform copolymer particles carrying both the carboxyl and hydroxyl groups were produced by the emulsion copolymerization of methyl methacrylate, 2-hydroxyethyl methacrylate, and methacrylic acid monomers by using ammonium persulfate and sodium dodecyl sulfate as the initiator and the stabilizer, respectively, within the aqueous continuous

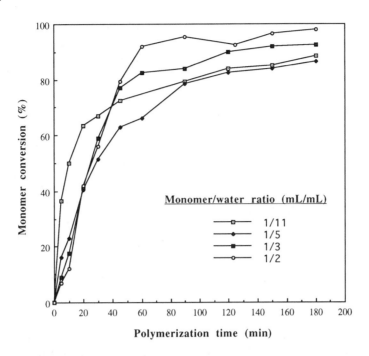

Figure 7 The effect of monomer-to-water ratio on the variation of monomer conversion by the polymerization time in the emulsion polymerization of styrene. KPS = 1.65 mM; SDS = 15.4 mM; reaction volume = 300 ml; stirring rate = 250 rpm; temperature = 70°C.

medium. Ethylene glycol dimethacrylate was used as the crosslinker and the synthesis of copolymer latices was carried out at 70°C for 1 h with quantitative conversion of the monomers. A typical recipe for the synthesis of uniform hydrophilic copolymer microspheres is given in Table 3. This recipe provided uniform hydrophilic copolymer microspheres 140 ± 11 nm in size.

4. Polymerization Temperature

The selection of the polymerization temperature for the emulsion polymerization system is strongly related to the initiation system. A polymerization temperature in which the initiator system exhibits its best performance should be selected.

Table 3 A Typical Recipe for the Synthesis of Uniform Hydrophilic Microspheres [66]

Ingredient	Weight percent (%)
2-Hydroxyethyl methacrylate	4.5
Methyl methacrylate	8.55
Methacrylic acid	1.5
Ethyleneglycol dimethacrylate	0.45
Sodium dodecyl sulfate	0.097
Ammonium persulfate	0.011

The effect of temperature on emulsion polymerization may be summarized as follows: (1) free radical production rate and the concentration of free radicals in the continuous medium increase with the increasing polymerization temperature, (2) the diffusion rate of free radicals into the forming particles increases with the increasing polymerization temperature, (3) the number of micelles increases with the increasing polymerization temperature. This leads to an increase in the number of forming particles, (4) the monomer diffusion through the aqueous phase to the forming particles increases with the increasing polymerization temperature. This leads to an increase in the monomer content of forming particles during polymerization.

All these effects increase the overall polymerization rate and decrease the degree of polymerization. The effect of polymerization temperature on the variation of monomer conversion with the polymerization time is exemplified in Fig. 8 for the emulsion polymerization of styrene.

E. Soapless Emulsion Polymerization

When the emulsion polymerization is conducted in the absence of an emulsifier, this process is termed emulsifier free or soapless emulsion polymerization [68–73]. In this case, the particle formation occurs by the precipitation of growing macroradicals within the continuous

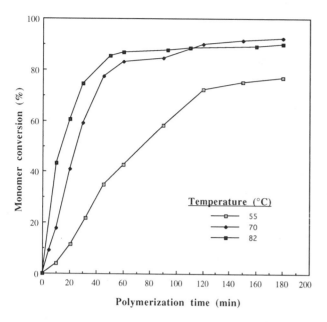

Figure 8 The effect of polymerization temperature on the variation of monomer conversion by the polymerization time in the emulsion polymerization of styrene. KPS = 1.65 mM; SDS = 15.4 mM; Styrene–water = 1/3, reaction volume = 300 ml; stirring rate = 250 rpm.

medium. The precipitation takes place when they reach a certain critical chain length, depending on the solvency of the continuous phase. The initially formed nuclei cannot be stabilized by the emulsifier molecules since no emulsifier is present. Therefore, the nuclei continue to collide to form larger particles. Note that the free sulfate radicals are formed by the decomposition of the peroxydic initiators and these radicals first interact with the monomer molecules before forming the growing polymer chains within the continuous phase. Due to this interaction, the precipitated polymer chains contain covalently attached and negatively charged sulfate groups. Therefore, the nucleated particles are stabilized by their own negative charges originated from the acidic sulfate groups in one end of the polymer chains. These hydrophilic groups are located on the surface of the growing particles and the number of sulfate groups in each particle increase with the increasing particle size. This leads to the better stabilization of particles by the progressing polymerization. However, Goodwin et al. [69] studied the stabilization of the positively charged latex particles by using amidinium initiators.

According to the other kinetic model proposed for the soapless emulsion process, the growing macroradicals may also form micelle structures at earlier polymerization times since they have both a hydrophilic end coming from the initiator and a hydrophobic chain [74].

Therefore, the polymerization progresses within the micelle structure by following the traditional mechanism of emulsion polymerization.

In the soapless emulsion process, the number of nuclei is lower, but the size of nucleated particles are larger than those obtained with conventional emulsion procedures. The emulsifier-free emulsion process usually provides uniform latex particles having larger final size. The average particle size values with the soapless emulsion process are in the range of 0.5–1.5 μm, depending on the polymerization conditions. The type and concentration of the monomer, the initiator concentration, ionic strength of the continuous medium, and polymerization temperature strongly affect the polymerization rate, final particle size, and average molecular weight of the latex particles in the emulsifier-free emulsion polymerization process [75]. However, the polymerization rate is usually slower than that of the conventional emulsion polymerization carried out under the same conditions due to the existence of a lower number of forming particles in the soapless emulsion medium.

The effects of polymerization variables on the soapless emulsion process may be summarized as follows: The increase in the polymerization temperature increases the free radical production rate and the concentration of free radicals. This leads to the formation of oligomers having shorter kinetic chain lengths, which increase the number of density of sulfate groups in each nucleated particle. This case provides the stabilization of relatively smaller particles due to the high surface charge. Therefore, the number of forming particles and, hence, the polymerization rate increases and the average particle size decreases with the increasing polymerization temperature. An effect on the particle size and on the polymerization rate in the same direction can also be obtained with the increasing initiator concentration since the free radical production rate increases. However, an increase in the initiator concentration increases the ionic strength of the continuous medium, which involves an increase in the average size. The apparent effect of initiator concentration can be observed as the summation of these two opposite effects. The monomer concentration in the continuous medium is roughly constant and determined by the ionic strength of medium until all the monomer droplets disappear. Therefore, the initial monomer–water ratio is not an effective parameter of the polymerization rate up to the particle formation step. After the particle formation, this becomes an important factor in controlling the final size of the product since the monomer absorption of forming particles is proportional to the amount of available monomer within the continuous phase.

Recently, Smigol et al. [75] extensively studied emulsifier-free emulsion polymerization of different monomers including styrene, methyl methacrylate, and glycidyl methacrylate in an aqueous medium by using potassium peroxydisulfate as the initiator. In this study, the average particle size increased with the increasing ionic strength and increasing monomer–water ratio and decreased with the increasing polymerization temperature.

III. DISPERSION POLYMERIZATION

Uniform polymeric microspheres of micron size have been prepared by dispersion polymerization. This process is usually utilized for the production of uniform polystyrene and polymethylmethacrylate microspheres in the size range of 0.1–10.0 μm.

A. General Mechanism

Monomer-soluble initiators are used in this polymerization technique. The monomer phase containing an initiator is dissolved in an inert solvent or solvent mixture including a steric stabilizer. The polymers or oligomer compounds having low solubility in the polymerization medium and moderate affinity for the polymer particles can be selected as the steric stabilizer [9,76]. The schematical representation of the dispersion polymerization process is given in Fig. 9.

In a typical dispersion polymerization system, there is initially only one phase where the initiator, the monomer, and the steric stabilizer molecules are dissolved. The polymerization is initiated in the homogenous solution. Upon heating, the initiator decomposes and free radicals react with the solubilized monomer molecules within the continuous phase to generate the oligomeric radicals. The polymerization medium is selected as a poor solvent for the polymer particles. The forming polymer chains become insoluble in the polymerization medium when they reach a certain critical length, depending upon the solvency of the medium. Therefore, a phase separation occurs at an early stage of polymerization by either the self- or aggregative nucleation mechanism [9].

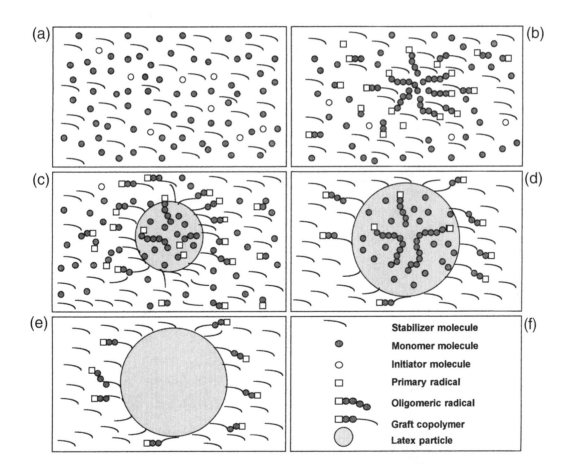

Figure 9 The schematical representation of dispersion polymerization process. (a) initially homogeneous dispersion medium; (b) particle formation and stabilizer adsorption onto the nucleated macroradicals; (c) capturing of radicals generated in the continuous medium by the forming particles and monomer diffusion to the forming particles; (d) polymerization within the monomer swollen latex particles, (e) latex particle stabilized by steric stabilizer and graft copolymer molecules; (f) list of symbols.

This step is termed nucleation and leads to the formation of primary particles. The nuclei, the smallest particles formed in the nucleation step, are considered to be formed by aggregation of growing polymer chains precipitating from solution as they exceed their critical chain length.

Therefore, the particle formation stage in the dispersion polymerization may be considered as a two-step process including the generation of nuclei and the aggregation of nuclei [77]. The oligomer chains grow in the continuous phase and precipitate to form nuclei that are unstable and rapidly aggregate with each other. In this period, the stabilizer and costabilizer are adsorbed from the continuous phase onto the precipitated oligomers to form the stable particle nuclei. Aggregation of these nuclei with themselves and their aggregates continues until mature and stable particles are formed. This occurs when sufficient stabilizer occupies the particle surface, which includes both the polymeric stabilizer and its graft copolymer which is created in situ [77].[a] After forming

stable particles in sufficient number, all the oligo-radicals and nuclei generated in the continuous phase are captured by the mature particles, no more particles form, and the particle formation stage is completed. The primary particles formed by the nucleation process are swollen by the unconverted monomer and/or polymerization medium. The polymerization taking place within the individual particles leads to resultant uniform microspheres in the size range of 0.1–10 μm. Various dispersion polymerization systems are summarized in Table 4.

B. Typical Recipe

Some typical dispersion polymerization recipes and the electron micrograph of the uniform polymeric particles with Recipe I are given in Table 5 and Fig. 10, respectively. As seen in Table 5, the alcohols or alcohol–water mixtures are usually utilized as the dispersion media for the dispersion polymerization of apolar monomers. In order to achieve the monodispersity in the final product, a costabilizer can be used together with a primary steric stabilizer, which is usually in the polymeric form as in

---

[a] Reproduced from Ref. 77, by the permission of John Wiley & Sons, Inc.

Table 4 Some Examples for Different Dispersion Polymerization Systems

Monomer	Initiator	Stabilizer/co-stabilizer	Medium	P. Size (μm)	Reference
MMA	AIBN	PMMA-g-OSA	Petroleum hydrocarbons	—	78
Styrene	AIBN	PAAc	Ethanol/water	2.0	79
MMA	AIBN	PVP/aliquat 336	Methanol	3.9	80
Styrene	AIBN	PVP/aliquat 336	Ethanol	2.0	80
Styrene	AIBN	PEI/aliquat 336	Ethanol	1.7	80
Styrene	AIBN	PAA/aliquat 336	Ethanol	2.4	80
Styrene	AIBN	PVME/aliquat 336	Ethanol	1.9	80
Styrene	BPO	HPC	Ethanol/2-methoxyethanol	3.0–9.0	81,82,83
Styrene	AIBN	PVP/aerosol OT	Ethanol	2.5–6.2	84
Styrene	AIBN	PVP/triton N-57	Ethanol	—	84
Styrene	AIBN	PVP/cetyl alcohol	Ethanol	—	84
Styrene	BPO	HPC	C_1–C_{10} alcohols	2.0–8.5	85
Styrene	ACPA	PVP/aerosol OT	Ethanol	1.9–4.5	86
Styrene	AIBN	PAAc	Ethanol/water	1.9	87
Styrene	BPO	HPC	Ethanol/isopropanol	2.1	88
Styrene	AIBN	PAAc	Isopropanol/water 1-Butanol/water	1.4–3.3	89
MMA	AIBN	PVP/aliquat 336	Methanol	2.5–8.0	90
DVB	AIBN	None	Acetonitrile	3.4–3.6	91
DVB	AIBN	PVP	Acetonitrile	0.9–2.6	92
Styrene	AIBN	PAAc	Isopropanol/water	1.2–2.8	93
ECA	H_3PO_4/HCl	PEO/PPO	Water	0.5–2.4	94,95
Acrolein	NaOH	PGA-NaHSO₃	Water	0.04–8.0	96
CMS	AIBN	PVP	Ethanol-DMSO	1.0	97

ACPA: azobis(4-cyanopentanoic acid); AIBN: azobis(isobutyronitrile); BPO: benzoyl peroxide; DVB: divinyl benzene, ECA: 2-ethylcyanoacrylate; HPC: hydroxypropyl cellulose; MMA: methyl methacrylate; PAAc: polyacrylic acid; PEI: polyethyleneimine, PEO/PPO: polyethylene oxide/polypyropylene oxide copolymer; PVME: polyvinylmethylether; PVP: polyvinylpyrrolidone K-30; DMSO: dimethylsulfoxide; PGA: polyglutaraldehyde; CMS: chloromethylstyrene; PMMA-g-OSA: polymethylmethacrylate grafted oligostearic acid.

Table 5 Typical Dispersion Polymerization Recipes Providing Uniform Latex Particles

	Recipe I (93)	Recipe II (84)	Recipe III (90)
Monomer:	Styrene (20 ml)	Styrene (12.5%)	MMA (2.50 g)
Medium:	90% IsoPrOH-10% Water (200 ml)	Ethanol (84.8%)	Methanol (21.475 g)
Initiator:	AIBN (0.14 g)	AIBN (1.0%)*	AIBN (0.025 g)
Stabilizer:	PAAc (2.0 g)	PVP K-30 (1.8%)	PVP K-30 (1.00 g)
Co-stabilizer:	—	Aerosol OT (0.5%)	—
Conditions:	70°C, 24 h, 150 rpm	70°C, 24 h, 30 rpm	60°C, 48 h, 32 rpm
Particle size:	2.3 μm	2.5 μm	4.0 μm

* The initiator concentration in recipe II was given as wt% based on monomer. The concentrations of other ingredients in recipe II were given as wt% based on total.
Source: Reproduced from Refs. 84 and 90, by the permisson of John Wiley & Sons, Inc.

Recipes II and III. The volumetric ratio of monomer-to-continuous medium in the dispersion polymerization is lower relative to that of emulsion polymerization for the synthesis of uniform polymeric particles. The requirements for the formation of uniform particles in the dispersion polymerization process are: (1) the nucleation step should be very short, (2) secondary nucleation should be prevented. In other words, all the oligomeric radicals generated in the continuous phase during the particle growth stage should be captured by the existing particles before they precipitate and form new particles, and (3) the coalescence between particles in the particle growth stage should be prevented by the use of proper steric stabilizer systems [86].[b]

---

[b] Reproduced from Ref. 86, by the permission of John Wiley & Sons, Inc.

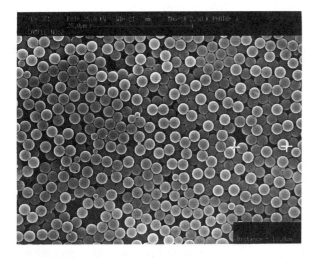

Figure 10 The electron micrograph of the uniform polymeric particles produced with Recipe I. Magnification: 2000×.

C. Effect of Process Variables

1. Initiator

The type and concentration of oil-soluble initiator are effective both on the polymerization rate and on the average size of the final product. The polymerization rate and the average size of the final product usually increase with the increasing initiator concentration.

Tseng and coworkers [84] tried different oil-soluble initiators in the dispersion polymerization of styrene carried out in the ethanol medium. They used poly(N-vinyl pyrollidone) and aerosol OT as the stabilizer and the costabilizer, respectively. Five different initiators, 2-2'-azobis(isobutyronitrile) (AIBN), 4,4'-azobis(4-cyanopentanoic acid) (ACPA), 2-2'-azobis(2-methylbutyronitrile) (AMBN), 2-2'-azobis(2,4-dimethylvaleronitrile) (ADVN), and benzoyl peroxide (BPO), were used by changing the initiator concentration between 0.5–4.0 wt% based on the monomer. By keeping the monodispersity of the final product, the average size of the final product increased from 2.0 μm to 4.5–5.0 μm with the increasing initiator concentration when AIBN, ACPA, or AMBN were used as the initiator. BPO and ADVN provided broad or bimodal size distributions. These results showed that the initiators having very slow (BPO) and very fast (ADVN) decomposition rates caused the formation of polidispersity in the final product [84]. Ober and Hair [98] used BPO to initiate the dispersion polymerization of styrene in the ethanol–methoxyethanol medium, using hydroxypropylcellulose as the stabilizer. The average size and the polymerization rate increased and the molecular weight of the final product decreased with the increasing BPO concentration [98].

We also studied the effect of initiator on the dispersion polymerization of styrene in alcohol–water media by using a shaking reactor system [89]. We used AIBN and polyacrylic acid as the initiator and the stabilizer, respectively. Three different homogenous dispersion media including 90% alcohol and 10% water (by volume) were prepared by using isopropanol, 1-butanol, and 2-

butanol and the initiator concentration was changed between 0.5–2.0 mol% based on the monomer. The average size increased with the increasing initiator concentration in all media [89]. In another study, the effects of AIBN concentration on the polymerization rate of styrene and on the average size of the polystyrene particles were investigated in a medium containing 90% isopropanol and 10% water by a stirred polymerization reactor [93]. The electron micrographs of the final products and the variation of the monomer conversion by the polymerization time at different initiator concentrations are exemplified in Fig. 11. As seen here, similar results to the other studies were obtained for the effects of initiator concentration on the polymerization rate and the average particle size.

The role of initiator concentration on dispersion polymerization can be summarized as follows. The in-

crease in the initiator concentration increases the free radical production rate, which in turn causes an increase in the propagation rate of oligomeric radicals. Therefore, an increase occurs in the concentration of precipitated oligomer chains leading to an increase in final particle size. However, depending on these effects, the polymerization rate increases and the average molecular weight of the final product decreases with the increasing initiator concentration. Further increase in the initiator concentration results in the formation of relatively larger but fewer particles during the first nucleation period. Therefore, the radical capturing ability of forming particles decreases and this leads to the formation of a new crop of particles by the repeated nucleations. This results in the formation of a wide size distribution in the final product.

2. Stabilizer

The stabilizer plays an important role in the dispersion polymerization process in which a block or graft copolymer including both soluble and insoluble polymer segments is usually selected as a stabilizer. The stabilizer determines both the particle stability during the particle formation step and the viscosity of the continuous medium. However, a graft copolymer can be formed in situ when a precursor polymer that contains proper active sites for chain transfer of the oligomeric radicals is used as a stabilizer [77]. Hydroxypropylcelulose, poly(acrylic acid), and poly(N-vinyl pyrollidone) (PVP) are the typical and most widely used stabilizers in the precursor form. The formation of graft copolymer during the dispersion polymerization is a complex process. The structure and concentration of the precursor polymer control the rate of formation, concentration, and the properties of the graft copolymer. The formed graft copolymer and the precursor polymer are both adsorbed competitively onto the surface of the polymer particles.

Shen and coworkers [77] performed an excellent study in which the role of stabilizers in the dispersion polymerization process was clearly identified. In their study, the dispersion polymerization of methyl methacrylate was carried out in methanol by using AIBN and PVP having α hydrogens as possible chain transfer sides as the initiator and the stabilizer, respectively. The generation and aggregation of nuclei during the polymerization were monitored by differential light scattering (DLS). They reported two different ways for the adsorption of stabilizer onto the polymeric particles. The adsorption of PVP was described as the physical adsorption, which was weak and reversible. The adsorption of PVP-g-PMMA was termed anchoring adsorption (including chemically bound and strongly entrapped graft copolymer molecules on the surface), which was irreversible. After the determination of physically adsorbed and chemically bound PVP, their results showed that the physical adsorption of the stabilizer was significant and its effect on the particle formation stage should not be

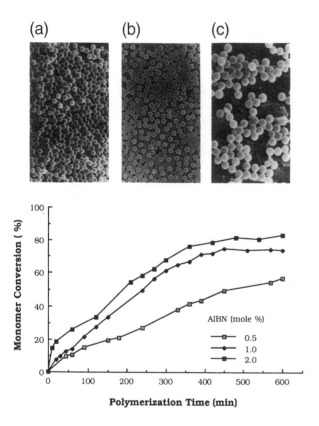

Figure 11 The electron micrographs of the final products and the variation of the monomer conversion with the polymerization time at different initiator concentrations in the dispersion polymerization of styrene. Initiator concentration (mol%): (a) 0.5, (b) 1.0, (c) 2.0. The original SEM photographs were taken with 2600×, 2000×, and 2600× magnifications for (a), (b), and (c), respectively, and reduced at a proper ratio to place the figure. (From Ref. 93. Reproduced with permission from John Wiley & Sons, Inc.)

neglected. Three types of PVP were used including PVP K-15 (M_W = 10.000), PVP K-30 (M_W = 40.000), and PVP K-90 (M_W = 360.000). The final particle size decreased with the increasing PVP concentration for all PVP types. The physical adsorption rate of PVP and anchoring adsorption rate of the PVP-g-PMMA copolymer increase with the increasing stabilizer concentration. These changes reduce the aggregation of nuclei, which leads to a smaller particle size. However, the increase in the molecular weight of the stabilizer increases the viscosity of the continuous medium, which in turn causes better stabilization of particles by the longer stabilizer chains. Therefore, the particle size can be reduced by using a higher molecular weight stabilizer at the same concentration with a lower molecular weight stabilizer. In such a case, the physical and anchoring adsorption rates of the stabilizer will be lower since the medium viscosity is high and the stabilizer concentration is low.

The same PVP series were also tried for the dispersion polymerization of styrene in the ethanol medium by using AIBN as the initiator and aerosol OT as the costabilizer [84]. PVP K-15 usually yielded polymeric particles with a certain size distribution and some coagulum. The uniform products were obtained with PVP K-30 and PVP K-90 in the presence of the costabilizer. The tendencies for the variation of the final particle size with the stabilizer concentration and with the molecular weight of the stabilizer were consistent with those obtained for the dispersion polymerization of methyl methacrylate [84].

Paine et al. [99] tried different stabilizers [i.e., hydroxypropylcellulose, poly(N-vinylpyrollidone), and poly(acrylic acid)] in the dispersion polymerization of styrene initiated with AIBN in the ethanol medium. The direct observation of the stained thin sections of the particles by transmission electron microscopy showed the existence of stabilizer layer in 10–20 nm thickness on the surface of the polystyrene particles. When the polystyrene latexes were dissolved in dioxane and precipitated with methanol, new latex particles with a similar surface stabilizer morphology were obtained. These results supported the grafting mechanism of stabilization during dispersion polymerization of styrene in polar solvents.

We have also examined the effect of stabilizer (i.e., polyacrylic acid) on the dispersion polymerization of styrene (20 ml) initiated with AIBN (0.14 g) in an isopropanol (180 ml)–water (20 ml) medium [93]. The polymerizations were carried out at 75°C for 24 h, with 150 rpm stirring rate by changing the stabilizer concentration between 0.5–2.0 g/dL (dispersion medium). The electron micrographs of the final particles and the variation of the monomer conversion with the polymerization time at different stabilizer concentrations are given in Fig. 12. The average particle size decreased and the polymerization rate increased by the increasing PAAc concentra-

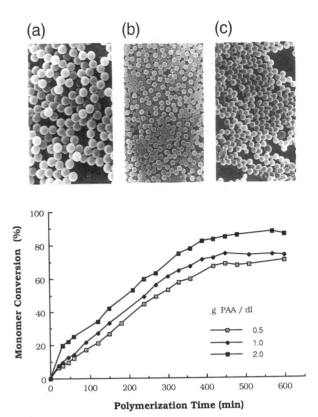

Figure 12 The electron micrographs of the final particles and the variation of the monomer conversion with the time at different stabilizer concentrations in the dispersion polymerization of styrene. Stabilizer concentration (g/dL): (a) 0.5, (b) 1.0, (c) 2.0. The original SEM photographs were taken with 2600×, 2000×, and 2600× magnifications for (a), (b), and (c), respectively, and reduced at a proper ratio to place the figure. (From Ref. 93. Reproduced with permission from John Wiley & Sons, Inc.)

tion. The increase in the PAAc concentration increases either the physical adsorption rate of PAAc or the chemical adsorption rate of PAAc-g-PS graft copolymer onto the forming particles. The better and rapid stabilization of forming nuclei reduces the aggregation of nucleated particles. Therefore, a large number of relatively small particles can be obtained by the increasing stabilizer concentration. The increase in the number of formed particles causes an increase in the apparent polymerization rate since it is directly proportional to the number of particles present in the polymerization medium based on the kinetic model proposed by Lu et al. [86].

3. Dispersion Medium

Uniform polymeric microspheres in the micron size range have been prepared in a wide variety of solvent combinations by dispersion polymerization. The polarity of the dispersion medium is one of the most important parameters controlling the average size and the mono-dispersity of the final product. A general relation that can explain the effects of medium polarity on the dispersion polymerization process has not yet been obtained. There are too many solvents or solvent mixtures that can be used as the continuous phase in the dispersion polymerization process. Linear alcohols, homogeneous alcohol–water or alcohol–ether solutions are usually preferred as the continuous medium in the dispersion polymerization process. The polarity of the dispersion medium is usually represented by the solubility parameter. This value is obtained by the summation of the dispersion, polar, and H-bonding forces of the solvent according to the following expression:

$$\delta^2 = \delta_D^2 + \delta_P^2 + \delta_H^2 \qquad (5)$$

Where, δ is defined as the solubility parameter of the solvent. δ_D, δ_P and δ_H are the dispersion, polar, and H-bonding forces, respectively. A homogenous mixture of polar solvents can also be used as the continuous phase. In this case, the solubility parameter of the homogeneous mixture is calculated according to the following expression [89]:

$$\delta_m = \sum (\chi_i \delta_i)^{1/2} \qquad (6)$$

Where, χ_i is the volume fraction of component i, δ_m, and δ_i are the solubility parameters of the homogenous solvent mixture and the component i, respectively. The solubility parameters of some solvents that are widely used as the continuous medium in the dispersion polymerization are given in Table 6.

Paine et al. [85] extensively studied the effect of solvent in the dispersion polymerization of styrene in the polar media. In their study, the dispersion polymerization of styrene was carried out by changing the dispersion medium. They used hydroxypropyl cellulose (HPC) as the stabilizer and its concentration was fixed to 1.5% within a series of n-alcohols tried as the dispersion media. The particle size increased from only 2.0 μm in methanol to about 8.3 μm in pentanol, and then decreased back to 1 μm in octadecanol. The particle size values plotted against the Hansen solubility parameters

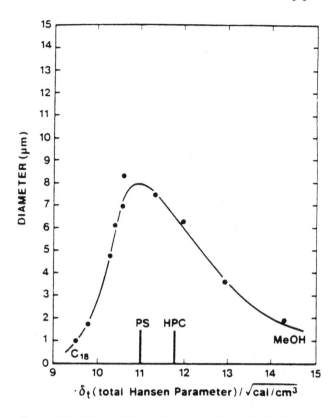

Figure 13 The variation of particle size with the Hansen solubility parameter of the n-alcohols. (Adapted from Ref. 85 with the permission of John Wiley & Sons, Inc.)

of the n-alcohols are given in Fig. 13. As seen here, the maximum particle size was obtained with n-pentanol. The Hansen solubility parameters of polystyrene and HPC were 11.0 and 11.8, respectively. These values are very close to the Hansen solubility parameter values of n-alcohols in which relatively larger particles were obtained. But, this graph was not a predictive one for the particle size value when a solvent system that is not included in the linear alcohol series is used as the disper-

Table 6 The Solubility Parameters of Some Solvents [100]

Solvent	$\delta(cal/cm^3)^{1/2}$	$\delta_D(cal/cm^3)^{1/2}$	$\delta_P(cal/cm^3)^{1/2}$	$\delta_H(cal/cm^3)^{1/2}$
t-Butanol	10.6	—	—	—
Ethanol	12.7	7.7	4.3	9.5
Isopropanol	11.5	—	—	—
Methanol	14.5	7.4	6.0	10.9
2-Methoxyethanol	11.4	7.9	4.5	8.0
Water	23.4	6.0	15.3	16.7
Styrene	9.3	9.1	0.5	2.0
Methyl methacrylate	8.8	—	—	—

sion medium. Therefore, a further analysis was also performed by considering the terms included in the definition of the Hansen solubility parameter (i.e., the hydrogen bonding term, δ_H, and the polarity term, δ_P) to obtain a predictive chart showing the dependency of particle size to the medium properties. A chart, the so-called Hansen solvency map, was obtained by plotting δ_P values of the solvents against their δ_H values. The particle size values obtained by different solvents were placed on the related points defined by the respective δ_P and δ_H values of the used solvents. This chart indicated that the largest particles could be obtained with the solvent systems having the closest δ_P and δ_H values to the respective values of HPC. Paine [101] also proposed a mathematical model for the prediction of particle size in the dispersion polymerization of styrene in polar media. Two different research groups used ethanol–methoxyethanol medium for the dispersion polymerization of styrene initiated with BPO and stabilized by HPC [81,85]. In both studies, the average particle size decreased and the polymerization rate increased with the increasing ethanol content of the dispersion medium. This tendency originated from the more polar character of ethanol relative to methoxyethanol. Lok and Ober [81] studied the effect of water content on the average particle size in the dispersion polymerization of styrene in ethanol–water mixture initiated by BPO and stabilized by HPC. They also obtained the same tendency for the variation of particle size with the water content of the dispersion medium, but with the smaller final particle size values. They also tried different alcohols including methanol, ethanol, isopropanol, and t-butanol as dispersion medium. In this series, the final size with isopropanol medium did not obey the correlation between final particle size and medium solubility parameter.

Almog et al. [80] studied the dispersion polymerization of styrene in different alcohols as the continuous medium by using AIBN and vinyl alcohol–vinyl acetate copolymer as the initiator and the stabilizer, respectively. Their results showed that the final particle size decreased with the alcohol type according to the following order:

$$t\text{-Butanol} > \text{Isopropanol} > \text{Ethanol} > \text{Methanol}$$

All these alcohols provided uniform final particles while a size distribution was observed in both n-propanol and sec-butanol media. The increase in the particle size as a function of the alcohol type was attributed to the swelling capacity of the alcohol. The degree of swelling of the precipitated polymer was inversely proportional to the difference between the solubility parameters of the dispersion medium and the polymer. The plotting of the volume of final particles against the solubility parameter difference gave a linear relation showing the validity of this consideration. However, the formation of larger particles with the increasing carbon number of the alcohol

was also explained by the effect of medium polarity on the nucleation step. According to this explanation, when the solvency of the medium increases (i.e., the polarity decreases), the critical degree of polymerization at which the oligomeric molecules nucleate will increase. This resulted in the retardation of particle formation, and, therefore, fewer and correspondingly larger particles were produced.

Okubo et al. [87] used AIBN and poly(acrylic acid) ($M_W = 2 \times 10^5$) as the initiator and the stabilizer, respectively, for the dispersion polymerization of styrene conducted within the ethyl alcohol/water medium. The ethyl alcohol–water volumetric ratio (ml:ml) was changed between (100:0) and (60:40). The uniform particles were obtained in the range of 100:0 and 70:30 while the polydisperse particles were produced with 35:65 and especially 60:40 ethyl alcohol–water ratios. The average particle size decreased form 3.8 to 1.9 μm by the increasing water content of the dispersion medium.

Shen et al. [90] included the effect of water content on the average particle size in the dispersion polymerization of methyl methacrylate in methanol–water medium by using AIBN and poly(N-vinylpyrollidone). The results indicated that relatively larger and uniform particles were obtained with the lower water contents. The polydispersity increased and the average size decreased by the increasing water content. According to the explanation given by the authors, the critical chain length decreases and the rate of adsorption of PVP-g-PMMA copolymer increases with the increasing water content of the polymerization medium. These led to an increase in the rate of nuclei formation resulting in smaller particles. If the water content is further increased, the generation rate of nuclei and higher adsorption rate of the graft copolymer make it more difficult for existing particles to capture all the nuclei and aggregates from the continuous phase before they become stable particles. Therefore, the particle formation stage is extended, resulting in broad size distributions [90].

We have also investigated the effect of alcohol–water ratio on the dispersion polymerization of styrene with AIBN (1.0% mol) as the initiator and poly(acrylic acid) (1.0 g/dL disp. medium) as the stabilizer [89]. We have used isopropanol, 1-butanol, and 2-butanol for the preparation of dispersion media. The alcohol–water ratio was changed between 100:0–70:30 for each alcohol type. The polymerizations were carried out in a shaking system at 70°C, for 24 h, with 100 cpm shaking rate, 5.0 ml of styrene was used in 50 ml of the dispersion medium in the polymerizations. The average particle size decreased with the increasing water content of the polymerization medium for each alcohol type. The average size values in the homogenous dispersion media plotted against the average solubility parameters of the dispersion media are given in Fig. 14. As seen here, a certain relation was obtained between the particle size and polarity of the medium when the alcohol–water mix-

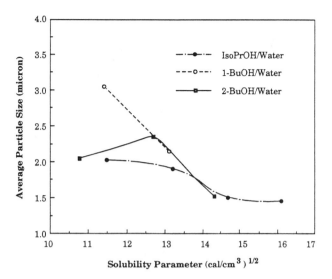

Figure 14 The variation of average size of the polystyrene particles by the average solubility parameter of the homogeneous alcohol–water dispersion medium. (From Ref. 89. Reproduced with the permission of John Wiley & Sons, Inc.)

tures were used as the dispersion medium. Notice that the particle size values with pure 1-butanol and 2-butanol deviated from this tendency.

We have also followed kinetics of the dispersion polymerization of styrene in a stirred reactor system by changing the isopropanol–water ratio at the same conditions [93]. The apparent polymerization rate did not change significantly with the water content at an earlier stage of polymerization up to the nucleation step. But, after this period, the polymerization rate clearly increased with the increasing water content. The tendency of decreasing final particle size with the increasing water content was also observed in these polymerizations. The formation of smaller final particles with the increasing water content was explained by the formation of smaller, but a large number of nuclei with the precipitation of lower molecular weight oligomers since the oligomer solubility and critical chain length decrease with the increasing water content. The rate of dispersion polymerization is directly proportional to the number of particles in the polymerization medium according to the proposed kinetic model [84]. Therefore, the increase in the number of particles with the increasing water content caused an increase in the apparent polymerization rate after completion of the particle formation step as was also experimentally observed.

4. Monomer Type and Concentration

The monomers that are apolar relative to the continuous media are usually used in the dispersion polymerization

process. The solvency of the medium for the polymer increases with the increasing monomer concentration since the monomer is a part of the initial medium. This allows an increase in the solubilities of the oligomeric molecules and the critical chain length value of the forming oligomers at the earlier stage of polymerization. The high-monomer concentration also causes an increase in the propagation rate of oligomer chains. Thus, prior to the precipitation stage, the oligomers can reach a higher molecular weight by remaining in the soluble form within the dispersion medium. The nucleation occurs with the oligomeric chains having higher molecular weight and higher concentration. However, high-monomer concentration causes a decrease in the adsorption rate of the stabilizer or costabilizer onto the forming particles, which in turn increases the possibility of aggregation. So, fewer nuclei, but nuclei that are larger in size are formed by the aggregation of these oligomeric materials. A reduction in the interfacial tension between forming polymer particles and continuous medium decreases in the presence of high-monomer concentration. The swelling of forming particles with the monomer becomes easier and a larger extent of monomer can be absorbed by the fewer nucleated particles. All these effects provide a clear increase in the average size of the final product with the increasing monomer concentration in the dispersion medium.

After completion of the nucleation period, an oligomer initiated in the continuous phase is captured by the stable nucleated particles before growing to the critical chain length value under appropriate conditions. Therefore, the initiation in the continuous phase is suppressed by the diffusion capture mechanism according to the model proposed by Fitch and Tsai [13]. When the monomer concentration is further increased, the growth rate and the concentration of oligomers will increase, but the number of nuclei will decrease. This will cause a decrease in the capturing possibility of oligomeric radicals by the existing nuclei, and the chance of secondary nucleation will increase. Therefore, a size distribution in the final product will be observed since the repeated nucleations in the continuous phase with the oligomeric chains have different molecular weights.

We have studied the effect of monomer concentration in the dispersion polymerization of styrene carried out in alcohol–water mixtures as the dispersion media. We used AIBN and poly(acrylic acid) as the initiator and the stabilizer, respectively, and we tried isopropanol, 1-butanol, and 2-butanol as the alcohols [89]. The largest average particle size values were obtained with the highest monomer–dispersion medium volumetric ratios in 1-butanol–water medium having the alcohol–water volumetric ratio of 90:10. The SEM micrographs of these particles are given in Fig. 15. As seen here, a certain size distribution by the formation of small particles, possibly with a secondary nucleation, was observed in the poly-

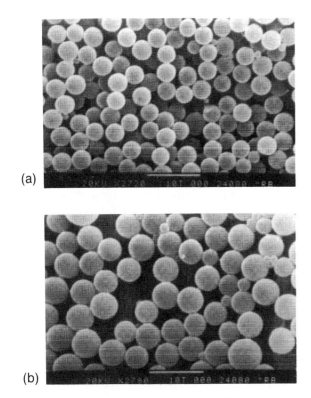

(a)

(b)

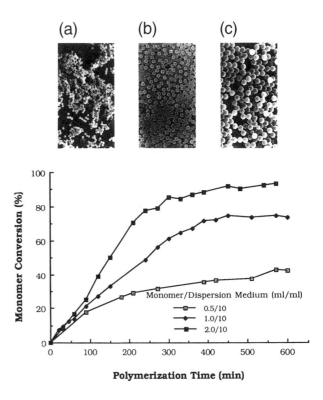

Figure 15 The SEM photographs of the polystyrene particles produced in 90% 1-butanol–10% water medium by the high monomer dispersion medium ratios. Monomer/dispersion medium volumetric ratio (mL/mL): (a) 2.0/10.0, (b) 2.5/10.0. (From Ref. 89. Reproduced with the permission of John Wiley & Sons, Inc.)

Figure 16 The variation of the monomer conversion by the polymerization time at different monomer to dispersion medium ratios in the dispersion polymerization of styrene. Monomer/dispersion medium (mL/mL): (a) 5/100, (b) 10/100, (c) 20/100. The original SEM photographs were taken with 2600×, 2000×, and 2600× magnifications for (a), (b), and (c), respectively, and reduced at a proper ratio to place the figure. (From Ref. 93. Reproduced with the permission of John Wiley & Sons, Inc.)

styrene particles produced by the highest monomer dispersion medium ratio of 2.5:10 mL/mL used in this study.

The kinetics of the dispersion polymerization was followed in a stirred reactor by using isopropanol–water mixture as the dispersion medium and by changing the monomer–water ratio [93]. The results are presented in Fig. 16. As seen here, the polymerization rate and the average particle size increased with the increasing initial monomer concentration. The increase in the monomer concentration increases the production and growth rate of the oligomeric radicals at the earlier course of polymerization. After the particle formation, the diffusion rate of monomer from the continuous phase to the forming particles will be higher since the magnitude of the driving force (i.e., concentration difference between the forming particles and continuous medium) is larger. After completion of the monomer diffusion, more monomer will be available for the polymerization reaction within the forming particles. These effects cause an increase in the apparent polymerization rate as shown in Fig. 16.

5. Temperature

Shen et al. [90] studied the effect of polymerization temperature on the final size of particles in the dispersion polymerization of methyl methacrylate carried out in methanol. They used AIBN and PVP. K-30 as the initiator and stabilizer, respectively. They observed a clear increase in the average particle size by changing the polymerization temperature between 50–60°C. The effect of polymerization temperature on the average size and on the molecular weight of the final product was investigated by Ober and Hair [98] for the dispersion polymerization of styrene conducted in ethanol-methoxyethanol medium. They observed that the polymerization rate increased and the average molecular weight decreased with the increasing polymerization temperature. Based on the results of these studies, the effect of polymerization temperature on the dispersion polymerization may be explained as follows. The increase in the polymerization temperature increases the free radical production rate that in turn increases the concentration of oligomeric radicals within the continuous phase. This leads to a decrease in the average molecular weight of the final particles and an increase in the apparent polymerization rate. However, the solvency of the medium for oligomeric radicals increases with the increasing polymerization temperature, which causes an increase in the critical length of the forming oligomeric chains. The adsorption rate of stabilizer decreases since the solubility of the stabilizer within the continuous medium increases. The latter two factors provide an increase in the final size of particles with the increasing polymerization temperature.

D. Kinetics of Dispersion Polymerization

A conceptual kinetic model on the dispersion polymerization process was recently proposed by Lu et al. [86]. In their study, the rate of dispersion polymerization of styrene in ethanol medium was compared with the rate of solution polymerization of styrene in cyclohexanol medium conducted under the same conditions with the dispersion polymerization process. Their results indicated that dispersion polymerization provided a higher polymerization rate and a higher molecular weight of polymer relative to those of the solution polymerization. A monomer partitioning study performed in the same dispersion system indicated that most of the styrene monomer retains in the continuous phase at the earlier stage of the polymerization. Therefore, most of the initiator is found within the continuous phase depending on the monomer distribution. This case involves the existence of two different polymerization loci in the dispersion polymerization system. The two different loci were defined as the continuous phase and the monomer swollen forming particles in their study. At the earlier stages of the dispersion polymerization process, the polymerization starts out and progresses within the continuous phase as a solution polymerization. Therefore, at relatively low conversion values in dispersion polymerization, the polymerization rate and the molecular weight of the polymer are very close to those obtained with the solution process under the same conditions. However, the forming particles after the nucleation step become important as a polymerization loci, and the rate of polymerization differs from that of the solution polymerization. It is known that, the polymerization rate and the final average molecular weight of the polymer produced by dispersion polymerization are lower than those produced with the emulsion polymerization. The reasons for this difference are the lower monomer concentration within the forming particles, lower number of forming particles, and higher number of oligomeric radicals in the forming particles in the dispersion polymerization system relative to the emulsion polymerization.

Lu et al. [86] also studied the effect of initiator concentration on the dispersion polymerization of styrene in ethanol medium by using ACPA as the initiator. They observed that there was a period at the extended monomer conversion in which the polymerization rate was independent of the initiator concentration, although it was dependent on the initiator concentration at the initial stage of polymerization. We also had a similar observation, which was obtained by changing the AIBN concentration in the dispersion polymerization of styrene conducted in isopropanol–water medium. Lu et al. [86] proposed that the polymerization rate beyond 50% conversion could be explained by the usual heterogenous polymer kinetics described by the following equation:

$$R_p = k_p(M_p)(\rho_a V_p/2k_t)^{1/2} \tag{7}$$

In this equation, M_p is the monomer concentration within forming particles, ρ_a is the adsorption rate of oligomeric radicals by the forming particles, V_p is the volume fraction of forming particles within the system, and k_p and k_t are the rate constants of propagation and termination, respectively.

E. Dispersion Polymerization with Different Monomers

Li and Stöver [91,92] produced uniform poly(divinylbenzene) microspheres by the dispersion polymerization of divinylbenzene in acetonitrile. The oil-soluble initiators 2,2′-azobisizobutyronitrile, benzoyl peroxide, and 2,2′-azobis-(2,4-dimethylvaleronitrile) were used in this study. This process did not require stabilizers of any type and produced uniform crosslinked poly(divinylbenzene) particles in the size range of 2.0–5.0 μm in which the dispersion polymerizations conducted at 70°C for 24 h. The particle formation and growth mechanism proposed in this study were very similar to that of conventional dispersion polymerization except that the forming particles were stabilized by their rigid and crosslinked surfaces against the coagulation rather than by steric stabilizers. The researchers tried different organic solvents including n-alcohols as the dispersion medium, but the spherical uniform particles were obtained only by using acetonitrile.

The dispersion polymerization of alkylcyanoacrylates provides degradable uniform polyalkylcyanoacrylate latex particles in submicron size range. These particles are termed as biodegradable nanoparticles in the common literature [102–107]. The general structure of alkylcyanoacrylates is:

Couvreur et al. [104,106] extensively studied the dispersion polymerization of alkylcyanoacrylate monomers. Polyalkylcyanoacrylate nanoparticles have attracted much attention as a colloidal drug carrier because of their ease of preparation, ability to absorb efficiently large amounts of drugs, biodegradability, and biocompatibility. These particles can be degraded into corresponding alcohol and formaldehyde in the alkaline medium. The dispersion polymerizations of alkylcyanoacrylates are usually carried out in the aqueous media since they are highly polar monomers that are soluble in water. The polarity and the water solubility of the monomer decreases with the increasing alkyl chain length. The polymerization rate is very high and is not suitable for the particle formation when the aqueous dispersion media having neutral pH values are used. Therefore, the formation of submicron size polyalkylcyanoacrylate nanoparticles is achieved within the aqueous dispersion media having acidic pH values. For this purpose, H_3PO_4 or HCl are usually used to control the particle formation and to regulate the polymerization rate. According to the proposed mechanism, the polymerization occurs via an anionic mechanism involving the initiation by nucleophilic attack on the β-carbon of an alkylcyanoacrylate monomer [107]. Polymeric stabilizers such as dextran or polyethyleneoxide-polypropyleneoxide copolymers (Pluronics) are usually preferred. The polymerizations are conducted at room temperature in a 24-h period by using the usual stirring regimes for conventional dispersion polymerization. The typical ingredients used in the dispersion polymerization of alkylcyanoacrylate monomers are listed in Table 7.

In our recent studies [94,95], we studied the dispersion polymerization of 2-ethylcyanoacylate monomer in an aqueous medium. In this approach, all ingredients, including the monomer, initially are completely soluble in the aqueous acidic dispersion medium. We selected a stabilizer system containing a polyethyleneoxide–polypropyleneoxide copolymer and dextran, and we used HCl and H_3PO_4 together in the polymerization recipe. These modifications increased the particle size from submicron-size range to the micron-size range and com-

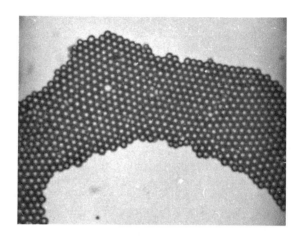

Figure 17 A typical optical micrograph of the uniform polyethylcyanoacrylate particles 2.1 μm in size. Magnification: 1000 ×.

pletely eliminated the particle size distribution [95]. The average particle size changed between 0.5–2.5 μm by changing the polymerization conditions. A typical optical micrograph of uniform polyethylcyanoacrylate particles 2.1 μm in size is given in Fig. 17. We have also studied the degradation of newly developed uniform PECA microspheres in micron-size range in the aqueous media having different pH values. The results indicated that, fast degradation rates were obtained in alkaline pH range while the degradation rate was very slow in the acidic region. A SEM study on the degradation process indicated that the degradation of uniform PECA particles mainly occurred by surface erosion [95].

Another group of uniform latex particles were prepared by the dispersion polymerization of acrolein in an aqueous medium by Margel and coworkers [96]. A typical polymerization was conducted in the aqueous alkaline medium having a pH of 10.5, which was obtained by the aqueous NaOH solution. Polyglutaraldehyde-$NaHSO_3$ conjugate was synthesized by the reaction of

Table 7 The Dispersion Polymerization Studies on Alkylcyanoacylates in the Aqueous Media

Monomer	Stabilizer	Acid type	Particle size (nm)	Reference
Methylcyanoacrylate	Tween-20	HCl	200	104
2-Ethylcyanoacrylate	Dextran-glucose	HCl	155	106
2-Ethylcyanoacrylate	Pluronic F88-dextran	HCl-H_3PO_4	400–2500	95
Isobutylcyanoacylate	Dextran-glucose	HCl	165	106
Isobutylcyanoacylate	Dextran	HCl	159–203	102
Isohexylcyanoacylate	Dextran-glucose	HCl	68	106
hexylcyanoacrylate	Pluronic F68-dextran	H_3PO_4	130	105

polyglutaraldehyde with NaHSO$_3$ in water and used as the stabilizer in the dispersion polymerization of the acrolein. Uniform poly(acrolein) microspheres in the range of 0.04–8.0 μm could be produced by changing the concentrations of acrolein and the stabilizer, and the pH of the dispersion medium. The authors concluded that stabilized nuclei, probably 0.01 μm in diameter, formed rapidly at the earlier polymerization times, and the diffusion of acrolein through the aqueous phase into the forming particles resulted in a size increase due to the polymerization of acrolein within the forming particles without formation of new nuclei and agglomeration. The microspheres produced contained free aldehyde groups on their surfaces. Uniform poly(acrolein) microspheres were also produced by the aqueous radical polymerization of acrolein by irradiation with a cobalt source [96].

Margel et al. [97] also produced uniform polychloromethylstyrene (PCMS) microspheres by the dispersion polymerization of chloromethylstyrene in ethanol-dimethylsulfoxide medium by using AIBN and polyvinylpyrollidone as the initiator and the stabilizer, respectively. In this study, PCMS microspheres were obtained in the uniform character with an average size of 1.0 μm with the dispersion polymerization conducted at 70°C. The presence of chloromethyl group on the surface of these micropheres provides significant advantages in the surface derivation studies for the attachment of specific ligands and functional groups onto the microsphere surfaces [97].

IV. MULTISTAGE EMULSION POLYMERIZATION

Multistage emulsion polymerization techniques are usually applied for (1) the synthesis of large uniform latex particles, (2) the introduction of functional groups into the uniform latex particles, or (3) the synthesis of macroporous uniform latex particles.

A. Synthesis of Large Uniform Latex Particles

Uniform polymeric microspheres in the size range of 0.01–1.0 μm can be produced by single stage standard emulsion polymerization techniques. It is reasonably difficult to produce uniform polymeric particles >1.5 μm by applying this techniques. However, dispersion polymerization is still an alternative single stage method for producing uniform polymeric particles in the size range of 0.1–10.0 μm. But, some defects in the monodispersity can appear when the final product size is >5 μm.

Multistage emulsion polymerization has been proposed by Ugelstad et al. [108,109] for the synthesis of large uniform latex particles. In general, the multistage emulsion polymerization techniques include two main

steps. The first one is the swelling of seed particles by a monomer with a subsequently high-swelling ratio, and the second step is the polymerization of the monomer within the individual swollen particles. Different methods may be applied for obtaining the monomer swollen form of the uniform seed particles prior to the repolymerization step. One of these methods is to introduce an oligomeric material (Y) into the seed particles before the swelling process that will be carried out with the selected monomer.

The introduction of oligomeric material into the seed particles may be performed by following two different ways [109]: (1) the seed particles are swollen by a proper monomer containing a chain transfer agent and the monomer is polymerized within the swollen particles to form an oligomeric material, (2) polymer seed particles are swollen by a monomer including an oil-soluble initiator in a proper concentration providing the formation of oligomeric material by the polymerization of the monomer within the individual beads.

Instead of the formation of oligomeric material within the uniform seed particles, a low-molecular weight, highly water-insoluble compound (Y) may be also introduced into the uniform seed particles before the swelling of the particles with the monomer [108]. The swelling of seed particles with the low-molecular weight compound is usually conducted in the aqueous emulsion of this agent to facilitate the diffusion process through the water phase to the particles. A water-soluble organic compound, such as acetone, may be added into the aqueous emulsion medium to increase the diffusion rate of the low-molecular weight, highly water-insoluble organic compound and is removed from the medium after the completion of the swelling process (i.e., prior to the monomer swelling step). When a step involving the introduction of a swelling agent (i.e., an oligomeric material or a low-molecular weight, highly water-insoluble organic compound) into the seed particles is applied, before the swelling of these particles by the monomer, this procedure is termed as "activated swelling method." The main feature of the new process proposed by Ugelstad et al. [108,109] is that one initially activates polymer seed particles, so that in aqueous dispersion they are capable of absorbing monomer in an amount which far exceeds that of pure polymer particles. The activation of the seed particles results from the presence of an oligomeric material or a low-molecular weight, highly water-insoluble organic compound within the uniform seed particles. In some cases, an oil-soluble initiator may be used as compound Y, which acts both as an initiator and as an activator for the monomer adsorption [109].

The role of a swelling agent in the activated swelling method may be explained by considering the theoretical basis of the process. The swelling of pure polymer particles with the monomer can be described by the Morton equation:

$$\ln \Phi_M + \Phi_P + \Phi_P^2 X_{MP} + 2\psi_M\gamma/rRT = 0 \qquad (8)$$

If the swelling process is carried out in the presence of a swelling agent, the Morton equation may be written as follows:

$$\ln \Phi_M + (1 - J_M/J_Y)\Phi_Y + (1 - J_M/J_P)\Phi_P + \Phi_Y^2 X_{MY}$$
$$+ \Phi_P^2 X_{MP} + \Phi_P\Phi_Y(X_{MY} + X_{MP} - X_{YP}J_M/J_Y)$$
$$+ 2\psi_M\gamma/rRT = 0 \qquad (9)$$

Where, the subscripts of M, P, and Y denote the monomer, the polymer, and the swelling agent, respectively; Φ_i is the volume fraction of the component i within the swollen seed particles at equilibrium swelling, ψ_M is the partial molar volume of the monomer, γ is the interfacial tension between the swollen seed particles and the continuous medium, r is the radius of the swollen particles at equilibrium swelling, and R and T are the ideal gas constant and the absolute temperature, respectively. The molar volume ratios of the monomer to the swelling agent and the monomer to the polymer are J_M/J_Y and J_M/J_P, respectively. For practical purposes, J_M/J_P is assumed to be very close to zero when the molecular weight of the polymer is sufficiently high. The Flory–Huggins interaction parameter per mole of compound i with compound j is X_{ij}. The particle radius at equilibrium swelling may be written in terms of initial radius of the seed particles by considering the following equation:

$$\left(\frac{V_P}{V_M + V_P + V_Y}\right) = \left(\frac{r_o}{r}\right)^3 \qquad (10)$$

The final form of the Morton equation for the activated swelling process may be written by expressing the equilibrium swelling radius of the seed particles in terms of initial radius by using Eq. (10).

$$\ln\left(\frac{V_M}{V_M + V_P + V_Y}\right) + (1 - J_M/J_Y)\left(\frac{V_Y}{V_M + V_P + V_Y}\right)$$

$$+ \left(\frac{V_P}{V_M + V_P + V_Y}\right) + \left(\frac{V_Y}{V_M + V_P + V_Y}\right)^2 X_{MY}$$

$$+ \left(\frac{V_P}{V_M + V_P + V_Y}\right)^2 X_{MP} + \left(\frac{V_P}{V_M + V_P + V_Y}\right)$$

$$\times \left(\frac{V_Y}{V_M + V_P + V_Y}\right)(X_{MY} + X_{MP} - X_{YP}J_M/J_Y)$$

$$+ \left(\frac{2\psi_M\gamma}{r_oRT}\right)\left(\frac{V_P}{V_M + V_P + V_Y}\right)^{1/3} = 0 \qquad (11)$$

This equation may be used for the estimation of the swelling capacity of the activated seed particles with the monomer. A typical graph sketched based on Eq. (11) is given in Fig. 18. This graph shows the variation of the swelling capacity of the seed polymer particles (V_M/V_P) with the ratio of interfacial tension–initial particle radius

(γ/r_o) for two different concentrations of the swelling agent within the swollen particles (V_Y/V_P). As seen here, the swelling capacity of the seed particles significantly increased with the increasing concentration of the swelling agent especially at low values of γ/r_o. Ugelstad and coworkers [108,109] have shown that monomer–polymer swelling ratio as high as 1:100 can be expected if a swelling agent is used prior to the swelling with the monomer. The polymerization of the highly swollen particles is usually achieved with the oil-soluble initiators. The use of stabilizers in the polymeric or anionic form helps to prevent the coagulation of the swollen particles. The presence of water-soluble inhibitor within the continuous phase is useful for reducing the possibility of nucleation leading to the formation of secondary particles during the repolymerization step.

Ugelstad and coworkers [108] used the activated swelling process to produce uniform polymeric particles starting from 0.63-μm polystyrene seed particles produced by the soapless emulsion polymerization by using sodium persulfate as the initiator. A low-molecular weight, highly water-insoluble compound, 1-chlorododecane was selected as the activator for the swelling of seed particles. The activated swelling procedure followed in this study is summarized in Table 8.

Highly uniform latex particles up to 5 μm in size were obtained by applying the activated swelling procedure. It is also possible to use larger beads as seed particles to achieve uniform final beads with larger sizes. By applying activated swelling method on 1-μm oligomeric seed latex, 10 μm uniform crosslinked poly(styrenedivinylbenzene) particles were also produced by Ugelstad et al. [109]. The relative standard deviation of the diameter was measured as <1%, with a high-precision Coulter instrument [109].

Jansson et al. [110] achieved the high levels of monomer swelling without the utilization of a swelling agent. The polystyrene seed particles with a diameter of 0.49 μm were swollen by the styrene, including benzoyl peroxide, in an aqueous emulsion medium prepared by using sodium dodecylsulfate as the emulsifier. They used acetone in the swelling medium to facilitate the diffusion of monomer through the aqueous phase into the seed particles. The polymerization of monomer within the swollen particles provided uniform latex particles with a diameter of 1.06 μm. But, the formation of small particles in the second stage polymerization was also observed. They also obtained about the same final size and the same swelling ratio by applying the same procedure in the absence of acetone and by using $NaNO_2$ as the water-soluble initiator [110].

Sheu and coworkers [111] produced polystyrene–polydivinylbenzene latex interpenetrating polymer networks by the seeded emulsion polymerization of styrene-divinylbenzene in the crosslinked uniform polystyrene particles. In this study, a series of uniform polystyrene latexes with different sizes between 0.6 and 8.1

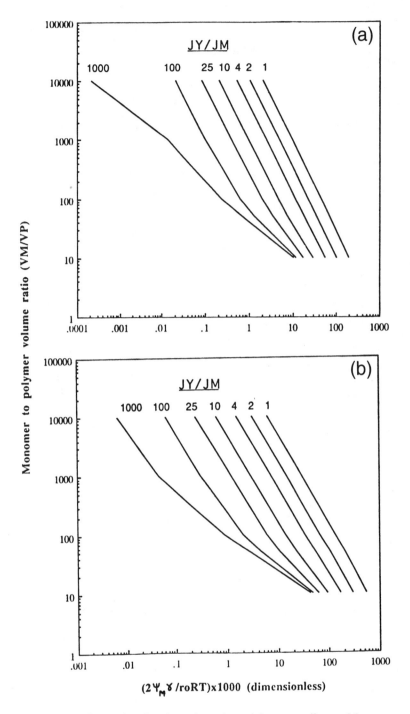

Figure 18   The variation of the swelling ratio of activated seed particles according to Morton equation (X_{MY} = X_{MP} = 0.4, X_{YP} = 0). (a) V_Y/V_P = 1.0, (b) V_Y/V_P = 3.0.

Table 8 The Steps of the Activated Swelling Process [108]

Step I: Preparation of aqueous emulsion of 1-chlorododecane containing benzoyl peroxide as the oil soluble initiator
  (i) Mixing of organic phase containing BPO in 1,2-dichloroethane (DCE) and 1-chlorododecane (CCD) with the aqueous SDS solution.
  (ii) Homogenization of the emulsion in a high pressure homogenizer.
Step II: Swelling of polystyrene seed particles with CDD
  (i) Mixing of the seed latex and CDD emulsion.
  (ii) Adding of acetone 10% volume of the water present in the resulting mixture.
  (iii) Swelling of seed particles with CDD and BPO with ordinary stirring at 35°C.
  (iv) Removal of acetone and DCE by evaporation in vacuum after the swelling is completed.
Step III: Swelling of activated seed particles by the monomer
  (i) An aqueous SDS solution was mixed with the emulsion of activated seed particles and styrene added into the mixture and emulsified.
  (ii) Swelling of activated seed particles with monomer at 30°C.
Step IV: Repolymerization of monomer within the swollen seed particles
  (i) Repolymerization of the monomer within the swollen beads.

μm and with different degrees of crosslinking were prepared by successive seeded emulsion polymerization and were used as seed latexes. The seed particles were swollen with the styrene-divinylbenzene mixture, and the monomer mixture within the swollen beads were polymerized at 70°C. Uniform nonspherical particles, e.g., ellipsoidal and egglike singlets, symmetric and asymmetric doublets, and ice-cream conelike and popcornlike multiplets were obtained as final particles. The nonspherical particles were formed by separation of the second stage monomer from the crosslinked seed network during the swelling and polymerization. The degree of phase separation increased with the increasing crosslinking density of seed particles, monomer–polymer swelling ratio, polymerization temperature, and seed particle size and with the decreasing divinylbenzene concentration in the swelling monomer.

Highly uniform large latex particles were also produced by Okubo et al. [112] by using a different multistage polymerization procedure. The method in their study is termed "dynamic swelling method," which makes it possible for a large amount of monomers to be absorbed into polymer seed particles. This method does not require a swelling agent and includes only one swelling step. This process is developed based on the swelling of the seed particles in a continuous medium having an increasing polarity value by the swelling time, which controls the diffusion rate of the monomer from the continuous phase to the seed particles. The polarity of the swelling medium is continuously increased with time by the slow addition of water with a constant flow rate. Therefore, the solubility of apolar monomer in the continuous phase decreases by the increasing water content. This accelerates the diffusion of monomer from the polar phase to the more apolar one (i.e., the seed particles). The absorption rate of monomer by the seed particles can be controlled by regulating the water flow rate and this leads to the uniformly swollen seed particles by monomer at the end of the process. The size and size distribution of the dynamically swollen seed particles were also compared with those of the swollen seed particles under standard conditions. This comparison revealed that the defects in the monodispersity formed during the conventional swelling process could be completely eliminated by applying the dynamic swelling method. This method was first used for the preparation of 6.1-μm uniform polystyrene particles starting from uniform seed latex 1.8 μm in size [112]. But a small particle fraction about 0.1 μm in size was obtained together with the desired product since a new crop of nucleating particles was formed during the repolymerization step.

B. Introduction of Functional Groups

The presence of functional groups on the surface of the uniform polymeric microspheres is required for the applications involving the chemical derivation of the microspheres (i.e., attachment of ligands with specific recognition abilities and immobilization of biological molecules onto the surface of microspheres, etc.). It is usually difficult to derive the plain polymeric surfaces in the absence of functional groups. Various emulsion and dispersion copolymerization recipes are available in the literature that aim to produce uniform and functional polymeric microspheres in micron-size range. The synthesis of these microspheres usually involves the copolymerization of a hydrophobic, apolar monomer like styrene with a polar comonomer containing a functional group. Acrylate-based functional comonomers are usually preferred in these copolymerizations. Carboxyl, hydroxyl, amide, amine, aldehyde, or chloromethyl functional groups can be incorporated onto the surface of polymeric microspheres by keeping the monodispersity of the final product [84,87,93,113–120]. The polarities and the reactivities of monomer and comonomer play an important role in the selection of the copolymerization method. The dispersion copolymerization procedures developed for the synthesis of uniform latex particles carrying functional groups on their surfaces are summarized in Table 9.

With some limitations, Single-stage soapless emulsion polymerization can be used for the direct copolymerization of a relatively apolar monomer with a polar

Table 9 The Dispersion Copolymerization Procedures Used for the Synthesis of Uniform Latex Particles

Method	Seed latex	Comonomers	Functional group	Reference
DDC	—	S/HEA	Hydroxyl	84
	—	S/DMAEM	Amine	84
	—	S/MAAc	Carboxyl	84
	—	S/AAm	Amide	84
	—	S/ATES	Silane	84
	—	S/VTES	Silane	84
TSDC	PS	S/CMS	Chloromethyl	87
	PS	S/AAc	Carboxyl	93
	PS	S/HEMA	Hydroxyl	93
	PS	S/DMAEM	Amine	93
	PS	GA	Aldehyde	119–120
DSEP	—	S/AAm	Amide	113
	—	S/HEMA	Hydroxyl	113
	—	S/AAc	Carboxyl	114
	—	S/AAm/AAc	Amide-carboxyl	114
	—	S/HEMA/AAc	Hydroxyl-carboxyl	114
SSEC	PS/PMAAc	S/MAAc	Carboxyl	115
	PS/PMMA/PMAAc	S/MMA/MAAc	Carboxyl	115
	PS/PHEMA	S/HEMA	Hydroxyl	116
	PS/PAAc	S/AAc	Carboxyl	117

DSEP: direct soapless emulsion polymerization, SSEC: seeded soapless emulsion copolymerization, DDC: direct dispersion copolymerization, TDSC: two-stage dispersion copolymerization, ATES: Allyl trietoxysilane, VTES: vinyl trietoxysilane, DMAEM: dimethylaminoethyl-
methacrylate, CMS: chloromethylstyrene, GA: glutaraldehyde, AAc: Acrylic acid; Aam: Acrylamide; HEMA: 2-hydroxyethylmethacrylate.

one. The main disadvantages of this method can be summarized as follows.

1.   The water solubilities of the functional comonomers are reasonably high since they are usually polar
compounds. Therefore, the initiation in the water phase
may be too rapid when the initiator or the comonomer
concentration is high. In such a case, the particle growth
stage cannot be suppressed by the diffusion capture
mechanism and the solution or dispersion polymerization of the functional comonomer within water phase
may accompany the emulsion copolymerization reaction. This leads to the formation of polymeric products
in the form of particle, aggregate, or soluble polymer
with different compositions and molecular weights. The
yield for the incorporation of functional comonomer into
the uniform polymeric particles may be low since some
of the functional comonomer may polymerize by an un-
desired mechanism.

2.   The size distribution of the final product may
become wide since the monodispersity cannot be controlled due to the formation of the copolymer particles
with a secondary nucleation.

3.   Too rapid initiation may lead to the agglomeration of the particles, which usually results in the formation of coagulum.

In the organic phase (i.e., monomer + comonomer)-
to-water ratio, the polar comonomer concentration in

the organic phase and the water-soluble initiator concentration are kept at sufficiently low levels, single-stage
soapless emulsion polymerization may be used for producing uniform copolymer microspheres containing
functional groups. But, it should be noted that this
method is usually suitable for the introduction of small
amounts of comonomers into the final uniform particles.
Uniform poly(styrene-acrylamide) [P(S/AAm)] and poly-
(styrene-hydroxyethylmethacrylate) [P(S/HEMA)] latexes were prepared by a direct emulsion copolymerization method by Shirahama et al. [113]. A similar procedure was also used for the synthesis of uniform poly(styrene-acrylic acid) [P(S/AAc)], poly(styrene-acrylic acid-
acrylamide), [P(S/AAm/AAc)], and poly(styrene/2-hy-
droxyethylmethacrylate/acrylic   acid)   [P(S/HEMA/
AAc)] copolymer microspheres in the another study of
the same group [114]. The sizes of the uniform copolymer latices prepared by direct soapless emulsion
polymerization are given in Table 10.

Styrene monomer was also copolymerized with a
series of functional monomers by using a single-step dispersion copolymerization procedure carried out in
ethanol as the dispersion medium by using azobisizobutyronitrile and polyvinylpyrrollidone as the initiator and
the stabilizer, respectively [84]. The comonomers were
methyl methacrylate, hydroxyethyl acrylate, methacrylic acid, acrylamide, allyltrietoxyl silane, vinyl polydimethylsiloxane, vinylsilacrown, and dimethylamino-

Table 10 The Size of the Uniform Copolymer Latices Prepared by Direct Soapless Emulsion Copolymerization [113,114]

Latex	Styrene/comonomer (M/M)	Size (μm)	Reference
P(S/AAm)	0.768/0.281	0.458	113
P(S/AAc)	1.0/0.02	0.560	114
P(S/AAm/AAc)	1.0/0.02/0.02	0.440	114
P(S/HEMA/AAc)	1.0/0.02/0.02	0.454	114

ethylmethacrylate. The comonomer was added into the dispersion polymerization medium as 1.0% based on the monomer. The uniform copolymer latices in the size range of 2.4–3.7 μm were achieved with a single-step process. The variation of the average size of the final product with the comonomer type was attributed to the formation of costabilizer in situ by the copolymerization reaction between styrene and comonomer.

Soapless seeded emulsion copolymerization has been proposed as an alternative method for the preparation of uniform copolymer microspheres in the submicron-size range [115–117]. In this process, a small part of the total monomer–comonomer mixture is added into the water phase to start the copolymerization with a lower monomer phase–water ratio relative to the conventional direct process to prevent the coagulation and monodispersity defects. The functional comonomer concentration in the monomer–comonomer mixture is also kept below 10% (by mole). The water phase including the initiator is kept at the polymerization temperature during and after the addition of initial monomer mixture. The nucleation takes place by the precipitation of copolymer macromolecules, and initially formed copolymer nuclei collide and form larger particles. After particle formation with the initial lower organic phase–water ratio, an oligomer initiated in the continuous phase is captured by an existing particle before it grows to some threshold degree of polymerization for aggregative or self-nucleation to take place.

Higher organic phase–water ratio and higher concentration of the comonomer in the polymerization medium increase the concentration of growing oligomers and growth rate of the oligomer chains, thus favoring secondary nucleation leading to the polidispersity in the final product. The rest of the monomer–comonomer mixture is continuously fed into the reactor, over a wide period of time after the nucleation step. The reactor is also kept at the polymerization temperature during this period. Therefore, the slow addition of the monomer mixture makes the complete absorption of monomer and comonomer from the aqueous medium by the seed particles possible. So, the copolymerization reaction progresses within the individual uniform beads without producing new particles. The formation of larger microspheres stabilized by the initiator residues also decreases the risk of coagulation in this process.

The soapless seeded emulsion copolymerization method was used for producing uniform microspheres prepared by the copolymerization of styrene with polar, functional monomers [115–117]. In this series, polystyrene-polymethacrylic acid (PS/PMAAc), polystyrene-polymethylmethacrylate-polymethacrylic acid (PS/PMMA/PMAAc), polystyrene-polyhydroxyethylmethacrylate (PS/PHEMA), and polystyrene-polyacrylic acid (PS/PAAc) uniform copolymer microspheres were synthesized by applying a multistage soapless emulsion polymerization process. The composition and the average size of the uniform copolymer latices prepared by multistage soapless emulsion copolymerization are given in Table 11.

The uniform polymeric microspheres in submicron- or micron-size range can also be prepared as seed particles by the soapless emulsion or dispersion polymerization of a hydrophobic monomer like styrene. The uniform seed particles are swollen with the organic phase including functional comonomer, monomer, and oil-soluble initiator at a low temperature in an aqueous

Table 11 The Composition and the Average Size of the Uniform Copolymer Latices Prepared by the Multistage Soapless Emulsion Copolymerization [115–117]

Latex	Initial composition (styrene/comonomer) (M/M)	Size (μm)	Reference
PS/PMAAc	1.62/0.085	0.578	115
PS/PMMA/PMAAc	1.08/0.54/0.087	0.488	115
PS/PHEMA	1.60/0.085	0.510	116
PS/PHEMA	1.48/0.165	0.491	116
PS/PAAc	1.67/0.035	0.541	117
PS/PAAc	1.65/0.085	0.515	117

The initial compositions were calculated based on the reported amounts of the ingredients.

emulsion medium. The initial weight ratio of the organic phase to the seed particles is fixed to a certain value in which the organic phase is completely absorbed by the seed particles from the emulsion medium to keep the monodispersity. The next step is the seeded polymerization or copolymerization of the functional monomer within the individual swollen seed particles. The copolymerization reaction proceeds smoothly without producing new particles, so the monodispersity in the final product is achieved.

A series of uniform polymer microspheres having different functional groups on their surfaces were prepared by Okubo et al. [87]. A two-step polymerization process was followed for the preparation of uniform polystyrene microspheres having chloromethyl group [87]. In this study, first, uniform polystyrene seed particles 1.9 μm in size were prepared by the dispersion polymerization of styrene in the ethanol–water medium by using azobisizobutyronitrile as the initiator. Polyacrylic acid having a viscosity average molecular weight of 2.0×10^5 was used as the stabilizer in the preparation of seed particles. Prior to the seeded copolymerization, the swelling medium was stirred at 0°C for 24 h for the absorption of monomers by the seed particles. The swelling medium was comprised of polystyrene seed emulsion, styrene, chloromethylstyrene, and AIBN. The 2.1-μm uniform polymer particles, having chloromethyl group on their surfaces, were obtained by the seeded copolymerization carried out at 70°C [87].

Okubo and coworkers [118] also produced uniform polymer microspheres having cationic groups by reacting the crosslinked uniform polystyrene particles having a chloromethyl group with polyamines. Triethylenetetraamine or ethylenediamine was selected to create the cationic groups on the surface of microspheres, which were prepared by following the two-step polymerization process previously described [87]. The produced microspheres exhibited positive zeta potential values within the acidic pH range due to the presence of cationic groups on their surfaces.

Submicron-size uniform polymer particles having aldehyde groups were synthesized by Okubo et al. [119,120] by using a two-stage polymerization method. In these studies, uniform polystyrene seed particles 0.47 μm in size were obtained by the soapless emulsion polymerization of styrene by using potassium persulfate as the initiator. Glutaraldehyde was polymerized on the surface of polystyrene seed particles within the aqueous-alkaline dispersion medium. Prior to the polymerization, polystyrene seed emulsion and glutaraldehyde were stirred at the acidic pH value to make glutaraldehyde absorb into polystyrene seed particles. The concentration of free aldehyde groups on the surface of polymeric microspheres was found as 4.8×10^{-6} mol/m^2 after cleaning of the microspheres by the serum replacement method [119].

We have also produced uniform polymeric particles containing functional groups on their surfaces. These latices were prepared by a two-step polymerization technique aiming the copolymerization of styrene with a series of acrylate comonomers [93]. The comonomers, acrylic acid (AAc), 2-hydroxyethylmethacrylate (HEMA), and dimethylaminoethylmethacrylate (DMAEM), were selected for the synthesis of uniform polystyrene/poly(acrylic acid) (PS/PAAc), polystyrene/poly(hidroxyethylmethacrylate) (PS/PHEMA), and polystyrene/poly(dimethylaminoethylmethacrylate)(PS/PDMAEM) copolymer microspheres, respectively. The uniform polystyrene seed latex 2.3 μm in size as prepared by the dispersion polymerization of styrene [93]. A typical procedure for the preparation of copolymer latices is summarized below [93]. A seed monomer solution containing styrene (2 ml), functional acrylate monomer (4 ml), and the initiator (0.12 g) was added into the seed latex dispersion prepared by the dilution of the seed latex (100 ml) with 100 ml of distilled water. The resulting mixture was stirred 24 h at room temperature for absorption of the monomer solution by the seed particles. At the end of the absorption period, the temperature was increased first to 75°C and kept at this value for 8 h for the copolymerization. The copolymerization was continued at 82°C for 16 h. The scanning electron micrographs of the copolymer latices are given in Fig. 19. As seen here, the uniform copolymer latex particles were obtained by applying a two-stage polymerization method and by using highly uniform polystyrene seed particles. The presence of functional groups of the comonomers on the microsphere surface was shown by FTIR-ATR and x-ray photoelectron spectroscopy.

In another study [2], uniform polystyrene latex particles containing hydroxyl groups were produced by the crosslinking of polyvinyl alcohol ($M_R = 14.000$, 100% hydrolized) on the surface of uniform polystyrene microspheres produced by the dispersion polymerization. The uniform seed latex 4.0 μm in size was produced by the dispersion polymerization of styrene (35 ml) in the ethanol (100 ml)–methoxyethanol (100 ml) medium. AIBN (0.75 g) and polyacrylic acid (3.5 g) were used as the initiator and the stabilizer, respectively. The polymerization was carried out at 75°C for 16 h and at 80°C for 8 h with 250 rpm stirring rate. The seed latex was washed with distilled water by the serum replacement method. Polyvinylalcohol (70 mg) was adsorbed onto the surface of the polystyrene microspheres (3.0 g) in the aqueous medium (100 ml) containing Na$_2$SO$_4$ (ionic strength of Na$_2$SO$_4$ = 0.2). The PVA adsorption was carried out at room temperature for 2 h with 200 rpm stirring rate. The amount of polyvinyl alcohol adsorbed onto the polystyrene microspheres was determined as 19.0 mg/g by the KI/I$_2$ method. At the end of the adsorption period, the final acid concentration of the medium was fixed to 0.1 M by the addition of aqueous HCl solu-

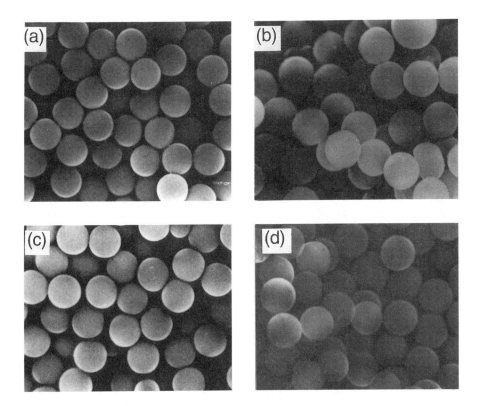

Figure 19 The scanning electron micrographs of the polystyrene seed latex and the copolymer latices carrying carboxyl, hydroxyl and amine functional groups. (a) PS/PAA, (b) PS/PHEMA, (c) PS/PDMAEM. The original SEM photographs were taken with 10,000× magnification and reduced at a proper ratio to place the figure. (From Ref. 93. Reproduced with the permission of John Wiley & Sons, Inc.)

tion. The aqueous solution of crosslinker terephatalaldehyde (TPA) (10 ml, 1 mg TPA/ml) was added into the resulting dispersion. The medium was stirred at 500 rpm at 25°C for 48 h. The temperature then was increased to 80°C and the dispersion was kept at this temperature for 4 h for completion of the crosslinking reaction. The presence of the hydroxyl group on the microsphere surface was shown by IR spectroscopy.

In another study, uniform composite polymethylmethacrylate/polystyrene (PMMA/PS) composite particles in the size range of 1–10 μm were prepared by the seeded emulsion polymerization of styrene [121]. The PMMA seed particles were initially prepared by the dispersion polymerization of MMA by using AIBN as the initiator. In this polymerization, poly(N-vinyl pyrollidone) and methyl tricaprylyl ammonium chloride were used as the stabilizer and the costabilizer, respectively, in the methanol medium. Seed particles were swollen with styrene monomer in a medium comprised of seed particles, styrene, water, poly(N-vinyl pyrollidone), Polywet KX-3 and aeorosol MA emulsifiers, sodium bicarbonate, hydroquinone inhibitor, and azobis(2-methylbu-

tyronitrile). The uniform composite particles were obtained by the polymerization carried out within the individual swollen particles. The unique morphology of the composite particles comprised three types of polystyrene domains embedded in a continuous PMMA matrix: the dispersed "internal" domains in the interior, the interconnected "subsurface" domains that form a crust beneath the surface, and the separated "surface" domains at the surface [121].

The uniform latices containing reactive groups, styrene-acrylonitrile (S/AN) [122] and styrene-glycidyl methacrylate (S/GMA) [123] were also prepared by the coreshell emulsion copolymerization and by the soapless emulsion copolymerization method, respectively. In the preparation of P(S/GMA) copolymer particles, S and GMA were copolymerized in an aqueous medium by using potassium peroxydisulfate as the initiator at 65°C. The average size was changed between 0.22–0.44 μm by changing the initiator concentration and ionic strength of the medium. The reactive oxine groups of the latex particles were modified later by hydrolysis, ammonolysis reaction with Na_2S, or periodic acid oxidation of the hydrolyzed or ammonolyzed form of the latex particles.

PS/PHEMA particles in micron-size range were also obtained by applying the single-stage soapless emulsion copolymerization method [124]. But, this method provided copolymer particles with an anomalous shape with an uneven surface. PS or PHEMA particles prepared by emulsifier-free emulsion polymerization were also used as seed particles with the respective comonomer to achieve uniform PS/PHEMA or PHEMA/PS composite particles. PS/PHEMA and PHEMA/PS particles in the form of excellent spheres were successfully produced 1 μm in size in the same study.

C. Uniform Macroporous Particles

Uniform macroporous polymer particles have been prepared in the size range of 5–20 μm by the multistage emulsion polymerization methods. Several methods are available in the literature describing the synthesis and the properties of macroporous uniform particles. The main steps of these methods may be summarized as follows.

1. Preparation of uniform seed particles: Soapless emulsion polymerization is usually preferred for the preparation of uniform seed particles since this technique provides emulsifier-free, larger, and highly uniform microspheres relative to those that can be obtained by the conventional emulsion recipes including emulsifiers and various additives. The size of uniform seed particles with the soapless emulsion procedure is in the range of 0.6–1.2 μm depending on the polymerization conditions [75,108].

The use of the dispersion polymerization method provides uniform seed particles in the size range of 2.0–8.0 μm. But, the conditions of dispersion polymerization should be selected carefully for the synthesis of a seed latex having highly uniform character. Since the multistep swelling procedure involves the systematic enlargement of primary latex particles, any defect present in the seed particles, such as a lack of size uniformity, will also be magnified by the swelling process. Therefore, it is desirable to use as a starting material with beads that are as large as possible [75]. The other advantage of larger seed particles is a reduction in the time required to swell them to the desired size and, therefore, a reduction in the occurrence of undesirable processes, such as coalescence, etc.

2. Swelling of seed particles by an organic phase: For this purpose, the seed particles are redispersed in the aqueous emulsion of the organic phase including an inert diluent, monomer, crosslinker, and oil-soluble initiator prepared by using an anionic or nonionic emulsifier. The swelling process is usually carried out in an aqueous emulsion at a low temperature value (i.e., room temperature or 0°C) with mild stirring regime [125]. The activated

swelling method is preferred by some researchers to achieve the swelling of seed particles [109].

3. Repolymerization of monomer and crosslinker within the swollen particles: The repolymerization step is started after transferring all the monomer, crosslinker, initiator, and diluent (solvent or nonsolvent) from the continuous medium into the seed particles by the diffusion process through the aqueous medium. The presence of monomer in the continuous medium after completion of the swelling of seed particles usually leads to broad-size distribution of the final product. Therefore, the initial weight ratio of monomer-to-seed particles should be fixed to a value by which the seed particles absorb all the monomer from the aqueous medium. In other words, this ratio should not exceed the equilibrium swelling value of the seed particles, which can be achieved by the studied conditions of the swelling medium. The use of high organic phase–seed latex ratios can lead to the coalescence and the agglomeration of particles during the repolymerization step due to the soft and sticky character of forming particles. The polymerization of monomer and crosslinker within the swollen seed particles is usually carried out at around 70°C with mild stirring or shaking regimes over a period of 24 h. The macroporous matrix builds up during the repolymerization step according to the mechanism described by Cheng et al. [125]. It is essential to use oil-soluble initiators for achieving the polymerization of monomer within the swollen seed particles. In some cases, a polymeric stabilizer can be added into the medium prior to the repolymerization step to prevent the coalescence and/or the agglomeration of forming macroporous particles [126,127,128]. The use of water-soluble inhibitors in this step reduces the formation of smaller particles by the nucleation of oligomers produced by the polymerization of the monomer remaining in the continuous phase of the swelling medium [127]. The repolymerization step provides macroporous uniform particles containing inert organic diluents and linear polymer.

4. Removal of diluent by an extraction process: To obtain the final stable macroporous structure, the liquid organic diluents and the linear polymer are removed from the crosslinked structure by extraction with a good solvent for the inert diluents and particularly for the linear polymer. Toluene or methylene chloride are usually preferred for the removal of linear polystyrene from the divinylbenzene crosslinked macroporous polystyrene particles [125,128]. The extraction is carried out within a Soxhelet apparatus at the boiling point of the selected solvent over a period usually more than 24 h.

A typical scanning electron micrograph of the uniform macroporous polystyrene–divinylbenzene particles is given in Fig. 20. First studies on the synthesis of macroporous uniform particles were started by Ugelstad et al. [109]. They used a two-step activated swelling method to obtain macroporous uniform particles in the

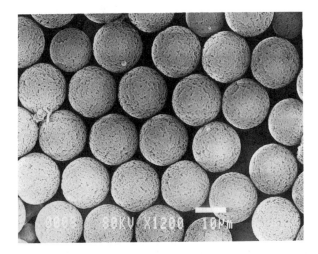

Figure 20 A typical scanning electron micrograph of the macroporous uniform poly(styrene-divinylbenzene) latex particles, Magnification: 1200×, (particle size = 16.0 μm average pore diameter = 200 nm).

size range of 1–100 μm. In this method, uniform polystyrene particles prepared by soapless emulsion polymerization are first swollen with a low-molecular weight organic compound like dodecyl chloride or dioctyl phatalate including an oil-soluble initiator. The oil-soluble initiator also acts as an activator for the adsorption of monomer in the next step by the swollen seed particles. The first step swelling process is performed in the aqueous emulsion of the organic compound prepared with an anionic emulsifier (i.e., sodium lauryl sulfate). An inertial organic solvent like 1,2-dichloroethane can also be added into the organic phase to increase the solubility of the initiator. Second step swelling is performed by the monomer including the crosslinker within the aqueous emulsion of the monomer phase. Porous matrices are obtained when polymerization and crosslinking take place in the presence of inert diluents, which leads to the formation of permanent pores in the material after removal of the diluent. The monomer type and reactivity, degree of crosslinking, amount of diluent, and diluent solvency for the polymer were found as the most important parameters controlling the pore structure of the uniform particles [126].

A research group in Lehigh University has extensively studied the synthesis and characterization of uniform macroporous styrene-divinylbenzene copolymer particles [125,126]. In their studies, uniform porous polymer particles were prepared via seeded emulsion polymerization in which linear polymer (polystyrene seed) or a mixture of linear polymer and solvent were used as inert diluents [125]. The average pore diameter was on the order of 1000 Å with pore volumes up to

0.9 ml/g and specific surface areas up to 200 m²/g. The uniform polystyrene latex 8.7 μm in size was used as a seed latex in the preparation of uniform macroporous particles. The ingredients of the successive seeded emulsion polymerization were the seed particles, styrene, divinylbenzene, 2-2′-azobis-(2-methylbutyronitrile) initiator, hydroquinone inhibitor, sodium bicarbonate buffer, and aerosol MA, Polywet KX-3, and polyvinylpyrrolidone emulsifiers. The swelling of seed particles was carried out in the aqueous emulsion of the ingredients at room temperature. The polymerization was conducted after the swelling step at 70°C for 24 h. A typical polymerization recipe used for the synthesis of macroporous uniform polymeric particles 11 μm in size is given in Table 12.

The physical characteristics of uniform macroporous particles produced in the existence of different diluents are exemplified in Table 13. Some important results obtained in this study are summarized as follows:

1. The macroporous particles prepared by using only linear polystyrene as diluent yielded lower pore volume and specific surface area values.
2. The pore volume and the specific surface area of the uniform macroporous particles increased and the average pore size decreased with the increasing divinylbenzene concentration within the monomer phase.
3. The shape of the pore size distribution curve strongly depends on the molecular weight distribution of the linear polymer. The narrowest pore size distributions were obtained with the linear polymers having the lowest polydispersity indices.

Table 12 Typical Recipe for 11-μm Diameter Uniform Macroporous Polymer Particles

Ingredients	Weight percent (%)
Polystyrene seed particles (8.7 μm)	7.5
Aerosol MA	0.005
Polywet KX-3	0.014
Poly-N-vinyl pyrrolidone (K-90)	0.8
Water	69.115
Monomers:	
Styrene	13-Variable
Divinylbenzene 55	Variable[a]
n-Hexane	9.5
2-2′-Azobis-(2-methylbutyronitrile) initiator	0.018
Sodium bicarbonate buffer	0.027
Hydroquinone inhibitor	0.021

[a] Different levels: 5, 8, 15, 25, and 33% divinylbenzene content based on total polymer and monomers.
Source: Ref. 125. Reproduced with the permission of John Wiley & Sons, Inc.

Table 13 Physical Characteristics of Uniform Porous Polymer Particles ($M_{w,LP} = 1.49 \times 10^6$, 15% DVB)

Diluent	$d(\mu m)$	$S(m^2/g)$	$Vp(mL/g)$	$\Phi(\%)$	$\rho_a(g/mL)$
LP/n-hexane	11	44	0.75	45	0.593
LP/n-heptane	11	54	0.73	44	0.600
LP/n-hexanol	11	35	0.72	43	0.604
LP	13	16	0.40	30	0.748
LP/toluene	12	32	0.72	43	0.604

LP: linear polystyrene; d: diameter; S: specific surface area; Vp: specific volume; Φ: porosity; ρ_a: apparent density.
Source: Ref. 125. Reproduced with the permission of John Wiley & Sons, Inc.

Cheng et al. [126] proposed a pore formation mechanism for the synthesis of uniform macroporous polymer particles. According to the proposed mechanism, the pore formation is a two-step process. The first step is termed as the formation and agglomeration of highly crosslinked gel microspheres within the particle structure. At the beginning of this step, linear copolymer chains with pendant vinyl groups are formed by the copolymerization reaction between divinylbenzene and styrene. The linear chains are converted into crosslinked structures and two different phases appear within the swollen particles by the progressing copolymerization. These phases are termed as the crosslinked copolymer rich phase and the diluent (i.e., linear polymer + nonsolvent) rich phase. The crosslinked copolymer chains are separated by forming gel microspheres with the increasing crosslinking density. The macrogelation of whole particles takes place at a certain conversion of the seed monomer and the formed structure contains the agglomerated gel microspheres. The second stage in the pore formation process is the binding and the fixation of microspheres and agglomerates within the particles. The precipitated gel microspheres are bound and fixed within the particle structure by the polymer chains including dominantly monovinyl groups by the progressing polymerization. The voids between the microspheres and agglomerates are filled with the porogen solution including linear polymer and nonsolvent. The macroporous structure is obtained after the removal of the diluent by an extraction process.

Macroporous uniform polymeric particles were also prepared by using a multistep-activated swelling procedure. The synthesis method followed in this study can be summarized as follows [128]. The 1.1-μm uniform polystyrene latex particles produced by soapless emulsion polymerization were first swollen by dibutyl phatalate and then with styrene or styrene–methyl methacrylate containing benzoyl peroxide as the oil-soluble initiator. These two swelling steps were carried out at room temperature in the aqueous emulsion media of the organic phases including sodium lauryl sulfate as the emulsifier. After the second swelling step, polyvinyl alcohol and sodium nitrite were added into the resulting emulsion as the stabilizer and the water-soluble inhibi-

tor, respectively. The next step was the repolymerization of styrene or styrene–methylmethacrylate mixture within the individual seed particles at 70°C for 24 h. After this stage, the swelling of uniform seed particles by the second monomer phase including styrene and divinylbenzene and benzoyl peroxide was conducted under similar conditions to those of the previous steps. The monomer mixture including the crosslinker was repolymerized within the individual beads after the swelling process was completed. After removal of the secondary small particles in the second stage polymerization by decantation, the uniform macroporous particles were washed with water and methanol and extracted with toluene for obtaining the stable pore structure.

In this study, dibutyl phatalate and the linear polymer obtained in the first repolymerization step were used as porogen. The number average molecular weight of the linear polystyrene was changed between 511.000 and 4100 by changing the BPO concentration within the monomer phase in the first repolymerization step. The results indicated that the median pore size decreased from 283 nm to 49 nm by the decreasing molecular weight of the linear polymer. No significant change was observed in the median pore size when poly(styrene-co-methylmethacrylate) was used as a porogen instead of linear polystyrene.

We prepared uniform poly(styrene-divinylbenzene) macroporous beads in the size range of 5.0–15.0 μm by following the activated swelling procedure [129]. The uniform polystyrene seed latices with different average molecular weights and in the size range of 1.9–7.5 μm were prepared by dispersion polymerization. The seed latices were swollen in the aqueous emulsion medium, first by dibutyl phatalate and then by the monomer phase including only divinylbenzene 55 or styrene-divinylbenzene 55 mixture and benzoyl peroxide. The crosslinker and monomer were polymerized within the swollen beads by using polyvinyl alcohol as the stabilizer. The isolated uniform macroporous beads were extracted with methylene chloride.

The developed method reduced the number of steps required for the synthesis of uniform macroporous beads since the larger uniform seed latices were prepared by the dispersion polymerization relative to those obtained with the soapless emulsion process. The results indicated that the average pore size was strongly dependent on the porogen viscosity. As a conclusion, an increase in the porogen viscosity resulted in an appreciable increase in the average pore size. The smaller seed particles (1.9–3.6 μm) with higher average molecular weights provided craterlike macroporous structures including macropores up to 2000 nm. The macroporous structures having average pore size values between 100–500 nm were obtained by the larger seed particles (5.5–7.5 μm) with relatively lower average molecular weights.

REFERENCES

1. M. T. Ercan, A. Tuncel, B. Caner, M. Mutlu and E. Pişkin, *Nucl. Med. & Biol., 18*: 253 (1991).
2. A. Tuncel, A. Denizli, D. Purvis, C. R. Lowe and E. Pişkin, *J. Chrom., 634*: 161 (1993).
3. E. Pişkin, A. Tuncel, A. Denizli and H. Ayhan, *J. Biomater. Sci., Polym. Ed., 5*: 451 (1994).
4. F. Hoffman, K. Dellbruck and K. Gottlob, German Patent, DRP 250, 609 (1909).
5. M. Luther, C. Heuck, German Patent, DRP 558, 890 (1927).
6. L. B. Bangs and M. T. Kenny, *Industrial Research, 18*: 46 (1976).
7. J. Ugelstad, P. C. Mork, K. H. Kaggerud, T. Ellingsen and A. Berge, *Adv. Colloid & Interface Sci., 13*: 101 (1980).
8. J. Ugelstad, P. C. Mork, A. Berge, T. Ellingsen and A. A. Khan, *Emulsion Polymerization* (I. Piirma, ed.), Academic Press, New York, Chapter 11, (1982).
9. K. E. J. Barret, *Dispersion Polymerization in Organic Media*, John Wiley & Sons, New York, (1975).
10. W. V. Smith and R. H. Ewart, *J. Phys. Chem., 16*: 592 (1948).
11. A. E. Alexander and D. H. Napper, *Prog. Polym. Sci., 3*: 145 (1971).
12. M. Nomura, M. Harada, W. Eguchi and S. Nagata, *Polym. Preprint Am. Chem. Soc., Div. Polym. Chem., 16*: 217 (1975).
13. R. M. Fitch and C. H. Tsai, *Polymer Colloid I* (R. M. Fitch, ed.) Plenum Press, New York, Chapter 5, (1971).
14. A. R. Goodal, M. C. Wilkinson and J. Hearn, *Prog. Colloid Interface Sci., 53*: 327 (1975).
15. C. Y. Chen and I. Piirma, *J. Polym. Sci., Polym. Chem. Ed., 18*: 1979 (1980).
16. H. J. Van Den Hull and J. W. Wanderhoff, *J. Electroanal. Chem. Interfacial Electrochem., 37*: 161 (1972).
17. R. H. Ottewill, *Emulsion Polymerization* (I. Piirma, ed.) Academic Press, New York, Chapter 1, (1982).
18. W. D. Harkins, *J. Am. Chem. Soc., 69*: 1428 (1947).
19. W. D. Harkins, *J. Polym. Sci., 5*: 217 (1950).
20. F. K. Hansen and J. Ugelstad, *J. Polym. Sci., Polym. Chem. Ed., 17*: 3069 (1979).
21. A. G. Parts, D. E. Moore and J. G. Watterson, *Makromol. Chem., 89*: 156 (1965).
22. M. Harada, M. Nomura, H. Kojima, W. Eguchi and S. Nagata, *J. Appl. Polym. Sci., 16*: 811 (1972).
23. B. M. E. Van Der Hoff, *Polymerization and Polycondensation Process*, Advances in Chemistry Series, American Chemical Society, Washington D.C. (1962).
24. A. Tuncel and E. Pişkin, *Polym. Plast. Technol. & Eng., 31*: 787 (1992).
25. F. K. Hansen and J. Ugelstad, *Emulsion Polymerization* (I. Piirma, ed.) Academic Press, New York, Chapter 2, (1982).
26. J. Ugelstad, M. S. El-Aasser and J. W. Wanderhoff, *J. Polym. Sci., Polym. Lett. Ed., 11*: 505 (1973).
27. J. Ugelstad and F. K. Hansen, *Rubber Chem. Technol., 49*: 536 (1976).
28. J. L. Gardon, *J. Polym. Sci., Polym. Chem. Ed., 6*: 623 (1968).
29. J. L. Gardon, *J. Polym. Sci., Polym. Chem. Ed., 6*: 643 (1968).
30. J. L. Gardon, *J. Polym. Sci., Polym. Chem. Ed., 6*: 665 (1968).
31. J. L. Gardon, *J. Polym. Sci., Polym. Chem. Ed., 6*: 687 (1968).
32. J. L. Gardon, *J. Polym. Sci., Polym. Chem. Ed., 6*: 2853 (1968).
33. J. L. Gardon, *J. Polym. Sci., Polym. Chem. Ed., 6*: 2859 (1968).
34. J. L. Gardon, *J. Polym. Sci., Polym. Chem. Ed., 9*: 2763 (1971).
35. M. Harada, H. Nomura, W. Eguchi and S. Nagata, *J. Chem. Eng. Jpn., 4*: 54 (1971).
36. H. Nomura, M. Harada, K. Nakagawara, W. Eguchi and S. Nagata, *J. Chem. Eng. Jpn., 4*: 160 (1971).
37. F. K. Hansen and J. Ugelstad, *Macromol. Chem., 180*: 2423 (1979).
38. R. M. Fitch, *Br. Polym. J., 5*: 467 (1973).
39. R. M. Fitch and L. B. Shih, *Prog. Colloid Polym. Sci., 56*: 1 (1975).
40. A. Tuncel, Doctoral Thesis, Hacettepe University, Institute of Pure and Applied Sciences, Ankara (1989).
41. A. Tuncel and E. Pişkin, *Polym. Plast. Technol. & Eng., 31*: 807 (1992).
42. I. M. Kolthoff, P. R. O'Connor and J. L. Hansen, *J. Polym. Sci., 15*: 459 (1955).
43. P. J. Flory, *J. Am. Chem. Soc., 59*: 241 (1937).
44. H. Warson, *Emulsion Polymerization* (I. Piirma and J. L. Gardon, eds.) p. 228–235 (1976).
45. A. S. Dunn, *Emulsion Polymerization* (I. Piirma, ed.) Academic Press, New York, Chapter 6, (1982).
46. W. C. Griffin, *J. Soc. Cosmet. Chem., 1*: 311 (1949).
47. W. C. Griffin, *Encyclopedia of Chemical Technology*, Kirk-Othmer, 3rd ed., *8*: p. 900 (1980).
48. W. C. Griffin, *J. Soc. Cosmet. Chem., 5*: 249 (1957).
49. J. T. Davies, *Proc. Int. Congr. Surface Activity, 1*, Butterworths, London, p. 426, (1957).
50. G. H. Greth and J. E. Wilson, *J. Appl. Polym. Sci., 5*: 135 (1961).
51. E. Bartholome, H. Gerrens, R. Herberck, and H. M. Weitz, *Z. Elektrochem., 60*: 334 (1956).
52. A. S. Dunn, W. A. Al-Shahib, *J. Polym. Sci. Polym. Chem. Ed., 16*: 677 (1978).
53. S. H. Maron, M. E. Elder and I. N. Ulevitch, *J. Colloid Sci., 9*: 89 (1954).
54. I. Piirma and S. R. Chen, *J. Colloid Interface Sci., 74*: 90 (1980).
55. T. R. Paxton, *J. Colloid Interface Sci., 31*: 19 (1969).
56. W. M. Sawyer and S. J. Rehfeld, *J. Phys. Chem., 67*: 1973, 1963.
57. S. S. Medvedev, A. V. Zuikov, I. A. Gritskova and V. V. Dudukin, *Polym. Sci., USSR, 13*: 1572 (1971).
58. E. G. Bobalek and D. A. Williams, *J. Polym. Sci., Polym. Chem. Ed., 4*: 3065 (1966).

59. H. J. Van Den Hull and J. W. Wanderhoff, *Brit. Polym. J., 2*, 121, 1970.
60. J. W. Vanderhoff, J. F. Vitkuske, E. B. Bradford and T. E. Alfrey, *J. Polym. Sci., 20*: 225 (1956).
61. F. A. Bovey, M. Kolthoff, A. I. Medalia and E. J. Meehan, *Emulsion Polymerization*, Vol. IX in High Polymer Series, Interscience Publishers Inc., New York (1955).
62. C. E. Schildknecht, *Vinyl and Related Polymers*, John Wiley & Sons Inc., New York (1961).
63. G. S. Whitby, *Synthetic Rubber*, John Wiley & Sons Inc., New York (1954).
64. C. E. Schildknecht, *Polymer Processes*, Interscience Publishers Inc., New York (1961).
65. G. Messwarb, E. Pasckhe and P. Seibel, *Angew. Chem., 71*: 604 (1959).
66. A. Rembaum, S. P. S. Yen, and W. Wolksen, *Chemtech, March*: 182 (1978).
67. R. S. Molday, W. J. Dreyer, A. Rembaum and S. P. S. Yen, *J. Cell Biol., 64*: 75 (1975).
68. R. H. Ottewill and N. J. Shaw, *Colloid & Polym. Sci., 218*: 34 (1976).
69. J. W. Goodwin, R. H. Ottewill, R. Pelton, G. Vionello and D. E. Yates, *Brit. Polym. J., 10*: 173 (1978).
70. J. W. Goodwin, R. H. Ottewill and R. Pelton, *Colloid & Polym. Sci., 257*: 61 (1979).
71. R. M. Fitch, *Macromolecules* (H. Benoit and P. Remp, eds.), Pergamon Press, New York, (1982).
72. Z. Sang and G. W. Poehlein, *J. Colloid Sci., 128*: 501 (1989).
73. Z. Sang and G. W. Poehlein, *J. Polym. Sci. Polym. Chem. Ed., 28*: 2359 (1989).
74. J. W. Goodwin, J. Hear, C. C. Ho and R. H. Ottewill, *Brit. Polym. J., 5*: 347 (1973).
75. V. Smigol, F. Svec, K. Hosoya, Q. Wang and J. M. J. Frechet, *Die Angewandte Macromol. Chem., 195*: 151 (1992).
76. R. Arshady, *Colloid & Polym. Sci., 270*: 717 (1992).
77. S. Shen, E. D. Sudol and M. S. El-Aasser, *J. Polym. Sci., Polym. Chem. Ed., 32*: 1087 (1994).
78. K. E. J. Barret, *Brit. Polym. J., 5*: 259 (1973).
79. T. Corner, *Colloids & Surfaces, 3*: 119 (1981).
80. Y. Almog, S. Reich and M. Levy, *Brit. Polym. J., 14*: 131 (1982).
81. K. P. Lok and C. K. Ober, *Can. J. Chem., 63*: 209 (1985).
82. C. K. Ober, K. P. Lok and M. L. Hair, *J. Polym. Sci., Polym. Lett. Ed., 23*: 103 (1985).
83. C. K. Ober, K. P. Lok and M. L. Hair, *Macromolecules, 20*: 268 (1987).
84. C. M. Tseng, Y. Y. Lu, M. S. El-Aasser and J. W. Vanderhoff, *J. Polym. Sci., Polym. Chem. Ed., 24*: 2995 (1986).
85. A. Paine, *J. Polym. Sci., Polym. Chem. Ed., 28*: 2485 (1990).
86. Y. Y. Lu, M. S. El-Aasser and J. W. Vanderhoff, *J. Polym. Sci., Polym. Phys. Ed., 26*: 1187 (1988).
87. M. Okubo, K. Ikegami and Y. Yamamoto, *Colloid & Polym. Sci., 267*: 193 (1989).
88. Y. Chen and H. W. Yang, *J. Polym. Sci., Polym. Chem. Ed., 30*: 2765 (1992).
89. A. Tuncel, R. Kahraman and E. Pişkin, *J. Appl. Polym. Sci., 50*: 303 (1993).
90. S. Shen, E. D. Sudol and M. S. El-Aasser, *J. Polym. Sci., Polym. Chem. Ed., 31*: 1393 (1993).
91. K. Li, D. Harald and H. Stöver, *J. Polym. Sci., Polym. Chem. Ed., 31*: 2473 (1993).
92. K. Li, D. Harald and H. Stöver, *J. Polym. Sci., Polym. Chem. Ed., 31*: 3257 (1993).
93. A. Tuncel, R. Kahraman and E. Pişkin, *J. Appl. Polym. Sci., 51*: 1485 (1994).
94. H. Çiçek, A. Tuncel, M. Tuncel and E. Pişkin, *J. Biomat. Sci., Polym. Ed., 6*: 845 (1994).
95. A. Tuncel, H. Çiçek and E. Pişkin, *J. Biomed. Mat. Res., 29*: 721 (1995).
96. S. Margel and E. Wiesel, *J. Polym. Sci., Polym. Chem. Ed., 22*: 145 (1984).
97. S. Margel, E. Nov and I. Fisher, *J. Polym. Sci., Polym. Chem. Ed., 29*: 347 (1991).
98. J. K. Ober and M. L. Hair, *J. Polym. Sci., Polym. Chem. Ed., 25*: 1395 (1987).
99. A. J. Paine, Y. Deslandes, P. Gerroir and B. Henrissat, *J. Colloid & Interface Sci., 138*: 170 (1990).
100. J. Brandrup and E. H. Immergut, *Polymer Handbook*, Wiley Interscience Publishers, Toronto (1975).
101. A. J. Paine, *Macromolecules, 23*: 3109 (1990).
102. S. J. Douglas, L. Illum, S. S. Davis and J. Kreuter, *J. Colloid Interface Sci., 101* (1984).
103. J. G. Kreuter, *Methods in Enzymology* Academic Press, New York, 112: pp. 129–138 (1985).
104. P. Couvreur, B. Kante, M. Roland, P. Guiot, P. Baudin and P. Speiser, *J. Pharm. Pharmacol., 31*: 331 (1979).
105. L. Illum, P. D. E. Jones, R. W. Baldwin and S. S. Davis, *J. Pharm. Exp. Therap., 230*: 733 (1984).
106. R. H. Muller, C. Lherm, J. Herbort and P. Couvreur, *Biomaterials, 11*: 590 (1990).
107. D. C. Pepper, *Polym. J., 12*: 629 (1980).
108. J. Ugelstad, K. H. Kaggerud, F. K. Hansen and A. Berge, *Macromol. Chem., 180*: 737 (1979).
109. T. Ellingsen, O. A. Sintef, J. Ugelstad and S. Hagen, *J. Chrom., 535*: 147 (1990).
110. L. H. Jansson, M. C. Wellons and G. W. Poehlein, *J. Polym. Sci., Polym. Lett. Ed., 21*: 937 (1983).
111. H. R. Sheu, M. S. El-Aasser and J. W. Vanderhoff, *J. Polym. Sci., Polym. Chem. Ed., 28*: 653 (1990).
112. M. Okubo and T. Nakagawa, *Colloid & Polym. Sci., 270*: 853 (1992).
113. H. Shirahama and T. Suzawa, *J. Colloid & Interface Sci., 126*: 269 (1988).
114. H. Tamai, M. Hasegawa and T. Suzawa, *J. Appl. Polym. Sci., 38*: 403 (1989).
115. T. Suzawa, H. Shirahama and T. Fujimoto, *J. Colloid & Interface Sci., 86*: 144 (1982).
116. H. Shirahama and T. Suzawa, *J. Appl. Polym. Sci., 29*: 3651 (1984).
117. H. Shirahama and T. Suzawa, *Polym. J., 16*: 795 (1984).
118. M. Okubo, Y. Iwasaki and Y. Yamamoto, *Colloid & Polym. Sci., 270*: 733 (1992).
119. M. Okubo, Y. Kondon and M. Takahashi, *Colloid & Polym. Sci., 271*: 109 (1993).
120. M. Okubo and M. Takahashi, *Colloid & Polym. Sci., 272*: 422 (1994).
121. S. Shen, M. S. El-Aasser, V. L. Dimonie, J. W. Vanderhoff and E. D. Sudol, *J. Polym. Sci., Polym. Chem. Ed., 29*: 857 (1991).
122. V. Dimonie, M. S. El-Aasser, A. Klein and J. W. Vanderhoff, *J. Polym. Sci., Polym. Chem. Ed., 22*: 2197 (1984).
123. E. Zurkova, K. Bouchall, D. Zdenkova, Z. Pelzbauer and F. Svec, *J. Polym. Sci., Polym. Chem. Ed., 21*: 2949 (1983).

124. S. Kamei, M. Okubo and T. Matsumoto, *J. Polym. Sci., Polym. Chem. Ed., 24*: 3109 (1986).

125. C. M. Cheng, F. J. Micale, J. W. Vanderhoff and M. S. El-Aasser, *J. Polym. Sci., Polym. Chem. Ed., 30*: 235 (1992).

126. C. M. Cheng, J. W. Vanderhoff and M. S. El-Aasser, *J. Polym. Sci., Polym. Chem. Ed., 30*: 245 (1992).

127. M. Galia, F. Svec and J. M. J. Frechet, *J. Polym. Sci., Polym. Chem. Ed., 32*: 2169 (1994).

128. Q. C. Wang, F. Svec and J. M. J. Frechet, *J. Polym. Sci., Polym. Chem. Ed., 32*: 2577 (1994).

129. A. Tuncel, H. Çiçek and C. Alagöz, *Proceedings of II. National Biomed. Sci. & Tech. Symp.*, METU, Ankara (1995).

16

Reaction Mechanism of Vinyl Polymerization with Amine in Redox and Photo-Induced Charge-Transfer Initiation Systems

Xin-De Feng, Kun-Yuan Qiu, and Wei-Xiao Cao
Peking University, Beijing, China

I. INTRODUCTION

Organic peroxide-aromatic tertiary amine system is a well-known organic redox system[1]. The typical examples are benzoyl peroxide(BPO)-N,N-dimethylaniline(DMA) and BPO-DMT(N,N-dimethyl-p-toluidine) systems. The binary initiation system has been used in vinyl polymerization in dental acrylic resins and composite resins [2] and in bone cement [3]. Many papers have reported the initiation reaction of these systems for several decades, but the initiation mechanism is still not unified and in controversy [4,5]. Another kind of organic redox system consists of organic hydroperoxide and an aromatic tertiary amine system such as cumene hydroperoxide(CHP)-DMT is used in anaerobic adhesives [6]. Much less attention has been paid to this redox system and its initiation mechanism. A water-soluble peroxide such as persulfate and amine systems have been used in industrial aqueous solution and emulsion polymerization [7–10], yet the initiation mechanism has not been proposed in detail until recently [5]. In order to clarify the structural effect of peroxides and amines including functional monomers containing an amino group, a polymerizable amine, on the redox-initiated polymerization of vinyl monomers and its initiation mechanism, a series of studies have been carried out in our laboratory.

A substantial number of photo-induced charge transfer polymerizations have been known to proceed through N-vinylcarbazole (VCZ) as an electron-donor monomer, but much less attention was paid to the polymerization of acrylic monomer as an electron receptor in the presence of amine as donor. The photo-induced charge-transfer polymerization of electron-attracting monomers, such as methyl acrylate(MA) and acrylonitrile (AN), have been recently studied [4]. In this paper, some results of our research on the reaction mechanism of vinyl polymerization with amine in redox and photo-induced charge transfer initiation systems are reviewed.

II. REDOX INITIATION SYSTEMS

A. Diacyl Peroxide-Amine Systems

1. Initiation Mechanism and Structural Effects of Amine and Peroxide

The diacyl peroxide-amine system, especially BPO-DMT or BPO-DMA, has been used and studied for a long time but still no sound initiation mechanism was proposed. Some controversy existed in the first step, i.e., whether there is formation of a charge-transfer complex of a rate-controlling step of nucleophilic displacement as Walling[1] suggested:

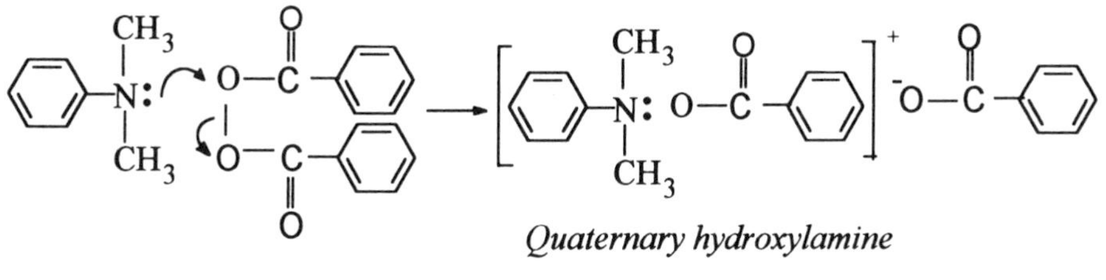

Quaternary hydroxylamine

227

Qiu et al. [11] reported that the aromatic tertiary amine with an electron-rich group on the N atom would favor nucleophilic displacement and thus increase the rate of decomposition of diacyl peroxide with the result of increasing the rate of polymerization (Table 1). They also pointed out that in the MMA polymerization using organic peroxide initiator alone the order of the rate of polymerization R_p is as follows:

LPO > BPO

with an activation energy of polymerization E_a for LPO 77.6 kJ/mol and E_a for BPO 85.8 kJ/mol. However, in the presence of amine DMT, the order of R_p changed to:

BPO-DMT > LPO-DMT

with E_a for BPO-DMT 35.9 kJ/mol and LPO-DMT 56.1 kJ/mol, it is explained that in nucleophilic displacement, the benzoic acid is more acidic than lauric acid so the anion of the stronger acid will be the more reactive leaving group. All of the above results seem to favor Walling's suggestion:

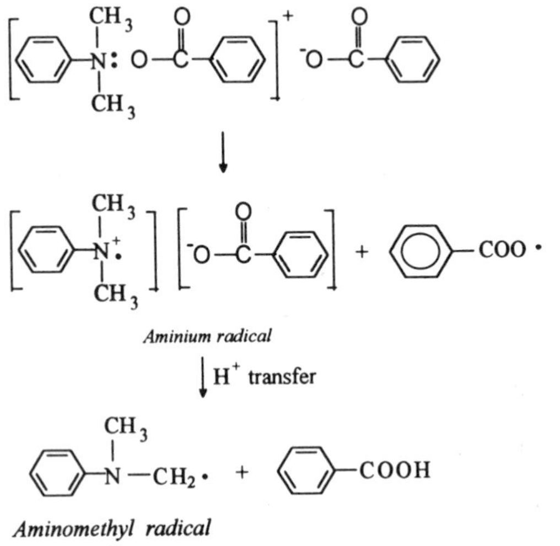

Aminium radical

↓ H^+ transfer

Aminomethyl radical

Although Otsu et al. [12] have studied the BPO-DMA system by electron spin resonance (ESR) technique and trapped the aminomethyl radical, there is still a lack of direct proof of the above second step, particularly concerning the behavior of the aminium radical salt. We [13] have proposed the aminium radical salt with purple color through this reaction of DMT with CCl_4 in the presence of O_2 following the displacement reaction as:

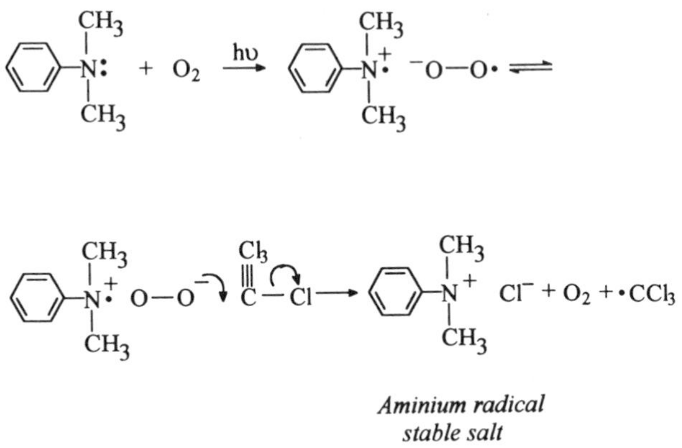

Aminium radical stable salt

This aminium radical salt in aqueous solution in the form of solvated radical salt is very stable and will not polymerize acrylonitrile even with C_6H_5COONa to form the corresponding benzoate. Therefore, we believe that in the nucleophilic displacement, there must be some intermediate step, such as intimate ion pair and cyclic transition state, which will then proceed the deprotonation to form the active aminium radical ion [14], as shown in Scheme 1. The presence of the above aminomethyl radical has also been verified [15] through ultraviolet (UV) analysis of this polymer formed such as PAN or PMMA with the characteristic band as the end group.

Table 1 R_p and E_a of MMA Bulk Polymerization Using Diacylperoxide-Amine Initiation System

Amine in BPO-amine system	DMT	HDMA	DMA	NDMA	DMAB
$R_p \times 10^5$ (mol/L·s)	71.0	56.0	29.2	9.4	9.0
R_r	2.43	1.92	1.00	0.32	0.31
E_a (kJ/mol)	35.9	39.4	44.7	54.4	50.9
Amine in LPO-amine system	DMT	HDMA	DMA	NDMA	DMAB
$R_p \times 10^5$ (mol/L·s)	21.5	15.8	11.7	9.8	9.2
R_r	1.84	1.35	1.00	0.84	0.79
E_a (kJ/mol)	56.1	56.6	67.0	71.6	71.6

[Diacyl peroxide] = [amine] = 1.0×10^{-2} mol/L, 45°C; HDMA: *p*-hydroxymethyl-*N*,*N*-dimethylaniline; NDMA: *p*-nitro-*N*,*N*-dimethylaniline; DMAB: *p*-dimethylaminobenzaldehyde.

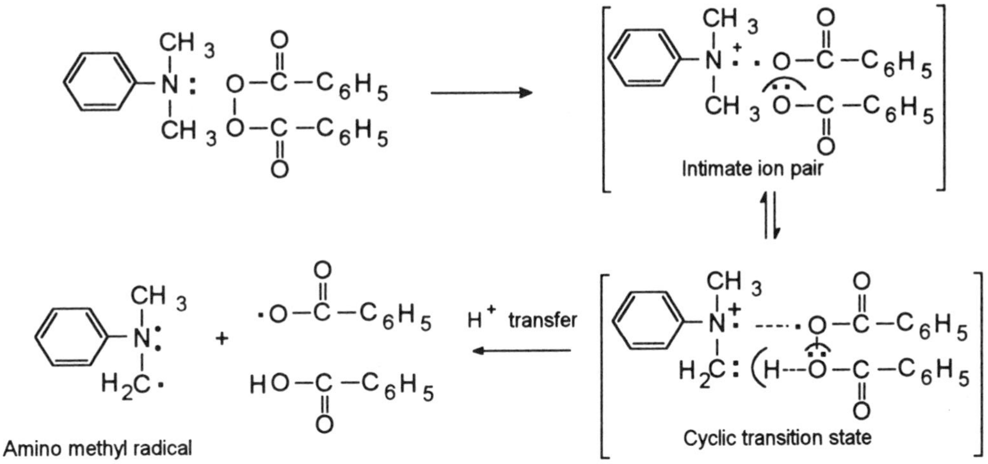

Scheme 1

2. Novel Peroxide-Amine Initiation Systems

We have also investigated [15,16] the following novel BPO-amine initiation systems:

BPO-N,N-di(2-hydroxyalkyl)-p-toluidine Systems

The typical systems are BPO-DHET(*N,N*-di(2-hydroxyethyl)-*p*-toluidine) system, BPO-DHPT(*N,N*-di(2-hydroxypropyl)-*p*-toluidine) system, BPO-HMA(*N*-2-hydroxyethyl-*N*-methyl-aniline), and BPO-HMT(*N*-2-hydroxylethyl-*N*-methyl-*p*-toluidine) system [17–19]. Their polymerization rate and overall activation energies of polymerization E_a are determined and the data are compiled in Table 2.

From ESR studies the formation of free radicals on the α-C atom of the amines attached to hydroxyalky group as:

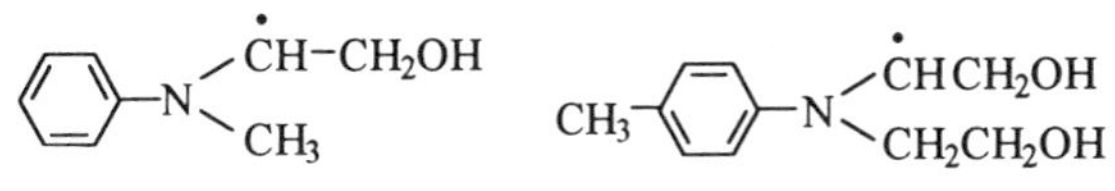

were verified. The UV spectra of the polymer solution revealed that the DHET, DHPT, HMA, and HMT moities were present as the end group of the polymer obtained.

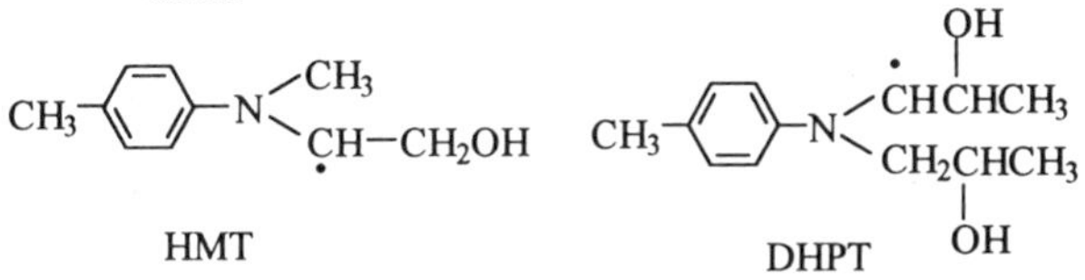

BPO-Heterocyclic Tertiary Amine Systems

Besides aromatic tertiary amines, the aliphatic cyclic tertiary amines such as *N*-methyl(or ethyl) morpholine can also be used in coupling with BPO to enhance the R_p of MMA polymerization [20]. Since the

system of peroxide and heterocyclic tertiary amine has not been received much attention, we have reported that a drug, such as pilocarpine, containing heterocyclic tertiary amine, i.e., imidazolyl ring, can couple with BPO to form a redox initiation system for vinyl polymerization at 40°C, such as MMA, HEMA, or NVP(*N*-vinyl-pyrrolidone) to form a controlled drug delivery device for the drug pilocarpine in film or hydrogel form [21]. This work has been extended to other drugs such as Rifampin in the BPO-Rifam system, Perpheminazium in the BPO-Perphem system, and physostigme in the BPO-Physos system.

Peroxide-Functional Monomer Containing Amino Group Systems

It is interesting to study the polymerization of functional monomer containing an amino group, so-called polymerizable amine by peroxide initiator, which could be anticipated to cause less pulpal irritation and toxic

Table 2 MMA Polymerization Initiated by BPO-Aromatic Tertiary Amine Systems

Initiation system	$R_p \times 10^4$ (mol/L·s)	R_r	E_a (kJ/mol)
BPO-DHET	9.30[a]	1.7	43.5
BPO-DHPT	8.31[a]	1.5	44.6
BPO-DMT	5.45[a]	1.0	39.4
BPO-DMA	2.19[a]	0.4	45.5
BPO-HMT	9.07[b]	1.6	41.3
BPO-DMT	5.78[b]	1.0	39.4
BPO-HMA	3.30[b]	0.6	43.4
BPO-DMA	2.34[b]	0.4	45.5

[a] [MMA]/[Toluene] = 1/1 at 45°C, [BPO] = [Amine] = 2.0×10^{-2} mol/L.
[b] MMA in bulk at 40°C, [BPO] = [Amine] = 2.0×10^{-2} mol/L.

reaction than the low-molecular weight amine during prolonged implantation of the polymeric materials. Therefore, such a system has received much attention recently. Li et al. [22–24] reported that 2,2'-azobisisobutyronitrile (AIBN) can polymerize these functional monomers normally. The structures of functional monomers are shown in Table 3.

3. Rate Equations of MMA Polymerization Using Peroxide-Amine Initiation Systems

Usually, the rate equation of redox initiated polymerization is shown as follows:

$$R_p = K[\text{Reductant}]^{0.5}[\text{Oxidant}]^{0.5}[\text{Monomer}]^{1.0} \tag{1}$$

The rate equations of MMA polymerization initiated by peroxide amine systems are listed in Table 4. It shows that No. 1, 2, 3, 8, 9 are in good agreement with the redox-initiated polymerization rate equation. However, there are deviations in the order of concentration of peroxide, amine, and MMA, respectively. It is interesting to note the rate equation in the LPO-amine initiation system. The dependence of R_p on the aliphatic tertiary amine concentration was very low, i.e., < 0.1 order (see No. 11), under a normal ratio for [NMMP]/[LPO]. While in a higher concentration of NMMP, it changed to 0.30 order as shown in No. 12. Similar results are observed in the LPO-NEP initiation system (see Nos. 13 and 14).

In addition, we found that all of the functional monomers having amino group would act as an amine component and with LPO form a redox system to initiate the polymerization of functional monomer itself with the rate of polymerization as:

$$R_p = k[\text{LPO}]^{0.5}[\text{Functional monomer}]^{1.5} \tag{2}$$

When the functional monomer is in low concentration, it can be used as an amine component of a redox initiation

Table 3 The Structure of Functional Monomers

Functional monomers	Structure	R=	Reference
Dimethylaminoethyl methacrylate (DMAEMA)	$CH_2{=}\overset{\underset{\displaystyle |}{CH_3}}{C}{-}COOCH_2CH_2N(CH_3)_2$		[25,26]
2-hydroxy-3-dimethylamino-propyl methacrylate (DMAHPMA) 2-hydroxy-3-diethylamino-propyl methacrylate (DEAHPMA)	$CH_2{=}\overset{\underset{\displaystyle |}{CH_3}}{C}{-}COOCH_2\overset{\underset{\displaystyle |}{OH}}{CH}CH_2NR_2$	CH_3 CH_3	[26]
N-acryl-N'-methylpiperazine (AMP) N-methacrylyl-N'-methyl-piperazine (MAMP)	$CH_2{=}\overset{\underset{\displaystyle |}{R}}{C}\,CO\,N{\bigcirc}N{-}CH_3$	H CH_3	[23]
N-(N'-methylene-morpholino) acrylamide (MMAA) N-(N'-methylene-morpholino) methacrylamide (MMMA)	$CH_2{=}\overset{\underset{\displaystyle |}{R}}{C}{-}CONCH_2N{\bigcirc}O$	H CH_3	[28]
N-(N',N'-dimethylaminophenyl) acrylamide (DMAPAA) N-(N',N'-dimethylaminophenyl) methacrylamide (DMAPMA)	$CH_2{=}\overset{\underset{\displaystyle |}{R}}{C}{-}CONH{-}\bigcirc{-}N(CH_3)_2$	H CH_3	[27]
4-dimethylaminobenzyl methacrylate (DMABMA)	$CH_2{=}\overset{\underset{\displaystyle |}{CH_3}}{C}{-}COOCH_2{-}\bigcirc{-}N(CH_3)_2$		[22,29]
N,N-di(methacryloyloxy-propyl)-p-toluidine ((MP)₂PT)	$(CH_2{=}\overset{\underset{\displaystyle |}{CH_3}}{C}{-}COOCHCH_2)_2N{-}\bigcirc{-}CH_3$		[30]
N-methacryloyloxyethyl-N-methyl aniline	$CH_2{=}\overset{\underset{\displaystyle |}{CH_3}}{C}{-}COOCH_2CH_2\,\overset{\underset{\displaystyle |}{CH_3}}{N}{-}\bigcirc$		[32]

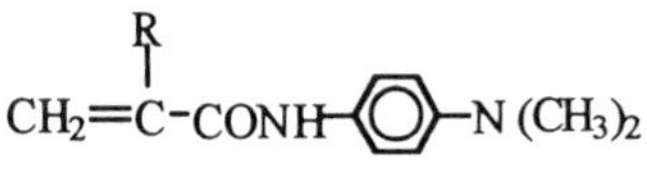

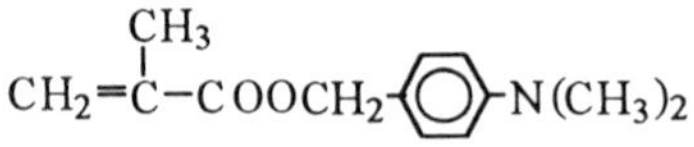

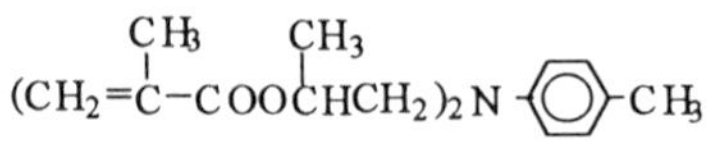

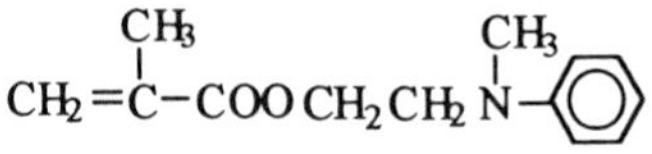

Table 4 Rate Equation of MMA Polymerization $R_p = $ k[Peroxide]^a[Amine]^b[MMA]^c

No.	Initiation system	a	b	c	Reference
1	BPO-DMT	0.47	0.47	1.07	[10]
2	BPO-DHPT	0.51	0.48	1.09	[15]
3	BPO-TMDAPM	0.48	0.44	1.1	[14]
4	BPO-NMMP	0.47	0.40	1.17	[20]
5	BPO-NEP	0.57	0.34	1.02	[29]
6	BPPD-DHET	0.67	0.77	1.39	[16]
7	BPPD-DHPT	0.65	0.60	1.52	[16]
8	LPO-DMT	0.5	0.5	1.0	[10]
9	LPO-DHPT	0.51	0.49	0.93	[15]
10	LPO-TMDAPM	0.47	0.43	1.07	[14]
11	LPO-NMMP	0.48	0.07	1.01	[20]
12	LPO-NMMP*	0.48	0.30	1.01	[20]
13	LPO-NEP	0.53	0.02	1.03	[29]
14	LPO-NEP*	0.53	0.39	1.03	[29]

* Polymerization under higher concentration of amine.

system. The rate equations for MMA polymerization by peroxide and functional amine systems are obtained as follows:

$$R_p = k[BPO]^{0.5}[DMAPAA]^{0.5}[MMA] \qquad (3) \quad [27]$$

$$R_p = k[BPO]^{0.5}[DMAPMA]^{0.5}[MMA] \qquad (4) \quad [27]$$

$$R_p = k[LPO]^{0.5}[MAMP]^{0.5}[MMA] \qquad (6) \quad [23]$$

$$R_p = k[BPPD]^{0.42}[(MP)_2PT]^{0.52}[MMA]^{1.1} \qquad (7) \quad [33]$$

$$R_p = k[LPO]^{0.52}[(MP)_2PT]^{0.46}[MMA]^{0.98} \qquad (8) \quad [33]$$

$$R_p = k[BPO]^{0.49}[(MP)_2PT]^{0.46}[MMA]^{1.1} \qquad (9) \quad [33]$$

These equations are in accordance with redox initiated poymerization rate Equation 1.

B. Organic Hydroperoxide-Amine Systems

1. Structural Effects of Amine

Less attention has been paid to the ROOH-amine system, but we [34] have investigated the effect of amine

structure on the R_p of MMA polymerization in bulk and found that in contrast to the diacyl peroxide amine system (Table 1), the change of R_p is much less sensible whether the substituent is an electron-donating group or electron-withdrawing groups with the following data (Table 5) for comparison:

In our laboratory, Sun et al. [35] reported that the terahydrofuran hydroperoxide (THFHP)-DMT system could initiate vinyl polymerization actively with very low E_a as 35.2 kJ/mol for MMA and 34.3 kJ/mol for AAM polymerization.

2. Initiation Mechanism of ROOH-Amine Systems

We have proposed an initiation mechanism for the ROOH-amine system in which some H-bond complex may be formed [36]. Then Sun et al. [37,38] thoroughly investigated the initiation mechanism of ROOH-Amine through IR spectra of TBH-triethylamine, TBH-DMT, and CHP-DMT. From the wideness of the shift of OH absorption bands at 3120, 3336, and 3257 cm^{-1} were

Table 5 R_p of MMA Initiated with ROOH-Amine System

TBH-Amine	DMT	DMA	NDMA	DMAB
$R_p \times 10^5$ (mol/L·s)	5.83	5.46	5.20	4.67
R_r	1.07	1.00	0.95	0.86
E_a (kJ/mol)	51.7	52.6	57.4	50.4
CHP-Amine	DMT	DMA	NDMA	DMAB
$R_p \times 10^5$ (mol/L·s)	10.0	7.57	8.11	3.85
R_r	1.32	1.00	1.07	0.51
E_a (kJ/mol)	38.9	45.3	55.7	62.4

TBH: *t*-butyl hydroperoxide; CHP: cumene hydroperoxide; [ROOH] = [Amine] = 2.0 × 10^{-2} mol/L 50°C.

obtained served to confirm the formation of an H-bond complex. We then proposed the first step as:

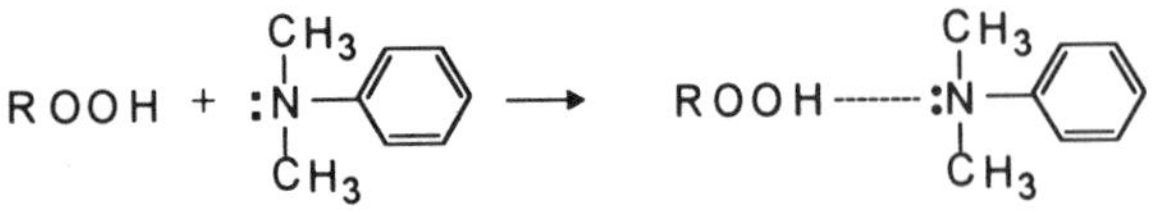

Through radical trapping and ESR spectrum the same radical, i.e., N-methyl-p-toluidine methyl radical, as in the BPO-DMT system was verified but with a weakened signal. Therefore, the above result favored our formerly proposed mechanism as follows:

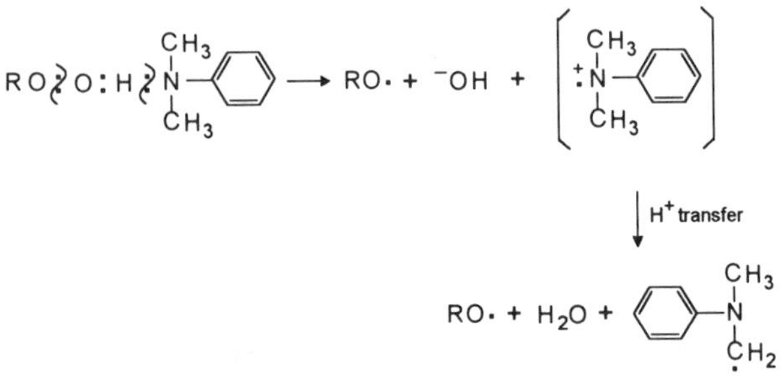

Formation of an intimate ion pair of OH^- and aminium radical cation was also proposed for the intermediate step before deprotonation. The presence of the above radical was verified through UV analysis of the polymer formed with the characteristic band on the end group. Through chromatographic analysis of the TBH-DMT reaction products, H_2O was detected as the above mechanism proposes after deprotonation.

Pavlinec and Lazar [39] reported that organic hydroperoxide and piperidine(PD) could be used as an initiator for MMA polymerization. In our laboratory, we also found that TBH-NMMP, TBH-NEMP [20], TBH-PD(piperidine) [31], TBH-NEP(N-ethylpiperdine) [31], TBH-TMDAPM (N,N'-tertramethyl-diaminodiphenyl-methane), and TBH-TMEDA($N,N,N'N'$-teramethylethylenediamine) [15] systems could initiate MMA to polymerize. The kinetic equation of MMA polymerization initiated with CHP-DMT system has been investigated in our laboratory and the rate equation of polymerization is shown as follows:

$$R_p = k[CHP]^{0.34}[DMT]^{0.45}[MMA]^{1.0}$$

The departure of dependence of R_p on the concentration of CHP from 0.5 order might be ascribed to induction decomposition of ROOH type to form ROO· radical, which has very low activity to initiate monomer polymerization [40], but can combine with the propagation chain radical to form the primary radical termination. For the same reason, the order of concentration of TBH was also lower than 0.5 when the TBH-DMT system was used as the initiator in MMA bulk polymerization. But in the BPO-DMT initiation system as shown in Table

4, the orders of dependence of R_p on concentrations of BPO and DMT were both 0.47–0.5, as there was not such a peroxy radical ROO·, but a benzoyloxy radical C_6H_5COO· formed.

C. Persulfate-Amine Systems

1. Persulfate-Aliphatic Monoamine Systems

Several articles [7,8] have reported that a persulfate-amine system, particularly persulfate-triethanol amine and persulfate-tetramethylethylenediamine (TMEDA) can be used as redox initiators in aqueous solution polymerization of vinyl monomers. Recently, we studied the effect of various amines on the AAM aqueous solution polymerization and found that not only tertiary amine but also secondary and even primary aliphatic amine and their polyamines can promote the vinyl polymerization as shown in Table 6 [40–42].

For aliphatic monoamine [43], it is shown that secondary amines R_2NH always possess a higher promoting effect for the polymerization of AAM and even the primary amine PA will enhance the polymerization with $R_r = 1.47$ and $E_a = 36.4$ kJ/mol, while the tertiary aliphatic amine TPA will not provide the polymerization due to some steric hindrance (Table 6). All of the data of cyclic amines listed in Table 7; are effective, i.e., NMMP with $R_r = 1.81$ and $E_a = 29.9$ kJ/mol showing the absence of steric hindrance.

2. Persulfate-Aliphatic Diamine Systems

For aliphatic diamines [40] it is shown that TMEDA is the well-known and most effective redox initiation system with APS. The data for the effects of the diamines on AAM polymerization with APS are compiled in Table 8. From the structural condition there are three generations, i.e.:

Tertiary amine>	Secondary amine>	Primary amine
TMEDA	as, sym-DMEDA	EDA
—CH$_2$CH$_2$— >	—(CH$_2$)$_2$—CH$_2$— >	—(CH$_2$)$_3$—CH$_2$—
TMEDA	TMPDA	TMBDA
Dimethylamino >	Diethylamino	
TMEDA	TEEDA	

3. ESR Studies and End Group Analysis

ESR studies on the initial free radicals were carried out by using MNP(2-methyl-2-nitrosopropane) or DMPO (5,5-dimethylpyrroline N-oxide) as the spin-trapping agent. The reactions are shown as:

Table 6 Effects of Monoamines on AAM Polymerization

Series	Initiation system	Structure of amine	$R_p \times 10^4$	R_r	E_a
	APS		2.33	1.00	62.0
	APS/PA[a]	$CH_3CH_2CH_2NH_2$	3.43	1.47	36.4
1	APS/BA[a]	$CH_3CH_2CH_2CH_2NH_2$	2.89	1.24	36.9
	APS/DPA	$(CH_3CH_2CH_2)_2NH$	6.25	2.68	26.6
	APS/DBA	$(CH_3CH_2CH_2CH_2)_2NH$	6.49	2.78	35.9
	APS		3.01	1.00	
2	APS/TEA	$(CH_3CH_2)_3N$	3.62	1.20	
	APS/TPA	$(CH_3CH_2CH_2)_3N$	3.01	1.00	
	APS/TBA	$(CH_3CH_2CH_2CH_2)_3N$	3.61	1.20	

Series 1: Polymerization in water solution at 45°C.
Series 2: Polymerization in mixed solvent H_2O: $CH_3OH = 10:1$ at 50°C.
$[APS] = [Amine] = 1.00 \times 10^{-3}$ mol/L, $[AAM] = 1.00$ mol/L.
[a] $[PA] = 1.92 \times 10^{-3}$ mol/L; $[BA] = 1.61 \times 10^{-3}$ mol/L.

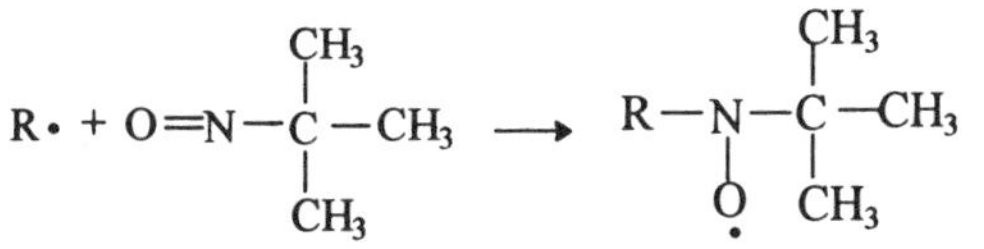

The spin adducts of free radicals and MNP or DMPO were observed by means of an ESR spectrometer. The data of hyperfine splitting constants were compiled in Tables 9 and 10 [40–42,44,45]. ESR studies on the initial free radicals revealed that the monoalkylamino radical RHN·, dialkylamino radical $R_2N·$, and aminomethyl radical $·CH_2N<$ or aminoethylidene radical $>N(·CHCH_3)$ were obtained from the corresponding primary, secondary, and cyclic tertiary amine. In case of a tertiary diamine such as TMEDA, formation of

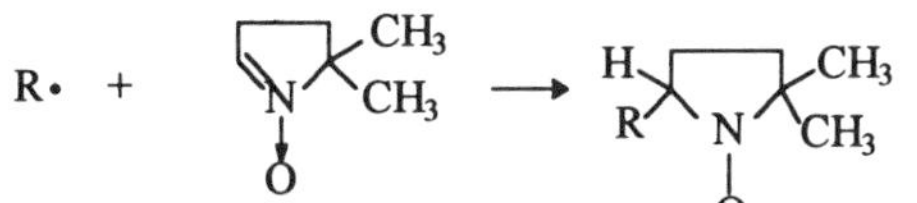

$·CH_2N(CH_3)R$ rather than $(CH_3)_2N$ ($·CH$ R) showed that the methyl group is the preferable group for substitution. Meanwhile, a secondary product was also formed and verified through ESR as $·CH_2CH_2N(CH_3)_2$ (N,N-dimethylaminoethylene radical) from TMEDA, which was considered to form from the scission of the primary radical as follows:

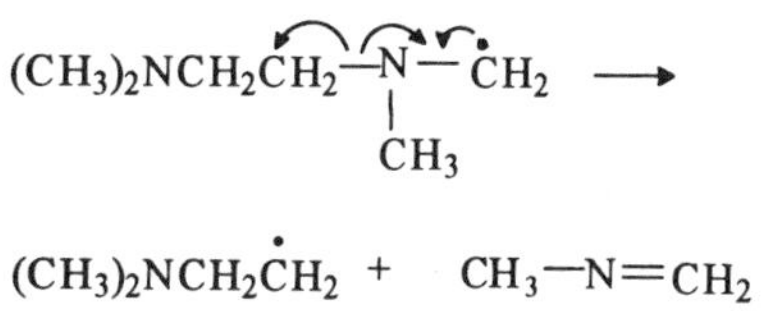

$$(CH_3)_2NCH_2\overset{\bullet}{C}H_2 + CH_3—N=CH_2$$

End group analysis on the charge transfer complex(CTC) method [40,41] using tetracyanoethylene (TCNE) as a strong electron charge transfer complex acceptor to react with a donor molecule such as aromatic or aliphatic amine forms a CTC or ion radical pairs exhibiting a characteristic absorption in UV or visible range. The CTC method can be used for the analysis of the amino end group of the polymer formed.

Table 7 Effects of Aliphatic Cyclic Amines on Polymerization

Initiation system	$R_p \times 10^4$ (mol/L·s)	R_r	E_a (kJ/mol)	MW of PAAM (10^{-6})
KPS	2.52	1.00	62.5	3.37
KPS/MP	4.11	1.63	34.3	3.10
KPS/PD	3.94	1.56	40.3	2.32
KPS/NMMP	4.55	1.81	29.9	2.94
KPS/NEMP	3.91	1.55	38.0	3.38
KPS/NEP	3.22	1.28	54.8	0.82

$[KPS] = [Amine] = 1.0 \times 10^{-3}$ mol/L; $[AAM] = 1.0$ mol/L, 45°C.

Table 8   Effects of the Diamines on AAM Polymerization

Initiation system	Structure of diamine	$R_p \times 10^4$ (mol/L·s)	R_r	E_a (kJ/mol)	MW of PAAM(10^{-6})
APS		1.97	1.00	62.0	2.14
APS/EDA	$H_2NCH_2CH_2NH_2$	2.63	1.33	36.9	1.99
APS/sy-DMEDA	$CH_3NHCH_2CH_2NHCH_3$	6.83	3.47	24.9	1.28
APS/as-DMEDA	$(CH_3)_2NCH_2CH_2NH_2$	9.10	4.62	16.2	1.70
APS/as-DEEDA	$(C_2H_5)_2NCH_2CH_2NH_2$	4.84	2.46	31.6	1.12
APS/as-DMPDA	$(CH_3)_2NCH_2CH_2CH_2NH_2$	5.87	2.98	19.7	0.72
APS/TMMDA	$(CH_3)_2NCH_2N(CH_3)_2$	4.10	2.08	35.6	0.74
APS/TMEDA	$(CH_3)_2NCH_2CH_2N(CH_3)_2$	9.74	4.94	22.0	1.64
APS/TEEDA	$(C_2H_5)_2NCH_2CH_2N(C_2H_5)_2$	6.83	3.47	24.2	0.83
APS/TMPA	$(CH_3)_2N(CH_2)_3N(CH_3)_2$	8.01	4.07	25.0	1.91
APS/TMBDA	$(CH_3)_2N(CH_2)_4N(CH_3)_2$	5.06	2.57	27.5	1.68

APS: ammonium persulfate; [APS] = [Amine] = 5.0×10^{-4} mol/L, [AAM] = 1.0 mol/L, 45°C.

4. Initiation Mechanism of Persulfate-Amine Systems

Based on the results discussed previously, the following initiation mechanism was proposed for a tertiary amine, N-methylmorpholine [46], which involves the formation of aminium radical as the intermediate step and deprotonation to methyl radical as one of the active initiators:

Table 9   Hyperfine Splitting Constants of Spin Adduct Formed APS/Amine/MNP or DMPO System

System	Radical trapped by MNP or DMPO	Hyperfine splitting constant (0.1 mT)			
		a_α^N	a_β^H	a_β^N	a_γ^H
PA/APS/MNP	$CH_3CH_2CH_2\dot{N}H$	15.46	1.33	1.33	0.67(2H)
BA/APS/MNP	$CH_3CH_2CH_2CH_2\dot{N}H$	15.47	1.33	1.33	0.67(2H)
EDA/APS/MNP	$H_2NCH_2CH_2\dot{N}H$	15.47	1.33	1.33	0.67(2H)
PDA/APS/MNP	$H_2NCH_2CH_2CH_2\dot{N}H$	15.47	1.33	1.33	0.67(2H)
EDA/APS/DMPO	$(CH_3CH_2)_2\dot{N}$	13.68	9.87	1.28	
DPA/APS/DMPO	$(CH_3CH_2CH_2)_2\dot{N}$	13.65	10.0	1.19	
MP/APS/MNP	 O⌷N·	18.72		0.88	0.88(2H)
PD/APS/MNP	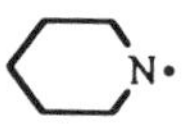 ⌷N·	18.61		0.93	0.93(2H)
MP/APS/DMPO	 O⌷N·	15.0	9.33	2.13	
PyD/APS/DMPO	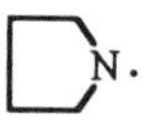 ⌷N·	15.25	17.6	2.29	
NMMP/APS/MNP	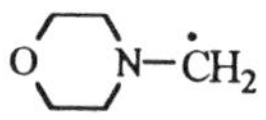 O⌷N–$\dot{C}H_2$	14.16	2.38(2H)	0.80	
NEMP/APS/MNP	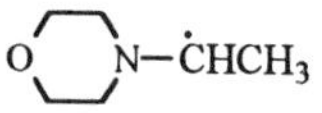 O⌷N–$\dot{C}HCH_3$	15.28	3.20	3.20	
NMP/APS/MNP	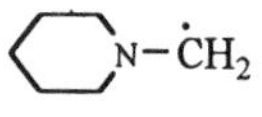 ⌷N–$\dot{C}H_2$	15.07	7.55(2H)	3.52	

Table 10 Hyperfine Splitting Constants of Spin Adducts Obtained from APS/Amine/MNP Systems

System	Radical trapped by MNP	Hyperfine splitting constant (0.1 mT)			
		a_α^N	a_β^H	a_β^N	a_γ^H
TMEDA	$(CH_3)_2NCH_2CH_2$—N—$\dot{C}H_2$ $\qquad\qquad\quad$ CH$_3$	15.73	9.84(2H)	2.08	
	$(CH_3)_2NCH_2\dot{C}H_2$	15.87	10.35(2H)		0.80(2H)
TMPDA	$(CH_3)_2N(CH_2)_3N$—$\dot{C}H_2$ $\qquad\qquad\qquad$ CH$_3$	15.0	8.0(2H)	3.63	
TMBDA	$(CH_3)_2N(CH_2)_4N$—$\dot{C}H_2$ $\qquad\qquad\qquad$ CH$_3$	15.0	8.0(2H)		
TEEDA	$(C_2H_5)_2NCH_2CH_2N\dot{C}HCH_3$ $\qquad\qquad\qquad$ C$_2$H$_5$	15.57	5.63		
	$(C_2H_5)_2NCH_2\dot{C}H$—$N(C_2H_5)_2$	15.73	19.5		
	$(C_2H_5)_2NCH_2\dot{C}H_2$	15.87	10.67		0.53
as-DMEDA	$H_2NCH_2CH_2N$—$\dot{C}H_2$ $\qquad\qquad\quad$ CH$_3$	15.60	9.73(2H)	2.13	
	$H_2NCH_2\dot{C}H_2$	16.0	10.8(2H)		0.53(2H)
as-DMPDA	$H_2N(CH_2)_3N$—$\dot{C}H_2$ $\qquad\qquad\quad$ CH$_3$	15.36	9.47(2H)	2.53	
DETA	$(H_2NCH_2CH_2)_2N\cdot$	14.93		2.13	
TETA	$R_1R_2N\cdot$	14.96		2.16	
TEPA	$R_3R_4N\cdot$	14.93		2.16	

$R_1 = R_3 = H_2NCH_2CH_2-$; $R_2 = H(HNCH_2CH_2)_2-$; $R_4 = H(HNCH_2CH_2)_3-$

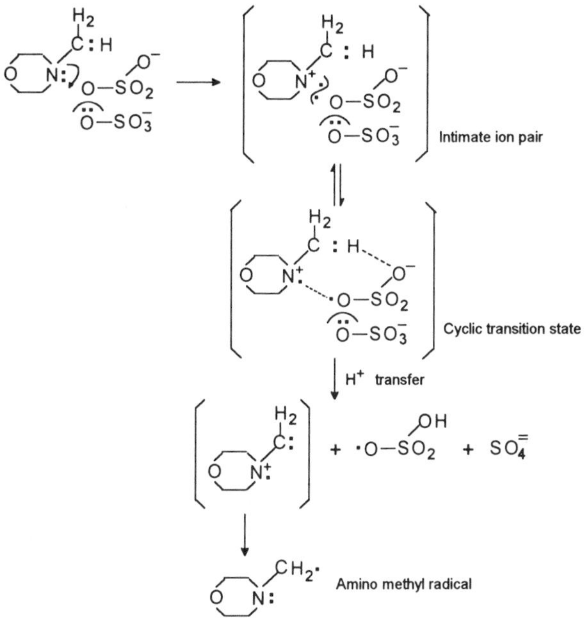

In case of secondary amine or primary amine it would form amino radical as follows:

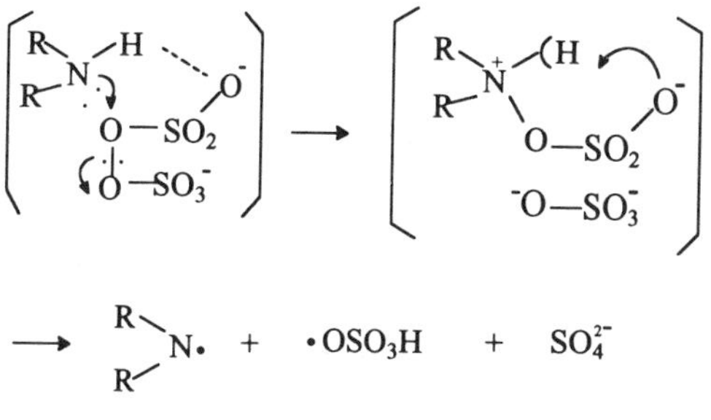

The ethylenediamine derivative [31] possesses higher promoting activities than other diamines. This phenomenon may be ascribed to the copromoting effect of the two amino groups on the decomposition of persulfate through a CCT (contact charge transfer complex) formation. So we proposed the initiation mechanism via CCT as the intimate ion pair and deprotonation via CTS (cyclic transition state) as follows:

Moreover, the initiation mechanism of APS/ethylene-disecondary-amine and APS/ethylene-diprimary-amine systems has also been proposed [47].

III. PHOTO-INDUCED CHARGE TRANSFER POLYMERIZATION OF VINYL MONOMERS

The charge transfer theory was established by Mulliken [48–53] in the 1950s. Since then a great number of organic reactions have been found to proceed through formation of charge transfer complexes (CTC) [54–57]. In the 1960s, this theory was applied to the polymerization field and successfully explained the mechanism of many polymerizations. Some of the earliest studies were published by Scott et al. [58] and Ellinger [59] independently in 1963. They reported the polymerization of N-vinylcarbazole (VCZ) as an electron-donating monomer in the presence of electron acceptors through the formation of CTC. Before that, Norrish et al [60,61] had reported the polymerization of styrene photoinitiated by anthracene and proposed the formation of an excited triplet complex as an intermediate, but did not mention whether the complex was of the charge-transfer type or not.

Charge-transfer polymerization has been developed within the last 30 years on the basis of the interaction between an electron donor (D) and an electron acceptor (A) involved in the initiating and/or propagating processes. Such polymerization has attracted a great deal of attention because it is processed through a novel type of initiating or propagating mechanism with a charge transfer interaction. This exists throughout in a wide variety of organic compounds as donors or acceptors and also is found with low energy due to the charge transfer interaction in the formation of the reactive centers. Therefore, the polymerization would be carried out under more moderate conditions.

A. Vinyl Monomers as Electron Donors

Charge-transfer photopolymerizations of electron-donating monomers initiated by electron-accepting initia-

tors were studied in great detail in the 1960s and 1970s. The most notable work in this area was the photopolymerization of VCZ, as studied by Shirota et al. [62–64] and the photopolymerization of α-methyl-styrene (α-MSt), as studied by Irie and Hayashi [65].

VCZ as a strong electron-donating monomer can form CTC with various strong electron acceptors in the ground state, such as chloranil (CA), bromanil (BA), tetracyanoethylene (TCNE), maleic anhydride (MAn), tetranitromethane (TNM), and trinitrobenzene (TNB), and can be polymerized in the presence of such acceptors. With many weak electron acceptors, however, usually no polymerization occurs at room temperature in the dark. Polymerization of these systems are induced or accelerated by photoirradiation since the charge-transfer interaction is greatly enhanced under irradiation and the excited charge-transfer complex is rather polar and readily gives rise to ion radicals.

Studies of the effect of the wavelength of the incident light on polymerization have shown that two possible processes produce the active initiating species: (1) a stable ground state CTC is excited, and (2) either a donor or an acceptor is excited, followed by charge transfer and electron transfer with a ground state acceptor or donor. While the photopolymerization of VCZ-thiobenzophenone in benzene or toluene is initiated through local excitation of thiobenzophenone [66], the polymerization of VCZ-p-quinoid compounds in benzene [67] and VCZ-trinitrofluorenone in nitrobenzene [68] has been shown to be initiated by selective excitation of the ground state CTC. Obviously, only the second process is operative for the system where no charge-transfer interaction exists in the ground state. For the VCZ-CA and VCZ-BA systems, the above two processes are operative, depending on the wavelengths of the incident light used [67,69].

Charge-transfer and electron-transfer processes in photopolymerization systems have been demonstrated by the measurement of fluorescence spectra and by means of flash photolysis. Typically, VCZ-dimethyl terephthalate (DMPT) [70], VCZ-FN, and VCZ-diethyl fumarate (DEF) [71] systems have been shown to form exciplexes by local excitation of VCZ. Dynamic quenching of the VCZ fluorescence occurred by adding acceptors. Broad and structureless exciplexes were observed at longer wavelengths in nonpolar solvents. Although no exciplex fluorescence was observed in the polar solvents, the quenching of the VCZ fluorescence occurred, indicating the occurrence of charge transfer or electron transfer in the excited singlet state of VCZ [72,73]. With flash photolysis and laser photolysis techniques, the photochemical formation of the transient VCZ cation radical and the electron-acceptor (A) anion radical (A·$^-$) has been confirmed [74–77]. The transient absorption spectra obtained for the VCZ-A system in various solvents consists of two band systems: one is due to A·$^-$ and the other is due to the VCZ cation radical (VCZ·$^+$).

These results indicate that a complete electron transfer is brought about from the charge-transfer interaction and can be summarized [62] as follows:

$$VCZ + A \xrightarrow{h\nu} A\overset{-}{\cdot} + VCZ\overset{+}{\cdot}$$

$$A\overset{-}{\cdot} + VCZ\overset{+}{\cdot} \quad \begin{array}{c} \xrightarrow{C_6H_5NO_2} \text{Cationic polymerization} \\ \xrightarrow{HMPA} \text{Radical polymerization} \end{array}$$

HPMA: Hexamethylphosphoramide

B. Vinyl Monomers as Electron Acceptors

1. Aromatic Tertiary Amine and Electron-Accepting Monomer

In 1982 Wei et al. [78,79] studied the quenching of N,N-dimethyltoluidine (DMT) fluorescence by adding the electron-accepting monomer MA or MMA and successfully observed broad and structureless exciplex fluorescences at longer wavelengths in nonpolar solvents for the first time.

It has been shown that the excited charge transfer complex can be formed through two different routes depending on the wavelengths of the light used [80–83]. Both routes can bring about polymerization. For example, AN can form CTC with DMT in the ground state, which was verified through UV spectra. This CTC will be photoactivated with 365-nm light to form an exciplex. If 313-nm light is used, the activated DMT will form an exciplex directly with the ground state AN molecule as follows:

$$
\begin{array}{ccc}
\text{DMT} & \xrightarrow[\;(1)\;]{h\nu,\, 313 \text{ nm}} & \text{DMT}* \\
\big\downarrow \text{AN} & & \big\downarrow \text{AN} \\
(\text{DMT} \cdot \text{AN}) & \xrightarrow[\;(2)\;]{h\nu,\, 365 \text{ nm}} & (\text{DMT} \cdot \text{AN})*
\end{array}
$$

Route (1) is referred to as "local excitation" and route (2) as "CTC excitation." It has been observed that the different routes bring about the polymerization of AN with different kinetic behaviors. A 365-nm light will irradiate the CTC only, and in this case the rate of polymerization for different aromatic tertiary amines descends in the following order:

$$p\text{-}CH_3C_6H_4N(CH_3)_2 > p\text{-}CH_3C_6H_4N(CH_2CH_2OH)_2$$
$$> p\text{-}HOCH_2C_6H_4N(CH_3)_2 > C_6H_5N(CH_3)_2$$

It is in the same order as the equilibrium constants of CTC of amine-FN. That is, the stronger the ability of an amine to form CTC with electron acceptors, the faster the rate of photopolymerization. However, under 313-nm irradiation, local excitation plays a principal role and the rate of polymerization is observed to descend in a different order [80]:

$$p\text{-}CH_3C_6H_4N(CH_3)_2$$
$$> C_6H_5N(CH_3)_2 > p\text{-}HOCH_2C_6H_4N(CH_3)_2$$
$$> p\text{-}CH_3C_6H_4N(CH_2CH_2OH)_2$$

This order agrees with that of quenching constant $(k_q\tau)$ values of the fluorescence of amines by AN. That is, the easier the reaction between an excited aromatic tertiary amine and the ground state AN, the faster the initiation.

Local excitation was also studied for primary and secondary amines under irradiation at 313 nm. The results are summarized in Table 11. In order to estimate the photoinitiating efficiency of the amines, the measurement was performed at a chosen constant absorbance (0.40) of the reaction mixture. The rates of polymerization were found to be in the following order:

$$C_6H_5N(CH_3)_2 > C_6H_5NH_2 > C_6H_5NHCH_3$$
$$> p\text{-}CH_3C_6H_4NH_2 > p\text{-}CH_3OC_6H_4NH_2$$
$$> m\text{-}CH_3C_6H_4NH_2$$

which is consistent with the order of the quenching rate constant k_q values calculated with the Stern–Volmer equation:

Table 11 The Photopolymerization of AN Initiated by Aromatic Amines (at 313 nm)

Amine	$\epsilon_{313}{}^a$	$C^b \times 10^4$	R_p (mol/L·s) $\times 10^5$	$k_q\tau^c$	$\tau(\text{ns})^d$	$k_q \times 10^{-8}$
$C_6H_5N(CH_3)_2$	1800	2.2	3.4	46.0	2.3	2.00
$C_6H_5NH_2$	860	5.2	2.6	44.5	2.3	1.93
$C_6H_5NHCH_3$	1760	2.3	2.3	48.0	2.7	1.78
$p\text{-}CH_3C_6H_4NH_2$	1240	3.0	2.1	44.0	2.5	1.78
$p\text{-}CH_3OC_6H_4NH_2$	2300	1.6	1.8	48.0	2.9	1.66
$m\text{-}CH_3C_6H_4NH_2$	1030	3.8	1.7	27.3	2.5	1.09

^{a} The molar extinction coefficient of amine at 313 nm measured in DMF-AN solution ([AN] = 3.8 mol/L).

^{b} The amine concentration (mol/L) used in the polymerization at which the absorbance of the system was 0.40.

^{c} Calculated with the Stern–Volmer equation, $I_0/I = 1 + k_q\tau[\text{AN}]$ in alcohol ([amine] = 2×10^4 mol/L).

^{d} Measured in alcohol ([amine] = 2×10^4 mol/L).

$$I_o/I = k_q\tau[\text{AN}] + 1$$

This result reveals that exciplex formation plays a principal role in the initiation of polymerization. Since the absorption band is broadened toward longer wavelengths as the result of formation of CTC between AN and aniline, a certain concentration of aniline can be chosen so that 365-nm light is absorbed only by the CTC but not by the aniline molecule. Therefore, in this case the photopolymerization may be ascribed to the CTC excitation selected. For example, a 5×10^{-2} mol/L aniline solution in AN could absorb light of 365 nm, while solutions in DMF or cyclohexane with the same concentration will show no absorption. Obviously, in this case the polymerization of AN is caused by CTC excitation. The rates of polymerization for different amines were found to be in the following order (Table 12):

$$C_6H_5N(CH_3)_2 > C_6H_5NHCH_3 > C_6H_5NH_2$$

The process for initiating radical formation in aromatic amine-vinyl monomer systems have been studied by Feng et al. [80–86] who proposed the formation of an aminium radical as the active state of an exciplex as intimate ion-pair and then a cyclic transition state which then would undergo a proton transfer process of deprotonation leading to the formation of active radical species for initiation as follows:

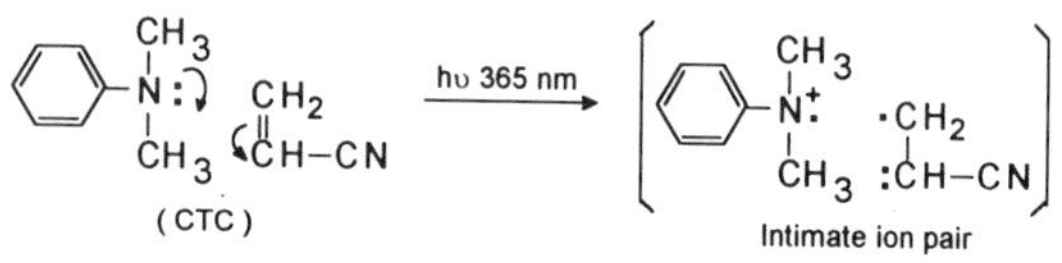

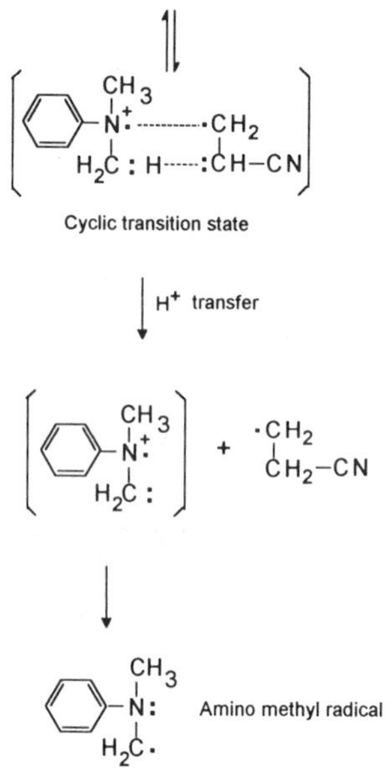

(CTC) → Intimate ion pair

Cyclic transition state

H^+ transfer

Amino methyl radical

2. Aromatic Secondary or Primary Amine and Electron-Accepting Monomer

Li et al. [87,88] found that aniline will process the photopolymerization of AN either in N,N-dimethylformamide (DMF) solution or in bulk with a fair rate of polymerization only next to DMT. From UV spectra it is proved that aniline will form a CTC with AN. Using 313-nm radiation that CTC is excited to an exciplex and polymerization proceeds. N-methylaniline will polymerize AN similarly. The following mechanism was proposed:

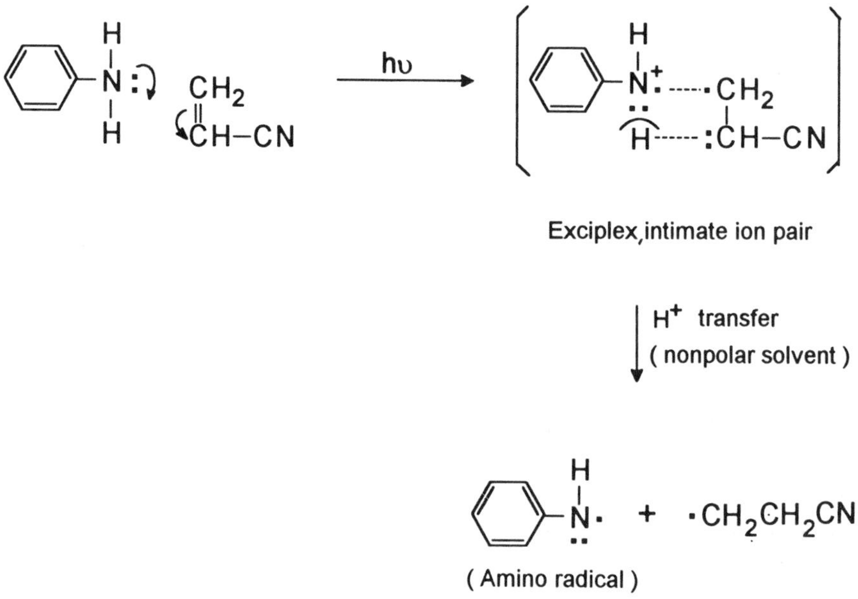

Exciplex, intimate ion pair

H^+ transfer
(nonpolar solvent)

(Amino radical)

Table 12 The Bulk Polymerization of AN Initiated by Aromatic Amines

Amine	Absorbance in DMF	Absorbance in AN	R_p (mol/L·s)
$C_6H_5NH_2$	0	0.03	1.7×10^{-4}
$C_6H_5NHCH_3$	<0.01	0.09	2.1×10^{-4}
$C_6H_5N(CH_3)_2$	<0.01	0.21	2.8×10^{-4}

([amine] = 5×10^{-2} mol at 365 nm)

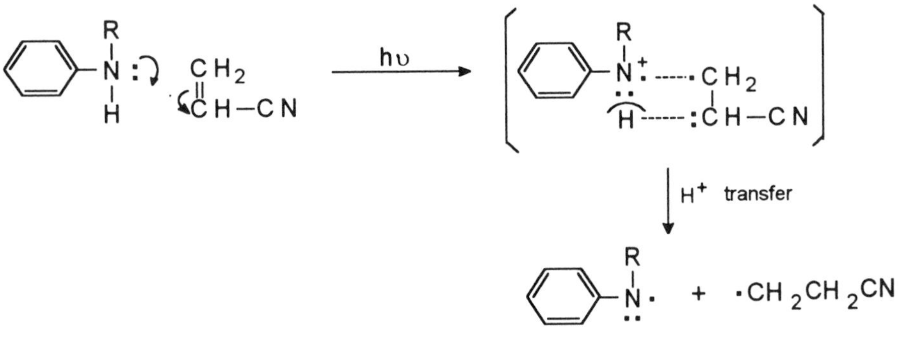

Thus, deprotonation of the aminium radical from a secondary or primary amine will at last form an amino radical instead of an aminoalkyl radical and a $\cdot CH_2CH_2CN$ radical. This amino radical will then serve as one of the active species for the initiation of polymerization.

C. Aromatic Secondary or Primary Amine and Benzophenone

The well-known photopolymerization of acrylic monomers usually involves a charge transfer system with carbonyl compound as an acceptor and aliphatic tertiary amine, triethylamine (TEA), as a donor. Instead of tertiary amine such as TEA or DMT, Li et al. [89] investigated the photopolymerization of AN in the presence of benzophenone (BP) and aniline (A) or N-methylaniline (NMA) and found that the BP-A or BP-NMA system will give a higher rate of polymerization than that of the well-known system BP-TEA. Still, we know that secondary aromatic amine would be deprotonated of the H-atom mostly on the N-atom so we proposed the mechanism as follows:

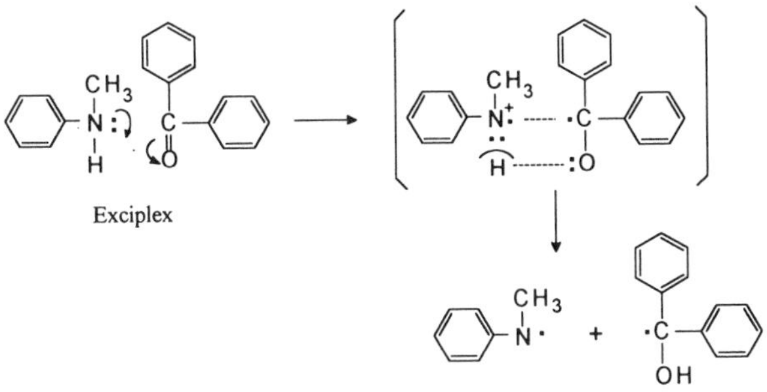

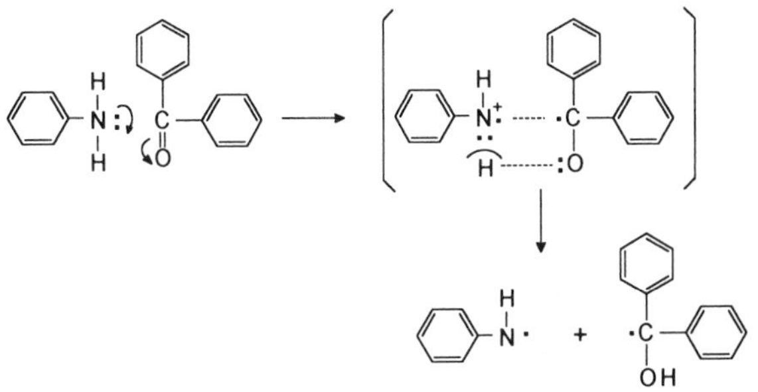

Thus, an aminium radical from primary or secondary amine will at last form an amino radical instead of an aminomethyl radical. This amino radical will then serve as the only active radical species to initiate the vinyl polymerization.

The end group of the polymers, photoinitiated with aromatic amine with or without the presence of carbonyl compound BP, has been detected with absorption spectrophotometry and fluororescence spectrophotometry [90]. The spectra showed the presence of tertiary amino end group in the polymers initiated with secondary amine such as NMA and the presence of secondary amino end group in the polymers initiated with primary amine such as aniline. These results show that the amino radicals, formed through the deprotonation of the aminium radical in the active state of the exciplex from the primary or secondary aromatic amine molecule, are responsible for the initiation of the polymerization.

The proton transfer mechanism described previously was confirmed somewhat by the influence of solvent polarity on polymerization. The rate of photopolymerization of AN initiated by aniline in an acetonitrile-cyclohexane mixture has been studied, and it was found that with an increase in the proportion of acetonitrile, the rate of polymerization first increases because the moderate polar solvent is favorable for exciplex formation, and then the rate falls as the polarity increases further. This is ascribed to the dissociation of the exciplex into solvated ion-radicals in a highly polar solvent. This result shows that the intimate radical-ion pair state of the exciplex, but not the solvated ion-radical state, is responsible for proton transfer:

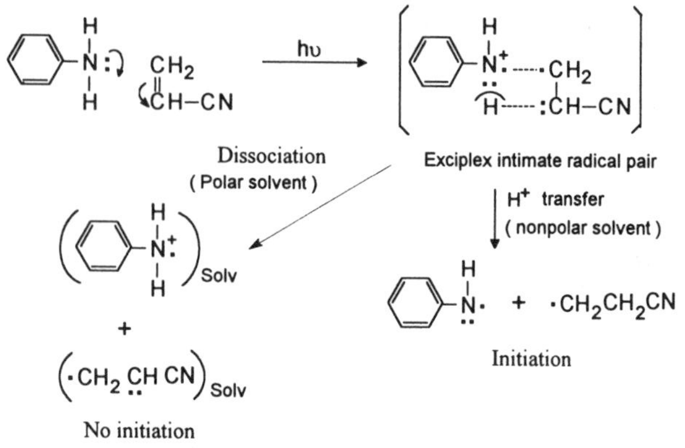

We have prepared a copolymer-bearing amino side group and used it either alone or in combination with BP to initiate the photopolymerization of MMA [89]. The gel permeation chromatography (GPC) plot of PMMA initiated by the former system showed a bimodal distribution of molecular weight because both the radicals produced initiate polymerization as follows:

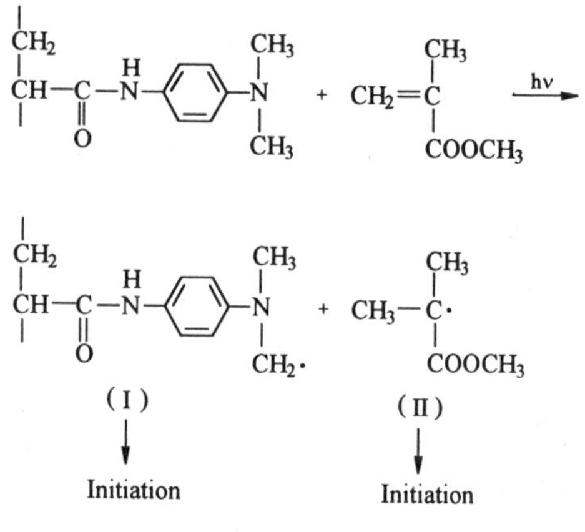

The high-molecular weight was assigned to the PMMA grafted to the copolymer chains and the low-molecular weight to the PMMA initiated by the MMA radical (II). However, only one molecular weight distribution peak was observed for the PMMA initiated by the latter system, i.e., in combination with BP, which implies that only aminomethyl radicals are capable of initiating the polymerization.

The polymers initiated by BP amines were found to contain about one amino end group per molecular chain. It is reasonable to consider that the combination of BP and such polymers will initiate further polymerization of vinyl monomers. We investigated the photopolymerization of MMA with BP-PMMA bearing an anilino end group as the initiation system and found an increase of the molecular weight from GPC and viscometrical measurement [91]. This system can also initiate the photopolymerization of AN to form a block copolymer, which was characterized by GPC, elemental analysis, and IR spectra. The mechanism proposed is as follows:

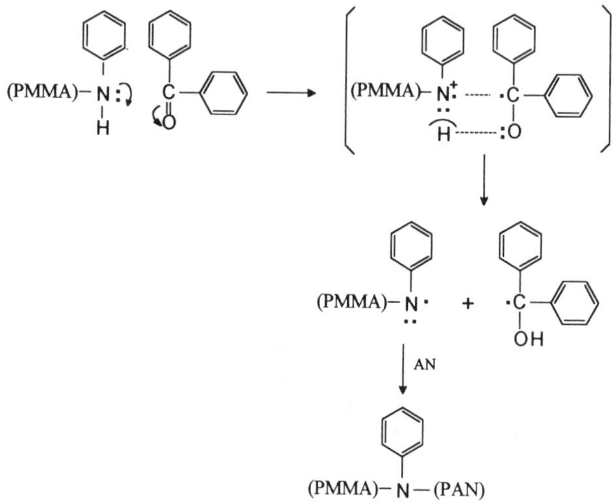

Thus, BP-aniline may serve as the photoinitiator in the establishment of some new methods of block copolymerization.

ACKNOWLEDGMENT

The authors are indebted to the National Natural Science Foundation of China for financial support of this work and also grateful to those coworkers whose research works have been cited.

REFERENCES

1. C. Walling, *Free Radicals in Solution,* John Wiley & Sons, Inc., New York, (1957), p. 590.
2. G. M. Brauer and H. Argentar, *Initiation of Polymerization,* F. E. Jr. Bailey, ed., American Chenical Society, Washington D.C., p. 359 (1983).
3. Q. R. Wu, D. F. Yen, and X. D. Feng (S. T. Voong), *Preprints, The 4th China-Japan Symposium on Radical Polymerization,* October 1–3, 1986, Chengdu, China, p. 156.
4. X. D. Feng, *Chinese J. Polym. Sci., 3*: 109 (1986).
5. X. D. Feng, *Makromol. Chem., Macromol. Symp., 63*: 1 (1992).
6. W. X. Cao, J. F. Chung, and S. T. Voong, *Polym. Commun., 2*: 80 (1978). (in Chinese).

7. G. E. Serniuk, U.S. Patent, 2,529,315, Nov. 7, 1950.

8. T. Tanaka, *Scientific American, 244(1):* 110 (1981).

9. E. Boschetti, U.S. Patent, 4,189,370, Feb. 19, 1980.

10. M. K. Gupta and R. Bansil, *Polym. Preprints, 22(2):* 375 (1981).

11. K. Y. Qiu, L. Shui, and X. D. Feng, *Polym. Commun.,* (Beijing), *1:* 64 (1984).

12. T. Sato, S. Kita, and T. Ostu, *Makromol. Chem., 176:* 561 (1975).

13. G. Yang, *Master Thesis,* Polymer Division of Chemistry Department, Peking University, (1984).

14. X. D. Feng, *Preprints, China-Japan Bilateral Symposium on the Synthesis and Materials Science of Polymers,* October 21–24, 1984, Beijing, China, p. 13.

15. K. Y. Qiu, J. Y. Zhang, Y. J. Hu, and X. D. Feng, *Polym. Commun.,* (Beijing), *1:* 76 (1985).

16. K. Y. Qiu and T. Zhao, *Acta Polym. Sin., 1:* 72 (1988).

17. J. Fu, X. Q. Guo, K. Y. Qiu, and X. D. Feng, *Preprints, Symposium on Polymer Syntheses, Polymerization reaction and Mechanism,* Nov. 7–11, 1988, Nanjing, China, p. 7; *Acta Polym. Sin., 1:* 67 (1990).

18. K. Y. Qiu, D. J. Guo, and X. D. Feng, *Acta Polym. Sin., 1:* 84, (1991).

19. K. Y. Qiu, Z. H. Zhang, and X. D. Feng, *Acta Polym. Sin., 2:* 178 (1993).

20. K. Y. Qiu, Y. B. Shen, and X. D. Feng, *Chinese J. Polym. Sci., 4:* 92 (1986).

21. Q. R. Wu, Y. L. Li, and X. D. Feng, *Preprints, Kunming International Symposium on Polymeric Biomaterials,* May 3–7, 1988, Kunming, China, p. 127.

22. F. M. Li, L. Wang, and X. D. Feng, *Polym. Commun.,* (Beijing), *2:* 144 (1987).

23. F. M. Li, L. Wang, Y. W. Ding, and X. D. Feng, *Polym. Commun.,* (Beijing), *4:* 316.

24. F. M. Li, L. Wang, Y. N. Ding, and X. D. Feng, *Polym. Commun.,* (Beijing), *5:* 393 (1985).

25. F. M. Li, *Preprints, The 1st Japan-China Symposium on Radical Polymerization,* Nov. 5–6, 1980, Osaka, Japan, p. 45.

26. F. M. Li, Z. W. Gu, W. P. Ye, and X. D. Feng, *Polym. Commun.,* (Beijing), *2:* 122 (1983).

27. F. M. Li, L. Wang, and X. D. Feng, *Polym. Commun.,* (Beijing), *3:* 237 (1984).

28. Z. Y. Zhang, H. R. Zhou, A. R. Wnag, and X. D. Feng, *Polym. Commun.,* (Beijing), *2:* 149 (1987).

29. F. M. Li, W. P. Ye, and X. D. Feng, *Polym. Commun.,* (Beijing), *5:* 397 (1982).

30. K. Y. Qiu and J. Fu, *Chinese J. Polym. Sci., 7(3):* 258 (1989).

31. J. Y. Zhang, K. Y. Qiu, and X. D. Feng, *Chem. J. Chinese Univ., 1(2):* 114 (1985).

32. K. Y. Qiu, D. J. Guo, and X. D. Feng, *Chinese J. Polym. Sci., 8:* 363, (1990).

33. K. Y. Qiu, J. Fu, X. Q. Guo, and X. D. Feng, *Chinese J. Polym. Sci., 8:* 188 (1990).

34. K. Y. Qiu, X. Q. Guo, X. L. Wang, and X. D. Feng, *Chem. J. Chinese Univ., 1(1):* 141 (1985).

35. Y. H. Sun, K. Y. Qiu, and X. D. Feng, *Kexue Tongbao (Science Bulletin), 28(4):* 475 (1983).

36. S. T. Voong and K. Y. Qiu, *Kexue Tongbao (Science Bulletin), 17(4):* 247, (1966).

37. Y. H. Sun, K. Y. Qiu, and X. D. Feng, *Scientia Sinica, B27(4):* 349 (1984).

38. Y. H. Sun, *Doctor Thesis,* Polymer Division, Department of Chemistry, Peking University (1984).

39. J. Pavlinec and M. Lazar, *Collect Czech Chem. Commun., 42:* 5023 (1977).

40. X. Q. Guo, K. Y. Qiu, and X. D. Feng, *Scientia Sinica, B30(9):* 897 (1987).

41. X. Q. Guo, *Doctor Thesis,* Polymer Division, Department of Chemistry, Peking University (1987).

42. X. D. Feng and K. Y. Qiu, *Advances in Science of China, Chemistry, 4:* 39 (1992).

43. K. Y. Qiu, X. Q. Guo, and X. D. Feng, *Kexue Tongbao (Science Bulletin), 31(11):* 736 (1986).

44. X. D. Feng, X. Q. Guo, and K. Y. Qiu, *Acta Polym. Sin., (2):* 95 (1988).

45. X. Q. Guo, K. Y. Qiu, and X. D. Feng, *Makromol. Chem., 191:* 577 (1990).

46. X. D. Feng, X. Q. Guo, and K. Y. Qiu, *Polymer Bulletin, 18:* 19 (1987).

47. X. Q. Guo, K. Y. Qiu, and X. D. Feng, *Preprints, The 5th Japan-China Symposium on Radical Polymerization,* August 8–10, 1988, Osaka, Japan, p. 33.

48. R. S. Mulliken, *J. Am. Chem. Soc., 72:* 600, 4493 (1950).

49. R. S. Mulliken, *J. Am. Chem. Soc., 74:* 811 (1952).

50. R. S. Mulliken, *J. Chem. Phys., 19:* 514 (1951).

51. R. S. Mulliken, *J. Chem. Phys., 23:* 397 (1955).

52. R. S. Mulliken, *Rec. Trav. Chim., 75:* 845 (1956).

53. R. S. Mulliken, *J. Phys. Chem., 56:* 801 (1952).

54. E. M. Kosower, *Prog. Phys. Org. Chem., 3:* 81 (1965).

55. R. Foster (ed.), *Organic Charge Transfer Complexes,* Eleck Science, London, 1974.

56. R. Foster (ed.), *Molecular Complexes,* Vol. 2, Eleck Science, London, 1974.

57. S. G. Cohen, A. Parola, and G. Parson, *Chem. Rev., 73:* 141 (1973).

58. H. Scott, G. A. Millerand, and M. M. Laber, *Tetrahedron Lett.,* 1073 (1963).

59. C. P. Ellinger, *Chem. Ind.* (London), 1982 (1963); *Polymer, 5:* 559 (1964); *6,* 549 (1964).

60. V. S. Anderson and R. G. W. Norrish, *Proc. R. Soc.* (London), *A251:* 1 (1959).

61. R. G. W. Norrish and I. P. Sirnous, *Proc. R. Soc.* (London), *A251,* 4 (1959).

62. Y. Shirota and H. Mikawa, *J. Macromol. Sci.-Rev. Macromol. Chem., C16:* 129 (1977–1978).

63. Y. Shirota and H. Mikawa, *Mol. Cryst. Liq. Cryst., 126:* 43 (1985).

64. Y. Shirota, in *Encyclopedia of Polymer Science and Engineering* (J. I. Kroschwitz et al., eds.), 2nd ed., Vol. 3, John Wiley and Sons, Inc., New York, (1985), p. 327.

65. M. Irie and K. Hayashi, *Prog. Polym. Sci. Jpn., 8:* 105 (1975).

66. A. Ohno, N. Kito, and N. Kwase, *J. Polym. Sci., Polym. Lett. Ed., 10:* 133 (1972).

67. M. Shimizu, K. Tanabe, K. Tada, Y. Shirota, S. Kusabayashi, and H. Mikawa, *Chem. Commun.,* 1628 (1970).

68. M. Yamamoto, S. Nishimoto, M. Ohaka, and Y. Nishijima, *Macromolecules, 3:* 706 (1970).

69. M. Shimizu, K. Tada, Y. Shirota, S. Kusabayashi, and H. Mikawa, *Makromol. Chem., 176:* 1953 (1975).

70. M. Yamamoto, T. Ohmichi, M. Ohoka, and Y. Nishijima, *Prog. Polym. Phys. Jpn., 12:* 457 (1969).

71. K. Tada, Y. Shirota, and H. Mikawa, *J. Polym. Sci., Polym. Lett. Ed., 10:* 691 (1972).

72. S. Tazuke, *J. Phys. Chem., 74:* 2390 (1970).

73. H. F. Kaufman, J. W. Breitenbach, and O. F. Olaj, *J. Polym. Sci., Polym. Chem. Ed., 11:* 737 (1973).

74. Y. Shirota, K. Kawai, N. Yamamoto, K. Tada, H. Mi-

kawa, and H. Tsubomura, *24th Annu. Meet. Jpn. Chem. Soc.,* p. 11232 (1971).

75. Y. Shirota, K. Kawai, N. Yamamoto, K. Tada, H. Mikawa, and H. Tsubomura, *Chem. Lett.,* p. 145 (1972).

76. Y. Shirota, K. Kawai, N. Yamamoto, T. Shida, H. Mikawa, and H. Tsubomura, *Bull. Chem. Soc. Jpn., 45*: 2693 (1972).

77. Y. Shirota, T. Tomikawa, T. Nogami, N. Yamamoto, H. Tsubomura, and H. Mikawa, *Bull. Chem. Soc. Jpn., 47*: 2099 (1974).

78. Y. Wei, W. X. Cao, and X. D. Feng, *Kexue Tongbao (Science Bulletin), 28*: 1068 (1983).

79. Y. Wei, W. X. Cao, and X. D. Feng, *Chem. Abstr., 98*: 143870t (1983).

80. T. Li, W. X. Cao, and X. D. Feng, *Polym. Commun.* (Beijing), *(2)*: 127 (1983).

81. T. Li, W. X. Cao, and X. D. Feng, *Chem. Abstr., 99*: 140446U (1983).

82. T. Li, W. X. Cao, and X. D. Feng, *Polym. Commun.* (Beijing), 4: 260 (1983).

83. T. Li, W. X. Cao, and X. D. Feng, *Chem. Abstr., 100*: 103923r (1984).

84. X. D. Feng, *Preprints of the 3rd Japan-China Symposium on Radical Polymerization,* Osaka, Japan, 1984, p. 5.

85. T. Li, W. X. Cao, and X. D. Feng, *Chem. J. Chin. Univ.* (Engl. Ed.) *3*: 52 (1987).

86. X. D. Feng, *Preprints, 2nd China-Japan Bilateral Symposium on Synthesis and Material Science of Polymers,* 1984, Beijing, China, p. 17.

87. T. Li, W. X. Cao, and X. D. Feng, *Scientia Sinica B, 30(12)*: 1260 (1987).

88. T. Li, W. X. Cao, and X. D. Feng, *Chem. J. Chinese Univ. 7(10)*: 953 (1986).

89. T. Li, W. X. Cao, and X. D. Feng, *Scientia Sinica B, 7*: 685 (1987).

90. T. Li, W. X. Cao and X. D. Feng, *Photographic Sci. and Photochemistry,* (Beijing), *2*: 49 (1986).

91. T. Li, W. X. Cao and X. D. Feng, *Chinese Science Bulletin, 34*: 1611 (1989).

17

Photoinitiation of Free Radical Polymerization by Organometallic Compounds

Salah M. Aliwi
Mustansiriya University, Baghdad, Iraq

I. INTRODUCTION

Photopolymerization, in general, can be defined as the process whereby light is used to induce the conversion of monomer molecules to a polymer chain. One can distinguish between true photopolymerization and photoinitiation of polymerization processes. In the former, each chain propagation step involves a photochemical process [1,2] (i.e., photochemical chain lengthening process in which the absorption of light is indispensable for each propagation step); whereas in the latter, only the initial step results from the interaction of light with a photosensitive compound or system [3–6]. This chapter deals only with the photoinitiation of polymerization process or photochemical postpolymerization. This may occur in two different ways:

1. Through energy transfer of an excited sensitizer molecule (S) to either a monomer (M) or foreign molecule (A) resulting in the formation of species capable of initiation (e.g., radical):

$$S \xrightarrow{h\nu} [S]^*_{excited} \begin{cases} \xrightarrow{M} S + \overset{*}{M} \\ \xrightarrow{A} S + \overset{*}{A} \end{cases} \longrightarrow R^{\bullet}\ (free\text{-}radical) \tag{1}$$

2. Through direct excitation of a monomeric or polymeric molecule or of a molecular complex (A) followed by a reaction producing an initiating species:

$$A \xrightarrow{h\nu} [A]^*_{excited} \longrightarrow R^{\bullet}\ (free\text{-}radical) \tag{2}$$

The radicals created in (1) and (2) interact with monomer molecules to produce macroradicals, and ultimately, by termination of these radicals, a polymeric chain is produced:

$$\begin{aligned} R^{\bullet} + M &\longrightarrow RM^{\bullet} & (initiation) \\ RM^{\bullet} + nM &\longrightarrow R(M)_n M^{\bullet} & (propagation) \\ 2R(M)_n M^{\bullet} &\longrightarrow R(M)_{n+1}-(M)_{n+1}R & (termination) \end{aligned} \right\} \tag{3}$$

The rate of formation of radical from the photoinitiator molecule (the only light-absorbing molecule in the system) $V_{R^{\bullet}}$ is generally given by Eq. (4):

$$V_R{}^{\bullet} = \propto I_o(1 - e^{\varepsilon_{In} l [In]}) \tag{4}$$

where I_o is the incident light intensity, $[In]$ and ϵ_{In} are the photoinitiator concentration and its molar extinction coefficient, respectively, and l is the cell path length, while α is the overall quantum yield of the initiating species ($R^{\bullet}$).

Not all initiating radicals ($R^{\cdot}$) succeed in initiating polymerization, recombination of these radicals in the solvent can decrease the efficiency (f) to a value lower than 1. Detailed kinetic treatment of photoinitiation processes are discussed by Oster and Yang [3].

Photoinitiation of polymerization has played an important role in the early developments of polymer chemistry. The main features of this type of initiation are:

1. The rate of formation of the initiating species and, accordingly, the molecular weight distribution can be controlled by variation of the incident light intensity.
2. Photoinitiation is an excellent method for studying the pre- and posteffects of free radical polymerization, and from the ratio of the specific rate constant (k_x) in non-steady-state conditions, together with steady-state kinetics, the absolute values of propagation (k_p) and termination (k_t) rate constants for radical polymerization can be obtained.
3. Photoinitiators provide a convenient route for synthesizing vinyl polymers with a variety of different reactive end groups. Under suitable conditions, and in the presence of a vinyl monomer, a block AB or ABA copolymer can be produced which would otherwise be difficult or impossible to produce by another polymerization method. Moreover, synthesis of block copolymers by this route is much more versatile than those based on anionic polymerization, since a wider range of a monomers can be incorporated into the blocks.
4. Photoinitiation with a high quantum yield of radical production in the visible light is of practical importance for photocuring processes [5,6].

One of the first methods of polymerizing vinyl monomers was to expose the monomer to sunlight. In 1845, Blyth and Hoffman [7] obtained by this means a clear glassy polymeric product from styrene. Berthelot and Gaudechon [8] were the first to polymerize ethylene to a solid form and they used ultraviolet (UV) light for this purpose. The first demonstration of the chain reaction nature of photoinitiation of vinyl polymerization was done by Ostromislenski in 1912 [9]. He showed that the amount of poly(vinyl bromide) produced was considerably in excess of that produced for an ordinary chemical reaction.

In more recent years, photoinitiation of polymerization proved to be of immense value in the understanding of the precise nature of polymerization. Several systems used for the initiation of radical polymerization were reviewed by Oster and Yang [3], Rabek [10], and Davidson [5,6].

Two types of photoinitiation processes may be considered:

1. A chain reaction polymerization of vinyl monomer, which is usually carried out by a photoinitiator to produce a primary radical ($R^{\cdot}$), which can interact with a monomer molecule (M) in a propagating process to form a polymer chain composed of a large number of monomer units (see Eq. [2] and reaction Scheme [3].
2. The photografting and photocrosslinking processes in which the photoinitiation is carried out by pre-existing polymer in the presence of vinyl monomer. The pre-existing polymer may itself be photoactive or, with the presence of photoinitiator, can interact with a polymer to produce a radical on the backbone of the polymer chain.

$$\text{In} \xrightarrow{h\nu} \quad \xrightarrow{nM} \quad (P^{\cdot}) \qquad (5)$$

If ($P^{\cdot}$) is terminated by a chain transfer to a solvent or a monomer, a graft copolymer is formed, or, if the termination is from a combination, a crosslinked network polymer is formed. If the pre-existing polymer (B) contains an end group that itself is photosensitive (or can produce a radical by interacting with photoinitiator) and in the presence of a vinyl monomer (A), block copolymer of type AB can be produced if the photosensitive group is on one end of the polymeric chain. Type ABA block copolymer can be produced if the polymer chain (B) contains a photosensitive group on both ends.

Free radicals capable of initiating polymerization of vinyl monomers can be produced photochemically from a wide variety of substances, such as organic, inorganic, and organometallic compounds in one or two component initiating systems.

The photoinitiation of vinyl polymerization by organic compounds (carbonyl, azo, peroxide, disulphide compounds, etc.) or inorganic salts (e.g., metal halides and their ion pairs, etc.) will not be discussed here, since these type of photoinitiators are beyond the scope of the present chapter.

Two types of organometallic photoinitiators for free radical vinyl polymerization are considered: (1) transition metal complexes with ligand, mainly carbon monoxide (metal carbonyl complexes), generally, but not invariably, in the presence of organic halide (e.g. CCl_4 or CBr_4), and (2) transition metal chelates. Systems of type (1) contain metal in a low, often a zero oxidation state, while those in type (2) contain metal in a high oxidation state.

II. TRANSITION METAL CARBONYL COMPLEXES AS PHOTOINITIATORS

The primary photochemical of transition metal carbonyls $[Me(CO)_n]$ involves the scission of carbon monoxide (CO) and the formation of coordinated unsaturated species:

$$Me(CO)_n \xrightarrow{h\nu} [Me(CO)_n]^* \Longrightarrow Me(CO)_{n-1} + CO \qquad (6)$$

$$excited$$

The first experimental evidence with the relevant theoretical background of this type of photochemical reaction was given by Koerner Von Gustorf and Grevels [11]. The product $Me(CO)_{n-1}$ can combine with carbon monoxide to regenerate the original carbonyl $Me(CO)_n$, or it may react by the addition of an n or an electron doner (S) (e.g., monomer) according to Blyth and Hoffman [7].

$$Me(CO)_{n-1} + S(M) \Longrightarrow (M)S...Me(CO)_{n-1} \qquad (7)$$

The species $(M)S\cdots Me(CO)_{n-1}$ arises from Eq. (7) and reacts with suitable halides (e.g., CCl_4) with a generation of free radicals:

$$(M)S...\overset{0}{Me}(CO)_{n-1} + CCl_4 \Longrightarrow (M)S...\overset{+1}{Me}(CO)_{n-1}Cl + \overset{\cdot}{C}Cl_3 \qquad (8)$$

Thus, a mixture of simple carbonyls $Me(CO)_n$ and halides should behave as a photoinitiator of free radical polymerization. Many such systems have been found to function in this way. Complexes formed by irradiation of $Fe(CO)_5$ in the presence of a vinyl monomer (M) (such as MMA, styrene, vinyl acetate, propylene, and vinyl ether) have been studied by Koerner Von Grustrof and colleagues [12,13] and shown to have the chemical structure $MFe(CO)_4$ in which the iron atom is coordinated to the vinyl double bond. Further, it has been demonstrated by the same authors that irradiation of $Fe(CO)_5$ in the presence of a vinyl monomer and an organic halide leads to photoinitiated polymerization in a suitable system.

The following reaction scheme is suggested for methyl methacrylate monomer and $CHBr_3$ halide as coinitiators:

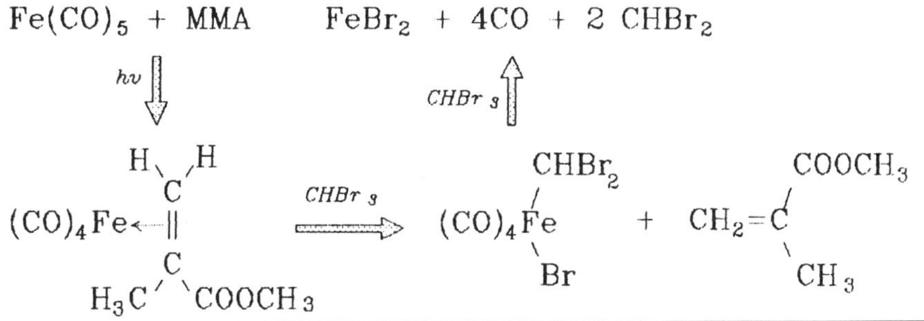

$$(9)$$

Strohmeier and Hartmann [14] first reported in 1964 the photoinitiation of polymerization of ethyl acrylate by several transition metal carbonyls in the presence of CCl_4. Vinyl chloride has also been polymerized in a similar manner [15,16] No detailed photoinitiation mechanisms were discussed, but it seems most likely that photoinitiation proceeds by the route shown in reaction Scheme (9).

The pioneering work of Bamford and coworkers in 1965 [17] and 1966 [18] has given a detailed kinetics study and a mechanism of the photoinitiation of polymerization of methyl methacrylate (MMA) by manganese and ruthenium carbonyls in the presence of carbon tetrachloride as the coinitiator. The long wavelength limits of absorption by these materials are approximately 460 nm and 380 nm, respectively, and photoinitiation occurs up to these wavelengths. For both photoinitiating systems, it was found that the rate of polymerization of MMA monomer is dependent on CCl_4 concentration (at constant absorbed light intensity). Quantum yields for initiation at 25°C for MMA polymerization in the presence of 0.1 mol/1 CCl_4 were determined at $\lambda = 435$ nm for Mn_2

$(CO)_{10}$ and $\lambda = 365$ for $Re_2(CO)_{10}$. The values reported are close to unity.

However, MMA polymerization photoinitiated by $Re_2(CO)_{10}$ ($\lambda = 365$ nm) shows a long-lived aftereffect, persisting for several hours at 25°C after irradiation has

been discontinued. Aftereffects of this kind are not shown by systems containing $Mn_2(CO)_{10}$ photoinitiator with the same monomer. According to the kinetics and spectroscopic observations, the following reaction, Scheme (10), was suggested for the photoinitiating system composed of $Mn_2(CO)_{10}/CCl_4$ system:

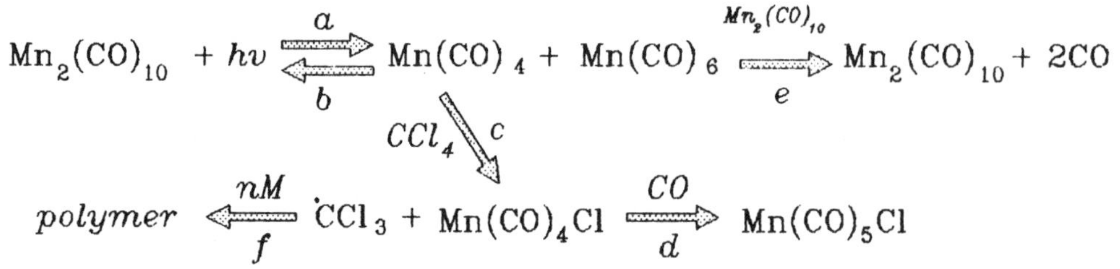

$$Mn_2(CO)_{10} + h\upsilon \underset{b}{\overset{a}{\rightleftarrows}} Mn(CO)_4 + Mn(CO)_6 \xrightarrow[e]{Mn_2(CO)_{10}} Mn_2(CO)_{10} + 2CO$$

$$CCl_4 \searrow c$$

$$polymer \underset{f}{\overset{nM}{\rightleftarrows}} \dot{C}Cl_3 + Mn(CO)_4Cl \xrightarrow[d]{CO} Mn(CO)_5Cl \tag{10}$$

Reaction 10e is relatively slow in the $Re_2(CO)_{10}$ initiating system, and the thermal reaction between $Re(CO)_6$ formed in 10a and CCl_4 generated $\dot{C}Cl_3$ radicals thermally in the dark and so is responsible for the aftereffect. However, $Mn_2(CO)_{10}$ reacts rapidly according to 10e and no aftereffect is observed for the $Mn_2(CO)_{10}/CCl_4$ photoinitiating system.

Bamford and coworkers [19] have shown that a prolonged aftereffect can be obtained with Mn-carbonyl in the presence of certain additives, notably cyclohexane and acetylacetone (S). It was suggested that the photochemical reaction between $Mn_2(CO)_{10}$ and (S) produces the active species (Z), which generates free radicals by interaction with halide and Z probably formed from $Mn(CO)_6$ species:

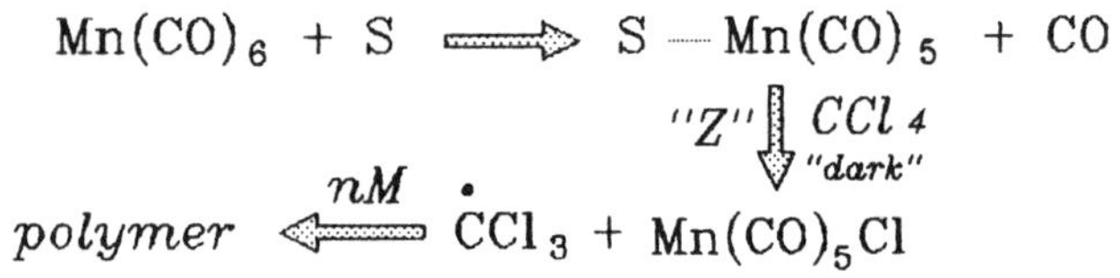

$$Mn(CO)_6 + S \Longrightarrow S\text{---}Mn(CO)_5 + CO$$

$$"Z" \Big\| \overset{CCl_4}{\underset{"dark"}{}}$$

$$polymer \overset{nM}{\Longleftarrow} \dot{C}Cl_3 + Mn(CO)_5Cl \tag{11}$$

Osmium carbonyl ($Os_3(CO)_{12}$) acts as a photoinitiator of vinyl polymerization [20], which can function without a halide additive. The mechanism of photoinitiation is by a hydrogen abstraction from monomer to photoexcited osmium carbonyl molecule. Addition of CCl_4 leads to only a moderate increase in the rate of initiation, but photoinitiation by the $Os_3(CO)_{12}/CCl_4$ system is markedly increased by the addition of dimethyl sulfoxide (DMSO) as the electron doner.

This has been explained by the formation of additional complexes with e-doner, which increases the rate of reaction with stabilization of incipient osmium cations.

Strohmeier and Grübel [21] have reported that some vanadyl carbonyls of $CpV(CO)_4$ (Cp = cyclopentadienyl) can photochemically induce the polymerization of vinyl chloride in the presence of CCl_4.

Bamford and Mullik [22], in 1973 demonstrated that managanese carbonyl is able to photosensize ($\lambda = 436$ nm) the polymerization of tetrafluoroethylene (TFE) in the absence of any other halide. The reaction occurs readily in bulk TFE or in carbon disulfide at -93°C; thus, $Mn_2(CO)_{10}/TFE$ can photosensitize the free radical polymerization of vinyl monomers such as MMA, styrene, and acrylonitrile at 25°C. TFE as halide in this system is considerably less active than CCl_4 and perfluorocyclohexone is completely inactive. Initiation by $Mn_2(CO)_{10}/TFE$ appears to be the result of electron transfer from the product of photolysis of $Mn_2(CO)_{10}$ (e.g., $Mn(CO)_4$) to TFE, producing first a radical anion and resulting ultimately in a species containing Mn—C bond:

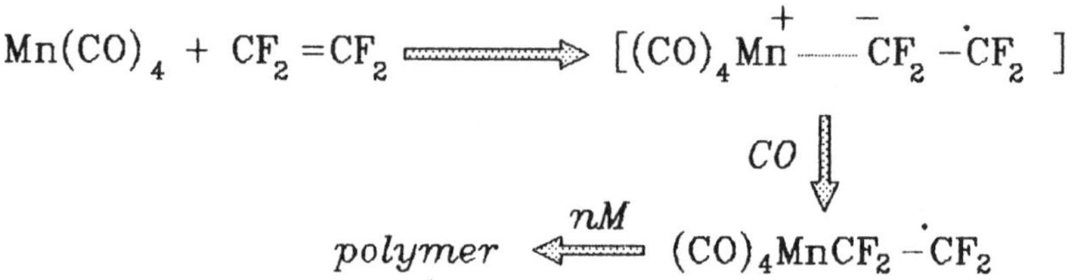

$$Mn(CO)_4 + CF_2{=}CF_2 \Longrightarrow [(CO)_4\overset{+}{Mn}\text{---}\overset{-}{CF_2}\text{-}\dot{C}F_2\,]$$

$$CO \Big\downarrow$$

$$polymer \overset{nM}{\Longleftarrow} (CO)_4MnCF_2\text{-}\dot{C}F_2 \tag{12}$$

Bamford and Mullik [23] have also investigated a new photoinitiating system composed of $Mn_2(CO)_{10}$ or $Re_2(CO)_{10}$ with acetylene, acetylene dicarboxylic acid, diethyl fumarate, diethyl maleate, or maleic anhydride. It was concluded that the primary radical responsible

for initiation comes from the addition of metal carbonyl fragments (e.g., $Me(CO)_4$ or $Me(CO)_6$) to multiple bond in the acetylinic or olefinic derivatives. The following initiating radicals photochemically produced are suggested:

$$(CO)_5 M_n - \overset{|}{\underset{|}{C}} = \overset{|}{\underset{|}{C}} \cdot \quad , \quad (CO)_5 Re - \overset{|}{\underset{|}{C}} - \overset{|}{\underset{|}{C}} \cdot \quad \text{and} \quad (CO)_5 Re - \overset{|}{\underset{|}{C}} = \overset{|}{\underset{|}{C}} \cdot$$

$Mn_2(CO)_{10}$ is only active in the presence of acetylene derivatives, where as $Re_2(CO)_{10}$ is active in the presence of both olefinic and acetylene derivatives.

Bamford and coworkers [24] also investigated the kinetics and mechanism of free radical polymerization of bulk MMA photoinitiated by $Mn_2(CO)_{10}$ or $Re_2(CO)_{10}$ in the presence of a series of fluoro-olefins such as:

$$CF_2 = CFCl \quad , \quad CF_2 = CHF \quad , \quad CF_2 = CH_2 \quad \text{and} \quad CHF = CH_2$$

The average quantum yield of an initiation process is 0.65, which is evidently less than the quantum yield in the $Mn_2(CO)_{10}/CCl_4$ or $Re_2(CO)_{10}$ photoinitiating systems under similar conditions. Two different mechanisms were proposed for the production of the initiating radicals, either by chlorine abstraction with formation of $CF_2 = \cdot CF$ primary radicals, or by the addition process, such as that occurring with the $Mn_2(CO)_{10}/C_2F_4$ initiating system, which would lead to initiating radicals of the type $(CO)_5 Mn\ CF_2 \cdot CF\ Cl$ (see reaction Scheme [12]).

Photolysis of several arene chromium tricarbonyls $(Ar\ Cr\ (CO)_3)$ $(Ar =$ benzene, toluene, or xylene) and photoinitiation by these derivatives have been studied by Bamford and Al-Lamee [25,26]. In the presence of active halide (e.g., CBr_4 or CCl_4) as coinitiator, the chromium carbonyl excited state interacts with the halide molecule through an electron transfer process to give the initiating radical derived from the halide molecule. Mechanisms similar to that shown in Schemes (10) and (11) are suggested, and the quantum yield of the photoinitiation process approaches unity.

In the polymerization of phenyl acetylene [27] by tungsten and molybdenum hexacarbonyls, high-polymer yields were obtained in CCl_4 solvent. The following reaction scheme was proposed, which is different from that reported by Bamford and coworkers [17–20]:

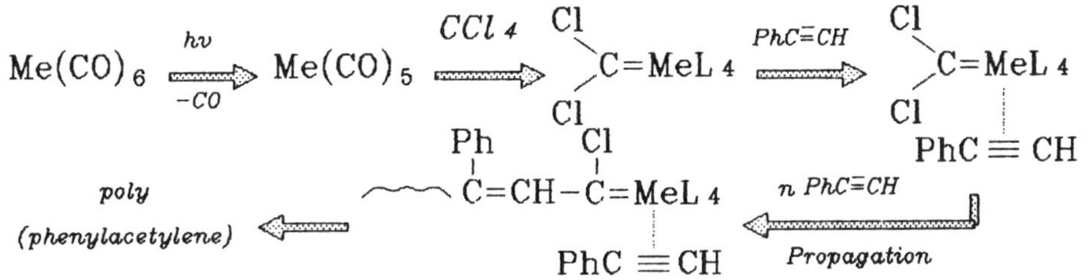

$$\tag{13}$$

Copolymerization of methacrylic acid with butadiene and isoprene was photoinitiated by $Mn_2(CO)_{10}$ without any halide catalyst [28,29]. The polymerization system is accompanied by a Diels-Alder additive. Cross propagation reaction was promoted by adding triethylaluminum chloride.

It has generally been concluded that the photoinitiation of polymerization by the transition metal carbonyls/halide system may occur by three routes; (1) electron transfer to an organic halide with rupture of C—Cl bond, (2) electron transfer to a strong-attracting monomer such as C_2F_4, probably with scission of-bond, and (3) halogen atom transfer from monomer molecule or solvent to a photoexcited metal carbonyl species. Of these, (1) is the most frequently encountered.

III. PHOTOINITIATION OF POLYMERIZATION BY TRANSITION METAL CHELATES

Compared with other metal coordination compounds, relatively little is known about chelate photochemistry.

The photochemistry of transition metal 1,3-diketone chelate complexes has been known for some time [30,31], and their photophysical and photochemical properties and photocatalytic activity in different chemical reactions were reviewed in 1990 by Marciniak and Buono—Core [32]. Further discussion on the photochemistry of metal chelate will not take place here since this subject is out of the scope of this chapter.

Studies in the photoinitiation of polymerization by transition metal chelates probably stem from the original observations of Bamford and Ferrar [33]. These workers have shown that Mn(III) tris-(acetylacetonate); (Mn(acac)$_3$) and Mn (III) tris-(1,1,1-trifluoroacetyl acetonate); (Mn(facac)$_3$) can photosensitize the free radical polymerization of MMA and styrene (in bulk and in solution) when irradiated with light of $\lambda = 365$ at 25°C and also abstract hydrogen atom from hydrocarbon solvents in the absence of monomer. The initiation of polymerization is not dependant on the nature of the monomer and the rate of photodecomposition of Mn(acac)$_3$ exceeds the rate of initiation and the initiation species is the acac radical. The mechanism shown in Scheme (14) is proposed according to the kinetics and spectral observations:

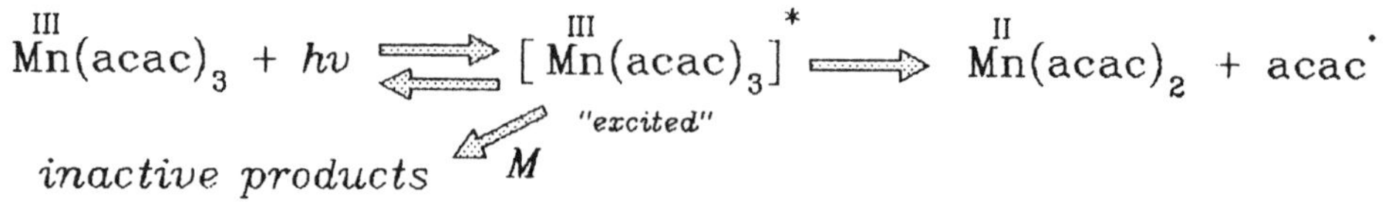

$$\overset{III}{Mn}(acac)_3 + h\upsilon \rightleftarrows [\overset{III}{Mn}(acac)_3]^* \longrightarrow \overset{II}{Mn}(acac)_2 + acac^\cdot \qquad (14)$$

$$\text{inactive products} \overset{M}{\longleftarrow} \quad \text{``excited''}$$

Mn(acac)$_3$ in the above mechanism undergoes an intramolecular photooxidation–reduction reaction arising from the ligand to metal charge transfer process (LMCT).

The quantum yield of the initiation process (ϕ_i) is quite low 8×10^{-3}, indicating the great stability of the chelate ring toward photolysis. However, the quantum yield of photodecomposition (ϕ_d) under similar condition is 2×10^{-2}, which is higher than (ϕ_i). It is clear, therefore, that not every molecule of Mn(acac)$_3$ that is decomposed initiates polymerization; apparently, ex-

cited molecules of the chelate may react with a monomer to form inactive products.

Photoinitiation of polymerization of MMA and styrene by Mn(facac)$_3$ was also investigated, and it was shown that the mechanism of photoinitiation is different [33] from that of Mn(acac)$_3$ and is subject to the marked solvent effect, being less efficient in benzene than in ethyl acetate solutions. The mechanism shown in Schemes (15) and (16) illustrate the photodecomposition scheme of Mn(facac)$_3$ in monomer-ethyl acetate and monomer-benzene solutions, respectively. (C = manganese chelate complex.)

In ethyl acetate (E):

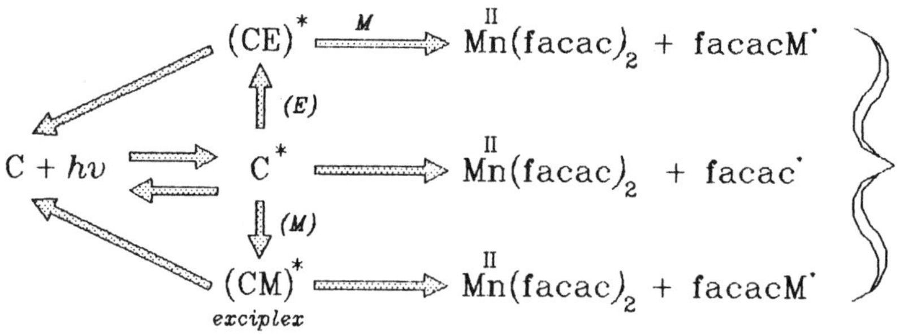

$$
\begin{array}{c}
(CE)^* \overset{M}{\Longrightarrow} \overset{II}{Mn}(facac)_2 + facacM^\cdot \\
\uparrow (E) \\
C + h\upsilon \rightleftarrows C^* \longrightarrow \overset{II}{Mn}(facac)_2 + facac^\cdot \\
\downarrow (M) \\
(CM)^* \longrightarrow \overset{II}{Mn}(facac)_2 + facacM^\cdot \\
\text{exciplex}
\end{array}
\qquad (15)
$$

In benzene (B)

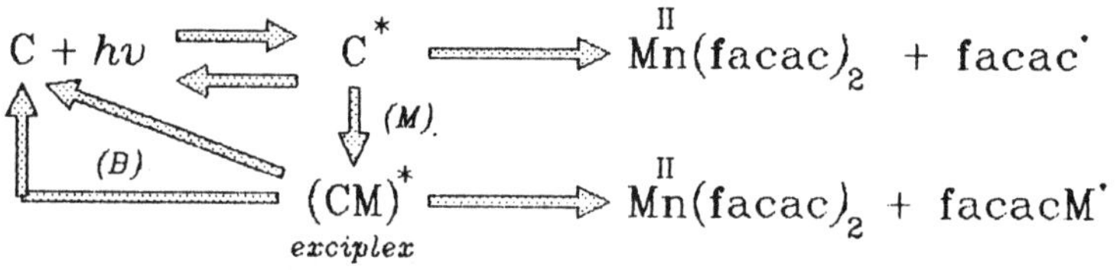

$$
\begin{array}{c}
C + h\upsilon \rightleftarrows C^* \Longrightarrow \overset{II}{Mn}(facac)_2 + facac^\cdot \\
(B) \quad \downarrow (M) \\
(CM)^* \Longrightarrow \overset{II}{Mn}(facac)_2 + facacM^\cdot \\
\text{exciplex}
\end{array}
\qquad (16)
$$

If the monomer is denoted by (M) and photoexcited chelate by (C)*, the proposed mechanism with ethyl acetate (15,16) involves formation of an exciplex (CM*), which may revert to (C + M) or decompose to facacCM$^\cdot$ radicals and Mn(facac)$_2$.

This alone is inadequate, since it predicates a linear relation between the rate of initiation and the monomer concentration, and, therefore, it is believed that monomer and ethyl acetate (E) are, to some extent, interchangeable in the reaction and that E may also form as exciplex (CE)*.

The corresponding reaction Scheme, Scheme (16) proposed for benzene solution is similar, the only additional reaction being the deactivation of the exciplex (CM*) by the banzene molecule (B).

Quantum yields for photoinitiation and photodecomposition are also low in Mn(facac)$_3$ (1.5×10^{-2}) and

are approximately equal. This is a different behavior that observed for the Mn(acac)$_3$ photoinitiator.

Kaeriyama and Shimura [34] have reported the photoinitiation of polymerization of MMA and styrene by 12 metal acetylacetonate complex. These are Mn(acac)$_3$, MoO$_2$(acac)$_2$, Al(acac)$_3$, Cu(bzac)$_2$, Mg(acac)$_2$, Co(acac)$_2$, Co(acac)$_3$, Cr(acac)$_3$, Zn(acac)$_2$, Fe(acac)$_3$, Ni(acac)$_2$, and (Ti(acac)$_2$) − TiCl$_6$. It was found that Mn(acac)$_3$ and Co(acac)$_3$ are the most efficient initiators. The intraredox reaction with production of acac radicals is proposed as a general route for the photodecomposition of these chelates.

Aliwi and coworkers have investigated many vanadium (V) chelate complexes as photoinitiators for vinyl polymerization [36–43]. The mixed ligand complex of chloro-oxo-bis(2,4-pentanedione) vanadium (V). VO(acac)$_2$ Cl is used as the photoinitiator of polymerization of MMA in bulk and in benzene solution (irradiation wavelength $\lambda = 365$ nm) [36].

The neutron activation analysis of the polymer reveals that initiation is effected predominantly by chlorine atoms. No retardation or inhibition were detected, but the rate of photodecomposition increases linearly with an increasing monomer concentration [M], and the rate of decomposition is equal to the rate of initiation at finite [M]. The following photo-redox reaction is suggested:

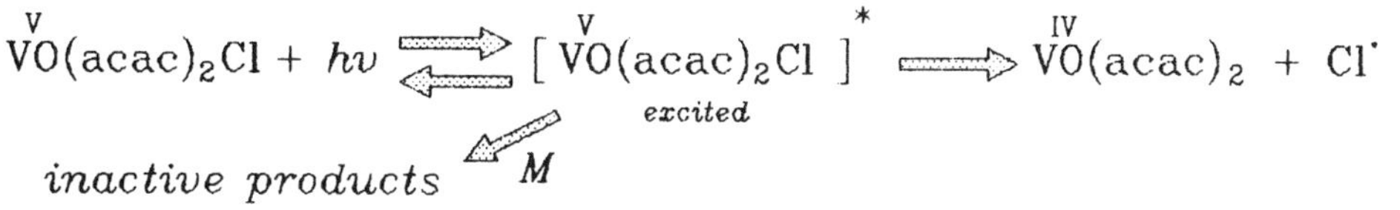

$$\overset{V}{VO}(acac)_2Cl + h\nu \rightleftharpoons [\overset{V}{VO}(acac)_2Cl]^* \underset{excited}{\Longrightarrow} \overset{IV}{VO}(acac)_2 + Cl^\cdot \tag{17}$$

$$inactive\ products \overset{M}{\swarrow}$$

It has been proposed that the monomer undergo an insertion reaction into the V—Cl bond of the photoexcited chelate molecule, resulting in decomposition of the chelate in the anon-radical route.

The quantum yield of initiation is also low, approximately 2.07×10^{-2}. This could be increased to 0.59 or 0.125 when 0.2 M of a strong electron doner (D) such as dimethyl sulfoxide (DMSO) or pyridine (Py) is used, respectively [36,37].

According to the UV visible spectral and conductivity changes after the addition of DMSO or Py to the MMA solution of the VO(acac)₂Cl, it has been suggested that an ion-pair complex of the type $\{[VO(acac)_2]^+Cl^-\}$ is formed, which is photosensitive at $\lambda = 365$ nm (25°C) and is able to polymerize MMA monomers.

Neutron activation analysis of a polymer suggests that when Py is used as the electron doner (D), the initiation proceeds through the Cl⁻ atom, but when D = DMSO, both Cl⁻ and DMSO residues are the primary radicals produced from the photoexcited ion-pair complex. The following reaction scheme is proposed:

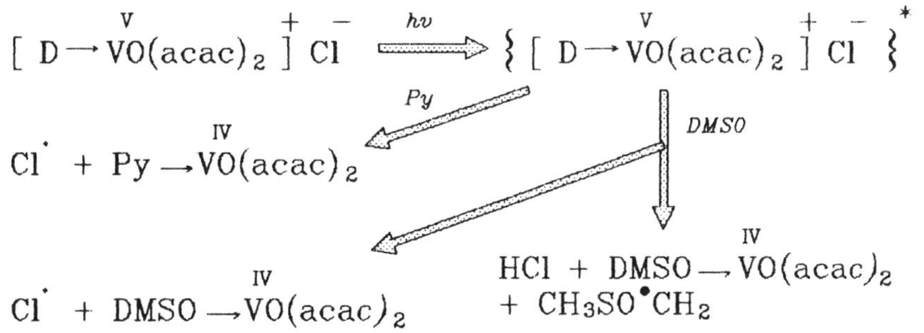

$$[D \rightarrow \overset{V}{VO}(acac)_2]^+ Cl^- \overset{h\nu}{\Longrightarrow} \{[D \rightarrow \overset{V}{VO}(acac)_2]^+ Cl^-\}^* \tag{18}$$

$$Cl^\cdot + Py \rightarrow \overset{IV}{VO}(acac)_2$$

$$Cl^\cdot + DMSO \rightarrow \overset{IV}{VO}(acac)_2$$

$$HCl + DMSO \rightarrow \overset{IV}{VO}(acac)_2 + CH_3SO^\cdot CH_2$$

The polymerization of MMA photoinitiated by alkoxo-oxo-bis(8-quinolyloxo) vanadium (V) complex [VOQ₂ OR] has also been studied [38,39]. The alkyloxo radical ($^\cdot$OR) formed from the photodecomposition of the chelate ($\lambda = 365$) nm at 25°C was found to be the initiating species:

$$\overset{V}{VOQ_2}OR \overset{h\nu}{\Longrightarrow} \overset{IV}{VOQ_2} + {}^\cdot OR \tag{19}$$

Spin trapping (e.s.r) and C¹⁴-labeling techniques were used to study the structure of the alkyloxy radicals produced and to show that these radicals isomerize to hydroxy radicals derivatives (e.g., $CH_3O^\cdot \leftrightarrow {}^\cdot CH_2OH$). Again the quantum yield of initiation is rather low being approximately equal to 2.26×10^{-3}, which reflects the importance of the deactivation of the photoexcited chelate molecule by usual photophysical processes.

The ion-pair complex formed by the interaction of hydroxobis(8-quinolyloxo) vanadium (V) [VOQ₂OH] and n-butyl amine is also effective in photoinitiation of polymerization of MMA in bulk and in solution [40]. The quantum yield of initiation and polymerization determined are equal to 0.166 and 35.0, respectively. Hydroxyl radical ($^\cdot$OH) is reported to be the initiating radical and the following photoreaction is suggested:

$$\{\overset{V}{VOQ_2}(n-BuNH_2)\}^+ OH^- \overset{h\nu}{\Longrightarrow} n-BuNH_2 \rightarrow \overset{IV}{VOQ_2} + nBuNH_2 + {}^\cdot OH \tag{20}$$

Several vanadium (V) complexes were also studied by Aliwi [41] in 1988 as possible photoinitiators for the radical polymerization of MMA. These complexes are oxo-tris(ethoxo) vanadium (V), VO(OC₂H₅)₃, oxo-tris-(triphenylsiloxy) vanadium (V), VO(Si(C₆H₅)₃)₃, oxo-tris(benzoyloxo) vanadium (V), VO(OCOC₆H₅)₃, n-propylthio-oxo-bis(8-quinlyoxo) vanadium (V) VOQ₂SCH₂CH₂CH₃, and oxo-tris-indenyl vanadium (V).

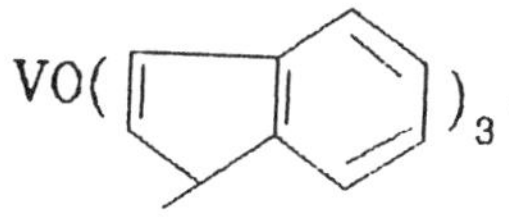

In no case was retardation or inhibition detected during the polymerization process; the radical responsible for the initiation step is derived from the complex ligand produced by the scission of the V—O, V—S, or V—C with reduction of V(V) to V(IV) derivatives. The following photochemical reactions for primary radical generation were suggested:

$$VO(OC_2H_5)_3 \xrightarrow{h\upsilon} VO(OC_2H_5)_2 + {}^{\cdot}OC_2H_5 \tag{21}$$

$$VO(OSi(C_6H_5)_3)_3 \xrightarrow{h\upsilon} VO(OSi(C_6H_5)_3)_2 + {}^{\cdot}OSi(C_6H_5)_3 \tag{22}$$

$$VO(OCOC_6H_5)_3 \xrightarrow{h\upsilon} VO(OCOC_6H_5)_2 + {}^{\cdot}OCOC_6H_5 \tag{23}$$

$$VOQ_2SCH_2CH_2CH_3 \xrightarrow{h\upsilon} VOQ_2 + {}^{\cdot}SCH_2CH_2CH_3 \tag{24}$$

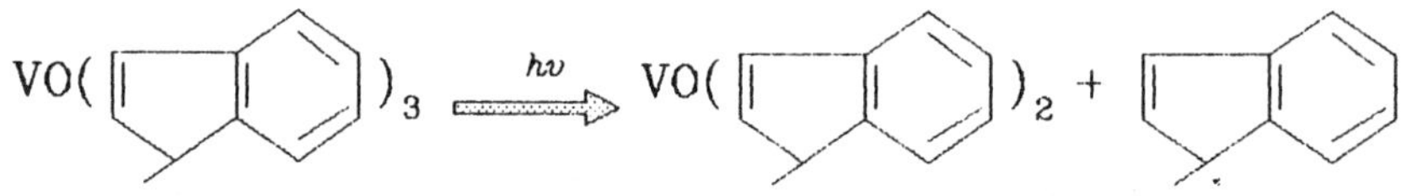

$$VO()_3 \xrightarrow{h\upsilon} VO()_2 + \tag{25}$$

Recently, Aliwi and Abdullah [42] have investigated the photoinitiation of styrene by oxo-tris(dimethyl dithiocarbomato) vanadium (V) ($VO(S_2CN(CH_3)_2)_3$) using light of $\lambda = 365$ nm. Spectroscopic and kinetic analyses show that initiation occurs predominantly through scission of N,N-dimethyldithiocarbomate ligand with reduction of V(V) to V(IV), and $VO(S_2CN(CH_3)_2)_2$ is the final photolytic product.

$$\overset{V}{VO}(S_2CN(CH_3)_2)_3 \xrightarrow{h\upsilon} \overset{IV}{VO}(S_2CN(CH_3)_2)_2 + {}^{\cdot}SC(S)N(CH_3)_2 \tag{26}$$

The quantum yield of polymerization is 6.72 and for photoinitiation $\phi_i = 2.85 \times 10^{-3}$. The polystyrene produced with this initiator shows photosensitivity when irradiated with UV light ($\lambda = 280$ nm). This polymer, which carries two photosensitive end groups of $-SC(S)N(CH_3)_2$, behaves as a telechelic polymer and it is useful for production of ABA block copolymer.

In another type of mixed ligand oxo-vanadium Schiff base complexes: chloro-oxo-bis[N(4-bromophenyl) salicylideneiminato] vanadium (V) chelate, VOL_2Cl, and methoxo-oxobis[N(4-bromophenyl)salicylideneimine vanadium (V) chelate, VOL_2OCH_3 Aliwi and Salih [43] show that these complexes are active photoinitiators for styrene and methyl methacrylate monomers. These initiators are also active in photocuring of unsaturated polyester, and VOL_2Cl could be incorporated into secondary cellulose acetate to produce a photosensitive polymer that upon irradiation in the presence of styrene a grafted and crosslinked (network) copolymer are obtained. A number of Cu(II) amino acid chelate CuL_2(L = glutamic acid or serine) have also been found to be effective photoinitiators for the polymerization of acrylamide [44,45]. The general sequence of events is outlined in the following scheme.

This was postulated to account for the results from flash photolysis of the Cu^{II}-glutamic acid and Cu^{II}-serine systems.

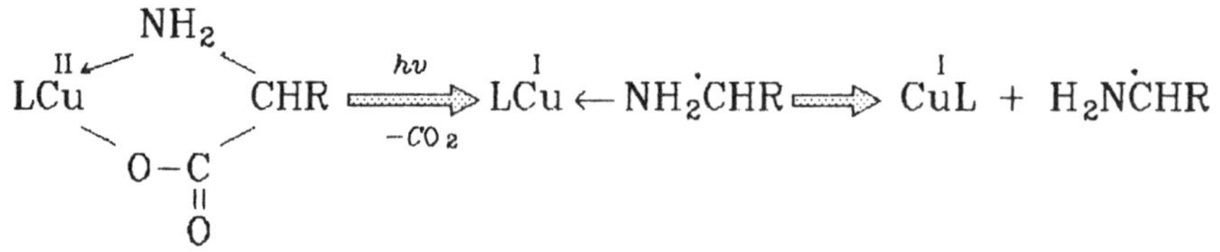

$$\overset{II}{LCu} \cdots \overset{NH_2}{CHR} \xrightarrow[-CO_2]{h\upsilon} \overset{I}{LCu} \leftarrow NH_2\dot{C}HR \Longrightarrow \overset{I}{Cu}L + H_2N\dot{C}HR \tag{27}$$

IV. OTHER PHOTOINITIATING SYSTEMS BASED ON METAL COMPLEXES

To complete the information regarding photoinitiation of radical polymerization by metal complexes, special photoinitiation systems will be discussed and reviewed in this section.

The participation of a monomer molecule in the initiation step of polymerization has not been required in the examples described so far. Tris(thiocyanato) tris(pyridine) iron(III) complex forms a complex with methyl methacrylate [46]. By subjecting the compound to UV radiation, the complex decomposes to give $^{\cdot}$SCN as the initiating radical.

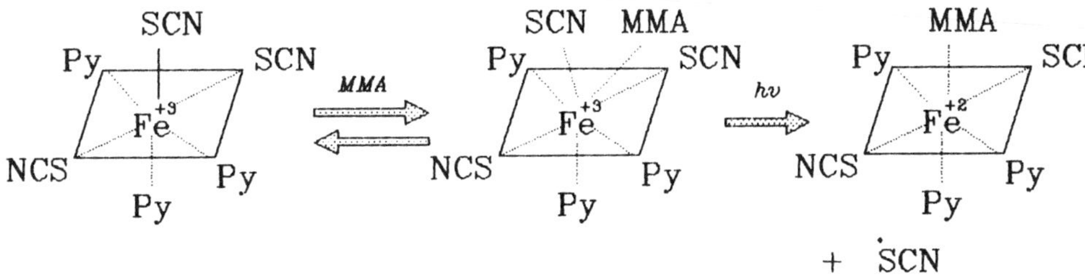

$$
\begin{array}{c}
\underset{NCS}{\overset{Py}{\diagdown}}\!\!\!\overset{\displaystyle SCN}{\underset{\displaystyle Py}{Fe^{+3}}}\!\!\!\overset{SCN}{\diagup}Py \quad \underset{\xleftarrow{}}{\xrightarrow{MMA}} \quad \underset{NCS}{\overset{Py}{\diagdown}}\!\!\!\overset{\displaystyle SCN\ MMA}{\underset{\displaystyle Py}{Fe^{+3}}}\!\!\!\overset{SCN}{\diagup}Py \quad \xrightarrow{h\nu} \quad \underset{NCS}{\overset{Py}{\diagdown}}\!\!\!\overset{\displaystyle MMA}{\underset{\displaystyle Py}{Fe^{+2}}}\!\!\!\overset{SCN}{\diagup}Py
\end{array}
\tag{28}
$$

$$+ \ \overset{\cdot}{S}CN$$

This photoinitiating system is also used for the polymerization of other vinyl monomers such as styrene (St), acrylonitrile (AN), and vinylacetate (VA). The efficiency of photoinitiation by this system follows the order:

$$MMA \geqslant St > AN \geqslant VA$$

The involvement of acrylamide monomer in the initiation step is also observed in the pentamine-aqua-cobalt(III) complex [47].

The initiating radical is derived from the monomer by addition of the H_2O molecule with a reduction of Co^{3+} to Co^{2+}. (reaction Scheme [29])

Acrylmide monomer in aqueous solution was suc-cessfuly photoinitiated by a similar type of Co^{3+} complex:

$$[COL(NH_3)_5]X_2(\lambda = 470 \text{ nm})$$
$$(L = N_3^-, SCN^- \cdot X = NO_3^-, SO_4^=)$$

The initiating radicals are assumed to be $\cdot SCN$, $\cdot ONO$ or $\cdot N_3$ free radicals. Tris oxalate-ferrate-amine anion salt complexes have been studied as photoinitiators ($\lambda = 436$ nm) of acrylamide polymer [48]. In this initiating system it is proposed that the $CO_2^{\overline{\cdot}}$ radical anion found in the primary photolytic process reacts with iodonium salt (usually diphenyl iodonium chloride salt) by an electron transfer mechanism to give photoactive initiating phenyl radicals by the following reaction machanism:

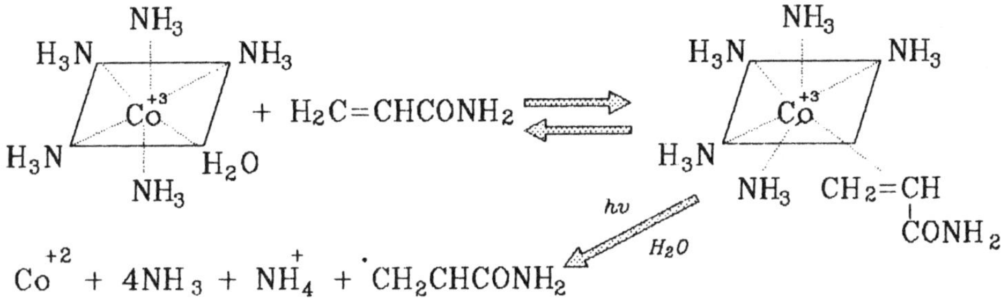

$$
\underset{H_3N}{\overset{H_3N}{\diagdown}}\!\!\!\overset{\displaystyle NH_3}{\underset{\displaystyle NH_3}{Co^{+3}}}\!\!\!\overset{NH_3}{\diagup}H_2O \ + \ H_2C{=}CHCONH_2 \ \underset{\xleftarrow{}}{\xrightarrow{\quad}} \ \underset{H_3N}{\overset{H_3N}{\diagdown}}\!\!\!\overset{\displaystyle NH_3}{\underset{\displaystyle NH_3\ \ CH_2{=}CH}{Co^{+3}}}\!\!\!\overset{NH_3}{\diagup}\underset{CONH_2}{}
\tag{29}
$$

$$\overset{+2}{Co} + 4NH_3 + \overset{+}{NH_4} + \overset{\cdot}{C}H_2CHCONH_2 \quad \underset{H_2O}{\overset{h\nu}{\nwarrow}}$$

$$[Fe(C_2O_4)_3]^{-3} \xrightarrow{h\nu} [Fe(C_2O_4)_2]^{-2} + C_2O_4^{\overline{\cdot}}$$

$$C_2O_4^{\overline{\cdot}} \xrightarrow{\quad} CO_2^{\overline{\cdot}} + CO_2 \tag{30}$$

$$(C_6H_5)_2\overset{+}{I} + CO_2^{\overline{\cdot}} \xrightarrow{\quad} CO_2 + (C_6H_5)_2\overset{\cdot}{I} \xrightarrow{\quad} C_6H_5I + \overset{\cdot}{C}_6H_5$$

Triphenylsulfonium tetrafluoroborate [$(C_6H_5)_3 S^+ BF_4^-$] is used instead of diphenyl iodonium chloride to give phenyl radical as the initiating species. Potassium [tris(oxalato) cobaltate) (III)] with diphenyl iodonium chloride also has been used as the photoinitiator of acryl-amide. The anion radicals $CO_2^{\overline{\cdot}}$ and $C_2O_4^{\overline{\cdot}}$ are probably the initiating species in this system [49].

Silver nitrate and/or cupric nitrate [50] can photoinitiate the polymerization of acrylonitrile in a dimethyl formamide (DMF) medium. The photoactive species is the complex formed between the monomer and salt molecules:

$$AgNO_3 + CH_2{=}CHCN \xrightarrow{\quad} CH_2{=}CHCN\text{-----}AgNO_3$$

$$\qquad\qquad\qquad\qquad\qquad\qquad h\nu \Big\downarrow \tag{31}$$

$$\overset{\cdot}{N}O_3 + CH_2{=}CHCN \xleftarrow{DMF} \overset{+}{N}O_3^{-}\overset{\cdot}{C}H_2CHCN + Ag^{\circ}$$

Okimato and coworkers [51] have introduced a new multicomponent photoinitiating system composed of the metal-ion-amine CCl$_4$ for the polymerization of MMA ($\lambda > 300$ nm). The most active system is FeSO$_4\cdot$7H$_2$O/

triethyamine tetramine/CCl$_4$ ($\approx 22\%$. Conversion was achieved for 4 h irradiation at 0°C). The following scheme was reported for the production of ˙CCl$_3$ initiating primary radicals:

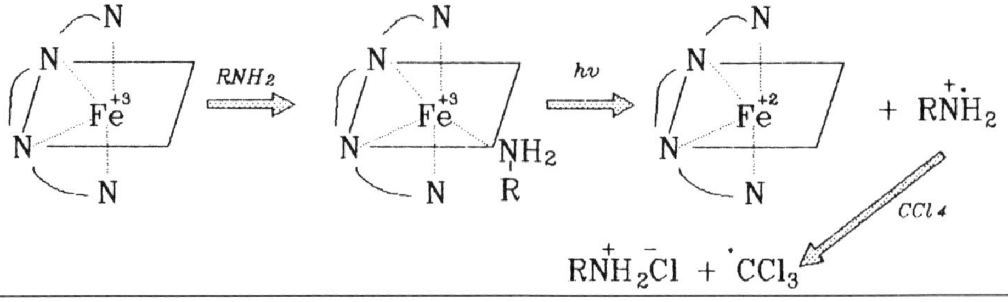

$$\tag{32}$$

In another type of multicomponent photoinitiating system, Okimato and colleage [52] have investigated the Fe^{+3}-salt/saccharide (SH) system for the photoinitiation

of acrylonitrile and acrylamide monomers in aqueous solution ($\lambda > 300$ nm). The initiating radical is derived from saccharide molecule:

$$Fe^{+3} + SH \rightleftharpoons [Fe^{+3}\cdots SH] \xrightarrow{h\nu} Fe^{+2} + \overset{˙}{S} + H^+$$

$$\overset{˙}{S} + nM \Longrightarrow Polymer \tag{33a}$$

The following order of decreasing efficiencies for different saccharides in the system was found:

glucose > fructose > lactose $\approx$ maltose.

The triplet photoexcited state of the complex found

between MMA and NiCl$_2$ in N,N-dimethylaniline solvent is responsible for the formation of (MMA) initiating radicals produced by H abstraction from MMA monomer (or solvent RH) by the triplet excited state molecule [53].

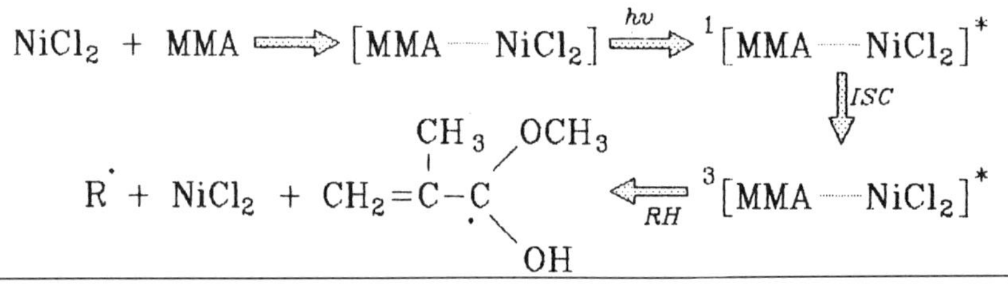

$$NiCl_2 + MMA \Longrightarrow [MMA\cdots NiCl_2] \xrightarrow{h\nu} {}^1[MMA\cdots NiCl_2]^*$$

$$\Big\downarrow ISC$$

$$\overset{˙}{R} + NiCl_2 + CH_2{=}\overset{\overset{\displaystyle CH_3}{|}}{\underset{\underset{\displaystyle OH}{˙}}{C}}{-}C\overset{\diagup OCH_3}{\diagdown} \underset{RH}{\Longleftarrow} {}^3[MMA\cdots NiCl_2]^* \tag{33b}$$

Both ˙R and MMA˙ radicals are found to be responsible for the photoinitiation process. Chaturvedi and coworkers [54,55] introduced phenyl dimethyl sulfonium-ylide cupric chloride and chromium thiophene carboxylate as the photoinitiator of styrene and MMA. No reaction mechanism was given for these systems.

Iwai and coworkers [56] have introduced a novel type of multicomponent photoinitiating system for water-soluble monomer (acrylamide, acrylic acid, acrylonitrile, etc).

This photoinitiator (or photocatalyst) is composed of an electron relay system in which ruthenium bipyridyl

chloride (Ru(BPY)$_3$Cl$_2$) is acting as the photosensitizer for the generation of triethanol amine (TEA) radical cation (TEA)$^+$ by electron transfer from the amine molecule to the photoexcited ruthenium complex. The reduced form of this complex {(Ru(BPY)$_3$$^+$} interacts with the acrylamide monomer (AA) to give the initiating monomer radical anion (AA$^-$) and, therefore, regenerates the original [Ru(BPY)$_3$]$^{2+}$ Sensitizer, which is not consumed through the process. The following reaction scheme shows the photosensitizing process of this electron relay initiating system:

$$\tag{34}$$

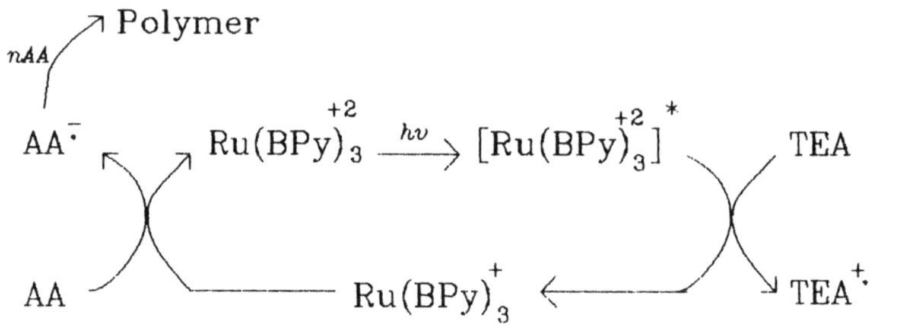

It was found that [Ru(1,10 phenanthroline)$_3$]$^{2+}$ and [Ru(2,2'-bipyrazine)$_3^{2+}$] complexes can photosenstize the polymerization acrylamide in aqueous solution [57]. A similar mechanism to that shown in Scheme (34) is suggested. However, maleic acid (MA) comonomer is used as a strong photoelectron doner in the [Ru(2,2'-bipyrazine)]-acrylamide/TEA relay system. The high-molecular weight, 7.7×10^6 g/mol, of the acrylamide-maleic anhydride random copolymer is obtained.

V. PHOTOINITIATED GRAFTING AND CROSSLINKING BY TRANSITION METAL COMPLEXES

The most important practical application of the organometallic complex photoinitiators is the possibility of using these types of initiators in modifying the pre-existing polymer chain, e.g., block, graft, and crosslinked copolymers preparation.

It is clear from the preceding discussion that organometallic photoinitiators (metal carbonyl or chelate derivatives) can provide a convenient route for synthesizing vinyl polymers with a variety of different reactive end group or photoreactive pendant groups or side chains through the polymer chain.

In the first type, a block AB or ABA copolymer, and in the second, a grafted or crosslinked copolymer could be produced under suitable conditions. Bamford [58] first reported that a block copolymer could be produced from the polymer chain A carrying CCl$_3$ end groups prepared by the Mn$_2$(CO)$_{10}$/CBr$_4$ photoinitiating system (see reaction Schemes [10] and [11]) and vinyl monomer (B) in the presence of the Mn$_2$(CO)$_{10}$ photoinitiator. Here CBr$_3$ ends in the macro molecules and is considered a halide coinitiator. Three types of block copolymer could be formed by this technique: (1) block copolymer type ABA, when the CBr$_3$ active groups located at one end of homopolymer (A) and the termination is exclusively by combination:

$$\sim\sim (A)CBr_3 \xrightarrow[Mn_2(CO)_{10}]{h\nu} \sim\sim A \sim\sim CBr_2 \xrightarrow[nB]{monomer} \sim\sim ACBr_2(B)_{n-1}B^{\cdot}$$

$$\sim\sim ACBr_2(B)_nCBr_2 \sim\sim A \sim\sim \xleftarrow{\qquad}_{termination}$$

$$\text{"ABA" block copolymer} \tag{35}$$

(2) if termination is mainly by disproportionation or the chain transfer process is significant, AB block copolymer is produced:

$$2 \sim\sim ACBr_2(B)_{n-1}B^{\cdot} \xrightarrow[or\ (RH)]{disproportionation} 2 \sim\sim ACBr_2B_n \sim\sim \tag{36}$$

$$\text{"AB" block copolymer}$$

(3) if the initial polymer carries two terminal CBr$_3$ groups per chain, the first product is: AB block copolymer; Br$_3$C $\sim$ ACBr$_2$(B) CBr$_2$ $\sim$ A $\sim$ C(Br)$_3$. Further reactions will activate CBr$_3$ groups in this copolymer, so that the next product will be the seven block copolymer having four blocks of A and three blocks of B and so on. Generally, therefore, the product will consist of chains of alternating blocks of A and B linked by CBr$_2$ units and with CBr$_3$ terminals. CBr$_2$ groups are proven to be much less reactive towards Mn$_2$(CO)$_{10}$ than a CBr$_3$ group. A grafted copolymer will not be considerable or not produced at all [58]. Moreover, since all primary radicals produced are macroradicals, the synthesis produces only a minimal (or nil) amount of B homopolymer (homopolymer of B is produced only by chain transfer process).

Using the above method, Bamford and Han [59] have succeeded in synthesizing and characterizing copolymers in which each block is an alternating copolymer. The block copolymers prepared are of general structure:

poly(A-alt-B-) block-poly(C-alt-D) block-poly (A-alt-B)

in which A = styrene, B = methylacrylate, C = isoprene, D = MMA.

or

A = isoprene, B = methylacrylate, C = styrene, D = MMA.

or

A = butadine, B = methylacrylate, C = styrene, D = MMA.

Using the same method a block copolymer of polypeptides and vinyl monomers was also prepared. As mentioned in Section II, Bamford and Mullik [22] introduced an interesting method of photoinitiation of vinyl monomers by the Mn$_2$(CO)$_{10}$ or the Re$_2$(CO)$_{10}$/C$_2$F$_4$ system. By these methods polymeric molecule with (CO)$_5$Mn—CF$_2$CF$_2$—terminals is produced (see Scheme [12]). If a polymer of this kind is heated to 100°C in the presence of vinyl monomer, a block copolymer AB or ABA with CF$_2$—CF$_2$ linkage is produced [60]:

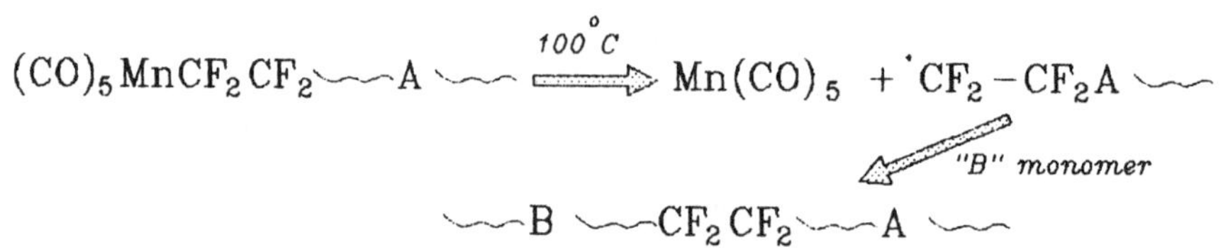

$$(CO)_5 MnCF_2CF_2 \smile A \smile \xrightarrow{\text{100}^{\circ}C} Mn(CO)_5 + {}^{\cdot}CF_2 - CF_2 A \smile$$

$$\smile B \smile CF_2CF_2 \smile A \smile \quad \xleftarrow{\text{"B" monomer}}$$

(37)

Among the pairs of monomers used in synthesizing block copolymers by this method are: MMA/AN, C_2F_4/MMA, and HEMA/DMA (HEMA = hydroxyethyl methacrylate and DMA = decyl methacrylate).

The block copolymer produced by Bamford's metal carbonyl/halide-terminated polymers photoinitiating systems are, therefore, more versatile than those based on anionic polymerization, since a wide range of monomers may be incorporated into the block. Although the mean block length is controllable through the parameters that normally determine the mean kinetic chain length in a free radical polymerization, the molecular weight distributions are, of course, much broader than with ionic polymerization and the polymers are, therefore, less well defined.

It is also important to note that this procedure of synthesizing block copolymers possesses an advantage in that homopolymers are avoided, since all the initiating radicals are attached to a polymer chain. This is not so with many free radical syntheses.

An analogous series of reaction gives rise to graft copolymer when the copolymer initiator carries a side chain with active halide groups (e.g., CBr_3, CCl_3, CHClCOOH or CF_2CF, etc.) photoinitiation of vinyl monomer (M) by this type of a preformed halide-containing polymer in the presence of metal carbonyls (e.g., $Mn_2(CO)_{10}$) then leads to synthesis of a graft and a cross-linked polymer, or a mixture of the two may be formed depending on the character of termination of $M^{\cdot}$ radicals. Reaction Scheme (38) represents the formation of a crosslink in a typical case:

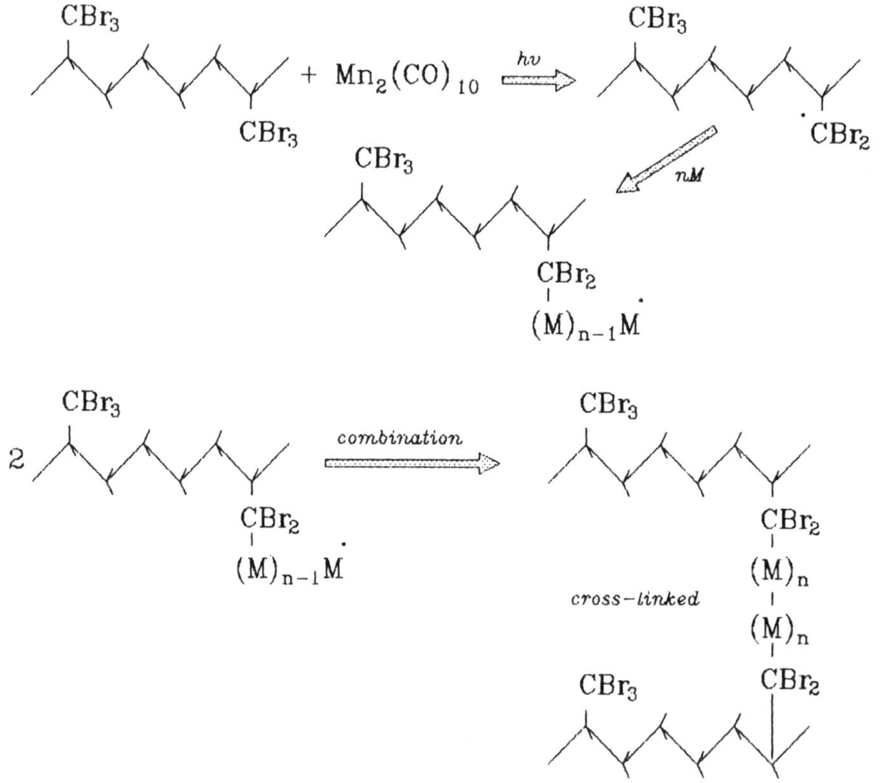

(38)

A wide range of polymer networks are constructed in this manner. Poly(vinyltrichloacetate) was used as the coinitiator with styrene, MMA and chloroprene as crosslinking units. Polycarbonates, polystyrene, N-halogenated polyamide, polypeptides, and cellulose acetate, suitably functionalized, have been used as a coinitiator

with metal carbonyl to produce polymeric networks that are not readily synthesized by other routes.

In an attempt to formulate a new photoresistant and presensitized lithographic plate, Wagner and Purbrick [61] have used poly(vinyl trichloroacetate) and styrene, which together with manganese carbonyl or phenyl chromiumtricarbonyl ($C_6H_5Cr(CO)_3$), were coated onto grained anodized aluminum foil. On exposure to light, the illuminated portions were crosslinked and insolubilized so that an image developed in a mixture of ethanol and cyclohexanone was formed.

Unfortunately, poly(vinyl trichloroacetate) does not have suitable mechanical properties for practical use in phototoresistant or lithographic printing materials and the use of the liquid monomer (styrene) is undesirable in this technique.

A polymer was, therefore, designed to overcome these shortcomings. Its chains with an average composition shown below carried both CBr_3 group and polymerizable metharylate units [61].

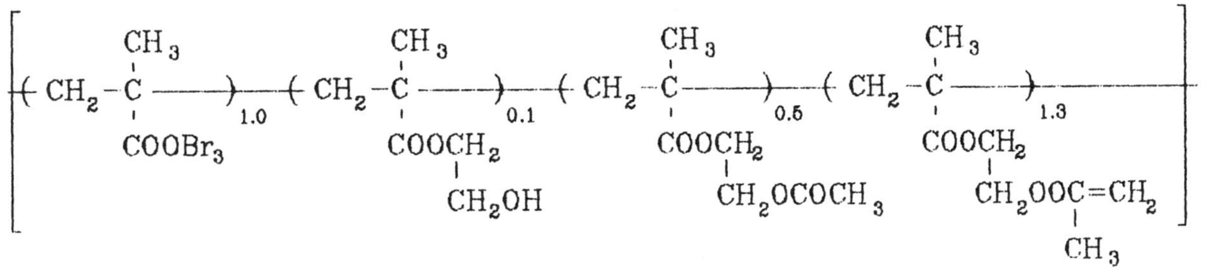

$$\left[\left(CH_2-\underset{\underset{COOBr_3}{|}}{\overset{\overset{CH_3}{|}}{C}} \right)_{1.0} \left(CH_2-\underset{\underset{\underset{CH_2OH}{|}}{COOCH_2}}{\overset{\overset{CH_3}{|}}{C}} \right)_{0.1} \left(CH_2-\underset{\underset{\underset{CH_2OCOCH_3}{|}}{COOCH_2}}{\overset{\overset{CH_3}{|}}{C}} \right)_{0.5} \left(CH_2-\underset{\underset{\underset{\underset{CH_3}{|}}{CH_2OOC=CH_2}}{COOCH_2}}{\overset{\overset{CH_3}{|}}{C}} \right)_{1.3} \right]$$

This system gave results that were further improved upon by the addition of a photosensitizer (e.g., a thiopyrilium, pyrilium, and selena-pyrilium salt) to bring the absorption of light into the visible region.

Bamford and Mullik [62] have succeeded in photografting a vinyl monomer onto a styrene–MMA copolymer using the $Mn_2(CO)_{10}/C_2F_4$ photoinitiating system in acetic acid. The following scheme was reported for this process:

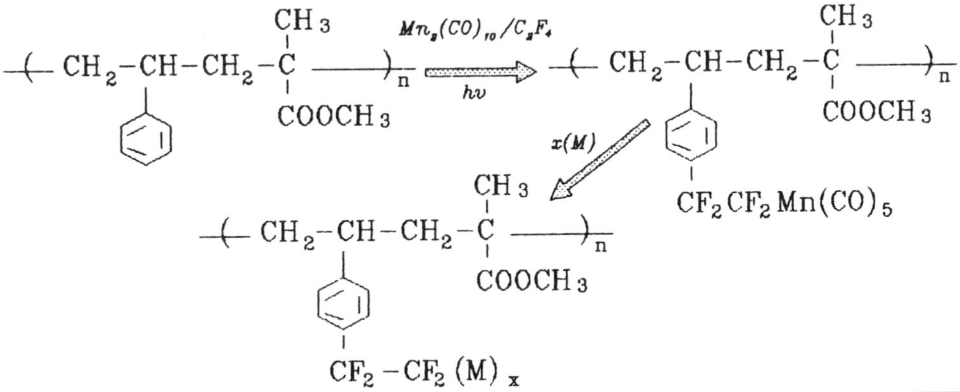

$$\tag{39}$$

Terminal bonding of a polymer chain to a surface is a subject of great academic and industrial interest. Eastmond et al. [63] used $Mn_2(CO)_{10}$ and $Re_2(CO)_{10}$ photoinitiators to graft poly(methyl methacrylate) onto a chlorinated glass surface.

Active halogen groups were bonded to a glass surface by nonhydrolyzable links using the reactions set out in Scheme (40):

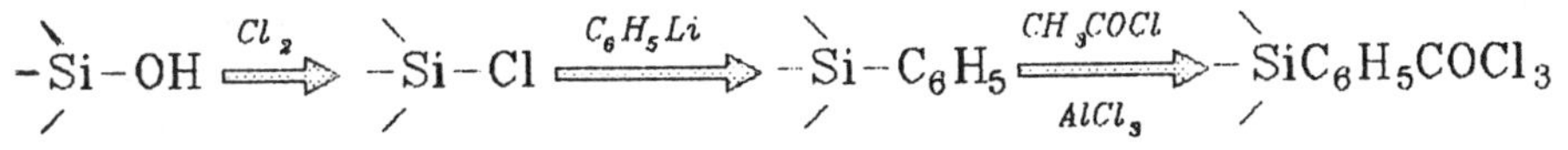

$$-Si-OH \xrightarrow{Cl_2} \overset{\diagdown}{-Si-Cl} \xrightarrow{C_6H_5Li} \overset{\diagdown}{-Si-C_6H_5} \xrightarrow[AlCl_3]{CH_3COCl} \overset{\diagdown}{-SiC_6H_5COCl_3} \tag{40}$$

The chlorinated surface of glass then acted as a co-initiator, and when photochemically treated with metal carbonyl in the presence of MMA monomer, a thin layer of 0.4 μm thickness of poly(methyl methacrylate) was obtained.

Grafting and modification of polymers have been found to have applications in the biomedical field. For example, poly(etherurethane), which has good elastomeric and often mechanical properties and a relatively high compatibility with blood, has been used in the man-

ufacture of prostheses, such as artificial hearts and arteries and extracorporeal circulatory systems.

The highest possible degree of hemocompatibility is essential for certain applications, including small-bore artificial prostheses.

Two general methods of improving polymer hemocompatibility are known: (1) grafting of hydrophilic chains, and (2) the chemical attachment of an antiplatelet agent. In this respect, many hydrophilic and other monomers are grafted on solid polyurethane surfaces such as

MMA, styrene, 2-hydroxy ethyl methyacrylate, 2-hydroxypropyl methacrylate, acrylic and methacrylic acids, acrylamide and N-vinyl pyrrolidone (NVP) [64]. By this process, polyurethane is partially converted to N-chloro or N-bromo derivatives by a short immersion in a dilute aqueous sodium hypochlorite solution at room temperature and vinyl monomer to produce polyurethane-grafted copolymer as shown in Scheme (41). This was subjected to irradiation with $\lambda = 365$ nm at 25°C in the presence of $Mn_2(CO)_{10}$.

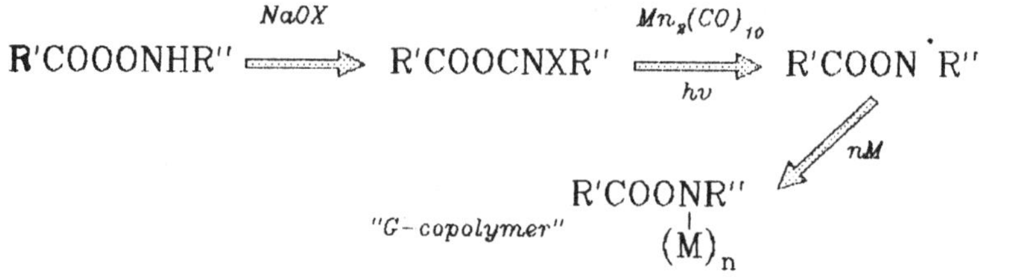

$$R'COOONHR'' \xrightarrow{NaOX} R'COOCNXR'' \xrightarrow[h\upsilon]{Mn_2(CO)_{10}} R'COON{}^\bullet R''$$

$$\underset{\text{"G-copolymer"}}{R'COONR''} \Big\uparrow {nM}$$
$$\underset{(\overset{|}{M})_n}{}$$

(41)

The photografting process was also done in dimethylformamide (DMF) solution. Grafting to the solid polyurethane has the advantage of leaving the interior and the mechanical properties effectively unchanged. Furthermore, the inner surface of a tube, such as a prosthetic artery, may be halogenated by allowing the hypochlorite solution to flow through the tube for a short time; grafting is subsequently confined to the halogenated surface.

Polymers with an antiplatelet agent of 5-(6-carboxylhexyl) 1-(3-cyclohexyl-3-hydroxyropyl) hydantoin as an end group were synthesized by esterifying the hydantoin molecule with a haloalcohol, such as CCl_3CH_2OH or $BrCH_2CH_2OH$, and using the product as coinitiator with

$Mn_2(CO)_{10}$ in a simple photoinitiated free radical polymerization. The antiplatelet activity of these polymers were compared with that of poly(ether-Wrethane) carrying the hydantoin residues in side chains.

Metal chelates have also been used in photografting and crosslinking of different types of polymers [61,65–67].

A novel type of grafting process was developed using a new photosensitive polymer containing vanadium (V) chelates. These polymers were generally synthesized by the condensation of a VOQ_2OH complex and a hydroxy-containing polymer to produce photoactive polymer (red in color) with pendant vanadium (V) chelate.

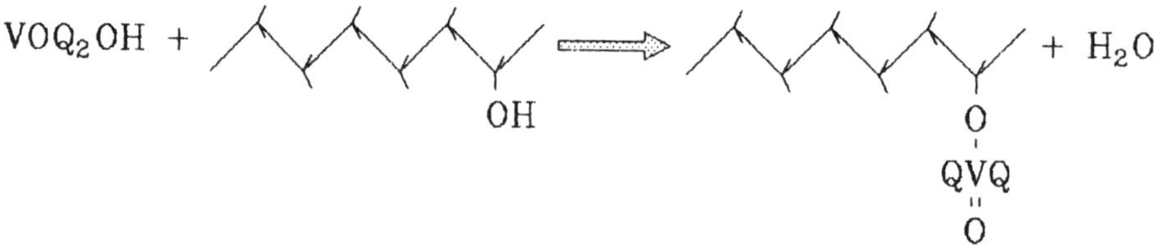

$$VOQ_2OH + \underset{OH}{\wedge\!\wedge\!\wedge\!\wedge} \Longrightarrow \underset{\underset{O}{\overset{||}{QVQ}}}{\wedge\!\wedge\!\wedge\!\wedge} + H_2O$$

(42)

Hydroxy-containing polymers such as poly(methylmethacrylate-co-hydroxyethyl methacrylate) [65,66] or secondary cellulose acetate [67,68] were used for this purpose. Vanadium (V) 8-hydroxy quinoline-hydroxyethyl methacrylate adduct, prepared by condensation of the latter with a VOQ_2OH complex, is polymerized to

produce a photoactive polymer similar to that shown in Scheme (42).

When these polymers are subjected to light of $\lambda = 365$ nm in bulk vinyl monomer, (MMA or styrene) grafted or extensive crosslinking polymers were produced. The photografting or photocrosslinking occurs through the macro-radicals photochemically generated on the backbone of the polymer:

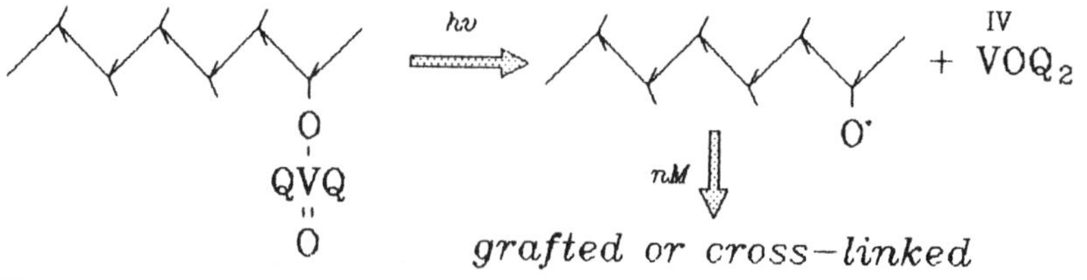

$$\underset{\underset{O}{\overset{||}{QVQ}}}{\wedge\!\wedge\!\wedge\!\wedge} \xrightarrow{h\upsilon} \underset{O^\bullet}{\wedge\!\wedge\!\wedge\!\wedge} + \overset{IV}{VOQ_2}$$
$$\underset{nM}{\Big\Downarrow}$$
$$grafted\ or\ cross-linked$$

(43)

This grafting technique proved to possess several advantages: (1) the initiating macroradicals [shown in

Scheme (43)] have high activity and no other initiating species, which rise to homopolymers, are formed (i.e., no chance for any homopolymer formation); (2) unreacted vanadium chelate attached to the polymer may be readily removed after polymerization by substitution reaction with methanol or ethanol:

$$\bigwedge\!\!\bigwedge\!\!\bigwedge\!\!\bigvee \underset{\substack{O \\ | \\ QVQ \\ \| \\ O}}{} + CH_3OH \Longrightarrow \bigwedge\!\!\bigwedge\!\!\bigwedge\!\!\bigvee_{OH} + VOQ_2OCH_3 \tag{44}$$

(3) The low quantum yield of the photografting process ($\Phi \approx 2 \times 10^{-3}$) provides a good opportunity to control the network formation (curing time control), and accordingly, the desirable properties of the crosslinked or grafted copolymer might be obtained.

Grafting of cotton with a styrene monomer is also possible after conversion of the cotton to a secondary cellulose acetate and after grafting, hydrolysis of acetate groups to cellulose grafted with styrene is possible [67].

A similar type of condensation between a hydroxyl-containing polymer (such as secondary cellulose acetate) with $VO(BrC_6H_4N\!=\!CHO\ C_6H_4)_2$ $Cl[VOL_2Cl]$ produces photoactive polymers [68]. When irradiated with UV light in the presence of styrene or MMA, grafted and crosslinked polymers were obtained:

$$\bigwedge\!\!\bigwedge\!\!\bigwedge\!\!\bigvee_{OH} + VOL_2Cl \Longrightarrow \bigwedge\!\!\bigwedge\!\!\bigwedge\!\!\bigvee_{\substack{O \\ | \\ QVQ \\ \| \\ O}} + HCl \tag{45}$$

$$grafted\ or\ cross-linked \overset{M}{\Longrightarrow}$$
$$copolymer$$

Photocuring of commercial unsaturated polyester–styrene mixture was effectively done in the presence of the VOL_2Cl photoinitiator complex. The chlorine atom produced by the scission of V—Cl bond in the VOL_2Cl complex is proven to be the initiating species for the photocuring process [68].

Polypyridine ruthenium (II) chelate complex [Ru(Bpy)$_3$]$^{2+}$ is known to participate in a photoredox reaction on excitation with visible light, coupled with the tendency of the intermediate ruthenium species to revert to the divalent state. Included in the reactions of these complexes are certain photoinitiating processes [56,57,69]. The reactions normally proceed via the lowest metal to ligand change transfer state, which is luminescent, and thus provides a valuable mechanistic prope. Persulfate ions are known to quench [Ru(Bpy)$_3$]$^{2+}$ luminescence by both dynamic and static processes with the following dominant reaction:

$$[Ru(BPy)_3]^{+2} + S_2O_8^{-2} \overset{h\nu}{\Longrightarrow} [Ru(BPy)_3]^{+3} + SO_4^{-\cdot} + SO_4^{-2} \tag{46}$$

This reaction is utilized by Burrows et al. [70] to photoinitiate grafting of acrylamide, acrylic acid, methacrylamide, and acrylonitrile on cellulose triacetate in acidic aqueous solution.

It was found that the sulfate radical anion SO_4^- produced photochemically in Scheme (46) is responsible for generating the cellulose derivative macroradicals by hydrogen abstraction, which added the vinyl monomer to produce the grafted copolymer. The main disadvantage of this method is the production of large quantities of undesirable homopolymers in addition to the grafted copolymers.

ACKNOWLEDGMENTS

The author would like to thank Dr. Najat J. Salah and Mr. Saad N. Farhan for assistance in preparing and typing the manuscript.

REFERENCES

1. F. C. DeSchryver, *Pure Appl. Chem., 34*: 213 (1973).
2. A. D. Jenkins, and A. Ledwith (eds.), *Reactivity and Structure in Polymer Chemistry,* John Wiley & Sons, New York, Chapter 14 (1974).
3. G. Oster, and N. Yang, *Chemical Reviews, 68*: 125 (1968).
4. G. Oster, *Encyclopedia of Polymer Science and Technology,* John Wiley & Sons, New York, *10*: 145 (1969).
5. R. S. Davidson, *J. Photochem. Photobiology, A: Chem., 73*: 81 (1993).
6. R. S. Davidson. *J. Photochem. Photobiology, A: Chem., 69*: 263 (1993).
7. J. Blyth, and A. W. Hoffmann, *Ann., 53*: 292 (1845).
8. D. Berthelot, and H. Gaudechon, *Compt. Rend., 150*: 1169 (1900).
9. I. Ostromislenski, *J. Russ. Phys. Chem. Soc., 44*: 204 (1912).
10. J. F. Rabek, *Photochem. Photobiology, 7*: 5 (1968).
11. E. Koerner Von Gustorf, and F. W. Grevels, *Fortschr. Chem. Forsch., 13*: 366 (1969).

12. E. Koerner Von Gustorf, M. C. Henry, and C. Dipietro, *Z. Naturforsch, 21B*: 42 (1966).
13. E. Koerner Von Gustorf, M. J. Jun, and G. O. Schensk, *Z. Naturforsch, 18B*: 503 (1963).
14. W. Strohmeier, and P. Hartmann, *Z. Naturforsch, 19B*: 882 (1964).
15. W. Strohmeier, and H. Grubel, *Z. Naturforsch, 22B*: 98 (1967).
16. W. Strohmeier, and H. Grubel, *Z. Naturforsch, 22B*: 553 (1967).
17. C. H. Bamford, P. A. Crowe, J. Hobbs, and R. P. Wayne, *Proc. Roy. Soc. A, 284*: 455 (1965).
18. C. H. Bamford, P. A. Crowe, J. Hobbs, and R. P. Wayne, *Proc. Roy. Soc. A, 292*: 153 (1966).
19. C. H. Bamford, and J. Paprotny, *Polymer, 13*: 208 (1972).
20. C. H. Bamford, and M. U. Mahmud, *Chem. Commun.* 762 (1972).
21. W. Strohmeier, and H. Grübel, *Z. Naturforsch, 22B*: 98 (1967).
22. C. H. Bamford, and S. U. Mullik, *Polymer, 14*: 38 (1973).
23. C. H. Bamford, and S. U. Mullik, *J. Chem. Soc., Faraday Trans I, 72*: 368 (1976).
24. C. H. Bamford, S. M. Aliwi, and S. U. Mullik, *J. Polym. Sci., Symp. 50*: 33 (1976).
25. C. H. Bamford, and K. K. Al-Lamee, *J. Chem. Soc., Faraday Trans. I, 80*: 2175 (1984).
26. C. H. Bamford, and K. K. Al-Lamee, *J. Chem. Soc., Faraday Trans. I, 80*: 2187 (1984).
27. T. Musuda, Y. Yamamoto, and T. Higashimura, *Polymer, 23*: 1663 (1982).
28. C. H. Bamford, and X. Z. Han, *J. Chem. Soc., Faraday Trans. I, 78*: 855 (1982).
29. C. H. Bamford, and X. Z. Han, *J. Chem. Soc., Faraday Trans. I, 78*: 869 (1982).
30. V. Balzani, and V. Carassiti, *Photo Chemistry of Coordination Compounds,* Academic Press, London Chapter 5 (1970).
31. A. W. Adamson, and P. D. Fleischauer (eds.) *Concepts of Inorganic Pohotochemistry* John Wiley and Sons, London, Chapter 7 (1975).
32. B. Marciniak, and G. E. Buono-Core, *J. Photochem. Photobiology, A: Chem. 52*: 1 (1990).
33. C. H. Bamford, and A. N. Ferrar *J. Chem. Soc. Faraday Trans. I, 68*: 1243 (1972).
34. K. Kaeriyama, and Y. Shimura, *Makromol. Chemie, 167*: 129, (1973).
35. T. Okimoto, *J. Polym. Sci., Part B, Polym. Letter, 12*: 121 (1974).
36. S. M. Aliwi, and C. H. Bamford, *J. Chem. Soc., Farady Trans. I, 70*: 2092 (1974).
37. S. M. Aliwi, and C. H. Bamford, *J. Chem. Soc., Farady Trans. I, 71*: 52 (1975).
38. S. M. Aliwi, and C. H. Bamford, *J. Chem. Soc., Farady Trans. I, 71*: 1733 (1975).
39. S. M. Aliwi, and C. H. Bamford, *J. Chem. Soc., Farady Trans. I, 73*: 776 (1977).
40. S. M. Aliwi and C. H. Bamford, *J. Photochem. Photobiology A: Chem., 47*: 353 (1989).
41. S. M. Aliwi, *J. Photochem. Photobiology A: Chem., 44*: 179 (1988).
42. S. M. Aliwi, and S. M. Abdullah, *Polym. Inter., 35*: 309 (1994).
43. S. M. Aliwi, and N. J. Saleh, *J. Photochem. Photobiology* (in press) (1995).
44. C. N. Namasivayan, and P. Natarajan, *J. Polym. Sci., Polym. Chem. Ed., 21*: 1371 (1983).
45. C. N. Namasivayan, and P. Natarajan, *J. Polym. Sci., Polym. Chem. Ed., 21*: 1385 (1983).
46. N. Sakota, K. Takahashi, and K. Nishihara, *Makromol. Chem., 161*: 173 (1972).
47. G. A. Delzenne, *IUPAC Symp., Macromol. Preprint* p. 145, Prague (1965).
48. H. Baumann, B. Strehmel, and H. J. Tempe, *Polym. Photochem., 4*: 223 (1984).
49. H. Baumann, B. Strehmel, H. J. Tempe, and U. Lammel, *J. Prakt. Chem., 326*: 415 (1984).
50. H. Schnecko, *Chimia, 19*: 113 (1965).
51. T. Okimoto, M. Takahashi, Y. Inaki, and K. Takemoto, *Angew. Makromol. Chem., 38*: 81 (1974).
52. T. Okimoto, Y. Inaki, and K. Takemoto, *J. Macromol. Sci., A7*: 1537 (1973).
53. J. Bartŏn, I. Leboc, I. Capek, and J. TKăc, *Makromol. Chem., Rapid Commun. 1*: 7 (1980).
54. B. Chaturvedi, and A. K. Srivastava, *J. Photochem. Photobiology, A: Chem., 64*: 183 (1992).
55. B. Chaturvedi, and A. K. Srivastava, *J. Photochem. Photobiology, A: Chem., 59*: 393 (1991).
56. K. Iwai, Y. Uesugi, and F. Takemura, *Polym. J. 17*: 1005 (1985).
57. K. Iwai, M. Uesugi, T. Sakabe, C. Hazama, and F. Takemura, *Polym. J., 23*: 757 (1991).
58. C. H. Bamford, *European Polym. J.,* (Suppl. 1) (1969).
59. C. H. Bamford, and X. Z. Han, *Polymer, 22*: 1299 (1981).
60. C. H. Bamford, and S. U. Mullik, *Polymer, 17*: 94 (1976).
61. H. M. Wagner, and M. D. Purbrick, *J. Photographic Sci., 29*: 230 (1983).
62. C. H. Bamford, and S. U. Mullik, *Polymer, 19*: 948 (1978).
63. G. C. Eastmond, C. Nguyen-Hun, and W. H. Piret, *Polymer, 21*: 598 (1980).
64. C. H. Bamford, and I. P. Middelton, *Europ. Polym. J. 19*: 1027 (1983).
65. S. M. Aliwi, and C. H. Bamford, *Polymer, 18*: 375 (1976).
66. S. M. Aliwi, and C. H. Bamford, *Polymer, 18*: 382 (1976).
67. S. M. Aliwi, and B. Y. Al-Banna, *Polym. Plast. Tech. Eng., 33*: 503 (1994).
68. N. J. Saleh, Ph.D. Thesis, Mustansiriah University, Baghdad (1995).
69. C. K. Gratzel, M. Jirousek, and M. Gratzel, *Langmuir, 2*: 292 (1986).
70. H. Burrows, P. J. Eliseu, M. Helena, and R. M. Freire, *J. Photochem. Photobipology A: Chemistry, 59*: 81 (1991).

18

Chemical Modification of Polystyrenes in the Presence of Cationic Catalysis and Their Industrial Applications

Ramazan Mirzaoglu, Refiga Kurbanova, and Mustafa Ersöz
Selçuk University, Konya, Turkey

I. INTRODUCTION

The synthesis of new polymeric materials having complex properties has recently become of great practical importance to polymer chemistry and technology. The synthesis of new materials can be prepared by either their monomers or modification of used polymers in industry. Today, polystyrene (PS), which is widely used in industrial applications as polyolefins and polyvinylchlorides, is also used for the production of plastic materials, which are used instead of metals in technology. For this reason, it is important to synthesize different PS plastic materials. Among the modification of PS, two methods can be considered, viz. physical and chemical modifications. These methods are extensively used to increase physico-mechanical properties, such as resistance to strike, air, or temperature for the synthesizing of new PS plastic materials.

In the present chapter, we will start to briefly summarize all the aspects of this subject, which have been covered in previous specialized reviews, allowing for a broad and general discussion on the subject. Then we will focus on the chemical modification of PS in the presence of cationic catalysts.

When the physical modification method is used, PS is modified by mechanical stirring with various synthetic rubbers such as polybutadiene, polybutadiene styrene, polyisoprene, polychloropropene, polybutadiene styrene-acrylonitrile copolymers. In the chemical modification, PS is modified with polyfunctional modificators in the presence of cationic catalysis.

New elastic polymeric materials (resistance to higher stroke or air) can be obtained by using physical modification methods, but using this method, two phases (PS and rubber) in the mixture were formed. Small rubber particles spread as a PS layer and, after awhile, the relationship between the layers decreases and rubber particles gather in the upper layer of the materials. This can be the cause of the loss of resistance of the materials. These material disadvantages have stimulated the polymer synthesis to increase the PS resistance to higher physico-mechanical properties, such as higher temperature and stroke for the chemical modification of PS with various functional modifiers.

The reaction capability of PS is weak, but the reaction capability can be improved by anchoring the functional group to the aliphatic chain or aromatic ring of PS using chemical or conversion reactions. Aliphatic chain reactions are: halogenation reactions, oxidation reactions, or unsaturated acids to bonded aliphatic chain of PS (in the presence of a radical catalysis).

II. REACTION OF POLYSTYRENE

A. Halogenation Reactions

The halogenation reaction is very important for the production of new polymeric materials [1]. If AlCl$_3$ was used, the halogenation reaction took place using the aromatic ring of PS [2]. With regard to photochemical halogenation reaction is occurred either the aliphatic or aromatic ring of PS [3,4].

Bevington and Ratti [5] reported that when CCl$_4$ was used as a solvent on the photochlorination of PS at 78°C, the reverse C atom was chlorinated. In this case, the halogenation reaction is very important for copolymer synthesis due to the C—Cl bond losing its resistance [6] as shown in Scheme (1).

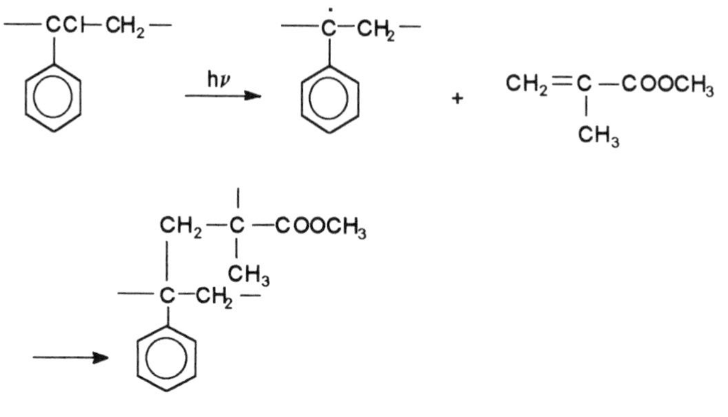

Scheme 1

B. Metallic Reactions

It is known that PS containing metal can be easily obtained by metallic reaction. When PS solution in benzene is stirred with potassium at 80°C, the potassium derivatives of PS are obtained. When the derivative compound was carboxylated, carboxyl groups (86–96%) were attached to the phenyl ring as L-type [7] (Scheme [2]).

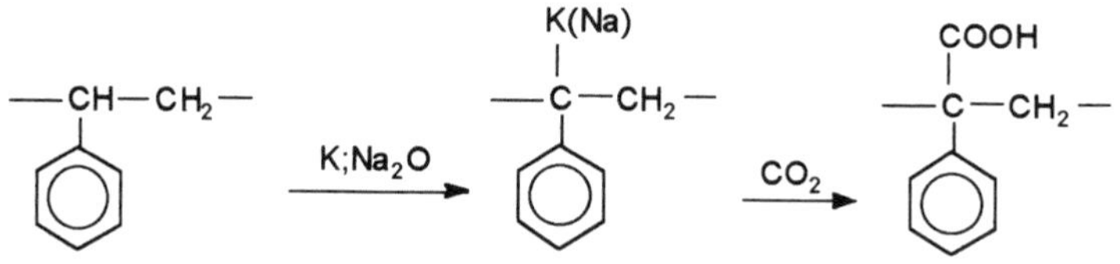

Scheme 2

It was shown that chemical reactions generally occur from the aromatic ring of PS. PS has high dielectric properties and high hardness, but is not resistant to higher temperature and stroke and has no adhesion capability to metals. For this reason, polymeric materials having physico-mechanical properties could be obtained by anchoring some functional groups to the aromatic ring. For example, ion exchange resins are prepared by amination after chloromethylation of PS and some cross-linking polymers are also used as a membrane by dehydrochlorination.

C. Chloromethylation Reaction

Generally, PS containing amine groups are synthesized by condensation of chlorinated PS with amines. These type of resins are widely used as anionic resins.[8] PSs containing imidazol rings have antistatic properties and are used as additives to make dyeing of synthetic fiber materials easy [9] (Scheme [3]).

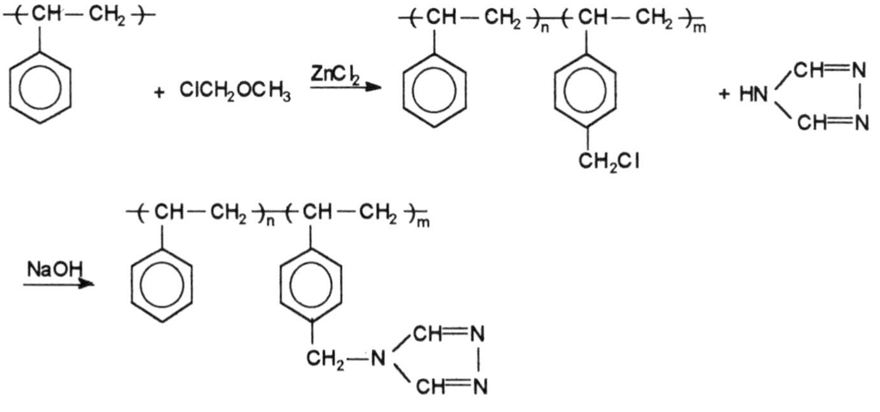

Scheme 3

D. Alkylation Reaction

The alkylation reaction of PS with different compounds in the presence of Lewis catalysis have been widely investigated [10–12. Polymeric materials that have a resistance to high temperature can be synthesized by bonding alkyl- or cycloalkyl groups to the aromatic ring of PS. The obtained polymer materials with C_{16}–C_{20} olefins of PS in the presence of Lewis catalysis at -12–$35°C$ are used as a depressing agent for oils [13]. The alkylation of PS with cyclohexylchloride in the presence of $AlCl_3$ catalysis at 20–$22°C$ has been reported, and the product has resistance to higher temperature of about 180–$185°C$ [14].

The alkylation of PS with N-methylolacetamide in the presence BF_3 catalysis at room temperature was prepared by Swiger et al. [15,16] who reported that the polymer has resistance to higher stroke and temperature as shown in Scheme (4).

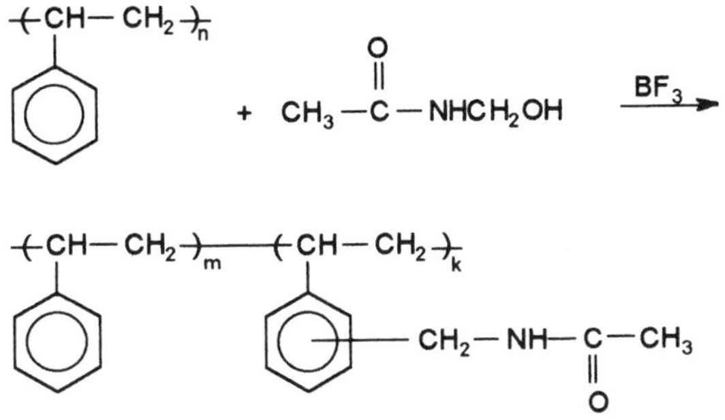

Scheme 4

The bonding of the phosphorus group to the aromatic ring of PS is an important element of these groups, which creates a higher resistance capability to fire [17–21]. The conversion of PS into its peroxide form cannot be made possible by oxidation. This property is normally obtained from the first alkylation of PS with isopropylchloride and then oxidation of the obtained product [22], as shown on Scheme (5).

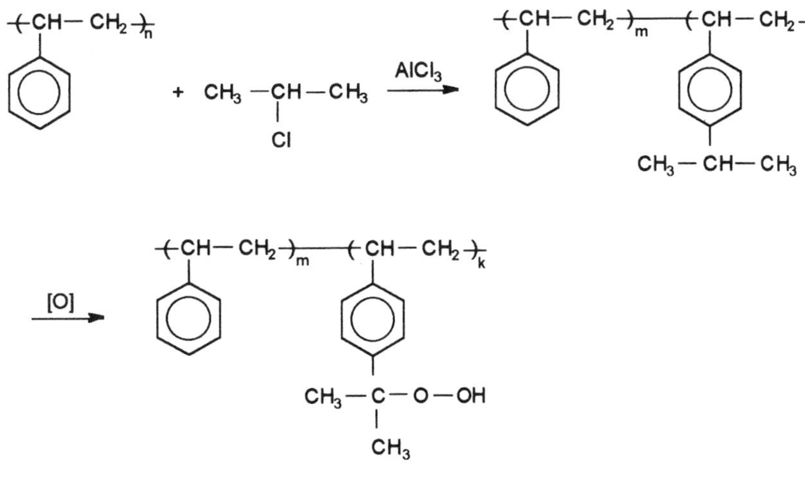

Scheme 5

These combination polymers can be used as a starting substance for methyl methacrylate polymerization.

Polymers containing epoxy groups have been used for polymer modification, either the synthesis of branched copolymers or their monomer having amine groups [23]. The PS containing epoxy group was first synthesized by Sembay et al. [24], who used styrene with epoxyperoxides (1,2-epoxy-3-tert-butylperoxypropan). However, this process is difficult and only about 0.16–0.54% of the epoxy group can be anchored.

E. Acylation Reaction

PS has apolar characteristics and, thus, it is difficult to form a bond with metals or polar materials. The adhesion capability of saturated polyhydrocarbons are dependent on the basis of polar properties of polymers [25]. Mitsuaki and Masyasu [26] investigated the chemical modification of PS for anchoring of the carboxyl group to PS macromolecules with maleic anhydride (MA) in the presence of radical catalysis at 90–150°C. These authors showed that bromulated PS is useful as a starting material for achieving more bonding of MA to the chain. The synthesis of acylated PS and its condensation with alde- hydes from the reaction of PS with acetic acids chloroan- hydride in the presence of AlCl₃ catalysis has been syn- thesized in Japan patents [27,28], as shown on Scheme (6).

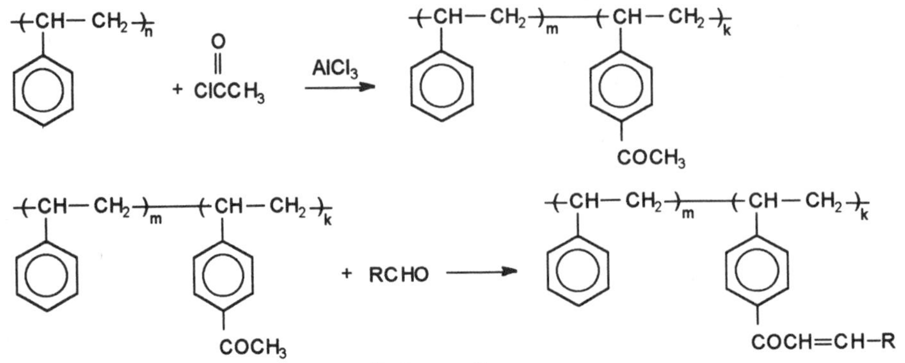

Scheme 6

Therefore, the obtained —CO—CH=CH—R group PS has a high photosensitivity property and can be cross- linked with the effect of light. The group is used in the field of photography [29,30].

F. Li-Polystyrene

In American patents [31,32, the acylation of PS with acryl, methacrylate chloride, and α and β-halogen pro- penylchloride has been done in the presence of AlCl₃ catalysis at 60–100°C. This process can be accomplished by using two methods in the presence of nonsolvent; either CH₂=CHCOCl, ClCH₂—CH₂—COCl and BF₃ catalysis passed from PS as vapor or acylated material passed from a mixture of PS and AlCl₃ as vapor. This method is not available due to the use of more catalysis (corresponding catalysis: PS ratio is 0.4–1/1 mol). The reaction of chloromethylated PS with furanacrylic acid or unsaturated aromatic acids are preferred for the an- choring of the carboxyl group to the aromatic ring of PS. The incorporation of various functional groups to the aromatic ring of PS has been given an American patent [33]. This incorporation is possible through the alkylation and acylation reaction of Li-PS with organic solvents. For instance, the various reactions are given in Scheme 7 (carbonylated PS with carboxyl acids; hydroxylated PS with ketones; epoxylated PS with epi- chlorohydrin).

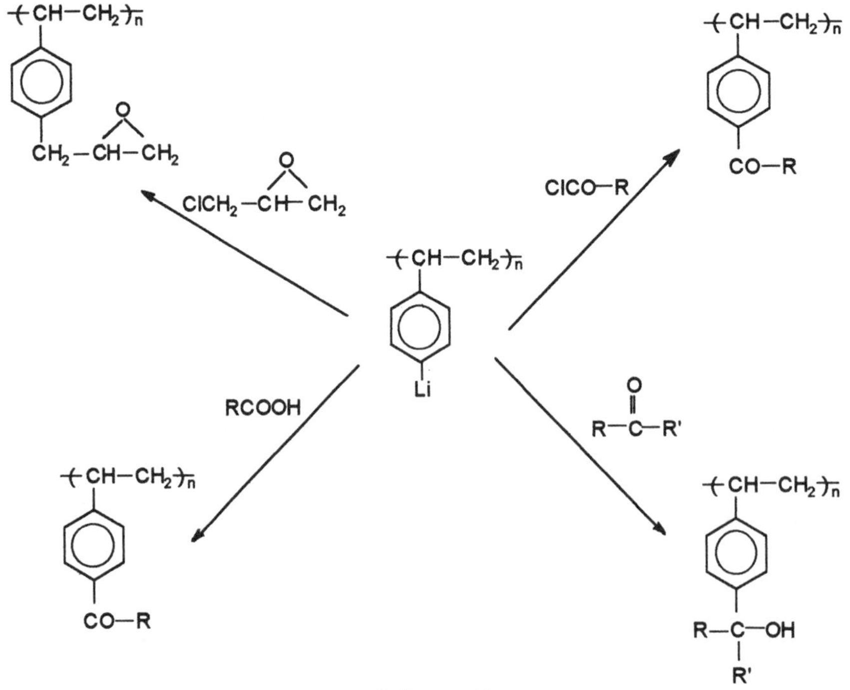

Scheme 7

The above methods occurred in 3 steps, therefore, these methods are not preferred. For instance, in the first step, o-, m-, and p-bromostyrene and its copolymer are synthesized. In the second step, Li-PS is synthesized from the reaction of copolymers with an organic compound containing Li. The abovementioned reactions are made with different compounds of Li-PS in the third step. These methods were also investigated by Ayres and Mann [34], who used the synthesis of PS containing chloro groups with chloromethylated PS as the first step. In the second step, formil resin was obtained by oxidation of chlorometylated PS. In the third step, carboxylated PS was obtained by the oxidation of formol resin with acetic acid at 20°C for 48 h. There are some disadvantages in this process. The oxidation process could take place from the aliphatic chain of PS.

G. Carboxylation Reaction

Another approach for anchoring the carboxyl groups to the aromatic ring of PS is to use a two-step method [35]. In this method, first, the reaction of PS with 2-chlorobenzoylchloride is carried out. The product can then be followed as shown in Scheme (8). If chloro atom is as o-position on benzoyl chloride the reaction occurred as (a). If there is no substitute group o-position on benzoyl chloride, the preferred is reaction (b).

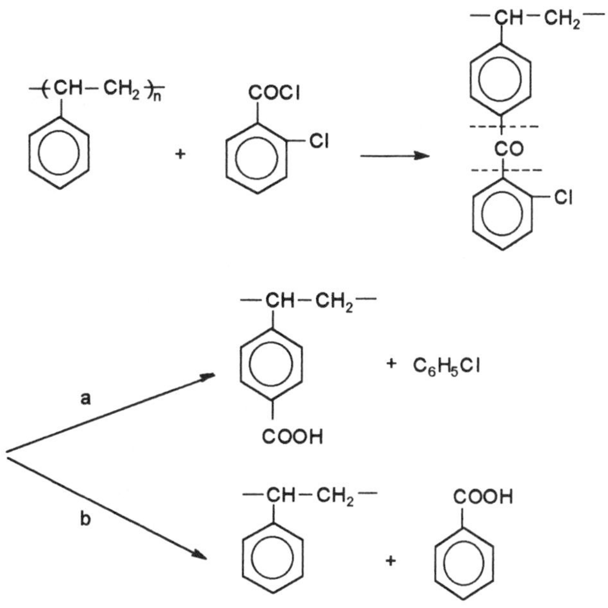

Scheme 8

Compared to the chemical modification reactions of PS, alkylation and acylation reactions are preferred to other reactions, such as halogenation, nitrolation, sulfonation, amination, and chloromethylation, etc. because the obtained polyfunctional PS has higher physico-mechanical properties.

III. CHEMICAL MODIFICATION REACTION OF POLYSTYRENE

The chemical modification of PS with epichlorohydrin (EC), maleic anhydride (MA), acetic anhydride (AA), butadiene, and isoprene in the presence of cationic catalysis such as $AlCl_3$, $FeCl_3$, $BF_3 \cdot O(C_2H_5)_2$, $ZnCl_2$, $TiCl_4$, and $SnCl_4$, have been extensively studied under various conditions for the last 15 years. We have also studied their kinetics, physico-mechanical, thermal, and dielec-

tric properties under various conditions and also their industrial applications. Moreover, the structure of modified PS and the nature and quantity of functional groups to anchored aromatic rings have also investigated [36–52].

A. Alkylation Reaction of Toluene as Model Compound with Epichlorohydrin

The mechanism of chemical modification reactions of PS were determined using toluene as a model compound with EC in the presence of $BF_3 \cdot O(C_2H_5)_2$ catalyst and the kinetics and mechanism of the alkylation reaction were also determined under similar conditions [53–55]. The alkylation reaction of toluene, with epichlorohydrin, underwent polymerization of EC in the presence of Lewis acid catalysis at a low temperature (273 K) as depicted in Scheme (9).

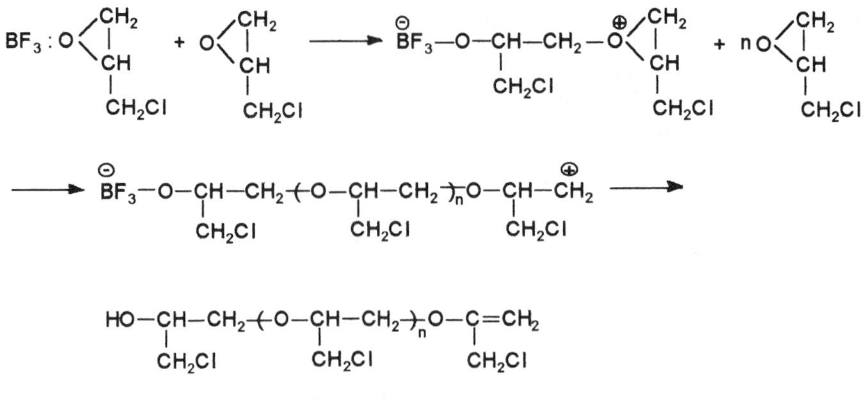

Polyepichlorohydrine (PEC)

Scheme 9

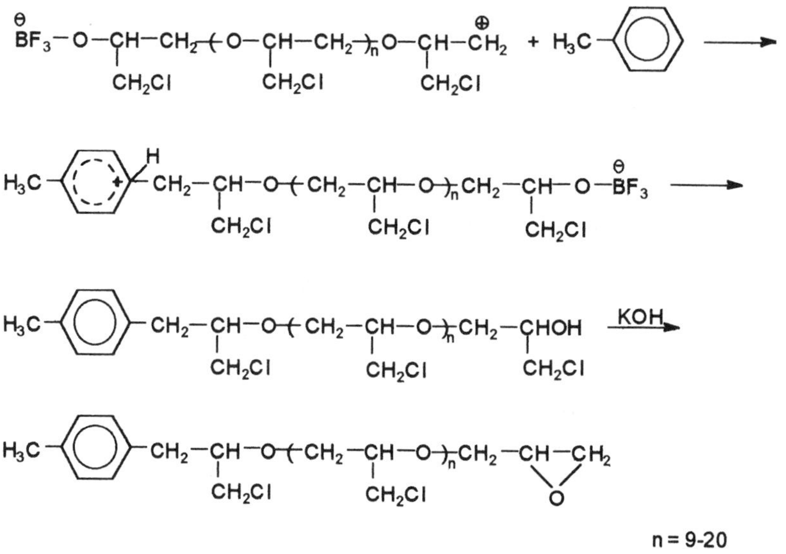

Epoxy toluene olygomer (ETO)

n = 9-20

Scheme 10

The epoxy toluene oligomer was also obtained by alkylation of toluene with PEC as shown in Scheme (10).

The structure of the products were characterized by two detector gel-chromatography (recractometer and UV). The direction of the alkylation reaction of toluene with EC was changed depending on the reaction condition as follows: if the ratio of toluene–EC is 1:5 mol, the alkylation reaction is toward obtaining the PEC at 273 K and if the ratio is reverse at 333 K, the reaction undergone to obtain epoxy toluene oligomer is shown in Fig. 1.

B. Alkylation of Polystyrene with Epichlorohydrin

It has been shown that the alkylation reaction can be used for the chemical modification of PS to obtain epox-

ylated and propylchlorhydrinated PS. The PS was alkylated at first with EC in the presence of Lewis catalysis [36–40]. In this method, the alkylation reaction is formed either by obtaining the EC homopolymer or alkylation of the aromatic ring of PS as shown in Scheme 11.

The alkylation reaction of PS was dependent on the temperature. When temperature was decreased from 353 K to 293 K, the following changes occur while the PEC is increased, the quantity of EC and anchored propyl chlorohydrin group to aromatic ring was decreased from 4.34 mol to 2.77 mol. As shown in Scheme 11, epoxylated PS was synthesized by dehydrochlorination of propylchloride group PS with NaOH. In addition, the alkylation reaction of PS with EC can occur by alkylation of the aromatic ring or destruction of macromolecules (decreasing molecular weight) in the presence of Lewis catalysis as depicted in Table 1.

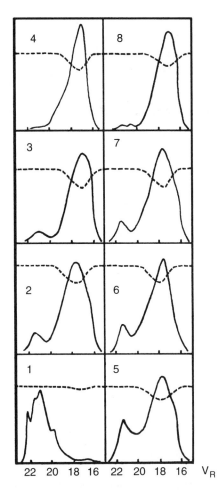

Figure 1 The gel-chromatogram of the obtained product from the alkylation reaction of toluene with EC under various reaction conditions; T, K = 273 (1), 293 (2), 313 (3), 333 (4). The ratio of toluene : epichlorohydrin; 0.5 : 1 (5); 1 : 1 (6), 2 : 1 (7), and 5 : 1 (8). ($-$) refractometer detector; (---) UV-detector.

Table 1 Effect of Lewis Catalysis on Molecular Weight and Quantity of the Obtained Products

Catalysis	Quantity of functional group (mol%)		$Mn \cdot 10^{-4}$
	Propylchlorohydrin	Epoxy	
$BF_3 \cdot O(C_2H_5)_2$	4.90	3.75	21.0
$AlCl_3$	4.70	3.53	17.6
$ZnCl_2$	0.5	0.32	20.4
$SnCl_4$	4.32	3.24	26.7
$FeCl_3$	1.02	0.61	20.3

The effect of catalysis on the quantity of anchoring functional groups and destruction of macromolecules is obtained in the following order:

$$BF_3 \cdot O(C_2H_5)_2 > AlCl_3 > SnCl_4 > FeCl_3 > ZnCl_2$$

The kinetics and activation parameter of the alkylation reaction of PS and toluene as a model compound with EC in the presence of $BF_3 \cdot O(C_2H_5)_2$ catalysis are given in Table 2. The initial rate and reaction rate constant was increased with increasing temperature as

Table 2 Kinetic Parameters of PS and Toluene as a Model Compound

T, K	$W_0 \cdot 10^5$ (mol/L·s)		$K \cdot 10^5$ (L/mol·s)		E_a, kJ/mol	
	PS	Toluene	PS	Toluene	PS	Toluene
293	0.1	0.3	2.06	8.0		
303	0.2	0.6	4.12	15.0	76.59	57.4
323	0.4	1.7	10.3	44.0		

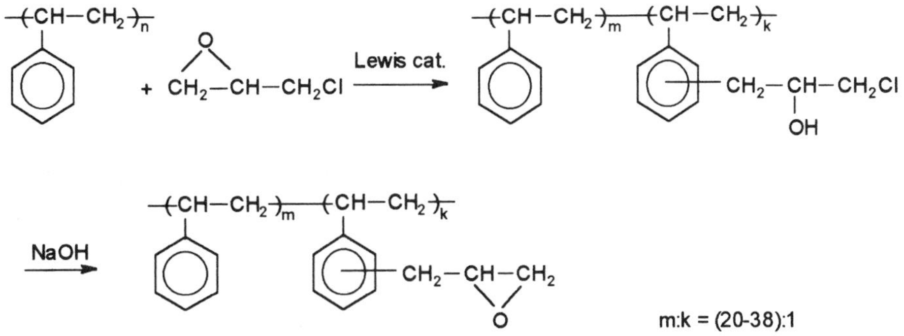

Scheme 11

shown in Table 2. Thus, 2-oxy-3-chloropropyl group PS and oligoepichlorohydrin (OECH) were obtained.

C. Alkenylation Reaction of Polystyrene

The chemical modification of PS with diene hydrocarbons in the presence of Lewis catalysis are important for synthesizing of higher resistance, elasticity, and adhesion-capable polymers. When polybutadiene or polyisopropene fragments were chemically anchored to the macromolecule of PS, the physico-mechanical and adhesion capability increased. For that reason, the alkenylation reaction of PS also occurred by use of diene hydrocarbons, butadiene, and isoprene in the presence of $BF_3 \cdot O(C_2H_5)_2$. The reaction can be formed by either the alkenylation of the aromatic ring or obtaining the diene homopolymer (Scheme 12).

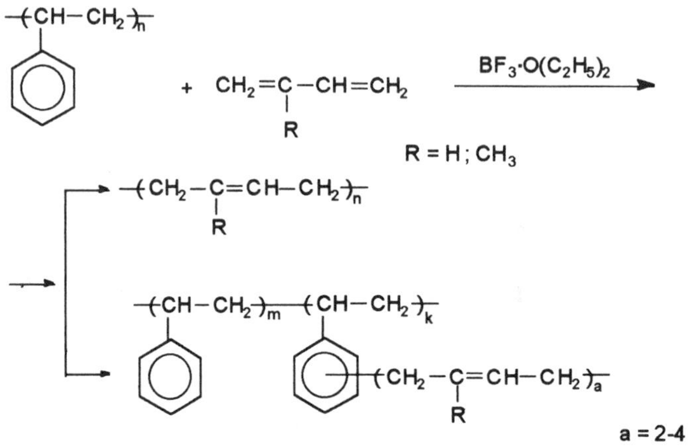

Scheme 12

D. Acylation Reaction of Polystyrene with Maleic Anhydride and Acetic Anhydride

The acylation reaction of PS with organic anhydrides, such as maleic and acetic anhydrides, are very important for synthesizing polyfunctional (carbonyl-, carboxyl-, keto-, olefinic) PS. The incorporation of these groups to PS caused an increase of adhesion capability, physico-mechanical properties, elasticity, and photosensitivity [41–46].

When MA was used as the reactive, the acylation reaction occurred on both sides of MA both from the olefinic bond or anhydride group. However, the acylation reaction occurred from the anhydride group as shown in Scheme (13). Therefore, to fully understand the acylation reaction mechanism of PS with MA, toluene or ethylbenzene as model compounds has been investigated under similar conditions. The structure of the acylation reaction product of ethylbenzene with MA was determined by chromatographic, spectral, and chemical analysis. The results show that the reaction occurred by opening the anhydride groups and anchoring the aromatic ring.

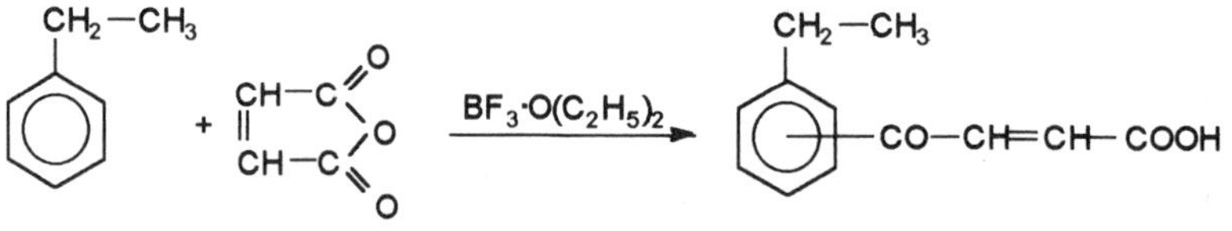

Scheme 13

The acylation reaction of PS with MA by using model compounds in the presence of Lewis catalysis can be explained as follows.

First, it should be taken into consideration that benzene is easily acylated with MA in the presence of $AlCl_3$ [56]. This fact can serve as a model system for the study of PS-MA-Lewis acid systems. Results obtained allow one to present the general scheme for the side chain modification of PS in the presence of Lewis acid in the following way:

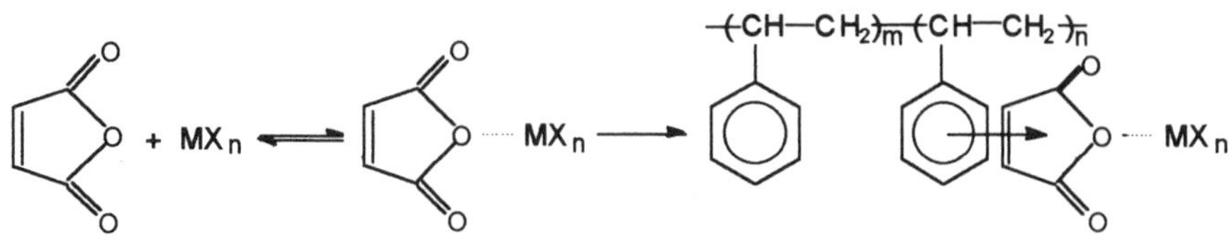

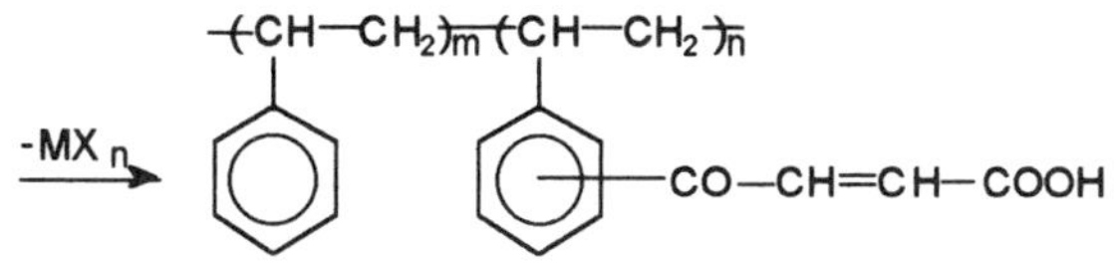

Scheme 14

In the case of use of $BF_3 \cdot OEt_2$ as the catalyst, the reaction scheme can be suggested as follows:

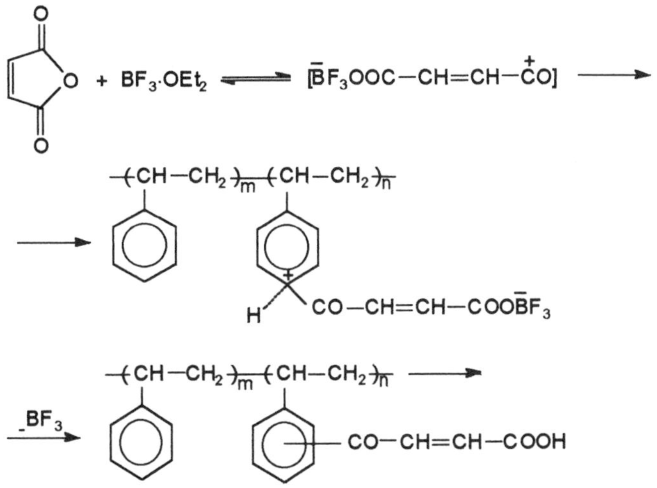

Scheme 15

The scheme consists of several stages: (1) complex-formation of MA with Lewis acid (MX_n), (2) addition of MA . . . MX_n complex to the phenyl ring of PS, and (3) the break of the hydrogen atom from p- or o-position of the phenyl ring and its addition to the maleate fragment. In this scheme, complex-formation plays a significant role in the acylation reaction. It can also be suggested that in the acylation reaction charge transfer complex between MA and phenyl ring of PS:

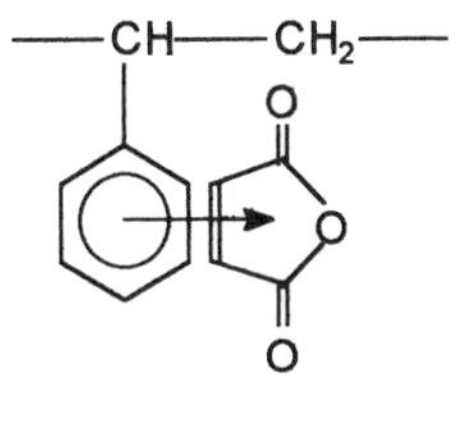

or between complexed MA and PS:

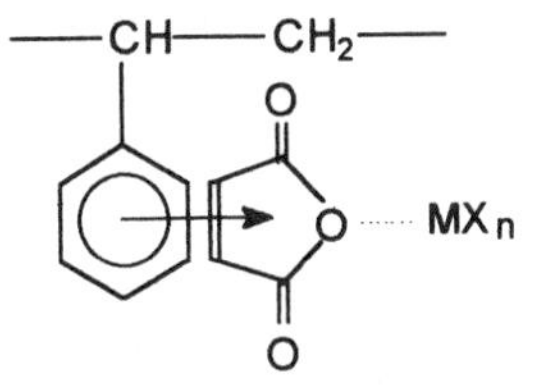

also can take place, which facilitate the processing of the acylation in the expected direction. Complex-formations between MA and aromatic carbohydrogens such as benzene and toluene were noted in earlier studies [57].

The results of analyses for carboxyl group contents, intrinsic viscosities, molecular weights of functionalized PSs, and the ratio of virgin and acylated PS fragments (m:n) are presented in Table 3. Based on the data in Table 3, it is clear that modification of the PS by MA occurred in all cationic catalyst media, but at different levels. The highest carboxyl group concentration is found to occur with $BF_3 \cdot OEt_2$ and $TiCl_4$. But in the case of $TiCl_4$, acylated PS has a relatively low value of viscosity, which indicates that the process of modification is accompanied by degradation of the main chain. Results of chemical analyses show that carboxyl group concentrations have increased from 4.5 to 20.8 mol% in the case of PS with $M_n = 2.5 \times 10^5$ and from 2.5 to 10.2 mol% for PS with $M_n = 5.5 \times 10^5$. If one compares the molecular weights of modified PS, the highest was obtained in the case of $BF_3 \cdot OEt_2$, which shows the ease of application of this catalyst in the modification reaction and hence activity of it.

The degree of acylation of PS essentially depends on the type of catalyst used and the molecular weight of initial virgin PS. As it is seen from the data in Table 3, virgin PS with higher molecular weight is acylated to a smaller degree in the equal conditions of modification.

Table 3 Results of Characterization of Modified PS Prepared by Using Various Catalysts

	Initial PS with $M_n = 2.5 \cdot 10^5$				Initial PS with $M_n = 5.0 \cdot 10^5$			
Catalyst	A.N. (mgKOH/g)	—COOH (mol%)	(m:n)[a]	$[\eta]$ (dL/g)	$M_n\ 10^{-5}$	—COOH (mol%)	(m:n)[a]	$M_n\ 10^{-5}$
BF$_3$·OEt$_2$	136	20.0	5:1	0.75	1.91	10.2	9:1	3.7
AlCl$_3$	132	16.6	6:1	0.51	0.96	8.7	11:1	3.1
TiCl$_4$	140	20.8	5:1	0.48	0.91	—	—	—
SnCl$_4$	106	14.3	7:1	0.53	1.16	4.4	22:1	4.05
FeCl$_3$	64	7.7	13:1	0.54	1.17	3.4	30:1	3.61
ZnCl$_2$	39	4.5	22:1	0.55	1.21	2.5	41:1	3.82

[a] Ratio of contents of virgin PS (m) and acylated PS (n) units in macromolecules.
Condition: PS/MA = 2.33, MA/Catalyst = 1.0, 20°C, 2 h.

In fact, as can be easily traced from Table 3, the activities of Lewis acids in the reactions studied are as follows:

$$BF_3 \cdot OEt_2 \geq TiCl_4 > AlCl_3 > SnCl_4$$
$$> FeCl_3 > ZnCl_2$$

Curves of precipitations (Fig. 2) of modified PS prepared in the presence of various Lewis acids also prove that BF$_3$·OEt$_2$ is the most convenient catalyst for the reaction studied.

In Table 4, intrinsic viscosities and carboxyl contents of different fractions of modified PS obtained at different catalyst media are presented. As evidenced from these data, all PS samples used were acylated. As the viscosity of the fractions decreases, the carboxyl contents in the product increase. The highest degree of acylation for all fractions is observed when BF$_3$·OEt$_2$ is used as the catalyst.

Existence of carboxyl groups in the structure of modified PS was also confirmed by IR studies (Fig. 3). IR spectra of acylated PS have characteristic bands at 1760, 1725, 1555, and 1410 cm^{-1} corresponding to $\nu_{C=O}$

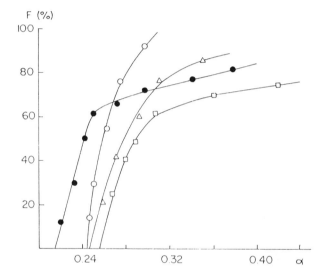

Figure 2 Curves of precipitation for functionalized PS prepared in the presence of (—○—) BF$_3$·OEt$_2$, (—●—) TiCl$_4$, (—△—) SnCl$_4$, and (—□—) FeCl$_3$. F = the content of fractions (%); α = the ratio of solvent(benzene)/precipitant(methanol).

Table 4 Effect of Lewis Acid on the Intrinsic Viscosity and Degree of Acylation for the Various Fractions of Modified Block PS

	Fractions									
	I		II		III		IV		V	
Catalyst	$[\eta]$[a]	—COOH[b]	$[\eta]$	—COOH	$[\eta]$	—COOH	$[\eta]$	—COOH	$[\eta]$	—COOH
BF$_3$·OEt$_2$	0.93	2.12	0.64	2.77	0.51	3.12	0.50	8.70	0.40	17.82
AlCl$_3$	1.08	2.38	0.86	5.26	0.55	9.10	0.32	12.52	0.27	16.15
TiCl$_4$	1.63	1.37	1.15	2.27	0.86	3.33	0.44	8.71	0.30	14.32
SnCl$_4$	1.30	1.85	0.83	3.12	0.47	7.14	0.35	10.26	0.27	15.15
FeCl$_3$	1.27	1.64	1.04	2.27	0.77	4.54	0.50	7.14	0.40	7.71
ZnCl$_2$	1.33	1.26	0.92	2.12	0.61	2.77	0.34	4.02	0.26	5.55

[a] $[\eta]$ in toluene at 250°C (dL/g).
[b] —COOH is content of carboxyl group in the modified PS (mol%).

of acyl fragments in the side chain, which are absent in the spectra of virgin PS. These bands are more intensive in the acylated PS prepared in the presence of $BF_3 \cdot OEt_2$ (spectra 5) and $TiCl_4$ (spectra 4).

The mechanism of the acylation reaction of PS with acetic anhydride in the presence of Lewis catalysts can be considered as follows:

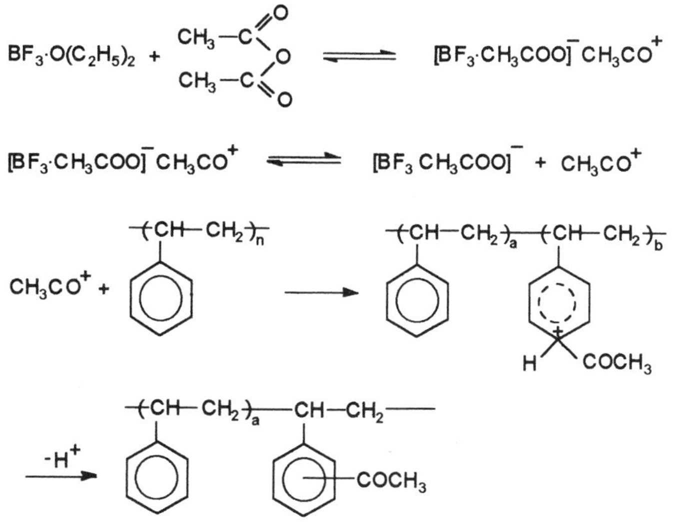

Scheme 16

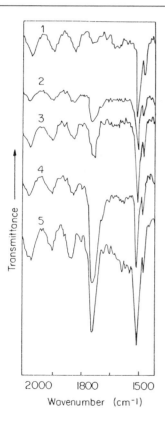

Figure 3 IR spectra of (1) virgin PS and (2–5) acylated PS synthesized by using various catalysts: (2) $FeCl_3$, (3) $SnCl_4$, (4) $TiCl_4$, (5) $BF_3 \cdot OEt_2$.

The anchoring of the CH_3—CO— group to the aromatic ring of PS was confirmed by spectral and chemical methods and the (CH)(OH)—CH_3 group was determined by hydrogenation with $LiAlH_4$.

Compared with the bonding groups (mol%) to aromatic ring of PS, the degree of acylation was observed when MA was used. These results was obtained by determination of kinetic parameters of PS with MA and AA under the same reaction conditions. As shown in Table 5, if the initial rate (W_0) and rate constant (K) of the acylation reaction between MA and AA are compared, the MA is almost 10–14 times higher than AA in the presence of $BF_3 \cdot OEt_2$ catalyst. This fact is due to the stretching structure of MA and the effect of the catalyst.

IV. PHYSICO-MECHANICAL PROPERTIES OF POLYSTYRENE

The physico-mechanic and thermal properties of polyfunctional PSs that were obtained by the chemical modification with MA, AA, EC, butadiene, and isoprene in the presence of [$BF_3 \cdot OEt_2$, $TiCl_4$, $AlCl_3$, $SnCl_4$, $FeCl_3$, $ZnCl_2$) catalysts have been investigated under various conditions. The adhesion and photosensitivity properties of new polyfunctional PSs were determined. They depended on the binding of functional groups to the aromatic ring, and especially increased with the binding of olefinic, epoxy, and carboxyl groups. In this study, we have investigated the physico-mechanical and thermal properties of synthesized PSs, which have different

Table 5 Kinetic Parameters of Acylation Reaction of PS with MA and AA in the Presence of $BF_3 \cdot OEt_2$ Catalyst

T, K	$W_0 \cdot 10^5$, mol/L·s		$K \cdot 10^4$, L/mol·s		E_a, kJ/mol	
	MA	AA	MA	AA	MA	AA
298	15.8	1.1	12.7	0.9		
323	21.6	2.1	17.1	1.6	11.3	17.1
343	28.3	3.1	22.7	2.3		

functional groups, and the nature and the quantity of binding functional groups to the aromatic ring. In addition, the adhesion capability and photosensitivity of polyfunctional PSs were also determined.

The physico-mechanical, thermal, and adhesion properties of the synthesized polyfunctional PSs are dependent on the nature of functional groups in the aromatic ring. In this case, the following are properties of the chlorohydrin and epoxy groups: highest elasticity, resistance to strike, and adhesion properties with carboxyl and olefinics. Furthermore, the —CO—CH=CH-—COOH group was provided new properties such as the photosensitive capability. Functionalized PSs obtained are characterized by their high thermostability, adhesion, and photosensitivity.

In the chemical modification of PS with MA, AA, EC, butadiene, and isoprene using cationic catalysis caused either destruction of macromolecules or the binding of functional groups to the aromatic ring.

In general, physico-mechanical properties of polymers depend on the molecular weight. However, the physico-mechanical properties of PSs decreased in the presence of cationic catalysis, but increased in the case of the binding of functional groups to the aromatic ring in spite of the destruction of PS. Therefore, new properties such as adhesion and photosensitive capability increase

in the PS. The results of physico-mechanic, thermal, and adhesion capability properties of the synthesized polyfunctional PSs are given in Table 6.

The physico-mechanic, thermal, and adhesion properties of functionalized PS are dependent on the anchored functional groups (Table 5) the following properties were observed: highest thermal properties in the —CH(OH)CH$_2$Cl, highest resistance against light in the —CO—CH=CH—COOH, highest adhesion capability in the —C=C—, $\overset{-C-C-}{\underset{O}{\diagdown\diagup}}$, and; highest resistance to stroke in the —C=C— and —CO—CH=CH—COOH groups.

When the molecular weight of PS was decreased from 5.0×10^5 to $(3.0–4.05) \times 10^5$, the abovementioned properties were also decreased in the presence of cationic catalysis after the destruction of PS. These predicted properties are related to the nature and the quantity of functional groups.

When —CO—CH=CH—COOH groups bonded to the aromatic ring of PS, the physico-mechanic, thermal, and adhesion properties increased from 4.5 mol% to 20.0 mol%. This caused the following changes: the resistance of PS increased from 14.0 to 19.2 kJ/m^2, the resistance to stretch polymer material itself increased from 35.0 to

Table 6 Physico-Mechanic, Thermal, and Adhesion Properties of Functional Polystyrenes

Functional group Group	mol %	σ (MPA)	α (KJ/m^2)	ϵ' (%)	T_v (°C)	H_R (N/mm^2)	A (MPA)	$Mn \cdot 10^{-4}$
$\overset{-C-C-}{\underset{O}{\diagdown\diagup}}$	3.7	60.0	15.3	12.5	120	200	5.3	30.0
—CH—CH$_2$Cl OH	4.9	48.5	13.6	10.1	150	228	3.8	30.7
—CO—CH$_3$	9.8	62.0	9.0	16.1	115	195	2.4	37.0
—CO—CH=CH—COOH	20.0	63.5	19.2	17.5	100	180	7.3	40.5
—CH=CH—	10.0	42.0	39.0	22.0	105	170	8.0	39.5
PS	—	35.0	13.0	1.5	80	180	0.2	50.0
After destruction of PS	—	30.0	10.5	2.1	76	125	0.5	30.0

Resistance to Stretch, (σ); resistance to stroke, (α); relative extension, (ϵ'); resistance to heat, (T_v); hardness, (H_B) A = adhesion.

63.5, and adhesion capability to metal surfaces increased from 0.2 to 7.3 MPA. Furthermore, the sensitivity of the polymer to light (87 cm²/J) and the higher thermostability (according to DTA and TGA analysis) is increased.

From the data of DTA and TGA analyses illustrated in Fig. 4, its evident that the visible exo-peaks in the range of 140–150°C appear in DTA curves (Fig. 4a), which can be taken as evidence of a crosslinking reaction of acylated macromolecules at the expense of side chain

fragments. However, the TGA curves in Fig. 4b show that there is a small weight loss (4–5%) at 110–140°C for modified PS, after which all TGA curves straighten up to 275°C. Observed weight loss probably is related to the decarboxylation reaction of unsaturated acyl groups with formation of vinylketone groups in the side chain, which additionally took place easily in the crosslinking reaction.

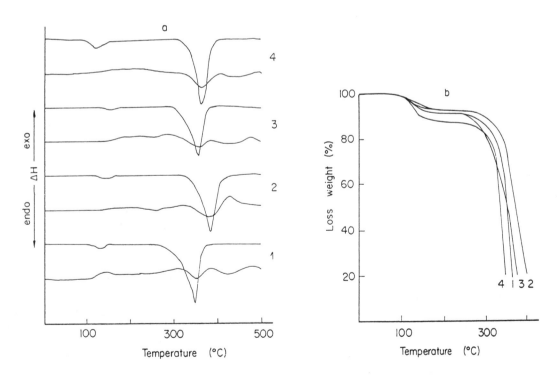

The possibility of processing the first reaction during thermotreatment of modified PS is also confirmed by the observed weight loss in the TGA curves (Fig. 4b) at the beginning stage. Crosslinking of macromolecules is confirmed by exo-peaks of DTA curves (Fig. 4a) and also by the rather high thermostability of studied polymer analogies observed. The temperature of decomposition for these polymers is 310–335°C. The characteristics of DTA and TGA curves also depend on the nature of the catalyst used in the synthesis of functionalized PS characteristics by the content of carboxyl groups in the side chain (Table 3).

Results of IR studies illustrated in Fig. 5 for modified PS films thermotreated at 75 and 150°C for 45 and

30 min, respectively, can also confirm the process of the crosslinking reaction. Figure 5 shows the dependence of the intensities of carbonyl bands (1760 and 1725 cm⁻¹) on the condition of thermotreatment. It is seen that the thermotreatment at 75°C does not affect the intensity of the carbonyl bands, while the thermotreatment at 150°C significantly decreases the intensity of these bands. This fact can be explained by the decarboxylation reaction of side chain unsaturated acyl groups and their further crosslinking.

Figure 6 shows the scans of virgin and modified PS synthesized by using various Lewis acids. The position of each transition is designated by arrows. These transitions, which are observed to occur above room tempera-

Figure 4 DTA and TG (a) and TGA (b) curves of modified PS in the presence of (1) $BF_3 \cdot OEt_2$, (2) $FeCl_3$, (3) $AlCl_3$, (4) $SnCl_4$. Heating rate 10°C/min in air.

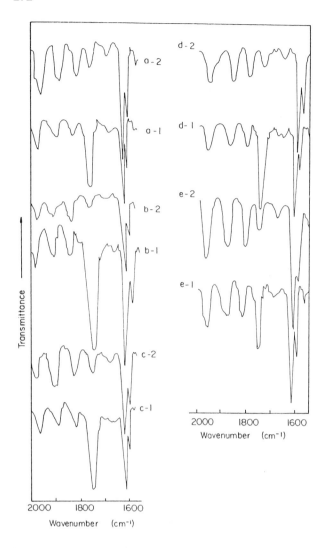

Figure 5 IR spectra of acylated PS films after thermo-treatment at 75°C during 45 min (1) and at 150°C during 30 min (2). Catalysts: (a) $BF_3 \cdot OEt_2$, (b) $TiCl_4$, (c) $SnCl_4$, (d) $ZnCl_2$, (e) $FeCl_3$.

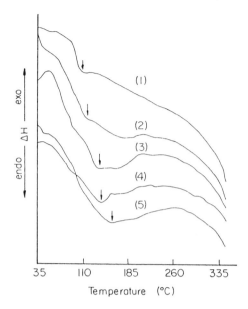

Figure 6 DSC scans of (1) PS ($M_n = 2.5.10^5$) and its acylated derivatives with following contents of carboxyl group: (2) 4.5, (3) 14.3, (4) 16.6, (5) 20.0 mol%, prepared by using $FeCl_3$, $SnCl_4$, $AlCl_3$, and $BF_3 \cdot OEt_2$, respectively.

ture, are probably associated with the melting of hard-segment domains. An increase of graft/carboxyl fragments in the macromolecules leads to the displacement of T_g to a higher field. For PS (virgin) with $M_n = 2.5 \times 10^5$, $T_g = 108°C$, while for its modified derivatives, T_g changes from 116 to 145°C, depending on the degree of acylation. Observed exo-effects in the field of 150–280°C, which are absent in the DSC curves of virgin PS, indicated the process of the chemical reactions of acylated macromolecules.

As expected, the introduction of polar and highly reactive unsaturated acyl groups in the PS introduce significant improvement of some very important properties. Thermal characteristics, adhesion, and photosensitivity of modified PS with different compositions are presented in Table 7.

Modified PS by use of $BF_3 \cdot OEt_2$ catalyst had better properties comparised with virgin PS and other modified polymers. High thermostability and photosensitivity of modified PS compared with virgin PS are explained by the crosslinked structure of macromolecules formed during the processes of thermo- and phototreatment.

High adhesion and sensitivity to UV irradiation of these polymer coatings allow one to use them as a base for the preparation of polymer resistance of a negative type.

The increase of the quantity of functional groups bonded in PS causes the increase in the resistance to stretch from 40 to 63.5 MPA for —CO—CH=CH—COOH group and 53.9 to 62.0 MPA for —CO—CH₃ group, and adhesion capability from 2.0 to 7.3 MPA for —CO—CH=CH—COOH group and from 0.5 to 2.4 MPA for —CO—CH₃ group, respectively (Figs. 7 and 8). As shown in Fig. 8, adhesion capability increased, on the contrary, a difference in the resistance to stretch in the carboxyl and acetyl groups was not observed.

When the chlorohydrine group was bonded from 2.65 to 4.9 mol% to the aromatic ring of PS, the following changes were obtained: hardness increased from 175 to 228 N/mm² and resistance to light increased from 100°C to 150°C. When this polymer was converted to epoxylated PS in the basic medium, the same mentioned above properties were also observed. Moreover, the stretch, breaking, and adhesion capabilities increased from 48.5–60.0 MPA and 3.8–5.3 MPA, respectively (Fig. 9 and Table 5).

Table 7 Thermostability, Adhesion, and Photosensitivity of Functionalized PS Prepared in the Presence of Various Lewis Acids

Lewis acid	A.N. (mgKOH/g)	Adhesion[a] (%)	Photosensitivity (cm²/J)	T_0 (°C)	Loss of weight (%) at (°C)		
					200	250	300
BF₃·OEt₂	136	98	87	145	7.0	9.0	12.0
AlCl₃	132	92	75	131	9.0	10.5	18.5
SnCl₄	106	90	68	128	12.5	13.5	17.0
FeCl₃	39	85	16	116	7.5	8.0	11.0
PS (virgin)	—	2–3	—	108	38	44	50

[a] Adhesion of polymer coatings (thickness 60–70 μk) formed on the glass surface is obtained by "lattice notch" method.

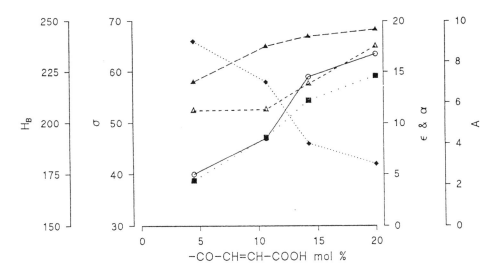

Figure 7 The physico-mechanical properties of —CO—CH=CH—COOH groups polystyrenes; (○, —) resistance to stretch, (σ); (▲----) resistance to strike, (α); (△----) relative extention, (ϵ); (◆----) hardness (H_B); (■----) adhesion, (A).

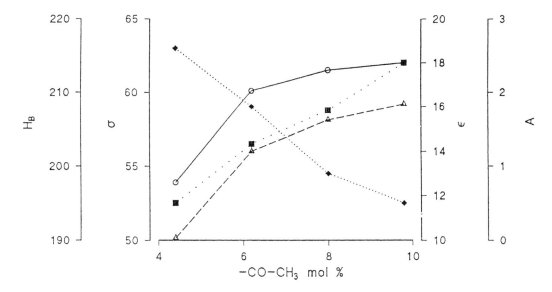

Figure 8 The physico-mechanical properties of —CO—CH₃ groups polystyrenes; (○, —) resistance to stretch, (σ); (△----) relative extention, (ϵ); (◆----) hardness (H_B); (■----) adhesion, (A).

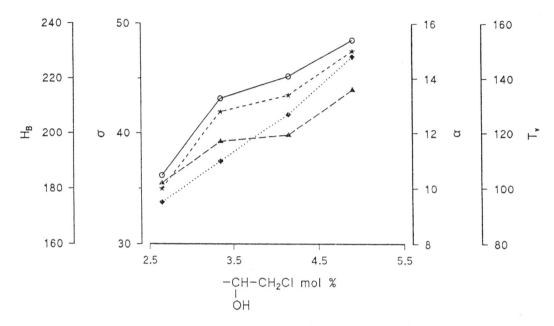

Figure 9   The physico-mechanical properties of —CH(OH)—CH₂Cl groups polystyrenes: (○,—) resistance to stretch, (σ); (▲----) resistance to strike, (α); (*----) resistance to heat, (T_v); (◆----) hardness (H_B).

When the quantity of olefinic increased from 2.5–10.0 mol% in the alkenylated PS with butadiene and isoprene, the following advantages were achieved: resistance to strike, 21.0–39.0 KJ/m²); elasticity, 13.0–22.0%; and adhesion capability, 4.9–8.0 MPA, (Fig. 10).

The investigation of physico-mechanic and adhesion properties of synthesized polyfunctional PS has

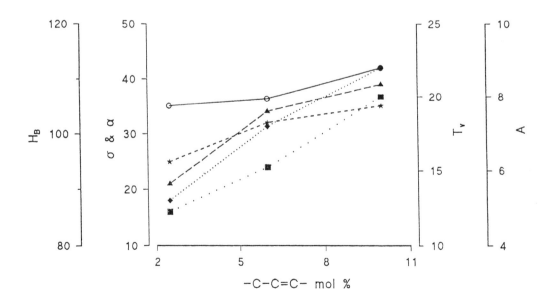

Figure 10   The physico-mechanical properties of —C—C=C— groups polystyrenes; (o,—) resistance to stretch, (σ); (▲, --) resistance to stroke, (α); (*, - -) resistance to heat, (T_v); (◆, ·····) hardness (H_B), and (■, · · ·) adhesion, (A).

Table 8 Glass Transition Temperatures and Thermomechanic Properties of Functional PSs

Functional groups	Glassing point temperature (Tg) K	Thermo-mechanic analysis (TMK), K

—CH—CH$_2$Cl
  |
  OH
	413	418

—CO—CH=CH—COOH	388	385
—C=C—	370	368
Polystyrene	358	360

shown that highest resistance to strike and highest adhesion capability were obtained on the alkenylated PS (Fig. 10). These advantages come from polybutadiene and polyisoprene fragments that were bonded to the aromatic ring of PS in the chemical modification.

The glass transition temperatures (T$_g$) of both modified and unmodified PSs were determined by DSC analysis, and thermomechanic analysis was controlled by TMK. The results are given in Table 8. It is seen from Table 8 that the highest glass transition temperature (410 K) was obtained with chlorohydrinated PS and that of the lowest (370 K) with olefinic PS. The lowest glass transition temperature in the alkenylated PS caused to elasticity properties on polybutadien and polyisopren fragments.

These observations demonstrate that the different functional groups can be attached to the aromatic ring of PS with various chemical modification conditions, and it is possible to obtain different technical properties for polymer materials.

REFERENCES

1. S. H. Usakov, and R. A. Matuzov, *J. Prikl. Chem., 17*: 538 (1944).
2. G. B. Bachman, *J. Org. Chem., 12*: 108 (1947).
3. H. M. Jones, and V. G. Robertson, *Nature, 174*: 78 (1954).
4. H. M. Jones, *Can. J. Chem., 34*: 948 (1956).
5. J. G. Bevington, and J. Ratti, *Europ. Polym. J., 8*: 1105 (1972).
6. K. Kalriyama, and J. Shimura, *J. Appl. Polym. Sci., 16*: 3035 (1972).
7. A. A. Marton, and L. D. Treylor, *J. Org. Chem., 24*: 1167 (1959).
8. J. D. Jones, *Ind. Engineer. Chem., 44*: 2865 (1952).
9. A. C. 366202 (USSR), T. A. Sokolova, D. L. Cijenko-B. I., 7 (1973).
10. Pat. 2713570 (USA), W. O. Kenyon and G. P. Waugh, (1955).
11. W. O. Kenyon, and G. P. Waugh, *J. Polym. Chem., 32*: 83 (1958).
12. J. A. Blanchette, and J. D. Gotman, *J. Org. Chem., 23*: 1117 (1958).
13. Pat. 704813 (Brit.).
14. S. M. Aliyev, and A. A. Sarkisyen, *J. Azerb. Chem., 2*: 43 (1971).
15. R. T. Swiger, *Amer. Chem. Soc., Polym. Prep., 15*: 267 (1974).
16. R. T. Swiger, *Amer. Chem. Soc., Polym. Prep., 17*: 504 (1976).
17. A. Magno, *J. Polym. Sci. Part A, 15(2)*: 513 (1977).
18. F. Propessu, *Rev. Rom. Chem., 14*: 1525 (1969).
19. F. Propessu, *Rev. Rom. Chem., 16*: 285 (1971).
20. F. Propessu, *Rev. Rom. Chem., 16*: 899 (1971).
21. Pat. 4007318 (USA), A. Magno, and J. L. Webb (1977).
22. D. Y. Metz, and R. B. Mersobian, *J. Polym. Sci., 16*: 345 (1955).
23. Pat. 3208980 (USA), J. Cruver, and G. Kraus (1965).
24. A. S. 382649 (USSR) E. I. Sembay, T. I. Elagina, and T. I. Yurjenko, *Bull.* 23 (1973).
25. Pat. 3450560 (USA) R. Backal (1963).
26. N. Mitsuaki, and A. Masyasu, *J. Chem. Soc. Japan, 70*: 1432 (1970).
27. Pat. 49-46396 (Japan), F. Hirosi (1974).
28. Pat. 49-46397 (Japan), F. Hirosi (1975).
29. Pat. 2731301 (USA), P. A. Allen (1957).
30. E. B. Kuznesov, I. P. Prochorov, and D. A. Fayzullina, *Visok. Molek. Soed., A, 3*: 1544 (1961).
31. Pat. 3304294 (USA), P. A. Garney, and F. G. Leavitt (1967).
32. Pat. 3299025 (USA), P. A. Garney, and F. G. Leavitt (1967).
33. Pat. 3234196 (USA) (1965).
34. J. T. Ayres, and C. K. Mann, *J. Polym. Sci., 3*: 433 (1965).
35. C. R. Harrison, and H. Philip, *Macromol. Chem., 176*: 267 (1975).
36. A. S. 417443 (USSR), MKL. C 0827/00, S. I. Sadikzade, R. A. Ismaylova, A. V. Ragimov, K. A. Aslanov, and R. I. Mustafayev, *Bull. 8*: 72 (1974).
37. S. I. Sadikzade, R. A. Ismaylova, A. V. Ragimov, and K. A. Aslanov, *Plastmassi., 6*: 66 (1974).
38. R. A. Kurbanova, A. V. Ragimov, K. A. Aslanov, and D. Y. Misiyev, *Visokomoleik. Seed., B, 18*: 542 (1976).
39. R. A. Kurbanova, A. V. Ragimov, K. A. Aslanov, and V. Y. Aliyev, *Lakokras. Mater., 6*: 19 (1979).
40. R. A. Kurbanova, A. V. Ragimov, and K. A. Aslanov, *J. Prikl. Chem., 1*: 2311 (1979).
41. R. A. Kurbanova, A. V. Ragimov, and D. N. Aliyeva, *Plastmassi, 9*: 70 (1976).
42. R. A. Kurbanova, and D. N. Aliyeva, *Preprint Inder. Sump. on Macromol.*, IUPAC, Dublin, Vol. 2, 195 (1977).
43. D. N. Aieva, R. A. Kurbanova, and A. V. Ragimov, *Dokl. AN Azerb. SSR, 39*: 43 (1983).
44. A. S. 713813 (USSR), MKI, C 08 F 212/08, R. A. Kurbanova, and A. V. Ragimov, *Bull., 5*: 76 (1980).
45. A. S. 704097 (USSR), MKI, C 08 F 212/08, A. V. Ragimov, and R. A. Kurbanova, D. N. Alieva (Closed).
46. R. A. Kurbanova, D. N. Alieva, and A. V. Ragimov, *Lakokras. Mater., 1*: 51 (1980).
47. R. A. Kurbanova, A. A. Mehraliev, T. M. Orucova, and K. A. Aslanov, *Lakokras. Mater., 3*: 34 (1981).
48. R. A. Kurbanova, A. A. Mehraliev, T. M. Orucova, K. A. *J. Plastmassi., 9*: 61 (1984).
49. R. A. Kurbanova, T. M. Orucova, and A. A. Mehraliev, *J. Lakokras. Mater., 3*: 53 (1985).

50. R. A. Kurbanova, A. A. Mehraliev, M. M. Gurbanov, and A. V. Ragimov, *J. Plast. Massi., 10*: 12 (1989).

51. R. A. Kurbanova, A. V. Ragimov, S. F. Sadikov, and M. M. Gurbanov, *J. Lakokras. Mater., 1*: 103 (1990).

52. A. M. Kroxmalniy, I. I. Zin, Y. M. Nagieva, and R. A. Kurbanova, *J. Fizikoximiceskaya Mexanika Mater., 5*: 114 (1990).

53. A. S. 684044 (USSR) MKI C 08 C 59/00, R. A. Kurbanova, A. V. Ragimov, and N. R. Bektasi, *Bull., 33,* 92 (1979).

54. A. V. Ragimov, N. R. Bektasi, and R. A. Kurbanova, *J. DAN Azerb. SSR., XYII, 10*: 37 (1987).

55. A. V. Ragimov, N. R. Bektasi, R. A. Kurbanova, and K. A. Aslanov, *Azerb. Xim. J., 1*: 84 (1988).

56. B. Pummer, *Ber., 69*: 1005 (1906).

57. E. Tsuchida, T. Tomana, and H. Sana, *Macromol. Chem., 151*: 245 (1972).

19

Performance of Polyethylenes in Relation to Their Molecular Structure

Ajit B. Mathur
Indian Petrochemicals Corporation Ltd., Baroda, India

Ishwar Singh Bhardwaj
R. S. Petrochemicals Ltd., Baroda, India

I INTRODUCTION

Polyethylene (PE) is the most versatile polymeric material having a major share of almost all areas of application of commodity plastics. The family of ethylene polymers is large and includes a range of homo- and copolymers. The technological advancement and use of a wide range of polar and nonpolar comonomers for ethylene-based polymers has made it possible to tailor the properties as per the specific end use applications. PEs can, therefore, be regarded more as speciality resins instead of commodity materials [1]. Their broad spectrum of properties, ranging from stiff to very soft and flexible (elastomers), can now be manufactured in a single reactor [2,3].

The type of manufacturing process, reaction conditions, and catalyst are the controlling factors for the molecular structure of the polymers [4–8]. The molecular features govern the melt processability and microstructure of the solids. The formation of the microstructure is also affected by the melt-processing conditions set for shaping the polymeric resin [9]. The ultimate properties are, thus, directly related to the microstructural features of the polymeric solid.

The primary molecular parameters affecting the processing and ultimate properties of PEs are type, content, and distribution of chain branching, molecular weight (MW), and molecular weight distribution (MWD).

The control of linearity and/or branching in PEs through the catalyst is very well known [4–8]; a highly branched PE is obtained by a free radical initiator, whereas a stereospecific catalyst (Ziegler–Natta catalyst) provides linear polymer. Use of nonpolar comonomers, i.e., alpha-olefins with ethylene, is made during polymerization to introduce chain branching in the polymer but in a controlled manner. The chain branching controls the crystallinity and, in turn, the density of PE. The density increases with the increase in crystallinity. Commercial grades of PEs are defined by their density, which generally ranges from 0.915–0.965 g/cc.

Being large molecules, the molecular parameters, i.e., MW and MWD, play a significant role in controlling the characteristics of the polymers. The polymerization reactions lead to synthesis of polymers with heterogeneity in MW, i.e., the polymer molecules differ in the number of repeating units. Hence, a polymer is a mixture of molecules of various MW and molecular size. The MW of the polymers is, therefore, expressed as average MW ($\overline{M}$). The heterogeneity of the MW in a polymer is termed as MWD or polydispersity.

The MW at values above a critical MW has the following relationship with melt viscosity (η):

$$\eta = \overline{M}^{3.4}$$

The melt viscosity, therefore, rapidly increases with the increase in MW. Whereas the MW below a critical value does not provide the required chain entanglement, and hence, the viscosity increase is slower in the low-MW range.

The different MW averages are derived by using the differential or frequency distribution curves and are reported as number average MW ($\overline{M}n$), weight average

MW ($\overline{M}$w), Z-average MW ($\overline{M}$z), Z + 1 average MW ($\overline{M}$z + 1), and so on. The measurement of the number of molecules, functional groups, or particles that are present in a given weight of sample allows the calculation of $\overline{M}$n. The $\overline{M}$n value is more dependent on the amount of the lower MW fraction available in the polymer. The experiments in which each molecule or chain makes a contribution to the measured result provide the value of $\overline{M}$w. The dependence of $\overline{M}$w is more on the high-MW fraction in the polymer. Higher MW averages, such as $\overline{M}$z and $\overline{M}$z + 1, are dependent on very high MW fractions.

The molecular structure and properties of polyolefins have been explained by several workers in the past [10–14]. This chapter deals with the primary molecular parameters and their effect on processability and ultimate properties of PEs. Since molecular parameters are closely interrelated, it is not possible to discuss one without referring to the other. Hence, in the section relating to the effect of chain branching, reference has also been made to MW and MWD and vice versa.

According to the end use application, PEs are processed by various techniques, which include injection moulding, blow moulding, rotomoulding, and film extrusion. However, since the bulk of the processed material is used as film in the area of packaging, the discussion in this chapter focuses mainly on processing behavior and the ultimate properties of tubular blown film.

II. CHAIN BRANCHING

According to the length, chain branching is defined as long chain branching (LCB) and short chain branching (SCB). The LCBs vary in length and can be as long as the main chain. They significantly affect the solution viscosity and melt rheology because of the molecular size reduction and entanglement [15]. The SCBs mainly control the morphology and solid state properties as they hinder the orderly arrangement of polymer molecules [16–18].

PE produced by a high-pressure polymerization process (pressure: 1000–3000 atm) using a free radical initiator is a highly branched material that contains both LCBs and SCBs. The polymer so produced is a low-density material (density up to about 0.925 g/cc) and is known as high-pressure low-density PE (HP LDPE). The LCBs are formed via intermolecular hydrogen transfer [19], whereas SCBs are formed by intramolecular hydrogen abstraction [16].

Using a low-pressure (pressure: 5–50 atm) polymerization process in the presence of Ziegler–Natta catalyst, the PE produced is generally linear with or without the trace of SCBs and has a high density (upto about 0.965 g/cc); the polymer, therefore, is known as high-density PE (HDPE).

Deliberate addition of alpha-olefin comonomer in an ethylene polymerization reactor leads to the formation of linear polymer with controlled SCBs. These PEs can be produced by almost the same process that is used for HDPE. They are linear and their densities are lower than the density range of HDPE. These ethylene-alpha-olefin copolymers are termed as linear low-density polyethylene (LLDPE) with densities around 0.92 g/cc.

To get an insight into the branched structure of PEs it is important to analyze them for their content and distribution in the macromolecules. The branching analysis of PEs have been reviewed by many workers [13,20,21].

A. Distribution of Chain Branching in HP LDPE

Various fractionation methods have been used to study the heterogeneity of LCBs and SCBs of PEs [13]. The average values of LCBs in the commercial grades of HP LDPE range from 0.07–12.9 per 1000 C atoms [13,20,21], and their concentration increases with the increase in MW [15]. The content of SCBs (mainly from 1–6 carbon atoms) ranges from 8.5–22.4 per 1000 C atoms in HP LDPE [22,23]. It is reported that paired and/or branched branches are also found in HP LDPE [23].

The SCB distribution (SCBD) has been extensively studied by fractionation based on compositional difference as well as molecular size. The analysis by cross fractionation, which involves stepwise separation of the molecules on the basis of composition and molecular size, has provided information of inter- and intramolecular SCBD in much detail. The temperature-rising elution fractionation (TREF) method, which separates polymer molecules according to their composition, has been used for HP LDPE; it has been found that SCB composition is more or less uniform [24,25]. It can be observed from the appearance of only one melt endotherm peak in the analysis by differential scanning calorimetry (DSC) (Fig. 1) [26]. Wild et al. [27] reported that HP LDPE prepared by tubular reactor exhibits broader SCBD than that prepared by an autoclave reactor. The SCBD can also be varied by changing the polymerization conditions. From the cross fractionation of commercial HP LDPE samples, it has been found that low-MW species generally have more SCBs [13,24].

B. Distribution of Chain Branching in LLDPE

In LLDPE, the type of alpha-olefin comonomer determines the length of the SCBs. While being incorporated into the polymer chains, two of the carbon atoms of the comonomer become part of the polymer backbone (Table 1).

Comonomers with four, six, and eight carbon atoms are commonly used for commercial LLDPE [28], whereas a comonomer having five carbon atoms has recently been used to achieve a typical balance of polymer properties [29].

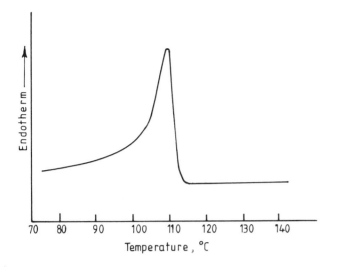

Figure 1 Typical DSC melting thermogram of HP LDPE.

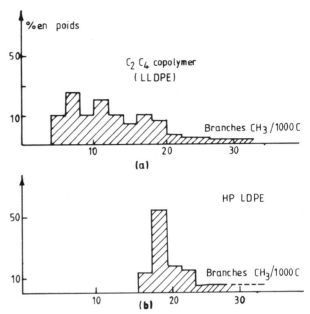

Figure 2 Short chain branching distribution in polyethylenes. *Source:* Ref. 31.

Contrary to HP LDPE, LLDPEs have heterogeneity in SCBD, i.e., in the latter case very wide SCBD exists [13,27,30–33]. The derivative SCBD in LLDPE and HP LDPE are represented in Fig. 2. This heterogeneity exists both at inter- and intramolecular levels [13,32] and is independent of comonomer content in LLDPE [32,34]. DSC traces have shown broad endotherm with more than one peak in the melting range due to heterogeneity in SCB [13,30,32,35]. Mathot [32] has explained the shape of DSC thermogram, i.e., the low-melt temperature peak due to a highly branched fraction and high-melt temperature peak due to a low-branched fraction (Fig. 3). Irrespective of the manufacturing process, commercially available LLDPEs have the same SCBD pattern [33].

The SCBD depends on the MW, i.e., low-MW fraction consists of a large number of SCBs as compared with high-MW fractions in LLDPE. The results of fractionation of two LLDPEs, differing in their type and content of comonomer but having the same density and

melt flow index (MFI), are shown in Fig. 4. These results indicate the dependence of SCBD on the MW of the polymer fractions [32]. Cross fractionation studies have shown the heterogeneity of SCBD both at inter- and intramolecular levels in LLDPE (Fig. 5).

The melting behavior of PE depends on the size, content, and distribution of SCBs as these molecular

Table 1 Comonomers Used for LLDPE and the Size of SCBs

Alpha–Olefin comonomer	Carbon content	Branch length
Propylene	C_3	C_1
1-Butene	C_4	C_2
1-Pentene	C_5	C_3
1-Hexene	C_6	C_4
4-Methylpentene-1	C_6	C_4
1-Octene	C_8	C_6

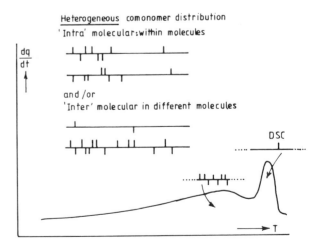

Figure 3 Two examples of heterogeneity of comonomer distribution and a possible explanation of the shape of the DSC curve. *Source:* Ref. 32.

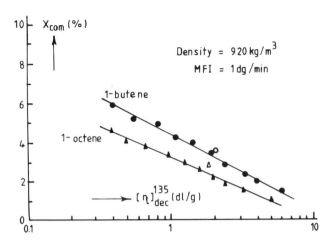

Figure 4 Mole percentage of comonomer and intrinsic viscosities of a 1-butene LLDPE (○) and an 1-octene LLDPE (△) and direct extraction fractions thereof (●) and (▲), respectively. *Source:* Ref. 32.

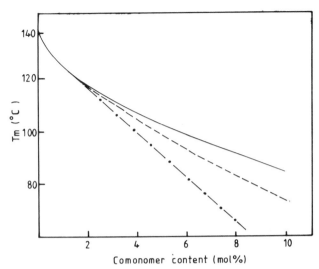

Figure 6 Dependence of Tm on comonomer contact for LLDPE containing: (———) 1-butene; (----) 1-octene; and (-·-) 1-octadecene. *Source:* Ref. 36.

parameters control the crystallization characteristics [13,32,34,36]. The homogeneous catalyst provides even distribution of SCBs in LLDPE as compared with those produced by a heterogenous catalyst. Therefore, in the former case, the polymer crystalline morphology is found to be more uniform [37]. Clas et al. [36] studied the melting behavior of LLDPEs synthesized by using comonomers, i.e., 1-butene, 1-octene, and 1-octade-cene, in the presence of $Et_3Al_2Cl_2/VOCl_2$ as the cata-lyst. The samples were more homogenous with respect to SCBD as compared with commercially available

LLDPEs. DSC analysis has shown that the melting point (Tm) of the samples is independent of branch length at lower concentrations, but at higher concentrations (5–9 mol% of comonomer), it decreases with the increase in branch length (Fig. 6). LLDPE containing 1-octadecene comonomer showed a broad DSC melting endotherm with no distinct peak for melting. Since the shape of the melt endotherm is indicative of the degree of crystallin-ity, it is clear that the sequences of ethylene and 1-octa-decene are greatly diluted by $C_{16}H_{33}$ branches, which prevent them from entering in somewhat thicker crystal-lites as are formed by copolymers of ethylene and 1-butene [36].

III. EFFECT OF MOLECULAR PARAMETERS ON RHEOLOGY AND PROCESSABILITY

PEs, as other polymers, exhibit nonlinear behavior in their viscous and elastic properties under practical pro-cessing conditions, i.e., at high-shear stresses. The MFI value is, therefore, of little importance in polymer pro-cessing as it is determined at a fixed low-shear rate and does not provide information on melt elasticity [38,39]. In order to understand the processing behavior of poly-mers, studies on melt viscosity are done in the high-shear rate range viz. 100–1000 s^{-1}. Additionally, it is important to measure the elastic property of a polymer under similar conditions to achieve consistent product quality in terms of residual stress and/or dimensional accuracy of the processed product.

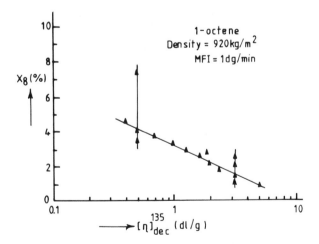

Figure 5 Cross fractionation of a 1-octene LLDPE by crystallization/dissolution treatment after direct extraction. *Source:* Ref. 32.

Several workers [38–45] have studied the effect of MW, MWD, and chain branching on the rheological properties of PEs. The melt viscosity at low-shear rate depends on $\overline{M}w$, but in the higher shear rate range the drop in the value (shear thinning) depends more on the molecular parameters, such as $\overline{M}n$ and MWD [44]. While comparing the rheological behavior of commercial HP LDPEs of similar MFI, it has been found [38] that shear thinning behavior is more of broad MWD sample. This is attributable to the easier disentanglement of macromolecules of broad MWD polymer under a shear motion as compared with narrow MWD polymer.

The viscoelastic properties of HP LDPE melt undergo shear modification due to LCBs [46–50]. This is purely a physical change in the interaction of polymer chains to one another. Therefore, the MW, MWD, and frequency of LCBs remain unaltered. The shear work on the melt is said to cause disentanglement of the polymer network, thus changing the density of entanglement couplings. An explanation for this phenomenon is that two types of entanglements [51] exist for HP LDPE: one presumably occurs between the LCBs, another is assumed to arise from looping and coiling of the main chains at the branching points. The entanglement and disentanglement is relatively easy in the former case while during shearing action [52], but the latter does not easily entangle or disentangle because of steric hindrance at the branching points. The attainable degree of modification depends to some extent on the MWD and the degree of LCB in the polymer molecules. Comparison of two different grades of HP LDPE showed that a higher degree of modification can be obtained at a lower energy input level for more highly branched grades [49].

Shear modification of HP LDPE has been found to be a completely reversible action (memory effect) [46–50]. This can be achieved by thermal treatment or solution precipitations [46,53]. The time required for complete recovery increases with the increase in $\overline{M}w$, which implies that the material with high $\overline{M}w$ forms the entanglement with more complicated steric hinderance. Figure 7 shows the effect of shear history on the swelling ratio of an HP LDPE sample and the recovery after solvent treatment. The disentangled network of macromolecules exists in the nonequilibrium transient rheological state of material. As a result of this, characteristics of the polymer melt, i.e., viscous and, especially, the elastic behavior, are altered. Thus, the change in properties, such as an increase in MFI, decrease in die swell, and reduction in surface roughness of the extrudates, takes place [46,49,50]. An increase in gloss and a decrease in haze of HP LDPE film are associated with the decrease in elasticity of the melt, which results in the reduction of the surface roughness [48,54,55]

While comparing the shear thinning behavior of HP LDPE and LLDPE having identical MFI (2.0 g/10 min), the value of zero shear viscosity (η_0) of the former has been found higher than the latter despite the lower $\overline{M}n$

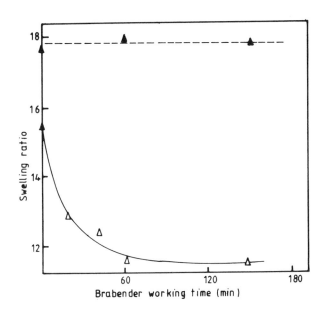

Figure 7 Effect of shearing history on swelling ratio for HP LDPE sample: ($\triangle$) Brabender worked at 190°C; ($\blacktriangle$) solvent treated sample. *Source:* Ref. 39.

and $\overline{M}w$ (Fig. 8) [39]. This is attributed to the greater amount of entanglement caused by the presence of LCB in HP LDPE [39,56]. In other words, the amount of energy required for disentangling the large molecules of HP LDPE, when they are virtually at the state of rest, is greater than that for LLDPE that is devoid of LCBs. Once the macromolecules are sufficiently disentangled under shearing motion, $\overline{M}n$ becomes the controlling factor in determining the amount of energy required for shearing the macromolecules. Hence, the rate of drop in shear viscosity of LLDPE is slower in the high-shear rate range than HP LDPE. Similar observations have been made during extrusion, and it is found that power consumption, pressure before the die, and temperature of the melt are characteristically higher for the LLDPE resins in comparison to HP LDPE [56].

Kalyan et al. [56] have also studied the effect of alpha-olefin comonomers on the rheological properties and processing of LLDPE. The characteristics of the resins are shown in Table 2. It is found that 1-octene-based LLDPE has the lowest shear viscosity as compared to 1-butene- and 1-hexene-based polymers (Fig. 9). Decrease in power consumption, pressure before the die, temperature in the die, and increase in output has also been found according to shear viscosities of the polymers during tubular film extrusion.

Figure 10 shows that upon cessation of shear flow of the melt, shear stress relaxation of LLDPE is much faster than HP LDPE because of the faster reentangle-

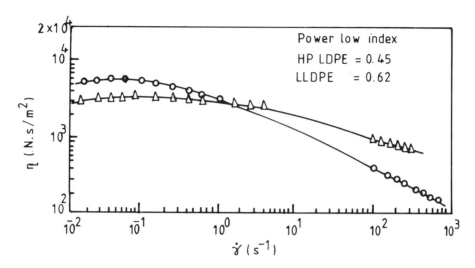

Figure 8 η vs. γ at 220°C for: (○) HP LDPE ($\overline{M}n - 1.39 \times 10^4$, $\overline{M}w - 1.12 \times 10^5$); (△) LLDPE ($\overline{M}n - 5.60 \times 10^4$, $\overline{M}w - 2.21 \times 10^5$).

Table 2 Properties of the PE Resins and Their Blow Film Samples (Blow up Ratio-2) [56]

| | | Resin | | | |
| | | LLDPE | | | | HP LDPE |
Property		A	B	C	D	E
Comonomer		1-Butene	1-Hexene	1-Hexene	1-Octene	—
Number of methyl groups/1000 C atoms		18–19	7–8	12–14	12–14	—
$\overline{M}w$		94336	95341	77280	78100	70222
Mn		25614	30243	22487	26184	19786
MWD		3.68	3.15	3.44	2.98	3.54
Melt index (g/10 min)		1.0	0.8	1.0	1.0	1.4
Density (g/cc)		0.918	0.926	0.919	0.920	0.923
		Film				
Tensile strength at yield, Mpa[a]	MD/TD	11.7/13.1	16.6/22.1	12.4/15.2	12.4/12.4	13.1/14.5
Tensile strength at break, Mpa[a]	MD/TD	40.7/29	46.2/40.7	51/35.2	49/40	35.2/18.6
Elongation at break, %[a]	MD/TD	810/1380	715/940	626/985	680/1410	230/835
Sceant modulus, Mpa[b]	MD/TD	238/339	425/590	332/467	253/332	290/394
Puncture resistance, in-lb		28.3	32	30	31	11
Tear resistance, g/mil	MD/TD	110/518	127/914	90/1005	223/1039	410/222
Impact strength, grf.		87	88	91	143	48

MD = machine direction; TD = transverse direction.
[a] Cross head speed 50 cm/min
[b] Cross head speed 2.5 cm/min

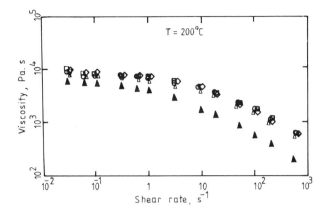

Figure 9 Shear viscosity as a function of shear rate at the wall at 200°C, (○) Resin A, (□) Resin B, (◇) Resin C, (△) Resin D, (▲) Resin E. (Refer to Table 2 for symbols code.) *Source:* Ref. 56.

ment of the molecules in the former due to the absence of LCBs [56]. Rate of shear stress relaxation of HP LDPE is slower due to the presence of LCBs, which hinders the reentanglement of macromolecules [57]. Among the LLDPEs, 1-octene-based polymer, having a relatively longer chain branching, shows the lowest relaxation but it is higher than HP LDPE. This relatively higher shear stress relaxation of LLDPE than HP LDPE resin results in lowest orientation, more isotropy, and better weld integrity in the processed product of the former.

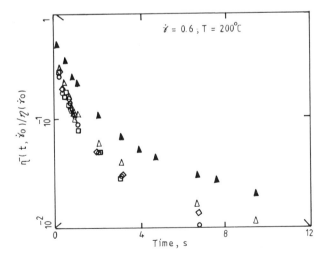

Figure 10 Relaxation of shear stress with time upon cessation of steady flow. (○) Resin A, (□) Resin B, (◇) Resin C, (△) Resin D, (▲) Resin E. (Refer to Table 2 for symbol code.) *Source:* Ref. 56.

The study on elongational viscosity (η_E) provides useful information on elasticity of the polymer melt. The melt elasticity of PEs fall in the typical order, i.e., HP LDPE > HDPE > LLDPE. η_E of HP LDPE being higher than LLDPE causes poor bubble stability but more downgauging in tubular blown film extrusion of the latter than the former. Han and coworkers [38,44] studied the elongational viscosity (η_E) of commercial HP LDPE samples. It has been found that η_E increases with the increase in elongational rate and that a steady-state approaches only at low-elongational rates when studied with respect to time. η_E is reported to increase with the broadening of MWD, which manifests the increase in strain hardening due to tensile stress while in film extrusion.

The major processing difficulties in the tubular film blowing operation, i.e., bubble instability and breakage of the tubular blown bubble can be overcome with a better understanding of molecular parameters and their effect on rheological properties of polymer. The bubble instability results in nonuniform thickness of the film, and the bubble breakage does not allow higher take-up speed and limits downgauging of the film.

The improvement in the take-up ratio with respect to bubble rupture of HP LDPE samples is found with the narrowing of MWD and lowering the degree of LCB since these molecular parameters help provide a balance to shear thinning and melt elasticity (reduction in strain hardening) during uniaxial stretching thus improving the blowability of the polymer [38,44]. Kwack et al. [39] in their study compared melt rheology and tubular blown film extrusion of three commercial grades of LLDPE with HP LDPE samples. The η_E of LLDPE samples showed a decreasing trend with the increase in elongation rate and the polymer melt exhibited strain softening behavior instead of strain hardening as was found in HP LDPE samples. Similar observations have been made by others [56]. It is reported that tensile stress on the tubular film bubble both in machine and transverse directions (S_{11F} and S_{33F}) increases with blow-up and take-up ratios for LLDPE but is relatively much slower compared with that of HP LDPE [39]. Figure 11 shows the change in ratio of S_{11F}/S_{33F} with respect to an increase in blow-up ratio of HP LDPE and LLDPE samples indicating superior blowability of the latter compared with the former.

The strain-softening behavior and slow rate of increase in S_{11F} and S_{33F} of LLDPE during tubular film extrusion facilitates extensive drawdown "downgauging" of the film [56]. On the contrary, bubble rupture occurs in the case of HP LDPE due to strain hardening and increase in melt tension during drawdown. Although an increase in the processing temperature brings down the elasticity and the tension of the melt in HP LDPE resin and, thereby, increases the possibility of downgauging. This can lead to instability of the bubble due to very low η_E.

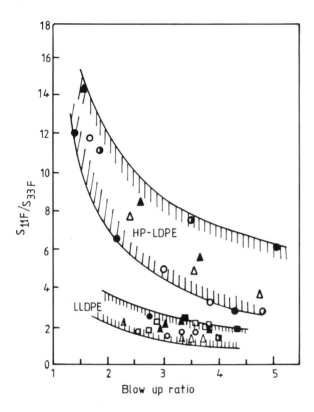

Figure 11 S_{11F}/S_{33F} ratio vs. blow up ratio for different HP LDPE and LLDPE samples. *Source:* Ref. 39.

The uniaxial η_E increases with the increase in length of SCB in LLDPE, which is implied to the increase in stability of the bubble in film processing [56,58].

IV. FORMATION OF MICROSTRUCTURE DURING SOLIDIFICATION

While cooling the melt to solidify in a processing operation, alignment of molecules controlling the crystallinity and overall morphology depends on the molecular structural parameters of PEs.

The formation of the microstructure involves the folding of linear segments of polymer chains in an orderly manner to form a crystalline lamellae, which tends to organize into a spherulite structure. The SCB hinder the formation of spherulite. However, the volume of spherulite/axialites increases if the branched segments participate in their formation [59]. Heterogeneity due to MW and SCB leads to segregation of PE molecules on solidification [59–65]. The low MW species are accumulated in the peripheral parts of the spherulite/axialites [63]. The low-MW segregated material is brittle due to a low concentration of interlamellar tie chains [65] and

fracture thus propagates preferentially through these regions [62,65]. The increase in MW and control of SCB and their distribution thus help to improve properties such as impact strength, environmental stress crack resistance (ESCR), low temperature brittleness, puncture resistance (due to the increase in concentration of tie molecules), and regulation of lamellar thickness and its distribution. The increase in length of SCB also decreases the crystallization of PEs, and thus decreases the stiffness and increases the impact and tear resistance. The higher alpha-olefin comonomers such as 1-octene give larger side branches, which disrupt the lamellar growth of crystallites [31]. More crystallites are thus joined by the molecules. These molecules in the amorphous phase, between crystallites, are responsible for higher levels of mechanical properties. This fits the observation that the proportion of higher alpha-olefins in the amorphous phase is higher than with 1-butene. Also, that 1-butene-based LLDPE needs more side chains to achieve a given density [31].

V. EFFECT OF CHAIN BRANCHING ON SOLID STATE PROPERTIES

The length and degree of chain branching significantly affect the solid state properties of PEs. In the case of LLDPE, the type and content of alpha-olefin comonomer determine the polymer properties. Several workers [30,38,66–72] have studied the structure properties correlation and reported the role of chain branching in the formation of the microstructure of PEs. A marked difference in the properties of LLDPE and HP LDPE is due to the absence of LCBs in the former, whereas the properties of the latter are governed by the degree of LCBs. The difference in the alignment of macromolecules during shearing and extensional flows in tubular film processing depends on length of SCBs in LLDPE [56]. This markedly affects the crystallization as well as the orientation of amorphous and crystalline structural units upon solidification [30,73,74]

In their study on LLDPE resins containing 1-butene, 1-hexane, and 1-octene comonomers, Kalyon and Moy [30] found a significant variation in their film thickness when measured around the circumference of tubular bubbles processed under identical conditions. The samples blown with a blow-up ratio of two, exhibited more significant variation in thickness than those prepared with a blow-up ratio of three. However, film processed at a higher blow-up ratio has been found to have less variation in thickness.

Circumferential variation in the orientation of the film bubble is not found to have any linear relationship with the content and length of SCBs. The resin containing a lower content of comonomer (SCBs) showed the highest value of orientation due to a higher degree of crystallinity [30].

The density of LLDPE resin is affected by the concentration of SCBs, whereby an increase in the degree of chain branching leads to a decrease in crystallinity and density [30]. Thus, by varying the amount of comonomer, it is possible to produce LLDPE with densities in the range of 0.900–0.940 g/cc. Use of very high amounts of comonomers can lead to the reduction of density even below 0.900 g/cc. The density as low as 0.86 g/cc is reported for polymers having C_2 units in the main chain as 76 wt%, indicating a very high amount of SCBs [34]. Therefore, the linear PEs in this range of density have been termed as ultra low-density PE (ULDPE). ULDPEs are characterized by extremely low crystallinity and low melting temperature; Table 3 provides a comparison with HDPE. The low crystallinity and, in turn, the density are due to the interruptive effect of the alpha-olefin units during crystallization [34]. While comparing 1-butene (C_4)-based polymer with 4-methylpentene-1 (C_6)-based polymer, it was found that the latter has lower crystallinity with correspondingly lower Tm as compared to the former. ULDPEs are reported to have higher impact strength and better heat sealability at low temperatures due to very low crystallinity and low melting temperature, respectively.

Ultimate properties of LLDPE are superior to HP LDPE resins [75,76]. The tubular blown film of LLDPE has higher tensile strength, elongation, and outstanding puncture resistance as compared to HP LDPE film. Mechanical properties data of LLDPE and HP LDPE blown film samples prepared under identical processing conditions are shown in Table 2 [30]. The higher values of tensile strength at break and elongation at break of LLDPE samples compared with HP LDPE also indicate superior toughness of the former. This is in agreement with higher puncture and impact resistance of LLDPE film than HP LDPE. The high value of tear resistance in machine direction and lower value in transverse direction of HP LDPE as compared to LLDPE, which has a relatively higher value of tear resistance in the transverse direction, indicates a better balance of film properties of the latter. Lowering of comonomer content leads to higher tensile strength at yield and secant modulus and lower dart impact of the LLDPE. A comparison can be made from the test data shown in Table 2 of the polymers containing different contents of 1-hexene comonomer. 1-Butene-based LLDPE exhibits a higher solid drawability, i.e., greater elongation at break in comparison to 1-hexene- and 1-octene-based LLDPE samples. This together with the observed greater drawability of the LLDPE resins as compared with HP LDPE is due to shorter branch length at equal crystallinity [30,77]. Under comparable processing conditions of tubular blown film extrusion, a more uniform tensile strength in machine and transverse direction is achievable with LLDPE resins as compared with HP LDPE resins [39].

The longer SCBs provide a better overall balance of LLDPEs toughness properties than smaller ones. Most of the commercially available LLDPE grades are based on 1-butene, 1-hexene, and 1-octane with 1-butene-based having the major market share. Their order of toughness with respect to branch length is found to be $C_6 > C_4 > C_2$. The toughness of LLDPE based on propylene comonomer is inferior to conventional HP LDPE. The rate of improvement in polymer properties begins to diminish at a point somewhere between C_6 and C_8 on the comonomer scale. A decline in ultimate properties appears to occur with 1-decane (C_{10}) comonomer [28]. The effect of comonomer type and increase of comonomer content (decrease in density) on dart impact strength of LLDPE film is shown in Fig. 12.

To improve the end use performance and make the processability easy, control of MW and MWD as well as the use of more than one comonomer has been reported for LLDPE [28]. Union Carbide's high MW-LLDPE with broad MWD is a 1-hexene-based resin, and its film provides superior (about 30–50% higher) tensile strength, puncture resistance, and dart impact strength than conventional 1-hexene-based resin, but with lower tear resistance in the transverse direction. The broad MWD makes the resin processability easy on the conventional extruder.

Use of 4-methylpentene-1 comonomer with ethylene provides LLDPE resin have film properties (i.e., tensile strength, modulus, transverse direction tear strength, and impact strength) superior to 1-butene-based LLDPE resin as has been claimed by B.P. Chemicals. 1-Butene has also been used as the second comonomer with 4-methylpentene-1 to tailor the properties of LLDPE resin [28]. The properties of 4-methylpentene-1-based LLDPE film are given in Table 4.

The optical properties of tubular blown film depends greatly on the surface irregularities and the size of crystallites domain in film, which, in turn, are dependent on

Table 3 Characteristics of ULDPE and HDPE [34]

Polymer	Comonomer	Tm, °C	Density, g/cc
ULD-PE	1-Butene	75.2–84.2	0.887–0.895
ULD-PE	4-Methylpentene-1	40.0–40.8	0.865–0.867
HDPE	—	129	0.956

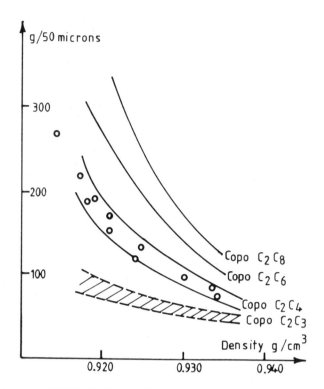

Figure 12 Dart impact (ASTM D1709) of different copolymers (at constant MW). *Source:* Ref. 31.

Table 4 Comparison of Film Properties of LLDPE, Based on 1-Butene and 4-Methylpentene-1 Comonomer [28]

Resin property[a]	1-Butene copolymer	1-Butene/4-methylpentene-1 terpolymer	4-Methylpentene-1 copolymer
Tensile at break, MPa			
M D[b]	24.7	33.5	42.1
T D[b]	18.7	25.8	31.7
Tensile at yield, MPa			
M D[b]	7.9	10.8	12.7
T D[b]	8.8	11.7	13.4
Elongation, %			
M D[b]	460	460	510
T D[b]	620	600	680
Modulus, MPa			
M D[b]	153.8	205.5	272.2
T D[b]	205.5	233.7	353.7
Elmendorf tear, G/mil			
M D[b]	200	240	250
T D[b]	470	540	720
Dart drop			
F-50 at 660 mm	140	161	180

[a] Extrusion conditions = 1.25 mil film blown at 2 and 1 blow up ratio using 63.5 mm, 20D smooth bore extruder, die gap 80 mil, melt temperature 193°C.
[b] MD = Machine direction; TD = transverse direction.
Source: U.S. Industrial Chemicals Co.

the processing conditions [78]. The LLDPE film exhibits higher value of haze and lower value of gloss as compared with HP LDPE sample [30,39,79,80]. This is suggested to be due to higher scattering of light from the surface of LLDPE film being more rough than HP LDPE material [80].

In another study [31] it has been reported that haze of LLDPE is mainly a consequence of light scattered by the spherulites. Thus, origin of poor clarity is not the same in LLDPE as in HP LDPE. In HP LDPE the haze is produced by surface defects coming from processing rheology [48,54,55]. Processing variables, therefore, do not have the same effect on LLDPE as they do on HP LDPE.

As described in an earlier section, the heterogenous catalysts (Ziegler–Natta catalyst) used for the manufacture of LLDPEs provide nonuniform distribution of comonomers at intermolecular and intramolecular levels in the polymer chains. This structural distribution correspondingly reflects in the distribution on crystalline levels such as lamella thickness distribution. The wide lamella thickness distribution is the characteristic of LLDPEs that strongly influences its basic mechanical properties. Contrary to the LLDPE produced by the Ziegler–Natta catalyst, polymer prepared with a homogenous catalyst system (metallocene catalyst) has characteristic narrow structural distribution, narrower MWD, and very low density [8,37]. Due to lower density than conventional LLDPEs, these polymers are termed as very low-density PEs (VLDPEs) [8]. As a consequence of narrow structural distribution, the polymer attains narrow lamella thickness distribution and morphology, resulting in superior properties (e.g., impact strength, transparency, etc.) than the polymer produced by the Ziegler–Natta catalyst. A comparison of properties of resins and their extruded film samples are given in Table 5.

B.P. Chemicals [81] have also reported the new metallocene-based LLDPE (with 1-hexene comonomer) having much superior properties than those produced by using the Ziegler–Natta Catalyst.

VI. EFFECT OF MOLECULAR WEIGHT AND MOLECULAR WEIGHT DISTRIBUTION

MW and MWD are very significant parameters in determining the end use performance of polymers. However, difficulty arises in ascertaining the structural properties relationship, especially for the crystalline polymers, due to the interdependent variables, i.e., crystallinity, orientation, crystal structure, processing conditions, etc., which are influenced by MW and MWD of the material. The presence of chain branches and their distribution in PE cause further complications in establishing this correlation.

The different values of MW averages of the polymer fall in an order, i.e., $\overline{M}z + 1 > \overline{M}z > \overline{M}w > \overline{M}n$; their approximate locations in a distribution curve are shown in Fig. 13. If all the polymer molecules in a sample were of the same molecular size, then all the MW averages would be identical. However, this is not the case with commercial polymers.

The MWD is the ratio of $\overline{M}w : \overline{M}n$ and is called the polydispersity index. This largely varies from one grade of polymer to the other, depending on the polymerization conditions and the type of catalyst used. Figure 14 shows different types of MWD for the polymers.

Table 5 Properties of VLDPEs Prepared with Homogeneous and Heterogeneous Catalyst System [8]

Comonomer catalyst	1-Butene homogeneous	1-Hexene homogeneous	1-Butene heterogeneous
Resin properties			
MFR (g/10 min)	1.7	1.8	1.4
Density (g/cc)	0.909	0.908	0.903
Tensile impact strength (Kg-cm/cm)	1132	2930	780
Blown film properties[a]			
Haze (%)	3.7	3.8	7.0
Gloss (%)	130	125	105
Dart impact strength (kg cm/mm)	2920	>4000	400
Tear strength (kg/cm) MD/TD	46/60	83/120	36/74
Ultimate strength (kg/cm²) MD/TD	550/410	580/560	370/290

[a] Melt temp = 170°C; Die = 125 ϕ–2.0t; Blow up ratio = 1.8; Film thickness = 30 μ Output = 20–25 kg/h.

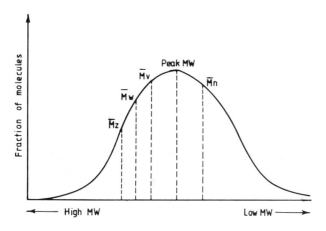

Figure 13 Approximate location of MW averages on a MWD curve.

Being polydisperse, characterization of low- and high-MW fractions has been very useful for obtaining a better understanding of their role in polymer performance. The quantitative estimation of the macromolecules based on their MW is, therefore, carried out by using different types of fractionation methods such as addition of nonsolvent to a polymer solution, cooling a solution of polymer, solvent evaporation, zone melting, extraction, diffusion, or centrifugation. Gel permeation

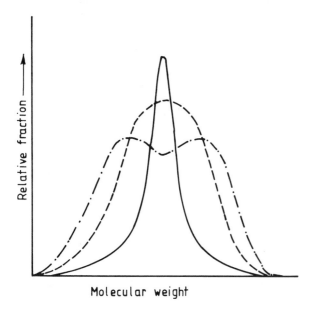

Figure 14 Representative differential weight distribution curve; (———) relatively narrow distribution curve; (---) relatively broad distribution curve; (-··-) bimodal.

chromatography involves separation of molecules based on hydrodynamic size by sieving action.

A. Molecular Weight Effect

The effect of MW and MWD on the solid state properties have been extensively studied [11,12,82]. These studies have been made both on fractionated and whole polymer samples. Attempts have also been made to correlate the solution viscosity, melt viscosity, MFI and other related parameters, which represent the MW and MWD of the polymers, with the solid state properties. Table 6 summarizes the results of various studies on effect of MW and MWD on the properties of PEs.

The different MW averages are found to provide better correlation with polymer properties viz., $\overline{M}n$ on brittleness, impact strength and flow properties; $\overline{M}w$ on tensile strength and hardness; and $\overline{M}z$ on flex life and stiffness [97,109–112]. Most of the studies have been limited to evaluate the effect of $\overline{M}w$ and $\overline{M}n$ on the properties of polymers. In a recent study [113], the role of higher MW averages, i.e., $\overline{M}z$ and $\overline{M}z + 1$ is reported in controlling the drop impact properties of moulded HDPE products. Three commercial moulding grades of HDPE used in this study had their densities, melt indexes, Tms, tensile strengths, and flexural moduli that were similar in magnitude. The $\overline{M}w$, $\overline{M}n$, and MWD values were also close to each other, but the performance in drop impact test was dissimilar. It was found [113] that the sample having the highest values of $\overline{M}z$ and $\overline{M}z + 1$ (i.e., long high MW tail) did not fail the drop impact test. This is in agreement with other studies [105,114] reporting that in crystalline polymers, brittle fracture decreases because of the increase in interlamaller "tie" molecules connecting hard crystalline and soft amorphous phases with the increase in MW.

The study on commercial HDPE samples could not provide a correlation of the izod impact test with the field performance test, i.e., drop impact resistance on moulded products [113]. It was found that the sample of highest density and lowest izod impact strength passed the drop impact test, but other samples of lower density and higher izod impact strength could not withstand shock loading by drop impact and failed in brittle manner. This may be due to the fact that velocities and modes of loading vary widely in different impact tests. It has been reported that even the qualitative agreement between the different impact tests is poor because the test bars and moulded products often have different orientation characteristics, particularly near the surface [115].

A typical balance of processability and end use performance is the general requirement of polymeric resins. The studies on the different polymer fractions have provided a great support in tailoring the MW and MWD in order to achieve the required properties and eliminating the unwanted molecular species. The increase in low-

Table 6 Effect of MW and MWD on the Solid State Properties of PE

Property	When MW increases [Ref.]	When MWD narrows down [Ref.]
Spherulite size	(−) [83]	
Density	(+)[a,b] [84–86]	
Resistance to indentation/scratch resistance	No effect up to MW >10^4 [87]	
Hardness		(+) [88]
Tensile strength[c,d]	(+) [12,87,89,90]	(+)[e] [91–93]
Elongation[c]	(+) [12,87,89,90]	(+) [91,92]
Young's modulus	No effect [89]	(+) [94,95]
Izod impact strength	(+) [84,86,93]	(+) [84,86,93,96]
Charpy impact strength[f]	(+) [97]	
Falling dart impact resistance[g]	(+) [96]	
Resistance to low temperature brittleness	(+) [12,91,96,98]	(+) [12,91,96,98]
Abrasion resistance	(−) [87]	
Resistance to creep	(+) [99–101]	
Folding endurance of film	(+) [84,102]	No significant effect [84]
ESCR	(+)[h] [12,84,97,102,103]	(+)[i,j] [104–106]
Softening point	(+) [89,102]	

+ = increase; − = decrease.

[a] Measurement made on fractionated samples of HDPE [84].
[b] Crystallinity decreases by increase in concentration of chain ends in low MW samples [85,86].
[c] Change in morphology and/or residual stresses due to MW and MWD may adversely effect the property (as reported for HDPE [87]).
[d] Results reported are of both fractionated and unfractionated samples.
[e] Broadening of MWD at constant $\overline{M}n$ improves the tensile strength [92].
[f] Increases with the increase in $\overline{M}n$ of LDPE [97].
[g] Determined for HDPE [96].
[h] Less effective in case of HDPE [99,107].
[i] Studied after removal of low MW fraction [104–106].
[j] Narrowing of MWD reduces the amount of low MW reject material while spherulite formation thus refines the crystalline structure and lowers the void content, as a result ESCR improves [108].

MW fraction increases the melt flow, thus improving the processability but at the cost of toughness, stiffness, and stress crack resistance. In addition, the improvement in performance through narrowing the MWD is restricted by the catalyst, the process hardware, and the process control limitations. Dow has developed a reactor grade HDPE of optimized breadth, peak, and shape of MWD by controlling the chain length frequencies for thin-walled moulded products with excellent balance of processability and performance [116]. Wherein, the polymers of 40 and 60 MFI can be processed as 65 and 95 MFI materials, respectively, requiring lesser energy because of lower processing temperature and pressure and can fill the large mould cavity with ease. This could be achieved due to greater number of short chains. The increase in the number of long chains and thereby, high-MW fraction could provide high toughness to the moulded product, and reduction in peak MW helped in balancing the properties (Fig. 15).

B. Molecular Weight Distribution Effect

It is evident from the foregoing discussion that MW is the fundamental characteristic of polymer, controlling the performance properties. However, simple correlation of this molecular parameter can be misleading without taking the MWD into consideration. Control of MWD provides a proper balance of polymer performance characteristics. The effect of change in MWD on the properties of PEs is given in Table 6.

It is well understood that low MW species lubricate the polymer melt and thus improve the flow and avoid frozen-in-orientation by fast molecular relaxation in the moulded product, whereas higher MW fractions control the mechanical properties in the solid [117,118]. Despite superior performance characteristics, film of high MW HDPE, therefore, requires more energy input in processing as compared with medium-MW HDPE. The high melt temperature built-up during processing of high-MW HDPE can also lead to polymer degradation. The high-MW fraction also causes more anisotropy, which results in an imbalance of tear resistance in machine and transverse direction of film. The narrowing of MWD im-

proves the performance characteristics of the polymer in solid state, but at the same time makes the melt less sensitive to shear rate and therefore the processing of high-MW polymer becomes more difficult. Although some of the processing limitations of high MW can be overcome by broadening the MWD, it is at the cost of some mechanical properties. Broadening of the MWD strongly decreases the melt viscosity at high shear rates [119,120], which in turn reduces the processing defects, such as melt fracture [121,122], and increases the melt strength [123–126].

The advantages of broad MWD polymer, both unimodal and bimodal are well known [127–129], but most of the commercially available grades of polymers are of unimodel MWD. The limitations of high-MW unimodal polymers can be overcome by achieving bimodal MWD, providing a better balance of end use performance and processability. A reactor blend of both low- and high-MW fractions during polymerization provides a polymer of bimodal MWD [130]. To get a product of very broad MWD with two distinct maxima for a 50:50 (by weight) blend, there must be a certain distance between the average MW of the low- and high-MW fractions. It has been reported that a homogenous polymer blend with broad bimodal MWD is impossible to achieve by mechanical mixing of two resins with an average MW ratio higher than 10 [131]. A typical properties comparison of the unimodal and bimodal high-MW HDPE resins and their films are given in Table 7. It can be well understood that

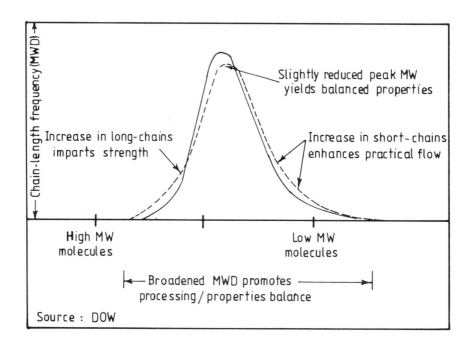

Figure 15 Typical gel permeation chromatograms of polyethylenes (HDPEs). (———) conventional high-flow resin; (----) new generation super flow resin. *Source:* Ref. 116.

Table 7 Properties of High MW-HDPE [132]

Property	Resin			
	Unimodal MWD		Bimodal MWD	
Flow index (I_{21})	9		8	
MFR (I_{22}/I_5)	23		25	
Density (kg/m³)	948		948	
	Film[a]			
Thickness (μ)	12	25	12	25
Dart impact (G)	190	310	420	570
Elmendorf tear strength (N/mm)				
MD	4.6	6.9	13.9	12.0
TD	20.1	44.8	32.0	21.2

[a] Film extruded on 50 mm ALPINE, 4:1 Blow up ratio, 80 mm die.

one can afford to increase the MW of high-MW fraction even more in bimodal MWD polymer than unimodal MWD polymer and thus further improve the end use performance without sacrificing the processability, which is taken care of by low-MW fraction. The major advantage to the processing industry is the downgauging by use of bimodal MWD-HDPE resin; film thickness could be reduced from 20–25 μm to 12–15 μm without loss or improvement in dart impact strength (Fig. 16) [132]. The tear strength of the film has also been found to be less effected by orientation.

Böhm et al. [130] have reported the incorporation of comomomer (1-butene, 1-hexene) in the high-MW fraction of bimodal PE during polymerization, obtaining a resin blend of low-MW homopolymer and high-MW copolymer. It was found that the long chain density in-

creased in comparison to a unimodal MWD polymer containing these comonomers, but it decreases linearly with the increase in comonomer content. The stress crack resistance (failure time) increased by approximately two orders of magnitude as compared with unimodal MWD homo- and copolymers due the SCB incorporated in polymer by alpha-olefin comonomer (Fig. 17). Yeh et al. [133] in their study reported a sharp increase in ESCR of linear PE (LLDPE) with the increase in

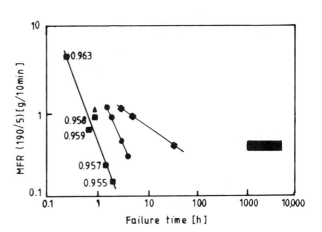

Figure 17 Correlation of failure time and melt flow rate MFR 190/5 of sharp notched bars under stress for unimodal homopolymer and copolymers, and bimodal copolymers. *Source:* Ref. 130.

Unimodal resin	Density (g/cc)
■ Homopolymer	see numbers
▲ C₃-copolymer	
● C₄-copolymer	0.940 ± 0.002
◆ C₆-copolymer	
▬ Bimodal resin	0.948 ± 0.002

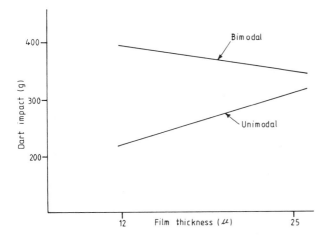

Figure 16 Dart impact strength of HMW-HDPE films. *Source:* Ref. 132.

length of SCBs. It is understood that the concentration of the tie molecules with the increase in $\overline{M}w$ improve the ESCR [134–136]. The chain diffusion within the crystalline lamellae occurs [137]. Under stress, crack propagation inside the semi-crystalline material is hindered by fibrillation (crazing) at the crack tip. Tie molecules stabilize the fibrils provided they are fixed in the crystalline regions. To reduce rapid lateral diffusion of the molecules through the crystallites, small number of SCBs must be present along the main chain. A blend of low-MW PE and high-MW ethylene-alpha-olefin copolymer, therefore, provides excellent stress crack resistance. The bimodal MWD PE have thus been found to have superior properties such as stiffness, impact strength, and long-term stress crack resistance [130] as compared with unimodal MWD PE.

ACKNOWLEDGMENTS

We wish to thank Dr. Shashikant and Dr. U. Dayal for helpful suggestions in preparing the manuscript. For the task of typing the manuscript we would like to thank Mr. A. D. Kharadi. The author, A. B. M. thanks Dr. R. C. Jain for his support and Dr. M. Ravindranathan for giving permission to publish this chapter. In addition, permission was generously granted by the copyright holders.

REFERENCES

1. *Mod. Plast. Int.*, p. 52, Sept. 1988.
2. *Mod. Plast. Int.*, p. 30, May 1986.
3. *Mod. Plast. Int.*, p. 4, Oct. 1988.
4. J. Boor, *Ziegler Natta Catalyst and Polymerizations*, Academic Press, New York, Chap. 1 (1979).
5. V. Zucchini, and G. Cecchin, *Adv. Polym. Sci., 51*: 101 (1983).
6. R. Spitz, *Recent Advances in Mechanistic and Synthetic Aspects of Polymerization*, Reidel Publishing Co., Japan, p. 485 (1987).
7. C. S. Speed, B. C. Trudell, A. K. Mehta, and F. C. Stehling, *Polyolefins, VII*: 45 (1991).
8. S. Hosoda, A. Uemura, Y. Shigematsu, I. Yamamoto, and K. Kojima, *Studies in Surface Science and Catalyst' 89, Catalyst Design for Tailor-Made Polyolefins*, (K. Soga, and M. Teruno, eds.), Elsevier Science BV., The Netherlands, p. 365 (1994).
9. R. M. Patel, T. I. Butler, K. L. Walton, and G. W. Knight, *Polym. Eng. Sci., 34(19)*: 1506 (1994).
10. N. Bikales (ed.), *Mechanical Properties of Polymer*, Wiley Interscience, New York (1971).
11. J. R. Martin, J. F. Johnson, and A. R. Cooper, *J. Macromol Sci.-Revs. Macromol. Chem.*, (G. B. Butler, K. F. O'Driscoll, and M. Shen, eds.) Marcel Dekker, New York, p. 57 (1972).
12. H. V. Boening, *Polyolefins: Structure and Properties*, Elsevier, Amsterdam and New York (1966).
13. T. Usami, *Handbook of Polymer Science and Technology*, Vol. 2, (N. P. Cheremisinoff, ed.), Marcel Dekker, Inc., New York, p. 437 (1989).
14. J. M. Dealy, and K. F. Wissbrun, *Melt Rheology and its Role in Plastics Processing*, Van Nostrand Reinhold, New York (1990).
15. F. W. Billmeyer, Jr., *J. Am. Chem. Soc., 75*: 6118 (1953).
16. M. J. Roedal, *J. Am. Chem. Soc., 75*: 6110 (1953).
17. M. J. Richardson, P. J. Flory, and J. B. Jackson, *Polymer, 4*: 221 (1963).
18. K. Shirayama, S. Kita, and H. Watabe, *Makromol. Chem., 151*: 97 (1972).
19. P. J. Flory, *J. Am. Chem. Soc., 69*: 2893 (1947).
20. P. A. Small, *Adv. Polym. Sci., 18*: 1 (1975).
21. T. G. Scholte, *Developments in Polymer Characterization*, Vol. 4 (J. V. Dawkins, ed.), Applied Science, London, p. 1 (1983).
22. Y. Sugimura, and S. Tsuge, *Macromolecules, 12*: 512 (1979).
23. H. Ohtani, S. Tsuge, and T. Usami, *Macromolecules, 17*: 2557 (1984).
24. K. Shirayama, T. Okada, and S. Kita, *J. Polym. Sci., A2(3)*: 907 (1965).
25. C. Bërgstom, and E. Avela, *J. Appl. Polym. Sci., 23*: 163 (1979).
26. T. Usami, Y. Gotoh, and S. Takayama, *Polym. Prep. (Am. Chem. Soc., Div. Polym. Chem.) 27*: 110 (1986).
27. L. Wild, T. Ryle, D. C. Knobeloch, and I. R. Peat, *J. Polym. Sci., Polym. Phys. Ed., 20*: 441 (1982).
28. R. D. Leaversuch, *Mod. Plast. Int.*, p. 66, Sept. 1986.
29. R. D. Leaversuch, *Mod. Plast. Int.*, p. 12, Dec. 1991.
30. D. M. Kalyon, and F. H. Moy, *Polym. Eng. and Sci., 28(23)*: 1551 (1988).
31. M. Hert, "Proceedings of Polycon'84-LLDPE," p. 19 (1984).
32. V. B. E. Mathot, "Proceeding of Polycon'84-LLDPE," p. 1 (1984).
33. T. Usami, Y. Gotoh, and S. Takayama, *Macromolecules, 19*: 2722 (1986).
34. B. K. Kim, M. S. Kim, H. M. Jeong, K. J. Kim, and J. K. Jang, *Die Angew Macromol. Chemie, 194*: 91 (1992).
35. F. Mirabella, Jr., and E. Ford, *J. Polym. Sci. Part B, Polym. Phys. Ed., 25*: 777 (1987).
36. S. D. Clas, D. C. Mcfaddin, K. E. Russell, and M. V. Scammell-Bullock, *J. Polym. Sci., Part A, Polym. Chem., 25*: 3105 (1987).
37. V. K. Gupta, S. Satish, and I. S. Bhardwaj, *J. Macromol. Sci. Rev. Macromol. Chem. Phys., C34(3)*: 439 (1994).
38. C. D. Han, Y. J. Kim, H. K. Chuang, and T. H. Kwack, *J. Appl. Polym. Sci., 28*: 3435 (1983).
39. T. H. Kwack, and C. D. Han, *J. Appl. Polym. Sci., 28*: 3419 (1983).
40. L. Wild, R. Ranganath, and D. C. Knobelock, *Polym. Eng. Sci., 16*: 811 (1976).
41. J. E. Guillet, R. L. Combs, D. E. Slonaker, D. A. Weems, and H. W. Coover, *J. Appl. Polym. Sci., 9*: 757 (1965).
42. E. W. Bagley, *J. Appl. Phys., 31*: 1126 (1960).
43. R. L. Combs, D. F. Slonaker, and H. W. Coover, *J. Appl. Polym. Sci., 13*: 519 (1969).
44. C. D. Han, and T. H. Kwack, *J. Appl. Polym. Sci., 28*: 3399 (1983).
45. C. D. Han, and C. A. Villamizar, *J. Appl. Polym. Sci., 22*: 1677 (1978).
46. M. Rokudai, *J. Appl. Polym. Sci., 23*: 463 (1979).
47. M. Rokudai, S. Mihara, and T. Fujiki, *J. Appl. Polym. Sci., 23*: 3289 (1979).

48. M. Rokudai, and T. Fujiki, *J. Appl. Polym. Sci., 26*: 1343 (1981).
49. G. Ritzau, A. Ram, and L. J. Izrailov, *Poly. Eng. and Sci., 29(4)*: 214 (1989).
50. J. H. Prichard, and K. F. Wissburn, *J. Appl. Poly. Sci., 13*: 233 (1969).
51. R. A. McCord, and B. Maxwell, *Mod. Plast., 38(9)*: 166 (1961).
52. W. W. Grassley, *J. Chem. Phys., 43*: 2696 (1965).
53. A. Ajji, P. J. Carreau, H. P. Schreiber, and A. Rudin, *J. Polym. Sci., Part B., Polym. Phys., 24*: 1983 (1986).
54. T. Fujiki, *J. Appl. Polym. Sci., 15*: 47 (1971).
55. F. Stehling, C. S. Speed, and L. Westerman, *Macromolecules, 14*: 698 (1981).
56. D. M. Kalyan, D. W. Yu, and F. H. May., *Polym. Eng. Sci., 28(23)*: 1542 (1988).
57. P. G. de Gennes, *J. Chem. Phys., 55*: 572 (1971).
58. S. Kurtz, L. Searola, and J. Miller, *SPE ANTEC Tech. Papers, 28*: 192 (1982).
59. U. W. Gedde, J-F. Jansson, G. Liljenstrom, S. Eklund, S. R. Holding, P.-L. Wang, and P-E. Werner, *Polym. Eng. Sci., 28(20)*: 1289 (1988).
60. U. W. Gedde, and J.-F. Jansson, *Polymer, 24*: 1521 (1983).
61. B. Underlich, and A. Mehta, *J. Polym. Sci., Polym. Phys. Ed., 12*: 255 (1975).
62. U. W. Gedde, S. Eklund, and J.-F. Jansson, *Polymer Bull., 8*: 90 (1982).
63. U. W. Gedde, and J.-F. Jansson, *Polymer, 25*: 1263 (1984).
64. U. W. Gedde, S. Eklund, and J.-F. Jansson, *Polymer, 24*: 1532 (1983).
65. U. F. Gedde, and J.-F. Jansson, *Polymer, 26*: 1469 (1985).
66. C. S. Speed, *Plast. Eng., xxxviii*: 39–42, (1982).
67. F. DeCandia, A. Perullo, and V. Vittoria, *J. Appl. Polym. Sci., 28*: 1815 (1983).
68. R. Bubeck, and H. Baker, *Polymer, 23*: 1680 (1982).
69. G. Attalla, and F. Bertinotti, *J. Appl. Polym. Sci., 28*: 3503 (1983).
70. D. LaMontia, A. Valenza, and D. Acierno, *Europ. Poly,. J., 22 (8)*: 647 (1986).
71. J. Pezzutti, and R. Porter, *J. Appl. Polym. Sci., 30*: 4251 (1985).
72. H. Ashizawa, J. Spruiell, and J. White, *Polym. Eng. Sci., 24(13)*: 1035 (1984).
73. D. M. Kalyon, F. H. Moy, V. Tan, and S. Bhakhuni, SPE ANTEC Papers, 32, 741 (1986).
74. D. M. Kalyon, V. Tan, S. Bhakuni, and F. H. Moy, *J. Plast. Film Sheeting, 2*: 310 (1986).
75. W. A. Fraser, L. S. Scarola, and M. M. Concha, "TAPPI Paper Synthetic Course Processings," Technical Association of the Pulp and Paper Industry, Inc., Atlanta, GA, p. 237 (1980).
76. W. A. Fraser, and G. S. Cieloszyk, U. S. Pat. 4, 243, 619 (1981).
77. F. DeCadia, A. Perullo, and V. Vittoria, *J. Appl. Polym. Sci. 28*: 1815 (1983).
78. N. D. Huck, and P. L. Clegg, *Soc. Plast. Eng. Trans., 1(3)*: 121 (1961).
79. J. Pezzutti, and R. Porter, *J. Appl. Polym. Sci., 30*: 4251 (1985).
80. H. Ashizawa, J. Spruiell, and J. White, *Polym. Eng. Sci., 24(13)*: 1035 (1984).
81. *Mod. Plast. Int.* p. 16, Dec. 1993.
82. P. E. Slade, Jr., *Techniques of Polymer Evaluation,* Marcel Dekker, New York (1975).
83. G. W. Bailey, *J. Polym. Sci., 62*: 41 (1962).
84. G. R. Williomson, B. Wright, and R. N. Haward, *J. Appl. Polym. Sci., 14*: 131 (1964).
85. S. Matsuoka, *Polym. Eng. Sci., 5(3)*: 142 (1965).
86. A. S. Kenyon, I. O. Salyer, J. E. Kurz, and D. R. Brown, *J. Polym. Sci., Part C, 8*: 205 (1965).
87. C. S. Myers, *Mod. Plast., 21(12)*: 103 (1944).
88. N. H. Shearer, J. E. Guillet, and H. W. Coover, *SPE. J., 17*: 83 (1961).
89. C. A. Sperati, W. A. Franta, and H. W. Starkweather, *J. Am. Chem. Soc., 75*: 6127 (1953).
90. L. E. Shalaeva, and N. M. Domareva, *Past. Massy., 9*: 10 (1961).
91. E. T. Darden, and C. F. Hammer, *Mater. Methods, 44*: 94 (1956).
92. E. V. Veselovskaya, M. D. Pukshanki, and L. F. Shalaeva, *Plast. Massy. 9*: 44 (1969).
93. L. H. Tung, *SPE J., 14(7)*: 25 (1958).
94. G. N. B. Burch, G. B. Field, F. H. Mc Tigue, and H. M. Spurlin, *SPE J., 13*: 34 (1957)
95. H. Kojima, and K. Yamaguchi, *Kobunshi, Kagaku, 19*: 715 (1962).
96. H. Grimminger, *Kunstoffe, 57*: 496 (1967).
97. L. Wild, J. F. Woldering, and R. T. Guliana, *SPE Tech. Pap., 13*: 91 (1967).
98. P. I. Vincent, *Polymer, 1*: 425 (1960).
99. L. L. Lander, and R. H. Carey, *Plastics* (London), *28 (310)*: 87 (1963).
100. R. H. Garey, *Ind. Eng. Chem., 50*: 1045 (1958).
101. G. R. Gohn, J. D. Cummings, and W. C. Ellis, *Amer. Chem. Soc. Test. Mater, Proc., 49*: 1139 (1949).
102. R. N. Haward, B. Wright, G. R. Williamson, and G. J. Thackray, *J. Polym. Sci., Part A, 2*: 2977 (1964).
103. A. N. Karasev, I. N. Andreeva, N. M. Domareva, K. I. Kosmatykh, M. G. Karaseva, and N. A. Domnicheva, *Vysokomol, Soedin., Ser. A, 12*: 1127 (1970).
104. A. A. Burriyat-Zade, and A. B. Azimova, *Plast. Massy., 6*: 37 (1970).
105. P. Hittmair, and R. Ullman, *J. Appl. Polym. Sci., 6*: 1 (1962).
106. J. B. Howard, *Polym. Eng. and Sci., 51*: 125 (1965).
107. R. McFedries, W. E. Brown, and F. J. McGarry, *SPE Trans., 2*: 170 (1962).
108. J. N. Herman, and J. A. Biesenberger, *Polym. Eng. Sci., 6*: 341 (1966).
109. Water Associates Data Module Model 730, Operator's Manual, Appendix E, Data Module: GPC Applications, Manual No. OM.82908, Sept. 1981, Rev. B.
110. F. W. Billmeyer, Jr., *Text Book of Polymer Science,* John Wiley & Sons Inc., New York, 1971.
111. P. J. Flory, *Principles of Polymer Science,* Cornell University Press, Itaca, New York, 1953.
112. J. M. G. Cowie, *Polymers: Chemistry and Physics of Modern Materials,* Intertext Books, Aylesbury, U.K. (1973).
113. U. Dayal, A. B. Mathur, and Shashikant, *J. Appl. Polym. Sci. 59*: 1223 (1996).
114. M. Fleissner, *Kunstoffe, 77*: 45 (1987).
115. R. A. Horsley, D. J. A. Lee, and P. B. Write. *The Physical Properties of Polymers* (SCI Monogr. No. 5) SCI, London, p. 63–79, (1959).
116. R. D. Leaversuch, *Mod. Plast. Int., Aug 14, (1991).*
117. M. Flissner, *Angew Makromol. Chem., 24*: 197 (1981).

118. M. Fleissner, *Angew Macromol. Chem., 105*: 167 (1982).

119. C. D. Han, *Rheology in Polymer Processing,* Academic Press, New York (1979).

120. H. Watanabe, and T. Kotaka, *Macromolecules, 17*: 2316 (1984).

121. G. V. Vinogradov, *Rheology Acta, 12*: 273 (1973).

122. E. Uhland, *Rheol Acta, 18*: 1 (1979).

123. C. D. Han, T. C. Yu, and K. Kim, *J. Appl. Polym. Sci., 15*: 1149 (1971).

124. W. W. Grassley, and M. J. Struglinski, *Macromolecules, 19*: 1754 (1986).

125. C. D. Han, and J. Y. Park, *J. Appl. Polym. Sci., 19*: 3291 (1975).

126. M. Fleissner, *Int. Polym. Proc., II*: 229 (1988).

127. J. P. Montfort, G. Marin, and P. Monge, *Macromolecules, 19*: 1979 (1986).

128. A. M. Birks, and L. E. Dowd, "Polyethylene Film," *SPE Technical Papers, 25*: 714 (1979).

129. H. H. Zabusky, and R. F. Heitmiller, *SPE Trans, 4*: 17 (1964).

130. L. L. Böhm, H.-F. Enderle, and M. Fleissner, *Studies in Surface Science and Catalyst'89, Catalyst Design for Tailor Made Polyolefins,* (K. Soga, and M. Terano, eds.) Elsevier Science BV. The Netherlands, p. 351 (1994).

131. R. Hayes, and W. Webster, *Plast. Int. Trans., 32*: 219 (1964).

132. K. C. H. Yi, and N. J. Maraschin, Bimodal HDPE Via Gas Phase Process, A New Frontier © 1990, Union Carbide Chemicals & Plastics Technology Corporation./ Mod. Plast. Int. January 1991, p. 12.

133. J. T. Yeh, J.-H. Chen, and H.-S. Hong, *J. Appl. Polym. Sci., 54*: 2171 (1994).

134. A. Lustiger, and R. L. Markham, *Polymer, 24*: 1647 (1983).

135. N. Brown, and I. M. Ward, *J. Mater. Sci., 18*: 1405 (1983).

136. Y.-L. Huang, and N. Brown, *J. Polym. Sci., B, Polym. Phys., 29*: 129 (1991).

137. K. Schmidt-Rohr, and H. W. Spiess, *Macromolecules, 24*: 5288 (1991).

20

The Crystallization of Polyethylene Under High Pressure

Rui Huang, Xiongwei Zhang, and Guangxian Li
Sichuan Union University, Sichuan, China

I. INTRODUCTION

The actual experimental moduli of the polymer materials are usually about only 1% of their theoretical values [1], while the calculated theoretical moduli of many polymer materials are comparable to that of metal or fiber reinforced composites, for instance, the crystalline polyethylene (PE) and polyvinyl alcohol have their calculated Young's moduli in the range of 200–300 GPa, surpassing the normal steel modulus of 200 GPa. This has been attributed to the limitations of the folded-chain structures, the disordered alignment of molecular chains, and other defects existing in crystalline polymers under normal processing conditions.

Continuous efforts have been made to search for an effective approach to realize the optimum alignment of the polymer chain to achieve the maximum properties that polymers theoretically possess. In 1964, Wunderlich and coworkers [2–4] reported that by crystallization from the melt at sufficiently high pressure, PE can form an extended-chain, crystal (ECC) rather than the normally observed folded-chain, that is several micrometers thick, much thicker than normal polymer crystal thickness (~10 nm). Fibers made in this way have a tensile strength of 2 GPa, which is even stronger than that of alloy steel wires [1]. The novel configurations and the outstanding physical performance of ECC polymer materials demonstrate the tremendous potential remaining in the polymer itself and have provoked great interest from many polymer scientists and engineers. A great deal of research has been done since the middle 1960s on the structure–property relationship of polymers under high pressure, this research was mainly concentrated on crystalline polymers, such as PE, polychloro-trifluoroethylene, polyamide, polyvinylidene fluoride, polyethylene terephthalate, etc. [2–38]. The pressure condition used was often within the range of 0–700 MPa. These investigations led to findings of some new crystal structures and an understanding of the ECC growth mechanism and crystalline dynamics. As a matter of fact, the study of the behavior of polymers under high pressure has become one of the significant fields of polymer morphology and polymer processing, and has formed a basis for developing a new approach, which parallels physical blending and chemical synthesis, toward the making of high-performance materials.

The aim of this chapter is to review the development of high-pressure crystallization of PE over the past three decades, based on the current literature. The following topics are discussed in this chapter: the ECC of polymers; the melting–crystallization of PE under high pressure; the high-pressure hexagonal phase of PE; the types of hexagonal phases; patterns of crystallization of PE under high pressure; theory of nucleation of ECC under high pressure; effect of molecular weight on the melting–crystallization of PE under high pressure; and the effect of other components on the melting–crystallization of PE under high pressure.

II. THE EXTENDED-CHAIN CRYSTALS OF POLYMERS

Crystalline polymers are pressure-sensitive materials due to their weak interchain potential. Therefore, they have different crystal structures under different conditions.

It is well known that flexible polymer chains usually crystallize at atmospheric pressure in a folded-chain manner, giving rise to lamellar crystals with a thickness in the range of 10–15 nm, which is much thinner than the length of a fully extended chain. However, this is not the most thermodynamically favorable conformation, but it occurs due to kinetic reasons. Crystallization can take place faster through chain folding where the chains need to continuously deposit only along a fraction of their lengths, provided that the resulting crystal is still stable when supercooled. It is believed that the most thermodynamically stable conformation should correspond to a crystal with extended chains, a state that normally is not realizable during primary crystal growth because of a high-activation barrier to such an extended-chain deposition, whether enthalpic (large surface free energy [39]) or entropic (low-deposition probability [40,41]), making such crystal growth prohibitively slow. For example, Anderson [42] discovered that the lamellar thickness of a PE fraction of 12,000 molecular weight (MW) crystallized for up to 10 days at 128°C in vacua was approximately equal to its molecular length.

A completely different mode of crystallization occurs at appropriately elevated pressures (hundreds of MPa). In 1964, Wunderlich and coworkers [2–4] discovered that PE samples crystallized from a melt at about 500 MPa had a density up to 0.994 g/cm³, with a degree of crystallinity close to 100%, and a maximum melting point at atmospheric pressure of about 140°C. The morphology of crystals, revealed by electron microscopy of fracture surface replicas of the samples, was striated, banded, and had very thick lamellae. The spectrum of thickness extended to 3 μm, with an average value of 250 nm. The C axis direction, identified by electron diffraction and optical birefringence, was parallel to a characteristic striation on fracture surfaces. These structures were accordingly referred to as extended-chain crystals (ECC) compared to the usual thinner folded-chain crystals (FCC) formed by atmospheric pressure crystallization. From the comparison between molecular length and crystal thickness distributions, determined by statistics of fracture surface of samples and fuming nitric acid degradation followed by gel-permeation chromatography (GPC) measurement of molecular weight distribution of the degraded samples, it has been found [5–9] that molecular length is usually longer than the crystal thickness of ECC in the chain axis direction, implying that molecular chains in ECC lamellae also may be folded with their chain ends included in the lamellae, except that chain ends are not turned in for molecular lengths <60 nm. Slower crystallization and the use of narrow polydisperse samples tend to cause the exclusion of chain ends from lamellae and a move toward full chain extension. Nevertheless, molecular length is not the upper limit of crystal thickness. In suitable conditions, molecules will fit end-to-end inside lamellae in fully extended conformation. Hatakeyama et al. [10] observed

that prolonged crystallization of a 50,000 MW PE sample at 237°C and 480 MPa for 200 h produced lamellae up to 40 μm thick, 20% thicker than 10 μm, whereas only 1.2% of molecules were longer than 10 μm.

There is no unanimity in regard to the exact mechanism of ECC formation under high pressure. Wunderlich et al. [11–18] suggested that when a flexible polymer molecule crystallizes from the melt under high pressure, it does not grow in the form of a stable extended chain, rather it deposits as a metastable folded chain.

Each new crystallizing polymer molecule has to go through a nucleation step during its packing into the crystal. The typical size of folded-chain nuclei along the molecular chain axis is 5–100 nm. The metastability due to such a chain folding would suggest that there exists a considerable driving force toward reorganization of chain conformation into a crystal with a larger fold length. Thus, the formation of ECC can be taken as a two-step process (Fig. 1). The first step is the transition of the molecule from mobile random conformation in the melt to a folded-chain conformation in the nucleus or crystal. The second step is the rearrangement of the folded chain into a longer fold length, resulting in lamellar thickening and the formation of ECC. The mechanism of nucleation at atmospheric pressure is also valid for high-pressure crystallization; the major difference is the relative ease with which the lamellar thickness can be increased at high pressures.

Bassett et al. [6,19–22] and others [23–25] suggested, however, that chain-folded and chain-extended growth are two distinct, rival processes. ECC does not form from FCC by lamellar thickening but arises directly through crystallization from the melt under high pressure. This viewpoint was based on the following experimental observations. These showed that at a cooling rate of ~1 K/min, crystallization of a 50,000 MW PE fraction occurred at characteristic supercoolings of ~16 K for chain-folded crystallization and ~12 K for chain-extended crystallization [22]. In the region where both ECC and FCC existed, two DTA peaks were detected having the characteristic supercoolings and widths measured on either side of this region [21]. Under particular isothermal, isobaric conditions ECC formed prior to FCC [21].

It should be noted that high-pressure crystallization is not the only means to obtain ECC [26]. It has been

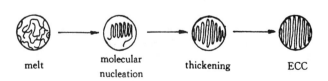

Figure 1 Schematic diagram of formation of ECC by crystallization under high pressure. (From Ref. 18.)

known that ECC can also be produced by solid-state extrusion under hydrostatic pressure [43] and ultra-drawing techniques [44,45]. Zone polymerization of caprolactam as described by Wunderlich et al. [46] led to the formation of ECC. Slower crystallization of low molecular weight PE and polytetrafluoroethylene (PTFE) [27,47,48] at atmospheric pressure has been found to also produce ECC. However, it is not the case that all of the polymers can form ECC by high-pressure crystallization. In spite of efforts made by many workers, so far only the following polymers have been reported to form ECC by crystallization from the melt under high pressure since the first discovery in PE: polyethylene (PE); polychlorotrifluoroethylene (PCTFE) [28], nylon-6 [29–33], 11 [34], 12 [35]; polyvinylidene fluoride (PVDF) [36]; and polyethylene terephthalate (PET) [37,38].

III. THE MELTING–CRYSTALLIZATION OF PE UNDER HIGH PRESSURE

It is known that elevating the crystallization temperature [42,49] or annealing above the crystallization temperature [50] of PE results in a thicker folded-chain lamella of up to ~200 nm. In addition to the higher temperature, if high pressure is applied, crystals can grow as thick as several micrometers in the chain axis direction [2–4,51].

The viewpoint of Wunderlich et al. [11–18] that the polymer molecules transfer from the melt to a folded-chain conformation and further undergo enhanced lamellar thickening and transform into ECC was once commonly accepted. In 1972, however, Bassett and Turner [20] reported that melting and crystallization of PE at pressures above ~400 MPa occurred in two stages. An additional high-temperature endothermic or exothermic peak in DTA curves appeared, accompanied by discontinuous changes in volume. The same phenomena were also reported by Yasuniwa et al. [52] and Maeda and Kanetsuna [53–55]. Although the actual results of the three groups of workers were in good agreement, the respective interpretations for the peaks were quite different. Yasuniwa et al. [52] ascribed the peaks to the melting and crystallization of an unknown structure, which are illustrated in Figs. 2–4. The processes of DTA can be classified into six groups. Run 1 shows the melting curve under high pressure of a polydisperse PE sample (M_W = 40,000) crystallized at atmospheric pressure (heating rate 6°C/min). Run 2 shows the crystallization curve under high pressure on slow cooling (cooling rate is shown in the figure). Run 3 shows the melting curve under high pressure of the sample crystallized under the same pressure by slow cooling (heating rate 6°C/min). Run 4 shows the crystallization curve under high pressure by rapid cooling (cooling rate is shown in the figure). Run 5 shows the melting curve under high pressure of the sample crystallized under the same pressure by

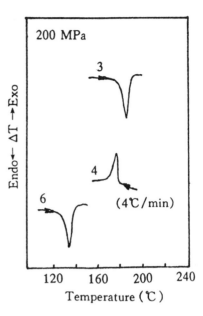

Figure 2 DTA curves of melting and crystallization of PE at 200 MPa. Numbers in the figure correspond to the experimental conditions described in the text. (From Ref. 52.)

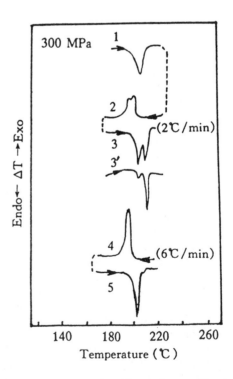

Figure 3 DTA curves of melting and crystallization of PE at 300 MPa. (From Ref. 52.)

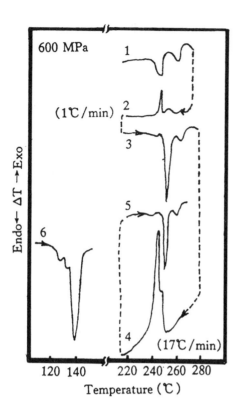

Figure 4 DTA curves of melting and crystallization of PE at 600 MPa. (From Ref. 52.)

slow cooling (heating rate 6°C/min). Run 6 shows the melting curve at atmospheric pressure of the sample crystallized under high pressure (heating rate 6°C/min). It is shown in Fig. 2 that the melting and crystallization curves are not complicated up to about 250 MPa, and only a single endothermic or exothermic peak appears, which corresponds to the melting or crystallization of FCC. As pressure is increased to 300 MPa, the melting and crystallization curves begin to be complex (Fig. 3). The melting curve of the sample crystallized at atmospheric pressure (run 1) exhibits a single endothermic melting peak of FCC, but the crystallization curve by slow cooling (run 2) shows two exothermic crystallization peaks corresponding to the ECC and FCC. The melting curve of this sample (run 3) shows two endothermic melting peaks corresponding to FCC and ECC. If the sample is crystallized by slow cooling (0.3°C/min), the melting peak of the ECC develops strongly (run 3′). On the contrary, the sample crystallized by rapid cooling (run 4) exhibits strong development of FCC (run 5). As pressure is further increased to 600 MPa, the melting and crystallization curves become more complicated (Fig. 4). An additional high-temperature endothermic peak appears in the melting curve of the sample crystallized at atmospheric pressure (run 1). The two endothermic peaks in run 1 were assigned to the melting of

ECC formed from the recrystallization of FCC and the so-called unknown structure formed from the ECC, respectively. The crystallization curve under high pressure on slow cooling (run 2) shows exothermic crystallization peaks of this unknown structure and an ECC. The melting curve under high pressure of the sample (run 3) shows a small melting peak of the FCC, a large melting peak of the ECC, and a melting peak of the unknown structure. If the sample is crystallized by rapid cooling (run 4), the crystallization peak of the FCC disappears. The melting curve of that sample (run 5) shows the same feature as that of run 3. The melting curve at atmospheric pressure by slow cooling (run 6) shows the feature of fractional crystallization, as mentioned by others [2,13,16,50]. Yasuniwa et al. [56], by using a diamond-anvil cell with an optical microscope, observed that the growth feature of the unknown structure crystal is similar to that of a liquid crystal, thus they presumed that the unknown structure may be a nematic liquid crystal. Maeda et al. [53–55] considered that the two stages of crystallization or melting represent the formation or melting of two kinds of ECC with different lamellar thickness, which they referred to as high ECC (500–800 nm) and ordinary ECC (200–500 nm), and assigned the high-temperature exothermic or endothermic peak and the related volume change to the crystallization or melting of high ECC and the low-temperature exothermic or endothermic peak and the related volume change to that of ordinary ECC. Bassett et al. [20,22] postulated from a number of thermodynamic criteria that a new high-pressure phase may exist between the normal orthorhombic phase and the melt at pressures above ~300 MPa, based on the volumetric and thermal measurements of melting and crystallization in the range of 0–600 MPa. The phase diagram of PE constructed from the observed melting peak temperatures of high-pressure crystallized PE of 5×10^4 molecular weight is shown in Fig. 5. Bassett et al. [20,22] suggested that the curious phenomena at high pressures, i.e., two stages of melting and crystallization, are due to the reversible first-order transitions into and out of the new high-pressure phase. The quite confused situation in which three groups of workers interpreted essentially the same facts but toward different conclusions was resolved by the discovery of the high-pressure intermediate phase postulated by Bassett et al. [20,22].

IV. THE HIGH-PRESSURE HEXAGONAL PHASE OF PE

In 1974, Bassett et al. [57] confirmed the existence of the predicted high-pressure phase of PE by using a high-pressure and high-temperature x-ray measurement employing the gasketed diamond-anvil cell. The new phase observed by one x-ray reflection was suggested to be hexagonal with lattice parameters of a = 8.46 Å and b = 4.88 Å (Fig. 6a and b). It is found, in the hexagonal

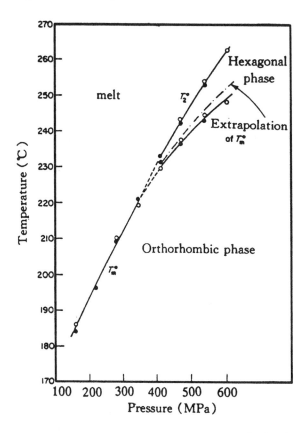

Figure 5 P–T phase diagram of a linear PE fraction of 50,000 molecular weight. (From Ref. 85.)

cell, that the specific volume is increased from 1.005 cm^3/g in the orthorhombic phase to 1.109 cm^3/g, an increase of only 8.5%, whereas the expansion in the cross-sectional area per chain is 18.2–20.6 Å^2, i.e., 13%. It is deduced, accordingly, that the average C axis spacing per ethylene unit has to contract from 2.53 Å in the orthorhombic phase to 2.45 Å. These facts imply that the molecules in the hexagonal phase seem to have lost the all-*trans* conformation and include a proportion of *gauche* bonds along the chain, so that the hexagonal phase is probably not a fully ordered one. Pechhold et al. [58] have used a perturbed helix model of alternating sequences of all-*trans, T,* and helical *TGTG** bonding (with *G* and *G** representing *gauche* bonds of opposite sense) to have a statistical mechanical calculation and obtained excellent agreement between observed and predicted thermodynamic functions of the hexagonal phase. The x-ray observations of Yamamoto et al. [59,60] further confirmed the deduction of Bassett et al. [57]. Three Bragg reflections, corresponding to the (100), (110), and (200) reflections of the hexagonal lattice, were observed on the equator. The line width of each of these three Bragg reflections is not very different, indicating that the two-dimensional lattice perpendicular to the chain axis is quite ordered, therefore, the hexagonal phase is not a nematic liquid crystal. However, there is no appreciable Bragg reflection on the nonequatorial line. Distinct diffuse scattering is seen instead, which indicates the disorder of molecular conformation along the chain axis. On the basis of calculations from the

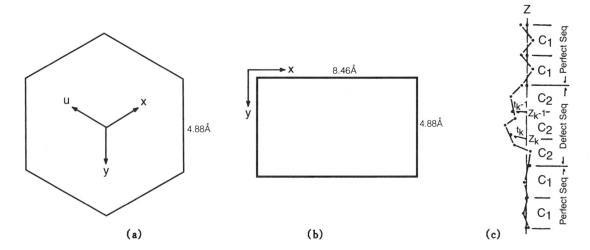

Figure 6 The lateral cell dimensions of the hexagonal cell at 500 MPa. (a) Referred to the hexagonal axes and (b) referred to orthohexagonal axes. (From Ref. 57.) (c) Structural model of the intramolecular defect in the hexagonal crystal. (From Ref. 60.)

intensities of diffuse scatterings, Yamamoto [60] proposed a locally perturbed all-*trans* chain model for the hexagonal phase as shown in Fig. 6c. The molecular chains are highly extended, oriented, and hexagonally packed in the direction perpendicular to the chain axis. Each unit cell contains only one molecule, while each molecular chain consists of one-third defect and two-thirds all-*trans* sequences with average sequence lengths of about 2 subunits and 4 subunits, respectively. The average C axis spacing is about 2.43 Å, in good agreement with the experimental results by dilatometry.

The spectroscopic studies [61–67] have provided further information about the structure of the hexagonal phase of PE. Far infrared spectroscopy [61] at high temperatures and pressures has shown that at 450 MPa and 237°C a new absorption band at 93.5 cm^{-1} was discovered and assigned to a lattice vibration of the hexagonal phase. This implies a well-ordered lattice over considerable distances. From the frequency that lies between the frequencies of the B_{1u} and B_{2u} lattice vibrations of the orthorhombic phase, it is deduced that the hexagonal phase consists mainly of helical sequences and that the lattice vibrations contribute little to the entropy difference of orthorhombic–hexagonal transition. High-pressure Raman spectroscopic studies [62–67] have confirmed the presence of *gauche* bonds in the hexagonal phase due to the observation of a medium-intensity band at 1090 cm^{-1} in the 950–1250 cm^{-1} skeletal optical modes region attributable to a C—C stretching vibration of a *gauche* bond. From calculations of the order parameter, $r = I(1130)/I(1090)$, the ratio of the intensity of the all-*trans* band at 1130 cm^{-1}, $I(1130)$, to the intensity of the *gauche* band at 1090 cm^{-1}, $I(1090)$, it is found that the population of *gauche* bonds in the hexagonal phase lies between 32% and 38%.

In thermodynamic terms, the condition for the existence of a hexagonal phase is that its specific free enthalpy, g (Gibbs's function per unit mass), should fall below those of the orthorhombic and melt phases [22,68]. The specific free enthalpy of the respective phases (orthorhombic, hexagonal, and melt phases) under high pressure as a function of temperature is schematically shown in Fig. 7. The temperature interval of existence of the hexagonal phase lies between the orthorhombic–hexagonal transition temperature, T_t, and the hexagonal melting point, T_m, recalling that $(\delta g/\delta T)_p = -s$, and can be written as:

$$T_m - T_t = \frac{\Delta g(s_m - s_o)}{(s_m - s_h)(s_h - s_o)} \tag{1}$$

A large number of experiments [22,68–71] have shown that the specific volume, v_h, and specific entropy, s_h, of the high-pressure hexagonal phase of PE are closer to the corresponding values v_m and s_m of the melt than they are to those of the orthorhombic phase, v_o and s_o. For instance, at 500 MPa, the ratio of entropies of the orthorhombic–hexagonal and hexagonal-melt transi-

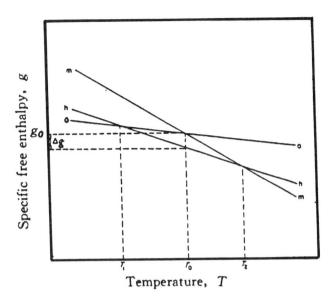

Figure 7 Schematic diagram of the free enthalpy–temperature relationships for (*o*) orthorhombic, (*h*) hexagonal, and (*m*) melt phases. (From Ref. 85.)

tions is $(s_m - s_o)/(s_m - s_o) = 0.72$ (indicating that the entropy of fusion of the hexagonal phase is only about one-fourth of the orthorhombic phase), while the ratio of specific volumes of these transitions is $(v_h - v_o)/(v_m - v_o) = 0.64$. Thus, the inequality is satisfied:

$$\frac{s_h - s_o}{s_m - s_o} > \frac{v_h - v_o}{v_m - v_o} \tag{2}$$

The hexagonal phase of PE has accordingly been termed a high-entropy phase. It is the high entropy that offsets the disadvantage of high volume and leads to its formation. X-ray measurements have shown that the compressibility of the hexagonal phase at 500 MPa is very large and has a value comparable to that of the melt. Ultrasonic studies [67,72] have found that the shear modulus of the hexagonal phase is of the same order in magnitude as that of the melt. Raman spectroscopic studies [63,65] have shown that the spectra of the hexagonal phase are nearly identical to those of the melt. There are no detectable differences with the exception of a minor change in the C—C stretching region. All of these facts seem to imply that the nature of the hexagonal phase resembles that of the melt. Therefore, the high-pressure hexagonal phase of PE has usually been described as a liquid crystal. From this context, taking the longitudinal conformational disorder and the lateral hexagonal packing into account, it would be placed in the smectic B category [71].

Besides the thermodynamic modelings, the statistical mechanical analysis has also been used to study the high-pressure hexagonal phase of PE. Priest [73–75] regarded the mechanical model of the hexagonal phase as a fixed bond angle chain confined to a cylindrical cavity formed by six neighboring chains and simulated by using a statistical mechanical approach the orthorhombic–hexagonal transition features. The predicted values such as the change of cross-sectional area per chain, the contraction of the chain axis; the differences of entropy, enthalpy, and volume; the phase boundary between the orthorhombic and hexagonal phases; etc., are shown in good agreement with observed ones. On the other hand, Yamamoto [76,77] treated the mechanical model of the hexagonal phase as a one-end fixed chain in a cylindrical potential produced by six neighboring chains and simulated by using a Monte Carlo method the molecular motion, conformational disorder in hexagonal phase, and the orthorhombic–hexagonal transition behaviors. The calculated values are also found in good agreement with experiments.

V. THE TYPES OF HEXAGONAL PHASES

In addition to the occurrence in PE at high temperature and pressure, the hexagonal phase is frequently found in paraffin substances under various circumstances. The analysis of crystal structure and order of these hexagonal phases is not only interesting, but also important for the understanding of their origins. The circumstances in which hexagonal phases arise in paraffin substances can be summarized as follows:

1. On heating lower n-paraffins the so-called rotator phases occur a few degrees below the melting point; they are either hexagonal or nearly so [78–82]. This is the earliest recognized example of a hexagonal phase and thus the best known.

2. In low-dose irradiated [83–86] or fuming nitric acid treated PE [87,88], hexagonal phases can be observed at low temperature and pressure.

3. A hexagonal phase appears at atmospheric pressure in ultra-drawn PE fibers containing highly extended chains when the fiber is heated above the normal temperature while the ends are held fixed to prevent retraction [89–92].

4. The presently described case of formation of PE hexagonal phase at suitably high temperatures and pressures.

5. On heating a sufficiently high-dose (>500 Mrad) irradiated PE, a hexagonal phase can be induced to form prior to melting at atmospheric pressure [93,94].

6. A hexagonal phase is found at room temperature and atmospheric pressure in some ethylene–propylene copolymers containing a small amount of diene component [86,93].

We shall now discuss the common features and the differences between the hexagonal phases in Cases 1–6.

Regarding Case 1, it is known that the rotator phases of n-paraffins are three dimensionally ordered solids with molecules in all-*trans* conformation, their specific volumes, v_h, and specific entropies, s_h, are closer to corresponding values v_o and s_o for the orthorhombic phases than they are to those for the melt v_m and s_m. In contrast, the high-pressure hexagonal phase of PE, ordered only in two dimensions due to the partial destruction of the longitudinal order resulting from the introduction of *gauche* bonds in the chains, like many liquid crystals, is closer to the melt in specific volume and entropy than to the orthorhombic phase. A further distinguishing feature between the rotator and high-pressure hexagonal phases is their different responses to the application of pressure. The temperature interval of existence of the high-pressure hexagonal phase widens with increasing pressure, while that for the rotator phase decreases. Rotator phases are not found at pressures exceeding 300 MPa, as shown in Fig. 8. This difference implies that rotator phases are ones of high volume that satisfy the inequality:

$$\frac{s_h - s_o}{s_m - s_o} < \frac{v_h - v_o}{v_m - v_o} \tag{3}$$

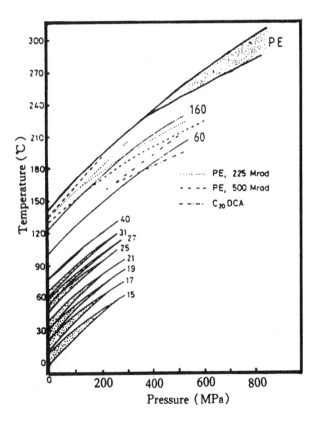

Figure 8 P–T phase diagrams of n-paraffins and PE. shaded areas indicate the regions of the stable hexagonal (or rotator) phase. (From Ref. 86.)

Conversely, the high-pressure hexagonal phase is one of high entropy for which the reverse inequality is true. In lower n-paraffins, the occurrence of the rotator phase at atmospheric pressure is due to the fact that the orthorhombic–hexagonal (or rotator) transition temperature, $T_{o\rightarrow h}$, is lower than the temperature of fusion of the orthorhombic phase, $T_{o\rightarrow m}$, as shown in Fig. 9b. While at high pressures (>300 MPa), the disappearance of the rotator phase is ascribed to the fact that $T_{o\rightarrow h}$ increases with increasing pressure so steeply as to exceed $T_{o\rightarrow m}$. In addition, it is known that $T_{o\rightarrow h}$ in lower n-paraffins increases with increasing chain length. The inability to form the hexagonal phase in higher n-paraffins and PE under ordinary conditions is due to the fact that $T_{o\rightarrow m}$ does not increase with increasing chain length as sharply as does $T_{o\rightarrow h}$. Thus, $T_{o\rightarrow h}$ is higher than $T_{o\rightarrow m}$ (Fig. 9a), i.e., for sufficiently long chains the orthorhombic crystals melt directly.

In Cases 2 and 3, $T_{o\rightarrow m}$ is raised beyond $T_{o\rightarrow h}$. Low-dose irradiation results in the crosslinking that takes place in the fold surface of the PE single crystal or in the amorphous phase of the bulk polymer [93]. On the other hand, one of the most important chemical changes induced by fuming nitric acid treatment is the replacement of end groups by —COOH and partly by —NO$_2$, which is confirmed by the IR spectrum [88]. Thus, the fuming nitric acid treated PE is a dicarboxylic acid. Carboxyl groups form strong hydrogen bonds. Molecules in the melt are joined by hydrogen bonds along the chain axis [88]. All these lead to the decrease in configurational entropy of the melt and, thus, the increase in $T_{o\rightarrow m}$ since $T_{o\rightarrow m} = \Delta H_{o\rightarrow m}/\Delta S_{o\rightarrow m}$. Consequently, hexagonal phases occur at low temperatures and pressures. In the case of constrained fiber (Case 3), keeping the polymer chains stretched also results in the reduction of configu-

rational entropy of the melt and an increase in $T_{o\rightarrow m}$, whereas $T_{o\rightarrow h}$ remains practically unaffected [92], $T_{o\rightarrow m}$ is, thus, raised above $T_{o\rightarrow h}$ enabling the hexagonal phase to appear even at atmospheric pressure.

In Cases 4, 5, and 6, $T_{o\rightarrow h}$ is again brought below $T_{o\rightarrow m}$, but this time it is achieved mainly by reducing $T_{o\rightarrow h}$ due to introduction of defects into the crystal lattice. In bulk PE at high temperatures and pressures (Case 4) the main defects are *gauche* bonds [62], and in high-dose irradiated PE (Case 5) and the ethylene–propylene copolymer (Case 6), they are crosslinks and methyl branches, respectively [93]. These crystal defects result in an increase of $\Delta S_{o\rightarrow h}$ and, thus, a decrease of $T_{o\rightarrow h}$ since $T_{o\rightarrow h} = \Delta H_{o\rightarrow h}/\Delta S_{o\rightarrow h}$. Although the crystal defects are primarily responsible for the occurrence of the hexagonal phases in Cases 4–6, another effect that favors their appearance is the reduction in the configurational entropy of the melt caused by the increase of intrachain regularity in the melt due to the application of high pressure (Case 4), the crosslinking of polymer molecules (Case 5), and the methyl branches (Case 6). This leads to a lowering of $\Delta S_{o\rightarrow m}$ with the consequence that $T_{o\rightarrow m}$ remains high. Thus, the constraint imposed on the melt, which is the primary factor in producing the hexagonal phase in Cases 2 and 3, is also operative to some extent in Cases 4–6.

In summary, a hexagonal phase occurs in PE for the following reasons: in Cases 2 and 3, because $T_{o\rightarrow m}$ increases (by constraining the melt) beyond $T_{o\rightarrow h}$; and in Cases 4–6 primarily because $T_{o\rightarrow h}$ decreases below $T_{o\rightarrow m}$ (by introduction of crystal defects).

In what follows the degree of order of the hexagonal phases in Cases 1–6 will be compared to see whether there are basic differences between them. For this purpose the heats of fusion for the hexagonal phases

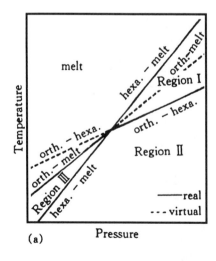

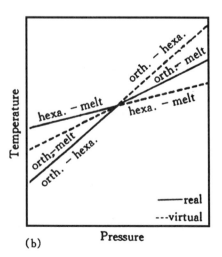

Figure 9 Schematic P–T phase diagrams (a) for PE including metastable and virtual phase boundaries. (From Refs. 98 and 99.) (b) For n-paraffins including virtual boundaries.

($\Delta H_{o \to m}$) will be compared together with the lattice parameters.

The heat of fusion of the rotator phase in n-paraffins is 38–40 cal/g, while $\Delta H_{o \to m}$ for PE irradiated with 1000 Mrad is about 29 cal/g. The corresponding value of $\Delta H_{o \to m}$ for hexagonal PE at high pressure (500 MPa) is only around 15 cal/g. Thermograms of constrained PE fibers showed similarly low heats of fusion of the hexagonal phase.

The cross-sectional area per chain in the hexagonal lattice of irradiated PE varies between 20.6 and 22.0 Å^2. It is, thus, always greater than the cross-sectional area in the rotator phase in paraffins (19.5–20.0 Å^2), but on average somewhat smaller than that in constrained PE fibers above $T_{o \to h}$ (21.4–22.7 Å^2). An ethylene–propylene diene copolymer with approximately 64%, 32%, and 4% by weight of each component, respectively, was found to contain hexagonal crystals with a cross-sectional area per chain of 20.3 Å^2.

From the above data it can be seen that the different hexagonal phases in Cases 1–6 under the conditions of different temperatures and pressures have different degrees of order, so that a knowledge of their expansion coefficients and specific heats is important. By all existing information, the following sequence in ascending degree of order emerges for different hexagonal phases: constrained PE and PE at high pressures (Cases 2–4), high-dose irradiated PE (Case 5), ethylene–propylene copolymer (Case 6), and paraffin rotator phase (Case 1).

VI. PATTERNS OF CRYSTALLIZATION OF PE UNDER HIGH PRESSURE

Bassett et al. [6,20,22,57,95,96] suggested that it is the intervention of the hexagonal phase of PE that is directly responsible for the formation of ECC by crystallization from the melt at high pressures. ECC forms by crystallization from the melt into the hexagonal phase, whereas FCC forms by crystallization from the melt into the orthorhombic phase. This viewpoint can well explain the crystallization behavior of PE in the low-pressure region (below ~200 MPa) and the high-pressure region (above ~350 MPa) in P–T phase diagram of Fig. 5. At pressure below ~200 MPa, crystallization from the melt gives a FCC with orthorhombic structure, a behavior similar to that at atmospheric pressure. At pressure above ~350 MPa, however, there are two possible crystallization patterns, depending on cooling rate. By rapid cooling from the melt region into the orthorhombic region without intermediate crystallization, once again, a FCC with orthorhombic structure will form directly from the melt, whereas at normal cooling rate the hexagonal phase will occur and subsequently transform to the orthorhombic phase, giving rise to a ECC. Some authors [25,53–55], however, did not agree with the above mentioned opinion of Bassett et al. Their objections primarily came from

the complicated crystallization behaviors in the intermediate pressure region (200–350 MPa) where ECC and FCC are often found to coexist. On cooling from the melt at intermediate pressure, the reflections of the hexagonal phase were not detected by x-ray measurement, and the exothermic peak of this phase was not monitored by DTA method either. These facts led them to conclude that ECC may grow directly on crystallization at intermediate pressures without passing through the hexagonal phase, and, thus, the hexagonal phase is not always necessary for the formation of ECC. Nevertheless Bassett [70,97] commented later that, at such pressures, PE crystallizing from the melt is only so fleeting in the hexagonal phase before undergoing the transition to the orthorhombic structure that the hexagonal phase would not be detected with currently available x-ray and DTA facilities over the time scale of the experiments reported. However, this is not the case for optical microscopic observation. The hexagonal crystals (cigar-shaped objects) and their transformation with alteration of their birefringence color to the orthorhombic crystals at pressures of ~300 MPa have been observed. This suggests that at such pressures, the hexagonal phase still intervenes the crystallization. The interpretation [70] for the occurrence of the hexagonal phase below the triple point is that there is a kinetic competition between formations of the (metastable) hexagonal and the (stable) orthorhombic phases and that the two rates only become equal at conditions sufficiently below the triple point to offset the characteristic 3–4 K difference of supercooling between the two processes. The recent studies by Hikosaka and Rastogi et al. [98–100] together with more precise optical microscopy and x-ray measurement have supported the suggestion of Bassett. They found that crystallization at pressures above 200 MPa always starts in the hexagonal phase (uniformly bright crystals) and crystals can only grow in the hexagonal phase, irrespective of whether in the hexagonal or orthorhombic region of the phase diagram. The hexagonal phase can appear as a metastable phase in the stable region of the orthorhombic phase and transforms into the stable orthorhombic phase (blotchy crystals) after a certain lapse of time at a certain stage of growth, irrespective of whether above or below the triple point. At pressures below the triple point, crystals in the hexagonal and orthorhombic phases melt individually without passing through any transition; the melting temperature of metastable hexagonal crystals is lower than that of the stable orthorhombic crystals. These results enabled them to map out an extended phase diagram, which is supplemented with metastable regions and virtual phase transitions, and represented by Fig. 9a in an idealized form. Region I above the triple point (in terms of P) in between hexagonal-melt and orthorhombic–hexagonal transition lines is a region of the stable hexagonal phase. Region II with upper bound of hexagonal-melt and orthorhombic–hexagonal transition lines below and above the triple point, respectively, is a region of the stable orthorhombic phase where the hexagonal phase can also exist as the metastable phase. Region III bounded by hexagonal-melt and orthorhombic-melt transition lines below the triple point is also a region of the stable orthorhombic phase, but is distinct from region II by the fact that here the hexagonal phase cannot persist even in the metastable form because it lies above the hexagonal-melt line. In region II, only the crystals in the metastable hexagonal phase can grow. On transformation into the stable orthorhombic crystals, all growth was found to stop. Region III is a nongrowth region, i.e., where crystals cannot grow, because the metastable hexagonal phase does not appear. The metastability and growth-controlling effect of the hexagonal phase reported by Hikosaka and Rastogi et al. [98–100], however, only extend down to the pressure of ~200 MPa and supercooling of ~10°C. Additional work is required in order to know how far they can extend into the stable orthorhombic region.

Direct optical microscopic observation of the crystallization processes of PE under high pressure can be performed using a diamond-anvil cell [56,57,98–101]. The optical appearance of FCC grown from the melt at low and intermediate pressures is the normal spherulitic texture, while that of ECC is the coarse spiky band structure [21,70]. The two can even be distinguished when they coexist in the same sample (grown at intermediate pressures) and can be identified according to their melting temperatures using hot-stage microscopy. On cooling at pressures above 350 MPa, when the temperature falls and reaches that of crystallization of the hexagonal phase, lens-like (or cigar-shaped) hexagonal crystals form gradually and develop in the direction of their lens diameters [56]. Bassett et al. [57] specifically observed that the hexagonal crystals grow outward behind a tapered growth edge. The lamellar thickness increases rapidly in a region several microns wide behind the growing edge. When the temperature decreases to that of hexagonal–orthorhombic phase transition, the lens-like (or cigar-shaped) hexagonal crystals are divided by many striations and turn into band structure. The width of striations corresponds to the band thickness revealed at the fracture surface of the sample by electron microscope. The subdivision of the hexagonal crystals into several bands is very rapid at this temperature, and the visual field of the microscope is filled by these bands in a short time. When the temperature further decreases to that of crystallization of FCC, a slight gap between bands is filled up rapidly by many fine spherulites, but the quantity of the fraction of these spherulites is very small (below ~5%).

Miyashita et al. [102] have proposed a possible model for the hexagonal–orthorhombic phase transition (Fig. 10). They proposed that since the hexagonal crystal contains many defects along the chain axis, such as kinks and jogs, on its phase transition to the orthorhombic phase, these defects are excluded from the crystal

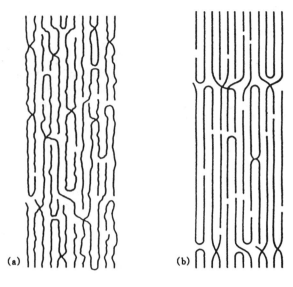

Figure 10 Model for the phase transition of PE from the hexagonal to the orthorhombic phase. (a) In the hexagonal phase, there are many defects along C-axis (kinks, jogs). (b) On the phase transition, these defects are excluded to form boundaries between bands. (From Ref. 102.)

by molecular motions and aggregated on a plane perpendicular to the chain axis to form the boundaries between the bands. This process could minimize the free energy of the system and cause a thick hexagonal crystal to split into several bands on the phase transition. This model agrees with the optical observations of Yasuniwa et al. [56]. The hexagonal phase has three equivalent crystallographic A axes. It is thus reasonable that (100) direction of the hexagonal phase turns out to be either the (010) or the (110) direction of the orthorhombic phase [102,103] (Fig. 11). Since the specific volume of the hexagonal phase is larger than that of the orthorhombic phase, the dilatational strain as well as shear strain in the phase boundary makes the orthorhombic crystals orientate in different directions during the phase transition, resulting in the twists of bands as often observed at the fracture surfaces of high-pressure crystallized samples by electron microscope. Bassett et al. [71] discovered by permanganic oxidizing technique that there are two principal types of subdivision of the hexagonal crystals. The first one is that boundaries are formed perpendicular, or nearly so, to lamellae, and the second and rarer type of subdivision is that parallel to the plane of lamellae. Our results [104] have showed that the transformation of the (100) direction of hexagonal phase to the (110) direction of orthorhombic phase indeed occurs on the phase transition. The spacing and diffraction intensity of the (110) plane of orthorhombic ECC increase with increasing crystallization pressure, indicating that this transformation is intensified by pressure.

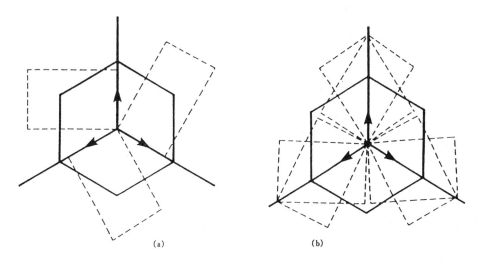

Figure 11 Schematic representations for the possible orientations from the hexagonal to the orthorhombic phase. (a) $(100)_{hexa} \to (010)_{orth}$; (b) $(100)_{hexa} \to (110)_{orth}$. (From Ref. 102.)

VII. THEORY OF NUCLEATION OF ECC UNDER HIGH PRESSURE

Concerning the nucleation of PE ECC under high pressure, so far two different theories have been proposed. Hikosaka et al. [105–109] suggested that there is no essential difference between the mechanisms of formation of FCC and ECC. At the primary nucleation stage, a FCC lamella forms first from the melt, while at the secondary nucleation stage, i.e., when crystallization proceeds into the disordered hexagonal phase, an ECC grows from the FCC by lamellar thickening via chain-sliding diffusion (Fig. 12). By extending Frank and Tosi's kinetic theory and using a generalized linear sequential process, Hikosaka derived a formula of the secondary nucleation rate j for both ECC and FCC, which

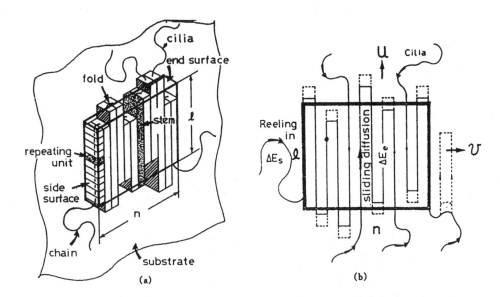

Figure 12 Monolayer nucleus with uneven end surface: (a) definition of terms; (b) definition of ΔE_e and ΔE_s, the activation free energies for chain sliding diffusion within a nucleus and for reeling from the melt or the solution, respectively. U and ν are the growth rates parallel and normal to the stem axis, respectively. (From Ref. 109.)

can be given as:

$$j = \eta_0 \left(\frac{kT}{h}\right) \left[\sum_{m=0}^{\infty} \exp\left\{\frac{G_m}{kT} + \frac{\Delta E_m}{k(T - T_g)}\right\}\right]^{-1} \quad (4)$$

where η_0 is a constant, kT is the thermal energy at a temperature T, h is Planck constant, G_m is the free energy for forming a nucleus of the mth stage, ΔE_m is the activation energy necessary for diffusion from the mth to the $(m + 1)$th stage, and T_g is the glass transition temperature. It is shown that j is determined by two competing factors, a thermodynamic factor G_m and a kinetic factor ΔE_m. The latter consists of two kinds of diffusion activation energy, namely $\Delta E_{s,m}$ for diffusion of a chain within the melt or solution and $\Delta E_{e,m}$ for sliding diffusion of a chain within a crystal (nucleus or lamella). In the two diffusion activation energies, $\Delta E_{e,m}$ plays an essential role in the mechanism of formation of FCC and ECC. When PE crystallizes from the melt into the disordered hexagonal phase, $\Delta E_{e,m}$ is small because chains can easily slide within crystals due to the high mobility of the hexagonal phase, the thermodynamic factor G_m becomes dominant, and ECC will develop by lamellar thickening caused by the thermodynamic driving force. On the contrary, when PE crystallizes into the ordered orthorhombic phase, $\Delta E_{e,m}$ is large because chain-sliding diffusion within orthorhombic crystals is difficult, thus the kinetic factor ΔE_m will become dominant and FCC will form because lamellar thickening is difficult. According to this theory, Hikosaka predicted that lamella of any polymer will tend to thicken if chains can slide to some extent within the crystal. Therefore, some polymers that crystallize from the melt into the hexagonal phase will show a continuous change from FCC to ECC just from a change in the crystallization temperature.

Sawada et al. [110] and the authors of this Chapter [104,111] have proposed another theory, the bundle-like nucleation theory, for the mechanism of ECC formation. Both groups of workers suggested that crystallization under high pressure starts from partially extended-chain nucleation rather than from the folded-chain nucleation as proposed by Hikosaka [103,104]. This theory was established on the basis of the following facts:

Extended- and folded-chain crystallization are mutually independent processes, and extended-chain crystallization can take place prior to folded-chain crystallization [6,19–25].

A rapid cooling of PE under high pressure has shown that the average length of the extended chain segments in the melt increases with increasing pressure [112]. High-pressure Raman spectroscopic study of the PE melt has shown that the ratio, $r = I(2890)/I(2850)$, of the intensities of the 2890 cm^{-1} and 2850 cm^{-1} bands in the C—H stretching region increases with increasing pressure, indicating that the concen-

tration of *trans* bonds and degree of intrachain order in the melt increase with increasing pressure [77]. These results provided experimental evidence for the extended-chain nucleation.

Most of the high-pressure crystallization kinetics studies have shown that ECC grows one dimensionally at high pressures [113–116].

Our results [104,117] have shown that the extended chains of ethyl cellulose liquid crystal at high pressures can act as the nucleus of PE ECC and induce the formation of ECC. (The details will be introduced in Section IX.)

The excess free energy of the nucleus surface, σ_e, plays an essential role in determining the nucleation mode. It is known that both folded-chain and extended-chain crystallization rates increase with increasing pressure [113–115,118]. Since the crystalization rate is sensitive to the value of σ_e, it may be concluded that both the excess free energy of the nucleus surface for folded-chain nucleation, σ_{ef}, and that for bundle-like nucleation, σ_{eb}, are reduced, but their reduction extents are different with increasing pressure. From the data of the isothermal crystallization rates, Sawada et al. [113] estimated that the value of σ_{ef} at 300 MPa is a half of that at atmospheric pressure. Nevertheless, it is natural that there should be a lower limit on the value of σ_{ef} corresponding to the lowest energy to form the fold surface, in other words, σ_{ef} will decrease to a constant value with increasing pressure. However, a limit on σ_{eb} has not been found. To illustrate these situations, the dependences of σ_{ef} and σ_{eb} on pressure are schematically shown in Fig. 13. It shows a continuous change from folded-chain to bundle-like nucleation with a change in crystallization pressure. At pressures below the critical value, P_T, $\sigma_{ef} < \sigma_{eb}$, the folded-chain nucleation mode is dominant and results in the formation of FCC, whereas when the pressure is raised above P_T, $\sigma_{eb} < \sigma_{ef}$, the bundle-like nucleation mode becomes dominant and ECC will form. In the case of the addition of liquid crystal polymer, the σ_{eb}–P curve shifts toward lower pressure, that is, the critical pressure P_T for bundle-like nucleation decreases to P_T' [104,117].

In the bundle-like nucleation, there are two possibilities: one is the attachment of a whole molecule to the crystal substrate and the other is the attachment of part of a molecule to the substrate. For long molecules, the former situation is unfavorable because of the high free energy created on forming a thick nucleus. It would be reasonable to consider that each molecule is partially incorporated into the crystal substrate. This bundle-like nucleation can be expressed by the model shown in Fig. 14.

Nearby the melting point, the formation or disappearance of a nucleus is actually a stochastic process. Thus, the nucleation process can be treated by using the method of the stochastic processes. This method has

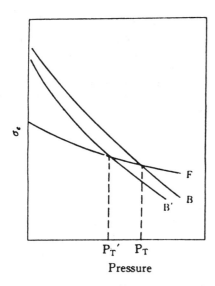

Figure 13 Schematic diagram of the dependence of σ_e on pressure. (F) Denotes folded-chain nucleus, (B) denotes bundle-like nucleus and (B') denotes addition of ethyl cellulose liquid crystal polymer. (From Refs. 104, 110, 111, and 117.)

been used to describe the nucleation rate for a folded-chain nucleus [119,120], similarly, it can be used to describe the surface nucleation rate for a bundle-like nucleus.

By denoting $\alpha_{\gamma-1}$ and β_γ as the forward and backward transition rates between the states $\gamma - 1$ and γ, respectively, the net transition rate J_γ between the states

$\gamma - 1$ and γ can be written as:

$$J_\gamma = \alpha_{\gamma-1}N_{\gamma-1} - \beta_\gamma N_\gamma \qquad (5)$$

where $N_{\gamma-1}$ and N_γ are the occupation numbers at the states $\gamma - 1$ and γ, respectively, N_γ is time-dependent and given by:

$$\frac{dN_\gamma}{dt} = J_\gamma - J_{\gamma+1} \qquad (6)$$

under steady-state conditions, $dN_\gamma/dt = 0$.

In this model, it is assumed that, except for the first step that has transition rate pairs α_0 and β_1, the following steps have the same transition rate pairs α_1 and β_2. The net transition rates are:

$$J_1 = \alpha_0 N_0 - \beta_1 N_1$$
$$J_2 = \alpha_1 N_1 - \beta_2 N_2 \qquad (7)$$
$$\cdots\cdots\cdots\cdots\cdots\cdots$$
$$J_\gamma = \alpha_1 N_{\gamma-1} - \beta_2 N_\gamma$$

For simplicity of treatment, steady-state conditions are imposed. Therefore, the net transition rates are all equal from Eq. (6) and can be denoted by J. Each equation in Eq. (7), except the first, is multiplied by (β_1/α_1) $(\beta_2/\alpha_1)^{i-2}, i = 2, 3, \ldots, \gamma$, and all equations are added to obtain:

$$J = A_0 N_0 - B_1 N_\gamma \qquad (8)$$

where:

$$A_0 = \frac{\alpha_0}{1 + (\beta_1/\alpha_1) \sum\limits_{k=1}^{\gamma-1} (\beta_2/\alpha_1)^{k-1}} \qquad (9)$$

$$B_1 = \frac{\beta_1(\beta_2/\alpha_1)^{\gamma-1}}{1 + (\beta_1/\alpha_1) \sum\limits_{k=1}^{\gamma-1} (\beta_2/\alpha_1)^{k-1}} \qquad (10)$$

The condition $\beta_2/\alpha_1 < 1$ has to be satisfied to assure nucleation growth, and if γ is larger, the steady-state solution can be obtained:

$$J = \frac{\alpha_0(\alpha_1 - \beta_2)N_0}{\alpha_1 + \beta_1 - \beta_2} \qquad (11)$$

According to Flory [121], Mandelkern [122], and Price [123], the free energy change in the formation of a bundle-like nucleus shown in Fig. 14 can be expressed as:

$$\Delta F = 2b_0 l\sigma + 2ma_0 b_0 \sigma_{eb} - ma_0 b_0 l\Delta g$$
$$+ kT\left[\frac{ml}{Lc_0} - m \ln \frac{L - n + 1}{L}\right] \qquad (12)$$

where $a_0, b_0,$ and c_0 are the dimensions of the molecules comprising the bundle-like nucleus, $m,$ the number of molecules in the nucleus, $n,$ the number of repeat units

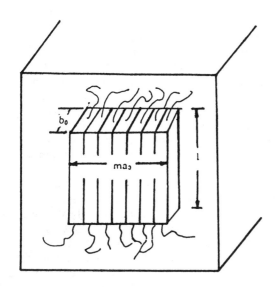

Figure 14 Model of monomolecular growth of bundle-like nucleus. (From Refs. 104, 110, and 111.)

in a crystalline sequence, $l = ncc_0$, the thickness of the bundle-like nucleus, L, the contour length of a molecule in number of repeat units, Δg, the bulk free energy of fusion, σ, the excess free energy of the lateral surface, T, the crystallization temperature, and k, Boltzmann constant. The bulk free energy of fusion Δg is given by $\Delta h \Delta T / T_m^\circ$, where Δh is the heat of fusion, ΔT, the supercooling, and T_m°, the equilibrium melting temperature.

In Eq. (12), the fourth term results from the increased volume available to the ends of the polymer chains on melting and the fifth term results mainly from the requirement that the ends of the molecules should stay out of the crystallites. Both terms are entropy terms giving the molecular weight dependence of the formation of bundle-like nucleus. Thus, the net transition rate J can be determined by the following equations:

$$\alpha_0 = \frac{kT}{h} \exp\left[- \left\{ 2b_0 l\sigma + 2b_0^2 \sigma_{eb} + kT \left(\frac{n}{L} - \ln \frac{L - n + 1}{L} \right) - 0.5 b_0^2 l \Delta g \right\} \middle/ kT \right] \tag{13}$$

$$\alpha_1 = \frac{kT}{h} \exp\left[- \left\{ 2b_0^2 \sigma_{eb} + kT \left(\frac{n}{L} - \ln \frac{L - n + 1}{L} \right) - 0.5 b_0^2 l \Delta g \right\} \middle/ kT \right] \tag{14}$$

$$\beta_1 = \beta_2 = \frac{kT}{h} \exp(-0.5 b_0^2 l \Delta g / kT) \tag{15}$$

where h is Planck constant. For simplicity, it is assumed that $a_0 = b_0$, and the free energy gain $b_0^2 l \Delta g$ on attaching a stem of l to the crystal substrate is equally shared in the forward and backward steps.

In the primary nucleation stage of crystallization at small supercoolings and high pressures, the growth rate G and net transition rate J can be correlated by the following relation:

$$G \propto b_0 J \tag{16}$$

Therefore, through Eqs. (11), (13), (14), (15), and (16),

the growth rate G is a function of the crystallization temperature, the crystal thickness, and the molecular weight. When the crystallization temperature and the molecular weight are held constant, $G(l)$ shows a maximum at the value of the thickness l^*, which satisfies the condition $\delta G / \delta l = 0$. The thickness l^* is defined as the crystal thickness, since it is expected that the fast crystal growth is experimentally observed [124].

For the bundle-like nucleation, the relation $\sigma_{eb} < \sigma_{ef}$ holds. Since the values of σ_{ef} are usually in the range of 80–100 erg/cm², a value of $\sigma_{eb} = 60$ erg/cm² was used in this calculation. By substituting the values of $b_0^2 = 18 \times 10^{-16}$ cm² and $c_0 = 1.27 \times 10^{-8}$ cm (from the study of PE unit cell by Bunn [125]) and $\Delta h = 2.8 \times 10^9$ erg/cm³ (by Mandelkern et al. [126]), the estimated growth rates $G(l^*)$ are shown against the temperature variable $T_m^\circ / T\Delta T$ for three cases in Fig. 15. It can be seen that at a constant high pressure, the increase in crystallization temperature will lead to an increase in the temperature variable $T_m^\circ / T\Delta T$ and thus a decrease in the growth rate G, that is, elevating the crystallization temperature is unfavorable for the growth of ECC, which agrees with the experimental results. On the other hand, at a given crystallization temperature, the increase in pressure will lead to a decrease in $T_m^\circ / T\Delta T$ due to the increase of T_m° with pressure, and thus an increase in G. Therefore, high pressure favors not only the reduction of σ_{eb}, but also the growth of ECC, which has been demonstrated by high-pressure crystallization kinetics studies [113–115,118].

In the secondary nucleation stage, the remaining amorphous portions of the molecule begin to grow in the chain direction. This is schematically shown in Fig. 16. At first, nucleation with the nucleus thickness l_1 takes place in the chain direction and after completion of the lateral deposition, the next nucleation with the thickness l_2 takes place, and this process is repeated over and over. The same surface nucleation rate equation as the primary stage can be used to describe these nucleation processes.

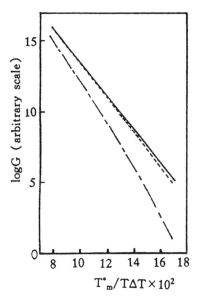

Figure 15 Plots of log G vs. $T_m^\circ / T\Delta T$ for (-··) L = 1000, (---) L = 10,000, and (—) L = 100,000. Calculated for σ = 10 erg/cm² and σ_{eb} = 60 erg/cm². (From Ref. 110.)

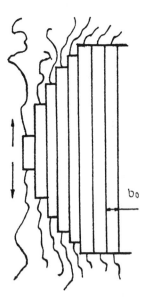

Figure 16 Model of the longitudinal growth of bundle-like nucleus seen from lateral direction. The arrows indicate the direction of longitudinal growth at a velocity u. (From Refs. 104, 110, and 111.)

By using Eq. (12), the transition rates were chosen in the following way.

$$\alpha_{0j} = \frac{kT}{h} \exp\left[-\left\{2b_0 l_j \sigma + kT\left(\frac{n_j}{L} - \ln\frac{L_j - n_j}{L_j}\right)\right.\right.$$
$$\left.\left. - 0.5 b_0^2 l_j \Delta g\right\} \middle/ kT\right] \tag{17}$$

$$\alpha_{1j} = \frac{kT}{h} \exp\left[-\left\{kT\left(\frac{n_j}{L} - \ln\frac{L_j - n_j}{L_j}\right)\right.\right.$$
$$\left.\left. - 0.5 b_0^2 l_j \Delta g\right\} \middle/ kT\right] \tag{18}$$

$$\beta_{1j} = \beta_{2j} = \frac{kT}{h} \exp(-0.5 b_0^2 l_j \Delta g/kT) \tag{19}$$

where $L_j = L - \sum_{k=0}^{j-1} n_k + 1$ and $n_j = l_j/c_0$, $j = 1, 2, \ldots$, and $n_0 = l^*/c_0$ is set. Subscript j denotes the jth nucleation process.

The growth rate for jth process can be expressed as:

$$u_j \propto l_j J_j \tag{20}$$

where J_j is the net transition rate for jth nucleation process. The longitudinal growth rates u vs. the temperature variable $T_m^\circ/T\Delta T$ are plotted in Fig. 15.

In fact, the further longitudinal growth of a bundle-like nucleus will be hindered by such factors as chain entanglement and deformation as crystallization proceeds. But such effects are not taken into account in this model.

VIII. EFFECT OF MOLECULAR WEIGHT ON THE MELTING–CRYSTALLIZATION OF PE UNDER HIGH PRESSURE

In the case of atmospheric pressure crystallization, the formation of ECC becomes difficult as the molecular weight of polymers increases [8]. But in the case of high-pressure crystallization, the situation is completely reversed, that is, the formation of ECC is much easier with increasing molecular weight. This results from the difference in the mechanism of the formation of ECC at atmospheric pressure and high pressure. The extended-chain crystallization at atmospheric pressure takes place in a folded-chain nucleation manner and has a mode of three-dimensional growth. The formation of ECC of high-molecular weight polymers at atmospheric pressure is difficult primarily because of the high free energy created on the formation of a large critical nucleus. However, the extended-chain crystallization under high pressure starts probably in a partially extended-chain manner and has a mode of one-dimensional growth [127]. According to the bundle-like nucleation model, the longer molecules should be kinetically favorable for producing ECC. Figure 15 shows that longer molecules give rise to higher growth rates at the primary nucleation stage. At low supercoolings, the dependence of the growth rate G on molecular weight is notable. For example, at $T_m^\circ/T\Delta T = 0.1$ the ratio of the growth rate for $L = 10,000$ to that for $L = 1000$ is about 20; however, at $T_m^\circ/T\Delta T = 0.16$, the ratio is about 4000. The dependence of the longitudinal growth rate u on molecular weight at the secondary nucleation stage is the same as in the case of the primary nucleation stage, as shown in Fig. 17. Therefore, at high pressures where the crystallization at low supercoolings is possible, crystallization for high-molecular weight is considerably favorable.

The triple point, the orthorhombic–hexagonal transition temperature and the melting temperature of hexagonal phase in the phase diagram of Fig. 6 are sensitive to molecular weight. Bassett et al. [6] have shown that the location of the triple point can be extrapolated to as low as 270 MPa for ultra-high molecular weight linear PE, whereas it is above 500 MPa for molecular weights below 10^4. Takamizawa et al. [128] have studied the melting and crystallization behaviors at 500 MPa by DTA for the samples of molecular weight from 6.5×10^3 to 1.3×10^5. According to their results, the hexagonal phase begins to appear at the molecular weight above 10^4, the transition temperature of the orthorhombic–hexagonal phase and the melting temperature of the hexagonal phase increase with molecular weight up to 1.3×10^5. Asahi [88] has studied the effect of molecular weight on the phase diagram of low-molecular weight PE (MW: 1000, 2000, 6500, 16,000) by x-ray diffraction

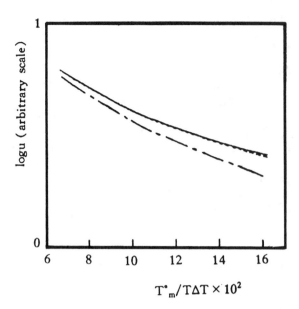

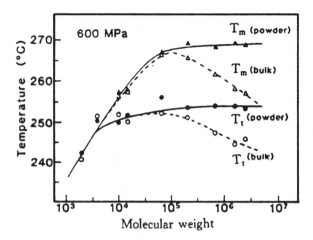

Figure 17 Plots of $\log u$ vs. $T_m^\circ / T \Delta T$ for (-··) L = 1000; (---) L = 10,000; (—); L = 100,000. Calculated for $\sigma = 10$ erg/cm². (From Ref. 110.)

Figure 18 Molecular weight dependencies of the phase transition temperature (T_t) from orthorhombic to hexagonal phase and the melting temperature (T_m) of the hexagonal phase of PE. $\bigcirc \bullet$ = phase transition from orthorhombic to hexagonal phase; $\triangle \blacktriangle$ = melting of the hexagonal phase. (From Ref. 131.)

with a high-pressure diamond-anvil cell. Phase diagrams of these samples from 20–300°C up to pressures of about 1 GPa has been reported. It was shown that the samples of MW 6500 and 16,000 exhibit the hexagonal phase, but those of MW 1000 and 2000 do not. Although the orthorhombic–hexagonal transition temperatures are almost the same for the samples of MW 6500 and 16,000, the melting temperature of the hexagonal phase of MW 16,000 is higher than that of MW 6500 by 10 or 15°C. Hoehn et al. [129,130] have studied the crystallization of ten linear PE samples ranging from $M_W = 4.9 \times 10^4$ to 4.6×10^6 at 510 MPa and 242°C. They found that with increasing molecular weight the degrees of crystallinity of these high-pressure crystallized samples decreases from 100% to 80%, and the content of ECC decreases from 98% to 85% over the indicated molecular weight range. Yasuniwa et al. [131–134] have determined the phase diagrams between about 400 and 600 MPa for as-polymerized powder and bulk PE samples ranging from low (2×10^3) to ultra-high molecular weight (UHMW 2.5×10^6). For both samples the orthorhombic–hexagonal transition temperature (T_t) and the melting temperature (T_m) of the hexagonal phase increase with the molecular weight in the same manner up to the molecular weight about 10^5. Above this molecular weight, T_m and T_t of the bulk samples decrease with increasing molecular weight, while those for the powder sample increase slightly (Fig. 18). The high-pressure hexagonal phase region for an as-polymerized powder sample of UHMW PE is higher in pressure and temperature than that of the bulk sample, while the high-pressure

phase region of the solution-crystallized sample of UHMW PE lies between those of as-polymerized powder and bulk samples. However, the morphology of high-pressure crystallized PE changes remarkably with increasing molecular weight. Three phases, namely a lamellar crystalline, a crystalline–amorphous interphase, and an amorphous phase are present in high-molecular weight samples. Solid-state ^{13}C NMR spectroscopy [135] show that the thickness of the crystalline–amorphous interphase reaches 8.0 nm and the mass fraction of the amorphous phase does not exceed 0.05. GPC measurements [136] also show that ECC in high-molecular weight samples may have a surface layer of about 20 ± 10 nm, including a completely disordered amorphous layer. Nevertheless, the amorphous phase is absent from low-molecular weight samples. The morphology of pressure-crystallized low- and medium-molecular weight PE is a closely spaced band structure formed by ECC lamellae, while the band in pressure-crystallized high-molecular weight PE is formed by the parallel arrangement of fibrils. There are many cracks between the bands. The morphology of pressure-crystallized powder sample of UHMW PE is fibrous band formed by the parallel arrangement of long fibrils whose diameters are less than about 500 nm. The width of the fibrous band is about 0.2 μm. The pressure-crystallized bulk sample of UHMW PE has three types in its morphology: isolated fibril, textile structure, and band structure. The textile structure is the network superstructure of the fibrils, and the band structure is also formed by the parallel arrangement of the fibrils [131–134]. Concerning the drastic change of the melting behavior and morphology of high-

and ultra-high molecular weight PE, Yasuniwa et al. [131–134] have given an interpretation as follows. As the molecular chain length of high- and ultra-high molecular weight PE is approximately the same as the width of a band (ca. 1 μm) or longer than it, during crystallization under high pressure it would be difficult for the chain disentangling to occur on the time scale of experiments and, therefore, the entanglements should be trapped in the sample. The entanglements hinder the crystal growth and cause defects in the crystal. Consequently, crystals become smaller and more disordered with molecular weight, resulting in a decrease of the melting temperature. In contrast, the melting temperature increases by superheating, which is also caused by the entanglements. Since the UHMW PE bulk sample, which was kneaded in the melt at atmospheric pressure, has too many entanglements, the former effect is large. Thus, the UHMW PE bulk sample shows the lowest melting temperature. Conversely, the latter effect is large in the UHMW PE powder sample, therefore the UHMW PE powder sample shows the highest melting temperature. The entanglement density in the solution-crystallized UHMW PE sample is the lowest among these samples, so these two effects are not prominent in the solution-crystallized UHMW PE sample. Consequently, it shows an intermediate high-pressure hexagonal phase region between the UHMW PE powder and bulk samples.

Fractional crystallization [6,53–55,70,71,136–142] features the high-pressure crystallization of polydisperse PE. When this has happened, the melting curves both at atmospheric pressure and high pressure for the pressure-crystallized polydisperse PE always exhibit quite complicated endothermic peaks, which indicate the melting of several types of crystals with a different morphology and different average lamellar thickness. A typical example [136] is shown in Fig. 19. The DSC melting curve at atmospheric pressure for a polydisperse PE sample (M_w 165,000, $M_w/M_\eta = 5.2$) crystallized at 513 MPa and 234°C for 235 h shows four endothermic peaks. The corrected nitration/GPC curve for this sample also shows four main peaks and corresponds well to the DSC melting pattern of the original sample. The average chain lengths of these peaks represent the average lamellar thickness of four types of crystals and are estimated to be 1.85 μm, 594 nm, 212 nm, and 59 nm. The first three types of crystals can be regarded as ECC, while the last one as FCC. By permanganic oxidizing technique the different populations can be identified within the overall morphology. This fractional crystallization behavior can be interpreted in terms of the movements of the triple point to higher pressures for shorter molecules, and to lower pressures for longer molecules. At a given crystallization temperature and pressure, the supercooling will be greatest, and the growth rate fastest for the longest molecules. Therefore, the longest molecules will be the first to form ECC via the hexagonal phase, while the longer molecules tend to form the hexagonal phase

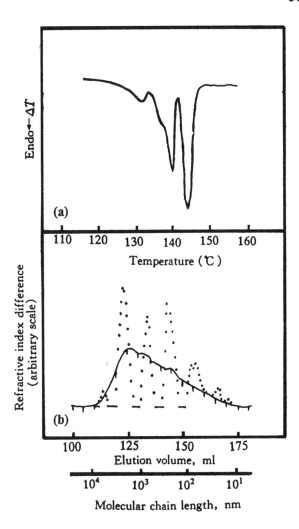

Figure 19 (a) DSC melting curve at atmospheric pressure for PE crystallized at 234°C and 513 MPa for 235 h; (b) matching GPC curve after degraded with fuming nitric acid for 10 h. (From Ref. 136.)

slowly. In addition, the shorter molecules may have to wait to form the hexagonal phase until the supercooling has reached a large enough value. Thus, they will only crystallize relatively late in the cooling process. The lamellar thicknesses of the later-formed crystals are influenced so strongly by the first-formed ECC that their values are relatively small. The shortest molecules are unable to grow as the hexagonal phase in the time available and eventually crystallize into the orthorhombic phase directly from the melt. For linear and methyl-branched PE, the lamellar thickness is generally of the order of micrometers, but the lamellar thickness is remarkably reduced for branched PE containing ethyl, butyl, or larger branches, as such branches are excluded from the hexagonal and orthorhombic lattices and the inter-branch separation gives an upper limit to the lamellar thickness [143].

IX. EFFECT OF OTHER COMPONENTS ON THE MELTING–CRYSTALLIZATION OF PE UNDER HIGH PRESSURE

In recent years, there have been increasing reports on the study of the melting and crystallization behaviors of PE under high pressure by addition of other components. Nakafuku et al. [144–147] have reported that some high melting temperature diluents such as 1-, 2-, 4-, 5-tetrachlorobenzene; 1-, 3-, 5-tribromobenzene; hexamethylbenzene; and tetracontane affect the melting and crystallization processes, the phase transition, and

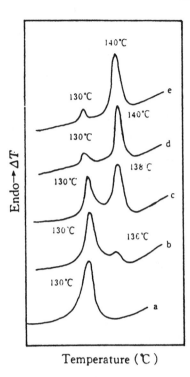

Figure 21 DSC melting curves at atmospheric pressure for PE–ethyl cellulose mixture crystallized at elevated pressure. (a) 100 MPa; (b) 150 MPa; (c) 300 MPa; (d) 400 MPa; and (e) 750 MPa. (From Ref. 117.)

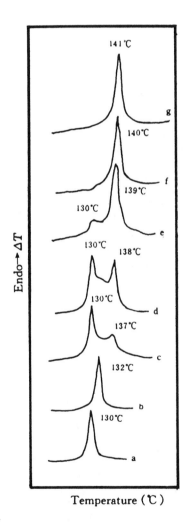

Figure 20 DSC melting curves at atmospheric pressure for PE crystallized at elevated pressure. (a) 0.1 MPa; (b) 300 MPa; (c) 440 MPa; (d) 750 MPa; (e) 1.00 GPa; (f) 1.21 GPa; and (g) 1.50 GPa. (From Ref. 117.)

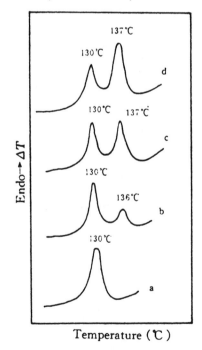

Figure 22 DSC melting curves at atmospheric pressure for PE–ethyl cellulose mixture crystallized at 150 MPa and various temperatures. (a) 150°C; (b) 170°C; (c) 190°C; and (d) 210°C. (From Ref. 117.)

the formation of ECC of PE under high pressure. The phase transition of medium- and ultra-high molecular weight (MMW and UHMW) PE to the hexagonal phase is impeded by the addition of tetracontane and does not occur below a weight fraction of PE (for MMW PE 0.7, for UHMW PE 0.4) at 500 MPa. The content of ECC of MMW and UHMW PE formed by high-pressure crystallization decreases with tetracontane, and below a weight fraction of PE (for MMW PE 0.5, for UHMW PE 0.35), ECC does not form even on crystallization at 500 MPa. Nakafuku et al. [144–147] considered that in the binary mixture of PE and tetracontane, the liquid tetracontane behaves like a lubricant and impedes the formation of entanglements or pressure-induced crosslinkage of PE molecules. Therefore, the hexagonal phase of PE disappears in the binary mixture of the weight fraction below the indicated critical value. In the binary mixtures of PE/1-, 3-, 5-tribromobenzene and 1-, 2-, 4-, 5-tetrachlo-robenzene, the phase transition from the orthorhombic phase to the hexagonal phase of PE does not occur under high pressure, but the transition did occur in the PE/hexamethylbenzene system. The formation of ECC of PE is largely impeded in the mixtures with 1-, 3-, 5-tribromobenzene and 1-, 2-, 4-, 5-tetrachlorobenzene.

A result different from that of Nakafuku et al. [144–147] was obtained by us from the study of a binary mixture of PE–ethyl cellulose liquid crystal under high pressure. We have reported [104,117] that addition of 1% ethyl cellulose by weight facilitates the formation of ECC of PE and moderates the conditions for the formation of ECC, that is, the pressure limit is lowered from 440 MPa to 150–200 MPa, and the temperature limit lowered from 200–245°C to 170°C. The DSC melting curves at atmospheric pressure for pure PE ($M_\eta = 1.06 \times 10^5$, $\rho = 0.9556$ g/cm^3) and PE–ethyl cellulose mixture crystallized at various pressures are shown in Figs. 20 and

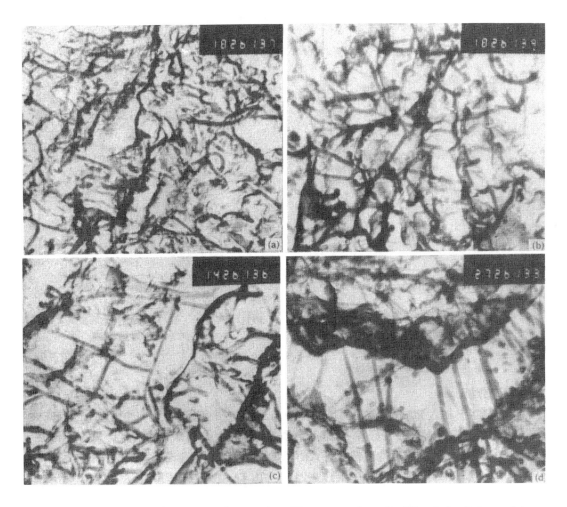

Figure 23 Transmission electron microscopy micrographs of fracture surfaces for PE–ethyl cellulose mixture crystallized at 170°C and various pressures: (a) 150 MPa, ×18,000; (b) 300 MPa, ×18,000; (c) 440 MPa, ×14,000; and (d) 750 MPa, ×27,000. (From Ref. 117.)

21, respectively. At crystallization pressures below 440 MPa, the DSC melting curves for pure PE show a single endothermic peak at 130°C corresponding to the melting of FCC. At 440 MPa, an additional high-temperature endothermic peak at 137°C appears in the DSC curve corresponding to the melting of ECC. This high-temperature peak develops strongly and moves to a higher temperature of 141°C with increasing crystallization pressure, reflecting the increase in content of ECC. These facts imply that ECC of pure PE cannot be produced until the crystallization pressure reaches 440 MPa. By mixing with 1% ethyl cellulose by weight, however, the high-temperature peak corresponding to the melting of ECC appears in the DSC curve for PE ethyl cellulose mixture crystallized at a pressure as low as 150 MPa. As the pressure is increased to 440 MPa, the high-temperature peak becomes the main one and moves to a higher temperature of 140°C, indicating a considerable increase in content of ECC at this pressure (440 MPa). Figure 22 gives the DSC melting curves at atmospheric pressure for PE–ethyl cellulose mixture crystallized at 150 MPa and various temperatures. At crystallization temperatures below 170°C, the DSC curves only show a melting peak of FCC at 130°C, indicating that it is impossible to obtain ECC at these temperatures. It can be accounted for in terms of the transition temperature of ethyl cellulose from the solid to the liquid crystal phase (170–180°C). Below 170°C, ethyl cellulose does not display the liquid crystal behavior, i.e., chain extension arrangements, thus it cannot induce the formation of ECC. At crystallization temperatures above 170°C, however, the DSC curves show the melting peak of ECC in addition to that of FCC, and the melting peak of ECC becomes strong with increasing crystallization temperature, indicating that the induction role of ethyl cellulose to the formation of ECC is enhanced by elevating crystallization temperature. Transmission electron microscopy reveals that ECC and FCC coexist in the PE–ethyl cellulose samples crystallized at 150 and 300 MPa, and the former becomes dominant in the sample crystallized at 440 MPa. Ethyl cellulose forms a fibrous network inlaid in ECC lamellae of PE (Fig. 23). Wide angle x-ray diffraction shows that the addition of ethyl cellulose results in the remarkable change of crystal lattice parameters of orthorhombic phase in high-pressure crystallized PE–ethyl cellulose samples. The fact that the addition of ethyl cellulose into the PE could relieve the ECC-forming condition may be explained in terms of the ordered alignment of molecular chains of ethyl cellulose in the liquid crystal state. This structure should be easy to initiate such a nucleation that on which the PE ECC can grow easily and rapidly under a relaxable temperature and pressure due to the reduction of excess free energy of the nucleus surface. Further work is required to explore the details of the mechanism. This finding provides a new way to obtain ECC and is important for the practical application of ECC.

REFERENCES

1. I. M. Ward, *Mechanical Properties of Solid Polymer,* Wiley-Inter Science, London (1971).
2. B. Wunderlich, and T. Arakawa, *J. Polym. Sci. Part A-2* 2: 3697 (1964).
3. P. H. Geil, F. R. Anderson, B. Wunderlich, and T. Arakawa, *J. Polym. Sci. Part A-2* 2: 2043 (1964).
4. P. H. Geil, F. R. Anderson, B. Wunderlich, and T. Arakawa, *J. Polym. Sci. Part A-2* 2: 3707 (1964).
5. D. V. Rees, and D. C. Bassett, *J. Polym. Sci. Part A-2* 9: 385 (1971).
6. D. C. Bassett, *Polymer 17*: 460 (1976).
7. G. Czornyj, and B. Wunderlich, *J. Polym. Sci. Polym. Phys. Edn.* 15: 1905 (1977).
8. R. H. Olley, and D. C. Bassett, *J. Polym. Sci. Polym. Phys. Edn.* 15: 1011 (1977).
9. D. C. Bassett, B. A. Khalifa, and R. H. Olley, *J. Polym. Sci. Polym. Phys. Edn.* 15: 995 (1977).
10. T. Hatakeyama, H. Kanetsuna, and T. Hashimoto, *J. Macromol. Sci. Phys. B7*: 411 (1973).
11. B. Wunderlich, and L. Mellilo, *Makromol. Chem 118*: 250 (1968).
12. B. Wunderlich, *J. Polym. Sci. Part A-2 5*: 7 (1967).
13. B. Wunderlich, and T. Davidson, *J. Polym. Sci. Part A-2 7*: 2043 (1969).
14. R. B. Prime, and B. Wunderlich, *J. Polym. Sci. Part A-2 7*: 2061 (1969).
15. R. B. Prime, and B. Wunderlich, *J. Polym. Sci. Part A-2 7*: 2073 (1969).
16. R. B. Prime, B. Wunderlich, and L. Mellilo, *J. Polym. Sci. Part A-2 7*: 2091 (1969).
17. B. Wunderlich, *Pure Appl. Chem. 31*: 49 (1972).
18. B. Wunderlich, *Macromolecular Physics, Vol. 2,* Academic Press, New York (1976).
19. D. C. Bassett, B. A. Khalifa, and B. Turner, *Nature (Phys. Sci.) 239*: 106 (1972).
20. D. C. Bassett, and B. Turner, *Nature (Phys. Sci.) 240*: 146 (1972).
21. D. C. Bassett, and B. Turner, *Phil. Mag. 29*: 285 (1974).
22. D. C. Bassett, and B. Turner, *Phil. Mag. 29*: 925 (1974).
23. H. Kanetsuna, S. Mitsuhashi, M. Iguchi, T. Hatakeyama, M. Kyotani, and Y. Maeda, *J. Polym. Sci. Part C 42*: 783 (1973).
24. P. D. Calvert, and D. R. Uhlmann, *J. Polym. Sci. Part A-2 10*: 1811 (1972).
25. M. Yasuniwa, R. Enoshita, and T. Takemura, *Jpn. J. Appl. Phys. 15*: 1421 (1976).
26. X. W. Zhang, and R. Huang, *Chinese Polym. Bull. 1*: 9 (1994).
27. D. C. Bassett, and R. Davitt, *Polymer 15*: 721 (1974).
28. Y. Miyamoto, C. Nakafuku, and T. Takemura, *Polym. J. 3*: 122 (1972).
29. S. Gogolewski, and A. J. Pennings, *Polymer 14*: 463 (1973).
30. S. Gogolewski, and A. J. Pennings, *Polymer 16*: 673 (1975).
31. S. Gogolewski, and A. J. Pennings, *Polymer 18*: 647 (1977).
32. S. Gogolewski, and A. J. Pennings, *Polymer 18*: 654 (1977).
33. S. Gogolewski, *Polymer 18*: 63 (1977).
34. S. Gogolewski, and A. J. Pennings, *Polymer 18*: 660 (1977).
35. J. E. Stamhuis, and A. J. Pennings, *Polymer 18*: 667 (1977).

36. K. Matsushige, and T. Takemura, *J. Polym. Sci. Polym. Phys. Edn. 16*: 921 (1978).
37. A. Siegmann, and P. J. Harget, *J. Polym. Sci. Polym. Phys. Edn. 18*: 2181 (1980).
38. N. Hiramatsu, and S. Hirakawa, *Polym. J. 12*: 105 (1980).
39. J. D. Hoffman, L. J. Frolen, G. S. Ross, and J. I. Lauritzen, *J. Res. Natl. Bur. Stand. (U.S.) A79*: 671 (1975).
40. D. M. Sadler, and G. H. Gilmer, *Polymer 28*: 242 (1987).
41. D. M. Sadler, *Nature 326*: 174 (1987).
42. F. R. Anderson, *J. Appl. Phys. 35*: 64 (1964).
43. R. S. Porter, *J. Mater. Sci. 19*: 1193 (1974).
44. I. M. Ward, *J. Polym. Sci. Polym. Phys. Edn. 12*: 635 (1974).
45. B. Smith and T. Lemstra, *J. Polym. Sci. Polym. Phys. Edn. 19*: 877 (1981).
46. F. N. Liberti, and B. Wunderlich, *J. Polym. Sci. Part A-2 6*: 833 (1968).
47. C. W. Bunn, A. J. Cobbold, and R. P. Palmer, *J. Polym. Sci. 9*: 385 (1958).
48. L. Mellilo, and B. Wunderlich, *Kolloid-Z. Z. Polym. 250*: 417 (1972).
49. L. Mandelkern, J. M. Price, M. Gopalan, and J. G. Fatou, *J. Polym. Sci. Part A-2 4*: 385 (1966).
50. C. L. Gruner, B. Wunderlich, and R. C. Bopp, *J. Polym. Sci. Part A-2 7*: 2099 (1969).
51. T. Arakawa, and B. Wunderlich, *J. Polym. Sci. Part C 16*: 653 (1967).
52. M. Yasuniwa, C. Nakafuku, and T. Takemura, *Polym. J. 4*: 526 (1973).
53. Y. Maeda, and H. Kanetsuna, *J. Polym. Sci. Polym. Phys. Edn 12*: 2551 (1974).
54. Y. Maeda, and H. Kanetsuna, *J. Polym. Sci. Polym. Phys. Edn. 13*: 637 (1975).
55. Y. Maeda, and H. Kanetsuna, *J. Polym. Sci. Polym. Phys. Edn. 14*: 2057 (1976).
56. M. Yasuniwa, and T. Takemura, *Polymer 15*: 661 (1974).
57. D. C. Bassett, S. Block, and G. J. Piermarini, *J. Appl. Phys. 45*: 4146 (1974).
58. W. Pechhold, E. Liska, H. P. Grossman, and P. C. Hägele, *Pure Appl. Chem. 96*: 127 (1976).
59. T. Yamamoto, H. Miyaji, and K. Asai, *Jpn. J. Appl. Phys. 16*: 1891 (1977).
60. T. Yamamoto, *J. Macromol. Sci. Phys. 16*: 487 (1979).
61. U. Leute, *Polym. Bull. 4*: 407 (1981).
62. S. L. Wunder, *Macromolecules 14*: 1024 (1981).
63. R. Eckel, H. Schwickert, M. Buback, and G. R. Stroble, *Polym. Bull. 6*: 559 (1982).
64. H. Tanaka and T. Takemura, *Polym. J. 12*: 355 (1980).
65. H. Tanaka and T. Takemura, *Jpn. J. Appl. Phys. 22*: 1001 (1983).
66. L. B. Shih and R. G. Priest, *Appl. Spectrosc. 38*: 687 (1984).
67. K. Matsushige and T. Takemura, *J. Cryst. Growth 48*: 343 (1980).
68. K. Asai, *Polymer 23*: 391 (1982).
69. U. Leute, W. Dollhopf, and E. Liska, *Colloid Polym. Sci. 256*: 914 (1978).
70. D. C. Bassett, *Principles of Polymer Morphology,* Cambridge University Press, London, (1981).
71. D. C. Bassett, *Developments in Crystalline Polymers, Vol. 1,* Applied Science Publishers, London, p. 115 (1982).
72. K. Nagata, K. Tagashira, S. Taki, and T. Takemura, *Jpn. J. Appl. Phys. 19*: 985 (1980).
73. R. G. Priest, *J. Appl. Phys. 52*: 5930 (1981).
74. R. G. Priest, *Macromolecules 15*: 1357 (1982).
75. R. G. Priest, *Macromolecules 18*: 1504 (1985).
76. T. Yamamoto, *Polymer 24*: 943 (1983).
77. T. Yamamoto, *Polymer 25*: 178 (1984).
78. J. D. Barnes, and B. M. Fanconi, *J. Chem. Phys. 56*: 5190 (1972).
79. G. Strobl, B. Ewen, E. W. Fischer, and W. Piesczek, *J. Chem. Phys. 61*: 5257 (1974).
80. J. Doucet, I. Denicolo, A. F. Craievich, and C. Germain, *J. Chem. Phys. 80*: 1647 (1984).
81. M. Maroncelli, S. P. Qi, H. L. Strauss, and R. G. Snyder, *J. Am. Chem. Soc. 104*: 6237 (1982).
82. T. Yamamoto, *J. Chem. Phys. 82*: 3790 (1985).
83. K. Takamizawa, Y. Urabe, and H. Hasegawa, *Polym. Prepr. Jpn. 28*: 1854 (1979).
84. H. Orth, and E. W. Fischer, *Makromol. Chem. 88*: 188 (1965).
85. A. S. Vaughan, G. Ungar, D. C. Bassett, and A. Keller, *Polymer 26*: 726 (1985).
86. G. Ungar, *Macromolecules 19*: 1317 (1986).
87. T. Yamamoto and K. Asai, *Polym. Prepr. Jpn. 27*: 1828 (1978).
88. T. Asahi, *J. Polym. Sci. Polym. Phys. Edn. 22*: 175 (1984).
89. S. B. Clough, *Polym. Lett. 8*: 519 (1970).
90. A. J. Pennings, and A. Zwijnenburg, *J. Polym. Sci. Polym. Phys. Edn. 17*: 1011 (1979).
91. N. A. J. M. Van Aerle, and P. J. Lemstra, *Polym. J. 20*: 131 (1988).
92. S. Tsubakihara, A. Nakamura, and M. Yasuniwa, *Polym. J. 23*: 1317 (1991).
93. G. Ungar, and A. Keller, *Polymer 21*: 1273 (1980).
94. D. T. Grubb, *J. Mater. Sci. 9*: 1715 (1974).
95. D. C. Bassett, and B. A. Khalifa, *Polymer 17*: 275 (1976).
96. R. B. Morris, and D. C. Bassett, *J. Polym. Sci. Polym. Phys. Edn. 13*: 1501 (1975).
97. D. C. Bassett, *High Temp. High Press. 9*: 553 (1977).
98. M. Hikosaka, S. Rastogi, A. Keller, and H. Kawabata, *J. Macromol. Sci. Phys. B31*: 87 (1992).
99. S. Rastogi, M. Hikosaka, H. Kawabata, and A. Keller, *Macromolecules 24*: 6384 (1991).
100. S. Rastogi, M. Hikosaka, H. Kawabata, and A. Keller, *Makromol. Chem. Macromol. Symp. 48/49*: 103 (1991).
101. J. F. Jackson, T. S. Hsu, and J. W. Brasch, *J. Polym. Sci. Part B 10*: 207 (1972).
102. S. Miyashita, T. Asahi, H. Miyaji, and H. Asai, *Polymer 26*: 1791 (1985).
103. D. C. Bassett, and B. A. Khalifa, *Polymer 17*: 291 (1976).
104. Q. Fu, Ph. D. Thesis, Sichuan Union University (in China) (1993).
105. M. Hikosaka, and S. Tamaki, *J. Phys. Soc. Jpn. 50*: 638 (1981).
106. M. Hikosaka, and T. Seto, *Jpn. J. Appl. Phys. 21*: 332 (1982).
107. M. Hikosaka, and T. Seto, *Jpn. J. Appl. Phys. 23*: 956 (1984).
108. M. Hikosaka, *Polymer 28*: 1257 (1987).
109. M. Hikosaka, *Polymer 31*: 458 (1990).
110. S. Sawada, K. Kato, and T. Nose, *Polym. J. 11*: 551 (1979).
111. Q. Fu, R. Huang and X. W. Zhang, *Science in China Series E 26:* 467 (1996).
112. D. J. Cultler, P. J. Hendera, and R. D. Sang, *Faraday Discuss. Chem. Soc. 68*: 320 (1979).
113. S. Sawada, and T. Nose, *Polym. J. 11*: 477 (1979).

114. T. Hatakeyama, H. Kanetsuna, H. Kaneda, and T. Hashimoto, *J. Macromol. Sci. Phys. B10*: 359 (1974).
115. M. Kyotani, and H. Kanetsuna, *J. Polym. Sci. Polym. Phys. Edn. 12*: 2331 (1974).
116. X. W. Zhang, and R. Huang, *Chinese Polym. Bull. 2*: 86 (1994).
117. Q. Fu, R. Huang, and H. Huang, *Science in China Series A 24*: 1218 (1994).
118. D. R. Brown, and J. Jonas, *J. Polym. Sci. Polym. Phys. Edn. 22*: 655 (1984).
119. J. I. Lauritzen, and J. D. Hoffman, *J. Appl. Phys. 44*: 4340 (1973).
120. I. C. Sanchez, and E. A. Dimarzio, *J. Chem. Phys. 55*: 893 (1971).
121. P. J. Flory, *J. Chem. Phys. 17*: 223 (1949).
122. L. Mandelkern, *J. Appl. Phys. 26*: 443 (1955).
123. F. P. Price, *Nucleation,* A. C. Zettlemoyer, ed.) Marcel Dekker Publishers, New York (1969).
124. I. C. Sanchez, and E. A. Dimarzio, *Macromolecules 4*: 677 (1971).
125. C. W. Bunn, *Trans. Faraday Soc. 35*: 482 (1939).
126. L. Mandelkern, J. G. Fatou, and C. Howard, *J. Phys. Chem. 68*: 3386 (1964).
127. S. Sawada, and T. Nose, *Polym. J. 11*: 227 (1979).
128. K. Takamizawa, Y. Urabe, A. Ohno, and T. Takemura, *Polym. Prepr. Jpn. 22*: 447 (1973).
129. H. H. Hoehn, R. C. Ferguson, and R. R. Hebert, *Polym. Eng. Sci. 18*: 457 (1978).
130. R. C. Ferguson, and H. H. Hoehn, *Polym. Eng. Sci. 18*: 466 (1978).
131. M. Yasuniwa, K. Haraguchi, C. Nakafuku, and S. Hirakawa, *Polym. J. 17*: 1209 (1985).
132. M. Yasuniwa, and C. Nakafuku, *Polym. J. 19*: 805 (1987).
133. M. Yasuniwa, S. Tsubakihara, and C. Nakafuku, *Polym. J. 20*: 1075 (1988).
134. M. Yasuniwa, M. Yamaguchi, A. Nakamura, and S. Tsubakihara, *Polym. J. 22*: 411 (1990).
135. R. Kitamaru, F. Horii, Q. Zhu, D. C. Bassett, and R. H. Olley, *Polymer 35*: 1171 (1994).
136. Y. Maeda, and H. Kanetsuna, *Polym. J. 13*: 357 (1981).
137. Y. Maeda, and H. Kanetsuna, *Polym. J. 13*: 371 (1981).
138. L. Kardos, H. M. Li, and K. A. Huckshold, *J. Polym. Sci. Part A-2 9*: 2061 (1971).
139. M. Hikosaka, S. Minomura, and T. Seto, *Jpn. J. Appl. Phys. 19*: 1763 (1980).
140. M. Hikosaka, *Jpn. J. Appl. Phys. 20*: 617 (1981).
141. Y. Maeda, H. Kanetsuna, K. Nagata, K. Matsushige, and T. Takemura, *J. Polym. Sci. Polym. Phys. Edn. 19*: 1313 (1981).
142. Y. Maeda, H. Kanetsuna, K. Tagashira, and T. Takemura, *J. Polym. Sci. Polym. Phys. Edn. 19*: 1325 (1981).
143. J. A. Parker, D. C. Bassett, R. H. Olley, and P. Jaaskelainen, *Polymer 35*: 4140 (1994).
144. C. Nakafuku, *Polym. J. 17*: 869 (1985).
145. C. Nakafuku, *Polymer 27*: 353 (1986).
146. C. Nakafuku, M. Yasuniwa, and S. Tsubakihara, *Polym. J. 22*: 110 (1990).
147. C. Nakafuku, H. Nakagawa, M. Yasuniwa, and S. Tsubakihara, *Polymer 32*: 696 (1991).

21

Structure, Stability, and Degradation of PVC

Mohammad Kazim Naqvi

Saudi Basic Industries Corporation (SABIC), Riyadh, Saudi Arabia

I. INTRODUCTION

The monomer, vinyl chloride, was discovered by Justus Von Liebig in 1835. On assignment from Liebig, Victor Regnault confirmed this reaction and was allowed to publish alone and gain credit for this discovery. In 1872, E. Baumann detailed the light-induced change of vinyl chloride monomer to solid products, which he thought were isomers of the monomer. The properties described by him are those we ascribe to poly(vinyl chloride) (PVC) [1].

On its own, PVC is an extremely unstable polymer, in fact, almost certainly the least naturally stable polymer in commercial use. As Grassie has noted, "Had this polymer been discovered at the present stage of development of the plastics industry, it would almost certainly have been eliminated as useless because of its general instability to all common degradative agents" [2].

PVC started gaining commercial significance in the 1930s only after the successful development and application of effective means of stabilization. With the help of modifying agents (stabilizers, plasticizers, lubricants, fillers, and other additives), PVC can be made to exhibit a wide spectrum of properties, ranging from the extremely rigid to the very flexible. In fact, it is unmatched by any other polymer in the diversity and range of its applications.

PVC is commercially one of the most important thermoplastics in the world today. The 1994 global capacity for PVC production stood at 23.8 million metric tons (Table 1). Its growth rate averaged 7% per annum in the 1970s. In 1980 it was the second largest volume thermoplastic used in the United States (the first being low-density polyethylene [LDPE]) and was the lowest priced among the five leading plastics: LDPE, PVC, high density polyethylene (HDPE), polypropylene (PP), and polystyrene (PS). The diversity of its applications and its low cost are responsible for its importance in the world market. PVC compositions have competed with and successfully displaced such materials as metals, wood, leather, rubber, cellulose and other natural and synthetic polymers, textiles, conventional paints and coatings, masonry and ceramics, glass, and paper. The manufacture of wire and cables, resilient flooring, and phonograph records represent a few examples where production was revolutionized by the commercial advent of vinyl chloride resins.

PVC demand in the United States by application is shown in Table 2.

From Table 2 it is clear that:

1. PVC pipe shows a slight increase in its share of the portfolio and remains by far the major use of PVC. This indicates that it may be winning the intermaterial competition in certain areas. Most likely areas are the sectors of underground drainage, waste, and vent applications where it has a better price/performance compromise than ABS or PP. Other construction sectors and related applications such as cladding and cable insulation retain their positions as large PVC users and stay in line with overall PVC growth.

Table 1 Poly(vinyl Chloride) Capacity, 1994 (Thousand Metric Tons)

Western Europe	5,838
United States	5,503
Japan	2,359

Table 2

Application	Demand % (1993)	Demand % (2000)
Pipe	42.6	44.2
Wire and cable	4.3	4.3
Siding	12.9	12.9
Film/sheet	7.4	7.7
Other extrusion	3.0	1.8
Calendaring	11.9	11.7
Molding	5.7	5.1
Coating	4.0	3.4
Other	8.2	8.9

2. Film and sheet refers more to the rigid packaging and construction areas and is, therefore, continuing to grow in contrast to calendered film, which falls slightly as a result of food packaging film losses. This is not so marked as in Europe, however, where public opinion and environmental pressure against PVC in end uses is more extreme.

3. Molding includes bottle blow molding, and the continued move toward PET for use in soft drink and water bottles depresses growth slightly, although the trend remains positive.

Current (September 1995) international prices (in U.S. dollars per ton) of various bulk polymer in different regions of the world are given in Table 3 for comparison.

From Table 3 it is clear that PVC is by far the cheapest among the five bulk polymers in the world today. Its unmatched versatility and low cost make PVC commercially one of the most important thermoplastics today. Even in applications in which it is in competition with some of the other bulk polymers its price–performance ratio gives it a slight edge. Despite the attacks from the environmental lobby, PVC continues to retain its commercial significance in the world market.

PVC on its own is extremely heat- and shear-sensitive and cannot be processed into finished goods, as it starts degrading at temperatures considerably lower than those required to process it. The processing of PVC requires a number of additives. These include heat stabilizers, impact modifiers, processing aids, and lubricants.

Table 3

	PVC	LDPE	HDPE	PP	PS
Europe	650	800	750	800	1000
USA	500	750	720	750	950
Far East	600	850	720	800	750
Southeast Asia	600	850	730	800	800

Each of these additives plays an important role in either the processing or service life of the finished goods. In addition, fillers and pigments are two larger scale additives that play an important role in product economics and aesthetics.

II. THERMAL STABILITY OF PVC

The polymerization of vinyl chloride monomer, in common with other vinyl monomers, proceeds by a free-radical mechanism involving the usual steps of initiation, propagation, and termination. Poly(vinyl chloride) is formed in a regular head-to-tail manner Eq. (1) [3–6].

$$\text{--C--C--}\left[\text{C--C}\right]_n\text{--} \tag{1}$$

Thermogravimetric analysis and other studies made on low-molecular weight "model" compounds such as 1, 3, 5,-trichlorohexane [7,8] corresponding to the idealized head-to-tail structure of PVC show these structures to be considerably more stable than the polymer. This "abnormal" instability of the polymer is attributed to structural irregularities or defects in the polymer chain, which serve as initiation sites for degradation.

Unstabilized PVC starts degrading even under mild heating conditions. In the initial stages of degradation, HCl starts "zipping-off" from the polymer backbone, resulting in the formation of polyene sequences, which can be observed as a gradual coloration and darkening of the polymer. On continued heating in the later stages of degradation dienes, aromatics, and other hydrocarbons are given-off from the polymer. The polymer undergoes chain scission, leading to a gradual deterioration of mechanical properties and chemical resistance, accompanied by crosslinking, which predominates in the later stages. The presence of oxygen leads to rapid oxidation of polyene sequences, resulting in chain scission and formation of carbonyl and hydroperoxide groups. The process of degradation is autocatalytic and quickly reduces the polymer to a worthless black char.

It is possible to suppress these undesirable degradation reactions by using stabilizers, but the exact mechanism of dehydrochlorination of PVC is still a subject of controversy. PVC is a highly complex polymer and its reactions with stabilizers are also quite complex involving side reactions and side products, which also influence the course of degradation of the polymer. Hence, the generalizations made regarding the reactions of stabilizers with PVC should be treated with a certain amount of caution. Investigation of the degradation and stabilization of PVC remains a very active area of research in polymer chemistry. Such interest reflects the worldwide economic importance of this unstable polymer.

A polymer is a complex mixture of molecules that is difficult to define and reproduce. The quality of the polymer is markedly affected by the conditions of preparation. Different degrees and types of branching, differences in the number and distribution of various irregular structures, along with the degree of purity of the finished product and conditions of further treatment all influence the thermal stability of the polymer and the course of its thermal degradation. This further complicates the study of this polymer and explains the differences be-

tween the results and conclusions obtained by different workers.

III. STRUCTURAL DEFECTS IN PVC

A. Unsaturation

Since it is well known that chloroalkenes are often much less stable than the corresponding alkanes, olefinic unsaturation may be an important source of thermal instability in PVC. Chain-end unsaturation could arise by disproportionation during bimolecular reaction of polymer radicals Eq. (2).

$$2\,(\text{\textasciitilde}CH_2\overset{\bullet}{C}HCl) \longrightarrow \text{\textasciitilde}CH_2CH_2Cl \;+\; \text{\textasciitilde}CH=CHCl \tag{2}$$

Chain transfer reactions involving the monomer could also result in unsaturation of the chain ends according to the following two reactions [Eqs. (3) and (4)].

$$\text{\textasciitilde}CH_2\overset{\bullet}{C}HCl \;+\; CH_2=CHCl \longrightarrow \text{\textasciitilde}CH_2CH_2Cl \;+\; CH_2=\overset{\bullet}{C}Cl$$

$$CH_2=\overset{\bullet}{C}Cl \;+\; CH_2=CHCl \longrightarrow CH_2=CClCH_2CHCl\text{\textasciitilde}^{\bullet} \tag{3}$$
$$\text{(A)}$$

$$\text{\textasciitilde}CH_2\overset{\bullet}{C}HCl \;+\; CH_2=CHCl \longrightarrow CH_3\overset{\bullet}{C}HCl \;+\; \text{\textasciitilde}CH=CHCl \tag{4}$$
$$\text{(B)}$$

(A) can react further to give a branched polymer with a tertiary chlorine at the branch point [Eq. (5)]. This is also an important structural feature from the point of

view of thermal stability of PVC and will be discussed in the following section.

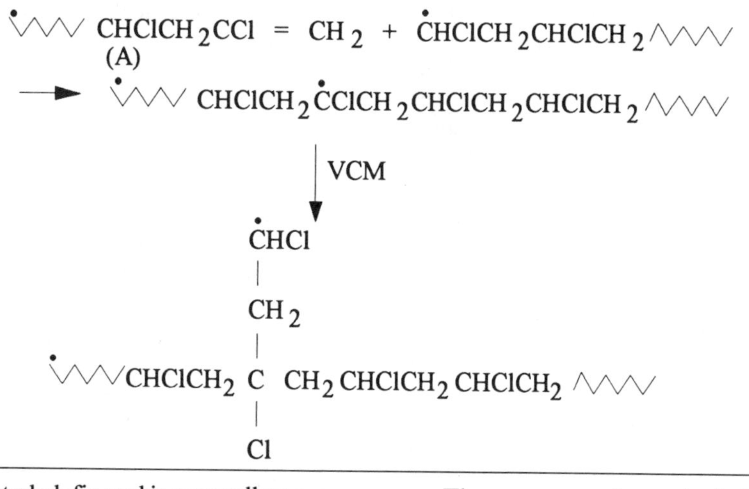

$$\overset{\bullet}{\text{\textasciitilde}}CHClCH_2CCl=CH_2 \;+\; \overset{\bullet}{C}HClCH_2CHClCH_2\text{\textasciitilde}$$
$$\text{(A)}$$
$$\longrightarrow \overset{\bullet}{\text{\textasciitilde}}CHClCH_2\overset{\bullet}{C}ClCH_2CHClCH_2CHClCH_2\text{\textasciitilde}$$

$$\Big\downarrow \text{VCM}$$

$$\overset{\bullet}{C}HCl$$
$$|$$
$$CH_2$$
$$|$$
$$\overset{\bullet}{\text{\textasciitilde}}CHClCH_2\;C\;CH_2\;CHClCH_2\;CHClCH_2\text{\textasciitilde}$$
$$|$$
$$Cl$$

$$\tag{5}$$

(B) is a 1,2-disubstituted olefin and is generally unreactive in free-radical reactions.

Braun and Schurek [9] assumed that during polymerization a reaction can occur between the polymer and free radicals that leads to the elimination of hydrogen chloride and formation of a double bond. The formation of HCl during the polymerization of vinyl chloride has been observed [10].

The presence of acetylenic impurities or butadiene in the system can also give rise to unsaturation in the polymer backbone.

Starnes et al. [11] appear to have evidence for the following mechanism of formation of a double bond next to the chain end [Eq. (6)]:

$$\text{\scriptsize WWW}\ CH_2\ \overset{\bullet}{C}HCl \xrightarrow[\substack{\text{head-to-head}\\ \text{addition}}]{VCM} \text{\scriptsize WWW}\ CH_2CHClCHCl\overset{\bullet}{C}H_2$$

$$\downarrow \text{1,2-Cl migration}$$

$$\text{Polymer with} \xleftarrow{VCM} \text{\scriptsize WWW}\ CH_2\ CHCl\overset{\bullet}{C}HCH_2Cl \qquad (6)$$
$$CH_2\ Cl\ \text{branch}$$

$$\downarrow -\overset{\bullet}{C}l$$

$$\text{\scriptsize WWW}\ CH_2\ CH = CHCH_2Cl$$

$$+\ \overset{\bullet}{C}l$$

There are many conflicting reports regarding chain-end unsaturation as initiation sites for thermal degradation. Arlman [13] found a linear dependence between the rate of dehydrochlorination and the reciprocal value be attributed to signal broadening for protons in conjugated polyene structures and, thus, seems to constitute direct evidence for the involvement of olefinic defect sites in the early stages of the degradation process and their incorporation into runs of conjugated double bonds in the later stages.

There are many conflicting reports regarding chain-end unsaturation as initiation sites for thermal degradation. Arlman [13] found a linear dependence between the rate of dehydrochlorination and the reciprocal value of the average molecular weight of several PVC samples with molecular weights that varied between 56,000 and 175,000. The same linear dependence was found by Bengough and Sharpe [14,15] and Talamini and Pezzin [16]. Bengough and Sharpe used four samples of PVC initiated by AIBN at different temperatures in bulk. Talamini and Pezzin's results were based on three samples obtained by fractionation of a commercial suspension-polymerized product, Sicron 548. The molecular weights in both cases lay within the range of about 12,000–60,000.

However, Bengough and Varma [17,18] affirm that there is no systematic dependence between the rate of dehydrochlorination and molecular weight for samples obtained by fractionation of Geon III.

Onazuka and Asahina [19] investigated the thermal decomposition of PVC divided into 30 fractions by measuring the HCl evolved and monitoring the changes in ultraviolet (UV) spectra. They concluded that fractions having lower and higher molecular weights were less stable than the medium fractions. This is in agreement with the findings of Crosato-Arnaldi and coworkers [20].

Studies on model compounds also suggest that unsaturated chain-end groups should not have an important influence on the thermal stability of PVC [21]. In conclu-

sion, it may be said that the effect of unsaturated end groups on the stability of PVC is minor. Definite effects have been demonstrated on low-molecular weight samples, and this is to be expected because reactions favoring the formation of unsaturated end groups favor low-molecular weight. The nature of this effect is not clear.

Braun [22] showed from ozonolysis that for fractions of bulk PVC the number of internal double bonds and the rate of thermal degradation, although dependent on each other, were independent of the molecular weight. This clearly demonstrated the role of internal unsaturation on the stability of the polymer. After careful chlorination of the double bonds, an increase in thermal stability was observed and the number of double bonds as determined by oxidation with potassium permanganate were reduced. It was also shown that one polyene sequence was formed from each isolated double bond.

The correlation of stability with the number of internal double bonds has been demonstrated by other workers [20,23,24]. In work reported by Lindeschmidt [25], this correlation was not observed.

Studies of thermal degradation on low-molecular weight model compounds have shown that the structure [Eq. (7)]

$$R'CH\!\!=\!\!CHCHClCH_2R'' \qquad (7)$$

$$(R' = \text{alkyl}, R'' = \text{H or alkyl})$$

in the gas phase [26–28] and in the liquid phase [29] is the most labile one. It is now generally accepted that chlorine atoms allylic to internal double bonds are the most labile and along with tertiary chlorines at branch points play the most significant role in the thermal degradation of PVC.

B. Branching

Branches in PVC can be formed by transfer to polymer during polymerization. Short branches in PVC have usually been considered to be formed by backbiting mechanisms similar to those occurring in polyethylene. Such branches should have a tertiary chlorine at the branch point [Eq. (8)]. Thus, in many earlier investigations the methyl content was taken as a measure of the number of tertiary chlorines present in the polymer. Another possible route to tertiary chlorines is copolymerization with unsaturated moieties at the ends of chains [30] [Eqs.

(3) and (5)]. However, the latter path requires terminal $CH_2=C(Cl)$-groups and, thus, is not consistent with the work of Enomoto [10], which suggests that the β-hydrogens of vinyl chloride are involved in the process of chain transfer to monomer. Thus, chain transfer to dead or growing polymer seems to be the most likely mechanism for the introduction of tertiary chlorine atoms [Eqs. (9a) and (9b)].

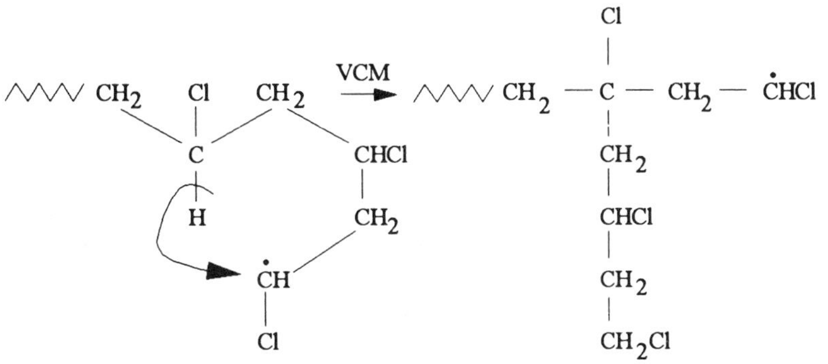

(8)

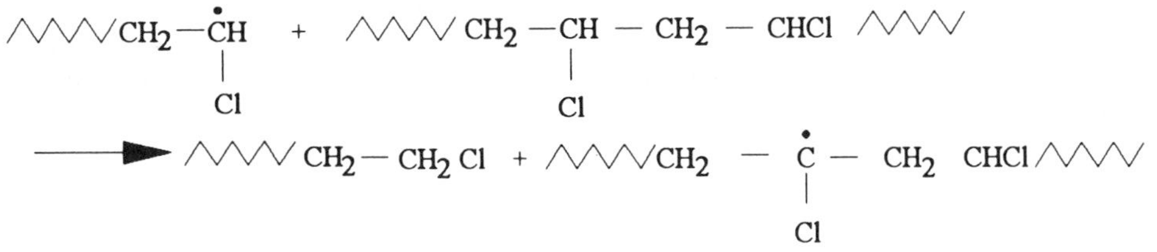

(9a)

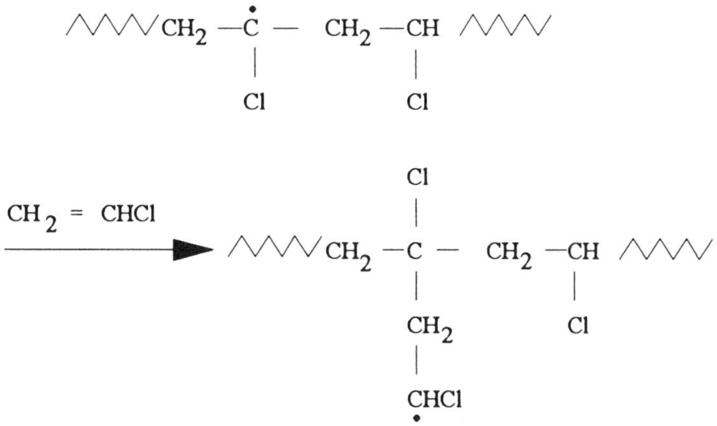

(9b)

Branching in PVC has been suggested as a possible source of instability for a long time. 2-methyl-2-chloropropane and 3-ethyl-3-chloropentane are much less stable than the corresponding secondary chlorine compounds [31]. Comparison of 3-chloro-3-ethylpentane with 3-chloro-4-ethylhexane showed that the latter was by far the most stable [32]. Macromodels containing ter-

tiary chlorine atoms have also been studied. Caraculacu [33] studied the degradation of a polymer obtained by copolymerization of vinyl chloride with 2-chloropropene and showed that tertiary chlorine atoms readily initiate degradation of temperatures lower than is usual for pure PVC. Copolymers of vinyl chloride and 2,4-dichloropent-1-ene [Eq. (10)] confirmed these findings [33].

$$
\begin{array}{c}
\mathrm{Cl} \\
| \\
-\mathrm{CH_2}-\mathrm{CHCl}-\mathrm{CH_2}-\mathrm{C}-\mathrm{CH_2}-\mathrm{CHCl} \\
| \\
\mathrm{CH_2} \qquad (10) \\
| \\
\mathrm{CHCl} \\
| \\
\mathrm{CH_3}
\end{array}
$$

Work with vinyl chloride/2-chloropropene copolymer has suggested that only 0.1–0.2 mol% of such groups would be needed to account for the instability of PVC [30]. Thus, the type of atom attached to the tertiary carbon at the branching sites in PVC should have a substantial effect on the thermal degradation of the polymer.

On the other hand, investigations with copolymers from vinyl chloride and 2,4-dichloropentene-1 led to the conclusion that because of steric reasons PVC could not contain branch points with tertiary chlorine atoms [34]. Braun and Wiess [35,36] confirmed these findings by further investigations with copolymers of vinyl chloride with 2-chloropropene. The thermal degradation of such copolymers, with the same content of methyl groups as of branch points in radically prepared PVC, was much faster. Also, the distribution of the formed polyene sequences of different length was quite different from that of PVC after the same degree of degradation. In copolymers, a remarkable shift to a shorter polyene sequences was observed.

Braun and Schurek [9] could not find any relationship between the number of branch points and the rate of degradation. Short branches in PVC are mainly chloromethyl groups with a hydrogen attached to the tertiary carbon. The methyl content is, thus, a measure of structures that are more stable than tertiary chlorines. For a model substance resembling the chloromethyl type of structure, Suzuki and Nakamura [29] reported a decomposition temperature exceeding 180°C. Abbas and Sorvik [23] found no obvious correlation between the dehydrochlorination rate and the methyl content in reduced PVC. There was a slight trend toward higher degradation rates at higher amounts of branching but the data were rather scattered. Suzuki et al. [37] found similar results. For very high degrees of branching they observed a much higher dehydrochlorination rate. The degradation rate was considered to be almost constant within the interval common for commercial polymers. During dehydrochlorination in oxygen, however, a greater dependence was observed.

Hjertberg and coworkers [38–41] were able to correlate the amount of labile chlorine, tertiary and internal allylic chlorine, to the dehydrochlorination rate. They studied PVC samples with increased contents of labile chlorine, which were obtained by polymerization at reduced monomer concentration. According to their results, tertiary chlorine was the most important defect in PVC. In agreement with other reports [42,43], the results also indicated that secondary chlorine was unstable at the temperatures in question, i.e., random initiation would also occur.

C. Determination of Labile Chlorines in PVC

The presence of allylic chlorines and tertiary chlorines and their influence on the thermal stability of PVC has now been established with some degree of confidence, and together they are considered to constitute the labile chlorine structures in the polymer. Numerous chemical modification methods involving the selective nucleophilic substitution of labile chlorines in PVC with other chemical moieties for identifying and quantifying labile structures have been reported in the literature.

Bengough and Onozuka [44] showed evidence of substitution of labile chlorines in PVC by acetate groups by treating the polymer with cadmium acetate under vigorous conditions and using IR spectroscopy. Naqvi [45] subjected PVC to Bengough and Onazuka's treatment and observed that the resulting polymer was less stable, as determined by the rate of thermal dehydrochlorination, than the untreated polymer. These observations have an important bearing on the Frye and Horst mechanism of PVC stabilization and the effect of polar and nonpolar groups on the thermal stability of the polymer and will be discussed in some detail in a later section.

Thane et al. [46] reported that in pentane suspension, alkylaluminum compounds efficiently alkylate labile chlorines in PVC, and alternatively, PVC carbonium ions could alkylate aromatic compounds to give rise to polymers of increased stability. The values of 2–3% for labile chlorines estimated by them were considerably higher than now generally believed.

In a series of articles, Caraculacu and coworkers described a method based on substitution of labile chlorines with phenol. IR [34] and UV [47] were used for determination of incorporated phenol. The published data indicate a detection limit of 2 and 0.5 labile chlorines per 1,000 monomer units.

Caraculacu et al. [48] also quantitatively determined allylic chlorines in PVC by isotopic exchange with $SO^{36}Cl_2$. The selective exchange of chlorine in the polymer was verified by experiments with model compounds. The number of allylic chlorines in PVC was found to be between 0.12 and 0.16 for 100 monomer units.

Using thiophenol instead of phenol, Michel et al. [49] found a new selective reaction that takes place exclusively with allylic chlorines and not with tertiary chlorines. A single product of thioether structure is formed [Eq. (11)].

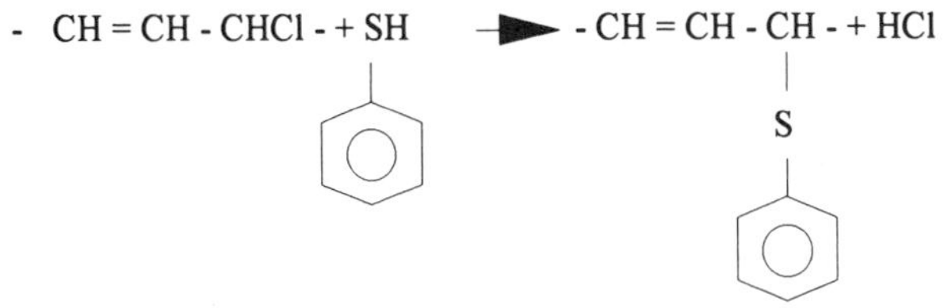

$$- CH = CH - CHCl - + \overset{\bigcirc}{SH} \longrightarrow - CH = CH - CH - + HCl$$

(11)

In usual PVC 1.53–2.54 allylic chlorines per 1,000 monomer units and in a benzene-soluble low-molecular weight PVC fraction, a higher value of 3.07 allylic chlorines per 1,000 monomer units was obtained. This method has come under certain criticism because it was found to give significantly higher values for labile (allylic) chlorines than do other methods, e.g., phenolysis [50].

Starnes and Plitz [51], in their studies of PVC stabilization with di(n-butyl)tin bis (n-dodecyl mercaptide) or with mixtures of the mercaptide and di(n-butyl)tin dichloride, found that the rate of dehydrochlorination was inversely related to the amount of sulphur content of approximately 0.9%. The following mechanism of allyl chloride substitution was proposed [Eq. (12)].

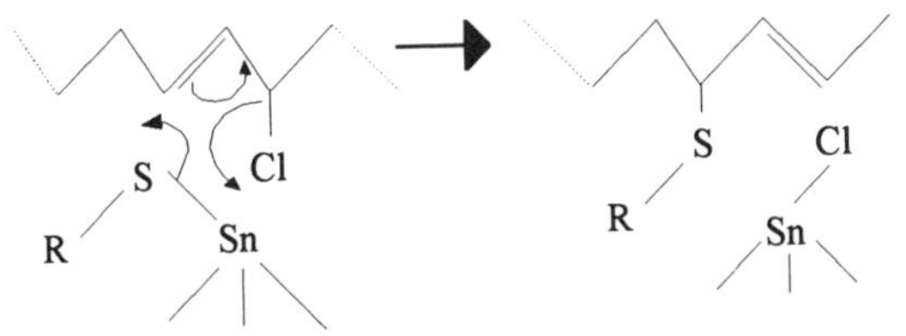

(12)

Lewis et al. [52,53] have also determined labile chlorines in PVC by a crown ether catalyzed acetoxylation of PVC and the thermal degradation characteristics of the modified polymer. The values were comparable with those obtained by the phenolysis method.

Rogestedt and Hjertberg [41] treated PVC with ethanol, trimethylaluminum, and dibutyltin maleate in order to substitute labile chlorines. The degradation behavior of the modified samples was compared with that of an ordinary suspension PVC and a PVC obtained by anionic polymerization [54–56]. The dehydrochlorination rate of anionic PVC, polymerized by using butyllithium as initiator, is substantially lower than that of ordinary PVC [57]. All modified samples and the anionic PVC showed the same behavior when degraded in pure nitrogen. Besides a decreased rate of dehydrochlorination, the polyenes became shorter. Degradation in an atmosphere containing HCl resulted in a higher dehydrochlorination rate and longer polyenes for all samples

with improved heat stability, except the sample treated with trimethylaluminum, which exhibited excellent thermal stability. The results showed that the polyene sequence distribution depends on the presence of HCl in the sample during degradation. The content of incorporated methyl groups in the alkylated sample was determined to be about 1 per 1,000 monomer units. Furthermore, the content of tertiary chlorine was reduced to less than 10% of that of unreacted PVC. It was concluded that the enhanced thermal stability was caused by removal of labile chlorines.

D. Initiator End Groups

The effect of the decomposition products of the polymerization initiator incorporated at the beginning of the chain is a controversial one. If the polymerization of vinyl chloride is initiated with organic peroxides, which decompose according to Eqs. (13) and (14):

$$(R-COO)_2 \longrightarrow 2R - CO\overset{\bullet}{O} \qquad (13)$$

$$R-CO\overset{\bullet}{O} \longrightarrow \overset{\bullet}{R} + CO_2 \qquad (14)$$

they lead to the structures of the type:

$$RCO\overset{\bullet}{O} + nCH_2 = CHCl \longrightarrow RCOO(CH_2CHCl)^{\bullet}_n$$

(15)

or,

$$\overset{\bullet}{R} + nCH_2 = CHCl \longrightarrow R(CH_2 - CHCl)^{\bullet}_n \qquad (16)$$

Where R is an alkyl or aryl group [58].

Peroxide initiators may also undergo primary radical transfer to produce unsaturated end groups, which may result in a less stable polymer. In the case of benzoyl peroxide, an additional possibility is initiation by phenyl radicals to give a polymer with terminal phenyl groups [Eq. (17)]:

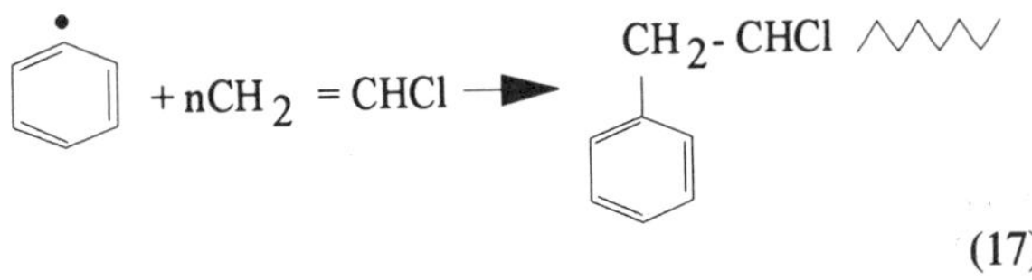

(17)

Instability at the chain end could then arise via the benzylic hydrogen atoms due to the possible resonance stabilization of the resulting radical [Eq. (18)]:

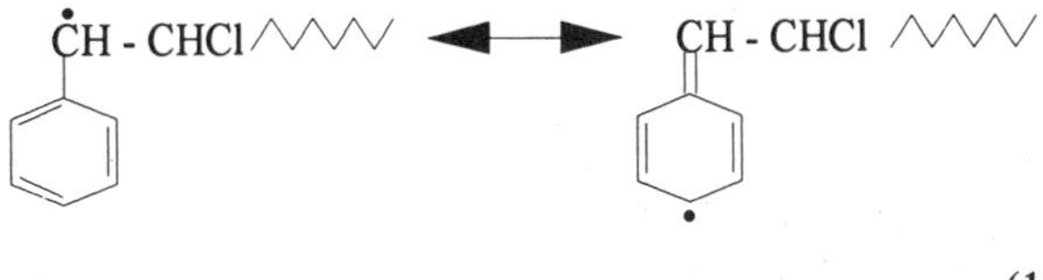

(18)

In earlier investigations chain ends were suggested to be important initiation sites for dehydrochlorination. Provided there are no transfer reactions during polymerization, at least half the polymer chain ends will carry initiator fragments. In practice, transfer reactions swamp the normal termination processes and <30% of the chain ends carry initiator residues [59].

Cittadini [60] and Corso [61] found that azo-initiated polymer liberates HCl more rapidly than peroxide-initiated material. Stromberg et al. [62] confirmed these findings and in addition found that PVC prepared by gamma irradiation was still more stable. Talamini and coworkers [63] kept the molecular weight relatively constant and compared the rate of dehydrochlorination for samples obtained with different initiators. Their results were consistent with those mentioned previously. Park and Smith [64] have shown that PVC prepared in solution by free-radical initiation contains 0.1–1.0 initiator fragments per polymer molecule. These polymers are more suitable for investigating the effects of initiator end groups than polymer prepared in suspension or emulsion systems where chain transfer to monomer is much more predominant.

Park and Skene [65] found that the effectiveness of the initiator end groups for causing dehydrochlorination was in the ratio lauroyl peroxide–isopropyl peroxidicarbonate–benzoyl peroxide–azoisobutyronitrile = 9.7:5.8:4.4:1.6. These values were obtained at 220°C, and the degradation rates were taken as the mean value for 0–20% dehydrochlorination.

However, the effect of the initiator has been suggested to be less important as model substances for such structures are much more stable than other possible structural irregularities already discussed [19,22,66,67].

E. Head-to-Head Structures

Head-to-head units formed in a molecule have not only been considered as initiation sites for the dehydrochlorination but also as termination points for the growing polyene sequences [19,66,68]. Head-to-head units can either be formed through termination by combination [Eq. (19)] or by head-to-head addition during propagation [Eq. (20)].

$$2 \text{ } \wedge\!\!\wedge\!\!\wedge\text{CH}_2-\overset{\bullet}{\text{C}}\text{H} \longrightarrow \text{ } \wedge\!\!\wedge\!\!\wedge\text{CH}_2-\underset{\underset{\text{Cl}}{|}}{\text{CH}}-\underset{\underset{\text{Cl}}{|}}{\text{CH}}-\text{CH}_2\text{ }\wedge\!\!\wedge\!\!\wedge \tag{19}$$

$$\wedge\!\!\wedge\!\!\wedge\text{CH}_2-\underset{\underset{\text{Cl}}{|}}{\overset{\bullet}{\text{C}}\text{H}} \xrightarrow[\substack{\text{Head-to-head} \\ \text{addition}}]{\text{VCM}} \wedge\!\!\wedge\!\!\wedge\text{CH}_2-\underset{\underset{\text{Cl}}{|}}{\text{CH}}-\underset{\underset{\text{Cl}}{|}}{\text{CH}}-\overset{\bullet}{\text{C}}\text{H}_2\text{ }\wedge\!\!\wedge\!\!\wedge \tag{20}$$

Shimizu and Ohtsu [69] have proposed a chemical method to determine head-to-head structures in PVC. Mitani et al. [70] found 2.5–7.0 head-to-head structures per 1,000 monomer units, increasing with the polymerization temperature. It has not been possible to detect internal head-to-head structure by ^{13}C-NMR spectroscopy with the detection limit of 2 per 1,000 monomer units [71]. Starnes et al. [71] found evidence for the absence of neighboring methylene groups by ^{13}C-NMR spectroscopy. However, the proposed rearrangement of head-to-head units at the radical chain ends resulting in chloromethyl branches [Eq. (6)] would partially explain their consumption during polymerization and their absence in the final product.

Hjertberg et al. [72] have studied various PVC samples using Shimizu and Ohtsu's method [69]. They concluded that this method mainly gave a measure of the content of saturated 1,2-dichloroethyl chain end groups, the presence of which has been conclusively demonstrated [73]. With caution they contend there are 0–0.2 head-to-head structures per 1,000 monomer units.

Head-to-head PVC prepared by adding chlorine to cis-polybutadiene has been found to be less stable than ordinary head-to-tail polymer [70].

From the results obtained by thermal decomposition of both low-molecular weight vicinal dichlorides in the gas phase [74,75] and of the copolymers of vinyl chloride and trans-1,2-dichloroethylene [72], it is not possible to attribute the cause of the thermal instability of PVC to the individual head-to-head structures. Crawley and McNeill [76] chlorinated cis-1,4-polybutadiene in methylene chloride, leading to a head-to-head, and a tail-to-tail PVC. They found, for powder samples under programmed heating conditions, that head-to-head polymers had a lower threshold temperature of degradation than normal PVC, but reached its maximum rate of degradation at higher temperatures.

The extent of head-to-head units in PVC and their effect on stability of the polymer is yet to be conclusively demonstrated, although it would seem that as compared to other structural defects their contribution to polymer instability is a minor one.

F. Oxygen-Containing Defects

Various oxygen-containing structures [68,77] arising either by a reaction with traces of oxygen already present during polymerization or by oxidation during storage or treatment of the finished polymer in air have also been considered to be a source of labile sites in the PVC molecule. Virgin PVC may contain up to 500 ppm peroxide [78], which together with hydroperoxides may act as initiation sites for dehydrochlorination. Popova et al. [79] reported a higher dehydrochlorination rate for the PVC polymerized in the presence of oxygen (0.1%). Decomposition temperatures were also 10–15°C lower. Garton and George [80] also observed decreased stabilities for polymers prepared in the presence of oxygen. This method of preparation introduces polyperoxide moieties that can decompose into HCHO, HCl, and CO [81]. However, the importance of oxygenated structures remains unclear for polymer prepared under normal conditions.

Sonnerskog [82] found that PVC could oxidize a number of phenols at room temperature in the absence of oxygen. In addition, a graft copolymer was formed when acrylonitrile was added to PVC. These results were taken as indications for the peroxy groups in PVC.

As far as oxidation of the polymer with oxygen of the air is concerned, the β-hydrogen atom in the neighborhood of the C=C double bond is the most likely one to be attacked by oxygen with the formation of hydroperoxide which undergoes further decomposition [19]. OH and CO groups have been detected spectroscopically in the polymer [67,83].

Minsker et al. [84,85] discussed the initiation of PVC degradation as a consequence of conjugated ketoallyl groups [Eq. (21)],

$$—CO—CH=CH—CHCl—CH_2— \qquad (21)$$

which should be expected to be more important in causing instability in PVC than chloroallyl groups.

Svetly et al. [86], in a very interesting study of the effect of cis- and trans-α,β-unsaturated ketones on the thermal stability of PVC, concluded that the propagation of polyene sequences occurs via migration of activating groups between polymer chains and that the activating groups are represented by oxygen-containing structures.

G. Tacticity and Crystallinity

Tacticity or stereochemical arrangement of atoms in three-dimensional space in relation to each other along the polymer chain cannot really be termed a structural defect. But researchers have shown that tacticity has an important bearing on the reactivity and thermal stability of PVC. For this reason tacticity is being discussed under this section.

Guyot et al. [87] studied the influence of regular structures on thermal degradation of PVC in an inert atmosphere and concluded that a higher degree of polymer crystallinity favors intermolecular condensation and, therefore, the acceleration of dehydrochlorination. They [88] also studied the thermal degradation of semicrystalline, low-molecular weight PVC fractions prepared by polymerizations catalyzed by tert-butyl magnesium chloride and observed that crystallinity of some fractions gave considerable thermal stability as long as the pyrolysis temperature remained below the melting point of the sample. In an interesting recent article, Behnisch and Zimmerman [89] relate the dehydrochlorination of PVC under various conditions to the crystalline structure observable by DSC. Absorption peak position in UV/visible spectra corresponding to different polyene sequence lengths was related to the different crystallite types described by Guerrero and Keller [90,91]. Thermal degradation above T_g produced long, conjugated sequences (21 repeat units in length), which were attributed to the degradation of solution-grown lamellar crystallites of higher melting temperature.

It was suggested that structural configuration of monomer units could also influence the course of thermal degradation. Millan and coworkers [92,93] have shown that rate of nonoxidative thermal dehydrochlorination of PVC is increased by the presence of syndiotactic sequences, although there seems to be no significant effect of tacticity in the overall energy of activation. These observations are consistent with a slow initiation step whose activation energy is tacticity independent, followed by rapid polyene growth that is facilitated by the syndiotactic arrangement.

In a series of studies on the effect of tacticity on the non-oxidative thermal degradation of PVC, Millan and coworkers (94–97) demonstrated that the higher the syndiotactic content of the polymer, the higher the rate of degradation and the length of polyene sequences formed. The number of oxidative scissions, by ozonolysis, of the degraded samples was found to be low for syndiotactic sequences and high for atactic sequences, which was accounted for by the increased clustering of double bonds to form long polyene sequences in the more syndiotactic polymers.

Millan (98) studied the effect of tacticity on the ionic dehydrochlorination and chlorination of PVC. For the dehydrochlorination reaction, both the reaction rate and the polyence sequence distribution depend markedly on the syndiotactic content. Chlorination appeared to be easier through heterotactic parts than through syndiotactic sequences as shown by ^{13}C-NMR.

Millan and coworkers (99–101) also studied the effect of tacticity on the nucleophilic substitution reactions of PVC. Sodium thiophenate and phenol were used for these reactions. The central chlorine in isotactic triads and, to a lesser extent, in heterotactic triads was found to be most reactive. It was concluded that initiation of degradation may occur by normal structures, and polyene build-up may be favored by syndiotic sequence. This is in favor of simultaneous participation of normal chlorines along with defect sites in the thermal degradation of PVC.

In recent years this view has received increased attention [102]. Razuvaev et al. [103] suggested that even in the early stages of degradation, dehydrochlorination of normal and of chloroallylic fragments proceeds simultaneously. Chain propagation stages consist of consecutive degradation of chloroallylic fragments of the macromolecules.

Abbas and Sorvik [21] studied the structural changes in PVC during degradation in nitrogen. They concluded that although allylic chlorine atoms seem to be the main points of initiation, other sites cannot be excluded as the number of initiation points increased appreciably during the early stages of degradation.

IV. FRYE AND HORST MECHANISM OF PVC STABILIZATION

The autocatalytic effect of HCl, the initial degradation product, on the degradation of PVC is a well-established phenomenon [104–107].

Neutralization of HCl by bases of sufficient strength according to Eq. [22] helps in stabilization.

$$MY_2 \xrightarrow{HCl} MYCl + HY \xrightarrow{HCl} MCl_2 + HY$$

$$(22)$$

Where $M = R_2Sn^{2+}$ (R is an alkyl group), Ba^{2+}, Cd^{2+}, Zn^{2+}, Pb^{2+}, and so on.
$Y = RS^-$, $RCOO^-$, or RO^- (where R is a hydrocarbon residue or an organic moiety containing one or more heteroatom functionalities).

As discussed earlier, the "abnormal" instability of PVC has been attributed to structural defects in the polymer chain, which serve as the initiation sites for degradation. The mechanism of stabilization of PVC was given by Frye and Horst in 1959 [108–112]. It proposed that organic metal salts can undergo selective reactions with PVC structural defects containing "particularly labile chlorine atoms," e.g., allylic and/or tertiary chlorine, converting them into thermally more stable structures. Thus, initiation of degradation from these defect sites is effectively prevented or retarded. Moreover, the pendant substituent groups tend to block the growth of any polyene sequences whose formation actually begins.

Organic metal salts retard the development of color in the thermal treatment of PVC, and their ability to react selectively with allylic and tertiary chlorine structures according to Eq. 23 has been demonstrated with model compounds [19,32,113,115].

$$RCl + MY_2 \longrightarrow RY + MYCl \xrightarrow{RCl} RY + MCl_2$$

$$(23)$$

$R = $ allyl, tert-alkyl, etc.

In some cases the products RY obtained from models have been shown to be more stable thermally than their halogenated progenitors [19,29,32,106,116,117]. The presence of allylic and tertiary chlorines in PVC as structural defects was earlier a subject of controversy. Only recently their presence in PVC in extremely low levels has been demonstrated [118]. It has been shown by the analytical studies of Frye et al. [108–112] and other workers [119–124] that the anionic moieties of organic metal salts are incorporated into PVC itself during thermal treatment. These results favor the Frye and Horst mechanism of PVC stabilization, which is widely accepted. Although the Frye and Horst mechanism of PVC stabilization has come to be widely accepted, time and again researchers have come up with results that could not be conclusively explained on its basis. Some controversy regarding this mechanism has always existed and will be discussed in the following section.

V. CRITICISM OF THE FRYE AND HORST MECHANISM

In the stabilization of PVC, the principal mode of action of the various stabilizer systems has been explained in terms of the Frye and Horst mechanism, i.e., substitution of labile chlorines by more stable groups. Evidence for other actions, such as HCl neutralization, addition to polyene sequences, and bimetallic complex formation have also been given. Despite the wide acceptance of the Frye and Horst mechanism, researchers have frequently contended that this could not be the dominant mechanism in the stabilization of PVC.

The presence of allylic and tertiary chlorines in PVC as structural defects has now been demonstrated, but their levels in the polymers have been determined to be extremely low [118]. In view of the limited length of the conjugated double bond sequences obtained from each defect structure (the zipper-like evolution of HCl continues up to the formation of 7–14 double bonds after which the gain in delocalization energy of the Π-electrons is no longer sufficient to drive the reaction further) once degradation has been initiated and the low initial concentration of defects, it is difficult to account for the later stages of degradation unless the reaction starts in the normal PVC structures.

As discussed in Section III.G, the possibility of simultaneous initiation of degradation from normal units in the PVC chain is receiving increased attention. The work of Millan and coworkers who showed that the normal secondary chlorines in isotactic triads are also labile and can act as points of initiation for thermal degradation was also discussed. This favors simultaneous participation of normal chlorines along with defect sites in the thermal degradation of PVC. The content of nondefect labile structures in PVC is much higher than that of the defect structures [125].

Thus, the role of the defect labile structures in the thermal degradation of PVC, which is the key to the Frye and Horst mechanism of stabilization, is itself debatable. The Frye and Horst mechanism—the substitution of labile chlorines by more stable groups—has also been criticized.

Organic metal salts have frequently failed to produce an appreciable chemical stabilization effect, either during dehydrochlorination induction periods or in later decomposition stages. While this does not rule out the occurrence of Frye and Horst substitution reactions, it does suggest that these reactions may not be responsible for the observed retardation of color developments [126–128].

Results of some attempts to stabilize PVC by pretreatment with "Frye-Horst metal salts" have been negative or inconclusive. Frye and Horst [108] reported a stabilization effect resulting from prior reaction of polymer with cadmium 2-ethylhexanoate, but they used color change rather than HCl evolution as the measure of stability; nor did they rigorously verify the removal of metal salts from the chemically pretreated resin. Morikawa [129] observed a rapid decomposition on reheating PVC, which was decomposed earlier with addition of a metal benzoate. Minsker et al. [130] found that the pretreatment of PVC with dibutyltin dicaprilate did not affect the thermal stability. This result is consistent with the work of Troitskaya and Troitskii [131] who reported a lack of correlation between the dehydrochlorination rate of PVC containing dibutyltin distearate and the number of stearate groups attached to the chain of this polymer at various dehydrochlorination levels. Czako et al. [132] studied the effect of barium and cadmium stabilizers on PVC stability. Using partly deuterated stearates, they showed that esterification of the polymer chain could not be regarded as the predominant mechanism of stabilization. Hoang et al. [133] studied the reaction of model compound 4-chloro-2-hexene with stearates of zinc and calcium. Mixtures of the two stearates exhibited synergistic esterification effects, which were ascribed, not to ligand exchange, but to the formation of bimetallic complexes whose ability to catalyze dehydrochlorination and diene oligomerization was less than that of zinc chloride.

Naqvi [45] substituted chlorines in PVC by acetoxy groups, using potassium acetate in the presence of a crown ether, and studied the thermal stability of the modified polymer. Contrary to expectation, initially the substitution of chlorines by acetoxy groups greatly destabilized the polymer. With further substitution the stability decreased linearly but more gradually. In the later stages substitution was accompanied by elimination (dehydrochlorination) as evidenced by the gradual coloration of the polymer. Naqvi et al. [53] also observed that substitution of labile chlorines in PVC by acetoxy groups resulted in a polymer less stable than unmodified PVC. These results are in direct contradiction to the Frye and

Horst mechanism of PVC stabilization. In fact, based on his observations Naqvi [134] has proposed an alternative model for the degradation and stabilization of PVC based on polar interactions within the polymer matrix, which will be discussed in the following section (Section VI).

In conclusion, it may be said that a lot of literature has been published that favors the Frye and Horst mechanism of stabilization. Most of this is based on studies done on low-molecular weight model compound for allylic chlorines in PVC, i.e., 4-chloro-2-hexene. Although the large contribution of these studies toward understanding the mechanism of stabilization of PVC cannot be denied, the extrapolation of these results to the processes involved in the actual stabilization of the polymer should be done with extreme care. The polymer represents a complex mixture of macromolecules, which in the melt is not only physically a very different system compared to the low-molecular weight model compound, but invariably contains, apart from stabilizers, other additives, such as plasticizers, lubricants, processing aids, etc., that further complicate the situation. The criticism of the Frye and Horst mechanism is also based on solid experimental evidence, and hence, the controversy is still very much alive.

VI. NAQVI's MODEL

Naqvi [134] has proposed an alternative model to the Frye and Horst mechanism for the degradation and stabilization of PVC. At room temperature, PVC is well below its glass transition temperature (about 81°C). The low thermal stability of the polymer may be due to the presence of undesirable concentrations of like-poles in the more or less frozen matrix with strong dipoles. Such concentrations, randomly distributed in the polymer matrix, may be considered to constitute weak or high energy spots in the polymer, the possible sites of initiation of thermal dehydrochlorination.

On heating PVC, as the glass transition temperature is approached, the tensions within the concentrations of like-poles may be released by atoms being pushed apart to an extent that some break from the polymer backbone, resulting in the initiation of dehydrochlorination of the polymer. Unstabilized PVC is known to start degrading at approximately its glass transition temperature [135].

Figure 1 is a schematic representation of a PVC matrix having clusters of chlorine atoms, which, being partially negatively charged, would greatly repel each other, and, energy permitting, would have a tendency to move apart. In the process, some of the chlorine atoms, which are highly electro-negative, may break from the polymer backbone as chlorine atoms or chloride ions, depending upon whether the C—Cl bond undergoes homolysis or heterolysis. The latter, which would involve charge sep-

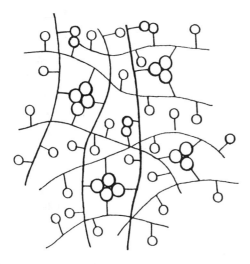

Figure 1 Schematic drawing of PVC molecules with the chlorine atoms represented by circles. Dark areas represent the weak spots formed by clusters of partially negatively charged chlorine atoms.

aration, would require more energy. Both these species could initiate degradation of the polymer by abstracting hydrogen atoms or ions giving rise to unsaturation in the polymer backbone and thereby activating the adjacent chlorine atom by making it allylic. Degradation initiated by chlorine atoms resulting from homolysis of the C—Cl clipole is shown in Eqs. (24) and (25).

$$-CH_2-CH-CH-CH-CH_2- \xrightarrow{\text{homolysis}}$$
$$\quad\quad | \quad\quad | \quad\quad |$$
$$\quad\quad Cl \quad\quad H \quad\quad Cl$$

$$-CH_2-CH-CH-CH-CH_2$$
$$\quad\quad \bullet \quad\quad | \quad\quad |$$
$$\quad\quad \overset{\bullet}{Cl} \quad\quad H \quad\quad Cl \quad\quad\quad (24)$$

$$-CH_2-CH-CH-CH-CH_2$$
$$\quad\quad \bullet \quad\quad | \quad\quad |$$
$$\quad\quad \overset{\bullet}{Cl} \quad\quad H \quad\quad Cl$$

$$\xrightarrow{\quad\quad} -CH_2-CH=CH-CH-CH_2$$
$$\quad\quad\quad\quad\quad\quad\quad\quad |$$
$$\quad\quad Cl-H \quad\quad\quad\quad Cl \quad (25)$$

(allylic)

Degradation initiated by chloride ions resulting from the heterolysis of the C—Cl dipole is shown in Eqs. (26) and (27).

$$-CH_2-CH-CH-CH-CH_2- \xrightarrow{\text{heterolysis}}$$
$$\quad\quad | \quad\quad | \quad\quad |$$
$$\quad\quad Cl \quad\quad H \quad\quad Cl$$

$$-CH_2-CH-CH-CH-CH_2-$$
$$\quad\quad + \quad\quad | \quad\quad |$$
$$\quad\quad \overset{-}{Cl} \quad\quad H \quad\quad Cl \quad\quad (26)$$

$$-CH_2-CH-CH-CH-CH_2-$$
$$\quad\quad + \quad\quad | \quad\quad |$$
$$\quad\quad \overset{-}{Cl} \quad\quad H \quad\quad Cl$$

$$\xrightarrow{\quad\quad} -CH_2-CH=CH-CH-CH_2-$$
$$\quad\quad\quad\quad\quad\quad\quad\quad |$$
$$\quad\quad Cl-H \quad\quad\quad\quad Cl \quad (27)$$

(allylic)

Recent research has conclusively demonstrated the presence of extremely low levels of tertiary and allylic chlorines in PVC as structural defects [118]. Such chlorines, which are extremely labile, when present in areas of like-charge concentration in the polymer matrix would be more susceptible to breaking from the chain than the ordinary secondary chlorines.

Introduction of highly polar groups into the PVC matrix may intensify the undesirable polar interactions described above and result in destabilization of the polymer. Conversely, the introduction of nonpolar groups may stabilize the polymer by deintensifying such polar concentrations by acting as a buffer or diluent. The stabilizing effect would be limited by the degree of compatibility between the polar and nonpolar phases. The results of much research in this area could be explained more conclusively on the basis of the above postulates put forward by Naqvi and will be discussed in the following section.

VII. EFFECT OF POLAR AND NON-POLAR GROUPS

A. Effect of Polar Groups

Grassie and coworkers [136,137] carried out a comparative study of the thermal degradation of vinyl chloride–vinyl acetate copolymers covering the entire composition range. They observed that <10% of vinyl acetate produced a marked decrease in stability as compared with PVC. Thereafter, there was a progressive decrease in stability as the amount of comonomer was increased with a minimum at 40–50% vinyl acetate. Stability progressively increased with further increase in vinyl acetate content, with poly(vinyl acetate) being more stable than poly(vinyl chloride). The thermal degradation behavior of copolymers of vinyl chloride and vinyl acetate was explained on the basis of assistance provided by acetate groups in the elimination of acids in the following manner [Eqs. (28) and (29)]:

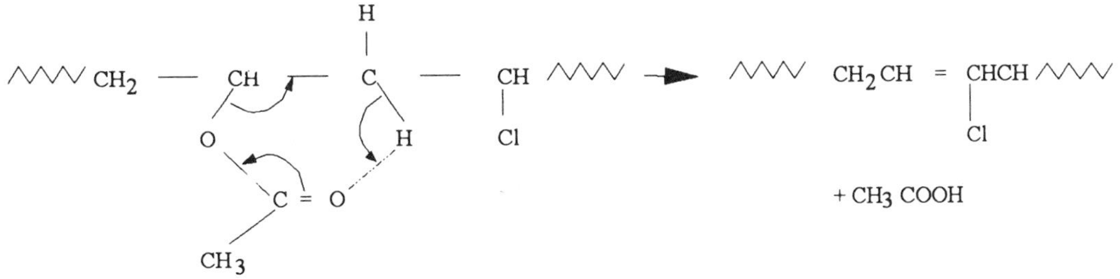

$$(28)$$

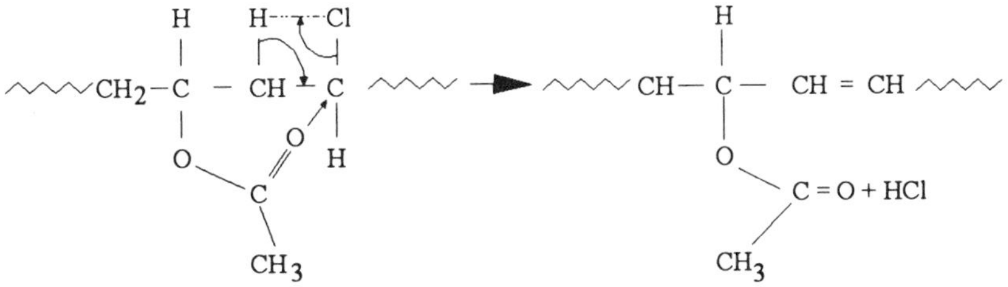

$$(29)$$

These results may be explained somewhat more convincingly on the basis of the ideas postulated by Naqvi. Replacement of chlorines by acetate groups in PVC may be expected to exert a dual effect on the stability of the polymer. The C—O bond is more stable than the C—Cl bond (the bond strength of C—Cl bond is 326.4 kJ/mol and C—O bond is 334.7 kJ/mol [138] also poly[vinyl acetate] is more stable than PVC) and this would contribute toward increasing the stability of the polymer. However, the introduction of a highly polar carbonyl group, with its π-electron cloud, would tend to intensify polar interactions within the PVC matrix by exerting a repelling influence on all chlorine atoms in its vicinity (both intra- and intermolecular), especially those labile chlorines that are present in undesirable likepole clusters (Fig. 1). This would contribute to destabilization of the polymer. Hence, the net effect on stability would be the result of these two opposing effects.

Up to the composition of 10–15% vinyl acetate monomer units in the copolymer, the destabilizing effect of the polar carbonyl group greatly predominates, resulting in a sharp decrease in stability. With further increase in vinyl acetate content, stability decreases, but more gradually. This may be due to the decrease in the T_g of the copolymer (the T_g of PVC is 80°C and of PVAc, 32°C). The temperature at which polar tensions within the matrix can be released by molecular motion decreases, and hence correspondingly, the degree of destabilization decreases. At about 40–50% vinyl acetate, the stabilization effect, due to the stronger bond strength of C—O compared with C—Cl, substantial reduction in the number of chlorines present in the copolymer, and the lower T_g that permits the release of polar tensions within

the matrix at considerably lower temperatures, starts to predominate and a gradual increase in stability is observed, with pure PVAc being more stable than pure PVC.

It may not be appropriate to compare the thermal stability characteristics of VC/VAc copolymer to that of a VC homopolymer (PVC). The copolymerization would involve different kinetics and mechanism as compared to homopolymerization resulting structurally in quite different polymers. Hence, copolymerization of VC with VAc cannot be regarded as a substitution of chlorines in PVC by acetate groups. To eliminate the possibility of these differences Naqvi [45] substituted chlorines in PVC by acetate groups, using crown ethers (18-crown-6) to solubilize potassium acetate in organic solvents, and studied the thermal stability of the modified PVC. Following is the mechanism of the substitution reaction:

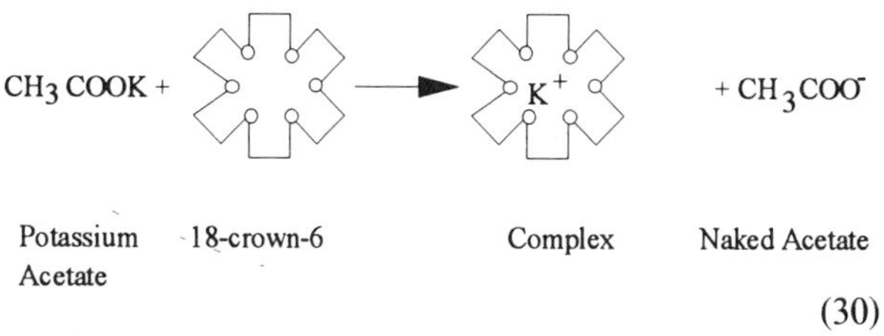

$$(30)$$

The K^+ fits into the cavity of the crown ether generating high unsolvated strongly nucleophilic acetate anions also termed "naked" anions.

The anionic moiety can substitute chlorines in PVC by an S_n^2 mechanism [Eq. (31)]. The reaction can also take place by an S_n^1 mechanism. This would involve the formation of a cationic center on the polymer backbone in the rate-determining step [Eq. (32)], which could then undergo substitution [Eq. (33)] or the competing elimination reaction [Eq. (34)].

$$\sim CH - CH \sim \longrightarrow \sim CH - CH \sim + \overset{\ominus}{Cl}$$

(31)

$$\sim CH - CH \sim \longrightarrow \sim CH - \overset{\oplus}{CH} \sim + \overset{\ominus}{Cl}$$

(32)

$$\sim CH - \overset{\oplus}{CH} \sim + \overset{\ominus}{OAc} \longrightarrow \sim CH \quad CH$$

(33)

$$\sim CH - \overset{\oplus}{CH} \sim \longrightarrow \sim CH = CH \sim + \overset{\oplus}{H}$$

(34)

The anionic moiety, depending on its basicity, could also facilitate the elimination reaction without undergoing substitution [Eqs. (35) and (36)].

$$\sim CH - CH \sim \longrightarrow \sim CH - CH \sim + AcOH$$

(35)

$$\sim CH - CH \sim \longrightarrow \sim CH = CH \sim + \overset{\ominus}{Cl}$$

(36)

Naqvi and Joseph [139] studied the kinetics of acetoxylation of PVC using 18-crown-6 to solubilize KOAc in organic solvents under mild reaction conditions. They concluded that the substitution of chlorines in PVC by acetoxy groups proceeded by an S_n^2 mechanism.

Thermal stabilities of modified PVC samples acetoxylated to varying degrees (reaction temperature 46°C) were determined [45]. Rate of thermal dehydrochlorination at 1% degradation was taken as a measure of thermal stability. The log of the degradation rate is plotted against the acetate content of the polymer in Fig. 2.

The interesting feature of the plot is the markedly discontinuous increase in thermal instability of the polymer with increase in the acetate content. The polymer was greatly destabilized at very low levels of acetate content, but further increase in instability with increasing acetate content is much more gradual and occurs at

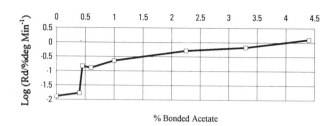

Figure 2 Variation of thermal stability of PVC with acetate content.

a more or less constant rate. These results are in agreement with what Grassie and coworkers [136,137] observed for VC/VAc copolymers containing varying amounts of vinyl acetate and can be explained quite convincingly on the basis of the ideas put forward by Naqvi to explain the thermal stability behavior of VC/VAc copolymers reported by Grassie.

Naqvi [45] also estimated the labile chlorine atoms in PVC from the degradation characteristics of PVC acetoxylated, under very mild conditions, to varying degrees. Double bond content of these samples was also determined and was found to remain unchanged with acetoxylation indicating that the competing elimination reaction did not occur under the conditions of study. The results showed that the substitution of labile chlorines in PVC by acetoxy groups resulted in a polymer that was less stable than the unmodified polymer. The estimation of labile chlorine atoms was based on the interpolation of data obtained for higher conversions as the analytical technique used (infrared spectroscopy) was not sensitive enough to quantify the low levels of labile chlorine atoms. But a separate estimation of labile chlorines by the phenolysis method was found to agree. Figure 3 shows the effect of substitution of labile chlorines by acetoxy groups on the thermal stability of the polymer.

From Fig. 3 it can be seen that substitution of labile chlorines by acetoxy groups resulted in destabilization that was at a maximum at 40–50% conversion of labile chlorines. Further substitution resulted in stabilization that was at a maximum at 100% substitution of labile

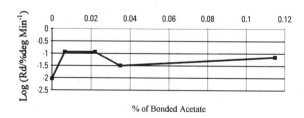

Figure 3 Variation of thermal stability with acetate content of PVC samples acetylated at 27°C.

chlorines, but was still less than that of the unmodified polymer. With further substitution the stability was found to decrease progressively. This interesting behavior was found to be reproducible.

The effect of substitution of labile chlorines by acetoxy groups on the thermal stability of polymer taken in isolation appears to correspond to the thermal degradation behavior of VC/VAc copolymers over the whole composition range [136,137].

This further interesting effect of very low levels of acetoxylation on the thermal stability of PVC may be explained on a basis similar to that applied to the copolymers, with specific reference to labile chlorines taken in isolation. But this can only be done with a certain amount of caution.

In view of the preceding discussion, the stabilization of PVC by metal carboxylates cannot be comprehensively explained on the basis of substitution of labile chlorines by the more stable carboxylate groups (Frye and Horst mechanism). We have seen that substitution of chlorines by acetoxy groups results in destabilization of the polymer. Substitution of chlorines in PVC by carboxylate groups in metal carboxylate-stabilized systems would involve the incorporation of highly polar carbonyl groups. The effect should be a destabilizing one, similar to one already discussed. But the fact is that metal carboxylates are one of the most effective commercial stabilizers for PVC. Naqvi and coworkers [140] have demonstrated that the length of the alkyl chain of the metal carboxylate stabilizers has an important bearing on the thermal stabilization of PVC. It was also shown that low-molecular weight crystalline polyethylenes stabilize PVC. This may be explained on the basis that, just as the introduction of polar groups destabilizes PVC, by intensifying the polar interactions within PVC, incorporation of nonpolar hydrocarbons may stabilize PVC by reducing such polar interactions by more or less acting as a diluent. They may also result in decreasing the glass transition temperature, which would make possible the easing off of tensions within concentrations of like-poles at lower temperatures. The effectiveness of such stabilizers would be influenced by their degree of compatibility with the polar PVC matrix.

Wirth and Andreas [141] studied the effect of octyltin chlorides on the thermal stability of PVC. They were found to significantly retard the dehydrochlorination in the following order:

$$Oc_4Sn > Oc_3SnCl > Oc_2SnCl_2 > OcSnCl_3$$

$$> \text{Pure PVC}$$

This shows a direct relationship of nonpolar n-alkyl groups with the stabilizing effectiveness of the tin compound. The larger the number of nonpolar n-alkyl groups attached to tin, the more effective it is in retarding dehydrochlorination.

In the case of metal carboxylates it may be the long n-alkyl chains, normally incorporated to reduce volatil-

ization during processing, and leaching out of stabilizer from the polymer during the service life of the product, that may be making the major contribution towards stabilization by acting as lubricants and reducing polar interactions within the polymer matrix. But since reactions of metal carboxylates with the PVC in the solid phase involve a series of complex reactions and the overall stability is also influenced by all these reactions and their products, oversimplifications regarding the mechanisms of stabilization should be made with a certain amount of caution.

Jamieson and McNeill [142] studied the degradation of poly(vinyl acetate) and poly(vinyl chloride) and compared it with the degradation of PVC/PVAc blend. For the unmixed situation, hydrogen chloride evolution from PVC started at a lower temperature and a faster rate than acetic acid from PVAc. For the blend, acetic acid production began concurrently with dehydrochlorination. But the dehydrochlorination rate maximum occurred earlier than in the previous case indicating that both polymers were destabilized. This is a direct proof of the intermolecular nature of the destabilizing effect of acetate groups on chlorine atoms in PVC. The effects observed by Jamieson and McNeill were explained in terms of acid catalysis. Hydrochloric acid produced in the PVC phase diffused into the PVAc phase to catalyze the loss of acetic acid and vice-versa.

The effect of hydrochloric acid on PVAc is shown in Eq. (37) and the effect of acetic acid on PVC in Eq. (38).

$$(37)$$

$$(38)$$

The results of Jamieson and McNeill cannot be accounted for by the intramolecular mechanism proposed by Grassie and coworkers [136,137] for the thermal degradation behavior of VC/VAc copolymers (Eqs. [28] and [29]). They can be accounted for much more convincingly by the alternative approach proposed by Naqvi based on polar interactions within the PVC matrix. Just like in copolymers even in blends, the polar carbonyl group of PVAc intensifies the concentration of like-poles in the PVC matrix resulting in destabilization.

Grassie and McGuchan [143] studied the thermal stability of acrylonitrile–vinyl chloride copolymers. They found the vinyl chloride structure in the copolymer was quite unstable and hydrochloric acid evolution occurred more readily in the copolymers than in PVC itself. It was thought that the neighboring acrylonitrile units exerted an activating influence on the elimination. Acrylonitrile units were found to be more stable in the copolymer as compared to pure poly(acrylonitrile). These observations can also be more convincingly explained on the basis of ideas put forward by Naqvi. The nitrile group, which is highly polar, can be expected to exert a destabilizing influence in exactly the same way as the polar carbonyl group in VC/VAc copolymers.

McNeill et al. [144] studied the thermal stability of PVC blends with poly(acrylonitrile) and found that poly(acrylonitrile) was stabilized by mixing with PVC, but PVC was destabilized by the presence of poly(acrylonitrile). Substantial destabilization of PVC was observed in the temperature range where poly(acrylonitrile) was stable in the blend. This is an interesting observation. Poly(acrylonitrile) by its mere presence without undergoing any change itself was exerting a destabilizing influence on PVC. The explanation of mutual acid catalysis put forward in the case of PVC/PVAc blends cannot be applied here. The authors suggested that the physical state of the blend may be such that hydrochloric acid diffusion is hindered, giving increased opportunity for catalysis. Here also Naqvi's hypothesis that introduction of polar groups into the PVC matrix would result in destabilization of the polymer by intensifying undesirable polar interactions seem more plausible. Poly(methacrylonitrile) was found to have a similar effect on the thermal stability of PVC.

McNeill and Neil [145,146] studied the thermal stabilities of PVC and poly(methylmethacrylate) blends and observed that chain scission of poly(methylmethacrylate) occurred at a much lower temperature in the blend and the PVC was slightly stabilized. Although at first sight this would seem to contradict the effect of highly polar groups on chlorines in PVC put forward by Naqvi, in reality it does not. The carbonyl of the methacrylate group would destabilize the chlorines in PVC and cause HCl to evolve at a lower temperature. The catalytic effect of HCl on degradation of PVC would be prevented by the removal of HCl from the blend by reaction with the ester groups of poly(methylmethacrylate) and would

contribute to the stability of PVC. This could be the reason for the observed overall increase in the stability of PVC in the blend.

Kovacic and coworkers [147] have studied the thermal degradation of blends of PVC and poly(α-methylstyrene-acrylonitrile) (PMSAN) over the whole composition range. These polymers are immiscible over the whole composition range as all the blends exhibit two T_gs corresponding to the T_g of pure polymers. When temperature at 5% weight loss (TGA) is taken as a measure in PMSAN content, thermal stability is reduced up to 10% PMSAN content. With further increase in PMSAN content, the thermal stability of the blend goes through a minimum and then continuously increases with pure PMSAN being much more stable than PVC. These observations can also be explained on the basis used earlier to describe the effect of polar comonomers and homopolymers on the thermal stability of PVC. Despite the fact that PVC/PMSAN blends are phase separated the polar nitrile groups of PMSAN present at the interface of the two phases exert a destabilizing influence on the PVC phase by intensifying the undesirable polar interactions within the PVC matrix. The reversal in thermal stability behavior of 20% PMSAN (increase in stability) is a bit difficult to explain. Some kind of phase inversion might be a factor contributing to this, but this is pure conjecture at this time.

McNeill and Basan [148] have studied the thermal degradation of blends of PVC with poly (dimethylsiloxane) (PDMS). Structure of PDMS is given in Eq. (39).

$$\begin{array}{ccccccc}
& CH_3 & & CH_3 & & CH_3 & \\
& | & & | & & | & \\
\sim\!\!\sim\!\!\sim & Si & -\!O-\!\! & Si & -\!O-\!\! & Si & -\!O \sim\!\!\sim\!\!\sim \\
& | & & | & & | & \\
& CH_3 & & CH_3 & & CH_3 &
\end{array}$$

$$(39)$$

Every alternating atom in the PDMS chain is the electronegative oxygen, which makes it a polar polymer. The authors observed that at low loadings of PDMS, PVC is destabilized. This is to be expected according to the alternative model for degradation and stabilization of PVC put forward by Naqvi. But for compositions with 50% or more PDMS, both polymers are stabilized. No possible explanation can be given for this reversal in thermal stability at this time.

Some polymers have both polar and nonpolar groups in the polymer chain. Thermal stability behavior of blends of some such polymers with PVC have also been studied. McNeill and coworkers [149] have studied the degradation of blends of PVC with poly (tetramethylene sebacate) (PTMS). The structure of PTMS is given in Eq. (40).

$$\begin{array}{ccccc}
& O & & O & \\
& \| & & \| & \\
\sim\!\!\sim\!\!\sim O & -\!C- & (CH_2)_8 & -\!C & -\!O-\!(CH_2)_4 \sim\!\!\sim\!\!\sim
\end{array}$$

$$(40)$$

The polymer contains both polar and nonpolar groups. Polar carboxylate groups are separated by nonpolar ethylene units. Degradation studies of blends of PVC and PTMS showed that incorporation of small amounts of PTMS into PVC leads to significant stabilization in the initial stages of degradation, but in the presence of a large excess of PTMS, PVC is subject to marked destabilization. These observations can also be explained on the basis of ideas put forward by Naqvi. At low levels of PTMS, PTMS may be miscible with PVC and the blends may be homogeneous. The nonpolar long ethylene units in PTMS may be reducing the undesirable polar concentrations in the PVC matrix by acting as a buffer or diluent thereby imparting stabilization to PVC. At very high loadings of PTMS the nonpolar ethylene units may phase-separate in which case the mechanism of stabilization will become inoperative and destabilization may result due to tensions of the interface of the polar and nonpolar phases. Also, the HCl evolved during degradation may become entrapped in the polar PVC phase due to poorer mobility in the nonpolar PTMS phase. This will result in autocatalysis of the degradation process by HCl, which would also cause destabilization. A similar reversal in thermal stability due to the occurrence of phase-separation has been observed when chlorines in PVC have been progressively substituted by hydrogen atoms using tri-*n*-butyltin-hydride. This will be discussed in some detail in the next section.

McNeill and Basan [150] have studied the blends of PVC with poly (ethylene adipate) (PEAD). Structure of PEAD is given in Eq. (41).

$$\sim\!\!\sim\!\!\sim O - \underset{\substack{\|\\O}}{C} - (CH_2)_4 - \underset{\substack{\|\\O}}{C} - O - (CH_2)_2 \sim\!\!\sim\!\!\sim$$

(41)

Structure of PEAD is similar to that of PTMS except that the number of nonpolar ethylene units between polar carboxylate groups is smaller. The thermal degradation behavior of PVC/PEAD blends was observed to be similar to that of PVC/PTMS blends and can be explained on a similar basis described earlier.

McNeill and Basan [151] studied the thermal degradation of blends of PVC with bisphenol-A polycarbonate. The structure of bisphenol-A polycarbonate is given in Eq. (42).

$$\sim\!\!\sim\!\!\sim \overbrace{\bigcirc} - \underset{\substack{CH_3\\|\\|\\CH_3}}{C} - \overbrace{\bigcirc} - O - \underset{\substack{\|\\O}}{C} - O \sim\!\!\sim\!\!\sim$$

(42)

As can be seen from the structure, the polymer contains both polar and nonpolar segments. The authors

observed that bisphenol-A polycarbonate has a stabilizing influence on PVC over the whole composition range. From these results it appears that the blends are miscible over the whole composition range, and the observations can be explained on the basis similar to that used for PV/PTMS and PVC/PEAD blends.

B. Effect of Nonpolar Groups

Copolymerization of a small amount of a second monomer with vinyl chloride has been an important preventive stabilization technique for PVC. By including 1–3% of ethylene or propylene in vinyl chloride polymerizations, copolymers of greatly improved thermal stability, compared with vinyl chloride homopolymers of equivalent intrinsic viscosity, have been produced with very little effect on the other properties characteristic of PVC [152–154]. The stabilization is believed to be due to the blocking of the progressive "unzipping" of the HCl from the PVC backbone by the second monomer. In my opinion this seems rather implausible in view of the fact that during the degradation of PVC, the HCl unzipping does not continue *ad infinitum* but tends to stop after the formation of about 7–14 conjugated double bonds as the gain in delocalization energy is no longer significant enough to keep the reaction going further. Hence, the presence of only 2–3% comonomer may not be expected to exert a very significant effect on the 'unzipping' of HCl from the polymer backbone. On the other hand, ethylene and propylene are nonpolar and when randomly dispersed in the PVC, matrix, may exert a stabilizing influence by deintensifying the undesirable like-pole interactions in the PVC matrix by acting as a buffer or diluent as postulated by Naqvi.

Naqvi and Sen [155] studied the thermal stability and thermal characteristics of PVC and *cis*-polybutadiene rubber (PBR) blends. The conclusions of their study may be summarized as follows:

1. PBR increases the inherent stability of PVC by acting as a nonpolar dispersant and reducing the undesirable polar tensions within the PVC matrix.
2. PBR increases the stability of the degradation process itself by absorbing the evolved HCl and preventing it from catalyzing further degradation.
3. Both the stabilizing mechanisms are limited by the degree of interaction and miscibility between the two unlike polar (PVC) and nonpolar (PBR) phases.
4. With increasing content of PBR (above 10%) the inherent tendency of the two unlike phases to separate out increases, resulting in reduced interaction and miscibility. The effectiveness of stabilization is accordingly reduced.

Kolawole and Olugbemi [156] have published a study of photo and thermal degradation of the two-phase system of PVC and poly(isobutylene). They have shown that poly(isobutylene) has a remarkable stabilizing influence on PVC. Polystyrene has also been found to exert a stabilizing influence on PVC in the blends of two polymers [157,158]. These results may also be explained on the basis of the effect the nonpolar matrix has on the polar interactions within PVC.

Braun et al. [159] have published an article on the preparation of vinyl chloride–ethylene copolymers of varying composition by partial reductive dechlorination of PVC with tri-*n*-butyltin-hydride and have reported the studies on the thermal stability of the copolymers. Naqvi [160] has interpreted the observations of Braun and coworkers [159] in the light of the alternative approach to thermal degradation and stabilization of PVC, based on polar interactions within the polymer matrix, proposed by him [134]. The vinyl chloride–ethylene copolymers obtained by reductive dechlorination of PVC showed a continuous increase in stability up to a substitution of about 25 mol% of chlorines by hydrogen. This stabilization effect cannot be explained merely on the basis of substitution of labile chlorines in PVC by more stable hydrogen, although this may make a nominal contribution. The levels of labile chlorines present as structural defects in PVC have been found to be extremely low—much less than even 1 mol%. The reversal in the stability of the copolymers at about 35 mol% of ethylene sheds some doubt on the mechanism of stabilization based on blocking of the progressive unzipping of HCl from the polymer backbone and will be discussed subsequently. The observed increase in the stability of the copolymer may be explained on the basis of the diluent effect the interdispersed nonpolar short ethylene sequences have on the undesirable concentrations of like-poles in the PVC matrix described earlier. The T_g of the copolymer would decrease with increasing ethylene content, making possible the relaxation of repulsive tensions between concentrations of like-charges by molecular motion at lower temperatures. This may also be expected to contribute to the stabilization of the copolymer.

The observed reversal in the thermal stability of the copolymer at a critical composition, which appears to be between 30 and 40 mol% of ethylene, may be explained on the basis of the emergence of phase-separation between the nonpolar ethylene and polar vinyl chloride blocks. Although crystallization of the ethylene blocks in the copolymer is only observed when more than 70 mol% ethylene units are present, the possibility of phase-separation occurring at lower contents of ethylene units cannot be excluded. Also, round about the critical copolymer composition, the T_g of the copolymer may be reduced to a level that would facilitate separation between the unlike phases by increased molecular mobility within the polymer matrix. As has been discussed earlier, occurrence of phase-separation in the copolymer would not only make the mechanism of stabilization due

to the dilution of polar-interactions in the PVC matrix inoperative but could result in its destabilization. After the critical composition at which phase-separation begins to appear in the copolymer matrix, the degree of phase-separation would progressively increase with increasing ethylene content until even crystallization of ethylene blocks becomes possible at levels of more than 70 mol% of ethylene, i.e., a further phase-separation in the already separated polyethylene phase. The stability of the copolymer would be expected to decrease correspondingly, which is what is observed.

The results of the studies of solution degradation of the copolymer samples, which showed all copolymers to be more stable than PVC, also strongly suggest that the physical state of the copolymers in the bulk may be playing a dominant role in the observed stabilization and destabilization of the copolymers. These observations strongly favor the ideas put forward by Naqvi.

The appearance of any appreciable degree of phase-separation in the copolymer should be reflected in different T_gs of the two phases. But Braun and coworkers [159] observed only single T_gs in the reported range of 0–57.0 mol% of ethylene in the copolymer. This may constitute a criticism of the ideas put forward here.

Bowmer and Tonelli [161] have also studied the thermal characteristics of the whole range of ethylene-vinyl chloride copolymers prepared by partial reductive dechlorination of PVC using tri-*n*-butyltin-hydride. Naqvi [162] has substantiated further his explanations for the thermal stability characteristics of ethylene–vinyl chloride copolymers reported by Braun et al. [159] using the results of Bowmer and Tonelli [161] as a basis.

Bowmer and Tonelli [161] have observed a single T_g (54.5°C) for copolymers containing up to 14.7 mol% of ethylene and two T_gs for copolymers containing 15.7 mol% (24°C, 60°C), 29.3 mol% (23°C, 72°C), and 38.5 mol% (0°C, 57°C) of ethylene, respectively. The lower T_g obviously characterizes the ethylene phase and the higher T_g the vinyl chloride phase. This is direct evidence of the presence of two phases in copolymers in the composition range 16–38 mol% of ethylene; that is, there is some degree of phase-separation between the nonpolar ethylene segments and the highly polar vinyl chloride segments. This is in agreement with the explanations put forward by Naqvi for the thermal stability behavior of the copolymers prepared in a similar manner, based on phase effects within the polymer matrix.

Bowmer and Tonelli [161] have also observed that the magnitude of the glass transition (ΔC_p) increases with the ethylene content of the copolymer, goes through a maximum at about 30 mol%, and then continually decreases until no glass transition is observed at more than 80 mol% of ethylene. This may constitute further evidence in favor of the explanations put forward by Naqvi for the thermal stability behavior of similar copolymers reported by Braun et al. [159]. Initially, with increasing content of nonpolar ethylene units in the copolymer, the T_g and cohesive energy density decreases and internal plasticization of vinyl chloride segments increases (the same factors contribute to increased thermal stability by dilution or reduction of undesirable polar interactions in the vinyl chloride matrix), resulting in increasing ΔC_p. A significant degree of phase-separation between ethylene and vinyl chloride segments appears at about 30 mol% of ethylene and with further increase in ethylene content, ΔC_p decreases. This is probably due to the fact that with an increasing degree of phase-separation between the ethylene and vinyl chloride segments, the degree of plasticization of vinyl chloride segments by ethylene segments is reduced. Also, the degree of order in the ethylene phase of the copolymer increases with increasing ethylene content and finally results in the appearance of crystallization of ethylene units when their content is more than 60 mol%. Increasing order in the ethylene phase would result in a decrease in the number of its internal degrees of freedom; that is, in the possible modes of motion of the molecules, and, correspondingly, ΔC_p would decrease. The reversal in ΔC_p at about 30 mol% of ethylene observed by Bowmer and Tonelli [161] and the reversal in thermal stability between 30 and 35 mol% of ethylene observed by Braun et al. [159] tend to suggest that a significant degree of phase-separation occurs in their respective copolymer samples at about the same composition, especially in view of the explanations that have been put forward for the observed reversals. This shows that the correlations made between the thermal stability characteristics and thermal transitions of two different sets of ethylene–vinyl chloride copolymers prepared in a similar manner are reasonably justified. They support the alternate explanations put forward by Naqvi [160], based on phase-effects within the polymer matrix, for the thermal stability behavior of the copolymers.

REFERENCES

1. M. Kaufman, *The Chemistry and Industrial Production of Polyvinyl Chloride: The History of PVC,* Gordon and Breach, New York (1969).
2. N. Grassie, *Encyclopedia of Polymer Science and Technology, 4:* John Wiley & Sons, New York, p. 647 (1966).
3. C. S. Marvel, J. H. Sample, and M. F. Roy, *J. Amer. Chem. Soc., 61:* 3241 (1939).
4. C. S. Marvel, G. D. Jones, T. W. Martin, and G. L. Schertz, *J. Amer. Chem. Soc., 64:* 2356 (1942).
5. C. S. Marvel, *Chemistry of the Large Molecules,* Interscience Publishers, New York, p. 219.
6. C. S. Marvel, and C. Horning, *Gilman's Organic Chemistry 1:* John Wiley & Sons, New York p. 754 (1943).
7. J. J. P. Staudinger, *Plast. Prog., 9:* (1953).
8. H. V. Smith, *Rubber J. Inst. Plast., 138:* 966 (1960).
9. D. Braun, and W. Shurek, *Angew. Macromol. Chem., 7:* 121 (1969).
10. S. Enomoto, *J. Polym. Sci., Part A1, 7:* 1255 (1969).
11. W. H. Starnes, F. C. Schilling, K. B. Abbas, R. E. Cais, and F. A. Bovey, *Macromolecules, 12:* 556 (1979).
12. A. Caraculacu, and E. Bezdadea, *J. Poly. Sci., Polym. Chem. Ed., 15:* 611 (1977).
13. E. J. Arlman, *J. Polym. Sci., 12:* 547 (1954).
14. W. I. Bengough, and H. M. Sharpe, *Macromol. Chem., 66:* 31 (1963).
15. W. I. Bengough and H. M. Sharpe, *Macromol. Chem., 66:* 45 (1963).
16. T. Talamini, and G. Pezzin, *Macromol. Chem., 39:* 26 (1960).
17. W. I. Bengough, and I. K. Varma, *Eur. Polym. J., 2:* 49 (1966).
18. W. I. Bengough, and I. K. Varma, *Eur. Polym. J., 2:* 61 (1966).
19. M. Onozuka, and M. Asahina, *J. Macromol. Sci.-Rev. Macromol. Chem., C3:* 325 (1969).
20. A. Crosoto-Arnaldi, G. Palma, E. Peggion, and G. Talamini, *J. Appl. Polym. Sci., 8:* 747 (1964).
21. K. B. Abbas, and E. M. Sorvik, *J. Appl. Polym. Sci., 19:* 2991 (1975).
22. D. Braun, *Pure Appl. Chem., 26:* 173 (1971).
23. K. B. Abbas, and E. M. Sorvik, *J. Appl. Polym. Sci., 20:* 2395 (1976).
24. D. Braun, and W. Quarg, *Angew Macromol. Chem., 29/30:* 163 (1973).
25. G. Lindeschmidt, *Angew. Macromol. Chem., 47:* 79 (1975).
26. V. Chytry, B. Obereigner and D. Lim, *Eur. Polym. J., Suppl:* 379 (1969).
27. M. Asahina and M. Onozuka, *J. Polym. Sci., A2:* 3515 (1964).
28. V. Chytry, B. Obereigner and D. Lim, *Eur. Polym. J., 7:* 1111 (1971).
29. T. Suzuki and M. Nakamura, *Jpn. Plast., 4:* 16 (1970).
30. A. R. Berens, *Polym. Eng. Sci., 14:* 318 (1974).
31. M. Asahina and M. Onozuka, *J. Polym. Sci., A2:* 3502 (1964).
32. T. Suzuki, I. Takakura and M. Yoda, *Eur. Polym. J., 7:* 1105 (1971).
33. A. A. Caraculacu, *J. Polym. Sci., Part A-1, 4:* 1839 (1966).
34. A. Caraculacu, E. C. Bezdadea, and G. Istrate, *J. Polym. Sci., Part A-1, 8:* 1239 (1970).
35. D. Braun, and F. Weiss, *Angew. Macromol. Chem., 13:* 55 (1970).
36. D. Braun, and F. Weiss, *Angew. Macromol. Chem., 13:* 67 (1970).
37. T. Suzuki, M. Nakamura, M. Yasuda and J. Tatsumi, *J. Polym. Sci., Part C, 33:* 281 (1971).
38. T. Hjertberg and E. M. Sorvik, *Polymer, 24:* 673 (1983).
39. T. Hjertberg and E. M. Sorvik, *Polymer, 24:* 685 (1983).
40. T. Hjertberg and E. M. Sorvik, *Polymer Stabilization and Degradation,* (P. P. Klemchuk, ed.), ACS Symposium Series No. 280, American Chemical Society, Washington, DC, p. 259, (1985).
41. R. Rogestedt and T. Hjertberg, *Macromolecules, 25:* 6332 (1992).
42. W. H. Starnes, *Developments in Polymer Degradation - 3,* (N. Grassie, ed.), Applied Science, London, p. 135, (1981).
43. B. Evan, J. P. Kennedy, T. Kelen, F. Tudos, T. T. Nagy, and B. Turcsanyi, *J. Polym. Sci., Polym. Chem. Ed., 21:* 2177 (1983).
44. W. I. Bengough, and M. Onozuka, *Polymer, 6:* 625 (1965).
45. M. K. Naqvi, PhD Thesis, University of Wales Institute of Science and Technology, UK (1980).

46. N. G. Thane, R. D. Lundberg, and J. P. Kennedy, *J. Polym. Sci., Poly. Chem. Ed., 10*: 2507 (1972).
47. G. Robila, E. C. Buruiana and A. A. Caraculacu, *Eur. Polym. J., 13*: 21 (1977).
48. E. C. Buriana, V. T. Barbinta, and A. A. Caraculacu, *Eur. Polym. J., 13*: 311 (1977).
49. A. Michel, P. Burille and A. Guyot, *Prepr. Makro Maintz*, p. 603 (1979).
50. B. Evan, F. Tudos, O. Egyed and T. Kelen, *Makromol. Chem., Rapid Commun., 3*: 727 (1982).
51. W. H. Starnes, Jr. and I. M. Plitz, *Macromolecules, 9*: 633 (1976).
52. J. Lewis, M. K. Naqvi and G. S. Park, *Makromol. Chem., Rapid Commun., 1*: 119 (1980).
53. J. Lewis, M. K. Naqvi and G. S. Park, *Polym. Prepr., ACS Div. Polym. Chem., 23*: 140, (1982).
54. V. Jisova, M. Kolinsky and D. Lim, *J. Polym. Sci., A1, 8*: 1525 (1970).
55. V. Jisova, M. Kolinsky, and D. Lim, *J. Polym. Sci., C42*: 467 (1973).
56. M. Kolinsky, V. Jisova and D. Lim, *J. Polym. Sci., C42*: 657 (1973).
57. T. Hjertberg, E. Martinsson and E. Sorvik, *Macromolecules, 21*: 603 (1988).
58. G. A. Razuvaev, G. G. Petukhov and V. A. Dodonov, *Tr. Khim. Khim. Tekhnol., 3*: 163 (1960).
59. G. A. Razuvaev, G. G. Petukhov and V. A. Dodonov, *Polym. Sci. USSR, 3*: 1020 (1962).
60. A. Cittadini and R. Paoillo, *Chim. Ind.* (Milan), *41*: 980 (1959).
61. C. Corso, *Chim. Ind.* (Milan), *43*: 8 (1961).
62. R. Stromberg, S. Straus and B. G. Achhammer, *J. Polym. Sci., 35*: 355 (1959).
63. G. Talamini, C. Cinzue and G. Palma, *Mater. Plast., 30*: 317 (1964).
64. G. S. Park and D. G. Smith, *Makromol. Chem., 131*: 1 (1970).
65. G. S. Park and C. L. Skene, *J. Polym. Sci., Part C, 33*: 269 (1971).
66. Z. Mayer, *J. Macromol. Sci.-Rev. Macromol. Chem., C10*: 262 (1974).
67. Baum and Wartman, *J. Polym. Sci., 28*: 537 (1958).
68. W. C. Geddes, *Rubber Chem. Technol., 40*: 177 (1967).
69. A. Shimizu and T. Ohtsu, *Kogyo Kagaku Zasshi, 67*: 966 (1964).
70. K. Mitani, T. Ogata, H. Awaya and Y. Tomari, *J. Polym. Sci., Polym. Chem. Ed., 13*: 2813 (1975).
71. W. H. Starnes, F. C. Schilling, K. B. Abbas, R. E. Cais, and F. A. Bovey, *Macromolecules, 12*: 556 (1979).
72. T. Hjertberg, E. Sorvik and A. Wendel, *Makromol. Chem., Rapid Commun., 4*: 175 (1983).
73. T. Hjertberg and E. M. Sorvik, *J. Macromol. Sci-Chem., A17*: 903 (1982).
74. F. V. Erbe, T. Grewer and K. Wehage, *Angew. Chem., 74*: 988 (1962).
75. A. Maccoll, *Chem. Rev., 69*: 33 (1969).
76. S. Crawley and I. C. McNeil, *J. Polym. Sci., Polym. Chem. Ed, 16*: 2593 (1978).
77. Z. Vymazalova and Z. Vymazal, *Chem. Listy, 61*: 1204 (1967).
78. Y. Landler and P. Lebel, *J. Polym. Sci., 48*: 477 (1960).
79. Z. V. Popova, N. V. Tikhova, and G. A. Rozuvavev, *Polym. Sci. USSR, 7*: 588 (1965).
80. A. Garton and M. H. George, *J. Polym Sci., Polym. Chem. Ed., 12*: 2779 (1974).
81. J. Bauer and A. Sabel, *Angew. Makromol. Chem., 47*: 15 (1975).
82. S. Sonnerskog, *Acta Chem. Scand., 13*: 1634 (1959).
83. B. G. Achhammer, *Anal. Chem. 24*: 1925 (1952).
84. K. S. Minsker, A. A. Berlin, V. V. Lisickij, S. V. Kolesov, and R. S. Korneva, *Dokl. Akad. Nauk SSSR, 232*: 1 (1977).
85. K. S. Minsker, A. A. Berlin, V. V. Lisickij, and S. V. Kolesov, *Plast. Massy, 10*: 69 (1976).
86. J. Svetly, R. Lukas, J. Michalkova, and M. Kolinsky, *Makromol. Chem., Rapid Commun., 1*: 247 (1980).
87. A. Guyot, P. Roux, and P. Q. Tho, *J. Appl. Polym. Sci., 9*: 1823 (1965).
88. A. Guyot, M. Bert, and P. H. Tho, *J. Appl. Polym. Sci., 12*: 639 (1968).
89. J. Behnisch and M. Zimmerman, *Makromol. Chem., 190*: 2347 (1989).
90. S. J. Guerrero and A. Keller, *J. Macromol. Sci.-Phys., B20*: 101 (1981).
91. S. J. Guerrero and A. Keller, *J. Macromol. Sci.-Phys., B20*: 167 (1981).
92. G. Martinez, J. Millan, M. Bert, A. Michel, and A. Guyot, "Second International Symposium on PVC," Lyon-Villeaurbanne, France, July 1976, pp. 293–296 (1976).
93. J. Millan, E. C. Madruga, and G. Martinez, *Angew. Makromol. Chem., 45*: 177 (1975).
94. G. Martinez, J. Millan, M. Bert, A. Michel, and A. Guyot, *J. Macromol Sci-Chem., 12*: 489 (1978).
95. J. Millan, G. Martinez, and C. Mijangos, *J. Polym. Sci., Polym. Chem. Ed., 18*: 505 (1980).
96. J. Millan, E. L. Madruga, M. Bert, and A. Guyot, *J. Polym. Sci., Polym. Chem. Ed., 11*: 3299 (1973).
97. J. Millan, M. Carranza, and J. Guzman, *J. Polym. Sci., Polym. Symp., 42*: 1411 (1973).
98. J. Millan, *J. Macromol. Sci.-Chem., A12*: 315 (1978).
99. J. Millan, G. Martinez, and C. Mijangos, *Polym. Bull., 5*: 407 (1981).
100. G. Martinez, C. Mijangos, and J. Millan, *J. Macromol. Sci-Chem., A17*: 1129 (1982).
101. C. Mijangos, G. Martinez, and J. L. Millan, *Eur. Polym. J., 18*: 731 (1982).
102. I. Tvaroska, "Second International Sympositum on PVC," Lyon-Villeurbanne, France, July 1976, p. 269 (1976).
103. G. A. Razuvaev, B. B. Toritskii, and L. S. Troitskaya, "Second International Symposium on PVC," Lyon-Villeurbanne, France, July 1976, p. 261 (1976).
104. C. David, *Compr. Chem. Kinet., 14*: 78 (1975).
105. S. Van der Ven, and W. F. de Witt, *Angew. Makromol. Chem., 8*: 143 (1969).
106. T. Morikawa, *Chem. High Polym., Japan, 25*: 505 (1968).
107. M. I. Abdullin, V. P. Malinskaya, S. V. Kolesov, and K. S. Minsker, "Second International Symposium on PVC," Lyon-Villeurbanne, France, July 1976, p. 273–276 (1976).
108. A. H. Frye, and R. W. Horst, *J. Polym. Sci., 40*: 2119 (1959).
109. A. H. Frye and R. W. Horst, *J. Polym. Sci., 45*: 1 (1960).
110. A. H. Frye, R. W. Horst, and M. A. Poliobagis, *J. Polym. Sci. (Pt A2), 2*: 1769 (1964).
111. A. H. Frye, R. W. Horst, and M. A. Poliobagis, *J. Polym. Sci. (Pt A2) 2*: 1785 (1964).
112. A. H. Frye, R. W. Horst, and M. A. Poliobagis, *J. Polym. Sci. (Pt A2) 2*: 1801 (1964).

113. G. Ayrey, R. C. Poller, and I. H. Siddiqui, *J. Polym. Sci. (Pt B)*, 8: 1 (1970).
114. G. Ayrey, R. C. Poller, and I. H. Siddiqui, *J. Polym. Sci. (Pt A1)*, 10: 725 (1972).
115. G. Ayrey, R. C. Poller, and I. H. Siddiqui, *Polymer, 13*: 299 (1972).
116. T. Suzuki, M. Yasuo, and T. Masuda, *Jpn. Plast.*, 6: 11 (1972).
117. D. F. Anderson, and D. A. McKenzie, *J. Polym. Sci. (Pt A1)*, 8: 2905 (1970).
118. M. K. Naqvi, *J. Macromol. Sci.-Rev. Macromol Chem. Phys.*, C25: 119 (1985).
119. E. N. Zilberman, A. E. Kulikova, S. B. Meiman, N. A. Okladnov, and V. P. Lebedev, *J. Polym. Sci. (Pt. A1)*, 8: 2631 (1970).
120. D. Braun, and D. Hepp, *J. Polym. Sci. (Pt C), 33*: 307 (1971).
121. A. E. Kulikova, S. B. Meiman, N. A. Okladnov, and E. N. Zilberman, *J. Appl. Chem. USSR (Engl. Transl.)*, 45: 686 (1972).
122. D. Braun, and D. Hepp, *Angew. Makromol. Chem., 32*: 61 (1973).
123. F. Alavi-Moghadam, G. Ayrey, and R. C. Poller, *Eur. Polym. J., 11*: 649 (1975).
124. K. Figge, and W. Findeiss, *Angew. Makromol. Chem., 47*: 141 (1975).
125. G. Martinez, C. Mijangos, and J. Millan, *Polym. Bull., 13*: 151 (1985).
126. L. D. Loan and F. H. Winslow, *Polymer Stabilization,* W. L. Hawkins, ed.) Wiley-Interscience, New York, pp. 125–140 (1972).
127. L. D. Loan, *ACS Polym. 11*: 224 (1970).
128. T. Kimura, *Enka Biniiru To Porima,* 5: 18 (1965).
129. T. Morikawa, *Kobunshi Kagaku,* 602 (1962).
130. K. S. Minsker, G. T. Fedoseyeva, T. B. Zavarova, and E. O. Krats, *Polym. Sci. USSR (Engl. Transl.), 13*: 1013 (1971).
131. L. S. Troitskaya and B. B. Troitskii, *Plast. Massy,* 12 (1968).
132. E. Czako, Z. Vymazal, K. Volka, I. Stibor, and J. Stepak, *Eur. Polym. J., 15*: 81 (1979).
133. T. V. Hoang, A. Michel, and A. Guyot, *Eur. Polym. J., 12*: 337 (1976).
134. M. K. Naqvi, *Polym. Deg. Stab., 13*: 161 (1985).
135. G. Palma and M. Carenza, *J. Appl. Polym. Sci., 14*: 1737 (1970).
136. N. Grassie, I. F. McLaren, and I. C. McNeill, *Eur. Polym. J., 6*: 679 (1970).
137. N. Grassie, I. F. McLaren, and I. C. McNeill, *Eur. Polym. J., 6*: 865 (1970).
138. I. L. Finar, *Organic Chemistry,* Vol. 1, ed. 6, p. 36.
139. M. K. Naqvi and P. Joseph, *Polym. Commun., 27*: 8 (1986).
140. M. K. Naqvi, P. A. Unnikrishnan, Y. N. Sharma, and I. S. Bhardwaj, *Eur. Polym. J., 20*: 95 (1984).
141. H. O. Wirth and H. Andreas, *Pure Appl. Chem., 49*: 627 (1977).
142. A. Jamieson and J. C. McNeill, *J. Polym. Sci., Polym. Chem. Ed., 12*: 387 (1974).
143. N. Grassie and R. McGuchan, *Eur. Polym. J., 9*: 507 (1973).
144. I. C. McNeill, N. Grassie, J. N. R. Samsom, A. Jamieson, and T. Straiton, *J. Macromol. Sci.-Chem., A12*: 503 (1978).
145. I. C. McNeill and D. Neil, *Eur. Polym. J., 6*: 143 (1970).
146. I. C. McNeill and D. Neil, *Eur. Polym. J., 6*: 569 (1970).
147. T. Kovacic, I. Klaric, A. Nordelli, and B. Baric, *Polym. Deg. Stab., 40*: 91 (1993).
148. I. C. McNeill and S. Basan, *Poly. Deg. Stab., 39*: 139 (1993).
149. I. C. McNeill, J. G. Gorman, and S. Basan, *Polym. Deg. Stab., 33*: 263 (1991).
150. I. C. McNeill and S. Basan, *Polym. Deg. Stab., 41*: 311 (1993).
151. I. C. McNeill and S. Basan, *Polym. Deg. Stab., 39*: 145 (1993).
152. R. J. Ireland, *SPE Tech. Papers, 14*: 75 (1968).
153. M. J. R. Cantow, C. W. Cline, C. A. Heiberger, D. Th. A. Huibers, and R. Phillips, *Mod. Plast., 46*: 6, 126 (1969).
154. C. A. Heiberger, R. Phillips, and M. J. R. Cantow, *SPE Tech. Papers, 15*: 306 (1969).
155. M. K. Naqvi and A. R. Sen, *Polym. Deg. Stab., 33*: 367 (1991).
156. E. G. Kalawole and P. O. Olugbemi, *Eur. Polym. J., 21*: 187 (1985).
157. B. Dodson and I. C. McNeill, *J. Polym. Sci.-Polym. Chem. Ed., 14*: 353 (1976).
158. I. C. McNeill, N. Grassie, J. N. R. Samson, A. Jamieson, and T. Straiton, *J. Macromol. Sci.-Chem., A12*: 503 (1978).
159. D. Braun, W. Mao, B. Bohringer, and R. W. Garbella, *Die Angew. Makromol. Chem., 141*: 113 (1986).
160. M. K. Naqvi, *Polym. Deg. Stab., 17*: 341 (1987).
161. T. B. Bowmer, and A. E. Tonelli, *Polymer, 26*: 1195 (1985).
162. M. K. Naqvi, *Polym. Deg. Stab., 22*: 1 (1988).

22

Solution State of Metal Complex Calixarenes and Polymeric Calixarenes

Mustafa Yilmaz
Selçuk University, Konya, Turkey

I. COMPLEXES WITH ALKALI AND ALKALINE EARTH CATIONS

The calixarenes are almost totally insoluble in water and only partially soluble in most organic solvents [1]. The first success in demonstrating their complex abilities was achieved by Izatt and coworkers [2,3]. Izatt, who has carried out extensive investigations on the complex behavior of crown ethers and various related types of compounds, perceived a structural resemblance between the crown ethers, cyclodextrins, and calixarenes, and proceeded to test the latter for their ability to transport cations across a liquid membrane. They also used an apparatus in which an aqueous source phase containing the host molecule (the carrier) and the cations were separated by an organic phase (e.g., chloroform) from an aqueous receiving phase. They discovered that although the calixarenes are ineffective cation carriers in neutral solution, they possess significant transport ability for Group I cations in a strongly basic solution. This is in sharp contrast to 18-crown-6 compounds, which are more effective in a neutral rather than in a basic solution. Group II cations, including Ca^{2+}, Ba^{2+}, and Sr^{2+}, are not effectively transported by these calixarenes. Control experiments with *p-tert*-butyl phenol, which shows little or no transport ability itself, support the idea that the macrocyclic ring plays a critical role, although it is not yet clear what the role is exactly. The diameters of the annuli of the calixarene mono anions are *ca.* 1.0 Å for the cyclic tetramer, 2.4 Å for the cyclic hexamer, and 4.8 Å for the cyclic octamer. Thus, for complexes in which the cation and oxygen are coplanar, the cyclic tetramer has too small an opening even for Li^+, whereas the cyclic octamer has too large an opening to fit snugly,

around even Cs^+. The cyclic hexamer seems to be the system best constituted to behave in crown ether-like fashion and, indeed, even the synthesis of *p-tert* butyl calix[6]arene may be influenced by a template effect. Extending their initial studies to include multicarbon systems comprised of equimolar mixtures of two, three, and four cations from NaOH, KOH, RbOH, and CsOH, Izatt and coworkers [3] found that selective transport of Cs^+ occurs in all cases. In another work in this field, Izatt and coworkers [4] have a patent that has been issued for a process that uses this technology, and was utilized to recover cesium from radioactive wastes.

Bocchi et al. [5] pointed out that while the tetraacetate of *p-tert*-butyl calix[4]arene fails to form complexes with guanidium ions or Cs^+, the octa-(3,6-dioxaheptyl)-ether of *p-tert*-butyl calix[8]arene forms strong complexes with these cations. The Parma group [6] later showed that the hexa-(3-oxabutyl)ether of *p-tert*-butyl calix[6]arene also forms complexes with these ions.

An extensive survey has been carried out by McKervey and coworkers [7], who prepared the carbo-alkoxymethyl ethers of *p-tert*-butyl calix[4]arene, *p-tert*-butyl calix[6]arene, *p-tert*-butyl calix[8]arene, calix[4]arene, calix[6]arene, and calix[8]arene, and measured their abilities to extract cations from the aqueous phase into the nonaqueous phase. They concluded the following general aspects for the phase-transfer experiments: (1) the calix[4]arene compounds show the greatest selectivity for Na^+; (2) phase-transfer of Li^+ is inefficient with all of the compounds; (3) the calix[6]arene compounds show less affinity for Na^+ than for K^+, with plateau selectivity for Rb^+ and Cs^+; (4) the calix[8]arene compounds are the least efficient of the cyclic oligomers, showing low levels of transport and low discrimination for all five cations; (5) the calix[6]arene compounds are significantly more effective than 18-crown-6 for Na⁺ and K⁺ and much more so for Cs⁺; and (6) the *tert*-butyl group appears to increase to some extent the selectivity of the calix[4]arene compounds for Na⁺.

Calestani et al. [8] prepared dietylamide of the tetra-carboxymethyl ether of *p-tert*-butyl calix[4]arene and reported alkali metal picrate extraction constants of 1.9

$\times 10^9$ for Na⁺, 2.8×10^7 for K⁺, and 1.3×10^7 for Li⁺. These constants were of a magnitude that persuaded the authors to liken the binding properties of these calixarenes to cryptands and spherands.

McKervey [7] and Chang and Cho [9] carried out similar studies. They prepared various ester derivatives of calixarene and tested their ion carrying capacities (Scheme 1).

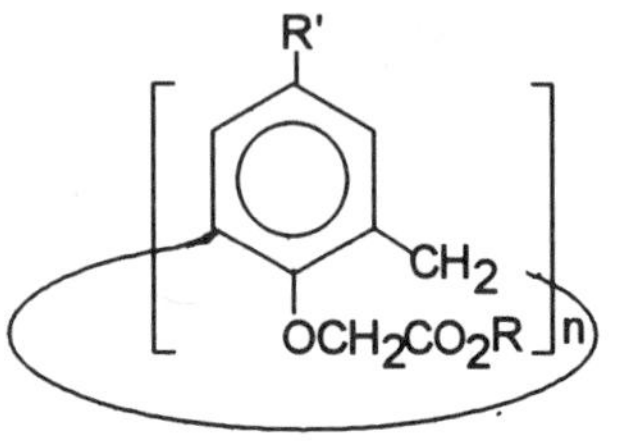

a	R'= tert-butyl, R=Et, n=4	g R'=H, R=Et, n=6
b	R'= tert-butyl, R=Me, n=4	h R'=H, R=Me n=6
c	R'=H, R=Et, n=4	i R'=tert-butyl, R=Et n=8
d	R'=H, R=Me, n=4	j R'=tert-butyl, R=Me, n=8
e	R'= tert-butyl, R=Et, n=6	k R'=H, R=Et, n=8
f	R'=tert-butyl, R=Me, n=6	l R'=H, R=Me, n=8

Scheme 1

Chang and coworkers [10] have synthesized amide derivatives of calixarenes and examined their ion binding properties with Group I and Group II cations. They observed that although the amides are much less effective than the esters for the complexation of Group I cations they are more effective for Group II cations.

Thus, it has been shown that calix[4]aryl esters exhibit remarkably high selectivity toward Na⁺ [11–14]. This is attributable to the inner size of the ionophoric cavity composed of four $OCH_2C=O$ groups, which is comparable to the ion size of Na⁺, and to the cone conformation that is firmly constructed on the rigid calix[4]arene platform (Scheme 2).

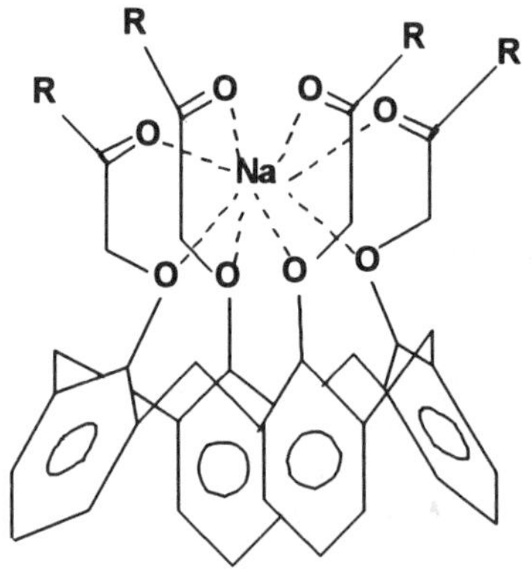

R = OC₂H₅

Scheme 2

Shinkai [15] concluded that *p-tert*-butyl calix[n]arene tetra esters form stable monolayers at the air–water interface and the metal responds, therein, quite differently from that in solution. They reported that examination of the metal "template effect" on the conformer distribution established that when the metal cation present in the base used serves as a "template," the cone conformer results are predominant [16]. Hence, Na⁺ in

NaH should serve as an efficient "template ion" to yield cone conformers. Conformer distribution for the reaction of *p-tert*-butyl calix[4]arene and ethyl bromoacetate is given in Table 1.

Kimura and coworkers [17], Diamond [18], and Damien et al. [19] have described that the polymeric calix[4]arenes have been used as ionophores in ion selective electrodes for Na⁺ (based on calixarene esters and amides) and for Na⁺ and Cs⁺ (based on *p*-alkylcalixarene acetates). The electrodes are stated to function as potentiometric sensors as well, having good selectivity for primary ion, virtually no response to divalent cations, and being stable over a wide pH range.

Other recent works in this field, studies on the transport of alkali and alkaline earth cations with *p-tert*-butyl calix[n]arene esters and amides, were carried out by Arnaud-Neu et al. [20] and Casnati et al. [21]. They prepared 1,3-alternate calix[4]arene-crown-6 as a new class of cesium selective ionophore.

In another work, ionophoric *bis*-calix[4]aryl esters in which the lower edges confront each other were syn-

Table 1 Conformer Distribution for the Reaction of *p-tert*-Butyl Calix[4]arene and Ethyl Bromoacetate [16]

		Distribution of *a* in Scheme 1		
Solvent	Base	Cone	Partial-cone	1,3-Alternate
THF	NaH	100	0	0
DMF	Li₂CO₃	100	0	0
DMF	Na₂CO₃	88	12	0
DMF	K₂CO₃	84	16	0
DMF	Cs₂CO₃	27	73	0
Aceton	Na₂CO₃	100	0	0
Aceton	K₂CO₃	96	3	0
Aceton	Cs₂CO₃	0	100	0

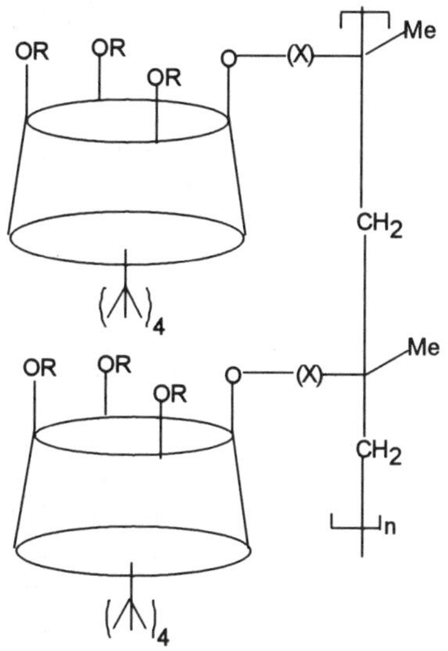

R=CH₂CO₂Et, X=CH₂CO₂CH₂CH₂OCO

Scheme 3

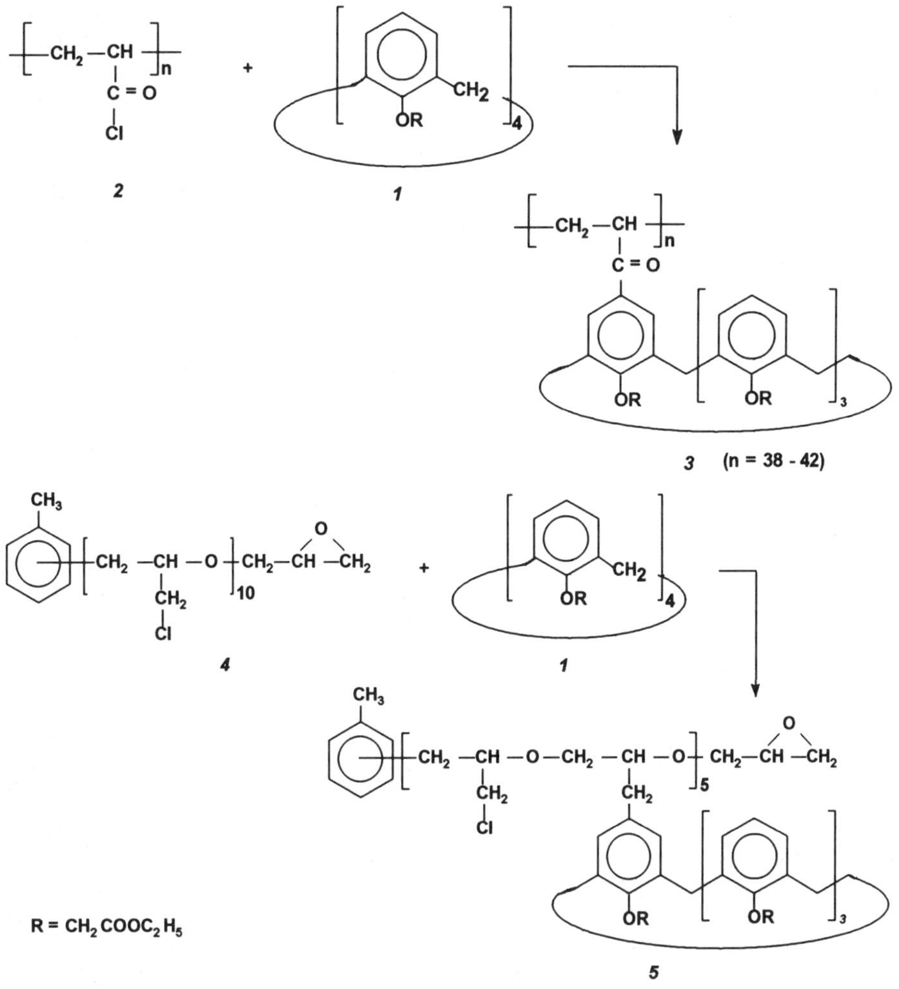

thesized by Ohseto, et al. [22]. The calixarenes are interesting compounds for which the unique intramolecular metal-hopping is observed using the ^{1}H-NMR spectral method.

The majority of the literature reports deal with the reaction of calixarenes with Group I and II cations. Polymeric calixarenes have been the subject of a more recent innovation. Harris et al. [23] have prepared a calix[4]arene methacrylate, its polymerization, and Na$^+$ complexation (Scheme 3). They concluded that both monomers and polymers form stable complexes with sodium thiocyanate.

Recently, Deligöz and Yilmaz [24,25] described the preparation of two polymeric calix[4]arene tetra esters (Scheme 4) and their Na$^+$-complexation. Based on phase-transfer experiments with these compounds using alkali picrates in water–dichloromethane, they confirmed that polymers are as Na$^+$ selective as monomers.

R = CH₂COOC₂H₅

Scheme 4

II. SELECTIVE IONOPHORES FOR METAL CATIONS

The most important contributions in this field were made Shinkai and coworkers [26–32]. They have embarked on an ambitious program that focuses on the uses to which calixarenes can be put. A good example, and one that represents an especially interesting study of cation complexation by calixarenes, deals with the extraction of uranium from sea water. The world's oceans contain a total of about 3 billion tons of uranium in the form of UO_2^{2+} associated with carbonate. Although this represents an enormous quantity of material, its concentration is only about 3 parts per billion, an amount that corresponds to less than 1 mg in a large-size backyard swimming pool. Also, the UO_2^{2+} is accompanied by numerous other cations, most of them present in a much larger concentration. Thus, the extraction of uranium from sea water poses a tantalizing challenge, which has been addressed by a number of chemists during the past decade. If the "greenhouse effect" proves to be responsible for adverse global changes in the climate, it may force greater attention to nuclear energy, and the recovery of uranium from sea water will become a more pressing problem in the future than it is at the present time. Early reports of UO_2^{2+} complexation came from the laboratories of Alberts and Cram [33] in 1976, where crown ethers chemistry was coming to fruition. Another work in this field came from Japan under the leadership of the late Iwao Tabushi, who found that the macrocyclic triamine-1 is particularly effective [34], having a K_{assoc} of $10^{20.7} M^{-1}$ for UO_2^{2+}. This material was tested in the ocean by placing it 10 m below the surface in the strong Kuorshio Current off the coast of Mikura, an island in Japan, where it absorbed 50 μg of uranium per gram of resin per day [35].

1

A polymer-bound analog of the *p*-sulfonatocalix[6]arenes is described in a Shinkai patent [31,32], which states that the hexakis(carbetoxymethyl)ether of *p*-sulfonatocalix[6]arene was partially nitrated, aminated, and fixed on crosslinked chloromethylated polystyrene. This resin is stated to absorb 108 μg of uranium

from sea water per 0.1 g of resin in 7 days at a flow rate of 30 mL/min. Another polymer-bound analog has also been reported by Shinkai et al. [36] who treated *p*-(chlorosulfonyl)calix[6]arene with polyethleneimine and obtained a gel-like product, which contained one calixarene unit for each 15 ethyleneimine units. It exhibited a similar binding power and selectivity for UO_2^{2+} as that of the parent calixarene, *p*-sulfonatocalix[6]arene.

Harrowfield et al. [37–39] have described the structures of several dimethyl sulfoxide adducts of homo bimetallic complexes of rare earth metal cations with *p*-tert-butyl calix[8]arene and *bis*-ferrocene derivatives of bridged calix[4]arenes. Ludwing et al. [40] described the solvent extraction behavior of three calixarene-type cyclophanes toward trivalent lanthanides La³⁺ (Ln = La, Nd, Eu, Er, and Yb). By using *p*-tert-butyl calix[6]arene hexacarboxylic acid, the lanthanides were extracted from the aqueous phase at pH 2–3.5. The extractability is: Nb, Eu > La > Er > Yb.

Yosida et al. [41] found that *p*-tert-butylcalix[6]arene can extract Cu^{2+} from the alkaline–ammonia solution to the organic solvent. Nagasaki and Shinkai [42] described the synthesis of carboxyl, derivatives of calix[n]arenes (n = 4 and 6) and their selective extraction capacity of transition metal cations from aqueous phase to the organic phase. Gutsche and Nam [43] have synthesized various substituted calix[n]arenes and examined the complexes of the *p*-bromo benzene sulfonate of *p*-(2-aminoethyl)calix[4]arene with Ni^{2+}, Cu^{2+}, Co^{2+}, and Fe^{2+}.

The titanium complexes of calixarene were obtained by Olmstead et al. [44] and Bott et al. [45], who examined their x-ray characteristics. Recent research in that field has been conducted by Rudkevich et al. [46]. They prepared calix[4]arene-triacids as receptors for lantanides.

Yilmaz and Deliğoz [47,48] prepared a calix[6]arene substituted with six aminoglyoxime groups (**2**), and two *vic*-dioxime compounds of calix[4]arene (**3,4**), and they examined their chelating ability with Co^{2+}, Cu^{2+}, and Ni^{2+}.

Yilmaz and Deligöz [49] have theorized that calixarenes might be utilized as selective ionophores for Fe³⁺. Thus, they studied the selective liquid–liquid extraction of Fe³⁺ cation from the aqueous phase to the organic phase by using *p*-tert-butyl calix[4]arene (**1**), calix[4]arene (**2**), *p*-nitro-calix[4]arene (**3**), calix[4]arene-*p*-sulfonic acid (**4**), *p*(diethyl-amino)methylcalix[4]arene (**5**), tetra-methyl-*p*-tert-butyl calix[4]arene-tetraketone (**6**), 25,27-dimethyl-26,28-di-hydroxy-*p*-tert-butyl calix[4]arene-diketone (**7**), calix[4]arene bearing dioxime groups on the lower rim (**8**), and a mono-oxime (**9**). The affect of varying pH upon extraction ability of calixarenes substituted with electron donating and electron with drawing groups at their *p*-position. Observed results were compared with those found for unsubstituted calix[4]arene. Compounds **3** and **4** were used as electron with-

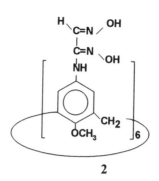

2

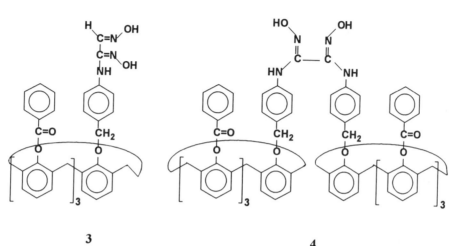

3

4

1 : R' = C(CH₃)₃ R = H

2 : R' = H R = H

3 : R' = NO₂ R = H

4 : R' = SO₃H R = H

5 : R' = CH₂N(C₂H₅)₂ R = H

6 : R' = C(CH₃)₃ R = CH₂COCH₃

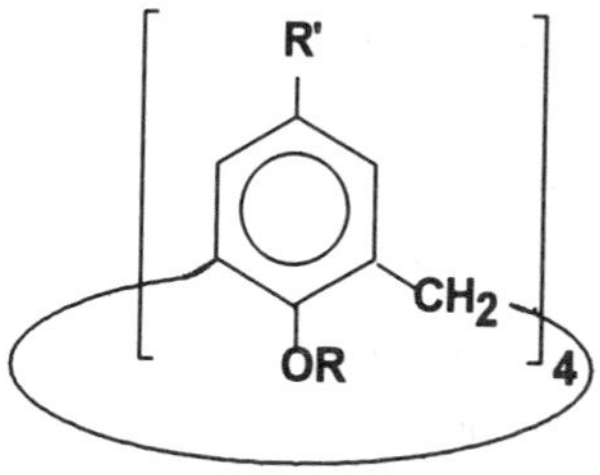

7 : R = CH₂COCH₃

N-OH
||
8 : R = CH₂CCH₃

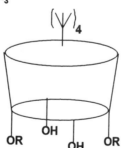

9 : H₃C—⬡—CH=N—OH

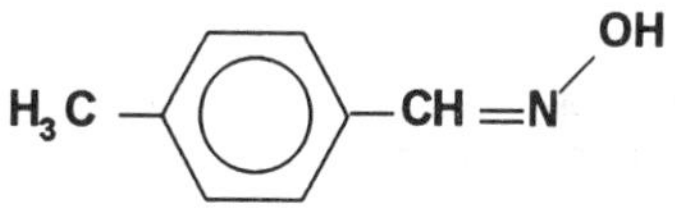

drawing and **1** and **5** as electron donating calixarenes (Table 2). The results of the extraction experiments indicated that the nature of *p*-substituted group has not significantly affected the extraction process. The extraction of Fe^{3+} ions was increased by increasing pH, but exhibited a decrease at pH 2.2; 51.0% extraction was accomplished at pH 5.4 with **6** in which all phenolic groups were substituted. When compound **7**, in which two phenolic groups were substituted, was used, the extraction ratio increased significantly (Table 2). The above observations evidently indicate the important role of phenolic oxygen in this procedure. Extraction experiments with the compound **8** yielded similar results as obtained with other compounds. The effect of pH on the extraction of **8** was observed to be lower; 62.8% extraction was accomplished even at pH 2.2. This result is due to the presence of two adjacent oxime groups (—C=N—OH) in the compound **8**. Since the extraction experiments performed with its monomer **9** the ratio was only 3.6% at pH 2.2. The above observations indicate that the cone conformation of calixarene bearing the oxime groups also plays an important role in the extraction process.

The ultraviolet spectrum of compound **2** in DMF did not exhibit an absorption maximum above 300 nm. Formation of such a complex was evident from the changing color of solution into brown and the appearance of an absorption maximum at 534 nm. The metal–ligand ratio determined at this wavelength by the Job method was 1:1.

The extraction reaction of the present systems can be expressed by Eq. (1).

$$M^{n+}_{(aq)} + [LH_m]_{(org)} \rightleftharpoons [MLH_{m-n}]_{(org)} + nH^+_{(aq)}$$
$$(1)$$

Where *aq* and *org* denote the species in the aqueous and the organic phase, respectively.

$$D = [MLH_{m-n}]_{org}/[M^{n+}]_{aq} \tag{2}$$

The extraction equilibrium constant (K_{ex}) is given by:

$$K_{ex} = \frac{[MLH_{m-n}]_{org} \cdot [H^+]^n_{aq}}{[M^{n+}]_{aq} \cdot [LH_m]_{org}} \tag{3}$$

$$\log D = n\,pH + \log K_{ex} + \log [LH_m]_{org} \tag{4}$$

Equation (4) indicates that the slope *n* of the log *D* versus the pH plot corresponds to the number of protons released upon extraction. If the logarithm of the ratio between Fe^{3+} content in aqueous phase and organic phase is plotted as a function of pH, a linear relation was obtained between pH 3.5–5.4, which deviated from linearity at lower pH values (2.2). The fact that the slope of the curves were very close to unity indicates that only one proton has separated from the ligand (Eq. [5]).

$$Fe^{3+} + HL \rightleftharpoons LFe^{2+} + H^+ \tag{5}$$

Since one proton separation will hardly occur at pH 2.2, extraction ratios of Fe^{3+} with compounds **1–7** are quite low. Yet compound **8**, which carries oxime groups, was able to extract a considerable amount of Fe^{3+} at pH 2.2.

The logarithmic extraction constants $\log K_{ex}$ (K_{ex} in mol/L) corresponding to Eq. (4) are as follows:

$$\log K_{ex} = 1.79 \pm 0.15 \ (\mathbf{1})$$

$$\log K_{ex} = 1.50 \pm 0.15 \ (\mathbf{4})$$

$$\log K_{ex} = 3.40 \pm 0.10 \ (\mathbf{5})$$

$$\log K_{ex} = 1.80 \pm 0.10 \ (\mathbf{7})$$

The solvent extraction mechanism with **1–5** and **7,8** is different from **6**, since **6** is a remarkably good ion-extracting compound for Na^+ ions [12–14]. In a two-phase solvent extraction of Fe^{3+} from the organic phase into the aqueous phase with **6**, can be explained in terms of the exchange of Fe^{3+} ions by the Na^+ ions, which is in an organic phase. The extraction processes are shown as in Fig. (1). In order to ascertain that the Na^+ salt of **6** indeed exists in the chloroform phase, an aqueous solution (25 mL) containing 0.01 M CH_3COOH/CH_3COONa buffer (pH 4.5) and a chloroform solution (5 mL) containing 5.3×10^{-4} M of **6** were mixed and shaken. After 12 h at 25°C the chloroform phase was separated and extracted with 0.1 M HCl solution.

The flame photometric analysis of this solution established that 70% of **6** is dissociated as the Na^+ salt.

Based on the above results they have concluded that the ligand groups circularly arranged on the lower rim of the calixarene cavity construct a novel cyclic metal receptor for selective extraction of transition metal cations. Results suggest that the fine tuning in molecular

Table 2 Extraction of Fe^{3+} Cation with Ligand (%) [49]

pH	L_1	L_2	L_3	L_4	L_5	L_6	L_7	L_8	L_9
2.2	7.0	8.4	27.1	15.7	22.0	12.0	20.7	62.8	3.6
3.8	22.4	56.0	42.1	28.2	40.5	18.5	48.3	72.0	18.5
4.5	46.5	57.5	77.0	65.0	72.0	20.4	87.8	88.6	28.1
5.4	66.0	90.0	77.1	77.8	84.5	51.0	92.1	89.4	40.7

Aqueous phase[Metal nitrate = 1.06×10^{-4} M]. Organic phase, [chloroform, (ligand) = 5.3×10^{-4} M]. pH: 2.2 (0.01 M $NaNO_3/HNO_3$, $\mu = 0.1$ with KCl), pH 3.8, 4.5 and 5.4 (0.01 M CH_3COONa/CH_3COOH, $\mu = 0.1$ with KCl), at 25°C for 12 h.

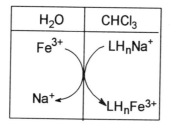

Figure 1 Extraction mechanism proposed for L_6.

design can be done by using functional groups arranged on the lower rim (closed side of the calixarene cavity) rather than by using those arranged on the upper rim (open side of the calixarene cavity).

Recently, Deligöz and Yilmaz [51] prepared three polymeric calix[4]arenes, which were synthesized by reacting chloromethylated polystyrene with 25,26,27-tribenzoyloxy-28-hydroxy calix[4]arene (**2a, 3a**) and polyacryloyl chloride with 25,26,27,28-tetraacetoxy calix[4]arene (**4a**). After alkaline hydrolysis of the polymers, they were utilized for selective extraction of transition metal cations from aqueous phase to organic phase.

To support a polystyrene onto the upper rim of calix[4]arene (phenolic-O- of calix[4]arene) and 25,26,27-tribenzoyloxy-28-hydroxy, calix[4]arene was treated with chloromethylated polystyrene in the presence of K_2CO_3 (Scheme 7). Polymeric calix[4]arene (**3a**) thus obtained was hydrolyzed in the benzoyl groups prior to use for the extraction process.

Selective extraction experiments were then performed to see transference of some transition elements (Cu^{2+}, Ni^{2+}, Co^{2+}, and Fe^{3+}) from the aqueous phase to the organic phase by the synthesized polymeric calixarenes. Phase-transfer studies in water–chloroform confirmed that polymer **2b** and **3b** were Fe^{3+} ion-selective as was its monomer (**1**). Extraction of Fe^{3+} cation with **2b** and **3b** was observed to be maximum at pH 5.4. Only trace amounts of other metal cations such as Cu^{2+}, Ni^{2+}, and Co^{2+} were transferred from the aqueous to the organic phase (Table 3). Furthermore, the extracted quantities of these cations remained unaffected with increasing pH. The effect of pH on the extraction of **3b** was lower and 56% extraction was accomplished even at pH 2.2. The extraction experiments were also performed with calix[4]arene (**1**); the ratio was 8.4% at pH 2.2. The polymeric calix[4]arenes were selective to extract Fe^{3+} from an aqueous solution, which contained Cu^{2+}, Ni^{2+}, Co^{2+}, and Fe^{3+} cations, and it was observed that the

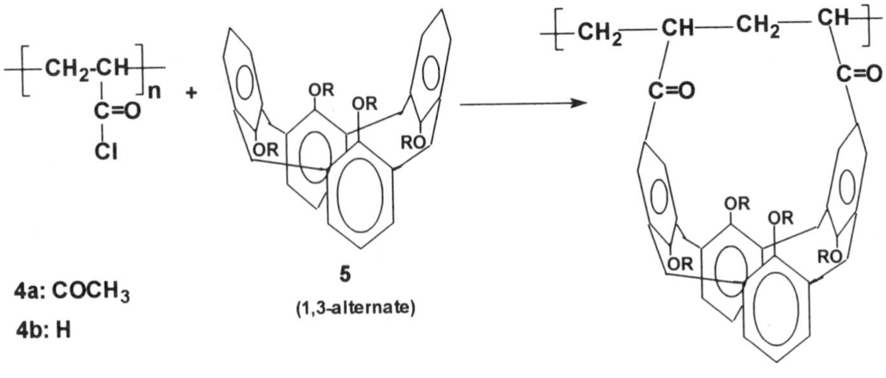

4a: COCH₃

4b: H

5

(1,3-alternate)

4

polymer supported onto the lower rim calix[4]arene is a more efficient carrier for Fe^{3+} in the extraction process.

The results of the extraction of metal cations are summarized in Table 3. The **4b** showed the selectivity toward Fe^{3+} cation and cannot act as an extractant for Fe^{3+} at pH 2.2. The extraction of Fe^{3+} cation with **4b** was increased with increasing pH, but the Fe^{3+} cation ratios at all pH decrease than that of **2b** and **3b**. The results suggested that the conformation of **4b** and the C=O groups in the polymer affected the extraction of Fe^{3+} in this process.

Based on the preceding results, we have observed that the extraction process with the polystyrene sup-

ported onto the lower rim calix[4]arene (**3b**) can act cooperatively, whereas those induced onto the upper rim (i.e., open *p*-position side) act rather independently.

In another study in this field, Deligöz et al. [50], synthesized a polymeric calixarenes by combining 25,26,27-tribenzoyloxycalix[4]arene with the oligomer **1** in the presence of NaH. Based on the chlorine analysis of this product, it was observed that compound **2** did not attach to each consecutive (CH₂-Cl) groups in a regular array.

The polymeric calixarene(**2**) thus obtained was hydrolyzed with ethanolic NaOH solution to remove the benzoyl groups prior to use for the extraction process.

Table 3　Extraction of Metal Cations with Ligands*

Ligand	Metal cations extracted (%)				pH
	Fe^{3+}	Ni^{2+}	Co^{2+}	Cu^{2+}	
1	8.4	<1.0	<1.0	4.6	2.2
1	56.0	<1.0	<1.0	7.3	3.8
1	57.5	<1.0	<1.0	8.9	4.5
1	90.0	<1.0	<1.0	9.6	5.4
2b	2.4	<1.0	<1.0	2.6	2.2
2b	50.6	<1.0	<1.0	4.6	3.8
2b	59.0	<1.0	<1.0	7.3	4.5
2b	98.0	<1.0	<1.0	7.8	5.4
3b	56.0	<1.0	<1.0	2.5	2.2
3b	95.0	<1.0	<1.0	4.7	3.8
3b	97.0	<1.0	<1.0	8.5	4.5
3b	98.0	<1.0	<1.0	9.1	5.4
4b	0.3	<1.0	<1.0	2.1	2.2
4b	38.7	<1.0	<1.0	5.2	3.8
4b	40.6	<1.0	<1.0	8.7	4.5
4b	69.7	<1.0	<1.0	8.9	5.4

* Aqueous phase [metal nitrate = 1.06×10^{-4} M]. Organic phase [chloroform (ligand) = **1** (0.23 g/L), **2b** (0.36 g/L), **3b** (0.36 g/L), or **4b** (0.32 g/L)]; 25°C, for 12 h. pH: 2.2 (0.01 M NaNO₃/HNO₃, μ = 0.1 with KCl), pH. 3.8, 4.5 and 5.4 (0.01 M CH₃COONa/CH₃COOH, μ = 0.1 with KCl), at 25°C for 12 h.

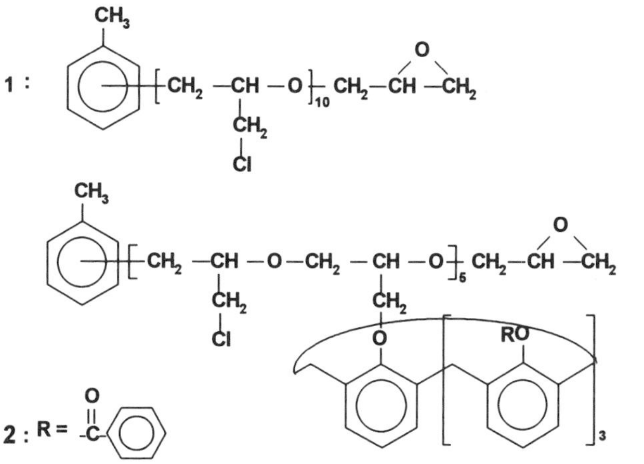

1:

2 : R = $-\overset{\overset{\displaystyle O}{\|}}{C}$

3 : R = H

Extraction of Fe^{3+} with calix[4]arene, which was 7%, has increased to 82% when the polymeric calix[4]arene

1 : n= 4	R = H
2 : n= 6	R = H
3 : n= 4	R = CH$_2$COOC$_2$H$_5$
4 : n= 4	R = CH$_2$COCH$_3$
5 : n= 4	R = CH$_2$CCH$_3$ (N=OH)
6 : n= 6	R = CH$_2$CCH$_3$ (N=OH)

7 :

8 : R = $-\overset{\overset{\displaystyle O}{\|}}{C}$

9 : R = H

was used (Table 4). Only trace amounts of other metal cations such as Cu^{2+}, Co^{2+}, and Ni^{2+} were transferred from aqueous to organic phase at pH 2.2. Furthermore, the extracted quantities of these cations remained unaffected with increasing pH.

Deligöz and Yilmaz [52] reported that the selective liquid–liquid extraction of various alkali and transition metal cations from the aqueous phase to the organic phase as carried out by using p-tert-butyl calix[4]arene (**1**), p-tert-butyl calix[6]arene (**2**), tetra-ethyl-p-tert-butyl calix[4]arene-tetra-acetate (**3**), tetra-methyl-p-tert-butyl calix[4]arene-tetraketone (**4**), calix[n]arenes (n = 4 and 6) bearing oxime groups on the lower rim (**5** and **6**) and a polymeric calix[4]arene (**8**). It was found that compounds **5** and **6** showed selectivity towards Ag^+, Hg^+, Hg^{2+}, Cu^{2+}, and Cr^{3+} and the order of the extractability was $Hg^+ > Hg^{2+} > Ag^+ > Cu^{2+} > Cr^{3+}$. The polymeric calix[4]arene (**8**) was selective for Ag^+, Hg^+, and Hg^{2+}, unlike its monomeric analog.

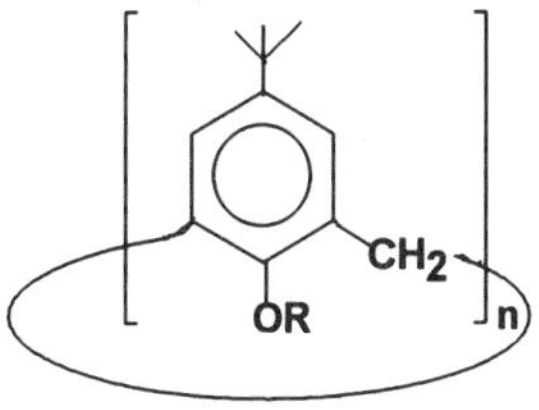

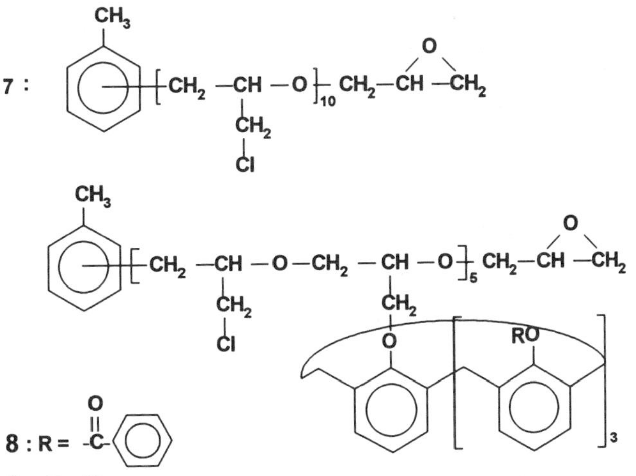

In this work, the effectiveness of the ester, ketone, and oxime derivatives of p-tert-butyl calix[4]arene (**1, 3, 4,** and **5**) and the oxime derivative of p-tert-butyl calix[6]arene (**2** and **6**) in transferring the alkali metal cations such as Na^+, K^+, and Li^+ and transition metals such as Ag^+, Hg^+, Hg^{2+}, Co^{2+}, Cu^{2+}, Ni^{2+}, Cd^{2+}, Fe^{3+}, and Cr^{3+} from aqueous phase into chloroform (Table 5) were studied.

From the data given in Table 5, it can be seen that ligands with ester and ketone groups (**3** and **4**) are more effective at extracting Na^+ and K^+ ions than other ligands. This conclusion is not new, however, and has been

Table 4 Extraction of Metal Cations with Ligands*

Ligand	Metal cations extracted (%)				pH
	Fe^{3+}	Ni^{2+}	Co^{2+}	Cu^{2+}	
3	82.0	<1.0	<1.0	3.7	2.2
3	93.0	<1.0	<1.0	6.5	3.8
3	94.0	<1.0	<1.0	9.2	4.5
3	96.0	<1.0	<1.0	9.8	5.4

* Results of metal extraction with calx(4)arene were given in Table 3. Aqueous phase[Metal nitrate = 1.06×10^{-4} M]. Organic phase, [chloroform, (ligand) = 5.3×10^{-4} M]. pH: 2.2 (0.01 M $NaNO_3$/HNO_3, μ = 0.1 with KCl), pH 3.8, 4.5 and 5.4 (0.01 M CH_3COONa/CH_3COOH, μ = 0.1 with KCl), at 25°C for 12 h.

previously reported in the literature [12]. Oxime derivatives of *p-tert*-butyl calix[4]arene (**5**) and *p-tert*-butyl calix[6]arene (**6**) do not possess the same properties. These ligands, which are very effective in transferring the transition elements, particularly Ag^+, Hg^+, Hg^{2+}, Cu^{2+}, and Cr^{3+}, do not extract the alkali metal cations to a significant extent as reported by Nomura et al. [53], who used *p*-phenylazocalix[6]arene as the ligand.

These phenomena can be explained by the (hard–soft) acid–base principal as follows: C=N-OH is a soft base, hence has stronger affinity towards soft basic metal cations than hard metal cations. The strong participation of the N-OH group in complex formation was further confirmed by the results shown for extraction experiments with **5** and **6**.

It was observed that while the ester (**3**) and ketone (**4**) derivatives of *p-tert*-butyl calix[4]arene extracted negligibly small amounts of Cu^{2+} and Cr^{3+} ions, its oxime derivative efficiently extracted these ions. Furthermore, **5** was found to be more effective than **6** in extracting Cr^{3+} ion.

The fact that all ligands failed to transfer Fe^{3+} ion from the aqueous into the organic phase was not unexpected, since this ion prefers to bind with picric acid more than the other ligands. This property is typical only for the Fe^{3+} ion [54]. Yet, our previous observations [49] indicated that, when $Fe(NO_3)_3$ was used instead of metal picrate, it was possible to efficiently extract Fe^{3+} into the organic phase by utilizing ligands **1**, **3**, and **4**.

Another interesting result of the present investigation was the fact that the oligomer **7** of higher calix[4]arene content exhibited a different property than that of **1**. While **1** extracted little Ag^+, Hg^+, and Hg^{2+}, **8** was capable of extracting these ions.

To investigate whether these results were caused by the oligomer (**7**) or calix[4]arene itself, experiments were performed with calix[4]arene, before it was reacted with **7**. Observations showed that when extraction was performed with unreacted calix[4]arene, the transfer of Ag^+, Hg^+, and Hg^{2+} ions was very close to unity. The compound (**8**) contains benzoyl groups. To understand the effect of this group on the extraction process, the

Table 5 Extraction of Metal Picrates with Ligands*

Ligand	Picrate salt extracted (%)											
	Li^+	Na^+	K^+	Ag^+	Hg^+	Hg^{2+}	Co^{2+}	Ni^{2+}	Cu^{2+}	Cd^{2+}	Fe^{3+}	Cr^{3+}
1	19.9	8.0	2.3	6.3	—	15.5	7.7	6.3	—	9.4	—	13.0
2	10.5	4.6	6.3	4.1	31.1	13.9	4.6	3.5	—	6.6	—	6.5
3	41.0	95.0	60.0	68.5	47.7	67.3	9.4	3.7	6.0	1.2	—	18.0
4	54.0	69.0	59.0	49.5	46.3	11.5	26.3	11.5	—	17.3	—	7.3
5	9.1	2.4	3.3	87.3	91.2	90.2	18.1	18.2	65.0	20.1	—	63.5
6	15.4	10.7	15.8	85.8	87.6	87.6	26.3	21.3	51.0	28.0	—	32.5
7	—	—	—	—	9.3	30.8	19.3	—	—	5.1	—	2.4
8	—	—	—	37.3	63.8	66.6	—	—	17.1	5.7	—	—
9	—	—	—	38.0	27.0	57.3	—	—	—	—	—	—

* H_2O/$CHCl_3$ = 10/10 (v/v); [Picric acid] = 2×10^{-5} M, [Ligand] = 1×10^{-3} [Metal nitrat] = 10^{-2} M for transition metals; 298 K for 1 h.

benzoyl groups of **8** were removed by hydrolysis with an NaOH/EtOH mixture of yield **9**. The extraction experiments performed with **9** have indicated that while the Hg$^+$ cation ratio decreased, the ratio of Ag$^+$ and Hg^{2+} cations remained unaffected. This result suggests that the C$=$O groups are the efficient groups for Hg$^+$ cation extraction. The present experimental data is not sufficient to explain the very high selectivity of **9** to Hg^{2+} cation. Investigations related to this aspect are in progress.

Based on these results, it was concluded that the transfer of ions such as Ag$^+$, Hg$^+$, and Hg^{2+} with polymeric calix[4]arene follows a different mechanism than that of calix[4]arene.

REFERENCES

1. C. D. Gutsche, *Calixarenes,* Royal Society of Chemistry, Cambridge, UK (1989).
2. R. M. Izatt, J. D. Lamb, R. T. Hawkins, P. R. Brown, S. R. Izatt, and J. J. Christensen, *J. Am. Chem. Soc., 105*: 1782 (1983).
3. S. R. Izatt, R. T. Hawkins, J. J. Christensen, and R. M. Izatt, *J. Am. Chem. Soc., 107*: 63 (1985).
4. R. M. Izatt, J. J. Christensen, and R. T. Hawkins, U.S. Patent 4,447,377, 16 Oct. 1984 (*Chem. Abstr., 101*: 21 833m).
5. V. Bocchi, D. Foina, A. Pochini, and R. Ungaro, *Tetrahedron 38*: 373 (1982).
6. R. Ungaro, A. Pochini, G. D. Andreetti, and P. Domiano, *J. Incl. Phenom., 3*: 35 (1985).
7. M. A. McKerwey, E. M. Seward, G. Ferguson, B. L. Ruhl and S. J. Harris, *J. Chem. Soc., Chem. Commun.,* 388 (1985).
8. G. Calestani, F. Ugozzoli, A. Arduni, E. Ghidini, and R. Ungaro, *J. Chem. Soc., Chem. Commun.,* 344 (1987).
9. K. Chang, and I. Cho, *Chem. Lett.,* 477 (1984).
10. S. K. Chang, S. K. Kwon, and I. Cho, *Chem. Lett.,* 947 (1987).
11. A. Arduini, A. Pochini, S. Reverberi and R. Ungaro, *Tetrahedron, 42*: 2089 (1986).
12. F. Arneud-Neu, E. M. Collins, M. Deasy, G. Ferguson, S. J. Harris, B. Kaitner, A. J. Lough, M. A. McKervey, E. Marques, B. L. Ruhl, M. J. S. Weill and E. M. Seward, *J. Am. Chem. Soc., 111*: 8681 (1989).
13. S. K. Chang and I. Cho, *J. Chem. Soc., Perkin Trans. I,* 211 (1986).
14. Arimura, M. Kubota, T. Matsuda, O. Manabe, and S. Shinkai, *Bull. Chem. Soc. Jpn, 62*: 1674 (1989).
15. S. Shinkai, *Tetrahedron, 49*: 8933 (1993).
16. K. Iwamoto, and S. Shinkai, *J. Org. Chem., 57*: 7066 (1992).
17. K. Kimura, M. Matsuo, and T. Shono, *Chem. Lett.,* 615 (1988).
18. D. Diamond, Anal. Chem. Symp. Ser., 1986, 25 (Chem. Abstr., 106:148452b); ''Idem. in Electrochemistry, Sensors, and Analysis,'' Proceedings of International Conference, Dublin, June 10–12 25: 155 (1986).
19. W. Damien, M. Arrigan, G. Svehla, S. J. Harris, and M. A. McKervey, *Analytical Proceeding, 29*: 27 (1992).
20. F. Arnaudneu, S. Fanni, L. Guerra, W. Megregor, K. Ziat. M. J. Schwingweill, G. Barrett, M. A. Mckervey, D. Marrs, and E. M. Seward, *J. Chem. Soc. Perkin Trans. II,* 113 (1995).
21. A. Casnati, A. Pochini, R. Ungaro, F. Ugozzoli, F. Arnaud, S. Fanni, M. J. Schwing, R. J. M. Egberink, F. Dejong, and D. N. Reinhoult, *J. Am., Chem., Soc., 117*: 2767 (1995).
22. F. Ohseto, T. Sakaki, K. Arak, and S. Shinkai, *Tetrahedron Lett., 34*: 2149 (1993).
23. S. J. Harris, G. Barrett and M. A. McKervey, *J. Chem. Soc., Chem. Commun.,* 1224 (1991).
24. H. Deligöz and M. Yilmaz, *Synth. React. Inorg. Met.-Org. Chem., 26*: 285 (1996).
25. H. Deligöz and M. Yilmaz, *J. Pol. Sci., Part A; Polymer Chemistry, 33*: 2851 (1995).
26. S. Shinkai, H. Koreishi, K. Ueda, and O. Manebe, *J. Chem. Soc. Chem. Commun.,* 233 (1986).
27. S. Shinkai, H. Koreishi, K. Ueda, and O. Manebe, *J. Am. Chem., 109*: 6371 (1987).
28. G. Deng, T. Sakaki, Y. Kawahara, and S. Shinkai, *Tetrahedron Lett., 33*: 2163 (1992).
29. G. Deng, T. Sakaki, Y. Kawahara, and S. Shinkai, *Chem. Lett.,* 1287 (1992).
30. S. Shinkai, Y. Shirahama, H. Satoh, O. Manabe, T. Arimura, K. Fujimoto, and T. Matsuda, *J. Chem. Soc., Perkin Trans. II.,* 1167 (1989).
31. S. Shinkai, O. Manabe, Y. Kondo, and T. Yamamoto, (Kanebo Ltd.), Jpn. Tokyo Koho, Jp 62, 136, 242, 1987 (Chem. Abstr., 108:644109).
32. Y. Kondo, T. Yamamoto, O. Manabe, and S. Shinkai, (Kanebo Ltd.) Jpn. Kokai Tokkyo Koho, Jp 62, 210, 055, 1987 (Chem. Abstr., 108:116380b).
33. A. H. Alberts, and D. J. Cram, *J. Chem. Soc., Chem. Commun.* 958 (1976).
34. I. Tabushi, Y. Kabuke, and A. Yoshizawa, *J. Am. Chem. Soc., 106*: 2481 (1984).
35. I. Tabushi, Y. Kabuke, N. Nakayama, T. Aoki, and A. Yoshizawa, *I & EC Prod. Res. Dev., 23*: 445 (1984).
36. S. Shinkai, H. Kawaguchi, and O. Manabe, *J. Pol. Sci., Part C; Polymer Letters, 26*: 391 (1988).
37. J. M. Harrowfield, M. I. Ogden, and A. H. White, *Aust. J. Chem., 44*: 1249 (1991).
38. J. M. Harrowfield, M. I. Ogden, and A. H. White, *Aust. J. Chem., 44*: 1237 (1991).
39. J. M. Harrowfield, M. I. Ogden, and A. H. White, *J. Chem. Soc. Dalton Trans,* 2625 (1991).
40. R. Ludwig, K. Inoue, and T. Yamato, *Solvent Extr. Ion. Exch., 11*: 311 (1993).
41. I. Yoshida, S. Fujii, Uneo, S. Shinkai, and T. Matsuda, *Chem. Lett.,* 1535 (1989).
42. T. Nagasaki, and Shinkai, *Bull. Chem. Soc. Jpn., 65*: 471 (1992).
43. C. D. Gutsche, and K. C. Nam, *J. Am. Chem. Soc., 110*: 6153 (1988).
44. M. M. Olmstead, G. Sigel, H. Hope, X. Xu, and P. P. Power, *J. Am. Chem. Soc., 107*: 8087 (1985).
45. S. G. Bott, A. W. Coleman, and J. K. Atwood, *J. Chem. Soc., Chem. Commun.,* 610 (1986).
46. D. M. Rudkevich, W. Verboon, E. Vandertol, C. J. Vanstaveren, F. M. Kaspersen, J. W. Verhoeven, and D. N. Reinhoudt, *J. Chem. Soc. Perkin Trans. II.* 131 (1995).

47. M. Yilmaz, and H. Deligöz, *Synth. React. Inorg. Met.-Org. Chem., 23*: 67 (1993).

48. H. Deligöz, and M. Yilmaz, *Synth. React. Inorg. Met.-Org. Chem., 3*: (1997).

49. M. Yilmaz, and H. Deligöz, *Separation Sci. and Technol., 31*: 2395 (1996).

50. H. Deligöz, M. Tavasli, and M. Yilmaz, *J. Pol. Sci., Part A., Polymer Chem., 32*: 2961 (1994).

51. H. Deligöz, and M. Yilmaz, *Reactive Polymers, 31*: 81 (1996).

52. H. Deligöz, and M. Yilmaz, *Solvent Extr. Ion. Exch., 13*: 19 (1995).

53. E. Nomura, H. Taniguchi, K. Kawaguchi, and Y. Otsuji, *Chem. Lett.*, 2167 (1991).

54. M. Yilmaz, and U.S. Vural, *Synth. React. Inorg. Met.-Org. Chem., 21*: 1231 (1991).

23

Thermodynamic Opportunity Due to the Degradation Reaction Initiation

E. F. Vainstein and G. E. Zaikov
Russian Academy of Sciences, Moscow, Russia

I. INTRODUCTION

The final power of any process, including chemical degradation, is the change of free energy, ΔG. The proceeding reaction becomes possible if:

$$\Delta G = G_e - G_{in} < 0 \qquad (1)$$

where the free energies of end (G_e) and initial (G_{in}) systems determine the formation. The condition of thermodynamic possibility of the reaction initiation, with performance defined by kinetic parameters, is set by the equation:

$$\Delta G = 0 \qquad (2)$$

Frequently conditions occur at which, with regularity, Eq. (1) is fulfilled at storage and exploitation of polymeric materials [1–9] as a result of the change in external parameters. Consider the conditions fulfilling Eq. (2) parameters for polymers, existing in different phases, at which degradation process proceeding can be initiated in the system. We can restrict through thermodynamics (formation free energy) the initial and end states, in which definite chemical substances are contained, because the end state at the defined initial one does not depend on how it is reached. That is, it is possible not to take into account the detailed mechanism of the process to clear up thermodynamic possibility. Consider the proceeding possibilities of the following reactions:

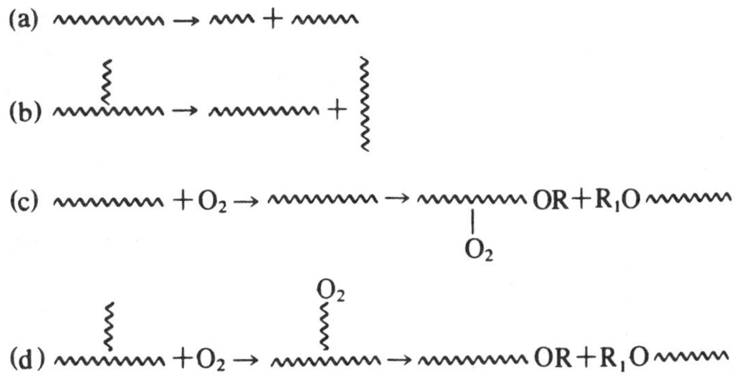

Scheme 1

Reactions (a) and (c) describe the process of a chemical bond break in the main chain, and (b) and (d) in the side chain. Reactions (a) and (b) are characteristic for thermodegradation, and (c) and (d) for thermooxidation

degradation. It is evident that reactions (a) and (b) differ sufficiently from (c) and (d) by the concrete mechanism of the proceeding reaction. In particular, for polymers under thermodynamic consideration, it is necessary to take into account the process of the first attachment of oxygen (or other chemical agents) only. Further on it will proceed the chemical transformation of groups, neighbor to attached oxygen, followed by a chain fracture. That is why we first consider the conditions of thermodynamic possibility of the initiation of (a) and (b) type reactions. Depending on the place of its proceeding, reaction (a) will occur with a sufficient decrease of molecular weight at the breaking of the chain middle. If the chain break occurs by the end, reaction (a) will be similar to that of the (b) type. We will consider that the components of free energy additive, which in some cases can lead to different parameters which differ from parameters at which the process occurs. The probability of such a consideration is connected with independence of the free energy of the formation of the end and defined initial system states, i.e., in this case as though it is composed the way of transition for each stage, during which only one component of free energy changes. First, consider the simplest case—thermodynamic opportunity of degradation reaction initiation in diluted solution.

A. Limited Sizes of Thermodynamically Stable Chains in Diluted Solution

Free energy of solution, containing chain molecules of a definite length (before the beginning of degradation), can be written as follows:

$$G_{in} = G_x + G_s + G_{id} + G_{ex} + G_{defl} + \Delta G_s + \Delta G_{ch-s} \tag{3}$$

where G_x is the free energy of chain molecule formation, representing "rigid bar"; G_s is the free energy of solvent formation; G_{id} is the free energy of ideal mixing; G_{ex} is the excessive free energy of different size molecules mixing (polymer and solvent); G_{defl} is the free energy of chain deflection, ΔG_s is the change of free energy of solvent molecule interaction with each other at transition from solvent volume to the chain (Fig. 1). This interaction is stipulated by the change of the interatomic distance in solvent molecules disposed in polymer volume among the chain. ΔG_{ch-s} is the free energy of the chain–solvent interaction—precisely, the change of free energy is the consequence of interaction between solvent molecules in bulk and chain molecules in mass (by local interactions of solvent molecules with the chain).

Free energy of solution formation, consisting of the same amount of solvent and degradated molecules, is presented as follows:

$$G_e = G_x + G_s + G'_{id} + G'_{ex} + G'_{defl} \\ + \Delta G'_s + \Delta G'_{ch-s} + \Delta G_x \tag{4}$$

Indexes in Eq. (4) mark the same components, as in Eq. (3), but relate to a new state.

$$\Delta G_x = \Delta H - T\Delta S$$

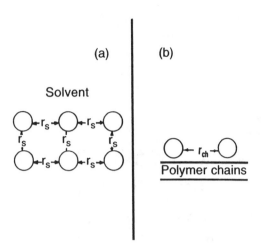

Figure 1 The distance between solvent molecules in volume (r_s) − (a). The distance between solvent molecules disposed neighboring the polymer chain (r_{ch}) − (b).

G_x is the change of free energy at new bond formation. We can neglect the contribution of bulk into free energy change, owing to its low change at degradation in diluted solution. We then obtain:

$$\Delta G = G_e - G_{in} = (G'_{id} - G_{id}) \\ + (G'_{ex} - G_{ex}) + (G'_{defl} - G_{defl}) \\ + (\Delta G'_s - \Delta G_s) + (\Delta G'_{ch-s} - \Delta G_{ch-s}) \\ + \Delta G_x \tag{5}$$

At a similar reaction with the participation of low-molecular compounds (LMC), only ΔG will be the combination of all components, except $(G'_{ex} - G_{ex}) + (G'_{defl} - G_{defl})$, values of $(\Delta G'_s - \Delta G_s) + (\Delta G'_{ch-s} - \Delta G_{ch-s})$ differing from the ones of the same components in Eq. (5) among the chain. In most cases of low-and high-molecular reactions, these terms possess similar values because they characterize local interactions, i.e., that between the closest atoms. Designing these terms as G_{LMC}, we obtain:

$$\Delta G = \Delta G_{LMC} + (G_{ex} - G'_{ex}) + (G'_{defl} - G_{defl}) \tag{6}$$

Consequently, the features of degradation proceeding in diluted solutions of chain molecules in comparison with similar reactions in low-molecular compositions will be stipulated by the change of chain flexibility and number of polymer–solvent contacts. The change of both components is stipulated by the change of chain length (by its shortening). The present consideration does not take into account the existence of rotary isomers in chains and their changes as a consequence of chain length change. If considered, the free energy of both initial and end states should be added by terms, taking into account their amount and probability of various disposition in the chain. The changes of flexibility and contact number should be taken into account at the change of correlation between conformers, in the process.

For infinitely long chains (polymers), terms $(G'_{ex} - G_{ex}) + (G'_{defl} - G_{defl})$ will be close to zero. Consequently, in infinitely long chains and polymers, free energy change in the process can be close to the one in similar reactions of low-molecular compounds (P. Flori principle) [10].

If ΔG_x values are sufficiently lower (greater by absolute value) than ΔG of the rest of the components of free energy change, the process can proceed at all the values of molecular weights from the point of view of thermodynamics. Consequently, the application ΔG in the definite conditions is unadvisable.

$(G'_{id} - G_{id}) < 0$ for selected mechanism, i.e., in the whole range of molecular weights. The change of free energy of an ideal mixing is favorable to degradation process proceeding.

$$G'_{id} - G_{id}$$
$$= -RT\left[\frac{1}{2X'_2 X_1}\ln\left(\frac{X_1}{X_2}\right) + \ln(1 + X_2) - X'_2\ln 2\right] \quad (7)$$

Here X_1, X_2, X'_1, X'_2 are the molar parts of solvent and chain molecules in initial and end (') states. In the calculation per mole of double mixture:

$$G_{id} = -RT\sum_i x_i \ln x_i$$

where x_i is the molar parts of components. It is necessary to introduce corresponding coefficients into equation (7) if changes in concentrations and numbers of moles are taken into account. The value of $(G'_{id} - G_{id})$ is constant at constant initial concentration and number of breaks of chain molecules. Values of $(G'_{id} - G_{id})$ for a definite number of chain breaks (>1) can be estimated in analog.

G_{ch-s}, characterizing free energy of polymer–solvent interaction for sufficiently long chains, can be expressed as follows:

$$G_{ch-s} = C\left[n_c q^c_{p-s} + \sum_i t_i q\right] \quad (8)$$

where C is the chain concentration; q^c_{p-s} is the molar free energy of local interaction of middle chain units; n_c is the number for end groups, also, if the chain is chemically inhomogenous, summation of free energies of all monomeric units is required. Chain inhomogeneity and solvent contacting with it may be stipulated by chemical defects as well as by rotary isometry. For sufficiently long homogenous chains, G_{p-s} becomes a linear function of molecule number; q_{ie} is the local free energy of interaction for different end groups t. If there are two ends, and they are chemically equal, then:

$$G_{ch-s} = C[n_c q^c_{p-s} + 2t q^G_{p-s}]$$

At small chain lengths, G_{ch-s} depends nonlinearly on the chain length (the number of monomeric units), as a consequence of its deflection (chain form does not represent "statistic tangle" or "rigid bar"). In the presence of conformers, it occurs as a consequence of conformatic composition depending on the chain length. This component can be written down similarly after degradation. In this case, it is necessary to take into account that the number of molecules becomes twice as high, but the total length of both chains, in the ranges of undertaken suppositions, differs from the initial degradating chain by one monomeric unit, and the number of ends becomes twice as great for each chain. For example, at degradation of complexes of AlR$_3$ polyethylene glycol end groups $\sim$OAlR$_2$ and R—CH$_3$ are formed in addition to two initial ones. They are formed according to the equation:

$$\begin{array}{c} OH \\ | \\ HO\text{\Large$\wedge\!\wedge\!\wedge$}CH_2\text{—}O\text{—}CH_2\text{\Large$\wedge\!\wedge\!\wedge$}OH \rightarrow CH_2OAlR_2 + RCH_2\text{\Large$\wedge$}OH \\ | \\ AlR_3 \end{array}$$

It is clear that this value, relating to a single break, does not depend on concentration and chain length and is similar to low-molecular compounds, beginning from a definite length of the initial chain, only. In this case, it was supposed that the molecules of solvent and chains do not interact with each other in the bulk. In a more general case, it is necessary to take into account the fact that when solving, G_{ch-s} is expressed by the following equation:

$$G_{ch-s} = n q_{s-s} + m q_{ch-ch} - l q_{ch-s}$$

where q_{s-s}, q_{ch-ch}, and q_{ch-s} are the free energies of interaction; n, m, and l are the numbers of contacts between them. $G'_{ch-s} - G_{ch-s}$ is the difference of free energies of contact formation because chains repesent diluted solution before and after degradation in accordance with accepted suppositions, i.e., they do not interact with each other. Consequently, $G'_{ch-s} - G_{ch-s}$ should include the change of free energy, stipulated by the interaction of solvent molecules with each other, which become free as a result of the process.

For relatively long chains, average value of interaction energies of solvent molecules surrounding the chain is independent of the chain length. One more additional component of free energy of chain formation occurs because the distance between solvent molecules is defined by both interaction with the chain and their disposition among the chain. Figure 1 shows directions of interaction of solvent molecules, disposed neighbor to the chain, with the chain and with each other. Similar interactions in processes with the participation of low-molecular components only can differ by their value. Based on the above mentioned facts, it may be decided that the ΔG_{in-in} term can differ insufficiently from similar value of the process, proceeding with the polymer participation. As equilibrium constants, determined from the change of Gibbs free energy during the process, are related to a mole of reacting substance, then the components of free energy change will be considered further on, taking this fact into account. Free energy of homogenous isolated chain deflection depends on its local rigidity and length and may be characterized by the correlation ($\overline{R}_o^2/\overline{R}_1^2$) of mean-squares of the chain distances at "free rotation" of bonds and the chain being studied [11]. The dependence of deflection entropy of isolated chain related to monomeric unit is shown in Fig. 2, and the dependence of $\overline{R}_o^2/\overline{R}^2$ on chain length is shown in Fig. 3. It is seen from Fig. 2 that accompanying the growth of isolated chain length this component, related to a mole, approaches the limit (R), stipulated by "free rotation" of bonds (by their statistic independence). The higher local rigidity is at larger lengths (molecular weights) this limit is reached. $S_{ch} = 0$ for very small chain lengths of simple structures, and $S_{ch}/n = R$ for very large ones. Local rigidity decreases as temperatures increase. Consequently, the higher the temperature is, at a shorter chain length the limit is reached.

Solvent disturbs chain deflection by interacting with it. At first, solvent molecules must be moved at chain deflection. In this case it may sufficiently decrease the distance between interacting atoms, which are closely disposed and valently disconnected. The model of a homogenous chain in solution is described [12]. The chain in the solvent possesses a higher observable local rigid-

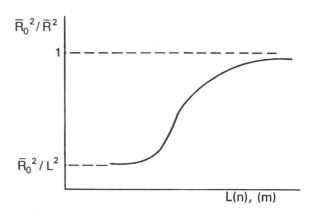

Figure 3 The dependence of $\overline{R}_o^2/\overline{R}^2$ on chain length $L(n),(m)$.

ity. Observable rigidity is defined as the local rigidity of the isolated chain, which possesses a contour length and a mean-square of the distance between chain ends, the same as in the solvent. Vainstein [13] showed the following value of observable rigidity of a homogenous chain, a_n:

$$a_n = (40a^2)/(40a - L^2p) \qquad (9)$$

Where, a is the local rigidity of isolated chain, determined by using the basis of a potential hole: L is its contour length, p is the interaction parameter, characterizing the influence of the solvent on chain deflection. Vainstein details the parameter p and its physical sense [12]. As seen from Eq. (9), the greater the chain length, the higher the a_n value is at the definite temperatures. It is evident that at a definite value of chain length (L_c) the denomimator becomes zero and a_n becomes infinite. It physically means that the chain cannot exist at the length greater than this value ($a_n < 0$). Consequently,

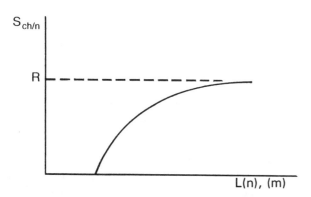

Figure 2 The dependence of deflection entropy of isolated chain related to monomeric unit (bond) S_{ch}/n on chain length $L(n),(m)$.

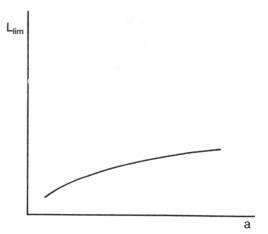

Figure 4 The dependence of limited length of thermodynamically stable chains L_{lim} on its local rigidity a.

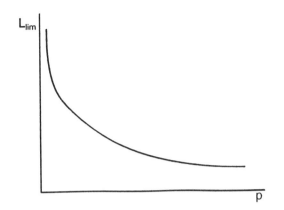

Figure 5 The dependence of limited length of thermodynamically stable chain L_{lim} on interaction parameter chain-solvent p.

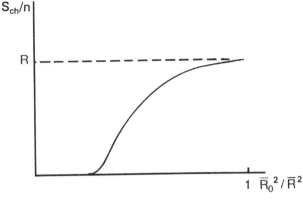

Figure 6 The dependence of deflection entropy related to monomeric unit, S_{ch}/n on $\overline{R}_o^2/\overline{R}^2$.

any solvent must show the chain length, larger which value the chain is thermodynamically instable. The value of L_c equals:

$$L_c = \sqrt{40a/p} \qquad (10)$$

It is seen from Eq. (10) that the larger the p value and the smaller the a value, the lower the critical chain length. The dependence of L_c on a and p is shown in Figs. 4 and 5. However, the value of free energy of molecule deflection in solution, should be considered because the length of thermodynamically instable chain is defined by total change of free energy. If the chain represents a "rigid bar," as previously mentioned, entropy of its deflection equals zero. If "free rotation" of bonds in the chain exists free energy of deflection, related to monomeric unit (bond), equals RT. As entropy is characterized by the number of chain states, its value in the first approximation can be characterized by correlation of mean-squares of distances between chain ends at "free rotation" of the chain in the solvent (precisely the correlation of distribution functions). If $\overline{R}_o^2$ can be calculated easily for the chains of any structure, $\overline{R}_s^2$ may be determined experimentally, for example, by viscosity. As the view of S_{ch}/n function in general cases is unknown, it may be written down as the expansion:

$$S_{ch}/n = a + b\frac{\overline{R}_o^2}{\overline{R}^2} + c\left(\frac{\overline{R}_o^2}{\overline{R}^2}\right)^2 + \cdots$$

Usually for noncyclic chains $\overline{R}_o^2/\overline{R}^2 \leq 1$.

Apparently, in most cases, consideration may be limited by the first three terms of the expansion with the

accuracy usually satisfying the experimental data:

$$S_{ch}/n = a + b\frac{\overline{R}_o^2}{\overline{R}^2} + c\left(\frac{\overline{R}_o^2}{\overline{R}^2}\right)^2$$

It is evident that this function is continuous (Fig. 6). Consequently, the following additional correlations may be used for the estimation of expansion coefficients for noncyclic chains:

$$a + b\left(\frac{\overline{R}_o^2}{L^2}\right) + c\left(\frac{\overline{R}_o^2}{L^2}\right)^2 = 0$$

$$R = a + b + c$$

For infinitely large chains $a = 0$ and $b + c = R$. Coefficients b and c can be separated, $\overline{R}^2$ value being determined in any experimental point. One can use the expression containing a greater number of terms in the expansion if required. Consequently the number of experimental determinations of coefficients will grow.

For the chain (homogenous) consisting of one conformer, osmotic forces are similar to the ones stretching the molecule by the ends. Then, labor of the distance being estimated at constant temperature T_n, one can estimate S_{ch} value from the condition $A_d = F \cdot \Delta \overline{R}^2 = T_n(\Delta S_{ch})$. If a more accurate estimation of the distance change value between the ends is required, one may calculate the $\overline{R}$ value, taking into account the distribution function of the distances between the ends $\overline{R}^2$. The value of the mean-square distances between the ends of the chain, being stretched by forces, applied to the ends equals [14]:

$$\overline{R}_d^2 = 2\cdot\int_0^L (L - l)\exp\left[-\left(RT/\sqrt{f_1 a_1}\right)\text{th}\left(1/2\cdot\sqrt{(f_1/a_1)}l\right) + RT/\sqrt{a_2 f_2}\cdot\text{th}\left(1/2\cdot\sqrt{(f_2 l/a_2)}\right)\right] dl$$

where l is the particular coordinate; a_1, a_2 are the values of coefficient a in different directions; f_1, f_2 are the values of the stretching force in different directions. The value of S_{ch}, characterizing the number of chain states, may be estimated also by knowing the distribution function of distances between the chain ends is known. Evidently, two new chains occur at the considered fracture:

$$L = 2L_1 + \Delta l$$

where L_1 is the single chain length; $L_1 + \Delta l$ the other chain length. If the initial chain is long enough, then $2L_1 \gg \Delta l$. Then it is possible to estimate the value of observable local rigidity of fractured parts a'_{OB} and to compare the decrease of local rigidity a_{OB}:

$$a'_{OB} - a_{OB} = \frac{120a^2L^2p}{(40a - L^2p)(160a - L^2p)}$$

$$= a_o \left[\frac{3L^2p}{160a - L^2p} \right] \quad (11)$$

where $40a > L^2$.

Equation (11) smoothly describes an increasing function, i.e., the value of differences of observable local rigidities increases with the chain length sufficiently faster than a_o of the initial chain (Fig. 7) (increasing coefficient $3L^2p/[40a - L^2p]$). It should be mentioned that entropy of isolated chain deflection must increase according to the applied model with the increase of the chain length as it approaches nRT. The value of local rigidity of initial and broken isolated chains are equal, and entropy change in the solution according to initial chain length is extreme function (Fig. 8).

At the consideration of this formula [11] it is supposed that both new chains are close in size. If the obtained chains differ sufficiently by size, their observable

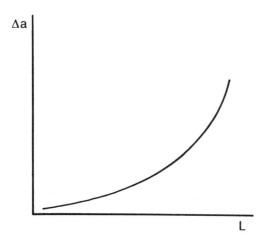

Figure 7 The dependence of changing of observable rigidities (Δa) of rupted and initial chains on initial chain length L.

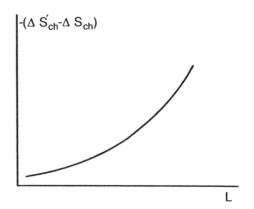

Figure 8 The dependence of entropy change ($\Delta S'_{ch} - \Delta S_{ch}$) on initial chain length L.

local rigidities are to be estimated by Eq. (9). The dependence of the chain entropy on its length, according to the Landau–Lifshitz model [12], is shown in Fig. 2 and observable local rigidities on p and L in Figs. 9 and 10. If $\overline{R}_o^2$ is reached, for an isolated chain at infinite length, then such a length cannot be reached in the solvent, as previously mentioned.

$$\frac{S'_{defl}}{2n}(n - 1) - \frac{S_{defl}}{n} \cdot n$$

$$= \left(\frac{S'_{defl}}{n - 1} - \frac{S'_{defl}}{n} \right) \cdot n - \frac{S'_{defl}}{n - 1} > 0$$

For relatively short chains:

$$G'_{defl} - G_{defl} = -T \left(n \left[\frac{S'_{defl}}{n - 1} - \frac{S_{defl}}{n} \right] - \frac{S'_{defl}}{n - 1} \right)$$

for infinitely long chains $n \gg 1$

$$G'_{defl} - G_{defl} = -Tn \left(\frac{S_{defl}}{n - 1} - \frac{S_{defl}}{n} \right) \quad (12)$$

As $(a_{OB} - a'_{OB})$ increases with the chain length, $(G'_{defl} - G_{defl})$ increases with higher rate (with higher derivative):

$$d(G'_{defl} - G_{defl})/dL$$

Using the Flori–Haggins formula [14,15] to estimate the change of mixture excessive entropy:

$$G_{ex} = -RT \left[x_1 \ln \frac{\phi_1}{x_1} + x_2 \ln \frac{\phi_2}{x_2} \right]$$

where $\phi_i = r_i x_i / \sum_i r_i x_i$ is the volumeric part of components where $i = 1, 2$; $r_2 = r$ — the number of places occupied by the chain molecule in condition, that solvent occupies one place in it ($r_1/r_2 \sim V_1^o/V_2^o$). The result, characterizing the physical sense of the process, is obtained also at the consideration of the Guggen-

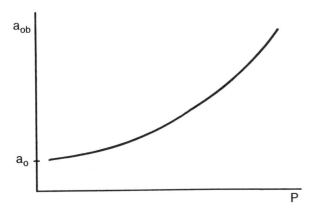

Figure 9 The dependence of observable local rigidities of chain (a_{ob}) on interaction parameter chain-solvent p.

game–Stavermann theory [16–18] and more accurate expressions of Prigojin's theory [19]. As we are just interested in the thermodynamic opportunity of the degradation reaction, the simplest model generally accepted for polymers was selected for the estimation of excessive entropy of different size molecule mixing. It is necessary to point out that Prigojin's model, applying con-RTtact number, is suitable for the chains of limited size molecule mixing, which are contained before the process initiation, equals:

$$G_{ex} = -RT\left\{\frac{N_1}{N_1 + N_2}\cdot\ln\left[\frac{N_1 + N_2}{N_1 + rN_2}\right]\right.$$
$$\left. + \frac{N_2}{N_1 + N_2}\cdot\ln\left[\frac{r(N_1 + N_2)}{N_1 + rN_2}\right]\right\}$$

After degradation the number of molecules becomes twice as high, and if it is supposed that the break occurred at approximately the middle of the chain, then it can be written:

$$G'_{ex} = -RT\left\{\frac{N_1}{N_1 + 2N_2}\cdot\ln\left[\frac{N_1 + 2N_2}{N_1 + (r/2\cdot2N_2)}\right]\right.$$
$$\left. + \frac{2N_2}{N_1 + 2N_2} \times \ln\left[\frac{r/2\cdot(N_1 + 2N_2)}{N_1 + rN_2}\right]\right\} \quad (13)$$

It was supposed at the derivation of Eq. (13) that the number of places in quasi-lattice decreases twice. Such supposition describes perfectly the relatively large chains, where $(r/2 + t) \simeq r/2$. Here t is the difference in the number of solvent molecules surrounding the chain. In comparison with $r/2$, t is stipulated by contacts of the solvent with the ends, forming at chain breaks. Then:

$$G'_{ex} - G_{ex}$$
$$= -RT[\ln(1 + x_2) + (1/2\cdot x_1 x_2'\ln r) - x_2\ln r] \quad (14)$$

It is seen from Eq. (14) that $(G'_{ex} - G_{ex})$ should increase by absolute value according to chain length (r increases) at constant concentration of the initial chains. In terms of free energy change, characterizing variation of chain nature of a molecule, decrease according to molecular weight, then independently on terms, characterizing the change of local interactions, molecular weight will be reached, at which $\Delta G = 0$ and, kinetic opportunities existing, degradation proceeding will start. But, if there is no possibility for the degradation proceeding according to kinetic causes, then according to thermodynamics the growth of molecules must be limited. The value of molecular weight at which the break occurs is defined by total selection of the components of the process free energy change, i.e., by the change of bond energies, temperature, and solvents. The weaker the bond change is, the higher molecular weight provides the occurrence of the chain degradation. Consider the present dependence on the example of degradation of simple ether polymeric complexes with AlR_3. Thus, according to calorimetric data, complexes of $Al(CH_3)_3$ with polyethylene glycol starts from $\overline{M}_n = 5000$, complexes of $Al(C_2H_5)_3$—from $\overline{M}_n \sim 6000$. For complexes of $Al(C_2H_5)_3$ with polyformaldehyde the degradation is observed at $\overline{M}_n \sim 800$, for complexes of $Al(CH_3)_3$—at $\overline{M}_n = 525$ [13]. The reaction was performed in benzene solution with polymers possessing sharp molecular weight distribution. At present, it may be established that for the chains of any structure, particularly, there exist extreme molecular weights of the chains in diluted solution. On the whole, these extreme molecular weights are the consequence of degradation processes, however, the limiting of their growth is probable. In this way it was shown [20] the occurrence of

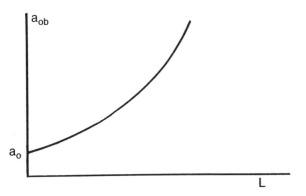

Figure 10 The dependence of observable rigidity of chain (a_{ob}) on chain length L.

$$\begin{array}{cc} H & H \\ \diagdown & \diagup \\ -C - C - \\ \diagup & \diagdown \\ H & H \end{array}$$

bond degradation. Growth limiting is usually characteristic for extremely rigid chains with large bond energies, for example, for the ones with the conjugation system [21]. Solvent influence and the value of the chain length at which thermodynamic opportunity of degradation process proceeding occurs can be considered on the example of influence on separate components of free energy change and their combinations. Evidently, the picture of the occurrence is very diversified, because the solvent influences practically all components (except $G'_{id} - G_{id}$). The influence on ΔG_x is stipulated by the change of the chain surrounding by solvent at new end group formation. Solvent molecule size influences most sufficiently ($G'_{ex} - G_{ex}$) because r decreases with the solvent molecule size growth. Consequently, the larger solvent molecules are, at greater chain lengths thermodynamic opportunity of polymer degradation occurs, in other similar conditions. Local interaction of the solvent with the chain is expressed by the change of depth, as well as by the form of potential pit ($\Delta G'_{ch-s} - \Delta G_{ch-s}$) (the change of observable local rigidity) and by the change of the interaction of solvent molecules that surround the chain with each other ($\Delta G'_s - \Delta G_s$). As ($\Delta G'_{ch-s} - \Delta G_{ch-s}$) and ($\Delta G'_s - \Delta G_s$) are connected linearly with the chain length (in the first approximation for sufficiently long chains), and the change of observable local rigidity leads to linear dependence of the change of this free energy component on the chain length, then their break is probable according to the chain length. Moreover, local interactions (their components) are usually stipulated preferably by energetic components. The change of deflection free energy is stipulated preferably by entropy change.

Temperature influence is reflected in all components of free energy change. The changes, related to ΔG_x, ($G'_{id} - G_{id}$), are similar to the ones observed in the processes with low-molecular compounds participation, only. They are described in detail in many manuals on physical chemistry, thermodynamics, and kinetics. At temperature increase the changes among molecules surrounding the chain are probable due to local interaction weakening. This may lead to the change of their interaction and to the decrease (increase) of the number of chain–solvent contacts, and, consequently, to the change of ($\Delta G'_s - \Delta G_s$) and ($\Delta G'_{ch-s} - \Delta G_{ch-s}$) components. If we accept the changes of the rest of the components to be insufficient at temperature change, which is wrong in many cases, even in these conditions there is observed nonlinear dependence of ΔG on temperature. Under extreme conditions, energies of local interactions will approach zero (solvent molecules will be separated by a sufficiently long distance from each other). If the interaction intensity equals zero, the system represents a rarefied gas. In this case, association is not taken into account. Usually these deviations are small, but the ones related to ($G'_{ex} - G_{ex}$) change may become the determi-

nant in ΔG change. The $G'_{ex} - G_{ex}$ changes in two ways by means of the T coefficient in Eq. (14), as well as by means of the r change. It is necessary to mention that if T changes in one direction, increases, for example, r, may decrease. Usually, r decreases smoothly according to T growth approaching constant value. At $T > 0$ (the case, which is not fulfilled in practice), the value of r is defined by the length of stretched chain and by sums of van der Waals radii of contacting atoms. At $T \to \infty$ (the case, which is also not practically fulfilled), r value is defined by minimally probable density of solvent, in the supposition, that distribution of solvent molecules in the volume is equally probable. If it is assumed that the rest of the parameters are constant, extrema may exist in the system, although they apparently occur very rarely. It is most probable that in the most number of cases thermodynamic opportunity of degradation occurs at temperatures below that which extremum is observed. Extremum values of ($\Delta G'_{ex} - \Delta G_{ex}$) can be observed in dependence on temperature at the correlation between T and r, expressed by the equation:

$$r = e^{(c/(T-\Delta))}$$

where c is the integrity constant determined from experimental data, $\Delta = \ln[(1 + x_2) - x_2\ln 2]/1/2 \cdot x_1 x'_2$. Thus, ($\Delta G'_{ex} - \Delta G_{ex}$) abruptly decreases with the temperature (increases by its absolute value), i.e., at temperature increase the contribution of this term into the change of free energy of the process increases. Temperature change influences ($G'_{ex} - G_{ex}$) also doubles. On one hand, at temperature increase, local rigidity increases from infinity at $T \to 0$ ("rigid bar") to the value defined by "free bond rotation" in the chain. After reaching the temperature at which "free rotation" of chain bonds is observed, local rigidity becomes constant. On the other hand, chain-solvent interaction is weakening, and the mobility of solvent molecules increases. This leads to an even sharper increase of the observable local rigidity.

If it is supposed that chain molecules are not associated with each other in the definite concentration range, then ($G'_{ex} - G_{ex}$) is a function of concentration according to the dependence. This term of free energy change decreases (increases by its absolute value) with the concentration increase [Eq. (14)]. Consequently, at a definite ratio of free energy change terms thermodynamic opportunity of degradation reaction intiation may occur, beginning at a definite concentration, i.e., if there exist kinetic opportunities for reaction performing, the proceeding of the chemical degradation reaction may then be observed at definite polymer concentrations in solution. Evidently, as it follows from Eq. (14), thermodynamic opportunity of initiation and display of the reaction at definite concentration depends on temperature and solvent molecule sizes. Considering the opportunity of a similar reaction proceeding by a chain end or side group (Scheme 1 (b)), we obtain the largest change, contributed by $(G'_{\text{defl}} - G_{\text{defl}})$ term into motive force [Eq. (5)].

As the number of chain molecules is the same before and after the reaction, then $(G'_{\text{ex}} - G_{\text{ex}})$ term practically does not change. Consequently, in this case there are no specific features of the process display different from similar reactions of low-molecular compounds and stipulated by their concentration. As $(G'_{\text{defl}} - G_{\text{defl}})$ depends on chain length and on the place of reaction proceeding, then thermodynamic probabilities of its initiation by end and middle of the chain will be different. The reaction, proceeding by the end, depends weakly on molecular weight. That is why the probability of its initiation will differ insufficiently from that for similar reactions with participation of low-molecular compounds only. For reactions proceeding by side groups, the decrease (increase by absolute value) of $(G'_{\text{defl}} - G_{\text{defl}})$ term will occur in the case of formation of a sphere possessing lower observable local rigidity during the reaction. As this dependence on chain length is of extremal character, initiation and proceeding of the reaction at definite molecular weight is then probable. Border values of molecular weights can be estimated based on the expression $\Delta G = 0$. The place of the reaction proceeding will be of a special meaning for such processes. Forming more rigid parts, the most probable place of the reaction proceeding will situate closer to the chain end; in more flexible ones—closer to the middle (Fig. 11). That is why in some cases chemical reaction is not probable for all side groups. This fact should be taken into account, calculating thermodynamic and kinetic parameters of the process. If more rigid parts are formed during the reaction, then the chain itself will be more stable thermodynamically and the reaction will be decelerated according to the chain length growth (in some cases it will stop fully).* At the formation of more flexible parts in the reaction it is necessary to take into account the probability of reaction proceeding on the chain itself according to the mentioned overall, i.e., it is also necessary to control the opportunity of the reaction initiation at the chain itself. Estimating thermodynamic opportunity of the reaction initiation, proceeding in accordance with Scheme (b), let us take into account the main demand of thermodynamics—about the independence of free energy change on the way of transfer from particular initial state to the end one.

According to all the abovementioned:

$$\Delta G = \Delta G_1 + \Delta G_2,$$

where ΔG_1 is the change of free energy of transfer from initial state to an intermediate one, ΔG_2 is the free en-

---

* At the opposite influence on the change of local rigidity and chain-solvent interaction change on $\bar{R}^2$ there may exist two probable places of the reaction, symmetric relative to the chain middle (Figs. 12–14).

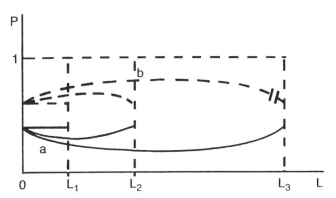

Figure 11 The dependence of probability of reaction proceeding P on the active center position for different chain lengths: L_1 = low-molecular chains, L_2 = oligomers; L_3 = polymers; a = the place after the reaction proceeding has higher observable local rigidity than in the initial chain; b = the place after the reactions proceeding has lower observable local rigidity than in the initial chain.

ergy change at the transfer from intermediate state to end one.

Parameter ΔG_2 is similar to ΔG in reactions (a) and (b) and consequently is described fully by the dependencies presented above:

$$\Delta G_1 = \Delta G_x + (G'_{\text{ex}} - G_{\text{ex}}) + (G'_{\text{defl}} - G_{\text{defl}}) \qquad (15)$$

As the number of chain molecules does not change in this reaction, then $(G'_{\text{ex}} - G_{\text{ex}})$ is small in analog to (b) type reactions, and in many cases it may not be taken into account. At the formation of more flexible chain sites comparable to the initial chain sites or intermediate complexes, the reaction will proceed preferably in the chain middle as in (b) scheme, and at the formation of more rigid parts–preferably by chain ends (Fig. 12). Remember that a chain fracture from a small number of breaks is mostly probable at the chain middle. It is precisely the formation of intermediate reaction products and/or chain degradation that stipulates reaction proceeding in the solution by the chain middle. Consequently, the feature of degradation reaction, as well as other chain ones, is the necessity of the reaction ability control of different reaction spheres. Evidently, the solvent can influence the correlation and disposition of conformers that may change the most probable place of the reaction (Figs. 13 and 14). As the change of the most probable place of the reaction causes the obtaining of other products, it is necessary to take into account the probability of reaction mechanism change in polymer reactions by means of the change of chemical acts order (in analog to low-molecular compound reactions), as well as the change of the reaction place at the same order of elementary chemical acts. Overall this is related to

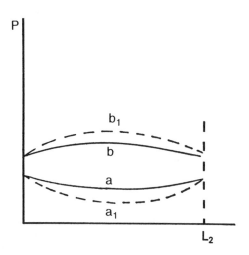

Figure 12 The dependence of probability of reaction proceeding on the active center position in the chain. a = the
place after the reaction proceeding has higher observable
local rigidity than initial chain; a_1 = the chain itself after
reaction proceeding has higher local rigidity than the initial
chain during the formation of a more rigid intermediate
chain-solvent interaction; b = the place after the reaction
proceeding has a lower observable local rigidity than in the
initial chain; b_1 = the chain itself after the reaction proceeding has lower observable local rigidity than the initial chain
during the formation of a more weak intermediate chain-
solvent interaction.

thermooxidation degradation as well as to any other
chemical reactions proceeding through the formation of
intermediate products. Thus, at complex formation of
AlR_3 with polyethylene glycole, at complex composition
of (1:1) per monomeric unit, further reaction of chain
break proceeds preferably by the chain middle (complex
formation and degradation coincides). At the proceeding
of the reaction:

$$\sim O\text{---}O\sim \; \rightarrow \; \sim\sim O\text{---}AlR_2 + R\sim\sim O\sim\sim$$
$$\diagdown\!\diagup$$
$$AlR_3$$

the break proceeds preferably by the end. Evidently the
difference is observed for the chains of limited size. For
polymers (infinitely long chains), break and complex formation becomes equally probable, i.e., the reaction proceeds according to chance law [13].

Association of macrochain molecules is displayed
strongly as usual as the association of similar low-molecular compounds. Free energies of deflection and mixing
(of ideal solution and excessive one) are lost during association, but free energy of intermolecular bond formation increases. It is clear that the longer the chain, the
higher the probability of bond formation. That is why if

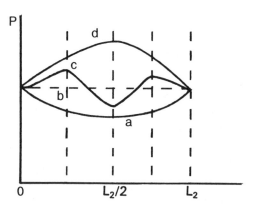

Figure 13 The dependence of probability of the reaction
proceeding on the position of active center in chain on different chain-solvent influence. The active center after the
reaction proceeding has a lower local rigidity than the initial
chain: a = influence of solvent on deflection of reacted and
unreacted place of the same chain; b = influence of solvent
on deflection is compensated by the influence of the local
rigidity; c = influences of solvent and local rigidities on
deflection are equal, but have opposite directions of action.
d = influence of solvent on the deflection is stronger than
the influence of the local rigidity.

one considers the initial state as the solution of associated initial molecules and the end one as the solution of
fractured molecules, the change of free energy of the
process will be lower than at isolated molecule break in
the solution because it is necessary to spend required

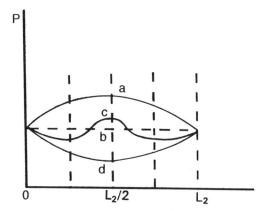

Figure 14 The dependence of probability of the reaction
proceeding on the active center position in the chain. The
reacted place has a higher local rigidity than the initial
chain: a = influence of reacted and unreacted places of
chain on deflection are the same; b = influences of solvent
and local rigidity on deflection are equal, but have opposite
directions of action; c = influence of solvent on deflection
is completely compensated by the local rigidity influence;
d = influence of solvent on the deflection is predominate
compared with the local rigidity influence.

free energy for an associate break to obtain initial isolated molecules.

Fracture of initial chains existing in association is less probable thermodynamically even when the end state represents the associate, including fragments, formed during the initial molecule break. This is stipulated by the circumstance that free energy of associates increases with the chain length growth. For example, if the initial associate consisted of two molecules and the associate in the end state contained three molecules (one of the chains broke into two parts), the associate is less stable in the end state than in the initial one because it can contain smaller amounts of intermolecular bonds. In addition, at associate fracture in the end state a great number of molecules is formed, and deflection entropy in the end state at any rate is not higher, than in the initial one (excessive and free energy of different size molecules mixing).

Evidently, in the intermediate case of simultaneous existence of isolated and associated parts of fractured molecules in the end state, the break will be less thermodynamically profitable in comparison with isolated molecule fracture.

Consequently, any association must decrease chain tendency to degradation. However, the existence of such intermediate particles at association, which possess lower height of the reaction barrier, may be probable. In this case, kinetic probabilities of the process performance increase. A sufficiently sharp increase of kinetic probabilities of the reaction must be observed in the case, if a low-molecular compound (oxygen, for example) participating in the reaction is highly stressed. But it is necessary to remember that even if kinetic probabilities of the process are increased, the reaction will also proceed in the case of its thermodynamic benefit. As association depends on macromolecule concentration, it should be taken into account at the calculation of kinetic and thermodynamic parameters of the process according to thermodynamics.

B. Limited Sizes of Cyclic Compositions

Free energy of solution formation containing cyclic large molecules is expressed before degradation as follows:

$$G_{in} = G_x + G_{id} + G_{ex} + G_{defl} + G_s + \Delta G_s + G_{ch-s}$$

Free energy of solution formation containing linear molecules after degradation equals:

$$G_e = G_s - G_x + G_{id} + G'_{ex} + G'_{defl} + \Delta G'_s$$
$$+ G'_{ch-s} + \Delta G_c$$

The main symbols of both equations are similar to the ones whose physico-chemical sense was explained previously. ΔG_c the change of free energy at chemical bond break in cycle and noncyclic bonds formation. G_{id} is equal for initial and end states.

It is assumed that solvent molecules and macrochain molecules, including cyclic ones, are not associated with each other. It is also assumed that if solutions are diluted, we can neglect the contribution of the volume change into the change of process free energy. As the number of molecules changes at cycle break, the change of free energy of the process equals:

$$\Delta G = G_e - G_{in} = (G'_{ex} - G_{ex})$$
$$+ (G'_{defl} - G_{defl}) + (\Delta G'_s - \Delta G_s)$$
$$+ (G'_{ch-s} - G_{ch-s}) + \Delta G_c \qquad (16)$$

It is evident, that all of the abovementioned relates to the reaction of cycle opening type and the formation of active centers (radicals) at molecule ends:

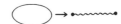

According to the P. Flori approximation, G_{ex} and G'_{ex} equals, respectively:

$$G_{ex} = -RT\left[\frac{N_1}{N_1 + N_2}\cdot\ln\frac{N_1 + N_2}{N_1 + rN_2}\right.$$
$$\left. + \frac{N_2}{N_1 + N_2}\cdot\ln\frac{r(N_1 + N_2)}{N_1 + rN_2}\right]$$

$$G'_{ex} = -RT\left[\frac{N_1}{N_1 + N_2}\cdot\ln\frac{N_1 + N_2}{N_1 + (r + t)N_2}\right.$$
$$\left. + \frac{N_1}{N_1 + N_2}\cdot\ln\frac{(r + t)(N_1 + N_2)}{N_1 + (t + r)N_2}\right]$$

Here t is the change of the number of contacts chain molecule–solvent at cycle break:

$$\Delta G_{ex} = G'_{ex} - G_{ex}$$
$$= -RT\left[\ln\frac{N_1 + rN_2}{N_1 + (t + r)N_2} + \frac{N_2}{N_1 + N_2}\cdot\ln\frac{r + t}{r}\right]$$
$$= -RT\left[\ln\left(1 - \frac{tN_2}{N_1 + (t + r)N_2}\right)\right.$$
$$\left. + \frac{N_2}{N_1 + N_2}\ln\left(1 + \frac{t}{r}\right)\right] \qquad (17)$$

In many practical cases t is sufficiently lower than r, and $tN_2/N_1 + (t + r)N_2$ is smaller than 1 in diluted solutions. So let's expand ΔG_{ex} in series by x_2:

$$\Delta G_{ex} = -RT\left\{\left[\frac{tx_2}{x_1 + (t + r)x_2} + \frac{1}{2}\left(\frac{tx_2}{x_1 + (t + r)x_2}\right)^2\right.\right.$$
$$\left. + \frac{1}{3}\left(\frac{tx_2}{x_1 + (t + r)x_2}\right)^3 + \cdots\right]$$
$$\left. + x_2\left[\frac{t}{r} - \frac{1}{2}\left(\frac{t}{r}\right)^2 + \frac{1}{3}\left(\frac{t}{r}\right)^3 + \cdots\right]\right\}$$

At very low concentrations we can neglect $tx_2/[1 + (t + r - 1)x_2]$ value, and the more so neglect all other terms of expansion

$$\ln\left(1 - \frac{tN_2}{N_1 + (t + r)N_2}\right)$$

In this case ΔG_{ex} decreases linearly with x_2 growth (increases by its absolute value). If $t < r$ (large cycles), one can restrict by the first term $\ln(1 + t/r)$, then $G_{ex} = -x_2(t/r)RT$. At the break of very large cycles at low concentrations (formation of infinitely long chains) the value of ΔG_{ex} approaches zero. It is more correct not to use the value of excessive free energy of mixing the molecules of different sizes calculated from I. Progojin's model [19] for thermodynamic estimation of the probability of relatively small size cycle break.

If cycles being broken are relatively large, $(G'_{ch-s} - G_{ch-s})$ and $(\Delta G'_s - \Delta G_s)$ will be close to similar values at the break of low-molecular compounds. Since the change of the polymer–solvent contact number depends weakly on the molecular weight of the cycle, then both $(G'_{ch-s} - G_{ch-s})$ and $(\Delta G'_s - \Delta G_s)$ will be constant. At relatively small sizes of the cycle, the number of macrochain molecule–solvent contacts can be connected nonlinearly with the chain length. Then, as in the previous paragraph:

$$\Delta G = C_{ch}^+(G'_{defl} - G_{defl}) + (G'_{ex} - G_{ex})$$

where $C_{ch}^+ = \Delta G_c + (G'_{ch-s} - G_{ch-s}) - (\Delta G'_s - \Delta G_s)$

As ΔG_{ex} value in diluted solution is small, in a number of practical cases:

$$\Delta G = C_c^+ + (G'_{defl} - G_{defl}) \tag{18}$$

The term $(G'_{defl} - G_{defl})$ characterizes labor spent for the displacement of linear chain ends to the distance equal to the length of the forming bond. Mean-square of the distance between linear chain ends increases with molecular weight growth. It is clear that the average distance between chain ends will also increase. Evidently, labor required for chain ends transferring will increase with molecular weight growth, and a moment will occur at which this labor gain will exceed free energy of the chemical bond formation. Thermodynamic opportunity of the cycle break appears at $G'_{defl} - G_{defl} = C_c^+$. Term $(G'_{defl} - G_{defl})$ is characterized by the deflection entropy change. Unlike chain degradation reaction of the linear chain critical length of the thermodynamically stable cycle will be observed for isolated chains, as well as for the chain in solution. The better solvent is from the point-of-view of chain hardening, at lower molecular weights thermodynamic opportunity of cycle break will be displayed under other equal parameters and conditions. It can be believed that the chain length at which cyclization reaction is still thermodynamically probable is at any rate not greater than the one at which

the linear chain becomes thermodynamically instable, i.e., it can be treated by degradation. Apparently, if one of the chemical bonds of the chain is weaker, it cannot exclude the possibility of its stretching with cycle size growth, and cycle break will occur at this bond. If all chemical bonds in the cycle are uniform, then the stretch of each bond will be lower than the similar bond would be the weakest in the cycle.

As the change of free energy is uniform for the bond break in infinitely long chains and low-molecular compounds, its value can be determined not only by the break of the largest cycle. As this value is constant in the first approximation, then the variation of free energy change of the recyclization process will define the change of free energy by the account of system flexibility change, i.e., it will define the increase of labor required for chain ends transfer from a particular distance to the one equal to the chemical bond length [Eq. (18)]. The distribution by bond lengths may not be taken into account at the analysis because fluctuation of bond lengths is low compared with that of the distances between chain ends. If G_{defl} value can be estimated according to experimental data or on the basis of a corresponding model, then by determining the C_c and molecular weight at which the cycle becomes unstable thermodynamically, one can calculate G'_{defl}, i.e., free energy of cyclic compound deflection. For the chain consisting of one conformer in the solvent, it is possible to estimate G_{defl} according to Vainstein [12] and for isolated chain, according to Landau and Lifshitz [14]. Maximum value of deflection free energy of isolated chain equals nRT by its absolute value, and minimum value of free energy of cycle deflection equals zero (the analog of "rigid bar"). That is why $C_c \geq nRT$. A typical example of bond weakening in the cycle (precisely the increase of its rigidity or flexibility) is complex formation [21]. In the presence of complex formation at the increase of cycle rigidity, its break can be observed (in the absence of complex formation the cycle can be thermodynamically stable). We have not considered the question about the influence of rotary isometry on the process of cycle degradation in the present case. At temperature growth, G'_{defl} increases, approaching the limit defined by the nRT expression (the chain approaches "free rotation of bonds"). As the length of the breaking bond changes insufficiently, then $(G'_{defl} - G_{defl})$ approaches nRT according to temperature growth. In general, the value of C_c decreases with the temperature according to thermodynamics. The deviation of the observed dependencies, according to the temperature from similar ones of the processes in which only low-molecular compounds participate, can be stipulated by the change of chain-solvent contacts according to temperature. If C_c decreases and $(G'_{defl} - G_{defl})$ increases with the temperature, minimum size of thermodynamically stable cycle then decreases. If C_c increases, which is rarely observed, then maximum

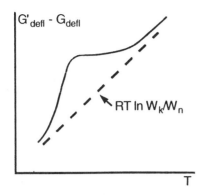

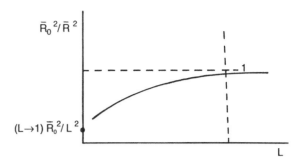

Figure 17 The dependence of $\bar{R}_o^2/\bar{R}^2$ on chain length L.

Figure 15 The dependence of deflection free energy changing during chain scission ($G'_{\text{defl}} - G_{\text{defl}}$) on temperature.

size of the thermodynamically stable cycle may depend on the extreme temperature. More complicated temperature dependencies of maximum size of the thermodynamically stable cycle are also probable. Extreme size of the thermodynamically stable cycle is connected linearly with the temperature at which ($G'_{\text{defl}} - G_{\text{defl}})/nRT$ changes weakly (very low or very high temperatures), and C_c changes in linear order (according to approximations of ΔH and ΔS constancy in narrow temperature ranges accepted in chemical thermodynamics). Let us agree that all chemical bonds are uniform before break, or one of them differs from the rest (naturally, the formation of a more complicated chemical composition of cycles is possible, too). Chemical bonds in chains will differ from the end groups after cycle break. That is why G'_{defl} of linear chains possesses uniform temperature dependencies in both cases. For initial cycles containing two types of bonds, the two values of local rigidities and parameters of chain-solvent interactions should be taken into account. Temperature dependencies of G'_{defl} and G_{defl} for constant n value, i.e., in the absence of chemical processes in cycles and linear chains, are shown in Fig. 15. Consequently, in this case, ($S'_{\text{defl}} - S_{\text{defl}}$) will possess extremal character (extremum is weakly ex-

pressed). The probability of the formation of cycles with small amount of bonds is stipulated by free energy gain at bond formation as well as by its loose in consequence of valent angles deformation (of bond lengths rarely). It is described in detail in many manuals on theoretical organic chemistry (see for example [22]). Thus, there may be observed two borders of thermodynamically stable cycle existence that depend on the chain length.

As the maximum size of the thermodynamically stable cycle depends preferably on flexibility change during the process, it will also depend on its disposition in chain molecule. To simplify the consideration let us examine a new cycle contained in chain molecule (Fig. 16). The formation of the cycle in the chain usually makes the latter more rigid (local rigidity of the cycle disposition place in the chain is higher than that of the rest of the chain). Estimating G'_{defl}, one can consider a chain containing three parts (two parts, if the cycle is disposed at the end of the chain), two of which (end ones) possess uniform local rigidity. The dependence of $\bar{R}^2$ for such chains in solvent and in isolated state is shown on Fig. 17. According to Fig. 17, chains possess higher formation free energy when the cycle is disposed in the middle of the chain or in the chain points at which $\bar{R}^2$ is maximum (the smaller $\bar{R}^2$, the higher chain entropy). After completing the reaction local rigidity of the chain part, where the break has occurred can be even lower than that of the rest of the chain parts. So maximum value of ($G'_{\text{defl}} - G_{\text{defl}}$) will depend on the place of initial cycle disposition. At close influence of solvent on local rigidities of various parts ($G'_{\text{defl}} - G_{\text{defl}}$) will possess extremal character in dependence on the reaction place (Fig. 18); maximum value will be observed when the initial cycle is disposed in the middle of the chain, and minimum value at the end. As the change of energy at cycle break and new bond formation depend weakly on the chain length, then extreme size of thermodynamically stable cycles will depend greatly on the place of disposition in the chain (maximum size of thermodynamically stable cycle will be observed at chain ends). It is clear that the solvent may change the present dependence (Figs. 12–14).

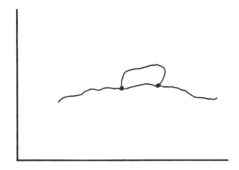

Figure 16 Scheme of cycle disposition in the chain.

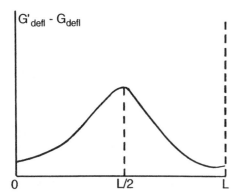

Figure 18 The dependence of deflection free energy changing on the reacted place disposition in the chain.

At the increase of the number of cycles in the chain, free energy of deflection in initial and end states will increase; however, $(G'_{defl} - G_{defl})$ will change extremely, i.e., the cycles will be more stable, disposing at the end of the chain. Cycle break is supposed in the present example. Maximum of $(G'_{defl} - G_{defl})$ absolute value will increase with local rigidity growth of the main chain as well as of the cycle. Consequently, the break of the cycle disposed in the middle of the chain will be the most profitable, the more rigid the chain, and, consequently, the more cycles dispose in it. Minimum $(G'_{defl} - G_{defl})$ absolute value will depend weakly on chain rigidity, including the number of cycles in it, because the change of $\overline{R}^2$ change varies weakly at the change of the local rigidity of the chain ends. On the whole, the dependence of thermodynamically stable cycle size on conditions (temperature, solvent, etc.) and on chain structure (local rigidities, length) is defined by "chain effect" [23], i.e., by entropy change at variation of the chain deflection in all cases. Thus, the solvent, interacting more strongly (stretching) with the chain, can change sufficiently the place of more stable cycle disposition in the chain (Figs. 12–14). As with the chain length growth, the rigidity of chains increases, containing cycles, as well as the chains after their break, then $(G'_{defl} - G_{defl})$ value decreases. For infinitely long chains, $(G'_{defl} - G_{defl})$ change becomes minimal, and the sizes of thermodynamically stable cycles are determined in analog in the case in which the cycle is formed at chain ends. It is clear that the limited size of the thermodynamically stable cycle decreases with the chain length growth and temperature rise. Flexibility of the initial and end chains increases according to temperature growth that leads to leveling of the extreme size of thermodynamically stable cycle. However, in many cases this size decreases as a result of parameter C_{ch} change, which value, as mentioned previously, does not depend on cycle disposition in the chain, but depends only on process conditions. If chemical bonds in the cycle are uniform, then cycle

break proceeds most probably by the place where both parts of side chain formed at the break possess maximum size (Fig. 19). If one or several chemical bonds are disposed in the main chain, the most probable place of the break in the chain will be defined by the change of distances between ends and will probably depend strictly on reaction conditions and chain structure. If the break occurred by the bond disposed in the chain, then not only local rigidity changes, but the increase of its contour length, too, and consequently, of observable flexibility. The break, occurred by bonds, disposed in the side chain, increases local rigidity of the reaction place without changing the contour length of the main chain. Correlation of these contributions into the change of deflection free energy defines the most probable place of the reaction and, consequently, the place at which the reaction proceeding becomes thermodynamically probable, reaching corresponding conditions. At temperature increase as well as at the increase of the main chain length, the break in various places of cycles approaches equally probable one. It should be noted that the different stability of various cycle sizes may lead to various cycle compounds (cycles) formed, for example, by complex formation, which are able to participate in other processes further on, for example, in degradation and catalysis [20]. The influence of solvent on the most probable place of the reaction proceeding is defined by its interaction with local places of the chain and corresponds strictly to "chain effect" as shown by Vainstein [13].

C. Limited Sizes of Chain Sites Between Network Points

Free energy of network formation before the first break is:

$$G_{in} = G_c + \sum_i G_{defl,i} + G_{im} + A + G_s$$
$$+ G_{ch-s} + \Delta G_{s-s} + G_{com}$$

Here G_{im} is the free energy of intermolecular bond for-

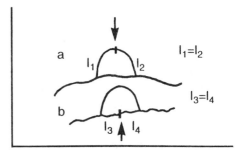

Figure 19 Scheme of the most probable place of cycle break in the chain: a = the break occurs at a bond of cycle; b = the break occurs at a bond of chain itself.

mation; $\sum_i G_{\mathrm{defl},i}$ is the total free energy of nonassociated chain parts deflection (it is supposed, that free energy of associated parts deflection is close to zero); A is the labor performed over the system or by it; G_s is the free energy of solvent formation, disposing in the network or contacting with it; G_{com} is the free energy, stipulated by combinatorial entropy of solvent molecule disposition in the network.

After the break of a single bond in the system, free energy equals:

$$G_e = G_c + \sum_j G_{\mathrm{defl},j} + G'_{\mathrm{im}} + A' + G_s$$
$$+ \Delta G'_{s-s} + G'_{ch-s} + G'_{\mathrm{com}} + \Delta G_c$$

where ($'$) index marks the values related to the end state; ΔG_c is the change of free energy at network break and formation of new bonds. As in the previously described cases, we neglect the change of system volume and conformation change:

$$\Delta G = G_e - G_{\mathrm{in}} = \left(\sum_j G_{\mathrm{defl},j} - \sum_i G_{\mathrm{defl},i} \right)$$
$$+ (G'_{\mathrm{im}} - G_{\mathrm{im}}) + (A' - A)$$
$$+ (\Delta G'_{s-s} - \Delta G_{s-s}) + (G'_{ch-s} - G_{ch-s})$$
$$+ (G'_{\mathrm{com}} - G_{\mathrm{com}}) + \Delta G_c \qquad (19)$$

The following model was considered when writing this thermodynamic equation [24]. The thermodynamic system contains 1 mole of elements of uniform structure, each representing a part of the chain between chemical points. Each part contains n_x bonds of uniform chemical structure. We neglect differences in the chemical structure of the point. Stress, which is induced by internet points, is calculated in $\sum_j G_{\mathrm{defl},j}$, $\sum_i G'_{\mathrm{defl},i}$ A', and A. Chain parts between crosslinkages may form intermolecular (physical) bonds, the amount of which is n_{ph} per element, and free energy of formation is G_{im}. If it is supposed that a single chain of a part can form one intermolecular bond with a chain of an opposite part, then $n_c \geq n_{\mathrm{ph}} \geq 0$. Labor is attached to network elements. This labor can be performed by means of external influence, as well as by means of "internal" forces, which occurred in the sample during production, storage, and exploitation. Moreover, the solvent exists in the network as a consequence of swelling. It stretches chain parts. If the network is swelling in a good solvent, n_{ph} approaches zero at the limit. Consequently, a slightly swelled polymer is considered here. Its slight swelling is stipulated by both chemical and physical points.

As in previous cases, the state is accepted as a standard one in which elements are stretched completely ("rigid bars"). The solvent amount in the system is constant. As we consider just the initial break, the number of molecules does not practically change then, and $(G'_{\mathrm{com}} - G_{\mathrm{com}})$ is close to zero, i.e., the role of the com-

binative term in the swelled network, which may be dominant in the reactions in solutions, is insufficient. If the term $\Delta G_{\mathrm{com}} + (G'_{ch-s} - G_{ch-s}) + (\Delta G'_s - \Delta G_s)$ is accepted as constant, then C_{con}. Under more accurate consideration the number of solvent molecules will be defined by the sizes of chains and solvent molecules, as well as by the amount and distribution of intermolecular bonds in the parts. The present supposition is correct for the case if n_c is high enough and the $(n'_{\mathrm{ph}} - n_{\mathrm{ph}})$ number is changed by the constant value. Then:

$$\Delta G = C_{\mathrm{con}} + \left(\sum_j G_{\mathrm{defl},j} - \sum_i G_{\mathrm{defl},i} \right)$$
$$+ (A' - A) + (G'_{\mathrm{im}} - G_{\mathrm{im}}) \qquad (20)$$

Consequently, ΔG is defined by C_c coefficient as well as by the change of element deflection, labor over the system, and the number of intermolecular bonds. The value of C_c approaches the ΔG value observed in similar reactions with the participation of only low-molecular compounds. As intermolecular bonds are distributed in elements according to Gibbs distribution, then chain parts between the molecular bonds and branching points possess different lengths in which the lengths of nonassociated parts are also different. Gibbs distribution is only performed in polymer equilibrium, which usually exists in so-called stationary states.

Then:

$$\sum_i G_{\mathrm{defl},i} = (n_x - n_{\mathrm{ph}})RT\alpha$$

where α is the coefficient characterizing a part of free energy of deflection rested from free energy at free rotation of bonds, fitted single bond. It is clear that, in general, case α is the averaged value, taking into account the sizes of nonassociated parts and their disposition in the element. $\sum_i G_{\mathrm{defl},i}$ characterizes averaging deflection free energy by sizes of nonassociated parts. In analog:

$$\sum_j G_{\mathrm{defl},j} = \alpha' (n_x - n_{\mathrm{ph}})RT$$

Then we obtain:

$$\sum_j G_{\mathrm{defl},j} - \sum_i G_{\mathrm{defl},i}$$
$$= RT[\alpha'(n_x - l - n_{\mathrm{ph}}) - \alpha(n_x - n_{\mathrm{ph}})] \qquad (21)$$

As free energy of deflection decreases total free energy, then it must enter the expression with a negative sign. According to the accepted approximation, G_{im} is defined by multiplying the free energy by the number of bonds because the principle used is, similar to the P. Flori one for chemical bonds. The energy of intermolecular interaction $E_{\mathrm{ph}} = \beta RT_{\mathrm{lim}}$ [25], where T_{lim} is the boundary temperature over which we can neglect intermolecular interaction (entropy losses are calculated by free energy of deflection); β is the coefficient, characterizing the part of heat energy accepted by the bond, $1 > \beta > 0$. Then.

$$G_{\mathrm{im}} = n_{\mathrm{ph}}\beta RT_{\mathrm{lim}};$$
$$G'_{\mathrm{im}} = \beta n_{\mathrm{ph}}RT_{\mathrm{lim}};$$
$$G'_{\mathrm{im}} - G_{\mathrm{im}} = \beta RT_{\mathrm{lim}}(n'_{\mathrm{ph}} - n_{\mathrm{ph}}) \qquad (22)$$

Free energy of intermolecular interaction also decreases the general free energy of system formation. The last component of a free energy change in this process is labor change. Labor can be positive and negative, according to the type and value of the stress. At swelling there occurs the stretch of nonassociated parts of the chain, leading to the decrease of their flexibility. That is why swelling is equivalent to stretching load.

Equation (20) may be rewritten as:

$$\Delta G = C_{\mathrm{con}} - RT[\alpha'(n_x - l - n'_{\mathrm{ph}}) - \alpha(n_x - n_{\mathrm{ph}})]$$
$$- \beta RT_{\mathrm{lim}}(n'_{\mathrm{ph}} - n_{\mathrm{ph}}) + (A' - A) \qquad (23)$$

As it is seen from Eq. (23), the thermodynamic opportunity of the reaction initiation ($\Delta G = 0$) is defined by network properties (C_{con}, n_x, T_{lim}), as well as by the conditions of production, storage, and exploitation (n_{ph}, α), and by external influence (T, A). As mentioned previously, the C_{con} value is the function of chemical structure of the network and the solvent (in a number of cases the solvent amount disposed in the network may depend on n_{ph} and intermolecular bonds distribution).

If the system is not swelled ($[\Delta G'_{\mathrm{s-s}} - \Delta G_{\mathrm{s-s}}] = 0$; $[G'_{\mathrm{p-s}} - G_{\mathrm{p-s}}] = 0$) and there is no labor influencing it, then $C_{\mathrm{con}} = \Delta G_{\mathrm{com}}$, and the temperature at which thermodynamic opportunity of reaction proceeding occurs is determined from the correlation of n_{ph} and α:

$$T = \frac{\Delta H_x + \beta RT_{\mathrm{lim}}(n_{\mathrm{ph}} - n'_{\mathrm{ph}})}{\Delta S_x + [\alpha'(n_x - l - n'_{\mathrm{ph}}) - \alpha(n_x - n_{\mathrm{ph}})]} \qquad (24)$$

The present conditions characterize material storing in the initial approximation. It is natural that the network structure must be stated.

If the polymer system was able to exist in an equilibrium state only, then a strictly defined correlation between (α, n_{ph}) and (α', n'_{ph}) would exist in particular conditions, according to minimum of free energy of system formation. Consequently, there would occur only one temperature at which process initiation is thermodynamically probable. In rare cases there may occur different correlations between (n_{ph}, α) and (n'_{ph}, α'), which display one and the same value of free energy minimum of system formation.

It is known that polymers may exist in various stationary states, which are defined by the amount and distribution of intermolecular bonds in the sample at definite network structure. The latter is defined by the conditions of storage, exploitation, and production of the network. That is why T values may be different. The highest value is observed in the equilibrium state of the system. In this case it is necessary to point out, that the n'_{ph} value becomes close to the n_{ph} one at n_x.

There are many proofs in the literature about the chemical reaction initiation in the network starting from a definite temperature. Thus, it was observed in the investigation of fracturing crosslinked butadiene copolymers with acrylonitrile that below a certain temperature an observed constant rate of one of the degradation reactions was several decimal degrees lower than that estimated according to the Arrhenius equation [26]. Although linear dependence of the Arrhenius equation anamorphosis was observed until a definite temperature (rate constant approached zero abruptly with the temperature decrease). If polymers exist in a highly elastic state, then it is necessary to take into account both the intermolecular bonds and the flexibility of chain parts. If polymers exist in a vitriolic state (plastics), the contribution of deflection entropy is close to zero, and system stability increases according to Eq. (24), i.e., it increases with the growth of the number of intermolecular bonds. Estimating the influence of n_{ph} and α on system stability, one should take into account that $0 < \alpha < 1$, and n_{ph} may change in very wide ranges ($0 \leqslant n_{\mathrm{ph}} \leqslant n_x$). If there are no intermolecular bonds in the system, we can determine temperature value at which process proceeding becomes probable as follows:

$$T = \frac{\Delta H_x}{\Delta S_x + (\alpha' - \alpha)n_x - \alpha'l}$$

We can neglect values of ($\alpha' - \alpha$) and $\alpha'l$ for sufficiently long chain parts between branching points. In this case the E value should not differ from the analog one for the process with the participation of low-molecular compounds only. In the rest of the cases, the T value is lower, i.e., T dependence on n_{ph} possesses extreme character (the minimum exists). These circumstances, namely, define sufficiently high chemical stability of low-molecular compounds compared with polymers possessing similar chemical structure. The load attached to the network can stretch nonassociated parts of chains between branching points or intermolecular bonds so that it promotes the increase of the number of intermolecular bonds (polymer network transites to vitriolic state from a highly elastic one). It also can create the stress on the intermolecular bonds, decreasing T_b (stretching them), and changing the valent angles and lengths of chemical bonds. As the labor necessary for these actions possesses different values, one of the effects can be predominant at various stress values. In general, under deformation of valent angles and bond lengths, stress value is more sufficient than at the change of flexibility and molecular bonds deformation. The influence of it on ΔH_x and ΔS_e is displayed at relatively large values only. As the stress contribution to the deformation of intermolecular bonds is the usually higher than to flexibility change, the numerator increases and denominator decreases according to stress attachment. Consequently, at low stresses, according to the increase of the intermolecular bond number, the temperature increases, at which point the process of initiation becomes probable. Further on, this temperature begins to decrease. The most abrupt decrease of it will then occur at valent angle deformation and chemical bond lengths. It is necessary to point out that in the present consideration it is supposed that the load influences are sufficiently higher at the initial state than at the end one. The present supposition is stipulated by the circumstance that in the initial state the ends of chain parts are influenced by stretching forces,

and in the end state this force decreases the least (the stress transits to neighboring parts)

Stress occurring during production, storage, and exploitation of the sample is explained by this very fact that one and the same temperature of a probable reaction may be obtained at the corresponding stationary state, as well as at the stress attachment to the equilibrium state. That is why at low stresses the transition may proceed from one state to another, accompanied by the increase of system entropy, and then the decrease will be observed. Energy change, according to the change of the intermolecular bond number at system entropy, increased under the stress influence is compensated by flexibility increase. The value of free energy change, depending on the amount of decrease of intermolecular bonds, becomes higher than that, connected with the flexibility decrease, only after reaching the definite stationary state. The number of intermolecular bonds then starts increasing.

All abovementioned facts relate to the stretching loads. However, we can similarly consider other ones, including more complicated stresses.

The solvent in which the polymer network swelled is able to change the number of intermolecular bonds, to decrease observable flexibility of chain parts between points, and to stretch the system, i.e., to perform labor over it. Its influence on the C_{con} parameter, according to accepted approximations, is similar to the influence on the process in which only low-molecular compounds participate. Some deviation can be observed as a consequence of suppressing the solvent molecules in the network. One can find that the change of observable local rigidity of the chain as a consequence of network swelling is similar by its first approximation to its change for analog chains in solutions. In some cases, the number of solvent molecules disposed in the network at its swelling will depend on both the chemical structure of the chain and the solvent and the number and distribution of intermolecular bonds in the sample as a consequence of observable local rigidity dependence on the length of nonassociated parts. For example, if it is necessary to break a single intermolecular bond disposed between associated and nonassociated parts, the change of the C_{con} parameter will be uniform in all cases, and the gain in deflection free energy will be different. Consequently, maximum amounts of the solvent in the network will depend on both chemical structure of the compositions and on sample prehistory. If the solvent amount is lower than the maximally possible one (incomplete swelling), it will be disposed in places where it is possible not to change the intermolecular interaction and where flexibility of the chain parts will minimally change (e.g., at chain stretching), and maximally, in rare cases of chain suppressing to the value at which observable local rigidity is close to free rotation. Consequently, the solvent will probably be distributed in the sample nonequally. First of all, it will be distributed between nonassociated parts of chains of optimum size (in analog to interphase plastification). Then it will dispose in the zones in which free energy of deflection will grow according to the increase (in equilibrium state). At the end, if it is possible, it will penetrate into other zones, breaking more and more bonds.

At the attachment of the stress, lower than it is required for deformation of valent angles and chemical bond lengths, in general, the maximal amount of the solvent in the network should decrease. It is evident that if the number of intermolecular bonds and their distribution decrease, temperature should decrease (if $C_{con} = $ constant), and at their break temperature will decrease more. In present consideration, it was also accepted that the influence of the solvent on the initial state is higher than on the end one (similar to the load influence). The increase of a particular temperature can occur as a consequence of the C_{con} decrease and at transition from one stationary state to another.

D. Limited Sizes of Chains on the Surface

Free energy of the formation of the chain disposed on the surface, is:

$$G_{in} = G_x + G_{id} + G_{ex,s} + \sum_i G_{defl,i} + G_{ch-s}$$
$$+ G_s + \Delta G_{com,s}$$

Free energy of chain formation on the surface after degradation is:

$$G_e = G_x + G'_{id} + G'_{ex,c} + \sum_j G_{defl,j} + G'_{ch-s}$$
$$+ G_s + \Delta G_x + G'_{com,s}$$

Here ΔG_x is the free energy of chain break and formation of new bonds; G_m is the free energy of chain surface bond formation; G_s is the free energy of the surface formation; $G'_{ex,s}$ is the excessive combinatorial free energy stipulated by different disposition of chain molecules on the surface; $\Delta G'_{com,s}$ is the combinatorial free energy stipulated by different disposition of intermolecular chain surface bonds on chain molecule. The rest of the G terms possess the abovementioned physical sense. Index (') relates to the end state of the system.

$$\Delta G = G_e - G_{in} = (G'_{id} - G_{id}) + (G'_{ex,s} - G_{ex,s})$$
$$+ \left(\sum_j G_{defl,j} - \sum_i G_{defl,i} \right) + (G'_{ch-s} - G_{ch-s})$$
$$+ (G'_{com,s} - G_{com,s}) + \Delta G_x \qquad (25)$$

We ignored the contribution into free energy of the initial and the end state, depending on the volume change. It is supposed that the degree of surface filling is low (similar to diluted solution). As at the consideration in the previous paragraph, we will accept that local rigidity of sites with intermolecular bond is nearly infinite. As a consequence of the distribution of intermolecular bonds along and between the chain, we use $\sum G_{defl}$ value. In this case:

$$\sum_i G_{defl,i} = \alpha(n_x - n_{ph})RT;$$

$$\sum_j G_{defl,j} = \alpha'(n_x - l - n'_{ph})RT$$

As in a previous case, the general number of adsorbed initial molecules is 1 mole. If intermolecular bonds are great enough, molecules in the volume cannot transit, and there is no motion at the surface. In this case, we can omit terms connected with combinatorial entropy of molecule disposition in the volume, and, even more so, their difference:

$$\Delta G = \Delta G_x + \left(\sum_j G_{defl,j} - \sum_i G_{defl,i} \right) + (G'_{im} - G_{im})$$

$$+ (G'_{com,s} - G_{com,s})$$

In analog to the previous paragraph:

$$G'_{ch-s} = \beta n'_{ph} RT_{lim}; \qquad G_{ch-s} = \beta n_{ph} RT_{lim}$$

As usual, n_{ph} represents the mean arithmetical number of intermolecular bonds, n'_{ph} being the total amount of bonds in both parts of fractured initial molecules:

$$G'_{ch-s} - G_{ch-s} = \beta RT_{lim}(n'_{ph} - n_{ph})$$

Terms G'_{ch-s} and G_{ch-s} are defined by combinatorial entropy, depending on the transposition of intermolecular bonds at the molecule, taking into account the geometry of the surface and the chain (some intermolecular bonds, for example, the ones at neighboring chain atoms cannot be performed as a consequence of the structure

of the chain itself or because of group disposing, which form these bonds on the surface). Therefore the thermodynamic opportunity of the initiation of the degradation reaction, as well as any other process, will be defined by properties of the surface and chain features. In the simplest case, intermolecular bonds can be formed at any place of the volume—combinatorial term is defined by the totality of the transposition number of a definite bond amount from 0 to the formation of intermolecular bonds on all active centers. This totality equals Newton binomial. According to the definition, the formula can be obtained in the case of equal probability of the formation of intermolecular bonds. In real cases, the number of states is stipulated by chemical and geometrical composition of the chain and the surface and can be sufficiently lower. Then:

$$G'_{com,s} = n'RT_{ex} + n''RT_{ex}; \quad G_{com,s} = nRT_{ex}$$

Here n' and n'' are the amounts of bonds formed into molecules.

$$G'_{com,s} - G_{com,s} = lRT_{lim}$$

If l is small, compared with n, then we can neglect this term. In this case:

$$\Delta G = C - \beta RT_{lim}(n'_{ph} - n_{ph})$$
$$- [\alpha'(n_x - n'_{ph} - l) - \alpha(n_x - n_{ph})]RT \quad (26)$$

Concluding from the last expression, the temperature at which reaction initiation is thermodynamically probable, is determined from the equation:

$$T = \frac{\Delta H_x + \beta RT_{lim}(n'_{ph} - n_{ph})}{\Delta S_x + R[\alpha'(n_x - n_{ph} - l) - \alpha(n_x - n_{ph})]} \quad (27)$$

Consequently, at a definite number and distribution of intermolecular bonds, the degradation process may begin at a definite temperature only if there are kinetic probabilities for the reaction proceeding at the present temperature. At a lower temperature, n_{ph} increases and α decreases, i.e., numerator does not change at least (grows more frequently), and denominator decreases, approaching ΔS_c. Moreover, it is necessary to point out, that at a low temperature n_{ph} is also high. That is why it is hard to remove formed chain fragments that especially promote their combination at radical break mechanism. If a molecule break occurs at the end of the chain, then the formed molecule can be removed into space. This promotes the break at relatively low temperatures. In the limit, $n_{ph} - n'_{ph} = l$. At higher temperatures, $(n_{ph} - n'_{ph}) > l$, and the break in the chain middle becomes more probable (denominator increases). At uniform length of nonassociated site it is more profitable, thermodynamically, for it to dispose at the end (number of states and, consequently, entropy will be higher for uniform length of nonassociated sites if the site is fixed by one end and not by both). This case promotes the reaction proceeding at chain ends at low temperatures.

We will consider this question in more detail in the following text.

Considering the influence of various parameters on T, it is necessary to take into account that α, n_{ph}, α', n'_{ph} depend on n_x and on each other. Parameters α and α' increase with the chain length at constant n_{ph} and n'_{ph} and decrease with n_{ph} and n'_{ph} growth at constant n_x value. As the change of α lies in relatively narrow ranges compared with n_{ph}, in most cases, the predominant influence on the degradation process is the amount and distribution of intermolecular bonds. With the chain length growth, the difference (the distribution width) between the maximum possible amount of the intermolecular bonds and the minimal amount of those necessary for the chain retaining on the surface increases. It follows from this fact, that the probability of the equilibrium formation, as well as of stationary states at adsorption, occurs in dependence on storage conditions. Apparently, different amounts and distributions of intermolecular bonds is one of the main causes of stationary state formation.

If $n_{\text{ph}} - n'_{\text{ph}}$ is (for example, molecule was fixed by both ends before the process, and then each fragment is fixed by one end), then α' is greater than α (flexibility of the chain, fixed by both ends, is usually lower than that of chains fixed by one end). In this case the denominator increases and T decreases with n_x growth.

Although it is clear that α grows with n_x, the molecule can be unstable at any definite temperature at definite length. If the chain is stretched sufficiently by the surface (intermolecular bonds are sufficiently removed from each other), then we can accept for the first approximation $\alpha \rightarrow 0$, and $\alpha' \rightarrow 1$. Then:

$$T = \frac{\Delta H_x}{\Delta S_x + R(n_x - 3)} \qquad (28)$$

At definite values of T, ΔH_x and ΔS_x from the equation (28), the length of nonassociated parts is determined at which degradation reaction is possible. At definite ΔH_x and ΔS_x, T will be thermodynamically lower than the temperature at which the break of chemical bonds in low-molecular compound proceeds. The value of α at definite n_x will be defined by the distance between engagement points (there exists optimal value of this distance, at which α possesses maximum value). The value of T will depend on engagement places, and apparently, on the disposition and distribution of surface active centers. In the limit, $\Delta H_x - T\Delta S_x = RTn_x$ where n_x for the first approximation equals the length of nonassociated sites of the chain. In this case it is necessary to take into account that if $T > T_b$ (T_b is the boiling temperature), molecules will evaporate, and if $T < T_b$, molecules will be decomposed until they form the chain fragments, being able to evaporate. Analyzing degradation it is necessary to take into account that at temperature increase the value of free energy change of bond break decreases, and the difference of $\sum_j G_{\text{defl},j}$ and $\sum_i G_{\text{defl},i}$ increases.

The first term may increase in consequence of the growth of α' and T, and the second one is a consequence of temperature growth only.

Considering the process proceeding on the surface, the process of chemical degradation itself should be taken into account as well as the break of intermolecular bonds. However, at break of all intermolecular bonds there exists a limited length of thermodynamically stable chain on the surface, which is stipulated by the impossibility of a large molecule evaporating. Thermodynamic instability is defined by the fact that if molecule occurs on the surface, all the states cannot be realized in it, as would occur in gas. The transition of molecules into the gas phase depends on the attraction force. A limited size of thermodynamically stable chain will change in dependence on this attraction. In the absence of the attraction, molecules are able to transit into the gas, and their stability will be the same even for infinitely large molecules as for low-molecular compounds. This can be easily shown by modeling isolated molecules. If a molecule is fixed by two points on the surface, it represents something like cyclic compositions. In this case, system stability depends on fixing places of the chain itself onto the surface (Fig. 20). The disposition will be mostly stable when engagement places are situated nearby and closer to the chain end. Then the system may be considered as fixed by single end. Free energy of formulation will decrease according to translocation of fixation places because longer ends possess high α value (for infinitely long chains α does not practically depend on their length). At the decrease of chain ends, i.e., broadening engagement places, free energy of the initial system formation will decrease and become minimal when both ends are fixed, disposed from each other at the distance equal to contour chain length. In this state, α will be close to zero, and α' will be sufficiently higher (in the limit $\alpha' \rightarrow 1$). As two ends are formed at the chain

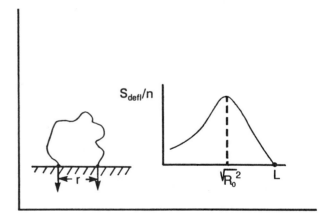

Figure 20 The dependence of deflection entropy related to monomeric unit (bond) S_{defl}/n on the distance of fixed points. The molecule is fixed by two points on the surface.

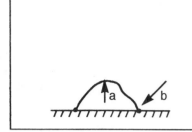

Figure 21 Scheme of the probable break of the molecule. Molecule is fixed by two points on the surface. Arrows show the break direction. a = in the middle of the cycle; b = on the chain-surface bond.

break and their free energy of deflection depends on length, then the break will occur at the middle of the "cycle" according to the highest probability (Fig. 21). If changes of free energy of chain–surface bond formation are lower than that of chemical bond formation, then with regard to higher free energy of longer sites fixed by one end, chain–surface bonds will break with higher probability. This phenomenon must be observed at the association with the formation of physical bonds.

If a molecule is fixed on the surface by three and more active centers, and if the energy of chain–surface bond is higher than bond energy in the macromolecule (chemical chain–surface bond), degradation may be considered for different cycles separately. In this case the break at the middle of the cycle will be more thermodynamically probable. If the bond energy in the chain is higher than that of chain–surface bond, then the reaction initiation proceeding by chain-surface bond is more probable. In this case, reactions forming longer nonassociated ends, are preferable, i.e., the break of the "end" chain–surface bonds is more thermodynamically profitable (Fig. 22). However, in the case of probable formation of large cycle (Fig. 23) the break of chain–surface bond will be thermodynamically profitable for "internal" bonds, too, in consequence of high benefit of free

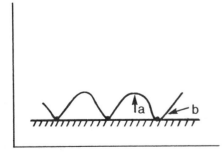

Figure 22 Scheme of the adsorbed molecule break. Arrows show direction of break: a = break of chain; b = detaching of the end of the chain-surface bond.

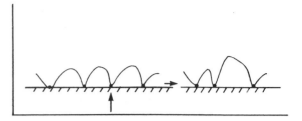

Figure 23 Scheme of the detaching of a chain-surface bond with increasing length between two fixed points.

energy by means of the increase of the length of nonassociated site in the cycle, comparing with the increase of the length of the end site. For example, if the length of the end site increases by one unit and the large cycle is formed at the break of the middle bond, the reaction of cycle formation is thermodynamically preferable. The place of reaction proceeding depends mainly on entropic components of free energy. Both reaction places are equally probable if free energy change is the same for an equal number of breaks of variously disposed chain–surface bonds. In this case, equal probability of the break may be observed at the reactions of cycle break and at the growth of the chain end, and at the reactions of cycle size increase. If lengths increase equally (for example, such a case may be displayed at equal probability of bond formation at any place of the chain), the increase of nonassociated end length is always preferable. That is why chain ends are less associated than its middle at adsorption, macromolecule association, and other similar processes. Calculation of combinatorial free energy change should be performed in the presence of relatively weak (physical, preferably) chain–surface bonds. The influence of this component change on the process is similar to that of excessive combinatorial entropy of different size molecules in solution. However, absolute value of excessive combinatorial entropy on the surface is usually lower than for the solution because the number of various dispositions of molecules in the solution is calculated for three dimensions, and on the surface for two dimensions. Solvent influence on the change of process free energy is displayed by practically all the terms, except $(\Delta G'_{\text{com,s}} - \Delta G_{\text{com,s}})$. Thus, estimating $(G'_{\text{ex,s}} - G_{\text{ex,s}})$, it is necessary to take into account combinatorial analysis, too, as a consequence of the surrounding of nonassociated sites of chain molecules and contact with nonassociated molecules. As previously mentioned, the solvent makes the chain more rigid in most cases. That is why in many cases the $(\sum_j G_{\text{defl},j} - \sum_i G_{\text{defl},i})$ value can decrease. Moreover, the solvent changes local free energy of chain–surface bond formation, because at its break there may occur the gain of free energy of surface formation. In the presence of a solvent in the first approximation, it may be taken that all local places of the surface able to form intermolecular bonds are occupied by these bonds or by solvent molecules. That is why $(G'_{id} - G_{id})$ will change weakly in comparison with their value in solvent absence, if the disposition of molecules on the surface is considered only. The $(G'_{id} - G_{id})$ will also depend on the system volume, as well as on the surface size, if general combinatorial analysis of the system is calculated with regard to solvent molecules disposed on the surface. However, in the last case it is necessary to take into account the transfer of initial and degradated polymer molecules as well as solvent ones to the volume (into the solution). Moreover, it is necessary to introduce components of free energy change connected with the change of local interaction of the molecules surrounding the chain with each other, and local interaction of associated and nonassociated parts of chains with the solvent. The influence of these components is similar to their influence on molecule degradation in the solution. Only $(G'_{com,s} - G_{com,s})$ will be practically constant because this term depends just on the structure of chain and surface. The exception is in the case of changing local conformation, as well as the whole chain.

The chapter did not consider the contribution of conformation variation into the change of the process of free energy. In general, case values of limited parameters at which there appears thermodynamic opportunity of reaction initiation decrease at the formation of thermodynamically unstable conformers.

Thus, as it follows from the consideration of this text thermodynamic opportunity of reaction initiation occurs at the change of conditions. The performance of this reaction is defined by kinetic opportunities of the process. As the conditions of production usually differ sufficiently from those of storage and exploitation of polymers, then as a consequence of all the abovementioned, there occur the opportunities of degradation. This chapter estimated probabilities of degradation initiation for comparatively simple processes. In general, a similar approach can be applied to the change of reaction mechanisms, which can prevail in various conditions of storage and usage of the sample [27–30].

REFERENCES

1. N. M. Emanuel and A. L. Buchahenko, *Chemical Physics of Polymer Degradation and Stabilization,* VNU Science Press, p. 340 (1987).
2. G. E. Zaikov, A. L. Iordanskii, and V. S. Markin, *Diffusion of Electrolytes in Polymers,* VSP Science Publisher Utrecht, p. 322 (1988).
3. Yu. A. Slyapnikov, S. G. Kiryushkin, and A. P. Mar'in, *Antioxidative Stabilization of Polymers,* Ellis Horwood, Chichester (W. Sussex), p. 246 (1994).
4. E. F. Vainstein, *The Role of Intermolecular Interaction in Polymer Degradation Processes, Polymer Yearbook,* (R. A. Pethrick and G. E. Zaikov, eds.) Gordon and Breach, London, vol. 9, pp. 69–79 (1993).
5. A. Popov, N. Rapoport, and G. Zaikov, *Oxidation of Stressed Polymers,* Gordon and Breach, London, p. 336 (1991).
6. K. S. Minsker, S. V. Kolesov, and G. E. Zaikov, *Degradation and Stabilization of Polymers on the Base of Vinylchloride,* Pergamon Press, Oxford, p. 526 (1988).
7. E. F. Vainstein, *The Limiting Sizes of the Large Molecules, Polymer Yearbook,* (R. A. Pethrick, and G. E. Zaikov, eds.) Gordon and Breach, London, vol. 8, pp. 85–113 (1992).
8. Yu. V. Moiseev and G. E. Zaikov, *Chemical Resistance of Polymers in Reactive Media,* Plenum Press, New York, p. 586 (1987).
9. R. M. Aseeva and G. E. Zaikov, *Combustion of Polymeric Materials,* Karl Hanser Verlag, München, p. 390 (1986).
10. N. Grassie, and G. Scott, *Polymer Degradation and Stabilization,* Cambridge University Press, Cambridge, p. 222 (1985).
11. E. F. Vainstein, A. A. Sokolovskii, and A. S. Kuzminskii, *Kinetics of the Changing Products Molecular-Mass Distribution in Thermodegradation of Associated Polymers, Polymer Yearbook,* (R. A. Pethrick and G. E. Zaikov, eds.) Gordon and Breach, London, vol. 9, pp. 79–101 (1993).
12. E. F. Vainstein, G. M. Gambarov, and S. G. Entelis, *Doklady Akademii Nauk SSSR, 113(2):* 375–378 (1974).
13. E. F. Vainstein, Thesis Dr. of Sci., Institute of Chemical Physics USSR Academy of Sciences, Moscow (1981).
14. L. D. Landau, and E. M. Lifshitz, *Statistical Physics* (in Russian), Nauka, Moscow (1964).
15. P. J. Flory, *J. Chem. Phys., 10(1),* 51 (1942).
16. H. L. Huggins, *Ann. N.-Y. Academy of Sciences,* 43 (Pt. 1): 1 (1942).
17. E. M. Guggenheim, *Mixtures,* Clarendon Press, Oxford, p. 272 (1952).
18. A. J. Staverman, *Rec. Trav. Chim. Phys.-Bas.,* 1 (1950).
19. I. Prigogine, *The Molecular Theory of Solution,* North-Holland Publishing Co., Amsterdam, p. 448 (1957).
20. A. D. Pomogailo, *Immobilized Polymeric-Metal Complex Catalysts* (in Russian), Khimiya, Moscow (1986).
21. I. Oshima Takumi, and Nagai Toshikazu, *J. Org. Chem., 56(2):* 673 (1991).
22. *Chemical Encyclopedian Dictionary,* (I. L. Knunyants, ed.) Sovetskaya Encyclopedia, Moscow, p. 791 (1983).
23. E. F. Vainstein and G. E. Zaikov *Processes with Oligomeres, Polymer Yearbook,* (R. A. Pethrick and G. E. Zaikov, eds.) Gordon and Breach, London, *10:* 231–235 (1993).
24. E. F. Vainstein, and A. A. Sokolovskii, *11:* 49–73 (1994).
25. O. F. Shlensky, A. A. Matyukhin, and E. F. Vainstein, *J. Thermal. Anal.* 34: 645 (1988).
26. A. A. Sokolovskii, E. F. Vainstein, K. M. Gubeladze, and A. S. Kuzminskii, *Vysokomolekulyarnye Soedineniya, 30B* (1): 244–248 (1988).
27. G. E. Zaikov, *Russian Chemical Review, 62(6):* 603–623 (1993).
28. G. E. Zaikov, *Intern. J. Polym. Mater. 24, (1–4):* 1–19 (1994).
29. V. V. Kharitonov, B. L. Psikha, and G. E. Zaikov, *Intern. J. Polym. Mater. 26 (3–4):* 121–176 (1994).
30. G. E. Zaikov, 208th American Chemical Society National Meeting and Exposition Program, Washington DC, August 21–25, p. 125 (1994).

24
Polymerization by Ylides

A. K. Srivastava

Harcourt Butler Technological Institute, Kanpur, India

I. INTRODUCTION

Polymers in one form or another are basic constituents of every kind of living matter, whether plant or animal. It is only in the present century, however, that as a result of a wide range of scientific studies, their existence as a coherent group has come to be recognized and understood. Following this recognition and this understanding, the possibility has arisen of actually producing polymers by means of suitable chemical reactions. Originally these synthetic products, regarded as substituents for existing natural polymers such as rubber or silk, led to the introduction of a vast range of entirely new compounds in the field of plastics, rubber, and fibers, many of which have properties different from those existing in natural materials. At present, the study of polymers, though it includes the original natural polymers, tends to be dominated by these synthetic materials. It is from the industrial development of such polymers that the main stimulus to scientific research has arisen.

Progress in the chemistry of macromolecules results not only from the introduction of an increasing numbers of novel monomers from known polymerization reactions, but also from the application and development of new initiators. Investigations devoted to the latter occupy an important place in the polymer field and are of great significance; as a result of this, many compounds such as peroxides, hydroperoxides, redox initiators, and organometallic compounds, including ylides, have been reported. Over the past two decades, ylide–polymer chemistry has been an area of great interest to chemists due to its theoretical implications and industrial significance.

The first ylide was prepared and isolated by Michaelis and Gimborn [1] in 1894. This was an isolated event, and the first flurry of activity in the field of ylides oc-curred approximately 50 years later, probably stimulated by the development of the Wittig synthesis of olefins. The term "ylide" was first introduced by Wittig [2] in 1944. It represents the ending of the name given to nitrogen organic derivatives, *trimethyl ammonium methylide;* "yl" is the ending of organic radicals (methyl, phenyl) and suggests a free valency, and "ide" implies anionic character (acetylide, phenacylide).

The ylides may be defined as dipolar compounds in which a carbanion is covalently bonded to a positively charged heteroatom. They are represented by the following general formula:

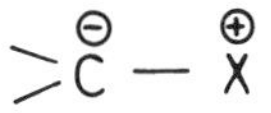

$$>\overset{\ominus}{C} - \overset{\oplus}{X} \tag{1}$$

where X may be P, As, Sb, S, N, Bi and Tl. According to this, an ylide structure is characterized by two adjacent charged atoms, the carbon atom having an orbital uninvolved in the covalent bond occupied by an electron pair.

A. Classification

The ylides have been classified on the basis of the heteroatom covalently bonded to the carbanion. Accordingly, they can be differentiated into nitrogen ylide (Scheme 2), sulfur ylide Scheme 3, phosphorus ylide Scheme 4, arsenic ylide Scheme 5, antimony ylide (Scheme 6), bismuth ylide (Scheme 7) and thallium ylide (Scheme 8).

$$\geq\overset{\oplus}{N} - \overset{\ominus}{C}< \tag{2}$$

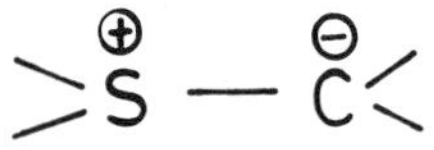

$$\overset{\oplus}{>\!S} - \overset{\ominus}{C}\!< \qquad (3)$$

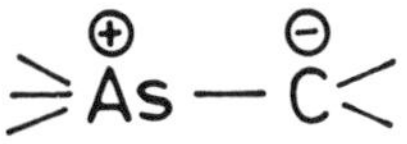

$$\overset{\oplus}{\geq\!\!P} - \overset{\ominus}{C}\!< \qquad (4)$$

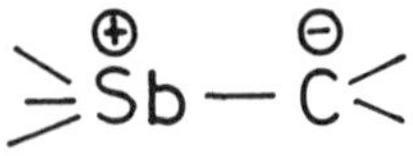

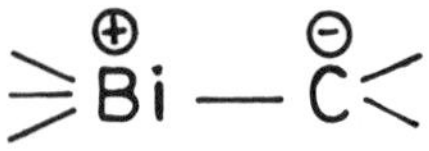

$$\overset{\oplus}{\geq\!\!As} - \overset{\ominus}{C}\!< \qquad (5)$$

$$\overset{\oplus}{\geq\!\!Sb} - \overset{\ominus}{C}\!< \qquad (6)$$

$$\overset{\oplus}{\geq\!\!Bi} - \overset{\ominus}{C}\!< \qquad (7)$$

$$\overset{\oplus}{\geq\!\!Tl} - \overset{\ominus}{C}\!< \qquad (8)$$

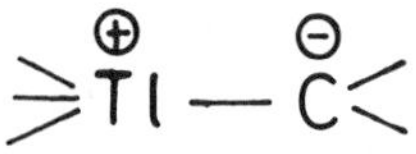

B. Stability

A comparative study on ylide stability as a function of the heteroatom type was carried out by Doering et al. [3,4]. They concluded that the phosphorus and sulfur ylides are the most stable ones. The participation of three-dimensional orbitals in the covalency determines the resonance stabilization of the phosphorus and sulfur ylides [5–8]. The nitrogen ylides are less stable from this point of view. The only stabilization factor involves electrostatic interactions between the two charges localized on adjacent nitrogen and carbon atoms [9].

II. APPLICATION OF YLIDES IN POLYMER SCIENCE

Many review have documented the utilization of ylides in synthetic organic chemistry [10]. This chapter focuses on recent developments and special topics of ylide polymer science. The major purpose of this review is to pro-

vide a comprehensive guide to the literature reported from 1966 to 1995, thereby indicating the advances made during the past three decades in the polymer–ylide arena for the synthesis of polymers and the scope of ylides in the polymerization field.

A. Nitrogen Ylide

The effectiveness of ylides in the field of polymer science was first described in 1966 by George et al. [11] who felt that 3- and 4-(bromo acetyl) pyridines, which contain both the α-haloketone and the pyridine nucleus in a single molecule, could be quaternized to polymeric quaternary salts and finally to polymeric ylides Schemes 9 and 10 by treating these polymeric salts with a base.

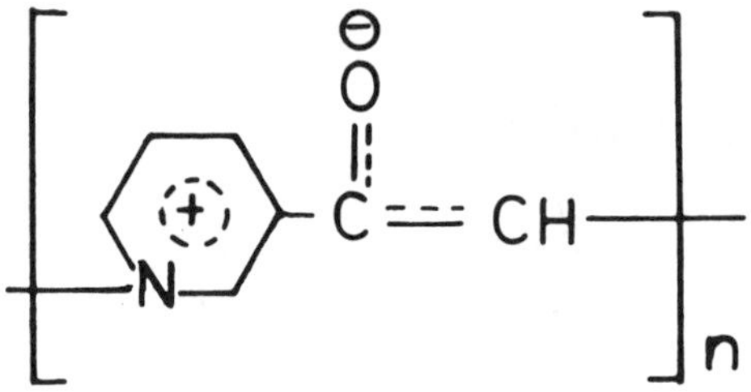

$$(9)$$

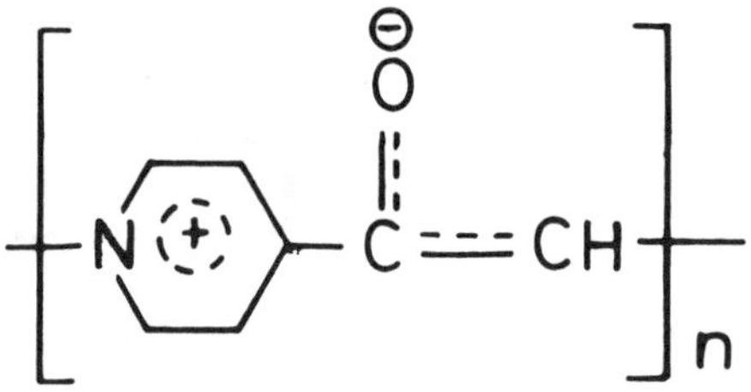

$$(10)$$

Similar results were obtained when 4,4'-bipyridyl was quaternized with p-bis(bromo-acetyl) benzene in refluxing benzene to give a 92% yield of the blue–green salt. Potassium carbonate treatment of their blue-green salt gave a greenish black polymeric ylide (Scheme 11), which was similar in appearance and properties to (Scheme 12).

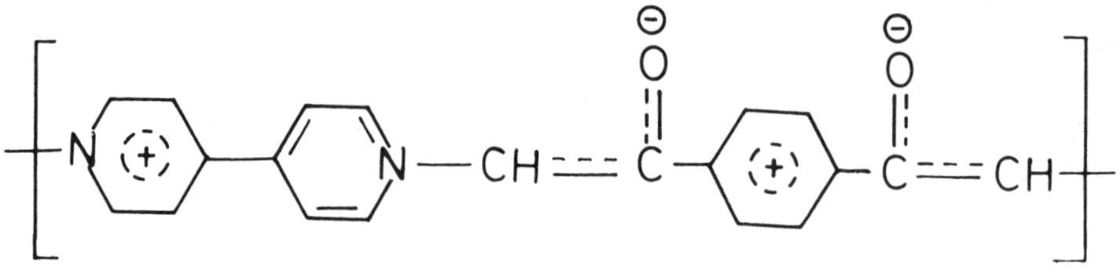

$$(11)$$

Pyridinium ylide is considered to be the adduct carbene to the lone pair of nitrogen in pyridine. The validity of this assumption was confirmed by Tozume et al. [12]. They obtained pyridinium bis-(methoxycarbonyl) methylide by the photolysis of dimethyl diazomalonate in pyridine. Matsuyama et al. [13] reported that the pyridinium ylide was produced quantitatively by the transylidation of sulfonium ylide with pyridine in the presence of some sulfides. However, in their method it was not easy to separate the end products. Kondo and his coworkers [14] noticed that this disadvantage was overcome by the use of carbon disulfide as a catalyst. Therefore, they used this reaction to prepare poly[4-vinylpyridinium bis-(methoxycarbonyl) methylide (Scheme 12) by stirring a solution of poly(4-vinylpyridine), methylphenylsulfonium bis-(methoxycarbonyl)methylide, and carbon disulfide in chloroform for 2 days at room temperature.

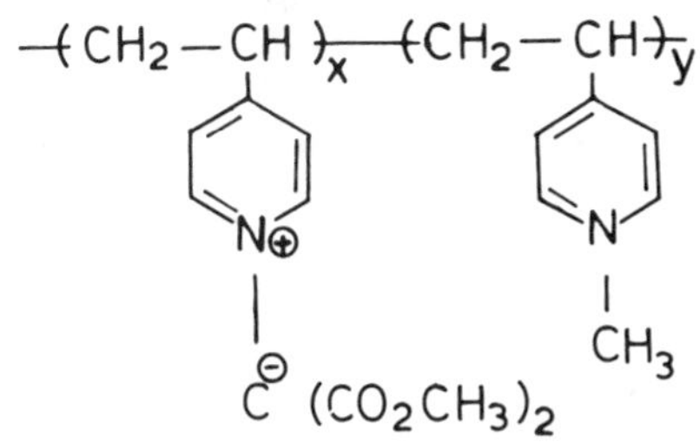

$$\text{+(CH}_2\text{—CH})_x\text{+(CH}_2\text{—CH)}_y \qquad (12)$$

On the other hand, poly(ethoxycarbonylimino-4-vinylpyridinium ylide) (Scheme 13) was prepared essentially by the same method from 1-ethoxycarbonyliminopyridinium ylide, as described by Hafner [15] from the reaction of poly (4-vinylpyridine) with nitrene, generated from the pyrolysis of ethyl azidoformate.

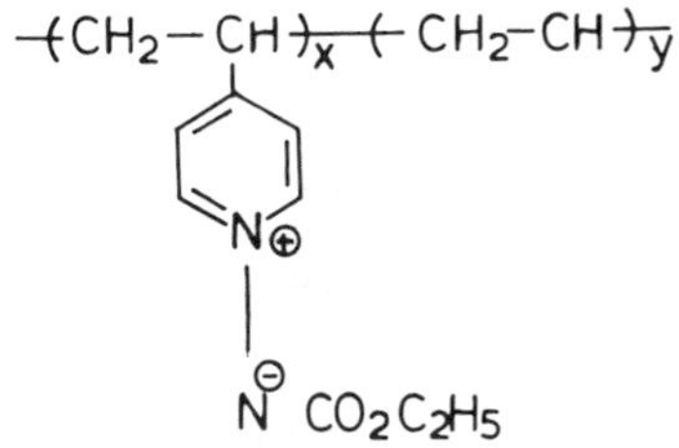

$$\text{+(CH}_2\text{—CH})_x\text{+(CH}_2\text{—CH)}_y \qquad (13)$$

Polymers in Schemes 12 and 13 were the first examples of the preparation of pyridinium and iminopyridinium ylide polymers. One of the more recent contributions of Kondo and his colleagues [16] deals with the sensitization effect of 1-ethoxycarbonyliminopyridinium ylide (IPYY) (Scheme 14) on the photopolymerization of vinyl monomers. Only acrylic monomers such as MMA and methyl acrylate (MA) were photoinitiated by IPYY, while vinylacetate (VA), acrylonitrile (AN), and styrene were unaffected by the initiator used. A free radical mechanism was confirmed by a kinetic study. The complex of IPYY and MMA was defined as an exciplex that served as a precursor of the initiating radical. This ylide is unique in being stabilized by the participation of a

nitrogen anion in the resonance of the pyridine ring [17] as shown below:

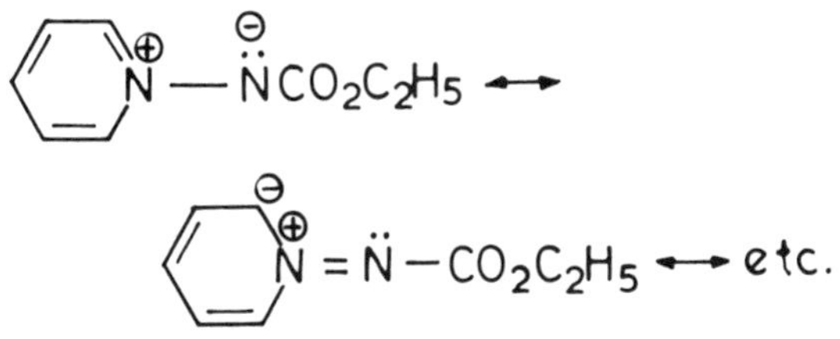

$$(14)$$

We have also used β-picolinium-p-chlorophenacylide (β-PCPY) [15], a cycloimmonium ylide, as the radical initiator for the polymerization of MMA [18,19] at 60–70°C. (See also Table 1.) The kinetics of the system has been studied.

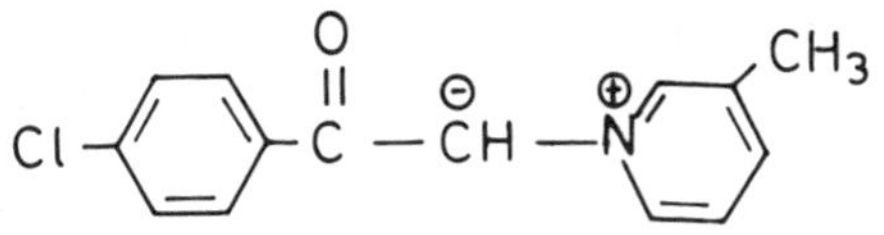

$$(15)$$

Photopolymerization of MMA was also carried out in the presence of visible light (440 nm) using β-PCPY as the photoinitiator at 30°C [20]. The initiator and monomer exponent values were calculated as 0.5 and 1.0, respectively, showing ideal kinetics. An average value of k_p^2/k_t was 4.07×10^{-2} L·mol^{-1}·s^{-1}. Kinetic data and ESR studies indicated that the overall polymerization takes place by a radical mechanism via triplet carbene formation, which acts as the sources of the initiating radical.

Autoacceleration in the polymerization of MA poses a serious problem [21–23]. Saini et al. [24] attempted to polymerize MA by using β-PCPY as the initiator with a view to minimize the difficulties experienced due to this phenomenon. The findings led to the conclusion that β-PCPY can be used to obtain 19.5% conversion of MA without gelation due to autoacceleration, which is nearly double the conversion obtained by using the conventional free radical initiator (AIBN) in the same experimental conditions.

According to Bevington [25], a substance that decreases the rate of polymerization as well as the average degree of polymerization is called a degradative chain transfer agent. In the case of polymerization of VA, β-PCPY was found to act as initiator-cum-degradative chain transfer agent [26]. The [β-PCPY] and [VA] exponents evaluated were 0.80 ± 0.15 and 1, respectively, in the temperature range 30°C and 40°C. The values of ΔE and k_p^2/k_t were calculated to be 90.3 kJ/mol and 0.37 $\times 10^{-2}$ L·mol^{-1}·s^{-1}, respectively, for this system.

β-PCPY does not initiate the polymerization of electron-donating monomers, such as styrene, although it

Table 1 Polymerization of MMA Initiated by β-Picolinium p-Chlorophenacylide[a] [18]

S.N.	[Ylide] $\times 10^3$ mol L^{-1}	$R_p \times 10^5$ mol L^{-1} S^{-1}	η_{int}(dL/g)	$\overline{M}v$	$\overline{P}n$	$\overline{M}w$	$\overline{M}w/\overline{M}n$
1.	0.5	0.875	—	—	—	—	—
2.	1.0	1.310	—	—	—	—	—
3.	1.5	1.750	—	—	—	—	—
4.	2.0	2.090	0.033	4083	46.98	—	—
5.	2.5	2.440	0.044	5994	64.42	—	—
6.	3.0	2.810	0.066	10280	101.90	9528	1.54
7.	3.25	3.120	0.078	12810	123.90	—	—

[a] MMA $= 36 \times 10^{-3}$ mol, 60°C, 4 h.

accelerates the AIBN-initiated polymerization of styrene [27]. The R_p was proportional to [AIBN]$^{0.33}$ $\alpha[\beta$-PCPY]$^{0.14}$ α[Styrene]$^{0.5-1.0}$. The $\overline{P}n$ values increased with increasing [β-PCPY]; however, the energy of activation decreased in the presence of β-PCPY. The accelerating effect was explained on the basis of a decrease in the rate of termination.

In continuation of this series, Saini et al. [24] studied the initiation effect of α-PCPY on the polymerization of MMA [28] and MA [29]. The systems followed nonideal kinetics in both cases, which was explained in terms of

degradative chain transfer agent to initiator and primary radical termination, respectively. A suitable mechanism was also proposed, according to which α-PCPY (Scheme 16) dissociates thermally into triplet carbene [30] (Scheme 17), which acts as a source of radicals. The triplet carbene reacts with monomer (MMA or MA) to form the diradical (Scheme 18) [31,32], which further decomposes into two radicals (Scheme 19) and (Scheme 20). The radical (Scheme 20) is resonance stabilized, and, therefore, polymerization is brought about only by the radical (Scheme 19).

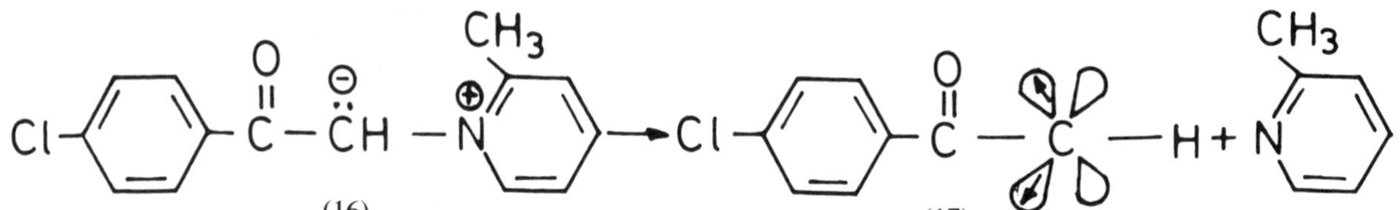

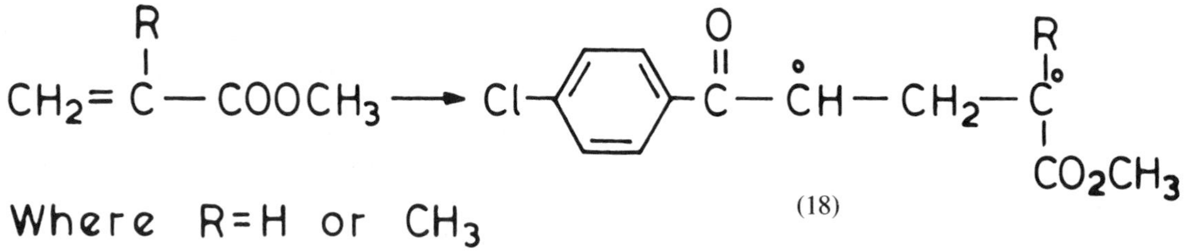

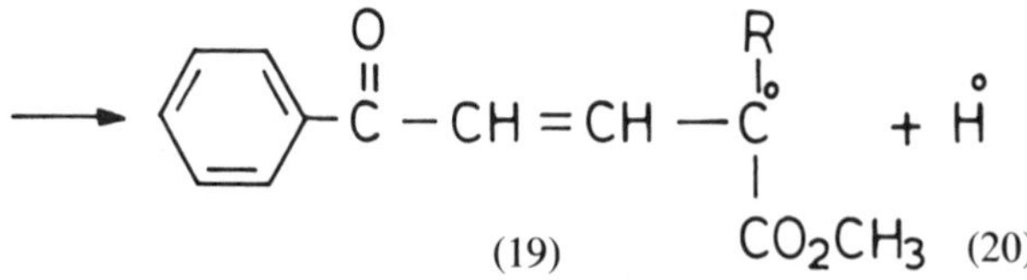

In an attempt to broaden the use of nitrogen ylides in polymerization, Saini et al. [33] tried using imidazolium-p-chlorophenacylide (ICPY) for the polymerization of styrene [33] and MA [34]. Unlike β-PCPY, ICPY

initiates the polymerization of styrene significantly. The kinetic rate equation of the system (R_p α[ICPY]$^{0.38}$ α[Styrene]$^{1.5}$) shows nonideality, which appears to be due to primary radical termination. The nonideal kinetics were also obtained with MA ($R_p \alpha [MA]^{0.36}$), which was reasonably explained on the basis of degradative chain transfer to initiator.

It was recently found that β-PCPY can also be used as a radical initiator to obtain an alternate copolymer of MMA with styrene [35], which was only possible in the presence of Lewis acids [36,37] in the past. The kinetics of the system has been formulated as $Rp \, \alpha [\beta\text{-PCPY}]^{0.5}$ $\alpha [MMA]^{1.0} (1/\alpha [Styrene]^{1.0})$ The values of k_p^2/k_t and ΔE were evaluated as 1.43×10^{-3} L·mol^{-1}·s^{-1} and 87 kJ/mol, respectively, for the system. NMR spectroscopy was used to determine the structure composition and stereochemistry of copolymers. Radical copolymerization of AN with styrene [38] by using β-PCPY as the initiator at 55–65°C also resulted in an alternate copolymer. Rp is a direct function of β-PCPY and AN, and is inversely related to styrene.

In contrast to β-PCPY, ICPY did not initiate copolymerization of MMA with styrene [39] and AN with styrene [40]. However, it accelerated radical polymerization by increasing the rate of initiation in the former case and decreasing the rate of termination in the latter case. The studies on photocopolymerization of MMA with styrene in the presence of ICPY has also been reported [41], β-PCPY also initiated radical copolymerization of 4-vinylpyridine with methyl methacrylate [42]. However, the ylide retarded the polymerization of N-vinylpyrrolidone, initiated by AIBN at 60°C in benzene [44]. (See also Table 2.)

B. Phosphonium Ylide

Mckinley and Rakshys [45] contributed significantly to our store of knowledge by their classical investigation on the preparation of phosphonium ylide polymer. They also synthesized [46] crosslinked resins having phosphorous ylide groups by dehydrohalogenating the corresponding haloalkyl phosphonium salt. These phosphonium ylide polymers have been used for converting ketons or aldehydes to olefins via the Wittig reaction.

Table 2 Copolymerization of Styrene with MMA [43] Initiated by Pyridinium Dicyanomethylylide[a]

S.N.	Ylide $\times 10^4$ mol/L^{-1}	Conversion percentage	$R_p \times 10^4$ mol L^{-1} s^{-1}	$\bar{P}n$
1.	2.7	3.12	3.9	745
2.	5.6	5.0	6.2	623
3.	4.8	5.8	7.3	293
4.	11.1	6.8	8.5	200
5.	14.0	4.3	5.4	—
6.	28.0	3.5	4.3	—
7.	42.0	3.5	4.2	—
8.	56.0	2.5	3.1	—

[a] {Styrene} = 1.38 mol L^{-1}; {MMA} = 1.44 mol L^{-1}, Time = 60 min; temperature = 85 ± 0.1°C.

The first important lead toward the application of ylides as an initiator came from the observations of Zweifel and Voelker [47] in 1972. In their experiment, these authors polymerized lactones or unsaturated compounds initiated by phosphorus ylides prepared directly from tertiary phosphines or similar compounds.

Meanwhile, it was found by Asai and colleagues [48] that tetraphenylphosphonium salts having such anions as Cl$^-$, Br$^-$, and BF$_4^-$ work as photoinitiators for radical polymerization. Based on the initiation effects of changing counteranions, they proposed that a one-electron transfer mechanism is reasonable in these initiation reactions. However, in the case of tetraphenylphosphonium tetrafluoroborate, it cannot be ruled out that direct homolysis of the p-phenyl bond gives the phenyl radical as the initiating species since BF$_4^-$ is not an easily photooxidizable anion [49]. Therefore, it was assumed that a similar photoexcitable moiety exists in both tetraphenyl phosphonium salts and triphenylphosphonium ylide, which can be written as the following resonance hybrid [17] (Scheme 21):

$$Ph_3P = C \overset{\diagup}{\diagdown} \longleftrightarrow Ph_3 \overset{\oplus}{P} - \overset{\ominus}{C} \overset{\diagup}{\diagdown} \qquad (21)$$

If direct homolysis occurs in the case of tetraphenylphosphonium tetrafloroborate, triphenylphosphonium ylide was expected to function as a photoinitiator of radical polymerization because of its similar structure. Therefore, another milestone was reached by Kondo and colleagues [50] who investigated the use of triphenylphosphonium ethoxycarbonylmethylide (TPPY) (Scheme 22) as an effective photoinitiator for the polym-

$$Ph_3 \overset{\oplus}{P} - \overset{\ominus}{C}HCO_2C_2H_5 \qquad (22)$$

erization of methyl methacrylate and styrene. The polymerization proceeded through a free radical mechanism. However, under thermal conditions at 60°C, TPPY did not initiate the polymerization of these monomers.

The proposed mechanism is based on the basis of the fact that ylides (Scheme 23 and Scheme 24) undergo bond fission between the phosphorus atom and the phenyl group in TPPY as reported by Nagao et al. [51] and between the sulfur atom and the phenyl group in POSY as observed in triphenylsulfonium salts [52–55] when they are irradiated by a high-pressure mercury lamp. The phenyl radicals thus produced participate in the initiation of polymerization.

C. Arsonium Ylide

Vasishtha [56] also reported p-acetyl benzylidene triphenyl arsonium ylide radically initiated bulk polymerization of styrene at $60 \pm 0.1°C$ for 20 h. The system followed ideal kinetics with bimolecular termination. The values of the activation energy and k_p^2/k_t were 64 kJ/mol and 0.10×10^{-2} L/mol sec^{-1}, respectively. The authors proposed that the ylide dissociated to produce phenyl radicals that led to polymer. The mechanism is as follows:

$$(C_6H_5)_3 - \overset{\oplus}{As} - \overset{\ominus}{CH} - \langle \text{ring} \rangle - COCH_3$$

$$CH_3 - \overset{O}{\overset{\|}{C}} - \langle \text{ring} \rangle - \overset{\ominus}{CH} - \overset{\oplus}{As} \overset{\displaystyle C_6H_5}{\overset{\displaystyle -C_6H_5}{\diagdown C_6H_5}}$$

$$\downarrow$$

$$CH_3 - \overset{O}{\overset{\|}{C}} - \langle \text{ring} \rangle - \overset{\ominus}{CH} - \overset{\oplus\bullet}{As} \overset{\displaystyle C_6H_5}{\diagdown C_6H_5} + \dot{C}_6H_5 \tag{23}$$

D. Sulphonium Ylide

In an attempt to prepare sulfonium–ylide polymer, Tanimoto and coworkers [57,58] carried out the reaction of a sulfonium salt polymer with benzaldehyde in the presence of a base and obtained styrene oxide. The reaction was considered to process via a ylide polymer formation (Scheme 24), which may be unstable and has not been isolated.

$$\begin{array}{c} +CH - CH_2 \rangle \\ | \\ \langle \text{ring} \rangle \\ | \\ \overset{\oplus}{S} - \overset{\ominus}{Me}CH_2 \end{array} \tag{24}$$

These pioneer studies laid dormant until 1977 and, influenced by Kondo and colleagues' [59] reports on the synthesis of poly(vinylsulfonium ylide) with a trivalent sulfur attached directly to the polymer chain, poly[ethylvinylsulfonium bis-(methoxycarbonyl) methylide] (Scheme 25) was prepared by irradiation of a benzene

$$\begin{array}{c} +CH_2 - CH \rangle_x + CH_2 - CH \rangle_y \\ | \qquad \qquad \qquad | \\ | \qquad \qquad \qquad SEt \\ Et - \overset{\oplus}{S} - \overset{\ominus}{C} - (CO_2CH_3)_2 \end{array} \tag{25}$$

solution of poly(ethyl vinyl sulfide) and dimethyl diazomalonate in a Pyrex tube by a high-pressure mercury lamp. In a similar manner, an attempt was made to prepare poly[phenylvinylsulfonium bis-(methoxycarbonyl) methylide] (Scheme 26), but it was not successful. (See also Table 3.) However, this compound was obtained by the thermal reaction of diazomalonate and poly(phenylvinyl sulfide) in the presence of cupric sulfate as catalyst in benzene.

$$\begin{array}{c} +CH_2 - CH \rangle_x + CH_2 - CH \rangle_y \\ | \qquad \qquad \qquad | \\ | \qquad \qquad \qquad SPh \\ \overset{\oplus}{S} - \overset{\ominus}{C} (CO_2CH_3)_2 \end{array} \tag{26}$$

The structures of these ylide polymers were determined and confirmed by IR and NMR spectra. These were the first stable sulfonium ylide polymers reported in the literature. They are very important for such industrial uses as ion-exchange resins, polymer supports, peptide synthesis, polymeric reagent, and polyelectrolytes. Also in 1977, Hass and Moreau [60] found that when poly(4-vinylpyridine) was quaternized with bromomalonamide, two polymeric quaternary salts resulted. These polyelectrolyte products were subjected to thermal decyanation at 7200°C to give isocyanic acid or its isomer, cyanic acid. The addition of base to the solution of polyelectrolyte in water gave a yellow polymeric ylide.

In a pioneering article, Farrall et al. [61] reported the preparation of fully regenerable sulfonium salts anchored to an insoluble polymer and their use in the preparation of epoxides by reaction of their ylides with carbonyl compounds. Their results clearly indicate that

Table 3 Effect of [p-ABTAY] on the Rate of Polymerization of Styrene Initiated by p-Acetylbenzylidene Triphenylarsonium ylide at $60 \pm 0.1°C^a$

S.N.	[p-ABTAY] $\times 10^4$ [mol L^{-1}]	Percent conversion	$R_p \times 10^6$ [mol L^{-1} s^{-1}]	$\overline{P}n$
1.	1.14	0.082	2.10	—
2.	3.42	0.100	3.20	305
3.	6.84	0.150	4.20	250
4.	9.13	0.160	5.30	190
5.	11.4	0.220	6.11	170
6.	13.6	0.298	6.94	146
7.	17.1	0.318	7.40	137
8.	22.2	0.500	9.00	—
9.	34.2	0.571	11.10	—
10.	45.6	0.582	12.00	—

^{a} Styrene 17.3×10^{-2} mol, time, 20 h.

phasetransfer catalysis is the method of choice for the generation of sulfonium ylides on insoluble resins from a polymeric sulfonium salt.

Kondo maintained his interest in this area, and with his collaborators [62] he recently made detailed investigations on the polymerization and preparation of methyl-4-vinylphenyl-sulfonium bis-(methoxycarbonyl) methylide (Scheme 27) as a new kind of stable vinyl monomer containing the sulfonium ylide structure. It was prepared by heating a solution of 4-methylthiostyrene, dimethyldiazomalonate, and *t*-butyl catechol in chlorobenzene at 90°C for 10 h in the presence of anhydride cupric sulfate, and Scheme 27 was polymerized by using α, α′-azobisisobutyronitrile (AIBN) as the initiator and dimethylsulfoxide as the solvent at 60°C. The structure of the polymer was confirmed by IR and NMR spectra and elemental analysis. In addition, this monomeric ylide was copolymerized with vinyl monomers such as methyl methacrylate (MMA) and styrene.

$$H_2C = CH \underset{}{\overset{}{\longleftrightarrow}} \overset{\overset{CH_3}{|}}{\underset{}{S}}{}^{\oplus} - \overset{\ominus}{C} - (CO_2CH_3)_2$$

$$\overset{\overset{CH_3}{|}}{C_2H - \underset{}{S}{}^{\oplus}} - \overset{\ominus}{C} - (CO_2CH_3)_2 \tag{27}$$

Previously, the same author [52] reported that compounds containing the tricoordinated sulfur cation, such as the triphenylsulfonium salt, worked as effective initiators in the free radical polymerization of MMA and styrene [52]. Because of the structural similarity of sulfonium salt and ylide, diphenyloxosulfonium bis-(methoxycarbonyl) methylide (POSY) (Scheme 28), which contains a tetracoordinated sulfur cation, was used as a photoinitiator by Kondo et al. [63] for the polymerization of MMA and styrene. The photopolymerization was carried out with a high-pressure mercury lamp; the orders of reaction with respect to [POSY] and [MMA] were 0.5 and 1.0, respectively, as expected for radical polymerization.

$$C_6H_5 - \overset{\overset{CH_3}{|}}{\underset{}{S}}{}^{\oplus} - \overset{\ominus}{C} - (CO_2CH_3)_2 \tag{28}$$

Recently, in connection with the use of sulfur ylides in polymerization, Kondo and his coworkers [64] attempted to use diphenylsulfonium bis(methoxycarbonyl)methylide (DPSY) (Schemes 27, 29) methylphenylsulfonium bis-(methoxycarbonyl)methylide (MPSY) (Scheme 30) and dimethylsulfonium bis (methoxycarbonyl) methylide (DMSY) (Scheme 31) as photoinitiators for the polymerization of MMA and styrene. They concluded that DPSY and MPSY are effective photoini-

tiators, but DMSY has little initiation ability. The differences were explained on the basis of ultraviolet spectra.

$$\overset{R^1}{\underset{R^2}{>}}\overset{\oplus}{S} - \overset{\ominus}{C}\overset{CO_2CH_3}{\underset{CO_2CH_3}{<}} \qquad R^1 = R^2 = C_6H_5 \tag{29}$$

$$R^1 = C_6H_5 , R^2 = CH_3 \tag{30}$$

$$R^1 = R^2 = CH_3 \tag{31}$$

Dimethyl sulfonium-2-pyridyl carbonyl methylide (Ypy-5) [65] initiated radical polymerization of styrene in dimethyl sulfoxide at 85 ± 0.1°C for 6 h under a nitrogen blanket. (See Table 4.)

III. CONCLUSIONS

1. Ylides can be used as new novel radical initiators, accelerators, chain transfer agents or retardants, and curing agents.
2. Low concentration (0.2–0.6 mg ylide in case of PDMY) of ylides is needed to achieve desirable percentage of conversion.
3. Polymerization takes place by triplet carbene formation.
4. Alternating copolymer can be synthesized using most of the ylides as radical initiators even in the absence of Lewis acids otherwise essential for other conventional radical initiators.
5. Ylides can be used as initiators to get a polymethacrylate of $\overline{P}n$ 571 at 14% conversion to minimize the autoacceleration effect.

IV. FURTHER SCOPE

It is clear from the achievements and prospects in polymer ylide science that this emerging field of polymer science merits new efforts and investigations. There is every reason to believe that during the course of such

Table 4 Effect of Ylide Concentration on the Rate of Polymerization of Styrene Initiated by Dimethyl Sulfonium 2-Pyridyl Carbonyl Methylide

S.N.	Ylide concentration $\times 10^{-3}$ (mol/L^{-1})	Conversion (%)	$R_p \times 10^7$ (mol/L^{-1} $-$ S^{-1})	$\overline{P}n$
1.	1.84	0.66	8.74	—
2.	5.52	1.85	11.20	—
3.	9.20	2.52	12.75	313
4.	11.00	2.88	13.33	268
5.	12.87	3.22	15.00	—
6.	14.70	3.66	15.55	186
7.	18.41	3.87	16.66	136

investigations new possibilities for the applications of ylides in other polymer fields will be reported. A few areas that need the attention of polymer chemists are:

1. To examine potentiality of other ylides and their metal complex containing Sb, As, P, Bi, and Se as new novel initiator in polymer synthesis via living radical polymerization.
2. Synthesis of metal ylide complex and their application in polymer chemistry.
3. Synthesis and polymerization of new monomers using ylides as starting materials.
4. Kinetics and mechanism of polymerization of vinyl monomers initiated by ylides.
5. Kinetics and mechanism of curing reactions using ylides as a new curing agent.

REFERENCES

1. A. Mischaelis and H. V. Gimborn, *Ber. Deut. Chem. Gest.* 27: 272 (1894).
2. G. Wittig and G. Felleteschiin, *Ann., 555*: 133 (1944).
3. W. E. Doering, K. L. Levy, and K. Schreiber, *J. Am. Chem. Soc.*, 77: 509 (1955).
4. W. E. Doering and K. Hoffmann, *J. Am. Chem. Soc.*, 77: 521 (1955).
5. D. P. Craig, A. Maccol, and L. E. Sutton, *J. Am. Chem. Soc.* 76: 332 (1954).
6. S. W. Kantov and C. R. Hauser, *J. Am. Chem. Soc.*, 73: 4122 (1951).
7. A. W. Johnson, *J. Org. Chem.*, 24: 252 (1959).
8. A. W. Johnson, *J. Org. Chem.*, 25: 183 (1960).
9. I. Zugravescu and C. Petrovanu, *N-Ylid Chemistry*, McGraw-Hill, New York (1976).
10. Yu. V Belkin and N. A. Plezhaeva, *Russ. Chem. Rev.*, 50: 481 (1981).
11. D. E. George, R. E. Putnam, and S. Selman, *J. Polym. Sci. Pt. A-1, 4*: 1323 (1966).
12. S. Tozume, T. Yagihara, S. Ando, and T. Migita, 21st Annual Meeting of the Chemical Society of Japan, p. 1791 (1969).
13. H. Matsuyama, H. Minato, and M. Kobayashi, *Bull. Chem. Soc. Jpn.*, 46: 2845 (1973).
14. S. Kondo, K. Takagishi, T. Obata, M. Senga, Y. Yamashita, and K. Tsuda, *J. Polym. Sci., Polym. Chem. Ed., 21*: 3597 (1983).
15. K. Hafner, *Tetrahedron Lett.*, p. 1733 (1964).
16. S. Kondo, S. Hashiya, M. Muramtsu, K. Tsudo, *Angew. Macromol. Chem. 126*: 19 (1984).
17. A. W. Johnson, *Ylide Chemistry*, Academic Press, New York (1966).
18. A. K. Srivastava and S. Saini, *J. Macromol. Sci. Chem., A22*: 43 (1985).
19. A. K. Srivastava and S. Saini. *Acta Polym., 35*: 667 (1984).
20. S. Saini and A. K. Srivastava, *Polym. Photochem., 7*: 179 (1986).
21. C. Walling, *J. Am. Chem. Soc.*, 70: 2591 (1948).
22. K. S. Bagdasaryan, *J. Phys. Chem., Moscow, 22*: 1181 (1948).
23. V. Mahadevan and M. Santhapa, *Makromol. Chem., 16*: 119 (1955).
24. S. Saini, A. K. Shukla, P. Kumar, and A. K. Srivastava, *J. Macromol. Sci. Chem., 23*: 1107 (1986).
25. J. C. Bevington, *Radical Polymerization*, Academic Press, New York, p. 102 (1961).
26. A. K. Shukla, S. Saini, P. Kumar, J. S. P. Rai, and A. K. Srivastava, *J. Polym. Sci., Polym. Chem. Ed., 227*: 43 (1989).
27. S. Saini, A. K. Shukla, and A. K. Srivastava, *Polym. J., 17*: 1117 (1985).
28. A. K. Shukla, S. Saini, P. Kumar, and A. K. Srivastava, *Angew. Makromol., 141*: 103 (1986).
29. A. K. Shukla, S. Saini, P. Kumar, and A. K. Srivastava, *Indian J. Chem., 24A*: 1054 (1985).
30. J. Streith and J. M. Cassal., C. R. Hebd, *Seancees Acad. Sci., Ser. C, 264*: 1307 (1967).
31. W. Kirmse, *Carbene Chemistry,* Academic Press, New York, p. 285 (1971).
32. R. J. Cvetanovic, H. E. Avery, and R. S. Irwin, *J. Chem. Phys., 49*: 1993 (1967).
33. S. Saini, A. K. Shukla, P. Kumar, and A. K. Srivastava, *Acta Polym., 38(7)*: 432, (1987).
34. S. Saini, S. K. Nigam, J. S. P. Rai, and A. K. Srivastava, *Eur. Polym. J., 23(11)*: 913 (1987).
35. S. K. Nigam, A. K. Shukla, S. Saini, P. Kumar, and A. K. Srivastava, *Angew. Makromol. Chem., 149*: 139 (1987).
36. N. G. Gaylord and B. K. Patnaik, *J. Polym. Sci., Part B 9*: 269 (1971).
37. H. Hirai, K. Takeuchi, and M. Komiyama, *J. Polym. Sci., Polym. Chem. Ed., 23*: 901 (1985).
38. S. K. Nigam, S. Saini, and A. K. Srivastava, *Indian J. Chem. 25A*: 944 (1986).
39. S. K. Nigam, and A. K. Srivastava, *Acta Polym., 38*: 244 (1987).
40. S. K. Nigam, S. Saini, and A. K. Srivastava, *Indian J. Technol., 24*: 743 (1986).
41. S. K. Nigam, S. Saini, and A. K. Srivastava, Short Lecture in the 31st IUPAC Symposium, Merseburg, (1987).
42. R. Vasishtha and A. K. Srivastava, *Br. Polym. J. 22(1)*: 53 (1990).
43. P. Shukla and A. K. Srivastava, *Polym. J. 20*: 941 (1988).
44. R. Vasishtha and A. K. Srivastava, *Colloids Polym. Sci., 268*: 645 (1990).
45. S. V. Mckinley and J. W. Rakshys Jr., *J. Chem. Soc., Chem. Commun.*, 134 (1972).
46. S. V. Mckinley and J. W. Rakshys Jr., U.S. Patent 3,725,365 (1973).
47. H. Zweifel and T. Voelker, *Chimia, 26*: 345 (1972).
48. D. Asai, A. Okada, S. Kondo, and K. Tsuda, *J. Macromol. Sci. Chem., 18*: 1011 (1982).
49. L. E. Orgel, *Q. Rev., 8*: 422 (1954).
50. S. Kondo, Y. Kondo, and K. Tsuda, *J. Polym. Sci., Plym., Lett. Ed. 21*: 217 (1983).
51. Y. Nagao, K. Shima, and H. Sakurai, *Kogyo Kagaku Zasshi, 72*: 236 (1969).
52. S. Kondo, M. Muramatsu, M. Senga, and K. Tsuda, *J. Macromol. Sci. Chem., A19*: 999 (1983).
53. J. W. Knapczyk and W. E. McEven, J., *Org. Chem., 35*: 2539 (1970).
54. S. L. Nikol and J. A. Kampmeier, *J. Am. Chem. Soc.*, 95: 1908 (1973).
55. J. W. Crivello and J. H. W. Lam, *J. Polym. Sci., Polym. Chem. Ed., 17*: 977 (1979).
56. R. Vasishtha and A. K. Srivastava, *Polymer, 31*: 150 (1990).
57. S. Tanimoto, J. Horikawa, and R. Oda, *Kogyo Kagaku Zasshi, 70*: 1969 (1967).

58. S. Tanimoto, J. Horikawa, and R. Oda, *Yuki Gosei Kagaku Kyokai Shi, 27*: 989 (1969).
59. S. Kondo and K. Tsuda, *J. Polym. Sci. Polym. Chem. Ed., 15*: 2797 (1977).
60. H. C. Hass and R. D. Moreau, *J. Polym. Sci. Polym. Chem. Ed., 15*: 1225 (1977).
61. M. J. Farrall, T. Durst, and J. M. J. Frechet, *Tetrahedron Lett.*, p. 203 (1979).
62. M. Senga, S. Kondo, and K. Tsuda, *J. Polym. Sci., Polym. Lett. Ed. 20*: 657 (1982).
63. S. Kondo, M. Muramstsu, M. Senga, and K. Tsuda, *J. Polym. Sci., Polym. Chem. Ed., 22*: 1187 (1984).
64. S. Kondo, S. Itoh, and K. Tsuda, *J. Macromol. Sci, Chem., A20*: 433 (1983).
65. U. Bhatnagar and A. K. Srivastva, *Polym. International 25*: 13 (1991).

25

Characterization and Properties of Thermotropic Polybibenzoates

Ernesto Pérez, José M. Pereña, Rosario Benavente, and Antonio Bello
Instituto de Ciencia y Tecnología de Polímeros (CSIC), Madrid, Spain

I. INTRODUCTION

In recent years the interest in thermotropic materials has grown extraordinarily. These compounds usually exhibit a mesophase at temperatures above the region of the crystalline solid and before the formation of an isotropic melt. This liquid crystalline phase represents a state of order between long-range, three-dimensionally ordered crystals and the disordered amorphous or isotropic state. Thus, mesophases are characterized by the absence of positional long-range order in at least one dimension [1], but a long-range orientational order is preserved.

Aromatic polyesters constitute an important class of main-chain liquid-crystalline polymers, but present the inconvenience of their reduced solubility and very high transition temperatures (sometimes not detected before the degradation of the sample). Their processability can be improved in several ways [2,3], e.g., reduction of the rigidity of the mesogen, lengthening of the spacer, or introduction of lateral substituents.

Polybibenzoates are a kind of thermotropic polyesters obtained by polycondensation of 4,4'-biphenyldicarboxylic acid (*p,p'*-bibenzoic acid) with a diol. These polyesters contain the biphenyl group, which is one of the simplest mesogens. They are synthesized by melt transesterification of the dimethyl or diethyl ester of *p,p'*-bibenzoic acid and the corresponding diol, using a titanium compound as catalyst, according to the following scheme:

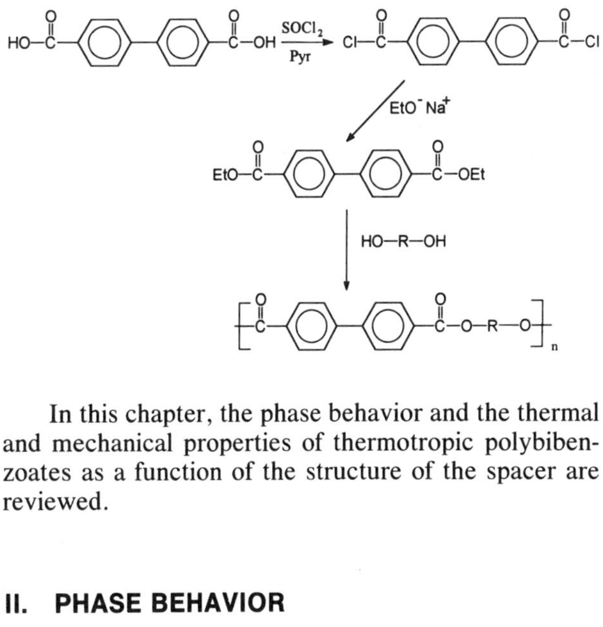

In this chapter, the phase behavior and the thermal and mechanical properties of thermotropic polybibenzoates as a function of the structure of the spacer are reviewed.

II. PHASE BEHAVIOR

Several works have been published about the phase behavior of polybibenzoates, showing the ability of the biphenyl group to produce mesophase structures. Different spacers have been used, and the results show that the structure of the spacer influences very much the transition temperatures and the nature and stability of the mesophases, as well as the ability to generate three-dimensional crystals.

383

The phase behavior of polybibenzoates has been investigated mainly by DSC, variable-temperature x-ray diffraction, and optical microscopy. However, only the first two techniques are useful in the case of polymers with the high-molecular weights required for materials with good mechanical properties and, in such cases, revealing textures are not usually observed by optical microscopy.

Both all-methylene and oxyalkylene spacers have been used. The presence of ether groups in the spacer has a profound influence on the phase behavior, as will be shown in this chapter.

A. Polybibenzoates with All-Methylene Spacers

Several articles [4–11] deal with the properties of polybibenzoates with all-methylene spacers where the number, m, of methylene groups ranges from 2–10. On cooling these polyesters from the melt, smectic-type mesophases are typically formed, which experience a subsequent transformation into a three-dimensional crystal. The behavior of this crystal on heating is, however, not the same for all polybibenzoates, since for those polymers with $m \geq 7$ it undergoes a monotropic transformation directly into the isotropic melt, while for the lower members this transformation is enantiotropic and the mesophase is also observed on heating. Thus, Fig. 1 shows the DSC curves corresponding to a sample of poly(heptamethylene p,p'-bibenzoate), P7MB. On cooling from the isotropic melt (lower curve), two exotherms, at 141 and 99°C, are obtained, while only one endotherm, at 168°C, is observed in the subsequent heating cycle (upper curve). The transitions involved were analyzed by variable-temperature x-ray diffraction. Figure

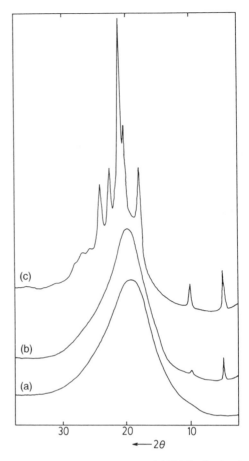

Figure 2 X-ray diffractograms of P7MB obtained at different temperatures [10]. Curve a: 182°C; curve b: 152°C; curve c: 72°C. Noise has been suppressed.

2 shows the diffractograms corresponding to three temperatures. The lower curve, obtained at 182°C, represents the isotropic melt without any indication of order. The middle diffractogram, however, taken at 152°C on cooling from the melt, shows a sharp peak (and its second order) at a spacing of about 1.7 nm. It was attributed to the spacing of the smectic layers [10]. In addition, the broad peak, centered at 0.43 nm (similar in shape to the one for the isotropic melt) is attributed to the lateral disorder within the layers in this smectic phase. Finally, the upper diagram in Fig. 2, taken at 72°C, presents sharp diffractions characteristic of three-dimensional order. Identical diffractograms were obtained both at lower temperatures and in the subsequent heating cycle, maintaining the same pattern until total isotropization. Thus, the crystal experiences a monotropic transition to the isotropic melt and the mesophase is not detected on heating [10].

It can be observed in Fig. 2 that the lower-angle peak appears in both the mesophase and the crystal at

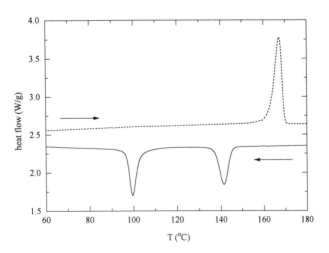

Figure 1 DSC curves of P7MB representing a cooling cycle (lower) starting from the isotropic melt, and the subsequent heating cycle (upper). Scanning rate: 10°C/min.

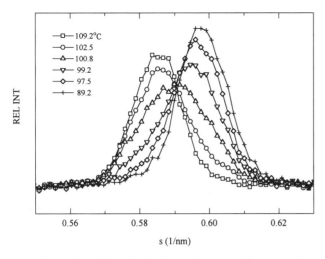

Figure 3 Plot of the relative intensity versus the scattering vector at the indicated temperatures for a sample of P7MB cooled from the melt at 10°C/min, analyzed in a synchrotron source [12].

approximately the same spacing. Real-time small-angle synchrotron experiments [12] have been performed in order to elucidate whether for the crystalline sample that peak represents a crystal repeat distance or indicates the coexistence of crystal and mesophase. The scattering profiles as a function of temperature in a sample of P7MB cooled from the melt show that only the isotropic melt, characterized by the absence of order, is present at high temperatures [12], while a sharp diffraction at about 1.69 nm, corresponding to the spacing of the smectic layers, was obtained at a lower temperature. This peak, which appears at temperatures around 140°C, corresponds to the first exotherm in Fig. 1. At subsequent lower temperatures, the peak at 1.69 nm gradually disappears and a crystal peak of a very close spacing (about 1.65 nm) is formed [12], as it is observed in Fig. 3. This figure shows that the crystal peak is obtained around 100°C, in perfect agreement with the second DSC exotherm in Fig. 1. Simultaneously, a long spacing at about 19 nm is obtained, indicating the formation of the three-dimensional crystal.

Therefore, these synchrotron experiments show that the repeat distance for the smectic layer lies very close to, but is distinct from, that of the crystal. Moreover, the smectic peak disappears totally when the transformation is complete, indicating that the mesophase has been destroyed in the semicrystalline sample.

In the case of monotropic behavior, the isotropization endotherm and the corresponding thermodynamic parameters for the mesophase-isotropic transition can be obtained by isolating the mesophase when cooling from the melt and holding the temperature in a region where the transformation into the crystal is very slow

[10,13]. Thus, Fig. 4 shows the heating curves of a sample of P7MB cooled from the melt to 135°C and maintained at this temperature for different periods of time. The lower curve in this figure corresponds to the untransformed mesophase [10] and the isotropization temperature is shown to be 160°C. The other curves in Fig. 4 show the appearance of a second melting endotherm at a temperature about 7 degrees higher, whose intensity increases with the crystallization time as the mesophase is being transformed into the crystal and the isotropization endotherm vanishes. This high-temperature peak represents the melting of the crystal directly into the isotropic melt and corresponds to the single endotherm in Fig. 1.

The variation of the transition temperatures of these polybibenzoates with the number of methylene units in the spacer is shown in the lower part of Fig. 5. Melting temperatures, T_m, (crystal-isotropic melt transition) are obtained [9] for $m \geq 7$ and $m = 3$ (monotropic behavior), while for the other members, T_m really represents the

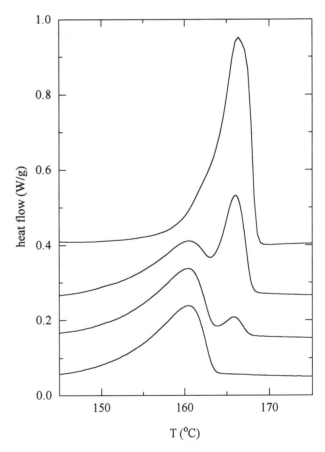

Figure 4 DSC melting endotherms of P7MB after isothermal crystallization at 135°C, starting from the isotropic melt [10]. The curves correspond to 0, 3, 6, and 35 min of crystallization time, from bottom to top. Scanning rate: 5°C/min.

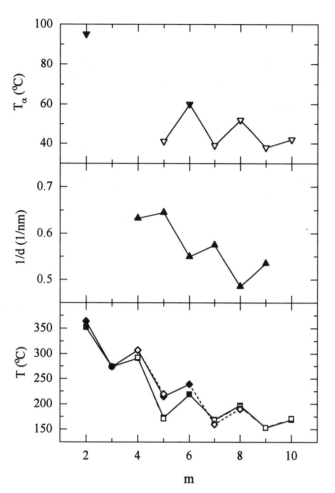

polymer chains between even and odd members, considering that the long-range orientational order in the mesophase will force the mesogenic groups to align approximately parallel to the director. Figure 6 shows an illustration of P7MB and P8MB in their all-*trans* conformation. A fairly extended conformation can be obtained in the even members, with the mesogenic groups approximately parallel to the chain axis, while for the odd members, the valence angles force the chain to adopt a less extended arrangement. In a real polybibenzoate it has to be considered that the molecular chains must adopt conformations that meet the dimensions of the smectic layer. However, the relatively low requirements of the mesomorphic order will allow more than a single conformation to intervene in the formation of the smectic layers [14].

The angular correlation between two consecutive mesogenic groups is also very different between odd and even members, as shown by conformational analyses [14–16] based on the rotational isomeric state model. The corresponding calculations for P7MB and P8MB are plotted in Fig. 7. It can be observed in P8MB that for nearly 28% of the conformers the consecutive rigid groups form an angle that lies in the range 0–10°, while for about 72% of the conformations this range lies between 100–115°. It is evident that only the first fraction will fulfill the arrangement required to adopt mesomorphic order. On the contrary, for P7MB about 76% of all conformers have correlation angles between 63–75°, and

Figure 5 Summary plot of the transition temperatures (lower), inverse of the smectic layer spacing (middle), and temperature of the α relaxation (upper) of polybibenzoates as a function of the number of methylene units in the spacer: ◆ T_i, [Ref. 9]; ■ T_m, [Ref. 9]; ▲ 1/d, [Ref. 7]; ▼ T_α, [Ref. 9]; open symbols: our results.

crystal-mesophase transition. Two facts, usual for thermotropic polyesters, are derived from Fig. 5. First, a very important decrease of the transition temperatures with an increase in the number of methylene units is observed as a consequence of the decrease of molecular rigidity. The second fact is the odd–even effect for the transition temperatures with considerably higher values for the polymers with even number of methylenes in the spacer. This odd–even effect has been also found for the transition entropies [2,3,7], for the smectic layer spacing (middle part of Fig. 5), and for the temperature of the α-relaxation corresponding to the glass transition temperature (upper part of Fig. 5).

In principle, the odd–even effect can be explained from the differences in the packing arrangement of the

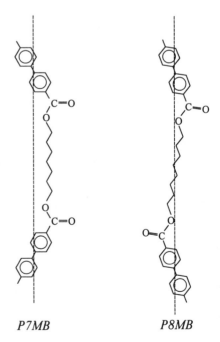

Figure 6 Extended-chain structures of P7MB and P8MB.

only 21% are in the range 171–180°. Moreover, conformers with correlation angles below 30° are not found. Therefore, the odd-membered mesophases will have two successive mesogens forming an angle in the range 60–80°, while the mesogens will be practically collinear with the chain axis for the even members, thus explaining the odd–even oscillations.

The odd–even character of the spacer in polybibenzoates is also reflected in the type of mesophase formed. Thus, the x-ray photographs of the oriented mesophase of P5MB show a broad outer reflection split into two intense portions lying above and below the equator [8,17] and indicating a tilting angle of about 25° between the mesogenic groups and the layer normal. This structure has been characterized as a smectic C_2 phase, similar to that of the regular smectic C mesophases but with the tilt direction of the mesogenic groups being opposite in adjacent layers [17]. On the contrary, a smectic A structure is deduced from the x-ray photographs of oriented polybibenzoates with an even number of methylenes in the spacer, and the mesogens are perpendicular to the layer planes [8,17]. The smectic phases of polybibenzoates have also been characterized by optical microscopy in samples where the degree of polymerization was controlled [17] to be in the range 20–40, since low-molecular weights favor the observation of optical textures.

The effect of lateral methyl groups in the spacer on the phase behavior has been studied in several polybibenzoates [18,19] derived from poly(tetramethylene p,p'-bibenzoate), P4MB. The branched polymers display transition temperatures significantly lower than P4MB. Moreover, the substituents have a clear effect on the kind of mesophase formed. Thus, P4MB displays a smectic A mesophase, while the lateral methyl groups

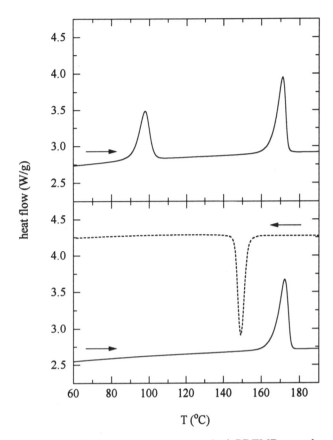

Figure 8 DSC curves for a quenched PDTMB sample (lower) representing a cooling cycle (dashed line) starting from the isotropic melt, and the subsequent heating cycle (continuous line), and the melting of a sample annealed at 70°C during 24 days (upper). Scanning rate: 10°C/min.

favor the obtention of smectic C or nematic structures, and, in some cases, no liquid crystalline behavior was observed [18,19]. The formation of smectic C mesophases is particularly interesting since chiral centers are present in some of these polymers and smectic C* chiral mesophases can be obtained.

B. Polybibenzoates with Oxyalkylene Spacers

The phase behavior of several polybibenzoates with oxyalkylene spacers has been reported [11,14,15,20–27]. These spacers include the dimer of trimethylene glycol and different ethylene oxide oligomers. The most noticeable characteristic of these polybibenzoates with ether groups in the spacer is the considerable decrease of the rate of the mesophase-crystal transformation. Thus, Fig. 8 shows the DSC curves corresponding to a sample of poly[oxybis(trimethylene)p,p'-bibenzoate], PDTMB, with a structure similar to that of P7MB but with the

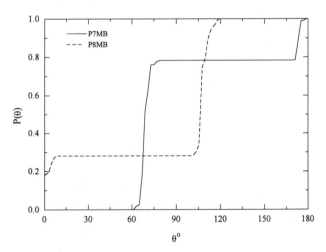

Figure 7 Integrated distribution curves for the angle θ between two successive mesogenic groups calculated [14,15] for P7MB (continuous line) and P8MB (broken line).

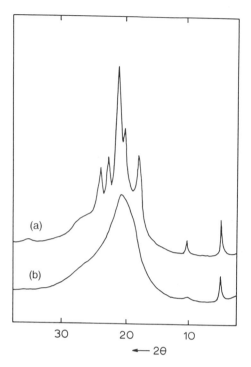

(a)

(b)

30   20   10

← 2θ

Figure 9 Ambient-temperature x-ray diffractograms of PDTMB samples: (a) annealed at 70°C for 24 days, and (b) freshly quenched from the melt [25,26]. Noise has been suppressed.

central methylene in the spacer replaced by an oxygen atom. The dashed curve in this figure represents the cooling from the isotropic melt, and only one exothermic peak at about 150°C is observed. The subsequent heating (lower curve in Fig. 8) led to a single endotherm at 172°C. The nature of this single first-order transition was analyzed by variable-temperature x-ray diffraction [20]. A diffractogram taken at 192°C presents no sharp features, since it corresponds to the isotropic melt. On the contrary, the one obtained at room temperature presents a sharp diffraction (and its second order) at about 1.7 nm, characteristic of a smectic mesophase (lower diagram in

Fig. 9). Moreover, the lateral disorder within the layers is represented by the broad peak, centered at 0.43 nm, similar in shape to that of the isotropic melt. The conclusion is that the mesophase of PDTMB is stable at any temperature (below its isotropization point) during a considerable time. However, at very long annealing times and temperatures above the glass transition (17°C), this mesophase undergoes the transformation to the crystal [15]. Thus, the top diffractogram in Fig. 9 was obtained after annealing a sample of PDTMB at 70°C during 24 days. This diffractogram is similar to that of the crystal of P7MB (Fig. 2), but showing a smaller crystallinity, which may be due to the fact that longer annealing times are needed to complete the mesophase-crystal transformation.

The DSC melting curve of this annealed sample of PDTMB is presented in the upper part of Fig. 8. Two endotherms are obtained: the first one, at about 100°C, corresponding to the transition crystal-mesophase [15], and the final isotropization at 172°C. In conclusion, the presence of the central ether group has two effects: first, the mesophase of PDTMB is stable at the time scale of the DSC measurements (in fact, several days of annealing at any temperature between the glass transition and the isotropization point are necessary to observe any appreciable crystallinity); second, the annealed PDTMB samples show enantiotropic behavior and the mesophase is observed on heating, contrary to the case of P7MB, the all-methylene analogue. However, a comparison of the thermodynamic parameters of the transitions of PDTMB and P7MB (Table 1), shows that the parameters for the mesophase are very similar for both polymers, i.e., the substitution of the central methylene in the spacer of P7MB by an oxygen atom to give PDTMB has a very little effect on the properties of the mesophase [15,20].

The different phase behaviors are evidenced in the corresponding free energy diagrams, which have been estimated for both polymers [15]. These diagrams are shown in Fig. 10 (due to the different approximations used in the calculation of the free energy differences, these diagrams are only semiquantitative [15]). It can be seen that the monotropic transition of the crystal in

Table 1 Thermodynamic Parameters of the Transitions Between the Crystal (c), Smectic (s), and Isotropic (i), Phases of Several Polybibenzoates [13–15]

Polymer	T/°C				ΔH/kcal mol^{-1}			ΔS/cal mol^{-1}K^{-1}		
	c→i	s→i	i→s	s→c	c→i	s→i	s→c	c→i	s→i	s→c
P7MB	168	160	135	95	2.6	1.5	1.1	5.9	3.5	3.0
PDTMB		172	154			1.5			3.5	
P8MB	195	190	178	160	3.9	2.3	1.6	8.3	5.0	3.4
PTEB		114	85			0.9			2.3	

P7MB is characterized by the crossing of the curves corresponding to the crystal and the mesophase at a temperature higher than the isotropization point, contrary to the case of PDTMB, with enantiotropic behavior. Evidently, this behavior is expected to depend on molecular weight and crystal perfection, since these parameters may shift the free energy curves of the various phases in a different way [28]. For instance, linear dimeric bibenzoate compounds, with spacers varying from 4–9 methylene units, all present enantiotropic behavior [29], even though they show odd–even effects and smectic mesophases similar to the case of polybibenzoates.

Regarding the layer thickness, Fig. 11 shows the synchrotron profile of a freshly quenched sample of PDTMB (dotted line), corresponding to the pure mesophase. A thickness of 1.618 nm is deduced for the smectic layer of PDTMB, somewhat smaller than that for P7MB [26]. The solid curve in Fig. 11 corresponds to the sample of PDTMB annealed at 70°C, and shows a peak centered at 1.631 nm, i.e., now the crystal peak appears at a longer spacing than that for the mesophase, contrary to the case of P7MB. The x-ray diffraction photographs of a stretched PDTMB sample present a diffuse outer halo split into two intense maxima lying above and

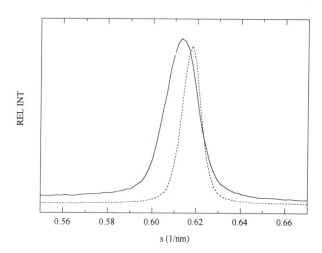

Figure 11 Intermediate-angle synchrotron profiles [26] of PDTMB samples: freshly quenched from the melt (dashed line) and annealed at 70°C for 24 days (continuous line).

below the equator [15] (both maxima form an angle of about 60°), similarly to the case of polybibenzoates with odd number of methylene groups in the spacer.

The phase behavior of several polybibenzoates with oxyethylene spacers has been also reported [14,21,22, 24,27]. The number of oxyethylene units in the spacer clearly affects the phase behavior in several ways. First, the isotropization temperature decreases greatly as the length os the spacer increases, changing from 192°C for poly(diethylene glycol p,p'-bibenzoate), PDEB, to only 114°C for poly(triethylene glycol p,p'-bibenzoate), PTEB. Moreover, in the case of poly(tetraethylene glycol p,p'-bibenzoate), PTTB, the mesophase is not obtained at all [21,24], and this polymer was found to be amorphous at room temperature.

Focusing attention on PTEB, it has been found that, similar to the case of PDTMB, the mesophase experiences a very slow transformation into the crystal. Thus, only the isotropization is observed in a sample freshly cooled from the melt [27]. However, after a long time at room temperature, the transformation mesophase-crystal is produced, owing to a glass transition temperature of about 14°C. Moreover, several endotherms were obtained before the final isotropization for a sample of PTEB annealed at 85°C for 12 days, i.e., PTEB shows enantiotropic behavior. The different endotherms may arise from polymorphism or melting-recrystallization phenomena [30].

The x-ray diffractograms of those three samples are shown in Fig. 12. The lower pattern corresponds to the quenched sample where only the mesophase is present (the layer line, appearing at lower angles is not shown). The other two diagrams corresponding to the annealed samples present several sharp diffraction peaks, which

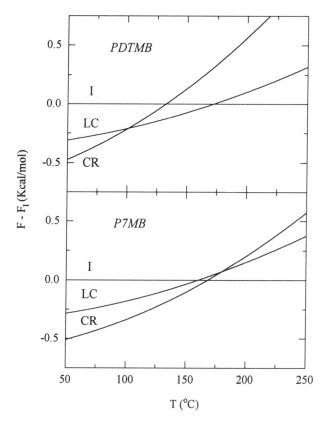

Figure 10 Free-energy diagrams [15] corresponding to PDTMB (upper) and P7MB (lower).

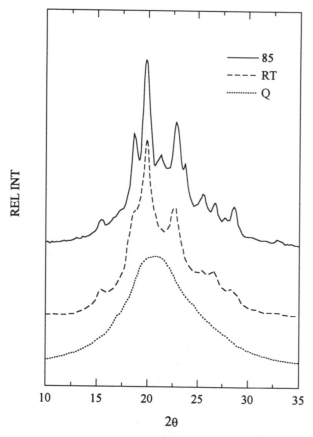

Figure 12 Ambient-temperature x-ray diffractograms [30] corresponding to three PTEB samples: freshly quenched from the melt (dotted line), annealed at room temperature for 12 months (dashed line), and annealed at 85°C during 12 days (continuous line). Noise has been suppressed.

leads to a considerable influence on the stability of the mesophase. Moreover, the type of mesophase is also different for PTEB. The x-ray diffraction photographs of the oriented polymer reveal a smectic C mesophase for this polymer [11,14], since the chain axes of the molecules are not perpendicular to the layer planes.

Solid-state ^{13}C NMR techniques have been applied to the characterization of the different phases of several polybibenzoates [25,30], including P7MB, PDTMB and PTEB. The last two polymers offer the advantage of the stability of the mesophase at room temperature. The spectra corresponding to the pure mesophase of these samples only exhibited a broad component, while the spectra of the annealed samples were separated into two components: crystal and noncrystal. The shapes of the mesophase and the noncrystal components are very similar, and only modest variations in the relaxation times were observed between these two components. The degree of crystallinity of these samples was determined

are characteristic of three-dimensional order. These samples have been also analyzed by small-angle x-ray diffraction, (SAXD), at room temperature in a synchrotron source [30]. The small angle scattering region is shown in Fig. 13. The quenched sample does not exhibit any defined long spacing (up to the experimental limit of 40 nm), supporting the idea of large coherence lengths in the liquid crystalline state. On the contrary, long spacings are clearly observed for the annealed samples, although the estimated crystal thicknesses are rather small [30]. The corresponding profiles in the region of the layer spacing show that the smectic layer peak (quenched sample) appears at 1.893 nm, while the annealed samples present diffraction peaks at very close spacings [30], similar to the cases mentioned previously for other polybibenzoates.

The thermodynamic parameters of the transitions in PTEB are compared in Table 1 with those of P8MB, the all-methylene homologue. It can be deduced from this table that the presence of two ether groups in the spacer

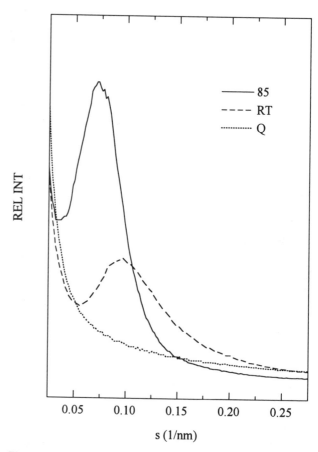

Figure 13 Small-angle synchrotron profiles [30] corresponding to three PTEB samples: freshly quenched from the melt (dotted line), annealed at room temperature for 12 months (dashed line), and annealed at 85°C during 12 days (continuous line).

from the apparent amounts of narrow and broad components corrected for $T_{1\rho}^H$ differences. These values are in reasonable agreement with those deduced from the x-ray diffractograms [25,30].

III. MECHANICAL PROPERTIES

The mechanical properties of polymers are of interest in all applications where they are used as structural materials. The analysis of the mechanical behavior involves the deformation of a material under the influence of applied forces, and the most important and characteristic mechanical property is the modulus. A modulus is the ratio between the applied stress and the corresponding deformation, the nature of the modulus depending on that of the deformation. Polymers are viscoelastic materials and the high frequencies of most adiabatic techniques do not allow equilibrium to be reached in viscoelastic materials. Therefore, values of moduli obtained by different techniques do not always agree in the literature.

The mechanical properties can be studied by stretching a polymer specimen at constant rate and monitoring the stress produced. The Young (elastic) modulus is determined from the initial linear portion of the stress–strain curve, and other mechanical parameters of interest include the yield and break stresses and the corresponding strain (draw ratio) values. Some of these parameters will be reported in the following paragraphs, referred to as results on thermotropic polybibenzoates with different spacers. The stress–strain plots were obtained at various drawing temperatures and rates.

A. Polybibenzoates with All-Methylene Spacers

Polybibenzoates with all-methylene spacers display a very fast mesophase-crystal transformation [10,13], and, consequently, their mechanical properties are akin to those of semicrystalline polymers. As a typical example, P7MB stretched at 23°C reaches very low draw ratios ($\lambda \le 1.02$). When the drawing temperature is 40°C the stretching process takes place through a neck. The intensity increases as the crosshead speed does. The draw ratio attained at this temperature is around 5 and the initial Young modulus (≈ 200 MPa) is higher than those obtained at 65 and 80°C. At these temperatures the P7MB specimens can be stretched to draw ratios 6 or 7 and the neck is less pronounced than those corresponding to the 40°C drawing process [31].

The homologous polybibenzoate with a central oxygen atom in the spacer, PDTMB, exhibits a liquid-crystal state at room temperature, owing to its low crystallization rate [15]. This fact is one of the main differences between P7MB and PDTMB, and is reflected on the mechanical properties of both polybibenzoates. The initial

Young moduli of PDTMB samples are much lower (≈ 20 MPa at 40°C) than those of P7MB, the necks are less pronounced, and the final draw ratios range from 9 (at 23 and 40°C) to 11.5 (at 80°C) [31].

These differences on the stress–strain behavior of P7MB and PDTMB show the marked influence of the mesomorphic state on the mechanical properties of a polymer. When increasing the drawing temperatures and simultaneously decreasing the strain rate, PDTMB exhibits a behavior nearly elastomeric with relatively low modulus and high draw ratios. On the contrary, P7MB displays the mechanical behavior typical of a semicrystalline polymer.

B. Polybibenzoates with Oxyethylene Spacers

A member of the polybibenzoate series containing oxyethylene spacers, PTEB, is well suited for studying the influence of the mesophase on the mechanical properties. A sample of this polymer obtained after melting at 150°C and quenching to room temperature (Q sample) only shows the glass transition (17–20°C) and an endotherm at 114°C, corresponding to the isotropization, T_i, of the mesophase [22]. This mesophase is stable at room temperature for several days, and, thus, PTEB-Q represents a liquid crystalline (LC) state. Stress–strain curves corresponding to room temperature drawing of this PTEB-Q sample (Fig. 14) show homogenous deformation and low-stress levels similar to an amorphous polymer [32].

A considerable change in the properties is observed when the sample is left at room temperature for a long time (PTEB-RT sample) because at this storing temperature (between T_g and T_i), the mesophase of PTEB changes to a crystalline state [33]. This transformation is very slow at any temperature and particularly at ambient conditions as the temperature is then only a few degrees above T_g. When the samples stored at room temperature are stretched at this temperature, both the yield stress and the Young modulus increase and the drawing process takes place through necking even though the stretching temperature is higher than T_g (Fig. 14). In general, a slight increase in the values of different parameters (yield stress, modulus, strain hardening) with the drawing rate is observed for this sample. The reason is the presence of a small degree of crystallinity in these specimens as revealed by the x-ray diffraction pattern (Fig. 12).

The influence of the annealing process at a higher temperature was also studied, and important changes in the properties of a sample annealed at 55°C during 5 days (PTEB-CR sample) were observed [33]. The PTEB-CR specimens were stretched at 1 and 10 cm/min, but the behavior was similar in both cases. Typical stress–strain plots are shown in Fig. 14 for samples stretched at 23°C

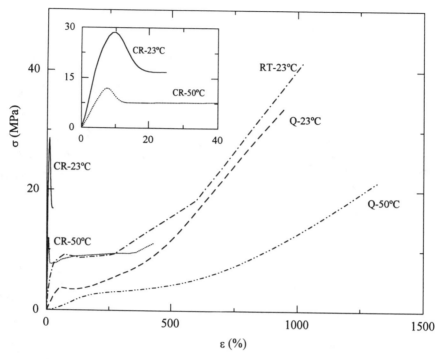

Figure 14 Stress–strain plots of several PTEB specimens, stretched at 1 cm/min and two drawing temperatures. The inset shows the two CR specimens at low deformations.

and 50°C. DSC and x-ray results indicate the presence of crystallinity in the samples, which has a clear effect on the mechanical properties. Thus, the Young modulus increases considerably but the break strain is small ($\lambda_B = 1.20$) at room temperature, similar to semicrystalline polymers. When raising the temperature of the drawing process it is possible to increase λ_B up to 6, but the values of the modulus decrease, although they remain higher than those of the LC samples (PTEB-Q).

A common feature of the three PTEB samples is that the yield stress decreases as the drawing temperature increases (Table 2), whereas it does not change significantly with the strain rate. The Young modulus does not change with the strain rate but it decreases and the break strain increases as the drawing temperature increases. The main conclusion is that the behavior of PTEB-RT is intermediate between the other two samples, with the advantage of a considerable increase in the modulus in relation to sample PTEB-Q and without much decrease in the break strain (Table 2).

IV. DYNAMIC MECHANICAL PROPERTIES

Viscoelastic phenomena always involve the change of properties with time and, therefore, the measurements of viscoelastic properties of solid polymers may be called dynamic mechanical. Dynamic mechanical thermal analysis (DMTA) is a very useful tool for studying

the motional processes taking place in the macromolecular chain along a wide interval of temperatures and a short range of frequencies, this latter limitation due to instrumentation drawbacks.

DMTA is based on the response of a viscoelastic material to an oscillatory excitation. Considering a fixed frequency, while the temperature is changed, there will

Table 2 Average Values[a] of the Modulus, Yield Stress, Yield Strain, and Strain at Break for Three Samples of PTEB Stretched at Different Temperatures and Deformation Rates

Sample	T_d (°C)	v_d (cm/min)	E (MPa)	σ_Y (MPa)	ϵ_Y (%)	ϵ_B (%)
PTEB-Q	23	1	9	2.8	31	1200
		10	9	3.2	36	1150
	50	1	2	—	—	1100
		10	4	—	—	1350
PTEB-RT	23	1	45	9	21	1050
		10	60	11	45	950
PTEB-55	23	1	470	28	12	30
		10	430	25	9	20
	50	1	250	14	9	450
		10	150	12	12	450

[a] Average of three tests.

be a temperature at which a motion will occur. This motion, involving a group, a chain segment, or the entire macromolecule, will produce an energy transfer from the oscillatory force to the moiety in motion that will be evident in a loss maximum and the corresponding drop (relaxation) in the storage modulus. The ratio between the loss and storage parts of the complex modulus, either elastic (E) or shear (G), is the tangent of the phase lag or loss tangent, tanδ, related to the mechanical damping, Λ, according to the expression:

$$\frac{\Lambda}{\pi} \sim \tan\delta = \frac{E''}{E'} = \frac{G''}{G'}$$

The relaxation processes defined in the preceding paragraph are also a function of the frequency of measurement. The narrow range of frequencies (around four decades) of DMTA instrumentation is sufficient for studying the shifting of damping maxima to higher temperatures with increasing frequencies. The secondary relaxations obey the Arrhenius law for frequency-temperature shift:

$$\ln f = \ln f_0 \exp \frac{-\Delta H}{RT}$$

where ΔH is the apparent activation energy of the motional process. On the other hand, the main relaxation (glass transition) is ruled by the Williams, Landel, and Ferry relationship.

In DMTA a sinusoidal strain (or stress) of known frequency is applied to the sample and the resulting stress (or strain) together with the phase lag between excitation and response are recorded. Depending on the geometry and thermal behavior of the sample studied different clamps can be used (shear, tensile cantilever, bending). The most accurate type of measurement in DMTA is to heat the sample in a temperature scan at different frequencies, using a heating rate compatible with the number and absolute values of the selected frequencies. With this type of scan it is possible to obtain in a few hours a complete picture of the relaxation motions of a macromolecular chain, together with the activation energies of the movements.

An alternative method of studying the molecular motions of a polymeric chain is to measure the complex permitivity of the sample, mounted as dielectric of a capacitor and subjected to a sinusoidal voltage, which produces polarization of the sample macromolecules. The storage and loss factor of the complex permitivity are related to the dipolar orientations and the corresponding motional processes. The application of the dielectric thermal analysis (DETA) is obviously limited to macromolecules possessing heteroatomic dipoles but, on the other hand, it allows a range of frequency measurement much wider than DMTA and its theoretical foundations are better established.

In the following section the results on dynamic mechanical relaxation of various polybibenzoate series are summarized. In general, all the polybibenzoates display three dynamic mechanical relaxations, called α, β, and γ in order of decreasing temperature.

A. Polybibenzoates with All-Methylene Spacers

Among the members of this polybibenzoate series we will report the results of the polymers containing spacers with seven [34] and eight [13] methylenic units, P7MB and P8MB, respectively.

The variations of storage modulus and loss tangent as a function of temperature are plotted in Fig. 15, where it can be seen that the α relaxation temperature increases with the length of the spacer (46°C to 52°C in passing from P7MB to P8MB). In principle, longer spacers should lead to lower temperatures, but the odd–even effect also has to be considered (see upper part of Fig. 5). The α relaxation is considered the glass transition of the two polybibenzoates owing to the high tanδ maximum value of the relaxation and the corresponding sharp

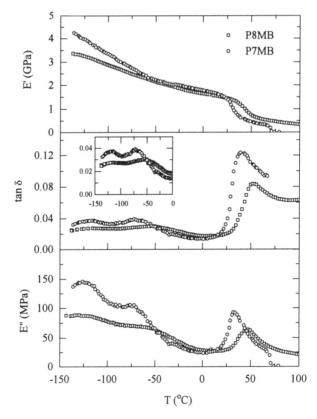

Figure 15 Storage modulus, (E'), loss tangent (tanδ), and loss modulus, (E''), as a function of temperature for P7MB and P8MB at 3 Hz.

decrease of the storage modulus. The apparent activation energy of the process is higher than 400 kJ mol^{-1} in both polymers; a result that confirms the assignation of the α relaxation as the glass transition of these polybibenzoates.

The β relaxation is very broad for the two polybibenzoates (Fig. 15) and appears at around $-60°C$. It has a complex origin, is characteristic of polyesters, and originates from movements of phenyl and carboxyl groups. The very broad peaks observed in Fig. 15 are the consequence of this complex character.

The γ relaxation takes place at the lowest temperature, overlaps with the β relaxation (Fig. 15), and coincides in location and activation energy with the typical γ relaxation of polyethylene [35,36], and also of polyethers [37], and polyesters [38] with three or more consecutive methylene units. It appears, for 3 Hz and tanδ basis, at $-120°C$ (P7MB) and $-126°C$ (P8MB), and its location and activation energy (35–45 kJ mol^{-1}) agree with the values of a similar relaxation associated with kink motions of polymethylenic sequences.

The effect on dynamic mechanical results of the introduction of a central oxygen atom in the spacer has been also analyzed by comparing the behavior of P7MB and the homologous PDTMB [34]. The α relaxation of PDTMB takes place at 20°C. The decrease of 26°C with respect to the α relaxation temperature of P7MB is attributed to the increased flexibility of the PDTMB chain, which lowers the glass transition temperature of this polybibenzoate containing an oxygen atom in the spacer. The other two relaxations of PDTMB, β and γ, do not change significantly when varying the chemical structure of the spacer. In the case of the γ relaxation, the practical constancy of the location, intensity, and activation energy of the relaxation suggest that the same type of motion is responsible for the γ relaxation of either all-methylene or oxymethylene chains.

B. Polybibenzoates with Oxyethylene Spacers

Although the liquid crystalline phase of most polybibenzoates usually undergoes a rapid transformation into a three-dimensional crystal, the introduction of oxygen atoms in the spacer of polybibenzoates has been used to prevent or to slow down this transformation. The dynamic mechanical behavior of polybibenzoates with 2, 3, or 4 oxyethylene groups in the spacer (PDEB, PTEB, and PTTB, respectively) is determined by the composition of the spacer [24], as discussed in this section.

Considering the α relaxation, the modulus drop occurring between the glassy and rubbery states ranges from two to three decades, and it increases as the length of the spacer does. Figure 16 shows that the glass–rubber relaxation shifts toward lower temperatures for longer spacers. Moreover, the intensity of the α peak,

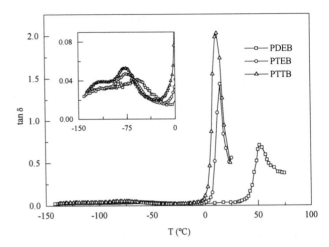

Figure 16 Loss tangent (tanδ) at 3 Hz, as a function of temperature for a series of main chain thermotropic polyesters having oxyethylene spacers.

considering tanδ values, increases with the length of the spacer.

The three polymers studied display remarkable physical aging effects [24], with a strong increase of modulus and changes in the location and intensity of the glass transition. This phenomenon is exemplified in Fig. 17 where the changes of dynamic moduli in a PDEB sample aged for 14 months are very apparent.

The β relaxation in the members of this polybibenzoate series takes place in the temperature interval around $-70°C$ and its temperature location decreases as the length of the spacer increases. This result seems to confirm the complex character of the β relaxation and the possibility of different steric hindrances of the reorganizational motions of the carboxyl and phenylene groups depending on the spacer length, because the activation energy of this relaxation increases slightly with the length of the spacer (and is smaller in the case of the amorphous polymer, PTTB).

The relaxation at the lowest temperature (γ relaxation) takes place below $-100°C$. The two polymers with shorter spacers (PDEB and PTEB) show weak relaxations overlapped with the β ones due to the low tanδ values (0.03 and 0.04, respectively). Notwithstanding this, the γ relaxation is clearly distinguished when using loss modulus plots, even in the case of PDEB, that shows the weakest maximum (see Fig. 16). For PTTB, tanδ values in the γ relaxation interval are of the order of 0.05.

It is usually considered that the γ relaxation arises from crankshaft and kink movements of polymethylenic sequences, but the clear maximum of tanδ and loss modulus for the three polybibenzoates here reported leads to the conclusion that the motion responsible of this relaxation also takes place when one of the methylenic

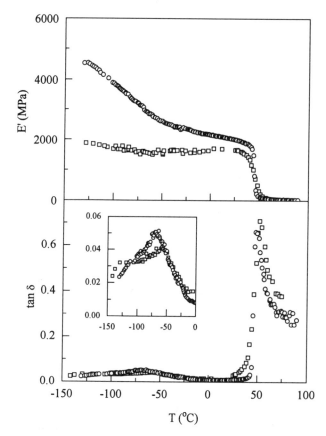

Figure 17 Variation of the storage modulus (E') and loss tangent (tanδ) at 3 Hz, for two PDEB specimens: □ freshly quenched, and ○ aged for 14 months.

(PTEB-Q) to the annealed ones, owing to the presence of the crystalline phase. Moreover, the temperature of the peak increases with the annealing, as well as the broadness of the relaxation. These results suggest that the liquid crystalline phase gives raise to an α relaxation similar to that of amorphous polymers despite the existence of the two-dimensional order characteristic of smectic mesophases, and it changes following the same trend than that of semicrystalline polymers.

The annealing process imposed to the samples increases the value of the storage modulus, which always shows a profound decrease in the vicinity of the glass transition (Fig. 18). This strong decrease is less marked in the annealed samples because of the presence of three-dimensional order. In addition to this general result, there are some meaningful differences among the annealed samples when compared with the freshly quenched one. Thus, the sample annealed at room temperature for several months can be considered as physically aged, while the sample annealed at 85°C for 12 days shows a higher three-dimensional crystallinity. Both the aging and the crystallization processes lead to an in-

units is substituted by an oxygen atom. The consideration of this relaxation as a γ one is further confirmed by its activation energy, which amounts to ~30 kJ mol^{-1}, i.e., the common value of the activation energy of the typical γ relaxation. In conclusion, we can state that the oxygen atom can cooperate in the crankshaft and kink movements responsible for the γ relaxation when it substitutes for a methylenic unit in the polymethylene chain [27,39].

Among these three polybibenzoates, PTEB has a smectic mesophase stable during several days at any temperature below its isotropization point, although the transformation into a three-dimensional crystal can be attained by annealing at the appropriate temperatures, thus making it possible to analyze the effect of the thermal history on the dynamic mechanical relaxations of PTEB [27].

The influence of the thermal history on the location and intensity of the α peak of PTEB can be observed in Fig. 18. It can be seen that the intensity of this relaxation measured on both E'' and tanδ bases decreases considerably on passing from the liquid crystalline sample

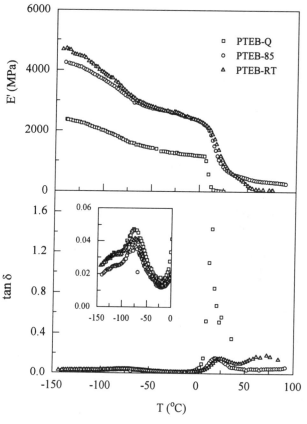

Figure 18 Storage modulus (E') and loss tangent (tanδ) as a function of temperature, at 3 Hz, for the indicated samples.

crease of the modulus at temperatures below T_g, as occurs in the present results.

On the contrary, the phase structure and the thermal history do not have important effects on the location and intensity of the β relaxation. This relaxation is very broad in all the samples and overlaps the γ relaxation. The activation energy of the β peak is about 85 kJ mol^{-1} for the three samples, of the same order of magnitude as that of other polyesters [38,40]. Finally, the γ relaxation is found in the three samples of PTEB with no remarkable influence of the thermal history.

In conclusion, the different thermal histories imposed to PTEB have a minor effect on the β and γ relaxations, while the α transition is greatly dependent on the annealing of the samples, being considerably more intense and narrower for the specimen freshly quenched from the melt, which exhibits only a liquid crystalline order. The increase of the storage modulus produced by the aging process confirms the dynamic mechanical results obtained for PDEB [24], a polyester of the same series, as well as the micro-hardness increase [22] (a direct consequence of the modulus rise) with the aging time.

C. Copolyesters Derived from Bibenzoic Acid

The presence of three oxyethylene units in the spacer of PTEB slows down the crystallization from the mesophase, which is a very rapid process in the analogous polybibenzoate with an all-methylene spacer, P8MB [13]. Other effects of the presence of ether groups in the spacer are the change from a monotropic behavior in P8MB to an enantiotropic one in PTEB, as well as the reduction in the glass transition temperature. This rather interesting behavior led us to perform a detailed study of the dynamic mechanical properties of copolymers of these two polybibenzoates [41].

The reduction in T_g (α relaxation temperature) is accompanied by a marked increase (1500%) of the intensity for the α relaxation measured on a tanδ basis as shown in the middle part of Fig. 19. Although a parallel sharpening of the α peak is obtained as the oxyethylene content increases (Fig. 19), the total area under this peak also experiences a very important increase on passing from P8MB to PTEB (it is about ten times higher for PTEB, on the tanδ plots). Therefore, the intensity of the α relaxation seems to be dependent on the proportion of oxygen atoms in the spacer, confirming the previous results on polybibenzoates with oxyethylene spacers [24]. In that work, the tanδ maximum for the α peak was found to be as high as 1.8 for the polymer with four oxyethylene units in the spacer (PTTB). Moreover, the fall of the storage modulus at the glass transition is also dependent on the oxyethylene content as shown in the upper part of Fig. 19.

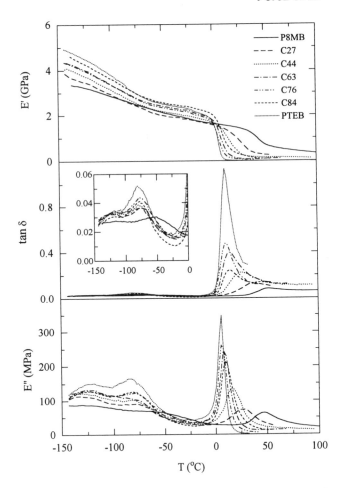

Figure 19 Storage modulus (E'), loss tangent (tanδ), and loss modulus (E''), as a function of temperature, at 3 Hz, for the indicated samples.

Figure 19 shows that the β relaxation increases in intensity with the oxyethylene content. The maximum peak of the β relaxation for P8MB appears at around $-55°C$, although the curve is broad, indicating a complex relaxation. In copolyesters, the intensity increases with the content in oxyethylene units and the peak is sharper, but the location of the maximum does not appreciably change with composition (Table 3). This table also lists the values of the apparent activation energy, showing a small decrease in the value when increasing the oxyethylene content.

The γ relaxation appears around $-120°C$ (Fig. 19). PTEB and the copolyesters show the γ maximum at a temperature that is practically constant, slightly lower than the one for P8MB. The low activation energy values (Table 3) are the usual ones for this relaxation.

The results for the γ relaxation, compared with those of the β relaxation, give a deeper insight about the influence of the spacer structure on the viscoelastic

Table 3 Temperature Location (tanδ basis, 3 Hz) and Activation Energies of the β and γ Relaxations for Different Samples with Varying Content in Oxyethylene Units (f_{TEG})

Sample	f_{TEG}	T_β (°C)	ΔH_β (kJ/mol)	$T\gamma$ (°C)	ΔH_γ (kJ/mol)
P8MB	0	−55	138	−116	41
C27	27	−76	124	−120	36
C44	44	−76	90	−124	33
C63	63	−77	86	−123	35
C76	76	−78	104	−123	36
C84	84	−79	84	−122	49
PTEB	100	−78	94	−122	42

properties. The substitution of methylenes by ether groups produces a significant increase of the intensity of the β relaxation, whereas the intensity increase of the γ relaxation is relatively small. These results hold for all pairs of polybibenzoates with the same number of groups (methylenes and/or oxygens) in the spacer [11,27,34]. Owing to these concurring features, the ratio of intensities of the β and γ relaxations shows a marked increase with the oxyethylene content for the samples studied here.

ACKNOWLEDGMENTS

The financial support of the CICYT (project no. IN89-0066), DGICYT (project no. PB94-1529), and Consejería de Educación de la Comunidad de Madrid is gratefully thanked. We also acknowledge the support of NATO (grant CRG 920094) and the assistance of Dr. W. Bras and the Daresbury Laboratory (UK) in the synchrotron experiments.

REFERENCES

1. H. Finkelmann and G. Rehage, *Adv. Polym. Sci. 60/61*: 99 (1984).
2. C. K. Ober, J. Jin, Q. Fhou, and R. W. Lenz, *Adv. Polym. Sci. 59*: 103 (1984).
3. C. Nöel, *Makromol. Chem., Macromol. Symp. 22*: 95 (1988).
4. P. Meurisse, C. Nöel, L. Monnerie, and B. Fayolle, *Br. Polym. J. 13*: 55 (1981).
5. W. R. Krigbaum, J. Asrar, H. Toriumi, A. Ciferri, and J. Preston, *J. Polym. Sci. Poly. Lett. Ed. 20*: 109 (1982).
6. W. R. Krigbaum and J. Watanabe, *Polymer 24*: 1299 (1983).
7. J. Watanabe and M. Hayashi, *Macromolecules 21*: 278 (1988).
8. J. Watanabe and M. Hayashi, *Macromolecules 22*: 4083 (1989).
9. W. J. Jackson and J. C. Morris, *ACS Symp. Ser. 435*: 16 (1990).
10. E. Pérez, A. Bello, M. M. Marugán, and J. M. Pereña, *Polymer Commun. 31*: 386 (1990).
11. A. Bello, J. M. Pereña, E. Pérez, and R. Benavente, *Macromol. Symp. 84*: 297 (1994).
12. E. Pérez, A. Bello, J. M. Pereña, R. Benavente, M. M. Marugán, B. Peña, Z. Zhen, M. L. Cerrada, and J. Pérez-Manzano, *Daresbury Annual Report* 113 (1993).
13. E. Pérez, Z. Zhen, A. Bello, R. Benavente, and J. M. Pereña, *Polymer 35*: 4794 (1994).
14. E. Pérez, E. Riande, A. Bello, R. Benavente, and J. M. Pereña, *Macromolecules 25*: 605 (1992).
15. A. Bello, E. Riande, E. Pérez, M. M. Marugán, and J. M. Pereña, *Macromolecules 26*: 1072 (1993).
16. A. Abe, *Macromolecules 17*: 2280 (1984).
17. J. Watanabe and S. Kinoshita, *J. Phys. II 2*: 1237 (1992).
18. J. Watanabe, M. Hayashi, S. Kinoshita, and T. Niori, *Polymer J. 24*: 597 (1992).
19. A. J. B. Loman, L. Van der Does, A. Bantjes, and I. Vulic, *J. Polym. Sci. Part A: Polym. Chem. 33*: 493 (1995).
20. A. Bello, E. Pérez, M. M. Marugán, and J. M. Pereña, *Macromolecules 23*: 905 (1990).
21. E. Pérez, R. Benavente, M. M. Marugán, A. Bello, and J. M. Pereña, *Polymer Bull. 25*: 413 (1991).
22. E. Pérez, J. M. Pereña, R. Benavente, A. Bello, and V. Lorenzo, *Polymer Bull. 29*: 233 (1992).
23. M. M. Marugán, E. Pérez, R. Benavente, A. Bello, and J. M. Pereña, *Eur. Polym. J. 28*: 1159 (1992).
24. R. Benavente, J. M. Pereña, E. Pérez, and A. Bello, *Polymer 34*: 2344 (1993).
25. E. Pérez, M. M. Marugán, and D. L. VanderHart, *Macromolecules 26*: 5852 (1993).
26. E. Pérez, M. M. Marugán, A. Bello, and J. M. Pereña, *Polymer Bull. 32*: 319 (1994).
27. R. Benavente, J. M. Pereña, E. Pérez, A. Bello, and V. Lorenzo, *Polymer 35*: 3686 (1994).
28. A. Keller, G. Ungar, and V. Percec, *ACS Symp. Ser. 435*: 308 (1990).
29. J. Watanabe, H. Komura, and T. Niiori, *Liq. Cryst. 13*: 455 (1993).
30. E. Pérez, R. Benavente, A. Bello, J. M. Pereña, and D. L. VanderHart, *Macromolecules 28*: 6211 (1995).
31. M. M. Marugán, J. M. Pereña, R. Benavente, E. Pérez, and A. Bello, *Progress and Trends in Rheology IV*, Steinkopff, Darmstadt, p. 332 (1994).
32. R. Benavente and J. M. Pereña, *Polym. Eng. Sci. 27*: 913 (1987).
33. R. Benavente, J. M. Pereña, A. Bello, and E. Pérez, *Polymer Bull. 34*: 635 (1995).
34. J. M. Pereña, M. M. Marugán, A. Bello, and E. Pérez, *J. Non-Cryst. Solids 131–133*: 891 (1991).
35. R. Benavente, J. M. Pereña, A. Bello, and E. Pérez, *British Polym. J. 23*: 95 (1990).
36. R. Benavente, J. M. Pereña, A. Bello, E. Pérez, C. Aguilar, and M. C. Martínez, *J. Mater. Sci. 25*: 4162 (1990).
37. J. M. Pereña and C. Marco, *Makromol. Chem. 181*: 1525 (1980).
38. C. C. González, J. M. Pereña, and A. Bello, *J. Polym. Sci., Polym. Phys. Ed. 26*: 1397 (1988).
39. R. Benavente, J. M. Pereña, E. Pérez, and A. Bello, *Macromol. Reports A31*: 953 (1994).
40. R. Benavente and J. M. Pereña, *Makromol. Chem. 189*: 1207 (1988).
41. R. Benavente, Z. Zhu, J. M. Pereña, A. Bello, and E. Pérez, *Polymer 37*: 2379 (1996).

26

Polymeric Ultraviolet Stabilizers for Thermoplastics

Jayant S. Parmar
Sardar Patel University, Gujarat, India

Raj P. Singh
National Chemical Laboratory, Pune, India

I. INTRODUCTION

There is a great interest at present in the photo-oxidative thermal degradation of polymeric materials because macromolecules have increasingly widespread commercial applications. All commercial organic polymers degrade in air when exposed to sunlight as the energy of sunlight is sufficient to cause the breakdown of polymeric bonds, moreover, degradation is due to the action of short-wavelength ultraviolet (UV) rays ($\lambda > 290$ nm) present in the solar spectrum. As a consequence of degradation, the resulting smaller fragments do not contribute effectively to the mechanical properties and the article becomes brittle. Thus, the life of thermoplastics for outdoor applications becomes limited due to weathering [1–5].

Weathering in a tropical climate causes polyethylene containers to crack, polypropylene (PP) ropes to rupture, and ABS telephones to fail. Polystyrene (PS) and cellulose acetate films used as packing materials also fail due to weathering.

II. STABILIZATION OF POLYMERS

The degradation and stabilization of polymers are very important problems from the scientific and industrial points of view. A better understanding of their mechanism is necessary to achieve better stabilization. A small amount of compounds, called stabilizers, are added into the polymer matrix to retard degradation and to impart long-term outdoor stability to polymeric systems. The photostabilization of polymers may be achieved in many ways. The following stabilizing systems have been developed, which depend on the action of stabilizer.

A. Light Screener

The light screeners are interposed as a shield between the radiation and the polymer. The light screener functions either (1) by absorbing the damaging radiation before it reaches the photoactive chromophoric species in the polymer, or (2) by limiting the damaging radiation penetration into the polymer matrix. The paints, coatings, and pigments show the screening activity and tend to stabilize the polymer. Generally, the pigments are used in dispersed form within the polymer bulk.

Schonhorn and Luongo [6] assumed that due to its low surface energy the pigment shows protective activity. The pigments also quench certain photoactive species in the polymer [7]. The carbon black is commonly used as a pigment, because it is the most effective light screen [8], especially at high temperatures. Several theories have been advanced to explain its technically important behavior in the polymer. The carbon black contains (1) a fair number of quinone groups, (2) phenolic groups, and (3) an occasional carboxylic group [9]. The quinone and the polynuclear aromatic structure would act as traps for free radicals that are produced during photo-oxidative thermal degradation and may form stabilized free radicals, which can terminate the kinetic chain. The phenolic groups usually act as antioxidants.

B. UV Absorber

UV absorbers have been found to be quite effective for stabilization of polymers and are very much in demand. They function by the absorption and harmless dissipation of the sunlight or UV-rich artificial radiation, which would have otherwise initiated degradation of a polymer material. Meyer and Geurhart reported, for the first time in 1945 [10], the use of UV absorber in a polymer. They found that the outdoor life of cellulose acetate film was greatly prolonged by adding phenyl salicylate (salol) [10]. After that, resorcinol monobenzoate, a much more effective absorber, was introduced in 1951 [11] for stabilization of PP, but salol continued to be the only important commercial stabilizer for several years. The 2,4-dihydroxybenzophenone was marketed in 1953, followed shortly by 2-hydroxy-4-methoxybenzophenone and other derivatives. Of the more commonly known UV absorbers, the 2-hydroxybenzophenones, 2-hydroxy-phenyl-triazines, derivatives of phenol salicylates, its metal chelates, and hindered amine light stabilizers (HALS) are widely used in the polymer industry.

C. Excited-State Quencher

Although for many years chain termination, UV absorption, and peroxide decomposition were the traditional methods for photo-, thermal, and radiation stabilization, metal chelates, and quenchers also are in current use. It has been reported that metal chelates with a variety of ligands are excellent quenchers for excited states [12,13]. In the solid state, transfer of energy occurs by resonance or dipole–dipole interactions. The photo-oxidative degradation is promoted by the electronically excited oxygen molecule (commonly called singlet oxygen $^1O_2$) formed in polymers. The $^1O_2$ can also be generated in polymers by energy transfer from electronically excited carbonyl group to dissolved (chemisorbed) molecular oxygen [7]. The quenching of $^1O_2$ is also necessary for effective stabilization, and nickel-chelates [14–16] have proved to be effective quenchers for excited states of $^1O_2$. The quenching may occur in one of the two ways.

1. Energy Transfer

The energy transfer may result in a reactive or nonreactive quencher molecule:

$$^1O_2^* + S \rightarrow O_2 + S^*$$

$$S^* \rightarrow S + h\nu$$

$$^1O_2^* + S \rightarrow SO_2 + h\nu$$

$$RH^* + S \rightarrow [RH—S]^* \rightarrow \text{photophysical processes}$$

where S^* is an excited singlet or triplet state of the stabilizer (quencher). The quencher must be capable of dissipating most of the accepted energy harmlessly, i.e.,

without destroying the polymer bonds or its own structure.

The various light-emission processes are best described with reference to the Jablonski diagram. The diagram for a simple carbonyl compound is given in Fig. 1. The absorption process leads to the formation of excited singlet states (S_1, S_2, etc.). A very rapid depopulation of the upper excited states occurs by internal conversion through vibrational relaxation processes. After deactivation to the first excited singlet state (S_1), several processes are possible for the molecule to reduce its excess energy. It may react chemically, resulting in the formation of a new chemical species or a free radical. The first excited singlet state may lose its excess energy by emitting a photon of light, and this emission is known as fluorescence. It is also possible for the first excited singlet state to lose its energy by some nonradiative process. Thus, internal conversion to the ground state may occur or, alternatively, since overlap of higher vibrational levels is possible, a nonradiative transition from S_1 to T_1 may occur by electron-spin reorientation. This transition from a singlet to a triplet state of a molecule is termed intersystem crossing.

The triplet molecule loses any excess vibrational energy to its surroundings and goes to the lowest vibrational level of the T_1 state. This molecule also loses its excess energy through chemical reaction or by a nonradiative process of intersystem crossing or by internal conversion. However, if the excited molecule in T_1 state has not returned to the ground state, it will emit a photon and return to some vibrational level of the ground state. This emission process ($T_1 \rightarrow S_0$) is known as phosphorescence and occurs at longer wavelengths than those at which fluorescence occurs.

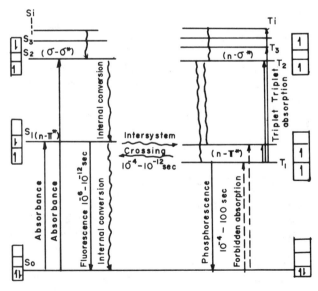

Figure 1 Energy level diagram of a carbonyl compound.

2. Formation of the Excited-State Complex

The most reasonable explanation for the quenching mechanism is the exciplex formation between excited chromophore (carbonyl or ketone) and the ground state of the quencher. The mechanism involves energy transfer from both the singlet and triplet excited states to the ground state of the quencher. Coupling of the excited carbonyl electronic energy and the quencher vibrational energy results in the "exciplex" intermediate [17] and leads to subsequent deactivation of the excited-state species to yield the observed products. These emissions occur from excimer and exciplex species [18,19]. The mechanism is illustrated in the following scheme below.

The excimer emission occurs from an excited associated complex (D*) formed between a species in the excited singlet state (S^*) and a similar ground-state (S_0) species. The excimer is also called a dimer and is short-lived.

$$S^* + S_0 \rightarrow (D^*) \rightarrow S + S + h\nu$$

This emission occurs at longer wavelengths than the normal fluorescence. The exciplex emission, on the other hand, occurs from an excited associated complex formed between an excited species and a different ground-state species.

$$S^* + B \rightarrow (SB)^* \rightarrow S + B + h\nu$$

The nature of the light emissions is influenced by the way in which the absorbed energy is transferred through the polymer matrix. In crystalline polymers, "exciton migration" is possible as all molecules lose their energetic individuality and all electronic and oscillation levels are coupled [20]. Thus, new "exciton absorption and emission bands" are formed and the excitation energy can move along the chain:

$$-A^*-A-A-A- \rightarrow -A-A-A-A^*-$$

The impurities may capture this migrating exciton and lose its excess energy. The mutual annihilation of two or more 'triplet excitons' occurs in the same polymer chain and delayed fluorescence is observed.

In addition to the above mechanism, the metal chelates can influence the process of quenching through electron-transfer:

$$ROO^{\cdot} + (\text{metal chelate}) \rightarrow ROO^- + (\text{metal chelate})^{+\cdot}$$

and formation of charge-transfer exciplex intermediate [21]:

$$[ROO^{\cdot} + (\text{metal chelate})]$$
$$\rightarrow [ROO^- —(\text{metal chelate})^+] \rightarrow \text{nonreactive products}$$

Thus, chelates are able to dissipate the radiation harmlessly as infrared radiations or heat through resonating structures. Carlsson and Wiles [22] have confirmed in their studies that the quencher slowly migrate through the solid polymer destroying the hydroperoxide group.

D. Free-Radical Scavenger

Since radicals are mostly responsible for the degradation reactions, scavenging of radicals is effective for protecting polymers against photo-, thermal, and radiation degradation. The mercaptans [23] and aromatic amines [24] are the most common class of radical scavengers. The scavenging by such compounds proceed through the donation of a hydrogen atom to radical sites in the polymer.

$$R^{\cdot} \text{ (or } ROO^{\cdot}) + SH \rightarrow RH \text{ (or } ROOH) + S^{\cdot}$$

Mercaptans first act as electron acceptors and then transfer the negative charge to the H atom.

Any substance capable of reacting with free radicals to form products that do not reinitiate the oxidation reaction could be considered to function as free-radical traps. The quinones are known to scavenge alkyl free radicals. Many polynuclear hydrocarbons show activity as inhibitors of oxidation and are thought to function by trapping free radicals [25]. Addition of $R^{\cdot}$ to quinone or to a polynuclear compound on either the oxygen or nitrogen atoms produces adduct radicals that can undergo subsequent dimerization, disproportionation, or reaction with a second $R^{\cdot}$ to form stable products.

E. Peroxide Decomposer

The salts of alkyl xanthates, N,N'-di-substituted dithiocarbamates and dialkyldithiophosphates [26] are effective peroxide decomposers. Since no active hydrogen is present in these compounds, an electron-transfer mechanism was suggested. The peroxide radical is capable of abstracting an electron from the electron-rich sulfur atom and is converted into a peroxy anion as illustrated below for zinc dialkyl dithiocarbamate [27]:

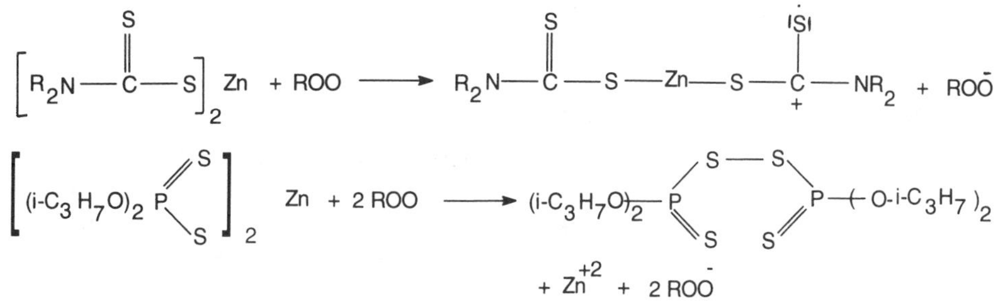

F. Combined Effect/Synergism

The combined effect of screeners, quenchers, UV absorbers, and stabilizers, which are synergistic toward one another, can provide protection against degradation in two possible ways, namely (1) the synergistic effects and (2) the antagonistic effects. The synergism is the cooperative action in which the total effect is greater than the sum of the several individual effects taken independently [28,29]. The opposite of synergism is termed as antagonism.

III. POLYMERIC UV ABSORBERS

Additives such as antioxidants and photostabilizers of low-molecular weight face two major problems (1) they may evaporate during high temperature moulding and extrusion process or (2) they may migrate to the surface of the plastic and get extracted. There are, in general, three ways of overcoming these problems.

1. By attaching the UV absorber moiety chemically to the polymer backbone [30] or by introducing as one of the comonomer:

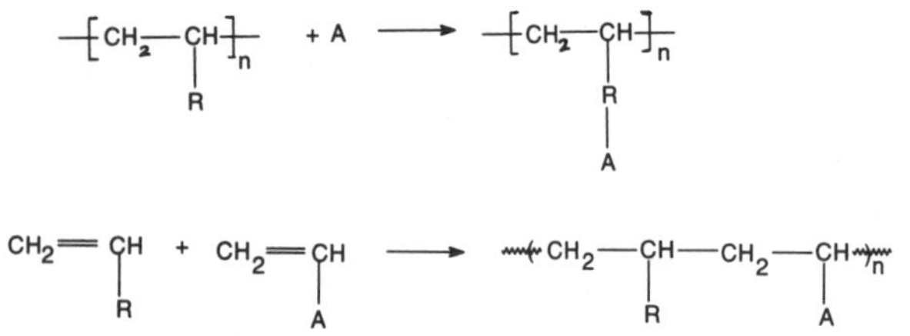

Where A is UV absorbing moeity.

2. By preparing polymeric UV absorbers and then blending them in proper proportions with commercial plastics:

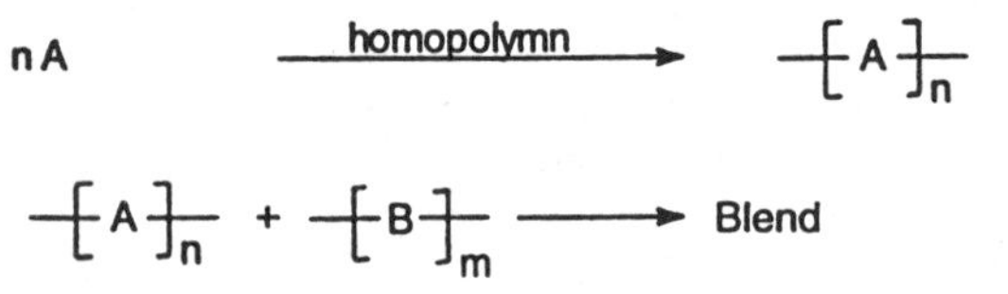

3. By using photorearranging polymers as additives [31].

The polymeric UV absorbers have potential use as stabilizers for fibers, films, and coatings. Until now several monomeric stabilizers containing a vinyl group are homopolymerized and used as stabilizers for PE, PVC, acrylates, polystyrene, cellulose acetate [32–34] and several vinyl polymers. 2-Hydroxybenzophenones and its 4-substituted derivatives, which are very effective photostabilizers, can be transformed to polymeric stabilizers of type 2 by reacting with formaldehyde in acidic condition [35,36]. Accordingly 2,4-dihydroxybenzophenone-formaldehyde resin [DHBF], 2-hydroxy-4-methoxybenzophenone-formaldehyde resin [HMBF], 2-hydroxy-4-butoxybenzophenone-formaldehyde resin [HBBF], and 2-hydroxy-4-octyloxybenzophenone-formaldehyde resin [HOBF] are prepared as shown below [37–39].

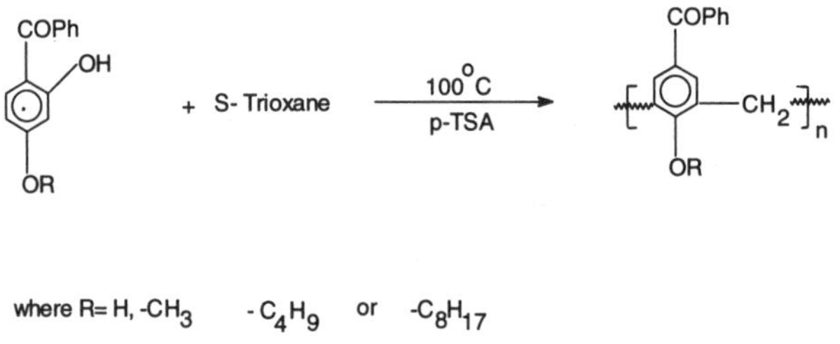

where R= H, -CH$_3$ -C$_4$H$_9$ or   -C$_8$H$_{17}$

The properties of these compounds are as shown in Table 1.

Table 1 Physical Properties of Benzophenone-Formaldehyde Resins

Resin	Color	Mol. wt. (Mn)	Thermal stability		T$_{max}$ (°C)
			T$_0$ (°C)	T$_{10}$ (°C)	
DHBP-F	Light brown	3000	45	330	675
HMBP-F	Yellow	3300	75	375	680
HBBP-F	Pale yellow	3900	75	370	680
HOBP-F	Light brown	4700	50	260	585

IV. POLYMER-BOUND STABILIZERS

A polymer-bound hindered amine light stabilizer [P-HALS] has been synthesized by terminating the living anionic polymerization of isoprene with 4(2,3-epoxypropoxy)-1,2,2,6,6-pentamethylpiperidine followed by hydrogenation of the resulting polymer to E-P copolymer using Zeigler type catalyst [40]:

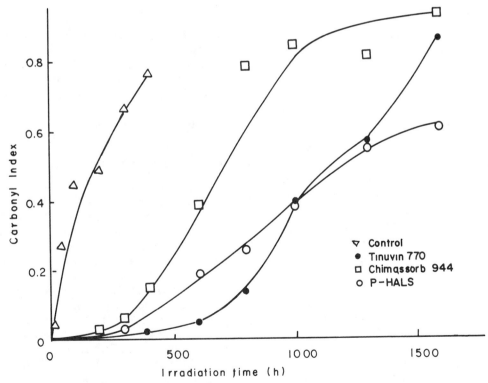

The photostabilizing efficiency of polymer-bound HALS in i-PP and E-P copolymers were studied and compared with commercial HALS (i.e., Tinuvin 770 and Chimassorb 944) by measuring the carbonyl index at 1720 cm^{-1}. Plots of carbonyl index versus irradiation time in i-PP and E-P copolymer samples stabilized with Tinuvin 770 and Chimmasorb 944 are shown in Figures 2 and 3. The unprotected i-PP and E-P copolymer films showed rapid increase in the carbonyl absorbance after 25 h while it showed its appearance after 40 h only in

Figure 2 Plot of carbonyl index versus irradiation time in i-PP films at 0.2 wt% concentration of stabilizers.

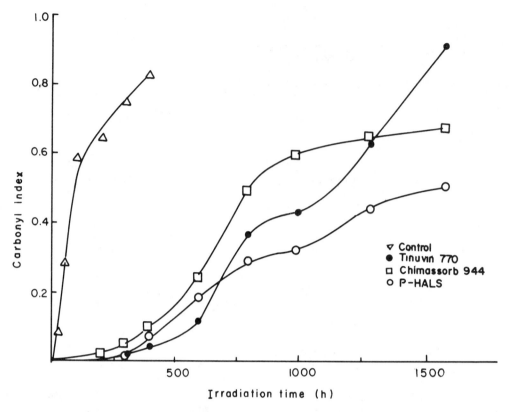

Figure 3   Plot of carbonyl index versus irradiation time in E-P copolymer at 0.2 wt% concentration of stabilizers.

all the stabilized films. The increase in carbonyl index is almost linear in P-HALS stabilized samples whereas in the case of Tinuvin 770 and Chimassorb 944 stabilized samples, the increase is nearly exponential and reaches an asymtotic value around 1000 h irradiation. This is understood based on the fact that low-molecular weight Tinuvin 770 undergoes migration and surface evaporation at longer irradiation time whereas Chimassorb 944 shows negligible diffusion and solubility in polyolefins due to its high-molecular weight and the polar nature of the backbone polymer. On the contrary P-HALS has significant diffusion compared to Chimassorb 944 and thus remains in the polymer for a long enough time to be able to exert its stabilizer behavior.

V.   APPLICATION OF POLYMERIC UV ABSORBERS

Protection of polymers against thermal and photo-oxidative degradation is achieved with appropriate stabilizers that ensure the desirable polymer properties throughout the entire service life of the polymer. Compatible and polymeric stabilizers usually give the best protection. In order to avoid migration and evaporation, polymeric stabilizers are used.

Since 2-hydroxy-4-alkoxybenzophenones are widely used to stabilize polystyrene, flexible and rigid PVC, celluloses, acrylics, and polyolefins such as PE and PP, the polymeric UV stabilizers shown in Table 1 are used with polystyrene, polymethylmethacrylate, and cellulose triacetate (CTA). The polymeric-HALS are used in polyolefins.

A.   Performance Evaluation

The films containing 1% UV stabilizer were prepared by a solution casting process using chloroform for PS and PMMA and acetone for CTA. Films of thickness 0.03 ± 0.01 mm were obtained. The polymer thin films (i-PP, E-P copolymer) were also made by pressing the material between Teflon sheets at 180°C by applying 150 kg cm^{-2} pressure for 45 s in a preheated hydraulic press. The films were quench cooled in tap water for 5 min. Performance evaluation is an empirical science. The fundamental approach is to examine the changes occurring in a polymer upon exposure to a weather-o-meter. The UV stability test was performed by irradiating the films at a low-dose rate using a UV tube (6 W) having an intensity of 3.36 Wm^{-2} and keeping the samples at a distance of 10 cm from the source for 0 to 100 h. The photo-oxida-

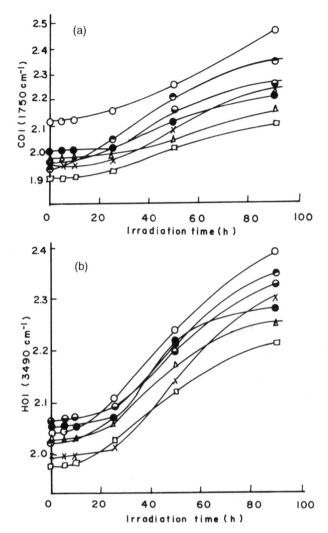

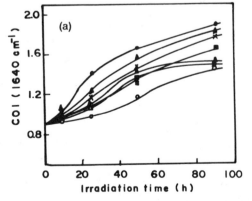

tion was performed in a SEPAP 12/24 chamber at 60°C. The unit consists of four 400 W "medium pressure" mercury sources filtered by a Pyrex envelope supplying radiations longer than 290 nm. The mercury sources were located at the four corners of a square chamber (50 × 50 cm). The samples were irradiated on a rotating support located at the center at a constant speed. Two fans on the walls of the chamber provided regulation of the sample temperature.

B. Characterization of Films

1. The films are characterized by IR for carbonyl index (COI) and hydroxyl index (HOI). The results are shown in Fig. 4, 5, and 6.
2. Ultraviolet spectra has been recorded between 200–400 nm to see the increase in number of chromophore by recording optical density (OD) (at 280 nm) versus irradiation time Fig. 7, 8, and 9.
3. Tensile strength and elongation at break has been determined by an Instron mechanical analyzer, and the results are as shown in Fig. 10, 11, and 12.
4. The evaporation test had been performed to see the loss of polymeric UV absorbers compared to their counterpart monomeric stabilizers, by placing the films in an air-circulating oven at 80°C for 200 to 1000 h (Fig. 13, 14, and 15).

From Table 1, it could be seen that the polymeric stabilizers prepared have 10 to 15 times higher molecular weight than respective monomeric compounds and possess very good thermal stability. In most of the polymers new functional groups, such as C=O, —OH, —OOH are formed during the degradation process and, as a result, an increase in their concentration shows the extent of degradation. From Figs. 4, 5, and 6, it can be seen that compared to unstabilized polymers, the rate of increase of COI and OHI in PS, PMMA and CTA suggests

Figure 4 Changes in (a) carbonyl index and (b) hydroxyl index versus irradiation time for polystyrene films. ○-control; ◐-2,4-DHBP; ●-2H-4MBP; ◓-2H-4BB; X-DHBP-F; △-HMBP-F; and □-HBBP-F.

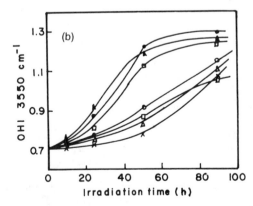

Figure 5 (a) Carbonyl index (at 1640 cm^{-1}) and (b) hydroxyl index (at 3550 cm^{-1}) versus irradiation time for PMMA films. X-control; ●-DHBP; ▲-HMBP; ■-HBBP; ○-DHBP-F; △-HMBP-F; and □-HBBP-F.

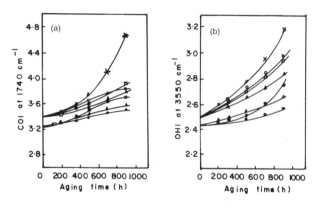

Figure 6 (a) Carbonyl index (at 1740 cm⁻¹ and (b) hydroxyl index (at 3550 cm⁻¹) versus aging time for CTA films: X-control; ○-DHBP; □-HMBP; △-HBBP; ●-DHBP-F; ■-HMBP-F; and ▲-HBBP-F.

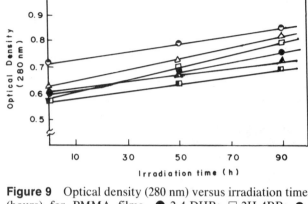

Figure 9 Optical density (280 nm) versus irradiation time (hours) for PMMA films. ◐-2,4-DHB; □-2H-4BB; ●-DHBP-F; △-2H-4MB; ▲-HMBP-F; and ◨-HBBP-F.

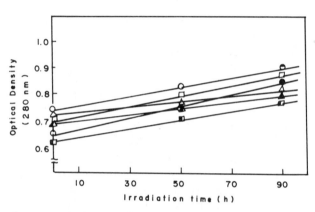

Figure 7 Optical density (280 nm) versus irradiation time (hours) for polystyrene films: ◐-2,4-DHB; □-2H-4BB; ●-DHBP-F; △-2H-4MB; ▲-HMBP-F; and ◨-HBBP-F.

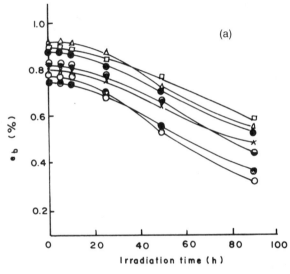

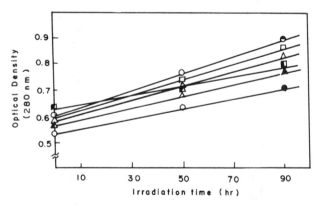

Figure 8 Optical density (280 nm) versus irradiation time (hours) for cellulose triacetate films: ◐-2,4-DHB; □-2H-4BB; ● DHBP-F; △-2H-4MB; ▲-HMBP-F; and ◨-HBBP-F.

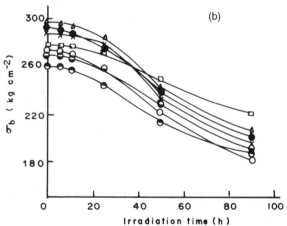

Figure 10 (a) Percent elongation at break (e_b), and (b) tensile strength (σ_b) versus irradiation time for polystyrene films: ○-control; ◐-2,4-DHBP; ●-2H-4MBP; ◐-2H-4BBP; X-DHBP-F; △-HMBP-F; and □-HBBP-F.

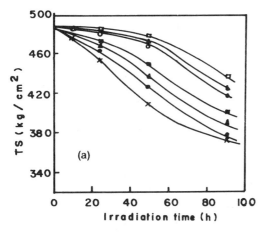

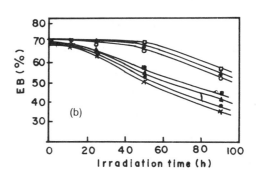

Figure 11 (a) Tensile strength (TS), and (b) elongation at break (EB) versus irradiation time for PMMA films: X-control; ●-DHBP; ▲-HMBP; ■-HBBP; ○-DHBP-F; △-HMBP-F; and □-HBBP-F.

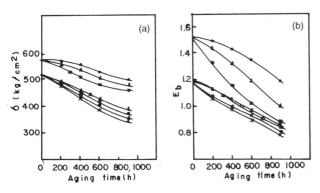

Figure 12 (a) Tensile strength (σ_b), and (b) elongation at break (E_b) versus aging time for CTA films: X-control; ○-DHBP; □-HMBP; △-HBBP; ●-DHBP-F; ■-HMBP-F; and ▲-HBBP-F.

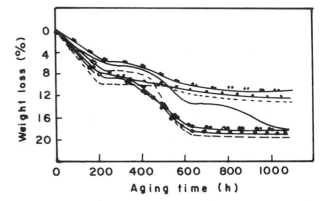

Figure 14 Weight loss of PMMA films versus aging time; (—) control, (——) DHBP, (—○—) HMBP, (—○○—) HBBP, (----) DHBP-F, (—●—) HMBP-F, and (—●●—) HBBP-F.

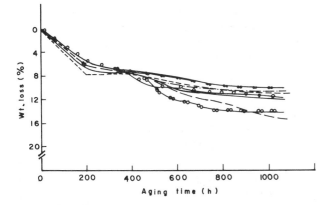

Figure 13 Percent weight loss versus oven aging time for polystyrene films incorporated with: (—) control, (——) 2,4-DHBP, (—○—) 2H-4MBP, (—○○—) 2H-4BBP, (----) DHBP-F, (—●—) HMBP-F, and (—●●—) HBBP-F.

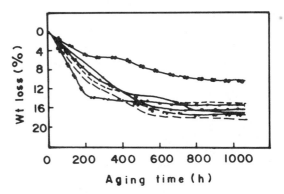

Figure 15 Weight loss of CTA films versus aging time; (—) control, (——) DHBP, (—○—) HMBP, (—○○—) HBBP, (----) DHBP-F, (—●—) HMBP-F), (—●●—) HBBP-F.

improved stabilization. The formation of chromophores, such as C=O can also be monitored by UV spectral study. The increase in OD at 280 nm, which is most probably due to the formation of new carbonyl chromophores [41,42], is measured to get a fair idea about the chemical changes occurring due to aging. The results as seen from Figs. 10, 11, and 12, show that compared to unstabilized polymers, the increase in OD for stabilized polymers is lower, but in all cases the rate of increase is almost same. Except for PMMA, the polymeric stabilizers show good effect in PS and CTA. This may be due to the inherent good stability of PMMA compared to PS and CTA.

Photolysis of polystyrene by light of wavelength 250–300 nm in the presence of oxygen causes chain-scission, some crosslinking, and yellowing with the evolution of water and CO_2 [37,38]. These changes are reflected in the reduction of mechanical properties as seen in Fig. 10. It can be seen that PS film with HBBP-F stabilizer has the lowest reduction in tensile strength at break (σ_b) and elongation at break (e_b), which suggests its good stabilizing capability. Similarly, photolysis of PMMA leads to several chemical changes like unzipping [39], formation of aldehyde groups [41], and conjugated olefinic groups. These changes are reflected in reduction of σ_b and e_b (Fig. 11) and among polymeric UV stabilizers studied, HBBP-F shows maximum σ_b even after 90 h of exposure. CTA films also are stabilized to a greater extent (Fig. 12) by HMBP-F and HBBP-F.

The weight loss on aging may occur due to several reasons. Loss of gaseous compounds formed does not contribute very much, but evaporation or exudation of low-molecular weight compounds are the most responsible. The results (Fig. 13, 14, and 15) indicate that the major weight loss occurs within 200 h. Further, it can be seen that for PS, 6–8%; PMMA, 6–10%, and for CTA 6–14% weight loss occurs in this region. All the figures clearly indicate that in comparison to monomeric stabilizers, the weight loss with polymeric stabilizer is relatively low at all stages and the difference is very high at longer periods of aging.

From various tests performed with polymeric stabilizers and after comparing with the respective monomeric ones, the following trend of effectiveness has been observed [43,44]:

HBBP-F > HMBP-F > DHBP-F > HBBP
> HMBP > DHBP

and for i-PP and E-P copolymers:

P-HALS > Chimassorb 944 > Tinuvin 770

VI. FUTURE DEVELOPMENTS

In recent years, developments in additives have been aimed at minimizing dose level, prolonging stabilizing action, preventing migration, and minimizing undesirable color of the polymer. Decreased volatility, increased heat stability, and increasing environmental considerations, during processing and in use, is another goal of new product developments.

Most of the recent excitement regarding UV stabilizers centers around polymeric UV stabilizers and polymer-bound HALS. A new benzotriazole, Topanox 100 BT has been introduced by ICI, which absorbs light in 290–380 nm range and melts at 128–132°C. It is compatible with PVC, styrenics, acrylics, unsaturated polyesters, and ABS. A liquid benzotriazole has been introduced by Ciba-Geigy (Tinuvin 571) primarily for polyurethane, flexible PVC, as well as nylon and PET fibers. Sometimes a flexible, rather short spacer group between the polymer backbone chain and the active stabilizer moiety is necessary. The polymers containing UV stabilizing groups did not diminish the activity of the stabilizers. Allyl groups have also been introduced in 3-position of 2,4-dihydroxybenzophenone. The 2(2-hydroxyvinylphenyl)2-benzotriazole has been used as a grafting monomer for atactic PP, E-P copolymers, ethylene/vinyl acetate, PMA, and PMMA.

Polymeric HALS show good compatibility with polyolefins and show synergism with many secondary antioxidants. New product forms and synergistic packages of HALS dominate the recent development. The free-flowing, nondusting, granular and liquid forms of HALS (Tinuvin 622 FB and Tinuvin 785), which are especially suitable for PS, PU, and polyolefins have been introduced by Ciba-Geigy. This field will probably attract a great deal of interest during the next few years.

ACKNOWLEDGMENTS

We thankfully acknowledge the facilities provided by Prof. M. N. Patel (Late) and Prof. V. S. Patel at Sardar Patel University, Department of Chemistry, V. V. Nagar, India, and Dr. S. Sivaram, Deputy Director and Head, Polymer Chemistry Divison, National Chemical Laboratory, Pune, India, for the necessary cooperation and the facilities for the present work.

REFERENCES

1. J. F. Rabek and B. Ranby, *Photodegradation, Photo-oxidation and Photostabilization of Polymers,* Interscience, New York (1975).
2. A. Davis, *Developments in Polymer Degradation* (N. Grassied, ed.), Applied Science, London (1977).
3. F. Gugumus, *Mechanism of Polymer Degradation and Stabilization,* (G. Scott, ed.), Elsevier, Amsterdam, pp. 169–210 (1990).
4. A. K. Kulshreshtha, *Handbook of Polymer Degradation,* (S. M. Hamid, M. B. Amir, A. G. Maadhah, eds.) Marcel Dekker, New York, pp. 55–91 (1991).

5. S. Sivaram and R. P. Singh, *Adv. Polym. Sci., 101*: 169 (1991).
6. H. Schonhorn and P. J. Luongo, *Macromolecules, 2*: 364 (1969).
7. M. Heskins and J. E. Guillet, *Macromolecules, 1*: 97 (1968).
8. P. J. Papillo, *Mod. Plastics, 4*: 31 (1967).
9. M. J. Astle and J. R. Shelton, *Organic Chemistry*, II ed., New York, p. 731 (1949).
10. L. W. A. Mcyer and W. M. Geurhart, *Ind. Eng. Chem., 37*: 232 (1945).
11. W. M. Geurhart, R. O. Hill, and M. H. Broyles, U.S. Pat. 2,571,703 (1951).
12. H. J. Heller, *Eur. Polym. J, (suppl.)* 105 (1969).
13. R. P. Fors, D. O. Cowan, and G. S. Hammond, *J. Phys. Chem., 68*: 3747 (1964).
14. R. G. Schmitt and R. C. Hirt, *J. Appl. Polym. Sci. 7*: 1565 (1963).
15. J. P. Guillory and C. F. Cook, *J. Polym. Sci, Polym. Chem. Ed., 11*: 1927 (1973).
16. P. J. Briggs and J. F. McKellar, *J. Appl. Polym. Sci., 12*: 1825 (1968).
17. H. C. Ng and J. E. Guillet, *Macromol., 11*: 937 (1978).
18. A. M. North *Brit. Polym. J., 7*: 119 (1975).
19. A. C. Somersall and J. E. Guillet, *J. Macromol Sci. Revs. Macromol. Chem. C13*: 135 (1975).
20. J. G. Calvert and J. N. Pitts, *Photochemistry*, John Wiley & Sons, New York (1966).
21. R. P. Singh, *Polym. Degradn. Stab., 13*: 313 (1985).
22. D. J. Carlsson and D. M. Wiles, *J. Polym. Sci. Polym. Chem. Ed., 12*: 2217 (1974).
23. H. G. Heine, H. J. Rosenkranz, and H. Rudolph, *Appl. Polym. Symp., 26*: 157 (1975).
24. A. Davis and D. Sims, *Weathering of Polymers*, Applied Sci Publishers, New York, p. 242 (1983).
25. V. T. Kozlov, Z. N. Tarasova, E. R. Klinshpont, V. K. Milinchuk, and B. A. Dogadkin, *Vysokomol Soed A9*: 1541 (1967).
26. A. S. Kuzminskii and M. A. Zakirova, *Radiats Khim Polim, Mater Simp,* Moscow, 388 (1966).
27. A. S. Kuzminskii, L. I. Lyubchanskaya, E. S. Yurtseva, and G. G. Yudina, *Radiats Khim Polim, Mater Simp,* Moscow, 301 (1966).
28. H. S. Olcott and H. A. Mattill, *J. Am. Chem. Soc., 58*: 2204 (1936).
29. H. S. Olcott and H. A. Mattill, *Chem. Rev., 29*: 257 (1941).
30. H. Kamogawa, M. Nanasawa, and Y. Uehara, *J. Polym. Sci., Polym. Lett. Ed., 15*: 675 (1977).
31. S. B. Maerov, *J. Polym. Sci. A-1, (3)*: 487 (1965).
32. J. Fertig, A. I. Goldberg and M. Skoultchi, U.S. Patent 3,141,986 (1964).
33. S. Tocker, *J. Makromol. Chem., 101*: 23 (1967).
34. K. Hiroyoshi, T. Yoko, and N. Masato, *J. Polym. Sci., Polym. Chem. Ed., 19*: 2947 (1981).
35. Susha. R. Menon, Chetan. G. Patel, and J. S. Parmar, *Angew. Makromol. Chem., 189*: 87 (1991).
36. S. R. Menon and J. S. Parmar, *Makromol. Chem., 210*: 61 (1993).
37. N. Grassie and N. A. Weir, *J. Appl. Polym. Sci., 9*: 999 (1965).
38. J. F. Rabek and B. Ranby, *J. Polym. Sci. Polym. Chem. Ed., 12*: 273 (1974).
39. R. B. Fox, L. G. Isaacs, and S. Stokes, *J. Polym. Sci. Part A., 1079* (1979).
40. R. Mani, R. P. Singh, S. Chakrapani, and S. Sivaram Polymer (in press).
41. R. B. Fox, L. G. Isaacs, S. Stokes, and R. E. Kagarise, *J. Polym. Sci. Part A., 2085* (1964).
42. P. R. E. Cowly and H. W. Molville, *Proc. R. Soc. A., 210*: 461 (1952).
43. S. Ramachandran and J. S. Parmar, *High Performance Polymers, 4*: 81 (1992).
44. D. K. Patel, S. R. Menon, M. M. Patel, and J. S. Parmar, *Angew. Makromol. Chem., 226*: 117 (1995).

27

High-Performance and Functional Materials from Natural Monomers and Polymers

Chennakkattu Krishna Sadasivan Pillai
Regional Research Laboratory (CSIR), Thiruvananthapuram, India

I. INTRODUCTION

Early humans used a variety of naturally occurring polymers to meet their material needs. They used them not with the perception of the chemistry and physics of modern high polymers, but for their survival, food, shelter, and clothing. A variety of materials made from wood, bark, animal skins, cotton, wool, silk, natural rubber, etc. were essentially playing key roles in early civilizations [1]. They could mechanically modify materials into useful tools (stone axes; wood carvings; animal skins; the twisting of cotton, wool, and flax to form threads; weaving; preparation of thin-skinned papyrus and vegetable tissues for writing, etc). Embalming of corpses was one of the earliest practices that involved chemical modification (crosslinking of proteins by formaldehyde). With the growth of human civilizations, the use and applications of materials extended to metals and ceramics, and, by the nineteenth century, there were eight classical materials—metals, stones, woods, ceramics, glass, skins, horns, and fibers—of which woods, skins, horns, and fibers are organic polymers. During the past one and one half centuries, two more materials were added to the list, rubber and plastics, both of which are polymeric in nature. However, by 1900, there were only a few plastics in use, e.g., shellac, gutta percha, ebonite, and celluloid [2]. Although a variety of chemical modifications came into vogue, four great discoveries formed the foundation of the industrial use of natural polymers, which even today form the basis of continued support and maintenance of these industries against stiff competition from synthetics. They are: the vulcanization of natural rubber; the mercerizing of cotton, hemp, and flax; the tanning of leather; and the loading of silk [1,3]. Even today, these discoveries are the basis of many modern industries based on natural polymers that produce standard materials with a definite specification.

The advent of synthetic polymers based on petrochemicals adversely affected many of the thriving industries based on natural polymers, but the time has come for regeneration. The spectacular growth and developments of synthetic polymers during this century revolutionized the world of materials and now polymers are classed along with advanced metals and ceramics. Apart from the host of polymers with unprecedented qualities, one of the greatest developmental outcomes is the generation of information necessary for tailor making polymers to meet specific property profiles for any conceivable application [4–6]. There have been many attempts to apply this information to natural polymer systems to achieve the desired property profiles with varying degrees of success, which have given and are giving it a new outlook as a possible alternative source for the production of polymers [3,7–20]. This chapter will briefly discuss a few of the significant developments in this area, with particular emphasis on certain natural monomers as a source for production of high-performance and functional polymers.

There are a number of factors that favor the use of natural polymers as a source for polymer production. The concerns of environmental pollution, tensions in the Persian Gulf countries regarding oil, the fear of a possible future depletion of oil, etc., give credence to the move toward a bio-based material policy [8,21]. The

Table 1 Carbon in Biosphere [23]

Form	Tonnes $\times 10^{12}$
Carbon in sediment	18,000
Organic carbon in sediment	6,800
CO^2 in atmosphere	0.65
Living matter on land	0.08
Dead organic matter on land	0.70
CO^2 in ocean	35.40
Living matter in ocean	0.008
Dead organic matter in ocean	2.7

overdependence of the chemical industry on petroleum resources could be reduced by the full utilization of renewable natural resources wherever possible. Ranging from algae to wood, its availability is limited only by the photosynthetic efficiency of plants [22]. Table 1 gives the availability of carbon over all of the earth, of which the total biomass is substantially high (more than the total production of polymers), which means that there will be sufficient agro by-products and related materials for utilization [23]. These materials are renewable and are also adaptable through genetic manipulations [24]. The availability of some materials is flexible through crop rotation, for example. Apart from naturally existing polymers, there are a variety of naturally occurring monomers that exist in a free or a combined form and could be obtained by extraction, cleavage, or depolymerization from the biomass [3,16,24–30]. The abundant availability and structural variety of agricultural and forest products lend support to the need to have a fresh look into their utilization [31]. The major limitations will be, unlike petrochemicals, the variation in properties from source to source, the presence of contaminants, lack of uniformity in composition, etc., which can be solved with a new approach based on the application of the information gained from the synthetics, as indicated earlier [32].

II. ADVANCED SYNTHETIC STRUCTURES VERSUS NATURAL POLYMERS

One may now ask whether natural systems have the necessary structural evolution needed to incorporate high-performance properties. An attempt is made here to compare the structure of some of the advanced polymers with a few natural polymers. Figure 1 gives the cross-sectional microstructure of a liquid crystalline (LC) copolyester, an advanced polymer with high-performance applications [33]. A hierarchically ordered arrangement of fibrils can be seen. This is compared with the microstructure of a tendon [5] (Fig. 2). The complexity and higher order of molecular arrangement of natural materi-

als are evident. Tendons and similar materials are examples of several orders of heirarchical orders of molecular arrangement that evolved over millions of years to specifically cater to specific functions. This can be explained by examining the structure of hemoglobin. There are four levels of structural heirarchy, increasing in complexity and improving specificity in function. The primary structure involves the covalent bonded polypeptide chain, which by hydrogen bonding and helical arrangement forms the secondary structure. The tertiary structure is formed by folding the resulted configuration into the disulphide bonds giving rise to a myoglobin; four units of which combine to form the quarternary structure of hemoglobin that carries oxygen without fail to its destination. It was not possible to find a manmade polymer with a similar structural complexity and functional specificity (the honeycombed structures of ablatives used in spaceships can be said to have a higher complexity in structure and function) [35]. It is well known that one of the main factors in the high strength of Kevlar® originates from the hydrogen bonded structure of the polyaramide, as given in Fig. 3 [36]. The source of the structural stability and strength of many natural polymers is hydrogen bonding, which need not be emphasized fur-

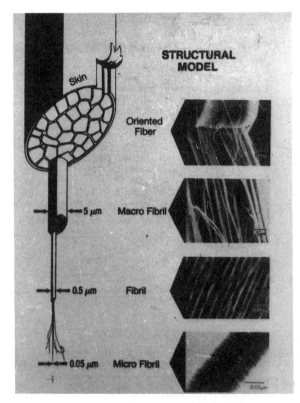

Figure 1 Schematic representation of the microstructure and cross-sectional view of a liquid crystalline copolyester fiber [33].

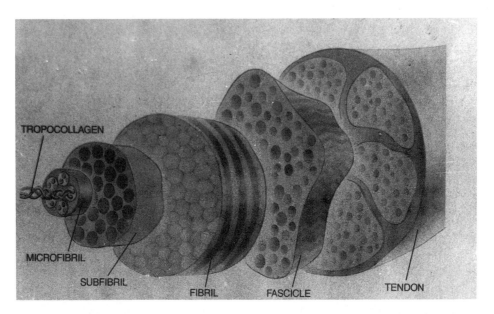

Figure 2 Schematic diagram of the cross-section of tendon showing the hierarchical structural arrangement [5].

ther. From a materials point of view, the strength of cellulose owes much to the hydrogen bonded structure made possible by the β-glycosidic linkage in comparison with that of starch, also made from glucose but with α-glycosidic linkage. Comparisons can be made with the helical structures of synthetic and natural polymers to emphasize the role of advanced structures on properties. A number of such comparisons can be made available to show that the natural polymers are much more complex than synthetics and, therefore, what is required is more imaginative chemical approaches to deal with natural systems.

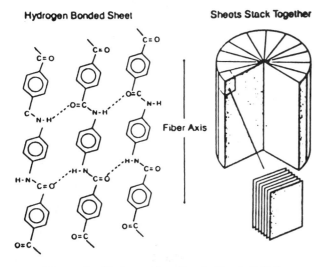

Figure 3 Hydrogen bonding in Kevlar® [34].

III. NATURAL POLYMERS

Biomass is generally made up of polymers, oligomers, monomers, and other nonpolymerizable simple organic compounds, including metallic salts and complexes [17,29,30]. Polymers are, of course, the major components and have been serving human civilizations from time immemorial. The literature on natural polymers is vast and only a few reviews and books are cited here for further reference [3,17,18,24,29–31,37–53]. The outstanding aspect of natural polymers is their wide variety, which provides innumerable opportunities for structural modifications and utilization.

A. Classification

Natural polymers can be classified in a variety of ways [17,30], but it appears that a classification based on structural heirarchy is most appropriate. Thus, depending on the nature of the hetero atom inserted in the main chain, the polymers can be classified into four major types as hydrocarbon polymers (e.g., natural rubber), carbon–oxygen (e.g., carbohydrates: cellulose, starch, etc.; phenolics: lignin, humus, etc.; and polyesters: shellac), carbon–oxygen–nitrogen/sulphur (e.g., proteins with the exception of phospho proteins) and carbon–oxygen–nitrogen–phosphorus-(e.g., nucleic acids) containing polymers. Of these polymers, polysaccharides, proteins, and nucleic acids are grouped as polymers having pronounced physiological activity. When material applications are considered, nucleic acids can be excluded, except some recent genetic engineering outlets which, of course, are a promising growth mode [17].

Lignin

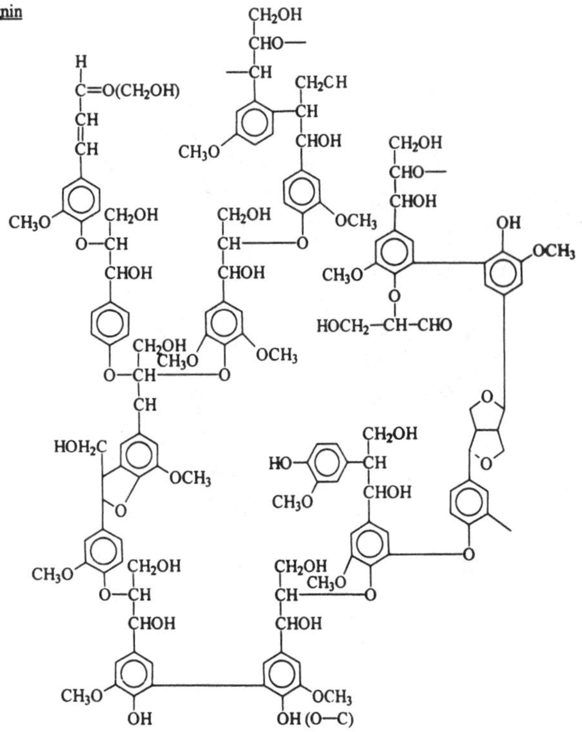

Cellulose

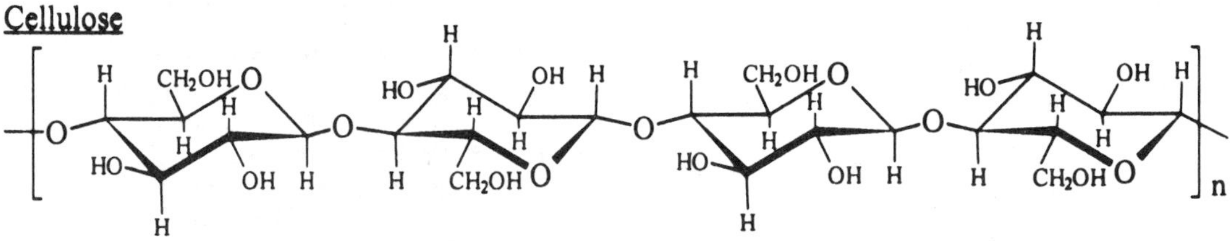

Starch

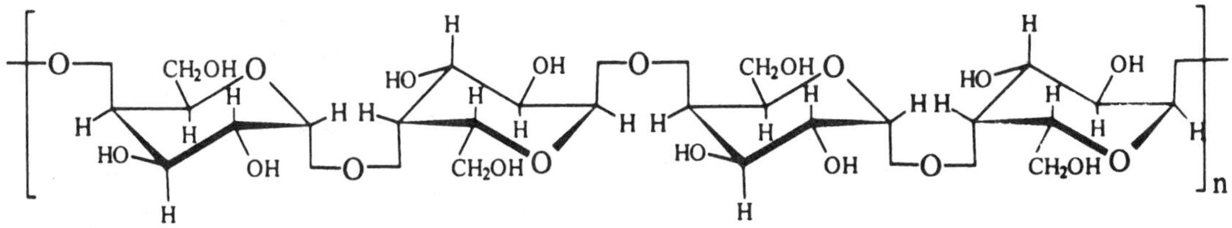

Natural Rubber

$$-\!\left[CH_2-CH=\overset{\displaystyle CH_3}{\overset{|}{C}}-CH_2\right]_n\!-$$

Protein

$$R\!\left[\overset{\displaystyle O}{\overset{\|}{C}}-\overset{\displaystyle H}{\overset{|}{N}}\right]_n\!R' \text{ (polypeptide repeat unit)}$$

Figure 4 Primary structures of some natural polymers.

Table 2 Approach to Modification of Polymers

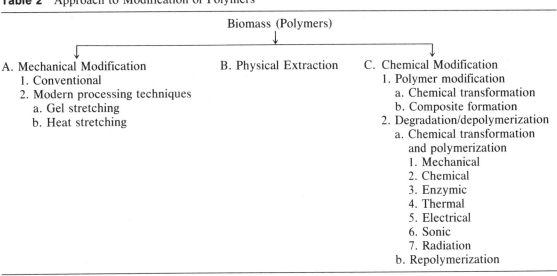

Thus, based on material applications, the following polymers are important: natural rubber, coal, asphaltenes (bitumens), cellulose, chitin, starch, lignin, humus, shellac, amber, and certain proteins. Figure 4 shows the primary structures of some of the above polymers. For detailed information on their occurrence, conventional utilization, etc., refer to the references cited previously.

B. Modification of Natural Polymers

Table 2 depicts an approach to the modification of biomass with a view to effecting necessary changes, structural or otherwise, in natural polymers. Conventionally, modifications [54–63] are effected mainly by two ways: (1) mechanical and/or chemical modifications without destroying the main structural backbone, and (2) cleavage of the polymer into smaller fragments/monomers and repolymerzation. Cleavage can be effected in a variety of ways, such as mechanical, chemical, enzymatic, thermal, electrical, sonic, or by irradiation methods. (As it is intended to give more emphasis to certain natural monomers in this chapter, the discussion on natural polymers will only be indicative of the pertinent points and not an exhaustive discussion.) This is exemplified in Table 3, where a possible fragmentation pattern of ligno-cellulosic material is shown as a typical sample for its utilization. The fact that structure and properties of polymers can be manipulated to meet a specific need and the availability of such information [4–7] to effect such a transformation suggest that speciality properties could be built into the existing natural polymers through appropriate chemical modifications/process design tech-

Table 3 Fragmentation of Lignocellulosic Materials by Hydrolysis

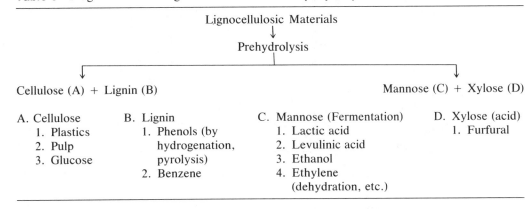

niques to ensure performance on par with synthetics. This can be illustrated by taking the scheme (Fig. 5) of chemical reactions proposed by Lindburg et al. [64] that can be performed on lignins [37,47,48,60] so that it can be converted to a series of speciality or even high-performance/functional polymers.

1. Cellulose

The first thermoplastics and synthetic fibers were derived from cellulose, but their markets were eroded by the rapid growth of petrochemically derived polymers. Even the present sustainment of the cellulosic fibers can be attributed to the property gain obtained by the introduction of viscose and other processes. This was followed by the development of various cellulose esters, ethers, crosslinked cellulose, graft copolymers, etc., some of which have limited markets. Whether cellulosics can expand further will depend upon the possibilities for further modifications to tailor-make specific end products and whether it is possible to accommodate the cost of such a transformation. The literature on cellulose

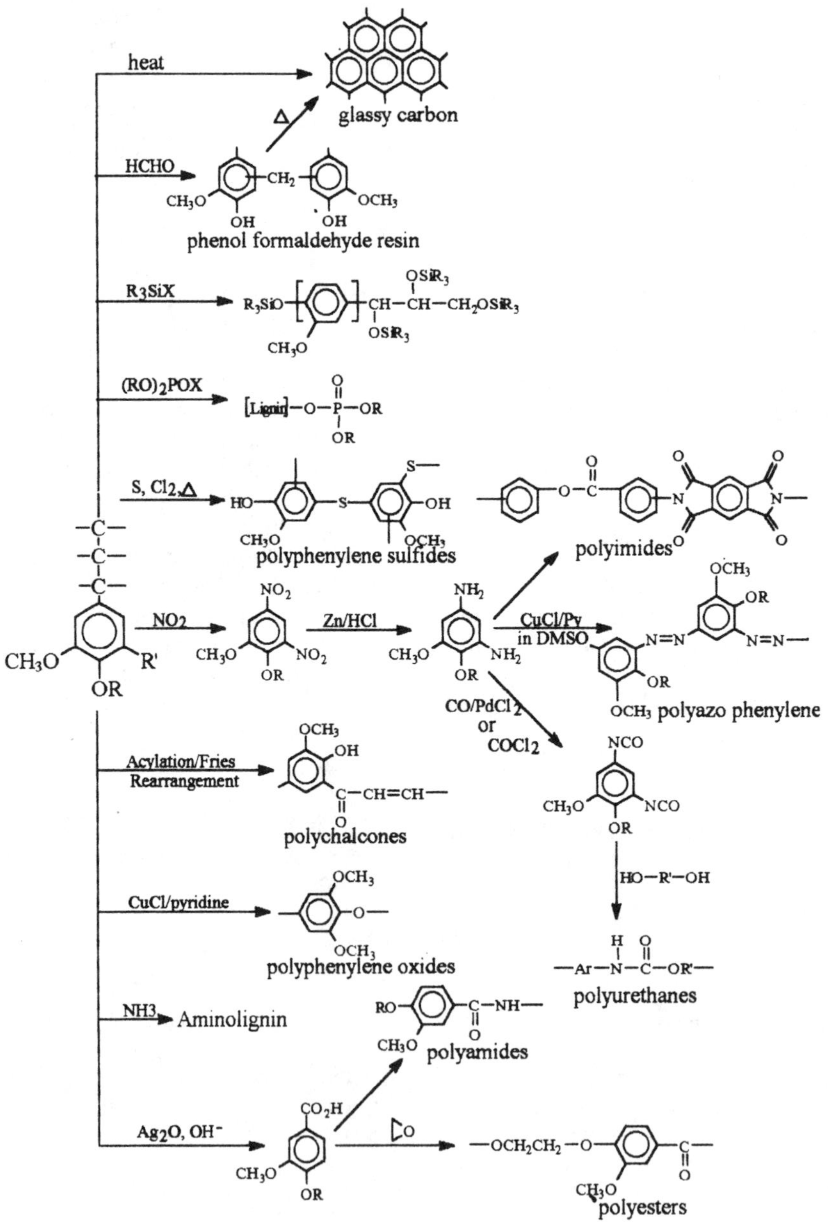

Figure 5   Speciality polymers from lignin.

and cellulose modifications are numerous [3,17,24,29, 65–98], and what is cited here is not exhaustive. Newer molecular design and process design techniques can be applied to effect such a transformation [4,5]. Cellulose, being the most abundant natural polymer in the earth requires much thought and imagination. A few suggestions to generate speciality/functional properties on the system can be projected as follows:

1. the strength of natural fibers, which are composites of cellulose and lignin, can be improved many times by applying heat/gel stretching techniques followed by effective crosslinking using low viscous cross-linking molecules or metal ions. (It should be noted here that while cellulose has a theoretical tensile moduli of 500–800 g/day, what is achieved is only 100–140 g/day.)
2. Graft reactions on cellulosics are well studied and are well known to incorporate desired properties in polymers [61,72,73,76,77,99–102], but commercialization of the processes on cellulosics are not increasing. (Table 4 gives a summary of the techniques of grafting.) A fresh imaginative approach is required to solve this problem.
3. Newer and cost-effective techniques of separation of cellulose from lignin need to be developed.

2. Starch and Other Polysaccharides

Cellulose and starch are related in structure by having the same repeat unit. The 1,4-α-glycosodic linkage of glucose units in starch, rather than the 1,6-β-glycosodic linkage in cellulose, gives rise to a branched structure with a concomitant change in properties. The high strength and rigidity of cellulose originate, thus, from the linear structure and the hydrogen bonding made possible. Apart from other uses as gums and adhesives, a promising application is the area of biodegardable polymers either after grafting with suitable polymers or after blending with other synthetic polymers [103,104].

A number of other polysaccharides, such as glycogen, dextran, chitin, etc., possess interesting structures for chemical modification [103,104]. Dextran has been used as a blood plasma substitute. Although it can be converted to films and fibers, chitin's relatively small resource restricts its commercialization.

3. Lignin and Related Polymers

Mention has been made of lignin modification in Section III B. Lignin is the "cement" that holds celluose fibrils together, giving wood much of its dimensional stability [37,47,48,60]. Constituting about 25–30% of wood, and next only to cellulose, lignin is abundantly available and as a by-product of the paper and cellulose industries, it has yet to reach its potential with regard to polymer applications [105–110]. The problems with lignins are: the structure of lignin varies greatly according to the source; during cellulose separation, there is degradation in lignin molecular weight (this decay in molecular weight is, however, used in the development of adhesives, etc.); it is a highly crosslinked polymer so that further processing requires degradation to smaller units or monomer structures, etc. New routes for controlled degradation of lignin might have to be developed to solve these problems.

Humus and Coal

Humus and coal are related to lignin in structure. Humus, which constitutes the organic component of soil, is presumed to originate from lignin [111,112]. Unlike lignin, humus contains carboxy groups, which makes it an excellent scavenging agent for toxic metals.

Coal is an extremely complex polymer with a structure analogous to those of lignin and humus. It is presently used as fuel, considered as an alternative to petroleum, and as a feed stock for petrochemicals, coal has great potential for processing into high-performance/functional materials. Kerogen is another material found in oil shale [113]. Structurally related to coal, it gives low-molecular weight fragments on thermal cracking.

4. Natural Rubber

Natural rubber was the only polymer for elastomer production until the advent of synthetics. Natural rubber, however, continues to maintain its competitive edge due mainly to the gain in properties such as high resilience, low hysteresis, low heat buildup, and excellent tack with mechanical properties achieved through the process of vulcanization [114–115]. The industry is said to be self-sufficient with a good technological base and is expected to compete successfully with synthetics because of the edge in properties mentioned above [116,117].

Table 4 Methods of Preparation of Graft Copolymers

A. Crosslinking reactions of two polymer chains of different types.
B. Initiation of sites on a polymeric backbone where a monomer can be grafted.
1a. Ionic grafting initiated by radiation.
1b. Ionic grafting initiated by chemical means.
2. Anionic grafting.
3. Free radical grafting.
3a. Free radical grafting initiated by chemical means.
3b. Free radical grafting by redox reactions: direct oxidation, chain transfer.
4. Grafting by chain transfer other than redox system.
5. Grafting by mechanical degradation.
5a. Plasma radiation-induced grafting.
5b. Photo-induced grafting.
5c. High-energy radiation induced grafting.
6. Free radical grafting without using a free radical initiator.

5. Wool, Silk, Gelatin, and Leather

The history of wool [118] and silk [119] for use in clothing goes back 5000 years. Both are proteins with considerable variation in amino acid composition. The high proportion of cystine in wool make it more chain coiling through disulphide linkages. Whereas wool exhibits an alpha-helical secondary configuration, silk assumes the extended chain beta configuration. Numerous chemical modifications of wool and silk have been reported [118,119]. The silk industry, however, is known to have its survival based on a reaction of some tin compounds to improve its stability and performance [119].

Collagen [120,121], the main constituent of skin and hides, is the main source for gelatin and leather. While gelatin is a degraded form of collagen, leather in produced by tanning of hides. Tanning of hides to produce leather is as old as 5000 years and is the oldest industry in continuous production [122]. The chemistry involves the cross-linking of protein molecules by gallic acid. The industry with improved methods of production and quality assurance is safe against newer materials.

Plastics and fibers have been produced from regenerated proteins obtained from a number of sources [17]. The process involves dispersing the proteins in dilute sodium hydroxide followed by extrusion through a spinneret into an acid bath to form the fibers that are then crosslinked with formaldehyde to improve strength. The fibers are used along with silk and wool.

6. Bitumen, Shellac, etc.

Another natural polymer that needs a fresh look into its structure and properties is bitumen [123], also called asphaltines, that are used in highway construction. Although a petroleum by-product, it is a naturally existing polymer. It primarily consists of polynuclear aromatic and cyclocaliphatic ring systems and possesses a lamellar-type structure. It is a potential material that requires more study, and high-performance materials such as liquid crystalline polymer (LCP) could be made from it.

Shellac [124,125] is a natural resin used in very old times for varnishes and moulding compounds. The resin secreted by the lac insect, *Kerriar paca*, is collected by scraping the shellac-encrusted trees found in southern parts of Asia. It consists of a complex mixture of cross-linked polyesters derived from hydroxy acids, principally aleuritic acid (9,10,16-trihydroxyhexadecanoic acid). From a structural point of view, it appears that this material can be used as a crosslinking agent and/or as a monomer for developing dentrite-like polymers. The question is whether it is possible to produce this material from shellac by controlled hydrolysis.

Biomass is a relatively inexpensive raw material. Since it is made by nature there is an enormous saving of energy. The main research areas include (1) isolation and purification of natural monomers and polymers, (2) modification of natural monomers and polymers, and (3) search for new natural monomers and polymers. In effecting modifications, both conventional modifications and modern methods as evolved from the recent concepts on molecular design and process design of polymers to meet specific end use applications could be applied. This requires another look into its potential vis-a-vis synthetics in terms of both properties and cost. The variety and maneuverability of structures present in natural polymers and the availability of information needed to effect transformation to high-value polymers stress the need to to have a long-term perspective to solve problems in the area.

IV. Natural Monomers

Unlike natural polymers, natural monomers have not received much attention from scientists and technologists. The abundant availability of monomers from petroleum sources has restricted research on natural monomers. Moreover, natural monomers required extensive purification as they were always found contaminated with other extraneous matter. They also showed variations in properties from source to source. It is, however, interesting to note that there are a variety of monomers existing in nature free or in the combined form that can be obtained by extraction, cleavage, or depolymerization from biomass [25–30]. There are also some naturally existing monomers/oligomers, such as the amber found in fossil resources. The monomers exhibit a variety of interesting structures that can be manipulated to obtain speciality/high-performance properties. Following is a discussion of (1) those monomers that have already been shown to be useful, and (2) those monomers that need further work to find applications. Particular attention will be given to the extensive work carried out in this laboratory on natural monomers having unsaturated long-chain hydrocarbon structures from which it was shown that speciality/high-performance/functional polymers could be derived by appropriate chemical modifications.

A. Chemically Modified Natural Monomers

There is a good amount of data on the transformations of natural polymers, but the literature concerning chemical modifications of natural monomers is rather scanty. However, there are a few natural monomers that have found applications. The monomers in these cases are either modified into other suitable monomers of industrial importance or are polymerized directly into polymers. A few examples are given below for illustration.

Furfural is a natural monomer obtained by the steam acid digestion of corncobs, bagasse, rice husks, oat hulls, or similar materials. It acts as the precursor for the preparation of two important monomers (Scheme 1), adipic acid and hexamethylene diamine, used in the

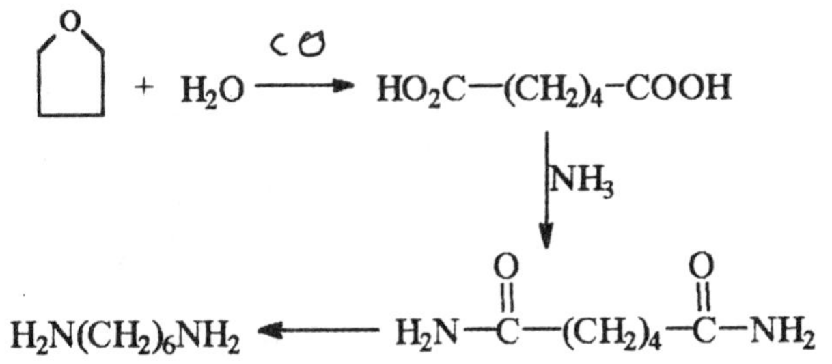

Scheme 1 Chemical modification of furfural.

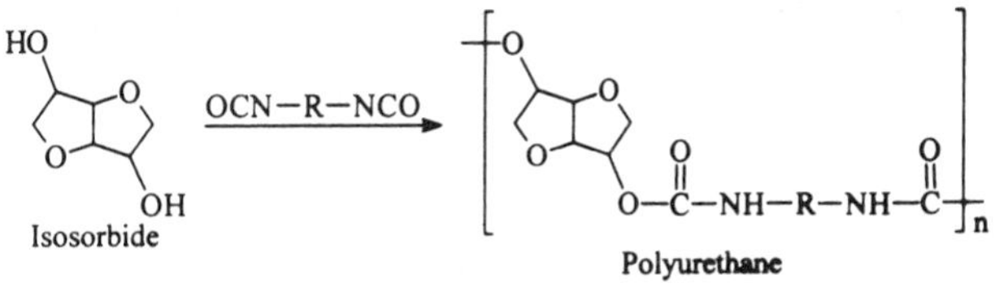

Scheme 3 Chemical modification of sorbitol.

production of nylon-66 [126]. Furfural and its derivative, furfuryl alcohol, are known to react with phenol to form thermosetting resins in the presence of alkaline/acid catalysts [25]. These resins are widely used in moulding compounds and in the manufacture of coated abrasives.

Terpene monomers are another class of interesting natural monomers because they give, on polymerization, hydrocarbon therplastic resins that exhibit a high degree of tackiness useful in pressure sensitive tapes [25]. They are also used for sizing paper and textile materials. Terpene-phenol resins are effective heat stabilizers for high-density polyethylene.

Fatty acids, both saturated and unsaturated, have found a variety of applications. Brassilic acid (1,11-undecanedicarboxylic acid [BA]), an important monomer used in many polymer applications, is prepared from erucic acid (Scheme 2), obtained from rapeseed and crambe abyssinica oils by ozonolysis and oxidative cleavage [127]. For example, an oligomer of BA with 1,3-butane diol-lauric acid system is an effective plasticizer for polyvinylchloride. Polyester-based polyurethane elastomers are prepared from BA by condensing with ethylene glycol-propylene glycol. Polyamides based on BA are known to impart moisture resistance.

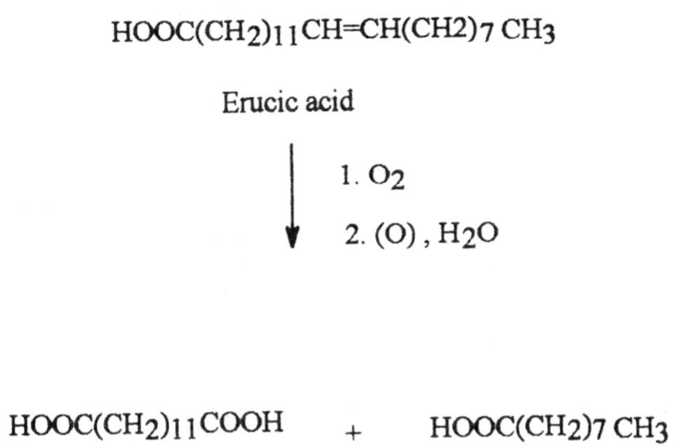

$$HOOC(CH_2)_{11}CH=CH(CH_2)_7 CH_3$$

Erucic acid

1. O_2

2. (O) , H_2O

$$HOOC(CH_2)_{11}COOH \quad + \quad HOOC(CH_2)_7 CH_3$$

Brassylic acid(BA)   Pelargonic acid

Scheme 2 Synthesis of brassilic acid and its conversion.

Sorbitol is an interesting natural monomer obtained from glucose (sugar waste is used for this purpose) by dehydrogenation. It easily dehydrogenates into isosorbide (Scheme 3), which is useful as a polyol for production of polyesters/polyurethane [128]. D-glucose, available from sugar waste, can be converted to D-glucose methacrylate whose polymer has found many potential applications in medicine, secondary ill recovery, etc [129].

Lactic acid and levulinic acid are two key intermediates prepared from carbohydrates [7]. Lipinsky [7] compared the properties of the lactide copolymers [130] obtained from lactic acid with those of polystyrene and polyvinyl chloride (see Scheme 4 and Table 5) and showed that the lactide polymer can effectively replace the synthetics if the cost of production of lactic acid is made viable. Poly(lactic acid) and poly(1-lactide) have been shown to be good candidates for biodegradeable biomaterials. Tsuji [131] and Kaspercejk [132] have recently reported studies concerning their microstructure and morphology.

B. Long-Chain Hydrocarbon Phenols

Long-chain hydrocarbon phenols are phenolic lipids present in plants from a number of families, notably the *Anacardiacae*, found in many parts of the world [133]. They are widespread in tropical and temperate climates in trees, shrubs, many small plants, and certain bacterial sources. Thyman [133] wrote an excellent review codifying the available data on these monomers. Table 6 provides a representaive sample of structures of a few phenols. They are mostly monohydric or dihydric phenols or phenolic acids with a hydrocarbon side chain at the meta position. In general, many of these monomers are found to be mixtures of four components varying in the degree of unsaturation of the side chain. Being unsaturated, these phenols can undergo oxidative coupling polymerization giving rise finally to crosslinked films and, hence, their applications in surface coatings. Thus, some of

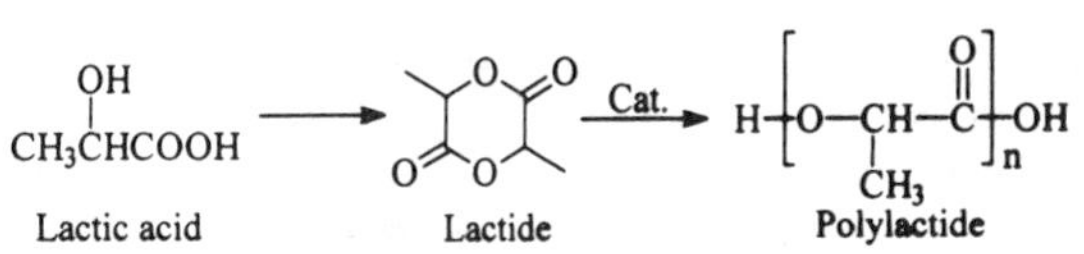

Scheme 4 Lactide polymers.

Table 5 Comparison of Properties of Lactide Polymers with Polystyrene and Poly(vinyl chloride)

Polymer/Property	Polystyrene	95/5 Lactide/Caprolactone	95/15 Lactide/Caprolactone	Flexible PVC
T.S. (psi)	7,000	6,900	3,200	1,500–3,000
Elongation (%)	2	1.6	6–500	200–450
Initial modulus (psi)	450,000	112,000	84,000	50,000–100,000
Impact strength (ft-lb/in.)	0.26	0.36	No break	0.4–7.0
Specific gravity	1.008	1.26	1.26	1.16–1.35

them have found applications in artistic uses, but there are others that have found technical uses [135–137].

From the point of view of chemistry, phenols are interesting because of their dual-aliphatic and aromatic character [133]. The aliphatic entity gives rise to hydrophobic behavior, whereas the phenolic moiety is hydrophilic in character. From an application point of view, the reactions are (1) that of the phenolic moiety undergoing the conventional polycondensation reaction with formaldehyde to give phenolic type resins, (2) the unsaturated side chain undergoing chain reaction polymerization to flexible/rubbery polymers, and (3) the auto-oxidation/oxidative coupling polymerization to give rise to crosslinked polymers. Apart from these possibilities, these lipids have numerous sites such as the hydroxyl position, the aromatic ring, and the side chain that are amenable to chemical modifications [16,32,137]. So, compared with simple phenols, they have more opportunities for chemical modifications and polymerizations for effecting structural changes for tailor making a polymer for a specific application. This will be illustrated taking the phenolic lipid available from the tree *Anacardium occidentalle L.* as an example. Among the phenolic lipids of nonisoprinoid origin, the two most important lipids of commercial significance are those available from the trees *Anacardium occidentalle* and *Rhus vernicifera* [133]. Further discussions will be centerd around these two monomer sources only.

1. Rhus vernicefera

Originally from China, *Rhus vernicefera* has been under cultivation in Japan since the sixth century AD. The latex is collected in the same way as the rubber plant *Hevea brasiliensis*. The product is known as urushiol, which consists mostly of dihydric phenols of structures (Fig. 6) and is used as lacquers.

Extraction

A crude process of extraction of urushiol from the tree *Rhus vernicifera* was used by the Chinese during the Chou dynasty of 1122–249 BC, and the process was systematised by the Japanese. The tree is tapped at about the 10th year of cultivation by a lateral sloping incision into the bark during June to September. The sap is white to grayish in color, but on exposure to air turns yellow-brown and then black. The crude sap contains approximately 70% urushiol, 4% gum, 2% albuminous materials, and 24% water. It is stirred and filtered and heated to reduce the moisture level [138] and finally stored in air-tight containers.

Structure

As is true in the case of other phenolic lipids, urushiol is also a mixture of components varying mostly in the degree of unsaturation. Thus, the urushiol from *Rhus vernicefera* has structures shown in Fig. 6 [139]. *Rhus toxicodendron* is also known to give urushiol, but its

Table 6 List of a Few Long-Chain Hydrocarbon Phenols [133]

Botanical Name	Common Name	Main Component	Source
Anacardium occidentalle L.	Cashew nut shell liquid	Anacardic acid, Cardanol, Cardol	Brazil, Tanzania, Kenya, Nigeria Mozambique, India
Rhus vernicefera	Japan lac, Chinese lac	Urushiol	Japan, China, Korea
Rhus toxicodendron	Poison oak	Urusiol	North America
Anacardium semecarpus	Bhilawen nut shell liquid (Bhilawainol)		India
Grevilla pyramidalis	A cardol monoene	5-tridecylresorcinol	N.W. Australia
Triticum vulgere	Wheat bran phenol	5-nonadecylresorcinol	N.W. Australia
Semicarpus vernicefera	Formosan lac	Laccol	Formosa
Rhus succedanea	Indo China lac		Vietnam
Melannorrhea usitata wall	Burmese lac	5-pentadecylcatechol	Burma

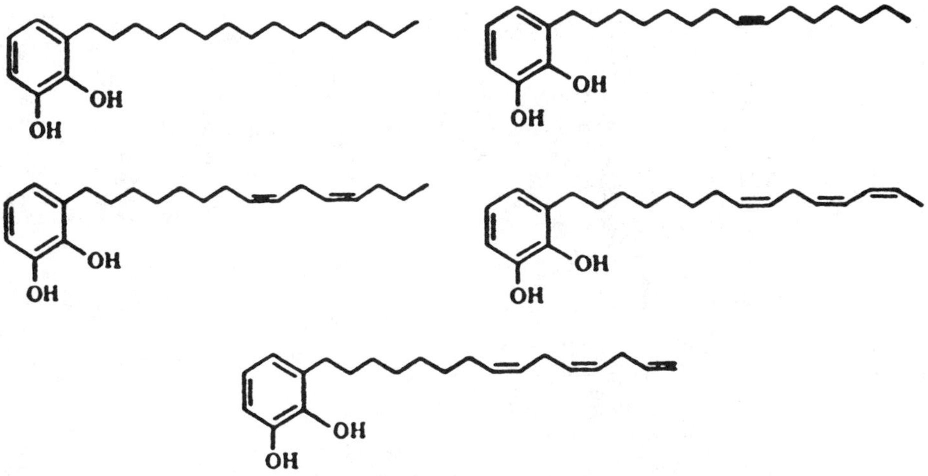

Figure 6 Structure of dihydric phenols of urushiol.

triene component has a vinyl end group unlike that of *Rhus vernicefera* [140].

Applications

Urushiol is used largely in Japanese lacquering, which is the single most application used in China and Japan and some other far eastern countries. In lacquering [141], the base object is cleaned and a series of coatings are applied over a period of weeks, with each coating applied only after drying, hardening, and polishing the previous coat. Overall, this is a very complex process, but the final coating has a high standard of artistic value. The hardening process is thought to involve oxidative polymerization and crosslinking [142]. Urushiol is also used in other surface coating applications. High adhesion, thermal, acid, and alkali-resistant epoxy resin paints have been made by adding the product of reaction between urushiol and hexamethylene diamine [143]. Black ointments for printing ink, plastics, and carbon paper have been obtained from urushiol and certain iron salts [144]. Salts of thiosulphate ester of urushiol have found use as lubricants and additives [145].

Although urushiol possesses an interesting structure for transformation into speciality polymers, no attempt has been reported. Notwithstanding its applications in a specified area, it appears that it is not properly put to use as it can be converted to polymers with better properties. The possibilities for such conversions into high-performance polymers are illustrated by cardanol, a phenolic lipid of related structure obtained from *Anacardium occidentale*.

2. Anacardium occidentale

This is one of the most widely distributed plants cultivated formally to obtain the well-known edible nut popularly known as a cashew nut. The phenolic lipid is only a by-product known under the name cashew nut shell liquid (CNSL). The kidney-shaped nut is attached to the base of the cashew apple (Figs. 7 and 8). It consists of an ivory colored kernel covered by a thin brown membrane (testa) and enclosed by an outer porous shell, the mesocarp of which is about 3-mm thick with a honeycomb structure where the reddish brown liquid (CNSL) (18–27% of raw nut weight) is stored [133,134,146]. In the natural form, CNSL is a mixture of four components with a phenolic acid, anacardic acid as the major component [133,134,147]. The other major ingredient is a dihydric phenol, cardol, with traces of a monohydric phenol, cardanol, and 2-methyl cardol [142,148–151].

Extraction

Originally considered a waste material, CNSL has attracted attention only when its potential as a source for phenol for formaldehyde polymerization was recognized and used in a compounded form as a friction dust in automotive brakes and clutches [135]. A number of processes exist for the extraction of CNSL, but most industries follow the hot oil "CNSL bath process" because it directly gives cardanol through the decarboxylation of anacardic acid effected by the hot conditions in the bath [134]. The process, in brief, consists of heating the raw nuts submerged in a pool of CNSL moving on a slowly traveling conveyor belt. The CNSL thus obtained is called technical grade and has a specification in India (IS: 840(19964) [152]). Since the main demand is for cardanol or cardanol-enriched CNSL, the hot oil bath process acquires significance. A solvent extraction process is practiced on a limited scale in Brazil, East Africa, and India [137]. There is also the expeller technique whereby the shell obtained after the removal of the kernel is subjected to an expeller under pressure to expell the CNSL. This and the solvent process have the disadvantage that they give CNSL rich in anacardic acid and, hence, it has

Figure 7 Cashew nut and apple.

to be subsequently decarboxylated. However, technical grade CNSL is contaminated by the polymer formed by heat polymerization.

Structure

The structures of anacardic acid, cardol, cardanol, and 2-methyl cardol are given in Fig. 9. Each component is a mixture of four constituents differing in structure in the degree of unsaturation of the side chain. The percentage content of the individual constituents of cardanol is given in Table 7. On heating to above 120°C anacardic acid gets decarboxylated to cardanol. The composition of natural and technical CNSL is given in Table 8. Cardanol is generally obtained by the distillation of CNSL under reduced pressure [153].

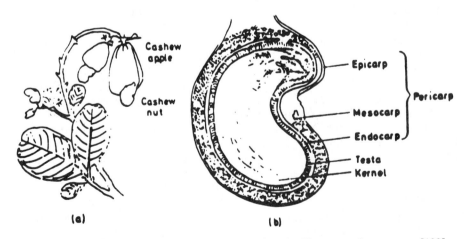

Figure 8 Cross-section of cashew nut showing the Honeycomb structure [133].

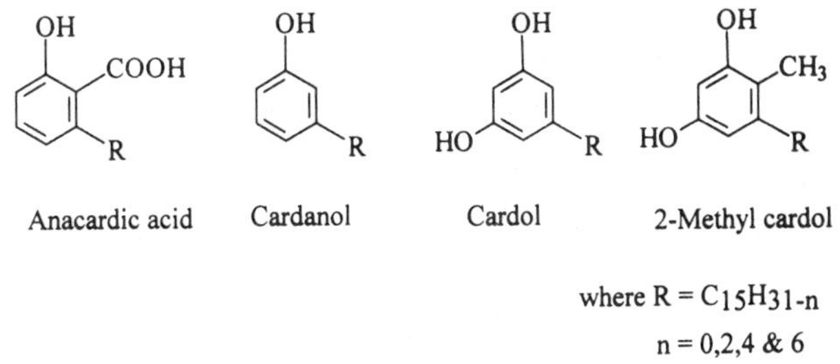

where R = $C_{15}H_{31-n}$

n = 0, 2, 4 & 6

Figure 9 Structures of CNSI components.

Table 8 Phenolic Composition of Natural and Technical CNSL

Component	Natural CNSL	Technical CNSL
Cardanol (%)	1.20	62.8
Cardol (%)	11.31	12.3
2-methylcardol (%)	2.04	2.1
Anacardic acid (%)	64.93	—
Polymer and minor materials (%)	22.82	23.8

Speciality Structural Features of the Components of CNSL

When compared to phenol, the structure of cardanol, a representative candidate among the components of CNSL, has many interesting features [16,140]. The presence of the C_{15} hydrocarbon side chain at the *meta* position provides additional opportunities for manipulation. For example, unlike phenol, cardanol can be polymerized by chain reaction mechanisms. This gives rise to opportunities for selection of control of polymerization for a particular product. For example, the preparation of resins for brake linings is reportedly made by making use of acid initiated side-chain polymerization followed by the well-known formaldehyde condensation [154]. This ensures build-up of enough flexibility to prevent brittle failures. Apart from the variety in polymerization reactions, cardanol can also undergo chemical modifications at the hydroxyl group, at the reactive sites of the aromatic ring, and on the side chain. Moreover, the presence of the hydrocarbon side chain imparts additional properties such as internal plasticization, flexibility, acid and alkali resistance, moisture resistance, etc. The anticipated reduction in mechanical properties can be compensated by using better curing agents for the system. The following discussion is oriented to bring to focus the potential of cardanol for high-performance applications vis-a-vis its conventional uses.

Utilization

Described often as "a versatile industrial raw material" CNSL is cited to have innumerable applications in many areas. The large number of patents [155–157], reports [135], monographs [146], reviews [136,137,153,

Table 7 Percentage Content of Individual Constitutents of Cardanol

Constituent	Percentage
Saturated	5–8
Monoene	48–49
Diene	16–17
Triene	29–30

158–162], and publications [16,133,134,137], on CNSL and its products provide ample evidence of their significance and applicability. This may, however, give the wrong information that the utilization of CNSL is at its maximum possible level. Contrary to expectations, it is generally reported, particularly in countries such as India, Africa, etc., that the demand for CNSL is far below production levels. Consequently, many factories have opted to stop production of CNSL, and they are content with the lucrative business of the kernels alone. In short, CNSL and its products have not found enough commercial opportunities for total utilization. For example, the total availability of CNSL in India and in the world is of the order of 15,000 tonnes and 125,000 tonnes, respectively. It is calculated based on the production of raw nuts that the potential availability of CNSL in India alone is 45,000 tonnes. This large gap is due to the lack of interest in successful commercial ventures that can offset the growth of synthetics. A study [32,163] carried out in this laboratory on CNSL utilization have brought out many significant points. A few are cited as follows: (1) the limitation in its total availability restricts high-volume productions; (2) variations in the properties of the raw materials from place to place are serious impediments to its development, particularly for products with specific property profiles; (3) CNSL and its distillation product cardanol are mixtures of a number of components and, hence, their polymer products may have variations in property profiles from batch to batch; (4) the components of CNSL, however, have very interesting structural features for transformation onto high-performance/functional polymers; (5) production of high-value low-volume materials will be within the total availability of CNSL, and, hence, the appropriate utilization of CNSL will be in the area of high-value polymers. This laboratory has, therefore, conducted extensive experiments to show the applicability of CNSL and cardanol for developing high-performance/functional polymers. This will be dealt with after giving a short account of the conventional industrial applications of CNSL.

Industrial Uses

Brake linings and clutch facings. CNSL and cardanol have found extensive uses in automotive brake lining applications [135,158]. Brake linings and clutch facings based on CNSL polymers show very low fade characteristics and very high recovery [164]. It probably acts by absorbing/dissipating the heat created during the braking action and, at the same time, the braking efficiency is improved. They also exhibit a low-noise level and give a quieter braking action. Thus, CNSL–formaldehyde resins and their crosslinked product processed into friction dust are extensively used in brake linings formulations.

Surface coatings. CNSL/cardanol-based surface coatings possess excellent gloss and a surface finish with a high level of toughness and elasticity [137,147, 158–162]. It is widely known that CNSL resin is added to synthetics by paint/varnish manufacturers to control property and to reduce cost. Its antitermite and antimicrobial properties are well known from ancient times. However, its dark color restricts its outlets to anticorrosion primers, black enamels, marine paints, etc. Photopolymerization under sunlight activated by copper sulphate has shown that CNSL is a good surface coating for extending the life of coconut leaf thatch [165,166].

Foundry core oil. CNSL resins are known to impart good scratch hardness to sand core after they are baked [133,134,137,167]. It also provided resistance to moisture and weathering and good green strength, and surface finish to molded articles.

Laminating resins. CNSL resins are added to laminates based on phenol-formaldehyde, epoxy, etc. to reduce brittleness and to improve flexibility of the product. The resins also exhibit better age hardening and improved bonding to the substrate [133,134,137,168].

Speciality/High-Performance Polymers from Cardanol

It has been indicated earlier that the components of CNSL have the structural features required for introducing speciality properties [16,32]. It required fresh ideas and approaches to introduce molecular design aspects of information gathered from developments in modern polymer science and technology. One of the impediments has been the lack of enough information on polymerization characteristics and structure property correlations on the system. Although there is a vast patent base for products based on cardanol, the gap between technology and scientific understanding on such systems is quite substantial [16]. Most of the processes were developed by trial-and-error methods to arrive at a process to make a particular product. There was not enough data on polymerization characteristics and structure–property relations of CNSL/cardanol polymers prior to 1990. The exception being the work of Misra and Pandey [169] on the kinetics of formaldehyde condensation of cardanol. Few attempts have been made to find out about the nature and mechanism of chain reaction polymerization through the double bonds in the side chain. Many of the patents cite polymerization involving acid catalysts such

as sulphuric acid, phosphoric acid, diethyl sulphate, metal oxides, or even metals [155–157]. Lewis acids such as aluminium chloride are reported to be used [134], but there is no information available as to kinetics or mechanism of polymerization. The first attempt toward creating some level of understanding of the nature of chain reaction polymerization of cardanol was done by Pillai and his group [16,25–28,32,137,170–173]. One of the intriguing aspects of free radical initiation of cardanol using initiators such as benzoyl peroxide or azo-*bis*-isobutyronitrile was the total absence of initiation. It was subsequently found that the inability to achieve polymerization by free radical initiators has its origin in the capability of the molecule to act as an antioxidant. Cardanol, however, was found to respond to cationic or anionic initiators. Manjula et al. [170] studied the kinetics and mechanism of oligomerization of cardanol using acidic initiators, such as mixtures of sulphuric acid–diethyl sulphate and phosphoric acid–diethyl sulphate, and a cationic initiator such as borntrifluorideetharate. Gel permeation chromatography (GPC) data of the polymer showed the formation of only oligomers consisting of dimer, trimer, tetramer, and some higher analogues in trace amounts. IR spectra of the products of oligomerization showed a decrease in the intensity of the double bond absorption band at 1630 cm^{-1} and the disappearance of the terminal vinyl band at 895 cm^{-1}. ^{1}H NMR spectra showed drastic changes in the unsaturated resonance signals at 5.5δ with respect to saturated protons at $0.2–2.25 \delta$. The ratio of resonance integrals of unsaturated to saturated protons decreased from $1:6.5$ to $1:20$ after oligomerization. The kinetic studies showed that the oligomerization reaction follows a first order kinetics with respect to the monomer concentration and the rate constant is $6.6 \times 10^{-5} \text{ s}^{-1}$. It appears that the oligomerization may involve carbo-cationic mechanism with the addition of a hydrogen ion to one of the double bonds of the side chain. Thyman [133] has proposed that the polymerization of cardanol under acidic conditions might involve the formation of an allylic carbonium ion from the more reactive triene component. The present studies indicate the formation of a carbo-cationic center as the initiation step, but no evidence could be obtained on the formation of an allylic carbonium ion. A kinetic scheme proposed to account for the oligomerization is given in Fig. 10.

When concentrated sulphuric acid alone was used as the initiator, the polymerization was found to follow a different path. It is well known that Bronsted acids can function as cationic/pseudocationic initiators in the oligomerization of olifins [174]. If the counter ion has a higher nucleophilicity as it forms cation-conjugate pairs, which collapse rapidly, polymerization will not take place. As the counter ion in the case of sulphuric acid is not very strong compared to the cation, oligomerization can take place, but may not be to a very high molecular weight. This, however, depends on the nature of the

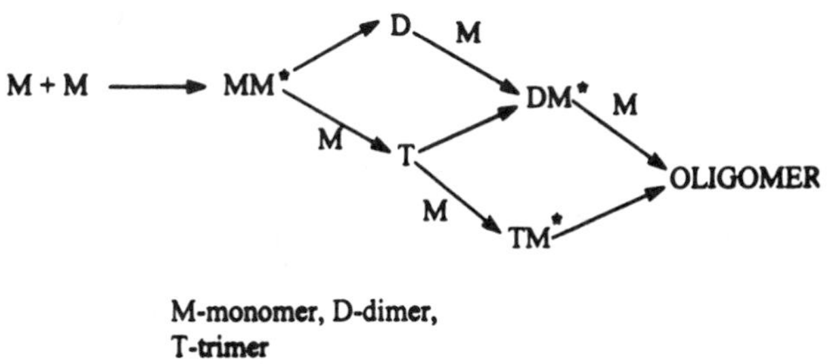

M-monomer, D-dimer,
T-trimer

Figure 10 Kinetic schemes of oligomerization of cardanol.

monomer. Oligomerization can only be expected from cardanol because of the long side chain. This has been demonstrated by the work of Manjula et al. [170]. However, when concentrated sulphuric acid was used as the initiator, the system was found to gel after a certain time and temperature. For example, at 1% concentrated sulphuric acid as initiator, sudden gelation occurs above 200°C and after 3 h. Addition of concentrated sulphuric acid at high concentrations was found to initiate gelation at the ambient temperature itself. A steep rise in viscosity was noted just prior to gelation [175].

The oligomerization of cardanol with boron trifluoride etharate as the initiator was studied in detail by Antony et al. [171]. The reaction conditions were optimized by using gel permeation chromatography as 140°C with an initiator concentration of 1%. GPC data indicate conversion of all monoene, diene, and triene components into polymer except the saturated component, indicating participation of all the unsaturated components in polymerization. It is possible that the initiation of po-

lymerization might start with the triene component because of the presence of the vinyl grouping. Later, John and Pillai [173] showed that cardanol gets grafted onto cellulose in the presence of BF₃-etharate and the mechanism of grafting has found to follow a cationic mechanism as given in Scheme 5.

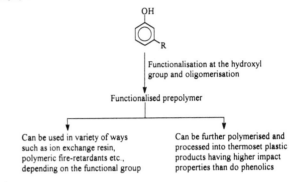

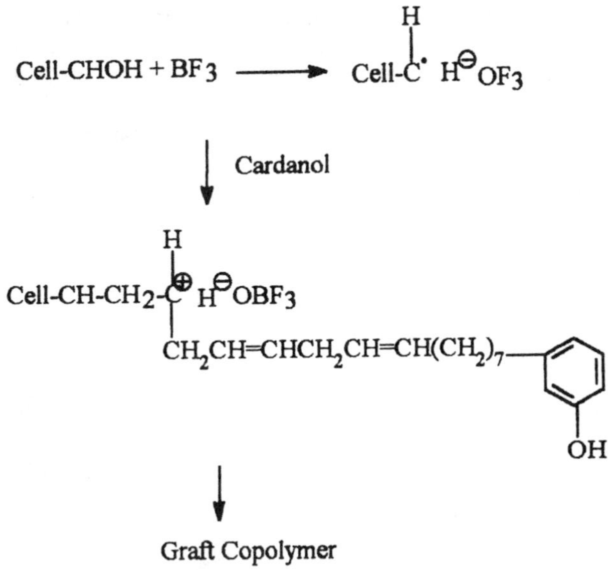

i. Prepolymer route:

Functionalisation at the hydroxyl group and oligomerisation

Functionalised prepolymer

Can be used in variety of ways such as ion exchange resin, polymeric fire-retardants etc., depending on the functional group

Can be further polymerised and processed into thermoset plastic products having higher impact properties than do phenolics

ii. Dimer route:

Dimerisation → Dimer

Copolymerisation with rigid and bifunctional monomers

Linear chain polymers with repeating sequences of hard and soft segments. Possibility of formation of liquid crystal polymers and thermoplastic elastomers.

iii. Free radically polymerisable monomer route:

Introduction of vinyl groups through hydroxyl group → Free radical polymerisation → Self-crosslinkable polymers

iv. Oxidation route :

Chemical modification at the hydroxyl group and oxidation

Copolymerisation with rigid rod monomers → Liquid crystal polymers

v. Reduction route:

Diazotisation and coupling → Azocompound → Reduction → New monomer:

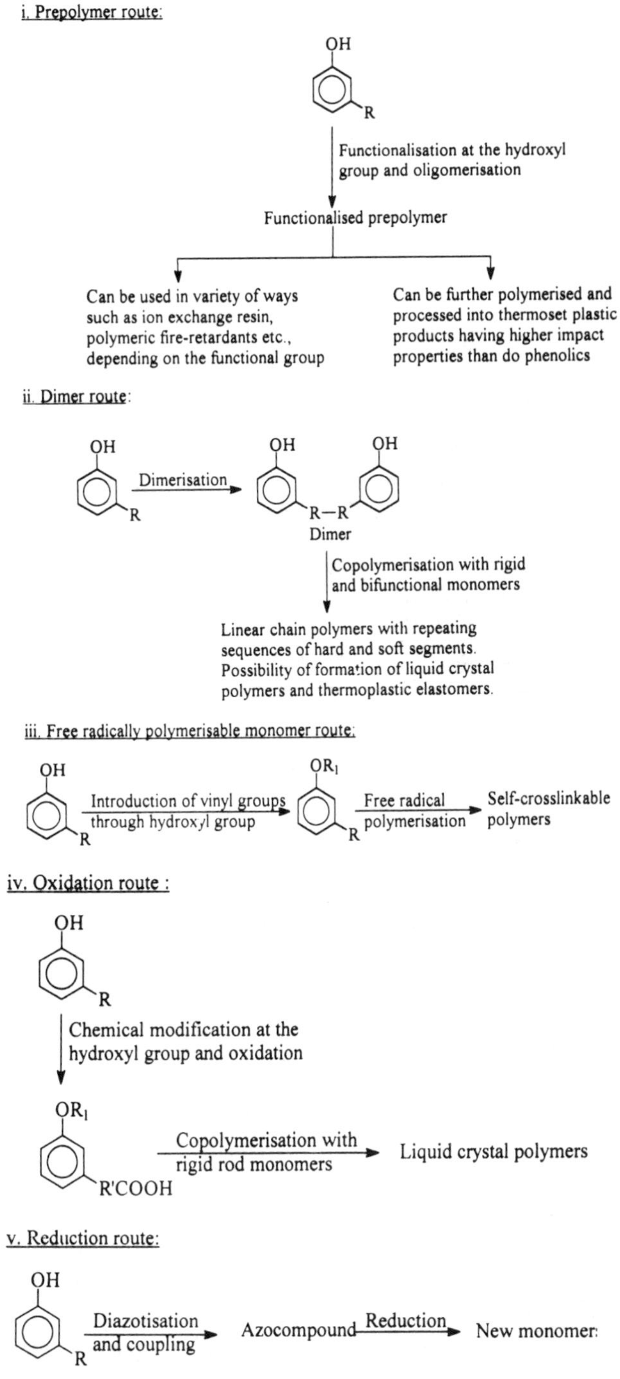

Figure 11 Novel strategies for value addition to cardanol polymers.

Cell-CHOH + BF₃ ⟶ Cell-Ċ H⊖OF₃

↓ Cardanol

Cell-CH-CH₂-C⊕ H⊖OBF₃

CH₂CH=CHCH₂CH=CH(CH₂)₇——⟨ ⟩

OH

↓

Graft Copolymer

Scheme 5 Mechanism of grafting of cardanol onto cellulose.

Table 9 Chemical and Physical Properties of PCP

Number	Property	Cardanol	PCP
1.	Color	Light yellow	Light brown
2.	Moisture content (%)	0.001	0.05–0.065
3.	Specific gravity at 29°C	0.9320	1.0590
4.	Viscosity at 30°C cps	450–520	35,000–45,000
5.	Hydroxyl number, mg of KOH	185–200	4–10
6.	Iodine value g/100 g of resin (Wijs method)	212 min.	90–100
7.	Unreacted phosphoric acid content	—	0.009–0.01
8.	Phosphorus content (%)	—	7.9

Novel Approaches to Synthesize High-Value Polymers from Cardanol

Considering the special structural features of cardanol and that it is still underutilized, especially in India, novel strategies were designed to develop methods to design speciality/high-performance polymers from cardanol [16,32]. Figure 11 gives a rough outline of some of the approaches adopted by the author and his group. All the methods mentioned do not lead to high-performance/functional polymers, but there is definitely a value addition that might be of use in ultimate utilization of cardanol. Hence, all the methods that give a value addition to the exising products/materials are discussed.

Flame retardants. Cardanol, although one may point out that the long side chain can act as fuel provider under actual flame conditions, has certain interesting structural features for developing into wide spectrum flame retardants (FRs). The presence of both hydrophobic and hydrophilic groups in the same molecule makes it an ideal material that can be compatible with a wide spectrum of polymers. Cardanol has reactive sites for incorporation of FR elements such as phosphorus and bromine. It can also be polymerized to a prepolymer so that problems, such as migration to surface, blooming, etc., faced by the conventional FRs can be overcome. The resulting wide spectrum FRs can be used for a variety of polymers (generally, for many of the applications a specific polymer has been found to require a specific FR) so that definite cost reductions can be achieved. Cardanol can be phosphorylated by simple reations with orthophosphoric acid/phosphoryl chloride/phosphorus pentoxide. Pillai et al. [16,163] studied the direct reaction of orthophosphoric acid with cardanol and noted that phosphorylation is simultaneously followed by oligomerization to give phosphorylated cardanol prepolymer (PCP). The product was light-brown in color (see Table 9 for chemical and physical properties); soluble in most of the common organic solvents, but insoluble in water, and semi-solid at room temperature, but starts to flow above 80°C. Although GPC showed the presence of only oligomeric species in the system, PCP exhibited extremely high viscosities to the level of 0.35–0.45 million cps. The high viscosity has been ascribed to hydrogen bonding [176] between the phosphate moieteis and not due to contributions from chain length.

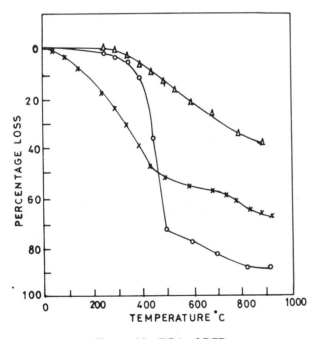

Figure 12 TGA of PCP.

Table 10 Effect of PCP on the LOI of Low-Density Polyethylene and Cellulose

Number	System	% PCP	LOI
1.	LDPE	0	16.9
2.		10	18.5
3.		20	18.5
4.	Cellulose	0	17.5
5.		10	21.5
6.		20	23.0
7.		30	23.5
8.		40	24.0

Table 11 Synergistic behavior of Nitrogen Compounds on Cellulose–PCP blends

System	Effect Due to	Change in LOI
Cellulose/Urea/30%PCP	PCP alone (P)	4.5
	Urea alone (N)	7.0
	Combined effect (P + N)	14.5
	Expected additive effect	11.0
Cellulose/Melamine/30%PCP	PCP alone (N)	4.5
	Melamine alone (N)	3.0
	Combined effect (P + N)	7.5
	Expected additive effect	7.3
Cellulose/HMTA/30%PCP	PCP alone (N)	4.5
	HMTA alone (N)	1.5
	Combined effect (P + N)	7.0
	Expected additive effect	6.0

The cationic polymerization of cardanol under acidic conditions has been referred to earlier [170,171]. NMR studies [16] indicated a carbonium ion initiated mechanism for oligomerization. PCP was found to be highly reactive with aldehydes, amines, and isocyates. Highly insoluble and infusible thermoset products could be obtained. Hexamine-cured PCP showed much superior thermal stability (Fig. 12) at temperatures above 500°C to that of the unmodified cardanol–formaldehyde resins. However, it was definitely inferior to phenolic resins at all temperatures. The difference in thermal stability between phenolic and PCP resins could be understood from the presence of the libile hydrocarbon segment in PCP.

Completely cured PCP was highly resistant to hydrolysis by water and in this respect it will have a score over phenolic resin. Studies on mechanical properties showed that PCP cured with HMTA gave a tensile strength of 17–22 MN/m^2 and an impact strength of 1.63–2.04 J compared with 24–48 MN/m^2 for phenol–formaldehyde resin (novolac) prepared under similar conditions [16,163]. It shows that the presence of the flexible segments in PCP allows it to have a higher impact property than that of phenolic resin. The use of cardanol-based epoxy resin for high-impact applications can be understood based on this information. One of the significant improvements in properties has been seen in the case of lap shear strength (lss) of PCP bonded wood pieces. The lss value for PCP was 400 ± 10 N/cm^2, whereas that for conventional cardanol–formaldehyde was only 60 ± 10 N/cm^2. So it appears that this material can have applications in bonding wood and similar materials.

With a phosphorus content of 7.9%, PCP showed FR behavior mainly in oxygenated polymers. This is in agreement with the well-known fact that phosphorus compounds are more effective in oxygenated polymers in bringing out flame retardancy than in other polymers [177]. Thus, Table 10 shows that at 20% loading, PCP

gives better performance with cellulose than with polyethylene. However, at 25% loading, PCP gave a LOI value of 24.5 [178] with LDPE indicating that there is some other mechanism involved in the flame retardation of LDPE. As CNSL is reported to exhibit an antioxidant property, it may act to prevent surface oxidation of LDPE, thereby enhancing the LOI value [179].

As nitrogen compounds are known to improve the LOI values of phosphorus FR positively, the effect of a few nitrogen compounds such as urea, melamine, HMTA, etc. were studied. Table 11 shows that nitrogen gives a synergistic effect with phosphorus in cellulose. Similar results have also been observed for LDPE.

Other applications of PCP. CNSL resins are known to provide better fade and quieter braking action in brake lining materials. Table 12 gives the properties

Table 12 Typical Properties of Anorin-38 Based Asbestos Short Fiber Reinforced Brake Linings

Property	Value
Tensile strength (MN/m^2)	17–22
Impact strength (ft-lb)	1.2–1.5
Water absorption (96 h, %)	2.9–3.5
Acetone extract (%)	3.0–4.0
Abration loss (g/mm^2)	1.12×10^{-4}–6.2×10^{-5}
Average coefficient of friction (dry)	0.335–0.472
Average coefficient of friction (wet)	0.335–0.419
Fade, dry (%)	19–40
Fade, wet (%)	40–50
Recovery, dry (%)	87–115
Recovery, wet (%)	80–100
Wear loss (cm^3/kwh)	0.691–1.479
Density (g/cc)	1.48–1.51
Maximum variation in the initial test for coefficient of friction (%)	3–11

Table 13 Flame Retardant Properties of NR-PBPCP System

Property	NR Alone	NR-PBPCP System (cured)
LOI	15	28
Time for self-extinction (s)	107–145	1–3
Weight loss (%)	35.5	1.6

of Anorin-38 [180] (phosphorylated CNSL prepolymer) based on brake linings with asbestos fibers as the reinforcing agent [137,181,182]. The fade value of 19% is the lowest reported in the literature.

Anorin-38 has also shown an interesting effect as a multifunctional additive (a single additive to replace many of the conventional additives) for natural rubber (NR). It showed excellent blending behavior and compatibility with NR. Aorin-38 enhances the tensile properties and percent elongation, decreases fatigue, acts as an antioxidant and antiozonant, and positively affects many of the other properties, apart from acting as a process aid and a cure enhancer [183–186].

The FR characteristics of PCP and Anorin-38 were improved substantially by introducing bromine. When five bronine atoms were introduced by controlled bromination, the resin (the product obtained was still fluid in nature and could be crosslinked by hexamine to get hard partially brominated PCP–PBPCP) showed excellent FR characteristics exhibiting self-extinguishing property and UL 94 V-0 grade when blended with polyethylene, NR, etc. There was excellent compatibility with polyolefins as well as with other polymers such as cellulose, polyurethane, etc. Table 13 shows the FR properties of NR–PBPCP.

One of the interesting properties of PBPCP [187] was its fast heat dissipation characteristics and so it was tested by the well-known oxy-acetylene panel test (ASTM 285-70) for ablative materials. Figure 13 shows the survival of a flower for 100 s. kept on the 6.35-mm asbestos fiber-reinforced hexamine-cured panel. The ablation rate value of this material was 3.2×10^{-3} in/s in comparison with 3.6×10^{-3} in/s for asbestos–phenolic. As the char content of PBPCP was only 27% compared with 60% for conventional phenolics, mechanisms involving transpiration processes rather than heat blocking by char formation might be playing a greater role in this case [188].

Figure 13 Oxy-acetylene panel test showing the survival of a flower kept on the asbestos–CNSL polymer composite of thickness 6.35×10^{-3} m.

Table 14 LOI Values, Char Yields, and Thermal Stability (in Air)

Polymer	LOI	Cy (%)	T_{50}, °C
CF	19	11	480
MCPAF	27	21	480
BrCF	45	18	430
BrMCPAF	49	27	425
PF	35	62	—
MPPAF	50	66	—
BrPF (linear)	48	2	400
BrMPPAF	56	26	325
PPF	46	58	—

CF = cardanol-formaldehyde resin (cured); MCPAF = monocardanyl phosphoric acid-formaldehyde resin (cured); BrCF = bromo derivative of CF; BrMCPAF = bromoderivative of MCPAF; PF = phenol-formaldehyde (cured); MPPAF = monophenyl phosphoric acid-formaldehyde (cured), BrPF = bromo derivative of PF; BrMPPAF = bromo derivative of MPPAF; PPF = phenol-formaldehyde resin phosphorylated (cured).

A series of phosphorus- and bromine-containing FRs were synthesized and studied to understand their role, especially their combined effects. Thus, monocardanyl phosphoric acid, its bromo derivatives and their formaldehyde condensates and crosslinked products [28,188] were prepared and their properties compared with analogous products made from phenol [28,189]. Table 14 gives the LOI values, char yields (Cy at 600°C), and thermal stability at 50% (T_{60}) decomposition.

The lower thermal stability of cardanol–formaldehyde resin and their derivatives were expected because of the presence of the libile side chain in the system. Although phenolics are superior in their properties, their bromo derivatives exhibit very low char yields. Oxidation of the char by a decomposition product is suspected. Evaluation of the LOI data with char yields individually for phosphorus and bromine suggests a positive interac-

tion between phosphorus and bromine in the case of the cardanol derivatives. No support for any synergestic effect nor for an additive effect between phosphorus and bromine could be found. The mechanistic contributions of phosphorus and bromine were found in the condensed phase and vapor phase, respectively [188,189].

As halogen-based FRs give rise to toxic fumes and gases, the search for alternate FRs with similar properties are now a topic of great interest. The efficiency of halogen FRs are such that they are very difficult to be substituted. In this connection, polyphosphate-based nonhalogen FRs based on CNSL and cardanol offer opportunities for development [190].

It has been shown earlier that cardanol possesses multiple sites for chemical modification. It has of late been pointed out that brominated compounds with bromine attached to both aliphatic and aromatic moieties of the same molecule would serve as better FRs as they can provide an uninterrupted supply of HBr over of a wide range of temperature. As the aromatic brominated compounds have melting points nearer to the decomposition temperature of many polymers, it is possible that the effect of the FR will not be felt at the initial stages of decomposition. This problem could be overcome by the use of compounds having bromine attached to both aliphatic and aromatic moieties. Cardanol with a structure having both aliphatic and aromatic moieties in it is the right choice. Moreover, the improved miscibility and compatibility of cardanol-based FRs are expected to enhance the functioning of such FRs. Thus, direct halogenation of cardanol gives FRs with very high halogen contents [190]. A series of FRs [191,192] were thus prepared by direct bromination of CNSL and cardanol by bromine/bromine water to get FRs with a varying number of bromine atoms in them. Of these FRs, 3-(tetrabromopentadecyl) 2,4,6-tribromophenol (TBPTP) [192] and 3-(tetrabromopentadecyl) 2,4,5,6-tetrabromophenol (TBPTP) [192] were further characterized and TBPTP has been found to be both cost effective and have better

Table 15 Properties of EVA–TBPTP Based Cable Material

Property	EVA–TBPT Cable	EVA–DBDPO Cable	Specification (Jacketing material)
LOI	30	29.5	30 min.
Oxygen index at 250°C	436	380	250 min.
% Smoke density rating	3.0	48.5	20–30 max.
% Light transmission	96	20	80 min.
Max. % light absorption	4	80	20 min.
Acid gas emission (%)	2	10.5	2 max.
Volume resistivity (ohm cm)	1.4×10^{10}	1.4×10^{10}	—
Dielectric constant (at 100 Khz)	2.08	5.76	—
Tensile strength (Mpa)	6.8	7.2	6.0
% Elongation	400	450	150 min
Aging at 100°C/168 hrs. TS/% Elongation	7.8/300	7.7/350	± 30% variation

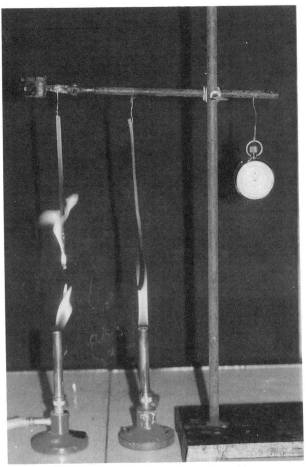

Figure 14 Vertical burning test of EVA–TBPTP cable material. The untreated control sample of EVA can be seen burning.

with PMMA did not alter the mechanical properties much, it showed that the PMMA decomposition was altered and a stabilization from 50% decomposition at 350°C to 15% decomposition was noted. However, an unusual increase in the T_g of CF from 128°C to 144°C was observed. This suggests a restriction in the segmental motion of the CF phase brought about by the mixing of another rigid polymer such as PMMA [27,193,194].

Free radically polymerizable monomers and polymers. It has been indicated that cardanol resists polymerization by free radical initiators. So introducing free radically polymerizable moieties would enable cardanol to become a potential monomer for better utilization. With this view in mind, the hydroxyl group of cardanol was chemically modified with acrylate/methacrylate groups [24,195,196]. The reaction schemes are given in Scheme 6. When prepared in solution, PCA on exposure to air was found to get crosslinked to a transparent film. This gives an opportunity to utilize CNSL for preparation of clear/transparent coatings and paints of all colors as it has been noted that it is almost impossible to change the dark-brown color of CNSL/cardanol polymers. A dream of CNSL manufacturers has come true with the successful preparation of this polymer! The formation of the transparent film was ascribed to autoxidation reaction taking place at the side chain of cardanol moiety of PCA [196]. This resin may find use in speciality coatings or as a polymer support (see below) or in

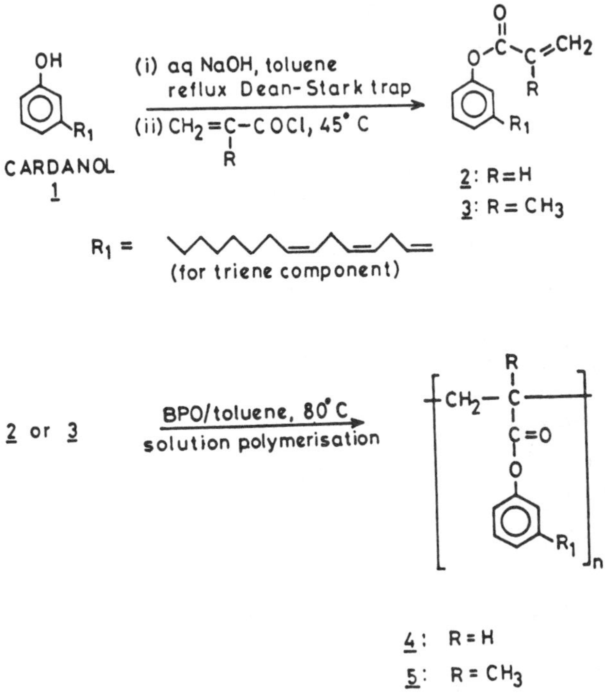

performance. The performance of this FR was established for use in ethylene–vinyl acetate (EVA)-based cable material. It was surprisingly found that TPBTP gives rise to extremely low smoke in the case of EVA and the mechanism by which it functions still needs to be elucidated. The formulation containing the FR for EVA could be categorized in the grade FRLS (flame retardant low smoke) (see Table 15 for the properties in comparison to decabromodiphenyloxide (DBDPO). The cable also exhibited self-extinguishing behavior and could be classified under UL 94 V-0 grade. Figure 14 depicts a vertical burning experiment in comparison to EVA alone without the FR. Formulations for EVA–polyethylene blend and polyethylene alone also gave good FR behavior with TBPTP.

Interpenetrating network polymer. In a separate study, it was shown that cardanol–formaldehyde resins form semi-interpenetrating networks with polymethylmethacrylate (PMMA). Although interpenetration of CF

Scheme 6 Synthesis of cardanyl acrylate and poly(cardanyl acrylate).

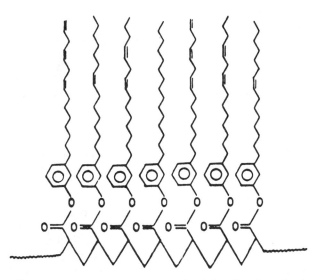

Figure 15 Schematic structure of poly(cardanyl acrylate).

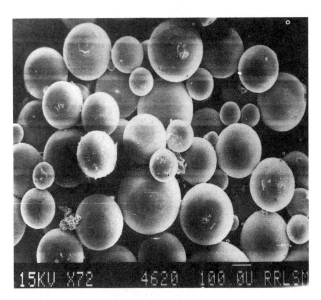

Figure 16 Scanning electron micrograph of poly(cardanyl acrylate) beads.

photoresistent material. The cross-linking reaction is thought to be favored by the acrylate backbone providing a template for the side chains coming in close proximity to form the crosslinks. Figure 15 depicts a schematic representation of PAC to favor such a reaction.

When the polymer was prepared by the suspension polymerization technique, the product was crosslinked beads of unusually uniform size (see Fig. 16 for SEM picture of the beads) with hydrophobic surface characteristics. This shows that cardanyl acrylate/methacrylate can be used as comonomers-cum-cross-linking agents in vinyl polymerizations. This further gives rise to more opportunities to prepare polymer supports for synthesis particularly for experiments in solid-state peptide synthesis. Polymer supports based on activated acrylates have recently been reported to be useful in supported organic reactions, metal ion separation, etc. [198,199]. Copolymers are expected to give better performance and, hence, coplymers of CA and CMA with methyl methacrylate (MMA), styrene (St), and acrylonitrile (AN) were prepared and characterized [196,197].

The beads were almost uniform in size and crosslinked. Effective crosslinking has occurred without addition of any cross-linking agents. Table 16 gives the swelling behavior of the beads in various solvents. Swelling is one of the factors that affect the accessibility of functional groups for chemical interactions. Swelling studies help to identify a good solvent for media to be selected for performing a reaction on polymer supports. It can be seen from Table 16 that swelling in chloroform is comparatively higher than in other solvents for St and MMA copolymers. This possibly suggests a hydrophobic character for the beads. Similarly, as AN copolymers show higher swelling in DMF than in other solvents, DMF might be a good solvent for the AN copolymer.

Cardanol grafted cellulose. One of the advanced techniques to improve the properties of a polymer is to graft another polymer onto it. Grafting of vinyl monomers onto cellulose has been the subject of extensive studies during the last decade or two. Grafted cellulose copolymers have been found to have improved proper-

Table 16 Swelling Behavior of the Copolymers of CA and CMA with MMA, St, and AN

Polymer	Weight of Beads (g)	Chloroform (g)	Toluene (g)	Ethyl Acetate (g)	Dimethylformamide (g)
CA:MMA	1	2.43	1.2	1.85	1.1
CA:St	1	2.51	1.31	1.92	1.05
CA:AN	1	1.7	1.02	1.3	1.85
CMA:MMA	1	2.39	1.32	1.41	1.10
CMA:St	1	1.72	1.39	1.52	1.02
CMA:AN	1	1.42	1.39	1.50	1.93

Table 17 Water Repellency of Cardanol-Grafted
Cellulose in Comparison to Other Copolymers

Copolymer	Extent of Grafting (%)	Time Required to Sink (s)
Untreated cellulose	0	1
Cellulose–n-butyl acrylate	29.4	720
Cellulose–isobutyl-acrylate	29.6	490
Cellulose–methylmethacrylate	28.9	120
Cellulose–styrene	30.7	240
Cellulose–cardanol	25.3	1600

ties in terms of water repellency, easy wear characteristics, crease resistance, toughness, flame retardance, etc. [101,102]. Cardanol was found to be a biomonomer, which has not been attempted at all for grafting reactions. The grafting was carried out using boron trifluorideetharate as the initiator [24]. The graft yield of 15–20% was obtained within 2 hours without significant gelation of the monomer. The grafted products exhibited excellent water repellency and resistance to acid aging. The grafted sample took 1600 s to sink in water, whereas the untreated sample sank in 10 s. Table 17 gives comparative evaluation of water repellency of various grafted copolymers. The superiority of the cardanol grafted product is also seen in the acid aging data. The polycardanol grafted cellulose took 24 h for complete degradation compared with the ungrafted one, which degraded within 10 min in 72% sulphuric acid. This was superior to a similar product of styrene grafted cellulose as reported by Jenkins et al. [200]. The superior performance of cardanol grafted cellulose might be due to the crosslinking of the side chain during grafting. This is exemplified by the fact that the film obtained after the acid hydrolysis is insoluble and infusible.

Liquid crystalline polymers. One of the remarkable achievements in converting cardanol into high-performance polymers was the successful synthesis of LC polymers [201–203]. They represent a new class of polymers that exhibit unusual properties and represent one of the most exciting developments in high-performance polymers. LC polymers are characterized by the existence of mesophase that exhibits pronounced anisotropy in shape. This generates organized fluid phases either on melting (thermotrotropic) or on dissolution (lyotropic). Making use of this anisotropy in the processing of LC polymers, it was possible to achieve impressive properties for the production of high-performance polymers (high strength, high modulus, high heat resistance that are close to theoretical values) fibers, films, etc. Such outstanding mechanical properties are characteristic of main chain LC polymers. In contrast, side-chain LC polymers find applications in display or imaging technol-

ogy because they orient themselves in electric or magnetic fields. Some of the LC polymers are being projected as the futuristic materials in nonlinear optics.

The total insolubility and intractability of the wholly aromatic LC homopolyesters and homopolyamides posed serious problems in their processability during the developmental stage. Introduction of disrupters (flexible species, rigid kinks, etc.) into the main chain or copolymerization with disrupter monomers reduces the perfectly arrayed chains structures and lowers their melting. Vectra® and Xyder® are two of the early LC commercial products [204–207].

These copolyesters are said to have melting points near or above 300°C and, hence, have processing problems. The incorporation of a molecule having both "kink" structure and a flexible segment built into the same molecule would be an ideal system for further reduction in melting point. Such a monomer was prepared from cardanol by phase transfer catalyzed permanganate oxidation. This gave rise to the novel monomer, 8-(3-hydroxyphenyl) ocatnoic acid (HPOA), which possesses both a flexible segment and a kink structure [201,208] (Scheme 7). (The synthesis of this monomer solves all the existing problems in the utilization of cardanol as described earlier. This also fits into the idea of high value–low volume product.) HPOA on copolymerization with p-hydroxy benzoic acid gave an LC polymer having a transition temperature at 256°C [209,210]. Figure 17 shows the DSC trace of the polymer. As its thermal stability and mechanical properties are expected to be lower than those of Vectra® and Xyder®, it may find applications possibly in electronics, for example, connectors.

Cardanol could also be converted to an azophenyl group containing LC polymer. This was possible by per-

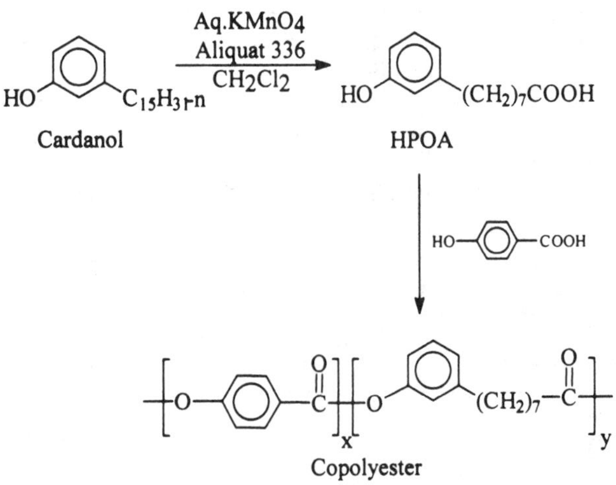

Scheme 7 Synthesis of HPOA and its polycondensation with *para* hydroxy benzoic acid.

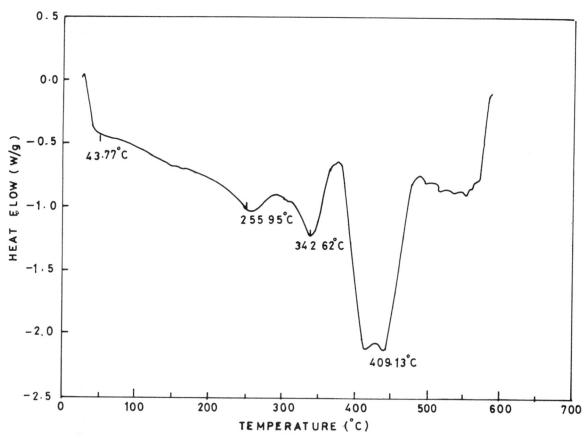

Figure 17 Differential scanning colorimetric trace of the liquid crystalline copolyester of 8-(3-phenyl hydroxy) octanoic acid and *para*-hydroxy benzoic acid.

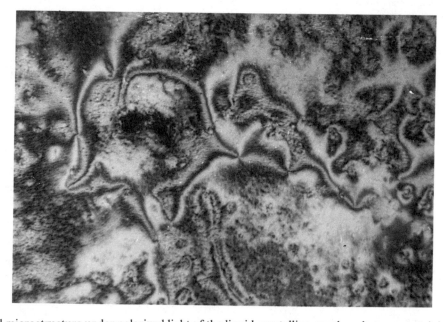

Figure 18 Optical microstructure under polarized light of the liquid crystalline azophenyl group containing polyester [211].

Table 18 Naturally Occurring Monomers [25]

Natural Monomer Already Polymerized		Natural Monomer Not Yet Polymerized	
Monomer	Source	Monomer	Source
Abeitic acid	Oleoresin	Acronylin	Acronichia
Adipic acid	Sugar beet	Alanosine	S. alanosinius
Anacardic acid	Anacardium occidentalle L	Anisoxide	Star anise oil
Ascaridole	Cheopodium oil	Betonicine	Betonica officinalis
Camphene	Oil of orange and lemon	-Bisabolone	Chrysanthamum
Cardanol	*Anacardium occidentalle L*	(3-carboxy-4-hydroxy phenyl) glycine	Dyer's weed
Cardol	Anacardium	Carnavaline	Cassia carnaval speg.
Cinnamic acid	Natural balsam, cocoa leaves, cinnamol oil	Carlina oxide	Carline aeaulis L.
Citral	Lemon grass oil, mandarin oil	Chelidonic acid	Asparagus, Celandine, anemone
Citronellol	rose oil, lemon oil	Chrysin	pinus species
Eugenol	Myrteceae and lauraceae	1,4-ciniole	Piper cubea
Furfuraldehyde	Enteromorpha, corncobs	Convolvulinic acid	Convolvulacea
Furfuryl alcohol	Oil of cloves	Coronaric acid	Chrysanthamum coronarium
Beta-furoic acid	Eronymus atropurpureus	Daizdein	Peuria thunbergia, berth
Fumaric acid	Tegestes erecta	Embelin	Embelia ribes brum
Glutamine	Sugar beet, lemon *Sweet orange*	Eremophelone	Eremophila mitcheli
Glutaric acid	Green sugar, beet juice	Eugenin	Eugeniacarruophyll ata
Isamic acid	Ongokea oil	Eparin	E. canabinum
Limonene	Lemon, orange oil	Farnesol	rose oil, Ambretta sed oil
Linalool	Mysore oil	Funtamine	Funtamia latifolia, stapf
Linoleic acid	Grape seed, tobacco seed, sun flower seed oil	Gorlic acid	Seed of Onsoba gorlic seed
Linolenic acid	Linseed oil, hemp seed oil, perilla oil	Holarrihidine	holarrhena antidy, senterica
Myrcene	Turpentine oil	Hordenine	Anhalonium species
Nerol	Neroli, bergamot oil	Jacobine	S. jacobaca L.
Ocimene	Ocimum bacilicum	5,6-dehydro-kawain	Aniba firmule, kawa root
Oleic acid	Olive oil, almond oil	Kojoic acid	Aspergilli
Ricinoleic acid	carstor oil. Argot oil	Meconic acid	Papaveraceaa
Sabinic acid	Juniperus sabina	Mimosine	Leucen glauca
Sebacic acid	Castor oil	Morin	Calico yellow
Succinic acid	Turpentine oil	Mycophenolic acid	Pencillium brevi, compactum
Squalene	Vegetable oil, olive oil, rice bran oil	Myrtenol	Myrtus cuninus
Teracasidin	Exudates from woody plants, tannin	3-methyl naphthalene 1,8-diol	Diopyrosmollis
Tartaric acid	shizardra, chimnsis, tamarindus	Neroli diol	Neroli oil
Urushiol	Rhus vernicefera, Rhus toxicodendron	O-orsellinic acid	Chaetomium-cochliodes
		Oryzoxymycin	Perilla malcinesis
		Perillene	Perilla citriodora, Makino
		Podocarpic acid	Resin of podocarpus cupressinium
		Pontica epoxide	Artemisia including A. pontica L.
		Prosopine	Prosopis afficana, Taub
		Quercetin	Aesculus
		5-nondecylresorcinol	Wheat bran
		5-tridecyl resorcinol	Grevillea robusta A.
		Sabinol	Oil of juniper
		Selinene	Celery seed oil
		Siamin	Salvia sclarea L.
		Solacongestidine	Solanium congestiflorum
		Stigmasterol	calaber been, Soya been oil
		Stizolobic acid	Stizolobium hassjoo
		trans-Epoxy succinic acid	Aspergillus fumigatus
		Tariric acid	Tariric seed, bitter bush oil
		Vaccenic acid	Vegetable fats
		Vernolic acid	Vernonia anthelmintica wild
		Verticine	Fertilleria verticcillata wild
		Zingiberene	Oil of ginger

forming a diazotization reaction between cardanol or pentadecyl cardanol and *para*-aminobenzoic acid. Polymerization of the resulting monomer gave poly 4 [-4-hydroxy 2-pentadecyl phenyl) azo benzoic acid] [202,203,211–213]. Figure 18 shows the optical micrograph under polarized light exhibiting the typical schlieren nemetic texture of liquid crystals. Azo-based LC polymers are, in general, insoluble, and introduction of the pentadecyl group has significantly improved its solubility. Moreover, azo-based LC polymers are well known to give nonlinear optical behavior and for such applications the polymers should not absorb in the ultraviolet (UV) range of 100–200 nm [203]. The UV spectra of the LC polymer prepared did not absorb in this region, indicating possibly that this polymer or related structures might show useful nonlinear optical properties. Side-chain LC polymers from cardanol are under development with a view to developing them for NLO applications.

A number of workers elsewhere have also been applying similar concepts to prepare speciality polymers from cardanol. Thus, Pillot et al. [214] reports the hydrosilylation of cardanol with $HMeSiCl_2$ followed by hydrolysis to get polysiloxane grafted phenolic resins with thermally stable Si-C linkages. Trivedi et al. [215] studied the dimerization of cardanol with a view to developing epoxy resins from the dimer. Ramsri et al. [216] have developed Mannich bases and polyurethane coatings for use in electrodeposition. Speciality coatings based on cardanol–formaldehyde resins copolymerized with toluene diisocyanate have been reported by Hu et al. [217]. Sitaramam and Chatterjee [218] prepared pressure sensitive adhesives from 3-pentadecylphenol (hydrogenated cardanol). It appears that the opportunities for chemical modification and transformation of cardanol into speciality/high-performance polymers are vast and a tangible solution may soon emerge.

C. Natural Monomers Not Yet Studied and Utilized

There exists a large number of natural monomers that needs further studies for their exploitation Pillai and Manjula have reviewed [25] the subject and a comprehensive list of potential monomers and their source is given in Table 18.

V. CONCLUSIONS

Natural monomers and polymers present a scenario where they have a structural diversity and complexity that, with appropriate chemical modifications, and taking information from modern techniques of molecular and process designs could be utilized for transforming them into high-value polymers. This was exemplified by showing the example of a natural monomer, cardanol, which was converted to polymers with high performance and functional behavior. Chemical modifications on natural monomers and polymers were reviewed.

ACKNOWLEDGMENTS

Thanks are due to Dr. A. D. Damodaran, Dr. George John, Sri M. Saminathan, Dr. Rosy Antony, Dr. S. Manjula, Dr. C. Pavithran, Sri A. R. R. Menon, V. S. Prasad, Smt. J. D. Sudha, and Sri P. Anandan for helpful discussions and assistance during the preparation of this chapter.

REFERENCES

1. R. B. Seymour, *History of Polymer Science and Technology,* Marcel Dekker, Inc., New York (1982).
2. C. K. S. Pillai, *Chem. Ind. News, 31*: 781 (1987).
3. C. E. Carraher, Jr. and C. H. Sperling (eds.), *Polymer Applications of Renewable Resource Materials,* Plenum Press, New York and London (1981).
4. H. Mark, *Polym. Plast. Technol., 26*: 1 (1986).
5. E. Baer, *Sci. Amer., 255A*: 157 (1986).
6. C. K. S. Pillai, *J. Scient. Ind. Res., 51*: 776 (1992).
7. E. S. Lipinsky, *Science, 212*: 1465 (1981).
8. J. L. Cawse, J. L. Stanford, and R. H. Still, *Macromol. Chem., 85*: 6 97 (1984).
9. M. Morita and I. Sakta, *J. Appl. Polym. Sci., 31*: 831 (1986).
10. N. Shiraishi and H. Kishi, *J. Appl. Polym. Sci., 32*: 3189 (1986).
11. A. Heibish, S. A. Abdel-Hafis, and F. L1-Sisi, *J. Appl. Polym. Sci., 36*: 191 (1988).
12. S. S. Kelly, W. G. Glasser, and T. C. Ward, *J. Appl. Polym. Sci., 36*: 759 (1988).
13. S. S. Ray, A. K. Kundu, and M. Maiti, *J. Appl. Polym. Sci., 36*: 1283 (1988).
14. C. Pavithran, P. S. Mukherjee, M. Brahmakumar, and A. D. Damodaran, *J. Mater. Sci., 26*: 455 (1991).
15. C. K. S. Pillai, K. Gopakumar, and P. K. Rohatgi, *J. Scient. Ind. Res., 40*: 159 (1981).
16. C. K. S. Pillai, V. S. Prasad, J. D. Sudha, S. C. Bera, and A. R. R. Menon, *J. Appl. Polym. Sci., 41*: 2487 (1990).
17. M. P. Stevens, *Polymer Chemistry: An Introduction,* Oxford University Press, New York, Chapter 17 553 (1990).
18. R. A. Northey, *Emerging Technology of Materials and Chemicals from Biomass,*, ACS Symp. Series *476*: Washington, DC (1992).
19. S. Manjula, J. D. Sudha, S. Bera, and C. K. S. Pillai, *J. Appl. Polym. Sci., 30*: 1767 (1985).
20. C. Decker, H. L. Xuan, and T. N. T. Viet, *J. Polym. Sci. Polym. Chem. Edn., 33*: 2759 (1995).
21. M. F. Baladrin, J. A. Kloke, E. S. Wurtelle, and W. H. Wollinger, *Science, 228*: 1154 (1985).
22. P. Vasudevan, D. Kumari, and T. S. Kumar, *J. Scient. Ind. Res., 42*: 14 (1983).
23. B. Bolin, *Sci. Amer, 223*: 174 (1970).
24. C. E. Carraher, Jr. and L. H. Sperling (eds.), *Renewable Resource Materials: New Polymer Sources,* Plenum Press, New York (1985).

25. C. K. S. Pillai and S. Manjula, *Polym. News, 12*: 359 (1987).

26. G. John, *Chemical Modification of Natural Polymers,* Ph.D thesis, Regional Research Laboratory, Trivandrum and Kerala University, Trivandrum, India (1992).

27. S. Manjula, *Studies on Polymerization Characteristics of Naturally Occurring (Renewable) Monomers,* Ph.D thesis, Regional Research Laboratory, Trivandrum and Kerala University, Trivandrum, India (1988).

28. R. Antony, *Synthesis, Characterization and Thermal Behaviour of Chemically Modified Phenolic and Substituted Phenolic Polymers,* Ph.D thesis, Regional Research Laboratory, Trivandrum and Kerala University, Trivandrum, India (1993).

29. P. Bernfeld (ed.), *Biogenesis of Natural Compounds,* Pergamon Press, New York (1963).

30. E. A. MacGregor and C. T. Greenwood, *Polymers in Nature,* John Wiley & Sons, New York (1980).

31. J. E. Glass and G. Swift (eds.) *Agricultural and Synthetic Polymers: Biodegradability and Utilization,* ACS Symp. Series, *433*: Washington, DC (1990).

32. C. K. S. Pillai, *Rubber Rep., 12*: 145 (1988).

33. M. Jaffe, *Polymers for Advanced Technologies,* (M. Levin, ed.) VCH Publishers, Jerusalem (1988).

34. *Kevlar: From Laboratory to Market Place Through Innovation,* Dupont Report-H 16261, p. 34 (1990).

35. D. L. Schmidt, *Ablative Polymers in Aerospace Technology,* (G. F. D'Alelio and J. A. Parker, eds.) *Ablative Plastics,* Marcel Dekker, Inc., New York (1971).

36. T. David, G. Vlodek and S. R. John, *Proceedings on Polymers for Advanced Technologies: IUPAC International Symposium,* (M. Lewin, ed.) VCH Publishers, Jerusalem (1988).

37. K. V. Sarkanen and L. H. Heregrt *Lignin: Occurrence, Formation, Structure and Reactions,* (K. Sarkanen and C. Ludwig, eds.,) Wiley Interscience, New York. (1971).

38. S. I. Fakehag et al., ACS Symp. Series, *Resources for Plastics,* Pennsylvania (1975).

39. R. M. Brown (ed.), *Cellulose and Other Natural Systems,* Plenum Press, New York (1982).

40. E. R. Yescumbe, *Plastics and Rubber,* Applied Science Publishers, Essex (1976).

41. H. L. Fischer, *Chemistry of Natural and Synthetic Rubbers,* Reinhold, New York (1957).

42. W. Pigman and D. Horton (eds.), *The Carbohydrates,* 2nd edn., Academic Press, London and New York (1970).

43. R. L. Whistler and E. F. Paschall (eds.), *Starch: Chemistry and Technology,* Vol. 2, Academic Press, New York (1967).

44. M. Poulicek, M. S. Voss-Foucart, and Ch. Jeuniaux, *Chitin in Nature and Technology* (M. Muzzarelli, Ch. Jeuniaux, and G. W. Gooday, eds.) Plenum Press, New York (1986).

45. H. Daly and S. Lee *Applied Bioactive Polyemric Materials,* (C. G. Gebelein, C. E. Carraher, Jr., and V. R. Foster, eds.) Polym. Sci. Technology, Vol. 38, Technomic, Westport (1987).

46. I. A. Pearl, *The Chemistry of Lignin,* Marcel Dekker, Inc., New York (1967).

47. J. Nakasno and G. Mashitsuka, *Methods in Lignin Chemistry,* (S. Y. Lin and C. W. Dence, eds.) Springer-Verlag, Berlin (1992).

48. T. Kent Kirk, T. Higuchi, and H. Chang, *Lignin Biodegradation: Microbiology, Chemistry, and Potential Applications,* Vol. 1, CRC Press, Boca Raton (1980).

49. H. Neurah (ed.), *The Proteins,* 3rd edn., Vols. 1–3, Academic Press, New York 1975–1977).

50. G. E. Means and R. E. Feeney, *Chemical Modification of Proteins,* Holden-Day, Inc., London (1971).

51. S. Fox, K. Harada, and D. Rohlfing, *Polyamino Acids, Polypeptides and Proteins,* (M. Stahman, ed.), U. Wisconsin Press, Maddison, Wisconsin (1952).

52. M. F. Perutz, *Proteins and Nucleic Acids,* Elsevier, Amsterdam (1962).

53. J. D. Watson, *Molecular Biology of the Gene,* W. A. Benjamin Inc., New York, Amsterdam, (1965).

54. C. E. Carraher, Jr. and J. A. Moore, *Modification of Polymers,* Plenum Press, New York (1983).

55. R. B. Seymour and G. S. Kirschenbaun, *High Performance Polymers: Their Origin and Development,* Elsevier, New York (1986).

56. D. C. Sherrington and P. Hogde (eds.), *Synthesis and Separation Using Functional Polymers,* Wiley, Chichester (1988).

57. E. Marechal, *Comprehensive Polymer Science,* Vol. 6 (G. Allen and J. C. Bevinton, eds.) Pergamon, Press, New York, (1989).

58. W. T. Ford (ed.), *Polymeric Reagents and Catalysts,* ACS Symp. Series, Washington, DC (1986).

59. O. B. Wurtzburg, Jr., *Modified Starches: Properties and Uses,* CRC Press, Inc., Boca Raton (1986).

60. H. Hatakeyama, S. Hirose, and T. Hatakeyama, *Lignin: Properties and Materials,* ACS Symp. Series, *397*: 205, (1989).

61. T. M. Garver, Jr. and S. Srakanen, *Renewable Resource Materials: New Polymer Sources,* (C. E. Crraher, Jr. and L. H. Sperling, eds.) Plenum Press, London (1986).

62. *Amino Acids, Pepetides and Proteins,* Specialist Periodical Report of Chemical Society, London (1983).

63. J. C. Arther, *Comprehensive Polymer Science,* (G. Allen and K. C. Bhevington, eds.), Pergamon Press, New York (1989).

64. J. J. Lindburg, T. A. Kuusela, and K. Levon, *Lignin: Properties and Materials,* ACS Symp. Series, *397*: 205 (1989).

65. E. Ott, H. M. Spurlin, and W. G. Grafflin (eds.), *Cellulose and Cellulose Derivatives, Part 1,* Interscience Publishers, New York (1954).

66. B. A. K. Andrews and I. V. deGruvy, *Kirk-Other Encyclopedia of Chemical Technology,* 3rd edn, Vol. 7, (M. Greyson eds.), Wiley-Interscience, New York (1979).

67. T. P. Nevell and S. H. Zeronian (ed.), *Cellulose Chemistry and its Applications,* Horwood, Chichester (1985).

68. C. F. Cross and E. J. Bevan, *Researches in Cellulose,* Longmans, London (1905–1911).

69. J. F. Kennedy, G. O. Philips, and P. A. Williams (eds.), *Wood and Cellulosics: Industrial Utilization, Biotechnology, Structure and Properties,* Ellis Horwood, Chichester (1987).

70. A. Hebeish and J. T. Guthrie, *The Chemistry and Technology of Cellulosic Polymers,* Springer-Verlag, Berlin (1981).

71. R. M. Brown, Jr. (ed.), *Cellulose and Other Natural Polymer Systems: Biogesis, Structure and Degradation,* Plenum Press, New York (1982).

72. E. M. Fettes, (ed.), *Chemical Reactions of Polymers,* Wiley-Interscience, New York (1964).

73. N. S. David (ed.), *Graft Copolymerization of Lignocellulosic Fibres,* ACs Symp. Series, *187*: Washington, DC (1982).

74. K. Ward Jr., *Chemical Modification of Paper Making Fibres,* Marcel Dekker, Inc., New York, (1973).

75. R. Rowell and R. Young, *Modified Cellulosics,* Academic Press (1978).

76. H. A. J. Battaerd and G. W. Tregear, *Graft Copolymers,* Interscience, New York (1967).

77. R. J. Ceresa, *Block and Graft Copolymerization,* Vols. 1 and 2, John Wiley & Sons, New York (1973).

78. R. Narayan and M. Shay, *Recent Advances in Anionic Polymerization,* (T. E. Hogen-Esch and J. Scmid, eds.) Elsevier, New York, (1986) p. 137.

79. R. J. Ceresa, *Block and Graft Polymers,* Ch. 5, Butterworths, London (1962).

80. C. I. Simionescu and M. M. Macoveanu, *Cellulose Chemistry Technol., 11*: 197 (1976).

81. Z. A. Rogovin, *Chemical Transformation and Modification of Cellulose,* Khimiya (1967).

82. G. Mino and S. Kaizerman, *J. Polym. Sci., 31*: 242 (1958).

83. F. R. Duke and R. A. Forist, *J. Amer. Chem. Soc., 71*: 2790 (1949).

84. J. P. Kennedy, *Recent Advances in Polymer Blends, Grafts and Blocks,* L. H. Sperling, Plenum Press, New York (1974).

85. V. Stannet and T. Memetea, *J. Polym. Sci. Polym. Symp. Edn., 64*: 57 (1978).

86. P. L. Nayak, *J. Macromol. Sci. Rev. Chem. Phys., 14*: 193 (1976).

87. R. Jerome, R. Tyat, and T. Quhadi, *Prog. Polym. Sci., 10*: 87 (1984).

88. M. K. Misra, *J. Macromol. Sci. Rev. Macromol. Chem. Phys., 22*: 741 (1983).

89. R. K. Samal, K. Sahoo, and H. S. Samantary, *J. Macromol. Sci. Rev. Macromol. Chem. Phys., 26*: 81 (1986).

90. S. Lemka, *J. Macromol. Sci. Rev. Macromol. Chem. Phys., 22*: 303 (1983).

91. S. N. Bhattacharrya and D. Maldas, *Prog. Polym. Sci., 10*: 171 (1984).

92. W. K. Walsh, C. R. Jin, and A. A. Amstrong, *Text. Res. J., 35*: 648 (1965).

93. V. Stannet and A. S. Hoffman, *Am. Dyest. Rep., 91*: (1968).

94. V. Stannet and J. L. Williams, *J. Macromol. Sci. Chem., A10*: 637 (1976).

95. J. C. Arther, Jr., *J. Macromol. Sci. Chem., A10*: 635 (1976).

96. J. P. Kennedy and A. Vidal, *J. Polym. Sci. Polym. Chem. Edn., 13*: 1765 (1975).

97. A. Takahashi, Y. Sugahara, and Y. Hirano, *J. Polym. Sci. Polym. Chem. Edn., 27*: 3817 (1989).

98. V. Y. Kabanov, *Radiat. Phys. Chem., 33*: 51 (1989).

99. N. W. Taylor and E. B. Bagley, *J. Appl. Polym. Sci., 21*: 1607 (1977).

100. G. John, C. K. S. Pillai, and A. Ajayaghosh, *Polym. Bull., 30*: 415 (1993).

101. M. Putterman and R. M. Fitch, *J. Appl. Polym. Sci., 40*: 162 (1990).

102. V. T. Stannet, W. M. Donne, and G. F. Fanta, *Absorbancy,* (P. K. Chatterjee, ed.), Elsevier, Amsterdam (1985).

103. R. I. Whistler and E. F. Paschall (eds.), *Starch, Chemistry and Technology,* Vols. 1 and 2, Academic Press, New York (1965).

104. R. I. Whistler and C. I. Smart, *Polysaccharide Chemistry,* Academic Press, New York (1953).

105. K. Freudenberg and A. C. Neish, *The Constitution and Biosynthesis of Lignin,* Springer-Verlag, Berlin (1968).

106. V. P. Saraf. W. G. Glasser, G. L. Wilkes, and J. E. McGrath, *J. Appl. Polym. Sci., 30*: 2207 (1985).

107. K. Freudenberg, *Science, 148*: 595 (1965).

108. D. Feldman M. Lacasse, and L. M. Benaczuk, *Prog. Polym. Sci., 12*: 271 (1988).

109. S. S. Kelley, W. G. Glasser, and T. C. Ward, *J. Appl. Polym. Sci., 36*: 759 (1988).

110. S. C. Bera, C. K. S. Pillai, and K. G. Sathyanarayana, *J. Scient. Ind. Res., 44*: 599 (1985).

111. F. J. Stevenson, *Humus: Chemistry, Genesis, Composition, Reactions,* Wiley-Interscience, New York (1982).

112. M. Schnitzer and S. U. Khan, *Humic Substances in the Environment.,* Marcel Dekker, Inc., New York (1972).

113. I. Gobaty and K. Ouchi (eds.), *Advances in Chemical Series, 192*: ACS Symp. Series Washington, DC (1981).

114. M. Morton, *Rubber Technology,* 2nd edn., Van Nostarnd, New York (1973).

115. C. M. Blow and C. Hepburn (eds.) *Rubber Technology and Manufacture,* Butterworth Scientific, London (1982) p. 2.

116. C. K. S. Pillai, K. G. Sathyanarayana, and A. G. Mathew, *Chem. Ind. News, 31*: 781 (1984).

117. E. R. Gill, B. Bagostini, and M. J. Hooft-welvaars *The World of Rubber Economy, Structure, Changes and Prospects,* Johns Hopkins University Press, London and Baltimore, (1980).

118. P. Alexander and R. F. Hudson, *Wool, Its Chemistry and Physics,* Reinhold, New York, (1954).

119. W. C. Wolfgang, *Encyclopedia Polym. Sci. Technol.,* H. Mark, N. G. Gaylord, and N. M. Bikales, Interscience Publishers, New York *12*: p. 578 (1970).

120. K. H. Gustavson, *The Chemistry and Reactivity of Collagen,* Academic Press, New York (1956).

121. A. Vies, *The Macromolecular Chemistry of Gelatin,* Academic Press, New York (1964).

122. R. B. Seymour, *J. Chem. Edn., 65*: 327 (1988).

123. J. G. Speight and S. E. Moschopedis, *Chemistry of Asphaltenes,* J. W. Bunger and N. C. Li, *Adv. Chem. Ser. 195*: ACS, Washington, DC, (1981).

124. P. K. Bose, Y. Sankaranarayanan, and S. C. Sen, *Chemistry of Lac,* Indian Lac Research Institute, Ranchi, India, (1963).

125. S. Maiti and M. D. S. Rahman, *J. Macromol. Sci., Rev. Macromol. Chem. Phys., C26*: 441 (1986).

126. K. D. Carlson, V. E. Sohns, R. B. Perkins, Jr., and E. L. Huffman, *Ind. Eng. Chem. Prod. Res. Dev., 16(1),* 95, (1977).

127. K. Gidanian and G. J. Howard, *J. Macromol. Sci. Chem. A10(7)*: 1415 (1976).

128. H. Rudolph, Ger. Offen, 3002762, 1981; *Chem. Abstr. 95*: 151439n, (1981).

129. J. Klein and W. E. Kulicke, *Polymer Pre-prints,* ACS Symp, Series Washington, DC (March 1982).

130. R. G. Sinclair, U.S. Patents, 4,045,418 and 4,057,537 (1977).

131. H. Tsuji, *Polymer, 36*: 2709 (1995).

132. J. E. Kaspercejk, *Macromolecules, 28*: 3937 (1995).

133. J. H. P. Thyman, *Chem. Soc. Rev., 8*: 499 (1979).
134. A. R. R. Menon, J. D. Sudha, C. K. S. Pillai, and A. G. Mathew. *J. Scient. Ind. Res., 44*: 324 (1985).
135. R. J. Wilson, *The Market for Cashew Kernels and Cashew Shell Liquid,* Report No. G91 Tropical Products Institute, London (1975).
136. C. K. S. Pillai, *Pop. Plast.,* Nov. Issue (1993).
137. W. Knop and A. Scheib, *Chemistry and Applications of Phenolic Resins-Polymer Properties and Applications,* Springer-Verlag, Berlin (1979).
138. W. F. Symes and C. R. Dawson, *J. Amer. Chem. Soc., 76*: 2959 (1954).
139. S. V. Santhkumar and C. R. Dawson, *J. Amer. Chem. Soc., 76*: 5070 (1954).
140. J. H. P. Thyman, and A. J. Mathews, *Chem. Ind.* (London), 740 (1977).
141. A. S. Dyer, *Interchem. Rev., 4*: 35 (1945).
142. J. H. P. Thyman, *J. Chromatog., 111*: 285 (1975).
143. S. Iwahashi, Japanese Patent 6142 (1957).
144. K. Ohashi, U.S. Patent 291978 (1959).
145. J. R. Morris, US Patent (to Texas Corp.) 2417562 (1947).
146. K. M. Nair, E. V. V. Bhaskara Rao, K. K. N. Nambiar, and M. C. Nambiar (eds.), *Cashew (Anacardium occidentalle L. -Monograph on Plantation Crops-1,* Central Plantation Crops Research Institute, Kasargod-Kerala, Kasargod (1979).
147. N. D. Ghatge and N. N. Maldar, *Rebber Rep., 6*: 139 (1981).
148. S. K. Sood, J. H. P. Thyman, A. Durrni, and R. A. Johnson, *Lipids, 21*: 241 (1986).
149. J. H. P. Thyman and V. Tychopoulos, *J. Planar. Chromatog, 1*: 227 (1988).
150. P. H. Gedam, P. S. Sampathkumaran, and M. A. Sivasamban, *Indian J. Chem., 10*: 338 (1972).
151. V. J. Paul and L. M. Yeddanappalli, *J. Amer. Chem. Soc., 78*: 5675 (1956).
152. Indian Standard Specification for Cashewnut Shell Liquid, *IS*: 840 (1964).
153. P. H. Gedam and P. S. Sampathkumaran, *Prog. Org. Coatings, 14*: 115 (1986).
154. B. G. K. Murthy, M. C. Menon, J. H. Aggarwal, and S. H. Zaheer, *Paint Manuf., 31*: 47 (1961).
155. *Cashewnut Shell Liquid Patents* U.S. Vol. 1 and UK, Indian and Japan, Vol. 2, Cashew Export, June, Cashew Export Promotion Vcouncil, Ernakulam (1964).
156. *Cashewnut Shell Liquid-Extraction and Uses—A Survey of World Patents upto 1976,* Cashew Export Promotion Council, Ernakulam (1978).
157. *Indian Cashew Nut Shell Liquid—A Versatile Industrial Raw Material of Great Promise—Regional Research Laboratory,* Trivandrum and Cashew Export Promotion Council, Ernakulam (1983).
158. B. G. K. Murthy and M. A. Sivasamban, *Cashew Causerie, 1*: 8 (1979).
159. B. G. K. Murthy and J. S. Aggarwal, *J. Clour. Sci., 11*: 2 (1972).
160. L. C. Anad, *Pop. Plast., June*: (1978).
161. A. Orazio, S. Z. Franco, P. Franco, and S. Alexandro, *Chemica Ind., Milan, 61*: 718 (1970).
162. J. S. Aggarwal, *J. Clour. Sci., 15*: 14 (1976).
163. *Development of Value Added Polymer Resins/Products from Cashew Nut Shell Liquid,* Project report No. RT.09/MO/50, Regional Research Laboratory, Trivandrum (1987).
164. M. G. Jacko, P. H. S. Tsang, and S. K. Rhee, *Wear, 100*: 503 (1984).
165. C. K. S. Pillai, M. A. Venkataswamy, K. G. Sathyanarayana, V. P. Sredharan, C. Indira, and P. K. Rohatgi, *J. Mat. Sci., 17*: 2861 (1982).
166. C. K. S. Pillai, M. A. Venkataswamy, K. G. Sathyanarayana, and P. K. Rohatgi, *Appropriate Technol., 12*: 249 (1985).
167. P. K. Biswas, N. Balagopal, and C. K. S. Pillai, *Indian J. Engg. Mater. Sci., 1*: 99 (1994).
168. J. D. Sudha, C. Pavithran, and C. K. S. Pillai, *Res. Ind., 34*: 139 (1989).
169. A. K. Misra and G. N. Pandey, *J. Appl. Polym. Sci., 30*: 969 (1985).
170. S. Manjula, V. G. Kumar, and C. K. S. Pillai, *J. Appl. Polym. Sci., 45*: 309 (1992).
171. R. Antony, K. J. Scria, and C. K. S. Pillai, *J. Appl. Polym. Sci., 41*: 1765 (1992).
172. C. K. S. Pillai, V. S. Prasad, J. D. Sudha, A. R. R. Menon, S. C. Bera, and A. G. Mathew, *Rubber Plast. Annual, 11*: 5 (1987).
173. G. John and C. K. S. Pillai, *Polym. Bull., 22*: 89 (1989).
174. J. P. Kennedy and E. Marechal, *Carbocationic Polymerization,* Wiley Interscience Publishers, John Wiley & Sons, New York (1982).
175. C. K. S. Pillai et al., unpublished results.
176. G. John, C. K. S. Pillai, K. G. Das, and K. Saramma, *Indian J. Chem., 29A*: 89 (1990).
177. C. K. S. Pillai, A. R. R. Menon, C. Pavithran, and A. D. Damodaran, *Metals Materials Processes, 1*: 151 (1989).
178. C. K. S. Pillai et al., unpublished results.
179. E. D. Weil, *Proc. Rec. Adv. Flame Retard. Polym. Matter,* (L. Menachem, ed.), *1*: Business Communication Co., Norwalk, Connecticut (1990).
180. C. K. S. Pillai, J. D. Sudha, V. S. Prsasd, S. C. Bera, A. R. R. Menon, A. D. Damodaran, S. Alwan, S. K. Lakshmidasn, and K. N. Govindaraman, Indian Patent, Provitional No. 1157, Del/88 (1988).
181. S. C. Bera, C. K. S. Pillai, P. N. Rangan, A. R. Arankale, and H. Chirmade, *Indian J. Technol., 27*: 393 (1989).
182. S. C. Bera, C. K. S. Pillai, A. R. R. Menon, P. N. Rangan, A. R. Arankale, H. Chirmade, and S. G. Gurav, Proceedings of the International Conference on Rubbers and Rubber Like Materials, Indian Institute of Technology, Kharagpur, November 6–8 (1986).
183. A. R. R. Menon, C. K. S. Pillai, and G. B. Nando, *J. Appl. Polym. Sci., 51*: 2257 (1994).
184. A. R. R. Menon, C. K. S. Pillai, and G. B. Nando, *Kaustschuk Gummi Kunststoffe, 45*: 708 (1992).
185. A. R. R. Menon, C. K. S. Pillai, and G. B. Nando, *Polym. Degrad. Stab., 52(3)*: 265 (1996).
186. A. R. R. Menon, C. K. S. Pillai, and G. B. Nando, *J. Adhesion Sci. Technol., 9*: 443 (1995).
187. C. K. S. Pillai, J. D. Sudha, V. S. Prsasd, S. C. Bera, A. G. Mathew, and A. D. Damodaran, Indian Patent No 171782 July 13 (1988).
188. R. Antony and C. K. S. Pillai, *J. Appl. Polym. Sci., 54*: 429 (1994).
189. R. Antony and C. K. S. Pillai, *J. Appl. Polym. Sci., 49*: 2129 (1993).
190. R. Antony, *J. Polym. Sci. Polym. Chem. Edn., 31*: 3187 (1993).

191. C. K. S. Pillai, C. Pavithran, A. R. R. Menon, V. S. Prasad, J. D. Sudha, V. G. Jayakumari, M. Brahmakumar, P. Anandan, and A. D. Damodaran, *Specialty Polymers from Long Chain Hydrocarbon Phenols,* Project Completion Report, Regional Research Laboratory, Trivandrum (1994).

192. C. K. S. Pillai, V. S. Prasad, J. D. Sudha, V. G. Jayakumari, A. R. R. Menon, M. Brahmakumar, C. Pavithran, and A. D. Damodaran, Indian Patents, Provisional Nos. 897/DEL/93, August 18 (1993); 898/DEL/93, August 18 (1993); 102/DEL/94, January 27 (1994); 1404/DEL/95, July 27 (1995); 1405/DEL/95, July 27 (1995) and 1405/DEL/95, July 27 (1995).

193. S. Manjula, V. G. Kumar, C. Pavithran, and C. K. S. Pillai, *J. Mater. Sci., 267:* 4001 (1991).

194. S. Manjula, V. G. Kumar, and C. K. S. Pillai, *Thermo Chemica Acta, 159:* 255 (1990).

195. G. John and C. K. S. Pillai, *Macromol. Chem. Rapid. Commun., 13:* 255 (1992).

196. G. John and C. K. S. Pillai, *J. Polym. Sci. Polym. Chem. Edn., 31:* 1067 (1993).

197. G. John, S. K. Thomas, and C. K. S. Pillai, *J. Appl. Polym. Sci., 53:* 1415 (1994).

198. A. Akelah and D. C. Sherrington, *Chem. Rev. 81:* 551 (1981).

199. R. Arshardy, *Macromol. Chem., 185:* 2287 (1984).

200. A. D. Jenkins and D. J. Wilson, *Polym. Bull., 20:* 101 (1988).

201. C. K. S. Pillai, D. C. Sherrington, and A. Sneddon, *Polymer, 33:* 3968 (1992).

202. M. Saminathan, PhD. Thesis, Synthesis and Characterization of Novel Liquid Crystalline Polymers Containing Azobenzene Mesogen, Regional Research Laboratory, Trivandrum and University of Kerala, Trivandrum (1995).

203. M. Saminathan, C. K. S. Pillai, and C. Pavithran, *Macromolecules, 26:* 7103 (1993).

204. W. J. Jackson, Jr. and K. F. Kuhfus, *J. Polym. Sci. Polym. Chem. Edn., 14:* 2046 (1976).

205. G. W. Calundan, *Amer. Chem. Soc. Polym. Prepr., 27:* 473 (1986).

206. C. K. Ober, J. J. Lin, and R. W. Lenz, *Adv. Polym. Sci., 13:* 103 (1984).

207. C. Noel and P. Navard, *Prog. Polym. Sci., 16:* 55 (1991).

208. C. K. S. Pillai, D. C. Sherrington, and A. Sneddon, Indian Patent No. 677/del/92, July 29 (1992).

209. C. K. S. Pillai, D. C. Sherrington, and A. Sneddon, Indian Patent, No. 678/del/92, July 29 (1992).

210. C. K. S. Pillai, D. C. Sherrington, and A. Sneddon, Indian Patent No. 679/del/92, July 29 (1992).

211. M. Saminathan, C. K. S. Pillai, and C. Pavithran, Indian Patent No. 12791/del/92, December 31 (1992).

212. M. Saminathan, C. K. S. Pillai, and C. Pavithran, Indian Patent No. 12972/del/92 December 31 (1992).

213. H. K. Hall, Jr., T. Kuoand, and T. M. Leslie, *Macromolecules, 22:* 3525 (1989).

214. J. P. Pillot, J. Dunogues, J. Gerval, M. D. The, and M. V. Thanh, *Eur. Polym. J., 25:* 285 (1989).

215. M. K. Trivedi, M. J. Patni, and L. Bindal, *Indian J. Technol., 27:* 281 (1989).

216. M. Ramsri, G. S. S. Rao, P. S. Sampathkumaran, and M. M. Shirsalker, *J. Appl. Polym. Sci., 39:* 1993 (1990).

217. Y. Hu and M. Gou, *Linchan Huaxue Yu Gongye, 9:* 43 (1989), *Chem. Abstr., 113:* 213896k (1991).

218. B. S. Sitaramam and P. C. Chatterjee, *J. Appl. Polym. Sci., 37:* 33 (1989).

28

Ionic Thermoplastic Elastomer Based on Maleated EPDM Rubber

Santanu Datta and S. K. De

Indian Institute of Technology, Kharagpur, India

I. INTRODUCTION

Polymers can be modified by the introduction of ionic groups [1]. The ionic polymers, also called ionomers, offer great potential in a variety of applications. Ionic rubbers are mostly prepared by metal ion neutralization of acid functionalized rubbers, such as carboxylated styrene-butadiene rubber, carboxylated polybutadiene rubber, and carboxylated nitrile rubber [2–5]. Ionic rubbers under ambient conditions show moderate to high tensile and tear strength and high elongation. The ionic cross-links are thermolabile and, thus, the materials can be processed just as thermoplastics are processed [6].

Ethylene propylene diene monomer (EPDM) rubber functionalized by sulfonation and maleation reactions form sulfonated and maleated EPDM rubber, respectively. Ionic rubbers based on sulfonated EPDM rubber have been studied by several workers [7–10]. Maleated EPDM rubber (m-EPDM), on the other hand, has not been exploited as a potential ionomer and has mostly been used as a component in reactive blending, that is, for the impact modification of nylon-66 or polybutadiene terephthalate [11,12]. Kresge [13] has reported studies on physical and mechanical properties of m-EPDM-based ionomers. Zinc stearate is known to act as a plasticizer for the ionic domains and facilitates processing of the ionic polymers above the melting point of zinc stearate ($>128°C$) [14,15]. Although reinforcing fillers in general are known to adversely affect the strength properties of TPEs [16], there are reports of reinforcement of ionic thermoplastic elastomers by carbon black [17–20].

This chapter reports the results of studies on the physical, dynamic mechanical, and rheological behavior of zinc oxide neutralized m-EPDM, particularly in the presence of stearic acid and zinc stearate, with special reference to the effects of precipitated silica filler.

II. EXPERIMENTAL

Formulations of different mixes are shown in Tables 1 and 2.

m-EPDM was mixed with zinc oxide and other ingredients in a laboratory size two-roll mill at a nip gap of 2 mm. Cold water was circulated inside the rollers in order to prevent excessive temperature rise during mixing. Mixing schedules were as follows:

(a) For formulations without fillers:

Mixing Step	Time (min)
Add rubber and band	2
Add ZnO	2
Add stearic acid	2
Add zinc stearate	10
Sheet out	2

(b) For formulations with fillers:

Mixing Step	Time of Mixing (min)
i. Add rubber and band	2
ii. Add ZnO	1
iii. Add stearic acid	1
iv. Add 1/3 zinc stearate	3
v. Add 1/2 filler	3
vi. Add 1/3 zinc stearate	3
vii. Add 1/2 filler	3
viii. Add 1/3 zinc stearate	3
ix. Sheet out	1

Table 1 Formulations of Mixes and Molding Conditions [33]

Ingredient	Mix Number						
	EP0	EP1	EP2	EP3	EP4	EP5	EP6
m-EPDM	100	100	100	100	100	100	100
ZnO	—	10	10	10	10	10	10
Stearic acid	—	—	1	1	1	1	1
Zinc stearate	—	—	—	10	20	30	40
DCP	2.0	—	—	—	—	—	—
Molding temperature, (°C)	150	170	120	120	120	120	120
Molding time, (min)	8	60	20	20	20	20	20

Mixes were compression molded to form 2-mm thick sheets at a pressure of 10 MPa at 120°C for 20 min.

The hardness was determined as per American Society for Testing and Material (ASTM D 2240, 1986) and expressed in Shore A. The stress–strain properties were measured using dumbbell-shaped specimens according to ASTM D 412 (1987) in a Zwick universal testing machine (UTM), model 1445, at a crosshead speed of 500 mm/min. Hysteresis studies were made according to ASTM D 412 (1980) by stretching the dumbbell-shaped specimens upto an extension of 200%. Tear strength was measured in the Zwick UTM 1445 using a 90° nick cut crescent sample according to ASTM D 624–86. The tension set at 100% extension was determined as per ASTM D 412 (1987).

Measurement of dynamic mechanical properties was carried out under tension mode using a viscoelastometer, (Rheovibron DDV-III-EP, M/s. Orientec Corp., Tokyo, Japan). Sample size was 3.5 cm × 6.5 mm × 2 mm. Testing was carried out at a low amplitude, 0.025 mm, over a temperature range of −100°C to +200°C. Heating rate was 1°C/min and frequency of oscillation was 3.5 Hz or 110 Hz.

Rheological and processability behaviors were studied in a Monsanto processability tester (MPT), which is an automatic high-pressure capillary viscometer. The entire barrel and capillary are electrically heated with a microprocessor-based temperature controller [14]. The

capillary length to diameter ratio was 30:1. Preheat time for each sample was 5 min. Reprocessability studies were made by repeated extrusions, up to four cycles, of the samples through MPT at different shear rates.

III. RESULTS AND DISCUSSION

A. Effect of Zinc Stearate

Physical properties of the different systems are listed in Table 3. The m-EPDM–DCP system (mix EP0) was cured to the same level as the m-EPDM–ZnO (mix EP1) by choosing molding times so that the torque increase (i.e., the difference between the maximum torque and the minimum torque) is the same in both cases. As compared with the m-EPDM–DCP system, the m-EPDM–ZnO system shows higher hardness, modulus, tensile strength, and tear resistance presumably due to the presence of ionic clusters, which will be discussed later. The higher elongation at break in the m-EPDM–ZnO system is believed to be due to the occurrence of stress-induced ion exchange, which causes lowering of stress concentration, resulting in high elongation [21]. Hysteresis loss follows the order EP0 < EP1 < EP2 < EP3. Hysteresis of the ionic system (EP1) is higher than the DCP cured system (EP0). Reinforcing fillers, in general, are known to cause an increase in hysteresis of rubbers. Accordingly, it is believed that the ionic aggregates in the m-EPDM–ZnO system act not only as crosslink sites, but also as fillers providing reinforcement to the matrix [22]. The effect becomes more pronounced on addition of stearic acid and zinc stearate. Incorporation of stearic acid facilitates formation of ionic aggregates. It has been reported that at ambient conditions crystalline zinc stearate acts as a reinforcing filler in metal oxide crosslinked carboxylic rubbers [23]. Results of measurements of physical properties at 70°C reveal that the reinforcing ability of zinc stearate diminishes at higher test temperatures, presumably due to the onset of melting of zinc stearate and the consequent plasticization.

Table 2 Formulation of Silica-Filled Mixes[a] [27]

Ingredients	Mix Number			
	S0	S1	S2	S3
m-EPDM	100	100	100	100
ZnO	10	10	10	10
Stearic acid	1	1	1	1
Zinc stearate	30	30	30	30
Silica	0	10	20	30

[a] All mixes were molded at 120°C for 20 min.

Table 3 Physical Properties[a] [33]

Properties	Mix Number						
	EP0	EP1	EP2	EP3	EP4	EP5	EP6
300% modulus (MPa)	0.70	1.70	2.20	2.60	2.80	2.90	2.90
	(0.76)	(1.80)	(1.90)	(1.90)	(2.06)	(1.80)	(1.50)
Elongation (%)	580	850	620	880	800	760	710
	(512)	(422)	(503)	(303)	(422)	(431)	(440)
Tensile strength (MPa)	0.89	4.40	5.80	7.15	8.54	7.67	6.34
	(0.80)	(2.24)	(2.20)	(1.91)	(2.34)	(2.42)	(1.88)
Tear strength (N/cm)	143	332	370	381	400	418	422
	(130)	(202)	(211)	(244)	(347)	(233)	(220)
Hysteresis loss (J/m^2) $\times$ 10^2	3.3	4.5	7.0	9.0	10.0	10.7	13.9
Permanent set at 100% extension (%)	8	11	10	11	13	13	15
Hardness (Shore A)	40	47	53	56	58	62	64

[a] Values in the parentheses are the results of tests done at 70°C.

Figure 1 shows the variation of dynamic storage modulus (E′) with temperature. As compared with the m-EPDM–DCP system, the m-EPDM–ZnO system shows a broad rubbery plateau, due to physical cross-links arising out of the ionic aggregates. Addition of zinc stearate causes a gradual increase in E′ at ambient temperatures, but a sharp fall in E′ at a temperature exceeding 100°C, indicating onset of transition from the rubbery state to the viscous flow state. It is also evident that the rate of fall of E′ increases with an increase in zinc stearate loading. It is believed that at room temperature zinc stearate strengthens the ionic aggregates causing reinforcement of the matrix resulting in higher E′. But at a temperature approaching its melting point (i.e., 128°C), zinc stearate plasticizes the ionic aggregates, presumably by solvation or exchange reaction. In the absence of zinc stearate, plasticization cannot occur and, accordingly, the transition from the rubbery state to the viscous flow state is not observable under the studied conditions of temperature range and strain levels. The absence of the transition results in a broad rubbery plateau in the m-EPDM–ZnO and m-EPDM–ZnO–stearic acid systems.

Figure 2 shows typical plots of tanδ versus temperature, from there it is evident that the glass to rubber transition in the m-EPDM–ZnO system occurs at a slightly higher temperature than the m-EPDM–DCP system. Furthermore, incorporation of stearic acid and zinc stearate shifts the glass–rubber transition temperature (T_g) of the m-EPDM–ZnO system to the higher temperature side. The results are summarized in Table 4. The value of tanδ at T_g (i.e., tanδ$_{max}$) decreases with the incorporation of stearic acid and zinc stearate. This indicates that with the introduction of stearic acid and zinc stearate, the chains become considerably tightened, and, therefore, the backbone relaxation related to T_g oc-

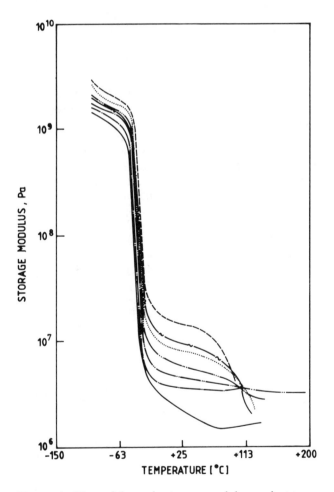

Figure 1 Plots of dynamic storage modulus against temperature at 3.5 Hz [33]. EP0 (———), EP1 (—·—), EP2 (—··—), EP3 (—···—) EP4 (·····), EP5 (—∼—), EP6 (------).

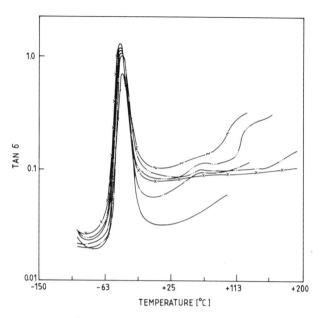

Figure 2 Plots showing variation of tanδ against temperature [33]. EP0 (———), EP1 (—·—), EP1D (—x—), EP2 (—··—), EP3 (—···—), EP5 (—~—).

The high temperature relaxation peak is believed to be due to onset of motion of chain segments firmly held by the ionic aggregates called clusters [24,25]. Figure 3 is a model proposed for the different types of ionic aggregates in ionomers and the region of restricted mobility in the chain segment causing a biphasic structure [25]. Restricted mobility region is separated from the "free" mobility region by a skin layer. Formation of the ionic clusters is favored by higher ion content, provided other factors remain the same. That the high temperature relaxation is due to ionic clusters and not by any other relaxation of the base polymer is confirmed by the absence of the peak in the m-EPDM–DCP system.

Thus, it can be concluded that in the present ionomer system, zinc stearate plays a dual role. First, below its melting point it reinforces the matrix and strengthens the ionic aggregates and, second, at a higher temperature it results in solvation of the ionic aggregates and plasticizes the system, thus, facilitating the transition from the rubbery state to the viscous flow state [23].

While studying melt rheology in the MPT, it was observed that the m-EPDM–ZnO and m-

curs at a slightly higher temperature with consequent lowering of tanδ at T_g. In the high-temperature region (i.e., above room temperature) a second relaxation occurs, which is, however, not prominent in the m-EPDM–ZnO and m-EPDM–ZnO–stearic acid systems. The high temperature relaxation becomes prominent in the presence of zinc stearate. Characteristic features of the second transition is that the transition temperature, designated as T_i, shifts to a higher temperature side with the incorporation of zinc stearate and the tanδ value at T_i increases. At higher loading of zinc stearate (i.e., 30 and 40 phr) tanδ at T_i gets overshadowed by the rise in tanδ due to an early onset of melting of zinc stearate.

Table 4 Results of Dynamic Mechanical Studies at 3.5 Hz [33]

Mix No.	Glass–Rubber Transition		Ionic Transition	
	T_g (°C)	tanδ at T_g	T_i (°C)	tanδ at T_i
EP0	−44.0	1.273	a	—
EP1	−41.8	1.192	38.0	0.094
EP2	−40.0	1.084	40.9	0.103
EP3	−38.6	0.944	46.3	0.112
EP4	−38.2	0.934	51.7	0.121
EP5	−36.8	0.837	52.6	0.128
EP6	−37.0	0.820	54.8	0.139

a = absent.

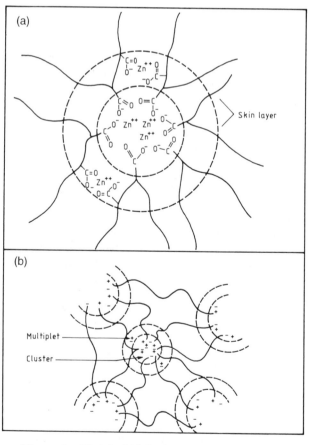

Figure 3 Model of biphasic structure in ionomer.

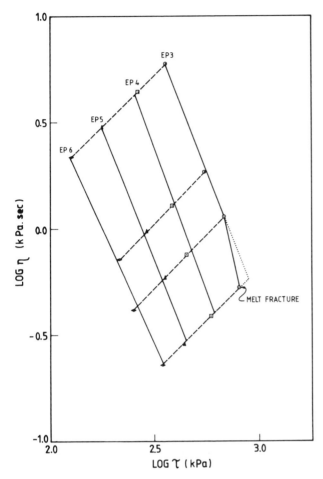

Figure 4 Plots showing dependence of apparent viscosity (η) on apparent shear stress (τ) [33].

loading. It is also apparent that the shear stress dependence of viscosity decreases with increasing zinc stearate loading, since higher loading of zinc stearate provides greater extent of plasticization and enhances the flowability of the system.

At high-processing temperatures molten zinc stearate is believed to solvate the ionic aggregates (Fig. 5) and high-shear force applied may facilitate exchange reactions. Due to the interchange reaction of the intermolecular carboxylate salts (both bridging and non-bridging types) with zinc stearate, the intermolecular metal carboxylate bonds can break and result in formation of a new structure with ionic groups attached to the stearate moiety at one end. Since intermolecular salt bridges cease to exist in the postexchange reaction stage, the chains become free to flow.

Since the system is processed as thermoplastics are processed, its reprocessability was studied under repeated cycles of extrusion in the MPT. Results of reprocessability studies are shown in Table 5. It is evident that after the first cycle viscosity increased slightly, which may be due to the orientation effect. In the subsequent cycles viscosity remained almost constant. Therefore, it is concluded that the zinc stearate plasticized zinc salt of m-EPDM is melt processable just as thermoplastics.

B. Effect of Precipitated Silica Filler

The physical properties of the different mixes are summarized in Table 6. As expected, both hardness and modulus increase with an increase in filler loading. Incorporation of silica causes a gradual increase in tensile strength up to 20 phr of the filler, when the strength is almost twice that of the unfilled system. At 30 phr of filler loading, the tensile strength comes down to the level of 10 phr, but it is still 1.5× the value of the unfilled system. The increase in elongation at break on filler incorporation is believed to be due to a slippage of the polymer chains over the silica surface [26]. Because of the thermo-reversible nature of the ionic crosslinks, tensile strength dropped down at an elevated test temperature (i.e., 70°C), but the silica filled system registered higher strength than the unfilled system even at 70°C.

EPDM–ZnO–stearic acid systems could not be extruded even at 190°C. This is not unexpected since the material, in the absence of zinc stearate, shows no transition from the rubbery state to the viscous flow state (Fig. 1). In the presence of 10 phr of zinc stearate, the m-EPDM–ZnO–stearic acid system could be extruded but melt fracture occurred at a lower temperature (150°C) at all shear rates. At 160°C and 170°C, however, the extrudates showed melt fracture only at high shear conditions. At 20 phr loading of zinc stearate, melt fracture of the extrudate occurred at high shear conditions at 150°C, but at higher temperatures no melt fracture occurred and the extrusion was smooth under all shear conditions. At 30 and 40 phr loadings of zinc stearate, the extrudates were smooth under all shear conditions at all temperatures.

Figure 4 shows the plots of apparent viscosity versus shear stress at 170°C for different mixes. It is evident that the materials behave as pseudoplastic fluids, and the viscosity decreases with increasing zinc stearate

Table 5 Results on Reprocessability Studies [33]

No. of Cycles	Shear Stress (KPa)	Apparent Viscosity (η) KPa-sec	300% Modulus of Extrudate (MPa)
1	232.6	1.5	2.69
2	253.3	1.7	2.74
3	259.1	1.7	2.87
4	263.2	1.8	2.74

Mix number, EP5; temperature, 170°C; shear rate, 159 s^{-1}.

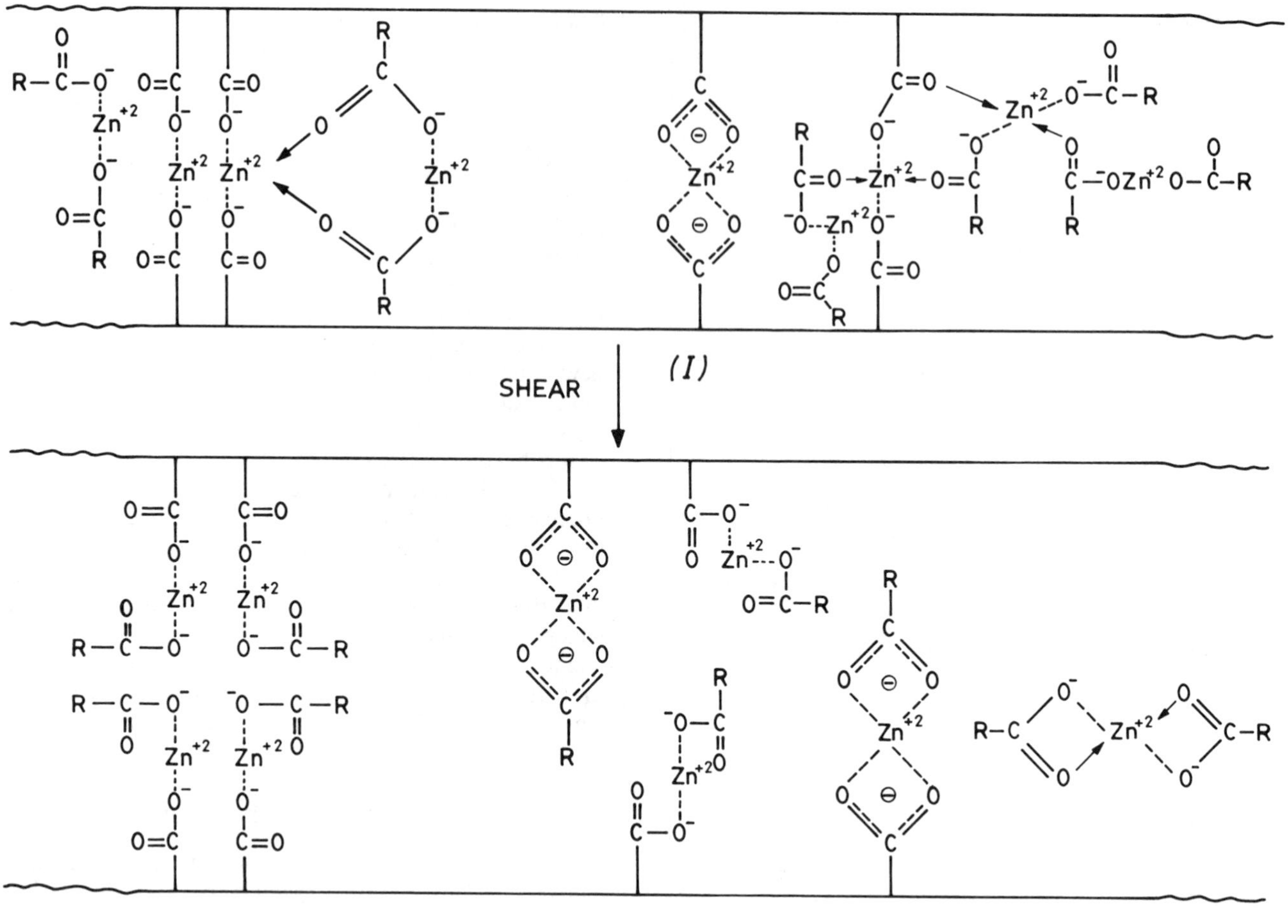

Figure 5 Probable mechanism of shear-induced exchange reactions during melt flow process [33]. (I) Interaction of zinc stearate, $(RCOO)_2 Zn$, with ionic aggregates before melt flow. (II) Exchange reactions during melt flow.

Table 6 Physical Properties[a] [27]

	Mix Number			
	S0	S1	S2	S3
Hardness (Shore A)	62	65	67	69
300% modulus (MPa)	2.90	3.26	3.64	4.49
	(1.82)	(2.58)	(2.35)	(3.13)
Tensile strength (MPa)	7.60	10.30	14.35	10.43
	(2.42)	(3.10)	(3.59)	(4.00)
Elongation at break (%)	760	1370	1340	1000
	(431)	(400)	(530)	(440)
Tear strength (N/cm)	418	436	449	424
	(233)	(256)	(296)	(283)
Hysteresis work, $(J/m^2) \times 10^3$	80.1	98.0	112.0	110.2
Tension set at 100% extension (%)	17	20	22	22

[a] Values in parentheses are the results of tests done at 70°C.

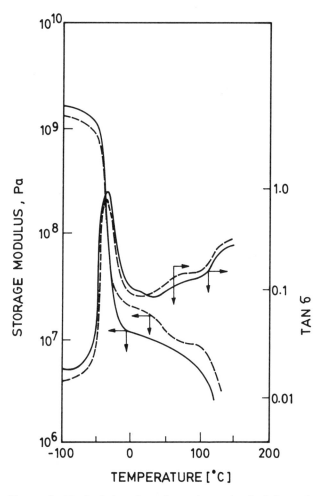

Figure 6 Typical plots from dynamic mechanical thermal analysis showing storage modulus and tanδ variation with temperature [27]. S0 (——), S2 (------).

Both tear resistance and hysteresis increase on incorporation of silica, but the effect is less pronounced as compared to the stress–strain properties. Tension set of the ZnO-neutralized m-EPDM system is low (around 20%) and incorporation of filler causes only a marginal increase in set due to chain slippage over the filler surface, as previously discussed. Measurement of physical properties reveal that there occurs an interaction between the filler surface and the polymer. Results of dynamic mechanical studies, subsequently discussed, support the conclusions derived from other physical properties.

Table 7 Results of Dynamic Mechanical Studies [27]

Mix Number	T_g, °C	tanδ at T_g	T_i, °C	tanδ at T_i
S0	−36.8	0.837	+52.6	0.128
S1	−37.0	0.814	+60.6	0.139
S2	−36.4	0.715	+62.9	0.148
S3	−35.5	0.692	+72.8	0.166

Typical plots of the storage modulus versus temperature and mechanical loss (tanδ) versus temperature are shown in Fig. 6. It is clear that, apart from the glass–rubber transition (T_g) occurring around −37°C, there occurred the ionic transition (T_i), at high temperature (>50°C), which, in the presence of silica, becomes prominent and is shifted to the higher temperature side. The high temperature transition is believed to be due to relaxation of the restricted chain segments arising out of the ionic aggregates called multiplets and clusters, as depicted in Fig. 3. It is shown that the skin layer separates the restricted mobility region from the free mobility region, which constitutes the bulk of the chain segments. Results of dynamic mechanical studies are summarized in Table 7.

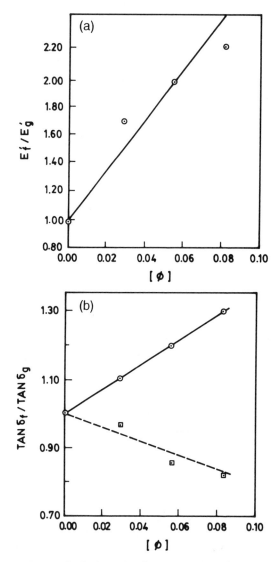

Figure 7 (a) Variation of E'_f/E'_g with volume fraction [8] of the filler at room temperature [27]. (b) Variation of $\tan\delta_f/\tan\delta_g$ with volume fraction [8] of filler [27]. T_g, (---□---) T_i (———○———).

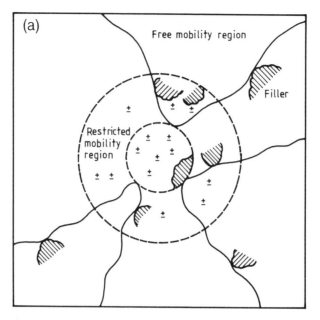

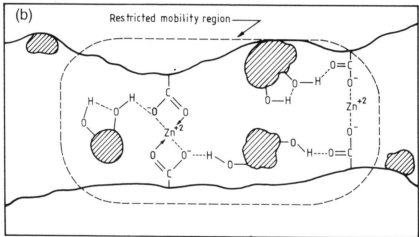

Figure 8 (a) Schematic diagram showing distribution of fillers in different parts of anionic elastomer [27]. (b) Proposed structural model showing interaction of silanol groups on silica surface with carboxylate groups [27].

The variation in room temperature storage modulus on filler incorporation is shown in Fig. 7. The results could be fitted into the following equation [27]:

$$\frac{E'_f}{E'_g} = 1 + 12.1\,[\theta] \tag{1}$$

where E'_f is the storage modulus of the silica filled system and E'_g refers to the modulus of the gum or unfilled system and $[\theta]$ is the volume fraction of silica. This is very similar to the relationship proposed by Smallwood [28] in the case of diene rubbers. The greater slope of the plot in Fig. 7, as compared with conventional rubber systems, is ascribed to the strong interaction between

the polymer chains containing ionic aggregates and the active sites on the filler surface.

Figure 7 also shows the plots of $(\tan\delta)_f/(\tan\delta)_g$ versus volume fraction of the filler (θ) at T_g and T_i. Here "f" stands for the silica filled system and "g" denotes the gum or unfilled system. The results could be fitted into the following relations [27]:

At T_g,

$$\frac{(\tan\delta)_f}{(\tan\delta)_g} = 1 - 2.6\,[\theta] \tag{2}$$

Equation (2) is similar to that obtained in the case of conventional rubber systems and depicts weak rub-

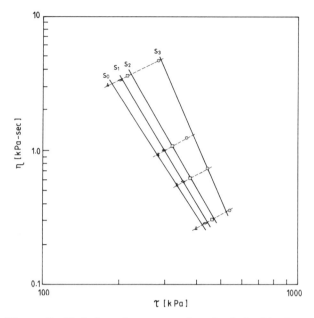

Figure 9 Variation of apparent viscosity [η] with shear stress [τ] [27].

ber–filler interaction [29,30], involving the backbone chains of the polymer.

At T_i,

$$\frac{(\tan\delta)_f}{(\tan\delta)_g} = 1 + 3.1\ [\theta] \tag{3}$$

This indicates that the occurrence of the high temperature relaxation of the restricted mobility zones of the chain segments is facilitated by the filler, thereby implying strong bonding between the ionic aggregates and the polar sites of the filler. From the above results, it can be inferred that the rubber–filler interaction, in the case of silica filled ionic rubber, is of two types: (1) the interaction between the filler particles and the nonionic segments of the polymer backbone, which is similar to the interaction involving diene rubbers and reinforcing fillers, as manifested in the lowering of tanδ at T_g; and (b) the interaction between the ionic groups of the polymer and the polar sites (or silanol groups) present on the filler surface, which is manifested in an increase in tanδ at T_i.

Table 8 Activation Energy (ΔE) for Viscous Flow (kJ/mol) [27]

Shear rate (sec^{-1})	Mix Number			
	S0	S1	S2	S3
61	16.0	28.1	31.8	39.5
305	12.2	20.5	24.3	28.2
610	9.36	15.6	16.9	18.6

Table 9 Results of Reprocessability Studies [27]

Cycle	Shear Stress (kPa)	300% Modulus of Extrudate (MPa)
1	285.71	3.82
2	293.27	3.98
3	296.40	4.00
4	297.29	3.91

Mix no., S2; temperature, 170°C; shear rate, 121 s^{-1}.

At T_i the slope in the plot of Eq. (3) is positive, thereby inferring that an increase in filler loading strengthens the high temperature relaxation, presumably by increasing the ionic cluster-induced chain rigidity. While the rubber–filler interaction involving the nonionic polymer backbone is of weak van der Waals type, the same due to ionic aggregates can be of both hydrogen bonded type involving the polar sites on the filler surface [13] and van der Waals type, as depicted in Fig. 8. Mondal et al. [31] made a similar observation in the case of zinc oxide neutralized carboxylated nitrile rubber filled with precipitated silica filler.

Figure 9 shows the variation of apparent viscosity with apparent shear stress. It is evident that the mixes are pseudoplastic in nature. Furthermore, as expected, viscosity increases with an increasing filler loading.

Activation energy (ΔE) for viscous flow was calculated according to the following equation [32].

$$\eta = A\ e\ \Delta E/RT \tag{4}$$

Here η is the apparent viscosity at temperature T, R is the universal gas constant, and A is an empirical constant, called frequency factor for melt flow. The activation energy values for different systems and at different shear rates are summarized in Table 8. It is evident that activation energy for flow increases with filler loading, but it decreases with an increase in shear rate.

At high-processing temperature, zinc stearate plasticizes the matrix and solvates the ionic clusters [28,29], thus facilitating the polymer flow. In the silica filled systems, however, the flow is hindered due to the presence of silica particles anchoring the chain segments both in the regions of free mobility and restricted mobility, thereby increasing the free energy of activation. Furthermore, under high-shear conditions, it is likely that the loosely held silica aggregates are broken down, causing a reduction in the apparent activation energy of flow.

Since the system exhibits thermoplastic behavior, its reprocessability was studied in the MPT and the results are given in Table 9. Viscosity during repeated extrusions and the modulus of the corresponding extrudates remain constant up to four cycles of extrusion, thereby confirming the thermoplastic elastomeric nature of the silica filled m-EPDM–ZnO–stearic acid-zinc stearate composition.

IV. CONCLUSIONS

Zinc salt of maleated EPDM rubber in the presence of stearic acid and zinc stearate behaves as a thermoplastic elastomer, which can be reinforced by the incorporation of precipitated silica filler. It is believed that besides the dispersive type of forces operative in the interaction between the backbone chains and the filler particles, the ionic domains in the polymer interact strongly with the polar sites on the filler surface through formation of hydrogen bonded structures.

REFERENCES

1. R. A. Weiss, J. A. Fitzerald, and D. Kim, *Macromolecules, 24*: 1071 (1991).
2. M. Pineri, C. Mayer, A. M. Levelut, and M. Lambert, *J. Polym. Sci., Polym. Phys., 12*: 115 (1974).
3. R. Jerome, J. Horrion, R. Fayt, and Ph. Tayssie, *Macromolecules, 17*: 2447 (1984).
4. U. K. Mondal, D. K. Tripathy, and S. K. De, *Polymer, 34*: 3832 (1993).
5. K. Sato, *Rubber Chem. Technol., 56*: 1942 (1984).
6. R. A. Weiss, W. J. MacKnight, R. D. Lundberg, K. A. Mauritz, C. Thies, and D. A. Brant, "Coulombic Interactions in Macromolecular Systems," A. C. S. Symposium Series, *302*: 1 (1986).
7. P. K. Agarwal and R. D. Lundberg, *Macromolecules, 17*: 1918 (1984).
8. P. K. Agarwal and R. D. Lundberg, *Macromolecules, 17*: 1928 (1984).
9. A. U. Paeglis and F. X. Oshea, *Rubber Chem. Technol., 60*: 223 (1988).
10. A. J. Oostenbrink and R. J. Gaymans, *Polymer, 33*: 3086 (1992).
11. D. R. Paul and J. W. Barlow, *Polymer, 33*: 268 (1992).
12. R. Greco, M. Malineonico, E. Martuscelli, G. Rogosta, and G. Scarizi, *Polymer, 28*: 1185 (1987).
13. E. N. Kresge, "Ionic Bonding in Elastomeric Networks," 18th Canadian High Polymer Forum, Hamilton, Ontario (1978).
14. U. K. Mondal, D. K. Tripathy, and S. K. De, *Polym. Engg. Sci. 36*:283 (1996).
15. I. Duvedevani, P. K. Agarwal, and R. D. Lundberg, *Polym. Eng. Sci., 22*: 500 (1982).
16. E. J. Quan, *Handbook of Fillers and Reinforcements for Plastics*, (H. S. Katz and J. V. Milewski, ed.), Van Nostrand Reinhold Co., New York (1978).
17. U. K. Mondal, D. K. Tripathy, and S. K. De, *Plastics Rubber and Composites—Processing and Applications, 24*: 19 (1995).
18. K. Sato, *Rubber Chem. Technol., 56*: 942 (1983).
19. A. U. Paeglis, and F. X. O. Sheha, *Rubber Chem. Technol., 60*: 228 (1988).
20. T. Kurian, P. P. De, D. Khastgir, D. K. Tripathy, S. K. De, and D. G. Peiffer, *Polymer 36*: 3875 (1995).
21. H. P. Brown and C. F. Gibbs, *Rubber Chem. Technol., 28*: 937 (1955).
22. B. Hird and A. Eisenberg, *Macromolecules, 25*: 6466 (1992).
23. M. R. Tant and G. L. Wilkes, *J. Macromolecular Sci. Rev., Pt. C, 28*: 1 (1988).
24. A. Eisenberg, B. Hird, and R. B. Moore, *Macromolecules, 23*: 4098 (1990).
25. W. J. MacKnight, W. P. Taggert, and R. S. Stein, *J. Polym. Sci., Polym. Symp., 45;* 113 (1974).
26. S. K. Chakraborty, and S. K. De, *Polymer, 24*: 1055 (1983).
27. S. Dutta, A. K. Bhattacharya, S. K. De, E. G. Kontos, and J. M. Wefer, *Polymer 37*: 2581 (1996).
28. H. M. Smallwood, *J. Appl. Phys., 15*: 758 (1944).
29. L. E. Nielsen, *J. Polym. Sci., Polym. Phys. Ed., 17*: 1897 (1979).
30. D. Roy, A. K. Bhowmick, and S. K. De, *J. Appl. Polym. Sci., 49*: 263 (1993).
31. U. K. Mondal, D. K. Tripathy, and S. K. De, *J. Appl. Polym. Sci., 55*: 1185 (1995).
32. A. A. Tager, *Physical Properties of Polymers*, Mir Publishers, Moscow (1978).
33. S. Dutta, S. K. De, E. G. Kontos, and J. M. Wefer, *J. Appl. Polym. Sci. 61*: 177 (1996).

29

Elastomeric Compound—The Importance of Consistency in Economic Component Manufacture

Robert Bond
Chicago Rawhide (Americas), Elgin, Illinois

I. INTRODUCTION

Today with the everincreasing global market place and the demands for customer satisfaction, the end user is placing more and more emphasis on product quality, reliability, and value. The result is that companies that produce components for the major original equipment manufacturers (OEMs), particularly the automotive industry, are being required to: improve quality, improve reliability, increase product life, improve cost effectiveness, and reduce response time.

To this end, manufacturing throughput time has to be minimized. This allows the flexibility required to keep the response time to customers down to a minimum. The introduction and use of electronic data interchange (EDI) for changing order schedules, in some cases on a daily basis, make it essential for component manufacturers to have responsive and flexible processes. One of the major contributing factors that enable these objectives to be achieved is consistency, whether it be the incoming raw material or the processes used in the manufacturing process. To get a more in-depth understanding of the issues relating to raw materials and processes, the manufacture of radial lip shaft seals will be used as the component under consideration.

II. THE RADIAL LIP SHAFT SEAL

The radial lip shaft seal is a dynamic sealing device. In general, its role is to seal fluids in and contaminants out when sealing against a rotating, reciprocating, or oscillating shaft. The seal design can take many forms, e.g., with or without a garter spring; with or without dirt exclusion lips; metal case, rubber covered, or half metal/half rubber.

Figure 1 shows a typical rubber covered radial lip shaft seal. It is made up of three basic components, the metal insert, the garter spring, and the elastomeric lip. Following is a description of the role of each component.

Metal insert: The metal insert acts as the stiffening element of the seal, which is essential when the seal is fitted into the bore or housing.

Elastomeric lip: The elastomeric lip forms half the dynamic sealing interface, the other half being the shaft. In the case of the rubber covered seal, the elastomeric compound also acts as half the static seal between the bore/housing and the seal; the bore itself being the other half of the static seal.

Garter spring: The garter spring generates the majority of the radial load the elastomeric lip exerts on the shaft, which is essential when the lip has to follow the shaft when operating under dynamic conditions.

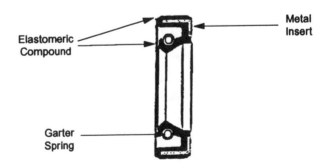

Figure 1 Rubber covered lip type rotary shaft seals.

It can be seen that the role of the radial lip seal and the tolerances to which it has to be manufactured make it an engineered product that is used in the engineering industry; a radial lip shaft seal is not a rubber component to use typical rubber tolerances that happen to be used by the engineering industry.

III. MANUFACTURING PROCESS

The manufacture of elastomeric radial lip seals centers around the molding operation. Figure 2 [1] illustrates the manufacturing flow diagram for the two primary processes used: compression molding and injection molding.

To minimize the cost of manufacture, one approach is to minimize the manufacturing "throughput" time, sometimes called "just in time" or "the Toyota lean manufacturing process."

The result will be minimum work-in-process (WIP) through the elimination of "queue time" and a significant reduction in the cost of quality when compared to the more conventional approaches to manufacturing used in recent years. However, the new approaches to manufacturing demand three things from the raw materials, purchased components, and the individual processes: high quality, consistency, reliability.

Therefore, every effort must be placed on ensuring the quality, consistency, and reliability of supply of the incoming raw materials and components. Since this chapter is about the need for consistent, high-quality elastomeric compound, emphasis will be placed on: incoming raw materials, compound mixing, extrusion (where appropriate), and molding.

A. Incoming Raw Materials

The requirement for consistent, high-quality incoming raw materials comes from the need for the elastomeric compounds to have consistent processing properties. Therefore, it is important to:

1. Understand the critical properties of the mixed compound that affect processability. It should be understood that the properties and acceptable operating values of those properties, which control the compound processability, are dependent on the processes under consideration, e.g., hot or cold feed extrusion, compression or injection molding.

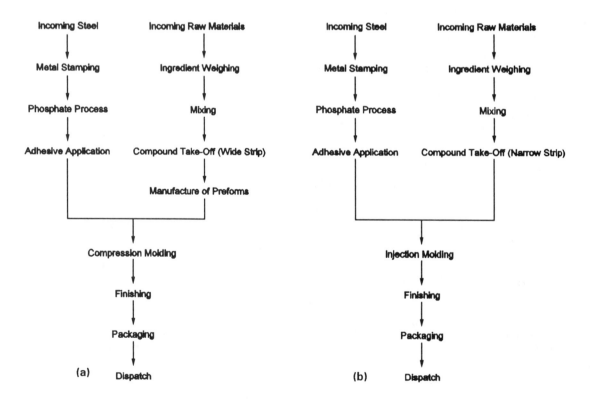

Figure 2 Typical process flow diagrams: (a) compression molding, (b) injection molding.

2. Establish the contribution the individual ingredients in the elastomeric compounds make to both the vulcanized compound properties and the uncured properties required for consistent processability.

3. Establish realistic, workable individual ingredient specifications.

At the present time, the difficulty is correlating incoming raw ingredient properties with processability. Elastomeric compounds can be considered to be made up of four components: polymers, fillers, process aids, and other chemicals.

During recent years a considerable amount of research has been undertaken to understand what in the makeup of a polymer affects the processability. In the late 1980s, the Rubber Manufacturers Association in the United States undertook a research project with the Department of Polymer Engineering at the University of Akron to evaluate the laboratory equipment available using specially made butadiene–acrylonitrile polymers with different acrylonitrile levels, molecular weights, and molecular weight distributions. The results from the study confirmed that, from the processing variables viewpoint, the major factors are: frequency (shear rate), temperature (temperature), and deformation (strain).

Therefore, if processability is to be measured on a regular basis, it would be extremely useful if a piece of equipment was available that could measure the dynamic properties under realistic operating conditions. Fortunately, one piece of test equipment has been developed, which is commercially available, the RPA 2000 (Monsanto Co.), which may meet the requirements. A considerable number of investigations have been reported on the RPA 2000 [2], that support the view that it may meet the requirements of an instrument that measures both polymer and compound processability. The work to date identifies differences in polymers and compounds. However, it is important to relate those differences to processing characteristics in the manufacturing environment.

An alternate approach was taken by a shaft seal manufacturer and a polymer producer working together [2]. The project involved making a series of butadiene-acrylonitrile polymers of known acrylonitrile contents with various molecular weights and molecular weight distributions then making seals under normal manufacturing conditions. There were four polymers evaluated:

A.	Bound acrylonitrile	34%
	Average molecular weight	275,500
	Intrinsic viscosity	1.05
	Compound Mooney	28
B.	Bound acrylonitrile	34%
	Average molecular weight	309,600
	Intrinsic viscosity	1.15
	Compound Mooney	37
C.	Bound acrylonitrile	34%

	Average molecular weight	379,700
	Intrinsic viscosity	1.44
	Compound Mooney	56
D.	Bound acrylonitrile	34%
	Average molecular weight	306,200
	Intrinsic viscosity	1.13
	Compound Mooney	37

Note: (1) Polymer D was made up of blending A and C (50:50) in the latex stage. The result was a bimodal molecular weight distribution having an average value close to B; (2) All compounds used the same formulation.

The performance of the different polymers was based on the molding scrap levels obtained for two categories of scrap:

Excess flash—Indicates the compound flow was excessive during the mold closing process (shelf loaded compression molding) [3].

Unfills—Indicates that the compound did not flow sufficiently during the mold closing process (shelf loaded compression molding) [3].

The variable used in the manufacturing process was the mold closure speed to investigate the effect of varying the shear rate (frequency). Figures 3, 4 and 5 illustrate the unfill scrap for the polymers A through D for mold closure times of 5 s, 8 s, and 11 s, respectively. Figure 6 shows the excess flash scrap for the same polymers with a mold closure time of 11 s. The results show that for butadiene-acrylonitrile polymers, molecular weight and molecular weight distribution have a significant effect on the compound flow during the molding process: In addition, it shows that shear rate (frequency) plays an important role.

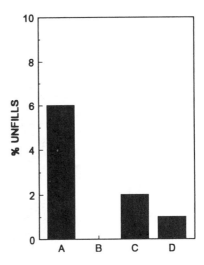

Figure 3 Percent unfills by elastomer type for 5-s mold closure.

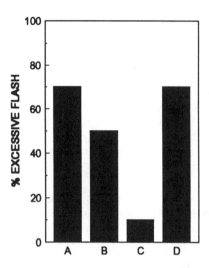

Figure 4 Percent unfills by elastomer type for 8-s mold closure.

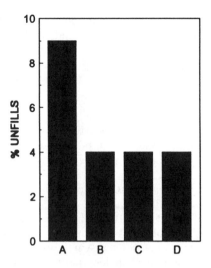

Figure 6 Percent excess flash by elastomer type for 11-s mold closure.

Unfortunately, at the present time there does not appear to be a detailed understanding of the relationship between the properties of the raw materials or the ingredients making up an elastomeric compound and the compound's processability. Therefore, the most widely used approach presently being adopted is to use the conventional characteristics with the standard specification limits issued by the supplier or special agreed upon limits with the supplier to accommodate specific requirements in the fabricators manufacturing process.

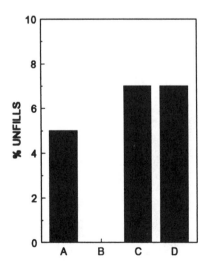

Figure 5 Percent unfills by elastomer type for 11-s mold closure.

B. Compound Mixing

The previous section has discussed the importance of understanding the role the individual ingredients play in the compound processability. The next major issue is the mixing of the elastomeric compound. In any manufacturing operation that uses rubber (elastomeric compound) as one of the major components in the product being produced, the consistency of the mixed elastomeric compound is critical not only for product reliability and performance, but also for keeping the cost of quality down and enabling the manufacturing throughput time to be minimized. Therefore, the major requirements from the mixing operation are: flexibility, high-quality mixed compound, compound consistency (within batch and batch to batch), compound delivery, and cost effectiveness.

There are two types of mixing processes available, continuous and batch. The nature of the component manufacturing business generally dictates that there is a need to mix a large variety of compounds. Therefore, the majority of component manufacturers use the batch mixing process to produce the different elastomeric compounds required. This process consists of four basic areas all of which are important for consistent mixing: ingredient weighing, batch mixing process, take-off system, and cooling and batch-off system.

1. Ingredient Weighing

The systems used for ingredient weighing can vary from a totally manual system to a totally automated system where the ingredients are untouched by human hands. The critical role of the ingredient weighing system, whether it is manual or fully automated, is to ensure that the correct weight of the correct ingredient arrives at the mixer hopper charging door in the correct order at the right time.

The least costly approach is to use a manual weighing system. However, this approach also has the highest risk for manual error, although with a very experienced mixing crew the problem can be minimized. The other extreme, a fully automated weighing system, is the most expensive but has the advantage that the possibility for human error is removed. This applies not only to the weighing process itself but also to the filling of the bulk containers, the critical ingredient containers, and the other chemical containers, including oils and greases if the incoming ingredients have the appropriate bar code label attached. The bar code labels are read prior to filling the container; if the bar code label corresponds to the correct ingredient then the container can be opened, if it is incorrect, the containers will not open.

The system generally considered to be the most appropriate is a combination of both approaches, a semi-automated computer prompted weighing system using the locked container/bar code labelling technique to minimize, hopefully eliminate, the possibility of getting the incorrect material in a given container. For consistent, high-quality compounds, both within batch and from batch to batch, it cannot be overstated that the correct weight (tight tolerances) of the correct ingredient must arrive at the mixer hopper charging door at the right time in the correct order.

2. Compound Mixing

A cross-section of a typical batch mixer can be seen in Fig. 7 [4]. Using a simple analogy, the batch mixing process can be said to be similar to mixing a fine fruit cake. The ingredients are placed in the mixing chamber then mixed together using two mixing rotors and finally discharged from the mixer.

There are two fundamental rotor designs in general use, the tangential rotor (Fig. 8a [5]) and the intermeshing rotor (Fig. 8b [5]) designs. In recent years, a number of in-depth studies have been completed comparing the mixing capability of the tangential rotor to the intermeshing design. Bond and Takada [5] in 1992 reported that in both cases the latest (at that time) intermeshing and two-wing tangential rotor mixers gave superior temperature control and mixing capability when compared to the conventional 3D Banbury* mixer used to establish the base line results. It also demonstrated the role the use of computer controls played in improving the mixed compound consistency, both within batch and from batch to batch. Therefore, there are additional important factors that have to be taken into account when purchas-

---

* Banbury is a trademark of Farrell Corporation.

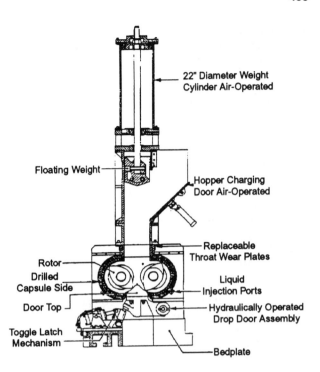

Figure 7 Typical batch mixer.

ing a new mixer, e.g., equipment reliability, spare part availability, and technical service.

3. Take-Off System

When the compound is dropped out of the mixer, it is essential to remove the heat from the batch of rubber as quickly as possible; this is particularly important when the curatives are in the compound. It is just as important to ensure that the batch is fully blended before it is finally cooled and stored prior to use.

There are three systems in use today:

1. A twin-screw extruder fitted with a "stuffer box." The twin-screw extruder continues to blend the compound as it passes through; although with the dryer,

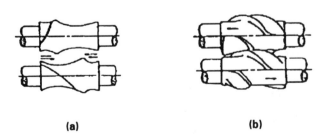

Figure 8 Internal mixer rotor designs: (a) tangential rotor, (b) intermeshing rotor.

more difficult compound formulations can have problems. If the heat is to be removed efficiently, the extruder must have extremely good cooling capabilities.

2. A single drop mill sized appropriately for the mixer. With the single, two-roll drop mill approach, the mill can be either beneath the mixer or alongside it and fed using a tub on rails beneath the mixer. This enables the heat to be taken out of the compound from the mixer. If time is not a critical issue in the mixing process, the single drop mill approach also accommodates the blending required.

3. A two-mill take-off system, the mills sized appropriately for the mixer. The two two-roll mill take-off system is made up of one mill beneath the mixer to take out the heat from the compound (the drop mill) and a second mill used for the final blending process. This system is recommended when time and productivity are major issues.

To ensure that each batch of compound gets the same amount of blending (energy input), the blending mill is generally fitted with a stock blender to improve the blending consistency. To ensure that each batch of compound gets the same amount of blending, automated controls can be incorporated on the blending mill to control the following variables:

Number of passes on the stock blender.
Individual roll speeds.
Individual roll temperature.
"Nip" setting (gap between rolls).

4. Cooling and Batch-Off System

The conventional system in general use takes the compound from the blending mill, through an antistick applicator onto a festoon system with air cooling, and finally through a "wig-wag" take-off system into storage baskets.

5. Optimized Mixing Line with a Computer Prompted Weighing System

Bond and Takada [5] showed how replacing a conventional, manual mixing line based on a 3D Banbury mixer with a mixing line comprised of a computer prompted weighing system using bar code labels and readers, the latest 80-L Banbury mixer, a two two-roll mill take-off system, and conventional cooling and batch-off improved compound consistency. Figure 9 schematically illustrates the improved mixing line.

Figures 10–14 illustrate the reductions obtained in hardness, 50% modulus, 100% modulus, elongation, and tensile strength, respectively, for a fluoroelastomer compound produced under normal production conditions. Table 1 shows the improvement obtained in detail.

C. Extrusion

In the two typical shaft seal manufacturing processes shown in Fig. 2, the extrusion process is only required if the compression molding process is being used. There are basically two types of extruders in general use, the cold feed and the hot feed. In both cases their role is to produce the preforms for use with compression molding. The critical issues with the preform are: the shape, the dimensions, and the weight.

1. Cold Feed Extrusion

Cold feed extruders use compounds in a strip form. The strip is generally made during the take-off process in mixing; a series of cutters are used on the blending mill

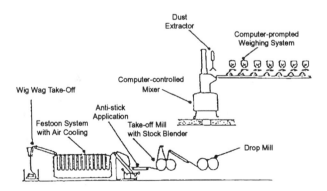

Figure 9 Flow diagram: improved mixing line.

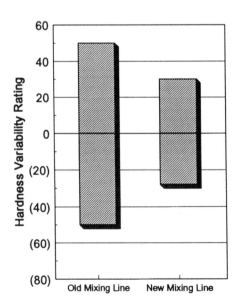

Figure 10 Comparison of hardness ±3 sigma variability ratings: fluoroelastomer compound.

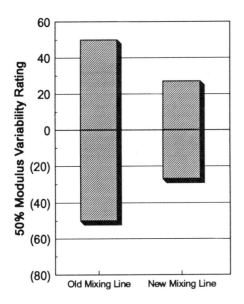

Figure 11 Comparison of 50% modulus ±3 sigma variability ratings: fluoroelastomer compound.

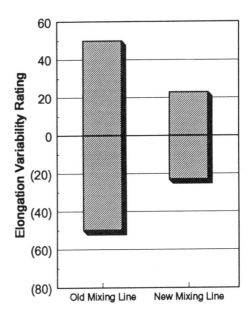

Figure 13 Comparison of elongation ±3 sigma variability ratings: fluoroelastomer compound.

to produce the required strip width, then passed through the cooling system and batched-off into either baskets or wound onto coils for an easier feed into the extruder.

There are two types of cold feed extruders, the standard screw design type (Fig. 15) and one with a screw feed using a ram action to feed the extruder. In both

cases the preforms are produced using a rotating blade system immediately after the extruder die.

If consistency is critical, it is important to remember that the preforms are only as consistent as the compound or rubber strip feeding the extruder. Because of the critical nature of the preform dimensions, extruder manufac-

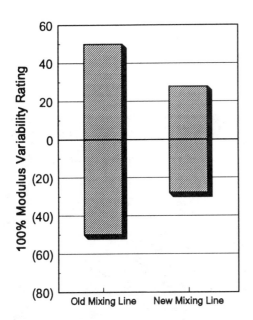

Figure 12 Comparison of 100% modulus ±3 variability ratings: fluoroelastomer compound.

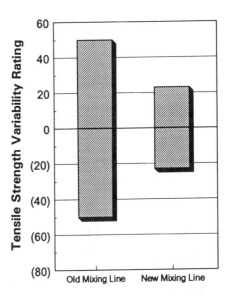

Figure 14 Comparison of tensile strength ±3 sigma variability ratings: fluoroelastomer compound.

Table 1 Comparison of Fluoroelastomer Compound Properties

Property	Old mixing line ± 3 sigma rating July–Sept. 1989	New mixing line ± 3 sigma rating July–Sept. 1990
Hardness (IRHD)	100	56.4
50% modulus	100	54.0
100% modulus	100	59.0
Elongation	100	46.0
Tensile strength	100	52.0

turers have developed machine controls that allow the compound to leave the extruder at a constant rate. However, the rate varies depending on the nature and type of compound being extruded. The need for this type of control system is to minimize any variation or inconsistency within the compound being extruded.

After leaving the extruder, the preforms are cooled. There are two types of cooling processes used. The choice depends on the compound type being extruded.

1. If the compound being extruded is not attacked by water, for example, conventional butadiene-acrylonitrile based compounds, the preforms are passed through a water cooling system. The water cooling system can be using either the spray technique or complete submersion.
2. If the compound can be attacked by water, for example polyacrylate and fluoroelastomer elastomer based compounds, the preforms are air cooled.

2. Hot Feed Extrusion

There are two types of hot feed extruders, one similar to the screw type cold feed extruder, except that the strip feeding the extruder is taken directly off the two-roll mill. The mill is used to further blend and heat the compound. It is then taken off the mill in strip form and fed directly into the extruder. The second type of hot feed extruder is based on the action of a ram being used to feed the compound into the extruder die (Fig. 16). The compound is placed on a two-roll mill to further blend and heat rubber. It is then taken off the mill in "pig" form to fill the extruder chamber.

In both cases, the preforms are produced using the rotating cutter system as described for the cold feed extruders. The preforms are also cooled the same way.

D. Molding

In the manufacture of shaft seals, the molding is the process where the compound and treated metal stamping (phosphated and adhesive applied) are brought together to form the conventional lip type shaft seal. Therefore, it is critical that the compounds used are formulated to suit the molding process. There are three fundamental molding processes in use today: compression, injection, and injection/transfer.

The choice of process is generally based on economic issues such as: volume of seals to be produced; elastomeric compound, cost and waste; and product design.

There are four critical issues required for all the molding processes listed previously, they are [6]:

Temperature distribution across the mold (top and bottom).

Pressure distribution across the mold.

Figure 15 Cold feed extruder.

Figure 16 Hot feed extruder.

Application of vacuum, whether directly through the mold or completely evacuating the chamber.

Press controls.

The variation in temperature across the top and bottom halves of the mold is a function of the press platen size, the flatness of the mold and platen surfaces at the two mold/platen interfaces, the mold construction, and the platen heating system. The larger the platen size, the more difficult it is to maintain a constant temperature across the platen. Therefore, one solution to the problem is to use a large number of molding presses with a relatively small platen size, for example, four presses with platens 355 × 355 mm will have approximately the same production capacity as one press with a 710 × 710 mm platen. In reality, for high-volume production, large presses are more economical from the cost point of view and the floor space required.

The two accepted methods used for heating platens are both electric: resistance heating and induction heating.

Experience has shown that the most efficient method for minimizing temperature variation across a larger platen is the use of multiinduction heating coils. With a three zone system on a 710 × 710 mm platen correctly balanced, it is possible to maintain the temperature variation across the platen to within ± 2.5°C.

The pressure distribution across the mold is a function of both the mold surface and the platen flatness. The larger the mold size and the platen, the more difficult it is to maintain the high degree of flatness required. The mold and platen flatness is also critical in the area of heat transfer from the platen to the mold itself.

During the shaft seal molding process, the elastomeric compound has relatively large distances to travel/flow in the mold, particularly when using shelf loaded compression molding (Fig. 17a) and injection molding (Fig. 17b). History has shown that without the use of a vacuum, either applied directly through the mold or completely evacuating the mold chamber, "air trapping" is a major problem. Therefore, the trend started in the United States of evacuating the mold/mold chamber has become the accepted way to mold lip type rotary shaft seals.

Press control is critical; it is essential that the elastomeric compound reaches the required cure state to optimize product performance yet remains in the press the shortest time period to maximize productivity. To meet this objective, both compression and injection presses now use microprocessor controls, which enable variations in platen temperatures and compound cure characteristics to be accommodated without sacrificing product performance or productivity.

One compound formulation is not suitable for all the different molding techniques used. It was reported by Berens and Born [7] and Horve [8] that molded lip seals produced using "shelf" loaded preforms, as shown in Fig. 17a, gave more consistent product performance than the "cavity" loaded approach, also shown in Fig. 17a. However, for "trimmed" lip seals, both "shelf" and "cavity" loaded preforms gave similar performance. When considering injection molding, various "gating" systems, as shown in Fig. 17b, have been successfully used for producing "molded" lip type rotary shaft seals. The reason being that the elastomeric compound formulation plays a major role in the compound flow and particle orientation. The same compound formulation is not suitable for all the different approaches to injecting the rubber into the mold.

1. Compression Molding

Compression presses are produced in various sizes throughout the world. In general, shaft seals are manufactured on presses with platen sizes in the range of 350 × 350 mm to 715 × 715 mm with clamp loads in the range of 1500 kN to 4000 kN.

Figure 18 shows a typical line of compression presses fitted with vacuum chambers. This type of press can be purchased with platen sizes in the range 355 × 355 mm to 710 × 710 mm. The choice of press size is dependent on: product size, product design, and volume.

In general, the mold is made up of two components: the two holding plates (top and bottom) and the mold cavity. The larger the platen size, the larger the mold size and the more cavities per mold. However, if the molds are designed correctly (holding plates and cavities), the same cavities can be used in the different size holding plates. This is a major issue in today's environment where flexibility is essential to be successful. As mentioned earlier, to manufacture lip type rotary shaft seals with a molded primary lip to a consistent high-quality standard using the compression molding process, it is necessary to use the "shelf" loading technique for the elastomeric compound preforms.

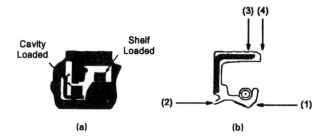

Figure 17 (1) "Diaphragm" gating system through primary lip. (2) "Diaphragm" gating system through dirt lip. (3) "Point" gating system through seal outside diameter. (4) "Fan" gating system through seal outside diameter.

Figure 18   A typical compression press molding line.

Figure 20   A horizontal reciprocating screw injection molding machine.

Figure 19 illustrates one type of compression mold suitable for molding lip type shaft seals with a metal outside diameter (O.D.). Figure 19a shows the compound preform resting on the "shelf" and the mold beginning to close. The mold continues to close until it gets to the position shown in Fig. 19b. Finally, the mold opens as shown in Fig. 19c. It can be seen that this molding technique produces a molded lip seal with the flash already removed. In general, when using a vacuum chamber on the press, to ensure that all the air and/or gases are evacuated from the mold after closure, one or

two "bumps" are applied between 5 and 10 s after the mold initially closes. The term "bump" is used when the mold is opened fractionally, approximately 1 or 2 mm for a fraction of a second, to allow all the gases to escape from the mold.

2. *Injection Molding*

Rubber injection molding machines used to manufacture shaft seals fall into three categories:

Horizontal reciprocating screw machines (Fig. 20).
Vertical screw/ram machines (Fig. 21).
Vertical reciprocating screw machines (Fig. 22).

In recent years, the trend has been toward an increasing use of the reciprocating screw machines. One of the most common reasons given for the change is that the "first in–last out" principle applies for screw/ram

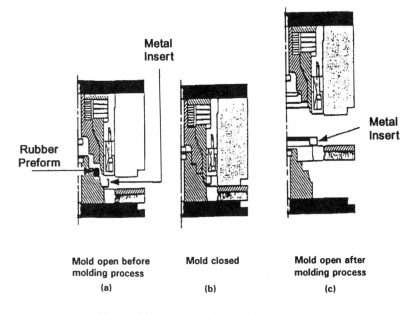

Metal Insert

Rubber Preform

Metal Insert

Mold open before molding process	Mold closed	Mold open after molding process
(a)	(b)	(c)

Figure 19   Compression molding process.

Figure 21 A screw/ram injection molding machine.

machines. The major difference between the two types of machines is as follows.

Reciprocating Screw Machine

The rubber compound is fed into the machine and the screw loads a known volume of rubber into the cylinder. When the cylinder is completely filled, the screw then acts as a ram to force the rubber compound into the runner system and the mold.

Screw/Ram Machine

In general, their are two components, a screw and a cylinder/ram assembled in a 90° "V" configuration. The rubber is fed into the screw, which forces the rubber into the cylinder. When the cylinder is full, the ram then forces the rubber into the runner system and the mold.

Figure 22 A vertical reciprocating screw injection molding machine.

Injection molding machines vary in size, but those most commonly used for shaft seal manufacture range in size as:

Platen size	250 × 250 mm to 500 × 500 mm
Shot size	50 to 220 cc
Load	800 to 3000 kN

To ensure consistency, the two critical areas for control specific to injection molding are shot size control and temperature control on the injection system, i.e., screw portion for the reciprocating screw machines, and both the screw and ram sections for the screw/ram machines. As mentioned previously, for temperature and cure control, the use of microprocessor controls has significantly improved the performance of injection molding machines, including control of the shot size.

An alternative approach for injection molding is the use of a rotary table injection machine. This type of machine uses the multistation concept, each station having a single cavity mold generally using a "point" gating/injection point as previously mentioned. Consider a rotary table machine with eight stations, station one would be the injection station and station seven would be the unload and load metal insert station. This type of process has been successfully used for different types of lip type rotary shaft seals where more expensive elastomers are used and waste needs to be kept to a minimum.

The trend toward automation has resulted in the increasing use of the vertical injection molding machines as shown in Fig. 22. This type of machine has horizontal platens and, as a result, the mold is in the horizontal plane making it easier to load and unload when using automated procedures, including "pick" and "place" robots. Figure 23 illustrates a cross section of half a four-cavity injection mold used to produce rubber covered lip type shaft seals. Figure 23a shows the mold in the closed position and Fig. 23b shows the mold in the open position highlighting the seal (component), the flash cap, and the compound runner. It can be seen that in this case, the "diaphragm" gating system has been used and the seal being removed after molding has had the flash removed during the molding cycle. Figure 24 illustrates the configuration of the four cavities in the mold as shown by the seals as they come out of the mold, including the flash caps and the runner system.

Over the years the trend has been toward an increasing use of the more expensive elastomers. Concern has been expressed about the cost of waste compound when using a multicavity mold and injection molding, primarily due to the amount of elastomeric compound in the runner system. There are three approaches to overcome the problem:

1. The use of small injection molding machines with a single cavity mold. This approach removes the need for a runner system and, when using a "dia-

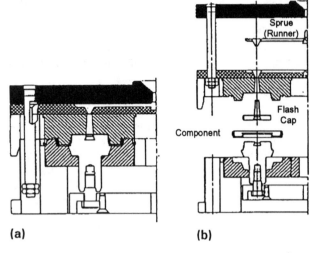

(a)   **(b)**

Figure 23 Four-cavity injection mold: (a) mold closed, (b) mold open.

phragm'' gating system, the only waste is the flash cap, and, for rubber covered seals, a small amount on the outside diameter of the seal that helps the rubber flow. The single cavity mold approach also lends itself more readily to automation using a ''pick'' and ''place'' robot.

2. The use of multicavity injection molds together with a ''cold'' runner system. The objective of the ''cold'' runner system is to prevent the runner system in the mold from curing during the molding process. The result is that the elastomeric compound held in the run-

ner system can be used to fill the mold during the next molding cycle. Therefore, the only waste compound generated during the molding process is the flash cap, and, for rubber covered seals, a small amount on the outside diameter of the seal that helps the rubber flow.

3. The use of multinozzle injection machines. In general, this type of machine has a horizontal reciprocating screw that feeds four injection nozzles each connected to a single cavity mold thus minimizing waste. However, one concern with this approach is the need to have the mold accurately lined up (the top and bottom surfaces of the mold will be in the vertical plane) to eliminate compound leakage at the nozzle/mold interface.

3. Injection/Transfer Molding

Injection/transfer molding is generally used for high-volume manufacture. In principle, the transfer ''pot'' replaces the runner system in a conventional injection mold and the transfer pot contains a ram that forces the uncured compound into the individual cavities making up the mold. However, like large platen injection molding using high-cavity count molds, there is an excessive compound waste that is very undesirable when using the higher cost elastomers. The large runner system from injection molding is replaced by a thin sheet of cured compound from the bottom of the transfer pot.

One approach to overcome the compound waste issue has been the development of the ''cold pot'' system, which has the same result as the ''cold runner''

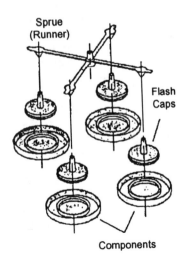

Figure 24 Seal/flash cap/runner system configuration from four cavity mold.

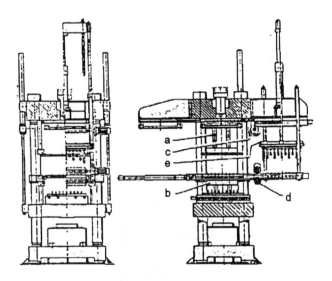

Figure 25 Schematic diagram of an injection/transfer molding machine [9]. (a) Hydraulic separation unit for upper mold plate. (b) Hydraulic separation unit for middle mold plate. (c) Shuttle system with automatic sprue nipple removal. (d) Brushing unit for cleaning middle mold plate. (e) Hydraulic ejector for automatic ejection.

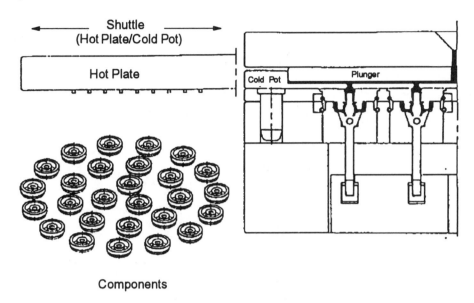

Figure 26 A typical mold/transfer pot layout, which can be used in the injection/transfer process and a typical component (seal) layout.

system on conventional injection molding that minimizes the amount of waste compound generated during the molding process. The use of a "cold pot" injection/transfer molding process has all the advantages of high-volume compression molding without having to make rubber preforms and then loading the preforms into each cavity.

Figure 25 shows a typical injection press that can be used for "cold pot" injection/transfer molding. It is important to note that the molding cycle is different than the conventional injection process. The major difference is that when the rubber compound has been moved from the transfer pot to the individual cavities, the transfer pot is removed and the "hot plate" moved in to take its place. Figure 26 illustrates a typical mold/transfer pot layout that can be used in the injection/transfer process. The layout also shows the "hot plate" ready to move into position when the transfer pot is removed and a typical layout for seals produced on the injection/transfer process. It can be seen that there is no flash or waste compound. The small amount of waste is, in fact, removed with the "hot plate." This system lends itself to high-volume automation.

SUMMARY

There is no one individual issue that ensures product and manufacturing process consistency. However, there are a number of elements that are essential if consistency is to be achieved:

1.   The component manufacturer and the raw material suppliers should work closely together to ensure that appropriate and realistic raw material specifications are agreed upon. The specifications should be developed to enable the elastomeric compounds produced with those ingredients to be consistent and processable in the manufacturing equipment used.

2.   The manual content involved in the compound mixing operation should be minimized. However, economic issues must be considered and the trend, therefore, is toward computer weighing systems (generally, a combination of full automation for critical ingredients and fillers combined with a computer prompted, manual system for the remainder), computer controlled mixers, and computer controlled blending mills.

3.   To ensure that consistent, high-quality parts can be molded on any of the three molding processes discussed: compression, injection, or injection/transfer, the following issues must be addressed:

a.   The temperature distribution across the mold (top and bottom) should be minimized.

b.   Every effort should be made to have a uniform pressure distribution across the mold.

c.   It is advisable to use presses fitted with either vacuum chambers or applying vacuum directly through the mold.

d.   Press control is critical and if one is to ensure that the elastomeric compound reaches the required cure state yet minimize the time the component spends in the mold, it is essential to use microprocessor controls.

e.   The elastomeric compound formulation is critical since product performance is dependent on the molding process used.

The choice of molding process is dependent on three primary issues: the volume of parts to be molded, the elastomeric compound to be used, and the product design. The injection and injection/transfer processes lend themselves more readily to total automation. For lower volume requirements, conventional smaller injection molding machines are appropriate; when the higher cost elastomers are being used, the preferred options appear to be either the use of single cavity molds using small injection machines or multinozzle injection machines or the use of a cold runner system with a multicavity mold. Both methods minimize the amount of compound waste. For high-volume requirements, the injection/transfer process using a "cold pot." For short production runs, the small press approach appears to be the most commonly used.

To minimize throughput time, component inventory, and work in process, it is necessary to use the "pull" system for manufacturing. The object being to have the ability to respond quickly to customer demands through efficient, fast response manufacturing techniques, not through supplying from finished goods inventory, which is costly.

The "pull" system requires that the suppliers to the component fabricator, the component fabricator, and the fabricator's customers work closely together. This is essential as there is little room for error that causes costly "line down" situations for the fabricators' customer. The trend is also for each part of the chain, supplier, fabricator, and customer, to take responsibility for the quality and performance of their own products. This has resulted in a term in general use today, "self-certified supplier."

It can be seen that in today's environment where quality and consistency are expected, the only way this can be achieved in an economic way is for the suppliers, component fabricators, and their customers to work closely together. Each should keep the other informed of issues of importance to ensure a constant flow of product and, where appropriate, work together on joint developments.

ACKNOWLEDGMENTS

The author wishes to thank Chicago Rawhide (Americas) for permission to present the information contained in this chapter.

REFERENCES

1. R. Bond, "Compound Ingredient Consistency—Its Role in Process Capability," Ninth Annual Meeting of the Polymer Processing Society, Manchester, England, April 5–8 (1993).
2. J. S. Dick, and H. A. Pawlowski, "Applications for the Rubber Process Analyzer." Rubber Division of the American Chemical Society, Nashville, Tennessee, November 3–6 (1992).
3. R. Bond, *J. Appl. Polym. Sci.: Appl. Polym. Symp. 53:* 219–232 (1994).
4. "F" Series Banbury Mixers. Farrel Corp. Bulletin No. 224-A.
5. R. Bond and Y. Takada, *J. Appl. Polym. Sci.: Appl. Polym. Symp. 50:* 175–184 (1992).
6. R. Bond, "Shaft Seal Manufacture—Alternate Molding Techniques." Polymer Processing Society European Regional Meeting, Prague, Czechoslovakia, September 21–24 (1992).
7. A. S. Berens and J. H. Born, "The Effect of Surface Microstructure on the Performance of Rotary Shaft Lip Type Oil Seals." Fourth Rubber and Plastics Conference, July 4–7 (1974).
8. L. A. Horve, "The Correlation of Rotary Shaft Lip Seal Service Reliability and Pumping Ability to Wear Track Roughness and Microasperity Formation." S.A.E. Paper 910530 (1991).
9. The New ITM Injection-Transfer Molding Technique. Maschinenfabrik J. Dieffenbacher GmbH & Co., Germany.

30

Advancement in Reactively Processed Polymer Blends

Mohini M. Sain and Claude Daneault
University of Quebec, Trois-Rivieres, Quebec, Canada

I. INTRODUCTION

This chapter discusses the processing technologies involved in preparing rubber–rubber and rubber–plastic blend vulcanizates and reviews recent developments in the analysis and performance of these materials. The capital intensive processes involving synthesis of new polymers with modified properties are becoming less interesting. This is because polymer blending has become a widely acceptable method of modification of polymer properties. A review of rubber blends was published by Ronald [1]. The emphasis was more on thermodynamics and properties of rubber–rubber blends. Coran [2] also summarized the properties of various rubber–plastic blends with special reference to technological compatibility.

In recent years, interest has been generated by polymer blends involving polyolefins and PVC in combination with natural and various new elastomers. Simple mixtures of elastomer and plastic were commercialized to prepare either impact resistant plastics or easily processible rubber [3] blends. Understanding that with technological compatibilization blends of polymers can have properties better than those of the component polymers, more development in this area has been made in recent years. However, technological difficulties are frequently faced in some type of mutually incompatible polymers. This incompatibility may be due to either viscosity difference or thermodynamic incompatibility. These problems are generally encountered with selectivity of elastomers and plastics. Addition of a phase compatibilizer or modification of component polymers usually improves product immissibility by reducing the surface energy difference. In few instances, optimization of the processing parameter improves compatibility.

A definite route to achieve compatibility in rubber or rubber–plastic blends is covulcanization or cocrosslinking. In elastomer–plastic blends covulcanization needs to satisfy stringent requirements of cure rate and inhibition period. When blend components are cured with the same curing agents, the equality of cure rate is hardly achieved and mostly poor cocuring results either from the solubility differences of curatives in each polymer or due to their unequal reactivity [4,5]. The controlled addition of curing additives in some cases give covulcanizates. For example, maleic anhydride modified EPDM showed improved covulcanizate properties with natural rubber (NR).

A new process to develop interface vulcanization is grafting of selective accelerators onto a polymer chain, which in the subsequent process of vulcanization acts as an effective cure accelerator for the second polymer component in the blend. Beniska et al. [6] prepared SBR-PS blends where the polystyrene phase was grafted with the accelerator for curing SBR. Improved hardness, tensile strength, and abrasion resistance were obtained. Blends containing modified polystyrene and cis-1,4-polybutadiene showed similar characteristics as SBS triblock copolymers.

Another area of recent interest is covulcanization in block copolymers, thermoplastic rubbers, and elasto–plastic blends by developing an interpenetrating network (IPN). A classical example for IPN formation is in polyurethane elastomer blended acrylic copolymers [7].

We have found that proper choice of curatives and the effective modification of polymer components improved covulcanizate properties in polymer blends. The focus of this chapter is on more recent advances in co-vulcanization of modified and unmodified rubber–rubber and rubber–plastic blends. The modification process adopted here is to add a controlled amount of special cross-linking agents before the final molding process, so that the two phases mutually interact to form chemical linkages between them. Such physically mixed blends are expected to resemble block copolymers and would show improvement of some properties of semi-rigid plastics, particularly, impact strength and yield stress, as well as improve elasticity of softer grade thermoplastic rubber vulcanizates. This chapter will deal with the methods of preparing these modified elastomer–plastic and rubber–rubber blends.

II. TECHNOLOGY FOR SOME PROCESS REACTIVE BLENDS PREPARATIONS

The processing technologies for elastomeric blends, thermoplastic elastomer-based on mechanical mixing, and elastomer–plastic vulcanizates are distinctly different. Depending on the type and nature of blend, size, and their final application, a wide range of processing equipment is now in use both industrially as well as in laboratory scale preparation.

A. Elastomer–Elastomer Blends with Dissimilar Cure Rate

The oldest technology involved in the elastomer blending and vulcanization process is essentially a temperature controlled two roll mill as well as internal mixers followed by an optimum degree of crosslinking in autoclave molds (compression, injection, etc.) in a batch process or in a continuous process such as continuously heated tube or radiated tubes. A few examples of laboratory scale preparation of special purpose elastomeric blends is cited here.

Rubber blends with cure rate mismatch is a burning issue for elastomer sandwich products. For example, in a conveyor belt composite structure there is always a combination of two to three special purpose rubbers and, depending on the rubber composition, the curatives are different. Hence, those composite rubber formulations need special processing and formulation to avoid a gross dissimilarity in their cure rate. Recent research in this area indicated that the modification of one or more rubbers with the same cure sites would be a possible solution. Thus, chlorosulfonated polyethylene (CSP) rubber was modified in laboratory scale with 10 wt% of 93% active meta-phenylene bismaleimide (BMI) and 0.5 wt% of dimethyl-di-(*tert*-butyl-peroxy) hexane (catalyst). Mixing was carried out in an oil heated Banbury-type mixer at 150–160°C. The addition of a catalyst was very critical. After 2 min high-shear dispersive melt mix-

ing of CSP and BMI, the catalyst was added and the mixing reaction was continued for another 3 min. The system temperature was controlled carefully to avoid any possible increase in temperature. The composition was then dumped and chopped into pieces to use further as a reactive cure site. This BMI-treated CSP composition is designated as "modified CSP" (MCSP) in the text.

All dissimilar elastomer compositions, whether unmodified or modified, were prepared in a Banbury mixer at 120–140°C. The recipes of unmodified masterbatches are given in Table 1. Four recommended and generally used curing systems were used for three selected elastomer blends. When ionomeric MgO cross-linking systems were widely used for separate crosslinking of CSP and CM rubbers, this same system was used to understand the features of cure-rate incompatibility in the CSP–CM blend. A typical curing system recommended for acrylic rubber only was also taken as an example to explain the need for high selectivity of common curing agents in elastomer blends. The compositions for the curing systems are given in Table 2.

B. Preparation of Elastomer–Plastic Blend by Melt Mixing Without Vulcanization

Elastomer–plastic blends without vulcanization were prepared either in a two roll mill or Banbury mixer. Depending on the nature of plastic and rubber the mixing temperature was changed. Usually the plastic was fed into the two roll mill or an internal mixer after preheating the mixer to a temperature above the melting temperature of the plastic phase. The plastic phase was then added and the required melt viscosity was attained by applying a mechanical shear. The rubber phase was then added and the mixture was then melt mixed for an additional 1 to 3 min when other rubber additives, such as filler, activator, and lubricants or softeners, were added. Mixing was then carried out with controlled shear rate

Table 1 Elastomer Masterbatch Compositions

Composition	CSP–MB	ACM–MB	CM–MB
Rubber	100	100	100
PEG 300	3.0	0	0
Stearic acid	2.0	1.0	0
HAF black	50	50	0
FEF black	0	0	55
Chlorinated paraffin	10	0	0
Ketone-amine condensate	0	0	0.25

MB-master batch; ACM-acrylic rubber; CM-chlorinated rubber; PEG-polyethylene glycol; MV-ML(4 + 1), 100°C for CSP, ACM, and CM masterbatches are 34, 30, and 32, respectively.

Table 2 Recommended Curing Systems

Composition	a	b	c	d	ab	ac	ad	bc
MgO	7.0	0	7.0	—	3.5	7.0	7.0	3.5
DPTS	2.0	—	—	—	1.0	1.0	1.0	—
TDD	—	—	2.5	—	—	1.25	—	1.25
NC	—	—	3.5	—	—	1.75	—	1.75
B-18	—	0.8	—	—	0.4	—	—	0.4
DOTG	—	0.5	—	—	0.25	—	—	0.25
Peroxide 14/40	—	—	—	5.0	—	—	2.5	—
TAC	—	—	—	1.0	—	—	0.5	—

DPTS-di-pentamethylene thiuram tetrasulfide; TDD-thiodiazole derivative; NC-fatty acid amide amine; B-18-special curative; DOTG-diortho-tolyl guanidine; peroxide 14/40-dicumyl peroxide; TAC-triallyl cyanurate.

and temperature, and the mixture was then dumped into a granulator to form granulates for further use. The difficulty of the system is the temperature sensitivity of the compound. As soon as the mixing is over it is essential that the molten mass be released from the mixer without much delay to avoid hardening of the plastic phase. The granulated product is then either compression molded or injection molded to obtain a test specimen. The compression mold as well as injection mold should be provided with heating and cooling arrangements and the products must be cooled at least 40°C below the melting point of plastic phase to avoid stress cooling.

C. Preparation of Elastomer–Plastic Blends by Static Vulcanization

NBR–PVC blend was one of the first invented statically cured elastomer–plastic systems to find industrial application [8]. The concept of vulcanized elastomer–plastic blend was introduced even earlier by Fischer [9]. In such blends, the rubber phase is crosslinked by conventional curing systems. However, crosslinking is carried out only after a medium- to high-speed shear mixing of the molten plastic phase with the heat soften rubber phase. We have made several statically cured elastomer–plastic blends including NBR–PVC, EVA–PE, PRP–EVA, NR–PP, NR–LDPE compositions. The blend preparation process is very similar to that used for unvulcanized blends. However, in this case respective curing agents are also added in the internal mixer 1 min prior to the unloading process. However, the mixing temperature is always kept at least 20–30°C lower than the curing temperature. Generally, a temperature of about 100–140°C is used for melt mixing, depending on the curative. For conventional sulfur and peroxide curable blends, the mixing temperature is usually restricted to 120°C, whereas for high-temperature curable blends the mixing temperature can go as high as 140°C. For both single phase crosslinking and cocrosslinking the curatives were added just before the end of the mixing cycle to avoid

premature vulcanization or covulcanization. After the mixing process the compositions are cooled and then granulated. They are finally compression molded for a given time at their optimum cure temperature determined by a Monsanto rheometer. Table 3 gives the stock mixing and molding temperatures.

D. Elastomer–Plastic Blends by Dynamic Vulcanization

Gessler [10] prepared semi-rigid rubber–plastic compositions containing a small amount of vulcanized rubber. Later, Fischer [9] and, more recently, Coran and Patel [11] used the same techniques to develop partially vulcanized blends, which are thermoplastic in nature with different end use properties. Coran and Patel [12] used various combinations of plastics and rubber with a large number of curing systems, and they have concluded that if enough plastic phase is present in the molten state, then the compositions are processible as thermoplastics. A similar method as used by Gessler [10] and Coran and Patel [12] was used in our study. The elastomer–plastic compositions were different from those described by others in that the thermoplastic part contained a highly nonpolar polyolefin phase and a highly polar vinyl polymer. The elastomers were either naturally occurring hy-

Table 3 Mixing and Curing Temperatures

Blend composition	Mixing temperature (°C)	Curing temperature (°C)
NBR–PVC	110	160–180
EVA–PRP	180–190	200
NR–LDPE	120	150–160
NR–PP	170	180
NR–PRP	180	170–190

drocarbon polymers or synthetic elastomers containing polar pendant groups. The process used was as follows. All preparations were carried out either in a Brabender internal mixer only or by using a combination of a Babury mixer and a twin-screw extruder. For blends prepared in the Brabender mixer, equipment was first preheated to the melting temperature of the plastic phase. The plastic phase was then added and mixing was continued at about 40 to 60 rpm for several minutes until the plastic phase melted completely. Rubber was then introduced slowly by applying ram pressure and mixing was continued for 2 min to disperse the plastic or the elastic phase, whichever was less in quantity. Other compounding ingredients were added slowly including compatibilizers, activator, fillers, and plasticizers and mixing was continued for an additional 2 min. At this point vulcanizing agents were added and the mixing speed was increased from 60 to about 90 rpm and mixing was continued until the torque began to decrease. The batches were then removed from the mixer, pressed, and sheeted out to approximately 3-mm slabs. Final molding was done by using the following cycle at a given temperature determined by the type of composition: preheating for 10 min, pressing between the platens for 15 min, and cooling for 15 min at 60°C.

In another case where the twin-screw extruder was used, the rubber and plastic were melt mixed with all ingredients in a similar manner as described in blend compositions for static vulcanizations. The product was then dumped, cooled, and granulated. The premixed granules were then fed into a twin-screw extruder where a very narrow temperature profile was maintained with a relative high compression (2:1), and the screw speed was adjusted depending on the final torque and the flow behavior of the extruded stock. The stock was cured by shear force and temperature enforced by the twin-screw extruder. The dynamically crosslinked blend was taken out in the form of a strip or solid rod to determine the physical and mechanical properties and the flow characteristics.

III. PROPERTIES OF REACTIVELY PROCESSED BLENDS

Unlike incompatible heterogeneous blends of elastomer–elastomer, elastomer–plastic, and plastic–plastic, the reactively processed heterogeneous blends are expected to develop a variable extent of chemical interaction. For this reason the material properties, interfacial properties, and phase morphology of reactively processed blends would differ significantly from heterogeneous mixtures.

A. Properties of Elastomeric Blends with Dissimilar Cure Rates

Table 4 demonstrates the cure rate differences among different elastomers with their respective recommended curing agents. Virtually, there is no match in the optimum cure time between CSP and CM or ACM. It is then obvious that a combination of respective curing systems for elastomers in a two phase blend will result in overcuring of CSP and/or undercuring of CM or ACM. Two different cure systems were tested on CM rubber. However, none of them appeared to be a closer cure match to unmodified CSP. Similarly, the mechanical properties of CSP–CM blends are extremely poor compared to that of CSP. Interestingly, it was found that the cure system "ab" was an excellent combination to achieve a good cure of the CSP–ACM blend as it exhibited a synergism in torque value. Some important mechanical and heat aging properties of those blends are given in Table 4. Even the CSP–ACM combination gave a high torque at optimum cure, but it failed to attain a good strength property. Apparently, the nature of chemical bonding

Table 4 Cure Characteristics and Properties of Unmodified Blends

Stock no.	Unit	1	2	3	4	5	6	7
CSP–MB	phr	100				50	50	
ACM–MB	phr		100			50		50
CM–MB	phr			100	100		50	50
Cure system		a	b	c	d	ab	ad	bc
Cure temp.	°C	155	165	155	155	160	155	160
Cure time	min	11	28	26	30	20	9	30
Max. tor.	Nm	37	24	29	16	41	30	22
Tensile st.	MPa	23.7	12.4	11.8	15.8	12.7	19.5	11.4
Elongation	%	244	375	684	740	157	405	510
Tension set	%	1.0	4.5	2.0	2.0	1.5	1.0	4.0
Hardness	Sh A	81	61	67	72	78	74	59
Aged, 150°C	Tensile	52	110	69	80	163	90	153
72 h, %reten	Elong.	4	50	12	4	67	10	81

during crosslinking of CSP–ACM blends is not identical to those formed during their separate vulcanization process. This is because the strength property improved dramatically after aging at 150°C for 72 h. It is possible that some slow intermolecular reaction continues between the two elastomer phases during heat aging of the blend, which makes the blend more resistant to thermooxidative degradation. No proof of such a reaction, if any, was documented. In contrary, the cure rate mismatch was very significant for the CSP–CM blend system. An incompatible CSP–CM system shows additive values of mechanical, and aging properties. No synergism was found in mechanical properties. The modification of CSP with inbuilt additional cure functionality showed a better cure compatibility with the CM system. Table 5 furnishes the cure characteristics and some useful properties of cure compatible blends after functional modification of CSP (MCSP). It is evident that the optimum cure time of MCSP is closer to that of the CM rubber. Moreover, a 50:50 blend of MCSP and CM rubber has a cure time of about 20 min, which is reasonably close to that obtained for each of the rubber components. The cure modified blends showed improved oil and heat resistant properties compared to a blend with high cure rate mismatch. Another advantage of BMI MCSP was its improved scorch safety. It is predicted that the complex degradation reaction of sulphonyl chloride groups generates free radicals to react with BMI. This MCSP rubber apparently bound with pendant reactive functionality and probably retarded the cure rate. It is further believed that the better cure compatibility gives rise to a better distribution of crosslink density in each rubber phase resulting in a better heat and oil resistance.

B. Properties of Conventionally Vulcanized Elastomer–Plastic Blends

1. Natural Rubber–Polyolefin Blends

The cure characteristics of NR–LDPE blends containing varying amounts of LDPE for two different cure systems

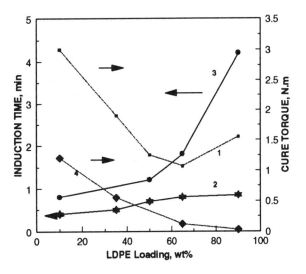

Figure 1 LDPE loading on inhibition period torque; 1, 2-peroxide; 3,4 = BMI. *Source*: Ref. 14.

are shown in Fig. 1. Evidently, increasing the polyethylene load decreases the rheometric torque monotonously for the BMI curing system, and for the peroxide cure system, it passes through a minimum. Moreover, the measured torque was always higher with peroxide. It is known that in the presence of peroxide, the LDPE phase also gets crosslinked in addition to the NR phase. It is also reported that BMI does not cure PE [13]. Thus, the lowering of torque with BMI is explained by the plasticization effect of PE and the changes of torque with LDPE in the peroxide cured system is a complex dependence of plasticization effect of PE and degree of crosslinking of both the phases. From Table 6, it is evident that the optimum crosslinking depends in a complicated manner on the cross-linking rate. The first order vulcanization rate constant (k) is higher for pure rubber than that of pure LDPE in a peroxide cured system. Further-

Table 5 Cure Behavior and Properties of Modified Blends

Stock No.	Unit	1	4	7	9	10
CSP–MB	phr	100		50		
MCSP–MB	phr				100	50
CM–MB	phr		100	50		50
Cure system		a	d	ad	a	ad
Op. Cure time	min	11	30	20	23	20
Max. tor.	Nm	37	16	22	32	29
Heat aged at 125°C, 168 h	% retention tensile st.	116	106	109	121	123
	% retention elongation	72	62	70	89	77
ASTM Oil 3 aged at	% retention tensile st.	67	78	59	69	74
100°C and 72 h	% retention elongation	70	72	62	71	68

Table 6 Vulcanization Rate Constant (k) and Optimum Cure Time (τ_{90})

NR–LDPE Composition	τ_{90} min (DCP)	k (DCP) min^{-1}	τ_{90} min (BMI)	k (BMI) min^{-1}
100:0	8.5	0.309	17	0.207
90:10	7.3	0.383	16	0.181
65:35	8.1	0.366	16	0.189
50:50	9.1	0.322	15.5	0.165
35:65	10.4	0.282	15.5	0.201
10:90	11.0	0.272	—	—
0:100	10.7	0.277	—	—

DCP = dicumyl peroxide; BMI = m-pheneylene bis-maleimide.
Source: Ref. 14.

more, a vulcanization rate synergism has been achieved with low PE content in the blend. A probable explanation of such a synergistic effect is the occurrence of an additional cross-linking reaction besides the crosslinking of the two individual phases. Depending on the blend composition, this additional reaction seems to be possible, assuming a large interpenetrating interfacial area formed during peroxide curing of the blend where cocrosslinking can take place at the interface through peroxide linkages. Thus, in contrast to elastomer–elastomer blends where a microphase covulcanization is best characterized by the activation energies and the induction periods rather than by the vulcanization rate [14], the vulcanization rate probably plays a dominant role over the induction period.

Evidently, the mechanical and processing properties of the conventionally vulcanized NR–LDPE blend would be guided by the nature of the curing agent. Table 7 gives some important properties of the NR–LDPE blend containing 65% elastomer phase for different curing systems. The reactivity of the sulfur system was very much restricted to the rubber phase only. BMI reacted preferably with the rubber phase as a curing agent, but at the same time it modified the LDPE by in situ free radical reaction with the formation of some weak interaction between cured rubber and the LDPE phase [13]. Such interaction, if any, would lead to better mechanical strength and high mixing torque. This is evident from Table 7. It is evident that neither sulfur nor the BMI system shows a tensile synergism (T.S. of NR = 12.6 MPa and of LDPE = 15.8 MPa). Peroxide vulcanized NR is seldom used commercially due to its poor mechanical properties [15]. Our earlier report [15] suggested that by optimizing the NR–LDPE ratio, cure additives, and stabilizers, a blend with mechanical properties comparable to carbon black reinforced NR vulcanizate can be achieved. It seems that the tensile synergism is a direct effect of crosslinking of the plastic phase. However, interfacial adhesion strength of the NR–LDPE blend cured with the peroxide–HQ system was found to be much higher than the corresponding sulfur cured system [15], suggesting a cocrosslinking of NR with LDPE. The disadvantage of such blends are their poor thermal stability. In fact, a peroxide cured blend could be easily degraded under an accelerated aging condition. Therefore, a suitable choice of antioxidant was found to be

Table 7 Effect of Various Curing Systems on NR–LDPE (65:35) Blend

Curing System	S/MBTS/ZnO/ Stearic Acid	BMI/MBTS	DCP (40%)	DCP/EGDM	DCP/Sulfur	DCP/HQ
Composition	2/1.3/5/1	3/0.8	5.0	2.0/5.0	3.5/0.35	5.0/0.6
Property						
Torque (Nm)	22.9	33.6	40.6			42.6
Mod(3) (MPa)	2.7	8.7	11.2	8.9	9.7	14.7
Tensile (MPa)	7.9	13.4	16.1	13.6	13.9	19.7
Hardness (A)	48	65	63			
Ten.set (%)	24	22	65			91
Elong. (%)	475	550	415	400	510	510

EGDM-ethylene glycol dimethacrylate; HQ-hydroquinone; Mod(3)-300% Modulus.

essential to improve the aging behavior of peroxide cured NR–LDPE blends. The best recommended antioxidant system is a suitable combination of 1,1,3-tris(2-methyl-4 hydroxy-5-*tert*-butyl phenyl)butane, dilauryl thio-dipropionate with DCP. The aging resistance of such a composition was greater than 60% at 100°C after 72 h. The other advantage of blending NR with PE is the improvement of electrical properties of the relatively poor electrical insulant, NR. Figure 2 shows the effect of the LDPE content on volume resistivity (ρ_v) and loss tangent (tanδ) values of the blends. The increase in the LDPE loading causes (1) a sharp increase in volume resistivity, and (2) a decrease in the loss factor. Blend permitivity remained practically unchanged, which is about 2.4 for any given vulcanizate containing minimum 10% PE and maximum 90% PE.

Unlike LDPE or HDPE, other polyolefins are difficult to crosslink by the peroxide curing system because polyolefins containing propylene as a monomer unit in the main chain are very prone to free radical decomposition with peroxide. Therefore, in a elastomer–plastic system containing PP or PRP (propylene–random ethylene propylene–propylene) copolymer as plastic phase peroxide system is not a successful reaction initiator. Technological compatibility was needed to process PP or PRP with NR in order to achieve better strength and processability. Coran and Patel [16] found that phenolic resin could be useful to modify PP in order to improve the technological compatibility of PP with other polar elastomers, such as NBR. A similar processing technology as described by Coran and Patel [16] was used to make PP or PRP as a curative for NR. It is well known that phenolic resin can vulcanize NR under certain experimental conditions. Therefore, a PP or PRP matrix containing pendant phenolic functionality would serve as a curative for NR–PP or NR–PRP blends for a result-

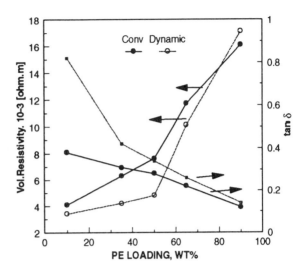

Figure 2 Electrical properties of NR–LDPE. *Source*: Ref. 13.

ing cocuring effect during the processing or molding operation. In our study [17], both PP and PRP were treated with 2–4 parts of phenolic curatives in the presence of $SnCl_2 \cdot 2H_2O$ and MgO to obtain phenolic modified PP or PRP. The modified polyolefin was then melt mixed with NR under an optimum shear condition and temperature to develop chemical bonding between NR and polyolefin via quinone methide formation [18]. About 1 min before dumping, the batch curatives were added to the mixer to disperse the curatives in a partly reacted blend. The total curing of the rubber phase in a compression mold then improved the properties of the compatibilized blend over unmodified PP or PRP blended and cured stocks. Table 8 gives the properties of the blends after

Table 8 Properties of NR–PP or PRP Blends (Modified with Phenolic Resin) at 50:50 Ratio

Preparation	NR–PRP				NR–PP			
Cure system	E	E	F	F	E	E	F	F
Cure temperature (°C)	180	180	180	180	180	180	180	180
Cure time (min)	20	20	20	20	20	20	20	20
Processing	R.M.	B.M.	R.M.	B.M.	R.M.	B.M.	R.M.	B.M.
Properties								
Modulus, 100% (MPa)	11.9	9.0	11.1	9.0	13.6	9.9	13.3	11.7
Tensile strength (MPa)	20.8	23.6	22.5	23.3	26.9	24.7	29.3	28.3
Elongation at Br. (%)	400	475	475	480	400	475	500	500
Hardness (Sh A)	87	86	88	88	86	88	89	88
Tension set (%)	28	27	28	26	53	37.5	37.5	28
True stress at break (MPa)	104	136	129	135	135	142	176	170
Cross. density (10^5)	9.0	11.8	5.8	7.5				
Reprocessibility					No	No	No	No

R.M. = roll mill; B.M. = Brabender mixer; E = DCP–HQ/4.0:0.5%; F = PF resin–$SnCl_2$/5.0:0.5%.
Source: Ref. 19.

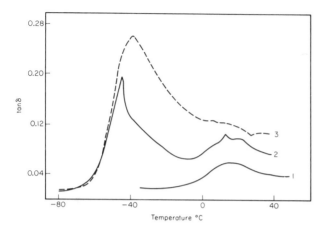

Figure 3 tanδ versus temperature plot for NR–PP blends (50:50); curve 1, purePP; curve 2, dynamically cured blend; curve 3, conventionally cured blend. *Source*: Ref. 19.

processing in two different machines. The vulcanizates have excellent mechanical properties. The mixing techniques, however, have no significant influence on the blend properties. The blends are not repressible. As mentioned before such blends are technologically compatibilized by reactive processing. The improved phase interaction of the conventionally vulcanized system is apparent from spectra shown in Fig. 3. A shift in the glass transition temperature of NR and the distinct shape of the original sharp PP glass transition peak confirms better interaction for the resin containing the conventionally cured blend.

2. NBR–PVC Blends

Our study was also extended to obtain the effect of a selective curing system on NBR–PVC compositions.

NBR–PVC is a very useful blend for the electrical cable sheathing application that demands high flexibility, strength, oil resistance, and low-temperature resistance. Table 9 summarizes some results of the effect of curing systems during blend processing for a 60:40 NBR–PVC blend. A marked change is visible on the mechanical properties depending on the curing systems. While the system containing PETA–DHBP develops the best strength properties before aging, which is associated with the high gel content of the vulcanizate, the mixture containing PNDA showed good strength only after aging. It is expected that this change in the mechanical properties in the presence of different curing systems is due to a change in the vulcanizate structure during the molding process. The peroxide covulcanization of NBR–PVC showed poor thermo-oxidative properties when sulfur curing of the same composition resulted in a single phase cure only.

Based on our previous experience, we have also studied the effect of selective cross-linking systems on NBR–PVC blends containing dominant PVC phase, 60% by weight. Some experimental results are in Table 10. Evidently, at a higher concentration of BMI the gel content was somewhat higher than for the gel content prorated for NBR content in the blend. It seems a partial crosslinking of PVC or NBR–PVC interface has occurred. The results are in conformity with the mechanical properties of the BMI cured blend. In most of these cases the measured values of mechanical strength has gone through a maximum, indicating a change in the reaction mechanism during the molding process with curatives.

In order to support our prediction that the change in mechanical properties with different curing systems is due to a change in the vulcanizate structure, results were compared from dynamic–mechanical measurements as shown in Figs. 4 and 5.

Table 9 Vulcanization Characteristics and Properties of NBR–PVC Blends

Curing system	Sulfur	Sulfur/PNDA	BMI/MBTS	PETA/DHBP
Cure characteristics				
Temperature (°C)	160	160	180	180
Time (min)	5	5	15	20
MV,ML(1 + 4)	36	35	—	—
Properties				
Modulus, 100%	1.8(3.7)	1.7(2.8)	3.2(4.3)	6.6(8.7)
Modulus, 300%	3.7(6.5)	3.2(6.6)	5.3(6.8)	17.1
Tensile strength (MPa)	7.1(13.3)	8.6(4.5)	6.9(7.7)	17.7(16.6)
Elongation (%)	740(700)	780(638)	475(375)	330(215)
Hardness (ShA)	54(55)	56(58)	64(68)	80(78)
Gel content	56(69)	67(84)	35.7(42)	93(95)

PNDA = *p*-dinitroso diphenyl amine; PETA-penta erythritol triacrylate; DHBP-peroxy hexane; values in parentheses are after aging at 100°C, 72 h.
Source: Ref. 20.

Table 10 Effect of Short-Time Aging on Mechano-Chemical Properties of PVC–NBR Blend

Cure system (wt%)	BMI–MBTS/5.0:1.7	PETA–TEA 10:0.2	S/TMT/MOZ/PNDA/ 0.1:0.33:0.44:1.1
Modulus 100% (MPa)	8.3(14.5)	8.2(13.1)	7.1(14.3)
Modulus 300% (MPa)	15.4(18.6)	—	13.3(20.3)
Ten. strength, (MPa)	16.3(16.3)	11.5(12.8)	15.0(19.8)
Elong. at break, (%)	362(282)	255(173)	405(385)
True stress at break (MPa)	75.6(65)	41(34.5)	75.1(95.4)
Hardness (ShA)	71(69)	73(73)	65(73)
Gel Content (%)	54(10.1)	52(71)	24(34)

TEA = triethylene amine. *Source*: Ref. 20.

The values of gain modulus, loss modulus, and tanδ show only one relaxation peak within the studied temperature range. A comparison of the maximum absorption peaks in Fig. 4 indicates glass transition temperature changes with the curing system, and the relative change is maximum for the PETA–DHBP system. This high T_g of the PETA–DHBP system is associated with poor fatigue resistance of the blend vulcanizate [21]. While the PNDA–S system shows maximum fatigue resistance with lowest T_g of the blend.

C. Properties of Unvulcanized Elastomer–Plastic Blends

1. Unmodified PRP–EVA Blend

In the preceding sections, our discussion has been limited to softer grade elastomer–plastic vulcanizates. Commercial interest, however, also centers on another major family of polymer blends, semi-rigid impact resistant polyolefins. Thus, we report some of our findings on PRP triblock copolymer and EVA rubber blends without

rubber phase vulcanization at first and then modified interfacial reaction.

As with NR–PP or NR–PRP blends, EVA–PRP blend is also incompatible. Therefore, technological compatibility was a must to obtain useful properties from this blend. In this case a reactive surfactant was introduced in the EVA–PRP blend during the processing operation. This reactive surfactant decreases the surface energy between the EVA and PRP interface by forming a chemical bond with one of the polymers and physically compatibilizing with the other phase. The interfacial surfactant was prepared by several methods. In one method used previously by Muras et al. [22], powdered PP was reacted with maleic anhydride (MAH) in solid phase in the presence of a peroxide catalyst below the melting temperature of PP. The grafted polymer was recovered by extraction and used as an interfacial surfactant for EVA–PRP blend preparation. The interfacial surfactant is designated here as MAH–PP. Depending on the grafted MAH content, they are classified as MAH–PP-

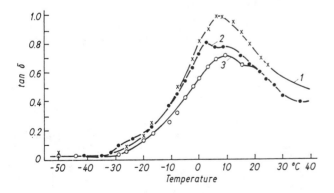

Figure 4 tanδ versus temperature for NBR–PVC blends; curve 1-BMI/MBTS; curve 2-S/PNDA; curve 3 = DHBP/ PETA. *Source*: Ref. 21.

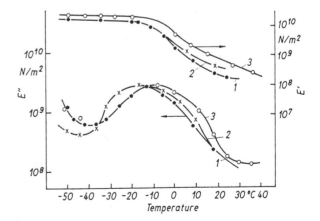

Figure 5 Gain (E') and Loss (E'') modulus (see Fig. 4). *Source*: Ref. 21.

Table 11 Effect of the Mixing Temperature and Processing Method on PRP–EVA Blends

| | Value of the property for the following EVA content and mix temperature | | | | |
| | 8 wt% | | 30 wt% | | 47 wt% |
Property units	170°C	195°C	170°C	195°C	170°C
Tensile st., MPa	17.7	18.9(19.5)	8.3	12.0(11.3)	7.4
Elong. at break, %	169	196(82)	16	75(60)	66
Hardness, ShA	63	64(65)	56	53(57)	48
MFI, g/10 min	5.8	5.7(6.2)	5.2	7.6(7.0)	−21
Brittle point, °C	−50	−56(−50)	−23	−28(−31)	9.4
Yield stress, MPa	19.0	18.7(−)	8.7	12.3	7.7

Source: Ref. 23.

I (0.98%), MAH–PP-II (1.41), and MAH–PP-III (2.95%) [23]. Table 11 gives the result of unmodified EVA–PRP blends at variable process conditions. The EVA polymer used was a rubber grade with 40% vinyl acetate content. The results in Table 11 for unmodified and uncured blends showed that blend composition has a maximum influence on the properties and in that case the optimum composition was 8% EVA. Generally the blends are weak in nature but are flowable. Higher temperature of mixing has a marginal effect on blend properties and the mixing method hardly influences the properties. Apparently the blends are grossly incompatible.

2. Reactivity of Maleic Anhydride Grafted PP on PRP–EVA Blend

To introduce some interfacial physico-chemical linkage between EVA and PRP, blends were made by adding different quantities of MAH–PP. Some results are demonstrated in Table 12. The physico-mechanical properties of the PRP–EVA compositions modified with MAH–PP showed that properties are influenced by MAH–PP concentration. Compositions with better impact strength and improved brittleness can be prepared by varying the modifier concentration. Tensile strength and elongation are not significantly influenced by the addition of a modifier. An increase in the modifier con-

centration is marked by an initial increase in low-temperature impact resistance and lowering of the brittle point of the studied composition. However, both impact strength and brittleness temperature deteriorated when the modifier level was very high. To elucidate the cause of this complex dependence of low–temperature brittleness and impact strength on the modified concentration, we have studied the dynamic mechanical properties as well as the flow behavior of unmodified and modified compositions. Figure 6 gives tanδ versus temperature correlation for various blends. The lowering of the glass transition temperature from −45 to −55°C of the PRP polymer in the PRP–EVA composition only suggests a possible phase interaction between the two phases. However, more light will be thrown on this subject during a morphological study discussed later in this text. It is also interesting to note that the flow behavior of the blend first marginally deteriorated and then, with further increase in the MAH–PP concentration, it was further improved. It also indicates that at higher concentrations of MAH–PP it might function as a flow promoter.

D. Properties of Dynamically Vulcanized Blends

The method of "dynamic vulcanization" was applied to mixtures of NR with LDPE, PP homopolymer, and PRP

Table 12 Effect of Concentration of MAH–PP-III on PRP–EVA 70:30 Blend

MAH–PP-III (wt%)	MAH$_g$ Content (%)	Tensile (MPa)	Elongation (%)	Impact st. (kJ/m²)	Brittle point (°C)	Hardness (ShA)	MFI (g/10 min)
0.5	0.015	15.0	81	31.0	−36	64	7.7
1.0	0.029	15.7	79	51.0	−48	63	7.3
1.5	0.044	16.1	98	36.5	−46	62	7.5
2.0	0.050	15.8	95	30.1	−46	63	8.5
3.0	0.088	15.5	88	28.1	−45	60	8.8

Source: Ref. 23.

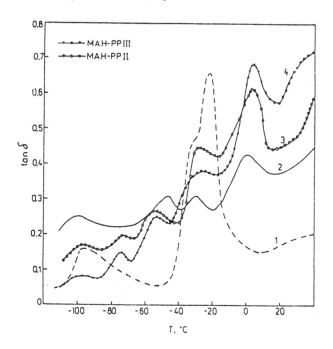

Figure 6 Variation of tanδ with temperature for (1) EVA, (2) PRP–EVA/92:8 blend, and (3 and 4) EVA–PRP blend 92:8 modified with MAH–PP-III. *Source*: Ref. 23.

Table 14 Properties of Dynamically Vulcanized NR–LDPE (65:35) Blends

Preparation		
Curing system	BMI/MBTS	Sulfur/MBTS
Cure Temp. (°C)	180	180
Cure time (min)	5	5
Properties		
Modulus, 100% (MPa)	4.1	0.5
Tensile st. (MPa)	9.5	4.0
Break elong. (%)	320	600
Hardness (Sh A)	57	32
Tension set (%)	12.5	45
True stress at break (MPa)	40	28
Permittivity	2.35	2.4
Vol. resistivity (ohm·m·10^{-13})	6.5	6.3

triblock copolymer. The main advantage of a dynamically cured blend is the reprocessibility. Therefore, these blends are preferred in many industrial applications where dimensional accuracy and product complexity leads to a high rejection rate. Industrial cables are one example that needs high precision and quality. In many cases such application does not need very high mechanical properties. Thus, various custom-designed thermoplastic elastomers can be used. A suitable blend of polyolefin and natural rubber is good for many applications. They are good insulants and are highly resistant to polar solvents, moisture, etc. After dynamic crosslinking, such blends could develop desirable strength without sacrificing processability. Mixing characteristics

during dynamic crosslinking of some NR–LDPE blends are given in Table 13. The mixing torque decreases with increase in plastic content. The Mooney viscosity of the dynamically cured blend passes through a maximum. However, the dynamically cured blend showed higher Mooney viscosity compared with a statically cured blend. High viscosity may be due to the formation of IPN in the dynamically cured blends as suggested by several coworkers [7,24]. It is predicted that the IPN formation depends on the blend compositions, and once formed, it strongly resists phase separation. This could be the reason for poor flow behavior of the dynamically vulcanized NR–LDPE blend at an intermediate composition as given in Table 13. Therefore, the dynamically cured blend of technical interest was found to be a composition with 35% LDPE content. Table 14 demonstrates some important properties of this 65:35 NR–LDPE blend for two different curing systems. Evidently, the mechanical properties of the dynamically cured blend are very much influenced by the nature of the curing agent; i.e., the nature of the chemical bond formed dur-

Table 13 Blend Composition and Dynamic Curing Characteristics for NR–LDPE[a]

Composition NR–LDPE	Dump temp. (°C)	Mixing torque (Nm)	MV, ML[4 + 1] 121	MFI (g/10 min)
90:10	165	30.4	28	8.1
65:35	162	29.4	44	1.7
50:50	158	27.4	35	10.2
35:65	156	22.6	26	21.6
10:90	155	19.6	9.5	44.7

[a] Stocks compounded with 3 phr BMI and 0.8 phr MBTS.

ing the dynamic curing process. Sulfur curing gives bonds that are flexible in nature and lead to softer composition. On the other hand, imide linkage formed during crosslinking with BMI is rigid in nature and mechanical properties are reasonably high. Moreover, the blend is reprocessible. Thus, a dynamically cured NR–LDPE blend with BMI is of commercial interest in electrical industry.

In the past several attempts to apply dynamic vulcanization techniques to NR and PP blends [25,26], for both PP homopolymer and PP copolymer, were less effective owing to the difference in their miscibility characteristics. Compositions with practical interest, however, have been prepared by modifying polyolefins and then dynamically curing them. PP homopolymer and PRP copolymer were pretreated with phenolic resin and SnCl$_2$ to improve blend compatibility; hydroquinone, a hydroperoxide radical scavenger, was used to prevent degradation of the PP chains. The physical properties of the blends cured by the dynamic process are given in Table 15. It is evident that reactive mixing of NR–PP and NR–PRP could lead to reprocessible thermoplastic compositions with excellent mechanical strength. A comparison of two curing systems suggests that phenolic resin curing by dynamic vulcanization probably leads to a highly compatible and miscible blend with reasonable crosslinking of the rubbery phase. In addition, both these blends have excellent electrical properties. Figure 3 demonstrates that loss tangent value is very reasonable for medium voltage electrical insulation. However, tanδ versus the temperature of the dynamically crosslinked blend with peroxide (Fig. 3) indicates less interfacial interaction between two phases as there is only a marginal shift in the glass transition values. The morphological changes with phenolic modified blends are discussed in the following section.

IV. PHASE MORPHOLOGY OF REACTIVE PROCESSED BLENDS

Dissimilar polymer blends show phase separation and poor mechanical properties. A suitable processing technique in combination with a technical compatibilization of blend could reduce the immiscibility significantly. Reactive mixing leading to graft copolymerization or co-crosslinking improved mechanical strength of various polymer blends discussed here. The differences in the mechanical properties between sulfur cured and peroxide cured NR–LDPE blends are very much reflected in their phase structure. Figure 7 shows SEM microphotographs of fractured and etched surfaces of NR–LDPE (65:35) compositions. The sulfur crosslinked blend has a higher particle size of the plastic phase, and the dispersion of those particles is also poor. Again, blends cured with peroxide have very good dispersion of the cured plastic phase and the particle size was much smaller. Further morphological investigation of the interfacial adhesion on the reactively cured blends was carried out by taking SEM microphotographs under externally applied predetermined strains. The SEM micrographs of a series of samples with a series of gradually increasing strains have been analyzed.

Figure 8 shows the SEM images with a low level of strain (50%). It is clear that even with a low-strain level defects are initiated in the sulfur cured system with the formation of large cracks at the boundary layer between the two phases. However, in the peroxide cured system the mechanism of crack initiation is very different. In the latter case the NR–LDPE interface is not the site for crack initiation. In this case, stress due to externally applied strains is distributed throughout the matrix by formation of fine crazes. Furthermore, such crazes are developed in the continuous rubber matrix in a direction

Table 15 Properties of Dynamically Cured NR–PP and NR–PRP Blends

Preparation	NR–MPP	NR–MPP	NR–MPRP	NR–MPRP
Curing system	DCP–HQ	PF–SnCl$_2$	DCP–HQ	PF–SnCl$_2$
Cure Temp. (°C)	180	180	180	180
Cure time (min)	5	5	5	10
Properties				
Modulus, 100% (MPa)	7.6	10.2	7.5	9.4
Tensile st. (MPa)	7.6	18.5	8.8	16
Break elong. (%)	100	280	150	300
Hardness (ShA)	88	89	88	90
Tension set (%)	34	37.5	26	27.5
True stress at break (MPa)	15.2	70.5	22	64
Flow property	Flowable	Flowable	Flowable	Flowable

MPP = phenolic resin (PF) modified PP; MPRP = phenolic resin modified PRP.

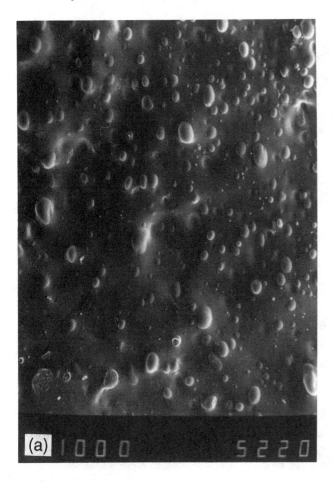

Figure 7 Optical micrographs for NR–LDPE (65:35) blends: (a) cured with sulfur; and (b) cured with peroxide. *Source*: Ref. 19.

perpendicular to the applied strain. This difference in the mechanism of stress distribution is one of the reasons for differences in the mechanical properties of the system studied. A similar morphological feature was also noticed for NR–PP blends. The phenolic modified NR–PP blends after BMI and phenolic curing showed a continuity between the two phases. In contrast, unmodified NR–PP blends either conventionally or dynamically cured by sulfur chemistry resulted in a grossly heterogeneous structure leading to interfacial separation during mechanical stress or repeated fatigue tests [19].

Effect of processing temperature as well as relative concentration of two phases in a blend were found to have a significant effect on the blend morphology and interfacial interaction. Results of the investigation of PRP–EVA blends with and without maleic modification give more insight into the morphological properties. The results of an investigation into the unmodified PRP–EVA blend in Fig. 9 illustrate that the mixing tem-

perature necessary to obtain the best blend property is the temperature at which the EVA rubber is best dispersed in the plastic matrix, i.e., the temperature at which the maximum interfacial area is developed. From an analysis of the SEM photographs and subsequent evaluation of the maximum interfacial area per unit volume of the blend, it is evident that for both the polymer compositions studied (8% EVA and 30% EVA), as illustrated in Fig. 9, the maximum calculated value of interfacial area was obtained at 195°C when optimum blend properties were also achieved. The calculated values of the interfacial area from SEM micrographs in Fig. 10 also confirmed that the blend with 8 wt% EVA developed a maximum specific interfacial area with the narrowest particle size distribution. This observation is in line with the conclusion drawn by Eliott and Tinker [26].

Maleic modified EVA–PRP blend showed some differences in the blend morphology from unmodified EVA–PRP blend. Addition of MAH–PP significantly in-

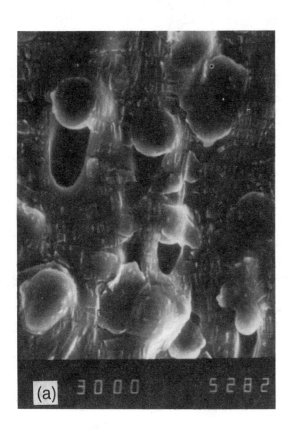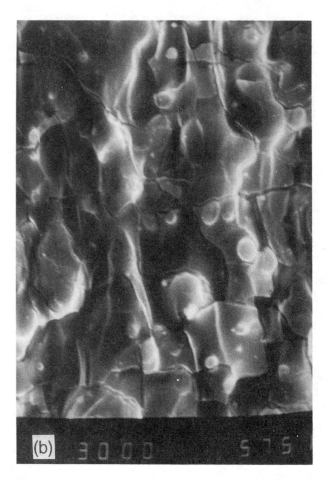

Figure 8 SEM images of etched surfaces of blends with 50% stretching; (a) sulfur cured (3000×); and (b) peroxide cured (3000×). *Source*: Ref. 27.

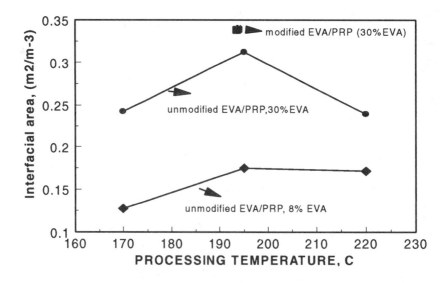

Figure 9 Effect of processing temperature on interfacial area.

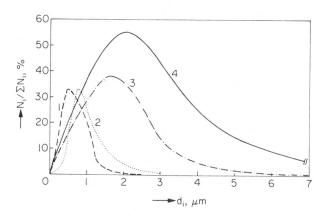

Figure 10 Particle size distribution of dispersed EVA rubber in PRP: (1) 1.5% EVA; (2) 8% EVA; (3) 15% EVA; and (4) 30% EVA. *Source*: Ref. 19.

creased the interfacial area of the dispersed phase as evident from the value shown in Fig. 9, indicating a stronger physical interaction between the polymeric phases. Figure 11 shows SEM images of the etched unfractured (a) and (b) unmodified and MAH–PP modified, respectively, and fractured surfaces (c) and (d) from unmodified and MAH–PP modified PRP/EVA blend, respectively; elastomer domains with a particle size of about 0.5 to 3.0 μm are visible in the form of dark holes. It is apparent from the fractured surface of the unmodified composition in Fig. 11c that fractures are brittle in nature with "dropping out" (dark holes) of elastomer particles. On the other hand, the modified composition resists phase separation by partly preventing dropping out of the elastomer phase.

Dispersed elastomeric micro-particles preferentially "break" (bright spots) with the development of crazes around the broken rubber domain as evident from Fig.

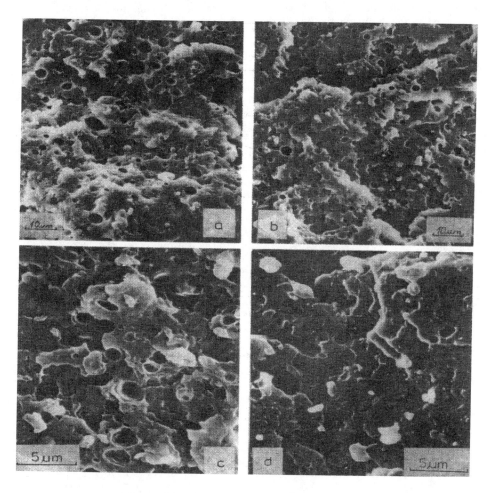

Figure 11 SEM images of PRP–EVA 92:8 blends. *Source*: Ref. 23.

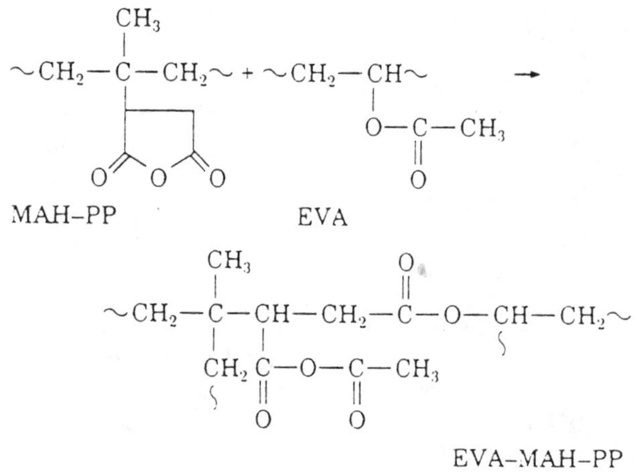

Scheme 1 *Source*: Ref. 23.

11d. Thus, it is expected that in the modified PRP–EVA blend, probably due to interface modification by reactive processing, a transesterification between the pendant MAH group in MAH–PP, and acetate groups in the EVA elastomer, as predicted in reaction Scheme 1, the dispersed rubber particles become more efficient in craze initiation.

REFERENCES

1. C. M. Ronald, *Handbook of Elastomers,* (A. K. Bhowmik and H. Stephens, eds.), Marcel Dekker, Inc., New York, pp. 183–214, 1988.
2. A. Y. Coran, *Handbook of Elastomers,* (A. K. Bhowmik and H. Stephens, eds.), Marcel Dekker, Inc., New York, pp. 249–312, 1988.
3. D. S. Cambell, D. J. Eliott, and M. A. Wheelans, *NR Technol., 9*: 21 (1978).
4. M. E. Woods and J. A. Davidson, *Rubber Chem. Technol., 49*: 112 (1976).
5. V. A. Shersnev, *Rubber Chem. Technol., 55*: 537 (1982).
6. J. Beniska, E. Staudner and E. Spirk, *Makromol. Chem. 161*: 113 (1974).
7. T. Tanaka and T. Yokoyama, *Polymer Sci., Part C: Polymer Symp. 23*: 865 (1968).
8. K. Mori and Y. Nakamura, *J. Polym. Sci., Polym. Chem. Ed. 16*: 2055 (1978).
9. W. K. Fischer, U.S. Patent 3,758,643, (Uniroyal Inc.) Sept. 11, 1973.
10. A. M. Gessler, (Esso Research & Engineering Co.) U.S. Patent 3,037,954, June 5, 1962.
11. A. Y. Coran and R. P. Patel, *Rubber Chem. Technol., 54*: 892 (1981).
12. A. Y. Coran and R. P. Patel, *Rubber Chem. Technol., 56*: 1045 (1983).
13. M. M. Sain, J. Lacok, J. Beniska, and V. Khunova, *Kautch. Gummi. Kunstst. 41*: (9) 895 (1988).
14. M. M. Sain, J. Lacok, J. Beniska, and J. Bina, *Acta Polymerica, 40*: (2) 136 (1989).
15. M. M. Sain, J. Beniska and P. J. Rosner, *Polym. Mater., 6*: 9 (1989).
16. A. Y. Coran and R. P. Patel, *Rubber Chem. Technol., 56*: 210 (1983).
17. M. M. Sain, I. Hudec, J. Beniska, and P. Rosner, *Rubber Chem. Technol.; 61*: 748 (1988).
18. Morris Morton, (ed.) *Rubber Technology,* VNR, New York (1973).
19. M. M. Sain, I. Hudec, J. Beniska, and P. Rosner, *Mat. Sci. Eng., A108*: 63 (1989).
20. M. M. Sain, J. Oravec, E. Sain, and J. Beniska, *Acta Polymerica, 43*: 51 (1992).
21. M. M. Sain, J. Lacok, J. Oravec, and J. Beniska, *Acta Polymerica 40*: (10) 649 (1989).
22. J. Muras and Z. Zamorsky, *Plasty Kaucuk,* (in Czech) *28*: (5), 137 (1991).
23. I. Hudec, M. M. Sain, and V. Sunova, *J. Appl. Polym. Sci., 49*: 425 (1993).
24. D. S. Cambell, D. J. Eliott, and M. A. Wheelans, *NR Technol., 9*: (2) 21 (1979).
25. W. K. Fischer, U.S. Patent 3,806,558, 1974.
26. D. J. Eliott and A. J. Tinker, *Int. Rubber Conf.,* Kuala Lumpur, Malaya (1985).
27. I. Hudec, M. M. Sain, and J. Kozankova, *Polymer Testing, 10*: 387 (1991).

31

Graft Copolymerization Onto Natural and Synthetic Polymers

Inderjeet Kaur and Bhupendra Nath Misra
Himachal Pradesh University, Shimla, India

I. INTRODUCTION

A polymer is a large molecule built up by the repetition of small, simple chemical units. In some cases the repetition is linear while in other cases the chains are branched or interconnected to form three-dimensional networks. The polymer can be formed not only through linear addition, but also through condensation of similar units as well.

If a polymer consists of two or more different monomers, then the resulting polymer is called a "copolymer." The copolymer is further classified into five types depending on the sequence of monomers in the polymer backbone.

1. Random copolymer—Distribution of the monomer units does not follow any definite sequence. These are produced in bulk, aqueous, suspension, or emulsion using free radical initiators of the peroxide type or redox systems.

2. Regular copolymer—These copolymers have an ordered sequence in the distribution of monomers. These are produced by controlled feeding of the monomers during copolymerization.

3. Alternating copolymer—In these copolymers, the monomers are arranged in an alternate sequence. Alternating copolymers are produced by special copolymerization processes.

4. Block copolymer—These copolymers are built of chemically dissimilar terminally connected segments. Block copolymers are generally prepared by sequential anionic addition or ring opening or step growth polymerization.

5. Graft copolymer—A graft copolymer consists of a polymeric backbone with covalently linked lateral side chains of different polymers. Graft copolymers are generally prepared by free radical, anionic or cationic addition, or by ring opening polymerization of a monomer onto the backbone polymer. Thus, the graft copolymer is a mixture of macromolecules consisting of a block of constitutional units of type A (backbone) having blocks of units of type B (grafts) attached to the backbone (Fig. 1).

In addition to the genuine copolymer macromolecules, the rough copolymer product usually contains grafted backbone and, according to the conditions of preparation, the attendant homopolymer. The part of the graft copolymer that consists of macromolecules bearing grafts attached to the backbone will be referred to as "true copolymer." The synthesis of a graft copolymer requires the formation of a reactive center on a polymer molecule in the presence of a polymerizable monomer.

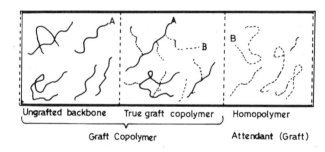

Figure 1 Components of rough copolymer products in graft copolymerization.

Most methods of synthesizing graft copolymers involve the use of radical polymerization, although ionic graft copolymerizations are also gaining importance. Graft copolymerization can be carried out either in homogeneous or heterogeneous systems depending on whether the polymer being grafted is soluble or insoluble in the monomer and in the solvent being used. Usually grafting can occur by either the "grafting from" or "grafting onto" method as illustrated by the following:

1. Grafting from: An active site generated along a polymer backbone starts to propagate monomer and thus produces branches.

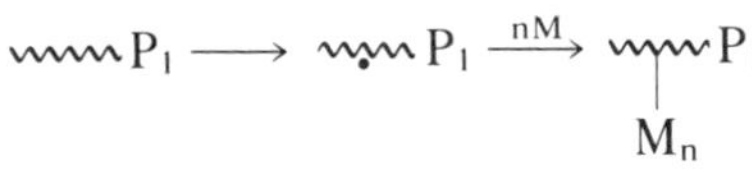

$$\sim\!\!\sim\!\!\sim P_1 \longrightarrow \sim\!\!\sim\!\!\sim P_1 \xrightarrow{nM} \sim\!\!\sim\!\!\sim P_1 \atop | \atop M_n \qquad (1)$$

2. Grafting onto: A growing polymer chain P_2 attacks another polymer P_1.

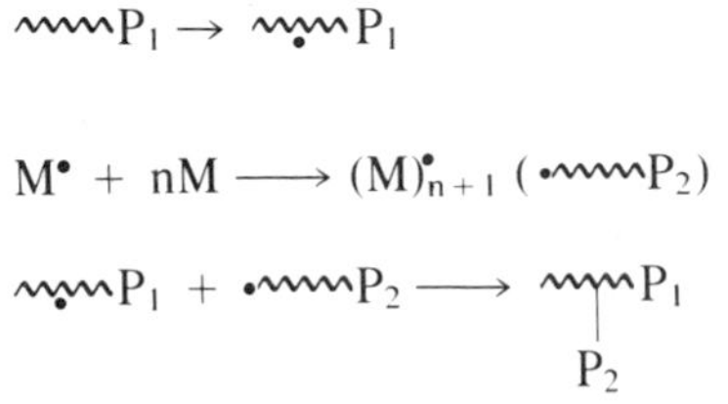

$$\sim\!\!\sim\!\!\sim P_1 \rightarrow \sim\!\!\sim\!\!\sim P_1$$

$$M^\bullet + nM \longrightarrow (M)^\bullet_{n+1} (\bullet\!\!\sim\!\!\sim P_2) \qquad (2)$$

$$\sim\!\!\sim\!\!\sim P_1 + \bullet\!\!\sim\!\!\sim P_2 \longrightarrow \sim\!\!\sim\!\!\sim P_1 \atop | \atop P_2$$

During formation of the graft copolymer the following assumptions are made:

1. The attachment of grafts to backbones is random [1–3]. Each segment of the backbone has the same probability to be grafted. Segment is defined as that part of the backbone to which just one graft can be attached. The probability of attachment of another graft to the backbone is not affected by the grafts already attached to the backbone.

2. Copolymer macromolecules are composed of a single backbone having simple grafts attached to it, i.e., the macromolecules are of the comb-like type. No further grafting of grafted chains is contemplated (4).

Usually a graft copolymer is a mixture of homopolymer, ungrafted backbone polymer, and graft copolymer. The graft copolymer is usually separated from the homopolymer by the solvent extraction method. Often graft copolymers prepared from water-soluble backbone polymers (such as starch, gelatin, poly(vinyl alcohol) [PVA]) and water-soluble vinyl monomers (such as acrylamide, acrylic acid, etc.) are also water-soluble. In such cases a rough copolymer is obtained that is not a true graft copolymer but a mixture of ungrafted polymer, grafted polymer, and the homopolymer. In such cases homopolymer cannot be removed from the mixture by the conventional solvent extraction method. Hence, such products are referred to as composites. Evidence for grafting in such cases has been obtained from turbidimetric studies [5–7].

II. IMPORTANCE OF GRAFTING

Grafting provides a convenient means for modifying the properties of numerous polymers. It is often required that a polymer possess a number of properties. Such diverse properties may not be easily achieved by the synthesis of homopolymers alone but can be achieved through the formation of copolymers or even terpolymers. The formation of graft copolymer with sufficiently long polymeric sequences of diverse chemical composition opens the way to afford speciality polymeric materials.

III. METHODS OF GRAFTING

The preparation of a graft (-g-) copolymer in most cases starts from a homopolymer A, which is appropriately activated, and then the growth of the graft of monomer B is initiated. The model must fulfill the following requirements.

1. The copolymer is of poly(A-g-B) type.
2. The probability that a graft will grow on a given unit of the backbone does not depend on the number or size of the other grafts in the molecule or on the size of the backbone to which it is attached.

The efficiency of grafting is dependent upon several competing reactions:

1. Competition between the various species present in the reaction mixture such as monomer, solvent, and backbone for the growing polymer radical, which means that there is competition between chain growth and various chain transferring steps.
2. Competition for the initiator radical between monomer and backbone.
3. Competition among the terminating processes such as disproportionation after the polymer radical has formed.
4. Competition among the terminating processes in the growing graft species.

If the termination occurs through recombination of growing polymeric chains, it should lower the grafting efficiency. But even though the efficiency of this process is low, it has received considerable use because the reactions are easily carried out. One needs merely to conduct polymerization of the monomer in the presence of the backbone polymer and a proper free radical initiator. A solvent may or may not be present. The addition of the good chain transferring agent, such as a mercaptan [8], CCl_4 or CH_3I [9], causes reduction in grafting efficiency. The other reaction parameters, such as concentration of the initiator, monomer, temperature, and the time of reaction, affect the grafting efficiency. The nature of the initiating radical appears to bear a strong influence on the efficiency of grafting. Smets and coworkers [10] found that when methylmethacrylate was polymerized in the presence of polystyrene initiated with benzoyl peroxide (BPO), an appreciable amount of grafting took place. When, however, the same reaction was initiated with azobisisobutyronitrile (AIBN) or ditertiarybutyl peroxide (DTBP), the amount of grafting was insignificant. Similar observations were made by Misra et al. [11] during grafting of vinyl acetate onto starch in aqueous medium using BPO and AIBN as radical initiators.

Various methods are available for effecting grafting. In most of them prior activation of the backbone polymer is involved to afford active sites where vinyl monomers can be grafted. Different methods of activation include: (1) physical activation, (2) chemical activation, and (3) radiation activation.

A. Physical Activation

Application of stress to polymeric backbone causes segmental motions and molecular flow, which may lead to bond scission and consequent formation of free radicals. Free radicals may also be generated by mastication and milling of the backbone polymer or can be accomplished by swelling it with suitable solvent. Penetration of the solvent through the polymer causes rupture of bonds leading to formation of free radicals. William and Stannett [12] utilized freezing and thawing of the polymer monomer mixture technique for effecting grafting of vinyl monomers onto wool. Grinding of polyethylene wax with styrene in hexane gave a grafted product having good flowability [13]. Heat and impact resistant polymer moulding materials for automobiles were prepared by the injection moulding procedure for grafting unsaturated carboxylic acid monomers onto ethylene-propylene block copolymer [14]. Thermosetting vinyl polymers grafted with hydrolyzable silanes onto high-density polyethylene (HDPE) and vinyl trimethoxysilane-forming radicals on mechanical treatment [15]. Production of a polymer with pendant basic moieties was performed by grafting dimethylaminoethyl-methacrylate to linear low-density polyethylene (LDPE) in the melt in a batch type internal mixer. The extent of grafting was determined by FTIR and nuclear magnetic resonance (NMR) spectroscopy, and the degree of crosslinking was determined by measuring the product melt index. The grafted material was found to be more stable suggesting an antioxidant effect of the grafted moieties [16]. Formation of graft copolymers has also been effected by carrying out the reaction at higher temperatures [17–19]. Grafting in melt [20] or treating the polymer-monomer mixture in a mixer also produces graft copolymer [21,22].

B. Chemical Activation

Conventional radical initiators, various redox systems, β-diketonates of transition metals alone and in conjunction with additives, Lewis acids, and strong bases for chemical grafting have been used successfully for a long time. Grafting by use of any of the previously mentioned initiators proceeds with the initial formation of the free radicals, which initiate the polymerization of the vinyl monomer. Simultaneously, active sites are generated on the polymeric backbone where grafting can take place.

Grafting by free radical methods is discussed in the following sections.

1. Radical Initiator

In the presence of radical initiators such as benzoyl peroxide (BPO), azobisisobutyronitrile (AIBN), persulfates ($S_2O_8^{2-}$), etc., grafting of vinyl monomers onto polymeric backbones involves generation of free radical sites by hydrogen abstraction and chain transfer processes as described below:

$$I_2 \rightarrow 2I^{\bullet}$$

$$I^{\bullet} + PH \rightarrow P^{\bullet} + IH$$

$$I^{\bullet} + M \rightarrow IM^{\bullet} \xrightarrow{nm} I - M_n - M^{\bullet}$$

$$PH + I - Mn - M^{\bullet} \rightarrow P^{\bullet} + I - (M)_{n+1} - H$$

$$P^{\bullet} + M \rightarrow PM^{\bullet} \xrightarrow{nm} P - M_n - M^{\bullet}$$

$$P^{\bullet} + I - M_n - M^{\bullet} \rightarrow P - (M)_{n+1} - I \tag{3}$$

where I_2 represents initiator molecule, PH is the polymer backbone and M is the vinyl monomer.

The extent of grafting is dependent on various reaction parameters, such as concentrations of initiator and monomer, reaction time, temperature, etc. The nature of the initiator and the monomer influences the reactivity of initiator and monomer toward grafting. It has been observed by Misra et al. that during grafting of vinyl monomers onto starch, wool, or rubber, conventional free radical initiators exhibit different reactivity toward grafting. When vinyl acetate was grafted onto starch using BPO and AIBN as the initiator, a lower percentage of grafting was obtained with AIBN [11]. The low reactivity of AIBN is explained on the basis that the $(CH_3)_2$—$C^{\bullet}$—$C\equiv N$ is resonance stabilized and less reactive. These radicals are, however, capable of initiating polymerization.

$$\tag{4}$$

Misra et al. [11] further observed that the molecular weight of poly(vinyl acetate) prepared by using either AIBN or BPO was higher than the molecular weights of poly(vinyl acetate) isolated from starch-g-poly(vinyl acetate) graft copolymer and also the molecular weights of poly(vinyl acetate) (PVAC) or isolated PVAC graft prepared by using AIBN as the initiator are higher than those prepared by using BPO. It is assumed that benzoyl radicals, being highly reactive, terminate the growing chains with greater frequency and with consequent decrease in viscosity average molecular weight [11]. Acrylonitrile (AN) and methylacrylate (MA) have been grafted onto starch using BPO as the initiator, and AN was found to be more reactive than MA. The observed reactivity was explained on the basis of polarity and solubility of the monomers in the reaction medium [23]. Modification of natural rubber has also been successfully attempted by grafting MA using BPO, potassium persulfate (KPS), and AIBN as radical initiators [24].

Ethylacrylate (EA) and butylacrylate (BA) was grafted onto natural rubber using BPO as the initiator [25]. KPS and AIBN were ineffective in the production of the graft and the failure to form the graft is attributed to the "initiator effect."

The reactivity of different monomers toward grafting onto natural rubber using different radical initiators was found to be:

$$EA > BA > MA \qquad (5)$$

When tertiary butyl hydrogen peroxide (TBHP) was used alone as the radical initiator, no grafting of methylmethacrylate (MMA) onto wool was observed. However, TBHP in conjunction with mineral acids, such as H_2SO_4, HNO_3, or $HClO_4$ afforded good results [26]. Protonation of TBHP by the acid aided in the dissociation of TBHP to yield free radicals, which initiated grafting reaction.

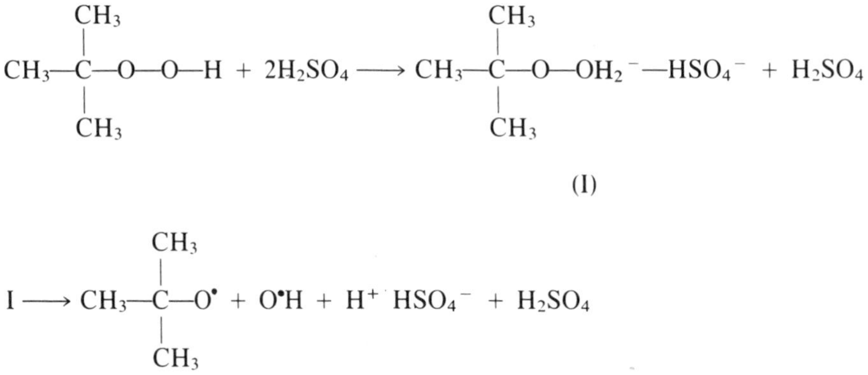

$$\underset{\underset{CH_3}{|}}{\overset{\overset{CH_3}{|}}{CH_3-C}}-O-O-H + 2H_2SO_4 \longrightarrow \underset{\underset{CH_3}{|}}{\overset{\overset{CH_3}{|}}{CH_3-C}}-O-OH_2{}^--HSO_4{}^- + H_2SO_4$$

$$(I) \qquad\qquad\qquad\qquad\qquad\qquad (6)$$

$$I \longrightarrow \underset{\underset{CH_3}{|}}{\overset{\overset{CH_3}{|}}{CH_3-C}}-O^\bullet + O^\bullet H + H^+ \; HSO_4{}^- + H_2SO_4$$

One molecule of acid first protonates TBHP to give a protonated complex (I), which is stabilized by the second molecule of H_2SO_4 involving hydrogen bonding in the following manner:

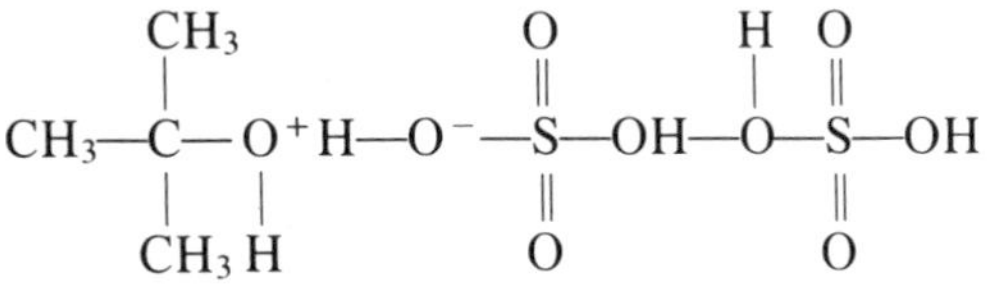

$$\underset{\underset{CH_3\;H}{|\quad|}}{\overset{\overset{CH_3}{|}}{CH_3-C}}-O^+H-O^--\overset{\overset{O}{\|}}{\underset{\underset{O}{\|}}{S}}-OH-O-\overset{\overset{H\;\;O}{|\;\;\|}}{\underset{\underset{O}{\|}}{S}}-OH \qquad (7)$$

The formation of the complex is expected to decrease the free energy of activation for the homolysis of the peroxide bond, and the decomposition of TBHP would occur at a lower temperature. It was further observed that at a higher concentration of mineral acid, the decomposition of TBHP occurs via an ionic pathway, as reported by Turner [27].

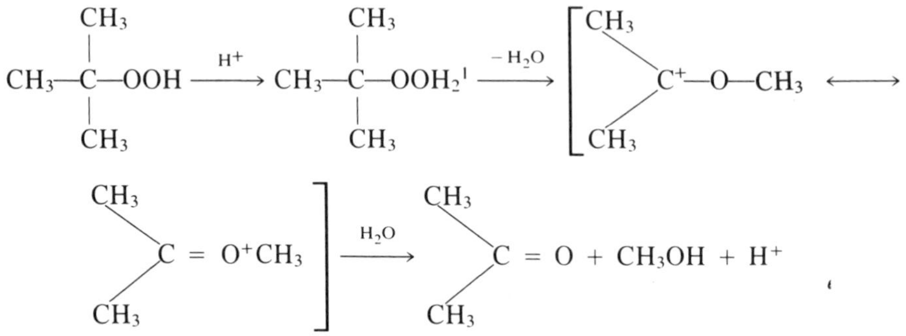

$$\underset{\underset{CH_3}{|}}{\overset{\overset{CH_3}{|}}{CH_3-C}}-OOH \xrightarrow{H^+} \underset{\underset{CH_3}{|}}{\overset{\overset{CH_3}{|}}{CH_3-C}}-OOH_2{}^1 \xrightarrow{-H_2O} \left[\underset{CH_3}{\overset{CH_3}{>}}C^+-O-CH_3 \longleftrightarrow \right.$$

$$\left. \underset{CH_3}{\overset{CH_3}{>}}C = O^+CH_3 \right] \xrightarrow{H_2O} \underset{CH_3}{\overset{CH_3}{>}}C = O + CH_3OH + H^+ \qquad (8)$$

Different mineral acids showed the following reactivity order toward grafting in the TBHP-mineral acid system:

$$H_2SO_4 > HClO_4 > HNO_3 \qquad (9)$$

Chandel and Misra [28] observed that benzoyl peroxide alone was ineffective in the production of styrene grafted wool, but good grafting was obtained when used in conjunction with pyridine and acetic acid mixture, which acted as a pH modifier. The addition of pH modifiers along with penetrating agents during BPO initiated grafting of acrylic acid onto PET fiber was also studied by Suzuki et al. [29]. Persulfates have also been used as the radical initiator for graft copolymerization of various vinyl monomers onto polymeric backbones. Grafting of

MMA onto cellulose was carried out by Hecker de Carvalho and Alfred using ammonium and potassium persulfates as radical initiators [30]. Radical initiators such as H_2O_2, BPO dicumylperoxide, TBHP, etc. have also been used successfully for grafting vinyl monomers onto hydrocarbon backbones, such as polypropylene and polyethylene. The general mechanism seems to be that when the polymer is exposed to vinyl monomers in the presence of peroxide under conditions that permit decomposition of the peroxide to free radicals, the monomer becomes attached to the backbone of the polymer and pendant chains of vinyl monomers are grown on the active sites. The basic mechanism involves abstraction of a hydrogen from the polymer to form a free radical to which monomer adds:

$$R—O—O—R \longrightarrow 2R\overset{\bullet}{O}$$

$$\text{\large\char`\~}CH_2—CH_2—CH_2\text{\large\char`\~} + R\overset{\bullet}{O} \rightarrow \text{\large\char`\~}CH_2—\overset{\bullet}{C}H—CH_2\text{\large\char`\~} + ROH$$

$$\text{\large\char`\~}CH_2—\overset{\bullet}{C}H—CH_2\text{\large\char`\~} + M \longrightarrow \text{\large\char`\~}CH_2—\underset{\underset{M^{\bullet}}{|}}{CH}—CH_2\text{\large\char`\~} \xrightarrow{nM} Graft \qquad (10)$$

Polyethylene grafted with a wide variety of monomers using inorganic and organic peroxides shows improvement in thermal stability, elasticity, modulus, melt viscosity, and water resistance [31,32]. Carboxylic ion exchange resins [33] and chemical and heat resistant cation exchange membranes [34] have also been prepared by grafting vinyl monomers onto polyethylene using chemical methods.

It was observed by Borsig and coworkers [35] that in the presence of a reactive radical initiator such as

dicumyl peroxide, polypropylene (PP) is not only cross-linked but also degraded. Cross-linking efficiency can be increased by polyfunctional monomers such as pentaerythritol diacrolein acetal (PEDAA) (a difunctional) or pentaerthritol tetraallyl ether (PETA) (a four functional monomer), and pentaerythritol difurfural acetal (PEDFA). The improved cross-linking efficiency of peroxides in the presence of polyfunctional monomers is due to the ability of the latter to add to radicals that are formed on the polymer chain via the double bonds of polyfunctional monomer.

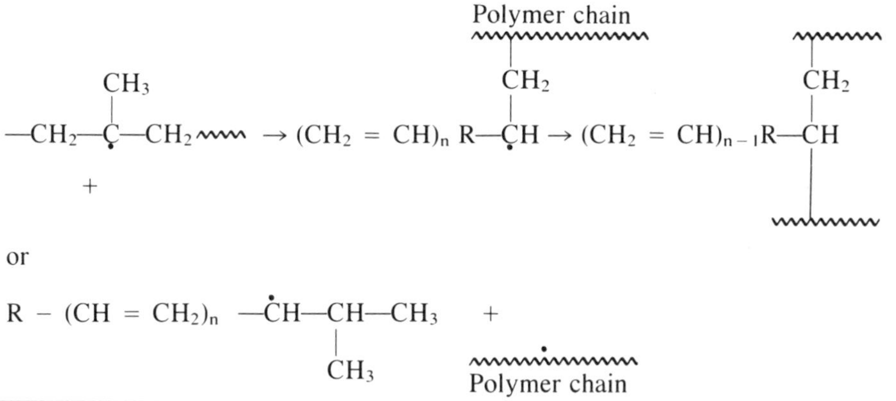

$$\underset{+}{\underset{\underset{CH_3}{|}}{—CH_2—\overset{}{C}—CH_2}\text{\large\char`\~}} \rightarrow (CH_2 = CH)_n R—\overset{\bullet}{C}H \rightarrow (CH_2 = CH)_{n-1}R—CH \qquad (11)$$

Polymer chain ... CH_2 ... CH_2

or

$$R - (CH = CH_2)_n \ \underset{\underset{CH_3}{|}}{—\overset{\bullet}{C}H—CH—CH_3} \ +$$

Polymer chain

2. Redox System

In redox initiation, the free radicals that initiate the polymerization are generated as transient intermediates in the course of redox reaction. Essentially this involves an electron transfer process followed by scission to give free radicals. A wide variety of redox reactions, involving both organic and inorganic components either wholly

or in part, may be used for this purpose. Because most of the redox initiators are ionic in nature, their reactions take place in aqueous media and occur rather rapidly, even at relatively low temperatures. The general mechanism of grafting involves an abstraction of the hydrogen atom from the backbone of the polymeric material by the transient radical formed during the redox reaction, thus generating macroradicals onto which the monomer molecules add to produce graft copolymer.

Commonly used oxidants include peroxides, persulfates peroxydisulfates, permanganate, etc., and the reducing agents may be salts of metals such as Fe^{2+}, Cr^{2+}, V^{2+}, Ti^{3+}, Cu^{2+}, oxyacids of sulfur, hydroxyacids, etc. There is no commonly accepted method of classifying redox initiators. They are subdivided in accordance with the nature of their main oxidizing component. There exists a large number of redox systems that can initiate polymerization of vinyl monomers. The following are some of the few redox systems utilized effectively for graft copolymerization reactions.

Peroxydisulfate System

The activation of persulfates by various reductant viz. metals, oxidizable metals, metal complexes, salts of various oxyacid of sulfur, hydroxylamine, hydrazine, thiol, polyhydric phenols, etc. has been reported [36–38]. Bertlett and Colman [39] investigated the effect of methanol on the decomposition of persulfates and proposed the following mechanism.

$$S_2O_8{}^{2-} + CH_3OH \rightarrow HSO_4{}^- + SO_4{}^-\cdot + \dot{C}H_2OH$$
$$SO_4{}^-\cdot + CH_3OH \rightarrow HSO_4{}^- + \dot{C}H_2OH$$
$$S_2O_8{}^{2-} + \dot{C}H_2OH \rightarrow HSO_4{}^- + SO_4{}^-\cdot + CH_2O$$
$$\dot{C}H_2OH + \dot{C}H_2OH \rightarrow CH_3OH + CH_2O \qquad (12)$$

The $Ag^4 - K_2S_2O_8$ is also a very useful redox agent and the following mechanism for its decomposition was proposed by Bawn and Margarison [40].

$$Ag^+ + S_2O_8{}^{2-} \rightarrow Ag^{2+} + SO_4{}^{2-} + SO_4{}^-\cdot$$
$$Ag^{2+} + S_2O_8{}^{2-} \rightarrow Ag^{3+} + SO_4{}^{2-} + SO_4{}^-\cdot$$
$$Ag^{3+} + 2OH^- \rightarrow Ag^+ + 2\dot{O}H \qquad (13)$$

Polyamide fiber was grafted with MMA using the peroxydisulfate Ag^+ redox system [41]. Other metals such as Cu^{2+}, Ti^{3+}, and CO^{2+} have also been coupled with persulfate to form an effective redox system. A variety of vinyl monomers have been grafted onto different polymeric backbones using persulfate reducing agents as initiators. The $K_2S_2O_8 - Na_2S_2O_3{}^-$-Cu ion-aqueous or organic ligand complex system was used as the initiator for grafting methacrylic acid onto polyamide-6 fiber. It was observed that the system was most active when the fiber was treated with Cu-phenanthroline complex prior to polymerization. The copper complex in combination with $Na_2S_2O_3$ and $K_2S_2O_8$ in 25:1 ratio fully inhibited polymerization. The initial rate of grafting was higher for unoriented fiber due to increased swelling, but the yield of the graft copolymer was higher for oriented fiber due to inhibited chain termination [42].

Potassium persulfate with ferrous sulphate was successfully used as the initiator for grafting vinyl monomers onto wool [43] and cellulose [44] by Misra et al. They observed that there exists a critical molar ratio of

(oxidant)/(reductant) at which the percentage of grafting is at its maximum and beyond which it decreases. This was explained on the basis that during Fe^{2+}-induced decomposition of $S_2O_8{}^{2-}$ excess of Fe^{2+} consumes $SO_4{}^-\cdot$ to produce Fe^{3+} and $SO_4{}^{2-}$. Fe^{3+} at a higher concentration terminates the process. In order to establish the site of grafting on wool, grafting of MA and VAC was carried onto reduced wool. Reduction was effected by use of thioglycolic acid and the disulphide groups were converted to thiol (—SH) groups. The disulphide linkage is known to form a redox system with persulfate in the following manner.

$$R—S—S—R + O_3{}^-S—O—O—SO_3{}^-$$
$$\rightarrow RS—O—SO_3{}^- + \dot{R}S + SO_4{}^-\cdot \qquad (14)$$

Both $\dot{R}S$ and $SO_4{}^-\cdot$ directly or indirectly initiate polymerization. It was observed that the percentage of grafting of MA onto reduced wool was higher than in wool. This suggested that the additional —SH groups provide sites for grafts [45].

$$R—SH \rightleftharpoons RS^- + H^+$$
$$RS^- + O_3S—O—O—SO_3{}^-$$
$$\rightarrow \dot{R}S + SO_4{}^{2-} + SO_4{}^-\cdot \qquad (15)$$

Peroxydiphosphate System

In conjunction with Ag^+, V^{5+}, CO^{2+}, and acid, peroxydiphosphate forms an efficient redox system for polymerization of vinyl monomers. $H_2P_2O_8{}^{2-}$ is assumed to be an active species of peroxydiphosphate. The initiating species are $\dot{O}H$ and $HPO_4{}^-\cdot$ and the termination is considered to be exclusively by mutual method. The following mechanism has been proposed for the redox reaction [46].

$$H_2P_2O_8{}^{2-} + Ag^+ \rightarrow HPO_4{}^-\cdot + HPO_4{}^{2-} + Ag^{2+}$$
$$HPO_4{}^-\cdot + Ag^+ \rightarrow HPO_4{}^{2-} + Ag^{2+}$$
$$HPO_4{}^-\cdot + HOH \rightarrow \dot{O}H + H_2PO_4{}^-$$
$$Ag^{2+} + HOH \rightarrow \dot{O}H + H^+ + Ag^+ \qquad (16)$$
$$HPO_4{}^-\cdot + M \rightarrow \dot{M}—HPO_4 \text{ (or } \dot{M}_1)$$
$$\dot{O}H + M \rightarrow \dot{M}—OH \text{ (or } \dot{M}_1)$$
$$\dot{M}_1 + nM \rightarrow M_1—M_{(n-1)}—\dot{M}$$

If the polymerization is carried out in the presence of the backbone polymer, grafting occurs. Both $\dot{O}H$ and $HPO_4{}^-\cdot$ radicals can generate an active site on the backbone polymer by abstracting hydrogen, where grafting can take place. Peroxydiphosphate–$VO_2{}^{2+}$ redox system has also been investigated [47]. This system gave better results than when Ag^+ was used as the reductant. The rate of polymerization was found to increase with increasing temperature in the $P_2O_8{}^{4-} - VO_2{}^{2+}$ system but not in the Ag^+ containing redox system. Nayak et al. [48] grafted MMA onto wool using peroxydiphosphate in acid medium and suggested that H_2PO_4, $\dot{O}H$, and

HPO$_4^-$• are responsible for grafting. These radicals are capable of interacting with wool giving wool macroradicals where grafting can occur. Nayak et al. have also successfully used peroxydiphosphate–thiourea [49], peroxydiphosphate–Fe^{2+} [50], and peroxydisphosphate–Mn^{2+} [50] redox systems for grafting of vinyl monomers onto wool, silk, and polyamide-6 fibers. Better results were obtained with the peroxydiphosphate–Mn^{2+} system.

Ce^{4+} Redox System

Following the findings of Mino and Kaizerman [51] that ceric ion can form a redox system with cellulose, grafting onto various natural polymers has been carried out by the ceric ion method. In the case of cellulose, the reaction between ceric ion and cellulose occurs to produce active sites on cellulose in the following manner:

$$Ce^{4+} + R_{cell}CH_2OH \rightarrow Complex \rightarrow Ce^{3+},$$
$$+ H^+ + R_{cell}\overset{\bullet}{C}HOH \qquad (17)$$

This is a highly selective process and very good results on cellulose and starch grafting have been observed. Ceric ion initiated grafting is usually carried out at lower temperatures and, therefore, wastage of monomer in chain transfer reactions is minimal.

Extensive work on grafting of vinyl monomers onto different polymeric backbones has been carried out using Ce^{4+} as the redox initiator. Misra et al. have successfully carried out Ce^{4+} initiated graft copolymerization of vinyl monomers onto wool [52,53] and gelatin [54–56]. In order to establish the site of grafting in wool, grafting has been attempted onto reduced wool and it was found that the —SH group of cystine of wool are involved in the grafting process [57,58]. Kinetic studies on ceric ion initiated grafting of AN onto sodium salt of partially carboxymethylated amylose have been reported by Patel et al. [59].

When ceric ammonium nitrate (CAN) was used as a source of ceric ion, the presence of nitric acid was found to play a significant role. Ceric ion in water is believed to react in the following manner:

$$Ce^{4+} + H_2O \rightarrow [Ce(OH)_3]^{3+} \quad (i)$$
$$2[Ce(OH)_3]^{3+} \rightarrow [Ce—O—Ce]^{6+} + H_2O \quad (ii) \qquad (18)$$

Thus, in aqueous medium, ceric ion exists as Ce^{4+} [Ce(OH)$_3$]$^{3+}$ and [Ce—O—Ce]$^{6+}$. No grafting was observed in the absence of nitric acid. This is explained on the basis that in the absence of nitric acid, [Ce—O—Ce]$^{6+}$, because of large size, is unable to form a complex with wool. However, with an increase in [HNO$_3$], the equilibria [Eqs. 18 (i) and (ii)] shift toward more and more of Ce^{4+} and [Ce(OH)$_3$]$^{3+}$, which easily form complexes with functional groups of wool. The complex decomposes to give free radicals on wool where grafting occurs. At higher concentrations, these species adversely affect grafting efficiency.

Misra et al. have utilized the ceric-amine redox system for grafting MMA onto wool [60] and gelatin [61]. The graft yield was explained in terms of basicity, nucleophilicity, and steric requirements of amines. A complex of ceric ion and amine (AH) decomposes to generate free radical species, which produce additional active sites onto the polymeric backbone where grafting can occur.

$$WH + Ce^{4+} \rightarrow Complex_1 \rightarrow \overset{\bullet}{W} + H^+ + Ce^{3+}$$
$$A–H + Ce^{4+} \rightarrow Complex_2 \rightarrow \overset{\bullet}{A} + H^+ + Ce^{3+}$$
$$\overset{\bullet}{A} + WH \rightarrow \overset{\bullet}{W} + AH \qquad (19)$$

where WH = wool.

When grafting of ethylacrylate was carried out onto gelatin using the ceric amine system [62], it was found that in the presence of more basic amines, which can form a complex with Ce^{4+} with greater facility, a higher percentage of grafting was obtained. Besides basicity and nucleophilicity, steric requirements of the amines also play a part in the complex formation and these affect grafting. Similar results were also observed by Patel et al. [63] during grafting of AN onto sodium salt of partially carboxymethylated starch using a ceric-amine redox system. The effect of radioprotecting agents (RPA) on ceric ion-initiated grafting of MA onto cellulose has been studied by Misra et al. [64]. Complex formation between the RPA and Ce^{4+} was assumed and it was found that threonine possessing —OH, —NH$_2$, and —COOH groups can easily form a complex with Ce^{4+} and, therefore, increases percentage of grafting. On the other hand, in the presence of 5-hydroxy tryptophan and 5-hydroxy tryptamine, the percentage of grafting was found to be suppressed. This was explained on the basis that the phenolic —OH groups present in these additives inhibit both polymerization and grafting.

Complex formation between Ce^{4+} and the functional groups of the backbone polymer was also suggested by Chinese workers [65] during grafting of acrylamide onto 1,4-butanediol-4,4-diphenyl methane diisocyanate. Polytetramethylene glycol copolymers capable of forming a complex between the carbamate groups and ceric ion lead to the generation of free radicals on the N-atom, which acts as the maximum site of grafting.

Metal Chelates Initiated Graft Copolymerization

A number of metal chelates containing transition metals in their higher oxidation states are known to decompose by one electron transfer process to generate free radical species, which may initiate graft copolymerization reactions. Different transition metals, such as Zn, Fe, V, Co, Cr, Al, etc., have been used in the preparation of metal acetyl acetonates and other diketonates. Several studies demonstrated earlier that metal acetyl acetonates can be used as initiators for vinyl polymerization. Few studies have been reported on the use of metal chelates as initiators for grafting. However, in recent years, attempts have been made by Misra et al. and Nayak et al. to effect grafting of different vinylmonomers onto starch, cellulose, and wool using various β-diketonates as initiators.

Metal chelates are known to decompose upon heating to generate free radicals, which can abstract hydrogen atoms from the polymeric backbone producing an active site where grafting can take place.

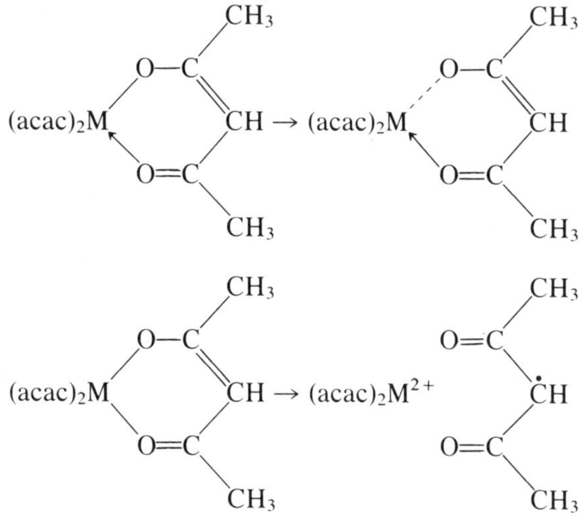

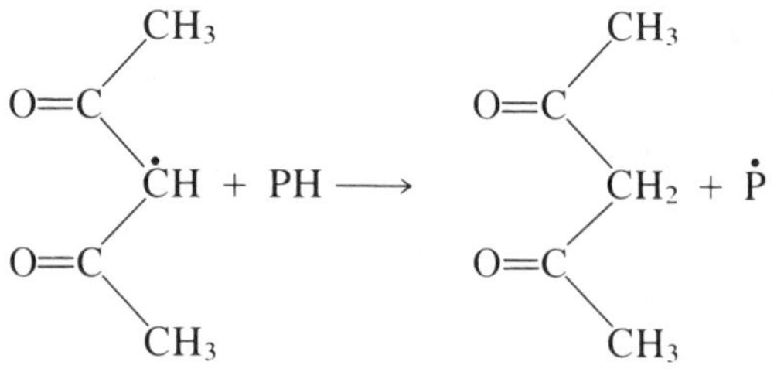

$$(20)$$

where PH is the polymer backbone, acac is the acetylacetone.

Metal chelates afford a better initiating system as compared to other redox systems since the reactions can be carried out at low temperatures, thus avoiding wastage reactions due to chain transfer. Homopolymer formation is also minimum in these systems. It was observed by Misra et al. [66,67] that the maximum percentage of grafting occurs at a temperature much below the decomposition temperature of the various metal chelates indicating that the chelate instead of undergoing spontaneous decomposition receives some assistance either from the solvent or monomer or from both for the facile decomposition at lower temperature. The solvent or monomer assisted decomposition can be described as:

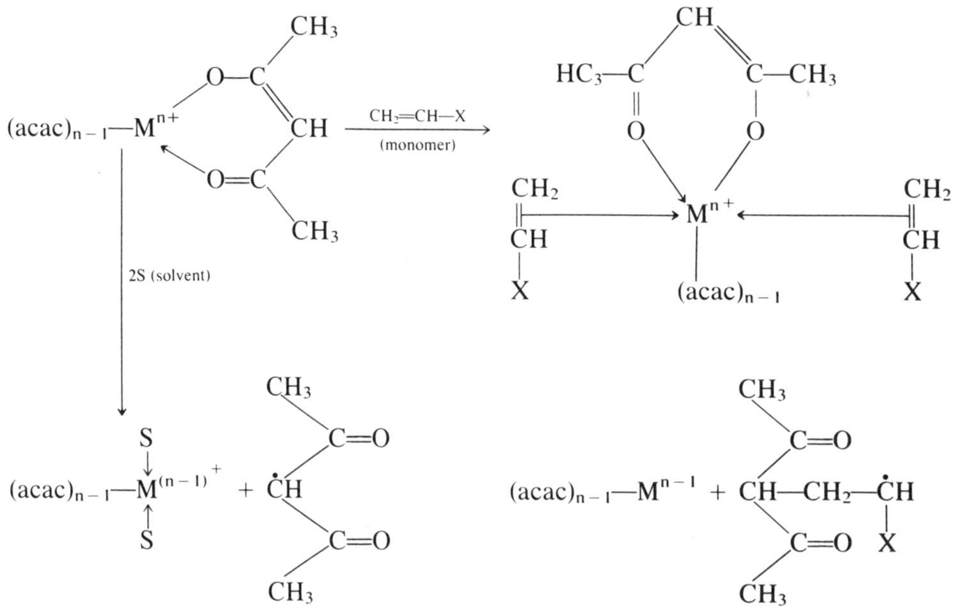

$$(21)$$

Evidence for the formation of such a complex was obtained from IR spectroscopy [68] of the solid residue obtained by the treatment of MA with Fe(acac)$_3$. Characterization of the complex by the ultraviolet (UV) spectral method also indicated the formation of the complex [68]. Misra et al. have carried out grafting of vinyl mono-

mers onto cellulose [68] and wool [69–74] using different metal chelates. The following reactivity order was established for grafting of MMA onto wool in the presence of various metal chelates as initiators.

Fe(acac)$_3$ > Cr(acac)$_3$ − TBHP > Mn(acac)$_3$ > Vo(acac)$_2$

Nayak et al. have also reported grafting of vinyl monomers onto wool [75] and silk [76] using metal chelates as initiators.

Hydrogen Peroxide—Reductant System

Combinations of H_2O_2 and Fe(II) salts, generally known as "Fenton reagents," have for many years been used in organic and polymer chemistry. Medalia and Kolthoff [77] summarized the principal features of this reaction.

1. In aqueous solution containing only Fe(II) salts, H_2O_2, stable inorganic salts and acids, the stoichiometric reaction is followed if the iron salt is in excess over the peroxide.

$$H_2O_2 + 2Fe^{2+} + 2H^+ \rightarrow 2Fe^{3+} + 2H_2O \quad (22)$$

2. When the peroxide is in large excess compared with the Fe(II) less iron is oxidized than would correspond to the H_2O_2 consumed if the above reaction was followed. The excess of the H_2O_2 is decomposed to oxygen and water.

3. In the presence of many organic compounds as well as of certain easily oxidized inorganic compounds such as the iodide ion, part of the oxidized H_2O_2 is added to the compound.

Following the discovery of Haber and Weiss [78] that Fe(II) salts react with H_2O_2 by one electron transfer process to give OḢ, Baxandale et al. [79] used the Fe^{2+} − H_2O_2 system for effecting polymerization of vinyl monomers.

Besides Fe^{2+}, other reducing agents that may be used in conjunction with H_2O_2 are aliphatic amines, $Na_2S_2O_3$ thiourea, ascorbic acid, glyoxal, sulfuric acid, $NaHSO_3$, sodium nitrite, ferric nitrate, peroxidase, $AgNO_3$, tartaric acid, hydroxylamine, ethylene sulfate, sodium phosphite, formic acid, ferrous ammonium sulphate, acetic acid, ferrous sulphate, and HNO_2, etc.

Fenton reagents have been used in numerous graft copolymerization reactions. Misra et al. have utilized Fenton reagent as an initiator for grafting vinyl monomer onto wool [80] and cellulose [81, 82] and observed that the molar ratio of $[Fe^{2+}]/[H_2O_2]$ influences percentage of grafting. Recently, Trivedi et al. [83] have grafted acrylonitrile onto sodium Alginate using Fenton reagent as an initiator and observed that there exists a critical concentration of Fe^{2+} and H_2O_2 at which maximum percentage of grafting was obtained. They also studied the kinetics of reaction and found that the plots of lnRp versus ln[M], lnRp versus $ln[H_2O_2]^{1/2}$ and lnRp versus $ln[Fe^{2+}]^{1/2}$ bear a linear relation. This confirms the validity of the proposed reaction [84].

C. Radiation-Induced Graft Copolymerization

Grafting by means of radiation is by far the most popular synthetic technique for modification of polymers, and the majority of the literature on graft copolymerization describes the use of a radiation method in some form or another. The growth and popularity of radiation as the initiating system stems from the improvement in the availability of ionizing radiation as a result of the introduction during recent years of more powerful nuclear reactors. Apart from being inexpensive, it is a very convenient method for graft initiation as it allows a considerable degree of control to be exercised over structural factors, such as the number and length of the grafted chains by careful selection of the dose and dose rate. Radiation is also unique in its ability to enable grafting to be carried out on prefabricated or shaped articles.

The following sections describe, in a general sense, the various methods of radiation grafting, their efficiency, applicability, and the mechanism of the grafting reaction. Among the various methods used for effecting grafting by using high-energy radiations, the following methods deserve special mention:

1. The direct radiation grafting of a vinyl monomer onto a polymer by the mutual method.
2. Grafting of vinyl monomers onto preirradiated polymers either in vacuum or air by the preirradiation method.
3. Grafting initiated by trapped radicals.
4. Intercrosslinking of two different polymers.
5. The double irradiation method involving the mutual method of grafting of vinyl monomers onto preirradiated polymer recently used by Kaur et al. [85].

Initiation by these methods can be carried out by either gamma rays or by electron beam. Usually grafting has been carried out either by the mutual or the preirradiation method using Co^{60} as a source of gamma rays.

1. Mutual Method

In its simplest form the direct grafting method involves the irradiation of polymeric substrate in the absence or presence of oxygen. Graft copolymerization of the monomer to the polymer is then initiated through the free radicals generated in the latter. The reaction can be schematically written as:

$$(23)$$

where Ap ∿∿∙ and Ap ∿∿∿∿Ap are polymeric free radicals derived from Ap, and Ṙ is a low-molecular weight radical or a hydrogen atom. Reaction (i) is expected to occur if Ap is a polymer of degrading type resulting in the formation of a block copolymer. If Ap crosslinks under irradiation, then reaction (ii) is more likely to occur leading to the formation of equal numbers of graft copolymer and homopolymer. In addition, the monomer B is also irradiated in the process to give the homopolymer:

$$B\!\!\sim\!\!\sim\!\!\rightarrow \dot{R} \xrightarrow{\ nB\ } Bq \qquad (24)$$

In order to reduce the amount of homopolymer it may be advantageous to reduce the concentration of the monomer in the polymer monomer combination. Thus,

much higher yields are obtained if the polymer Ap is slightly swollen by the monomer B than if Ap is dissolved in a large excess of monomer B [86]. Good results are obtained by swelling the polymer with the monomer vapor [87]. A reduction in the concentration of monomer may further influence a number of other factors involved in the reactions such as viscosity of the reaction medium, the rate of chain propagation, the lengths of grafted branches, etc.

If the grafting is carried out in air, the active sites on the polymeric backbone is attacked by atmospheric oxygen leading to the formation of macroperoxy radical, which might abstract the hydrogen atom from the backbone polymer by an inter- or intramolecular process to give hydroperoxide groups as shown.

$$(25)$$

The hydroperoxide groups thus formed undergo decomposition during irradiation producing macroxy radicals that offer sites for grafting.

(macroxy radical)

$$(26)$$

The radiation sensitivity of a substrate is measured in terms of its GR value or free radical yield, which is the number of free radicals formed per 100 eV energy absorbed per gram. The highest grafting yields will occur for polymer monomer combinations in which the free radical yield of the polymer is much greater than for the monomer. It also follows that the grafting yield will increase at a lower monomer concentration.

Graft copolymerization of different vinyl monomers has been successfully attempted onto both natural [88–90] and synthetic polymers [91–93] by the mutual method. Percentages of grafting have been studied as a function of various reaction parameters that seem to influence the grafting reaction. Total dose used for grafting determines the number of grafting sites while the dose rate determines the length of the grafted branches. These can also be controlled by other factors such as presence of chain transfer agent, monomer concentration, reaction temperature, and amount of the solvent used. Misra et al. have extensively studied the modification of natural polymers such as rayon [94], wool [95–97], and gelatin [98–99], as well as synthetic polymers such as poly(vinyl alcohol) PVA [100–102], isotac-

tic polypropylene (IPP) [103], and polyamide-6 [104] by mutual method. In order to avoid chain transfer reaction, grafting was carried out in aqueous medium since water has zero chain transfer constant. It was observed that a critical value of the total dose produce a maximum percentage of grafting beyond which the grafting percentage decreased. At a higher value of total dose, bond scission or crosslinking of the backbone polymer may take place, which decreases the extent of grafting. A plausible mechanism of this reaction based on conventional oxidative degradation is outlined as follows:

$$(27)$$

In aqueous medium, excessive generation of hydroxyl radicals from radiolysis of water preferentially initiate homopolymerization of vinyl monomers leading to a lower percentage of grafting. Homopolymer, thus formed, if soluble in the reaction medium, increases the viscosity of the system, which results in the restriction of the mobility of the growing polymeric chains to the active site thereby decreasing the grafting level. In a medium of high viscosity, the primary radicals experience difficulty in adding to the active sites of the backbone polymer. With increasing total dose, recombination of primary radicals is accelerated leading to the formation of homopolymer at the expense of the graft.

Another factor that affects the extent of grafting is the nature and the concentration of the monomer. Generally, maximum graft copolymer is observed at an optimum monomer concentration beyond which grafting percentage decreases. At higher concentrations, homopolymer formation becomes the preferred process. In addition, monomer chain transfer reactions are also accelerated at higher concentrations. If the GR values of the monomer are higher as compared to that of the backbone polymer, homopolymer formation would be a preferred process. Use of excess of water decreased grafting percentage. This is because the excessive hydroxyl radicals produced as a result of radiolysis of water initiate polymerization at the expense of grafting. Also, if the monomer is soluble in water, an increased amount of water would entrap the monomer molecules and would decrease the rate and extent of grafting.

The presence of alcohols in the aqueous medium generally decreases grafting. This is expected since the addition of alcohol breaks the tetrahedral hydrogen bonded structure of water and thus disturbs the association of active sites with water. This will lead to a decrease in grafting. In the presence of alcohols, chain transfer reactions are accelerated leading to wastage of monomer and hence a decrease in grafting is observed.

2. Preirradiation Method

In this method the polymeric substrate (PH) is irradiated in air or oxygen to produce peroxide bonds [PH(O$_2$)] or hydroperoxide bonds [PH—OOH]. Irradiation of hydrocarbon using a low-dose rate usually leads to the formation of hydroperoxide on the polymer backbone. Macrohydroperoxides upon heating decompose giving a hydroxyl radical, which can initiate the graft copolymerization of the added monomer. Polypropylene in particular appears to form a considerable amount of hydroperoxide groups on irradiation in air. The efficiency of the grafting reaction will depend directly on the kinetics of the radiation peroxidation process. The peroxide yield will necessarily depend on both the GR value of the backbone polymer and the stability of the resulting peroxide at the irradiation temperature. The following processes have been suggested for peroxidation reactions:

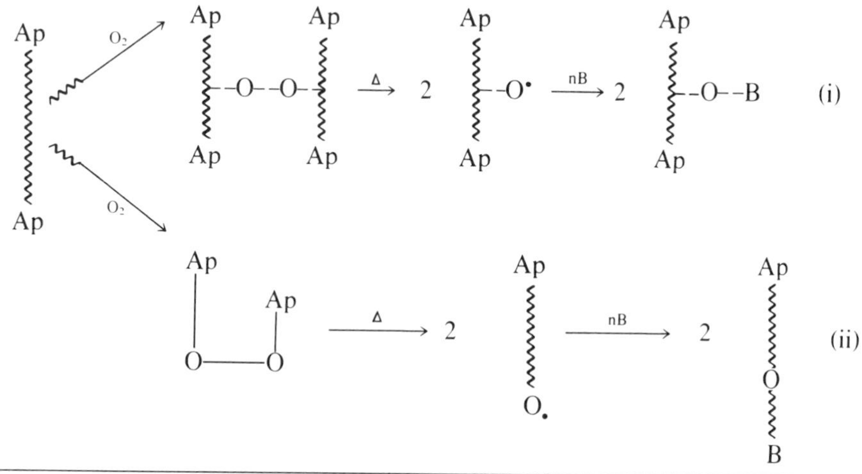

$$(28)$$

Graft or block copolymers are expected to be formed via process (28) (i) and (ii), respectively, depending on whether the polymer Ap is of a crosslinking type or a degrading type. No homopolymerization occurs in this reaction other than by chain transfer to the monomer or by thermal initiation.

For polymer containing labile hydrogen atoms, chain peroxidation can lead to hydroperoxide formation. The following reactions are expected to take place.

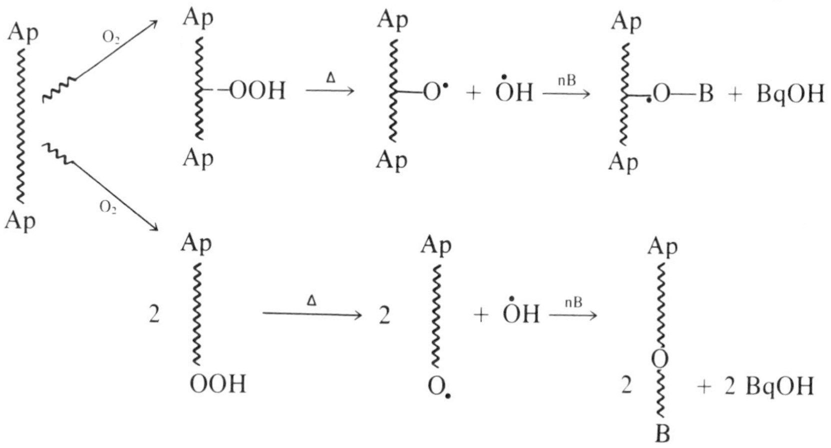

$$(29)$$

Crosslinking and degradation reactions can predominate when the rate of diffusion of oxygen within the polymer is such that it cannot cope with the polymer radical formation. The concentration of polymer peroxides will increase with radiation dose. For low doses the peroxide concentration has been shown to build up linearly until for very high doses radiolysis of the peroxide groups becomes significant and reduction in the apparent overall rate of peroxidation is noticed. This may be the reason that at higher doses, the percentage of grafting was found to decrease after attaining the maximum value at an optimum total dose. The advantage of this method over the mutual method is the avoidance of direct radiation polymerization of the monomer and, hence, a higher percentage of grafting with minimum homopolymer formation is obtained. Misra et al. studied

modification of various polymers by graft copolymerization using the mutual as well as the preirradiation methods [105–109] and found that the preirradiation method produces better results [110].

3. Double Irradiation Method

Another method has recently been utilized by Kaur et al. [85] involving grafting of a vinyl monomer onto pre-irradiated backbone polymer by the mutual method. This method offers certain advantages over the conventional mutual method of grafting.

Irradiation of the polymer in air prior to grafting introduces hydroperoxide groups in the backbone polymer by the intermolecular, intramolecular, or hydrogen abstraction process.

Intermolecular

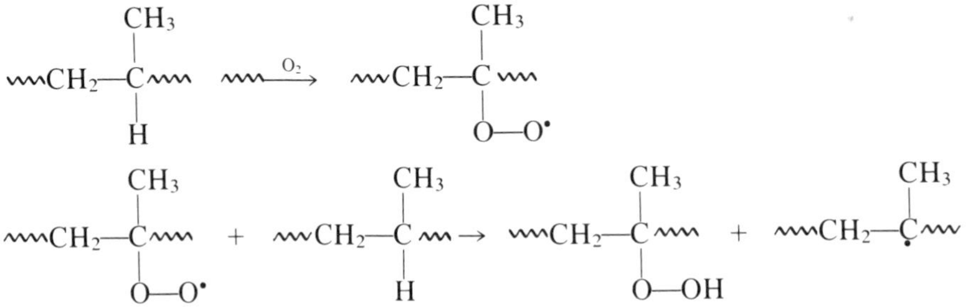

$$(30)$$

Intramolecular

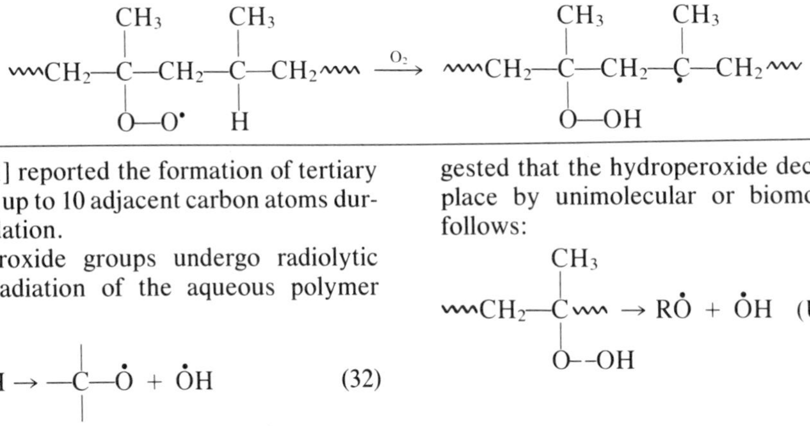

$$(31)$$

Chien et al. [111] reported the formation of tertiary hydroperoxy groups up to 10 adjacent carbon atoms during oxidative degradation.

These hydroperoxide groups undergo radiolytic cleavage during irradiation of the aqueous polymer monomer mixture.

$$(32)$$

Photolytic cleavage of hydroperoxy groups to give macroxy radical has been reported [112–114]. They sug-

gested that the hydroperoxide decomposition may take place by unimolecular or biomolecular reactions as follows:

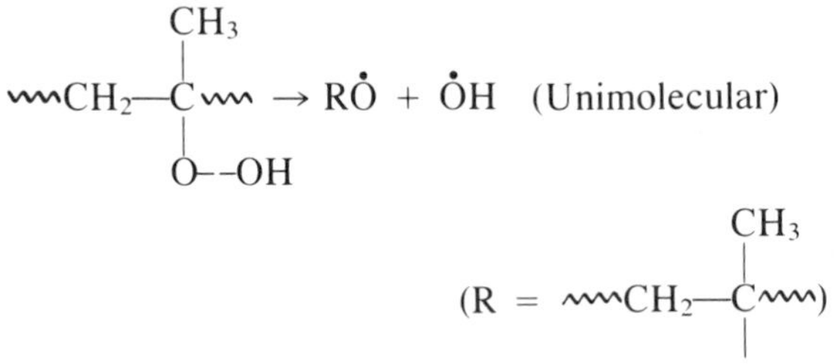

$$(33)$$

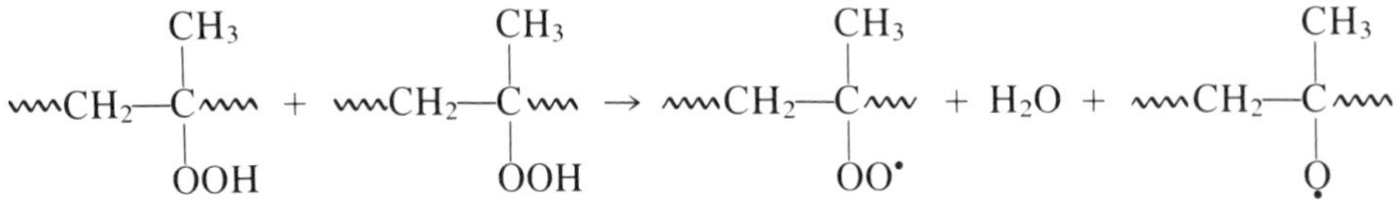

$$(R = \text{wwCH}_2-\overset{CH_3}{\underset{|}{\overset{|}{C}}}\text{www})$$

$$(34)$$

(Bimolecular)

However, reaction (32) is expected to be more facile since it required minimum energy. Gray et al. [115] have found that alkoxy radicals formed by the decomposition

of macrohydroperoxide groups can decompose by β-scission reaction (35).

$$\underset{\substack{\text{O} \\ \text{|} \\ \text{CH}_3}}{\text{wwCH}_2\text{-}\overset{\text{CH}_3}{\underset{\text{O}^{\cdot}}{\text{C}}}\text{-}\overset{\text{H}}{\underset{\text{H}}{\text{C}}}\text{-}\overset{\text{CH}_3}{\underset{\text{H}}{\text{C}}}\text{ww}} \rightarrow \text{wwCH}_2\text{-}\overset{\text{CH}_3}{\underset{\text{O}}{\text{C}}} + \overset{\text{CH}_3}{\dot{\text{C}}\text{H}_2\text{-}\underset{\text{H}}{\text{C}}\text{ww}}$$

(35)

Upon photolysis of polypropylene hydroperoxide (PP—OOH) a major absorption at 1726 and 1718 cm^{-1} has been observed in the IR spectrum, which is attributed to the carbonyl groups. Sometimes the macroradical having free radical site reacts with a neighboring newly born hydroperoxide causing the formation of a macroalkoxy radical [116].

$$\underset{\text{OOH}}{\overset{\text{CH}_3}{\text{wwC}}}\text{-CH}_2\text{-}\underset{\text{OOH}}{\overset{\text{CH}_3}{\text{C}}}\text{-CH}_2\text{-}\overset{\text{CH}_3}{\dot{\text{C}}\text{ww}} \rightarrow \underset{\text{OOH}}{\overset{\text{CH}_3}{\text{wwC}}}\text{-CH}_2\text{-}\underset{\text{O}^{\cdot}}{\overset{\text{CH}_3}{\text{C}}}\text{-CH}_2\text{-}\underset{\text{OH}}{\overset{\text{CH}_3}{\text{Cww}}}$$

(36)

Such a reciprocal motion of the kinetic chain (or back reaction) results in the decomposition of hydroperoxide groups without any interruption of kinetic chains and leads to the decrease in hydroperoxy groups. It has

been suggested by many workers [117,118] that β-scission of tertiary alkoxy radicals is the major source of chain backbone scission during photodegradation of polypropylene.

$$\underset{\text{O}^{\cdot}}{\overset{\text{CH}_3}{\text{wwCH}_2\text{-C}}}\text{-CH}_2\text{-}\overset{\text{CH}_3}{\underset{}{\text{CH}}}\text{-CH}_2\text{ww} \rightarrow \underset{\text{O}}{\overset{\text{CH}_3}{\text{CH}_2\text{-C}}} + \dot{\text{C}}\text{H}_2\text{-}\overset{\text{CH}_3}{\underset{}{\text{CH}}}\text{-CH}_2\text{ww}$$

(37)

$$\underset{\text{O}^{\cdot}}{\overset{\text{CH}_3}{\text{wwCH}_2\text{-C}}}\text{-CH}_2\text{-}\overset{\text{CH}_3}{\underset{}{\text{CH}}}\text{-CH}_2\text{ww} \rightarrow \underset{\text{O}}{\overset{\text{CH}_3}{\text{wwCH}_2\text{-C}}}\text{-CH}_2\text{-}\overset{\text{CH}_3}{\underset{}{\text{CH}}}\text{-CH}_2\text{ww} + \dot{\text{C}}\text{H}_3$$

(38)

The OH radical formed during the decomposition of hydroperoxy groups is very reactive and can either terminate a reaction (39) or may attack the weak bond of the backbone polymer resulting in the formation of a new radical (40).

$$\underset{}{\overset{\text{CH}_3}{\text{wwCH}_2\text{-}\dot{\text{C}}\text{-CH}_2\text{ww}}} + \dot{\text{O}}\text{H} \rightarrow \underset{\text{OH}}{\overset{\text{CH}_3}{\text{wwCH}_2\text{-C}}}\text{-CH}_2\text{ww}$$

(39)

$$\underset{\text{OOH}}{\overset{\text{CH}_3}{\text{wwCH}_2\text{-C}}}\text{-CH}_2\text{ww} + \dot{\text{O}}\text{H} \rightarrow \underset{\text{O-O}^{\cdot}}{\overset{\text{CH}_3}{\text{wwCH}_2\text{-C}}}\text{-CH}_2\text{ww} + \text{H}_2\text{O}$$

(40)

On the basis of the above findings, grafting of vinyl monomers onto irradiated polypropylene has been attempted successfully by the mutual method. Upon irradiation hydroperoxide groups are introduced, which provide sites for grafting. During mutual irradiation in the presence of the monomer in aqueous medium, these hydroperoxide groups and water undergo decomposi-

tion to yield macro hydroperoxy radical and hydroxyl radical. The hydroxyl radical initiates the polymerization of the monomer.

$$\text{M} + \dot{\text{O}}\text{H} \rightarrow \text{OH-}\dot{\text{M}} \overset{n\text{M}}{\longrightarrow} \text{OH} - (\text{M})_n - \dot{\text{M}} \quad (41)$$

The monomer can also directly polymerize onto the macroperoxy radical as follows:

$$\underset{\text{O}^{\cdot}}{\overset{\text{CH}_3}{\text{wwCH}_2\text{-C}}}\text{-CH}_2\text{ww} + n\text{M} \rightarrow \underset{\text{O-M}_{n-1}\text{-M}^{\cdot}}{\overset{\text{CH}_3}{\text{wwCH}_2\text{-C}}}\text{-CH}_2\text{ww}$$

(42)

It was observed that beyond optimum total dose the percentage of grafting decreased. This may be due to the fact that at higher total doses beyond optimum, chain degradation by β-scission (reaction 35, 37, 38) occurs. Further at higher doses, hydroxyl radicals arising from

the radiolysis of the hydroperoxide and water, kills the active sites [39,40]. Homopolymer formation may also become the preferred process at higher total doses.

Use of a large amount of water also affects the extent of grafting. Termination of the growing polymeric chains or killing of the active sites may become facile due to the excessive generation of hydroxyl radicals. Also in the presence of excess of water, a considerable

amount of $H\overset{\bullet}{O}_2$ may be formed that may interact with the macroperoxide radical in the following manner leading to a decrease in the percentage of grafting.

$$\text{wCH}_2\!\!-\!\!\underset{\underset{\displaystyle OO^\bullet}{|}}{\overset{\overset{\displaystyle CH_3}{|}}{C}}\!\!-\!\!CH_2\text{w} + H\overset{\bullet}{O}_2 \rightarrow \text{w}\overset{\bullet}{C}H_2\!\!-\!\!\underset{\underset{\displaystyle OOH}{|}}{\overset{\overset{\displaystyle CH_2}{\|}}{C}} + CH_2\text{w} + O_2 \qquad (43)$$

Baxendale and Thomas [119] and Alexander and Charlesby [120] also observed such β-scission involving the reaction between $H\overset{\bullet}{O}_2$ and $P\overset{\bullet}{O}_2$.

Polyethylene films were preirradiated with gamma-rays and again irradiated in aqueous poly(vinyl alcohol) solutions to give water lubricated antifriction films [121].

D.   Mosaic Grafting

In recent years, mosaic grafting has attracted much attention for the preparation of bipolar membranes. In this process, a polymer film is sandwiched between two lead plates, one of which is perforated. This assembly is placed in the gamma-rays or electron beam source, and grafting of the monomer is allowed to proceed by the mutual method in vacuum or in air. During this process, monomer M_1 will be grafted onto the exposed area of the film. After the grafting of M_1 is over, the grafted film is placed upside down between the perforated lead plates and the grafting is done by the mutual method using monomer M_2. By this method it is possible to prepare a membrane having both the monomers M_1 and M_2 grafted onto the opposite sides of the same film. If M_1 and M_2 are polar monomers, this will lead to the preparation of a bipolar membrane.

Alternatively, the film is placed between two perforated lead plates and grafted with monomer M_1 by the mutual method. After grafting, the film is now placed between the two perforated lead plates taking precautions that only the ungrafted portions of the grafted film are exposed to gamma radiations in the presence of monomer M_2. Schematically, the preparation of the mosaic membrane is shown in Fig. 2.

Preparation of charge mosaic membranes by radiation-induced graft copolymerization of polyethylene (PE) films with styrene and 4-vinyl pyridine was carried out by Chuanyins [122]. Effects of preirradiation dosage (electron beam), atmosphere storage time, and storage temperature of grafting of 4-vinyl pyridine onto PE film in the vapor phase were studied. Bifunctional membranes containing both carboxylic and pyridine groups are obtained by successive grafting of acrylic acid and 4-vinyl pyridine onto thin films of polytetrafluoroethylene (PTFE) [123]. Mosaic membranes containing geometrically well-defined acid and basic zones are made by preirradiation with gamma-rays using two absorbing shields. The equilibrium and transport properties of these monomers have been determined. They can either result from the addition of the properties of each

component in the case of conductance or give rise to a new property (e.g., abnormal negative osmosis). Whatever the structure of the bifunctional membrane (homogeneous or mosaic), this negative osmosis occurs within a pH range in which the membrane potential varies greatly, i.e., when the ionic character of the membrane changes.

E.   Successive Grafting

Another recent method of radiochemical grafting involves grafting of a vinyl monomer M_1, onto already grafted polymeric backbone with vinyl monomer M_2. During this process M_2 may be grafted onto the main chain of the backbone polymer and/or on the grafted polymer chain depending upon the GR values of the backbone polymer as well as grafted polymer. Grafting of MMA vapors was carried out onto preirradiated PE already grafted with styrene by Omichi and Araki [124]. The second step focussed mainly in the PE portion. On the other hand, when styrene vapors was introduced onto PE grafted with MMA, only radicals of poly(MMA) decreased. In this case second step grafting occurred in PMMA portion, which covered the whole surface of PE powder. When monomer vapors were alternatively introduced onto preirradiated PE powder, the second step of grafting occurred at the growing chain end of the first monomer.

F.   Grafting by Trapped Radicals

The presence of trapped radicals has been detected in many irradiated polymers. In some cases radicals can

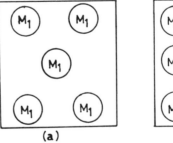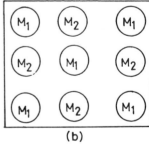

Figure 2   Preparation of mosaic membrane: (a) film grafted with monomer M_1; and (b) M_1 grafted film further grafted with monomer M_2.

remain trapped for extremely long periods (several days or even several months) and as such the reactivity of these radicals is drastically reduced if these radicals are embedded in viscous medium. Free radicals created in irradiated polymers may become even more rigidly trapped when the substance is kept at a temperature below its glass transition point. The reaction medium is then a glass, which is more viscous than a swollen gel. If the irradiated polymer is partly crystalline, the free radicals formed within the crystalline region are still more firmly trapped, since the mobility of polymeric segments is much lower if these are involved in an organized structure. This effect has been particularly studied in the case of PE. It was found that the free radicals that are formed in the amorphous regions can be readily utilized in a number of chemical processes, such as crosslinking [125] or grafting [126], whereas those radicals that are located within the crystalline areas only react at elevated temperatures when most of the crystallites have melted. Oxygen and ethylene can, however, diffuse into the crystalline regions and react with the trapped radicals.

The trapped radicals, most of which are presumably polymeric species, have been used to initiate graft copolymerization [127,128]. For this purpose, the irradiated polymer is brought into contact with a monomer that can diffuse into the polymer and thus reach the trapped radical sites. This reaction is assumed to lead almost exclusively to graft copolymer and to very little homopolymer since it can be conducted at low temperature, thus minimizing thermal initiation and chain transfer processes. Moreover, low-molecular weight radicals, which would initiate homopolymerization, are not expected to remain trapped at ordinary temperatures. Accordingly, irradiation at low temperatures increases the grafting yield [129].

Oxygen is known to destroy trapped radicals, because it converts radicals into peroxidic radicals that can abstract hydrogens from surrounding polymer molecules. Much higher radical yields are indeed found if irradiation is carried out in the absence of oxygen. If grafting is only desired at the surface of the shaped polymer articles, such as films, the yield of the formation of trapped radicals can be increased by irradiating the polymer surface in the presence of a properly selected fluid in which "active" free radicals are formed under irradiation. These radicals (e.g., $\overset{\bullet}{O}H$, $\overset{\bullet}{C}l$, etc.) are then capable of abstracting hydrogens or other atoms from the polymer surface, thus creating polymeric radicals by indirect effect. Grafting experiments along these lines have been carried out with various films, using water and aqueous solutions of H_2O_2 as activating fluids [130,131].

G. Plasma Grafting

Plasma is a state of matter consisting of neutral excited radicals and ionic particles or fragments of molecules and also comprising electrons and photons. If a solid substrate is exposed to a plasma, the action of all these reactive species results in distinct modifications, either in the substrate or in its surface depending on the kind of plasma gas used for modification of polymer materials. Only nonthermal plasmas are generally used. These are generalized by electrical glow discharge under reduced pressure of gas (≈ 5 m. bars) and the application of a high-frequency electrical field. Factors that influence the distribution of polymer deposition in an electrodeless glow-discharge system were investigated for acetylene and ethylene by Yasuda and Hirotsu [132] under the conditions in which full glow is maintained. The distribution of polymer deposition from pure monomer flow systems is nearly independent of flow rate of monomer or of the system pressure in discharge but is largely determined by the characteristic (absolute) polymerization rates (not deposition rates) of the monomer. Acetylene has high tendency to deposit polymer near the monomer inlet, whereas ethylene deposits polymer more uniformly in wider areas in the reaction. The addition of a carrier gas, such as argon, or a partially copolymerizing gas, such as N_2, H_2, and CCl_2F_2, was found to narrow the distribution of polymer deposition. The distribution of polymer deposition is also influenced by a glow characteristic that is dependent on flow rate and discharge power.

Polyethylene films were evaporated in glass plasmas of Ar, N_2, O_2, and H_2O. The deposits were analyzed with IR spectroscopy. The deposits prepared in Ar atmosphere had a rather high concentration of —CH_3 groups and many double bonds were produced in this film. The deposits prepared in Ar and N_2 plasma showed similar spectra, which showed twice the concentration of —CH_3 groups than the deposits in Ar atmosphere and also contained a few carbonyl and hydroxyl groups. The deposits treated in O_2 plasma contained the largest amount of carbonyl groups and the lowest number of double bonds among the plasma treated deposits [133].

Polyethylene powder pretreated with the plasma of electric discharge was evacuated and then exposed to vapors of MMA, acrylic acid (AAC), or methacrylic acid and reevacuated. The amount of grafted polymer was determined [134]. A porous HDPE film treated with a low-temperature plasma was graft copolymerized with acrylonitrile. The grafted film was treated with NH_2OH to give a polymer with low swelling in water [135]. Graft copolymerization was observed for vinyl trimethylsilane in contact with the surface of HDPE and LDPE subjected to glow discharge treatment with H_2O_2 or Ar plasma. The grafting occurred by an anionic mechanism and was initiated by anionic active centers formed on the polymer surface by the glow discharge treatment [136]. Tensile strength loss upon grafting was reduced if the PE fibers were first treated with plasma of air, N_2, Ar, and/or He *in vacuo* and then grafted with vinyl monomers [137]. MMA or EA was grafted onto a PP surface by a low-pressure application of monomer vapors after high-frequency Ar plasma exposure to PP. Even the very small amount of oxygen could stop the grafting, and the discharge in monomer vapor produced mainly the acrylic homopolymer and caused just a slight increase of the copolymer content [138]. Durable membranes with good permeability are prepared by plasma treatment of hydrophobic porous films, grafting with hydrophillic monomers in the absence of plasma, and complexing with complementary polymers [139].

1. Characterization of the Graft Copolymer

Two parameters such as percentage of grafting (%G) and grafting efficiency are usually determined as functions of different variables that influence graft copolymerization. Percent grafting is usually determined from the increase in weight of the backbone polymer after grafting after complete removal of the homopolymer. It is expressed as follows:

$$\%G = \frac{W_2 - W_1}{W_1} \times 100 \qquad (44)$$

where W_2 and W_1 are, respectively, the weights of the grafted polymer after complete removal of the homopolymer and the original polymer.

Grafting efficiency is determined from the total conversion of monomer into homopolymer and graft copolymer and is expressed as:

$$\% \text{ Efficiency} = \frac{B}{A + B} \Big/ C \times 100 \qquad (45)$$

where B = weight of the graft copolymer
     A = weight of the homopolymer
     C = weight of the monomer charged

These two parameters give the quantitative measurement of the grafting reaction. The general characterization of the graft copolymer is based on the following:

1. Spectroscopic measurements including IR, NMR, ESCA, and x-ray diffraction studies.
2. Thermal analysis.
3. Isolation of the grafted polymer.
4. Scanning electron microscopy (SEM).

2. Spectroscopic Measurements

The formation of the graft copolymer can be established by taking the IR spectra of the original backbone polymer and the grafted polymer. The grafted polymer would show the additional peaks due to the pendant groups present in the grafted polymer. Thus, the IR spectra of IPP-g-poly (4-VP) or IPP-g-poly-(MAN) showed peaks at 1580 cm^{-1} and 2940 cm^{-1} assigned to —C≡N and —CH stretching of poly(4-VP) and at 2230 cm^{-1} assigned to —C≡N of poly(MAN) grafted to IPP. The IR spectrum of IPP did not show these peaks [107].

IR analysis of powdered IPP containing acrylic acid grafts with acidic groups esterified with epichlorohydrin revealed that the carbonyl content gradually increased when the esterification is carried out at 90–120°C [140]. IR spectra for AAC-grafted PP fiber cation exchanger in both acidic and salt form was studied that revealed dimerization of the carboxylic groups [141]. The surface of granules of grafted PE were examined by ESCA. Spectral peaks were assigned for pure PE, polyacrylic acid (PAAC) and various grafted samples. Evaluation of the adhesion of grafted PE showed that the attachment of a very small amount of AAC leads to considerable improvement without significant improvement in the quality of insoluble material [142].

The structure of PP-g-polystyrene was studied by wide and narrow angle x-ray diffractometry. The polystyrene component formed discrete amorphous structures, which were identified after chloromethylation [143]. Structure of Rayon–Styrene graft copolymer was characterized by IR, wide angle x-ray, and SEM, grafting occurred mainly in the amorphous regions of rayon [144].

Raman spectroscopy allowing the determination of global variation of the ionization of the membrane as a function of pH was studied. Experimental and theoretical Pk$_a$ value was found to be 5.2 from which the average number of five carboxylic groups/graft was determined [145].

3. Thermal Analysis

Thermal studies of the graft copolymers and the backbone polymer showed a marked difference in their thermal behavior, which suggests that upon grafting the properties of the grafted polymer are changed. Thermogravimetric analysis (TGA), differential scanning calorimetry (DSC), differential thermal analysis (DTA), and differential thermogravimetric analysis (DTG) are different methods of characterizing the graft copolymers. DTA and TG were used to study the thermal stability of PP fibers grafted with acrylic acid for cation exchangers. The thermal stability of cation exchanger in acid form and rare earth ions form was higher than that of cation exchanger in alkali metal ion form [146]. Misra et al. [107] studied the thermal behavior of graft copolymers of IPP, PVA [100], wool [147], and cellulose [44] and found that the thermal stability of the grafted polymer is better than that of the original polymer. However, in the case of gelatin and grafted gelatin, thermal stability of the grafted gelatin was higher, up to a 40% weight loss beyond which the decomposition temperature of the original gelatin is higher than the decomposition temperature of the grafted gelatin. Polyamide-6-g-poly (4-VP) showed poorer thermal stability than the original polyamide-6 fiber [104]. Strength–strain measurement and DSC analysis were carried out to analyze the physical properties of methacrylamide grafted silk fibers [148].

4. Isolation of the Grafted Polymer from Graft Copolymer

Different methods have been used for removing the grafted polymeric chains from the graft copolymer depending upon the nature of the backbone polymer. To isolate the α-methyl styrene grafted chains for molecular weight measurements, the graft copolymers were oxidized by m-chloroperbenzoic acid to cleave the polybutadiene backbone quantitatively [149]. The grafted chains from poly (ethylene oxide)-g-poly (styrene) were separated by immersing the graft copolymer in H_2O_2 solution of $CuCl_2$ at room temperature for 2 d [150]. The same method has been used by Kaur et al. [107] for separating the graft polymers of polypropylene.

The selective degradation of the preformed polymer component (polycarbonate/polystyrene) of graft copolymers by two phase alkaline hydrolysis provides a clean system for quantitative recovery and subsequent characterization of the graft [151].

A procedure for photochemical detachment, followed by isolation and characterization of MMA grafts from glass surfaces is described. The molecular weights and molecular weight distribution can be varied by control of kinetics of grafting. Thermodynamics studies are used to explain the immiscibility of grafts with free homopolymer at high surface coverage by grafts, which is entropy driven [152].

Starch and cellulose graft copolymers have been hydrolyzed using 6N HCl. The grafted polymer, separated as a resinous mass, is separated out while starch and cellulose go into the solution as glucose. This method has been used by Misra et al. [11,44] and many other workers [153] for separating polymer from graft copolymers of starch and cellulose. Grafted wool samples were hydrolyzed with a 1:1 mixture of benzene-sodium hydroxide solution. All wool goes into the solution. The presence of benzene in the hydrolyzing mixture solubilizes the polymer (poly MMA, poly MA, and poly EA) and thus assists in the removal of the polymer from the backbone polymer. The molecular weight of the isolated polymer was determined by the viscosity method [147].

5. Scanning Electron Microscopy

Deposition of the polymer grafted into the preformed polymer backbone can be characterized by scanning electron microscopy (SEM). Misra et al. have studied the scanning electron micrographs (SEM) of wool, grafted wool [71,154], cellulose, grafted cellulose starch [44], and grafted starch [155] and found that a considerable amount of the grafted polymer is deposited on the respective polymer backbones.

6. Use of Graft Copolymers

Generally, a graft copolymer should offer a clear advantage over the physical blend when a high degree of incompatibility exists between the component parts. It is difficult to make a well-dispersed blend of polymers that have poor solubility in one another. Even when this can be done, migration and separation will occur under stress because of poor adhesion and the combination will have low-strength properties. The mutual solubility of components will be of the same order in graft or physical mixture. In a graft copolymer, components are tied by chemical bonds, however, the components will necessarily be in a highly dispersed state and will be prevented from separating or may, in turn, give rise to an organized structure. In general, grafting means addition of side chains to the backbone polymer. Such side chains may be located at the surface or may be deeply penetrating. If the option is such that the grafting does not encompass the far interior of the backbone polymer, it may be envisaged that this will cause little perturbation in the molecular property of the backbone polymer. The properties of grafts will, of course, vary according to the degree of segregation of monomer units, i.e., according to the number and length of the segments. The addition of a side chain will impart its own characteristics to the grafted backbone polymer. On the other hand, if the penetration of the side chains is deeper, major changes in the properties of the graft copolymer will develop. Therefore, some properties may be linked with grafting on the surface only, whereas others may be grouped due to the bulk grafting.

In general, grafting of hydrophillic monomers have been found to lead to an increase in wettability, adhesion, dyeing, and rate of release of oil stains by detergent solution. On the other hand, if the monomer is hydrophobic, the result will be decreased wetting by all liquids including oil stains. If grafting is not restricted to surface alone but encompasses the bulk of the backbone polymer, then the properties such as flame resistance, water sorption, crease resistance, etc. will be affected.

Thus, it is evident that while synthesizing a graft copolymer, one can, according to the requirement, tailor-make the graft copolymer for a particular use or a combination of different uses.

REFERENCES

1. J. Vorliček and P. Kratochvil, *J. Polym. Sci., Polym. Phys. Ed., 11*: 1251 (1973).
2. Y. Ikada and F. Horii, *Makromol. Chem., 175*: 227 (1974).
3. T. Kotaka, N. Donkai, and T. I. Min., *Bull. Inst. Chem. Res. Kyoto Univ., 52*: 332 (1974).
4. A. R. Shultz, *Polym Prepr. Amer. Chem. Soc. Div. Polym. Chem., 20(2)*: 179 (1979).
5. I. Kaur, B. N. Misra, R. Barsola, and K. Singla, *J. Appl. Polym. Sci., 47*: 1165 (1993).
6. I. Kaur, B. N. Misra, S. Chauhan, M. S. Chauhan, and A. Gupta, *J. Appl. Polym. Sci. 59*: 389 (1996).
7. I. Kaur, S. Maheshwari, A. Gupta, and B. N. Misra, *J. Appl. Polym. Sci. 58*: 835 (1995).
8. R. Hayes, *J. Polym. Sci. 11*: 531 (1953).

9. A. A. Berlin, L. V. Stupin, B. I. Fedoseyera, and D. M. Yanosky, *Dokl. Akad. Nauk USSR, 121*: 644 (1958).
10. G. Smets, J. Roovers, and W. Van Humbeek, *J. Appl. Polym. Sci., 5*: 149 (1961).
11. B. N. Misra, R. Dogra, I. Kaur, and D. Sood, *J. Polym. Sci. Polym. Chem. Ed., 18*: 341 (1980).
12. J. L. Williams and V. Stannett, *J. Polym. Sci., B-8*: 711 (1970).
13. V. Tokashi, A. Toshihiro, I. Yoshikuni, I. Toshiyuki, and H. Naoshi, *Jpn. Kokai Tokkyo Koho JP* 63, 191, 817 (88, 191, 817), (Cl 08 F 255/02); 09 Aug. 1988, Appl. 87/23, 552, 05 Feb. 1987; 12 pp.
14. M. Kazuo and K. Hiroyuki, *Jpn. Kokai Tokkyo Koho Jp* 63, 238, 149 (88, 238, 149) (Cl C 08 L51/06); 04 Oct 1988, Appl. 87/70, 494, 26 Mar 1987; 7 pp.
15. P. Bastin, G. DeLaunois, J. Polart, and J. J. Versluijs, *Belg. BE* 905, 048, (ClCO8F), 03 Nov. 1986, Appl. 2, 168, 04 Jul. 1986; 16 pp.
16. A. Simmons and W. E. Baker, *Polym. Eng. Sci., 29(16)*: 1117 (1989).
17. A. Romanov, I. Novak, L. M. Dutaj, *Dieter Czech CS,* 227, 606, 1987; 7 pp.
18. V. F. Zheltobryukhov; USSR SU 1, 381, 123 (Cl C 08 G69/48), 15 Mar. 1988 (101), 98.
19. Suren Co. Ltd. Jpn. Kokai Tokkyo Koho JP 58, 191, 279 (83, 191, 279) (Cl DO6 M17/OO) 08 Nov. 1983, Appl. 82/71, 288, 30 Apr. 1982; 4 pp.
20. M. Zhang, Y. Cai, *Suliao, 17(2)*: 23 (1988).
21. W. E. Baker and M. Saleem, *Polymer, 28(12)*: 2057 (1957).
22. C. A. Strait, G. M. Lancaster, R. L. Tabor, U. S. US4, 762, 890 (Cl. 525–527; C 08F 255/02), 09 Aug. 1988, Appl. 905, 099, 05 Sept. 1986; 5 pp.
23. B. N. Misra, R. Dogra, I. Kaur, and D. Sood, *Ind. J. Chem., 17A*: 390 (1979).
24. B. N. Misra and J. Kaul, *Ind. J. Chem., 21A*: 922 (1982).
25. B. N. Misra and J. Kaul, *Ind. J. Chem., 22A*: 601 (1983).
26. R. K. Sharma and B. N. Misra, *J. Macromol. Sci. Chem. A20(2)*: 225 (1983).
27. J. O. Turner, *Tetrahedron Lett.*, 887 (1971).
28. P. S. Chandel and B. N. Misra, *J. Polym. Sci. Polym. Chem. Ed., 15*, 1549 (1977).
29. S. Kimihiro, K. Iichiro, and N. Kiyoji, *Kobunshi Kagaku, 29(10)*: (1973).
30. L. Hecker de Carvalho and A. Rudin, *J. Appl. Polym. Sci., 29(9)*: 2921 (1984).
31. L. P. Krul, Yu. I. Matusevich, L. Yu. Brazhnikova, N. P. Prokopchuk, *Vysokomol, Soedin, Ser. B, 30(9)*: 695 (1988).
32. Hitachi Cable Ltd. Jpn. Kokai Tokkyo Koho JP 6037, 604, (85, 37, 604) (Cl. HOL B 3/44), 27 Feb 1985, Appl. 83/144, 677, 08 Aug. 1983; 3 pp.
33. A. Dimov and I. Aleksandrova, *God. Vissh Khim. Tekhno. Inst. Prof. d-r. As Zlataror, gr. Burgas 1985, 20(2)*: 55 (1986).
34. Hitachi Cable Ltd., Jpn. Kokai Tokkyo Koho JP 5838, 710 (83 38, 710) (Cl.C08 F289/00), 07 Mar 1983, 81/138, 021, 02 Sep. 1981; 4 pp.
35. E. Borsig, J. Masler, and M. Lazar, *IUPAC, 28th Macromol, Symp.*, July 1982, pp. 597.
36. S. R. Palit and T. Guha, *J. Polym. Sci., A-I*: 1877 (1963).
37. S. Okamura and T. Motoyama, *J. Polym. Sci., 58*: 221 (1962).
38. D. S. Verma and R. K. Sarkar, *Angew Makromol Chem., 37*: 167 (1974).
39. P. D. Bertlett and J. D. Colman, *J. Amer. Chem. Soc., 71*: 1419 (1949).
40. C. E. H. Bawn and D. Margarison, *Trans Faraday Soc., 51*: 925 (1955).
41. S. Lenka, *J. Appl. Polym. Sci., 27(6)*: 2295 (1982).
42. Bogoeva Gatseva, G. A. Gabrielyan, and L. S. Gal'braikh, *Khim Volokna, 25(4)*: 260 (1988).
43. B. N. Misra, R. Dogra, I. K. Mehta, and A. S. Singha, *Die Ang. Makromol. Chemie, 90*: 83 (1980).
44. B. N. Misra, I. K. Mehta, and R. C. Khetarpal, *J. Polym. Sci. Polym. Chem. Ed., 22*: 2767 (91984).
45. B. N. Misra, R. Dogra, I. K. Mehta, and Kiran Dip Gill, *J. Appl. Polym. Sci. 26*: 3789 (1981).
46. S. S. Hariharan and A. Meenakshi, *J. Polym. Sci. Polym. Lett., 15(1)*: 1 (1977).
47. S. S. Hariharan and A. Meenakshi, *Curr. Sci., 46(20)*: 708 (1977).
48. P. L. Nayak, S. Lenka, and M. K. Mishra, *J. Appl. Polym. Sci., 25(1)*: 63 (1980).
49. P. L. Nayak, S. Lenka, and M. K. Mishra, *J. Polym. Sci., 18(7)*: 2247 (1980).
50. S. Bhusan Dash, A. K. Pradhan, N. C. Pati, and P. L. Nayak, *J. Macromol. Sci. Chem., A19(3)*: 343 (1983).
51. G. Mino and S. Kaizerman, *J. Polym. Sci., 31*: 242 (1958).
52. B. N. Misra, I. K. Mehta, and R. Dogra, *J. Appl. Polym. Sci., 25*: 235 (1980).
53. B. N. Misra, D. S. Sood, and I. K. Mehta, *J. Macromol. Sci. Chem., A18(2)*: 209 (1982).
54. B. N. Misra and I. K. Mehta, *J. Macromol Sci. Chem., A15*: 457 (1981).
55. R. C. Khetarpal, K. D. Gill, I. K. Mehta, and B. N. Misra, *J. Macromol Sci. Chem. A18(3)*: 445 (1982).
56. B. N. Misra and R. C. Khetarpal, *J. Polym. Matr., 1*: 7 (1984).
57. B. N. Misra, I. K. Mehta, and R. K. Sharma, *Polym. Bull. 3*: 115 (1980).
58. B. K. Patel, V. K. Sinha, and H. C. Trivedi, *J. Polym. Matr., 8*: 321 (1991).
59. B. N. Misra, I. K. Mehta, and R. K. Sharma, *Polym. Bull., 4*: 635 (1981).
60. B. N. Misra and I. Kaur, *J. Polym. Sci., Polym. Chem. Ed., 18*: 1911 (1980).
61. B. N. Misra and P. S. Chandel, *J. Polym. Sci., Polym. Chem. Ed., 18*: 1171 (1980).
62. R. C. Khetarpal, J. Kaul, I. K. Mehta, and B. N. Misra, *Ind. J. Chem., 23A*: 983 (1984).
63. B. K. Patel, V. K. Sinha, C. P. Patel, and H. C. Trivedi, *Starke, 45(5)*: 178 (1993).
64. B. N. Misra, I. K. Mehta, M. P. S. Rathore, and S. Lakhanpal, *J. Appl. Polym. Sci., 49*: 1979 (1993).
65. X. D. Feng, Y. H. Sun, and K. Y. Qui, *Macromolecules, 18(11)*: 2105 (1985).
66. B. N. Misra, J. K. Jassal, and R. Dogra, *J. Macromol Sci. Chem., A16(6)*: 1093 (1981).
67. B. N. Misra, J. K. Jassal, R. Dogra, and D. S. Sood, *J. Macromol. Sci. Chem., A14(7)*: 1061 (1980).
68. B. N. Misra, J. K. Jassaal, and R. Dogra, *J. Macromol Sci. Chem., A16(6)*: 1093 (1981).
69. D. S. Sood, A. S. Singha, and B. N. Misra, *J. Macromol. Sci. Chem. A20(2)*: 237 (1983).
70. B. N. Misra and B. R. Rawat, *J. Macromol. Sci. Chem. A21(4)*: 495 (1984).
71. D. S. Sood and B. N. Misra, *J. Macromol. Sci. Chem., A21(10)*: 1267 (1984).

72. D. S. Sood, B. R. Rawat, and B. N. Misra, *J. Appl. Polym. Sci., 30*: 135 (1985).

73. B. N. Misra, I. K. Mehta, and D. S. Sood, *J. Macromol Sci. Chem., A14(8)*: 1255 (1980).

74. B. N. Misra, D. S. Sood, and R. K. Sharma, *Die Ang. Makromol. Chem., 59*: 102 (1982).

75. A. K. Tripathy, S. Lenka, and N. C. Pati, *J. Polym. Sci., Polym. Chem. Ed., 17*: 3405 (1979).

76. P. L. Nayak, S. Lenka, and N. C. Pati, *J. Polym. Sci., Polym. Chem. Ed., 17*: 3405 (1979).

77. A. I. Medalia and I. M. Kolthoff, *J. Polym. Sci. 4*: 377 (1949).

78. F. Haber and J. Weiss, *Naturwiss,* (a) *20*: 948 (1932), (b) Proc. R. Soc. A.147, 332 (1934).

79. J. H. Baxendale, M. G. Evans, and G. S. Park, *Trans. Faraday Soc., 42*: 155 (1946).

80. B. N. Misra, P. S. Chandel, and R. Dogra, *J. Polym. Sci. Polym. Chem. Ed., 16*: 1801 (1978).

81. B. N. Misra, R. Dogra, I. Kaur, and J. K. Jassal, *J. Polym. Sci. Polym. Chem. Ed., 17*: 1861 (1979).

82. B. N. Misra, R. Dogra, and I. K. Mehta, *J. Polym. Sci. Polym. Chem. Ed., 18*: 749 (1980).

83. S. B. Shah, C. P. Patel, and H. C. Trivedi, *J. Appl. Polym. Sci., 51*: 1421 (1994).

84. S. B. Shah, C. P. Patel, and H. C. Trivedi, *J. Appl. Polym. Sci., 52*: 857 (1994).

85. I. Kaur, B. N. Misra, and R. Barsola, *J. Appl. Polym. Sci., 48*: 575 (1993).

86. A. Chaprio, M. Magat, and J. Sebban, French Patent 1,130,099 (1956) to Centre National de La Recherche Scientifique.

87. Societe des Usines Chimiques Rhone–Poulenc. French Patent 1,181,893 (1959).

88. F. Fazilat and S. H. Rostamie, *J. Macromol. Sci. Chem., A13(8)*: 1203 (1979).

89. Y. Schigeno, K. Konda, K. Takemoto, *J. Macromol. Sci. Chem., A17(4)*: 571 (1982).

90. C. H. Ange, J. L. Garnett, R. Levot, A. Mervyn Long, *J. Appl. Polym. Sci., 27(12)*: 4893 (1982).

91. N. H. Rao and K. N. Rao, *Radiat. Phys. Chem., 26(6)*: 669 (1989).

92. J. Dobo, *Makromol. Chem. Makromol. Symp., 28*: 107, (1989).

93. L. N. Grusheukaya, R. E. Alieu, V. Ya. Kabanov, *Vysokomol Soedin, Ser., A31(7)*: 1398 (1989).

94. B. N. Misra, I. Kaur, B. Kapoor, and S. Lakhanpal, *Ind. J. Fibre Textile Res., 17*: 107 (1992).

95. R. K. Sharma and B. N. Misra, *Polym. Bull., 6*: 183 (1981).

96. B. N. Misra and B. R. Rawat, *J. Polym. Sci. Polym. Chem. Ed., 23*: 307 (1985).

97. B. N. Misra, B. R. Rawat, and G. S. Chauhan, *J. Appl. Polym. Sci., 42*: 3223 (1991).

98. I. Kaur, R. Barsola, A. Gupta, and B. N. Misra, *J. Appl. Polym. Sci., 54*: 1131 (1994).

99. I. Kaur, B. N. Misra, S. Chauhan, M. S. Chauhan, and A. Gupta, *J. Appl. Polym. Sci. 59*: 389 (1996).

100. B. N. Misra, J. Kishore, M. Kanthwal, and I. K. Mehta, *J. Polym. Sci., Polym. Chem. Ed., 24*: 2209 (1986).

101. B. N. Misra, I. K. Mehta, M. Kanthwal, and S. Panjloo, *J. Polym. Sci. Polym. Chem. Ed., 25*: 2117 (1987).

102. I. Kaur, S. Maheshwari, and B. N. Misra, *J. Appl. Polym. Sci. 58*: 835 (1995).

103. I. Kaur, S. Kumar and B. N. Misra, "Grafting of 2-VP and styrene onto isotactic polypropylene by mutual method" (unpublished result).

104. I. Kaur, B. N. Misra, and R. Barsola, *Ange. Makromol. Chem., 234*: 1–12 (Nr. 3786) (1996).

105. B. N. Misra, S. K. Gupta, and G. S. Chauhan, *Polym. Appl. Sci. Engg., 61*: 900 (1989).

106. I. Kaur, G. S. Chauhan, and B. N. Misra, *J. Appl. Polym. Sci., 54*: 1171 (1994).

107. I. Kaur, R. Barsola, and B. N. Misra, *J. Appl. Polym. Sci., 51*: 329 (1994).

108. I. Kaur, R. Barsola, and B. N. Misra, *Polym. Preprints, 32*: 109 (1990).

109. B. N. Misra, G. S. Chauhan, and I. Kaur, *Collection of Czech Chem. Common 61*: 259 (1996).

110. I. Kaur, B. N. Misra, and S. Kumar, *J. Appl. Polym. Sci.* (In Press).

111. J. C. W. Chien, E. J. Vandenberg, and H. Jobloner, *J. Polym. Sci., A-1(6)*: 381 (1968).

112. D. J. Carlsson and D. M. Wiles, *J. Macromol. Sci. Rev., Macromol. Chem., C-14(i)*: 65 (1976).

113. N. V. Zolotova and E. T. Denisov, *J. Polym. Sci., A-1(9)*: 3311 (1971).

114. D. Emanuel and M. Emanuel, *Liquid Phase Oxidation of Hydrocarbons,* Plenum Press, NY (1967).

115. P. Gray, R. Shaw, and J. C. J. Thynne, *Progr. Reaction Kinetics, 4*: 63 (1967).

116. S. Kiryushkin and Yu. Shlyapnikov, *Dokl. Akad. Nauk, USSR, 220*: 1364 (1975).

117. D. J. Carlsson, F. R. S. Clark, and D. M. Wiles, *Text. Res. J., 46(8)*: 590 (1976).

118. P. Blais, D. J. Carlsson, F. R. S. Clark, P. Sturgean, and D. M. Wiles, *Text Res. J., 46(9)*: 641 (1976).

119. J. H. Baxendale and J. K. Thomas, *Trans Faraday Soc., 54*: 1515 (1958).

120. P. Alexander and A. Charlesby, *J. Polym. Sci., 23*: 355 (1957).

121. Idemitsu Petrochemical Co. Ltd., Jpn. Kokai Tokkyo Koho JP 5840, 323 (8340, 323) (cl. C 08 G81/00), 09 Mar 1983, Appl. 81/137, 709, 03 Sep. 1981; 6 pp.

122. D. Chuanyin, *Desalination, 62*: 275 (1987).

123. A. M. Jendry Chowska-Bonamour and J. Millequant, *Eur. Polym. J., 16*: 39 (1980).

124. H. Omichi and K. Araki, *J. Polym. Sci., Polym. Chem. Ed., 15*: 1833 (1977).

125. E. J. Lawton, J. S. Balwit, and R. S. Powell, *J. Polym. Sci., 32(257)*: 277 (1958).

126. D. Ballantine, A. Glines, G. Adler, and D. J. Metz., *J. Polym. Sci., 34*: 419 (1959).

127. L. A. Wall and D. W. Brown, *J. Research Natle. Bur. Standards, 57*: 131 (1956).

128. J. C. Bevington and D. E. Eaves, *Nature, 178*: 1112 (1956).

129. E. T. Cline and D. Tanner, Belg. Patent 549,388, French Patent 1, 155,579 (1957); Brit. Patent 801,531 (1958) to E. I. Dupont de Nemours and Co.

130. Le Clair, Belg. Patent 549,388, French Patent 1,155,579 (1957); Brit. Patent 801,531, (1958) to E. I. Dupont de Nemours and Co.

131. N. G. Gaylord, Belg. Patent 549, 387 (1957) to E. I. DuPont De Nomours and Co.

132. H. Yasuda and T. Hirotsu, *J. Polym. Sci., 16*: 229 (1978).

133. M. Ashida, E. Ikada, Y. Ueda, and H. Aizawa, *J. Polym. Sci., Polym. Chem. Ed., 20*: 3107. (1982).

134. A. N. Ponomarev, *Teplo–Massoobmen Plazmokhim Protsessakh. Mater. Mezhdunar. Shk. Semin., 1*: 154 (1982).

135. Kan Nakajima, Koichi Kono, Kenji Miyasaka, Masato

Komatsu, Shigeo Fujii, Joichi Tabuchi, Jpn. Kokai Tokkyo Koho JP 61, 157, 344 (86, 157, 344) Cl B01 J.20/26),
17 Jul. 1986, Appl. 84/274, 686, 28 Dec. 1984; 5 pp.

136. A. A. Kalachev, S. Yu. Lobanov, M. V. Shorokhova,
N. A. Plate, *Dokl. Akad. Nauk SSSR, 302*(2): 338 (1988).

137. Taku Tokita, Hajime Inagaki, Jpn. Kokai Tokkyo Koho
JP 63, 203, 882 (88, 203, 882) (Cl. DO 6p 5/20), 23 Aug.
1988 Appl. 86/176, 720, 29 Jul. 1986; 5 pp.

138. M. Mastihuba, J. Blecha, V. Kamenicka, L. Lapcik, I.
Topolsky, *Acta Phys. Solvaca, 35*: 355 (1985).

139. Kanebo Ltd. Kanebo Synthetic Fibers Ltd. Jpn. Kokai
Tokkyo Koho JP 59, 160, 505 (84, 160, 505) (Cl BO 1 D
13100), 11 Sep. 1984, Appl. 83/34, 848, 02 Mar. 1983, 9
pp.

140. P. Citovicky, J. Majer, E. Staudner, V. Chrastova, J.
Mejzlik, *Eur. Polym. J., 21(1)*: 89 (1985).

141. I. N. Ermolenko, K. F. Paraskeovova, L. G. Vlasov,
R. B. Virnik, *Zh. Prikl. Spektrosk., 44(2)*: 312 (1986).

142. J. Villoutreix, P. Nogues, R. Berlot, *Eur. Polym. J.,
22(2)*: 147 (1986).

143. P. O. Nizovtseva, A. A. Shunkevich, G. Yu. Zonov.,
P. L. Krul, S. V. Soldatov, *Vysokomol, Soedin Ser. A.,
24(8)*: 1582 (1982).

144. H. Zhou, W. Yu, T. Sun, *Fangzhi Xuebao*, 6(1): 13
(1985).

145. J. L. Bribes, J. Maillors, M. E. I. Baukari, and R.
Gaufres, *C. R. Acad. Sci., Ser. 2, 308(14)*: 1199
(1989).

146. I. N. Ermolenko, K. F. Paraskeovova, L. G. Vlasov, N.
V. Kachurina, and R. B. Virnik, *Veskski Akad. Navuk,
BSSR, Ser. Khim Navuk, 5*: 68 (1985).

147. B. N. Misra, I. K. Mehta, and D. S. Sood, *J. Appl.
Polym. Sci., 34*: 167 (1987).

148. M. Tsukada, A. Aoki, *Nippon Sanshigaku Zasshi, 54(5)*:
354 (1985).

149. R. J. Ambrose and J. J. Newell, *J. Polym. Sci., Polym.
Chem. Ed., 17*: 2129 (1979).

150. J. Omichi and K. Araki, *J. Polym. Sci., Polym. Chem.
Ed., 17*: 1401 (1979).

151. C. G. Eastmond and L. W. Harvey, *Br. Polym. J., 17(3)*:
275 (1985).

152. C. G. Eastmost, G. Mucciariello, *Br. Polym. J., 16(2)*:
63 (1984).

153. G. F. Fanta, R. C. Burr, C. R. Russell, and C. E. Rist.
J. Appl. Polym. Sci., 14: 2601 (1970).

154. D. S. Sood, J. Kishore, and B. N. Misra, *J. Macromol.
Sci. Chem., A22(3)*: 263 (1985).

155. B. N. Misra and R. Dogra, *J. Macromol. Sci. Chem.,
A14(5)*: 763 (1980).

32

Methods, Characterization, and Applications of Grafting

Elsayed Mohamed Abdel-Bary
Mansoura University, Mansoura, Egypt

Eman Mohamed El-Nesr
National Center for Radiation Research and Technology, Cairo, Egypt

I. INTRODUCTION

Polymers are classified according to their chemical structures into homopolymers, copolymers, block copolymers, and graft copolymers. In a graft copolymer, sequences of one monomer are grafted onto a backbone of the other monomer and can be represented as follows:

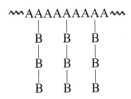

where A and B are two different monomers. The properties of a graft copolymer differ from those of either a random copolymer ~ABABBAB~ or a physical mixture consisting of two homopolymers from (A)n and (B)n.

The relation between structure and properties of macromolecules interested many researchers who were devoted to developing new materials capable of competing with natural polymers and possessing new characteristics. In this regard, chemical modifications have been devised to impart certain desirable properties of both natural and synthetic macromolecules that would lead to new applications. Various chemical modifications such as change of functionality, inter- and intramolecular gelation, and graft copolymerization, have been tried to add improved properties to the base polymers. However, modification of polymers using graft copolymer-

ization created increasing interest due to the varieties of both monomers and substrates participating in such a process. In addition, the wide field of applications makes such a modification method important for industrial and biomedical applications.

Grafting reactions alter the physical and mechanical properties of the polymer used as a substrate. Grafting differs from normal chemical modification (e.g., functionalization of polymers) in the possibility of tailoring material properties to a specific end use. For example, cellulose derivatization improves various properties of the original cellulose, but these derivatives cannot compete with many of the petrochemically derived synthetic polymers. Thus, in order to provide a better market position for cellulose derivatives, there is little doubt that further chemical modification is required. Accordingly, grafting of vinyl monomers onto cellulose or cellulose derivatives may improve the intrinsic properties of these polymers.

Graft copolymerization of vinyl monomers onto polymeric materials, including cellulose and cellulose derivatives, has been the subject of extensive studies for about four decades [1,2]. In spite of the huge number of published papers and patents and the interesting results obtained, there has been comparatively little commercialization of the grafting process. The reasons for the lack of industrialization on a large scale have been partly economic. Among the technical problems that still remain to a considerable extent are the concurrent formation of homopolymers in most cases and the lack of reproducibility in these largely heterogeneous reactions.

In addition, there is the difficulty of controlling the grafted side chains in molecular weight distribution. In the context of this chapter, the authors hope to give a systematic, comprehensive, up-to-date review on the various methods of grafting, the methods used for characterization of grafted substrates, and their possible end use applications.

There are a considerable number of methods available for effecting graft copolymerization onto preformed polymers, each with its own particular advantages and disadvantages.

Graft copolymerization is effected, generally, through an initiation reaction involving an attack by a macroradical on the monomer to be grafted. The generation of the macroradical is accomplished by different means such as: (1) a decomposition of a weak bond or the liberation of the unstable group present in side groups in the chemical structure of the polymer, (2) chain-transfer reactions, (3) redox reaction, (4) photochemical initiation, and (5) gamma radiation-induced copolymerization.

In the following sections, we give some examples of each of these methods as it is very difficult to account for all methods and all techniques. For chemical grafting using different initiating systems, cellulose was used as an example as it is the most extensively investigated substrate. This holds true for photo-induced grafting of cellulose and cellulose derivatives. Grafting using gamma radiation is concentrated on polyolefins and

some vinyl polymers and elastomers, which are usually difficult to graft by chemical means without prior chemical modification of the substrate.

II. GRAFTING OF CELLULOSE USING CHEMICAL METHODS

A. Decomposition of Azo Groups

The presence of an azo group as a side group in the polymer structure can be used to produce a macroradical via its decomposition under the effect of metallic ions [3,4]. This macroradical reacts with vinyl monomers leading to grafting with a minimum amount of homopolymers. This reaction was first applied to a polymer by Chapman et al. [5]. The reaction proceeds as follows: (See structure below.)

The NH_2 groups can be diazotized and reduced in the presence of thiosulphates and different metal ions. The effect of some metal ions, namely Fe^{2+}, Sn^{2+}, Cu^{1+}, and Co^{2+} on the graft yield of cotton modified with aryl diazonium groups via its reaction with 2,4-dichloro-6-(p-nitroaniline)-s-triazine in the presence of alkali and followed by reduction of nitro group was studied [4].

B. Chain Transfer Initiation

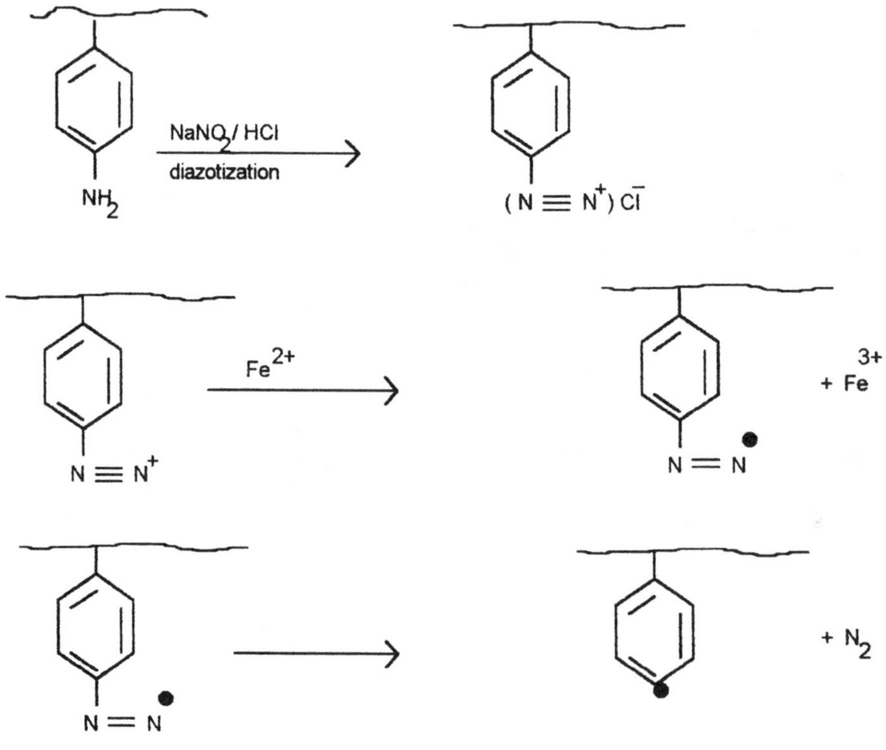

Chain-transfer reactions take place during vinyl polymerization involving abstraction of an atom such as hydrogen or halogen by the growing chains of M units. Polymerization can take place at these newly formed reactive sites leading to the formation of a branched polymer. This mode of synthesis was first suggested by Adkin et al. [6] and Flory [7] between 1935 and 1937.

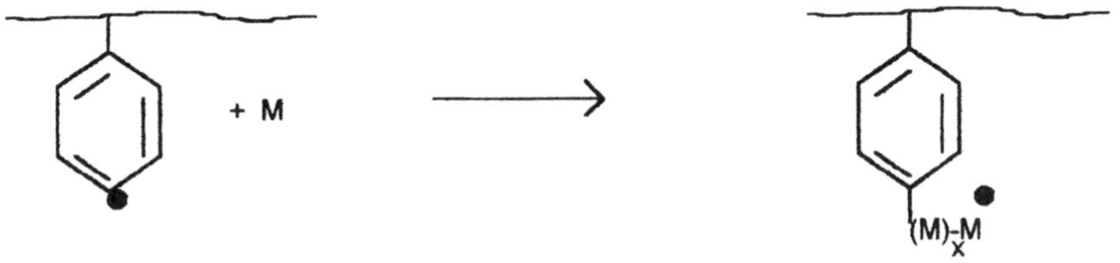

The general steps are as follows:

$$R^{\cdot} + \text{\small ⋀⋀}PH\text{\small ⋀⋀} \rightarrow RH + \text{\small ⋀⋀}P^{\cdot}\text{\small ⋀⋀}$$

$$\text{\small ⋀⋀}P^{\cdot}\text{\small ⋀⋀} + CH_2{=}CH(X) \rightarrow \text{\small ⋀⋀}P\text{\small ⋀⋀}$$
$$\overset{|}{CH_2{-}CH(X)}$$

$$\underset{X}{\overset{\text{\small ⋀⋀}P\text{\small ⋀⋀}}{\overset{|}{CH_2{-}CH}}} + nCH_2{=}\underset{X}{\overset{|}{CH}} \rightarrow \underset{X}{(\overset{\text{\small ⋀⋀}P\text{\small ⋀⋀}}{\overset{|}{CH_2{-}CH}})_{n+1}} \text{ branched Polym.}$$

R may be a radical formed by the decomposition of an initiator or a growing radical chain. Similarly, grafting by the chain-transfer mechanism occurs when the branched part consists of another monomer. Since cellulose is a poor transfer agent [8], the efficiency of grafting is quite poor. Incorporation of —SH groups into cellulose enhances the probability of chain transfer. This can be achieved as follows:

$$CellOH + \overset{\diagdown\diagup}{\underset{S}{CH_2{-}CH_2}} \rightarrow Cell{-}O{-}CH_2CH_2SH$$

$$Cell{-}O{-}CH_2{-}CH_2SH + R^{\cdot}$$
$$\rightarrow Cell\ O{-}CH_2{-}CH_2\ S^{\cdot} + RH$$
$$Cell{-}O{-}CH_2{-}CH_2{-}S^{\cdot} + M$$
$$\rightarrow Cell{-}O{-}CH_2{-}CH_2{-}S\ M^{\cdot}$$

The hydrogen abstraction from —SH groups is faster than from —OH groups. Hebeish et al. [9] and Misra et al. [10,11] reported the chain-transfer method of initiation of graft copolymerization onto cellulosic substrates with azobisisobutyronitrile (AIBN) and benzoyl peroxide (BPO) as initiators.

C. Redox Method

1. Ceric Ion Initiation

The use of ceric ions to initiate graft polymerization was first discussed by Mino and Kaizerman in 1958 [12]. Schwab and coworkers [13] were among the first to extend this method to the grafting of cellulose. Following their work, numerous papers have appeared in the literature on the grafting of vinyl monomers onto cellulose by this technique.

Mino and Kaizerman [12] established that certain ceric salts such as the nitrate and sulphate form very effective redox systems in the presence of organic reducing agents such as alcohols, thiols, glycols, aldehyde, and amines. Duke and coworkers [14,15] suggested the formation of an intermediate complex between the substrate and ceric ion, which subsequently is disproportionate to a free radical species. Evidence of complex formation between Ce(IV) and cellulose has been studied by several investigators [16–19]. Using alcohol the reaction can be written as follows:

$$Ce(IV) + RCH_2OH \rightleftarrows [\text{Ceric - alc. complex}]$$
$$CeIII + H^{+} + R{-}CH{-}OH$$
$$\text{or } [R{-}CH_2O.]$$

If polyvinyl alcohol is used as the reducing agent and the oxidation is conducted in the presence of vinyl monomer, grafting occurs. This method of grafting yields substantially pure graft copolymers since the free radicals are formed exclusively on the polymer backbone.

By using this technique acrylamide, acrylonitrile, and methyl acrylate were grafted onto cellulose [20]. In this case, oxidative depolymerization of cellulose also occurs and could yield short-lived intermediates [21]. They [21] reported an electron spin resonance spectroscopy study of the affects of different parameters on the rates of formation and decay of free radicals in microcrystalline cellulose and in purified fibrous cotton cellulose. From the results they obtained, they suggested that ceric ions form a chelate with the cellulose molecule, possibly, through the C_2 and C_3 hydroxyls of the anhydroglucose unit. Transfer of electrons from the cellulose molecule to Ce(IV) would follow, leading to its reduction

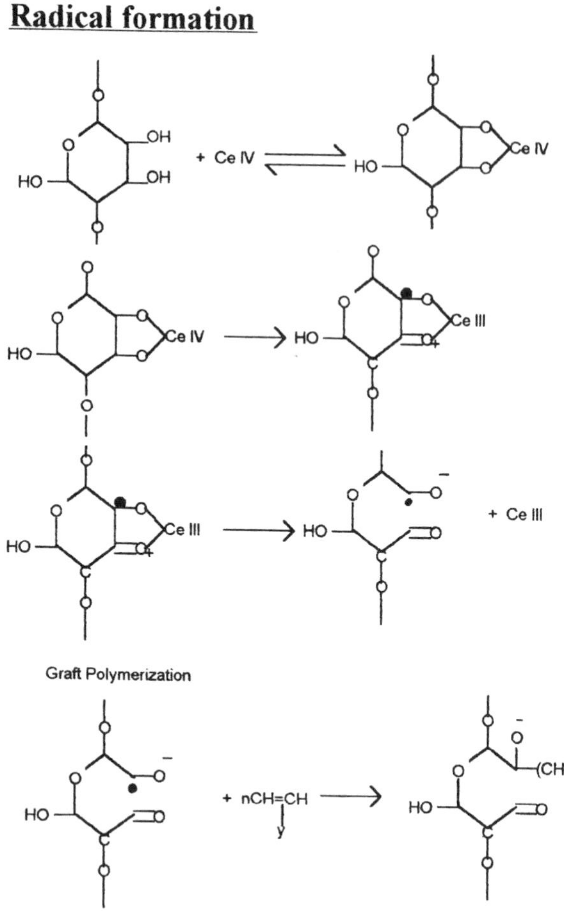

Radical formation

Graft Polymerization

to Ce(III), breakage of the C_2—C_3 bond, and formation of a free radical site.

Grafting of acrylic acid and methacrylic acid onto cellulose is very interesting due to the possibility of developing enhanced ion exchange and water sorbency properties. Unfortunately, it is difficult to realize such grafting due to their sensitivity, which leads to homopolymerization and irreproducibility. In view of the wide application of acrylic acid grafted fibers, ceric ion-initiated grafting of acrylic acid onto cellulosic fibers has been reported by Richards and White [22]. Gagneux et al. [23] developed a method for grafting acrylic acid onto cellulose powder, solka floc, for use in textile waste treatment. The cellulose was treated with ceric ion in aqueous solution prior to its reaction with acrylic acid. A benzene-acrylic acid solution was used for grafting to reduce homopolymer formation. They obtained grafting yields up to approximately 70% accompanied by 45% homopolymer.

Graft copolymerization of acrylonitrile with various vinyl comonomers such as methyl acrylate, ethyl acrylate, vinyl acetate, and styrene onto cellulose derivatives using ceric ion was studied [24]. The results showed that

for polymerization in aqueous medium, total conversion was greatly increased in the order.

methyl acrylate > ethyl acrylate > vinyl acetate > styrene.

This is the same order of the solubility of the monomer in water. The increase in acrylonitrile concentration increases the percent grafting and grafting efficiency.

Another binary mixture, namely, (acrylic acid/ acrylamide) was grafted onto carboxy methyl cellulose (CMC) [25].

Stannett et al. [26] used the ceric ion method to graft acrylic acid onto rayon filament. They used a special design to optimize grafting yields and minimize homopolymer formation. Thus, they pretreated the rayon filaments with an aqueous solution of ceric ammonium sulphate followed by washing to remove the excess of ceric ion at the fiber surface. The grafting reaction was carried out in toluene to reduce homopolymer formation. Several solvents were used to reduce the concentration of ceric ammonium sulphate and the aqueous medium at the fiber surface. The most suitable rinsing agent was found to be a mixture of 10:90 methanol:toluene by volume whereby high grafting yield resulted.

Attempts have been also made to use ceric ion initiation for grafting vinyl monomers onto lignocellulosic fibers. Lin et al. [27] grafted MMA and AN onto bambo.

2. Vanadium V(V) Systems

Oxidation of organic substances such as alcohols, acids, aldehydes, sugars, and mercaptans by using vanadium V(V) has been reported by Waters and Littler [28]. Most such reactions were found to proceed through a free radical path that effectively initiates vinyl polymerization. Macromolecules containing pendant groups such as (OH, CHO, NH_2, SH, and —OOH) interact with V(V) to form free radicals on the backbone. The macroradicals attack monomers in the immediate vicinity, resulting in the formation of a graft copolymer. Such grafting has been reported by Ikada et al. [29] in the reaction of polyvinyl alcohol with V(V). Also, the hydroxy group of the anhydroglucose unit in polysaccharides reacts with V(V) to produce macrocellulosic radicals and, consequently, grafting occurs in the presence of vinyl monomers. Morin et al. [30] reported that V(V) was the most efficient metal ion compared to Cr(IV) and Fe(III) where a high rate of grafting was achieved involving negligible homopolymer formation. Nayak et al. [31] observed that an increase of V(V) concentration to <0.0025 M increases the graft yield of MMA onto cellulose. Beyond this value, the graft yield decreases. Besides, an increase of acid concentration had an adverse effect on graft yield. As in the case of Ce(IV), the grafting proceeds via the formation of an intermediate complex between the substrate and vanadium ion, which is subsequently disproportionate to a free radical species.

3. Mn(III) Initiating System

Mn(III) is able to oxidize many organic substrates via the free radical mechanism [32]. The free radical species, generated during oxidation smoothly initiate vinyl polymerization [33–35]. Mn(III) interacts also with polymeric substrates to form effective systems leading to the formation of free radicals. These radicals are able to initiate vinyl polymerization and, consequently, grafting in the presence of vinyl monomers.

The pyrophosphate complex of Mn(II) ion was found to initiate grafting of acrylonitrile onto cellulose and its derivatives [36,37].

4. Co(III) Initiating System

As in the case of ceric and vanadium ions, the reaction of organic compounds with Co(III) proceeds via formation of an intermediate complex. Such a complex decomposes and produces free radicals capable of initiating vinyl polymerization. However, only a few reports on Co(III) ion-initiated grafting onto cellulose fibers are available [38].

5. Metal Complex System

It has been known for more than 30 years that certain metal chelates with —O—O— donor atoms in the ligand molecule yield free radicals on thermal decomposition in both aqueous and nonaqueous media [39] with polymeric ligands. The net reaction results in the formation of graft copolymers. This type of grafting has been reviewed by Samal [40]. Metal complex-initiated graft copolymerization was first observed by Shirai et al. [41]. Mishra et al. [42] reported graft copolymerization of MMAs onto cellulose using Fe(III)-acetyl acetonate, Al(III)-acetyl acetonate, and Zn(II)-acetyl acetonate complexes. The efficiency of these complexes can be arranged as follows:

$$Fe(II) > Al(III) > Zn(II) \text{ complexes.}$$

D. Persulphate Systems

In aqueous solutions the persulphate ion is known as a strong oxidizing agent, either alone or with activators. Thus, it has been extensively used as the initiator of vinyl polymerization [43–47]. However, only later, Kulkarni et al. [48] reported the graft copolymerization of AN onto cellulose using the $Na_2S_2O_3/K_2S_2O_8$ redox system.

Gaylord et al. [49] reported the dilution and matrix effects in grafting of the styrene/AN binary mixture onto cellulose with $K_2S_2O_8$ as the initiator. Titledman and coworkers [50] reported the effect of hydroxypropylmethyl cellulose on the course of $(NH_4)_2S_2O_8$ decomposition and claimed a route for grafting of vinyl monomers onto the polymer backbone. The decomposition of the peroxo salt, under the catalytic influence of the polymer, has been studied at variable reaction conditions.

The reaction proceeds by a free radical mechanism. The mechanism involves the decomposition of persulphate ions to sulphate radical ions either when heated alone or in the presence of a reducing agent as described below:

a) $S_2O_8^{2-} \rightleftarrows 2\ ^-SO_4^{\cdot}$

b) $S_2O_8^{2-} + Fe(II) \rightarrow Fe(III) + \ ^-SO_4^{\cdot} + \ ^-SO_4^{\cdot}$

$\overline{SO_4^{\cdot}} + H_2O \rightarrow \ ^{\cdot}OH + HSO_4^{\ -}$

Generation of macroradicals

a) $Cell—H + \ ^-SO_4^{\cdot} \rightarrow Cell^{\cdot} + HSO_4^{\ -}$

b) $Cell—H + \ ^{\cdot}OH \rightarrow Cell^{\cdot} + H_2O$

c) $Cell—H + Fe(III) \rightarrow Cell^{\cdot} + H^+ + Fe(II)$

Initiation

$Cell^{\cdot} + M \rightarrow Cell—M^{\cdot}$

$R^{\cdot} + M \rightarrow RM^{\cdot}\ (R = \ ^{\cdot}OH, \ ^-SO_4^{\cdot})$

Propagation

$Cell\ M^{\cdot} + (n - 1)\ M \rightarrow Cell\ M_n$

$RM^{\cdot} + (m - 1)\ M \rightarrow RM_m$

Termination

$Cell\ M^{\cdot}_n + Fe(III) \rightarrow$ Graft copolymer

$Cell\ M^{\cdot}_n + Cell\ M^{\cdot}_n \rightarrow$ Crosslinked graft copolymer

$RM^{\cdot}_m + Fe(III) \rightarrow$ Homopolymer

Oxidation termination

$Cell^{\cdot} + Fe(III) \rightarrow$ Oxidation products

$Cell^{\cdot} + \ ^-SO_4^{\cdot} \rightarrow$ Dehydrocellulose + $HSO_4^{\ -}$

E. Fenton's Reagent (Fe(II)/H₂O₂) System

The use of Fenton's reagent (H_2O_2) as an effective system for grafting vinyl monomers onto preformed macromolecules was reported for the first time by Rogovin and coworkers [51–53]. Rogovin [53] and Ogiwara et al. [54] suggested that the reaction of hydrogen peroxide and Fe^{2+} is a typical redox reaction.

$$HO—OH + Fe(II) \rightarrow Fe(III) + \ ^{\cdot}OH + \ ^-OH$$

The OH radicals may initiate grafting via abstraction of the hydrogen atom from cellulose or may also initiate homopolymerization.

Grafting of methyl methacrylate on cellulose containing sulphonic acid groups [55] and carboxy methyl groups [56] using the Fe^{2+}/hydrogen peroxide redox system was investigated. Also, the rates of vinyl graft copolymerization of AN and α-methyl styrene onto partially carboxymethylated cotton having different concentrations of carboxyl groups were studied [57]. In this case, the Fe^{2+}—H_2O_2 redox system was used for initiation. The presence of sulphonic and carboxylic groups enables the iron ions to be in the vicinity of the cellulose backbone chain. In this case, the radicals formed can easily attack the cellulose chain leading to the formation of a cellulose macroradical. Grafting of methyl methacrylate on tertiary aminized cotton using the bi-sulphite-hydrogen peroxide redox system was also investigated [58].

The effect of Fe(II) on grafting of 2-hydroxyethyl methacrylate onto polyester fibers in the presence of benzoyl peroxide was investigated [59]. It was found that increasing the iron ion concentration decreases the graft yield. This suggest that excess Fe(II) ions participate in the generation of free radical species and the iron ions seem to contribute to the termination and, consequently, decrease the graft yield.

Thus, the reaction is not specific for initiation of grafting. Another disadvantage is that Fe(II) ions formed—if not carefully removed—cause discoloration of the resulting product. The addition of Fe(II) sulphoxylate is claimed to increase the rate of grafting, the yield of grafted polymer, and the conversion of monomer to polymer [60,61]. The mechanism of grafting can be represented as follows:

Cell—H + ˙OH → Cell˙ + H_2O

Cell˙ + M → Cell M˙

OH + M → HO M˙

Cell M˙ + (n − 1) M → Cell M˙$_n$

HOM˙ + (m − 1) M˙ → HO M˙$_m$

Cell M˙$_n$ + Fe(III) → Graft copolymer

M˙$_m$ + Fe(III) → Homopolymer

Cell + ˙OH → Oxidation product

F. Hydroperoxide and Peroxy Acid Redox Systems

Peracetic acid formed in situ from acetic acid and hydrogen peroxide in acid aqueous solution was found to initiate graft copolymerization of vinyl monomers [62,63]. In the presence of Fe^{2+} and Fe^{3+} with peracetic acid, an optimum concentration was found to reach a high conversion of MMA, acrylamide, and acrylic acid, and a high yield of grafted cellulose [64]. The addition of a complexing agent, such as ethylenediamine tetraacetate (EDTA) to the Fe(II) ions makes it possible to operate the grafting process at high pH. The following reaction mechanism was suggested:

AcOH + H_2O_2 → AcOOH + H_2O

AcOOH → AcO˙ + ˙OH

AcOOH + Fe^{2+} → AcO˙ + ˉOH + Fe^{3+}

Cell H + AcO˙ → Cell˙ + AcOH

Cell˙ + nM → Cell M˙$_n$

In this case, two kinds of free radicals are formed leading to the formation of homopolymer and graft copolymer. The latter is due to the formation of cellulosic macroradicals.

III. XANTHATE METHOD

The xanthate method [62] is considered as one of the most promising methods for industrial chemical modification. The principal involved in the xanthate method of grafting is that cellulosic xanthate either ferrated or in acidic conditions reacts with hydrogen peroxide to produce macroradicals. The following reaction mechanism has been proposed:

a) Cell—CH_2—O—C(S)—S—Fe $\xrightarrow[-Fe(OH)_2]{H_2O_2}$ Cell—CH_2—O—C(S)—S˙

Cell—O—C(S)—S˙ + n CH_2=CH(X) → Cell—CH_2—O—C(S)—(CH_2CH˙X)n

b) Cell—CH_2—O—C(S)—S˙ $\xrightarrow{-CS_2}$ Cell—CH_2—O˙ or Cell—˙CH—OH

The reaction of any one of these two radicals with vinyl monomer leads to the formation of a grafted copolymer.

IV. PHOTO-INDUCED METHOD

The principal of photo-induced grafting is the absorption of electromagnetic radiation in the visible and ultraviolet region. In this case, the molecule is said to be in an excited state. This requires that the polymer or the molecule should contain chromophoric groups and should absorb the radiation in this region. The energy-rich molecule can either dissociate into reactive free radicals or dissipate this energy by fluorescence, phosphorescence, or collisional deactivation. The dissociation of polymer energy-rich molecules into reactive free radicals can be used to initiate block and graft copolymerization. This process can be promoted by the addition of photosensitizers. Such compounds may decompose easily into reactive free radicals or transfer their energy to other molecules in the system and, consequently, promote the graft copolymerization reaction. The photosensitizers may exist freely in the reaction mixture or as pendant groups on the polymeric backbone. In the latter system they undergo facile homolysis to free radicals as reported by Norrish et al. [65, 66].

The photo-induced process of modification of cellulose and its derivatives was reported by Geacintov and coworkers [67,68]. Thus, acrylonitrile, vinyl acetate, styrene, MMA, and the binary system of styrene and AN were grafted onto cellulose and cellulose derivatives. In this case, photosensitive dyes, such as the disodium salt of anthraquinone (2,7-disulphonic acid), were used. Many reports are available on free radical generation in cellulose and photochemical grafting of various monomers in the presence of a large variety of initiators onto cellulose and cellulose derivatives [69–71].

Photo-induced grafting onto wood cellulose with several vinyl monomers using phenyl acetophenone and benzophenone derivatives as photosensitizers has been reported [72]. A fast increase in grafting was observed for short irradiation times, whereas longer ones resulted in a decreased percent grafting.

Hon and Chan [73] examined the photo-induced free radical-generated cellulose and cellulose derivatives (methyl cellulose, ethyl cellulose, acetyl cellulose, hydroxyethyl cellulose, and carboxymethyl cellulose) using ESR spectroscopy. They reported that a high concentration of free radicals was generated in all samples irradiated with ultraviolet of wavelength $\gamma > 254$ and 280 nm. Several types of free radicals were generated in cellulose due to chain scission, dehydrogenation, and decarboxymethylation reactions, whereas free radicals generated in cellulose derivatives were, by and large, due to the cleavage of the substituted side chains. Accordingly, grafting reactions of cellulose took place at the main backbone, whereas grafting reactions of cellulose derivatives took place at the substituted side chains. A higher degree of grafting is always achieved in a homogeneous medium than in a heterogeneous medium.

The mechanism that explains the free radical generation on the cellulose derivatives can be represented as follows:

$$\text{Cell—OCH}_3 \xrightarrow{h\nu} \text{Cell—O}^\bullet + {}^\bullet\text{CH}_3$$

$$\text{Cell—O—CH}_2\text{—CH}_3 \xrightarrow{h\nu} \text{Cell—O}^\bullet + {}^\bullet\text{CH}_2\text{—CH}_3$$

The reaction between Cell—O radical and vinyl monomers leads to the formation of grafted cellulose. In the presence of photosensitizers generally used as photoinitiators, such as benzophenone and phenylacetophenone derivatives, the photoinitiator absorbs the UV radiation and transforms to its singlet (S*) and then triplet (T*) states. After that it may decompose into free radicals or transfer its energy to cellulose or any other molecules in the system. Take benzophenone as an example:

$$\text{Ph—}\overset{\overset{\displaystyle O}{\|}}{\text{C}}\text{—ph} \xrightarrow{h\nu} S^* \longrightarrow T^*$$

$$T^* + \text{CellOH} \rightarrow \text{Ph—}\underset{\underset{\displaystyle OH}{|}}{\overset{+}{\text{C}}}\text{—Ph} + \text{Cell}^\bullet$$

$$\text{Cell}^\bullet + M \rightarrow \text{graft}$$

If 2,2′-dimethyoxy-2-phenyl acetophenone was used as the photoinitiator, the exited molecule decomposes into a free radical as follows:

$$\text{Ph—}\underset{\underset{\displaystyle OMe}{|}}{\overset{\overset{\displaystyle O \quad OMe}{\| \quad |}}{\text{C—C}}}\text{—Ph} \rightarrow S^* \rightarrow T^* \rightarrow \text{Ph—}\overset{\overset{\displaystyle O}{\|}}{\text{C}}{}^\bullet + {}^\bullet\underset{\underset{\displaystyle OMe}{|}}{\overset{\overset{\displaystyle OMe}{|}}{\text{C}}}\text{—Ph}$$

The formed radicals may abstract hydrogen from cellulose or a cellulose derivative leading to the formation of grafted cellulose or may initiate homopolymer formation.

Grafting of MMA onto dialdehyde cellulose proceeded more easily than onto mono-carbonyl cellulose [74]. In this case, the grafting reactions start with the photolysis of carbonyl groups.

The mechanistic details of photo-induced grafting of styrene onto cellulose in the presence of mineral acids with methanol and dioxane as solvents was investigated [75]. It was found that acid enhances grafting and homopolymer formation. Analysis of homopolymers shows that acid reduces the chain length but increases the number of grafted chains.

Akira and coworkers [76] reported photo-induced grafting onto cellulose in the presence of powerful oxidizing agents such as Ce(IV) and NaIO$_4$ and free radicals generating aliphatic azo compounds (AIBN). The authors reported that grafting was initiated by radicals via redox reactions between Ce(IV) and OH groups in cellulose, the grafting rate was affected by UV radiation. Photolysis of carbonyl groups formed as a result of oxidation in the presence of NaIO$_4$ enhances the grafting. In all cases the degree of grafting and the apparent number of grafted chains increased linearly with an increase in the concentration of the chemical oxidants and the time of irradiation. Photo-induced homogeneous copolymerization of ethyl acrylate onto cellulose acetate using a series of photosensitizer combinations, such as aliphatic amines [77] and benzophenone and AIBN, benzoin and bromine-pyridine system [78], was studied. In addition, the kinetics of the grafting of ethyl cellulose with isobutyl methacrylate in homogeneous medium was studied using the conventional photosensitizers [79]. It was found that benzoin is more efficient than biacetal, benzil, and phenyl acetophenone, respectively.

V. GAMMA RADIATION-INDUCED GRAFTING

The growth in popularity of radiation as the initiating system for grafting arises from the improvement in availability and cost of ionizing radiation. This is due to the introduction of more powerful nuclear reactors. Apart from its inexpensiveness, radiation is a very convenient method for graft initiation because it allows a considerable degree of control to be exercised over such structural factors as the number and length of the grafted chains by careful selection of the dose and dose rate. Thus, the advantages of radiation-chemical methods are: (1) the ease of preparation as compared to conventional chemical methods, (2) the general applicability to a wide range of polymer combinations (due to the relatively unselective absorption of radiation in matter), and (3) the more efficient (and, thus, more economical) energy transfer provided by radiation compared to chemical methods requiring heat.

A. Mechanism of Gamma Radiation Grafting

The theory of radiation-induced grafting has received extensive treatment. The direct effect of ionizing radiation in material is to produce active radical sites. A material's sensitivity to radiation ionization is reflected in its G value, which represents the number of radicals in a specific type (e.g., peroxy or allyl) produced in the material per 100 eV of energy absorbed. For example, the G value of poly(vinyl chloride) is 10–15, of PE is 6–8, and of polystyrene is 1.5–3. Regarding monomers, the G value of methyl methacrylate is 11.5, of acrylonitrile is 5.6, and of styrene is >0.69.

During gamma radiation of polymer/monomer systems, numerous radiation chemical processes involving macromolecules could, in principle, lead to the production of graft copolymers. The typical steps involved in free radical polymerization are also applicable to graft copolymerization including initiation, propagation, and chain transfer. However, the complicating role of diffusion prevents any simple correlation of individual rate constants to the overall reaction rate. Among the various methods of radiation grafting, four have received special attention; these include: (1) the direct radiation of grafting of a vinyl monomer onto a polymer, (2) grafting on radiation-peroxidized polymer, (3) grafting initiated by trapped radicals, and (4) the intercrosslinking of two different polymers.

B. Direct Method

In this method of grafting, the trunk polymer (Am) is irradiated while in contact with monomer B, existing either as a liquid, in a solution or vapor state. In this case, irradiation leads to the formation of active sites in the polymer (Am) or on the monomer (B). Polymer radicals either directly formed or via transfer reaction with the growing homopolymer radicals are directly used to initiate the graft copolymerization process. The resulting product is either a block or a graft copolymer depending on whether Am is a degrading polymer or can be crosslinked under the effect of radiation [80].

This can be explained as follows:

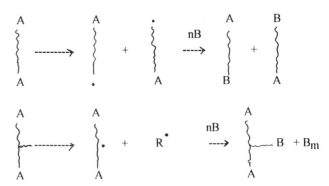

R is a low-molecular weight radical or hydrogen atom. A and B are monomers. The homopolymer $(B)_m$ arises from initiation by small radicals R and also by radiolysis of the monomer B.

$$B \rightarrow R^\cdot \xrightarrow{nB} R(B)_m$$

The rate of free radical production from Am and B are dependent on the G value of both:

$$Ri\ Am = G(Am)\ [I]\ [Am]$$
$$Ri\ B = G(B)\ [I]\ [B]$$

where Ri is the rate of initiation of the reaction, G(Am) and G(B) are the radiation chemical yields for free radical production of polymer (Am) and monomer (B), respectively, and [I] is the dose rate.

From the basic kinetics of radiation grafting, one of the most important properties in determining the efficiency of the process is the radiation sensitivity of the polymer, i.e., G(Am) relative to the monomer G(B) [81]. Thus, if G(Am) ≫ G(B), grafting is favored. The mutual technique works well if the reaction yields a little homopolymer. A good example is the polyethylene/styrene system. If the contrary is the case, mainly homopolymer is formed and the polymer substrate remains largely unaffected. A good example for this case is the PE/acrylic acid system where mainly polyacrylic acid is formed. However, methods have been worked out where a polymerization inhibitor is added to the monomer solution, thus greatly reducing the homopolymer formation [82].

If a solvent or swelling agent needs to be added, these should be chosen carefully, otherwise solvent radicals may predominate leading again to initiate homopolymerization. Dose effects are also important with the mutual method. If high-dose rates are used, the growing chains will rapidly terminate and graft yield and graft efficiency may result. For this reason, gamma sources such as gamma ^{60}Co or ^{137}Cs are preferred. Solvent or swelling agents are normally added to increase monomer diffusion and enhance the efficiency and uniformity of the grafting. This can be achieved also with the vapor phase technique. However, nonsolvents have also been used to restrict the grafting close to the polymer substrate surfaces. This could be useful, for example, to improve adhesion or biocompatibility [83].

C. Preirradiation Method

In this method, the backbone polymer (Am) is irradiated in the absence of oxygen prior to exposure to the monomer (B), which may be in a vapor or liquid form. Irradiation of the polymer produces relatively stable trapped radicals. Graft copolymers resulted when the diffused monomers react with these active sites. The amount of formed homopolymers is very little by using this technique. However, the success of this method depends largely on the crystallinity of the substrate. Also, it depends on the relative rates of reaction of monomers with trapped radicals as well as the thermal decay of radicals at the higher temperature required for grafting. Thus, the grafting efficiency decreases with increasing the decay rate of the radicals. Accordingly, it is advisable to conduct the grafting by irradiating the polymer at a low temperature and to allow the monomers to diffuse at the same temperature to prevent radical–radical combination on warming. If the polymer is semi-crystalline or below the glass transition temperature, the radical formed by radiation will remain trapped and able to react with the monomer to form a grafted side chain. One advantage of this method is that an electron beam accelerator can be used for the irradiation step and the irradiated substrate can be reacted with the monomer in a continuous operation. This is valid for all vinyl monomers.

One disadvantage of this method, is that, unlike the mutual technique, there is no protection of the polymer by the monomer or swelling agent. This can lead to degradation, crosslinking, or other changes in the polymer. With most grafting systems, however, this has not proved to be a major drawback. A second disadvantage is that the process is sensitive to the presence of air. This can be turned into a plus factor if the resulting peroxides are also used to initiate grafting [84].

D. Peroxidation Method

Instead of using the polymer directly for grafting, it is possible to modify them by introducing a new function group via irradiation in air or oxygen. In this case, stable diperoxide or hydroperoxides are formed, depending on the nature of the polymer and the irradiation conditions. These peroxy polymers can be removed from the radiation source and stored at room or lower temperature until ready for reaction. The reaction takes place in contact with monomer (B) in solvent and, if necessary, in air or vapor at an elevated temperature to form a graft copolymer. Catalysts can be added to accelerate free radical formation from the peroxy polymers and thus enhance grafting efficiency.

The advantage of this technique is that the intermediate peroxy polymers may be kept for long periods of time before performing the final grafting step. The grafting reactions using this technique can be described as follows:

a) Formation of diperoxide

$$PH + O_2 \xrightarrow{\gamma\text{-rays}} P—OO—P \xrightarrow{heat} 2PO^{\bullet}$$
$$PO^{\bullet} + n(B) \longrightarrow PO—(B)_n$$

if diperoxide formed at the chain ends, block copolymers are formed.

b) formation of hydroperoxide

$$PH + O_2 \xrightarrow{\gamma\text{-rays}} POOH \xrightarrow{heat} PO^{\bullet} + {}^{\bullet}OH$$
$$PO^{\bullet} + nB \longrightarrow PO(B)_x$$
$$HO^{\bullet} + n(B) \longrightarrow HO(B)_y$$
$$x + y = n$$

Similarly, if the hydroperoxide is formed at the end group, a block copolymer is obtained.

Experimental evidence is available to show that at room temperature diperoxides are formed in polyethylene, whereas polypropylene and polyvinyl chloride generate hydroperoxides. The temperature at which the peroxides decompose and initiate grafting depends on the type of polymer used.

E. Factors Affecting Radiation Grafting

There are a number of important factors that must be considered before applying gamma radiation-induced grafting. These factors include the radiation sensitivity of the polymer/monomer system, radiation dose and dose rate, type and concentration of inhibitor, type of solvent or diluent, and monomer concentration. The effect of such parameters on the grafting efficiency during mutual grafting is given below.

1. Radiation Sensitivity of the Polymer/ Monomer System

Ionizing radiation is unselective and has its effect on the monomer, the polymer, the solvent, and any other substances present in the system. The radiation sensitivity of a substrate is measured in terms of its G value or free radical yield G(R). Since radiation-induced grafting proceeds by generation of free radicals on the polymer as well as on the monomer, the highest graft yield is obtained when the free radical yield for the polymer is much greater than that for the monomer. Hence, the free radical yield plays an important role in grafting process [85].

2. Dose and Dose Rate Effects

The dose and dose rate of irradiation are important factors in any radiation grafting system. In the direct method the total dose determines the number of grafting sites, while the dose rate determines the number of grafting sites, and the dose rate determines the length of the grafted branches. The length of the branches is also controlled by other factors, such as the presence of chain-transfer reaction, the concentration of the monomer, the reaction temperature, the viscosity of the reacting medium, diffusion phenomena, etc. Diffusion of the monomer plays an important role in the direct radiation method. It is the way by which the monomer reaches the active sites within the polymer. It would be expected that the rate of graft polymerization would be directly proportional to the radiation dose rate. In some cases, however, the diffusion of monomer cannot satisfy the increased rate of initiation within the polymer [86,87].

3. Effect of Temperature

The effect of temperature on the kinetics of the direct radiation method is quite complex. Increase in temperature increases the monomer diffusion rate but also increases transfer and termination reaction rates of the growing chains and reduces the importance of the gel effect. Solubility and radical mobility may also change as the temperature is varied [88,89].

4. Effect of Inhibitors

During mutual graft copolymerization, homopolymerization always occurs. This is one of the most important problems associated with this technique. When this technique is applied to radiation-sensitive monomers such as acrylic acid, methacrylic acid, polyfunctional acrylates, and their esters, homopolymer is formed more rapidly than the graft. With the low-molecular weight acrylate esters, particularly ethyl acrylate, the homopolymer problem is evidenced not so much by high yields as by erratic and irreproducible grafting.

To avoid homopolymer formation, it is necessary to ensure true molecular contact between the monomer and the polymer. Even if this is initially established, it needs to be maintained during the radiation treatment while the monomer is undergoing conversion. Several methods are used for minimizing the homopolymer formation. These include the addition of metal cations, such as Cu(II) and Fe(II). However, by this metal ion technique, both grafting and homopolymerization are suppressed to a great extent, thus permitting reasonable yield of graft with little homopolymer contamination by the proper selection of the optimum concentration of the inhibitor [83,90,91].

5. Effect of Solvent and Monomer Concentration

The correct choice of solvents is essential to the success of radiation-induced graft copolymerization. Their influence on radiation grafting has been the subject of many studies. It has been established that solvents play an important role in grafting because of the significance of polymer swelling. The grafting patterns to PP are solvent-dependent, thus, it is essential to examine the role of solvents. It was found that grafting in the alcohol is better than when other polar solvents such as dimethyl-

Table 1

Substrate	Monomer	References
PE	Acrylamide	96–98
	Acrylic acid	99–102
	AN	104, 105
	Styrene	92–94, 103, 105–109
PP	Acrylamide	110, 111
	AN	112
	Acrylic acid	113
	Methacrylic acid	114
	Acrylates	115
	Styrene	106, 107
PVC	Styrene	116
	HEMA, NVP	117
	4-Vinyl pyridine	118

sulphoxide, dimethylformamide, acetone, or dioxane are used. The other important feature is the presence of an accelerated grafting effect in the alcohol, in dimethylsulphoxide, and dimethylformamide but not in acetone and dioxane. The accelerating effect of methanol observed in the radiation grafting of styrene to polyethylene has been attributed to the Thromsdorff effect [92–94] similar to that found in the radiation grafting of styrene to cellulose [95]. In the case of cellulose, the wetting and swelling properties of the solvents as well as their radiation chemical characterization were shown to be important in decreasing gel effects found with grafting of this trunk polymer. The first two properties are necessary to permit access of monomer to cellulose, since only hydrophilic solvents such as methanol, ethanol, and propanol were found to be useful for grafting, whereas butanol was not.

In Table 1 a collection of some general polymer/monomer systems in radiation grafting is given. These references are only representatives as the number of references in this area is very huge.

VI. CHARACTERIZATION

A. Gravimetric Method

The graft products are usually characterized by different methods. The first method is the calculation of graft parameters known as the grafting percentage (GP), grafting efficiency (GE), and weight conversion percentage (WC). These parameters can be calculated as follows:

$$GP = \frac{A - B}{B} \times 100$$

$$GE = \frac{A - B}{C} \times 100$$

$$WC\% = \frac{A}{B} \times 100$$

where A, B, and C are the weights of the extracted graft product, substrate, and monomer, respectively. This gravimetric method gives a direct and rapid indication about the graft reaction.

Other characterization methods are usually used to detect the changes in physical properties, which usually result from the changes in the morphology and structures of the substrates due to grafting.

B. Infrared

Proof of grafting has generally been indicated by the changes in the infrared spectra. The IR spectra of grafted cellulose has been examined by several investigators. This is valid for many other substrates. For example, the IR spectra of acrylonitrile grafted jute did not show a carbonyl peak that was present in the physical mixture, indicating that grafting had occurred at this point [119]. Kulkarni and Mehta [120], while examining the IR spectra of acrylonitrile grafted chains after removal of the cellulose backbone, found additional peak characteristics of a —O—CO— ester group indicating the presence of cellulose fragments.

The amount of acrylonitrile grafting onto PE powder using an electron beam was determined from the absorbance of a nitrile group at 2240 cm^{-1} after extraction of homopolymer. In order to minimize the effects of a weighing error, an internal reference method utilizing the methylene absorbance of PE at 730 cm^{-1} was adopted. Thus, the mass of PAN in a sample was correlated to the ratio of the absorbance A_{2240}/A_{730}, and the weight percent graft defined before was computed from the mass of PAN.

Another technique often used to verify actual grafting is to observe the change in the insolubility of the grafted cellulose in solvents such as cuene and cupriethylene diamine, which otherwise dissolve cellulose. This change in solubility was found to be dependent upon the grafting level. In other solubility studies, where the cellulose substrate and homopolymer were both soluble, the grafted cellulose was completely dissolved in cuprammonium solution. The solution was acidified, precipitating the cellulose and grafted polymer and leaving the homopolymer in solution.

C. Thermal Analysis

In polymers having a certain degree of crystallinity, a differential scanning calorimeter is used to determine the heat of fusion and, consequently, the changes in the degree of crystallinity in the grafted and ungrafted samples. The changes in the crystallinity of PE found after grafting include a small, 2.5°C, drop in the location of the maximum in the melting curve and a significant decrease in the area under the melting peak [121]. Similar results were observed in case of grafting PP and PE/EVA blends [98]. While the decrease in the melting temperature, represented by the shift in the melting curve,

indicates there is some change in the crystallinity caused by grafting, comparison of the areas before and after grafting indicates this may be a small effect. By assuming the difference in areas is due only to a difference in the amount of PE or PP present, in other words no difference in the degree of crystallinity, the percent graft can be calculated from the equation:

$$\%G = \frac{(A_1 - A_2)}{A_2} \times \frac{\rho PAN}{\rho PE} \times 100$$

where A_1 = area before grafting and A_2 = area after grafting, ρ is the density.

D. Scanning Electron Microscope

The scanning electron microscope is generally used to detect the topography of the grafted surface, which usually changes due to grafted monomers onto the surface. Besides, this method can be used also to detect the depth of grafting into the matrix. If binary monomer mixture was used for grafting, the scanning electron micrographs helps to detect the grafted monomer distribution by comparing with micrographs of each grafted monomer separately.

E. Swelling Measurements

Equilibrium swelling of grafted samples in the proper solvent helps to detect the presence of grafted monomer. For instance, polyethylene does not swell practically in water. However, if polyethylene was grafted with water-soluble polymers such as polyacrylic acid or polyacrylamide, the equilibrium swelling of the product obtained markedly increases. Accordingly, the increase in swelling is an evidence of grafting. In contrast, the ability of natural rubber or styrene butadiene rubber vulcanizates to swell in gasoline or benzene markedly decreases due to grafting with polyacrylonitrile. This decrease in swelling, again, is an evidence of grafting.

F. Molecular Weight and Molecular Weight Distribution

Molecular weight distribution of the grafted part is essential in order to precisely design the functional polymeric membranes by application of radiation-induced graft polymerization and to control such a process. For example, the length and density of the polymer chains grafted onto cellulose triacetate microfilteration membrane will determine the permeability of the liquid through and the adsorpitivity of the molecules on the functionalized microfilteration membrane. Thus, the molecular weight distribution of methyl methacrylate grafted onto cellulose triacetate was carried out by the acid hydrolysis method of the substrate. From the gel-permeation chromatogram, the molecular weight distribution was determined [122]. This method is valid only when it is possible to degrade the substrate. In the case of grafted natural rubber, for example, ozonolysis is a very convenient process used to destroy the natural rubber segments while leaving the plastomer chains intact [123]. Alternatively, oxidation with perbenzoic acid can be used [124]. Osmometry or solution viscosity may then be used to determine the molecular weight of the isolated nonrubber fraction.

G. Dielectric Relaxation

Dielectric relaxation measurements of polyethylene grafted with acrylic acid(AA), 2-hydroxyethyl methacrylate (HEMA) and their binary mixture were carried out in a trial to explore the molecular dynamics of the grafted samples [125]. Such measurements provide information about their molecular packing and interaction. It was possible to predict that the binary mixture used yields a random copolymer PE—g—P(AA/HEMA), which is greatly enriched with HEMA. This method of characterization is very interesting and is going to be developed in different polymer/monomer systems.

VII. APPLICATIONS

Since the changes in physical properties are often the impetus for grafting, it is necessary to briefly touch on this, in this section. A number of general reviews on grafting have also included some discussion on the changes in physical properties [126–129] that usually determine the field of applications. Some other reviews deal with certain properties and applications, such as sorbency [70] and ion exchange properties [130] of cellulose.

Grafting has often been used to change the moisture absorption and to transport properties of cellulosic materials when hydrophilic monomers, such as acrylamide, acrylic acid, and methacrylic acid, were grafted. Grafting of acrylic acid onto cellulose or starch can impart super water absorbing materials for use in sanitary napkins, diapers, and for soil stabilizer and other agricultural uses [131,132]. The dyeability of cellulose was found to be affected by grafting [133] and has been examined along with the corresponding lightfastness and washfastness [119]. Thus, for practical applications, continuous and semi-continuous methods for grafting of partially carboxy methyl cellulose were investigated [134].

Several authors have discussed the ion exchange potentials and membrane properties of grafted cellulose [135,136]. Radiation grafting of anionic and cationic monomers to impart ion exchange properties to polymer films and other structures is rather promising. Thus, grafting of acrylamide and acrylic acid onto polyethylene, polyethylene/ethylene vinyl acetate copolymer as a blend [98], and waste rubber powder [137,138], allows

a new product with reasonable ion exchange capacity to be obtained.

Applications of radiation grafting in the coating industry for improving adhesion and other properties has been an active field. For instance, grafting of styrene onto polyester fibers was found to improve the interfacial adhesion between grafted chopped polyester fibers and polystyrene used as a matrix [139].

A number of possible uses of radiation grafting are being explored for microlithography, diazo printing, and various copying and printing processes.

Radiation grafting for various biomedical applications remains an extremely active field of development. The grafted side chains can contain functional groups to which bioactive materials can be attached. These include amine, carboxylic, and hydroxyl groups, which can be considered as a center for further modifications.

Improvement in the solvent and oil resistance of rubbers can be achieved via grafting of acrylonitrile onto rubber [140–142] and rubber blends [143]. The careful control of the degree of grafting allows vulcanized rubber with high-mechanical properties compared with ungrafted vulcanized rubber to be obtained. Also, acid resistance [144] and resistance to microbiological attack [145,146] was improved for cellulose grafted with acrylonitrile, and increases in base resistance were also noted for MMA and a mixture of MMA and ethyl acrylate [13].

Photodegradation of polyethylene waste can be markedly accelerated via its grafting with acrylamide [98]. In contrast, photostabilization of polyethylene and polypropylene can be achieved as a result of the grafting of 2-hydroxy-4-(3-methacryloxy-2-hydroxy-propoxy) benzophenone using gamma radiation [147]. In this case, the grafted compound acting as a UV stabilizer is chemically bound to the backbone chain of the polymer and its evaporation from the surface can be avoided.

REFERENCES

1. S. N. Ushakov, *Fiz. Mat. Nauk.*, *1*: 35 (1946).
2. G. Landells and C. S. Whowell, *J. Soc. Dyes Colour*, *67*: 338 (1951).
3. Yu. G. Kryazhev, Z. A. Rogovin, and V. V. Chernaya, *Vysokomol Soedin. Tsellylozaziee Proizvodnye, Sb. Statei*, p. 94, 1963.
4. A. Hebeish, E. M. Abdel-Bary, A. Waly, and M. S. Bedeawy, *Angew. Makromol. Chem.*, *86*: 47 (1980).
5. C. R. Chapman and L. Valentine, *Ric. Sci.*, *25* (suppl.) 278 (1955).
6. H. Adkin and R. A. Houtz, *J. Am. Chem. Soc.*, *55*: 1609 (1933).
7. P. J. Flory, *Am. Chem. Soc.*, *59*: 241 (1937).
8. G. Machells and G. N. Richards, *J. Chem. Soc.*, *1961* 3308 (1961).
9. A. Hebeish, M. I. Khalil, and M. H. El-Rafie, *Makromol. Chem.*, *37*: 149 (1979).
10. B. N. Misra, J. K. Jassal, and C. S. Pande, *J. Polym. Sci., Polym. Chem. Ed.*, *16*: 295 (1978).

11. B. N. Misra, J. K. Jassal, and C. S. Pande, *Indian J. Chem., 16A(12)*: 1033 (1978).
12. G. A. Mino and S. A. Kaizerman, *J. Polym. Sci., 31*: 242 (1958).
13. E. Schwab, V. Stannett, D. H. Rakowitz, and J. K. Magrani, *Tappi 45*: 390 (1962).
14. F. R. Duke and R. A. Forist, *J. Am. Chem. Soc., 71*: 2790 (1949).
15. F. R. Duke and R. F. Bremer, *J. Am. Chem. Soc., 73*: 5179 (1951).
16. V. I. Kurlyankina, V. A. Molotkov, O. P. Koz'mina, A. K. Khripunov, and I. N. Shetennikova, *Eur. Polym. J. (suppl.)*: 441 (1969).
17. A. Y. Kulkarni and P. C. Mehta, *J. Appl. Polym. Sci., 12*: 1321 (1968).
18. A. Y. Kulkarni and P. C. Mehta, *J. Polym. Sci. Part B, 5*: 509 (1967).
19. Y. O. Ogiwara, Y. U. Ogiwara, and H. J. Kubotu, *Polym. Sci. Part A-1*: 1489 (1968).
20. G. N. Richards, *J. Appl. Polym. Sci., 5*: 545 (1961).
21. J. C. Arthur, Jr., P. G. Baugh, and O. Hinojosa, *J. Appl. Polym. Sci., 10*: 1591 (1966).
22. G. N. Richards and E. F. T. White, *J. Polym., Sci. Part C, 4*: 1251 (1963).
23. A. Gagneux, D. Wattiez, and E. Marechel, *Eur. Polym. J., 12*: 535 (1976).
24. A. Takahashi, S. Takahashi, and E. Naguchi, *Sen't Gakkaishi, 32(7)*: T. 307 (1966).
25. T. Shiro and T. Akira, *Kogakuin Daigaku Kenkyu Hokoku, 43*: 232 (1977).
26. V. Stannett, B. S. Gupta, and D. J. McDowall, *ACS, Symp. Ser., 187*: 45 (1982).
27. K. C. Lin, Y. S. Lin, and C. T. Hus, *J. Chin. Chem. Soc., 27*: 83 (1980).
28. W. A. Waters and J. S. Littler, *Oxidation in Organic Chemistry, Part A* (K. B. Wiberg, ed.) 186–240 (1965).
29. Y. Ikada, Y. Nishizaki, T. Uyama, T. Kawahara, and I. Sakurada, *J. Polym. Sci. Polym. Chem. Ed., 14*: 2251 (1976).
30. B. P. Morin and Z. A. Rogovin, *Polym. Sci. USSR, A18*: 2451 (1976).
31. P. L. Nayak, S. Lenka, and M. Mishra, *J. Appl. Chem. Sci., 25(7)*: 1323 (1980).
32. W. A. Waters and J. S. Littler, *Oxidation in Organic Chemistry Part A*. (K. B. Wiberg, ed.), Chap. III (1965).
33. R. K. Samal, P. L. Nayak, and N. Baral, *J. Macromol. Sci.-Chem., A11*: 1071 (1977).
34. R. K. Samal, P. L. Nayak, and M. C. Nayak, *Eur. Polym. J., 14(4)*: 290 (1978).
35. R. K. Samal, M. C. Nayak, C. N. Nada, and D. P. Das, *Eur. Polym. J., A18(6)*: 999 (1982).
36. B. Ranby and I. Gadda, *ACS Symp. Ser., 187*: 33 (1982).
37. B. Ranby and H. Sundstrom, *Eur. Polym. J., 19*: 1062 (1983).
38. V. L. Kurliankina, V. A. Molotkov, S. L. Kleinen, and S. Ya. Liubina, *J. Polym. Sci., 18(2)*: 3369 (1980).
39. E. A. Arnett and M. A. Mendelson, *J. Am. Chem. Soc., 84*: 3821 (1962).
40. R. K. Samal, *J. Macromol. Sci.-Chem. A18(5)*: 719 (1982).
41. N. Hojo and H. Shirai, *Nippon Kagaku Kaishi 8*, 1316 (1972).
42. B. N. Misra, J. K. Jassal, and R. Dagra, *J. Macromol. Sci., A(16)6*: 1093 (1981).
43. R. G. R. Bacon, *Trans. Faraday Soc., 42*: 140 (1946).
44. L. B. Morgan, *Trans. Faraday Soc., 42*: 169 (1946).
45. I. M. Kolthof, A. I. Medalia, and H. P. Rain, *J. Am. Chem. Soc., 73*: 1733 (1951).
46. C. S. Marvel, *J. Polym. Sci., 3*: 181 (1948).
47. R. W. Rainwards, *J. Polym. Sci., 2*: 16 (1947).
48. A. Y. Kulkarni, A. G. Chitale, B. K. Vaidya, and P. C. Mehta, *J. Appl. Polym. Sci., 7*: 1581 (1968).
49. N. G. Gaylord and T. Tomano, *J. Polym. Sci., Polymer Lett. Ed., 13(11)*: 698 (1975).
50. G. I. Titledman, V. A. Volkov, A. A. Pantilov, and E. N. Zilberman, *Izv. Vyssh. Uchebn. Zaved. Khim. Khim. Technol., 25(10)*: 1258 (1982).
51. A. A. Gulina, R. M. Livchits, and Z. A. Rogovin, *Khim. Volokna, 3*: 29 (1965).
52. A. Y. Korotkova and Z. A. Rogovin, *Vysokomol. Soedin., 7*: 1571 (1965).
53. A. A. Gulina, R. M. Livchits, and Z. A. Rogovin, *Polym. Sci., USSR, 7(7)*: 1247 (1965).
54. H. Kubota and Y. Ogiwara, *J. Appl. Polym. Sci., 13*: 1569 (1969).
55. A. Hebeish, A. Waly, E. M. Abdel-Bary, and M. S. Bedeawy, *Cell. Chem. Technol., 15*: 505 (1981).
56. A. Hebeish, E. M. Abdel-Bary, M. H. El-Rafie, and A. El-Hussini, *Cell. Chem. Technol., 14*: 159 (1980).
57. M. A. Morsi, E. M. Abdel-Bary, M. A. El-Tamboly, and A. Hebeish, *Cell. Chem. Technol., 15*: 193 (1981).
58. A. Hebeish, A. Waly, E. M. Abdel-Bary, and M. S. Bedeawy, *Cell. Chem. Technol., 15*: 441 (1981).
59. E. M. Abdel-Bary, A. A. Sarhan, and H. H. Abdel-Razik, *J. Appl. Polym. Sci., 35*: 439, (1988).
60. B. P. Morin, Z. A. Rogovin, *Polymer Sci., USSR, A18*: 2451 (1976).
61. B. P. Morin, I. P. Breusova, G. I. Stanchenko, M. P. Bereza, T. N. Koptelnikova, and Z. A. Rogovin, Russian Patent, 401,675 (1973).
62. B. Ranby and H. Hatakeyama, *Cell Chem. Technol., 9*: 583 (1975).
63. B. Ranby, *New Methods for Graft Copolymerization onto Cellulose and Starch* (R. M. Rowell and R. A. Young, eds.), Academic, New York, p. 171 (1978).
64. S. Hirose and H. Hatakeyama, 178th ACS, Natl. Meeting, p. 199 (1979).
65. J. E. Guillet and R. G. W. Norrish, *Nature, 173*: 625 (1954).
66. J. E. Guillet and R. G. W. Norrish, *Proc. R. Soc. London, Ser. A. 233*: 172 (1956).
67. N. Geacintov, E. W. Abrahamson, and V. Stannett, *Macromol. Chem., 36*: 52 (1959).
68. N. Geacintov, E. W. Abrahamson, J. J. Hermans, and V. Stannett, *Appl. Polym. Sci., 3*: 54 (1960).
69. R. M. Reinhardt and J. A. Harris, *Text. Res. J., 50*: 139 (1980).
70. A. Hebeish and J. T. Guthrie, *The Chemistry and Technology of Cellulosic Copolymers*, Springer-Verlag, New York (1981).
71. V. T. Stannett and T. Memeteat, *J. Polym. Sci., Polym. Symp., 64*: 57 (1978).
72. N. P. Fouasster, *ACS Symp., Ser. 187*: 83 (1982).
73. N. S. Hon and H. C. Chan, *ACS Symp., Ser. 187*: 101 (1982).
74. T. Akira and T. Shiro, *ACS Symp., Ser. 187*: 119 (1982).
75. C. H. Ang, J. L. Garnett, and S. V. Jankiewicz, *ACS Symp., Ser. 187*: 141 (1982).
76. T. Akira, S. Yasusato, and T. Shiro, *Kogakuin Daigaku Kenkyu Hokoku, 50*: 71 (1981).
77. E. A. Abdel-Razik, M. M. Ali, M. I. Ahmed, and E. M. Abdel-Bary, Proceedings 3rd Arab International Confer-

ence on Polymer Science and Technology, 4–7 September, Mansoura, Egypt (1995). VII, p. 295.

78. E. M. Abdel-Bary, E. A. Abdel-Razik, M. M. Ali, and M. I. Ahmed, Proceedings 3rd Arab International Conference on Polymer Science and Technology 4–7 September, Mansoura, Egypt (1995). VII, p. 265.

79. E. A. Abdel-Razik, M. M. Ali, M. I. Ahmed, and E. M. Abdel Bary, Proceedings of 2nd Arab International Conference on Advances in Material Science and Engineering, Cairo, Egypt, p. 429. April (1993).

80. A. Chapiro, *Radiat. Chem., 9*: 55 (1977).

81. J. L. Garnett, *Radiat. Phys. Chem., 14*: 79 (1979).

82. E. M. Abdel-Bary, A. M. Dessouki, E. El-Nesr, and A. A. El-Miligy, *J. Appl. Polym. Sci., Appl. Polym. Symp., 55*: 37 (1994).

83. V. Y. Kabanov, *Radiat. Phys. Chem., 33*: 51 (1989).

84. V. T. Stannett, *Radiat. Phys. Chem., 35*: 82 (1990).

85. A. K. Mukharjee and B. D. Gupta, *J. Macromol. Sci.-Chem., A19(7)*: 1069 (1983).

86. R. F. Stamm, E. F. Hosterman, C. D. Felton, and C. S. Chen, *J. Appl. Polym. Sci., 7*: 723 (1963).

87. V. Y. Kabanov, H. Kubota, and V. Stannett, *J. Macromol. Sci.-Chem., A13*: 807 (1979).

88. M. H. Rao and K. V. Rao, *Polymer Bulletin, 1*: 727 (1979).

89. A. Furuhashi and M. Kadonega, *J. Appl. Polym. Sci., 10*: 127 (1966).

90. E. A. Hegazy, N. H. Taher, and H. Kamal, *J. Appl. Polym. Sci., 38*: 539 (1989).

91. E. M. Abdel-Bary, A. M. El-Dessoki, E. El-Nesr, and A. A. El-Miligy, *J. Polymer-Plastic Technol. Engineering 34*: 383 (1995).

92. G. Odian, A. Rossi, and E. N. Trachtenberg, *J. Polym. Sci., 52*: 575 (1960).

93. G. Odian, T. Acker, and M. Sobel, *J. Appl. Polym. Sci., 7*: 245 (1963).

94. M. B. Huglin and B. L. Johnson, *J. Polym. Sci., A-1, 7*: 1379 (1969).

95. S. Dilli, J. L. Garnett, E. C. Martin, and D. H. Phuoc, *J. Polym. Sci., Polym. Symp., 37*: 57 (1972).

96. L. N. Grushevskaya, R. E. Aliev, V. Y. Kabanov, *Vysokomolek. Soedin., Seria A, 31(7)*: 1398 (1989).

97. E. A. Hegazy, I. Shigaki, J. Okamato, *J. Appl. Polym. Sci., 26(9)*: 3117 (1981).

98. E. M. Abdel-Bary and E. M. El-Nesr, *Radiat. Phys. Chem.*, (in press).

99. Wang-Zhongyang, Liu-Yjian, Chen-Zuliang, Zhang-Yuehong, *Nuclear-Techniques, 12(5)*: 267 (1989).

100. I. Ishigaki, T. Sugo, T. Takayama, T. Okaola, J. Okamoto, and S. Machi, *J. Appl. Polym. Sci., 27*: 1043 (1983).

101. S. E. Kobiela, *Radiat. Phys. Chem., 16*: 329 (1980).

102. M. I. Aly, K. Singer, N. A. Ghanem, and M. A. El Azmirly, *Eur. Polym. J., 14*: 545 (1978).

103. L. P. Krul, Deposited, Doc, Vinit, 37,84,420 Availl. Viniti (1979).

104. K. S. Maeng, T. S. Hwang, J.-H. Baek, Industrial–Technology–Research–Inst., Chungnam–University V. 5(1) p. 17 (1990).

105. Y. Shinohara and K. Tomioka, *J. Polym. Sci., XIIV*: 195 (1960).

106. J. L. Garnett and N. Y. Yen, *J. Polym. Sci., Polym. Lett. Ed., 12*: 225 (1974).

107. C. H. Ang, J. L. Garnett, R. Levot, M. A. Long, *J. Polym. Sci., Polym. Lett. Ed., 21*: 257 (1983).

108. A. Rabi and G. Odian, *J. Polym. Sci., Polym. Chem. Ed. 15*: 469 (1977).

109. T. Matsuda, K. Hayakawa, B. Eda, and K. Kawase, *Kobonshi, Kagaku, 18*: 639 (1961); C.A. 13086 (1962).

110. I. Y. Prazdnikova, R. R. Shifrina, S. A. Pavlov, M. A. Bruk, and E. N. Teleshov, *Vysokomolek. Soedin., Seriya A, 31(8)*: 1631 (1989).

111. I. V. Bykova and I. Y. Prazdnikova, *Vysokomolek. Soedin., Seriya, B32(1)*: 67 (1990).

112. N. Kabay, A. Katakai, and T. Sugo, *J. Appl. Polym. Sci., 42(3)*: 599 (1993).

113. J. Pruzinec and A. Manova, *J. Radioanalyt. and Nucl. Chem. Lett., 153(1)*: 29 (1991).

114. N. Taher, E. A. Hegazy, A. M. Dessouki, M. El-Arnaouty, *Radiat. Phys. Chem., 33*: 129 (1989).

115. P. A. Dworjanyn, J. L. Garnett, *Radiat. Process. Polym., 393*: 93 (1992).

116. K. Friese and F. Tannert, Proceedings of the 7-Tihany Symposium on Radiation Chemistry. Budapest (Hungary/Hungarian Chemistry Society) p. 367 (1991).

117. V. K. Krishnan and A. Jayakrishnam, *J. Mat. Sci. Mat. Med., 2(1)*: 56 (1991).

118. E. A. Hegazy, A. M. Dessouki, M. M El-Dessouky, and N. M. El-Sawy, *Radiat. Phys. Chem., 26*: 143 (1985).

119. I. M. Trivedi and P. C. Mehta, *Cell. Chem. Technol., 7*: 401 (1973).

120. A. Y. Kulkarni and P. C. Mehta, *J. Appl. Polym. Sci., 12*: 1321 (1968).

121. P. W. Morgan and J. C. Corelli, *J. Appl. Polym. Sci., 28*: 1879 (1983).

122. H. Yamagishi, K. Saito, S. Furusaki, T. Sugo, F. Hoson, and J. Okamoto, *J. Membrane Sci., 85*: 71 (1993).

123. P. W. Allen, G. Ayrey, C. G. Moore, and J. Scanlan, *J. Polym. Sci., 36*: 55 (1959).

124. J. A. Blanchette, L. E. Nielson, *J. Polym. Sci., 22*: 317 (1956).

125. A. A. Mansour, E. M. Abdel-Bary, E. M. El-Nesr, *J. Elastomers Plastics, 26*: 355 (1994).

126. R. B. Philips, R. B. Que're', J. G. Guiroy, and V. T. Stannett, *Tappi, 55*: 858 (1972).

127. J. C. Arthur, Jr., *Adv. Macromolec. Chem., 2*: 1 (1970).

128. J. C. Arthur, Jr., *Adv. Chem. Ser., 99*: 321 (1971).

129. D. J. Mcdowall, B. S. Gupta, and V. T. Stannett, *Prog. Polym. Sci., 10*: 1 (1984).

130. R. Chatelin, *Inf. Chem., 312*: 181; C.A. 95: 99519x, 1981.

131. A. H. I. Zahran, J. L. Wiliams, and V. T. Stannett, *J. Appl. Polym. Sci., 25*: 535 (1980).

132. V. T. Stannett, G. F. Fanta, and W. M. Doane, *Absorbency* (P. K. Chatterjee, ed.), Elsevier Science Inc., New York (1985).

133. M. H. El-Rafie, E. M. Abdel-Bary, A. El-Hussini, and A. Hebeish, *Angew. Makromol. Chem., 88*: 193 (1980).

134. A. Hebeish, M. H. El-Rafie, E. M. Abdel-Bary, and A. El-Hussini, *Angew. Makrom. Chem., 88*: 89 (1980).

135. Y. Iwakura, T. Kurosaki, K. Uno, and Y. J. Imai, *Polym. Sci., Part C4*: 673 (1963).

136. G. N. Richards and E. F. T. White, *J. Polym. Sci., Part C4*: 1251 (1963).

137. E. M. Abdel-Bary, M. A. Dessoki, E. M. El-Nesr, and M. Hassan, Proceedings 3rd Arab International Conference on Polymer Science and Technology, Sept. 4–7, Mansoura, Egypt (1995); *Polym.-Plastic Technol. Engineering, 36* (2), 241 (1997).

138. G. Adam, A. Sebenik, U. Oserdkar, Z. Veksli, F. Ranogajec, and Veksij, *Rubber Chemistry and Technology, 64*: 133 (1991).

139. H. Käufer and E. M. Abdel-Bary, *Colloid and Polymer Sci., 260*: 788 (1982).

140. A. K. Pikaev, L. V. Shemenkova, V. F. Timofeeva, and P. Y. Glazumov, *Radiat. Phys. Chem., 35(13)*: 132 (1990).

141. A. A. Katbab, R. P. Burford, and J. L. Garnett, *Radiat. Phys. Chem., 39(3)*: 293 (1992).

142. Haddadi-Asl, R. P. Burford, and J. L. Garnett, *Radiat. Phys. Chem., 45(2)*: 191 (1995).

143. E. M. Abdel-Bary and E. M. El-Nesr, (unpublished work).

144. A. Nagaty, S. E. Shakra, S. T. Ibrahim, and O. Y. Mansour, *Cell. Chem. Technol. 14*: 177 (1980).

145. S. Kaizerman and G. Mino, *Text. Res. J., 32*: 136 (1962).

146. R. S. Rao and S. L. Kapur, *J. Appl. Polym. Sci., 13*: 2649 (1969).

147. F. Ranogajec, M. Mlinac, I. Dvornik, *Radiat. Phys. Chem. 18*: 511 (1981).

33

Surface Properties of Modified Polymers

Jayasree Konar, Bhola Nath Avasthi, and Anil K. Bhowmick
Indian Institute of Technology, Kharagpur, India

I. WHAT ARE SURFACE PROPERTIES?

Most polymers have very good bulk properties, but many industrial applications require them to have special surface properties as well. Surface treatments are used to change the chemical composition of the surface, increase the surface energy, modify the crystalline morphology and surface topography, and remove contaminants and weak boundary layers. For example, good adhesion requires that the adhering surfaces be free of contaminants and that the substrate adequately wet the adherend. The removal of contaminants from the surface is necessary to ensure good adhesion. Surfaces of adherends of low-surface energy, such as polymers, are treated to introduce certain functionally reactive groups necessary for the adhesive to be bonded. The surfaces may also be treated by the incorporation of a surfactant to change the surface tension of the adherend. The modification also helps to eliminate physical handling problems associated with static electricity and to enhance printing, decorating, wetting, and laminating qualities.

II. MEASUREMENT OF SURFACE PROPERTIES

A number of techniques are available for determining the composition of a solid surface. Since the surface plays an important role in many processes, such as oxidation, discoloration, wear, and adhesion, these techniques have gained importance. The choice of a surface analysis technique depends upon such important considerations as sampling depth, surface information, analysis environment, and sample suitability. Different techniques provide different, and sometimes complementary, information.

The most widely used techniques for surface analysis are Auger electron spectroscopy (AES), x-ray photoelectron spectroscopy (XPS), secondary ion mass spectroscopy (SIMS), Raman and infrared spectroscopy, and contact angle measurement. Some of these techniques have the ability to determine the composition of the outermost atomic layers, although each technique possesses its own special advantages and disadvantages.

A. Spectroscopic Techniques

1. Attenuated Total Reflectance Spectroscopy (ATR)

An infrared beam is directed through a crystal of refractive index (n_1) onto a sample of smaller refractive index (n_2). The intensity of the reflected beam is monitored as a function of the wavelength of the incident beam. These absorptions are used to identify the chemical structure. ATR has a sampling depth of about 0.3–3.0 microns.

2. X-Ray Fluorescence Spectroscopy (XFS)

This method allows us to characterize the first several hundred nanometers of depth of a solid. This may be an attachment to a scanning electron microscope (SEM). As the energetic electrons bombard the sample, ionization takes place. Ions with an electron vacancy in their atomic core rearrange to a lower energy state resulting in the release of electromagnetic energy of a specific wavelength. Analysis of the wavelengths of the X-radiation emitted identifies the atomic species present.

3. Auger Electron Spectroscopy (AES)

This is a two-step process: an electron drops into a core vacancy and a secondary Auger electron escapes. The energy of the Auger electron depends on the chemical bonding state of the element from which it escaped. The maximum depth from which Auger electrons can escape is only about 0.3–0.6 μm for most materials. Thus, Auger spectroscopy is a technique that truly characterizes the surface region of the irradiated specimen. AES uses a low energy, 1–5 KeV, electron beam gun for surface bombardment to minimize surface heating.

4. Electron Spectroscopy for Chemical Analysis (ESCA) or X-Ray Photoelectron Spectroscopy (XPS)

In this method the surface is bombarded with low-energy x-rays, which are less disruptive than an electron beam. The energy is absorbed by ionization, resulting in the direct ejection of a core level electron, that is, a photoelectron. These electrons have an escape depth of less than a nanometer. Although XPS is less sensitive than AES, it provides a direct measure of the binding energy of core level electrons and gives simpler spectral line shapes than AES. The following relationship holds:

Energy of emitted photoelectron
= Incident x-ray energy
    − Binding energy of ionized core-level electron

5. Ion-Scattering Spectroscopy (ISS)

A solid sample is bombarded with a stream of inert gas ions. Some of these ions are backscattered with some loss of energy after colliding with the surface atoms. Analysis of the scattered ion energies is done for identification of the surface atoms present.

6. Secondary Ion Mass Spectroscopy (SIMS)

The surface is bombarded with a stream of inert gas ions of energy, E_0, and the sputtered target secondary ions of energy, E, are monitored, rather than the backscattered primary beam ions. Mass analysis of the secondary ions is carried out. The intensity and the energy are also determined. Each element has a characteristic value of E/E_0. This allows the elemental analysis of the surface.

    Among the techniques mentioned previously, XPS has the greatest impact on polymer surface analysis. A major additional source of chemical information from polymers comes from IR and Raman spectroscopy methods. These vibrational data can be obtained from the bulk and the surface region, although the information depth is much greater than with AES, XPS, or ISS.

    A very simple, though indirect, method of surface analysis is the measurement of the angle of contact that a liquid makes with the solid surface being analyzed. This method has been widely used to study changes introduced in a polymer surface by various treatments.

Being sensitive to the chemical composition of the outermost layers of the surface, contact angle measurement is widely used for characterizing polymer surfaces. Surface characterization for polymers using contact angle measurement and XPS will now be described in detail, as these are the most widely used methods.

B. Contact Angle Measurement

Measurement of the contact angle at a solid–liquid interface is a widely used method for the determination of the surface energy of solid polymers. Fowkes [1] first proposed that the surface energy of a pure phase, γ_a, could be represented by the sum of the contribution from different types of force components, especially the dispersion and the polar components, such that:

$$\gamma_a = \gamma_a^d + \gamma_a^p \tag{1}$$

where γ_a^d is the dispersion force component and γ_a^p is the polar force component. Fowkes [2] then suggested that the geometric mean of the dispersion force components was a reliable prediction of the interaction energies at the interface caused by dispersion forces. Thus, the interfacial free energy, γ_{ab}, between phases a and b may be given by:

$$\gamma_{ab} = \gamma_a + \gamma_b - 2(\gamma_a^d \gamma_b^d)^{1/2} \tag{2}$$

where γ_a and γ_b are the surface free energies of phases a and b, respectively. Owens and Wendt [3] and Kaelble and Uy [4], by analogy with the work by Fowkes, proposed another energy term, $2(\gamma_a^p \gamma_b^p)^{1/2}$, to account for the effect of the polar forces (γ_a^p and γ_b^p are the polar components of the surface free energy of phases a and b, respectively). Thus,

$$\gamma_{ab} = \gamma_a + \gamma_b - 2(\gamma_a^d \gamma_b^d)^{1/2} - 2(\gamma_a^p \gamma_b^p)^{1/2} \tag{3}$$

    Considering a solid–liquid system, this relationship may be combined with the well-known Young's equation to eliminate the interfacial free energy. Hence,

$$\cos\Theta = [-1 + 2(\gamma_s^d \gamma_l^d)^{1/2}/\gamma_l + 2(\gamma_s^p \gamma_l^p)^{1/2}/\gamma_l] \tag{4}$$

assuming the spreading pressure to be negligible and s and l to represent solid and liquid, respectively. The thermodynamic work of adhesion could also be obtained from the equilibrium contact angle Θ and the surface tension of the liquid as follows:

$$W_A = \gamma_l(1 + \cos\Theta) \tag{5}$$

Combining Eqs. (4) and (5):

$$W_A = 2(\gamma_s^d \gamma_l^d)^{1/2} + 2(\gamma_s^p \gamma_l^p)^{1/2} \tag{6}$$

Various workers have used these equations extensively to understand the surface energetics of polymeric solids.

1. Contact Angle Hysteresis

For a three phase system of solid–liquid–vapor in thermodynamic equilibrium, an angle Θ_e, is subtended at the triple line by the tangents onto the solid–liquid and liquid–vapor interfaces. In thermodynamic terms, this system parameter Θ_e depends, via Young's equation, on the interfacial tensions. Instead of the equilibrium angle Θ_e, two other angles are obtained on real surfaces: the advancing angle Θ_a after enlarging the drop (wetting) and the receding angle Θ_r after reducing the contact area of the drop with the solid (dewetting). The difference between the two angles is widely known as hysteresis:

$$\Delta\Theta = \Theta_a - \Theta_r$$

Both the advancing and receding angles can be observed with one drop when the substrate is tilted [5].

The contact angle hysteresis of a polymer–liquid combination gives further insight into the wetting behavior of a liquid onto a solid surface. Surface roughness and surface heterogeneity are the common causes of contact angle hysteresis. It is found that hysteresis becomes negligible when the roughness is 0.5–0.1 μm and when the heterogeneous phase is smaller than 0.1 μm.

2. Measurement of Contact Angles

The most commonly used techniques for contact angle measurements are the sessile drop method and the Wilhelmy plate method. Results obtained from these two methods are in good agreement.

3. Tangent Method

The tangent method is generally used for measuring the contact angle. It involves direct observation of the profile of a liquid drop resting on a plane solid surface. The contact angle is obtained by measuring the angle made between the tangent to the profile and the solid surface at their point of contact. This can be done on a projected image or a photograph of the drop profile, or directly using a telescope fitted with a goniometer eyepiece.

The polar and dispersion components of the surface energy are generally obtained using two liquids, for example water and formamide. To calculate γ_s^p and γ_s^d, the following values for γ_l^p and γ_l^d were taken [3]:

Water: $\gamma_l = 72.8$, mN/m; $\gamma_l^d = 21.8$, mN/m;

$\gamma_l^p = 51.0$ mN/m

Formamide: $\gamma_l = 58.2$, mN/m; $\gamma_l^d = 39.5$, mN/m;

$\gamma_l^p = 18.7$ mN/m

C. X-Ray Photoelectron Spectroscopy

Its ability to distinguish among different elements and different chemical bonding configurations has made XPS the most popular surface analytical technique for providing structural, chemical bonding, and composition data for polymeric systems. All elements, except hydrogen, are readily identifiable by XPS, since the different core level binding energies are highly characteristic. The concentration of different elements on a surface may be found by measuring the relative peak intensities and dividing them by appropriate sensitivity factors. Further, small shifts in the binding energy of a core level can be corroborated by considering the presence of different functional groups. For example, when a carbon atom is bonded to different groups of atoms of increasing electronegativity, a systematic shift in the binding energy of the C1s peak is observed: the higher the electronegativity of the group, the higher the binding energy of the C1s peak.

1. The XPS Experiment

A basic XPS spectrometer consists of an x-ray source, an electron analyzer, a detector that counts the number of photoelectrons, and a data acquisition and processing system. XPS is conducted in a very high-vacuum environment ($< 10^{-9}$ torr). A high vacuum is also needed to prevent surface contamination and minimize scattering between the photoelectrons and gaseous molecules. A practical problem with using polymer in a high-vacuum environment is that low-molecular weight components, additives, and water, may volatilize.

In the majority of spectrometers, Al and Mg are commonly used as x-ray target materials. With two anodes, Al and Mg, it is possible to resolve overlapping photoelectron and Auger electron peaks. This is because in an XPS spectrum the position of the Auger peaks changes if Al K$_\alpha$ radiation is replaced by Mg K$_\alpha$ radiation, but the positions of the photoelectron peaks are unaltered.

III. SURFACE PROPERTIES OF MODIFIED POLYMERS

Over the past several decades, many surface modification methods for polymers have been developed. Pretreatment with corona discharge, gas plasma, flame, electron beam, and oxidizing solutions, and modification by blending, grafting, and copolymerization are common. Though many of these processes were developed primarily for treating polyethylene, they have been used with varying degrees of success for other polyolefinic materials, such as polypropylene, ethylene–propylene copolymers, and high-density polyethylene. The surface properties of polymers modified by these techniques are now discussed briefly.

A. Modification by Chemical Methods

This method is one of the oldest and most widely used techniques of improving the adhesion of polyolefins to metal and other substrates. Chemical treatment is usually used for irregular and particularly large articles when other treatment methods are not applicable. It involves immersion of the article in the etchant solution.

Many types of chemical treatment are used in industry. Chromic, permanganic, sulphuric, and chlorosulphonic acids are often used as the oxidants. It has been shown that the adhesion of polyethylene to substrates, such as cellophane, steel, aluminium, and epoxy adhesives, improves upon pretreatment with any of the etchants mentioned previously.

Chromic acid etching has a significant effect upon surface topography of the etched polymer as shown by Blais et al. [6]. They studied chromic-sulphuric acid-etched low- and high-density polyethylene, as well as polypropylene, using transmission electron microscopy and found the etching rates to be independent of time upto 6 h at 70°C with the trend PP > LDPE ≫ HDPE. Reflection IR studies revealed extensive chemical changes on the surface in the case of LDPE but not HDPE or PP. New bonds corresponding to the introduction of $-OH$, $>C=O$ and $-SO_3H$ groups were detected. XPS studies of similarly treated LDPE and PP also confirmed both oxidation and sulphonation [7]. Yet another report on XPS studies revealed that the surface oxidation of PP rapidly reaches an equilibrium state and involves only a thin layer of the polymer (<100 Å), whereas LDPE oxidation increases in degree and in depth with time. These results confirm the hypothesis of Blais et al. [6] that rapid attack at tertiary carbon atoms results in material loss of oxidized PP at an earlier stage in the reaction compared with LDPE, the equilibrium oxidized layer thickness being insufficient for detection by reflection IR.

In chromic acid etching using H_2SO_4, all surface sulphur is in the form of $-SO_3H$ groups. Comparison of XPS and adhesion results shows that there is a good correlation between adhesion level and degree of surface oxidation, but no correlation can be obtained between adhesion and $-SO_3H$ concentration [8].

Oxidation of polyethylene by sulphuric acid and potassium chlorate [9,10] improves its adhesiveness. The free energy of adhesion of the polymer is found to increase linearly with the surface density of the hydrophyllic sites created by oxidation.

The surface tension, polarity, wettability, and bondability of fluoro-polymers are improved by sodium etching [11,12]. The etchant solutions are the equimolar complexes of sodium and naphthalene dissolved in tetrahydrofuran. XPS shows complete disappearance of a fluorine peak, the appearance of an intense oxygen peak, and broadening and shifting of the C1s peak to a lower binding energy. A good number of functional groups, such as carbonyl, carboxyl, and C=C unsaturation are introduced.

The oxidation methods described previously are heterogeneous in nature since they involve chemical reactions between substances located partly in an organic phase and partly in an aqueous phase. Such reactions are usually slow, suffer from mixing problems, and often result in inhomogeneous reaction mixtures. On the other hand, using polar, aprotic solvents to achieve homogeneous solutions increases both cost and procedural difficulties. Recently, a technique that is commonly referred to as phase-transfer catalysis has come into prominence. This technique provides a powerful alternative to the usual methods for conducting these kinds of reactions.

In the light of the success of phase-transferred permanganate (purple hydrocarbon) as an oxidant in nonpolar media, Konar et al. [13–15] have oxidized several polyolefins, such as LDPE, HDPE, PP, and ethylene–propylene copolymer, with the help of tetra-butyl ammonium permanganate in hydrocarbon media. Characterization of the oxidized polyolefins confirmed the introduction of polar functional groups onto the polymer surface [16–18]. The contact angle values and the corresponding surface energies of LDPEs oxidized by purple hydrocarbon using different phase transfer catalysts such as tetrabutyl ammonium bromide (TBAB), tetrapentyl ammonium iodide (TPAI), di-cyclohexyl-18-crown-6 (DC-18-C-6), and benzyl triphenyl phosphonium chloride (BTPC) are given in Table 1. It reveals that on oxidation with different phase transfer catalysts the equilibrium contact angles of water and formamide decreased from 93° to 58° and from 76° to 38°, respectively [19]. At the same time the surface energy of PE increased on oxidation. One important point to be noted is that the LDPEs oxidized using DC-18-C-6 and BTPC catalysts have a relatively greater polar contribution to

Table 1 Contact Angles (Θ) and Surface Energies of LDPEs Oxidized Using Purple Hydrocarbon as Oxidant

Catalyst used for oxidation of LDPE	Θ (H$_2$O) (degrees)	Θ (HCONH$_2$) (degrees)	$\gamma_s{}^d$ (mJ/m^2)	$\gamma_s{}^p$ (mJ/m^2)	γ_s (mJ/m^2)
None	93	76	18.16	4.18	22.34
TBAB	71	41	38.62	7.72	46.32
TPAI	68	38	39.50	8.39	47.89
DC-18-C-6	58	42	24.17	21.01	45.18
BTPC	60	38	26.08	18.55	44.63

Table 2 Functionalities Produced by Oxidation with Purple Hydrocarbon

Catalyst used for oxidation of LDPE	Carboxyl present [20] (%)	Hydroperoxide present [16] (%)	Carbonyl present [17] (%)	Epoxy present [16] (%)
TBAB	8.0	7.8	3.09	—
TPAI	5.5	8.9	1.95	0.059
DC-18-C-6	15.0	22.2	2.22	0.016
BTPC	20.0	15.2	1.45	0.045

the total surface free energies than the LDPEs oxidized using TBAB and TPAI catalysts. From Table 2 it is seen that the total polarity measured in terms of various oxo-groups is certainly higher in the first two instances.

As a result of oxidation, the adhesion strength between PE and Al is found to increase by about 8 to 28 times. The adhesion strength increases with an increase in carboxyl content, total amount of oxo- groups, and the combined carboxyl and carbonyl content.

A similar reaction carried out on PP is interesting. The C1s and O1s spectra of PP and oxidized PP are shown in Fig. 1. The binding energies and areas calculated from the C1s and O1s spectra are given in Table 3. The XPS spectra of untreated PP indicate a low level of oxygenated species. The C1s spectrum of the untreated sample shows a single peak at about 285 eV, which indicates the presence of only C—C bonds in PP. The deconvoluted C1s spectrum of oxidized PP shows four peaks that can be assigned to —CH$_2$ at 285.0 eV, —CH$_2$O— (e.g., alcohol, ether, ester, or hydroperoxide) at about 286.5 eV, >C=O (e.g., aldehyde or ketone) at 288.0 eV, and —COO— (e.g., carboxylic acid or ester) at 289.5 eV [21]. Some of these functional groups were also found in the infrared spectrum of oxidized PP [14]. The O1s spectrum of the control sample shows a peak at about 531.5 eV, whereas the O1s peak for oxidized PP is shifted slightly toward the high-binding energies (to about 533 eV). The shift of the O1s peak toward high-

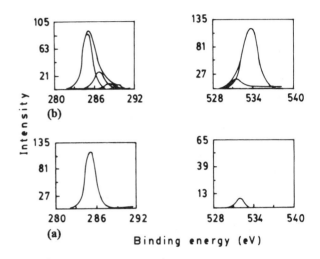

Figure 1 C1s (left) and O1s (right) peaks for (a) PP and (b) oxidized PP.

binding energy confirms that the relative concentration of carboxyl groups increases in oxidized PP. It is known that most oxygen functional groups give O1s binding energies at about 532 eV, whereas the ester oxygen in carboxyl groups shows a binding energy at about 533.5 eV [21].

Table 3 Binding Energies and Areas of C1s and O1s Spectra of PP and Oxidized PP

Sample	C1s Peak position (eV)	C1s Area (eV/ms)	O1s Peak position (eV)	O1s Area (eV/ms)
PP	285.0	230.5	531.5	20.8
Oxidized PP		254.4 (total)		255.4 (total)
	285.0	165.8	531.5	7.0
	286.5	65.2	533.7	248.4
	288.2	16.9		
	289.3	6.5		

Table 4 Contact Angle and Contact Angle Hysteresis of Modified Polyethylene

Polymer	Degree of grafting (mmol/100 g)	Θ_a (H$_2$O) (degree)	Θ_r (H$_2$O) (degree)	$\Delta\Theta$ (H$_2$O) (degree)	Θ_a (HCONH$_2$) (degree)	Θ_r (HCONH$_2$) (degree)	$\Delta\Theta$ (HCONH$_2$) (degree)
LDPE	—	83	73	10	75	69	6
PEgVTMS$_1$	20	80	68	12	73	65	8
PEgVTMS$_2$	25	78	65	13	70	58	12
PEgDBM$_1$	13	67	55	12	61	53	8
PEgDBM$_2$	17	64	49	15	55	44	11
PEgDBM$_3$	20	57	40	17	50	36	14

B. Modification by Graft Copolymerization

Functionalization of polyolefins through graft copolymerization of unsaturated monomers containing polar groups has received much attention in recent years. This type of functionalization is generally carried out for modification of some of the properties, such as adhesion and dyeability. Sometimes further crosslinking is done on these functionalized polymers. Graft copolymerization of an unsaturated polar monomer onto a preformed polymer is also widely recognized as a potential practical route to generate a novel class of interaction promoters (compatibilizers) in the case of multicomponent polymer systems. Vinyl silanes are among the most effective classes of such organofunctional monomers. They can be grafted easily onto the backbone chains of polyolefins through melt processing. The technology of making moisture-curable polyolefins through silane grafting by bulk processing has been developed [22,23]. The method of making crosslinkable polyolefins through silane grafting has gained attention because of the various advantages, such as easy processing, low cost and capital investment, and favorable properties in the processed materials.

Until recently, maleic anhydride has been used mostly for the functionalization of polyolefins due to the high reactivity of the anhydride group toward sucessive

reactions [24–26]. The maleic anhydride, although more reactive and effective than the corresponding ester, is very volatile, toxic, and corrosive. Hence, the ester dibutyl maleate has been used in a few applications as it is more compatible with polyolefins than maleic anhydride [27].

LDPE has been functionlized in bulk through dicumyl peroxide (DCP)-initiated grafting of dibutyl maleate (DBM) and vinyltrimethoxy silane (VTMS) in the temperature range from 140–200°C [28]. The advancing and receding contact angles of water and formamide on VTMS grafted PE (PEgVTMS) and DBM grafted PE (PEgDBM) are tabulated in Table 4. The corresponding values of contact-angle hysteresis are also presented in that table. The advancing and the receding angles for all the grafted PEs decrease when compared with the values of the control PE. PEgDBM shows lower Θ_a and Θ_r than those of the silane system. These differences are more marked when the plane of PE is tilted at 35°. The change in Θ(H$_2$O) value is 12° for DBM-grafted PE1 as compared with PE. The contact-angle hysteresis value for grafted PEs increases with an increase of the degree of grafting.

It is observed that the total surface energy increases with grafting (Table 5). The values become almost double for PEgDBM3. The increased surface energy of the grafted polymer comes mostly from the polar compo-

Table 5 Surface Energy of Grafted Polyethylene

Polymer	Degree of grafting (mmol/100 g)	γ_s^d (mJ/m^2)	γ_s^p (mJ/m^2)	Total surface energy, γ (mJ/m^2)
LDPE	—	12.18	10.02	22.20
PEgVTMS$_1$	20	12.69	10.89	23.58
PEgVTMS$_2$	25	13.35	12.29	25.64
PEgDBM$_1$	13	13.63	19.59	33.22
PEgDBM$_2$	17	15.36	21.23	36.59
PEgDBM$_3$	20	15.46	25.76	41.22

nent γ_s^p. γ changes more than twofold for PEgDBM3. Owing to the presence of polar groups on the surface, the surface energies of grafted PEs are expected to increase; the extent of this enhancement will depend on how many polar groups have been incorporated.

The x-ray photoelectron (XP) spectra of pressed polyethylene indicates a low level of oxygenated species. The C1s and O1s regions of the XP spectra of PEgDBM and the C1s, O1s, and Si2p regions of the XP spectra of PEgVTMS are shown in Fig. 2. The C1s spectrum of polyethylene shows a single peak at 285.2 eV, indicative of the carbon–carbon bonds in PE. The C1s spectra of grafted polyethylene show new peaks at high-binding energy (near 288–289 eV). The shift in the C1s peak indicates the presence of carbon-oxygen functionalities. To obtain atomic ratios, O1s, C1s, and Si2p peak areas have been corrected for different photoemission cross sections of these core electrons. These data are summarized in Table 6. It is observed that the O:C ratio increases with grafting and, surprisingly, it is more for silane-grafted polymer. There is a four- to twelve-fold increase depending on the level of grafting and DCP contents.

An examination of Table 6 also confirms that increasing DCP concentration leads to increased O:C ratio for PEgVTMS. The increased O:C ratio for PEgDBM might be due to the presence of DBM groups on the polymer backbone. The appearance of a new peak at the higher binding energy region (288–289 eV) of the C1s

Table 6 XPS Data of Grafted Polyethylene

Polymer	Degree of grafting (mmol/100 g)	% O to total C	% Si to total C
LDPE	—	0.65	—
PEgVTMS$_1$	20	6.95	4.2
PEgVTMS$_2$	25	7.52	6.5
PEgDBM$_1$	13	2.30	—
PEgDBM$_2$	17	4.12	—
PEgDBM$_3$	20	6.90	—

spectra can be ascribed to the carboxylic ester group [21].

The higher O:C and Si:C ratios for PEgVTMS can be explained by considering the incorporation of the -Si-O-C group onto the polymer backbone during grafting.

At higher values of the O:C ratio, lowering of the contact angle is expected. Higher O:C ratio means more functional groups are present and, naturally, one can expect an enhanced surface energy of the solid. Lowering of the contact angle suggests better wetting and, therefore, enhancement of surface energy. However, it is interesting to note here that the surface energy of PEgVTMS is lower, although the O:C ratio is higher than that of PEgDBM. It is well known that the silicones

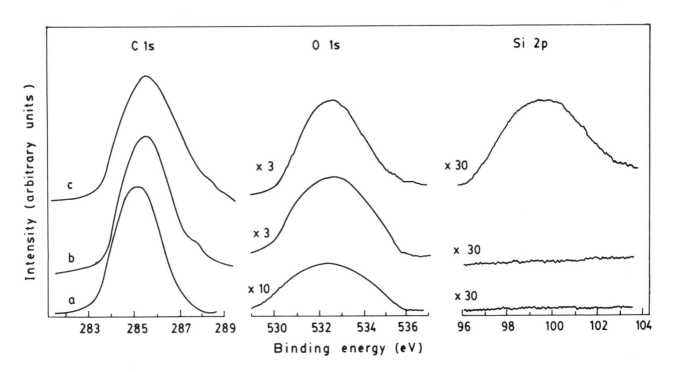

Figure 2 Binding energies (in eV) of C1s, O1s, and Si2p peaks from surfaces of (a) LDPE, (b) PEgDBM, and (c) PEgVTMS.

are release agents and have very low surface energy [29]. Apparently, the expected increase in γ_s due to increased O:C ratio is partially nullified by the presence of silicon atoms.

C. Modification by Electron Beam Treatment

Electron beam-initiated modification of polymers is a relatively new technique with certain advantages over conventional processes. Absence of catalyst residue, complete control of the temperature, a solvent-free system, and a source of an enormous amount of radicals and ions are some of the reasons why this technique has gained commercial importance in recent years. The modification of polyethylene (PE) for heat-shrinkable products using this technique has been recently reported [30,31]. Such modification is expected to alter the surface properties of PE and lead to improved adhesion and dyeability.

The equilibrium contact angles for PE and triallyl cyanurate (TAC) grafted PE with water and formamide are presented in Table 7. The grafting of TAC onto PE increases the surface energies of modified PEs [32].

Table 7 reveals that the grafting of TAC onto PE decreases the equilibrium contact angles of water and formamide from 92° to 65° and from 75° to 53°, respectively. This decrease is a function of the monomer level and the irradiation dose. At a fixed irradiation dose of 15 Mrad, variation of the TAC level from 0.5 to 3 parts causes a reduction in the contact angles of water by 13° (from 88° to 75°) and of formamide by 11° (from 72° to 61°). This is due to the fact that the concentration of

polar groups on the surface increases as a result of grafting of the polyfunctional unsaturated monomer. These polar groups are evident from the IR and XPS studies [32]. With increasing irradiation dose at a constant TAC level of 1 part, the contact angles of water and formamide are lowered by 7° up to an irradiation dose of 10 Mrad, and then increased for up to a 20 Mrad irradiation dose. The results can be explained in terms of the level of surface modification described later.

The effect of irradiation of samples in the absence of TAC on the contact angles is also reported in Table 7. Modification of the surface takes place, as is evident from the decrease in the contact angles of water and formamide. The change, which is maximum at an irradiation dose of 10 Mrad, is due to the generation of polar functionalities on the surface. This is also corroborated from the IR/XPS studies described later. The contact angles are lowered further when TAC is incorporated in the system (compare T0/5 with T1/5, T0/15 with T1/15, etc.)

The results of surface energy of grafted PEs are reported in Table 7. It is observed that the dispersion component of the surface energy, γ_s^d does not increase with the incorporation of TAC, although there is an approximately two- to three-fold increase in the polar component of the surface energy, resulting in an overall increase of the surface energy value. A change in the total surface energy of the samples without TAC due to irradiation (i.e., T0/2, T0/5, T0/10, etc.) is also evident.

With the increase in irradiation dose, however, there is an optimum value of the surface energy at 10 Mrad irradiation. In order to explain these results, the grafting levels calculated from the IR spectra [30] are

Table 7 Contact Angle Θ and Surface Energies of PE and TAC Grafted PEs

Sample	Grafting level (mmol/100 g)	Θ (H$_2$O) (degrees)	Θ(HCONH$_2$) (degrees)	γ_s^d (mJ/m^2)	γ_s^p (mJ/m^2)	γ_s (mJ/m^2)
T0/0	—	92	75	20	4	24
T0.5/15	4.8	88	72	20	6	26
T1.5/15	5.1	82	67	20	8	28
T2/15	15.0	79	64	21	10	31
T3/15	25.0	75	61	20	12	32
T1/2	5.7	72	60	19	15	34
T1/5	6.7	69	58	18	17	35
T1/10	7.4	65	53	20	19	39
T1/15	4.6	70	60	17	17	34
T1/20	2.6	72	62	17	16	33
T0/2	—	88	73	18	6	24
T0/5	—	82	72	15	10	25
T0/10	—	78	68	16	12	28
T0/15	—	80	69	16	11	27
T0/20	—	84	69	19	7	26

T1/15 for example, means, sample with 1 part TAC irradiated at 15 Mrad.

given in Table 7. It is clear that the changes in surface energy values are generally in accord with the changes in the grafting level. As the grafting level increases, the surface energy values increase.

IR studies of irradiated samples without TAC indicate absorbance peaks at 1140 and 1732 cm^{-1} due to the generation of —O—CH$_2$— and >C=O functionalities. The ratio of the peak at 1732 cm^{-1} with respect to that at 1470 cm^{-1} was calculated for all samples. On introduction of TAC, the ratio increases by four- to eight-fold depending on the amount of TAC and the irradiation dose.

Figure 3 shows the high-resolution spectra for PE in the O1s, N1s, and C1s regions and illustrates several points of interest. Three representative samples, that is, control PE and grafted PEs at 15 Mrad irradiation dose using 2 parts TAC (T2/15) and at 10 Mrad irradiation dose with 1 part TAC (T1/10), are shown in the figure. The C1s spectrum for the control PE film surface shows a peak at 284.5eV indicative of the carbon bonds in PE with a small shoulder, probably related to the presence of an O1s peak. This control sample shows a small percentage of oxygen but no nitrogen on the surface before grafting. After grafting, several percentages of oxygen and a small percentage of nitrogen are incorporated into the PE surface, and the C1s spectrum shows new peaks toward higher binding energy, indicating the formation of carbon-oxygen and carbon-nitrogen functionalities. A simple deconvolution of the spectrum for C1s in Fig. 3c is shown. The high-binding energy region of the C1s spectrum can be fitted with three peaks corresponding to carbon atoms with a single bond to oxygen or nitrogen at 287.0 eV, carbon atoms with two bonds to oxygen at 288.0 eV, and carbon atoms with three bonds to oxygen at 290.0 eV.

The O1s spectrum of the control sample shows a peak at about 532.2 eV, while the O1s peak for grafted PEs shifts slightly toward higher binding energies (532.9 eV) (Table 8). This shift confirms that the relative concentration of carboxyl and carbonyl groups increases in grafted PE. Also, the O1s peak area increases three times on modification. The grafted samples show a N1s

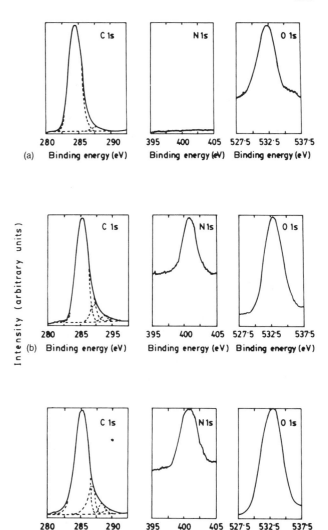

Figure 3 (a) Core level spectra of polyethylene (T0/0). (b) Core level spectra of polyethylene grafted with 2 parts TAC at an irradiation dose of 15 Mrad (T2/15). (c) C1s, N1s, and O1s peaks for polyethylene grafted with 1 part TAC at an irradiation dose of 10 Mrad (T1/10).

Table 8 XPS Details of C1s, N1s, and O1s Spectra from TAC Grafted PE

	C1s		N1s		O1s	
Sample	Peak center (eV)	Area (eV/ms)	Peak center (eV)	Area (eV/ms)	Peak center (eV)	Area (eV/ms)
T0/0	284.5	20.34	—	—	532.2	2.02
T2/15	285.3	20.29	400.7	0.89	532.9	7.30
T1/10	285.5	20.31	400.9	0.91	533.0	7.41
T0/10	284.9	20.35	—	—	532.4	2.23
T0/15	285.0	20.28	—	—	532.6	2.64

T2/15 for example, means, sample with 2 parts TAC irradiated at 15 Mrad.

peak at 400.7 eV, and the incorporation of a nitrogen peak suggests a grafting reaction. It must be pointed out here that the samples even without TAC show a surface concentration of oxygen on irradiation in line with the IR studies. The O/C ratio, however, increases significantly for the TAC grafted PE.

D. Modification by Blending

Multiphase polymers, such as block and graft copolymers and polymer blends, show distinct chemical and physical properties at the surface [33,34]. The low surface energy component generally comes to the surface because the thermodynamic driving force reduces the total free energy of the system. Siloxane polymers are commonly used in copolymers or blends to reduce surface energy and thereby reduce their coefficient of friction. The surfaces of most block copolymers and blends of siloxane polymers are rich in siloxane [35]. Similar behavior is apparent in the blend of equal parts by weight of silicone and EPDM, where silicone diffuses onto the surface during aging [36]. Surface segregation of these polymers occurs only at very low concentrations. Hence, this method has been recently proposed to modify the surface without affecting the bulk properties of polymers.

The equilibrium contact angles and surface energies of silicone, EPDM, and a 50:50 blend of Si:EPDM are reported in Table 9. On aging, at 175°C, the contact angle of water gradually decreases with aging time and then increases at 48 h of aging. For all the rubbers, the surface energy increases with an increase of time of aging in the initial stage and then decreases. The peak value after which there is a decrease in the surface energy is 9 h for the silicone and the blends and 24 h for the EPDM. In the blend of silicone and EPDM (50:50 by weight) the surface energy values in the initial stages of aging are similar. But these are lowered for the blends at longer times of aging. It is clear that the polar contribution increases from 7.95 mJ/m^2 for the unaged sample of EPDM to 50.55 mJ/m^2 for a 24 h aged sample, a six-fold increase; from 1.86 mJ/m^2 for the unaged sample of silicone rubber to 16.40 mJ/m^2 for the 9 h aged sample, a nine-fold increase; and from 8.73 mJ/m^2 for the blend to 17.19 mJ/m^2 for the 9 h aged sample. These values decrease on continued aging. The dispersion component, on the other hand, shows a decrease in general before the final increase. There is a clear indication of increased polar functionalities, mainly due to >C=O function, from IR measurements. For example, during aging at 175°C for 48 h, there is about a 25% increase of >C=O functionalities for the control EPDM, while a 15% increase of the same group is registered for the EPDM in the blend. Clearly, the EPDM in the blend is protected from oxidation by the presence of silicone rubber.

From the XPS measurements of 50:50 silicone–EPDM blend, the concentrations of carbon, oxygen, and silicone have been calculated [36]. It is observed that the relative area percent, which is proportional to the concentration of a particular chemical species, changes due to oxidation. The total peak area of O1s increases due to oxidation (Fig. 4). Silicone peaks Si2p and Si2s appear at about 102 and 153.5 eV, respectively, in the blend. It is interesting to note that the peak area of silicone increases from 31.74 to 46.6 units (47% increase) on aging for 9 h at 175°C. It can be inferred that the silicone diffuses onto the surface during aging. The total concentration of silicone is 18.5% after aging, as compared with 12.4% on the unaged surface. Concomitantly, there is approximately a 7% increase in the oxygen concentration. The C1s concentration, however, is lowered by about 13%. These results are in general agreement with the contact angle values and IR observations [36].

E. Modification by Corona Treatment

This technique is designed to generate a sufficiently high-voltage electrical discharge at the surface of the moving substrate, which may be in the form of sheet or film. The corona treatment functions at atmospheric pressure and relatively high temperature. It works with both electrically conducting and insulating substrates. There is substantial roughening of the surface following corona discharge treatment [37]. Very significant surface oxidation also occurs [38]. XPS techniques identify

Table 9 Contact Angle of Water and Surface Energies of EPDM, Silicone and Their 50:50 Blend

Status of sample	$\Theta(H_2O)$ (degrees)			γ_s (surface energy) (mJ/m^2)		
	EPDM	Silicone	50:50 blend	EPDM	Silicone	50:50 blend
Unaged	87	105	95	23.00	15.47	15.47
Air aging at 175°C						
For 3 h	80	94	90	25.08	15.74	18.73
For 9 h	70	90	90	36.08	18.96	19.30
For 24 h	67	104	102	51.22	11.34	11.41
For 48 h	86	100	107	21.07	13.65	9.88

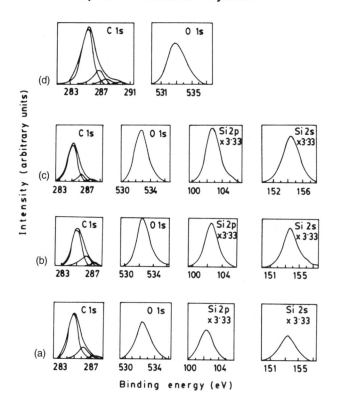

Figure 4 C1s, O1s, Si2s, and Si2p peaks of: (a) unaged 50:50 silicone–EPDM blend; (b) 50:50 silicone–EPDM blend aged for 9 h at 175°C; (c) silicone rubber aged for 9 h at 175°C; and (d) EPDM aged for 9 h at 175°C.

the presence of hydroxyl, ether, ester, hydroperoxide, aldehyde, carbonyl, or carboxylic groups in corona-discharge treated polyolefins. Briggs and Kendall [39] have provided some insight into the adhesive interaction between discharge-treated LDPE and a commercial nitrocellulose based printing ink. Using reagents that block specific functional groups, they showed that if enolic OH was eliminated from the discharge treated surface, then ink adhesion dropped to zero. On the other hand, elimination of carboxylic acid functions brought about only a slight loss of ink adhesion. By analogy with autohesion findings, it was concluded that hydrogen binding between enolic OH groups and an acceptor moiety of the printing ink was mainly responsible for the adhesion.

F. Modification by Gas Plasma Treatment

Gas plasma treatment operates at low pressure and relatively low temperature. While the corona treatment is applicable to substrates in sheet or film form, the gas plasma process can treat objects of virtually any shape. The gases most widely used to generate plasma by free-radical reactions include air, argon, helium, nitrogen, and oxygen. All these, with the exception of oxygen,

behave as inert gases. Oxygen plasma is normally used to clean surfaces by oxidizing them, resulting in the formation of —CO—, —OH, or —O— groups. Argon, helium, and nitrogen plasmas improve surface wettability and bondability by means of surface activation and crosslinking.

The diffusion of electrons plays a major role in the surface treatment of polymers in a gas plasma apparatus. The adhesion of PE tapes increases to a maximum by this treatment at 0.20 to 0.30 Å for 20 to 30 s [40]. IR absorption at 1600–1750 cm^{-1} increases 1.5-fold on discharge treatment owing to the formation of polar >C=O and —COOH groups, which increase tape adhesion.

Brosse et al. [41] modified isotactic polypropylene and other polyolefins by a cold plasma. In isotactic polypropylene, plasma treatment results in a polypropylene crystallization of paracrystalline or smectic form into a α-crystalline form. Further, the active films are susceptible to react with monomers in a postgrafting reaction.

Tetrafluoromethane plasma modified polyethylenes indicate two mechanisms: degradation and fluorination. These reactions are competitive and parallel [42].

G. Modification by Flame Treatment

Flames are also plasmas, characterized by electron densities of about 10^8/cm^3 and electron energies of about 0.5 eV. Many excited species are present in the flame, namely free radicals, ions, excited atoms and molecules, and electrons [43]. Excited species that have been observed include O, OH, NH, NO, and CH [44].

Flame treatment is predominantly used with articles of relatively thick section, such as blow moulded bottles, although it has been applied to polyolefin films as well. The most important variables in the process are the air–gas ratio and their rate of flow, the nature of the gas, the separation between burner and surface, and the exposure time.

H. Modification by Other Methods

UV irradiation on a polymer surface produces chemical modification as well as wettability and bondability improvement. It causes chain scission and oxidation on polymer surfaces even in the presence of an inert gas [45]. Carbonyls are found to be introduced onto polyethylenes on UV irradiation. Sivram et al. [46] have used photochemical treatments for surface modification of polymers. They have generated surfaces of vaying surface energies by simple organic reactions.

Thermal aging is another simple pretreatment process that can effectively improve adhesion properties of polymers. Polyethylene becomes wettable and bondable by exposing to a blast of hot (~500°C) air [47]. Melt-extruded polyethylene gets oxidized and as a result, carbonyl, carboxyl, and hydroperoxide groups are introduced onto the surface [48].

The effect of thermal aging on polyethylene and iso-tactic polypropylene have been studied by Konar et al. [49]. They used contact angle, contact angle hysteresis, and XPS to characterize the modified surfaces of the polymers. Hysteresis increased with aging temperature. In the case of polyethylene, thermal aging led to a significant increase in adhesion strength of polyethylene with aluminium, but the increase in the case of polypropylene was much less marked.

Oxidative degradation of nitrile and hydrogenated nitrile rubber was studied using IR, XPS, and contact angle measurement [50]. The contact angle of rubbers decreased with aging time. XPS studies indicated that the oxidation of hydrogenated nitrile rubber takes place through $-C\equiv N$, whereas the double bonds are attacked in nitrile rubber. Surface modification by plasma, laser and ozone treatment has been discussed in a recent book [51].

REFERENCES

1. F. M. Fowkes, *J. Phys. Chem.*, 67: 2538 (1963).
2. F. M. Fowkes, *Treatise on Adhesion and Adhesives* (R. L. Patrick, ed.), Marcel Dekker, Inc., New York, Vol. 1, p. 344, (1967).
3. D. K. Owens and R. C. Wendt, *J. Appl. Polym. Sci.*, 13: 1740 (1969).
4. D. H. Kaelble and K. C. Uy, *J. Adhesion*, 2: 50 (1970).
5. S. Wu, *Polymer Interface and Adhesion*, Marcel Dekker, Inc., New York p. 15, (1982).
6. P. Blais, D. J. Carlsson, G. W. Csullog, and D. M. Wiles, *J. Colloid Interface Sci.*, 47: 636 (1974).
7. H. A. Willis and V. J. I. Zichy, *Polymer Surfaces* (D. T. Clark and W. J. Feast, eds.), Wiley Interscience, New York p. 287, (1978).
8. D. Briggs, D. M. Brewis, and M. B. Konieczko, *J. Mat. Sci.*, 11: 1270 (1976).
9. A. Baszkin, L. Ter-Minassian-Saraga, and C. R. Lisbeth, *Acad. Sci., Paris, Ser C*, 268: 315 (1969).
10. C. Foresca, J. M. Perena, J. G. Faton, and A. Bello, *J. Mat. Sci.*, 20: 3283 (1985).
11. D. W. Dwight and W. M. Riggs, *J. Colloid Interface Sci.*, 47: 650 (1974).
12. E. H. Andrews and A. J. Kinloch, *Proc. R. Soc. Lond.*, A332: 385 (1973).
13. J. Konar, S. Ghosh, A. K. Banthia, and R. Ghosh, *J. Appl. Polym. Sci.*, 34: 431 (1987).
14. J. Konar and P. Maity, *J. Mat. Sci. Letts.* 13: 197 (1994).
15. J. Konar, G. Samanta, B. N. Avasthi, and A. K. Sen, *Polym. Degradation Stability*, 43: 209 (1994).
16. J. Konar and R. Ghosh, *Polym. Degradation Stability*, 21: 263 (1988).
17. J. Konar and R. Ghosh, *J. Appl. Polym. Sci.*, 40: 719 (1990).
18. J. Konar, R. Ghosh, and S. K. Ghosh, *Polym. Degradation Stability*, 22: 43 (1988).
19. J. Konar and R. Ghosh, *J. Adhesion Sci. Technol.*, 3: 609 (1989).
20. J. Konar, R. Ghosh, and A. K. Banthia, *Polym. Commun.*, 29: 36 (1988).
21. D. Briggs, V. J. I. Zichy, D. M. Brewis, J. Comyn, R. H. Dahm, M. A. Green, and M. B. Konieczko, *Surf. Interface Anal.*, 2: 107 (1980).
22. H. G. Scott, U.S. Patent 3,646,155 (Feb. 29, 1972).
23. P. Swarbrick, W. J. Green, and C. Maillefer, U.S. Patent 4,117,195 (Sep. 26, 1978).
24. N. G. Gaylord and M. Mehta, *J. Polym. Sci. Polym. Lett.*, 20: 481 (1982).
25. N. G. Gaylord, M. Mehta, and R. Mehta, *J. Appl. Polym. Sci.*, 33: 2549 (1987).
26. Y. Minoura, M. Ueda, S. Mizunuma, and M. Oba, *J. Appl. Polym. Sci.*, 13: 1625 (1969).
27. R. Greco, G. Maglio and P. V. Musto, *J. Appl. Polym. Sci.*, 33: 2513 (1987).
28. J. Konar, A. K. Sen, and A. K. Bhowmick, *J. Appl. Polym. Sci.*, 48: 1579 (1993).
29. K. E. Polmanteer, *Handbook of Elastomers—New Developments and Technology* (A. K. Bhowmick and H. L. Stephens, eds.), Marcel Dekker, Inc., New York, p. 551, (1988).
30. T. K. Chaki, S. Roy, A. B. Majali, R. S. Despande, V. K. Tikku, and A. K. Bhowmick, *J. Appl. Polym. Sci.*, 53: 141 (1994).
31. T. K. Chaki, A. B. Majali, R. S. Despande, V. K. Tikku, and A. K. Bhowmick, *Angew. Makromol. Chem.*, 217: 61 (1994).
32. J. Konar and Anil K. Bhowmick, *J. Adhesion Sci. Technol.*, 8: 1169 (1994).
33. D. W. Fakes, M. C. Davies, A. Brown, and J. M. Newton, *Surf. Interface Sci.*, 13: 233 (1988).
34. H. Inoue, A. Matsumoto, K. Matsukawa, A. Ueda, and S. Nagai, *J. Appl. Polym. Sci.*, 41: 1815 (1990).
35. N. M. Patel, D. W. Dwight, J. L. Hedrick, D. C. Webster, and J. E. McGrath, *Macromolecules*, 21: 2689 (1988).
36. Anil K. Bhowmick, J. Konar, S. Kole, and S. Narayanan, *J. Appl. Polym. Sci.*, 57: 631 (1995).
37. K. Rossman, *J. Polym. Sci.*, 19: 141 (1956).
38. C. Y. Kim and D. A. I. Goring, *J. Appl. Polym. Sci.*, 15: 1357 (1971).
39. D. Briggs and C. R. Kendall, *Surf. Interface Anal.*, 1: 189 (1980).
40. V. I. Bukhgatler, R. I. Belova, N. V. Evdokimova, and A. L. Goldenberg, *Plast. Massy.*, 2: 56 (1981).
41. F. Poncin-Epaillard, B. Chevet, and J. C. Brosse, *J. Appl. Sci.*, 53: 1291 (1994).
42. F. Poncin-Epaillard, B. Pompui, and J. C. Brosse, *J. Polym. Sci. Polym. Chem., Pt. A*. 31: 2671 (1993).
43. J. R. Hollahan and A. T. Bell, *Techniques and Applications of Plasma Chemistry*, Wiley Interscience, New York (1974).
44. A. G. Gaydon, *The Spectroscopy of Flames*, Chapman and Hall, London (1974).
45. N. J. DeLolis, *Rubber Chem. Technol.*, 46: 549 (1973).
46. S. Sivram, Lecture at the Indo-French Symposium at P&M Curie University, Paris (1995).
47. W. H. Kreidl and F. Hartmann, *Plast. Technol.*, 1: 31 (1955).
48. D. Briggs, D. M. Brewis, and M. B. Konieczko, *Eur. Polym. J.*, 14: 1 (1978).
49. J. Konar, A. K. Bhowmick, and M. L. Mukherjee, *J. Surface Sci. Technol.*, 8: 331 (1992).
50. S. Bhattacharya, Anil K. Bhowmick, and B. N. Avasthi, *Polym. Degradation Stability*, 31: 71 (1991).
51. K. L. Mittal, Polymer Surface Modification: Relevance to adhesion, VSP, Utrecht, Netherlands, 1996.

34

Grafting of Cellulose

Abd-Alla M. A. Nada
National Research Centre, Cairo, Egypt

Mohamed Adel Yousef
Helwan University, Cairo, Egypt

I. INTRODUCTION

Grafting copolymerization of different monomers onto cellulose is clarified. Several oxidant systems have been used to generate free radicals on cellulose molecules to initiate graft copolymerization as photoactive quinones, ozone, hydroxyl radicals, ceric ammonium nitrate, potassium bromate, and potassium perminganate. Partially, xanthate cellulose is also used for the preparation of grafting. On the other hand, the effects of different physical treatments, e.g., beating, grinding, and cellulose swelling with ethylene diamine, on the reactivity of cellulose toward grafting are studied. Also the effect of partial carboxymethylation and acetylation with a different degree of substitution (DS) on the grafting of cellulose has been studied. Finally, the effect of grafting medium (aqueous or solvent) on the cellulose graftability with monomer is discussed.

II. GRAFT COPOLYMERIZATION OF CELLULOSE

Grafting presents a means of modifying the cellulose molecule through the creation of branches of synthetic polymers, which impart to the cellulose certain desirable properties without destroying the properties of cellulose. The polymerization of vinyl monomers may be initiated by free radicals or by certain ions. Depending on the monomer, one or the other type of initiation may be preferred. The grafting process depends on the reactivity of the monomer used, the type of initiation, and cellulose accessibility [1,2].

Most of the grafting methods involve the creation of free radicals on the cellulose molecule to initiate the graft polymerization. Free radicals are produced by γ-radiation [3,4] or by chemical means [5]. The most important way of producing radicals and initiating grafting is the utilization of redox systems [6]. Hydrogen peroxide and ferrous salts have been used to initiate polymerization [7]. The possibility to use thiocarbonate (xanthate method) for the preparation of graft copolymers has been recognized by Faessinger and Conta [8]. A graft copolymer is formed, whose composition and properties depend on the type of monomer and on the conditions of the reaction. Briekman [9] stated that the xanthate grafting process does not require a special atmosphere. For optimum efficiency of the grafting, the presence of some solvents is usually necessary.

III. INITIATION BY FREE RADICALS

Free radicals can be generated on the cellulose chain by hydrogen abstraction, oxidation, the ceric ion method, diazotization, introduction of unsaturated groups, or by γ-irradiation.

If a vinyl monomer is polymerized in the presence of cellulose by a free radical process, a hydrogen atom may be abstracted from the cellulose by a growing chain radical (chain transfer) or by a radical formed by the polymerization catalyst (initiator). This leaves an unshared electron on the cellulose chain that is capable of initiating grafting. As cellulose is a very poor transfer agent [10], very little copolymer results from the abstraction of hydrogen atoms by a growing chain radical. The only way in which a chain-transfer reaction between a growing chain radical and cellulose can be made to succeed is by first introducing into the cellulose molecule certain atoms, or groups, that are readily abstracted by radicals. Therefore, in most cases a radical produced by the initiator is responsible for the formation of the graft copolymer [11,12].

This technique is based in the fact that when cellulose is oxidized by ceric salts such as ceric ammonium nitrate $\{Ce(NH_4)_2(NO_3)_6\}$ free radicals capable of initiating vinyl polymerization are formed on the cellulose. However, the possibility remains that the radical formed is an oxygen radical or that the radical is formed on the C-2 or C-3 instead of the C-6 carbon atom. Another mechanism, proposed by Livshits and coworkers [13], involves the oxidation of the glycolic portion of the anhydroglucose unit. Several workers [14,15], however, have found evidence for the formation of some homopolymer. In the ceric ion method free radicals are first generated and are then capable of initiating the grafting process [16–18].

A. Grafting of Cellulose with Ceric Ions

Grafting of pulp with methylmethacrylate monomer using a ceric salt redox system is carried out. The effect of different variables for example, monomer dose, reaction time, reaction temperature, acid concentration, initiator concentration, and liquor ratio/is studied. The effect of the presence of different amounts of residual lignin in pulp on the grafting process is demonstrated. Residual lignin percentage in the pulp plays an important role in determining the grafting rate.

1. Effect of Monomer Concentration

Figure 1 illustrates the effect of monomer concentration (0.5–3 mL/g pulp) at 25°C for 1 h, 0.1% initiator, and 1% acid concentration at 30:1 liquor ratio on the total conversion and grafting efficiency percentage. Total monomer conversion percent is increased from 34% to 126% by increasing the monomer concentration from 0.5 to 3 mL/g cellulose. On the other hand, it is evident that grafting efficiency increases as the methylmethacrylate (MMA) concentration increases from 0.5 to 1 mL. Enhancement of the grafting efficiency due to the increase of the monomer concentration can be interpreted in terms of immobility of the cellulose macroradicals. Hence, availability of MMA molecules in the proximity of these radicals was a prerequisite for graft initiation and propagation. The increase in grafting efficiency can also be attributed to the fact that the increase of monomer concentration from 0.5 to 1 mL/g pulp allows the graft formation to grow, and long graft chains are formed since the active sites are the same in all monomer concentrations. By increasing the monomer concentration more than 1 mL, the grafting efficiency is decreased, since this increase allows the formation of more homo-

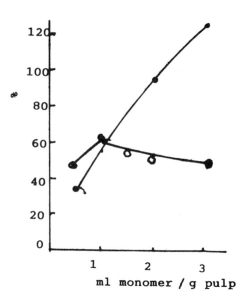

Figure 1 Effect of monomer concentration on the total conversion percent and grafting efficiency. LR 30:1, acid concentration 1%, initiator concentration 0.1%, grafting time 1 h, and reaction temperature 27°C ●—● = total conversion (%); ○—○ = grafting efficiency (%).

polymers. It seems that both grafting and polymerization are favored at high-monomer concentration, but homopolymerization prevails over grafting.

2. Effect of Reaction Time

Total monomer conversion percent is increased by increasing the reaction time with a slow rate, reaching the maximum value (60.6%) after a reaction time of 1 h. Increasing the reaction time more than 1 h causes the reaction rate to nearly level off. The leveling off can be ascribed to the depletion in monomer and initiator as well as to the shortage of available grafting sites as the reaction proceeds. However, the grafting efficiency has the same trend as the total monomer conversion percent; that is, it is increased by increasing the reaction time and reaches its maximum value after 1 h. By increasing the grafting time still further, grafting efficiency is slightly decreased or nearly leveled off, indicating that the formation of homopolymer predominates. The maximum grafting efficiency (59.6%) is obtained after a reaction time of 1 h.

3. Effect of Acid Concentration

The effect of acid concentration on the rate of grafting has been studied. It is clear that increasing the acid concentration from 0.5% to 1.0% increases not only the total monomer conversion percent from 60.1% to 61.5%, but also the grafting efficiency from 58.0% to 59.6%. This can be explained as follows: at 1% acid concentration (H^+) is sufficient to establish a balance between the suppression of the rate of active species formation and the rate of primary radical generation. By increasing the acid concentration more than 1%, the total monomer conversion and grafting efficiency are decreased. Increase of acid concentration over 1% may cause oxidation of the formed free radicals and thereby suppress the rate of initiation.

4. Effect of Liquor Ratio

The effect of changing the grafting liquor ratio on the total monomer conversion and grafting efficiency ranging from 20:1 to 60:1 was studied. Total monomer conversion percent as well as grafting efficiency is increased by increasing the liquor ratio from 20:1 to 40:1. Increasing the liquor ratio up to 40:1 allows more homogeneous distribution of the chemicals and increases the mobility of the monomer, which, in turn, increases its penetration velocity onto cellulose macroradicals and, consequently, increases the grafting efficiency.

5. Effect of Initiator Concentration

Figure 2 shows the effect of ceric ammonium nitrate concentration on the total monomer conversion and grafting efficiency of the produced grafted pulp. From this figure, it is clear that the total conversion percent increases with an increasing initiator concentration, reaching a maximum value of 64.9% using a concentration of 0.15%, and grafting efficiency reaches its maximum value (59.6%) at 0.1% initiator concentration. This increase in grafting efficiency can be discussed in terms

of the increasing rate of the termination step as ceric ion increases the redox process, which generates more radical sites on the cellulose, and then propagates by the addition of monomer.

6. Effect of Reaction Temperature

Total conversion percent is increased by increasing the reaction temperature from 10°C to 50°C. A sharp increase is observed up to 30°C, then it is increased slightly by increasing the temperature up to 50°C. On the other hand, the grafting efficiency increases with increasing reaction temperature and reaches its maximum value (66.9%) at 30°C. It is well known that the rising reaction temperature, up to 30°C, enhances the grafting efficiency through one or more of the following factors: (1) increased mobility of the monomer and, hence, a higher rate of monomer diffusion from the reaction medium into cellulose; (2) reaction of the already formed homopolymer chain with cellulose macroradicals; (3) increased cellulose swellability; and (4) enhancement in the rate of initiation and propagation of the grafting process. Increasing the grafting temperature more than 30°C decreases the grafting efficiency.

7. Effect of the Residual Lignin in Pulp

Graft polymerization of methylmethacrylate monomer onto pulps of different residual lignin contents using ceric ammonium nitrate as the initiator was carried out to study the influence of this residual lignin on the graftability of these pulps (Fig. 3). From this figure one can

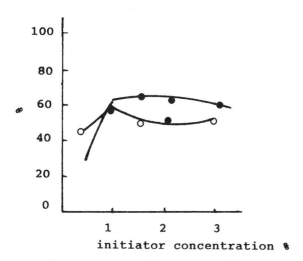

Figure 2 Effect of initiator concentrator on total conversion percent and grafting efficiency. LR 30:1, reaction time 1 h, reaction temperature 27°C, monomer concentration 1 mL/g pulp, and acid concentration 1% ●—● = total conversion (%); ○—○ = grafting efficiency (%).

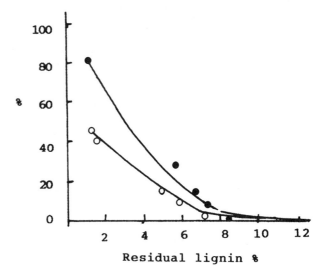

Figure 3 Effect of residual lignin percent in the pulp on total conversion percent and graft yield percent. LR 40:1, reaction time 1 h, reaction temperature 27°C, monomer concentration 1 mL/g pulp, initiator concentration 1%, and acid conversion 1%; ●—● = total conversion (%), ○—○ = graft yield (%).

notice that decreasing the residual lignin percent in the pulp enhances not only the total conversion percent but also the graft yield percent. A sharp decrease in total conversion and graft yield percent is noticed upon increasing the residual lignin percent in pulp from 1.68% to 4.97%. The graft yield and total conversion percent reach nearly a zero value at 12% residual lignin in the grafted pulp. This can be attributed to the fact that increasing the residual lignin in pulp increases the consumption of ceric ions due to the oxidizing effect of ceric ammonium nitrate on the polyhydric alcohol of lignin; this results in the formation of quinonoid structure and, hence, significantly retards and/or inhibits the grafting reaction, which lowers the propagation of the reaction [19]. Moreover, the presence of lignin, which acts as a cement to the cellulose fibers, decreases the penetration of the chemicals through the cellulose fibers, and, consequently, decreases the grafting reaction.

To sum up, the optimum conditions for methylmethacrylate grafting onto pulp by the ceric ion redox system can be summarized as follows: the grafting is done at 30°C for a 1-h reaction time, using liquor ratio 40:1, acid concentration 1%, initiator concentration 0.1%, and monomer 1 mL/g pulp.

Free radicals can also be formed on the cellulose molecule by using H_2O_2 [20–23], ozone–oxygen mixture [21,22], perminganate [22], and bromate [24].

B. Grafting by Xanthate Method [25]

Cotton linters and viscose grade wood pulp were partially xanthated under different conditions to study the effect of the degree of substitution on the acrylamide grafting of these pulps. Sodium hydroxide solutions of 2%, 4%, and 6% were used and the vapor phase xanthation process was applied for 0.5, 1.0, 1.5, and 2.0 h for each concentration. Grafting of the partially xanthated cellulose samples was carried out under the same conditions. The results obtained show that there is no direct relation between the degree of substitution and the grafting yield. The most important factors that affect the grafting process by the xanthate method are sodium hydroxide concentration and time of xanthation.

Cotton linters and viscose grade wood pulp were grafted with acrylamide using the xanthate method. The effects of monomer concentration, reaction time, hydrogen peroxide concentration, reaction temperature, and liquor ratio on the grafting process were studied. Optimum conditions for the grafting reaction were established [26].

$$
\begin{array}{ccc}
\text{Cell—OH} + \text{C} = \text{S} \rightarrow \text{Cell} - \text{OH} + \text{CS}_2 \\
| \qquad\qquad | \qquad\qquad\quad | \\
\text{O}^\pm \qquad \text{S}^{\cdot\cdot} \qquad \text{O}^\cdot \\
\text{Cell—OH} + \text{Monomer} \rightarrow \text{Cell} - \text{OH} \qquad (1) \\
| \qquad\qquad\qquad\qquad\qquad\qquad | \\
\text{O}^\pm \qquad\qquad\qquad\qquad\qquad \text{O Monomer}
\end{array}
$$

1. Effect of Monomer Concentration

The effect of acrylamide monomer concentration on the grafting percentage, grafting efficiency, polymer loading, and polymerization efficiency are shown in Fig. 4. Increasing the monomer concentration from 0.5 to 3 g/g cellulose increases the grafting from 21.3% to 59.5% and from 12.8% to 37.5% for cotton linters and wood pulp, respectively. On the other hand, the grafting efficiency decreases by increasing the monomer concentration. The grafting and polymerization efficiency of the grafted cotton linters were higher than those of wood pulp. This can be explained by the differences in the chemical and physical structures of the two pulps rather than the degree of substitution of the partially xanthated pulps.

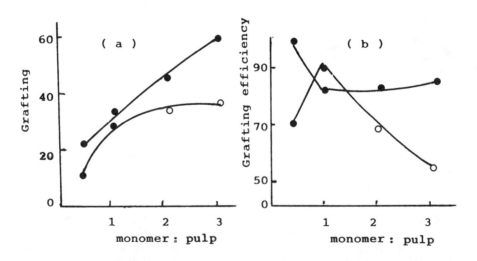

Figure 4   Effect of monomer: pulp ratio on grafting yield and grafting efficiency. ● = cotton linters; ○ = wood pulp.

2. Effect of Grafting Time

Polymer loading, polymerization efficiency, and grafting of cotton linters and viscose wood pulp were increased with increasing the grafting time. In the first 30 min the reaction was slow, then its rate increased until 60 min; after 60 min the reaction rate acquires a leveling value. This can be ascribed to the depletion of the monomer and initiator concentration as well as a shortage of the available grafting sites as the reaction proceeds. The grafting efficiency also increases with time for up to 1 h, then decreases.

3. Effect of Hydrogen Peroxide Concentration

Figure 5 shows the effect of hydrogen peroxide concentration. The grafting percentage and grafting efficiency as well as polymer loading were increased by increasing the dose of hydrogen peroxide added up to 0.5 mL/g cellulose. The same result was obtained for grafting of wood pulp applying the same method [27]. Upon increasing the hydrogen peroxide concentration over 0.5 mL/g cellulose, the grafting percentage, grafting efficiency, and polymerization efficiency are decreased. The initial increase in the graft yield is due to the increase in the number of xanthate groups that have reacted with hydrogen peroxide, thus creating more sites (free radicals) that can initiate polymerization.

4. Effect of Grafting Temperature

The effect of the reaction temperature on the grafting process is discussed. Upon increasing the reaction temperature from 15°C to 40°C the grafting decreases from 30.7% to 29.1% and from 30.0% to 28.5% for cotton linters and viscose wood pulp, respectively. This can be

attributed to the exothermic character of the polymerization reaction; also increasing reaction temperature may cause the oxidation of some of the created free radicals. Therefore, both the grafting and polymer loading were decreased by increasing the grafting temperature.

C. Grafting of Cellulose by Using γ-Radiation

Low-energy irradiation, such as ultraviolet rays through a mercury lamp [28,29], high-energy irradiation, such as γ-rays from a Co-60 source [30,31], and mechanical degradation by ball-milling of cellulose acetate in the presence of a monomer [32] are capable of generating free radicals on the cellulose molecules. Grafting of cellulosic material with monomer can be achieved by the free radicals mechanism, which can be produced by physical or chemical treatment. γ-irradiation is considered one of these treatments. The effect of ionizing radiation on the cellulosic fibers' properties has attracted considerable interest. It is generally accepted that oxidation and cleavage are the principal reactions that cellulose chains undergo on the exposure to ionizing radiation [33]. The possibility of modifying cellulosic fibers' properties by graft copolymerization using this radiation has received considerable attention. Graft copolymerization is based on the existence of free radicals in an existing cellulose chains. The free radicals can be induced by various initiation methods, such as ultraviolet and ionizing radiation [34]. Grafting of irradiated cellulose has low-grafting efficiency, i.e., high quantity of homopolymer. Figure 6 shows the effect of γ-radiation dose on the graft yield percent of the grafted cotton linters with MMA monomer. It is clear that the graft percent has a

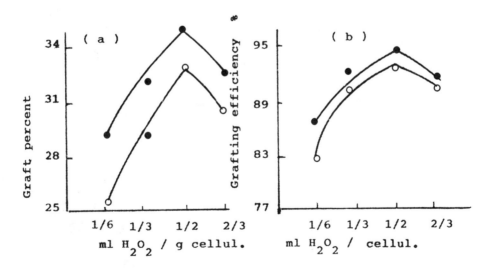

Figure 5 Effect of hydrogen peroxide concentration on grafting yield and grafting efficiency. ● = cotton linters; ○ = wood pulp.

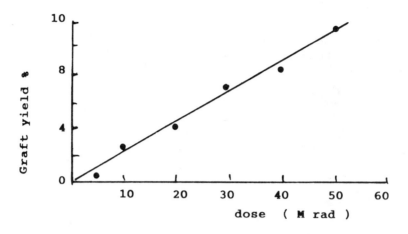

Figure 6   Effect of radiation dose on the graft yield (%) of grafted cotton linters.

linear relationship with the increase of the γ-radiation dose from 5–52 Mrad.

D.   Grafting of Cellulose by Using KBrO₃ or KMnO₄ Redox System

Paper wood pulp was grafted with methylmethacrylate using the potassium bromate redox system. The effects of monomer concentration, reaction time, initiator concentration, and liquor ratio on the graft yield and polymer loading of the grafted samples were studied. The optimum conditions of the grafting reaction were determined [24]. Cotton linters were grafted with methylmethactylate using the KMnO₄ redox system. The effect of type and concentration of acid as the activator for

KMnO₄ and different parameters affecting the grafting process, e.g., reaction temperature, reaction time, and monomer concentration, were investigated to ascertain the optimum conditions of the grafting process [23].

1.   Amount of Consumed KMnO₄

Figure 7 shows the effect of $KMnO_4$ concentration on the consumed $KMnO_4$. The amount of $KMnO_4$ consumed by the cotton linters increases with increasing $KMnO_4$ normality, i.e., MnO_2 deposit content enhances with increasing $KMnO_4$ concentration. In the presence of acid, free radicals are formed as a result of acid action on the deposited MnO_2. These created free radicals depend on the nature and normality of the acid and produce

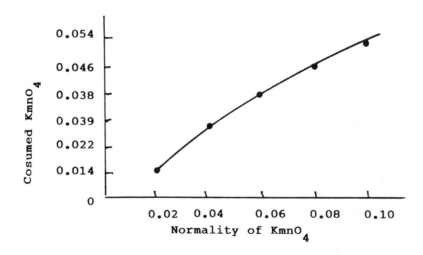

Figure 7   Effect of $KMnO_4$ concentration on the amount of consumed $KMnO_4$. Liquor ratio 1:50; t = 30 min; T = 50°C.

cellulose macroradicals as follows:

$$Cell—OH + R• \rightarrow Cell—O• + RH \qquad (1)$$

In presence of the monomer cellulose, macroradicals are added to the double bond of the monomer.

2. *Effect of Acid Concentration*

The effect of acid concentration on the polymer loading and grafting efficiency is illustrated in Fig. 8. The polymer loading increases with increasing acid concentration. The efficiency of MMA grafting onto cotton linters increases until the acid concentration of 0.05 N is reached. By increasing the normality of the acid more than 0.05 N, the grafting efficiency decreases. This can be interpreted by the fact that higher acid normality establishes a balance between the suppression of the rate of formation of active species and the rate of generation of primary radicals.

On the other hand, the decrement in grafting observed with oxalic acid is higher than that observed with sulfuric acid. This can be due to the fact that oxalic acid is more active than sulfuric acid for the formation of free radicals and, consequently, the grafting in the presence of oxalic acid is greater than in the presence of sulfuric acid.

IV. ANIONIC GRAFTING

Strongly acidic monomers, such as nitroethylene, can only be grafted under anionic conditions. Also, anionic grafting is useful for conventional monomers such as acrylonitrile and methylmethacrylate. In addition to the fact that certain monomers cannot be grafted by free radical methods, ioic grafting has some advantages, e.g., it is fast, it can be carried out at low temperatures, and homopolymer formation is excluded. Using this method, cellulose is treated with a Lewis acid, such as boron trifluoride, aluminium chloride, or stannic chloride [35]. Initiation of graft copolymerization of vinyl monomers onto cellulose by formation of free radicals using ceric ions is the most widely used method.

V. ACTIVATION OF CELLULOSE TOWARD GRAFTING

Since the cellulose hydroxyl groups may not be available for reaction, because of crystallinity or insolubility of the cellulose, this hinders the access of the reagent to the innermost hydroxyl groups. Nonuniform products can result due to the differences in accessibility of different portions of the cellulose. It is important, therefore, to have the cellulose in such a form that allows a maximum number of hydroxyl groups available for reaction. This can be brought about by decreasing the hydrogen bonding between the chains using various activation processes.

Cellulose may be activated by various mechanical or physical treatments such as, beating, grinding, and swelling in different swelling agents. Different chemical treatments, involving the introduction of new groups into the cellulose chains, e.g., acetylation or carboxymethylation, may also be used for activating cellulose as a result of the opened crystalline structure of the low-substituted derivatives.

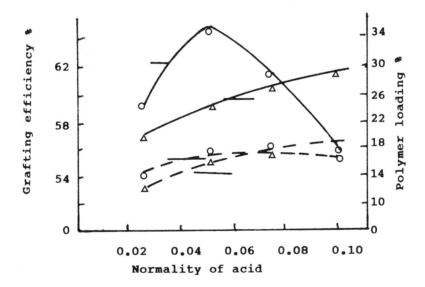

Figure 8 Effect of acid concentration on the grafting of cotton linters. $[KMnO_4]$ = 0.06 N; [monomer] = 1 mol/g cell; liquor ratio 1:50; T = 50°C; t = 1 h. (—) Oxalic acid; (----) sulfuric acid.

A. Mechanical or Physical Activation

1. Beating

The principal effect of beating is a physical one that causes breakage of some intrafiber hydrogen bonds and their replacement of fiber by water hydrogen bonds [36].

In addition, the graft yield and the number of grafted chains are greatly affected by the beating degree of the pulp [37]. The effect of beating wood and bagasse pulps at various beating degrees on their reactivity toward grafting has been studied [38]. The beating process has a great effect on the properties of cellulosic fibers. It increases the fiber flexibility and increases the surface area by fibrillating the fibers and, consequently, increases the area of contact between the fibers. Beating causes some reduction in the size of the crystalline zones. Also, it can be noticed that minor chemical changes that occur in the beaten pulp may be due to the large increase in surface activity and the changes in some other physical properties of the fibers.

By comparing the effect of the degree of beating on the graft yield of pulp using a 50:1 liquor ratio, 3 mL MMA:1 cellulose, 1% acid for 2 h at room temperature, it is apparent that increasing the degree of beating up to 37°SR is accompanied by a remarkable increase in the graft yield from 39% to 78%. This indicates that the reactivity of the pulp toward grafting increases as the beating degree increases until reaching this optimum value. The reactivity tends to decrease when increasing the beating degree more than 37°SR. The observed increase in the pulp reactivity caused by low-beating degrees is due to the significant increase in the surface area of the fibers and improved fiber flexibility. Beaten pulp of 61°SR shows the lowest grafting yield at all monomer concentrations. Total monomer conversion is increased by increasing the monomer concentration, and it decreases according to the following order:

$$\text{beaten sample for } 37°SR > 20°SR > 48°SR >$$
$$\text{unbeaten sample} > 61°SR \qquad (2)$$

2. Grinding

In the case of grinding, the cellulose fibers go over a state of fine fibrillation into a more or less powdery substance. This mechanical severance of cellulose may break main valence bonds and will, therefore, decrease its degree of polymerization. In addition, the crystal structure of cellulose fibers is nearly lost [32]. Grinding of the cellulose fibers also, appreciably increases its surface area.

Effect of Grinding on the Graftability

Compared with untreated pulp, a relatively high liquor ratio is necessary to attain maximum graft yield (43.5%) because of the small particle size of the ground pulp. For this reason, ground pulp needs a higher liquor ratio than the untreated pulp to help the mobility of the particles. On the other hand, comparing the graft yield

for ground untreated pulps at a liquor ratio 50:1, it is clear that the graft yield attained by the ground pulp is higher than that of untreated pulp. This can be attributed to the high surface area of the ground pulp, which allows better contact between the pulp and chemicals.

3. Swelling

The accessibility of cellulose for reaction is improved by using a swelling agent, which expands the crystalline lattice and permits reagents to penetrate to all cellulose molecules. The action of the swelling agents in disrupting the intermolecular hydrogen bonds may be either reversible or irreversible. Water, amines, and amine solutions have a reversible effect [40–44].

Bleached cotton stalk pulp is treated with different concentrations of ethylene diamine (50–100%) for 20 min. It is clear that the crystallinity index (Crl) of these treated pulps is decreased by increasing the concentration of ethylene diamine; that is, the decrystallization increases. The degree of polymerization is nearly the same, but some increase is shown in the sample treated with 100% ethylene diamine. This indicates that 100% ethylene diamine may act as a dissolving agent for low degree of polymerization (DP) of cellulosic chains and hemicellulose.

Of samples swollen with ethylene diamine, the graft yield at a 50:1 liquor ratio increases as the concentration of ethylene diamine increases. This is due to the increase of decrystallization of swollen samples, which helps the penetration velocity of the chemicals through the cellulosic chains. Graftability of the samples treated with 100% ethylene diamine is lower that of the sample treated with 75%. This is due to the dissolution of low DP chains and some of the hemicelluloses, which is detectable by the increase in DP of the sample teated with 100% ethylene diamine.

From the preceding results, it is seen that there is a large difference in graft yield between grafted swollen pulp treated with ethylene diamine and untreated pulp at a liquor ratio of 50:1 at all monomer concentrations.

Summing up these results, it is clear that beating and swelling of cotton stalk pulp have important effects on graftability. At high-monomer concentrations, the graftability is higher in the swollen samples than in the beaten samples.

B. Chemical Activation

It has been found that substitution of cellulose hydroxyls of cotton by acetyl groups, greatly affects the graft yield [45].

The grafting reaction depends upon the degree of substitution as well as the kind of pulp used. Introducing acetyl groups in the cellulose chains (high substitution) causes a large reduction of its swellability, which reduces the diffusion of the reactants. Thus, acetylation lowers the graftability of the cellulose.

Partial carboxymethylation of wood pulp significantly increases its susceptibility toward grafting with acrylonitrile using the ceric ion as the initiator [46]. Studies dealing with grafting of various vinyl monomers, such as acrylonitrile, methylmethacrylate, and acrylamide, onto partially carboxymethylated cotton cellulose using tetravalent cerium as the initiator have been reported [47].

Cotton linters and viscose grade wood pulp were partially acetylated and carboxymethylated. The samples were grafted under the same conditions with acrylamide. Grafting yield and efficiency depend on several factors, such as the kind of pulp, chemical and physical structure, type of the introduced substituent, and degree of substitution (DS).

1. Grafting of Partially Acetylated Cellulose

Wood pulp and cotton linters were partially acetylated to different acetyl contents ranging from approximately 6%–13% (0.23–0.54 degree of substitution [D.S.]). The increase of the DS of partially actylated cotton linters from 0.26 to 0.56 causes the percent graft of partially acetylated cotton linters to decrease slightly from 0.93% to 0.6%. In the case of wood pulp, the effect is somewhat different—the percent graft increases with an increasing DS until about 0.48, then the percent graft decreases.

Grafting of the unacetylated samples gives higher values in the case of wood pulp than in cotton linters. This is explained by the crystallinity of both pulps, which is 80% and 56.8% for cotton linters and wood pulp, respectively.

2. Grafting of Partially Carboxymethylated Cellulose

In the case of cotton linters, the percent graft slightly decreases with increasing D.S., then it begins to increase. This behavior can be explained by the opposite effects of introducing carboxyl groups. These effects are the blocking of some of the cellulose hydroxyl groups and the opening of the cellulose structure due to decrystallization. At an early stage of carboxymethylation of cotton linters, some of the OH groups are blocked, whereas the opening of the cellulose structure is very low or negligible, therefore, the graft yield tends to decrease. By increasing the D.S., the effect of the carboxyl groups in opening the cellulose structure begins to predominate and, hence, the graft yield increases gradually.

VI. EFFECT OF GRAFTING MEDIUM

Graft copolymerization of methylmethacrylate onto paper wood pulp using ceric ammonium nitrate as the initiator has been studied. Different experimental conditions have been used, including both water and water-organic solvent systems. The effects of composition of the grafting medium and grafting temperature on the grafting process are examined.

A. Effect of Different Wetting Solvent–Water Systems

Table 1 shows that alcohols, e.g., methanol, ethanol propanol, butanol, and isobutanol, have a distinct effect of chain length on the grafting efficiency. There is a decrease in the grafting yield with an increasing chain length and degree of branching of the alcohols. Copolymerization is the most pronounced reaction in methanol–water and is decreased in isobutanol–water medium. A small molecule, such as methanol, is not only capable of swelling the trunk polymer but also miscible in all proportions with methylmethacrylate and water, thus simplifying access and diffusion to the grafting sites. By contrast, butanol and isobutanol are relatively poor swelling agents for cellulose and the grafting is correspondingly low (Table 1).

By further increasing the methanol in the grafting medium, the graft yield decreases. This can be related to the lower solubility of the initiator in the grafting medium and a reduced formation of free radicals, which

Table 1 Effect of Ratio of Different Solvents in Water System on Polymer Loading and Grafting Yield

Solvent/water ratio	0/30		1/29		2/28		4/26		8/22		12/18		29/1	
Type of solvent	p.l. %	g.y. %	p.l. %	g.y. %	p.l. %	g.y. %	p.l. %	g.y. %	p.l. %	g.y. %	p.l. %	g.y. %	p.l. %	g.y. %
Methanol	—	—	40	34	49	40	52	42	54	44	34	27	11	8
Ethanol	—	—	35	30	42	32	46	34	45	34	32	21	10	7
Propanol	—	—	30	28	36	32	43	35	16	13	8	7	—	—
Butanol	—	—	28	25	25	20	22	18	10	8	8	7	—	—
Isobutanol	—	—	25	21	22	18	15	14	10	9	8	6	—	—
Water	43	33												

p.l. = polymer loading; g.y. = grafting yield.

decrease the initiation and propagation of the grafting process.

The grafting in aqueous medium is higher than in case of the 97% methanol medium. This can be attributed to the lower dissolution of the initiator in the solvent medium. On the other hand, grafting in methanol–water medium (8:22) produces higher grafting than in water medium.

B. Effect of Mixed Solvent–Water Systems on Grafting

A mixture of methanol and butanol or isobutanol in water with a ratio of 4:4:22 in the grafting process produces a higher graft yield than a mixture of butanol or isobutanol in water with a ratio of 8:22.

Partial replacement of ethanol by methanol has nearly no effect. In the case of propanol an increase in grafting is visible. This can be attributed to the mixing of higher carbon alcohols, e.g., butanol and isobutanol, with the active solvent methanol, which increases the miscibility of the monomer in these grafting systems and, consequently, increases the penetration of monomer to the active sites on the cellulose chains.

C. Effect of the Addition of Acetone on the Grafting Process

The effect of the addition of acetone to the grafting medium has been investigated. The acetone–water system produces a high-grafting yield, especially at low concentrations of acetone in the grafting medium. This is explained by the inhibition of the formation of homopolymer. By increasing the ratio of acetone–water up to 8:22, the grafting yield is lower than that in case of the 8:22 methanol–water medium.

D. Grafting in a Nonwetting Solvent

Grafting of methylmethacrylate onto cellulose using ceric ammonium nitrate (Ce^{4+}) as the initiator in a benzene–water system is also demonstrated. The grafting yield in a benzene–water system is much lower than in the case of the methanol–water system and decreases by increasing the ratio of benzene to water. This can be due to the lower polarity and wetting power of benzene, which leads to poor swelling of the cellulose.

ACKNOWLEDGMENTS

The authors wish to express their gratitude to Dr. A. Ibrahem for his kind cooperation and his helpful discussion of some parts of this work. The authors gratefully acknowledge the financial assistance received from cellulose of paper Det., National Research Centre, Dokki; Cairo, Egypt.

REFERENCES

1. A. Hebeish, and P. C. Mehta, *Textile Res. J., 37*: 9115 (1967).
2. A. Hebeish, and P. C. Mehta, *J. Appl. Polym. Sci., 12*: 1625 (1968).
3. H. A. Krassto, and V. T. Stannett, *Adv. Polym. Sci., 4*: 111 (1965).
4. J. J. Guthrie, and Z. Haa, *Polymer, 15*: 133 (1974).
5. R. B., Phillips, J. Ouere, Z. Quiroy, and V. T. Stannett, *Tappi 55*: 858 (1972).
6. G. Mino, and S. Kaizerman, *J. Polym. Sci. 31*: 242 (1958).
7. J. H. Richard, *J. Soc. Dyers Colocirists, 80*: 640 (1965).
8. R. W. Faessinger, and J. S. Conta, U.S. Patent 3,359,224 (December 19, 1967).
9. W. J. Briekman, *Tappi, 56*: 97 (1973).
10. G. Machell, and G. N. Richards, *J. Chem. Soc.,* 3308 (1961).
11. S. Kaizerman, G. Mino and C. F. Meinheld, *Textil. Res. J., 32*: 136 (1962).
12. E. Schwob, V. T. Stannett, W. H. Rakowiz, and J. K. Magrane, *Tappi, 45*: 390 (1962).
13. R. M. Livshits, V. P. Alachev, M. V. Prokofeva, and Z. A. Pogovin, *Vysekomolekul. Soedin, 6*: 655 (1964).
14. S. Kimura, and H. Imoto. Makromole Chem. 42, 140 (1960).
15. Y. Iwakura, T. Kutopaki, and Y. Imai, *J. Polym. Sci., (A) 3*: 1185 (1965).
16. A. M. A. Nada, M. A. Yousef, and A. A. Ibrahem, *Acta Polymercca, 38*: 93 (1987).
17. A. A. Shabaka, M. A. Yousef, and A. M. A. Nada, *Polym. Plast. Techn. Eug., 29(1, 2)*: 167 (1990).
18. A. M. A. Nada, S. F. El-Kalyoubi, and I. El-Roweing, *Polym. Plast. Techn. Eug., 28(4)*: 439 (1989).
19. Q. Y. Mauson, Nagaty, A. Beshay, M. Nosier, *J. Polym. Sci., 21*: 715 (1983).
20. A. Hebeish, N. Abore-Zeid, A. Waly, and E. El-Rofie, *J. Appl. Polym. Sci., 23*: 306 (1979).
21. Y. Londier, and P. Label, German Patent 1,100,286 (Appl. Oct. 11, 1958).
22. A. A. Ibrahem and A. M. A. Nada, *Acta Polymercca, 36*: 342 (1985).
23. A. M. A. Nada, and M. A. Yousef, *Acta Polymerica, 40*: 69 (1989).
24. M. A. Yousef, and A. M. A. Nada, *Acta Polyanerica, 41*: 581 (1990).
25. A. A. Ibrahem, and A. M. A. Nada, *Bull., NRC, Egypt, 10*: 206 (1985).
26. A. M. A. Nada, and A. A. Ibrahem, *Acta Polymercca, 38*: 28 (1989).
27. S. El-Meadawy, and A. El-Ashmawy, *Acta Polymerica, 34*: 229 (1983).
28. I. Sakurad, T. Gkada, and K. J. Kaji, *Polymer Sci., C37*: 1 (1972).
29. J. J. Harris, J. H. Garra, I. V. Dogray, and J. C. Arthur, Jr. *Text. Res. J., 42*: 14 (1994).
30. T. Akira, Y. Sugahara, and Y. Honkawa, *Transaction, 43(7)*: 362 (1987).
31. A. A. Shabaka, A. El-Agramy, and A. M. A. Nada, *Isotopenpraxis, 27*: 251 (1991).
32. W. Deters, and D. Huang, *Fasuforsch. Textiltech, 19(S)*: 183 (1963).
33. J. G. Guthrie, M. B. Huang, and G. O. Phillips, *Eur. Polymer J., 8*: 47 (1982).
34. K. D. Lawrence, and D. Rades. *Appl. Polym. Sci., 17*: 2653 (1973).
35. G. Rauseng, and S. Sumner, *Tappi, 45(1)*: 203 (1962).

36. H. G. Higgurs, and J. de Yong, *Transaction of Oxford Symposium, 2*: 51 (1962).

37. Y. Ogrwara, and H. Kubota, *Kogyo Kagaku Zasshi 71(1)*: 171 (1968).

38. S. Heileol, and S. El-Kalyoubi, *J. Appl. Polym. Sci., 27*: 3691 (1982).

39. S. Lipatou, D. V. Zharhovskii, and M. Zagraevskaya, *Kollod Zh. 21*: (1959). Absk Bull. Inst. Paper. Chem. 32(3), 1333 (1961).

40. N. I. Klenkova, *Zur. Priki-Chim, 40*: 2191 (1967).

41. A. Koura, and A. A. Ibrahem, *Faserfersch. Textiltech., 25*: 57 (1974).

42. A. M. A. Nada, A. A. Shabaka, M. A. Yousef, and K. N. Abd El-Nour, *J. Appl. Polym. Sci., 40*: 731 (1990).

43. A. M. A. Nada, and M. Z. Sefain, *Cell. Chem. Tech., 20(2)*: 209 (1986).

44. A. M. A. Nada, S. F. El-Kalyoubi, and I. El-Roweing, *Polym. Plast. Tech. Eng., 29(182)*: 27 (1990).

45. A. Kantouch, A. Hebeish, and M. El-Rafie, *Appl-Polym. Sci., 15*: 11 (1971).

46. N. El-Shinawy, A. M. A. Nada, and S. F. El-Kolyorbi, *Cell. Chem. Techn., 20*: 333 (1986).

47. A. A. Ibrahem, and A. M. A. Nada, *Acta Polymerica, 37(5)*: 320 (1986).

48. A. Kantoueh, A. Hebieoh, and M. El-Rofie, *Eur. J. Polym., 6*: 1575 (1970).

49. A. M. A. Nada, M. D. Bodry, and K. N. Abd El-Nour, *Polym. Degrad. Stab., 36*: 201 (1992).

35

Graft Copolymerization of Vinyl Monomers Onto Macromolecules Having Active Pendant Group via Ceric Ion Redox or Photo-Induced Charge-Transfer Initiation

Kun-Yuan Qiu and Xin-De Feng
Peking University, Beijing, China

I. INTRODUCTION

Graft copolymers have been synthesized from the grafting vinyl monomer copolymerization onto polymers by radical, anionic, and cationic initiation processing. In general, there are three ways for synthesis of graft copolymers by radical initiation reaction: (1) chain-transfer reaction; (2) redox initiation with ceric salt, Ce(IV) ion, and photo-induced charge-transfer reaction; and (3) copolymerization of vinyl monomer with macromer [1,2]. However, these methods yield a mixture of homopolymer, graft copolymer, and ungrafted original polymer. Some systems such as high-impact polystyrene (HIPS), acrylonitrile-butadiene-styrene (ABS) polymer, and methyl methacrylate-butadiene-styrene (MBS) polymers have been commercialized. For characterization, the purification of graft copolymer is necessary and even the purification process is tedious. Furthermore, the free radical grafting reactions of polymers can proceed either along the main chain (backbone) of the polymer or from the pendant groups (side chain), these make it difficult to realize the grafting site in a graft copolymer. In our laboratory on the basis of investigation of vinyl radical polymerization initiation systems [3–7], i.e., Ce(IV) ion redox and amine-benzophenone photo-induced charge-transfer initiation systems, we have developed some techniques for graft copolymerization via initiation reaction of active pendant groups of polymers either by redox or photo-induced charge-transfer reaction. The results of our research work concerning the model compounds of active functional group for Ce(IV) ion redox

initiation systems and their initiation mechanism; the synthesis of macromolecules having active pendant group and its grafting reaction; and the grafting mechanism as well as the grafting site are reviewed in this chapter.

II. CERIC ION REDOX INITIATION SYSTEMS

Since the 1950s, ceric salt, Ce(IV) ion alone, or Ce(IV) ion redox systems consisting of Ce(IV) ion and alcohol, aldehyde, ketone, amine, and carboxylic acid were investigated to initiate vinyl monomers to polymerize [8–12]. Although some papers reported the grafting reaction of polymer initiated by Ce(IV) ion [13–16], much less attention was paid to the grafting reaction of macromolecules having active pendant groups initiated with Ce(IV) ion. Lin et al. [17,18] have reported the graft copolymerization of acrylamide (AAM) onto macromolecules having active pendant groups such as 3,4-dihydroxycyclohexylethyl acrylate and 2,3-dihydroxypropenyl acrylate initiated with Ce(IV) ion. In order to develop the scope of the active pendant group, a series of works concerning the effect of model compounds on the rate of polymerization in the presence of Ce(IV) ion were carried out in our laboratory, and some novel redox initiation systems consisting of Ce(IV) ion-acetanilide [3,4], Ce(IV) ion-alkyl tolylcarbamate [19], Ce(IV) ion-1,3-dicarbonyl [20,21], Ce(IV) ion-triketone [22], Ce(IV) ion-actoacetanilide [23], and Ce(IV) ion-N-acetyl-N'-4-

tolylurea [24] have been established. The initiation mechanism was proposed based on the end group analysis of the resulting polymer and electron spin resonance (ESR) studies. Moreover, the active functional group mentioned previously can serve as an active site for the grafting reaction of polymers having such an active pendant group. Grafting reactions of macromolecules having 4-tolylcarbamoyl, 1,3-diketone, and 4-tolylurea pendant groups initiated with Ce(IV) ion were performed successfully and are discussed in Section V.

III. NOVEL CERIC ION REDOX INITIATION SYSTEMS

Recently, in our laboratory the following novel Ce(IV) ion redox initiation systems have been investigated for vinyl radical polymerization.

A. Ce(IV) Ion–Phenylcarbamoyl Compound Systems

Samal et al. [25] reported that Ce(IV) ion coupled with an amide, such as thioacetamide, succinamide, acetamide, and formamide, could initiate acrylonitrile (AN) polymerization in aqueous solution. Feng et al. [3] for the first time thoroughly investigated the structural effect of amide on AAM polymerization using Ce(IV) ion, ceric ammonium nitrate (CAN) as an initiator. They found that only acetanilide (AA) and formanilide (FA) promote the polymerization and remarkably enhance R_p. The others such as formamide, N,N-dimethylformamide (DMF), N-butylacetamide, and N-cyclohexylacetamide only slightly affect the rate of polymerization. This can be shown by the relative rate (R_r), i.e., the rate of AAM polymerization initiated with ceric ion-amide divided by the rate of polymerization initiated with ceric ion alone. R_r for CAN-anilide system is approximately 2.5, and the others range from 1.04–1.11.

1. Ce(IV) Ion–Acetanilide Systems

Feng et al. [3] have studied the structural effect of acetanilide on the AAM polymerization either in water-formamide [3], water-acetonitrile [4], and water-DMF [26] mixed solution using Ce(IV) ion-acetanilide and its substituted derivatives as the initiator. The results showed that an electron donating substituent on the phenyl group would enhance the R_p, while an electron withdrawing group would decrease it, as shown in Table 1 [26].

2. Ce(IV) Ion–Alkyl Phenylcarbamate Systems

We have investigated Ce(IV) ion-carbamates, such as methyl and butyl 4-methylphenyl-carbamate, (MTC and BTC, respectively), or methyl, ethyl, and butyl phenylcarbamate, (MPC, EPC and BPC, respectively), systems for AAM polymerization [19]. It was found that the presence of carbamate compounds can promote the polymerization and enhance the rate of AAM polymerization (R_p) in descending order as:

$$MTC \approx BTC \gg MPC \approx EPC \approx BTC \qquad (1)$$

The structural effect of alkyl groups such as methyl, ethyl, and n-butyl on the R_p is small. Alkyl 4-methylphenylcarbamate can be chosen as a model compound for the hard segment of poly(ether-urethane) (PEU). This group can initiate grafting reaction with Ce(IV) ion and the grafting site was proposed at the hard segment of PEU [3,15] as shown in Scheme (1).

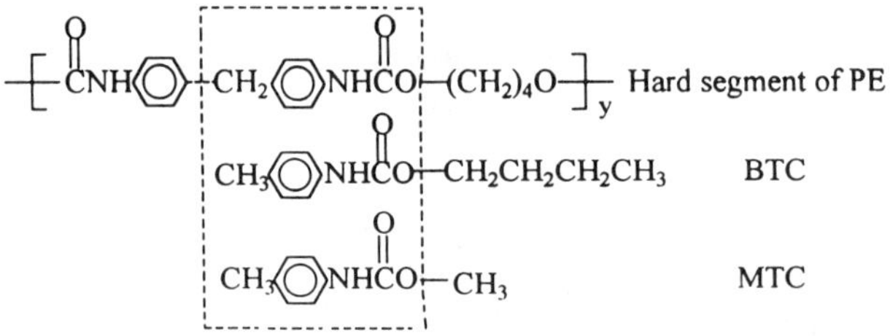

Scheme 1

Table 1 Structural Effect of Acetanilide on AAM Polymerization

Compound	Substituent (X)	$R_p \times 10^5$ (mol/L·s)	Relative rate R_r	R(X/H)
—		8.75	1.00	
AA	H	16.8	1.92	1.00
p-APT	p-CH$_3$	26	2.97	1.54
m-AABA	m-COOH	15.3	1.75	0.91
m-AAl	m-CH$_2$OH	18	2.06	1.07
m-AAe	m-CH$_2$OCOCH$_3$	17.5	2	1.04
o-AABA	o-COOH	19.3	2.2	1.14
o-AAl	o-CH$_2$OH	13	1.48	0.77
o-AAe	o-CH$_2$OCOCH$_3$	12.8	1.46	0.76

AA = acetanilide; p-APT = N-(p-methylphenyl) acetamide; m-AABA = m-acetamino benzoic acid; m-AAl = m-acetamino phenylmethanol; m-AAe = o-acetaminophenylmethyl acetate; o-AABA = o-acetaminobenzoic acid; o-AAl = o-acetaminophenylmethanol; o-AAe = o-acetaminophenylmethyl acetate.

B. Ce(IV) Ion–1,3-Dicarbonyl Compound Systems

Although carbonyl compounds, such as formaldehyde [27,28], can couple with Ce(IV) ion to initiate acrylonitrile (AN) or methyl methacrylate (MMA) polymerization, the remarkable activity of aliphatic aldehyde had not been noticed until the paper of Sun et al. [29] was published. They found that aliphatic aldehydes always possess higher activity than alcohol compounds, in other words, the redox reaction between aldehyde and Ce(IV) ion is more favorable than alcohol. Therefore, based on the investigation of the effect of *cis*- and *trans*-1,2 cyclohexane-diol on the AN polymerization initiated with Ce(IV) ion, Li et al. [30], and Yu et al. [31] commented that the following Mino's mechanism (1956) of 1,2-diol with Ce(IV) ion as a fashion for the grafting reaction of cellulose and starch is questionable:

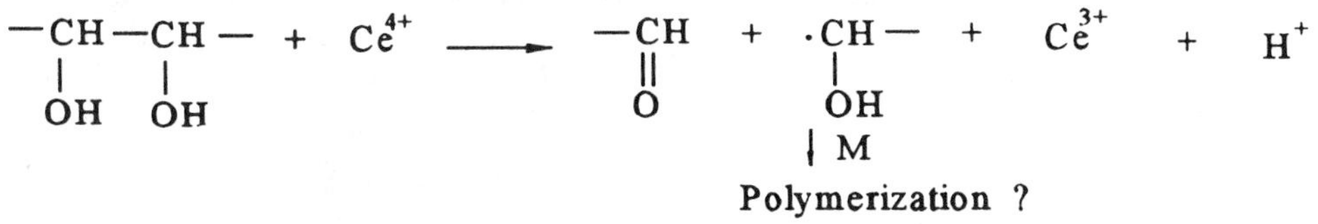

$$-\overset{|}{\underset{OH}{CH}}-\overset{|}{\underset{OH}{CH}}- + Ce^{4+} \longrightarrow -\overset{\|}{\underset{O}{CH}} + \cdot\overset{|}{\underset{\underset{\downarrow M}{OH}}{CH}}- + Ce^{3+} + H^+$$

Polymerization ?

Yu et al. [31] pointed out that hydroxymethene radical will react with Ce(IV) ion to form a carbonyl group very quickly, and then it can further oxidize to form carbonyl radical. Therefore, the grafting polymerization would take place on the carbonyl radical as the following:

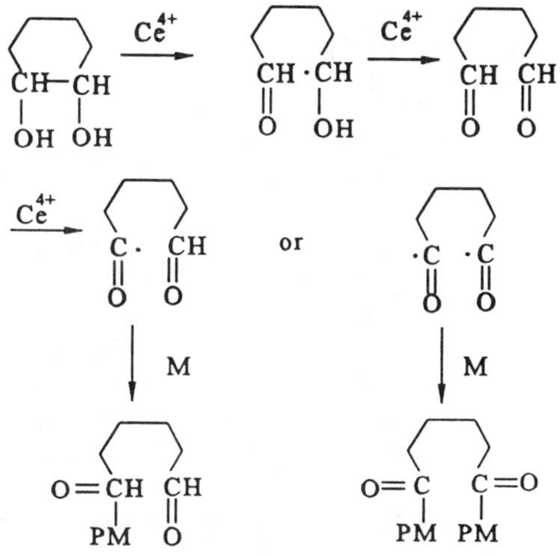

Scheme 2

Voong et al. [32] have investigated the effect of ketone compounds on AN polymerization in the presence of Ce(IV) ion and found that acetone, 2-butanone, 3-pentanone, cyclopentanone, and cyclohexanone were effective in promoting polymerization. Much less attention has been paid to the Ce(IV) ion-1,3-diketone compound initiation system. In a patent [33], an initiator consisting of Ce(IV) ion-acetylacetone(2,4-pentanedione, AcAc) was used for AAM polymerization. To date no detailed information has been published. Recently, we [20,21,34–36] have thoroughly investigated a series ceric redox initiation system containing 1,3-diketone compounds and found that 1,3-diketones, such as AcAc, benzoylacetone (BzAc), 3-benzyl acetylacetone(3-benzyl-2,4-pentanedione, [BzyAcAc]), dibenzoylmethane (DBzM), and triketone, such as 3-benzoyl acetyl acetone (BzAcAc), and 1,3-ketone ester, such as ethyl acetoacetate (EAcAc), coupled with Ce(IV) ion can form a redox initiator for AAM polymerization. The data are compiled in Table 2.

It can be seen that all diketones, even the substituted diketone such as BzyAcAc, can remarkably enhance the R_p. BzyAcAc can be chosen as a model of poly[3-(4-vinylphenylmethyl)-2,4-pentanedione] [P(St-

Table 2 Effect of 1,3-Diketones or Triketone on AAM Polymerization Initiated by Ce(IV) Ion

Reductant	Structure	$R_p \times 10^4$ (mol/L·s)	E_a (kJ/mol)
None	—	1.2	41.2
2-Butanone	$CH_3COCH_2CH_3$[a]	2.37	—
AcAc	$CH_3COCH_2COCH_3$	9.43	24.7
BzyAcAc	$CH_3COCH(CH_2C_6H_5)COCH_3$	9.08	12.7
BzAc	$C_6H_5COCH_2COCH_3$	11.98	23.9
DBzM	$C_6H_5COCH_2COC_6H_5$	12.46	22.5
BzAcAc	$CH_3COCH(COC_6H_5)COCH_3$	11.66	4.5

[AAM] = 1.0 mol/L; [CAN] = 2.5 × 10^{-3} mol/L; [Reductant] = 1.0 × 10^{-3} mol/L.
[a] [AAM] = 2.0 mol/L; [CAN] = 5.0 × 10^{-3} mol/L; [Monoketone] = 5.0 × 10^{-3} mol/L.
H_2O/CH_3CN = 3:1 (v/v) at 40.0°C.
CAN = ceric ammonium nitrate.

CH$_2$-AcAc)], the grafting reactions of this polymer initiated with Ce(IV) ion will be discussed in Sec. VI.B. Furthermore, the order of the promoting activity of diketone is always greater than that of monoketone.

C. Ce(IV) Ion–Acetoacetanilide Systems

As we have mentioned previously, 1,3-diketone and anilide were very effective reducing agents for vinyl polymerization initiated by ceric ion, respectively. Acetoacetanilide (AAA), a compound having a 1,3-diketone and an anilide structure as well:

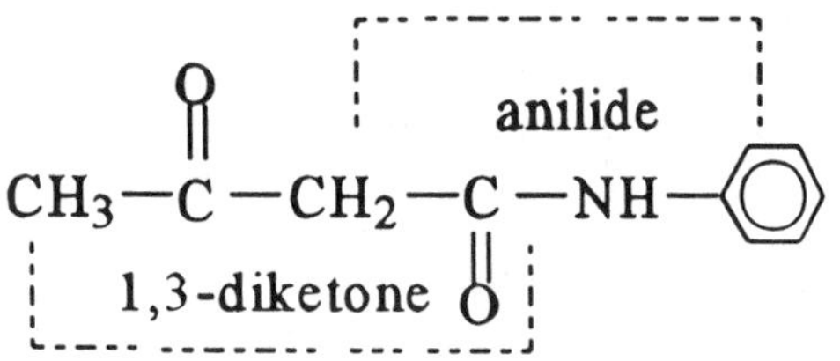

Can it promote vinyl polymerization initiated with Ce(IV) ion? Dong et al. [37–39] for the first time reported that AAA and its derivatives such as o-acetoacetotoluidine (AAT), o-acetoacetanisidide (AAN), and 2-benzoyl acetanilide (BAA) possess very high reactivity toward Ce(IV) ion in initiating the polymerization of vinyl monomer. The results are tabulated in Table 3.

From Table 3, it can be seen that the reactivity of acyl acetanilide, such as BAA or AAA, is higher than that of the other reductant reported from our laboratory, i.e., acetanilide (AA), N-acetyl-p-methylaniline (p-APT), acetylacetone (AcAc), and ethyl acetoacetate (EAcAc). Moreover, the promoting activities of derivatives of acetoacetanilide were affected by the ortho substituent in benzene ring, and the relative rate of polymerization (R_r) decreased with the increase of the bulky ortho substituent to the redox reaction between Ce(IV) ion and substituted acetoacetanilide.

D. Ce(IV) Ion-*N*-Acyl-*N'*-4-tolylurea Systems

Urea has no promoting effect on the AAM polymerization initiated with Ce(IV) ion. Recently, Qiu et al. [24] have studied the effect of *N*-acryloyl-*N'*-4-tolylurea (ATU), *N*-methacryloyl-*N'*-4-tolylurea (MTU), and *N*-acetyl-*N'*-4-tolylurea (AcTU) on AAM polymerization initiated with Ce(IV) ion and found that these three urea compounds have a high promoting effect on the polymerization of AAM. The data are cited in Table 4.

It can be seen from Table 4 that the descending order of R_p is as follows:

$$\text{CAN-AcTU} > \text{CAN-ATU} > \text{CAN-MTU} \qquad (2)$$

It reveals that either the steric effect of the bulky *N*-substituent on the interaction of Ce(IV) ion and urea reductant, or the electron withdrawing effect of the vinyl group, will reduce the coordination of Ce(IV) ion with the carbonyl group, thus resulting in a decrease of the R_p.

IV. INITIATION MECHANISM

To clarify and reveal the initiation mechanism of the Ce(IV) ion redox system in vinyl polymerization, we have thoroughly determined the initial radical structure by means of ESR spectrum analysis and end group analysis of the resulting polymer. The ESR spectra can reveal the structure of the free radicals directly. Sometimes, however, it is difficult to trap a nitrogen center radical in an amide, such as CH$_3$CON·(C$_6$H$_5$) with 2-methyl-2-nitrosopropane (MNP), with the result that no signal is observed in the ESR spectrum of CAN/AA/MNP system. End group analysis is a unique technique to reveal the presence of reductant moiety as an end group in the resulting polymer initiated with the Ce(IV) ion reductant system. This method also can demonstrate the structure of the initial radical indirectly—as the radical formed, it will initiate monomer polymerization and will enter into the polymer chain as an end group.

Table 3 Effect of Acetoacetanilide and Its Derivatives on AAM Polymerization Initiated by Ce(IV) Ion in Aqueous Media at 25.0°C

Reductant	Structure	$R_p \times 10^4$ (mol/L·s)	R_r	E_a (kJ/mol)
—	—	1.6	1.0	58.3
AAA	C$_6$H$_5$NHCOCH$_2$COCH$_3$	48.4	30.3	9.18
AAT	o-CH$_3$C$_6$H$_4$NHCOCH$_2$COCH$_3$	38.4	24.0	—
AAN	o-OCH$_3$C$_6$H$_4$NHCOCH$_2$COCH$_3$	35.6	22.3	−1.26
BAA	C$_6$H$_5$NHCOCH$_2$COC$_6$H$_5$	55.9	34.9	—

[CAN] = 1 × 10^{-3} mol/L; [Red.] = 0.5 × 10^{-3} mol/L; [AAM] = 1.0 mol/L.

Table 4 Effect of N-acyl-N'-tolylurea on AAM Polymerization Initiated by Ce(IV) at 35°C in H$_2$O–THF (v/v = 1:1)

Reductant	Structure	$R_p \times 10^4$ (mol/L·s)
—	—	No polym. during 1 h
ATU	CH$_2$=CHCONHCONH-C$_6$H$_4$CH$_3$-p	0.86
MTU	CH$_2$=C(CH$_3$)CONHCONH-C$_6$H$_4$CH$_3$-p	0.78
AcTU	CH$_3$CONHCONH-C$_6$H$_4$CH$_3$-p	1.24

[CAN] = 5.0 × 10^{-4} mol/L; [Red.] = 1.0 × 10^{-4} mol/L; [AAM] = 0.2 mol/L.

The initial radicals formed from the reaction of Ce(IV) ion and reductant systems can be trapped by MNP. The spin adducts of the initial radicals and MNP were observed by means of an ESR spectrometer. The structure of the initial radicals and the hyperfine splitting constants of the spin adduct of the radical with MNP and α-phenyl-N-$tert$-butylnitrone (PBN) are compiled in Tables 5 and 6, respectively.

The formation of spin adducts from radicals and MNP is as follows:

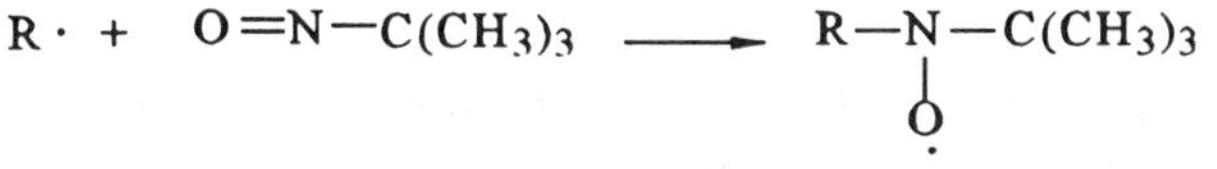

$$R\cdot \ + \ O=N-C(CH_3)_3 \longrightarrow R-\underset{\underset{O}{|}}{N}-C(CH_3)_3$$

The formation of spin adducts from radicals and PBN is as follows:

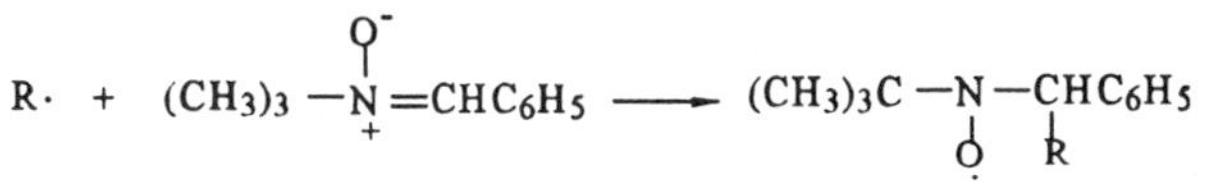

$$R\cdot \ + \ (CH_3)_3-\overset{\overset{O^-}{|}}{\underset{+}{N}}=CHC_6H_5 \longrightarrow (CH_3)_3C-\underset{\underset{O}{|}}{N}-\underset{\overset{|}{R}}{C}HC_6H_5$$

The initial radicals formed from the Ce(IV) ion redox system can initiate a monomer to polymerize and form an end group of the resulting polymer. When the reductant exhibits a carbonyl group, the amide group can be conveniently detected by the FT-IR spectrum of the polymer, such as polyacrylonitrile (PAN). The FT-

Table 5 Hyperfine Splitting Constants of CAN-Reductant–MNP Systems [20,36,38,40]

Reductant	Radical trapped by MNP	Hyperfine splitting constant (mT)		
		a_α^N	a_β^N	a_γ^N
AcAc	(CH$_3$CO)$_2$CH·	1.36	0.36	—
BzyAc	(CH$_3$CO)$_2$C·(CH$_2$C$_6$H$_5$)	1.44	—	—
BzAc	CH$_3$COCH·(COC$_6$H$_5$)	1.41	0.46	—
DBzM	(C$_6$H$_5$CO)$_2$CH·	1.41	0.63	—
BzAcAc	CH$_3$COCH·(COC$_6$H$_5$)[a]	1.4	0.48	—
EAcAc	CH$_3$COCH·(COOC$_2$H$_5$)	1.39	0.36	—
BAA	C$_6$H$_5$NHCOCH·(COC$_6$H$_5$)	1.45	0.49	—
AAA	C$_6$H$_5$NHCOCH·(COCH$_3$)	1.43	0.3	0.09
AAT	o-CH$_3$C$_6$H$_4$NHCOCH·(COCH$_3$)	1.44	0.25	0.09
AAN	o-OCH$_3$C$_6$H$_4$NHCOCH·(COCH$_3$)	1.43	0.22	0.09

[a] A triketone compound BzAcAc when it reacts with Ce(IV) ion, the free radical observed is obtained by splitting an acetyl group in BzAcAc.

Table 6 Hyperfine Splitting Constants of CAN-reductant–PBN Systems [38]

Reductant	Radical trapped by PBN	Hyperfine splitting constant (mT)		
		a_α^N	a_β^N	a_γ^N
BAA	C$_6$H$_5$COCH$_2$CON·(C$_6$H$_5$)	1.44	0.49	0.07
AAA	CH$_3$COCH$_2$CON·(C$_6$H$_5$)	1.43	0.3	0.09
AAT	CH$_3$COCH$_2$CON·(C$_6$H$_4$CH$_3$-o)	1.44	0.25	0.09
AAN	CH$_3$COCH$_2$CON·(C$_6$H$_4$OCH$_3$-o)	1.43	0.23	0.09

IR spectra of PAN obtained from the CAN-1,3-diketone systems revealed that, besides the characteristic absorption band of CN group at 2244 cm^{-1}, the characteristic absorption bands at 1727 and 1700 cm^{-1} (keto form of 1,3-diketone) and 1625 cm^{-1} (enol form of 1,3-diketone) were observed simultaneously. Therefore, the 1,3-diketone moiety is present as an end group in PAN.

Although the initial radical formed from the reaction of Ce(IV) ion and acetylanilide (AA) and N-p-tolylacetamide (PTA) has never been observed in the ESR studies, the presence of AA, PTA moieties in the end group of PAN obtained from initiating the CAN-AA, CAN-PTA system have been detected by the FT-IR spectra analysis method. Similar results were observed in the end group analysis of CAN-phenylcarbamate, CAN-N-acyl-N'-tolylurea initiation systems.

Based on the ESR studies and the end group analysis, the initiation mechanism of Ce(IV) ion redox systems is proposed as:

Ce(IV) ion–acetanilide system:

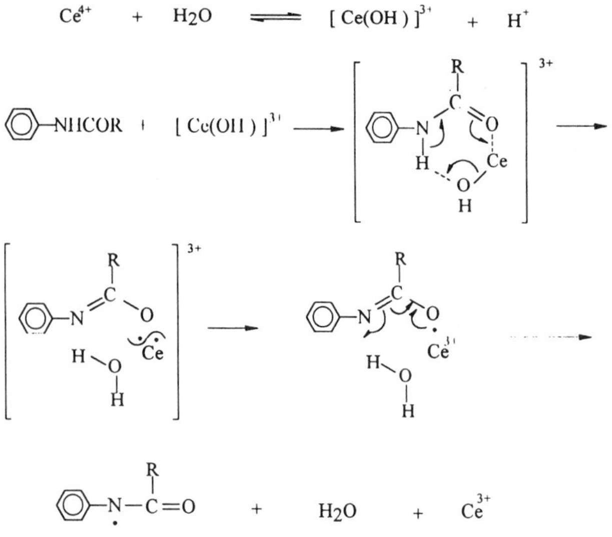

Scheme 3

Ce(IV) ion–1,3-diketone system:

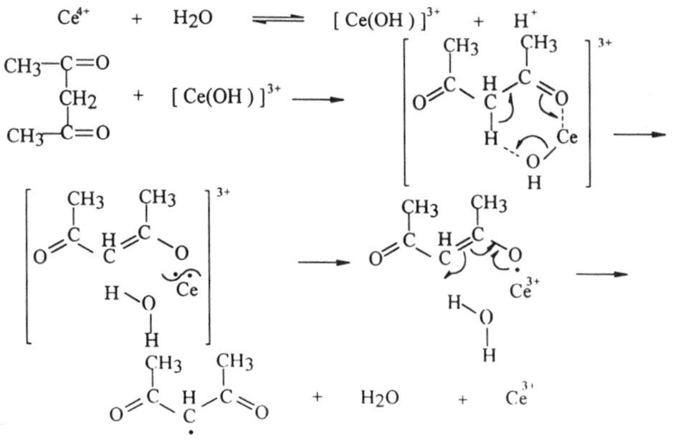

Scheme 4

Ce(IV) ion–acetoacetanilide system:

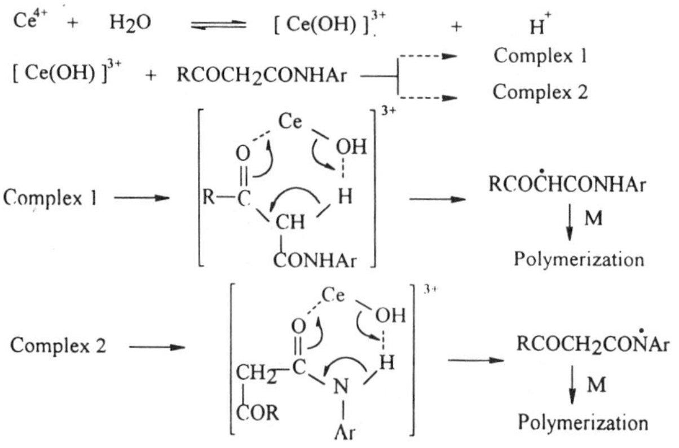

Scheme 5

V. GRAFTING REACTION OF MACROMOLECULES

Among the transition metal ions, Ce(IV) ion is the most widely used initiator for grafting reaction of cellulose. Many papers reported on the grafting copolymerization of vinyl monomers onto cellulose by using the Ce(IV) ion as the initiator. This technique is based on the fact that when cellulose is oxidized by the Ce(IV) ion, free radicals capable of initiating vinyl polymerization are formed on the cellulose. Since the radicals on the cellulose are produced more quickly than the homopolymerization of monomer initiated by the Ce(IV) ion alone, the result is that less homopolymers are obtained. Some review articles [13,41,42] have summarized the graft copolymerization using the ceric ion method. Although the mechanism of grafting copolymerization is still not completely confirmed, it is important to note the work of Yu et al. [31] on the studies of grafting mechanism of 1,2-diols compound, as pointed out in Section III.B.

VI. GRAFTING REACTION OF MACROMOLECULES HAVING AN ACTIVE PENDANT GROUP INITIATED WITH CERIC ION

Lin et al. [17,18] have reported the grafting copolymerization of AAM onto macromolecules having active pendant groups using the Ce(IV) ion as the initiator. Few studies in the similar field of grafting reaction are reported so far. Based on the investigation of the novel Ce(IV) ion redox system, we have developed a new technique for the grafting reaction of macromolecules having an active pendant group, such as p-tolylcarbamoyl(4-

methylphenyl carbamoyl) and 1,3-diketone pendant group, initiated with Ce(IV) ion.

The synthetic methods of macromolecules having an active pendant group include: (1) the transformation reactions of polymer and copolymers, and (2) polymerization and copolymerization of functional monomers having active pendant groups. The macromolecules, either in the shape of film or microbeads, can be used as the substrate. As we have mentioned previously, the rate of polymerization initiated with the Ce(IV) ion redox system is much faster than that initiated by Ce(IV) ion alone, as expressed in $R_r \gg 1$. Therefore, the graft copolymerization is favored much more than the homopolymerization, and the yield of the homopolymer was depressed to a very low level.

A. Grafting Reaction of Macromolecules Having 4-Tolylcarbamoyl Pendant Group

Macromolecules having 4-tolycarbamoyl and phenylcarbamoyl pendant groups can be synthesized by the polymerization and copolymerization of functional monomers. Qiu and Song [43] have reported that the copolymers having 4-tolylcarbamoyl and phenylcarbamoyl pendant groups were synthesized by the copolymerization of functional monomers, i.e., N-4-tolyl-methacrylamide (NTMAAM), N-4-tolyl-acrylamide (NTAAM), and N-phenyl-methacrylamide (NPMAAM) with methyl acrylate (MA) and methyl methacrylate (MMA), respectively. The copolymers are represented as follows:

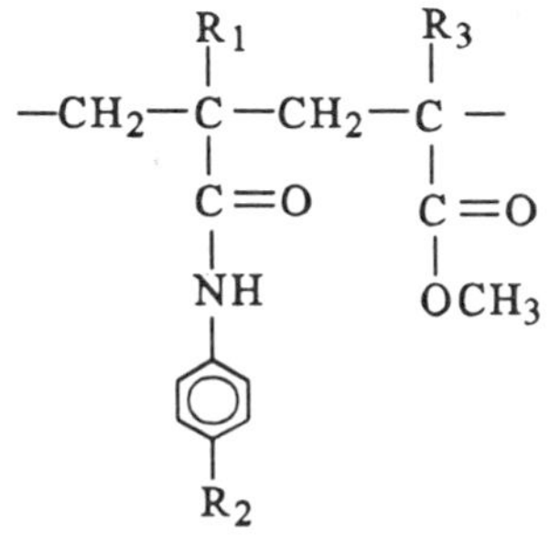

					R_1	R_2	R_3
poly(NTAAM-co-MA)	1				H	CH_3	H
poly(NTAAM-co-MMA)	2				H	CH_3	CH_3
poly(NTMAAM-co-MA)	3				CH_3	CH_3	H
poly(NTMAAM-co-MMA)	4				CH_3	CH_3	CH_3
poly(NPMAAM-co-MA)	5				CH_3	H	H
poly(NPMAAM-co-MMA)	6				CH_3	H	H

As mentioned previously, the 4-tolylcarbamoyl group is highly reactive toward the Ce(IV) ion. The graft copolymerization of AAM onto the films of copolymers 1 ~ 6 in the presence of Ce(IV) ion would take place, and the order of grafting percentage (G%) was observed as:

$$1 > 2 \gg 3 > 4 \gg 5 > 6 \qquad (3)$$

Obviously, the copolymers containing a more reactive functional monomer, i.e., monomer having a 4-tolylcarbamoyl group, have a higher grafting percentage. The grafting mechanism is proposed similar to the initiation mechanism of the Ce(IV) ion and acetanilide system, and the grafting reaction occurred at the pendant group

of the functional monomer unit via the hydrogen abstraction of amide to form a nitrogen center radical capable of initiating graft copolymerization.

Other types of copolymers having a 4-tolylcarbamoyl pendant group, poly(VAc-co-MAMT) 7 and poly(MA-co-MAMT) 8, were synthesized from the reaction of 4-toluidine with poly(vinyl acetate-co-maleic anhydride) and poly(methyl acrylate-co-maleic anhydride), respectively [44]. Graft copolymerization of AAM onto films of copolymers 7 and 8 were successfully performed by using the Ce(IV) ion as the initiator. The grafting reaction created from the reaction of the tolylcarbamoyl group with the Ce(IV) ion to form the nitrogen radical active site to initiate AAM graft copolymerization was proposed as follows (Scheme 6):

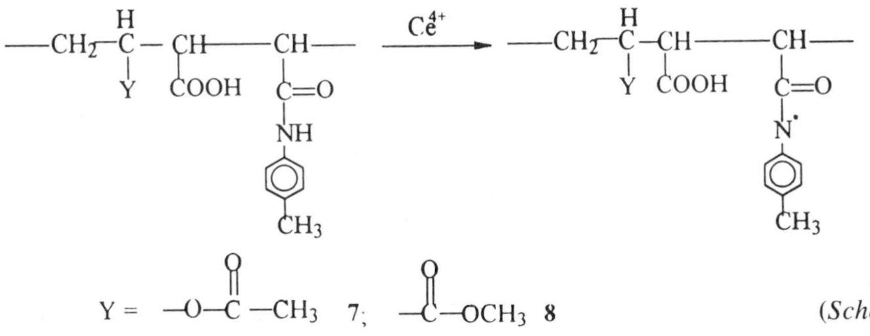

(Scheme 6 continues)

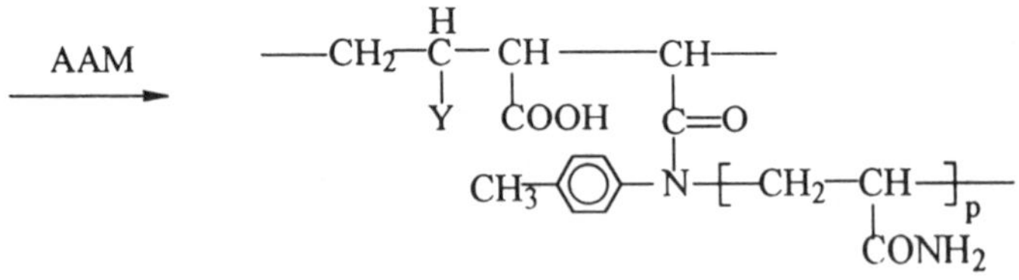

AAM →

$$—CH_2—\overset{H}{\underset{Y}{C}}—\overset{}{\underset{COOH}{CH}}——\overset{}{\underset{\overset{C=O}{|}}{CH}}—$$

$$CH_3—\text{⬡}—N\!-\!\!\left[CH_2—CH\right]_p$$
$$\underset{CONH_2}{|}$$

Scheme 6

Recently, poly(itaconamide) with 4-tolylcarbamoyl pendant groups have been synthesized in our laboratory. The polymer **9** and copolymers **10** and **11** were synthesized via aminolysis of poly(N-4-methyl-phenylitaconi-mide) and its copolymers, respectively. Graft copolymerization of AAM on the surface of the polymer films using Ce(IV) ion as the initiator has been investigated by us [45]. The mechanism of graft copolymerization was proposed as shown in Scheme 7.

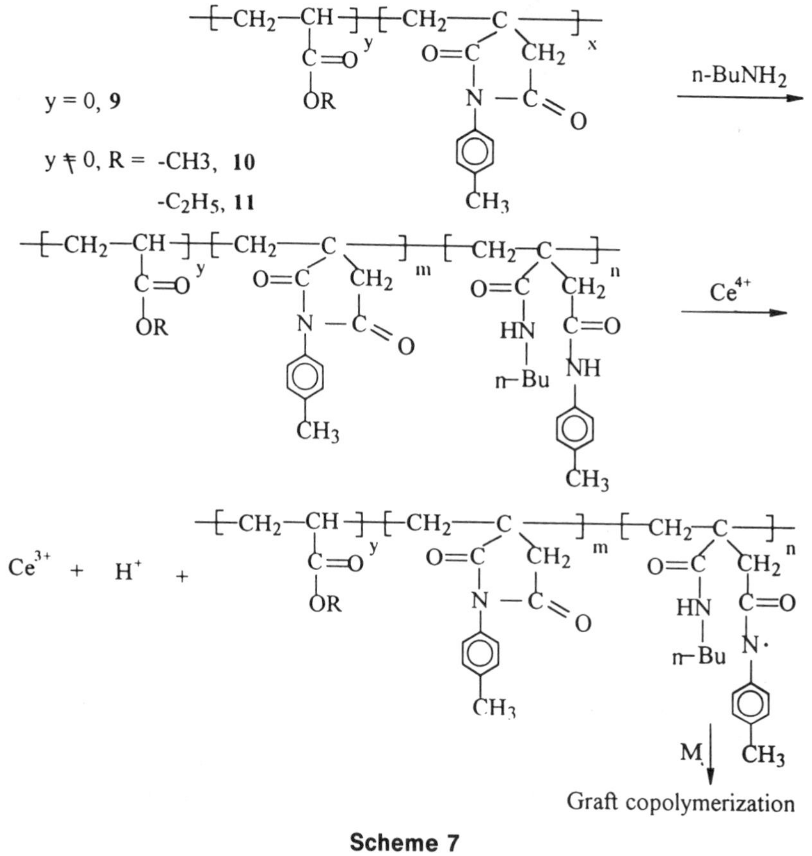

$y = 0,$ **9**

$y \neq 0, R = -CH3,$ **10**

$-C_2H_5,$ **11**

n-BuNH₂ →

Ce⁴⁺ →

Ce³⁺ + H⁺ +

M↓

Graft copolymerization

Scheme 7

The mechanism proposed that the nitrogen radical was formed in the 4-tolylcarbamoyl groups rather than in the butylcarbamoyl groups, which were both pendant groups of the polymers. In our previous work, as mentioned in Section III, the promoting activity of the 4-tolylcarbamoyl group (in APT R_r = 17) is much higher than the butylcarbamoyl group (in N-butylacetamide R_r = 1.10). Therefore, we proposed that graft copolymerization took place predominantly on the 4-tolylcarbamoyl pendant group.

Although we have proposed the grafting reaction mechanism of polymers having 4-tolylcarbamoyl pendant group initiated with ceric ion and the grafting site of the copolymer at the N-H of the 4-tolylcarbamoyl linkage via the oxidation of H by Ce(IV) ion as shown in Scheme (3), there is still lack of evidence for the grafting site, due to the difficulties of analyzing the nitrogen radical signals in the initiation system by ESR measurement.

Recently, Zhao and Qiu [46] have designed two

kinds of polymers having 4-tolylcarbamoyl pendant groups, i.e., poly(4-acetaminostyrene), P(St-NHCOCH$_3$), poly(NTMAAM) **12**, poly(NTMAAM-*co*-MA) **13**, poly(NTMAAM-*co*-EA) **14**, and poly(NT-MAAM-*co*-VAc) **15**, and they were chosen as the original polymer in the grafting reaction with Ce(IV) ion. Poly(4-acetaminostyrene) was prepared from cross-

linked polystyrene beads by nitration, reduction, and acetylation of the three-step reactions. Acrylamide graft copolymerization took place onto poly(4-acetaminostyrene) beads initiated with Ce(IV) ion, and the graft copolymer P[(St-NHCOCH$_3$)-*g*-AAM] was characterized. The mechanism of graft copolymerization is shown as Scheme 8.

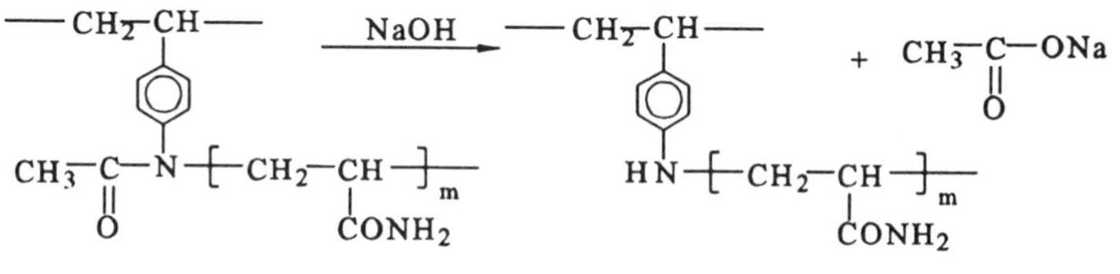

Scheme 8

When P[(St-NHCOCH$_3$)-*g*-AAM] was hydrolyzed in the basic solution no PAAM was released. The scanning electron microscopy (SEM) micrograph of the copolymer shows that the hydrolyzed grafted beads are still covered with PAAMs with salient micrographs. The results reveal that AAM graft copolymerization is initiated by the nitrogen radical rather than any other radical.

In a comparable experiment, AAM graft copolymers were prepared by the AAM graft copolymerization

on the surface of polymer films of polymers **13–16** using the Ce(IV) ion as the initiator. Hydrolysis of the grafted copolymer films in NaOH aqueous solution caused the grafted PAAM macromolecules to separate from the substrate films. The hydrolysis of the grafted copolymer provides some evidence for the grafting site and the mechanism of graft copolymerization of polymers having a 4-tolylcarbamoyl group. The reactions are shown in Scheme (9).

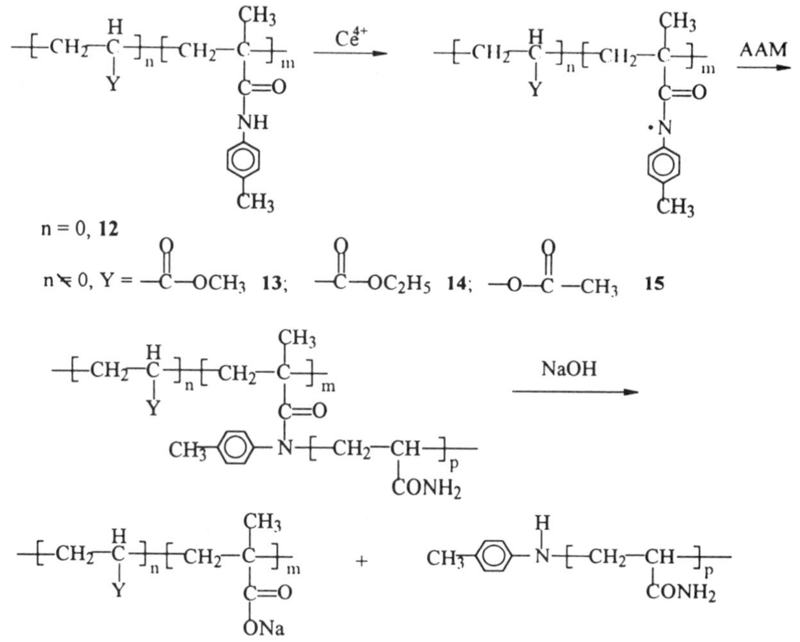

Scheme 9

B. Grafting Reaction of Macromolecules Having a 1,3-Diketone Pendant Group

Recently, Zhao et al. [34,47], and Qiu et al. [21] have reported the graft polymerization of AAM onto macro-

molecules having a 1,3-diketone pendant group such as poly[3-(4-vinylphenylmethyl)-2,4-pentanedione], P(St-CH$_2$-AcAc) **16** and poly[*N*-(4-acetoacetylphenyl)-methacrylamide], systematic name: poly{1-[4-(2-methyl-1-oxo-2-propenylamino]phenyl-1,3-butadione} (PMPAPB)

17, were initiated by the Ce(IV) ion, respectively. The formation of the grafted copolymer was revealed by x-ray photoelectron spectrometry (XPS) spectra, FT-IR spectra, and photomicrographs.

Based on the ESR studies of Ce(IV) ion-BzyAcAc-MNP, Ce(IV) ion BzAc-MNP systems as mentioned before, the grafting reaction of P(St-CH₂-AcAc) will take place on the methene carbon of 1,3-dikeone via the abstraction of hydrogen by the Ce(IV) ion to form radicals and then initiate monomer graft copolymerization. The initiation mechanism of graft copolymerization is proposed in Scheme (10).

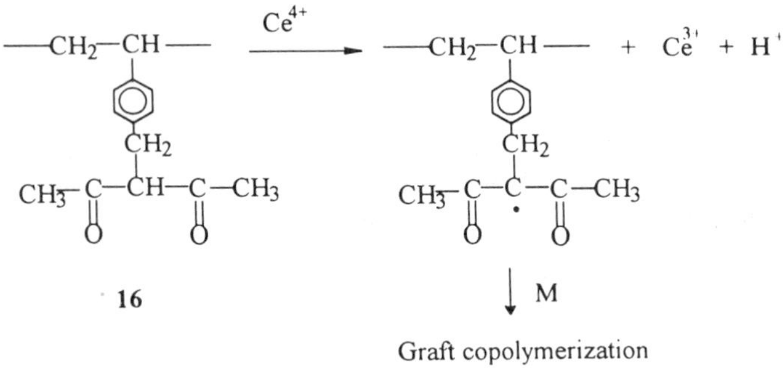

Graft copolymerization

Scheme 10

In P(MPAPB) there are two types of functional groups, i.e., 1,3-diketone group and methacrylanilide group, which can probably react with the Ce(IV) ion to form radicals to initiate the graft copolymerization. From the kinetic studies, we have found that 1,3-diketone group has a much higher activity than that of acetanilide group containing electron withdrawing substituent acetyl at 4-position of the benzene ring. Therefore, the grafting reaction of P(MPAPB) will take place on the

methene carbon of 1,3-diketone by the abstraction of hydrogen with Ce(IV) ion as shown in Scheme (11).

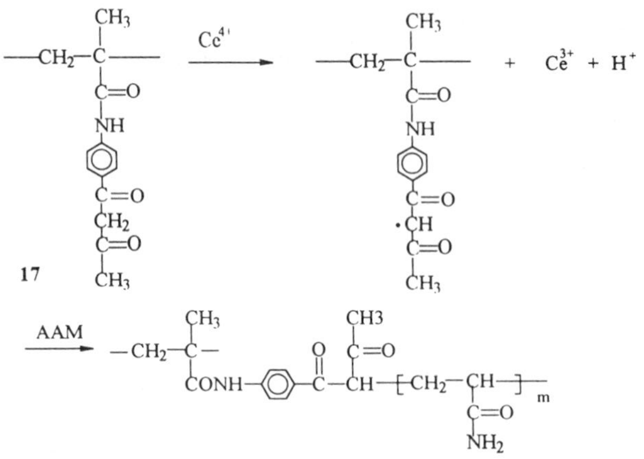

Scheme 11

C. Grafting Reaction of Macromolecules Having 4-Tolylureido Pendant Groups

Qiu et al. [24] have reported the synthesis of macromolecules having 4-tolylureido pendant groups, such as poly(N-acryloyl-N'-4-tolylurea-co-ethyl acrylate) [poly(ATU-co-EA)] 18, and poly(N-methacryloyl-N'-4-tolylurea-co-EA) [poly(MTU-co-EA)] 19, from the copolymerization of ATU and MTU with EA, respectively. Graft copolymerization of acrylamide onto the surface of these two copolymer films took place using the Ce(IV) ion as initiator. The graft copolymerization is proposed as Scheme (12).

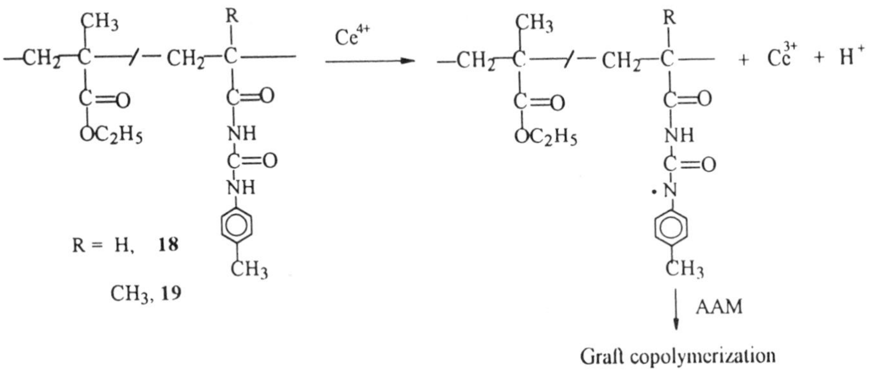

Graft copolymerization

Scheme 12

1. Characterization of Graft Copolymers

When the AAM monomer was grafted onto polymer films, the grafted films showed much higher water absorption than the original ungrafted films. This is due to

the higher water absorbability of PAAM. XPS determination of the surface composition of polymer films would give the higher ratio value of N/C on the surface of grafted films than that on the surface of the original films. SEM photographs of grafted films always showed salient micrographs, showing the presence of the PAAM macromolecules chain on the surface of the films, while the original films were smooth and essentially featureless.

2. Advantage of Graft Copolymerization Initiated with Ce(IV) Ion

As we have previously mentioned, the graft copolymerization of vinyl monomer onto macromolecules having an active pendant group initiated with the Ce(IV) ion will give the following features: (1) Since the radicals capable of initiating vinyl graft copolymerization are produced almost exclusively on the active pendant group, grafting is favored much more than homopolymer formation as compared with other grafting reaction systems; (2) The redox reaction of an active pendant group with Ce(IV) ion is much faster than the Ce(IV) ion initiating vinyl homopolymerization, and graft copolymerization will usually take place rather than homopolymerization; (3) The homopolymer yields can be depressed to very low levels and the separation of homopolymer from the grafted films is very easy.

D. Mechanism of Graft Copolymerization of Chitosan Model Compounds

Chitosan, having a similar chemical backbone as cellulose, is a linear polymer composed of a partially deacetylated material of chitin [(1-4)-2-acetamide-2-deoxy-β-D-glucan]. Grafting copolymer chains onto chitosan can improve some properties of the resulting copolymers [48–50]. Yang et al. [16] reported the grafting reaction of chitosan using the Ce(IV) ion as an initiator, but no detailed mechanism of this initiation has been published so far.

Recently, Li et al. [30], Yu et al. [31] reinvestigated the mechanism of graft copolymerization of vinyl monomers onto carbohydrates such as starch and cellulose initiated by the Ce(IV) ion with some new results as mentioned in Section II. Furthermore, they investigated the mechanism of model graft copolymerization of vinyl monomers onto chitosan [51]. They chose the compounds containing adjacent hydroxyl-amine structures, such as D-glucosamine, *trans*-2-amino-cyclohexanol, 2-

amino-3-phenyl-1-propanol, 1-amino-2-propanol, as the model for chitosan reacted with the Ce(IV) ion. Based on experimental results, they suggest that there are two ways for the initiation reaction to take place, depending on the reaction temperature. At lower temperatures (40°C), the adjacent hydroxyl-amine linkage was oxidized to aldehyde and —CH=NH groups. At higher temperatures (90°C), however, the —CH=NH group could be hydrolized to form another aldehyde group. In each case the aldehyde group was further oxidized to an acyl radical, therefore, one or two acyl radicals were responsible for initiating polymerization at 40°C and 90°C, respectively.

From this initiation mechanism, the important role of the aldehyde group in the reaction mechanism of 1,2-diol and 1-amino-2-hydroxy compound and Ce(IV) ion initiation systems can again be seen.

VII. GRAFTING REACTION OF MACROMOLECULES HAVING AN ACTIVE PENDANT GROUP BY PHOTO-INDUCED CHARGE-TRANSFER INITIATION

A convenient method for synthesizing macromolecules that have amino pendant groups is by means of radical polymerization and copolymerization of the functional monomer containing an amino group, the so-called polymerizable amine. A well-known photo-induced charge-transfer initiation system is the benzophenone (BP)-amine system, which can be used as the initiator for photopolymerization of vinyl monomers [52–58]. Li et al. [52] investigated the macromolecules having an amino pendant group, i.e., copolymer of N-(4-dimethyl-aminophenyl) acrylamide and MMA[poly(DMAPAA-*co*-MMA) **20**]. In the presence of BP under irradiation of ultraviolet light, the active amino group would react with BP to generate two kinds of radicals, i.e., the anilomethyl radical, which would initiate MMA for graft copolymerization, and the semipinacol radical, which would undergo primary termination. Therefore, the graft copolymer was obtained without contamination of homopolymer PMMA (Scheme 13).

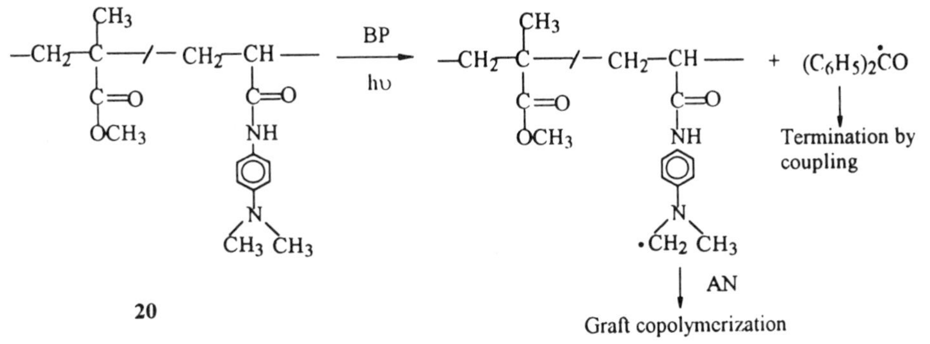

20

Scheme 13

Similar results were obtained in the photografting reaction of *N*-methyloyloxethyl-*N*-methyl-*p*-toluidine

polymer[(PMEMT) **21**] and its copolymer, poly(MEMT-*co*-MA) **22**, using BP as the initiator [53], as shown in Scheme (14).

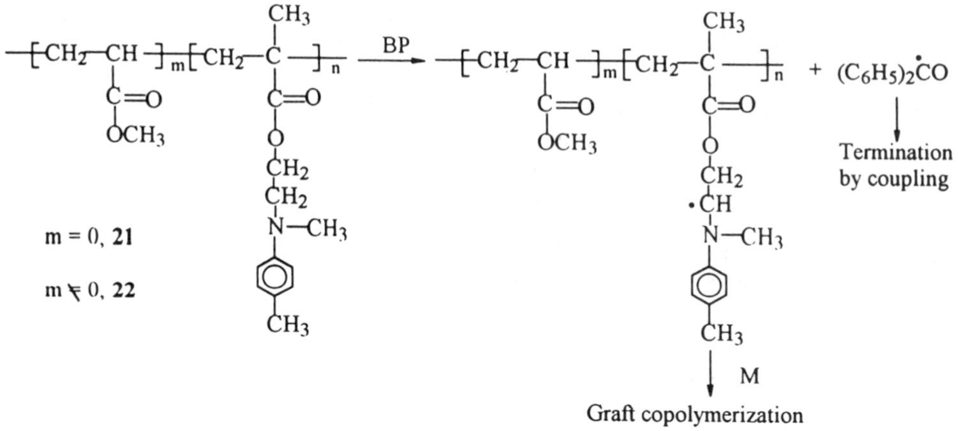

m = 0, **21**

m ≠ 0, **22**

Scheme 14

Recently, Si et al. [59,60] have investigated the synthesis of polymerizable amines, such as *N*-(3-dimethylaminopropyl) acrylamide(DMAPAA) and *N*-(3-dimethylaminopropyl) methacrylamide (DMAPMA), and their copolymerization reaction. DMAPAA or DMAPMA in conjunction with ammonium persulfate was used as a redox initiator for vinyl polymerization. Copolymers having amino pendant groups, such as copolymer of

MMA and DMAPMA poly(MMA-*co*-DMAPMA) **23**, obtained by radical copolymerization, can produce a photografting reaction with acrylonitrile (AN) using BP as the initiator [61]. The formation of a graft copolymer, poly[(MMA-*co*-DMAPMA)-*g*-AN] was confirmed by FT-IR spectrophotometry. Based on ESR studies and end group analysis, the mechanism of grafting reaction is proposed as follows:

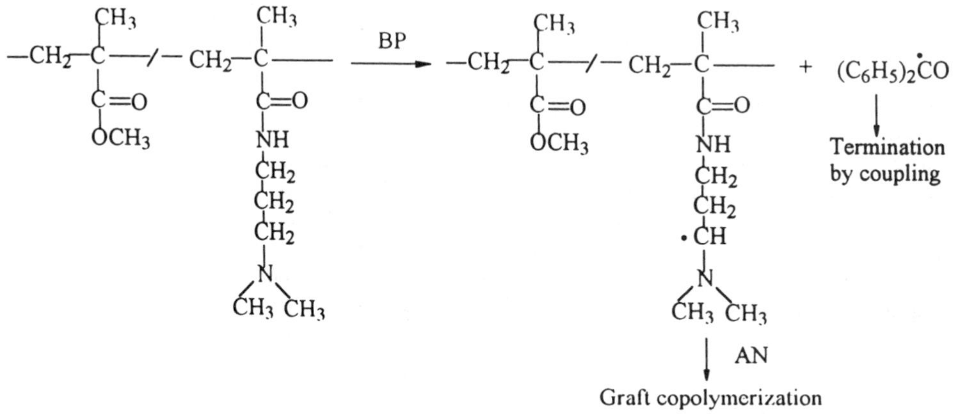

Scheme 15

Therefore, the graft copolymerization of vinyl monomers onto macromolecules having active an pendant group can be achieved either by redox initiation with a Ce(IV) ion or by photo-induced charge-transfer initiation with BP, depending on the structure of the active groups.

ACKNOWLEDGMENTS

This reported research was supported by the National Natural Science Foundation of China. The authors are grateful to those coworkers whose research works have been cited.

REFERENCES

1. H. A. J. Battaerd and G. W. Tregear, *Graft Copolymers*, Wiley, New York (1967).
2. R. J. Ceresa, ed., *Block and Graft Copolymerization*, Vol. 1, Wiley, New York (1973).
3. X. D. Feng, Y. H. Sun, and K. Y. Qiu, *Macromolecules*, *18*: 2105 (1985).
4. K. Y. Qiu and W. Wang, *Acta Polym. Sin.*, *3*: 355 (1989).
5. W. Wang, K. Y. Qiu, and X. D. Feng, *Acta Polym. Sin.*, *2*: 219 (1991).
6. Y. H. Sun, K. Y. Qiu, and X. D. Feng, *Sci. Sinica*, *B27(4)*: 349 (1984).
7. T. Li, W. X. Cao, and X. D. Feng, *Sci. Sinica, B31(3)*: 294 (1988).
8. J. Saldick, *J. Polym. Sci.*, *17*: 73 (1956).
9. G. Mino and S. Kaizerman, *J. Polym. Sci.*, *31*: 242 (1958).
10. P. L. Nayak and S. Lenka, *J. Macromol. Sci.-Rev. Macromol. Chem., C19*: 83 (1980).
11. G. S. Misra and U. D. N. Bajapai, *Prog. Polym. Sci.*, *8*: 61 (1982).
12. K. Y. Qiu, *Makromol. Chem. Macromol. Symp.*, *63*: 69 (1992).
13. O. Y. Mansour and A. Nagaty, *Prog. Polym. Sci.*, *11*: 91 (1985).
14. Y. Y. Chen, R. Oshima, K. Hatanaka, and T. Uryu, *J. Polym. Sci. Pt. A Polym. Chem.*, *24(7)*: 1539 (1986).
15. X. D. Feng, Y. H. Sun, and K. Y. Qiu, *Makromol. Chem.*, *186*: 1533 (1985).
16. J. X. Yang, D. Q. He, J. Wu, H. Yang, and Y. Gao, *J. Shandong College Oceanogr*, *14(4)*: 58 (1984).
17. Y. Q. Lin, and G. B. Butler, *J. Macromol. Sci.-Chem.*, *A26(4)*: 681 (1989).
18. Y. Q. Lin, H. Pledger Jr., and G. B. Butler, *J. Macromol. Sci.-Chem.*, *A25(8)*: 999 (1988).
19. W. Wang, Y. H. Sun, Y. Song, K. Y. Qiu, and X. D. Feng, *Acta Polym. Sin.*, *(2)*: 213 (1990).
20. K. Y. Qiu, X. Q. Guo, D. Zhang, and X. D. Feng, *Chinese J. Poly. Sci.* (English Ed.), *9*: 145 (1991).
21. K. Y. Qiu, J. B. Zhao, and J. H. Dong, *Polym. Bull.* (*Berlin*), *32*: 581 (1994).
22. J. B. Zhao, K. Y. Qiu, *Chem. J. Chinese Universities* *17(2)*: 315 (1996).
23. J. H. Dong, K. Y. Qiu, and X. D. Feng, *Macromol. Chem. Phys.*, *195*: 823 (1994).
24. K. Y. Qiu, Z. Y. Li, and T. Zhao, *Macromol. Chem. Phys.*, *196*: 2483 (1995).
25. R. K. Samal, M. C. Nayak, G. Panda, G. V. Syrryinaryana, and D. P. Das, *J. Polym. Sci., Polym. Chem. Ed.*, *20*: 53 (1982).
26. K. Y. Qiu, Z. H. Lu, A. J. Tang, and X. D. Feng, *Chinese J. Polym. Sci.*, *10*: 366 (1990).
27. M. Snatappa and V. S. Anathanarayanan, *Proc. Indian Acad. Sci.*, *A62*: 159 (1965).
28. S. V. Subramanian and M. Santappa, *Makromol. Chem.*, *112*: 1 (1986).
29. Y. H. Sun, X. N. Chen, X. D. Feng, *Makromol. Chem.*, *191*: 2093 (1990).
30. W. Li, C. Yu, J. J. Zhang, and X. D. Feng, The 6th China-Japan Symposium on Radical Polymerization, Oct. 8–12, Guilin, China, p. 27 (1991).
31. C. Yu, W. Li, and X. D. Feng, IUPAC International Symposium on Olefin and Vinyl Polym. & Functionalization, Oct. 14–18 Hangzhou, China, p. 55 (1991).
32. S. T. Voong (X. D. Feng) and K. Y. Chiu (K. Y. Qiu), *Kexue Tongbao* (English Ed.), *17*: 399 (1966).
33. K. Odagawa (Mitsubishi Chem. Ind.) Japan Kokai 75,46794, Chem. Abstr., 83, 98407 (1975).
34. J. B. Zhao and K. Y. Qiu, Xi'an International Conference of Reactive Polymers, Oct. 8–11, Xi'an China, p. 119 (1994).
35. J. B. Zhao and K. Y. Qiu, *Ion Exch. and Absorp.*, *10*: 338 (1994).
36. J. B. Zhao, *Doctoral Thesis*, Dept. of Chem., Peking Univ., Beijing, China (1994).
37. J. H. Dong, K. Y. Qiu, and X. D. Feng, The 6th China-Japan Symposium on Radical Polymerization, Oct. 8–12, Guilin, China, p. 68 (1991).
38. J. H. Dong, K. Y. Qiu, and X. D. Feng, *J. Macromol. Sci. Macromol. Rept.*, *A31*: 499 (1994).
39. J. H. Dong, C. Deng, K. Y. Qiu, and X. D. Feng, *Chem. Research in Chin. Univ.*, *10*: 327 (1994).
40. K. Y. Qiu, W. Wang, X. Q. Guo, D. Zhang, and X. D. Feng, *Acta Polym. Sin.*, *(4)*: 496 (1990).
41. S. B. Bhattacharryya and D. Maldas, *Prog. Polym. Sci.*, *10*: 171 (1984).
42. M. K. Mishra, *J. Macromol. Sci.-Rev. Macromol. Chem., C19(2)*: 193 (1980); *C22(3)*, 471 (1982–1983).
43. K. Y. Qiu and Y. Song, *Chinese J. Polym. Sci.* (English Ed.), *6*: 75 (1988).
44. W. Wang, K. Y. Qiu, and X. D. Feng, *Chinese J. Polym. Sci.*, *12(2)*: 137 (1994).
45. K. Y. Qiu and T. Zhao, *Polymer Inter.*, *38(1)*: 71 (1995).
46. T. Zhao and K.Y.Qiu, Symposium on Polymers, Polym. Div. of Chinese Chem. Soc., Guangzhou, China, p. 146 (1995).
47. J. B. Zhao and K. Y. Qiu, *J. Macromol. Sci.-Pure Appl. Chem. A33(11)*: 1675 (1996).
48. R. S. Slagel and G. D. M. Sinkoitz, U.S. Patent 3,709,780 (1993).
49. K. Kojima, M. Yoshikuni, and T. Suzuki, *J. Appl. Polym. Sci.*, *24*: 1587 (1979).
50. K. Shigeno, K. Kondo and K. Takenmoto, *J. Macromol. Sci. Chem.*, *A17(4)*: 571 (1982).
51. W. Li, Z. Y. Li, W. S. Liao and X. D. Feng, *J. Biomater. Sci. Polymer Edn.*, *4(5)*: 557 (1993).
52. T. Li, W. X. Cao, and X. D. Feng, *Scientia Sinica, B30(7)*: 685 (1987).
53. Z. H. Zhang, K. Y. Qiu, X. D. Feng, *Acta Polym. Sin.*, *(2)*: 246 (1993).
54. H. Block, A. Ledwith and A. R. Tarlar, *Polymer, 12*: 271 (1971).
55. M. R. Sadner, C. L. Oshorn, and D. J. Trecher, *Polymer, 14*: 250 (1973).
56. S. R. Clarke and R. A. Sanks, *J. Macromol. Sci.-Chem.*, *A14*, 69 (1980).
57. T. Li, W. X. Cao, and X. D. Feng, *J. Macromol. Sci., Rev. Chem. Phys.*, *C29*: 153 (1989).
58. X. D. Feng, K. Y. Qiu, and W. X. Cao, *Handbook of Engineering Polymeric Materials* (N. P. Cheremisinoff, ed.) Marcel Dekker Inc., NY Chapt. 16 (1997).
59. K. Si, K. Y. Qiu, *J. Macromol. Sci. Macromol. Reports*, *A32(8)*: 1139 (1995).
60. K. Si, X. Q. Guo, K. Y. Qiu, *J. Macromol. Sci. Macromol. Reports*, *A32(8)*: 1149 (1995).
61. X. D. Feng, K. Y. Qiu, and W. X. Cao, *Handbook of Engineering Polymeric Materials*, (N. P. Cheremisinoff ed.) Marcel Dekker Inc., NY, Chapt. 16 (1997).

36
Modification of Properties of Nitrile Rubber

Susmita Bhattacharjee
Philipps Universität, Marburg, Germany

Anil K. Bhowmick and Bhola Nath Avasthi
Indian Institute of Technology, Kharagpur, India

I. INTRODUCTION

Nitrile rubber (NBR) was first commercialized by I.G. Farbindustry, Germany, in 1937, under the trade name of Buna N. Its excellent balance of properties confers it an important position in the elastomer series. Nitrile rubber, a copolymer of butadiene and acrylonitrile, is widely used as an oil-resistant rubber. The acrylonitrile content decides the ultimate properties of the elastomer. In spite of possessing a favorable combination of physical properties, there has been a continuous demand to improve the aging resistance of NBR due to the tougher requirements of industrial and automotive applications.

Although, the heat resistance of NBR is directly related to the increase in acrylonitrile content (ACN) of the elastomer, the presence of double bond in the polymer backbone makes it susceptible to heat, ozone, and light. Therefore, several strategies have been adopted to modify the nitrile rubber by physical and chemical methods in order to improve its properties and degradation behavior. The physical modification involves the mechanical blending of NBR with other polymers or chemical ingredients to achieve the desired set of properties. The chemical modifications, on the other hand, include chemical reactions, which impart structural changes in the polymer chain.

Bhattacharjee et al. [1] recently did an exhaustive review on properties and degradation of NBR. The improvement of physical properties of NBR by both chemical and physical methods have been thoroughly described. They examined the available literature on various modified NBRs and discussed in detail the refurbishment of properties in comparison with NBR. The present chapter focuses on one of the most important chemical modifications: the hydrogenation of NBR.

Hydrogenated nitrile rubber (HNBR) is the latest high-performance elastomer that has emerged as an important product in the last decade. The principal driving force behind the development of HNBR is the increasing performance demanded of elastomers by the automotive and oil drilling industries. Even the conventional oil-resistant rubbers, namely, nitrile rubber, chloroprene rubber, and chlorinated polyethylene are reaching their performance limits. Fluoroelastomers, which are the possible substituents in these applications, are very expensive and possess processing difficulties. Hence, HNBR has been developed to bridge the price performance gap between general purpose oil-resistant rubbers and fluoroelastomers. It has high heat and aging resistance, good resistance to swelling in technical fluids, including those containing aggressive additives, and outstanding wear resistance in extremely adverse conditions. HNBR has recently been marketed by Nippon Zeon Co., (Japan) under the tradename of 'Zetpol' [2]; and Bayer A-G (Germany) and Polysar Rubber Co. (Canada) under the tradename of Therban [3] and Tornac [4], respectively.

The double bond present in the diene part of the elastomer is generally more susceptible to thermal and oxidative degradation. The selective hydrogenation of olefinic unsaturation in NBR imparts significant improvements in resistance to degradation and other properties, such as permeability, resistance to ozone and chemicals, and property retention at high temperature.

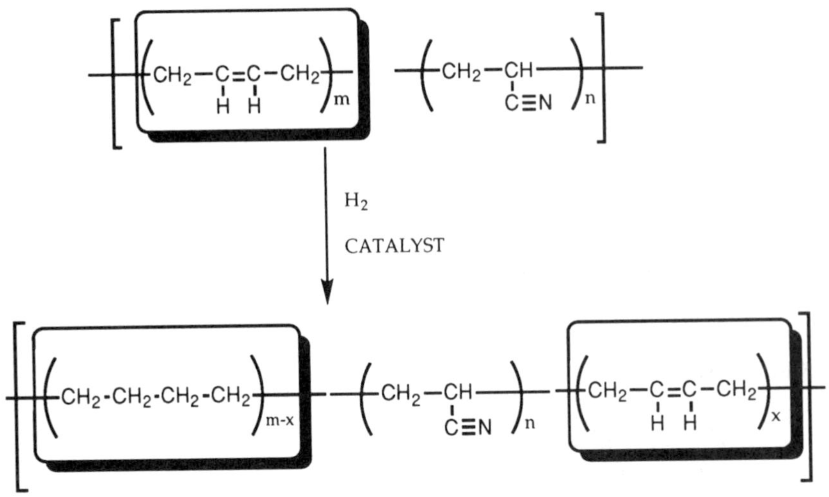

This chapter mainly aims at describing the various methods and processes developed for hydrogenation of nitrile rubber. The characterization, physical properties, and application of hydrogenated nitrile rubber are also discussed. Another small section on hydroformylation of nitrile rubber has been included.

II. HYDROGENATION OF NITRILE RUBBER

Nitrile rubber hydrogenation is one of the most significant developments in the field of scientific and technological interest. Modification by hydrogenation of olefinic unsaturation in NBR offers an economic route for the synthesis of a new specialty elastomer, which is not accessible by standard polymerization techniques. The hydrogenation of various diene polymers has been reviewed previously [5–7]. The recent advancement on the chemical modification of polymers, with special reference to homogeneous catalytic hydrogenation, has been extensively discussed by McManus and Rempel [8]. McGrath et al. [9] have described the functionalization of polymers by metal-mediated processes.

The survey of the available literature reveals that conversion of nitrile rubber to hydrogenated nitrile rubber has been achieved by both catalytic and noncatalytic methods [5–7]. In catalytic hydrogenation, polymer is reacted with hydrogen in the presence of either a heterogeneous or a homogeneous catalyst. Noncatalytic hydrogenation is carried out in the presence of a suitable reducing agent. The important points to be considered in these reactions are the extent of conversion, side reactions, and selectivity. The recent citations suggest that the most popular method for noncatalytic hydrogenation is by diimide reduction using hydrazide reagents [5–7]. The advantage of noncatalytic methods over catalytic

methods is that a special hydrogenation apparatus is not required. Despite this advantage, there are a few side reactions, such as isomerization, attachment of hydrazide fragment to the polymer chain, depolymerization, and cyclization, during diimide reduction. However, under carefully controlled conditions, side reactions can be minimized [10]. The studies on catalytic hydrogenation of nitrile rubber are much larger in number as compared with noncatalytic hydrogenation. The scope of the present chapter is to analyze all these techniques critically. The use of heterogeneous and homogeneous catalysts in selective reduction of olefinic bonds in the butadiene unit of nitrile rubber is summarized separately in the following sections. The initial reports of this vast body of literature on preparation of HNBR mainly covers the heterogeneous catalysis. But the more recent advancements are in the field of homogeneous catalysis, since it offers a better understanding of quantitative hydrogenation.

A. Thermodynamics of Nitrile Rubber Hydrogenation

The thermodynamic feasibility of a chemical reaction can be assessed before the experimental studies. Though the thermodynamic data alone are not sufficient to indicate the favorable conditions of the reactions, they definitely give some indications about the probable range of temperature and pressure under which the reaction can be carried out. The estimation of heat of reaction from thermodynamical data gives the magnitude of the heat effects during the reaction. It also allows the calculation of the equilibrium constant from the standard free energies of the reacting materials. The expected maximum attainable yield of these products of the reaction can be estimated from the equilibrium constant.

Bhattacharjee et al. [11] have calculated the thermodynamic parameters for hydrogenation of acrylonitrile–butadiene copolymer.

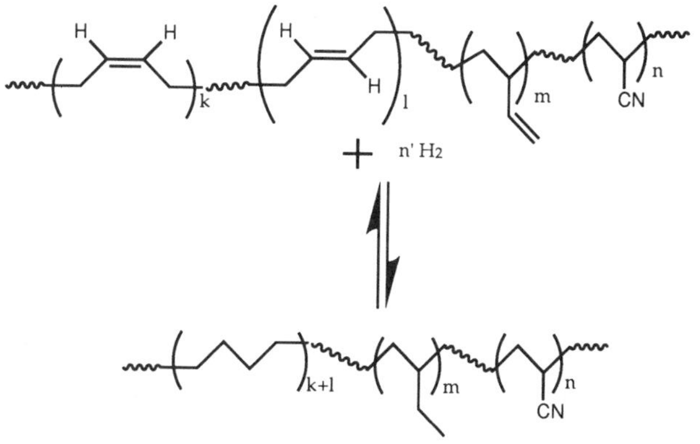

where k is the number of cis-1,4-unit, l is the number of $trans$-1,4-unit, m is the number of 1,2-unit, and n is the number of acrylonitrile unit in NBR and:

$$n' = k + l + m \qquad (1)$$

The nitrile rubber used for the experimental studies contained 40% acrylonitrile and the values of k, l, m, and n for the elastomer were calculated from spectroscopic studies as 359, 559, 215, and 770, respectively.

The relationship used for the calculation of $\Delta G°$ at room temperature (298 K) is given by:

$$\Delta G° = \Delta H° - T\Delta S° \qquad (2)$$

where $\Delta H°$ and $\Delta S°$ are the standard enthalpy change and standard entropy change of the reaction, respectively. The standard heats of formation and entropies for hydrogenation are available from the literature [12], $\Delta S°$ and $\Delta H°$ of the nitrile rubber and hydrogenated nitrile rubber were calculated from group contributions [12]. $\Delta G°$ values at higher reaction temperatures were calculated according to the relationship:

$$\Delta G° = \Delta H_0 - \Delta aT \ln T - (1/2) \Delta bT^2 \\ - (1/6) \Delta cT^3 + IT \qquad (3)$$

Table 2 Values of $K_p°$ and $\Delta G°$ of NBR Hydrogenation at Different Temperatures

	Temperature (K)				
	298	323	343	363	373
$\ln K_p° \times 10^{-4}$	4.4	4.0	3.7	3.4	3.3
$\Delta G° \times 10^{-4}$ (kJ/gmol)	−11.1	−10.8	−10.6	−10.4	−10.3

Source: Ref. 11.

and

$$\Delta H° = \Delta H_0 + \Delta aT + (1/2) \Delta bT^2 + (1/3) \Delta cT^3 \qquad (4)$$

The coefficients a, b, and c for hydrogenation were obtained from the literature [13] and those for nitrile and hydrogenated nitrile were calculated from a group contribution method reported by Rihani and Doraiswami [14]. All the necessary data are listed in Table 1. The integration constant l and ΔH_0 have been calculated by incorporating the values of $\Delta G°$ and $\Delta H°$ at 298 K in Eqs. (3) and (4). The equilibrium constant at atmospheric pressure and various temperature has been calculated according to the relationship:

$$\Delta G° = -RT \ln K_p° \qquad (5)$$

The value of $\Delta G°$ at various reaction temperatures and corresponding $K_p°$ values are reported in Table 2. It can be understood from Table 2 that the formation of hydrogenated nitrile rubber from nitrile rubber is thermodynamically feasible.

III. HETEROGENEOUS CATALYTIC HYDROGENATION

This section discusses the processes for nitrile rubber hydrogenation developed during past years using a heterogeneous catalyst. The published reports suggest that mostly the palladium catalyst has been used for nitrile

Table 1 Thermodynamic Parameters at Room Temperature (298 K)

Sample	$\Delta H°$ kJ/gmol	ΔS kJ/(gmol)(K)	a	$b \times 10^2$	$c \times 10^4$
NBR 40 mol% ACN	−36657.07	569.33	578.90	13593.02	−818.03
H$_2$	8.46	0.10	6.42	0.10	-7.81×10^{-4}
HNBR 40 mol% ACN	−96993.84	576.36	2055.53	14653.92	−858.16

Source: Ref. 11.

Table 3 Heterogeneous Catalytic Hydrogenation of Nitrile Rubber

Sl. No.	Catalyst	Solvent	Pressure (MPa)	Temperature (°C)	Time (h)	Degree of hydrogenation (%)	Ref.
1.	Pd/C	Isobutyl-methyl ketone	—	—	—	47	16
2.	Pd/C	Acetone	6.0	60	5	83	17
3.	Pd-Ca/C	Acetone	5.0	50	4	95.3	18
4.	Pd-Al/C	Acetone	5.0	50	4	92.2	19
5.	PdCl$_2$/C	—	—	—	—	—	20
6.	Pd/Silica	Acetone	5.0	50	5	89.5	21,22
7.	Pd/C	Isobutyl-methyl ketone	—	—	—	Iodine value = 31	23
8.	Pd/Silica + lithium acetate	Tetrahydrofuran	5.0	50	4	84.0	24
9.	Pd/CaCO$_3$	Acetone	5.0	50	4	98.5	25
10.	Pd/C	—	—	—	—	—	26
11.	Pd/TiO$_2$	Acetone	5.0	50	4	96.8	27
12.	Pd/C	Tetrahydrofuran	5.0	70	2	93.0	28
13.	PdO	Acetone	5.0	50	6	87.8	29

rubber hydrogenation. However, heterogeneous catalysts based on different metals have been used for hydrogenation of many other diene polymers [15]. The selection of an optimum catalyst system is determined by the requirement for the selectivity in hydrogenation. Heterogeneous catalytic hydrogenation of NBR is generally performed by stirring a solution of NBR with an insoluble catalyst under an atmosphere of hydrogen gas. The reduced product is usually isolated simply by filtration of the catalyst, followed by precipitation of the rubber and drying. The selection of the solvent and hydrogen pressure for a given hydrogenation is dependent on the catalyst chosen, its surface area, and on its inert support where it is deposited. In many cases the catalyst can be recovered and reused.

Nippon Zeon Co. has published many patents on heterogeneous catalytic hydrogenation of NBR. Initially, they used 5% palladium on carbon as the catalyst for hydrogenation of NBR in iso-butyl-methyl ketone to obtain only 47% conversion [16]. However, the product on compounding and vulcanizing showed better tensile strength, elongation, hardness, ozone cracking, and high-temperature oil resistance than the corresponding NBR. Later they changed the reaction condition and particle size of the carbon carrier to improve the degree of hydrogenation [17]. The reaction condition and catalysts used for heterogeneous catalytic hydrogenation of NBR [16–29] is arranged in Table 3. Supported palladium catalysts have been widely used for hydrogenation of conjugated diene polymers containing the nitrile group. Most of the reactions were carried out at 5.0 MPa hydrogen pressure, 50°C temperature, and 4–6 h reaction time [16–29]. A high degree of hydrogenation could be obtained by changing the active porous support and using other metals along with palladium.

Researchers from Nippon Zeon Co. have used palladium and aluminum or palladium on carbon for preparing highly saturated nitrile rubber [18,19]. For the similar reaction in the presence of only palladium on carbon, the extent of reaction was quite low [19]. Palladium supported on silica showed higher catalytic activity during hydrogenation of NBR and could easily be removed from hydrogenation mixtures by simple filtration [21,22]. The variation of average particle size of powdered silica support changed the degree of hydrogenation. Acetone is the commonly favored solvent for these systems. Tetrahydrofuran and isobutyl-methyl ketone have been used in some reactions. It is important to note that all the reactions mentioned in Table 3 proceed selectively without reduction of the nitrile group. Kubo et al. [24] hydrogenated carbon–carbon double bond in NBR in the presence of silica-supported palladium and lithium acetate promoter. Buding et al. [25] suggested that acrylonitrile–butadiene copolymers can be hydrogenated to high conversion by using palladium on calcium carbonate. The catalyst could be easily recovered from the reaction mixture and reused. Kubo et al. [27] developed another catalyst for hydrogenation of diene polymers that had high activity and could be easily recovered for reuse. It consisted of 1% palladium supported on titanium dioxide and saturated 97% of the double bonds present in nitrile rubber containing 35% acrylonitrile. Recently, Kubo [30] discussed the development of palladium catalyst on a special silica carrier for selective hydrogenation of NBR. He reviewed [31,32] the manufacture of hydrogenated NBR in the presence of supported palladium catalysts and its improved physical properties. Takahashi et al. [29] developed a polymer-supported catalyst for hydrogenation of conjugated diene polymer. They stirred a solution of NBR with the catalyst (PdO) at room temperature for 20 h, filtered the solution, and then conducted the hydrogenation. The pretreatment of the catalyst with the polymer enhanced its activity.

It can be summarized from the available data in Table 3 that supported palladium catalysts selectively hydrogenated carbon–carbon double bonds in the presence of the nitrile group in NBR. However, there is no detailed fundamental study on heterogeneous catalytic hydrogenation of nitrile rubber in the literature that can provide an insight into the reaction. The available information is limited since most of the literature is patented.

IV. HOMOGENEOUS CATALYTIC HYDROGENATION

In the last 15 years, there has been a tremendous explosion in the literature on the homogeneous catalytic hydrogenation of NBR. This section discusses the available reports on the development of various methods and processes for preparation of hydrogenated NBR in the presence of homogeneous catalysts. The principle of homogeneous catalytic hydrogenation is the activation of molecular hydrogen by a transition metal complex in solution and subsequent hydrogen transfer to an unsaturated substrate. The most important advantage offered by homogeneous catalysts is their selectivity as compared with heterogeneous systems. The homogeneous catalysis also extends the opportunity to tailor ligands so as to enhance the reactivity and selectivity of the catalyst. Since the reaction occurs in the homogeneous phase, it also facilitates the study of the mechanistic steps involved in the catalytic cycle. However, it has the disadvantage that the catalyst remains in the product due to difficulty in removal. Even then, most of the previous literature suggests that the homogeneous catalytic methods for hydrogenation of unsaturated polymers are preferable to heterogeneous ones.

The hydrogenation of functional unsaturated copolymers presents a special problem because many of the functional groups coordinate with the catalyst and poison its activity or are themselves reduced. These problems are particularly acute for highly coordinating functionalities such as nitrile, carboxyl, amino, hydroxyl, etc. Moreover, when the polymer contains two or three functional groups that are all prone to reduction, careful selection of the catalyst system and reaction condition becomes important if regioselectivity during hydrogenation is desired. The choice of the catalyst also

depends on the cost of the metal and stability of the complex or ease of handling the complex.

The hydrogenation procedure for most of the catalyst systems reported follows the same route. At first the polymer is dissolved in a suitable organic solvent. The solution is purged with hydrogen and then the requisite amount of catalyst is added. The mixture is subjected to the desired temperature and hydrogen pressure. After the completion of the reaction, the mixture is cooled and the polymer is precipitated out and finally dried. The reaction can be controlled up to the desired level of hydrogenation by proper combinations of the reaction pressure, temperature, time, and catalyst concentration. The homogeneous catalyst systems developed so far for NBR hydrogenation are discussed in the following sections. They have been divided into three broad classes based on the metals.

A. Rhodium Catalyst

The major breakthrough in the history of selective hydrogenation is the discovery of the Wilkinson's catalyst [33], i.e. tris(triphenylphosphine) chloro rhodium(I) [RhCl(PPh$_3$)$_3$]. This catalyst offers remarkable activity toward hydrogenation of carbon–carbon double bond in the presence of other reducible functional groups. Several other similar complexes have been introduced in the homogeneous catalyst series. Rhodium (Rh) complexes have been extensively used for hydrogenation of carbon-carbon double bond (C=C) in NBR. The major advantage offered by Rh complexes is their selectivity to hydrogenate C=C without causing any reduction of nitrile (CN) groups. The hydrogenation of CN groups, if any, can be considered negligible. In the presence of a nitrile group, which inhibits the catalytic activity during hydrogenation, Rh complexes are capable of retaining their high activity without any pronounced difference. The main criteria of NBR hydrogenation is the selectivity toward reduction, in order to maintain the oil resistance and other physical properties of the hydrogenated product.

It can be noted that possibly the most effective catalyst for selective hydrogenation of olefinic unsaturation in NBR is the Wilkinson's catalyst [15]. There are quite a few reports of fundamental studies on this catalyst system. One of the reasons for the increasing interest in this complex is the simple method to synthesize this in high yield and the ease of handling due to its great air stability as a solid. McManus and Rempel [8] have very systematically reviewed the commercial application of Wilkinson's catalyst in preparation of HNBR. Generally, the desirable reaction condition for a commercial process to hydrogenate NBR using precious metal catalyst is the use of minimum catalyst concentration with respect to the polymer. Hence, an ideal operating condition in an industrial scale would involve temperatures greater than 100°C, pressure greater than 2.7 MPa, and

Table 4 Rhodium Catalysts Used for NBR Hydrogenation

Sl. No.	Catalyst	Solvent	Pressure (MPa)	Temperature (°C)	Time (h)	Degree of hydrogenation (%)	Ref.
1.	RhCl(PPh$_3$)$_3$	Chlorobenzene	7.0	100	8	92	34
2.	RhCl(PPh$_3$)$_3$	Chlorobenzene	6.0	100	5.5	Iodine value = 14	35
3.	RhCl(PPh$_3$)$_3$	Chlorobenzene	0.240–0.275 6.9	70–75 100	12	99.7	36
4.	RhCl(PPh$_3$)$_3$	Chlorobenzene	19.0	120	5	99.7	37
5.	RhCl(PPh$_3$)$_3$	2-Butanone	<0.101	40		100	38
6.	RhCl(PPh$_3$)$_3$	Chlorobenzene	5.6	100	11	100	11
7.	RhCl(PPh$_3$)$_3$	Chlorobenzene	15.0	120	—	99.2	39
8.	RhCl(PPh$_3$)$_3$	Chlorobenzene	5.6	100	11	93 (L-XNBR)	40
9.	RhH(PPh$_3$)$_4$	Chlorobenzene	1.4–2.8	87–130	2–15	51–99.95	41
10.	RhH(PPh$_3$)$_4$	Chlorobenzene	0.09	55	19	91	42
11.	RhH$_2$(dppb)$_2$ (1,3-diphenyltriazenido)	—	—	—	—	—	43
12.	RhH(dibenzophosphole)$_3$	—	—	—	—	—	44
13.	Rh(COD) (PPh$_3$)$_2$ OTs	—	4.9	100	10	99.5	45
14.	RhCl(sodium diphenyl-phosphino-benzene-m-sulfonate)$_3$	Water	0.101	75	—	60	46

polymer concentration around 2–10% by weight. On the basis of these criteria, Bhattacharjee et al. [11] have optimized the reaction condition of NBR hydrogenation, which might be considered useful for a commercial process. It is also necessary to add an excess of free triphenyl phosphine (PPh$_3$) to the reaction mixture to achieve quantitative hydrogenation of C=C in the presence of RhCl(PPh$_3$)$_3$. The choice of solvent is also important for efficient hydrogenation. The butadiene unit of NBR has cis-, *trans*-, and vinyl double bonds. The preferential reduction of these double bonds are different for various solvents used. Also the catalyst does not show its proper activity in the presence of many solvents.

The continued interest in the use of the Wilkinson's catalyst in commercial application and fundamental studies [11,34–47] has been chronologically organized in Table 4. The first patent on NBR hydrogenation, in the presence of RhCl(PPh$_3$)$_3$ appeared in 1972 [34]. Oppelt et al. [35] hydrogenated NBR in chlorobenzene solution using RhCl(PPh$_3$)$_3$ to obtain a polymer with complete saturation of vinyl and *trans* double bonds. Weinstein [36] prepared elastomeric tetramethylene–ethyl ethylene acrylonitrile copolymer by selective hydrogenation of butadiene acrylonitrile rubber in the presence of RhCl(PPh$_3$)$_3$ and excess of triphenyl phosphine. Almost complete hydrogenation was obtained at both low- and high-hydrogenation pressures. However, at low-hydrogenation pressure appreciable excess of triphenyl phosphine was required. Weinstein [36] observed that physi-

cal and chemical properties of modified polymer gums and their sulfur and peroxide cured vulcanizates were resistant to attack by oxygen and showed unexceptionably high tensile strength, which was attributed to the crystalizability of the gums on stretching. Peroxide-cured polymer vulcanizates prepared from green stock containing magnesium oxide had a combination of resistance to heat aging in air, hydrogen sulfide, and oil, suggesting their applicability to oil drilling components.

The industrial application of RhCl(PPh$_3$)$_3$ has stimulated a few academic studies to obtain a better understanding of hydrogenation of NBR substrate. Thermodynamics and kinetics of this reaction have been thoroughly elucidated. This has given valuable insight into the possible mechanism for the hydrogenation process. Mohammadi and Rempel [38], for the first time, conducted a detailed study involving the hydrogenation of acrylonitrile–butadiene copolymer in the presence of RhCl(PPh$_3$)$_3$. Under mild reaction conditions, this catalyst provided quantitative hydrogenation of carbon–carbon unsaturation without any reduction of the nitrile functionality. The reactions were carried out in a ketone solvent (2-butanone) at 40°C under a hydrogen pressure < 101.32 KPa. The selectivity of the catalyst for terminal versus internal double bonds present in the polymer was markedly influenced by the nature of the solvent. In 2-butanone, the catalyst did not exhibit selectivity for the hydrogenation of 1,2- over 1,4-units of the copolymer, whereas a distinct preference toward saturation of the 1,2-unit was observed when the reaction medium was chlorobenzene. They also observed that, under this mild condition, there is no need to add an excess of PPh₃ to maintain the catalytic activity. During the hydrogenation, an excess of PPh₃ prevents the formation of a binuclear Rh complex that is inactive as a hydrogenation catalyst. Under relatively mild conditions, the nitrile group of the polymer might inhibit the formation of inactive catalytic species, whereas the contrast is observed at higher reaction temperature and pressure [11]. At elevated reaction conditions, it is necessary to add an excess of PPh₃ to stabilize the rhodium complex and prevent the reduction of the complex to Rh metal.

The reaction kinetics showed significant effects on the concentration of olefinic substrate and hydrogen gas as well as solvent. A first order dependence of the hydrogenation rate on the catalyst concentration was obtained. The reaction was first order at low-hydrogen pressure and zero order at higher hydrogen pressure. The proposed mechanism based on the kinetic studies and spectral observation is illustrated in Scheme 1. Mohammadi and Rempel [38] suggested that the active catalyst species, $RhClH_2(PPh_3)_2$ interacted with $C=C$ in the rate-determining step. The nitrile functionality present in the copolymer also reacted with the active catalyst and inhibited the rate of hydrogenation to some extent. The effect of temperature on the reaction rate provided the activation parameters associated with the rate constants. The activation energy of the reaction had a value of 87.28 kJ/mol and at 40°C the activation enthalpy and activation entropy of the reaction were estimated to be 84.67 kJ/mol and −8.3 J/(mol K), respectively.

Bhattacharjee et al. [11] optimized the reaction conditions for hydrogenation of NBR in the presence of $RhCl(PPh_3)_3$. In order to identify the important factors

that influenced the nature and extent of the reaction, parameters such as temperature, pressure, time, catalyst concentration, and solvent were varied. When the reaction was carried out in chlorobenzene, there was no conversion at room temperature. The extent of the reaction increased with an increase in temperature, hydrogen pressure, and catalyst concentration. The authors found that the reaction of NBR at a catalyst concentration of 0.02 mmol, under 5.6 MPa hydrogen pressure, at 100°C in chlorobenzene for 11 h was optimum for complete hydrogenation. The amount of catalyst required under this condition is much lower as compared to that reported by Mohammadi and Rempel [38]. This supports the practical importance of the reaction optimized at the elevated condition.

Bhattacharjee et al. [11] also attempted to study the kinetics of the NBR hydrogenation catalyzed by $RhCl(PPh_3)_3$ at higher reaction pressure and temperature. The experiments were conducted at various intervals of time and corresponding hydrogenated polymer samples were characterized to know the extent of reaction. The experimental data were examined to understand the dependence of reaction variables on the rate of hydrogenation. The observations made were the same as that reported earlier by Mohammadi and Rempel [38]. The activation energy calculated for the reaction at 100°C was 22 kJ/mol. The value of activation energy is significantly lower than that reported by Mohammadi and Rempel [38]. The formation of the active catalyst precursor $RhClH_2(PPh_3)_2$ might be more favorable at elevated temperatures. Bhattacharjee et al. [11] characterized the hydrogenated product and studied its physical properties. They found that no degradation or side reaction occurred during the hydrogenation process and

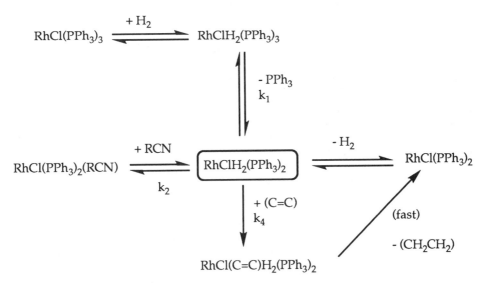

Scheme 1 Mechanism of NBR hydrogenation catalyzed by RhCl(PPh₃)₃. *Source*: Ref. 8.

the molecular weights of the polymer did not change significantly.

RhH(PPh$_3$)$_4$ has been also used for NBR hydrogenation [41,42]. It is claimed that the analogous complexes of RhCl(PPh$_3$)$_3$ containing 5-phenyl-5H phenylphosphole ligand [44], RhH$_2$(dppb)$_2$(1,3-diphenyltriazenido) [43] and Rh(COD)(PPh$_3$)$_2$OTs [45] offer excellent activity for NBR hydrogenation. Recent patents have described the hydrogenation of low-molecular weight NBR [47] and butadiene-isoprene–(meth)acrylonitrile copolymers [39] using RhCl(PPh$_3$)$_3$.

The rhodium complexes are excellent catalysts for hydrogenation of NBR. At low temperature and pressure, high catalyst concentrations are used to obtain a better rate of reactions. Due to higher selectivity of the reaction, pressure and temperature can be increased to very high values. Consequently the rhodium concentration can be greatly reduced, which leads to high turnover rates. The only practical drawback of Rh complex is its high cost. This has initiated the development of techniques for catalyst removal and recovery (see Section VII), as well as alternate catalyst systems based on cheaper noble metals, such as ruthenium or palladium (see Sections IV.A and B).

B. Ruthenium Complexes

Hydrogenation of olefinic unsaturation using ruthenium (Ru) catalyst is well known. It has been widely used for NBR hydrogenation. Various complexes of Ru has been developed as a practical alternative of Rh complexes since the cost of Ru is one-thirtieth of Rh. However, they are slightly inferior in activity and selectivity when compared with Rh catalyst.

Reports on NBR hydrogenation catalyzed by Ru complexes [48–74] are summarized in Table 5. A variety of Ru complexes, such as RuCl$_2$(PPh$_3$)$_3$ [48–52], RuH(O$_2$CR)(PPh$_3$)$_3$ [54], RuHCl(CO)(PPh$_3$)$_3$ [57], and RuCl(O$_2$CR)(CO)(PPh$_3$)$_3$ [58] have been used as the catalysts for hydrogenation of NBR. The easy accessibility of these complexes is a great advantage in commercialization of the process. Almost all the ruthenium–phosphine complexes offer adequate activity for reduction of carbon–carbon double bonds in NBR. However, the main drawback of these systems is the gel formation during the hydrogenation process, unless the reaction is carried out in ketone solvents.

Table 5 indicates that RuCl$_2$(PPh$_3$)$_3$ has been frequently used for selective hydrogenation of C=C in NBR [48–52]. This is commercially available and is also easy to synthesize. In most of the patented processes, low-molecular weight ketone solvents are used to avoid the gel formation. The activity of the catalyst can be enhanced by the use of certain additives, such as triethylamine [59], isopropanol [52], and ammonium hexafluorophosphate [50] in the reaction system. This might be

attributed to the fact that the formation of the active catalytic species: RuHCl(PPh$_3$)$_3$ is promoted by these additives. The nonketone solvents can be used for certain Ru complexes containing organic phosphine ligands when RCO$_2$H and PPh$_3$ are the additives [57,74].

Rempel et al. [67,68] have reported that the complexes of the form RuXY(CO)L$_2$Z (X and Y may be a variety of anionic ligands, L is a bulky phosphine such as tricyclohexyl or triisopropyl phosphine, and Z may be any neutral coordinating ligand) offer quantitative hydrogenation of C=C in NBR in nonketone solvents without any gel formation. Besides the cost factor, these catalyst systems are also more practical in commercial fields due to higher operating temperature and pressure. Quantitative reduction of olefinic unsaturation in NBR can be achieved at elevated temperature and pressure without addition of any excess ligands (such as PPh$_3$) for maintaining the catalytic activity. However, these complexes cause crosslinking of the elastomer during hydrogenation. This was observed from the unusually high viscosities of the resultant hydrogenated polymer [61]. Furthermore, it was also found that the Ru complexes catalyzed little reduction of nitrile groups in the elastomer. This might be a reason for the high viscosities of the polymer. It was later found that the nitrile group was converted to secondary amines. This could be prevented by the addition of small amounts of primary amines to the reaction system [61]. They might react with the imines, which are the primary product of nitrile hydrogenation, thus preventing the crosslinking. The aqueous solutions of carboxylic or mineral acid or transition metal salts can inhibit crosslinking caused by the ruthenium catalyst [73,74]. They probably avert crosslinking by inducing hydrolysis of imine groups produced by hydrogenation of nitriles. The crosslinking and gel formation are the major drawbacks of the Ru catalyzed hydrogenation process. Consequently, many answers have arisen to surmount this problem.

Recently, Guo et al. [67,73] have studied the hydrogenation of NBR in the presence of RuCl(PhCO$_2$)-(CO)(PPh$_3$)$_3$. They conducted the reaction at relatively low temperatures (65–120°C) and pressures (approximately 0.1 MPa) and obtained only 40% hydrogenation. The activity of this catalyst was poorer than RhCl(PPh$_3$)$_3$, under similar conditions. This complex also catalyzes the isomerization of C=C in the polymer backbone. Due to positional isomerization, C=C shifted conjugated to nitrile groups of the elastomer. Such C=C are difficult to hydrogenate as they are stabilized by electron delocalization due to conjugation. Later, they could achieve more than 97% hydrogenation, at low-polymer concentration, using higher pressure and temperature. The gel formation in this case was also less.

Martin et al. [69] undertook a study of the kinetics and mechanism of NBR hydrogenation using various Ru complexes. They examined the activity of RuXCl-(CO)L$_2$ (X = H, Ph, or CH=CHPh; L = PCy$_3$, P^{i}Pr$_3$,

Table 5 Hydrogenation of NBR using Ruthenium Complex

Sl. No.	Catalyst	Solvent	Pressure (MPa)	Temperature (°C)	Time (h)	Degree of Hydrogenation (%)	Ref.
1.	$RuCl_2(PPh_3)_3$	Butanone	14.0	150	—	99.1	48
2.	$RuCl_2(PPh_3)_3$	—	4.9	145	—	93	49
3.	$RuCl_2(PPh_3)_3$	—	—	145	6	100	50
4.	$RuCl_2(PPh_3)_3$	Acetone	4.9	145	10	95.9	51
5.	$RuCl_2(PPh_3)_3$	2-Butanone	14.0	135	5	99.8	52
6.	$RuCl_2(PPh_3)_3/NH_4PF_6/Et_3N$		4.9	—	—	—	53
7.	$RuH(MeCO_2)(PPh_3)$	Butanone	14.0	145	4	99	54
8.	$RuX[(L_1)(L_2)n]$, X = halo, $SnCl_2$, H L_1 = halo, hydrido, indenyl, fluoroenyl, L_2 = phosphine, bisphosphine or arsene, (n = 1 or 2)	Low molecular weight ketone	—	—	—	—	55
9.	$RuH_4(PPh_3)_3$	2-Butanone	14.0	130	4	95	56
10.	$RuHCl(CO)(PPh_3)_3 + CH_3COOH$	Chlorobenzene	1.5	140	1.8	99.5	57
11.	$RuH(OAc)(PPh_3)_3$	Acetone	—	145	6	90.0	58
12.	$RuCl_2(PPh_3)_3 + Et_3N$	Acetone	—	145	4	92.0	59
13.	$RuCl[(CO) (tricyclohexyl-phosphine)_2]$	Chlorobenzene	5.6	140	2	99	60, 61
14.	$RuCl_3·3H_2O + PPh_3 + Et_4NCl$	Acetone	9.8	145	5	94.5	62
15.	$RuCl_3·3H_2O + PPh_3$	Acetone	9.8	145	4	92.2	63
16.	$RuCl_3·3H_2O + PPh_3 + CaCl_2·2H_2O$	Acetone	—	145	3	85	64
17.	$RuCl_3·3H_2O + PPh_3$	Acetone	—	145	—	94	65
18.	$RuCl_3·3H_2O + PPh_3 + AcOH$	Acetone	—	155	6	97.2	66
19.	$RuCl [(CO)(COPh) (PPh_3)_2]$	—	—	—	—	—	67,68
20.	$Ru(X)Cl(CO)L_2$, X = H or styryl L = tricyclohexyl or triisopropyl phosphine	—	—	—	—	—	69
21.	$RuXY(CO)ZL_2$ or $RuX(NO)(CO)L_2$ X = halogen, carboxylate or phenyl vinyl group, Y = halogen, H, Ph, phenyl vinyl, carboxylate, Z = CO, pyridine, PhCN, $(MeO)_3P$, L = Hindered alkyl phosphine	Chlorobenzene	5.0	140	0.5	99	70
22.	$RuCl_2(styryl) (tricyclohexyl phosphine)_2$	2-Butanone	—	—	—	99	71
23.	$RuCl_2(PPh_3)_3/FeSO_4/(NH_4)SO_4$	2-Butanone	—	—	—	99	72,73
24.	Ru(II) complex + HOAc	Organic solvent + water	—	—	—	99.8	74

or PPh_3) toward selective hydrogenation of NBR. They observed that the Ru complexes with PPh_3 ligand cause gel formation, whereas the complexes with PCy_3 and P^iPr_3 ligands steer efficient hydrogenation without gel formation. They also concluded that the gel formation in the Ru catalyzed hydrogenation of NBR in nonketone solvents is only due to reduction of nitrile groups during the reaction process. The kinetics of hydrogenation using $RuCl(CH=CHPh)(CO)(PCy_3)$ was also examined by Martin et al. [69]. The kinetics of hydrogenation showed a first order dependence on C=C concentration,

hydrogen pressure, and catalyst concentration and an inverse dependence on nitrile concentration. They postulated a mechanism (Scheme 2) for hydrogenation using this catalyst. The coordination of the nitrile group to the metal center is the reason for hydrogenation of these groups. The mechanism is slightly different from the Rh catalyst where no ligand is displaced from the metal center.

Table 5 lists other ruthenium complexes that could catalyze selective hydrogenation in NBR. However, their activity could not be properly compared as they

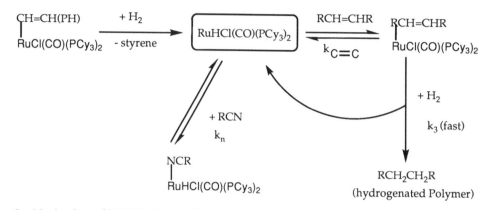

Scheme 2 Mechanism of NBR hydrogenation catalyzed by RuCl(CH=CHPh)(CO)(PCy$_3$)$_2$. *Source*: Ref. 8.

occur under different reaction conditions. In spite of possessing certain disadvantages, the ruthenium catalyst is an attractive alternative for selective and quantitative hydrogenation of nitrile rubber.

Guo et al. [70,71,73] recently attempted to hydrogenate NBR in emulsion form using Ru-PCy$_3$ complexes. However, successful hydrogenation can only be obtained when the emulsion is dissolved in a ketone solvent (2-butanone). A variety of Ru–phosphine complexes have been studied. Crosslinking of the polymer could not be avoided during the reaction. The use of carboxylic acids or first row transition metal salts as additives minimized the gel formation. The reactions under these conditions require a very high catalyst concentration for a desirable rate of hydrogenation.

C. Palladium Catalysts

In the preceding section, it has been shown that considerable attention has been devoted to palladium as a heterogeneous catalyst. The present section describes the homogeneous palladium catalysts developed for hydrogenation of NBR. The main drive behind the development of various catalyst systems is to find suitable substituents of the Rh catalyst. Palladium complexes are much cheaper as compared with Rh and exhibit comparable activity and selectivity to Rh and Ru complexes.

The most widely used homogeneous palladium catalyst is palladium carboxylate. Table 6 lists the palladium-based catalyst systems developed for NBR hydrogenation [75–87]. They are regarded as homogeneous cata-

Table 6 Hydrogenation of Nitrile Rubber in Presence of Palladium Catalyst

Sl. No.	Catalyst	Solvent	Pressure (MPa)	Temperature (°C)	Time (h)	Degree of hydrogenation (%)	Ref.
1.	Pd[(OAc)$_2$]$_3$	Acetone	4.9	50	3	97.2	75
2.	Pd[(OAc)$_2$]$_3$ reduced with hydrazine	Acetone	4.9	—	5	98.5	76
3.	Pd[(RCO$_2$)$_2$]$_3$ R = Me or Pr	—	—	—	—	—	77
4.	Pd[(OAc)$_2$]$_3$	Acetone	2.7	60	1	96	78
5.	[Pd(OAc)(2-benzoyl pyridine)]$_2$	Acetone	5.6	75	6	80	79
6.	[Pd(OAc)(2-benzoyl pyridine)]$_2$	Acetone	2.7	60	1	68	40,80
7.	[PdCl$_2$(MeCN)$_2$]	Acetone	4.9	50	6	84.9	81
8.	PdO(H$_2$O)$_n$/MeCN/PhPO$_4$	Tetrahydrofuran	2.41	75–85	—	70.0	82
9.	PdO complexed with NBR	—	—	—	—	—	83
10.	Pd(2-ethylhexanoate)/methyl aluminoxane	2-Butanone	—	—	—	84	84
11.	Pd(2-ethylhexanoate)/Et$_3$Al	2-Butanone	6.2	60	2	90	85
12.	Pd(benzoate)$_2$	Benzene	3.0	50	6	90.2	86
13.	Pd[(OAc)$_2$]$_3$/HNO$_3$	Water	3.0	50	6	78.2	87

lysts because they are dissolved in the polymer solution during hydrogenation. It is presumed that they are reduced to colloidal Pd metal under the hydrogenation condition. In a patent, Kubo et al. [75] mentioned that the catalyst can be removed from the polymer solution by filtration or centrifugation. This might be attributed to the formation of a uniform dispersion of very finely divided metal throughout the substrate solution when the catalyst is subjected to the reaction. The palladium complex should be properly soluble in the reaction mixture in order to incur a high degree of hydrogenation. This is supported by the studies on PdCl$_2$ catalyzed hydrogenation of NBR [75]. PdCl$_2$ is not a readily soluble complex in the polymer solution and, hence, offers inferior levels of conversion as compared to palladium carboxylates.

Acetone is the best solvent for NBR hydrogenation in the presence of palladium carboxylates. No hydrogenation is achieved when chloroform or chlorobenzene are the solvents. Since it is understood that palladium is reduced to colloidal metal in the presence of hydrogen, attempts have also been made to reduce the palladium by hydrazine [76], methylaluminoxane [84], and trialkyl aluminum [85] to improve the catalytic activity.

Bhattacharjee et al. [78] have conducted a fundamental study on the catalytic activity offered by palladium acetate toward hydrogenation of C=C in nitrile rubber. Palladium acetate is a simple, commercially available, air stable, and relatively less expensive complex of palladium. It exhibited excellent selectivity and hydrogenation efficiency. With increases in reaction time, pressure, temperature, and catalyst concentration, the extent of hydrogenation increased. A maximum conversion of 96% could be achieved at 60°C under 2.7 MPa pressure and 0.54 mmol catalyst for 1 h in acetone. The reaction is very rapid even at room temperature. From the spectroscopic analysis, they confirmed that the CN group was not hydrogenated during the reaction. A kinetic study of NBR hydrogenation suggests a first order dependence on the olefinic substrate. The rate of hydrogenation increases with increases in catalyst concentration and hydrogen pressure. The activation energy of the reaction calculated from the Arrhenius plot is 29.9 kJ/mol. This is comparable to that obtained for the Rh catalyst [11]. The apparent activation enthalpy and activation entropy are 27.42 kJ/mol and −0.20 kJ/(mol-K). Gel permeation chromatographic measurements show that there is no change in the molecular weight of the polymer after hydrogenation. The glass-transition temperature reduces gradually with the increase in the level of hydrogenation. They also studied some of the physical properties of the hydrogenated product.

Bhattacharjee et al. [79] introduced another new catalyst based on a Pd complex containing both acetate and benzoyl pyridine ligands (Table 6). This was developed to hydrogenate liquid carboxylated nitrile rubber (L-XNBR) [80]. Selective hydrogenation of C=C in L-

XNBR is more intricate due to the presence of two additional functional groups: carboxyl and nitrile. Palladium acetate does not hydrogenate L-XNBR, whereas cyclopalladate complex of 2-benzoyl pyridine effects measurable amounts of hydrogenation [40]. The reduction was strictly restricted to C=C in the presence of carboxyl and nitrile groups. RhCl(PPh$_3$)$_3$, on the other hand, cause significant amounts of decarboxylation along with hydrogenation. Hydrogenation of solid nitrile rubber with different acrylonitrile content and solid carboxylated nitrile rubber (XNBR) has also been accomplished by using this catalyst [79]. The physical properties of

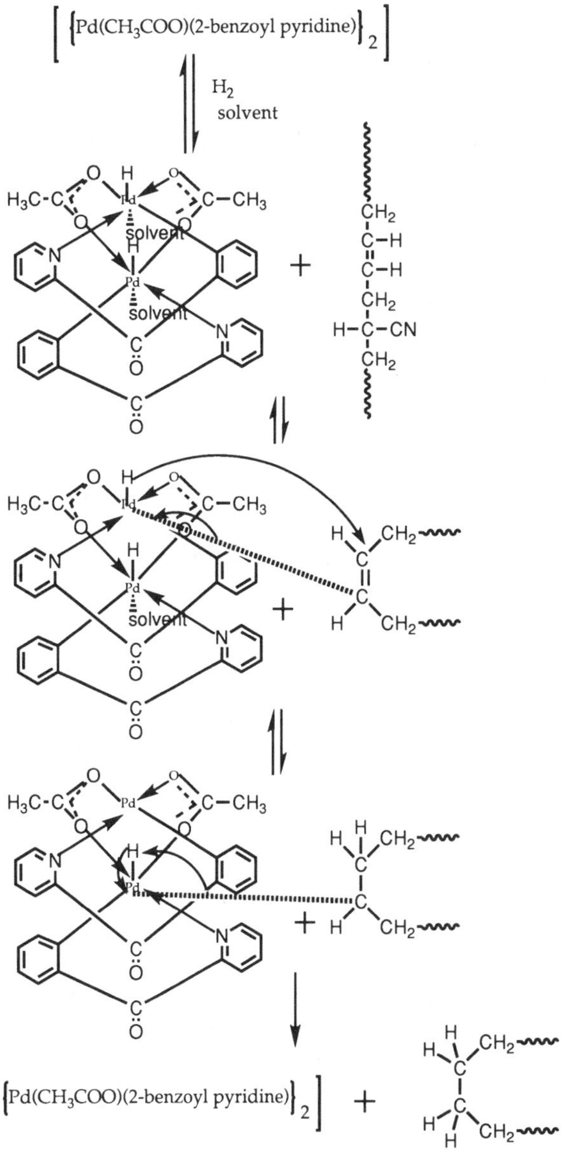

Scheme 3 Mechanism of hydrogenation of NBR catalyzed by [Pd(OAc)(2-benzoyl pyridine)]$_2$. *Source*: Ref. 79.

the modified rubbers were also examined. In contrast to palladium acetate, chloroform and chlorobenzene could also be used as solvents. The authors have proposed a mechanism for hydrogenation in the presence of cyclo-palladate complex (Scheme 3), which is similar to that reported for Rh and Ru catalyst (Schemes 1 and 2).

The first attempt to hydrogenate NBR in emulsion form was carried out in the presence of palladium benzoate [86]. However, an organic solvent and high concentration of catalyst were required to attain a reasonable level of hydrogenation. A similar observation was experienced later for Ru catalyzed hydrogenation in emulsion form. A very recent patent from Nippon Zeon Co. reveals that NBR latex can be hydrogenated by using palladium acetate in the presence of nitric acid [87]. This is a very significant development as it does not require any organic solvent.

The available data in Table 6 reveal that palladium complexes are excellent catalysts for selective hydrogenation of C=C in NBR. Recent attempts to recover the catalyst (see Section VII) after hydrogenation and lower the cost of the metal make it an attractive supplement in the industrial production of HNBR.

V. HYDROGENATION OF LIQUID RUBBER

The liquid nitrile rubbers are generally used as nonvolatile and nonextractable plasticizers. They also function as binders and modifiers for epoxy resins. Their moderate heat resistance limits their ability to meet industrial requirements. Hence, attempts have been made to improve their thermal and oxidative resistance by saturating the polymer backbone.

Hashimoto et al. [88] have hydrogenated liquid nitrile rubbers for the betterment of the thermo-oxidative resistance and retention of properties at higher temperature. Studies on hydrogenation of liquid rubbers are quite challenging, as most of them contain more than one reducible functionality. Hydrogenation of liquid carboxylated nitrile rubber (L-XNBR) complicates the method, because both the carboxyl and nitrile functionality interfere in the double bond reduction. Palladium acetate, which evolved as an efficient catalyst system for selective hydrogenation of solid NBR, could not hydrogenate L-XNBR. Wilkinson's catalyst offered reduction of C=C along with a significant amount of decarboxylation and, under certain conditions, hydrogenation of nitrile groups [40]. Bhattacharjee et al. [40,80] have developed a new catalyst based on the palladium complex for quantitative hydrogenation of C=C in L-XNBR. Successful hydrogenation of C=C was obtained without disturbing the carboxyl and nitrile content of the polymer. These groups also did not restrict the catalytic activity of the complex. The degree of hydrogenation at a particular reaction condition depended mainly on the

ratio of catalyst to that of the diene unit of the liquid rubber. The reaction kinetics exhibit a first order dependence with respect to the olefinic substrate [80]. This is in line with the earlier observation on Rh [11,38] and palladium acetate [78] catalysts. An increase in the amount of nitrile and carboxyl groups decreased the rate constants of the reaction. The activation energy of the L-XNBR hydrogenation was 20.2 kJ/mol. The glass-transition temperature of the saturated liquid polymer decreased significantly.

VI. HYDROGENATION OF NITRILE RUBBER LATEX

The idea of hydrogenating nitrile rubber in latex form is of high practical importance in the commercial production of HNBR. This could lead to a direct preparation of HNBR from the monomers without isolation of NBR if hydrogenation could be introduced at the emulsion stage. The expense of using an organic solvent and subsequent removal could also be curtailed. Attempts have been made to hydrogenate NBR emulsion using Ru [71–73] and Pd [86] catalysts. However, they all require an organic solvent to swell the emulsion and a high concentration of catalyst for a sufficient degree of hydrogenation.

Recently there was a report of NBR latex hydrogenation in the presence of palladium acetate which does not require any additional use of organic solvent [87]. Researchers from Nippon Zeon Co., have developed a novel method for selective hydrogenation of C=C in NBR latex by using palladium acetate and nitric acid. The introduction of an equimolar amount of nitric acid along with palladium acetate helps to avoid the organic solvent, which is normally used to dissolve the catalyst and swell the latex particles. However, the investigations in this area are not sufficient to generalize the method. Singha et al. [46] developed a water-soluble analog of Wilkinson's catalyst for hydrogenation of NBR in the latex stage. RhCl (sodium diphenyl phosphino benzene-m-sulfonate)₃ effected a moderate degree of hydrogenation of C=C in NBR. However, hydrogenation was accompanied by an increased gel content of the latex. The noncatalytic methods have also been used for latex stage hydrogenation of NBR (see Section VIII).

The research in this area has a great potential from scientific and technological aspects and requires further exploration. However, the reported attempts are a welcoming endeavor to hydrogenate nitrile rubber in latex form.

VII. REMOVAL OF THE CATALYST

The only weakness of the homogeneous catalytic hydrogenation of elastomer is the removal of the catalyst from the polymer mixture. This is the main reason for the high cost of hydrogenated nitrile rubber. If the expensive catalyst can be recovered and reused, the difference in the cost of NBR and HNBR can be brought down to a minimum level. The service life and resistance to high-temperature degradation might be enhanced further when these active metals are removed from the polymer substrate. Several researchers have come up with new ideas to remove the catalyst from the hydrogenated polymer.

Ahlberg et al. [89] reported the removal of $RhCl(PPh_3)_3$ catalyst residues from HNBR. They kneaded the rubber with 20–500 phr methanol in a mixing apparatus at 60°C. The rubber was later separated from methanol. The polymer sample, which initially contained 116 ppm Rh and 1.46 wt% PPh_3, after extraction showed 86 ppm Rh and 1.18 wt% PPh_3. Osman et al. [90] developed another procedure for removal of Rh residues from HNBR solution. The reaction mixture (200 mL solution of chlorobenzene containing 6 g of HNBR), after catalytic hydrogenation, was treated with 10 g of starch solution in water, refluxed at 110–120°C for 6–7 h, and cooled to ambient temperature. A precipitate, formed upon cooling, was separated by filtration. The rubber was coagulated with methanol and dried. The residual Rh content of the elastomer decreased to 5 ppm from 85 ppm. Madgavkar et al. [91] lowered the level of catalyst residues below 5 ppm in polymer solutions. They oxidized the polymer solution in the presence of oxygen and then treated with carbon black to remove the catalyst. Panster et al. [92] reported the recovery of Rh and Ru catalysts from solutions of HNBR. More than 80% of the catalyst could be recovered by absorption on siloxanes containing tertiary amino groups, sulphide, phosphine, thiourea, or urea. However, the research is still in progress and probably a more convenient method will soon be developed to completely recover and reuse the expensive catalysts.

VIII. NONCATALYTIC HYDROGENATION

The available studies indicate that diimide has been used as a reducing agent for the preparation of HNBR. It has been used mainly as an alternative for hydrogenation of nitrile rubber latex. The use of diimide to hydrogenate low-molecular weight olefines is well known in the organic literature [93]. Diimide can be conveniently generated in situ by thermal treatment of solutions of p-toluenesulfonyl hydrazide or oxidation of hydrazine.

Wideman [94] developed a method for reducing the C=C of NBR in the latex stage. The process involved refluxing NBR latex with hydrazine, an oxidizing agent, and a metal ion activator. The latex was later filtered and coagulated. During the in situ generation of diimide,

there is formation of some by-products. The reaction of these by-products with unsaturated sites of the polymer leads to molecular weight degradation. Apart from this, hydrazine can remain attached to the polymer chain. A novel process for preparation of highly saturated nitrile rubber latex has been introduced by Parker et al. [10]. It involves the diimide reduction that is generated in situ within the latex by the oxidation of hydrazine hydrate in the presence of a catalyst. The reaction operates at ambient pressure and at 40–70°C temperature. Nitrogen and water are the by-products of the reaction. The process is quite efficient for selective hydrogenation of C=C in NBR and has a high commercial interest because it converts NBR latex into HNBR latex in a one-step reaction. The HNBR latex produced by this method has been used in many latex-related applications. Scientific studies on diimide reduction of low-molecular weight polymers by p-toluene sulfonyl hydrazide are also reported [95].

Schiessl [96] attempted to scavenge hydrazine attached to the latex following the reduction procedure. NBR latex was reduced with hydrazine and peroxide at a 60–70°C temperature for 1–5 h in the presence of copper sulfate or ferrous sulfate initiators. The hydrogenated polymer latex was then treated with compounds consisting of various functional groups, such as isocyanates, acrylates, acids, ketones, diketones, or aldehydes in large excess to eliminate the free hydrazine in the system. In another process to minimize residual hydrazine in HNBR latex [97], they treated HNBR latex containing 3.76% hydrazine with oxygen at 0.28 MPa in the presence of 0.2% hydroquinone at 50°C for 20 h and lowered the hydrazine level significantly to 100 ppm.

Noncatalytic hydrogenation is promising from a commercial point of view if the side reactions during the process can be carefully excluded. Moreover, this is of particular importance for online reduction of NBR emulsion.

IX. HYDROFORMYLATION OF NITRILE RUBBER

Hydroformylation of nitrile rubber is another chemical modification that can incorporate a reactive aldehyde group into the diene part and further open up new synthetic routes to the formation of novel nitrile elastomers with a saturated backbone containing carboxyl or hydroxyl functionalities.

Bhattacharjee et al. [98] reported that nitrile rubber can be selectively hydroformylated in the presence of $RhH(CO)(PPh_3)_3$ and $RhCl(CO)(PPh_3)_2$ under high pressure of carbon monoxide and hydrogen. They found that $RhH(CO)(PPh_3)_3$ was more efficient than $RhCl(CO)(PPh_3)_2$. However, both catalysts offered se-

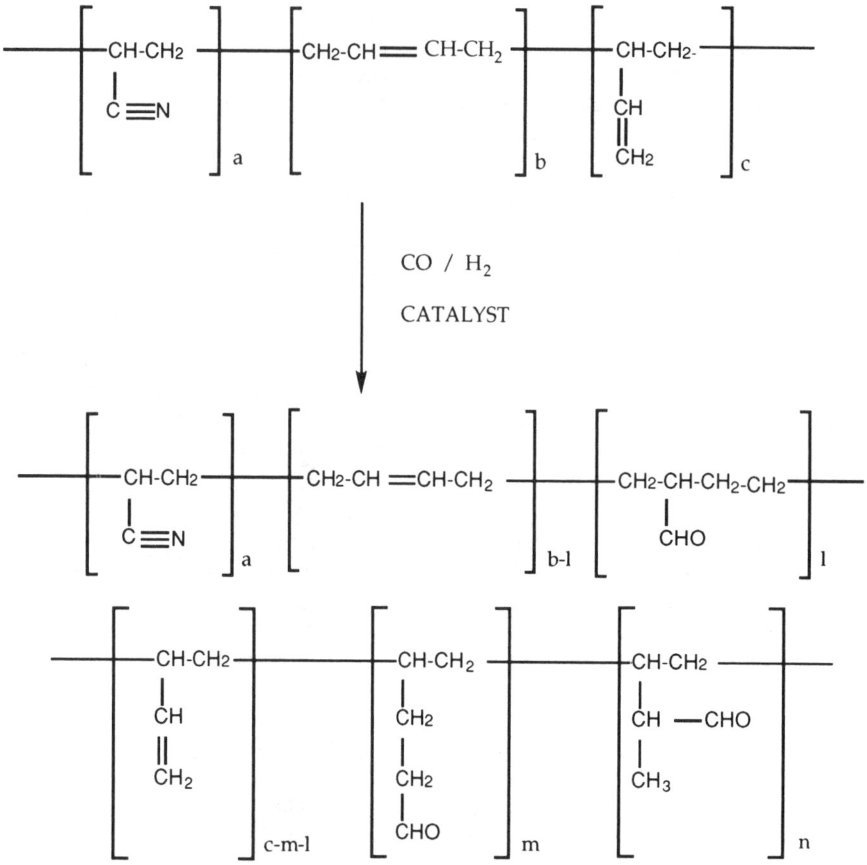

$$\left[\begin{array}{c} \text{-CH-CH}_2 \\ | \\ \text{C}\equiv\text{N} \end{array}\right]_a \left[\begin{array}{c} \text{-CH}_2\text{-CH}=\text{CH-CH}_2 \end{array}\right]_b \left[\begin{array}{c} \text{-CH-CH}_2\text{-} \\ | \\ \text{CH} \\ \| \\ \text{CH}_2 \end{array}\right]_c$$

CO / H₂

CATALYST

$$\left[\begin{array}{c} \text{-CH-CH}_2 \\ | \\ \text{C}\equiv\text{N} \end{array}\right]_a \left[\begin{array}{c} \text{-CH}_2\text{-CH}=\text{CH-CH}_2 \end{array}\right]_{b\text{-}l} \left[\begin{array}{c} \text{-CH}_2\text{-CH-CH}_2\text{-CH}_2\text{-} \\ | \\ \text{CHO} \end{array}\right]_l$$

$$\left[\begin{array}{c} \text{-CH-CH}_2\text{-} \\ | \\ \text{CH} \\ \| \\ \text{CH}_2 \end{array}\right]_{c\text{-}m\text{-}l} \left[\begin{array}{c} \text{-CH-CH}_2\text{-} \\ | \\ \text{CH}_2 \\ | \\ \text{CH}_2 \\ | \\ \text{CHO} \end{array}\right]_m \left[\begin{array}{c} \text{-CH-CH}_2\text{-} \\ | \\ \text{CH} \; \text{-CHO} \\ | \\ \text{CH}_3 \end{array}\right]_n$$

lectivity toward the addition of aldehyde group to C=C in the presence of nitrile functionality; 30% conversion could be achieved at 90°C under 5.6 MPa pressure (CO:H₂ = 1:1) in presence of a 0.43 mmol/L catalyst. When the hydroformylation was more than 30%, gel formation occurred. The rate constant was reduced by the higher partial pressure of carbon monoxide. With the increase in the acrylonitrile concentration of the copolymer, there was a significant decrease in the extent of hydroformylation as well as the rate constants of the reaction under certain reaction conditions. Characterization of the products by spectroscopic techniques indicated the formation of internal aldehyde from a 1,4-unit and terminal aldehyde from a 1,2-unit of the diene fraction of the copolymer. The intrinsic viscosity of the modified polymer decreased with an increase in aldehyde incorporation. The glass-transition temperature increased with the extent of hydroformylation. The thermal stability of the hydroformylated product was, however, lower than that of nitrile rubber.

Hydroformylated nitrile rubbers might have potential as binders for coupling with other rubbers due to the presence of polar aldehyde group. This might also open up the still greater potential of NBR containing saturated backbone in composite materials [1].

X. CHARACTERIZATION OF MODIFIED NITRILE RUBBER

Characterization and understanding of the microstructure become important after hydrogenation and hydroformylation of the nitrile rubber since the amount and distribution of the residual double bonds influence the properties of modified rubber. The conventional analytical tools have been used to characterize the elastomers. Spectroscopy is the most useful technique for determination of the degree of hydrogenation in nitrile rubber.

Brück [99,100] described the infrared (IR) spectrophotometric methods for determination of the proportion of acrylonitrile, butadiene, and hydrogenated butadiene in hydrogenated acrylonitrile–butadiene rubbers of unknown composition. In addition, it was also demonstrated that the cis/*trans* ratio of the 1,4-butadiene structure in NBR and the residual double bonds in HNBR could be calculated from the IR measurements without a knowledge of the corresponding absorption coefficients. These calculations are based on the fact that the nitrile group is not reduced during the hydrogenation process. This can be supported by the IR spectrum of a HNBR sample, which does not show any additional peaks due to hydrogenation of nitrile functionality. The selectivity

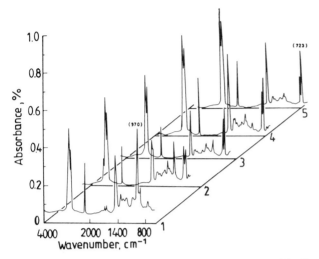

Figure 1 IR spectra of: 1-NBR and 2–5-HNBR with 58, 71, and 99.9 mol% hydrogenation, respectively. *Source*: Ref. 11.

of the catalyst can also be established from the IR studies.

Figure 1 illustrates the IR spectra of NBR and HNBR samples with different concentrations of residual double bonds [11]. The CN stretching vibration is observed at 2222 cm^{-1} in NBR and HNBR. The peak at 1440 cm^{-1} is for C—H deformation of —CH$_2$— groups. The =C—H out of plane deformations of trans, vinyl, and cis double bonds are observed at 970 cm^{-1}, 920 cm^{-1}, and 730 cm^{-1}, respectively. These peak absorbances decrease gradually, and a new peak at 723 cm^{-1} appears on the spectrum of HNBR for —CH$_2$— rocking vibration [when (CH$_2$)$_n$, n > 4] [11,78]. The CN stretching vibration is taken as an internal standard and the

ratio of absorbances at 970 cm^{-1}, 920 cm^{-1}, and 730 cm^{-1} with respect to 2222 cm^{-1}, is compared with the peak absorbance ratio at 723 cm^{-1} for an estimation of the level of hydrogenation. The degree of hydroformylation [98] was calculated from the relative decrease in the absorbance ratio due to =C—H and increase in the absorbance ratio at 723 cm^{-1}.

Marshall et al. [101] developed a method for quantification of the level of hydrogenation in HNBR by nuclear magnetic resonance spectroscopy (NMR). The measurements are based on relative amounts of olefinic and aliphatic signals in proton NMR (^{1}H NMR spectra). Figure 2 shows the ^{1}H NMR spectra of NBR and two HNBR samples. The signals due to olefinic protons appear as a broad peak between 5.0 to 5.8 ppm. All the protons in the —CH$_3$, —CH$_2$, and —CH microstructure are present between 0.5 to 2.8 ppm. Upon hydrogenation, the integral of the peaks due to olefinic protons decreases and all the aliphatic protons shift in the region of 0.5 to 2.0 ppm. The —C—H attached to the CN group can be clearly seen at 2.6 ppm. The degree of hydrogenation calculated from the ratio of olefinic to aliphatic protons is comparable to that of IR spectroscopic method [11,78]. Figure 3 represents the carbon-13 NMR (^{13}C NMR) spectra of NBR and HNBR [11]. The decrease in the concentration of the olefinic carbons can be clearly understood from the spectra. The selectivity of the reaction can further be confirmed from the ^{13}C NMR spectra of a completely hydrogenated polymer sample [11].

The classic chemical technique for measuring the degree of unsaturation in diene polymers is iodometry (iodine value) [102]. Kubo et al. [103] extensively measured the iodine value to determine the amount of residual double bonds present in the HNBR. However, this method exhibited significantly poorer precision as compared with IR and NMR spectroscopies [99–101]. Acid

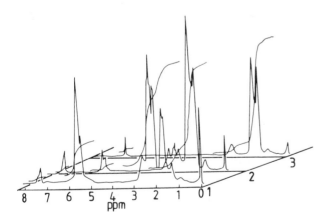

Figure 2 ^{1}H NMR spectra of: 1-NBR and 2–3-HNBR with 71 and 99.9 mol% hydrogenation, respectively. *Source*: Ref. 11.

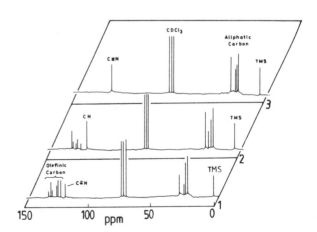

Figure 3 ^{13}C NMR spectra of: 1-NBR and 2–3-HNBR with 80 and 99.9 mol% hydrogenation, respectively. *Source*: Ref. 11.

values were calculated by Bhattacharjee et al. [40,80] to monitor any loss of carboxyl functionality during hydrogenation of L-XNBR.

The structural characterization of HNBR by pyrolysis gas chromatography together with IR and NMR spectroscopies was carried out in details by Kondo et al. [104]. The characteristic peaks on the program by pyrolysis gas chromatography, specific absorption bands in the IR spectra, and resonance peaks in the NMR spectra were interpreted in terms of the microstructure of the hydrogenated polymer. ^{13}C NMR spectra gave insight into the sequence distribution and hydrogenation mechanism. Kondo et al. [104] also indicated that the butadiene units next to the acrylonitrile units in the polymer chains were more likely to be hydrogenated than those next to the butadiene unit. The preferential reduction of the 1,2-unit over the 1,4-unit was understood from IR and NMR spectra [11,78,80,104]. Bhattacharjee et al. [78] used electron spectroscopy for chemical analysis (ESCA) as an additional tool to prove the selectivity offered by palladium acetate toward hydrogenation of C=C in the presence of the CN group.

There is no significant change in the molecular weight of the elastomer after hydrogenation [11,78,103] as observed from gel permeation chromatography. Bhattacharjee et al. [40,98] measured the intrinsic viscosity of hydrogenated and hydroformylated polymers. They studied solution properties of HNBR and determined the second virial coefficient and Huggins constant [105]. However, the solubility parameter values changed very little upon hydrogenation.

The modified NBR samples were characterized by differential scanning calorimetry [11,78–80,98]. The glass-transition temperature (T_g) decreased with the level of hydrogenation. In the case of HFNBR, T_g increased with an increase in the addition of aldehyde groups to the polymer chain. Thermogravimetric analysis of the modified polymers have also been carried out [15].

XI. PHYSICAL PROPERTIES OF HYDROGENATED NITRILE RUBBERS

The properties of HNBR are mainly determined by the degree of saturation and acrylonitrile content. The degree of saturation is responsible for the heat resistance, rheological behavior, and processability of the elastomer. The residual unsaturation present in the rubber gives a favorable balance between crosslinking by the sulfur–accelerator system and loss of heat, ozone, and chemical resistance. However, it has been found that superior performance is obtained when peroxide curing is used rather than the traditional sulfur curing system [106–108]. For completely saturated nitrile rubber, only peroxide curing is done. The physical properties of HNBR in comparison with NBR have been discussed

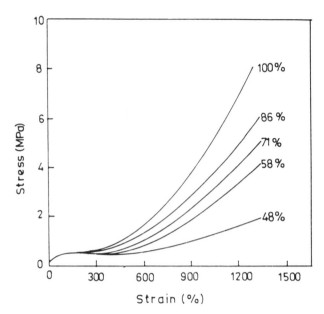

Figure 4 Stress–strain curves of hydrogenated NBR containing 40 mol% ACN with different degree of hydrogenation. *Source*: Ref. 11.

in detail by Bhattacharjee et al. [1]. The production and properties of different commercial grades of HNBR have been reviewed by Kovshov et al. [109].

The gum strength of HNBR is much higher in comparison to NBR [11] (Fig. 4). With the increase in degree

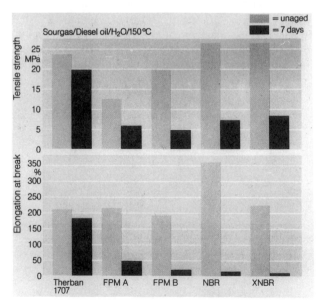

Figure 5 Comparison of HNBR with fluoroelastomer (FPM), NBR and XNBR in "Sour Crude Oil" (sour gas = 20% H_2S + 65% CH_4 + 15% CO_2). *Source*: Bayer AG, Germany.

of hydrogenation, the stress–strain properties improve. HNBR vulcanizates have much higher tensile strength. They can achieve tensile strength values up to 40 MPa, which is much superior to NBR and fluoroelastomers. HNBR is able to retain high levels of physical properties even under a combination of many adverse conditions. It exhibits very good abrasion resistance and low compression set values at high temperature. The wear of HNBR vulcanizates have been examined by Thavamani et al. [110,111]. The various physical properties of HNBR vulcanizates have been extensively studied [112–116].

Low temperature flexibility of HNBR is better than NBR. With the increase of the degree of hydrogenation,

the glass-transition temperature (T_g) of the elastomer decreases [11,78,79]. T_g can be further lowered without noticeable loss of heat resistance by adding plasticizers [117]. The improved performance of HNBR at low temperature is investigated by Jobe et al. [118].

Rubber to metal bond strength of HNBR vulcanizates exceeds its inherent tear strength in presence of suitable rubber to metal bonding agent. Bond strength remains almost the same after hot air aging of metal–rubber composite [119]. Suitably formulated HNBR offers good adhesion to different reinforcing fabrics and cords.

The high strength and high-elasticity retention over a wide range of temperatures imparts HNBR with a good retention of dynamic behavior. Stein et al. [120] have

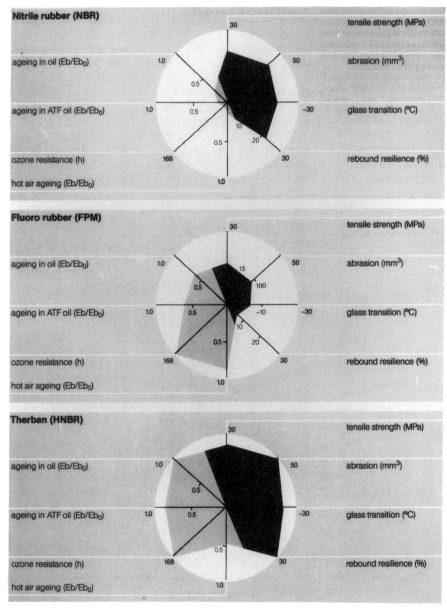

Figure 6 Performance profile of HNBR compared to NBR and FPM. *Source*: Bayer AG, Germany.

compared HNBR under low temperature and high temperature dynamic conditions. Thavamani and Bhowmick [121] studied the dynamic mechanical properties of HNBR and observed the effect of crosslink density, curing system, filler, and resin. HNBR is slightly superior to NBR in gas and water vapor permeation [122,123].

HNBR is able to maintain the physical properties under a combination of hot air, ozone, steam, oils, chemicals, and high-energy radiation, which can singly cause rapid failure of other elastomers. The most outstanding property of HNBR is its ozone resistance under extreme exposure conditions [124,125]. The oil and fuel resistance of HNBR is inherited from the parent NBR. The performance characteristics of HNBR are much better than NBR in retaining these properties at a higher temperature for a prolonged period [126,127]. It also shows higher resistance to various oil contaminants and lubricating oil additives [128]. Oil field applications involve exposure to amines, hydrogen sulfide, and carbon dioxide. This kind of "sour crude oil" resistance of HNBR is illustrated in Fig. 5, which shows its better accomplishment than fluoroelastomers, XNBR, and NBR [129]. HNBR is now used for military and aerospace equipment due to its excellent wear resistance. Blends of HNBR are used in tank tracks and exhibits superior performance as compared with traditional styrene–butadiene rubber compositions [130–133]. A relative performance profile of HNBR in comparison with NBR and fluoroelastomer is represented in Fig. 6.

The thermal degradation behavior of raw HNBR has been investigated by Bhattacharjee et al. [134] and the observations have been compared with raw NBR. Studies on low-temperature degradation (75–150°C) of HNBR and NBR in the presence of air reveal that the molecular weight of saturated rubber decreases initially with aging time and finally decreases with prolonged aging. In the case of NBR, the molecular weight decreases with aging time. The contact angle of both the samples decreases with aging time and temperature. The enhanced wettability of the polymer surface is attributed to the oxidation of the surface and the incorporation of carboxyl, peroxide, and other oxygen containing groups. However, HNBR samples show more resistance to surface changes with aging time and temperature than NBR. NBR undergoes gel formation on aging, whereas there is no gel formation in HNBR even after prolonged aging time. Both the saturated and unsaturated nitrile rubbers generate carbonyl and ester functionalities on thermal oxidation. The degradation in saturated polymers takes place through nitrile groups, which might be converted to imide groups.

Hydrogenated nitrile rubber is much more stable than NBR at elevated temperature. Bhattacharjee et al. [134] performed a high-temperature degradation (30–800°C) of HNBR and NBR in nitrogen atmosphere

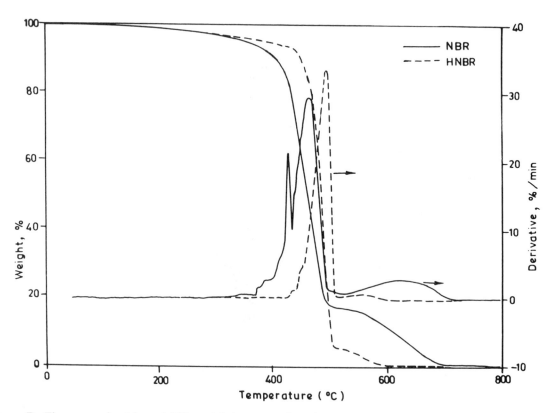

Figure 7 Thermogravimetric and differential thermogravimetric curves of NBR and HNBR. *Source*: Ref. 134.

(thermogravimetric analysis). The onset of degradation for NBR (40% ACN) is at 357°C and that for corresponding HNBR is at 419°C. The maximum degradation in HNBR takes place at 495°C. On the other hand, the corresponding NBR shows two temperature maxima at 433°C and 476°C (Fig. 7). The degradation in HNBR is a first order reaction, whereas in NBR, it is of second order. The activation energies of degradation are 350 and 175 kJ/mol for HNBR and NBR, respectively. This indicates the better thermal stability of HNBR over NBR. The thermo-oxidative stability of L-XNBR can be improved to a considerable extent by hydrogenation of olefinic bonds [80]. There are a few citations in the current literature regarding the degradation behavior of HNBR vulcanizates [135–137].

XII. APPLICATION

The versatility of hydrogenated nitrile rubber is reflected by its wide range of industrial application. HNBR finds high recommendation in the automotive, mechanical engineering, and oil industries [3]. The service life of HNBR is much longer than other conventional rubber articles. The admirable combination of its physical properties along with the capability of property retention under aggressive conditions have extended its domain as a specialty elastomer. Due to better performance characteristics, HNBR has replaced many conventional elastomers that were traditionally used. The various fields where HNBR finds application are grouped together in the following table.

Applications	Properties responsible for the application
Automotive Industry	
Hydraulic hoses, valves, seals, expansion hoses, radial shaft sealing ring, and axle boots	Heat, hot oil, and swelling resistance
Bearings	Heat resistance
Rotary vibration dampers, synchronous and transmission belts, and fluid blocks in steering linkages	Heat, hot oil resistance, and dynamic strength
Water pump seals, hoses for air-conditioning systems, seals in cooling systems and covers in diesel injection systems	Heat, hot oil, swelling, and ozone resistance
Chain tensioners and skids for chain-controlled camshafts	Heat, hot oil, swelling, and wear resistance
Oil Field Industry	
Drill pipe protectors and packer seals	Heat, hot oil, swelling, and wear resistance
Pump stators and blow-out preventors	Heat, hot oil, swelling and wear resistance, and dynamic strength
Electrical Industry	
Cable sheathings for special heavy-duty cables	Heat, hot oil, wear, and ozone resistance
Marine cables and drill-lock cables for deep wells	Hot oil, heat, and ozone resistance
Mechanical Engineering Industry	
Various parts for cellulose manufacture	Heat, swelling and wear resistance, and high dynamic strength
Lamination and pressure rollers for paper, steel, and printing industries; rollers and wheels for heavy-duty transport; cable railway guide rolls	Heat, hot oil, swelling and wear resistance, and dynamic strength
Ship Building Industry	
Seals for hatch covers for oil tankers, prow and stern pipe seals, and couplings for ship propulsion units	Heat, hot oil, and swelling resistance

XIII. CONCLUSION

Hydrogenation of nitrile rubber is an excellent example of tailoring the properties of an existing elastomer in order to compete with the demands in the field of application. Today HNBR replaces NBR and other conventional rubbers in many of their traditional operations. Hydrogenation has modified and improved all the major drawbacks of NBR and widened its range of potential uses. This vast body of literature suggests that considerable attention has been attributed toward the development of hydrogenation processes for NBR modification. The only drawback of HNBR is its high cost. Research is in progress to reduce the cost of HNBR production by using inexpensive metal catalysts and subsequently recovering them. The latex stage hydrogenation is another promising development. However, the versatility offered by HNBR during the designing of end products actually outclasses its possible disadvantages. Efforts to overcome the existing detriments are still continuing. Hydroformylation of NBR is another step toward development of specialty elastomers. However, the commercial viability of HFNBR is yet to be explored. Nevertheless, these modifications can be considered as a great revolution in nitrile rubber technology.

REFERENCES

1. S. Bhattacharjee, A. K. Bhowmick and B. N. Avasthi, *Elastomer Technology Handbook* (N. P. Cheremisinoff, ed.), CRC Press, Florida, p. 519 (1993).
2. K. Hashimoto and Y. Todani, *Handbook of Elastomers—New Developments and Technology* (A. K. Bhowmick and H. L. Stephens, eds.), Marcel Dekker, Inc., New York, p. 741 (1988).
3. Bayer AG, *Manual for the Rubber Industry,* Bayer AG, Leverkusen Germany, p. 119 (1993).
4. A. Rawlinson and N. Djuricis, "Polysar Technical Literature," Polysar Technical Centre/Jugoauto, Sarajevo, 8–10 Nov., (1988).
5. D. N. Schulz, S. R. Turner, and M. A. Golub, *Rubber Chem. Technol.,* 55: 809 (1982).
6. D. N. Schulz, *Encyclopedia of Polymer Science and Engineering,* (H. F. Mark, N. Bikales, C. C. Overberger, and G. Menges, eds.), vol. 7, p. 807 (1986).
7. D. N. Schulz, *Handbook of Elastomers—New Developments and Technology* (A. K. Bhowmick and H. L. Stephens, eds.), Marcel Dekker, Inc., New York, p. 75 (1988).
8. N. T. McManus and G. L. Rempel, *J.M.S.-Rev. Macromol. Chem. Phys.,* C 35(2): 239 (1995).
9. M. P. McGrath, E. D. Sall, and S. J. Tremont, *Chem. Rev.,* 95: 381 (1995).
10. D. K. Parker, R. F. Roberts, and H. W. Schiessl, *Rubber Chem. Technol.,* 65: 245 (1992).
11. S. Bhattacharjee, A. K. Bhowmick, and B. N. Avasthi, *Ind. Eng. Chem. Res., 30*: 1086 (1991).
12. O. A. Hougen, K. M. Watson, and R. A. Regatz, *Chemical Process Principles, Part II, Thermodynamics,* Wiley, New York, p. 982 (1960).
13. D. D. Wagman, W. H. Evans, I. Halow, V. B. Parker, S. M. Baily, and R. H. Schumm, *CRC Handbook of Chemistry and Physics* (R. C. Weast, ed.), CRC Press, Florida, p. D-61 (1979).
14. D. N. Rihani and L. K. Doraiswami, *Industrial Engineering Chemistry; Fundamental, 4*: 17 (1965).
15. S. Bhattacharjee, Ph.D. Thesis, Indian Institute of Technology, Kharagpur, India (1992).
16. A. Maeda, K. Hashimoto, S. Yagishata, M. Ingami, and H. Fukushima, (Nippon Zeon Co. Ltd., Japan) Ger Offen 2,913,992 (1979); CA 92:7695.
17. Nippon Zeon Co. Ltd., Japan, Jpn Kokai Tokkyo Koho, JP 56,28,803 (1981); CA 95:134184.
18. Nippon Zeon Co. Ltd., Japan, Jpn Kokai Tokkyo Koho, JP 56,81,305 (1981); CA 95:151487.
19. Nippon Zeon Co. Ltd., Japan, Jpn Kokai Tokkyo Koho, JP 56,28,306 (1981); CA 95:151980.
20. Nippon Zeon Co. Ltd., Japan, Jpn Kokai Tokkyo Koho, JP 57,202,305 (1982); CA 98:180880.
21. Nippon Zeon Co. Ltd., Japan, Jpn Kokai Tokkyo Koho, JP 58,17,103 (1983); CA 99:6737.
22. Nippon Zeon Co. Ltd., Japan, Jpn Kokai Tokkyo Koho, JP 57,205,404 (1982); CA 99:39641.
23. N. Watanabe and Y. Kubo (Nippon Zeon Co. Ltd., Japan) Ger Offen 3,418,357 (1984); CA 102:63445.
24. Y. Kubo and K. Ora (Nippon Zeon Co. Ltd., Japan) Jpn Kokai Tokkyo Koho, JP 61,130,303 (1986); CA 106:19226.
25. H. Buding and R. Casper (Bayer AG, Germany) Ger Offen 3,514,403 (1986); CA 106:343926.
26. T. Sawanobori and N. Furuki (Mitsubishi Kasei Corp.) Jpn Kokai Tokkyo Koho JP 01,121,303 (1989); CA 112:8607.
27. Y. Kubo and K. Ohura (Nippon Zeon Co. Ltd., Japan) Ger Offen DE 3,905,432 (1989); CA 112:100468.
28. N. Furuki and K. Sone (Mitsubishi Kasei Corp., Japan) Jpn Kokai Tokkyo Koho JP 03,81,302 (1991); CA 115:9701.
29. K. Takahashi and Y. Kubo (Nippon Zeon Co. Ltd., Japan) Ger. Offen. DE 4,110,349 (1991); CA 115:258099.
30. Y. Kubo, *Sekiyu Gakkaishi (Japan) 33*: 189 (1990); CA 113:99140.
31. Y. Kubo, *Shokubai (Japan) 32*: 218 (1990); CA 113:133928.
32. T. Muroi, *Shokubai (Japan) 33*: 566 (1991); CA 117:27173.
33. J. A. Osborn, F. A. Jardine, J. F. Young, and G. Wilkinson, *J. Chem. Soc., A*: 1711 (1966).
34. A. M. T. Finch Jr., U.S. Patent 3,700,637 (1972); CA 78:5206.
35. D. Oppelt, H. Schuster, J. Thoermer, and R. Braden, British Patent 1,558,491 (1976); CA 87:168798.
36. A. H. Weinstein, *Rubber Chem. Technol., 57*: 203 (1984).
37. H. Buding, H. Koenigshofen, Z. Szentivanyi, and J. Thoermer, (Bayer A-G) Ger Offen DE 3,329,974 (1985); CA 102:221994.
38. N. A. Mohammadi and G. L. Rempel, *Macromolecules, 20*: 2362 (1987).
39. H. Buding, J. Thoermer, J. Oppenheimer-Stix, and F. Liebrandt, Eur. Pat. Appl. EP 471,250 (1992); CA 116:196005.
40. S. Bhattacharjee, A. K. Bhowmick and B. N. Avasthi, *Makromol. Chem., 193*: 659 (1992).
41. G. L. Rempel and H. Azizian (Polysar Ltd.) Eur. Pat. Appl. EP 111,412 (1984); CA 101:92603.

42. G. L. Rempel and H. Azizian (Polysar Ltd.) Fr. Demande FR 2,540,503 (1984); CA 102:26127.
43. K. Kato, K. Kishimoto and T. Kamada, Japanese Patent JP 01,289,806 (1989); CA 112:200382.
44. G. L. Rempel and H. Azizian, U.S. Patent, U.S. 4,503,196 (1985); CA 102:7723.
45. H. Wada and Y. Yoshinori, Japanese Patent, JP 01,045,402 (1989); CA 112:8598.
46. N. K. Singha, S. Sivaram, and S. S. Talwar, *Rubber Chem. Technol., 68*: 281 (1995).
47. H. Buding, S. Zsolt, and J. Thoermer, U.S. Patent, U.S. 4,647,627 (1986); CA 104:44494.
48. H. Buding, P. Fiedler, H. Koenigshofen, and J. Thoermer (Bayer A-G), Ger Offen DE 3,433,392 (1986); CA 105:98996.
49. H. Wada and Y. Hara, Japanese Patent, JP 01,045,404 (1989); CA 111:59379.
50. H. Wada and Y. Hara, Japanese Patent, JP 01,045,403 (1989); CA 112:8599.
51. H. Wada and Y. Hara, Japanese Patent, JP 01,256,501 (1989); CA 113:42313.
52. H. Buding, J. Thoermer, W. Nolte, J. Fiedler, C. Paul, and T. Himmler, Eur. Pat. Appl., EP 405,266 (1991); CA 114:187325.
53. Y. Hara, K. Endo, H. Inagaki, and K. Wada, Res. Dev. Rev., Mitsubishi Kasei Corp. (Japan) 6, 22 (1992); CA 119:10234.
54. P. Fiedler, H. Buding, R. Braden, and J. Thoermer (Bayer A-G), Ger Offen DE 3,529,252 (1987); CA 106:177942.
55. T. Himmler, P. Fiedler, R. Braden, and H. Buding (Bayer A-G) Ger Offen DE 3,541,689 (1987); CA 107:135642.
56. P. Fiedler, H. Buding and R. Braden (Bayer A-G) Ger Offen DE 3,540,918 (1987); CA 107:116764.
57. G. L. Rempel, N. A. Mohammadi and R. Farwaha, Eur. Pat. Appl. EP 298,386 (1989); CA 111:135794.
58. H. Wada and Y. Hara (Mitsubishi Kasei Corp., Japan) Jpn Kokai Tokkyo Koho JP 01,045,402 (1989); CA 111:59376.
59. Y. Hara (Mitsubishi Kasei Corp., Japan) Jpn Kokai Tokkyo Koho JP 02,147,605 (1990); CA 113:116885.
60. G. L. Rempel, N. T. McManus, and N. A. Mohammadi, Eur. Pat. Appl., EP 455, 154 (1991); CA 116:256315.
61. G. L. Rempel and N. T. McManus, U.S. Patent, U.S. 5,075,388 (1991); CA 116:130954.
62. N. Murai, M. Aoki, H. Hata and T. Takeuchi (Mitsubishi Kasei Corp., Japan) Jpn Kokai Tokkyo Koho JP 04,202,403 (1992); CA 117:214299.
63. N. Murai, M. Aoki, H. Hata and T. Takeuchi (Mitsubishi Kasei Corp., Japan) Jpn Kokai Tokkyo Koho JP 04,202,408 (1992); CA 117:173186.
64. N. Furuki and K. Mizushima (Mitsubishi Kasei Corp., Japan) Jpn Kokai Tokkyo Koho JP 04,323,201 (1992); CA 118:193451.
65. N. Murai, M. Aoki and H. Hata (Mitsubishi Kasei Corp., Japan) Jpn Kokai Tokkyo Koho JP 04,253,706 (1992); CA 118:61337.
66. N. Murai, M. Aoki and H. Hata (Mitsubishi Kasei Corp., Japan) Jpn Kokai Tokkyo Koho JP 04,323,205 (1992); CA 118:192552.
67. X. Y. Guo and G. L. Rempel, *Stud. Surf. Sci. Catal., 73*: 135 (1992).
68. G. L. Rempel and X. Y. Guo, *Chem. Ind. (Dekker), 47*: *(Cat. Org. React.),* 105 (1992).
69. P. Martin, N. T. McManus and G. L. Rempel, *Stud. Surf. Sci. Catal., 73: (Prog. Catal.),* 161 (1992).
70. G. L. Rempel and X. Y. Guo, U.S. Patent, U.S. 5,210,151 (1993); CA 119:119260.
71. G. L. Rempel and X. Y. Guo, U.S. Patent, U.S. 5,208,296 (1993); CA 119:97733.
72. G. L. Rempel, N. T. McManus, and N. Mohammadi, U.S. Patent, U.S. 5,057,581 (1991); CA 116:42943.
73. G. L. Rempel, N. T. McManus, and X. Y. Guo, U.S. Patent, U.S. 5,241,013 (1993); CA 120:56476.
74. G. L. Rempel, N. T. McManus, and X. Y. Guo, U.S. Patent, U.S. 5,258,467 (1993); CA 120:79184.
75. Y. Kubo, K. Takaaki, and K. Oura (Nippon Zeon Co. Ltd., Japan) Ger. Offen. DE 3,346,888 (1984); CA 101:172839.
76. Y. Kubo and K. Oura (Nippon Zeon Co. Ltd., Japan) Jpn Kokai Tokkyo Koho JP 62,218,687 (1987); CA 108:9549.
77. V. V. Moiseev, Yu. S. Kovshov, I. P. Zornikov, G. N. Kirei, T. P. Zharkivh, G. Filv, V. V. Kur'min, and V. M. Lisov, USSR Patent, SU 1,574,610 (1990); CA 113:213682.
78. S. Bhattacharjee, A. K. Bhowmick, and B. N. Avasthi, *J. Polym. Sci., Polym. Chem., 30*: 471 (1992).
79. S. Bhattacharjee, A. K. Bhowmick, and B. N. Avasthi, *J. Appl. Polym. Sci., 41*: 1357 (1990).
80. S. Bhattacharjee, A. K. Bhowmick, and B. N. Avasthi, *J. Polym. Sci., Polym. Chem., 30*: 1961 (1992).
81. K. Takahashi and Y. Kubo (Nippon Zeon Co. Ltd., Japan) Ger. Offen. DE 4,106,466 (1991); CA 116:61341.
82. R. T. Patterson, Eur. Pat. Appl., EP 467,468 (1992); CA 116:153614.
83. Yu. S. Kovshov, V. V. Moiseev, I. P. Zornikov, G. N. Kirei, and A. S. Molchadskii, *Pr-vo i Ispol'z. Elastomer, 5*: 12 (1990).
84. R. J. Hoxmeier and L. H. Slaugh (Shell Oil Co.) U.S. Patent, U.S. 4,876,314 (1989); CA 112:99565.
85. R. J. Hoxmeir (Shell Oil Co.) U.S. Patent, U.S. 4,892,928 (1990); CA 112:199426.
86. Y. Kubo and K. Oura (Nippon Zeon Co. Ltd., Japan) Jpn Kokai Tokkyo Koho JP 02,178,305 (1990); CA 113:212911.
87. K. Oora and O. Mori (Nippon Zeon Co. Ltd., Japan) Jpn Kokai Tokkyo Koho JP 06,287,219 (1994); CA 122:163245.
88. K. Hashimoto, M. Oyama, T. Nakagawa, K. Murakata, T. Saya, and Y. Aimura, 136th Meeting of the Rubber Division, A.C.S., Detroit, Michigan, Oct. 17–20 (1989).
89. D. T. Ahlberg, D. K. Padliya, and J. T. Reed (Polysar Ltd.), U.S. Patent, U.S. 4,857,632 (1989); CA 112:37892.
90. A. Osman and W. G. Bradford (Polysar Ltd.), Eur. Pat. Appl. EP 354,413 (1990); CA 112:218553.
91. A. M. Madgavkar, D. W. Daum, and C. J. Gibler (Shell Oil Co.) U.S. Patent, U.S. 5,089,541 (1992); CA 116:195155.
92. P. Panster, S. Wieland, H. Buding, and W. Obrecht (Bayer AG), Ger. Offen. DE 4,032,597 (1992); CA 116:113320.
93. S. Hunig, H. R. Muller, and W. Thier, *Angew. Chem. Int. Ed. Eng., 4*: 271 (1965).
94. L. G. Wideman (Goodyear Tire Co.) U.S. Patent, U.S. 4,452,950 (1984); CA 101:56322.
95. Y. Luo, *J. Appl. Polym. Sci., 56*: 721 (1995).
96. H. W. Schiessl (Goodyear Tire and Rubber Co.) U.S. Patent, U.S. 5,302,696 (1994); CA 121:109986.

97. H. W. Schiessl and S. A. Manke (Goodyear Tire and Rubber Co.) U.S. Patent, U.S. 4,954,614 (1990); CA 113: 192623.

98. S. Bhattacharjee, A. K. Bhowmick, and B. N. Avasthi, *Angew. Makromol. Chem., 198*: 1 (1992).

99. D. Brück, *Kautsch. Gummi Kunstst., 42*: 107 (1989).

100. D. Brück, *Kautsch. Gummi Kunstst., 42*: 194 (1989).

101. A. J. Marshall, I. R. Jobe, T. Dee, and C. Taylor, *Rubber Chem. Technol., 63*: 144 (1990).

102. A. R. Kemp and H. Peters, *Rubber Chem. Technol., 17*: 61 (1944).

103. Y. Kubo, K. Hashimoto, and N. Watanabe, *Kautsch. Gummi Kunstst., 40*: 118 (1987).

104. A. Kondo, H. Ohtani, Y. Kosugi, S. Tsuge, Y. Kubo, N. Asada, H. Inaki, and A. Yoshioka, *Macromolecules, 21*: 2918 (1988).

105. S. Roy, S. Bhattacharjee, A. K. Bhowmick, B. R. Gupta, and R. A. Kulkarni, *Macromol. Rep., A30*: 301 (1993).

106. J. Thoermer, J. Mirza, Z. Szentivanyi, and W. Obrecht, *Kautsch. Gummi Kunstst., 41*: 1209 (1988).

107. K. C. Smith, Can. Patent Appl. CA 2,081,707 (1993); CA 120:32714.

108. U. Eisele, Z. Szentivanyi, and W. Obrecht, *J. Appl. Polym. Sci., Appl. Polym. Symp., 50*: 185 (1992).

109. Yu. S. Kovshov, V. V. Moiseev, T. P. Zhasrkikh, and I. P. Zornokov, *Int. Polym. Sci. Technol., 18*: T6 (1991).

110. P. Thavamani and A. K. Bhowmick, *J. Mater. Sci., 28*: 1351 (1993).

111. P. Thavamani, D. K. Khastgir, and A. K. Bhowmick, *J. Mater. Sci., 28*: 6318 (1993).

112. W. Obrecht, H. Budding, U. Eisele, Z. Szentivanyi, and J. Thoermer, *Angew. Makromol. Chem., 145/146*: 161 (1986).

113. R. Weir and G. C. Blackshaw, *Eur. Rubber J., 3*: 2 (1989).

114. J. R. Dunn, G. C. Blackshaw, and J. Timar, *Elastomerics, 2*: 20 (1986).

115. A. Nomura, J. Takano, A. Toyoda, and T. Saito, *Nippon Gomu Kyokaishi, 66*: 830 (1993); CA 121:37396.

116. N. Nagata, T. Saito, T. Fujii, and Y. Saito, *J. Appl. Polym. Sci., Appl. Polym. Symp., 53*: 103 (1994).

117. S. Hayashi, H. Sakakida, M. Oyama, and T. Nakagawa, *Rubber Chem. Technol., 64*: 534 (1991).

118. G. J. Arsenault, T. A. Brown, and I. R. Jobe, *Kautsch. Gummi Kunstst., 48*: 418 (1995).

119. K. Mori, H. Hirahara, Y. Sasaki, T. Kono, H. Sasaki, and T. Komukai, *Nippon Gomu Kyokaishi, 65*: 181 (1992); CA 117:92027.

120. G. Stein, A. Von, and M. Ernst, *Kautsch. Gummi Kunstst., 44*: 928 (1991).

121. P. Thavamani and A. K. Bhowmick, *J. Mater. Sci., 27*: 3243 (1992).

122. R. P. Campion and G. J. Morgan, *Plast. Rubber Compos. Process. Appl., 17*: 51 (1991).

123. R. Forte and J. L. Leblanc, *J. Appl. Polym. Sci., 45*: 1473 (1992).

124. T. Nakagawa, T. Toya, and M. Oyama, *J. Elastomers Plast., 24*: 240 (1992).

125. D. K. Parker, PCT Int. Appl. WO 92 17,512 (1992); CA 118:23592.

126. W. A. Wiseman and J. J. Ridland, *Kunstst. Rubber, 43*: 12 (1990).

127. K. Hashimoto, N. Watanabe, and A. Yoshioka, *Rubber World, 190*: 32 (1984).

128. M. Oyama, H. Shimoda, H. Sakakida, and T. Nakagawa, *Mech. Eng. (Marcel Dekker) 80*: 765 (1993).

129. J. Thoermer, G. Marwende, and H. Budding, *Kautsch. Gummi Kunstst., 36*: 269 (1983).

130. P. Touchet, P. E. Gatza, G. Rodriguez, D. P. Butler, D. M. Teets, H. O. Feuer, and D. P. Flanagen, Eur. Pat. Appl., EP 326,394 (1989); CA 112:160238.

131. P. Thavamani, D. K. Khastgir, and A. K. Bhowmick, *Plast. Rubber Compos. Process. Appl., 19*: 245 (1993).

132. P. Thavamani, A. K. Bhowmick, and D. K. Khastgir, *Wear, 170*: 25 (1993).

133. P. Thavamani and A. K. Bhowmick, *Plast. Rubber Compos. Process. Appl., 18*: 35 (1992).

134. S. Bhattacharjee, A. K. Bhowmick, and B. N. Avasthi, *Polym. Degrad. Stab., 31*: 71 (1991).

135. W. M. Rzymaki and J. Jentzsch, *Plaste. Kautsch., 39*: 269 (1992).

136. T. Kleps, D. Jaroszynska and L. Slusarski, *Polymery (Warsaw), 38*: 258 (1993); CA 120:272797.

137. P. Thavamani, A. K. Sen, D. K. Khastgir, and A. K. Bhowmick, *Thermochim. Acta, 219*: 293 (1993).

37

Industrial Perspective of Cellulosics in Thermoplastic Composites

Debesh Maldas*
University of Quebec, Trois-Rivieres, Quebec, Canada

I. INTRODUCTION

Cellulose is globally very important because billions of tons are being produced annually by photosynthesis of plants. Historically, cellulosic filled plastic systems have been dominated by most of the common thermoset resin matrix systems. The low viscosity of thermoset resins, which allowed ease of cellulosic fiber wet-out in prepregnation, their good fiber–matrix bonding capability, decreased shrinkage in the mold, and good mechanical properties of the finished products are the main factors that led to the popularity of cellulose-thermoset composite technology. During the last two decades, emphasis has been placed on the production of cellulosic-thermoplastic composite materials. Major driving forces for this development are the potential advantages, compared with thermoset matrix-based systems, in impact resistance (damage tolerance), lower cost of processing, and recyclability. Moreover, since the last decade the problems of recycling plastic materials and utilization of renewable resources have received enormous attention from academic, industrial, and governmental authorities due to both economic and environmental factors. These major incentives played the key role in gaining momentum in the production of cellulosic-thermoplastic composite materials. The utilization of cellulosic materials in thermoplastic composites can be classified into five categories: (1) plastic-cellulose networks; (2) composite sheet laminates; (3) panel products; (4) wood-polymer composite (WPC); and (5) synthetic wood.

II. PLASTIC-CELLULOSE NETWORKS

Since the mid-1960s synthetic fibers or fiber bundles (e.g., polyesters, acrylics, and nylons) have been blended with cotton and rayon (i.e., regenerated cellulose) in the form of flock or fabric. The conventional method to prepare synthetic polymer-containing cellulosic networks in sheet or mat form is the wet-laid technique. The first moldable wood product using the wet-laid method was developed in 1946 by Deutche Fibrit [1]. However, the wet-laid technique is well suited for lightweight, lofty, absorbent products, although a disadvantage is the amount of water that needs to be dried off (often approximately equivalent to the weight of the dry product produced), as well as a certain binder loss. For these reasons, an attractive alternative is to keep the plastic-cellulose networks and consolidation process completely dry [2]. A variety of cellulosic and synthetic fibers can be assembled into a web or mat using air-forming or nonwoven web technology. This is possible by uniformly mixing thermoplastic fibers with cellulosic fibers forming a web from this mixture. The fibers are initially held together by mechanical interlocking. The bonding potential of the thermoplastic fibers is then activated by applying heat and pressure over a certain time interval. Sheets of such products can be thermoformed to complex shapes, painted, and laminated with plastic sheets. The interest in dry-form polymer-bonded networks of cellulose fibers that yield paper products with relatively high bulk and high softness, compared with those of conventional materials, has evidently increased during recent years [3,4]. Generally, the use of 10–20% by weight of polyolefins containing a small amount of polar groups results in these fiber networks having im-

* *Current affiliation*: Michigan State University, East Lansing, Michigan.

proved mechanical properties. Moreover, thermoplastics, thermoset, and elastomers are often used in the pulp and paper industries to improve the sheet properties of the paper. In 1982, a patent was issued to Sciaraffa et al. [5] for producing a nonwoven web. Bither [6] reported that effective binding ability of nonwoven products can be achieved by polyolefin pulps. Youngquist and Rowell [7] reviewed the opportunities for combining wood fiber with synthetic fibers. Brooks [1] reviewed the history of technology development for the production and use of moldable wood products and air-laid nonwoven mat processes and products.

Recently, a few new polymer composite sheets, i.e. Templex™ (Temple-Eastex, Inc., Silsbee, Texas) and All-seasonware™ (Signode Corporation, Glenview, Illinois), were introduced in the marketplace [8,9]. Polymer composite sheet, neither a paperboard nor a plastic, is a synergistic product resulting from a combination of cellulosic fiber and polyolefin resin. Conversion of the sheet is completed through a thermopressing operation that heats and densifies the sheets into specific shapes. It has a specific gravity of approximately 0.6 unconsolidated, and can be consolidated (compacted) to a specific gravity of 1.2. Polymer composite sheet is manufactured on a 120 inch wide multiply cylinder type machine capable of producing approximately 41,000 metric tons annually. The unconsolidated sheet has extremely low strength properties, especially tensile and stiffness, compared with equal caliper paperboard or plastic. It has been pointed out that dry-formed structures are structurally very inhomogeneous, i.e., the fibers are curled, wrinkled, may be slightly damaged, etc., which may yield rather high local stress concentrations and, thus, reduce the strength of the sheet. An alternative explanation of the low strength (and stiffness) is that the stress transfer between fibers is poor in dry-formed networks of cellulose fibers. However, once consolidated, the product is very strong in all physical aspects as well as being extremely rigid. The thermopress system has six modular sections [8]: roll stand, feed-cutter, heat conveyor, press, mold/die assemly, and a counter/stacker. The thermopressing line is fully automatic, and it is capable of consolidating the sheet to approximately one-half of the original thickness to provide a broad range of product applications. The thermopressing action yields a moisture-resistant and smooth surfaced product. From the standpoint of heat resistance, there is a pronounced synergetic effect of the components. Polymer composite sheet has a high temperature capability that is at least equivalent to the nonmetallic high-heat packaging alternatives.

A new manufacturing technology that recycles plastic and paper is being used by Sonoco Products Co., Hartsville, S.C., to produce Edgeboard™, a protective industrial packaging product [10]. Edgeboard is a right-angled edge and surface protector made from comingled paper and plastic that are passed through a two-stage extruder. This is the first extruded product of its kind to incorporate high percentages of plastic, i.e., 50% by weight. This product provides users with moisture resistance and better product surface protection during shipping and distribution.

In order to provide additional interfiber bonding, however, synthetic (polymer-based) binding materials, e.g., maleated polyolefins, thermosetting resins, and latex, are incorporated in the web [11–14]. In practice, the amount of synthetic binder used for dry-formed paper sheets can be of the order of 20%, and for other nonwovens it may even be higher [15,16]. For economic reasons, it would be desirable to substantially decrease this level. For instance, Youngquist et al. [14] reported the bonding of air-formed wood-polypropylene fiber composite in conjunction with maleated polypropylene. The process for producing a flexible mat using thermoplastic fiber in combination with a thermosetting resin have been developed by two different companies [17,18]. Generally, the mat is fed through an oven to melt and set the thermoplastic fiber without affecting the setting of the thermosetting resin component. Most binders used are latex polymers that depend on cross-linking during the curing stage to develop the required high-molecular weights. Although the curing operation is rather simple, in practice it is one of the most misunderstood and difficult steps of the entire process [11]. Moreover, the application of a synthetic binder gives rise to one of the more serious complications of the drying and curing operations. The cost of the binders represents one of the significant cost factors in the air-laid process. For this reason, great care is given to the adhesive application with the goal of minimizing the quantity to be applied for a maximum of web strength.

Once again, the mechanical properties of the bonded network are determined by a number of factors in addition to the corresponding properties of the fibers and the binder, e.g., the amount of binder, the binder distribution, the network density, the adhesion between the fiber and the binder, etc. The type of polymer used may influence the properties of the sheet. In general, the softer the polymeric binder, the softer and more ductile is the nonwoven product [19]. Sometimes the glass transition temperature (T_g) of the polymer is used as a rough guideline for estimating the degree of softness. This may be of value in some cases, but the absolute value of the modulus of the polymer in the rubbery state is also an important factor. Many polymeric binders are also thermally crosslinked after impregnation to enhance the strength properties.

The processing flexibility of plastic-cellulose networks gives rise to products in various thickness, from a material only a few millimeters thick to structural panels up to several centimeters thick, that include [14]:

Storage bins for crops or other commodities.

Temporary housing structures.

Furniture components, including flat and curved surfaces.

Automobile and truck components.

Paneling for interior wall sections, partitions, and door systems.

Floor, wall, and roof systems for light-frame construction.

Packages, containers, cartons, and pallets.

Filters for air and liquids.

III. COMPOSITE SHEET LAMINATES

Lamination is the process of bringing together two or more webs/films and bonding them with an adhesive or by heating under pressure. Laminates are one of the finest packaging materials where flexible packaging is required. As with coating, laminating produces a multilayer film for packaging. A synthetic binder can be added to the dry-formed paper (or any other nonwoven structure) in the form of latex, powder, fiber, or a solution. Typical lattices used in this connection are those based on styrene–butadiene rubbers, vinyl acetate copolymers, acrylic copolymers, and vinyl chloride copolymers. The latex systems can be applied to the dry-formed structures by means of impregnation (saturation) or spraying. In practice, the latter technique is usually preferred, although both spraying and impregnation appear to be most efficient. Composite sheet lamination techniques can be classified as: wet adhesive bonding, dry adhesive bonding, dry bonding (without adhesive), and extrusion coating.

A. Wet Adhesive Bonding Process

Water-based polymer latex binders (e.g., styrene-butadiene rubbers and copolymers based on vinylacetate, acrylates, or vinylchloride) are normally sprayed onto both sides of the dry-formed web. Adhesion of the layers is achieved by wet bonding, which requires evaporation of a solvent through the pores of one layer. Wet bonding using aqueous-based adhesives is not normally successful with polyolefin–paper laminates because of the poor wetting of the surface of polyolefin film. Moreover, spraying of the water-based dispersions can have a negative effect on the structure and properties of the network. The cellulosic fibers swell in the presence of water and, upon subsequent drying, the local inhomogeneity of the structure may increase, which results in a deterioration of the mechanical properties [20,21]. Furthermore, from the processing point of view, the spraying of water-based dispersions is somewhat complicated and less desirable. With the spray technique, problems associated with nonuniform coverage of the web surface and insufficient penetration of the binder through the thickness of the cellulosic network can be encountered. Solvent-based adhesives pose fewer problems but can cause odor problems in the finished laminates, particularly in food-packaging applications.

B. Dry Adhesive Bonding Process

Polymeric binder can be added to the network either as an aqueous latex dispersion or as a solution that should be dried prior to lamination in this process. In either case, the polymer should form a film and join adjacent fibers together and thus improve the stress transfer characteristics of the fibrous network. Provided that the proper film forming conditions are available, the property profile of the bonded network is determined to a significant degree by the properties of the polymeric binder at the temperature of use [20,22]. For example, if a softer type of product is desired, a binder with a relatively low glass transition temperature (T_g) is often chosen.

A decorative laminate suitable as a floor or wall covering is made by coating a substrate, such as PVC resin-impregnated web of cellulosic fibers, with vinyl plastisol or organosol, and heating the coated substrate to gel the vinyl plastisol and firmly bond the PVC resin particles [23]. However, this technique has a number of drawbacks, which are mainly due to the presence of a plasticizer that undergoes migration and evaporation, causing damage to the product [24]. Moreover, this technique involves a series of operations, e.g., application of the adhesive, drying, recovery of the solvents and shaping, which substantially increases the cost of the articles thus produced [25].

C. Dry Bonding Without Adhesives and Extrusion Coating Techniques

Lamination of thermoplastics (such as polyethylene [PE] or polypropylene [PP]) in the form of film, powder, or fibers are accomplished by means of heat, pressure, and oxidation of the laminating surfaces in these processes. However, for dry bonding (without adhesive) process one possible way to combine cellulose fibers with thermoplastics is to use cellulose fibers in the form of paper, and laminates can be prepared by interleaving thermoplastic polymer films and paper sheets on off rolls and then hot pressing. The main requirement is that the polymer melts at a temperature that is lower than that at which structural changes occur in the cellulosic substrate, e.g., by crosslinking or degradation reactions. It may be noted that under the bonding conditions used here, a yellowing of the cellulosic fibers was noted after a heat treatment of about 175–180°C. These methods, of course, completely eliminate any problem of odor associated with the presence of an adhesive.

A high-quality extrusion-coated paper for high-quality printing was produced by the extrusion of PE onto a highly smoothed base sheet of paper and passing the coated base through the nip of two rolls. The roll is a low-friction gloss chill roll with a surface finish roughness between that of high-gloss and a mat finished roll [26].

A high-strength printable paperlike web used in protective wrapping materials was made by forming a furnish including three different polyester fibers, PP fibers, and wood pulp; sheeting out the furnish by wet paper making methods; and hot-calendering so as to bond the sheet via melting and solidification of the lowest melting polyester [27]. Good barrier properties and excellent adhesion between the polymer film and the substrate were obtained after hot pressing polymer powder (i.e., PP and biopolyester, e.g., poly-beta-hydroxyalkanoates) that had been electrostatically deposited on the paper surface using an electrostatic spray applicator [28].

The basic requirements for any laminate will usually include mechanical strength, machine handling qualities, a heat seal medium, and barrier properties (water vapor, gases, odor, etc.). In addition, the outer material must normally be easily printable. The degree of mechanical strength and machine handleability required are normally dictated by the type of packaging machine used and the distribution hazards encountered, while the barrier properties required depend on the product. Moreover, one of the first questions to be answered is whether the product to be packaged requires a transparent or an opaque wrapper. Transparency may be required in order to give sales appeal or simply to show the type or amount of the contents. On the other hand, opacity may be dictated by the sensitivity of the product to visible or ultraviolet light.

The nature, amount, and distribution of the adhesives within the sheet also strongly affect the physical properties of the network structure, and binder migration during drying or curing can also present a problem. There is a great deal of overlapping of properties among various base materials and a large number of different laminates can be made, giving similar performance. The choice will then be determined by cost. Laminate costs are difficult to arrive at theoretically because of the numerous factors involved, such as varying prices for the component plies, adhesion costs, and lamination costs. Perhaps the overriding factor in laminate choice at present is availability, and here customers is really on their own.

IV. PANEL PRODUCTS

Wood that is known as a natural composite is one of the oldest and most widely used materials. In fact, wood has been used as a construction material for thousands

of years. Among the abundant sources of cellulosic materials only a limited number of wood species are suitable for lumber and furniture. Moreover, prime trees in forests required for sawmilling are becoming scarce. As a result, reconstitution of wood from smaller and smaller entities (such as wood waste, various nonwoody plants, e.g., sugarcane bagasse, corn husks, and other wastes from agricultural industries, as well as different common or uncommon wood species) into woodlike composite materials, e.g., particleboard, medium-density fiberboard (MDF), flakeboard, waferboard, oriented-strandboard (OSB), plywood, etc., is a recent trend.

In 1907, phenolic resins made by Baekeland were first combined with wood flour as an additive before molding [29]. The adhesives used for bonding in cellulosic materials are many and varied, for example, formaldehyde condensation products from urea (UF), melamine (MF), phenol (PF), resorcinol, and a mixture of these resins; polyvinyl acetate emulsion or white glues; thermoplastic and thermosetting types; rubber-based and mastic types; diisocyanates, polyurethanes, epoxies, and hot melts [30–36]. However, the great bulk of adhesives used in woodlike composite materials are mainly based on aminoplastic thermosetting resins (e.g., urea formaldehyde (UF) and urea melamine formaldehyde (UMF) resins) [37,38]. On the other hand, all of the formaldehyde-based resins have a tendency to release some formaldehyde vapor if they are improperly manufactured or used over prolonged periods. This causes criticism from the public and environmental protection agencies. As a result, nonformaldehyde-based adhesives are emerging in the market, although most of them remain in the developmental stage.

A typical method for making particleboard or the like is provided by the means for combining the wood chips/wafers/particles and binder in a predetermined desired ratio [39]. Both binder and particles are fed separately to a mixing station with the amount of binder being fed remaining constant. The particles are fed to the mixing station by a continuous moving belt that passes under a rake assembly, which meters the thickness of the particle layer on the belt. The bulk density of the particles on the belt is measured continuously and senses volumetrically fluctuations in the bulk density. The sensing is used to control the height of the rake assembly and, hence, the amount of particles being fed to the mixing station.

In another method, the liquid resin is sprayed onto wood chips. The condition for rapid and complete impregnation of the wood chips or sawdust in resins is to dilute the resins either in monomer [40] or in swelling agents [41]. Dilution of the resin with monomers of roughly 10% permits complete impregnation within a very short period. The mixed or impregnated product was subjected to press under pressure and temperature to produce a board with a natural appearance. Percentage of thickness recovery after hot pressing decreased rapidly with an increase in pressing time (critical pressing time being about 25 min for 130°C and about 15 min for 150–190°C). Actually, the density distribution of boards pressed within the temperature range of 130–190°C is essentially the same. Moreover, density distribution is affected by the presence of a spacer between the pressing platens.

A typical method of preparation of 5- and 7-ply plywood panels is by pressing at room temperature and the simultaneous use of a one-step exothermic heat gluing/monomer curing technique. Prior to monomer impregnation, all veneer samples were conditioned at room temperature. The assembly was compressed to 1.03 MPa (150 psi) between steel plates in a hydraulic press, and the steel plates fastened with a pair of bolts and nuts at both ends.

V. WOOD-POLYMER COMPOSITES (WPC)

Wood is a traditional construction and decorative material, with many excellent properties, but, as a natural product, it cannot be expected to be ideal for every purpose. Wood is readily degraded by variety of microorganisms, insects, such as termites, and other elements, e.g., fire, ultra-violet light, etc. Moreover, because of its hygroscopic nature, wood absorbs/desorbs moisture with changing relative humidity of the atmosphere, which causes alternate swelling and shrinkage, resulting in physical degradation, such as deterioration of mechanical properties. As a result, numerous types of treatments have been developed to improve its dimensional stability, mechanical strength, fire endurance, and durability against microorganisms. On the basis of treatments WPC can be classified into two distinct types: chemical-modified wood, which comprises ethers, acetals, and ester bonds produced in the wood by the chemicals [42]; and polymer-impregnated wood, which is prepared by the impregnation of wood with a monomer followed by curing [43]. Obviously, the basic experimental conditions of preparation of these two types of composite materials widely differ.

From 1930 to 1960, a number of new wood treatments were introduced: acetylation of the hydroxy groups, crosslinking of the cellulose with formaldehyde, cynoethylation, ethylene oxide addition to the hydroxy groups, ozonolysis, propiolactone grafting of side chains, polyethylene glycol (PEG) bulking of the cell walls and, of course, phenol-formaldehyde treatments [44]. However, acids and base catalysts used with some of those treatments degrade the cellulose chain and cause brittleness of the composite.

During the early 1960s, various vinyl monomers (e.g. styrene, t-butyl styrene, chlorostyrene, vinyl chloride, vinyl acetate, acrylonitrile, acrylates, especially methyl methacrylate) were used to treat wood, and po-

lymized those in situ into the solid polymer by means of a free radical catalyst [43,45]. Vinyl monomers have a large range of properties from soft rubber to hard, brittle solids depending upon the groups attached to the carbon–carbon backbone. Moreover, vinyl polymerization was an improvement over the condensation polymerization reaction because the free radical catalyst did not cause wood to degradate. In addition, the reaction does not leave behind a reaction product that must be removed from the final composite.

The two most widely used methods for the preparation of WPC, however, are the solvent exchange techniques of impregnation and the graft polymerization of monomers onto cellulose [43]. Generally, wood contains about 50–70% voids that are interconnected in a very tortuous manner. As a result, permeation of a monomer is difficult [44]. To overcome this problem, the wood is evacuated and monomers are added, followed by pressure applied using an inert gas. Under these conditions, the wood is allowed to soak for several hours and monomers slowly fill the void spaces in the wood. A dilution of the monomers in swelling agents (penetration promoter) enhances the impregnation process. In fact, the monomer content in the wood linearly increases with the increase in the monomer concentration. The monomer-impregnated wood is then removed and subjected to catalyst heat or radiation curing until polymerization of the monomer is essentially complete (90–100%) conversion. Finally, WPC is dried in the air or in a kiln.

WPCs technology was praised as one of the ten important scientific achievements in the world in 1964. It entered the world's markets in 1970 and has been in use ever since [47]. The major advantages of WPCs are the enhancement of certain physical and mechanical properties, e.g., decay-resistance, microbia-resistance, hardness, compression, bending and shear strengths, modulus, static bending, internal bond, dimensional stability, durability, diffusion coefficient, and lower water absorption [45–51]. In general, hardness increases 2–8 times, antiimpact strength increases 3–8 times, curving strength increases 4–5 times, even strength of wearability being 5 times as high as marble. Once again, dry wood swells by about 45% based on volume. On the other hand, impregnated wood reaches only about a 25% absorption level, i.e., as much water as the corresponding untreated wood. This percentage of water uptake decreases with the increase of polymer loading percentage. The swelling or deswelling of different magnitudes implies the corresponding dimensional change. The dimensional change can also be expressed by the average value of antishrink efficiency (ASE).

The improved physical and mechanical properties of the wood-plastic composites lead to a diversity of applications, e.g., automotive parts, furniture, construction (e.g., building panel, flooring veneers), toys, cutlery handles, industrial pattern, sports equipment, musical instruments, as well as construction for soundproofing, heat insulation, or as a barrier against atomic radiation (neutrons or gamma-rays), etc. [50,52–54]. Hills et al. [53] refer to two curious applications especially suited to the damp climate of England. One is for golf club heads, which are often stored under moist conditions with minimum ventilation, and the second is for organ consoles in old cathedrals and churches where the atmosphere has become more variable since the recent introduction of central heating. In fact, one wood-plastic composite console is said to have been installed in an English cathedral.

Lawniczak [50] proved the usability of WPC for floors in freight railway cars and sleepers. After several years long service, car floors constructed of WPC demonstrated good performance without symptoms of wear due to abrasion, tightness in all seasons of the year, and resistance to chemical or biotic erosion.

VI. SYNTHETIC WOOD

In the recent years, an urgent need of the utilization of cellulosic materials as reinforcing fillers for both virgin and recycled thermoplastics has been established because of environmental pollution issues and because of the various potential advantages they offer compared with those of conventional reinforcing fillers, e.g., calcium carbonate, talc, mica, and glass. The advantages are: low cost, renewability, low density, nonabrasiveness, biodegradability, low-energy consumption, and high specific properties. Moreover, the nonabrasive nature of cellulosic materials permits the use of high-fiber loadings without the extensive damage to compounding and molding equipment that can occur with much harder mineral fillers. Although the challenge and opportunity of using cellulosic materials as reinforcing fillers for thermoplastic composites is very appealing, the industrial success of using cellulosic-thermoplastic composites is still limited. The general problems of using cellulosic materials in thermoplastics are: noncompatibility with resins, thermal instability at temperature above 200°C, low-bulk density, high-moisture absorption, difficulty of dispersion in plastic processing equipment, and poor microbial resistance. These disadvantages are minimized in composites by proper measure and by selection of applications where these drawbacks are not of prime consideration. In the early 1980s the increasing cost of oil triggered interest in achieving up to 60% cellulose loadings in PP extruded sheets [55]. In 1986, however, declining oil and resin costs knocked the bottom out of the market of cellulose-filled composites.

Cellulosic materials, such as wood, in their different forms (i.e., wood flour and wood pulp), cotton, shell flours, ground corn cobs, and other vegetable by-products or agro-wastes are used as the source of cellulosic raw materials for the plastic industry [29,56], at least as cost-cutting fillers. However, the demand for unmodified fillers that reduce costs but contribute little to property improvement has risen only marginally because of difficulties associated with surface interactions between hydrophilic cellulosics and hydrophobic thermoplastics [57]. Such divergent behavior results in difficulties in the compounding of these materials and poor mechanical properties. Moreover, commonly used coupling agents, i.e., that are being used for mineral fillers, do not function efficiently in such composite systems and may be too expensive. The most successful, as well as profitable, methods are: the modification of the polar cellulose surface by grafting with compatible thermoplastic segments or coating with a compatibilizer or a coupling agent prior to the compounding step, the addition of a compatibilizer or a coupling agent in the compounding step, and the modification of the matrix polymer with a polar group [58,59].

Composites of polyethylene and polypropylene with up to 50% saw dust, called Woodstock,™ was developed by I.C.M.A. (Milan, Italy) [60]. The materials are combined in a high-intensity mixer, and then extruded in a twin-screw extruder and calendered in sheets 0.32 cm (⅛ inch) thick. The sheet can be thermoformed to complex shapes, painted, laminated with vinyl or polyurethane foam, or embossed. Alternatively, the composite can be injection-molded. This material outperforms wood in workability and has superior technical characteristics. For example, the elastic modulus is 243 MPa (352,00 psi), and water absorption after 24 h of immersion is only 2.5%. Heat resistance is also excellent. No distortion resulted after this composite material was exposed to 40°C at 100% relative humidity for 70 h. Dry heat (100°C for 24 h) had no measurable effect. The U.S. licensee for this technology is American Woodstock, Sheboygan, WI. This company manufactured approximately 26 million pounds of Woodstock panel in 1991 for the automotive interior trim market [61,62]. However, the same company is also trying to replace wood flour by waste newspaper in their extruded sheet. According to a statistical forecast, if the annual North American production of 12 million passenger and light truck vehicles were each fitted with 25 pounds of Woodstock, some 300 million pounds of panel would be required, and 120–150 million pounds of waste newspaper would be used, that is, roughly 36–46% of the recovered waste newspaper.

In the United States, Advanced Environmental Recycling Technologies (Bioplaste™), Springdale, AR Mobil Chemical Comp. (Trex™), Norwalk, CT, and Phoenix, Inc. (Strandex™), Madison, WI have launched cellulose–polyolefins composites in the marketplace [63,64] (D. Maldas, personal communication, 1995.) In fact, 50–70% virgin or recycled plastics, waste wood flour, and bonding agents are being continuously proccessed in a single/twin-screw extruder. For instance, Mobil's process uses continuous extrusion for throughput exceeding 1000 lb/h. Sales of Trex lumber in the U.S. now exceed 10 million lb/year. The product is priced between top-grade, treated marine lumber ($600/1000 board ft) and cedar ($1000/1000 board ft). In general, these materials are highly resistant to moisture, insect, fungus, and ultraviolet radiation. Similar to wood, they can be sawed, routed, sanded, nailed, drilled, turned on a lathe, and painted. These materials hold fasteners tighter than wood (A. U. Ferrari, Mobil Chemical Comp., personal communication, 1994.). Above all, these materials are environmentally friendly, i.e., made of mostly recycled plastics and reclaimed wood, and are recyclable.

During 1989–1990, the Research Institute for Petrochemistry, Czechoslovakia developed extrudable PVC-wood (25 wt.%) composites using a twin-screw extruder, which produce 442 lb/h and about 0.33 million lb/year hardboards with approximately 0.25–0.85 cm ($\frac{1}{10}$–$\frac{1}{3}$ in) thickness and 101.6 cm (40 in) wide (D. Maldas, personal communication, 1992.)

The cellulosic-reinforced composite materials (i.e., synthetic wood) are suitable for a wide range of applications including [62] (A. U. Ferrari, Mobil Chemical Comp., personal communication, 1994):

Automotive interior trim.
Window and door frames.
Decking, landscaping, and planter boxes.
Marine docks/decking, broadwalks/walkways, and industrial flooring.
Playground equipment, site amenities, and fencing.
Offset blocks, noise barriers, and right-of-way fencing.

However, synthetic wood is not a one-for-one structural replacement for wood in all applications. Moreover, it is not intended for use as columns, beams, joists, stringers, and other primary load-bearing elements.

VII. CONCLUSION

Potential advantages of both thermoplastics and cellulosic materials combined with the economic and environmental viewpoint have lead to a promising utilization of both these materials in various forms of composites. Although various branches of cellulosic-thermoplastic composites' industries are booming in recent years, their growth rate is very slow. In order to achieve the full potential of such valuable materials as various engineering materials and commodity products more incentives from academic, industrial, and governmental authorities are needed.

REFERENCES

1. S. H. Brooks, *Proc. Tappi Nonwoven Conference,* Atlanta, Georgia, p. 87 (1990).

2. K. Moller, *Tappi Eng. Conference,* Atlanta, Georgia, p. 639 (1981).

3. B. Westerlind, M. Rigdah, and A. Larson, *Composite System from Natural and Synthetic Polymer* (L. Salmen, A. de Ruvo, J. C. Seferis, and E. B. Stark, eds.), Elsevier Science Publishers, Amsterdam, London, New York, p. 83 (1986).

4. R. Mansfield and Associates (1989), *Nonwoven Ind., 20(11)*: 40 (1989).

5. M. A. Sciaraffa, D. G. Thome, and C. M. Bogt, U.S. Patent 4,333,979 (1982).

6. P. Bither, *Proceedings Air-Laid and Advanced Forming Conference,* Hilton Head Island, S.C., Nov. 16–18 (1980).

7. J. A. Youngquist and R. M. Rowell, Proceedings 23rd Washington State University International Particle Board/Composite Materials Symposium, Pullman, WA, p. 141 (1989).

8. M. D. Kennedy, *Polymers, Laminations and Coating Conference,* San Franscisco, California, Book 1, p. 135 (1987).

9. B. R., Mitchell, G. W. Alexander, and B. J. Conrad, *Tappi J., 70*: 47 (1987).

10. Anonymous, *Plastics Eng., 47(4)*: 7 (1991).

11. J. A. Villalobos, *Tappi, 64(9)*: 129 (1981).

12. A. de Ruvo, H. Hollmark, S. Hartog, and C. Fellers, *Seven. Papperstidn., 85(7)*: 16 (1982).

13. J. R. Starr, Tappi Engineering Conference, Atlanta, Georgia, p. 417 (1981).

14. J. A. Youngquist, A. M. Krzysik, and J. H. Muehl, *Wood Fiber/Polymer Composites: Fundamental Concepts, Processes, and Material Options* (M. P. Wolcott, ed.), U.S. Forest Products Society, Wisconsin, p. 79 (1993).

15. J. R. Starr, *Pulp Paper Int., 22(Aug.)*: 38 (1980).

16. T. Miller, B. Westerlind, and M. Rigdah, *J. Appl. Polym. Sci., 30*: 3119 (1985).

17. P. E. Caron and D. G. Allen, U.S. Patent 3,367,820 (1968).

18. R. P. Doere and J. T. Karpik, U.S. Patent 4,474,846 (1984).

19. R. A. Gill, T. J. Drenner, E. J. Sweeney, and L. Mlynar, *Tappi, 55*: 761 (1972).

20. M. Rigdahl, B. Westerlind, H. Hollmark, and A. de Ruvo, *J. Appl. Polym. Sci., 28*: 1599 (1983).

21. M. Rigdahl and A. de Ruvo, *J. Appl. Polym. Sci., 29*: 187 (1984).

22. R. D. Athey, *Tappi, 60*: 118 (1977).

23. W. J. Kauffman, T. D. Colyer, and M. Dees Jr., U.S. Patent 4,950,500 (1990).

24. P. Vaccari, A. Palvarini, and L. Sinatra, British Patent 1,243,254 (1971).

25. P. Georlette and R. Bouteille, U.S. Patent 4,356,226 (1982).

26. E. Avni, S. Salama, and E. Turi, Canadian Patent 2,013,618 (1990).

27. J. A. Goettmann and J. R. Boylan, U.S. Patent 5,133,835 (1990).

28. R. H. Marchessault, P. Rioux, and I. Saracovan, *Nord. Pulp Pap. Res. J. 8(1)*: 211 (1993).

29. B. M. Walker, *Handbook of Fillers for Plastics* (H. S. Katz and J. V. Milewski, eds.), Van Norstrand Reinhold Company, New York, London, Toronto, Chap. 20, p. 420 (1987).

30. R. H. Gillespie, *J. Adhesion, 15*: 51 (1982).

31. P. Zadorecki and A. J. Michell, *Polym. Comp., 10*: 69 (1989).
32. D. Maldas and B. V. Kokta, *Biores. Technol., 35*: 251 (1991).
33. P. Zadorecki and P. Flodin, *Polym. Comp., 7*: 170 (1986).
34. G. S. Han and N. Shiraishi, *Mokuzai Gakkaishi, 37*: 39 (1991).
35. W. J. Groah, *ACS Sym. Ser., 316*: 17 (1986).
36. M. A. Arle and A. J. Bolton, *Holzforschung, 42*: 53 (1988).
37. D. Maldas and B. V. Kokta, *Trends in Polym. Sci., 1(6)*: 174 (1993).
38. D. Maldas, *The Polymeric Materials Enclopedia: Synthesis, Properties and Applications* (J. C. Salamone, ed.), CRC Press, Florida, June (1996).
39. B. Greten, U.S. Patent, 4,065,030 (1977).
40. R. Schaudy and E. Proksch, *Holzforschung, 30(5)*: 164 (1976).
41. Y., Nakayama, M. Ogawa, and T. Iwamoto, Japanese Patent 31314/74 (1974).
42. S. Kumar and K. Kohli, *Polymer Science and Technology* (E. Carraher, Jr. and L. H. Sperling, eds.), Plemum Press, New York, London, Vol. 33, p. 147 (1986).
43. J. A. Meyer, *Wood Sci., 14(2)*: 49 (1981).
44. J. A. Meyer and W. E. Loos, *Forest Prod. J., 19(12)*: 32 (1969).
45. S. Manrich and J. A. Marcondes, *J. Appl. Polym. Sci., 37*: 1777 (1989).
46. A. J. Stamm, *Wood and Cellulose Science,* Ronald Press New York, (1964).
47. Q. W. Yang, Int. Symp. on Wood and Pulp. Chem., p. 203 (1989).
48. D. J. Fahey, *Forest Prod. J., 26(7)*: 32 (1976).
49. S. Yashizawa, T. Handa, M. Fukuoka and T. Nakamura, *Rept. Prog. Polymer Phys., Japan, 23*: 227 (1980).
50. M. Lawniczak, *Holzforschung Holzverwertung, 30*: 25 (1978).
51. I. Santar, J. Urban, D. Pinicha and J. Martinek, Czechoslovakian Patent 195, 967 (1982).
52. J. A. Manson and L. H. Sperling, *Polymer Blends and Composites* Plenum Press, New York and London, Chap. 11, p. 336 (1976).
53. P. R. Hills, R. L. Barrett, and R. J. Pateman, *U.K. Atomic Energy Agency Report* (1969).
54. F. M. L. Verdson, French Patent 2,038,434 (1971).
55. R. Leaversuch, *Modern Plastics, 64(5)*: 51 (1987).
56. F. J. Washabaugh, *Modern Plastics Encyclopedia, 63*: 144 (1986–1987).
57. D. Maldas and B. V. Kokta, *Comp. Interf., 1(1)*: 87 (1993).
58. D. Maldas and B. V. Kokta, *Wood Fiber/Polymer Composites: Fundamental Concepts, Processes, and Material Options* (M. P. Walcott, ed.), U.S. Forest Products Society, Wisconsin, p. 112 (1993).
59. Maldas and B. V. Kokta, *Macromolecules '92* (J. Kahovec, ed.), VSP International Publishers, The Netherlands, p. 349 (1992).
60. Anonymous, *Plastics Eng., 30(2)*: 7 (1974).
61. G. E. Myers and C. M. Clemons, *Final Report for Solid Waste Reduction and Recycling Demonstration,* U.S. Dept. of Agriculture, Madison, Wisconsin, Project No. 91–5 (1993).
62. K. P. Gohr, Wood Fiber-Plastic Composite Conf., Madison, WI (1993).
63. Anonymous, *Modern Plastics, 67(4)*: 52 (1990).
64. Anonymous, *Plastics Week, June 29*: 4 (1992).

38

Liquid Crystalline Polymer Composite—Preparation and Properties

Yongsok Seo
Korea Institute of Science and Technology, Seoul, Korea

I. INTRODUCTION

Liquid crystalline polymers may be thermotropic or lyotropic [1,2]. A lyotropic material, small molecule or polymer, is one that forms an ordered solution when dissolved in an appropriate solvent. Thermotropic liquid crystalline polymers (TLCPs), featuring mesomorphic melts (anisotropic melts) within a defined temperature range, have attracted considerable attention in the past two decades [3,4]. For the most part, the interests in TLCPs lie in their excellent balance of mechanical performance and melt processability in the liquid crystalline state. When blended with a conventional thermoplastic resin, the melt processable TLCP phase not only can serve as a processing aid by lowering the melt viscosity of the material, but is also a prime candidate for improving the mechanical performance of the host thermoplastic matrix by forming fibrous reinforcements in situ, hence, the name in situ composites. The concept of generating TLCP reinforcing fibrils on an in situ base has a number of potentially attractive advantages over the use of the more commonly used glass-reinforced composites [5,6]. The potential advantages include a wider range of processing options, improved surface appearance, recyclability, and lower processing requirements. Most of these advantages, as well as their general processing conditions, have been discussed in many excellent reviews [3–8]. Recently, Handlos and Baird [5] reviewed overall processing conditions and associated properties of in situ composites.

A large number of studies have been conducted to elucidate the rheology and thermodynamics of binary TLCP/thermoplastic blends, as well as the processing variables controlling the morphology and mechanical properties [8]. It is realized that the mechanism of forming an elongated TLCP fibrous structure in situ is controlled by factors such as the relative composition, the interfacial tension, the viscosity ratio, and the processing conditions. This is also related to the fact that most of TLCPs are generally immiscible with general thermoplastics. Even though the enhancement of mechanical properties of in situ composites can be achieved by the formation of fine TLCP fibrils having a large aspect ratio in flow direction, their adhesion at the phase boundary are quite weak due to immisicibility. How to improve the poor mechanical properties in transverse direction while keeping high performance in the flow direction, that is how to overcome mechanical anisotropy, has been the main concerns of recent studies [9–14]. In spite of numerous studies of in situ composites and TLCP blends with commercial thermoplastics, not all TLCP blends with commercial thermoplastics can produce in situ composites. (A TLCP starting with a high degree of nematic order in the melt is the best candidate [9].) Melt spinning of such a polymer produces fibers with a high degree of nematic order "frozen" in. This develops high-chain continuity, highly extended-chain structures, and high-mechanical properties, especially high-tensile modulus). For deformation of the TLCP phase into fibrils in polymer processing operations, some conditions should be met such as processing conditions, component rheological properties, etc. [15,16].

Since many good reviews have appeared in the literature [1–8], this chapter focuses on the mechanism of

TLCP droplet deformation in processing equipment and fibrillation, and recent advances in the fibrillation techniques.

II. DISPERSED TLCP DROPLET DEFORMATION AND MORPHOLOGY DEVELOPMENT

To understand how the dispersed phase is deformed and how morphology is developed in a two-phase system, it is necessary to refer to studies performed specifically on the behavior of a dispersed phase in a liquid medium (the size of the dispersed phase, deformation rate, the viscosities of the matrix and dispersed phase, and their ratio). Many studies have been performed on both Newtonian and non-Newtonian droplet/medium systems [17–20]. These studies have shown that deformation and breakup of the droplet are functions of the viscosity ratio between the dispersity phase and the liquid medium, and the capillary number, which is defined as the ratio of the viscous stress in the fluid, tending to deform the droplet, to the interfacial stress between the phases, tending to prevent deformation:

$$N_{Ca} = \frac{\dot{\gamma}\eta_0}{\left(\frac{\sigma}{b}\right)} \tag{1}$$

where η_0 is the viscosity of the suspending liquid, $\dot{\gamma}$ is the shear rate, σ is the interfacial tension, and b is the initial diameter of the droplet. The classical analysis of Taylor on the elongation of droplets in different suspensions shows that in simple shear flow of Newtonian fluids, the initial droplet is modified according to the relationship:

$$\frac{b_1 - b_2}{b_1 + b_2} = \frac{\dot{\gamma}\eta_0 b}{\sigma} \frac{19\,\delta\,+\,16}{16\,\delta\,+\,16} \tag{2}$$

where b_1 and b_2 are the principal semiaxes of the deformed drop, δ is the ratio between the viscosity of the dispersed liquid and the viscosity of the suspending liquid (η_d/η_0) [21]. Taylor suggested that when the maximum value of the pressure difference (which tends to disrupt the drop) across the interface between a suspending liquid and a droplet exceeds the force due to surface tension (which tends to hold it together), the drop will burst. This occurs when:

$$4\dot{\gamma}\eta_0 \frac{19\,\delta\,+\,16}{16\,\delta\,+\,16} > \frac{2\,\sigma}{b} \tag{3}$$

Taking into account the viscosities of TLCP and the matrix, when δ is small, this equation can be reasonably simplified to:

$$\frac{2\dot{\gamma}\eta_0 b}{\sigma} > 1 \tag{4}$$

indicating that if the ratio of shear stress is larger than half of the interfacial energy, the droplet will elongate. On the other hand, if δ is much greater than 1, then elongation will be simplified to:

$$\frac{2\dot{\gamma}\eta_0 b}{\sigma} > \frac{16}{19} \tag{5}$$

not much different from Eq. (4). Eqs. (4) and (5) show that droplet deformation happens when the transferred stress is greater than the interfacial stress between the phases, tending to prevent deformation. Hence, dispersed TLCP phase can be deformed or not, depending on the strength of transferred stress. The lack of good alignment in shear flow is ascribable to the high-interfacial tension, too small particle diameter for deformation, and poor adhesion at the interface. Since shear flow is a weak flow, more strong flow is required for easy deformation [2,5,22]. Uniaxial extension most strongly aligns molecules along a single axis, because stretching occurs in one direction only [22]. Elongational deformation is necessary to produce fibrils because the condition of Eq. (3) is not always satisfied if the polymer melt is non-Newtonian. In immiscible polymer blends, both the dispersed phase and the matrix phase may often show increasingly complex non-Newtonian and viscoelastic melt behavior. Elasticity of the fluid takes a very important role in the particle deformation [23,24]. Its explicit expression for the deformation of TLCP particles, of which elastic behavior is important, is different from usual polymeric melts. Even for simple steady shear flow, polymeric melts generally show shear-thinning behavior in the usual processing condition, i.e., their viscosities decrease two or three orders of magnitude with the shear rate.

Relatively few contributions on drop behavior in a viscoelastic system have been published. Viscoelasticity, and especially a yield stress, may severely slow down or completely prevent the deformation of the drop and the occurrence of thread breakup [25]. Van Oene approaches the deformation and breakup of the particle including the difference in "elasticity" across the interface under laminar flow conditions [17,18]. Even though there is a question about his thermodynamic equilibrium state assumption, he demonstrated the dominant influence of "elastic" terms on composite morphology. According to his theory and expression, low-shearing conditions, such as occur in an internal mixer, are usually sufficient, especially when the components stratify. However, when the difference in elasticity is large and the disperse phase forms droplets neither low shear nor high mixing is satisfactory, since the droplets resist deformation. Even in this case, he concluded that a liquid droplet may still be drawn into a fiber by an elongational flow, subsequent shearing may cause the fiber breakup. Elmendorp et al. [19] showed that in model liquids exhibiting distinct yield stresses no breakup occurs if yield stress is larger than the pressure difference that is generated in the thread by the different radii of curvature. In his experimental result, he showed that the solution having higher normal stresses exhibited the smaller deformations, although the viscosities of the two components of the pair were equal. The droplets having higher normal stresses (higher "elasticity") appeared the most stable. Recently, there appeared some articles about the viscoelastic droplet deformation in thermoplastic melts [26–30]. Theoretically and experimentally none of them is thorough yet, but they generally agree that it is difficult to deform the droplet size when the normal stress of the dispersed phase is large and the drop size should not change even if the shear rate is increased. However, TLCP melts show a unique phenomenon of negative first normal stress difference where a positive first normal stress difference is generally common to polymeric systems, which can affect the capillary number that controls the droplet deformation [24]. Still there is much dispute over the general rheological characteristics of TLCPs. It can be easily conjectured that the viscosity of the TLCP melt will be quite low at shear rates commonly seen in processing equipment because of the highly shear thinning viscosity. Even though elongational deformation that normally applied outside die exit is believed to be the major cause of fibril generation in in situ composites, the viscosity ratio, which is shown to affect drop deformation, may be significantly different than determined in standard rheological tests. Preshearing can cause a reduction in the TLCP viscosity and viscosity may be much lower during processing than normally observed using rheometers, which affects the capillary number. Based on Eq. (1), droplet deformation is more favorable if high-shear stress is transferred. High-matrix viscosity provides high-shear stress. Most in situ composite studies used thermoplastics whose viscosities at the processing condition are much higher than those of TLCPs [8,23,24]. Also, this was the reason common engineering plastics such as nylon 6 or polyesters could not produce in situ composites [23,31,32].

Until now, we have considered mostly rheological properties of TLCP phase and matrix thermoplastics. As briefly mentioned, flow type also plays a decisive role in the TLCP deformation. Previous studies revealed that shear flow tends to break up the polydomain structure, while extensional flow may consolidate and orient it [33]. The initial polydomain structure of TLCP is affected differently by shear flow than by extensional flow. After the destruction of the polydomain structure into smaller domains, they will be extended, form aggregates, or rotate [33,34]. In the initial compounding, TLCP particles are dispersed in the matrix through elongation into cylindrical shapes that break up into smaller droplets in quiescent regions. These drops are elongated again in high-shear regions, break up again, and so on until viscous stress would equilibrate the interfacial

stress. As the droplets get more shearing action, dispersed particles are aggregated by the high-energy input. The final average particle size of the dispersed phase is the result of a dynamic equilibrium between the dispersive mixing and coalescence in the melt [10].

Coalescence arises from forced collision of dispersed particles during the mixing process. This collision takes place in random directions. As more mixing takes place, the concentration of domains will progressively increase, and, consequently, the rate of collision of these particles. There might be tiny particle coalescence around the big particles (Ostwald ripening). This is negligible in nonequilibrium mixing process. Recently, Seo et al. [33] have investigated deformation of a TLCP (Vectra B950, poly(ester amide) [Hoechst Celanese Co., U.S.A.]) in different die geometries. Varying the converging angle of extrusion die, they tested the effect of shearing on TLCP particle deformation in the matrix. Blends of Vectra B950 with a thermoplastic elastomer (EPDM [ethylene-propylene-diene terpolymer]) were made in a twin-screw extruder. Experimental results regarding the flow of TLCP in extrusion dies show that fibril shape of TCLP phase can be formed by the shearing action without strong elongation in the outer region of the extrudate where most shearing action happens.

Macrofibril generation by shearing action only can be seen in Fig. 1. No drawing was applied. As shown in Figure 1, more oriented material exists at the outer region, which has undergone mostly shear flow. Shear flow is effective for dispersed large TLCP particle deformation. Some extended TLCPs have fiber shape, but others having a "Vienna-sausage" structure can also be observed. Based on their size, they can be classified as oriented fiber or macro fibrils, but few of them can be in the category of microfibrils (having a diameter less than 1 μm). On the basis of microscopic examination, the process in the die and out of the die exit can be presented schematically as shown in Fig. 2. This is similar to the Tsebrenko et al.'s [35] schematic figure for the offering an explanation for the occurrence of fibrillation of dispersed particles (polyoxymethylene) in the matrix of copolyamide. However, there are some differences. They investigated the fibrillation process in the entrance to the extrusion orifice. This study is concerned about the flows in the die and out of the die exit. As the fluid enters the die entrance, dispersed droplets formed initially during the mixing operations are elongated at the outer region due to shearing action and collide with each other more frequently in the core region (region A, Fig. 2).

The tensile stresses acting in the direction of converging stream lines can ellipsoidally deform the big particles, but not so much as to form fine fibrils from small particles (region B). The matrix are also elongated in the converging section. As they pass the die exit (region C), recoil of the matrix occurs to release the stored energy

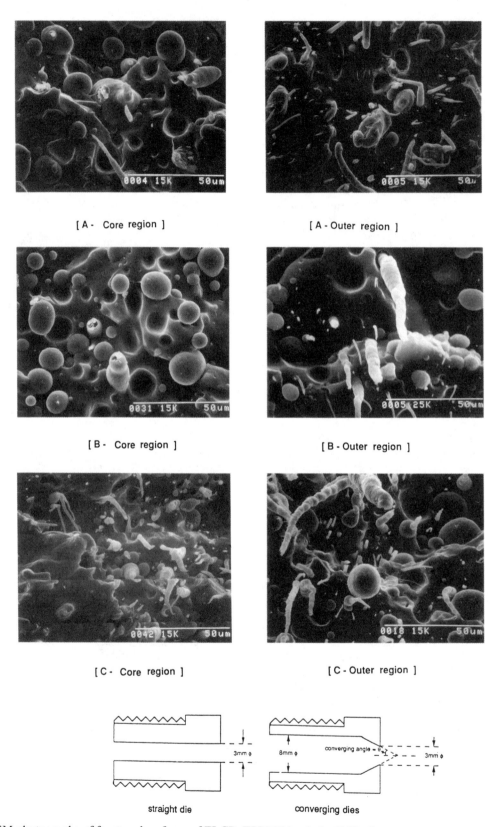

Figure 1 SEM photographs of fractured surfaces of TLCP–EPDM blends (×1000). Screw speed was 10 rpm. (A) Straight die, (B) 7.5° converging die, (C) 45° converging die. *Source*: Ref. 33.

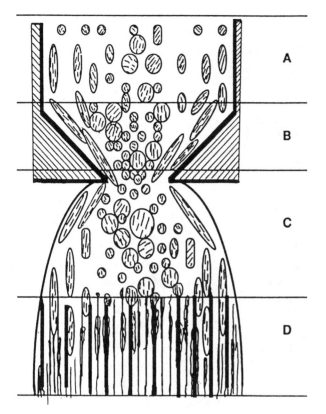

Figure 2 Representation of TLCP deformation process in die exit zone (micro scale). *Source*: Ref. 33.

(extrude swell) and compression at the center region reforms elongated particles while later there occurs the elongation on the free surface region. If there is a strong drawing on the extrudate in region D, then dispersed particles would be deformed into threadlike fibrils. Very high drawing can suppress the swelling and even very small particles can be deformed into fine fibrils [33]. This fibril formation starts as soon as the fluid comes out from die exit. The effect of shear rate can be observed in Fig. 3, which shows clearly more fibril formation at high-extrusion rate. As expected, more macrofibrils appear in the outer section, because of more severe shearing action. The size of dispersed particles are also reduced. Seo et al.'s [33] study shows a clear experimental evidence that shearing effect also takes an important role in TLCP droplet deformation.

The mechanism of droplet deformation can be briefly summarized as follows. The factors affecting the droplet deformation are the viscosity ratio, shear stress, interfacial tension, and droplet particle size. Although elasticity takes an important role for general thermoplastics droplet deformation behavior, it is not known yet how it affects the deformation of TLCP droplet and its relationship with the processing condition. Some of

these factors are correlated with each other. By studying the morphology development in the processing, we find that shearing action can produce fibrils when the viscosity of the matrix is larger than that of discrete TLCP phase. This is clearly demonstrated in Figs. 1 and 3. However, regardless of the system, a strong extensional flow field is much more effective for very small droplet deformation. It is worth noting that Eq. (4) shows that interfacial tension and droplet size can affect the deformation. Both of these factors can be affected by miscibility or compatibility between the matrix and TLCP phases. Hence, the next section is concerned with the effect of miscibility or compatibility on the droplet deformation and properties of in situ composites.

III. THE ROLE OF BLENDS MISCIBILITY IN THE GENERATION OF IN SITU COMPOSITES

The maximum enhancement of the mechanical properties of short fiber composites are achieved by very fine fibrils with a large aspect ratio and by strong interfacial adhesion between the fibers and the polymeric matrix. Most of thermoplastics studied so far are incompatible with TLCPs. This incompatibility between the matrix polymers and reinforcing TLCPs brings poor interfacial adhesion. Thus, the reinforcing effect is less than that obtained from the miscible system. For an example, good mechanical properties have been achieved for immiscible blends of poly(ether imide) and TLCPs in flow direction [36]. However, mechanical properties in transverse direction was quite poor. This can be improved if they are miscible or compatible. As we pointed out, the morphology and properties of in situ composites are largely affected by the interaction between the matrix and TLCP phase [10,11]. Miscibility or compatibility can significantly contribute to the interaction between these two phases and, hence, to the physical properties of in situ composites.

In order to get the enhanced properties of the in situ composites, compatibility between the matrix polymer and the reinforcing TLCPs has been sought [9–16,37–49]. Compatibilization has been known to overcome the problems of poor dispersion and poor adhesion in blends of other incompatible polymers [37]. To provide the compatibility, miscible TLCP systems with the matrix polymer have been investigated by many researchers. Baird and coworkers [14] studied the miscibility between various TLCPs with PEI (Ultem 1000, G.E., U.S.A.). In the blending system of TLCPs (HX1000 and HX4000, DuPont) and PEI, they observed that morphologies of the partially miscible system were quite different from that of an immiscible system (which showed pull-out of TLCP microfibrils and large voids where the microfibrils were pulled out of the matrix, revealing poor matrix/fibril interfacial adhesion). In con-

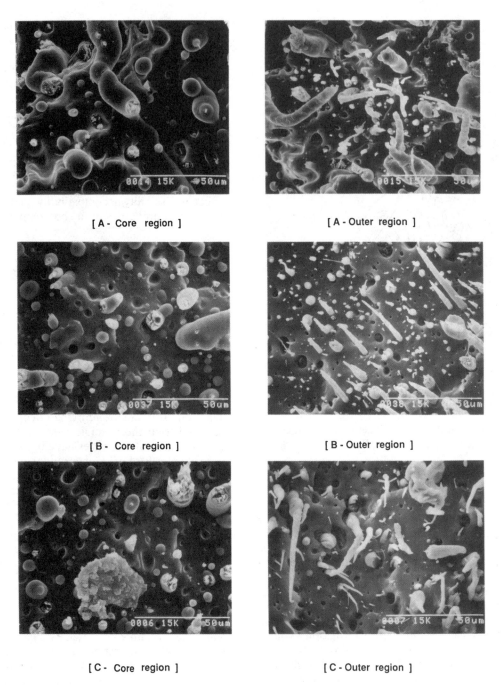

[A - Core region]   [A - Outer region]

[B - Core region]   [B - Outer region]

[C - Core region]   [C - Outer region]

Figure 3   SEM photographs of fractured surfaces of TLCP–EPDM blends (× 1000). Screw speed was 50 rpm. (A) Straight die, (B) 7.5° converging die, (C) 45° converging die. *Source*: Ref. 33.

trast, partially miscible blends did not show the pulled out microfibrils. Using this partial miscibility between HX4000 and Ultem, Bretas and Baird [38] further investigated ternary blend system of TLCP (HX4000)/PEI (Ultem)/PEEK (Victrex 450G I.C.I., England). The measurement of tensile properties showed that ternary blends with high modulus can be obtained at high TLCP

loadings, while composition with high-ultimate tensile strength can be obtained with high loadings of PEI or PEEK. Studies dealing with the miscibility of blends of a TLCP consisting of ethylene terephthalate and *p*-hydroxybenzoate (PET/PHB) with various thermoplastics can be found in the literature [39–43]. The effects of miscibility of the matrix and reinforcing phase on the development of in situ composites have been studied by Zuhang et al. [41]. The authors concluded that both polycarbonate (PC) and PET were partially miscible with PET/PHB60 (60 mol% PHB), PET being miscible to a higher degree than PC. The miscibility was checked by glass-transition temperature (T_g) movement. Little attempt was made to explain the reason for the observed morphology or correlate it with miscibility. Recently, these have been studied by many other researchers. Lee et al. [44,45] also investigated the blend system of PET/PHB60 with PET and PBT to correlate the partial miscibility between the matrix and dispersed phase to chemical reaction (transesterification) and formation of copolymers (initially they form a block copolymers but after some reaction time they were changed as random copolymers), which was checked by Nuclear Magnetic Resonance (NMR) spectroscopic analysis. They were also confirmed by others [46–48]. Produced copolymers act as the compatibilizer at the interface to reduce the size of dispersed phase and increase the mechanical properties by good adhesion at the interface [49,50].

Another way of compatibilizing an immiscible system is to use the third component as a compatibilizer or coupling agent. DeBenedetto and coworkers [16] investigated the feasibility of introducing a second TLCP as a compatibilizer, or coupling agent, in order to improve the adhesion and dispersion between components of incompatible TLCP/thermoplastic blends. They used a wholly aromatic copolyester (K161) and an aliphatic-containing TLCP (PET/PHB60) as a dispersed phase and a coupling agent. Polycarbonate (PC) and poly(ethylene terephthalate) (PET) were used as the matrix. They found that the adhesion between the reinforcing K161 and the matrix PET depends on both the concentration of the coupling PET/PHB60 phase and the relative composition of two TLCPs. Improved adhesion was also observed for the PC blend system. Mechanical properties of the composite system were improved. Baird et al. [12,49] used a functionalized polypropylene (PP) as a compatibilizer for TLCP–PP blends. The mechanical properties of the compatibilized PP–TLCP (Vectra B) blends compared favorably with those of the glass-filled PP. Miller et al. [51] also investigated the effect of acrylicacid-functionalized polypropylene (PP-AA) added as a compatibilizer on the properties of polypropylene–TLCP blends. They found that the functional compatibilizer improved interfacial adhesion and, thus, fiber properties as well as it significantly enhanced the thermal stability of the fiber. Incorporation of PP-AA increased fiber crystallinity and orientation but with just a slight enhancement of fiber properties, which can be attributed to the promotion of specific polar interactions between the blend components.

As recognized by others in the case of TLCP blends, it may be detrimental to fiber formation if the matrix and TLCP are too compatible [12,13,49–51]. This can

be easily understood based on Eq. (4). Compatibility can reduce not only the interfacial tension but also the size of the dispersed phase, whereas good interfacial adhesion is favorable to the stress transfer across the interface. Hence, it has both pros and cons for the fibril formation. Recently, Seo et al. [10,11] investigated the effect of compatibilizer (poly[ester imide]) (PEsI) on the properties and morphology of the in situ composite of poly(ether imide) (Ultem) and a TLCP (poly[ester amide], Vectra B950). Morphological evidence demonstrated that addition of the proper amount of PEsI reduces the TLCP particle size and induces a fine distribution. These can be corroborated by morphological observation. Figure 4 shows the fractured surfaces of the noncompatibilized and compatibilized Ultem–Vectra B950 blends (70:30 wt%) blends at a draw ratio of 1, which is defined as the extrudate diameter ratio at die exit to far down stream. TLCP domains are relatively large in noncompatibilized blends indicating a poor dispersion. The micrographs also demonstrate the poor adhesion between the two phases (Fig. 4A), which leads to an open ring hole around the TLCP domain and whole TLCP being pulled out during the fracturing of the samples. In contrast, the fracture is seen to occur within the fibrils in the compatibilized blends with a low content of PEsI (Fig. 4B and C), and there is no open ring around the TLCP domain reflecting better bonding on adhesion between the two phases. Furthermore, TLCP fibrils are more evenly distributed and finer than noncompatibilized ones. For blends containing more than 2.2 phr of PEsI, TLCP fibrils are poorly distributed and thicker than that of 1.5 phr PEsI added composite. This is because of the droplets flocculation by excess compatibilizer. The same trend is observed for the fiber of draw ratio equal to 4, as shown in Fig. 5. When draw ratio is 4, fine fibril formation is clearly observed for the composites containing PEsI less than 1.5 phr and it is hampered by more addition of the compatibilizer. This is similar to the behavior of the surfactant in an emulsion system [52]. From emulsion studies, flocculation of the dispersed phase is known to happen by strong interparticle interactions. The quantity of surfactant required to fully cover an interface is related to many variables. For the case shown in Figs. 4 and 5, 1.5 phr PEsI is the maximum quantity beyond which significant size reduction of the dispersed droplet no longer occurs. Instead of size reduction, excess compatibilizer tends to coalesce the dispersed TLCP phase (Fig. 4D, E, and F). The flocculation and coalescence of the TLCP phase result in poor dispersion of the TLCP. The interfacial area between the matrix and TLCP is reduced. An excess amount of the compatibilizer coalesces the TLCP particles before fibril is formed. Particle coalescence prevents fine fibril formation. Even fibril domains are not homogeneous. They may include part of compatibilizer in the fibril, which acts as a defect in the fiber. As

[A]   [B]   [C]

[D]   [E]   [F]

Figure 4   SEM photographs of fractured surfaces of PEI–TLCP blend fibers at the draw ratio of 1 (×3000). The samples were fractured after freezing in liquid nitrogen. The amount of PEsI in the blends are (A) 0 phr, (B) 0.75 phr, (C) 1.5 phr, (D) 2.25 phr, (E) 3.75 phr, and (F) 7.5 phr. *Source*: Ref. 11.

a result, tensile properties of the drawn fibers became even worse than pure PEI/TLCP blend when excess amounts of compatibilizer were added. This can be clearly seen in Fig. 4F and Fig. 5F. When 7.5 phr PEsI is added, the TLCP phase does not even form fibril shape at the draw ratio of 4. It remains like a stubby particle. Plochocki et al. [53] similarly observed the minimum of dispersed phase size with the compatibilizer amount in a blending system of low-density polyethylene/polystyrene.

Figure 6 shows SEM micrographs of the fracture surface after tensile test. For the immiscible Ultem/Vectra B blends, numerous microfibrils are observed to be pulled out from the surface (Fig. 6A). Many matrix voids generated by the pull-out of microfibrils reveal the poor

matrix/fibril interfacial adhesion for this system. In the case of the compatibilizer-added composites, matrix voids are barely observed. The interfacial adhesion looks good, suggesting the compatibilization existence at the interface. However, TLCP domains grow with the amount of compatibilizer due to the flocculation and coalescence. This is clearly seen in Fig. 6F. Diminished interaction between the TLCP phase and the matrix can be seen from the pulled-out big holes.

In order to see the effect of the compatibilizer more clearly, SEM (scanning electron microscopy) micrographs of the peeled back exposed surface of the spun fibers are shown in Fig. 7. In a noncompatibilized blend, the long TLCP fibrils are bundled together (Fig. 7A). The fibril surface looks quite clean and smooth along the flow direction, which indicates poor adhesion between TLCP and the matrix. In contrast, the compatibilized composite containing less than 1.5 phr of PEsI exhibits much finer fibrils with rough and rugged surfaces. Definitely this shows a good adhesion between the TLCP fibrils and the matrix phase. The tendency of the TLCP to coalesce, however, proceeds with more compatibilization as shown in Fig. 7C. Thick fibril bundles appear with more PEsI addition, but their surfaces are still rough due to the strong interaction. Figure 8 shows an enlarged micrograph of a fibril surrounded by the matrix, which manifests the adhesion of the matrix on the fibril surface.

Therefore, we can conclude that there exists an optimum amount of compatibilizer for the best dispersion of TLCP phase and for the most improvement of the in situ composites with high fibrillation. Excess amounts of PEsI coalesce the TLCP droplets. The adhesion at the interface was good due to chemical reaction and strong interaction by the compatibilizer with the dispersed phase and the matrix [11]. Tensile properties showed maximum at the optimum amount of compatibilizer and poor values with excess compatibilizer (Fig. 9), while impact strength also showed maximum, but always higher values compared to the noncompatibilized system (Fig. 10). This is ascribed to the different failure modes between these two tests [11]. In the impact strength test, propagating stress is transmitted to the TLCP phase through the compatibilizer, which deforms the TLCP phase. Excess energy is consumed by plastic deformation of the TLCP particles or fibril shape. In a noncompatibilized system, propagating stress passes around the TLCP phase since they are immiscible and the phases are separated. Hence, good adhesion enables the compatibilized system to always have higher impact strength. On the other hand, compatibilizer is included

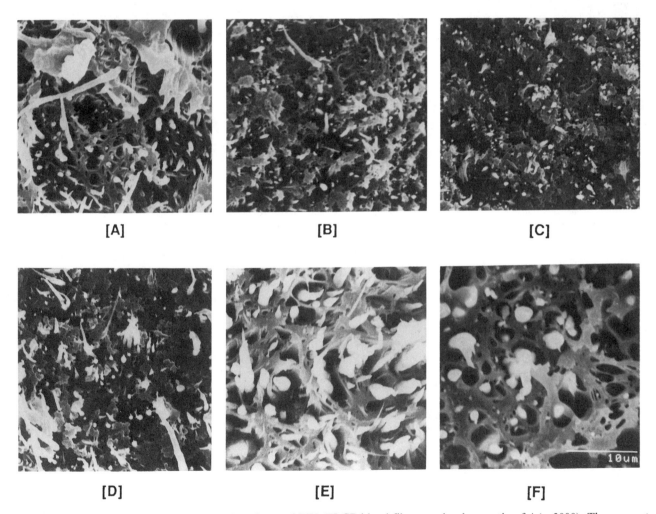

Figure 5 SEM photographs of fractured surfaces of PEI–TLCP blend fibers at the draw ratio of 4 (×3000). The amount of PEsI in the blends are (A) 0 phr, (B) 0.75 phr, (C) 1.5 phr, (D) 2.25 phr, (E) 3.75 phr, and (F) 7.5 phr. *Source*: Ref. 11.

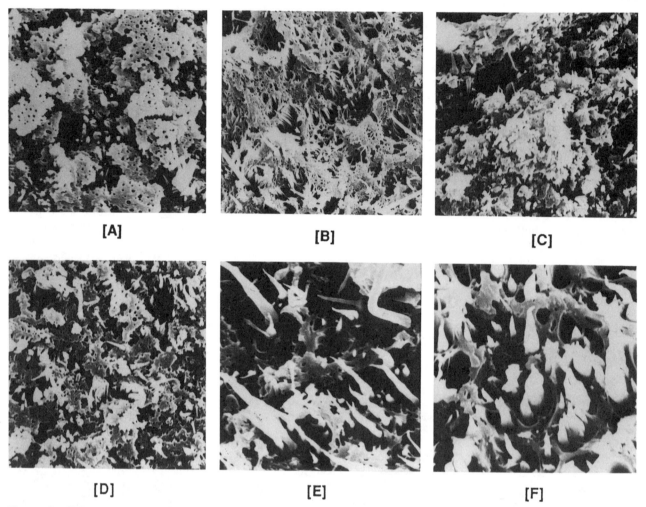

Figure 6 SEM photographs of fractured surfaces of PEI–TLCP blend fibers after tensile test (×3000). Draw ratio is 4. The amount of PEsI in the blends are (A) 0 phr, (B) 0.75 phr, (C) 1.5 phr, (D) 2.25 phr, (E) 3.75 phr, and (F) 7.5 phr. *Source*: Ref. 11.

in the coalesced TLCP phase (Fig. 6D–F), which decreases the tensile modulus and tensile strength of the TLCP phase. Under the tensile stress, breaking can happen at the TLCP particles contacting surface area that is occupied by the compatibilizer. Also, as we noted, excess compatibilizer brought poor dispersion of TLCP phase. Hence, the excess compatibilized system always has lower tensile properties than noncompatibilized system.

If compatibilizing action is too much promoted, even in the compatibilizer concentration range of noncoalescence or no self-phase formation, it may be detrimental to the formation of fibrils because deformation of small droplets into fibrils is more difficult. However, this is not detrimental, as long as the melt strength of the matrix is strong enough to sustain strong extensional deformation and elongation is applied to the blends. High viscosity of the matrix with extensional deforming

action can overcome the difficulties of small particle deformation. Fine microfibrils under submicron range can be easily observed in Figs. 4 and 5. When the two conditions mentioned previously are not met, fibril formation can be hampered by too much fine dispersion of the TLCP phase by the compatibilizer. This is especially true when the matrix viscosity is not so high compared to TLCP viscosity [54,55]. Recently Seo et al. [56,57] investigated the effect of compatibilizer on the physical properties of PBT–poly(ester) TLCP (Vectra A 950) and nylon 6–Vectra B950 blends. As noted by others, nylon 6 and PBT are not good as the matrix for in situ composite production since fibril formations from the crystalline phase are not easily proceeded due to the lower viscosities of the matrices than TLCP droplets under the processing condition [48,54]. Seo [56] noticed that processing temperature is above the glass-transition temperature of the TLCP phase, hence, droplets are de-

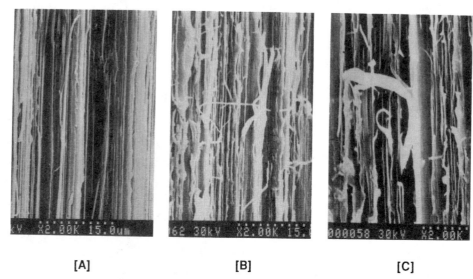

[A]   **[B]**   **[C]**

Figure 7 SEM photographs of peeled back surfaces of PEI–TLCP blend fibers. Draw ratio is 4 ($\times$2000). The amount of PEsI in the blends are (A) 0 phr, (B) 0.75 phr, and (C) 3.75 phr. *Source*: Ref. 11.

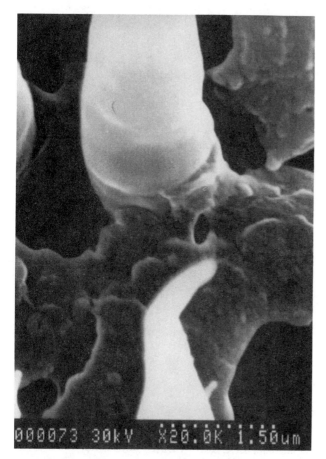

Figure 8 SEM photographs after tensile test showing the adhesion between the TLCP fibril and the matrix (PEI) ($\times$2000). PEsI content is 7.5 phr and draw ratio is 4. *Source*: Ref. 11.

formable if enough stress is transferred. They used a functionalized elastomer (maleic anhydride grafted EPDM, SA-g-EPDM), which can act as a compatibilizer at the interface, as well as a toughening agent in the matrix. Polarized microscope pictures of nylon 6–Vectra B–SA-g-EPDM blend and PBT–Vectra A–SA-g-EPDM blend are shown in Fig. 11. PBT blend exhibits a deformed structure of Vectra A. At the extrusion shear rate (about 20 s^{-1}) and temperature of 290°C, the viscosity of PBT was at least two orders lower than that of TLCP. Although the addition of the elastomer phase gave a rise to blend viscosity (10 times higher than that of PBT), this viscosity was still lower than that of Vectra A. At a similar shear rate, PBT of low viscosity could not by itself deform and break the spherical particles of the dispersed phase during flow. A more surprising feature is that no drawing was applied to the extrudate. Most previous results were consistent with the finding that when the TLCP was the minor component, the viscosity ratio of the TLCP to that of the isotropic polymer was a decisive factor determining the deformation and structure development of the TLCP phase. Elongation and orientation of the TLCP phase to a fibrillar structure takes place when the viscosity of the matrix polymer is higher than that of the suspended TLCP phase. Therefore, a high-shear rate processing should be performed to lower the viscosity of the TLCP phase. The shear rate of this process was low (about 20 s^{-1}), and it should be emphasized that the TLCP viscosity is higher than that of PBT. More details of the morphology can be observed in the SEM picture of the blend shown in Fig. 12A. We can clearly see the TLCP phase is in fibril shape. Some appear to be coalesced into a large fiber shape, but fine fibrils can be also observed.

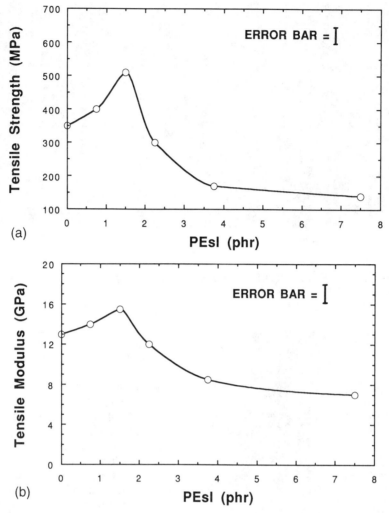

Figure 9   (a) Tensile strength versus PEsI content for PEI–TLCP in situ composite at a draw ratio of 4. (b) Tensile modulus versus PEsI content for PEI–TLCP in situ composite at a draw ratio of 4. *Source*: Ref. 11.

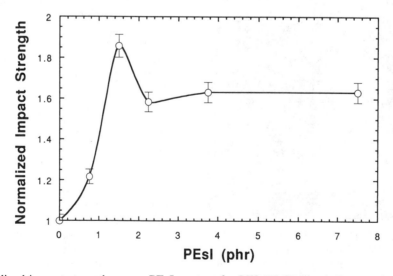

Figure 10   Normalized impact strength versus PEsI content for PEI–TLCP blend. Draw ratio is 4. *Source*: Ref. 11.

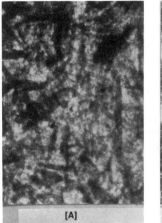 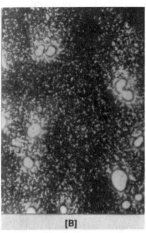

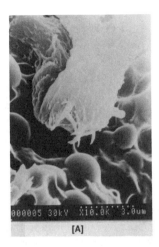 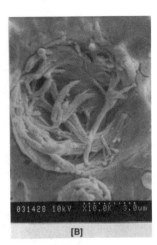

Figure 11 Polarized microscope photographs (×800) (A) PBT–TLCP–elastomer blend (60:25:15 wt ratio). Dark phase is the TLCP phase. (B) Nylon 6–TLCP–elastomer blend (60:25:15 wt ratio). Dark phase is the TLCP phase and large white one is the elastomer phase. *Source*: Ref. 56.

Figure 13 Higher magnification of SEM photographs of nylon 6–TLCP–elastomer blend (×10,000) (A) Dispersed fine microfibrils of TLCP, (B) cluster of TLCP microfibrils. *Source*: Ref. 56.

As shown in Fig. 11B, dispersion morphology for the nylon 6/Vectra B/SA-g-EPDM blend was totally different from that of the PBT–Vectra A–SA-g-EPDM blend. TLCP phases were very uniformly and finely dispersed in the nylon 6–Vectra B–SA-g-EPDM blend and a large fibril shape observed in the PBT–Vectra A–SA-g-EPDM blend could not be seen under polarized microscope. It should be noted that the size of the dispersed TLCP phase is very small (submicron size). This small size of the TLCP phase in the nylon 6/elastomer matrix was not observed by any others [4,54,55,58]. A closer look by SEM more clearly revealed the dispersion of Vectra B in the matrix (Fig. 12B). TLCP phases are very

finely and uniformly dispersed with elongated shapes. Higher magnification showed dispersed microfibrils and their clusters in the TLCP phase (Fig. 13). Again it should be emphasized that no drawing was applied. We believe that this happened due to the compatibilizing action and good stress transfer through good adhesion at the interface. If the compatibilizing action is too vivid to reduce the size of TLCP droplets to very small size, tiny droplet deformation would be very difficult even if strong elongation is applied since transferred stress would not be greater than the resisting stress, i.e., interfacial tension divided by the radius of droplet [Eq. (1)].

Seo et al. [57] also investigated nylon 46–Vectra B950 and functionalized elastomer (SA-g-EPDM) blends. Matrix viscosity was again much lower than that of TLCPs. Fibril formation could not be obtained so much as nylon 6–Vectra B950 case. Still some fibril formation was observed, however, especially in the extrudate outer region where the shearing action is most active. Local elongational action on the extrudate outside to set the surface velocity from zero at die exit to average value in a short distance may account for the deformation [59], but it is not so manifest as we can see the droplet shape in the binary blend (Fig. 14A and B) [57]. We should emphasize again that the key factor for the droplet deformation is enough shear stress transfer through good adhesion at the interface by compatibilizing action since no compatibilizer-added system shows no fibril formation nor droplet deformation under similar processing condition (Fig. 14A and B). Tensile strength and modulus of the blends increased significantly compared to the noncompatibilized binary blends. Tensile modulus increased rapidly with draw ratio and then reached a plateau after a draw ratio of 6 (Fig. 15). Fibril

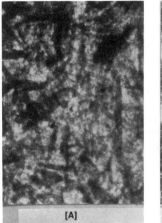 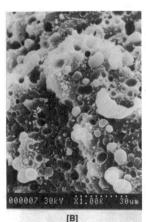

Figure 12 SEM photographs of fractured surfaces (×1000). (A) PBT–TLCP–elastomer blend. (B) Nylon 6–TLCP–elastomer blend. *Source*: Ref. 56.

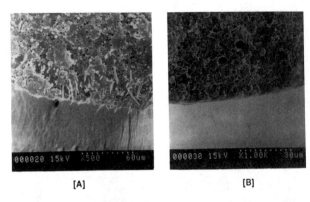

[A]	[B]

Figure 14 SEM photographs of fractured surfaces of non-elongated nylon 46–Vectra B (75:25 wt ratio) blend fibers. (A) 2.7 wt% SA-g-EPDM was added. (B) No compatibilizer was added. *Source*: Ref. 57.

amount of crystalline phase present as the filler. The miscibility between PET–PHB60 and PC enables the TLCP phase to act as a plasticizer, which induces depression of the glass-transition temperature of the matrix in the blend, and, therefore, the rate of shrinkage for the blend increases because of the increasing difference between the temperature of the experiment and the glass transition temperature of the matrix. On the other hand, if enough to form a second phase is added, it can be oriented during the drawing process and will act as short fibers, strongly reducing the shrinkage of the drawn objects for $T_m > T > T_g$, where T_g is the glass-transition temperature of the matrix and T_m is the melting temperature of the filler. This reinforcing effect of the TLCP phase can be preserved up to the temperature close to the melting zone of the crystals, where the filler loses its shape and effectiveness. This is the reason why strong flow such as elongational flow field after die exit is most effective for fibril formation. We already saw

formation in the outer region and good adhesion between TLCP and the matrix result in the increase of tensile strength. It seems that compatibilizer action was not as active in nylon 46 matrix as in nylon 6 (end group concentration would be low) [11,57]. As shown in Fig. 12, simultaneous increase of tensile strength and modulus is interesting since it is contrary to the general fact that addition of the elastomer reduces the blend modulus [60]. Regional fibrillation and good adhesion are ascribed to be the reason. This suggests a new method to produce strong and tough plastic blends.

Miscibility or compatibility provided by the compatibilizer or TLCP itself can affect the dimensional stability of in situ composites. The feature of ultra-high modulus and low viscosity melt of a nematic liquid crystalline polymer is suitable to induce greater dimensional stability in the composites. For drawn amorphous polymers, if the formed articles are exposed to sufficiently high temperatures, the extended chains are retracted by the entropic driving force of the stretched backbone, similar to the contraction of the stretched rubber network [61,62]. The presence of filler in the extruded articles significantly reduces the total extent of recoil. This can be attributed to the orientation of the fibers in the direction of drawing, which may act as a constraint for a certain amount of polymeric material surrounding them.

As Carfagna et al. [61] suggested, the addition of a mesophasic polymer to an amorphous matrix can lead to different results depending on the properties of the liquid crystalline polymer and its amount. If a small amount of the filler compatible with the matrix is added, only plasticization effect can be expected and the dimensional stability of the blend would be reduced. Addition of PET–PHB60 to polycarbonate reduced the dimensionality of the composite, i.e., it increased the shrinkage [42]. This behavior was ascribed to the very low

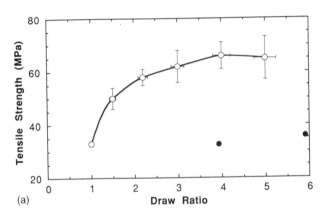

(a)

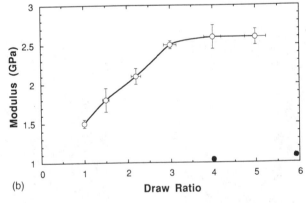

(b)

Figure 15 (A) Tensile strength versus draw ratio of nylon 46–Vectra B (75:25 wt ratio) and (B) tensile modulus of the blends when 2.7 wt% of SA-g-EPDM was added. Lines are guides for eyes. Closed symbols are mechanical properties of the binary nylon 46–Vectra B blend (75:25 wt ratio) *Source*: Ref. 57.

that fibrils can be formed in compatible blends under high-shear stress, but if the residence time in the extruder or die is longer than relaxation time of TLCP, then fibril structure formed in the melt can break up. Thus, using a short die can be more effective for in situ composite preparation, since it provides more extrudate swelling as well as short residence time. This extrudate swell will hinder heat transfer from the inside to the surface, which would give more fluidity for the TLCP phase to be deformed into fibril shape when it is under strong elongational flow.

In conclusion, miscibility or compatibility between TLCPs and matrices can have a great effect on droplet deformation and fibrillation as well as their dispersion and adhesion. It is necessary to point out that use of highly viscous matrix polymer is not the only way for high-stress transfer. Processing of highly viscous matrix requires more energy. For instance, fibrillation in the nylon 6 matrix can be possible even without a strong elongational field provided there is good adhesion at the interface. Since low surface tension by compatibility does exhibit the same effect, more understanding of the droplet deformation mechanism is necessary for choosing the most appropriate processing condition. Mechanical properties of produced in situ composites can be much higher than incompatible ones.

IV. CONCLUSIONS

Deformation and fibrillation of TLCP with polymeric materials is an intriguing phenomena. Reduction of the resistant stress against TLCP droplet deformation by lowering interfacial tension can bring the same effect as if matrix polymers of high viscosity were used to increase the viscous stress for droplet deformation. In fact, good adhesion at the interface increases transferred stress, which enables the droplet deformation. The amount of transferred stress is in some cases greater than expected, noticeably producing fine fibrils without strong elongation in the matrix polymers whose viscosity is smaller than that of the TLCP phase at low shear rate. This is in clear contrast to the traditional in situ composite preparation method, which always needs matrix polymers that have higher viscosity at the processing condition than TLCPs and also needs a strong elongational flow to deform very small particles. Fine fibril formation is especially important to obtain improved mechanical properties for in situ composites.

With respect to good adhesion, reduced interfacial tension, fine distribution of TLCP phase, and the use of a compatibilizer can be very effective for this purpose. Remarkably improved mechanical properties (good impact properties as well as tensile properties) can be obtained with optimum amounts of the compatibilizer. Excess amounts of the compatibilizer causes the emulsifying effect to coalesce the dispersed TLCP phase, hence, deteriorate dispersion and mechanical properties. Due to different failure modes, the impact strength of the compatibilized system shows still improved values. Addition of a functionalized elastomeric phase, which can promote the chemical reaction at the interface to produce some copolymers acting as the compatibilizer, suggests a new method that can be applicable to produce tough and strong materials. Also, generation of in situ composites having microfibrils in the matrix polymers, whose processing temperature is lower than the TLCP nematic transition temperature, can help to solve the nonisotropy problem common to all in situ composites. Dimensional stability provided by fibril formation or deformation of TLCP phase will not be affected by the compatibilizer unless excess amounts of compatibilizer are used to make the TLCP droplet coalesce. In the compatibilized system, the proper amount of TLCP should be used to get the dimensional stability from fibril formed in the in situ composites.

Strong elongational deformation and use of matrix polymers whose viscosity is higher than that of TLCP phase are better to ensure uniform and fine fibril formation. But application of compatibilizing techniques to in situ composite preparation can be useful to get the most desirable products. These can reduce the high costs of the liquid crystalline polymers and expensive special engineering plastics used for the in situ composite preparation and reduce the processing cost, whereas they can increase the performance of produced in situ composites, hence, their applications, too.

ACKNOWLEDGMENTS

The author appreciates Drs. Soon Man Hong, Seung Sang Hwang, Sang Mook Lee, and Eunwon Han for their help with the experiments and fruitful discussion, as well as stimulating suggestions. Special thanks are going to Mr. Youngwook Seo and Ms. Youngin Seo for their support.

REFERENCES

1. N. A. Plate (ed.), *Liquid Crystal Polymers,* Plenum Press, New York (1993).
2. A. Cifferi, W. R. Krigbaum, and R. B. Meyer (eds.), *Polymer Liquid Crystals,* Academic Press, New York (1982).
3. A. I. Isayev and T. Limtasiri, *International Encyclopedia of Composites* (S. M. Lee, ed.), *Vol. III,* VCH Publishers, New York (1990).
4. D. Acierno and M. R. Nobile, *Thermotropic Liquid Crystal Polymer Blends,* (F. P. La Mantia, ed.), Technomic Publishing, Lancaster (1993).
5. A. A. Handlos and D. G. Baird, *J. Macromol. Sci., Rev. C 35(2):* 183 (1995).
6. W. Brostow, *Polymer, 31:* 979 (1990).
7. F. N. Cogswell, B. O. Griffin, and J. B. Pose, U.S. Patent to I.C.I. (England) 4,386,174 (1983), 4,433,083 (1984), 4,438,236 (1984).

8. F. P. LaMantia (ed.), *Thermoplastic Liquid Crystal Blends,* Technomic Publishing, Lancaster, PA (1993).

9. C. Ryu, Y. Seo, S. S. Hwang, S. M. Hong, T. S. Park, and K. U. Kim, *Intern. Polym. Proc., IX*: 266 (1994).

10. Y. Seo, S. M. Hong, S. S. Hwang, T. S. Park, K. U. Kim, S. Lee, and J. W. Lee, *Polymer, 36*: 515 (1995).

11. Y. Seo, S. M. Hong, S. S. Hwang, T. S. Park, K. U. Kim, S. Lee, and J. W. Lee, *Polymer, 36*: 525 (1995).

12. A. Datta and D. G. Baird, *Polymer, 36*: 505 (1995).

13. S. S. Bafna, T. Sun, and D. G. Baird *Polymer, 36*: 259 (1995).

14. S. S. Bafna and D. G. Baird, *Polymer, 34*: 708 (1993).

15. A. Datta, H. H. Chen, and D. G. Baird, *Polymer, 34*: 759 (1993).

16. W. Lee and A. T. DiBenedetto, *Polymer, 34*: 684 (1993).

17. H. Van Oene, *J. Colloid Interf. Sci., 40*: 448 (1972).

18. H. Van Oene, *Polymer Blends* (D. R. Paul, and S. Newman, eds.) Academic Press, Vol. 1, Ch. 7 pp 295–352, (1978).

19. J. J. Elmendorp and R. J. Maalcke, *Polym. Eng. Sci., 25*: 1041 (1985).

20. W. K. Lee and R. W. Flumerfelt, *Int. J. Multiphase Flow, 7*: 363 (1981).

21. G. I. Taylor, *Proc. Royal. Soc. (A)., 146*: 501 (1934).

22. R. G. Larson, *Constitutive Equations for Polymer Melts and Solutions,* Butterworths, Boston (1988).

23. D. G. Baird, *Polymeric Liquid Crystals* (A. Blumstein, ed.) Plenum Press, New York, p. 119 (1985).

24. F. N. Cogswell, *Recent Advances in Liquid Crystalline Polymers* (L. L. Chapoy, ed.), Elsevier Applied Science, London (1985).

25. F. P. La Mantia, A. Valenza, M. Paci, and P. L. Magagnini, *Rheol. Acta, 28*: 417 (1989).

26. P. Ghodganokar and U. Sundararaj, *Polym. Eng. Sci., 36*: 1656 (1996).

27. L. Levitt and C. W. Macosko, *Polym. Eng. Sci., 36*: 1647 (1996).

28. I. Delaby, B. Ernst, D. Froelich, and R. Muller (*Macromol. Symp., 100*: 131 (1995).

29. I. Delaby, B. Ernst, D. Froelich, and R. Muller *Polym. Eng. Sci., 36*: 1627 (1996).

30. U. Sundararaj and C. W. Macosko *Macromolecules, 28*: 2647 (1995).

31. F. P. La Mantia, A. Valenza, M. Paci, and P. L. Magagnini, *Polym. Eng. Sci., 30*: 7 (1990).

32. F. P. La Mantia, M. Saiu, A. Valenza, M. Paci, and P. L. Magagnini, *Eur. Polym. J. 26*: 323 (1990).

33. Y. Seo, S. S. Hwang, S. M. Hong, T. S. Park, and K. U. Kim, *Polym. Eng. Sci., 35*: 1621 (1995).

34. G. G. Viola and D. G. Baird, *J. Rheol., 30*: 601 (1986).

35. M. V. Tsebrenko, A. V. Yudin, T. I. Ablazova, and G. V. Vinogradov, *Polymer, 17*: 831 (1976).

36. S. Lee, S. M. Hong, Y. Seo, T. S. Park, S. S. Hwang, K. U. Kim, and J. W. Lee, *Polymer 35*: 519 (1994).

37. L. A. Utracki, *Polymer Alloys and Blends,* Carl Hanser Verlag, New York (1989).

38. R. E. S. Bretas and D. G. Baird, *Polymer, 34*: 759 (1993).

39. A. Nakai, T. Shiwaku, H. Hasegawa, and T. Hashimoto, *Macromolecules, 19*: 3010 (1986).

40. M. Paci, C. Barona, and P. L. Magagnini, *Polym. Sci. Polym. Phys. Edn., 25*: 1595 (1987).

41. P. Zuhang, T. Kyu, J. L. White, *Polym. Eng. Sci., 28*: 2817 (1988).

42. M. R. Nobile, E. Amendola, L. Nicolais, D. Acierno, and C. Carfagna, *Polym. Eng. Sci., 29*: 244 (1989).

43. D. Acierno, F. P. LaMantia, G. Polizzotti, A. Cifferi, and B. Valenti, *Macromolecules, 15*: 455 (1982).

44. J. Lee, Y. Seo, S. M. Hong, and S. S. Hwang, *Intern. Polym. Process.* (in press) (1997).

45. J. Lee, Y. Seo, S. M. Hong, and S. S. Hwang, *J. Appl. Polym. Sci. Korea Polym. J. 4*: 198 (1996).

46. M. Kimura and R. S. Porter, *J. Polym. Sci. Polym. Phys. Ed., 22*: 1697 (1984).

47. F. P. La Mantia, F. Cangialosi, U. Pedretti, and A. Roggers, *Eur. Polym. J., 29*: 671 (1993).

48. R. S. Porter, J. M. Jonza, M. Kimura, C. R. Desper, and E. R. George, *Polym. Eng. Sci., 29*: 55 (1989).

49. H. J. O'Donnel and D. G. Baird, *Polymer, 36*: 3113 (1995).

50. A. I. Isayev and M. Modic, *Polym. Compos., 8*: 158 (1987).

51. N. M. Miller, J. M. G. Cowie, J. G. Tait, D. L. Brydon, and R. R. Mather, *Polymer, 36*: 3107 (1995).

52. I. Piirma, *Polymeric Surfactants,* Marcel Dekker, New York (1992).

53. A. P. Plochocki, S. S. Dagli, and R. D. Andrews, *Polym. Eng. Sci., 30*: 741 (1990).

54. F. P. LaMantia, A. Valenza, M. Paci, and P. L. Magagnini, *J. Appl. Polym. Sci., 38*: 583 (1989).

55. D. G. Baird and T. Sun, *Liquid Crystalline Polymers, ACS Symp. Ser. 435* (R. A. Weiss and C. K. Ober, eds.), ACS Washington, (1990).

56. Y. Seo, *J. Appl. Polym. Sci.* (in press) (1997).

57. Y. Seo, S. M. Hong and K. U. Kim, *Macromolecules* (in press) (1997).

58. G. Kiss, *Polym. Eng. Sci., 27*: 410 (1987).

59. Y. Seo and E. H. Wissler, *J. Appl. Polym. Sci., 37*: 1159 (1989).

60. F. Coppola, R. Grew, E. Martuscelli, H. W. Kammer, and C. Kummerlowe, *Polymer, 28*: 47 (1987).

61. C. Carfagna, E. Amendola, and M. R. Nobille, *International Encyclopedia of Composites* (S. M. Lee, ed.), Vol. II, VCH Publishers, New York (1990).

62. J. Tunnicliffe, D. J. Blundell, and A. M. Windle, *Polymer, 21*: 1259 (1980).

39

Self-Organization of Core-Shell Type Polymer Microspheres and Applications to Polymer Alloys

Reiko Saito and Koji Ishizu
Tokyo Institute of Technology, Tokyo, Japan

I. INTRODUCTION

It is well known that block copolymers and graft copolymers composed of incompatible sequences form the self-assemblies (the microphase separations). These morphologies of the microphase separation are governed by "Molau's law [1]" in the solid state. Nowadays, not only the three basic morphologies but also novel morphologies, such as ordered bicontinuous double diamond structure, are reported [2–6]. The applications of the microphase separation are also investigated [7–12]. As one of the applications of the microphase separation of AB diblock copolymers, it is possible to synthesize core-shell type polymer microspheres upon crosslinking the spherical microdomains [13–16].

These core-shell type microspheres have very interesting structural features in that the cores are hardly crosslinked and the shell chains are fixed on the core surface with one end of the shell chains. The other end of the shell chains is free in good solvents for the shell chains. As the result of such a specific structure, the solubilities of the core-shell type polymer microspheres are governed by, not the core, but by the shell sequences, and the core-shell structures do not break even in the dilute solution [9,10].

Generally, the number of the shell chains in a microsphere ranges from a few hundred to a few thousand. The range of the diameter of the core is from 10–100 nm. Such a core-shell structure is very similar to the (AB)n type star block copolymers, which have many arms and spherical polymer micelles of the block or graft copolymers formed in selective solvents that are good for the corona sequence and bad for the core sequence. In fact, many theoretical investigations of the chain con-

formations of the (AB)n type star block copolymers have been reported by using the core-shell type microspheres as the models [17–19].

Among them, the detail conformations of the shell chains were proposed using a scaling theory by Doaud and Cotton [17]. Based on their theory, the conformation of the shell chains takes three regions from the center to the outside: (1) the perturbed region, (2) the semidilute region, and (3) the dilute region. When the molecular weights of the core-shell type polymers and linear polymers are the same, the core-shell type polymers have lower viscosity and stronger repulsive force than the linear polymers in the highly concentrated solution. These are due to the higher density of the shell sequences at region I and II. From these structural peculiarities, Witten and coworkers [20] theoretically expected the self-organization of the (AB)n type star block copolymers, such as super lattice, in the good solvents above the overlap concentration (C*) from the scaling theory. Birshtein et al. [21] also expected the self-organization of them from the viewpoint of the mean field theory. In fact, the self-organization (the super lattice formation) was confirmed for the poly(styrene) star block copolymers and the (polystyrene-b-isoprene)n type star block copolymers by the small angle x-ray scattering measurement above the C* concentration in good solvents [22,23]. For the core-shell type polymer microspheres, the self-organization in the solvents was also obtained above the C* [24,25].

The feature of the core-shell type polymer microspheres that differentiates them the most from the (AB)n type star block copolymers is size. The external diameters of the core-shell type polymer microspheres are generally from about 20–200 nm in the good solvents instead

Table 1 Characteristics of the Block Copolymers [24,25,36,37]

Name	Blocks	Mn ($\times 10^{-4}$)[a] PS	Block	PS content (mol%)[b]	Domain Type[c]	Size (nm)[d]
BC1	P(S-b-4VP)	8.0	11.1	72	Sp$_{4VP}$	48.4
BC2	P(S-b-4VP)	3.6	4.2	86	Sp$_{4VP}$	17
B1	P(S-b-4VP)	1.6	2.4	68	Lm	24
B2	P(S-b-IP)	2.3	3.2	72.5	Sp$_{IP}$	10
B3	P(S-b-2VP)	2.0	2.8	70	Sp$_{2VP}$	18

[a] Number-average molecular weights determined from GPC and ^{1}H NMR.
[b] Estimated from ^{1}H NMR.
[c] Estimated from TEM. Sp = sphere; Lm = lamellar.
[d] Diameter for sphere and thickeness for lamellar.

of 30 nm for the (AB)n type star block copolymers. The diameter of the core-shell type polymer microspheres is in the region of the usual microphase separation. Each core-shell type microsphere can be assumed to be the smallest composing unit of the microphase separation. Therefore, the architectural effect of the core-shell type microspheres on the aggregation behavior would be different from their precursor block copolymers. In this chapter, we will focus on the poly(4-vinyl pyridine) core-polystyrene-shell type polymer microspheres and discuss the details of their self-organization and the application to polymer alloys with other block copolymers from the viewpoint of the controlling of the microphase separation.

II. EXPERIMENTAL MATERIAL

A. Precursor Block Copolymers

As previously described, all microspheres discussed in this chapter were synthesized from AB type diblock copolymers. Precursor block copolymers, poly(styrene-b-4-vinyl pyridine) (P[S-b-4VP]) diblock copolymers, were synthesized using the additional anionic polymerization technique [13]. The basic properties of the block copolymers were determined elsewhere [24,25] and are listed

in Table 1. In order to obtain the polystyrene (PS) matrix type morphologies, according to Molau's law [1], PS contents of all block copolymers were greater than 65 mol%.

As these block copolymers were synthesized using the anionic polymerization technique, their molecular weight distributions were narrow. The microspheres with narrower size distribution are better for well-ordered self-organization. Actually, all block copolymers synthesized for these works formed poly(4-vinyl pyridine) (P4VP) spheres in the PS matrices with narrow size distributions.

The poly(styrene-b-isoprene) (P(S-b-IP)) and poly(-styrene-b-2-vinyl pyridine) (P(S-b-2VP)) block copolymers with narrow molecular weight distributions for blending with the microspheres were also synthesized using the additional anionic polymerization technique. The number-average molecular weights ($\overline{Mn}$s) and PS contents are also shown in Table 1.

B. Synthesis of the Microspheres

The core-shell type polymer microspheres were synthesized upon the chemical crosslinking of the spherical microdomains in the microphase separated films. The block copolymers were dissolved in 1,1,2-trichloroeth-

Table 2 Characteristics of the Microspheres

Name	Block type	Diameter (nm) P4VP core[a]	Internal[a]	External[b]	Crosslink density (mol%)[c]	C*[d] (wt%)
MC1 [24]	BC1	49.7	77.8	180.5	20.8	4.4
MC2 [25]	BC2	16	36	66	35.4	3.9

[a] Determined by TEM in solid state.
[b] Measured by DLS in benzene at 20°C.
[c] Measured by Volhard titration.
[d] Overlap concentration in benzene calculated from the external diameter in benzene.

ane at 5 wt%. The polymer solution was cast on the Teflon sheet and the solvent was gradually evaporated at room temperature. The P4VP sequences were cross-linked with 1,4-dibromobutane gas at 80°C for a certain number of hours. The details of the synthesis and characteristics have been reported elsewhere [24]. Among them, the important characteristics for the self-ordering are listed again in Table 2. External structures, sizes, and inner textures of the core-shell type polymer microspheres were determined by transmission electron microscopy (TEM; Hitachi H-500) at 75 kV and dynamic light scattering measurements (DLS; Otsuka Denshi, DLS-700) at an angle of 90 degrees). Figure 1 shows the specific TEM micrographs of the microspheres (MC1) dispersed on the carbon substrate. The microspheres shown in this chapter are true sphericals and can be dispersed separately on the carbon substrate when they are cast from dilute solution (< 0.1 wt%). The C* values were calculated using the following equation: ($C^* = 3 MW/4\pi Rh^3 Na$) with the hydrodynamic radii of the microspheres in the solvent measured by DLS at 0.1 wt/vol% of polymer concentration.

C. Characteristics of the Self-Organization of the Microspheres

1. Transmission Electron Microscopic Observations

The block copolymer and the microsphere were cast from polymer–benzene solution on a Teflon sheet. The solution was gradually dried at room temperature. Film was microtomed vertically at 80 nm thick by the UltraCut-N (Reichert Nissei). In order to obtain enough contrast for TEM observation, the P4VP microdomains in the film were stained with OsO_4. The film was observed by TEM (JEOL CX-100) at 100 kV.

2. Small Angle X-Ray Scattering (SAXS)

The SAXS intensity distribution was measured with a rotating anode x-ray generator (Rigaku Denki, Rotaflex, RTP 300 RC) operated at 40 kV and 100 mA. The x-ray source was monochrolmatized to CuK_α ($\lambda = 0.154$ nm) radiation. The SAXS patterns were taken with a fine-focused x-ray source using a flat plate camera (Rigaku Denki, RU-100). In the measurement of the solution sample, we used a glass capillary ($\phi = 2.0$ mm; Mark-Rohrchen Ltd.) as a holder vessel.

III. RESULTS AND DISCUSSION

A. Self-Organization of Microsphere in Two- and Three-Dimensions

As described in the Introduction, because of the structural similarity of the core-shell type microspheres to the (AB)n type star block copolymers with many arms

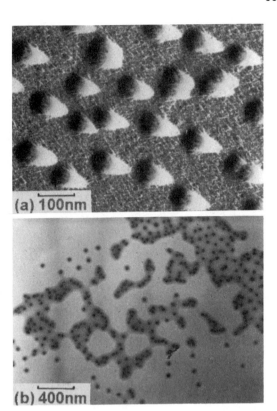

Figure 1 The transmission electron micrographs of the crosslinked products of MC1 cast from benzene, (a) at a 0.05 wt% polymer concentration and shadowed with Cr at an angle of 20°, and (b) at a 0.05 wt% concentration [24].

[26], the hierarchical super structure formation, such as macrolattice formation in solid state, was expected. First, the two-dimensional macrolattice formation of the microspheres was geometrically shown. Figure 2 shows the TEM microspheres of the as-cast films of the MC1 cast from benzene and their precursor block copolymer BC1 cast from 1,1,2-trichloroethane. It is clear that the spherical microdomains (P4VP domains) of the microspheres were arranged in one layer with a hexagonal arrangement better than the block copolymers. According to Thomas et al. [27], the hexagonal lattice is the most favorable state for two-dimensional packing of the spherical microdomains. Figure 3 shows the degree of ordering of these P4VP spherical domains, quantitatively by using the radius distribution function between the centers of the P4VP spherical microdomains. P4VP domains of the block copolymer MC1 were not orderly arranged, as no clear peaks are observed. On the other hand, the three clear peaks (79.9, 139.4, and 159.7 nm) can be observed for the microsphere as-cast film. The ratio of these distances for the peaks observed for MC1 were 1 : 1.75 (139.4/79.9), :2.00(159.7/79.9), which agreed well with the theoretical ratio of hexagonal packing of

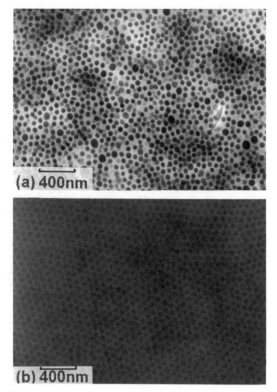

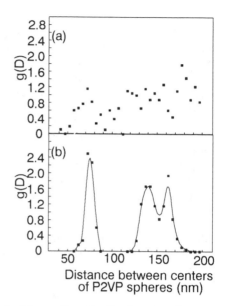

Figure 3 The radius distribution function between the centers of the P4VP domains of (a) the block copolymer BC1, and (b) the microsphere MC1 [24].

Figure 2 The transmission electron micrographs of samples cast from solution containing 1 wt% of polymer, (a) the block copolymer BC1, and (b) the microsphere, MC1 [24].

the microsphere, $1:\sqrt{3}:2$ (Fig. 4). Therefore, it was concluded that the microspheres were more easily arranged completely hexagonally than block copolymers in two dimensions.

Next, the three-dimensional packing structures of the microspheres are shown. In general, when spheres from a highly ordered microphase separated structure, the capable packing structures of the spheres are simple (SC), face-centered (FCC), and body-centered (BCC) cubic. There are two major experimental methods to investigate the arrangement of the microspheres three-dimensionally. One is to combine the TEM images obtained for essentially two-dimensional projections [28,29]. Figure 5 shows the TEM micrographs of the cross-section of the film of the microspheres MC1 that had been cast from benzene and tilted to the y-axis. Figure 5b is the tilted profile of Fig. 5a at an angle of 45 degrees. From the movement of the two-dimensional address of certain P4VP spheres on the micrographs, the exact addresses of the spheres in three dimensions could be determined. Figure 5c shows the schematic arrangement of the P4VP spheres calculated from each three-dimensional address. The average distances between the

centers of the microspheres (D) are 57.3 and 80 nm for the a-axis and b-axis, respectively. As the ratio of these distances ($80/57.3 = 1.4 \approx \sqrt{2}$) agreed well with the packing pattern of the (110) plane of the FCC, the structure of the crystal lattice was concluded FCC.

Another method for the determination of the structure of the crystal lattice is SAXS [30,31]. Figure 6 shows the specific SAXS profiles of microsphere film (MC2). The cubic packing values (dl/di) are listed in Table 3. Three clear peaks appeared at 0.35, 0.42, and 0.66 degrees in Fig. 6. The dl/di values of the second and third peaks are $\sqrt{4/3}$ and $\sqrt{11/3}$, respectively. These values are peculiar to the FCC structure. Thus, the lattice structure of the microspheres is an estimated FCC. As both

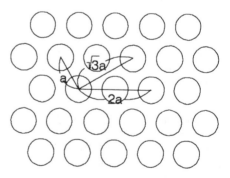

Figure 4 Schematic representation of microspheres showing a two-dimensional hexagonal packing arrangement [24].

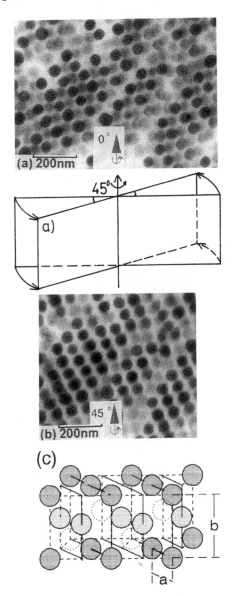

Figure 5 The transmission electron micrographs of cross-section of MC1 (a) without any tilt, (b) tilted at an angle of 45 degrees of the y-axis, and (c) schematic representation of the arranged microspheres after tilting [24].

results of the titled TEM micrographs and SAXS profiles agreed well, the lattice structure of the recast microspheres in the solid state was the concluded FCC.

In three dimensions, Ohta and Kurokawa [32] reported that a BCC arrangement was only slightly more favored than the FCC arrangement. In fact, many BCC structures have been reported for AB type block copolymers and the blends of homopolymer–block copolymer systems [27,33–35]. However, the lattice structure of the core-shell type polymer microspheres was FCC. This FCC formation resulted in the lower viscosity of

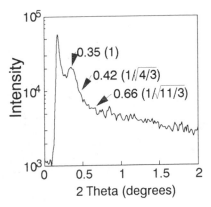

Figure 6 SAXS intensity distributions for MC2 film in the small-angle region. The arrows show the scattering maxima. The values in parentheses indicate the cubic packing (dl/di) [25].

the microspheres' solution than the precursor block copolymer solution at high-polymer concentration. Consequently, the macrolattice was formed by the core-shell type polymer microsphere and its structure was FCC instead of BCC for AB type block copolymers.

B. Alloy of the Core-Shell Type Polymer Microspheres and Block Copolymers

As described in the Introduction, the core-shell type polymer microspheres can be considered as the smallest unit of the microphase separation with the spheres in a matrix. By using the high ordering ability of the microspheres, the novel microphase-separated structures can be obtained by alloying with other copolymers. We succeeded in obtaining the novel morphologies by blending the core-shell type polymer microspheres and AB type block copolymers with lamellar [36] and spherical morphologies [37]. In order to prevent the macrophase separation of the core-shell type polymer microspheres and the block copolymers, one sequence of the AB type block copolymer and the shell sequences of the microspheres are polystyrene.

Figures 7 and 8 show the original morphologies of the block copolymers observed by TEM selectively stained P4VP, P2VP, and polyisoprene (PIP) sequences

Table 3 SAXS Data for MC2 Film [25]

n	2Theta	di(nm)	d1/di
1	0.35	25.2	1
2	0.42	21.0	$\sqrt{4/3}$
4	0.66	13.4	$\sqrt{11/3}$

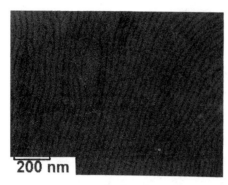

Figure 7 The transmission electron micrograph of the block copolymer B1 for blend [36].

with OsO$_4$. Based on Molau's law [1], the morphologies of the microphase separation of the block copolymers were P4VP lamellar and PIP and P2VP spheres in the PS matrices for B1, B2, and B3, respectively. The $\overline{Mns}$ of the PS sequences of these block copolymers were small enough (smaller than that of the shell sequences of the core-shell type polymer microspheres) for good solubility of them. Benzene was used as the good common solvent for both block copolymer and core-shell type polymer microspheres. The morphologies of the microphase separation in two dimension were observed by TEM on the carbon substrate.

First, Fig. 9 shows the TEM micrographs of interesting novel morphologies observed on the blend of MC1 and B1 [36]. The blend ratio of B1 (r) was varied from 0.33 to 0.66. At any blend ratio, the macrophase separation of MC1 and B1 did not appear. Additionally, the diameter of the P4VP core (48 nm) and the thickness of the P4VP layer (11 nm) did not change by blending at any weight fraction of blending. In general, the morphologies of blends of A homopolymer and lamellar AB diblock copolymer can be divided into two cases: (1) the AB diblock copolymer forms a spherical microdomain of B in the blend when the AB diblock copolymer and A homopolymer are mixed homogeneously and the composition of the B block in the blend is less than 25 vol%, or (2) the AB diblock copolymer forms the onion-ring structure in the blend [38]. In this study, neither the usual onion-ring structure nor the spherical microdomain of P4VP from B1 was observed at any blend ratio.

At $r = 0.5$ (Fig. 9b), the most interesting and novel morphology can be observed. This morphology can be described as follows. The P4VP cores of the microspheres form a regular structure, and a P4VP bilayer surrounds each microsphere with a honeycomb-like structure, similar to a cell wall, as the number of the microsphere surrounded by the P4VP wall (K) was 1.08. Similar structures have been observed for ABC triblock copolymers [39]. Our honeycomb-like novel structure, however, is different from that of the ABC triblock co-

polymer. From a detailed observation of Fig. 9b, the cell of the microspheres seems hexagonal, and the AB type block copolymer forms a bilayer between the microspheres. The hexagonal surrounding with the P4VP bilayer indicates that the unit cell of the microsphere was hexagonal in two dimensions. This agreed well with the assumptions proposed by Thomas et al. [27] and Birshtein et al. [21] that the unit cell of the spherical microdomain of the microphase-separated film is hexagonal in two dimensions.

The chain arrangement of this morphology was schematically proposed as in Fig. 10. The cell of the microsphere has a hexagonal surface, and the AB diblock copolymers form a bilayer between the microspheres. From this schematic arrangement, the optimal blend ratio of the AB block copolymer in this system was calculated as 0.46. This value was very close to the blend ratio of the AB type block copolymer 0.5 at which the blend showed the hexagonal packed honeycomb-like structure.

When r was less than 0.46 (Fig. 9a), the P4VP layer surrounded some microspheres in groups ($r = 0.33$). The K at $r = 0.33$ was 2.48. This indicates that the amount of B1 block copolymer was insufficient to surround each microsphere separately. When r was larger than 0.46 (Fig. 9c), the wide dark regions of P4VP were also observed. These regions were horizontally oriented lamellar microdomains of B1, resulting in a minimization

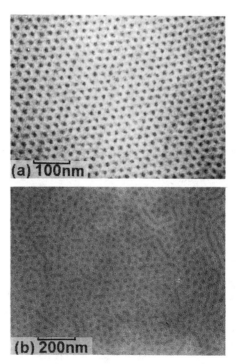

Figure 8 The transmission electron micrographs of block copolymers for blends, (a) B2, and (b) B3 [37].

of the air–polymer surface tension [40]. It was confirmed that the B1 was in excess to the microsphere at 0.66 of the blend ratio.

The degree of ordering of the microspheres was estimated by using the radial distribution function g(D) of the P4VP cores of the microspheres (Fig. 11). As previously described, for hexagonal packed spheres, the ratio of the peaks of the distances between the centers of the cores would be $1:\sqrt{3}:2$. For the film at $r = 0.5$, the narrow three peaks appeared at D = 88 nm, 158 nm, and 178 nm. As the ratio of these distances, 1:1.80 (158/88): 2.02 (178/88) agreed well with the theoretical values, it was concluded that the microspheres were hexagonal packed even in the blends. For the other blend ratio, the peaks became wider and appeared without special

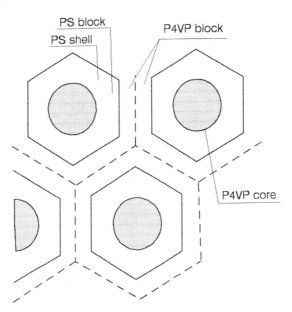

Figure 10 Schematics of two-dimensional chain conformations of the block copolymer and the microsphere in a binary blend [36].

packing patterns. Thus, it was also found that there was an optimal blend ratio for complete hexagonal packing of the novel morphology.

The morphology obtained from the blend of the core-shell type microspheres and AB type block copolymers with spherical morphology is shown next [37]. Figure 12 shows the typical morphologies of the blend ob-

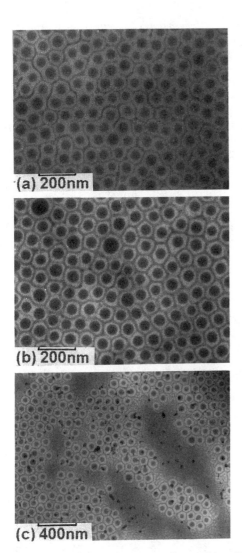

Figure 9 The transmission electron micrographs of binary blend of MC1/B1 on a carbon substrate (a) $r = 0.33$, (b) $r = 0.5$, and (c) $r = 0.66$ [36].

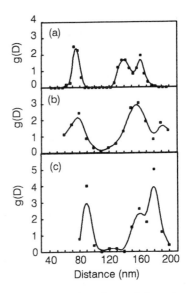

Figure 11 Distribution functions of the microspheres of the MC1/B1 blend. (a) MC1, (b) binary blend at $r = 0.33$, and (c) binary blend at $r = 0.5$ [36].

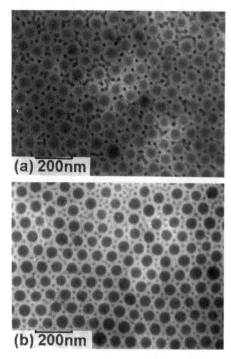

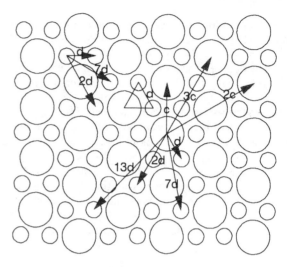

Figure 13 Schematic arrangement of the microdomains of microsphere and block copolymer [37].

Figure 12 The transmission electron micrographs of the blend of MC1 with (a) B2, and (b) B3 [37].

served by TEM. The blend ratio of the AB block copolymers was set to 0.5. The dark regions in the micrographs are P4VP, P2VP, and PIP stained with OsO$_4$. On both of the micrographs, common interesting morphologies were observed. Six small spheres were ar-

ranged around each large sphere. The larger spherical microdomains seemed to be hexagonal packed in two dimensions. From the results of the diameters of each microdomain, the larger spheres with 50-nm diameter were P4VP cores of the microspheres, and the smaller spherical microdomains were owing the spherical microdomains of the block copolymers. This morphology could be observed only when the molecular weights of the PS sequences of the block copolymers were smaller than the PS shell sequence of the microsphere.

The schematic arrangement of this morphology is shown in Fig. 13. Here, c is the distance between the

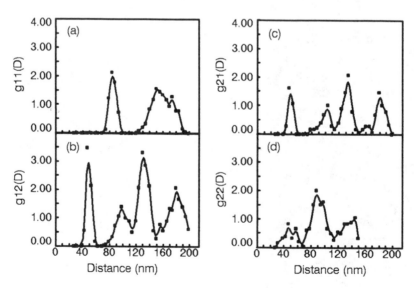

Figure 14 Radius distribution functions (a) between the P4VP cores of the microspheres, (b) from P4VP core to P2VP spherical microdomains, (c) from P2VP spherical microdomain to P4VP core, and (d) between P2VP spherical microdomains [37].

centers of the nearest large spheres, and d is the distance between the centers of the nearest small spheres. Figure 14 shows the radius distribution functions gmn(D) of the blend of MC1 and B3 in two dimensions. For the gmn(D), 1 and 2 indicate the P4VP core and the P2VP spherical microdomain, respectively. First, the degree of ordering between P4VP cores, g11(D), was estimated. In the g11(D), three peaks are observed at D = 85.2, 146.4, and 170.9 nm and the ratio of distance was 1:1.72:2.01. This packing pattern indicates the complete hexagonal packing of the P4VP cores. By comparison with Fig. 3b, it was also found that the distance between the P4VP cores was increased from 79.9 to 85.2 nm with blending the AB diblock copolymer. Consequently, the packing state of the P4VP cores was not destroyed but expanded.

Between the P4VP core and P2VP sphere, two types of radius distribution functions, g12(D) and g21(D), were obtained. For both functions, the patterns of the peaks appeared at 49, 98, 130, and 177 nm. From the schematic arrangement (Fig. 13), the peaks should appear at d, 2d, $\sqrt{7}d$ (= 2.645d), and $\sqrt{13}d$ (= 3.606d). As the experimental ratio of the peaks, 1:2:2.65:3.61, agreed well with the schematic ratio, not only the P4VP cores but also the spherical microdomains of the AB block copolymers were orderly arranged as shown in Fig. 13.

IV. CONCLUSIONS

The core-shell type microspheres can form many self-ordering structures with or without blending other polymers. The structures of itself without blending are hexagonal and a face-centered cubic in two and three dimensions, respectively. The degree of ordering of the microspheres was better than the precursor block copolymers. By blending with the block copolymers with lamellar or spherical morphologies, the hexagonal packing of the microspheres was kept, and the phase-separated structures of the blend block copolymers appeared around each microsphere. Consequently, well-ordered novel morphologies could be obtained. These microspheres can be viewed as an interesting material for controlling the microphase-separated structures.

REFERENCES

1. G. E. Molau, *Block Copolymers* S. L. Aggarwall, (ed.), Plenum Press, New York, p. 79 (1970).
2. D. B. Alward, D. J., Kinning, E. L. Thomas and L. J., Fetters, *Macromolecules, 19*: 215 (1986).
3. D. J. Kinning, E. L. Thomas, D. B. Alward, L. J. Fetters and D. L. Handlin, Jr., *Macromolecules, 19*: 1288 (1986).
4. E. L. Thomas, D. B. Alward, D. J. Kinning, D. C. Martin, D. L. Handlin Jr. and L. J. Fetters, *Macromolecules, 19*: 2197 (1986).
5. H. Hasegawa, H. Tanaka, K. Yamasaki and T. Hashimoto, *Macromolecules, 20*: 1651 (1987).
6. D. S. Herman, D. J. Kinning, E. L. Thomas and L. J. Fetters, *Macromolecules, 20*: 2940 (1978).
7. R. Saito, S. Okamura and K. Ishizu, *Polymer, 33*: 1099 (1992).
8. K. Ishizu, Y. Yamada, R. Saito, T. Yamamoto and T. Kanbara, *Polymer, 33*: 1816 (1992).
9. K. Ishizu, Y. Yamada, R. Saito, T. Kanbara and T. Yamamoto, *Polymer, 34*: 2256 (1993).
10. Y. Ng Cheong Chan, R. R. Schrock and R. E. Cohen, *Chem. Mater., 4*: 24 (1992).
11. C. C. Cummins, R. R. Schrock and R. E. Cohen, *Chem. Mater., 4*: 27 (1992).
12. Ng Cheong Chan Y., Craig G. S. W., Schrock R. R., and Cohen R. E., *Chem. Mater., 4*: 885 (1992).
13. Ishizu K. and Fukutomi T., *J. Polym. Sci. Polym. Lett. Ed., 26*: 281 (1988).
14. Ishizu K. and Onen A., *J. Polym. Sci., Polym. Lett. Ed., 27*: 3721 (1989).
15. Saito R., Kotsubo H., and Ishizu K., *Eur. Polym. J., 27*: 1153 (1991).
16. Saito R., Kotsubo H., and Ishizu K., *Polymer, 33*: 1073 (1992).
17. Doaud M. and Cotton J. P., *J. Phys., 43*: 531 (1982).
18. Halperin A., *Macromolecules, 20*: 2943 (1987).
19. Halperin A. and Alexander S., *Macromolecules, 22*: 2403 (1989).
20. Witten T. A., Pincus P. A., and Cates M. E., *Eur. Phys. Lett., 2*: 137 (1986).
21. Birshtein T. M., Zhulina E. B., and Borisov O. V., *Polymer, 27*: 1078 (1986).
22. Ono, T., Uchida, S., Saito, R., Ishizu, K., *Polym. Prep. Jpn., 45*: 2550 (1996).
23. Uchida, S., Saito R., Ishizu, K., *Polym. Prep. Jpn., 45*: 2548 (1996)
24. Saito R., Kotsubo H., and Ishizu K., *Polymer, 35*: 1747 (1994).
25. Ishizu K., Sugita M., Kotsubo H., and Saito R., *J. Colloid & Interface Sci., 169*: 456 (1995).
26. De la Cruz M. O. and Sanchez I. C., *Macromolecules, 19*: 2501 (1986).
27. Thomas E. L., Kinning D. J., Alward D. B., and Henkee C. S., *Macromolecules, 20*: 2934 (1987).
28. Spontak R. J., Williams M. C., and Agard D. A., *Polymer, 29*: 387 (1988).
29. Ma G. H. and Fukutomi T., *Macromolecules, 25*: 1870 (1992).
30. Plestal J. and Baldrian J., *Makromol. Chem., 176*: 1009 (1975).
31. Hashimoto T., Nagatoshi K., Toda A., Hasegawa H., and Kawai H., *Macromolecules, 7*: 364 (1974).
32. Ohta T. and Kawasaki K., *Macromolecules, 19*: 2621 (1986).
33. Hashimoto T., Fujiura M., and Kawai H., *Macromolecules, 13*: 1660 (1980).
34. Roe R. J., Kishkis M., and Chang V. C., *Macromolecules, 14*: 1091 (1981).
35. Richards R. W. and Thomason J. L., *Polymer, 22*: 581 (1981).
36. Saito R., Kotsubo H., and Ishizu K., *Polymer, 35*: 1580 (1994).
37. Siato R., Kotsubo H., and Ishizu K., *Polymer, 35*: 2296 (1994).
38. Molau, G. E. and Wittbrodt, W. M., *Macromolecules, 1*: 260 (1968).
39. Gido, S. P., Schwark, D. W., Thomas, E. L., and Goncalves, M. C., *Macromolecules, 26*: 2636 (1933).
40. Ishizu, K., Yamada, Y., and Fukutomi, T., *Polymer, 31*: 2047 (1990).

40

Flow Behavior of Polymer Blends as Affected by Interchain Crosslinking

Chapal K. Das
Indian Institute of Technology, Kharagpur, India

I. INTRODUCTION

Flow behavior of the polymer blends is determined by their structure, which is governed by the degree of dispersion of the component and by the mode of their distribution. For blends having identical compositions, it is possible to produce systems in which one and the same component may be either a dispersion medium or a dispersed phase [1]. This behavior of the polyblend systems depends on various parameters, the most important of which is the blending sequence. It is, therefore, difficult to obtain a uniform composition property relationship for the polymer blends even though the composition remains identical.

The flow behavior of the polymer blends is quite complex, influenced by the equilibrium thermodynamic, dynamics of phase separation, morphology, and flow geometry [2]. The flow properties of a two phase blend of incompatible polymers are determined by the properties of the component, that is the continuous phase while adding a low-viscosity component to a high-viscosity component melt. As long as the latter forms a continuous phase, the viscosity of the blend remains high. As soon as the phase inversion [2] occurs, the viscosity of the blend falls sharply, even with a relatively low content of low-viscosity component. Therefore, the S-shaped concentration dependence of the viscosity of blend of incompatible polymers is an indication of phase inversion. The temperature dependence of the viscosity of blends is determined by the viscous flow of the dispersion medium, which is affected by the presence of a second component.

The treatment of blends as a two phase system opened up an interesting field of modifying the composite properties by the use of a (third component within the interface boundaries, which is termed as compatibilizers [1]. Such modifications are still being extended to the formation of microgel out of the interaction between the two blend partners having a reactive for functionalities. This type of interchain crosslinking does not require any compatibilizer to enhance the blend properties and also allows the blends to be reprocessed by further addition of a curative to achieve still further improved properties [3,4]. Such interchain crosslinking is believed to reduce the viscoelastic mismatch between the blend partners and, thus, facilitates smooth extrusion [5,6].

The study of composition dependence of the flow behavior and rheological parameters of the polymer blends give more insight into the microstructural changes that occur during the process [7–9]. This study can be successfully correlated with a scanning electron microscopy (SEM) study of the phase morphology. The inflexion points obtained in the composition dependence of rheological parameters suggest phase inversion [10–12], which may be affected by the introduction of interchain crosslinking. As a result, all the properties of the blend change. Here we have chosen the system blends of: (1) acrylonitrile–butadiene rubber (NBR) and chlorosulphonated polyethylene (CSPE), (2) NBR and polyacrylic rubber (ACM), (3) NBR and polysulphide rubber (Thiokol), (4) carboxylated NBR and polyacrylic rubber (ACM), and (5) carboxylated NBR and polyurethane elastomer (AU) for the detailed discussion with special reference to the effect of interchain crosslinking on their flow behavior.

II. EXPERIMENTAL

Two types of blends were studied: (1) virgin polymers, well blended (preblends), and (2) preblended virgin polymers heated in the mold at 150°C for 40 min (preheated blends). Before studying the rheological behavior, both types of blends were kept at ambient for 24 h to attain equilibrium. Blending was performed in a two-roll open mixing mill (keeping the nip gap and friction ratio constant for each set of blends) at ambient temperature. Entire blend ratios were studied including 100% at the extreme ends. Rheological measurements were studied with the help of capillary rheometry at different shear rates and at different temperatures. The non-Newtonian index (n) and consistency index (k) were determined by regression analysis. Extrudate samples were used to determine the swelling ratio (α); viscoelastic parameters such as stored elastic energy (W), shear modulus (G), maximum recoverable deformation (γ_m); and relaxation time (t_R) were calculated following the mathematical model as developed by Das et al. [13]:

$$\gamma_m = \sqrt{(1/2c)(\alpha^4 + 2\alpha^{-2} - 3)}$$

where, $c = (3n + 1)/4(5n + 1)$ and γ_m is the maximum recoverable deformation.

$$W = c\,\gamma_m\,\tau_m$$

where, τ_m is the maximum shear stress.

$$G = 2\,W/(\alpha^4 + 2\alpha^{-2} - 3)$$

$$t_R = [1/\gamma^{(1-n)/n}][(nk^{1/n})/G(1 - n)] \times \{\exp[(1 - n)/n] - 1\}$$

Melt fracture of the extrudate was studied using M-45 wild photoautomat. Blend morphology was studied by SEM after differential solvent swelling.

III. DISCUSSION

A. Blends of NBR–CSPE (Hypalon)

Variation of apparent viscosity with the blend ratio for both preblends and preheated blends is shown in Fig. 1. Comparing preblends and preheated, the viscosity of preheated 50:50 (NBR–Hypalon) blends becomes maximum, whereas the prebends show a continuous decrease in viscosity from 100% Hypalon to 100% NBR in all shear rates studied. This decrease is explained by the difference in viscosity between two virgin polymers. Preheating of the blends may result in interchain crosslinking and it seems to be maximum at a 50:50 ratio.

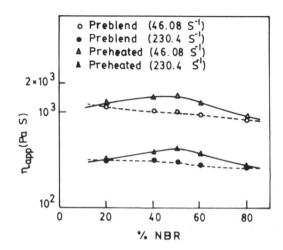

Figure 1 Variation of apparent viscosity with the percent NBR content in NBR–CSPE blend.

The non-Newtonian index "n" is plotted against the blend ratio in Fig. 2. There are three distinct stages of the change of "n" value with %NBR in the preblend. First, a decrease of up to 40% of NBR, a rapid rise of up to 60% of NBR, and beyond this ratio a further decrease are observed. Heating of blends shows the only minimum at 50:50 ratio. It is obvious that 60–50% of Hypalon in the NBR–Hypalon blend is an optimum range where maximum extent of interchain crosslinking reaction is expected, and this blend is supposed to be more pseudoplastic.

Representative plots of extrudate swelling ratio as a function of NBR content are shown in Fig. 3. Shear rate increases the die-swell in all blends. The change of die-swell with NBR content exhibits a decreasing trend up to 60% of NBR, and beyond this level it shows a saturation in die-swell. Preheating of blends exhibits the minimum at 50:50 ratio irrespective of shear rates. We

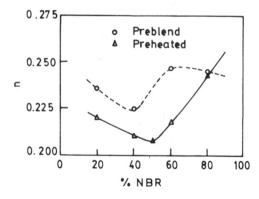

Figure 2 Variation of power law index (n) with the percent NBR content in NBR–CSPE blend.

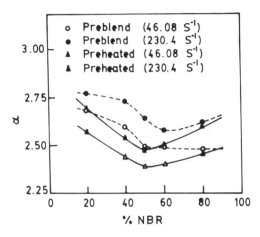

Figure 3 Variation of die-swell with the percent NBR content in NBR–CSPE blend.

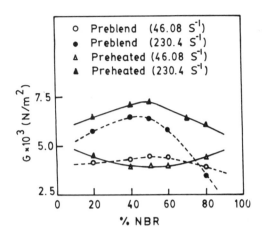

Figure 5 Variation of shear modulus (G) with the percent NBR content in NBR–CSPE blend.

believe that interchain crosslinking plays a major role in decreasing the die-swell in the preheated blend, and that 50:50 ratio is the effective region of such phenomena.

The rheological parameter variations with blend compositions are shown in Figs. 4, 5 and 6. The variation of relaxation time (Fig. 4) seems to show an increasing pattern up to 40% of NBR, then there is a sharp decrease until 60% of NBR, and finally there is a further increase of relaxation time in the preblends. Preheating of blends, however, results in a longer relaxation time than in the preblends and a maximum at 50:50 ratio at all shear rates.

Shear modulus also changes with the blending ratio (Fig. 5), such as the relaxation time. Preblends show an

inflection point around 50–60% of NBR in shear modulus. Preheating of blends provides higher shear modulus, particularly at higher shear rates, than the preblends.

Stored elastic energy (Fig. 6) decreases with NBR content at lower shear rates. But the "W" values are likely to be independent of shear rate at higher levels of Hypalon in the blend, and "W" slightly decreases in NBR-rich blend. Here again we notice that preheating of the blends slightly changes the stored elastic energy as compared with preblends.

From the rheological parameters it appears that there is a phase inversion around 50% of NBR in the blend. Preheating of the blends seems to show the same ratio as an optimum region for interchain crosslinking. The rheological parameters change appreciably with the blend ratio, particularly at higher shear rate regions. This may be due to shear-induced structural changes in

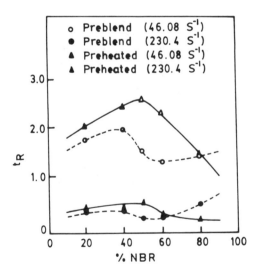

Figure 4 Variation of relaxation time (t_R) with the percent NBR content in NBR–CSPE blend.

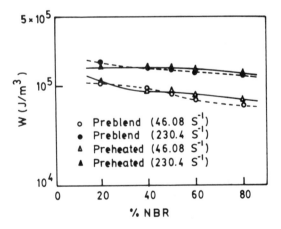

Figure 6 Variation of stored elastic energy (W) with the percent NBR content in NBR–CSPE blend.

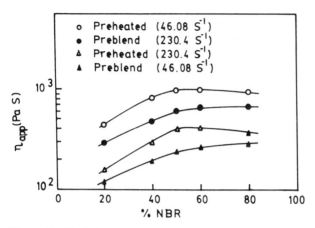

Figure 7 Variation of apparent viscosity with the percent NBR content in NBR–ACM blend.

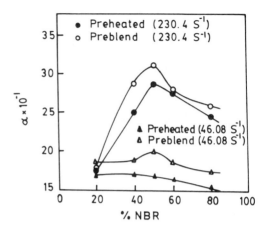

Figure 9 Variation of die-swell with the percent NBR content in NBR–ACM blend.

the blend. It may be assumed that preheating induces microgels in the blend, depending on the blend ratio, which are prone to shear rate effects at its higher level.

Earlier studies [14,15] clearly reveal that there is a reaction between two polymers and that the extent of reaction depends on the blend ratio. As 50:50 ratio has been found to the optimum (from rheological and infrared studies) ratio for interchain crosslinking, the higher heat of reaction for the NBR-rich blend may be attributed to the cyclization of NBR at higher temperatures. There is an inflection point at 50:50 ratio where maximum interchain crosslinking is expected. Higher viscosity, relaxation time, and stored elastic energy are observed in the preheated blends. A maximum 50–60% of Hypalon in NBR is supposed to be an optimum ratio so far as processibility is concerned.

B. Blends of NBR–ACM

Variations of melt viscosity with the blend ratio for both the preblends and preheated blends are shown in Fig.

7. There are two distinct stages in viscosity change with the addition of NBR in the blend. First, there is a rapid rise of viscosity with the initial addition of NBR, then beyond 50% of NBR, viscosity increases slowly, except for the preheated blends where viscosity decreases marginally at the higher level of NBR.

The non-Newtonian index "n" is plotted against the blend ratio in Fig. 8. We see that for both the preblends and the preheated blends, the "n" value increases with NBR content up to about 50% of NBR. Beyond this while the preblend shows a continued increase in "n" value at a slower rate, the preheated blends seem to show a saturation in "n" value. The low values of "n" may be attributed to the interchain crosslinking at higher levels of NBR in the blends.

Representative plots of the extrudate swelling ratios with NBR content are shown in Fig. 9 as a function of

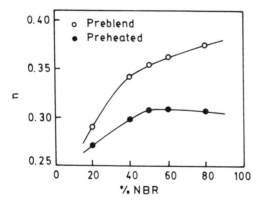

Figure 8 Variation of power law index (n) with the percent NBR content in NBR–ACM blend.

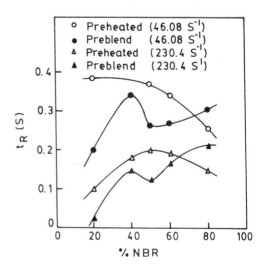

Figure 10 Variation of relaxation time (t_R) with the percent NBR content in NBR–ACM blend.

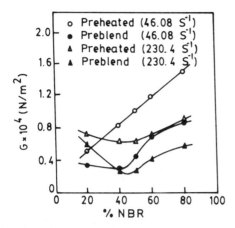

Figure 11 Variation of shear modulus (G) with the percent NBR content in NBR–ACM blend.

shear rate for both types of blends. At a high-shear rate, the die-swell increases as the NBR content increases, attains a maximum at 50% NBR, and then decreases on further increase in NBR. However, at low-shear rates, the swelling ratio tends to decrease marginally with the increase in NBR content in the blend. We see that preheating of blends decreases the swelling in all shear rates. Here again there is an inflection point at 50:50 level for the preblends (without heating) at low-shear rate.

The rheological parameters with blend compositions are shown in Figs. 10–12. The relaxation time increases with the increase in NBR content up to a 50% level for both the preblends and preheated blends. Comparing the preblends and preheated blends, we see that

the relaxation time decreases from 40–50% of NBR and then increases gradually up to 100% of NBR for the preblends at all the shear rates studied; but there is a clear inflection point around 50:50 ratio for the preheated blends. Preheating of the blends increases the relaxation time.

Shear modulus is also changed with the blending type (Fig. 11). Preheated blends provide higher modulus than the preblends. It is obvious that shear modulus either decreases or remains the same at the initial level of NBR for preblends, but beyond 45% of NBR there is a further rise in the "G" value irrespective of shear rates; whereas at the lower shear rate, preheated blends show the continuous rise in the shear modulus for the entire composition range.

Stored elastic energy (Fig. 12) also increases with shear rate both for preblends and preheated blends. Here again, we see that the "W" values increase sharply with NBR, attain a maximum at 50:50 level, and beyond 50% NBR the stored elastic energy decreases.

From the previously mentioned rheological parameters, it appears that there is a phase inversion at around the 50:50 blend ratio of NBR and polyacrylic rubber (ACM). However, preheating does not seem to change the position of this inversion point. These rheological parameters change appreciably with the blend ratio for the preheated blends, particularly at a high-shear rate region. This may be due to shear-induced structural changes in the blend. It may be assumed that preheating induces microgels in the blend [16,17], depending on the blend ratio, that are prone toward shear rate at its higher level. This type of shear sensitivity may be absent in the preblends.

C. Blends of NBR–Thiokol

Variation of melt viscosity with blend ratio for both the preblends and preheated blends is shown in Fig. 13.

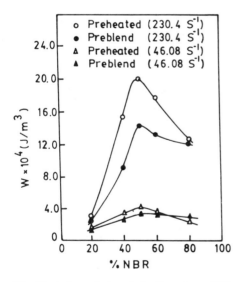

Figure 12 Variation of stored elastic energy (W) with the percent NBR content in NBR–ACM blend.

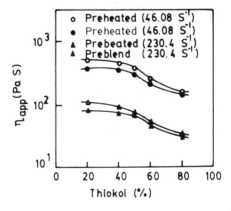

Figure 13 Variation of apparent viscosity with the percent, Thiokol rubber in NBR–Thiokol blends.

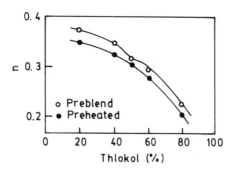

Figure 14 Variation of power law index (n) with the percent Thiokol rubber in NBR–Thiokol blends.

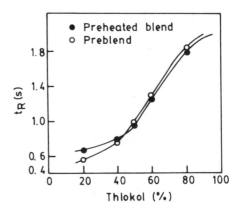

Figure 16 Variation of relaxation time (t_R) with the percent Thiokol rubber in NBR–Thiokol blends.

Comparing preblends and preheated blends, the viscosity of 80:20 (NBR–Thiokol rubber) preheated blends is higher. Addition of Thiokol rubber decreases the viscosity in both preblends and preheated blends. The difference in viscosity between the two plots narrows at high Thiokol rubber–NBR levels in both types of blends (Fig. 13), suggesting the negligible extent of polymer–polymer interaction crosslinking.

The non-Newtonian index "n" is plotted against the blend ratio in Fig. 14 for all blends. As observed from Fig. 14, the addition of Thiokol rubber in NBR lowers the "n" values for both preblends and preheated blends. The preheating decreases the non-Newtonian index throughout the entire region studied. Around a 50:50 blend ratio there appears to be an inflection point in both cases. The decrease in "n" values is higher in the high NBR region of preheated stock, which may be a manifestation of interchain crosslinking, which occurs to a limited extent in the high Thiokol rubber region.

A representative example of the extrudate swelling behavior is shown in Fig. 15 as a function of blend ratio

for both types of blends. Die-swell increases with shear rate irrespective of blend type. The plot shows a slow rise of die-swell up to 45% Thiokol, then a rapid rise up to 60% Thiokol rubber, and finally a slow increase in die-swell until the end. However, preheating of the blends decreases the swelling ratio. Here again there is an inflection point at around the 50:50 ratio where the die-swell increases rapidly with Thiokol rubber. The greater decrease in die-swell for the NBR-rich blend due to heating may be attributed to network formation due to interchain crosslinking [18]. This type of interchain crosslinking is also present at higher Thiokol rubber blends, although to a lesser extent.

Rheological parameters, such as relaxation time, shear modulus, and stored elastic energy, are determined from the extrudate swell and stress–strain data as previously described. Representative examples of the variation of these parameters with blend ratios for both blends are shown in Figs. 16–18. Figure 16 shows that relaxation time for both preblends without heating and

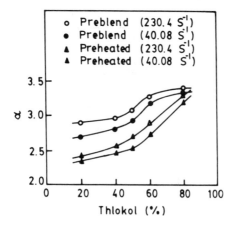

Figure 15 Variation of die-swell with the percent Thiokol rubber in NBR–Thiokol blends.

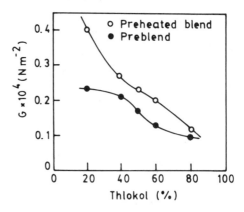

Figure 17 Variation of shear modulus (G) with the percent Thiokol rubber in NBR–Thiokol blends.

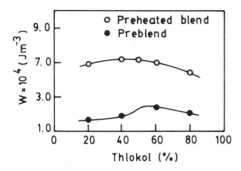

Figure 18 Variation of stored elastic energy (W) with the percent Thiokol rubber in NBR–Thiokol blends.

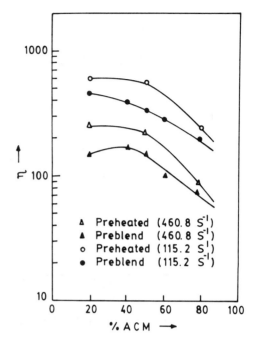

Figure 19 Variation of apparent viscosity with the percent ACM in XNBR–ACM blends.

preheated blends increases with a decrease of the NBR–Thiokol rubber ratio. In both cases, there is a point around 45–50% of Thiokol rubber beyond which the relaxation time shoots up. Preheating increases the "t_R" values, especially for NBR-rich blends; however, at higher levels of Thiokol rubber these two curves tend to merge.

Shear modulus decreases rapidly in the preheated blends (Fig. 17) with an increase in Thiokol rubber. In the higher NBR region the two plots diverge, but they tend to converge at higher Thiokol levels. There is also a remarkable change in the trend observed at around 45–50% of Thiokol rubber. Preheating of the blend is accompanied by an increase in the shear modulus values.

Stored elastic energy also increases up to 45–50% of Thiokol rubber and then decreases gradually until the end in the preheated blends (Fig. 18). A similar phenomenon is also noticed in the preblends without heating where the inflection point shifts toward the slightly higher level of Thiokol rubber.

From the rheological parameters it appears that there is a phase inversion at around 50–55% in the rubber blend. However, preheating does not seem to change the position of this inversion point. These rheological parameters change appreciably with the blend ratio for the preheated blends, particularly in NBR-rich blends. Hence, it may be logical to assume that preheating induces some crosslinks resulting in microgels in the blend, depending on the blend ratio, and this may be predominant at higher NBR–Thiokol rubber ratios.

D. Blends of XNBR–ACM

Variation of melt viscosity for both the preblends and preheated blends with the blend ratio are shown in Fig. 19. There are two distinct regions in viscosity change with the addition of polyacrylic rubber (ACM) in the blends. First, in the higher shear rate region, the viscosity increases with the addition of the ACM (up to 40% ACM) in the blend and then it decreases. In the lower

shear rate region, the viscosity gradually decreases with the addition of ACM in the blend. In the case of preheated blends, the viscosity decreases very slowly up to a 50:50 blend ratio in both the shear rate regions, then it drops faster with further addition of ACM in the blend. As ACM content in the blend is increased, the difference in viscosity between the preblend and preheated blends decreases.

The non-Newtonian index "n" is plotted against the blend ratio in Fig. 20. It is observed that for both the

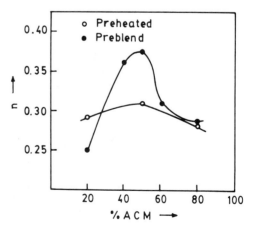

Figure 20 Variation of power law index (n) with the percent ACM in XNBR–ACM blends.

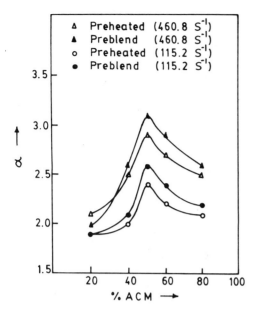

Figure 21 Variation of die-swell with the percent ACM in XNBR–ACM blends.

preblends and preheated blends, the "n" value increases with the addition of the ACM in the blend and then decreases. The rise is sharper in the case of the preblends in the ACM-rich blend. The "n" values are not changed as much as in the case of preheated blends.

The plot of the extrudate swelling ratios with ACM content are shown in Fig. 21 as a function of shear rate for both types of blends. The die-swell increases as the ACM content in the blends is increased and reaches maximum at a 50:50 blend ratio and then decreases with the further addition of ACM in the blend. The increase is more at higher shear rate regions. Preheating of the blends decrease the swelling ratio for both the cases.

The plot of the rheological parameters (relaxation time, shear modulus, and stored elastic energy) are shown in Figs. 22–24. The relaxation time increases as the ACM content is increased to attain a maximum at 60:40 = ACM:XNBR blend ratio for the preblends. For lower shear rate the rise is sharp and after 60:40 blend ratio, "t_R" remains almost constant, whereas for the higher shear rate region the rise is not sharp and after 60:40 blend ratio "t_R" decreases as ACM percent increased in the blend. In the case of the preheated blends the "t_R" increases up to 50:50 blend ratio and then decreases with the addition of ACM in the blend. The preheating increases the "t_R" in both shear rate regions.

The shear modulus, "G" varies with the blend ratios. In the preblend system, the shear modulus decreases sharply with the addition of ACM up to the 50:50 blend ratio and then it remains constant with the further addition of ACM in the blend. In the case of preheated blend, the shear modulus decreases gradually with the addition of ACM. In both the preblend and preheated blends it is observed that the shear modulus at higher shear rate and lower shear rate gradually converge as the ACM content in the blend is increased.

The stored elastic energy, "W" (Fig. 24) increases with the increase of ACM content in the blend up to 50:50 blend ratio and then it decreased with the further

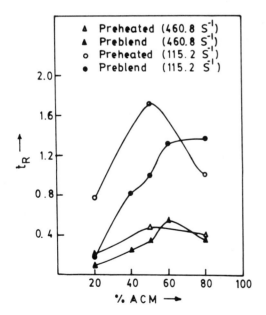

Figure 22 Variation of relaxation time (t_R) with the percent ACM in XNBR–ACM blends.

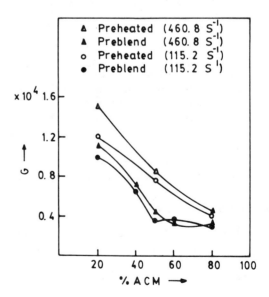

Figure 23 Variation of shear modulus (G) with the percent ACM in XNBR–ACM blends.

addition of ACM in the blend for both the preblend and the preheated blends and at both the shear rates. However, the increase is drastic for the preheated blends and at higher shear rates.

From the above rheological parameter, it appears that there is phase inversion at around 50–60% ACM content in the blends. These rheological parameters change appreciably with the blend ratio for the preblends and preheated blends. Preheating changes the rheological parameters of the blends, which is more prominent at the higher shear rates. This is due to the shear-induced structural changes in the blends. Preheating of the blends leads to microgel formation depending on the blend ratio, which are prone toward shear rate at its higher level [19].

E. Blends of XNBR–AU

Variation of melt viscosity for both the preblends and preheated blends with the blend ratio are shown in Fig. 25. The viscosity of the blends gradually decreases with the addition of the AU in the blends. This decrease is more drastic in the higher shear rate region. In the case of the preheated blends, viscosity does not change much with the addition of AU in the blend up to a 50:50 blend

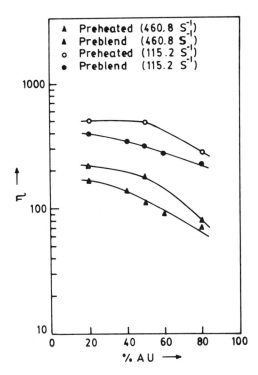

Figure 25 Variation of apparent viscosity with the percent polyurethane rubber in XNBR–AU blends.

ratio, but with further addition of AU in the blends the viscosity falls rapidly. Again the lowering of viscosity is found to be more sharp at higher shear rate.

The non-Newtonian index "n" is plotted against the blend ratio in Fig. 26. It is observed that for the preblends the "n" values gradually decrease with the addition of AU in the blends, and the decrease becomes more

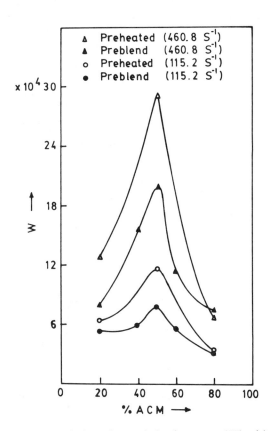

Figure 24 Variation of stored elastic energy (W) with the percent ACM in XNBR–ACM blends.

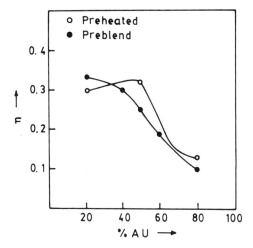

Figure 26 Variation of power law index (n) with the percent polyurethane rubber in XNBR–AU blends.

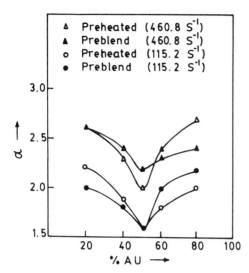

Figure 27 Variation of die-swell with the percent polyurethane rubber in XNBR–AU blends.

drastic after 40:60 AU–XNBR blend ratio. Whereas for the preheated blends the "n" values first increase with the addition of AU in the blends and then fall drastically with a further addition of AU in the blends.

The plot of the extrudate swelling ratios with AU content are shown in Fig. 27 as a function of shear rate for both types of blends. The die-swell gradually decreases with the addition of AU in the blend up to the

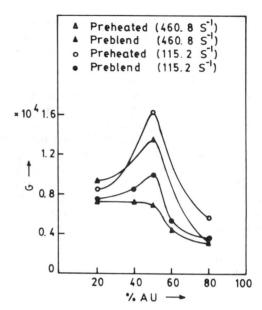

Figure 29 Variation of shear modulus (G) with the percent polyurethane rubber in XNBR–AU blends.

50:50 blend ratio and then it increased with a further addition of AU in the blends. The swelling is more at higher shear rates. It has been observed that at higher shear rate the die-swell of the preheated blends is lower than the preblends up to 50:50 blend ratio and beyond that the trend is reversed. Whereas in a lower shear rate

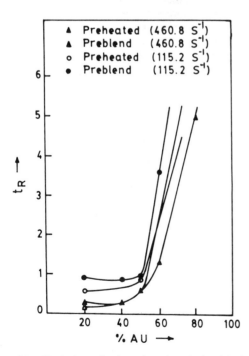

Figure 28 Variation of relaxation time (t_R) with the percent polyurethane rubber in XNBR–AU blends.

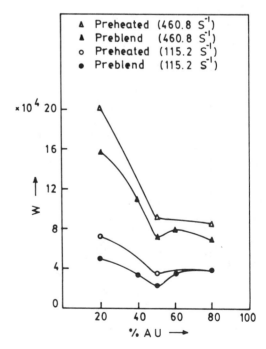

Figure 30 Variation of stored elastic energy (W) with the percent polyurethane rubber in XNBR–AU blends.

region, the die-swell of preheated blend is higher than the preblends up to 50:50 blend ratio, but after that the die-swell of the preheated blend is lower compared to the preblends.

The plot of the rheological parameters (relaxation time, "t_R"; shear modulus, "G"; and stored elastic energy, "W") are given in Figs. 28–30. The relaxation time of both preblends and preheated blends remains almost constant up to 50:50 blend ratio and then shoots up drastically at both shear rates. Up to 50:50 blend ratio it is observed that the relaxation time is more at lower shear rate. Preheating of blends lowers the "t_R" values.

The shear modulus varies with the blend ratios in the preblend system, "G" remains constant up to 50:50 blend ratio and then it decreased with the further addition of AU in the blend at higher shear rate. At lower shear rate, "G" increased up to the 50:50 blend ratio and then it fell. Preheating of the blends resulted in increasing the "G" values. At both shear rates "G" increased up to 50:50 blend ratio and then decreased with further addition of AU in the blends.

The stored elastic energy, "W", decreases drastically with the addition of AU in the blends up to 50:50 blend ratio for both the preblends and preheated blends at both shear rates. After 50:50 blend ratio at both shear rates the "W" does not change for preheated blends. For preblends the "W" increase up to 60:40 AU–XNBR blend ratio for both the shear rates and then remain constant at lower shear rate, whereas at higher shear rates it decreased further with the addition of AU in the blend. At lower shear rates, the "W" gradually converge to the same value as the AU is added to the blend.

IV. CONCLUSIONS

From the previous rheological parameters, it appears that the changes mostly occurred at around 50:50 blend ratio. These rheological parameters change appreciably with blend ratio for both the preblends and preheated blends. Preheating changes the rheological parameters of the blends and is more prominent at higher shear

rates. This is due to the shear-induced structural changes in the blends. Preheating of the blends results in the interchain crosslinking reaction [20], which leads to microgel formation depending on the blend ratio that are prone toward shear rate on its higher level.

REFERENCES

1. G. V. Uinogradov and A. Y. Malkin, *Rheology of Polymers*, Mir. Publs., Moscow (1980).
2. R. S. Lenk, *Polymer Rheology*, Allied Publs. Ltd., London (1978).
3. L. A. Utracki, *Polymer Alloys and Blends*, Hanser Publs., Munich (1989).
4. C. K. Das and M. K. Ghosh, *Polym. Adv. Technol.*, 5: 390 (1994).
5. M. K. Ghosh and C. K. Das, *J. Polym. Eng.*, 13: 4 (1994).
6. C. K. Das and A. R. Tripathy, *Proc. Polyblend–95* October, Boucherville, Canada (1995).
7. P. Mukhopadhyay and C. K. Das, *J. Appl. Polym. Sci.*, 40: 1833 (1990).
8. P. Mukhopadhyay and C. K. Das, *J. Appl. Polym. Sci.*, 39: 49 (1990).
9. P. Mukhopadhyay and C. K. Das, *J. Polym. Plast. Technol. Engg.*, 28(56): 537 (1989).
10. M. K. Ghosh and C. K. Das, *J. Reinforced Plast. Comp.*, 11/12: 1376 (1992).
11. M. K. Ghosh and C. K. Das, *J. Thermoplastics Comp. Mate.*, 6: 275 (1993).
12. M. K. Ghosh and C. K. Das, *J. Appl. Polym. Sci.* 51: 646 (1994).
13. C. K. Das, D. Sinha, and S. Bandyopadhyay, *Rheologica Acta*, 25: 507 (1986).
14. A. R. Tripathy and C. K. Das, *Kauts. Chuk. Gummi Kunsts* 45: 626 (1992).
15. A. R. Tripathy and C. K. Das, *Polym. Plast. Tech. Eng.*, 33(2): 195 (1994).
16. A. R. Tripathy and C. K. Das, *J. Polym. Eng.*, 13(1): 49 (1994).
17. A. R. Tripathy and C. K. Das, *J. Appl. Polym. Soc.*, 51: 245 (1994).
18. A. R. Tripathy and C. K. Das, *Plast. Rubb. Comp. Proc. Appln.*, 21(5): 8283 (1994).
19. S. K. Sinha Roy and C. K. Das, *J. Elast. Plast.*, 27: 239 (1995).
20. S. K. Sinha Roy and C. K. Das, *Polym. Polym. Compsts.*, 3(5): 1 (1995).

41
Tough Composites Based on Premixed PP/LCP Blends

Markku T. Heino* and Tommi P. Vainio†
Helsinki University of Technology, Espoo, Finland

I. INTRODUCTION

Thermotropic main-chain liquid crystalline polymers (LCPs) consist of rigid rodlike molecules that are capable of aligning to a very high degree during melt flow, and thus form a highly ordered melt phase in a certain temperature range above the melting temperature. Owing to the relatively long relaxation time, the orientation is retained when the polymer melt is cooled, resulting in a highly ordered fibrous structure and anisotropic properties in the solid state. The good mechanical strength and dimensional and thermal stability of thermotropic LCPs are based on this fibrous structure [1,2]. Because of the highly oriented melt phase, thermotropic LCPs exhibit lower viscosities and are typically more shear thinning than isotropic polymers [3]. The degree of orientation of the LCP fibers can be modified through the choice of processing conditions, especially by applying high shear or elongational forces to the molten polymer. The close relationship between the orientation during processing and the flow-induced morphology makes the processing of LCPs important [4–8].

Materials with totally new property combinations may be achieved by blending two or more polymers together. Through blending of thermotropic main-chain LCPs with engineering thermoplastics, the highly ordered fibrous structure and good properties of LCPs can be transferred to the more flexible matrix polymer. LCPs are blended with thermoplastics mainly in order to reinforce the matrix polymer or to improve its dimensional stability, but LCP addition may modify several

other properties of thermoplastics as well. Owing to the relatively low-melt viscosity of thermotropic LCPs, often a small amount of LCP decreases the blend viscosity significantly and renders the matrix thermoplastics easier to process [9–21].

The blends of thermotropic LCPs and thermoplastics are generally two-phase systems where the dispersed LCP phase exists as small spheres or fibers within the thermoplastic matrix. Often a skin/core morphology is created with well-fibrillated and oriented LCP phases in the skin region and less-oriented or spherical LCP domains in the core.

The primary factors determining the size, shape, and distribution of the LCP domains in the matrix are the LCP content, processing conditions, and rheological characteristics of the blend components, in particular the viscosity ratio [22,23]. The morphology is also affected by the interfacial adhesion between the components, which may be modified by the addition of suitable compatibilizers [24,25]. The effects of shear stress, viscosity ratio, and interfacial tension are related to each other by the Weber number [26–29]. Besides the above, additional drawing greatly enhances the fibrillation and orientation of the LCP phases, resulting in improved strength and stiffness in the fiber direction [22,30–37].

The effect of viscosity ratio on the morphology of immiscible polymer blends has been studied by several researchers. Studies with blends of LCPs and thermoplastics have shown indications that for good fibrillation to be achieved the viscosity of the dispersed LCP phase should be lower than that of the matrix [22,38–44].

In the work reported here, we produced highly fibrillar polypropylene (PP)–LCP blends that were subsequently processed by injection molding without melting the LCP fibers again in order to create tough PP–LCP

Current affiliations:
* Nokia Cables Ltd., Espoo, Finland.
† Valmet Corporation, Jyväskylä, Finland.

composites. Hence, a thermotropic main-chain LCP was first blended with PP in a twin-screw extruder with the take-up speed varied to achieve blends with different LCP fiber dimensions. In a second stage, these blends were processed both above and below the melting temperature of the LCP by extrusion and injection molding. The object of the study was to ascertain the feasibility of processing PP–LCP blends without melting the LCP and to compare the morphology and properties of these and conventional melt processed blends.

II. BACKGROUND

In an earlier study (44) on the effect of viscosity ratio on the morphology of PP–LCP blends we found that the viscosity ratio is a critical factor in determining the blend morphology. The most fibrillar structure was achieved when the viscosity ratio ($\eta_{LCP} : \eta_{PP}$) ranged from about 0.5–1. At even lower viscosity ratios the fiber structure was coarser, while at viscosity ratios above unity, the LCP domains tended to be spherical or clusterlike (Fig. 1).

In addition, it was found that the blends with highly fibrillar structure exhibited a significantly lowered viscosity. Increased shear rate caused slight changes in the blend morphology but did not enhance the fiber formation. Thus, in addition to shear, elongational forces are needed to achieve a well-fibrillated blend structure and significant mechanical reinforcement.

Since the processing conditions and mixing equipment have a crucial effect on the morphology of immiscible polymer blends [45], experiments were carried out in four different types of extruders to find optimal conditions for blend preparation and fibrillation. Nevertheless, the morphologies of PP–LCP blends produced by

different mixing equipment differed only slightly. The greatest variation in the mixing efficiency was found for blends whose components had totally dissimilar melt viscosities. The slight differences in morphology due to melt blending in dissimilar equipment were decreased after injection molding, whereas the differences in morphology due to dissimilar viscosity ratios were still evident in the injection molded blends.

In manufacturing and processing polymer blends, it is thus important that the viscosity ratio be within the optimal range in the actual processing conditions. Not only the polymers to be blended but also the temperature and processing conditions (shear, elongation) should be carefully selected. Other factors, such as interfacial tension [46,47] and elasticity of the blended polymers, may also influence the blend morphology.

III. EXPERIMENTAL

A. Materials

The matrix polymer was PP (VB1950K, Neste) and the dispersed LCP was Vectra® A 950 (Hoechst Celanese). Vectra is a totally aromatic polyester-type thermotropic main-chain LCP, consisting of p-hydroxybenzoic acid (HBA) and 2,6-hydroxynaphthoic acid (HNA). The following properties are given by the manufacturer: density 1.40 g/cm³, melting point 280°C, tensile strength 165 MPa, elastic modulus 9700 MPa, and elongation at break 3.0% [48]. The PP grade was selected on the basis of earlier studies [44] so that the viscosity ratio ($\eta_{LCP} : \eta_{PP}$) was in the optimum range (between 0.5–1) for good fibrillation of the LCP phase. In all experiments the LCP content was 20 wt%. For preliminary experiments another, less viscous PP grade (VB6511B, Neste) was used as well.

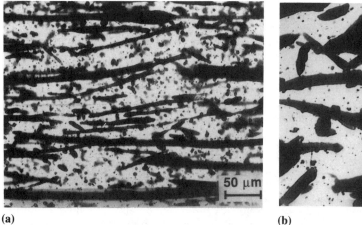
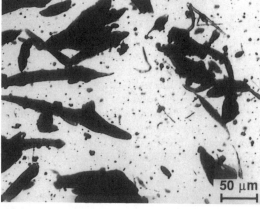

(a)   (b)

Figure 1 Optical micrographs in the flow direction of the extruded strands of the PP–LCP blends exhibiting viscosity ratios of (a) $\eta_{LCP} : \eta_{PP} = 0.6$, and (b) 2.8 [44].

B. Blending

Prior to blending, the LCP was dried at 155°C for 5 h. The melt blending of the materials was carried out with a Berstorff ZE 25 × 33D corotating twin-screw extruder at a melt temperature of 290°C, with a screw speed of 200 rpm, and an output of 6.4 kg/h. The extrudate was immediately quenched in a water bath and repelletized.

The blends were prepared with two take-up speeds to achieve morphologies with different LCP fiber dimensions. The lower speed (L), leading to a draw ratio of 1:1, represents the normal blending procedure, which was what we used in our previous work [44]. The higher speed (H) resulting in a draw ratio of 6:1 was applied to form long highly oriented LCP fibers. The draw ratio for each strand was determined as the ratio between the die and strand cross-sections (S_0/S_s).

C. Processing

1. Preliminary Tests

In preliminary tests, melt mixed blends of PP and LCP were processed at six different temperatures (T_{cyl}: 230, 240, 250, 260, 270, and 280°C) with a Brabender Plasti-Corder PLE 651 laboratory single-screw extruder. The measured melt temperatures were about 10°C higher than the cylinder temperatures (T_{cyl}). The objective was to study the influence of temperature on the size and shape of the dispersed LCP phase. Two different polypropylenes were used to ascertain the effect of the viscosity of the matrix on the final morphology. Different draw ratios were obtained by varying the speed of the take-up machine.

2. Extrusion

The blends prepared by twin-screw extruder with two different draw ratios were extruded with the Brabender single-screw extruder at temperatures ranging from 180–280°C. The sample designation and specific processing conditions are given in Table 1.

Table 1 Designation and Temperatures of Subsequent Processing Steps for the Two Melt Mixed Blends

	Low-draw ratio (1:1) in melt blending (L)		High-draw ratio (6:1) in melt blending (H)	
Extrusion				
Code			He1 He2 He3 He4 He5	
T (°C)[a]			180 200 230 250 280	
Injection molding:				
Code	Li1	Li2	Hi1	Hi2
T (°C)[a]	180	280	180	280

[a] Measured melt temperatures were about 10°C higher.

3. Injection Molding

The same two blends as in the extrusion tests were injection molded at two different melt temperatures with an Engel ES 200/40 injection molding machine (Table 1). Prior to the injection molding the blends were dried overnight at 80°C.

D. Characterization

The morphology of the blends and composites was investigated with an optical microscope (Olympus BH-2) equipped with a hot stage. The samples were cut in the flow direction. The fiber formation and dispersion of LCP were studied by melting the PP matrix at 180°C between thin glass plates. At this temperature, PP was transparent, while the LCP domains were solid and unchanged and could easily be inspected with transmitted light. This technique appeared to be advantageous in describing the changes in the size and shape of the dispersed LCP phase.

Tensile and flexural properties were studied with an Instron 4204 testing machine. Tensile tests were performed on the drawn strands at a test speed of 3 mm/min, while three-point-bending tests (ISO 178) at a speed of 5 mm/min were applied to the injection molded specimens. Charpy impact strength was measured of the unnotched samples with a Zwick 5102 pendulum-type testing machine using a span of 70 mm. The specimens (4 × 10 × 112 mm) used for three-point-bending tests were also used for the impact tests. It should be noted that neither the tensile tests for the strands nor the impact tests were standard tests. The samples were conditioned for 88 h at 23°C (50% r.h.) before testing.

Viscosities of the blends and composites were measured in shear flow with a Göttfert Rheograph 2002 capillary viscosimeter. The shear rate was investigated from 100–10000 s^{-1}. The L:D ratio of the capillary die was 30 mm:1 mm. Rabinowitch correction was made to the measurements, but Bagley correction was not applied.

IV. RESULTS AND DISCUSSION

A. Morphology

Preliminary tests were made on melt mixed blends of PP and LCP to study the effect of processing temperature on the shape and size of the dispersed LCP phase. Extrusion experiments were made below as well as above the melting temperature of the LCP. Two different polypropylenes were tested to determine the effect of the viscosity of the matrix on the final morphology.

According to these experiments there were three processing temperature regions with three different morphologies. At high temperatures both the PP and the LCP were molten, and in situ fibrillation of the LCP component took place during cooling of the oriented

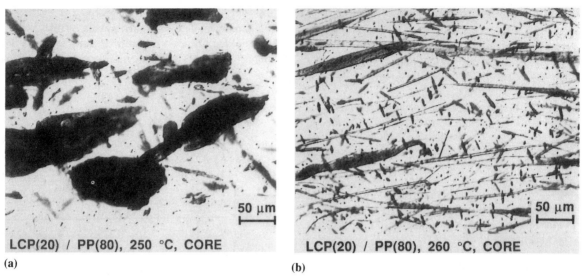

Figure 2 Optical micrographs of melt mixed PP–LCP blends single-screw extruded at melt temperatures of (a) 250°C, and (b) 260°C.

melt phase (Fig. 2b). Owing to viscous heat generation in the extruder, the LCP component also remained molten at cylinder temperatures slightly below the melting temperature of the LCP. At moderate temperatures (240–250°C) the LCP component was in solid form, but it could be deformed and agglomerated due to softening. This led to an undesirable, clusterlike morphology without reinforcing fibers (Fig. 2a). At still lower temperatures the pregenerated LCP fibers remained unchanged in fibrillar form, resulting in a composite morphology. The differences in mechanical properties were related to the different morphologies, the poorest properties being associated with the clusterlike structure.

Effects were similar with the two polypropylenes as matrix. Nevertheless, as in our earlier studies [44], the in situ fibrillation was more pronounced for the more viscous PP (VB1950K) and this PP grade was selected for the additional experiments.

The morphologies of the PP–LCP melt blends prepared by a twin-screw extruder at 290°C with low (L) and high (H) drawing speed are presented in Fig. 3 and the corresponding mechanical properties in Table 2. At the lower draw ratio a clear skin/core morphology was generated, whereas the higher take-up speed resulted in an extremely highly fibrillated morphology and there were virtually no differences between skin and core. Especially with the higher draw ratio, it was expected that the fibers in the strand might be several millimeters long. In actual pellets, however, the length did not naturally exceed that of the pellet (about 2 mm).

The two blends depicted in Fig. 3 were used in all further processing steps. The aim of the further studies was to investigate the morphologies after extrusion and

injection molding at various temperatures both above and below the melting temperature of the LCP. Special attention was paid to the LCP fiber dimensions to determine whether the differences in the fibrous structure still are significant after these processing steps.

Table 2 Tensile Properties of Extruded Blends and Composites

Material	E (MPa)	σ (MPa)	ϵ (%)	d (mm)	D.R.[a]
Reference					
Polypropylene	1050	24	>100	3.1	1.1
Twin-screw blending (Fig. 3)					
Melt blend (L)	1550	34	4.1	3.1	1.1
Melt blend (H)	3630	54	2.0	1.3	6.1
Subsequent single-screw extrusion (Fig. 4)					
He1	1310	25	6.0	4.2	0.6
He2	1570	26	4.4	3.6	0.8
He3	1200	20	8.0	3.8	0.7
He4	1160	19	>10	3.7	0.7
He5 (less drawn)	1810	34	6.4	3.3	0.9
He5 (highly drawn)	2530	33	1.7	1.9	2.8

[a] D.R. = draw ratio = S_0/S_s
E = elastic modulus; σ = maximal strength; ϵ = elongation; d = diameter of the strand.

1. Extrusion

The morphology of the highly drawn blend after extrusion at different temperatures is depicted in Fig. 4. The effect of processing temperature on the morphology was similar to that found in the preliminary tests. At the highest temperature (280°C Fig. 4e) both blend components were molten during processing and a fibrous morphology with fine well-oriented fibrils was formed in situ during the rapid cooling of the ordered melt. Elongational drawing of the solidifying extrudate increases the orientation of the LCP phases. Additional drawing at temperatures between the glass transition (T_g) and melting temperature (T_m) of the LCP phase, i.e., cold drawing, may further improve the fibrillation and orientation. At moderate temperatures (230–250°C; Figs. 4c,d) the LCP fibers were softened and agglomerated to clusters. At low temperatures (180–200°C; Figs. 4a,b) the fiber structure created during the previous melt blending step was fairly well maintained, as expected, but fiber orientation was poorer than in the blend processed at 280°C. Additional drawing could not be applied to the composites generated at the lower temperature owing to the poor melt strength of the material. Thus, the enhanced fiber orientation and reinforcing effect achieved with the melt blends could not be achieved with the composites.

2. Injection Molding

The morphology of the injection molded blends is shown in the optical micrographs of Fig. 5.

After injection molding at 180°C, just as before, the

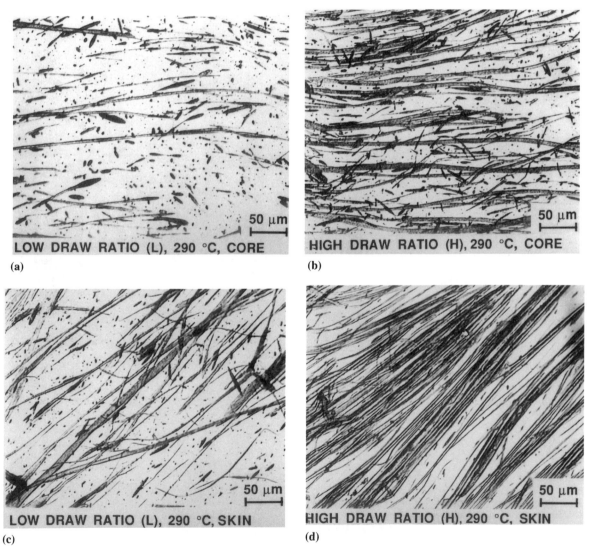

Figure 3 Twin-screw extruded PP–LCP blend processed at a melt temperature of 290°C with low- (left) and high-draw ratio (right). Upper micrographs are taken from the core and lower ones from the skin region.

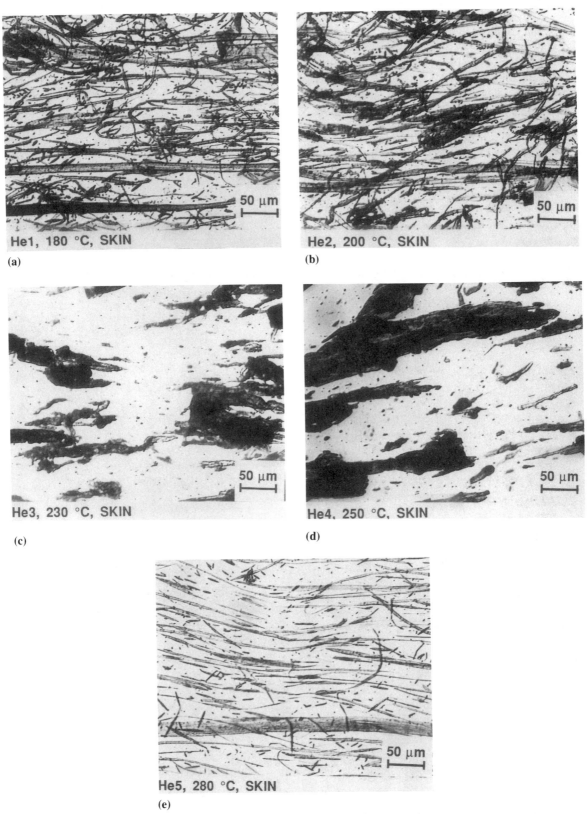

Figure 4 Optical micrographs from the skin region of the single-screw extruded strands processed at cylinder temperatures of (a) 180°C, (b) 200°C, (c) 230°C, (d) 250°C, and (e) 280°C.

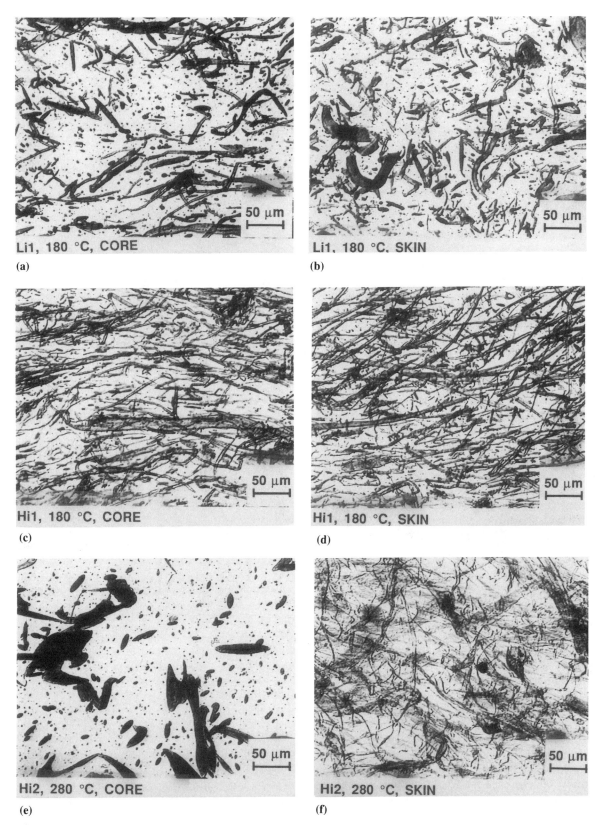

Figure 5 Optical micrographs of specimens injection molded at 180°C (a–d) and 280°C (e,f). Samples were taken from core (left) and skin region (right). (Sample codes as in Table 1.)

blend prepared with the higher draw ratio exhibited clearly thinner, longer, and more oriented LCP fibers than the less drawn blend (Fig. 5a,c). Thus, the difference in fiber length was maintained when the material was processed at a temperature clearly below the T_m of the LCP. From the highly drawn blend a composite structure with long highly oriented fibers was achieved.

At the higher temperature (280°C) a distinct skin/core morphology was formed with thin fibers in the skin region and dispersed LCP particles in the core (Figs. 5e,f). At this temperature both the blend components were molten during injection molding and the previously created fiber structure was destroyed. However, in the melt stage the LCP phases were still at least partly elongated owing to the long relaxation time of the LCP. The fiber morphology was recreated during cooling in the mold, taking place as in situ fibrillation in the skin region due to the high-shear forces near the mold wall. Thus, the highly ordered morphology created during the injection stage could be locked by rapid cooling in the mold (low mold temperature). It should be noted that very high mold temperatures could cause relaxation of the oriented LCP phases.

The composites injection molded at the lower temperature (180°C) did not exhibit any skin/core effect, but rather contained fibers throughout.

B. Mechanical Properties

1. Extruded Strands

The tensile properties of the extruded blends and composites are presented in Table 2. Compared to the neat PP, a clear reinforcement was achieved after twin-screw blending. The reinforcing effect was even more pronounced with the higher take-up speed (H), evidently due to the extremely fibrillar morphology, as seen in Fig. 3.

The changes in the morphology of the highly drawn blends after single-screw extrusion (Fig. 4) were reflected in the mechanical properties, which also varied with the processing temperature regions. A good reinforcing effect was achieved with the composites processed well below the T_m of the LCP (He1 and He2, in Table 2) and with the melt blend (He5). However, the composites processed at moderate temperatures (He3 and He4) showed poor mechanical properties due to the clusterlike morphology.

The melt blends exhibited the best mechanical properties, which could be still further improved with additional drawing. The composites He1–He4 could not be drawn to improve the mechanical properties. In the case of the melt blends, even higher draw ratios than used in this study will increase the fibrillation and orientation of the LCP phase leading to significant improvements in strength and modulus [21,30].

2. Injection Molded Specimens

The mechanical properties of the injection molded blends and composites are shown in Table 3.

The differences in fiber length and orientation in the composites prepared at 180°C (Fig. 5a–d) are reflected in the strength and modulus. The composite based on the highly drawn blend showed better strength and modulus than the composite based on the less drawn one. After injection molding, both melt blends (Fig. 5e,f) exhibited a higher level of modulus than the composites, but the effect of the preceding drawing diminished. Nevertheless, the elastic modulus of the highly drawn melt blend was slightly better owing to the more elongated LCP phases. By contrast, the value of the impact strength was almost twice as large for the composites as for the melt blends. This significant difference is explained by the totally fibrillar structure of the composites. Although the thin fibrils in the skin region of the melt blends were able to carry most of the load applied in the flexural test, they could not resist the impact load to the same extent.

In both the blends and composites, the addition of LCP reinforced the PP matrix considerably. On the basis of the fibrillar morphology throughout the specimens, even better mechanical properties were expected for the composites than for the blends. The poorer than expected reinforcement was primarily due to the lack of adhesion between fiber and matrix.

C. Rheology

Shear viscosities of the twin-screw blended materials were measured at 190°C and 290°C (Fig. 6), the same temperatures as the melt temperatures during processing: 190°C for the composites and 290°C for the melt blends.

The rheological behavior of the blends and composites was totally different. Addition of LCP reduced the

Table 3 Flexural and Impact Properties of Injection Molded Specimens

Material	E (MPa)	σ (MPa)	ϵ_Y (mm)	Charpy impact strength (kJ/m²)[a]
PP (180°C)	1390 (50)	39 (1)	12 (1)	not broken
Li1 (180°C)	1580 (40)	40 (1)	12 (1)	41 (5)
Hi1 (180°C)	1770 (40)	45 (1)	12 (1)	39 (5)
Li2 (280°C)	1900 (150)	41 (1)	11 (1)	22 (3)
Hi2 (280°C)	1970 (110)	42 (1)	10 (1)	22 (1)

[a] Dimensions of the unnotched test bar = $4 \times 10 \times 112$ mm; span = 70 mm. Standard deviations in parentheses.

E = elastic modulus; σ = maximal strength; ϵ_Y = elongation at yield.

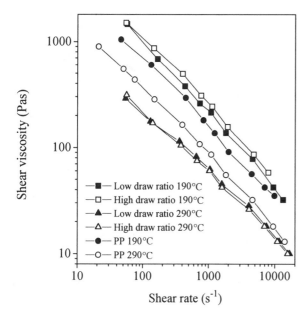

Figure 6 Shear viscosities of the premixed PP–LCP blends measured at 190°C and 290°C.

viscosity of PP in the melt blends, as reported earlier [24,44], but increased it in the composites. Nevertheless, the increase in viscosity caused by the solid LCP fibers was surprisingly small and the material was still easily processable at the temperature normally used for PP. A probable explanation is the small dimensions of the LCP fibers relative to conventional glass fibers for example. The differences in fiber length induced by drawing did not affect the viscosity markedly.

V. CONCLUSIONS

Blends of polypropylene (PP) and liquid crystalline polymer (LCP) processed without melting the LCP were compared with conventional melt processed blends. In a first stage, PP was blended with 20 wt% of LCP in a twin-screw extruder with the take-up speed varied to achieve blends with different LCP fiber dimensions. In the second stage, these blends were processed both below and above the T_m of the LCP by extrusion and injection molding.

At lower temperatures (180–200°C) the material was processed without melting the LCP and a real composite structure with solid LCP fibers in the PP matrix was formed. When processing was done above the T_m of the LCP (280°C), all the material was molten during processing and a compositelike blend morphology was created in situ during cooling of the oriented melt phase.

The composite and blend morphologies were nevertheless different in the solid state. In the composite there

were LCP fibers throughout the sample, whereas the blends showed a clear skin/core morphology with long oriented LCP fibers in the skin region and dispersed LCP domains in the core. The differences in fiber length induced by different drawing speeds in the blending stage were maintained in the composites, but vanished during the melt processing at higher temperatures. At processing temperatures 230–250°C the LCP fibers did not melt but were softened and agglomerated and an undesirable morphology with LCP clusters was obtained.

The twin-screw mixed blend with the higher draw ratio showed significant reinforcement compared to the neat PP and to the less drawn blend, and this was promising for the additional experiments. The melt blends exhibited the highest level of modulus, but the effect of the preceding drawing seen in composites disappeared after injection molding. In comparison to the melt blends the composites exhibited significantly higher impact strength.

The addition of LCP sharply decreased the viscosity of PP in the melt blends, but increased it in the composites. The increase in viscosity effected by the solid LCP fibers was nevertheless surprisingly small.

This work led us to conclude that preblended PP–LCP blends can be processed without melting the LCP fibers. The resulting composite showed interesting morphology. In particular, the impact strength, which often is poor for conventional uncompatibilized LCP melt blends, was significantly higher for the composites. Other mechanical properties of the composites were not on a satisfactory level perhaps due to the lack of adhesion between fiber and matrix. By the addition of appropriate coupling agents the adhesion could be improved. On the other hand, the length of the pregenetrated fibers might be increased simply by increasing the length of the pellets.

The main benefits achieved by melt processing PP–LCP blends are decreased viscosity and increased strength and modulus, which can be further improved by elongational drawing. Owing to the in situ fibrillation of melt blends, the morphology ultimately depends on the final processing conditions. The processing method leading to composites is advantageous in achieving good impact strength for moldings. Nevertheless, as typical for all composites, the poor melt strength of the PP–LCP composites does not allow high-speed drawing in extrusion. Although the added LCP does not act as a processing aid as it does in the melt blends, it does not dramatically increase the viscosity either.

REFERENCES

1. W. J. Jackson, Jr. and H. F. Kuhfuss, *J. Polym. Sci., Polym. Chem. Ed., 14*: 2043 (1976).
2. T.-S. Chung, *Polym. Eng. Sci., 26*: 901 (1986).
3. K. F. Wissbrun, *J. Rheol., 25*: 619 (1981).

4. T. Schacht, Doctor's thesis, Rheinisch-Westfälischen Technischen Hochschule, Aachen, West Germany (1986).
5. Y. Ide and Z. Ophir, *Polym. Eng. Sci., 23*: 261 (1983).
6. S. Kenig, *Polym. Eng. Sci., 29*: 1136 (1989).
7. Z. Ophir and Y. Ide, *Polym. Eng. Sci., 23*: 792 (1983).
8. L. C. Sawyer and M. Jaffe, *J. Mat. Sci., 21*: 1897 (1986).
9. F. N. Cogswell, B. P. Griffin and J. B. Rose, *U.S. Patent* 4,386,174, to Imperial Chemical Industries (ICI) PLC, UK May 31, 1983.
10. E. G. Joseph, G. L. Wilkes and D. G. Baird, *Polymeric Liquid Crystals,* (A. Blumstein, ed.), Plenum Press, New York & London, p. 197 (1985).
11. A. Siegmann, A. Dagan and S. Kenig, *Polymer, 26*: 1325 (1985).
12. G. Kiss, *Polym. Eng. Sci., 27*: 410 (1987).
13. Isayev, A. I. and Modic, M., *Polym. Compos., 8*: 158 (1987).
14. Weiss, R. A., Huh, W., and Nicolais, L., *Polym. Eng. Sci., 27*: 684 (1987).
15. Apicella, A., Iannelli, P., Nicodemo, L., Nicolais, L., Roviello, A., and Sirigu, A., *Polym. Eng. Sci., 26*: 600 (1986).
16. Brostow, W., Dziemianowicz, T. S., Romanski, J., and Werber, W., *Polym. Eng. Sci., 28*: 785 (1988).
17. Carfagna, C., Amendola, E., and Nobile M. R., *International Encyclopedia of Composites,* Vol. 2 (S. M. Lee, ed.), VCH Publishers, New York, p. 350 (1990).
18. Dutta, D., Fruitwala, H., Kohli, A., and Weiss, R. A., *Polym. Eng. Sci., 30*: 1005 (1990).
19. Seppälä, J. V., Heino, M. T., and Kapanen, C., *J. Appl. Polym. Sci., 44*: 1051 (1992).
20. Heino, M. T. and Seppälä, J. V., *Int. J. Mater. Product Technol.* 7: 56 (1992).
21. Heino, M. T. and Seppälä, J. V., *Polym. Bull., 30*: 353 (1993).
22. Blizard, K. G. and Baird, D. G., *Polym. News, 12*: 44 (1986).
23. Blizard K. G. and Baird, D. G., *Polym. Eng. Sci., 27*: 653 (1987).
24. Heino, M. T. and Seppälä, J. V., *J. Appl. Polym. Sci., 48*: 1677 (1993).
25. Datta, A., Chen, H. H., and Baird, D. G., *Polymer, 34*: 759 (1993).
26. Taylor, G. I., *Proc. Roy. Soc., A138*: 41 (1932).
27. Taylor, G. I., *Proc. Roy. Soc., A146*: 501 (1934).
28. Wu, S., *Polym. Eng. Sci., 27*: 335 (1987).
29. Plochocki, A. P., Dagli, S. S., and Andrews, R. D., *Polym. Eng. Sci., 30*: 741 (1990).
30. Heino, M. T. and Seppälä, J. V., *J. Appl. Polym. Sci., 44*: 2185 (1992).
31. Amano, M. and Nakagawa, K., *Polymer, 28*: 263 (1987).
32. Ramanathan, R., Blizard K., and Baird, D., *SPE ANTEC, 34*: 1123 (1988).
33. Nobile, M. R., Amendola, E., Nicolais, L., Acierno, D., and Carfagna, C., *Polym. Eng. Sci., 29*: 244 (1989).
34. Malik, T. M., Carreau, P. J., and Chapleau, N., *Polym. Eng. Sci., 29*: 600 (1989).
35. Blizard, K. G., Federici, C., Federico, O., and Chapoy, L. L., *Polym. Eng. Sci., 30*: 1442 (1990).
36. Carfagna, C., Amendola, E., Nicolais, L., Acierno, D., Francescangeli, O., Yang, B., and Rustichelli, F., *J. Appl. Polym. Sci., 43*: 839 (1991).
37. Crevecoeur, G. and Groeninckx, G., *Polym. Eng. Sci., 33*: 937 (1993).
38. Done, D., Sukhadia, A., Datta, A., and Baird, D. G., *SPE ANTEC, 36*: 1857 (1990).
39. Beery, D., Kenig, S., and Siegmann, A., *Polym. Eng. Sci., 31*: 451 (1991).
40. Brinkmann, T., Höck, P., and Michaeli, W., *SPE ANTEC, 37*: 988 (1991).
41. Carfagna, C., Amendola, E., and Nicolais, L., *Int. J. Mater. Product Technol., 7*: 205 (1992).
42. Incarnato, L., Nobile, M., and Acierno, D., *Makromol. Chem., Macromol. Symp., 68*: 277 (1993).
43. Kato, S., Naito, T., and Nakamura, K., *Japanese Patent 05305626 A2,* to Toppan Printing Co., Ltd., Japan November 19 (1993).
44. Heino, M., Hietaoja, P., Vainio, T., and Seppälä, J., *J. Appl. Polym. Sci., 51*: 259 (1994).
45. Vainio, T. P. and Seppälä, J. V., *Polym. Polym. Compos., 1*: 427 (1993).
46. Kirjava, J., Rundqvist, T., Holsti-Miettinen, R., Heino, M., and Vainio, T., *J. Appl. Polym. Sci., 55*: 1069 (1995).
47. Holsti-Miettinen, R., Heino, M., and Seppälä, J., *J. Appl. Polym. Sci., 57*: 573 (1995).
48. ®Vectra Flüssigkristalline Polymere (LCP), brochure from Hoechst Celanese, September 1989/1. Auflage.

42

Compatibilization of Thermoplastic Elastomer Blends

R. Asaletha*
Rubber Research Institute of India, Kerala, India

Zacharia Oommen and Sabu Thomas
Mahatma Gandhi University, Kerala, India

I. INTRODUCTION

During the last several years researchers all over the world have been trying to produce new polymeric materials with specific properties for specific applications. Since the field of new materials is highly exposed, scientists have turned their interest to modified forms such as polymer blends, polymer composites, interpenetrating networks, etc. As a new and important challenge for researchers, polymer blends have gained much interest and, of course, have become a new branch of macromolecular science. The blending technique is quite attractive due to the fact that already existing polymers can be used and, thus, the costly development of new polymers via copolymerization or polymerization of new monomers is avoided. Since there is no generally accepted definition of polymer blends, they are generally considered as a combination of two or more polymers. Polymer blends can be obtained by mixing the two polymers in the molten state, casting from common solvent, or latex blending, etc., and these methods do not involve the formation of chemical bonds. Blending is the simplest and cheapest process of combining the properties of different polymeric materials.

We can classify blends into three categories: miscible, partially miscible, and immiscible. Miscibility can be defined in thermodynamic terms. For a binary blend to be miscible the following two conditions should be satisfied:

$$\Delta G_m < 0 \tag{1}$$

$$\frac{\delta^2(\Delta G_m)}{\delta^2(\phi_2)^2} > 0 \tag{2}$$

where ΔG_m is the free enthalpy of mixing per unit volume and ϕ_2 is the volume fraction of component 2. In miscible polymer blends, molecular level mixing of the components is obtained. They are characterized by single-phase morphology. Immiscible blends do not satisfy either of the above criteria and will show a two-phase morphology. In the case of partially miscible blends, the second criteria is not satisfied and will show either two-phase or single-phase morphology.

II. INCOMPATIBILITY

A. Problems and Solutions

Only a very few polymers form truly miscible blends. These include poly(phenylene oxide) (PPO/PS), poly(vinyl chloride) (PVC)–polymeric plasticizers [1]. The rest of the blends, which are either partially miscible or immiscible, may undergo micro- or macrophase separation, leading to heterophase polymer blends. This heterogeneity is an unfavorable one, and this often leads to problems that reflect in the overall performance of the resultant material. Blending can give rise to morphologies that lead to certain specific characteristics. It is expected that this process can give rise to a material with the proper balance of properties than would be obtainable with single polymers. Practically, it is difficult to get the expected combination of properties due to the fact that many of the polymers are thermodynamically immiscible and it is difficult to get a homogeneous product. In an immiscible blend, the situation at the blend interface is critical, i.e., a high-interfacial tension and poor adhesion between the phases are observed. The high

* *Current affiliation*: Mahatma Gandhi University, Kerala, India.

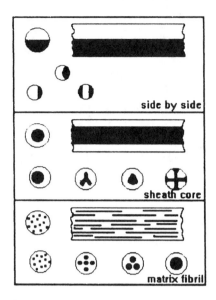

Figure 1 Various phase configurations upon the extrusion of binary immiscible blend.

viscosities associated with such systems are responsible for poor dispersion and lack of stability to gross phase segregation. The various phase configurations of a binary blend as suggested by Paul et al. [2] are given in Fig. 1. The low-intermolecular force between the component phases, which is responsible for the poor properties of incompatible blends, can be improved by increasing the interfacial area and adhesion between the phases.

The term "compatibility" is used extensively in the blend literature and is used synonymously with the term "miscibility" in a thermodynamic sense. Compatible polymers are "polymer mixtures that do not exhibit gross symptoms of phase separation when blended" or "polymer mixtures that have desirable chemical properties when blended." However, in a technological sense, the former is used to characterize the ease of fabrication or the properties of the two polymers in the blend [3–5].

III. THERMOPLASTIC ELASTOMER BLENDS (TPEs)

TPEs are materials that possess, at normal temperatures, the characteristic resilience and recovery from the extension of crosslinked elastomers and exhibit plastic flow at elevated temperatures. They can be fabricated by the usual techniques such as blow molding, extrusion, injection molding, etc. This effect is associated with certain interchain secondary valence forces of attraction, which have the effect of typical conventional covalent crosslinks, but at elevated temperatures, the secondary

bonds dissociates and the polymer exhibits thermoplastic behavior.

TPEs have many processing advantages over the vulcanized rubbers. The complex vulcanization techniques are not required and very little compounding is needed. Thermoplastic processing techniques, such as blow molding, heat welding, etc., that are unsuitable for conventional rubbers can be applied successfully to TPEs. The short processing cycle and mixing will result in very low-energy consumption. Processing of scraps that are considered as a waste in a conventional system can be recycled here. Besides all these advantages, certain disadvantages were also observed, i.e., they show high creep and set on extended use and will melt only at elevated temperatures. Thermoplastic elastomers have become very important. The estimated use of TPEs world wide in 1985 was 500,000 metric tonnes. They are replacing many of the conventional rubbers as well as thermoplastics.

TPEs can be mainly classified into five groups. These include:

1. TPEs from rubber–plastic blends.
2. Polystyrene–elastomer block copolymers.
3. Polyurethane–elastomer block copolymers.
4. Polyamide–elastomer block copolymers.
5. Polyether–elastomer block copolymers.

TPEs from blends of rubber and plastics constitute an important category of TPEs. These can be prepared either by the melt mixing of plastics and rubbers in an internal mixer or by solvent casting from a suitable solvent. The commonly used plastics and rubbers include polypropylene (PP), polyethylene (PE), polystyrene (PS), nylon, ethylene propylene diene monomer rubber (EPDM), natural rubber (NR), butyl rubber, nitrile rubber, etc. TPEs from blends of rubbers and plastics have certain typical advantages over the other TPEs. In this case, the required properties can easily be achieved by the proper selection of rubbers and plastics and by the proper change in their ratios. The overall performance of the resultant TPEs can be improved by changing the phase structure and crystallinity of plastics and also by the proper incorporation of suitable fillers, crosslinkers, and interfacial agents.

IV. COMPATIBILIZATION CONCEPTS

Although blending is an easy method for the preparation of TPEs, most of the TPE blends are immiscible. Very often the resulting materials exhibit poor mechanic properties due to the poor adhesion between the phases. Over the years different techniques have been developed to alleviate this problem. One way is to alter the blending technique so that the interfacial area between the component phases can be increased. By the proper selection of the processing technique either a co-continuous or

Table 1 Compatibility Through Nonreactive Copolymers

Major component	Minor component	Compatibilizer
PE or PS	PS or PE	S-B, S-EP,S-I-S, S-I-HBD, S-EB-S, S-B-S, PS–PE graft copolymers
PP	PS or PMMA	S-EB-S
PE or PP	PP or PE	EPM, EPDM
EPDM	PMMA	EPDM-g-MMA
PS	PA-6 or EPDM or PPE	PS–PA-6 block copolymers or S-EB-S
PET	HDPE	S-EB-S

Source: Ref. 1.

Table 2 Compatibility Through Reactive Copolymers

Major component	Minor component	Compatibilizer
ABS	PA-6–PA-6,6 copolymer	SAN–MA copolymer
PP or PS-6	PA-6 or PP	EPM–MA copolymer
PE	PA-6 or PA-6,6	Ionomers, carboxyl functional PEs
PP or PE	PET	PP-g-AA, carboxyl functional PE

Source: Ref. 1.

Table 3 Compatibility Through Low-Molecular Weight Reactive Compounds

Major component	Minor component	Compatibilizer
Fluororubber, FPM	NBR or CHR	Triazine dithiol complex
PVC or LDPE	LDPE or PVC	Polyfunctional monomers plus peroxide
NBR	PP	NBR curative and interchain copolymer
PVC or PP	PP or PVC	Cholrinated paraffin
PPE	PA-6,6	Amino silane

Source: Ref. 1.

interpenetrating phase morphology can be obtained that results in direct load sharing without the need for stress transfer across the phases. The second way is by the addition of a third component that is able to have interaction with blend components, (e.g., block and graft copolymers and low-molecular weight materials). The third technique is to blend suitably functionalized polymers capable of specific interaction or chemical reactions. This functionalization can be done in solution or in a compounding extruder [6] and may involve reactions such as halogenation, sulfonation, hydroperoxide formation, and the in situ formation of block and graft copolymers.

The "in situ" formed copolymers act as very good compatibilizers in many systems. These are formed during compounding, mastication, polymerization of one monomer in the presence of another polymer, etc., and have segments that are chemically identical to the homopolymers. Hajian et al. [7] reported the in situ formation of styrene–ethylene graft copolymers during the mixing of PS and PE. Anderson [8] studied the compatibilizing action of in situ-formed EPDM-g-MMA during the melt extrusion of EPDM and methylmethacrylate (MMA). The in situ formation of compatibilizers using functionalized polymers form the subject of several studies. Ide and Hasegawa [9] studied the effect of maleic anhydride grated polypropylene in PP–polyamide-6 blend.

Block or graft copolymers that act as compatibilizers are of two types, reactive and nonreactive. Nonreactive ones have segments capable of specific interaction with each of the blend components. In reactive copolymers, segments are capable of forming stronger covalent or ionic bonds with the blend components. Copolymers of both A–B type and A–C type can act as efficient compatibilizers in the A–B system, provided C is miscible with B. Tables 1 and 2 contain a few examples of polymer systems that are compatibilized through reactive and nonreactive and nonreactive copolymers, respectively. Systems with low-molecular weight compounds as compatibilizers have been subjected to various studies in the field of blends. Co-crosslinking, crosslinking, and grafting reaction may be involved in such systems and may lead to the formation of certain copolymers. Table 3 deals with such a system involving low-molecular weight compounds as compatibilizers.

V. COMPATIBILIZERS

A. Requirements

The efficiency of the copolymers, either block or graft, acting as the compatibilizer depends on the structure of the copolymers. One of the primary requirements to get maximum efficiency is that the copolymer should be located, preferentially at the blend interface (Figs. 2a, b, and c). There are three possible conformations, as shown in the figure. Many researchers [10–12] found that the actual conformation is neither fully extended nor flat (Fig. 2c). A portion of the copolymers penetrates into the corresponding homopolymer and the rest re-

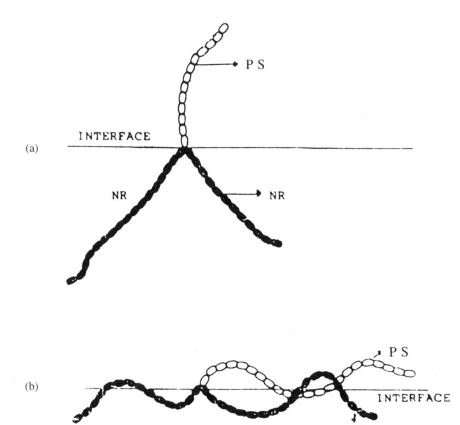

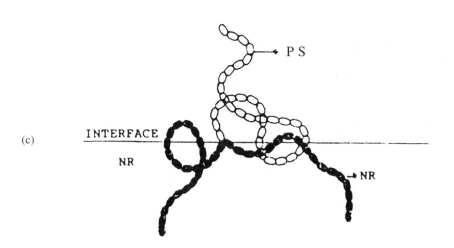

Figure 2 Conformation of copolymer at the blend interface: (a) completely extended, (b) completely flat, and (c) neither completely extended nor completely flat.

mains at the interface. In this respect, a block copolymer is supposed to be superior to a graft copolymer because a less-complicated conformation is expected in the former case [13,14]. In the case of graft copolymer, multiple branches should be avoided because they would restrict the penetration of the backbone into the homopolymer phases, and a diblock copolymer will be more effective than a triblock copolymer. Fayt et al. [15] demonstrated that a tapered diblock is ore efficient than a pure diblock with the same composition and molecular weight. Pure diblock copolymer contains highly incompatible sequences. These sequences segregate into domains and no mixing occurs. But tapered block copolymers do not form domains of their own and therefore provide a strong adhesion. Compared with diblock, the tapered block copolymers can be easily dispersed due to their low viscosity. Chemical identity of the copolymer segment with the homopolymer phase is important. Even though there is no chemical identity between the copolymer segment and the homopolymer, copolymers can be kept equally efficient provided the segment is miscible with the homopolymer.

Another important requirement is that the copolymer should have the propensity to segregate into two phases. In addition, the copolymer, both block and graft, should not be miscible as a whole in one of the homopolymer phases.

The amount of the copolymer to be added into a binary blend depends on several factors. Since the copolymer preparation is an expensive process, it should be used to its maximum efficiency. The amount of copolymer (m) required to saturate a unit volume of blend is given by:

$$m = 3\phi_A M/aRN \qquad (3)$$

where ϕ_A is the volume fraction of polymer A, R is the radius of dispersed particle A in a matrix B, N is the Avogadro's number, and M is the molecular weight of the copolymer. For a copolymer to be fully efficient, its molecular weight (M) should be higher than the molecular weight of the homopolymers. Riess and Jolivet [14] studied the effect of molecular weight on solubilization. When the homopolymer molecular weight is larger than that of the corresponding block segment, the homopolymer forms a separate phase and is not solubilized into the domains of the block copolymer. In the case of high-molecular weight copolymers, the long segments are able to anchor the immiscible phases firmly. Besides all these arguments, there should be an optimum molecular weight when the cost-benefit of the resultant material is a concern. The effect of the molecular weight of poly-(styrene-butadiene) copolymer in PS–PB system has been explained by Legge et al. [16]. They have shown the interfacial profiles of two systems: (1) with block copolymer having molecular weight less than that of the homopolymer, (2) with block copolymer having molecular weight higher than that of the homopolymer. It was found that the copolymer with high-molecular weight has greater concentration in the interfacial region. The emulsifying effect of block and graft copolymers in polystyrene–polyisoprene (PS–PI) blends was demonstrated by Riess and Joliviet [14]. Mechanical properties of the incompatible blends can be enhanced by controlling the dispersed phase size and adhesion between the components, which can be achieved by the addition of a suitable copolymer. Molecular weight and composition are two important parameters that determine whether the copolymer will locate at the blend interface, in the continuous phase, or in the dispersed phase. Block copolymers of equal segmental mass are more effective as a compatibilizer than those of unequal segmental mass. There are various techniques for locating the copolymer. These include the use of a copolymer with a fluorescent group, x-ray scanning microanalysis, analysis of gel formed after crosslinking the elastomer phase by γ-radiation, etc.

In addition to the abovementioned parameters, various factors such as viscosity of the copolymer and its interaction with the homopolymers also play a major role in the compatibilization process.

VI. THEORETICAL ASPECTS OF COMPATIBILIZATION

Leibler [17] and Noolandi et al. [18,19] developed thermodynamic theories concerning the emulsification of copolymers (A-b-B) in immiscible polymer blends (A–B).

The theory of Leibler holds for mainly compatible systems. Leibler developed a mean field formalism to study the interfacial properties of two polymers, A and B with an A–B copolymer. An expression for interfacial tension reduction was developed by Noolandi and Hong [18] based on thermodynamics to explain the emulsifying effect of the A-b-B in immiscible A–B blends (A–A-b-B–B). [18,19]. The expression for interfacial tension reduction (Δr) in a binary lend upon the addition of divalent copolymer is given by:

$$\Delta r = d\phi_c(\tfrac{1}{2}\chi + 1/Zc) - 1/Zc \exp(Zc\chi/2) \qquad (4)$$

where d is the width at half height of the copolymer reduced by Kuhn's statistical segment length, ϕ_c is the bulk volume fraction of the copolymer in the system, χ is the Flory-Huggin's interaction parameter between A and B segment of the copolymer, and Zc is the degree of polymerization of the copolymers. Although this expression was derived for the action of symmetrical diblock copolymers (A-b-B) in homopolymer blend (A–B), i.e., A–A-b-B–B, this theory can be successfully applied to A–A-g-B–B system.

According to this theory, the interfacial tension reduction should decrease linearly with copolymer content at low concentrations followed by a leveling off at higher concentrations. The theory of Noolandi and Hong [18] has been testified to by Koberstein and coworkers [19a] and Thomas [10].

VII. COMPATIBILIZATION OF TPEs

A. Specific Cases

There are a large number of studies related to the compatibilization of TPEs by the addition of copolymers (both block and graft) and by dynamic crosslinking.

Park et al. [20] reported on the synthesis of poly-(chloroprene-co-isobutyl methacrylate) and its compatibilizing effect in immiscible polychloroprene–poly(isobutyl methacrylate) blends. A copolymer of chloroprene rubber (CR) and isobutyl methacrylate (iBMA) poly[CP-Co-(BMA)] and a graft copolymer of iBMA and polychloroprene [poly(CR-g-iBMA)] were prepared for comparison. Blends of CR and PiBMA are prepared by the solution casting technique using THF as the solvent. The morphology and glass-transition temperature behavior indicated that the blend is an immiscible one. It was found that both the copolymers can improve the miscibility, but the efficiency is higher in poly(CR-Co-iBMA) than in poly(CR-g-iBMA).

Oommen and Thomas [21] studied the interfacial activity of natural rubber-g-poly(methylmethacrylate) in incompatible NR–PMMA blends. Graft copolymer of NR and PMMA was prepared using a redox initiator consisting of cumene hydroperoxide and tetraethylene pentamine. Mechanical and morphological analyses of the blends with and without the compatibilizer were studied, and it was found that the mechanical properties increase with increasing loading of the graft copolymer concentration (Table 4). Morphological analyses are in agreement with the mechanical data. Figure 3 gives the scanning electron micrograph (SEM) of 60:40 NP–PMMA blends containing different levels of the compatibilizer. They further studied the effect of casting solvent, mode of addition of compatibilizer, molecular weight of homo- and copolymers, etc., on the morphological and mechanical properties. The experimental results were compared with the theoretical predictions of Noolandi and Hong [18].

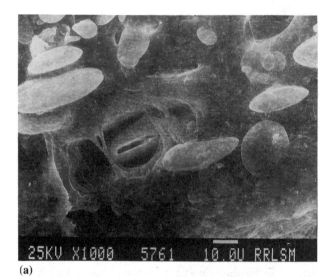

(a)

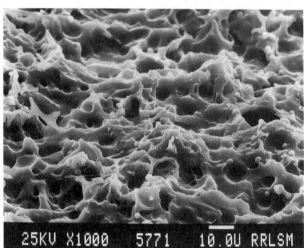

(b)

Figure 3 Scanning electron micrographs of 60:40 NR–PMMA blend: (a) 0% and (b) 3.5% graft copolymer.

Asaletha and coworkers [12,22] further studied the compatibilizing effect of NR-g-PS in NR–PS blends. NR–PS blend is incompatible and immiscible and its compatibility can be improved by the addition of the

Table 4 Effect of Compatibilizer Loading on Mechanical Properties

Percent of graft copolymer	Stress at 50% elongation (N/mm)	Stress at 100% elongation (N/mm)	Stress at 130% elongation (N/mm)	Tensile strength (N/mm²)	Elongation at break (%)
0	1.05	1.41	1.70	3.98	362
1	1.07	1.54	1.96	6.97	429
2.5	2.55	3.92	4.55	7.43	262
3	1.18	1.95	2.56	7.81	381

Source: Ref. 21.

Table 5 Mechanical Properties of 50:50 NR–PS Blends

Wt% of graft polymer	Stress at 15% elongation (MPa)	Stress at 30% elongation (MPa)	Stress at 50% elongation (MPa)	Tensile strength (MPa)	Elongation at break (%)	Tensile impact strength (J/m^2)
0	1.24	1.78	2.37	3.60	454	0.30×10^5
1.5	1.54	1.82	2.45	3.86	194	1.64×10^5
3	1.96	2.05	2.75	4.50	190	2.10×10^5
4.5	1.99	2.28	3.07	10.10	252	1.63×10^5
6	2.56	2.78	3.24	13.24	247	1.39×10^5
7.5	3.20	3.47	3.88	13.15	241	1.37×10^5

Source: Ref. 12.

graft copolymer. Up to a particular concentration of the compatibilizer, the mechanical properties of the blend increased, thereafter the properties leveled off at higher graft loading (Table 5). The compatibilizing effect de-

pends upon the solvent used for the preparation of the blend (Fig. 4), which in turn, depends upon the solubility parameter of the concerned solvent. The molecular weight of copolymer and homopolymers, mode of prepa-

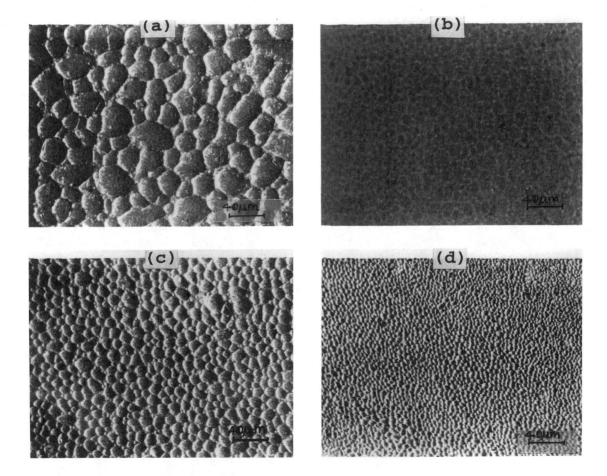

Figure 4 Effect of casting solvent on compatibilization studies of NR–PS blend. (a) 0% graft (CHCl$_3$), (b) 1.2% graft (CHCl$_3$), (c) 0% graft (CCl$_4$), and (d) 1.2% graft (CCl$_4$).

ration of the blend, composition of the blend, etc. are other parameters that control the compatibilization process. Morphology and mechanical properties were studied in detail and the area occupied by the compatibilizer in each case was calculated and correlated with the theoretical predictions of Noolandi and Hong [18].

Characterization and control of interfaces in the incompatible polymer blends were reported by Fayt et al. [23]. They used techniques such as electron microscopy, thermal transition analysis, and nonradiative energy transfer (NRET), etc. They have illustrated the exciting potentialities offered by diblock copolymers in high-performance polymer blends.

Chu et al. [24] correlated viscosity-morphology and compatibility of PS–PB blends. The effect of styrene–butadiene triblock copolymer in PS–PB was studied, and it was found that the domain size decreases with an increase of compatibilizer loading. The blending methods influenced the morphology due to the difference in the extent of mixing.

Dynamic crosslinking as a means to improve the impact strength and other mechanical properties of polypropylene–elastomer blends has been discussed in detail by Inoue [25]. All these blends contain 80% PP and 20% elastomer. Elastomers include EPDM, styrene–butadiene–styrene (SBS) and styrene–isoprene–styrene (SIS) and the crosslinking system comprised of N,N'-m-phenylene-bis-maleimide and 6-ethoxy 2,2,4-trimethyl-1,2-dihydroquinoline or poly(2,2,4-trimethyl-2,2-dihydroquinoline). Impact strength and other mechanical properties, such as tensile strength at yield, ultimate elongation, flexural modulus, etc., showed remarkable increase after crosslinking. This is due to the increase of interfacial adhesion caused by the PP graft elastomers located at the blend interface.

Interfacial adhesion and, thereby, compatibility can be enhanced by the selective crosslinking reaction in polymer blends. Inoue and Suzuki [26] reported the properties of blends dynamically crosslinked PP–EPDM blends. The crosslinking agent was N,N'-m-phenylene-bismaleimide - poly(2,2,4 - trimethyl - 1,2-dihydroquinoline) system. Increase in interfacial adhesion leads to

an improvement in Izod impact strength. Various other mechanical properties such as tensile strength at yield, ultimate elongation, and flexural modulus were also studied before and after the crosslinking reaction.

Krulis et al. [27] also described the dynamic crosslinking as a route to improve the mechanical properties. It was found that a high-impact strength is obtained in PP–EPDM blends by slow curing with sulfur. Thiuram disulfide N-(cyclohexylthio)phthalimide was used as an inhibitor of curing, and its effect on the impact strength of dynamically cured PP–EPDM blends was studied (Table 6). It was also found that the one-step method of blend preparation also has a favorable effect on the impact strength of the resultant blend system.

Compatibilization along with dynamic vulcanization techniques have been used in thermoplastic elastomer blends of poly(butylene terephthalate) and ethylene propylene diene rubber by Moffett and Dekkers [28]. In situ formation of graft copolymer can be obtained by the use of suitably functionalized rubbers. By the usage of conventional vulcanizing agents for EPDM, the dynamic vulcanization of the blend can be achieved. The optimum effect of compatibilization along with dynamic vulcanization can be obtained only when the compatibilization is done before the rubber phase is dispersed.

Ha [29] has shown that in PP–HDPE (of high-density PE) dynamically cured EPDM, the cured EPDM act as a compatibilizer to the HDPE–PP system. Blending was done in two ways. EPDM was cured first and then blended with PP and HDPE. In the second case, EPDM was cured in the presence of PP and HDPE using dicumyl peroxide (DCP) as the vulcanizing agent. In EPDM-rich composition, mechanical properties were increased by increasing the concentration of DCP, whereas in PP-rich composition, the reverse was the case.

Dynamic vulcanization as a method to improve the mechanical properties of NR–PE blends has been discussed in detail by Choudhary et al. [30]. The physical properties of unvulcanized and vulcanized NR–HDPE blends are given in Table 7 where notations A, B, and C indicate 70:30, 50:50, 30:70 NR–HDPE blends, respectively. The subscripts C and D denote blends containing DCP and high abrasion furnace (HAF) black (40 phr), respectively.

In all the compositions, the DCP-cured blends showed better properties than the corresponding unvulcanized samples. Choudhary et al. [30] further demonstrated the use of EPDM, chlorinated PE, chlorosulfonated PE, maleic anhydride modified polyethylene, and blends of epoxidized natural rubber–sulfonated EPDM as compatibilizers in NR–LDPE (low-density PE) blends.

It was found that PP–EPDM blends [30] with a slow curing EPDM have a high-impact strength. Effect of tetramethyl thiuram disfulfide (TMTD) concentration on the Charpy notched impact strength of PP–EPDM blend is given in Table 8.

Table 6 Effect of Cure Inhibitor Concentration on the Charpy Notched Impact Strength (a_k) at 0°C

CTPI (phr)	E_{min} (min)	t_{max} (min)	$t_{max} - t_{min}$ (min)	a_k, kJm^{-2} 85/15	a_k, kJm^{-2} 80/20
0	0.3	1.7	1.4	13.9	24.1
0.3	0.6	2.9	2.3	14.2	38.2
0.6	1.0	3.3	2.3	17.0	39.4
1.0	2.0	4.3	2.3	19.1	37.7

Source: Ref. 27.

Table 7 Physical Properties of NR–HDPE Blends

Property	HDPE	A	A_c	B	B_c	C	C_c	C_b	C_{bc}	NR
Young's modulus (MPa)	29.4	13.4	14.5	5.8	6.1	1.4	2.4	2.1	4.5	0.4
Modulus 100% (MPa)	—	—	—	7.2	8.7	2.2	4.6	4.1	5.7	0.8
Tensile strength (MPa)	32.1	13.5	14.6	10.9	11.3	9.5	13.0	13.7	12.7	2.3
Elongation at break (%)	30.0	65.0	24.0	420	430	960	500	470	300	450
Tear strength (kNm^{-1})	118.0	72.5	74.4	47.2	55.7	28.1	40.8	43.6	36.4	4.35
Tensile set (%)	—	—	—	92	>100	72	64	70	68	>100

Source: Ref. 30.

A binary blend of polypropylene–ethylene propylene diene monomer rubber and a ternary blend of PP–EPDM–poly(ethylene-co-methacrylic acid) ionomer was prepared by Kim et al. [31]. The rheological, mechanical, and morphological properties of these blends were analyzed. Two kinds of ionomers neutralized with different metal ions (Na^+ and Zn^{2+}) were used and their concentration varied from 5–20% based on the total amount of PP and EPDM. It was found that rheological and morphological properties of the binary and ternary blends showed much variation due to the compatibilizing effect of the ionomer. Na neutralized ionomer (ionomer A) showed a compatibilizing effect, whereas the Zn neutralized ionomer (ionomer B) did not, and the effect was prominent at 5 wt% of the ionomer concentration. The effect of ionomer concentration on the storage modulus of PP–EPDM blends is given in Fig. 5. It was seen that the ternary blends with 5 wt% of ionomer A showed better properties than other blends.

The effect of ionomer concentration on the mechanical properties of PP–EPDM blends is given Table 9. It is seen that the tensile strength and modulus show a maximum at 5 wt% of both ionomer A and B, thereafter, it decreases at higher ionomer loading. The properties are higher for ternary blends containing ionomer B than these containing ionomer A. On the other hand, addition of ionomer will reduce the elongation at break regardless of the ionomer type.

The structure–property relationship of graft copolymers based on an elastomeric backbone poly(ethyl acrylate)-g-polystyrene was studied by Peiffer and Rabeony [32]. The copolymer was prepared by the free radical polymerization technique and, it was found that the improvement in properties depends upon factors such as the number of grafts/chain, graft molecular weight, etc. It was shown that mutually grafted copolymers produce a variety of compatibilized ternary component blends.

Coran and Patel [33] selected a series of TPEs based on different rubbers and thermoplastics. Three types of rubbers EPDM, ethylene vinyl acetate (EVA), and nitrile (NBR) were selected and the plastics include PP, PS, styrene acrylonitrile (SAN), and PA. It was shown that the ultimate mechanical properties such as stress at break, elongation, and the elastic recovery of these dynamically cured blends increased with the similarity of the rubber and plastic in respect to the critical surface tension for wetting and with the crystallinity of the plastic phase. Critical chain length of the rubber molecule, crystallinity of the hard phase (plastic), and the surface energy are a few of the parameters used in the analysis. Better results are obtained with a crystalline plastic material when the entanglement molecular length of the

Table 8 Effect of Composition of TMTD Curing System on the Charpy Notched Impact Strength (a_k) for 85/15 and 80/20 PP/EPDM Blends

ZnO (phr)	TMTD (phr)	MBT (phr)	t_{min} (min)	t_{max} (min)	$t_{max} - t_{min}$ (min)	a_k, kJm^{-2} 85/15	a_k, kJm^{-2} 80/20
1.5	0.5	0.5	2.0	12.1	10.1	12.4	24.7
3	1	1	1.5	7.2	5.7	15.6	27.3
3	3	1	0.4	3.5	3.1	12.0	17.8
3	3	2	0.4	4.0	3.6	13.8	23.7
3	3	3	1.5	7.0	5.5	17.6	23.7

Source: Ref. 30.

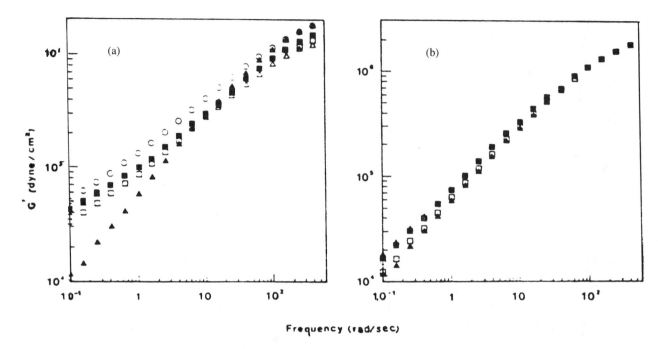

Figure 5   Effect of ionomer concentration on storage modulus of PP–EPDM blends. (a) PP50-EP50/IA, and (b) PP50-EP50/
IB (▲) 0 wt%, (○) 5 wt%, (□) 10 wt%, (△) 15 wt%, and (■) 20 wt%.

rubber is low. The other two conditions that are to be satisfied are: (1) the surface energies of the plastic and rubber should not be very high, and (2) both rubber and plastic should not decompose in the presence of each other during melt mixing.

In the case of NBR–nylon blends it was reported that the addition of the curative *m*-phenylene-bis-maleimide, improved the strength and stiffness of the blend. An attempt to compare the effect of different curative

Table 9   Effect of Ionomer Concentration on the Tensile Properties of Blends of PP–EPDM

Sample	Modulus, E (Kgf/cm^2)	Tensile strength (Kgf/cm^2)	Elongation at break (%)
PP50–EP50	575.82	105.57	234.75
PP50–EP50/IA5	630.21	109.23	138.60
PP50–EP50/IA10	497.35	88.08	170.02
PP50–EP50/IS15	487.49	74.68	237.00
PP50–EP50/IA20	421.72	56.23	190.43
PP50–EP50/IB5	719.63	112.62	195.18
PP50–EP50/IB10	680.21	93.07	206.33
PP50–EP50/IB15	640.81	85.54	215.45
PP50–EP50/IB20	611.58	65.50	176.66

Source: Ref. 31.

m-phenylene bismaleimide and dimethylol phenolic compounds was also reported. Nylon–NBR graft molecules formed during the crosslinking induce better homogenization in the system, which leads to overall enhancement of the blend performance.

In addition to dynamic vulcanization, the technological compatibilization technique was also adopted by Coran and Patel [34] to obtain thermoplastic vulcanizate having good mechanical integrity and elastic recovery.

Swelling of thermoplastic elastomeric vulcanizates using a model EPDM–PP blend in various solvents, such as cyclohexane, butyl acetate, methyl ethyl ketone (MEK), etc. was studied by Coran and Patel [35]. Blends were vulcanized both by dynamic and by static means and the mechanical properties, such as ultimate tensile strength, Young's modulus, ultimate elongation, hardness etc., were determined in each case. All the properties except elongation break (EB) are higher for dynamically vulcanized samples. Static samples are not processable with the typical thermoplastic processing techniques, whereas the dynamic samples can be molded. The amount of swelling of the thermoplastic vulcanizates was found to be less than the average swelling of the rubber and plastic. It was also noted that the vulcanizates prepared by the dynamic technique will swell less than the composition prepared by static means.

Legge et al. [36] have discussed the PS–PI blend and the corresponding PS-b-PI as a model system for rubber-modified thermoplastics. The emulsifying effect of the block copolymer is evaluated by checking the transparency of the polymer blend. The transparency of an incompatible PS–PI system having different refractive indices can be obtained by reducing the particle size of the dispersed phase below a certain level. This may be possible by the compatibilizing action of the block copolymer.

Frounchi and Burford [37] studied the effect of styrene block copolymer as a compatibilizer in isotactic PP–ABS blends. It was found hat in PP-rich blends a marginal improvement in mechanical properties was obtained. However, in acrylo nitrile butadiene styrene (ABS) rich blends no improvement was obtained. The effects of four different block copolymers, SBS, SIS,

Table 11 Effect of Different Block Copolymers on the Mechanical Properties of ABS-Rich Blends

Composition of blend (ABS-rich)	Yield tensile strength, σ_y (MPa)	Elongation at break, EB (%)
ABS80–ABS20	19	5
ABS80–PP20–SEBS5	22	10
ABS80–PP20–SEB5	24	15
ABS80–PP20–SIS5	22	17
ABS80–PP20–FSEBS5	25	23

Source: Ref. 37.

Table 10 Effect of Different Block Copolymers on the Mechanical Properties of PP-Rich Blends

Composition of blend (PP-rich)	Yield tensile strength, σ_y (MPa)	Elongation at break, EB (%)
PP80–ABS20	19	5
PP80–ABS20–SEBS5	22	10
PP80–ABS20–SEB5	24	15
PP80–ABS20–SIS5	22	17
PP80–ABS20–FSEBS5	25	23

Source: Ref. 37.

styrene–ethylene–butylene–styrene (SEBS), and a slightly maleated functionalized SEBS were compared. Mechanical properties of the PP–ABS blends with different block copolymers are given in Tables 10 and 11.

The effect of ABS wt% on the Charpy impact strength is given in Fig. 6.

Elliot [38] has reported that interfacial adhesion in the NR–PP blend can be enhanced by the addition of small amounts of HDPE. Addition of HDPE does give some improvement in the notched Izod impact strength of NR–PP blend (Fig. 7). The effect of HDPE on the impact modification of NR–PP is associated with the improved crystallinity of PP, enhanced by HDPE. During the mill mixing of NR and PP, chain scission may occur to give polymeric radicals that, on reaction with

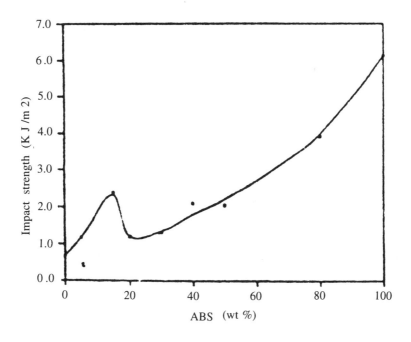

Figure 6 Effect of ABS weight percent on Charpy impact strength.

Izod impact strength (J m⁻¹)

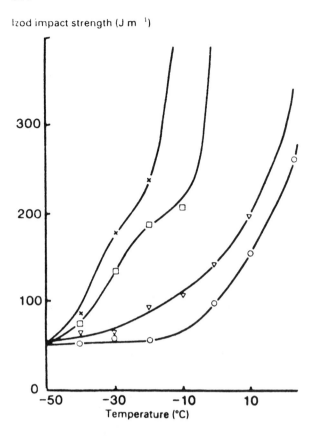

Figure 7 Effect of addition of HDPE on Izod impact strength of NR–PP blend. (○) 20:80 NR–PP homopolymer, (▽) 20:67:13 NR–PP–homopolymer–HDPE, (□) 15:85 NR–PP copolymer grade, and (x) 15:75:10 NR–PP–copolymer–HDPE.

the added multifunctional radical acceptor, can give graft copolymer and, in turn, can act as a compatibilizer in a NR–PP system. The effect of such compatibilizers on the BS-notched Izod impact strength is given in Fig. 8.

Compatibility and various other properties such as morphology, crystalline behavior, structure, mechanical properties of natural rubber–polyethylene blends were investigated by Qin et al. [39]. Polyethylene-b-polyisoprene acts as a successful compatibilizer here. Mechanical properties of the blends were improved upon the addition of the block copolymer (Table 12). The copolymer locates at the interface, and, thus, reduces the interfacial tension that is reflected in the mechanical properties. As the amount of graft copolymer increases, tensile strength and elongation at break increase and reach a leveling off.

Morphological studies of these blends revealed that the compatibilization was very effective in decreasing the interfacial tension and increasing the adhesion between the two phases (Fig. 9).

Izod impact strength (J m⁻¹)

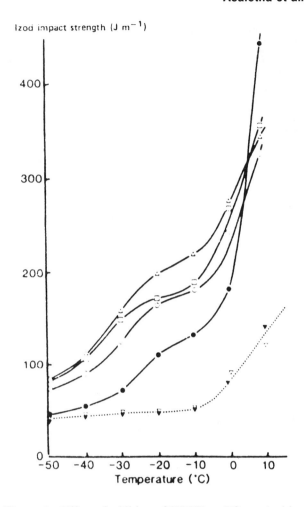

Figure 8 Effect of addition of HDPE on BS-notched impact strength of NR–PP blend. (▼) PP, (▽) PP + 0.5 p.p.h.p. HVA2, (●) blend, (○) blend + 0.2 p.p.h.p. HVA2, (□) blend + 0.5 p.p.h.p. HVA2, and (△) blend + 1 p.p.h.p. HVA2.

Table 12 Mechanical Properties of NR–LLDPE Compatibilized with PE-b-PI

Sample	Tensile strength (kg/m⁻²)	Elongation at break (%)
0:100:0	240	1387
50:50:0	84	975
49:50:1	99	965
47:50:3	112	1190
45:50:5	132	1089
41:50:9	113	1089
37:50:13	110	1188
37:50:13[a]	96	935

[a] $[\eta]$ PE-b-PI = 12.8 dlg⁻¹; 18.9 dlg⁻¹ for other blends.
Source: Ref. 39.

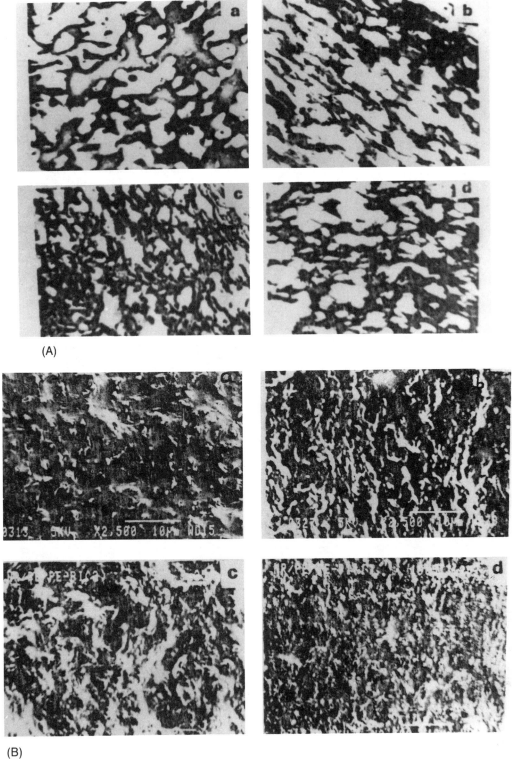

Figure 9 (A) SEM micrographs of NR–LLDPE–PE-b-PI: (a) 50:50:0, (b) 45:50:5, (c) 37:50:13, and (d) 37:50:13; (B) TEM micrographs of NR–LLDPE–PE-b-PI: (a) 50:50:0, (b) 45:50:5, (c) 37:50:13, and (d) 37:50:13.

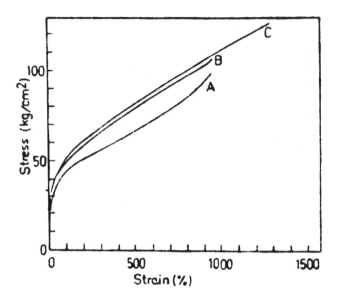

Figure 10 Effect of compatibilizer molecular weight on mechanical properties. (A) NR–LLDPE–PE-b-PI = 50:50:0, (B) NR–LLDPE–PE-b-PI = 37:50:13 (M_n = 1.29 × 10^5), and (C) NR–LLDPE–PE-b-PI = 37:50:13 (M_n = 2.79 × 10^5).

The molecular weight of the copolymer (PE-b-PI) has a profound influence on the mechanical properties of these blends (NR–LLDPE) (Fig. 9). High-molecular weight copolymer is a more effective compatibilizer compared with a low-molecular weight one.

Improvement in the compatibility of NBR–PP blends by various compatibilizers was reported by Rader and Sabet [40]. The effect of different compatibilizers on the properties of NBR–PP blends is given in Table 13.

Wang and Chen [41] studied the compatibility problems of incompatible NBR–PVC blends. Poly(vinylidene chloride-covinyl chloride) is reported to act as an efficient interfacial agent. Blends of PVC, NBR, and the copolymer were prepared by the solution casting technique using THF as a solvent. Improvement in mechanical properties can be achieved in NBR–PVC blend by the addition of different types of rubbers [42]. Different rubbers include NR, styrene butadiene (SBR) and butadiene (BR). Replacement of a few percent of NBR by other rubbers will improve the mechanical properties and at the same time reduce the cost of the blend.

Compatibility of immiscible PP–NBR blends was improved by the reactive compatibilization technique using various modified polypropylenes. In this study,

Table 13 Effect of Different Compatibilities on NBR–PP Compositions

	Composition number							
	1	2	3	4	5	6	7	8
Recipe								
PP	50	0	50	0	50	45	50	49
Chlorine-modified PP	0	50	0	0	0	0	0	0
ATBN masterbatch	50	50	0	0	0	0	0	0
Maleic/TETA-modified PP masterbatch	0	0	0	50	0	0	0	0
Carboxylated NBR	0	0	0	0	50	50	0	0
Nylon-6/BR block copolymer	0	0	0	0	0	10	0	0
NBR	0	0	0	0	50	45	50	50
NBR/PP block copolymer	0	0	0	0	0	0	0	1.0
Phenolic resin	3.75	3.75	3.75	3.75	2.5	2.5	3.75	3.75
Tin(II) dichloride dihydrate	0.5	0.5	0.5	0.5	0.5	2.5	0.5	0.5
Properties								
UTS (MPa)	8.5	11.9	8.2	19.2	8.7	10.1	8.8	18.3
Stress at 100% strain (MPa)	—	11.8	—	14.2	—	10	—	12.7
Young's modulus (MPa)	231	239	180	137	233	170	209	195
Ultimate elongation (%)	8	160	13	250	19	140	19	330
Tension set (%)	—	46	—	39	—	—	—	43
True stress at break (MPa)	9	31	9	67	10	24	10	78

Source: Ref. 35.

glycidyl methacrylate (GMA), 2-hydroxyethyl methacrylate (HEMA), 2-hydroxypropyl methacrylate (HPMA), *t*-butylaminoethyl methacrylate (TBAEMA), dimethylaminoethyl methacrylate, and 2-isopropenyl-2-oxazoline (IPO) were used as the modifiers. It was found that IPO and GMA are effective in compatibilizing the PP–NBR blends [43]. The compatibilization of NBR–PP [44] and NBR–HDPE [45] blends has been reported by Thomas and coworkers. High-impact PS can be obtained from PS and EPDM by the coupling of EPDM and PS in the mixing chamber of a HAAKE plastograph. Lewis acids were added to the melt, and it was found that rubber became crosslinked or was coupled with the PS molecules and improvement in mechanical properties was observed [46]. Santra et al. [47] reported the in situ compatibilization of low-density polyethylene and polydimethyl siloxane rubber blends using ethylene–methyl acrylate copolymer as a chemical compatibilizer. Ethylene methacrylate (EMA) reacted with the rubber to form EMA-grafted rubber during the melt mixing, which acts as the compatibilizer. They have conducted dynamic mechanical analysis, adhesion studies, and phase morphology and found that 6 wt% of the compatibilizer was the optimum quantity required.

Els and McGill [48] reported the action of maleic anhydride on polypropylene–polyisoprene blends. A graft copolymer was found in situ through the modifier, which later enhanced the overall performance of the blend. Scott and Macosko [49] studied the reactive and nonreactive compatibilization of nylon–ethylene-propylene rubber blends. The nonreactive polyamide-ethylene propylene blends showed poor interfacial adhesion between the phases. The reactive polyamide-ethylene propylene-maleic anhydride modified blends showed excellent adhesion and much smaller dispersed phase domain size.

Greco et al. [50] studied the effect of the reactive compatibilization technique in ethylene propylene rubber–polyamide-6 blends. Binary blends of polyamide-6–ethylene propylene rubber (EPR) and a ternary blend of polyamide-6–EPR–EPR-g-succinic anhydride were prepared by the melt mixing technique, and the influence of the degree of grafting of (EPR-g-SA) on morphology and mechanical properties of the blends was studied.

VIII. CONCLUSION

Thermoplastic elastomers are materials that have the properties of vulcanized rubbers but can be processed by techniques associated with thermoplastics. The commercial importance of TPEs is due to their superior processing properties and economic advantages over conventional rubbers and plastics. TPEs from rubber–plastic blends became important because they combine the superior processability of thermoplastics and the

very good mechanical properties of the elastomers. Both natural and synthetic elastomers can be blended with thermoplastics either by melt blending or by the solution-casting technique for the preparation of TPEs. Most of these TPEs are immiscible and incompatible and exhibit poor properties. This problem can be alleviated by the proper addition of a suitable compatibilizer, which may be a block copolymer or graft copolymer or homopolymer or low-molecular weight materials. The dynamic vulcanization of the rubber phase is widely used as a means to improve the compatibility. There are many structural and technical parameters that control the overall performance of the compatibilized blend system. A few theoretical approaches were also discussed. The changes in mechanical and morphological properties of the blends upon the addition of the copolymer were discussed in detail.

REFERENCES

1. M. Xanthos, *Polym. Eng. Sci., 28*: 1392 (1988).
2. D. R. Paul and J. W. Barlow, *J. Macromol. Sci. Rev. Macromol. Chem., C18(1)*: 109 (1980).
3. D. R. Paul, C. E. Vinson, and C. E. Locke, *Polym. Eng. Sci., 12*: 157 (1972).
4. A. J. Yu, "Multicomponent Polymer Systems" (N. A. J. Platzer, ed.), *Advanced Chemistry Series,* Vol. 99, p. 2, Amer. Chem. Soc., Washington, DC (1971).
5. N. G. Gaylord, "Copolymers, Polyblends and Composites" (N. A. J. Platzer, ed.), *Advanced Chemistry Series,* Vol. 142, p. 76, Amer. Chem. Soc., Washington, DC (1975).
6. C. S. Tucker and R. J. Nicholas, *S. P. E. ANTEC Tech. Papers, 33*: 117 (1987).
7. M. Hajian, C. Sadrmohaghegh, and G. Scott, *Eur. Polym. J., 20*: 135 (1984).
8. P. G. Anderson, U.S. Patent, 4, 476, 283 (1984).
9. F. Ide and A. Hasegawa, *J. Appl. Polym. Sci., 18*: 963 (1974).
10. S. Thomas and R. E. Prud'homme, *Polymer, 33*: 4260 (1992).
11. Z. Oommen and S. Thomas, *Polym. Eng. Sci. 36*: 151 (1996).
12. R. Asaletha, M. G. Kumaran, and S. Thomas, *Rubber Chem. Technol. 68*: 671 (1995).
13. G. Riess, J. Kohler, C. Tournut, and A. Banderet, *Makromol. Chem., 101*: 58 (1967).
14. G. Riess and Y. Jolivet, "Copolymers, Polyblends and Composites, (N. A. J. Platzer, ed.), *Advanced Chemistry Series,* Vol. 142, p. 243, Amer. Chem. Soc., Washington, DC (1975).
15. R. Fayt, R. Jerome, and Ph. Teyssie, *J. Polym. Sci. Polym. Phys. Edn., 20*: 2209 (1982).
16. G. Riess, P. Bahadur, and G. Hurtrez, *Encyclopedia of Poly. Sci. Eng. 2*, 234 (1985), Second Edn., John Wiley and Sons, New York.
17. L. Leibler, *Makromol. Chem. Symp., 16(1)*: 17 (1988).
18. J. Noolandi and K. M. Hong, *Macromolecules, 15*: 482 (1982).
19. J. Noolandi, *Polym. Eng. Sci., 24*: 70 (1984).

19a. S. H. Arastasiadis, I. Gancarz, and J. T. Koberstein, *Macromolecules 22*: 1149 (1989).

20. C. K. Perk, C. S. Ha, J. K. Lee, and W. J. Cho, *J. Appl. Polym. Sci., 50*: 1239 (1993).

21. Z. Oommen and S. Thomas, *Polymer Bulletin, 31*: 623 (1993).

22. R. Asaletha, M. G. Kumaran, and S. Thomas, *Polym. Plast. Technol. Engg., 34*: 633 (1995).

23. R. Fayt R. Jerome, and Ph. Teyssie, *Polym. Eng. Sci., 27*: 328 (1987).

24. L. H. Chu, S. H. Guo, and H. C. Tseng, *J. Appl. Polym. Sci., 49*: 179 (1993).

25. T. Inoue, *J. Appl. Polym. Sci., 54*: 723 (1994).

26. T. Inoue and T. Suzuki, *J. Appl. Polym. Sci., 56*: 1113 (1995).

27. Z. Krulis, I. Fortelny, and J. Kovar, *Collect. Czech. Chem. Commun., 58*: 2642 (1993).

28. A. J. Moffett and M. E. J. Dekkers, *Polym. Eng. Sci., 32*: 1 (1992).

29. C. S. Ha, *J. Appl. Polym. Sci., 37*: 317 (1989).

30. N. R. Choudhuary, P. P. De, and A. K. Bhowmick, *Thermoplastic Elastomers from Rubber-Plastic Blends,* Ellis Horwood, England, Chap. 3, p. 79 (1990).

31. Y. Kim, C. S. Ha, T. Kang, Y. Kim, and W. Cho, *J. Appl. Polym. Sci., 51*: 1453 (1994).

32. D. G. Peiffer and M. Rabeony, *J. Appl. Polym. Sci., 51*: 1283 (1994).

33. A. Y. Coran and R. Patel, *Rubber Chem. Technol., 54*: 892 (1981).

34. A. Y. Coran and R. Patel, *Rubber Chem. Technol., 55*: 116 (1982).

35. A. Y. Coran and R. Patel, *Rubber Chem. Technol., 55*: 1063 (1982).

36. G. Riess, P. Bahadur, and G. Hurtrez, *Encyclopedia of Poly. Sci. Eng.* 2, 234 (1985). Second Edn., John Wiley and Sons, New York.

37. M. Frounchi and R. P. Burford, *Iranian J. Polym. Sci. Technol., 2*: 59 (1993).

38. D. J. Elliot, *Thermoplastic Elastomers from Rubber-Plastic Blends,* Ellis Horwood, England, Chap. 4, p. 121 (1990).

39. Chuan Qin, J. Yin, and B. Huang, *Polymer, 31*: 663 (1990).

40. C. P. Rader and S. A. Sabet, *Thermoplastic Elastomers from Rubber-Plastic Blends,* Ellis Horwood, England, Chap. 6, p. 177 (1990).

41. Y. Wang and S. Chen, *Polym. Eng. Sci., 21*: 47 (1981).

42. K. E. George, Rani Joseph, D. Joseph Francis, and K. T. Thomas, *Polym. Eng. Sci., 27*: 1137 (1987).

43. M. C. Liu, H. Q. Xie, and W. E. Baker, *Polymer, 34*: 4680 (1993).

44. S. George, R. Joseph, K. T. Varughese, and S. Thomas, *Polymer 36*: 4405 (1996).

45. J. George, R. Joseph, K. T. Varughese, and S. Thomas, *J. Appl. Polym. Sci., 57*: 449 (1995).

46. E. Mori, B. Pukanszky, T. Kelen, and F. Tudos, *Polymer Bulletin, 12*: 157 (1984).

47. R. N. Santra, B. K. Samantaray, A. K. Bhowmick, and G. B. Nando, *J. Appl. Polym. Sci., 49*: 1145 (1993).

48. C. Els and W. J. McGill, *Plast. Rubb. Comp. Proc. Appl., 21*: 115 (1994).

49. C. E. Scott and C. W. Macosko, *International Polym. Proc., 10*: 1 (1995).

50. R. Greco, M. Malinconico, E. M. Celli, G. Ragosta, and G. Scarinzi, *Polymer, 28*: 1185 (1987).

43

Polymer Blends and Alloys

B. S. Kaith, A. S. Singha, and Sunil
Regional Engineering College, Hamirpur, India

I. INTRODUCTION

Most materials, be they natural or synthetic, have limited utility. However, technical ingenuity has increased the utility of these materials beyond anyone's wildest imagination. The enormous range of steel that can be produced by adding carbon or other elements to give it the required balance of properties, such as strength and hardness, related to changes in their microstructure [1–3] is just one example.

An analogy can be drawn in the case of pure polymers, i.e., on their own they exhibit extremely limited utility. In fact, many pure polymers are difficult to process into satisfactory products. In order to make them useful, certain additives are incorporated to give ease of processibility, and useful mechanical and load-bearing properties. Based on the broad principles governing the science and technology of compounding, the skill and experience of the compounder plays a vital role in delivering the right material for a particular application [4–25].

One of the major areas of thrust in science and technology for the decade has been the field of materials, of which polymer blends and alloys (PBAs) have the major share and are one of the most dynamic sectors in the polymer industry. Considering the worldwide scenario in the field of PBAs and their growing commercial applicability, this field is probably the right and timely choice for raw material and polymer educators, scientists, and technologists. PBAs are gaining importance in academic and industrial research. These blends have some unique properties that are different from the basic polymers from which these have been produced. To improve the processing behavior for end use, one polymer blending with another polymer is a common practice. The exploitation of certain unique sets of properties of the individual polymer for the benefit of the overall properties of a multicomponent system form the basis of the polymer blending [26–28].

Blending of polymers is the most versatile way for producing new materials with tailor-made properties, which are not possible via copolymerization, for demanding applications in the polymer technology. Block and graft copolymerization share many common features and purposes as blends, but these materials differ from blends by only a few chemical bonds. In block and graft copolymers, the thermodynamics of phase separation of the constituent polymers is different from that of common polymer blends. The existence of covalent bonds greatly influence or control the phase behavior and improve the miscibility of two dissimilar polymeric phases, thereby altering and optimizing their physical and mechanical properties. However, in PBAs proper selection of blend ratios, processing techniques, and use of a selected compatibilizer may be tried to derive different levels of improved physical and mechanical properties; selective chemical modifications may also offer enhanced property balance in making tailor-made polymers of special significance [29–38].

In a fundamental sense, the miscibility, adhesion, interfacial energies, and morphology developed are all thermodynamically interrelated in a complex way to the interaction forces between the polymers. Miscibility of a polymer blend containing two polymers depends on the mutual solubility of the polymeric components. The blend is termed "compatible" when the solubility parameter of the two components are close to each other and show a single-phase transition temperature. However, most polymer pairs tend to be immiscible due to differences in their viscoelastic properties, surface-tensions, and intermolecular interactions. According to the terminology, the polymer pairs are incompatible and show separate glass transitions. For many purposes, miscibility in polymer blends is neither required nor desirable. However, adhesion between the components is one of the most important factors. Optimum adhesion between the dispersed phase (domain) and the matrix are gainfully attained through the use of a third component known as the reactive compatibilizer, which has a chemical structure matching those of the component polymers [9,10,13,39–53].

Following are some of the important terms used in the field of polymer blends and alloys [54].

1. Polymer Blends: Mixture of chemically different polymers or copolymers with no covalent bonding between them.
2. Polymer Alloys: A class of polymer blends, heterogeneous in nature with modified, controlled interfacial properties or morphology.
3. Homologous Polymer Blends: A subclass of polymer blends limited to mixtures of chemically identical polymers differing in molecular mass.
4. Miscible Polymer Blends: A subclass of polymer bends encompassing those blends that exhibit single-phase behavior.
5. Immiscible Polymer Blends: A subclass of polymer blends referring to those blends that exhibit two or more phases at all compositions and temperatures.
6. Partially Miscible Polymer Blends: A subclass of polymer blends including those blends that exhibit a "window" of miscibility, i.e., they are miscible only at certain concentrations and temperatures.
7. Compatible Polymer Blends: A term indicating commercially useful materials, mixture of polymers with strong repulsive forces that is homogeneous to the eye.
8. Interpenetrating Polymer Network (IPN): A subclass of PBs reserved for the mixture of two polymers where both components form a continuous phase and at least one is synthesized or crosslinked in the presence of the other [8].

The PBAs [55], however, must be distinguished from polymeric components that are defined as follows:

1. Composites consisting of two or more physi-

cally distinct and mechanically separable materials.
2. Composites can be made by mixing the separate materials in such a way that the dispersion of one material in the other can be done in a controlled way to achieve optimum properties [10–22].
3. The properties are superior and possibly unique in some specific respects to the properties of the individual components.

II. INTERNATIONAL SCENARIO

The world production of plastics in 1995 is projected at 76 million metric tons (mT) with an annual growth rate (AGR) of 3.7%. The expected AGR of PBAs is 12% and that of composites 16%. In 1987, 21% of polymers were used in blends and 29% in composites and filled plastics [56]. If this trend continues, by 1995 all manufactured resins will be used in multiphase polymeric systems. Two factors moderating the tendency are:

1. A need for single-phase polymers in some applications, e.g., polytetrafluoroethylene, VHMWPE, light-sensitive polymers, etc.
2. Use of polymer alloys and blends as matrices for composites, reinforcements, and foams.

There has been a rapid growth of the demand for plastics from less than 20 billion pounds in 1970 to nearly 50 billion pounds consumed in the United States in 1986, mostly due to the substitution of traditional raw materials. All over the world, plastics have replaced metals, glass, ceramics, wood papers, and natural fibers in a wide variety of industries including packaging, consumer products, automobiles, building and construction, electronics and electrical equipment, appliances, furniture, piping, and heavy industrial equipment [57–121]. Consumption patterns of PBAs in some countries are shown in Tables 1 and 2.

The market study conducted by Frost and Sullivan [122] forecasted that sales of PBAs would continue strongly in the United States with consumption growing from the 1990s total of 363,000 MT to 466,000 MT in 1995.

Table 1 PBA Global Consumption Patterns (in kMT)

PBAs	USA		W. Europe		Japan	
	1986	1996	1986	1996	1986	1996
High-volume PBAs	109	270	76	141	23	7
Engineering plastics (PBAs)	179	409	105	225	88	216
Speciality plastics (PBAs)	8	36	0.5	1	—	—
Growth rate	9.5%		8.2%		10.3%	

Table 2 Consumption of Six Main Families of PBAs in Western Europe (MT)

PP–EPDM:	90,000	Automobile applications:	56%
PC–ABS:	49,000	Appliances:	19%
PS–PPE:	43,000	Electronics:	15%
PC–PBT:	31,000	Industrial uses:	10%
Nylons:	18,000	Construction:	7%
PVC–ABS:	16,000	Consumer applications:	4%

III. ADVANTAGES OF BLENDING

The combination of two or more commercially available polymers through alloying or blending represents an inexpensive route to product differentiation for suppliers. For a processor and end user, alloying and blending technology permits tailoring of a polymer compound to their specific application requirements, often at a lower cost than the current material and over a shorter developmental period. Alloy and blend development is typically market driven and requires an ongoing dialogue between supplier and customer to enable commercial success [123]. The major advantages of blending are as follows:

1. High-impact strength
2. Easy in processibility
3. High-tensile strength
4. Modulus/Rigidity
5. Head deflection temperature
6. Flammability
7. Solvent resistance
8. Thermal stability
9. Dimensional stability
10. Elongation
11. Gloss

There are number of reasons that justify the positive prospects for not only the established PBAs but also for the new ones:

1. PBAs close technological and economic gaps among the thermoplastics.
2. The production of PBAs means that a successful combination of properties can be achieved to almost perfectly fit the respective requirements of an application.
3. The development time needed for PBAs before a product can be launched into the market is shorter when compared with new materials.
4. The blends can be manufactured without high additional capital investment. There is familiarity with the processing of such materials.
5. Comprehensive know-how in PBAs technology has been developed and is constantly expanded.
6. PBAs technology provides an opportunity for

added value for the recycling of plastics, thus, it is an environment-friendly mechanism for plastic wastes management.

IV. SELECTION OF BLENDING COMPONENTS

Selection of blending components is a difficult but important task for polymer technologists. This task can be achieved by the selection of blending components in such a way that the principal advantages of the first polymer will compensate for deficiencies of the second polymer and vice-versa. At the beginning, it is essential to highlight the advantages and disadvantages of the polymers (Table 3) that are to be used and then make combinations to suit the requirements for a particular set of properties. All this should be done while keeping in mind the processibility of the materials, the limitation of the equipment, and the stability of the materials. Moreover, in the selection of the blend components, the economic factor must also be taken into consideration. This can be illustrated by the following examples.

A. Golf Balls

Originally golf balls were made by stuffing feathers under high pressure into a stitched leather container. Subsequently, gutta-percha or balata was used. With the passage of time, the use of gutta-percha became obsolete. Around 1900, a ball with a core made from winding rubber threads was prepared. However, it was a tedious process and the winding techniques differed. With advancements in technology, this process has been re-

Table 3 Formulation of a ''O'' Type of Mount in a Load Deflection Operation

	Phr	
NR	100.0	25.0
Bromobutyle rubber	0.0	75.0
ZnO	5.0	5.0
Acid	1.5	1.5
Polybutadiene nitrile-acrylate	1.0	1.0
IIPD	1.0	1.0
FEF	10.0	10.0
Black pine tar	10.0	10.0
Chlorobutyl-costyrene	1.5	1.5
Sulphur	1.5	1.5
TMTD	0.2	0.2
Physical Properties		
Hardness	40.0–41.0	40.0
Tensile strength (MPa)	14.4	12.7
Mod. 300% MPa	4.9	5.0
Percent elongation	650.0	630.0

Table 4 Typical Thread Recipe

	Percentage weight
cis-Polybutadiene	100.0
Aldehyde/amine	0.5
Zinc oxide	3.0
4,4′ Dithiodimorpholine	1.0
Sulphur	1.0

Cure time at 135°C is 2 h.

placed by a different method of preparing golf balls, i.e., a two-piece ball. One is the "inner core" and the other is the "outer cover"; both are made of separate materials.

1. Inner Core

Since 1960, the inner core has been made from cis-polybutadiene by the compression moulding technique. This replaced the earlier material made from a suspension of barytes or bentonite clay in water and glycerine or the winding of rubber threads made from cis-polyisoprene, either from latex or a dry rubber compound. A typical thread recipe is given Table 4.

This recipe gives an elongation of about 100% with high-tensile strength. But this threaded core has been replaced by a solid moulded core, which is based on the use of a crosslinkage system. The monomers preferred are metal salts of polymerizable organic acids, such as zinc diacrylate and zinc dimethacrylate. These may be formed separately or in situ during compounding.

Generally, metal-containing polymerizable monomers produce cores of higher resilience, for which an extra protection cover is needed. A typical formula is shown in Table 5.

2. Cover Material

Since 1960, gutta-percha or synthetic balata has been replaced by "ionomer" for use as a cover material in golf balls. It is based on a combination of ethylene and a carboxylic acid. It can be processed at a reasonable temperature to produce good, tough ball covers. Because of higher processing temperature of the ionomer

Table 5 Formula for Higher Resilience Cores

	Percentage weight
cis-Polybutadiene	100.0
Zinc diacrylate	35.0
Zinc oxide	18.0
Dicumyl peroxide (40% active)	5.0

Cure time at 160°C is 20 min.

cover material (160°C), separate moulding procedures are adopted for the cover and the inner core.

Technology

Cover compounds with desired properties are obtained by blending two or more different ionomers or by blending with other thermoplastic materials. Half shells produced by injection moulding are linked together for ease of handling. Since ionomers require high-moulding temperatures (160°C), core/cover assemblies can be put into a hot mould, closed rapidly, and then immediately cooled. Special mould coatings are necessary for good release property. Finally, the equator sprue line is, accurately trimmed after moulding.

In a particular application involving a typical "O" type mount, the nitrile rubber (NR) compound causes a resonance frequency of 28 Hz, whereas resonance was previously tolerable only well below 20 Hz. In this case, a blend of NR and bromobutyl rubber is more suitable. A comparison is shown in Table 3.

Similarly, in the case of a cup type mount, an initial attempt to restrict resonance amplification featured the use of white factice in a NR formulation. However, the successful formulation was a blend of predominantly bromobutyl with NR, as shown in Table 6.

B. Automobile Panels and Bumpers

Generally, 85–90% of rubber modified polypropylenes (PP) are used in the automobile industry. It allows for a better way of reducing the weight of the vehicle, increased longevity, improved impact resistance, a glossy appearance, and reduction in cost. Suitable blends contain 10–35%, by weight, of ethylene propylene diene rubber (EPDM). Their specific gravity varies from 0.87–0.89 g/cm^3. Impact strength is retained at normal temperatures down to −30°C and in stable to ultraviolet radiations and heat aging. Recently, the blends have

Table 6 Formulas for a Cup Type Mount

	Phr	
NR	100.0	25.0
Bromobutyl rubber	0.0	75.0
FEF	5.0	5.0
Black pine tar	10.0	10.0
White factice	20.0	20.0
Chlorobutyl-costyrene	1.5	1.5
Sulphur	1.5	1.5
Physical Properties:		
Hardness	35.0	35.0
Tensile strength (MPa)	13.3	11.1
Mod. 300% MPa	4.5	4.7
Percent elongation	650.0	675.0
Amplification at resonance	8.5	3.5

been reinforced with fibrous fillers, such as carbon and glass fibers, for higher impact strength. The PP–EPDMS blends have been replaced by other polymer blends such as poly(2,6 dimethy 1:4 phenylene ether)–polybutylene terephthalate (PPE/PBT), polycarbonates–styrene-comaleic-anhydride (PC/SMA), poly(2,6 dimethyl 1:4 phenylene ether)–polyamide (PPE/PA), and acrylonitrile butadiene styrene–polycarbonates (ABS/PC), etc. [89].

V. CLASSIFIED PBAs

A. Thermoplastic Elastomers (TPEs)

TPEs are a new class of materials, combining the processibility ease of thermoplastics and the functional performance of conventional thermoset elastomers. TPEs need no vulcanization and can be processed using conventional techniques, such as injection moulding, blow moulding, extrusion, etc. The morphology of the TPE provides unique properties to it, and it is a phase-separated system. The first phase is hard and solid, while the second one is soft and rubbery at room temperature. The properties depend on the nature and amount of the hard phase present. The temperature range of applications is determined by the combined effect of the glass-transition temperature of the soft phase and the melting point of the hard phase. There are two categories of TPEs: (1) Block copolymers of two or more monomers, and (2) Simple blends of thermoplastics and elastomers.

Plastics, such as PE, PP, polystyrene (PS), polyester, and nylon, etc., and elastomers such as natural rubber, EPDM, butyl rubber, NR, and styrene butadiene rubber (SBR), etc., are usually used as blend components in making thermoplastic elastomers. Such blends have certain advantages over the other type of TPEs. The desired properties are achieved by suitable elastomers/plastic selection and their proportion in the blend.

B. Elastomeric Alloys (EAs)

EAs are generated from the synergistic interaction of two or more polymers possessing properties better than those of a simple blend and are a special class of thermoplastic elastomers. In EAs the elastomer is crosslinked and dispersed in a continuous matrix of thermoplastic under a dynamic condition resulting in a fine dispersion of fully crosslinked rubber particles in EA. Two-phase EAs differ from the common thermoplastic elastomers containing two-phase morphology by: (1) the presence of a high degree of crosslinked elastomer chains, and (2) the degree of interaction between the thermoplastic and elastomeric phases. The alloying process between the two polymers ultimately attains the following characteristics:

1. Better tensile strength.
2. Lesser tension and compression set.

3. Greater resistance toward solvents and greater flex resistance.
4. Retention of properties at elevated temperatures.
5. More consistent processibility.

Conventional crosslinking agents, such as sulphur, accelerators, and peroxide, etc., used in dynamic vulcanization and melt-mixing of two polymers is the first step in the preparation of an EA. The dynamic vulcanization is done by dynamic shear at a high temperature to activate the process.

A compatibilizer is sometimes used to overcome the interfacial tension between the two phases of dissimilar polymers. It enables a fine dispersion of highly crosslinked rubber particles. The function of the compatibilizer is to provide greater, but not total, thermodynamic compatibility between the two polymers [8].

VI. PREPARATION OF PBAs

PBAs are generally prepared by three commercial methods [57,124]: latex mixing, solution mixing, and melt mixing.

A. Latex Mixing

Latex mixing is the most important technique for the preparation of commercial PBAs. In this process two important factors are to be taken into consideration:

1. pH of latex: It should be approximately equal for better mixing of two polymers.
2. particle size: If small particle size latex is to be blended with a latex of large particle size, stabilizer should be added to the latex with the larger particle size prior to blending the two latexes.

Polymers are suspended as microparticles in the latex and interactions between these microparticles are prevented by the presence of adsorbed suspending agent and soap molecules. Blending results in a random suspension of dissimilar particles in the mixture of latexes, each unaffected by the other. Rate of flocculation depends entirely on the stabilizer and not on the polymer characteristics as such. Coagulated mass contains an intimate mixture of the polymers. Acrylonitrile butadiene styrene (ABS) polymers [23–25] may be prepared by this method.

Compounding and palletizing of the latex blended material is sometimes done by melt mixing. In such cases precautions must be taken to avoid thermal and shear degradation of the blended polymer.

B. Solution Mixing

This method is used for polymers when they are not amenable to melt processing. In this technique, a diluent is added in order to lower the temperature and shear-force requirement for satisfactory mixing of polymers without any degradation. However, attempts to remove the diluent from the mixture may lead to a change in the domain sizes in the blend, which in severe cases may cause gross-phase separation [124,125].

One of the most important solution blend polymers is high-styrene resin, which is manufactured by several companies worldwide. This is a latex blend of high-styrene rubber and normal styrene butadiene rubber. The different high-styrene master batches are available in the world as:

Company name	Product code
1. Japan synthetic rubber	SBR 0051,0061
2. Kubochem	AF POL 537, 539
3. BF Goodrich Tire Group Ameripol Tire Div.	Ameripol 1901, 1904
4. Chemische Werke Huels A.G.	DURANIT B
5. Nippon Zeon Co. Ltd., Japan	NIPOL HS 850, HS 860
6. Synthetics & Chemicals Ltd., India	SYNAPRENE 1958
7. Apar Ltd., India	POWERENE 958
8. Apar Ltd., India	APARENE 198

C. Melt Mixing

In this technique, the melting of constituent polymers is of utmost importance for the generation of a uniform-phase morphology, which finally controls the performance of the blended materials. In general, melt mixing is used for a system in which thermal degradation does not ordinarily occur. To prepare the commercial PBAs, uniform blending of immiscible polymer pairs is often done through the use of a reactive compatibilizer. This method involves the preparation of a master batch of the compatibilizer and the dispersed phase, which is then blended with the matrix through the melt-mixing technique.

In this mixing process, contaminants such as solvent and/or diluents as well as their removal problems can be avoided. Degradation of the polymers is avoided by proper maintenance of the viscosity and shearing rates.

Other methods of blending include: (1) fine powder mixing, and (2) monomer as a solvent for other components of the blend, followed by polymerization for making an interpenetrating network (IPN) [15].

VII. COMPATIBILIZATION OF PBAs

There are many definitions of polymer compatibility. On one hand, compatible polymers are the polymer mixtures that have desirable physical properties when blended, while, on the other hand, those mixtures that do not exhibit gross symptoms of phase separation when blended are said to be compatible blends. However, the most used definition of polymer compatibility is the total miscibility at the molecular level of blending components. Only a few blends form complete miscible systems, which are characterized by a single T_g and homogeneity of 5- to 10-nm scale. Most of the blends are immiscible, possessing a phase-separated morphology. In such cases, important physical and mechanical properties are related to the finely dispersed phase and resistance to gross-phase separation. Moreover, it is reasonable to believe that PBAs exhibiting no gross symptoms of phase separation on blending and having desirable properties show at least some mixing of polymer segments on a microscopic scale. This implies the establishment of some degree of thermodynamic compatibility or the existence of physical constraints such as crosslinking, graft or block copolymer sequences, which are expected of a single-phase material.

The following example provides a clear understanding of the term miscibility. If one refluxes PS on a mill with polyphenyl oxide (PPO) as the second component, a single-phase polymer system results that is thermodynamically stable and is an example of a miscible blend, i.e., polyvinyl chloride (PVC)–NBR system. (already commercialized under the tradename of Noryl by General Electric Co., USA. It has been observed that if PS is refluxed with an equal amount of PVME on a mill at 80°C, a clear one-phase mixture results, indicating a miscible system at that temperature. However, if the temperature of the system is raised to 140–150°C, two phases appear in the system, indicating the development of immiscibility. Thus, the driving forces for the transition from the miscible state (single phase) to immiscible state (two phase) are thermodynamic in origin and do not depend on the extent or intensity of mixing. As a result of recent advances in the field of continuous reactive processing, particularly in extrusion compatibilization of polymer blends through reactions during compounding is becoming increasingly important. The versatility of twin-screw extruders as the reactor combined with wider availability of polymers functionalized with a variety of reactive groups appear to be among the reasons contributing to the growth in the field of reactive compatibilization [126,127]. Methods depending on reactive compatibilization can be classified [126] as follows:

1. Those involving the use of functionalized blend components that produce in situ compatibilizing interchain copolymers: some common reactive functionalities include carboxyl, epoxy, isocyanate, and anhydride, etc.
2. Methods involving the addition of a third high-molecular weight component capable of reacting with at least one blend component. In this

case, compatibilization may result through covalent or ionic bonding.

3. Through the addition of a low-molecular weight component that may promote copolymer formation and crosslinking or co-crosslinking reactions.

Blend of (1) and (2) type categories mostly include the modification of engineering thermoplastics with another thermoplastic or rubber. PS–EPDM blends using a low-molecular weight compound (catalyst) Lewis acid have been developed [126]. Plastic–plastic blends, alloys of industrial importance, thermoplastic elastomers made by dynamic vulcanization, and rubber–rubber blends are produced by this method.

VIII. CHARACTERIZATION OF PBAs

Complexity within homopolymers as well as that of PBAs have made the task of analysis and characterization a difficult one. Basically, the task of analysis and characterization of PBAs is not different from that of simple low-molecular weight polymers, provided adequate solubility and sites are available for accepting artificial stimulation responses to those stimuli that may be used as functional tools for characterization. Properties of the blend mainly depend on the homogeneity of blends. The processes that are used for characterization of the PBAs are discussed in the following sections [128–131].

A. Microscopy

Electron and optical microscopes are being used to see blend homogeneity. Elastomer–plastic blends are somewhat easier to identify than elastomer–elastomer blends because normal staining techniques, e.g., osmium tetraoxide, can be used in the case of plastic–elastomer blends. Normally, there are two methods that are followed for examining the blend surface by electron microscopy.

1. Ebonite Method
The blends are cured with a sulphur–sulphonamide zinc stearate mixture so that the elastomer phase is hardened and then microtomed.

2. Cryogenic Method
The blends are frozen in liquid nitrogen and then microtomed and stained with osmium tetraoxide, which stains only unsaturated elastomers.

B. Solubility Method

Solubility differences among components of polymer blends can be utilized in different ways to identify the polymer blends.

C. Optical Properties

Light scattering of polymer blends also helps in characterizing the different phases of PBAs.

D. Thermal and Thermomechanical Analysis

Thermal analysis helps in measuring the various physical properties of the polymers. In this technique, a polymer sample is subjected to a controlled temperature program in a specific atmosphere and properties are measured as a function of temperature. The controlled temperature program may involve either isothermal or linear rise or fall of temperature. The most common thermoanalytical techniques are: (1) differential scanning analysis (DSC), (2) thermomechanical analysis (TMA), and (3) thermogravimetry (TG).

These techniques help in providing the following information: specific heat, enthalpy changes, heat of transformation, crystallinity, melting behavior, evaporation, sublimation, glass transition, thermal decomposition, depolymerization, thermal stability, content analysis, chemical reactions/polymerization linear expansion, coefficient, and Young's modulus, etc.

DSC helps in determining the glass-transition temperature, vulcanization, and oxidative stability. TG mainly is applied for the quantitative determination of major components of a polymer sample. TMA or DLTMA (dynamic load thermomechanical analysis) measures the elastic properties viz. modulus.

Thermal analysis of homopolymer samples are simpler than those of blends. Separate thermal analysis of individual polymer components are made before doing the same for a blend in order to get more accurate and proper information on thermal characteristics.

It is possible to distinguish between SBR and butyl rubber (BR), NR and isoprene rubber (IR) in a vulcanizate by enthalpy determination. In plastic–elastomer blends, the existence of high T_g and low T_g components eases the problems of experimental differentiation by different types of thermal methods. For a compatible blend, even though the component polymers have different T_g values, sometimes a single T_g is observed, which may be verified with the help of the following equation:

$$\frac{1}{T_g} = \frac{w_1}{T_{g_1}} + \frac{w_2}{T_{g_2}}$$

where w_1 and T_{g_1}, w_2 and T_{g_2} are the weight fractions and the glass-transition temperature of the two polymer components 1 and 2, respectively [129]. Glass-transition temperature (T_g) of the blend can also be determined by the differential scanning calorimeter, which gives indication about the nature of the blend.

E. IR and NMR Spectroscopy

The miscibility of two or more polymers in the solid state can be investigated by this technique.

F. X-Ray Diffraction Method

For the structure evaluation this method is one of the most suited methods.

G. Gel Permeation Chromatography

It is basically a fractionation process that depends not only on molecular size, but also on chemical composition, stereo-configuration, branching, and crosslinking. For multicomponent systems, fractionation with different ion polymolecularity, chemical heterogeneity and sequence length distribution, solubility or elution fractionation is of primary importance. Therefore, gel permeation chromatography or size exclusion chromatography is used as an important tool for the characterization of PBAs.

IX. BLEND MORPHOLOGY/PROPERTY RELATIONSHIP

One of the basic criteria for the process choice is the optimum size and shape of the dispersed phase leading to better overall properties. As a brief guideline, Table 7 can be used to choose the correct manufacturing process.

In multiphase polymeric systems, the properties of the end products do not solely depend on the properties of the pure components, but other various parameters also have a great impact (Fig. 1). In order to emphasize these factors, the following systems are taken into consideration: (1) elastomer toughened styrene system, (2) elastomer toughened polycarbonate blends, and (3) direct reactive blend processing.

A. Elastomer Toughened Styrene System

In the late 1940s, the demand for styrene homopolymers (PS) and styrene–acrylonitrile copolymers (SAN) was drastically reduced due to their inherent brittleness. Thus, the interest was shifted to multiphase high-impact polystyrene (HIPS) and rubber-modified SAN (ABS). In principle, both HIPS and ABS can be manufactured by either "bulk" or "emulsion" techniques. However, in actual practice, HIPS is made only by the bulk process, whereas ABS is produced by both methods [132,133].

1. Bulk Process

Initially, this process consists of a rubber dissolved in a monomer(s). During polymerization, the phase homogeneous system transforms into the heterogeneous system over a wide conversion range. The heterogeneous phase originates due to the formation of a subemulsion [134] consisting of polystyrene droplets in a rubber solution and can be correlated with the viscosity conversion profile [135].

Table 7 Effect of Morphological and Molecular Aspects on the Technological Properties

Technological properties	Morphological aspects (dispersed phase)		Molecular aspects (continuous phase)	
	Variation due to increasing			
	Phase–volume ratio	Particle size and its distribution	Reactive compatibilization	Molecular weight and its distribution	Processing aids
Impact strength	Increases sharply	Increases, reaches an optimum followed by a decrease	Increases, reaches an optimum followed by a decrease	Increases sharply	Increases marginally
Modulus	Decreases sharply	Decreases slowly	Increases slowly	Does not change	Decreases slowly
Flow properties	Becomes inferior	Improves marginally	Improves marginally	Becomes inferior	Improves
Solvent crazing resistance	Enhances	Enhances	Initially remains unchanged then deteriorates	Enhances	Deteriorates
Heat deflection temperature	Lowers	Remains unchanged	Initially remains constant followed by an increase	Remains unchanged	Lowers

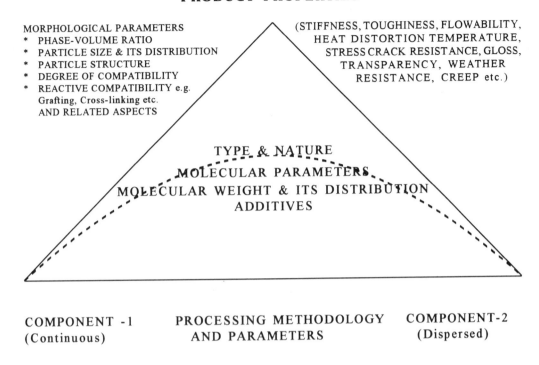

PRODUCT PROPERTIES

MORPHOLOGICAL PARAMETERS
* PHASE-VOLUME RATIO
* PARTICLE SIZE & ITS DISTRIBUTION
* PARTICLE STRUCTURE
* DEGREE OF COMPATIBILITY
* REACTIVE COMPATIBILITY e.g.
  Grafting, Cross-linking etc.
AND RELATED ASPECTS

(STIFFNESS, TOUGHINESS, FLOWABILITY,
HEAT DISTORTION TEMPERATURE,
STRESS CRACK RESISTANCE, GLOSS,
TRANSPARENCY, WEATHER
RESISTANCE, CREEP etc.)

TYPE & NATURE
MOLECULAR PARAMETERS
MOLECULAR WEIGHT & ITS DISTRIBUTION
ADDITIVES

COMPONENT -1
(Continuous)

PROCESSING METHODOLOGY
AND PARAMETERS

COMPONENT-2
(Dispersed)

MANUFACTURING - BULK/EMULSION etc.
METHODOLOGY - MELT/ SOLUTION etc.

Figure 1   Molecular and morphological parameters influencing technological properties of multiphasic polymer blends.

Molecular Parameters

The molecular weight of the continuous phase is an important parameter that affects the mechanics and the melt flow of the end product. It can be controlled by the use of a suitable chain transfer agent (e.g., *tert*-dodecyl mercaptan C_T = 4.0) or their combinations (e.g., primary mercaptans C_T = 26.0 and dimeric α-methyl styrene C_T = 0.1) [132].

The particle size of the dispersed phase depends upon the viscosity of the elastomer–monomer solution. Preferably the molecular weight of the polybutadiene elastomer should be around 2×10^5 and should have reasonable branching to reduce cold flow. Furthermore, the microstructure of the elastomer provides an important contribution toward the low-temperature impact behavior of the final product. It should also be emphasized that the use of EPDM rubber [136] or acrylate rubber [137] may provide improved weatherability. It has been observed that with an increase in agitator speed the mean diameter of the dispersed phase (D) decreases, which subsequently levels out at high shear [138–141]. However, reagglomeration may occur in the case of bulk

ABS at high shears [142]. $\overline{D}$ also increases with the increase of the viscosity ratio of the phases [143].

Any additive, such as different surfactants and grafted copolymers that lower the interfacial tension, decrease the mean particle size of the dispersed phase (Fig. 2). In order to enhance compatibility of the two phases, the length of the block copolymers, acting as a surfactant, must be greater than the critical entangled chain length. In general, particle size <1 μm is not effective in toughening. Particle size between 1–2 μm is satisfactory in the toughening process. However, submicron particle size can act as a good toughner if the dispersed phase is crosslinked. It is interesting to note that blending of two different grades of high–impact polystyrene (HIPS) having widely different dispersed phase particle sizes can easily be used for optimization of impact strength and gloss (Table 8) [143].

Some common structural features of the dispersed phases that can be observed in HIPS are shown in Fig. 3 [144].

The importance of the morphological aspects of the HIPS on their Izod impact strength and gloss has been

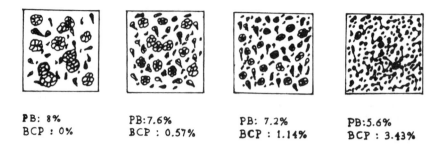

PB: 8%   PB:7.6%   PB: 7.2%   PB:5.6%
BCP : 0%   BCP : 0.57%   BCP : 1.14%   BCP : 3.43%

Figure 2   Influence of added styrene–butadiene block copolymer (BCP) on the particle size.

highlighted for some commercially available HIPS in Fig. 4 [132].

Another problem faced during cooling of the melt is the mismatch of the shrinkage behavior of the continuous and the dispersed phases. Triaxial stress thus developed around the dispersed phase will lead to craze nucleation under load. Void formation within the system can be avoided by the incorporation of a minimum amount of crosslinks, which will help in retaining the elasticity, load modulus, and glass-transition temperature. Those thermally activated crosslinking will significantly affect the swelling index of HIPS–ABS [145].

Morphological Parameters

Phase–value ratio depends upon the following factors: dispersed phase contents, number and size occlusion, and degree of reactivity.

During the manufacturing process, if the grafting increases during early stages of the reaction, the phase volume will also increase, but the size of the particles will remain constant [146–148]. Furthermore, reactor choice plays a decisive role. If the continuous stirred tank reactor (CSTR) is used, little grafting takes place and the occlusion is poor and, consequently, the rubber efficiency is poor. However, in processes akin to the discontinuous system (e.g., tower/cascade reactors), the dispersed phase contains a large number of big inclusions.

Dispersed Phase Particle Size

The size of the dispersed phase is effectively decided within a small conversion zone (between

phase–inversion and high viscosity). However, within this zone it depends upon the agitator speed, resulting shear stress, viscosity ratio of phases, and nature of the interphase. It can be qualitatively summarized in the fol-

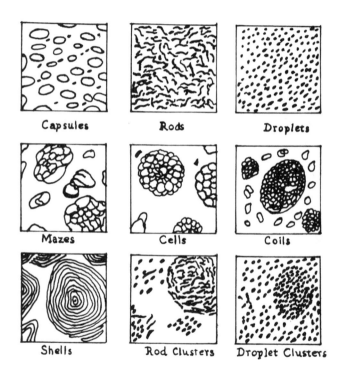

Capsules   Rods   Droplets

Mazes   Cells   Coils

Shells   Rod Clusters   Droplet Clusters

Figure 3   Particle structure in HIPS.

Table 8   Impact Strength and Gloss of HIPS Blends

Rubber particle size (μm)	Large particles in blends (%)	Rubber phase volume fraction	Notched Izod impact strength ft. lbs/in.	Gloss
0.6	0.0	0.200	0.90	100.0
3.0	100.0	0.450	1.85	50.0
0.6	5.0	0.213	1.85	96.3
3.0	30.0	0.275	1.90	81.3

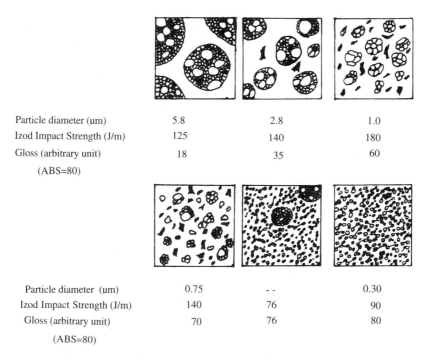

Particle diameter (μm)	5.8	2.8	1.0
Izod Impact Strength (J/m)	125	140	180
Gloss (arbitrary unit)	18	35	60
(ABS=80)			

Particle diameter (μm)	0.75	- -	0.30
Izod Impact Strength (J/m)	140	76	90
Gloss (arbitrary unit)	70	76	80
(ABS=80)			

Figure 4 Correlation between morphology, impact strength, and gloss of HIPS.

lowing equation [149]:

$$D_c = \frac{K\lambda\sigma\eta_d^{\beta-1}}{\eta_d^{\beta}} = K' \frac{\sigma}{\tau}(\eta_d/\eta_c)^{\beta-1}$$

where

D_c = critical drop break-up diameter
K, K' and β = constants
λ = time constant
η_c and η_d = viscosity of the continuous and dispersed phases, respectively.
τ = shear stress.

Transparent Heterogeneous Blends

In general, the multiphasic heterogenous nature of the impact grade styrene-based polymers is the root cause of their opaque–turbid nature. In determining the transparency of the blends, size and the size-distribution pattern of the dispersed phase along with the refractive index difference between the continuous and the dispersed phases are two very important criterion [133].

High-Heat Distortion Grade Blends

High-distortion temperature (HDT) of the styrenic materials is around 100°C. Blending with polymers of high HDT may lead to blends of high HDT. Both compatible and noncompatible polymers can be used judiciously for this purpose, e.g., PS is compatible with PPE (HDT = 220°C) [150] on a molecular level. Therefore, it is not surprising that a blend of HIPS with PPE will exhibit the same morphology as those of the original HIPS. However, the size of the dispersed phase is smaller than the conventional systems. Thus, the proper formation will lead to the blends exhibiting high-heat resistance, high-impact strength, high-dimensional stability, high-hydrolytic stability, and halogen-free flame retardant compositions.

B. Elastomer Toughened Polycarbonate Blends

Heterogeneous compatible blends of preformed elastomers and brittle plastics are also an important route for the development of blends of enhanced performance with respect to crack or impact resistance. Polycarbonate blends with preformed rubber particles of different sizes have been used to provide an insight into the impact properties and the fracture modes of these toughened materials. Izod impact strength of the blends having 5–7.5 wt% of rubber particles exhibits best overall product performance over a wide range temperature (RT to −40°C) [151–154].

C. Direct Reactive Blend Processing

Recently, it has been proposed that direct extrusion processing may be a far better alternative for the blend preparation. It has various advantages, such as a one-step fabrication process, a reduction of the production cost, and it minimizes the thermal degradation of the polymer blend components by reducing their thermal history from compounding to moulding.

Sakai and coworkers [155] have developed a direct compounding and extrusion system, i.e., a hybridized processing system for manufacturing PBAs with excellent impact strength. The main purpose of this direct reactive blend processing technology is reduction of moulding pressure and temperature, utilization of low-melt viscosity thermosetting resins, improvement in physical properties of resultant thermosets, such as impact strength, to obtain blends/alloys equivalent and/or superior to corresponding thermoplastics by incorporation of rubbery. The process is composed of the following three distinct stages:

First Stage

Mixing and kneading of the thermosetting resin, rubbery components and other additives.

Second Stage

At the downstream of the extrusion process a suitable reaction catalyst, reactive diluent, e.g., crosslinking monomer, is fed into the molten polymer mix.

Third Stage

Fabrication of end product in the moulding press.

X. INDUSTRIAL PERSPECTIVES

The United States, Western Europe, and Japan consume almost 88% of the performance plastics of the total world production. Speciality plastics, such as PEEK, LCP, polyimides, conductivity plastics, and polyetherimides, etc., are exclusively consumed by them. In the era of the globalization of business, global players have little motivation to give up their dominant position in the field of engineering plastics, thus restricting the flow of the technology. Further technological trends are, therefore, likely to be driven toward the knowledge-based materials rather than capital-intensive large-volume materials. Speciality engineered plastics are, therefore, the front-runner of plastic materials for the next decade. The major challenges at the present state of the engineering plastics industry may be identified as building up of own technological base for negotiating on equal footing with technology owners, encouraging the development of speciality plastics and their applications to encase intrinsic technical strength, establishing joint research and development (R&D) ventures with the engineering plastics manufacturers for exploiting bigger markets, and developing a culture of cooperative teamwork in order to achieve the goals.

A. Commercial PBAs

Commercial thermoplastics are the engineering materials containing two or more compatibilized polymers that are chemically bounded in a way that creates a controlled and stable morphology with a unified thermodynamic profile. In view of multiplicity and contradictory requirements of various properties for most of the applications, almost all the commercial PBAs are made of two or more thermoplastics, elastomeric modifiers along with a series of compatibilizers with modifiers compounded together. A considerable number of blends have been appearing in the market regularly, some of which are listed in Table 9.

Generally, in each of the PBAs a specific characteristic is being imparted by the dispersed phase and the resultant material has certain properties that are an optimized combination of both the components. Some of the resultant features of the PBAs and the contribution of each phase is shown in Table 10.

Table 9 List of the Engineering PBAs

Materials	Trade name	Manufacturers
PC–ABS	Bayblend	Bayer
	Pulse	Dow Chemicals
	Cycloy	GE Plastics
	Triax-2000	Monsanto
	Aruloy	ARCO
PC–ASA	Geloy-xp 4001	GE Plastics
PC–PBT	Makroblend	Bayer
	Sabre	Dow Chemicals
	Ektar	Eastman Performance
	Xenoy	GE Plastics
Nylon–PPE	Noryl GTX	GE Plastics
	Ultranyl	BASF
	Dimension	Allied Signal
	MPPE	Ashahi
Nylon–ABS	Triax-1000	Monsanto
	Elemid	Borg-Werner
Nylon–PP	Orgalloy	Atochem
	Dexlon	D & S Plastics
	Enphite	Chisso
Nylon–Polyolefins	Akulon	DSM engineering
	Selar	Dupont
Nylon–Polyarylate	Bexloy	Dupont
Nylon–PC	Dexcarb	D & S Plastics
PBT–POM	Duralloy	Celanese
ABS–PBT	Cycolin	GE Plastics
SMA–PBT	Dylark	ARCO
	Steron	Firestone
	K-Resin KRO3	Phillips
Polysulfone–ABS	Mindel A & B	AMOCO

ASA = acrylate-costyrene-coacrylonitrile; POM = polyoxymethylene. For other abbreviations see text.

Table 10 Commercially Available Engineering PBAs: Contribution from Each Component

Trade Name	Polymer -I	PBAs	Polymer-II	
Noryl-GTX GE Plastics Nylon Nylon/PPE Blends	Processing Impact Strength Crystallinity *(adv.)* / Water Absorption *(disadv.)*	High Heat Resistance Chemical Resistance Dimensional Stability Ductility Less Moisture Abs.	Heat Distortion temp. Rigidity *(adv.)* / Processing Impact Resistance *(adv.)*	PPE/ PPO
Xenoy PC GE Plastics PC/PBT Alloy	Low temperature Toughness *(adv.)* / Stress crack sensitivity Chemical Resistance *(disadv.)*	Balance of chemical, Resist. Low Temp. Impact strength Toughness Rigidity, Shrinkage Stress crack Resistance	Chemical Resistance Crystallinity *(adv.)* / Shrinkage Toughness Processibility *(adv.)*	PBT/ PET
Bayblend Mobay PC PC/ABS Alloy	Low Temperature Toughness HDT *(adv.)* / Stress Crack Sensitivity Chemical Resistance *(disadv.)*	Processibility Impact strength at Broad Temp. range Heat Resistance Stiffness	Impact Resistance Processing Weatherability *(adv.)* / HDT *(disadv.)*	ABS/ ASA
Noryl PPO GE Plastics PPO/HIPS Alloys	HDT Rigidity Flame Resistance *(adv.)* / Processing Impact Strength *(disadv.)*	Improved Processing Combination of Heat Resist. with dimensional stability Toughness with Impact resistance	Processing Impact Resistance *(adv.)* / HDT *(disadv.)*	HIPS

☐ Advantages
▣ Disadvantages

The properties profile of some of the PBAs is illustrated in Table 11.

B. R&D Management of PBAs Development

Novel copolymerization and alloying technology were predicted to be the critical forces for shaping the development of new materials in the 1990s. However, the reality of this technology indicates a period of cautious reassessment of the daunting technical and economic hurdles for the development of PBAs in view of the challenges of compatibilization, cost performance profile in actual applications, and competition from recycling of single resins.

Table 11 Properties Profile of Some of the Engineering PBAs

Properties	Materials					
	Nylon–ABS	Nylon–PPE	Nylon–PP	PC–PBT Alloy	PC–ABS Alloy	Polysulfon–ABS
Specific gravity	1.06	1.08	1.04–1.13	1.17–1.22	1.09–1.12	1.13
Tensile-strength-psi	6800–7400	8000	6500–7500	6500–7600	6600–8000	7200
Elongation %	270–330	—	—	—	5	4
Izod impact str. ft. lb/in.	16–19	10	18	2.4–18	6.9–12	7.00
Flex. modulus $\times 10$ E3	600	275	250–300	250–300	330–370	316
HDT-°F @264 psi	167	250	160	135–190	205–220	300
Moisture abs. 24 h	1.2	0.85	1.05	0.15	0.05	—

The technology required for PBAs is not less complex than that needed to develop a totally new polymer. However, the cost-risk of developing alloys from existing material stocks is still less than that for inventing new polymers.

R&D management of PBAs development is an important and complex task as the time, money, and technical expertise required for development of a successful PBA virtually mirrors the commitments necessary to invent a new polymer with market application considerations. The following flow diagram (see below) briefly illustrates the practices and designs of PBAs, particularly of engineering plastics.

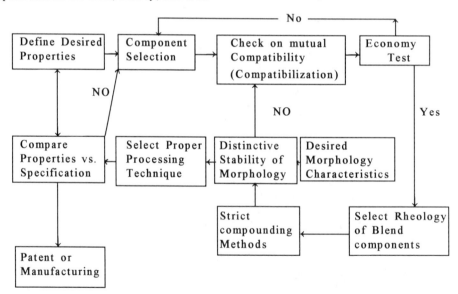

Table 12 Polymeric Modifiers Used for Various Matrices in Engineering

Property	Matrix	Modifiers
Impact strength toughness	PVC, PP, PE PC, PA, PPE, PET	ABS, ASA, SEBS, Block copolymer, SEBS, EPR, EPDM, NBR, MBA, etc.
Heat distortion temp. stiffness	PC, PA, ABS, SAN	PET, PPE, Arylates, PC, PSO
Chemical resistance	Amorphous polymers	Crystalline polymers
Barrier properties	Polyolefins	PA, PVOH, PVDC
Processibility	PVC, PP, PE, PS, etc.	Liquid crystal polymers

PET = polyethylene terephthalene; SEBS = styrene–ethylene butylene–styrene; PSO = polysulfone. For other abbreviations see text.

PBAs are designed explicitly to meet the needs of specific applications on the basis of their property-processing-cost performances. One polymer is incorporated into the matrix of other polymers to impart specific characteristics as per the requirement along with the appropriate compatibilizer to ensure stress transfer in between phases. The polymer blend constituents and composition must, therefore, be selected on the basis of the compensation of properties, considering the advantages and disadvantages associated with each phase. Table 12 indicates some of the components used as modifiers.

XI. APPLICATIONS OF PBAs

PBAs technology established itself in the late 1980s as a premier field for the creation of new engineering materials and has promised virtually unlimited combinations of resins that could target specific market applications while taming the processing difficulties and prohibitive cost of exotic polymers. During the 1990s novel copolymerization and alloying technology came into the field for fabrication of the materials with desired properties. PBAs are used commercially in various fields, such as the aerospace industry [58,59,156] and the military as well as in industries and domestic articles, such as adhesives [60–83], development of new materials with improved properties, such as gas and moisture barrier, resistance to heat and cold, chemicals, corrosion, abrasion and moisture, etc. [84–121,156–164].

ACKNOWLEDGMENT

The authors acknowledge with a profound sense of gratitude and obeisance the generous guidance and constant encouragement extended by the esteemed and erudite principal, Dr. R. L. Chauhan, Regional Engineering College, Hamirpur, India, during the preparation of this chapter.

REFERENCES

1. C. W. Bert and P. H. Francis, *Trans. N. Y. Acad. Sci. Ser.*, II, *36(7)*: 663 (1974).
2. J. E. Gordon, *The New Science of Strong Materials*, 2nd. ed., Penguin Books Ltd.
3. H. Worth, UK, *An Excellent Historical Perspective on Structural Materials and Fine on Stiffness, Strength and Toughness* (1976).
4. P. J. Fenelon, *Plastics Eng., 37(a)*: 37 (1981).
5. R. L. Jalbert, *Alloys in MPE*, 127 (1982).
6. H. F. Mark et al. (eds.), *Composites, Encyclopedia of Polymer Science and Engineering*, 3rd ed., 3 (1985).
7. H. F. Mark et al. (eds.), *Composites, Encyclopedia of Polymer Science and Engineering*, 4th Ed. 4 (1986).
8. F. W. Billmeyer, Jr., *Text Book of Polymer Sciences*, 3rd ed., A Wiley-Interscience Publication, John Wiley and Sons, New York, p. 470.
9. E. Rofael and H.-A. May, *Proceedings of the Seventeenth International Symposium on Particle Board Composite Materials*, (T. M. Maloney, ed.) Washington State University, Pullman, Washington, pp. 283–295 (1983).
10. D. B. Agarwal and L. J. Broutman, *Analysis and Performance of Fibre Composites*, Wiley-Interscience, New York (1980).
11. P. Beardmore, et al., *Science, 208*: 833 (1980).
12. M. R. Piggott, *Load-Bearing Composite Materials*, Pergamon Press, New York (1980).
13. S. W. Tsai and H. T. Hahn, *Introduction to Composite Materials*, Technomic, Westport, Connecticut (1980).
14. J. Delmonte, *Technology of Carbon and Graphite Fibre Composites*, Van Nostrand Reinhold, New York (1981).
15. L. H. Sperling, *Interpenetrating Polymer Networks and Related Materials*, Plenum Press, New York (1981).
16. D. Hull, *Introduction to Composite Materials*, Cambridge University Press, Cambridge (1981).
17. F. J. McGarry, *ECT, 13*: 968 (1981).
18. J. S. Hearsons, *MPE*, 153 (1982).
19. G. Lubin, *Handbook of Composites*, Van Nostrand Reinhold, New York (1982).
20. L. H. Miner, *MPE*, 152 (1982).
21. H. L. Peterson, *MPE*, 152 (1982).
22. R. P. Sheldon, *Composite Polymeric Materials*, Elsevier, New York (1982).
23. C. H. Baskedis, *ABS Plastics*, Reinhold, New York (1964).
24. G. A. Morneau et al., *ECT, 1*: 442 (1978).
25. J. M. Lantz, *MPE*, 6 (1982).
26. J. A. Manson and L. H. Sperling, *Polymer Blends and Composites*, Plenum Press, New York (1976).
27. D. R. Paul and Seymour Newman (eds.), *Polymer Blends*, Academic Press, New York (1978).
28. O. Olabisi, *ECT, 18*: 443 (1982).
29. W. M. Saltman, *EPST, 2*: 678 (1965).
30. J. P. Kennedy and E. Törnquist (eds.), *Polymer Chemistry of Synthetic Elastomers*, Wiley-Interscience, New York, Part I (1968) and Part II (1969).
31. I. D. Rubin, *Poly (1-Butene): Its Preparation and Properties*, Gordon and Breach, New York (1968).
32. R. G. Bauer, *ECT, 8*: 608 (1979).
33. L. J. Kuzma and W. J. Kelly, *ECT, 8*: 546 (1979).
34. J. E. McGrath, *ECT, 8*: 446 (1979).
35. J. A. Brydson, *Rubber Chemistry*, Applied Science, London (1978).
36. H. W. Robinson, *EPST, 6*: 505 (1967).
37. H. L. Hsieh et al., *Chem. Tech., 11*: 626 (1981).
38. D. C. Blackley, *Synthetic Rubbers: Their Chemistry and Technology*, Elsevier, New York (1983).
39. M. G. Arora et al., Industrial Polymers, *Polymer Chemistry*, 1st ed., Anmol Publications Pvt. Ltd., New Delhi, p. 277.
40. R. W. Lang, et al. *Polym. Eng. Sci., 22*: 982 (1982).
41. J. C. Seferis and L. Nicolais, (eds.), *The Role of the Polymeric Matrix in the Processing and Structural Properties of Composite Materials*, Plenum, New York (1983).
42. J. P. Trotignon et al., *Polymer Composites, 3*: 230 (1982).
43. L. J. Buckley et al., *SAMPE, Q*: 16 (1984).
44. T. Mathews, *Polymer Mixing Technology*, Elsevier, North-Holland, Inc, New York, Applied Science Publishers, Barking, U.K. (1982).
45. L. Mascia, *The Role of Additives in Plastics*, Edward Arnold Publishers, Ltd., London (1974).

46. J. A. Brydson, *Plastic Materials*, 4th Ed., Butterworth Publishers, Ltd., London (1982).
47. D. V. Rosato, *Handbook of Composites*, (G. Dubin, ed.), Van Nostrand Reinhold Co., New York (1982).
48. K. M. Prewo, *Frontiers in Materials Technology*, (M. A. Myers, ed.), Elsevier, Amsterdam (1985).
49. W. K. Fischer, U.S. Patent 3806, 558, 3835, 201 (1979).
50. H. L. Morris, *J. Elastomer Plast.*, *6*: 1 (1974).
51. R. J. Crawford, *Plastic Engineering*, 2nd ed. (1987).
52. M. O. W. Richardson, *Polymer Engineering Composites*, Applied Science Publications, London (1977).
53. F. Rodriguez, *Principles of Polymer Systems*, 2nd ed., McGraw Hill, International Ed., p. 258 (1983).
54. D. Mull, *An Introduction to Composite Materials*, Cambridge Solid-State Science Series, Cambridge University Press, Cambridge, Massachusetts, 3 (1981).
55. J. M. G. Couri, *Encyclopedia Polymer Science Eng.* (J. I. Krotschwitz, Exe. ed.), 2nd. Editing Supplement Volume, Wiley Guter Science Publication, John Wiley and Sons, New York, pp. 455–480 (1989).
56. L. A. Utracki et al., *Polymer Alloys, Blends and Ionomers: An overview in Multiphase Polymers: Blends and Ionomers* (L. A. Utracki and R. A. Weiss, eds.), ACS Symposium Series No. 395, 1–35 (1989).
57. S. Y. Kienzle, *Advances in Polymer Blends and Alloys Technology* (Kohudic, ed.), Technomic Publishing Co. Inc., Lancaster, 1, 1–11 (1988).
58. B. C. Hoskin and A. A. Baker, *Composite Materials for Aircraft Structures*, AIAA Publ. Inc., New York (1986).
59. S. M. Lee (ed.), *International Encyclopedia of Composites*, 1–4, VCH, New York (1990).
60. P. J. Corish, *Science and Technology of Rubber* (F. R. Erich, ed.), Academic Press, New York (1978).
61. B. S. Gesner, *Encyclopedia of Polymer Science and Technology*.
62. S. N. Angove, *Rubber J. 149(3)*: 37 (1967).
63. D. C. Blackley and R. C. Charnock, *J. Inst. Rubber Ind. 7*: 60 (1973).
64. D. C. Blackley and R. C. Charnock, *J. Inst. Rubber Ind. 7*: 113 (1973).
65. A. King, *RAPRA Res. Bull.*, 127 (1964).
66. M. H. Walters and D. N. Keyte, *Trans. Inst. Rubber Ind.*, *38*: 40 (1962).
67. L. Bohn, *Rubber Chem. Tech.*, *41*: 495 (1968).
68. K. A. Burgess et al., *Rubber Age, 97(9)*: 85 (1965).
69. D. J. Angier and W. F. Watson, *J. Polym. Sci.*, *18*: 129 (1955).
70. D. J. Angier and W. F. Watson, *Trans. Inst. Rubber Ind.*, *33*: 22 (1957).
71. W. M. Hess, et al., *Rubber Chem. Tech.*, *66*: 329 (1993).
72. N. Tokita, *Rubber Chem. Tech.*, *50*: 292 (1977).
73. G. N. Avgeropoulos, et al. *Rubber Chem. Tech.*, *49*: 93 (1976).
74. W. M. Hess, et al., *Rubber Chem. Tech.*, *47*: 64 (1974).
75. L. Evans, et al., *Rubber Age, 94*: 272 (1963).
76. W. M. Hess, et al., *Rubber Chem. Tech.*, *50*: 301 (1977).
77. W. M. Hess et al., *Rubber Chem. Tech.*, *58*: 350 (1985).
78. P. A. Marsh et al., *Rubber Chem. Tech.*, *40*: 359 (1967).
79. P. A. Marsh et al., *Rubber Chem. Tech.*, *43*: 400 (1970).
80. H. E. Trexler et al., *Kautsch Gummi Kunsts.*, *40*: 945 (1987).
81. A. Y. Coran and R. Patel, *Rubber Chem. Tech.*, *54*: 892 (1981).
82. A. Y. Coran and R. Patel, *Rubber Chem. Tech.*, *55*: 116 (1982).
83. D. K. Setua, et al., *Kautsch Gummi Kunsts.*, *44*: 821 (1991).
84. R. R. Goddard, *Monitor, 31*: 1 (1990).
85. F. S. Proste, *Monitor, 31*: 11 (1990).
86. W. J. Kores, *Barrier Polymers and Structures: Overview in Barrier Polymers and Structures* (W. J. Kores, ed.) *ACS*, Washington, 1 (1990).
87. S. A. Stlrn and S. Tohalaki, *Fundamentals of Gas Diffusion in Rubbery and Glassy Polymers, ACS*, Washington, 22 (1990).
88. P. M. Subramaniam, *Polymer Blends: Morphology and Solvent Barriers, ACS*, Washington (1990).
89. R. Bell et al., *Automotive Polymers and Design*, 13 (1991).
90. W. M. Hess, et al., *Rubber Chem. Tech. 66*: 329 (1993).
91. A. Y. Coran and R. Patel, *Rubber Chem. Tech.*, *53*: 141 (1980).
92. A. Y. Coran, *Rubber Chem. Tech.*, *63*: 599 (1990).
93. A. Y. Coran and S. Lee, *Rubber Chem. Tech.*, *65*: 231 (1992).
94. T. R. Wolfe, *Thermoplastic Elastomers* (N. R. Legge et al., eds.) Hanser, New York, Chap. 6 (1987).
95. C. P. Roder, *Handbook of Thermoplastic Elastomers* (B. M. Walker, ed.) Van Nostrand-Reinhold, New York, Chap. 4.
96. L. H. Sperling, *IPN and Related Materials*, Plenum Press, New York (1981).
97. E. T. McDonel et al., *Polymer Blends*, II, (D. R. Paul et al., eds.) Academic Press, New York, Chap. 19.
98. P. Ghosh et al., *Polymer Science*, Sivaram, Tata McGraw Hill, p. 871.
99. L. M. Glanville and P. W. Milner, *Rubber Plast. Age*, *48*: 1059 (1967).
100. J. F. Svetlik et al., *Rubber Age, 96(4)*: 570 (1965).
101. S. M. Hirschfield, U.S. Patent, 3,281,289 (1966).
102. C. W. Snow, U.S. Patent, 3,280,876 (1966).
103. V. Abagi et al., *Mater-Plast. (Buchanest), 6(2)*: 84 (1969).
104. Chemische Werke Hüls A. G. French Patent 1, 499, 094 (1966).
105. W. C. Flanigan, *Rubber Age, 101(2)*: 49 (1969).
106. R. N. Kienle et al., *Rubber Chem. Tech.*, *44*: 996 (1971).
107. A. Springer, *RAPRA* Transl., 1170 (1964).
108. M. S. Sutton, *Rubber World, 149(5)*: 62 (1964).
109. K. Satake et al., *J. Inst. Rubber Ind.*, *4*: 102 (1970).
110. M. S. Banerji et al., *International Conference Polymer '94, India IV*, BCN–29.
111. D. D. Dunnom and H. K. de Decker, *Rubber World, 151(6)*: 108 (1965).
112. B. Topcik, *Mater. Plast. Elastomeri, 35*: 201 (1969).
113. S. A. Banks et al., *Rubber World, 151(3)*: 62 (1964).
114. A. K. Bhowmik et al., *Rubber Products Manufacturing Tech.*, Marcel Dekker Inc., 693.
115. M. S. Banerji et al., *International Rubber Conference*, China (1992).
116. M. S. Banerji et al., IRMRA, 16th Rubber Conference (1993).
117. X. Zhang and Y. Zhang, *International Rubber Conference*, India, 29 (1993).
118. M. S. Banerji et al., *International Rubber Conference*, India, 39 (1993).
119. M. S. Banerji et al., *International Rubber Conference*, USA (1993).
120. C. K. Das et al., IRMRA, *15th Rubber Conference*, India (1990).

121. H. Domininghaus, *Gummi Gasern Kauststoffe, 7*: 352 (1992).
122. Y. P. Singh, *Proceedings of QIP STC, PBA-94*, pp. 116–130 (1994).
123. L. A. Utracki, *Inter. Polymer Processing, 2*: 3 (1987).
124. M. Xanthos and S. S. Dagli, *Polym. Engg. Sci., 31*: 929 (1991).
125. Y. P. Singh and R. P. Singh, *Eur. Polym. J., 19*: 533 (1983).
126. M. Xanthos (ed.), *Reactive Extrusion, Principles and Practice,* Hanser Publishers, Munich (1992).
127. S. Shaw and R. P. Singh, *J. Appl. Polym. Sci., 38*: 1677 (1989).
128. E. Schroder et al., *Polymer Characterization,* Hanser Publisher, New York (1989).
129. A. K. Sircar, *Proceedings of the International Conference on Structure—Property Relation of Rubber,* India (1980).
130. D. J. Walsh et al. (eds.), *Polymer Blends and Mixtures,* Martinus Nijhoff Publisher, Netherlands (1985).
131. F. R. Eirich (ed.), *Science and Technology of Rubber,* Academic Press, London (1978).
132. A. Echte, *Rubber Toughened Plastics,* (C. K. Riew, ed.), Advances in Chemistry, C S, Washington, DC, p. 222 (1989).
133. B. J. Schmitt, *Angew. Chem., 91*: 286 (1989).
134. G. Riess and P. Gallard, *Polymer Reaction Engineering,* (K. H. Reichert and W. Geiseler, (eds.), Hanser, Munich, p. 221 (1983).
135. G. E. Molau et al., *J. Appl. Polym. Sci., 13*: 2735 (1969).
136. G. Freund et al., *Angew. Macromol. Chem., 58/59*: 199 (1977).
137. H. Haaf et al., *J. Sci. Ind. Res., 40*: 659 (1981).
138. A. Echte, *Angew, Makromol. Chem., 58/59*: 175 (1977).
139. J. L. Amos, *Polym. Eng. Sci., 14*: 1 (1974).
140. G. F. Freeguard, *Br. Polym. J., 6*: 205 (1974).
141. F. Ide and I. Sasaki, *Kobunshi Kagaku, 27*: 607 (1970).
142. H. J. Karam and J. C. Bellinger, *Ind. Eng. Chem. Fundam., 7*: 576 (1968).

143. R. E. Lavengood, *Monsanto,* U.S. Patent 4,214,056 (1980).
144. A. Echte, *Chemische Technologie,* 4th ed. (K. Harnisch, R. Steiner, and K. Winnacker, (eds.), Hanser, Munich, 6: 373 (1982).
145. K. McCreedy and H. Keskkula, *Polymer, 20*: 1155 (1979).
146. C. B. Bucknall et al., *J. Mater. Sci., 21*: 301 (1986).
147. C. B. Bucknall et al., *J. Mater. Sci., 22*: 1341 (1987).
148. C. B. Bucknall et al., *J. Mater. Sci., 21*: 307 (1986).
149. R. W. Flumerfelt, *Ind. Eng. Chem. Fundam., 11*: 312 (1972).
150. J. Jelenic et al., *Makromol. Chem., 185*: 129 (1984).
151. C. K. Riew and R. W. Smith, *Rubber Toughened Plastics* (C. K. Riew, ed.), Advances in Chemistry, *ACS,* Washington, DC, p. 222, Chap. 9 (1989).
152. C. K. Riew et al., *Toughness and Brittleness of Plastics* (R. D. Deanin and A. M. Crugnola, eds.), *Advances in Chemistry, ACS,* Washington, DC, *154*: 326 (1976).
153. A. F. Yee et al., *Toughness and Brittleness of Plastics,* (R. D. Deanin and A. M. Crugnola, eds.), *Advances in Chemistry, ACS,* Washington, DC, 154, 97 (1976).
154. A. F. Yee, *J. Mater. Sci., 12*: 757 (1977).
155. T. Sakai, *Adv. Polymer. Tech., 11(2)*: 99 (1991–92).
156. E. M. Trewin et al. (eds.), *Proceedings of the Intl. Conf. on Composite Materials,* American Institute of Mining, Metallurgical and Petroleum Engineers, New York, pp. 1462–1473 (1978).
157. E. E. Hardesty, *Automation Cleveland* (1970).
158. K. M. Prewo, *Frontiers in Materials Technology,* (M. A. Myers, ed.), Elsevier, Amsterdam (1985).
159. H. Schnell, *Chemistry and Physics of Polycarbonates,* Wiley-Interscience, New York (1964).
160. L. Bottonbruch, *EPST, 10*: 710 (1969).
161. D. W. Fox, *ECT, 18*: 479 (1982).
162. S. L. Page, *MPE*: 60 (1982).
163. N. L. Hancox, (ed.), *Fibre Composite Hybrid Materials,* McMillan, New York (1981).
164. J. A. Radosta, *Plastics Eng. 33(9)*: 28 (1977).

44

Reactive Compatibilization of Immiscible Polymer Blends

Snooppy George, Josephine George, and Sabu Thomas
Mahatma Gandhi University, Kerala, India

I. INTRODUCTION

Blending of polymers has resulted in the development of polymeric materials with desirable combinations of properties. In most cases, simple blending will not result in the attainment of desirable combinations of properties because of some inherent problems. Frequently, the two polymers are thermodynamically immiscible and the product will not exhibit homogeneity. Often a low level of inhomogeneity is preferred as both the components can retain their identity in the blend and thereby contribute synergistically toward blend properties. However, the interfacial situation in an immiscible blend is very critical in determining the blend properties. A typical incompatible blend exhibits high interfacial tension and poor adhesion between the phases. The high interfacial tension results in poor dispersion during mixing and the subsequent lack of stability during later processing or use. Often such blends show poor mechanical properties, particularly those related to ductility, that can preclude their commercial utilization. Moreover, the weak interfacial interaction in these blends leads to premature failure under stress as a result of the usual crack opening mechanism [1].

II. COMPATIBILIZATION CONCEPTS

The two generic terms found in the blend literature are "compatibility" and "miscibility." Components that resist gross phase segregation and/or give desirable blend properties are frequently said to have a degree of compatibility even though in a thermodynamic sense they are not miscible. In the case of immiscible systems, the overall physicomechanical behavior depends critically on two demanding structural parameters [2]: a proper interfacial tension leading to a phase size small enough to allow the material to be considered as macroscopically "homogeneous," and an interphase "adhesion" strong enough to assimilate stresses and strains without disruption of the established morphology.

Various techniques have been reported to combat the problem of incompatibility [3]. One way is to organize the phases and to increase their interfacial areas so that they are able to effectively transfer applied forces despite having low natural interfacial adhesion. Techniques of blending [4] and blending methods combined with certain block copolymers to stabilize the blend morphology [5] have been used successfully to form interpenetrating co-continuous networks of phases. Such systems have improved mechanical properties because the continuous nature of the phases allow direct load sharing of the components without the need for shear force transfer across phase boundaries. It is widely known that the presence of certain polymeric species, usually suitably chosen block or graft copolymers, can alleviate to some degree these problems as a result of their interfacial activity [6–13]. The segments of these copolymers can be chemically identical with those in the respective phases [8,13–15] or miscible with or adhered to one of the phases [16–18]. As pointed out by Paul [19] this type of surface activity should: (1) reduce the interfacial energy between the phases, (2) permit finer dispersion during mixing, (3) provide a measure of stability against gross segregation, and (4) result in improved interfacial adhesion [20].

There have been many experimental investigations that studied the interfacial and surfactant properties of block copolymers. In addition to documenting the compatibilizing activity in polymer blend dispersion, these investigations have also provided evidence of the interfacial activity in polymeric systems [21–25], surface activity [26–29], and dispersed efficiency [30,31] of block copolymers. For example, Gailard and coworkers [21,24] demonstrated the surface activity of block copolymers by studying the interfacial tension reduction in demixed polymer solutions.

Addition of poly(styrene-block-butadiene) block copolymer to the polystyrene–polybutadiene–styrene ternary system first showed a characteristic decrease in interfacial tension followed by a leveling off. The leveling off is indicative of saturation of the interface by the solubilizing agent.

III. REACTIVE COMPATIBILIZATION

Addition of block copolymers, crosslinking of rubber phase, or the generation of interpenetrating networks, which are being used for controlling morphology and mechanical properties of incompatible polymer blends, are not always preferred. Therefore, more attention has been focused on reactive polymer processing in which superior polymer alloys are produced during melt processing, making it an attractive cost effective alternative.

The basic principle underlying reactive compatibilization is that, by making use of the functionalities present in one or more polymers, one can form graft or block copolymers in situ during processing. These copolymers act as compatibilizers by reducing the interfacial tension and increasing the adhesion between the phases and thereby allowing finer dispersion and more stable morphology. Several review articles [32–34] have appeared in the literature highlighting the significance of reactive polymer processing.

A. Polyolefin–Polyamide Blends

Blends based on polyolefins have been compatibilized by reactive extrusion where functionalized polyolefins are used to form copolymers that bridge the phases. Maleic anhydride modified polyolefins and acrylic acid modified polyolefins are the commonly used modified polymers used as the compatibilizer in polyolefin–polyamide systems. The chemical reaction involved in the formation of block copolymers by the reaction of the amine end group on nylon and anhydride groups or carboxylic groups on modified polyolefins is shown in Scheme 1.

One of the earliest references on compatibilizing a nylon-6–polypropylene blend using maleic anhydride grafted PP (PP-g-MAH) was the work of Ide and Hasegawa published in 1974 [35]. In their study, the formation of a graft copolymer was confirmed by DSC after solvent extraction of the PP component. Blends with PP-g-MAH

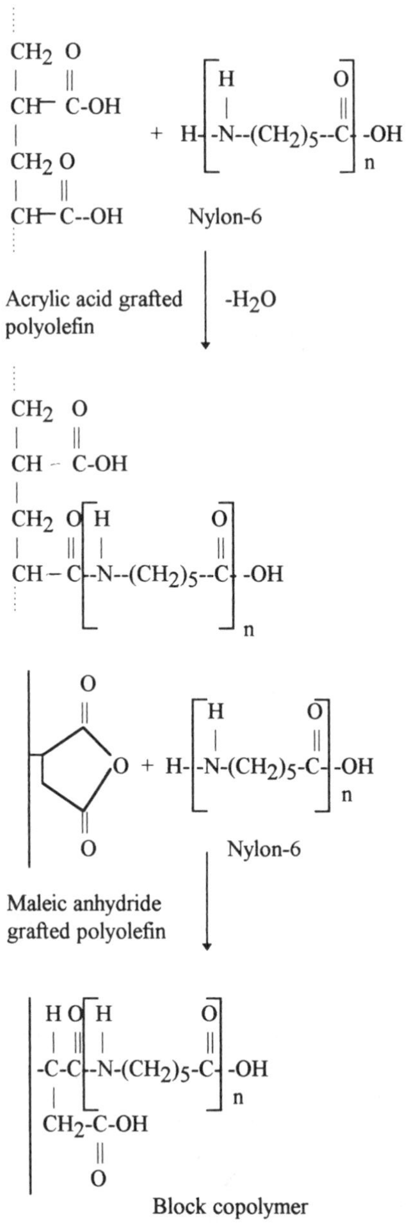

Scheme 1 The proposed acid/amine and anhydride/amine reaction between carboxylic groups/anhydride groups on modified polyolefins and amine end groups on Nylon-6.

showed a minimum in melt flow rate versus composition curve (varying the amount of N6), whereas blends without PP-g-MAH first showed a monotonic increase. There are several Japanese publications [36–38] and patents maintaining similar compatibilization of a nylon-6–PP blend using PP-g-MAH. A few studies [39–42] have been reported on laminates of N6 and PP and, in

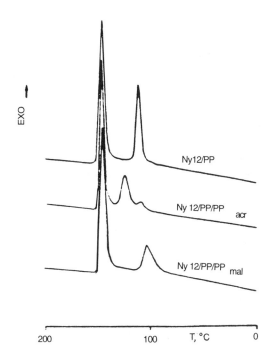

EXO

Ny12/PP

Ny 12/PP/PP acr

Ny 12/PP/PP mal

200						100				T, °C				0

Figure 1 DSC thermograms of Nylon–12–PP blends. *Source*: Ref. 43.

behavior to the greater reactivity of maleic groups than acrylic groups to the amine end groups of nylon to form block copolymers.

In the PP–PA system, the DSC thermograms showed two peaks corresponding to nylon and PP. For the compatibilized system the crystallization peak of nylon remains unaltered, while that of PP shifted toward a higher temperature in the case of PPacr, and for PPmal, the shift was to lower temperatures (Fig. 1). This may be due to the fact that PPacr was acting as the nucleating agent. The average crystallinity of the blend was also decreased by the incorporation of compatibilizer. The mechanical properties of these blends was improved by the addition of PPmal and PPacr as compatibilizers (Table 1).

Generally, chemical reaction between the components of a blend increases the viscosity. Thus, an increase in viscosity in the presence of a compatibilizer may indicate that there are strong interactions at the interface. In the PP–PA system, Valenza et al. [43] reported such an increase in viscosity in the presence of a compatibilizer due to the increased interaction at the interphase. Miettinen and coworkers [44] also reported such an increase in viscosity of the PP–PA-6 system by the addition of SEBS-g-MA as a compatibilizer. Here also the reaction at the interface is in between anhydride and amine groups. The authors have evaluated the effectiveness of the SEBS-g-MA in PP–PA system in terms of impact property, morphology, and dynamic mechanical properties. In these blends, the addition of SEBS-g-MA leads to a very interesting morphology. In the ternary blends, in which PA–PP ratios are 20:80, 40:60, a well-dispersed phase of combined SEBS-g-MA and polyamide is seen in the polypropylene matrix on TEM investigation. A clear change in morphology is observed in a PA–PP ratio of 80:20 with 10% SEBS-g-MA. In this case the morphology shows a fine dispersion of size of about 0.04 μm and coarse dispersion of 0.1–1 μm in the continuous polypropylene matrix. Miettinen et al. [44] interpreted this morphology as follows. In the blends of PA–PP 60:40 with 10 wt% SEBS-g-MA, the PA/SEBS-g-MA agglomerates and begins to overlap when the amount of PA is increased to 80%. Therefore, the phase inversion inside the agglomerates extends to the whole

particular, on improving the adhesion of these laminates by the use of modified PP or other modified polar thermoplastics in the packaging and automotive industries.

The morphology, rheology, and crystallization behavior of reactive compatibilized blends of polyolefin–polyamide blends were widely investigated. Polypropylene–polyamide compatibilization through functionalized polypropylenes is one of the systems studied recently. Valenza et al. [43] investigated the effectiveness of maleic anhydride grafted PP (PPmal) and acrylic acid grafted PP (PPacr) as the compatibilizer in the PP–PA system. In these systems, the average diameter of the inclusion is decreased by the addition of modified polypropylenes and this reduction is more pronounced in the case of PPmal. They attributed this

Table 1 Mechanical Properties

	PP	Nylon-12	Nylon-12–PP	Nylon-12–PP–PPmal	Nylon-12–PP–PPacr
Stress at yield (MPa)	30	35	—	36	—
Elongation at yield (%)	15	25	—	18	—
Stress at break (MPa)	40	55	22	30	26
Elongation at break (%)	400	200	60	150	110
Izod impact (Jm^{-1})	37	95	35	80	65

Source: Ref. 43.

sample. This results in a morphology with a bimodel dispersion of SEBS-g-MA as the continuous phase and PA-6 as a coarse dispersion of 0.4–1 μm. This morphology leads to exceptional impact properties (Fig. 2).

The rheology and morphology of polyacrylamide–polypropylene blends compatibilized through MA-g-PP was studied by Haddout et al. [45]. In the compatibilized blend, the viscosity shows two types of behavior, i.e., at low shear rates up to 10^3 s^{-1} a positive deviation is observed. The viscosity of compatibilized blends is higher than the uncompatibilized ones. The compatibilized blend also shows a decrease in domain size of the dispersed phase. The higher values of viscosity of compatibilized blends compared with uncompatibilized ones are due to the reduction of the average size of the dispersed nodules and also due to the better adherence between the two phases in the presence of compatibilizer. The in situ compatibilization of LDPE–polyamide-11 was carried out by Lambla and Seadan [46]. The graft copolymer was formed by the thermal decomposition of selected peroxides in the melt. The reactivity could further be enhanced by the addition of monomers, such as undecenal or maleic anhydride, both of which lead to significant changes in morphology and macroscopic properties in comparison with blends containing no such monomers.

The kinetics of the reactive compatibilization of nylon-6–PP by acrylic acid modified PP was investigated by Dagli et al. [47]. The compatibilization reaction in this system involved the reaction between the acid group of acrylic acid modified PP and the amine group of nylon-6. A typical intensive batch mixer torque (τ) vs time (t) trace for a ternary blend showing an increase in mixing torque upon the addition of PP-g-AA to a binary PP–NBR (85:7.5) blend is shown in Fig. 3. The kinetic

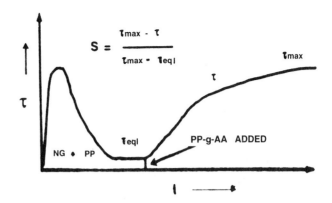

Figure 3 Torque vs time trace for the PP–N6–PP-g-AA ternary blend. *Source*: Ref. 47.

studies revealed that the compatibilization reaction follows a second order kinetics as shown in Fig. 4. Interestingly, acid-amine nylon polymerization reaction also follows second order kinetics [48–49]. Also, they have investigated the effect of process variables on the compatibilization reaction. It was observed that changes in the rotor speed affected the reaction kinetics significantly, whereas temperature changes in the 15–20°C range have no effect. The experiments in the twin-screw extruder had revealed that the screw speed, presence of venting, and sequence of feeding had a noticeable effect on the reactive compatibilization process. Venting and

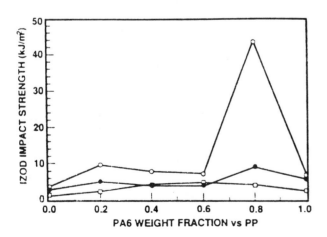

Figure 2 Notched Izod impact strength of ternary PA–PP–SEBS-g-MAH blends with 0 wt%, 5 wt%, and 10 wt% SEBS-g-MAH. *Source*: Ref. 44.

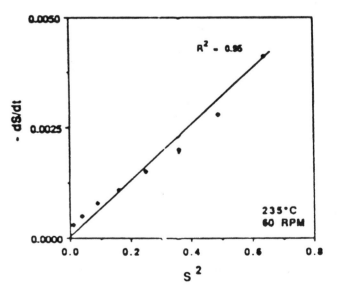

Figure 4 Reaction kinetics plot showing the use of a differential method of rate determination of PP–N6–PP-g-AA ternary blend. *Source*: Ref. 47.

residence time had a more significant effect on the grafting reaction. Screw speed was important from the distribution of the compatibilizer point of view. However, feed rate and temperature were not found to be significant in the range tested.

Ionomers are sometimes used to compatibilize N6 and polyolefins blends. Carbonyl groups from the ionomer were found to interact with the amine groups of N6 by forming hydrogen bonds. Ionomer concentrations as low as 0.5 wt% were found to be effective. Willis and Favis [50] have carried out extensive studies to determine the emulsifying effect of an ionomer (polyethylene-methacrylate acid-isobutyl acrylate terpolymer) on the morphology of polyolefin–polyamide blends. The addition of the ionomer to polyolefin–polyamide blends was found to increase the adhesion between the matrix and dispersed phase. Much finer dispersion of the minor phase in the matrix polymer as well as a reduction in the polydispersity of the particle size were also observed. One important finding was that as little as approximately 0.5% ionomer by weight of the blend was sufficient to produce a maximum reduction of the dispersed phase size. Larger quantities of ionomer yielded aggregates of the dispersed polyamide, possibly due to the clustering phenomenon of ionomers.

B. PS-Based Blends

Baker and Saleem [51] have reported on the reactive compatibilization of oxazoline modified PS and carboxylated polyethylene. The coupling reaction results in amide-ester linkages at the time of melt mixing. A schematic representation of the reaction is shown in Scheme 2.

The graft polymer formed during melt mixing results in good interfacial adhesion between the phases. The

intermolecular reaction between the two polymers is further substantiated by the rise in mixing torque, FTIR, and DSC studies.

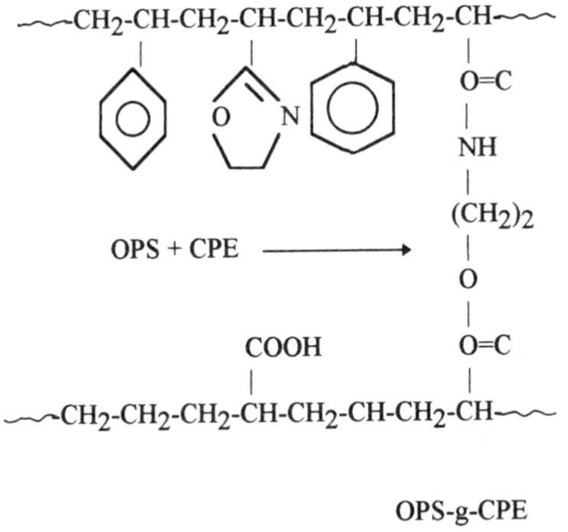

The Lewis acid initiated melt reactions of polystyrene and EPDM rubber to obtain high impact polystyrene were reported by Mori et al. [52]. It was found that among the Lewis acids investigated, the NaCl AlCl$_3$ double salt has the best initiating effect. The greatest number of crosslinks and the smallest extent of degradation was observed in this reaction. The extent of coupling increases with increasing Lewis acid and rubber content. As a result of coupling, the solubility of the elastomer particles in the continuous PS matrix increases. Increased solubility results in increased adhesion and mechanical properties.

Willis et al. [53] have reported on the morphology and impact properties of polystyrene-maleic anhydride–bromobutyl rubber blends as a function of interfacial modification and melt processing conditions. It was found that dimethyl amino ethanol (DMAE) serves as a reactive compatibilizing agent for these blends. The number average and volume average diameters as a function of DMAE concentration for the PS–bromobutyl rubber (80:20) blend is shown in Fig. 5. It is evident that the size of the dispersed rubber phase and polydispersity decrease as the extent of interfacial modification increases. This gives rise to a corresponding decrease in the interparticle distance. The particle size tends to equilibrate toward a larger concentration of DMAE. The impact strength values obtained for 80:20 PS–bromobutyl rubber blend as a function of DMAE concentration are shown in Fig. 6. A three-fold increase in the notched Izod impact strength is observed over the range of DMAE concentrations studied. It is very interesting to note that by increasing the concentration of DMAE from 2–5% (i.e., by saturating the interface), the impact strength is significantly enhanced while the particle size, polydispersity, and interparticle distance remain almost constant. This illustrates the importance of saturating the interface with the reactive compatibilizing agents.

Teh and Rudin [54] have reported on the compatibilization of the PS–PE blend through reactive processing in a twin-screw extruder. In fact, this study [54] was aimed at generating the polystyrene radicals at the interface so that the coupling reaction could occur at the interphase between the PS and PE. Here compatibilization was achieved through reactive processing with styrene monomer, dicumyl peroxide, and TAIC in a twin-screw extruder by simultaneous melt blending, polymerization, grafting, and coupling. The polymerization, grafting/coupling and degradation reactions possible in a mixture of PS, PE, and styrene monomer (ST) in the presence of peroxide and the coupling agent are shown in Scheme 3.

From thermal, molecular weight, and FTIR studies on the blends and isolated PE and PS fractions, it was deduced that a polyalloy of PE-TAIC-PS was formed.

Scheme 2 Reaction scheme for the formation of block copolymer between OPS and CPE.

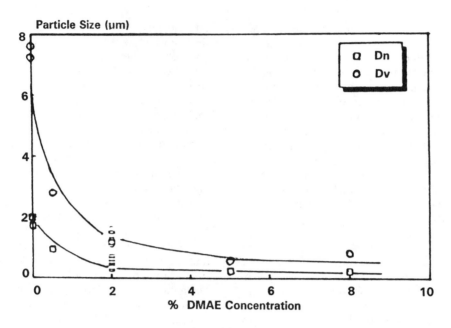

Figure 5 Dependence of particle size of dispersed bromobutyl rubber on the DMAE concentration. *Source*: Ref. 53.

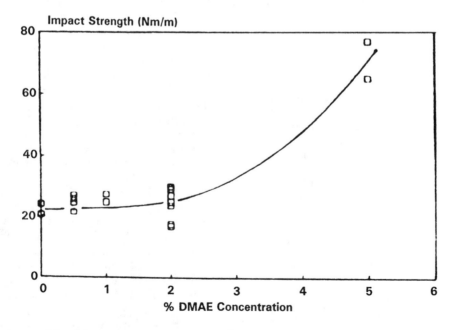

Figure 6 Dependence of Izod impact strength on the DMAE concentration in 80:20 PS–bromo butyl rubber blends. *Source*: Ref. 53.

Radical generations:

$$ROOR \longrightarrow 2RO\cdot$$

$$ROOR + TAIC \longrightarrow 2ROH + \cdot TAIC\cdot$$

$$RO\cdot + PE \longrightarrow ROH + PE\cdot$$

$$RO\cdot + PS \longrightarrow ROH + PS\cdot$$

$$RO\cdot + ST \longrightarrow ST\cdot$$

Polymerization:

$$ST\cdot + nST \longrightarrow PS\cdot_n$$

Grafting/Coupling:

$$PE_x\cdot + PE\cdot_y \longrightarrow PE_x - PE_y$$

$$PS_x\cdot + PS_y\cdot \longrightarrow PS_x - PS_y$$

$$PE\cdot + PS\cdot \longrightarrow PE - PS$$

$$2PS\cdot + \cdot TAIC\cdot \longrightarrow PS - TAIC - PS$$

$$2PE\cdot + \cdot TAIC\cdot \longrightarrow PE - TAIC - PE$$

$$PE\cdot + \cdot TAIC\cdot + PS \longrightarrow PE - TAIC - PS$$

Scheme 3 Reactions in a mixture of PS, PE, and styrene monomer in the presence of a peroxide and the coupling agent.

The formation of the polyalloy results in improvement in the performance of the blends. This system is similar to the production of high-impact polystyrene (HIPS) where a rubber is dissolved in styrene monomer and then polymerized in the usual way. Even though the impact strength of the compatibilized PS–PE blend was higher than that of PS, it was much less than that of HIPS. In another study, Van Ballegooie and [55] have confirmed

the improvement in properties of PS–PE blends via the reactive extrusion process in the presence of dicumyl peroxide and triallyl isocyanurate (TAIC). In that case, the distribution of the radical generator and coupling agent was likely to be uniform throughout the whole melt rather than concentrated at the interphase boundaries.

The reaction between the oxazoline group and the carboxylic group has also been exploited in the case of rubber toughening of polystyrene using acrylonitrile butadiene rubber having carboxylic groups [56]. The concentration of the reactive oxazoline groups was varied by mixing polystyrene with a copolymer of styrene and vinyloxazoline. The reactive blends showed significant improvement in impact properties. The 5% OPS blends showed an increase of 73% in impact strength over the nonreactive blend and then decreased with increasing oxazoline content (Table 2).

The important factors that affect the rubber toughening are: (1) interfacial adhesion, (2) nature of the matrix, (3) concentration of the rubber phase, and (4) shape and size of the rubber particles. In the PS–XNBR blend containing OPS, due to the reaction between oxazoline groups of OPS and carboxylic groups of XNBR, the interfacial adhesion increases and as a result, the minor rubber phase becomes more dispersed. The immiscible blend needs an optimum interfacial adhesion and particle size for maximum impact property. In PS–XNBR, a very small concentration of OPS provides this optimum interfacial adhesion and particle size. The interfacial adhesion beyond this point does not necessarily result in further toughening.

C. PET- and PBT-Based Blends

Reactive compatibilization of engineering thermoplastic PET with PP through functionalization has been reported by Xanthos et al. [57]. Acrylic acid modified PP was used for compatibilization. Additives such as magnesium acetate and *p*-toluene sulfonic acid were evaluated as the catalyst for the potential interchange or esterification reaction that could occur in the melt. The blend characterization through scanning electron microscopy, IR spectroscopy, differential scanning calorimetry, and

Table 2 Impact Results of Unnotched Charpy Specimens of PS–OPS–XNBR Blends

XNBR (wt%)	PS (wt%)	OPS (wt%)	Impact energy (J)	Peak load (N)	Ductile ratio
0	100	0	0.86	1900	0.22
20	80	0	2.05	2120	0.29
20	75	5	3.54	2380	0.41
20	70	10	2.74	2180	0.38
20	60	20	1.21	1690	0.23

Source: Ref. 56.

mechanical properties revealed that functionalized PP promotes fine dispersed phase morphology, improves processability and mechanical properties, and modifies the crystallization behavior of the polyester component. The presence of the additives that are used to catalyze interchange or esterification reactions does not, in general, improve any further the blend morphology, properties, or its processability.

Kanai et al. [58] recently reported the impact modification of various engineering thermoplastics (PPS, POM, and PBT). Functionalized elastomers were used as impact modifiers. They have correlated the impact strength with the particle size and interparticle distance of the modifiers. In the case of PBT with maleic anhydride functionalized ethylene olefin rubber (EOR), the plot of impact strength vs the particle diameter did not have any significant relationship, however, the plotting of impact strength vs interparticle distance gave a good correlation (Fig. 7). A critical interparticle distance approximately of 0.4 μm was obtained for PBT–EOR blends.

D. EPDM-Based Blends

Amine–anhydride reaction has also been exploited in the impact modification of polyamide with maleic anhydride grafted EPDM rubber and with ethylene-ethyl acrylate-maleic anhydride terpolymer [59,60]. Greco et al. [61] have reported on the degree of grafting of functionalized EPR on mechanical properties of rubber modified PA6. The reaction between succinic anhydride grafted EPR and PA6 resulted in the formation of a graft PA6-g-(EPR-g-SA) copolymer. The degree of grafting was varied from 0.6–4.5 wt%. The results reveal that the morphology and impact properties were better when the degree of grafting was higher.

The Charpy resilience (R) as a function of the test temperature for PA6 homopolymer and for ternary blends at an increasing degree of grafting is shown in Fig. 8. From the figure, it is clear that pure PA6 exhibits very brittle behavior with very low resilience values, which remain unchanged over the whole investigated temperature range. This type of behavior is further substantiated from morphological observations. However, for blends of the same compositions but containing EPR-g-SA as the rubbery phase, a large enhancement of impact properties is observed with respect to pure PA6. From the figure, it is clear that R increases with the increasing degree of grafting of the EPR-g-SA. Furthermore, all these blends show a marked variation in R values over the temperature range in which the behavior of the material changes from a brittle to a ductile mode of failure. The location of this transition temperature is a function of degree of grafting and blend composition. A shift toward lower temperature is observed with the increasing degree of grafting value of EPR-g-SA.

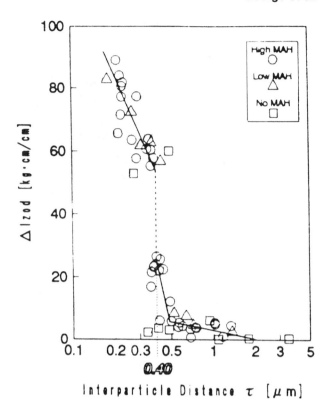

Figure 7 Notched impact strength vs. interparticle distance for PBT–maleic anhydride grafted EOR blends. *Source*: Ref. 58.

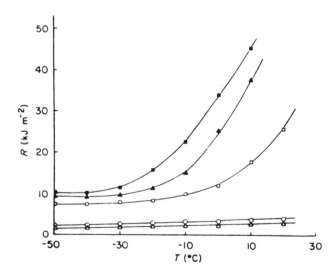

Figure 8 Charpy resilience (R) as a function of test temperature for PA-6 homopolymer and PA-6–EPR binary blends with increasing degree of grafting (DG), ○-0%, □-0.6%, ▲-2.4%, and ■-4.5% DG. *Source*: Ref. 61.

The morphology development during reactive and nonreactive blending of ethylene propylene rubber with nylon and polystyrene was investigated by Scott et al. [62]. For the reactive blends, MA-EPDM is used in nylon–EPM and oxazoline-g-PS, and maleic anhydride-g-EPM are used in PS–EPM blends. The reduction in domain size with the time of mixing was investigated. They reported that the major reduction in domain size occurs at short mixing times. In the case of the reactive system the volume average particle diameter of the dispersed phase is reduced from ~4 μm to 1 μm within the first 90 s of mixing time. For the reactive blend systems the mixing torque and temperature are higher than that of nonreactive systems. This is due to the chemical reaction occurring at the interface. In PA–EPM-MA, the reaction is in between the amine end groups of PA and anhydride groups on EPM-MA, and in PS-OX–EPM-MA the reaction is in between the oxazoline groups of PS and anhydride groups of EPM-MA. The chemical reaction between the polymers increases the molecular weight of the polymer, and, as a result, the mixing torque is increased. The increase in torque and temperature is less in PS-OX–EPM-MA compared to PA–EPM-MA, since the oxazoline-anhydride reaction is slow compared to amine-anhydride reaction. The morphology of the blends also shows a reduction in domain size for reactive blends than nonreactive blends. The reduction in size is also more pronounced in the PA–EPM-MA system due to the more reactivity of this system.

The possibility of improving the mechanical and impact properties of poly(D(-)-3-hydroxybutyrate) (PHB), which is highly prone to brittle failure, by melt mixing with functionalized rubber has been investigated by Abbate et al. [63]. It has been found that the best results were obtained when succinic anhydride grafted EPR was used. Here the chemical interactions taking place between the two components during the blending process result in good adhesion and thereby improved mechanical properties.

E. Natural Rubber-Based Blends

The reactive extrusion of polypropylene–natural rubber blends in the presence of a peroxide (1,3-bis(t-butyl peroxy)benzene) and a coagent (trimethylol propane triacrylate) was reported by Yoon et al. [64]. The effect of the concentration of the peroxide and the coagent was evaluated in terms of thermal, morphological, melt, and mechanical properties. The low shear viscosity of the blends increased with the increase in peroxide content initially, and beyond 0.02 phr the viscosity decreased with peroxide content (Fig. 9). The melt viscosity increased with coagent concentration at a fixed peroxide content. The morphology of the samples indicated a decrease in domain size of the dispersed NR phase with a lower content of the peroxide, while at a higher content the domain size increases. The reduction in domain size

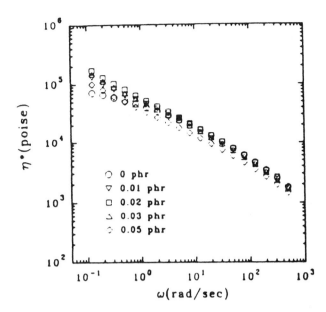

Figure 9 Complex viscosity of PP–NR blends vs. peroxide content of PP–NR blends at 220°C. *Source*: Ref. 64.

has been correlated with the compatibilizing effect of the interpolymers, which were formed in situ during melt extrusion, and to the increased viscosity of the PP matrix by crosslinking. The mechanical properties, such as yield strength and impact strength, also increase up to 0.03 phr peroxide content and then decrease.

The extrusion of PP and polyisoprene in the presence of maleic anhydride was reported as a compatibilization method by Els and McGill [65]. They isolated the block copolymer formed during mixing using the solvent extraction technique and characterized by means of TGA, DSC, and IR spectra. The IR spectra of the xylene extracted fraction at 100°C showed the presence of peaks characteristic of PP, polyisoprene, and maleic anhydride. The DSC scan also had the evidence for the formation of a graft copolymer. The WAXS investigations show the formation of the β-crystalline form of polypropylene along with the α-form for the blends extruded in the presence of MA. However, the addition of MA had little effect on the percentage crystallinity in the PP phase. The impact properties of the blends were also found to improve by extrusion in the presence of MA.

F. SEBS Rubber-Based Blends

The influence of maleic anhydride modified styrene–(ethylene-co-butylene)–styrene (SEBS) triblock copolymer as a reactive compatibilizer in a nylon-6–SEBS blend was investigated by Wu et al. [66]. When the maleated SEBS was incorporated into the PA-6–SEBS blend, the domain size of the dispersed phase decreased and became more uniform. The xylene extracted samples of the uncompatibilized blend showed two large bare cavities on SEM investigation. With the addition of 2 wt% of maleated SEBS in the blends, the morphology showed holes with a peculiar shape, volcano-like and protruding from the plane of the continuous phase. The presence of the protruding holes reveals the presence of better interfacial interactions. The protrusion of holes arises from the release of SEBS chains of the discrete phase from entanglements formed from the reaction of maleated SEBS chains with the continuous polyamide phase. The impact strength of the blend was also significantly improved on the addition of MA-g-SEBS (Table 3).

G. Dimethyl Siloxane Rubber (PDMS)-Based Blends

Santra et al. [67] investigated the effect of ethylene methylacrylate copolymer as a chemical compatibilizer in the low-density polyethylene–polydimethyl siloxane rubber blends. The IR investigation of the individual blend components and a blend containing 6 wt% EMA showed evidence for the formation of a graft copolymer between EMA and PDMS rubber. They had suggested a mechanism for the formation of the graft copolymer. The lap shear adhesion strength of the blends using EMA in one of the phases shows substantial improvement with an increase in the concentration of EMA (Table 4). The tensile impact strength of the blends also improved significantly with the incorporation of EMA up to 6 wt%, and after that the increase is marginal.

Table 4 Adhesion strength of the LDPE and PDMS Rubber Blends Containing EMA

Sample code[a]	Adhesion strength (N/cm^2)
PE_0	2.6
PE_2	3.2
PE_4	4.1
PE_6	4.3
PE_{10}	4.5
Si_0	2.6
Si_4	4.2
Si_6	7.0
Si_{10}	7.3

[a] The subscripts represent the proportion of EMA in each phase.
Source: Ref. 67.

The SEM investigation shows that the particle size of the dispersed domain size decreased from 3.3 to 1.1 μm with the incorporation of 6 wt%, EMA, and this indicates the increased surface area of the dispersed phase morphology. The increase in surface area led to effective compatibilization and is responsible for the increased adhesion strength and tensile impact strength of compatibilized blends.

H. NBR-Based Blends

Most of the commercially available reactive compatibilized systems contain acidic functional groups. Reactive

Table 3 The Formulation and the Impact Strength of the PA-6–SEBS-g-MA Blends

Recipe				
PA6 (wt%)	SEBS (wt%)	MA-g-SEBS (wt%)	MA graft ratio (gMA/gSEBS)	Impact strength (J/m)
100	—	—	—	42.6 ± 3.14
85	15	—	—	103.1 ± 16.40
85	14	1	0.09	112.0 ± 9.41
85	13	2	0.09	154.6 ± 15.30
85	12	3	0.09	1157.8 ± 75.20
85	11	4	0.09	1100.6 ± 50.90
85	14	1	0.11	117.6 ± 22.50
85	13	2	0.11	202.6 ± 33.30
85	12	3	0.11	1101.5 ± 71.60
85	11	4	0.11	804.6 ± 85.40
85	14	1	0.14	107.7 ± 12.90
85	13	2	0.14	149.7 ± 14.60
85	12	3	0.14	424.1 ± 56.50
85	11	4	0.14	202.2 ± 21.10

Source: Ref. 66.

compatibilization of immiscible polymer blends via the incorporation of basic functional groups have also been developed. A well-known example is the reactive PS developed by Dow Chemical Company [68]. Basic functional groups such as glycidyl methacrylate (GMA) have been grafted onto LDPE [69] and PP [70]. Recently, a novel process of making primary amine functionalized polyolefins by copolymerization of masked functional monomers has been patented [71,72].

Recently, Baker et al. [73] have reported on the effectiveness of different basic functional groups for the reactive compatibilization of polymer blends. In this study, glycidyl methacrylate (GMA), 2-hydroxy ethyl methacrylate (HEMA), 2-hydroxy propyl methacrylate (HPMA), *t*-butylaminoethyl methacrylate (TBAEMA), dimethylaminoethyl methacrylate (DMAEMA), and 2-isopropenyl-2-oxazoline (IPO) were melt grafted onto PP homopolymer. The effectiveness of these functionalized PPs as compatibilizers for PP–acrylonitrile co-butadiene-co-acrylic acid rubber (NBR) blends was evaluated in terms of impact properties and blend morphology. The effect of the functionalized PPs on the impact energy of the PP–NBR blends is shown in Fig. 10. There were minor improvements in impact energy in the blends when HEMA and TBAEMA grafted PPs were introduced into the matrix phase. These improvements

may be due to possible hydrogen bonding interactions between either the hydroxy group of HEMA or the secondary amine group of TBAEMA with the carboxylic acid group in the NBR rubber phase. The impact energies of PP–NBR blends having HPMA and DMAEMA grafted PPs in the matrix phase were actually lower than blends without any functionalized PPs. This is due to the fact that the tertiary amine functionality cannot form an amide linkage by reacting with carboxylic acid and the secondary hydroxy group in HPMA is much less active than a primary hydroxy group in esterification reactions.

However, a substantial improvement in impact energies of PP–NBR blends with GMA or IPO functionalized PPs is observed. The PP–NBR blends went through a brittle-ductile transition as the concentration of the functionalized PPs in the matrix phase reached a leveling off at 13 wt% in the case of IPO functionalized PP and 25 wt% in the case of GMA functionalized PP. Up to a ten-fold improvement in impact energy was observed when the brittle-ductile transition was reached.

A typical force–displacement impact curve obtained using the instrumental impact tester for PP and PP–NBR blends with various amounts of GMA grafted PP in the PP matrix is shown in Fig. 11. The maximum loads for both PP and the nonreactive PP–NBR blend and their total displacement were small. The force decreased rapidly after the maximum force was reached, indicating brittle failure. As GMA grafted PP (13 wt%) was introduced into the PP matrix phase the maximum load and the total displacement were significantly improved. However, the force dropped quickly to zero after the maximum force was reached, again indicating brittle failure. Only when 25 wt% of GMA grafted PP

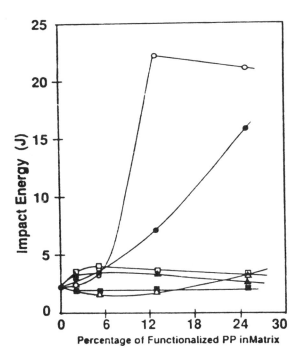

Figure 10 Effect of different functionalized PPS on the impact energy of PP–NBR blends, ■-HPMA, ●-GMA, ▲-TBAEMA, □-HEMA, ○-IPO, and △-DMAEMA. *Source*: Ref. 73.

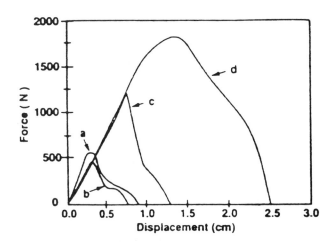

Figure 11 Force displacement curves of a-PP, b-nonreactive PP–NBR blend, c-reactive blend containing 13 wt% GMA functionalized PP, and d-reactive blend containing 25 wt% GMA functionalized PP. *Source*: Ref. 73.

Table 5 Effects of Various Phenolic Modified Polypropylenes on the Properties of NBR-Polypropylene Compositions Containing Amine Terminated NBR

	1	2	3	4	5
Recipe					
Polypropylene	50	—	—	—	—
Phenolic modified polypropylene	—	50(A)	50(B)	50(C)	50(D)
NBR	45	45	45	45	45
ATBN	5	5	5	5	5
SP-1045	5	5	5	5	5
SnCl$_2$-2H$_2$O	0.5	0.5	0.5	0.5	0.5
Mixing time before curative addition (min)	5	5	5	5	30
Properties					
Tensile strength, σ_B (MPa)	9.6	9.8	15.3	16.5	20.1
Stress at 100% strain σ_{100} (MPs)	—	—	10.2	9.9	10.1
Young's modulus, E (MPa)	149	105	107	106	106
Ultimate elongation, ϵ_B (%)	36	79	390	420	450
Tension set, ϵ_s (%)	—	—	54	54	54
True stress at break, σ_B^* (MPa)	13.1	17.5	75.0	85.8	110.55

Source: Ref. 74.

was added to the PP matrix phase did the blend become ductile with the force decreasing slowly in a stable manner after the maximum force was achieved, while at the same time the total displacement increased significantly.

Coran and Patel [74] investigated the reactive compatibilization of PP–NBR and HDPE–NBR blends using phenolic modified polyolefin, maleic anhydride modified polyolefin, and amine terminated nitrile rubber as reactive components. Dynamic vulcanization was also inves-

tigated as a way of reactive compatibilization. The mechanical properties and percentage oil swelling was improved by the reactive compatibilization of these blends (Table 5). The improvement in mechanical properties and the decrease in oil swelling in reactive blends are attributed to the increase in interfacial adhesion caused by the graft copolymer formed in situ during mixing. The graft copolymer is formed by the reaction between amine end groups of NBR and the anhydride and

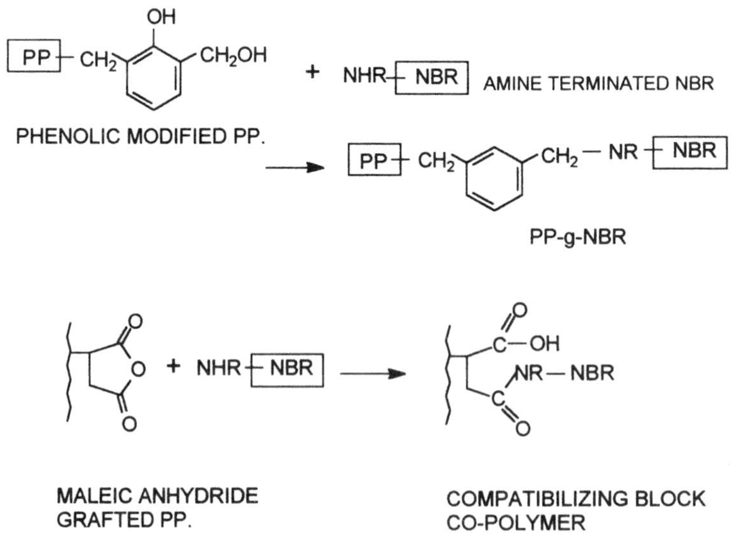

Scheme 4 Reaction scheme for the formation of graft copolymer between amine terminated NBR and modified polypropylenes.

phenolic groups on the modified polyolefins as shown in Scheme 4.

The reactive compatibilization of HDPE–NBR and PP–NBR blends has been studied by Thomas and coworkers [75,76]. The maleic anhydride modified polyolefins and phenolic modified polyolefins are used as compatibilizers. The effect of the concentration of these compatibilizers on the compatibility of these blends was investigated in terms of morphology and mechanical properties. It was found that in these blends an optimum quantity of the compatibilizer was required to obtain maximum improvement in properties, and after that a leveling off was observed. The domain size of the dispersed NBR phase in these blends is decreased up to a certain level and then increases (Fig. 12 and 13). The reduction in domain size is attributed to the increase in interfacial adhesion and to the reduction in interfacial tension. The phenolic modified polyolefin reacts with the NBR particles to form a block copolymer as shown in Scheme 5.

This block copolymer acts as an emulsifying agent in the blends leading to a reduction in interfacial tension and improved adhesion. At concentrations higher than the critical value, the copolymer forms micelles in the continuous phase and thereby increases the domain size of the dispersed phase.

The applicability of Noolandi and Hong's theory of compatibilization of immiscible blends using block copolymers has been extended to the reactive compatibilization technique by Thomas and coworkers [75,76]. According to Noolandi and Hong [77], the interfacial tension is expected to decrease linearly with the addition

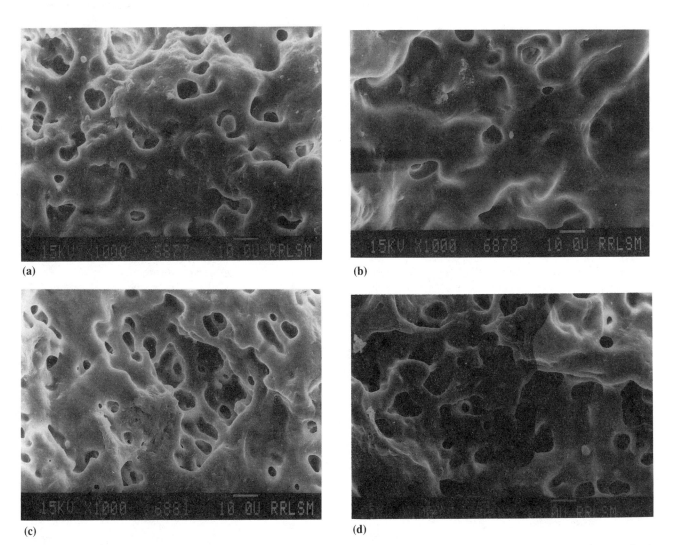

(a)　　　　　　　　　　　　　　　(b)

(c)　　　　　　　　　　　　　　　(d)

Figure 12 Scanning electron micrographs of 70:30 HDPE–NBR blend with (a) 0 wt%, (b) 1 wt%, (c) 5 wt%, and (d) 10 wt% maleic anhydride modified polyethylene. *Source*: Ref. 75.

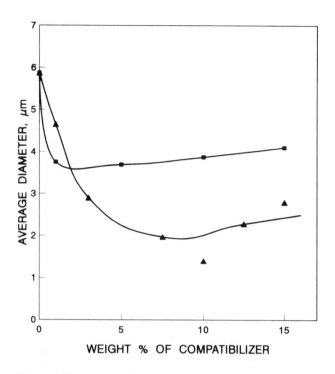

Figure 13 Average diameter of dispersed NBR domains vs. phenolic modified polypropylene and maleic anhydride modified polypropylene concentration. *Source*: Ref. 75.

of compatibilizer below CMC, and a leveling off is expected after that. The expression for interfacial tension reduction ($\Delta \nu$) in a binary blend upon the addition of divalent copolymer A-b-B is given by

$$\Delta \nu = d\phi_c(1/2\chi + 1/Zc) - 1/Zc \exp(Zc\chi/2) \qquad (1)$$

where d is the width at half height of the copolymer profile by Kuhn statistical segment length, ϕ_c is the bulk volume fraction of the copolymer in the system, χ is the Flory Huggin's interaction parameter between A and B segments of the copolymer, and Zc is the degree of polymerization of the copolymer. Based on this equation, the plot of interfacial tension reduction versus ϕ_c should yield a straight line. Since interfacial tension reduction is directly proportional to the particle size reduction, the interfacial tension reduction term in Eq. (1) can be replaced by the term particle size reduction. Therefore,

$$\Delta d = Kd\phi_c (1/2\chi + 1/Zc) - 1/Zc \exp(Zc\chi/2) \qquad (2)$$

where K is the proportionality constant. In HDPE–NBR and PP–NBR blend the particle size reduction (Δd) varies linearly with blend composition at low compatibilizer concentrations, whereas at higher concentrations it levels off, in agreement with the theories of Noolandi and Hong (Fig. 14).

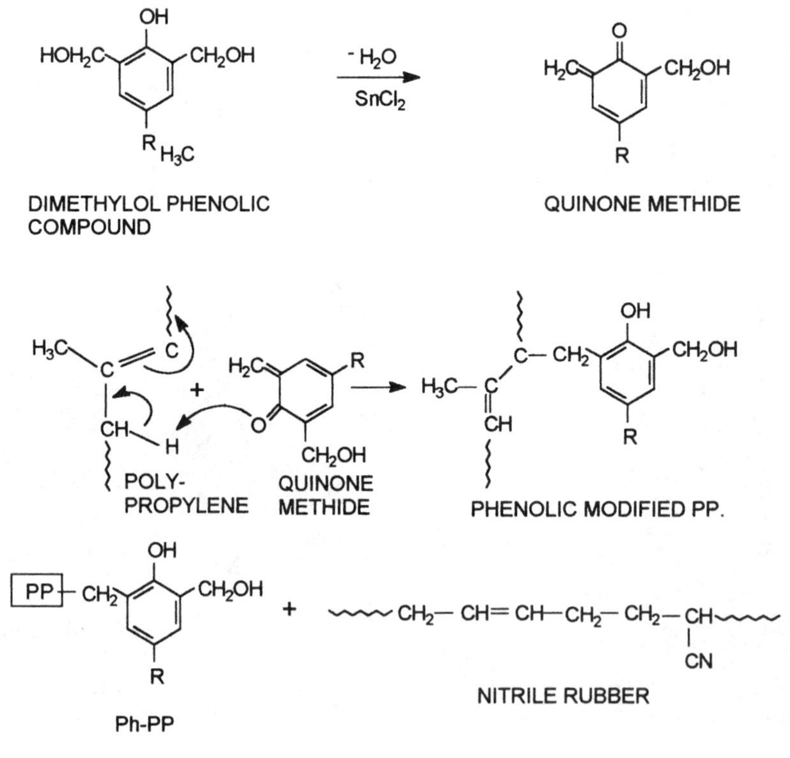

Scheme 5 Reaction scheme for the formation of graft copolymer in PP–NBR blends in the presence of phenolic modified PP.

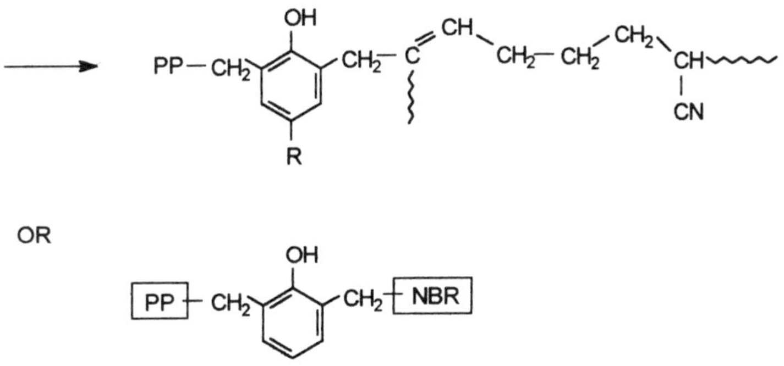

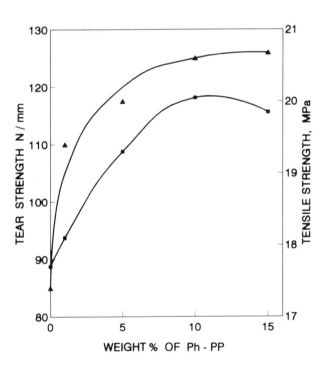

Scheme 5 *Continued*

The mechanical properties of these blends are also increased with an increase in compatibilizer concentration. The improvement in tensile strength and tear strength on the addition of modified polyolefins are shown in Fig. 15. The mechanical properties also show a leveling off after the optimum concentration. The im-

provement in strength with compatibilizer concentration is due to the improvement in interfacial adhesion, which leads to better stress transfer between the matrix and the dispersed phase. Ramesh and De [78] have investigated the effect of carboxylated nitrile rubber as a reactive compatibilizer for immiscible blends of poly(vinyl chloride) and epoxidized natural rubber. In PVC–ENR binary blends there is crosslinking due to interchain reactions. These blends are immiscible at various blend ratios. The incorporation of XNBR into these systems increased the compatibility of these blends. The IR in-

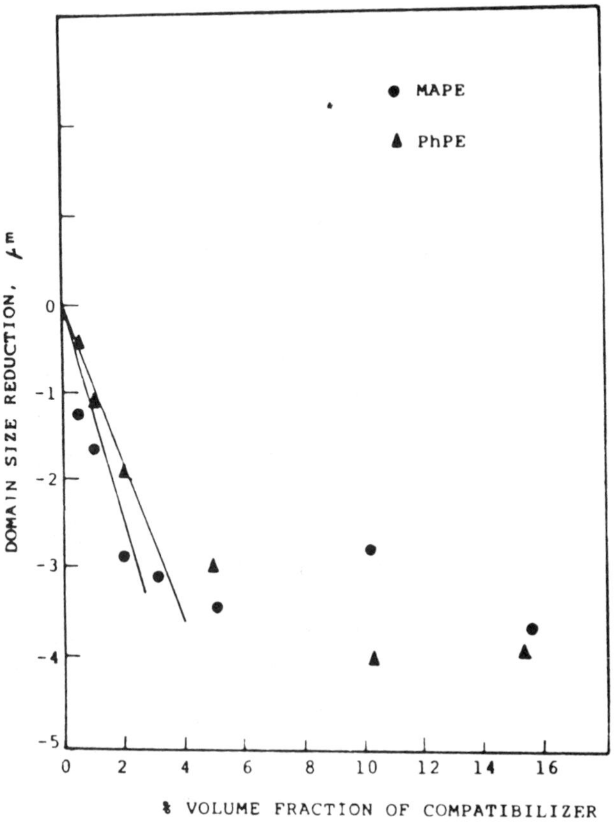

Figure 14 Effect of volume fraction of maleic modified PE and phenolic modified PE on the particle size reduction of 70:30 HDPE–NBR blend. *Source*: Ref. 75.

Figure 15 Effect of phenolic modified PP concentration on tensile strength and tear strength of 70:30 PP–NBR blend. *Source*: Ref. 76.

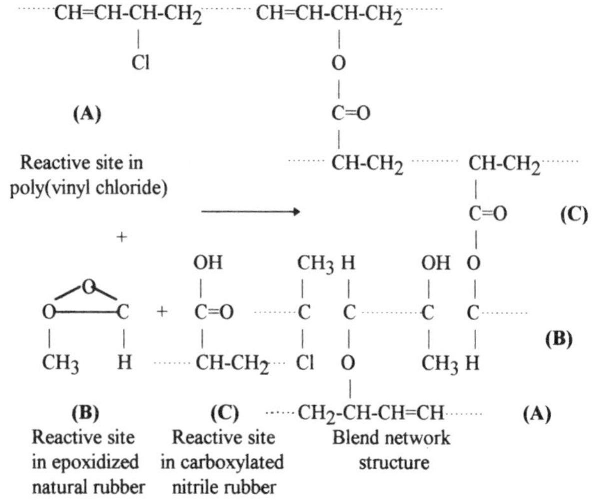

CH=CH-CH-CH₂   CH=CH-CH-CH₂
        | |
       Cl   O
                                    |
   (A)   C=O
                                    |
Reactive site in   CH-CH₂ ⋯⋯ CH-CH₂
poly(vinyl chloride) |
                                              C=O   **(C)**
   ────────→ |
                                              O
      +

                    OH   CH₃ H   OH   O
                     | | | | |
   O──────O | | | | |
 O──────C   + C=O ⋯ C ── C ⋯⋯ C ── C   **(B)**
   | | | | |
  CH₃   H   CH-CH₂ Cl   O   CH₃ H

   (B)   **(C)**   ⋯CH₂-CH-CH=CH⋯   **(A)**
Reactive site   Reactive site   Blend network
in epoxidized   in carboxylated structure
natural rubber nitrile rubber

Scheme 6 Possible mechanism of crosslinking between PVC, ENR, and XNBR.

vestigation of the PVC–ENR–XNBR ternary blends showed evidence for the formation of PVC-ENR, PVC-XNBR, and ENR-XNBR reactions. The probable crosslinking reaction is shown in Scheme 6. The amount of XNBR required to make the system miscible increased as the PVC content in the rubber blend increased (Table

Table 6 T_gs of the Ternary Blends

Blend composition PVC–ENR–XNBR	T_g (°C) from tan δ vs temperature plots
100:0:0	101
0:100:0	−5
0:0:100	−8
25:75:25	−1
25:75:50	3
25:75:75	9
50:50:25	16, 85
50:50:50	13, 73
50:50:75	Broad
50:50:100	5
50:50:125	3
50:50:150	3
75:25:25	11, 99
75:25:50	6, 87
75:25:75	−1, 87
75:25:100	−9, 75
75:25:125	−7 and broad
75:125:50	−8 and broad
75:25:200	Broad
75:25:250	3

Source: Ref. 78.

6). The dynamic mechanical analysis of the ternary blends with 75:25 PVC–ENR blend showed single T_g at all levels of XNBR concentrations, which indicates the miscibility of the system. In the 50:50 PVC–ENR blend, when the concentration of XNBR increased, the blend becomes progressively miscible.

IV. SUMMARY

To meet the increasing demand of new polymeric materials with improved properties, it is possible to use a wide range of combinations via reactive compatibilization. Here there is no need to prepare the expensive block or graft copolymer separately. The only thing to be kept in mind is to introduce appropriate functional groups to the existing polymers, thereby making them amenable to react at the time of processing which leads to the in situ formation of the compatibilizer. Finally, it is also important to mention that reactive compatibilization is more economical than any other technique.

GLOSSARY

ATBN - amine terminated nitrile rubber
χ - Flory Huggins interaction parameter
CPE - carboxylated polyethylene
d - width at half height of the copolymer profile given by Kuhn statistical segment length
DMAE - dimethyl amino ethanol
Δr - interfacial tension reduction
Δd - particle size reduction
DSC - differential scanning calorimetry
EMA - ethylene methyl acrylate copolymer
ENR - epoxidized natural rubber
EOR - ethylene olefin rubber
EPDM - ethylene propylene diene monomer
EPM - ethylene propylene monomer rubber
EPR - ethylene propylene rubber
EPR-g-SA - succinic anhydride grafted ethylene propylene rubber
ϕ_c - bulk volume fraction of the copolymer
FTIR - Fourier transform infrared spectroscopy
GMA - glycidyl methacrylate
HDPE - high density polyethylene
HEMA - 2-hydroxy ethyl methacrylate
HIPS - high impact polystyrene
HPMA - 2-hydroxy propyl methacrylate
IPO - 2-isopropenyl-2-oxazoline
IR - infrared spectroscopy
K - proportionality constant
LDPE - low density polyethylene
MA-EPDM - maleic anhydride grafted ethylene propylene diene monomer rubber
MA-g-PP - maleic anhydride grafted polypropylene
μm - micrometer

MPa - Megapascal
N6 - nylon-6
NBR - acrylonitrile-co-butadiene rubber
NR - natural rubber
OPS - oxazoline modified polystyrene
PA - polyamide
PBT - polybutylene terephthalate
PDMS - polydimethyl siloxane rubber
PE - polyethylene
PET - polyethylene terephthalate
PHB - poly[D(-)]-3-hydroxy butyrate
PP - polypropylene
PP-g-AA - acrylic acid grafted polypropylene
PP-g-MAH - maleic anhydride grafted polypropylene
PPacr - acrylic acid grafted polypropylene
PPmal - maleic anhdride grafted polypropylene
PPS - polyphenylene sulfide
PS - polystyrene
PS-OX - oxazoline grafted polystyrene
PVC - polyvinylchloride
R - resilience
SEBS-g-MA - maleic anhydride grafted styrene (ethylene-co-butylene)styrene copolymer
$SnCl_2$ - stannous chloride
SP-1045 - dimethylol phenolic resin
ST - styrene monomer
t - time
τ - torque
TAIC - triallyl isocyanurate
TBAEMA - t-butyl amino ethyl methacrylate
T_g - glass-transition temperature
TGA - thermogravimetric analysis
WAXS - wide angle x-ray scattering
XNBR - carboxylated nitrile rubber
Zc - degree of polymerization of the copolymer

REFERENCES

1. D. R. Paul and S. Newman (eds.), *Polymer Blends*, Academic Press, New York, Vol. 2 (1978).
2. S. A. Anaslsiadis, I. Gancarz and J. T. Koberstein, *Macromolecules, 22*: 1449 (1989).
3. J. W. Barlow and D. R. Paul, *Polym. Eng. Sci., 24*: 525 (1984).
4. G. N. Avgeropoulos, F. C. Weissert, P. H. Biddison, and G. G. A. Bohm, *Rubber Chem. Technol., 49*: 93 (1963).
5. W. P. Gergen and S. Davison, *US Patent*, 4,107,130, Aug. 15, 1978, assigned to Shell Oil Company.
6. N. G. Gaylord, *US Patent*, 3,485,777, Dec. 23, 1969.
7. N. G. Gaylord, *Copolymers, Polyblends and Composites* (N. A. J. Platzer, ed.), Advances in Chemistry 142; American Chemical Society, Washington, DC, p. 76 (1975).
8. G. Riess and Y. Jolivet, *Copolymers, Polyblends and Composites* (N. A. J. Platzer, ed.), Advances in Chemistry 142; American Chemical Society, Washington, DC, p. 243 (1975).
9. G. Riess, J. Kohler, C. Tournut, and A. Banderet, *Makromol. Chem., 58*: 101 (1967).
10. J. Kohler, G. Riess and A. Banderet, *Eur. Polym. J., 4*: 173 (1968).
11. J. Kohler, G. Riess and A. Banderet, *Eur. Polym. J., 4*: 187 (1968).
12. G. Riess, J. Periard, and Y. Jolivet, *Angew Chem. Int. Ed. Engl., 11*: 339 (1972).
13. R. Fayt, R. Jerome, and Ph. Teyssie, *J. Polym. Sci., Polym. Lett. Ed., 24*: 25 (1986).
14. A. R. Ramos and R. E. Cohen, *Polym. Eng. Sci., 17*: 639 (1977).
15. W. M. Barentsen, D. Heikens, and P. Piet, *Polymer, 15*: 119 (1974).
16. T. Ouhadi, R. Fayt, R. Jerome, and Ph. Teyssie, *Polym. Commun., 27*: 212 (1986).
17. T. Ouhadi, R. Fayt, R. Jerome, and Ph. Teyssie, *J. Polym. Sci., Polym. Phy. Ed., 24*: 973 (1986).
18. T. Ouhadi, R. Fayt, R. Jerome, and Ph. Teyssie, *J. Appl. Polym. Sci., 32*: 5647 (1986).
19. D. R. Paul, *Polymer Blends* (D. R. Paul and S. Newman, eds.), Academic Press, New York, Vol. 2, p. 35 (1978).
20. G. E. Molau, *Block Copolymers* (S. L. Aggarwal, ed.), Plenum Press, New York, p. 79 (1970).
21. P. Gailard, M. Ossenbach-Sauter, and G. Riess, *Makromol. Chem., Rapid Commun., 1*: 771 (1980).
22. R. Cantor, *Macromolecules, 14*: 1186 (1981).
23. G. Riess, J. Nervo, and D. Rogez, *Polym. Eng. Sci., 17*: 634 (1977).
24. P. Gailard, M. Ossenbach-Sauter, and G. Riess, *Polymer Compatibility and Incompatibility: Principles and Practice* (K. Solc, ed.), MMI Symposium Series 2, Harwood, New York (1982).
25. H. T. Patterson, K. H. Hu, and T. H. Grindstaff, *J. Polym. Sci., Part C, 34*: 31 (1971).
26. H. B. Gia, R. Jerome, and Ph. Teyssie, *J. Polym. Sci., Polym. Phys., 18*: 2391 (1980).
27. M. J. Owen and T. C. Kendrick, *Macromolecules, 3*: 458 (1972).
28. G. L. Gaines, Jr. and G. W. Bender, *Macromolecules, 5*: 82 (1972).
29. J. J. O'Malley, H. R. Thomas, and G. N. Lee, *Macromolecules, 12*: 996 (1979).
30. D. J. Meier, *J. Phys. Chem., 71*: 1861 (1967).
31. D. H. Napper, *Polymeric Stabilization of Colloidal Dispersion*, Academic Press, New York (1983).
32. R. W. Holfeld, *Plast. World, 43(8)*: 85 (1985).
33. J. A. Sneller, *Mod. Plast., 63(7)*: 56 (1985).
34. Z. N. Frund, *J. Plast. Compounding, 9(5)*: 24 (1986).
35. F. Ide and A. Hasegawa, *J. Appl. Polym. Sci., 18*: 963 (1974).
36. H. Terada, T. Kono, M. Kawajima, and A. Hasegawa, *Japanese Patent*, 7,138,022 (1971).
37. F. Ide, T. Kodama and A. Hasehawa, *Kobunshi Kagaku, 29(4)*: 259, 265 (1972) [English abstract].
38. F. Komatsu and A. Kaeriyama, *Muroran Kogyo Daigaku Kenkya Hohoku, 7(3)*: 719 (1972) [English Abstract].
39. S. Kasura, Mitsui Petrochemical Industries Ltd., *Japanese Patent*, 6,128,539 (1986).
40. M. Sakuma, Y. Yagi, N. Yamamoto, and T. Yokokura, Tonen Sekiyu Kaglen, Inc., *Japanese Patent*, 62,158,739 (1987).
41. J. M. Wills and B. D. Favis, *Polym. Eng. Sci., 28*: 1416 (1988).
42. P. Scholtz, D. Froelich, and R. Muller, *J. Rheol., 33*: 481 (1989).
43. A. Valenza and D. Acierno, *Eur. Polym. J., 30(20)*: 1121 (1994).

44. R. M. H. Miettinen, J. V. Seppala, O. T. Ikkala, and I. T. Reima, *Polym. Eng. Sci., 34*: 395 (1994).
45. A. Haddout, J. Villoutreix, and G. Villoutreix, *Inter. Polym. Process, 10*: 1 (1995).
46. M. Lambla and M. Seadan, *Polym. Eng. Sci., 32*: 1687 (1992).
47. S. S. Dagli, M. Xanthos, and J. A. Beissenberger, *Polym. Eng. Sci., 34*: 1720 (1994).
48. D. Heikens, P. H. Hermans, and G. M. Van der Want, *J. Polym. Sci., 44*: 437 (1960).
49. M. I. Kohan (ed.), *Nylon Plastics,* Wiley, Interscience, New York (1973).
50. J. M. Willis and B. D. Favis, *Polym. Eng. Sci., 28*: 1416 (1988).
51. W. E. Baker and M. Saleem, *Polymer, 28*: 2057 (1987).
52. E. Mori, B. Pukanszky, T. Kelen, and F. Tudos, *Polymer Bulletin, 12*: 157 (1984).
53. J. M. Willis, B. D. Favis, and J. Lunt, *Polym. Eng. Sci., 30*: 1073 (1990).
54. J. W. Teh and A. Rudin, *Polym. Eng. Sci., 32*: 1678 (1992).
55. P. Van Ballegooie and A. Rudin, *Polym. Eng. Sci., 28*: 1434 (1988).
56. M. W. Fowler and W. E. Baker, *Polym. Eng. Sci., 28*: 1427 (1988).
57. M. Xanthos, M. W. Young, and J. A. Beissenberger, *Polym. Eng. Sci., 30*: 355 (1990).
58. H. Kanai, V. Sullivan, and A. Auerbach, *J. Appl. Polym. Sci., 53*: 527 (1994).
59. B. N. Epstein, *US Patent,* 4,174,358 (1979), assigned to E. E. DuPont de Nemours and Co.
60. N. Yamaguchi, J. Nambu, Y. Toyoshima, K. Mashita, and T. Ohmae, 5th Ann. Meet. PPS, Japan (1989).
61. R. Greco, M. Maliconico, E. Martuscelli, G. Ragosta, and G. Scarinzi, *Polymer, 28*: 1185 (1987).
62. C. E. Scott and C. W. Macosko, *Polymer, 35(25)*: 5422 (1994).
63. M. Abbate, E. Martuscelli, G. Ragosta, and G. Scarinzi, *J. Mater. Sci., 26*: 1119 (1991).
64. L. K. Yoon, C. H. Choi, and B. K. Kim, *J. Appl. Polym. Sci., 56*: 239 (1995).
65. C. Els and W. J. McGill, *Plast. Rubb. Comp. Process. Appln., 21*: 2 (1994).
66. C. J. Wu, J. F. Kuo, and C. Y. Chen, *Polym. Eng. Sci., 33*: 1329 (1993).
67. R. N. Santra, B. K. Samantaray, A. K. Bhowmick, and G. B. Nando, *J. Appl. Polym. Sci., 49*: 1145 (1993).
68. "Reactive Polystyrene," Form No. 171-012-85, Dow Chemical Corporation, Midland, MI, USA (1985).
69. R. R. Gallucci and R. C. Going, *J. Appl. Polym. Sci., 27*: 425 (1982).
70. P. Citovicky, V. Chrastova, J. Majer, J. Mejzlik, and G. Benc, *Collect. Czech. Chem. Commun., 45*: 2319 (1980).
71. S. Datta and E. N. Kresge, *US Patent,* 4,987,200 (1991).
72. S. Dutta, *High Performance Polymers* (A. Fawcett, ed.), Royal Society of Chemistry, London, Chap. 2, p. 33 (1990).
73. N. C. Liu, H. Q. Xie, and W. E. Baker, *Polymer, 34*: 4680 (1993).
74. A. Y. Coran and R. Patel, *Rubber Chem. Technol., 56*: 1045 (1983b).
75. J. George, R. Joseph, K. T. Varughese, and S. Thomas, *J. Appl. Polym. Sci., 57*: 449 (1995).
76. S. George, R. Joseph, K. T. Varughese, and S. Thomas, *Polymer 36*: 4405 (1995).
77. J. Noolandi, K. M. Hong, *Macromolecules, 17*: 1531 (1984).
78. P. Ramesh and S. K. De, *J. Appl. Polym. Sci., 50*: 1369 (1993).

45

Polymer Blends Containing Thermotropic Liquid Crystalline Polymer

Xiao-Su Yi

Zhejiang University, Hangzhou, China

I. INTRODUCTION

Polymer blends containing a thermotropic liquid crystalline polymer (TLCP) have received considerable attention in the past years. By blending TLCP with some thermoplastic (TP) resins, significant self-reinforcement is obtained in tensile strength, flex strength, and modulus. In addition, the processing is considerably easier. This is in contrast to the traditional chopped glass composites, which are more difficult to process than the base matrix resins and can lead to problems because of the abrasive nature of the filler (e.g., machine wear, process drift, and contamination of the polymer). The detriment is, however, a reduction in the elongation to break and, in some cases, poorer weldline performance.

This chapter focuses on recent developments in rheological and mechanical examinations of TLCP blends. Although most of the data presented are taken from actual works done at the Institute of Polymers and Processing of Zhejiang University, Hangzhou, China, this is merely a matter of accessibility and in no way implies that this data is the best or only data of this nature available.

II. FLOW BEHAVIOR AND A VISCOSITY FUNCTION

A. Introduction

One of the most attractive features of TLCPs is their ability to alter the rheology of bulk thermoplastic polymers. Most reports in the academic literature are concerned with viscosity reduction. For example, Siegmann et al. [1] observed a steep viscosity drop when a TLCP

was blended with an amorphous polyamide. The greatest reduction was seen at 5% TLCP loading. With the addition of more TLCP the blend viscosity increased. A review of this problem by Dutta et al. [2] is available.

Nobile et al. [3] reported that viscosity of a polycarbonate–TLCP blend can increase or decrease in the same system at the same temperature, depending on the shear condition. At very low shear rates the viscosity was found to increase with TLCP loading, whereas at high shear rates a significant drop was observed. But in all of these cases, the way in which the TLCPs alter the bulk polymer flow is not yet well understood.

The purpose of our study was to model the steady-state (capillary) flow behavior of TP–TLCP blends by a generalized mathematical function based on some of the shear-induced morphological features. Our attention was primarily confined to incompatible systems.

B. Theoretical Analysis

To study the flow behavior of an incompatible TP–TLCP binary system through a capillary, we first take into consideration some experimental observations and theoretical assumptions:

1. There is a steady-state laminar flow of concentric layers in a capillary.
2. During a steady-state capillary flow, several shear-induced effects emerge on blend morphology [4–6]. It is, for instance, frequently observed that TLCP domains form a fibrillar structure. The higher the shear rate, the higher the aspect ratio of the TLCP fibrils [7]. It is even possible that fibers coalesce to form platelet or interlayers.

3.   Under the steady-state flow conditions, there is an increasing tendency of this fiberlike structure moving toward the capillary wall as shear stress, flow flux, and radial position increase. In fact, we often obtained extrudates with a very thin TLCP-rich skin layer from the capillary test [8].

4.   It is believed that the thin TLCP-rich skin layer or interlayer may be responsible for a pluglike flow (i.e., a continuous velocity profile), due to a composition-dependent interfacial slippage [9], and, therefore, for the improved fluidity of this binary system.

Starting from these observations and assumptions, we introduce two parameters:

1.   Critical interface shear stress τ_c (Pa). This is a shear stress that causes a relative slide on the phase interface of the two components.
2.   Interface slip factor α (m^{-1}). This factor is defined as a phenomenological parameter characterizing the lubrication behavior on the phase interface as a slide occurs.

An interfacial lubrication or a slippage can be considered a composition-dependent phenomenon. Therefore, we postulate a relationship between α and the weight fraction ϕ:

$$\alpha = \frac{\alpha_0}{\phi^\epsilon(1 - \phi)} \tag{1}$$

where ϕ is the weight fraction of TLCP, the exponent ϵ depicts the influence of ϕ on the slip factor α, and α_0 is a proportional factor.

Because the interfacial slippage is assumed to be caused by the thin TLCP-rich interlayers, we can form a stress equilibrium relation as:

$$\Delta V \alpha \eta_0 = \tau_w \frac{2r}{R_0} - \tau_c \tag{2}$$

where ΔV (m/s) represents the interface slide velocity, η_0 (Pas) the matrix viscosity, r(m) the distance of the TLCP-rich interlayer to the capillary axis, R_0(m) the capillary radius, and τ_w (Pa) the shear stress on the capillary wall.

The shear-stress distribution is uneven in a capillary. Since an interfacial slippage takes place only at a point where the shear stress exceeds a critical value, a critical radius r_c can be defined as:

$$r_c = \frac{R_0}{2} \times \frac{\tau_c}{\tau_w} \tag{3}$$

If $r < r_c$, there is no interfacial slippage. Thus, the improved fluidity, i.e., the increased volume flux under a constant pressure through a capillary, can only be attributed to the TLCP-rich interlayers formed in the area where $r \geq r_c$.

A TLCP-rich interlayer can be visualized as a layer consisting of numerous TLCP fibrils. The layer may

have a thickness of d, which is assumed to be exactly equal to the average fibril diameter. The fibril number, n, within a layer can be calculated by the equation:

$$n = 4\phi \frac{R_0}{d} \tag{4}$$

where τ_c/τ_w in Eq. (3) defines a critical stress ratio: $\lambda = \tau_c/\tau_w$. If $\lambda > 1.0$, the volume flux through the capillary should follow the values predicted by the simple additivity rule. Otherwise, if $\lambda \leq 1.0$, there must be an additional volume flux of a quantity of ΔQ due to the interfacial slippage and/or interfacial lubrication under the constant capillary pressure.

The flux increase should depend on the fibril number n. When each fibril causes a flux increase of ΔQ_1 of the TLCP component, an argument of continuity suggests that there must also be a flux contribution of $\Delta Q_1 (1 - \phi)/\phi$ coming from the matrix component. Thus, the whole flux increase produced by each TLCP-rich interlayer is:

$$\begin{aligned} \Delta Q_n &= n\Delta Q_1/\phi = 4\Delta Q_1 R_0/d = 4\Delta V \pi dR_0 \\ &= 4\pi dR_0(\tau_w - \tau_c)/(\alpha\eta_0) \\ &= 4\pi dR_0\tau_w(1 - \lambda)\phi^\epsilon(1 - \phi)/(\alpha_0\eta_0) \\ &= \frac{128Qd\phi^\epsilon(1 - \phi)(1 - \lambda)}{\alpha_0R_0^2} \end{aligned} \tag{5}$$

Furthermore, when m interlayers exist within a capillary flow, where $m = kR_0/d$ and k is a linear factor, the total flux increase of ΔQ caused by the slippage of these m interlayers can be written as:

$$\begin{aligned} \Delta Q &= 128kQ\phi^\epsilon \frac{(1 - \phi)(1 - \lambda)}{\alpha_0R_0} \\ &= KQ\phi^\epsilon(1 - \phi)(1 - \lambda) \end{aligned} \tag{6}$$

$$K = 128 \frac{k}{\alpha_0R_0}$$

If TLCP is well dispersed and fibrillated, R_0/d in Eq. (4) is larger than 10^3 for most cases. When $R_0 = 1$ mm and $\phi = 25\%$, n will be more than 1,000. These fibrils with their average diameter less than 1 μm may just form only one interlayer near the capillary wall, or only one thin TLCP-rich skin layer, as mentioned previously. This phenomenon has been observed by us [7,8] and reported by many other researchers [9].

Finally, a generalized viscosity function in the form of a weight fraction-dependent viscosity ratio, η_0/η, could be derived as follows:

$$\frac{\eta_0}{\eta} = 1 - \phi + \frac{\phi}{\delta} + K\phi^\epsilon(1 - \phi)(1 - \lambda) \tag{7}$$

where η is the blend viscosity, δ is the viscosity ratio of η_1/η_0, and η_1 is the TLCP viscosity.

When $\phi = 0$, Eq. (7) reduces to $\eta_0 = \eta$; while when $\phi = 1$, it becomes $\eta_1 = \eta$. When $K \to 0$, Eq. (7) reduces to the inverse additivity rule as:

$$\frac{1}{\eta} = \frac{1}{\eta_0} + \frac{1}{\eta_1} \tag{8}$$

which often gives the lower limit of composite properties. $K \to 0$ corresponds to a situation where $m = 0$, indicating that there exists no interlayer that promotes the blend melt flow.

When other parameters are fixed, the viscosity function, Eq. (7) is a reducing function of TLCP weight fraction. Figure 1 shows the patterns. Generally, the blend viscosities are lower than the matrix viscosity (i.e., $\eta/\eta_0 < 1.0$), even when a small amount of TLCP is added to the system and even when the TLCP itself has a viscosity higher than that of the matrix (e.g., when $\delta = 10$). The smaller the ratio δ of the TLCP viscosity η_1 to the matrix viscosity η_0, the lower their blend viscosity η.

The influence of the viscosity ratio δ on the flow behavior in a capillary was discussed by Rumscheidt and Mason [10]. They pointed out that when the viscosity ratio is small, the dispersed droplets are drawn out to great lengths but do not burst, and when the viscosity ratio is of the order of unity, the extended droplets break up into smaller droplets. At very high viscosity ratios, the droplets undergo only very limited deformations. This mechanism can explain our observations and supports our theoretical analysis assumptions, summarized previously as points 2, 3, and 4.

Equation (7) is log-normal symmetric in weight fraction for $\epsilon = 1$ (Fig. 2). In this case, it describes systems in which there are no concentration-dependent phase transitions of any kind. In other words, Eq. (7) should

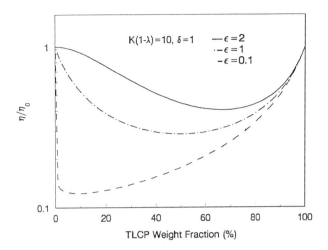

Figure 2 Theoretical viscosity function patterns by varying the exponent ϵ.

not be used for a system with partial solubility and change of morphology. This symmetric dependence of blend viscosity on weight fraction was observed, for instance, on a PC–TLCP blend and reported by Nobile et al. [3]. Otherwise, when $\epsilon < 1$, Eq. (7) describes the viscosity reduction giving a minimum value in the approximate range of $\phi < 50\%$, whereas when $\epsilon > 1$, the opposite is true. Incompatible systems, such as most TP–TLCP blends, belong to the first kind of flow behavior. But it was also reported that some TP–TLCP blends have composition-dependent viscosity indicating two minima below and above the 50 wt% of TLCP. An example for this behavior was a polysulfone–TLCP blend, its flow curves were measured at 240°C [9]. In all the cases, they must be able to describe their special flow behaviors by applying Eq. (7).

The viscosity drop becomes more pronounced in the whole range of ϕ (0% < ϕ < 100%) when $K(1 - \lambda)$ increases, indicating that the value of parameter K considerably affects the flow behavior of the blend (Fig. 3). Note that $K(1 - \lambda)$ is related to the fibrillation and migration of TLCP-rich interlayer.

As demonstrated, Eq. (7) gives complete information on how the weight fraction influences the blend viscosity by taking into account the critical stress ratio λ, the viscosity ratio δ, and a parameter K, which involves the influences of the phenomenological interface slip factor α or α_0, the interlayer number m, and the d/R_0 ratio. It was also assumed in introducing this function that: (1) the TLCP phase is well dispersed, fibrillated, aligned, and just forms one interlayer; (2) there is no elastic effect; (3) there is no phase inversion of any kind; (4) $\lambda < 1.0$; and (5) a steady-state capillary flow under a constant pressure or a constant wall shear stress.

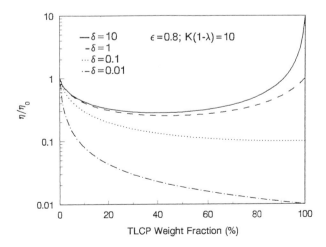

Figure 1 Theoretical viscosity function patterns by varying the viscosity ratio δ.

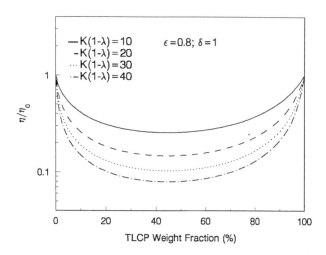

Figure 3 Theoretical viscosity function patterns by varying the parameter $K(1 - \lambda)$.

C. Experimental Results and Discussions

The TLCP used was KU 9231 produced by Bayer AG, Germany. The matrix material was an engineering plastic polyethersulfone (PES) manufactured by Jilin University, China. In an earlier article [11] we reported that KU 9231 was incompatible with the PES.

The materials were dried in a vacuum oven at 115°C for 24 h. They were then melt blended by using a domestic twin-screw extruder (ϕ35) [screw diameter = 35 mm]. The weight ratios of PES–TLCP were 90:10 and 70:30, respectively [12].

Figure 4 shows viscosity versus shear rate results for the two original components and their blends, re-

spectively. Rheological experiments were conducted on an Instron Capillary Rheometer, Mode 3211. The data were systematically corrected according to the Bagley and Rabinowitsch procedures. The test temperature was 310°C. Prior to testing, the blended materials were dried again in the vacuum at 125°C for 4 h.

As mentioned previously, and as expected, the mixtures containing a TLCP component exhibited shear viscosities lower than those of their original components, whether the TLCP weight fraction was 10% or 30%. The higher the TLCP weight fraction, the lower the blend viscosity. The viscosity curves of the two pure components crossed each other at a shear rate about 25 s^{-1}, i.e., at this point the viscosity ratio was 1.

The effect of the blending ratio on the melt viscosity of the two-phase system can best be seen from the cross plots at different shear stress. For this purpose, the rheological data were first fitted to Eq. (7) by a computer programming, and then, for $\tau = \eta\gamma$ being fixed, both the theoretical blend viscosity curves and the experimental values were plotted against TLCP weight fraction in Fig. 5. It was found that the computer-fitted curves ran perfectly through the experimental data points. The calculated parameters $K(1 - \lambda)$, ϵ, and δ are listed in Table 1, according to dependence on the shear stress.

It is obvious that all values of $K(1 - \lambda)$ were found to be larger than 30. They increased with increase of the shear stress (i.e., shear rate), and it has been well known that high shear stress leads to the formation of dispersed TLCP fibrils with a higher aspect ratio. Experimental evidence is given for the same PES–TLCP system in Yi et al. [7], where a larger average fibril aspect ratio was microphotographed at a higher shear rate. The average value of the exponent was approximately $\epsilon = 1$, indicating that the viscosity curves were nearly symmetric.

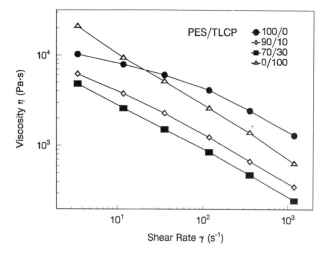

Figure 4 Experimental viscosity curves of PES, TLCP, and their blends.

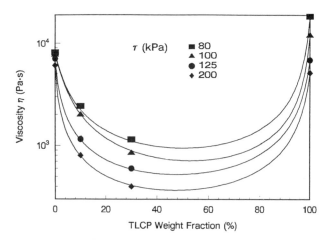

Figure 5 Comparison of theoretical viscosity curves with experimental value points of PES/TLCP blends for fixed shear stress τ.

Table 1 Parameters Provided by Fitting the Experimental Data to Eq. (7) for PES–TLCP Blends

τ(kPa)	$K(1 - \lambda)$	ϵ	δ
80	32.13	1.03	2.259
100	43.43	1.14	1.583
125	45.94	0.90	0.962
200	55.78	0.88	0.838

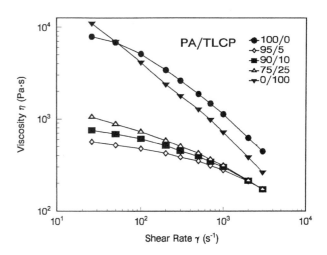

Figure 7 Experimental viscosity curves of PA, TLCP, and their blends of different blending ratios [1].

The influence of viscosity ratio δ on blend viscosity reduction can be well described by Eq. (7). As an example of $\epsilon < 1$, it is seen from Fig. 6 that the theoretical viscosity curves have two ranges, both are completely below the order of unity. In the range of $\delta < 1$, particularly when < 0.1, there is a considerable increasing tendency of η/η_0 against δ, suggesting that the lower viscosity component (TLCP) may migrate to the capillary wall, controlling the flow behavior of the blends. The higher the shear stress and rate, the stronger this viscosity reduction effect. At δ ratios greater than 1, i.e., at lower shear stress, the matrix becomes the lower viscosity component and may begin to control the blend flow. Thus, the η/η_0 ratio will be less affected by δ. Our investigation was performed just near the turning range of δ.

Equation (7) depicts the viscosity decrease independent of the chemical features of materials. Also for fixed τ, Figs. 7 and 8 demonstrate a further example of a polyamide–TLCP blend with different weight ratios. The rheological data in Fig. 7 were taken from Siegmann et al. [1]. It is obvious that the lowest blend viscosity is obtained at a TLCP loading of only 5%. This result is

somewhat different than the PES–TLCP blend previously reported. However, the weight fraction dependence of blend viscosity also followed Eq. (7) quite well (Fig. 8).

The parameters were also evaluated by fitting the rheological data to Eq. (7) and are listed in Table 2. It appears to be natural that both the ϵ and δ values were significantly lower than those of the PES–TLCP blend, previously discussed.

It is interesting to note that our derivation of the viscosity function Eq. (7) was not restricted within TP–TLCP systems. The negative deviation flow behav-

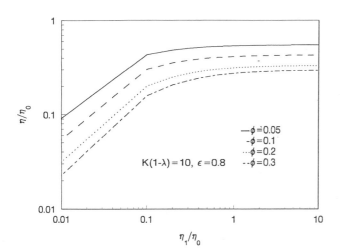

Figure 6 Theoretical viscosity function versus viscosity ratio $\delta = \eta_1/\eta_0$ at different blending ratios ϕ.

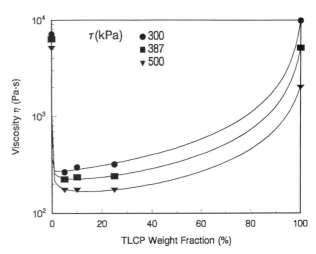

Figure 8 Comparison of theoretical viscosity curves with experimental values of PA–TLCP blends for fixed shear stress τ.

Table 2 Parameters Provided by Fitting the
Experimental Data to Eq. (7) for PA–TLCP Blends

τ(kPa)	$K(1 - \lambda)$	ϵ	δ
300	29.00	0.03	1.384
340	32.90	0.06	1.100
387	38.39	0.10	0.814
440	43.73	0.14	0.391
500	47.14	0.15	0.391

ior from the log-additivity rule has been also widely observed on different blends, for example polycarbonate–polyolefin blends reported by Huang et al. [14]. They found that in both PC–HDPE and PC–PP systems the addition of 5% polyolefins to 95% PC considerably reduces the blend viscosity to a value lower than that of both parent polymers. Utraki and Sammut [15] examined the morphology of a LLDPE [linear LDPE] blend containing 25% PC and reported that the low-viscosity component migrates to the capillary wall and controls the flow behavior. As a result, they also observed the viscosity reduction. In any case, an interfacial lubrication or an interfacial slippage of these incompatible systems is often a reasonable concept for understanding the viscosity reduction. It is hoped that the function in Eq. (7) can serve as a possible basic tool to deal with the viscosity reduction of the two-phase systems.

III. FIBRILLATION BEHAVIOR AND RHEOLOGICAL DETERMINATION

A. Introduction

Theoretically, the fibrillation behavior of a dispersed phase in a matrix is dictated by many characteristic factors such as Weber number W_e, viscosity ratio δ, breakup time t_b, first normal stress difference, and the flow types. Under certain conditions [16–18], when discrete domains of a well-dispersed minor phase are deformed, elongated ellipsoidal domains may form to nearly cylindrical domains. These domains, in turn, disintegrate into smaller drops, if a breakup time of t_b is exceeded. So, the disintegration and coalescence determine the morphology. Furthermore, a necessary condition for formation of fiberlike structures in the non-Newtonian range is that the first normal stress difference of the matrix phase should be larger than that of the minor phase at the same shear stress [19]. According to this model, the deformation behavior of fiberlike drops in simple shear flow has been demonstrated with many, mostly incompatible, blends [20].

In this section, we examine fibrillation behavior of a polycarbonate (PC)–TLCP blend by injection molding.

B. Theoretical Analysis

Concerning a liquid droplet deformation and drop breakup in a two-phase model flow, in particular the Newtonian drop development in Newtonian median, results of most investigations [16,21,22] may be generalized in a plot of the Weber number W_e against the viscosity ratio δ (Fig. 9). For a simple shear flow (rotational shear flow), a U-shaped curve with a minimum corresponding to $\delta = 1$ is found, and for an uniaxial extentional flow (irrotational shear flow), a slightly decreased curve below the U-shaped curve appears. In the following text, the U-shaped curve will be called the Taylor-limit [16].

Rumscheidt and Mason [10] studied the drop fracture of Newtonian fluids according to the value of a dimensionless δ. They found that $0.1 < \delta < 1.0$ is most favorable for extensional deformation and subsequently

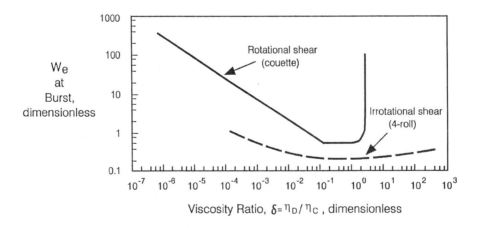

Figure 9 $W_e - \delta$ plot: comparison of effect of viscosity ratio δ on critical shear $W_{e,cri}$ in rotational and irrotational shear fields [18].

breakup. The exact $W_e - \delta$ graphs vary somewhat from author to author, but the general criteria is the same. Intermediate flow types between the two "regular" flow types give critical Weber numbers in between the curves [21].

Both dimensionless Weber number and viscosity ratio are defined by:

$$W_e = \frac{\eta_D \gamma d}{\sigma} = \frac{\tau_D d}{\sigma} \qquad (9)$$

$$\delta = \frac{\eta_D}{\eta_C} = \frac{\eta_1}{\eta_0} \qquad (10)$$

where η_D and η_C stand for the viscosity of the dispersed and the continuous phase, respectively, in our case $\eta_D = \eta_1$ and $\eta_C = \eta_0$. Representing the shear rate, shear stress, droplet diameter, and interfacial tension coefficient are γ, τ_D, d, and σ, respectively.

Based on a lot of experimental observations, criteria for the drop stability can be defined as below the U curve, namely $W_e < W_{e,cri}$, the interfacial stress can equilibrate the shear stress, and the drop will only deform into a stable prolate ellipsoid. Above this curve, the viscous shear stress becomes larger than the interfacial stress. The drop is at first extended and finally breaks up into smaller droplets.

The droplet deformation is also time-dependent. A practicable dimensionless breakup time t_b^* can be defined as:

$$t_b^* = \frac{t_b \sigma}{2 d \eta_D} \qquad (11)$$

Grace [18] evaluated t_b^* as a function of δ for viscoelastic matrices. His results are shown in Fig. 10. The reduced time increases as the viscosity ratio increases. If the residence time during a processing is less than

the breakup time, for the irrotational shear or rotational shear, the elongated drops will keep their form.

C. Experimental

The TLCP used was a widely investigated one, Vectra A950, manufactured by Hoechst-Celanese. The PC was produced by Enichem under the tradename Sinvet 303. Table 3 lists the main properties of these materials according to the manufacturers. The materials were dried in a vacuum oven at 120°C for 4 h before processing. Prior to injection molding of the composite samples, two components were melt mixed by a ZSK, W&P, Stuttgart, extruder with a constant weight ratio of PC–TLCP $= 80:20$ [23].

The standard injection molding machine used had a screw diameter of 30 mm and the aspect ratio of 23.70. The barrel temperature profile was 270, 280, 290, and 295°C. The mold temperature was about 90°C. The injection molded tensile samples were processed according to the CAMPUS specification (Computer Aided Materials Preselection by Uniform Standards) [24] and DIN 53455 Form 3. To obtain the different flow conditions, four groups of samples were injection molded by varying melt

Table 3 Main Properties of Experimental Materials

	PC (Sinvet 303)	TLCP (Vectra A 950)
ρ (g/cm³)	1.20	1.40
E (MPa)	2500	13000
σ_b (MPa)	65	156
HDT/A (°C)	130	168
Elongation at break (%)	>150	2.6

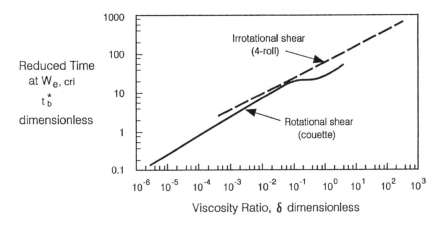

Figure 10 $t_b^* - \delta$ plot: comparison of effect of viscosity ratio δ on reduced breakup time t_b^* at critical $W_{e,cri}$ limit for rotational and irrotational shear [18].

Table 4 Processing Condition and Mechanical Properties of PC–TLCP Composite Samples Injection Molded

No.	Processing condition			Tensile properties	
	T_m (°C)	γ (1/s)	Q (cm³/s)	E_t (MPa)	σ_b (MPa)
1	280	116	8	4922	112
2	320	116	8	4648	104
3	280	1158	80	4092	95
4	320	1158	80	3704	81

Mold temperature = 90°C; maximum clamping force = 700 kN; maximum injection pressure = 225 MPa.

temperature T_m at the die and injection volume flux Q (Table 4). Hence, these samples will be identified in the following text by sample groups 1, 2, 3, and 4, respectively. The average shear rate was calculated using the Eq.: $6Qe/(bh^2)$, where Q is the volume flux, $e = 0.772$ [8], b and h are the width ($b = 10$ mm) and thickness ($h = 4$ mm) of the cross-section of the dumbbell samples.

Similar to prepared metallographic samples, the injection molded samples were cut along the flow direction, smoothed, and polished in order to expose their internal surface. After proper etching, the treated surfaces of the flank cross-section were photographed using a polarized light optical microscopy. Based on the color differences between the TLCP and matrix, volume fraction and aspect ratio of the TLCP fibers were measured [23].

The mechanical properties were obtained using a tensile machine at room temperature and for a strain rate of 1000%/h. Each reported value of the modulus was an average of five tests. The tensile modulus E_t was taken as the slope of the initial straight line portion of the stress-strain curve.

D. Morphological and Tensile Results

The results of viscosity versus shear rate are reported in Fig. 11 for the two pure components and their blend, respectively. The temperatures were the same for the viscosity measurements and for the injection molding. At temperatures of 280°C and 320°C, the viscosities of the blend are found to be values between the limits of the two pure components. In both cases, the TLCP still

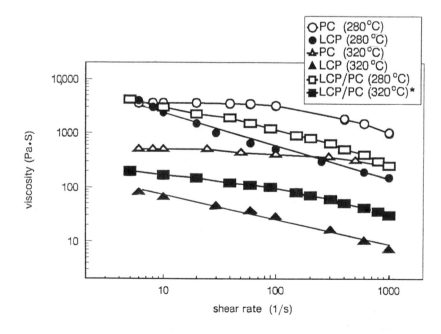

Figure 11 Viscosity curves of PC, TLCP, and their blends with respect to the test temperatures (* from WLF equation).

had the lowest viscosity. It was also observed that the pure TLCP exhibited a pronounced die shrinkage in contrast to the die-swell of most known thermoplastics.

Figure 12 shows the dependence of the average aspect ratio and the TLCP volume fraction on the relative sample thickness for the four processing conditions in the core layer, transition layer and skin layer, respectively, by a morphological examination [13]. Generally, the aspect ratio increases from core to skin layer, whereas the situation is reversed for the volume fraction. An average volume fraction about 20% can be clearly seen.

Table 4 also reports dependence of the mechanical tensile properties of the samples on the processing conditions. The highest tensile properties of sample 1, injection molded with a lower melt temperature and a lower volume flux, are attributed to the highest degree of fibrillation of the TLCP fibers, as shown in Fig. 12, by so-called in situ reinforcement.

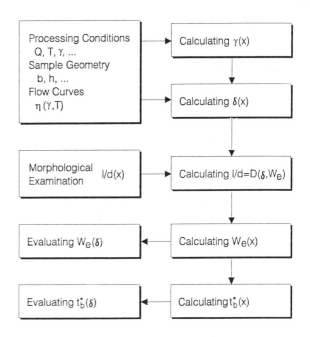

Figure 13 Calculation procedure.

E. Calculation and Discussion

The purpose of our calculation was to quantitatively evaluate the deformational behavior of the TLCP droplets and their fibrillation under the processing conditions, and finally, to establish a relationship among the calculated Weber number, the viscosity ratio, and the measured aspect ratio of the fibers. Figure 13 illustrates this procedure. All calculated results were plotted as

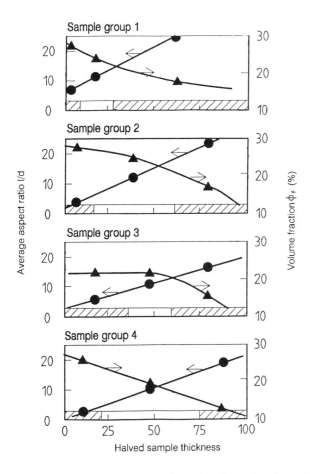

Figure 12 Average aspect ratio and volume fraction of TLCP fibers as functions of halved relative sample thickness for four processing conditions.

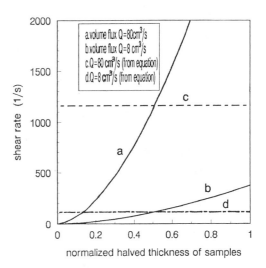

Figure 14 Calculated shear rate as function of the halved sample thickness for two injection volume fluxes Q.

functions of halved relative sample thickness with respect to the four processing conditions.

At first, the theoretical shear rate curves $\gamma(x)$ were calculated according to:

$$\gamma(x) = \frac{(m + 2) Q \left(\frac{2}{h}\right)^{m+1} x^m}{bh} \qquad (12)$$

for two injection volume fluxes of $Q = 80$ cm^3/s and 8 cm^3/s, respectively. Where m is the flow index of the viscosity curves, and it was an average value measured at 280°C and 320°C, and x is the thickness variable. This expression is only valid when assuming a steady or

quasi-steady flow state, constant thermal physical properties, homogeneous melt, and isothermal flow conditions [8]. As shown in Fig. 14, the shear rate increases in both cases from central to outside, and its gradient strongly depends on the volume flux Q. (The dashed lines in the figure represent the average volume flux.)

From the $\gamma(x)$ functions and the two melt temperatures used, and by using the viscosity curves from rheological examinations (Fig. 11), viscosity distributions $\eta(x)$ of the two pure components were easily determined, as shown in Figs. 15a and 15b. Subsequently, the viscosity ratio functions $\delta(x)$ were also calculated (Fig. 16). All four curves fall slightly from the core to the outside.

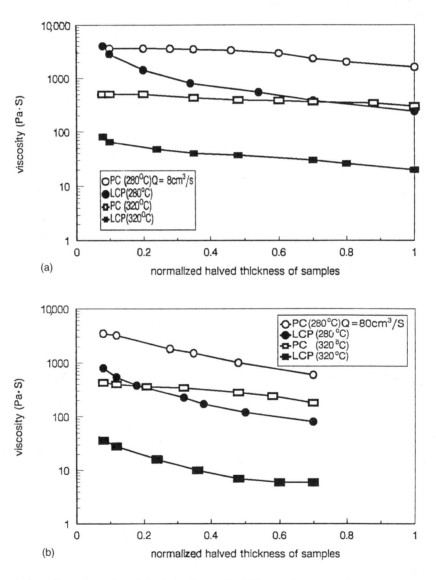

Figure 15   Calculated viscosity as function of the halved sample thickness for two melt temperatures and for: (a) injection volume flux of 8 cm^3/s; and (b) injection volume flux of 80 cm^3/s.

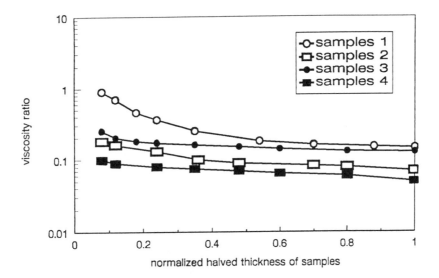

Figure 16 Calculated viscosity ratio as function of the halved sample thickness for the four sample groups.

To constitute the W_e number, characteristic values such as the drop diameter, d, and particularly the interfacial tension, ω, must be experimentally determined. However, the W_e number can also be obtained by deduction from mathematical analysis of droplet deformational properties assuming a realistic model of the system. For a shear flow that is still dominant in the case of injection molding, Cox [25] derived an expression that for Newtonian fluids at not too high deformation has been proven to be valid:

$$D = \frac{5(19\delta + 16)}{4 (\delta + 1) [(19\delta)^2 + (20W_e)^2]^{1/2}} = D(\delta, W_e) \quad (13)$$

in which, according to Taylor [16]:

$$D = \frac{1 - d}{1 + d} \quad (14)$$

where l and d are the long and short axis of a deformed droplet.

This expression can be rewritten as follows

$$\frac{l}{d} = \frac{1 + D(\delta, W_e)}{1 - D(\delta, W_e)} \quad (15)$$

where l/d is the aspect ratio that, in our study, had already been experimentally determined as $l/d = f(x)$, in Fig. 12.

With a computer program, which solved Eqs. (5) and (7) for W_e by numerical method, Weber numbers were obtained from layer to layer for each injection molding operation. The calculated points were then connected to curves, depending on the normalized thickness. Figure 17 shows these theoretically deduced $W_e(x)$

curves based on the known aspect ratio and the known viscosity ratio data. In general, the W_e curves have an increasing tendency, whereas for $\delta(x)$, the opposite is true.

To evaluate the fibrillation behavior of dispersed TLCP domains according to the $W_e - \delta$ relation discussed previously, different $W_e - \delta$ graphs were calculated by eliminating the thickness variable x. The result is reported in Fig. 18. It is obvious that all the points obtained are found to be relatively close to the critical curve by Taylor. The Taylor-limit is also shown in the figure with a solid curve. One finds that all the values calculated on sample 1 are completely above the limit, while all those determined on sample 4 are completely below the limit. The other two samples, 2 and 3, have the $W_e - \delta$ relation just over the limit.

According to the criteria, the dispersed phase embedded in the matrix of sample 1 must have been deformed to a maximum aspect ratio and just began or have begun to break up. By observing the relative position of the experimental data to the critical curve, the deformational behavior of the other samples can be easily evaluated. Concerning the fibrillation behavior of the PC–TLCP composite studied, the Taylor–Cox criteria seems to be valid.

More general information of the fibrillation behavior in the form of average aspect ratio against both the viscosity ratio and the W_e number is illustrated in Fig. 19 by using a mathematical smoothing operation. Generally, in the given range of W_e and δ, a higher W_e number combined with a lower δ value leads to a maximum average aspect ratio. It can be expected that if the W_e is equal to 1.0 by extrapolation, and δ is in a range of 0.1–0.3, the corresponding aspect ratio may be higher, indicating

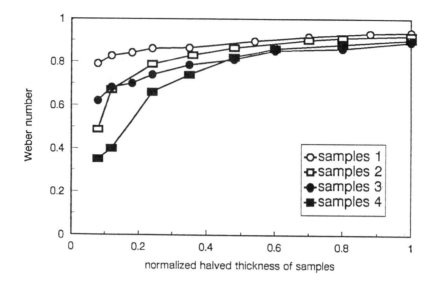

Figure 17 Calculated W_e number as function of the halved sample thickness.

a parameter area where the composite may be effectively reinforced in situ by a higher aspect ratio. In this presentation, one can also find the known $W_e - \delta$ relations of the four groups of samples under different processing conditions. This three-dimensional plot exhibits a clear picture of parameters that characterize the processing-dependent fibrillation process of the TLCP phase. Therefore, it may offer guidance of practical significance for the preparation of these types of composites.

Since the $W_e(x)$ and $\gamma(x)$ were known, the critical breakup time, t_c, could be also calculated by:

$$t_b(x) = 2t_b^* \frac{W_e(x)}{\gamma(x)} \qquad (16)$$

here t_b is the unit of time, namely $t_c (t_b = t_c)$. The dimensionless critical time, t_b^*, was taken directly from the Grace criteria according to Eq. (3) as a mathematical function in the δ range related. As shown in Fig. 10, the critical time for a drop fracture can be fitted approximately by a curve in the whole range of δ, both for rotational and irrotational shear.

It is seen from Fig. 20 that the injection flux significantly affects the critical burst time. If the samples are

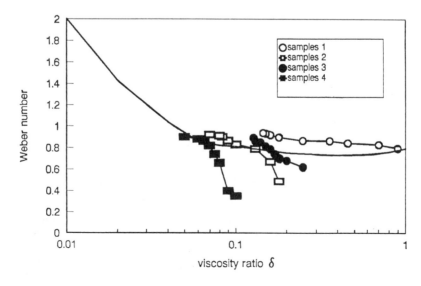

Figure 18 $W_e - \delta$ plot: comparison of the experimental values with the theoretical limit.

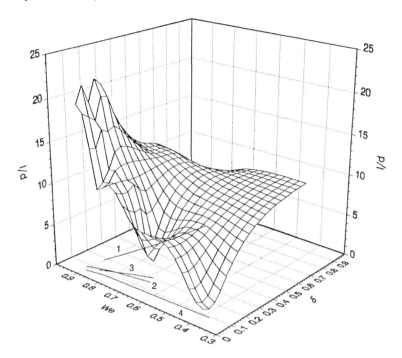

Figure 19 $(l/d) - W_e - \delta$ relation for the processing conditions studied.

injection molded with a lower flux of 8 cm³/s, the burst time is 10 times longer than those processed with a larger flux of 80 cm³/s. Obviously, the relative smaller average degree of fibrillation in samples 3 and 4, may be due to this breakup effect. In the core area the flow is more stable, therefore, the life time of deformed droplets is longer.

By plotting the t_c against δ, we obtained the curves reported in Fig. 21. The pattern seems similar to that by Grace, but the absolute values are significantly lower.

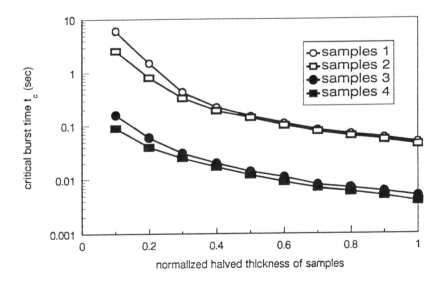

Figure 20 Calculated critical burst time t_c as function of the halved sample thickness.

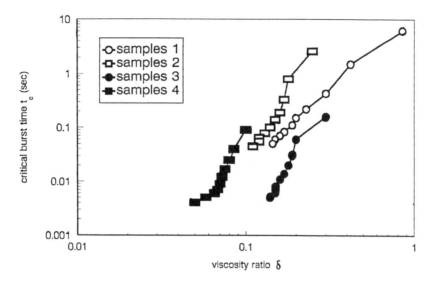

Figure 21 $t_c - \delta$ plot for the four sample groups.

The big scatter in the range of our experimental conditions and the lower values are not yet well understood.

IV. MODULUS OF DISPERSED TLCP FIBERS

A. Introduction

The formation of a fibrillar structure in TLCP blends makes the mechanical properties of this kind of composites similar to those of conventional fiber reinforced thermoplastics [11,26]. However, because the molecular orientation and fibrillation of TLCPs are generally flow-induced, the formation, distribution, and alignment of these droplets and fibers are considerably more processing-dependent. We do not know:

1. How to measure the mechanical constants of these dispersed fibers and deformed droplets that are always embedded in a matrix.
2. If these mechanical data are materials constants or processing- and morphology-dependent.

Kohli et al. [27], for instance, showed that the tensile modulus of a highly drawn PC–TLCP composite could be modeled effectively by the simple additivity rule of mixtures, while the compression molded composite samples with a spherical TLCP morphology had moduli according to the inverse rule. In both cases, the tensile modulus of the TLCP (E_{LC}) itself was assumed to be a constant value determined from a tensile test of the pure TLCP samples. But whether or not the dispersed TLCP fibers and deformed droplets have the same modulus as the bulk TLCP samples remains a question.

Blizard and coworkers [28] reported a calculated average modulus of 24.6 GPa for TLCP Vectra A950, as

microfibril dispersed in the extruded and drawn PC–TLCP composite strands, based also on the rule of mixtures. As known, this simple equation is only valid for unidirectional fiber composites. There is no parameter in the equation that is concerned with different morphological features.

It is, therefore, the intention of this section to present the tensile modulus of dispersed TLCP fibers and pure TLCP bulk materials, respectively, both processed by injection molding.

B. Experimental

The polymers studied and the experimental procedure were described in the last section. The samples were made of the same PC–TLCP composite and the pure TLCP Vectra A950, respectively. To study the influence of sample forms and geometries on mechanical constants, plate samples ($80 \times 80 \times 2$ mm^2) were also injection molded with a film gate. Tensile samples were then cut from the plate in the longitudinal (//) and transverse ($\_/\_$) direction, respectively. In the following text, the samples with their cross-section of 10×2 mm^2 are called "thin" samples, whereas the standard tensile samples are called "thick" samples.

The injection and holding pressures were 100 MPa and 50 MPa, respectively. For thick (DIN 53455 Form 3) and thin (cut from the plates) dumbbell-shaped samples, an average shear rate of 230 s^{-1} was kept constant.

C. Results and Discussion

Table 4 shows that tensile modulus E_t and strength σ_b decreased with increase of the melt temperature T_m and shear rate γ. From fractured surfaces, TLCP domains were found by microscope to be differently deformed, fibrillated, and distributed.

In all cases of the processing conditions, TLCP domains were well dispersed and deformed to droplets in the core layer, but there was only a narrow distribution of their aspect ratio (about $l/d \leq 6$) and less orientation. In both transition and skin layers, the domains were also well dispersed, but more oriented and fibrillated in the flow direction. From this reason, we give the distribution of aspect ratio (l/d) and fiber number (N) versus fiber length class in Fig. 22, only for skin and transition layers, respectively.

From the difference of these morphological characteristics, a simplified three-layer model could be set up and the inhomogeneous structure could be classified into these three layers. Figure 23 illustrates the relative thickness of the core, transition, and skin layer for the four sample groups, depending on the processing condition.

Within the conditions, the extremely thick skin layer was produced with a lower T_m of 280°C combined with a lower γ of $116\,s^{-1}$. In this case, the major cross-section of these samples was filled with highly oriented TLCP fibers. This led to the highest composite modulus and strength of 4922 MPa and 112 MPa, respectively. In-

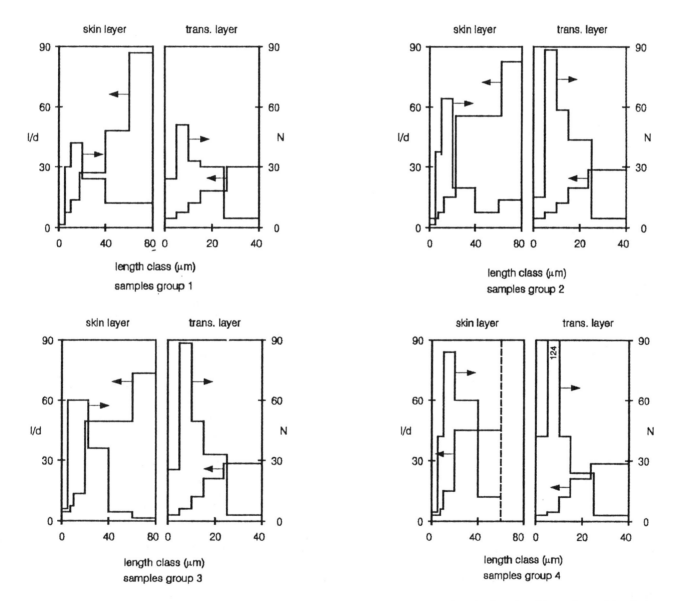

Figure 22   Distribution of fiber aspect ratio l/d and fiber number N versus fiber length class for skin and transition layer of the four groups of samples injection molded.

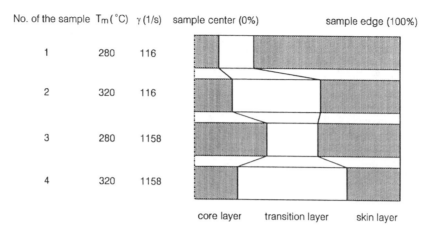

No. of the sample T_m (°C) γ (1/s)   sample center (0%)   sample edge (100%)

core layer   transition layer   skin layer

Figure 23 Distribution of average layer thickness within a sample in dependence on melt temperature T_m and shear rate γ by injection molding.

creasing the melt temperature and shear rate caused a decrease of degree of the fibrillation, and, therefore, the thickness of the skin and transition layer, due to the breakup time limit. This behavior is believed to be responsible for the lower mechanical values.

According to the composite theory, tensile modulus of fiber reinforced composites can be calculated by knowing the mechanical constants of the components, their volume fraction, the fiber aspect ratio, and orientation. But in the case of in situ composites injection molded, the TLCP fibrils are developed during the processing and are still embedded in the matrix. Their modulus cannot be directly measured. To overcome this problem, a calculation procedure was developed to estimate the tensile modulus of the dispersed fibers and droplets as following.

The three layers can be treated as an iso-strain parallel model in the flow direction. The total cross-section A_{total} and the total tensile modulus E_{total} (E_t) of a composite sample can be determined as follows:

$$A_{total} = \sum_{i=1}^{3} A_i$$

$$E_{total} = \sum_{i=1}^{3} E_i \frac{A_i}{A_{total}} \tag{17}$$

where E_i stands for the moduli of the three different layers in the sample ($i = 1 - 3$), while A_i refers to the corresponding areas in the cross-section, respectively. On the cross-section, the thickness of each layer in the sample width and thickness direction can be assumed to be the same. The effect of the edge areas cannot be considered.

According to the Cox–Darlington model [29,30], modulus E_i of each layers is expressed as below:

$$E_i = E_f \times \phi_i \times n_1 + E_m \times (1 - \phi_i) \qquad i = 1, 2, 3 \tag{18}$$

$$n_1 = \sum_{j=1}^{6} h_j \left(1 - \frac{\tanh x}{x} \right) \tag{19}$$

$$x = \sqrt{ \frac{2}{\ln\left(\frac{2}{\phi_1} \times \frac{\pi}{\sqrt{3}} \right)} \times \frac{E_m/E_f}{1 + v_m} \times \left(\frac{l}{d} \right)_j }$$

where E_f is the modulus of dispersed fibers, namely the E_{LC} to be determined in our case. The modulus of matrix, E_m, is in our composites equal to 2500 MPa, ϕ_i represents fiber volume fraction in each layer, and n_l is a so-called fiber length modifying factor that depends on the fiber aspect ratio (l/d) and volume fraction. Where h_j is equal to N_j/N_i, where N_j is the number of fibers at j class in each layer, $j = 1$ to 6, and N_i is the total number of fibers at layer i ($i = 1,2,3$), And v_m represents Poisson's ratio of matrix, here $v_m = 0.35$.

Equation (18) is an implicit function of E_f, which is given in Fig. 24 by a computer simulation. It indicates that E_t increases monotonously with increasing E_f. The E_f can be determined by replacing the experimental results of E_t (Table 4) of the composite samples in the function. It is interesting to note that the so-obtained four moduli were found approximately to be a constant of 24 GPa, independent of the deformational difference of the TLCP domains. In other words, the elastic tensile properties of the in situ formed TLCP may be changed less by varying the flow condition during the injection molding. The modulus value seems to be a material constant of these fibers and droplets.

For the calculation, the values of l/d and ϕ_i were take from Fig. 12. These values were an average of the

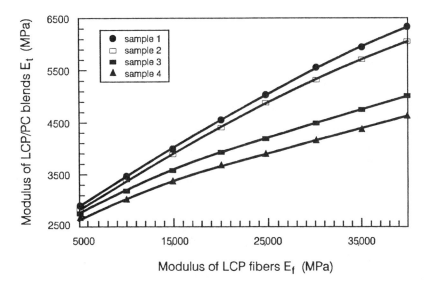

Figure 24   Total modulus E_t of PC–TLCP composite sample as function of the fiber modulus E_f for the four sample groups.

distribution functions in Fig. 22 combined with the results in Fig. 23.

The tensile modulus calculated previously is in fair agreement with that reported in the literature. Blizard et al. [28] also extruded pure TLCP Vectra A950 and its mixtures with PC with a single-screw extruder equipped with a capillary die having a diameter of 3.17 mm and a length of 9.6 mm to obtain the pure and composite strands, respectively. The extrudates were then taken up at different speeds after a quenching. Because of this stretching operation, the TLCP domains were mostly fibrillated in all cases of the composite strands and oriented nearly unidirectional in the machine direction. They reported a tensile modulus of 29 GPa for the extruded pure TLCP strand and an average modulus of 24.6 GPa for the individual TLCP microfibril by a calculation based on the additivity rule of mixtures. As known, the rule of mixtures is based on the assumption of continuous fibers and therefore establishes an upper limit.

Table 5 compares the tensile properties of Vectra A950 in the form of dispersed fibers and droplets in the matrix by injection molding, microfibril by extrusion and drawing [28], injection molded pure thick sample and pure thin sample, and the pure drawn strand [28]. As exhibited, our calculated fiber modulus with its average of 24 GPa is much higher than that of the thick and thin pure TLCP samples injection molded. It can be explained that in cases of pure TLCP samples the material may only be fibrillated in a very thin skin layer owing to the excellent flow behavior in comparison with that in the blends. However, this modulus value is lower than that of the extruded and drawn pure strand. This can be

viewed as a result of the drawing. As is well known, drawing results in a high orientation and fibrillation. Fiber spun Vectra has been, for example, reported to have moduli of 41 GPa at a draw ratio of 6 [31] and more than 62 GPa at draw ratios of 50 or more [32]. It appears to be natural that our calculated modulus was just between the two limits of the pure samples injection molded and extruded/drawn.

This constant value can be taken in turn into the composite functions to calculate the composite properties. A calculation result is illustrated in Fig. 25. For the four sample groups, the calculated layer moduli E_i are uneven in the cross-section within a composite sample group. The lowest value is still located in the core layer due to the lower deformation and, therefore, the lower

Table 5   Tensile Properties of Vectra A950 as Fibers and as Pure Bulk Materials

Tensile properties	Modulus (GPa)	Strength (MPa)
TLCP fibers in injection molded thick composite sample	24.0	?
TLCP fibrils in extruded and drawn composite strand [28]	24.6	—
Pure TLCP thick sample	13.5	175
Pure TLCP thin sample (//)	9.5	130
Pure TLCP thin sample (—/—)	1.7	53
Extruded and drawn pure TLCP strand [28]	28	—

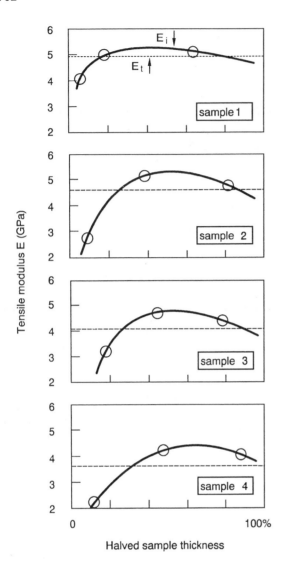

Figure 25 Calculated layer modulus E_i as function of the halved sample thickness for the four sample groups and a comparison with the total modulus E_t.

perimentally demonstrated, and a generalized function has been derived to evaluate the relationship between the blend viscosity reduction and TLCP weight fraction. Interfacial slippage of the TLCP-rich skin layer is the mechanistic explanation of the observed effect. Computer fitting and calculation of the experimental data can give property-related parameters characterizing the change manner of the blend viscosity with the TLCP weight fraction. Comparison of the theoretical curves with the experimental data gave satisfactory results. Therefore, this function has been proved to be a useful tool in dealing with the viscosity reduction behavior. Its applicability has been found to be independent of the special chemical nature of the materials.

The fibrillation behavior of a dispersed TLCP domains in a thermoplastic matrix can be described by using the $W_e - \delta$ relation initially given by Taylor and Cox. The flow-induced deformation is time-dependent. If the flow time exceeds the critical time, the resulting fibers break up. The life time of a TLCP fiber can be evaluated by Grace criteria. Based on morphological and rheological examinations, a calculation procedure was developed to predict the fibrillation behavior of a TLCP phase in a thermoplastic matrix during injection molding. In particular, the $W_e - \delta - l/d$ relation (Weber number of the melt, respectively for $W_e - \delta - ld$, viscosity ratio of TLCP to matrix aspect ratio of the fibrils) may give complete information for processing optimization in order to achieve highly oriented TLCP fibrous structure.

The variation in the tensile modulus of a dispersed TLCP phase depends primarily on the degree of orientation and fibrillation, or in other words, on the fiber–aspect ratio. Injection molding or extrusion offer samples that give only an integrated value. It can not be taken as the modulus of the in situ formed TLCP domains directly. In the case of injection molded blend samples, the modulus of dispersed TLCP domains can be estimated by a calculation procedure described in this chapter. For Vectra A950 tested, as it dispersed and deformed in the PC matrix, the average tensile modulus was found to be approximately constant, independent of the processing conditions. For most composites containing a deformed inclusion, the proposed calculation procedure may offer a useful way to estimate the modulus of the inclusion that is usually unknown due to testing problems.

aspect ratio of the TLCP droplets. In the transition layer, the modulus obtains its maximum. The experimentally measured total modulus E_t is approximately the average of E_i. Thus, the calculation procedure has been proven as valid for the in situ composites to estimate the modulus values.

V. CONCLUSIONS

A distinct viscosity reduction occurs as a small quantity of TLCP is added to a thermoplastic matrix. This negative deviation effect from the log-additivity rule was ex-

ACKNOWLEDGMENTS

This study was supported by the National Natural Science Foundation of China (NNSFC, Grants No. 58903427 and 59183016), the National Advanced Materials Committee of China (NAMCC, Grant No. 86371527), and the Deutsche Akademische Austauschdienst (DAAD, Germany) Foundation.

GLOSSARY

A_i = layer cross-section in three layers model ($i = 1$–3)

A_{total} = total cross-section of a tensile sample

b = width of a dumbbell-shaped sample

d = layer thickness, average fibril diameter, short axis of a deformed droplet

E = tensile modulus

$E_{f,m}(= E_{LC})$ = modulus of dispersed TLCP fibers (f) and of matrix (m)

E_i = layer modulus in three layers model ($i = 1$–3)

E_{total} ($= E_t$) = modulus of a composite sample experimentally obtained

h = thickness of a dumbbell sample

k = linear factor

l = long axis of a deformed droplet

l/d = fiber aspect ratio

m = interlayer number within a capillary flow, flow index of viscosity curve

n = fibril number within a layer

n_l = fiber length modifying factor

N_i = number of fibers at layer i ($i = 1 - 3$)

N_j = fiber number at j class in each layer ($j = 1 - 6$)

Q = volume flux

r = distance of a TLCP-rich interlayer to the capillary axis

r_c = critical radius

R_0 = capillary radius

t_b or $t_c (t_b = t_c)$ = breakup time

t_b^* = dimensionless breakup time

T_m = melt temperature

ΔV = interface slide velocity

W_e = Weber number

x = thickness variable

α = interface slip factor

α_0 = proportional factor

γ = shear rate

δ = viscosity ratio of η_1/η_0

ϵ = exponent

η = blend viscosity

η_0 = matrix viscosity or η_C viscosity of continuous phase ($\eta_C = \eta_0$)

η_1 = TLCP viscosity or η_D viscosity of dispersed phase ($\eta_D = \eta_1$)

$\lambda (= \tau_c/\tau_W)$ = critical stress ratio

v_m = Poisson's ratio of matrix

ρ = density

σ = interfacial tension coefficient

σ_b = tensile strength

τ_c = critical interface shear stress

τ_D = shear stress

τ_W = shear stress on the capillary wall

ϕ = weight fraction of TLCP

ϕ_i = fiber volume fraction in each layer (three layers model)

REFERENCES

1. A. Siegmann, A. Dagan, and S. Kenig, *Polym.*, 26: 1325 (1985).
2. D. Dutta, H. Fruitwala, A. Kohli, and R. A. Weiss, *Polym. Eng. Sci.*, 30: 1005 (1990).
3. M. R. Nobile, E. Amendola, L. Nicolais, D. Acierno, and C. Carfarna, *Polym. Eng. Sci.*, 29: 244 (1989).
4. L. A. Utracki, M. M. Dumoulin, P. Toma, *Polym. Eng. Sci.*, 26: 34 (1986).
5. N. Chapleau, P. J. Carreau, C. Peleteiro, P. A. Lavoie, and T. M. Malik *Polym. Eng. Sci.*, 32: 1876 (1992).
6. A. I. Isayev and M. J. Modic, *Polym. Compos.*, 8: 158 (1987).
7. X.-S. Yi, L. Wei, and H. Wang, *Cailiao Kexue Jingzhan, (Chin. J. Mat. Resear.)*, 6: 256 (1992), (Chinese).
8. X.-S. Yi, *Preparation and Processing of Plastics*, Zhejiang University Press, Hangzhou (1995) (Chinese).
9. V. G. Kulichiklin, O. V. Vasil'eva, I. A. Litinov, E. M. Antopov, I. L. Pasamyan and N. A. Plate, *J. Appl. Polym. Sci.*, 42: 363 (1991).
10. F. D. Rumscheidt and S. G. Mason, *J. Colloid Sci.*, 16: 238 (1961).
11. X.-S. Yi, G. Zhao, and F. Shi, *Polym. Intern 39(1)*: 11 (1996).
12. X.-S. Yi and G. Zhao, *J. Appl. Polym. Sci.*, 61: 1655 (1996).
13. X.-S. Yi, and L. Shen, *Polym.-Plast. Technol. Eng. 36(1)*: 153 (1997).
14. J.-C. Huang, H.-F. Shen, and Y.-T. Chu, *Adv. Polym. Techn.*, 13: 49 (1994).
15. L. A. Utraki and P. Sammut, *Polym. Eng. Sci.*, 30: 1027 (1990).
16. G. I. Taylor, *Proc. Roy. Soc. (London), Ser. A146*: 501 (1934); and *Ser. A138*: 41 (1932).
17. C. D. Han, *Multiphase Flow in Polymer Processing*, Academic Press, New York (1981).
18. H. P. Grace, *Chem. Eng. Commun.*, 14: 225 (1982).
19. J. J. Elmendorp, *Polym. Eng. Sci.*, 26: 418 (1986).
20. M. V. Tsebrenko, G. P. Danilova, and A. YA. Malkin, *J. Non-Newton. Fluid Mech.*, 31: 1 (1989).
21. B. J. Bentley and L. G. Leal, *J. Fluid Mech.*, 241 (1967).
22. N. Chapleau, P. J. Carreau, C. Peleteiro, P. A. Lavoie, and T. M. Malik, *Polym. Eng. Sci.*, 32: 1876 (1992).
23. Th. Brinkmann, PhD. Thesis, RWTH Aachen (1992).
24. J. Schmitz, E. Bornschlegel, G. Dupp, and G. Erhard, *Plastverarbeiter 39*: 50 (1988).
25. R. G. Cox, *J. Fluid Mech., 37(3)*: 601 (1969).
26. G. Kiss, *Polym. Eng. Sci.*, 28: 1248 (1988).
27. A. Kohli, N. Chung, and R. A. Weiss, *Polym. Eng. Sci.*, 29: 573 (1989).
28. K. G. Blizard, C. Federici, O. Fererico, and L. L. Chapoy, *Polym. Eng. Sci.*, 30: 1442 (1990).
29. X.-S. Yi and L. Shen, *Gaofenzi Xuebao (Acta. Polym. Sinica)*, 5: 621 (1994) (Chinese).
30. H. L. Cox, *Brit. J. Appl. Phys.*, 3: 72 (1952).
31. S. Kenig, *Polym. Eng. Sci.*, 29: 1136 (1989).
32. G. W. Calundann and M. Jaffe, *Proc. Robert Welch Conf. XXVI, Synthetic Polymers*, 247 (1982).

46
Energetic Composites

M. B. Khan*

CHEMTEC & Prime Glass, Jhelum, Pakistan

I. INTRODUCTION

Energetic composites, commonly referred to as composite propellants, are ignitable assemblies containing energetic particles (inorganic oxidizer and metal) that are embedded in a viscoelastic matrix. The primary function of these composites is to supply the energy needed for the propulsion of space vehicles and tactical missiles. The composite is normally contained in a casing, lined with an ablative thermal insulation to provide protection from the ultra-high temperatures (3000–3600°K) encountered in the combustion chamber. In the missile motor (Fig. 1), the sensitive heat available from the combustion reaction is converted into the kinetic energy of the gases at the nozzle exit plane, providing the thrust needed for vehicle propulsion.

From the ingredients to feed preparation to final fabrication in the casing, the solid fuel composite passes through several manufacturing stages, during which it is strictly scrutinized to achieve the requisite mechanical and ballistic properties in the end. Unlike other strategic technologies, energetic composites is a well-published discipline. The analytical and experimental efforts that have been undertaken to formulate, fabricate, characterize, and preserve these systems for long-term storage have all been well described. While abundant literature is available on the various aspects of the subject [1–10], a comprehensive treatment from the materials processing standpoint seems to be lacking. Processing constitutes an important sector in the production of energetic composites. It is during this phase that new material interfaces are formed, and the viscous prepolymer slurry is converted to a viscoelastic composite.

This chapter discusses the technical features of the production sequence, within the general domain of vis-

cous reactive processing, with special emphasis on multiphase rheology, the chemorheology and the associated morphology of the reactive polymer system, interfacial phenomena, and the equipment configuration. The discussion largely relies on the elucidation of the governing principles, with recourse to experimental data wherever necessary. The main discussion is preceded by a brief overview on some of the key performance parameters that influence the choice of the basic constituents. The chapter concludes with a technical description of the principal processing equipment used in the production of energetic composites.

II. OVERVIEW

A. Ingredients

An energetic composite is basically a fuel oxidizer assembly containing several important additives to perform specific functions. The fabricated system derives

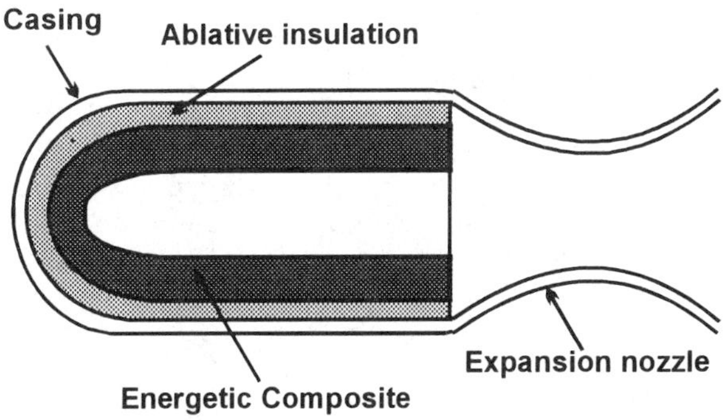

Figure 1 Energetic composite assembly showing various components.

---

* Former consultant, Pakistan Air Force.

its propulsion force from the combustion reaction between the oxidizer and the fuel components. The oxidizer together with the various ingredients are incorporated in a hydrocarbon fuel, a processable liquid prepolymer that upon curing transforms into an elastomeric matrix. The composition of the system is dictated by the often conflicting ballistic and processing constraints. To meet only the minimum performance requirements, the solid content (oxidizer plus metal) in the composite should be greater than 80%. This high level of solids loading limits the flexibility of the chemist in adjusting the content and rheology of the composition. Thus, matching performance and processability is often the real difficulty.

Aside from the oxidizer component, metal particles are added to augment the heat of combustion; ballistic catalysts may be used to enhance the burning rate; crosslinking and interfacial agents may serve to improve the mechanical properties; and polymerization catalysts may help to control the formation of the crosslinking network at a predetermined rate. Plasticizers and wetting agents are almost always used to aid in the processing, as they tend to reduce the viscosity of the mix to acceptable levels for the subsequent pouring and casting operations. Finally, the addition of an antioxidant improves shelf-life by arresting bond cleavage and/or crosslinking reactions at sites of unsaturation in the otherwise crosslinked polymer.

B. Material and Performance Parameters

In formulating energetic composites, the primary aim is to pack maximum thermal energy per unit volume. A principal performance characteristic of the fuel is the specific impulse, defined as the thrust delivered per unit mass rate of fuel burnt. From the ballistic standpoint, it is desirable to maximize the specific impulse of the propulsion unit, simply expressed as:

$$I_{sp} = v_e/g = f(\sqrt{T/M}) \tag{1}$$

The solid fuel composition is, therefore, concerned with the pursuit of higher I_{sp} involving both flame temperature elevation and molecular weight reduction. Concomitantly, it is desirable to have high heat of combustion, which also dictates the combustion chamber temperature. The actual composition is, therefore, set on the basis of the energy quantum inherent in the various constituents.

The fuel combustion enthalpy is primarily determined by the polymer binder and the metal heats of combustion. The polymer, which binds the particulates together, is an important constituent of the reactive composite: it contributes to the overall enthalpy to the tune of $30\text{--}40 \times 10^3$ kJ/kg, apart from conferring vital mechanical properties, such as tensile strength and elastic modulus. Al is the preferred choice as the metallic fuel (30 MJ/kg) because of its ease of incorporation and much higher specific gravity (2.70) compared with other more thermally active metals such as Be (63 MJ/kg and 1.81). A higher specific gravity is synonymous with higher mass ratios, which augments the vehicle burnout velocity.

Modern propellants use polybutadiene type binders whose main chain terminates with either hydroxyl

(HTPB: HO—[CH$_2$—CH=CH—CH$_2$)$_{.2}$—(CH$_2$—CH)$_{.2}$—(CH$_2$=CH—CH$_2$)$_{.6}$]$_{50}$—OH
$$\overset{|}{\underset{CH=CH_2}{}}$$

or carboxyl (CTPB: HOOCR—(CH$_2$—CH=CH—CH$_2$)$_n$— RCOOH)

end groups. For obvious reasons, polymers containing a high hydrogen content and higher heats of formation (those absorbing less heat on combustion) yield more energetic compositions. A comparison is shown in Table 1 between two commonly used prepolymers, namely HTPB and PPG (polypropylene glycol) [9].

The higher heat of formation and less oxygen in the HTPB molecule implies a higher heat output with greater oxidizer loading capacity. However, more oxygen in the

Table 1 Comparison of Prepolymer Heats of Formation and Oxygen Content

	ΔH_f (kJ/kg $\times$ 10^3)	O$_2$%
HTPB	20.92	5.0
PPG	−3740	26.0

PPG would allow more metal at the expense of the oxidizer to further energize the composition. In addition, HTPB is frequently preferred over PPG because of its superior mechanical properties, better aging characteristics, and lower glass-transition temperature (T_g). The latter is especially desirable because at low temperature, the higher strain rates produced by motor ignition decrease the elongation of the composite rather markedly.

C. Composition

Several techniques are available in the literature for evaluation of the flame temperature, exit temperature, equilibrium composition of combustion products, and performance parameters of energetic composites [11–13]. The optimum combination of the composite ingredients is determined by thermodynamic means, so as to arrive at a composition having maximum performance capabilities. The optimization procedure is, however, no longer dependent on trial and error methods. Thermochemical computer codes are now available [14,15] in which the motor thrust is expressed as a function of the reactants and optimized subject to linear or nonlinear constraints. Table 2 lists the various ingredients along with the possible range of composition for a typical energetic composite.

D. Performance Enhancement

The major emphasis in the development of new formulations has been on composites where an energetic solid is dispersed in an energetic polymer matrix. The polymer system in contention is the glycidyl azide polymer (GAP). Its curing properties are not ideal, but it was really designed to demonstrate that the energetic polymer approach is feasible [16]. Other advanced high-energy systems use nitramine (RDX or HMX) crystals in the so-called crosslinked double-base (XLDB) propellants, where a conventional prepolymer (PPG, HTPB, CTPB, PBAN, etc.) is mixed with double-base ingredients (nitrocellulose and nitroglycerine) together with a relatively large amount of RDX or HMX and crosslinked to provide a highly energized, HCl-free system, which also obviates a major ecological drawback of AP (ammonium percholorate)-based composites [17].

The inclusion of energetic solids in high-energy polymers also increases their vulnerability to detonation. The search, therefore, continues for new energetic solids that exceed the energy and density of RDX, but with less thermal and shock sensitivity. Polycyclic and bridged nitramines are families of new energetic materials with much promise. Plasticizers are also key ingredients because of their influence on energy, mechanical properties, and processing characteristics. Plasticizers containing nitric ester groups combine the ease of processing with higher energy content, but pose compati-

bility and sensitivity problems. A promising approach is the use of molecular modeling software to elucidate polymer–plasticizer interactions and evaluate the potential of new compounds. New and emerging materials exploit the nitramino and azado moieties for superior energy and reduced sensitivity relative to nitrato moieties [18].

Finally, there is active interest in developing catalyst systems, both ballistic and polymerization, that would promote combustion stability at high pressures (especially in metal-free systems for smokeless applications) and allow processing lattitude for relatively large motors. The ferric-based systems currently being used fall short of these performance measures. Compounds that form complex structures with the metal chelate to reduce its activity to acceptable levels seem to be most promising. Interestingly, the use of an antibiotic has been cited in this context [19].

III. CONCENTRATED PARTICLE PROCESSING

A. Multiphase Rheology

Rheology is the science of flow and deformation of matter. It is of direct importance to the production of energetic composites; processability heavily depends on the flow properties of the composite slurries. The process fluid in these systems is of a multiphase character, with a high percentage of solids suspended in a viscous and reactive liquid. The degree of particle loading, their size distribution, interfacial effects, and polymerization may substantially influence the rheological behavior of the system. However, the primary parameter that determines the bulk viscosity is particle concentration. This section will describe the rheology of suspensions and highlight the nature of particle assemblies that may be accommodated in the liquid, according to a particular size distribution. The influence of polymerization kinetics on the rheology will be discussed in a subsequent section.

1. Estimation of Suspension Viscosity

An estimation of the multiphase viscosity is a preliminary necessity for convenient particle processing. For particle-doped liquids the classical Einstein equation [20] relates the relative viscosity to the concentration of the solid phase:

$$\eta_r = 1 + 2.5\phi \qquad (2)$$

Equation (2) is valid only for very dilute suspensions of nondeformable, smooth, uniform spheres. It assumes a Newtonian liquid phase and neglects interaction between particles, a plausible condition when the volume of the solid phase is small compared with the liquid phase.

Table 2 Composition Range of a Typical Energetic Composite

Ingredient	Type	Wt%
Oxidizer	Perchlorate;	60–88
	Nitramine	20–40
Metal fuel	Al; Be; BeH$_2$	0–21
Prepolymer	Polyglycol; PB	8–12
Polymerizer	NCO; Epoxide	1–2
Crosslinker	Triol	0–3
Chain extender	Diol	0–3
Plasticizer	Alcohol derivative	1–2
Bonding agent	Polyalkylamine	0–1
Antioxidant	Phenol/amine derivative	0–1
Ballistic catalyst	Metal oxide	0–1
Curing catalyst	Metallic derivative	Trace

For relatively higher volume fractions (<0.15), a modification of the Einstein relation has been proposed [21]:

$$\eta_r = 1 + 2.5\phi + k_2\,\phi^2 + k_3\,\phi^3 \qquad (3)$$

to allow for both hydrodynamic and particle pair interactions. Experimental estimates of k_2 and k_3 vary enormously. A rigorous hydrodynamic solution is lacking due to the difficulty in handling multibody collisions. However, by appropriate extrapolation techniques, the scatter is reduced to fit a smooth curve of the form [22]:

$$\eta_r = 1 + 2.5\phi + 10.05\phi^2 + 0.00273 \exp(16.6\phi) \qquad (4)$$

This fitted the data well up to volume fractions of 0.55 and was so successful that theoretical considerations were tested against it. However, as the volume fraction increased further, particle–particle contacts increased until the suspension became immobile, giving three-dimensional contact throughout the system; flow became impossible and the viscosity tended to infinity (Fig. 2). The point at which this occurs is the maximum packing fraction, ϕ_m, which varies according to the shear rate and the different types of packings. An empirical equation that takes the above situation into account is given by [23]:

$$\eta_r = [1 - (\phi/\phi_m)]^{-[\eta]\phi_m} \qquad (5)$$

This equation has the advantage of having two adjustable parameters, $[\eta]$ and ϕ_m, which fit the data for both low- and high-shear rates, as ϕ_m is a function of shear rate. Since the viscosity of the suspension decreases the faster it is sheared, ϕ_m is evidently higher for higher shear rates. In practice, higher loadings also impart a finite yield stress, which must be overcome before the mixture will shear. In polydisperse systems, a large fraction of fines tend to produce dilatancy at low-shear rates.

The foregoing equations all express the multiphase viscosity as a function of the solids content, without any recourse to liquid parameters. A more realistic portrayal of the physical situation would include the fluid dynamic picture that compensates for entanglement and "absorbed liquids" carried along with the solid phase, thus effectively decreasing the liquid volume. An equation applicable to this case is [24]:

$$\eta_r = 1 - [k\phi/(1 - S\phi)] \qquad (6)$$

S is the so-called sediment volume. The volume of free liquid is $(1 - S)$. Typically, $k = -25$, and $S = 1.22$ for most high-energy propellants.

B. Particle Fluid Assemblies

State-of-the-art energetic composites use a high solids concentration in high fuel value polymer matrices. In a stoichiometric amount a typical oxidizer–metal combination, such as AP–Al, would constitute as much as 78 vol% of the composition. Special efforts are needed to attain such high loading levels without sacrificing fluidity in mixing, or even casting discontinuities in the binder, which may lead to poor mechanical properties. This may be accomplished by optimum packing of the particles in the continuous phase, according to the widely recognized principle that the maximum packing fraction of a monodisperse system can be increased by broadening the particle size distribution.

The essence of this idea is that there is a limit to which particles of like-size can occupy a given space, even when arranged in closely packed arrays (e.g., cubic or tetrahedral arrays). The voids that are left are usually smaller than the parent particles and may be filled by particles of smaller size to increase the concentrations of particles in space. Thus, polydispersity can give a lower viscosity at the same volume fraction or permit higher volume loading at the equivalent monodisperse viscosity.

Polydisperse particle processing is a close similarity to the arrangement of atoms in a crystal structure. This is so because smaller spheres can be placed between the closest packed spheres to produce various arrays. It turns out that the atomic arrays of most inorganic crystals are of this type. The largest atoms form the closest packing, while the smaller atoms are distributed among the available voids. Fig. 3a–c compares the three common geometrical structures comprised of closest packed arrays of one kind of particle, usually the larger particles in the system, with the smaller particles dispersed among the available voids. It is seen that the interparticle spatial arrangement depends on the relative diameters of the constituent particles. A pertinent question to ask from the processing angle is: what fine frac-

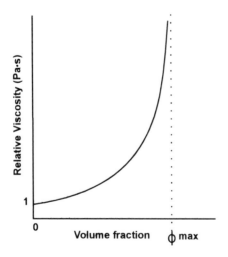

Figure 2 Variation of relative viscosity with particle volume fraction.

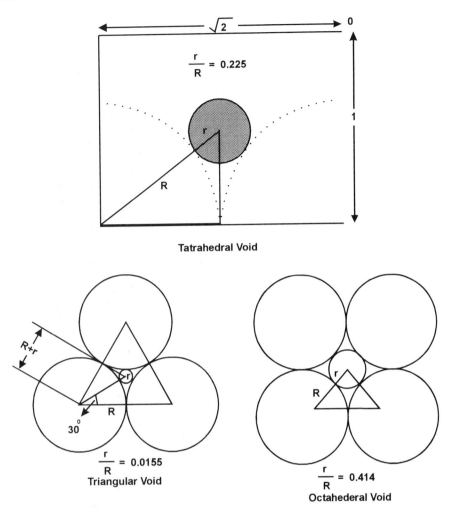

Figure 3 Schematic of model particle assemblies.

tion would allow the available space to be occupied most efficiently for a given d/D?

An analytical solution to this has already been attempted [25]. According to this model, the minimum concentration of fines would be that quantity required to coat each coarse particle with a monolayer of fines. Treating the particles as perfect spheres, the fractional change in combined particle volume due to additional film of fines is then:

$$\Delta V \approx 3[d/D + (d/D)^2] \qquad (7)$$

Since ΔV is composed of interstitial voids (between the fine particles), the weight fraction of fines in the total assembly is:

$$x_d = 3\phi[d/D + (d/D)^2]/1 + 3\phi[d/D + (d/D)^2] \qquad (8)$$

Thus, for a given particle loading, ϕ, the fine fraction in the bimodal assembly is a function of only the diameter ratio, d/D; a surprisingly simple result for an otherwise complex phenomenon. However, the inclusion of the liquid phase to the above particle assembly should modify the result somewhat. This can be seen very simply by expressing the interstitial particle population of the fines as a function of the continuous phase:

$$N_d = 6x_d/\pi d^3/[x_d + x_1(\rho_d/\rho_1)] \qquad (9)$$

Clearly, the inclusion of the liquid phase (x_1) will tend to reduce the number of fines in the system. Thus, a diminution in particle size must be effected to provide an equivalent particle population in the fluid–particle assembly. The particle size ratio normally used in practical systems tends to be somewhat lower than the one computed from Eq. (8).

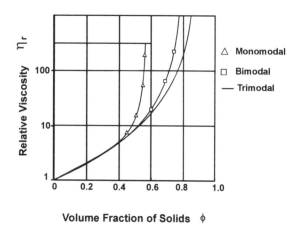

Figure 4 The effect of modality on the relative viscosity.

C. Multimodal Assemblies

The idea of adding smaller and smaller particles to fill in the interstices left by the larger particles can be continued. The viscosity of a multimodal suspension may be predicted from unimodal data based on the premise that the viscosity of the mixture of smaller fractions is the medium viscosity for the next largest fraction. This is an effective medium theory and basically assumes that the smaller particles act as a medium toward the larger particles. This was assuming at least an order of magnitude difference in size between successive fractions [26]. Thus, the viscosity of the ith component is:

$$\eta_r(\Pi_i) = \eta_{r1} \times \eta_{r2} \times \, ... \, \times \eta_{ri} \qquad (10)$$

However, the largest increase in volume fraction is observed when changing from a monomodal to a bimodal system, with successive systems giving a less significant reduction in viscosity, as depicted in Fig. 4.

IV. REACTIVE PROCESSING

A. Polymer Morphology

In the production of energetic composites, a viscous prepolymer slurry is reacted with an appropriate curative to yield a solid viscoelastic matrix. The reaction scheme for the commonly used polyaddition polymerizations is depicted in Fig. 5. All three systems use partially polymerized prepolymer (PPG, HTPB, or CTPB) because it cures with a minimum of shrinkage and heat release. The viscoelastic polymeric matrix normally consists of hard (rigid) and soft (flexible chain) segments, intercalated with adjoining bridges of chain-extender and/or crosslinking molecules. The two types of segments are thermodynamically incompatible and, thus, microphase segregated at the angstrom level. These materials, thus,

naturally are comprised of phases of intrinsically different chemical and mechanical properties. The reaction and processing parameters such as temperature, reactant ratio, rate constant, and the degree of mixing are often seen to influence the extent of microphase segregation or the matrix morphology [27–29].

In the multiphase system, structure buildup is either by localized phase separation to give a linear segmented polymer or crosslinking to yield a three-dimensional network. The final morphology of the segmented polymer is determined by a competition between the kinetics of reaction and microphase separation within the polymer building blocks. If the reaction is too slow (lower temperature or catalyst level), phase separation will dominate, effective crosslinking is ruled out, and a relatively brittle structure is obtained. On the other hand, too fast a reaction (higher temperature and/or catalyst level) will cause premature gelling, with the result that the system will loose its fluidity for proper mixing and casting. These ideas are important for optimizing the reaction conditions that are conducive for the attainment of viable physical morphology and mechanical properties.

Energetic composites are processed to attain a high polymer with a low glass-transition temperature, high-molecular weight, and a narrow molecular weight distribution; properties that usually result from an improved reactant mixing and a tightly crosslinked structure. For polymers the bulk viscosity is known to depend on molecular weight [30]:

$$\eta = KM^n \qquad (11)$$

The proportionality term, K, depends on interchain friction [31]. For crosslinking polymerizations, n lies between 1.0 and 2.6. Figure 6 gives the viscosity rise as a function of molecular weight at three different temperatures for a typical urethane binder [32]. The rather phenomenal rise in the viscosity of the polymerizing network over that of the prepolymer is evident. These data indicate that higher temperatures would lead to relatively high-molecular weight, high-modulus polymers. In crosslinked polymers, the latter is primarily a function of the crosslink density of the three-dimensional network [33–35]:

$$\begin{aligned} \mu &= C[(2r\alpha^2 - 1)/r\alpha^2]^3 \\ &= Kw_g^2/w_s \\ &= (\rho RT/M_c)\, v_r^{0.3} \end{aligned} \qquad (12)$$

in which the crosslink density has been expressed in terms of the reaction and swelling parameters. Apart from the reactant parameters (C, r, α), the sol fraction, w_s, may be used to determine the crosslink density, μ, and the optimum curing conditions of the composite [34]. An important observation is that μ is an inverse function of the molecular weight between crosslinks, M_c.

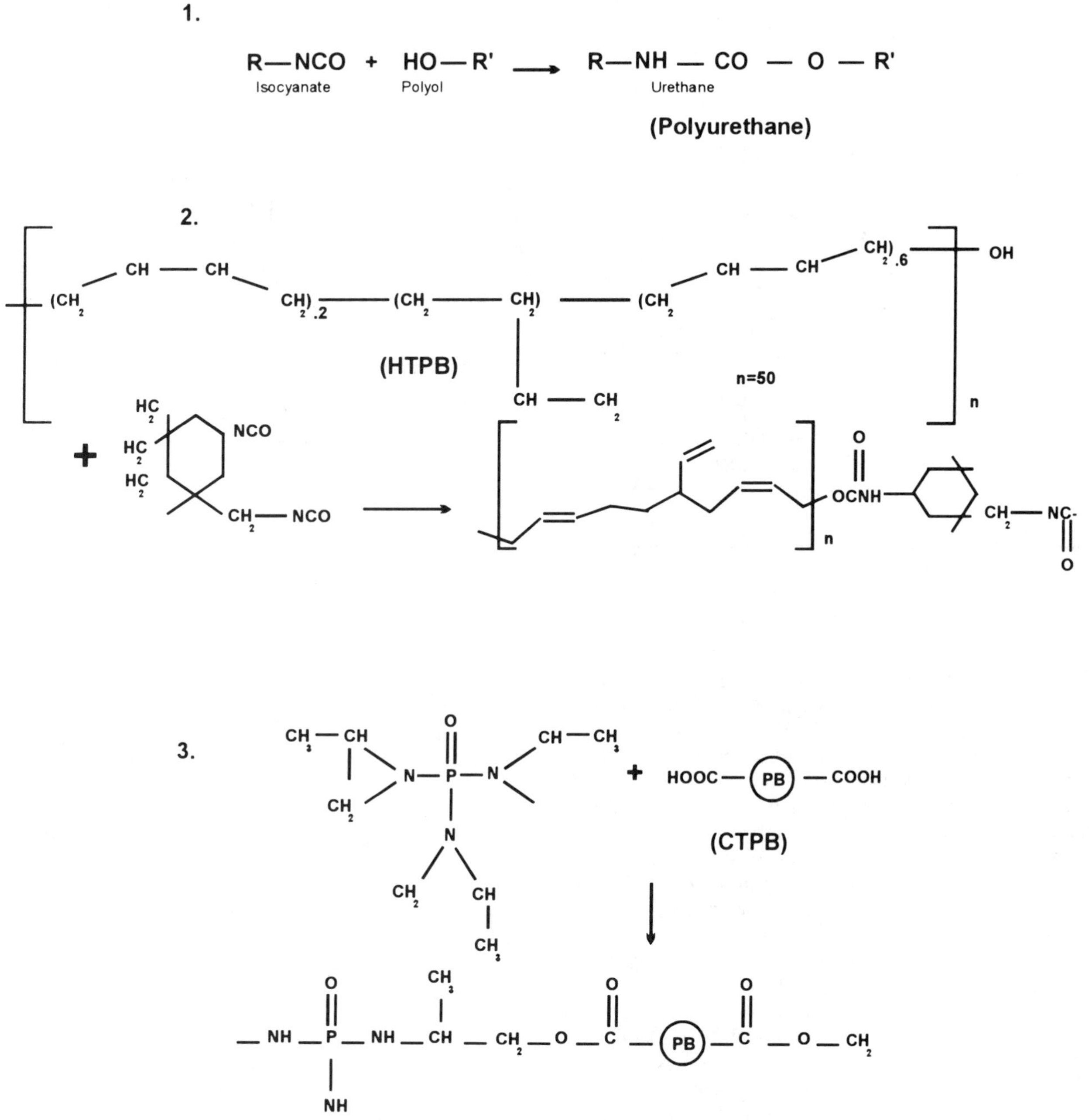

Figure 5 Urethane polymerization for three prepolymer systems.

In the case of the segmented polymers, the domains that form during phase separation will lead to the rapid buildup of viscosity and gelation, much like a crosslinking urethane, although these polymers are linear. An analogous expression for the viscosity rise in these systems is given by [36]:

$$\eta_r = [\alpha_g/(\alpha_g - \alpha)]^{C_1 + C_2\alpha} \tag{13}$$

Although Eq. (13) has been reported to fit the data well for $C_1 = 3.5$, and $C_2 = -2.0$, it provides no information on the phase separation process. In fact, there is little understanding about how the physical morphology and mechanical properties evolve with polymerization and time. The effect of various process parameters on the phase separation and morphology is obtained implicitly via final properties of the polymers. This is illustrated

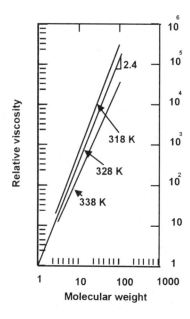

Figure 6   Evolution of MW and viscosity for a polymerizing network. *Source*: Ref. 32.

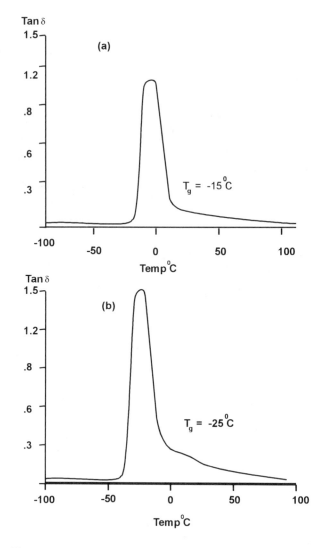

Figure 7   DMTA scan for (a) normal, and (b) shear gradient processed polymer.

in Fig. 7 by the dynamic mechanical analysis (DMTA) scans of the same polymer system that was subjected to different processing histories [29]. The scan in Fig. 7a is for the normal reaction injection molding (RIM) processed specimen, while the scan in Fig. 7b represents the specimen subjected to shear gradient processing. An enhanced microphase segregation in the latter specimen is reflected in the higher energy absorption or damping at the loss tangent peak, which also shifts the T_g of the polymer to a lower temperature (toward the higher phase-separated soft segment region).

Energetic composites are seldom processed via the phase separation route, they are almost always crosslinked through stable primary valence bonds to minimize deformation and creep of the grain. The foregoing discussion serves to emphasize that processing these composites is intimately related to the chemorheology and formulation parameters. A proper integration among the kinetics, microphase morphology, and the material and process parameters (temperature, catalyst level, reactant ratio, extent of mixing, particle size distribution and loading, etc.) is required for successful processing of these composites.

B.   Chemorheology

Chemorheology is concerned with the chemical kinetics and the associated flow properties of a model reacting system. Energetic composite rheology is a continuously evolving process. The initial slurry viscosity is determined by the system temperature, plasticizer content

(Fig. 8), and the solids loading and their distribution as discussed in some detail in the preceding sections. However, once the curing agent has been added, the polymerization and crosslinking reactions commence, with a concomitant rise in the viscosity of the system, as indicated by the temporal viscous flow curves (Fig. 9), for a relatively fast resin system [37]. The corresponding response of the reactive composite slurry is usually recorded by measuring the relative torque (Monsanto Rheometer) or the change in the amplitude of a vibrational instrument (Wallace Curimeter UK). Representative curing curves for the CTPB and HTPB systems appear in Fig. 10. The torque–viscosity curve for the production mixer is normally calibrated using these instruments.

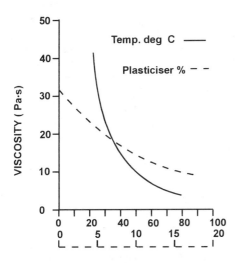

Figure 8 CTPB viscosity as a function of temperature and plasticizer content.

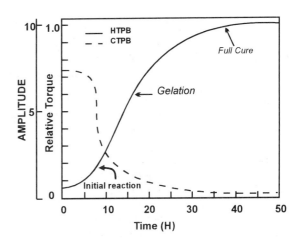

Figure 10 Curing curves for the HTPB and CTPB composite systems.

Referring to Fig. 9, the effect of the shear is to catalyze the reaction, presumably through suppression of the interfacial barrier by stretching the flow. The latter is believed to reduce the diffusion path, promoting the reaction rate, and hence the rate of increase in the viscosity. A similar effect is produced with temperature as a parameter, which also augments the reaction rate. The modified reaction rate constant in case of any external stimulus or perturbation acting on the system may be computed from the scalar K, where:

$$K = k/k_s = t_G/t_s \qquad (14)$$

With a knowledge of t_G and t_s from rheology data and k from reaction kinetics, the value of k_s may be readily computed, and hence the kinetics of the system is redefined. Means of obtaining the relative time scales from model rheology data are illustrated in Fig. 9.

The kinetic rate constant may be computed from the adiabatic temperature rise [38] or the isothermal heat release [37]. For a second order reaction:

$$k_{ad} = [\Delta T_{ad}/C_o(T_{max} - T)^2] \, dT/dt \qquad (15)$$

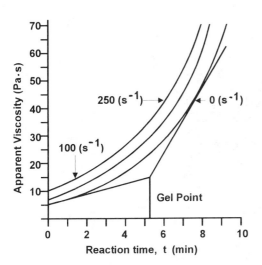

Figure 9 Temporal viscous flow curves for urethane polymerization.

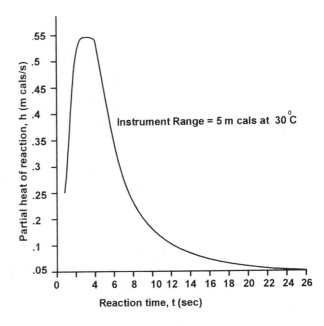

Figure 11 DSC isotherm for a typical urethane polymerization.

$$k_{iso} = \Delta H \ (dh/dt)/C_o(1 - h/\Delta H) \qquad (16)$$

The various terms appearing in these equations are self-evident. The differential heat release, dh/dt, data are computed from differential scanning calorimetry (DSC). A typical DSC isotherm for a polyurethane reactive system appears in Fig. 11. Energetic composite processing is normally conducted under isothermal conditions so that Eq. (15) is more applicable.

As mentioned previously, in the processing of reactive polymeric systems, the critical time domain is the gel point. A formal definition of the gel point would be when the system goes from a viscous character to a viscoelastic response. At this point the affine nature of the system is sufficiently well developed, and it is no longer capable of dissipating energy under viscous flow (infinite viscosity). Effective processing requires a gelling number or pot fraction (processing time/gel time) of much less than unity for proper mixing and casting of the slurry. Prediction of the chemorheological behavior for a given set of parameters is facilitated by the construction of a viscosity verses conversion curve using the independent kinetic and rheology data.

V. INTERFACIAL PROCESSING AND PROPERTIES

Energetic composites can be considered as particle-dispersed polymeric matrices. The mechanical properties of these systems are crucial to their dimensional and ballistic stability. Their ability to deform without rupture and to recover is important for successful performance.

The fabricated composite often experiences a variable loading environment, which includes handling and vibration, thermal cycling, ignition pressurization, and acceleration. To this may be added the weight of the grain, which may assume enormous proportions in relatively big motors. Apart from the stresses imposed due to self-loading, the composite weight influences the shear stress generated during rapid boost operations ($\tau = m\ a/\pi\ d\ l\ g$).

If structural failure of the fabricated composite occurs by cracks in the matrix, fracture of the particle–matrix interface, or case debonding under any applied load, the extra and exposed surface will cause an enormous rise in the motor pressure during the combustion event with disasterous consequences.

A. Interface Engineering

Special attention is required to provide sufficient fracture resistance to the highly loaded composite structures. Due to the viscoelastic nature of the binder matrix, the fabricated composite possesses significant time-dependent properties, implying that the response of the material to the applied load is a function of the strain rate. The strain rate dependence of tear energy is a fine example of such behavior (Fig. 12) [39]. High-strain rates are synonymous with low temperature, elastic properties; the reverse is true for lower strain rates, which often extract viscoelastic response. Both extremes are normally encountered by the composite during its service life, e.g., the high rate of ignition pressurization [40] and the slow cool down upon completion of

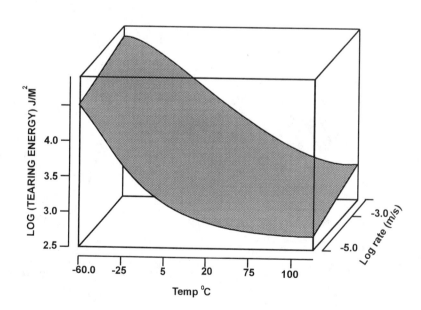

Figure 12 Three-dimensional plot of the tear energy of a viscoelastic composite as a function of temp. and strain rate. *Source*: Ref. 39.

cure. Interface engineering is concerned with the fabrication of resilient composite structures capable of efficiently dissipating energy at the material interface.

The mechanical properties of the composite are influenced largely by the volume fraction, morphology (shape) and size of the particles, the viscoelastic properties of the elastomeric matrix (molecular weight, cross-link density, degree of microphase segregation, etc.), and more importantly, the adhesion between the binder and filler particles in these otherwise nonreinforcing composites. From a mechanical standpoint, it is desirable to provide a high-fracture resistance in the polymer matrix and at the polymer–particle interface. However, for given properties of the polymer matrix, the entire edifice of the composite structure lies in the strength of the interface.

An important consideration is the effect of filler and its degree of interaction with the polymer matrix. Under strain, a weak bond at the binder–filler interface often leads to dewetting of the binder from the solid particles to formation of voids and deterioration of mechanical properties. The primary objective is, therefore, to enhance the particle–matrix interaction or increase debond fracture energy. A most desirable property is a narrow gap between the maximum (e_m) and ultimate elongation (e_b) on the stress–strain curve. The ratio, $e_m : e_b$, may be considered as the interface efficiency, a ratio of unity implying perfect efficiency at the interfacial junction.

B. Processing Techniques

1. Surface Activity

Several methods are practiced to achieve enhanced interfacial interaction. Primary physical interaction between the particles and the resin system may be simply increased by suitable roughening of the particle surface. Not only does this increase equilibrium wetting, it reduces dewetting as the irregularities allow them to key into the matrix, so that they are not easily dislodged along any broken or unsealed surface. Unfortunately, irregular morphology tends to increase the slurry viscosity, creating processing problems that often more than offset the advantage gained through reduced debonding. Alternatively, certain processing aids reduce the surface tension of the resin for its facilitated wetting on the substrate, thereby, increasing the intimacy of the polymer-substrate physical junction [41]. However, this is not very attractive thermodynamically. Decreasing the surface tension of the resin also reduces the thermodynamic work of adhesion:

$$W = \gamma_1 + \gamma_2 - \gamma_{12} \tag{17}$$

where W is energy that is required to separate the interface comprising the polymer with surface energy γ_1, and the substrate with surface energy γ_2; γ_{12} being the interfacial free energy. Clearly, reducing γ_1 will cause a corresponding reduction in the bond strength. A further re-

duction in the adhesion is also possible by the migration of the low-energy species to the discrete interface [42].

2. Interfacial Agents

Incorporation of bonding agents that exhibit bimaterial affinity is a proven means of producing localized changes in the character of multiphase interfacial systems [43–45]. These are compounds of usually ionic or even zwitterionic character that increase the concentration of polar groups in the system to promote adsorption or chemisorption through H-bonding, normally at the oxidizer–matrix interface, as illustrated in Fig. 13a,b for aziridine- (anionic) and ferrocene-based (zwitterionic) compounds. Physically, a bonding agent may homopolymerize together with the binder to encapsulate the particles, forming a tough polymeric shell at the material interface [46].

The efficiency of a particular bonding agent may be judged by the stress–strain capability of the composite. Figure 14 provides typical tensile loading data for an 80% solids. AP–urethane composite using a series of bonding agents [47]. Curve 1 is for the untreated specimen, while curve 5 represents the best bonding situation, as indicated by the superior stress–strain capability. The efficacy of a bonding agent in improving the interfacial bond quality may also be assessed by mechanical spectroscopy. According to this approach, the ratio of loss tangent of the specimens is related to the difference in void concentration (debonding), ΔC, between the two specimens [48]:

$$ln \ (\tan \delta_1 / \tan \delta_2) = -k \ \Delta C \tag{18}$$

Thus, increased energy dissipation at the weaker polymer–particle interface (through relative motion and friction) leads to higher mechanical energy absorption (higher tan δ value) close to the T_g [49]. This hypothesis has been tested by Hebeish and Bendak [48] where an increase in the loss tangent for the weaker interface is accompanied by the corresponding increase in the $e_b : e_m$ ratio, as expected. It is interesting to note that a similar effect is produced when the energy dissipation characteristics vary in the binder-phase, as described in Fig. 7. This shows that the mechanical properties of the composite may be varied both by varying the interfacial character as well as binder-phase morphology. The former is often more difficult to achieve. For this reason, it is relatively easier to develop a composite with high modulus and tensile strength than one with high ultimate elongation.

The bonding agent technique is usually not applicable to the metal particles in the composite. However, the surface of the metal is almost invariably covered by a thin (40–80 Å) oxide layer [50]. The free energy of oxide surfaces is normally quite large (10^3 mJ/m^2) to allow quick wetting by most organic polymers (40–60 mJ/m^2). Additionally, the metal surface may provide two

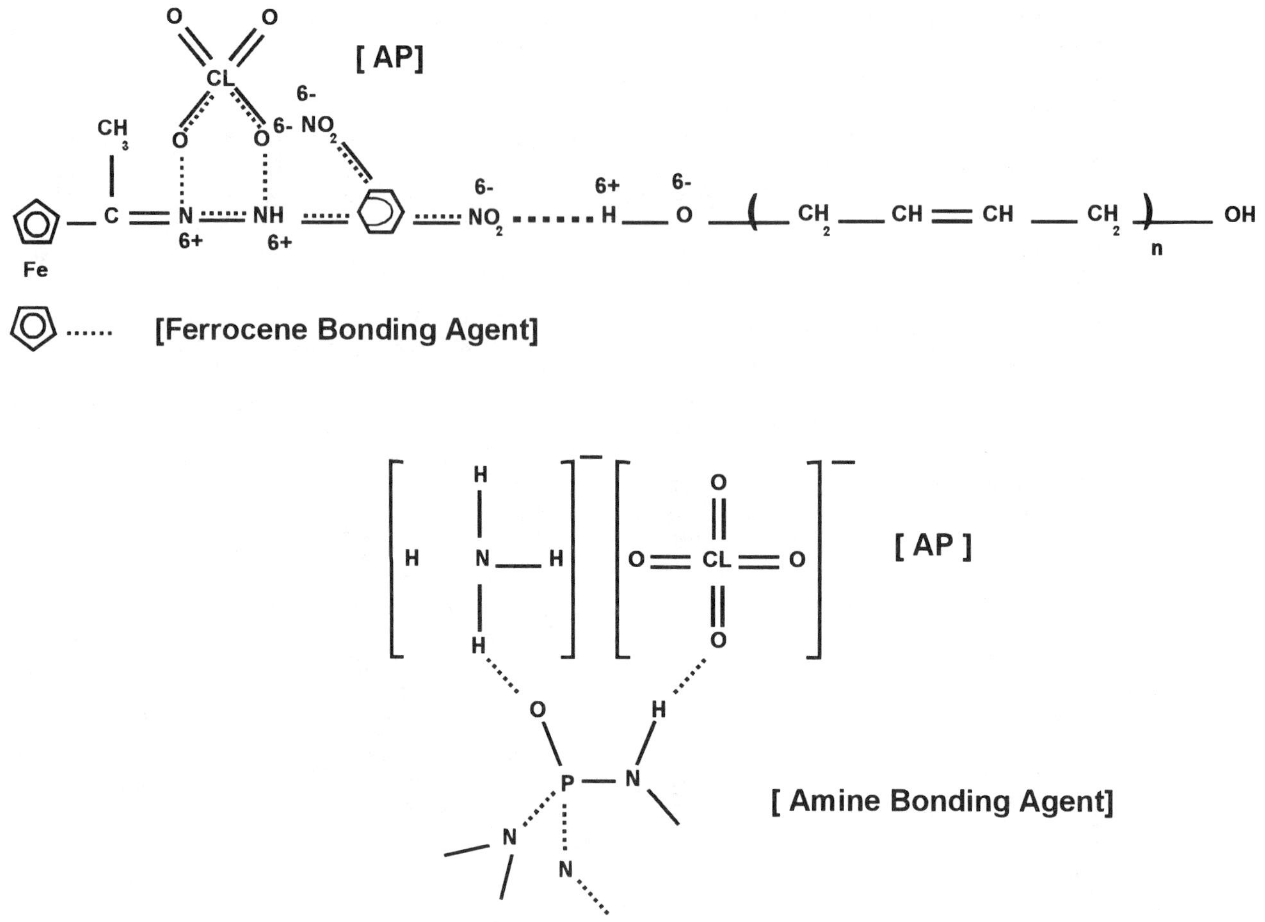

Figure 13   Interfacial bonding mechanism of the ferocene and aziridine based compounds.

possible modes of chemical bonding to a urethane: reaction of the ionic metal–hydroxide directly with the isocyanate, not too likely, or via hydrogen bonding with hydroxyl groups of the polyol. The possible existence of an ether linkage between the hydroxyl groups of the polyol and the oxide groups of the metal surface

(M—O | —(CH$_2$)$_4$ O C=O) has been suggested as another source of chemical interaction between the urethane and the metal counterface [51].

3. Transversal Modulus Gradient

A unique but not yet widespread technique that may influence the interfacial bond quality relies on the localized variation in the polymer modulus normal to the polymer–substrate junction in the composite assembly, as illustrated schematically in Fig. 15 [41,52]. The transversal modulus variation may be accomplished by interposing a tertiary interphase between the substrate and

the matrix. The presence of the interlayer alters the stress field in the matrix and near the polymer–substrate interface upon loading. For a flat simulation geometry (e.g., the blister test), the bond strength of the modified interfacial assembly expressed in terms of the debond pressure becomes:

$$P_c = P_o[(1 - c\,h/E)^{0.25}]^{-1} \qquad (19)$$

Evidently, the critical pressure to cause failure decreases with a stiffer interphase modulus, E, or a reduced interlayer thickness, h, or both. This hypothesis has been tested on several simulation systems, which confirm that increased adhesion is possible with a negative transversal modulus gradient at the material interface.

VI. PROCESSING EQUIPMENT

The processing sequence for the production of energetic composites mainly consists of the feed preparation, mix-

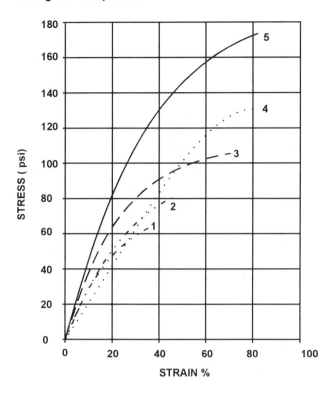

Figure 14 Instron tensile data for a urethane composite using several bonding agents. *Source*: Ref. 47.

ing, casting, curing, and quality control modules. The successful production of stress-free, smooth grains with homogeneous properties hinges upon the proper control of the mixing, casting, and curing operations. Since the thrust of this presentation is on interfacial reactive processing, emphasis will be placed on mixing, which can be considered as the most prominent part of the processing sequence, and it is the one that has received the least

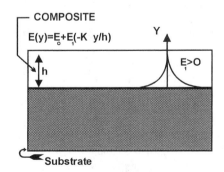

Figure 15 Shematic representation of transversal modulus gradient at the material interface.

attention in the past. The physical and chemical interfaces formed during the mixing process by and large determine the final mechanical properties (tensile strength, modulus, and elongation) of the composite. These mechanical properties are chiefly responsible for the structural integrity of the composite.

In a processing facility, the primary aim is to impart effective phase-contacting in the constituents within the safety and temporal constraints prescribed by the particular application. The basic design features of the equipment described in this section help to discern the performance capabilities of candidate geometries in this context.

A. Mixing Duty

Fluid mixing commonly requires the dispersion of one or more components into the body of the continuous phase. A mixing process essentially consists of two simultaneous steps: (1) the subdivision and (2) the uniform spatial distribution of the nondispersed phase in the continuum. From a reactive processing standpoint, the ultimate objective is to augment the intermaterial contact area for rapid rate of exchange. When the only nondispersed components are low-viscosity liquids, turbulence and fast diffusion provide easy mixing. In the case of energetic composites, the system viscosities may be as high as 1000 Pa.s. Neither diffusion nor turbulence can assist very much in mixing. Viscous systems, therefore, require laminar mixing, which relies on the protracted deformation of streamlines in creeping flow to achieve mixing.

An important parameter characterizing the laminar regime is the intersegment spacing over which the species must diffuse to be able to react. For kinetic control, as opposed to diffusion control of the reaction, this spacing must not exceed a certain scale of segregation given by [53]:

$$L = \sqrt{(2Dt)} \tag{20}$$

with D of the order of 10^{-12} m²/s and t of the order of tens of seconds, the reactant spacing, L, is only of the order of a few microns. It follows that viscous mixing requires a rather fine scale of segregation for kinetically controlled reactions. The mixing process must be highly efficient to achieve this type of mixing morphology. This is normally accomplished by special types of mixing elements as described in the subsequent sections.

B. Batch Processing

1. Planetary Mixer

In batch processing, the various additives are mixed with the prepolymer component in a jacketed horizontal (Z-blade) or vertical (helicone or spiral) mixer. Prior to this, however, a high-speed liquid mixer is used to disperse the minor additives (interfacial and crosslinking agents, ballistic and curing catalysts) in the prepolymer. Currently, most of the dedicated production of energetic composites is carried out in batch units; the low-shear, high-volume planetary mixer being the industry standard. As depicted in Fig. 16, the planetary has two sets of mixing elements: a central element that rotates around its axis and a side element that also rotates in orbit. The orbital motion of the side blade sweeps the contents around the mixer periphery, periodically passing through a narrow clearance with the edges of the central blade, providing effective mixing in the multiphase system.

The clearance between the walls of the blade is kept small to minimize dead spots (unmixed regions) in space. The shear rates prevailing close to the wall and at the gap between the two sets of blades are in the $50\text{--}70 \text{ s}^{-1}$ range. As the peripheral velocity is a strong function of blade size, in scale-up design, the rpm and the gap width (clearance) is adjusted to provide equivalent shear rates in the device. As opposed to the horizontal mixers, the vertical position of the blades in the planetary completely prevents the bearings and seals from having any contact with the slurry, thus avoiding contamination of the gear box and a higher level of safety is attained.

The primary objective of the mixing process is the thorough dispersion of the additives, both solid and liquid, in the prepolymer. The homogeneity is promoted by the wetting and spreading of the liquid-phase onto the solid particulate, during the phase-contacting operation. This is in essence a kinetic process that proceeds according to the minimization of the total free energy of the system, whereby two free surfaces combine to form a single interface. Enough residence time (1–2 h) is allowed in the mixer to attain equilibrium wetting (contact

angles) to ensure intimate contact at the solid–liquid interface.

It is this mixing process, assisted by appropriate reagents, that results in the required mechanical properties in the end. For the highly loaded, state-of-the-art systems, the solid particulate is introduced sequentially over a period of time in a cut of several sizes (in bimodal or trimodal distribution) to attain the maximum packing density together with reduced viscosity levels as described in Section III. The particle size distribution also influences some of the important ballistic properties (e.g., burn rate) not discussed in this chapter. The heat of polymerization is dissipated by cooling through the jacket to control the rate of reaction within acceptable limits. The extent of the reaction is normally indicated by the torque on the mixer shaft. The mixing is performed at moderate temperatures and under dynamic vacuum pressure of less than 30 mmHg to keep low overall humidity and to remove volatiles and entrapped air.

2. Casting

The final mix is of pastelike consistency (ca. 1000 Pa.s) and is cast by vacuum-assisted gravity transport to a pre-evacuated isothermic mold: a thermally protected case covered inside with a liner that bonds to the composite. For grains with a central port (for radial burning), the mold is equipped with a mandrel and the material is cast in the space between the mandrel and the case. The mandrel also confers the internal architecture to the grain, designed to conform to certain ballistic requirements. Figure 17 shows the profile of an eight-point star mandrel element: a most common configuration that maintains a steady deflagration surface for constant motor thrust. The material flow rate during casting may lie in the 10–100 kg/min range, depending on the size of the mold. This corresponds to low overall shear rates of $0.1\text{--}1 \text{ s}^{-1}$. The last processing step is the curing of the cast material, which renders it solid and fixed into the casing.

C. Continuous Processing

The mixing action of the batch devices is so slow that the interval between consecutive batches would generate an interface in the reactive composite to the detriment of structural properties. Continuous processing offers a viable alternative with cost and safety advantages, but adequate control of this mode poses a challenge. Continuous processing is distinguished from batch mixing by much shorter residence time (a few minutes as compared with a few hours) and less material holdup (a few pounds as compared with a few tons). A principal safety advantage over batch mixing is that only a relatively small amount of material is being processed at any given time. Consequently, an ignition or detonation at a continuous-processing plant is expected to be less dangerous to personnel and less costly than a comparable incident involv-

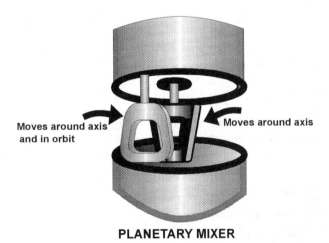

PLANETARY MIXER

Figure 16 A view of the mixing elements of the planetary mixer.

Figure 17 Viewgraph of the geometrical configuration of an eight-point star mandrel.

ing a large production batch, especially when more energetic (impact or friction sensitive) materials are used for advanced interceptor applications.

1. Plasticating Extruder vs. Kokneader

Baker-Perkins Kokneaders and Rotofead Deaerators (Fig. 18a) are the leading candidates for a continuous-processing facility [54–56]. These are low-pressure, low-shear, interrupted flight screw conveyors, as opposed to the high-pressure, friction-dominated extruders used by the polymer processing industry. In the plain barrel used in conventional extrusion, the degree of mixing, S, is expressed in terms of the total shear strain imparted to the mix [57]:

$$S = v/H \cdot f(Q_p/Q_d) \qquad (21)$$

since v/H is proportional to the shear rate and S is a direct function of the pressure flow rate, Q_p, an improvement in mixing relies on increased shear rate as well as back pressure; something unaffordable in the processing of energetic materials. The shear affects the degree of viscous heating in the fluid mass. In the flow of fluids, the internal fluid friction can cause self-heating under conditions of high viscosity and high-shear rates. The dimensionless group that measures this tendency is known as the Brinkman number, Br [58]:

$$\mathrm{Br} = \eta v^2/\lambda \, \Delta T \qquad (22)$$

The Br is a measure of the extent to which viscous heating is important relative to an impressed temperature difference. This can be of some concern in the scale-up design, v usually increasing, with other properties remaining constant. A comparison of the Br for a pilot scale (0.05-m screw) and an industrial (0.15-m screw) unit yields values of ca. 0.65, and 5.73, respectively, for the Br with $\lambda = 0.5$ w/m K and $\eta = 500$ Pa.s at 60 rpm. The numbers suggest that viscous dissipation will be important and will be much more pronounced in the case of an industrial unit.

The idea of the interrupted flights is to impart accentuated mixing in the multiphase system by bisecting the flow and, hence, maintaining low overall shear rates and system pressures. As indicated in the schematic of Fig. 18b, the bulk flow is split into several segments downstream of the mixers. These segments are radially deflected by the protrusions in the barrel, being simultaneously swept around the mixer periphery by screw rotation. Thus, phase-contacting relies on suitable subdivision of the fluid elements by mechanical means, rather than on continuous unidirectional shearing as in a plain barrel.

In continuous mixing, the numerous thermocouple wells on the barrel may be used as injection ports for the introduction of separate additive streams (crosslinker, curing agent, etc.) on-line. Figure 19 shows the cross-section of a hollow flighted-screw shaft together with a

(a)

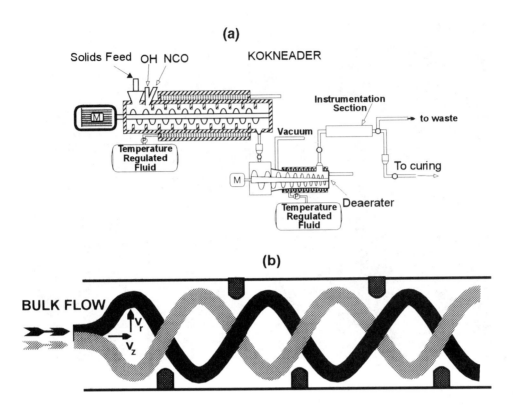

(b)

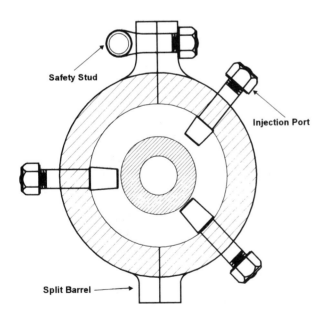

Figure 18 (a) Designer's view of a continuous processing facility showing kneading and deaeration modules. *Source*: Ref. 54. (b) Two-D representation of flow bifurcation in the Kokneader zone.

Figure 19 Cross-section of split barrel showing a series of injection ports.

series of injection ports situated around a split barrel. The ports also facilitate the on-line deaeration by exposing a thin film of the composite slurry being transported. It has been established in certain trials [59] that a vacuum of the order of 1.0 kPa would be practically required to mitigate the presence of microvoids in continuous processing.

Following the addition of the curing agent, the mixing morphology is controlled by the following behavior:

$$S(\text{OH}) = S(\text{NCO}) \times \eta_{\text{NCO}}/\eta_{\text{OH}} \quad (23)$$

since the viscosity ratio, $\eta_{\text{NCO}}:\eta_{\text{OH}}$, is less than unity, the total shear imparted to the more viscous component (OH) is always less than its low-viscosity counterpart (NCO), implying an increase in the mixing duty. The Kokneader design is especially suited to such on-line mixing requirements. Through its split barrel design, it also provides an arrangement for the immediate separation of the barrel halves in the event of undesirable pressure buildup within the mixer. This serves as a quick way of evacuating the mixer so as to prevent possible explosion.

2. Twin-Screw Extrusion

The modern trend is to use twin-screw extruders (Fig. 20). It is of interest to note that single-screw mixing depends on material friction between the screw and barrel to affect mixing, whereas the mixing action of the twin-screw occurs mostly between the screws, which results in less overall friction. This implies that the twin-screw configuration may be operated at a higher rpm throughput. A rigorous comparison of the mixing performance of the various laminar geometries has been presented [60], including the one between the corotating and counter-rotating type twin-screw machines. As inferred from the growth of the interfacial area as a function of the number of turns down channel or the number of twist sections, the superior mixing action of the counterrotating device is evident.

In the above investigation, the machine design was not discussed, however, the exponential nature of improvement in mixing performance was attributed to the generation of new interfaces every turn along the screw by some sort of redistribution process. On the other hand, there is little tendency for the velocity field produced by the corotating device to reorient the material being mixed. A logical explanation from the machine design point of view is that in the corotating device, the flight of the two sets of screws are parallel, while these are disposed at an angle in the counterrotating case. The latter design helps to reorient the material points for effective mixing. However, of the two common designs available [61], the one with intermeshing flights (conjugated) is ruled out due to safety considerations. The nonconjugated version that allows more room for material between flights is, therefore, the logical choice.

3. Crammer Feeder

The success of continuous processing heavily hinges on the speed and continuity of the feed materials into the process. A main problem with the continuous processing of energetic composites is the surging in the solids feed, making the process sequence hazardous and product

specification a difficult task. The flow of solids feed from conventional weight-loss feeders is such as very little pressure is likely to be transmitted, which may result in troublesome bridging and in some cases complete cessation of feed [62]. This may cause surging and instabilities in the downstream processing. Surging of the solids feeders is believed to have led to one or both of the disasters at the continuous-processing plants for the Polaris and Minuteman energetic composite programs in the United States [55].

Of course, the best remedy to minimize downstream instabilities is to minimize the input disturbance. Pressure initiation is thus incorporated, and the simplest and most effective way of achieving this is by vertical screw

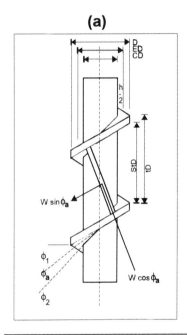

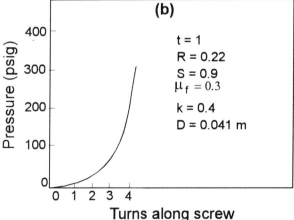

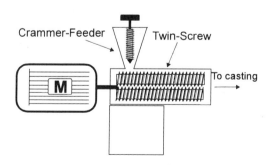

Figure 20 Schematic of a continuous feed twin-screw assembly.

Figure 21 (a) Geometry of the crammer screw. (b) Analytical pressure profile along the length of crammer screw.

feeders installed right above the extruder feedport, as shown in Fig. 20. Following the general principles of screw transport [63], the pressure buildup is directly dependent on local pressure and gravity terms, see Fig. 21a:

$$dP/dz = A/D_s \cdot P + B s \qquad (24)$$

A and B are complicated functions of the screw geometry, friction coefficients, and flow rate. A numerical solution to Eq. (24) yields the pressure, P, at the tip of the crammer screw and hence the extruder inlet. Solutions for one such case are indicated in Fig. 21b. It can be seen that pressures as high as 1.35 MPa would be generated by just four turns on a 0.041-m screw starting with feed gravity pressure. The analysis proves the efficacy of the forced feeding in continuous processing. However, the rate of feed through the crammer must be controlled so as to avoid excessive pressures and/or friction in the Kokneader/twin-screw feed zone.

4.   Reactive Injection Molding (RIM)

RIM generally involves the turbulent mixing (Re $\approx$ 500) of the reactant streams as they are made to impinge in a small mixing cavity [32]. Since this process involves the total consumption of kinetic energy by viscous dissipation to produce new interfaces, it is not suitable for the highly viscous, friction-sensitive propellant slurries. However, the recent development of a shear comminution theory and a device for minimizing energy input has made it possible to use low-pressure injection in creeping flow [64]. The device uses multiple transversal injection in an existing shear field as shown schematically in Fig. 22. A synergistic combination of a large initial interface (multiple injection ports) with radial flow and an optimum orientation of the shear displacement vec-

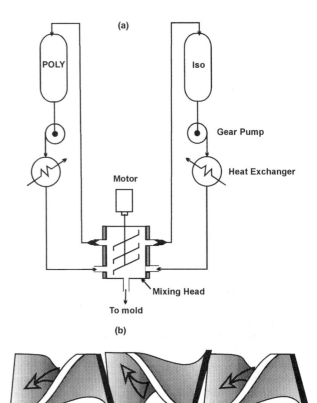

Figure 23 (a) Low-pressure RIM employing rotating helical elements. (b) Flow through assembly of helical elements.

tors is a promising factor for the modified RIM's use in the production of energetic composites.

Another promising contrivance within the domain of low-pressure RIM appears in Fig. 23a [65]. The premixes are heated and stirred in the hoppers, while they are drawn simultaneously by two pumps into a static mixer assembly. The latter originates from the plastic industry and is shown schematically in Fig. 23b. Flow through the elements is typically three-dimensional. A potential improvement lies in rotating the entire assembly of the elements relative to the confinement to augment the tangential flow component, just like a screw. Sample handling and material preparation are still cited as outstanding problems, but the low-pressure RIM program has shown that new processing methods should, in the future, enable RIM energy composites to be viable for system designers.

GLOSSARY

a = Vehicle acceleration (m/s^2)
C = Reactant concentration (mol/L)
C_0 = Initial concentration (mol/L)

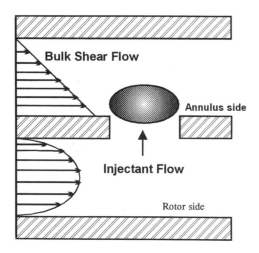

Figure 22 Illustration of interface production in a shear injection system.

C_1, C_2 = Empirical constants

d = Diameter of fine particles (m); grain diameter (m)

D = Diameter of coarse particles (m)

D_s = Screw diameter (m)

D = Self-diffusion coefficient (m²/s)

E = Interlayer modulus (Pa)

E_0 = Surface modulus (Pa)

E_1 = Bulk modulus (Pa)

h = Interlayer thickness (m); partial heat of reaction (J)

H = Overall heat of reaction (J/kg)

ΔH_f = Heat of formation (J/Kg)

I_{sp} = Specific impulse (s)

k = Rate constant [(L/mol)$^{n-1}$s]

k_{ad} = Adiabatic reaction rate constant [(L/mol)$^{n-1}$s]

k_{iso} = Isothemal reaction rate constant [(L/mol)$^{n-1}$s]

k_s = modified rate constant [(L/mol)$^{n-1}$s]

k_1, k_2 = Empirical constants

K = Gel times scaler; proportionality constant

L = Scale of segegation (m)

m = Mass of grain (Kg)

M = Molecular weight (Kg/mol)

M_c = Molecular weight between crosslinks (Kg/mol)

N_d = Number of fine particles

P = Pressure along screw (Pa)

P_c = Critical debond pressure (Pa)

P^o = Critical debond pressure without interlayer (Pa)

Q_d = Drag flow rate (m³/s)

Q_p = Pressure flow rate (m³/s)

r = Reactant ratio; fine particle radius (m)

R = Gas constant (J/mol K); coarse particle radius

s = Specific weight of feed

S = Degree of mixing

t = Time scale (s)

t_G = Gel time (s)

t_s = Modified time scale (s)

T = Flame temperature (K); system temperature (K)

ΔT_{ad} = Adiabatic temperature rise (K)

T_{max} = Maximum adiabatic temperature (K)

T_g = Glass transition temperature (K)

v = Screw peripheral velocity (m/s)

v_r = Volume fraction of composite in swollen phase

w_g = Gel fraction

w_s = Sol fraction

W = Thermodynamic work of adhesion (J/m²)

x_d = Weight fraction of fines

x_l = Weight fraction of liquid

Greek Symbols

α = Extent of reaction

α_g = Gel conversion

η = Dynamic viscosity (Pa.s)

η_r = Relative viscosity

$[\eta]$ = Intrinsic viscosity

ρ_l = Liquid density (Kg/m³)

ρ_p = Particle density (Kg/m³)

ϕ = Particle volume fraction

ϕ_m = Maximum particle volume fraction

γ_1 = Liquid surface free energy (J/m²)

γ_2 = Solid surface free energy (J/m²)

γ_{12} = Interfacial free energy (J/m²)

λ = Thermal conductivity (w/m K)

μ = Crosslink density (Mol/g)

τ = Shear stress (Pa)

REFERENCES

1. *Ind. Eng. Chem., 59(9)*: 754 (1960).
2. F. A. Williams, M. Barrere, and N. Huang, *AGARDO-GRAPH No. 116,* Technivision Services, Slough, England (1969).
3. *Propellants Manufacture, Hazards and Testing, Advances in Chemistry Series, No. 88,* (C. Boyers and K. Klager, eds.), ACS, Washington, D.C. (1969).
4. A. G. Chritiansen, L. H. Layton, and R. L. Carpenter, *J. Spacecraft, 18(3)*: 211 (1981).
5. *Fundamentals of Solid Propellant Combustion, Progress in Astronautics and Aeronautics,* Vol. 90, (K. K. Kuo and M. Summerfield, eds.), AIAA New York (1984).
6. R. Kent and R. Rat, *J. Electrostatics,* 17: 299 (1985).
7. G. P. Sutton, *Rocket Propulsion Elements,* Wiley, New York (1986).
8. Y. M. Timnat, *Advanced Chemical Rocket Propulsion,* Academic, New York (1987).
9. *Solid Rocket Propulsion Technology,* (A. Devenas, ed.), Pergamon, Oxford (1993).
10. M. B. Khan, *J. PIChE, Papers 1–6,* 23: 1 (1995).
11. W. B. White, S. N. Johnson, and G. B. Dantzig, *J. Chem. Phys.,* 28: 751 (1958).
12. V. Swaminathan and S. Rajagopalan, *Prop. Expl.,* 3: 150 (1978).
13. M. B. Khan, *J. PIChE,* 23: 1 (1995).
14. S. Gordan and B. J. McBride, *NASA SP-273* (1971).
15. D. R. Cruise, *NWC-TP-6037,* Naval Weapons Center, China Lake, California (1979).
16. N. Kubota, *Pop. Expl. Pyrotech.,* 16: 287 (1991).
17. A. Gany and Y. M. Timnat, *Space Technol., 13(1)*: 33 (1993).
18. *Solid Rocket Motor Technology,* AGARD CP-259, 6-5, (1979).
19. Z. Hussain, 13th Intl. Sem. Def. Sci. Technol., Wah Cantt. (1992).
20. A. Einstein, *Annalen der Physik,* 34: 591 (1911).
21. R. Simha, *J. Appl. Phys.,* 23: 1020 (1952).
22. D. G. Thomas, *J. Coll. Sci.,* 20: 267 (1965).
23. I. M. Kieger and T. J. Dougherty, *Trans. Soc. Rheol., III*: 137 (1959).
24. F. H. Strossand P. E. Porter, *ECT,* 2nd ed., Interscience, Wiley, New York, p. 483 (1970).
25. H. Trawinski, *Chem. Eng. Technol.,* 25: 229 (1953).
26. A. B. Metzner, *J. Rheol,* 29: 739 (1985).
27. C. S. Sung and N. S. Schneider, *Polymer Alloys* (D. Kemper and K. S. Frisch, eds.), Plenum, New York, p. 261 (1977).

28. I. D. Friedman, E. L. Thomas, L. J. Lee, and C. W. Macosko, *Polymer, 21*: 393 (1980).

29. B. J. Briscoe, M. B. Khan, and S. M. Richardson, *J. Adhesion Sci. Technol., 3(6)*: 475 (1989).

30. S. D. Lipshitz and C. W. Macosko, *Poly. Eng. Sci., 16*: 803 (1976).

31. G. Berry and T. Fox, *Adv. Polym. Sci., 5*: 261 (1968).

32. C. W. Macosko, *Fundamentals of Reaction Injection Molding,* Hanser, New York (1989).

33. F. N. Kelly, *Appl. Polym. Symp., 1*: 229 (1965).

34. L. R. G. Treloar, *The Physics of Rubber Elasticity,* Oxford Press (1975).

35. R. F. Fedors and R. F. Landel, *Polymer, 19*: 1189 (1978).

36. J. M. Castro, S. J. Perry, and C. W. Macosko, *Polym. Comm., 25*: 82 (1984).

37. B. J. Briscoe, S. M. Richardson, and M. B. Khan, *Plast. Rubb. Proc. Appl., 10*: 65 (1988).

38. R. E. Camargo, V. M. Gonzalez, C. W. Macosko, and M. Tirrell, *Rub. Chem. Technol., 56*: 773 (1983).

39. K. J. Min, PhD Thesis, University of Akron (1987).

40. K. K. Kuo, J. Moreci, and J. Manzaras, *J. Propulsion, 3(1)*: 19 (1987).

41. M. B. Khan, *J. Polym. Comp., 15(1)*: 83 (1994).

42. M. B. Khan, *Polym-Plast. Technol. Eng., 32(5)*: 467 (1993).

43. K. Hori, A. Iwama, and T. Fakuda, *Prop. Expl. Pyrotech., 10*: 176 (1985).

44. K. Kishore and P. Rajalingam, *J. Appl. Polym. Sci., 37*: 2845 (1989).

45. T. Suzuki and A. Kasuya, *J. Adhesion Sci. Technol., 3(6)*: 463 (1989).

46. A. E. Oberth and R. S. Bruenner, *Adv. Chem. Ser., 88*: 84 (1969).

47. A. E. Oberth and R. S. Bruenner, U.S. Patent No. 4000,023 (1976).

48. A. Hebeish and A. Bendak, *J. Appl. Polym. Sci., 18*: 1305 (1974).

49. M. B. Khan, *J. Composites, 26(3)*: 223 (1995).

50. N. J. DeLollis, *Adhesives for Metals: Theory and Technology,* Industrial Press, New York (1970).

51. E. H. Andrews and N. E. King, *J. Mat. Sci., 11*: 2004 (1976).

52. M. L. Williams and R. L. Chapbis, *3rd U.S. Nat. Cong. Appl. Mech.,* 281 (1958).

53. N. P. Suh and C. L. Tucker, *Poc. Intl. Conf. Polym. Process.* (N. P. Suh and N. Sung, eds.), MIT Press, Cambridge, Massachusetts and London (1982).

54. G. A. Fluke et al, U.S. Patent No. 3,296,043 (1967).

55. D. M. Husband, *Chem. Eng. Sci., May*: 55 (1989).

56. M. B. Khan, *Proc. 4th Intl. Symp. Adv. Mat.,* (A. Q. Khan and M. Anwar ul Haq, eds.) Islamabad (1995).

57. Z. Tadmor and I. Klein, *Engineering Principles of Plasticating Extuder,* Van Nostrand, New York, p. 349 (1970).

58. S. Middleman, *Fundamentals of Polymer Processing,* McGraw-Hill, New York (1977).

59. T. S. Lundstrom and B. R. Gebart, *J. Polym. Comp., 15(1)*: 25 (1994).

60. D. Bigio and W. Stry, *Polym. Eng. Sci., 30(3)*: 153 (1990).

61. R. J. Crawford, *Plastics Engineering,* 2nd Ed., Pergamon, Singapore, p. 169 (1989).

62. K. Schnieder, *Technical Report on Plastics Processing in the Feeding Zone of an Extruder,* Institute of Plastics Processing (IKV)TH, Aachan (1969).

63. J. G. A. Lovegrove and J. G. Williams, *Polym. Eng. Sci., 14(8)*: 589 (1974).

64. B. J. Briscoe, M. B. Khan, and S. M. Richardson, *Polym. Eng. Sci., 30(3)*: 162 (1990).

65. N. Giffiths, 11th Intl. Sem. Def. Sci. Tech., Wah Cantt. (1991).

47
Macrointermediates for Block and Graft Copolymers

Baki Hazer
Zonguldak Karaelmas University, Zonguldak, Turkey

I. INTRODUCTION

To insert a different polymer segment into a polymer backbone is important for obtaining new materials. These types of polymers are called block or graft copolymers [1–6]. Block copolymers are macromolecules comprised of chemically dissimilar, terminally connected segments. Their sequential arrangement can vary from A-B structures, containing two segments only, to A-B-A block copolymers with three segments, to multiblock -(-AB-)-$_n$ system possesing many segments. Starblock copolymers have star like different blocks bounded in a central point. These are shown in Fig. 1.

A graft copolymer is a polymer that is comprised of molecules with one or more species of blocks connected as side chains to the backbone, having constitutional or configurational features different from those in the main chain. The simplest case of a graft copolymer can be represented as follows:

$$\sim M\ M\ M\ M\ X\ M\ M\ M\ M\ M\ M\sim$$
$$|$$
$$G$$
$$G$$
$$G$$
$$G$$
$$G$$
$$G$$

where a sequence of the M monomer unit is the backbone, a sequence of the G-unit is the side chain (graft), and X is the unit in the backbone to which the graft is attached.

Block and graft copolymers have many similar characteristics. Thus, both graft and block copolymers behave in various respects as two immiscible polymers while at the same time having some properties that would be expected for a statistical polymer. Both block and graft copolymers can form two-phase morphologies, the phase separation depending on the length of the individual polymer sequences. Either of the two phases can form the continuous or matrix phase (which phase is continuous depends not only on the proportions of the two phases but also on the melting temperatures of the phases, their density, their surface tensions, and, of course, the method used to process the final product).

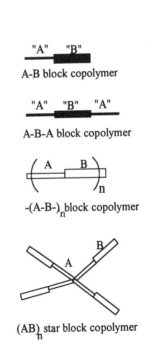

"A" "B"
A-B block copolymer

"A" "B" "A"
A-B-A block copolymer

$\left(\begin{array}{cc} A & B \end{array}\right)_n$
-(A-B-)$_n$ block copolymer

A B
(AB)$_n$ star block copolymer

Figure 1 Types of block copolymers.

725

In contrast to two-phase physical blends, the two-phase block and graft copolymer systems have covalent bonds between the phases, which considerably improve their mechanical strengths. If the domains of the dispersed phase are small enough, such products can be transparent. The thermal behavior of both block and graft two-phase systems is similar to that of physical blends. They can act as emulsifiers for mixtures of the two polymers from which they have been formed.

In terms of structural control, block copolymers have considerable advantages over graft copolymers. The segment length and sequence are generally more easily controlled for block copolymers than for graft copolymers.

A very special type of ABA block copolymer where A is a thermoplastic (e.g., styrene) and B an elastomer (e.g., butadiene) can have properties at ambient temperatures, such as a crosslinked rubber. Domain formations (which serves as a physical crosslinking and reinforcement sites) impart valuable features to block copolymers. They are thermoplastic, can be easily molded, and are soluble in common solvents. A domain structure can be shown as in Fig. 2.

Several macrointermediates to obtain this kind of copolymer were used via free radical, ionic, and/or free radical-ionic coupling polymerization. In this manner, macroinitiators, macromonomers, and macromonomeric initiators will be discussed in this chapter.

A. Macroinitiators

Macroinitiators are macromolecules having peroxygen and/or azo groups that can thermally initiate a vinyl polymerization to obtain block copolymers in one step. They can be classified as macroperoxyinitiators (MPI), macroazoinitiators (MAI), and macroazo-peroxyinitiators.

1. Macroperoxyinitiators

Several types of macroperoxyinitiators: macrobisperoxides [7,8] and oligoperoxides [8] have been reported in the literature.

Macrobisperoxides

Macrobisperoxides contain peroxide groups in both ends of the main chain. The first example of macro-

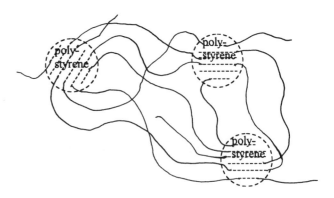

Figure 2 Schematic representation of the domain structure of styrene–butadiene–styrene block copolymer.

bisperoxides was reported by Bamford and Jenkins [7] via vinyl polymerization using 4,4′-azobiscyanopentanoic acid (ACPA) in the following reaction scheme:

$$HOOC-CH_2CH_2-\overset{\overset{\displaystyle CH_3}{|}}{\underset{\underset{\displaystyle CN}{|}}{C}}-N=N-\overset{\overset{\displaystyle CH_3}{|}}{\underset{\underset{\displaystyle CN}{|}}{C}}-CH_2CH_2-COOH \xrightarrow{+\ styrene,\ \Delta}$$

$$HOOC-CH_2CH_2-\overset{\overset{\displaystyle CH_3}{|}}{\underset{\underset{\displaystyle CN}{|}}{C}}\text{\small www}\ polystyrene\ \text{\small www}\overset{\overset{\displaystyle CH_3}{|}}{\underset{\underset{\displaystyle CN}{|}}{C}}-CH_2CH_2-COOH$$

$$\downarrow \begin{array}{l} 1)\ SOCl_2 \\ 2)\ t\text{-}BuOOH \end{array}$$

t-BuOO--www polystyrene www---OOBu-t

active-polystyrene

Polystyrene bisperoxides can be prepared by the termination of polystyryl anion with bromo methyl benzoyl peroxide [9]:

Styrene

$$\downarrow \begin{array}{l} 1,5\text{-di(2-3,3-dimethyl buthyl-1-lithio) naphtalene} \\ (DDBLN) \end{array}$$

⊖www polystyrene www⊖

$$\downarrow\ \begin{array}{l} +\ BrCH_2-\underset{}{\bigcirc}-\overset{\overset{\displaystyle O}{\|}}{C}-OO-\overset{\overset{\displaystyle O}{\|}}{C}-\underset{}{\bigcirc}-CH_2Br \\ (excess) \qquad\qquad (BBP) \end{array}$$

$$BrCH_2-\underset{}{\bigcirc}-\overset{\overset{\displaystyle O}{\|}}{C}-OO-\overset{\overset{\displaystyle O}{\|}}{C}-\underset{}{\bigcirc}-CH_2\ \text{\small www}\ polystyrene\ \text{\small www}\ CH_2-\underset{}{\bigcirc}-\overset{\overset{\displaystyle O}{\|}}{C}-OO-\overset{\overset{\displaystyle O}{\|}}{C}-\underset{}{\bigcirc}-CH_2Br$$

In this manner, growing polytetrahydrofuran cations can be terminated by using sodium salt of cumenehydroperoxide [10].

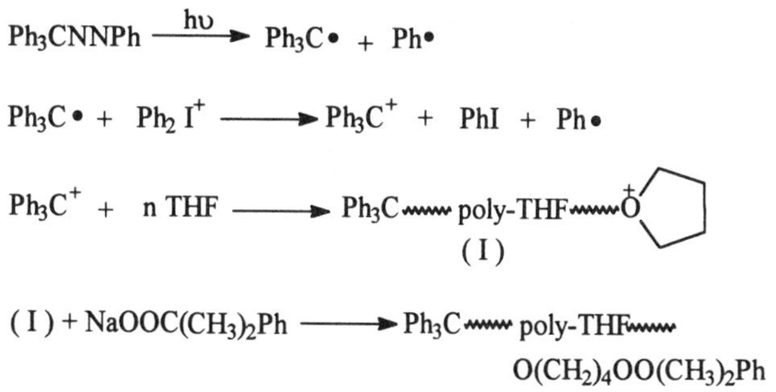

$$Ph_3CNNPh \xrightarrow{h\upsilon} Ph_3C\bullet + Ph\bullet$$

$$Ph_3C\bullet + Ph_2 I^+ \longrightarrow Ph_3C^+ + PhI + Ph\bullet$$

$$Ph_3C^+ + n\ THF \longrightarrow Ph_3C\text{\small www}\ poly\text{-}THF\text{\small www}\overset{+}{O}$$
(I)

$$(I) + NaOOC(CH_3)_2Ph \longrightarrow Ph_3C\text{\small www}\ poly\text{-}THF\text{\small www}$$
$$O(CH_2)_4OO(CH_3)_2Ph$$

Coupling reactions of diisocyanates with polyether glycols or polyesters lead to prepare macrobisperoxides [11,12]:

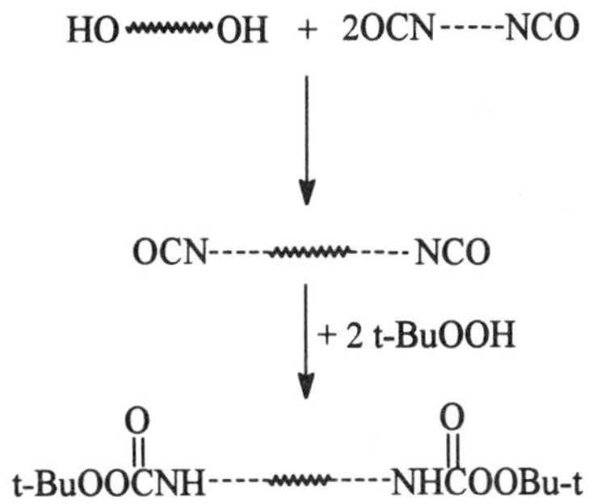

Cationic polymerization of tetrahydrofuran with the pairs of 4,4'-bromomethyl benzoyl peroxide and AgPF$_6$

or bis(3,5-di-bromomethyl) benzoyl peroxide and AgPF$_6$ yield active poly-THF having peroxide group in the main chain:

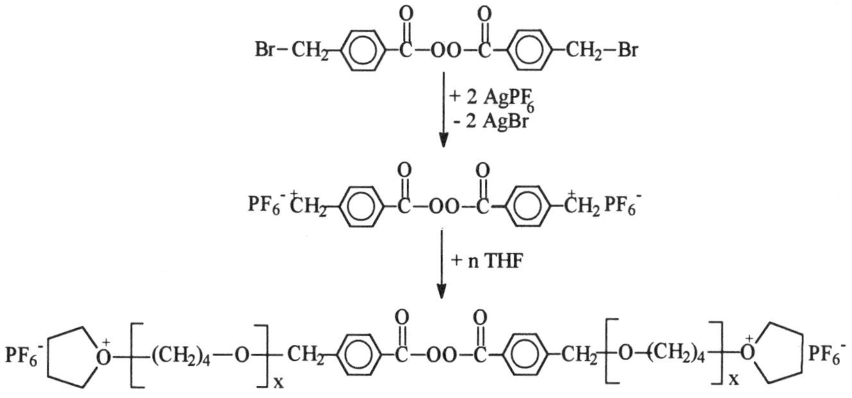

active-poly-THF

Oligoperoxides

Oligoperoxides [13–17] contain more than two peroxide groups, which can be prepared by the condensation reactions of aliphatic or aromatic diacid chlorides (e.g., adipoyl-, terephtaloyl-dichloride) and sodium peroxide or a dihydroperoxide (2,5-dimethyl 2,5-dihydroperoxy hexane). Oligoperoxides, until 1989, have been reviewed in detail by Hazer [1]. Oligoperoxy compounds containing peroxy groups of different activities [18] exhibit superiority as polymerization initiators. It was found that the acylperoxy group decomposes earlier than the peroxyester group: $E_a = 30$ kcal mol^{-1} and 36 kcal mol^{-1}, respectively:

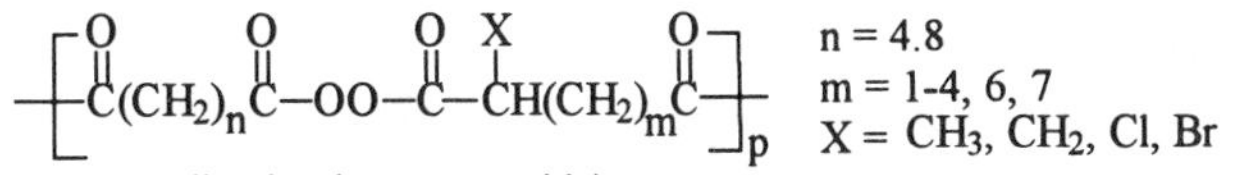

oligo(acylester peroxide)

A different kind of oligoperoxide, polymeric peroxycarbamate [19], was synthesized by reacting equimolar amounts of an aliphatic dihydroperoxide with an aliphatic diisocyanate.

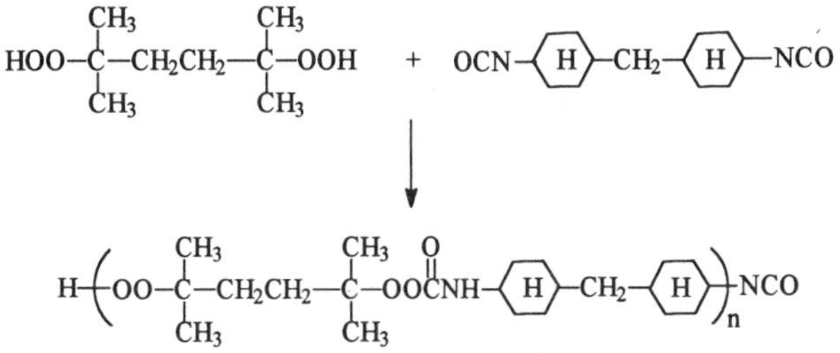

polymeric peroxycarbamate

An alternating copolymer of α-methyl styrene and oxygen as an active polymer was recently reported [20]. When α-methyl styrene and AIBN are pressurized with O$_2$, poly-α-methylstyreneperoxide is obtained. Polymerization kinetic studies have shown that the oligoperoxides mentioned above were as reactive as benzoyl peroxide, which is a commercial peroxidic initiator. Table 1 compares the overall rate constants of some oligoperoxides with that of benzoyl peroxide.

Active polymers from oligoperoxides. Active polymers can be obtained by the free radical polymerization of vinyl monomers with oligoperoxides as the initia-

Table 1 Overall Rate Constants, k, of Some Peroxidic Initiators for Styrene Polymerization at 80°C

Peroxide initiator	k (L/mol)$^{1/2}$sec^{-1}	Number of peroxide bond	Reference
Oligo(adipoyl-5-peroxy-2,5-dimethyl n-hexyl peroxide)	1.81×10^{-4}	~5	16
Oligododecandioyl peroxide	5.56×10^{-4}	~5	17
Polymeric peroxycarbamate	2.24×10^{-4}	~6	19
Poly α-methyl-styrene peroxide	0.60×10^{-4}	~50	20
Benzoyl peroxide	3.32×10^{-4}	1	19

tor. Lower polymerization temperature and a shorter polymerization period of time yield an active polymer having undecomposed peroxygen groups in the main chain but with low conversion. As the number of peroxy groups increases in oligoperoxide compounds, the amount of inactive polymer in the block copolymer decreases. To produce inactive polymer formation in the synthesis of active polymers, it is necessary to prepare oligoperoxide having as many peroxy groups as possible.

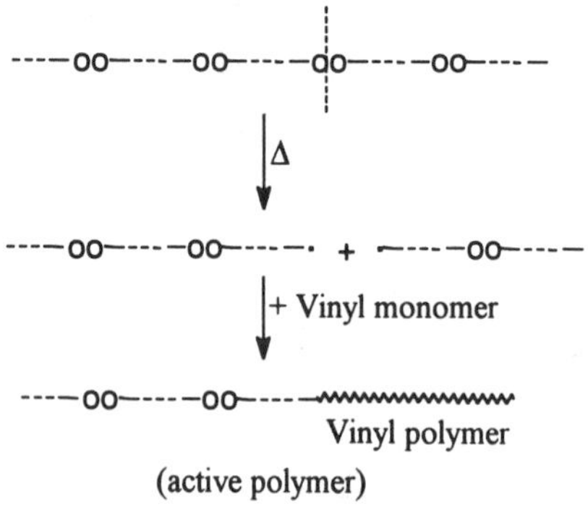

2. Macroazoinitiators

Macroazoinitiators have generally been obtained from a polyethylene glycol (PEG) and an azoinitiator: 4,4'-azobisisobutyronitril (AIBN) [21–23] and 4,4'-azobiscyanopentanoyl chloride (ACPC) [24].

Nitriles react with alcohols in the presence of hydrochloric acid to form iminoester hydrochlorides, which are hydrolyzed to the esters (Pinner synthesis). Heitz and coworkers [21–23] published several fine papers on the polyazoester synthesis from the reaction of a series of poly(oxyethylene) glycol or poly(oxypropylene) glycol and AIBN in the presence of dry hydrochloric acid at 0–5°C according to Pinner synthesis. Condensation reactions of ACPC and dihydroxy terminated poly(oxyethylene) glycol yield polyazoesters [24,25].

In a recent work of Haneda et al. [26], ACPC was reacted with several poly-diols: poly(ethylene adipate), poly(tetramethyleneadipate), poly(caprolactone), aliphatic poly(carbonate) to prepare various polyazoesters.

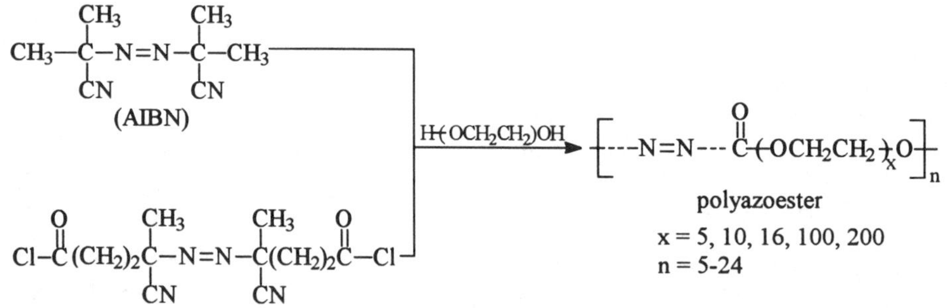

Interfacial polycondensation between a diacid chloride and hexamethylenediamine in the presence of small amounts of ACPC also yield polymeric azoamid, which is a macroazo initiator.[27] In this manner, azodicarboxylate-functional polystyrene [28], macroazonitriles from 4,4'-azobis(4-cyano-n-pentanoyl) with diisocyanate of polyalkylene oxide [29], polymeric azo initiators with pendent azo groups [3] and polybutadiene macroazoinitiator [30] are macroazoinitiators that prepare block and graft copolymers.

3. Macroperoxyazoinitiators

Azoperoxydic initiators are particularly important due to their capacity to decompose sequentially into free radicals and to initiate the polymerization of vinylic monomers. The azo group is thermally decomposed first to initiate a vinyl monomer and to synthesize the polymeric initiator with perester groups at the ends of polymer chain (active polymer) [31,32].

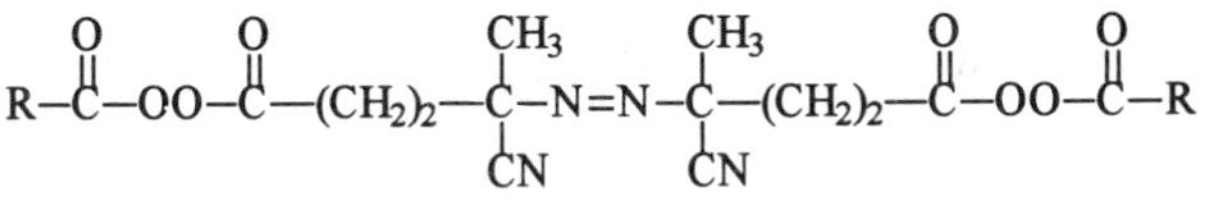

(Azoperoxydic initiator)

R: CH_3-, C_6H_5-, p.$NO_2-C_6H_4-$

Active block copolymers are obtained by using macroperoxyazoinitiators [33,34] having polyethylene glycol or polyester segments. Hydroxyl end groups of polyethylene glycol are transformed to peroxide groups by the capping reaction of excess terephtaloylchloride and *t*-butylhydroperoxide, respectively [34].

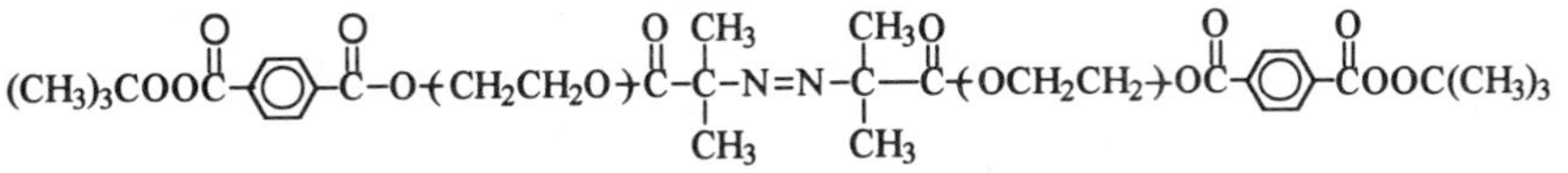

a macroperoxyazo initiator

Decomposition in the presence of styrene at 60°C or with a tertiary amine in the presence of methyl methacrylate gives the corresponding ABA active block copolymer or ABBA active block copolymer, respectively. When both active block copolymers are used as polymeric initiators in another vinyl polymerization, an ABCBA type multiblock copolymer is obtained [34].

B. Macromonomers

Macromolecular monomers, macromers, are, short polymer molecules carrying an end-standing unsaturation, which is polymerizable in turn. Copolymerization of a vinylic (or acrylic) monomer with a macromer should result in a graft copolymer the macromer units constituting grafts. A fine review article on macromer was published by Rempp and Franta [35]. In 1974, Milkovich and Chiang [36] demonstrated the syntheses and applications of a variety of macromers by the termination reactions of living anionic polystyrene with allyl chloride or with a halogen containing several more vinyl compounds [37].

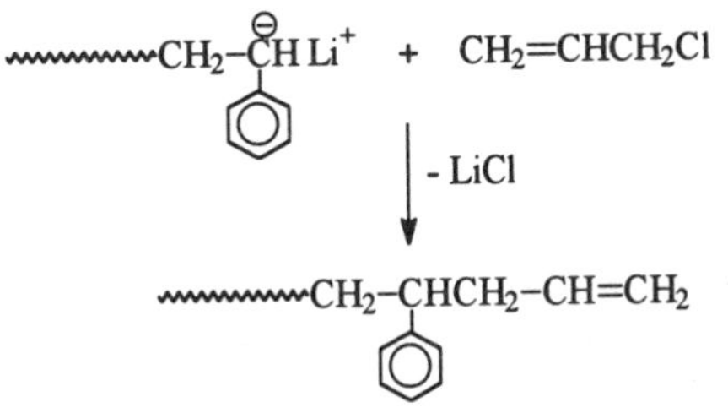

(macromer)

By using the Milkovich method, Asami et al. [38] prepared (*p*-vinyl benzyl) polystyrene containing no ether linkage by the direct reaction of living polystyrene with *p*-vinyl benzyl chloride. In a similar manner, living poly tetrahydrofuran or living poly(1-tert-butyl aziridine) can be coupled with sodium *p*-vinyl phenoxide [39], *p*-isopropenyl benzyl alcohol [40], methacrylic acid sodium salt [41].

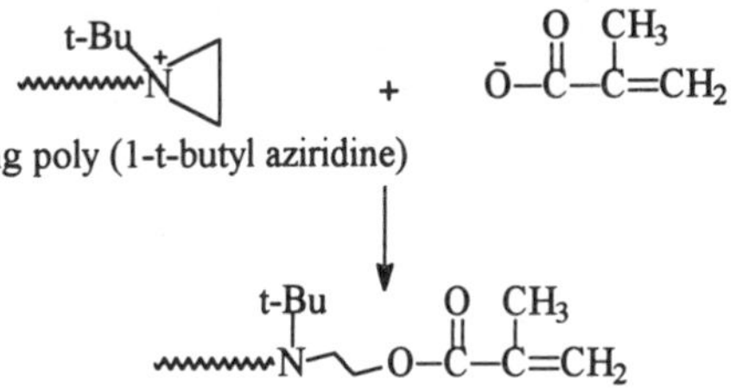

(macromer)

The cationic ring opening polymerization of oxolane (THF) or of N-substituted aziridines can be initiated by oxocarbenium salts [42]. The methacrylic ester unsaturation is insensitive to cationic sites, and polyoxolanes (poly-THF) macromonomers are obtained in good yields.

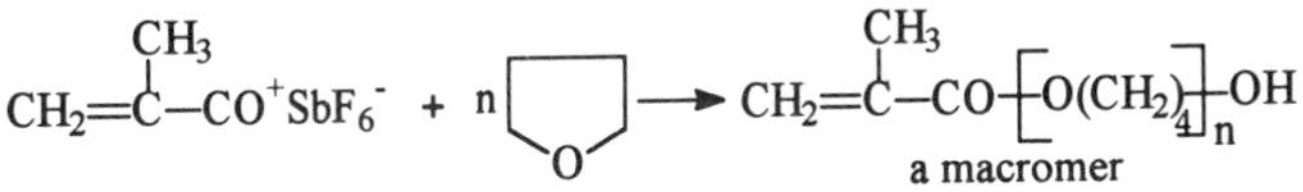

a macromer

The use of an unsaturated anionic initiator—such as potassium *p*-vinyl benzoxide—is possible for the ring opening polymerization of oxirane [43]. Although initiation is generally heterogenous, the polymers exhibit the molecular weight expected and a low polydispersity. In this case, the styrene type unsaturation at chain end cannot get involved in the process, as the propagating sites are oxanions.

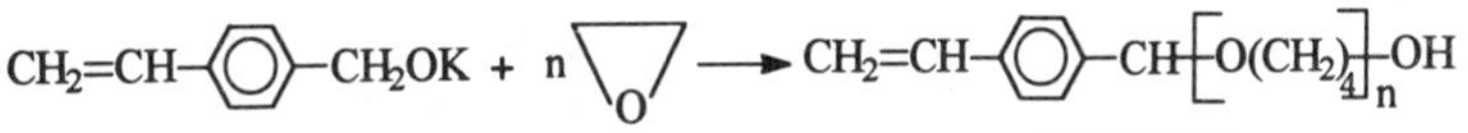

a macromer

Radical polymerizations of macromonomers are greatly influenced by the diffusion control effect [44]. Segmental diffusivity and translational diffusivity of the growing chains of macromonomers are strongly affected by the feed concentration and the molecular weight of the macromonomers. Furthermore, there is little difference in the degree of polymerization between macromers having methacrylate or vinyl benzyl group. The methacryloyl-terminated macromonomers are more reactive than the corresponding vinyl–benzyl-terminated macromonomers [45]. Macromonomers can be anionically prepared by the H-transfer polymerization [46]. H-transfer polymerization of *p*-vinyl benzamide yields a polymer with one vinyl end group.

$$CH_2=CH-\langle O \rangle-CONH_2 \xrightarrow{K^+O'Bu-t} CH_2=CH-\langle O \rangle-CONH(CH_2CH_2-\langle O \rangle-CONH)_n$$

C. Macromonomeric Initiators (Macroinimer)

A macromonomeric initiator called a ''macroinimer'' is a combination of a macromonomer and a macroinitiator [47]. They contain both vinyl and azo (or peroxygen) groups attached with polymeric blocks. Therefore, they can thermally homopolymerize by themselves or copolymerize with a vinyl monomer. In both cases, crosslinked or branched copolymers occur while macroinitiators give only linear block copolymers [48]. The vinyl end groups of a macromerinitiator are effective to obtain crosslinked or branched block copolymer. Macromerazoinitiator was reported first by Hazer [49] in 1991 by capping reactions of hydroxyl end groups of polyazoester [23] with isocyanatoethyl methacrylate:

$$CH_2=\overset{CH_3}{\underset{}{C}}-\overset{O}{\underset{}{C}}-OCH_2CH_2NH\overset{O}{\underset{}{C}}-O-PEG---N=N---PEG---O\overset{O}{\underset{}{C}}NHCH_2CH_2O\overset{O}{\underset{}{C}}-\overset{CH_3}{\underset{}{C}}=CH_2$$

Polymerization of styrene or methyl methacrylate by macroazoinimers having two vinyl groups (MIM-2v) resulted in crosslinked block copolymers, while macroazoinimers with one vinyl end (MIM-1v) group to polymerize vinyl monomers yielded branched block copolymers.

The macroazoinimer obtained by the end capping reaction of polyazoester with a diizocyanate and hydroxyethyl methacrylate was used in wood impregnation leading to the one-shot polymerization of styrene thermally [50].

Following is an ester type macroazoinimer [51] starting from the poly(ethylene glycol), 4,4'-azobiscyano pentanoyl chloride, and methacryloyl chloride.

$$2HO(CH_2CH_2O)H \quad + \quad \overset{O}{\underset{Cl}{C}}-CH_2CH_2-\overset{CH_3}{\underset{CN}{C}}-N=N-\overset{CH_3}{\underset{CN}{C}}-CH_2CH_2-\overset{O}{\underset{Cl}{C}} \quad \xrightarrow[\text{in benzene}]{+(C_2H_5)_3N}$$

PEG-200, 400, 1000

$$HO(CH_2CH_2O)_n\overset{O}{\underset{}{C}}-CH_2CH_2-\overset{CH_3}{\underset{CN}{C}}-N=N-\overset{CH_3}{\underset{CN}{C}}-CH_2CH_2-\overset{O}{\underset{}{C}}(OCH_2CH_2)_nOH$$

polyazoester

$$\downarrow \overset{CH_3\ O}{+2CH_2=C-C-Cl}$$

$$CH_2=\overset{CH_3}{\underset{}{C}}-\overset{O}{\underset{}{C}}-O(CH_2CH_2O)_n\overset{O}{\underset{}{C}}-CH_2CH_2-\overset{CH_3}{\underset{CN}{C}}-N=N-\overset{CH_3}{\underset{CN}{C}}-CH_2CH_2-\overset{O}{\underset{}{C}}(OCH_2CH_2)_nO-\overset{O}{\underset{}{C}}-\overset{CH_3}{\underset{}{C}}=CH_2$$

an ester type macroazoinimer

Homopolymerization of macroazoinimers and copolymerization of macroinimers with a vinyl monomer yield crosslinked polyethyleneglycol or polyethyleneglycol-vinyl polymer-crosslinked block copolymer, respectively. The homopolymers and block copolymers having PEG units with molecular weights of 1000 and 1500 still showed crystallinity of the PEG units in the network structure [48] and the second heating thermograms of polymers having PEG-1000 and PEG-1500 units showed that the recrystallization rates were very fast (Fig. 3).

The crosslinked polymers also show endothermic peaks in the range of 130–140°C. These endotherms can be attributed to the decomposition of the residual azo group and the polymerization of vinyl monomers (Fig. 1).

The swelling properties of these crosslinked copolymers in either toluene or water varied with the molecular weight of the PEG component and with the polymerization time. As polymerization time increased, the degree of swelling became smaller; and the degree of swelling increased with an increase in the molecular weight of PEG units in the macroazoinimer [48,51]. Figure 4 shows the change of the degree of swelling of crosslinked block copolymers obtained by macroazoinimers having PEG-200 (MAIM-200), PEG-400 (MAIM-400), PEG-600 (MAIM-600), PEG-800 (MAIM-800), PEG-1000 (MAIM-1000), and PEG-1500 (MAIM-1500).

Macroazoinimers were used in the dispersion polymerization of styrene in water without using any additional emulsifier or initiator [52,53]. Sulphonation of the styrene-PEG crosslinked block copolymer can be used

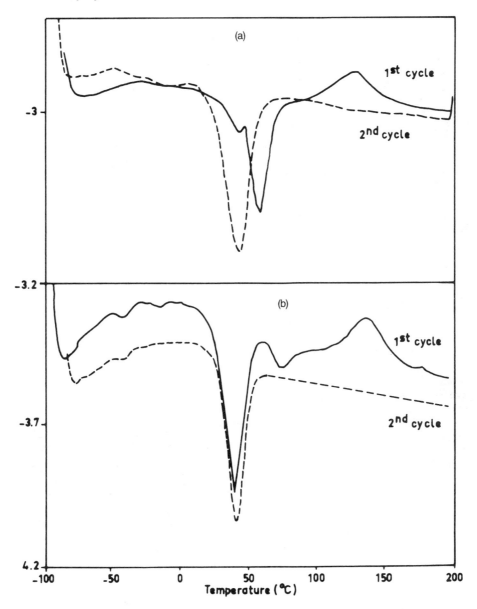

Figure 3 DSC thermograms for (a) homopolymer of an ester type macroazoinimer, MAIM-1000, having PEG-1000 units; and (b) styrene-PEG-1000 crosslinked polymer with MAIM-1000. *Source*: Ref. 50.

as an ion exchange resin [52]. In very recent works of Savaşkan and Hazer [54,55] macromerperoxy initiator having polytetrahydrofuran units has been reported. Macromerperoxyinitiator (MMPI) is synthesized by the reaction of 2,5-dimethyl 2,5-dihydroperoxide with isophoran diisocyanate, polytetrahydrofuran diol-1000 (poly-THF), and izocyanatoethyl methacrylate, respectively. Vinyl polymerization initiated by MMPI at 80°C in bulk gives crosslinked poly THF-*b*-polystyrene block copolymers. Since crosslinked block copolymers still contained undecomposed peroxygen groups, they initiated a vinyl polymerization in order to obtain interpenetrating polymers [55].

D. Block and Graft Copolymers By Macrointermediates

Vinyl polymerization initiated by macroinitiator yield AB, ABA, or (AB)ₙ types block copolymers. Macroinitiators such as macrobis peroxides, polyazoesters, and

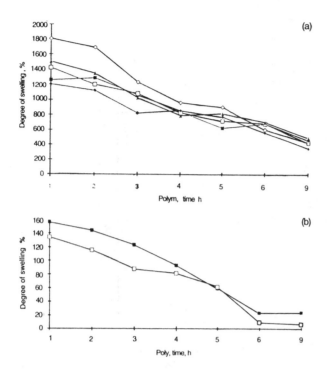

Figure 4 (a) Polymerization time versus degree of swelling (in toluene) curves for styrene-PEG crosslinked copolymers initiated by macroinimers: (◆) MAIM-200, (□) MAIM-400, (◇) MAIM-600, (△) MAIM-1000, and (■) MAIM-1500. (b) Polymerization time versus degree of swelling (in water) curves for styrene-PEG crosslinked copolymers initiated by (■) MAIM-1500 and (□) MAIM-400. *Source*: Ref. 50.

active polymers having peroxide groups obtained by the vinyl polymerization with oligoperoxides or azoperoxidic initiators thermally produce macroradicals by cleavage of their azo or peroxy groups.

Depending on the termination reaction of the vinyl monomer, termination by disproportionation or termination by combination occurs. As a result, AB or ABA block copolymers might be obtained.

Macromonomers always lead to the formation of graft copolymers. For example, the vinyl-terminated polystyrene can be copolymerized with ethylene to produce a graft copolymer of polyethylene, whereby the vinyl moiety of polystyrene is integrally polymerized into the linear polyethylene backbone:

Radical polymerizations of macromonomers are greatly influenced by a diffusion control effect. Segmental diffusivity and translational diffusivity of the growing chains of macromonomers are strongly affected by the feed concentration and the molecular weight of macromonomers [44]. Furthermore, the polymerization rate of macromonomers is readily decreased upon vitrification of the polymerization system even in the presence of a polymerization solvent. Copolymerization of a vinyl monomer with a macromonomer initiator yield a crosslinked block copolymer. In the beginning of the polymerization, the macroinimer behaves as a macrocrosslinker, then peroxygen decomposition proceeds and the crosslinking lattice gets smaller:

Grafting reactions onto a polymer backbone with a polymeric initiator have recently been reported by Hazer [56–60]. Active polystyrene [56], active polymethyl methacrylate [57], or macroazoinitiator [58,59] was mixed with a biopolyester: polyhydroxynonanaate [60] (PHN) or polybutadiene to be carried out by thermal grafting reactions. The grafting reactions of PHN with polymer radicals may proceed by H-abstraction from the tertier carbon atom in the same manner as free radical modification reactions of polypropylene or polyhydroxybutyratevalerate [61,62].

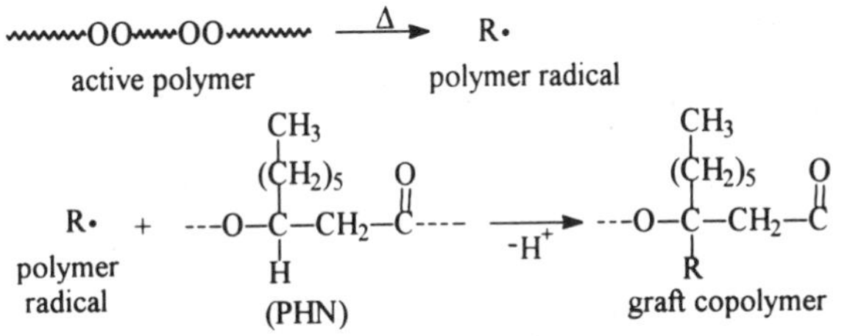

$$\text{\wavy OO\wavy OO\wavy} \xrightarrow{\Delta} R\cdot$$

active polymer → polymer radical

R· + (PHN) $\xrightarrow{-H^+}$ graft copolymer

Grafting reactions of polybutadiene with macrazoinimers or polyazoesters produced polyethylene glycol–polybutadiene crosslinked graft copolymers. Macroradicals thermally formed from macroazoinimers or polyazoesters attack 1,2-linked vinyl pendant groups of polybutadiene:

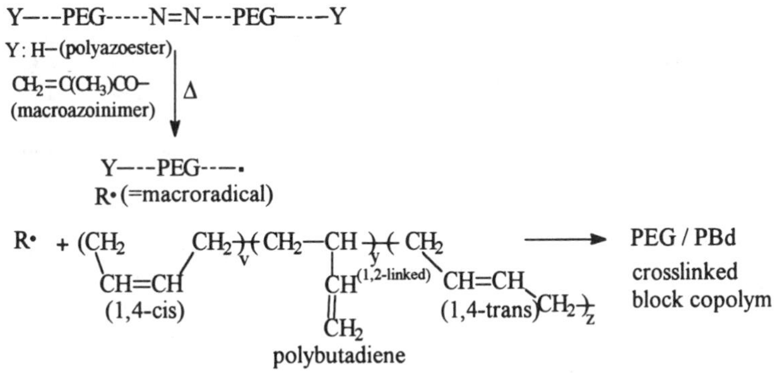

Y---PEG-----N=N---PEG----Y

Y: H–(polyazoester)
CH$_2$=C(CH$_3$)CO– (macroazoinimer) $\Big\downarrow \Delta$

Y---PEG---·
R· (=macroradical)

R· + polybutadiene (1,4-cis, 1,2-linked, 1,4-trans) $\longrightarrow$ PEG / PBd crosslinked block copolym

As an energetic polymer, poly(glycidyl azide) (PGA) was grafted onto poly(butadiene) to obtain PGA-g-poly-(butadiene), which is a high-energetic and high-performance solid propellant binder [63,64]. For this purpose, poly(glycidyl azide) macroazoinitiator synthesized from PGA and ACPC was mixed with poly(butadiene) and thermally grafted [64].

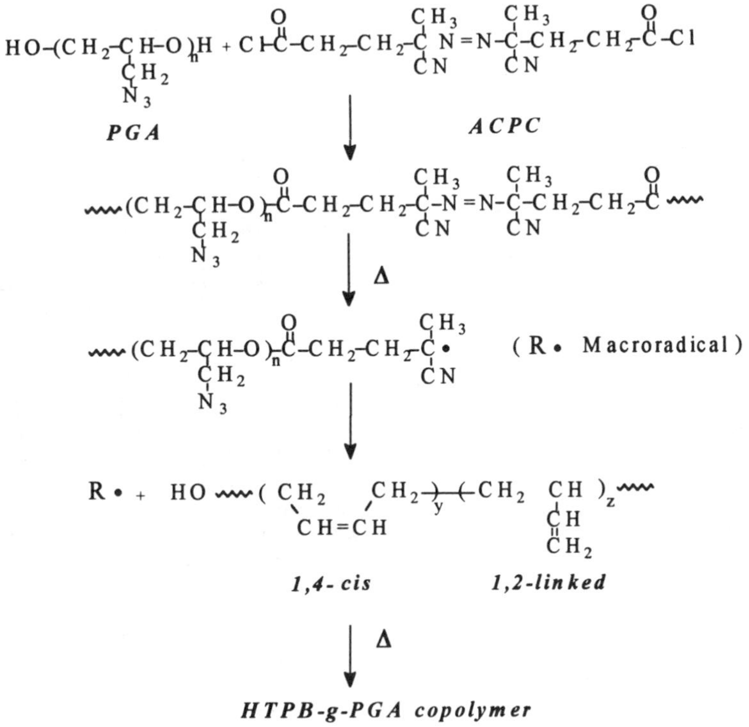

PGA + *ACPC*

$\downarrow$

$\downarrow \Delta$

(R· Macroradical)

$\downarrow$

R· + HO\wavy(CH$_2$ CH$_2$)$_y$(CH$_2$ CH)$_z$\wavy

1,4-cis *1,2-linked*

$\downarrow \Delta$

HTPB-g-PGA copolymer

REFERENCES

1. B. Hazer, *Handbook of Polymer Science and Technology* (N. P. Cheremisinoff, ed.) Vol. 1, Marcel Dekker Inc., New York, pp. 133–176 (1989).
2. R. P. Singh, *Prog. Poly. Sci., 17*: 1 (1992).
3. O. Nuyken and R. Weidner, *Adv. Polym. Sci., 73–74*: 145 (1986).
4. N. R. Legge, S. Davison, H. E. De La Mare, G. Holden, and M. K. Martin, *ACS Symp. Ser., 285*: 175 (1985).
5. R. Jerome, R. Fayt, and T. Ouhadi, *Prog. Polym. Sci., 10*: 87 (1984).
6. A. Noshay, J. E. McGrath, *Block Copolymers,* Academic Press, New York (1977).
7. C. H. Bamford and A. D. Jenkins, *Nature, 176*: 78 (1995).
8. A. V. Tobolsky and A. Rembaum, *J. Appl. Polym. Sci., 8*: 307 (1964).
9. B. Hazer, I. Çakmak, S. Küçükyavuz, and T. Nugay, *Eur. Polym. J., 28*: 1295 (1992).
10. B. Hazer, *Eur. Polym. J., 27*: 775 (1991).
11. I. Yilgör and B. M. Baysal, *Makromol. Chem., 186*: 463 (1985).
12. B. Hazer, *Eur. Polym. J., 26*: 1167 (1990).
13. H. V. Pechmann and L. Vanino, *Beriche, 27*: 1510 (1894).
14. N. S. Tsvetkov and E. S. Beletskaya, *Ukr. Khim. Zh., 29*: 1072 (1963).
15. Yu L. Zherebin, S. S. Ivancher, and N. M. Domareva, *Vysokomol. Sovedin., A16*: 893 (1974).
16. B. Hazer, *J. Polym. Sci. A, Polym. Chem., 25*: 3349 (1987).
17. B. Hazer and A. Kurt *Eur. Polym. J., 31*: 499 (1995).
18. T. A. Tolpygina, V. I. Galibei, and S. S. Ivanchev, *Vysokomol. Soyedin, A14*: 1027 (1972).
19. B. Hazer and B. M. Baysal, *Polymer, 27*: 961 (1986).
20. K. S. Murthy, K. Kishore and V. K. Mohan, *Macromolecules, 27*: 7109 (1994).
21. J. J. Laverty and Z. G. Gardlund, *J. Polym. Sci. A Polym. Chem., 15*: 2001 (1977).
22. H. R. Dicke and W. Heitz, *Macromol. Chem., Rap. Commun., 2*: 83 (1981).
23. R. Walz, B. Bomer and W. Heitz, *Makromol. Chem., 178*: 2527 (1977).
24. P. S. Anand, H. G. Stahl, W. Heitz, G. Weber and L. Bottenbruch, *Makromol. Chem., 183*: 1685 (1982).
25. B. Hazer, B. Erdem, and R. W. Lenz, *J. Polym. Sci. A. Polym. Chem., 32*: 1739 (1994).
26. Y. Haneda, H. Terada, M. Yoshida, A. Ueda, and S. Nagai, *J. Polym. Sci. A: Polym. Chem., 32*: 2641 (1994).
27. A. Ueda and S. Nagai, *J. Polym. Sci. A: Polym. Chem., 22*: 1783 (1984).
28. D. S. Campbell, D. E. Loeber and A. J. Tinker, *Polymer, 25*: 1141 (1984).
29. J. Furukawa, S. Takamori, and S. Yamashita, *Angew. Makromol. Chem., 1*: 92 (1967).
30. J. M. G. Cowie and M. Yazdani-Pedram, *Br. Polym. J., 16*: 127 (1984).
31. S. Dumitriu, A. S. Shaikh, E. Comanita, and C. R. Simionescu, *Eur. Polym. J., 19*: 263 (1983).
32. I. Piirma and L. H. Chou, *J. Appl. Polym. Sci., 24*: 2051 (1979).
33. B. Hazer, *Angew. Makromol. Chem., 129*: 31 (1985).
34. B. Hazer, A. Ayas, N. Beşırli, N. Saltek, and B. M. Baysal, *Makromol. Chem., 190*: 1987 (1989).
35. P. Rempp and E. Franta, *Recent Advances in Anionic Polymerization,* (T. E. Hogen-Esch and J. Smid, eds.), Elsevier Science Publishing Co., Inc., New York (1987).
36. R. Milkovich and M. T. Chiang, U.S. Patent 3, 786, 116 (1974).
37. R. Milkovich and M. T. Chiang, U.S. Patent 3, 842, 050 (1974).
38. R. Asami, M. Takaki, and H. Hanahata, *Macromolecules, 16*: 628 (1983).
39. R. Asami, M. Takaki, and K. Kyuda, and E. Asakura, *Polym. J., 15*: 139 (1983).
40. J. S. Vargas, P. Masson, G. Beinert, P. Rempp, and E. Franta, *Polym. Bull., 7*: 277 (1982).
41. E. J. Goethals and M. A. Vlegals, *Polym. Bull., 4*: 521 (1981).
42. J. S. Vargas, J. G. Zilliox, P. Rempp, and E. Franta, *Polymer Bulletin, 3*: 83 (1980)
43. P. Masson, G. Beinert, E. Franta, and P. Rempp. *Polymer Bulletin, 7*: 17 (1982).
44. Y. Tsukahara, K. tsutsumi, Y. Yamashita, and S. Shimada, *Macromolecules, 23*: 5201 (1990).
45. I. Capek and M. Akashi, *J. M. S. -Rev. Makromol. Chem. Phys., C33*: 369 (1993).
46. H. Jung and W. Heitz, *Makromol. Chem., Rap. Commun., 9*: 373 (1988).
47. B. Hazer, *The Polymeric Materials Encyclopedia* (J. C. Salomone, ed.), CRC Press, Inc., Florida (1995) Vol. 6. p. 3911–3918.
48. B. Hazer, B. Erdem, and R. W. Lenz, *J. Polym. Sci. A: Polym. Chem., 32*: 1739 (1994).
49. B. Hazer, *J. Macromol. Sci.- Chem., A28*: 47 (1991).
50. B. Hazer, Y. Örs, and M. H. Alma, *J. Appl. Polym. Sci., 47*: 1097 (1993).
51. B. Hazer, *Makromol. Chem., 193*: 1081 (1992).
52. S. Savaşkan, N. Beşırli, and B. Hazer, *J. Appl. Polym. Sci. 59*: 1515 (1996).
53. U. Yildiz, B. Hazer, and I. Capek, *Angew. Makromol. Chem. 231*: 135 (1995).
54. S. Savaşkan and B. Hazer, *239*: 13 (1996).
55. B. Hazer and S. Savaşkan, (submitted).
56. B. Hazer, *J. Macromol. Sci.- Chem., A32*: 81 (1995).
57. B. Hazer, *J. Macromol. Sci.- Chem., A32*: 477 (1995).
58. B. Hazer, *J. Macromol. Sci.- Chem., A32*: 679 (1995).
59. B. Hazer, *Macromol. Chem. Phys., 196*: 1945 (1995).
60. B. Hazer, *Polym. Bull., 33*: 431 (1994).
61. P. Cavallaro, B. Immirzi, M. Malinconico, E. Martuscelli, and M. G. Volpe, *Macromol. Rap. Commun., 15*: 103 (1994).
62. C. H. Bamford and K. G. Al-Lamee, *Macromol. Rap. Commun., 15*: 379 (1994).
63. M. S. Eroğlu, B. Hazer, O. Güven, and B. M. Baysal, *J. Appl. Polym. Sci. 60*: 2141 (1996).
64. M. S. Eroğlu, B. Hazer, and O. Güven, *Polym. Bull. 36*: 695 (1996).
65. M. K. Mishra and Y. Yağci, *Macromolecular Design: Concept and Practice;* Chapter 14, Polymer Frontiers Int., Inc., pp. 499–506 (1994).
66. B. Hazer, *Eur. Polym. J., 27*: 975 (1991).

48

Azo Initiators as Transformation Agents for Block Copolymer Synthesis

Yusuf Yagci and Ivo Reetz

Istanbul Technical University, Istanbul, Turkey

I. INTRODUCTION

Block copolymers have become increasingly important in recent decades. This importance is due to the fact that their special chemical structure yields unusual physical properties, especially as far as solid-state properties are concerned. Block copolymers are applied in various fields, they are used as surfactants, adhesives, fibres, thermoplastics, and thermoplastic elastomers.

A number of techniques for the preparation of block copolymers have been developed. Living polymerization is an elegant method for the controlled synthesis of block copolymers. However, this technique requires extraordinarily high purity and is limited to ionically polymerizable monomers. The synthesis of block copolymers by a radical reaction is less sensitive toward impurities present in the reaction mixture and is applicable to a great number of monomers.

In this chapter techniques of block copolymer preparation involving thermally labile azo compounds are reviewed. Upon heating, aliphatic azo compounds evolve nitrogen thus forming two carbon centered free radicals.

$$R_1-N{=}N-R_2 \xrightarrow{\Delta} \overset{\bullet}{R_1} + \overset{\bullet}{R_2} + N_2$$

The activation energies for this reaction are, supposing R_1 and R_2 are aliphatic, between 60 and 160 kJ mol^{-1}. Azo groups are, therefore, a highly suitable radical source with respect to the energy required. Despite being thermally labile, azo groups are photoactive, too. Aliphatic azo groups evolve nitrogen with quantum yields as high as $\phi = 0.44$ (AIBN [1]). However, the absorption is relatively low, being dependent on the nature of the two substituents, $\epsilon \approx 12. . .20$ at $\lambda_{max} \approx 350$ nm.

To obtain block copolymers, the initiating azo compound must be at least bifunctional—it must carry one or more reactive sites other than the azo function itself.

The thermal (or photochemical) decomposition of the azo group gives rise to a radically initiated polymerization. The reactive site F, the transformation site, however, can, depending on its chemical nature, initiate a condensation or addition type reaction. It can also start radical or ionic polymerizations. F may also terminate a polymerization or even enable the azo initiator to act as a monomer in chain polymerizations.

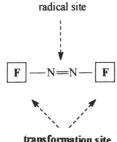

The azo initiator initially present therefore has to be classified as a transfer agent—it is able to combine monomers polymerized by different polymerization modes with each other. Three different modes of block copolymer synthesis via azo transfer agents can be distinguished:

$$\boxed{F}—N=N—\boxed{F} \;+\; nA \;\longrightarrow\; MAI \;\xrightarrow{\Delta,\,nB}\; block\;copolymer \qquad (1)$$

$$\boxed{F}—N=N—\boxed{F} \;+\; A_n \;\longrightarrow\; MAI \;\xrightarrow{\Delta,\,nB}\; block\;copolymer \qquad (2)$$

$$\boxed{F}—N=N—\boxed{F} \;+\; nA \;\xrightarrow{\Delta}\; \boxed{F}—A_n \;\xrightarrow[nB]{activation,}\; block\;copolymer \qquad (3)$$

Figure 1   F = Transformation site; MAI = macro-azo-initiator, having at least one azo group in the main chain.

The bifunctional azo initiator can initiate a polymerization of a monomer A forming a polymeric compound containing at least one azo group in its main chain [Figure 1, Scheme (1)]. Naturally, the monomer A has to have suitable sites for reacting with the transformation site F. The azo-containing polymers obtained in Scheme (1), so-called macro-azo-initiators (MAIs), are crucial in block copolymerization. By heating them in the presence of a second monomer B, the macroradicals formed initiate the polymerization thus leading to blocks of A and B sequences. Alternatively, in Scheme (2), the initially present bifunctional azo compound can react with a preformed polymer A forming MAIs. These can, by a treatment similar to Scheme (1), be transferred to a block copolymer. It is important to note that during the reaction of the transformation site F (in routes (1) and (2)), mild conditions regarding reaction temperature have to be applied to keep the azo function intact. The synthesis of block copolymers following reaction pathways (1) and (2) is dealt with in Section II of this chapter.

In a third type of block copolymer formation, Scheme (3), the initiator's azo group is decomposed in the presence of monomer A in a first step. The polymer formed contains active sites different from azo functions. These sites may, after a necessary activation step, start the polymerization of the second monomer B. Actually, route (3) of block copolymer formation is a *vice versa* version of type (1). It has been shown in a number of examples that one starting bifunctional azo compound can be used for block copolymer synthesis following either path. Reactions of type (3) are tackled in detail in Section III of this chapter.

Although most examples reported in the literature used azo initiators with two transformation sites F per molecule, in a few cases block copolymer synthesis was accomplished with transfer agents containing only one reactive site. The resulting macro-azo-initiators did contain exclusively terminal azo groups.

In this chapter the different transformations of the reactive site F will be dealt with separately. Due to the potential variability in the chemical nature of this second reactive site it was possible to combine a wide variety of chemically unlike monomers with each other thus designing novel block copolymers.

II. MACRO-AZO-INITIATORS (MAIs)—KEY ELEMENTS IN BLOCK COPOLYMER SYNTHESIS

The incorporation of thermally labile azo groups into polymer backbones was first reported in the early 1950s [2]. Since then, numerous techniques for synthesizing azo-containing polymers have been developed. The effort to create new azo-containing polymeric materials has been reviewed by several authors [3–8].

The number and location of azo functions in the polymeric azo initiator is substantial regarding its application for block copolymer synthesis. The cases illustrated in Table 1 will be discussed here.

It is possible, mainly by living cationic polymerization methods, to synthesize MAIs with exactly one azo function per polymer main chain (I). In this case, the azo functions are located then in the center of the chain [9–11]. Type I MAIs are very useful for the controlled synthesis of block copolymers with uniform block length. Most examples of MAIs reported refer to a type with more than one azo function per macromolecule (II). The azo function can be distributed statistically over the polymer backbone or occur in every repeating unit. Upon heating, polymer fragments that often differ in length are formed. Polymeric azo initiators with terminal azo groups (III and IV) are formed when the initial azo compound possesses only one reactive site F. However, for block copolymer synthesis, azo-terminated polymers are not very suitable. Being heated, they form low-molecular weight radicals R· giving rise to an immense quantity of homopolymer B_n. Polymeric compounds with azo groups in the side chain (V) undoubtedly play an important role in polymer chemistry but are not precursors for block copolymers. Heating them in the presence of a second monomer results in the formation of graft copolymers [7].

A. Formation of MAIs by Condensation and Addition Reactions

Bifunctional azo initiators with groups enabling them to participate in condensation or addition reactions can be classified as condensation radical and addition radical

Table 1 Types of MAIs

Location of the azo bond	Structure
I Central azo bond	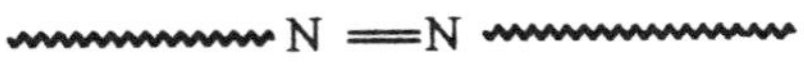 ⌇⌇⌇⌇⌇ N ═ N ⌇⌇⌇⌇⌇
II Azo groups in the main chain	 ⌇⌇⌇ N ═ N ⌇⌇⌇ N ═ N ⌇⌇⌇
III One terminal azo bond	 R — N ═ N ⌇⌇⌇⌇⌇
IV Two terminal azo bonds	 R — N ═ N ⌇⌇⌇⌇⌇ N ═ N — R
V Side chain azo bonds	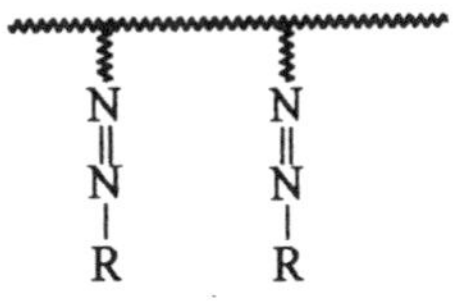 (side chain azo structure with N‖N–R groups)

transfer agents, respectively. They are able to combine radically polymerizable monomers with low-molecular weight compounds reacting in polycondensation or addition reactions thus generating block copolymers. In a large number of publications, acid or acid chloride derivatives of the well-known radical initiator AIBN were condensed with monomers or preformed polymers possessing alcohol, amine, acid chloride, or acid end groups. Regarding polyaddition, mostly dialcohols containing one central azo group were reacted with diisocyanates. In Table 2 the bifunctional azo compounds most frequently used in polycondensation and addition reactions are listed.

The carboxyl terminated ACPA, 4,4′-azobis-(4-cyanopentanoic acid), turned out to be a suitable reagent in condensation reactions. This compound can be prepared by Strecker's synthesis from levulinic acid following the method of Haines and Waters [12]. Regarding the formation of polymeric azo initiators, Matsakuwa et al. [13] reported on the condensation of ACPA with various diols and diamines in the presence of a condensation agent, 1-methyl-2-chlorpyridinium iodide, and a cata-

Table 2 Bifunctional Azo Compounds Used Frequently in Polycondensation and Addition Reactions

Formula	Abbreviation	Reference synthesis	Reference block copolymerization
$\left[HO-\overset{O}{\overset{\|}{C}}+CH_2)_2-\overset{CH_3}{\underset{CN}{\overset{\|}{C}}}-N= \right]_2$	ACPA	[12]	[13–15]
$\left[Cl-\overset{O}{\overset{\|}{C}}+CH_2)_2-\overset{CH_3}{\underset{CN}{\overset{\|}{C}}}-N= \right]_2$	ACPC	[16,17]	[14,15,17,20–38,40,41]
$\left[HO-(CH_2)_3-\overset{CH_3}{\underset{CN}{\overset{\|}{C}}}-N= \right]_2$	ACPO	[60]	[51–53,57–59]

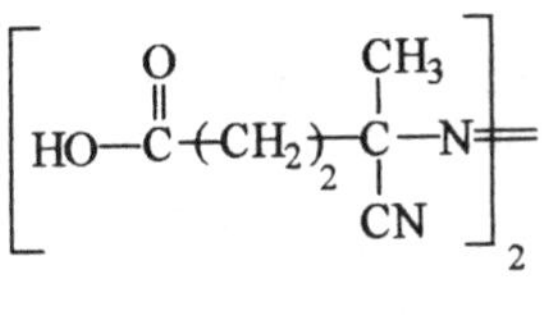

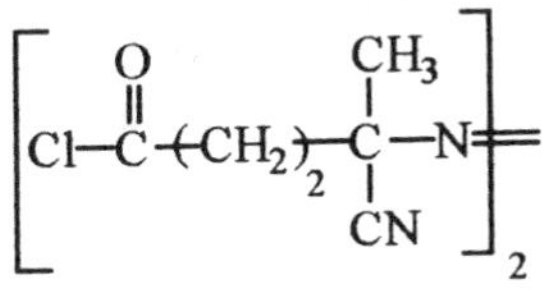

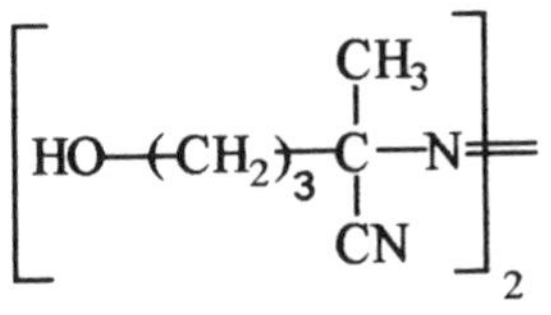

lyst, $(CH_3CH_2)_3N$) (see Scheme 4). Since the reaction was carried out at room temperature, the azo functions of ACPA remained unaffected. However, the disadvantage of using ACPA is the relatively low degree of polymerization; DP values between 8 and 15 were found in these reactions. The poly(ester)- and poly(amine)-based macro-azo-initiators obtained were used for the thermal polymerization of various vinyl monomers, such as styrene, methyl methacrylate, and vinyl acetate yielding block copolymers.

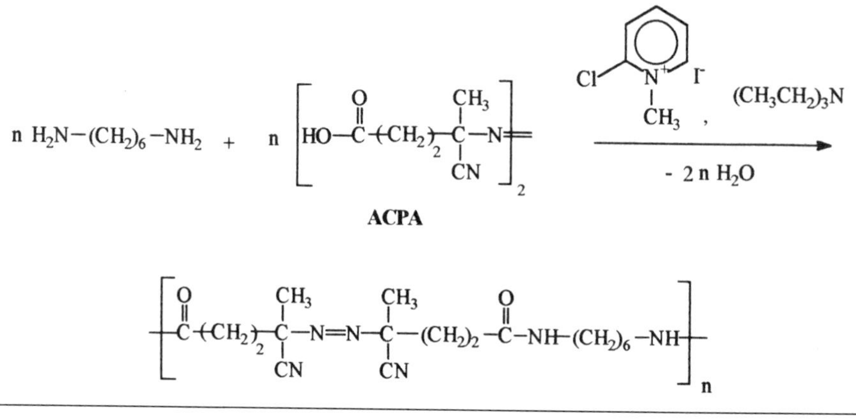

$$(4)$$

As far as polycondensations involving low-molecular weight azo compounds terminated with acid chloride groups are concerned, ACPC, 4,4'-azobis-4-cyanopentanoyl chloride, is the most prominent example. However, other acid chloride azo compounds were also used [18,19]. The enormous popularity of ACPC stems from the fact that it is both very reactive and easy to synthesize. ACPC can be obtained by reacting ACPA with PCl_5 [16] or $SOCl_2$ [17]. Due to its high reactivity, ACPC was used for the condensation with apparently all types of diols and diamines yielding thermally labile poly(ester)s and poly(amide)s [8]. The polycondensation of ACPC with dialcohols and diamins was first described by Smith [16].

Concerning the reaction of ACPC with diols, the frequent use of poly(ethylene glycol) has to be mentioned [20–24]. Ueda et al. ([22–24]) reacted preformed poly(ethylene glycol) (M_n between 6×10^2 to 2×10^4) with ACPC. In this case, unlike the reaction of ACPA with diols (*vide ante*), no additional condensation agent was needed. The ethylene glycol-based thermally labile polymers were used to produce blocks with poly(vinyl chloride) [22], poly(styrene) [23], poly(methyl acrylate), poly(vinyl acetate), and poly(acrylonitrile) [24].

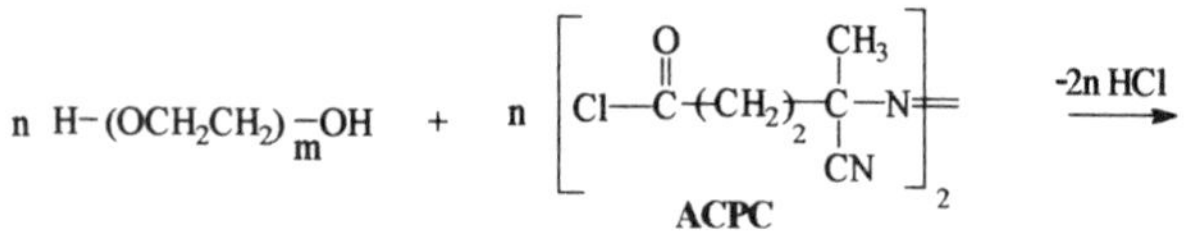

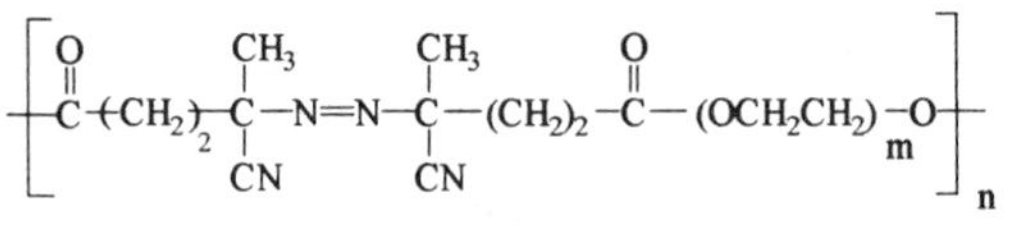

$$(5)$$

Hazer [20,25] reported on the reaction of a poly(ethylene glycol)-based azoester with methacryloyl chloride in the presence of $(CH_3CH_2)_3N$. In this reaction double bonds were attached to the chain ends of the poly(ester) thus obtaining a macroinimer. Being used for the thermal polymerization of styrene, the material formed an insoluble gel [20]. Probably, both the C=C double bonds and the azo bonds reacted in the course of the thermal treatment. The macroninimer in a later work [25] was used for thermally polymerizing poly(butadiene) thus leading to poly(ethylene glycol-b-butadiene) block copolymers.

Apart from poly(ethylene glycol), other hydroxyl-terminated polymers and low-molecular weight compounds were condensed with ACPC. An interesting example is the reaction of ACPC with preformed poly(butadiene) possessing terminal OH groups [26]. The reaction was carried out in chloroform solution and $(CH_3CH_2)_3N$ was used as a catalyst. MAIs based on butadiene thus obtained were used for the thermally induced block copolymerization with styrene [26] and dimethyl itaconate [27].

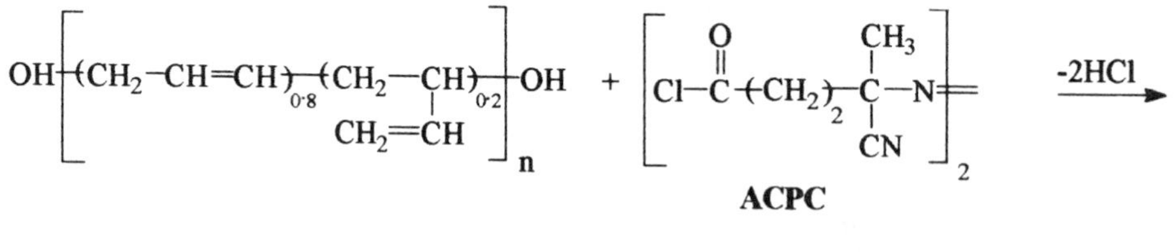

$$-O \left[(CH_2-CH=CH)_{0.8} (CH_2-CH)_{0.2} \atop \qquad\qquad\qquad\quad | \atop \qquad\qquad\qquad CH_2=CH \right]_n O-\overset{O}{\overset{||}{C}}(CH_2)_2\overset{CH_3}{\overset{|}{\underset{CN}{C}}}-N=N-\overset{CH_3}{\overset{|}{\underset{CN}{C}}}(CH_2)_2\overset{O}{\overset{||}{C}}- \qquad (6)$$

The solution polycondensation of ACPC with hydroxyl-terminated poly(arylate)s, consisting of bisphenol A and a mixture of terephathalic and isophathalic acid, was reported by Ahn et al. [28]. The azo-containing poly(arylate) was block copolymerized with styrene in order to improve the processability of the poly(arylate) without loss of its ultraviolet UV and thermal stability. The block copolymers obtained were furthermore utilized as compatibilzing agent in poly(arylate)–poly(styrene) blends [29]. Simionescu et al. [30] investigated the interfacial condensation of ACPA with bisphenol A and succinyl chloride yielding thermally labile copoly(ester)s. They were subsequently used to produce various poly(ester-*b*-alkylacrylate) block copolymers with different block lengths. The same authors [31] condensed a preformed hydroxyl-terminated poly(dimethyl siloxane) ($M_n = 7 \times 10^3$) with ACPC thus producing a suitable precursor for block copolymer synthesis. These investigations were aimed at the production of novel thermoplastic elastomers: vinyl polymers with incorporated poly(siloxane) blocks. In two recent papers, Ueda et al.

[32,33] reported on the condensation of ACPC with preformed poly(isoprene diol), poly(ethylene adipate), poly(caprolactone), and aliphatic poly(carbonates). The azo-containing prepolymers (M_n between 5×10^3 and 1×10^4) were successfully used for preparing block copolymers with either poly(styrene) or poly(methyl methacrylate). Galli et al. [34] condensed ACPC with bis(4-hydroxybenzoate) forming a polymeric azo inititiator for the syntesis of main- and side-chain liquid crystalline polymers. Together with ACPC, an excess of a second acid chloride was used for the condensation thus diminishing the number of thermally labile sites in the MAI.

The reaction of ACPC with linear aliphatic amines has been investigated in a number of Ueda's papers [17,35,36]. Thus, ACPC was used for a interfacial polycondensation with hexamethylene diamine at room temperature [17] yielding poly(amide)s. The polymeric material formed carried one azo group per repeating unit and exhibited a high thermal reactivity. By addition of styrene and methyl methacrylate to the MAI and heating, the respective block copolymers were formed.

$$n \; NH_2(CH_2)_6NH_2 \quad + \quad n \left[Cl-\overset{O}{\overset{||}{C}}(CH_2)_2\overset{CH_3}{\overset{|}{\underset{CN}{C}}}-N= \right]_2 \quad \xrightarrow{-\,2n\,HCl}$$

ACPC

$$\qquad\qquad\qquad\qquad\qquad\qquad\qquad\qquad\qquad\qquad\qquad\qquad\qquad (7)$$

$$\left[NH-(CH_2)_2-NH-\overset{O}{\overset{||}{C}}(CH_2)_2\overset{CH_3}{\overset{|}{\underset{CN}{C}}}-N=N-\overset{CH_3}{\overset{|}{\underset{CN}{C}}}(CH_2)_2\overset{O}{\overset{||}{C}} \right]_n$$

When ACPC was condensed with hexamethylene diamine in the presence of two other acid chlorides (sebacoyl or adipoyl chloride), poly(amide)s 6.6 and 6.10 with various numbers of azo groups per repeating unit (between 0.14 and 1.0, depending on the ratio of the acid chlorides used) could be obtained [35,36]. Thus, block copolymers with a controlled segment length of the poly(amide) blocks were attainable.

Macro-azo-initiators containing crown ether units were successfully synthesized by Yagci et al. [37,38] condensing ACPC with the *cis* or *trans* forms of 4,4'-diaminodibenzo-18-crown-6 (Scheme 8). The polymeric

initiators formed were used for the thermal polymerization of poly(methyl methacrylate) and poly(styrene) blocks. The incorporation of crown ether units into polymers opens new pathways for the utilization of the crown ether's complexation ability for cation exchange. Polymeric membranes, containing covalently bond crown ethers may easily be regenerated and thus be used many times [39]. A second advantage is that crown ethers bond to polymer matrices readily form sandwich type complexes enabling the exchange of relatively bulky cations.

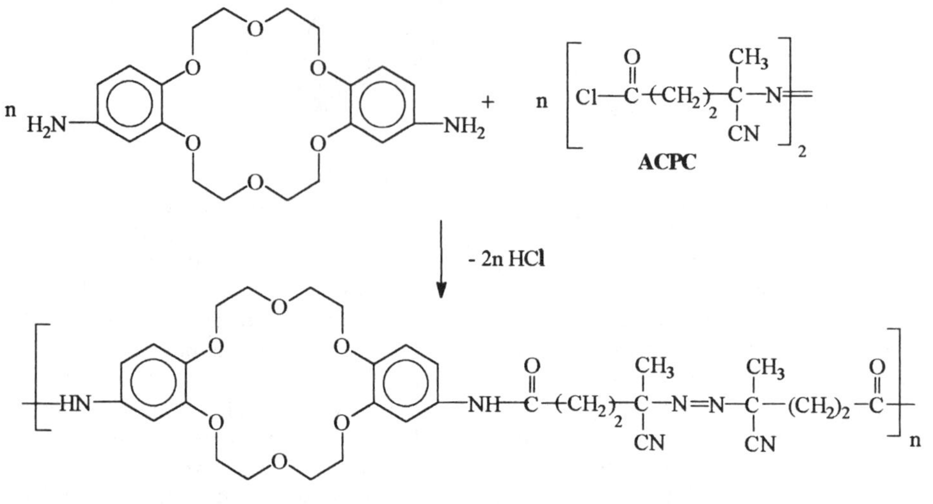

(8)

In a later work, Tunca and Yagci [40,41] used two other acid chlorides (adipoyl and terephthaloyl chloride) along with ACPC. By changing the ratio of the different acid chlorides the number of thermally labile azo bonds in the polymer backbone could be regulated.

Furthermore, macro-azo-initiators have also been synthesized by direct condensation of AIBN with diols (Pinner reaction, [42,43]) or formaldehyde [44,45]. The

reaction of AIBN with both polymeric (poly(ethylene glycol)) and low-molecular weight diols has been thoroughly studied by Heitz et al. [46–48]. The azo-containing reaction products, polyazo esters (degree of polymerization DP < 10), were used for partial decomposition in the presence of a first monomer and a subsequent thermal treatment forming blocks of the second monomer [46,49] (see Section II. D.)

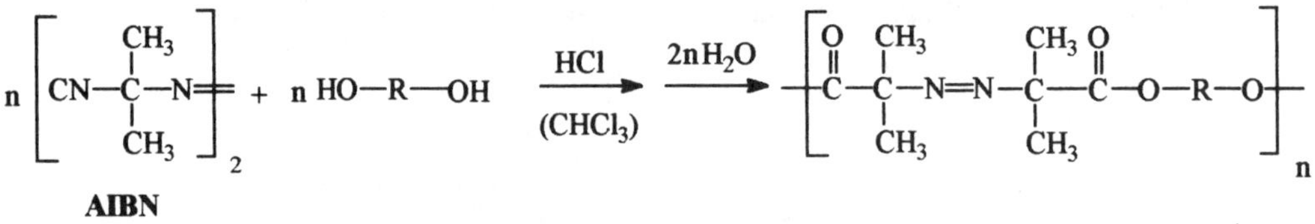

(9)

The reaction of AIBN with formaldehyde leads to polyazo amides. Block copolymers consisting of amide

units, styrene, and methyl methacrylate blocks were obtained by partial decomposition of the azo amide prepolymers [44,45].

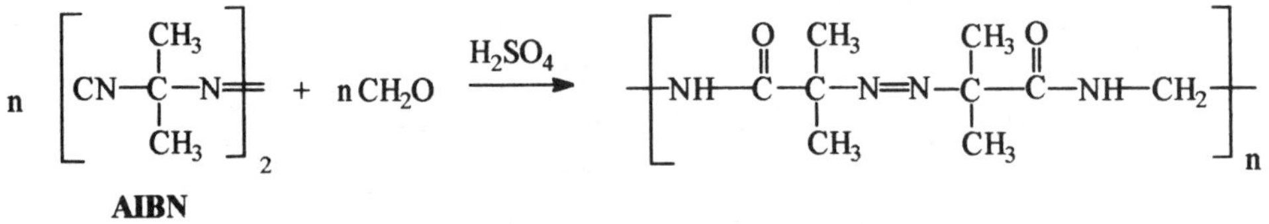

(10)

By means of a ring-opening polymerization of the condensation type Vlasov et al. [50] synthesized polypeptide based MAIs with azo groups in the polymeric backbone. The method is based on the reaction of a hydracide derivative of AIBN and a N-carboxy anhydride. Containing one central azo group in the polymer main chain, the polymeric azo initiator was used for initiating block copolymerizations of styrene and various methacrylamides.

Addition reactions were frequently used to create MAIs capable of forming block copolymers. Thus, one possible pathway is to react preformed polymer-contain-

ing terminal isocyano groups with 4,4'-azobis(4-cyano-n-pentanol) (ACPO). Yürük and Ulupinar [51] treated commercial poly(ethylene glycol) ($M_n \approx 4 \times 10^3$) and poly(propylene glycol) ($M_n \approx 2 \times 10^3$) with a low-molecular weight diisocyanate obtaining polymeric diisocyanates (Scheme 11). After a reaction with ACPO, macroazocarbamates were formed, materials bearing a central poly(alkylene oxide) block and two azo groups per molecule. These macroinitiators were used for a free radical polymerization of styrene. A similar strategy has also been described in other papers [52,53].

$$\text{OH}\text{-}(\text{CH}_2\text{-}\text{CH}_2\text{-}\text{O})_n\text{-}\text{H} \ + \ 2 \ \text{OCN}-\text{C}_{10}\text{H}_{18}-\text{NCO} \longrightarrow$$

$$2 \left[\text{OH}\text{-}(\text{CH}_2)_3\text{-}\overset{\overset{\displaystyle CH_3}{|}}{\underset{\underset{\displaystyle CN}{|}}{C}}\text{-}N \right]_2$$

$$\text{OCN}-\text{C}_{10}\text{H}_{18}-\text{NH}-\overset{O}{\overset{||}{C}}\text{-}(\text{CH}_2\text{-}\text{CH}_2\text{-}\text{O})_n\text{-}\overset{O}{\overset{||}{C}}\text{-}\text{NH}-\text{C}_{10}\text{H}_{18}-\text{NCO} \longrightarrow \qquad (11)$$

$$\text{OH}\text{-}(\text{CH}_2)_3\text{-}\overset{\overset{\displaystyle CH_3}{|}}{\underset{\underset{\displaystyle CN}{|}}{C}}\text{-}N=N\text{-}\overset{\overset{\displaystyle CH_3}{|}}{\underset{\underset{\displaystyle CN}{|}}{C}}\text{-}(\text{CH}_2)_3\text{-}O\text{-}\overset{O}{\overset{||}{C}}\text{-}\text{NH}-\text{R}-\text{NH}-\overset{O}{\overset{||}{C}}\text{-}O\text{-}(\text{CH}_2)_3\text{-}\overset{\overset{\displaystyle CH_3}{|}}{\underset{\underset{\displaystyle CN}{|}}{C}}\text{-}N=N\text{-}\overset{\overset{\displaystyle CH_3}{|}}{\underset{\underset{\displaystyle CN}{|}}{C}}\text{-}(\text{CH}_2)_3\text{-}\text{OH}$$

Polyaddition reactions based on isocyanate-terminated poly(ethylene glycol)s and subsequent block copolymerization with styrene monomer were utilized for the impregnation of wood [54]. Hazer [55] prepared block copolymers containing poly(ethylene adipate) and poly(peroxy carbamate) by an addition of the respective isocyanate-terminated prepolymers to polyazoesters. By both bulk and solution polymerization and subsequent thermal polymerization in the presence of a vinyl monomer, multiblock copolymers could be formed.

Moreover, polyaddition reactions of two low-molecular weight compounds can yield MAIs. Ueda et al.

[56] described the reaction of various azodiols, such as, 2,2'-azobis(2-cyanopropanol) with toluene diisocyanate (Scheme 12). MAIs with molar masses of $M_n \approx 1 \times 10^3$ and relatively small polydispersity (1.3) were formed and used for block copolymerization with styrene. Azo-containing poly(urethanes), with M_n between 2×10^3 and 6×10^3, were also synthesized by a ternary polyaddition using ACPO, diols containing ether bonds, and hexamethylene diisocyanate [57,58]. These macroinitiators were applied in the block copolymerization of various vinyl monomers.

$$n \left[\text{HO}-\text{CH}_2\text{-}\overset{\overset{\displaystyle CH_3}{|}}{\underset{\underset{\displaystyle CN}{|}}{C}}\text{-}N \right]_2 \ + \ n \ \text{(aromatic ring with CH}_3\text{, OCN, NCO)} \longrightarrow$$

$$(12)$$

$$\text{H}\left[\text{O}-\text{CH}_2\text{-}\overset{\overset{\displaystyle CH_3}{|}}{\underset{\underset{\displaystyle CN}{|}}{C}}\text{-}N=N\text{-}\overset{\overset{\displaystyle CH_3}{|}}{\underset{\underset{\displaystyle CN}{|}}{C}}\text{-}CH_2\text{-}O\text{-}\overset{O}{\overset{||}{C}}\text{-}\text{NH}-\text{(CH}_3\text{-aryl)}-\text{NH}-\overset{O}{\overset{||}{C}}\right]_n\text{-}\text{OH}$$

B. Formation of MAIs by Cationic Chain Polymerization—Cation Radical Transfer

Low-molecular weight azo compounds have frequently been used in cationic polymerizations producing azo-containing polymers. Thus, the combination of ionically and radically polymerizable monomers into block copolymers has been achieved. Azo compounds were used in all steps of cationic polymerization without any loss of azo function: as initiators, as monomers and, finally, as terminating agents.

With respect to the initiation of cationic chain polymerizations, the reaction of chlorine-terminated azo compounds with various silver salts has been thoroughly studied. ACPC, a compound often used in condensation type reactions discussed previously, was reacted with Ag^+X^-, X^-, being BF_4^- [10,61] or SbF_6^- [11,62]. This reaction resulted in two oxocarbenium cations, being very suitable initiating sites for cationic polymerization. Thus, poly(tetrahydrofuran) with M_n between 3×10^3 and 4×10^4 containing exactly one central azo group per molecule was synthesized [62a]. Furthermore, N-vinyl carbazol, *n*-butyl vinylether, and cyclohexene oxide have been polymerized following this procedure,

However, the latter yielded low-monomer conversions [10].

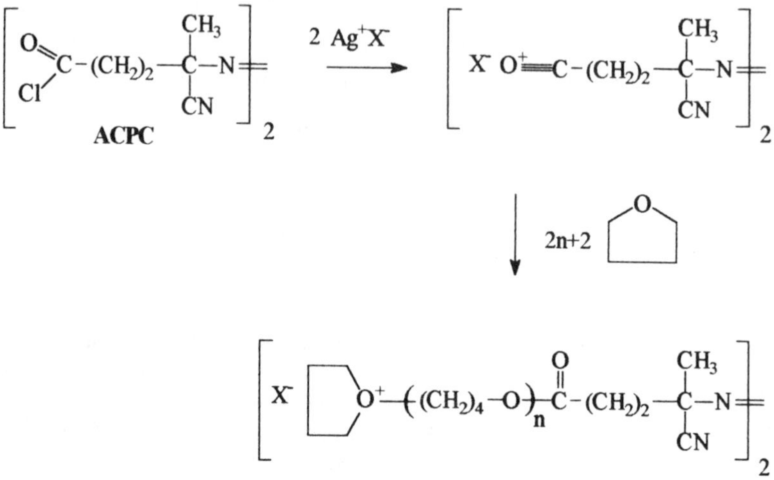

(13)

Denizligil et al. [62] terminated the cationic polymerization of tetrahydrofuran with pyridinium N-oxide derivatives (Scheme 14). Possessing both thermally labile azo sites and photolytically decomposable pyridinium ions, the resulting polymer is a bifunctional initiator being a suitable precursor for ABC triblock copolymers. The photodecomposition of pyridinium ions has to be performed at the wavelength of its maximum absorption $\lambda_{max} = 300$ nm. At this wavelength, the azo bond does not absorb light thus insuring sure that azo bonds remain unaffected in the step of pyridinium ion decomposition.

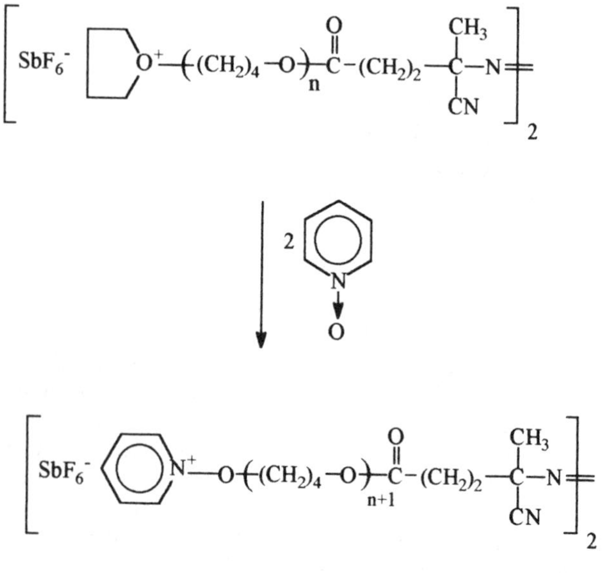

(14)

MAIs based on cyclohexene oxide, besides the previously mentioned reaction of acid chlorides with silver compounds, were prepared by a promoted cationic polymerization with high yields [63]. By means of a promoted cationic polymerization, azo-containing poly(epichlorhydrin) also was synthesized [63a]. Azo-containing poly(tetrahydrofuran) and poly(cyclohexene oxide) have been used for preparing novel liquid crystalline block copolymers [63,64,65]. The liquid crystalline blocks were obtained by polymerizing different acrylates containing substituted biphenyl mesogenes by means of azo functionalized prepolymer. Moreover, poly(tetrahydrofuran)-based azo initiators were used for the copolymerization of styrene and divinyl benzene preparing macroporous beads with good swelling properties [66].

An interesting pathway in synthesizing polymeric azo initiators is the living polymerization of azo-containing monomers. Nuyken et al. [67] have investigated a system consisting of monomer, HI and a coinitiator (tetrabutylammonium perchlorate, TBAP). Since a α,ω-divinyl ether containing one central azo function was used for reacting with the initiator in the very first polymerization step, polymeric initiators (M_n between 1.3×10^3 and 1.6×10^4) with one thermo labile function per repeating unit were obtained (see Scheme 15). If the azo-containing divinyl ether was used not only in the initiation step but also in the subsequent polymerization, insoluble crosslinked polymer would be formed due to the monomer's bifunctionality [68]. The macro-azo-initiator prepared by Nuyken et al. according to Scheme 15 was used for thermally polymerizing methyl methacrylate. It was found that the polymerization rates of the macro-azo-initiator are identical with those of AIBN.

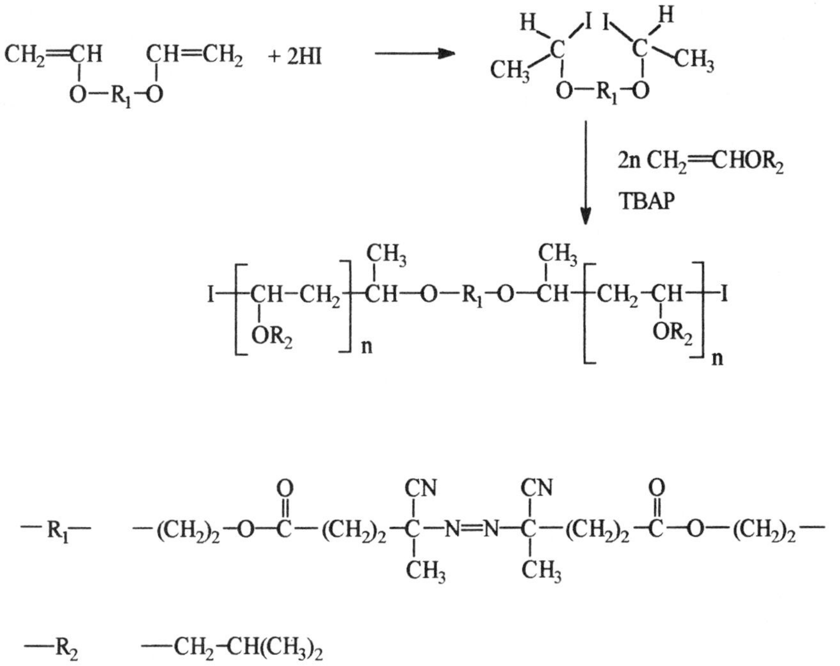

$$CH_2=CH \quad CH=CH_2 \quad + 2HI \longrightarrow \quad \underset{O-R_1-O}{\overset{\overset{\displaystyle H}{\underset{\displaystyle CH_3}{C}} \quad \overset{\displaystyle I \ I}{} \quad \overset{\displaystyle H}{\underset{\displaystyle CH_3}{C}}}{}}$$

$$\text{O—R}_1\text{—O}$$

$$\Big\downarrow \ \ 2n\ CH_2=CHOR_2$$
$$\text{TBAP}$$

$$I\!\!-\!\!\left[\underset{OR_2}{CH\!-\!CH_2}\right]\!\!-\!\!\underset{}{\overset{CH_3}{CH}}\!-\!O\!-\!R_1\!-\!O\!-\!\overset{CH_3}{CH}\!\!-\!\!\left[\underset{OR_2}{CH_2\!-\!CH}\right]_n\!\!-\!\!I \tag{15}$$

$$-R_1-\quad -(CH_2)_2-O-\overset{O}{\overset{\|}{C}}-(CH_2)_2-\overset{CN}{\underset{CH_3}{C}}-N=N-\overset{CN}{\underset{CH_3}{C}}-(CH_2)_2-\overset{O}{\overset{\|}{C}}-O-(CH_2)_2-$$

$$-R_2\quad -CH_2-CH(CH_3)_2$$

Besides being used as initiators and monomers, azo compounds may also be used for terminating a cationic polymerization. Thus, the living cationic polymerization of isobutylvinyl ether initiated by the previously mentioned system HI/coinitiator has been terminated by an azo-containing alcohol in the presence of ammonia [69].

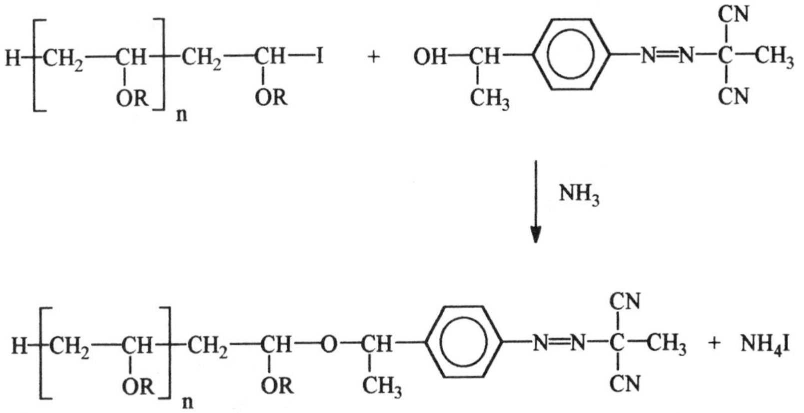

$$H\!\!-\!\!\left[\underset{OR}{CH_2\!-\!CH}\right]_n\!\!-\!\!\underset{OR}{CH_2\!-\!CH}\!-\!I \ \ + \ \ OH\!-\!\underset{CH_3}{CH}\!-\!\!\bigcirc\!\!-\!N\!=\!N\!-\!\overset{CN}{\underset{CN}{C}}\!-\!CH_3$$

$$\Big\downarrow \ \ NH_3 \tag{16}$$

$$H\!\!-\!\!\left[\underset{OR}{CH_2\!-\!CH}\right]_n\!\!-\!\!\underset{OR}{CH_2\!-\!CH}\!-\!O\!-\!\underset{CH_3}{CH}\!-\!\!\bigcirc\!\!-\!N\!=\!N\!-\!\overset{CN}{\underset{CN}{C}}\!-\!CH_3 \ \ + \ \ NH_4I$$

Another method to terminate cationic polymerizations was described by D'Haese et al. [70]. The authors reacted a living poly(tetrahydrofuran) chain with either azetidinium or thiolanium (see Scheme 17) thus stabilizing the cationic sites. In a second reaction step, the poly(tetrahydrofuran) was treated with the sodium salt of ACPA, disodium 4,4'-dicyano-4,4'-azodivalerate, yielding poly(tetrahydrofuran)-based polymeric initiators with either one azo group in the middle of the backbone or several azo groups at regular distances in the chain. Both initiator types were successfully used for the thermal synthesis of poly(styrene) and poly(methyl methacrylate) blocks.

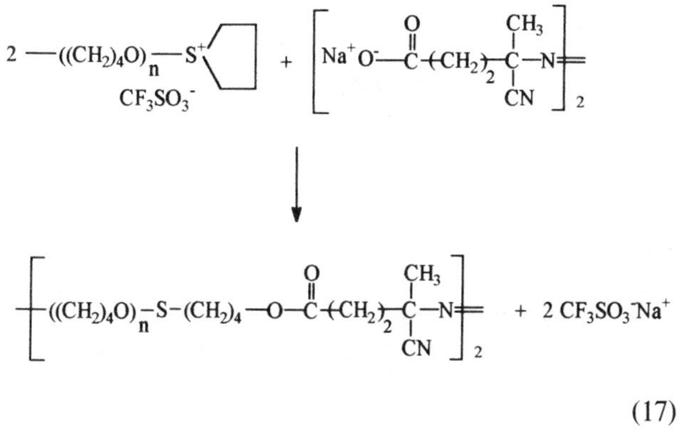

$$2\ -\!\!\underset{CF_3SO_3^-}{((CH_2)_4O)_n}\!\!-\!\overset{+}{S}\!\!\bigcirc \ \ + \ \ \left[Na^+O^-\!-\!\overset{O}{\overset{\|}{C}}\!\!-\!\!\overset{CH_3}{\underset{CN}{(CH_2)_2\!-\!C}}\!\!-\!N\!=\!\right]_2$$

$$\Big\downarrow$$

$$\left[-\!((CH_2)_4O)_n\!-\!S\!-\!(CH_2)_4\!-\!O\!-\!\overset{O}{\overset{\|}{C}}\!\!-\!\!\overset{CH_3}{\underset{CN}{(CH_2)_2\!-\!C}}\!\!-\!N\!=\right]_2 \ \ + \ \ 2\ CF_3SO_3^-Na^+ \tag{17}$$

C. Formation of MAIs By Anionic Chain Polymerization-Anion Radical Transfer

Regarding anion radical transfer, low-molecular weight azo compounds were used as terminating agents in anionic polymerizations. An interesting example is the addition of a living polystyrene chain to one nitrile group of AIBN [71]. The terminal styryl anion is likely to form a pseudocycle with the azo compound making the addition to the second nitrile group impossible (Scheme 18). Therefore, polystyrene with one terminal azo group is obtained. However, if the living poly(styrene) chain was first derivatized with α-methylstyrene or with 1,1-diphenyl ethylene, the dimer possessing a central azo group was isolated.

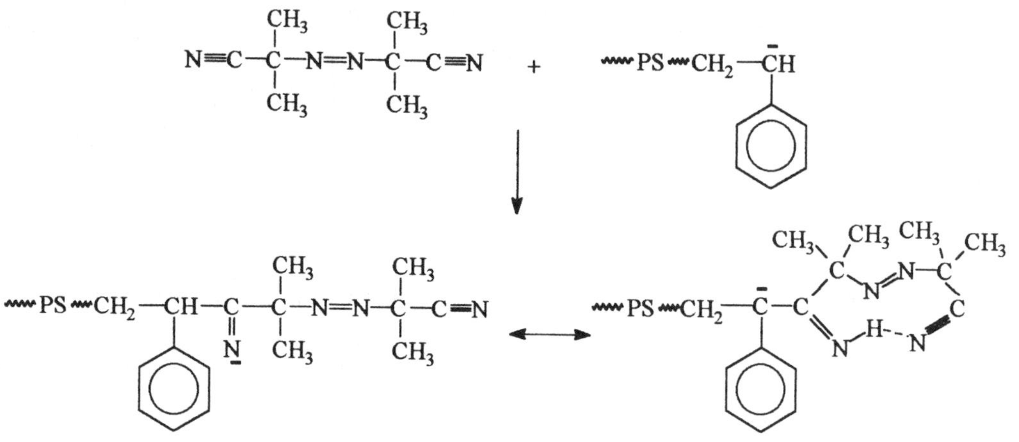

(18)

The addition of living poly(styrene) to AIBN leads finally, especially for high coupling efficiencies, to the elimination of one nitrile group [72]. More recently, Ren et al. [73] have used bis(2-chloroethyl)2,2'-azodiisobutyrate (see scheme 19) to terminate anionically initiated poly(butadiene) chains. Since the azo transfer agent possesses two functional groups (Cl) that are able to terminate anionic chains, a butadiene-based polymer (M_n between 1.2×10^3 and 3.2×10^3) with one central azo group has been obtained. This prepolymer was applied in a second stage to generate block copolymers poly(butadiene-b-methyl methacrylate). The block copolymer thus obtained consisted of a mixture of diblock, AB (A, poly(butadiene), B, poly(methyl methacrylate)), and triblock, ABA copolymer.

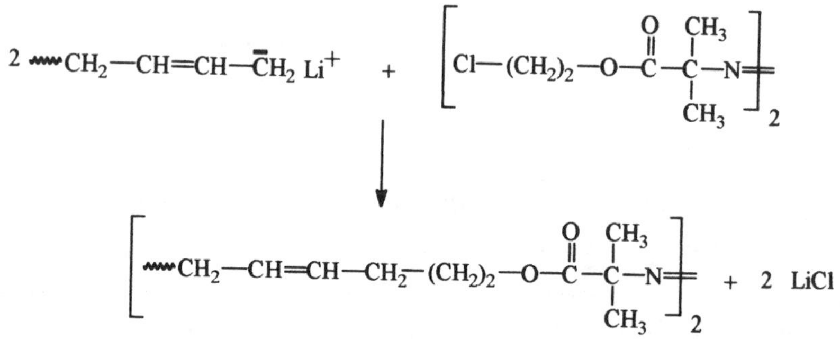

(19)

Similarly, two living polystyrene chains were terminated with the azobisacid chloride ACPC yielding poly(styrene) with one central azo group [74]. The yield of dimerization was found to be higher when the living chain was treated with 1,1-diphenyl ethylene before reacting it with ACPC.

D. Formation of MAIs by Radical Chain Reactions—Radical–Radical Transfer

As was explained, block copolymer formation by azo initiators always involves at least one radically polymerizable monomer. This is due to the fact that upon azo group decomposition radicals are formed starting chain polymerizations. However, there are numerous papers showing that a combination of two radically polymerized monomers to a block copolymer using azo initiators is also possible. In the examples referred to here, mostly initiators carrying two different radical forming sites were used. The use of polymeric azo initiators having a large number of identical azo sites in two subsequent radical polymerizations will also be discussed.

Frequently, initiators bearing simultaneously azo and peroxy groups were used as bifunctional initiators.

The activation energies of thermal decomposition was usually found to be higher for peroxy groups than for azo groups [75–77] (*vide infra*). Therefore, if azoperoxy bifunctional initiators are subjected to a thermal treatment in the presence of the first monomer, peroxy-containing polymeric initiators are formed. Peroxy macroinitiators thus formed were often used for block copolymer synthesis. This synthetic path, not involving MAIs but starting from azo-containing initiators, is discussed in detail in Section III. D. of this chapter.

However, under certain conditions peroxy functions decompose selectively prior to azo groups. If the decomposition of peroxy groups is carried out in the presence of a first monomer A, azo-containing polymer is formed. It was found that the excitation of azoperoxy compounds by UV light ($\lambda < 280$ nm) brings about a cleavage of peroxy links, leaving the azo groups intact [78]. The absorption maxima of aliphatic perester groups, usually present in azoperoxy initiators, are at wavelengths below 280 nm [79], whereas the $n \to \pi^*$ band of aliphatic azo groups occurs at c. 350 nm. Other methods to selectively promote the decomposition of peroxy groups, such as the addition of heavy metal ions or various aliphatic amines, have also been tried [78]. The latter was found to be most promising [80,81].

Moreover, free radical block copolymerization has been performed by means of low-molecular initiators containing two azo groups of different thermal reactivity. The first thermal treatment at a relatively low temperature in the presence of a monomer A results in a polymeric azo initiator. The more stable azo functions being situated at the end of A_n blocks can be subjected to a second thermal treatment at a higher temperature in the presence of monomer B.

Two low-molecular weight initiators with azo groups of different stability were used by Simionescu et al. [82,83] for polymerizations: phenylformamidoethyl 4-t-butylazo-4-cyanovalerate (Scheme 20) and N,N'-bis[(4-t-butylazo-4-cyanovaleryl)-oxoethyl]azo-bis-formamide (Scheme 21) [84–86].

$$(20)$$

$$(21)$$

In both compounds there are type (I) azo functions surrounded by alkyl groups and one cyano group. Upon heating, tertiary alkyl radicals and cyano alkyl radicals are formed. These radicals are relatively stable due to hyper conjugation and, in the case of cyano substituted alkyl radicals, to resonance. Therefore, azo groups (I) have a high proneness to thermal decomposition.

The second type azo function (II) is resonance stabilized by a phenyl ring and a carbonyl group or by two carbonyl groups in the case of Scheme 20 and Scheme 21, respectively. It was found that at temperatures below 90°C, nitrogen evolved stems almost exclusively from azo sites of type (I) [82,85]. Below this temperature, the Arrhenius plot of nitrogen evolution shows a straight line, the activation energy ($E_A = 99.6$ kJ mol^{-1} (for Scheme 20) being almost identical with the activation energy of the corresponding monofunctional azo initiator with type (I) azo functions. Hence, the polymerization of the first monomer A should be performed at tem-

peratures below 90°C. The MAI thus formed carries predominantly the thermally relatively stable azo sites (II). Obviously, care has to be taken that both B and the preformed block A_n does not undergo chemical alterations due to the high temperatures necessary.

MAIs may also be formed free radically when all azo sites are identical and have, therefore, the same reactivity. In this case the reaction with monomer A will be interrupted prior to the complete decomposition of all azo groups. So, Dicke and Heitz [49] partially decomposed poly(azoester)s in the presence of acrylamide. The reaction time was adjusted to a 37% decomposition of the azo groups. Surface active MAIs ($M_n > 10^5$) consisting of hydrophobic poly(azoester) and hydrophilic poly(acrylamide) blocks were obtained (see Scheme 22) These were used for emulsion polymerization of vinyl acetate—in the polymerization they act simultaneously as emulsifiers (surface activity) and initiators (azo groups). Thus, a ternary block copolymer was synthesized fairly elegantly.

hydrophobic   hydrophilic

$$(22)$$

Yagci and Denizligil [44] applied the method of partial decomposition of MAIs introducing styrene and methyl methacrylate blocks into poly(amide)s. The poly(amide)-based MAI had been prepared by a reaction of AIBN with formaldehyde (see Scheme 10). Evidently, since each unit of the preformed MAI carries one azo group, there are enough azo sites in every MAI molecule for a controlled and adjustable partial decomposition.

In hydroxyl-terminated azo initiators terminal free radicals may be formed in a redox reaction as was shown by Hazer et al. [87]. By means of Ce(IV), salts radicals were generated in poly(ethylene glycol)-based azo initiators (Scheme 23). In an other example [88, hydroxyl functions of a low-molecular weight azo initiator, 4,4'-azobis(4-cyano-n-pentanol) (ACPO), were, after a reaction with Ce(IV) ammonium nitrate, used to polymerize acryl amide. The resulting MAI contained one central azo function per molecule. In two recent papers, Tunca reported on the polymerization of acryl amide in aqueous medium using the redox pair Ce(IV)—methyloyl functional azo initiator [89,90]. The MAIs obtained were used for block and graft copolymerization with styrene and methyl methacrylate. Besides being thermally labile, the polymeric azo initiators were also suitable for photopolymerizations—due to the phenyl ring attached to one side of each azo group they possess a relatively high UV absorption.

The oxidizing capability of Ce(IV) has also been used for block copolymer synthesis starting from hydroxyl functional azo compounds, but not proceeding via the formation of MAIs (*vide infra*).

$$(23)$$

In a few cases, azo polymers were synthesized photochemically. Azobenzoin compounds have photo cleavable benzoin groups. Being irradiated with UV light ($\lambda = 350$ nm) 4,4'-azo-bis(4-cyanopentanoyl)-bis benzoin, ACPB, undergoes α-scission forming two free radicals per initiator molecule (Scheme 24).

ACPB

$$(24)$$

Hepuzer et al. [91] have used the photoinduced homolytical bond scission of ACPB to produce styrene-based MAIs. These compounds were in a second thermally induced polymerization transferred into styrene-methacrylate block copolymers. However, as Scheme 24 implies, benzoin radicals are formed upon photolysis. In the subsequent polymerization they will react with monomer yielding nonazofunctionalized polymer. The relatively high amount of homopolymer has to be separated from the block copolymer formed after the second, thermally induced polymerization step.

E. Formation of Block Copolymers Via MAIs

1. Thermal Decomposition of MAIs Azo Bond

Upon heating, azo compounds release nitrogen giving rise to the formation of two carbon centered radicals, provided the azo group is bonded at both sides to carbon atoms. The proneness to decomposition does, however, very much depend on the substituents. Thus, azobenzene is stable at 60°C [92], whereas azobis(triphenylmethane) has only a transient existence even at $-40°C$ [93]. AIBN, the compound mostly used for polymerizations, shows a high tendency to evolution of nitrogen and radical formation at temperatures above 60°C. The activation energy for thermolysis is 128 kJ mol^{-1} (in benzene [94]). Notably, the activation energies for macro-azo-initiators derived from AIBN or its derivatives do not show large deviations from the E_A value of AIBN. Furthermore, the decomposition rate constants are only slightly smaller than for AIBN [42]. The thermal decomposition of the MAIs derived from AIBN follows first-order kinetics.

However, when MAIs are thermolyzed in solution, the role of the cage effect has to be taken into account. The thermolytically formed macroradicals can, due to their size, diffuse only slowly apart from each other. Therefore, the number of combination events will be much higher for MAIs than for low-molecular weight AIBN derivatives. As was shown by Smith [16], the tendency toward radical combination depends significantly on the rigidity and the bulkiness of the chain. Species such as cyclohexyl or diphenylmethyl incorporated into the MAI's main chain lead to the almost quantitative combination of the radicals formed upon thermolysis. In addition, combination chain transfer reactions may also cause a lower initiation efficiency of MAIs as compared with low-molecular weight azo compounds. Studying the block copolymerization of styrene initiated by an poly(ethylene glycol)-based MAI, Furukawa et al. [52] observed that the initiator efficiency decreases by the factor of 3 and 2 due to combination and chain transfer, respectively. In other words, only as few as one radical in six reacted with the monomer present starting a chain polymerization.

2. Termination Governs Block Copolymer Structures

The mode of chain termination affects the type of block copolymer formed. For example, if a MAI (based essentially on the first monomer A) possessing one central azo bond is decomposed in the presence of monomer B, the growing chain B_n can terminate either by disproportionation or combination, leading to AB and ABA type copolymers, respectively.

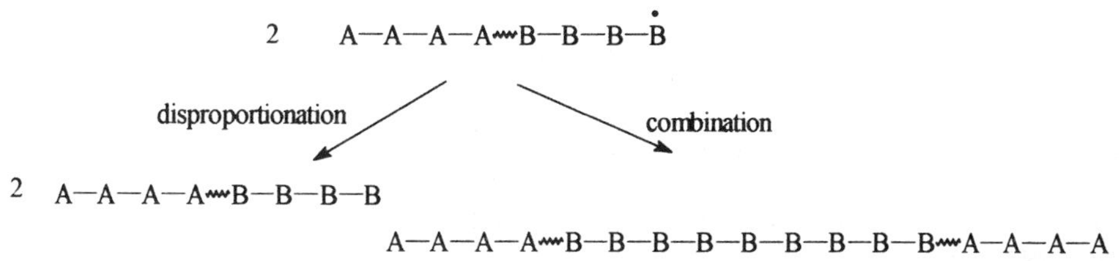

$$2 \quad A{-}A{-}A{-}A{\sim}B{-}B{-}B{-}\overset{\bullet}{B}$$

disproportionation   combination

$$2 \quad A{-}A{-}A{-}A{\sim}B{-}B{-}B{-}B$$

$$A{-}A{-}A{-}A{\sim}B{-}B{-}B{-}B{-}B{-}B{-}B{-}B{\sim}A{-}A{-}A{-}A$$

In Table 3 the types of block copolymers generated starting from different sorts of MAIs (see also Table 1) are summarized.

Which mechanism of termination will be preferably applied depends largely on the monomer used. Thus, methyl methacrylate chains terminate to a large extent by disproportionation, whereas styrene chains tend to termination by combination. The ratios of termination rate constants $\delta = k_{td}/k_{tc}$ (for disproportionation, k_{td}, combination, k_{tc}) are $\delta \approx 0$ and $\delta = 2$ for styrene [95] and methyl methacrylate [96], respectively. In the case of styrene, however, the values of δ reported in the literature are at variance. Berger and Meyerhoff [97] found $\delta = 0.2$, at 52°C. Therefore, it is possible that a fraction of styrene terminates by disproportionation.

In this context, methods of terminating radical chains by means of additives should be mentioned. An artificial termination of growing radical chains is a convenient tool for controlled synthesis of block copolymer. As is well known in polymer chemistry, radical chains may easily be terminated with thiols (see Scheme 25). Regarding the block copolymer formed, one would obtain AB blocks, provided that a growing block of monomer B is attached to a block of A and no other blocks are present. Generally speaking, the termination by additives gives exactly the same type of block copolymer as if all growing chains would terminate by disproportionation (see Table 3).

$$A{-}A{-}A{-}A{\sim}B{-}B{-}B{-}\overset{\bullet}{B} + R{-}SH \longrightarrow A{-}A{-}A{-}A{\sim}B{-}B{-}B{-}B{-}H + R{-}\overset{\bullet}{S} \quad (25)$$

However, upon terminating chains with thiols, sulphur centered low-molecular weight radicals are formed that are able to start a polymerization of the remaining monomer B. Therefore, formation of homopolymer consisting of B is inevitable if thiols are used. A suitable alternative to the classical transfer additives are degradative chain transfer agents, such as allylmalonic acid

diethylester [98]. The radical formed by abstracting one hydrogen atom from this compound is stabilized by allyl resonance (see Scheme 26) and does, therefore, not posses the reactivity necessary to start a polymerization. Consequently, no homopolymer is formed. Degradative chain termination was applied, e.g., to terminate growing styrene chains, that would, without transfer agents, predominantly undergo combination [99].

$$CH_2{=}CH{-}\overset{\bullet}{C}H{-}CH{\overset{COOCH_2CH_3}{\underset{COOCH_2CH_3}{<}}} \longleftrightarrow \overset{\bullet}{C}H_2{-}CH{=}CH{-}CH{\overset{COOCH_2CH_3}{\underset{COOCH_2CH_3}{<}}} \quad (26)$$

Table 3 Types of Block Copolymers

Type MAI	Disproportionation	Combination
One central azo bond	AB	ABA
Azo bonds in the main chain	BAB, AB	$(AB)_n$
One terminal azo bond	AB	ABA
Two terminal azo bonds	BAB	$(AB)_n$

III. BLOCK COPOLYMER SYNTHESIS WITH LOW-MOLECULAR WEIGHT AZO COMPOUNDS

As reported, much effort has been undertaken to chemically introduce azo groups to a macromolecule. However, the thermally labile azo groups of a bifunctional low-molecular weight azo initiator can also be decomposed initiating a free radical polymerization of a monomer A in the first step. Remaining with the so-formed blocks of A, the second reactive site can, after an activation, react with monomer B or preformed blocks of B (see reaction 3 in Scheme 1). Depending upon the chemical nature of the second reactive site, this reaction could potentially be either condensation or an ionic or radical chain reaction. However, no examples of block copolymer synthesis starting with the thermal decomposition of low-molecular weight azo compounds and applying subsequently anionic polymerization (radical anion transfer) have, to the best knowledge of the authors, been reported up to now.

A. Polycondensation

As pointed out in Section II. A., the AIBN derivatives ACPC and ACPA, bearing, respectively, acid chloride and acid end groups, show high reactivities in condensation reactions. When the azo group is used for a thermal polymerization of a vinyl monomer, —COCl or —COOH, terminated polymers are formed, being able to participate in condensation reactions. Thus, Ueda et al. [100] reported on the polymerization of styrene with ACPC. The resulting polymer ($M_n \approx 9 \times 10^4$) contained acid chloride groups at both chain ends. A subsequent condensation with 1,6-hexamethylene diol yielded block copolymers of the (AB)$_n$ type, n being between 3 and 4. In the case of a polycondensation with preformed poly(ethylene glycol), (AB)$_n$ blocks with even smaller n values (n $\approx$ 1.9) were observed. Recently, Van de Velde et al. [101] reported on interesting studies aimed at the synthesis of liquid crystalline block copolymers. Treating 4,4'azobis(4-cyanovaleric acid), ACPA, subsequently with diazomethane and hydrazine, 4,4'-azobis(4-cyanovalerohydrazide), ACPH, was obtained. ACPH was used for the thermal polymerization of a methyl methacrylate derivative-obtaining polymer with hydrazide end groups (Scheme 27). The methyl methacrylate monomer used carried mesogenic diphenyl units. The hydrazide-terminated methyl methacrylate-based

polymer was finally condensed with a preformed, carboxyl-terminated polystyrene thus obtaining a liquid crystalline block copolymer with mesogenic side chains.

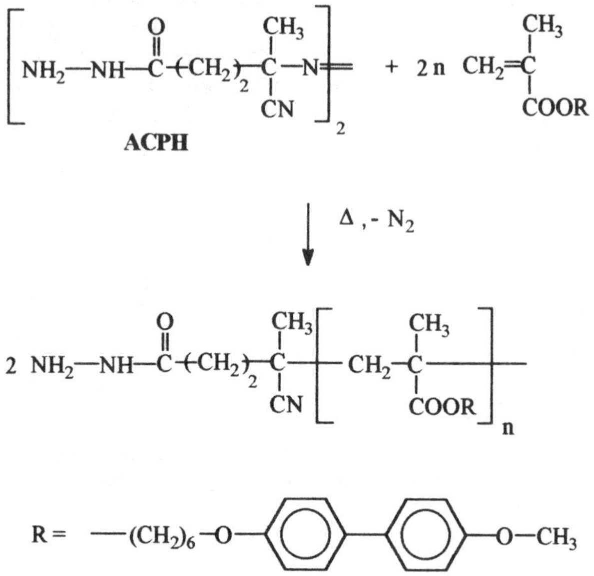

$$(27)$$

B. Cationic Techniques—Radical Cation Transfer

Sites suitable to start cationic polymerizations may be produced by reacting acid chlorides, such as ACPC, with silver salts (*vide ante,* Scheme 13). Following this procedure, azo-containing poly(tetrahydrofuran) has been prepared [10]. However, the azo group of ACPC may likewise be decomposed in the presence of a monomer in the first step, yielding polymer with terminal acid chloride functions. Thus, poly(styrene) was prepared starting from ACPC. Being terminated with acid chloride groups at both ends of the molecule, in a second step the polymer was treated with AgSbF$_6$ and used to generate poly(tetrahydrofuran) blocks [11,102]. Poly(styrene)-bearing photolabile benzoin groups was prepared starting with a benzoin-terminated azo initiator [103]. The alkoxy radicals formed upon irradiation of this polymer were, by means of pyridinium salts, oxidized to the corresponding carbocation capable of initiating the cationic polymerization of cyclohexene oxide. Thus, block copolymers of BAB type were formed, A and B being poly(styrene) and poly(cyclohexene oxide), respectively.

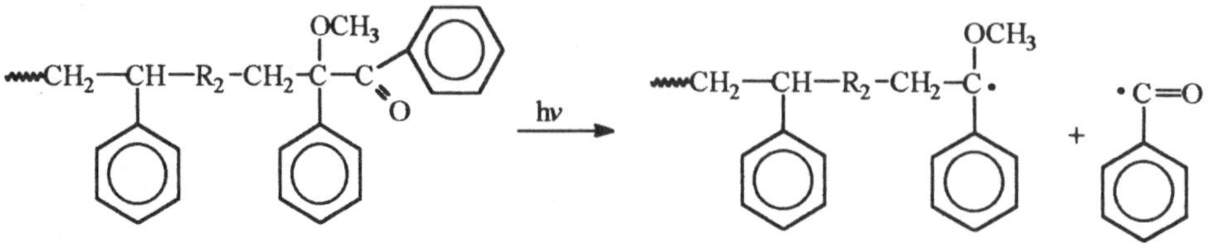

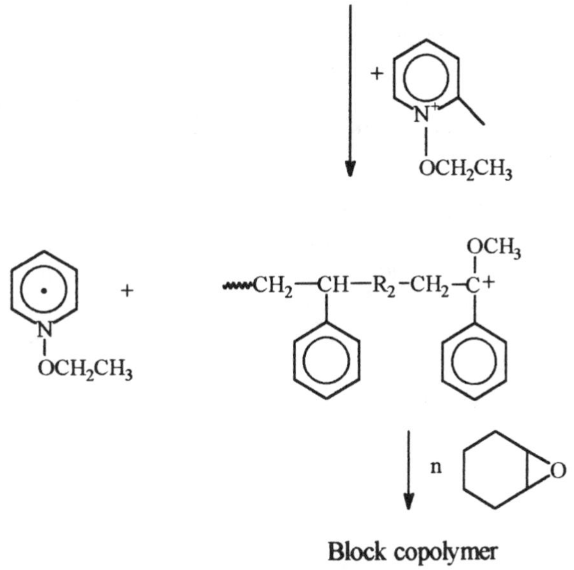

$$ \text{(28)} $$

Block copolymer

C. Radical Chain Reactions— Radical–Radical Transfer

As mentioned in Section II. D., azoperoxy bifunctional initiators play a significant role regarding block copoly-

mer synthesis. Mostly initiators of the general structure (Scheme 29) with R and n given in Table 4 were used. However, unsymmetrical low-molecular weight azoperoxy initiators have also been tried [104].

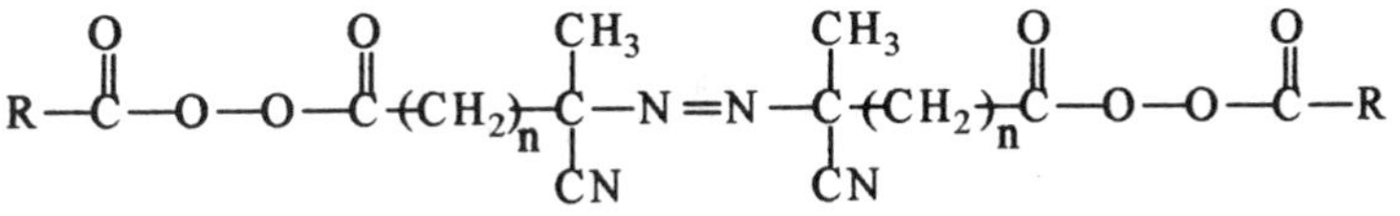

$$ \text{(29)} $$

The activation energies for the decomposition of the azo and peroxy groups compiled in Table 4 indicate that

the nature of the substituent R and also the number of CH_2-groups (n) between the azo and peroxy group af-

Table 4 Azoperoxy initiators

R	n	E_a^{per} in kJ mol^{-1}	E_a^{azo} in kJ mol^{-1}	Reference
CH_3—	2	189	155	[105]
$(CH_3)_3 C$—	2	not determined	not determined	[106]
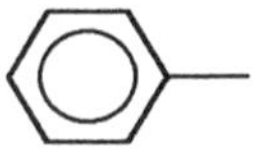	2	207	174	[76,105]
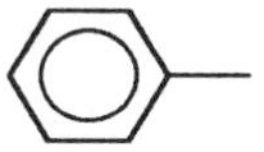	3	114	97	[76]
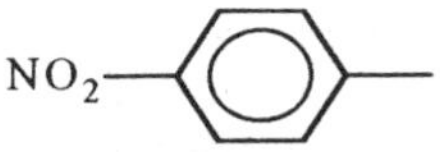 NO_2—	2	170	101	[75]
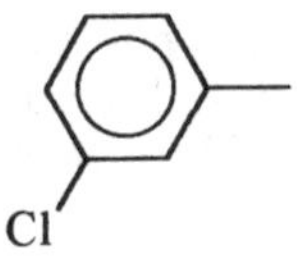	2	91	75	[75,107]

fects the thermal reactivity dramatically. In addition, it can be seen that the activation energy for decomposition of peroxy groups is higher than the corresponding E_a values for the decomposition of azo sites. Therefore, upon heating, the azo groups are going to break prior to the peroxy groups. If this bond scission is utilized for initiating a chain polymerization, the polymer will carry the more stable peroxy groups. In Scheme 30 [108], one example of a block copolymer synthesis using 4,4'-azo-bis-(4-cyanovaleryl)-benzoyl diperoxide is given.

In the example depicted in Scheme 30, the polymer-

ization of methyl methacrylate was carried out at temperatures between 60 and 80°C. Since methyl methacrylate growing chains predominantly terminate by disproportionation, polymer ($M_w \approx 5 \times 10^5$) with mostly one peroxide end group per molecule was formed. This peroxy-containing macroinitiator was subsequently used for copolymerizing styrene. In this step a reaction temperature of 95°C was applied. The resulting block copolymers are overwhelmingly of ABA type, since combination is the favored termination mechanism in the case of styrene.

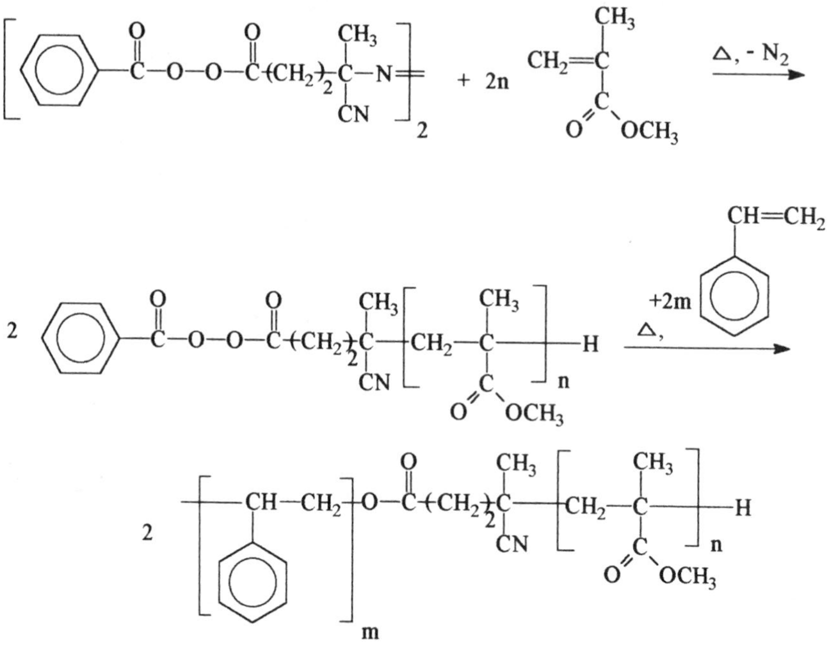

(30)

The formation of a polymeric initiator containing azo and peroxy groups has been reported by Hazer et al. [80]. In this paper, poly(ethylene glycol) (M_w $4 \times 10^2 \ldots 3 \times 10^3$) was condensed with AIBN (*vide ante*)

forming a MAI. Having OH-terminates these MAIs were functionalized with terephthaloyl chloride and subsequently treated with *t*-butyl hydroperoxide giving bifunctional initiator species (Scheme 31).

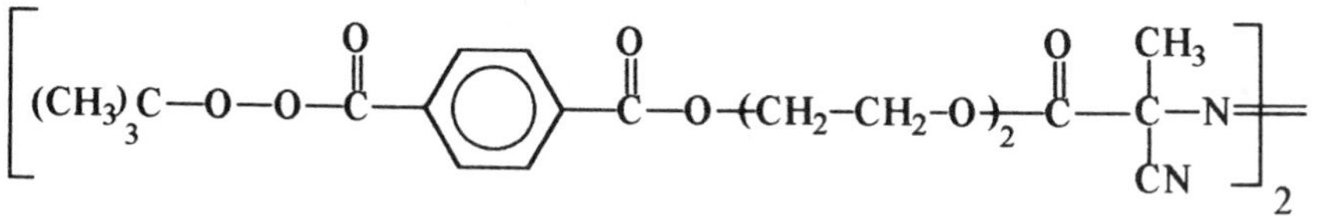

(31)

The initiator (Scheme 31) was decomposed at 60°C in the presence of styrene resulting in an ABA block copolymer, A and B being poly(ethylene glycol) and poly(styrene), respectively. In a second thermal treatment, at 80°C, the peroxy groups bound to the ABA block were decomposed in the presence of methyl methacrylate. Finally, a block copolymer of ABCBA type, C being poly(methyl methacrylate), was obtained. With respect to azo peroxy bifunctional initiators it must be noted that at a thermal treatment both groups will always undergo decomposition, although with different decom-

position rates. Therefore, the initially present peroxy groups may be partly decomposed in the first reaction step giving rise to formation of homopolymer A_n.

Moreover, block copolymers with two radically polymerizable monomers can be synthesized with a combination of thermal and photochemical polymerizations. Regarding their utilization in block copolymer synthesis, azocompounds with photoactive benzoin [103,109–111] and azyloximester groups [112] have been described. Two low-molecular weight azo benzoin initiators of the general formula (Scheme 32) were synthesized by a pyridine-catalyzed condensation of 4,4-azo-bis-(4-cyanopentanoic acid) (ACPA) with benzoin or

α-methylobenzoin methyl ether in the case of ACPB and ABME, respectively [110].

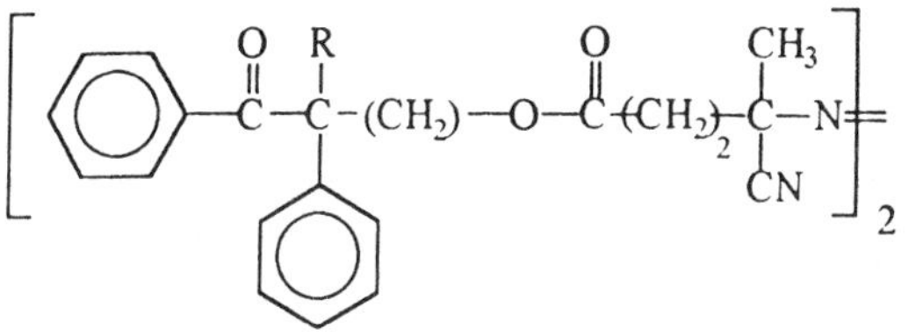

$$\left[\text{(structure)} \right]_2 \quad \begin{array}{l} \text{ACPB: R = H, n = 0} \\ \\ \text{ABME: R = -OCH}_3, \text{ n = 1} \end{array} \qquad (32)$$

As mentioned in Section II. D., benzoin groups undergo α-cleavage upon exposure to UV light (λ = 350 nm). Yagci and Önen [109] have used azobenzoin initiators for the thermal polymerization of styrene. The benzoin-terminated polystyrene was in a second reaction step photolyzed in the presence of methyl methacrylate monomer yielding block copolymers. Interestingly, the same compounds were used for synthesizing poly-(styrene-b-methyl methacrylate) block copolymers by photolyzing the benzoin groups in the first step (*vide ante*). It has to be noted that the conversion of the second monomer, methyl methacrylate, was higher for the photochemical initiation. This is most probably due to

the fact that the photolytically formed alkoxy benzyl radicals possess a higher initiation efficiency toward methyl methacrylate than cyanoalkyl radicals formed upon the decomposition of the central azo group.

Furthermore, photochemically induced homolytical bond cleavage can also be applied when the prepolymer itself does not contain suitable chromophoric groups [113–115]. Upon thermolysis of ACPA in the presence of styrene, a carboxyl-terminated polystyrene is formed. This styrene-based prepolymer was reacted with lead tetraacetate and irradiated with UV light yielding free radicals capable of initiating the polymerization of a second monomer (Scheme 33) [113].

$$Pb^{IV}(CH_3COO)_4 \;+\; R{-}COOH \;\rightleftharpoons\; R{-}COOPb^{IV}(CH_3COO)_3 \;+\; CH_3COOH$$

$$R{-}COOPb^{IV}(CH_3COO)_3 \;\xrightarrow{\;h\nu\;}\; \overset{\bullet}{R} \;+\; CO_2 \;+\; Pb^{III}(CH_3COO)_3 \qquad (33)$$

A bifunctional azo initiator containing terminal CCl_3 groups was obtained by an addition type reaction using ACPA and trichloroacetyl diisocyanate [114]. In this

case, free radicals were generated by a reaction of the CCl_3 terminates with photolytically formed $Mn(CO)_5$. The carbon centered radicals obtained were used for initiating a chain polymerization.

$$Mn_2(CO)_{10} \;\xrightarrow{\;h\nu\;}\; 2\; Mn(CO)_5$$

$$R{-}CCl_3 \;+\; Mn(CO)_5 \;\longrightarrow\; R{-}\overset{\bullet}{C}Cl_2 \;+\; Mn(CO_5)Cl \qquad (34)$$

The great advantage of reactions like Scheme 33 and 34, as compared with the direct attachment of a photolabile group to the polymer (see Scheme 24) is that in the former systems only polymer bound radicals are formed upon photolysis, whereas in the latter, additionally isolated small radicals are generated. Therefore, less homopolymer is produced in the photolytic step following reactions 33 and 34.

Generation of radicals by redox reactions has also been applied for synthesizing block copolymers. As was mentioned in Section II. D. (see Scheme 23), Ce(IV) is able to form radical sites in hydroxyl-terminated compounds. Thus, Erim et al. [116] produced a hydroxyl-terminated poly(acrylamid) by thermal polymerization using 4,4-azobis(4-cyano pentanol). The polymer formed was in a second step treated with ceric (IV) ammonium nitrate, hence generating oxygen centered radicals capable of starting a second free radical polymeriza-

tion. Block copolymers obtained following this procedure showed excellent flocculation properties as demonstrated in the example of tincal concentrate suspensions [117].

ACKNOWLEDGMENTS

The generous support of Alexander von Humboldt-Stiftung, Bonn, Germany, is gratefully acknowledged. One of the authors (I. R.) was granted a scholarship.

REFERENCES

1. P. S. Engel, D. J. Bishop, M. A. Page, *J. Am. Chem. Soc., 100*: 7009 (1978).
2. J. W. Hill (to E. I. du Pont de Nemours & Co., Inc.), U. S. Patent No. 2,556,876 (June 12, 1951), C.A. *45*: 9915 (1951).

3. O. Nuyken, *Encyclopedia of Polymer Science and Engineering*, 2nd ed., New York, Wiley & Sons, Vol. 2, p. 158 (1985).

4. G. S. Kumar, *Azo Functional Polymers—Functional Group Approach in Macromolecular Design*, Technomic Publ., Lancaster (1989).

5. A. Ueda and S. Nagai, *Nippon Setchaku Gakkaishi, 26*: 112 (1990).

6. K. Kinoshita and N. Araki, *Kobunshi-kako, 41*: 336 (1992).

7. O. Nuyken and B. Voit, *Macromolecular Design, Concept and Practice* (M. K. Mishra, ed.) Polymer Frontiers Int. Inc., New York, p. 313 (1994).

8. A. Ueda and S. Nagai, *Macromolecular Design, Concept and Practice* (M. K. Mishra, ed.) Polymer Frontiers Int. Inc., New York, p. 265 (1994).

9. M. Miyamoto, M. Sawamoto, and T. Higashimura, *Macromolecules, 18*: 123 (1985).

10. Y. Yagci, *Polym. Commun., 26*: 8 (1985).

11. Y. Yagci, *Polym. Commun., 27*: 21 (1986).

12. R. M. Haines and W. A. Waters, *J. Chem. Soc.,* 4256 (1955).

13. K. Matsukawa, A. Ueda, H. Inous, S. Nagai, *J. Polym. Sci., Part A: Polym. Chem., 28*: 2107 (1990).

14. O. Nuyken, J. Dauth, and W. Pekruhn, *Angew. Makromol. Chem., 187*: 207 (1991).

15. O. Nuyken, J. Dauth, and W. Pekruhn, *Angew. Makromol. Chem., 190*: 81 (1991).

16. D. A. Smith, *Macromol. Chem., 103*: 301 (1967).

17. A. Ueda, Y. Shiozu, Y. Hidaka, and S. Nagai, *Kobunshi Ronbunshu, 33*: 131 (1976).

18. B. Hazer, B. Erdem, and R. W. Lenz, *J. Polym. Sci., Part A: Polym. Chem., 32*: 1739 (1994).

19. D. Jayaprakash and M. J. Nanjan, *J. Polym. Sci., Polym. Chem. Ed., 20*: 1959 (1982).

20. B. Hazer, *Makromol. Chem., 193*: 1081 (1992).

21. O. S. Kabasakal, F. S. Güner, A. T. Erciyes, and Y. Yagci, *J. Coat. Techn., 67*: 47 (1995).

22. Y. Kita, A. Ueda, T. Harada, M. Tanaka, and S. Nagai, *Chem. Express, 1*: 543 (1986).

23. A. Ueda and S. Nagai, *J. Polym. Sci., Part A: Polym. Chem., 24*: 405 (1986).

24. A. Ueda and S. Nagai, *J. Polym. Sci. Polym. Chem. Ed., 25*: 3495 (1987).

25. B. Hazer, *Makromol. Chem. Phys., 196*: 1945 (1995).

26. A. Ueda and S. Nagai, *Kobushi Robunshu, 43*: 97 (1986).

27. J. M. G. Cowie and M. Yazdani-Pedram, *Brit. Polym. J., 16*: 127 (1984).

28. T. O. Ahn, J. H. Kim, J. C. Lee, H. M. Jeong, and J-Y. Park, *J. Polym. Sci., Part A: Polym. Chem., 31*: 435 (1993).

29. T. O. Ahn, J. H. Kim, H. M. Jeong, S. W. Lee, and L. S. Park, *J. Polym. Sci., Part B: Polym. Phys., 32*: 21 (1994).

30. C. I. Simionescu, E. Comanita, V. Harabagiu, and B. C. Simionescu, *Eur. Polym. J., 23*: 921 (1987).

31. C. I. Simionescu, V. Harabagiu, E. Comanita, V. Hamciuc, D. Giurgiu, and B. C. Simionescu, *Eur. Polym. J., 26*: 565 (1990).

32. H. Terada, Y. Haneda, A. Ueda, and S. Nagai, *Macromol. Reports, A32*: 173 (1994).

33. Y. Haneda, H. Terada, M. Yoshida, A. Ueda, and S. Nagai, *J. Polym. Sci., Part A: Polym. Chem., 32*: 2641 (1994).

34. G. Galli, E. Chiellini, M. Laus, M. C. Bignozzi, A. S. Angeloni, and O. Francescangeli, *Macromol. Chem. Phys., 195*: 2247 (1994).

35. A. Ueda and S. Nagai, *J. Polym. Sci., Polym. Chem. Ed., 22*: 1783 (1984).

36. A. Ueda and S. Nagai, *J. Polym. Sci., Polym. Chem. Ed., 22*: 1611 (1984).

37. Y. Yagci, Ü. Tunca, and N. Bicak, *J. Polym. Sci., Part C: Polym Lett., 24*: 491 (1986).

38. Y. Yagci, Ü. Tunca, and N. Bicak, *J. Polym. Sci., Part C: Polym. Lett., 24*: 49 (1986).

39. Ü. Tunca and Y. Yagci, *Prog. Polym. Sci., 19*: 233 (1994).

40. Ü. Tunca and Y. Yagci, *J. Polym. Sci., Part A: Chem., 28*: 1721 (1990).

41. Ü. Tunca and Y. Yagci, *Polym. Bull., 26*: 621 (1991).

42. R. Kalz, B. Bömer, and W. Heitz, *Makromol. Chem., 178*: 2527 (1977).

43. W. Heitz, H.-G. Stahl, and R. Dicke, (to Bayer AG) German, Offenl. 3,005,889 (Sep. 3, 1981), C.A. *95*: 151474v (1982).

44. Y. Yagci and S. Denizligil, *Eur. Polym. J., 27*: 1401 (1991).

45. S. Denizligil and Y. Yagci, *Polym. Bull., 22*: 547 (1989).

46. P. S. Anand, H. G. Stahl, W. Heitz, G. Weber, and L. Bottenbruch, *Makromol. Chem., 183*: 1685 (1982).

47. C. Oppenheimer and W. Heitz, *Angew. Makromol. Chem., 98*: 167 (1981).

48. R. Walz and W. Heitz, *J. Poly. Sci., Polym. Chem. Ed., 16*: 1897 (1978).

49. H. R. Dicke and W. Heitz, *Makromol. Chem., Rapid Commun., 2*: 83 (1981).

50. G. P. Vlasov, G. D. Rudkovsaya, and L. A. Ovsyannikova, *Makromol. Chem., 183*: 2635 (1982).

51. H. Yürük and S. Ulupinar, *Angew. Makromol. Chem., 213*: 197 (1993).

52. J. Furukawa, S. Takamori, and S. Yamashita, *Angew. Makromol. Chem., 1*: 92 (1967).

53. H. Yürük, A. B. Ozdemir, and B. M. Baysal, *J. Appl. Polym. Sci., 31*: 2171 (1986).

54. B. Hazer, Y. Örs, and M. H. Alma, *J. Appl. Polym. Sci., 47*: 1097 (1993).

55. B. Hazer, *Angew. Makromol. Chem., 129*: 31 (1985).

56. A. Ueda, Y. Agari, S. Nagai, N. Minamii, and T. Miyagawa, *Chem. Express, 4*: 193 (1989).

57. H. Kinoshita, M. Ooka, N. Tanaka, and T. Araki, *Kobunshi Ronbunshu, 50*: 147 (1993).

58. H. Kinoshita, N, Tanaka, and T. Araki, *Makromol. Chem., 194*: 829 (1993).

59. H. Yürük, S. Jamil, and B. M. Baysal, *Angew. Makromol. Chem., 175*: 99 (1990).

60. C. H. Bamford, A. D. Jenkins, and R. Wayne, *Trans. Faraday Soc., 56*: 932 (1960).

61. Y. Yagci, G. Hizal, A. Önen, and I. E. Serhatli, *Macromol. Symp., 84*: 127 (1994).

62. S. Denizligil, and A. Baskan, and Y. Yagci, *Macromol. Rapid Commun., 16*: 387 (1995).

62a. G. Hizal and Y. Yagci, *Polymer, 30*: 722 (1989).

63. I. E. Serhatli, G. Galli, Y. Yagci, and E. Chiellini, *Polym. Bull., 34*: 539 (1995).

63a. Y. Yagci, I. E. Serhatli, P. Kubisa, and T. Biedron, *Macromolecules, 26*: 2226 (1993).

64. E. Chiellini, G. Galli, A. S. Angeloni, M. Laus, M. C. Bignozzi, Y. Yagci, and E. I. Serhatli, *Macromol. Symp., 77*: 349 (1994).

65. E. Chiellini, G. Galli, E. I. Serhatli, Y. Yagci, M. Laus, and A. S. Angeloni, *Ferroelectrics, 148*: 311 (1993).

66. A. Akar, A. C. Aydogan, N. Talinli, and Y. Yagci, *Polym. Bull., 15*: 293 (1986).

67. O. Nuyken, H. Kröner, and S. Aechtner, *Macromol. Chem., Rapid Commun., 9*: 671 (1988).

68. B. Vollmert and H. Bolte, *Makromol. Chem., 36*: 17 (1960).

69. O. Nuyken, H. Kröner, and S. Achtner, *Macromol. Chem., Macromol. Symp., 32*: 181 (1990).

70. F. D'Haese, E. J. Goethals, Y. Tezuka, K. Imai, *Makromol. Chem., Rap. Commun., 7*: 165 (1986).

71. Y. Vinchon, R. Reeb, and G. Riess, *Eur. Polym. J., 12*: 317 (1976).

72. G. Riess, and R. Reeb, *J. Am. Chem. Soc. Symp. Ser., 166*: 477 (1981).

73. Q. Ren, H. J. Zhang, X. K. Zhang, B. T. Huang, *J. Polym. Sci., Part A: Polym. Chem., 31*: 847 (1993).

74. M. J. M. Abadie, D. Ourahmoune, and H. Mendjel, *Eur. Polym. J., 26*: 515 (1990).

75. C. Simionescu, K. G. Sik, E. Comanita, and S. Dumitriu, *Eur. Polym. J., 20*: 467 (1984).

76. A. S. Shaikh, E. Comanita, B. Sumitriu, and C. Simionescu, *Angew. Makromol. Chem., 100*: 147 (1981).

77. M. Schulz, G. West, and S. Ourk, *J. Prakt. Chem., 317*: 463 (1975).

78. M. Schulz, G. West, S. Ourk, and I. Strunz, *J. Prakt. Chem., 322*: 295 (1980).

79. P. D. Bartlett and J. M. McBride, *Pure Appl. Chem., 15*: 89 (1967).

80. B. Hazer, A. Ayas, N. Besirli, N. Saltek, and B. M. Baysal, *Makromol. Chem., 190*: 1987 (1989).

81. J. S. N. Su and I. Piirma, *J. Appl. Polym. Sci., 33*: 727 (1987).

82. C. I. Simionescu, E. Comanita, M. Pastavanu, and A. A. Popa, *Polym. Bull., 18*: 13 (1987).

83. C. I. Simionescu, A. A. Popa, E. Comanita, M. Pastravanu, and S. Dumitriu, *Eur. Polym. J., 24*: 515 (1988).

84. C. I. Simionescu, E. Comanita, M. Pastravanu, and A. A. Popa, *Bull. Polish Acad. Sci. Chem., 36*: 11 (1988).

85. C. I. Simionescu, A. A. Popa, E. Comanita, and M. Pastravanu, *Brit. Polym. J., 23*: 347 (1990).

86. C. I. Simionescu, A. A. Popa, E. Comanita, S. Manolache, and B. Comanita, *Acta Polym., 42*: 665 (1991).

87. B. Hazer, I. Cakmak, S. Denizligil, and Y. Yagci, *Angew. Makromol. Chem., 195*: 121 (1992).

88. Ü. Tunca, I. E. Serhatli, and Y. Yagci, *Polym. Bull., 22*: 483 (1989).

89. Ü. Tunca, *J. Appl. Polym. Sci., 54*: 1491 (1994).

90. Ü. Tunca, *Eur. Polym. J., 31*: 785 (1995).

91. Y. Hepuzer, M. Bektas, S. Denizligil, A. Önen, and Y. Yagci, *Macromol. Rep., A30*: 111 (1993).

92. M. T. Jaquis and M. Szwarc, *Nature (London), 170*: 312 (1952).

93. D. H. R. Barton, R. K. Haynes, G. Leclerc, P. S. Magnus, and I. D. Menzies, *J. Chem. Soc., Perkin Trans., 1*: 2055 (1975).

94. C. E. H. Bawn, and S. F. Mellish, *Trans. Faraday Soc., 47*: 1216 (1951).

95. G. C. Eastmond, *Comprehensive Chemical Kinetics* C. H. Bamford and C. F. H. Tipper, (eds.) Vol. 14a, Elsevier, Amsterdam p. 64 (1975).

96. G. C. Eastmond *Makromol. Chem., Macromol. Symp., 10/11*: 71 (1987).

97. K. C. Berger and G. Meyerhoff, *Makromol. Chem., 176*: 1983 (1975).

98. K. Ishizu, *J. Polym. Sci., Part A: Polym. Chem., 28*: 1887 (1990).

99. F. S. Erkal, A. T. Erciyes, and Y. Yagci, *J. Coat. Techn., 65*: 37 (1993).

100. A. Ueda, Y. Hidaka, and S. Nagai, *Kobunshi Ronbunshu, 36*: 123 (1979).

101. K. Van de Velde, M. Van Beylen, R. Ottenburgs, and C. Samyn, *Macromol. Chem. Phys., 196*: 679 (1995).

102. G. Hizal, H. Tasdemir, and Y. Yagci, *Polymer, 31*: 1803 (1990).

103. Y. Yagci, A. Önen, and W. Schnabel, *Macromolecules, 24*: 4602 (1991).

104. C. I. Simionescu, A. A. Popa, E. Comanita, and B. Comanita, *Polym. Plast. Technol. Eng., 31*: 451 (1992).

105. S. Dumitriu, A. S. Shaikh, E. Comanita, and C. I. Simionescu, *Eur. Polym. J., 19*: 263 (1983).

106. I. Piirma and L. H. Chou, *J. Appl. Polym. Sci., 24*: 2051 (1979).

107. K. G. Sik, S. Dumitriu, E. Comanita, and C. Simionescu, *Polym. Bull., 12*: 419 (1984).

108. A. S. Shaikh, S. Dumitriu, E. Comanita, and C. I. Simionescu, *Polym. Bull., 3*: 363 (1980).

109. A. Önen and Y. Yagci, *Angew. Makromol. Chem., 181*: 191 (1990).

110. A. Önen and Y. Yagci, *J. Macromol., Sci.-Chem., A27*: 743 (1990).

111. Y. Yagci and A. Önen, *J. Macrom. Sci., Chem., A28*: 129 (1991).

112. A. Önen, S. Denizligil, and Y. Yagci, *Angew. Makromol. Chem., 217*: 79 (1994).

113. Y. Yagci and A. Önen, *Macromol. Rep., A28*: 25 (1991).

114. Y. Yagci, M. Müller, and W. Schnabel, *Macromol. Rep., A28*: 37 (1991).

115. Y. Tagci, G. Hizal, and Ü. Tunca, *Polym Commun., 31*: 7 (1990).

116. M. Erim, A. T. Erciyes, E. I. Serhatli, and Y. Yagci, *Polym. Bull., 27*: 361 (1992).

117. A. T. Erciyes, M. Erim, B. Hazer, and Y. Yagci, *Angew. Makromol. Chem., 200*: 163 (1992).

49

Block Copolymers Derived from Macroinitiators: Recent Advances of Synthesis, Properties, and Applications

Akira Ueda
Osaka Municipal Technical Research Institute, Osaka, Japan

Susumu Nagai
Plastics Technical Society, Osaka, Japan

I. INTRODUCTION

Block copolymers have been synthesized on an industrial scale mainly by anionic or cationic polymerization, although monomers for block components are limited to ones capable of the process. Intensive academic and technological interest in radical block copolymerization using macroinitiators is growing. This process can be implemented in plants with easier handling of materials, milder conditions of operation, and a variety of materials to give various kinds of block copolymers to develop a wide application area [1–3].

A macroinitiator is a polymer or an oligomer having number of active sites for radical initiation at the ends or the middle part of the main chain or in the side chains. For the active site, peroxy group is used in oligomeric peroxy compounds or macroperoxyinitiator (MPI), while a scissile azo group is used in a macroazoinitiator (MAI), such as polyazocarbamate, polyazoester, and polyazoamide.

The principle of synthesizing block copolymers by macroinitiators is as follows:

1. Monomer A is polymerized initiated with a pair of radicals formed by thermolysis of an active site of macroinitiator. Since growing chain A propagates from the residual segment of the initiator, polymer A thus formed retains unreacted active sites in the chain end.
2. Polymer A is then decomposed in the presence

of monomer B, initiating the propagation of chain B from the chain end of A to form a block copolymer A-b-B.

In detail, the structure of a macroinitiator with active sites in the main chain is classified into two types that derive different types of block copolymers, as shown in Fig. 1.

Type I is composed of a linear combination of short repeating units containing an active site (◆) in every unit. Type II is a linear combination of oligomers or prepolymers linked with active sites. Starting from a type I macroinitiator, monomer A is first polymerized and then monomer B is polymerized, so as to give block copolymers of addition polymer segments A and B. In the case of type II macroinitiator, only one-step polymerization is needed for block copolymerization, since the oligomer or prepolymer originally linked in the macroinitiator is considered as block component A. Here, a block copolymer of a polycondensation polymer A and an additional polymer B is formed. In both cases, addition of more steps of polymerization of the third or fourth monomers will give multicomponent block copolymers, if desired.

For MAI, both types I and II have been synthesized, while type I has been the major item developed for MPI. The initiation efficiency is assumed as low as approximately 0.3 for an active site of MAI [4–6], but block efficiency is expected to be much higher, because even if a pair of radicals failed initiation, they tend to recombine themselves and then another active site of the same initi-

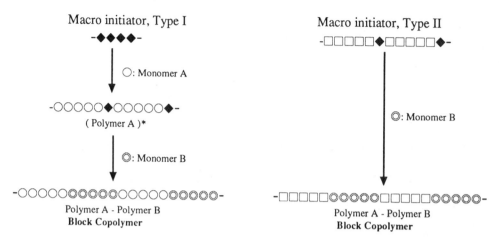

Figure 1 Principle of synthesis of block copolymer by the use of macro-initiator. ◆ = azo or peroxy group; ◯ = monomer A; ◎ = monomer B; ▢▢▢▢▢ = oligomer A.

ator takes a chance at the next initiation. Thus, it is considered that more than three active sites linked in a MAI will statistically realize a high block efficiency as 1.

For handling, MAI is safe under its decomposition temperature, but in the case of MPI, caution should be practiced against the possibility of induced explosion.

Many research papers, patents, and books have been published on this kind of block copolymerization. Recent topics will be reviewed in this chapter.

II. SYNTHESIS OF MACROINITIATORS

A. Macroazo Initiators

To incorporate a labile azo group as the essential active site to MAI, a series of azo compounds such as 2,2′ azobisisobutyronitrile (AIBN), 4,4′-azobis(4-cyanopentanoyl chloride) (ACPC), 2,2′ azobis (2-cyanopropanol) (ACPO), 2,2′ azobis [2-methyl-N-(2-hydroxyethyl)propionamide] (AHPA), etc., were used as starting materials for polycondensation with various diols, diamines, diacids, or diisocyanates.

$$\begin{array}{c} CH_3 \quad\quad CH_3 \\ | \quad\quad\quad | \\ CH_3C-N=N-CCH_3 \\ | \quad\quad\quad | \\ CN \quad\quad\quad CN \end{array}$$

AIBN

$$\begin{array}{c} O \quad\quad CH_3 \quad\quad CH_3 \quad O \\ || \quad\quad | \quad\quad\quad | \quad || \\ ClC(CH_2)_2C-N=N-C(CH_2)_2CCl \\ | \quad\quad\quad\quad | \\ CN \quad\quad\quad CN \end{array}$$

ACPC

$$\begin{array}{c} CH_3 \quad\quad CH_3 \\ | \quad\quad\quad | \\ HOCH_2C-N=N-CCH_2OH \\ | \quad\quad\quad | \\ CN \quad\quad\quad CN \end{array}$$

ACPO

$$\begin{array}{c} O \quad CH_3 \quad\quad CH_3 \; O \\ || \quad | \quad\quad\quad | \quad || \\ HO(CH_2)_2NHC-C-N=N-C-CNH(CH_2)_2OH \\ | \quad\quad\quad\quad | \\ CH_3 \quad\quad\quad CH_3 \end{array}$$

AHPA

A typical example of synthetic methods of macroazoester type I [7] or macroazoamide type II [8], respectively, is shown in Eqs. (1) or (2).

Type I

$$ACPC \;+\; HO(CH_2)_6OH \longrightarrow$$

$$\left[\begin{array}{c} O \quad\quad CH_3 \quad\quad CH_3 \quad O \\ || \quad\quad | \quad\quad\quad | \quad || \\ C(CH_2)_2C-N=N-C(CH_2)_2CO(CH_2)_6O \\ | \quad\quad\quad\quad | \\ CN \quad\quad\quad CN \end{array}\right]_n \tag{1}$$

In this case, ACPC is preferable to 4,4′ azobis(4-cyanopentanoic acid) (ACPA), since the reaction at room temperature proceeds with the labile azo group retained.

Type II

$$
ACPC \ + \ H_2NCH_2CH_2CH_2 \left[\overset{\overset{\displaystyle CH_3}{|}}{\underset{\underset{\displaystyle CH_3}{|}}{Si}} - O \right]_n \overset{\overset{\displaystyle CH_3}{|}}{\underset{\underset{\displaystyle CH_3}{|}}{Si}} - CH_2CH_2CH_2NH_2 \longrightarrow
$$

$$
\left[\overset{\overset{\displaystyle O}{\|}}{\underset{}{C}}(CH_2)_2\overset{\overset{\displaystyle CH_3}{|}}{\underset{\underset{\displaystyle CN}{|}}{C}} - N = N - \overset{\overset{\displaystyle CH_3}{|}}{\underset{\underset{\displaystyle CN}{|}}{C}}(CH_2)_2\overset{\overset{\displaystyle O}{\|}}{\underset{}{C}}NHCH_2CH_2CH_2 \left[\overset{\overset{\displaystyle CH_3}{|}}{\underset{\underset{\displaystyle CH_3}{|}}{Si}} - O \right]_n \overset{\overset{\displaystyle CH_3}{|}}{\underset{\underset{\displaystyle CH_3}{|}}{Si}} - CH_2CH_2CH_2NH \right]_m
$$

(2)

For achieving improved effective initiation for each step of block copolymerization, type I MAI having dual decomposition temperatures was developed [Eqs. (3) and (4)] [9,10].

$$
ACPC \ + \ ACPO \longrightarrow
$$

$$
\left[\overset{\overset{\displaystyle O}{\|}}{\underset{}{C}}(CH_2)_2\overset{\overset{\displaystyle CH_3}{|}}{\underset{\underset{\displaystyle CN}{|}}{C}} - N = N - \overset{\overset{\displaystyle CH_3}{|}}{\underset{\underset{\displaystyle CN}{|}}{C}}(CH_2)_2\overset{\overset{\displaystyle O}{\|}}{\underset{}{C}}OCH_2\overset{\overset{\displaystyle CH_3}{|}}{\underset{\underset{\displaystyle CN}{|}}{C}} - N = N - \overset{\overset{\displaystyle CH_3}{|}}{\underset{\underset{\displaystyle CN}{|}}{C}}CH_2O \right]_n
$$

(3)

$$
ACPC \ + \ AHPA \longrightarrow
$$

$$
\left[\overset{\overset{\displaystyle O}{\|}}{\underset{}{C}}(CH_2)_2\overset{\overset{\displaystyle CH_3}{|}}{\underset{\underset{\displaystyle CN}{|}}{C}} - N = N - \overset{\overset{\displaystyle CH_3}{|}}{\underset{\underset{\displaystyle CN}{|}}{C}}(CH_2)_2\overset{\overset{\displaystyle O}{\|}}{\underset{}{C}}O(CH_2)_2NH\overset{\overset{\displaystyle O}{\|}}{\underset{}{C}} - \overset{\overset{\displaystyle CH_3}{|}}{\underset{\underset{\displaystyle CH_3}{|}}{C}} - N = N - \overset{\overset{\displaystyle CH_3}{|}}{\underset{\underset{\displaystyle CH_3}{|}}{C}} - \overset{\overset{\displaystyle O}{\|}}{\underset{}{C}}NH(CH_2)_2O \right]_n
$$

(4)

Here, instead of an ordinary diol to react with ACPC, an azo-containing diol such as ACPO or AHPA was used to introduce two kinds of azo groups in the repeating unit. The decomposition temperature of newly incorporated azo groups were higher than that of the ACPC unit, and thus MAI shows dual decomposition temperatures so that a more labile azo group is specifically used for the first step polymerization and another one is effectively utilized for the second step polymerization. The effect of such structural modification was experimentally confirmed in the block copolymerization using methyl methacrylate (MMA), styrene (St), and vinyl acetate (VAc) as comonomers.

Another kind of dual activity was given to type I MAI by incorporating hydroxy-phenyl-propanedione-dioxime (HPO) group alternatively with ordinary azo group [Eq. (5)] [11].

$$
HO - \underset{HPO}{\overset{\displaystyle}{\bigcirc}} - \overset{\overset{\displaystyle O}{\|}}{\underset{}{C}} - \overset{\overset{\displaystyle CH_3}{|}}{\underset{}{C}} = NOH \ + \ ACPC \longrightarrow
$$

$$
\left[O - \bigcirc - \overset{\overset{\displaystyle O}{\|}}{\underset{}{C}} - \overset{\overset{\displaystyle CH_3}{|}}{\underset{}{C}} = N\overset{\overset{\displaystyle O}{\|}}{\underset{}{C}}(CH_2)_2\overset{\overset{\displaystyle CH_3}{|}}{\underset{\underset{\displaystyle CN}{|}}{C}} - N = N - \overset{\overset{\displaystyle CH_3}{|}}{\underset{\underset{\displaystyle CN}{|}}{C}}(CH_2)_2\overset{\overset{\displaystyle O}{\|}}{\underset{}{C}} \right]_n
$$

(5)

HPO group is sensitive to light, but stable to heat. Using this MAI, St was thermally polymerized at the first step, and then MMA was photopolymerized at the second step [12]. Block efficiency was 40–55% and the amount of PSt homopolymer decreased, while that of PMMA homopolymer increased, presumably due to chain transfer reaction.

Polycondensation of ACPC with triphenol gave a multibranched MAI with which a star block copolymer could be derived [13].

One of the most noticeable characteristic of MAI is the easy derivation of block copolymers involving poly-condensation polymer segments. Previously, polyurethane (PU), polyethyleneglycol (PEG), polybutadiene (PBd), polydimethylsiloxane (PDMS), polyamide, etc. were used for the members.

Recently, various polyesters such as poly(ethylene adipate), poly(tetramethylene adipate), poly(caprolactone), and poly(aliphatic carbonate), having terminal hydroxyl groups, were reacted with ACPC to give corresponding macroazoesters and their thermal behaviors were observed by DSC [14]. The block copolymers of these polycondensation polymers with addition polymers such as PSt and PMMA were synthesized [14].

The bis-amino derivative of PDMS was polycondensed directly with ACPA to give MAI by using 1,1'-dicarbonylimidazol as a condensation agent [15–18]. No formation of strong acid in this process will be acceptable for the plant.

A new polyazourethane was synthesized by the reaction of hexamethylenediisocyanate (HMDI) with diols composed of ACPO, dipropyleneglycol (DPG), and butylethylhydroxylpropanol (BEHP) [19,20]. Since the reaction between the hydroxyl group of ACPO and isocyanate group is slow, it is necessary to first react large excesses of HMDI with ACPO and then introduce DPG or BEHP to proceed polyaddition [Eq. (6)].

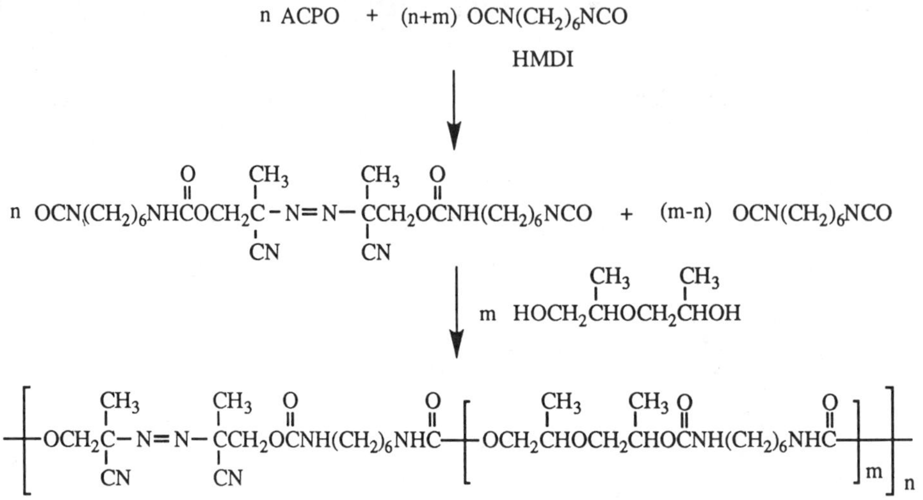

$$n \text{ ACPO} \quad + \quad (n+m) \text{ OCN(CH}_2)_6\text{NCO}$$

HMDI

$$n \text{ OCN(CH}_2)_6\text{NHCOCH}_2\overset{CH_3}{\underset{CN}{C}}{-}N{=}N{-}\overset{CH_3}{\underset{CN}{C}}\text{CH}_2\text{OCNH(CH}_2)_6\text{NCO} \quad + \quad (m{-}n) \text{ OCN(CH}_2)_6\text{NCO}$$

(6)

$$m \text{ HOCH}_2\overset{CH_3}{CHOCH_2}\overset{CH_3}{CHOH}$$

$$\left[\text{OCH}_2\overset{CH_3}{\underset{CN}{C}}{-}N{=}N{-}\overset{CH_3}{\underset{CN}{C}}\text{CH}_2\text{OCNH(CH}_2)_6\text{NHC}\left[\text{OCH}_2\overset{CH_3}{CHOCH_2}\overset{CH_3}{CHOCNH(CH}_2)_6\text{NHC}\right]_m\right]_n$$

B. Macroperoxy Initiators

For the case of MPI, Eq. (7) is a typical example of the preparative method [21].

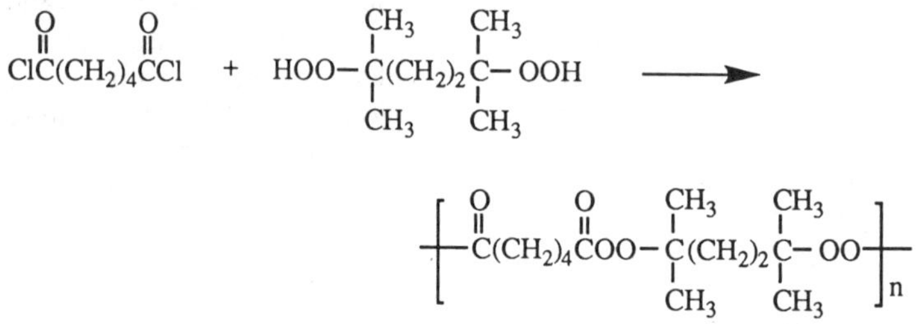

$$\text{ClC(CH}_2)_4\text{CCl} \quad + \quad \text{HOO}{-}\overset{CH_3}{\underset{CH_3}{C}}\text{(CH}_2)_2\overset{CH_3}{\underset{CH_3}{C}}{-}\text{OOH} \longrightarrow$$

(7)

$$\left[\text{C(CH}_2)_4\text{COO}{-}\overset{CH_3}{\underset{CH_3}{C}}\text{(CH}_2)_2\overset{CH_3}{\underset{CH_3}{C}}{-}\text{OO}\right]_n$$

MPI synthesized from the St–MMA–O$_2$ system was confirmed in its chemical structure as [——St—O—O—)$_x$—(—MMA—O—O—)$_y$—]$_n$— by means of NMR [22]. Alpha-methyl styrene was polymerized in the presence of pressurized oxygen to form an alternative copolymer with —O—O—. The product was used as an MPI for the block copolymerization of St and MMA [23]. However, initiation efficiency was as low as 0.02–0.04. It was explained that the radical of MPI decomposed into formaldehyde and acetophenone even in the presence of vinyl monomers [Eq. (8)].

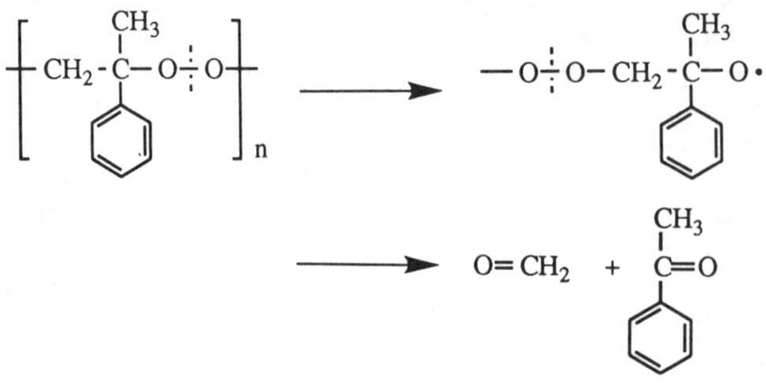

$$\left[\text{CH}_2\text{-}\overset{CH_3}{\underset{}{C}}\text{-O}{\vdots}\text{O}\right]_n \longrightarrow {-}\text{O}{\vdots}\text{O}{-}\text{CH}_2\text{-}\overset{CH_3}{\underset{}{C}}\text{-O}\cdot$$

(8)

$$\longrightarrow \text{O}{=}\text{CH}_2 \quad + \quad \overset{CH_3}{\underset{}{C}}{=}\text{O}$$

An MPI-containing peroxycarbonate group was synthesized from diethyleneglycolbis(chloroformate) re-acted with 2,5-dimethyl-2,5-dihydroperoxyhexane or 2,5-dimethyl-2,5-dihydroperoxyhexyne-3 [Eq. (9)] [24].

$$
\begin{array}{c}
\text{CH}_3 \qquad\quad \text{CH}_3 \qquad\quad\quad\; O \qquad\qquad\qquad\qquad O \\
\mid \qquad\qquad\;\; \mid \qquad\qquad\quad \parallel \qquad\qquad\qquad\qquad \parallel \\
\text{HOOC}-\text{CH}_2\text{CH}_2-\text{COOH} \;+\; \text{ClCOCH}_2\text{CH}_2\text{OCH}_2\text{CH}_2\text{OCCl} \longrightarrow \\
\mid \qquad\qquad\;\; \mid \\
\text{CH}_3 \qquad\quad \text{CH}_3
\end{array}
\tag{9}
$$

$$
\left[\!\!\begin{array}{c}
\qquad\quad \text{CH}_3 \qquad\qquad \text{CH}_3\; O \qquad\qquad\qquad\qquad O \\
\qquad\quad \mid \qquad\qquad\quad \mid \quad\; \parallel \qquad\qquad\qquad\qquad \parallel \\
\text{OOC}-\text{CH}_2\text{CH}_2-\text{COOCOCH}_2\text{CH}_2\text{OCH}_2\text{CH}_2\text{OC} \\
\qquad\quad \mid \qquad\qquad\quad \mid \\
\qquad\quad \text{CH}_3 \qquad\qquad \text{CH}_3
\end{array}\!\!\right]_n
$$

An MPI with lower active oxygen concentration was synthesized by inserting the esters of triethylenediol and tetramethylene- or heptamethylenedicarboxylic acid to the chain [25,26]. This MPI shows improved safety against shock and higher solubility to vinyl monomers.

A new ionic polymeric polycarbamate was synthesized after steps of polyurethane chemistry using 3-isocyanatemethyl-3,5,5-trimethylcyclohexyl isocyanate, 2,5-dimethyl-2,5-dihydroperoxyhexane, 1,6-butanediol, 2,4-tolylenediisocyanate, and N,N'-bis(β-Hydroxyethyl)piperazine [27]. Modification of the nitrogen of the piperazine ring into quaternary ammonium salt by treatment with methyliodide gave the MPI high electroconductivity.

$$
\begin{array}{c}
\qquad\qquad \text{CH}_3 \qquad\qquad\qquad\qquad\quad \text{CH}_3 \\
\qquad\qquad \overset{\oplus}{} \quad \text{CH}_2\text{CH}_2 \quad \overset{\oplus}{} \\
-\text{CH}_2\text{CH}_2\text{N} \qquad\qquad\qquad \text{NCH}_2\text{CH}_2- \\
\qquad\qquad \overset{\ominus}{I} \;\; \text{CH}_2\text{CH}_2 \quad I^{\ominus}
\end{array}
$$

Piperazine ring

III. BLOCK COPOLYMERIZATION— SYSTEMS, AFTERTREATMENT, AND ANALYSIS

A. Polymerization Systems

Block copolymerization is carried out by thermolysis of the macroinitiator in bulk, solution, suspension, or emulsion system. Further, it is possible to apply photolysis of azo group. In another case, an ionic active site coupled with an azo group is utilized [3].

Depending on the chemical structure of the MAI, a suitable solvent is sometimes needed to get a homogenous state of reaction mixture. Even if using the same combination of comonomers, for example, to prepare PMMA-b-poly(butyl acrylate) (PBA), the selection of the using order of comonomers for the first step or second step would affect the solvent selections, since PMMA is not easily soluble to BA monomer, while PBA is soluble to MMA monomer [28].

Suspension block copolymerization using MPI was reported elsewhere, but that of using MAI was recently reported [29]. Starting with type II MAI composed of poly(caprolactone), PBd, or PDMS, one-step suspension polymerization of St or MMA was successfully car-ried out using polyvinylalcohol as the suspension agent to give higher molecular weight block copolymers in comparison with the solution system, while it was rather difficult to obtain stable dispersion for the case of second-step polymerization with type I MAI.

Utilization of another function of the macroinitiator was tried in emulsion polymerization [30]. An MAI composed of PEG (molecular weight of a segment is 1000) linked with ACP units was confirmed to be usable as a surface active initiator (Inisurf) for preparing PSt-b-PEG [30]. A higher molecular weight block copolymer was obtained in comparison with the case of solution copolymerization.

B. Control of Block Sequences

Among the steps of initiation, propagation, chain transfer, and termination of radical polymerization, how to terminate determines the sequences of block, i.e. AB diblock, ABA triblock, or (AB)n multiblock. St, which dominates recombination, gives ABA or (AB)n type block. It also gives AB diblock, but only by a chance of chain transfer reaction. MMA, which prefers disproportionation at an ordinary temperature of thermal polymerization, mainly gives ABA or AB type block. Starting with type II MAI composed of polyisoprene or its hydrogenated one, block copolymer with PMMA was prepared, which was expected mainly of ABA type of sequences [31]. To prepare AB diblock, polyethyleneglycol monomethylether was reacted with ACPC to obtain MAI with the structure of PEG-ACP-PEG. The polymerization of MMA with the MAI resulted in diblock PEG-b-PMMA [32]. In contrast, photopolymerization initiated with a MAI at room temperature was found to give (AB)n type block copolymer of MMA system, since recombination of growing PMMA chain radicals becomes dominant at such a low temperature [33].

C. Isolation of Block Copolymer

Chain transfer reaction during propagation gives homopolymers as well as block copolymers. Separation of the homopolymers is performed by extraction with suitable solvents. Homopolymer A together with a small amount of block copolymer rich in component A are extracted at the same time, while homopolymer B is extracted together with a small amount of block copolymer rich in component B. Thus, block copolymer A-b-B is isolated as residue. In the polymerization of butyl methacrylate (BMA) initiated with type II MAI composed of PU, products were PU-b-PBMA (70–85%), PU homopolymer (10–28%), and PBMA homopolymer (0–4%) [34].

D. Analysis of Components

In turbidimetric titration using a precipitant that is added in increasing amounts, solutions of a block copolymer and that of a homopolymer blend show different patterns in their titration curves. The curve for a blend of homopolymers shows a distinct shoulder at each end point for precipitation of the component, but for a block copolymer, the changing mode is continuous. A fractional precipitation curve of polyarylate (PAR)-b-PSt was reported [35]. Fractionation by adsorption chromatography was also tried. Using a nucleosil column, PU-b-PSt was separated from PSt homopolymer by a gradient solvent from cyclohexane to tetrahydrofuran methanol, suggesting the usefulness of chromatography for the system [19].

E. Block Efficiency

In the polymerization of St initiated with type II MAI composed of polyvinylpyrrolidone (PVP), block efficiency was kept to 80% when feed concentration was above 3 mol/L, but it drastically decreased below 3 mol/L (Fig. 2) [36,37]. AIBN, the typical low-molecular weight azo initiator, shows a drastic decrease in its initiation efficiency below a critical feed monomer concentration, i.e., 0.5 mol/L. In the case of MAI, it seems that a similar decrease in initiation efficiency occurs at much higher critical monomer concentration due to immobility of macroinitiating radicals.

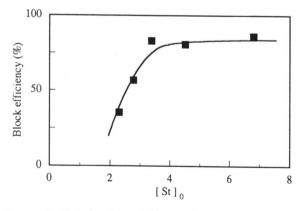

Figure 2 Relationship of block efficiency versus $[St]_0$ [36].

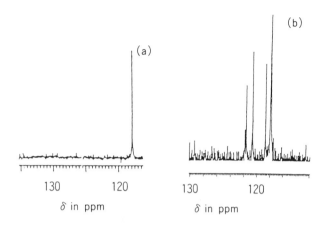

Figure 3 ^{13}C-nuclear magnetic resonance spectra of polyurethane type II macroazo initiator before (a) and after (b) decomposition of azo group [20].

In the emulsion polymerization of vinyl chloride (VC) initiated with type II MAI composed of PDMS, block efficiency was as high as 95–97%, and solid particles with a narrow range of particle size, 0.1–50 microns, were obtained in high yield [15].

F. Analysis of Active Sites

Quantitative analysis of the peroxy group of macroinitiators is performed by iodometry [38] and that of the azo group is done by ultraviolet (UV) spectrometry. Recently, type II MAI composed of PU was determined of its azo concentration by UV [20]. When the UV absorption spectral peak of the azo group overlaps other peaks, DSC is available by determining the azo group from the exothermal peak area [11].

The activation energy of thermolysis of the azo group was measured by DSC [14]. Type II MAIs, which are composed of various prepolymers such as aliphatic polyester, poly(caprolactone), and aliphatic poly (carbonate), showed almost the same activation energy irrespective of difference in prepolymer structure, suggesting that the neighboring group only affects the active site.

^{13}C of nitrile group neighboring to the azo group was detected by ^{13}C-NMR spectrum at 119 ppm, which was found to shift to 121–122 ppm after decomposition of the azo group (Fig. 3) [20]. During the preparation of MAI, the azo group was observed to retain above 95%.

IV. PROPERTIES AND APPLICATIONS OF BLOCK COPOLYMERS

Different polymers that are incompatible with each other do not form a stable homogeneous phase. In the block copolymer, however, several incompatible components are linked to each other by chemical bonds, and the bulk of the copolymers form a stable structure with two phases separated, showing the character or function of each component independently. By using macroinitiators, more than two kinds of components can be linked together to give multicomponent block copolymers, which are expected to exert multifunctions. The potential applications of block copolymers in relation to each function of the block component are summarized in Table 1. Various block copolymers have been synthesized for many applications. As for newly developed block copolymers, their potential uses are described in the following sections.

A. Compatibilizers

Polyarylate (PAR)-b-PSt and PAR-b-PMMA for compatibilizers are described [35,39,40]. The addition of PAR-b-PSt (1–10 parts) to 100 parts of a blend of PAR–PSt (7w–3w) resulted in improvement of the tensile and flexural modulus (Fig. 4), and PSt dispersed particles were diminished from 1–5 microns to an order that is undetectable by SEM, indicating the excellent, compatibilizing effect of the block copolymer. The alloy thus formed exert the characteristic of PAR, an engineering plastic, as well as easy processability of PSt. Addition of PAR-b-PMMA (3 or 8 parts) to 100 parts of a blend of PAR–polyvinylidenefluoride (PVDF) (7w–3w) resulted in improved microdispersed state of PVDF due to compatibility of PMMA with PVDF, while segregation of PVDF onto the surface was controlled.

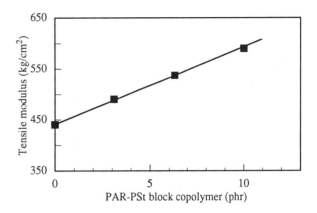

Figure 4 Tensile modulus of polyarylate–polystyrene (70:30) blend added with polyarylate-b-polystyrene (0–10) [39].

B. Impact Modifiers, Flexible Materials

PMMA-b-PBA shows improved izod impact strength compared to PMMA homopolymer [41]. Polyisobutylene (PIB) or its hydrogenated one (PIB-H) also acts as an impact modifier [31]. PSt-b-PIB, PSt-b-PIB-H, and PMMA-b-PIB-H derived from MAI have high- and wide-range molecular weight and show high flexibility and flow property [42]. The improved flexibility of PMMA-b-PEG synthesized as an elastomer, was confirmed by dynamic viscoelastic measurement [43].

Table 1 Relationship of Segments and Functions

	SEGMENT A — function A	SEGMENT B — function B	SEGMENT C — function C
Application	Function A	Function B	Function C
Compatibilizer	Solubility with matrix A	Solubility with matrix B	
Impact modifier	Dispersivity in matrix	Shock absorbance (flexibility)	
Thermoplastic elastomer	Elasticity	Physical crosslinking	
Coating	Weatherability, hardness, adhesiveness	Pigment dispersive	
Surface modifier to hydrophobic	Dispersive in matrix no immigration	Hydrophobicity surface localization	
Surface modifier to hydrophilic	Dispersive in matrix no immigration	Hydrophilicity	
Antistatic agent	Dispersive in matrix no immigration	Hydrophilicity, antistaticity	Surface localization, modification into hydrophobic, surface

C. Coatings

Flexibility of PU unit is also utilized. PU-b-(MMA-co-methacrylic acid) gives a transparent, flexible and impact-modified coating film [44], and PU-b-(ethylacrylate-co-dimethylaminoethylmethacrylate) gives a translucent and flexible one [44]. They are converted to coating materials that are water soluble and fire proof by changing to ammonium salt or acetate form. PU-b-(acrylate copolymer) has excellent dispersability on magnetic recording mediums, forming smooth surfaces, with durability and resistance to abrasives. It is applicable for coating materials for magnetic tapes, cards, and discs [45]. PU-b-(acrylate copolymer) or polyester-b-(acrylate copolymer) in which hydroxymethylmethacrylate is used as a component of acrylate copolymer can be reacted with methacryl isocyanate or glycidyl methacrylate to give a curable coating material with excellent appearance and low shrinkage [46].

PDMS-b-(acrylate copolymer) was synthesized for utilizing surface water repellency of PDMS segment [47]. It gives coating film with excellent adhesiveness onto copper plate and slate, pigment dispersability and water repellency. Incorporation of methacrylic acid to the acrylate copolymer unit results in giving a stable microdispersion for anionic electrodeposition, after neutralization with triethylamine followed by addition of water [48]. Hard, nonglaze adherent coating film with resistance to chemicals and high weatherability was obtained. Incorporation of dimethylaminoethyl methacrylate to the acrylate copolymr unit results in giving a dispersion for cationic electrodeposition for weatherable and antistain coatings, after neutralization with acetic acid, followed by addition of water [49]. PDMS-b-PMMA was synthesized initiated with MAI composed of PDMS via radical polymerization [50]. Using a mixture of solvents, n-hexane–isopropanol (3:1), which was selected to keep the SP value = 7.77, monodisperse polymer particles with a 240-nm diameter composed of a PMMA core–PDMS shell were obtained.

D. Surface Modifiers to Give Hydrophobicity

PDMS macromonomer was used as a component of block segment to obtain a graft block copolymer with PMMA (Scheme 1) [51–53]. This graft block copolymer is characteristic of surface water repellence, easy peeling, and weatherability superior to simple graft copolymers of the same members. PDMS-b-PVC film also shows long life surface water repellency with weatherability and very low coefficiency of abrasion [18,54].

E. Surface Modifiers to Give Hydrophilicity or Antistaticity

PSt or PMMA, respectively, was coupled with polymethacrylate having a PEG side chain or methylammoniumchloride side chain to prepare a block copolymer for giving a hydrophilic surface [55]. Also, PSt-b-PVP [36,37], PSt-b-(hydrophilic vinyl copolymer) [56], PSt-b-poly(sodium acrylate) (PNaA) [57], and PSt-b-PNaA-b-(polyperfluoroacrylate) (PFA) [58] were synthesized for the same application.

Another application of hydrophilic block copolymers is antistatics. Block copolymers having a PEG side chain or ammonium salt give antistaticity to PSt simply by blending. In the case of PSt-b-PVP film, however, antistatic function was demonstrated only after treating with KSCN [59]. PSt was given antistaticity abruptly when blended with PSt-b-PNaA over 15% (Fig. 5), where it was considered that the microdispersed state of the block copolymer in PSt matrix changed from spheres to continuous rods that exert antistaticity. The wet surface state, which was given by blending the above block copolymer, was improved by modifying the binary block copolymer to a ternary one by incorporating a PFA segment, which tends to localize onto the surface.

PSt-b-poly (N-acetyliminoethylene) (PAIE) and PMMA-b-PAIE were synthesized from a water-soluble

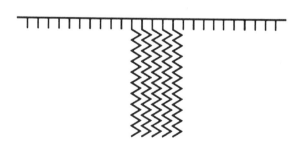

Scheme 1 Schematic structure of graft block copolymer [51]; ⊤⊤⊤⊤⊤ = polymethyl methacrylate segment; ∿∿∿ = polydimethylsiloxane segment.

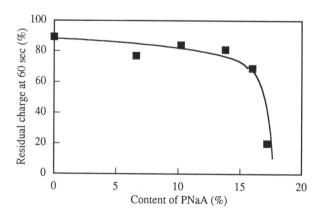

Figure 5 Corona discharge behavior of polystyrene added with polystyrene-b-poly(sodium acrylate) [57].

MAI composed of PAIE for application to emulsifiers, drug carriers, and ion-exchange resins [60,61]. Poly(perfluoropropyleneglycol)-b-poly(acrylic acid) was synthesized initiated with an MPI having fluoroalkyl group in the main chain [62]. It was soluble to water, ethanol, and THF, and showed water repellency, oil repellency, antifouling, and resistance to chemicals.

F. Other Applications

Poly(hydroxyphenyl maleimide)-b-PBA was added to thermosetting phenol resin to improve heat resistance [63]. PVC blended with poly(vinyl copolymer having cyclohexyl maleimide group)-b-PVC showed improved heat resistance and tensile strength with thermal stability during processing [64].

To prepare an interpenetrating polymer network (IPN) structure, PU networks having ACPA units were immersed with MMA and polymerized. PU–PMMA semi-IPN thus formed was given improved interfacial strength between PU and PMMA phases and showed flexibility with enforced tear strength [65,66].

A block copolymer composed of liquid crystalline polymer (LCP) segments or that composed of segments having an LCP unit in their main chain or side chain was synthesized [67,68]. The latter showed partial compatibility and second-phase separation even when in a melt liquid crystalline state.

An MAI having a cyclodiborazane group was synthesized by the reaction of thexyl borane, adipoyldinitrile, and AIBN. It was block copolymerized with St [69].

$$(CH_3)_2CHC(CH_3)_2 \diagdown \quad \diagup H$$
$$- CH = N \diagup^{\diagup B \diagdown}_{\diagdown B \diagup} N = CH -$$
$$H \diagup \quad \diagdown C(CH_3)_2CH(CH_3)_2$$

Cyclodiborazane group

In addition to a block copolymer, a microcapsule was made from suspension interfacial polycondensation between diacid chloride having aromatic–aliphatic azo group and aliphatic triamine [70,71]. The capsule was covered with a crosslinked structure having an azo group that was thermally stable but sensitive to light so as to be applicable to color photoprinting materials.

An MAI composed of polycaprolactam was investigated on its thermal decomposition behavior to develop easy decomposable polymers, the addition of oxidative deterioration was needed [72].

V. CONCLUSIONS

The utilization of macroinitiators for producing block copolymers provides the following advantages:

1. Easier handling of materials and milder operating conditions of the radical polymerization process than living ionic polymerization, although copolymers show polydispersity in molecular weight as well as in compositions.

2. Selection of optimum conditions makes block efficiency high enough to be practically acceptable.

3. Realizing a wide range of selection of composing members including vinyl polymers, polycondensation polymers, and polyaddition polymers, which opens a variety of application areas in the polymer manufacturing and processing industries.

REFERENCES

1. O. Nuyken, *Encyclopedia of Polymer Science and Engineering,* (Ed. J. K. Kroschwitz, et al.), John Wiley and Sons, New York, Vol. 2, p. 158 (1985).
2. O. Nuyken, and R. Weidner, *Advanced Polymer Science, Chromatographies-Forms-Copolymers,* (Ed. H. J. Cantow), Springer Verlag, Berlin, Vol. 73–74, p. 145 (1986).
3. A. Ueda, and S. Nagai, *Macromolecular Design, Concept and Practice,* (Ed. M.K. Mishra), Polymer Frontiers International, New York p. 265 (1994).
4. M. H. George and J. R. Ward, *J. Polym. Sci. Polym. Chem. Ed., 11*: 2909 (1973).
5. R. Walz and W. Heitz, *J. Polym. Sci. Polym. Chem. Ed., 16*: 1807 (1978).
6. Y. Oshibe and T. Yamamoto, *Koubunshi Ronbunshu, 44*: 73 (1987).
7. A. Ueda and S. Nagai, *Koubunshi Ronbunshu, 44*: 469 (1987).
8. H. Inoue, A. Ueda, and S. Nagai, *J. Polym. Sci. Part A Polym. Chem., 26*: 1077 (1988).
9. H. Takahashi, S. Nagai, and A. Ueda, *J. Polym. Sci. Part A Polym. Chem., 53*: 69 (1997).
10. S. Nagai, K. Takemoto, H. Takahashi, M. Yoshida, and A. Ueda, *IPST 94 IUPAC International Symposium Functional and High Performance Polymer,* p. 641 (Taipei, 1994).
11. A. Onen, S. Denizligil, and Y. Yagci, *Angew. Makromol. Chem., 217*: 79 (1994).
12. T. Imamoglu, A. Onen, and Y. Yagci, *Angew. Makromol. Chem., 224*: 145 (1995).
13. K. Yanagawa, S. Nagai, M. Shimada, and A. Ueda, *Preprints of 32nd Joint Meeting, Kyushu Sections of Chemistry Related Societies, Japan,* p. 127 (Fukuoka, 1995).
14. Y. Haneda, H. Terada, M. Yoshida, A. Ueda, and S. Nagai, *J. Polym. Sci. Part A Polym. Chem., 32*: 2641 (1994).
15. Y. Suenaga, and Y. Sugiura, *Japan Kokai Tokkyo Koho,* H7-2914 (1995).
16. Y. Suenaga, H. Kitagawa and Y. Sugiura, *Japan Kokai Tokkyo Koho,* H7-2915 (1995).
17. Y. Suenaga, H. Kitagawa, and Y. Sugiura, *Japan Kokai Tokkyo Koho,* H7-18139 (1995).
18. Y. Miyaki, Y. Sugiura, and Y. Suenaga, *Japan Kokai Tokkyo Koho,* H7-102210 (1995).
19. H. Kinoshita, M. Ooka, N. Tanaka, and T. Araki, *Kobunshi Ronbunshu, 50*: 147 (1993).
20. H. Kinoshita, N. Tanaka, and T. Araki, *Makromol. Chem., 194*: 829 (1993).
21. B. Hazer, *J. Polym. Sci. Part A Polym. Chem., 25*: 3349 (1987).

22. S. Jayanthi, and K. Kishore, *Macromolecules, 26*: 1985 (1993).
23. K. S. Murthy, K. Kishore, and V. K. Mohan, *Macromolecules, 27*: 7109 (1994).
24. J. Meijer, and P. J. T. Alferink, *PCT Int. Appl. WO, 94*: 03525 (1994).
25. T. Yamamoto, H. Ohmura, Y. Moriya, N. Suzuki, Y. Oshibe, M. Sugihara, *Nippon Kagaku Kaishi, 1992*: 1269 (1992).
26. S. Suyama, T. Watanabe, and H. Ishigaki, *Japan Kokai Tokkyo Kho,* H5-155849 (1993).
27. Y. Yagci, H. Yildirim, E. Altun, and B. M. Baysal, *J. Appl. Polym. Sci, 44*: 367 (1992).
28. A. Ueda and S. Nagai, *Kagaku To Kogyo, Osaka, 64*: 446 (1990).
29. N. Yokoyama, S. Nagai, and A. Ueda, *Preprints of 32nd Joint Meeting, Kyushu Sections of Chemistry Related Societies, Japan,* p. 128 (Fukuoka, 1995).
30. T. Nagamune, S. Nagai, and A. Ueda, *J. Appl. Polym. Sci., 62*: 359 (1996).
31. H. Terada, Y. Haneda, A. Ueda, and S. Nagai, *Macromol. Reports, A31*: 173 (1994).
32. S. Nagai, H. Terada, and A. Ueda, *Chem. Express, 8*: 157 (1993).
33. D. Fukuda, A. Ueda, and S. Nagai, *Preprints of Kobe Regional Conference on Polymers,* p. 17 (Kobe, 1995).
34. H. Kinoshita, N. Tanaka, and T. Araki, *Makromol. Chem., 194*: 2335 (1993).
35. T. O. Ahn, J. H. Kim, J. C. Lee, H. M. Jeong, and J. Y. Park, *J. Polym. Sci, Part A Polym. Chem., 31*: 435 (1993).
36. A. Ueda, Y. Agari, K. Kosai, N. Nishioka, and S. Nagai, *Chem. Express, 7*: 725 (1992).
37. A. Ueda, M. Shimada, Y. Agari, and S. Nagai, *Kobunshi Ronbunshu, 51*: 453 (1994).
38. H. Yildrim, A. Yilmazturk, B. M. Baysal, and Y. Yagci, *Polym. J., 21*: 253 (1989).
39. T. O. Ahn, J. H. Kim, H. M. Jeong, S. W. Lee, and L. S. Park, *J. Polym. Sci, Part B Polym. Physics., 32*: 21 (1994).
40. T. O. Ahn, J. C. Lee, H. M. Jeong, and K. W. Cho, *Eur. Polym. J., 30*: 353 (1994).
41. A. Ueda and S. Nagai, *Kagaku To Kogyo, Osaka, 66*: 385 (1992).
42. S. Suyama, T. Nakamura, M. Takei, and H. Nagai, *Japan Kokai Tokkyo Koho,* H6-9715 (1994).
43. H. Yuruk and S. Ulupinar, *Angew. Makromol. Chem., 213*: 197 (1993).
44. M. Fujii, *Japan Kokai Tokkyo Koho,* H4-89808 (1992).
45. H. Kinoshita, G. Iwamura, M. Ooka, *Japan Kokai Tokkyo Koho,* H4-209665 (1992).
46. H. Kinoshita, G. Iwamura, M. Ooka, *Japan Kokai Tokkyo Koho,* H4-279616 (1992).
47. T. Noguchi, T. Mise, H. Yoshikawa, H. Inoue, and A. Ueda, *Japan Kakai Tokkyo Koho,* H4-372675 (1992).
48. H. Yoshikawa, T. Mise, and T. Noguchi, *Japan Kokai Tokkyo Koho,* H5-179178 (1993).
49. H. Dezima, T. Noguchi, and T. Mise, *Japan Kokai Tokkyo Koho,* H7-179795 (1995).
50. K. Nakamura, K. Fujimoto, H. Kawaguchi, *Polymer Preprints, Japan, 44*: 1186 (1995).
51. H. Inoue, K. Matsukawa, A. Ueda, and K. Matsumoto, *J. Adhesion Society Japan, 30*: 146 (1994).
52. H. Inoue, K. Matsukawa, A. Ueda, and K. Matsumoto, *Kagaku To Kogyo, Osaka, 68*: 447 (1994).
53. T. Noguchi, T. Mise, H. Inoue, and A. Ueda, *Japan Kokai Tokkyo Koho,* H6-16756 (1994).
54. H. Murakami, S. Hirai, Y. Shikoku, A. Ueda, H. Inoue, and S. Nagai, *Japan Kokai Tokkyo Koho,* H4-39315 (1992).
55. H. Ohmura, Y. Oshibe, M. Nakayama, and T. Yamamoto, *Japan Kokai Tokkyo Koho,* H5-41668 (1993).
56. A. Ueda, Y. Agari, and S. Nagai, *Kagaku To Kogyo, Osaka, 68*: 98 (1994).
57. A. Ueda, M. Shimada, Y. Agari, and S. Nagai, *Polymer Preprints, Japan, 43*: 1820 (1994).
58. A. Ueda, M. Shimada, Y. Agari, and S. Nagai, *Polymer Preprints, Japan, 44*: 127 (1995).
59. A. Ueda, Y. Agari, K. Hayashi, and S. Nagai, *Kagaku To Kogyo, Osaka, 69*: 364 (1995).
60. C. I. Simionescu, G. David, A. Ioanid, V. Paraschiv, G. Riess, and B. C. Simionescu, *J. Polym. Sci. Part A Polym. Sci., 32*: 3123 (1994).
61. G. David, V. Bulacovschi, O. Ciochina, and B. C. Simionescu, *J. Macromol. Sci. Pure Appl. Chem., A32(8 & 9)*: 1649 (1995).
62. H. Sawada, H. Ishigaki, and S. Suyama, *Japan Kokai Tokkyo Koho,* H7-138362 (1995).
63. A. Matsumoto, A. Ueda, K. Hasegawa, and A. Fukuda, *J. Adhesion Society Japan, 29*: 504 (1993).
64. K. Kato, K. Toiuti, and M. Yoshida, *Japan Kokai Tokkyo Koho,* H4-332713 (1992).
65. M. Roha and B. Wang, *J. Appl. Polym. Sci., 45*: 1367 (1992).
66. M. Roha and F. Dong, *J. Appl. Polym. Sci., 45*: 1383 (1992).
67. G. Galli and E. Chiellini, *Macromol. Chem. Phys., 195*: 2247 (1994).
68. G. Galli, E. Chiellini, M. Laus, A. S. Angeloni, and M. C. Bignozzi, *Mol. Cryst. Liq. Cryst. Sci. Technol. Sect. A, 254*: 429 (1994).
69. P. Fritz and Y. Yagci, *Macromol. Rep., A32 (Suppl. 1 & 2)*: 75 (1995).
70. O. Nuyken, J. Dauth, and J. Stebani, *Angew. Makromol. Chem., 207*: 65 (1993).
71. O. Nuyken, J. Dauth, and J. Stebani, *Angew. Makromol. Chem., 207*: 81 (1993).
72. Y. Shimura and D. Chen, *Macromolecules, 26*: 5004 (1993).

50
Poloxamers

Iain F. Paterson, Babur Z. Chowdhry, and Stephen A. Leharne
University of Greenwich, London, England

I. INTRODUCTION

The term poloxamer is widely used to describe a series of ABA block copolymers of polyethylene oxide and polypropylene oxide, extensively used in industry as antifoams, emulsifiers, wetting agents, rinse aids, and in numerous other applications [1–5]. Poloxamers are amphiphilic in character, being comprised of a central polypropylene oxide (PO) block, which is hydrophobic, sandwiched between two hydrophilic polyethylene oxide (EO) blocks as shown below:

$$CH_3$$
$$HO(CH_2 CH_2O)_m(CHCH_2O)_n(CH_2 CH_2O)_mH$$

For the central PO block to serve as an effective hydrophobe, the value of n must be at least 15; the value of m in commercially manufactured poloxamers is such that the EO blocks constitute between 10–80% of the total polymer mass. The absolute and relative masses of the hydrophilic and hydrophobic blocks, on which the physico-chemical properties of the polymers depend, can be controlled during manufacture, enabling the production of poloxamers tailored to specific applications.

A. Nomenclature

The poloxamers manufactured by BASF (Cheadle, UK) are known as Pluronic PE block copolymers; those manufactured by ICI (Cleveland, UK) as Synperonic PE nonionic surfactants.

In the most commonly used system of nomenclature, each polymer is represented by a letter, which iden-

tifies its physical state (F = flake, P = paste, L = liquid), and a series of two or three numbers, the last of which, when multiplied by 10, gives the approximate weight percentage of EO in the polymer. The remaining numbers represent the mass of the PO block on an arbitrary scale.

In an alternative nomenclature, each poloxamer is represented by the letter P followed by three numbers. The third number, when multiplied by 10, again gives the approximate weight percentage of EO in the polymer, while the first two numbers, when multiplied by 100, give the approximate molecular mass of the PO block. A grid illustrating both systems of nomenclature is given in Fig. 1.

II. SYNTHESIS

The hydrophobic central segment is first synthesized by sequential addition of propylene oxide to a propylene glycol initiator:

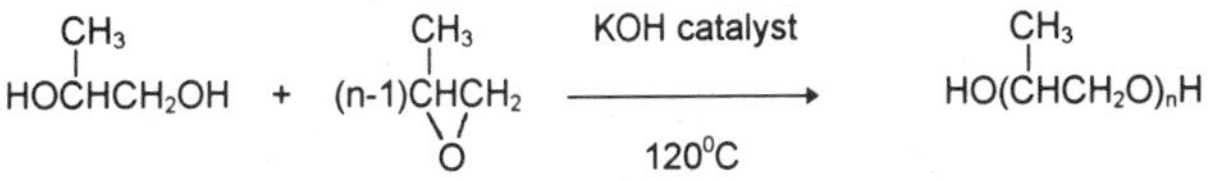

$$CH_3 \qquad\qquad CH_3 \qquad \xrightarrow[120^{\circ}C]{KOH\ catalyst} \qquad CH_3$$
$$HOCHCH_2OH \ + \ (n\text{-}1)CHCH_2 \qquad\qquad\qquad HO(CHCH_2O)_nH$$
$$O$$

The hydrophilic end segments are then formed by the addition of ethylene oxide:

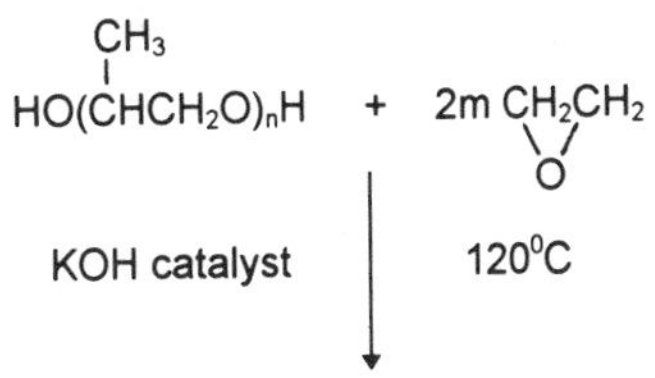

$$CH_3$$
$$HO(CHCH_2O)_nH \ + \ 2m\ CH_2CH_2$$
$$O$$

KOH catalyst $\qquad$ 120°C

765

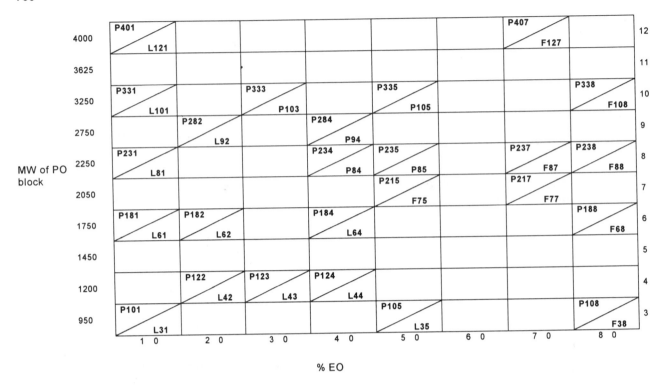

Figure 1 The poloxamer grid.

$$
\begin{array}{c}
CH_3 \\
| \\
HO(CH_2CH_2O)_m(CHCH_2O)_n(CH_2CH_2O)_mH
\end{array}
$$

The initiator usually constitutes less than 1% of the final product, and since starting the process with such a small amount of material in the reaction vessel may be difficult, it is often reacted with propylene oxide to produce a precursor compound, which may be stored until required [6]. The yield of poloxamer is essentially stoichiometric; the lengths of the PO and EO blocks are determined by the amount of epoxide fed into the reactor at each stage. Upon completion of the reaction, the mixture is cooled and the alkaline catalyst neutralized. The neutral salt may then be removed or allowed to remain in the product, in which case it is present at a level of 0.5–1.0%. The catalyst may, alternatively, be removed by adsorption on acidic clays or with ion exchangers [7]. Exact maintenance of temperature, pressure, agitation speed, and other parameters are required if the products are to be reproducible, thus poloxamers from different suppliers may exhibit some difference in properties.

A. Impurities

As propylene oxide is introduced into the reactor, a portion of it is converted to allyl alcohol and propenyl alcohol via a rearrangement [5]:

These unsaturated alcohols act as monofunctional initiators, giving rise to terminally unsaturated PO–EO diblock impurities, which may be quantified by determining the degree of unsaturation in the final product.

Condensation of ethylene oxide with any water present in the reactor (arising from, for example, the catalyst, initiator, epoxides, or leaks in the reactor cooling system) produces polyethylene glycols. The other main impurities are propionaldehyde, acetaldehyde, and their connate acids, which are capable of causing acceleration of peroxidation, depolymerization, and acidification in flaked products [8]. Pretreatment with $NaBH_4$ or addition of butylated hydroxytoluene (BHT) as an antioxidant is reported to eliminate the problem. Other materials that may be present are residual monomers and low-molecular weight by products such as 1,4 dioxane (from ethylene oxide) and allyl alcohol and dioxolanes (from propylene oxide) [9]. Any volatile impurities present may be removed by heating under vacuum or blowing an inert gas or steam through the product at elevated temperatures.

III. ANALYSIS

It is important to recognize that the following analytical methods essentially determine EO–PO ratio (¹H NMR, IR, cleavage methods) or even simply alkylene oxide content (compleximetric methods) of the analyte, and as such are not specific quantitative or qualitative methods for poloxamers, since EO–PO copolymers of a different structure (for instance, random copolymers, or PO-EO-PO block copolymers) may respond to the methods in a way indistinguishable from poloxamers. The principal technique that permits definitive identification of a sample as a poloxamer is ¹³C NMR, which allows structural details, such as the distribution of EO and PO units along the polymer chain, to be elucidated [10].

A. Spectroscopic Methods

¹H NMR is the method of choice for determining the ratio of EO units to PO units since the analysis is simple to perform, does not require calibration, and is applicable over a wide range of EO–PO ratios. The spectrum contains only two resonances: a doublet due to the methylene groups of the PO units, and a composite band, which is due to the CH$_2$O groups of the PO and EO units, the CHO of the PO units, and the terminal hydroxl groups. The hydroxyl resonance may be moved to a low field by the addition of trifluoroacetic acid, and comparison of the areas of the remaining signals yields the EO–PO ratio [11].

IR analysis can also be used quantitatively to determine the EO–PO ratio [12]. Using mixtures of polyethylene glycol and polypropyene glycol as calibration standards, the ratio of two absorbances, one due to the methyl group of the PO unit (e.g., the C-H stretch band at 2975 cm^{-1}) and one due to the methylene group (e.g., the C-H stretch band at 2870 cm^{-1}), are plotted against percent of PO content. The ratio of the same two absorbances taken from the IR spectrum of a poloxamer may then be used to determine its percent of PO content by interpolation.

B. Cleavage Methods

The EO–PO ratio may also be determined by methods in which the polymer chains are cleaved by chemical methods. The EO and PO segments each yield different products that may be separated and quantified by gas chromatography [9,13]. Reagents used to effect this cleavage and the principal products obtained are given in Table 1 [14–17].

Table 1 Cleavage Methods for the Analysis of Poloxamers

Reagent	Principal cleavage products		Comments
	From EO	From PO	
Hydrogen bromide	Dibromoethane	Dibromopropane	Yield of dibromopropane is less than yield of dibromoethane, so an appropriate calibration procedure is required [13].
Hydriodic acid	Diiodoethane	Diiodopropane	In an alternative method the diiodoalkanes are allowed to decompose, yielding iodine which may be titrated with sodium thiosulfate to obtain total alkylene oxide content [14]. Yield of diiodopropane lower than that of diiodoethane [13].
Phosphoric acid	Acetaldehyde	Propionaldehyde	Involves pyrolysis at 500°C. Good accuracy is obtained if the method is calibrated using mixtures of polyethylene glycol and polypropylene glycol of molecular weight >600 [15]. Products of cleavage with phosphoric acid undergo a characteristic color change [13].
Acetyl chloride	2-Chloro-1-methylethyl acetate, 2-chloropropyl acetate	2-Chloroethyl acetate	For EO content <30%, the ratio of chloroethyl acetate to total chloroethyl and chloropropyl acetates may be used to obtain percent of EO. A calibration is required for values of EO >30% [16].
Anhydrides of acetic acid and paratoluene sulfonic acid	Ethylene glycol diacetate	Propylene glycol diacetate	Values for the EO–PO ratio in good agreement with NMR data can be obtained [17].

C. Compleximetric Methods

Poloxamers are used primarily in aqueous solution and may be quantified in the aqueous phase by the use of compleximetric methods. However, a major limitation is that these techniques are essentially only capable of quantifying alkylene oxide groups and are by no means selective for poloxamers. The basis of these methods is the formation of a complex between a metal ion and the oxygen atoms that form the ether linkages. Reaction of this complex with an anion leads to the formation of a salt that, after precipitation or extraction, may be used for quantitation. A method reported to be rapid, simple, and consistently reproducible [18] involves a two-phase titration, which eliminates interferences from anionic surfactants. The poloxamer is complexed with potassium ions in an alkaline aqueous solution and extracted into dichloromethane as an ion pair with the titrant, tetrakis (4-fluorophenyl) borate. The end point is defined by a color change resulting from the complexation of the indicator, Victoria Blue B, with excess titrant. The Wickbold [19] method, widely used to determine nonionic surfactants, has been applied to poloxamer type surfactants [20]. Essentially the method involves the formation in the presence of barium ions of a complex between alkylene oxide groups and iodobismuthate ions, and determination of the bismuth content of the resulting precipitate by potentiometric titration.

D. Separation and Purification Methods

Nonionic surfactants, including EO–PO block copolymers, may be readily separated from anionic surfactants by a simple batch ion exchange method [21]; analytical separation of EO–PO copolymers from other nonionic surfactants is possible by thin-layer chromatography (TLC) [22,23] and paper chromatography [24], and EO–PO copolymers may themselves be separated into narrow molecular weight fractions on a preparative scale by gel permeation chromatography (GPC) [25].

Purification of poloxamers has been extensively investigated due to their use in medical applications, the intention often being to remove potentially toxic components. Supercritical fluid fractionation and liquid fractionation have been used successfully to remove low-molecular weight impurities and antioxidants from poloxamers. Gel filtration, high-performance liquid chromatography (HPLC), and ultrafiltration through membranes are among the other techniques examined [5].

Table 2 Physical Properties of the Poloxamers

		L31	L35	F38	L42	L43	L44	L61	L62	L64	F68	P75
Color[a]	Hazen (max.)	100	100	20[b]	100	100	100	100	100	100	20	—
Water content[d]	% w/v (max)	0.4	0.4	0.75	0.4	0.4	0.4	0.4	0.4	0.4	0.75	0.4
Cloud point[e]	(10% aq.)	29	80	(>100)	28	33	71	17	24	59	(>100)	0.4
Hydroxyl value[f]	mg KOH/g	102.0	59.3	24.0	67.0	60.0	51.0	53.0	46.8	39.1	13.5	87
pH[g]	(2.5% aqueous solution)	6.3	6.3	6.3	6.3	6.3	6.3	6.3	6.3	6.3	6.3	27.0
Molecular weight	(approx.)	1100	1900	4800	1650	1900	2200	2100	2400	2900	8350	6.3
Density	g/cm³	1.02	1.05	1.06	1.03	1.04	1.05	1.02	1.03	1.04	1.06	4150
Pour point/ Melting point	°C	−30	12	48	−18	−1	16	−30	−9	16	55	1.06
Kinematic Viscosity[h] (mm²s⁻¹)	at 20°C	170	355	n/a	280	360	435	360	420	720	n/a	27
	at 40°C	85	158	n/a	129	185	235	156	196	246	n/a	453
	at 100°C	14	27	595	21.5	31	39.2	25.6	32	41	544	72
Surface Tension at 20°C	mN/m (0.1% w/v)	44.2	38.6	41.6	40.0	37.5	34.5	36.6	39.6	40.0	43.8	34.2
Ross-Miles foam heights (mm) at 25°C (0.1% w/v)	initial	7	83.5	90	34	55	76	5	30	84	90	68
	after 5 min.	0	47	13	3	10	40	0	5	57	77	23
Cotton wetting power (s)	(0.1% w/v at 25°C Draves)	>360	>360	>360	>360	>360	>360	Ins.	>360	>360	>360	>360
Solubility guide at 25°C[i]	Water	S	S	S	S	S	S	I	S	S	S	S
	Ethanol	S	S	S	S	S	S	S	S	S	S	S
	Kerosene	P	I	I	I	I	I	P	I	I	I	I
	Ethylene glycol	I	P	I	I	I	I	I	I	P	I	I
	Toluene	S	S	P	S	S	S	S	S	S	P	P

[a] Measured according to ISO 2211.
[b] Measured on 10% w/w aqueous solution.
[c] Measured on 25% w/w aqueous solution.
[d] Measured according to ISO 4317.
[e] Measured according to ISO 1065.
[f] Measured according to ISO 4327.
[g] Measured according to ISO 4316.
[h] Measured according to ASTM D445.
[i] S-soluble; P-partially soluble; I-insoluble.

IV. PROPERTIES [26]

The physico-chemical properties of poloxamers, and thus their areas of application, are closely related to the relative and absolute sizes of the hydrophobic PO and hydrophilic EO blocks. However, the hydrophilic/lipophilic balance (HLB) scale, used widely to characterize the hydrophobicity or hydrophilicity of surfactants, and especially to predict their emulsifying behavior, is not considered to be useful in describing poloxamers, whose PO hydrophobes exhibit different lipophilic behavior from the hydrocarbon hydrophobes of the nonionic surfactants for which the HLB system was originally developed. Instead, the cloud point, the temperature at which a given concentration of poloxamer will "cloud out" of solution, may be used as a guide to hydrophobicity. The cloud point increases with increasing percent of hydrophile; as a rule, those poloxamers having low cloud points (10% or 20% EO) do not foam well and have the highest wetting capability, while those having high cloud points (70% or 80% EO) are best suited to stabilizing colloids of hydrophobic materials in aqueous solution.

A summary of the physical properties of poloxamers follows, with the emphasis on those properties most relevant to commercial applications. An extensive review (277 references) [4] provides a wealth of specific examples from the academic literature. Physical properties data is given in Table 2.

A. Foaming

The cloud point has a significant bearing on the foaming properties of poloxamers. Below the cloud point, foam production remains fairly constant. As the temperature is raised above the cloud point, foam generation drops sharply. Poloxamers are considered low-to-moderate foam-forming surfactants. At room temperature, as the EO content and hydrophobe molecular weight are increased, the foaming ability of the polymers increases to a maximum and then decreases. Poloxamers having a medium-to-high EO content exhibit maximum foam formation, though poloxamers are rarely used for their foam forming capability and are far more important as antifoams. The foaming properties of the poloxamers have been outlined in detail elsewhere [27].

B. Antifoaming

Poloxamers suppress the formation of foam by forming an insoluble monolayer, thus the antifoaming action of poloxamers depends on the cloud point, above which the polymer becomes insoluble. For the poloxamer to

F77	L81	P84	P85	F87	F88	L92	P94	L101	P103	P105	F108	L121	F127
20c	100	—	—	30c	20c	100	—	100	—	—	20c	100	20c
0.75	0.4	0.4	0.4	0.8	0.75	0.4	0.4	0.4	0.4	0.4	0.75	0.4	0.75
(>100)	16	71	86	(>100)	(>100)	16	73	11	51	90	(>100)	10	(>100)
17.0	41.0	25.0	23.5	14.6	9.8	32.1	25.0	29.5	22.7	17.3	7.7	25.5	10.0
6.3	6.3	6.3	6.3	6.3	6.3	6.3	6.3	6.3	6.3	6.3	6.3	6.3	6.3
6600	2750	4200	4650	7700	11800	3450	4600	3800	4950	6500	14000	4400	12000
1.04	1.02	1.04	1.04	1.04	1.05	1.03	1.04	1.02	1.04	1.05	1.06	1.02	1.05
50	−27	18	29	52	58	7	27	−9	30	30	59	−6	56
480	n/a	n/a	n/a	n/a	740	n/a	800	n/a	n/a	n/a	1600	n/a	
n/a	227	537	620	n/a	n/a	306	684	376.4	580	1315	n/a	413	n/a
234.5	36.7	86.9	99	254	1178	50.5	107	59.8	93.3	177.6	191.33	66.9	1032
42.0	35.0	36.8	37.4	40.7	39.0	33.5	33.5	32.4	31.2	34.0	38.0	32.5	40.8
69	9	56	76	87	66	16	78	17	98	83	67	24	65
47	4	23	66	66	60	4	72	9	95	78	60	17	59
>360	Ins.	260	>360	>360	>360	86	337	Ins.	24	213	>360	Ins.	>360
S	I	S	S	S	S	S	S	I	S	S	S	I	S
S	S	S	S	S	S	S	S	S	S	S	S	S	S
I	P	I	I	I	I	I	I	I	I	I	I	I	I
I	I	I	I	I	I	I	I	I	I	I	I	I	I
S	S	S	S	S	P	S	S	S	S	S	P	S	S

Source: Adapted from Ref. 1. Information in Table 2 is reproduced from technical literature by permission of ICI Surfactants subject to the following disclaimer: "The information and recommendations in this publication are believed to be accurate and are given in good faith, but the Customer should satisfy itself of the suitability of the contents for a particular purpose. ICI gives no warranty as to the fitness of the Product information and recommendations for any particular purpose and any implied warranty or condition (statutory or otherwise) is excluded except that such exclusion is prevented by law. Freedom under Patent, Copyright and Designs cannot be assumed."

be an effective antifoam, its cloud point at the concentration at which it used must be lower than the ambient temperature. Those poloxamers found to be most effective as antifoams contain 10% EO and have a low cloud point.

C. Wetting

A wetting agent may be defined as any substance that increases the ability of water to displace air from a liquid or solid surface. In practical terms, the ability of a surfactant to wet a particular surface may be evaluated by determining the spreading coefficient from surface and interfacial tension data, or, where information regarding the capacity of the surfactant to wet textiles is required, by using the Draves wetting test, in which the time taken for a skein of weighted waxed cotton yarn to completely sink in a cylinder of surfactant solution is measured. Optimum wetting ability is found in the high-molecular weight, low EO content poloxamers.

D. Emulsification

An emulsion is a suspension of one liquid in a second, immiscible, liquid. Emulsifiers are agents that facilitate the formation of emulsions and play a role in stabilizing the emulsion so formed. There are two main types of emulsions: oil in water and water in oil. Predominantly hydrophilic emulsifiers (for example, P188) will stabilize the former; predominantly hydrophobic emulsifiers (for example, P181) will stabilize the latter.

E. Dispersion

A dispersion consists of solid particles distributed in a liquid continuous phase. The function of a dispersant is to stabilize the dispersion by reducing the tendency of the particles to agglomerate. Nonionic surfactants, including poloxamers, achieve this by a process known as steric stabilization [28]. The PO segment adsorbs onto the surface of the dispersed particle, while the hydrophilic segments extend into the solution and form hydrogen bonds with the water molecules. The dispersion is stabilized by the surfactant physically keeping apart the dispersed particles and limiting the extent of the attractive van der Waals forces operating between them. Poloxamers with a medium-to-high EO content are reported to provide optimum dispersion, the precise application determining the actual poloxamer selected.

F. Detergency

Detergency may be defined as the removal of dirt from solid surfaces by surface chemical means [29], and may be related to several surfactant properties, including wetting and rewetting ability, foam generation, and surface and interfacial tension. It has long been observed that there is an optimum hydrophile–hydrophobe ratio for maximum detergency [30]: the poloxamers found to be most effective as detergents are found around the center of the poloxamer grid.

G. Gel Formation

When particles flocculate to form a continuous network structure, which extends throughout the available volume and immobilizes the dispersion medium, the resulting semi-solid system is called a gel [29]. Poloxamer gels are thought to be formed by the aggregation of micelles [31] and stabilized by hydrogen bonding between the water molecules of the aqueous dispersion medium and the ether oxygens of the polyalkylene oxide chains. P407 is the most effective gelling agent of the poloxamers, gelling at a concentration of 20% by weight in water at 25°C. Many poloxamers having a PO molecular weight of 1750–2750 form gels in water in the concentration range 40–80%, most having a PO molecular weight of 3250–4000 gel in the concentration range 30–90% [4]. The minimum poloxamer concentration required for gel formation is dependent on temperature and may be affected by other compounds added to the formulation. Gel strength for a given concentration of poloxamer may be increased by the addition of additives containing hydroxyl groups [26].

H. Solubilization

In common with other surfactants, poloxamers may be used to increase the apparent aqueous solubility of hydrophobic compounds through micelle formation. Micelles are aggregates of surfactant molecules orientated in such a way that the hydrophobic portions form a central core capable of incorporating hydrophobic solutes and the hydrophilic portions form an outer shell providing aqueous solubility. Aqueous solutions of poloxamers have been shown to substantially enhance the aqueous solubility of organic compounds [32–34], and their use to remove pollutants form water sources and effluent streams has been proposed [34]. Poloxamers having a high-molecular weight PO block and a low EO content have been shown to be most effective in solubilizing some hydrophobic compounds [34].

I. Aggregation Behavior

The aggregation behavior of poloxamers has been the subject of numerous studies, using techniques including dynamic and static light scattering, gel permeation chromatography, small-angle neutron scattering, high-sensitivity scanning calorimetry, and surface tension depression. Unusually, plots of surface tension against the logarithm of surfactant concentration for these surfactants often show two distinct changes in slope, or "breaks." The significance of this has been a matter of

Table 3 Techniques Used to Characterize the Aggregation Behavior of Poloxamers

	Abbreviation	References
Scattering techniques		
Laser light scattering	LLS	36,37
Static light scattering	SLS	36–47
Dynamic light scattering	DLS	39,40,45,45a,48–53
Rayleigh Brillouin scattering		49
Small angle x-ray scattering	SAXS	52,54
Small angle neutron scattering	SANS	45,45a,48,55–58
Other light-related techniques		
Light transmittance		59
Transient electric birefringence	TEB	37,45,45a
Fluorescence quenching and excimer formation		41,50,52,60,
Dye solubilization		61
Spectroscopic techniques		
IR and Raman spectroscopy		43
NMR		36,39,45,45a,50,59,60,62–66
Other techniques		
Differential scanning calorimetry	DSC	39,44,65,67–69
Viscosity and rheological measurements		36,41–43,45,45a,51–53,70,71
Sedimentation		53
Ultrasound velocimetry		43,44,53,54
Gel permeation chromatography	GPC	72
Adsorption by ellipsometry		73,74
Vapor pressure osmometry	VPO	36,71
Surface tension		38,45,45a,52,75–77
Vapor sorption		71
Microscopy		42,70,78

Source: Adapted from Ref. 31.

some debate. It has been proposed that the low-concentration break in the surface tension curve indicates the formation of monomolecular aggregates or that the existence of two breaks is due to the broad molecular distribution of the compounds. Recent work using high-sensitivity differential scanning calorimetry would appear to suggest that the low-concentration break occurs upon the formation of aggregates of 2–3 molecules, aggregates that coalesce to form larger micelles at higher concentrations, thereby producing the second break. Despite being the subject of extensive study using a broad range of techniques, the aggregation properties of poloxamers are still far from fully understood. Two recent comprehensive reviews have detailed the current state of knowledge in this field [31,35]. A guide to sources of information regarding the aggregation behavior of poloxamers is given in Table 3 [36–78].

V. APPLICATIONS

Poloxamers are most important commercially as antifoams, wetting agents, and emulsifiers, but have also been found to have numerous potential applications in the medical field.

A. Industrial Applications

In general terms, the commercial applications of poloxamers may be summarized as follows: poloxamers containing 10% EO are used as antifoams, poloxamers containing 20% EO are used as wetting agents, and poloxamers containing 80% EO are used as emulsifiers and gelling agents. Poloxamers containing intermediate amounts of EO are used as emulsifiers and demulsifiers. Table 4 illustrates the range of industrial applications of poloxamers.

Numerous uses of poloxamers have been proposed in the literature [4]. These have included using poloxamers as gelling agents in toothpaste, stabilizers in mouthwash, dispersants in aerosol antiperspirants and in contact lens cleaning preparations [79]. A major use of poloxamers is in formulations for hard-surface cleaning agents, often in dishwashing products. The function of the poloxamer in such formulations is to promote even draining of water from the washed surface to prevent "spotting" or the appearance of an aesthetically unacceptable residue on the surface. Traditionally, poloxamers having low cloud points have been used in this application since it was thought that rinse agents operated by adsorbing onto the surface at or above the cloud point,

Table 4 Applications of the Poloxamers

	L31	F38	L61	L62	L64	F68	F77	L81	P85	F87	F88	L101	F108	L121	F127
Agriculture															
Emulsifier/dispersant for aqueous systems						*	*		*	*	*		*		*
Cattle antibloat					*										
Detergency															
Autodishwash rinse aid			*	*				*				*		*	
Dairy and brewery bottle cleaning			*	*				*				*		*	
Low foam, hard surface cleaners			*	*				*				*		*	
Cleaning in place			*	*				*				*		*	
Emulsion polymerization															
Emulsion stabilizer		*				*					*		*		
Industrial antifoams															
Sugar beet processing			*					*				*		*	
Fermentation, antibiotics			*					*				*		*	
Citric acid and yeast			*					*				*		*	
Water treatment			*					*				*		*	
Paper and pulp			*					*				*		*	
Personal care															
Emulsifier/dispersant for aqueous systems (shampoos/ mouthwashes)						*	*		*	*	*		*		*
Gelling agent (hair treatment, etc.)															*
Thickening/creaming agent (antiperspirant, suntan, shaving, etc.)						*				*					
Wetting lotion (shaving, suntan, antiperspirant, bath oils)														*	
Blood plasma substitute						*									
Textiles															
Spin finish lubricants	*				*			*							
Wet end processing			*			*				*					

Source: Adapted from Ref. 1. Information in Tables 2 and 4 is reproduced from technical literature by permission of ICI Surfactants subject to the following disclaimer: ''The information and recommendations in this publication are believed to be accurate and are given in good faith, but the Customer should satisfy itself of the suitability of the contents for a particular purpose. ICI gives no warranty as to the fitness of the Product information and recommendations for any particular purpose and any implied warranty or condition (statutory or otherwise) is excluded except that such exclusion is prevented by law. Freedom under Patent, Copyright and Designs cannot be assumed.''

thereby reducing the solid–liquid interfacial energy and the contact angle leading to the formation of an evenly draining sheet. However, a recent patent [80] claims that poloxamers having high cloud points may be used successfully in such applications if combined with a suitable defoaming agent. A rinse additive composition containing a poloxamer and an insoluble zinc salt, which inhibits glassware corrosion caused by automatic dishwasher detergents, has also been recently described [81], as has a hard-surface detergent composition containing a po-

loxamer claimed to exhibit excellent spotting characteristics [82].

B. Clinical Applications

The potential use of poloxamers in the medical field has been extensively investigated [5,35]. The use of poloxamers in drug delivery, as gels in the controlled release of drugs and in solid form in the targetting of drugs at specific sites in the body, has received significant attention. Other potential clinical applications include the use of poloxamers to stimulate immune response, to protect human vein grafts during storage, and to cleanse wounds [83].

VI. ECOLOGY

An increasingly important drawback of poloxamers, when compared with many other surfactants, is their lack of biodegradability. The rate and extent of biodegradation of the EO chain has been shown to be inversely dependent on chain length [84]. It has been suggested that the mechanism of degradation of EO involves the progressive oxidation and hydrolysis of terminal glycol groups [85], and that the long chain length of the poloxamers thus accounts for their relatively poor biodegradability [86]. It is also the case that the high-molecular weight of poloxamers makes transport through bacterial cell walls difficult, so limiting the extent of intracellular degradation [86]. Introduction of PO into the EO chain reduces biodegradability, and the rate at which poloxamers are broken down by biological action decreases as their PO content increases [87]—an effect that has been explained as a consequence of the pendant methyl group [88]. PO–EO block coploymers have been shown to undergo only partial degradation [89]. A widely used method of assessing the biodegradability of surfactants is the OECD screening test, in which a dilute solution of the surfactant is mixed with a bacterial sample from a sewage plant and several inorganic nutrients. The test solution is agitated for up to 19 days to allow biodegradation to occur, and a surfactant passes the test if its concentration is reduced by 80% or more. Most major surfactants used in household products pass this test [90]; poloxamers do not. The lack of biodegradability of poloxamers has already resulted in their being excluded from some applications [27,91] and may further effectively restrict their use as environmental legislation becomes ever more stringent.

Although poloxamers show poor biodegradability, they exhibit very low acute toxicity [92] and are reported as having low potential for causing irritation and skin sensitization [26]. Toxicity decreases as ethylene oxide content increases, and the least toxic poloxamers are approved as food additives [80].

REFERENCES

1. The 'Synperonic' PE range of nonionic surfactants (ref. 200-20E), ICI Surfactants, Cleveland, U.K.
2. Pluronic PE types (ref. TI/ES 1026e), BASF, Cheadle, U.K.
3. I. R. Schmolka, *J. Amer. Oil Chem. Soc., 54*: 110 (1977).
4. I. R. Schmolka, Block and Graft Copolymerisation Vol. 2, (R. J. Ceresa, ed.), Wiley-Interscience, London (1976).
5. I. R. Schmolka, Poloxamers, The Versatile Surfactants for Medical Investigations. Conference Paper, 2nd International Conference on Clinical Hemorheology, Big Sky, Montana, (1995).
6. A. S. Davidsohn and B. Milwidsky, *Synthetic Detergents*, 7th edition, Longman, U.K. (1987).
7. J. Falbe, (ed.), *Surfactants in Consumer Products—Theory, Technology and Application*, Springer Verlag, Berlin (1986).
8. Th. F. Tadros, (ed.), *Surfactants*, Academic Press, London (1984).
9. T. M. Schmitt, *Analysis of Surfactants*, Surfactant Science Series Vol. 40, Marcel Dekker, New York (1992).
10. W. Gronski, G. Hellman, and A. Wilsch-Irrgang, *Makromol. Chem., 192*: 591 (1991).
11. A. Mathias and N. Mellor, *Anal. Chem., 38*: 472 (1966).
12. T. Uno, K. Miyajima, and Y. Miyajima, *Chem. Pharm. Bull., 15*: 77 (1967).
13. J. Cross, (ed.), *Nonionic Surfactants Chemical Analysis*, Surfactant Science Series Vol. 19, Marcel Dekker, New York (1987).
14. S. Siggia, A. C. Starke, Jr., J. J. Garis, Jr., and C. R. Stahl, *Anal. Chem., 30*: 115 (1958).
15. I. Zeman, L. Novak, L. Mitter, J. Stekla, and O. Holendova, *J. Chromatogr., 119*: 581 (1976).
16. P. Kusz, J. Szymanowski, K. Pyzalski, and E. Dziwinski, *LC-GC, 8*: 48 (1990).
17. K. Tsuji and K. Konishi, *J. Am. Oil Chem. Soc., 51*: 55 (1974).
18. R. D. Hei and N. M. Janisch, *Tenside, 26*: 288 (1989).
19. R. Wickbold, *Tenside, 9*: 173 (1972).
20. J. Chlebicki and W. Garncarz, *Tenside, 15*: 187 (1978).
21. M. J. Rosen, *Anal. Chem., 29*: 1675 (1957).
22. F. J. Ludwig, *Anal. Chem., 40*: 1620 (1968).
23. H. Brueschweiler, V. Sieber, and H. Weishaupt, *Tenside, 17*: 126 (1980).
24. G. L. Selden and J. H. Benedict, *J. Am. Oil Chem. Soc., 45*: 652 (1968).
25. M. Zgoda and S. Petri, *Chem. Anal.* (Warsaw), 1986, *31*: 577 (1986).
26. *Pluronic and Tetronic Surfactants*, BASF Technical Literature (1989).
27. M. J. Schick, (ed.), *Nonionic Surfactants Physical Chemistry*, Surfactant Science Series Vol. 23, Marcel Dekker Inc., New York (1987).
28. P. W. Houlihan, D. Fornasiero, F., Grieser, and T. H. Healy, *Colloids Surfaces, 69*: 147 (1992).
29. D. J. Shaw, *Introduction to Colloid and Surface Chemistry*, 2nd ed., Butterworths, London (1970).
30. T. H. Vaughn, H. R. Suter, L. G. Lundsted, and M. G. Kramer, *J. Am. Oil Chem. Soc., 28*: 294 (1951).
31. B. Chu, *Langmuir, 11*: 414 (1995).
32. J. H. Collett and E. A. Tobin, *J. Pharm. Pharmacol., 31*: 174 (1979).
33. A. Tontisakis, R. Hilfiker, and B. Chu, *J. Colloid Interface Sci., 135*: 427 (1990).
34. P. Hurter and T. A. Hatton, *Langmuir, 8*: 1291 (1992).
35. P. Alexandridis and T. A. Hatton, *Colloids Surfaces A: Physicochemical and Engineering Aspects, 96*: 1 (1995).
36. G.-W. Wu, Z.-K. Zhou, and B. Chu, *Macromolecules, 26*: 2117 (1993).
37. G.-W. Wu, Z.-K. Zhou, and B. Chu, *J. Polym. Sci., Part B: Polym. Phys., 31*: 2035 (1993).
38. N. K. Reddy, P. J. Fordham, D. Attwood, and C. Booth, *J. Chem. Soc., Faraday Trans., 86*: 1569 (1990).
39. G.-E. Yu, Y. Deng, S. Dalton, Q.-G, Wang, C. Price, and C. Booth, *J. Chem. Soc., Faraday Trans., 88*: 2537 (1992).

40. W. Brown, K. Schillen, M. Almgren, S. Hvidt, and P. Bahadur, *J. Phys. Chem., 95*: 1850 (1991).

41. M. Almgren, J. Alsins, and P. Bahadur, *Langmuir, 7*: 446 (1991).

42. P. Bahadur, K. Pandya, M. Almgren, P. Li, and P. Stilbs, *Colloid Polym. Sci., 271*: 657 (1993).

43. W.-D. Hergeth, I. Alig, J. Lange, J. R. Lochman, T. Scherzer, and S. Wartewig, *Makromol. Chem. Macromol. Symp., 52*: 289 (1991).

44. I. Alig, R.-V. Ebert, W.-D. Hergeth, and S. Wartewig, *Polym. Commun. 31*: 314 (1990).

45. G. Wanka, H. Hoffmann, and W. Ulbricht, *Colloid Polym. Sci., 268*: 101 (1990).

45a. G. Wanka, H. Hoffmann, and W. Ulbricht, *Macromolecules, 27*: 4145 (1994).

46. G.-W. Wu and B. Chu, *Macromolecules, 27*: 1766 (1994).

47. Z.-K. Zhou and B. Chu, *Macromolecules, 27*: 2025 (1994).

48. K. Mortensen and W. Brown, *Macromolecules, 26*: 4128 (1993).

49. K. Schillen, W. Brown, and C. Konak, *Macromolecules, 26*: 3611 (1993).

50. M. Almgren, P. Bahadur, M. Jansson, P. Li, W. Brown, and A. Bahadur, *J. Colloid Interface Sci., 151*: 157 (1992).

51. W. Brown, K. Schillen, and S. Hvidt, *J. Phys. Chem., 96*: 6038 (1992).

52. P. Bahadur and K. Pandya, *Langmuir, 8*: 2666 (1992).

53. K. Pandya, P. Bahadur, T. N. Nagar, and A. Bahadur, *Colloids Surfaces A: Physicochemical and Engineering Aspects, 70*: 219 (1993).

54. K. Schillen, O. Glatter, and W. Brown, *Prog. Colloid Polym. Sci., 93*: 66 (1993).

55. K. Mortensen, *Europhys. Lett., 19*: 599 (1992).

56. K. Mortensen and J. K. Pedersen, *Macromolecules, 26*: 805 (1993).

57. K. Mortensen, *Prog. Colloid Polym. Sci., 91*: 69 (1993).

58. B. Chu, G.-W. Wu, and D. K. Schneider, *J. Polym. Sci., Part B: Polym. Phys., 32*: 2605 (1994).

59. M. H. G. M. Penders, S. Nilsson, L. Piculell, and B. Lindman, *J. Phys. Chem., 98*: 5508 (1994).

60. M. Almgren, J. Stam, C. Linblad, P. Li, P. Stilbs, and P. Bahadur, *J. Phys. Chem., 95*: 5677 (1991).

61. P. Alexandridis, J. F. Holzwarth, and T. A. Hatton, *Macromolecules, 27*: 2414 (1994).

62. M. Malmsten and B. Lindman, *Macromolecules, 25*: 5446 (1992).

63. G. Fleischer, *J. Phys. Chem., 97*: 517 (1993).

64. G. Fleischer, P. Blosz, and W.-D. Hergeth, *Colloid Polym. Sci., 271*: 217 (1993).

65. N. Mitchard, A. Beezer, N. Rees, J. Mitchell, S. Leharne, B. Chowdhry, and G. Buckton, *J. Chem. Soc., Chem. Commun.*, 900 (1990).

66. A. E. Beezer, J. C. Mitchell, N. H. Rees, J. K. Armstrong, B. Z. Chowdhry, S. Leharne, and G. Buckton, *J. Chem. Res. (S)*, 254 (1991).

67. A. E. Beezer, N. Mitchard, J. C. Mitchell, J. K. Armstrong, B. Z. Chowdhry, S. Leharne, and G. Buckton, *J. Chem. Res. (S)*, 236 (1992).

68. J. K. Armstrong, B. Z. Chowdhry, A. E. Beezer, J. C. Mitchell, and S. Leharne, *J. Chem. Res. (S)*, 364 (1994).

69. A. E. Beezer, W. Loh, J. C. Mitchell, P. G. Royall, D. O. Smith, M. S. Tute, J. K. Armstrong, B. Z. Chowdhry, S. A. Leharne, D. Eagland, and N. J. Crowther, *Langmuir 10*: 4001 (1994).

70. M. Malmsten and B. Lindman, *Macromolecules, 26*: 1282 (1993).

71. P. Blosz, W.-D. Hergeth, C. Wohlfarth, and S. Wartewig, *Makromol. Chem., 193*: 957 (1992).

72. Q.-G. Wang, C. Price, and C. Booth, *J. Chem. Soc., Faraday Trans., 88*: 1437 (1992).

73. F. Tiberg, M. Malmsten, P. Linse, and B. Lindman, *Langmuir, 7*: 2723 (1991).

74. M. Malmsten, P. Linse, and T. Cosgrove, *Macromolecules, 25*: 2474 (1992).

75. W. Saski and S. G. Shah, *J. Pharm. Sci., 54*: 71 (1965).

76. R. A. Anderson, *Pharm. Acta Helv., 47*: 304 (1972).

77. K. N. Prasad, T. T. Luong, A. T. Florence, J. Paris, C. Vaution, M. Sellier, and F. Puisieux, *J. Colloid Interface Sci., 69*: 225 (1979).

78. A. A. Samii, G. Karlstrom, and B. Lindman, *Langmuir, 7*: 1067 (1991).

79. Flow Pharmaceuticals, Inc., Palo Alto, Calif., U.S.A. U.S. Patent No. 3,882,036.

80. Ecolab Inc., St Paul, MN, U.S.A International Patent Application No. PCT/US94/02504.

81. Procter & Gamble Co., Cincinnati, Ohio, U.S.A. European Patent Application No. 90301308.4.

82. Proctor & Gamble Co., Cincinnati, Ohio, U.S.A. European Patent Application No. 93202757.6.

83. Calgon Corporation, Robinson Township, PA, U.S.A. New Zealand Patent No. 236229.

84. R. N. Sturm, *J. Am. Oil Chem. Soc., 50*: 159 (1973).

85. P. Schoberl, E. Kunkel, and K. Espeter, *Tenside, 18*: 64 (1981).

86. D. R. Karsa and M. R. Porter (eds.), *Biodegradability of Surfactants*, Blackie, Glasgow, U.K. (1995).

87. C. G. Naylor, F. J. Castaldi, and B. J. Hayes, *J. Am. Oil Chem. Soc., 65*: 1669 (1988).

88. J. Karpinska-Smulikowska, M. Pawlaczyk-Szpilowa, and J. Plucinski, Proceedings of the VII International Congress of Surface Active Agents (Moscow) *4*: 23 (1976).

89. D. H. Scharer, L. Kravetz, and J. B. Carr, *TAPPI, No. 10* (1979).

90. M. R. Porter, Handbook of Surfactants, 2nd Edn., Blackie, Glasgow, U.K. (1994).

91. A. Hettche and E. Klahr, *Tenside, 19*: 127 (1982).

92. P. Schoberl and K. J. Bock, *Tenside, 25*: 86 (1988).

51
Recent Advances in Cationic Resins

Metwally Shafik Metwally
Al-Azhar University, Cairo, Egypt

I. INTRODUCTION

The extensive possibilities of the practical application of synthesis, and the study of the properties of ion-exchange resins have aroused widespread interest in chemistry. This chapter discusses some theoretical problems with cationic resins as catalysts in hydrolysis reactions. New types of cationic resins have been examined and some important generalizations on ion-exchange reactions have been formulated.

II. ION-EXCHANGE RESIN CATALYSIS IN HYDROLYSIS REACTIONS

In recent years, the rate of information available on the use of ion-exchange resins as reaction catalysts has increased, and the practical application of ion-exchanger catalysis in the field of chemistry has been widely developed. Ion-exchangers are already used in more than twenty types of different chemical reactions. Some of the significant examples of the applications of ion-exchange catalysis are in: hydration [1,2], dehydration [3,4], esterification [5,6], alkylation [7], condensation [8–11], and polymerization, and isomerization reactions [12–14]. Cationic resins in H^+ form, also used as catalysts in the hydrolysis reactions, and the literature on hydrolysis itself is quite extensive [15–28]. Several types of ion exchange catalysts have been used in the hydrolysis of different compounds. Some of these are given in Table 1.

Other possibilities for practical application of resin catalysis include some organic reactions involving: addition, cyclization, and structural rearrangement. Increased stability and specific control of structure has led to the increased use of cation exchange resins as catalysts. As in the case of cation exchange resins many

reactions that are base catalyzed are in a similar manner catalyzed by anion exchange resins. Increased stability and specific control of structure has led to the increased use of these resins as catalysts. Some of the molecular weight reactions catalyzed by both strong and weak basic anion exchangers are listed in Table 2.

There are several apparent advantages to the use of ion-exchange resins as either acid or base catalysts, several of which are as follows:

1. By a simple filtration step, catalyst-free products can be obtained.
2. The catalyst can be recovered frequently by means of a simple filtration step.
3. Continuous reactions can be obtained by the passage of the reactants through beds of ion-exchange catalysts.
4. Unusual selectivity effects are possible.
5. Side reactions can be kept at a minimum.
6. Special corrosion-resistant equipment is not as necessary as in the case of some soluble catalysts.

The catalytically active counter-ions of ion exchangers are in a solvated state, and in this respect are similar to the so-called free ions in ordinary electrolyte solutions. For that reason, both the reactions catalyzed in solution by ions and those catalyzed by counter-ions of ion-exchange resins are able to proceed in accordance with the same mechanism, and are of the same kinetic order. The rate of a given reaction depends, however, on whether a given amount of catalytically active ions is present in the form of a soluble electrolyte or in the form of an ion-exchanger. As suggested by Haskell and Hammett [16], the ratio of the hydrolysis rate ($q = K_{resin}/K_h$) is called efficiency. The efficiency of the exchanger, q, may be equal to [15,56], larger than [18,57],

Table 1 Several Types of Ion-Exchange Catalysts

Resin type	Compound hydrolysis	Ref.
KU-1	Ethyl acetate	29
KU-2	Esters of dicarboxylic acids	30
KU-2 and SG-50	Simple vinyl esters of aliphatic and aromatic alcohols	31
Duolite-C 25	Methyl acetate, methyl lactate	32
Amberlite IR-120	Allyl acetate	33–35
Amberlite IR-113	Phenylalanine peptides	36
Dowex-50W	Ethyl malonate	37
Amberlite IRA 410	Saccharose	38
KMT	Protein fraction of molecular weight 600–2000	39
Sulfonated coal	Starch	40
IR-120, DVB-20	Ethyl acetate and methyl acetate	41–43
IR-100		
Wolfatit KPS	Isobutyl acetate, mixtures of esters	25,44
Various anion exchangers	Ethyl iodide	45

or smaller than [16,58,59] unity and depends on the properties of the catalyst itself, as well as on those of the solvent. For the decomposition of the first-order reaction of:

$$AB \rightarrow A + B$$

Using Q_{AB} moles of starting compound, the reaction rate in a homogeneous system is:

$$dQ_{AB}/dt = k_h Q_{AB}$$

where k_h is the velocity constant. For the corresponding heterogeneous reaction carried out in the presence of an ion exchanger:

$$dQ_{AB}/dt = k_{\text{het}}.Q_{AB}$$

The relationship of the efficiencies of an ion exchanger for two compounds belonging to the same chemical class, is termed as specificity ratio [19]. If the reactants, e.g., esters, are simultaneously placed in contact with the catalyst, the value $r = q_1/q_2$ is a measure of the selectivity of the transformation of one component with

respect to the other. This value is quantitatively related to the standard free energies (F) of four intermediate states:

$$RT \ln q_1/q_2 = (FR_1 - FH_1) - (FR_2 - FH_2)$$

where the subscripts of H and R correspond to the reaction under homogeneous and heterogeneous conditions of catalysis, respectively. The efficiency of ion-exchangers depends on the following factors: (1) the properties of the catalyst and reactants, (2) the distribution constant of reactants between the solution and the cationic resin, (3) the nature of the solvent, and (4) the activation energy of the reactions.

III. FACTORS AFFECTING RESIN CATALYZED HYDROLYSIS OF ESTERS

A. Effect of Amount of Catalyst

The relation between the amount of catalyst and the efficiency of a new macroreticular sulfonated acrylonitrile

Table 2 Reactions Catalyzed by Anionic Exchangers

Reaction type	Reactants	Resin form	Ref.
Hydrolysis	Ethyl acetate	Weak base	46
Reduction	Nicotinonitrite	Strong base	47
	Various	Weak and strong	48
	Various	Strong base	49
	Aliphatic aldehyde	Weak base	50
Aldol condensation	Cyclic ketone	Weak base	51
	Furfural + aldehyde	Strong base	52
Methanolysis	Triglyceride	Strong base	53
Diacetone alcohol	Acetone	Strong base	54
Condensation	Glucose	Strong base	55

butadiene styrene copolymer previously crosslinked with phenol formaldehyde resin was studied by Metwally et al. [26]. The efficiency, q, was shown to be a function of resin concentration [26] and increases with an increasing amount of the resin catalyst (Table 3). The same observation was shown by Haskell and Hammett [16], Affrossman and Murray [23] and George and Robert [60].

The rate of hydrolysis in the presence of resins increases with the number of catalytically active ions. In some reactions, the reaction rate is a linear function of the quantity of catalyst added [26,34]. Figure 1 shows the effect of varying catalyst concentration on the rate of hydrolysis of ethyl acetate. Higher values of q are shown with the larger amount of catalyst.

B. Stability of Resins

The stability of the resins in various operations is of the utmost importance to the theory and practice of ion-exchange catalysis. The commonly observed loss in resin activity during a reaction is frequently due to a gradual decrease in the number of catalytically active ions. Polyanskii and Tulupov [59,61] and Fang [62] studied the mechanism of ion-exchange structural changes. However, it would be wrong to ascribe all changes to desulphonation, dephosporization, or degradation effects. In some reactions that take place under particularly mild conditions, the activity of sulfonated resins is found to decrease rapidly, although desulfonation under the experimental conditions seems unlikely. It was found that in these systems the pores of resin are blocked by products, and reactant molecules are, therefore, unable to reach the active sites [63–65].

C. Steric Hindrance

Strict hindrance of the counter-ions present in the cationic resins has been shown to be high enough to prevent exchange reactions [64,66]. Few studies was reported in the literature for the good operating lives of the cation exchange catalysts. Hydrolytic reactions are usually carried out at relatively low temperatures where the stability of resins is high. Granular sulfonated styrene–divinylbenze copolymer, operated intermittently during 3 years for hydrolysis and esterification reac-

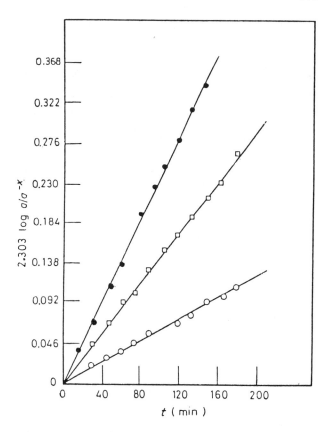

Figure 1   The effect of catalyst concentration on the rate constant. (●) 9 g, (◇) 6 g, and (○) 3 g.

tions, has retained its activity for 500 h without regeneration [67]. Mixed resins containing 50 wt% of Amberlite-IR 120 showed no detectable change after 5000 h operation [68]. On the other hand, Dowex-50 WX8 cation exchanger suffered blocking effects, particularly at elevated temperatures, when used as an inversion catalyst for concentrated saccharose solutions [69].

D. Hydrolysis Processes

Because hydrolytic reactions are reversible, they are seldom carried out in batchwise processes [26,28,36,70]. The reactor is usually a double jacket cylindrical flask fitted with a reflux condenser, magnetic stirrer, and thermometer connected with an ultrathermostat. The catalyst is added to the reaction mixture when the desired temperature has been reached [71,72]. A nitrogen atmosphere is used when the reactants are sensitive to atmospheric oxygen [36]. Dynamic methods require more complicated, but they have been widely used in preparative work as well as in kinetic studies of hydrolysis [72–74]. The reaction usually consists of a column packed with a layer of the resin and carrying a continuous flow of the reaction mixture. The equilibrium can

Table 3   Effect of Varying Catalyst Concentration on the Hydrolysis of Ethyl Acetate (0.2 M) at 45°C

	Weight of resin (q)		
	3	6	9
$k_r(10^6 \text{ s}^{-1})$	10.97	24.00	41.30
Efficiency, "q"	0.34	0.40	0.45

be strongly displaced toward hydrolysis by distilling the volatile reaction products [31] by extraction [29,67,68].

E. Particle Size Effect

Many workers [69,75–77] have examined the influence of the resin particle size on the reaction kinetics in order to throw light on the role of diffusional limitation. The hydrolysis rate constants of ethyl acetate [72] and acetamide [78] are practically independent of particle size over quite a wide range. The rate-limiting step in these reactions is the chemical act itself. Saccharose under static [70] or dynamic [41,75] conditions is much faster when smaller resin particles are used. Reed and Dranoff [75] have observed a linear relationship between the rate constant and the catalyst particle size. In the hydrolysis of sarin [34], the kinetic curves become steeper as the diameter of the catalyst particles decreases and the temperature increases. The rate of hydrolysis in the presence of ion-exchange resins is influenced by the bond strength being attacked. However, and in contrast to homogeneous catalysis, the steric factor (defined as the ratio of the size of the reactant molecules to the distance between the counter-ions of the nearest active groups in the catalyst) plays an important role in ion-exchange catalysis [17,37,74]. Ordyan and coworkers [79] found that methyl acetate is hydrolyzed 5 times faster than butyl acetate, and the latter only 1.2 times faster than methyl valerate, indicating that the reaction is retarded by branched chains. Similar effects have also been observed with n- and isopropyl acetate [72] and the hydrolysis of disaccharides [44]. Odioso et al. [80] and Smith and Steele [81] observed that hydrolysis rates decrease with an increasing chain length in simple aliphatic esters and the same is true for esterification in alcohol solution [82]. Helfferich [71,83] suggested that the variation of distribution coefficient might be the prime cause of the relationship between the efficiency of the resin catalyst and the structure of the reactant. However, few data are available on the study of the kinetics of resin-catalyzed

hydrolysis of esters substituted in the alkyl group [23]. The rate of hydrolysis of seven acetate esters substituted in the alkyl group have been studied in the presence of a sulfonated cation exchange resin in 70% aqueous acetone solution by Abdel Razik and others [84]. The slower hydrolysis rates with increasing size of substituent are related to increasing steric influence as determined by decreasing values of the steric substituent parameter, E_s. The efficiency of the resin catalyst is related to the entropy of substituents. The influence of steric hindrance on reaction rates accounts satisfactorily for observed variations of the enthalpies and entropies of activation with alkyl group substituents. The entropy Δs^* values increase in the following order:

$$—octyl— < ethyl— < methyl\ acetate$$

which represents the sequence of probability for the formation of activated complex on the exchange resin. For acid-catalyzed hydrolysis, the susceptibility of the reaction to polar effects is assumed to be almost negligible and the relative rates are determined by steric factors alone [85,86]. The steric substituent constant E_s gives a nearly quantitative measure of the steric factors associated with these substituents in rates of such reactions. Table 4 gives the rate coefficients and the Taft steric constant [87] E_s for esters substituted in the alkyl group [84]. The steric effect of substituents can affect the rate coefficient of the reaction by steric hindrance of internal molecular motions and/or by increasing strain in the transition state due to increasing nonbonded interactions. Both of these factors are expected to increase with increasing size and position of substituent groups. The order of decreasing E_s is in accord with this conclusion.

The slower rate of hydrolysis of alkyl substituted esters in the presence of the cation exchange resin can be explained by the assumption that the alkyl groups interfere more in the formation of the intermediate complex on the resin surface than in the homogeneous system. The efficiency of the resin q was less than unity

Table 4 Rate Coefficients, Efficiency "q," and Steric Substituent Constant

Alkyl substituent	E_s at 318 K	$K_H \times 10^5$ (s^{-1}) at 318 K	$k_r \times 10^5$ (s^{-1})			q at 318 K
			298 K	308 K	318 K	
Methyl	0.00	10.29	2.63	4.71	6.45	0.63
Ethyl	−0.12	8.34	2.15	3.22	4.86	0.58
Isopropyl	−0.15	7.78	2.10	2.73	4.61	0.56
n-Butyl	−0.25	5.85	1.68	2.11	3.61	0.55
Pentyl	−0.38	4.95	1.20	2.01	2.71	0.51
Isopentyl	−0.45	3.98	1.15	1.71	2.30	0.56
n-Octyl	−0.47	3.79	1.20	1.30	2.20	0.53
Cyclohexyl	−0.58	2.91	0.90	0.91	1.71	0.57

$E_s = (\log k_r/k_{Me})$, k_H, k_r are specific rate constants of acid and acid resin catalyst [84].

(Table 4), which indicates slower rates of hydrolysis of the alkyl acetates in the presence of the resin catalyst.

F. Effect of Solvent

A number of articles have appeared on studies of the hydrolysis rate of esters in binary solvent mixtures [88–97]. Little work has been done on the acidic resin catalysis of esters in aqueous organic solvents [18,23,27,37,83,98]. Predictions of Ingold [99] and Laidler and Landskoener [100] show that in a system of increasing dielectric constant, the rate of a reaction is expected to rise. However, in many instances, Parker [101] and Roberts [89] found that the rate constants decrease under similar conditions of increasing dielectric constant of the medium. The rate constants for the hydrolysis of ethyl acetate [29] and acetamide [78] in aqueous acetone solutions decrease as the mole fraction of acetone increases. We must assume that the inhibition of the reaction by increased concentration of organic solvent is due not only to the lowering in dielectric constant but also to the different distribution of reactant molecules between the solid and the liquid phase. A kinetic study of solvent effect on sulfonated resin-catalyzed hydrolysis of ethyl propionate has been done by Metwally [27] on water–acetone, water–dioxane, water–DMF, and water–DMSO solvent mixtures of varying compositions at different temperatures. Specific rate constants (k_r) were found to be first order and decreased with an increasing amount of organic cosolvent in the media (Table 5).

Metwally et al. [28] also studied the resin-catalyzed hydrolysis of ethyl formate in acetone–water mixtures at different temperatures. The experimental results indicated a linear dependence of the logarithm of rate constant on the reciprocal of the dielectric constant (Fig. 2). The decrease of dielectric constant may lower the concentration of the highly polar transition state and thereby decrease the rate [28].

G. Activation Energy

The calculated activation energy for the hydrolysis of sarin [43] on Amberlite-IR100 is relatively low (22,000

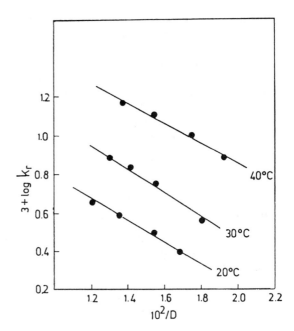

Figure 2 Log k_r for the resin catalyzed hydrolysis of ethyl formate as a function of the reciprocal of the dielectric constant.

J mol^{-1}). This is additional evidence in favor of rate limitation by inner diffusion. However, the same reaction in the presence of Dowex-50, which has a more open three-dimensional network, gave an activation energy of 44800 J mol^{-1}, and closely similar values were obtained for the hydrolysis of ethyl acetate [29] and dimethyl sebacate [30]. The activation energy for the hydrolysis of ethyl acetate on a macroreticular sulphonated cationic exchanger [93] is 3566 J mol^{-1}. For the hydrolysis of ethyl formate in a binary system, the isocomposition activation energy (E_c) [28,92] tends to decrease as the solvent content increases, while for solutions of the same dielectric constant, the iso-dielectric activation energy (E_D) increases as the dielectric constant of the solvent increases (Table 6).

The trends of variation of the activation parameters are correlated with the solvation mechanism and dielectric behavior of the medium. Thus, ΔH^*, ΔG^* and ΔS^* for the acidic resin-catalyzed hydrolysis of isopropyl acetate were calculated using the Wynne-Jones and Eyr-

Table 5 Specific Rate Constant ($k_r \times 10^4$ min^{-1}) of the Resin Catalyzed Hydrolysis of Ethyl Propionate in Aqueous Solvents at 25°C

| Solvent | Composition percent (V/V) | | | | |
	25	35	45	55	65
Water–acetone	40.06	35.18	32.47	26.85	20.91
Water–dioxane	46.08	40.46	36.61	32.79	24.79
Water–DMF	52.48	45.17	40.87	36.25	31.83
Water–DMSO	74.47	63.46	55.17	43.44	40.06

Table 6 Iso-Dielectric Activation Energy (E_D) for the Hydrolysis of Ethyl Formate in Acetone–Water Systems [28]

Dielectric constant values	55	60	65	70	75
E_D (KJ mol^{-1})	24.49	25.62	26.19	27.04	23.47

ing equations [104]. The variation of ΔG^* is considered to be small and can be taken as nearly constant in all solvents studied. The same behavior of ΔG^* was shown by Singh et al. [92] and others [28,98,105]. In addition, it is observed from Table 7 that there is a decrease in ΔS^* values for ester hydrolysis in water–dioxane, water–acetone, and water–DMF systems. This means that conversion of the initial state into the transition state is accompanied by an entropy decrease, which confirms the assumption that the transition state becomes more and more solvated compared with the initial state as the concentration of organic solvent increases.

H. Stirring Rate

The rate of hydrolysis of sarin on Dowex-50 cation exchange resin is insensitive to the stirring rate. However, with a more active catalyst (Amberlite-IRA 400), the rate constant at 20°C was 5.3, 7.5, and 8.5 h^{-1} at 60,800 and 1000 revolutions/min^{-1}, respectively, suggesting that film diffusion was the rate-limiting step. Thus, the mechanism of the rate-limiting step depends on the nature of the catalyst [34].

IV. NEW MACRORETICULAR CATIONIC RESINS

The majority of these exchangers are based on a reaction of synthesized linear resin or natural products with another synthesizing resin. Upon this chemically and physically robust structure a very wide range of functional groups may be bound, each corresponding to a new type of exchanger. A degree of flexibility in the design of new exchangers is thus introduced, which in principle allows us to design particular exchangers for particular purposes. As polymer chemistry and preparative techniques develop, these types of exchangers may become very important.

A. Macroreticular Cationic Resins in Analytical Chemistry

There are many applications of ion-exchange resins in analytical chemistry in both quantitative and qualitative

analysis. Ion-exchange resins have been used for the detection of esters [106], amides, imides, and anilides [107], nitriles [108], aliphatic and aromatic aldehydes [109], and are also used for the determination of aliphatic unsaturated amides and esters [110]. The resin spot technique was first developed by Fujimotu [111] in the detection of microgram compounds. Resin beads loaded with indicators have also been used for the titration of acids and bases [112–115]. Honda [116] used such resin beads to estimate pH in the resin phase. Qureshi et al. [117] used resin beads in Fe^{II} or p-dimethylaminobenzylidenerhodamine form as the indicator in precipitation reactions with potassium ferrocyanide. El-Hadi and Metwally [118] prepared new macroreticular cationic resins and used the synthesized resins as the indicator in acid-base titration both in aqueous and aqueous organic medium. Thus, phosphoric acid cation exchange resin was prepared by phosphorization of acrylonitrile–butadiene styrene copolymer crosslinked with phenol-formaldehyde resin (resol). First, the copolymer was dissolved in benzene at 70°C, and resol was then added to the polymer emulsion with continuous stirring. The phosphorylated resins were prepared by refluxing the crosslinked polymer with PCl$_3$ [118]. Nitric acid was used to oxide the phosphinic acid group to a phosphonic group. A comparison was made between the synthesized resin indicator and other available standard indicators. The optimum pH range for the function of the resin as indicator was found to be 8–10. The capacity of the phosphorylated resins were determined by two methods.

1. Direct titration [119,120]: In order to know the number of exchangeable hydrogen ions at different dissociation stayes, various salts of weak acids were used. Data observed in Table 8 shows an increase in the exchange capacity with an increase in the pH of the solution, indicating the presence of "weak acid" capacity [118].

2. Potentiometric titration curves: The procedure involves the addition of a salt of a weak acid to the resin and the determination of the pH of the equilibrated solution. Table 9 shows the pK values of the OH groups and dissociation constants of the studied resin. The first ionization occurs at a pH slightly higher than that of sul-

Table 7 Activation Parameters for the Resin Catalyzed Hydrolysis of Some Esters in Water–Solvent Mixtures

Ester	Solvent mixtures	Composition % (V/V)	ΔG^* KJ mol^{-1}	ΔS^* KJ mol^{-1}	ΔH^* KJ mol^{-1}
Ethyl propionate [27]	Water–acetone	35	62.57	−122	25.00
	Water–dioxane	35	62.16	−131	21.68
	Water–DMF	35	60.52	−114	25.45
	Water–DMSO	35	59.70	−135	18.08
Ethyl formate [28]	Water–acetone	30	59.84	−71.18	38.27
Ethyl acetate [91]	Water–dioxane	40	101.36	−191.32	39.2

Table 8 Experimentally Determined Capacities at Various pH Values

pH	Capacity (mEq/g)
6.8	0.51
9.0	0.99
10.3	1.12
12.7	1.31

Table 9 Dissociation Constants and pKa Values for Phosphoric Acid Resin

pKa value	Dissociation constant
$pK_1 = 2.192$	$K_1 = 1.523 \times 10^{-12}$
$pK_2 = 4.900$	$K_2 = 5.012 \times 10^{-7}$
$pK_3 = 7.199$	$K_3 = 3.980 \times 10^{-12}$

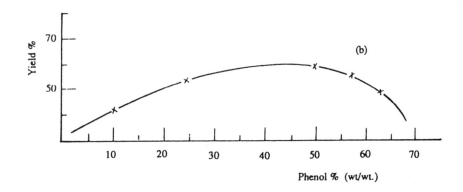

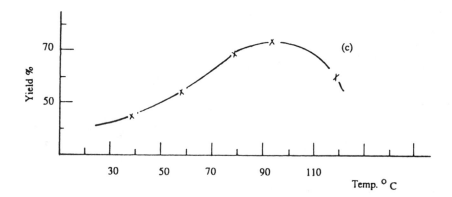

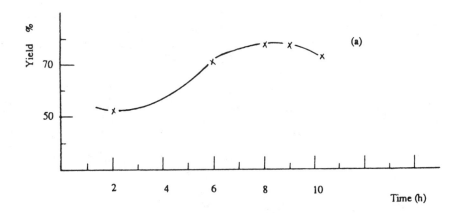

Figure 3 Effect of polycondensation time (a), phenol percentage (b), and reaction temperature (c), on the yield for the polycondensation of corncob–phenol resin.

phonic acid resin, while the second and third groups are at a pH intermediate between the carboxylic and phenolic types.

B. New Macroreticular Cationic Resins from Natural Products

1. Cationic Resins from Corncobs

Attempts have been made by Dheiveegan and Krishnamoorthy [121] and others [122–125] to prepare less expensive cationic resins from natural products. Earlier work has revealed that coke [126], sawdust [127], or lignite [128] could be substituted up to 30–40% in phenolic cationites. The production of phenolic resins from wood, lignin, and other cellulosic materials was studied by some workers [129–131], especially for the synthesis of polycondensate corncob–bisphenol resins [132]. Sulfonated cationic exchangers were prepared from polycondensation of Egyptian plants' by-products of corncobs with phenol in the presence of paraformaldehyde as a crosslinking agent by Metwally and Metwally [133]. Thus, phenol was react with corncobs by a polycondensation method in the presence of hydrochloric acid. The resulting product was treated with paraformaldehyde as a crosslinking agent. Sulfonation of the crosslinked resins was completed with sulfuric acid in the presence of silver sulfate as a catalyst. Figure 3 shows the effect of polycondensation time, phenol percentage, and reaction temperature on the yield of polycondensation of corncob–phenol formaldehyde resin. The optimum overall reaction time necessary for completion of the reaction is 8 h. Shorter reaction periods normally cause a decrease in the yield. However, a longer reaction time, as expected, did not affect the yield of the reaction product. The yield is decreased with an increase in the phenol percentage. The results of the investigation of the effect of reaction temperature on the yield show that maximum yield of the polycondensation product is obtained at a reaction temperature of 95°C. The relation between the paraformaldehyde percentage in the resins and the capacity is shown in Table 10.

The capacity of the sulfonated cationic resin is decreased with increases in the paraformaldehyde content.

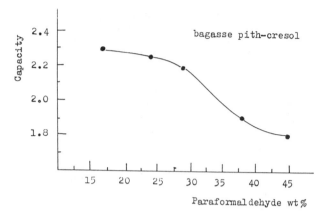

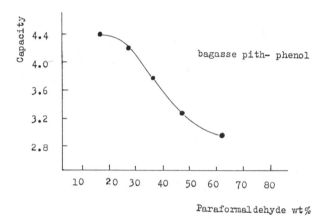

Figure 4 Effect of formaldehyde content on the capacity of cationic resins.

The average swelling percentages are decreased upon increasing the paraformaldehyde in the matrix. The rigidity of the matrix is proved from swelling measurements. These cationic resins with more crosslinking agents show a much more rigid shape; however, these resins may be useful where they are required to stand large osmotic shocks. Also, the swelling percentages are

Table 10 Capacities, Average Swelling Percentages, and Absolute Densities for Sulfonated Corncob Cationic Resins

Paraformaldehyde content (% wt/wt)	Capacity (mEq g^{-1})	Weight swelling (%)		Absolute density (gm^{-1})	
		H ± form	Na ± form	Hydrated state	Dehydrated state
10	3.95	76	79	1.10	0.99
20	3.79	72	75	1.12	1.01
30	3.22	68	70	1.12	1.15
45	2.81	55	58	1.16	1.19
60	1.83	49	52	1.18	1.20

calculated for the resins in the Na-form, however, higher values of swelling are observed for all the crosslinked samples studied. Generally, the absolute densities are increased as the paraformaldehyde content increases (Table 10).

2. Cationic Resins from Bagasse Pith

Bagasse pith is generally used as a furnace fuel of sugar mills[134] and for production of furfural [135]. The synthesis of polycondensate bagasse pith phenolic resins was also studied by some workers [129,130]. Egyptian plant by-products of bagasse pith are used for preparing cationic resins [136,137] by condensation the bagasse pith, first with cresol [136] or phenol [137], in the presence of HCl, second, by treating the resulting soluble polycondensate resins with various percentages of paraformaldehyde as a crosslinking agent, and, finally, by sulfonation of the products with sulfuric acid in the presence of a catalyst. The relationship among the various amounts of paraformaldehyde in the cationic resins and the capacities are shown in Fig. 4. For the same paraformaldehyde contents, the cationic resins of bagasse pith–phenol has a higher capacity then the bagasse pith–cresol cationic resins, and in general, the capacity is decreased with an increase in the paraformaldehyde content. The higher the amount of bagasse pith in the matrix, the larger is the capacity of cationic resins for the two types. The average swelling percentage (α) and also the capacities of the cationic resins with various amounts of bagasse pith are shown in Table 11. The weight swelling percentages are increased with an increasing amount of bagasse pith in the matrix. The values of the weight swelling are higher with resins of bagasse pith–phenol. The rigidity of the resin matrix is proved from swelling measurements. Those cationic resins with lower swelling show a much higher rigid shape.

Table 11 Swelling Percentages and Capacities of Bagasse Pith Cationic Resins

Resin	Bagasse pith (wt%)	Capacity mEq g^{-1} resin	Weight swelling % (H $\pm$ form)
Bagasse pith	0.00	3.00	112
with	22.72	3.42	124
phenol	37.04	3.65	132
	46.88	3.85	143
	54.05	3.92	152
Bagasse pith	0.00	1.81	57
with	14.29	2.19	93
Cresol	25.00	2.50	108
	33.33	2.68	117
	40.00	2.91	129

REFERENCES

1. J. R. Kaiser, H. Beuther, L. D. Moore, and R. C. Odioso, *Ind. Eng. Chem., Process Des. Devel., 1*: 296 (1962).
2. N. G. Polyanskii and T. I. Kazlova, *Zhur. Prikl. Khim., 39*: 1788 (1966).
3. J. Manassen and Sh. Khalif, *J. Catalysis, 7*: 110 (1967).
4. N. A. Ghanem and Z. H. Abd El-Latif, *Chem. Ind.,* 1650 (1965).
5. L. K. Maros and J. V. Szmercsangi, *Makromol. Chem., 78*: 224 (1964).
6. R. Kunin and E. Metzner, *J. Am. Chem. Soc., 84*: 305 (1962).
7. J. Kamis, *Coll. Czech. Chem. Comm., 29*: 923 (1964).
8. R. W. Fulmer, *J. Org. Chem., 27*: 4115 (1962).
9. H. C. Malhotra and S. Banerjee, *J. Indian Chem. Soc., 67*: 117 (1990).
10. R. E. Beal, D. E. Anders, and L. T. Black, *J. Amer. Oil Chem. Soc., 44*: 55 (1967).
11. G. M. Christensen, *J. Org. Chem., 27*: 1442 (1962).
12. S. Aoki, T. Otsu, and M. Imoto, *Chem. Ind.,* 1761 (1965).
13. S. Aoki, T. Otsu, and M. Imoto, *Makromol. Chem., 99*: 133 (1966).
14. A. Ozaki and S. Tsuzhiya, *J. Catalysis, 5*: 537 (1966).
15. G. G. Thomas and C. W. Davies, *Nature, 159*: 372 (1947).
16. V. C. Haskel and L. P. Hammett, *J. Am. Chem. Soc., 71*: 1284 (1949).
17. S. A. Bernhard and L. P. Hammett, *J. Am. Chem. Soc., 75*: 5834 (1953).
18. S. A. Bernhard and L. P. Hammett, *J. Am. Chem. Soc., 75*: 1798 (1953).
19. S. A. Bernhard, E. Garfield, and L. P. Hammett, *J. Am. Chem. Soc., 76*: 991 (1954).
20. P. Riesz and L. P. Hammett, *J. Am. Chem. Soc., 76*: 992 (1954).
21. C. H. Chen and L. P. Hammett, *J. Am. Chem. Soc., 80*: 1329 (1958).
22. S. Affrosman and J. P. Murray, *J. Chem. Soc., (B)*: 1015 (1966).
23. S. Affrosman and J. P. Murray, *J. Chem. Soc., (B)*: 579 (1968).
24. K. Rajamani, S. C. Shenoy, M. S. Rao, and M. G. Rao, *J. Appl. Chem. Biotechnol., 28*: 699 (1978).
25. A. B. Zaki, M. M. Abu Sekkina, and Y. A. El. Sheikh, *Egypt. J. Chem., 24*: 267 (1981).
26. M. S. Metwally, M. F. El-Hadi, M. A. El-Wardany, and A. Abdel Razik, *J. Mater. Sci., 25*: 4223 (1990).
27. M. S. Metwally, *React. Kinet. Catal. Lett., 47*: 319 (1992).
28. M. S. Metwally, A. Abdel Razik, M. F. El-Hadi, and M. A. El-Wardany, *React. Kient. Catal. Lett., 41*: 151 (1993).
29. Kh. R. Rustamov, L. F. Fatkalina, and K. A. Agzamor, *Vzb. Khim. Zhur., No. 2*: 32 (1961).
30. L. M. Gol'dshten and G. N. Freidlin, *Zhur. Prikl. Khim., 37*: 2540 (1964).
31. M. F. Shostakovskii, A. S. Atavin, B. A. Trofimova, and A. V. Gusarov, *Zhur. Vses. Khim. Obsheh. Imeni Mendeleeva,* 599 (1964).
32. L. Alexandru, F. Butaciu, and J. Ballint, *J. Prakt. Chem., 16*: 125 (1962).
33. L. M. Reed, L. A. Wenzel, and B. O'Hara, *Ind. Eng. Chem., 48*: 205 (1956).

34. M. G. Chasanov, and J. Epstein, *J. Polymer. Sci., 31*: 399 (1958).
35. S. Afrossman and J. P. Murray, *J. Chem. Soc. Ser, (B)*: 1015 (1966).
36. M. Pohm, *Naturwiss, 48*: 551 (1961).
37. M. J. Astle and I. A. Oscar, *J. Org. Chem., 26*: 1713 (1961).
38. M. Takeda and T. Imura, *Suisan Daigaku Kenky U. Hokuku, B*: 108 (1964).
38a. M. Takeda and T. Imura, *Chem. Abs., 62*: 9340g (1965).
39. G. V. Samsonov, M. V. Glikina, L. R. Gudkin, and A. D. Morozova, Symposium "Ionoobmennaya Tekhnologiya" (Ion Exchange Technology) Izd. Nauka, Moscow (1965).
40. P. K. Banerjle, K. S. Anand, B. P. Das, and A. N. Basu, *J. Proc. Inst. Chemists (India), 36*: 18 (1964).
41. C. W. Davies and G. G. Thomas, *J. Chem. Soc., Part II*: 1607 (1952).
42. A. B. Sidney and P. H. Louis, *J. Am. Chem. Soc., 75*: 1793 (1953).
43. S. Afrossman and J. P. Murray, *J. Phys. Org.,* 579 (1968).
44. H. Noller and P. E. Gruber, *Z. Phys. Chem., 38*: 2031 (1963).
45. Kh. Rustamov, A. Yuldashev, and R. Usmanov, *Uzb. Khim. Zhur., No. 2*: 24 (1960).
46. E. Mariani and F. Baldass, *Ricerea Sci., 20*: 324 (1950).
47. A. Galat, *J. Am. Chem. Soc., 70*: 3945 (1948).
48. G. V. Austerweil and R. Palland, *Bull. Soc. Chim. France,* 678 (1953).
49. K. Ueno and Y. Yamaguchi, *J. Chem. Soc. (Japan), 55*: 234 (1952).
50. G. Durr and P. Mastagli, *Compt. Rend, 235*: 1038 (1952).
51. G. Durr, *Compt. Rend., 236*: 1571 (1953).
52. P. Mastagli, A. Eloch, and G. Durr, *Compt. Rend.,* 1402 (1952).
53. H. Schlenk and R. Holman, *J. Am. Oil Chemists Soc., 30*: 103 (1953).
54. C. Schmidle and R. Mansfield, U.S. Patent No. 2,658,070 (1953).
55. H. Jenny, *J. Colloid Sci., 1*: 33 (1946).
56. C. W. Davies and G. G. Thomas, *J. Chem. Soc., II*: 1607 (1952).
57. S. Sussman, *Ind. Eng. Chem., 38*: 1228 (1946).
58. M. Marian, *Ann. Chim. Appl., 39*: 717 (1949).
59. N. G. Polyanskii and P. E. Tulupov, *Zhu. Prikl. Khim., 36*: 2244 (1963).
60. B. George and K. Robert, *J. End. Eng. Chem., 43*: 1082 (1951).
61. N. G. Polyanskii and P. E. Tulupov, *Zhur. Prikl. Khim., 37*: 2686 (1964).
62. F. T. Fang, Proceedings of the Third International Congress on Catalysis, Amsterdam, *2,* p. 90 (1964).
63. N. G. Polyanskii and N. L. Potudina, *Neftekhimiya, 3*: 706 (1963).
64. H. Noller and P. E. Gruber, *Z. Phys. Chem., 38*: 184 (1963).
65. H. Noller and P. E. Gruber, *Z. Phys. Chem., 38*: 2031 (1963).
66. W. Bauman, *J. Amer. Chem. Soc., 69*: 2830 (1947).
67. M. I. Balashov, L. A. Serafimov, K. M. Saldadze, and S. V. L'Vov, *Plast. Massy, No. 5*: 56 (1967).
68. H. Spes, *Chem.-Ztg., 90*: 443 (1966).
69. N. Lifshutz and J. S. Dranoff, *Ind., Eng. Chem., Process Des. Devel., 7*: 266 (1968).
70. K. J. Steinbach, K. S. Grunert, and K. Taufel, *Nahrung, 5*: 617 (1961).
71. F. Helfferich, *Ion Exchange* McGraw-Hill, New York (1962), p. 524–594.
72. R. Tartavelli, G. Nensetti, and M. Baccaredda, *Ann. Chim. (Italy), No. 5–6*: 1108 (1966).
73. H. Saito and F. Shimamoto, *Kogyo Kogekuzasshi, 64*: 1733 (1961).
73a. H. Saito and F. Shimamoto, *Chem. Abs., 57*: 28986 (1962).
74. R. Tartarelli, *Ann. Chim. (Italy), 56*: 156 (1966).
75. R. W. Reed and J. S. Dranoff, *Ind. Eng. Chem. Fund., 3*: 304 (1969).
76. F. Andeas, *Chem. Tech. (Berlin), 11*: 24 (1959).
77. D. Seletan and R. White, *Chem. Eng. Progr., 48*: 59 (1952).
78. P. D. Bolton and T. Henshall, *J. Chem. Soc., II*: 1226 (1962).
79. M. B. Ordyan, E. T. Eidus, P. A. Sarkisyan, and A. E. Akopyan, *Arm. Khim. Zhur., 19*: 632 (1966).
80. R. C. Odioso, A. M. Henke, and J. K. Frech, *Ind. Eng. Chem., 53*: 209 (1961).
81. H. A. Smith and J. H. Steele, *J. Am. Chem. Soc., 63*: 3466 (1941).
82. H. A. Smith and C. H. Reichardt, *J. Am. Chem. Soc., 63*: 605 (1941).
83. F. Hellferich, *J. Am. Chem. Soc., 76*: 5567 (1954).
84. A. Abdel Razik, M. S. Metwally, M. E. El-Hadi, and M. A. El Wardany *React. Kinet Catal. Lett., 48(1)*: 279 (1992).
85. K. J. Laidler, *Chemical Kinetics,* 11th reprint, p. 25a, TATA, McGraw-Hill Publishing Co. Ltd., New Delhi, (1985), p. 238–253.
86. C. K. Ingold, *Structure and Mechanism in Organic Chemistry,* 2nd ed. Bell, London (1969).
87. P. R. Wells, S. Ehrenson, and R. W. Taft, *Prog. Phys. Org. Chem., 6*: 147 (1968).
88. D. D. Roberts, *J. Org. Chem., 30*: 3516 (1965).
89. D. D. Roberts, *J. Org. Chem., 31*: 4037 (1966).
90. Sic. Rakshit and M. K. Sarkar, *J. Indian Chem. Soc., 48*: 605 (1971).
91. L. Singh, R. T. Singh, and R. C. Jha, *J. Indian Chem. Soc., 57*: 1089 (1980).
92. L. Singh, R. T. Singh, and R. C. Jha, *J. Indian Chem. Soc., 58*: 966 (1981).
93. I. M. Sidahmed, S. M. Salem, and F. M. Abdel Halim, *J. Indian Chem. Soc., 59*: 1139 (1982).
94. R. C. Jha, A. K. Gupta, R. Kumar, B. Singh, and L. Singh, *J. Indian Chem. Soc., 62*: 157 (1985).
95. B. Singh, A. K. Gmpta, D. B. Pathak, V. K. Singh, Y. P. Singh, and R. T. Singh, *J. Indian Chem. Soc., 66*: 377 (1989).
96. K. P. Singh, *J. Indian Chem. Soc., 67*: 463 (1990).
97. D. P. Singh and S. Prasad, *J. Indian Chem. Soc., 67*: 114 (1990).
98. M. S. Metwally, *Intern. J. Chem. (India), 2*: 37 (1991).
99. C. K. Ingold, *Structure and Mechanism in Organic Chemistry,* Cornell University Press, Ithaca, New York (1967).
100. K. J. Laidler and P. A. Landskoener, *Trans Faraday Soc., 52*: 200 (1956).
101. A. J. Parker, *Chem. Rev., 69*: 1 (1969).
102. F. Zidan, M. S. Metwally, and M. Abd El-Zahir, *J. Chem. Tech. Biotechnol, 56*: 151 (1993).
103. R. K. Wolford, *J. Phys. Chem., 67*: 632 (1963).

104. W. F. K. Wynne-Jones and H. Eyring, *J. Chem. Phys.*, *3*: 492 (1935).
105. R. K. Hudson, and J. Saville, *J. Chem. Soc.*, 4114 (1955).
106. M. Qureshi and S. Z. Qureshi, *Anal. Chim. Acta, 34*: 108 (1966).
107. P. W. West, M. Qureshi, and S. Z. Qureshi, *Anal. Chim. Acta, 47*: 97 (1969).
108. M. Qureshi, S. Z. Qureshi, and N. Zehra, *Anal. Chim. Acta, 47*: 169 (1963).
109. S. Z. Qureshi, M. S. Rathi, and S. Bono, *Anal. Chem.*, *46*: 1139 (1974).
110. M. Qureshi, S. Z. Qureshi, and S. C. Sihghal, *Anal. Chem., 40*: 1781 (1968).
111. M. Fujimoto, *Chem. Anal. (London), 49*: 4 (1960).
112. P. W. Miller, *Anal. Chem., 30*: 1462 (1958).
113. L. Legrodi, *Mayy. Kem. Foly., 76*: 66 (1960).
114. J. P. Rawat and P. S. Thind, *Cand. J. Chem., 54*: 1892 (1976).
115. P. S. Thind and S. S. Sandhe, *J. Indian Chem. Soc., 56*: 260 (1979).
116. M. Honda, *J. Chem. Soc. Jpn., 72*: 638 (1951).
117. M. Qureshi, S. Z. Qureshi, and N. Zehra, *Talanta, 19*: 377 (1972).
118. M. F. El-Hadi and M. S. Metwally, *J. Indian Chem. Soc., 62*: 774 (1985).
119. R. Kunin, *Ion-Exchange Resins*, John Wiley, New York, p. 341 (1958).
120. H. P. Gregor, F. Gultoff, and J. I. Pregman, *J. Colloid Sci., 9*: 245 (1951).
121. T. Dheiveegan and S. Krishnamoorthy, *J. Indian Chem. Soc., 65*: 731 (1988).
122. S. Ramachandran and S. Krishnamoorthy, *Indian J. Tech., 22*: 355 (1984).
123. J. Ragunathan and S. Krishnamoorthy, *J. Indian Chem. Soc., 61*: 911 (1984).
124. M. B. Chandrasekaran and S. Krishnamoorthy, *J. Indian Chem. Soc., 64*: 134 (1987).
125. N. Duraiswamy and S. Krishnamoorthy, *J. Indian Chem. Soc., 64*: 701 (1987).
126. N. L. N. Shrma, M. Joseph, and P. Vasudevan, *Res. Ind. (New Delhi), 21*: 173 (1976).
127. P. Vasudevan and N. L. N. Sharma, *J. Appl. Polym. Sci., 24*: 1443 (1979).
128. A. Havranek, *Polym. Bull., 8*: 133 (1982).
129. S. M. Saad, S. M. Sayyah, N. E. Metwally, and A. A. Mourad, *Acta Polymerica, 39*: 568 (1988).
130. A. B. Mostafa, S. M. Sagyah, N. M. Gawish, and A. A. Mourad, *Egypt. J. Chem., 27*: 737 (1984).
131. W. Fuchs, *J. Am. Chem. Soc., 58*: 673 (1936).
132. S. M. Sayyah, A. I. Sabry, I. A. Sabbah, and N. E. Metwally, *Acta Polymerica, 42*: 670 (1991).
133. M. S. Metwally and N. S. Metwally, *Polym. Plast. Technol. Eng., 31*: 773 (1992).
134. R. R. Naffziger and H. I. Mahon, *J. Agr. Good Chem., 1*: 847 (1953).
135. S. M. Saad, A. M. Naser, M. T. Zimaity, and H. F. Abdel-Maged, *J. Oil Colour Chem. Assoc., 61*: 43 (1978).
136. M. S. Metwally, N. E. Metwally, and T. M. Samy, *Die Angewandte Makromolekulare Chemie, 218*: 1 (1994).
137. M. S. Metwally, N. E. Metwally and T. M. Samy, *J. Appl. Polym. Sci., 52*: 61 (1994).

52
Natural Fiber Reinforced Plastics

A. K. Bledzki and J. Gassan
University of Kassel, Kassel, Germany

I. SURVEY OF TECHNICAL APPLICATIONS OF NATURAL FIBER COMPOSITES

Seventy years ago, nearly all resources for the production of commodities and many technical products were materials derived from natural textiles. Textiles, ropes, canvas, and paper were made of local natural fibers, such as flax and hemp. Some of them are still used today. In 1908, the first composite materials were applied for the fabrication of big quantities of sheets, tubes, and pipes in electrotechnical usage (paper or cotton as reinforcement in sheets made of phenol- or melamine-formaldehyde resins). In 1896, for example, airplane seats and fuel tanks were made of natural fibers with a small content of polymeric binders [1].

Because of low prices and steadily rising performance on the part of technical and standard plastics, the application of natural fibers came to a near halt. The current critical discussion about preservation of natural resources and recycling led to a reflection about natural materials, the focus centuring on regrowing raw materials [2].

Compared with Western Europe, India continued using natural fibers, mainly jute fibers, as reinforcement for composites. Pipes, protruded semi-products, for example, and panels with a polyester matrix were produced with these fibers [3]. The government of India promoted large projects where jute-reinforced polyester resins were used for buildings, e.g., the Madras House 1978 [4], or grain elevators. Certainly, applications of natural fibers as construction material for buildings were known long before. For centuries, mixtures of straw and loam dried in the sun were used as construction composites, e.g., in Egypt [1].

Today the renaissance of these materials as reinforcing fibers in technical applications is taking place mainly in the automobile industry and for packaging (e.g., egg-cartons), particularly in Germany. In this area, textile waste has been used for years to reinforce plastics in cars, especially in the Trabant [1]. At present, a K-car series is being developed by Mercedes, where the "K" stands for "kraut" and "compost" [2]. Locally grown fibers, such as flax and hemp, were used for these cars. Ramie fibers were also examined because of their specific properties [5]. As the following components, for example, were developed for these applications [2,6]:

- Door panels (molded wood, natural fiber moldings, laminated)
- Car roofs (composites made of the natural fiber fleece (flax) with epoxy resins or PUR composites)

A survey about the possible applications of natural fibers in automobiles, as presently developed, is shown in Fig. 1.

The use of flax fibers in car disc brakes as a replacement of asbestos fibers is another example of an application of this type of material [7].

II. NATURAL FIBERS

A. Technical Applications of Natural Fibers

In general, natural fibers are subdivided as to their origin, coming from plants, animals, or minerals (Fig. 2). Plant fibers usually are used as reinforcement in plastics. The plant fibers may be hairs, fiber sheafs of dicotylic plants, or vessel sheafs of monocotylic plants (bast and hard fibers).

The basic demand for technical application of these fibers is a defined quality, e.g., according to a specification sheet. These qualities have to be available for the

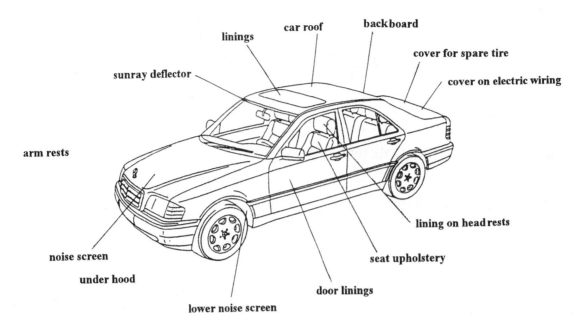

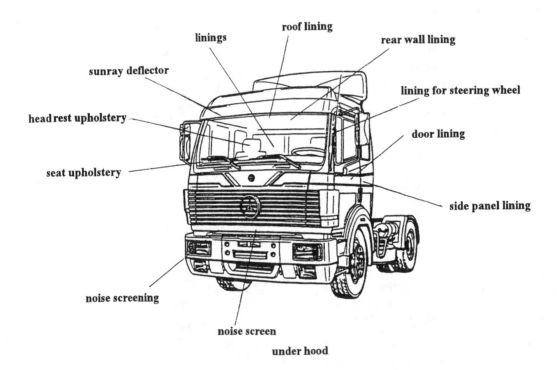

Figure 1 Possibilities of use for natural fibers in automobiles [2].

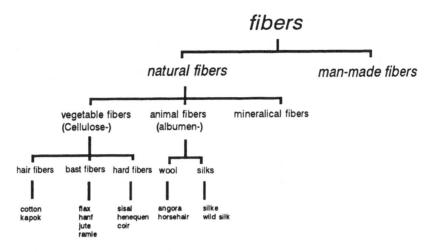

Figure 2 Classification of natural fibers [12].

industry over long periods in sufficient quantities. For several applications the fibers should be prepared or modified with respect to [8]:

- Homogenization of the properties of the fibers
- Degree of elementarization and degumming
- Degree of polymerization and crystallization
- Good adhesion between fiber and matrix
- Moisture repellence
- Flame-retardant properties

These properties can be partly produced by fiber disgestion (steam explosion, steam pressure, thaw roasting process, water roasting process [8,9]), as well as by a direct application to the fiber.

Presently, the demand for plant fibers for technical applications can be partially met (Table 1).

In recent years, prices for natural fibers were not stable, especially for flax fibers. Flax fibers, showing the highest values for strength (Table 2), are about 30% more expensive than glass fibers. Additionally, is price depends on the fiber preparation. Usually, glass fibers are delivered pretreated, i.e., treated with different sizes

for applications in composites. Natural fibers have to be specially treated, depending on different applications, thus raising the costs.

For these economic reasons, a substitution of glass fibers by natural fibers seems not to be easily realized. But natural fibers offer several advantages compared with glass fibers. Plant fibers are renewable raw materials, which could be available in nearly unlimited amounts. Additionally, natural fibers balance the CO_2 in a household. The plant releases the same amount of CO_2 during a combustion process, that was assimilated during growth. In addition, the problems of having to recycle glass fiber reinforced composites would be lessened. A method to compost natural fibers in composites was examined by Gatenholm et al. [10,11] However, this method seems to be practical only for special types of natural fiber composites.

B. Mechanical Properties of Natural Fibers

Natural fibers are principally suitable for reinforcing plastics because of their relatively high values of strength and stiffness (Table 2) [12]. The level of the characteristic values of flax and soft-wood craft fibers nearly reaches the values of glass fibers (types E). Nevertheless, as shown in Table 2, the dispersion of the characteristic values is remarkably higher than those of glass fibers. These values are in part determined by the structure of the fiber. The fiber structure is influenced by several conditions and varies by area of growth, climate, and age of the plant. The technical *disgestion* of the fiber also is an important factor, this determines the structure of the fibers and characteristic values.

As is the case with glass fibers, the tensile strength of natural fibers depends on the test length of the specimens. Examinations were made by Kohler et al. [8],

Table 1 Production of Plant Fibers in Comparison with Glass Fibers (1993) [60]

Fiber	Price in comparison to glass fibers (%)	Production (1000 t)
Jute	18	3600
E-glass	100	1200
Flax	130	800
Sisal	21	500
Banana	40	100
Coir	17	100

Table 2 Mechanical Properties of Natural Fibers in Comparison to Conventional Reinforcement Fibers

Fiber	Density (g/cm³)	Elongation (%)	Tensile strength (MPa)	Youngs modulus (GPa)	Reference
Cotton	1.5–1.6	7.0–8.0	287–597	5.5–12.6	[80–82]
Jute	1.3	1.5–1.8	393–773	26.5	[63,80,81,83]
Flax	1.5	2.7–3.2	345–1035	27.6	[80]
Hemp	—	1.6	690	—	[84]
Ramie	—	3.6–3.8	400–938	61.4–128	[80,82]
Sisal	1.5	2.0–2.5	511–635	9.4–22.0	[80,81,85]
Coir	1.2	30.0	175	4.0–6.0	[81,85]
Viscose (cord)	—	11.4	593	11.0	[82]
Soft wood kraft	1.5	—	1000	40.0	[86]
E-glass	2.5	2.5	2000–3500	70.0	[87]
S-glass	2.5	2.8	4570	86.0	[81,87]
Aramide (normal)	1.4	3.3–3.7	3000–3150	63.0–67.0	[87]
Carbon (standard)	1.4	1.4–1.8	4000	230.0–240.0	[87]

Mieck et al. [13], and Mukherjee et al. [14] on different types of flax and pineapple fibers. Their results show that tensile strength of flax fibers is remarkably more dependent on the length of the specimen than are usual glass fibers (Fig. 3). In contrast, tensile strength of pineapple fibers is less dependent on the length. The dispersion of the measured values is located mainly in the range of the standard deviation.

Tensile strength of the fibers is also determined by the refinement of the fiber [14] (Fig. 4). Hydrophilic properties are a major problem for all cellulose fibers. The moisture content of the fibers amounts to 10 wt% at standard atmosphere. Their hydrophilic behavior influences the properties of the fiber itself (Table 3) as well as the properties of the composite at production [15].

Table 3 Strength and Elongation at Break of Dry and Wet Fibers [88]

Fiber	Relative tensile strength (wet) (%)	Relative elongation (wet) (%)
Cotton	105–110	110–116
Jute	100–105	100
Flax	102–106	125–133
Ramie	115–125	100
Sisal	90–120	100
Glass	75–100	100
Aramide	78–80	—
Carbon	100	100

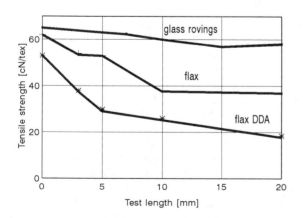

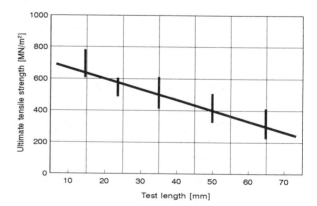

Figure 3 Dependence of tensile-strength on test length, flax fibers [70], and pineapple fibers [88] compared with textile glass-fibers [8].

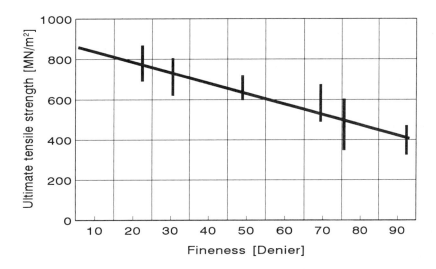

Figure 4   Dependence of tensile-strength on fiber refinement [14].

C.   Natural Fibers and Their Chemical Compositions

Climatic conditions, age, and the digestion process influence not only the structure of fibers but also the chemical composition. Mean values of components of plant fibers are shown in Table 4. With the exception of cotton, the components of natural fibers are cellulose, hemi-cellulose, lignin, pectin, waxes, and water-soluble substances.

1.   Cellulose

Cellulose is the essential component of all plant fibers. It is an isotactic β-1,4-polyacetal of cellubiose. The basic

unit, cellubiose, is composed of two molecules of glucose. As a result, cellulose is often called a polyacetal of glucose. The summation formula for cellulose is (P-degree of polymerization):

$$C_{6P} H_{10P + 2} O_{5P + 1}$$

elementary composition:

   44.4% carbon

   6.2% hydrogen

   49.4% oxygen

with the molecular weight of $m_0 = 162$.
The constitutional formula is given by [16]:

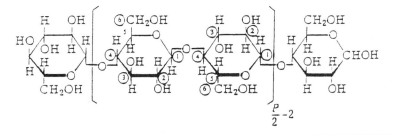

The degree of polymerization shows that the length of the polymer chains varies (Table 5).

Table 4   Components of Plant Fibers [12]

	Cotton	Jute	Flax	Ramie	Sisal
Cellulose	82.7	64.4	64.1	68.6	65.8
Hemi-cellulose	5.7	12.0	16.7	13.1	12.0
Pektin		0.2	1.8	1.9	0.8
Lignin	—	11.8	2.0	0.6	9.9
Water soluble	1.0	1.1	3.9	5.5	1.2
Wax	0.6	0.5	1.5	0.3	0.3
Water	10.0	10.0	10.0	10.0	10.0

Table 5   Degrees of Polymerization of Different Natural Fibers [89]

Fiber	P_n
Cotton	7,000
Flax	8,000
Ramie	6,500

Solid cellulose forms a microcrystalline structure with regions of high order, i.e., crystalline regions, and regions of low order that are amorphous. Naturally occurring cellulose (cellulose I) crystallizes monoclinic sphenodic. The molecular chains lay in the fiber direction:

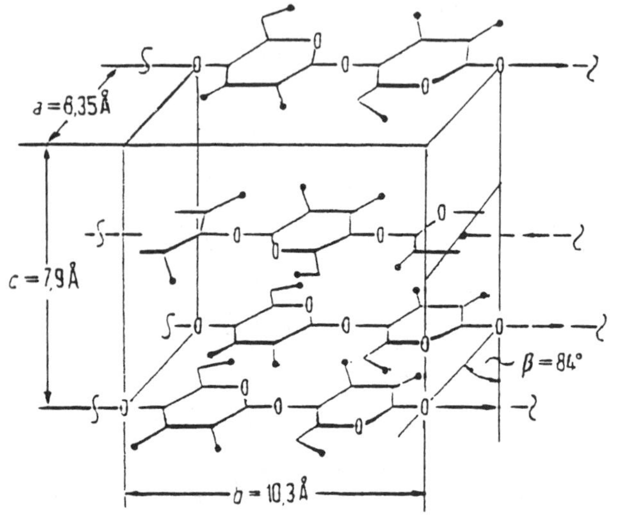

The geometry of the elementary cell is dependent on the type of cellulose (Table 6).

The mechanical properties of natural fibers depend on cellulose type because each type of cellulose has a specific cell geometry and the geometrical conditions determine the mechanical properties.

Fink et al. [17] correlated measurements from different authors and test methods to compare Young's modulus for cellulose of type I and II. Most of the authors determined higher characteristic values for type I than for type II (Table 7).

2. Additional Components

Lignin is a high-molecular, mainly three-dimensional aromatic substance, that, in contrast to polysaccharides, can be hydrolyzed only in small partitions [16]:

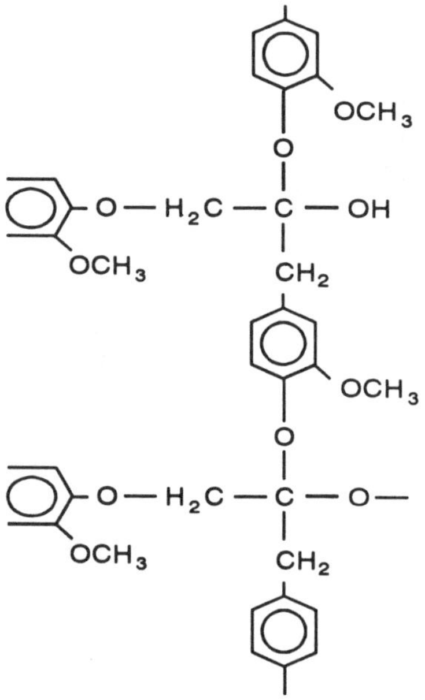

Table 7 Axial Young's Modulus for Cellulose [17]

Method	Young's modulus of cellulose (GPa)		Material
	Cellulose I	Cellulose II	
X-ray		70–90	Fortisan
X-ray	74–103		Flax, Hemp
X-ray	110		Flax
X-ray	130	90	Ramie
X-ray	120–135	106–112	Ramie
Calculated	136	89	
Calculated	168	162	
	134		Wood

The mechanical properties are distinctly lower than those of cellulose. At the value of 4 GPa for Young's

Table 6 Lattice Parameters of Elementary Cells in Different Types of Cellulose [90]

Type	Source	Dimensions (nm)			
		a	b	c	β (°)
Cellulose I	Cotton	0.821	1.030	0.790	83.3
Cellulose II	Cotton,	0.802	1.036	0.903	62.8
	Mercerized Viskose	0.801	1.036	0l.904	62.9
Cellulose III		0.774	1.030	0.990	58.0
Cellulose IV		0.812	1.030	0.799	90.0

modulus, the mechanical properties of isotropic lignin are distinctly lower than those of cellulose [18].

Pectin is a collective name for heteropolysaccharides, which consist essentially of polygalacturon acid. Pectin is soluble in water only after a partial neutralization with alkali or ammonium hydroxide [18].

Waxes make up the part of the fibers that can be extracted with organic solutions. These waxy materials consist of different types of alcohols, which cannot be solubilized in water as well as in several acids (palmitic acid, oleaginous acid, stearic acid) [18].

D. Physical Structure of Natural Fibers

The filaments of all plant fibers consist of several cells. These cells form crystalline microfibrils (cellulose), which are connected together into a complete layer by amorphous lignin and hemi-cellulose. Multiple layers stick together to form multiple layer composites, filaments. A single cell is subdivided into several concentric layers, one primary and three secondary layers. Figure 5 shows a jute cell. The cell walls differ in their composition and in the orientation of the cellulose microfibrils whereby the characteristic values change from one natural fiber to another.

The angle of the fibrils and the content of cellulose determine the properties of the plant fibers. The Hearle et al.'s model [19] considers only these two structure parameters. For the description of stiffness, solely, the S_2 layers were considered because the properties of these fibers were decisively dominated by the amount of these layers.

1. Proportion of Crystallinic Fibrils and Their Orientation

Many authors have tried to describe the mechanical characteristic values of natural fibers based on their structure, for example, Hearle in 1943. Thereafter followed models from Cowdrey and Preston in 1966, Page et al. in 1971, Brinson in 1973, Gordon and Jeronimides in 1974, Jeronimides in 1976, and Mc Laughlin and Tait in 1980. All these theories solely considered two parameters: content of cellulose and angle of fibrils [18].

The model described by Hearle et al. [19] allows a simple presentation of these relations. The basic idea is shown in Fig. 6.

This model was applied by Mukherjee et al. [20] for various natural fibers. By considering diverse mechanisms of deformation they arrived at different calculation possibilities for the stiffness of the fiber. According to Eq. (1), the calculation of Young's modulus of the fibers is based on an isochoric deformation. This equation sufficiently describes the behavior for small angles of fibrils ($<45°$) [19].

$$E = [X_{1C}E_{1C} + (1 - X_{1C})\, E_{NC}] \cos^2 X_2 := E^* \cos^2 X_2 \tag{1}$$

A springlike deformation of the fibrils overweighs at angles $>45°$. That means the length of the fibrils remains constant and Young's modulus of the fibers can be given by:

$$E_1 = \frac{K_{NC}}{1 - X_{1C}} (1 - 2 \cot^2 X_2)^2 = K(1 - 2 \cot^2 X_2)^2$$

It is possible to formally describe the stiffness of the fiber across the range of fibril angles with a suitable combination, i.e., a connection in the series of equations already mentioned [18].

$$E_1 = \frac{E^* \cos^2 X_2 \, [K(1 - \cot^2 X_2)^2]}{E^* \cos^2 X_2 + K(1 - 2 \cot^2 X_2)^2}$$

Nevertheless, the diagram of the characteristic values of plant fibers as dependent on structure parameters

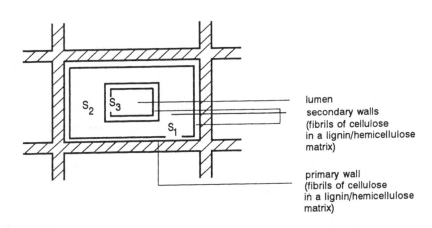

Figure 5 Constitution of a jute cell (diagrammatic drawing) [80].

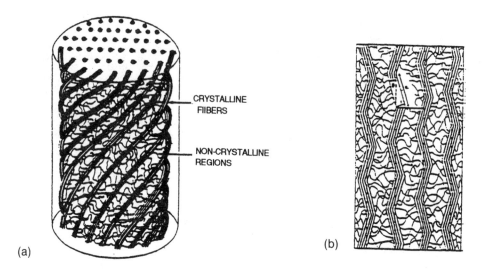

Figure 6   Model for the description of the stiffness of the fibers [19]. (a) S_2-layers in a 3-D view, (b) S_2-layers projected into a 2-D view.

show that the mechanical behavior (strength) of the fibers cannot be described sufficiently (Fig. 7).

The mechanical and physical properties of plant fibers are also influenced by the following structure parameters [17,21]:

- Degree of polymerization (DP)
- Crystal structure (type of cellulose and defects)
- Supramolecular structure (e.g., degree of crystallinity)

- Orientation of chains (noncrystalline and crystalline regions)
- Void structure (content of voids, specific interface, void size)
- Fiber diameter

2.   Cell Dimensions

The empiric relation developed by Murkherje et al. [20] shows a proportionality between mechanical and physi-

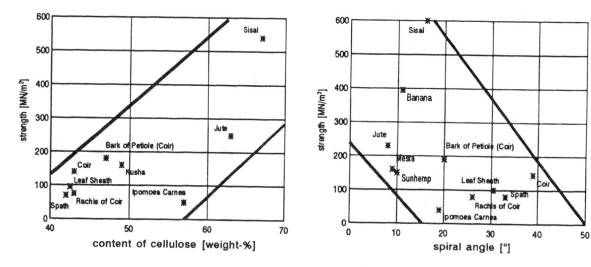

Figure 7   Dependence of fiber strength on the content of cellulose and on the angle of fibrils [91].

Table 8 Structure Parameters of Diverse Plant Fibers [20]

Fiber	Cellulose content (wt%)	Spiral angle (%)	Cross-sectional area $A \times 10^{-2}$ (mm^2)	Cell length "L" (mm)	"L/D" ratio (−)
Jute	61	8.0	0.12	2.3	110
Flax	71	10.0	0.12	20.0	1687
Hemp	78	6.2	0.06	23.0	960
Ramie	83	7.5	0.03	154.0	3500
Sisal	67	20.0	1.10	2.2	100
Coir	43	45.0	1.20	3.3	35

D-cell diameter.

cal properties Y, to the content of cellulose X_1, the angle of fibrils X_2, and the size of cells X_3.

$$Y \propto X_1^{a_1} X_2^{a_2} X_3^{a_3}$$

Table 8 also shows that the different plant fibers vary in the geometry of their cells.

III. METHODS FOR SURFACE MODIFICATION OF NATURAL FIBERS

The quality of the fiber matrix interface is significant for the application of natural fibers as reinforcement fibers for plastics. Physical and chemical methods can be used to optimize this interface. These modification methods are of different efficiency for the adhesion between matrix and fiber.

A. Physical Methods of Natural Fiber Modification

Reinforcing fibers can be modified by physical and chemical methods. Physical methods, such as stretching [22], calandering [23,24], thermotreatment [25], and the production of hybrid yarns [26,27] do not change the chemical composition of the fibers. Physical treatments change structural and surface properties of the fiber and thereby influence the mechanical bondings in the matrix.

Electric discharge (corona, cold plasma) is another method of physical treatment. Corona treatment is one of the most interesting techniques for surface oxidation activation. This process changes the surface energy of the cellulose fibers [28]. In the case of wood surface activation it increases the amount of aldehyde groups [29].

The same effects are reached by cold plasma treatment. Depending on the type and the nature of the used gases, a variety of surface modification can be achieved.

Surface crosslinkings could be introduced, surface energy could be increased or decreased, and reactive free radicals [28] and groups [30] could be produced.

Electric discharge methods are known [31] to be very effective for "nonactive" polymer substrates such as polystyrene, polyethylene, polypropylene, etc. They are successfully used for cellulose-fiber modification to decrease the melt viscosity of cellulose-polyethylene composites [32] and to improve the mechanical properties of cellulose-polypropylene composites [28].

An older method of cellulose fiber modification is mercerization [22,33–36], which has been widely used on cotton textiles. Mercerization is an alkali treatment of cellulose fibers. It depends on the type and concentration of the alkalic solution, its temperature, time of treatment, tension of the material, and the additives used [33,36]. At present there is a tendency to use mercerization for natural fibers as well. Optimal conditions of mercerization ensure the improvement of the tensile properties [33–35,37] and absorption characteristics [33–35], which are important in the composing process.

B. Chemical Methods of Modification for Natural Fibers

Strongly polarized cellulose fibers [38] inherently are rarely compatible with hydrophobic polymers [28, 39–41]. When two materials are incompatible, it is often possible to bring about compatibility by introducing a third material that has properties that are intermediate between those of the other two. There are several mechanisms [42] of coupling in materials:

1. Weak boundary layers—coupling agents eliminate weak boundary layers.
2. Deformable layers—coupling agents produce a tough, flexible layer.

3. Restrained layers—coupling agents develop a highly crosslinked interphase region with a modulus intermediate between that of the substrate and the polymer.

4. Wettability—coupling agents improve the wetting between polymer and substrate (critical surface tension factor).

5. Chemical bonding—coupling agents form covalent bonds with both materials.

6. Acid–base effect—coupling agents alter acidity of substrate surface.

The development of a definitive theory for the mechanism of bonding by coupling agents in composites is a complex problem. The main chemical bonding theory alone is not sufficient. Consideration of other concepts, including the morphology of the interphase, acid–base reactions at the interface, surface energy, and wetting phenomena, appears to be necessary.

1. Change of Surface Tension

The surface energy of fibers is closely related to the hydrophilicity of the fiber [38]. Some investigations are concerned with methods to decrease hydrophilicity. The modification of wood cellulose fibers with stearic acid [43] hydrophobizes those fibers and improves their dispersion in polypropylene. As can be observed in jute-reinforced unsaturated polyester resin composites, treatment with polyvinylacetate increases the mechanical properties [24] and moisture repellency.

Silane coupling agents may contribute hydrophilic properties to the interface, especially when amino functional silanes, such as epoxies and urethane silanes, are used as primers for reactive polymers. The primer may supply much more amine functionality than can possibly react with the resin at the interphase. Those amines that could not react are hydrophilic and, therefore, responsible for the poor water resistance of bonds. An effective way to use hydrophilic silanes is to blend them with hydrophobic silanes such as phenyltrimethoxysilane. Mixed siloxane primers also have an improved thermal stability, which is typical for aromatic silicones [42].

2. Impregnation of Fibers

A better combination of fiber and polymer is achieved by an impregnation of [44] the reinforcing fabrics with polymer matrixes compatible with the polymer. Polymer solutions [40,45] or dispersions [46] of low viscosity are used for this purpose. For a number of interesting polymers, the lack of solvents limits the use of the method of impregnation [44]. When cellulose fibers are impregnated with a bytyl benzyl phthalate plasticized polyvinylchloride (PVC) dispersion, excellent partitions can be achieved in polystyrene (PS). This significantly lowers the viscosity of the compound and the plasticator and results in cosolvent action for both PS and PVC [46].

3. Chemical Coupling

An important chemical modification method is the chemical coupling method. This method improves the interfacial adhesion. The fiber surface is treated with a compound that forms a bridge of chemical bonds between fiber and matrix.

4. Graft Copolymerization

An effective method of NVF chemical modification is graft copolymerization [34,35]. This reaction is initiated by free radicals of the cellulose molecule. The cellulose is treated with an aqueous solution with selected ions and is exposed to a high-energy radiation. Then, the cellulose molecule cracks and radicals are formed. Afterwards, the radical sites of the cellulose are treated with a suitable solution (compatible with the polymer matrix), for example vinyl monomer [35] acrylonitrile [34], methyl methacrylate [47], polystyrene [41]. The resulting copolymer possesses properties characteristic of both fibrous cellulose and grafted polymer.

For example, the treatment of cellulose fibers with hot polypropylene–maleic anhydride (MAH–PP) copolymers provides covalent bonds across the interface [40]. The mechanism of reaction can be divided in two steps:

1. Activation of the copolymer by heating ($t = 170°C$) (before fiber treatment), and

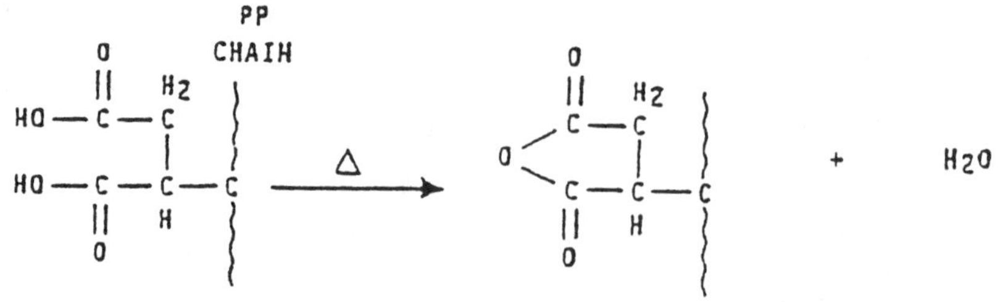

2. Esterification of cellulose.

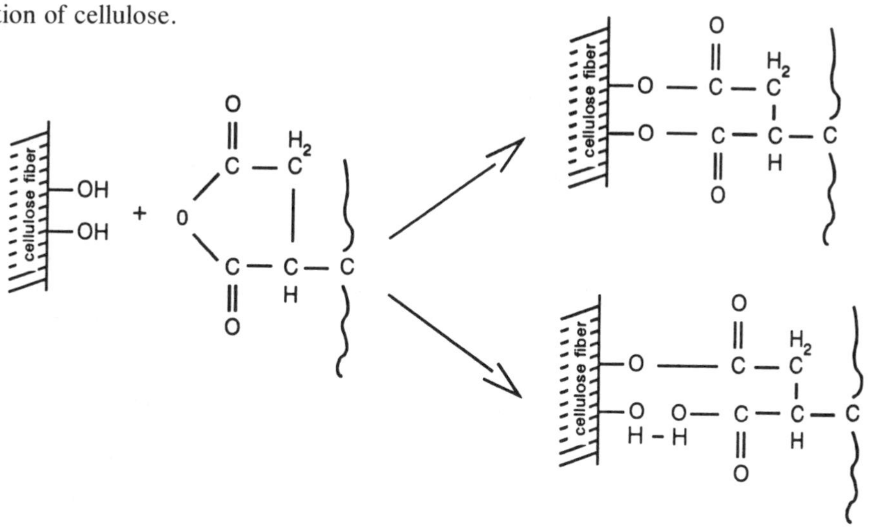

After this treatment the surface energy of the fibers is increased to a level much closer to the surface energy of the matrix. Thus, a better wettability and a higher interfacial adhesion are obtained. The polypropylene (PP) chain permits segmental crystallization and cohesive coupling between modified fiber and PP matrix [40]. The graft copolymerization method is effective, but complex.

5. Treatment with Compounds That Contain Methylol Groups

Chemical compounds that contain methylol groups ($-CH_2OH$) form stable, covalent bonds with cellulose fibers. Those compounds are well known and widely used in textile chemistry. Hydrogen bonds with cellulose can be formed in this reaction as well. The treatment of cellulose with methylolmelamine compounds before forming cellulose unsaturated polyesters (UP) composites decreases the moisture pickup and increases the wet strength of reinforced plastic [48,49].

6. Treatment with Isocyanates

The mechanical properties of composites reinforced with wood fibers and PVC or PS as resin can be improved by an isocyanate treatment of those cellulose fibers [41,50] or the polymer matrix [50]. Polymethylene-polyphenyl-isocianate (PMPPIC) in pure state or solution in plasticizer can be used. PMPPIC is chemically linked to the cellulose matrix through strong covalent bonds (Fig. 8).

Both PMPPIC and PS contain benzene rings, and their delocalized II electrons provide strong interactions. As a result, that there is an adhesion between PMPPIC and PS (Fig. 8). Comparing both methods, treatment with silanes or treatment with isocyanates, it is obvious that the isocyanatic treatment is more effec-

tive than the treatment with silane. Equal results are obtained when PMPPIC is used for the modification of the fibers or the polymer matrix [41].

7. Triazine Coupling Agents

Following is a schematic illustration of how triazine derivatives form covalent bonds with cellulose fibers:

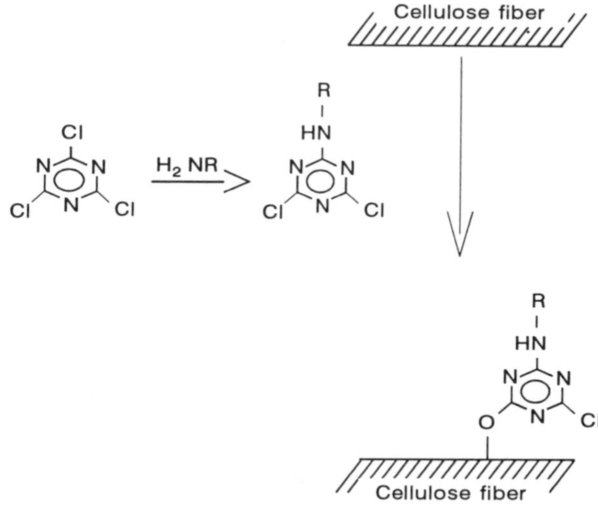

The reduction of the moisture absorption of cellulose fibers and their composites treated with triazine derivates is explained by [51,52]:

1. Reducing the number of cellulose hydroxyl groups that are available for moisture pickup.
2. Reducing the hydrophilicity of the fiber's surface.
3. Restraining the swelling of the fiber by creating a crosslinked network, due to covalent bonding, between matrix and fiber.

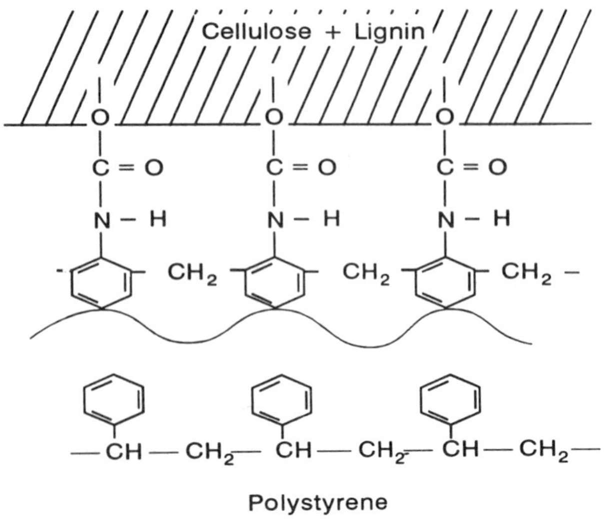

Figure 8 Hypothetical chemical structure of the cellulose-PMPPIC-PS interface area [41].

8. Organosilanes as Coupling Agents

Organosilanes are the main group of coupling agents for glass fiber-reinforced polymers. They have been developed to couple virtually any polymer to the minerals that are used in reinforced composites [42].

Most of the silane coupling agents can be represented by the following formula:

$$R - (CH_2)_n - Si (OR')_3$$

where $n = 0$–3, $OR' = $ hydrolyzable alkoxy group, and R = functional organic group.

The organofunctional group (R) in the coupling agent causes the reaction with the polymer. This could be a copolymerization and/or the formation of an interpenetrating network. This curing reaction of a silane-treated substrate enhances the wetting by the resin (Table 9).

The general mechanism of how alkoxysilanes form bonds with a fiber surface that contains hydroxyl groups is depicted in Figure 9.

Alkoxysilanes undergo hydrolysis, condensation (catalysts for alkoxysilane hydrolysis are usually catalysts for condensation), and a bond formation stage under base as well as under acid catalyzed mechanisms. In addition to this reaction of silanols with hydroxyls of the fiber surface, the formation of polysiloxane structures also can take place.

Analog-to-glass fibers silanes are used as coupling agents for natural fiber polymer composites. For example, the treatment of wood fibers with product A–175 improves wood dimensional stability [53]. In contrast, a decrease of mechanical properties was observed for coir-UP composites after a fiber modification with dichloromethylvinyl silane [54]. The treatment of mercer-

Table 9 Characteristics of Representative Commercial Silane Coupling Agents [38,91]

Organofunctional group	Chemical structure	OSi-Specialties Germany GmbH product	Critical surface tension of glass with silane treatment [dyne/cm]	Applied polymers (abbreviations according ASTM 1600)
Vinyl	$CH_2 = CHSi (OCH_3)_3$	A-171	25.0	UP
Vinyl	$CH_2 = CHSi (OC_2H_5)_3$	A-151	30.0	UP, PE, PP, DAP, EPDM, EPM EP
Chloropropyl	$ClCH_2CH_2CH_2 Si (OCH_3)_3$	A-143	40.5	
Epoxy	$\underset{CH_2CHCH_2O(CH_2)_3Si(OCH_3)_3}{\overset{O}{\triangle}}$	A-187	38.5–42.5	EP, PA, PC, PF, PVC, PUR
Methacrylate	$\underset{CH_2 = C\text{-}COO(CH_2)_3Si(OCH_3)_3}{\overset{CH_3}{\vert}}$	A-174	28.0	UP, PE, PP, DAP, EPDA, EPM
Primary amine	$H_2N(CH_2)_3Si(OC_2H_3)_3$	A-1100	35.0	UP, PA, PC, PUR, MF, PF, PI, MPF
Diamine	$H_2N(CH_2)_2NH(CH_2)_3Si(OCH_3)_3$	A-1120	33.5	
Cationic styryl	$CH_2 = CHC_6H_4CH_2N^+H_2(CH_2)_3Si(OCH_3)_3Cl^-$	—	—	all polymers
Phenyl	$C_6H_5Si(OCH_3)_3$	—	40.0	PS, addition to amine silanes
Mercapto	$HS(CH_2)_3Si(OCH_3)_3$	A-189	41.0	EP, PUR, SBR, EPDM

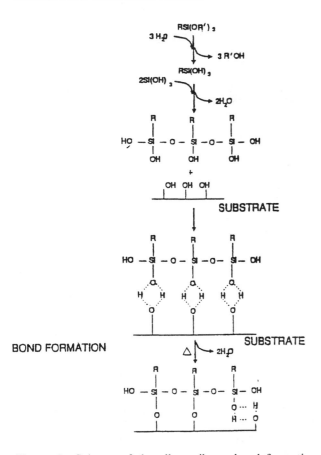

Figure 9 Scheme of the alkoxysilanes bond formation with hydroxyl groups contained by the fiber surface.

ized sisal fiber with Aminosilane–1100 [55] before forming sisal-epoxy composites markedly improves moisture repellency of the composite. These examples show that theories used for the silane treatment of natural fibers are not perfect and are contradictory, therefore, further studies are necessary.

C. Influence of Coupling Agents on the Mechanical Properties of Composites

As shown in Table 10, the surface modifications cause noticeable increases of the characteristic values of composites, depending on the fiber, matrix, and type of surface treatment used.

In addition to improving the mechanical properties, it is possible to minimize moisture sensivity by using, for example, silanes as coupling agents. For unsilanized composites, such as jute composites (Fig. 10), at a maximum humidity of 5.2 wt%, the strength values reduced to about 65% of the strength of dry composites were measured. By silanizing the fiber surface of jute, it was possible to reach a strength of the composite almost independent of humidity. Similarly, a slower decline of the dynamic characteristic values of the composite in respect to humidity could be noticed (Wöhler chart) when using silane coupling agents (Fig. 11). Silanized jute epoxy-resin composites with an average humidity of 4.5 wt% showed Wöhler chart values corresponding to results of unsilanized composites at a humidity of 2.1 wt%. No significant changes on the Sharpy-impact test values were found when silanizing the fiber surface [12].

Table 10 Influence of Coupling Agents on Natural Fiber Reinforced Plastics [12]

Fiber/matrix	Coupling agent	Increase in properties [%]			
		Tensile strength	Young's modulus	Compression strength	Impact energy
Thermosets					
Jute / EP	Acrylic acid	const.	—	—	100
Jute / UP and EP	Polyesteramid polyol	10	10	—	—
Sisal / EP	Silane	25	—	30	—
Cellulose / UP	Dimethylolmelamine	const.	—	—	100
Thermoplastics					
Cellulose / PS	Isocyanate	30	const.	—	50
Cellulose / PP	Stearinic acid	30	15	—	50
	Maleinanhydrid-PP-copolymer	100	const.	—	—
Flax / PP	Silane	const.	50	—	—
	Maleinanhydrid	50	100	—	—

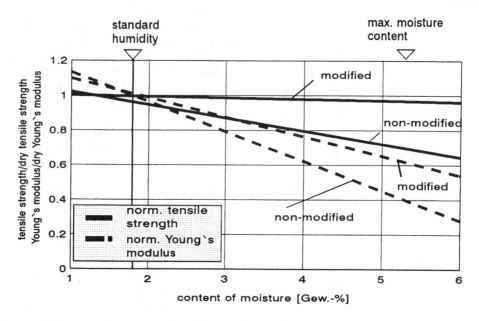

Figure 10   Influence of silane coupling agents on the strength of jute reinforced epoxy-resin composites at different moisture contents [12].

Aside from type of claim (fiber dominant or interfacet matrix dominant), the efficiency of surface treatments depends noticeably on the fiber content within the composite. At a fiber content of 30 vol%, tensile strength increases by 10% and shear strength increases by about 100% (Fig. 12). In contrast to modified fibers where shear strength rises with increasing fiber content, the chart, after having reached a maximum, shows a decreasing tendency on untreated fibers with rising fiber content because high-fiber contents facilitate a slipping of fiber and matrix [13].

As has been extensively discussed, the efficiency of fiber treatment is mainly interdependent with the adhesion between matrix and fiber. This has been shown, for example, in examinations of wood fiber HDPE composites by Raj et al. [56]. Similar results, with definite

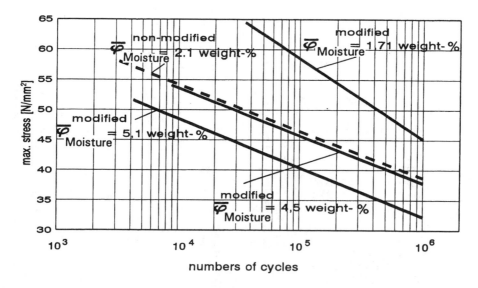

Figure 11   Wöhler chart of silanized and unsilanized jute reinforced epoxy-resin composites at different moisture contents (R = 0.1; f_{test} = 10 Hz; fiber content = 40 vol%) [12].

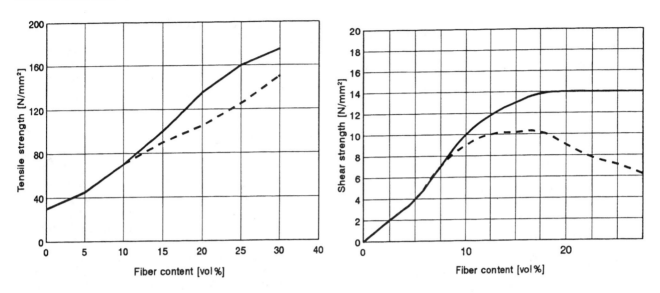

Figure 12 Dependence of tensile strength and shear strength of PP-flax composites on fiber content and on surface treatment. (------) untreated flax; (——) flax pretreated with hostaprime HC5.

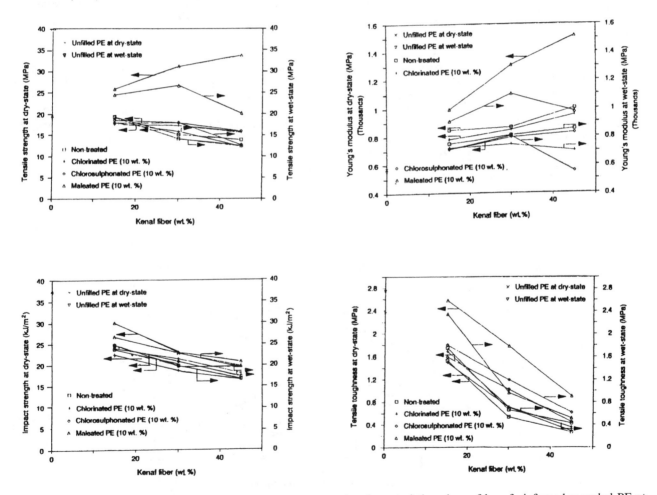

Figure 13 Influence of coupling agents and fiber content on the characteristic values of kenafreinforced recycled PE at room temperature (dry state) and after exposure in boiling water (wet-state) [57].

differences according to the coupling agents used, have been found by Maldas et al. [57] on kenaf-reinforced recycled PE (Fig. 13). Where there is a simultaneous influence of humidity, the examined coupling agents show an improvement only up to a fiber content of about 30 wt%.

IV. PROCESSING OF NATURAL FIBER REINFORCED PLASTICS

A. Influence of Humidity on the Processing of Natural Fiber Composites

Drying of fibers in advance of processing is an important factor because water on the fiber surface acts like a separating agent on the fiber matrix interface. Additionally, because of the evaporation of water during the reaction process, pores appear in the matrix (most of the thermosets have a reaction temperature over 100°C; processing temperature of thermoplastics lies distinctly over the evaporation temperature of water). Both phenomena lead to a decrease of mechanical properties. For jute epoxy resin composites, tensile strength of maximally predried fibers (moisture content: 1 wt%) rises about 10% compared with minimally dried fibers (moisture content: 10 wt%); the increase of stiffness of 20% is remarkably higher (Fig. 14).

Fiber drying can be done in a vacuum stove at different temperatures. This results in different degrees of loss of humidity (Fig. 15).

B. Natural Fiber Reinforced Thermosets

The economically most attractive glass fiber-reinforced plastics for high technical use are, next to RTM and winding technology, etc., semi-products made of SMC and BMC systems.

1. SMC

In the automobile and electronic industries, large amounts of pressed parts from SMC or BMC are used, for example, for bumpers, trunk covers, and spoilers.

In Fig. 16, pressed SMC materials based on flax fibers are compared with those made of glass fibers. It shows that the glass fiber-reinforced material attains higher characteristic values, except for tensile strength. However, if the measured values are regarded in reference to density, the results of the flax fiber-based SMC molded plastics are located in the same range as the glass fiber SMC molded plastics. As previously mentioned, the characteristic values of natural fiber materials are clearly dependent on their moisture content (Fig. 17). After the fibers are dried, they reach similar values as before the storage in the damp stage [58].

2. BMC

Aside from SMC molded plastics, BMC molded natural fiber plastics with good mechanical properties can be produced. Owolabi et al. [59] made such plastics with coir fibers, their basic recipe is shown in Table 11.

In this examination, glass fibers were replaced by coir fibers. A proper treatment of the fibers improved adhesion and, therefore, tensile strength was increased (Fig. 18).

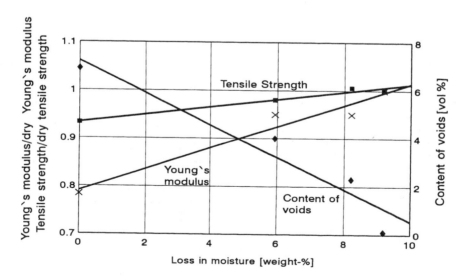

Figure 14 Influence of fiber drying on the characteristic values of jute reinforced epoxy-resin composites [12].

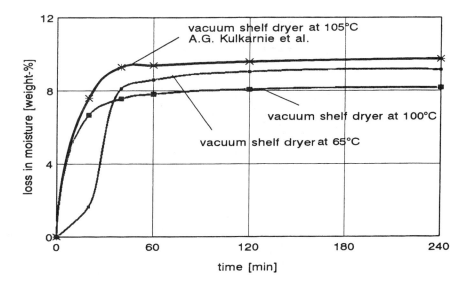

Figure 15 Loss of humidity during drying of jute fibers as dependent on the temperature in the vacuum stove [12].

C. Natural Fiber-Reinforced Thermoplastics

The lower thermal stability of natural fibers, up to 230°C, the thermal stability is only small, which limits the number of thermoplastics to be considered as matrix materials for natural fiber composites. Only those thermoplastics whose processing temperature does not exceed 230°C are usable for natural fiber reinforced composites. These are, most of all, polyolefines, such as polyethylene and polypropylene. Technical thermoplastics, such as poyamides, polyesters, and polycarbonates, require

Table 11 BMC—Recipe on the Basis of Coir Fibers [59]

Materials	Parts by weight
Unsaturated polyester resin	100
$CaCo_3$ filler	75
MgO	3
Styrene	12
Zinc–Stearate	2.5
Tert-Butyl perbenzoate	1.25
Chopped fibrous reinforcement	100

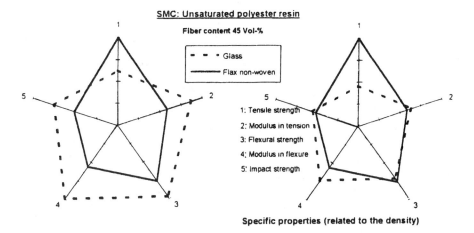

Figure 16 Characteristic values of glass fiber and flax fiber SMC molded plastics (absolute values and in reference to density) [58].

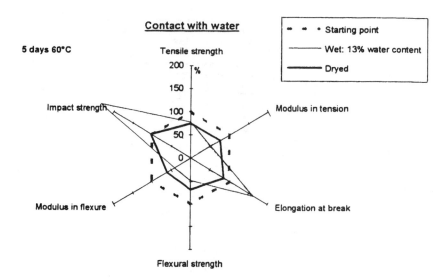

Figure 17 Influence of humidity on the characteristic values of flax fiber SMC molded plastics [58].

processing temperatures >250°C and are, therefore, not usable as a thermoplast matrix for natural fibers [60].

Other than the processes mentioned here, natural fibers are used as construction units, by applying hybrid nonwovens, i.e., natural fiber staple fiber fleece [61].

1. Natural Fiber Mat-Reinforced
Thermoplastics (NMT)

With the largest turnover, GMT is certainly the most important semi-product in the group of reinforced thermoplastics. A special production process for natural fiber-reinforced PP semi-products (NMT) has been de-

veloped by BASF AG [60]. For this process, natural fibers must be available in form of fiber mats. Mats are produced by stitching together layers of fibers that have been crumbled. Continued production of this semi-product is done by melt-coating the fiber mats in a double coil coating press, which is furnished with a heat- and cool-press zone. In such a coil coating press, the fiber mats are brought together with the polypropylenic melt between circulating steel bands. Wetting of the mats with the thermoplast melt takes place in the hot-press zone. Then the laminate is cooled under pressure in the cooling-press zone.

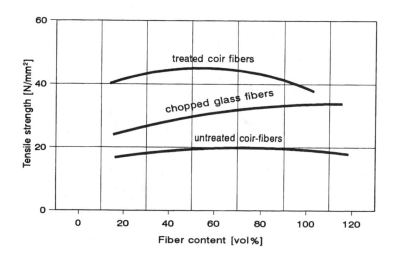

Figure 18 Tensile strength of B.M.C. molded plastics as dependent on their fiber content. %fibers = g fibers/100 g UP resin.

Table 12 Tensile Strength and Young's Modulus of Sisal, Flax, and Glass Fiber MTs with a Fiber Content of 40% (weight) [60]

Materials	Tensile strength (MPa)	Young's modulus (MPa)
PP–Wood flour	19	2500
PP–Sisal nonwoven	38	3600
PP–Sisal nonwoven with surface treatment	55	4800
PP–Flax nonwoven	47	5100
PP–Flax nonwoven with surface treatment	67	6700
PP–Glass nonwoven	100	6000

Sisal, flax, and glass fiber MTs can be classified by their mechanical properties, tensile strength, and Young's modulus (Table 12).

2. "Express" Processing

Extrusion *press* processing (express processing) was developed for the production of flax fiber-reinforced PP at the research center of Daimler Benz (Ulm, Germany) [62]. In this processing, natural fiber nonwovens and thermoplastic melt-films are alternatively deposited in a tempered molding tool and molded afterwards. The thermoplastic melt-films are laid on by a mobile extruder. If this process is optimally adapted to the element, a single passage by the extruder suffices. The structural order consists of three layers: two layers of nonwovens on the bottom and one on top, and in between the melt-film. In reference to density, the values for tensile strength of flax fibers PP composites exceed the characteristic values of sisal fiber-reinforced PP composites (Fig. 19).

D. Influence of Fiber Content on the Mechanical Properties of Natural Fiber-Reinforced Plastics

As is known of glass fiber-reinforced plastics, the mechanical and physical properties of composites, next to the fiber properties, and the quality of the fiber matrix interface, as well as the textile form of the reinforcement primarily depend on the volume content of fibers in the composite.

Tests by Roe et al. [63] with unidirectional jute fiber-reinforced UP resins show a linear relationship (analogous to the linear mixing rule) between the volume content of fiber and Young's modulus and tensile strength of the composite over a range of fiber content of 0–60%. Similar results are attained for the work of fracture and for the interlaminate shear strength (Fig. 20). Chawla et al. [64] found similar results for the flexural properties of jute fiber-UP composites.

E. Hybrid Composites Made of Natural Fibers and Glass or Carbon Fibers

Generally, the mechanical and physical properties of natural fiber-reinforced plastics only conditionally reach the characteristic values of glass fiber-reinforced systems. By using hybrid composites made of natural fibers and carbon fibers or natural fibers and glass fibers, the

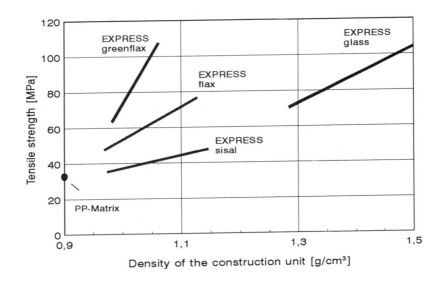

Figure 19 Tensile strength of different types of fiber reinforced composites, produced by the express-processing [62].

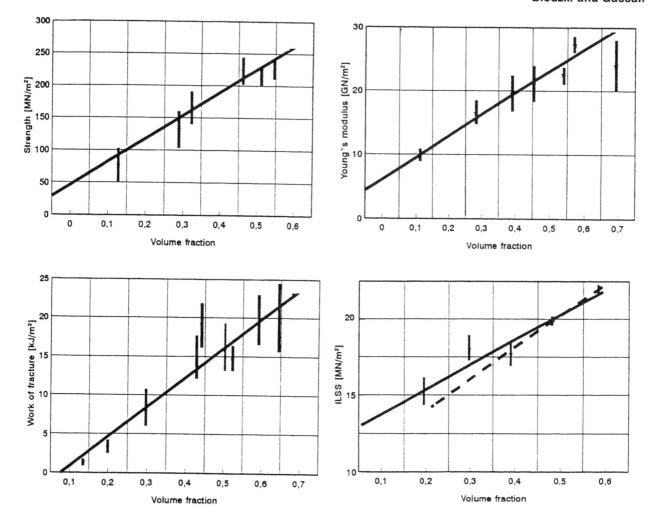

Figure 20 Influence of fiber content by volume on tensile strength, Youngs modulus, work of fracture, and interlaminate shear strength of one–dimensional jute fiber-reinforced UP resins [63].

properties of natural fiber-reinforced composites can be improved further [65–69], as Fig. 21 exemplifies for compression strength. To this, examinations were made by Mohan et al. [67,69], Philip [70], and Clark et al. [68] on jute-glass hybrid composites; Pavithran et al. [65] on coir-glass hybrid composites; Chand et al. [66,71,91] on sunhemp-carbon hybrid composites.

Additionally, the dependence of the mechanical properties on humidity is clearly reduced because of the moisture repellency of glass fibers [69]. The moisture absorption of composites is clearly smaller when natural fibers are replaced by glass fibers [65].

F. Biologically Degradable Composite Materials

For ecological reasons, more activities in the area of biologically degradable composite materials, i.e., natu-

ral fiber-reinforced biologically degradable polymers, were recognized. Applied as a matrix are the most readily available polymers as Biopol, PHB–HV coploymer, Bioceta, Mater Bi, or Sconacell A [72]. Tests with different flax fiber-reinforced biologically degradable polymers (Table 13) by Hanselka et al. [72] show that tensile strength and Young's modulus of these composites are clearly influenced by the particular matrix and the adhesion between fiber and matrix.

Tests by Gatenholm et al. [8,10] on PHB–HV copolymers containing cellulose fibers (for example, the tradenamed Biopol) show that the mechanical properties of these systems are determined by the fiber and the fiber matrix interface on the one hand, and on the other hand by the composition of the matrix, that is, of HV proportion in the matrix. At an increased proportion of HV, the stiffness of the composite is reduced up to 30%, whereas elongation at break increases until about 60%.

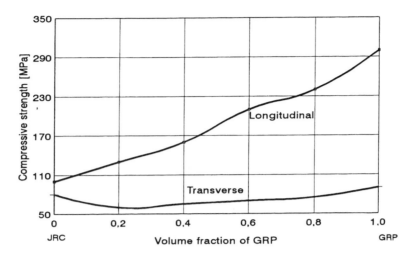

Figure 21   Compression strength as dependent on the content of GRP in jute fiber-reinforced hybrid-composites [67].

Table 13   Mechanical Properties of Flax Fiber
Reinforced Biologically Degradable Polymers (BDG) [72]

Fibers–BDP	Tensile strength (MPa)	Young's modulus (MPa)
Flax–Bioceta	65.7	1,400
Flax–Sconacell A	106.9	8,180
Flax–Mater Bi	124.3	10,580

Fiber content = 50% by vol., linen weave, cable direction = testing direction.

But the processing parameters are being greatly influenced by the proportion of HV. Tests made by Avella et al. [73] on straw fiber-reinforced PHB generally lead to expect good mechanical properties of such composites. The mechanical properties of extruded flax fiber-reinforced thermoplastic starch (structured with water or glycerin), clearly show increased values, especially for tensile strength and Young's modulus because of the addition of green-flax fiber rovings. Experiences mentioned in the literature on this subject suggest that from the point of view of the mechanical properties such biocomposites are suitable construction materials. At pres-

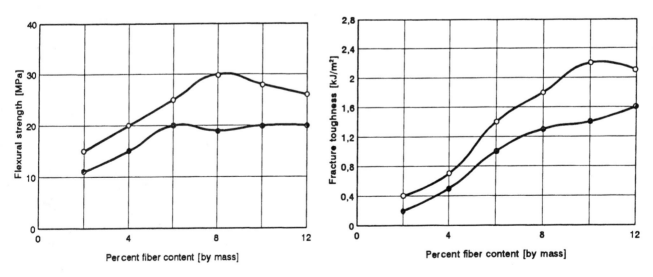

Figure 22   Influence of fiber content on flexural strength and fracture toughness of (○) softwood-cement composites and (●) hardwood-cement composites (air-cured) [78].

Table 14 Influence of Humidity on Flexural Strength and Fracture Toughness of Cellulose Fiber Reinforced Cements [78]

Fiber content (wt%)	Flexural strength (MPa)		Fracture toughness (kJ/m²)	
	r.h.	wet	r.h.	wet
2	10.6	8.6	0.25	0.33
4	14.2	10.5	0.51	1.00
6	20.9	10.4	1.06	1.61
8	20.3	8.4	1.37	1.49
10	20.1	9.6	1.46	1.83
12	20.6	9.3	1.68	1.79

r.h. = test samples were conditioned before testing at $50 \pm 5\%$ r.h. and $22 \pm 2°C$; wet = test samples were soaked in water for 48 h, dried with a cloth and then tested while wet.

Table 15 Influence of Chemicals on the Change of Bending Strength of Sisal–PP Composites [79]

Chemicals/time	Bending strength (MPa)	Bending modulus (GPa)
Reference samples	30	1.5
NaOH/50 h	24	1.1
NaOH/500 h	18	1.05
HCl/50 h	20	1.35
HCl/500 h	15	1.4

ent, limitations must be seen where excessive environmental conditions exist. Major prospects for these material systems are, therefore, lining elements with support function in the automobile, rail car, and furniture industries [72].

G. Additional Applications

When used as substitutes for asbestos fibers, plant fibers and "manmade" cellulose fibers show comparable characteristic values in a cement matrix, but at lower costs. As with plastic composites, these values are essentially dependent on the properties of the fiber and the adhesion between fiber and matrix. Distinctly higher values for strength and stiffness of the composites can be achieved by a chemical modification of the fiber surface (acrylic and polystyrene treatment [74]), usually produced by the Hatschek-process [75–77]. Tests by Coutts et al. [76] and Coutts [77,78] on wood fiber cement (soft-, and hardwood fibers) show that already at a fiber content of 8–10 wt%, a maximum of strengthening is achieved (Fig. 22).

Cellulose cement composites show a similar sensibility to humidity (Table 14), as do plastic composites, that is, they show a decrease of mechanical properties. Yet, formerly reached values can again be achieved by a drying process on the composite [75].

V. ENVIRONMENTAL EFFECTS

Whichever application of natural fiber or natural fiber-reinforced plastics will be used depends on the different environmental conditions, which are likely to add to the aging and degrading effects. On the other hand, such effects are often desirable, as is the case with com-

postable materials. Natural fibers are subject to degradation in acids and in alkaline solutions, as well as under UV rays. These effects, however, can be minimized by using suitable modifications. Unmodified cellulose fiber are normally degraded by enzymes after about 6–12 months; this can be altered through suitable treatments, so that no significant changes of mechanical properties can be noticed for 2 years. Within a period of 2.5 years, dry-stored fibers show only little changes in their mechanical properties. This is especially true with regard to strength and elongation at break. In this respect, sisal fibers are comparably more stable than Henequen and Abaka.

Lower temperatures, such as $-70°C$, clearly result in lesser strength, but this effect can be minimized by previous drying. Higher temperatures, such as 100–130°C, lead, in the case of cotton, to a noteable degradation after 80 days. Their strength is thereby reduced to 68%, that is 10% of the original value. Depending on temperature applied, these values are reduced to 41% and 12%, for flax fibers, and to 26% and 6%, for ramie fibers. In composites, moisture content results in a decrease of mechanical properties. This effect is greater with seawater than with freshwater. With freshwater, it is more likely that bacteria and fungi will appear. Against such influences, ramie, jute, and kopak fibers, are more resistant than other plant fibers [1]. Said lower resistants of natural fibers, against environmental factors, decisively effects the mechanical properties of the composites (Table 15) [79].

VI. CONCLUSIONS

The mechanical and physical properties of natural fibers vary considerably, as it is with all natural products. These properties are determined by the chemical and structural composition, which depend on the fiber type and growth circumstances. With this cellulose, the main component of all natural fibers varies from fiber to fiber.

The moisture sensibility is remarkable, certain natural fibers can easily be influenced by environmental effects. Generally speaking, the mechanical properties are lowered with rising moisture content.

Natural fibers compete with technical fibers, such as glass fibers or carbon fibers, as reinforcements for plastics. The advantages of technical fibers are their good mechanical properties, which vary only a little, but their recycling is difficult.

The mechanical properties of composites are mainly influenced by the adhesion between matrix and fibers of the composite. As it is known from glass fibers, the adhesion properties could be changed by pretreatments of fibers. So special process, chemical and physical modification methods were developed. Moisture repellency, resistance to environmental effects, and, not at least, the mechanical properties are improved by these treatments. Various applications for natural fibers as reinforcement in plastics are encouraged.

Several of these natural fiber composites reach the mechanical properties of glass fiber composites, they are already being used in the automobile and furniture industries. Up to now, the most important natural fibers were jute, flax, and coir.

Yet, the development of processing and modification methods is not finished. Further improvements need to be expected so that it might be possible to substitute technical fibers in composites even more widely. Natural fibers are reusing raw materials and they are recyclable. When recognizing the need for recycling and preserving natural resources, such a substitution is very important.

REFERENCES

1. A. K. Bledzki, J. Izbicka and J. Gassan, *Kunststoffe-Umwelt-Recycling*, Stettin (Poland) September 27–29 (1995).
2. W. Wittig, *Kunststoffe im Automobilbau*, VDI-Verlag, Düsseldorf (1994).
3. P. K. Pal, *Plastics and Rubber Processing and Applications, 4*:215, (1984).
4. A. G. Winfield, *Plastics and Rubber International, 4(1)*: 23 (1979).
5. N. N., *EUWID* 34, p.15.
6. K. P. Mieck and T. Reβmann, *Kunststoffe, 85(3)*: 366 (1995).
7. N. N., *Ingenieur-Werkstoffe 4(9)*: 18 (1992).
8. R. Kohler and M. Wedler, *Landinfo, 3*: 33 (1995).
9. R. W. Kessler, U. Becker, R. Kohler and B. Goth, *Biomass and Bioenergy*, (in press).
10. P. Gatenholm, J. Kubát and A. Mathiasson, *J. Appl. Polym. Sci., 45*: 1667 (1992).
11. P. Gatenholm and A. Mathiasson, *Polymeric Materials Science and Engineering, 67*: 361 (1992).
12. J. Gassan and A. K. Bledzki, *Die Angewandte Makromolekulare Chemie* (in press).
13. K.-P. Mieck, A. Nechwatal and C. Knobelsdorf, *Melliand Textilberichte, 11*: 892 (1994).
14. P. S. Mukherjee and K. G. Satyanarayana, *J. Mat. Sci., 21*: 51 (1986).
15. J. Gassan and A. K. Bledzki, *6.Internationales Techtexil Symposium 1994*, Frankfurt July 15–17 (1994).
16. G. E. Kritschewsky, *Chemische Technology von Textilmaterialien*, Legprombitisdat Moskau (1985).
17. H. P. Fink, J. Ganster and J. Fraatz, *Akzo-Nobel Viskose Chemistry Seminar "Challenges in cellulosic man-made fibers"*, Stockholm, May 30–June 3 (1994).
18. J. Gassan and A. K. Bledzki, *7. Internationales Techtexil Symposium 1995*, Frankfurt, June 20–22 (1995).
19. J. W. S. Hearle and J. T. Sparrow, *J. Appl. Polym. Sci., 24*: 1857 (1979).
20. P. S. Mukherjee and K. G. Satyanarayana, *J. Mat. Sci., 21*: 51 (1986).
21. J. Gassan and A. K. Bledzki, *9th International Conference on Mechanics of Composite Materials*, Riga (Latvia), October 17–20 (1995).
22. S. H. Zeronian, H. Kawabata and K. W. Alger, *Text. Res. Inst., 60(3)*: 179 (1990).
23. M. A. Semsarzadeh, *Polym. Comp., 7(2)*: 23 (1986).
24. M. A. Semsarzadeh, A. R. Lotfali and H. Mirzadeh, *Polym. Comp., 5(2)*: 141 (1984).
25. P. K. Ray, A. C. Chakravarty and S. B. Bandyopadhyay, *J. Appl. Polym. Sci., 20*: 1765 (1976).
26. A. N. Shan and S. C. Lakkard, *Fibre Sci. Techn., 15*: 41 (1981).
27. B. Wulfhorst, G. Tetzlaff and R. Kaldenhoff, *Techn. Text., 35(3)*: S. 10 (1992).
28. M. N. Belgacem, P. Bataille and S. Sapieha, *J. Appl. Polym. Sci., 53*: 379 (1994).
29. I. Sakata, M. Morita, N. Tsuruta and K. Morita, *J. Appl. Polym. Sci., 49*: 1251 (1993).
30. Q. Wang, S. Kaliaguine and A. Ait-Kadi, *J. Appl. Polym. Sci., 48*: 121 (1993).
31. S. Goa and Y. Zeng, *J. Appl. Polym. Sci., 47*: 2065 (1993).
32. S. Dong, S. Sapieha and H. P. Schreiber, *Polym. Eng. Sci., 32*: 1734 (1992).
33. T. P. Nevell and S. H. Zeronian, *Cellulose Chemistry and its Applications*, John Wiley & Sons, New York (1985).
34. S. C. O. Ugbolue, *Text. Inst., 20(4)*: 1 (1990).
35. J. I. Kroschwitz, *Polymers: Fibers and Textiles*, John Wiley & Sons, New York (1990).
36. V. V. Safonov, *Treatment of Textile Materials*, Legprombitizdat, Moscow (1991).
37. S. H. Zeronian, *J. Appl. Polym. Sci., 47*: 445 (1991).
38. B. S. Westerlind and J. C. Berg, *J. Appl. Polym. Sci., 36*: 523 (1988).
39. M. J. Schick, *Surface Characteristics of Fibers and Textiles. Part II*, Marcel Dekker, Inc., New York (1977).
40. J. M. Felix and P. Gatenholm, *J. Appl. Polym. Sci., 42*: 609 (1991).
41. D. Maldas, B. V. Kokta and C. Daneaulf, *J. Appl. Polym. Sci., 37*: 751 (1989).
42. K. L. Mittal, *Silanes and Other Coupling Agents*, VSP BV, Netherlands (1992).
43. R. G. Raj, B. V. Kokta, F. Dembele and B. Sanschagrain, *J. Appl. Polym. Sci., 38*: 1987 (1989).
44. W. Geβner, *Chemiefasern/Textilind.* (Ind.-Text.), *39/91(7/8)*: S.185 (1989).
45. A. C. Khazanchi, M. Saxena and T. C. Rao, *Text. Comp. Build. Constr.* 69 (1990).
46. P. Gatenholm, H. Bertilsson and A. Mathiasson, *J. Appl. Polym. Sci., 49*: 197 (1993).
47. P. Ghosh, S. Biswas and C. Datta, *J. Mater. Sci., 24*: 205 (1989).
48. L. Hua, P. Flodin and T. Rönnhult, *Polym. Comp., 8(3)*: 203 (1987).
49. L. Hua, P. Zadorecki and P. Flodin, *Polym. Comp., 8(3)*: 199 (1987).

50. D. Maldas, B. V. Kokta and C. Daneault, *J. Vinyl Techn.*, *11(2)*: 90 (1989).
51. P. Zadorecki and T. Rönnhult, *J. Polym. Sci.: Part A: Polym. Chem.*, *24*: 737 (1986).
52. P. Zadorecki and P. Flodin, *J. Appl. Polym. Sci.*, *31*: 1699 (1986).
53. M. H. Schneider and K. I. Brebner, *Wood Sci. & Techol.*, *19*: 67 (1985).
54. D. S. Varma, M. Varma and I. K. Varma, *J. Reinf. Plast. Comp.*, *4(10)*: 419 (1985).
55. E. T. N. Bisanda and M. P. Ansell, *Comp. Sci. Techn.* 165 (1991).
56. R. G. Raj and B. V. Kokta, *Polymer Engineering and Science*, *31(18)*: 1358 (1991).
57. K.-P. Mieck, A. Nechwatal and C. Knobelsdorf, *Die Angewandte Makromolekulare Chemie 225*: 37 (1995).
58. R. Kohler and M. Wedler, *6.Internationales Techtexil Symposium 1994*, Frankfurt, June 17–20 (1995).
59. O. Owolabi, T. Czvikovszky and I. Kovács, *J. Appl. Polym. Sci.*, *30*: 1827 (1985).
60. H. Baumgartl and A. Schlarb, *2.Symposium "Nachwachsend Rohstoffe—Perspektiven für die Chemie,"* Frankfurt, May 5–6 (1993).
61. R. Lützkendorf, K. Mieck and Th. Reußmann, *7.Internationales Techtexil Symposium 1995*, Frankfurt, June 20–22 (1995).
62. Th. Schlößer and Th. Fölster, *Kunststoffe, 85(3)*: 319 (1995).
63. P. J. Roe and M. P. Ansell, *J. Mat. Sci.*, *20*: 4015 (1985).
64. K. K. Chawla and A. C. Bastos, *3 International Conference on Mechanical Behaviour of Materials*, Cambridge (England), Aug. (1979).
65. C. Pavithran, P. S. Mukherjee and M. Brahmakumar, *J. Reinf. Plast. Comp.*, *10*: 91 (1991).
66. N. Chand and P. K. Rohatgi, *J. Mat. Sci. Letters, 5*: 1181 (1986).
67. R. Mohan, M. K. Shridhar and R. M. Rao, *J. Mat. Sci. Letters, 2*: 99 (1983).
68. R. A. Cark and M. P. Ansell, *J. Mat. Sci.*, *21*: 269 (1986).
69. R. Mohan and J. Kishore, *J. Reinf. Plast. Comp.*, *4*: 186 (1985).
70. A. R. Philip, *Engineering Materials and Design, 8*: 475 (1965).
71. N. Chand and P. K. Rohatgi, *Polymer Communication, 28(5)*: 146 (1987).
72. H. Hanselka and A. S. Herrmann, *7.Internationales Techtexil Symposium 1994*, Frankfurt, June 20–22 (1995).
73. M. Avella and R. dell'Érba, *Proceeding of the 9th International Conference on Composite Materials* Vol. II, Madrid, July 12–16 Vol. 9, p. 864 (1993).
74. A. C. Khazanchi, M. Saxena and T. C. Rao, *Textile Composites in Building Construction 1*: 69 (1990).
75. Y. M. Mai, M. I. Hakeem and B. Cotterell, *J. Mat. Sci.*, *18*: 2156 (1983).
76. R. S. P. Coutts and A. J. Michell, *J. Appl. Polym. Sci.: Appl. Polym. Symp.*, *37*: 829 (1983).
77. R. S. P. Coutts, *Composites, 15(2)*: 139 (1984).
78. R. S. P. Coutts, *J. Mat. Sci. Letters*, *6*: 955 (1987).
79. R. Selzer, *SAMPE, The Materials and Processes Society*, Kaiserslautern, March 28 (1995).
80. M. K. Sridhar and G. Basavarajappa, *Indian J. Text. Res.*, *7(9)*: 87 (1982).
81. E. T. N. Bisanda and M. P. Ansell, *J. Mat. Sci.*, *27*: 1690 (1992).
82. S. H. Zeronian, *J. Appl. Poly. Sci.*, *47*: 445 (1991).
83. A. N. Shan and S. C. Lakkard, *Fiber Sci Techn.*, *15*: 41 (1981).
84. S. C. O. Ugbolue, *Textile Institute 20(4)*: 1 (1990).
85. A. C. Khazanchi, M. Saxena and T. C. Rao, *Textile Composites in Building Constructions* 69 (1990).
86. A. J. Michell and D. Willis, *Appita, 31(3)*: 347 (1978).
87. H. Saechtling, *International Plastics Handbook*, Hanser, München (1987).
88. E. W. Wuppertal, *Die Textilen Rohstoffe*, Dr. Spohr-Verag, Frankfurt (1981).
89. E. Treiber, *Die Chemie der Pflanzenzellwand*, Springer-Verlag, Berlin (1957).
90. J. Warwicker, *J. Appl. Poly. Sci.*, *1*: 41 (1969).
91. S. M. Lee and R. M. Rowell, *International Encyclopedia of Composites*, VCH-Publishers Inc., New York (1991).
92. S. K. Pal, D. Mukhopadhyay, S. K. Sanyal and R. N. Mukherjee, *J. Appl. Polym. Sci.*, *35*: 973 (1988).
93. E. P. Plueddemann, *Interfaces in Polymer Matrix Composites*, Academic Press, New York (1974).
94. M. A. Semsarzadeh and D. Amiri, *Polym. Eng. Sci.*, *25(10)*: 618 (1985).
95. N. N., *ATZ, 97(5)*: 293 (1995).
96. G. Wacker and A. K. Bledzki, *Kompozyty I Kompozycje Polimerowe*, Stettin (Poland), June 22–24 (1994).
97. W. Aichholzer, *14. Stuttgarter Kunststoff-Kolloquium*, Stuttgart. March 22–23 (1995).

53
Short Fiber-Reinforced Plastic Composites

Jayamol George and Sabu Thomas
Mahatma Gandhi University, Kerala, India

I. INTRODUCTION

The name polymer and its applications have now reached the life style of the masses. Polymer is a generic name with its main permanent members being elastomers, plastics, and fibers. When considering materials for load bearing applications, designers are increasingly examining the advantages of using plastic materials, both thermosets and thermoplastics. The advantages of thermoplastics over others are their low specific gravities and low cost of fabrication, particularly when using the injection molding technique. The term composite is used in material science to mean a material made up of a matrix substance containing reinforcing agents. By definition, composite is a material composed of two or more distinct components. Composites are divided into two basic forms, composite materials and composite structure. Composite materials are composed of reinforcing structures surrounded by a continuous matrix, whereas composite structure exhibits a discontinuous matrix.

II. CLASSIFICATION OF COMPOSITES

The composites have been classified in general under the following heads: particulate composites, fibrous composites, and laminated composites [1].

A. Particulate Composites

Particulate composites consist of particles dispersed in a matrix. These particles are divided into two classes, skeletal and flakes. The first one consists of continuous skeletal structures filled with one or more additional materials. Flakes consist generally of flat flakes oriented parallel to each other. These particles may have any shape, configuration, or size. Concrete and wood particle boards are two examples of particulate-reinforced composites.

B. Fiber-Reinforced Composites

Fiber-reinforced composites consist of fibers of high strength and modulus embedded in a matrix with distinct interfaces between them. In this form, both matrix and fibers retain their physical and chemical properties and exhibit synergism. Fibrous composites are divided into two broad areas, continuous fibers and short fibers. Fibers of length less than 50 mm are generally accepted to be short fibers, whereas fiber lengths greater than 50 mm are regarded as continuous. Incorporation of short fibers into polymer matrices offer many attractive features, such as ease of fabrication and better economics at both the incorporation and fabrication stages, i.e., production of a complex-shaped article is possible with short fibers, but is completely impractical with long fibers.

C. Laminated Composites

In laminated composites, the individual layers of materials are bonded together to form an element or plate. When the constituent materials in each layer are the same, the laminated layers are called a laminate (e.g., plywood, papers, etc.). If the layers are of different constituent materials or of the same material with different reinforcing patterns, the laminate is said to be a hybrid laminate.

D. Hybrid Composites

Composites containing more than one type of fiber are commonly known as hybrid composites. The term "hy-

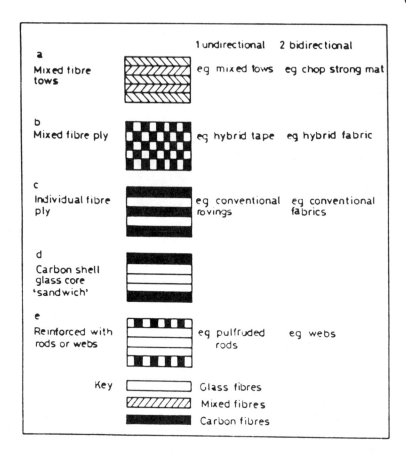

Figure 1 Different hybrid configurations.

brid'' is generally used to denote the incorporation of two different types of materials into one single matrix. The level of mixing can be either on a small scale (fibers, tows) or on a large scale (layers). The purpose of hybridization is to construct a new material that will retain the advantages of its constituents. There are several types of hybrid composites, characterized according to the way in which the constituent materials are arranged [2]. Figure 1 shows the various types of hybrid constructions. These include sandwich hybrids also known as core-shell, in which one material is sandwiched between two layers of another, interply or laminated where alternate layers of the two (or more) materials are stacked in a regular manner. In intraply, tows of two or more constituent types of fibers are mixed in a regular or random manner, and in intimately mixed hybrids, the constituent fibers are made to mix as much as possible so that no concentration of either type is present in the material.

III. FIBERS–MATRICES INTERFACE

A. Types of Fibers Used for Reinforcement

Fibers are mainly classified into two categories: natural and man-made fibers [3,4]. These two categories are further classified into various types, as shown in Fig. 2.

1. Man-Made Fibers

The man-made fibers are classified into two different categories, regenerated fibers and synthetic fibers, depending on the way in which they are prepared.

2. Regenerated Fibers

The fibers that are regenerated either from organic or inorganic systems are classified in this category. Polymers from natural sources are dissolved and regenerated after passage through a spinneret to form fibers (e.g.,

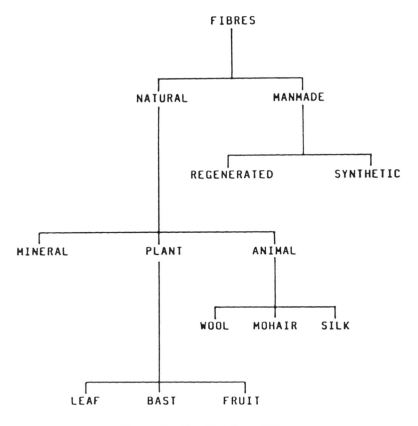

Figure 2 Classification of fiber.

rayon). These fibers are used for reinforcing plastics and rubbers.

Glass fibers are the most common of all reinforcing fibers for polymer matrix composites. The principal ingredient in all glass fibers is naturally occurring silica (SiO_2). The two types of glass fibers commonly used in the FRP industry are E and S glass. Other types are C glass, which is used in applications requiring greater corrosion resistance to acids, D glass for electrical application, A glass for making window glass and bottles, and L glass or lead glass for applications requiring radiation protection.

3. Synthetic Fibers

Fibers in which the basic chemical units have been formed by chemical synthesis, followed by fiber formation, are called synthetic fibers. Examples include nylon, carbon, boron fibers, organic fibers, ceramic fibers, and metallic fibers. Among all commercially available fibers, Kevlar fibers exhibit high strength and modulus. (Kevlar is a DuPont trademark for poly [p-phenylene diamine terephthalamide].) It is an aromatic polyamide (aramid) in which at least 85% of the

amide linkages are attached directly to two aromatic rings. Electron diffraction and electron microscopy studies revealed that Kevlar 49 fibers exhibit a radially oriented crystalline organization with a uniform distribution of ordered crystalline material throughout the fiber. Properties of some typical fibers are given in Table 1.

4. Natural Fibers

Fibers that come from natural sources, such as minerals, animals, and plants, are classified as natural fibers [5,6].

Mineral Fibers

Fibers in this category are composed of naturally occurring materials. A good example is asbestos. The most common type is chrysotile, representing more than 95% of world asbestos production. Chemically it is magnesium silicate ($Mg_6(OH)_4 Si_2O_5$). Today, use of this fiber is limited because long exposure to it may cause bronchial cancer.

Animal Fibers

Fibers obtained from living organisms are known as animal fibers, e.g., wool, which is obtained from domestic sheep; silk fiber, which is produced by the silkworm

Table 1 Properties of Kevlar and Other Competitive Materials

	Nylon	Kevlar 29	Kevlar 49	E-Glass	Steel
Specific gravity	1.44	1.44	1.45	2.55	7.86
Tensile strength (kNm^{-2})	999	2758	2758	1717	1965
Tensile modulus (MNm^{-2})	5.52	82.74	131.0	68.95	200
Elongation (%)	18	5	2.4	3	2

Source: Ref. 1.

in making its cocoon. Silk, in contrast to all other natural fibers, such as cotton, flax, and wool, etc., does not have a cellular structure. In this respect and in the way it is formed, silk closely resembles synthetic fibers.

Plant Fibers

These fibers are classified into three types depending on the part of the plant from which they are extracted:

1. Bast or stem fibers (jute, mesta, banana)
2. Leaf fibers (sisal, pineapple, screw pine)
3. Fruit fibers (coir, cotton, arecnut)

Extraction methods, amounts, and lengths of various natural fibers and their physical and chemical properties are given in Tables 2 and 3.

B. Matrices

The key difference between thermoplastic and more traditional thermoset-based matrices lies in the behavior of the matrix during processing. The thermoplastic matrix is not required to undergo a cure process to achieve its final mechanical properties. All that is required is melting, shaping, and subsequent solidification. The range of conventional thermoplastic composites and their resultant property advantages depend on the average length of fibers on molding.

The most important matrix material among common polymeric matrices are polyester and epoxy resins. Cured epoxy resins from the prime matrix for high-performance glass, aramid, and carbon fiber composites [7,8], whereas the unsaturated polyester resins are primarily used in industrial applications. The advantages of the former include resistance to water, a variety of chemicals, weathering, and aging. Polyimides represent the largest class of high-temperature polymers in use in composites today [9,10]. They have service temperatures between 250–300°C. Thermosetting resins are crosslinked with peroxide to obtain high modulus, strength, and creep resistance. However, they exhibit extreme brittleness. Thermoplastic resins are easier to fabricate than thermosetting resins [11], and they can be recycled. PEEK is a semi-crystalline aromatic thermoplastic. Thermoplastic resins such as PMMA have higher fracture energies of about 1 kJ/m^2. The typical properties of thermoplastic and thermosets are listed in Table 4.

C. Interface

The fiber-reinforced composite materials include three phases: surface of fiber side, the interface between fiber and matrix, and the interphase. These phases are collectively referred to as the interface [12]. The characteris-

Table 2 Extraction Methods, Amount, and Length of Various Natural Fibers

Fiber	Method	Amount	Length (mm)
Banana	Manual/raspador	1.5 wt% of stem	300–900
Coir	Retting/mechanical	8% of nut (this weighs 0.1 kg)	75–150
Jute	Retting and beating/chemical	3–4% of stem	1500
Linseed	Retting/dry scratching	20–25% of dry straw	—
Mesta	Retting and beating/chemical	Same as jute	—
Palmyrah	By hand (by beating)	0.5 kg per stalk	300–600
Pineapple	By hand/decorticator	2.5–3.5 wt% of green leaves	900–1500
Ramie	Decorticator	2.5–3.5 wt% of bark	900–1200
Sisal	Manual (beating)/microbial retting/decorticator	3–4% of green leaves	900–1200
Sunhemp	Manual/retting	2–4% of green stalk	—

Source: Ref. 5.

Table 3 Physical and Mechanical Properties of Some Natural Fibers

Fiber	Initial modulus (GN/m^2)	Ultimate tensile strength (MN/m^2)	Elongation at break $(\%)$	Flexural modulus (MN/m^2)
Sisal	9–22	568–640	3–7	12.5–17.5
Pineapple	34–82	413–1627	0.8–1	0.25–0.40
Banana	7–20	54–754	1–4	2–5
Jute	18	226	1.3	0.3–0.5
Mesta	—	—	1–2	0.35–0.65
Flax	—	780	2–4	0.18–0.25
Sunhemp	—	760	2–4	12.5–17.5
Palmyrah	4–6	180–215	7–15	—
Cotton	—	200–400	6–7	0.03–0.10

Source: Ref. 5.

tics of the interface are dependent on the bonding at the interface, the configuration, the structure around the interface, and the physical and chemical properties of constituents. As a result, the interface has a strong influence on the property of the composite material. These interface problems are seen as a type of adhesion phenomenon and are often interpreted in terms of the surface structure of the bonded material, i.e., surface factors such as wettability, surface free energy, the polar group on the surface, and surface roughness of the material to be bonded are often discussed as means of improving the bonding strength.

A method for the estimation of composite material performance from the characteristics of fillers and the matrices and from the configuration of filler is generally called the law of mixture. In the most basic form of the law of mixture, the characteristics of a composite material are represented as a function of characteristics of constituent components and their volume fractions, as shown in Fig. 3. For a composite material (characteristics:X_c) that consists of component A (characteristics: X_A, volume fraction: ϕ_A) and component B (characteristics: X_B, volume fraction: ϕ_B), the basic formulae of the law of mixture are as follows:

$$X_c = \phi_A X_A + \phi_B X_B \tag{1}$$

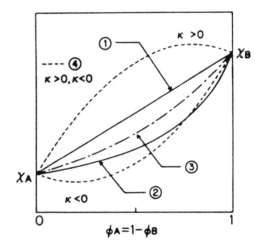

Figure 3 Relation between the properties of composites and various laws of mixture.

$$\frac{1}{X_c} = \frac{\phi_A}{X_A} + \frac{\phi_B}{X_B} \tag{2}$$

The parallel and series models (curves 1 and 2) are given in Fig. 3.

Table 4 Representative Properties of Some Polymeric Matrix Materials

	Epoxy	Polyimide	PEEK	Polyamideimide	Polyetherimide	Polysulfone sulfone	Polyphenylene	Phenolics
Tensile strength (MPa)	35–85	120	92	95	105	75	70	50–55
Flexural modulus (MPa)	15–35	35	40	50	35	28	40	—
Density (g cm^{-3})	1.38	1.46	1.30	1.38	—	1.25	1.32	1.30
Water absorption (24 h%)	0.1	0.3	0.1	0.3	0.25	0.2	0.2	0.1–0.2

The two curves exhibit theoretical upper and lower limits, respectively, based on a simple composite effect in general. A basic formula that generalizes Eqs. (1) and (2) is:

$$X_c{}^n = \phi_A X_A{}^n + \phi_B X_B{}^n \qquad (3)$$

where n ($-1 \leqslant n \leqslant 1$) represents the properties of the combination mode, i.e., the parallel model is predominant when n is close to 1 and the series model is predominant when n is close to -1.

$$\log X_c = \phi_A \log X_A + \phi_B \log X_B \qquad (4)$$

This function is intermediate between the parallel model and the series model and referred to as the logarithmic law of mixture shown in curve 3. The law of mixture is valid for a composite system when there is no interaction in the interface. However, it is natural to consider that interaction will occur in the interface due to contact between A and B. Then considering the creation of interfacial phase C, different from A and B, the following equation can be presented:

$$X_c = \phi_A X_A + \phi_B X_B + K\phi_A\phi_B \qquad (5)$$

This is referred to as quadratic law of mixture shown in curve 4. The parameter K involves an interaction between components A and B and provides an expression for the interfacial effect.

IV. PROCESSING TECHNIQUES

The choice of manufacturing technology for the fabrication of fiber-reinforced plastics or composite materials is intimately related to the performance, economics, and application of the materials. It also depends upon a number of factors, such as component numbers required, item complexity, number of molded surfaces, and type of reinforcement.

A. Injection Molding

One of the most common processing methods for thermoplastics is injection molding. Injection molding offers the advantage of rapid processing into complex shapes. The list of artifacts manufactured using this process is endless and include electric drill casings, gear wheels, telephones, brief cases, etc.

In injection molding, the polymer is fed from a hopper into a heated barrel where it softens and becomes a viscous melt. It is then forced under high pressure into a relatively cold mold cavity where the polymer has sufficient time to solidify. Then the mold is opened and the fabricated part is ejected. The cycle of operation is then repeated.

The processing of discontinuous fiber-filled thermoplastics owes much to the rheology of the system. Shear thinning and adiabatic heating associated with the broad molecular weight polymers in these compounds offer attractive processing conditions for engineering applications. During injection molding the shear dominated flow near the surface tends to align the fibers along the flow (shell region), whereas if the flow is extensional in the plane of the part, the fibers near the midplane of the cavity tend to align transverse to the flow (core region) [13–16]. It has been experimentally observed by many researchers in the past that the relative extent of core-shell regions across the thickness of the injection molded parts strongly depends upon the processing conditions during injection molding and the properties of the polymer used. Sanon et al. [13] found that cavity thickness also had an appreciable effect on fiber orientation. During injection molding, a thin layer of solidified polymer is developed near the cold mold walls. The thickness of the cold layer increases near the entrance and diminishes near the advancing front. This growing–diminishing thermal boundary layer results in a gap-wise converging–diverging flow even in a uniformly thick cavity [17]. This can affect the fiber orientation in the injection molded parts.

The fiber length distribution is determined by the following three stages in processing:

1. The plasticization stages of extrusion compounding
2. Remelting during injection molding
3. The higher shear during injection and mold filling

The plasticization stage dominates the fiber attrition process. The flow into the mold cavity during injection dictates the fiber orientation of the final molded part and, hence, the properties of the component. The addition of fibers makes polymer systems more shear thinning and thermally conductive. These factors influence the velocity and shear profiles of the melt front as it passes through the narrow mold channels. This also influences the rate of solidification. The central region is last to solidify and is free to relax into random in-plane orientations constrained by the decreasing channel thickness. This results in skin-core structure. The fiber damage in injection molding is severe. In plunger injection molding, the degradation is also significant. Increasing the diameter of the nozzle and gates and lowering the melt viscosity and back pressure reduces the extent of degradation for both thermoplastics and thermosets [18–21].

B. Hand-Lay or Contact Molding

Here the mold is coated with a proper release agent to prevent the item sticking on the mold surface. Gel coat resin is applied by painting or spraying to a thickness of 0.5 mm. This is a pigmental resin that will give the required quality of finish to the component to provide a degree of environmental protection and to prevent the fiber pattern from showing through the external part surface. The laminating resin and reinforcement are worked onto the surface by roller and brush to the required thickness. The objective of this is to ensure that all the reinforcing fibers are fully coated by the resin. The overall resin–fiber ratio must be controlled such that the mechanical properties of the composite materials are optimized. The item is allowed to cure before being removed from the mold. Curing is carried out at room temperature and can be varied depending on the size and complexity of the item. The cure rate can be adjusted by controlling the resin chemical formulation, i.e., by varying the catalyst and accelerator levels. The curing process can be accelerated by carrying out a postcure for several hours.

C. Spray Molding

Spray molding is the modification of the hand-lay process where the resin and glass fiber are deposited simultaneously on the molding tool. The fibers are mixed with the resin at the spray head before being deposited on the mold surface. Subsequent consolidation of the laminate is achieved by rolling in a similar manner to the hand-lay process. This method is suitable for large components. Here the capital cost is higher and the process is very operator sensitive.

D. Resin Transfer Molding

In this process, resin is injected into a closed mold containing the reinforcement preform. The resin can be injected either under pressure [22] or under vacuum [23]. The potential advantages of this process are: (1) low mold cost, (2) inserts can be incorporated, (3) low pressure requirements, (4) accurate fiber orientation, (5) automation possibilities, and (6) versatility. The resin formulation and process variables are selected so that no significant polymerization occurs until the mold cavity has been completely filled. This is achieved by the ad-

justment of supply pressure or flow rate mold temperature and catalyst/accelerator levels [24]. The overall cycle time is often limited by the time required to heat and cure the resin at the injection point.

E. Flexible Resin Transfer Molding

Flexible resin transfer molding (FRTM) is an innovative composite manufacturing process developed based on detailed cost analysis [25]. FRTM is a hybrid process that combines the technical characteristics and favorable economics of diaphragm forming and resin transfer molding. Separate sheets of fiber and solid resin are placed between elastomeric diaphragms and heated so that the resin liquefies. The fiber and resin are then compacted by drawing a vacuum between the diaphragms and formed to shape by drawing the diaphragm assembly over hand tooling (Fig. 4). The process control system combines an empirical resin polymerization model, a fluid flow model, and dielectric sensing of in situ resin properties in order to determine the optimum time for compaction and forming. In the resin polymerization model a sheet of resin is placed between rubber diaphragms along with the reinforcing fiber material during the FRTM process, the temperature is increased to an isothermal hold at the resin's cure temperature. As the temperature increases, the resin viscosity decreases until the polymerization reaction is activated at the beginning of the isothermal hold. As the polymer molecules begin to grow and molecular weight increases, the viscosity of fluid increases. In the polymerization model, the degree of cure α is zero at the time of the viscosity minimum. Thus, the effects of resin aging and resin heat-up rate on initial degree of cure are neglected.

In the fluid flow model, simulation is based on Darcy's law for the steady flow of Newtonian fluids through porous media. This law states that the average

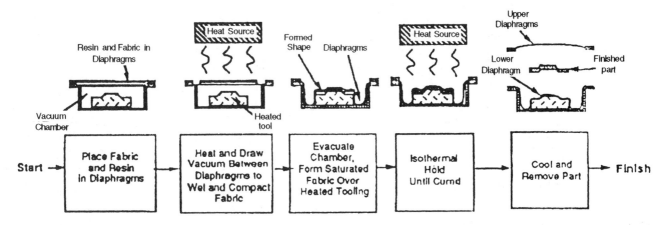

Figure 4 Schematic representation of FRTM process. (From Ref. 25.)

fluid velocity U is proportional to the average pressure gradient (∇P).

$$U = -K/\mu \; \nabla P \tag{6}$$

The constant of proportionality is K/μ, where K is the permeability and μ is Newtonian viscosity. The dielectric properties of the resin are also measured using sensors. These measurements were correlated with viscosity and used as a part of the FRTM control system.

F. Cold Press Molding

This process is carried out at low pressure and at ambient temperature. The fiber is placed in the tool in the form of a mat and the resin is poured in. The mold is closed, and the resin spreads through the reinforcement thereby impregnating and fully wetting out the fiber strands. The process is done on an hydraulic press. Pressure levels required for molding are typically around 10^5 MPa. Gel coat can be applied to the mold faces before the molding operation commences.

G. Hot Press Molding

A premixed sheet or dough material containing resin, fillers, and fibrous reinforcement is inserted between matched metal tools and the items are compression molded at elevated temperature and high pressure [26]. Different types of molding compounds used in hot press molding are dough molding compound (DMC), sheet molding compound (SMC), and bulk molding compound (BMC).

In DMC the material is supplied to the molder as a dough, which has been precompounded in a mixer. Great care must be exercised during the mixing stage to ensure an effective mixing. At the same time, the shear rate should not be severe to minimize the fiber damage. In SMC the molding materials are produced in a sheet form where the fiber is sandwiched between two resin layers, and the entire structure is then homogenized by "kneading" between specially countered rolls. This sheet is produced continuously and is supplied to the molder between two thermoplastic carrier films to prevent adhesion and contamination. BMC is similar in formulation to DMC, which consists of thermosetting polyester resin, mineral filler, and chopped fiber stands up to 12-mm in length. Pigments, catalysts, etc. are often included in formulations according to application and production requirements. Hydraulic presses are generally used in hot press molding. The molding pressure is in the range of 3.5–15 MPa.

H. Reinforced Reaction Injection Molding (RRIM)

Reaction injection molding (RIM) describes [27–29] the process whereby two liquid reactants are rapidly mixed at a mixing head and injected into a mold at relatively low pressures. RRIM refers to the composite processing route in which the reinforcement, generally in the form of particulates or very short fibers, is included in one of the monomer streams. For the most common polymer, the polyurethanes, the polyol would contain the filler. In SRIM, the reinforcement is placed in the mold in the form of a preform.

Simple pumps and static mixers can be used to disperse the monomers, when the rate of reaction is a lot slower than the fabrication cycle, but effective mixing of fast reacting reagents is best achieved with a self-creaming impingement mix-head, which permits the reactants to continuously circulate in their separate systems until an injection shot is required. The opening and closing of the valve needs to be controlled to accurately deliver the exact quantity of reactants to the mold, otherwise overfilling will cause mold flash and underfilling will cause an imperfect molding. Mixing occurs during mold filling and, more specifically, the dispersion of reinforcement must occur effectively. Orientation of fibers with higher aspect ratio occurs within the flow fields.

For RRIM-based composites, the main benefit is the case with which the properties can be varied across a wide range. Decorative finishes can be applied to RRIM molding by painting or, where color matching between components is less important, by self-coloring. However, color coating of the mold surface prior to RIM provides a better finish.

I. Pultrusion

This technique is used to manufacture sectional products in continuous lengths and is one of the most economically attractive methods for processing thermoset composite materials [30]. Here the fibrous reinforcement is impregnated with thermosetting resin and pulled through a heated steel die, often made from ultra high-molecular weight polyethylene, which shapes the product section and initiates the crosslinking reaction in the resin to form a solid product.

Most pultrusion dies are between 0.4–1 m long. The processing speed in pultrusion is generally in the range $1–5$ m/min^{-1}. Some improvement in speed may be obtained by preheating the material prior to entering the die. The product is pulled through the die by a powerful haul-off unit that may be either of the reciprocating type, where the pultrudate is gripped and pulled in a "hand over hand fashion," or of the caterpillar type. After the haul-off, the product is generally cut by a saw to the required length. Complex sections can be produced by pultrusion. Tubular and hollow shapes require mandrels located inside the die and anchored upstream of it. Any reinforcement available in continuous form may be processed by pultrusion. Glass fiber is most commonly used often as alternative layers of unidirectional rovings and continuous strand mat.

V. FACTORS INFLUENCING COMPOSITE PROPERTIES

A. Fiber Volume Fraction

One of the important factors affecting composite properties is the amount of fiber it contains, i.e., percentage by volume.

Fiber volume fraction (V_f) is defined as the ratio of fiber volume (V_f) to the total composite volume (V_c).

$$V_f = \frac{V_f}{V_c} \tag{7}$$

Matrix volume fraction (V_m) is defined as:

$$V_m = \frac{V_m}{V_c} \tag{8}$$

Using the shear lag theory, the minimum spacing between fibers is determined upon which the maximum volume fraction obtained is reported [31]. For a composite made of given fiber and matrix materials, there is an optimal spacing between fibers at which the fiber tensile strength will be fully exploited. This optimal spacing is the minimum allowable spacing between fibers, below which the structure will start to disintegrate under loading before the tensile failure. This minimum spacing then defines a maximum volume fraction allowable for a composite.

B. Strength, Modulus, and Chemical Stability of Fiber and Resin Matrix

The mechanical properties of fiber-reinforced composites very much depend on the strength and modulus of the reinforcing fiber [32–36]. Choice of the matrix resin depends on final requirements of the product in addition to low cost, ease of fabrication, environmental conditions, and chemical resistance of the matrix. The function of the resin matrix in a fiber composite will vary, depending on how the composite is stressed. For compressive loading, the matrix prevents the fibers from buckling and also provides a stress transfer medium, so that when an individual fiber breaks, it does not lose its load carrying capability. The physical properties of the resin influencing the behavior of shrinkage during cure are modulus of elasticity, ultimate elongation, tensile and flexural strength, compression, and fracture toughness.

C. Influence of Fiber Orientation

Orientation of fibers relative to one another has a significant influence on the strength and other properties of fiber-reinforced composites. With respect to orientation three extremes are possible as shown in Fig. 5. Longitudinally aligned fibrous composites are inherently anisotropic, in that, maximum strength and reinforcement are

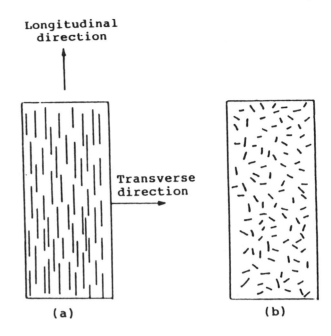

Figure 5 Schematic representation of (a) aligned and (b) randomly oriented fiber reinforced composites.

achieved along the direction of fiber alignment. In the transverse direction, fiber reinforcement is virtually nonexistent. Fracture usually occurs at very low tensile stress, which may be less than the strength of the matrix. In randomly oriented composites, strength lies between these two extremes.

Uniaxial fiber-filled composites can have very high longitudinal tensile strength, but the longitudinal compressive strength is generally less because of the buckling of the fiber [37,38]. Transverse compressive strength is limited by the strength of matrix and so is less than the longitudinal compressive strength [38]. By randomly orienting fibers in a plane or by making multilayered laminates in which the fibers in the various layers have different orientation directions, composites can be constructed that are essentially isotropic in plane, i.e., such composites have desirable properties in all directions in a plane. If fibers are aligned in all three directions, desirable properties can be achieved in three dimensions. However, to achieve good properties in two or three dimensions, there must be a sacrifice compared with the longitudinal direction of a uniaxially oriented fiber composite. Figures 6a and 6b show the effect of fiber orientation on the tensile strength and modulus of PALF–LDPE composites [39].

D. Influence of Fiber Length

The strength of a fiber-reinforced composite depends not only on the tensile strength of the fibers, but on the

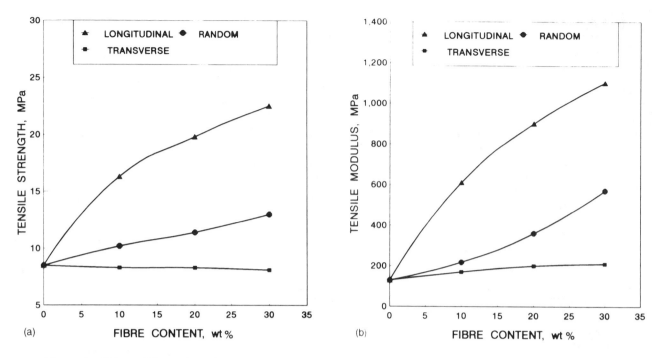

Figure 6   Effect of fiber orientation on (a) tensile strength and (b) tensile modulus of PALF–LDPE composites.

degree to which an applied load is transmitted to the fibers. The extent of load transmittance is a function of fiber length and the magnitude of the fiber–matrix interfacial bond. The critical aspect ratio that would result in fiber fracture at its midpoint can be expressed as:

$$(l/d)_c = \frac{S_f}{2\tau} \qquad (9)$$

where l = length of fiber,
   d = diameter of fiber,
   $(l/d)_c$ = critical aspect ratio,
   S_f = tensile stress at the fiber, and
   τ = fiber–matrix interfacial shear strength.

The rule of mixtures for discontinuous fiber composites may be expressed as:

$$S_c = V_f S_f \left[l - \frac{lc}{2l} \right] + V_m S_m \qquad (10)$$

where S_c = tensile strength of composite,
   S_f = tensile strength of fiber,
   V_f = volume fraction of fiber,
   V_m = volume fraction of matrix, and
   S_m = tensile strength of matrix.

More detailed discussion on the interface of fiber length is given in Section X, Theory and Mechanics of Reinforcement.

E. Coupling Agents

Studies on the composite materials have shown that the bonding between the reinforcing fiber and the matrix has a significant effect on the properties of the composite. Good bonding at the interface can be achieved by modifying the fiber–matrix interface with various surface reactive additives or coupling agents.

An important technique for improving compatibility and dispersibility between filler and matrix is to develop a hydrophobic coating of a compatible polymer on the surface of the filler before being mixed with the polymer matrix. Generally, coupling agents facilitate the optimum stress transfer at the interface between filler and matrix. The selection of a coupling agent that can combine both strength and toughness to a considerable degree is important for a composite material.

The most common coupling agents are silane-isocyanates—and titanates-based compounds whose chemical composition allows them to react with the surface of fiber. The structural formulae and properties of major silane and titanate coupling agents are given in Tables 5a and 5b. The coatings are applied to the fiber before they are compounded with the polymer matrix. Coupling action will be explained in Section VII. Peroxide treatment also has significant influence on improving the interaction between fiber and matrix [39].

Table 5a Structures and Physical Properties of Silane Coupling Agents

Chemical name	Structure	Molecular weight	Specific gravity at 25°C
Vinyltrichlorosilane	$CH_2=CHSiCl_3$	161.5	1.26
Vinyltriethoxysilane	$CH_2=CHSi(OC_2H_5)_3$	190.3	0.93
Vinyltri(β-methoxy-ethoxy)silane	$CH_2=CHSi(OC_2H_4OCH_3)_3$	280.4	1.04
γ-Glycidoxypropyl-trimethoxysilane	$CH_2 - CHCH_2OCH_2CH_2CH_2Si(OCH_3)_3$ $\underset{O}{\diagdown\diagup}$	236.1	1.07
γ-Methacryloxy-propyltrimethoxysilane	$CH_2=\overset{\overset{\displaystyle CH_3}{\vert}}{C}-\underset{\underset{\displaystyle O}{\parallel}}{C}-O-C_3H_6Si(OCH_3)_3$	248.1	1.04
N-(β-Aminomethyl) γ-aminopropyl-trimethoxysilane	$H_2NC_2H_4NHC_3H_6Si(OCH_3)_3$	222.1	1.03
N-(β-Aminomethyl) γ-aminopropylmethyl-dimethoxysilane	$H_2NC_2H_4NHC_3H_6\underset{\underset{\displaystyle CH_3}{\vert}}{Si}(OCH_3)_2$	206.1	0.98
γ-Chloropropyltri-methoxysilane	$ClC_3H_6Si(OCH_3)_3$	198.5	1.08
γ-Mercaptopropyltri-methoxysilane	$HSC_3H_6Si(OCH_3)_3$	196.1	1.06
γ-Aminopropyl-triethoxy silane	$H_2NC_3H_6Si(OC_2H_5)_3$	221.0	0.94

Source: Ref. 40.

F. Influence of Voids

During the incorporation of fibers into the matrix or during manufacture, air or some volatile may be trapped in the material. The most common cause of voids is the incapability of the resin to displace all the air that had entered within the rising or yarn as it passed through the resin impregnator. The rate at which the reinforcement passes the resin, the viscosity of the resin, the wettability or contact angle between the resin and reinforcement surface, and the mechanical working of the reinforcement will affect the removal of entered air. A high void content usually leads to lower fatigue resistance, greater susceptibility to water diffusion, and increased variation in mechanical properties. The amount of trapped air or volatiles that exist in the cured composite can be estimated by comparing the theoretical density with its actual density.

$$V_c = \frac{\rho_c - \rho}{\rho_c} \tag{11}$$

where V_c = volume fraction of voids,
  ρ_c = theoretical density, and
  ρ = actual density.

The voids represent a degree of unreproducibility that would be desirable to eliminate because they can limit the performance of these materials by providing paths for environmental penetration into and through the composite [42].

Table 5b Structures and Physical Properties of Titanate Coupling Agents

Chemical name	Structure	Molecular weight	Specific gravity at 25°C
Isopropyltriiso-stearoyltitanate	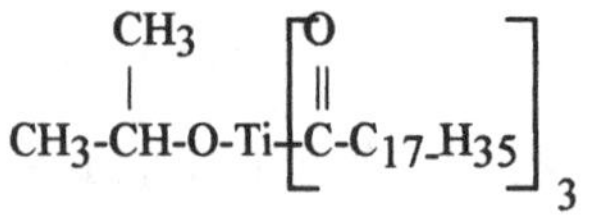 $CH_3-CH-O-Ti[-C-C_{17}H_{35}]_3$ (with CH_3 branch, $C=O$)	957	0.94
Isopropyltridodecyl-benzenesurfonil titanate 9S (2-3252)	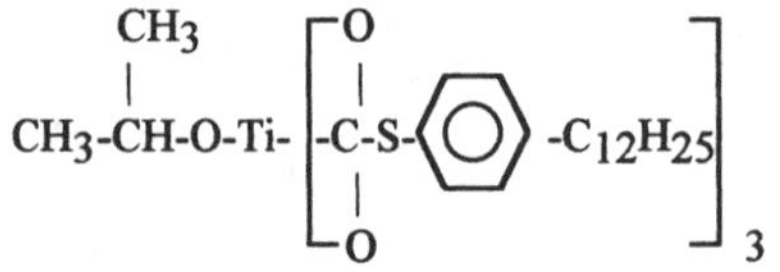 $CH_3-CH-O-Ti-[-C-S-\bigcirc-C_{12}H_{25}]_3$ (with CH_3 branch, O above and O below S)	1083	1.09
Isopropyl-tri(dioctyl-pyrophosphate)-titanate 38S	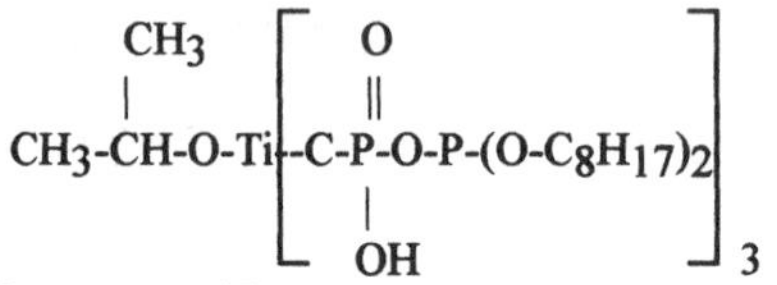 $CH_3-CH-O-Ti[-C-P-O-P-(O-C_8H_{17})_2]_3$ (with CH_3 branch, $C=O$, OH)	1311	1.10
Tetraisopropyl-bis-(dioctylphosphite) titanate 41B	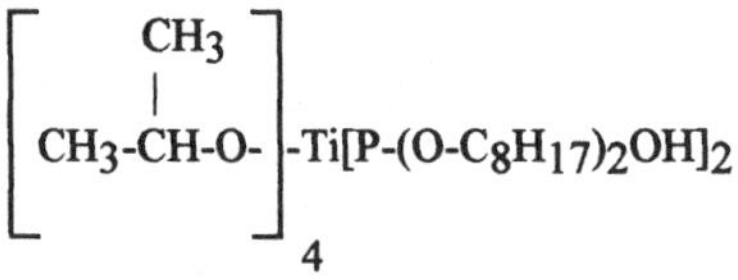 $[CH_3-CH-O-]_4-Ti[P-(O-C_8H_{17})_2OH]_2$ (with CH_3 branch)	897	0.94
Tetraoctyl-bis-(ditridecylphosphite) titanate 46B	$(C_8H_{17}-O-)_4-Ti.[P-(O-C_{13}H_{27})_2OH]_2$	1458	0.94
Tetra(2,2-diallyloxy-methyl-1-butyl)-bis-(ditridecylphosphite) titanate 55	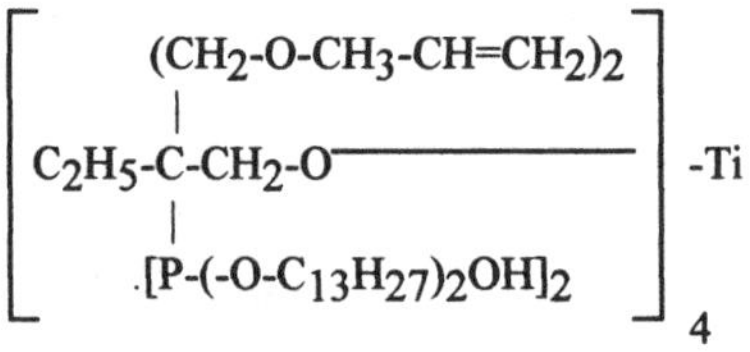 $[C_2H_5-C-CH_2-O$ ... $-Ti]_4$ with $(CH_2-O-CH_3-CH=CH_2)_2$ branch and $.[P-(-O-C_{13}H_{27})_2OH]_2$	1794	0.97
Bis(dioctylpyro-phosphate)-oxyacetate titanate 138 S	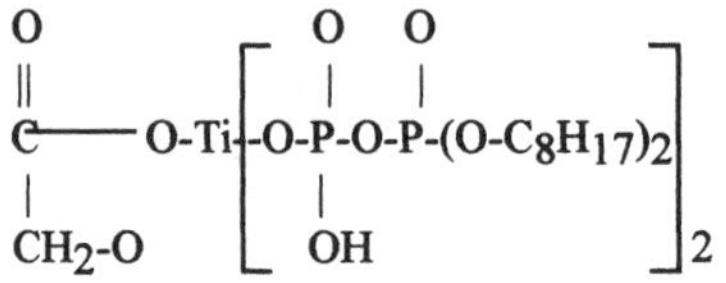 C(=O)$-O-Ti[-O-P-O-P-(O-C_8H_{17})_2]_2$ with CH_2-O branch, O O above, OH below	925	1.09
Bis(dioctylpyro-phosphate)-ethylene titanate 238 S	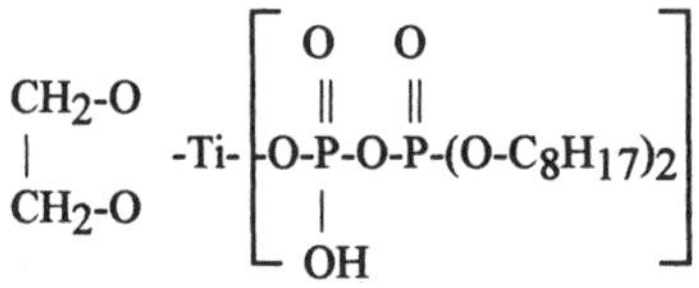 CH_2-O, CH_2-O branches $-Ti-[O-P-O-P-(O-C_8H_{17})_2]_2$ with O O above, OH below	911	1.08

Source: Ref. 41.

VI. THEORIES OF ADHESION

The mechanisms of adhesion are explained by four main theories: mechanical theory, adsorption theory, diffusion theory, and electrostatic theory.

A. Mechanical Theory

According to this theory, the adhesive interlocks around the irregularities or pores of the substrate. A rough surface will have a larger potential bonding area than a smooth one. The metal plating of the polymer acrylonitrile–butadiene styrene (ABS) is an example of mechanical adhesion. To obtain satisfactory adhesion, it is necessary to first pretreat with chromic acid, which dissolves rubber particles near the polymer surface, leaving a porous structure. A metal may then be deposited from solution into the porous structure providing a mechanical key. Geethamma and coworkers [43] have performed some electron microscopy studies to support this mechanism. Figures 7a and 7b show the untreated and alkali-treated coir fibers. During alkali treatment, certain holes are produced on the fiber surface after the removal of lignin. These holes will provide an anchoring effect and result in a better interaction between fiber and matrix.

B. Adsorption Theory

The adhesive macromolecules are adsorbed on to the surface of the substrate and are held by various forces of attraction. The adsorption is usually physical, i.e., due to van der Waals forces. However, hydrogen bond-

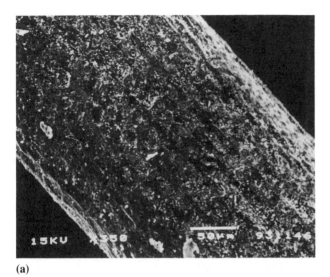

(a)

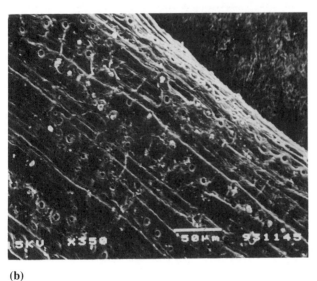

(b)

Figure 7 SEM of (a) untreated and (b) alkali-treated coir fiber. (From Ref. 43.)

ing and primary bonding (ionic or covalent) are involved in some cases. If primary bonds are involved, the term chemisorption is used. This theory assumes a definite interface between the adhesive and adherent. Hydrogen bonding is believed to be important in the bonding of tire cords to rubber. There is good evidence that hydrogen bonding is involved in the self-adhesion of corona-treated polyethylene [44,45].

The direct evidence of chemical bonding in adhesion is observed in silane coupling agents. Mablas et al. [46] have provided strong evidence for chemical bonding between isocyanate coupling agents and natural fiber-reinforced polyethylene systems. Although the exact nature of the interactions at the interface may be uncertain, the adsorption theory of adhesion is the most widely accepted mechanism.

C. Diffusion Theory

The adhesive macromolecules diffuse into the substrate, thereby eliminating the interface. It requires that the macromolecules of the adhesive and adherent have sufficient chain mobility and that they are mutually soluble. Diffusion will also take place when two pieces of the same plastic are heat sealed. Voyutskii [47] provided experimental evidence for this theory based on autohesion experiments, i.e., bonding experiments when the adhesive and substrate are identical. He studied the bonding of rubbers at elevated temperatures and found that joint strength increases with increasing period of contact, increasing temperature, increasing pressure, and decreasing molecular weight. The strength is also found to be affected by addition of plasticizers and with crosslinking.

D. Electrostatic Theory

In this theory, the adhesion is due to electrostatic forces arising from the transfer of electrons from one material of an adhesive joint to another. Evidence in support of this theory includes the observation that the parts of a broken adhesive joint are sometimes charged [48]. It has been shown that peeling forces are often much greater than can be accounted for by van der Waals forces or chemical bonds.

VII. INTERFACE MODIFICATION

A. Surface Modification of Polymers

1. Chemical Treatment

When a polymer is soaked in a heavily oxidative chemical liquid, such as chromic anhydride–tetrachloroethane, chromic acid–acetic acid, and chromic acid–sulfuric acid, and treated under suitable conditions, polar groups are introduced on the polymer surface and the surface characteristics are improved [49,50]. The surface of the polymer is heavily oxidized by nascent oxygen generated during reaction, i.e.,

$$K_2Cr_2O_7 + 4H_2SO_4 \rightarrow Cr_2(SO_4)_3 + K_2SO_4 \quad (12)$$
$$+ 4H_2O + 3[O]$$

The surface of polyolefine is activated by treating it with the liquid through the formation of polar groups such as $>C{=}O$, $-OH$, $-COOH$, and $-SO_3H$. Rasmussen et al. [51] qualitatively determined these polar groups in detail. The following mechanism for the formation of oxygen-containing polar groups has been proposed [52].

$$
\begin{array}{c}
R \\
| \\
-CH_2\text{-}C\text{-}CH_3 \\
| \\
H
\end{array}
\xrightarrow{\text{Chromic acid}}
\left[
\begin{array}{c}
R \\
| \\
CH_2\text{-}C\text{-}CH_3 \\
| \\
O \\
| \\
Cr(IV) \\
| \\
(OH)_3
\end{array}
\right]
\xrightarrow{H_2O}
\begin{array}{c}
R \\
| \\
-CH_2\text{-}C\text{-}CH_2\text{-} \\
| \\
OH
\end{array}
$$

$$
\longrightarrow \quad
\begin{array}{cc}
H & H \\
| & | \\
-C{=}O, & O{=}C\text{-}CH_2\text{-}
\end{array}
, \quad
\begin{array}{c}
O \\
\| \\
-CH_2\text{-}C\text{-}CH_2\text{-}
\end{array}
\quad (13)
$$

2. Corona Discharge Treatment

Corona discharge treatment results in the formation of high-polarity functional groups, such as carbonyl, at the polymer surface. Various mechanisms have been proposed for the improvement of the adhesive properties

of polyethylene by corona discharge treatment, some attributing it to electric formation [44], others to hydrogen bonding [53]. Figure 8 represents an ESCA analysis of PE treated by corona discharge [54]. The spectrum of C_1s, shown in Fig. 8 consists of carbon corresponding to functional groups of ethers, alcohols, peroxides, ketones, aldehydes, acids, and esters, which are not detected in untreated material. Chemical shifts related to the bonding conditions of C_1s are $+4.1$ eV for $R*COOH$, $+3.9$ eV for $R*COO{-}R$, $+3.2$ eV for $R*CHO$, $+3.1$ eV for $R{-}C*O{-}R$, $+3.3$ eV for $R{-}C*O{-}NHR$, $+1.5$ eV for $R*CH_2OH$, $+1.6$ eV for $RC*H_2{-}O{-}R$, and 0 eV for $R{-}C*H_2{-}CH_2{-}R$. By means of these values, peaks in the figure can be correlated with various functional groups.

3. Ultraviolet Irradiation

Some of the advantages of ultraviolet (UV) irradiation on polymer matrices are: (1) reaction occurs at ordinary temperature and pressure, (2) selective reaction is possible, and (3) light energy can be focused on the surface of thematrix. Modification is carried out by introducing functional groups: (1) by applying UV light to oxidize the material surface or alloying the material to contact a gas or sensitizer to cause a photochemical reaction or (2) by alloying UV irradiation graft polymerization to occur at the material surface. Adhesive strength increases with the increasing degree of treatment. To modify a polymer surface, a carboxylic acid group is produced at the surface by introducing double bonds through irradiation in acetylene followed by photo oxidation [55].

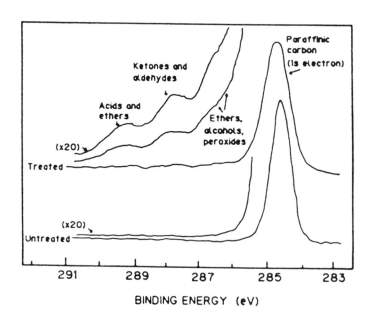

Figure 8 ESCA spectra of untreated and corona-treated PE surfaces. (From Ref. 54.)

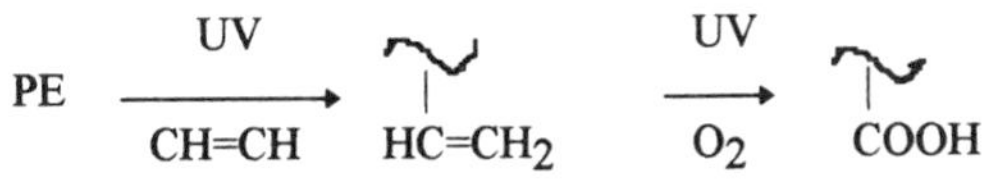

$$PE \xrightarrow[\text{CH=CH}]{\text{UV}} \text{HC=CH}_2 \xrightarrow[\text{O}_2]{\text{UV}} \text{COOH} \quad (14)$$

In addition, there are many surface modification processes that use triplet sensitizers to permit oxidation reactions. In a typical process, polyisocyanate is applied on a polyolefin together with a sensitizer such as benzophenone and then irradiated with UV light. As shown in Eq. (15) the sensitizer has an oxidizing effect to produce hydroxyl groups over the polymer surface. These hydroxyl groups finally react with isocyanate to provide a functional polymer [56,57].

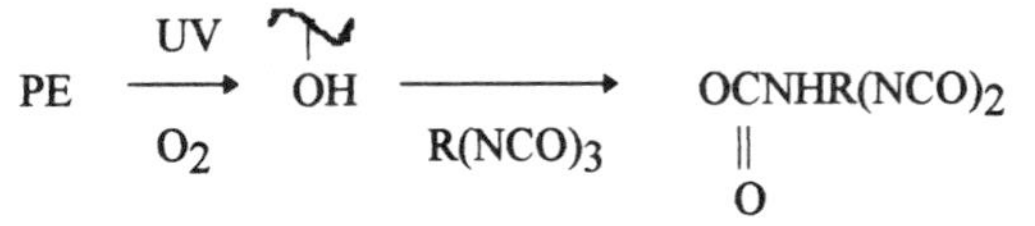

$$PE \xrightarrow[\text{O}_2]{\text{UV}} \text{OH} \xrightarrow{\text{R(NCO)}_3} \underset{\underset{O}{\|}}{\text{OCNHR(NCO)}_2} \quad (15)$$

In some cases adhesive properties can be improved by UV light irradiation alone [58].

4. Plasma Treatment

The plasma utilized for polymer treatment is generally called nonequilibrium low-temperature plasma [59]. In low-temperature plasma for polymer treatment, relatively few electrons and ions are present in the gas. Here, energy of electrons are in the range of 1–10 eV. This energy causes molecules of gas A to be ionized and excited. As a result radicals and ions are produced.

Ionization:	$A + e_f \longrightarrow A^+ + 2e$	(16)
Excitation:	$A + e_f \longrightarrow A^* + e$	(17)
Radical dissociation:	$A^* \longrightarrow A_1\cdot + A_2\cdot$	(18)
Luminescence:	$A^* \longrightarrow A^+ \, h\nu$	(19)
Electron addition:	$A + e \longrightarrow A_1^-$	(20)

The activated particles react with polymeric materials so that polymeric radicals are produced on the surface layer of materials. This causes the surface layer to be oxidized, crosslinked, or decomposed. On the other hand, $A\cdot$s are produced from molecules of the gas and are polymerized, so that the resultant polymers of A coat the surface of the material.

Crosslinking

In the presence of an inert gas, such as He or Ar, crosslinking can be introduced into the surface layer of material by plasma treatment. Hansen and Schonhorn [60] named this "Crosslinking by Activated Species of Inert Gases" (CASING). As a result, bond strength is enhanced because crosslinking strengthens the surface layer.

B. Surface Modification of Fillers

1. Coupling Agents

The structure of the silane coupling agent is expressed by the general formula $+(RO)_3$—Si—R', where the RO group represents functional groups, which hydrolyzed to give a silanol group (e.g., methoxy and ethoxy), and R' is for those groups that have an affinity for and display reactivity to the matrix materials. A large number of studies have been reported on the influence of silane coupling agents in polymeric materials reinforced with natural as well as inorganic fibers [61–65].

The possible reaction between silane coupling agents and inorganic fillers are [66–69]

1. The silane coupling agents undergo chemical reaction with the surface of inorganic substances to form an SiOM bond (M: Si atom in glass).
2. The silane coupling agent is physically adsorbed on the inorganic surface.
3. The Si—OH group on the glass surface forms a hydrogen bond with the silanol group derived from the silane coupling agent.
4. The silane coupling forms a sheathlike structure around the glass fibers.
5. A reversible equilibrium reaction takes place between the hydroxyl group on the surface and the silanol group derived from the silane coupling agent. The mechanism of action of silane coupling agents can be explained in terms of the chemical bonding theory.

As in Fig. 9, the hydrolyzable group in a silane coupling agent is first hydrolyzed to give a silanol group, which then undergoes a condensation reaction with the silanol group on the surface of the inorganic material (i.e., glass, silica, etc.) to form covalent bonding between the coupling agent and the material. However, the nonhydrolyzable functional group in the coupling agent bonds to the matrix through a chemical reaction. Thus, the coupling agent serves to increase the strength of the composite by providing chemical bonding to connect the glass surface and the matrix.

The structure of a titanate coupling agent is expressed by the general formula RO—Ti (OXRY)$_3$. These titanate coupling agents are useful for improving flexibility and processability [70]. Joseph and coworkers [71]

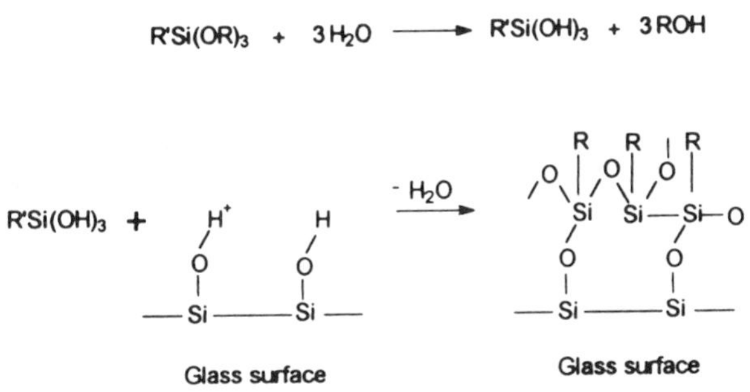

Figure 9 Chemical bonding theory.

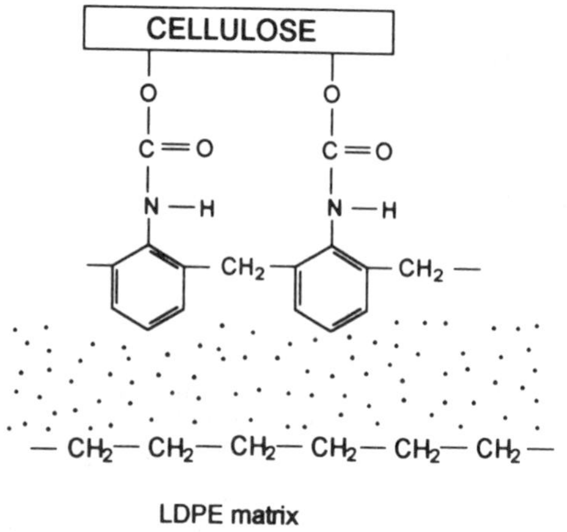

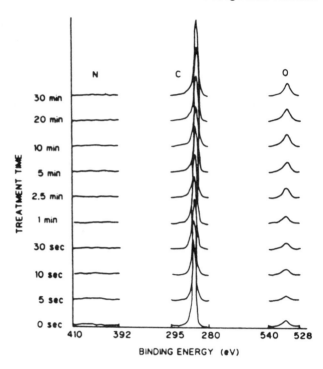

Figure 10 Schematic representation of the interfacial region of PMPPIC treated fiber and LDPE.

have reported the use of isocyanates (toluene diisocyanate, poly[methylene] poly[phenyl isocyanate] in natural fiber-reinforced thermoplastic composites. Urethane derivative of cardanol wastreated with sisal fibers to improve the compatibility between fiber and matrix. It was found that PMPPIC treatment has significant influence on the properties of composites, i.e., increased thermal stability, reduced water absorption, etc. The —N=C=O group in PMPPIC is highly reactive with the –OH group of cellulose and, therefore, a urethane linkage is formed.

$$-N=C=O + HO\text{-cellulose} \longrightarrow \underset{\overset{|}{H}}{-N}\overset{\overset{O}{\|}}{C}-O\text{-cellulose}$$

(21)

A hypothetical chemical structure in the interfacial area of the PMPPIC-treated composite [72] is shown in Fig. 10. The long-chain molecules present in PMPPIC interact with polyethylene leading to van der Waals type of interaction.

VIII. CHARACTERIZATION OF INTERFACE

The type of interaction along the interface will exert a great influence on the various properties of the composite materials. Therefore, to improve the performance of a composite material, it is absolutely necessary to characterize the structures of the interface. Some of the methods for analysis of the interface are ESCA, AES, IR-FTIR, SIMS, and SEM, etc. At present, ESCA is widely used in the surface analysis of elements and the qualitative analysis of functional groups. Figure 11 shows the ESCA spectrum of polyethylene treated with

Figure 11 Relation between ESCA spectra and Ar$^+$ treatment time of PE. (From Ref. 73.)

Ar plasma for various periods of time [73]. It is clear that there is a spectrum due to an oxygen atom even for an untreated sample having zero time. If the treating time is prolonged, the combined oxygen atoms increase and the shape of the carbon spectrum changes. Nitrogen atoms do not appear before and after treatment. It is possible to know how carbon is combined with oxygen from the C$_1$s spectrum, especially from the skirts on the high-binding energy side. Figure 12 shows an analysis by AES spectrum of the interface of a titanium composite material reinforced with W-SiC fiber [74]. The cross-section of a fiber in its center line is analyzed with a beam less than 50 mm in diameter. The results have revealed that a reaction phase of 7–12 μm width is formed on the interface between the SiC fiber and the Ti matrix.

Figure 13 shows the plot of the interface in the carbon fiber–A1 composite by SIMS [75]. Here measurements by SIMS indicate that the interface reaction phase consists of Al, Na, Mg, and Sn. The wettability of carbon fiber to Na is good, resulting in good penetration of Na into the carbon fiber. The penetrating phenomenon is believed to be the result of intergranular diffusion or the formation of an intercalation compound such as C$_{64}$Na. Carbon fiber is first covered by Na–SN intermetallic compound, and then by Mg–Sn intermetallic compound. This phenomenon is supposed to be due to the geometric

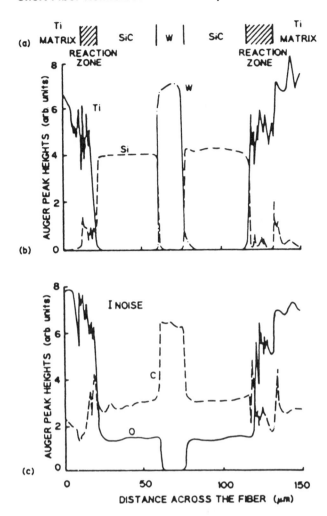

Figure 12 AES spectra of the W-SiC composite sample. (a) Schematic diagram of the sample (the shaded regions represent the reaction zone). (b) C and O line-scan profiles. The maximum PE noise is indicated by an error bar. (From Ref. 74.)

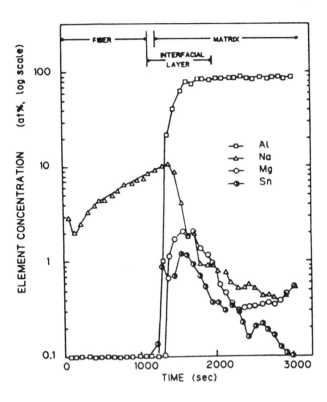

Figure 13 SIMS plot of Na, Sn, Mg, and Al concentrations as a function of time in graphite aluminium composite prepared by the sodium process. (Analysis starts in graphite fiber and proceeds into the aluminium matrix.) (From Ref. 75.)

effect of the irregular surface of carbon fiber or to a decrease of concentration on the interface. The thickness of the interface reaction phase is 0.45–0.55 μm.

The information obtained from an application of IR spectroscopy to a surface investigation includes the molecular structure, orientation, chemical reaction, conformation, crystallinity, and so on.

Figure 14 shows the ATR spectrum of the etched polyethylene surface treated with a chronic acid group [76]. Absorption bands due to surface treatment appear at 3300, 1700, 1260, 1215, and 1050 cm^{-1}. The band at 3300 cm^{-1} represents the absorption due to the hydroxyl group and that at 1700 cm^{-1} is due to the carbonyl group. The bands at 1260, 1215, and 1050 cm^{-1} are all due to the alkyl sulfonate group.

$$CH_3\text{-}O\text{-}\overset{\overset{O}{\|}}{\underset{\underset{O}{\|}}{S}}\text{-}O\text{-}CH_2\text{-}\overset{\overset{O}{\|}}{\underset{\underset{O}{\|}}{S}}\text{-}O^-$$

The formation of these polar groups contributes increased adhesion. Observation of disappearing vinyl groups in the silane coupling agent and of the formation of polystyrene in the silica by FTIR analysis (Fig. 15) have confirmed the occurrence of a reaction between the polymer and the silane coupling agent [77].

The observation of the spectrum for styrene polymerized on the surface of silane-treated silica and of the difference spectrum of polystyrene adsorbed on the surface of silica have revealed that there are absorption bands of atactic polystyrene at 1602, 1493, 1453, 756, and 698 cm^{-1}. The absorption bands at 1411 and 1010 cm^{-1} are related to vinyl trimethoxy silane, and C of the difference spectrum is below the base line. This indicates that the vinyl groups of silane react with styrene to form a copolymer.

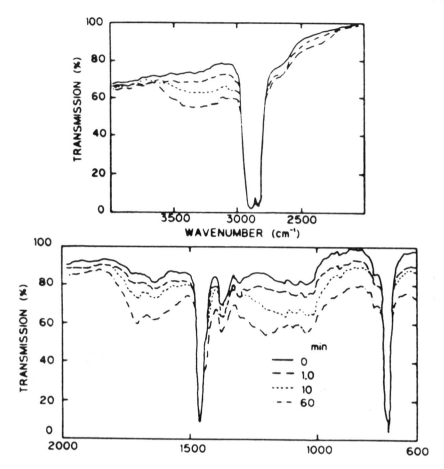

Figure 14 Surface IR spectra of etched LDPE-ATR spectra recorded with a KRS-5 reflection element, at 45° angle of incidence. Times refer to chronic acid itch duration. (From Ref. 76.)

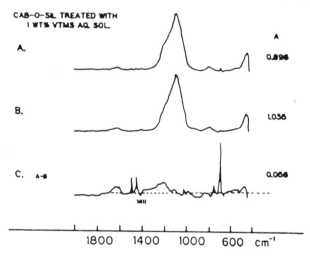

Figure 15 High-surface area silica treated with aqueous solution of 1 wt% vinyltrimethoxy silane. A silica was polymerized with styrene and washed with CS_2 three times. Polystyrene produced in experiment A was deposited with B silica and the silica washed with CS_2 three times. (From Ref. 77.)

SEM can also be used for the interface analysis of composites. Figures 16a and 16b show the SEM of PMPPIC-treated and untreated pineapple fiber-reinforced LDPE composites. Strong adhesion between fiber and matrix is evident from Fig. 16a, whereas Fig. 16b indicates fiber pullout [78].

Figure 17 shows the interface interaction between silane-treated E glass and polystyrene examined by a Raman spectrum [79]. Comparison of B in Fig. 17 with C indicates that the polymerization of styrene is proceeding on the silane-treated glass, and a comparison of C with D indicates the interaction between the silane coupling agent and styrene and homopolymerization of the styrene are taking place following the shift of absorption from 1718 cm^{-1} to 1702 cm^{-1}, as carbonyl stretching vibration of the silane has revealed.

IX. INTERFACIAL BOND STRENGTH DETERMINATION

Optimum mechanical properties in composite materials are strongly related to the efficiency of load transfer.

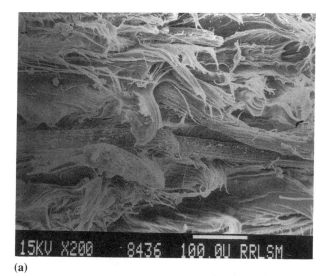

(a)

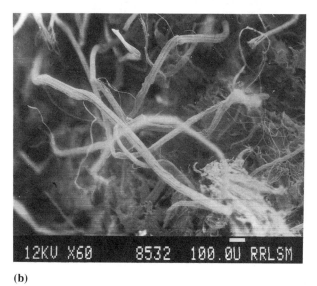

(b)

Figure 16 SEM of tensile fracture surface of (a) PMPPIC treated and (b) untreated PALF–LDPE composites.

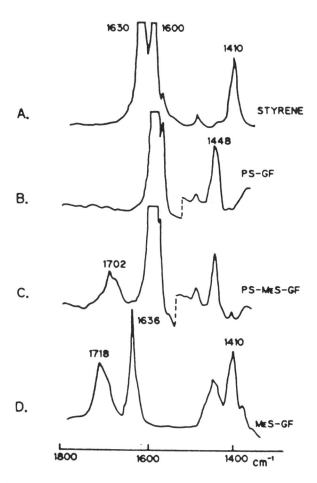

Figure 17 Raman spectra of a glass fiber/matrix interfaces. (A) styrene monomer; (B) untreated E-glass fiber coated with polystyrene, (C) E-glass fiber treated with γ-methacryloxy propyl trimethoxy silane.

When an external stress is applied to the specimen parallel to the fiber axis, the tensile load will be transmitted from the matrix to the fiber through the interfacial shear stress. An increase in the applied stress will result in a progressive breakup of the fibers into shorter fragments and a knowledge of the distribution of fragment length gives a value for the critical fiber length and, hence, interfacial shear strength to be determined. On the basis of the Kelly-Tyson model [80] the tensile stress (σ_x) of the fiber increases linearly from nearly zero at the fiber ends to a maximum value limited by the ultimate tensile strengths of fiber (σ_f). When this limit is reached, the fiber should break. However, the interfacial shear stress

(τ_u), which has a maximum value at the fiber ends, decreases to nearly zero toward the middle of the fiber. If the external stress is increased, this fiber breakage process should continue until all fiber fragments are less than a critical length, in which case the tensile stress in the fibers cannot reach σ_f. The minimum fiber length for which the tensile stress can reach the ultimate tensile strength of the fiber is termed the critical length (l_c).

The Kelly-Tyson model predicts that the fiber fracture process should result in a distribution of fragment lengths from $l_c/2$ to l_c. Often the distribution of fragment length obtained is broader than the ratio 2:1 predicted. This has been attributed to the existence of flaws in the fiber, thereby causing its strength to depend on length [80,81]. Many kinds of interfacial bond test methods have been proposed [82–85] to study the load transfer including the embedded single fiber tension test, the embedded single fiber compression test, the microdebond test, the single fiber pullout test, the bead pullout test, the short beam shear test, the transverse tension test, the transverse flexure test, and so on. Among these, the most popular are the fragmentation test, fiber pullout test, and micro indenter pushout tests. In the fragmentation test, a single fiber is embedded in a specimen of pure resin, which should have a high strain to failure. The specimen is loaded in tension parallel to the fiber. When the failure strain of the embedded fiber is smaller than the failure strain of the matrix, the fiber will begin to fracture. During the fragmentation test, the strain in the specimen is gradually increased, the strain in the fiber will also increase and reach a new critical fiber strain characteristic of the new fiber fragment length. In this way, the fiber fragments will be fractured again and the fiber is introduced by shear stresses along the fiber surface.

Thus, during a fragmentation test, a saturation fragment length or critical fiber length, (l_c) is reached at a certain applied strain, i.e., an equilibrium is reached between interface shear stress and fiber length at a particular fiber fragment length. The fragmentation process can be followed with a light microscope, and the fragment length as well as other features (debonding, matrix cracking, yielding) can be monitored as a function of the applied strain or load. The acoustic emission method can be used if the matrix is not transparent, but does need careful calibration.

The interface shear stress can be derived from the observed fragment length if one assumes that its absolute value is constant over the whole fragment length. Based on simple equilibrium of length, one can derive that in each point along the fiber

$$\tau_i = \frac{-df d\sigma_f}{4\, dx} \tag{22}$$

where df = fiber diameter.

If τ_i is constant, integration over the fiber length, (L) yields:

$$\tau_i = \frac{\sigma_{\text{fmax}} df}{2\, L} \tag{23}$$

If we assume that saturation during the fragmentation test just occurs when the maximum fiber stress does not equal the fiber length, the above formula where L = Lc and $\sigma_{fmax} = \sigma_{ufr(Lc)}$, the strength of a filament of length Lc, can be used to calculate the interface shear stress. As the interface shear stress is assumed to be constant and characteristics of the fiber–matrix interface, the obtained value is taken to be the interface shear strength [86]. The fabrication of resin block embedded with a single fiber is very time consuming. It is possible to make a specimen in the form of thin tapes using an extruder coupled with a slit die [87].

Because of the viscoelastic interactions in polymers, the fragmentation test may not always yield the correct information concerning the efficiency of load transfer or interfacial shear strength of polymer matrices, i.e., it does not take into account matrix viscosity and strain rate. The saturation fragment length is itself a reliable parameter, which for fibers with equal strength gives a good indication for the stress transfer capability of the interface. Interface bond strength and interface friction stress can be derived from the fragment length and the debond strength [88]. Absolute values of the two interface strengths, i.e., debond strength and frictional stress, can only be obtained when the actual stress is taken into account. Using Eq. (23), one should be aware that the basic assumptions are violated in reality, and the obtained interface strength will lie somewhere between the real interface debond strength and the interface frictional stress.

A. Single Fiber Pull Out Test

In the single fiber pull out test (SFPO), a small portion of the fiber is embedded in the bulky matrix and the interfacial strength is calculated from the peak load when the fiber is pulled out of the composite.

In this method [89], a single fiber is taken and partially embedded in a drop of uncured resin placed on a holder. The resin is then cured with the fiber held upright. The holder, with resin and fiber, is held in a grip attached to the crosshead and then pulled out from the resin. The force pulling the fiber out of the resin is balanced by shear stress at the resin–fiber interface holding the fiber in place. The maximum shear stress occurs as the embedded length tends to zero and is given by:

$$\tau_{\text{max}} = \tau_{av}\alpha le \coth (\alpha le) \tag{24}$$

where $\tau_{av} = F/2r_f le$ and $\alpha = (2Gi/br_f E_f)^{1/2}$

where F is the pull out force, r_f the fiber radius, le the embedded length, Gi the shear modulus of interface, b the effective width of interface, and E_f the tensile modulus of fiber. The embedded length has a significant effect on the value for τ_{max}. As the embedded length becomes smaller, the value of τ increases so the widest possible range of embedded lengths should be measured.

Depending on the fiber–matrix system, a wide range of information may be obtained from this test. Four possible regions on a load displacement trace have been identified that may be associated with different interfacial phenomena: an elastic region, a plastic deformation region, a region in which dynamic decoil of the fiber occurs after fiber matrix debonding, and a region where frictional forces predominate and a stick-slip mechanism may occur. In systems where strong stiff fibers have a strong interfacial bond, failure is often catastrophic and only a value for τ_{max} may be obtained. At long embedded lengths, fiber failure may occur within the embedded length resulting in a high apparent pull out load for a short embedded length. This will distort the calculated value for τ_{av}. As the bond strength becomes greater, the ability to successfully pull a fiber from the matrix drop without the fiber failing becomes increasingly harder, and problems may be encountered when testing systems with very high interfacial shear strengths. In single fiber pull out from a micro composite (SFPOM) test, another technique can be used for studying the effect of fiber volume fraction on interfacial shear strength that is not obtained from the SFPO test.

B. Microdebond Test

The single filament pull out test, sometimes called the microdebond test, has received attention for some years as a way to assess the adhesion between fibers and matrices in fiber composite [90,91]. It provides a direct measure of interfacial adhesion and can be used with both brittle and ductile matrix resins.

Figure 18 shows a widely used test configuration where the matrix is a sphere of resin deposited as a liquid onto the fiber and allowed to solidify. The top end of the fiber is attached to a load-sensing device, and the matrix is contacted by load points affixed to the crosshead of a load frame or another tensioning apparatus. When the load points are made to move downward, the interface experiences a shear stress that ultimately causes debonding of the fiber from the matrix.

The analysis depends on whether the interfacial failure occurs by yielding or by crack propagation. The simplest analysis is based on interfacial yielding where the shear stress is assumed to be distributed uniformly over the interface from top to bottom. According to this analysis, the interfacial shear stress increases uniformly until every location in the interface gives way simultaneously.

For this, failure interfacial shear strength (τ) is obtained by dividing the maximum load P_m by interfacial area A.

Sometimes the failure occurs by propagation of a crack that starts at the top and travels downward until the interface is completely debonded. In this case, the fracture mechanics analysis using the energy balance approach has been applied [92] in which P_m relates to specimen dimensions, elastic constants of fiber and matrix, initial crack length, and interfacial work of fracture (W_i).

$$P_m = \frac{2\pi r_f \sqrt{r_f W_i E_f}}{1 + \cos \text{ech2}\left[\left(\dfrac{E_m}{E_f (1 + V_m) \ln (R/r_f)}\right)^{1/2} \left(\dfrac{le-a}{r_f}\right)\right]} \tag{25}$$

where r_f = fiber modulus, R = matrix radius, E_f = fiber modulus, E_m = matrix modulus, V_m = matrix Poisson ratio, W_i = interfacial work of fracture, le = original embedded length of fiber, and a = area of small initial crack at the top of the interface.

In the push out test [93], the fiber is pushed into the matrix rather than being pulled out. The test allows the measurement of two quantities, F_{deb} (the force at which debonding occurs) and F_{Pi} (the force needed to push the fiber through the matrix sample if it is thin enough). The bond shear strength τ_{deb} is calculated using the shear lag theory:

$$\tau_{deb} = \frac{\beta F_{deb}}{4\pi r_f^2} \tag{26}$$

and

$$\beta = \sqrt{\frac{E_m}{E_f (1 + V_m) \ln (R/r_f)}} \tag{27}$$

where E_f, E_m are the elastic moduli of the fiber and the matrix, respectively, and R represents a distance far away in the matrix so the stress or strain at this point is equal to the applied stress or strain. If the test is performed on a thin specimen, then:

$$\tau_{avg} = \frac{F_{Pi}}{2\pi r_f \Delta x} \tag{28}$$

where Δx is the sample thickness.

X. THEORY AND MECHANICS OF REINFORCEMENT

A discontinuous fiber composite is one that contains a relatively short length of fibers dispersed within the matrix. When an external load is applied to the composite, the fibers are loaded as a result of stress transfer from the matrix to the fiber across the fiber–matrix interface. The degree of reinforcement that may be attained is a function of fiber fraction (V_f), the fiber orientation distribution, the fiber length distribution, and efficiency of

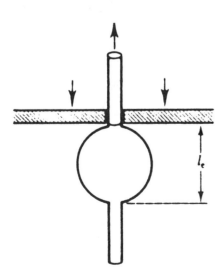

Figure 18 Geometry of the microdebond test.

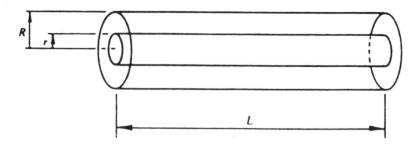

Figure 19 The representative element used with Cox analysis. The miner cylinder represents the fiber and outer as the matrix.

stress transfer at the interface. In general, the reinforcement is more effective when V_f is high, the fibers are long, the fibers are aligned in the principal stress direction, and the interface is strong.

There are two well-accepted models for stress transfer. In the Cox model [94] the composite is considered as a pair of concentric cylinders (Fig. 19). The central cylinder represents the fiber and the outer region as the matrix. The ratio of diameters (r/R) is adjusted to the required V_f. Both fiber and matrix are assumed to be elastic and the cylindrical bond between them is considered to be perfect. It is also assumed that there is no stress transfer across the ends of the fiber. If the fiber is much stiffer than the matrix, an axial load applied to the system will tend to induce more strain in the matrix than in the fiber and leads to the development of shear stresses along the cylindrical interface. Cox used the following expression for the tensile stress in the fiber (σ_f) and shear stress at the interface (τ):

$$\sigma_f = E_f \in_m \left[1 - \frac{\cos h\, \beta\, (R_a - X_r)}{\cos h\, \beta R_a} \right] \quad (29)$$

$$\tau = E_f \in_m \left[\frac{G_m}{2E_f \ln V_{f^{-1/2}}} \right]^{1/2} \frac{\mathrm{Sin}\, h\, \beta(R_a - X_r)}{\cos h\, \beta\, R_a} \quad (30)$$

$$\beta = \left[\frac{2G_m}{2E_f r^2 \ln V_f{}^{-1/2}} \right]^{1/2} \quad (31)$$

where E_f and G_m are tensile modulus of the fiber and shear modulus of the matrix, m is the applied strain, R_a is the aspect ratio of the fiber, $L/2r$ and Xr are the distances from the fiber end measured in terms of fiber diameter. The stress distribution along the fiber is shown in Fig. 20. The tensile stress in the fiber rises from zero at the end of the fiber to a maximum at the center. The rate of stress buildup from the fiber ends is greater when the parameters G_m/G_f and r/R (which is related to volume fraction) are greater. The overall reinforcing efficiency is given by the ratio of the area under the curve and the enclosing triangle. This tends to unity as the fiber aspect ratio tends to infinity.

The shear stress is greatest at the ends of the fiber and decays to zero somewhere along it. The tensile

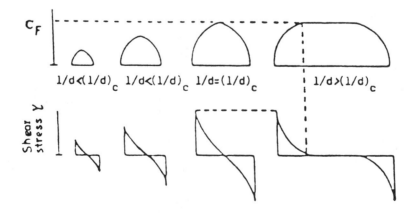

Figure 20 The variation of shear stress, τ, and the tensile stress along a short fiber in a matrix for short fibers of varying l/d ratio.

stress is zero at each end of the fiber and reaches a maximum at the center.

If the fiber is just long enough, the maximum tensile stress reaches the tensile stress in the matrix. The ratio $(l/d)_c$ that occurs under this condition is called the critical elastic aspect ratio [94], i.e.:

$$(l/d)_c = \frac{\sigma_f}{2\tau_i} d_f \tag{32}$$

For values of l/d less than $(l/d)_c$, the tensile stress in the fiber is always less than that in the matrix. The transfer of load from the matrix to the fiber is poor and the mechanical properties of the fiber are not fully utilized. If $l/d > (l/d)_c$, the tensile stress at the interface remains at a maximum over a greater proportion of fiber length. Here, the transfer of stress from the matrix to the fiber is very efficient, but the average tensile stress in the fiber is always less than that in the matrix because of reduced tensile stress at the end of the fiber.

By increasing or decreasing concentration of the coupling agent $(l/d)_c$ can be controlled. A coupling agent may increase τ_i, which in turn decreases the $(l/d)_c$.

When a fiber breaks, the normal stress at each of its broken ends become zero. Over a distance of $l_c/2$ from each end, stress builds back up to the average value by shear stress transfer at the fiber–matrix interface. Also, the stress state in a region close to the broken ends contain the following:

1. Stress concentration at the void created by the broken fiber.
2. High shear stress concentration in the matrix near the fiber ends.
3. An increase in the average normal stress in adjacent fibers.

When a transverse tensile load is applied, the fibers act as hard inclusions in the matrix instead of load carrying members. Although the matrix modulus is increased by the presence of fibers, local stresses and strains in the surrounding matrix are higher than the applied stress.

The second model introduced by Kelly and Tyson [95] is based on the concept of frictional stress transfer at the interface. It is considered that a constant shear stress is induced from the fiber ends. This results in a linear stress buildup. The frictional stress may be regarded as the interface shear strength (τ_i). This concept is often used for the experimental estimation of interface shear strength by fragmentation and the pull out test. This model gives a precise definition of the transfer aspect ratio R_t, or length. It should be noted that this transfer region increases as the applied strain is increased. If the fiber aspect ratio exceeds $2R_t$, there will be a plateau region along the central portion of the fiber. The reinforcement efficiency is obtained from a similar ratio of areas. Most discontinuous fiber composites contain fibers with a wide distribution of lengths distributed in a semi-random orientation, most often close to planar

random. Short fiber systems are more complex than the models due to fiber length distribution (FLD) and fiber orientation distribution (FOD). If the fiber aspect ratio distribution and the interfacial shear strength are known, the efficiency of each fiber may be determined using the Kelly-Tyson model [95]. The simplest procedure is to use the aspect ratio and orientation data to produce two constants, which may be applied to the Voigt equation:

$$E_c = \eta_o \eta_1 E_f V_f + E_m V_m \tag{33}$$

where η_1 and η_o are the length and orientation constants, respectively. They have a maximum value of unity, when the reinforcement efficiency is equal to that of continuous fibers.

The Bowyer and Bader [96] methodology can be used to predict stress–strain response of short fiber-reinforced plastics. The stress on the composite (σ_c) at a given strain (ϵ_c) can be computed by fitting the response to a form of Eq. (4) with two parameters, the fiber orientation factor (C_θ) and interfacial shear strength (τ_i).

$$\sigma_c = C_\theta \left[\sum_0^{R_c} V_x R_x \tau_i + \sum_{R_c}^{\infty} \epsilon_c E_f V_g \left[1 - \frac{\epsilon_c E_f}{4 R_y \tau_i} \right] \right] \tag{34}$$
$$+ \epsilon_c E_m (1 - V_f)$$

where $R_c = L_c/df$ − critical aspect ratio $R_y > R_c > R_x$ and L_c is the critical fiber length. Since the fiber distribution is larger and R_c is a function of ϵ_c,

$$\sum_{n=o}^{R_c} R_x V_x \tau_i$$

is the contribution from the subcritical fiber population and

$$\sum_{R_c}^{\infty} \epsilon_c E_f V_y \left[1 - \frac{\epsilon_c E_f}{4 R_y \tau} \right]$$

the contribution from the supercritical fiber population.

XI. VARIOUS FIBER-REINFORCED COMPOSITES AND APPLICATIONS

Fiber-reinforced plastics have been widely accepted as materials for structural and nonstructural applications in recent years. The main reasons for interest in FRPs for structural applications are their high specific modulus and strength of the reinforcing fibers. Glass, carbon, Kevlar, and boron fibers are commonly used for reinforcement. However, these are very expensive and, therefore, their use is limited to aerospace applications.

Fibrous fillers are now gaining more importance over particulate fillers due to the high performance in mechanical properties. The influence of fiber diameter on the tensile behavior of short glass fiber on polyimide was reported [95]. At higher concentrations thick fibers seem to be more advantageous probably because of the higher possibility for deformation process in the matrix between fibers with larger separation as in the situation when thick fibers are used.

Composite laminates made by reinforcing PF with glass and asbestos find some commercial applications, while those with carbon fiber are being widely evaluated. Current applications include barrier, spacing, and stacking functions in high-voltage transformers. The most important is in aircraft applications. Some of the typical fibers and their unidirectional composite properties are given in Table 6. Glass-reinforced nylon is replacing metals in many applications due to their high rigidity, creep resistance, low coefficient of friction, and high heat deflection temperature. Carbon fiber-reinforced nylon has found use in aerospace and tennis racket applications. The most widely used polyester fiber glass combinations have exceptional strength at low temperature. When combined with specific resins, polyester-reinforced glass fiber can function in temperatures exceeding 400°C. They provide mechanical properties equal to or exceeding those of most metals.

Thermoplastic materials are being used increasingly due to their lower costs, light weight, and better flexibility in part design. The use of cellulosic materials in thermoplastic composites is highly beneficial because the strength and toughness of plastics can be improved. The potential advantages of these natural fiber composites are low cost, light weight, flexibility, reduced wear of processing machinery, and no health hazards. Also, the hollow nature of vegetable fibers impart acoustic insulation properties to certain types of matrices. Different natural fibers are available throughout the world, and these are used as good reinforcements. These composites find immense applications as building materials. The effects of sisal, coir, bamboo, banana, hemp, and wood fibers as reinforcing agents have been studied in an attempt to produce reinforced building products. It was found that pineapple, banana, and bamboo fiber-reinforced cement composites were successful in building applications [97,98].

In 1981, Belmares and his research group in Mexico together with Parfoort and his research group in Belgium carried out an investigation on the reinforcement of polyester resin with palm, sisal, and henequene [99]. Their work explored the reinforcement of polyester resin with palm fibers, a low-cost fiber abundant in Mexico, by improving its mechanical properties, reducing water absorption, and biodegradation in order to make them competitive with glass-reinforced polyester resin. Palm fibers coated with a poly(vinyl acetate)–fumarate emulsion also increased the flexural properties of composites with polyester resin. Another important aspect of these composites is the prevention of biodegradation of cellulose fibers in the polyester. When a microbicidal agent ($ZnCl_2$) was added to the laminated samples, the tensile strength remained almost the same even after 45 days of soil biodegradation in palm fiber-reinforced polyester composites [99].

Polyethylene-henequene-sand laminates has been reported to have a relatively low density, high weathering resistance, and low water absorption [100]. PVC–ixtle–sand laminates [101] show good mechanical properties and are suitable for use as construction materials. In general, building and especially roofing applications are expected for these composites. Using pine craft pulp sheets as a basis for polyethylene laminates that show good flexural properties at low humidity have been produced for application in building and packaging. Pre-

Table 6 Comparative Fiber and Unidirectional Composite Properties

Fiber/composite	Elastic modulus (GPa)	Tensile strength (GPa)	Density (g/cm³)	Specific stiffness (MJ/kg)	Specific strength (MJ/kg)
E-glass fiber	72.4	2.4	2.54	28.5	0.95
Epoxy composite	45	1.1	2.1	21.4	0.52
S-glass fiber	85.5	4.5	2.49	34.3	1.8
Epoxy composite	55	2.0	2.0	27.5	1.0
Boron fiber	400	3.5	2.45	163	1.43
Epoxy composite	207	1.6	2.1	99	0.76
High strength graphite fiber	253	4.5	1.8	140	2.5
Epoxy composite	145	2.3	1.6	90.6	1.42
Aramid fiber	124	3.6	1.44	86	2.5
Epoxy composite	80	2.0	1.38	58	1.45

grafted and ungrafted pulp fibers from aspen and spruce have been combined with polystyrene by hot pressing, [102] and as a result, a general improvement of mechanical properties was observed at high-fiber contents.

Extruded composites of plasticized PVC and short cellulose fibers have been investigated by Goettler [103]. Pronounced increases in tensile modulus, yield, and ultimate tensile strength are observed. Single step processing of reinforcement and polymer with good product performance are key characteristics of the material whose field of application lies in the vinyl hose industry.

The major drawback associated with the use of natural fibers as reinforcement in the thermoplastics matrix to achieve composite material with improved mechanical properties and dimensional stability are the poor wettability and weak interfacial bonding with the polymer due to inherently poor compatibility. Kokta and others [104–109] have extensively studied the effect of different chemical modifications such as poly(methylene) poly(phenyl isocyanate) (PMPPIC), silane, and monomer grafting on the mechanical properties and dimensional stability of natural fiber-reinforced polymer composites (polyethylene, polystyrene, and PVC). They have reported that chemically modified cellulose fiber filled thermoplastic composites offer superior physical and mechanical properties. Felix and Gatenholm [110] reported the effect of compatibilizing agent and nature of adhesion in composites of cellulose fibers and polypropylene. Recently, from our laboratory, Thomas and coworkers [111–114] have reported on the use of sisal fiber as a potential reinforcing agent in polyethylene, thermosets (epoxy resin, phenol-formal-dehyde, polyester), and natural rubber. Among various natural fibers pineapple leaf fibers exhibit excellent mechanical properties. George et al. [39,72,78,115] reported the mechanical and viscoelastic properties of pineapple fiber–LDPE composites. Mechanical properties of LDPE filled with pineapple fiber are given in Table 7. The very close viscosity values at high shear rates of these composites for filled and unfilled thermoplastics indicate the successful exploitation of these materials in injection molding technology since very little additional power will be required to mold the filled materials.

XII. CONCLUSION

At the present time, high strength, high toughness, and long-term durable products can be made from natural fibers and polymers using high technology. By developing low-cost chemical pretreatment of natural fibers and low-cost manufacturing processes, the integeneous fibers of undeveloped nations will soon be available to produce inexpensive building materials.

GLOSSARY

AES = auger electron spectroscopy
ATR = attenuated total internal reflection
BMC = bulk molding compound
DMC = dough molding compound
e = electron
e_f = energy of electron
E_f = tensile modulus of fiber
ESCA = electron spectroscopy for chemical analysis
FRTM = flexible resin transfer molding
FTIR = Fourier transform infrared spectroscopy
G_i = shear modulus
IR = infrared spectroscopy
K = permeability
l/d = aspect ratio
LDPE = low-density polyethylene
μ = Newtonian viscosity
∇P = pressure gradient
PALF = pineapple leaf fiber
PEEK = poly(ether ether ketone)
PMMA = poly(methyl methacrylate)
PMPPIC = poly(methylene) poly(phenyl) isocyanate
PVC = poly(vinyl chloride)
ρ = density
r_f = fiber radius
RIM = reaction injection molding
RRIM = reinforced reaction injection molding
S_c = tensile stress of composite
SEM = scanning electron microscopy
σ_f = tensile stress
S_f = tensile stress of fiber
SIMS = secondary ion mass spectroscopy

Table 7 Mechanical Properties of PALF-LDPE Composites

Fiber content (wt%)	Tensile strength (MPa)	Young's modulus (MPa)	Elongation at break (%)	Tear strength	Hardness (shore-D) (kN/m)	Tension set (%)	Density (g/cm³)
LDPE	8.5	130	110	63	45	166	0.90
10	16.3	610	11	72	55	2	0.95
20	19.8	720–900	9	81	60	2	0.99
30	22.5	1095–1100	4	97	65	1	1.03

Source: Ref. 39.

Sm = tensile stress of matrix
SMC = sheet molding compound
SRIM = structural reaction injection molding
τ = shear strength
V_c = volume fraction of composite
V_f = volume fraction of fiber
V_m = volume fraction of matrix
W_i = interfacial work of fracture

REFERENCES

1. M. H. Datoo, *Mechanics of Fibrous Composites,* Elsevier Applied Science Publishers (1991).
2. G. Kretsis, *Composites, 18:* 13 (1987).
3. D. W. Clegg and A. A. Collyer, *Mechanical Properties of Reinforced Thermoplastics,* Elsevier Applied Science Publishers, London and New York (1986).
4. P. K. Mallick, *Fibre Reinforced Composites,* Marcel Dekker, Inc., New York (1988).
5. J. G. Cook, *Handbook of Textile Fibre and Natural Fibres,* 4th Ed. Morrow Publishing, England (1968).
6. K. G. Satyanarayana, B. C. Paw, K. Sukumaran, and S. G. K. Pillai, *Hand book of Ceramics and Composites,* Vol. 1, (N. P. Cheremisinoff, ed.), Marcel Dekker, New York (1990).
7. R. G. Weatherhead, *Fibre Reinforced Resin System,* FRP Technology, Applied Sciences, London (1980).
8. G. E. Green, *Composite Materials in Aircraft Structure,* (D. H. Middleton, ed.), Longman, Harlon, UK (1990).
9. B. Sillion, *Comprehensive Polymer Science,* Vol. 5, (G. C. Eastmend, A. Ledwith, S. Russo and P. Sigwalt, eds.), Pergamon Press, Oxford (1989).
10. J. W. Verbicky, *Encyclopedia of Polymer Science and Engineering,* 2nd ed., Vol. 12, (H. F. Mark, N. M. Bikales, C. G. Overberger, and G. Menges, eds.), John Wiley and Sons, New York (1988).
11. K. K. Chawla, *Composite Materials, Science and Engineering,* Springer-Verlag, New York, Berlin, Heidelberg (1987).
12. R. Yosomiya, K. Morimoto, A. Nakajima, Y. Ikada, and T. Suzuki, *Adhesion and Bonding in Composites,* Marcel Dekker Inc., New York (1989).
13. M. Sanon, B. Chang and C. Cohen, *Polym. Eng. Sci., 25:* 1008 (1985).
14. M. Vincent and J. F. Agassant, *Polym. Compos., 7:* 76 (1986).
15. M. W. Darlington and A. C. Smith, *Polym. Compos., 8:* 16 (1987).
16. P. Singh and M. R. Kamal, *Polym. Compos., 10:* 344 (1989).
17. M. Gupta and K. K. Wang, *Polym. Comp., 2292/ ANTEC/93.*
18. R. Bailey and H. Kraft, *Int. Polym. Process, 2:* 94 (1987).
19. M. J. Follers and D. Kells, *Plastics Rubber Proc. Appl., 5:* 125 (1965).
20. R. A. Schweizer, *Polym. Plast. Technol. Eng., 18:* 81 (1982).
21. J. B. Shortall and D. Pennington, *Plast. Rubber Proc. Appl., 2:* 33 (1982).
22. C. F. Johnson, *Resin Transfer Molding and Structural Reaction Injection Molding in Engineering Materials Handbook,* Vol. 2, ASM, International, USA (1988).
23. T. M. Gotch and P. E. R. Plowmin, 10th BPF Reinforced Plastics Congress, Brighton, UK (1978).
24. C. D. Rudd, *Hand book of Polymer Fibre Composites,* (F. R. Jones, ed.), Polymer Science and Technology Series (1994).
25. M. F. Foley, *SAMPE J., 28:* 15 (1992).
26. R. W. Mayer, *Polyester Molding Compounds and Molding Technology,* Chapman and Hau, New York (1987).
27. F. R. Jones, *Reinforced Reaction Injection Molding, Handbook of Polymer Fibre Composites,* (F. R. James, ed.), Polymer Science and Technology Science, New York (1994).
28. F. M. Sweeney, *Introduction to Reaction Injection Molding,* 2nd Ed., Technomic Publishers, Westport (1990).
29. C. W. Macosko, *Reaction Injection Molding Fundamentals of RIM,* Hanser, Munich (1989).
30. J. A. Quinn, *Metals and Materials, 5:* 270 (1989).
31. N. Pan, *Polym. Composites, 14:* 85 (1993).
32. L. J. Broutman and R. H. Krock, *Composite Materials,* Vol. 6, Academic Press, New York (1974).
33. L. J. Broutman and R. H. Krock, *Modern Composite Materials,* Addison-Wesley Publishing Co., London (1967).
34. W. D. Callister Jr., *Materials Science and Engineering,* John Wiley and Sons, Inc., New York (1987).
35. J. V. Milewski and H. S. Katz, *Handbook of Reinforcements for Plastics,* van Nostrand Reinfold Company, Inc., New York (1987).
36. C. Zweben and H. T. Hahn, *Tsu-Weichou, Mechanical Behaviour and Properties of Composite Materials,* Vol. 1, Technomic Publishers, Westport (1989).
37. B. W. Rosen, *Fiber Composite Materials,* Amer. Soc. for Metals, Metal Park, Ohio (1965).
38. L. J. Broutman and R. H. Krock, *Modern Composite Materials,* Addison-Wesley Reading, Massachusetts (1967).
39. J. George, S. S. Bhagawan N. Prabhakaran, and S. Thomas, *J. Appl. Polym. Sci. 57:* 843 (1995).
40. Catalogs of Shinetsu Silicone Ltd. and Torcy Silicone Ltd.
41. K. Kubo, M. Koishi, and T. Tsunoda, *Composite Materials and Interface,* Sogo Gitutsu, Shuppan, Tokyo, p. 13 (1986).
42. J. T. Paul and J. B. Thompson, Proc. 20th Conf. SPI. Reinf. Plast. Div., 1988.
43. V. G. Geethamma, R. Joseph, and S. Thomas, *J. Appl. Polym. Sci., 55:* 583 (1995).
44. D. K. Owens, *J. Appl. Polym. Sci., 19:* 265 (1975).
45. D. Briggs and C. R. Kendall, *Polymer, 20:* 1053 (1979).
46. D. Maldas, B. V. Kokta, and C. Daneault, *J. Appl. Polym. Sci., 37:* 751 (1989).
47. S. S. Voyutskii, *Autohesion and Adhesion of High Polymers,* Interscience, New York (1963).
48. B. V. Derjaguin and V. P. Smilga Proc. 3rd Int. Congress of Activity, 11, 349 (1960).
49. D. Briggs, D. M. Brewis, and M. B. Konieczo, *J. Mater. Sci., 11:* 1270 (1976).
50. K. Nakao and M. Nishiuchi, *J. Adhes. Sci. Jpn., 2:* 239 (1966).
51. J. R. Rasmussen, E. R. Stedronsky, and G. M. Whitesides, *J. Am. Chem. Soc., 99:* 4736 (1977).
52. P. Blais, D. J. Carlsson, G. W. Csullog, and D. M. Wiles, *J. Colloid, Int. Sci., 47:* 636 (1974).
53. C. Y. Kim, J. Evans, and D. A. I. Goring, *J. Appl. Polym. Sci., 15:* 1365 (1971).

54. H. L. Spell and C. P. Christenson, *Tappi, 62*: 77 (1979).

55. H. Kimuraand and H. Nakayama, *Surface Treatment of Plastics, Color Mater. Jpn., 54*: 149 (1981).

56. R. A. Bragole, *J. Elastomers Plast., 6*: 213 (1974).

57. C. D. Storms, *Plast. Des. Process, 17*: 57 (1977).

58. K. M. I. Ali, M. K. Uddin, M. I. U. Bhuiyan, and M. A. Khan, *J. Appl. Polym. Sci., 54*: 303 (1994).

59. H. V. Boening, *Plasma Science and Technology*, Cornell University Press, New York (1982).

60. R. H. Hansen and H. Schonhorn, *J. Polym. Sci. Polym. Lett. Ed., 4*: 203 (1966).

61. R. G. Raj, B. V. Kokta, D. Maldas, and C. Daneault, *J. Appl. Polym. Sci., 37*: 1089 (1989).

62. R. G. Raj, B. V. Kokta, G. Grouleau, and C. Daneault, *Polym. Plast. Technol. Eng., 29*: 339 (1990).

63. W. E. Jones, *J. Adhesion, 15*: 59 (1982).

64. F. D. Osterholz, *Modern Plastics Encyl., 63*: 126 (1987).

65. A. T. D. Benedetto, G. Haddad, C. Schilling, and F. Osterholtz, *Interfacial Phenomena in Composite Material* (F. R. Jones, ed.), UK (1989).

66. H. Ishida and J. L. Koenig, *J. Colloid Interface Sci., 64*: 555 (1978).

67. S. Wu, *Polymer Interface and Adhesion*, Marcel Dekker Inc., New York (1982).

68. B. M. Vanderbilt, *Mod. Plast., 37*: 125 (1959).

69. H. A. Clark and E. P. Plueddemann, *Mod. Plast., 40*: 133 (1963).

70. T. Nakao, *Finechem. Jpn., 4*: 3 (1984).

71. K. Joseph, C. Pavithran, B. Kuriakose, C. K. Premalatha, and S. Thomas, *Plast. Rubb. Comp. Proc. Appl., 21*: 237 (1994).

72. J. George, S. S. Bhagawan, and S. Thomas, *J. Thermal Analysis 47*: 1121 (1996).

73. Y. Ikada, *Surf. Jpn., 22*: 119 (1984).

74. E. P. Zironi and H. Poppa, *J. Mater. Sci., 16*: 3115 (1981).

75. D. M. Goddard, *J. Mater. Sci., 13*: 1841 (1978).

76. P. Blais, D. J. Carlsson, G. W. Csullog, and D. M. Wiles, *J. Colloid Interface Sci., 47*: 636 (1974).

77. H. Ishida and L. J. Koenig, *J. Polym. Sci. Polym. Phys. Ed., 17*: 615 (1979).

78. J. George, R. Janardhan, A. R. Anand, S. Bhagawan, and S. Thomas, *Polymer 37*: 5421 (1996).

79. H. Ishida and J. L. Koening, *Polym. Eng. Sci., 18*: 128 (1978).

80. S. H. Morrel, *Plast. Rubber Process. Appl., 1*: 179 (1981).

81. S. Z. Lu, 35th Annu. Tech. Conf. Reinforced Plastic/ Composite Institute, The Society of the Plastic Industry Inc., (1980).

82. M. R. Piggot, Proc. 36th Int. SAMPE Symp., p. 1173 (1991).

83. M. Narkis, E. J. H. Chen, and R. B. Pipes, *Polym. Composites, 9*: 245 (1988).

84. I. Verpoest, M. Desaeger, and R. Keunings, *Composites, Controlled Interphases in Composite Materials*, (H. Ishida, ed.), Elsevier Science Publishing Co. Inc., p. 653 (1990).

85. M. R. Piggot and S. R. Dai, *Polym. Eng. Sci., 31*: 1256 (1991).

86. M. Desaeger, T. Lacroix, B. Tilmans, R. Keunings, and I. Verpoest, *J. Comp. Sci. Tech., 43*: 379 (1992).

87. M. J. Folkes and W. K. Wong, *Polymer, 28*: 1309 (1987).

88. M. Desager, T. Lacroix, B. Tilmans, R. Keunings, and I. Verpoist, *J. Comp. Sci. Technol., 43*: 379 (1992).

89. L. B. Greszezule, *Interfaces in Composites*, ASTM, STP, p. 452 (1969).

90. P. S. Chua and M. R. Piggot, *Comp. Sci. Tech., 22*: 107 (1985).

91. C. T. Chou and L. S. Penn, *J. Adhesion, 35*: 127 (1991).

92. L. S. Penn and S. M. Lee, *J. Comp. Tech. Res., 1*: 23 (1989).

92a. L. S. Penn and S. M. Lee, *J. Comp. Tech. Res., 12*: 164 (1990).

93. M. K. Tse, *SAMPE J., 21*: 11 (1985).

94. H. L. Cox, *Brit. J. Appl. Phys., 3*: 72 (1952).

95. A. Kelly and W. R. Tyson, *J. Mech. Phys. Solids, 13*: 329 (1965).

96. W. H. Bowyer and M. G. Bader, *J. Mater. Sci., 7*: 1315 (1972).

97. F. Ramsteiner and R. Theysohn, *Comp. Sci. Technol., 24*: 231 (1985).

98. R. S. P. Coutts, Y. Ni, and B. C. Tobias, *J. Mater. Sci. Lett., 13*: 283 (1994).

99. R. S. P. Coutts, *J. Mater. Sci. Lett., 9*: 1235 (1990).

100. M. A. Semsarzadeh, *Polym. Plast. Technol. Eng., 24*: 323 (1986).

101. A. Padilla and A. Sanchez, *J. Appl. Polym. Sci., 29*: 2405 (1984).

102. A. Padilla, A. Sanchez, and M. Castro, *Advances in Rhelogy 1x*, Vol. 3, (B. Mena, A. G. Regon and C. R. Nafaile, eds.) Elsevier Science Publishing Co. Inc., New York (1985).

103. L. A. Goittler, *Polym. Composites, 4*: 249 (1983).

104. B. V. Kokta, R. Che, C. Daneault, and J. L. Valde, *Polym. Plast. Technol. Eng., 4*: 229 (1983).

105. R. G. Raj, B. V. Kokta, D. Maldas, and C. Danealt, *Makromol. Chem. Macromol. Symp., 28*: 187 (1989).

106. D. Maldas and B. V. Kokta, *J. Comp. Mater., 25*: 375 (1991).

107. D. Maldas and B. V. Kokta, *J. Appl. Polym. Sci., 40*: 917 (1990).

108. R. G. Raj, B. V. Kokta, and C. Daneault, *Intern. J. Polymeric Mater., 12*: 235 (1989).

109. D. Maldas and B. V. Kokta, *Comp. Sci. Technol., 36*: 167 (1989).

110. J. M. Felix and P. Gatenholm, *J. Appl. Polym. Sci., 42*: 609 (1991).

111. K. Joseph, C. Pavithran, and S. Thomas, *Eur. Polym. J. 32*: 1243 (1996).

112. K. Joseph, S. Thomas, C. Pavithran, and M. Brahmakumar, *J. Appl. Polym. Sci., 47*: 1731 (1993).

113. S. Varghese, B. Kuriakose, S. Thomas, and A. T. Koshy, *Ind. J. Nat. Rubb. Res., 4*: 55 (1991).

114. S. Varghese, B. Kuriakose, and S. Thomas, *J. Adh. Sci. Technol., 8*: 235 (1994).

115. J. George, K. Joseph, S. S. Bhagawan, and S. Thomas, *Mater. Lett., 18*: 163 (1993).

54

Peculiarities of the Fine Structure of PET Fibers and the Relationship to Their Basic Physical Properties

Grzegorz Urbańczyk and Andrzej Jeziorny
Technical University of Lodź, Lodź, Poland

I. INTRODUCTION

Polyethylene terephthalate (PET) fibers and filaments are currently the most important kind of man-made fibers. The manufacture of these fibers on an industrial scale, initiated in England in 1947 and then in the United States, originated with J. R. Whinfield and J. T. Dickson's patent (British Patent 578079 submitted in Great Britain on July 29, 1941). The PET fiber production, which in the 1940s and 1950s fell in the shadow of the then dominating (in regards to the rate of production) polyamide fibers, increased dynamically in the course of time. This dynamic increase was confirmed by the proportion indices of PET fibers in the world production of synthetic fibers: 34% in 1970, 47% in 1980, and 54% in 1990. At the beginning of the 1970s the volume of PET fibers production was greater than the quantity of polyamide fibers produced, which placed them in first position not only among synthetic fibers but among all chemical fibers.

In 1994, the proportion of PET fibers in the world production of synthetic fibers was 62.9% and of chemical fibers was 55.3%, while in the total volume of all kinds of fibers it was 27.4%. Out of PET fibers presently produced, 38% are staple fibers and 52.5% are filament yarns, with a marked tendency toward an increase in the latter. A 55% proportion is anticipated in the year 2000. At present, about 75% of PET fibers are used for textile purposes and 25% for nontextile purposes.

This dynamic increase in production was accompanied by the qualitative development of PET fibers, which manifested itself in the widening of assortment of the fibers being produced (e.g., staple microfibers and filament yarns of the POY, MOY, FOY, and HOY type) and in the manufacture of second-generation fibers on a large scale. As for the latter, some unfavorable properties of standard fibers were minimized and some new favorable ones were selectively obtained. Fibers less susceptible to pilling, with improved dyeability, highly shrinkable fibers, or capillary porous fibers of increased hygroscopicity are a few examples. The favorable functional qualities of PET fibers result from their specific physical properties (i.e., mechanical, thermal, frictional, and electrical) and are particularly strongly conditioned by the specificity of their fine structure. This fact, as well as the fact that the fine structure of these fibers can be shaped during the manufacturing process, over a wide range justify the consideration of the basic aspects of their fine structure.

II. FINE STRUCTURE OF PET FIBERS

PET fibers in final form are semi-crystalline polymeric objects of an axial orientation of structural elements, characterized by the rotational symmetry of their location in relation to the geometrical axis of the fiber. The semi-crystalline character manifests itself in the occurrence of three qualitatively different polymeric phases: crystalline phase, intermediate phase (the so-called mesophase), and amorphous phase. When considering the fine structure, attention should be paid to its three fundamental aspects: morphological structure, in other words, super- or suprastructure; microstructure; and preferred orientation.

A. Superstructure

PET fibers and filaments are characterized by a fibrillar superstructure that corresponds to the general concept of the fibrillar structure of synthetic fibers. The fibrillar structure concept was established by Peterlin [1,2], developed further by Brestkin [3], Gojchman [4], Pechhold [5], and finalized by Prevorsek and Kwon [6]. According to this concept, the fiber in macroscopic scale is formed by a system of interconnected fibrils, sometimes referred to as *macrofibrils*. Particular fibrils are connected to one another to form a coherent whole by means of *tie molecules* passing from one to another fibril. The structure of the fibrils is determined by their substructure. Most often this is a *microfibrillar substructure*, also known as *longitudinal substructure*. In the case of this type of substructure, the fibril is made up of a bundle of (3–7) smaller fibrillar elements laid in parallel, so-called *microfibrils*. The connections between microfibrils are ensured by tie molecules passing from one to another microfibril.

The microfibril, as in the fibril, is an elongated object of a lateral dimension of 25–80 A, and a longitudinal dimension of 1000–10,000 A. Along the microfibrils there occur—regularly, periodically, and alternately—*crystalline blocks*, also called *crystallites*, and noncrystalline areas between them, referred to as *separating layers*. The longitudinal periodic regularity in the structure of microfibrils manifests itself in the fact that in each fragment of its length, the sum of the height of the crystalline region and the separating layer are constant and equal to the value of the *long period* or *mean long spacing* (L). For PET fibers, this long period is within the range of 90–170 A. In the long period, about 50–100 A falls to the height of the crystalline block, while 40–70 A falls to the thickness of the separating layer. The principle of structure of microfibrils is shown in Fig. 1. The periodic regularity of the microfibril structure causes its degree of crystallinity to be constant, and when two-thirds of the value of L falls to the crystalline layer thickness, it amounts to 0.66 (volume degree of crystallinity) or 0.75–0.8 (mass degree of crystallinity).

Crystallites in microfibrils of PET fibers are created in their central part of the set of parallel, straightened, and spatially ordered segments of molecules to form a crystalline region. The developed space lattice of the crystalline region is characterized by a triclinic crystallographic system. On both ends the crystalline region adjoins so-called *crystallite surface layers*. The surface layers are formed mainly by regular and irregular chain folds. The length of the regular folds, also called adjacent reentry folds, is much smaller than that of the irregular folds and also the smallest possible fold. The surface layers are characterized by relatively high ordering of the regular chain folds. As shown by one of the coauthors [7], this ordering can be confirmed by the relatively low value of the surface free energy of these layers.

Noncrystalline separating layers of microfibrils are formed from two crystallite surface layers that are side by side in the microfibril and from a central part called the *intermediate zones*. The latter are mainly made up of tie molecules among which are distinct straightened

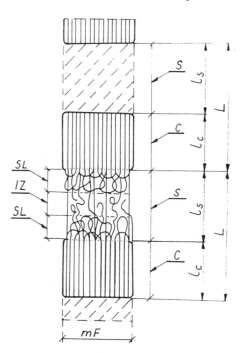

Figure 1 The structure of a microfibril. C - crystallite; S - separating layer; SL - surface layer; IZ - intermediate zone; mF - border of microfibril; l_c - crystallite length; l_s - separating layer length; L - mean long period (spacing).

and stressed molecules (so-called *taut tie molecules* [TTM]) joining adjacent crystalline blocks in the microfibril. The intermediate zones also comprise the ends of macromolecule chains, so-called *cilia*, and coiled ends of molecules.

Microfibrils are formed in PET fibers during the stretching of a solidified polymer stream. With an increase in the draw ratio, microfibrils become increasingly slender. The microfibril length is assumed to be proportional to the draw ratio (R), their lateral dimension proportional to $\sqrt{R}$, while the length-to-lateral dimension ratio is proportional to $R^{3/2}$.

Another type of fibril substructure in PET fibers, besides the microfibrillar type already discussed, is the *lamellar substructure*, also referred to as the *lateral substructure*. The basic structural unit of this kind of substructure is the *crystalline lamella*. Formation of crystalline lamellae is a result of lateral adjustment of crystalline blocks occurring in neighboring microfibrils on the same level. Particular lamellae are placed laterally in relation to the axis of the fibrils, which explains the name—*lateral substructure*. The principle of the lamellar substructure is shown in Fig. 2.

The distinction between the types of substructures of the microfibrils occurring in the PET fiber is possible based on the value of the *substructure parameter* (Λ).

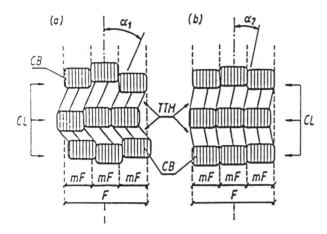

Figure 2 The lamellar substructure of a fibril. (a) Reciprocal positions of crystalline lamellae as a result of fiber annealing. (b) The situation after relaxation of stress affecting TTM. $\alpha_{1,2}$ - average angle of orientation of TTM; CL - crystalline lamellae; CB - crystalline blocks (crystallites); mF - border of microfibrils; and F - fibril. In order to simplify it was assumed that: (1) there are the taut tie molecules (TTM) only in the separating layers, (2) the axis of the fibril is parallel to the fiber axis.

(For details see Section III. A., Mechanical Properties.) For values of this parameter less than 1, we can conclude that a substructure of the microfibrillar type occurs, whereas for the values equal or close to 1 a substructure of the lamellar-type occurs.

While preserving its essence, the superstructure of PET fibers can be subject to modification, depending on the manufacturing conditions of the fiber, and to additional refining treatment. This modification first changes the value of the long period (L). The change in the value of L is due to an increase in the height of the crystalline blocks and reduces the thickness of the separating layers. The values, listed in Table 1, are an illustration of such changes.

Table 1 Values of the Long Period of Differently Drawn PET Fibers

	Long period L (nm)	
Draw ratio	Before annealing	After annealing (200°C, Lmin, silicon bath)
3.0×	11.2	14.0
3.5×	12.0	16.0
4.0×	13.4	17.0
4.5×	14.0	18.0
5.2×	16.6	19.0

B. Microstructure

Characteristic of the microstructure of PET fibers in their final production form is the occurrence of three types of polymer phases: crystalline, mesomorphous, and amorphous. The first phase is the result of crystalline aggregation of PET molecules, the second phase—of mesomorphous or, in other words, paracrystalline aggregation, the third phase—of amorphous aggregation. The mesomorphous and amorphous phases together form a noncrystalline part of the fiber.

1. Crystalline Phase

The crystalline phase is formed by a set of crystalline regions that occur in the fiber. In particular, crystalline regions have to do with a spatial regular system of fragments of molecules that form a crystalline space lattice. In the case of PET fibers, it is determined by a *triclinic crystallographic system*. Developing a crystalline phase is tantamount to the formation of crystallites, which is the result of the process of the fiber polymer crystallization. In the case of PET fibers, stress-induced crystallization, which occurs in the course of stretching the solidified fiber, is decisive as is thermal crystallization, which is the thermal treatment of the fiber while manufacturing or refining it. In order for crystallization to occur there must be a sufficiently high value of elongation imparted while stretching and a heat input temperature above the glass temperature (T_G). The process of stress-induced crystallization starts at a threshold value of elongation of about 250–300%. The uncoiling, straightening, and parallelization of molecules, occurring during the stretching of a freshly solidified fiber, is accompanied by the conformation of the transformation of the dioxymethylene fragment of the chains. This transformation consists of the transition of the initial stereoisomer of the gauche-type of this group into the stereoisomer of the *trans*-type (Fig. 3). The quantitative effect of crystallization also depends on the average molecular weight

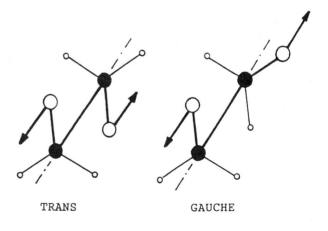

Figure 3 Trans and gauche isomers of PET.

of the polymer used for fiber production. When the remaining production parameters are on the same level, a smaller average molecular weight usually leads, to stronger crystallization of the fiber.

A univocal confirmation of the development of crystalline aggregation in the fiber is the occurrence of layer reflexes $01\bar{1}$, $11\bar{1}$, $\bar{1}11$, and 101 on the textural x-ray diffraction pattern. The details of organization of the space lattice are defined by the parameters of the unit cell and the number of polymers falling into one cell. The data, established by different authors, are presented in Table 2. Daubenny and Bunn's [8] pioneer findings are considered the most probable for space lattices occurring in PET fibers.

A specific attribute of unit cell building is the inclination of the axis of macromolecule chains, in relation to the normal, to the plane of the base of the cell (ab). According to Yamashita [11] this inclination is within the range of 25–35° (Fig. 4). Against the background of space lattices of other types of fibers, the lattice of crystalline regions in PET fibers is characterized by a number of specific features. These are:

1. The perfection of the geometrical structure. This is confirmed by a great number of x-ray reflexes, including, and in particular, the occurrence of layer reflexes and reflexes of a higher order than the first one (Fig. 5 and Table 3).

2. The translational direction of the lattice (c), which is the direction that the crystallite axis is not, because of the triclinic crystallographic system, is perpendicular to the plane of the unit cell base (ab).

3. The space lattice does not undergo polymorphous transformation. As with other kinds of fibers, no transformation of the space lattice under the effect of any physical or chemical treatment of PET fibers has yet been found.

Crystallites occurring in PET fibers can assume two kinds of *morphological forms*. The first form represents the crystallite formed by molecules of folded conformation, while the other is formed from molecules of extended-chain conformation. The first form is sometimes called a flexural morphological form, whereas the other is called a straightened morphological form. The flexural form is the typical and prevailing morphological form in PET fibers. However, it should be stressed that no

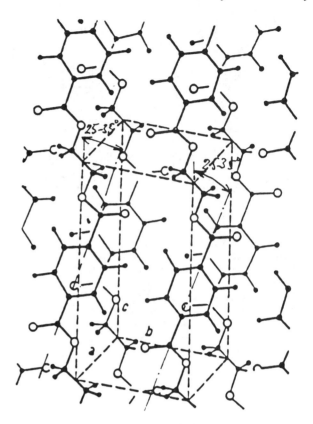

Figure 4 The scheme of the unit cell of PET.

occurrence of the spherolitic morphological form of crystalline aggregation has been observed in PET fibers. Recognition of the kind of a morphological form of the crystallite can be achieved by means of x-ray, IR spectroscopic, or thermal analysis. In the first case, it consists of studying the low-angle scattering of x-radiation. The occurrence of long period reflections is the evidence of the occurrence of the first morphological form. In the case of IR spectroscopic analysis [12,13], the presence of an absorption band of 988 cm^{-1} confirms the occurrence of this form. According to Baranova et al. [14] and Pakszwer et al. [15], the occurrence of the 853-cm^{-1} band can be considered the criterion for the presence of crystallites of the first form, while a 846-cm^{-1} band points to the second form.

Table 2 Parameters of the Unit Cell of PET

a (A)	b (A)	c (A)	α (°)	β (°)	γ (°)	d_c (g/cm³)	Reference
4.56	5.94	10.75	98.5	118	112	1.455	[8]
4.52	5.98	10.77	101.0	118	111	1.476	[9]
4.48	5.85	10.75	99.5	118.4	111.2	1.515	[10]

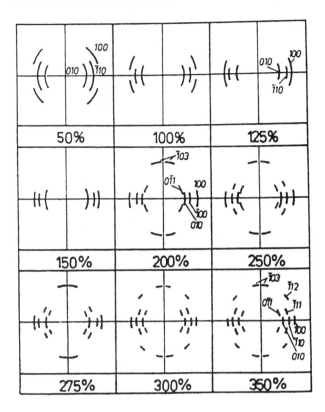

Figure 5 X-ray reflexes of PET fibers differentiated by draw ratio (in %).

treatment. The decisive factors are the draw ratio of the fiber, and the temperature and the duration of action of heat. Similarly, the average size of crystallites is also diversified, and as with the degree of crystallinity, it depends on production conditions and the additional refining treatment.

The numerical data, as determined by the authors, characterizing the dependence of the degree of crystallinity and the size of crystallites on the draw ratio of the fiber, are presented in Table 4.

2. Mesomorphous Phase

The mesomorphous phase, also called an intermediate phase or a mesophase, is formed by molecules occurring in surface layers of the crystallites. It can be assumed that the mesophase is made up largely by regularly adjacent reentry folds. However, it cannot be excluded that the mesophase is also composed of some irregular chain folds, which are characterized by a long length and run near the crystal face in the direction perpendicular to the microfibril axis.

The mesomorphous phase is characterized by a paracrystalline arrangement of segments of molecules, analogous to the one occurring in nematic liquid crystals. The paracrystalline manner of the arrangement of the chains manifests itself in the development of a pseudo-lattice. According to Asano and Seto [16], this lattice for PET fibers is described by a monoclinic crystallographic system and a unit pseudo-cell of this lattice of the parameters: a = 4.3 A, b = 9.2 A, c = 10.5 A, and $\alpha = 100^0$. The density of such a pseudo-lattice is $1.376 - 1.392$ g/cm^3.

The occurrence of the mesophase in the fiber is confirmed by x-ray diffraction examination. The occurrence of three equatorial reflections 010, 110, and 100, the absence of layer and meridional reflections, and the manifestation of the intensity maximum of diffusively scattered radiation at $2\Theta = 19^0$ in the fiber diffraction pattern are the criterion for the presence of the mesophase. The

The quantitative proportion of crystalline aggregation in the total polymer mass of the fiber, i.e. *degree of crystallinity*, is strongly diversified for PET fibers. The degree of crystallinity is within the range of the values 0.10–0.70. Most often, however, it is within the range of 0.45–0.65. The degree of crystallinity depends on production conditions and the additional refining

Table 3 X-Ray Diffractions in Texture Diffractograms of PET Fibers

Reflex type	Relative intensity	Diffraction indices (hkl)	Interplanar spacing d$_{hkl}$ (A)	Glancing angle for CuK$_\alpha$-2Q$_{hkl}$ (°)
Equatorial	strong	010	5.02	17.50
Equatorial	strong	110	3.94	22.27
Equatorial	very strong	100	3.48	25.53
Layer Refl.	strong	011	5.46	16.20
Layer Refl.	strong	111	4.14	21.43
Layer Refl.	medium	111	3.21	27.77
Layer Refl.	weak	101	2.72	32.90
Layer Refl.	weak	112	3.90	22.80
Meridional	very weak	103	3.58	24.83
Meridional	medium	105	2.15	41.10

Table 4   Degree of Crystallinity and Average Crystallite Size of Differently Drawn PET Fibers

| Draw ratio | Degree of crystallinity | | Lattice disorder coefficient (k) | Average crystallite size perpendicular to the crystallographic plane (hkl) D_{hkl} (nm) | |
	X-ray method (X^a)	Density method (X)		(010)	(100)
2.0×	0.12	—	16.5	4.8	3.2
3.0×	0.22	0.26	2.8	4.6	3.2
3.5×	0.27	0.29	2.4	3.9	3.2
4.0×	0.31	0.31	3.1	3.7	3.0
4.5×	0.33	0.32	3.3	3.7	2.9
5.2×	0.35	0.40	4.5	3.6	2.8

[a] Ruland's [42] and Vonk's [43] methods.

broadening of the base of the curve of the azimuthal intensity distribution of the reflected radiation $J = f(\delta)$ for reflexes 010, $\bar{1}$10, and 100 is an additional confirmation of the occurrence of the mesophase. The x-ray identification of the presence of the mesomorphous phase is, however, hindered by the simultaneous occurrence of a crystalline phase with it. The latter is a source of equatorial reflexes that partly overlap the equatorial reflexes from the mesophase and dim the presence of the latter ones.

The quantitative proportion of the mesomorphous phase in PET fibers is strongly diversified and depends on the conditions of fiber manufacture. According to Lindner [17], it is estimated to be in the range 0.21–0.36. For poorly stretched fibers (below 300% of draw ratio), the mesophase constitutes almost exclusively an ordered part of the polymer in the fiber. With an increase in the draw ratio of the fiber, the proportion of the mesophase in the whole fiber mass increases, mainly at the cost of the amorphous phase. It is assumed that the mesophase constitutes about 0.5 of the amorphous phase mass for poorly stretched fibers and about 1.2 of this mass for highly stretched fibers.

3. *Amorphous Phase*

The amorphous phase is formed by fiber areas in which the chains of macromolecules or their fragments are characterized by a very low degree of their spatial ordering. The amorphous phase is first formed by taut tie molecules (TTM) in the separating layers of microfibrils and placed between crystalline lamellae in the fibrils. Tie molecules occurring between microfibrils and fibrils are another component of the amorphous phase. The amorphous phase is also made up of irregular chain folds occurring in crystallite surface layers and of cilia. It is assumed that, depending on the degree of chain ordering, the amorphous phase is strongly diversified with respect to the molecular cohesion. The amorphous

phase regions of the smallest cohesion will be characterized by the absence of intermolecular bonds. This fact justifies referring to polymer material of such regions as nonbonded amorphous material. In turn, for regions of a greater cohesion the intermolecular bonds will occur, which is the case for bonded amorphous material.

The amorphous phase differs from the mesophase and the crystalline phase by a clearly lower value of density. The amorphous phase density depends on the internal orientation of the fiber. Its value is in the range 1.335–1.357 g/cm³. In the case of a very high orientation, it can even reach the value 1.363 g/cm³.

The existence of the amorphous phase of the fiber is confirmed in x-ray examination by the occurrence of a distinct intensity maximum of the radiation scattered diffusively at $2Q = 21.6^0$. The fraction of the amorphous phase in the fiber depends on manufacturing conditions and a possible further refining treatment. It is estimated to vary from 0.25 to 0.60. With an increase of the draw ratio and following the thermal treatment of the fiber, the proportion of the amorphous phase only reaches the lower values of this interval.

C. Orientation

The development of the internal orientation in formation in the fiber of a specific directional system, arranged relative to the fiber axis, of structural elements takes place as a result of fiber stretching in the production process. The orientation system of structural elements being formed is characterized by a rotational symmetry of the spatial location of structural elements in relation to the fiber axis. Depending on the type of structural elements being taken into account, we can speak of crystalline, amorphous, or overall orientation. The first case has to do with the orientation of crystallites, the second—with the orientation of segments of molecules occurring in the noncrystalline material, and the third—with all kinds of structural constitutive elements.

1. Crystalline Orientation

The parallelization of crystallites, occurring as a result of fiber drawing, which consists in assuming by crystallite axes-positions more or less mutually parallel, leads to the development of *texture* within the fiber. In the case of PET fibers, this is a specific texture, different from that of other kinds of chemical fibers. It is called axial-tilted texture. The occurrence of such a texture is proved by the displacement of x-ray reflexes of paratropic lattice planes in relation to the equator of the texture diffractogram and by the deviation from the rectilinear arrangement of oblique diffraction planes. With the preservation of the principle of rotational symmetry, the inclination of all the crystallites axes in relation to the fiber axis is a characteristic of such a type of texture. The angle formed by the axes of particular crystallites (the translation direction of space lattice [c]) and the

fiber axis is called a tilting or inclination angle of the texture (φ), (Fig. 6). The direct cause for the formation of the axial-tilted texture in PET fibers is the triclinic crystallographic system of the space lattice and the specific arrangement of chain molecules within the space lattice (Fig. 4).

The value of the angle of tilting of the texture can be determined from the analysis of x-ray fiber diagrams. As Urbańczyk noted [18], the position of layer reflexes $0\bar{1}1$ and $\bar{1}11$ or the position of equatorial reflexes 010 and 100 can be analyzed. In the first case, the tilting angle of the texture (φ) can be determined from the equation:

$$a \times \cos^2(\varphi) + b \times \cos(\varphi) + c = 0 \tag{1}$$

where:

$$a = 351.9,$$
$$b = -451.2 \times \sin\delta_{\bar{1}11} - 512.8 \times \sin\delta_{0\bar{1}1}, \text{ and}$$
$$c = 291.8 \times \sin^2\delta_{\bar{1}11} + 336.1 \times \sin^2\delta_{0\bar{1}1} + 435.1 \times \sin^2\delta_{\bar{1}11} \times \sin\delta_{0\bar{1}1} \times 132.8$$

where $\delta_{\bar{1}11}$ and $\delta_{0\bar{1}1}$ are azimuthal angles determining the position measured from the equator intensity maximum of the reflexes $0\bar{1}1$ and $\bar{1}11$.

In the second case, from the equation:
$$\sin(\varphi) = \frac{\cos(\theta)_{100} \times \sin(\delta)_{100}}{\sin(\alpha)} \tag{2}$$

where α is an auxiliary angle defined by the relationship:

$$ctg(\alpha) = \frac{\cos(\theta)_{010} \times \sin(\delta)_{010} - \cos(59.5°) \times \cos(\theta)_{100} \times \sin(\delta)_{100}}{\sin(59.5°) \times \cos(\theta)_{100} \times \sin(\delta)_{100}} \tag{3}$$

where δ_{100} and δ_{010} are azimuthal angles determining the position of the maximum intensity of the reflexes 100 and 010 measured from the equator.

It is more convenient to use the procedure of the first kind for higher drawn fibers (over 350% of elongation) and the second procedure for fibers of smaller drawing. The value of the angle of inclination of the texture depends on the magnitude of the drawing applied. As a result of the investigations of one of the authors

[18], it is known that with an increase in the drawing, the value of the angle φ causes the axial tilted in the PET fiber's texture to be similar to the ordinary axial fibrous texture occurring in the majority of chemical fibers. As a consequence, the fiber diagram of PET fibers is practically indistinguishable from the kinds of texture diffractograms of other chemical fibers.

The ordering of crystallites in the fiber texture is best and univocally described in the quantitative manner

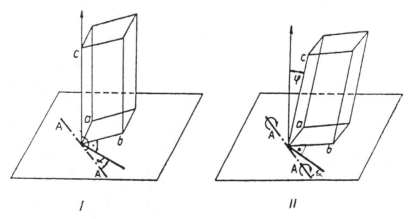

I   *II*

Figure 6 Schematic presentation of the ideal axial-fibrous texture (I) and ideal axial-tilted texture (II).

by the value of Hermans' axial orientation function (f_c) determined by the dependences:

$$f_c = \frac{1}{2}(3 \times \overline{\cos^2\phi} - 1)\, 1 - \frac{3}{2} \times (\overline{\sin^2\phi}) \qquad (4)$$

where ϕ is the angle formed by the crystallite axis (the translational direction of the space lattice [c]) and the geometrical axis of the fiber, i.e., called crystallite axis orientation angle.

As a rule, to determine the value of f_c, x-ray diffraction examination is used, which consists in the quantitative analysis of the azimuthal intensity distributions of the x-ray reflexes of the texture diffractogram. Two procedures are widely used. One of them consists of considering the meridional reflex $\overline{1}05$, the other in the analysis of the equatorial reflexes 010, $1\overline{1}0$, and 100. In the first procedure, which is considered theoretically to be more proper, the value of f_c is generally determined by the dependence in Eq. (4). In this case, the angle Φ value depends on the angle parameter of crystallographic ($\overline{1}05$) planes ($\epsilon_{\overline{1}05}$) (the angle between the "c" axis of a crystallite and a normal when considering crystallographic planes) and on the directional angle of the ($\overline{1}05$) planes ($\rho_{(\overline{1}05)}$). The equation connecting the mentioned angles is as follows:

$$\phi = \epsilon_{\overline{1}05} - \rho_{\overline{1}05} \qquad (5)$$

Using Eq. (5), the expression in Eq. (4) can be written in the form:

$$f_c = 1 - \frac{3}{2} \times (\overline{\sin^2} \times [\epsilon_{\overline{1}05} - \rho_{\overline{1}05}]) \qquad (6)$$

The correct determination of f_c depends largely on using the correct value of the ϵ_{105} angle. In light of the analysis of a coauthor [19], the value of ϵ_{105} is equal to ϵ_{105} 35.8^0. Assuming this ϵ_{105} value and the occurrence of the unit cell proposed by Daubeny and Bunn [8], after calculation of values of trigonometric functions, the expression in Eq. (6) may be written in the form [19]:

$$f_c = 0.487 - 0.474 \times \overline{\sin^2}\,\rho_{\overline{1}05} + 1.424$$
$$\times \sqrt{\overline{\sin^2}\,\rho_{\overline{1}05} \times (1 - \overline{\sin^2}\,\rho_{\overline{1}05})} \qquad (7)$$

where $\sin^2\rho_{\overline{1}05}$ is the variation in the parameter of the orientation of crystallographic ($\overline{1}05$) planes derived by Wilchinsky [20].

When using the procedure with equatorial reflections, the value of f_c is determined from the relationship deduced by Gupta and Kumar [22]:

$$f_c = 1 - 0.534 \times \overline{\cos^2}\,\rho_{010} - 1.150 \times \overline{\cos^2}\,\rho_{1\overline{1}0}$$
$$- 1.315 \times \overline{\cos^2}\,\rho_{100} \qquad (8)$$

where $\cos^2\rho_{010}$, $\cos^2\rho_{1\overline{1}0}$, and $\cos^2\rho_{100}$ are the variation parameters of the orientation of crystallographic (010), ($1\overline{1}0$), and (100) planes.

The quantitative assessment of the degree of crystallite orientation by x-ray examination is not free of ambiguity. From a comparative analysis [23] in which results obtained from the consideration of (105) and from three different variations of equatorial reflection were compared, the conclusion was that the first procedure can lead to underrated results, i.e., to the underestimation of the orientation. However, it can be assumed that this does not result from an incorrect procedure, but from ignoring the fact that the adjacent (105) reflex can overlap. The absence of the plate effect of the orientation is characteristic of the orientation of crystallites in PET fibers. The evidence of this absence is the nearly identical azimuthal intensity distributions of the diffracted radiation in the reflexes originating from different families of lattice planes. The lack of the plate effect of orientation in the case of PET fiber stretching has to do with the rod mechanism of the crystallite orientation.

The orientation of crystallites in PET fibers can also be assessed quantitatively by means of IR spectrographic examination. In this case, the basis for the assessment are the values of dichroic ratio (R) of the "crystalline" absorption bands in the fiber spectrogram. The determination of the values of f_c is made using Fraser's dependence [24,25] modified by Chranowski [26]:

$$f = \frac{2 \times (1 - R)}{(1 + 2R) \times (3 \times \cos^2\alpha - 1)} \qquad (9)$$

where α is the angle of transition moment for the absorption group correlated with the absorption band taken into consideration.

In an investigation [27], the determination was based on the consideration of crystalline bands 875 cm^{-1}, 1343 cm^{-1}, and 1473 cm^{-1}. In the case of absorption groups corresponding to these bands, i.e., C_6H_4, CH_2, and CH_2, respectively, the values of the angles of the transition moment (α) are equal to $\alpha_{875} = 71.0^0$, $\alpha_{1343} = 43.0^0$, and $\alpha_{1473} = 43.5^0$.

The crystallite orientation in PET fibers depends first and basically on the applicated draw ratio and second on the stretching rate. The values of f_c characteristics for PET fibers as established by the authors are in Table 5.

2. Amorphous Orientation

The amorphous orientation is understood as an arrangement of kinetic segments of macromolecules that form the amorphous phase in relation to the geometrical axis of the fiber. The development of an amorphous orientation in PET fibers initiated at the stage of melt spinning is intensified in the process of stretching the solidified fiber. Stretching in the production process is a decisive factor in the amorphous orientation in its final form. Developing the amorphous orientation starts at the beginning of stretching, then increases rapidly to reach the draw ratio of about 3.5×. For a higher draw ratio, i.e.,

Table 5 Crystallites Orientation Function (f_c) of Differently Drawn PET Fibers

| Draw ratio | X-ray method | | IR spectroscopy method | | |
	Equatorial reflexes (010), (110), (100)	Meridional reflex (105)	Absorption band 875 cm^{-1}	Absorption band 1343 cm^{-1}	Absorption band 1473 cm^{-1}
2.0×	—	—	0.176	0.269	0.309
3.0×	0.817	0.912	0.418	0.615	0.566
3.5×	0.820	0.916	0.571	0.691	0.626
4.0×	0.831	0.930	0.585	0.817	0.703
4.5×	0.842	0.945	0.713	0.870	0.851
5.2×	0.844	0.932	0.727	0.921	0.875

greater than 4.5×, an effect of an apparent decrease in the orientation index value can be observed [28]. Such an effect is consistent with the fact that at highest draw ratios part of the amorphous phase is transformed into the crystalline phase as a result of stress-induced crystallization. Consequently, in the amorphous phase only molecules of worse oriented kinetic segments remain.

The mechanism of development of the amorphous orientation largely consists of uncoiling and straightening molecules under the effect of internal tensile stress in the fiber that accompanies stretching. The attempts at a theoretical description of the amorphous orientation development process consisting in its approximation by the affine deformation of the polymer lattice stretched unidirectionally presented by Kuhn and Grun [29] and Roe and Kringbaum [30,31] do not yield results adequate to the effects obtained for stretching PET fibers. The main cause is that PET fiber stretching is done with pseudo-affine deformation, not with affine deformation as suggested by Ward [32].

The amorphous orientation is considered a very important parameter of the microstructure of the fiber. It has a quantitative and qualitative effect on the fiber deformability when mechanical forces are involved. It significantly influences the fatigue strength and sorptive properties (water, dyes), as well as transport phenomena inside the fiber (migration of electric charge carriers, diffusion of liquid). The importance of the amorphous phase makes its quantification essential. Indirect and direct methods currently are used for the quantitative assessment of the amorphous phase.

Intermediate methods include the earliest procedure based on Stein's equation [33] and one based on Samuels' equation [34]. Among the direct methods is an IR spectroscopic method based on the measurement of the dichroic ratio (R), of "amorphous" absorption bands. In the investigations [35], the amorphous bands 898 cm^{-1} and 1368 cm^{-1}, for which the angles of transition moment are $\alpha_{898} = 39^0$ and $\alpha_{1368} = 80^0$, respectively, were used. Other methods are spectroscopy of polarized fluorescent radiation [35,36], measurement of color di-

chroismus of the fiber dyed according the test methods [37–39], the x-ray method, which consists of an analysis of the azimuthal distribution of intensity of the x-ray scattered diffusively on the fiber [40,41].

The values of the amorphous orientation index in the form of Hermans' function of orientation (f_a), determined by the authors for PET fibers are listed in Table 6. The differences in the values of f_a quoted in Table 6 and referring to particular investigation methods can result from the fact that in some methods the orientation of the amorphous and mesophase are considered jointly. Consequently, in such a case the values of f_a will be overrated.

3. Overall Orientation

Overall orientation is understood as the joint arrangement of all the structural elements of the crystalline phase and noncrystalline part of the fiber in relation to the geometrical axis of the fiber. In its essence, the overall orientation of PET fibers, as a result of the crystalline and amorphous orientation, will be characterized by smaller values of the quantitative index of orientation than for the crystalline phase and by greater ones for the amorphous phase.

The quantitative assessment of the overall orientation of PET fibers is generally made on the basis of fiber optical anisotropy measurements, i.e., measurements of the optical birefringence of the fiber. The determination of the value of optical birefringence makes it possible to determine the value of Hermans' function of orientation based on the equation:

$$f_0 = \frac{\Delta n}{\Delta n_{ideal}} \times \frac{d_c}{d} \qquad (10)$$

where Δn_{ideal} is the optical birefringence of an ideal fiber, i.e., the fiber of a crystalline density and of ideal orientation, d_c is the density of the crystalline region, Δn is the birefringence, and d is the density of the fiber under investigations.

The overall orientation of PET fibers depends on the production conditions and on further refining treatment.

Table 6 Amorphous Orientation Function (f_a) of Differently Drawn PET Fibers

| Draw ratio | X-ray method | IR spectroscopy method | | Optical method | | Fluorescent polarization method | D R S D F method[c] |
		Absorption band 898 cm^{-1}	Absorption band 1368 cm^{-1}	a	b		
2.0×	—	0.220	0.208	—	—	0.082	0.047
3.0×	0.645	0.274	0.350	0.496	0.605	0.282	0.089
3.5×	0.744	0.379	0.362	0.522	0.715	0.336	0.156
4.0×	0.816	0.683	0.416	0.642	0.784	0.350	0.162
4.5×	0.834	0.703	0.510	0.681	0.834	0.323	0.130
5.2×	0.886	0.730	0.599	0.780	0.898	0.489	—

[a] $\Delta n_a = 0.253$, $\Delta n_c = 0.220$ [44].
[b] $\Delta n_a = 0.216$, $\Delta n_c = 0.210$ [45].
[c] Dichroic ratio of standard dyed fiber.

The factor having the strongest effect is the elongation imparted in the process of production stretching. Second, the overall orientation is affected by the stretching rate. For the same draw ratio, the overall orientation grows with an increase in the stretching rate. The effect of the draw ratio on the value of Hermans' function of orientation is illustrated by the values of f_0, established by the authors and depicted in Table 7.

III. PHYSICAL PROPERTIES

A. Mechanical Properties

Among the basic mechanical properties of fibers are their *deformability* and *tenacity*. When an axial stretching force is applied to the fiber, the principal quantitative indices of deformability are the *axial elastic modulus* (E)

Table 7 Overall Orientation Function (f_0) of Differently Drawn PET Fibers

| Draw ratio | Hermans method | | Cunningham method[c] |
	a	b	
0	0.0164	0.0172	0.0149
2.0×	0.1581	0.1657	0.1422
3.0×	0.6709	0.7028	0.5924
3.5×	0.7256	0.7601	0.6264
4.0×	0.8169	0.8558	0.7140
4.5×	0.8501	0.8906	0.7417
5.2×	0.9057	0.9488	0.7918

[a] Hermans method by assuming $\Delta n_{ideal} = 0.220$ after Dumbleton [44].
[b] Hermans method by assuming $\Delta n_{ideal} = 0.210$ after Okajima [45].
[c] Cunningham method [60].

(GPa) and the *relative elongation at break* (ϵ_r) (%). The first of the indices characterizes the resistance of the fiber to deformation under the affect of the acting force and determines the value of the tensile strength causing the formation of the unitary relative elongation (ϵ). The modulus E characterizes the susceptibility of the fiber to deform based on the principle of inverse proportionality. The other parameter, i.e., ϵ_r, characterizes the maximum ability of the fiber to deform under the effect of the acting tensile force.

As for the tenacity, which describes the fiber's resistance to the action of the tensile force, the quantitative index assumed most often is the *tensile strength* (σ_s) (cN/tex). For the quantitative assessment of tenacity an index of *mechanical long life* (τ) is used, proposed by Zurkow and Abasow [46,47]. This index defines the time after which, when applying a definite tensile strength and a definite temperature, the fiber breaks.

1. Axial Elastic Modulus (E)

The axial elastic modulus of PET fibers (E) depends, as with other kinds of fibers, on the value of the elastic modulus of crystalline material (E_c) and amorphous fiber material (E_a). When approximating their fine structure by the phenomenological model of Takayanagi, justified for PET fibers, it can be assumed that the elastic modulus (E) for PET fibers is described by the equation for the Takayanagi model [48]:

$$E = \Lambda \times \left[\frac{\phi}{E_c} + \frac{1 - \phi}{E_a} \right]^{-1} + (1 - \Lambda) \times E_a$$

(11)

where E_c is the elastic modulus of the crystalline region. This is a constant value, characteristic for a given kind of space lattice of crystalline regions. For PET fibers, $E_c = 140$ GPa according to Pierepelkin [49], $E_c = 110$

GPa according to Sakurada and Kaji [50], and $E_c = 108$ GPA according to Kunugi et al. [51]. The elastic modulus of amorphous material is E_a, Λ and Φ are parameters correspondingly proportional to the fraction of crystalline phase in the fiber. The product $(\Lambda \cdot \Phi)$, is defining the volume degree of fiber crystallinity (X); $X = \Lambda \cdot \Phi$. As stated by one of the authors [52], the parameters Λ and Φ can be defined by the following expressions:

$$\phi = \frac{l}{L} \qquad (12)$$

$$\Lambda = \frac{L \times X}{l} \qquad (13)$$

where l is the length of the crystallite; L is the mean long period; and X is the volume degree of fiber crystallinity.

The parameter Λ determined by the above relationship can be called a *substructure parameter*. Its numerical value determines the type of substructure of PET fibers. For $\Lambda < 1$ we have to do with the substructure of microfibrillar type, while for $\Lambda = 1$ with the substructure of lamellar type.

Contrary to widespread opinion, the value of E_a is not a constant quantity. As was proved previously [52], the value of E_a is variable, since it depends on the ordering of macromolecules in the amorphous material of the fiber. At the same time, one can suppose that this ordering will be affected by the specificity of the fine structure of the fiber, and particularly by the type of substructure of the fiber. The relationship determining the modulus E_a appropriate for a definite type of fiber substructure can be derived from Eq. (11) when appropriate values of Λ are assumed. In the case of the microfibrillar substructure, i.e., for $\Lambda < 1$, typical of PET fibers stretched, but not subjected to annealing, this equation has the form [52]:

$$a \times E_a^2 + b \times E_a + c = 0 \qquad (14)$$

where:

$$a = \frac{l}{L} - X,$$

$$b = E_c \times \left[X + 1 - \frac{l}{L} \right] - E \times \frac{l}{L},$$

$$c = E \times E_c \times \left[1 - \frac{l}{L} \right]$$

The value of E_a in this case is a positive value of the root of the above equation. In the case of the lamellar substructure, i.e., for $\Lambda = 1$, typical of stretched and then annealed fibers, the equation has the form:

$$E_a = \frac{E \times E_c \times (1 - X)}{E_c - X \times E} \qquad (15)$$

In the studies carried out by one of the authors [52], the values of E_a and E were determined for PET fibers of the microfibrillar and of the lamellar substructure. The results have been presented in Tables 8 and 9. The results obtained show that for both types of substructure the resistance to deformation, that is, the value of E, depends on the degree of molecular orientation of the amorphous material of the fiber (f_a) and the density of this amorphous phase of the fiber (d_a). However, this dependence assumes a different form for the microfibrillar and for the lamellar substructure. In the first case, it has the form:

$$E = A \times \left(\frac{f_a}{d_a} \right)^B \qquad (16)$$

In the case of the lamellar substructure:

$$E = C \times \frac{d_a}{f_a} + D \qquad (17)$$

where A, B, C, D are corresponding constant values.

2. Elongation at Break (ϵ_r)

The elongation at break of PET fibers depends decisively on the quantitative fraction and on the manner of the arrangement of the amorphous material of the fiber. In light of the results of investigations [53] that are presented in Table 10, one can assume that the value of the

Table 8 Fine Structure Parameters and Axial Elastic Moduli of PET Fibers of Microfibrillar Substructure

Draw ratio	Density of the amorphous material (d_a) (g/cm³)	Amorphous orientation function (f_a)	Crystallite length (l_c) (nm)	Long period (L) (nm)	Degree of crystallinity (X^a)	Substructure parameter (Λ)	Axial elastic modulus Fiber (E) (GPa)	Axial elastic modulus Amorphous phase (E_a) (GPa)
3.0×	1.333	0.645	7.9	11.2	0.36	0.51	9.1	4.4
3.5×	1.337	0.744	8.0	12.0	0.36	0.54	9.2	4.7
4.0×	1.338	0.816	7.8	13.4	0.37	0.64	9.7	5.5
4.5×	1.337	0.834	7.8	14.0	0.38	0.68	9.9	5.6
5.2×	1.338	0.886	7.9	16.6	0.39	0.82	10.8	6.5

a Determined by dsc method.

Table 9 Fine Structure Parameters and Axial Elastic Moduli of PET Fibers of Lamellar Substructure

Annealing temperature (°C)[a]	Density of the amorphous material (d_a) (g/cm³)	Amorphous orientation function (f_a)	Crystallite length (l_c) (nm)	Long period (L) (nm)	Degree of crystallinity (X[b])	Substructure parameter (Λ)	Axial elastic modulus	
							Fiber (E) (GPa)	Amorphous phase (E_a) (GPa)
120	1.337	0.310	4.6	10.4	0.44	0.99	9.1	5.3
140	1.330	0.265	5.1	11.0	0.46	0.99	9.9	5.6
180	1.333	0.185	5.9	11.8	0.50	1.00	11.0	5.8
220	1.345	0.115	7.6	13.8	0.56	1.00	14.0	6.6

[a] The fibers of draw ratio 1.5× were annealed unstressed in the dryer for 2 h in air.
[b] Determined by dsc method.

elongation at break (ϵ_r) increases with an increase in the fraction of the amorphous phase and a decrease in the degree of its orientation. It seems that at the same time both parameters act in the cumulated manner. Studies have shown the revealed straight line dependence between the elongation (ϵ_r) and the value of the cumulated index: $[(1 - X)/f_a]^4$, where $1 - X$ is the fraction of the noncrystalline phase, while f_a is the amorphous orientation function. In light of the above equations, the empirical dependence between the elongation at break and the fraction amorphous phase and its degree of orientation assumes the form:

$$\epsilon_r = F \times \left[\left(\frac{1 - X}{f_a} \right)^4 \right] + G \tag{18}$$

where F and G are constant values.

3. Tensile Strength

The tensile strength of PET fibers depends on their superstructure and internal orientation. The results from the investigations by one of the authors [54], show that the value of the tensile strength (σ_s) is affected by the fraction of taut tie molecules (β), the crystallites orienta-

tion function (f_c), and the amorphous orientation function f_a, Table 10. At the same time it appears that the enumerated parameters of the fiber fine structure, in the case of PET fibers, determine their tensile strength in the cumulative manner. The straight line dependence of the tensile strength on the product of β, f_c, and f_a, determined for PET fibers can testify to this fact [53]. The straight line correlation dependence being disclosed leads to the following empirical dependence of the tensile strength as a function of β, f_c, and f_a:

$$\sigma_s = H \times (\beta \times f_c \times f_a) + K \tag{19}$$

where H and K are constant values.

4. Mechanical Long Life (τ)

According to Żurkow and Abasow [46,47], authors of this τ index, it is determined by the relationship:

$$\tau = \tau_0 \exp \left[\frac{U_0 - \gamma \times \sigma}{R \times T} \right] \tag{20}$$

where:

τ_0 = a constant characterizing the thermal vibrations of atoms inside the molecules;

Table 10 Structural Determinants of Elongation, Tensile Strength, and Mechanical Long Life of PET Fibers

Draw ratio	Fraction of noncrystalline material ($1 - X$)	Amorphous orientation function (f_a)	Crystallites orientation function (f_c)	Fraction of tie molecules (β)	Elongation at break (ϵ_r) (%)	Tensile strength (σ_s) (cN/tex)	Structural parameter of the index τ (γ) (kJ/molMPa)	Mechanical long life (τ) (s)
3.0×	0.64	0.645	0.817	0.027	52.9	19.2	0.380	0.94
3.5×	0.64	0.744	0.820	0.030	26.7	30.8	0.242	0.98
4.0×	0.63	0.816	0.831	0.040	17.3	42.7	0.174	1.05
4.5×	0.62	0.834	0.842	0.043	16.8	55.3	0.134	1.10
5.2×	0.61	0.886	0.844	0.055	14.5	77.0	0.096	1.19

U_0 (kJ/mol) = the activation energy of the breakdown of the fiber corresponding to chemical bond energy for PET fibers $U_0 = 175$ (kJ/mol);

R_0 = gaseous constant = 8.314×10^{-3}(kJ/mol × °K);

T (°K) = the fiber temperature;

γ (kJ/mol × MPa) = the structural parameter of the index τ, and

σ = applied tensile strength.

The results of determinations of the τ index for PET fibers obtained [54], have been listed in Table 10. In light of these results, one can state that the mechanical long life of PET fibers is determined by the specificity of their amorphous regions. The degree of molecular orientation and the density of packing of macromolecules in the amorphous material of the fiber must be considered the most important parameters. The mechanical long life (τ) increases with an increase in the degree of molecular orientation of the amorphous phase. The increase in τ is also promoted by looser packing of the macromolecules in this part of the fiber; i.e., its smaller density. These dependences are reflected in the empirical straight line relationship described by the equation:

$$\gamma = -M \times \left(\frac{f_a}{d_a}\right) + N \tag{21}$$

where M and N are constant values.

B. Selected Thermal Properties

2. Thermal Volume Expansivity

PET fibers, among basic synthetic fibers, distinguish themselves by a relatively low thermal volume expansivity. The value of the coefficient of the thermal volume expansivity (α) is dependent on the fiber temperature. For temperatures below the glassy temperature, the values of α are lower and, depending on the draw ratio of the fiber, are within the range $7.9–15.9 \times 10^{-5}$ K^{-1}, while above the glassy temperature in the range $11.8–21.1 \times 10^{-5}$ K^{-1}. The thermal expansivity anisotropy occurring in the fibers causes the α to be the resultant of the linear expansivity $\alpha_\parallel$ and $\alpha_\perp$, which quantita-

tively characterize the expansivity in the direction of the fiber axis and perpendicular to it:

$$\alpha = \alpha_\parallel + 2 \times \alpha_\perp \tag{22}$$

In the case of PET fibers, $\alpha_\parallel$ assumes the values close to 0, often even negative ones, $\alpha_\perp$, which is perpendicular, is several times greater and always positive. According to Choy, et al. [55], $\alpha_\parallel = 0.5–0.3 \times 10^{-5}$ K^{-1}, and $\alpha_\perp = 4.2–7.8 \times 10^{-5}$ K^{-1} for the temperatures T < T_g, and $\alpha_\parallel = -0.5–0.5 \times 10^{-5}$ K^{-1}, and $\alpha_\perp = 6.2–10.3 \times 10^{-5}$ K^{-1} for the temperatures T > T_g.

In the investigations carried out by one of the authors [56], an attempt was made to examine the conditions for the thermal volume expansivity of PET fibers. Within the framework of these investigations, α_{25} was determined from the hydrostatic weight measurements using n-heptan as a liquid. The sought α_{25} values have been calculated from the equation:

$$\alpha_{25} = \frac{Q_0 - Q_1}{Q_1 \times (t - t_0)} \tag{23}$$

where Q_0 and Q_t refer to fiber density at $t_0 = 19$°C and $t = 25$°C, respectively.

Fibers of a diversified draw ratio in the range 2.0–5.2× were considered, determining the following parameters of their fine structure: the crystalline and amorphous orientation functions, f_c and f_a, degree of crystallinity, X, and critical dissolution time (CDS) in seconds. The results obtained are listed in Table 11.

The α values obtained show that the thermal volume expansivity of PET fibers changes only slightly as a function of their stretching degree. At the same time, it can be observed that at the initial stage of stretching, i.e., for the draw ratio up to 3.5×, the value of α deceases. For stretching above 3.5×, they are higher and practically constant. It can be assumed that the variation of α is directly related to the changes in the coefficients of the linear expansivity, $\alpha_\parallel$ and $\alpha_\perp$. As was stated by Choy et al. [55], at the initial stage of drawing of PET fibers, i.e., for the draw ratio below 3.5×, there occurs a rapid decrease in $\alpha_\parallel$ while simultaneously there is a very slow increase in $\alpha_\perp$. Thus, the resultant outcome must lead to a decrease in the value of α. In turn, for advanced

Table 11 Thermal Volume Expansivity (α_{25}) and Related Fine Structure Parameters of Differently Drawn PET Fibers

Draw ratio	$\alpha \times 10^{-4} \times$ K^{-1}	f_c	f_a	x	TTM fraction	CDT[a] (s)
3.0×	3.2	0.817	0.496	0.22	0.027	28
3.5×	2.9	0.820	0.522	0.27	0.030	76
4.0×	3.1	0.831	0.642	0.31	0.040	214
4.5×	3.0	0.842	0.681	0.33	0.043	138
5.2×	3.0	0.844	0.760	0.35	0.055	162

[a] CDT = critical dissolution time in mixture phenoltetrachlorethan 2:1 at 60°C.

drawing, i.e., for the draw ratio above 3.5×, the rates of decrease in $\alpha_{\parallel}$ and increase in $\alpha_{\perp}$ compensate each other, which will lead to the stabilization of the value of α.

It can be noted that the stated changes in α do not correlate with the variation of any separate considered fine structure parameter. This leads to the conclusion that the ascertained alteration in α must be evoked by a complex change in different parameters of the fiber substructure, microstructure, and orientation. The thermal volume expansivity of PET fibers undergoes an essential alteration resulting from annealing. The influence of annealing of PET fibers performed in air in a loose state at different temperatures and differentiated durations of heat treatment was studied [56]. The results obtained are presented in Fig. 7. As can be seen, the change in α depends on the annealing temperature and the duration of the heat treatment. The character of the alterations is attributed first to the level of the annealing temperature. Annealing time modifies only the range of changes but does not alter the character of the variation. The results obtained give evidence that there may be three distinguished temperature intervals of annealing in which α values attain different levels and undergo different alternations with an increasing annealing temperature. The first interval refers to the temperature range of 130–150/160°C within which the α values are larger than those of unannealed fibers and monotonically decline with the growth of the annealing temperature. The second temperature interval refers to 170–190°C

within which the α values are lower than those of unannealed fibers and continuously diminish reaching a minimum value at 190°C. The third temperature interval pertains to the temperature range of 200–210°C. Within this zone α increases with the annealing temperature reaching a level equal to or even higher than that of unannealed fibers.

Comparing the alteration in α values with the established changes in fine structure parameters presented in Table 12, it can be seen that the alteration in α values does not correlate with variations in any particular structure parameter. This leads to the conclusion that the ascertained alternation in α must be evoked by a complex change in different substructure, microstructure, and orientation parameters of the fiber.

2. Heat Capacity

PET fibers distinguish themselves among basic synthetic fibers by a relatively small value of heat capacity. The value of the heat capacity coefficient at constant pressure lies in the range of $C_p = 0.24$–0.44 (cal K^{-1} g^{-1}). It seems that the relatively low C_p value of PET fibers reflects the facility in enhancing the fibers internal energy by conveying heat to them. The heat capacity of PET fibers undergoes essential alternation as a result of the applied heat treatment of the fiber. The relationship between the c_p and the annealing conditions of PET fibers has been studied [56]. The heat capacities of a 4.5× drawn filament were estimated at constant pressure from DSC diagrams using the relationship:

$$C_p = \frac{H_s \times m_a \times C_{ap}}{H_a \times m_s} \tag{24}$$

where H_s and H_a refer to the amplitude of enthalpy change occurring at the temperature increase from 20 to 25°C, respectively, for fiber sample and reference substance (saphire); m_s and m_a are weights of sample and reference substance; and $c_{ap} = 775.43$ (J/kGK) is the heat capacity of the reference substance.

The obtained c_p values and the established fine structure parameters for the PET fibers examined are presented in Fig. 8 and Table 13.

The results obtained show a certain general regularity. The c_p values in the range of an annealing temperature of 130–160°C are higher than for the unannealed fiber and increase with an increase in temperature. For the range of temperatures exceeding 160°C, the c_p values monotonically decrease and assume smaller values than for the unannealed fibre. Comparing the alternation in c_p with changes in the established fine structure parameters, it will be noted that the change in c_p does not correlate with the variation of any separate structure parameter. Closer examination of the relations enables us to infer that the c_p values of annealed PET fibers are determined by a counteracting, mutually competitive influence of particular structure parameters. The level

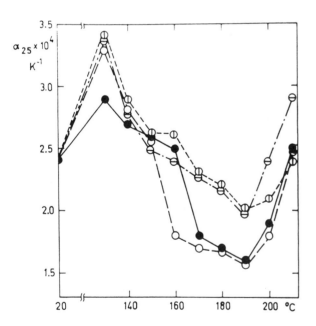

Figure 7 Thermal volume expansivity coefficient α_{25} of annealed PET fibers.

Table 12 Fine Structure Parameters of Annealed PET Fibers

Annealing temperature (°C)	Annealing time (min)	Birefringence (Δn)	$\Delta n/d$	Volume crystallinity (x_α) (%)	TTM fraction	Critical dissolve time (s)	Amorphous orientation function (f_a)
Without	—	0.1286	0.1296	36.7	0.107	9	0.50
130	2	0.1681	0.1209	45.6	0.048	19	0.50
	5	0.1615	0.1165	45.8	0.047	25	0.50
	10	0.1614	0.1184	45.4	0.045	41	0.47
	30	0.1653	0.1189	49.2	0.036	55	0.46
140	2	0.1645	0.1184	48.1	0.045	40	0.49
	5	0.1607	0.1156	48.6	0.041	35	0.49
	10	0.1593	0.1146	49.0	0.043	55	0.46
	30	0.1670	0.1202	48.8	0.036	74	0.45
160	2	0.1628	0.1169	50.3	0.043	103	0.47
	5	0.1572	0.1129	50.8	0.041	108	0.44
	10	0.1628	0.1170	51.3	0.043	106	0.44
	30	0.1622	0.1163	52.8	0.034	156	0.44
180	2	0.1592	0.1139	53.5	0.041	160	0.46
	5	0.1610	0.1154	53.1	0.038	150	0.44
	10	0.1606	0.1150	54.3	0.040	222	0.44
	30	0.1650	0.1180	55.6	0.029	245	0.43
200	2	0.1637	0.1171	55.5	0.038	240	0.45
	5	0.1601	0.1145	55.4	0.038	245	0.44
	10	0.1634	0.1167	57.8	0.039	323	0.44
	30	0.1663	0.1187	57.9	0.029	383	0.43
210	2	0.1623	0.1158	58.7	0.021	329	0.44
	5	0.1611	0.1150	58.1	0.031	405	0.43
	10	0.1573	0.1122	59.5	0.018	703	0.43
	30	0.1692	0.1203	62.3	0.021	1417	0.42

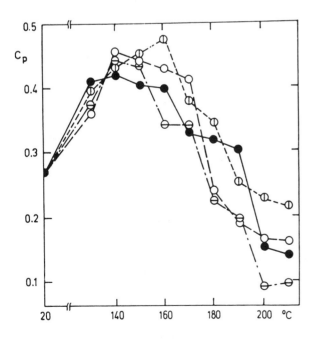

Figure 8 Heat capacity c_p of annealed PET fibers.

of c_p is always set by mutual opposite effects involved by the change of taut tie molecules fraction and the overall orientation on one hand, and the change in crystallinity on the other.

Table 13 Heat Capacity of Annealed PET Fibers

Annealing temperature (°C)	Time (min)			
	2	5	10	30
130	0.412	0.363	0.396	0.376
140	0.421	0.453	0.429	0.441
150	0.405	0.442	0.452	0.433
160	0.398	0.428	0.474	0.344
170	0.332	0.415	0.380	0.340
180	0.323	0.241	0.346	0.227
190	0.305	0.192	0.254	0.194
200	0.151	0.167	0.226	0.090
210	0.140	0.163	0.216	0.097
Without	0.269	—	—	—

C. Selected Electrical Properties

1. Electrical Conductivity

The electrical conductivity of PET fibers as compared with other main synthetic fibers is relatively low. This explains why PET fibers are often utilized in the manufacture of textiles as electroisolating materials. The value of the electrical resistivity characterizing reciprocal conductivity is of the order 10^{-13} ($\Omega \cdot$cm). The mechanism of the electrical conductivity of PET fibers is still a matter of controversy. According to results attained [57], there are convincing arguments that in the case of PET objects the electrical conductivity is due to the ionic mechanism.

The low electrical conductivity of PET fibers depends essentially on their chemical constituency, but also to the same extent on the fiber's fine structure. In one study [58], an attempt was made to elucidate the influence of some basic fine structure parameters on the electrical resistivity of PET fibers. The influence of crystallinity (x) the average lateral crystallite size (Λ), the mean long period (L), and the overall orientation function (f_0) have been considered. The results obtained are presented in the form of plots in Figs. 9–12.

The reported results give evidence that the electrical conductivity decreases with increasing crystallinity and with enhancement of the mean long period of the fiber. In addition, the conductivity simultanously decreases with the diminishing of the overall orientation. It seems that this relationship can be explained on the basis of the nature of electrical conductivity, i.e., on the migration of carriers of free electrical charges. The increase in crystallinity accompanied by the increase in L create more hindrances for the migration of carriers of electrical charges and, therefore, impedes the conductivity. The opposite result is caused by the improvement of the overall orientation.

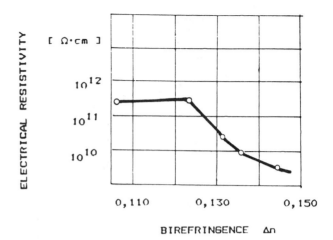

Figure 10 Electrical resistivity of PET fiber versus birefringence of the fiber.

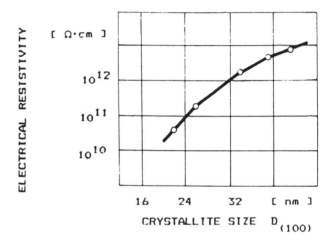

Figure 11 Electrical resistivity of PET fiber versus average crystallite size perpendicular to the chain direction.

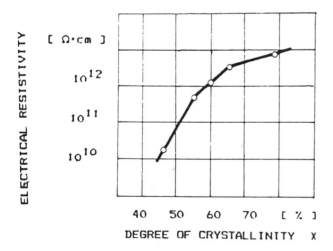

Figure 9 Electrical resistivity of PET fiber versus degree of crystallinity of the fiber.

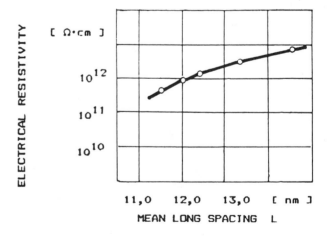

Figure 12 Electrical resistivity of PET fiber versus mean long spacing.

2. Electrostatic Charges

PET fibers are characterized by a strongly pronounced liability to electrostatic charging. Their high susceptibility to electrostatic charging is determined by a polar character of the polymer, and their low electrical conductivity. Investigations carried out by the authors [58] indicate that the effect of electrostatic charging of PET fibers depends additionally on the fine structure of the fiber. In these investigations an assessment was made of a charge in the surface density of electrostatic charges (F) (C/cm^2), generated during the rubbing of the fiber against a brass cylinder as a function of the crystallinity degree, average lateral crystallite size, mean long period, and overall orientation. The results obtained have been presented in the form of plots in Figs. 13–16. The dependences shown indicate that F increases with an increase in the crystallinity degree. The dependence on the average lateral crystallite size and the overall orientation is of the curvilinear character. Based on the results attained, it can be concluded that the electrostatic charging of PET fibers will be particularly strong for fibers of a high degree of crystallinity, coarse-grained crystalline structure, and low overall orientation.

D. Optical Properties

The basic optical properties of PET fibers consist of the anisotropy of light transmission and the anisotropy of light absorption. The first one manifests itself by the light double refraction, the other by the pleochroic effect of the fiber coloring. The double refraction causes the refractive indices of the ordinary and extraordinary light beams (n_0, n_e) originated by light transmission through

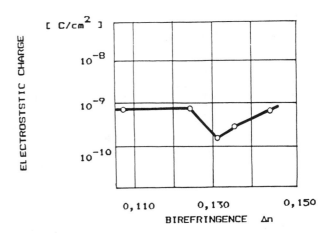

Figure 14 Electrostatic charge of PET fiber versus birefringence of the fiber.

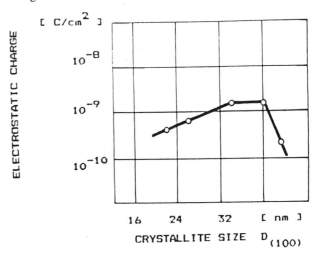

Figure 15 Electrostatic charge of PET fiber versus average crystallite size perpendicular to the chain direction.

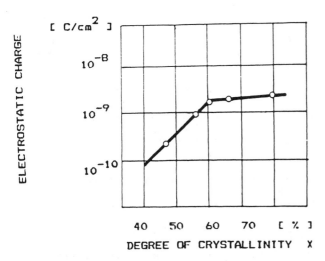

Figure 13 Electrostatic charge of PET fiber versus fiber degree of crystallinity.

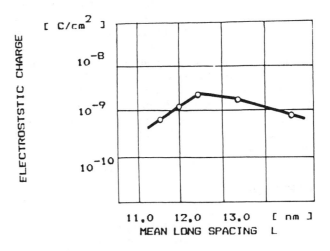

Figure 16 Electrostatic charge of PET fiber versus mean long spacing.

Table 14   Directional Refractive Indices ($n_\parallel$, $n_\perp$) and Birefringence (Δn) of Differently Drawn PET Fibers

Draw ratio	$n_\parallel$	$n_\perp$	Δn
1.0×	1.5732	1.5696	0.0036
2.0×	1.5934	1.5648	0.0286
3.0×	1.6866	1.5470	0.1396
3.5×	1.7037	1.5444	0.1593
4.0×	1.7119	1.5409	0.1710
4.5×	1.7172	1.5382	0.1790
5.2×	1.7224	1.5343	0.1881

the fiber to differ in their values. In view of the fact that PET fibers are optically positive, n_e is always greater than n_0. Considering that n_e is the value of the refraction index for the wave vibrating parallel to the fiber axis, it is additionally denoted by the symbol $n_\parallel$. In turn, n_0 being the value of the refractive index for the wave vibrating perpendicular to the fiber axis is also denoted by $n_\perp$. The measure of the optical anisotropy of light transmission of the fiber is optical *birefringence* (Δn).

$$\Delta n = n_e - n_0 \equiv n_\parallel - n_\perp \qquad (25)$$

In the case of PET fibers, the main directional refractive indices $n_\parallel$, $n_\perp$ and the optical birefringence Δn are dependent on the magnitude of the draw ratio at which the fiber was obtained and to a lesser extent on its crystallinity degree. The values of $n_\parallel$, $n_\perp$, and Δn for fibers of a diversified draw ratio are shown in Table 14. The ability to refract light independent of the direction of the light wave vibration is presented by the isomorphic refractive index (n_{iso}). For PET fibers $n_{iso} = 1.598$. In the case of PET fibers, we have to deal with the occurrence of internal radial differentiation of the birefringence. Results from investigations performed by one of the authors [59] show that the radial differentation of the Δn values varies for fibers of different stretching. For nondrawn fibers, the following are characteristic: a low value of Δn for the central axial part of the fiber and a low value for the surface layer. The remaining part of the fiber is distinguished by higher and, at the same time, variable values of Δn. Starting with the draw ratio 2.0×, the value of Δn for the surface layer and the central part increases. For the highest draw ratio (4.5–5.2×), the higher values of Δn occur only for the surface layer.

REFERENCES

1. A. Peterlin, *J. Polym. Sci., C9:* 61 (1965).
2. Peterlin, *Koll. Z., 216/217:* 129 (1966).
3. Jo. W. Brestkin, *Vysokomol. Sojed., A13:* 1794 (1971).
4. A. Sz. Gojchman, *Vysokomol. Sojed., B14:* 706 (1972).
5. W. Pechhold, M. E. T. Hauber, and E. Liska, *Koll. Z. Z. Polym., 250:* 1017 (1972).
6. D. C. Prevorsek and Y. D. Kwon, *J. Macromol. Sci., 852:* 447 (1976).
7. A. Jeziorny, *Acta Polymerica, 41:* 590 (1990).
8. R. T. Daubeny and C. W. Bunn, *Proc. Roy. Soc., A226:* 531 (1954).
9. Ju. Ja. Tomaszpolskij and G. S. Markova, *Vysokomol. Sojed., 6:* 274 (1964).
10. S. Fakirov, W. Fischer, and G. E. Schmidt, *Makromol. Chem., 176:* 459 (1975).
11. C. Yamashita, *J. Polym. Sci., A3:* 81 (1965).
12. J. L. Koenig and M. L. Hannen, *Macromol. Sci.-Phys., B1:* 617 (1971).
13. K. K. Mocherla and J. P. Bell, *J. Polym. Sci.-Phys. Edn., 11:* 1779 (1973).
14. S. A. Baranova et al., *Vysokomol. Sojed., A22:* 537 (1980).
15. S. L. Pakszwer et al., *Vysokomol. Sojed., A25:* 37 (1984).
16. T. Asano and T. Seto, *Polymer J., 5:* 72 (1973).
17. W. L. Lindner, *Polymer, 14:* 9 (1973).
18. G. Urbańczyk and I. Waszkiewicz, *Polimery, 25 (253):* 440 (1980).
19. A. Jeziorny, *Zesz. Nauk. P. L.-Inz.Wlok., 2/16:* 195 (1993).
20. Z. W. Wilchinsky, *Adv. X-Ray Anal., 6:* 231 (1963).
21. J. H. Dumbleton and B. B. Bowles, *J. Polym. Sci., A2-4:* 951 (1966).
22. V. B. Gupta and S. Kumar, *Text. Res. J., 49:* 405 (1979).
23. G. Urbańczyk, *J. Appl. Polym. Sci., 51:* 201 (1994).
24. R. D. B. Fraser, *J. Chem. Phys., 21:* 1511 (1953).
25. R. D. B. Fraser, *J. Chem. Phys., 28:* 1113 (1958).
26. W. A. Chranowski, *Vysokomol. Sojed., B25:* 96 (1983).
27. G. Urbańczyk, *Polimery, 21:* 26, 69 (1976).
28. G. Urbańczyk, *J. Polym. Sci.- Symp., 58:* 311 (1977).
29. W. Kuhn and F. Grún, *Koll. Z., 101:* 848 (1942).
30. R. I. Roe and W. R. Kringbaum, *J. Appl. Phys. 6:* 32 (1962).
31. R. I. Roe and W. R. Kringbaum, *J. Appl. Phys. 35:* 2215 (1964).
32. I. A. Ward, *Structure and Properties of Oriented Polymers,* Appl. Sci. Publ. Ltd., London, p. 34 (1975).
33. R. S. Stein and F. H. Norris, *J. Polym. Sci., 21:* 381 (1956).
34. R. I. Samuels, *J. Polym. Sci., A2-3:* 1741 (1965).
35. G. Urbańczyk, *Zesz. Nauk. P. L.,* nr.646, Wlokien., *50:* 83,55, (1993).
36. G. E. McGraw, *J. Polym. Sci., A2-8:* 1323 (1970).
37. J. M. Preston, *J. Soc. Dyers Colour, 66:* 361 (1950).
38. J. G. Blacker and D. Patterson, *J. Soc. Dyers Colour, 85:* 598 (1968).
39. B. Lipp-Symonowicz and G. Urbańczyk,- *Polimery, 27:* 198 (1982).
40. H. J. Biangardi, *J. Polym. Sci.-Phys., 18:* 903 (1980).
41. A. Jeziorny, *Polimery, 39:* 234 (1994).
42. W. Ruland, *Polymer, 5:* 89 (1964).
43. G. Vonk, *J. Appl. Crystallogr., 6:* 148 (1973).
44. J. H. Dumbleton, *J. Polym. Sci., A-2/6:* 795 (1968).
45. S. Okajima and K. Kayama, *Sen.I.Gakkaishi, 22:* 51 (1976).
46. S. N. Żurkow and S. A. Abasow, *Vysokomol. Sojed., 3:* 441 (1961).
47. S. N. Żurkow and S. A. Abasow, *Vysokomol. Sojed., 4:* 1703 (1962).
48. M. Takayanagi, K. Imada, and T. Kajima, *J. Polym. Sci., C15:* (1966).
49. K. E. Pieriepielkin, *Chim. Volok., 2:* 3 (1966).
50. I. Sakurada and K. Kaji, *J. Polym. Sci., C31:* 57 (1989).

51. T. Kunugi, A. Suzuki, and M. Hashimoto, *J. Appl. Polym. Sci., 26*: 1951 (1981).
52. A. Jeziorny, *Przegl. Wlokien., 47(10)*: 235 (1993).
53. A. Jeziorny, *Przegl. Wlokien., 48(10)*: 3 (1994).
54. A. Jeziorny, *Przegl. Wlokien., 49(4)*: 3 (1995).
55. C. L. Choy, M. Ito, and R. S. Porter, *J. Polym. Sci-Phys. Ed., 21*: 1427 (1983).
56. G. Urbańczyk and G. Michalak, *J. Appl. Polym. Sci., 32*: 4787, 3841 (1986).
57. G. Urbańczyk and W. Urbaniak-Domagala, *Polimery, 33*: 84 (1988).
58. G. Urbańczyk and A. Jeziorny, *Faserf. U. Textil., 24*: 151 (1973).
59. G. Urbańczyk, *Zeszyty Nauk. P. 1.646, Wlok. 50*: 55 (1993).
60. F. Cunningham and G. R. Davies, *Polymer, 15*: 743 (1974).

Index